2nd International Conference on Energy, Environmental and Information System (ICENIS 2017)

E3S Web of Conferences Volume 31 (2018)

Semarang, Indonesia
15 – 16 August 2017

Part 1 of 2

Editors:

Hadiyanto
Sudarno
Maryono

ISBN: 978-1-5108-5877-0

Printed from e-media with permission by:

Curran Associates, Inc.
57 Morehouse Lane
Red Hook, NY 12571

proceedings.com

Some format issues inherent in the e-media version may also appear in this print version.

Printed by Curran Associates, Inc. (2018)

For additional information, please contact EDP Sciences – Web of Conferences at the address below.

EDP Sciences – Web of Conferences
17, Avenue du Hoggar
Parc d'Activité de Courtabœuf
BP 112
F-91944 Les Ulis Cedex A
France

Phone: +33 (0) 1 69 18 75 75
Fax: +33 (0) 1 69 28 84 91

contact-edps@webofconferences.org

Additional copies of this publication are available from:

Curran Associates, Inc.
57 Morehouse Lane
Red Hook, NY 12571 USA
Phone: 845-758-0400
Fax: 845-758-2633
Email: curran@proceedings.com
Web: www.proceedings.com

TABLE OF CONTENTS

PART 1

PART 2

Preface

Population growth has significant impacts to sustainability of the natural and energy resources. Some problems arose from the uncontrolled exploitation of natural and energy resources. Several substantial efforts have been conducted by managing the environment through the instruments. The major efforts have yielded positive results in several aspects; however, some aspects have not improved. Innovation efforts, by academics, researchers, bureaucrats and entrepreneurs, are still needed to ensure a sustainable environmental management.

The 2nd International Conference on Energy, Environmental and Information System **(ICENIS)** 2017 organized by School of Postgraduate Studies, Universitas Diponegoro (UNDIP) has been conducted 15-16th August 2017. The conference has successfully enabled the exchangeof research results of researchers on the fundamentals and applications of energy, environment, and information system. More than 250 participants and presenters from several countries i.e. Indonesia, Malaysia, Germany, Sudan, Nepal, Australia, Japan, Libya have attended the conference to share their significant contributions in research related to Energy, Environment and Information System. This proceeding contains of 202 selected papers from the conferences.

We would like to express our gratitude to all authors and the members of scientific committee, reviewers and also organizing committee for their contribution to the success of the conference.

Guest Editors

Prof. Hadiyanto

Dr. –Ing. Sudarno

Dr. Eng. Maryono

sciences

Statement of Peer review

In submitting conference proceedings to E3S Web of Conferences, I certify to the Publisher that I adhere to the **Policy on Publishing Integrity** of the journal in order to safeguard good scientific practice in publishing.

1. All articles have been subjected to peer review administered by the proceedings editors.
2. Reviews have been conducted by expert referees, who have been requested to provide unbiased and constructive comments aimed, whenever possible, at improving the work.
3. Proceedings editors have taken all reasonable steps to ensure the quality of the materials they publish and their decision to accept or reject a paper for publication has been based only on the merits of the work and the relevance to the journal.

Title, date and place of the conference

The 2nd International Conference on Energy, Environmental and Information System (ICENIS) 2017

15 – 16 August 2017
Semarang, INDONESIA

Proceedings editor(s):

Prof. Dr. Hadiyanto, MSc   Dr.-Ing. Sudarno, MSc   Dr. Eng. Maryono, MT

Date and editor's signature. **January 20, 2018**

https://doi.org/10.1051/e3sconf/20183101001

$CaSO_4$ Scale Formation on Vibrated Piping System in the Presence Citric Acid

W. Mangestiyono[1,*], *J.* Jamari[2], *S.* Muryanto[3], *A. P.* Bayuseno[2]

[1]Mechanical Engineering, School of Vocation, Diponegoro University, Semarang 50275, Indonesia
[2]Graduate Program of Mechanical Engineering, Diponegoro University, Semarang 50275, Indonesia
[3]Chemical Engineering, Untag University, Semarang 50233, Indonesia

Abstract. Vibration in many industries commonly generated by the operation mechanical equipment such as extruder, mixer, blower, compressor, turbine, generator etc. Vibration propagates into the floor and attacks the pipe around those mechanical equipment. In this paper, the influence of vibration in a pipe on the $CaSO_4$ scale formation was investigated to understand the effect of vibration on the kinetics, mass of scale, crystal phases and crystal polymorph. To generate vibration force, mechanical equipment was prepared consisted of electrical motor, crankshaft, connecting rod and a vibration table at where test pipe section mounted. Deposition rate increased significantly when the vibration affected to the system i.e. 0.5997 and 1.6705 gr/hr for vibration frequency 4.00 and 8.00 Hz. The addition 10.00 ppm of citric acid declined the deposition rate of 8 Hz experiment from 3.4599 gr/hr to 2.2865 gr/hr.

1 Introduction

Vibration in piping system encounters in most industries because of the existing two factors i.e. internal and external factors. Chategorized as internal factor is fluctuation of fluid velocity which promotes the vibration mean while vibration produced by the operation of mechanical equipment such as blower, extruder, mixer, turbine, generator, pump, compressor are chategorized as external factors [1]. At industrial building, vibration propagates throughout the floor and wall, subsequently vibrates the pipes where are mounted around. Vibration is suspected affects the scaling process take place in the inner pipe.

In industries, pipe transports the fluid from one section to others. The fluid could be the water which comprises some impurities such as Ca, Na, SO_4, Cl, Mg salts [2]. When the saturation of those salt are exceeded the scale are formed through either homogeneous and heterogeneous crystallization [3].

In the current research, the formation $CaSO_4$ scale be the focus of the experiment by the consideration that $CaSO_4$ scale take place in almost of all industrial piping system [4]. $CaSO_4$ scale formation in piping system has been investigated by many researchers such as Muryanto, Hoang, Chunfang, Uchymiak [5, 6, 7, 8]. They reported that the scale forms a layer in the inner surface of the pipe and elevate thermal resistance. The existence of the scale affects the disturbance to industrial processes [9] i.e. decreases overall heat transfer coefficient [10]; prolongs the time of process [11]; waste the energy and increases the production cost [12]. Even $CaSO_4$ scale formation has been widely investigated but its formation in vibrate piping system has not been reported yet and supposed will be the novelty of this study. To complete the study, citric acid was added to the solution to inhibit scale growth. Indeed, citric acid has been used as inhibitor widely but its performance to inhibit $CaSO_4$ scale growth in vibrate piping system is still needed to investigate.

2 Methods

2.1 Material

$CaSO_4$ scale formation was carried out experimentally by mixing $CaCl_2.2H_2O$ and Na_2SO_4 powder which was supplied by Merck® according to the reaction such as shown in equation (1).

$$CaCl_2.2H2O_{(aq)} + Na_2SO_{4(aq)} \rightarrow CaSO_{4(s)} + 2NaCl_{(aq)} \quad (1)$$

Concentration of calcium solution was determined as 4.000 ppm, prepared by dissolved 147 gr of $CaCl_2.2H_2O$ into 10,000 ml of demineralized water.

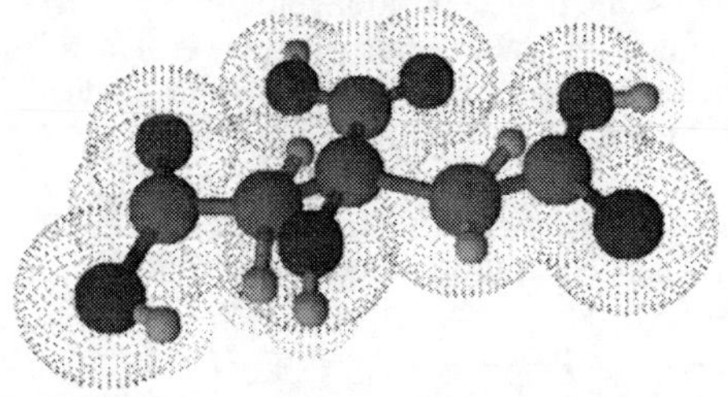

Fig. 1. Citric acid with three carboxylates form

*Corresponding author: rhombo_hedral@yahoo.co.id

Based on calculation of Ca^{2+} concentration in the solution, sodium solution was also prepared by dissolved 142 gr of Na_2SO_4 into the 10,000 ml. The solution was subsequently filtrated two times by 0.22 µm micropore® paper to ensure no unwanted particles remained in the solution. Each solution was then stored in a covered vessel to avoid from dusk and other dirty material.

Inhibitor that been used in the current research was citric acid which set at 10.00 ppm. Citric acid has been known as green inhibitor which was deliberately chosen because this acid had three carboxylathes form (COOH) as shown in Fig. 1. The carboxyl the was supposed to vary the ion and destructed the crystal lattice to inhibit its growth [13].

2.2 Deposition rate

Deposition rate in flowing and vibrating system must be defined earlier because of the existing new phenomenon in the system. (i).Vibration inserts reciprocal force to the precipitation processes and might affects the stretching and bending of molecule dissociation and association which can promotes more deposition rate. (ii). Vibration has potential to remove the younger crystals which not strong enough adheres in the pipe surface and being transported by fluid flowing, furthermore declines the deposition rate. (iii). Vibration also suspected will be the barrier of the scale to attach in the surface and brought by flowing force which such phenomenon often named as homogenous crystallization.

In fact, too difficult to distinct the crystals between the removal and homogeneous crystallization but both of crystals could be measured by seizure them through smooth strainer and was then named as deposition in the strainer (Wstrain).

To accommodate such phenomenon, the formula suggested by Sheikholeslami will be used as the basic. As it has published by Sheikholeslami that deposition rate has known as particulate, precipitation and removal [14] which expressed in equation (2)

$$W_{net} = A.\ W_{precip} + B.\ W_{part} - W_{rem} \qquad (2)$$

Where W_{net} is net deposition rate; W_{precip} is deposition rate through precipitation process; W_{part} is deposition rate because of the existing particulate; W_{rem} is the deposition rate of the removal and A, B are the constant.

Particle scaling is deposition process of particles carried by the fluid which the term particle is general and may refer to particulate matter or corrosion product. In this work, by filtrating the solution with strain paper 0.22 µmtwoo times, all particles in the solution were separated and deposition rate of the particulate could be neglected. Then, equation (1) was modified into equation (3) as follow:

$$W_{net} = W_{precip} + W_{strain} \qquad (3)$$

The transpose minus sign of -Wrem to +Wstrain was considered that all the deposition of the strainer stand as a whole part of reaction that must be calculated and inserted as the product of reaction, addressed to investigate the real deposition.

2.3 Variable

In this experiment, in depend variable was defined as vibration frequency which was chosen in 0.00; 4.00 and 8.00 Hz. The experiment under various vibration frequency was conducted for either blank and antiscalant experiment. Depend variable was defined as deposition rate; induction time; morphology; crystal phase of blank and antiscalant experiment. Otherwise displacement of the vibration; temperature; flow rate and calcium concentration were set as fix variable. All of test was conducted in atmospheric pressure. The summary of experiment variables was given in Table 1.

Table 1. Experiment variable

Variable	Test I	Test II	Test III
Frequency (Hz)	0	4.00	8.00
Displacement (mm)	0	4	4
Flow rate (ml/min)	30	30	30
Caconcent. (ppm)	4,000	4,000	4,000
Temperature (^{0}K)	318	318	318

Experiment of the addition citric acid to the solution in amount of 10 ppm was done in Test II and Test III and namely antiscalant experiment.

2.4 Phreeqc prediction

Scale potential prediction was analyzed by Phreeqc program. To operate this program, firstly, composition of the mineral in the solution was calculated based on the concentration of calcium, i.e. Ca: 2.00 ;Cl: 3.5499; H2: 0.20; O_2 : 1.600; Na : 2.0125 and SO_4 : 4.800 in the unit gr/kg water as the default unit in the program. After insert individual composition to the program, the program was run. The time needed to run the program was reported in 0.33 second only and the prediction of scale potential was ready. Specific conductance was predicted 15,222 µS/Cm, density 1.0049 gr/cm3, saturation index (SI) was predicted 0.24 and 0.54 for $CaSO_4$ and $CaSO_4.2H_2O$. According to the number of saturation index, production of dehydrate was higher than anhydrate. Crystal polymorph will be predominated by $CaSO_4$dihydrate.

2.5 Experimental process and characterization

To investigate the influence of vibration on $CaSO_4$ scale formation, an experimental rig was employed. This build in house experimental rig schematically illustrated in Fig. 1. Vessel (1) contained solution of $CaCl_2.2H_2O$ and vessel (2) contained solution of Na_2SO_4. An electrical heater was employed in each vessel (3) to control solution temperature and agitated by a stirrer (4) which had set at 30 rpm to provide homogeneous solution. Solution in vessel (1) and (2) was pumped by dosing pump CHEM FEED Ca-92683 (6) similarly in flowrate 30.00 mL/min and was met in test pipe section (8). In this test pipe section solution of $CaCl_2.2H_2O$ and Na_2SO_4 was reacted and precipitated as $CaSO_4$ scale. Four pair of coupons made from copper with 12.5 mm in inner diameter and 30 mm long was inserted in to the test pipe section where scale was needed to form. The pipe was mounted at the table (7) and vibrated by electrical motor (9) which the frequency was set by computer program (12). Vibration meter Lutron VT-8204(10) sensed the vibration parameter such as displacement peak to peak (m), velocity (m/s) and acceleration(m/s2) in RMS.

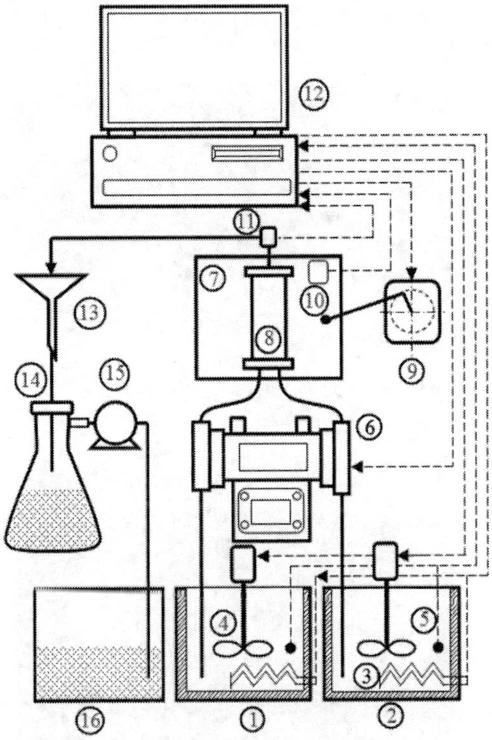

Fig. 2. Build inhouse experimental rig

Conductivity of the solution after left the coupon was measured by TDS meter (11). The data of conductivity and vibration was acquisitioned by computer program (12) to fullfill the accuracy. A strain paper 0.22 μm (13) was used to seizure the scale brought by the flow. A flask tube (14) and a vaccum pump (15) were used to help in sucking the water from the strainer and sent to vessel (16).

Scale of the coupon was heat in 60^0C as long as four hours to waste water molecule in the crystal. The weighing scale mass was done include coupon mass after dryer processing. Mass of scale was found by calculate the mass of coupon with scale deminished by the mass of coupon without scale. Furthermore, scale was labeled as the experiment i.e. frequency 0.00; 4.00 Hz and 8.00 Hz and kept in the dry cup.

Crystal polymorph was analyzed through scanning electron microscopy (SEM) to study the different shape as the presence of vibration. Crystal component was investigated through SEM EDX to show the alteration of crystal structure as the presence of vibration. Crystal phase was also investigated through X-Ray Diffractometer and then was confirmed to Powder Diffraction File to identify the crystal phase. Ion-molecule interaction was characterized through FTIR analysis.

3 Results and discussion

3.1 Induction time

Conductivity of the solution after left the coupon was measured by TDS meter and recorded by computer program every one minute during process of crystallization. Measurable change of conductivity showed that a number of ions in solution has been reacted and precipitated as the scale, so the amount of ions decrease and either conductivity of the solution. The time of first measurable change of conductivity called as an induction time [15] and shows the initial time of the nucleation as the start of crystallization. The induction time found were 31 ; 27; and 23 minute for frequency vibration of 0.00; 4.00 and 8.00 Hz of blank experiment.

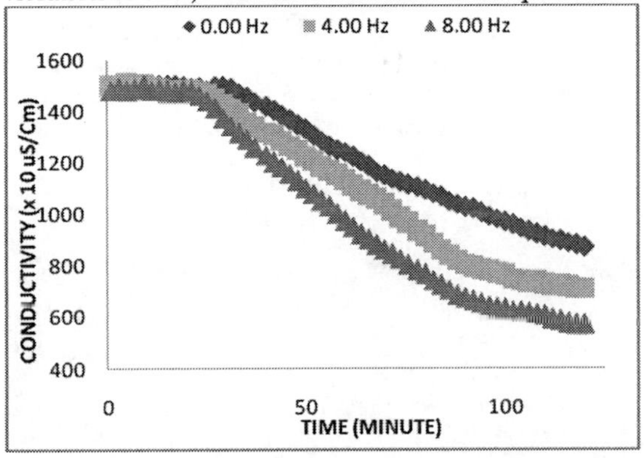

Fig. 3. Conductivity of the solution over the time

According to the data found, conductivity of solution at first measuring is 14,821 μS/Cm and after 31 minute TDS meter showed the change to be 14,076 μS/Cm for experiment without vibration. Thus, induction time was recorded at 31st minute as shown in plot of conductivity over time in Fig. 2 which shows the graph of induction time of the experiment when reaction process was affected by the vibration of 0.00; 4.00 and 8.00 Hz. The data prove that vibration affects $CaSO_4$ scale formation i.e. decrease nucleation time and shows that crystallization process was done more faster.

Unfortunately, no references found to explain the mechanism of vibration in affecting the nucleation time but it might be caused by the increase ionic collision which vibration of 8.00 Hz in this system would insert external force 57,600 times per hour or as much as

8,592kJ during four hours, for such experimental that has been done.

3.2 Deposition rate

The concept of deposition rate has been discussed in section two that scale formed was classified into deposition rate of precipitation (Wprecip), deposition rate in the strainer (Wstrain) and net deposition rate (Wnet) as the influence of vibration.

Table 2. Deposition rate in various vibration frequency

No	Vib (Hz)	W_{precip} (gr/hr)	W_{Strain} (gr/hr)	W_{net} (gr/hr)
1	0	0,1774	1,612	1,7894
2	4	0,3745	2,0146	2,3891
3	8	0,6657	2,7942	3,4599
4	4	0,2084	1,6379	1,8963
5	8	0,3306	1,9559	2,2865

Deposition rate of precipitation was measured in the coupon in respect to the time (hour) and named as grow rate of precipitation. Deposition rate in the strainer was measured in the strainer for every running. Net deposition rate (Wnet) as the presence of vibration was calculated as equation (3) for all vibration frequency. The result was listed in Table 2 and graphically depicted in Fig. 4. Number 4 and 5 of Table 2 shows the deposition of experiment with the presence citric acid in 10 ppm.

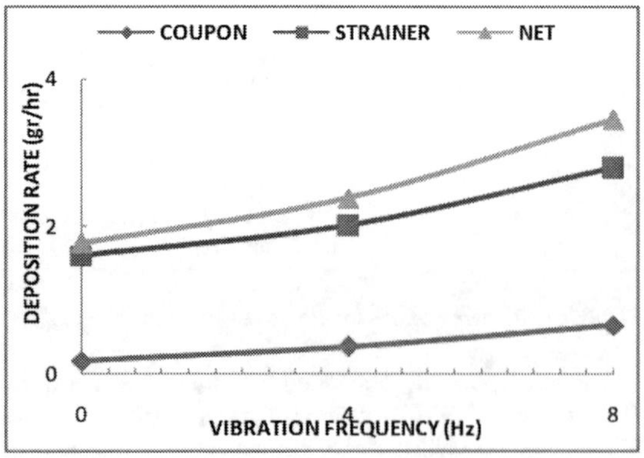

Fig. 4. Deposition rate of precipitation and strainer

Deposition rate increased significantly when the vibration affected to the system i.e. 0.5997 and 1.6705 gr/hr for vibration frequency 4.00 and 8.00 Hz. The addition 10.00 ppm of citric acid declined the deposition rate of 8 Hz experiment from 3.4599 gr/hr to 2.2865 gr/hr.

3.3 Crystal morphology

The phases of $CaSO_4$ crystal are anhydrate ($CaSO_4$), hemyhydrate ($CaSO_4.0.5H_2O$) and dihydrate ($CaSO_4.2H_2O$) [16]. All polymorph has no differences in shape so it phases could not be identified through the SEM. Even though, SEM could gives the information about the amount of needle and crystal size [17] as the influence of research parameter.

Crystal polymorph was investigated after it was characterized by Scanning Electron Microscopy FEI Inspect S50 to the scale found in the coupon of unvibrated-blank sample (Fig. 4a); 8.00 Hz vibrated-blank sample (Fig. 4b) and 8.00 Hz vibrated-antiscalant sample (Fig. 4 c) where all of them were in needle shape or orthorhombic [18]. The difference between those three crystals mainly emphasize at the size of needle which in experiment of frequency 8.00 Hz had larger needle compared to the other. This phenomenon could be caused by nucleation process of the experiment in which performed faster as the data found by induction time investigation.

Fig.5. Crystal morphology of: a). Unvibrated blank sample; b). Vibrated blank sample; c). Vibrated antiscalant sample

Fig. 5c) shows crystal polymorph of deposition of the addition citric acid in 10.00 ppm and vibrated at 8.00 Hz. The shape identified in irregular shape and fewer dimension as the evident that citric acid successfully destructed crystal lattice as promoted by Ersen [13].

3.4 Crystal structure

Composition of the crystal was characterized by SEM EDX FEI Inspect S50 to the specific area had been chosen and the result shown in Fig. 6 which the composition is given in % weight i.e. 50.91; 20.37 and 25.54 for O, S and Ca respectively. The peak shown in Fig. 6 prove that $CaSO_4$ crystal was pure. Its indicate that downstream process of this research such as washing, filtrating, drying and powder storaging was done very well [6]. This spectrum was then compared by the one had been published by Helalizadeh that the composition found not any differences [19].

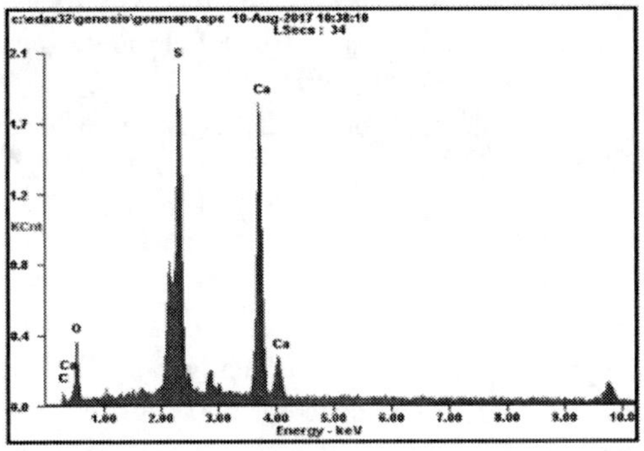

Fig. 6. Composition of $CaSO_4$ anhydrate crystal running in four hours and frequency 8.00 Hz

The differences between theoretical and actual value in % weight of crystal composition are 3.91; 2.63 and 3.6 for O, S and Ca respectively. The composition shows that the crystal was $CaSO_4$dihydrate. This phenomenon similar to the prediction by phreeqc program which has been predicted before that polymorph $CaSO_4$dihydratepredominated the scale formed with number saturation index 0.54 even $CaSO_4$ anhydrate was predicted only 0.24.

3.5 Crystal Phase

Crystal phase was analyzed through X-Ray Diffractometer which was operated in 40 kV, 30 mA and incident beam 100 < 2θ < 800, step size 0.02 degree (2θ). Start position at 10.0100 even end position at 79.9900, scan type continuous with goniometer radius 240 mm and measurement temperature at 25^{0}C.

The data found was then analyzed through Rietzfeld quantification method which operated under FullProf program to investigate crystal phase distribution. Three crystal of the experiment i.e. blank; vibrated in 8 Hz and citric acid addition-vibrated in 8 Hz experiment were characterized. AMCSD 0001807 was used as basic standard to analysis $CaSO_4$ $2H_2O$ crystal phase. All of peaks identified as $CaSO_4.2H_2O$ or recognized in popular name as gypsum.

The spectrum of XRD analysis is completed by miller indices of every peak to show the alteration each plane of every experiment (Fig. 7). Plane of (020); (121); (-141); (051); (042); (-233); (-204) and (-264) in Fig. 5 increases when the experiment was influenced by the vibration. The data shows not contrary to the data of deposition rate which vibrated experiment produced more scale and shows the agreement to the previous research which $CaCO_3$ was also increase when affected by vibration [20].

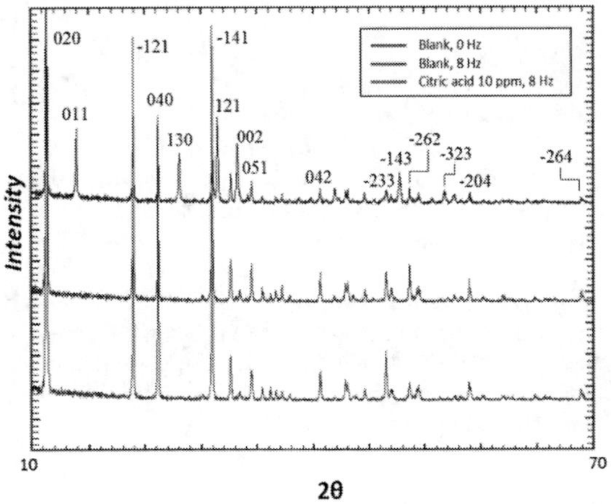

Fig. 7.XRD analysis of blank; vibrated and citric acid – vibrated experiment.

Planes of (011); (130); (002); (121) and (-323) did not arise anymore when the experiment was added by 10 ppm of citric acid and vibrated at 8 Hz mechanically. The loses planes shows the agreement to the deposition rates of the experiment which scale deposition of the addition citric acid decreased in significant value.

3.6 FTIR analysis

Fourier Transform Infra Red (FTIR) analysis (Fig. 6) was used to investigate ion interaction and scale formation among the reaction of unvibrated (a); vibrated reaction (b) and antiscalant reaction (c). Bands of 3,549 Cm^{-1} and 3,402 Cm^{-1} are assigned as O-H stretching [21]. Band of 1,681 $Cm^{-1}$and 1,620 Cm^{-1} assigned as stretching of water molecule in the crystal. Band of 671 and 574 are assigned as symmetric bending of sulfate ion [22]. The band 3,549 Cm^{-1}; 1,150 Cm^{-1} and 671 Cm^{-1} of blank experiment changed as the experiment was influenced by the vibration to be 3,543 Cm^{-1}; 1,148 Cm^{-1} and 665 Cm^{-1} respectively.
These alteration was also accompanied by the increase of scale deposition which unvibrated experiment produced larger scale than unvibrated. The phenomenon shows the agreement to Badge Rule that the lower band indicates the reaction takes place in easier [23].

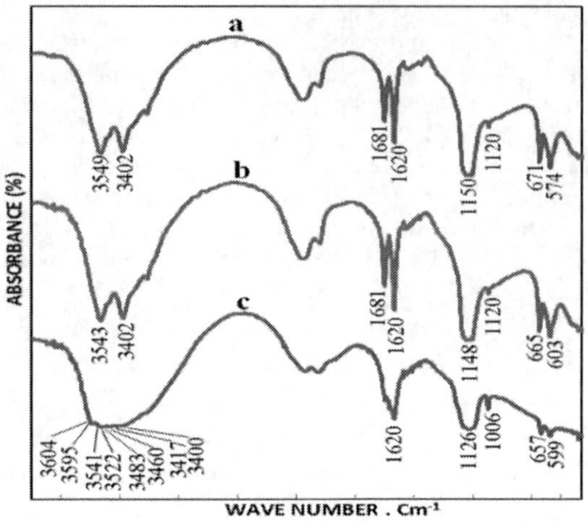

Fig. 8. FTIR analysis of $CaSO_4$ crystal. a). Blank-unvibrated experiment. b). Blank-vibrated experiment and c). Citric acid 10 ppm and vibrated experiment.

In the experiment with additive citric acid, more bands in the region 3,400 Cm-1 to 3,604 Cm^{-1} arised that assigned as O-H stretching. The arising peak indicated O-H stretching as the addition citric acid [23] which would be the reason of scale declination.

4 Conclusion

The influence of vibration on $CaSO_4$ scale formation especially in kinetical, morphology, microanalysis and deposition rate has been investigated systematically. Vibration has positive correlation to the kinetic and deposition rate of scale formation. Increasing vibration frequency would increase speed of reaction and increase mass of scale. Morphology crystal was needle shape or orthorhombic and crystal polymorph predominated by $CaSO_4$dihydrate .The addition citric acid in 10.00 ppm to the solution declined scale deposition and affected crystal polymorph in irregular shape.

5 Acknowledgement

Author would like to thank to laborant staff of Instrumentation and Metrology Laboratory, School of Vocation Diponegoro University, for their helps in device calibration.

Refferences

1. C.Ruiz, V.H.Jaramillo, D.Mba, J.R.Ottewil, Y.Cao, MechSyst Signal Pr, **66-67** (2015)
2. M.Euvrard, F.Membrey, C.Filiatre, A.Foissy, J Cryst Growth, **265** (2004)
3. A.Martinoda, MEuvrard, A.Foissy, A.Neville, Desalination, **220** (2008)
4. B.Bansal, X.D.Chen, H.M.Steinhagen, ChemEng Process, **47** (2008)
5. S.Muryanto, A.P.Bayuseno, Sutrisno, W. Manges-tiyono, W. Sediono, Ed ChemEng, **7** (2012)
6. T.A.Hoang, H.M.Ang, A.L.Rohl, Powder Technol, **179** (2007)
7. C.Fan, H.Teng, ChemGeol, **245** (2007)
8. M.Uchymiak, E.Lyster, J.Glater, Y.Cohen, J Membrane Sci, **314** (2008)
9. F.A.Setta, A.Neville, Desalination, **281** (2011)
10. R.Ketrane, B.Saidani, O.Gil, L.Leleyter, F.Baraud, Desalination, **249** (2009)
11. M.Jamialahmadi and H.M.Steinhagen, ChemEng Res Des, **85** (2007)
12. Z.Belarbi, J.Gambya, L. Makhloufi, B.Sotta, B.Tribollet, J Cryst Growth, **386** (2014)
13. A.Ersen, A.Smith, T.Chotard, J Mater Sci, **41** (2006)
14. R.Sheikoleslami, Heat Transfer Eng, **21**(3) (2000)
15. R.Isopescu, ChemEng Res Des, (2009)
16. M.G.Mwaba, G.Junjie, G.Mohammad, J Cryst Growth, **303** (2007)
17. G.O.Cakal, I.Eroglu, S.Ozkar, J Cryst Growth, **290** (2006)
18. A. Quddus, L.M. Al-Hadhrami, Desalination, **246** (2009)
19. A.Helalizadeh, H.M.Steinhagen, M.Jamialahmadi, ChemEngSci, **61** (2006)
20. W. Mangestiyono, S. Muryanto, J.Jamari, A.P. Bayuseno, Matec Web Conferences, **58** 01027 (2016)
21. L. Bin, L. S. Mei, L.J.Hua, Y.Mei, App Surf Sci**315** (2014)
22. M.A.Dabbas, M.Y.Eisal, W.H. Kadhim,Iraqi Journal of Science, **55**(2014)
23. Badger, Richard, J Chem Phys. **2**: 128(1934)

Utilization of Indonesia's Hot Spring Sources for Electricity using Kalina Cycle and Organic Rankine Cycle

Grano Prabumukti[1*], *Widodo* Wahyu Purwanto[1]

[1]Department of Chemical Engineering, Faculty of Engineering, Universitas Indonesia, Depok - Indonesia

Abstract. Indonesia posses 40% of the world's geothermal energy sources. The existence of hydrothermal sources is usually characterized by their surface manifestations such as hot springs, geysers and fumarole. Hot spring has a potential to be used as a heat source to generate electricity especially in a rural and isolated area. Hot springs can be converted into electricity by binary thermodynamic cycles such as Kalina cycle and ORC. The aim of this study is to obtain the best performances of cycle configuration and the potential power capacity. Simulation is conducted using UNISIM software with working fluid and its operating condition as the decision variables. The simulation result shows that R1234yf and propene with simple ORC as desired working fluid and cycle configuration. It reaches a maximum thermal efficiency up to 9.6% with a specific turbine inlet pressure. Higher temperature heat source will result a higher thermal efficiency.. Cycle thermal efficiency varies from 4.7% to 9.6% depends on source of hot spring temperature. Power capacity that can be generated using Indonesia's hot spring is ranged from 2 kWe to 61.2 kWe. The highest capacity located in Kawah Sirung and the least located in Kaendi.

1 Introduction

Indonesia is one of the world's richest geothermal reserves, with 40% of the world's geothermal potential. However, its current utilization is only 4.1% of its total potential. Geothermal systems in Indonesia are generally hydrothermal systems with temperatures greater than 225°C, only a few locations have temperatures between 125-225°C. The presence of hydrothermal sources in the sub-surface of the earth is usually characterized by surface manifestations such as hot springs, Geysers, fumaroles, mud pools, steaming ground and altered rock. [1]

Hot spring is a kind of heat manifestations which can be used as heat source to generate electricity. Hot spring in Indonesia usually has temperatures below 100°C. So far, geothermal manifestations in the form of hot spring in Indonesia, only used for use as a hot water bath. A hot spring source can be utilized to build a small-scale power plants by using binary cycles. [1]

Organic Rankine Cycle (ORC) power plant in commercial scale using the hot springs has been conducted in several countries. Holdmann was reporting ORC in Chena Hot Spring, Alaska, geothermal heat sources with temperatures 73°C, the working fluid R134a, water cooling fluid temperature 4.4°C, can produce a capacity of 210 kWe and an efficiency of 8.2% [2].

The working fluid used in binary generation systems, is a working fluid with a lower boiling point than water. The heat source with low temperature causes the thermodynamic efficiency of this cycle to be low. [3]

The purpose of this study is to obtain the maximum cycle efficiency for typical hot spring temperature range and its power capacity. The cycles assessed in this study are Kalina cycle with KCS11 type, simple ORC cycle and dual pressure ORC. The working fluids which is considered in this study are propane, propene, R1234yf, R407afor ORC and ammonia 85% for Kalina cycle. Heat source used is a heat source with a temperature of 60°C - 99°C. The temperature and flow rate of the hot spring use in the simulation are the average values [1].

2 Modeling the cycle

For geothermal systems with low temperatures, the thermal efficiency is ranged between 5% -12%. Even the energy efficiency is about 25% -50%. Some studies are done to improve the efficiency of binary cycle, depending on the working fluid or a combinations two or more working fluids is employed [4].

* Corresponding author: widodo@che.ui.ac.id

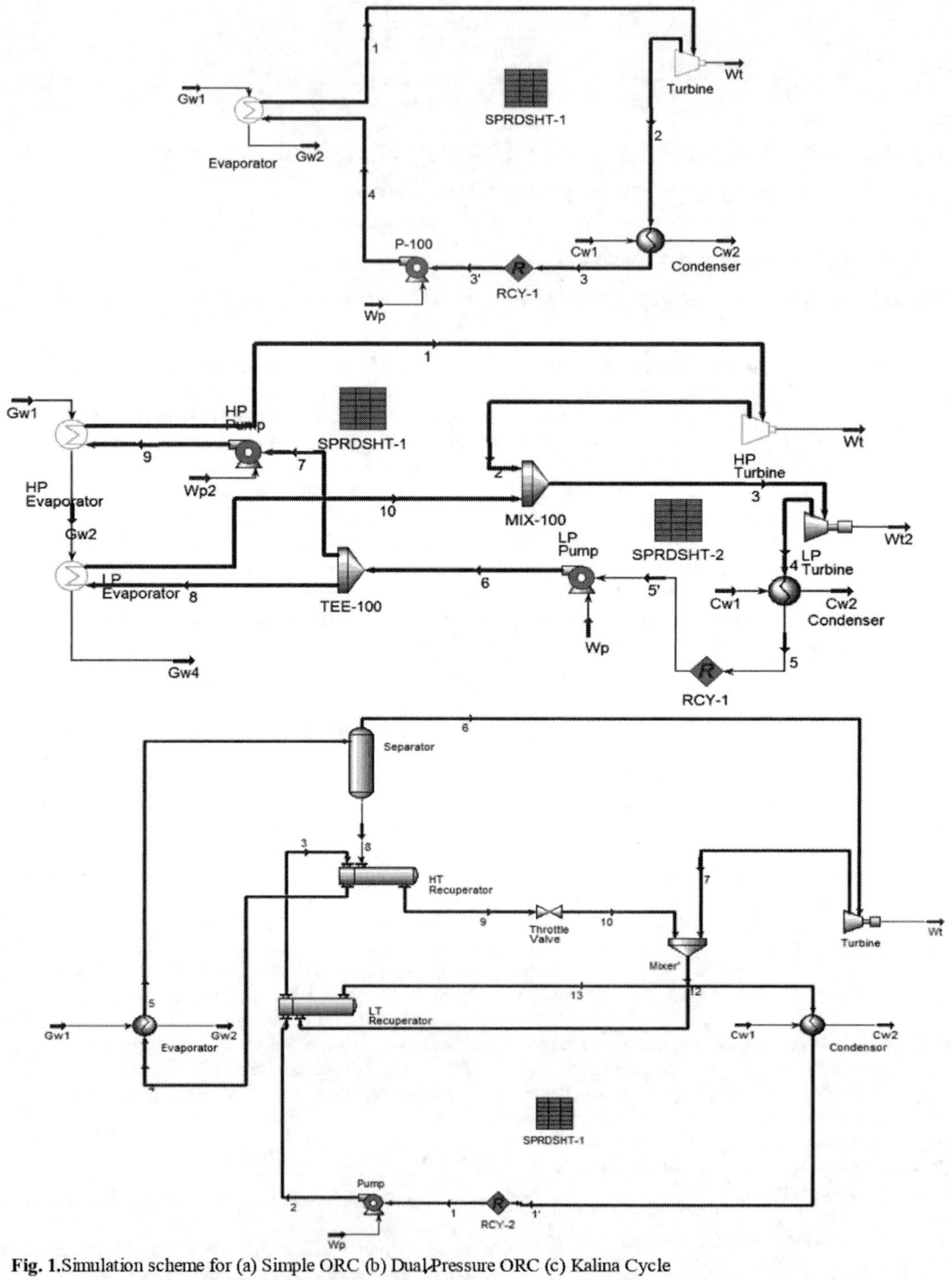

Fig. 1. Simulation scheme for (a) Simple ORC (b) Dual-Pressure ORC (c) Kalina Cycle

A simple binary cycle has a low thermal efficiency because of the low temperature difference between the heat source and the temperature at the heat sink. Carnot efficiency is the maximum possible efficiency of cycles that work in two temperature ranges [5].

In a thermodynamic cycle, cooling water is required for the condensation process in the condenser. Cooling water at a source is usually obtained from natural sources such as river or sea water. Normal average air temperature is usually around 25°C [5].

In a binary cycle, a heat exchanger is used to exchange heat between heat sources and working fluids. In addition, a heat exchanger is also required on the condenser to condense the working fluid that has passed through the turbine. To support the existence of irreversibility, a heat exchanger is usually having a certain temperature minimum approach. For low temperature heat sources, the temperature change in the heat exchanger is usually around 5°C. [6,7]

One type of thermodynamic cycle commonly used to generate electricity from heat sources with low temperatures is ORC.ORC uses the working fluid with low boiling point, it can convert the working fluid into its vapor phase at a low temperature of heat source. In ORC, there are four main components: Steam Turbine, Pump, Evaporator and Condensor. The main difference between a Rankine cycle and a ORC cycle is in the Rankine cycle use water as the working fluid to produce steam. In ORC, the system use organic working fluid such as a refrigerant or hydrocarbon ORC system use an evaporator as heat absorption equipment so that the cycle does not require burning process, hence it does not produce air pollution. [3]

Figure 1 shows simulation scheme for each cycle. Table 1 shows the assumptions which is used to simplified the modelling of each cycle. The pump efficiency, turbine efficiency and generator efficiency value which is used is a typical efficiency value. The assumptions for cooling water is due to limitation of real world condition. Certain pinch temperature is also used to keep the second law efficiency at certain level.

Table 1. Assumption used in the simulation

Variable	Simple ORC & Dual- Pressure ORC	Kalina Cycle
Tcooling water (°C)	27	27
Pump Efficiency (%)	80	80
Turbine Eficiency (%)	85	85
ΔT Pinch Evaporator (°C)	5	5
Generator Efficiency (%)	96	96
Working fluid	propane, propene, R1234yf, R407a	ammonia 85%

2.1Simple ORC

In simple ORC, the pressure of the working fluid as it exits the condenser reaches the evaporator pressure with the help of the pump. The working fluid gets heat from the geothermal fluid in the evaporator. After passing the evaporator, the working fluid expands in the turbine to the condenser pressure and is finally cooled by cooling water. The scheme of a Simple ORC can be seen in Figure 1a. A simple ORC is relatively simple and requires less equipment than other cycles. However, this cycle has a great thermodynamic irreversibility due to the large temperature difference in evaporators and condensers. [8]

Table 2 shows energy balance equations of each component of Simple ORC.

Table 2. Equation models of Simple ORC

Component	Energy Balance Equation
Turbine	$W_t=m_1(h_1-h_2)$
Pump	$W_p=m_3\cdot(h_3-h_4)$
Condenser	$m_2(h_2-h_3)=m_{cw1}(h_{cw1}-h_{cw2})$
Evaporator	$m_4(h_4-h_1)=m_{gw1}(h_{gw1}-h_{gw2})$
Cycle Efficiency	
$\eta=(W_t-W_p)/m_{gw1}(h_{gw1}-h_{gw2})$	

2.2 Dual pressure ORC

The process in a dual pressure ORC system is the same as in the simple ORC. The difference is the overall mass flow rate in the case is divided into two different evaporation pressure level. In series configuration of a dual pressure ORC system this separation occurs after Low Pressure (LP) Pump. Furthermore, the LP flow is evaporated on the low pressure (LP) evaporator while the remaining mass flow rate is pumped to high pressure (HP) pump. The working fluid which passes through the HP Pump is evaporated on high pressure (HP) Evaporator. High pressure working fluid is expanded in HP Turbine and mixes with low pressure fluid output of LP evaporator. Both fluids mix and expand in LP turbine.

The process scheme for dual pressure ORC can be seen in Figure 1c. In this system no recuperator or pre-heater is used for the cycle. The energy balance for a dual pressure ORC can be seen in the Table 3.

Table 3. Equation models of Dual Pressure ORC

Component	Energy Balance Equation
HP Turbine	$W_t=m_1(h_1-h_2)$
LP Turbine	$W_{t2}=m_3(h_3-h_4)$
HP Pump	$W_{p2}=m_7(h_7-h_9)$
LP Pump	$W_p=m_5\cdot(h_5-h_6)$
Condenser	$m_4(h_4-h_5)=m_{cw1}(h_{cw1}-h_{cw2})$
HP Evaporator	$m_9(h_9-h_1)=m_{gw1}(h_{gw1}-h_{gw2})$
LP Evaporator	$m_8(h_8-h_{10})=m_{gw1}(h_{gw1}-h_{gw2})$
Mixer	$m_2h_2+m_{10}h_{10}=m_3h_3$
Splitter	$m_6h_6=m_8h_8+m_7h_7$
Cycle Efficiency	
$\eta=(W_t-W_p)/m_{gw1}(h_{gw1}-h_{gw2})$	

2.3 Kalina cycle

For Kalina cycle, the ammonia-water mixture enters the separator after obtaining heat coming from geothermal fluid in the preheater and evaporator. In the separator, the mixture with the saturated vapor enters the turbine and expands to the condenser pressure. Meanwhile, the fluid mixture with the saturated liquid enters High Temperature Recuperator (HTR) to release heat to the

Low Temperature Recuperator (LTR) fluid. After that, the fluid mixture that has released heat in the HTR, enters the expansion valve and mixes with the turbine output fluid. The mixed fluid is exploited by heat in the LTR to heat the condenser output fluid.

Separator in a Kalina cycle is used to separate the vapor and liquid phase of working fluid that will enter the turbine. The separator output with the liquid phase will have a lower ammonia concentration than the separator output with the vapor phase. The output in the liquid phase will flow into the recuperator to absorb the heat. After passing the recuperator, the liquid phase fluid will enter the expansion valve to lower the pressure before mixed with the working fluid that has passed through the turbine. After mixing, the two working fluids will enter the condenser to produce a liquid phase.

The efficiency of the Kalina cycle is 2.17% points lower than the ORC baseline, 1.28% points of the ORC dual-fluid but 0.11% points higher than the dual-pressure ORC. [10]

In general, with the use of mixed ammonia and water working fluids, the Kalina cycle has an unstable evaporation and condensation temperature. This causes the Carnot efficiency of the Kalina cycle to be better than the ORC cycle. Table 4 shows the energy balance for each equipment in the Kalina cycle.

Table 4. Equation models of Kalina Cycle

Component	Energy Balance Equation
Turbine	$W_t = m_8(h_8 - h_9)$
Pump	$W_p = m_1(h_2 - h_1)$
Condenser	$m_5(h_{10} - h_1) = m_{cw1}(h_{cw1} - h_{cw2})$
Evaporator	$m_3(h_4 - h_3) = m_{gw1}(h_{gw1} - h_{gw2})$
Recuperator	$m_5(h_6 - h_5) = m_{11}(h_{11} - h_{12})$
Separator	$m_7 h_7 = m_8 h_8 + m_{11} h_{11}$
Mixer	$m_9 h_9 + m_{13} h_{13} = m_{10} h_{10}$
Throttling Valve	$m_{12} h_{12} = m_{13} h_{13}$
Cycle Efficiency	
$\eta = (W_t - W_p) / m_{gw1}(h_{gw1} - h_{gw2})$	

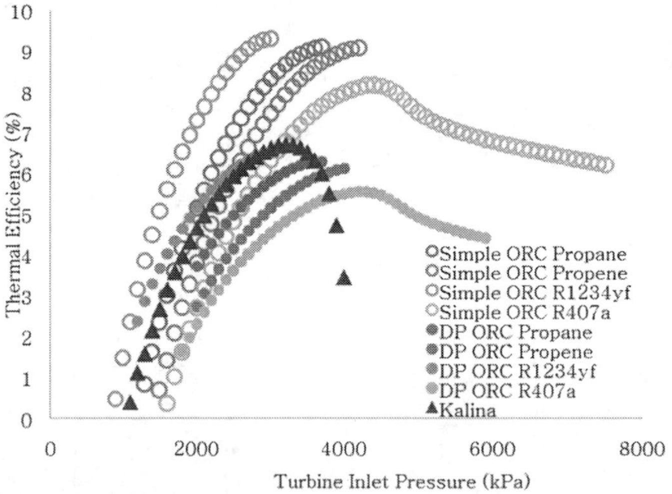

Fig.2. Cycle Efficiency for different Turbine Inlet Pressure

3 Result and discussion

3.1 Turbine Inlet Pressure Variations

The base case for the simulation is a heat source temperature value of 95°C with 50kg/s mass flow rate. The minimum approach used in the evaporator is 5°C, so the turbine inlet temperature value is 90°C. Figure 2 shows the influence of turbine input variation on thermal efficiency of the cycle. In general, an increase in turbine inlet pressure will result in increased efficiency. This is because of the greater difference between the enthalpy value of turbine inlet conditions and turbine output. This condition results in a larger W_{net} value, so the thermal efficiency value will rise. However, at some point, the efficiency will decrease as the pressure continues to increase. This is because the increase of enthalpy difference value between turbine inlet and outlet condition is not significant.

For each heat source temperature, an optimum value of turbine inlet pressure is obtained where the highest efficiency value is reached. The value of this optimum pressure depends on the saturation pressure value of each working fluid at the evaporator outlet temperature. Working fluids with a low saturation pressure will have an optimum pressure value at low pressure. In contrast, working fluid with a higher saturation pressure will have higher optimum pressure value.

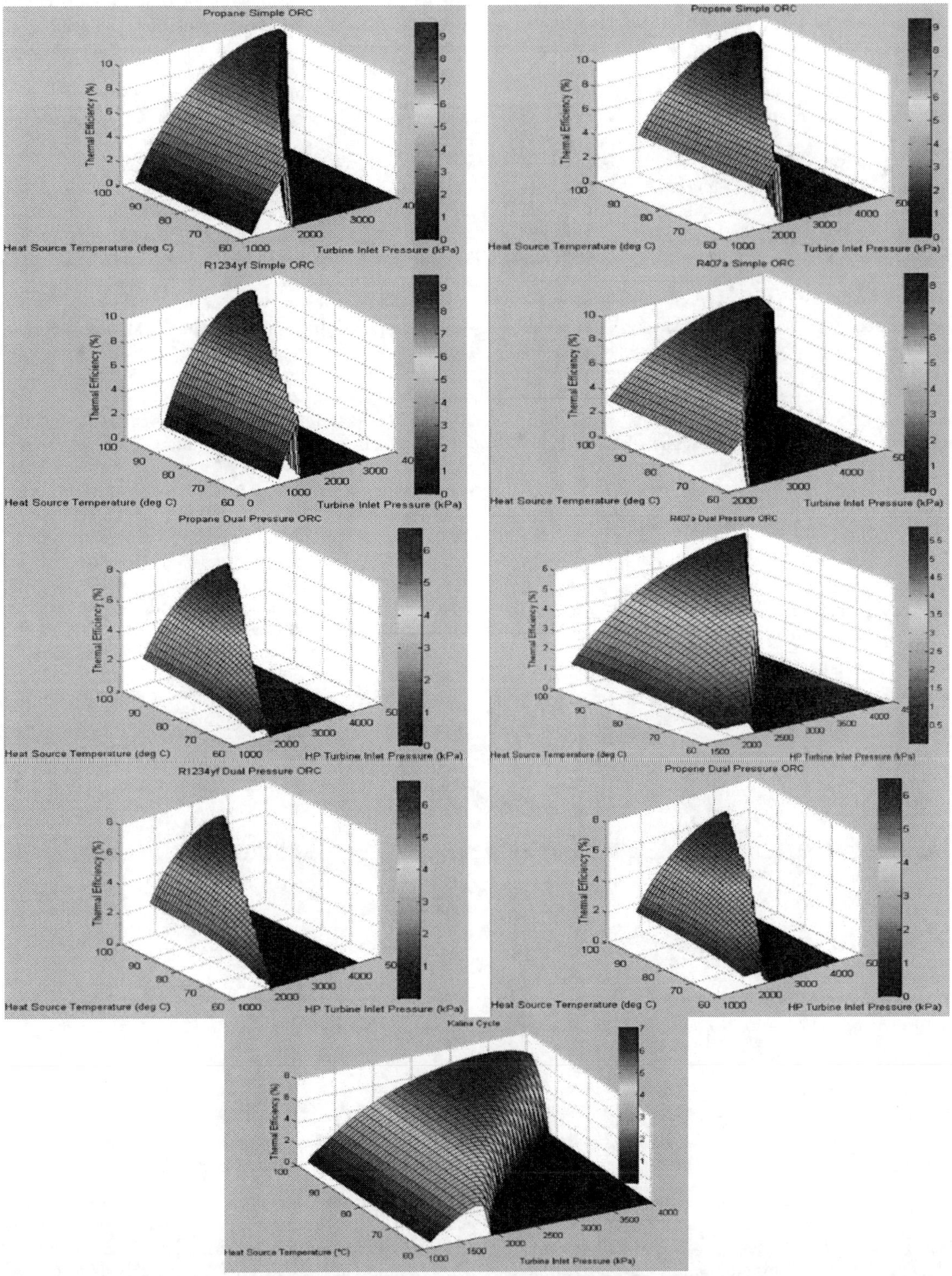

Fig.3. Cycle Efficiency for different Turbine Inlet Pressureand Heat Source Temperature

For heat source temperature of 95°C, the highest efficiency value is obtained on ORC simple configuration followed by Kalina and Dual-Pressure ORC. For ORC, the working fluid with the best efficiency value is R1234yf, followed by propane, propene and R407a.

For propane, propene and R1234yf working fluids, there is no decrease in efficiency when turbine inlet pressure continue to increase. The increase will be limited by the vapor fraction of the working fluid after passing through the evaporator. For case of propane, propene and R1234yf, the optimum pressure is reached when the fluid vapor fraction value is 1. As for the Kalina cycle and the working fluid of R407a with both ORC configuration, there is an optimum pressure when the efficiency value reaches the peak which is 4400 kPa for R407a Simple ORC, 4200 kPa for Dual Pressure ORC and 3100 kPa for Kalina Cyle.

Cycles with ORC simple configurations have greater efficiency values than Kalina cycle or dual-pressure ORC. This is because in the simple ORC, the fluid expansion occurs in a high pressure. High pressure expansion causes arise in power generated. High pressure expansion, in general, results in greater maximum cycle efficiency due to higher delta enthalpy between the inlet and outlet turbine condition.

3.2 Heat Source Temperature Variations

Figure 3 shows a three-dimensional graph which illustrates the effect of the temperature of the heat source and the turbine inlet pressure on the thermal efficiency of each working fluid of each cycle. Variations are performed to produce the maximum efficiency.

Each temperature level of the heat source will produce a maximum efficiency at a certain pressure. This pressure, in each configuration, in general, will be worth the highest pressure that can be achieved at a temperature level with a pinch temperature value in the evaporator of 5°C. With a set of 5°C pinch temperatures, the evaporator output temperature will be less 5°C from the temperature of the heat source for each heat source temperature level. Saturated pressure at the evaporator exit temperature of each working fluid will affect the maximum pressure that the working fluid can achieve in order to become vapor phase. The value of this maximum pressure will affect the efficiency, the lower the saturation pressure, the higher the efficiency can be achieved from the working fluid. Efficiency value of the cycle will increase as the temperature of the heat source increases. However, the increase in efficiency that is influenced by the rise of heat source temperature is not significant compared to the increase of efficiency due to the increase of pressure. This is because arise in the temperature of the working fluid will only result in a shrinking dome distance between the saturated liquid point and the saturated vapor in the T-s diagram, so the energy required to evaporate the fluid will be less. However, it is also followed by more heat coming into the system, thus making the efficiency of the cycle not rising significantly.

The temperature of a hot spring ranged from 60°C to99°C. Each hot spring temperature level with pinch temperature value of 5°C causes turbine inlet temperature difference. In general, turbine inlet temperatures will be 5°C lower than the temperature of the heat source due to pinch temperature value constraints. As shown in Figure 4,each temperature level of heat source will reach certain maximum efficiency for each configuration and each working fluid. The differences in maximum efficiency that can be achieved for each temperature level is due to difference in saturation pressure for each working at turbine inlet temperature. However, there are some performance ranking differences for certain level of heat source temperature. The simple ORC has the best efficiency value for each temperature range. The kalian cycle has

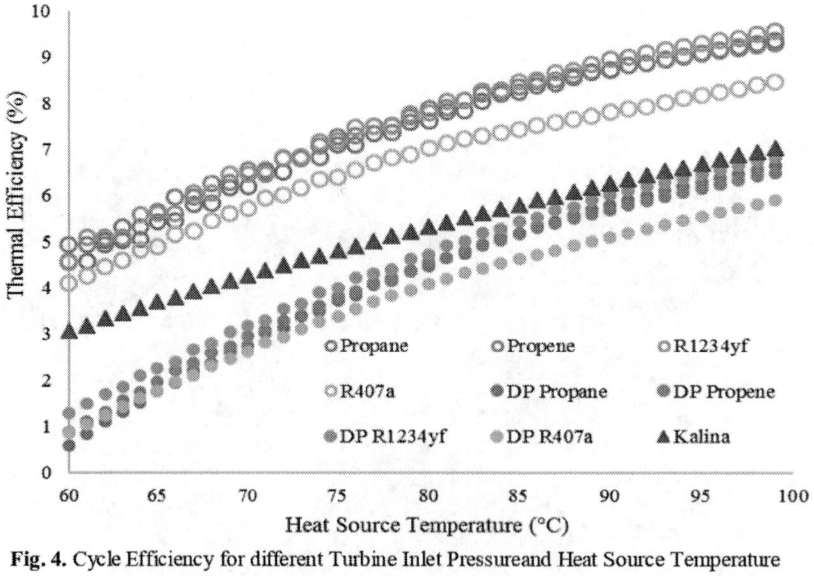

Fig. 4. Cycle Efficiency for different Turbine Inlet Pressureand Heat Source Temperature

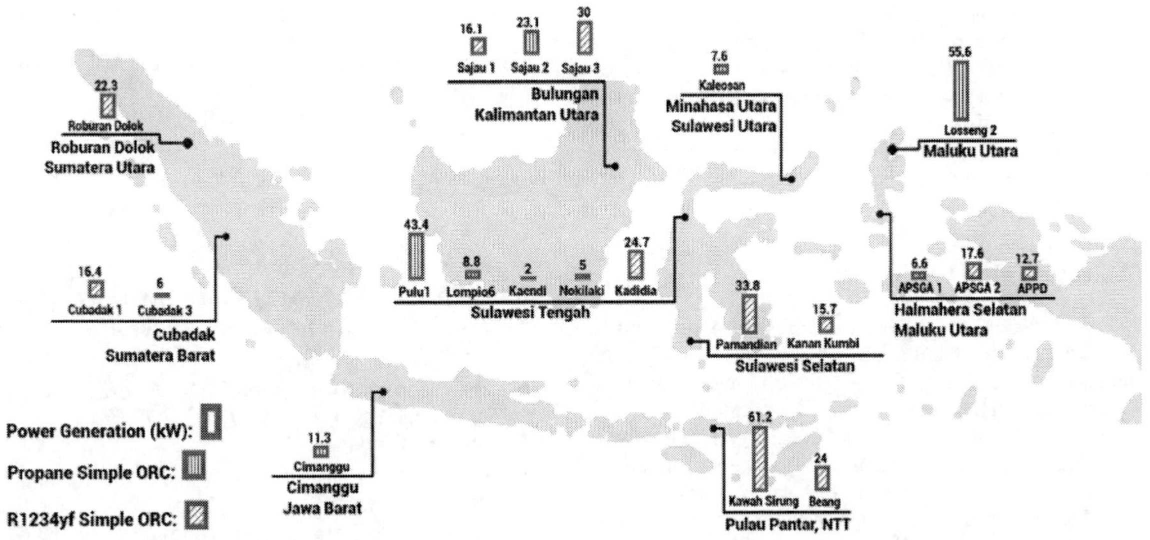

Fig. 5. Power Generation Potential for Hot Spring in Indonesia?

thermal efficiency under simple ORC, followed by a dual-pressure ORC configuration. However, the working fluid with the greatest efficiency for each temperature range is not fixed. In general, R1234yf has a greater efficiency than other working fluids. The propene produces the second highest efficiency, the third-order Propane, and R407a in fourth. But, at the temperature range of 91°C-99°C propane efficiency is better than propene.

Working fluid R1234yf with Simple ORC configuration can produce better efficiency than configuration and other working fluid, especially in temperature range 83°C to 99°C. Depends on the heat source temperature level, in general, the best working fluids are R1234yf and propene. Detail performance on each heat source temperature level can be seen in Figure 4.

3.3 Potential Power Capacity of Indonesia's Hot Spring Sources

In Indonesia there are various locations of hot spring characteristic varies depends on location, temperature and flow rate or heat duty. Table 5 shows hot spring location, temperature, mass flow rate and heat data. '

The majority of hot spring sources which is potential to be converted into electricity are located in Sulawesi and Maluku. This is because Sulawesi and Maluku region lies in the ring of fire which means many tectonic plates and volcanoes. The existence of a volcano contains a source of heat under the earth surface. The heat source beneath the surface of the earth brings about a geothermal reservoir, which results in the many geothermal manifestations that appear on the surface. Heat duty of hot spring sources ranged from 69 kWt to 1,030 kWt.

Table 5. Indonesia's Hot Spring Data and Electricity Generation Potential

Location	Temp (°C)	Flow (kg/s)	Heat Duty (kWt)	Electricity (kWe)
Robutran Dolok	91,0	1,3	320,3	22.3
Cubadak 1	74,8	2,0	399,0	16.4
Cubadak 3	72,7	1,0	190,7	6
Lompio 6	77,0	1,0	197,4	8.8
Sajau 1	74,5	2,0	392,3	16.1
Sajau 2	80,3	2,0	441,0	23.1
Sajau 3	85,8	2,0	487,2	30
Nokilalki	91,0	0,3	69,3	5
Kadidia	81,6	2,0	458,6	24.7
Kaendi	73,0	0,5	98,7	2
Kaleosan	80,0	0,7	161,7	7.6
Losseng 2	77,4	5,5	1032,6	55.6
Kawah Sirung	98,9	3,0	849,1	61.2
Beang	81,1	2,0	408,0	24
APSGA 1	99,0	0,5	145,7	6.6
APSGA 2	99,0	1,0	296,5	17.6
APPD	88,0	0,8	186,2	12.7
Cimanggu	78,0	1,2	308,7	11.3
Pulu 1	78,9	4,0	905,5	43.4
Pamandian	74,4	4,0	816,5	33.8
Kanan Kumbi	62,2	4,0	625,0	15.7

The efficiency value of the power generator at heat source is the function of the temperature level and the technology used. Figure 5 shows mapping of hot spring electricity production potential in Indonesia for each hot spring location with best performing technology. To estimate electricity generated in each location, the best performance cycle for each hot spring temperature is used. Using Figure 4, the cycle and the generated power can be estimated. Therefore, cycle efficiency in each hot spring location is different.

The power generated at hot spring range from 2 kWe to 61.3 kWe. The resulting power depends on temperature and heat. The largest power value in Kawah Sirung, East Nusa Tenggara. At the location of Kawah Sirung, a power generated of 61.3 kWe. The least power generated located in Kaendi, Central Sulawesi with only 2.1 kWe

4 Conclusion

An increase in turbine inlet pressure will result in an increase in the thermal efficiency value. However, the increase in efficiency will reach the optimum pressure at a certain point. After reaching the optimum value at a certain turbine inlet pressure, thermal efficiency will decrease. An increase in the temperature of the heat source results in increase in thermal efficiency value. The Increase in efficiency due to increased turbine inlet pressure is more significant than the increase in heat source temperature.

Hot spring temperature is ranged from 60°C to 99°C. Each temperature level has its own maximum efficiency. The thermal efficiency is ranged between 4.7% to 9.6%. Higher temperature will have higher thermal efficiency and vice versa. Each temperature level has its own optimum turbine inlet pressure.

Generally, R1234yf and propene with ORC simple configuration has the best performance compared to other working fluid and configurations. However, there are some differences in specific temperature range. No cycle with a specific working fluid always gives the best performance for each temperature level. Indonesia's hot spring sources have potential power capacity ranged between 2.1 kWe to 61.3 kWe.

Acknowledgement

This study is supported by Hibah PITTA DRPM Universitas Indonesia Contract No:849/UN2.R3.1/HKP.05.00/2017

References

1. G. Pikra, N. Rohmaha, R.I. Pramanaa, A.J. Purwanto, Energy Proced.**68**, 12 – 21 (2015)
2. G. Holdmann, Final Report: *400kW Geothermal Power Plant at Chena Hot Springs, Alaska* (Chena Power, LLC, 2007)
3. R. Di Pippo, *Geothermal Power Plants: Principles, Applications, Case Studies and Environmental Impact* (Elsevier, 2012)
4. C. E. C. Rodríguez, J. C. E. Palacio, O. J. Venturini, E. E. S. Lora, V. M. Cobas, D. M. dos Santos, F. R. L. Dotto, V. Gialluca,Appl. Therm. Eng.**52**, 109-119 (2013)
5. R. Di Pippo, Geothermics **33**, 565–586 (2004)
6. Z. Shengjun, W. Huaixin, G. Tao, Applied. Energy **88**, 2740–2754 (2011)
7. Walraven, D., Laenen, B., D'haeseleer, W. Energy Conv& Mgmt. **66**, 220–233 (2013)
8. N, Oscar, F. Cideos. *Power Production Using Low-Temperature Heat Sources In El Salvador* (Geothermal Training Programme, Iceland: 2012)
9. Y.Z. Wang, J. Zhao, Y. Wang, Q.S. An, Appl. Therm. Eng., (to be published, 2016)
10. N. Shokati, F. Ranjbar, M. Yari. Renew. Energy **83**, 527-542 (2015)

Managing Campus Energy: Compromising between Rapid Needs and Environmental Requirement

Ambariyanto Ambariyanto[1,2,*], *Yos* J. Utama[3],and *Purwanto*[4],

[1]Faculty of Fisheries and Marine Science, Diponegoro University, 50275 Jl. Prof.Soedarto SH, Tembalang Campus, Semarang, Indonesia
[2]Coastal Disaster Rehabilitation and Mitigation Center (COREM), Integrated Laboratory, Diponegoro University, 50275 Jl. Prof. H. Soedarto, SH, Tembalang, Semarang, Indonesia
[3]Faculty of Law,Diponegoro University, 50275 Jl. Prof.Soedarto SH, Tembalang Campus, Semarang, Indonesia
[4]School of Postgraduate Studies, Diponegoro University, 50275 Jl. Prof.Soedarto SH, Tembalang Campus, Semarang, Indonesia

Abstract. The utilization of energy, especially electricity at Diponegoro University campus continues to increase in line with the development of the university. This increase has a direct impact on the increased costs to be paid by the university. Some of the causes of increased utilization of electrical energy is the construction of new buildings to meet the needs, increased learning activities and education, research activities in the laboratory, and various other activities. On the other hand, the increase of energy utilization is considered not good from the environment point of view, especially the utilization of electrical energy coming from non sustainable resources. Efforts to compromise on both are to develop policies in developing environmentally friendly buildings, efficiency in utilization of electrical energy, and development of sustainable energy sources.

1 Introduction

Increased energy demand is now become an international issue [1,2,3]. The utilization of energy especially electricity continues to increase sharply in various sectors, both government, private and public (urban) sector, and in line with economic growth [4,5]. Increased utilization of energy is in line with the needs of the development of modern society around the world. Various fields and activities by all sectors, continue to move up, both in order to improve the ability, capacity and achievements and targets, including within the university campus, all require energy.

The development of campus around the world both in terms of technology and educational process to become a modern campus, also requires electrical energy continues to increase. This can not be avoided because of high energy demands of the modern campus development [6]. But on the other hand, the campus is also one place to be able to develop technology for new power plants through their research and development activities.

The issue of increasing the utilization of electrical energy is considered contrary to environmental principles, especially in relation to emission [7]. This is particularly aimed at electrical energy sourced from non-sustainable natural resources, such as oil resources, coal, and so on [8]. So that the increased use of electricity as it is considered to also encourage environmental damage.

A further impact is the emergence of renewable energy development issues that are growing all over the world. Various efforts continue to be done to produce sustainable energy. Increased utilization of science and technology also requires high electrical energy, but on the other hand with the advancement of science and technology are also able to efficiently utilize the energy, as well as generate new generating technologies. Currently, many power plants are derived from sustainable natural resources [9], among others, by utilizing wind, sunlight, currents or waves, etc.

This condition is also faced by various universities in Indonesia, including Diponegoro University (Undip) in Semarang. Increased utilization of electrical energy continues to increase in line with the policy of Undip to accelerate in achieving the vision of becoming a research university [10]. This condition has a direct impact with the increased cost of electricity that must be provided by the university. In addition, university management is also fully aware if this increase in the utilization of electrical energy is in conflict with environmental principles and sustainable university [11]. This brief paper will convey the compromise policy undertaken by Undip management in addressing the problem of increasing the utilization of electrical energy and efforts to become a sustainable environmental campus.

*Corresponding author: ambariyanto.undip@gmail.com

2 Causes of increased utilization of electrical energy

Increased utilization of electricity at Diponegoro University is mainly due to the acceleration programs to achieve the Undip vision. Educational and teaching activities, and research are the main causes, in addition to the construction of new buildings and rooms. Besides that, various other extracurricular activities run by faculty with students also influence the increase of electric energy usage. In more detail, the causes of increased utilization of electrical energy are as follows:

2.1. Development of new buildings

Construction of new buildings and rooms continues to be done at Diponegoro University because of the need due to the establishment of new units and because the old building is no longer feasible for certain faculties. Besides the effort to develop an ideal class, where the number of students is not too large in one class, also need new rooms. Some new buildings to meet these needs include for a newly formed vocational school, as well as new study programs, and construction of a new building for Psychology and other faculties.

By increasing the economics capabilities of lecturers, and employees, many of them use cars to work. This is worsen by the number of students who also use the car to campus. The problem is that some of them park their cars on the street because the parking space is not enough. This causes the disruption of traffic within the campus. To overcome the problem Undip provides a new parking building because the old parking lot is not able to accommodate the increasing number of cars.

Dormitory student demand continues to increase, especially by new students who come from outside the city of Semarang. To fulfill this high demand, nowadays Undip is building a new student dormitory that will accommodate many students. One of the newly completed buildings is the ICT II building which will serve as a training place as well as other commercial activities.

The existence of these new buildings requires electrical energy both for the construction process and for the operation of the building, for example for lighting, air conditioning, and for other needs. This causes an increase in the need for electrical energy at Diponegoro University.

2.2. Learning and education activities

One indicator of the success of a higher education is determined by the quality and competence of its graduates, especially after acting in the community and workplace. All universities continue to work to improve the quality of graduates in line with the development of science and technology.

Improving the quality and competence of graduates also continues to be done Diponegoro University through various programs, such as training activities, tutorials, discussion forums and so forth. These activities are also linked to efforts to accelerate the achievement of Undip's vision into a research university.

Education and learning activities are also increasing in line with the newly established study programs. For example, the opening of Pharmacy and Dentistry courses, as well as other programs. In addition, with the development of science and technology, encourage the process of education and learning on existing courses also experienced an increase in both quantity and quality.

2.3 Research activities

Research is one of the main activities of the university that has a vision to become a research university. The success of the research activities is not only determined by adequate human resources, but also the availability of laboratories and various other equipment and facilities [12]. Important policies ofDiponegoro University is to provide facilities and infrastructure to accelerate research activities for students and lecturers.

The year 2018 is the "laboratory year" in Undip. A special program is undertaken to address the problem of adequate laboratory availability for research that not only generates articles for international journals but is also expected to produce innovation products ready for commercialization. In turn these commercial products can support the university's finances.

By increasing the availability of laboratories, the research activities in Undip also continue to increase both in quantity and quality. It is also supported by the policy of allowing students and lecturers to conduct research in the laboratory for 24 hours 7 days. So with the increasing amount of equipment in the laboratory and research activities led to an increase in electrical energy demand [13, 14].

2.4 Other activities

There are many other activities conducted at Diponegoro University in addition to education, research and community service which is the core activities of universities. Various activities are undertaken by the entire academic community of Undip, both students, lecturers, and other employees. These activities are in some ways related to educational processes or skills and capacity improvements, such as summer courses, seminars, trainings, or discussions. In addition, there are also other activities related to sports such as futsal, tennis, badminton, basketball and so on. Other activities related to artistic skills such as classical and modern dance, choir, band, and so on.

Other activities are also carried out by student activity units both at the university level and at the faculty and department level. These activities of course require electrical energy both for stadium lighting, lighting and air conditioning, and related toequipments usage.

3 Compromise policies

Increased utilization of electric energy in Undip basically can be understood, because it is really needed, both for educational and research and other activities. On the other hand, Undip intends to develop sustainable university especially related to the application of environmentally friendly principles. Therefore, a policy compromise is required that minimizes the utilization of electricity at the same time encourages increased activity to achieve the university's vision. Some of the compromise policies are as follows:

3.1 Environmental friendly buildings

The construction of the building at Diponegoro University must meet environmental requirements and principles. Besides expected to have an artistic and functional design, the building must be built in accordance with the principle of environment, with one of the goals for the electricity needs to be minimal. There is a need an environmental assessment tools fof sustainable construction [15]. For example, the design of the building should be able to utilize sunlight for natural lighting in the building room, so it will not require lights during the day [16, 17]. The use of translucent walls is encouraged in the construction of buildings to exploit and use the availability of sunlight outside the building.

On the other hand, given that Semarang which is in tropical climate there is a tendency of high temperatures in a closed room [18]. The high temperature in the room will affect the performance of students and lecturers. In general, this is overcome by providing an air conditioner to increase comfort [19]. Therefore, the design of the building should be able to provide arrangements regarding airflow in the form of ventilation so that the room will be relatively cool, and the use of air conditioning is also getting minimal.

3.2 Efficiency energy utilization

In general, people awareness in conserving electricity, somewhat is still need to be improved [20]. In the case of our university, the use of electricity in the campus of Undip during this time is "less regulated" so freely utilized by all units. As a result there are things that cause excessive use of electricity, such as lighting, air conditioning and so on, which is not limited. Many found the room air conditioner was not turned off at night even though the office or classroom was closed and nobody was working in the room. But for special rooms where there are facilities or research equipment, or there is ongoing research that requires air conditioning then it is understandable and allowed. Our awareness to turn off the air conditioner when the room is not in used is still very low.

In comparison, our awareness on the use of lighting in rooms is much better. Generally the room lights are turned off when the office is closed at night. However, since this has not yet taken place in all parts of the campus, there is still room with lights on at night even though no one works in it. So the policy is to encourage all people, students, lecturers and other staff to turn off the air conditioner and lighting the room if it is no longer needed. In addition, the duty officers are also given the responsibility to do so, when at night they perform the activities of unit or campus security monitoring. There is currently a study on the development of automation to turn off air conditioning and lighting the room if it is not needed. There is a need to establish energy consumption information system to improve the awareness of reducing energy usage [21,22]. Through these policies, it is expected that efficiency of electricity utilization will be applied throughout the campus area of Undip.

3.3 Development of sustainable energy sources

Policies on the sustainable use of energy sources began to be implemented at Diponegoro University. Currently electrical energy is still obtained from the office of PLN (State Electricity Company) owned by the government, and costs continue to increase in line with increased utilization of electricity. Another policy taken today is the more sustainable use of electricity.

Currently Undip is working with the Japanese to develop electricity with three alternative sources i.e. solar power, wind power and micro hydro [23, 24]. These three alternatives are under studied at their efficiency level if implemented on the Undip campus. One consideration is whether the use of this power source will be able to reduce the cost burden that has been paid to PLN. Even though until now has not been decided, but Undip policy is to increase the use of electricity more sustainable and more environmentally friendly.

4 Conclusion

The rapid increase of electricity utilization on the Diponegoro University campus is understandable, given the good development of Undip to become a research university. However, considering that the use of this energy is considered not in line with environmental policy, it is necessary to compromise the efficiency of electricity usage policies but also encourage various activities aimed at accelerating the achievement of Undip vision to become a research university.

References

1. Asif, M. and Muneer, T. Renewable and Sustainable Energy Reviews, **11**.7: 1388-1413,(2007).
2. Pérez-lombard, L., Ortiz, J. and Pout, C. Energy and Buildings, **40**.3: 394-398, (2008)

3. Payne, J.E. Journal of Economic Studies, **37**,1: 53-95, (2010).
4. Santamouris, M.Cartalis, C., Synnefa, A. andKolokotsa, D. Energy and Buildings, **98**: 119-124,(2015).
5. Yoo, S-H. Energy Policy, **33**.12: 1627-1632, (2005).
6. Moreno-Muñoz, A.,Redel, M.D. and González, M. Power quality in high-tech campus: A case study. Proceedings of the Institution of Mechanical Engineers, Part A: *Journal of Power and Energy*, **220**.3: 257-269,(2006).
7. Chan, W.W., and Lam, J.C. International Journal of Hospitality Management, **21**.4: 381-391,(2002).
8. Sims, R.E.H.,Rogner, H-H. and Gregory, K. Energy Policy,**31**,13: 1315-1326,(2003).
9. Ahmad, S. and Tahar, R.M. Renewable energy,**63**: 458-466,(2014).
10. Utama, Y.J. and Ambariyanto, A. Achieving Research University: Indonesian Case. In *IOP Conference Series: Earth and Environmental Science*,**55**, 1: 012072, (2017).
11. Utama, Y.J., Purwanto, Ambariyanto, A. Developing Environmentally Friendly Campus at Diponegoro University. *Advanced Science Letters*, 23(3): 2584-2585,(2017).
12. Tobin, K. School science and Mathematics,**90.**5: 403-418,(1990).
13. Salpanio, R.,Warsito, A. and Winardi, B. Transmisi, **9**.2: 181-190. (2007).
14. Badruz zaman, JTET, **1**.2: 50-59,(2012).
15. Ding, G.K. Journal of environmental management, **86**(3):451-64, (2008).
16. Rosemann, A., Mossman, M. and Whitehead, L. Solar Energy, **82**.4: 302-310, (2008).
17. Duguay, M.A. and Edgar, R.M. Applied Optics, **16**.5: 1444-1446,(1977).
18. Hirunlabh, J., Kongduang, W., Namprakai, P. and Khedari, J. Renewable Energy, **18**.1: 109-119, (1999).
19. Ostro, B., Rauch, S. Green,R. and Malig, B. American Journal of Epidemiology, **172**.9: 1053-1061, (2010).
20. Mcmakin, A.H., Malone, E.L. and Lundgren, R.E. Environment and Behavior, **34**.6: 848-863, (2002).
21. Ueno, T.,Sano, F., Saeki, O, and Tsuji, K. Applied Energy, **83**.2: 166-183. (2006).
22. Vikhorev, K., Greenough, R. and Brown, N. Journal of Cleaner Production, **43**: 103-112,(2013).
23. Bakos, G.C. Applied Energy, **72**.3: 599-608,(2002).
24. Singh, G.K. A review. Energy, **53**: 1-13, (2013).

https://doi.org/10.1051/e3sconf/20183101004

Developing Energy Technology Course for Undergraduate Engineering Management Study Program in Lake Toba Area with Particular Focus to Sustainable Energy Systems in Development Context

Yosef Manik[1,*], *Rizal* Sinaga[1], and *Hadi* Saragi[1]

[1]Engineering Management Study Program, Institut Teknologi Del, Laguboti - Indonesia

Abstract. Undergraduate Engineering Management Study Program of Institut Teknologi Del is one of the pioneers for its field in Indonesia. Located in Lake Toba Area, this study program has a mission to provide high quality Engineering Management education that produces globally competitive graduates who in turn will contribute to local development. Framing the Energy Technology course—one of the core subjects in Engineering Management Body of Knowledge—in the context of sustainable development of Lake Toba Area is very essential. Thus, one particular focus in this course is sustainable energy systems in local development context that incorporates identification and analysis of locally available energy resources. In this paper we present our experience in designing such course. In this work, we introduce the domains that shape the Engineering Management Body of Knowledge. Then, we explain the results of our evaluation on the key considerations to meet the rapidly changing needs of society in local context. Later, we present the framework of the learning outcomes and the syllabus as a result of mapping the road map with the requirement. At the end, the summary from the first two semesters of delivering this course in academic year 2015/2016 and 2016/2017 are reported.

1 Introduction

The Undergraduate Engineering Management Study Program (UEMSP) in Institut Teknologi Del is one of the pioneers for its field in Indonesia. Founded in 2014, UEMSP is currently having a student body of 210 students. The core objective of UEMSP is to prepare students for careers that involve the complex interplay of technology, data, organizations and economics. Through an integrated curriculum, the Engineering Management bachelor's degree program combines a strong engineering core with training in data analysis and visualization, process and project management, engineering design, systems integration, and managerial economics.

The interdisciplinary of UEMSP faculties includes those with expertise in wide range of topics as engineering economics; decision and risk analysis; economic management and marketing; and management of people and organizations that affect technology. The program is designed to give students the knowledge to measure and evaluate technical, business, and human performance processes in engineering environments. In the required courses, students are expected to develop their perspectives on the core of technology in industry and to gain insight into other management issues critical to leading or managing a technological organization.

Energy Technology Course (ETC) is listed as a compulsory course during sophomore year. This course of 3 credit-hours per week is essentially aimed to provide UEMSP students with strong foundation in scientific and technological factors, as well as the nature of the energy in nature and society. This will enable them to see the big-picture perspective that comes with in the real world. Designing ETC for a non-traditional engineering sub-discipline in the context of developing country is unique and challenging. In this paper, the authors describe the design process, implementations and evaluations of ETC in terms of aims, philosophy, structure and outcomes from some of this initiative.

2 Methodology for Course Design

Various models have been presented in the engineering educational literature for the development of course for study programs in engineering. The model suggested here is a simple three-stage model that effectively encapsulates most features inherent in a logical and structured approach to course design that has been adapted from that described by Kruse and Keil [1] and by Dick et al. [2].

Essentially the model identifies the following stages in the design and development of ETC:

Stage 1. Analysis of some Key Considerations
Stage 2. Structuring the Course
Stage 3. Implementation and Evaluations

Each stage has a set of inputs which must be evaluated and then analyzed to produce the output from that stage.

* Corresponding author: yosef.manik@del.ac.id

However, the process is not linear as this would suggest but is highly iterative with multiple secondary interactions and revisions before moving on to the next stage. The process is illustrated in Fig. 1.

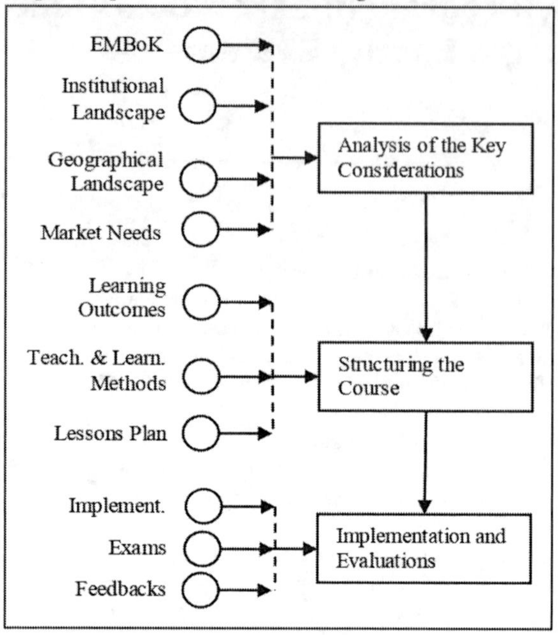

Figure 1. Course Design Process

3 Key Considerations of Course Design

3.1. Engineering Management Body of Knowledge (EMBoK)

Engineering Management (EM) is a relatively young discipline which combines a knowledge of both academic and practice topics. Over the last 25 years, there have been a number of articles which analysed EM curricula and helped define an EMBoK. The most prominent of these authors was Kocaoglu [3,4] and Farr [5] who analysed virtually all the EM related undergraduate and graduate programs using a consistent set of categories.

The ABET Engineering Criteria [6] requires that the Engineering Management curriculum must prepare the graduate "to understand the engineering relationships between the management tasks of planning, organization, leadership, control, and the human element in production, research, and service organizations; to understand and deal with the stochastic nature of management systems. The curriculum must also prepare graduates to integrate management systems into a series of different technological environments."

Simply put, in order to support the core competencies for EM, the following sets of courses must be provided:

[1] Science and Engineering domain: Physics, Chemistry, Thermodynamics, Energy, etc.
[2] Quantitative domain: Math, Statistics, Operation Research, Programming, Simulations, etc.
[3] Qualitative Conceptual domain: Management Theory, System Thinking, Organizational Behaviours, Leadership, Entrepreneurship, etc.
[4] Functional Technical Management domain: Logistics, Quality Management, Marketing, etc.
[5] Project-related domain: Engineering Design, Project Management, Capstone, Internship, etc.
[6] Finance and Accounting domain: Engineering Economics, Cost Accounting, etc.

It is clear that in order to comply with EMBoK, UEMSP shall equip its curricula with energy course, which is an essential element of its core competence in terms of Science and Engineering domain.

3.2 Institutional Landscape: Institut Teknologi Del

Institut Teknologi Del (IT Del) is the home institution for UEMSP. IT Del was founded in 2001 by Gen. (ret.) Luhut Pandjaitan as a polytechnic consisting of three study programs that mainly focus in informatics. The campus is located in Laguboti town (2°23'08.6"N 99°08'53.7"E), a remote area in Toba Highland (255 km away from Medan, the provincial capital of North Sumatra), just on the shore of Lake Toba. The initial goal of establishing IT Del was to provide access to quality education for outstanding students who face economic hardship in remote areas of Toba. In 2013 it transformed into institute of technology with eight study programs, including UEMSP.

At the moment, IT Del has about 1,500 students. Being a boarding school, all of these students live in the campus dormitory. Although admission is now open nationwide, demographically the student body is dominated by those come from around Lake Toba Area. As a matter of fact, generally they have not been exposed to urban life. Nevertheless, they have high motivation and sufficient intellectual capacity because they have passed a rigorous selection process. With dynamic and motivated student characters, it is appropriate that course materials are designed in a challenging, full of novelty and future-oriented forms.

3.3 Geographical Landscape: Lake Toba Area

Lake Toba Area is the area of which IT Del is located and from which most of the students originated. Physically, the area of Lake Toba is an area that is located around Lake Toba with delineation of boundaries based on the delineation of the Water Catchment Area. Lake Toba is the largest freshwater lake in Indonesia and Southeast Asia (90 x 30 km^2) that contains a lot of potential in many facets. In the context of geology, the Toba Caldera and its paroxysmal eruption 74,000 years ago have gained scientific notoriety for the past century [2]. In the context of socio-cultural, Lake Toba is a heritage to the civilization history of Batak tribe, one of the largest ethnic groups in Indonesia. In the context of economics, this lake is the source of livelihood of people living in its surrounding, such as fresh water source, irrigations, aquaculture, hydropower, and geothermal [8].

Understanding of the physical potential of Lake Toba Area, especially the potential of energy generation from hydropower and geothermal, is a huge plus in designing a contextual ETC material.

3.4 Market Needs: Trends and Changes in the Global Energy Business

Another consideration in designing ETC is the trends and changes in the global energy business. The renewable energy industry is growing rapidly amidst rising concerns about oil depletion and climate change [9]. Renewable energy is seen by many as part of the appropriate response to these concerns and some national Governments have put programs in place to support the wider use of sustainable energy systems.

Traditional engineering study programs nowadays are not trained to use these renewable energy technologies and most are not aware of the principles of sustainability. There is therefore an urgent need to develop and implement new courses that prepare engineers, scientists and energy planners to work with renewables to produce sustainable energy generation systems.

4 Structuring the Course

4.1 Learning Outcomes

Embarking from the key considerations discussed earlier, we agreed to define the learning outcomes, the understanding and competence of which the students will have after completing the ETC, as follow:

[1] Students understand the principles of energy flow in the biosphere, the principles of photosynthetic and global primary production, and the techniques in energy provision during the pre-industrial and industrial era.

[2] Students understand the basic concepts in chemistry and properties of hydrocarbon fuels, the techniques in extraction of selected fossil fuels, the sciences in combustion of selected fossil fuels, and the principles of selected utilization of fossil fuels in society.

[3] Students are able to develop understanding and construct arguments on selected topics through writing up a connection paper.

[4] Students understand the principles and emerging trends of renewable energies and the economics of energy in terms of cost and environmental issues

4.2 Teaching- and Learning Methods

This course is design for students at the sophomore year. As prerequisite, they need to have passed the Fundamental Physics (8 credit-hours) and Fundamental Chemistry (4 credit-hours). Thermodynamics course (3 credit-hours) is set as co-requisites.

In general, this course is delivered through classroom lecture. Presentation slides and whiteboards are used as the media. All electronic course materials are provided through a web-based academic information system <cis.del.ac.id> which can be downloaded by students at any time.

The lectures are provided by three faculties as a team, whose specific expertise is given in Table 1.

Table 1. Instructors for ETC

Instruct-or	Highest academic degree	Field of expertise
#1	PhD	Sustainability, bioenergy, combustion technology
#2	Masters	Physics, geothermal, thermodynamics
#3	Masters	Electrical power engineering

In the middle of the semester, two field visits are arranged. These field visits are intended to give students insight into the energy generation process in its true form. The visits are held interchangeably at energy plants located within Lake Toba Area, such as Sigura-gura Hydropower, Labuhan Angin Steam Power Plant, Sibayak Geothermal Plant, or Sarulla Geothermal Plant. As deliverable outcomes from the visits, students are assigned to write connecting papers about the findings and lesson learned, and to present the papers before the faculty panels.

A visiting lecture is also held. This lecture is delivered a practitioner, a government officer, or a decision maker in the field of energy business. Having a specialist in this particular subject area will help students to interact with the real actors in the energy business. It also benefits them because closer links can be forged with the expert's home institution in case they need it for internship or future employment.

Main reference for ETS are "Energy: Beginners' Guide" [10] and "Energy in the Nature and Society: General Energetics of Complex Systems" [11], both published by MIT press. Vaclav Smil is Distinguished Professor Emeritus at the University of Manitoba. He is a very productive author in the fields of energy, environmental and population change, food production and nutrition, technical innovation, risk assessment, and related public policy that Bill Gates wrote on his website in 2013:

> *"[...] there is no author whose books I look forward to more than Vaclav Smil."*

In these two books, systematic and exhaustive analysis of all the major energy sources, storages, flows, and conversions that have shaped the evolution of the biosphere and civilization are provided, so that we found it useful as the reference.

4.3 Lessons Plan

After the learning outcomes have been determined, the teaching- and learning method have been defined; the next step is to select the content and to organize in a manner that will facilitate the learning of it by the student. At this stage it is necessary to identify the elements of the curriculum that can serve as organizing threads, which, in engineering, are often concepts, percepts and skills. Learning is designed to take place in 7 consecutive weeks, followed by mid-term exam, and continued to another 7 consecutive weeks, followed by final exam. All learning materials are presented as a series of topics in 14 classes, the sequence of which is given Table 2.

Table 2. ETC Structure

Week	Topic	Sub-topic
1	Fundamentals and outlook	• Science of energy: origins and abstracts • Fundamental concepts: energies, conversion, efficiencies
2	Recitations on selected energy conversion problems	
3	Energy in the biosphere (1)	• Sun: the star and its radiation • Hydrosphere and atmosphere: thermal and mass fluxes
4	Energy in the biosphere (2)	• Water and air in motion: kinetic fluxes • Geo-energetics: heat, plate tectonics, volcanoes, earthquake
5	Bioenergetics of primary production	• Photosynthetic pathways • Global primary production • Cropping intensification
6	Fossil fuels (1)	• Hydrocarbon fuel chemistry and properties • Extraction and distribution
7	Field visit to energy plants	
--------------- Mid-term Exam ---------------		
8	Fossil fuels (2)	• Combustion and applications
9	Fossil fuels (3)	• Utilization in prime movers and electricity generation
10	Recitations on fossil fuel problems	
11	Renewables (1)	• Photovoltaic • Hydropower
12	Renewables (2)	• Geothermal • Biomass
13	Visiting lecture	
14	Energy economics	• Energy cost and valuation • Externality analysis
--------------- Final Exam ---------------		

The grading scheme is as follow:

[1] Assignments and quizzes : 10%
[2] Connecting paper : 15%
[3] Post field visit presentation : 15%
[4] Mid-term exam : 30%
[5] Final exam : 30%

In order to pass the course, the student must earn not least than 30% of the weighted score average.

5 Implementation and Evaluations

ETS has been implemented in two consecutive academic years: odd semester 2015/2016 and odd semester 2016/2017. The numbers of participating students for the academic year 2015/2016 and 2016/2017 are 30 and 60, respectively. All students are sophomore from Engineering Management major, with age range of 19 - 20 years old.

Participation and enthusiasms of the students were very high. This was reflected in their attendance and participation in each class and also in the interaction among them in group discussions. Classes were held according to lessons plan, including field visits and guest lecture. For the field visits, the students are brought to Labuhan Angin steam power plant and Sigura-gura hydropower. Guest lecture was given by practitioners in micro-hydropower. For the academic year 2015/2016, 100% of the students passed the exams. However, for the academic year 2015/2016, only 77% did.

The end-of-semester surveys show that student satisfaction level is high in both academic years. Some feedback suggests that field visits are a great help for them to understand what is really happening in the real world in the area of energy conversion and generation. As for things that still need to be improved, the survey suggests, is the exercise in solving quantitative analysis problems.

In the coming semesters such a pattern will continue while improving the things that still need to be improved. For this, the instructors will continue to improve their capacity and to extend collaboration with external partners who are expected to help improving the quality of this course.

References

1. K. Kruse, & J. Keil, , *Technology-based training: The art and science of design, development, and delivery.* San Francisco: Jossey-Bass (2000)
2. W. O. Dick, L. Carey, and J. O. Carey, *The systematic design of instruction.* Boston: Allyn & Bacon (2004)
3. D.F. Kocaoglu, *The Emergence of Engineering Management Discipline,* ASEE Conference Proceedings pp.240-247 (1986)
4. D.F. Kocaoglu, *Education for Leadership in Management of Engineering and Technology* pp.78-83 IEEE (1992)
5. J. Farr, and B.A. Bowman, Engineering Management Programs: Contemporary and Future Issues, *Eng. Mgt. J.,* **11** (1999)
6. ABET Engineering Accreditation Commission, *2015-2016 Criteria for Accrediting Engineering Programs* pp.11, (2014)
7. C.A. Chesner, *Quaternary International.* **258**, pp.5-18 (2012)
8. H. Moedjodo, P. Simanjuntak, P. Hehanussa, and Lufiandi. *Lake Toba: Experience and Lessons Learned Brief.* Lake Toba Heritage Foundation (2003)
9. REN21, *Renewables 2015 Global Status Report* Paris: REN21 Secretariat (2015)

10. V. Smil, *Energy : Beginners' Guide.* MIT Press (2006)
11. V. Smil, *Energy in Nature and Society: General Energetics of Complex Systems.* MIT Press (2008)

 https://doi.org/10.1051/e3sconf/20183101005

An effect of humid climate on micro structure and chemical component of natural composite (*Boehmeria nivea-Albizia falcata*) based wind turbine blade

S. Sudarsono[1*], *P. Purwanto*[2,3], *Johny* W Sudarsono[4]

[1]Department of Mechanical Engineering, Institut Sains dan Teknoogi AKPRIND, Yogyakarta
[2] Department of Chemical Engineering, Faculty of Engineering, Universitas Diponegoro
[3]Doctoral Program of Environmental Science, School of Postgraduate Studies, Universitas Diponogero, Semarang, Indonesia
[4]Depart. of Material and Metallurgy, Faculty of Engineering, Universitas Indonesia, Jakarta

Abstract. In this work, wind turbine blade NACA 4415 is fabricated from natural composite of Boehmeria nivea and Albizia falcate. The composite fabrication method used is hand lay up method. The aim of the work is to investigate an effect of humid climate of coastal area on micro structure and chemical composition of composite material of the blade. The wind turbine is tested at Pantai Baru, Bantul, Yogyakarta for 5.5 months. The micro structure scanning is performed with Scanning Electron Microscope (SEM) and material component is measured with Energy Dispersive X-ray spectrometer (EDS). The samples are tested before and after the use within 5.5 month at the location. The results show that composite material inexperienced interface degradation and insignificant change of micro structure. From EDS test, it is observed that Na filtration reduces C and increases O in composite material after 5.5 months.

1 Introduction

Wind energy conversion system is very attractive and have been developed worldwide. Composite based wind turbine blade is likely have good performance for low wind speed. Composite material form different natural material have been investigated by various researchers. Sandwich composited made from Albaizia wood and Kenaf fiber is fabricated and investigated by Diharjo et al [1]. The impact properties improves as the thickness of the core increased. Xu, *et al.[2]* studied the life cycle assessment of natural fibers in wood-fiber-reinforced polypropylene composites and introduced the term "material service density" (MSD), defined as the volume of material to meet the specific power requirements. When MSD is used as a functional unit, the wood fiber-reinforced composite is more environmentally friendly than polypropylene. Impact strength of laminating composite increases twice with the use of Ramie fiber [3]. Other natural material used for making composite is waste of starch fiber. This starch fiber waste was used for thermal insulator and sound absorber [4].

Composite material of wind turbine blade which exposed to coastal area will experience extreme weather condition. Regarding environmental factor, temperature and humidity may affect the performance of composite material. Increasing temperature from 50 to 100^{0}C lowers inter-laminar crack energy of laminar composite 25-30% [5]. Due to high humidity in particular location, water vapor content in air may infiltrate to the composite material. The infiltration may reduce inter matrix bounding [6], hairline crack [7], [8] and also may reduce temperature of glass transition [9].

In this work, wind turbine blade NACA 4415 is fabricated from natural composite of Boehmeria nivea and Albizia falcate. The composite fabrication method used is lay hand up method. The aim of the work is to investigate an effect of humid climate of coastal area on micro structure and chemical composition of composite material of the blade.

2 Material and method

Wind turbine blade NACA 4415 is fabricated from natural composite of Boehmeria nivea and Albizia falcate. The composite fabrication method used is hand lay up method. The wind turbine is tested at *Pantai Baru*, Bantul, Yogyakarta for 5.5 months. The micro structure scanning is performed with Scanning Electron Microscope (SEM) and material component is measured with EDS. The samples are tested before and after the use within 5.5 month at the site. The SEM unit used in this work as shown in Figure 1.

3 Results and discussion

Figure 2 presents photo of cross sectional area of 2 layers composite from the SEM test. The interface between Sengon wood core, resin, and ramie fiber indicated has good bonding due to polar resin used. SEM photo of cross sectional wind turbine after 5.5 months installed at Pantai Baru shown in Figure 3.

* Corresponding author: sudarsono1574@akprind.ac.id

https://doi.org/10.1051/e3sconf/20183101005

Fig. 1. Scanning Electron Microscope (SEM)

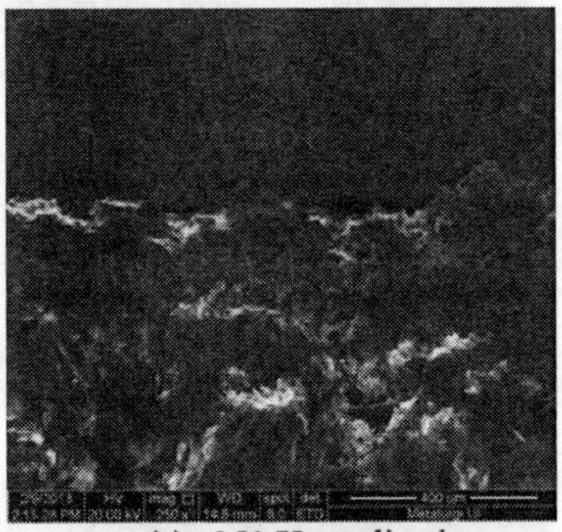

(a) 250 X amplication

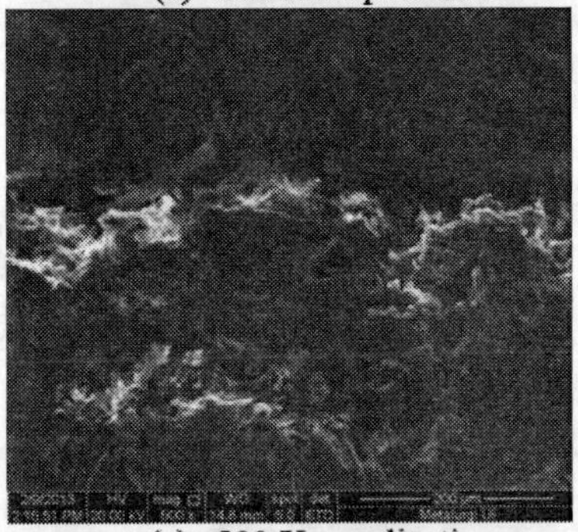

(a) 500 X amplication

Fig 2. SEM photo of composite before used as wind turbine blade

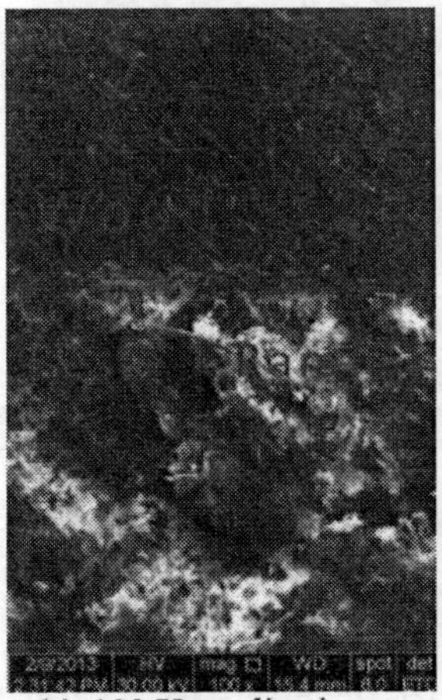

(a) 100 X amplication

(b) 250 X amplication

Fig.3. SEM photo of composite after 5.5 months

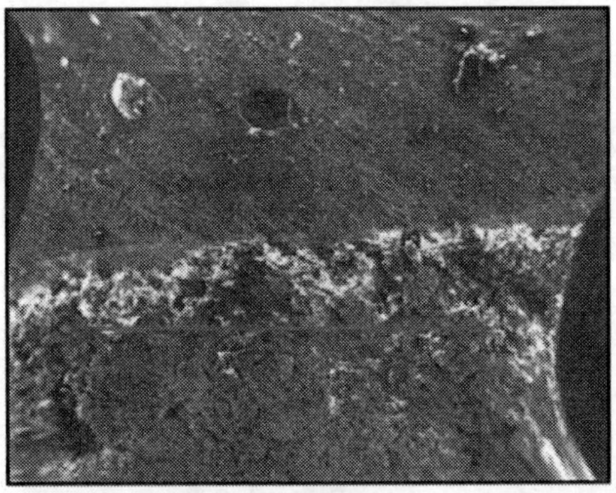

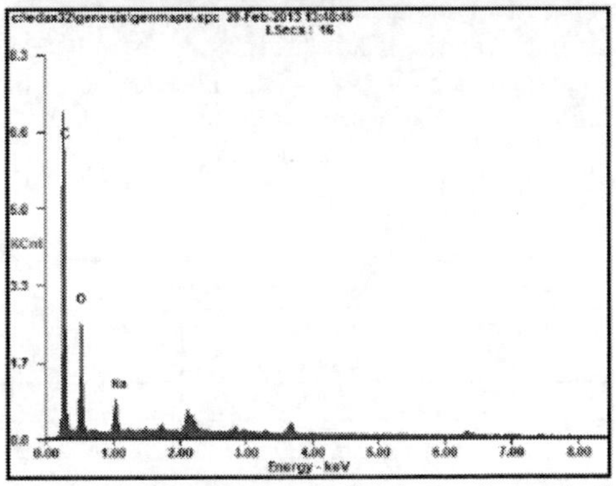

(a) SEM & EDS test of resin-ramie part before field test

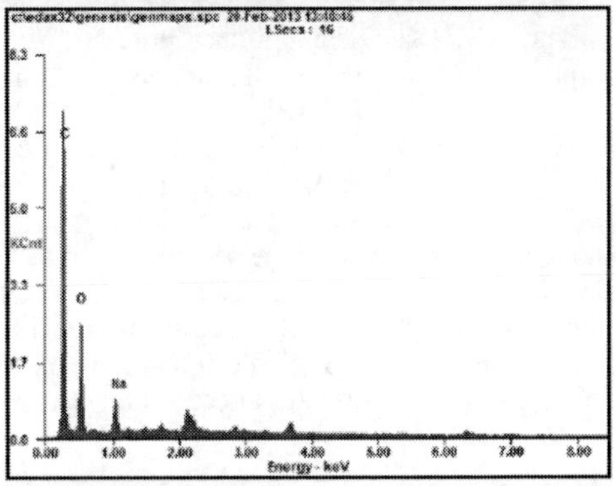

(b) SEM & EDS test of resin-ramie part after 5.5 months field test

Fig. 4. SEM & EDS test of resin-ramie fiber of the composite before and after field test

After 5.5 months, the micro structure of the composite inexperienced degradation. It indicates that the method of lay hand up is good enough for composite wind blade fabrication.

Figure 4 shows the results of SEM and EDS test of resin and ramie part of composite wind turbine blade. Meanwhile, Table 1 show chemical component of (C, O, Na) of resin-ramie part before and after field test.

Wood is mostly composed by three elements, Carbon, Hydrogen and Oxygen. These elements come from CO_2 in the air and from H_2O from the ground. However, the wood also contains other elements such as Nitrogen, Phospor, Kalium, Calcium, Magnesium, Silikon, Alumina and Natrium. After 5,5 months installation of model, EDS test is conducted. The test result indicates that Carbon decreases by 4.01%, while Oxygen increases by 7.85%. This phenomenon is caused by high humidity at observation site, about 69.98%. This high humidity increases a diffusion of oxygen air into the propeller material.

Table 1. C, O, Na composition of resin-ramie of blade before and after 5.5 months used

Component	Before being used (Wt%)	After 5.5 months (Wt%)
C	59.90	54.95
O	35.77	44.13
Na	4.33	0.91

Figure 4 and Table 2 show the results of SEM and EDS test of wood resin-ramie interface of composite wind turbine blade before and after the field test.

Table 2. C, O, Na composition of wood-resin-ramie of blade before and after 5.5 months used

Component	Before being used (Wt%)	After 5.5 months (Wt%)
C	59.69	55.68
O	36.17	44.02
Na	4.14	1.30

In the operational period of time, windmill propellers may be exposed to an extremely temperatures condition. In addition to accelerating the process of water absorption, the temperature also affects the composite resin.

The average weight of all three blades before being installed is 3025 grams. After 5,5 months installation of this composite as wind turbine blades, the weigth is found to be 3.106 grams. So the percentage content of water vapor diffused into the blade material during 5,5 months is 0.44%. This result is not much different from the result obtained by EDS. Percentage of oxygen in composite increases by 0.22% after 5,5 months.

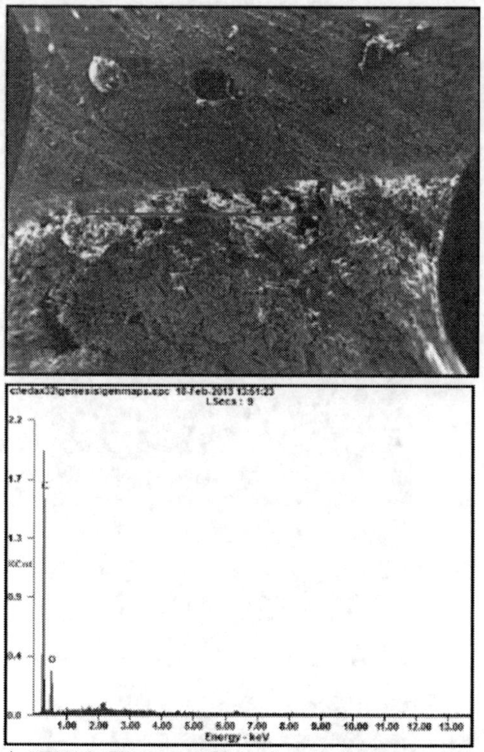

(a) SEM & EDS test of interface between wood and resin-ramie before field test

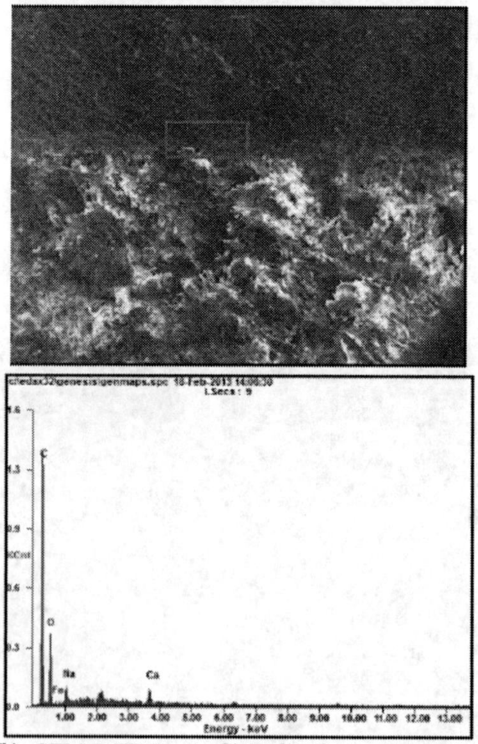

(b) SEM & EDS test of interface between wood and resin-ramie after 5.5 months field test

Fig. 5. SEM & EDS test of interface between wood and resin-ramie before and after field test

 https://doi.org/10.1051/e3sconf/20183101005

After being tested for 5.5 months at Pantai Baru-Bantul-Yogyakarta, EDS test shows that Carbon (C) content in composite decrease 4.01%, Oxygen (O_2) increases 7.85%. This is due to high humidity of the site during the field test. The humidity at the site during 5.5 months test is about 69.98%. In high moist air, oxygen content in the air infiltrates to composite wind turbine blade.

4 Conclusions

From EDS test, it is observed that Na diffusion reduces C and increases O in composite material after 5.5 months. The results of SEM test show that composite material inexperienced interface degradation and insignificant change of micro structure. It can be concluded that hand lay-up method give good performance of composite blade for encountering the effect of humidity on composite material.

References

1. K. Diharjo, Jamasri, Soekrisno, H. S. B. Rochardjo, *Proceeding of The Malaysia-Japan International Symposium On Advanced Technology (MJISAT 2007)*, (2007)
2. X. Xu, K. Jayaraman, C. Morin and N. Pecquex: Journal of Materials Processing Technology, Vol. **198** Issue 1 (2008), p. 168-177
3. N. Suizu, T. Uno, K. Goda and J. Ohgi: Journal of Materials Science, Vol **44** Issue 10 (2009), p. 2477-2482
4. Sudarsono, Saiful Huda, Murni Yuniwati, and Purnawan. Applied Mechanics and Materials Vol. **749** (2015) pp 227-230
5. G. Marom, 1989, *Application of Fracture Mechanics to Composite Materials, edited by Friedrich, K.*, pp. 397-423.
6. R. C. Schultheisz, G.W. McDonough, S. Kondagunta, L.C. Schutte, S.K. Macturk, Mc M. Auliffe, L.D. Hunston, *Effect of Moisture on E-Glass/Epoxy Interfacial and Fiber Strengths, Composite Materials: Testing and Design,* Thirteenth Volume.
7. L.C. Schutte, *Environmental Durability of Glass Fiber Composites, Polymer Composites Group, Polymers Division,* NIST, (1994)
8. S.T. Grant, , L.W. Bradley, Journal of Composite Materials, Vol.**29**, No.7 / 1995, pp. 853-867
9. H.F. Brinson, *Accelerated Life Prediction, Volume **2*** Engineered Materials Handbook, ASM International, pp.788-796, (1987)

The Effect of Acid Pre-Treatment using Acetic Acid and Nitric Acid in The Production of Biogas from Rice Husk during Solid State Anaerobic Digestion (SS-AD)

Winardi Dwi Nugraha[1,*], *Syafrudin*[1], *Cut* Fadhila Keumala [1], *Hashfi* Hawali Abdul Matin[3], and *Budiyono*[4]

[1]Environmental Engineering Department, Engineering Faculty, Diponegoro University – Indonesia
[2]Master Program of Environmental Science, School of Postgraduate Studies, Diponegoro University – Indonesia
[3]Chemical Engineering Department, Engineering Faculty, Diponegoro University – Indonesia

Abstract. Pretreatment during biogas production aims to assist in degradation of lignin contained in the rice husk. In this study, pretreatment which is used are acid and biological pretreatment. Acid pretreatment was performed using acetic acid and nitric acid with a variety levels of 3% and 5%. While biological pretreatment as a control variable. Acid pretreatment was conducted by soaking the rice straw for 24 hours with acid variation. The study was conducted using Solid State Anaerobic Digestion (SS-AD) with 21% TS. Biogas production was measured using water displacement method every two days for 60 days at room temperature conditions. The results showed that acid pretreatment gave an effect on the production of biogas yield. The yield of the biogas produced by pretreatment of acetic acid of 5% and 3% was 43.28 and 45.86 ml/gr.TS. While the results without pretreatment biogas yield was 29.51 ml/gr.TS. The results yield biogas produced by pretreatment using nitric acid of 5% and 3% was 12.14 ml/gr.TS and 21.85 ml/gr.TS. Results biogas yield with acetic acid pretreatment was better than the biogas yield results with nitric acid pretreatment.

1 Introduction

According to Minister of Energy and Mineral Resources, the dependence on fossil fuels, particularly oil in fulfilling domestic consumption is still high at 96% (petroleum 48%, gas 18% and coal 30%) of the total consumption and to maximize the utilization of energy renewable not capable of running as planned. Indonesian government's efforts in dealing with the energy issue is with the issuing of Law No. 5/2006 on the use of new and renewable energy.

Indonesia is an agricultural country with a large agricultural sector. One of the major agricultural sector is rice. Agricultural waste is containing lignocellulosic biomass. Rice straw is composed of three types of main components, cellulose, hemicellulose, and lignin. Cellulose and hemicellulose can be hydrolyzed to monosaccharides or simple sugars and can be fermented to biogas [1]. However, one of the problems in the production of biogas from biomass lignin is resistance to biodegradation or hydrolysis [2]. A key success factor for biogas production is to increase the efficiency of hydrolysis of biomass with increasing accessibility enzyme through pretreatment method [3]. Pretreatment using organic acids have several desirable characteristics, hydrolysis effective, less degradation products, and produce more sugar [4].

Pretreatment is a method to reduce or eliminate impurities or undesirable compounds in the raw material to be made into a product. In the process of biogas production from rice straw using a pretreatment in the form of delignification [5]. Chemical pretreatment refers to the use of chemicals, such as acids, bases, and ionic liquids to change the physical and chemical characteristics of lignocellulosic biomass [6]. Delignification process chemically induced by mixing and soaking acidic or alkaline compounds in the raw material before the fermentation process into biogas so that the raw material becomes softer and decreased levels of lignin [7]. Early treatment will reduce the lignin compound in the lignocellulose due to the destruction matrix structure of lignocellulose [8]. Lignin is a substance which, together with the cellulose found in wood. Lignin is a polymer with aromatic structure formed from units Pénil related propane jointly by several different types of bonds. Lignin is difficult to degrade because of the complex and heterogeneous structure that binds to

*Corresponding author: winardi_punk@yahoo.com

cellulose and hemicellulose in plant tissue. More than 30% of the crop is composed of lignin, which provides a solid form and provide protection against insects and pathogens [9].

Delignification process is done with the addition of microbes that can digest lignin thus decreasing the lignin contained in the raw material [10]. Pretreatment biology are considered to have the advantage and simplicity in a low capital investment more attractive [11]. Microbial consortium consisting of *Streptomyces sp.*, *Geobacilus sp.*, and *Trichoderma*. The function of microorganisms is doing delignification, reduce the degree of polymerization of the cellulose and hemicellulose hydrolysis. The addition of microbial consortium accelerate the degradation of cellulose, hemicellulose, and lignin into compounds needed by the microorganism biogas so that biogas production increased [7].

2 Materials and methods

This research is a kind of experimental research laboratories. The study was conducted at the Laboratory Waste Processing, Chemical Engineering, University of Diponegoro during the month of December 2016 - March 2017.

2.1 Content test TS rice straw

Analysis of total solids content (total solid / TS) by standard methods APHA

- The plate was dried at a temperature of 103-105 ° C for 1 hour, then cooled and stored in a desiccator until the cup is used.
- Weight the cup is weighed and recorded.
- Sample inserted into the cup as much as 25-50 g and weighed, then dried in an oven at a temperature of 103-105 ° C for 1 hour.
- Sample that has been dried and then cooled in a desiccators and weighed to the reduced weight 4% or 50 mg.

$$\%\, total\ solids = \frac{(A - B) \times 100}{C - B} \quad (1)$$

A = weight sample that has been dried + cup (mg)
B = weight cup (mg)
C = weight of the wet sample + cup (mg)

2.2 Pretreatment

- Took organic rice straw waste (biowaste), then conducted an examination of the total composition of solids and water content.
- The addition of acid is done by calculating the acid levels multiplied by the volume of the reactor is added with half the volume of water immersion chaff.
- Rice straw soaking for 24 hours.
- After soaking, chaff neutralized until the rice straw is no longer acidic
- Rice straw dried in the sun to dry.

2.3 Operational

- Rice straw has done acid pretreatment process is fed into the reactor in accordance with Table 1.
- Add water, enzymes, urea and rumen in accordance with Table 1.
- Wait for the fermentation process so that biogas formed.
- Measure the volume of biogas which is formed every two days until the biogas produced is not returned.

Table 1. Materials research

Variable	Microbial Consortium 5% v/v (ml)	Volume of water (ml)	Volume of Rumen (ml)	Rice Straw (gr)
1	10	76.61	76.61	46.77
2	10	76.61	76.61	46.77
3	10	76.61	76.61	46.77
4	10	76.61	76.61	46.77
5	10	76.61	76.61	46.77
6	10	76.61	76.61	46.77
7	10	76.61	76.61	46.77
8	10	76.61	76.61	46.77
9	0	76.61	76.61	46.77
10	0	76.61	76.61	46.77

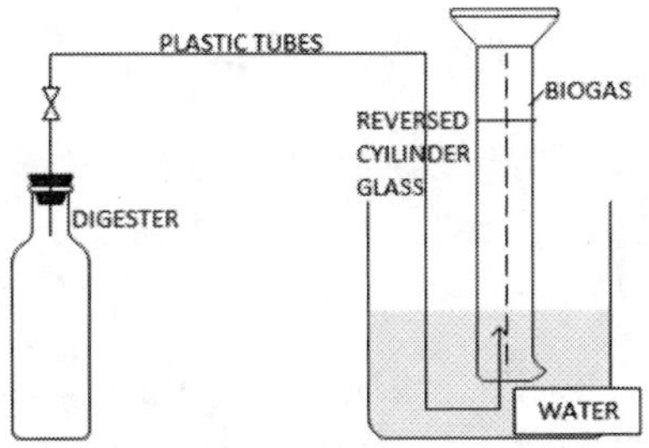

Fig 1. Schematic diagram of series laboratory batch assessment of SS-AD

3 Result and discussion

3.1 The Effect of Biological Pretreatment

Biological pretreatment aims to produce more hemicellulose and cellulose that help can be more widely available and easily digestible [6]. In this research using fixed variables microbial consortium of 5% of the total volume of liquid in the digester (200 ml). Then we get a volume of 10 ml. Biological pretreatment carried out after acid pretreatment. Biological pretreatment control variables used in the study, so there is no differences in the treatment of each study variable in terms of biological pretreatment.

3.2 The Effect of Acetic Acid Pretreatment

This study aimed to compare the biogas production between rice straw given early treatment with acidified with a solution of CH_3COOH with rice straw is not given early treatment. In addition, this study also compared two different levels of CH_3COOH to see the levels of acetic acid is optimum in the process of lignin degradation. Used CH_3COOH levels of 3 and 5% as a benchmark levels of acetic acid.

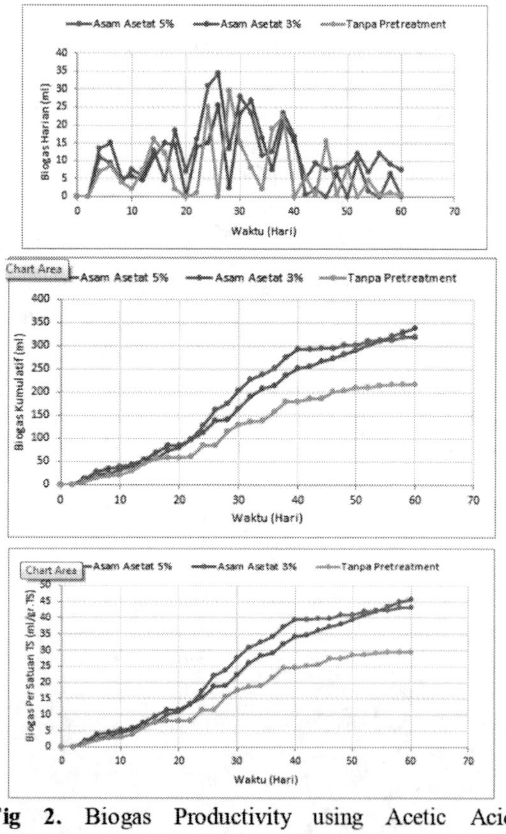

Fig 2. Biogas Productivity using Acetic Acid Pretreatment, (a). Daily Measurement, (b).The Cumulative Yield Biogas, (c). Cumulative/ TS

Acetic acid content of 5 and 3% with a concentration ratio of TS 21% compared to the levels of 3% tend to have a higher biogas yield than 5%. Total cumulative biogas yield per unit TS 3% acetic acid, 5% acetic acid, and without pretreatment respectively by 45.86; 43.28; and 29.51 ml.

Having an organic acid which is good potential to eliminate the inhibitory compounds such as lignin and cellulose structures are modified to be more susceptible to enzyme degradation [12]. The use of acid pretreatment mixture of acetic acid and propionic acid can significantly increase the production of methane up to 35.84% and approximately 34.19% can make the process of degradation of lignin, in addition as many as 21, 15% degree of hydrolysis obtained with 0.75% mol / L in the pretreatment for 2 hours with the liquids versus solids ratio of 1:20 (w / v) [13]. Treatment using a weak acid such as acetic acid at low concentrations can reduce the lignin content during the immersion process lasts a weak acid solution.

3.3 The Effect of Nitric Acid Pretreatment

This study aimed to compare the biogas production between rice straw given early treatment with acidified with HNO_3 solution with rice straw is not given early treatment. In addition, this study also compared two different levels of HNO_3 to see the levels of nitric acid which is optimum in the process of lignin degradation. Used HNO_3 content of 3% and 5% as a comparison of the levels of nitric acid.

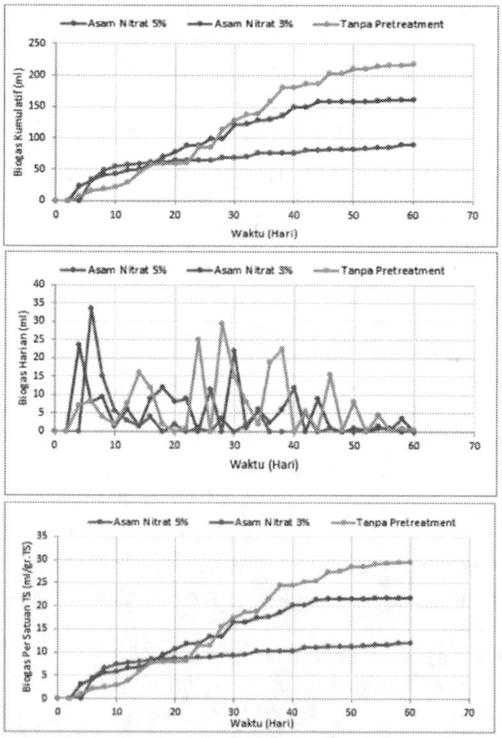

Fig 3. Biogas Productivity using Nitric Acid Pretreatment, (a). Daily Measurement, (b).The Cumulative Yield Biogas, (c). Cumulative/ TS

Nitric acid content of 5 and 3% with a concentration ratio of TS 21% compared to the levels of 3% tend to have a higher biogas yield than 5%. Total cumulative biogas yield per unit TS 3% acetic acid, 5% acetic acid, and without pretreatment respectively by 21.85 ml; 12.14 ml; and 29.51 ml.In research Tutt et al [7] suggest that nitric acid excellent in removing cellulose and hemicellulose, but generates byproducts that are difficult to be removed so that it

 https://doi.org/10.1051/e3sconf/20183101006

has a greater negative impact on fermentation. Another study by Helle et al (14), states that the formation of by-products such as acetic acid during pretreatment which then inhibit the fermentation process. In the previous research explains that the process of delignification, the acid will cause the structure of lignin and hemicellulose which binds cellulose will be condensed and settles so that the cellulose will be free structure that is separate from the lignin and hemicellulose then will facilitate the access of enzymes or microorganisms to convert cellulose to be monomers sugar as a nutrient for fermentation biogas[15]. Liginin The decline is also accompanied by a decrease in hemicellulose and cellulose is the main nutrient that will use microorganisms to produce biogas that biogas obtained results will be slightly[13].

3.4 The Rate of Biogas Productioan from Rice Straw with SS-AD Methods

Assuming that the rate of biogas production in the biodigester is proportional to the specific growth rate of methanogenic microorganisms inside the biodigester. The rate of biogas production will follow Gompertz Equation[16]. This equation is a mathematical model for the observation time series, which is the slowest growth at the beginning and end of the period of observation that has the general shape as the following equation.

$$P = A.\exp\left\{-\exp\left[\frac{Ue}{A}(\lambda - 1) + 1\right]\right\} \tag{2}$$

P = cumulative biogas production, liter
A = maximum biogas production, liter
U = constants the maximum rate of biogas production (liter/hari)
λ =lag phase, day
t =cumulative time for biogas productivity, day
e = Euler Number (e = 2.7182...)

There is a constant influence on the kinetics of biogas production from acid pre-treatment using acetic acid 3 and 5%. Kinetic constant biogas formed by the pretreatment given 5% acetic acid are as follows, daily biogas production (A), biogas production rate (U), and the minimum time of formation of biogas (λ) respectively 45.5338 (ml / g TS); 1.5057 (ml / g TS.days); 11.3514 days. While the rice husk given acid pre-treatment using 3% acetic acid produces biogas as follows kinetic constant, daily biogas production (A), biogas production rate (U), and the minimum time of formation of biogas (λ) respectively 54.0824 (ml / g TS); 1.0593 (ml / g TS.days); 8,3152 days.

Table 2. Kinetic constants on Acetic Acid Pretreatment Effect on Biogas Production

Variable	A (ml/g TS)	U (ml/g TS.days)	λ (days)
CH_3COOH 5%	45,5338	1,5057	11,3514
CH_3COOH 3%	54,0824	1,0593	8,3152
Without Pretreatment	33,7254	0,8121	8,9317

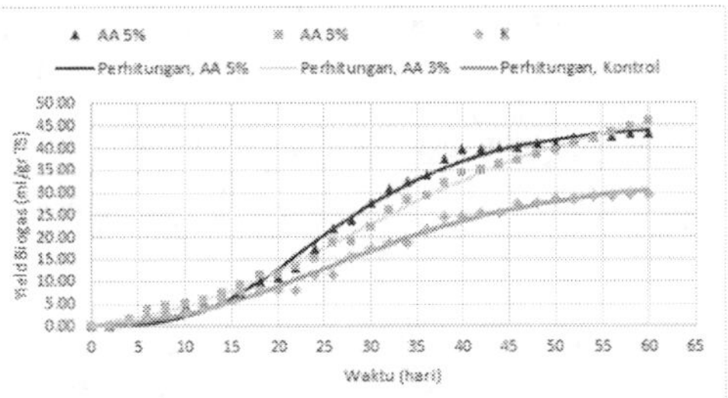

Fig 4. Relationship between Experiment data and calculation results on Effect of Pretreatment Research on Biogas Production Acetic Acid

There is a constant influence on the kinetics of biogas production from acid pretreatment using nitric acid 3% and 5%. Kinetic constant biogas formed with nitric acid pre-treatment given 5% is as follows, daily biogas production (A), biogas production rate (U), and the minimum time of formation of biogas (λ) respectively 10.7996 (ml / g TS); 0.6097 (ml / g TS.days); 0.1929 days. While the rice husk given acid pre-treatment using 3% nitric acid produces biogas as follows kinetic constant, daily biogas production (A), biogas production rate (U), and the minimum time of formation of biogas (λ) respectively 24.0036 (ml / g TS); 0.5759 (ml / g TS.days); 1.3080 days.

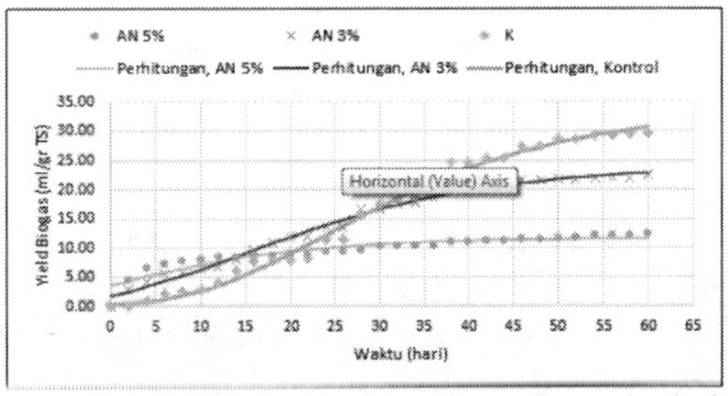

Fig 5. Relationship between Experiment data and calculation results on Effect of Pretreatment Research on Biogas Production Nitric Acid

After calculation by using the Polymath 6.0, known biogas production rate and can be seen on a day to stop what the biogas to produce biogas production. At 4:14 the picture shows the graphical form of biogas production rate with a variable yield acid pretreatment.

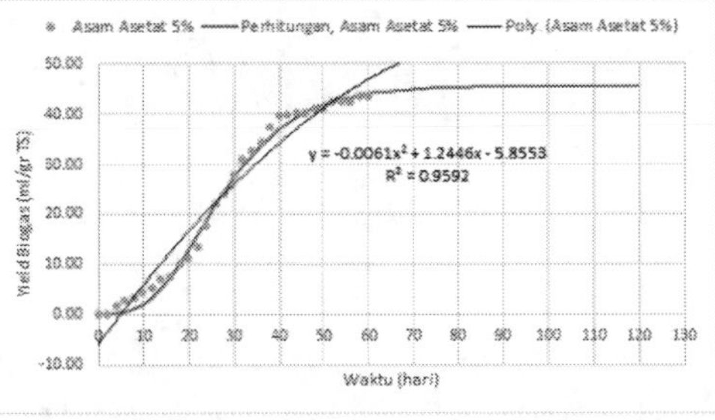

Fig 6. Calculation Result Data on Effects of Pretreatment Research Nitric Acid 5% of the Biogas Production

Calculation Result Data on Effects of Pretreatment Research Nitric Acid Production Biogas 5% against the trendline results obtained highest biogas yield on day 102 is equal to 57.63 ml / gr.TS. Then biogas productio further decline. While this is very different from the results of calculations using the Polymath 6.0, can be seen in Figure 4.15 that day all 118 results biogas yield reached the highest volume of 45.53 ml / gr.TS and constant or it can be said is not formed biogas.

Once the modeling is done on pre-treatment, namely by adding 3% CH_3COOH, obtained biogas yield of 45.86 ml / gr.TS. When there is a rice husk waste as much as one ton, it can produce biogas as much as 45.86 million ml or 45.860 m^3. If converted into other energy forms, obtain 1 m^3 of biogas equivalent to 4.7 kWh if it is converted into electrical energy. Then it can generate electricity capacity amounted to 215.54 kWh / day.

4 Conclusion

1. The preliminary treatment with acid 3% CH_3COOH produce biogas yield higher than 5% CH_3COOH and without pretreatment.
2. preliminary treatment without pretreatment acid generating biogas yield is higher than the pre-treatment HNO_3 HNO_3 3% and 5%.
3. The results of the biogas yield by pretreatment of organic acid (CH_3COOH) is better than the biogas yield results with pretreatment inorganic acid (HNO_3).
4. The biogas production rate in CH_3COOH 5% by value as follows: (A) as much as 45.5338 (ml / g TS); biogas production rate (U) of 1.5057 (ml / g TS.days); and the minimum time of formation of biogas (λ) is 11.3514 days.

References

1. H. Blanch, B. Simmons, D. Klein-Marcuschamer, Biotechnol. **6**, (2011), 1086–1102
2. V. Agbor, N. Cicek, R. Sparling, A. Berlin, D. Levin. Biotechnol. Adv. **29**, (2011), 675–685
3. A. Chandel, O. Singh. Appl. Microbiol. Biotechnol. **89**, (2011), 1289–1303
4. N. Khorshidi, B. Arikan. *Experimental Practice in Order to Increasing efficiency of Biogas Production by Treating Digestate Sludge.* University College of Boras (2008)
5. M. Kootstra, H. Beeftink, E. Scott, J. Sander, Biotechnol. Biofuels **2**, (2009), 31
6. Q. Zhang, J. He, M. Tian, Z. Mao, L. Tang, J. Zhang, H. Zhan. *Enchancement of Methane Production from Cassava by Biological Pretreatment Using a Constructed Microbial Consortium* (2011)
7. M. Tutt, T. Kikas, J. Olt. *Influence of Different Pretreatment Methods on Bioethanol Production From Wheat Straw*, 269-27 (2012)
8. A. Nopharatana, Pullammanappallil, C. Prata C, W.P Clarke. Waste Management, **27**. (2007), 595-603
9. A.T.W.M Hendriks, G. Zeeman. Bioresource Technology. **100,** (2009), 10–18
10. J. S. Eun, K. A. Beauchemin, S. H. Hong, M.W. Bauer. Animal Feed Science and Technology, (2006), 86-101
11. Z. Lei, J. Chen, Z. Zhang, N. Sugiura. Bioresource Technology, (2010), 4343-4348
12. Amnuaycheewa, Plaimein., Hengaroonprasan, Rotchanaphan., Rattanaporn, Kittipong., Kirdponpattara, Suchata., Cheen kachorn, Kraipat., Sriariyanun, Malinee., *Enchancing Encymatic Hydrolysis and Biogas Production from Rice Straw by Pretreatment with Organic Acids*, (2016)
13. S. Wahyuni. *Panduan Praktis Biogas*. Jakarta: Penebar Swadaya (2013)
14. S. Helle, D. Cameron, J. Lam, B. White, S. Duff. *Effect of Inhibitory Compounds Found in Biomass Hydrolystaes on Growth and Zylose Fermentation by A Genetically Engineered Strain of S. Cerevisiae* (2003)
15. F. PSari, Budiyono. *Enhanced Biogas Production From Rice Straw with various pretreatment : a review*, (2014), 17-25
16. A.MMshandete, L. Björnsson, A.KKivais, M.S.TRubindamayugi, B. Mattiasson. Afr. J. Biochem. Res. **2**, (2008), 211–218
17. W. Kenneth. *Handbook of Pulp and Paper Technology.* New York: Nab Nostrand Company, revise and enlarged, second edition (1970)
18. R. Zhao, Z. Zhang, R. Zhang, M. lo, Z. Lei, M. Utsumi, N. Sugiura. *Methane Production drom Rice Straw Pretreated by a Mixture of Acetic-Propionic Acid* (2010)
19. Y. Zheng, Z. Pan, R. Zhang, Int J Agric & Biol Eng **Vol. 2** No.3 (2009)
20. W. Zhong, Z. Zhang, Y. Luo, S. Sun, W. Qiao, X. Meng. *Effect of Biological Pretreatments in Enchancing Corn Straw Biogas Production*, (2011), 11177-111

E3S Web of Conferences **31**, 01007 (2018) https://doi.org/10.1051/e3sconf/20183101007
ICENIS 2017

Using a Fuzzy Light Sensor to Improve the Efficiency of Solar Panels

Suryono Suryono[1, 2,*], *Jatmiko* Endro Suseno[1, 2], *Ainie* Khuriati Riza Sulistiati[2], Tahan Prahara[3]

[1]Magister of Information System, School of Postgraduate Study, Diponegoro University, Semarang - Indonesia
[2]Department of Physics, Faculty Science and Mathematics, Diponegoro University, Semarang - Indonesia
[3]Department of Informatics Education, Veteran University, Semarang - Indonesia

Abstract. Solar panel efficiency can be increased by improving the quality of photovoltaic material, the effectiveness of electronic circuit, and the light source tracking model. This research is aimed at improving the quality of solar panels by tracking light source using a fuzzy logic sensor. A fuzzy light sensor property is obtained from two LDR (light dependent resistor) light sensors installed in parallel to each other and is given a light separator in between them. Both sensors are mounted on a solar panel. Sensor output is acquired using a 12 bit ADC from an ATSAM3XE microcontroller and is then sent to a computer using WIFI radio. A PID (Proportional-Integral-Derivative) control algorithm is used to manage the position of the solar panel in line with the input given by the fuzzy light sensor. This control mechanism works based on the margin of fuzzy membership from both sensors that is used to move a motor DC that in turn moves the solar panel. Experimental results show a characteristically symmetrical fuzzy membership of both sensors with a reflected correlation of R=0.9981 after gains from both sensors are arranged with a program. Upon being tested in the field, this system was capable of improving the performance of solar panels in gaining power compared to their original fixed position. The discrepancy was evident when the angle of incoming sunlight approached both 0° and 180°. Further calculations of data acquired by the fuzzy light sensor show increased solar panel power efficiency by up to 5.6%.

1 Introduction

Energy crisis is one of the biggest problems in most countries now. This issue has great effects on humanity. The underlying reason for this is dependency on fossil fuels with its dwindling available resources. Attempts have been done to generate electricity from other natural resources of energy such as the wind, geothermal, ocean wave, water, and sunlight. It is expected that in 2050 all countries in the world will have used 100% renewable energy [1].

Solar energy is a form of renewable energy available in its abundance. In tropical countries, solar energy can be harvested all year long. The technology required to convert solar energy into electric power is the photovoltaic solar panel (solar cell) made of semiconductor materials. Solar cell is advantageous from both environmental and efficiency point of views in transmitting power as the electric source can be placed near the facilities using it [2]. The efficiency in converting solar energy into electric power in solar panels is still a hot issue today. Efforts have been carried out to improve this efficiency by among others improving the photovoltaic material, using converter circuits, and tracking sunlight to maximize the power gained by a solar cell.

Monitoring the performance of solar cells connected to one another can improve their efficiency. This monitoring is carried out by mounting a sensor on each solar cell and subsequently reporting their individual conditions. These monitoring results are then sent to a monitoring station via a wireless network [3]. The use of artificial tracking algorithm has been widely used and has been proven to improve the efficiency of solar panels. The use of fuzzy logic algorithm to control solar panels in tracking sunlight has been simulated and has shown its effectiveness in improving the efficiency of solar panels it controlling [4].

Light sensors serve as information input in an artificial intelligent system to control light. Passive Infrared Sensors (PIRs) can be utilized to control lights on the streets as to lower their electric energy consumption [5]. Light sensors can also be used to control the brightness of LCD monitors in order to improve the efficiency of electric power consumption and add to the comfort of the eyes. These two systems can smartly combine ambient light intensity and set point values in order to control brightness, from the lowest to the highest levels [6].

Using a light sensor to know exact positions of a light source (sunlight) can significantly improve the light intensity taken by solar panels. A sensor's ability to track sunlight has been proven to improve efficiency in energy management. The sun changes its orientation over time. Therefore, a light sensor must adjust its position to the direction of incoming sunlight. A light sensor should be capable of directing itself to the position of apparent movement of a light source (the sun) on a daily to a monthly basis as to ensure maximum energy gained by solar panels. This means that angular inclination should be perpendicular to the light source as to ensure maximum light intensity exposed to the solar panel [7]. Once this sensor is connected to an automatic solar panel movement system, the performance of the later can improve by up to 25-30% [8].

Improving the efficiency of energy absorption by solar panels requires a supporting decision system that controls movements of the light sensor. The Opaque Tube technique was developed for sunlight tracking. This system consists of four photocells. It has been proven to improve tracking accuracy by high resolution [9]. The drawback of this technique is during its initial deployment whenever the incoming sunlight does not fall exactly on the sensor's position. This means that when the sunlight only hit the tube, instead of the sensor, it is not detected by the system. Therefore, there is a room for improvement on its construction. The addition of a computer algorithm is expected to minimize this problem and hence, improve its efficiency and effectiveness. This research focuses on the implementation of fuzzy characteristics of two sensors mounted in parallel with a light separator in between them. Output from this fuzzy sensor serves as input for the microcontroller that in turn moves the motor attached on the solar panel.

2 Research Method

The system built in this research consists of two sensors, a data acquisition system, and a computer program of proportional–integral–derivative control to move the solar panel. A solar panel coefficient largely depends on the incident angle of sunlight, despite the fact that other factors such as dust, snow, and some other external factors also have their roles. Maximum effectiveness is obtained when the incident angle of the sunlight has an angular coefficientt of $\eta_\theta = \sin 90°$. The correlation of light intensity and incident angle is shown in Figure 1.

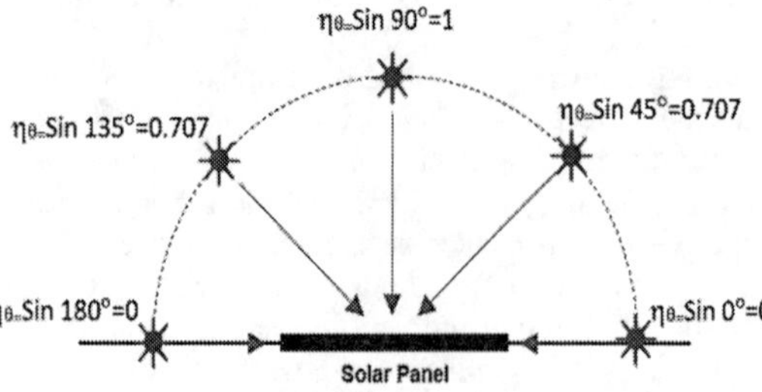

Fig. 1. Light intensity against incident angle of sunlight

This research makes use of a Light Dependent Resistor (LDR), an electronic component of changing resistance value upon contact with light. An LDR's maximum electric resistance is around 10MΩ in the dark, whereas its minimum is less than 1KΩ under bright light. The characteristics of an LDR light sensor are given in Figure 2a. In this system, sensors are configured as a voltage divider. In order to produce output voltage equal to that of light intensity, an LDR sensor must be connected in a series with a 1 KΩ resistor and given a voltage of 3.3 Volt. Zener diodes are installed at the sensor output to anticipate voltage hikes, so as to protect the other circuit connected to it. A circuit of LDR sensor is depicted in Figure 2b.

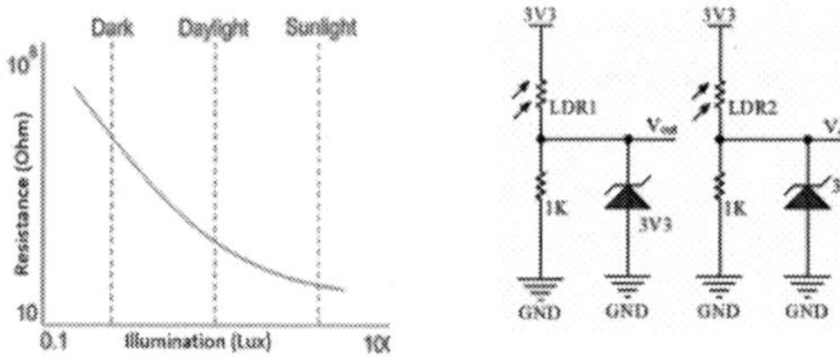

(a) (b)

Fig 2. (a) Characteristics of LDR sensor, and
(b) an LDR sensor circuit

The LDR sensor configuration yields maximum values when it is perpendicular to the incoming sunlight, and these values gradually reduce as the sensor is tilted. These values are used to generate fussy data. Sensor placement at certain angles represents weight against angle. A configuration of fuzzy light sensor is shown in Figure 3a, while its side view is given in Figure 3b. These sensors generate input fuzzy compilation data in the form of angle readings to be processed to obtain faster and more précised data. An ATSAM3X8E is used to acquire fuzzy sensor output and send it to a remote computer.

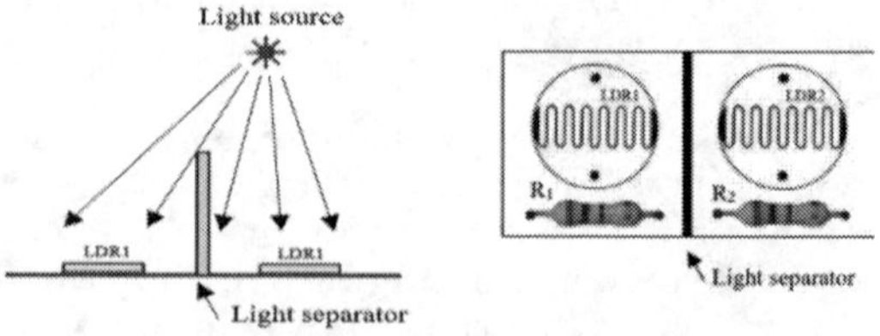

Fig. 3. (a) Configuration of a fuzzy light sensor, and
(b) side view of the fuzzy light sensor

They fuzzy light sensor was applied to a 100 watt solar panel. A 32 bit ATSAM3X8E microcontroller was used to acquire data and communicate those to a control computer via the TCIP/IP protocol. A circuit of that data acquisition and communication system is given in Figure

4. A WIFI radio was utilized as a data communication medium that connects ATSAM3X8E microcontroller to the computer's database system. Fuzzy light sensor output was acquired by a 12 bit ADC of the ATSAM3X8E microcontroller at channels A0 and A1. These acquired data were then sent to the control computer unit via an Ethernet module set at an IP address of the same class with the WIFI radio and the computer. The computer itself was programmed to using the client-socket and server-socket programming in order to access sensor data in line with the set port and these data were saved in the database every five (5) minutes. Stored data were then processed using the fuzzy logic algorithm according to the characteristics of their membership. Next, a de-fuzzy-fication process was carried out by the computer. This process changes fuzzy values into discreet values. Output values were obtained by adding multiplication results of input compilation membership for each input with the already set output values. The results of these were then divided with the number of multiplication of compilation membership for each input. De-fuzzy-fication transforms fuzzy units into numerical units. These unites were then sent to the microcontroller and set the basis for the Gearbox DC motor movement to position solar panels.

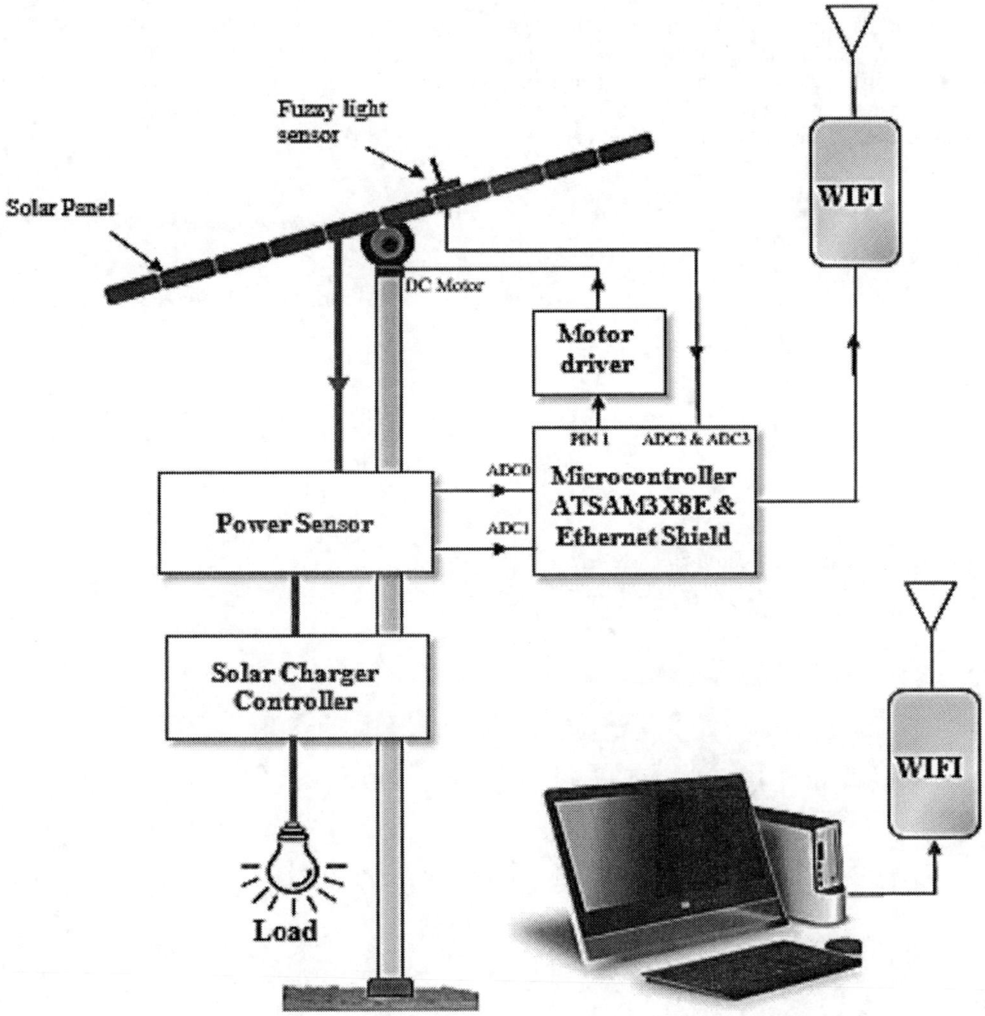

Fig. 4. Instrument system configuration and data acquisition and communication

https://doi.org/10.1051/e3sconf/20183101007

3 Analysis Results

Characteristics of the fuzzy light sensor were tested by moving a light source above the system. Data acquisition result was recorded by the computer. This result covered angles from 0° to 180°, adjusted with the position of the sun when the sensor was tested. In this research, the DC motor was set at 1° accuracy based on the delay time provided by the microcontroller. Results of this characterization are given in Figure 5. Characterization results in Figure 5 show that both sensors have logarithmic characteristics. Linear changes in scanning angles do not affect output voltage linearly but logarithmically instead. It is also shown that both sides of the sensors are not symmetrical. This is due to differences in intrinsic characteristics of the materials used for both sensors. Theoretically, fuzzy membership for both right and left sensors should be symmetrical.

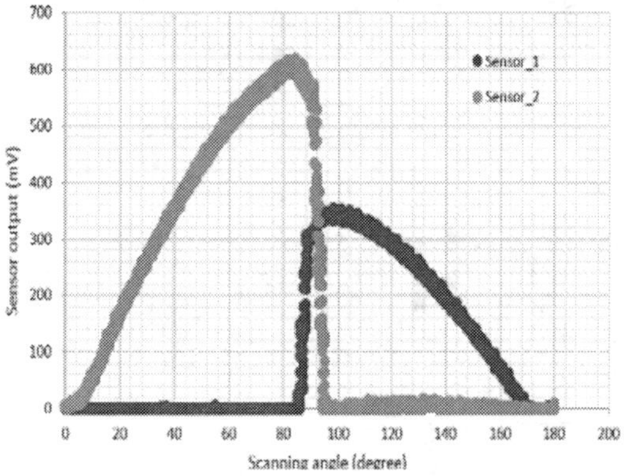

Fig 5. Characterization results for the fuzzy light sensor

Hence, this fuzzy system makes adjustment by carrying out a gain operation for both sensor characteristics with the help of a software. Subsequently, symmetrical values were obtained, with gains of 0.76 on sensor 1, and 1.34 on sensor 2. Adjustment results for both sensors are given in Figure 6, the reflected correlation of which is R=0.9981.

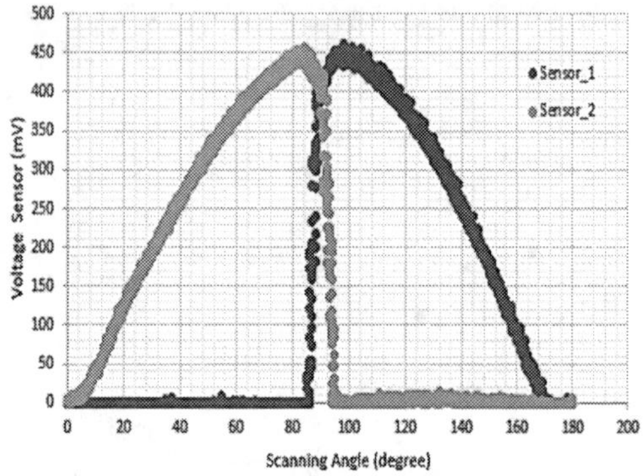

Fig. 6. Results of gain adjustments for both sensors

Fuzzy membership obtained from characterization of both sensors is obtained to control movements of the solar panel. Output from both sensors is used as the basis for solar panel movements. This research made use of the PID (Proportional-Integral-Derivative) control system algorithm for its computer user interface, as shown in Figure 7.

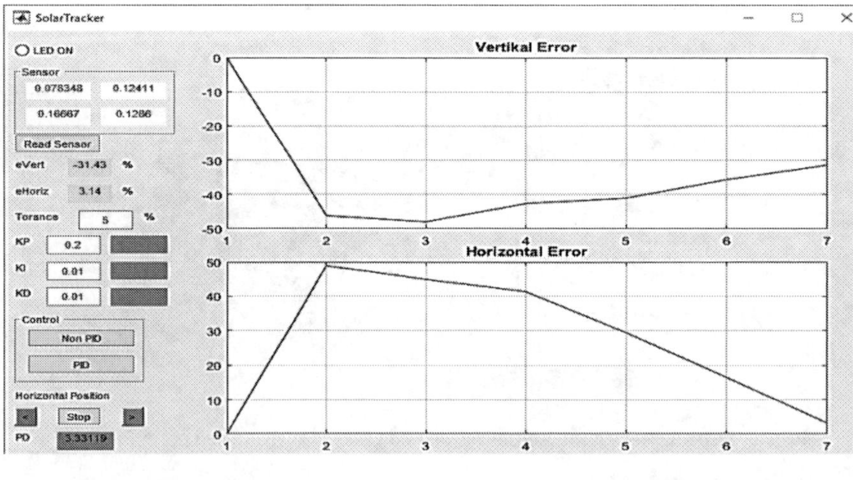

Fig. 7. Computer user interface to show control process for the light fuzzy sensor input in real-time

During each process, differences in output voltage from readings of sensor 1 and sensor 2 are displayed. These differences yield errors that are also displayed in real-time, as shown by the graph above. The less the differences in readings from the two sensors, the more optimum the solar panel position is. The number of iterations toward error =0 shows the speed of the control process. Less iterations mean faster control process.

This control system was then applied in the field. In order to figure out the effect of fuzzy sensor control system, electric power was acquired from two positions. The first was when the solar panel was lying horizontal at 0° and the second one was during tracking with the fuzzy sensor. Both situations were for the same solar panel and were observed during a slight time-lapse. Readings from both situations are depicted in Figure 8.

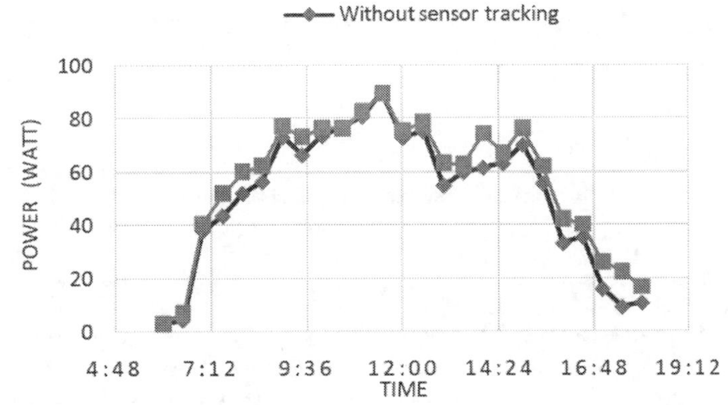

Fig. 8. Results of data acquisition by the solar panel using both fuzzy sensor and fixed position panels

Results of that data acquisition reveal that the power generated by the system employing the fuzzy light sensor is always on top. This means that fuzzy light sensor positively affects solar panel productivity. The effects are evident when the sunlight's angle approaches both 0° and 180°. This is further supported by the fact that the fixed position solar panel could not get enough sunlight, whereas the fuzzy light sensor solar panel was still capable of positioning itself toward the perpendicular line. Mathematically, it was then calculated that the fuzzy light sensor system improves power gain by up to 5.6%.

4 Conclusions

Sensors made of Light Dependent Resistors (LDRs) can be configured to be a light sensor with the properties of a fuzzy logic when they are installed parallel and given a light separator. The characteristics of this LDR fuzzy sensor is exponential with mirroring symmetry of correlation coefficient up to R=0.9981. Fuzzy membership of this sensor output can be utilized as input for the PID (Proportional-Integral-Derivative) control that in turn, is used to control the positioning of a solar panel. Differences in the fuzzy membership of both sensors are the foundation that decides the movement of the DC motor that moves the solar panel.

Fuzzy light sensor as input for PID control is capable for improving power performance of a solar panel compared to its fixed position counterpart. This fact is very evident when the angle of incoming sunlight approaches both 0° and 180°. Calculation results of data obtained from the fuzzy light sensor show power gaining improvement for solar panels by up to 5.6%.

References

1. T. N. Ngoc, Q. N. Phung, L. N. Tung, E. R. Sanseverino, P. Romano, F. Viola, Solar Energy, **150** pp. 325–334, (2017).
2. T. Sueyoshi, and D. Wang, Energy Economics, **65** pp. 389–398, (2017).
3. Vivek Kaundal, Amit Kumar Mondal, Paawan Sharma, Kamal Bansal, Engineering Science and Technology, **18** pp. 475-484, (2015).
4. S. Zhang, T. Wang, C. Li, J. Zhang, Y. Wang, *"Maximum power point tracking control of solar power generation systems based on type-2 fuzzy logic"*, 2016 12th World Congress on Intelligent Control and Automation (WCICA), pp. 770 – 774, (2016).
5. Y. M. Jagadeesh, S. Akilesh, S. Karthik, P. Prasanth, Procedia Technology **21** pp. 547 – 551, (2015).
6. K. Lee, S. Moon, B. Berkeley, S. Kim, Sensors and Actuators A: Physical, **130–131**, pp. 214-219, (2006).
7. Jeongjae Yoo and Yeonsik Kang, Journal of Automation and Control Engineering Vol. **2**, No. 4, December (2014), pp. 417 – 421
8. Stamatescu, I. Făgărăşan, G. Stamatescu, N. Arghira, and S. S. Iliescu, Procedia Engineering : 24th DAAAM International Symposium on Intelligent Manufacturing and Automation **69** pp. 500 – 507, (2014).
9. D. Fontani, P. Sansoni, F. Francini, D. Jafrancesco, L. Mercatelli, and E. Sani, International Journal of Photoenergy **2011**, (2011), pp. 1-9.

The Increase of Energy Consumption and Carbon Dioxide (CO_2) Emission in Indonesia

Hadi Sasana [1,*] and *Annisa* Eka Putri [2]

[1]Economics and Business Faculty , Diponegoro University, Semarang - Indonesia

Abstract. In the last decade, the increase of energy consumption that has multiplied carbondioxide emissions becomes world problems, especially in the developing countries undergoing industrialization to be developed ones like Indonesia. This aim of this study was to analyze the effect of fossil energy consumption, population growth, and consumption of renewable energy on carbon dioxide emission. The method used was multiple linear regression analysis with Ordinary Least Square approach using time series in the period of 1990 - 2014. The result showed that fossil energy consumption and population growth have a positive influence on carbon dioxide emission in Indonesia. Meanwhile, the consumption variable of renewable energy has a negative effect on the level of carbon dioxide emissions produced.

1 Introduction

The development of science and technology initially aiming to facilitate human work has caused new unrest and fear for human life as it has created negative implications to environment such as greenhouse effect that affects extreme climate changes on earth caused by increased carbon dioxide emissions. This effect disrupts the ecosystems of forest and others that function to absorb and reduce carbon dioxide.

Yang and Li [1] identified that technological advances lead to a rebound effect on carbon dioxide emissions in 30 provinces in China. Liberty, et al [2] who conducted a study by analyzing the impact of the natural resources used on the environment in Nigeria, a rich country in mineral resources like Indonesia, found that the natural resources greatly contribute to the national wealth and the people socially and economically, but, at the same time, a common problem of inadequate infrastructure emerges. Therefore, they suggested that both the government and the private sector should minimize waste through sorting, recycling, and pollution control. In addition, the government had to provide appropriate regulations, established law enforcement, and policy implementation. Meanwhile, the growing of the population as energy consumers results in the increased consumption of energy in aggregate, which drives a country to produce and provide more energy in order to meet their needs. The increase in the number of population can also lead to urbanizationthat has a positive and significant influence on energy consumption[3].

The increased energy consumption has resulted in the declining of the environmental quality. Impaired environmental quality can be measured by the increasing emissions or pollutants caused by energy exploration. In the long run pollutants can cause negative effects in the form of water, soil, and air contamination. One indicator that can be used to identify the amount of air pollution is to measure the levels of carbon dioxide emissions in the air due to energy exploration. According to U.S Energy Information Administration [4], human lifestyle triggers carbon dioxide as the largest emission or pollutant causing green house effect. The study of Ruijven et al [5] used CO_2 indicator because this gas is produced from production processes in various industrial sectors, including chemical and mining industries. In the cement industry, for example, every kg of cement produced contributes 0.5 kg of CO_2 emissions into the air. Therefore, CO_2 emissions becomes an indicator used by all sectors to measure air quality.

Meanwhile, Indonesia, the developing countries with abundant natural resources, according to World Resource Institute, is the 6th largest greenhouse gas emission-producing country. Some of the Indonesia's natural resources are potentially energy sources, both fossil and renewable, which might boost its economic development. Gerelmaa and Kotani [6] stated that abundant natural resources have a positive impact on economic growth. They identified countries having abundant natural resources in 1990 experienced good economic growth in the period of 1990 to 2010.

Liberty et al. [2] stated that the use of natural resources is essential in order to sustain life. Human being cultivates natural resources, either renewable or non-renewable, to produce sustainably supporting system to maintain population wellbeing by creating them for economic stability and social development. All life on earth is supported by energy and the sun. Plants and animals can store energy and some of them still exist even when the plants and animals have perished. The remains of ancient animals and plants form fossil energy.

* Corresponding author: hadisasana@live.undip.ac.id

The fossil energy is an unrenewable one that will run out at some point in time, and burning it produces greenhouse gases; therefore, fossil energy cannot be relied upon for energy generation. Consequently, renewable and sustainable energy that can be used repeatedly are needed.

Carbon dioxide (CO_2) is a chemical compound consisting of two oxygen atoms bonded covalently with a carbon atom. CO_2 is gaseous at the state of standard temperature and pressure and is present in the Earth's atmosphere. In addition, CO_2 is a colorless and odorless gas that can be produced by all animals, plants, fungi, and microorganisms. CO_2 can also be produced from the side effects of burning fossil energy.

Thao and Chon [7] stated that energy consumption has a positive impact on economy, but not to the environment. Energy consumption is widely known as the main reason for global warming and climate change to happen, particularly the consumption of fossil energy. The environmental adverse effects of such energy used are not only coming from the energy consumption but also from the exploitation process. Meanwhile, the renewable energy consumption has a negative relationship to CO_2 emissions, which means that an increase in the consumption of renewable energy will reduce CO_2 emissions.

Shi [8] argued that population growth can increase total carbon dioxide emissions, both in developed and developing countries. Thus, the larger the population is, the higher the human activity will be, which leads to the more energy is demanded. Shi [8] also concluded that if there is an increase in population by 1 percent, the emissions will increase by 1.28 percent in average. In addition, the impact of population growth on CO_2 emissions is significant in developing countries such as Indonesia than in developed countries. The study of Fei Li et al [9] using panel data from 28 Provinces in China found that energy consumption and economic growth, in long-term, affect CO_2 emissions, but CO_2 emissions and long-term economic growth have an impact on energy consumption.

Meanwhile, Ito [10] found that fossil energy consumption has a negative impact on economic growth in developing countries, and renewable energy consumption has a positive effect on economic growth. In this case, the consumption of fossil energy can cause pollution and environmental damage because the remaining burning of fossil energy is harmful to the environment; while, the renewable energy residue is considered more environmentally friendly. Moreover, Shafei and Ruhul [11] who conducted a study on OECD countries on the Kuznets Curve Hypothesis (EKC) between urbanization and CO_2 emissions found that non-renewable energy consumption has a positive relationship to CO_2 emissions, which means that an increase in non-renewable energy consumption will increase CO_2 emissions. In contrast, renewable energy consumption has a negative relationship to CO_2 emissions, which means that an increase in the consumption of renewable energy will reduce CO_2 emissions.

Given this description, this study aims to analyze the effect of fossil energy consumption, population growth, and renewable energy consumption on carbon dioxide emissions in Indonesia.

2 Research Method

This research used secondary data with Ordinary Least Square (OLS) regression analysis technique to investigate the influence of fossil energy consumption, population growth, and renewable energy consumption as the independent variable on carbon dioxide emission in Indonesia as the dependent variable. The data used were time series data from 1990 to 2014. In the time-series data, Logarithm Natural Equation (Ln) model was used to facilitate the estimation. In addition, Gujarati [12] proposed that the reason for choosing Logarithm Natural Function (Ln) is to close the data scale and to transform the data in case of deviation of classical assumption takes place. Thus, the estimated basic model is as follows:

$$CO_2 = (Pop, Fosil, R) \quad \text{......} 1$$

$$lnCO_{2t} = \alpha_0 + \alpha_1 Pop_t + \alpha_3 Fosil_t + \alpha_4 R_t + \mu_t \quad \text{......} 2$$

Note:

CO_2	= Total emissions of carbon dioxide (CO_2) generated from energy consumption in Indonesia
Pop	= Population growth
Fossil	= Fossil energy consumption
R	= Share of renewable energy consumption to total energy consumption
α_0	= Intercept
α	= Value of variable coefficients
t	= 1,2,3, ..., 25 (time series data of 1990 - 2014)
ln	= Natural log (LN)
μ	= Error term

3 Results And Discussion

Indonesia, located in Southeast Asia, is an archipelago country having approximately 17,000 islands with 1,922,570 km2 land and 3,257,483 km2water. Indonesia is the fourth most populous country in the world after China, India, and the United States. In 2016,its population was 263,846,946 million that make Indonesia has advantages as well as challenges for the future. Amid the imbalanced global economic recovery, the economic performance of Indonesia has continued to improve since 2010 marked by the economy growth in all sectors. In 2012, Indonesian economic growth was 6.5 percent; the highest figure in the last ten years[13]. The high rate of the economic growth was followed by the escalation in industrialization. Along with the growth of industrialization and population, the primary energy consumption in Indonesia also enlarged.

The higher consumption of energy led to greater CO_2 emissions or pollutants. CO_2 would continue to increase along with the increasing rate of industrialization in many developing countries, especially in Indonesia. In 1990 – 2014, there was an increase of carbon dioxide emissions as many as 0.045 million metric tons or 4.5% per year.

3.1 Description of the Variable

Table 1 exhibits that the average emissions of carbon dioxide in Indonesia is 5.7%, the average of the population growth is 1.4%, the average consumption of renewable energy is 45.5%, and the average consumption fossil energy is 62.1%.

Table 1. Descriptive Statistics of Research Variables

Var	Mean	Median	Max	Min	S.D
Ln CO_2	5.6626	5.666427	6.10256	5.0563	0.3186
POP	1.4241	1.352808	1.78145	1.2602	0.1526
R	45.458	44.37160	58.5976	38.066	6.5434
FOSSIL	62.109	62.44202	67.1548	53.431	3.6272

Source: Secondary Data, processed

3.2 Research Result

The result of the regression estimation of the variable of fossil energy consumption, population growth, and consumption of renewable energy to carbon dioxide emission in Indonesia during 1990 – 2014 by using regression method of Ordinal Least Square is as follows.

Table 2. Regression results

Variable	Coeffi	Std. Err	t-Statistic	Prob.
C	6.449592	1.324913	4.867938	0.0001
POP	1.417481	0.421336	3.364249	0.0029
R	-0.074606	0.012339	-6.046490	0.0000
FOSSIL	0.009431	0.015036	0.627216	0.5373
R-squared	0.964246	Mean dep.var		5.66257
Adjusted R2	0.959138	S.D. dep var		0.31862
S.E. of regr	0.064408	Akai criterion		-2.5015
Sum squ resid	0.087115	Schwarz criter		-2.3065
Log likelihood	35.26900	HQ criter.		-2.4474
F-statistic	188.7812	DW stat		1.31667
Prob(F-stat	0.000000			

The result of the regression estimation reveals that R2is equal to 0.964246, meaning that 96.4% of the dependent variable could be explained by the independent variable, and the rest of the 3.6% was explained by the variable of the outside model. At the significance level of $\alpha = 5\%$ with F prob = 0.00000, it could be concluded that simultaneously the variable of the population growth, renewable energy consumption, and fossil energy consumption influenced the amount of carbon dioxide emission in Indonesia. Begum et al. [14], on the effect of energy consumption and population growth on carbon dioxide emissions in Malaysia, reported that energy consumption has proven to have a positive effect on carbon dioxide emissions; while, population growth does not significantly affect the emissions of carbon dioxide.

Further analysis of this study found that the population growth had a significant positive effect on the amount of carbon dioxide emissions in Indonesia with a coefficient of 1.417481, which meant that if the population growth increases, the emissions of the carbon dioxide will also increase linearly. The population growth is one aspect that greatly affects the increase of carbon dioxide emissions around the world in the last two decades (Shi, 2001), in particular, the effect of the population growth in China positively and significantly affect carbon dioxide emissions [15].In addition, Wang et al [16] also stated that the population growth has a strong influence on the increase of carbon dioxide emissions in 30 provinces in China.

Furthermore, Yazdi, et al. [17] who conducted a research in Iran from 1975-2011 accounted that the population growth in Iran has a positive and significant relationship to CO_2 emission with the elasticity ratio of the CO_2 emissions to population growth is 0.97. Meanwhile, Yeh and Liao [18] found that the population growth has a positive impact on carbon dioxide emissions; the more the number of the population is, the more carbon dioxide emission are generated. Communities consume energy for various purposes; so that, the more the number of the population growth, the more energy consumption is demanded that result in the increased carbon dioxide emissions.

Further findings of this study was that the renewable energy consumption negatively affected the amount of carbon dioxide emissions in Indonesia. With the coefficient of -0.074606, the figure meant that if the consumption of renewable energy increases, the carbon dioxide emissions will be reduced. Similarly, Fathinah and Djoni [19] found that the estimation result has a negative and significant relationship between the proportion of renewable energy consumption and carbon dioxide emissions in ASEAN. Another finding by Bilgili et al. [20] also showed that the amount of renewable energy consumption has a negative and significant impact on the carbon dioxide emissions. This finding might indicate that the consumption of renewable energy could significantly reduce the carbon dioxide emissions. In this case, the increasing consumption of renewable energy was expected that reducing the dependence on fossil energy consumption would reduce the carbon dioxide emissions.

The study of Paramati et al. [21] in G20 countries concluded that the consumption of renewable energy reduces CO_2 emissions and increases the countries' economic output. A similar study conducted by Zoundi [22] found that renewable energy consumption has a negative effect on CO_2 emissions, and it is considered more environmentally friendly than fossil energy. In addition, in the long-term, the consumption of renewable energy will increase as it is efficient to replace fossil energy. Moreover, Liu et al [23] obtained similar result for renewable energy consumption negatively affects carbon dioxide emissions in 4 ASEAN countries (Indonesia, Malaysia, Philippines, and Thailand). The estimation results indicated that an increase in renewable energy consumption could reduce carbon dioxide emissions; so that, the use of the renewable energy was encouraged to achieve a cleaner and healthier environment.

The latest finding of this study was that the effect of the fossil energy consumption on the amount of carbon dioxide emissions in Indonesia was positive but not significant. This result was similar to the one of Bulut [24] who argued that the non-renewable energy or fossils had a positive effect on carbon dioxide emissions in Turkey in 1970 – 2013. Meanwhile, Shafiei and Ruhul [11] stated that the increased non-renewable or fossil energy consumption led to the increase of CO_2 emissions in OECD countries during 1980 to 2011. Similarly, Dogan and Fahri [25] who conducted a research in European countries stated that the consumption of nonrenewable energy will increase CO_2 emissions, and there is an indirect causal relationship between CO_2 emissions and non-renewable energy consumption. Zheng-Xin and De-Jun [26] proved that the consumption of fossil energy causes an increase in carbon dioxide emissions; while, Danish et al. [27] stated that the consumption of fossil energy has a positive effect on carbon dioxide emissions in Pakistan. Fossil energy consumption was the main cause of carbon dioxide emissions, and the combustion of fossil fuels are carbon dioxide gas that might damage the environment and people's health. Chibueze et al. [28] also stated that in the short and long term, the fossil energy consumption proved to have a positive and significant influence on the carbon dioxide emissions in Nigeria in the period 1971-2009.

4 Conclusion

Given the results and discussion of this research, some conclusion can be drown:

1. Based on the results of the OLS regression, the population growth variable has a positive and significant effect on carbon dioxide emissions in Indonesia in 1990-2014.
2. The variable of the consumption of the non-renewable fossil energy has positive but not significant effect on carbon dioxide emission in Indonesia in 1990-2014.
3. The variable of the consumption of the renewable energy has a negative and significant effect on Indonesia's carbon dioxide emissions in 1990-2014.

The results of the research suggested that the increased consumption of the non-renewable energy or fossil will increase the amount of carbon dioxide emissions; on the contrary, the consumption of the renewable energy can reduce the amount of the carbon dioxide emissions. Therefore, a developing country like Indonesia where industrialization is in progress, it is very important to reduce the use of the fossil energy and shift to the renewable one. In addition, the government is also expected to provide legislation and strict law enforcement if a production produces excessive emissions thereby lowering the quality of the environment.

Reference

1. Yang, Lisha and Li, Zhi. Science Direct *Energy Policy* Vol. **101** (2017) page 150-161.
2. J.T. Liberty, B.O. Ugwushiwu, G. IBassey, and V.N Eke. International Journal of Scientific & Engineering Research, Vol. **4**, No. 8, (2013) page 2115-2122.
3. Yang, Yingcun, Jianghua Liu and Yutao Zhang. An Journal of Cleaner Production Vol.**161**. (2017) page 1251-1262.
4. U.S Energy Information Administration. Download, 10 Juli 2017. https://www.eia.gov/
5. Rujiven, Bas J. Van, Detlef P. Van Vuuren, Willem Boskaljon, Maarten L. Neelis, Deger Saygin, and Martin K. Patel. *Resources, Conservation and Recycling*. Vol. **112**, (2016) page 15-36.
6. Gerelmaa, Lkhagva and Koji Kotani. Resources Policy Journal Vol. **50**, (2016) page 312-321.
7. Thao, Nguyen Thi Ngan and Le Van Chon. *Nonrenewable, renewable energy consumption and economic performance in OECD countries: A stochastic distance function approach*. (2016). http://veam.org/wp-content/uploads/2016/08/70.-Nguyen-Thi-Ngan-Thao.pdf
8. Shi, Anqing. *Population Growth and Global Carbon Dioxide Emissions*. Development Research Group The World Bank (2001).

   http://archive.iussp.org/Brazil2001/s00/S09_04_Shi.pdf
9. Fei Li, Suocheng Dong, Xue Li, Quanxi Liang, and Wangzhou Yang. Energy Policy.Vol. **39**, (2010) page 568-574.
10. Ito, Katsuya. International Economics.Vol. **151**. (2017) page 1-6.
11. Shafei, Sahar and Ruhul A. Salim. Energy Policy. Vol. **66**, (2013) page 547-556.
12. Gujarati, N.D. and Porter, D.C. *Basic Econometrics*. International Edition McGraw-Hill/Irwin, A Business Unit of The McGraw-Hill Companies, Inc., New York (2009).
13. Badan Pusat Statistik. *Ekonomi dan Perdagangan* (2012). https://bps.go.id/
14. Begum, RawshanAra, Kazi Sohag, Sharifah Mastura Syed Abdullah and MokhtarJafar. Renewable and Sustainable Energy Review. Vol. **41**, (2015) page 594-601.
15. Hang, Guo and Jiang Yuan-Sheng. Procedia Environmental Sciences. Vol. **11**, (2011) page 1183-1188.

16. Wang, Yanan, Yanqing Kang, Juan Wang and LinanXu. Ecological Indicators. Vol. **78**, (2017) page 322-330.
17. Yazdi, Soheila Khoshnevis, Bahman Khanalizadeh, and Nikos Mastorakis. Renewable, Non-Renewable Energy Consumption, Economic Growth and CO_2 emission: Evidence for Iran. *Advances in Environmental Sciences, Development and Chemistry*. (2014) page 399-404.
18. Yeh, Jong-Chao and Chih-hsiang Liao. Sustainable Environment Research .Vol. **27** (1), (2017) page 41-48.
19. Fathinah, Atikah and Hartono, Djoni. *Hubungan Antara Emisi Karbondioksida, Efisiensi Energi dan Konsumsi Energi Terbarukan di Asean (2000-2011)* (2014). http://www.lib.ui.ac.id/naskahringkas/2016-05/S56649-Atikah%20Fathinah
20. Bilgili, Faik, EmrahKocak and Umit Bulut. Renewable and Sustainable Energy Reviews Vol. **54**, (2016) page. 838-845.
21. Paramati, Sudharshan Reddy, Di Mo, and RakeshGupta. Energy Economics. Vol.**66**. (2017) page 360-371.
22. Zoundi, Zakaria. Renewable and Sustainable Energy Reviews. Vol. **72**, (2017) page 1067-1075.
23. Liu, Xuyi, Shun Zhang dan Junghan Bae. Journal of Cleaner Production. Vol.**164**. (2017) page 1239-1247.
24. Bulut, Umit. Environmental Science and Pollution Research Vol. **24**, (2017) page 15416-15426.
25. Dogan, Eyup and Fahri Seker. Renewable Energy. Vol. **94**, (2016) page 429-439.
http://www.sciencedirect.com/science/article/pii/S0960148116302622
26. Zheng-Xin, Wang and De-Jun Ye. Journal of Cleaner Production Vol. **142**, (2017) page 600-612.
27. Danish, Bin Zhang and Zhaohua, Wang. Journal of Cleaner Production.Vol. **156**, (2017) page 855-864.
28. Chibueze, E. Nnaji, O. Chukwu Jude and Nnaji Moses. International Journal of Energy Economics and Policy Vol. **3**, No. 3, (2013) page 262-271.

 https://doi.org/10.1051/e3sconf/20183101009

CFD Analysis to Calculate the Optimal Air Velocity in Drying Green Tea Process Using Fluidized Bed Dryer

Eflita Yohana[1,*], *Afif* Prasetya Nugraha[2], *Ade* Eva Diana[2], *Ilham* Mahawan[2], and *Sri* Nugroho[1]

[1] Department of Mechanical Engineering, Faculty of Engineering, Diponegoro University, Semarang - Indonesia
[2] Undergraduate of Mechanical Engineering Department, Faculty of Engineering, Diponegoro University, Semarang - Indonesia

Abstract. Tea processing is basically distinguished into three types which black tea, green tea, and oolong tea. Green tea is processed by heating and drying the leaves. Green tea factories in Indonesia are generally using the process of drying by panning the leaves. It is more recommended to use the fluidization process to speed up the drying process as the quality of the tea can be maintained. Bubbling fluidization is expected to occur in this research. It is a process of bubbles are formed in the fluidization. The effectiveness of the drying process in a fluidized bed dryer machine needs to be improved by using a CFD simulation method to proof that $u_{mf} < u < u_t$, where the average velocity value is limited by the minimum and the maximum velocity of the calculation the experimental data. The minimum and the maximum velocity value of the fluidization is 0.96 m/s and 8.2 m/s. The result of the simulation obtained that the average velocity of the upper bed part is 1.81 m/s. From the results obtained, it can be concluded that the calculation and the simulation data is in accordance with the condition of bubbling fluidization in fluidized bed dryer.

1 Introduction

Tea is a beverage form widely consumed in many countries. This beverage contains numerous active ingredients that act as both anti-oxidants and antimicrobials. Essentially, tea is processed into three various kinds which are black tea, green tea, and oolong tea. The amount of tea in the world that is processed into black tea is more than three quarters part of it. The process of black tea processing includes leaves chopping, and then drying the chopped ones in the sun until there is a chemical change that turns the leaves colour to brown and also provides a distinctive taste in the tea. For green tea, its leaves are heated only the directly dried so that the leaves colour remain the same. Oolong tea itself is a transitional type between black tea green tea. There is a fermentation occur in its processing that creates different flavour and characteristics to oolong tea. Also, the tea has health benefits and potentials that is as same as other tea [1].

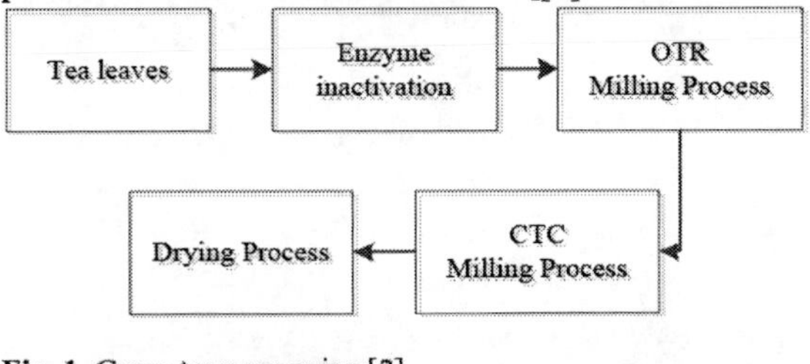

Fig. 1. Green tea processing [2].

A number of studies have found that catechins are such compound that is most responsible for the healthful effects that tea could provide. For instance, green tea is not only used as beverage products, but it also has an important component that can be used for foods, pharmaceutical, body and beauty products. The process scheme in Figure 1 shows the latest technology of green tea processing that works in to obtain a higher cathecin content than the one that is generated from conventional processing. In order to produce such green tea is by inactivating the enzyme polyphenol oxidize enzyme through steaming process that is applied evenly throughout the leaves' surface. Then proceed with a milling process which is a combination of OTR (Open Top Roller) and CTC (Crushing Tearing Curling) to increase the dissolved soils. The last process is namely drying process. This process uses fluidized bed dryer to get the tea leaves that contain water content of 4% - 5% and to avoid over burnt tea leaves in the drying process [2].

The fluidization drying method is used to speed up the drying process and maintain the leaves quality. This drying method is widely applied to dry such particulate or granular materials used for chemical, food, ceramic, pharmaceutical, agricultural, polymer and waste industries. There is a materials blowing occurring in this condition, thus it expands the drying area, increases the convection heat transfer coefficient and the rate of vapor diffusion. In order to achieve a fluidization condition, the

* Corresponding author: efnan2003@gmail.com

airflow velocity must be greater than the minimum fluidized velocity in which the dried materials can still be fluidized properly. When the airflow velocity is increased during this process, the static pressure of the drying air increases and the dried material is lifted to a certain height (floating) and causes the material be fluidized. In this condition, the material is uniformly stirred and the air cushion that holds the material at a certain height is in a minimum fluidized state [3].

The minimum fluidization velocity $\mathcal{U}_{mf}$ is used to determine the fluidization starting point so that the compressed lost value at the beginning state of the fluidization can be calculated. The minimum velocity fluidization equation is generally written as following equation :

$$\mathcal{U}_{mf} = \frac{(\psi\, d_p)^2}{150\mu}\,[g(\rho c - \rho g)]\,\frac{\varepsilon_{mf}^3}{1-\varepsilon_{mf}} \qquad (1)$$

where ψ is sphericity, d_p is particle diameter, g is gravity, ρ_c is the mass of the particle type, ρ_g is the vapor type mass, ε_{mf} is the porosity at the minimum state of fluidization [4]. There are several phenomenons that occur in the fluidization process, which they are [3] :

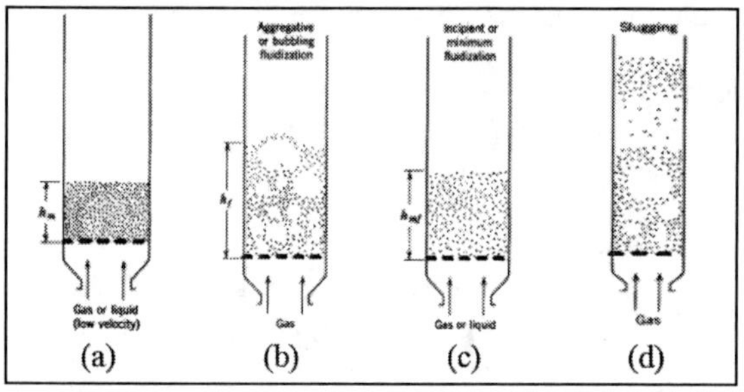

Fig. 2. Occurring phenomenon in fluidization process (a) fixed bed $\mathcal{U}<\mathcal{U}_{mf}$ (b) Minimum Fluidization $\mathcal{U}=\mathcal{U}_{mf}$ (c) Bubbling Fluidization $\mathcal{U}>\mathcal{U}_{mf}$, $\mathcal{U}<\mathcal{U}_t$ (d) Slugging [3].

a. The fixed bed phenomenon will occur when path for the fluid flow is less than the minimum rate that is required for the initial fluidization process. In this particular condition, the solid particles will remain stationary. The aforementioned phenomenon is shown in Figure 2(a).
b. The minimum phenomenon or incipient fluidization occurs when the fluid flow rate reaches the minimum flow rate that is required for the fluidization process. The solid particles begin to expand or float in this condition. This phenomenon is shown in Figure 2(b).
c. The bubbling fluidization phenomenon will occur when the density and distribution of the particles are not homogeneous, thus resulting in bubbles in the fluidization process. This phenomenon can be seen in Figure 2(c).
d. The slugging phenomenon will occur when the large bubbles that reach the column diameter width are formed in the solid particles. In this condition, there is a rampage so that the solid particles are lifted up. The just aforementioned phenomenon can be seen in Figure 2(d).

In the present study, fluidization can be used to speed up the drying process as well as to maintain the quality of the tea. Bubbling fluidization is expected to occur. It is a process of bubbles are formed in the fluidization. The effectiveness of the drying process in a fluidized bed dryer machine can be improved by using a CFD simulation method.

(1)

2 Research methods

2.1 Flow chart of research

The flow chart used for this study is shown in Figure 3 below.

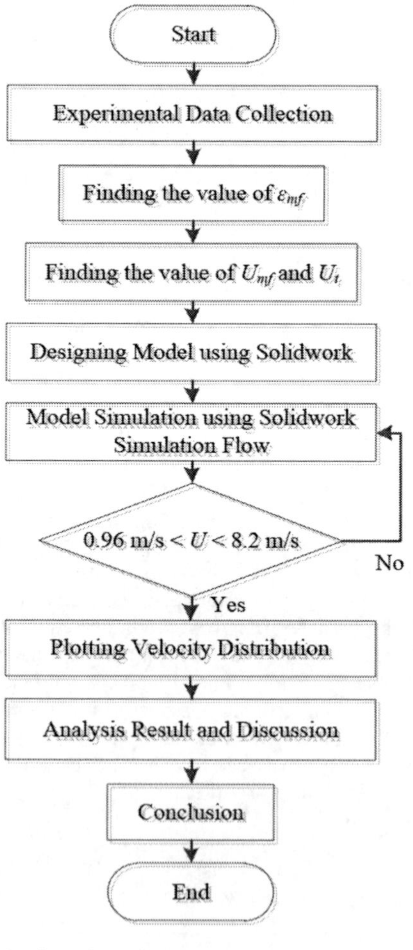

Fig. 3. Flow chart of research.

2.2 Experimental data

The data was collected at a tea factory located in Cibuni Estate area, West Java on a fluidized bed dryer with the following parameters in Table 1.

Table 1. Parameters and operating conditions in numerical simulations.

Parameter	Values
Temperature inlet	130^0 C
Velocity inlet	2.6 m/s
Particles diameter	0.5 mm
Particles weight	0.0016 gr
Pressure	2.5506 bar

2.3 Data calculation as parameter form

This calculation was done to look for the minimum porosity value of fluidization(ε_{mf}). It is used to find the minimum fluidized velocity value, as well as the maximum velocity value of fluidization.

$$\varepsilon_{mf} = 0.586\ \psi^{-0.72} \left(\frac{\rho_g}{\rho_c}\right)^{0.021} \left(\frac{\mu^2}{\rho_g \eta d_p^3}\right)^{0.029} \quad (2)$$

where ε_{mf} is the porosity at the minimum of fluidization, ψ is sphericity, μ is viscosity, η is gravitation term, d_p is particle diameter, ρ_g is density of the fluid, ρ_c is density of the particle [5]. The value are $\psi = 0.33$, $\rho_g = 0.00078$, $\eta = 239700.38$, $\mu = 0.282$, $\rho_c = 24.46$ and ε_{mf} is 0.888.

2.4 Velocity analyze simulation

The problem to be discussed in this study is the velocity analysis on fluidized bed dryer. The models are made using Solidworks. The assumptions used in the simulation are seen in few following points, which are :

1. Fluid velocity inlet of 2.6 m/s
2. The fluid used is in steam
3. Simulation is in steady state condition
4. The absence of heat transfer
5. The flow appears as turbulent flow
6. The pressure used is environmental stress of 101325 Pa

2.5 Fluidized bed geometry

The geometry on fluidized bed design is modeled by Solidworks software. This model is shown in Figure 4 below, which consists of some models that are assembled together into one single unit.

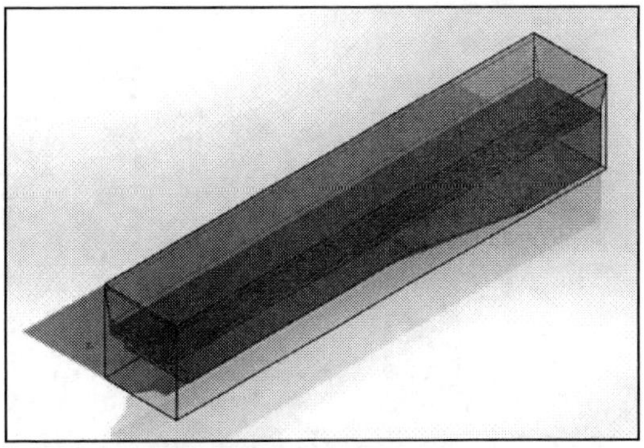

Fig. 4. The overall looks of assembly model.

3 Analysis results

3.1 Calculation of minimum fluidization velocity

The calculation of minimum fluidization velocity resulted as follows :

$$\mathcal{U}_{mf} = \frac{(\psi\, d_p)^2}{150\mu} \left[g(\rho c - \rho g)\right] \frac{\varepsilon_{mf}^3}{1-\varepsilon_{mf}} \quad (3)$$

where $\eta = g(\rho c - \rho g)$ [4]. The value are $d_p = 0.5$ and $\mathcal{U}_{mf} = 0.96\ {}^{m}/_{s}$.

3.2 Calculation of maximum fluidization velocity

The calculation of maximum fluidization velocity resulted as follows [3]:

$$\mathcal{U}_t = \left(\frac{1.78 \times 10^{-2} \eta^2}{\rho_g \mu}\right)^{\frac{1}{3}} d_p \quad (4)$$

where the value are $\mathcal{U}_t = 8.2\ {}^{m}/_{s}$.

3.3 Simulation result analysis

The simulation result on fluidized bed can be seen in Figure 5.

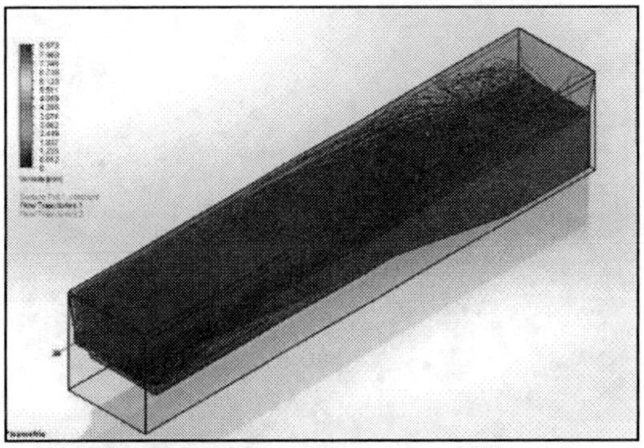

Fig. 5. The overall look of simulation result.

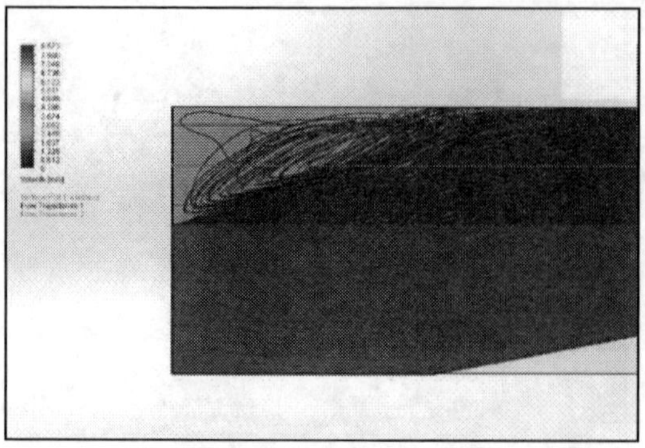

Fig. 6. The above view is zoomed particularly in the inlet area.

Figure 6 shows that when fluid enters the column side, the velocity increases as the lower bed cross-sectional area diminishes. We can also see in it that the fluid passing through the velocity filter will later decrease.

Fig. 7. The view is zoomed-in in the middle area of fluidizied bed.

Figure 7 shows that the lower bed velocity is increased due to the smaller cross-sectional area caused by the shape of the lower bed base leading to this that causes the fluid passing through the velocity filter is also increasing. We can also see in the upper bed that the fluid velocity increases too that is caused by the shape of the lower bed base that shrinks and points upward.

Fig. 8. The view is enlarged around outlet area.

Figure 8 shows that the fluid trapped in the lowerbed end makes a turbulence that serves to lift the falling particles from the filter so that the particles will rise back upward. On the upper bed, it indicates that the velocity of the outlet starts to accelerate according to the law of continuity; this is due to the fluid insistence from the lower bed that is getting more toned once it gets toward the end.

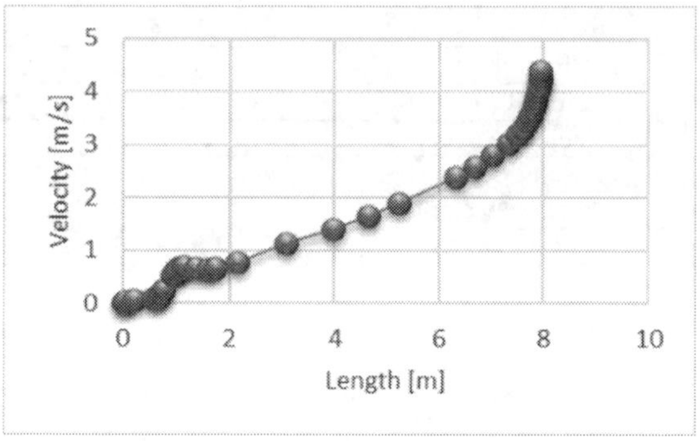

Fig. 9. Graphs of plot result in upperbed area.

Figure 9 above is a graph of the plot result in upper bed area. We can know from the graph that the fluid velocity is increasingly close to the oulet end of the bed. The average speed that occurs along the upper bed area is 1.81 m/s.

4 Conclusion

Prior calculating all of the data manually and conducting the simulation with Solidworks, the minimum fluidization velocity ($\mathcal{U}_{mf}$) was 0.96 m/s, the average fluidization velocity value ($\mathcal{U}$) was 1.81 m/s, and the maximum fluidized velocity value ($\mathcal{U}_t$) was 8.2 m/s. Overall, it can be concluded that from the results obtained, it can be expressed by following condition $\mathcal{U}_{mf} < \mathcal{U} < \mathcal{U}_t$; that the minimum fluidization velocity of 0.96 m/s is less than the average fluidization velocity of 1.81 m/s, and the average fluidization velocity is less than the maximum fluidization velocity of 8.2 m/s. This expression indicates that the calculation and simulation data is already in accordance with the condition of fluidization rule with such bubbling fluidization phenomenon occurs on fluidized bbed dryer. Whereas the bubbling fluidization phenomenon turns to be the best phenomenon to achieve a fluidization condition on the fluidized bed dryer machine.

References

1. A.D. Ananda, *Aktivitas Antioksidan dan Karakteristik Organolaptik Minuman Fungsional Teh Hijau (Camellia Sinensis) Rempah Instan, Bogor :* Fak Pertanian Institut Pertanian Bogor, (2009).
2. J. Towaha, *Kandungan Senyawa Kimia Pada Daun Teh (Camellia Sinensis), Warta Penelitian dan Pengembangan Industri,* **19** no 3, (2013).

https://doi.org/10.1051/e3sconf/20183101009

3. M. Sadeghi, *Vibration Effect on Particle Bed Aerodynamic Behavior and Thermal Performance of Black Tea in Fluidized Bed Dryers.* Iran : Tarbiat Modares University, (2012).
4. W. Widayati, *Fenomena dan Kecepatan Fluidisasi (U_{mf}),* **10**, (2010).
5. Pabis, *Measurements of porosity of gas – solid fluidized bed,* (2002).

A Case Study: The Potential of Energy Efficiency in Senior High School of Semarang Regency, Central Java, Indonesia

Ana Yustika[1,*], *P* Purwanto[2], and *H* Hermawan[3]

[1]Master Program of Environmental Science, School of Postgraduate Studies, Diponegoro University, Semarang - Indonesia
[2]Chemical Engineering Department, Faculty of Engineering, Diponegoro University, Semarang – Indonesia
[3]Electrical Engineering Department, Faculty of Engineering, Diponegoro University, Semarang – Indonesia

Abstract. The increasing of energy supply trend in Indonesia seems to be a serious problem in the implementation of sustainable development. This study case research aimed to determine the potential of energy efficiency in school environment. The subject of this research was SMA N 1 Ambarawa, located on Semarang Regency of Central Java, Indonesia. The data collection was done by used documentation, observation and interview method. The results showed that the average of electrical energy consumption in this school reached 11022.008 kWh/month, which resulted in the emergence of secondary emissions of CO_2 by 9644.257 kg CO_2/month. Overall, the consumption of electrical energy in this school was very efficient, with an Intensity of Energy Consumption (IEC) average 1.7957 kWh/m^2/month. In this case, the implementation of short-term no cost, long-term no cost, middle-cost, short-term high cost and long-term high-cost recommendation could save electricity energy sequent by 3.159%; 7.536%; 9.499%; 35.278% - 36.626%; and 42.084%. In conclusion, the school environment had a big potential of energy efficiency that could reduce the energy consumption and CO_2 gas emissions.

1 Introduction

Energy was one of the key factors to increase the economic growth level in a region [1]. Therefore, energy efficiency became one of the great programs that were currently developed by many countries in the world to support the implementation of sustainable development [2]. According to the Indonesian Law Number 32 of 2009 [3], the Indonesian Government also participated in implementing of efficiency energy. There were three pillars of sustainable development was be targeted in implementation of energy efficiency program were economic, social, and environmental aspect [4].

The high population growth significantly gave impact in energy consumption [5]. Impact of this, people would always demand an increase of energy supply to fulfill the economic and social needs [6]. In accordance with this, the trend of electrical energy supply in Indonesia showed an increase of 8.85% per year [7], with the largest percentage of electric energy source was from Steam Power Plant. The used of coal continuously as a primary energy source of Steam Power Plant caused the depletion of the coal quantity as one of the non-renewable natural resources on earth [8]. In addition, the use of coal for electric energy also caused of CO_2 emissions as the main cause of global warming phenomenon [9]. However, in reality, the burning activity of fossil fuel still happened every day in every country in the world [10]. Therefore, it should be emphasized the importance of improving energy efficiency in Indonesia to fulfill the economic and social needs of society, but without harm the environment [11]. In this case, the government was demanded to make a policy as soon as possible to reduce the amount of CO_2 emissions produced in the region [12].

The construction sector was expected to have a significant impact on implementation of energy efficiency program [13]. About of 42 percent of energy supply in the world was used for operational building [14]. Therefore, the Indonesian government also adopts a policy of energy efficiency in this sector. One of them is through by Adiwiyata program for school environment which expected being 'Adiwiyata school' or 'green school'.

Adiwiyata program has already begun since 2006 as a form of cooperation between the Ministry of Environment and The Education & Culture Department in Indonesia. As a result, in year 2006-2011, only 272 of the total 251,415 schools in Indonesia that successfully received a predicate as Adiwiyata School (MoE, 2012). This number indicates that the Adiwiyata program is still difficult to be implemented, so the stakeholders in 2013 revised the guidelines for the implementation of Adiwiyata program which contained in the Minister of Environment Rule Number 05 of 2013. However, it seems that the guidelines not bring significant impact in increase the number of Adiwiyata school. This is

* Corresponding author : anayustika13@gmail.com

because the guidelines not completed with the ways that can be done to determine the potential of energy efficiency which is owned by school. Based on this, this study was specifically conducted to investigate the potential of the energy efficiency in school environment.

2 Materials & Methods

This research was a study case, which aimed to determine the potential of energy efficiency in school environment. As for the high school which used as the research subject was SMA N 1 Ambarawa, located in Semarang Regency, Central Java. The data collection was done by using documentation, observation and interview method. Data analysis was done by test the statistic hypothesis by using t-test analysis [15] and SWOT analysis [16].

Planning strategy is needed to determine the steps which to be taken by a person or an agency to achieve the goal. In this case, SWOT analysis was done to determine the appropriate recommendation to improve the potential of energy efficiency in SMA N 1 Ambarawa. There are four key words in a SWOT analysis, which Strengths, Weakness, Opportunities, and Threats. Then, it can be formulated in SWOT matrix consist of four kinds planning strategies are S-O, W-O, S-T, and W-T.

In addition, data analysis was also done by calculating the Intensity of Energy Consumption (IEC), which referred to SNI 03-6196-2000 [17] and calculated the amount of secondary emissions of CO_2 which arose from the use of electrical energy in school which noticed the emissions produced by Steam Power Plant with used.

The formula calculation of CO_2 emission [18] :

$$CO_2 \text{ emission} = \Sigma A_i \times EF \quad \text{(i)}$$

With :

ΣA_i = amount of electricity consumed (kWh)

EF_i = emission factor of electricity in Central Java, Indonesia (0.857 kg CO_2/kWh)

3 Results & Discussion

3.1 Energy consumption

Electrical energy is the primary energy source used in SMA N 1 Ambarawa to support the learning activities. With an area of 6138 m^2 and the number of people in school was 1100 persons, the average of electrical energy consumption was 11022.008 kWh/month. The electricity consumption in SMA N 1 Ambarawa during past year can be seen in Table 1.

Table 1. Electricity consumption

No	Month – Year	The Number of Effective Days	Energy Subsidies from the Government (50%) (IDR)	Electricity Consumption (kWh)
1	Jan-2015	24	5,278,656	11730.35
2	Feb-2015	23	5,410,748	12023.88
3	March-2015	25	5,361,994	11915.54
4	April-2015	25	5,449,562	12110.14
5	May-15	23	5,397,015	11993.37
6	June-2015	14	3,469,587	7710.19
7	July-2015	8	3,014,779	6699.51
8	Aug-2015	26	5,086,717	11303.82
9	Sept-2015	25	5,357,083	11904.63
10	Oct-2015	27	5,459,791	12132.87
11	Nov-2015	25	5,500,139	12222.53
12	Dec-2015	17	3,774,551	8387.89
13	Jan-2016	24	5,458,071	12129.05
14	Feb-2016	24	5,419,960	12044.36

Over the past year, the highest electricity consumption amounted to 12129.047 kWh which occured in November 2015, while the lowest power consumption of 6699.509 kWh which occurred in July 2015.

Table 2. Light intensity & air temperature on each room

No	Room	Result			Standard			
		Air Temperature (°C)	Light Intensity (Lux)	Temperature Color	Air Temperature (°C) Minister of Health	Light Intensity (Lux) SNI	Light Intensity (Lux) Minister of Health	Temperature Color SNI
1	Class Room	27.5	277	Cool white	18 - 28	250	200-300	Cool white/Day light
2	Teacher Room	28.3	291	Cool white	18 - 28	-	200-300	-
3	Counseling Room	27.5	256	Cool white	18 - 28	-	200-300	-
4	UKS Room	27.6	234	Cool white	18 - 28	-	200-300	-
5	Around the Stairs	28.8	134	Cool white	-	-	100	-
6	Laboratory Room	23.7	503	Cool white	18 - 28	500	200-300	Cool white/Day light
7	Library Room	27.9	327	Cool white	18 - 28	300	200-300	Cool white/Day light
8	Canteen	30.3	319	Cool white	-	200	100	Warm white/Cool white
9	Toilet	26.6	114	Warm White	18 - 28	-	100	-
10	Praying Room	28.3	138	Cool white	18 - 28	-	100	-

The result of t-test indicated that there was no significant difference between electricity consumption in the odd semester and the second semester. This result consistent research result which showed that there is no difference in power consumption between the odd semester and the second semester on Semarang State Polytechnic [19]. This was because the effective number of days in each semester was relatively same. Meanwhile, there were significant differences between the electricity consumption on the months in early and mid-term semester than the other months at the end of semester. This was because there were the end of semester holiday which caused the number of effective days at the end of semester to be less than the other months.

3.2 Lighting & air temperature

The procurement of air conditioner and lamps in the room according the light intensity and temperature of room that can be seen in Table 2. The light intensity and temperature color on each room in SMA N 1 Ambarawa accordance with SNI 03-6197-2000 [20] about energy conservation lighting systems of buildings in Indonesia and also the Minister of Health of Republic of Indonesia Decision Number 1429/Men.Kes/SK/XII/2006 [21] about The Implementation Guidelines of Environmental Health School. The air temperature in each room, except the teacher's room and praying room was still considered normal based on the Minister of Health of Republic of Indonesia Decision Number 1405/Men.Kes/SK/XI/2002 [22] about the Environmental Health Work Requirements in the Office and Industrial. so no circulation of air tool which needed such as fan or air conditioner. Thus, the air conditioner in SMA N 1 Ambarawa only used in computer lab and language lab as temperature, humidity, and cleanliness control and also air circulation simultaneously because these rooms were always used in tightly closed [23].

3.3 Intensity of energy consumption

Calculation of Intensity of Energy Consumption (IEC) in SMA N 1 Ambarawa which conducted in accordance with SNI 03-6196-2000, the result showed that IEC average of 1.7957 kWh/m^2/month. In overall, building in SMA N 1 Ambarawa categorized as not air-conditioner building because the extent of it's less than 10% of the total area of building. Thus, based on the criteria in The Minister Energy & Mineral Resources Rule Number 13 of 2012 [24] which presented in Table 3, the electricity consumption in SMA N 1 Ambarawa classified as very efficient.

Table 4. Electricity consumption on each room

No	Room	Tipe of Building	Amount	Area (m^2)	Energy Consumption (kWh/month)	IEC	Criteria
1	Old Class Room	Non-AC	21	2217	1672.5	0.754397835	Very Efficient
2	New Class Room	Non-AC	9	950	900	0.947368421	Very Efficient
3	Cross-Interest Room	Non-AC	2	334	200	0.598802395	Very Efficient
4	Chemistry Laboratory	Non-AC	1	274	78.8	0.287591241	Very Efficient
5	Physic Laboratory	Non-AC	1	150	42.2	0.281333333	Very Efficient
6	Biology Laboratory	Non-AC	1	150	42.2	0.282666667	Very Efficient
7	Language Laboratory	Non-AC	1	120	51.6	0.43	Very Efficient
8	Computer Laboratory I	With-AC	1	150	2662.9	17.75266667	Quite Efficient
9	Computer Laboratory II	With-AC	1	120	2268.9	18.9075	Extravagant
10	Library	Non-AC	1	144	156.275	1.085243056	Very Efficient
11	Multipurpose Room/Hall	Non-AC	1	450	234	0.52	Very Efficient
12	UKS Room	Non-AC	1	16	8.4	0.525	Very Efficient
13	Cooperative Room	Non-AC	1	16	106.5	6.65625	Quite Efficient
14	Counseling Room	Non-AC	1	62	96.05	1.549193548	Very Efficient
15	Principal Room	Non-AC	1	40	164.795	4.119875	Efficient
16	Teacher Room	Non-AC	1	170	948	5.576470588	Efficient
17	Archive Room	Non-AC	1	30	507.4	16.91333333	Extravagant
18	Administration Room	Non-AC	1	48	380.71	7.931458333	Extravagant
19	Music Room	Non-AC	1	60	4.2	0.07	Very Efficient
20	OSIS Room	Non-AC	1	12	5	0.416666667	Very Efficient
21	Scout Room	Non-AC	1	12	5	0.416666667	Very Efficient
22	Principal Room's Toilet	Non-AC	1	8	16.8	2.1	Very Efficient
23	Teacher's Room Toilet	Non-AC	1	8	3	0.375	Very Efficient
24	Administration Office Toilet	Non-AC	1	8	3	0.375	Very Efficient
25	Counseling Office Toilet	Non-AC	1	8	3	0.375	Very Efficient
26	Boy Student Toilet	Non-AC	20	114	66	0.578947368	Very Efficient
27	Girl Student Toilet	Non-AC	20	114	66	0.578947368	Very Efficient
28	Warehouse	Non-AC	1	50	8.4	0.168	Very Efficient
29	Kitchen	Non-AC	1	21	15	0.714285714	Very Efficient
30	Praying Room/Chapel	Non-AC	1	120	101.4	0.845	Very Efficient
31	Canteen	Non-AC	6	156	28.95	0.185576923	Very Efficient
32	Security Post	Non-AC	1	6	0.375	0.0625	Very Efficient
33	(School Facility)	Non-AC	-	6138	1197	0.195	Very Efficient

Table 3. Criteria of Energy Used in Office Building

Criteria	Specific Energy Consumption (kWh/m²/month)	
	AC Building	Non AC Building
Very efficient	Smaller than 8,5	Smaller than 3,4
Efficient	8,5 up to smaller than 14	3,4 up to smaller than 5,6
Quite Efficient	14 up to smaller than 18,5	5,6 up to smaller than 7,4
Extravagant	Bigger than with 18,5	Bigger than with 7,4

The used of electrical energy and IEC in each building in SMA N 1 Ambarawa was varies widely, it depended the kinds of electronic equipment in it. The amount of electricity consumption in any room in this school in detail can be seen in Table 4. The room which used the highest electrical energy was computer lab, while the room which used the lowest of electrical energy was security post. High and low consumption of electrical energy in any room was influenced by the type and amount of equipment, the electric power used on any equipment, also the long of equipment operation. Based on IEC, almost all of rooms in SMA N 1 Ambarawa were classified as very efficient. But, unfortunately, computer laboratory and cooperatives were quite efficient. The teachers' room and the principal's office room were relatively efficient. Meanwhile, the second computer lab, storage room, and administration room were classified as wasteful.

Computers, lamps, and air conditioner were top of equipment list that used the highest electrical energy. This was supported by the results which stated that the use of electrical energy by the servers, computers, monitors, data communication equipment, and cooling systems has increased sharply in each time [25]. Additionally, a study of energy efficiency of the world also focused on building lighting systems because they had a major role of energy consumption in the world [26].

Table 6. SWOT Matrix

Internal Factor / External Factor	*Strengths* : - Number of Human Resources - The school community responded towards energy consumption was very high - The curriculum used - Energy subsidy	*Weakness* : - Environmental leadership which not yet strong - Lack of the school community was related the energy saving - Electrical power was not used as consideration when bought equipment - School had not policy which completed with strict sanction about energy efficiency - School had not electricity network area
Opportunities : - The funding source from government - Big support from the Ministry of Environment and the Department of Education & Culture	S-O : - Doing some energy efficiency efforts based school community	W-O: - Replacing the extravagant equipment with the energy efficient equipment - Cooperating with relevant agencies - Adding tools to control energy saving. The tool can be in human or sensor form - Make electricity network plan
Threats : - The scarcity of energy - Degradation of Environmental quality - Carbon tax	S-T : - Include the energy efficiency principles in learning curriculum - Streamline the use of long the equipment use which uses electricity - Do checking and improvement towards network electricity that get problem	W-T : - School makes the proper policy energy efficiency - School doing socialization related the importance of saving energy - Using electric power as a material consideration when buying the equipment

3.4 Climate change impact

Consumption of electrical energy in SMA N 1 Ambarawa indirectly caused the secondary emissions of CO_2. This was because the electricity production process from Steam Power Plant results the CO_2 gas emissions were the cause of global warming [27]. During the past year, secondary emission of CO_2 which resulted from electricity consumption at the school reach by

135,019.603 kg of CO_2. The highest emission occurred in November 2015, which amount by 10694.715 kg of CO_2, while the lowest emission occurred in July 2015, which amount by 5862.070 kg CO_2. The average emissions were reached by 9644.257 kg CO_2/month. Thus, it is known that the school environment give substantial contribution to the global warming phenomenon.

3.5 Energy savings

SWOT analysis was done to determine the appropriate recommendation to improve the potential of energy efficiency in SMA N 1 Ambarawa. The SWOT matrix completely can be seen in Table 6.

Based on this, the recommendations which offered were divided into three levels; they were no-cost, middle-cost and high-cost. No-cost recommendation which offered was school should make some policies about energy efficiency. Beside of according the SWOT analysis result, the recommendations which offered was also based on research which stated that government in many countries was always make every effort to implement the various policies related to energy efficiency in their own region [28]. The calculation basis of energy saving opportunities was long usage of lamps, LCD and speakers by the school community. Based on the data in Table 5, no-cost recommendation was divided into two kinds that called as short-term no-cost recommendation and long-term no-cost recommendation. Short-term recommendation aimed to calculate the energy saving in the rooms were wasteful of energy, but long-term recommendation aimed to calculate the energy saving in overall of SMA N 1 Ambarawa. As a result, the electrical energy which can be saved with short-term no cost recommendation was reached 3.159% and long-term no cost recommendation was reached 7.536%.

Middle-cost recommendation which offered was the school should add a controller for efficient used of time for all equipment which uses electrical energy. The calculation basis of energy saving opportunities was long usage of lamps, LCD and speakers by the school community also the duration of using the wifi router and access point, and water pump. As a result, electrical energy that can be saved was reached by 9.499%.

High-cost recommendation which offered were replacing the equipment which were still using high electric power with efficient equipment. This recommendation was supported by the research result stated that the used of LED light and lighting sensor system installation was good energy efficiency [29]. Moreover, the LED lights were known to save energy as much as 70% of electrical energy. Additionally, the concept of green-computing was one of the principles which emphasize in the importance of minimization of energy consumption in computer system [30]. The calculation basis of energy saving opportunities was the replacement of equipment become energy efficient equipment. As a result, the electrical energy that can be saved with short-term high cost recommendation was reached 35.278% - 36.626% and long-term high cost recommendation was reached 42.084%.

4 Conclusions

The school environment had a big potential of energy efficiency that can reduce the energy consumption and CO_2 gas emissions. The consumption of electrical energy in SMA N 1 Ambarawa was very efficient, with an IEC average 1.7957 kWh/m^2/month. The average of electrical energy consumption reached 11022.008 kWh/month, which resulted in the emergence of secondary emissions of CO_2 by 9644.257 kg CO_2/month. In this case, computers, lamps, and air conditioner were top of equipment list that used the highest electrical energy.

5 Acknowledgement

The authors are thankful to the Indonesia Endowment Funds for Education (LPDP) Scholarship of Indonesian Government for giving financial support to study at Diponegoro University.

References

1 Samiullah. Journal of Economics and Suistainable Development, **5**, 82-92 (2014)
2 Ozge K, & Zeynep K, International Journal of Energy Economics and Policy, **3**, 62-73 (2013)
3 Republic of Indonesia. Indonesian Law Number 32 of 2009 : Perlindungan dan Pengelolaan Lingkungan Hidup (Protection and Management of Environment). Republic Indonesia's Sheet Number 140 of 2009. Secretariat State, Jakarta, Indonesia, 1-71 (2009)
4 H. Ralph, AM. Harald, F. Peter, International Journal of Sustainable Development & World Ecology, **19**, 451-459 (2012)
5 S. Iwan, M. Hiroshi, S. Lusi, K. Ryushi, International Journal of Energy Economics and Policy, **5**, 360-373 (2015)
6 B. Cuma, A. Yusuf, International Journal of Energy Economics and Policy, **4**, 484-494 (2014)
7 Elinur, DS Priyarsono, T. Mangara, F. Muhammad, Indonesian Journal of Agricultural Economics, **2**, 97-119 (2010)
8 I. Joseph, M. Abraham, International Journal of Energy and Environment, **1**, 705-714 (2010)
9 A. Hasan, H. Mario, B. Helmut International Journal of Social Sciences, **3**, 12-21 (2014)
10 NH. Pham, HC, Tran, KL. Dong, NB. Duong, TVA. Pham, TML. Luong, TTH. Pham, KL. Nguyen, Environment Asia, **6**, 45-50 (2013)
11 O. Richard, S. Hannah, International Journal of Environmental & Sciences Education, **6**, 161-172 (2011)
12 I. Nuchjarin, P. Apirath, J. Chawalit, D. Aussadavut, Y. Pisal, Y. A Muliti-Objective Unit Commitment Model for Setting Carbon Tax to Reduce CO_2

Emission : Thailand's Electricity Generation Case, **8**, 9-17 (2015)
13 United Nations Development Programme. Promoting Energy Efficiency in Buildings : Lessons Learned from International Experience. UNDP, New York, 1-60 (2010)
14 AU. Usman, FK. Mohd, AS. Suleiman, M. Umar. Bonfring International Journal of Industrial Engineering and Mangement Science, **3**, 13-19 (2013)
15 Sugiyono. Statistika untuk Penelitian (Statistics for Research). 25thed. Alfabeta, Bandung, Indonesia, 137-142 (2014)
16 J. Arefeh, SN. Sajjadi, H. Habib, International Journal of Academic Research in Business and Social Sciences, **2**, 106-113 (2012)
17 National Standardization Institution. SNI 03-6196-2000 : Prosedur Audit Energi pada Bangunan Gedung (Energy Audit Procedure of Building). National Standardization Institution, Jakarta, 1-8 (2000)
18 SM. Affan, PS. Rulli, Jurnal Teknik Pomits, **3**, 5-7 (2014)
19 SP. Daeng, Jurnal Teknik Elektro, **1**, 1-7 (2012)
20 National Standardization Institution. SNI 03-6197-2000 : Konservasi Energi pada Sistem Pencahayaan (Energy Conservation in Lighting System). National Standardization Institution, Jakarta, 1-13 (2000)
21 Republic of Indonesia. The Rule from Minister of Health Number 1429/Men.Kes/SK/XII/2006 : Pedoman Penyelenggaraan Kesehatan Lingkungan Sekolah (The Implementation Guidelines of Environmental Health School). The Ministry of Health, Jakarta. 2002; 1-13.
22 Republic of Indonesia. The Rule from Minister of Health Number 1405/Men.Kes/SK/XI/2002 : Persyaratan Kesehatan Lingkungan Kerja Perkantoran dan Industri (The Environmental Health Requirements in Office and Industrial). The Ministry of Health, Jakarta, 1-35 (2002)
23 I. Shahid, B. Shameer, Azzizuddin, & K Vijaya KR, International Journal of Innovative Research in Science, Engineering and Technology, **2**, 7460-7464 (2013)
24 Republic of Indonesia. The Rule from Minister of Energy and Mineral Resource Number 13 of 2012 : Penghematan Pemakaian Tenaga Listrik (Saving Electricity Consumption). The Official Republic of Indonesia Number 556 of 2012. The Ministry of Energy and Mineral Resource, Jakarta, 1-10 (2012)
25 S. Biswajit, Green Computing. International Journal of Computer Trends and Technology, **14**, 46-50 (2014)
26 R. Sagar, KJ. Vijay, International Journal of Innovative and Applied Research, **3**, 28-33 (2015)
27 S. Tapan, C. Roberto, Z. Muslima, International Journal of Energy Economics and Policy, **3**, 30-42 (2013)
28 FG. Jaume, PV. Ignasi, International Journal of Energy Economics and Policy, **5**, 69-79 (2015)
29 AB. Jayashri, S. Arvind, International Journal of Scientific & Engineering Research, **3**, 1-6 (2012)
30 P. Kiruthiga, K. Vinoth, International Journal of Advanced Research in Computer and Communication Engineering, 3, 6318-6321 (2014)

Compressed Natural Gas Technology for Alternative Fuel Power Plants

Isworo Pujotomo[1,*]

[1] Electrical Engineering Department, School for Engineering of PLN's Foundation for Education & Welfare PT. PLN (Persero) Jakarta, Indonesia

Abstract. Gas has great potential to be converted into electrical energy. Indonesia has natural gas reserves up to 50 years in the future, but the optimization of the gas to be converted into electricity is low and unable to compete with coal. Gas is converted into electricity has low electrical efficiency (25%), and the raw materials are more expensive than coal. Steam from a lot of wasted gas turbine, thus the need for utilizing exhaust gas results from gas turbine units. Combined cycle technology (Gas and Steam Power Plant) be a solution to improve the efficiency of electricity. Among other Thermal Units, Steam Power Plant (Combined Cycle Power Plant) has a high electrical efficiency (45%). Weakness of the current Gas and Steam Power Plant peak burden still using fuel oil. Compressed Natural Gas (CNG) Technology may be used to accommodate the gas with little land use. CNG gas stored in the circumstances of great pressure up to 250 bar, in contrast to gas directly converted into electricity in a power plant only 27 bar pressure. Stored in CNG gas used as a fuel to replace load bearing peak. Lawyer System on CNG conversion as well as the power plant is generally only used compressed gas with greater pressure and a bit of land.

1 Introduction

Electricity consumption per capita is an index of living standard of a country. In Indonesia, with the increasing industrial activity and population, the need for electrical power also increased. Due to the increase in fuel prices today's world, PT. PLN (Persero) as one of the state electricity company should think businesses operating cost savings, of which 75% are in fuel costs. One effort that can be taken by PT. PLN (Persero) is the main power plant fuel switching from fuel oil (HSD and MFO) into natural gas.[1] [2]

Basically PLN operate Power Gas and Steam (Power Plant) as *peaker* plants (peak load) because operating costs are more expensive than base load generation (base load). Oil energy crisis that resulted in soaring oil prices caused the gas to be used as an alternative PLN for plant outside of peak load. During this time, when the load is low (outside of peak load), the supply of gas for the power plant is not absorbed optimally, as appropriate loading pattern in the Java-Bali system more filled by many ordinary coal plant production is cheaper. Nevertheless, the unabsorbed gas must still be paid, the gas supply contract is a take or pay. While at peak load gas supply is insufficient, so that some plants have to be operated using the fuel.[2]

Presence technology Compressed Natural Gas (CNG) is expected to be a solution to the problem mentioned above. Gas supply flow rate remain while outside of peak load, partially compressed into CNG tube pressurized to 250 bar for 10 hours. CNG use of technology for the generation of a rock thing, not in other countries. Much of the world uses Liquid Natural Gas (LNG) technology.

2 Gas Power Plant and Steam

Power Plant is a combination of the working principle of the power plant by gas and steam power plant or so-called combined cycle. Power plant using a gas turbine unit driving the generator, so that the working principle of the power plant following the working principle of the gas turbine. The gas turbine is designed and made for converting heat energy from burning fuel into mechanical energy. The system uses the principle of the Brayton cycle gas power plant. While the working principle of the power plant is converting the chemical energy in the fuel is converted into thermal energy in the form of vapor pressure and high temperature, the steam then changed to mechanical energy to drive generator. The system uses the principle of the Rankine cycle power plant. Figure 1 is a Grati power plant in East Java with a capacity of 526.850 MW which use the principle of combined cycle.[1] [3]

* Corresponding author: isworop@yahoo.com

Fig. 1. Grati Power Plant in East Java (Pasuruan)

Grati Power Plant in East Java (Pasuruan) consists of: Gas Turbine: 112.450 MW x 3; Steam turbine: 189.500 MW; Block output 526.850 MW.

2.1 Combined cycle

In improving the efficiency of gas turbines do combined cycle gas turbine with a steam turbine cycle in order to obtain the combined cycle which is commonly referred to as "co-generation". Meanwhile, to improve the thermal efficiency of the gas turbine is used combined cycle, forming a so-called "Combined Cycle" or Steam Gas Power Plant (Power Plant). Cycle power plant applying the Brayton cycle, whereas the ideal cycle Rankine cycle power plant apply.

2.2 Brayton cycle;

Brayton cycle (Brayton Cycle) shown in Figure 2 consists of isentropic compression process that ended with the release of heat at constant pressure. In Bryton cycle of each state processes can be analyzed as follows:[4]

1-2 (isentropic Compression).

Work required by the compressor;

2-3: Entry of the fuel at a constant pressure.

3-4: isentropic expansion in the turbine.

4-1: Discharge heat at constant pressure into the air.

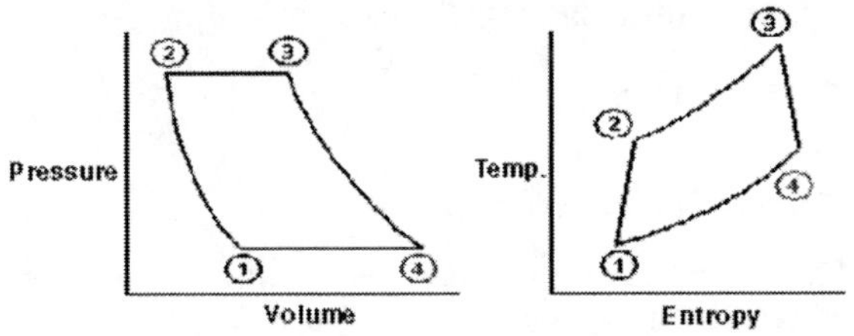

Fig. 2. *p-v and t-s Brayton Syklus Diagram*[3]

From Figure 2 on p-v diagram and t-s, it can be seen that the inclusion of heat takes place at a constant pressure:

$Q_{in} = m.c_p. (T_3\text{-}T_2)$ .... (1)

- Spending too hot at constant pressure:

$Q_{out} = m.c_p. (T_4\text{-}T_1)$ .... (2)

- Thus, useful work can be formulated as follows:

$W_u = Q_{in} - Q_{out}$ .... (3)

c. Rankine cycle;

Rankine cycle in Figure 3 is used in the steam turbine power plant system. The sequence steps as follows:

a - b: Water is pumped from the pressure p2 be p1. This step is a compression step isentropis, and this process occurs at the water pump filler.

b - c: Air pressure is increased the temperature until it reaches boiling point. Occurred in the LP heater, HP heater and Economizer.

c - d: Water transformed into saturated steam. This step is called evaporation with isobars isothermis process, occurred in the boiler is in the wall tube (riser) and the steam drum.

d - e: Steam is heated further until the vapor reaches a temperature further work into hot steam (superheated vapor). The move occurred in the boiler superheater with the isobars.

e - f: Steam does work so that the pressure and the temperature dropped. This step is a step isentropis expansion, and occur within the turbine.

f - a: Disposal of latent heat of the steam that turns into condensate water. This step is isothermis isobars, and occurs in the condenser.

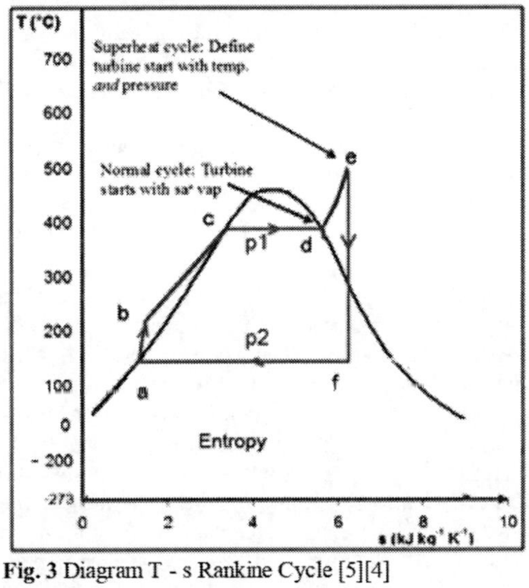

Fig. 3 Diagram T - s Rankine Cycle [5][4]

3 Compressed natural gas

Compressed Natural Gas (CNG) is a fluid gas that has been processed into a high-pressure natural gas compression in a tube. In general, the major components containing CNG methane (CH_3) and ethane (C_2H_8) with fraction of about 90%. CNG is made by compressing methane (CH_4), which is extracted from natural gas. CNG is stored and distributed through the packaging in the tank (pressure vessel or pressure vessel). Ideally, the pressure on the gas pipeline is 11 bar, while the CNG requires a pressure of 200 bar, or 197 atm, 197 times normal air pressure. With a pressure of 200 bar, charging the equivalent of 130 liters of premium gas can be done

within 3-4 minutes. Figure 4 shows an example of CNG tubes.[6] [7]
Calorific value is the amount of heat generated by the combustion of materials or fuels. Measured in units of energy per amount of material, for example kJ / kg. Calorific value of the fuel is divided into two kinds, namely: a calorific value above (HHV) and lower calorific value (LHV).

Fig. 4. pressure of 200 bar CNG Tubes

Higher Heating Value (HHV) or above is the calorific value calorific value derived from the combustion of 1 kg of fuel, taking into account the vapor condensation heat (water resulting from combustion are in liquid form). HHV gas = 1089 Btu / ft3 = 52.225 MJ / kg.

Lower Heating Value (LHV) or lower calorific value is obtained calorific value of combustion heat regardless vapor condensation (water produced from burning was on gas / steam). LHV gas = 983 Btu / ft^3 = 47.141 MJ / kg.

A fuel flash point is the lowest temperature at which the fuel can be heated so that steam output flame briefly when passed a flame. Flash point -187.80C CNG temperature. Burning point is the lowest temperature at which sufficient oxygen conditions, spontaneous combustion can occur. Burning point unit is degrees (^{0}) or degrees Celsius (^{0}) Fahrenheit. CNG point at a temperature of 540^{0}C. Octane number is a number that shows the number as a percentage (%) volume of isooctane in a mixture that consists of n-heptane issoctane and that does not cause outbreaks in the fuels being tested in a compression chamber of a combustion chamber.[8] [9]

According to the science of thermodynamics, the higher the ratio / percentage of compression used, the higher the efficiency of combustion in the combustion chamber. CNG octane value of the percentage of 130%.
Tube specifications for CNG fuel is as follows:[8]
Capacity = 50-100 (m^3)
Dimensions = 11657 x 2428 x 3028
Massa Netto = 10760 kg
Nominal Diameter = DN80
Material = 16 MNR
The selected specification is commercial fleet refuel CNG compressor station, with a pressure of 1.4-250 bar and a capacity of 1.3 m^3 / h.

4 Output Power Cng Capacity Analysis

4.1 Technical Analysis

Grati power plant that has the capacity of 3.5 MMSCF CNG with gas turbine efficiency of 45% and CF = 0.8 / year in operation for 6 hours output power can be determined as follows:
Production MWh = (SCF x 922 x η) / 3413
= (3.5 x 1,000,000 x 922 x 0.45) / 3413
= 425.476 MWh
Output power = (Production MWh) / (operating time)
= (425.476 MWh) / (6 h)
= 70.91 MW

Fuel is needed to generate power output of 70.91 MW are as follows:
BBG = (3413 x Power Output) / η
= (3413 x 70.91) / 0.45
= 538 MMBTU
= 512,380.9524 ft3
= 14,508,579.05 lt
with:
1 ft^3 = 1050 btu
1 ft^3 = 28.316 lt = 28.316 dm^3

4.2 Economical analysis [6]

4.2.1 Fixed costs

According to the above matter at Grati power plant installed two units of CNG with a capacity of 2 x 70.91 MW. If the interest rate i of 12% by age 25 years plants can be seen:
CRF = $(i\ (1 + i)^n) / ((1+ i)^{n-1}) = (0.12\ (1 + 0.12)^{25}) / ((1 + 0\ ,\ 12)^{25-1}) = 0.127$
Development costs PLT-CNG = 5.94864 $ / kWh
CF = 80%
CC = (cost of construction of generating capacity x CRF) / W
= $(5.94864 \times 400 \times 10^3 \times 0.127) / (400 \times 10^3 \times 0.8 \times 6 \times 365)$
= 4.4×10^{-4} $ / kWh

4.2.2 Operating and maintenance costs

O & M = (total cost of O & M) / (CF x time x cap) = (20347106.98 ($)) / (0.8 x 6 x 365 x 400 x 10^3 (kWh)) = 0.03 $ / kWh

4.2.3 The cost of fuel

Fc = (Ui 860) / 0.45 = (860 x 7.313 x $10^{(-7)}$) / 0.45 = 1.4 x 10^{-3} $ / kWh

4.2.4 Total cost

Tc = CC + O & M + Fc = 4.4 x 10-4 1.4 x 10-3 + 0.03 = 0.03184 $ / kWh

5 Conclusion

From the analysis of the utilization of CNG fuel at Grati power plant can be concluded that:

1. The use of CNG is useful to reduce system losses on gas purchases at gas power plant (power plant) in take or pay that lead to wasted unused gas.
2. Costs required by CNG stations is equal to 0.03184 $ / kWh with details:
   a) Construction of 4.4×10^{-4} Cost $ / kWh
   b) Operation and Maintenance Costs $ 0.03 / kWh
   c) Cost of Fuel 1.4×10^{-3} $ / kWh

CNG has a lower density than air, so it is not flammable.

References

1. Ahmad, A. *Meningkatkan Efisiensi Steam Turbine ketika Menggunakan Bahan Bakar Minyak. Pasuruan : PLTGU Grati,* (2011)
2. Mohamad, S. *Pengoperasian Preheater Dalam Mengoptimalkan Pemakaian Air Kondensat. Pasuruan : PLTGU Grati,* (2015).
3. Bo Yan Xu, Mikio Furuyama., JSAE Review **18**, 57-82, (1997)
4. Mikio Furuyama, Bo Yan Xu. *Mixing Flow Phenomena of Natural Gas and Air in the Mixer of a CNG Vehicle. SAE 981391.* Chiba University. Japan, (1998).
5. Heywood, J.B., *Internal Combustion Engine Fundamental,* McGraw-Hill, Singapore, (1988)
6. K. Kadirgama, M. M. Noor, A. R. N. A. Rahim, R. Devarajan, M. R. M. Rejab, N. M. Zuki N. M, *Design and Simulate Mixing of a Compressed Natural Gas With an Air Mixing Device,* MUCET (2008).
7. S. Shiga, S. Ozone, H.T.C. Machacon, T. Karasawa, H. Nakamura., *A Study of the Combustion and Emission Characteristic of Compresed Natural Gas Direct Injection Stratified Combustion Using a Rapid Compression Machine,* Combustion Institute Gunma University, Japan, (2002)
8. Pulkrabek, W.W., *Engineering Fundamentals of The Internal Combustion Engines,* 2nd Edition, Pearson Prentice-Hall, Wisconsin. (2004)
9. Willard. W. Pulkrabek, *Enginering Fundamental of the Internal Combustion Engine,* Prentice Hall, Upper Sadle River, New Jersey, (2003)

Impact of Training and Mentoring Activities Which are Given to The Level of Interest And Capability Industrial Target Group In Adopting SNI ISO 50001

Dwi Apriyanti[1,2]*, *Aris* Ika Nugrahanto[1], *Sanjaya* Shrestha[1]

[1]Industrial Energy Efficiency Project, United Nations Industrial Development Organization (UNIDO), Indonesia
[2]Master Program of Energy, School of Post Graduate Studies, Diponegoro University, Semarang - Indonesia

Abstract. Energy consumption in the industrial sector in Indonesia is increasing as a result of population and economic growth. The government is aware of this and seeks the answer to improve industrial competitiveness and increase energy security through energy efficiency programs. Some industries have implemented energy efficiency programs as ad-hoc, but have not applied systematically, so the results are not optimal. Through the cooperation of the Ministry of Energy and Mineral Resources (ESDM) with the United Nations Industrial Development Organization (UNIDO), there has been training and mentoring activities for industry on Energy Management System (EnMS) based on SNI ISO 50001. Based on the results of identification through survey conducted to 226 industry that has attended the training and 64 industries that have been trained and received assistance, obtained data that 45% of industries have fully adopted EnMS SNI ISO 50001, 17% industry has adopted a part of EnMS SNI ISO 50001 and 38% industry does not adopt EnMS SNI ISO 50001 altogether.

1 Introduction

Industrial growth contributes very important to economic growth in Indonesia. The industry's average energy demand grew by more than 2.7% per year from 2011 to 2035. This is driven by the shift of industrial-type growth from labor-intensive activities to energy-intensive industries [1]. However, on the other hand, the industry as the largest energy user is a significant greenhouse gas emitter. Increased greenhouse gas emissions from burning fossil fuels and rising world fuel prices are a serious threat to environmental sustainability and Indonesia's economic sustainability [2]. Beside, one of the cause of high energy consumption is inefficiency in energy use. Thus, the energy costs incurred by the industry become very high resulting in financial inefficiency and declining product competitiveness.

The Government of Indonesia is fully aware of the threat, so that various policy instruments and programs are implemented, including issuing Government Regulation (PP) no. 70 of 2009 on energy conservation. In this regulation, organizations that consume energy greater than 6000 TOE (Ton Oil Equivalent) are required to apply the Energy Management System (EnMS).

Energy management is an integrated activity to control energy consumption in order to achieve effective and efficient energy utilization to produce maximum output through technical action in a structured and economical way to minimize energy utilization including energy for production process and minimize consumption of raw materials and supporting materials. Energy Management is done by :

1. Appointed an energy manager
2. Developed energy conservation programs
3. Performed energy audits on a regular basis
4. Implemented recommendations of energy audit results
5. Reported the implementation of energy management every year to ministers, governors, or regents/ mayors in accordance with their authority [3]

An international standard on energy management systems was introduced in 2011 with the issuance of ISO 50001 which was adopted by the National Standardization Agency (BSN) to become SNI ISO 50001 in 2012. ISO 50001, Energy Management System, is a standard used to manage energy performance, including energy efficiency and consumption. The ISO 50001 concept uses a Management System model with Plan Do Check Action (PDCA) cycle approach for continuous improvement. Indonesia as an ISO member identically adopted the standard become SNI ISO 50001 [4]

This SNI has many benefits, there are to develop policies related to energy efficiency usage , set goals and targets in line with energy policy, data usage to make objective decisions about energy consumption, measure the results to identify areas of energy efficiency improvement, review the effectiveness of energy policy and result from energy efficiency improvement, and implement sustainable improvements in energy

* Corresponding author : aprigreen@gmail.com

https://doi.org/10.1051/e3sconf/20183101012

management practices. Improved in energy performance helps companies maximize the use of energy sources and energy-related assets so as to reduce energy consumption and energy costs.

The United Nations Industrial Development Organization (UNIDO), one of the United Nations (UN) agency participated in the making of ISO 50001 and in cooperation with the Government of Indonesia, namely the Ministry of Energy and Mineral Resources (MEMR), the Ministry of Industry (MOI) and the National Standardization Agency (BSN) in socializing and supporting the implementation of SNI ISO 50001 Energy Management System through the cooperation program "Promoting Industrial Energy Efficiency through System Optimization and Energy Management Standard in Indonesia ". This cooperation program runs from 2012 to 2017 (UNIDO, Project Document of Promoting Industrial Energy Efficiency Through System Optimization and Energy Management Standard in Indonesia, 2011), [5].

In its implementation, UNIDO in cooperation with MEMR has conducted several trainings on Energy Management System and System Optimization for industry and advanced training to produce national experts in the field. These trainings which was run from 2012 to 2015, have been done only for 6 sectors industry, there are chemical, textile, food and beverages, pulp and paper , ceramics, and others. The training was held for 2 days, where the participants who are representatives of the industry would be shared on how to apply energy management system in their own industries, in accordance with the clauses contained in SNI ISO 50001. In addition, this project also provided Energy Management System and System Optimization assistance to industries by making it a pilot company [6]. There are 28 industries which were become pilot companies on Energy Management System and 36 industries which are become pilot companies on System Optimization. The pilot company was implemented an energy management system accompanied by national experts, during their expert training periode, and under international expert supervision. Some of the pilot companies have been certified of ISO 50001.

Identification is conducted to determine the impact of training and mentoring activities provided by MEMR with UNIDO on the level of interest and capability of industrial target group in adopting SNI ISO 50001.

2 Methodology

The identification is done by conducting surveys to industries that have attended the training organized by MEMR together with UNIDO in the period of 2012 to 2015. The industries are divided into 6 sectors: chemical, textile, food and beverages, pulp and paper, ceramics, and others.

The survey was conducted by sending questionnaires via e-mail and conducting interviews with industry representatives, either by telephone or directly meeting at the factory.

The score of the questions given is between 0 to 5 [7]. Score 0 means that the company does not implement EnMS. A score of 3 means that the company 50% executes or implements EnMS. While score 5 means that the company implement 100% steps of EnMS. The evidence is the documentation of the system and the improvement of sustainable energy performance in each industry.

3 Result and Discussion

The survey was conducted for 226 industries, excluding UNIDO's pilot company (28 industries for Energy Management System and 36 industries for System Optimization). From questionnaires sent via e-mail and communications made via e-mail and telephone or directly meeting with industry representatives at the factory, feedback was obtained from 108 industries.

The questionnaire submitted contains questions about EnMS implementation in each industry regarding awareness of the importance of energy savings, management commitments, tasks and responsibilities, Significant Energy Users (SEU), Energy Performance Indicators (EnPI = Energy Performance Indicators), objectives and targets, action plans, and internal audits that have been implemented [8].

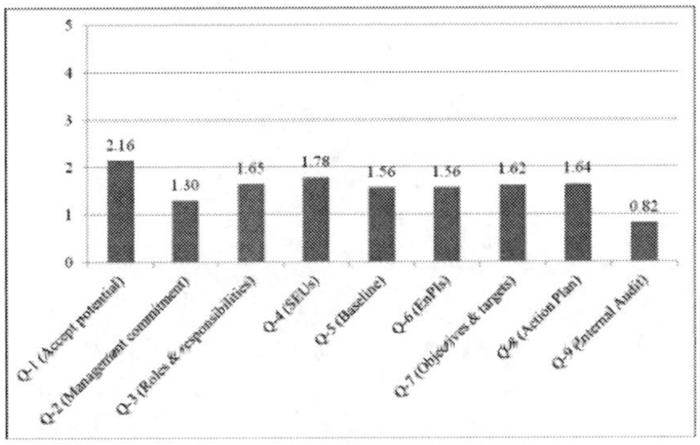

Fig.1. The score of all industries sector for each part of EnMS SNI ISO 50001

The expected result is the average score of each question reach 1.5. The results of each question based on the survey are shown in figure 1.

From the figure 1, the average is seen that for the question 1 (accept potential) obtained a score of 2.16, it means that the industry are already aware that energy efficiency needs to be done. Question 2 (management commitment) obtained score 1.3, meaning that only a few top management are committed to energy efficiency. Question 3 (roles & responsibility) obtained score of 1.65, meaning that industries which have implemented energy efficiency have made appointments to the energy team and their respective assignments. Question 4 (SEU's) obtained score of 1.78, meaning that the industries were already able to identify their significant energy users in their company. Question 5 (baseline) and question 6 (EnPIs) obtained score of 1.56, meaning that industries have determined and have baseline and energy performance indicators to measure their energy efficiency level. Question 7 (objective and target) obtained score of 1.62, meaning that industry has determined objectives and targets related to energy efficiency. Question 8 (action Plan) obtained score of 1.64, meaning that the industry has made the action plan in their energy efficiency implementation. Question 9 (internal audit) obtained score of 0.82, meaning that there are still few industries conducting internal audits to evaluate overall energy efficiency implementation based on SNI ISO 50001.

Based on the results of each given question, it is recommended to conducted advanced training on how to identify and determine the baseline and EnPI (ISO 50006). In addition, advance training on internal audit systems for these industries is required. Related to commitment from top management, further research is needed to identify factors that can increase commitment from top management in EnMS SNI ISO 50001 implementation.

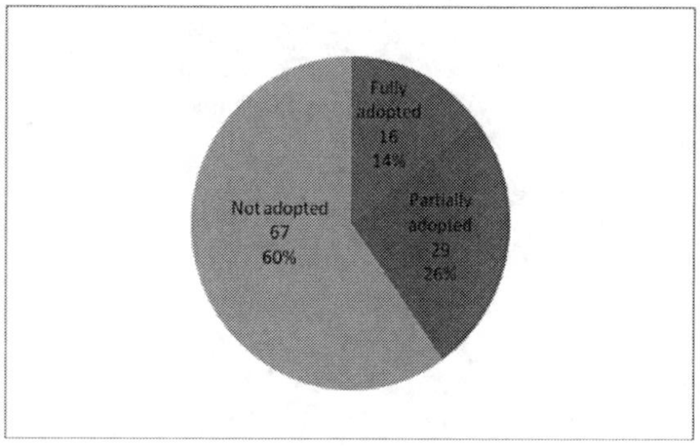

Fig.3. The survey result

Tabel 1. The number of adopted industries for EnMS

Item	Number of industries
UNIDO's pilot companies on Energy Management System (fully adopted)	28
UNIDO's pilot companies on System Optimization	36
Surveyed industries which was partially adopted on EnMS	29
Surveyed industries which was fully adopted on EnMS	16
Total	109

Based on the survey (excluding UNIDO's pilot company), it was found that the industries that have attended the training and implemented EnMS by fully adopting SNI ISO 50001 as many as 16 companies. The industry has implemented EnMS for 1 cycle of all stages to be implemented. Industries that partially adopt are 29 companies. That is, this industry has implemented EnMS but not yet completed 1 cycle, or just performing several stages only. While industries that do not adopt at all as many as 67 companies. That is, this industry does not implement EnMS, or has implemented energy savings but not systematically.

Taking into account the number of pilot companies that have attended training and obtained assistance, it can be obtained the number of industries that have implemented EnMS and adopted SNI ISO 50001 as many as 109 industry as shown in table 1 and figure 4.

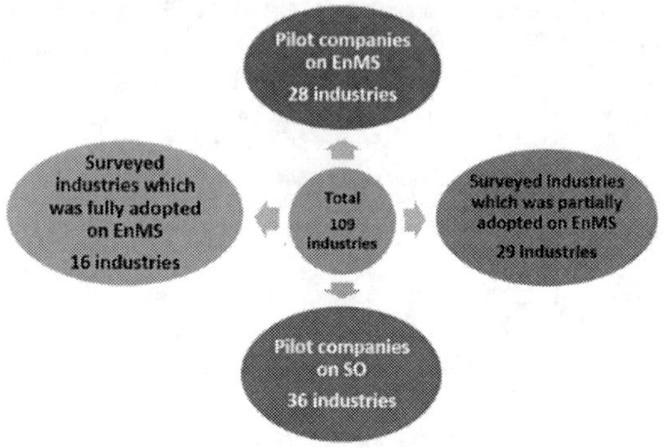

Fig.4. The number of adopted industries for EnMS

Based on the survey results and considered the pilot companies on EnMS and SO, it is found that 62 % of industries has adopted SNI ISO 50001 compared to industry which was attended the Energy Management System and System Optimization training. However, some of the clauses of SNI ISO 50001 have not been fulfilled maximally.

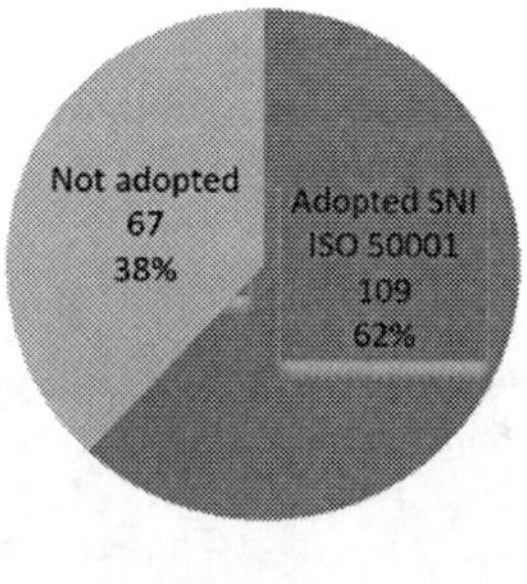

Fig.5. Adoption status

In order to have for more industries adopting SNI ISO 50001 and implementing sustainable development, it is necessary to conduct a more intensive approach between government and industry in EnMS implementation. By this effort, energy security can be achieved and negative impact of greenhouse gas emissions can be reduced.

4 Conclusion

Based on the identification result, the industries participating in the training and received mentoring of

 https://doi.org/10.1051/e3sconf/20183101012

SNI ISO 50001 Energy Management System organized by MEMR and UNIDO, and have fully adopted the system are 80 companies. While still partially adopted are 29 companies, and who do not adopt at all are 67 companies.

Advanced training on ISO 50006 is required in determining the baseline and EnPI, as well as advanced training on the internal audit system. In addition, further research is needed to identify factors that can increase commitment from top management in implementing EnMS SNI ISO 50001

References

1. Wijaya, D. E., Tumiwa, F., & Nugrahanto, A. I. (2017). In *Buku Panduan Pelatihan Pembiayaan Proyek Efisiensi Energi bagi Lembaga Jasa Keuangan*. Jakarta: UNIDO.
2. MEMR. (2016). *Handbook of Energy and Economic Statistic of Indonesia.* Jakarta: Ministry of Energy and Mineral Resources.
3. MEMR. (2012). *Peraturan Menteri ESDM No. 14.* Jakarta: MEMR (Ministry of Energy and Mineral Resources).
4. BSN. (2013). *Konservasi Energi dengan Menerapkan SNI ISO 50001.*
5. UNIDO. (2011). *Project Document of Promoting Industrial Energy Efficiency Through System Optimization and Energy Management Standard in Indonesia*. Jakarta: UNIDO.
6. Nugrahanto, A. I., Kamilya, N., Apriyanti, D., & Hutasoit, N. (2017). *Project Progress Report in Steering Committee Meeting.* Jakarta: UNIDO.
7. UNIDO. (2013). *Practical Guide for Implementing an Energy Management System.* Vienna: UNIDO.
8. BSN. (2012). *Sistem Manajemen Energi-Persyaratan dengan Pedoman Penggunaan, SNI ISO 50001.* Jakarta: Badan Standardisasi Nasional.
9. ESDM, D. E. (2016). *Statistik EBTKE 2016.* Jakarta: Dirjen EBTKE Kementerian ESDM.
10. GEF-UNIDO. (2011). *Project Council Document.* GEF-UNIDO

Performance Analysis of Isolated Hybrid Power Plant Model with Dynamic Load Conditions – Morning, Noon and Afternoon Transitions

Rina Irawati*

Research and Development Center for Electricity, New, Renewable Energy, and Energy Conservation Technology – Research and Development Agency of Ministry of Energy and Mineral Resources Republic of Indonesia, Jakarta – Indonesia

Abstract. Diesel Generator with Photovoltaic Hybrid Power Plant is one of the solutions for supply electric demand to isolated area. The energy sources that can be used for hybrid system are such as photovoltaic, wind turbine, and biomass or biogas, because these sources are almost available in every isolated area. This research used a model of hybrid system from diesel generator and 1.28 kWp photovoltaic power plant. The reliability and some of power quality of this system tested by 1300VA house hold daily load characteristic effectively 24 hour. Power quality and some electricity parameters during transition mode for each resource will be analyzed. Furthermore the power quality analyze will be conducted and evaluated base on Electrical Engineers' Association (EEA).

1 Introduction

Indonesia is an archipelagic country that has thousands of islands both populated and unpopulated. Along with the increasing of population, the need for electrical energy is also increasing. Most of the islands in Indonesia today still use diesel generator as electricity sources. With the high price of fuel oil, the price of electricity in the archipelago becomes very high. To reduce this dependence on fossil energy sources, it is necessary to develop a hybrid power plant system that combines diesel generators with electricity generated from renewable energy resources whose potential is in the area. So that the need for electric energy in isolated areas can be fulfilled with local potential energy sources.

Indonesia's geographical position crossed by the equator provides benefits for Indonesia, where renewable energy potential especially solar energy is very abundant because almost all regions in Indonesia can be exposed to the sun throughout the year. Nevertheless, electricity from solar energy does not necessarily directly replace electricity from fossil energy because the nature of renewable energy is generally intermittent depending on weather conditions. Therefore, the development of diesel generator hybrid system with solar power plant or Photovoltaic (PV) in archipelago areas is necessary[1].

The PV-Diesel hybrid system is potential to be developed in isolated islands but technical analysis in terms of the power of this system has not been done. Therefor we need to do research on the power quality of the PV-Diesel hybrid system.

This study aims to analyze the reliability and certain power quality parameters in an isolated PV-Diesel hybrid system, and provide an overview of the conditions that need to be anticipated on an isolated PV-Diesel hybrid system to maintain the reliability and quality of electricity.

The scope of work in this research including:

a. The analysis was carried out on a laboratory-scale isolated PV-Diesel hybrid system with a load profile in accordance with the 1300 VA household customer's daily load curve.
b. The power quality standard used is the Power Quality Guidelines Electricity Engineers' Association.
c. The power quality parameters evaluated are relevant parameters and can be detected by the Fluke 435 Power Quality Analyzer and Sunny Webbox form SMA.

2 Methodologies

The research conducted base on literature study, observation and testing method. The performance analysis conducted to lab-scale PV-Diesel hybrid system model based on literatures and several scenarios. All electrical data connected to power quality observed and recorded during test. The power quality parameters then evaluated and compared with standard from Electrical Engineer's Association (EEA).

* Corresponding author: rinairawatibm@gmail.com, rina.irawati@esdm.go.id

3 Literature Review

A hybrid power system is a system consist of two or more power sources, operated simultaneously, including (but not necessarily) a storage system and connected to a local (mini/microgrid) network, equipped with a multi-function inverter to convert DC and AC power, regulate generation and storage, and adjust the voltage and frequency of the system. Local mini/micro grid can also be connected to a large grid or national grid, thereby increasing the reliability and power quality of the system[2].

The advantages of using renewable energy sources for generating power in remote islands are obvious such as the cost of transported fuel are often prohibitive fossil fuel and that there is increasing concern on the issues of climate change and global warming. The disadvantage of standalone power systems using renewable energy is that the availability of renewable energy sources has daily and seasonal patterns which results in difficulties in regulating the output power to cope with the load demand. Combining the renewable energy generation with conventional diesel power generation will enable the power generated from renewable energy sources to be more reliable and affordable. This kind of electric power generation system, which consists of renewable energy and fossil fuel generators together with an energy storage system and power conditioning system, is known as a hybrid power system[3].With a hybrid system, power quality problems from renewable energy sources can be overcome.

To know whether power quality of the system is god or not, its need to be tested refers to existing standard. One of the standards governing the quality of power is the standard Power Quality Guidelines issued by Electricity Engineers 'Association. The standards set by the EEA are as follows[4]:

1. Steady-state voltage: The test is performed by data retrieval interval at least every ten minutes.
2. Voltage dips/sags: trigger level is set to 90% of nominal voltage.
3. Swell voltage: trigger level is set 110% of nominal voltage
4. Harmonic voltage: testing is successful when 95% of data retrieval done at intervals every ten seconds does not cross the boundary.
5. Frequency: follow the IEC 61000-4-30 standardi.e. data retrieval at least every ten seconds. Therefore for data retrieval on the power meter is set every ten seconds.

Performance analysis is conducted on hybrid power generation system installed in Renewable Energy Integration Laboratory at Mochtar Riady Plaza Quantum, Department of Electrical Engineering Faculty of Engineering University of Indonesia. This system represents a microgrid with two power sources that supply home loads but has considerable flexibility for other test scenarios such as on-grid and off-grid. The test was performed with a daily load profile of R-1300 VA that was simulated using a Programmable AC Load Controller as seen on Fig. 1[1].

4 Analysis

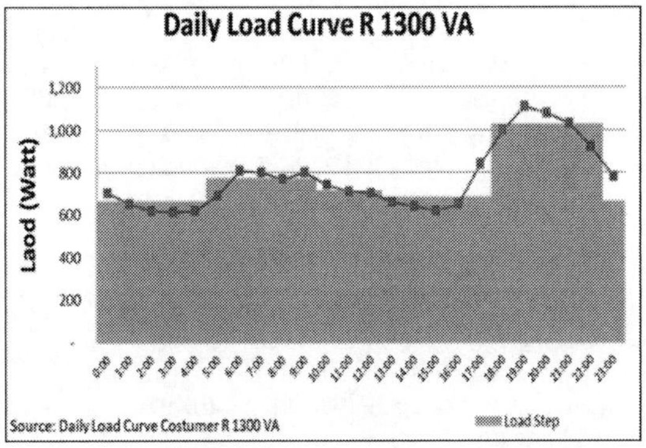

Fig. 1. Daily Load Profile.

The scheme in Fig. 2 is a simple representation of the installed test system. This system has two sources of electricity with one energy storage system. The first power source is a solar module whose output is converted into AC power and maximized with Maximum Power Point Tracker (MPPT) through Sunny Boy as a PV inverter and connected directly to the load and Sunny Island as a bidirectional inverter. The inverter PV output is connected to the bidirectional inverter and loads so when the energy production from PV larger then loads, PV inverter will supply the load and charges the batteries. However when the load power exceeds the power generated by PV, the load will be supplied by PV and batteries. The Bidirectional inverter itself is connected to an external power source in the form of a PLN grid or a diesel generator, and an energy storage system that is a battery bank. In the other hand, when the PV has no power output the PV inverter will detach from the system and the load will be supplied by the bidirectional inverter with the energy source of the battery. Then when the battery is within the specified State of Charge (SOC) limits, the diesel generator or PLN gird will supply the load and charge the battery through bidirectional inverter[1].

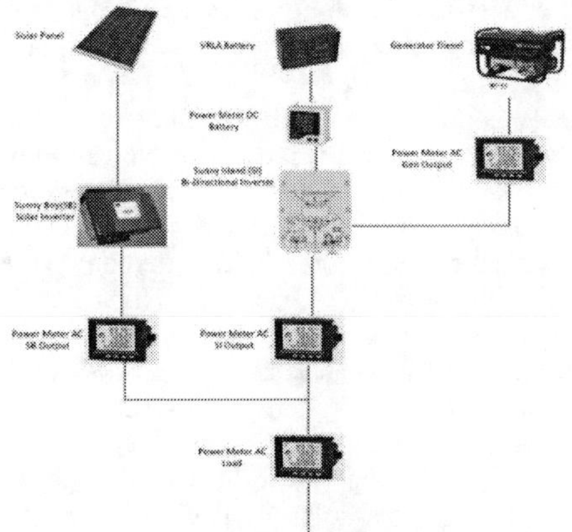

Fig. 2. Hybrid system test scheme.

In this test system the load used is a programmable AC Load Chroma 63803 with a maximum load capacity of 3.5 kW. The Power Factor value to be maintained is 0.9 lagging because the home load on the remote area is relatively linear load and if there is non-linear load it is inductive loads. Maximum current was set at 6 Ampere because the peak load is 1200 Watt which means current at system voltage 230 Volt is 5.2 Ampere.

In this study an analysis was conducted to the test results data along with the feasibility analysis of system power quality tested. The power quality parameters that are analyzed as follows:

- Steady-state voltage
- Harmonic voltage and Harmonic Flow
- Voltage dips / sags
- Swell voltage
- Frequency deviation

These parameters represent five of the eight power quality parameters found in the Electrical Engineer's Association (EEA) Power Quality Guideline. The power quality parameters analyzed are relevant to a low voltage (low voltage) test system and the availability of a measuring instrument that can measure the power quality parameters. The limits of these power quality parameters are as follows:

- Steady-state voltage: ± 6% of nominal voltage (216 Volts to 243.8 Volts for nominal voltage 230 Volt)
- Dip/sag voltage: 10% less than nominal voltage for a maximum of one minute (207 Volts for a nominal voltage of 230 Volts)
- Swell voltage: 10% more than nominal voltage for a maximum of one minute (253 Volts for a nominal voltage of 230 Volts).
- Harmonic voltage: maximum Voltage Total Harmonic Distortion is 8% for low voltage network
- Harmonic current: maximum Current Total Harmonic Distortion is 5% for low voltage network with current ratio less than 20.
- Frequency deviation: ± 1.5% of nominal voltage (49.25 Hz to 50.75 Hz for nominal frequency 50 Hz)

The evaluated parameters are measured in the same place i.e. Point of Evaluation (POE). POE in this system is Power Meter of Load. Power meter of load is selected as POE because it is the node closest to the load as can be seen in Fig. 2.

Analysis of power quality parameters is done on every circumstance and time so that we can understand more deeply the factors that cause the occurrence of a phenomenon or problem. The analysis is divided into six parts: Morning Transition, Noon, Afternoon Transition, Pre-Generator Night, Generator Night and Post-Generator Night.

Morning transition analysis is the analysis at 05.45 to 06.45. At this time the system supplies the load by using the battery even though the sun has shone until the inverter is able to supply the load to the system. This happens because the solar irradiation received by the solar cell does not produce enough voltage to initiate the inverter. We can verify when the inverter is on with bidirectional inverter Sunny Island data received in the Webbox. This means that in this analysis will be observed whether the initiation of inverters raises the problem of power quality on the system.

The noon analysis is the analysis at 08.00 to 14.30. In this analysis the irradiation of the solar cell is high enough so that the solar cell charges the battery and supplies the load for most of the time in this analysis. When irradiation is down and the solar cell output power is not sufficient to supply the load and charge the battery, then battery bank will supply the load. In this analysis can be observed the effect of solar cell and irradiation on the power quality parameters tested.

The Afternoon transition analysis is an analysis from 16.45 to 18.00. At 14.30 until 17.00 the sun is still shining but decreasing. At 17:00 until 18:00 solar irradiation is not enough to encourage solar cells to supply power to the system. Thus we can analyze the power quality parameters when the inverter changes the state of the supplying state until the inverter turns the operating status into deactivation.

The Pre-Generator night analysis is an analysis at 19.45 to 21.00. At this time all load is supplied purely by the battery due to deactivated PV inverters and State of Charge (SOC) batteries that have not reached the lower limit to initiate the generator to work. Thus in this analysis we can analyze the effect of the battery on the power quality parameters tested.

The Generator night analysis is an analysis with test time from 21:00 to 00:00. In this analysis the battery that previously supplied the load will pass the lower limit of the SOC that can initiate the generator until the generator supplies the load and charges the battery. That means we can analyze the quality parameters tested from the moment of the generator starts syncing in order to supply power to the system until the system is already running with the generator as the main supply.

The Post-Generation night analysis is analysis at 00.00 to 05.45. In the test period the batteries and loads previously supplied by the generator will be ordered to stop due to SOC batteries that have exceeded their upper limits. The data recorded in this analysis includes the post-stopping state of the generator system, this is so that we can analyze whether there is an effect when the generator is running that remains when the generator has stopped working. However, the main highlight of this analysis is the effect of transitions when the generator stops working.

Because each transition time has different effects on power quality, this paper will only cover morning, noon and afternoon transitions analysis. For night analysis, pre and post generators will be discussed in other papers.

4.1 Morning Transition Analysis

The morning transition analysis is intended to look deeper into the system condition when switching from the battery supply to PV supply due to the rising of solar irradiation. At 06:05 the status of the PV inverter changes to Waiting (Warten) which means the inverter is in Stand By condition waiting for the minimum voltage to supply the network achieved from the PV input. The status changes when the solar irradiation reaches the value of 9 W/m^2 to Mpp which means the inverter operates using MPPT to optimize the output power of the inverter. It should be noted that although the inverter is capable of exerting power but mostly still supplied by the battery, it can be seen in Fig. 3 where the battery SOC gradient still shows a negative value which means the battery is still decreasing.

Table 1. Webbox Data at 05:45 – 06:45

Time	Solar Irradiation	PV Inverter Output Power (Watt)	PV Inverter Status
5:45	0.67		
5:50	0.82		
5:55	0.95		
6:00	0.96	0	3: Warten
6:05	3.15	0	3: Warten
6:10	9	2.63	7: Mpp
6:15	13.63	4.63	7: Mpp
6:20	21.48	10.88	7: Mpp
6:25	32.87	21.03	7: Mpp
6:30	44.77	30.03	7: Mpp
6:35	55.1	44.03	7: Mpp
6:40	73.07	75.03	7: Mpp
6:45	87.88	91.23	7: Mpp

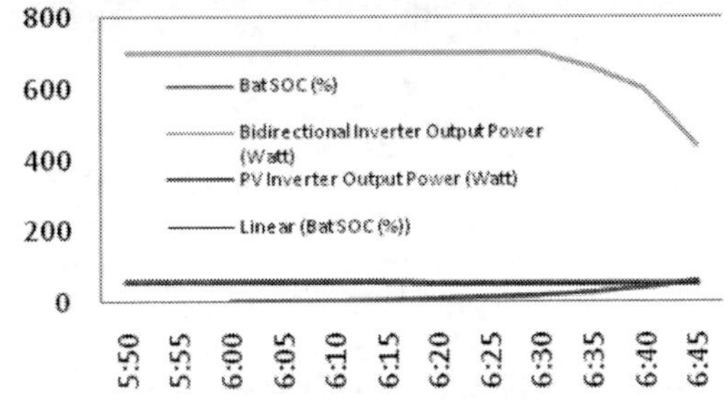

Fig. 3. Operation condition at 06:00 - 06:45

4.1.1 Steady-state Voltage, Dips / Sags, Swell

The active transition inverter does not seem to affect the voltage significantly. At 06:10 where the inverter status changed to Mpp the voltage increased only 0.2 Volt. The measured voltage has a range from the lowest 231.29 Volts to 231.58. With such a voltage range it can be concluded at this morning's transition the system is quite stable.

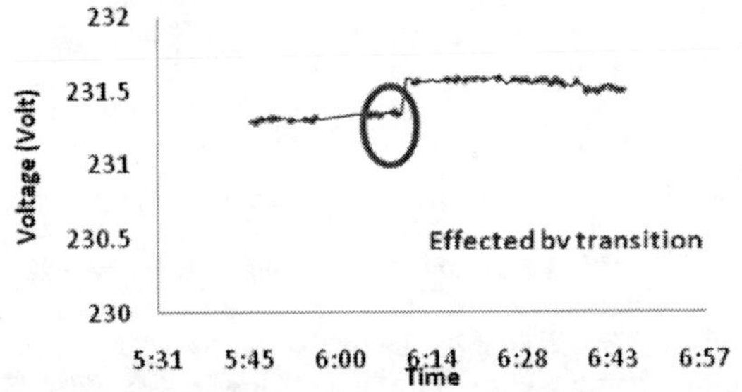

Fig. 4. Voltage at 05.00-06.45

4.1.2 Voltage Harmonic

The voltage harmonic at Fig. 5are seen to increase during the transition period. When the PV inverter Sunny Boy active voltage harmonic changed from 5.16% to 6.01%. However when the power output of PV inverter Sunny Boy began to increase and reduce the output of Sunny Island voltage harmonics also decreased. The dynamics that occur in this PV inverter Sunny Boy is observed to affect the voltage harmonics but these parameters are still within the safe limits of power quality.

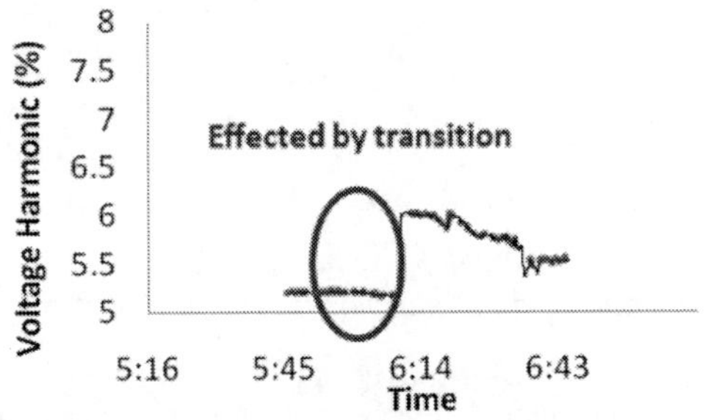

Fig. 5. Voltage harmonic at 05.45-06.45

4.1.3 Frequency Deviation

The advent of the PV inverter as a new supply at observations at 05.45-06.45 looks totally unaffected. The frequency remains stable until the end of the observation because the load and supply remain balanced. When the PV inverter Sunny Boy experiences increased power output bidirectional inverter Sunny Island can adjust well in reducing its output so that there is no overload or supply in the system.

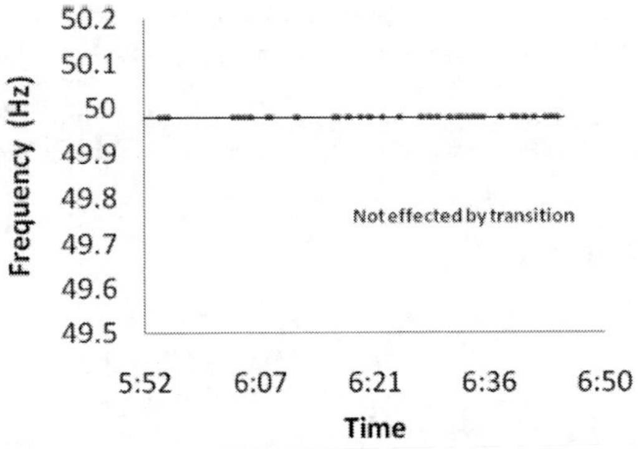

Fig. 6. Frequencies at 05.45-06.45

4.1.4 Current Harmonic

In the current harmonic parameter there is no active inverter effect. The current harmonic in this analysis remain stable at a range of values of 54%. The current harmonic parameters generally in the connected network are influenced by the load characteristics and in this experiment the load power is relatively fixed.

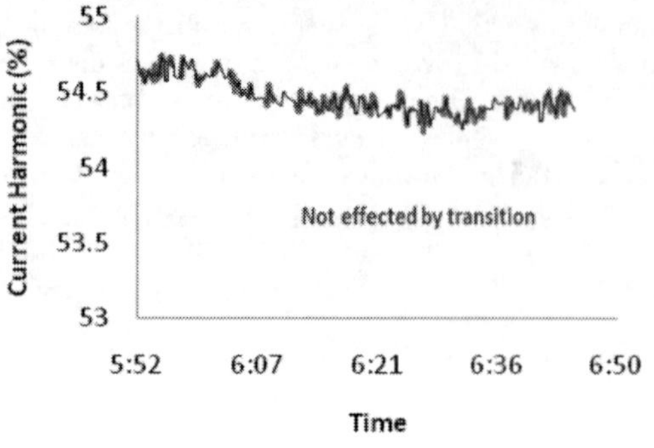

Fig. 7. Current Harmonic at 05:45-06.45

4.1.5 Conclusion of Morning Transition Analysis

From the data analysis above the active switching of the PV inverter Sunny Boy does not cause any meaningful problems either in terms of power quality or reliability but current harmonics are well over the limit. When the PV inverter Sunny Boy is active and starts supplying power to the system it only affects voltage and voltage harmonics with not really significant value. The frequency is not changed at all while the PV inverter Sunny Boy is active. The current harmonic parameters that should have a 5% limit on this analysis far exceed the limit to 54%.

4.2 Noon Analysis

In this analysis the PV inverters already able to supply power into the load or charge a battery depends on solar irradiation. This can be seen in the graphic shown in Fig. 8which is the result of footage at the time the system works during the day where during sun irradiation and Sunny Boy output is less than the load then Sunny Island which takes power from the battery helps to supply the load.

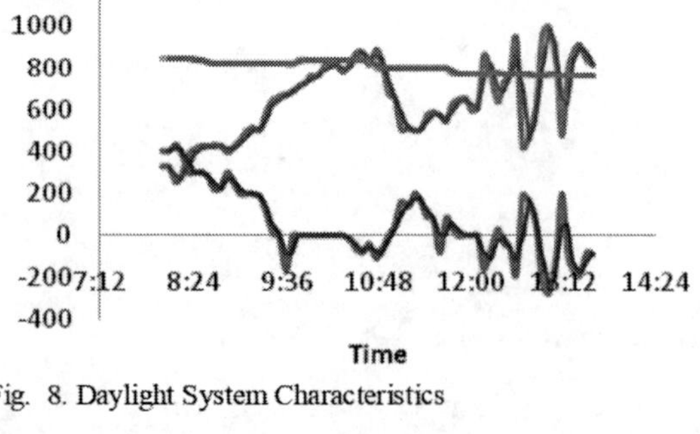

Fig. 8. Daylight System Characteristics

However, when the sun's irradiation and Sunny Boy's output is larger than the load, the Sunny Island output is negative, which means charging through Sunny Island. In order to understand in more detail in the next analysis it should be noted that the battery charging is done at 12:10 to 12:35; 12:55 to 13.05; And 13.15 to 13.35. At that time other than that time Sunny Island helped supply the load or output equal to zero.

4.2.1 Steady-stateVoltage, Dips / Sags, Swell

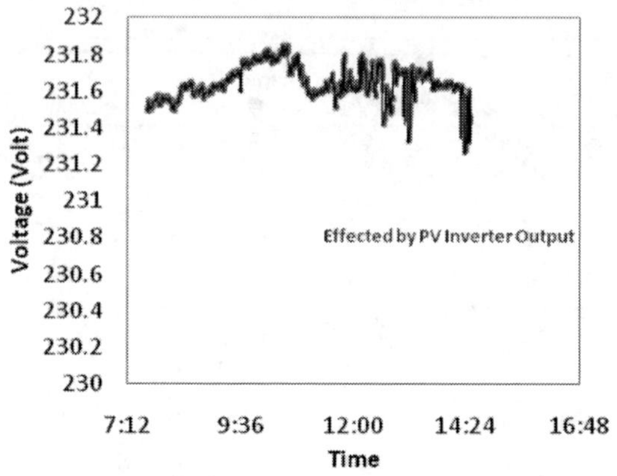

Fig. 9. Figure 11. Figure 4.8 System Voltages 08.00-14.30

The voltages in this analysis do not deviate too far and still within EEA standards. The system voltage should be 230 Volt but it rises up to the 231 Volt range. When compared to the voltage on the graph of Fig. 9, the sunny boy output, and the sun irradiation it can be seen that the ups and downs of the voltage follow the upward trend of sunny boy output and solar irradiation which means the system voltage is affected by the voltage supplied by the PV inverter. The highest voltage deviation is 231.85 around 10:30 and the lowest deviation is 231.25 at around 14:24.

4.2.2 Voltage Harmonic

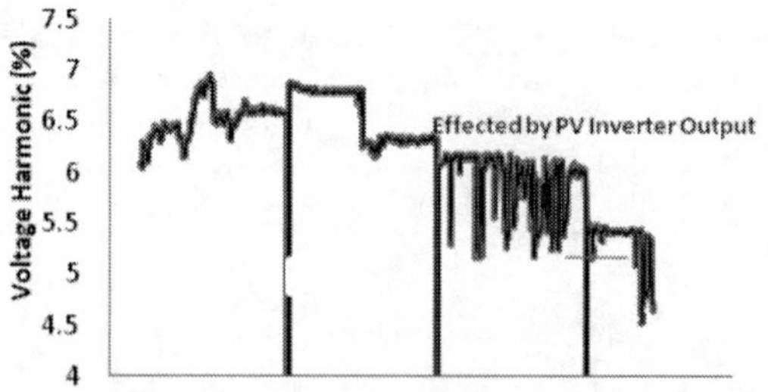

Fig. 10. Voltage Harmonic at 08.00-14.30

The voltage harmonics in this analysis do not exceed the 8% standard limit. The average voltage harmonic at this time period is in the range of 6%. The highest voltage harmonic is 6.93% and the lowest is 4.62%. The active components of the afternoon analysis are PV

Inverter Sunny Boy, Bidirectional Inverter Sunny Island and battery while the system works on a power factor of 0.9 lagging. This means the voltage harmonics are generated with the greatest contribution of the inverter as well as the bidirectional inverter. Although the voltage harmonic generator is theoretically inverted but the resulting data trend of voltage harmonics has a gradient that is more comparable to the load and output of bidirectional inverter Sunny Island than the PV inverter Sunny Boy.

4.2.3 Frequency Deviation

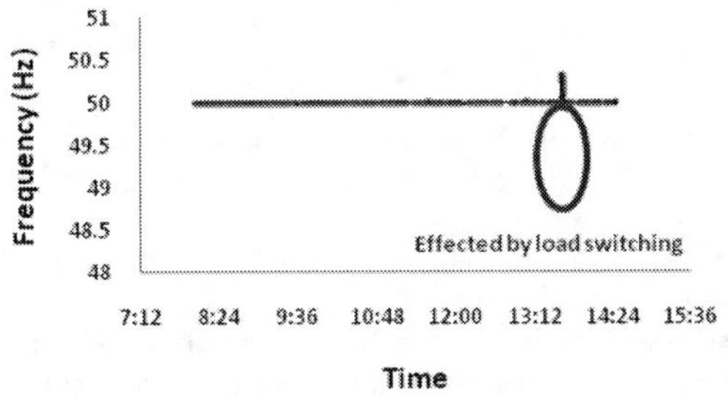

Fig. 11. Frequencies at 08.00-14.30

Recorded frequencies indicate that day-to-day operating systems exhibit stable frequencies without significant frequency deviations and are within the standard range. The largest frequency deviation occurs at 13:38 which is 50.33 Hz while the smallest appears at 13.40 of 49.92. Once traced, the cause of the frequency deviation between 13.38 and 13.40 is the instantaneous load change due to the load settings performed on the Programmable Load Chroma AC.

4.2.4 Current Harmonic

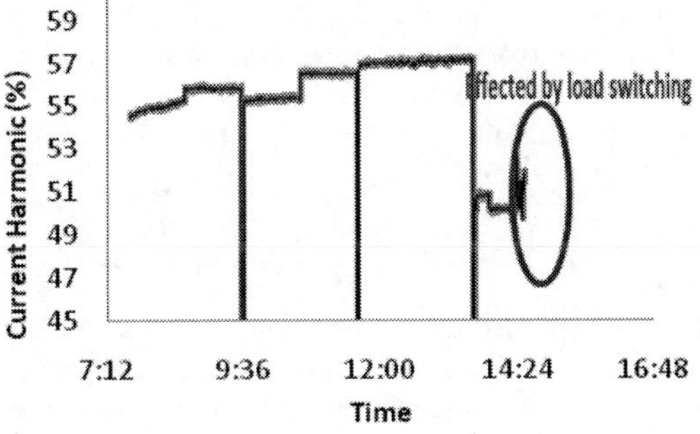

Fig. 12. Current harmonic at 08.00-14.30

The current harmonic at 08.00-14.30 appears to be affected by system dynamics in the afternoon calibration experiments. The most obvious change is when the current harmonics decrease by more than 7% at 13.38. These changes coincide with step load changes that are also down so this phenomenon becomes second evidence of harmonic currents in isolated tissue remains affected by the load. However, from 08:00 to 13:38 the harmonic currents tend to increase with the increase of the Sunny Boy inverter operation instead of the Sunny Island Inverter so that in this experiment it is seen that the current harmonics are affected not only by the load but also the Sunny Boy inverter.

4.2.5 Conclusion of Noon Analysis

In the noon analysis can be concluded that there is no problem of power quality or reliability other than the harmonic current parameters. Unobservable power quality limits are exceeded but note that voltage harmonics reach 6.93% and the graph indicates the proportionality of voltage harmonics with allowable load power when during the day there are higher loads or larger Sunny Island inverter output can cause voltage harmonics to pass through limit. The harmonic flow once again passes the limit with a much higher value of up to 57.13%. In terms of system reliability goes 100% without any trip at all.

4.3 Afternoon Analysis

From the Webbox data in Table 2 it can be observed how the system being tested operates affected by the start of sunset irradiation. At 16:45 to 17:00 where the lowest solar irradiation is 9.72 W/m^2 status Sunny Boy Inverter shows Mpp which means the inverter is working in MPPT mode. However in 17.05 the status of the inverter shows "Search-Mpp" (Mpp-Such).

Table 2. Webbox Data at16:45 to 18:00

Time	Solar Irradiation W/m^2	Bidirectional Inverter Current A	PV Inverter Power W	PV Inverter Status
16:45	17.03	3.26	11.53	7: Mpp
16:50	14.07	3.3	9.72	7: Mpp
16:55	11.6	3.3	6.77	7: Mpp
17:00	9.72	3.3	4.97	7: Mpp
17:05	8	3.37	2.56	6: Mpp-Such
17:10	6.14	3.32	0.04	2: Netzueb.
17:15	4.5	3.3	0	0: Offset
17:20	2	2.37	0	3: Warten
17:25	1.08	3.3	0	3: Warten
17:30	1	3.3	0	3: Warten
17:35	1	3.3	0	3: Warten
17:40	1.08	3.2	[illegible]	[illegible]
17:45	1	3.3	[illegible]	[illegible]
17:50	1	3.3	[illegible]	[illegible]
17:55	1	3.3	[illegible]	[illegible]
18:00	1.1	3.3	[illegible]	[illegible]

This means at that time the inverter is considering whether MPPT is still usable or not. At 17:15 the emerging status is off-set where in this state the inverter adjusts or arranges the measuring instrument to make it more sensitive to respond to smaller values. At 17:20 to

 https://doi.org/10.1051/e3sconf/20183102001

17:35 Warten status appears or in the Indonesian language means wait, the intended wait here is the inverter in stand-by mode in some time until the minimum voltage to supply the network is reached. However, in the afternoon transition the most common is the minimum voltage will not be reached and the inverter will break away from the system or go into deactivation mode as seen in the table at 17.30 onwards.

4.3.1 Steady-state Voltage, Dips / Sags And Swell

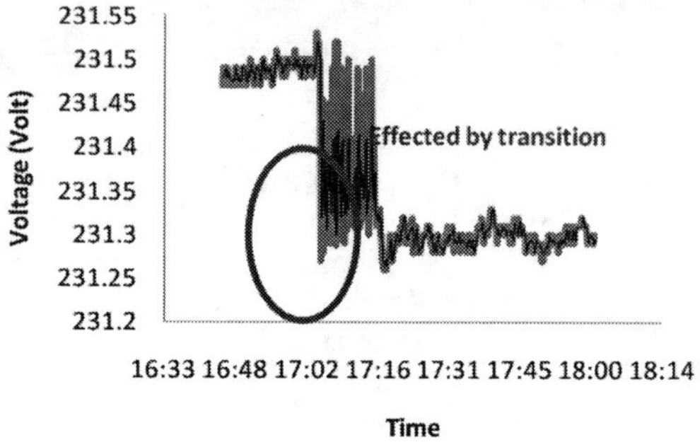

Fig. 13. System Voltages at 16:45 to 18:00

The system voltage at 16.45 to 18.00 is quite stable and is not far from its nominal voltage. The highest voltage observed at 17.03 is 231.5 V and the lowest is 231.26 V at 17.17. Changing the voltage can be seen at 17.02 where the inverter has just changed from the status of mpp to 17.17 where the status of the inverter will turn into stand-by. This shows that the inverter's inactive transition affects the voltage but is not in an alarming state because it does not approach the limits of steady-state, dip or swell.

4.3.2 Voltage Harmonic

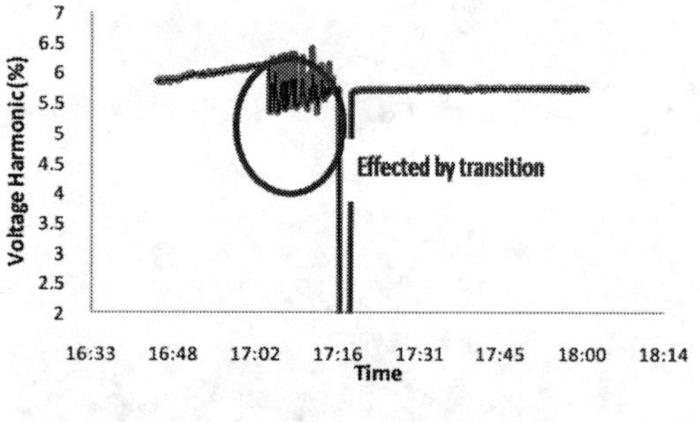

Fig. 14. Voltage Harmonic at 16:45 to 18:00

Harmonic stress in the analysis period that is analyzed is relatively stable with the harmonic quantity range is 5.8%. The greatest harmonic is 6.39% at 17.10 and the lowest harmonic is 5.33% at 17.13. In the evaluation of the voltage harmonic parameters it was observed that the value of this parameter varies when the inverter goes into its transitional period of inactivity but the effect of the transition is insignificant.

4.3.3 Frequency Deviation

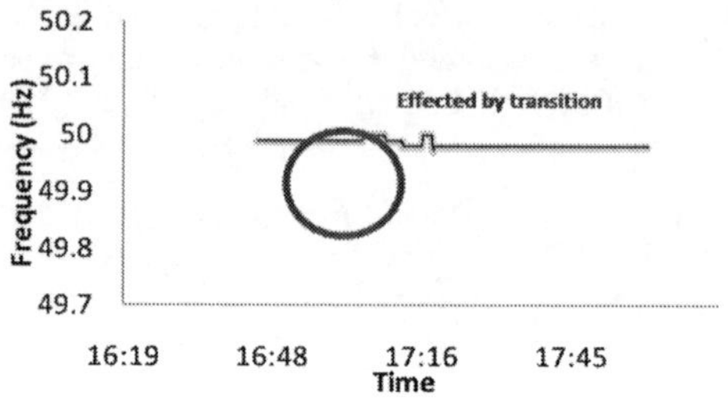

Fig. 15. System Frequencies at 16:45 to 18:00

The system frequency in this analysis ranges around 49.99 Hz and can be seen on the graph that the value is not exceeded and stable except at 17.02 to 17.16. The lowest frequency value is at 17:19 which is 49.97 and the highest is 50 Hz at some time. The graph shows that the inverter's inactive transition greatly does not affect the frequency that only moves in the 2 numbers behind the comma.

4.3.4 Current Harmonic

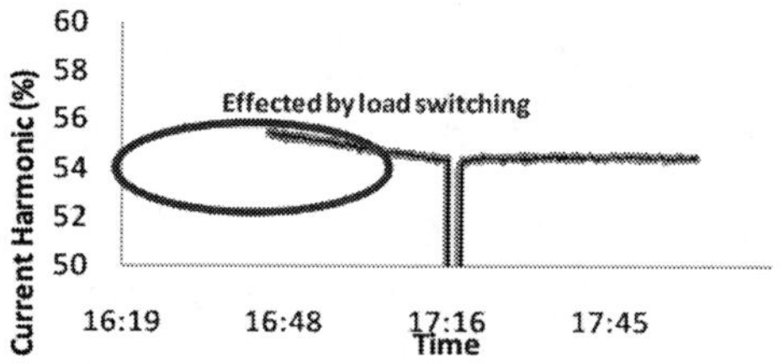

Fig. 16. Current Harmonic at 16.45-18.00

In the afternoon transition experiments the loads used are programmed to ensure the current harmonic trend against the load but only up to 17.17. The programmed load is linearly decreasing and from the above graph it can be seen that the harmonic current follows the load which tends to decrease. Similar to the previous analysis the current harmonics cross the limit with a significant value that is above 54%.

4.3.5 Conclusion of Evening Transition Analysis

In the analysis of Transition Evening analysis found that the transition of non-active inverter affect the parameters of power quality but not significant. The biggest effect of this transition is on the voltage harmonic where the transition period observed a shift of harmonic voltage value of 13.25% of harmonic voltage limits. Similar to the previous analysis the harmonic currents also exceed the limit of up to 55.34%. In terms of reliability this transition does not lead to reduced reliability of the system in supplying the load.

5 Conclusions

From the analysis of research conducted to evaluate the reliability and some power quality parameters of isolated hybrid system in dynamic conditions, it can be concluded several things as follows. Power quality problems arise in several conditions:

1. The peak load voltage harmonic condition exceeds the standard.
2. The condition of the system with the harmonic generator supply exceeds the standard.
3. The post-inactive condition of the bi-directional generator inverter frequency exceeds the standard limit.
4. Specification of bi-directional inverter showing the voltage harmonic will not exceed 3% not achieved.

The authors say thank you very much to Ryan Adilardy Prakoso and Brian Oscarino, undergraduates of Electrical Engineering UI, who have assisted authors in testing and retrieving data in the study of hybrid system at University of Indonesia Campus.

References

1. R. Irawati and E. A. Setiawan, "Analisis Keandalan Jaringan Listrik Mikro Cerdas Arus Bolak Balik Terisolasi," *Makara Univ. Indones.*, pp. 1–19, 2014
2. G. Léna, *Rural Electrification with PV Hybrid Systems*, Juli 2013. IEA, 2013.
3. A. Setiawan and C. Nayar, "Design of Hybrid Power System for a Remote Island in Maldives," ... *HOMER Webcast-NREL USA*, pp. 1–5, 2006.
4. N. Watson, V. Gosbell, S. Perera, S. Elphick, and S. Hardie, *Power Quality Guidelines*, Rev 3.8., no. January. Wellington: Electricity Engineers' Association, 2013
5. PT PLN (Persero), *Solar PV Development Plan.* (2013).
6. Anwari, M. I. Hamid & Makbul, *Single-Phase Photovoltaic-Inverter Operation Characteristic in Distritibuted Generation System.*. s.l. : InTech, Distributed Generation Vol. , (2010).
7. I. Muchsin, *PUSAT PENGEMBANGAN BAHAN AJAR-UMB*.Mesin Sinkron. s.l.
8. SMA Solar Technology AG.*Sunny Boy 1200/1700 User Manual.* Niestetal, Jerman : s.n.
9. Chroma ATE INC.,*Programmable AC/DC Electronic Load 63800 Series Operation and Programming Manual.* (2012).
10. Accuenergy,*Acuvim II Series Power Meter User's Manual.* (2012).
11. Endeavour Energy Power Quality & Reliability Centre, University of Wollongong.*Small Scale Domestic Rooftop Solar Photovoltaic System.* (2011).
12. *IEEE and IEC Harmonic Limits.* Haplin, Mark.(2007).
13. SMA Solar Technology AG.*Off-Grid Inverter Sunny Island 5048.*

Enhancement of Biogas Production from Rice Husk by NaOH and Enzyme Pretreatment

Syafrudin[1,*], *Winardi* Dwi Nugraha[1], *Shandy* Sarima Agnesia[1], *Hashfi* Hawali Abdul Matin[2], and *Budiyono*[3]

[1]Department of Environmental Engineering, Faculty of Engineering, Diponegoro University, Semarang - Indonesia
[2]Master Program of Environmental Science, School of Postgraduate Studies, Diponegoro University, Semarang - Indonesia
[3]Department of Chemical Engineering, Faculty of Engineering, Diponegoro University, Semarang - Indonesia

Abstract. Biogas is a renewable energy source that can be used as an alternative fuel to replace fossil fuel such as oil and natural gas. This research aims to analyze the impact of NaOH (Sodium hydroxide) and enzyme usage on the production of rice husk biogas using Solid State Anaerobic Digestion (SS-AD). Generally, SS-AD occurs at solid concentrations higher than 15%. The waste of rice husk are used as substrate with a C/N ratio of 25% and the total of solid that are used is 21%. Rice husk contains high lignin, therefore it is handled with chemical and biological treatment. The chemical preliminary treatment was using NaOH with various concentrations from 3%, 6% and 9% while the biological preliminary treatment was using enzyme with various concentration from 5%, 8%, and 11%. The biogas that is produced then measured every two days during 60 days of research with the biogas volume as a parameter observed. The result of the research shows that preliminary treatment with NaOH and enzyme can increase the production of biogas. The highest biogas production was obtained by the NaOH pretreatment using 6% NaOH which was 497 ml and by enzyme pretreatment using 11% enzyme which was 667,5 ml.

1 Introduction

Energy has become a principal needs to modern society nowadays The need of energy use is increasing along with population growth and the increasing consumption of society, as they use a wide range of tools and equipment to ease their life [1]. Nationally, the utilization of energy is still dominated by conventional energy sources, which are non-renewable energy such as crude oil, coal, and natural gas The contribution of non renewable energy for the national power plants has reached 85% [2]. One of the ways to save oil fuel and other unrenewable energy is to find another alternative that can be renewable [3]. That is why, to lessen the dependency for oil fuel, the government has established the presidential decree of Republic of Indonesia number 5 year 2006 on the issue of developing an alternative energy source as a replacement of fuel oil, microbial consortium and pretreatment. That policy highlights on the matter of energy source that can be renewable as an alternative to fossil fuel.

Biogas is a renewable energy that is able to be used as an alternative fuel to replace fuels that are originated from fossil such as petrol and natural gas [4]. Biogas is also one of a bioenergy that could be define as a gas that is released when organic materials such as human waste, manure, straw, rice husk and leaves from sorted out vegetables that are fermented or under a process called methanisation [5]

Indonesia can be considered as a country with the richest biomass yet has a low utilization of biomass potential [6]. Rice husk consists of organic elements such as cellulose, hemicellulose, and lignin [7]. The decomposition of organic material that contains cellulose, hemicellulose, and lignin transpire slowly. Thus, to fasten the degradation process of organic material that contains, the pretreatment of basic materials is necessary to occur [8].

Chemical pretreatment process can be done by using NaOH, KOH, calcium, and other types of alcalines, From those three alcaline type (NaOH, KOH and calcium) that have been tested for rice husk pretreatment, NaOH is the most effective to deprive lignin from the production of biogas[9]. This process is usually known as the delignification, which is the release of lignin from a complex compound.

The biologically pretreatment process with the help of microorganisms could also be used to process lingo cellulose and increase enzymeatic hydrolysis. Biological pretreatment for the increase of biogas production on AD included fungus pretreatment microbial consortium pretreatment and enzymeatic pretreatment [10].

2 Methods

This research is an experimental laboratory research. This research was done at a waste treatment laboratory, chemical engineering, University of Dipenegoro from December 2016 to February 2017.

* Corresponding author: udin_syaf@yahoo.com

 https://doi.org/10.1051/e3sconf/20183102002

2.1Test of rice husk TS contents

Analysis of Total Solid (TS) with the standard method of APHA

- Chemical cup is dried on 103-105°C temperatures for 1 hour, chilled and then stored in a desiccators until the chemical cup is going to be used
- The chemical cup's weigh is being measured and noted
- Insert 25-50 g sample to the chemical cup and then weigh, after that dried inside an oven at 103-105°C for 1 hour
- The samples that have been dried afterwards are cooled on the desiccators later weighed until its heaviness decreased on the amount of 4% or 50 mg

$$\% \, Total \, solids = \frac{(A-B)x \, 100\%}{C-B} \quad (1)$$

Note:
A = The weight of samples that have been dried + chemical cup (mg)
B = The weight of chemical cup (mg)
C = The weight of wet sample + chemical cup (mg)

2.2 Stage of Research

2.2.1 Preparation Stage

- Tools are prepared and planned in line with the variable of the research
- Take rice husks to be weighed and put into the vessel as many as the amount of the variable (14 variables)
- Do pretreatment to each corresponding variable

2.2.2 The Operation Stage

- Put rice husks and cow's rumen inside the biodigester with each biodigester filled with both rice husks and cow's rumens that are in line with determined variation
- Put the micronutrients as needed
- Input water into the biodigester
- Wait on the fermentation process until the biogas is formed
- Measure the volume of newly formed biogas every two days so that the biogas doesn't regenerated.

3Result and Discussion

3.1PretreatmentNaOH

In this research, the concentration variations that were used as a chemical pretreatment were the amount of 3%, 6% and 9%. This research aims to analyze the impact of NaOH on rice husks in biogas production using SS-AD methods. The data of biogas volume were presented as a daily cumulative biogas yield during 60 days of research, as seen on Figure 1.

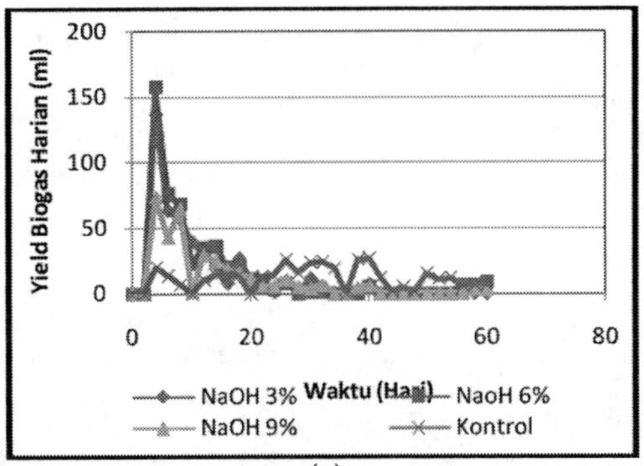

(a)

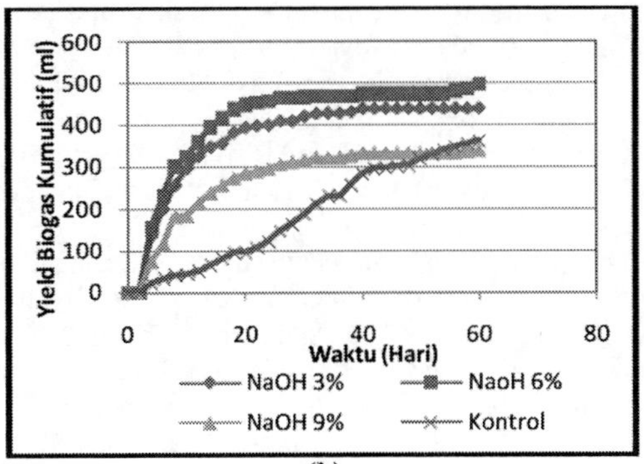

(b)

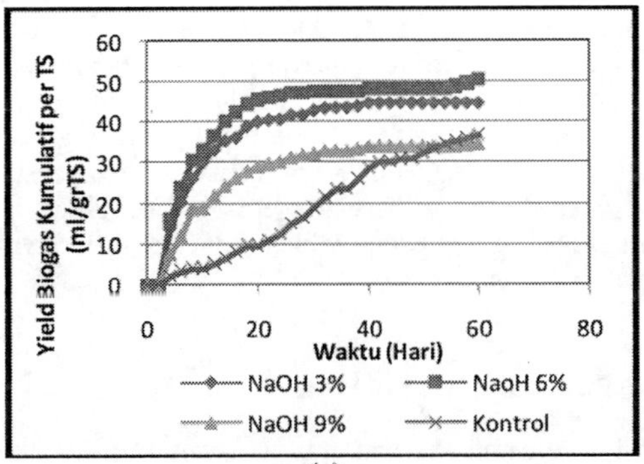

(c)

Fig. 1. The Impact of NaOH to Biogas Production

Figure1 (a) shows the 6% increase of NaOH, and at the measuring session on day 2 it reached the highest peak with the amount of 152 ml and 163 ml. The average daily biogas yield reached 157,5 ml. Then, on the next day of research the biogas production declined gradually. However, with a 6% NaOH reactor, it produced biogas yield again on the 56th, 58th and 60th day. Each with the amount of 6,5 ml, 7 ml, and 9 ml.

Figure1 (b) shows the impact of NaOH pretreatment on cumulative biogas production. Result showed that the reactor with 6% NaOH produce the highest biogas yield of 497 ml. Whereas, 3 % and 9 % concentrations of NaOH resulted 439 and 338,5 biogas yields. Whilst, a reactor that wasn't apply with pretreatment, resulted 361,5 ml biogas.

Figure 1 (c) shows cumulative biogas yield per TS unit result as the impact of using NaOH. The highest number of cumulative biogas yield per TS unit was resulted by a reactor filled with 6% NaOH which was 50,60 ml/gr TS. Furthermore, 3% NaOH resulted 44,70 ml/gr TS, while 9% NaOH and control reactor gave confiding result of 34,46 ml/gr and 36,81 ml/gr TS.

These sequence of elucidations showed that biogas yield increased after preliminary treatment with added NaOH and enzyme was done This points out that doing chemical and biological pretreatments influence biogas production significantly.

Whilst, for other reactors such as 3% and 9% each also used 5% enzyme, resulted 44,70 ml/gr TS and 34,46 ml/gr TS This states that the usage of 6% NaOH as a chemical pretreatment for rise husk wastes produce the most optimal biogas yield, compare with 3% and 9% NaOH. According to the research of He et al said that the utilization of 6% NaOH is the optimal dose for a pretreatment of rise husk [11]. The utilization of 6% NaOH on rice husks wastes indicated the increasing biogas production with the amount of 27,3-64,5% than the ones that were not using NaOH. Hosseini et al reported that with the increasing of sodium hydroxide (NaOH) from 15 to 30 gr/L it increased the rice stras biodegradability and so it also increased the production of biogas[12].

The usage of NaOH on pretreatment basically is a process of delignification that cause lignin to dissolved in a massive amount. Nevertheless, too much usage of NaOH will allow a high alkali atmosphere to substrate. Thus it will disturb the microorganisms's activity [13]. The application of 9% NaOH produced the lowest result of biogas yield than the ones without pretreatment. This shows that 9% NaOH does not increases biogas result on rice husks, This is because 9% NaOH gives alkali atmosphere to the reactor which influence the activity of microorganism.

3.2 Pretreatment Enzyme

In this research the concentration variation of enzyme that was used as a biological pretreatment were 5%, 8% and 11%. This research aims to analyze the impact of enzyme on rice husks in biogas production using SS-AD methods. The data of biogas volume are presented as a daily cumulative biogas yield during 60 days of research, as seen on Figure 2.

Figure 2 (a) shows the difference of daily biogas yield on rice husks that were on biological pretreatment with variety of enzyme amount such as 5%, 8%, and 11%, each also was given chemical pretreatments using 3% of NaOH and compared to the ones that didn't have chemical or biological pretreatments. Reactor that had 11% enzyme addition, on the 4th day of measure, the 11% enzyme reactor reach the highest number of 256 ml and 289 ml with biogas yield daily average production is 272,5 % ml. Moreover, on the next day of research, the biogas production decreased. Reactor 2 with 11 % enzyme produced yield biogas again on the 56th and 58th days each with the amount of 21 ml and 14 ml.

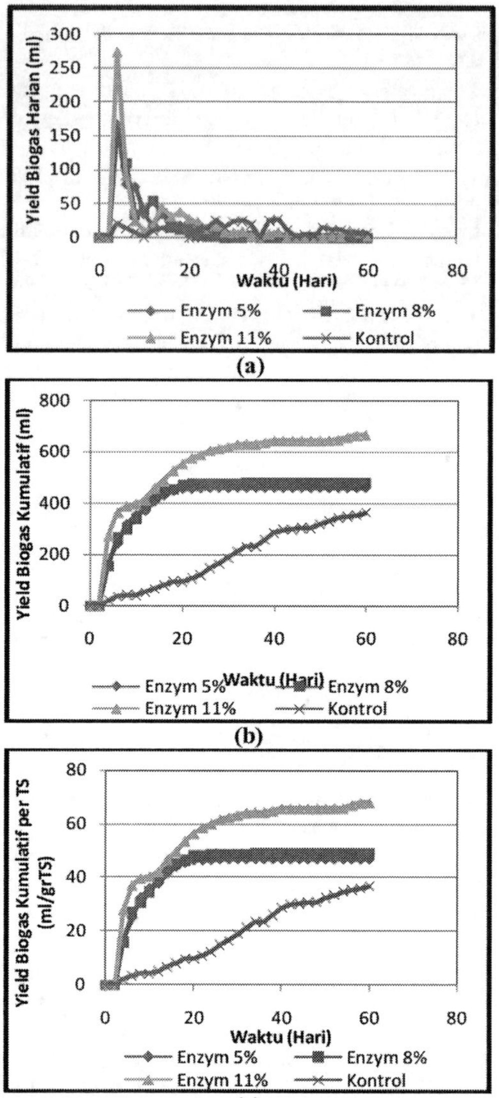

(c)

Fig. 2. The Impact of NaOH to Biogas Production

Figure2 (b) shows the influence of enzyme pretreatment on the cumulative biogas production. Result showed that reactor that had 11% enzyme generated the highest biogas yield which was 667,5% ml. Whereas with a 5 % and 8% enzyme concentrations each generated 462 ml and 480,5 ml biogas yields. While reactor that was not given pretreatment, resulted 361,5 ml biogas

Figure 2 (c) shows the production of biogas yield cumulative result per TS unit on the influence of enzyme on the biogas production on rice husk. The highest biogas yield cumulative result per TS unit was generated by reactor filled with 11% enzyme with the amount of 67,96 ml/gr TS. Moreover, for 5% and 8 % enzyme each produced each 47, 04 ml/gr TS and 48,92 ml/gr TS, while the control reactor produced 36,81 ml/gr TS biogas.

https://doi.org/10.1051/e3sconf/20183102002

Those series of explanations indicated that the biogas yield increased after the preliminary treatment using addition of enzyme and NaOH. On a reactor that used 11% enzyme and 3% NaOH resulted the highest biogas yield which was 67,96 ml/gr TS. Whilst for control reactor that didn't use enzyme or NaOH produced biogas yield of 36,81 ml/gr TS.

Other reactors that used 5% and 8% enzyme, both used 3% of NaOH and generated 47,04 ml/gr TS and 48,92 ml/gr TS. This confirms that the usage of 11% as a biological pretreatment for rice husks waste generated the most optimal biogas yield compares 5% and 8% enzyme.

Zhang et al. (2011) used tapioca dregs and did a microbial consortium pretreatment in a batch reactor with 55oC temperature could result methane yield of 259,46 mL/g VS this was obtainable on 12 hours of preliminary treatment where 96,63% was higher than control which was 131,95 mL/g VS) [14].This proofs that the use of microbial consortium on rice husks waste pretreatment can help degrade lignocellulose so the production of biogas is increased.

On the 11% enzyme research showed the best concentration on increasing biogas production rather than 5% and 8% enzyme. The reactor that was filed with 5 % enzyme, it resulted the lowest biogas between 8% and 11% enzyme, This is in line with what Purwoko et al. had researched which stated that the higher the utilization of microbial consortium with EM4 trademark then the higher the biogas production will increased [15].

3.3 The production rate of biogas from rice husks

In this research the objective is to analyze the production rate of biogas on rice husk wastes using SS-AD method. The data that were obtained then were solved in a numerical way with regression non linear using Polymath 6.0. Polymath 6.0 will gained biogas production rate constants (U), maximum production of biogas (A), and the minimum time of biogas creation (λ).

3.3.1 Production on the effect of NaOH

Table 1 and Figure 3 show that the usage of 6 % NaOH has given a real impact on the kinetic constant of biogas production. From the Polymath 6.0 result, 6% NaOH has the highest production rate compares to 3% NaOH, 9% NaOH, and control. The kinetic constant of 6% NaOH is avowed as the constant production rate of biogas rate (U), maximum biogas production (A), and the minimum time of biogas formation (□) each with a value of 3,97 ml/(gr TS per days), 48,06 ml/gr TS; and 0,75 per days. This means that rice husks that were treated with 6% NaOH pretreatment had a production rate constant of 3,97 ml/(gr TS days) resulting biogas production of 48,06 ml/gr TS and produce the first time on the 0,75 day.

The result of biogas yield with 6% pretreatment generated the highest biogas yield compared to 3% NaOH and 9% NaOH and control. The 3% NaOH had second order of biogas yield and biogas rate production each on the amount of 3,22 ml/gr TS and 43,77 ml/(gr TS days) with the initial formation of biogas on the 0,46 day. Further, 9% NaOH had the lowest biogas which was 33, 44 ml/gr and biogas production rate of 1,97 ml/(gr TS days) with the initial formation of biogas on the 0,47 day. On the other hand, reactor without pretreatment (control) had a biogas yield of 43,32 ml/gr TS and the lowest production rate which was 0,78 ml/(gr TS days) with the longest initial formation of biogas on the 7,83 days. This showed that the usage of 9% NaOH on rice husks caused lower biogas yield compares to rice husks that didn't have preliminary treatment.

Table 1. Kinetic constant in effect variation NaOH

Constant	NaOH 3%	NaOH 6%	NaOH 9%	Control
A, ml/gr TS	43,77	48,06	33,44	43,32
U, ml/(gr TS day)	3,22	3,97	1,97	0,87
λ, day	0,46	0,75	0,47	7,83

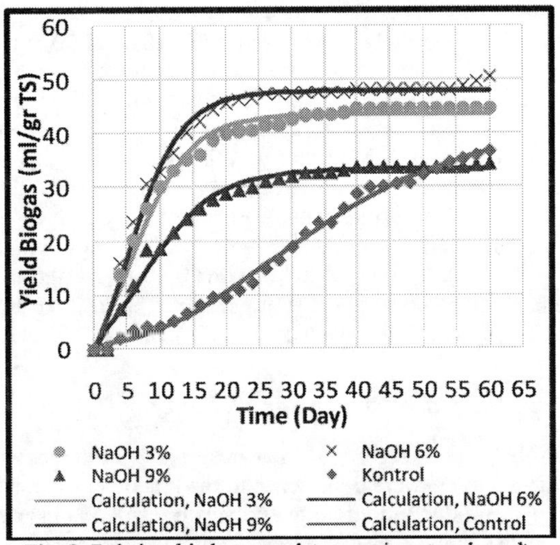

Fig. 3. Relationship between data experiment and results calculation on effect variation NaOH

From the data above, it can be concluded that the usage of 3% NaOH and 6% NaOH resulted biogas yield that was higher than the pretreatment (control) ones with faster production rate and faster biogas formation time. Moreover, the concentration variation of 6% NaOH gave the highest biogas yield of 48,06 biogas ml/gr TS, 6% NaOH had the highest rate production compare to 3% NaOH and 9% NaOH and control which is 3,97 ml/(gr TS day). This is because the 11% NaOH reactor biogas result is constant.

Figure 4 shows the calculation from the biogas result and 6% NaOH using Polymath 6.0. Thus, the maximum of biogas yield production can be found and the termination time of biogas production.

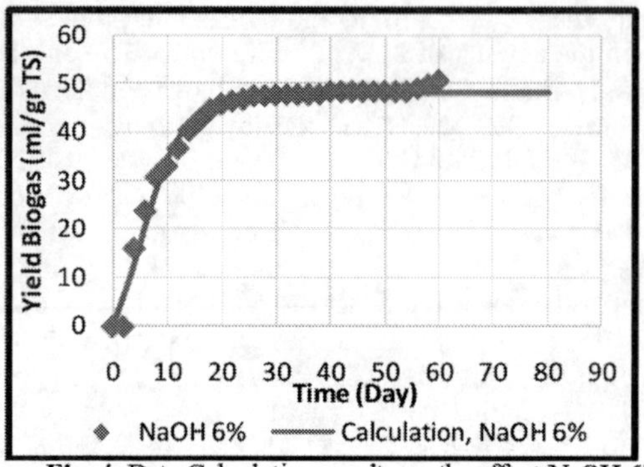

Fig. 4. Data Calculation results on the effect NaOH 6%

Based on Figure 4, it can be seen that the biogas yield maximum on 6% NaOH was 48,06 ml/gr TS and this was reached on the 50th day. This demonstrates the rice husks biogas with 6% NaOH is stable and can not result any more biogas. As seen from the data in the 50th day the biogas production has stopped. This phenomenon can be used to design SS-AD biodigester from rice husk with the continuation of NaOH preliminary treatment during 50 days.

Once the modeling preliminary treatment was done, by adding 6% of NaOH, then 48,06 ml/grTS biogas yield was gained. When there are 1 ton rice husks wastes, thus those can result biogas as much as 48.060.000 ml or 48,06 m3. If this is converted to other form of energy, then it will result 1 m3 biogas equals with 4,7 kWh so the electrical energy generated by 225,88 kWh. According PT. PLN (Persero) if the potential of electric energy is converted into electricity cost in March 2017 for Rp. 1.467,28/kWh, then the cost can be reduced to Rp. 331.432,1.

This can turn on the lights with 60-100 watt of power for 303 hours, can also run motor for 1 hp for 101 hour and produce 237,82 kwh amount of electricity.

3.3.2Production on the effect of enzyme

Table 2 and Figure 5 indicates that 11% enzyme application gave real influence to the kinetic constants of biogas production. From the polymath result 5% enzyme had the highest rate of biogas production compared to 8% enzyme and 11 % enzyme, and control. Kinetic constant of 5% enzyme declared as a constant rate of biogas rate production (U), maximum biogas production (A), and the maximum time of biogas creation (□) each on the value of 4,78 ml/(gr TS day), 46,80 ml/gr TS; and 1,16 day. This means that rice husk that had been given 5% enzyme pretreatment had constant rate of 4,78 ml/(gr TS day) with maximum biogas production result of 46,80 ml/gr TS delivered biogas for the first time on the 1,16 day.

In the biogas yield results with 11% pretreatment performed produced the highest biogas yield compared with 5% enzymes, 8% enzymes, and control. 11% enzyme reactor had the highest biogas yield which was 64.86 ml/g TS and was the lowest biogas production rate compared with the 5% and 8% enzyme which was 3.80 ml/(gTS-day) with the initial formation of biogas at day 0.99. In the enzyme reactor of 5%, it had the lowest biogas yield of 46.80 ml/g TS and the highest biogas production rate of 4.78 ml/(g TS-day) with the initial formation of biogas on the 1.16 day. In the enzyme reactor of 8%, it had a higher biogas yield than the yield of biogas enzyme 5% on the amount of 64.96 ml/g TS and a faster production rate compared with 11% enzyme that was 4.63 ml/(g TS-day) with the initial the formation of biogas on 1.13 day. While the reactor without pretreatment (control), had the lowest biogas yield of 43.32 ml/g TS and the lowest biogas production rate of 0.78 ml/(g TS-day) with the initial formation of biogas longest on the day 7.83.

Table 2. Kinetic constant in effect variation enzyme

Constant	*Enzyme* 5%	*Enzyme* 8%	*Enzyme* 11%	Control
A, ml/gr TS	45,05	43,72	64,86	42,55
U, ml/(gr TS day)	4,42	3,99	3,80	0,87
λ, day	1,31	1,51	0,99	7,83

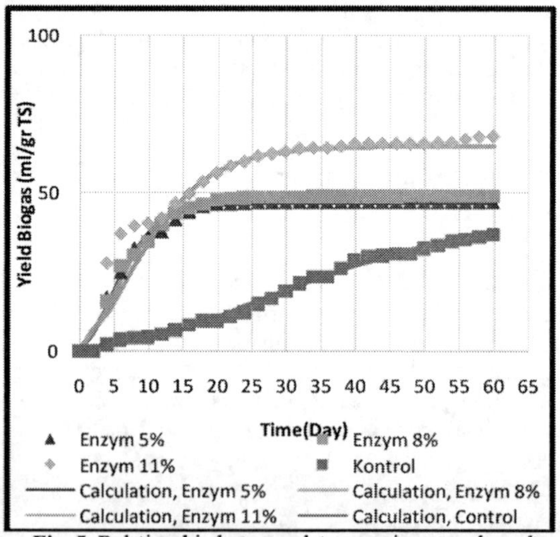

Fig. 5. Relationship between data experiment and results calculation on effect variation enzyme

From the data above it can be concluded that the use of variety of enzyme produce biogas yield higher than without pretreatment (control) with a faster production rate and faster time formed of biogas. Furthermore, the variation of enzyme concentration of 11% gave the highest biogas yield of 64.96 ml/(g TS). At 11% the enzyme reactor had the lowest production rate compared with the 5% and 8%, enzyme which was 3.80 ml/(g TS-day). This was because the enzyme reactor 11% still produced biogas at the longest period of time until the 60th day so that the biogas was neither stable nor constant.

Figure 6 shows the calculation results of the enzyme 5% biogas using Polymath 6.0 so the maximum yield of biogas which can be produced and the termination time to produce biogas.

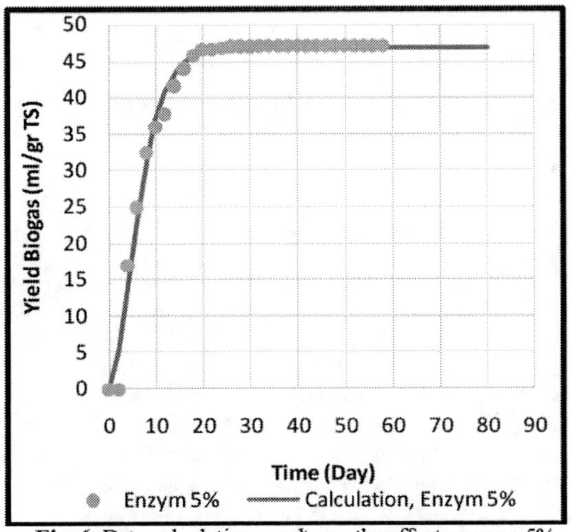

Fig. 6. Data calculation results on the effect enzyme 5%

Based on the data in Figure 7 it can be seen that the maximum biogas yield on the 5% enzyme was 46.80 ml/g TS can be achieved on the 36th day. This shows that the results of biogas from waste rice husks that had enzyme pretreatment of 5% is likely to be stable and does not produce more biogas. From the data it also can be acknowledged hat on the 36th day basically biogas production has stopped. This could be used to design biodigester SS-AD from rice husk with a continuous enzyme pre-treatment with a residence time of 36 days.

Once the modeling on the preliminary treatment is done, with adding 5% then 46,80 ml/gr TS biogas is obtained. When there are rice husks wastes as much as 1 ton, it can result to 46.800.000 ml or 46,80 m3 biogas. If it is converted into another form of energy, it will gained 1 m3 biogas equals with 4,7 kWh so the electrical energy generated by 219,96 kWh. According PT. PLN (Persero) if the potential of electric energy is converted into electricity cost in March 2017 for Rp. 1.467,28/kWh, then the cost can be reduced to Rp. 332.742,91.

4 Conclusion

The conclusions derived from this research are:

1. Pretreatment using three variation of NaOH can be concluded that, the usage of 6% NaOH generated the highest biogas yield compares to 3% NaOH and 9% NaOH.
2. From three variation of enzyme pretreatment, can be concluded that the usage of 11% enzyme resulted the highest biogas yield compared to 5% and 8% enzyme.
3. The rate of biogas production with the resulted rate maximum which is 6% NaOH with values as A, U, and λ each with the amount of 48,06 ml/gr TS; 3,97 ml/(gr TS.days); and 0,75 day. On the other hand, for enzyme pretreatment, the maximum rate result on 5% enzyme condition with A, U, and λ each on the amount of 46,80 ml/gr TS; 4,78 ml/(gr TS days); 1,16 day

References

1. Y. F.Saputri, T.Yuwono, and S.Mahmudsyah, *Pemanfaatan Kotoran Sapi untuk Bahan Bakar PLT Biogas 80 KW di Desa Babadan Kecamatan Ngajum Malang*. 1-6 (2014)
2. E.Mufrizon, and P. Subekti, *Kebijakan Energi Baru-terbarukan Serta Peluang Pemanfaatan Biogas dan Biomasa Limbah Pengolahan Kelapa Sawit untuk Pembangkit Tenaga Listrik di Propinsi Riau*(2013)
3. Febriyanita, Wahyu. *Pengembangan Biogas Dalam Rangka Pemanfaatan Energi Terbarukan di Desa Jetak Kecamatan Getasan Kabupaten Semarang* [Skripsi]. Semarang: Universitas Negeri Semarang (2015)
4. L.Houdkova, J. Boran., J. Pecek and P. Sumpela. Journal of Thermal Science 12(**4**) : 27 -33 (2008)
5. Ichsan, H. Hadiyanto and R. Hendroko. *Energy Procedia*. 47, 143-148 (2014)
6. Hasan M.H., Mahlia T.M.I., Nur H. *A review on energy scenario and sustainable energy in Indonesia*. May. Renew Sustain Energy Rev;16(**4**):2316–28 (2012)
7. H. W. Hsu, and B.S. Luh, *Rice Hull*. Rice Product And Utilization. Editor: Bor Shiun Luh. New York: Avi Publishing Company Inc. Hal. 736-740 (1980)
8. M. J. Taherzadeh, and Karimi, K. International Journal of Molecular Sci , **9**, 1621-1651 (2008)
9. D. Yang, Y. Zheng, R. Zhang, *In: An ASABE International Meeting*, Reno, Nevada, USA, 21–24 June, American Society of Agricultural Engineers. Paper No.: 095685 (2009)
10. M. Tuomela, M. Vikman, A. Hatakka, and M. Itavaara, Bioresource Technology **72**: 169-183 (2000)
11. Y. He, Y. Pang, Y. Liu, X. Li, K. Wang, Beijing: Departement of Enviromental Science and Engineering (2008)
12. S. M. Hosseini, H. A. Aziz, S. Meisam, V. Kiamahalleh, Urban Waters: resource or risks, Arcueil : France (2011)
13. M. Gunamantha, W. N. Yuningrat, Jurnal Sains dan Teknologi, Vol. **3**, No.1 (2014)
14. Q. Zhang, J. He, M. Tian, Z. Mao, L. Tang, J. Zhang, Bioresour Technol Vol. **102**:8899-906 (2011)
15. Purwoko, M. Romli, Suprihatin, L. Haditjaroko, Jurnal Teknologi Industri Pertanian, Vol. **26**, No. 2, 134-142 (2016)

Geomagnetic Survey to Explore High-Temperature Geothermal System in Blawan-Ijen, East Java, Indonesia

Yunus Daud[1,2*], *Syamsu* Rosid[1,2], *Fikri* Fahmi[1], *Faris* Maulana Yunus[2] and *Reza* Muflihendri[2]

[1]Master Program in Geothermal Exploration, Graduate Program of Physical Science, Universitas Indonesia, Jakarta, Indonesia
[2]Study Program of Geophysics Faculty of Mathematics and Natural Science, Universitas Indonesia, Depok, Indonesia

Abstract. Ijen geothermal area is high-temperature geothermal system located in Bondowoso regency, East Java. It is categorized as caldera-hosted geothermal system which is covered by quaternary andesitic volcanic rocks with steep topography at the surrounding. Several surface thermal manifestations are found, such as altered rocks near Mt. Kukusan and a group of Blawan hotsprings in the northern part of the caldera. Geomagnetic survey was conducted at 72 stations which is distributed inside the caldera to delineate the existence of hydrothermal activity. Magnetic anomaly was obtained by reducing total magnetic measured on the field by IGRF and diurnal variation. Reduction to pole (RTP) method was applied with geomagnetic inclination of about -32^0. In general, the result shows that high magnetic anomaly is distributed at the boundary of study area, while low magnetic anomaly is observed in the centre. The low anomaly indicates demagnetized rock that probably caused by hydrothermal activity. It has a good correlation with surface alteration observed close to Mt. Kukusan as well as high temperature reservoir drilled in the centre of caldera. Accordingly, the low magnetic anomaly also presents the possibility of geothermal reservoir in Ijen geothermal area.

1 Introduction

Ijen Volcanic Complex is located in the Bondowoso, West Java, Indonesia. Expectation of geothermal system appearance in the volcanic complex is indicated by the occurrence of Blawan hotspring in the northern part of caldera rim and steam heated extremely acid sulphate water on Ijen crater as the only surface manifestation in this area. The hot spring express an outflow of the system, while the Ijen crater indicates a young volcano which is uneconomical to be developed. Therefore, it is necessary to do further research to find an appropriate location to be targeted in such challenging hidden geothermal system.

In 2017, magnetic survey had been conducted over Ijen geothermal prospect area at 72 stations. Magnetic method is carried out to measure magnetic field variations in the earth's surface. The studies of magnetic anomalies are often useful for investigating high temperature geothermal systems hosted by Quaternary volcanic rocks [1]. These anomalies represent the demagnetization of reservoir rocks caused by hydrothermal processes due to thermal fluid and rock interactions. In this study, we used a reduce to pole (RTP) technique for imaging total force of magnetic anomalies. RTP transformation delineates the hydrothermally demagnetised rock which is shown as negative anomalies [2]. This paper presents the results of magnetic anomaly, RTP anomaly and the upward-continuation anomaly which were then interpreted to get a conclusion of the geothermal prospect zone in the Ijen Volcanic Complex.

2 Regional Geology

The Ijen Volcanic Complex (IVC) is located in East Java. The Old Ijen volcano is thought to form around the Pleistocene [3]. The caldera itself is formed by Plinian eruption, ejecting 80 km3 of volcanic material, back to the 0,2-0,05 Ma [4]. Currently, the product is revealed partly at the northern part. Meanwhile, in the southern part, it has been covered by younger volcanoes.

The post-caldera volcanoes are classified into two groups: Caldera Rim Volcanoes and Intra Caldera Volcanoes [5]. Some fault are found in the caldera and extents right in the centre of the caldera, strikes the remains caldera rim in the north and tear down the wall into a steep creek. Here in IVC, several structures that occurred are Blawan fault, Kawahwurung fault, Krepekan fault, Cemara-Kukusan fault, Kalipahit-Banyulinu fault, Djampit fault, Rante fault, Pawenan-Blau fault and Kendeng-Merapi fault (Fig.1).

The deformation may caused by tectonic activity, magma refilling in the chamber, or collaboration of those two that reactivate the caldera floor structures. Hot springs in Blawan area indicating that secondary permeability will affect the geothermal manifestation occurrences.

* Corresponding author: ydaud@sci.ui.ac.id

https://doi.org/10.1051/e3sconf/20183102003

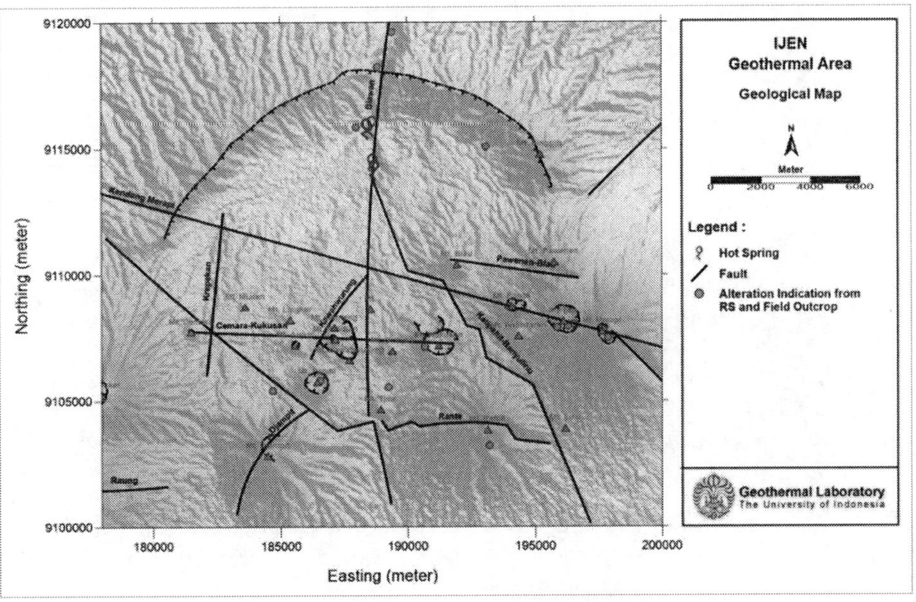

Fig. 1. Geological Map of Ijen Geothermal Area

3 Method

3.1 Acquisition and Data Processing

The acquisition of magnetic data had been carried out in the Ijen geothermal prospect area in 2017. The magnetic measurements were conducted using a Proton Precision Magnetometer (PPM). A total of 72 magnetic stations were measured with 1 to 1.5 km grid spacing and distributed inside the caldera rim.

Furthermore, several corrections had been applied to the measured magnetic field data set in order to get the anomaly reflected from the geological target. At the first step, the contour map of the observed magnetic field was reduced for diurnal variation. Then the magnetic data was corrected for IGRF to remove the normal geomagnetic field. By reducing the total magnetic field with diurnal variation and IGRF correction then the magnetic anomalies associated with local magnetic variations of rocks can be obtained.

Moreover, the reduction to the pole (RTP) technique was carried out by reducing the magnetic data to the pole with the angle of inclination -32°. It is used to get the response of geothermal reservoir located just below the low magnetic anomaly. The upward continuation was also conducted to reduce the effects of shallow/local anomaly. In this case, we performed the continuation to a height of 250 meters.

4 Magnetic Signature of Blawan-Ijen Geothermal System

4.1 Magnetic Anomaly

The magnetic anomaly map of Ijen geothermal prospect area is shown in the Fig. 2 with the highest value of about 1984 nT, while the lowest magnetic anomaly value of about -1206 nT.

The low magnetic anomalies were found covering almost the entire study area and they were surrounded by high magnetic anomalies. Several bipolar anomalies were identified in the three locations. In the southeast area, it is probably associated with the existence of Mt Kukusan while anomaly in the southwest area could be correlated with Mt Pendil. The other anomaly in the northern part of the study area has not been clearly determined regarding possible correlation with the geological condition in this area.

4.2 Reduce to Pole (RTP)

The result of reduction to the pole (RTP) can be seen in Fig. 3. The focus in the investigation for the location of the geothermal reservoir is low (negative) anomaly area due to the hydrothermal demagnetization. The negative magnetic anomalies occur over the northern part of the study area and extend to the southeast area.

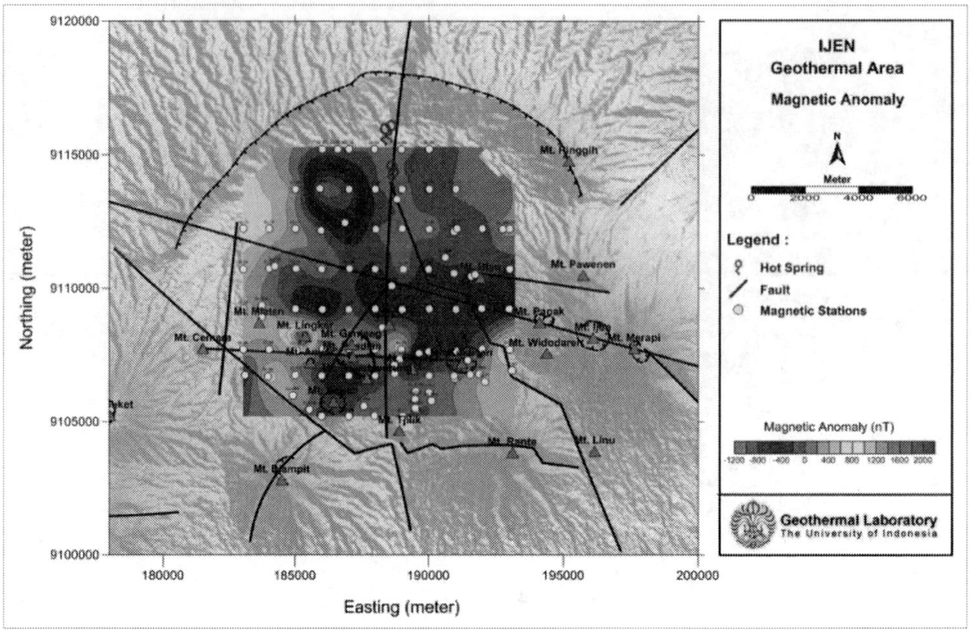

Fig. 2. Magnetic Anomalies over Ijen Geothermal Area

The pattern of negative anomaly found in the centre of the study area is probably associated with the presence of hydrothermal demagnetised rocks of Mt Kukusan. It is supported by the presence of surface alteration observed close to Mt. Kukusan.

The location of the prospect zone is also strengthened by the intersection of several faults; Blawan fault, Cemara-Kukusan fault, and Kawahwurung Fault which could form a high permeable zone.

4.3 Upward Continuation

Fig. 4 shows the upward continuation anomaly map. Generally, the pattern of the high and low magnetic anomaly in the upward continuation map yields the same pattern with the RTP anomaly.

The upward continuation anomaly represents a regional anomaly. As shown in the result, it further strengthens that the negative magnetic anomalies area occured in the northern part of the study area and extended to the southeast which could become the prospect zone in Blawan Ijen geothermal area.

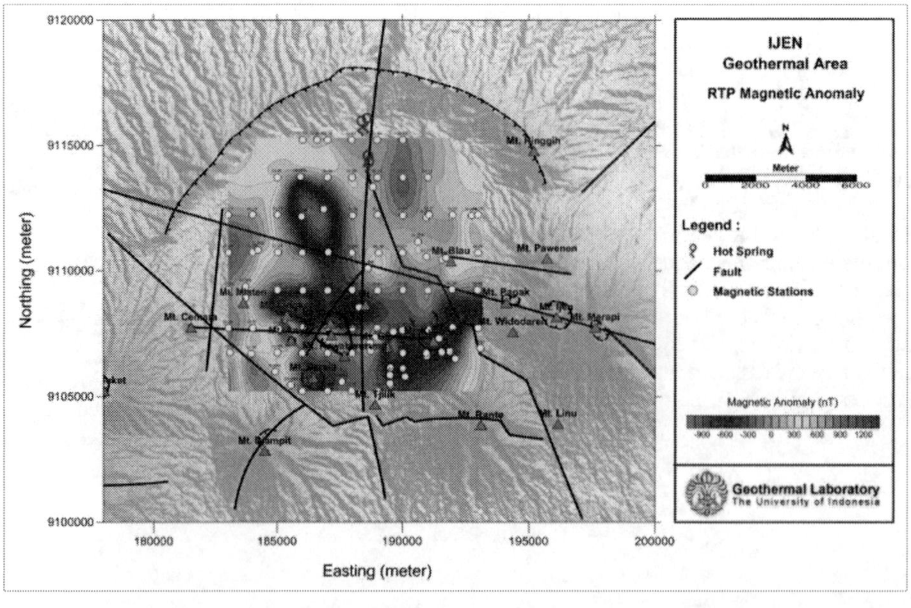

Fig. 3. The result of Reduce to Pole (RTP) of Magnetic Anomalies of Ijen Geothermal Area

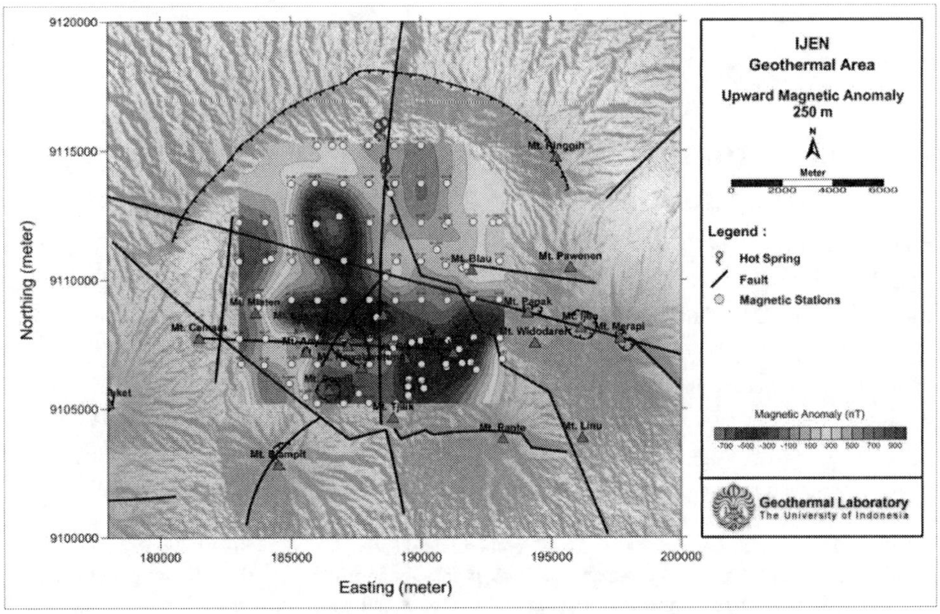

Fig. 4. The Map of Upward Continuation Anomalies of Ijen Geothermal Area

Conclusion

This study was carried out in the Ijen geothermal prospect area. The presence of surface manifestation is very rarely, only Blawan hotspring found in the northern part of study area and extremely acid fluid on the top of Mt. Ijen. Therefore, the magnetic method was conducted to determine the location of geothermal prospect zone between Blawan area and Mt. Ijen. A negative anomaly is identified in the center (North to southeast) of the study area and it was interpreted as the hydrothermal demagnetised rocks of Mt Kukusan. This negative anomaly has good correlation with the presence of altered rock found near Mt Kukusan. The intersection of several faults; Blawan fault, Cemara-Kukusan fault and Kawahwurung Fault in the center of study area supports the possibility of geothermal prospect zone location.

Acknowledgement

We greatly appreciate Directorate of Research and Community Engagement of Universitas Indonesia (DRPM UI) for funding this project through PITTA program.

References

1. S. Soengkono New Zealand: Geothermics **30**, 443-459 (2001)
2. S. Soengkono, *Proceeding 36th New Zealand Geothermal Workshop* 24-26 November 2014
3. M. Ben Rabha, M.F. Boujmil, M. Saadoun, B. Bessaïs, Eur. Phys. J. Appl. Phys. (to be published)
4. R.W. Van Bemmelen, The Geology of Indonesia Vol. IA: General Geology of Indonesia and Adjacent Achipelagoes (Government Printing Office, The Hague, 1949)
5. K. Sitorus, Stratigrafi dan Geokimia Kaldera Idjen,
6. Jawa Timur, Indonesia. Proceedings PIT XIX Ikatan Ahli Geologi Indonesia, Bandung (1997)
7. H.K. Handley, C.G. Macpherson, J.P. Davidson, K. Berlo, Journal of Petrology **48**, 1155-1183. Oxford University, (2007)

Geothermal Potential Based on Physical Characteristics of the Region (Case Study: Mount Karang, Pandeglang Regency and Banten Province)

Fhillipo Russel[1,*], *Astrid* Damayanti[2], and *Tjiong* Giok Pin[2]

[1]Bachelor Program of Geography, University Of Indonesia, Depok - Indonesia
[2]Department of Geography, Faculty of Mathematic and Natural Science, University Of Indonesia, Semarang - Indonesia

Abstract. This research is about geothermal potential of Mount Karang, Banten Province which is based on the characteristics of the region. This research method used is geochemistry sample of hot springs and integrated with GIS method for spatial of geothermal potential. Based on the geothermal potential, Mount Karang is divided into three regions, ie high potential, normal potential, and low potential. The high geothermal potential region covers an area of 24.16 Km^2 and which there are Cisolong and Banjar 2 hot springs. The normal potential covers Kawah hot spring. Index of the fault of Mount Karang region is one of the significant physical characteristics to determine geothermal potential.

1 Introduction

Indonesia is surrounded by active volcanoes in almost every region. This makes Indonesia the most potential country to have renewable energy in the form of geothermal. Mount Karang is a volcano that has volcanic activity. The hot springs in Mount Karang become a manifestation that the location has geothermal potential.
According to [1] to determine whether a mainstream has a commercial geothermal potential or not, a study of the type of hot springs is required. Testing this type of hot springs using geochemical methods by looking at chemical elements Na, K and Mg. Testing the element will produce three types of hot springs namely Immature Water, Partial Equilibrium and Full Equilibrium. Important to note that a geothermal resource is part of a geologic system where geologic factors such as lithology, faults, fractures, stress field, diagenesis, rock mechanics, fluid chemistry and geochemistry control key parameters [2].

The geothermal fluid contained in the hydrothermal reservoir of a manifestation derived from surface water includes rainfall that enters the surface and is heated [3,4]. The reservoir temperature is obtained by using geochemical method. Processed data consist of chemical data of hot water manifestation, isotope data, soil chemical and soil gas data [5]. The data is used to assess the possible development of geothermal resources that include various parameters such as resource size, temperature estimates, and formation permeability [6]. The concept of permeability of rock formations in relation to geothermal, is oriented in a proportional manner, namely primary permeability and secondary permeability [7]. Primary permeability includes rocks, rock layer contacts, misalignment and ancient solution features that are also overall known as stratigraphic approaches. Secondary permeability includes fractures, hydraulic fracturing, joints and hydrothermal leaching. The more fault or fault there will be many hot springs that appear to the surface [8]. The concept of permeability is also an indication of how the geothermal process occurs. According to [9] meteoric water in permeable and porous rocks is heated and hot water is trapped to form a geothermal reservoir.

The geological structure that controls Mount Karang in Pandeglang and Serang regencies is the same structure that is Maramang fault structures that trails northwest-southeast, which is a continuation of the regional fault structure in southern Lampung [10] but according to [11] it directed west-east. The purpose of this research is to analyze the geothermal potential areas and the relationship between the physical characteristics of the region as a reference for geothermal exploration.

2 Methodology

The parameters used are physical characteristics of geothermal area and geothermal potential distribution area. Physical characteristics of the area included in the study are the surface temperature of the soil, altitude and slope, type of parent rock (lithology) and fracture density. Geothermal potential is determined by the distance between the manifestations, the type of hot springs and the reservoir temperature. Distribution of geothermal potential areas is analyzed by overlay and scoring methods.

* Corresponding author: astrid.damayanti@ui.ac.id

https://doi.org/10.1051/e3sconf/20183102004

The distance between manifestation is obtained from field survey, then plotted and processed using ArcGIS 10.2 software and analyzed using buffer with range <500 meter, 500 - 3000 m and> 3000 m [12].

The reservoir temperature was obtained from the calculation using the [1,13] reservoir temperature equation based on the Na, K element contained from the hot springs sample.

$$T=[1390/(1.750 + \log (Na/K))] - 273.15 \quad (1)$$

Hot springs were analyzed from the percentages of Na, K and Mg elements treated with the following [1,14] equation:

$$S = ([Na] / 1000) + ([K] / 100) + [Mg]^{1/2} \quad (2)$$
$$\%Na = [Na] / 10.S \quad (3)$$
$$\%K = [K] / S \quad (4)$$
$$\%Mg = [Mg]^{1/2} / S \quad (5)$$

The percentage value of these elements is plotted into Ternary Diagram using software Rockware 14.

The surface temperature of the soil is processed from Landsat 8 (band 11) and Landsat 5 (band 6) image using Envi 4.7. Temperature data used 1996-2016 and divided into three periods namely in 1996, 2007 and 2016. Equations used in data processing is

$$T = K2 / \ln (K1/CV_{R2} + 1) \quad (6)$$

Legend :
T = Land Surface Temperature
K1 & K2 = Coeffiecient
CV = The radiance value on the thermal band

Altitude and slope data were obtained from DEM image, and processed using Envi software. Litologi data obtained from geological map sheet with scale 1: 100.000. Fraction density is calculated using the Fault Fracturing Density method in ArcGIS 10.2. Bands used to analyze fault and straightness are composites of bands 5, 6 and 7 on Landsat 8 image.

3 Geothermal Manifestation on Mount Karang

There are seven geothermal manifestations in the form of hot springs that appear on Mount Karang. The hot springs have temperatures ranging from 400C to 740C. The results of elemental analysis of Na, K and Mg from hot springs sample showed reservoir temperature manifestation ranged from 117.910C to 528.180C. Based on the interval [2,15] the temperature range includes a low classification (<1250C), medium (125-2250C), high (> 2250C) and the full information on the table 1.

Table 1. Identification of Geothermal Manifestations in Mount 4

Name of Hot Spring	Chemical Elements (ppm)			Hot spring Temperature °C	Reservoir Temperature °C
	Na	K	Mg		
Crater 1	8.2	4.76	8.6	40°C	426.68°C
Crater 2	7.74	4.15	9.17	74°C	414.73°C
Crater 3	3.05	3.16	1.7	71°C	528.18°C
Cisolong 1	45.29	31.39	25.91	55°C	454.90°C
Cisolong 2	44.34	35.71	26.19	63°C	480.64°C
Banjar 1	34.42	0.54	2.86	33°C	117.91°C
Banjar 2	45.84	3.68	3.45	32°C	225.36°C

The determination of the type of hot springs provides information on how the hydrothermal system is in a manifestation, so it can know the equilibrium of water in the rock type. Based on Ternary Diagram (Figure 1), hot springs belonging to the Immature Water type are hot springs Craters 1, 2 and 3 and Cisolong 1 and 2. Immature Water processes occur because rocks that store heat from beneath the Earth's surface experience the pressure and temperature Very high but before the vapor flow up to the top, the water vapor is already mixed with surface water (Meteoric Water). It is in accordance with opinion [16] G. Karang area is underlain by a concealed outflow of thermal water that has been diluted by groundwater infiltration. Banjar 2 hot springs is a type of Partial Equilibrium hot springs. This type of hot springs is very little affected by surface water flow. Banjar 1 hot springs include Full Equilibrium type which indicates that the reservoir temperature is not completely mixed with surface water.

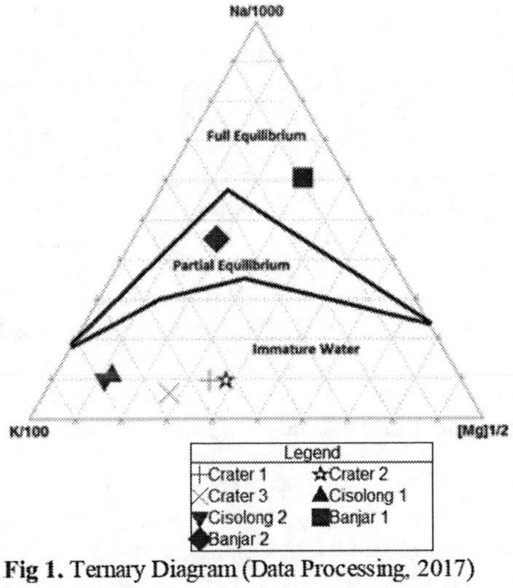

Fig 1. Ternary Diagram (Data Processing, 2017)

The distance between the manifestations indicates the geothermal potential. The closer the geothermal potential is higher [12]. The area of high geothermal potential exists between Crater 1 hot springs with Crater 2 and Cisolong 1 with Cisolong 2. The area between Cisolong hot springs with Banjar becomes a low geothermal potential region.

4 Physical Characteristics of Geothermal Potential Areas in Mount Karang

The highest mean surface temperature is 20.68^{0}C in Banjar 1 hot springs and the lowest is 15.13^{0}C in hot springs Crater 1. Banjar 1 hot springs has low reservoir temperature while hot springs Crater 1 has a reservoir temperature With high classification. This is different from the results of research [17]. Their research on Mount Salak indicates that geothermal-manifested regions typically have high surface temperatures. The anomaly occurs because in the hot springs Crater 1 is surrounded by dense forest and its water type is also Immature Water. While in Banjar 1 hot springs the anomaly can occur due to Full equilibrium eye type (Figure 1).

Discussion of morphological units with geothermal potential is closely related to the accessibility of geothermal exploration. Mount Karang has steep mountainous morphology units, steep mountains, mid ramps, mid-wavy and lowlands. The steep mountain morphology unit is a difficult area to reach, so it costs a lot to explore. On the other hand, in lowland areas with high accessibility and low exploration costs but low geothermal potential. So a morphological unit that has high geothermal potential to be explored is a mid-wavy and mid ramp morphological unit. In Mount Karang area which has mid-wavy and mid-ramped morphology units located in Banjar 2 hot springs and Cisolong 1 and 2 hot springs.

The hot springs emerging to the soil surface are caused by fractures due to fractures in areas that have hydrothermal activity. The water flow beneath the surface will be heated by volcanic activity so that there is a flow of convection currents. Water that has been heated will come to the surface through fractures and water still in normal temperature will flow down to be heated. Fracture density at Mount Karang is divided into two classes, namely low and medium. Low density areas are found in hot springs Craters 1, 2 and 3 and Banjar hot springs 1 and 2 with a density of 0-0.246 km^2. The densely populated region is in the hot springs of Cisolong 1 and 2 with a density of 0.246-0.722 km^2 (Figure 3.).

The characteristics and extent of the ore deposits and the geothermal fluids depend on the geometry and distribution of faults, fractures and porous media in the caldera and its basement [18,19,20,21]. Low fault density has low reservoir temperature and low permeability of rocks and low number of hot springs [14]. However, in hot springs Craters 1,2 and 3 whose low fault density has a high reservoir temperature with a sufficient number of hot springs. (Figure 3). This happens because the existing faults are not captured by the landsat image, and the fault has been covered by lake swamp sediments [22]. The anomaly is also seen from the rose diagram (Figure 2). Figure 2 is a rose diagram showing the direction and dominance of straightness and fracture on Mount Karang. Areas with high permeability are visible in the southwest-Northeast. This is different from the statement from [11,14] as stated in the introduction.

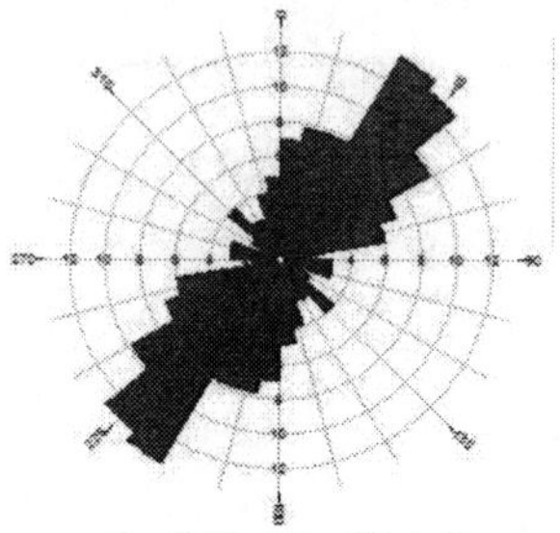

Fig 2. Rose Diagram Fault Density (Data Processing, 2017)

5 Region of Geothermal Potential Based on Physical Characteristics

Region of geothermal potential in the coral mountains are divided into three based on physical characteristics, ie low, medium and high. The division of potential areas is based on the results of spatial overlay analysis using 6 variables. These variables are distances between manifestations, hydrothermal activity, ground surface temperature, morphology, lithology and rock permeability.

5.1 Low Geothermal Potential Region

Region with low potential are almost all of the research area. The Region of low geothermal potential is 151.13 km^2, the geothermal potential area is low based on physical characteristics scattered near the crater and located near the Banjar 1 hot springs. This area has a surface temperature of 19,99^{0}C. The fault density index in this region belongs to a low classification or if translated in the numbers 0 - 0.246 km / km^2. The temperature of hot springs in Banjar 1 is 31^{0}C. Types of rocks that exist in hot springs Banjar 1 has a rock formation of Karang volcanic products with volcanic rock sediment type with the age of Pleitocene rock - Holocene.

The morphology of Banjar 1 hot springs is in low flat classification. The chemicals present in hot spring 1 have 34.42 ppm of Sodium element, 0.54 ppm of Potassium element, 2.86 ppm Magnesium element and 12.22 Ppm Sulfur element with reservoir temperature 117,91^{0}C. With the reservoir temperature included in the low classification but on the plot of Ternary triangle diagram showing that this hot springs are in Full Equilibrium it can be concluded that in the flow of hydrothermal fluid at Banjar 1 hot springs is not so hot, because the reservoir temperature is in the eye Hot water is not high enough so it can be concluded that the volcanic activity that occurs in Banjar 1 hot springs is not so active as Crater and Cisolong.

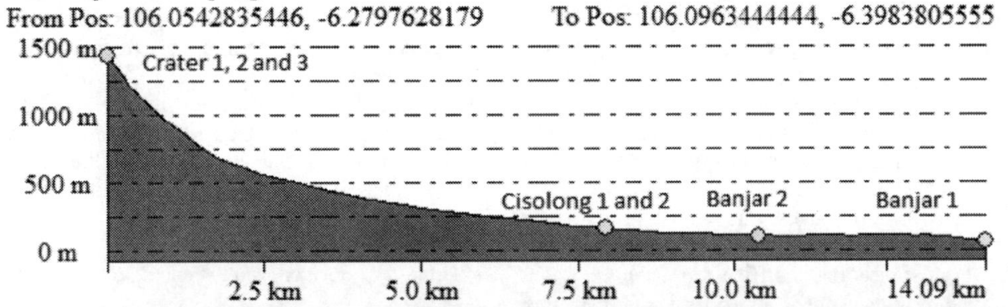

No	1. (Crater 1,2 & 3)	2. (Cisolong 1 & 2)	3 (Banjar1)	4 (Banjar2)
Lithology	Young Lava G Karang	Volcano ProductF		
Morphology	Steep mountains	Medium Wavey	Medium Wavey	lowland
Land Soil Temperature	15.13 – 16.06^{0}C	20.20 – 20.27^{0}C	19.99^0 C	20.68^0 C
Fault Density	Low	Medium	Low	Low
Reservoir Temperature	414.73 – 528.18^0 C	454.9 – 480.64^0 C	117.91^0 C	225.36^0 C
Type Of Hot Spring	Immature Water	Immature Water	Partial Equilibrium	Full Equilibrium

Fig. 3. Profile Karang Mount (Data Processing, 2017)

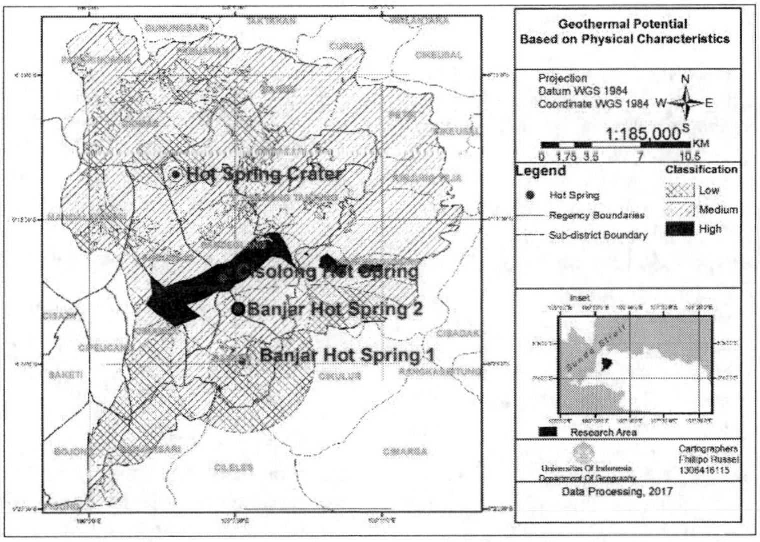

Fig 4. Maps Of Geothermal Potential Based on Physical Characteristics (Data Processing, 2017)

5.2 Medium Geothermal Potential Region

Region with geothermal potential are located in Crater hot springs. The area of geothermal potential with medium classification is 386.78 km^2. The ground surface temperature in this region is 15,60^{0}C. The index of fault density and faults in this region has a low classification with an interval of 0 - 0.246 km / km^2. Types of rocks that exist in the hot springs Crater was the same as Banjar 1 which has a young volcano lava formation Karang with volcanic rock sediment type with the age of Pleitocene rock - Holocene. Morphology in this region

also includes mid-wavy. The shape of the terrain had started steep and began to steep slopes.
Temperature reservoir in hot springs Craters 1,2 and 3 with temperature 414.73 to 528.18^{0}C. In reservoir temperatures, hot springs Craters 1,2 and 3 have reservoir temperatures that are included in the high classification in terms of geothermal potential. With a steep slope, an altitude above 1000 mdpl with a form of steep mountain terrain makes this crater hot springs difficult to be explored and utilized into a power plant because of its existence that is difficult to reach.

This type of hot springs is also a type of hot springs Immature Water where the water on the surface will enter and unite with water vapor so that the steam heat will be reduced. This fact is evidenced by field evidence showing that the ground surface temperature in the area is only 15.6^{0}C. It can be said that the soil surface temperature is low when compared with other geothermal potential areas that have a high ground surface temperature with a minimum of 25^{0}C.

5.3 High Geothermal Potential Region

Geothermal geothermal potential areas with high classification are located in hot springs Banjar 2 and also Cisolong hot springs 1 and 2. The area with geothermal potential is 24.16 km^2. The surface temperature of the land in this potential area is divided into two areas, namely in the area of the crater memilliki ground surface temperature 20,68^{0}C and for the Cisolong area has a surface temperature of 20.24^{0}C. It is interesting there is a ground surface temperature that is in the crater has the lowest temperature than other regions.

This spring type is different, the Banjar 2 hot springs have the Partial Equilibrium hot spring type indicating that the reservoir temperature obtained is still hindered by a small mix of surface water while the Cisolong hot spring type is the Immature water type where the reservoir temperature Is mixed with the surface water so that the actual reservoir temperature of the Cisolong spring is higher than expected in the absence of mixed water and precludes the calculation of reservoir temperature at Cisolong springs.

The density of fault and fault faults and this region is also divided into two, because the area located in Banjar has a low fault density while the density of the area in Cisolong has a moderate fracture density. This moderate density causes reservoir temperatures in Cisolong region to be high due to the reservoir temperature 467.77^{0}C with high geothermal potential classification based on reservoir temperature. With its high reservoir temperature, Cisolong area is a high potential area for geothermal energy. The results will be illustrated in Figure 4.

5.4 Correlation of Geothermal Potential Statistics with Physical Characteristics

The correlation of geothermal potential with physical characteristics using multiple linear regression method to determine the dominant variable statistically or visible number, thus the number validates the relationship between the geothermal potential and the physical characteristics. Dependent variable is Reservoir Temperature, whereas the independent variables are Rock Type, Fault Density, Morphology and Surface Temperature. Table 2 shows that the summary of the model has been made using multiple linear regression method.

Table 2. Correlation Non Parametric (Data Processing, 2017)

Correlation Non Parametric (Spearman's Rho)					
Reservoir Temperature		RS	FD	LST	M
	Correlation Coefficient	1	.820*	-0.273	-0.54
	Sig. (2-tailed)		0.024	0.553	0.211
	N	7	7	7	7
	*. Correlation is significant at the 0.05 level (2-tailed).				

Table 2 will provide information on the model equations created with surface temperature variables, fracture density and morphology, then other information which variables most influence the reservoir Temperature are the three variables between ground surface temperature, fault density and morphology. The equations created in this study are

$$Y = 1210.5 - 56.3\ LST + 147.8\ FD - 22.9\ M \quad (7)$$

Notes : LST= Land Surface Temperature, RS = Reservoir Temperature, FD = Fault Density, M = Morphology.

Other information to be seen is the influence of independent variables on the dependent variable in this study, to know it then things that must be seen is the value of sig in table 5.13. The limit value so that the variable has a real effect is <= 0.05 sig value. Land surface temperature variable has sig value 0.496 and sig value exceeds 0.05, it shows that ground surface temperature has no significant effect to reservoir temperature. Fault density variables have sig value 0.016 and sig value less than 0.05, it indicates that fracture density variables have a significant effect on the reservoir temperature variable. Morphological variables have sig value 0.952 and signya value exceeds 0.05 it shows that morphological variables have no significant effect on reservoir temperature. With that information it can be concluded that in the geothermal potential region the physical variable that influences is the fault density.
Y = 1210.5 – 56.3 LST + 147.8 FD – 22.9 M has meaning in this research. The meaning is that if the reservoir temperature rises 1 degree Celsius then the ground surface temperature will decrease by 56.34 degrees Celsius, the fault density will increase by 147,750 and the morphology will decrease by 22,917 with the constant as reference in the determination in equation of 1210.480. It shows that if reservoir temperature rises in research area, soil surface temperature and morphology value will decrease while

the fault density value will increase. These geothermal potential based on physical characteristics in Cisolong with reservoir temperature 467.77 have a mean fracture density value.

6 Conclusion

Region with geothermal potential in Mount Karang and surrounding areas based on their physical characteristics are divided into 3 ie with low potential, medium and high. The area with low geothermal potential based on the characteristics of its territory has an area of 151.13 km^2 with a ground surface temperature of 19.99^{0}C, a volcanic lowland morphological unit, a low fault density and a reservoir temperature of 117.91^{0}C and has manifestations in Banjar Springs 1. Medium geothermal potential region Ground level 15.13 - 16.06^{0}C with medium fault density, morphology unit high volcanic steep mountains with reservoir temperature 414.73 - 528.18^{0}C and manifestations are in Crater hot springs. Region with high geothermal potential are located in Banjar 2 and Cisolong with an area of 24.16 km^2, the physical characteristics of the area has a moderate fault density, surface temperature of 20.2 - 20.67^{0}C with mid-wavy morphological units and has a temperature reservoir of 225.36 - 528.18^{0}C.

The relationship between physical characteristics of the region with geothermal potential has a strong relationship, it is reinforced by spatial analysis and statistical analysis. Based on spatial analysis, areas with high geothermal potential have a moderate fault density value with high reservoir temperature and low geothermal potential having low fault density and low reservoir temperature, whereas based on statistical analysis of the four physical characteristic variables ie surface temperature, Type of rock, fracture density and morphological units only the fault density that affects the value of the reservoir temperature due to its value <0.05 which indicates that the fault density is closely related to the reservoir temperature.

The determination of the direction of fault and the dominant fracture in this study directed southwest and northeast, it is different from the research opinion with [16] that directed northwest and southeast and [3] research directed west-east. This shows that in determining the direction of fault and fault not only through remote sensing it is necessary to do field survey to show the direction of fault and fault that exist in a region.

Reference

1. W.F. Giggenbach, Geochemica Acta **52**. pp. 2749 – 2765. (1988).
2. I. S. Moeck, Elsevier. (2014)
3. 3 Nenny Miryani Saptadji. Karakterisasi reservoir panas bumi. Institut Teknologi Bandung, Bandung. (2009).
4. J. G. White, *Economic Geology*. pp. 48 – 65. (1967).
5. P. Stelling, L. Shevenell., N. Hinz., M. Coolbaugh., G.Melosh., and W. Cumming, Elsevier, Science Direct. (2016).
6. . Lawless, (eds). Lecture Note. Kingston Morrison. New York. pp. 15- 35. (1996).
7. Browne.P.R.L. *Geothermal exploration technology pre-advanced class*. Inhouse Training, Cirebon. pp. 1-30. (1996).
8. J.B. Purnomo and T. Pichler, Elsevier. (2014)
9. F.V.D. Meer, C. Hecker, F.V. Ruitenbeek, H.V.D. Werff, C.D. Wijkerslooth, C. Wechsler, Elsevier. (2014)
10. Puslitbang Geologi Provinsi Banten. Kondisi Geologi dan Litologi Wilayah Gunung Karang. Dinas Pertambangan dan Energi Provinsi Banten. Banten. (2013).
11. C. Endyana, F. Hirnawan, Hendarmawan, Mardiana. U, Buletin Sumber daya Geologi **Volume 6**. (2011).
12. Supriyanto Suparno. *Energi panas bumi: A present from the heart of the earth*. Universitas Indonesia, Depok. pp. 13 -26. (2009).
13. Yiman Lia, Zhonghe Panga, Fengtian Yang, Lijuan Yuan, Pinghui Tang, western China. Elsevier. (2017)
14. S.W.Y.D. Pambudi, M. Sakur, K. Ismail, D.I. Fajri, D.L. Setijadji, Delineasi daerah prospek panas bumi berdasarkan kelurusan citra landsat dan digital elevation model (DEM) daerah Gunung Lawu, Provinsi Jawa Tengah dan Jawa Timur. Universitas Gadjah Mada, Yogyakarta. (2014).
15. M. P. Hochstein, UNITAR/UNDP Centre for Small Energy Resources. Rome, Italy. (1990).
16. 16 M. P. Hochestein and S. Sudarman, History of geothermal exploration in Indonesia from 1970 to 2000. Elsevier. (2008)
17. Sukendar Putri., Mariasari Bandi., Sasmito. and Wijaya Arwan Putra. Analisis sebaran kawasan potensial panas bumi Gunung Salak dengan suhu permukaan, indeks vegetasi dan geomorfologi. Universitas Diponegoro, Semarang. (2016).
18. L. Guillou Frottier, E.B. Burov, J.-P. Milesi, J. Volcanol. Geotherm. Res. **102**, pp. 339–361. (2000)
19. D.W. Pals and P.G. Spry, Petrol. **79**,pp. 285–307.(2003)
20. J. Stix, B. Kennedy, M. Hannington, H. Gibson, R. Fiske, W. Mueller, J. Franklin, Geology **31**, pp. 375–378. (2003)
21. W.U. Mueller, J. Stix, P.L. Corcoran , R. Daigneault, Ore Geol. Rev. **35**, pp. 4–46. (2009)
22. BEICIP. Geothermal study, Banten area, West Java. Hydrogeochemical Reportprepared for Pertamina. (1979).

Sea Water Characterization at Ujung Kulon Coastal Depth as Raw Water Source for Desalination and Potential Energy

Dan Mugisidi[1], and *Okatrina* Heriyani[1*]

[1]Faculty of Engineering, Muhammadiyah Prof. Dr. HAMKA University, Jakarta - Indonesia

Abstract. Fresh water is basic need for life while the source is limited. Therefore, sea water is used as fresh water through desalination process. Sea water has different physical and chemical properties ranging from the surface to the seabed. The energy potential that can be obtained from the hydrostatic pressure also changes according to the depth. As part of the research of the utilization of sea water into fresh water, the aim of this study is to know the characteristics of sea water in the depth that can be utilized as source of fresh water. The sea water samples were taken at 11km from Ujung Kulon beach with depth of 0m, 20m, 40m, 60m, 80m, and 100m under the surface. The results showed that the physical properties at every depth were below the maximum allowable drinking water except for the amount of dissolved solids. Chemical characteristics at any depth above allowable level were fluoride, hardness (CaCo3), chloride, sodium, sulphate, and (KMnO4). In addition to the properties, pressure is one of the considerations in this study to determine the depth of sea water as sources for desalination. Pressure increased by 36.11% as the depth of the sea increased.

1 Introduction

Fresh water is a necessity for every creature on earth to stay alive. Population growth in Indonesia has a direct impact on the use of freshwater in the soil. This situation is completed with the reducing of water absorption area in various places due to the increase of buildings and factories. This resulted in the availability of clean water become threatened. UNICEF reports that 1 in 4 children will live in rare areas of water[1]. The condition is compounded by the decline in the soil surface due to the increasing amount of clean water taken from the soil[2]. Globally, water usage has tripled in the last 50 years[3], while the amount of fresh water on the surface of the earth is only 2.8% and the rest of 97.2% is sea water[4]. Therefore, the utilization of sea water as clean water becomes very potential. The problem is that seawater contains a greater amount of dissolved salt than fresh water, especially NaCl. Table 1 shows the dissolved salt content in different oceans and oceans in different parts of the world. Therefore, efforts to convert sea water into fresh water and efforts to eliminate salt content from seawater are widely practiced.

Various desalination technologies and their combination with renewable energy are shown in Figure 1. The utilization of solar energy for the desalination process is the least costly way, especially in tropical regions such as Indonesia. Direct heating method with solar heat is the most suitable way to produce fresh water up to 200 m^3/day[5]. For larger scales, the use of solar energy coupled with other energy sources such as heat collectors, waves, and geothermal is sufficient[6].

Table 1. Mean Value of Total Dissolved Salts in Various Seas and Oceans[4]

Sea or Ocean TDS (ppm)	Sea or Ocean TDS (ppm)
Baltic Sea 7000	Indian Ocean 33,800
Caspian Sea 13,500	Mediterranean near Crete 37,900
Black Sea 20,000	Mediterranean 39,000
White Sea 28,000	Red Sea 43,000
North Atlantic 29,000	Arabian Gulf 50,000
Pacific Ocean 33,600	Australian Shark Bay 70,000
Atlantic South 36,000	Kara Bogas Gulf 164,000
North Atlantic 36,200	(in Caspian Sea)
Adriatic Sea 31 ,400	Dead Sea 220,000

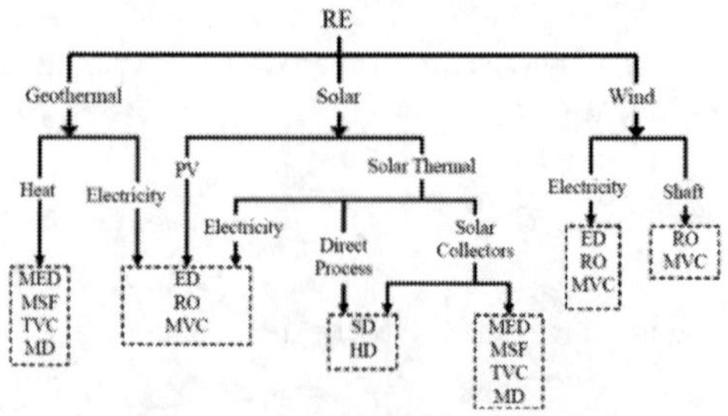

Fig. 1. Possible technological combinations of the main renewable energies and desalination methods[18]

* Corresponding author: oktarina@uhamka.ac.id

Renewable energies are energy sources that are continually replenished by nature and derived directly from the sun (such as thermal, photo-chemical, and photo-electric), indirectly from the sun (such as wind, hydropower, and photosynthetic energy stored in biomass), or from other natural movements and mechanisms of the environment (such as geothermal and tidal energy)[7]. The solar energy collector is one of the tools that developing in renewable energy. The various types of collectors such as flat-plate, compound parabolic, evacuated tube, parabolic trough, fresnel lens, parabolic dish, and Heliostat field collector (HFC) widely used as a tool for collecting solar energy[8]. Wind, solar, and other renewable technologies that can be used for desalination are rapidly emerging with the promise of economic and environmental viability on a large scale[9]. Renewable energy provides solutions to lower energy costs. One of the energy that can be utilized for the desalination process is hydrostatics pressure10. Hydrostatic pressure reduces the electricity consumption for the reverse osmosis process from 10 kwh/m^3 to 2.98 kwh/m^3[11]. Efficiency of hydrostatic usage as energy source is 80% in desalination process with RO[12].

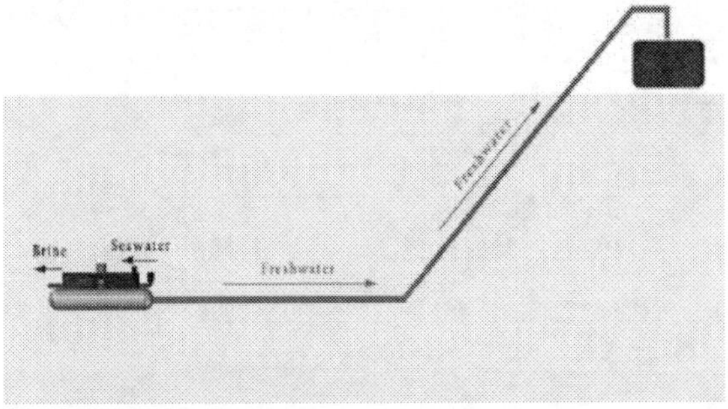

Fig. 2. Deep sea reverse Osmosis scheme[19]

At the depth of the ocean, the available pressure and power generated by the turbine can be estimated using the following equations[13]:

$$P = \rho \cdot g \cdot H \quad (1)$$

$$P_T = Q \cdot \rho \cdot g \cdot H \quad (2)$$

Where

P is the pressure due to the height difference ($\frac{N}{m^2}$),

P_T is the power generated turbine (*watt*),

Q is Flow Rate ($\frac{m^3}{s}$),

ρ is the density ($\frac{kg}{m^3}$),

g is gravity 9,81($\frac{m}{s^2}$), and

H is head difference (m).

Utilization of hydro energy storage, where the water is taken from the depth of the ocean, the water will be different from the water on the surface, such as temperature. Since ocean's temperature decreases steadily with depth, although the ocean's surface temperature varies from 40°C to -2°C, the average temperature in ocean's depth is almost constant and around 3 - 4°C[14]. In addition, sea water content will also change where salinity increased 4.52%, pH decreased an average of 1.46% and the conductivity increase an average 0.03% in line with the addition of depth[15].

In this study, the sample of sea water content was taken from depth 0*m*, 20*m*, 40*m*, 60*m*, 80*m*, and 100*m*. Tests were performed using drinking water standards. In addition, the potential power at each depth to be obtained will also be calculated

2 Method

The sea water used in this study was taken from offshore of Ujung Kulon, West Java, Indonesia. The depth of sampling is 0*m*, 20*m*, 40*m*, 60*m*, 80*m*, and 100*m*. Samples at each depth were taken as many as 40 *liters* and stored in jerrycan. Sampling is done by using 2 *inch* hose that closed at the end by using valve. The sample test was conducted at the Mechanical Engineering Laboratory, Prof. Dr. HAMKA University and at the Regional Health Laboratory (LABKESDA) of DKI Jakarta, in Rawasari. The comparative standard used is the drinking water standard applicable in the Republic of Indonesia, Permenkes 492 Tahun 2010[16].

3 Result and Discussion

In this study, the testing of sea water content taken from a depth of 0 m, 20 m, 40 m, 60 m, 80 m, and 100 m. The test is performed using the Drinking Water Standard as shown in Table 2.

Table 2. Physical Properties according to 492/2010

Parameter	Satuan	Kedalaman						Kadar Maksimum yang Diperbolehkan untuk Air Minum
		0 M	20 M	40 M	60 M	80 M	100 M	
Smell	-	Tidak berbau	Tidak berbau	Tidak berbau	Tidak berbau	Tidak berbau	Tidak berbau	Tidak berbau
TDS	mg/l	29,535	30,125	29,425	29,928	30,076	30,927	500
Turbidity		0.36	0.27	1.84	0.33	77.1	0.34	5
Temperature	Skala NTU	25.6	25.3	25.8	25.5	25.9	25.8	Suhu udara ±3^0
Color	Skala TCU	tt<2	tt<2	tt<2	tt<2	tt<2	tt<2	15

Table 2 is the result of testing of seawater physical properties that follow to the standard on drinking water testing. The test results showed that TDS (Total Dissolved Solid) was in the range of 30,000 *mg/l*, while the standard allowed for drinking water was 500 *mg/l*. In addition, turbidity at a depth of 80*m* differs from other

depths. At a depth of 0*m*, 20*m*, 40*m*, 60*m*, and 100*m* average turbidity is 0.628 below the drinking water standard, 5. Turbidity at a depth of 80*m* is different because at that depth there is undersea water flow. This is evidenced by the carrying of a hose by the water current at the depth. Furthermore, the chemical properties of seawater are shown in **Table 3**.

In Table 3, it appears that parameters that exceed the maximum permissible levels of drinking water are Flouride, Hardness, Chloride, Sodium, Sulfate, and Organic. This is consistent with that stated by Vassilis Belesiotis, et. Al[4].

Table 3. Chemical Properties according to 492/2010

Parameter	Satuan	Kedalaman						Kadar Maksimum yang Diperbolehkan untuk Air Minum
		0 M	20 M	40 M	60 M	80 M	100 M	
Mercury	mg/l	tt<0.0005	tt<0.0005	tt<0.0005	tt<0.0005	tt<0.0005	tt<0.0005	0.001
Alumunium	mg/l	tt<0.0127	tt<0.0127	0.0586	tt<0.0127	0.0851	tt<0.0127	0.2
Arsen	mg/l	tt<0.0021	tt<0.0021	tt<0.0021	tt<0.0021	tt<0.0021	tt<0.0021	0.01
Barium	mg/l	tt<0.0023	tt<0.0023	tt<0.0023	tt<0.0023	tt<0.0023	tt<0.0023	0.7
Ferrous	mg/l	tt<0.0178	tt<0.0178	tt<0.0178	tt<0.0178	tt<0.0178	tt<0.0178	0.3
Fluoride	mg/l	1.65	2.14	1.62	1.8	1.73	1.9	1.5
Cadmium	mg/l	tt<0.0003	tt<0.0003	tt<0.0003	tt<0.0003	tt<0.003	tt<0.0003	0.003
Hardness (CaCO^3)	mg/l	6313	6653	6473	6693	6633	7515	500
Chloride	mg/l	20621	20783	20358	20648	20855	20771	250
Chromium Total	mg/l	tt<0.0004	tt<0.0004	tt<0.0004	tt<0.0004	tt<0.0004	tt<0.0004	0.05
Manganese	mg/l	tt<0.0007	tt<0.0007	tt<0.0007	tt<0.0007	tt<0.0007	tt<0.0007	0.4
Sodium	mg/l	5380	5141	4718	4662	4526	4449	200
Nitrate	mg/l	tt<0.211	tt<0.211	tt<0.211	tt<0.211	tt<0.211	tt<0.211	50
Nitrite	mg/l	tt<0.012	tt<0.012	tt<0.012	tt<0.012	tt<0.012	tt<0.012	3
pH	-	7.67	7.63	7.3	7.61	7.17	6.87	6.5 – 8.5
Selenium	mg/l	tt<0.0021	tt<0.0021	tt<0.0021	tt<0.0021	tt<0.0021	tt<0.0021	0.01
Zinc	mg/l	tt<0.0335	tt<0.0335	tt<0.0335	tt<0.0335	tt<0.0335	tt<0.0335	3
Cyanide	mg/l	tt<0.0015	tt<0.0015	tt<0.0015	tt<0.0015	tt<0.0015	tt<0.0015	0.07
Sulfate	mg/l	2584	2580	2549	2546	2561	2548	250
Copper	mg/l	tt<0.0004	tt<0.0004	tt<0.0004	tt<0.0004	tt<0.0004	tt<0.0004	2
Lead	mg/l	tt<0.0013	tt<0.0013	tt<0.0013	tt<0.0013	tt<0.0013	tt<0.0013	0.01
Ammonia	mg/l	tt<0.02	tt<0.02	tt<0.02	tt<0.02	tt<0.02	tt<0.02	1.5
Organic ($KmnO_4$)	mg/l	111.76	119.46	145.67	116.38	141.04	148.75	10
Nickel	mg/l	tt<0.0003	tt<0.0003	tt<0.0003	tt<0.0003	tt<0.0003	tt<0.0003	0.07

Furthermore, data that is above the maximum level shown in Figure 3. In Figure 3, TDS that is above the maximum level increases with increasing depth. An average TDS increase of 10% for every 20m depth increase. Turbidity at depths of 0m, 20m, 40m, 60m, and 100m below permissible levels except at a depth of 80m.

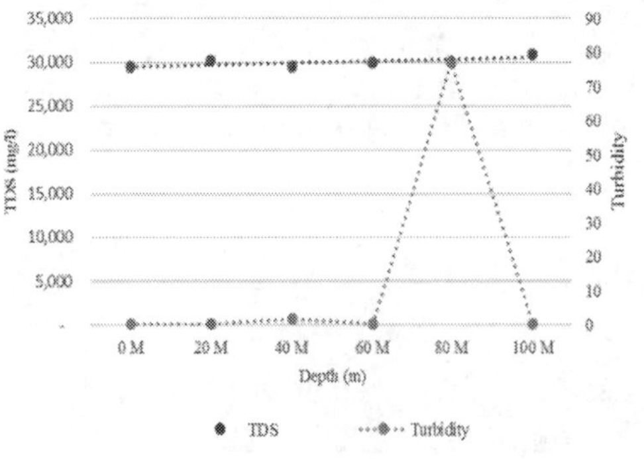

Fig. 3. Physical Properties TDS and Turbidity

In Figure 4, it appears that organic increases by 30% in line with increasing ocean depth. Mean while, on the average of flouride is no increase with the depth of the sea. This is in line with that proposed by Bewers[17] that there is no increase in levels of fluoride along with the addition of the depth of the ocean.

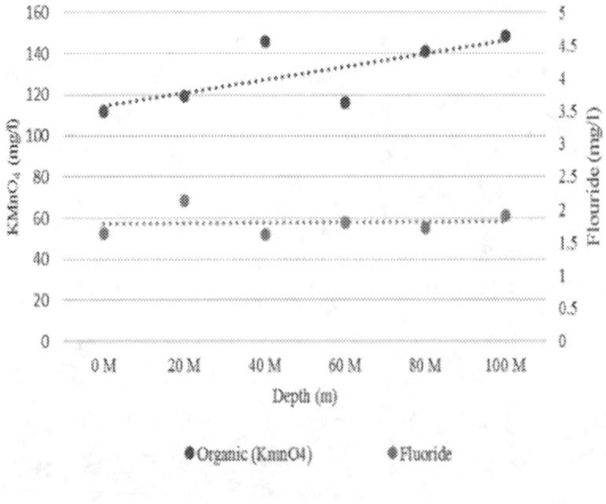

Fig. 4. Chemical Properties Organic and Flouride

The increase also does not occur in Chloride and Sulphate, as can be seen in Figure 5. In contrast to other elements, sodium has decreased by 20% with increasing sea depth, while Hardness is up 23% along with increasing depth of sea depth.

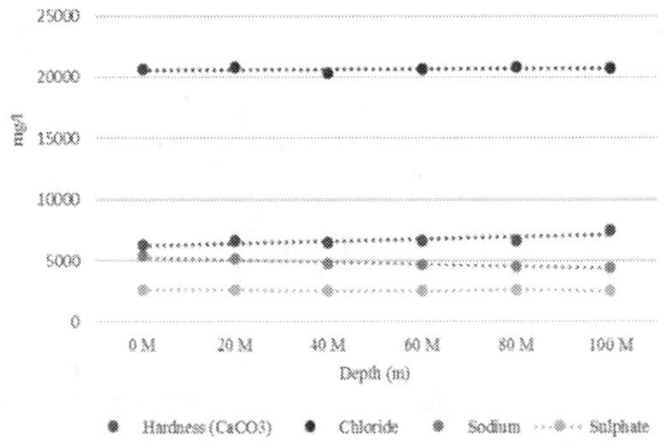

Figure 5. Chemical Properties ; Hardness, Chloride, Sodium and Sulphate

In addition to sea water testing, the potential power obtained by the pressure due to the depth of the sea is calculated, assuming the water requirement is 200 *liters/person/day* with the population on one island is 100 people, then the required debit is 20,000 *liters/day* or 0.8 m^3*/hour*. The gravity of sea water is 1.025 *kg/*m^3 and gravity is 9.81 *m/*s^2, the increase in hydrostatic pressure is 36.11% with increasing sea depth.

4 Conclusions

Based on the results of the test, it can be concluded that TDS, hardness, sodium, and organic have changed sequentially by 10%, 23%, -20%, and 30%. Flouride, chloride, and sulphate do not change with increasing sea depth. Potential pressure increased by 36.11% with increasing sea depth.

References

1. "UNICEF. 1 dari 4 Anak akan Tinggal di Daerah Langka Air" | Republika Online. http://www.republika.co.id/berita/internasional/global/17/03/23/on8xa8384-unicef-1-dari-4-anak-akan-tinggal-di-daerah-langka-air. Accessed June 29, (2017).
2. "Krisis Air Bersih di Jakarta Berdampak pada Penurunan Muka Tanah" | Republika Online. http://nasional.republika.co.id/berita/nasional/jabodetabek-nasional/16/10/26/ofnh6z335-krisis-air-bersih-di-jakarta-berdampak-pada-penurunan-muka-tanah. Accessed June 29, (2017).
3. Vigotti R, Hoffman A. The water scarcity and Water Scurity: po;itical and Social implications. In: *Technology and Water,* Rome: IEA Working Party, (2009).
4. V. Belessiotis, S. Kalogirou, E. Delyannis, 1St ed. (Convey M, ed.), ELSEVIER, (2016).
5. Garcia-Rodriguez L. Seawater desalination driven by renewable energies: a review, *Desalination,* (2002).
6. M. Shatat, S. Riffat, S. Ghabayen, SIn: *The 4th International Engineering Conference –Towards Engineering of 21st Century,* **1 -16**, (2012).
7. O. Ellabban, F. Blaabjerg, *Renew Sustain Energy Rev,* 39(August), **748-764**, (2014). doi:10.1016/j.rser.2014.07.113.
8. SA. Kalogirou, *Solar Thermal Collectors and Applications,* Vol **30**, (2004). doi:10.1016/j.pecs.2004.02.001.
9. S. Kalogirou, *Prog energy Combust Sci.* (2005). doi:10.1016/j.pecs.2005.03.001.
10. C. Charcosset, *Desalination*, (2009).
11. C.Charcosset. *Renew Energy*, 34 (12), **2878-2882**, (2009). doi:https://doi.org/10.1016/j.renene.2009.02.026.
12. M. Reali, M. de Gerloni, A. Sampaolo, *Desalination*, 109(3), 269-275, (1997). doi:10.1016/S0011-9164(97)00073-8.
13. F. Dietzel, *Turbin, Pompa Dan Kompresor*, Ketiga, Erlangga, (1992).
14. PR. Pinet *Invitation to Oceanography*, Jones and Bartlett Publishers, (2009).
15. D. Mugisidi, O. Heriyani, *Prosiding Seminar Internasional Universitas Muhammadiyah Purwokerto*, (2016).
16. Permenkes RI. PMK No. 492 ttg Persyaratan Kualitas Air Minum.pdf, (2010).
17. JM. Bewers, *Deep Sea Res Oceanogr Abstr*,18 (2), **237-241**, (1971). doi:10.1016/0011-7471(71)90113-6.
18. AE. Kabeel, SA. El-Agouz, R *Desalination*, 276 (1-3),**1-12,** (2011). doi:10.1016/j.desal.2011.03.042.
19. R. Dashtpour, SN. Al-Zubaidy, *International Proceedings of Chemical, Biological and Environmental Engineering*. Vol 33, **154-162,** (2012). http://ipcbee.com/vol33/031-ICEEB2012-B20001.pdf.

https://doi.org/10.1051/e3sconf/20183102006

Home Photovoltaic System Design in Pangkalpinang City

Wahri Sunanda[1,*]

[1]Department of Electrical Engineering, Faculty of Engineering, Universitas Bangka Belitung , Bangka - Indonesia

Abstract. This research aims to obtain the design of home photovoltaic systems in Pangkalpinang and the opportunity of economic savings. The system consists of photovoltaic with batteries. Based on electricity consumption of several houses with installed power of 1300 VA and 2200 VA in Pangkalpinang for one year, the daily load of photovoltaic system is varied to 40%, 30% and 20% of the average value of the daily home electricity consumption. The investment costs, the cost of replacement parts and the cost of electricity consumption accrued to PLN during lifetime of systems (25 years) are also calculated. The result provided that there are no economic saving opportunities for photovoltaic systems with batteries at home with installed power of 1300 VA and 2200 VA in Pangkalpinang. The most economical is the photovoltaic system with the daily load of 20% of the average value of the daily home electricity consumption. The configuration of photovoltaic system for 1300 VA home consists of 10 modules of 200 wattpeak and 4 batteries 150 AH, 12 Volt while photovoltaic system for 2200 VA home consists of 15 modules of 200 wattpeak and 6 batteries 150 AH,12Volt.

1 Introduction

According to the statistical data of PT. PLN (Persero) Bangka Belitung region in 2016, currently there are some power plants connected with 607,226 kVA and the amount of energy generated 839.68 GWh. The total electrical energy is supplied 60.1% by high speed diesel (HSD), 10, 6% of biodiesel, 27.3% of coal and 1% of biomass, biogas and solar. The number of customers in 2016 was 388.752 with 357.093 household, 21.657 business, 6.065 social and 3.448 government. Therefore, the diversification of energy sources needs to be developed in the Bangka Belitung Islands Province.

Solar cell technology can be applied in Indonesia especially in Province Bangka Belitung because sunlight is accepted almost throughout the year. With solar cells, sunlight is converted into electric current. Then, the resulting energy can be stored in storage media (battery). The use of PLN's electrical energy combined with a photovoltaic system can help solve the problem of electricity especially when applied to households because this sector is the largest consumer of electrical energy.

Previously there has been a study related to the hybrid system of solar power plants with PLN electricity using batteries as storage of electrical energy (storage system) for urban homes, photovoltaic system supplies electrical energy about 30% of the overall load of household electrical appliances while the remaining 70% of electricity from PLN [1]. Then photovoltaic system planning with batteries as a medium for storing electrical energy has also been done to meet the electrical energy needs of 10 houses in housing complexes in Banda Aceh [2]. There has also been research related to the concept of regulating the flow of power for the photovoltaic system connected to the grid system at residential, grid as a storage or temporary lender to meet the demand of the load so that only the investment cost of the photovoltaic system without batteries and the cost of leasing the network needed in the home system with photovoltaic system [3].

Then a research to design a standalone solar photovoltaic system for residential application with total load power 1981 watt for some direct current (DC) appliances and alternative current (AC) appliances [4]. Then a system to supply electrical demand of a rural house without connection to national grid, using day to day necessary appliances, by photovoltaic system [5]. Also the design of a grid connected photovoltaic (PV) system for typical house will be discussed in terms of sizing of P-V units and battery storage [6]. There is also design of solar PV system and a case study based on cost analysis of 1.0 kW off-grid photovoltaic energy system and costs of energy produced installed at Jamia Millia Islamia, New Delhi India [7]. However, the prior related researches have not conducted an economic analysis of the designed photovoltaic system. The research is expected to provide a description of household electrical energy consumption patterns and become a consideration in the use of solar cell technology for households that are integrated with electricity from PLN (grid connected) in the framework of developing renewable energy utilization.

2 Method

Collect data of household electricity consumption in Pangkalpinang city with installed power of 1300 VA (30

* Corresponding author: wahrisunanda@ubb.ac.id

https://doi.org/10.1051/e3sconf/20183102006

homes) and 2200 VA (30 homes) starting in April 2015 through March 2016. Then, perform data processing of household electrical energy consumption so it can be known the average value of daily electricity consumption. The daily load for the designed photovoltaic system in household is varied to three ie 40%, 30% and 20% of the average daily electricity consumption.

Make the design of photovoltaic systems with batteries connected to the household network/grid. The main components of photovoltaic systems with batteries are solar modules, BCR, batteries and inverters [12][13].

$$Number\ of\ solar\ modules\ required = \frac{daily\ load\ amount}{power\ module\ x\ system\ efficiency\ x\ long\ irradiation\ x\ \frac{daily\ intensity}{1000}} \quad (1)$$

$$Capacity\ of\ battery = \frac{daily\ load\ amount\ x\ autonomous\ days}{battery\ efficiency\ x\ DOD\ x\ battery\ voltage} \quad (2)$$

$$Capacity\ of\ BCR = Number\ of\ modules\ x\ I_{sc} \quad (3)$$

$Capacity\ of\ inverter =$ greater than the load power

$$load\ power = \frac{daily\ load\ amount}{load\ fulfillment\ time} \quad (4)$$

Finally, calculate total cost of photovoltaic system covering investment cost, component replacement cost during lifetime (25 years) and electricity consumption cost accrued to PLN during the system lifetime. Then, compare the total cost spent during the lifetime for the home with photovoltaic system and without photovoltaic system in order to know the feasibility and economic savings opportunities derived from the installation of the system photovoltaic in the house.

For calculation of component replacement cost during the lifetime of the photovoltaic system, it is assumed that the component price declines (8% per year) as a result of technological and industrial developments causing declining production costs [8]. Also assumed that the cost of household electricity consumption paid to PLN fixed during lifetime system because the electricity tariff not increase significantly each year.

3 Results

To determine the daily load of photovoltaic systems connected to the power grid, collected data of electric energy usage for 1 year from April 2015 to March 2016 in 30 households with installed power of 1300 VA and 30 households with installed power of 2200 VA in Pangkalpinang city. The average daily household power consumption with 1300 VA installed power is 9.51 kWh while the average daily household consumption with 2200 VA installed is 14.68 kWh. The average value is then used to determine the daily load of photovoltaic systems connected to the household power grid.

In this paper, the design is photovoltaic system using battery. The total daily load of photovoltaic system is 40%, 30% and 20% of the average daily household electricity consumption. Daily consumption of 30 household electric power customers with installed power 1300 VA can be seen in figure 1 and with installed power 2200 VA in figure 2. The daily load of photovoltaic systems in households also can be seen in Table 1.

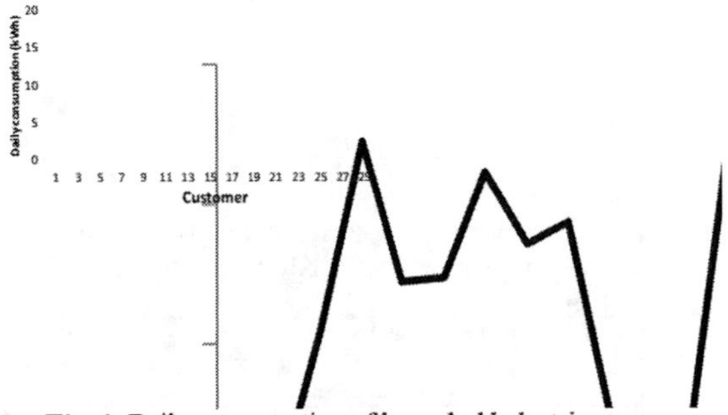

Fig. 1. Daily consumption of household electric power customers with installed power 1300 VA (kWh)

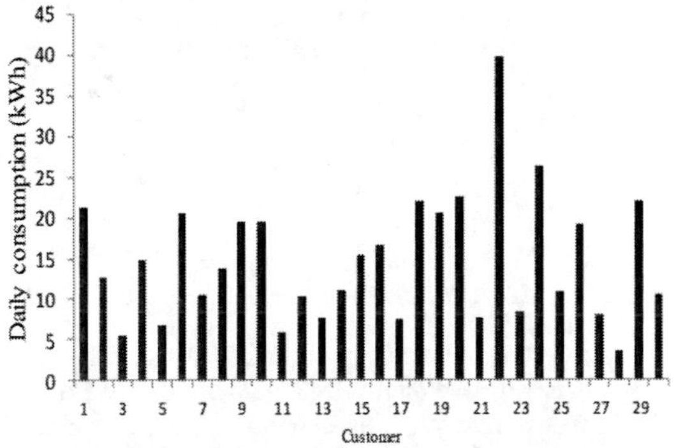

Fig. 2. Daily consumption of household electric power customers with installed power 2200 VA (kWh)

a. Photovoltaic system using storage media (battery)
Photovoltaic systems use storage media (batteries) so that the electrical energy generated can be stored in advance to meet the electricity needs at night. The design is carried out at installed power of 1300 VA and 2200 VA by varying the amount of photovoltaic system loads to 40%, 30% and 20% of the average daily consumption amount to know which photovoltaic system is most likely to be applied.

Table 2 shows the number of modules, inverter capacity, BCR capacity and battery capacity used for 1300 VA installed power with load variations of 40%, 30% and 20% with daily load of 3.8 kWh, 2.85 kWh and 1,902 kWh respectively.

Table 1. Total daily load of photovoltaic systems

Installed Power Capacity	Average Daily Electricity Consumption	Percentage of Daily Load	Daily Load
		40%	3,800 kWh
1300 VA	9,51 kWh	30%	2,850 kWh
		20%	1,902 kWh
		40%	5,870 kWh
2200 VA	14,68 kWh	30%	4,400 kWh
		20%	2,936 kWh

Table 2. Photovoltaic systems using batteries connected with the power grid for homes with 1300 VA installed power

System daily load	Number of modules	Capacity of inverter	Capacity of BCR	Battery capacity
3,8 kWh	20, @200 Wp	300 W, 24V (2 units)	80A, 24V (2 units)	150Ah, 12V (8 units)
40% of the average value daily electricity consumption				
2,85 kWh	15, @200 Wp	300 W, 24V (1 unit)	100A, 24V (1unit)	150Ah, 12V (6 units)
30% of the average value daily electricity consumption				
1,902 kWh	10, @200 Wp	300 W, 24V (1 unit)	80A, 24V (1unit)	150Ah, 12V (4 units)
20% of the average value daily electricity consumption				

Table 3 shows that for home of 2200 VA with load 40%, 30% and 20% which is 5.87 kWh, 4.4 kWh and 2,936 kWh respectively, the number of modules, inverters, BCR and batteries used are different compared to 1300 VA installed power

b. Economic analysis of designed photovoltaics systems

The economic analysis of the designed photovoltaic system is done by determining the assumption that the PLN electricity tariffs remain for the lifetime while the price of photovoltaic components decreases (8% per year). The calculation of the investment cost of the designed photovoltaic system, the cost of replacing the system components during the lifetime (25 years) and the accrued electricity cost to PLN during the lifetime of the system (25 years) is performed to determine the total cost required to meet household electricity needs during 25 years

The amount of electrical energy that must remain supplied by PLN, if reduced by the electric energy supplied by photovoltaic to meet the average daily electrical energy is shown in table 4.

Table 3. Photovoltaic systems using batteries connected with the power grid for homes with 2200 VA installed power

System daily load	Number of modules	Capacity of inverter	Capacity of BCR	Battery capacity
5,87 kWh	30, @200 Wp	300 W, 24V (2 units)	100A, 24V (2 units)	150Ah, 12V (12 units)
40% of the average value daily electricity consumption				
4,4 kWh	22, @200 Wp	300 W, 24V (2 units)	80A, 24V (2 units)	200Ah (4 units) , 150Ah (4 units)
30% of the average value daily electricity consumption				
2,936 kWh	15, @200 Wp	300 W, 24V (1 unit)	100A, 24V (1unit)	150Ah, 12V (6 units)
20% of the average value daily electricity consumption				

Table 4. Total daily household electricity needs still to be met by PLN

Installed power (VA)	Energy (kWh)
1300	5,71
	6,66
	7,608
2200	8,81
	10,28
	11,744

Table 5. The cost of a photovoltaic system using batteries connected with power grid on a house with installed power of 1300 VA

Cost Component	Total daily load of photovoltaic systems (kWh)		
	3,8	2,85	1,902
Investment costs (Rp)	124.716.000	91.578.000	62.358.000
Cost of component replacement (Rp)	33.027.720	22.808.230	16.513.860
PLN's electricity consumption costs for 25 years (Rp 1410,12/kWh)	72.466.067	84.522.593	96.553.737
Total cost (Rp)	230.209.787	198.908.823	175.425.597

Table 6. Savings opportunity of the installation of photovoltaic systems using batteries connected to the network/grid at home with 1300 VA installed power

	Cost of electricity consumption (Rp)		
Home without photovoltaic system for 25 years	120.692.171		
Home with photovoltaic system for 25 years	40% PV Load	30% PV Load	20% PV Load
	230.209.787	198.908.823	175.425.597
Difference cost for 25 years	-109.517.616	-78.216.652	-54.733.426

Table 5 shows that the cost of electricity consumption of house with photovoltaic systems connected to power lines using batteries for 25 years is still much larger when compared to the cost of electricity consumption if all the electricity needs of the house is met by PLN. Then table 6 shows the total cost for the smallest system lifetime is Rp 175.425.597 when the photovoltaic system meets 20% of the household's daily power supply (1,902 kWh). For this photovoltaic system, the cost of electricity consumption per month is more expensive Rp 182.445 than if the house does not use photovoltaic.

Table 7. The cost of a photovoltaic system using batteries connected with power grid on a house with installed power of 2200 VA

Cost Component	Total daily load of photovoltaic systems (kWh)		
	5,87	4,4	2,936
Investment costs (Rp)	183.156.000	136.956.000	91.578.000
Cost of component replacement (Rp)	45.616.460	35.823.525	22.808.230
PLN's electricity consumption costs for 25 years (Rp 1410,12/kWh)	111.808.415	130.464.302	149.044.044
Total cost (Rp)	340.580.875	303.243.827	263.430.274

Table 7 shows that the cost of electricity consumption of homes with photovoltaic systems connected to power lines by using batteries for 25 years is still much larger when compared to the cost of electricity consumption if all the electricity needs of the house is met by PLN. Then table 8 shows the lowest total cost during the system lifetime is Rp 263.430.274 when the photovoltaic system meets 20% of the household's daily electrical load (2,936 kWh). For this photovoltaic system, the cost of electricity consumption per month is more expensive Rp 257.084 than if the house does not use photovoltaic.

Table 8. Savings opportunity of the installation of photovoltaic systems using batteries connected to the network/grid at home with 2200 VA installed power

	Cost of electricity consumption (Rp)		
Home without photovoltaic system for 25 years	186.305.054		
Home with photovoltaic system for 25 years	40% PV Load	30% PV Load	20% PV Load
	340.580.875	303.243.827	263.430.274
Difference cost for 25 years	-154.275.821	-116.938.773	-77.125.220

4 Conclusions

From the research that has been done, it is known that there is no economic savings for photovoltaic systems connected to PLN power grids with battery-mounted system at home with 1300 VA and 2200 VA installed power in Pangkalpinang city. From some variations of photovoltaic systems that have been designed, the most economical is the photovoltaic system using batteries connected to the power grid with daily load of 20% of average daily household electricity consumption.

Photovoltaic system yet shows the feasibility in the economy, but the photovoltaic system is still feasible to be considered and implemented on a frequent home experienced a power outage, especially during the day.

Acknowledgement

This research was supported by Universitas Bangka Belitung year 2017.

References

1. L. E. B. L. E. Bien, I. Kasim, W. Wibowo, Jurnal Teknik Elektro, Vol. **8**, No 1, Universitas Trisakti, Jakarta (2008).
2. Suriadi, M. Syukri., Jurnal Rekayasa Elektrika, Vol. **9**, No. 2, Universitas Syiah Kuala, Banda Aceh (2010).
3. K. Kananda, R. Nazir., Jurnal Nasional Teknik Elektro,Vol. **2**, No. 2, Universitas Andalas, Padang (2013).

4. U.H.Ibrahim, D.A. Aremu, J.I. Unwaha., International Journal Of Scientific & Technology Research Vol. **2**, Issue 12, December (2013).
5. M. Aminy, N. Barhemmati,A. Hadadian, F.Vali., *Design Of A Photovoltaic System For A Rural House*, Second Iranian Conference on Renewable Energy and Distributed Generation (2012).
6. A.T. Elsayed, T.A. Youssef, A. Mohamed, O. A. Mohammed, *Design, Control and Management of P-V System for Residential Applications with Weak Grid Connection,* Eleventh LACCEI Latin American and Caribbean Conference for Engineering and Technology, Mexico (2013).
7. S. Ahsan , K. Javed, A. S. Rana, M. Zeeshan, Design And Cost Analysis Of 1 kW Photovoltaic System Based On Actual Performance In Indian Scenario, Perspectives in Science Vol. 8, pp 642—644 (2016).
8. G. Parkinson., *Cost Reduction (Example:USA)*, http://reneweconomy.com.au, accessible date on September 6th (2015).
9. Anonim, *Penetapan Penyesuaian Tarif Tenaga Listrik Bulan Agustus 2016*, http://www.pln.co.id, accessible date on August 18th (2016).
10. G. Boyle., *Renewable Energy, Power for a Sustainable Future,* Oxford University Press, Inc., New York (1996).
11. R. Boylestad, L. Nashelsky., *Electronic Devices and Circuit Theory,* Prentice Hall International, Inc., New York (2002).
12. M. Buresch., *Photovoltaic Energy Systems, Design and Installation,* McGraw-Hill, Inc., New York (1983).
13. H. Djojodihardjo., *Pengantar Ringkas Sistem Listrik Tenaga Surya,* Institut Teknologi Bandung, Bandung (2001).
14. R.J. Kodoatie., *Analisis Ekonomi Teknik,* Andi, Yogyakarta (2002).
15. Available online on : https://solar.schneider-electric.com/solutions/residential/

https://doi.org/10.1051/e3sconf/20183102007

Biogas Production from Rice Husk Waste by using Solid State Anaerobic Digestion (SSAD) Method

Hashfi Hawali Abdul Matin[1,*] and *Hadiyanto*[1,2]

[1]Master Program of Environmental Science, School of Post Graduate Studies, Diponegoro University - Indonesia
[2]Chemical Engineering Department, Engineering Faculty, Diponegoro University – Indonesia

Abstract. An effort to obtain alternative energy is still interesting subject to be studied, especially production of biogas from agriculture waste. This paper was an overview of the latest development of biogas researches from rice husk waste by Solid State Anaerobic Digestion (SSAD). The main obstacle of biogas production from rice husk waste was the lignin content which is very difficult degraded by microbes. Various pretreatments have been conducted, either physically, chemically as well as biologically. The SSAD method was an attractive option because of the low water content of rice husk waste. The biogas yield by SSAD method gave more attractive result compared to Liquid Anaerobic Digestion (LAD) method. Various studies were still conducted in batch mode laboratory scale and also has not found optimum operating conditions. Research on a larger scale such as bench and pilot scale with continuous systems will be an increase trend in the future research.

1 Introduction

Among of the global problems today is the issue of energy availability. The Indonesian government's efforts in addressing to cover this problem is by releasing the Government Regulation No 5/2006 about the use of new and renewable energy. Under this regulation, the Indonesian government targets to utilize new and renewable energy sources of 17% of the national energy mix in 2025 from the current of 5%. Along with technological developments and the potential of natural resources, this target is raised to 25% in the new vision referred to as the 25/25 vision. This effort can be classified in energy diversification and energy intensification.

Energy diversification means exploring new energy sources and renewable energy so as to obtain diverse types of which is energy from biomass which is often known as biogas, while intensifying the energy intended to make use of new energy as much as possible. As an agrarian country that also advanced in the field of industrialization, Indonesia has a very abundant biomass waste, including all industrial waste farms and livestock Biogas is an energy produced by organic matter fermented under anaerobic conditions [1]. Biogas is produced from renewable energy sources [2]. Biogas can be used as fuel for vehicles as well as for power and heat generation [3].

Biogas technology is actually not a new technology and this technology has been developed in Indonesia a few decades ago which is called first-generation biogas. Implementation technology of biogas production from livestock waste has also been done many researchers for decades. However, biogas production by anaerobic biodigester made from agricultural waste with a high lignocellulose content of Solid State Anaerobic Digestion (SS-AD) methods or called second generation biogas with pretreatment physical, chemical and biological methods with rumen bacterial biocatalyst is a breakthrough and the novelty of this biogas research.

Fig 1. Rice Husk

Biogas production research from agricultural waste is rice husk is based on the abundance of rice husk existence especially in rural area and low utilization done. So far the utilization is done only as a mixing material of red bricks and planting media. Based on data collected from the Directorate of Food and Agriculture (2013), it can be seen that rice plants produce grain (50%) and straw (50%). Then from the grain can be produced brown rice (80%) and husk (20%). According

* Corresponding author: hawalihashfi@gmail.com

to Central Bureau of Statistics of Central Java Province in 2014, rice crop productivity of 9,648,104.45 Ton. So the total rice production in one year reaches 964,810.44 Ton in 2014. It has excellent potential to be utilized as biogas. In addition, rice husk is very abundant also because of its difficulty to decompose because of high lignin content so it cannot be degraded by microorganisms easily. Through the preliminary treatment is expected to break down the content of lignin present in rice husks. The preliminary treatment may be physical, chemical and biological pretreatment [4].

In handling biogas production, utilizing anaerobic bacteria to produce methane gas (CH4). In order to obtain the maximum biogas production results required the right conditions for methane-producing microorganisms to live and grow. Some of the things that affect the growth of microorganisms are temperature, pH, alkalinity, F/M ratio and C/N ratio. The prospect of biogas technology development is very large, especially in rural areas where most of the people work in livestock and agriculture [2]. Biogas production technology, including SS-AD of rice husk waste biomass, should consider several parameters to produce optimal biogas.

The purpose of this article is to review the previous biogas research with rice husk waste materials through SSAD methods and to offer in-depth discussion on further research in order to obtain optimum operating conditions in batch or continuous scale.

2 Characteristics of Rice Husk

Agricultural waste is a biomass containing lignocellulose. Lignocellulose consists of cellulose, hemicellulose, lignin, extractive, and some inorganic materials. Table 1 shows the lignocellulosic components of various agricultural wastes. Cellulose, hemicellulose, and lignin are the three main components that are important sources for producing useful products, such as sugars from the fermentation process, chemicals and liquid fuels [5]. Lignocellulose can be obtained from wood, straw, rice husk, grasses, agricultural/forest wastes, industrial waste (wood, paper) and other fibrous materials [6].

Rice husk is one of agriculture waste containing lignocellulose which consists of 58.852% cellulose, 18.03% hemicellulose, lignin 20.9% and ash 0.16% - 1%. With a high amount of lignocellulosic content it is difficult to be degraded by microorganisms by itself. Usually farmers overcome rice husk waste by being burned and dumped. In fact, the way it can harm farmers and the surrounding environment because the combustion made to produce CO2 gas that is harmful to health. Meanwhile, landfill in the soil can be a contributing factor to the disease for the next crop. Therefore, in order to produce biogas from rice husk waste there is a need for special pretreatment before fermentation in the digester [7, 8].

Table 1. Lignocellulosic Components of Some Biomasses [9]

Biomass	Cellu-lose (%)	Hemi-cellulose (%)	Lignin (%)	Ash (%)
Rice husk	58.852	18.03	20.9	0.16-1
Rice straw	28-36	23-28	12-16	15-20
Oil palm empty bunches	36-42	25-27	15-17	0.7-6
Baggasse	32-44	27-32	19-24	1.5-5
Bamboo	26-43	15-26	21-31	1,7-5
Hardwood	40-45	7-14	26-34	1
Softwood	38-49	19-20	23-30	1

3 Pretreatment Method

3.1 Chemical Pretreatment

Chemical pretreatment refers to the use of chemicals, such as acids, bases, and ionic liquids to alter the physical and chemical characteristics of lignocellulosic biomass [10]. Despite many chemical pretreatment methods, only a few can be applied to the production of biogas in the anaerobic digestion process. The search of prior research with chemical pretreatment for biogas production is summarized in Table 2.

3.2 Biological Pretreatment

Microorganisms can be used for lignocellulosic pretreatment and increase enzymatic hydrolysis. Microorganisms are usually applied to degrade lignin and hemicellulose but very few parts of cellulose, because cellulose is more resistant than other parts of lignocellulose with biological treatment. The previous research search with biological pretreatment for biogas production is summarized in Table 3.

3.3 Physical Pretreatment

Pretreatment can physically increase surface area, reduce crystallinity and degree of cellulose polymerization [8]. Various types of physical processes such as grinding (eg ball milling, hammer milling, colloid milling) and irradiation (eg by gamma rays, electron beam or microwave) can be used to improve enzymatic hydrolysis or biodegradation of lignocellulosic waste materials. The previous research search with physical pretreatment for biogas production is summarized in Table 4.

4 Solid State Anaerobic Digestion

AD can be operated at different concentrations of different solids. AD processes with TS less than 15% are classified as liquid anaerobic digestion (L-AD), which is suitable for high water content wastes such as sewage and food waste [11]. So the L-AD needs a lot of water to handle low water waste such as lignocellulosic-containing biomass. In contrast, solid state anaerobic digestion (SS-AD) operates on TS more than 15% making it more suitable for digesting lignocellulose in biomass[11,12].

https://doi.org/10.1051/e3sconf/20183102007

Table 2. Chemically Pretreatment on lignocellulosic materials for biogas production

Chemical Pretreatment	Biomass	Treatment variables	Results
Base	• Agricultural residues: wheat straw, rice straw, corn stover, sugar beet root, corn, ensiled straw, bagasse, rapeseed, sunflower, pomace grape, and empty palm oil bunches	• Chemicals: NaOH, $Ca(OH)_2$, CaO, KOH, dan $NH_3.H_2O$ • Chemical loading = 1-10% (g/g dry matter) • Temperature = -15-170 °C • Time = 1 h to 10 days	Positive effects in many cases, namely increasing the yield of methane 3.2% to 2.3 times.
Acid	• Agricultural residues: sunflower oil cakes, greenhouse residues, sugar cane bagasse, herbal-extraction process residue (HPR), sunflower stalks, coconut fiber, OPEFB, rapeseed, sunflower meals, rice straw, and bracken	• Chemicals: H_2SO_4, HCl, HNO_3, H_3PO_4, acetic acid, maleic acid • Chemical loading = 1-4% (g/g dry matter) (Organic acids can be much higher, such as acetic acid = 35-80%)temperature = ambient temperature to 170 °C • Time = several minutes to hours, even 30 days	Positive effects in many cases by increasing the 20-200% methane yield.
Catalyzed steam-explosion	• Agricultural residues: digested biofibre, wheat straw • MSW: paper tube residuals	• Chemicals: H_2SO_4, SO_2, H_3PO_4, NaOH with and without H_2O_2 • Chemical loading = 0,5-4% (g/g dry matter) • temperature = 155-220 °C • Time= 5 menit-1 jam	increase methane yield 18-107%

Table 3. Biological Pretreatment on lignocellulosic materials for biogas production

Biological Pretreatment	Microbial and enzym	Biomass	Treatment variables	Results
Fungal pretreatment	White-, brown-, and soft-rot fungi (eg Ceriporiopsis subvermispora, Auricularia auricula-judae, Trichoderma reesei), and basidiomycete fungi (eg Ischnoderma resinosum and Fomitella fraxinea)	• Agricultural residues: sweet chestnut leaves / straw and sisal leaf decortications residue (SLDR) • Wood: Japanese cedar wood chip	28-37 °C for 12 days to 8 weeks, Autoclave or non autoclave raw material prior to inoculation, aerobic conditions	Increase yield of methane 15% to 5 times
Microbial consortium	Complex microbial agents containing yeast and cellulolytic bacteria, heat treatment of sludge, Clostridium thermocellum, and mixture of fungi and compost microbes	• Agricultural residues: corn straw, corn stalks, cotton stalks, cassava residues, and dirt biofibers	20-55 °C for 12 hours-20 days, Autoclave or no autoclave raw material prior to inoculation, aerobic conditions	Increased methane yield 25-96.63%
Enzymatic pretreatment	Laccase; Pectinase; Mixed cellulase and hemisellulase; Mixed cellulase, hemisellulase and β-glucosidase; And the Trichoderma complex enzyme	• Agricultural residues: beet pulp, spent hops, and biofiber impurities • MSW: pulp and paper sludge • Grass: jose tall wheatgrass	37 °C for 4-24 hours, no sterilization before enzyme loading, aerobic conditions (if pretreatment separated from AD) or anaerobic (if enzyme is loaded directly to AD reactor)	Increases 0-34% methane yield

Table 4. Physical Pretreatment on lignocellulosic materials for biogas production

Physical Pretreatment	Biomass	Treatment variables	Results
Comminution	Agricultural residue: wheat straw, rice straw, oats, clover, bagasse, coconut husk, hemp, banana leaf, cauliflower leaves, and digested biofibers	Particle size: 0,003-30 mm	Increase methane yield up to 30%.
Steam-explosion	Agricultural residues: wheat straw, corn stalks, corn straw, orange waste, potato slurry, and digested biofibers	160-260 °C, 0,7-4,8 MPa, During several second – minutes	Positive effect by increasing biogas production by 40%
Liquid hot water (LHW)	Agricultural residues: wheat straw, rice straw, empty palm oil bunches, sunflower stems, and bagasse	100-230 °C (0,1-2,8 MPa) for several second – hours	Increase methane yield 7-220%
Extrusion	• Agricultural residues: barley straw, corn, and solid dung fractions • Grass: Lolium multiflorum and peletted straw	0.45-3.5 MPa for several minutes (eg 4-12 minutes) and typical temperature 60-90 °C	Increase methane yield 8-70%
Irradiation	• Agricultural residues: wheat straw, barley straw, spring wheat, winter wheat, wheat straw, and rice stalks • Grasses: switch grass and hybrid grass	115-300 °C for several minutes to hours	Increase methane yield up to 4-28. Combination with acid or alkali will increase biogas yield.

Compared to L-AD, SS-AD has several advantages such as higher methane volume productivity, low heating energy requirements and little waste water [11, 13]. In America the SS-AD method has been used to convert organic waste that has a high solid content into biogas. Potential raw materials for SS-AD are lignocellulosic feedstocks, for example: food crops and forest products waste. In Europe the SS-AD system was used since the 1990s to cultivate municipal solid waste [14]. Here are the parameters that affect the SS-AD system.

4.2 pH

The gas can be produced if the temperature is between 4-60 °C and the temperature is kept constant. The higher the reaction temperature will also be faster but the bacteria will decrease. Some bacteria can survive over a certain temperature range.

Temperature affects the anaerobic digestion process of organic matter and gas production. The digestion takes place either at a temperature of 30-40 ° C for mesophilic conditions and at a temperature of 45-60 ° C for thermophilic conditions. The speed of fermentation decreases at temperatures below 20 ° C. The optimal temperature of most mesophilic bacteria is achieved at 35 ° C, but for thermophilic bacteria at 55 °C [15].

Generally SS-AD can be operated under mesophilic conditions with temperatures between 20-45 ° C or thermophilic conditions at temperatures of 45-65 °C [16]. At low temperatures, microorganisms in the fermentation process will produce fatty acids and alcohols because the methanogenic microorganisms are deactivated and thus cannot digest all the fermentation products that are formed. This results in a decrease in pH and the biogas production process stops [17]. While at high temperatures result in decreased biogas yield due to the formation of volatile gases such as ammonia which inhibits methanogenic activity [18, 19]

Comparison of thermophilic temperature with mesophilic is the activity of microorganism that produced 25-50% higher so as to produce more biogas yield. However, microorganisms in thermophilic conditions are more sensitive to temperature disturbances whereas at mesophilic temperatures microorganisms are more resistant to changes in environmental conditions [20]. Another disadvantage to thermophilic conditions is that it requires higher maintenance costs and less stable thermophilic digesters [21].

4.1 Temperature

The pH range of 7-7.5 is a neutral condition favored by most microorganisms. In the biogas process there are different types of microorganisms and pH for optimal growth of each microorganism [17]. The pH between 6.8-7.6 is the optimal condition for methanogenic bacteria and at pH below 6.6 the growth of methanogen bacteria will decrease, while the bacteria hydrolysis and acetogenesis can work at pH 5.5-6.6 [22]. If the pH value falls below 6.5, then organic acid production by hydrolytic bacteria can lead to a more drastic decrease in pH and the fermentation process will stop [23]. The addition of sodium hydroxide or sodium bicarbonate can control the decrease in pH value [24]. If a pH value above 8.5 would inhibit the digestion process [25]. Methanogenic bacteria are bacteria sensitive to acidic conditions and bacterial growth may be inhibited under acidic conditions [26]. Based on these conditions, it was determined that the optimal pH of anaerobic digestion was between 6.8-8.0 [21].

4.3 Alkalinity

In the anaerobic system, there are two large groups of working microorganisms are acid-forming bacteria and methane forming bacteria. Both of these bacteria have very different in duplicate capabilities that is 3 hours for acid-forming bacteria and 3 days for bacteria forming methane. On the other hand, methane-forming bacteria are very sensitive to acidity or very sensitive with low pH. This is very contradictory because at high feed load conditions tend to accelerate the occurrence of acid formation, thus lowering the pH. This decrease in pH will disrupt the performance of methane-forming bacteria that have not yet developed [3].

Alkalinity is the number of base amounts in the reactor to neutralize the acidic conditions in the digester so that the pH of the digester is stable [20]. The presence of alkalinity in a reactor with a certain concentration can be a buffer (Buffer) to keep the pH in neutral conditions in the event of acid addition, so that the overall process balance can still run normally. When the pH is allowed to fall then the system will stop altogether because the methane-forming bacteria will die [3].

4.3 Alkalinity

In the anaerobic system, there are two large groups of working microorganisms are acid-forming bacteria and methane forming bacteria. Both of these bacteria have very different in duplicate capabilities that is 3 hours for acid-forming bacteria and 3 days for bacteria forming methane. On the other hand, methane-forming bacteria are very sensitive to acidity or very sensitive with low pH. This is very contradictory because at high feed load conditions tend to accelerate the occurrence of acid formation, thus lowering the pH. This decrease in pH will disrupt the performance of methane-forming bacteria that have not yet developed [3].

Alkalinity is the number of base amounts in the reactor to neutralize the acidic conditions in the digester so that the pH of the digester is stable [20]. The presence of alkalinity in a reactor with a certain concentration can be a buffer (Buffer) to keep the pH in neutral conditions in the event of acid addition, so that the overall process balance can still run normally. When the pH is allowed to fall then the system will stop altogether because the methane-forming bacteria will die [3].

4.4 C/N Ratio

Generally, 2/3 of carbon is regenerated into CO2 and the other 1/3 is combined with nitrogen inside the cell. Microorganisms use N, P and other nutrients to make cell components but reduce organic N to organic acid and ammonia. Carbon from organic material not used in protein cells will be released especially to produce methane gas and less CO2 [27]. Nitrogen in the substrate has several advantages including providing essential elements for amino acid synthesis, neutralizing volatile acids produced by fermentation bacteria and also helping to attain neutral pH. The amount of nitrogen should be within sufficient limits as a nutrient. While the optimum C/N ratio for methane production is 20-35 [28].

The activity of microorganisms that play a role during the fermentation process depends on the C/N ratio. In anaerobic digestion the optimal C/N ratio is about 20-30 [11]. A high C/N ratio in organic matter will result in lower methane production. This is because the material with high C/N contains low levels of nitrogen. It causes rapid nitrogen consumption by methanogenic bacteria and results in biogas yields down due to the lack of nitrogen available for bacterial growth [8]. Therefore, to increase the level of nitrogen in the material can be done by adding organic materials containing high nitrogen such as livestock manure.

Meanwhile, if a very low C/N ratio causes ammonia to accumulate and may raise the pH of the digester by 8.5 which is toxic to methanogenic bacteria [8]. Therefore, it is necessary to add other materials containing carbon or high fiber, such as grass, straw, husk and foliage. How to mix organic material with carbon-containing material to achieve the ratio of C/N is determined. Here table C/N ratio contained in some organic materials.

Table 5. C/N Ratio some Biomass [4]

Biomass	C/N Ratio
Man manure	8
Duck dung	8
Chicken dung	10
Goat dung	12
Pig dung	18
Sheep dung	19
Buffalo/Cattle dung	24
water hyacinth	25
Elephant dung	43
Corn Straw	60
Rice Straw	70
Wheat Straw	90
Saw dust	> 200

4.5 TS Content

Water content is known as one of the important parameters that affect the anaerobic digestion process. Therefore, the total solid content is used to define two types of anaerobic digestion processes: wet digestion for TS <15% and dry digestion for TS> 15% [29]. The total solid content affects the pH value and the effectiveness of microorganisms in the digestion process [30]. The total solid concentration in the SS-AD system is about 20-30%. The total solid content (TS) of more than 30% can reduce biogas production by 17% due to the accumulation of organic acids and can decrease the substrate conversion [29, 31].

4.6 Feedstock and Inoculum

The substrate (feedstock) is converted to methane by anaerobic bacteria. Bacterial growth in the anaerobic digestion process can be accomplished by the addition of an inoculum or nutrient. This is because microorganisms need energy source to move on biogas production process. Nutrients required are macronutrients such as carbon, nitrogen, hydrogen, phosphorus, potassium and sulfur as well as micronutrients such as iron, copper, nickel, molybdenum, tungsten, cobalt and zinc [20].

Feedstock ratio to inoculum (F/I) is a parameter affecting SS-AD system in both batch and continuous process. In industrial SS-AD systems the F/I ratio is very low to avoid the risk of failure and maximize reaction kinetics [20]. The ratio of F/I used in the SS-AD system ranges from 2-6 [13].

5 Conclusions

Biogas production research from rice husk waste by SSAD method has been done by several researchers in laboratory scale. In the several study, an effort to increase biogas production were conducted the preliminary treatment of physics, chemistry and biology for the process of delignification. The activity aims to accelerate the fermentation process in the reactor by microorganisms. SSAD method is very suitable for agricultural waste, especially rice husk in terms of utilization as biogas because TS content is high so it does not require much water. According to some researchers, SSAD is able to produce more biogas than LAD. Through this overview, further research is needed on biogas from rice husk waste by SSAD method on a larger scale such as bench and pilot scale with continuous system.

References

1. Budiyono, I. Syaichurrozi, & Sumardiono, S. Waste Tech, **1-5** (2013)
2. A. Saedi, Rutz, D. Prassl, H. Kottner, M. Finsterwalder, T. Volk, S. et al. Biogas. Denmark (2008)
3. D. Padmono, *Kemampuan Alkalinitas Kapasitas Penyangga (Buffer Capacity) Dalam Sistem Anaerobik Fixed Bed*, **8(2)**, 119-127 (2007)
4. B. R. Saragih, Magister Teknik Elektro, Fakultas Teknik Universitas Indonesia (2010)
5. A.T.W.M. Hendriks, G. Zeeman, Bioresource Technology, **100**, 10–18 (2009)

https://doi.org/10.1051/e3sconf/20183102007

6. L. Appels, J. Lauwers, J. Degrève, L. Helsen, B. Lievens, K. Willems, Renew Sustain Energy Rev, **15**, 4295-301 (2011)
7. N. Khorshidi, B. Arikan, Applied Biotechnology 181-300 ECTS credits No. 3 (2008)
8. M.J. Taherzadeh, K. Karimi, International Journal of Molecular Sciences, **9**, 1621–1651 (2008)
9. M. Mulder, Basic Principles of Membrane Technology. 2nd Edition. Dordrecht Kluwer Academic (1996)
10. Y. Zheng, J. Zhao, F. Xu, Y. Li, Progress in Energy and Combustion Science pp. 35-53 (2014)
11. Y. Li, S.Y. Park, J. Zhu, Renewable and Sustainable Energy Reviews, **15(1)**, 821–826 (2011a)
12. L. Yang, Y. Li, Bioresource Technology, **171**, 233–239 (2014)
13. F. Xu, J. Shi, W. Lv, Z. Yu, Y. Li, Waste Management, **33(1)**, 26–32 (2013)
14. Y. Li, J. Zhu, C. Wan, S.Y. Park, American Society of Agricultural and Biological Engineers, **54(4)**, 1415–1421 (2011b)
15. S. Sahirman, *Kajian Pemanfaatan Limbah Cair Pabrik Kelapa Sawit untuk Memproduksi Gasbio*. Program Pascasarjana IPB : Bogor (1994)
16. L. Yu, P.C. Wensel, J. Ma, S. Chen, Bioremediation & Biodegradation (2013)
17. A. Schnürer, Å. Jarvis, Swedish Waste Management and Swedish Gas Centre (2010)
18. B. Fezzani, C.R. Ben, Bioresource Technology, **101(6)**, 1628–34 (2010)
19. A. Khalid, M. Arshad, M. Anjum, T. Mahmood, L. Dawson, Waste Management, **31(8)**, 1737–44 (2011)
20. A. Teghammar, Chalmers University of Technology (2013)
21. J.C. Akwaka, D.T. Kukwa, S.S. Mwekave, International Journal of Science and Technology, **3(4)**, 222–228 (2014)
22. A.K. Jha, J. Li, L. Nies, L. Zhang, African Journal of Biotechnology, **10(65)**, 14242–14253 (2011)
23. D. Deublein, A. Steinhauser, Wiley-VCH Verlag GmbH dan Co. KGaA, Weinheim (2008)
24. P. Weiland, Appl Microbiol Biotechnol, **85**, 849–860 (2010)
25. Zeshan. Asian Institute of Technology, School of Environment, Resources and Development (2012)
26. K.M. Kangle, S.V. Kore, V.S. Kore and Kulkarni, G.S. Universal Journal of Environmental Research and Technology, **2(4)**, 210–219 (2012)
27. Gotaas, B. Harold, Sanitary Disposal and Reclamation of Organic Wastes (1956)
28. D. Enri. Jurnal Ilmiah dalam Seminar PAU Bioteknologi ITB (1990)
29. J.-C. Motte, R. Escudié, N. Bernet, J.-P. Delgenes, J.-P. Steyer, C. Dumas, Bioresource Technology, **144**, 141–148 (2013)
30. N. Boontian, International Journal of Biological, Veterinary, Agricultural and Food Engineering, **66(9)**, 960–964 (2014)
31. A. Guendouz, A. Trably, E. Hamelin, J. Dumas, C. Steyer, J.P. Delgenès, J. Escudié, R. Bioresource Technology, **133**, 256–262 (2013)

The Effect of COD Concentration Containing Leaves Litter, Canteen and Composite Waste to the Performance of Solid Phase Microbial Fuel Cell (SMFC)

Ganjar Samudro[1,*], *Syafrudin*[1], *Winardi* Dwi Nugraha[1], *Endro* Sutrisno[1], *Ika* Bagus Priyambada[1], *Hilma* Muthi'ah[1], *Glory* Natalia Sinaga[1], and *Rahmat* Tubagus Hakiem[1]

[1] Department of Environmental Engineering, Faculty of Engineering, Diponegoro University, Semarang - Indonesia

Abstract. This research is conducted to analyze and determine the optimum of COD concentration containing leaves litter, canteen and composite waste to power density and COD removal efficiency as the indicator of SMFC performance. COD as the one of organic matter parameters perform as substrate, nutrient and dominating the whole process of SMFC. Leaves litter and canteen based food waste were obtained from TPST UNDIP in Semarang and treated in SMFC reactor. Its reactor was designed 2 liter volume and equipped by homemade graphene electrodes that were utilized at the surface of organic waste as cathode and in a half of reactor height as anode. COD concentration was initially characterized and became variations of initial COD concentration. Waste volume was maintained 2/3 of volume of reactor. Bacteria sources as the important process factor in SMFC were obtained from river sediment which contain bacteroides and exoelectrogenic bacteria. Temperature and pH were not maintained while power density and COD concentration were periodically observed and measured during 44 days. The results showed that power density up to 4 mW/m^2 and COD removal efficiency performance up to 70% were reached by leaves litter, canteen and composite waste at days 11 up to days 44 days. Leaves litter contain 16,567 mg COD/l providing higher COD removal efficiency reached approximately 87.67%, more stable power density reached approximately 4.71 mW/m^2, and faster optimum time in the third day than canteen based food waste and composite waste. High COD removal efficiency has not yet resulted in high power density.

1 Introduction

SMFC has become an alternative technology of solid waste treatment converting solid phase material to eco-friendly product and directly electrical energy. In recent years, the application of SMFCs have attracted the highest attention on research although the value of current output from freshwater SMFCs was not higher than MFCs [1]. Various factors cover material type, size and volume, nutrient, water content, electrode type and configuration, design configuration, electricity circuit, pH, and temperature are affect the performance of SMFC. The problems were found where COD concentration containing material type affect the COD removal efficiency and electrical production. Buffiere et al. [2] stated that COD measurement is a very important to know the value of biodegradability from waste to produce energy. Output electrical power depend on complex organic type in waste which is hard degradable so it restrict electrical production [3]. High COD concentration means high COD removal efficiency and electrical production but it was the problem. Higher organic content, higher COD removal efficiency [4,5]. But many research found that high COD concentration was not necessarily producing high electricity [5-7. According to Song et al. [1] that he suggested the biomass addition with appropriate proportion could increase the contact opportunity between matrix (anode and the biomass addition), increase the organic waste content and increase the cellulose activities, then as an important method to increase power output in SMFCs. He also stated that due to high internal resistance and lower oxidation rate changes, where generally lower output power in SMFCs than MFCs, so SMFCs could be operated on higher external resistance to increase power output. Leaves litter, canteen and the composite of leaves litter-canteen have known containing high COD concentration that can be utilized as substrate in SMFC process. COD concentration is used as a main part of Dissolved Organic Matter (DOM) or Biological Oxygen Demand (BOD) that means organic biodegradable. High COD concentration means high DOM or BOD, removal efficiency, and electrical production. Because solid phase materials contain high COD concentration, the correlation of it is made to show the new observed contribution of high COD concentration in SMFC. The aim of this study is to analyze and determine the optimum of COD concentration containing leaves litter, canteen and composite waste to power density and COD

* Corresponding author: ganjarsamudro@gmail.com

https://doi.org/10.1051/e3sconf/20183102008

removal efficiency as an indicator of SMFC performance.

2 Methods

Leaves litter, canteen and composite waste used in this study were collected from integrated solid waste treatment facility or it was called "TPST UNDIP" located on Diponegoro University in Indonesia. Waste samples were collected using Indonesia National Standard number 19-3964-1994 [8] about methods of sampling and measurement of generation and composition of urban waste. The collected wastes were identified as follows C-organic, N-Total, P-Total, C/N Ratio, pH, temperature, water content, and COD concentration using standard method analysis. River sediment as bacteria source were taken from Tanah Mas Semarang in Indonesia.

2.1. Reactor Design

SMFC has one compartment consists of catode and anode which were connected by electricity circuit and the distance between catode and anode is a half of reactor height. Its reactor was designed by 2 liter volume. Waste volume was maintained 2/3 of reactor volume along with water tap remaining to maintain moisture content and to enhance anolyte (substrate) conductivity [9]. del Campo [7] stated that 2/3 of waste volume was the optimum condition for metabolism bacteria. Sampling point was placed under the reactor which equipped by valve. Configuration of SMFC reactor is described in Figure 1 as follows:

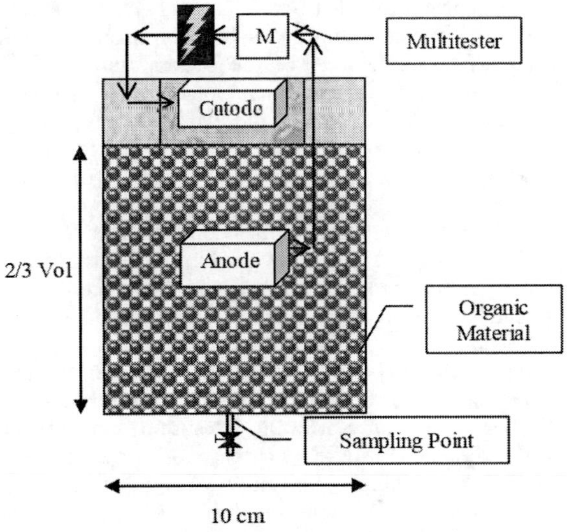

Fig. 1. SMFC Reactor Configuration.

2.2 Tools and Materials

The materials needed for this research consist of river sediment bacteria, chopped leaves litter 1.5 cm, fresh canteen based food waste, aquadest, mechanical pencil 2B, polyurethane, phosphate acid 85%, HCl 0.1 N, NaOH 0.1 N, wood glue, and sulphate acid 96% PA. The tools needed for this research consists of spectrophotometer, high digestion COD, low digestion COD, multimeter digital, pH meter, and thermometer.

2.3 Research Procedure

Research steps cover preparing step which were reactor construction and characterization test of leaves litter, canteen and composite waste, and also bacteria identification of river sediment that were used. Second step was main research which is started from seeding process to batch-running process. Its step was conducted during 44 days. At a certain period, sampling was conducted 15 times with the data taken cover COD in-out, voltage, ampere, resistance, pH, and temperature.

Research analysis covered COD removal efficiency and power density. In order to measure the COD, the samples were first diluted 10 times. Then 2 ml of the diluted samples were mixed with a digestion solution of a high-range COD reagent before heated at 150°C for 2 h in a thermo reactor which was analyzed using a spectrophotometer. The performance of COD removal efficiency in SMFC was calculated by the COD removal using the following equation: $\eta = \frac{C_{inlet} - C_{outlet}}{C_{inlet}} x\ 100\%$; where η is the percentage of BOD and COD removal, C_{inlet} is initial BOD and COD concentration (mg/L) and C_{outlet} is BOD and COD concentration in certain of time. The power density was calculated using the following equation: P = I x R; where, R is the applied external resistance (ohm) and I is the current (amps). The current was measured using the following equation: I = V/R; where, I is the current (amps), V is the voltage (volt), and R is the applied external resistance (ohm) [10]. Approximation of power density formulation can be followed equation: $P = \frac{V x I}{A}$; where A (m^2) is surface area of graphene anode. The voltage was measured using a digital multimeter, and the power density curve was obtained.

3 Result and Discussion

Initial characteristics of leaves litter, canteen and composite waste showed that canteen waste has higher COD concentration than leaves litter and composite wastes. Food based waste generated from canteen operations has highly biodegradable solid waste [9]. The following is presented in Table 1.

Table 1. Initial Characteristics of Leaves Litter, Canteen and Composite waste.

Parameter	Unit	Value					
		LL		Cant.		Comp.	
C-Organic	%	49.46	48.40	71.60	46.11	89.61	16.90-19.60
N-Total	%	1.56	1.24	0.94	3.19	1.02	1.30
P-Total	%	0.02		0.01	10.40	0.01	
C/N Ratio		31.70	39.03	76.26	20-26	87.85	14-36.14
pH		6.30		7.37	6-6.5	7.37	6.0-8.0
Temperature	°C	26.20		25.90		25.9	
Water Content	%	14.12	59.90	70.00	70.00	45.9	40.00-52.90
COD	mg/l	16,567		14,900	330,000-380,000	19,567	

Note: LL = Leaves Litter; Cant. = Canteen; Comp. = Composite

Based on the Table 1 that higher C-organic percentage and C/N ratio are not directly proportional with COD concentration. Water content affect COD concentration of all three materials which depend on C-organic percentage. Composite waste has particulary different result on COD concentration because of mixed materials between leaves litter and canteen waste. It can be said that higher C-organic percentage and C/N ratio with certain water content of composite waste provide different COD concentration result compared towards leaves litter and canteen waste. Because of this research based on an ideal condition so that the value of water content is avoided.

Only C-organic test result of two materials in accordance with the result of C-organic canteen waste reached above 46.11% [11], and C-organic composite waste reached in range 16.90-19.60% [12], except C-organic leaves litter test result reached below 48.4% [13]. N-Total and P-Total are not provide the same result that leaves litter reached 1.24%N-Total [13], canteen waste reached 3.19%N-Total and 10.4%P-Total [11], and composite waste reached 1.3%N-Total [12]. C/N ratio was also not provide the same result. pH and temperature are the physical standard parameter that were not significantly different or provide the same result. Water content of leaves litter result test was not provide the same result reached 59.90% [13], but water content of canteen and composite waste were provide the same result reached 70.00% [11,15] and reached 40.00-52.9% [12], respectively. COD concentration of canteen waste was only tested and reached 330,000-380,000 mg COD/l [14-15]. Meanwhile COD concentration of leaves litter and composite waste was not yet tested and experimented in SMFC.

Due to characterization of bacteria from natural river sediment is very important that become the main support on whole process of SMFC, then analysis of the existence of bacteria was obtained. The results were founded Escherichia coli as dominant and existence species in river sediment. The existence of bacteria in river sediment are identified in Proteobacteria phylum Escherichia, Salmonella, Desulfovibrio, Desulfobacteria, Acidovorax genus [16-19], Nitobacteria genus [20], and a half of Acidobacteria and Acinobacteria phylum [17], Nitrospirae phylum Nitrospira genus [17], Bacteroides phylum Flavobacterium genus [18], Firmicutes phylum Clostridium genus [19]. Proteobacteria phylum is a gram-negative type bacteria and facultative anaerobic which means that bacteria could live in available oxygen or not. This is accordance to running process on anaerobic SMFC. In the performance system of SMFC, Escherichia coli has a role to reduce glucose contained waste which was needed for reduction reaction with oxygen [21]. Proteobacteria phylum have the highest proportion to produce electricity [22]. According to research of Zhang et al. [23] that power density reached 100-600 mW/m^2 using Escherichia coli.

Mitov et al. [19] compared the performance of SMFC using river sediment and soil that river sediment showed the best performance of SMFC because of more organic substances and also the existence of exoelectrogenic bacteria with specific capability. Xing et al. [24] also stated that the river sediment has exoelectrogenic and fermentation function which known have good COD removal efficiency and electricity production.

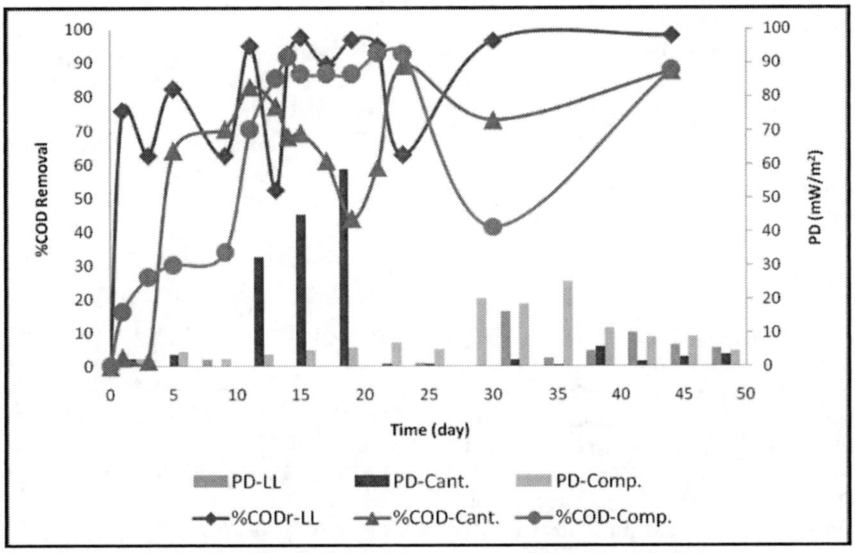

Fig. 2. Correlation of Waste Type to The Performance of SMFC.

An ideal condition that leaves litter, canteen and composite waste are adjusted by certain %C, %N-Total, %P-Total, C/N Ratio, pH, temperature, water content, and COD concentration provides the different results. Leaves litter contain 16,567 mg COD/l providing higher COD removal efficiency, more stable power density, and faster optimum time than 14,900 mg COD/l containing canteen based food waste and 19,567 mg COD/l containing composite waste as follows in Figure 2. Other reason is a statistical measure of how close the data are to the fitted regression or logarithmic or exponential line that means R-squared of percentage of COD efficiency, showed higher value of R-squared in canteen waste, composite waste and leaves litter, respectively. Based on scoring method, leaves litter also provide higher score than canteen and composite waste. Three statements based on the performance of SMFC result, R-squared, and the scoring method provide multiple argument, so the appropriate statement of leaves litter is selected as the optimum material. All of the statements can be stated that the optimum range of COD concentrations is 16,567 mg COD/l to 19,567 mg COD/l. This information opens up opportunities that the other of optimum range of COD concentration between 16,567 mg COD/l to 19,567 mg COD/l might be founded. Correlation of COD concentration with percentage COD removal efficiency and power density are described in Figure 3 as follows:

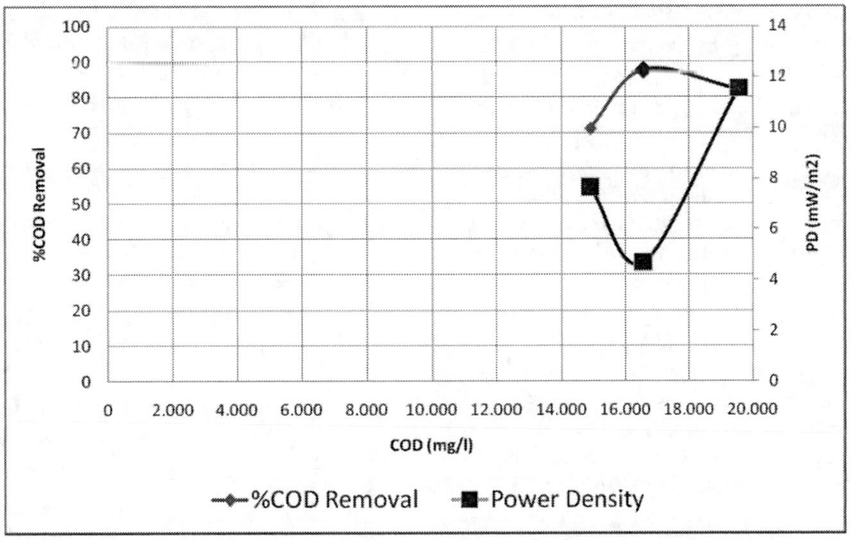

Fig. 3. Correlation of COD Concentration, COD Removal Efficiency and Power Density.

Correlation of COD concentration with percentage COD removal efficiency and power density are important that be high COD concentration has not yet high COD removal efficiency and power density. High COD concentration of 16,567 mg COD /l is an optimum COD concentration but power density performs lower value than low COD concentration of 14,900 mg COD /l and high COD concentration of 19,567 mg COD /l. It was same result with del Campo et al. [7] that the optimum result was not the highest COD concentration. It can be stated that high performance in COD removal efficiency has not yet resulted in high power density meanwhile low COD concentration performs the stability of COD removal efficiency and power density. Based on the explanation above that COD concentration between 16,567 mg COD/l to 19,567 mg COD/l might be founded is reliable. The increasing of COD concentration from 16,567 mg COD/l to 19,567 mg COD/l could increased the power density, but the COD removal efficiency were not decreased significantly.

4 Conclusion

High COD removal efficiency has not yet resulted in high power density. The stability result showed in COD concentration of 16,567 mg COD/l. The increasing of COD concentration from 16,567 mg COD/l to 19,567 mg COD/l could increase the performance of SMFC. Leaves litter contain 16,567 mg COD/l mg COD/l providing higher COD removal efficiency, more stable power density, and faster optimum time than canteen based food waste and composite waste.

The authors wish to thank the Department of Environmental Engineering, Faculty of Engineering, Diponegoro University for funding this research and helping the publication.

References

1. Song Tian-Shun, Wang De-Bin, Han Shuo, Wu Xia-yuan, C. Zhou Charles, International Journal of Hydrogen Energy **39**, 1056-1062 (2014)
2. P. Buffiere, S. Frederic, B. Marty, & J. Delgenes, *A comprehensive method for organic matter characterization in solid wastes in view of assessing their anaerobic biodegradability,* 1783–1789. http://doi.org/10.2166/wst.2008.517 (2008)
3. Y. Tian, W. Zuo, J. Zhang, X. Pan, L. Li, and X. Su, Bioresource technology, **205**, pp.104-110 (2016)
4. Y. Feng, Q. Yang, X. Wang, Y. Liu, H. Lee, N. Ren, Bioresour. Technol. **102**, 411–415 (2011)
5. GA. del Campo, J. Lobato, P. Cañizares, M. A. Rodrigo, F. J. F. Morales, *Short-term effects of temperature and COD in a microbial fuel cell.* http://doi.org/10.1016/j.apenergy.2012.02.064 (2012)
6. H. Liu, S. Cheng, BE. Logan, Environ Science & Technology, **39**(2):658–662 (2005)
7. GA. del Campo, J. F. Perez, P. Ca, M. A. Rodrigo, F. J. Fernandez, & J. Lobato, *Study of a photosynthetic MFC for energy recovery from synthetic industrial fruit juice wastewater,* 9, 0–8. http://doi.org/10.1016/j.ijhydene.2014.07.055 (2014)
8. Anonym, Indonesia National Standard number 19-3964-1994, *Methods of sampling and measurement of generation and composition of urban waste* (1994)
9. S.V. Mohan, and K. Chandrasekhar, Bioresource technology, **102**(14), pp.7077-7085 (2011)
10. J. Kim, K. Kim, H. Ye, E. Lee, C. Shin, P.L. McCarty, J. Bae, Environ. Sci. Technol. **45** (2), 576e581 (2011)
11. Y. Li, Y. Jin, J. Li, Energy, **102**, pp.139-147 (2016)
12. Lopez-Real, J. M. and Vere, A., Composting control parameters and compost product characteristics. In D. V. Jackson, J. M. Merillot and P. L'Hermite (eds.) Composting and compost quality assurance criteria (1992)
13. S. S. Hla, and D. Roberts, Waste Management, **41**, 12–19. (2015)
14. S.V. Mohan, K. Amulya, dan, K. Chandrasekhar, Waste Management, **45**, pp.57 (2015)
15. S.V. Mohan, R.K. Goud, P.S. Babu, International journal of hydrogen energy, **36**(10), pp.6210-6218 (2011)
16. Q. Wu, R. Zhang, S. Huang, H. Zhang, China:Journal Of Environmental Science, Volume **20**, Pages 404-412 (2008)
17. Y. Wu, K. Guan, Z. Wang, B. Xu, F. Zhao, PloS one 8:e73442 (2013)
18. G. Sun, A. Thygesen, A. S. Meyer, Appl. Microbiol, Biotechnol. **99** 4905–4915. 10.1007/s00253-015-6513-5 (2015)
19. M. Mitov, I. Bardarov, P. Mandjukov, Y. Hubenova, *Chemometrical assessment of the electrical parameters obtained by long-term operating freshwater sediment microbial fuel cells,* Bioelectrochemistry (2015)
20. H.L. Jiang, Y.L. Zhou, H.F. Wu, Z.S. Yan, H.Y. Cai, Ecological Engineering, **87**, pp.254-262 (2015)
21. KA. Kennedy, EG. Gachelet, B. Traxler, Journal of Biological Chemistry **279** (32), 33290-33297 (2004)
22. Fitrinaldi, *Microbial Fuel Cell Sebagai Energi Alternatif Menggunakan Bakteri Escherichia Coli.* Padang: Universitas Andalas (2011)
23. T. Zhang, C. Cui, S. Chen, H. Yang, P. Shen, *Electrochem.* Commun., 2008, 10(2): 293–297 (2008)
24. D. Xing, J. Jia, Y. Tang, B. Liu, D. Wu, N. Ren, Bioresource technology, **144**, pp.94-99 (2013)

Roles of K_2O on the CaO-ZnO Catalyst and Its Influence on Catalyst Basicity for Biodiesel Production

Luqman Buchori, *I.* Istadi*, *P.* Purwanto, *Louis Claudia* Marpaung, and *Rahmatika Luthfiani* Safitri

Department of Chemical Engineering, Faculty of Engineering, Diponegoro University, Semarang - Indonesia

Abstract. This research aimed to study the effect of K_2O impregnation on the basicity of the CaO-ZnO catalyst and its effect on biodiesel production. The effect of mole ratio of CaO to ZnO catalyst and %wt K_2O were also studied. The mole ratio of CaO to ZnO catalyst was varied at 1:1, 1:1.5, 1:2, 1:3, and 3:1, while the %wt K_2O was varied at 1, 3, and 5 %. The catalyst basicity was determined by titration method. The basicity of the catalyst increased after the CaO-ZnO catalyst was impregnated with K_2O in all mole ratios of CaO-ZnO catalyst. The addition of K_2O as a promoter also increase the basicity. The highest basicity was obtained at the CaO-ZnO mole ratio of 3:1 and 5%wt K_2O. The tranesterification process was carried out in a batch reactor at a methanol to oil mole ratio of 15:1, a reaction temperature of 60°C, a reaction time of 4 h, and a catalyst loading of 5%wt oil. The FAME yields obtained were 41.33%. These results proved that K_2O plays a role in enhancing the catalyst basicity. In addition, K_2O also serves as a binding agent to improve the mechanical properties of the catalyst.

1 Introduction

Biodiesel is an alternative fuel produced from vegetable oils or animal fats. Biodiesel consists of fatty acid methyl ester (FAME) which is a free fatty acid from vegetable oils and animal fat [1]. Biodiesel has lower emissions than diesel oil [2]. Biodiesel can be produced from sources of free fatty acids (feedstock) such as palm kernel oil, sunflower seed oil [3], soybean oil [4], rapeseed oil [5] and some other oils.

Biodiesel can be produced by transesterification using the homogeneous catalyst (acid or base), heterogeneous (solid catalyst) or enzyme [6]. Transesterification using solid heterogeneous catalysts has several advantages such as easier separation and purification processes due to different phases of the product, no water in the neutralization process, non-toxic, non-corrosive, low-cost, and easily regenerated [7–9]. Solid-base heterogeneous catalysts have higher effectiveness than acid and enzyme catalysts. This is due to the reaction rate of biodiesel production using heterogeneous catalyst is faster than acid catalyst [10]. However, the basic heterogeneous catalysts have some weaknesses such as lower yield, less activity and catalytic selectivity [11], leaching or dissolving the Ca active component in methanol [12, 13].

One of the catalysts developed is a metal oxide [2, 7, 14]. Potentially oxide base catalysts for the production of FAME include ZrO_2, TiO_2, ZnO, CaO, and SrO [15]. The higher the basicity of the catalyst, the higher the yield of FAME. The catalytic activity of the catalyst can be increased with the support of the catalyst [16]. Alkaline earth metal oxide is one of the catalysts used as a support, including ZnO, MgO, and BaO.

Istadi *et al.* [17] have prepared a basic solid catalyst with CaO as an active component. The catalyst is combined with ZnO by co-precipitation to increase the catalyst surface area and promoted with potassium oxide (K_2O) by impregnation to enhance its basicity. The results show that ZnO and CaO-ZnO catalysts have lower catalytic activity than K_2O/CaO-ZnO catalysts. However, in this research has not been an investigation to determine the effect of CaO, ZnO, and K_2O composition on catalyst activity and its effect on catalyst basicity. The mole ratio of CaO and ZnO on the catalyst and K_2O impregnation are thought to affect the crystal structure which results in the catalyst basicity so as to influence the yield of obtained biodiesel.

This research focused on studying the role of K_2O in CaO-ZnO catalyst at various mole ratios of CaO:ZnO and the effect of K_2O composition in K_2O/CaO-ZnO catalyst on the catalyst basicity and its influence in the transesterification process of soybean oil into biodiesel.

2 Materials and method

2.1 Materials

As the raw material in this research were soybean oil and methanol (99.9% Merck). Soybean oil was purchased from the local market. The chemical for the preparation of the catalyst comprises calcium nitrate tetrahydrate ($Ca(NO_3)_2.4H_2O$) (Merck, 99%), zinc nitrate hexahydrate ($Zn(NO_3)_2.6H_2O$) (Merck, 98,5%), potassium nitrate (KNO_3) (Merck, 99%), sodium

* Corresponding author: istadi@live.undip.ac.id

carbonate (Na_2CO_3) (Merck, 99%), and sodium hydroxide (NaOH) (Merck, 99%).

2.2 Preparation of K_2O/CaO-ZnO catalyst

The preparation of the catalyst was carried out by co-precipitation method followed by impregnation. CaO-ZnO catalysts with mole ratios of CaO:ZnO 1:1, 1:1.5, 1:2, 1:3, and 3:1 were prepared by dissolving calcium nitrate ($Ca(NO_3)_2$) 1 M and zinc nitrate ($Zn(NO_3)_2$) 1 M into the distillation water. Then, these solutions were mixed inside the beaker glass. Furthermore, 2 M Na_2CO_3 solution was drop wise into the mixed solution with constant stirring to form a white suspension. pH was adjusted by adding 1 M NaOH solution. Stirring was done for 24 hours at a temperature of 60°C. The suspension was then filtered and washed with distillation water until it was alkaline free. The resulting solid was dried in the oven at a temperature of 110°C overnight. The result was then calcined in box furnace (Ney Vulcan 3-550) at a temperature of 800°C for 3 hours resulting in a CaO-ZnO catalyst. The CaO-ZnO catalyst was impregnated with 1 M KNO_3 solution while stirred and then dried in the oven at 110°C overnight. Furthermore, dry solids were calcined at 300°C for 5 hours in box furnace (Ney Vulcan). The obtained results were K_2O/CaO-ZnO catalyst.

2.3 Testing of the catalyst basicity

The catalyst basicity was determined using titration method as conducted by Tanabe and Yamaguchi [18]. Half grams of catalyst were dissolved in 20 ml of benzene then dropped 3 drops of bromothymol blue indicator. The mixture was stirred for 30 minutes to form a suspension. The catalyst sample suspension was changed from yellow dye to green-blue color. The catalyst suspension was then titrated with benzoic acid while stirring at a fixed rate. The green color of the solid particle gradually disappeared. Titration was stopped when all the green color disappeared. Basicity could be calculated by equation (1)

$$Basicity\ \left(mmol\ .g^{-1}\right) = \frac{(VxN\ )_{benzoic\ acid}}{W_{catalyst}} x100\ \% \qquad (1)$$

2.4 Testing of catalysts on the biodiesel production

The experimental apparatus for the transesterification process of soybean oil over a 5%K_2O/CaO-ZnO catalyst was shown in Figure 1.

Methanol and oils with a mole ratio of 15:1 and a catalyst of 5% by weight of oil were heated to 60°C and kept constant for 4 hours. The transesterification process takes place inside this three neck flask. After the transesterification reaction was complete, the three neck flask was cooled. The resulting product was then inserted in a separating funnel and held for 24 hours to form 3 layers consisting of a top layer (methanol), a middle layer (methyl ester), and a bottom layer (glycerol and catalyst). The middle layer was taken and analyzed using Gas Chromatography-Mass Spectrometry (GC-MS).

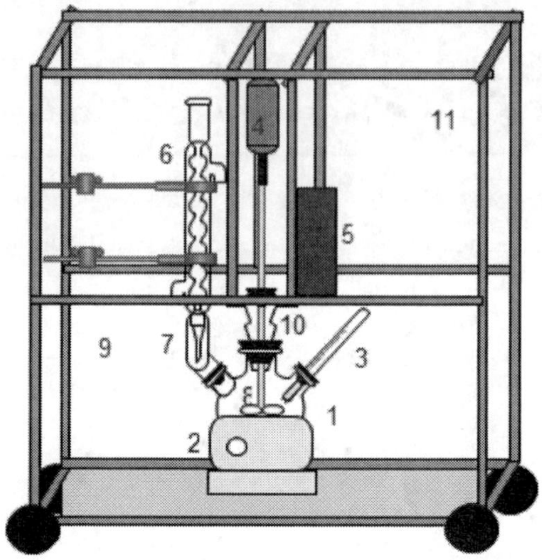

Fig. 1. The experimental apparatus for the transesterification process: (1) three-neck flask, (2) heating mantle, (3) thermometer, (4) driver motor, (5) controller motor, (6) condenser reflux, (7) elbow, (8) stirrer, (9) stative and clamps, (10) shell, and (11) portable frame

3 RESULTS AND DISCUSSION

3.1 The effect of mole ratio of CaO:ZnO on the catalyst basicity

The catalyst basicity was performed by titration method using benzoic acid and benzene solvent with bromothymol blue (BTB) as an indicator. The results of the catalyst basicity before and after impregnation showed in Table 1.

Table 1. The catalyst basicity before and after impregnation with 5%K_2O

Mole ratio of CaO:ZnO	Catalyst basicity (mmol/g)	
	Before impregnation	After impregnation
1:1	0.344	0.504
1:1.5	0.260	0.316
1:2	0.164	0.280
1:3	0.088	0.230
3:1	0.662	1.090

Table 1 shows that the larger mole of CaO in the CaO-ZnO catalyst leads to an increase in the catalyst basicity. This is due to the calcium oxide (CaO) has higher basic strength than ZnO [12]. In the periodic table, the Ca atoms are to the left of the Zn atom so that the radius of the Ca atom is larger and its base strength is also higher than the Zn atom. As a result, the greater the mole ratio of CaO in CaO-ZnO mixed metal oxide, catalyst basicity would increase.

Based on Table 1, the increase of ZnO mole in a CaO-ZnO catalyst would decrease the basicity. This is due to ZnO is a buffer/support component in the catalyst. ZnO plays a role in increasing surface area, porosity, mechanical properties, and catalyst stability, while the increase of basicity is the role of an active component of catalyst that is CaO [19]. Thus, the increase in CaO mole in the catalyst can improve alkalinity, however, the addition of ZnO leads to a decrease in the catalyst basicity. Table 1 shows that the catalyst with the mole ratio of CaO:ZnO 3:1 has the highest basicity, while the catalyst with the mole ratio of CaO:ZnO 1:3 has the lowest base.

3.2 Effect of K_2O impregnation on catalyst basicity

In this study, impregnation of the CaO-ZnO catalysts with K_2O was also investigated. The impregnation was performed by varying %wt K_2O. This study aims to determine the effect of the addition of K_2O to the catalyst basicity. Variation of % wt K_2O was performed on the CaO-ZnO catalyst with the mole ratio of CaO:ZnO 3:1 because previously the highest basicity was achieved in the CaO:ZnO mole ratio of 3:1. The K_2O impregnation on the CaO-ZnO catalyst is varied from 1, 3, and 5%. The catalyst basicity was determined by titration method using benzoic acid and benzene solvent with bromothymol blue (BTB) as an indicator [18]. The result of catalyst base test with the variation of wt% K_2O is presented in Table 2.

Table 2. The basicity catalyst of CaO(3)-ZnO(1) with variable of %wt K_2O

%wt K_2O	Basicity (mmol/ g)
1%	0.48
3%	0.856
5%	1.09

Table 2 shows that the greater the K_2O component in the catalyst, the catalyst basicity will increase. The K_2O component has a primary function as a binding agent which aims to improve the mechanical properties of the catalyst. K_2O as a binder serves to bind the catalyst components of the CaO-ZnO to be made as a pellet catalyst and also increase the activity, selectivity and desired stability effect [13, 17]. Table 2 also shows that the higher the content of K_2O in the catalyst, the higher the catalyst basicity. This is due to K_2O is a promoter designed to help the active components of CaO so that the increase of K_2O in the catalyst may enhance the presence of CaO as the active site of the catalyst. Therefore, the addition of K_2O component as a promoter in CaO-ZnO catalyst is necessary because it has a dual function that is to improve the mechanical properties (as a binder) and to increase the catalyst base.

3.3 Testing of 5%K_2O/CaO-ZnO catalyst for biodiesel production

The catalyst test was performed with the following process parameters: mole ratio of methanol:soybean oil of 15:1, catalyst loading of 5%wt oil, 4 hours reaction time, 60°C reaction temperature, and constant stirring. The biodiesel product is then analyzed by GCMS to determine the composition of FAME. Yield of FAME and biodiesel are calculated using equations (2) and (3). From the results of GCMS analysis, the yield of FAME and biodiesel were presented in Table 3.

$$Yield_{FAME} = \frac{\%\,FAME\ \ GC\ Area\ x\,\rho_{biodiesel}\ x\,V_{biodiesel}}{weight\ \ of\ soybean\ \ oil\ feed} x100\ \% \quad (2)$$

$$Yield_{biodiesel} = \frac{\rho_{biodiesel}\ x\,V_{biodiesel}}{weight\ \ of\ soybean\ \ oil\ feed} x100\ \% \quad (3)$$

Table 3. The results of 5%K_2O/CaO-ZnO catalyst testing for biodiesel production

Mole ratio of CaO:ZnO	% FAME GC Area	Yield of FAME (%)	Yield of biodiesel (%)
3:1	97.88	41.327	42.222

According to SNI, the minimum %FAME in biodiesel is 96.5%. Table 3 shows that the obtained biodiesel contains %FAME of 97.88%. This result fulfills the SNI standard [20]. The biodiesel product is also analyzed for viscosity, density, and acid number. The results are indicated in Table 4.

Table 4. Result of characterization of biodiesel product

Mole ratio of CaO:ZnO	Kinematic viscosity (mm^2/s)	Density (kg/m^3)	Acid number (mg KOH/gr sample)
3:1	4.597	874	0.4937
SNI of Biodiesel	**2.3-6**	**850-890**	**<0.5**

Table 4 indicates that all parameters, including kinematic viscosity, density, and an acid number, fulfill biodiesel SNI standards. However, the results of FAME and biodiesel obtained are still low at 41.33% and 42.22%, respectively. The biodiesel yield is strongly influenced by the density and volume of biodiesel produced. In this research, the volume of biodiesel produced is still low so that further research is required to obtain a high yield of FAME and biodiesel.

4 Conclusions

The K_2O/CaO-ZnO catalysts with a variation of mole ratio of CaO:ZnO and %wt K_2O was prepared and tested through the transesterification process to produce biodiesel from soybean oil. The results demonstrated that increasing the mole ratio of CaO to CaO-ZnO catalysts would enhance the catalyst basicity, while the addition of K_2O as a promoter would increase the basicity. The results showed that the basicity of the catalyst increased after the CaO-ZnO catalyst was impregnated with KNO_3 or K_2O in all mole ratios of CaO-ZnO catalyst. The highest catalyst basicity was obtained in the mole ratio of CaO:ZnO 3:1. The addition of %wt K_2O to the CaO-ZnO catalyst also enhances the catalyst basicity. The addition of K_2O component as a promoter in CaO-ZnO catalyst is necessary because it has a dual function that is to improve the mechanical properties (as a binder) and to increase the catalyst basicity. Testing of 5%K_2O/CaO-ZnO catalyst on biodiesel production showed that the biodiesel fulfills the SNI standard even though the biodiesel yield is still low.

Acknowledgment

The authors thank to the Directorate General of Research and Development, Ministry of Research, Technology, and Higher Education, the Republic of Indonesia for the financial support received under the research project of Hibah Kompetensi Year 2015-2017.

References

1. C.Y. Lin, H.A. Lin, L.B. Hung, Fuel, **85**, 1743–1749 (2006).
2. C.Y. Chen, K.L. Yeh, R. Aisyah, D.J. Lee, J.S. Chang, Bioresour Technol, **102**, 71–81 (2011)
3. M.L. Granados, M.D.Z. Poves, D.M. Alonso, R. Mariscal, F.C. Galisteo, R. Moreno-Tost, J. Santamarı´a, J.L.G. Fierro, Appl Catal B Environ **73**, 317–326 (2007)
4. M. Kouzu, T. Kasuno, M. Tajika, Y. Sugimoto, S. Yamanaka, J. Hidaka, Fuel, **87**, 2798–2806 (2008)
5. S. Gryglewicz, Bioresour Technol, **70**, 249–53 (1999)
6. L. Buchori, I. Istadi, P. Purwanto, Bull Chem React Eng Catal, **11**, 406–430 (2016)
7. P-L. Boey, G.P. Maniam, S.A. Hamid, Bioresour Technol, **100**, 6362–6368 (2009)
8. A.M. Dehkhoda, A.H. West, N. Ellis, Appl Catal A Gen, **382**, 197–204 (2010)
9. D.W. Lee, Y.M. Park, K.Y. Lee, Catal Surv from Asia, **13**, 63–77 (2009)
10. M. Zabeti, W.M.A. Wan Daud, M.K. Aroua, Fuel Process Technol, **90**, 770–777 (2009)
11. A.Z. Fadhel, P. Pollet, C.L. Liotta, C.A. Eckert, Molecules, **15**, 8400–8424 (2010)
12. Y.H. Taufiq-Yap, H.V. Lee, M.Z. Hussein, R. Yunus, Biomass and Bioenergy, **35**, 827–834 (2011)
13. I. Istadi, U. Mabruro, B.A. Kalimantini, L. Buchori, D.D. Anggoro, Bull Chem React Eng Catal, **11**, 34–39 (2016)
14. S. Semwal, A.K. Arora, R.P. Badoni, D.K. Tuli, Bioresour Technol, **102**, 2151–2161 (2011)
15. S.J. Yoo, H. Lee, B. Veriansyah, J. Kim, J-D. Kim, Y-W. Lee, Bioresour Technol, **101**, 8686–8689 (2010)

17\. C.H. Yulianti, R. Ediati, D. Hartanto, T.E. Purbaningtias, Y. Chisaki, A.A. Jalil, C.K.N.L.C.K. Hitam, D. Prasetyoko, Bull Chem React Eng Catal, **9**, 100–110 (2014)

17\. I. Istadi, S.A. Prasetyo, T.S. Nugroho, Procedia Environ Sci, **23**, 394–399 (2015)

18\. K. Tanabe, T. Yamaguchi, J Res Insitute Catal, **11**, 179–184 (1964)

19\. Ž. Kesić, I. Lukić, D. Brkić, J. Rogan, M. Zdujić, H. Liu, D. Skala, Appl Catal A Gen, **427–428**, 58–65 (2012)

20\. SNI 7182-2015, (2015)

Potential of Electric Power Production from Microbial Fuel Cell (MFC) in Evapotranspiration Reactor for Leachate Treatment Using *Alocasia macrorrhiza* Plant and *Eleusine indica* Grass

Badrus Zaman [1,*] and *Irawan* Wisnu Wardhana[1]

[1] Environmental Engineering Department, Engineering Faculty, Diponegoro University, Semarang-Indonesia

Abstract. Microbial fuel cell is one of attractive electric power generator from nature bacterial activity. While, Evapotranspiration is one of the waste water treatment system which developed to eliminate biological weakness that utilize the natural evaporation process and bacterial activity on plant roots and plant media. This study aims to determine the potential of electrical energy from leachate treatment using evapotranspiration reactor. The study was conducted using local plant, namely *Alocasia macrorrhiza* and local grass, namely *Eleusine Indica*. The system was using horizontal MFC by placing the cathodes and anodes at different chamber (i.e. in the leachate reactor and reactor with plant media). Carbon plates was used for chatode-anodes material with size of 40 cm x 10 cm x1 cm. Electrical power production was measure by a digital multimeter for 30 days reactor operation. The result shows electric power production was fluctuated during reactor operation from all reactors. The electric power generated from each reactor was fluctuated, but from the reactor using *Alocasia macrorrhiza* plant reach to 70 µwatt average. From the reactor using *Eleusine Indica* grass was reached 60 µwatt average. Electric power production fluctuation is related to the bacterial growth pattern in the soil media and on the plant roots which undergo the adaptation process until the middle of the operational period and then in stable growth condition until the end of the reactor operation. The results indicate that the evapotranspiration reactor using *Alocasia macrorrhiza* plant was 60-95% higher electric power potential than using *Eleusine Indica* grass in short- term (30-day) operation. Although, MFC system in evapotranspiration reactor system was one of potential system for renewable electric power generation.

1 Introduction

Leachate generation is a major problem for municipal solid waste (MSW) landfills as a liquid that passes through a landfill and has extracted dissolved and suspended matter from it. Leachate results from precipitation entering the landfill from moisture that exists in the waste when it is composed [1]. The most critical aspect is related to several high concentrations pollutants that can be divided into four main groups namely: dissolved organic material, inorganic compounds, heavy metals and xenobiotic organic substances [2,3]. Evapotranspiration system which using plant and microbial activity on plant root and in planting media was promising for leachate treatment [4].

Meanwhile, the cleaner generation of energy is a vital concept to ensure the survival of our current lifestyle past the depletion of the Earth's fossil fuel supply, where MFC systems are recognized as one of energy production systems with great potentials [5]. MFC was believed as a promising technology that can be used to produce bioenergy in the form of hydrogen and/or electricity directly from various oxidation process of organic and inorganic compounds [6,7,8,9]. MFC can generates electrical power while accomplishing simultaneous treatment of biodegradable contaminants in wastewater by utilizing microorganisms [10,11].

Most microbes can produce current if active oxidation- reduction (Redox) mediator was added into the system or settled on the electrode. In practice, the system was designed with self-mediated or directly transferable electron to anode through contact between membrane- anode (extracellular transfer Electrons through a protein membrane and/or a bacterial nano cable) [12].

MFC technology in wetlands in rice plants was using rhizodeposition substrate to be oxidized by microbes, thus generating electrical energy [13]. MFC system utilization using waste residue at wetland forest as substrates that are oxidized by microbes, whereas in sediments it in eutrophication lake condition which generates maximum electricity was 294 mW/m^2 at the same time can remove of nitrate more than 90% [14,15]. MFC operated with continuous flow can remove of 50% COD from wastewater with the result of electrical energy was 464 mW/m^2 [16]. Constructed Wetlands (CW) to COD from textile waste when the system was combined with MFC. Thus CW-MFC system is capable

* Corresponding author: badruszaman2@gmail.com

to remove 75% COD and generates a maximum power was 15.73 mW/m^2 with current Maximum was 69.75 mA/m^2 [17].

This study aims to determine the potential of electrical energy from leachate treatment using evapotranspiration reactor. The study was conducted using local plant, namely *Alocasia macrorrhiza* and local grass, namely *Eleusine Indica*. The system was using horizontal MFC by placing the cathodes and anodes at different chamber (i.e. in the leachate reactor and reactor with plant media).

2 Materials and Methods

2.1 Raw Leachate, plants and soil

Raw leachate collected from the Jatibarang landfill, Semarang City, Indonesia and store in the tank container with 4000lt capacity. Two plants species were used in the experiment namely *Eleusine indica* and *Alocasia macrorrhiza*, where thus species are growing in around the landfill. Soil media obtained from the Jatibarang landfill in Semarang City, Indonesia.

2.2 Reactors Set up

Main evapotranspiration-MFC reactor consists of two plastic containers with the size 80 gallons (h: 26 cm, d: 70 cm) and 70 gallons (h: 20 cm, d: 60 cm) each. Smaller container used for plants media which had maked ±1 cm^2 hole at the bottom and then put into large container. It's system design for 15 lt liquid volume. Electrode (carbon plate: w:10 cm, l:40 cm, t:1 cm) were put in the soil media as anode and under smaller container (in the large container) as cathode with horizontal position each (fig. 1).

2.3 Plants cultivation

After evapotranspiration reactors were ready, the gravel put into container as a bottom layer. Then, soil added with 20 cm high layer as plants cultivation media and put electrode. In each container was cultivated with 2 individual for *Eleusine indica* and Alocasia macrorrhiza. Then, put in the large container. All plants were maintained for 20 days, where each set reactor was made in duplo.

2.4 Experiment Set up

The reactors was operated at 30 days and the leachate sample was taken every 3 days from sample point above in accordance with the residence time of main reactor. Electricity was measure continuously using multimeter.

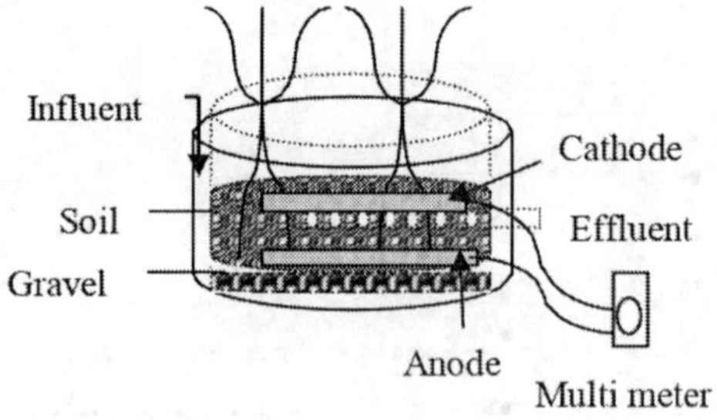

Fig. 1. Evapotranspiration-MFC reactor

3 Results and Discussion

3.1 Using *Alocasia macrorrhiza* Plant

A long 30 days reactor operation, electric power has produce from the reactor. It's shows electric power production was fluctuation during reactor operation from all reactors. Electric power generated from reactor were using *Alocasia macrorrhiza* plant at at the first day was reached 30 μwatt average. It's condition was similarity with control pattern. Thus shows at first operation (about 3 days) the plant and bacteria on plant roots are still in the adaptation stage condition by leachate presence, where more bacteria in the soil media was treat it's leachate.

Furthermore, the electric power generated has increased and reach to 60 μWatt and then decrease less than control, but during overall reactor operation, electric power generated has increase were up to 100 μWatt an average. Thus, pattern is influenced by *Alocasia macrorrhiza* leaves growth pattern. Where every 4-7 days will grow new leaves but old ones begin to wither. This condition causes its performance will decrease and then increase, but after leaf growth was stagnant, its' will reach maximum performance condition. The pattern will be repeated but in line with the growth of leaves and also roots develop and increase the number of bacteria (fig.2). It's bacteria was simultaneously Organic matter removal carried out by means of aerobic respiration, denitrification, sulphate reduction, fermentation or methanogenesis [19]

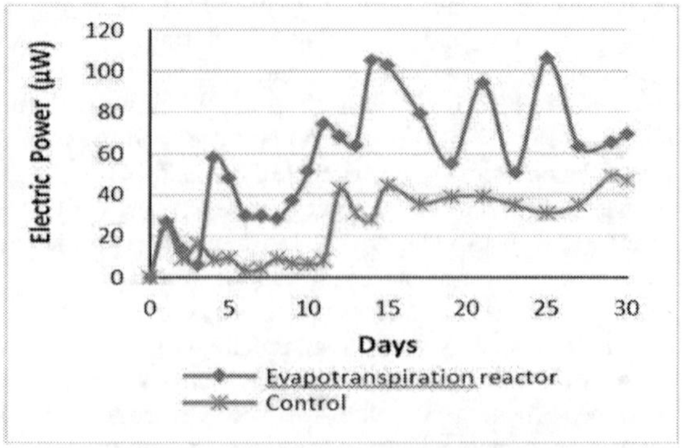

Fig.2. Electric power production from reactor using *Alocasia macrorrhiza Plant* and control

3.2 Using *Eleusine indica* Grass

From the reactor using *Eleusine Indica* grass on the first day was resulting electric power reach to 50 μwatt average and decreased until the 9th day was 10 μwatt average, but then gradually increased until 30th day was reached 60 μ watt average. This electric power production fluctuation is related to the bacterial growth pattern in the soil media and on the plant roots which undergo the adaptation process until the middle of the operational period and then in stable growth condition until the end of the reactor operation (fig.3).
The overall results indicate that the evapotranspiration reactor using *Alocasia macrorrhiza* plant was 60-95% higher electric power potential then using *Eleusine Indica* grass in short-term (30-day) operation. It's plants influence reactors performance by their ability to release oxygen or easily biodegradable substrates through the root system, also influence the redox conditions within the treatment bed due their ability to evapotranspirate water and influencing MFC performance. It's performance also influenced by biological, chemical and electrical factors [20].

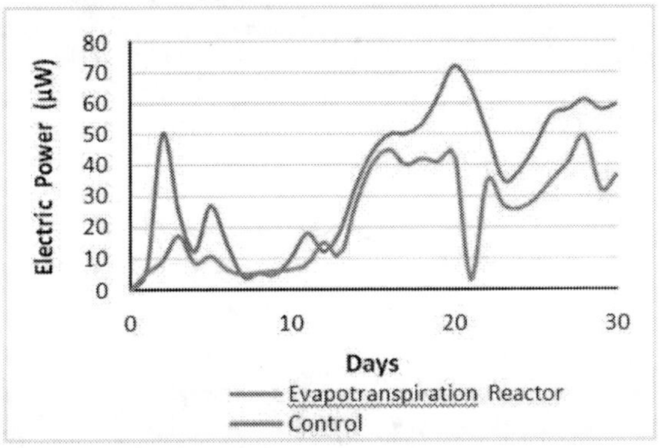

Fig. 3. Electric power production from reactor using *Eleusine indica* grass and control

Parameters defining MFCs performance are consists of substrate conversion rate, over potentials at the anode, over potentials at the cathode, proton exchange membrane related factors, and internal resistance of the MFC [21]. Also, concentration of chemical oxygen demand (COD) pH or temperature, distance of electrode [22].

Electric power generation from evapotranspiration system for leachate treatment was potential technology in the future but need to improve treatment and electric power performance according to parameters performance factors.

4 Conclusions

The results indicate that the evapotranspiration reactor using *Alocasia macrorrhiza* plant was 60-95% higher electric power potential then using *Eleusine Indica* grass in short-term (30-day) operation. MFC system in evapotranspiration reactor system was one of potential system for renewable electric power generation.

References

1. S.M. Raghab, Ahmed M. Abd El Meguid, Hala A. Hegazi. HBRC Journal **9**, 187–192 (2013)
2. Tengrui Long, A.F. Al-Harbawi, L.M. Bo, Z. Jun, X.Y. Long, Am. J.of Appl.Sci. **4(4)**: 211-214, (2007).
3. F. Kargi, M.Y. Pamukoglu, J. Enzyme & Microbial Tech. **33**:588- 595 (2003).
4. B. Zaman, P. Purwanto, S. Mangkoedihardjo, A Case Study for Jatibarang Landfill, Central Java. *J. Appl. Environ. Biol. Sci.*, **2(7):** 320-325 (2012).
5. C. Ozansoy and Heard R. Microbial Conversion of Biomass: A Review of Microbial Fuel Cells *School of Engineering and Science, Victoria University Melbourne, Victoria, Australia,* www.intechopen.com (2011).
6. S. Oh, B. E. Logan, *Water Res.* **39(19)**:4673-4682 (2005).
7. YA. Gorby, S. Yanina, JS. McLean, *et al.* Proc Natl Acad Sci USA; **103(30)**: 11358-11363 (2006).
8. IS. Chang, H. Moon, 0. Bretschger. *J Microbiol Biotechnol*, **16(12)**: 163-177 (2006)
9. Erable, B. Féron, D., Bergel, A., Chem Sus Chem, **5(6)**: 975-987(2012).
10. D. Pant, A. Singh, G. Van Bogaert, SI. Olsen, PS. Nigam, L. Diels, K. Vanbroekhoven, Rsc Adv **2(4)**:1248–1263 (2012).
11. M. Rahimnejad, A. Ghoreyshi, G. Najafpour, H. Younesi, M. Shakeri, Int. J. Hydrogen Energy. **37(7)**:5992-6000 (2012).
12. Biffinger, C. Justin and Ringeisen, R. Bradley Paper **22**: 150- 155 (2008).
13. L. De Schamphelaire, A. Cabezas, M. Marzorati, M.W. Friedrich, N. Boon, W. Verstraete, Appl Env Microbiol, **76 (6)**:2002– 2008 (2010).
14. J. Dai, *Application of microbial fuel cells in a forested wetland.* Master Theses. Graduate School of Clemson University (2010)
15. Y. Zhang, *Energy recovery from waste streams with microbial fuel cell (MFC)-based technologies.* PhD theses. DTU Environment Department of Environmental Engineering Technical University of Denmark (2012).
16. Y. Yuan, S. Zhou, L. Zhuang, Journal of Soils and Sediments, **10(7)**, 1427-1433, (2010).
17. A.K. Yadava, P. Dasha, A. Mohantya, R. Abbassi, B.K. Mishra, Ecol Eng, **47**:126– 131, (2012).
18. J. García, D. P. L. Rousseau, J. Morató, E. Lesage, V. Matamoros, J.M. Bayona, Crit Rev Env Sci Tec **40**:561-661 (2010).
19. C. Corbella, J. Puigagut, Alcalá de Henares. Spain. Abstracts, p 173 (2014)

20. C. Corbella, J. Puigagut, contributions To sci. **11**:127-134 (2015)
21. K. Rabaey, W. Verstraete, Trends Biotechnol **23**:291-298, (2005).
22. GS. Jadhav, MM. Ghangrekar, Bioresource Technol **100**:717-723 (2009)

https://doi.org/10.1051/e3sconf/20183102011

Technical Analysis Feasibility Study on Smart Microgrid System in *Sekolah Tinggi Teknik* PLN

Heri Suyanto

Lecturer, College of Engineering – PLN (Foundation for Education & Welfare PT. PLN (Persero). PLN tower. Jl. Outer West Lingkar, Duri Kosambi, Cengkareng, Indonesia 11750

Abstract. Nowadays application of new and renewable energy as main resource of power plant has greatly increased. High penetration of renewable energy into the grid will influence the quality and reliability of the electricity system, due to the intermittent characteristic of new and renewable energy resources. Smart grid or microgrid technology has the ability to deal with this intermittent characteristic especially if these renewable energy resources integrated to grid in large scale, so it can improve the reliability and efficiency of the grid. We plan to implement smart microgrid system at Sekolah Tinggi Teknik PLN as a pilot project. Before the pilot project start, the feasibility study must be conducted. In this feasibility study, the renewable energy resources and load characteristic at the site will be measured. Then the technical aspect of this feasibility study will be analyzed. This paper explains that analysis of ths feasibility study.

1 Introduction

The conditions of electricity in Indonesia is still inadequate, which the ratio of electrification is still low at 87% as of October 2015 (PT. PLN Persero). This shows that people access the energy is still limited. And also with the development of energy infrastructure in Energy for rural / remote and island - the outermost islands in general do not get adequate energy access. Dependence on electricity industry is more limited. Utilization of new and renewable energy and implementation of energy conservation are also not set optimal [1].

As the one of the way fulfil local needs for electricity, and in accordance with the policies of Indonesian government, in the past few years has been started the development of power systems of new and renewable energy (EBT) [2]. Due to the nature intermittent characteristic of new and renewable energy, we need a system to resolve the issue. Smart microgrid system or smart grid is one technology that can effectively and efficiently address the integration of new and renewable Energy in the electrical system for a large scale.

Since 2015 STT PLN has been conducting some research of smart microgrid, with stage: study potentials stats with (2015) and design of the smart microgrid laboratorium scale model (2016) [3], As the development of smart microgrid model and for application in the field, so as a first step of the implementation to start to establish the pilot project of smart microgrid system with in at the laboratory scale in STT PLN start in 2017

2 Technical Review

Research and development of renewable energy especially for Solar Power Generation based on smart-grid technology has been listed on RIP STT PLN 2013 - 2018. One of strategies to fulfill the electric energy needs is to utilize the potential of new and renewable energy (EBT) at the local site, In order to reduce transmission and distribution costs. Potential of Renewable Energy (ET) in local area is used to meet the needs of electrical energy in the area [4]. This will lead high concentration of renewable energy in into the grid. High concentrations of renewable energy sources can cause problems in stability, reliability and power quality on the main grid [5]. Smart microgrid is the key of the future grids where high concentration of renewable energy does not affect or disrupt with the quality of the grid.

A smart microgrid system can be interpreted as an electrical system consists of several Distributed Generations (DG) where the source is usually from EBT, such as photovoltaic panels, wind turbines, microturbons, with the addition of a storage system, load control, and an energy regulation system (Energy Management System - EMS). EMS allows smart microgrid systems to make its own free adjustment regardless grid positive in standalone mode (islanding). This microgrid smart system can operate in two modes, On-Grid / Grid-connected and Off-Grid / Standalone / Islanding. Off-Grid / Standalone / Islanding operation mode is when circuit breaker isolates the system from the main grid (PLN), so power generator equipment, storage, load regulator, power quality controller and other system

* Corresponding author: heri.suyanto@yahoo.com

https://doi.org/10.1051/e3sconf/20183102011

operation regulator are applied only to smart microgrid system [6]. In Off-Grid mode the load is supplied applied by DG, Diesel Generator (it backup) and Battery [7]. In an on-grid / grid connected mode, the microgrid smart system is also a controllable load, or the power generated from the grid is a backup / supplemental energy source. In On-Grid / Grid Connected mode this load is supplied by DG, grid (as backup) and Battery [8].

3 Research Methods

The expected outcome of this research program is the synergies which mutually beneficial and supportive synergies between university (STT PLN) and industrial partners to produce hybrid power plant products designed and manufactured in a micro-distributed (micro-grid structure), and Supported by high technology, so it can be operated in local/ remote island to fulfil needs of housing and also support the operation of NKRI territorial integrity [9].

Methodology and research stages to be used follow:

a. Measure New and Renewable Energy potential, the wind and solar energy at Sekolah Tinggi Teknik PLN.
b. Analysis of irradiation data on selected area (local / remote island), refers to data from BMKG and Lapan and other agencies.
c. Measure the load characteristics at the smart micro grid site.
d. Conducted an analysis of data on potential and characteristics of the load.
e. Conduct the analysis of feasibility for smart micro grid system as the technical specifications of solar cells, and determining the electrical power control system based on "power electronics" based on the study of literature [10].

4 Results And Discussion

Sekolah Tinggi Teknik PLN is located at Lingkar Barat, Duri Kosambi, Cengkareng, Jakarta Barat, Jakarta 11750. Applied load in the Campus main building of 100KW, and the load in the Laboratory of STT PLN is 15 KW.

After the discussions among colleagues and professors of Sekolah Tinggi Teknik PLN and by direct observation at the location it was determined that the project smart microgrid system will be built right at Laboratory at Research and Training Center For Renewable Energy Power System. Presence of solar Laboratory will save the electricity bill from PT. PLN

Blueprint New and Renewable Energy STT PLN as follows according to Figure 1 among others as follows:

1. Solar Power Plant (PLTS) capacity 14.5 kWp
2. Bayu / wind power plant (PLTB) capacity of 20 kW
3. Biomass Power Plant 16 kVA consisting of:
   - Waste Power Plants T gasifer 14 kVA
   - Waste Power Plants digester of 2 kVA
4. With KWH Export and Import KWH meter system (Exim) from PT. PLN (Persero) and Battery

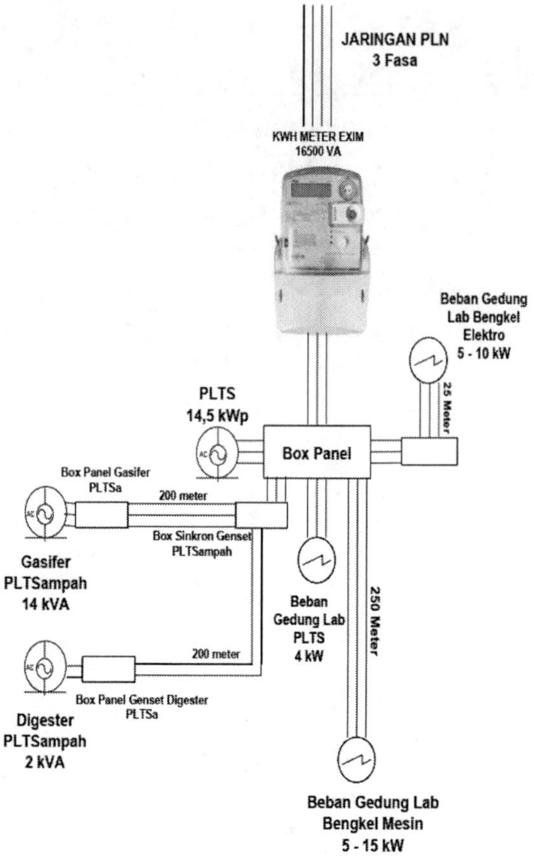

Fig. 1. Blueprint STT PLN Smart microgrid System.

This Installation that has been installed with a capacity of 14.5 kW with microgrid smart system On Grid and Off Grid.

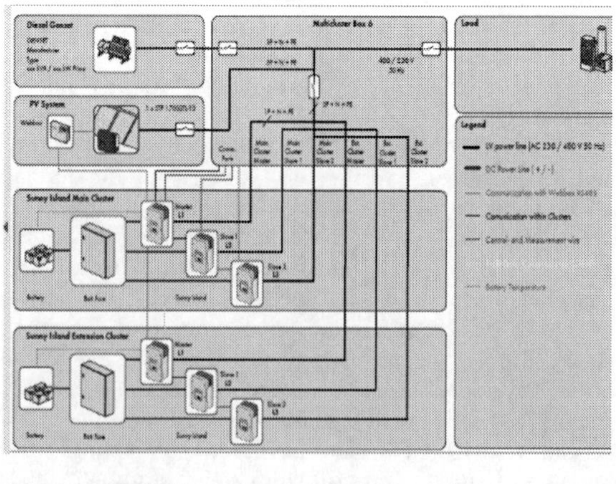

Fig. 2. Smart grid system design of Microgrid system for off grid at STT PLN

Our Solar Power Plants Capacity of 17 kW / peak has already Integrated KWH Meter Exim Export and Import and has installed with advanced technology monitoring system. This design is made as a Research by knowing new and renewable energy technologies especially Solar Power Generation and as a saver of electricity usage from PT. PLN (Persero) and the electricity cost down. Figure of generated electricity from PLTS monitoring system.

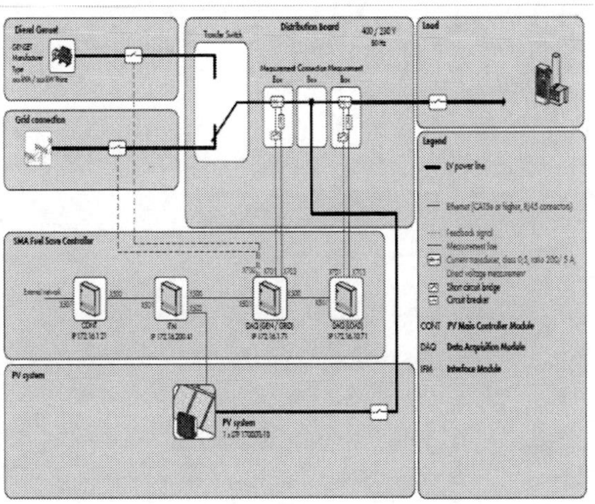

Fig. 3. Smart System Design of Microgrid system on grid at STT PLN

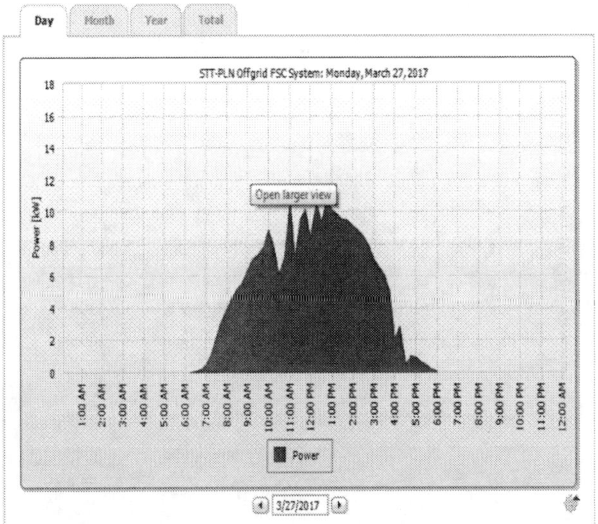

Fig. 4. Daily electricity curve of PLTS

It shows the maximum peak at 11:00 to 14:00 hours to reach 10 kW so that one can reach 70 kW in the position of maximum results Daily and Daily Average reaches 40 kW.

From the curve above, electricity generated from Solar Power Generation can reach 1.4 MWh.and Biomass Power Plant is in the design stage For Bayu / Wind Power Generation and Biomass Power Plant is still under costumer.

This Microgrid Project Smart will be applied as Research and Learning on New and Renewable Energy at the Research and Training Center Laboratory for Renewable Energy Power System STT PLN.

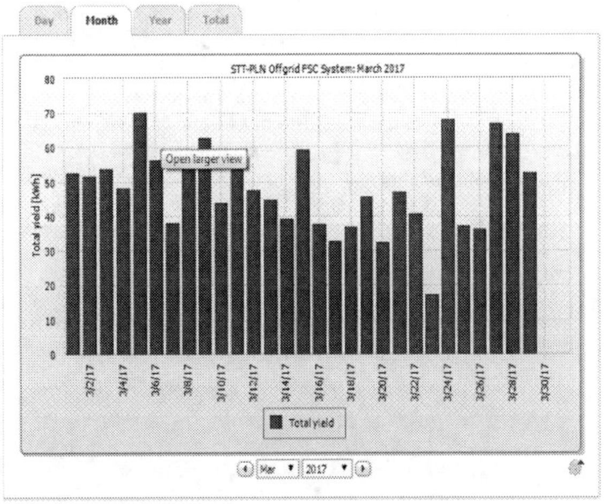

Fig.5. Monthly Electricity Curve of PLTS

5 Conclusion

- From the analysis conducted on the technical aspects of the feasibility study of the application of Smart Microgrid System Project in Sekolah Tinggi Teknik PLN is feasible with high potential availability and installation area of equipment as we expected.
- Microgrid Smart System Project in Sekolah Tinggi Teknik PLN Research and Training Center Laboratory for Renewable Energy Power system STT PLN as New and Renewable Energy Technology Research and as a saver of electricity usage from PT. PLN.
- Blueprint of smart microgrid system for New and Renewable Energy STT PLN consists of PLTS capacity 14.5 kWp, PLTB capacity 20 kW, PLTbiomass with 14 kVA Gasifer PLTS garbage and 2 kVA of waste PLTS Digester) and Battery.

References

1. CERTS *Microgrid Symposium*, Northern Power Systems
2. Clarke,S., *"Electricity Generation Using Small Wind Turbines At Your Home Or Farm"*, Fact Sheet, September (2003)
3. Jay Stuller, *"An Electric Revolution"*, Galvin Electricity Initiative
4. J. B. (Secretary of U. S. D. of C. P. D.G. (Director of N. I. of S. and Technology), *NIST Framework and Roadmap for Smart Grid Interoperability Standards , NISTSpecial Publication 1108R2 NIST Framework and Roadmap for Smart Grid Interoperability Standards*, NIST Speci. Energy Independence and Security Act of 2007, (2012).
5. Kurt Yeager, *"The Microgrid Revolution"*, Oktober (2010)

6. *Smart Houses interacting with Smart Grids to achieve next generation efficiency and sustainability*, Dr. Anke Weidlich, SAP Research 11.02.2009
7. IEC *Smart Grid Standardization Roadmap*, Prepared by SMB Smart Grid Strategic Group (SG3) June 2010; Edition 1.0
8. R. Ramakumar, *"Integrated Renewable Energy System – Micro Grid (IRES-MG) For Sustainable Development"*, ICSUNIDO, Trieste, Italy, April (2012)
9. *"The Smart Grid: Coming Of Age"*, Electric Power Research Institute Journal, EPRI,Inc. (2009).
10. T.Nakata, K. Kubo and A. Lamont, *"Design For Renewable Energy System With Application To Rural Area In Japan"*, white paper.

Solar Water Heating System for Biodiesel Production

Syaifurrahman[1,*], *A Gani* Usman[1], and *Rakasiwi* Rinjani[2,*]

[1]Department of Electrical Engineering, Faculty of Engineering, Tanjungpura University, Pontianak- Indonesia
[2]Department of Chemical Engineering, Faculty of Engineering, Tanjungpura University, Pontianak - Indonesia

Abstract. Nowadays, electricity become very expensive thing in some remote areas. Energy from solar panels give the solution as renewable energy that is environment friendly. West Borneo is located on the equator where the sun shines for almost 10-15 hours/day. Solar water heating system which is includes storage tank and solar collections becomes a cost-effective way to generate the energy. Solar panel heat water is delivered to water in storage tank. Hot water is used as hot fluid in biodiesel jacked reactor. The purposes of this research are to design Solar Water Heating System for Biodiesel Production and measure the rate of heat-transfer water in storage tank. This test has done for 6 days, every day from 8.30 am until 2.30 pm. Storage tank and collection are made from stainless steel and polystyrene a well-insulated. The results show that the heater can be reach at 50°C for ±2.5 hours and the maximum temperature is 62°C where the average of light intensity is 1280 lux.

1 Introduction

Energy always becomes important issue. Renewable energy of the sun is generated from the nuclear fusion of its hydrogen into helium, with resulting mass depletion rate of approximately 4.7 x 106 tons per second and solar radiation reaches the earth on a continuous basic amount to 120,000 terawatts [1].Solar energy is fast becoming a welcome alternative source and hazard free heat source, especially in the tropics [2].West Borneois in Equator line where the power of sunlight reaches ±1000 W/m^2[3]. The solar radiation incident on the surface of the earth can be conveniently utilized for the benefit of human society. One of the popular devices that harness the solar energy is solar hot water system [4]. Biodiesel as renewable energy is made by trans-esterification reaction at 40-65°C[5].In 2014, West Kalimantan has product 1.9 billion ton/year of crude palm oil which can produce become biodiesel.

Electricity generate from solar only 0.88 MW from 1.1 x 109 MW of solar energy potential in Indonesia. Solar water heating has been around for many years because it is the easiest way to use the sun save energy and money. Solar water heater do not pollutes. Solar water heater help to avoid carbon dioxide, nitrogen oxides, sulphur dioxides and the other air pollution and wastes created when the local utility generates power or fuel is burned to heat domestic water. When a solar water heater replaces an electric water heater, the electricity displaced over 20 years represents more than 50 tons of avoided carbon dioxide emissions alone [6].

Implementation for the idea is by design biodiesel unit using solar heater with no electricity needed. Heat of trans-esterification reaction can fill using hot water from solar heater. Mini unit of biodiesel production using solar water heating system is a simple and environment friendly technology. Hot water is used as hot fluid in biodiesel jacked reaction. Some areas which developing countries have used solar water heating system for trans-esterification reaction and to evaporate drying water in $4m^2$ solar collections for 19 litres each day of biodiesel[7].

Solar water heating system consists of storage tank and solar collection. There are two types of solar water heating system: active and passive. Active types have circulation pump and control, and passive, which don't. Passive solar water heaters rely on gravity and the tendency for water to naturally circulation as it is heated. The systems contain no electrical components are generally more reliable, easier to maintain, and possibly have a longer work life than active solar water heater systems. Most solar heaters require a well-insulated storage tank and have an additional outlet and inlet connected to and from the solar collector[6].

Solar water heating system can save 40-80% on electric or fuel. A solar energy incident on the absorber panel coated with selecting transfer to the riser pipes underneath the absorber panel. The water passing through the risers get heated up and is delivers the storage tank. The choice of system depends on heat requirement, weather condition, heat transfer fluid quality, space availability, annual solar radiation The solar water heater system are pollution free and easy for operation. A 100 litres capacity

*Corresponding author:rinjani_s@yahoo.com

solar water heater can replace an electric geyser for residential uses and saves 1500 units of electricity annually. The use of 1000 solar water heater of 100 litres capacity each can contribute to a peak load saving of 1 MW. A solar water heater of 100 litres capacity can prevent emission of 1.5 tonnes of carbon-dioxide per year [8].

This technology can apply in some remote area where there is no electricity. In West Borneo border side area, it is hard to reach some accommodation and transportation. Fuel becomes very expensive and rare things. The objective of this paper is to design and test solar water heating for biodiesel production. Solar water heater is design using generous material. Therefore we can produce renewable fuel with no electricity needs.

2 Methodology

There are two main steps that have been done in this research.

1. Designing solar water heating system
2. Testing the solar water heating system

Solar water heating system is made by using economy materials. Solar water heating system for biodiesel production consists of solar collection and storage tank. Type of solar heater is passive solar water heater with Thermosyphon system. Type of collector systems is called evacuated tube solar collector

The materials that have been used are:

1. Stainless steel storage tank
2. Stainless steel tubes
3. Stainless steel pipes
4. Acrylic glass
5. Foam
6. Aluminium foil
7. Valve
8. Black color
9. Nut-bolt
10. Elbow
11. Glass glue

Measuring instrument that have been used are:

1. Thermometer
2. Luxmeter

Water flows into the stainless steel pipe absorbed heat from solar. Hot water will be up stream into storage. Hot water then flow to biodiesel jacket-reactor.

Solar water heating system then is tested on the second floor of laboratory building which has outdoor space. Water flows into the inlet of collector solar water heater fixed to and the temperature in tube, atmosphere temperature and light intensity are measured.

The test was held in the Basic Electrical Laboratory, Engineering Faculty of Tanjungpura University for 5 hours each day. Solar water heating system is tested on clear sunny days for 6 day.

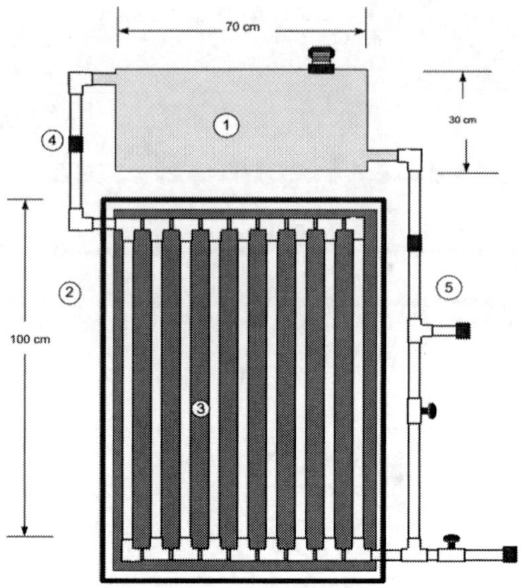

Fig.1. Design of Solar Water Heating System

Solar water heating system for biodiesel production parts: (1) Storage tank; (2) Solar collection; (3) Tubes; (4) Up-stream; (5) Down-stream.

3 Results and Discussion

3.1 Design of Solar Water Heating System

Solar collector is design for commercial application. Design for evacuated-tube collectors is small capacity and collecting surface. In a thermosyphon system there is no circulation pump and controller. Potable water flows directly to the tank on the roof. Solar heat water flows from the rooftop tank to the trans-esterification reactor. The result of design shows at Fig.1. The details of the design are:

1. The capacity of storage tank is proximally 50 litres. Storage tank is covered by foam and aluminium foil to save the heat well.
2. Solar collecting surface are proximally 7m^2.
3. They are 8 stainless steel-tubes which are colored black. The tubes are designed to deliver higher temperature. Hot and cool water flow using valve as control utility.
4. Acrylic glass is using as absorber tube.
5. Polystyrene as an isolator to inhibits radiation heat loss. Side of polystyrene face to the tube is colored black to increasing of solar absorption to the water in the tube.

Fig.2. Solar water heating system for biodiesel production

Hot water will be up stream into storage tank because of different density between hot and cool water. As the sun shines on the collector, the water inside the collector flow tubes is heated. As it heats, this water expands slightly and becomes lighter than cold water. Gravity then pulls heavier, cold water down from the tank and into the collector inlet. The cold water pushes the heated water through the collector outlet and the top of the tank, thus heating the water in the tank [4]. Acrylic glass is using because can trapped sun radiation inside the box as the solar water heater is able to provide the hot water even in cloudy days [5].

Fig 2 shows that water flows directly to the tank on the roof. Solar heated water flows from the rooftop tank to the reactor. Hot water will be up stream into storage. Hot water then flow to biodiesel jacket-reactor. Energy from hot water is being used for trans-esterification reaction to produce biodiesel from crude palm oil. The water that has been used than flows to the tube of collector and get heats again. Water temperature increases along with increasing of its density. Hot water than flows to the storage tank and cold water down to the collector and so on until becomes a circle.

3.2 Performance of Solar Water Heating System

Solar water heating system is tested for 6 days, every day from 8.30 am until 2.30 pm. The heater can be reach at 50°C for ±2.5 hours and the maximum temperature is 62°C where the average of light intensity is 1280 lux. The different temperature between outside and inside tube is ±20°C shows in **Table 1**.

Table 1. Solar water heating system temperature

Day	Heating time (hr)	T_{max}Tube (°C)	T_{max}atm (°C)	Light Intensity $_{max}$ (Lux)
1	6	50.3	34.4	1186
2	6	54.7	35.0	1280
3	6	54.4	35.9	1172
4	6	55.0	36.1	1164
5	6	53.4	36.3	1123
6	6	62.0	37.8	1020
6	6	62.0	37.8	1020

Solar water heating system fixed to the roof of the building where the sun heat can absorbed optimally as shows at **Fig 3**. The sun shines unstable for each day and therefore reduces temperature of water in tubes.

Fig.3. Solar water heating system position when was being tested

Solar collectors capture the sun's electromagnetic energy and convert it to heat energy. The efficiency of a solar collector depends not only on its materials and design but also on its size, orientation and tilt. Solar energy is its maximum at noon, when the sun is its highest point in its daily across the sky. Certain material absorbs more insulation then others. More absorptive material are generally dark with a matte finish, while more-reflective materials are generally lighter coloured with a smooth or shiny finish. The best tilt angle will vary not only with collector's geographical location but also with seasonal function. Solar water systems are designed to provide heat year-round.

Solar water heating is now a mature technology which can be attained 60-80°C. Solar water heating system can reach at 62°C shows that hot water from the system can be use as hot fluids for trans-esterification reaction.

Though the initial investment for a solar water heater is high compared to available conventional alternatives, the return on investment has become increasingly attractive with the increase in price of conventional energy.

Conclusion

Solar water heating system that has been design using thermosyphon system with evacuated-tube collectors can reaches 50°C for ±2.5 hours and maximum temperature is 62°C that can be used for biodiesel production.

The efficiency of a solar collector depends not only on its materials and design but also on its size, orientation and tilt

We are very grateful to the Ministry of research, technology and higher education in funding the research.

References

1. H. Bradke, C. Doetsch, H. Huhn, et al., Researching: energy, brochure, Aliance of Scientific Organizations, Germany, 2011.
2. Okafor, Basil. *Thermo Siphon Solar Water Heater*. IJET, Vol3 No3 (2013).
3. C. Philibert. *Solar Energy Perspectives, Renewable Energy Technologies*, French: International Energy Agency. (2014).
4. Gangane, S.D, Bhere, S.H, et al. *Economical Solar Water Heater*. IOSR-JMCE, PP 68-71 (2017).
5. C. Clifford. *The Reaction of Biodiesel: Trans-eterification*.The Pennsylvania State University. (2017). [7] Wasley. *Design of a Small-Scale Biodiesel Plant*. Bellairs Reasearch Institute: McGrill University. (2015).
6. C. Homola. *Solar Domestic Hot Water Heating Systems Design, Installation and Maintenance*. Solar Rating and Certification Corporation.(2014).
7. Verma, Abhishek, Kumar, Vishal. *Solar Water Heating System*. IJRAME. **Vol 3 Issue 1**, Pgs 53-63 (2015).
8. Ogie, Nosa. Oghogho Ikponmwosa, Jesumirewwhe. Design and Construction of a Solar Water Heater Based in the Thermosyphon Principle. JFREA. **Vol.3**, 8 (2013).

https://doi.org/10.1051/e3sconf/20183102013

Preliminary Study on the Location Selection of Microalgae Cultivation In Nusa Tenggara Region As A Potential Feedstock For Bioavtur

Rr. Citra Permata Kusuma Anggraini[1,*], *Nugroho* Adi Sasongko[2,3], *Yanif* Dwi Kuntjoro[2]

[1]Master Program of Energy Security, Faculty of Defense Management, Indonesia Defense University, Bogor - Indonesia
[2] Department of Energy Security, Faculty of Defense Management, Indonesia Defense University, Bogor - Indonesia
[3] Center of Energy Technology and Chemical Industry, The Agency for Assessment and Application of Technology (BPPT), Indonesia

Abstract. NTT is a province located in strategic areas between Bali and South Sulawesi which has economic growth 5,08% in 2016. This causes air transportation in NTT to grow rapidly so the need for avtur is increased by 6% per year. To meet the needs of avtur in NTT would require energy diversification with bioavtur development in which one of them comes from microalgae. The content of lipid and hydrocarbon in microalgae can be used as a source of bioavtur feedstock. The suitability of location for cultivation will influence the success of microalgae cultivation that will be used as a source of bioavtur feedstock. The purpose of this research is to choose the best location for microalgae cultivation in NTT by AHP method. The criteria used in this research are nutrient, water and technology. Sub criteria of nutrient elements are coal power plant emission, cement industry emission and synthetic fertilizers, sub criteria from water that is sea water, brackish water and fresh water, while sub criteria of technology are Photobioreactor, Open Raceway Pond and membrane. The result of AHP analysis shows the selection of microalgae cultivation location in Kupang with the weight of 0.308, with the source of nutrient derived from coal power plant emission, the type of water used is sea water and the technology used is Photobioreactor. Microalgae species used were *Nannochloropsis sp* with a lipid content of 31-68%. Based on the author assumption, microalgae have the productivity for bioavtur manufacture which amount of 24.489kL/ha/ yr. That can be used to meet the needs of 2% avtur in NTT which amount of 1.052,22 kL/yr and the area requirement for microalgae cultivation is 2,14 hectare.

1 Introduction

1.1 General

The province of East Nusa Tenggara is a provincial archipelago with 1.192 islands in total, with land area 47.931,54 km2 [1]. NTT is one of the provinces that has the most number of airports in Indonesia, to date there are 14 airports in operation. NTT has economic growth 5,18% in 2016[2]. As a developing region, many tourists who come to NTT using air transportation causing avtur demands to increase by 6% per year [3]. The government requires the use of bioavtures beginning in 2016 with a 2% mixture that could reduce dependence on imports while reducing air emissions resulting from exhausted fumes[4]. Microalgae have high lipid content, are easy to cultivate, produce high biomass and do not compete with food commodity so it can be developed as biofuel [5]. The location for microalgae cultivation is a major factor to be considered in the effort of providing bioavtur feedstock. In NTT there are 3 developing regions namely Kupang, Ende and Labuan Bajo which seen from the increase of flight passenger number in April 2017 are El Tari airport 53,52%, Komodo 12,34% and Ende 6,17% [6]. NTT has many tourism destination such as Komodo island, Kelimutu Lake, Diving spot in Labuan Bajo, and soon. NTT government has agenda to improve the infrastructure one of them is transportation sector. NTT has the best weather to cultivate microalgae as a bioavtur feedstock. Kupang has annual relative humidity 72,1%, rainfall 1144,89 mm and sun radiation 22,92 20,76 MJ/m^2/d. Ende has annual relative humidity 76%, rainfall 1367,43 mm y-1sun radiation 20,76 MJ/m^2/d. That locations is a strategic area in NTT so it can be used as a microalgae cultivation site. Figure 1 is a map of microalgae cultivation location.

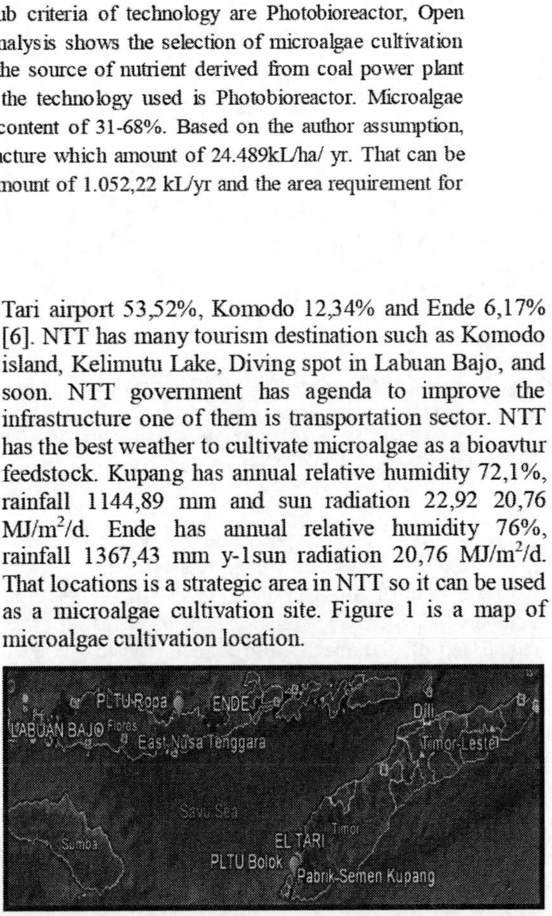

Fig. 1. Map of locations for microalgae cultivation

* Corresponding author: rr.citrapermata@gmail.com

1.2 Objective

The objective of this research is to choose the best location for microalgae cultivation in NTT by using AHP and to know the potential of microalgae to fulfill 2% of avtur requirement in NTT.

2 Methodologies

Analytic Hierarchy Process (AHP) is developed by Saaty (1980) and then it is used widely as an efficient multi criteria decision making (MCDM) method for ranking alternatives.

AHP is used on three principles :structure of the model, comparative judgment of the alternatives and criteria, synthesis of the priorities [7].

The main steps of AHP are given in the following: First of all, criteria and alternatives of the problem are defined. Then problem is organized hierarchically. The overall goal of this decision making problem is at the highest level and the alternatives are at the lowest level. Criteria and sub-criteria are placed between them [8]. The goal of the problem is located at level 1. Level 2 houses the major criteria. Level 3 houses sub criteria. Finally, the alternatives are located at the last level of the hierarchy. The hierarchy show in the Figure 2.

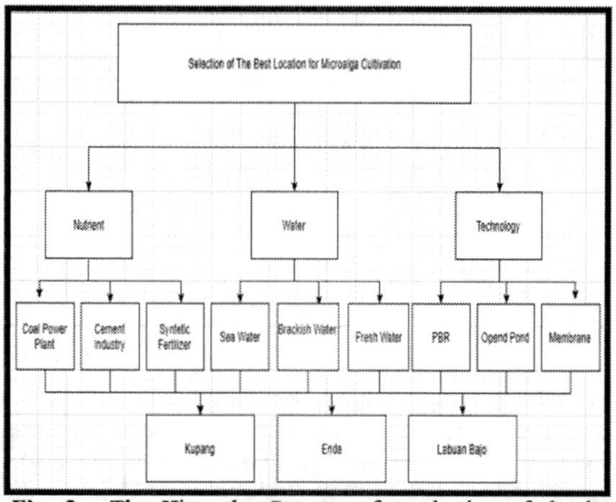

Fig. 2. The Hierarchy Structure for selection of the best location for microalgae cultivation.

After modelling the decision problem as shown above, data were collected from the decision maker by pairwise comparisons. In each level, the criteria are compared pairwise according to their levels of influence and based on the specified criteria in the higher level [7],[9]. These main criteria dan sub criteria for selection the best location for microalgae cultivation are listed in Table 1. Hierarchy structured with criteria, sub criteria and alternatives location for microalgae cultivation is given in Figure 2. There are three levels in the hierarchy structure for selection for the best location of microalgae cultivation. Saaty's 1-to-9 scale shown in Table 2 is used while comparing n criteria in the same level.

Table 1. Main criteria and sub criteria for selecting location for microalgae cultivation

Criteria	Sub criteria	Explanation
Nutrient	Coal power plant emission	CO_2 emission from coal power plant can be used as a nutrient for microalgae growth. NTT has 3 coal power plants there are PLTU Ende 2x7 MW, PLTU Kupang 2x15 MW and PLTU Rote 2x3 MW.
	Cement industry emission	CO_2 emission cement factory can be used as a nutrient for microalgaee growth. NTT has 2 cement industries in Kupang
	Fertilizer	Sufficient nutrient for microalgae growth.
Water source	Sea water	Salinity 30-35 ppt
	Brakish water	Salinity 6-29 ppt
	Fresh water	Salinity 0-5 ppt
Technology cultivation	Photobioreactor	In closed systems, algae growth conditions can be regulated much easily and precisely by engineers. Photobioreactor are also optimized to maximize algae cultivation.
	Opend Raceway Pond	The biggest advantage of these open ponds is their simplicity, resulting in low production costs and low operating costs. While this is indeed the simplest of all the growing techniques.
	Membrane	Growing microalgae on membrane minimizes water use and reduce harvesting cost.

Table 2. Saaty's 1-9 scale

Intensity of importance	Definition
1	Equal importance
3	Moderate importance
5	Strong importance
7	Very strong importance
9	Extreme importance
2,4,6,8	Intermediate values

In this research we use 3 experts. After the calculation of is finished pairwise matrix then it is continued by the arrangement of pairing matrix for doing normalization of interest level value of each elements on each hierarchy.

Table 3. Pairwise Comparison

Criteria/ Alternative	1	2	3	N
1	1	W12	W13	W1n
2	W21	1	W23	W2n
3	W31	W32	1	W3n
4	Wn1	Wn2	Wn3	1

Assessment in comparing between one criteria and the other criteria is independent of each other, and this may lead to inconsistency. Saaty [10] has proved that the consistency index of the ordered matrix can be obtained by the formula:

$$CI = (\lambda maks\text{-}n) / (n\text{-}1) \ldots\ldots\ldots\ldots(1)$$

Where:

CI= Consistency index

Λmaks = The largest eigen value of the ordered matrix

The limit of inconsistency is calculated using consistency ratio (CR) that is by comparison of consistency index (CI) with consistency random index value (RI). Consistency ratios can be formulated:

$$CR=CI/RI \ldots\ldots\ldots\ldots(2)$$

If the CR value is still less than 10%, then the inconsistency of opinion is still acceptable. The following is a list of random inconsistency indexes presented in Table 4.

Table 4. Index list of Random Inconsistency (RI)

n	1	2	3	4	5
RI	0.00	0.00	0.58	0.90	1.12

After AHP is finished then it is continued by modeling the productivity of *Nannochloropsis* sp as a bioavtur feedstock based on its productivity.

3 Result And Discussion

Based on the analysis of AHP, the obtained results are in accordance with the following figure.

Based on the above results, it is seen that the water source is the main criteria in the cultivation of microalgae with a weight value of 0.588. It is then followed by nutrient with a weight value of 0.159 and the last is technology with a weight value of 0.251.

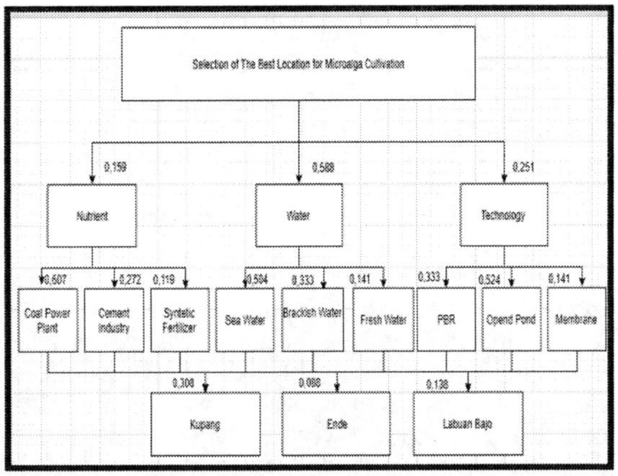

Figure 3. Result of selection for the best location for microalgae cultivation

Water is an absolute requirement of life for microalgae because it is a microalgae living habitat. Water criteria have priority of weighting result on sub-sea water criteria with value 0,524. Sea water has a salinity of 30-35 ppt which is a tolerable salinity for microalgae life. Sea water has a natural nutrient content that is used as a source of nutrients for microalgae life. The nutrient content of seawater is chloride (55%), sodium (31%), sulphate (8%), magnesium (4%), calcium (1%), potassium (1%) and the rest (less than 1%) comprises of bicarbonate, bromide, borax acid, strontium and fluoride[11]. The three main sources of salt in the ocean are from weathering of rocks on land, volcanic gases and the circulation of hydrothermal vents in the deep ocean.

Nutrient criteria have sub criteria of coal power plant emission with weight value of 0,607. In general, any source of CO_2 can be used for cultivating algae, however some options are more advantageous than others. Using CO_2 (~ 0.04 w%) and more must be supplied to facilitate optimal algae growth. The coal power plant's emissions can be used as a source of nutrients for microalgae growth because it contains high CO_2 content [12]. Microalgae require CO_2 to do photosynthesis so as to produce biomass. Biomass produced by microalgae can be used as a source of biofuel. According to the Ministry of Energy and Mineral Resources Regulation No 32nd 2008 [13] the definition of biofuels is that fuel is derived from vegetable materials or produced from other organic materials. In general, it takes an approximate mass ratio of 2.0 CO_2 to produce a quantity of algae biomass, e.g., around 2g of CO_2 are needed to produce 1g of ash-free dry algae. CO_2 purity is not required in microalgae culture so CO_2 emissions from coal power plant can be channelled directly to microalgae cultivation. Another advantage in utilizing CO_2 emissions from coal power plants in microalgae cultivation microalgae is that serves as carbon capture so as to reduce CO_2 emissions in the air. Water and nutrient management are used to tailor the biochemical composition of the produced algae biomass towards an economically viable biofuels feedstock.

In the criteria of technology, the sub-criteria of photobioreactor have weighting result of 0.524. The advantage of using photobioreactor is that the biomass produced is bigger than open pond, the cultivation condition can be regulated by the researcher, with low contamination risk and suitable for microalgae development using CO_2 emission as nutrient source.

Based on the results of AHP analysis it is seen that the priority of microalgae cultivation location is in the first position of Kupang city with a weighting of 0.308. This is supported by the existence of coal power plant located in Kupang. The existence of a coal power plant adjacent to the sea is in accordance with the weighting results stating that sea water is a priority in the selection of water sources. Then the technology used is photobioreactor which is considered more effective for mass scale cultivation of microalgae.

From the results it is assumed that microalgae used is *Nannochloropsis* sp which is a marine microalgae species with lipid content of 35-68% of its body so it can be used as source of bioavtur feedstock. Recent studies have reported that marine species *Nannochloropsis* sp are a potential microalgae for biofuel production owing to the fact that they are capable of producing lipid content as high as 27% (dry weight) and high biomass productivity [14].The screening of microalgae (*Chlorella vulgaris*, *Spirulina maxima*, *Nannochloropsis* sp., *Neochloris oleabundans*, *Scenedesmus obliquus* and *Dunaliella tertiolecta*) was done in order to choose the best one(s), in terms of quantity and quality as oil source for biofuel production. *Nannochloropsis* sp. proved to be suitable as raw materials for biofuel production, due to their high oil content (29.0 and 28.7%, respectively) [14].

The National Renewable Energy Laboratory (NREL) has extensive experience cultivating and manipulating microalgae to produce lipids or oils. NREL is now looking to re-establish its microalgae oil research in partnership with oil refiners, with a particular view towards jet fuel production [15].

Bio-jet fuel production by the blending process will be blended with Jet A (or JP5) and diesel to produce bio-jet fuel blend that can be used on a diesel engine. The US Navy has accepted bio-JP-5 as an alternative to JP-8 fuel [16].

Based on the assumption of microalgae biomass productivity cultivated with photobioreactor that is 24.489KL/Ha/yr [12,17]. The need for Avtur in NTT is 52.611 KL/Year. According to the KESDM appeal, the use of bioavtur 2% starts from 2016, which can be done using microalgae. The need for bioavtures 2% in NTT is 1.052,22 KL/Year which can be fulfilled from microalgae with productivity of 24.489kL/ha/yr and the area requirement for microalgae cultivation is 2.14 hectare.

4 Conclusions

Based on AHP results it is known that the best location for microalgae cultivation is in Kupang with nutrient source comes from steam power plant emission, type of water used is sea water and technology used is Photobioreactor. The Need of 2% bioavtur in ENT of 1052,22 kL/yr can be met by microalgae with productivity of 24489kL/ha/yr and the area requirement for microalgae cultivation is 2.14 hectare.

Acknowledgment

Faculty of Defense Management, Indonesia Defense University.

References

1. BPS-Statistic of Nusa Tenggara Timur Province, *Statistic of Nusa Tenggara Timur Province*, Catalog, 511 (2016).
2. H. Bifel, Gubernur: Presiden apresiasi pertumbuhan ekonomi NTT. Available online on http://www.antaranews.com/berita/613176/gubernur-presiden-apresiasi-pertumbuhan-ekonomi-ntt, Accessed on 10th July 2017.
3. Agency For The Assessment and Application of Technology, *Outlook Energy Indonesia 2016*, Catalog, 108 (2016).
4. The Ministry of Energy and Mineral Resourses, available online on http://ebtke.esdm.go.id/post/2014/08/26/656/pemerintah.dorong.pemanfaatan.avtur, Accessed on 10th July 2017.
5. Y.Chisti, Biodiesel From Microalgae, Biotechnology Advances 25, 294-306 (2007)
6. Ulum, M. Penumpang Pesawat Tujuan NTT Meningkat, available online on http://bali.bisnis.com/read/20170603/16/66860/penumpang-pesawat-tujuan-ntt-meningkat, Accessed on 10th July 2017.
7. M. Dagdeviren, S. Yavuz, N. Kilinc, *Weapon Selection Using the AHP and Topsisi methods under fuzzy environment, Expert system with Applications* **36**, 8143-8151 (2009).
8. L. Wang, J. Chu, J. WU, Int. J. of Production Economics, 107, 151-163 (2007).
9. E. Albayrak, Y. C. Erensal, Journal of Intelligent Manufacturing **15**, 491-503 (2004).
10. TL. Saaty, Int. J. Services Sciences, **Vol. 1**, No. 1 (2008)
11. K. Romimohtarto, and S. Juwana. *Biologi Laut*. Djambatan, 540 (2007).
12. International Energy Agency, *State of Technology Review-Algae Bioenergy*. 155 (2017).
13. The Ministry of Energy and Mineral Resources Regulation, No 32nd 2008 about Provision, Utilisation And Commerce Biofuel As Other Fuels
14. L. Gouveia and A. C. Oliveira, Journal of Industrial Microbiology and Biotechnology, **Vol. 36**, No. 2, PP. 269–274, (2009)

15. A. Darzins, *Algal Biofuel Technologies States Biomass/Clean Cities Web Conference*, National Renewable Energy Laboratory (2008)
16. M. Elmoraghy, I. H. Farag, International Journal of Engineering and Science, **Vol. 1**, PP 22-30 (2012).
17. H. Hadiyanto, and M.M.A. Nur. World Applied Sciences Journal, Vol **31 (5)**, 959-967 (2014)

Enhancing Biodiesel from Kemiri Sunan Oil Manufacturing using Ultrasonics

Slamet Supriyadi[1,2]; *Purwanto* Purwanto[3]; *Didi* Dwi anggoro[3]; *Hermawan*[4]

[1]Doctoral Program of Environmental Science, School of Postgraduate Studies, Diponegoro University, Semarang, Indonesia
[2]Department of Mechanical Engineering, Universitas PGRI Semarang, Semarang, , Indonesia
[3]Department of Chemical Engineering, Faculty of Engineering, Diponegoro University, Semarang, Indonesia
[4]Department of Electrical Engineering, Faculty of Engineering, Diponegoro University, Semarang, Indonesia

Abstract. Kemiri Sunan (*Reutalis trisperma* (Blanco) Airy Shaw) is a potential plant to be developed as biodiesel feedstock. The advantage of Kemiri Sunan seeds when compared to other biodiesel raw materials is their high oil content. This plant is also very good for land conservation. Due the increasingly demand for biodiesel, research and new methods to increase its biodiesel production continue to be undertaken. The weakness of conventional biodiesel manufacturing process is in the mixing process in which mechanical stirring and heating in the trans-esterification process require more energy and a longer time. A higher and stronger mixing process is required to increase the contact area between the two phases of the mixed substance to produce the emulsion. Ultrasonic is a tool that can be useful for a liquid mixing process that tends to be separated. Ultrasonic waves can cause mixing intensity at the micro level and increase mass transfer, so the reaction can be performed at a much faster rate. This study is to figure out the effect of ultrasonic irradiation on the transesterification process of biodiesel from Kemiri Sunan Oil.

1 Introduction

Indonesia, which is currently a net importer of fossil fuels, should make biofuels the backbone of fuel provision as a substitute for fossil fuels whose reserves are depleted. Sources of biofuels in Indonesia are abundant where more than 30 species of plants can be a source of biofuels. One of the potential plants to be developed as biodiesel feedstock is Kemiri Sunan (*Reutealis trisperma* (Blanco) Airy Shaw). In addition to its high oil content, Kemiri Sunan also has specific oil characteristics, relatively rapid growth, wide area development (from lowlands up to 1,000 m above sea level), high productivity, and very suitable for land conservation [1, 2].

Biodiesel is a fuel consisting of a mixture of mono-alkyl esters of long chain fatty acids, obtained from vegetable / vegetable oils or animal fats. As a fuel, biodiesel has several advantages: renewable energy, low CO_2 and SOx emissions, biodegradable, non-toxic, high Cetane number and does not require engine modification [3,4]. However, if biodiesel is made with raw materials derived from edible sources, it will compete with food requirements and is still not competitive compared to fossil fuel oil [5]. Therefore, non-food raw materials are more likely to be developed. Kemiri Sunan (*Reutalis trisperma* (Blanco) Airy Shaw) is a potential crop to be developed as biodiesel feedstock in Indonesia, whose seeds contain 52% of the oil content [2]. The productivity of Kemiri Sunan trees is 4-5 times greater than Jatropha and 2 times higher than Palm Oil [6,7,8] The process of making biodiesel includes esterification, transesterification and leaching which requires a strong mixing to make sufficient contact between vegetable oils with catalysts or alcohols. The mixing and stirring process between the biodiesel feedstock is carried out at the reactor, where the main raw material and the catalyst undergo a transformation. Reactor performance is very important in the operation of biodiesel production, because it will affect the efficiency and productivity of production process.

Conventional processing of biodiesel is performed manually in reactors by mechanically mixing and heating in the transesterification process which requires considerable energy, labor and a longer processing time [9,10] Therefore, the design of the reactor which is in accordance with the type of biodiesel feedstokes becomes important. Ultrasonic waves can be an alternative to be applied to biodiesel production processes to replace stirring and heating functions in transesterification. The waves can serve to increase mixing intensity at the micro level and increase mass transfer, so that the reaction process can be done at a much faster rate [11.12]. Since the transestrification process with the use of ultrasonic waves has not been done in processing of Kemiri Sunan for biodiesl, this paper will discuss the effectiveness of ultrasonic wave in processing biodiesel from Kemiri Sunan.

* Corresponding author: slametsupriyadi@upgris.ac.id

2 The Potential of Kemiri Sunan (*Reutealis trisperma* (Blanco) Airy Shaw) in Indonesia

Kemiri Sunan originally from the Philippines, has grown widely in Indonesia. The population of Kemiri Sunan crop is concentrated in Garut and Majalengka regency, West Java and in 2008 started to spread to various places that have various agro-ecosystem such as Central Java, East Java, East Nusa Tenggara, West Nusa Tenggara, Riau, Jambi, Bangka, And East Kalimantan. The data of Kemiri Sunan tree distribution in Jakarta are 3,500 trees, Bekasi as many as 30,000 trees, Kuningan as many as 10,000 trees, Majalengka as many as 10,000 trees, Jati Gede as many as 10,000 trees, Bandung as many as 3,000 trees, Ngawi as many as 40,000 trees, Lamongan as many as 13.000 trees, Nusa Penida -Bali as many as 15,000 trees, Lombok as many as 14,500 trees and Timor as many as 20,000 trees.

Bogor Agricultural University has prepared a land area of 200 hectares belonging to the University to be a pilot development project of Kemiri Sunan crops. Head of Department of Energy and Mineral Resources of West Java Province has data of land groups that are potential for planting of kemiri Sunan in West Java province: 591,000 hectares of critical land, 268,000 hectares of mine reclamation, 810,000 hectares of CSR from geothermal developers, 224,000 hectares Non- Active land, 16,000 km border of the main river in West Java, 2,000 km of roads along the province of West Java.

Currently, it is planned that around 5,000 hectares of area in Production Forest that belongs to Landak, Mempawah and Kubu Raya districts, will be planted with Kemiri Sunan plant. This area will be made forest area with special purpose which is managed by Tanjungpura University.

Chemical and Physical properties of Crude Oil from Kemiri Sunan seeds are: Acid Numbers, 13,26 mg KOH/ g; Fat Free Acid Content, 6,63%; Water content 9,6%; Density 985,49 kg / m; Kinematic Viscosity 26,57mm2 / s (c St), rendement 43,33% [8]. Furthermore, the oil composition consists of 10% palmitic acid, 9% stearic acid, 12% oleic acid, 19% linoleic acid, and 51% a-elaeostearic acid. Acid content a-elaeostearic explains the presence of toxins in oil [13]

Development of Kemiri Sunan in post mining reas is expected to be made as:

1. Land conservation facilities to reforest critical lands to improve the quality of the environment;
2. The source of diversified supply of raw materials to produce environmentally friendly biodiesel oil (one hectare of land with 100-150 Kemiri Sunan trees can produce 6-8 tons of biodiesel per year) instead of fuel oil and reduce dependence on imports of fuel oil;
3. Increasing the economy of the community by creating employment and business development, investment in the country, the development of downstream agricultural sector, as well as increased value-added products in the country.
4. Improving environmental quality by reducing greenhouse gas emissions, reducing air pollution levels, and improving air quality, public health, and community welfare.
5. Increasing national energy security through the provision of biodiesel from plants that do not compete with food and industrial raw materials.

3 Biodiesel Processing

The basic process of making biodiesel is the transesterification process. Transesterification is a chemical reaction of vegetable/animal fats with alcohols with the aid of a catalyst to produce methyl methyl ester (biodiesel) and glycerin. The transesterification process includes three stages: (1) mass transfer between oil and alcohol; (2) transesterification reaction; And (3) equilibrium formation [14]. Methanol is used as an alcohol in the reaction process because it is cheaper and easily available on the market. Several parameters affect biodiesel yields, among others the amount of catalyst, reaction temperature, alcohol to oil ratio and mixing intensity [10,14, 15]

A study on the effect of the amount of alcohol (methanol) on the production of biodiesel from Kemiri Sunan oil was conducted The results concluded that the mixed alcohol was about 1.6 times the theoretical amount, while more than 1.75 the theoretical amount did not speed up the reaction, even complicating the separation of glycerin from biodiesel [15]..

A two-stage transesterification method using oil extracted from Kemiri Sunan seed (50 l), methanol (11.5 l), KOH (96 g), water (10 l), chloroform and phenolphtalein indicator was conducted. The results showed that 44 liters of Kemiri Sunan biodiesel and 6 liters of glycerol were obtained. Using this method, yields of biodiesel from the Kemiri Sunan crude oil was about 88% [7].

The catalyst used in the production of biodiesel serves to decrease the activation energy of the reaction, so that the reaction can take place more quickly. Two commonly used catalysts are transesterification with alkaline catalyst and transesterification with acid catalyst. However, alkaline catalysts can provide better performance than acid catalysts, ie faster reaction, lower reaction temperature (at room temperature) and greater oil conversion [7, 10,15, 16] If not using a catalyst, high temperatures, long pressures and reaction times are required, with the consequence of which is a relatively large cost.

The most commonly used basic catalysts are Sodium Hydroxide (NaOH), Potassium Hydroxide (KOH), Sodium Methoxide ($NaOCH_3$) and Potassium Methoxide (CH_3OK). The use of lipase (enzyme) as a catalyst in the biodiesel production process which requires more complex processes and tools, has begun to be developed and aiming to overcome the problem of by-products at the end of the process. However, the constraint faced in the use of enzyme catalysts is expensive lipase prices.

The production of biodiesel from Kemiri Sunan oil, with 2 stages of reaction was examined, ie esterification with H_2SO_4 catalyst and transestrification using NaOH catalyst. In this study, the effect of catalyst concentration on biodiesel products and their characteristics was observed. The results showed that the yield of biodiesel increased with the increase of catalyst concentration from 0.5-1.0 wt%, then with the increase of catalyst concentration from 1.5-2.0 wt% made the yield decrease. The optimum yield was achieved at 84.7%, with a catalyst concentration of 1 wt% under reaction conditions of 65 ° C., a reaction time of 1 hour and a methanol / oil ratio of 1: 2 (wt / wt) [17].

The effect of catalyst concentration of NaOH on reaction time with at temperature 50 and 60°C with raw material of sunflower oil was examined by Meka et al. [18]. The molar ratio of methanol / oil is maintained at 6: 1. The study concluded that the reaction time decreased proportionally with increasing catalyst concentration from 1% to 2%. However, soap was formed when the catalyst concentration was more than 2%. Ataya et al. [19] conducted a transesterification study of Canola oil with results showing that the conversion of oil to biodiesel increased when the concentration of NaOH catalyst increased from 1-3%. The findings of Rashid and Anwar [20] showed that the production of biodiesel from rapeseed, increased at a KOH catalyst concentration between 0.25-1.5 %, with an optimum value at 1% concentration. Results are consistent with the results of the research by Meher et al. [21].

In biodiesel production, adequate mixing is required to create sufficient contact between oils with alcohol and catalysts. The use of mechanical stirring (mechanical stirring) is not yet maximized to produce the mixing process between the reactants, therefore ultrasonic gets attention in the production of biodiesel.

Ultrasonic energy transfers to the liquid and create a great vibration, which forms a cavitation bubble. As the bubble bursts, a sudden contraction in the liquid occurs, causing mixing in the bubble area. Ultrasonic waves cause intense mixing so that the reaction can continue at a much faster rate without involving high temperatures.

The ultrasonic irradiation method proved to be efficient, time-saving and economically functional to produce biodiesel fuel. Colucci et al. investigated the feasibility of ultrasonic use in mixing to produce biodiesel from soybean oil. The results showed that the reaction rate constants were three to five times higher than those reported in the literature for mechanical stirring. This was explained by the fact that the interface and microscopic and macroscopic areas of the bubble increased when the 20 kHz ultrasonic wave was applied to a two-stage system reaction [22,23].

The effect of ultrasonification on the droplet size of biodiesel blends was investigated. The result showed that ultrasonic mixing resulted in dispersion with an average droplet size 42% smaller than that produced using standard mechanical stirring implants [24]. Stavarache et al. [25] also conducted extensive studies on the application of ultrasonic energy in transesterification of commercial edible oil. The researchers concluded that by using ultrasonic, reaction time is much shorter than with mechanical stirring. Thus it can be concluded that using ultrasonic irradiation for transesterification of vegetable oils in biodiesel production is more efficient than using conventional methods.

Topare et al [26]. synthesized biodiesel from soybeans by using ultrasonic waves. The results showed that the amount of catalyst and alcohol required could decrease substantially compared to conventional processes. The ultrasonic technique resulted in short reaction time, shorter glycerin phase separation time (1-2 hours), and the glycerine obtained had a better purity level than that obtained in conventional process. Likewise the process was more energy efficient than heating and stirring in conventional processes [26].

Cancela et al.[27] conducted research on the effect of ultrasonic conversion of used cooking oil into biodiesel. In this study, with different process parameters such as reaction time (30-90 min), the amount of catalyst (0.5 - 1% NaOH weight) and temperature (20-40 °C) were also analyzed to obtain higher conversion. The ratio of methanol to 6: 1 molar oil, 0.5% of the amount of catalyst and 30 °C was sufficient to complete the process in 60 minutes. The results obtained in this study confirmed that ultrasonic assisted transesterification is a fast and efficient method of producing biodiesel from waste cooking oil despite low reaction temperatures [27].

4 Comparison between ultrasonic technique and mechanical stirring on Kemiri Sunan oil processing

The process of making biodiesel was carried out in Laboratory to establish optimum conditions. The transesterification treatment of mechanical stirring used mixture of methanol-catalyst by10%, 15%, 20% (v/v), with KOH and NaOH 0.4%, 0.6% (w/v). The process of transestrification was conducted in a way by heating until the temperature reaching 60° C and stirring for 1hour. The next process is the emulsion was keeping in a separation tank for a minimum of 3 hours, Further step was separating between biodiesel and glycerol. The ultrasonic experiment was carried out using the same amount of catalyst, reactant and molar ratio between methanol and crude oil as performed on transesterification using a mechanical stirrer. The frequency ultrasonic was 20 kHz [26] and reaction times were conducted at, 10 and 15 minutes. After completion of the reaction, the emulsion was transferred into a separation tank for phase separation.

The physical and chemical properties of biodiesel from the raw material of Kemiri sunan seed oil which was observed in this research were: water content, density, kinematic viscosity, cetane number. Biodiesel Standard according to Standar Nasional Indonesia (SNI) 2015 was used.

Density denotes the weight ratio of unit volume of a liquid at a certain temperature. Biodiesel with density value exceeding provisions standard requirements of biodiesel will improve engine wear, have high emission values, and damage engine components, otherwise this density value can be used to determine cetane number.

Viscosity is the most important biodiesel property because it affects the system of pressurized burning.

Lower biodiesel viscosity produces easier pumping and better spray pattern. The viscosity of biodiesel is affected by the amount of trigeliserides does not react with Methanol, the fatty acid composition of methyl constituents esters, as well as compounds between monoglycerides and Diglycerides having quite high polarity and the molecule weight is quite high

The value of cetane number is a measure of quality burning quality or burning waiting time. This is related to the time required for liquid fuel to burn after pumping to the combustion engine. The higher the Cetane number, the faster the burning time. This leads to burning More effective and efficient, smoke, CO and HC emission [28]. This number will be increasing with the increase of long carbon chain fatty acids and degrees of saturation.

Water content in vegetable oil will cause the occurrence of hydrolysis which raise the levels of free fatty acids in vegetable oil. The presence of excess water can cause some of the reaction to change into Saponification reaction between free fatty acid resulting from Hydrolysis of oil with the base catalyst to produce soap. Soap will reduce efficiency of the catalyst thereby increasing the viscosity, forming gel, and complicating the separation of glycerol with methyl esters.

Table 1 represents the result obtained on the conversion of Kemiri sunan oil, with mechanical stirring and untrasonic methods Yield of biodiesel for mechanical stirring method was ranged from 78.44 to 85.5%. while for ultrasonic method was ranged from 90.2 to 98.5 %. The yield of biodiesel produced was affected by raw materials on crude oil manufacture, origin of raw material (where to grow), timeliness of harvest,, seed storage, genetic factors, and moisture content [8]. Besides, it was also strongly influenced by the reaction between the oils with a mixture of methanol and base catalysts [23].

The high conversion obtained on the ultrasonic method can be caused by the acceleration and high mass transfer between methanol and oil, as well as the formation of micro-emulsions resulting from the ultrasonic cavitation phenomenon [23]. The ultrasonic method yielded 98.5% at 10 minutes while the mechanical stirring resulted in a much lower conversion extension of 85.5% at 60 minutes. The indication showed that the reaction under the ultrasonic reactor was much faster than the mechanical stirring process.

The transesterification process either by mechanical stirring or ultrasonic irradiation showed that it was capable of changing the saturated Kemiri Sunan oil into a diesel fuel as can be seen oin Table 2. The transesterification process is aimed to transforming [tri, di, mono] glycerides with high molecular weight and viscosity dominating the composition of vegetable oils into Fatty acid methyl ester (FAME) having lower molecule and viscosity.

Table 2 shows that properties of biodiesel from Kemiri sunan oil fulfilled the biodiesel requirement standard (SNI-2015). Density of biodiesel 867 kg/m^3 (mechanical Stirring) and 856 kg/m^3 (ultrasonics) have fulfilled the requirements set of 850-890 kg/m^3.

Table 1. Biodiesel yield from Kemiri Sunan (Aleurites trisperma) seed oil

Treatment	Biodiesel Yield (%)	
	Mechanical Stirring	Ultrasonic
Methanol 10%-KOH 0.4%	82.7	97.4
Methanol 15%-KOH 0.4%	85.3	98.5
Methanol 20%-KOH 0.4%	85,5	98.4
Methanol 10%-KOH 0.6%	83.1	96.3
Methanol 15%-KOH 0.6%	84.1	97.4
Methanol 20%-KOH 0.6%	85.2	97.9
Methanol 10%-NaOH 0.4%	79.4	91.5
Methanol 15%-NaOH 0.4%	78.4	90.2
Methanol 20%-NaOH 0.4%	80.2	92.5
Methanol 10%-NaOH 0.6%	78.4	96.5
Methanol 15%-NaOH 0.6%	79.3	97.4
Methanol 20%-NaOH 0.6%	80.3	96.2

Kinematic Viscosity of biodiesel produced of 5.42 mm^2 / s (cSt) (Mechanical Stirring) and 4.82 mm^2 / s (cSt (ultrasonics) indicated that the value of this viscosity already meet the requirements of biodiesel standards ie.
ranging from 2.3-6.0 mm^2 / s (cSt). The high of biodiesel oil viscosity is due to the bond of intermolecular hydrogen in the outer acid of the carboxyl group. The viscosity value is proportional to density, the higher the viscosity then the density will be higher [3].

Table 2. Properties of biodiesel from Kemiri Sunan (*Aleurites trisperma*) seed oil

Parameter	Methods	
	Mechanical Stirring	Ultrasonics
Kinematic viscosity mm^2/s (cSt)	5.42	4.82
Density (kg/m3)	867	856
Cetane number	56.3	59.8
Moisture content (%)	0.05	0.05

The biodiesel production results show that Cetane numbers of biodiesel has been fulfilled the minimum requirement of biodiesel which is 51.

5 Conclusion

The ultrasonic technique prooved to be much better than mechanical stirring in producing biodiesel from Kemiri Sunan. The Kemiri Sunan crude oil conversion was 98.5% compared to mechanical stirring 85.3%, with the methanol to oil molar ratio of Methanol 15%v/v/-KOH 0.4% w/v. The reaction time of ultrasonic irradiation required only 15% compared to reaction time required by mechanical stirring method. Properties of biodiesel meet the requirements of biodiesel standards. It can be concluded that ultrasonic is a more valuable tool for the transesterification of Kemiri Sunan oil to biodiesel compared to the mechanical stirring process.

References

1. Syafaruddin and A. Wahyudi, Perspektif Vol. **11** No. 1 /Juni 2012. Hlm 59 – 67 (2012).
2. S. Supriyadi, P. Purwanto, D.D. Anggoro, Hermawan, The Potential of Kemiri Sunan as Feedstock for the Production of Biodiesel, *ASL*, Volume **23**, number 3, March (2017).
3. A. Demirbas, *Ener Conv Mang* **50** 14–34(2009)
4. M. Canakci, J H. Van Gerpen, *Trans of the ASAE* Vol. **44** (6): 1429–1436 (2001)
5. M. K. Lam, K. T. Lee, M. Abdul Rahman, *Bio Adv* **28**, Issue 4, July–August 2010, Pages 500–518 (2010)
6. L. Lathifan., Cahyadi N., Kusmaningrum, *Reka Integra No.**04** ,Vol.02, Oktober* (2014)
7. A, Aunillah, and Pranowo D., *buletin Ristri 3(3):193-200 November* (2012).
8. D. Hendra, *Penelitian Hasil Hutan Vol. **32** No. 1,: 37-45 Maret* (2014)
9. M. J Haas, J. M. Andrew, C. Y.Winnie, A. F Thomas., *Bio.Tech.* **97** (2006) 671–678 (2006)
10. P. Chand,. Iowa State University Digital Repository @ Iowa State University (2008)
11. L. B Soon, Z. M. R Anika, H.Sulaiman, *International Journal of Automotive and Mechanical Engineering (IJAME)*, Volume **8**, pp. 1396-1405, July-December (2013)
12. H. Brian and J. H. Van Gerpen, *Biofuels* **3** (4), 479–488 (2012)
13. H.A.M. Vossen, and B.E. Umali, *Plant resources of South-East Asia No* **14.** Prosea Foundation. Bogor, Indonesia. (2002)
14. L. T. Thanh, K. Okitsu, L. V. Boi, and Maeda, *Catalysts* 2012, *2*, 191-222 (2012)
15. D. Pranowo, S. Muhammad, P. Bambang, Maman ., Asif ., Sumanto, , *Pembuatan Biodiesel dari Kemiri Sunan (Reutealis Trisperma (Banco) Airy Shaw) dan Pemanfaatan hasil Samping*, IAARD Press, Jakarta. (2014)
16. Z. Helwani, M.R. Othman, N. Aziz, W.J.N. Fernando, J. Kim, *Fuel Process*. Technol., **90**, 1502–1514(2009).
17. Holilah, T. P. Utami, D. Prasetyoko, *Jurnal MIPA* **36** (1): 51-59 (2013).
18. P.K. Meka, V. Tripathi, and R.P. Singh, *J Oleo Sci*, **56**(1), 9-12,(2007).
19. .F.Ataya, M. A Dube, M. Ternan, *Energy Fuels*, **22** (5), pp 3551–3556 (2008),
20. Rashid, U., & Anwar, F. Energy & Fuels, **22**, 1306–1312. (2008).
21. L.,C Meher, V.S. Dharmagadda, and S.N. Naik, *Bioresour Technol*, **97**(12), 1392-1397(2006)
22. H.D. Hanh, N.T. Dong, C. Starvarache, K. Okitsu, Y. Maeda, and R. Nishimura, *Energy Convers Manage*, 49, 276-280(2008).
23. J.A. Colucci, E.E. Borrero, and F. Alape, *J Am Oil Chem Soc*, **82**(7), 525-530. (2005).
24. P. Wu, Y. Yang, J.A. Colucci, and E.A. Grulke, *J Am Oil Chem Soc*, **84**, 877-884. (2007).
25. C. Stavarache, M. Vinatoru, and Y. Maeda, *Ultrason Sonochem*, **14**, 380-386(2007)
26. N S. Topare, D. P. Kiran, N. Parikshit, S. Akash, J. Pratik, *Emerging Trends in Chemical Engineering* Volume **2**, Issue 1, pp.1-8 (2015)
27. Á. Cancela, R. Maceiras, V. Alfonsín, Á. Sánchez, *European Journal of Suste Dev* **4,** 2, 401-406 (2015)
28. G. Antolín, F. V. Tinaut, Y. Briceño, V. Castaño, C. Pérez, and AI. Ramírez. *Bio Tech* **83**: 111–114. (2002).

Bioenergy Potential Based on Vinasse From Ethanol Industrial Waste to Green Energy Sustainability

Nani Harihastuti [1,*], and *Bekti* Marlena [1]

[1]Centre of Industrial Pollution Prevention Technology, Jl. Ki Mangunsarkoro No. 6 Semarang 50136 - Indonesia

Abstract. The waste water from alcohol industry is called vinasse has a high organic content, with BOD5 = 109.038 mg / l, COD = 353.797 mg / l and TSS = 7200 mg / l, pH 4-5 with a temperature of around 40-50°C. The current treatment of alcohol waste water, most still using facultative anaerobic technology with open ponds that are only covered with HDPE plastics. This technology produces less optimal biogas and has a weakness that is the hydraulic residence time (HRT) for long (40-50 days), wide land needs, low COD reduction efficiency as well as high risk of fire and leakage of biogas release high to trigger the occurrence of greenhouse gas and global warming effects. Development of technology with innovation reactor integration model Fixed Dome-Hybrid Anaerobic Filter aims to expand the contact area between the substrate and microbial with modification of the substrate flow system and the area of the filter and integrate with the gas accumulator. The design of this Fixed Dome-Hybrid Anaerobic filter integration model technology, has the advantage of producing optimal bioenergy with CH4 more than 50% content with decrease of COD more than 85% and hydraulic residence time of about 10 (ten) days, bioenergy result is renewable energy made from raw material vinasse from alcohol industrial waste which can be utilized for fuel substitution on the distillation process or boiler process of the industry in a sustainable and cleaner environment.

1 Introduction

Indonesia will face the threat of energy crisis in 2020-2030 [1]. One sector that will be affected by the energy crisis is industrial sector. The industrial sector that has been rely on fossil fuel sources must seek alternative energy sources such as renewable energy.

The Ministry of Industry ensuring the sustainability of industrial development through savings and energy diversification policies and greater attention to the development of renewable energy sources through Indonesia Government Regulation No. 14 year 2015 on the National Industrial Development Master Plan (RIPIN) of 2015-2035. Development of renewable energy sources is one of the government policies in the provision of energy needs for industry [2]

One potential renewable energy sources that has not been explored to its full potential is the energy potential from waste of alcohol industry. According to the Indonesian Ethanol Association (ASENDO), there are 16 large industries and some alcohol centers of 150 – 200 Small-Medium Enterprises (SMEs). The problem faced by both large and SMEs was environmental pollution from wastewater that is not managed properly.

Some researchers have proposed some treatments of this wastewater [3-8]. Generally anaerobic treatments [9-10] is chosen to treat this wastewater, especially anaerobic membrane [11], UASB [12-13], anaerobic Baffled Reactor [14], Anaerobic Sequencing Baffled Reactor [15]. Nevertheless, there are also several studies using ultra filtration [16] and catalytic thermolysis [17].

Large scale alcohol industries in Indonesia has utilized their wastewater to produce methane and organic fertilizer, but SMEs still discharged directly into environment. One of the industrial centers of alcohol located in Polokerto district, Sukoharjo regency, Central Java province. So far, almost all of the existing alcohol SMEs in Polokarto region discharge their wastewater directly to local irrigation channel towards Samin river. It makes water in irrigation channels unfit for use, even contaminate the well water that people use for cooking and washing [18].

In other hand, SMEs can utilize alcohol waste into bioenergy sources as an alternative energy sources as well as prevent environmental problems.

2 Method

This review summarizes field study of ethanol production, vinasse characteristics and production in Polokarto center, and literatures study on vinasse treatments and potential usages. Experience in treating organic wastewater, especially Palm Oil Mill Effluent (POME) into biogas as energy gives contribution in thought in vinasse treatment. Design of Fixed Dome High Rate Anaerobic Reactor (FDHRAR) is used as a model to convert vinasse into biogas. aims to expand the

* Corresponding author: nanisoeharto@yahoo.com

contact area between the substrate and microbial with modification of the substrate flow system and the area of the filter and integrate with the gas accumulator.

Primary attention is focused on the calculation of bioenergy potential obtained from digestion vinasse to methan.

3 Results and discussion

3.1 Characterization of vinasse

Vinasse is the residual remaining from the fermentation and distillation of alcoholic liquors. (Merriam-Webster Dictionary). Characteristic of vinasse SMEs in Polokarto are summarized in table 1.

Table 1. Typical characteristic of Polokarto vinasse

No	Parameter	Unit	Concentration
1	BOD_5	mg/l	109,038
2	COD	mg/l	353,797
3	TSS	mg/l	7,200
4	pH	-	4 - 6
5	Temperature	°C	40-50

There are about 115-150 business units SMEs in Polokarto center, produce an average of 1 m3 alcohol per day each units from 4 m3 fermented molasses. It is much smaller than distillery industry in India, which is reaches 1 : 4-10 L molasses [8]. The result of analysis of vinasse in table 1, the organic content is very high , look at the parameters of BOD and COD. The summarized that the waste water of vinasse is potential as renewable energy source from biomass.

In general equation the reaction of methane formation as follow [19] :

$$C_xH_yO_2 + (x - \tfrac{1}{4} y - \tfrac{1}{2} z) H_2O \longrightarrow (\tfrac{1}{2} x - 1/8\, y + \tfrac{1}{4} z) CO_2 + (\tfrac{1}{2} x - 1/8\, y + \tfrac{1}{4} z) CH_4 \quad (1)$$

3.2 Treatment by Modified of FDHRAR

BBTPPI has developed the Fixed Dome High Rate Anaerobic Reactor (FDHRAR) that has been applied in the palm oil industry [19]. The reactor is designed to produce bioenergy with the innovation of contact expansion between substrate and microorganism, to 30% filter volume. It would lead to increase the degree of degradation of complex compounds into simple organic compounds with a faster residence time (designed around 10-15 days). Calculated efficiency of COD removal will increase around 74,97% and bioenergy production increased with the content of CH_4 reached 47,5 %.

Development of FDHRAR by filter addition with innovation reactor integration model Fixed Dome-Hybrid Anaerobic Filter aims to expand the contact area between the substrate and microbial with modification of the substrate flow system and the area of the filter around 60-70 % and integrate with the gas accumulator to accommodate and regulate the flow of bioenergy production, thus minimizing the loss or "losses" of the biogas produced, and improve production efficiency, controlling the effects of gases and global warming.

It is in accordance with the opinion of Borja R, Banks CL and Sanches E, [20] that anaerobic treatment aimed to accumulating and utilizing the reformed gas must be performed on a reactor that has a special shape instead of an open pool as it is today.

Figure 1 is a picture of Fixed Dome Hybrid Anaerobic Filter technology innovation model of bioenergy generator from raw material of industrial waste of ethanol.

Through this modification, the efficiency of bioenergy production from alcohol waste will increase from 47.56% [19] become more than 50%, and the decrease in COD load can increase from about 74.97% become more than 85%. Hydraulic retention time will be shorter from 10-15 days to 6-8 days.

Seeing the characteristic properties of the waste water alcohols having organic content higher, then the stored energy potential of biomass is high, so before it is processed in the WWTP needs to discover the potential of this energy by passing on a reactor biogas digester to accumulate and utilize gases.

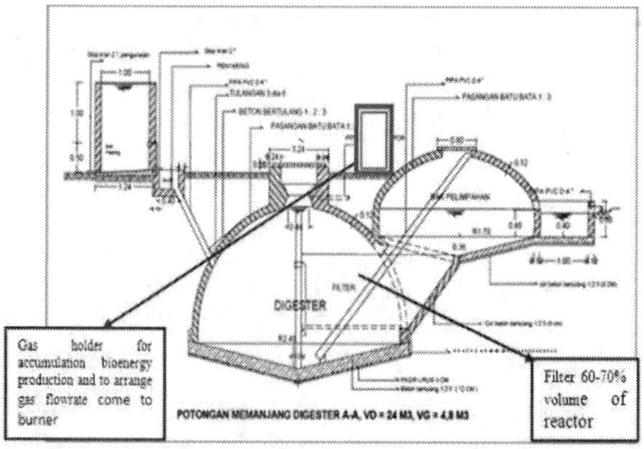

Fig. 1. Rector Design for the process of Vinasse Digestion

3.3 Bioenergy potential

Considering the characteristic properties of the vinasse with high organic content, it is necessary to extract the energy potential by passing it on a biogas digester reactor to accumulate and utilize the anaerobic reform gas before being processed in WWTP. As a fuel that has a high enough calorific value and the number is quite a lot.

According to Benefield, Larry.H [21] on each 1 kg reduction of COD in the anaerobic process will produce 350 liters of methane (CH4). The reshuffle process will reduce the COD value by 80-90%, so the next processing will be lighter load, and reduce the need for land.

Assumption of 75% COD is processed per day, then the potential of methane production in Polokarto's alcohol industry center is 307,045 m3 / day. If the calorific value of biogas approximately 4,500 - 6,300 kcal / m3, it will get energy of 1.381.703 - 1934.383.5 kcal / hari.

The average energy requirement in the distillation process is 2700 kcal / l of alcohol production [22](Navarro, Del C. Sepúlveda, & Rubio, 2000) so that the resulting bioenergy will be used to produce alcohols of approximately 511.7-716.4 m3.

It is expected that the bioenergy gained from the process of alcohol waste digestion can be used again as a renewable alternative energy in the process of separation (distillation). The separation process is the most energy-intensive process [22] and currently the energy used is firewood. Decreased availability of timber is currently in line with the decreasing forest area due to changing land use.

The scarcity and high price of firewood make the craftsmen start to switch to LPG fuel, but because LPG price is increasing every day it makes many craftsmen close their business. Looking at the energy potential contained in large biogas (theoretical calculations), the resulting bioenergy can be utilized as a renewable energy source (EBT) for use in the continuous separation / distillation process in the alcohol industry

From the use of such raw materials, the volume of waste generated by 2.7 m3 / industry / day x 150 industries, the amount of waste generated by industrial alcohol centers amounts to approximately 405 m3 / day. The quality of wastewater produced has BOD5 = 109.038 mg / l, COD = 353.797 mg / l and TSS = 7200 mg / l (BBTPPI, 2016). From these characteristics both in quality and quantity, industrial waste of alcohol is very potential to be used as raw material of biogas (bioenergy) energy sources (source: initial assessment of BBTPPI industry team).

4 Conclusion

Liquid vinasse wastes generated from the industrial production process of alcohol have high organic content, large amount and continuous production, so it is potential as raw material for anaerobic digestion process to produce renewable energy source from sustainable biomass (bioenergy).

The digestion process is carried out in a specially designed reactor with innovative Fixed Dome Hybrid Anaerobic Filter technology that will produce optimal bioenergy with fast retention time and impact on land use savings.

The company will get added value from renewable bioenergy sources that can substitute on other energy usage at factory operations.

The surrounding environment will become cleaner and healthier control of greenhouse gases and global warming.

5 Acknowledgments

Acknowledgments are shared with colleagues of the BBTPPI especially water laboratory Semarang, who have assisted in field surveys, sampling and analyzing the characteristics of vinasse.

References

1. A. Hidayat, Indonesia terancam krisis energi (2016) Retrieved from https://m.tempo.co/read/news/2016/05/24/090773660/pada-2020-2030-indonesia-terancam-krisis-energi
2. Pusat Komunikasi Publik Kementerian Perindustrian. (2015). Rencana Induk Pembangunan Industri Nasional 2015 - 2035.
3. G. J. Sheehan and P. F. Water Research, **14**, 257-277 (1980)
4. S. Mohana, B. K. Acharya, D. Madamwar, Journal of Hazardous Materials, **163**, 12–25 (2009)
5. S. R. Lekshmi, IJEEM, **4**(4), 339–344 (2013)
6. S. Ayub and S. Usmani, IJRET, 3 (2), 204-214 (2014).
7. M.P. Wags and P. D. Nemade, IJIRAE, **2**, 30-40 (2015)
8. D. Ariyanti and H.Hadiyanto, *Bulletin of Chemical Reaction Engineering and Catalysis*, **7**(3), 179-184 (2013)
9. H.Hadiyanto, D. Ariyanti, A.P.Aini, D.S.Pinundi, Energy Procedia, 47,108-112(2014)
10. X. L. Melamane, P. J. Strong and J. E. Burgess, S. Afr. J. Enol. Vitiv, 28, 1, 25-35 (2007)
11. H. Harada, S. Uemura, A. C. Chen and J. Bioresource Technology, **55** , 215-221 (1996)
12. B. Wolmarans, and G. H. De Villiers, Water SA, **28**(1), 63-68 (2002)
13. W. P. Barber , and D. C. Stuckey, Wat. Res. **33**, 7, 1559–1578 (1999).
14. M. L. Tansengco, D. L. Herera, J. C. Tejano. eJBio, **12**(4), 367–373 (2016)
15. Z. V. P. Murthy & L. B. Chaudari, Global NEST Journal. 11,2, 235-240 (2009)
16. A. Kadam, and K. Upahdhyay. Jr. Of Industrial Pollution Control, **28**(1),51-56 (2012)
17. Wicaksono, B. E. (2016). Tercemar Limbah Ciu, Kali Samin Sukoharjo Hitam Pekat Sukoharjo. Retrieved from http://www.solopos.com/2016/10/31/tercemar-limbah-ciu-kali-samin-sukoharjo-hitam-pekat-764880
18. N. Harihastuti and B. Marlena, Proceeding of Energy Seminar, UNDIP (2016)
19. R. Borja, C. L. Banks, and E. Sanches, Jurnal of Biotechnology, 45, 125-135 (1996)
20. L. D. Benefild, and C. W. Randal, Prentice Hall Inc. Englewood Clifts (2001)
21. A.R. Navarro, M. D. C. Sepulveda and M.C. Rubio, Waste Management, 20 (7), 581-585 (2000)

Optimization of Methane Gas Formation Rate with The Addition of EM4 Starter-made from Tofu Liquid Waste and Husk Rice Waste Using Biogas Reactor-Fixed Dome in Langensari West Ungaran

Fahmi Arifan[1] , *Fuad* Muhammad[2], *Sri* Winarni[3], *Hafizh* Rama Devara[1], *Latifah* Hanum[1]

[1]Departmen of Chemical Engineering, Faculty of Engineering, Diponegoro University
[2]Departmen of Biology, Faculty of Science and Mathematics, Diponegoro University
[3]Department of Biostatistic and Population study, Public Health, Diponegoro University

Abstract. Indonesia is a country that has abundant energy resources, namely oil, gas, coal, geothermal, and so forth. Biogas is an alternative fuel that can be used as a substitute for primary fuel. The term biogas is already familiar to the people, it is because biogas has usefulness as a vehicle fuel, domestic (cooking), and generate electricity. Cow dung has a value of C / N ratio is large enough that 18. Rice husk has a C / N ratio is sufficient High temperatures of 38.9. EM-4 (effective microorganism) is a bacterial culture which is usually used as an activator. In the manufacture of biogas from waste fluids out and chaff has the advantage because the content of the C / N is high enough. The composition of the raw materials used are liquid wastes out of 5 kg and 1 kg of husk-em with the addition of 4500 ml and the resulting calorific value of 1047.9 A fermentation time for 9 days. Ph maintained in neutral or alkaline conditions, namely 7-7.5, because the effectiveness of the methane formation is highly dependent on pH wherein the microorganism will grow and thrive in neutral. The test results has been done is the color of the flame and the time at yield is good enough where the color of the flame produced at day to9 blue with time for 40 seconds.

1 Introduction

Based on data from the Directorate General of Oil and Gas that shows the fuel oil consumption last 5 years start from 2006-2010, as many as 211 345 tons with an average increase of 42 269 per year and for gas consumption by 93,282 tons with an average increase of 18 656 tonnes per year and at 2014 is projected to reach 167 906 tonnes of fuel gas consumption in Indonesia [1]. With the increasing need for fuel for the population in Indonesia is not possible sometime in the future energy reserves exist in nature will be exhausted in view of an increased need for fuel oil and gas is a matter that can not be avoided and will continue to occur due to the increased population of the total population and the emergence of new industries and technologies that will continue to grow [2]. On the other hand, that the oil and gas reserves are owned by Indonesia increasingly limited because it is a product that can not be renewed. Therefore it is necessary for efforts to seek alternative fuels and environmentally friendly. Biogas is an alternative fuel that can be used as a substitute for primary fuel. The term biogas is already familiar to the people, it is because biogas has usefulness as a vehicle fuel, domestic (cooking), and generates electric power [3]. Biogas in the process usually requires considerable time to produce methane gas which is about 60-90 days. The main content of the biogas is methane and carbon dioxide. One of the things that affect the production of CH4 in biogas is the relationship between the amount of carbon (C) and nitrogen (N) contained in the organic material is expressed in terms of the ratio of C-N and PH also greatly affects the growth rate of biogas [4]. Waste know has a value of C / N ratio is large enough that 18. Rice husk has a C / N ratio is sufficient High temperatures of 38.9. With the C / N ratio is high is owned by rice husk waste out and expected to get quite a lot of methane gas from the second material is that by mixing the two. EM-4(effective microorganism) is a bacterial culture which is usually used as an activator. EM-4 may be added to a mixture of both ingredients so as to accelerate the formation of methane in Biogas Reactor Type Fixed Dome.

2 Method

Biogas production with waste materials out of the waste material consists out of 5 kg and EM-4500 ml. Biogas production process consists of two phases: preliminary and mixing the ingredients. To control the results obtained to be checked pH 3 days once within 17 days. Analysis of the results is done by connecting a gas stove on a biodigester through the tank.

* Corresponding author: fahmiarifan80@gmail.com

 https://doi.org/10.1051/e3sconf/20183102016

Table 1. Observations

Days to-	T Reactor (°C)	Increase in Gas Level (cm)	Gas volume (cm^3)	pH	Time light test (second)	Light Colour	T_1 (°C)	T_2 (°C)	ΔT (°C)	Calor (J)
1	25	0	-	7,5			-	-	-	-
2	25	0	-	7,5			-	-	-	-
3	25	0	-	7			-	-	-	-
4	25	0	-	6,5			-	-	-	-
5	25	0	-	6,5			-	-	-	-
6	26	0	-	6			-	-	-	-
7	26	0	-	6			-	-	-	-
8	26	0	-	5,5			-	-	-	-
9	26	0,5	830,53	5,5	2	Orange Yellow	30	30	0	-
10	27	1	1661,06	5,5	2	Orange Yellow	30	30	0	-
11	27	1	1661,06	6	10	Blue Yellow	30	30	0	-
12	27	1,5	2491,59	6,5	17	Blue Yellow	30	30,5	0,5	209,58
13	27	1,8	2989,908	6,5	23	Blue Yellow	30	30,5	0,5	209,58
14	27	2,2	3654,332	6,5	29	Blue	30	31	1	419,16
15	27	3	4983,18	6,5	34	Blue	30	31,5	1,5	628,74
16	27	3,3	5481,498	7	37	Blue	30	32	2	838,32
17	27	3,9	6478,134	7	40	Blue	30	32,5	2,5	1047,9

Trial of making biogas from waste out coupled with bacterial Effective Microorganism (EM). These bacteria function as decomposers and shorten the adaptation phase or phase lag of the safety. Make sure the connection is completely secure. Heats 250 ml of water and made observations flame colors produced.

3 Results and Discussion

Manufacture of biogas from waste out with the addition of EM-4 processed in batches, to determine the biogas production capacity and caloric value. Parameters measured were level at the height of the reactor biogas, biogas formed by the test flame test and the magnitude of the heat produced by the biogas in heating 100 ml of water. Data retrieval is done during the 17-day observation process of degradation. [5]

In experiments that have been carried out, the increase in the level of the gas cylinder holder be the first indicator of the formation of gas. The result of gas formation is one sign that the addition of bacterial inoculum (EM) as a bio-activator started working decipher the ingredients. Nutrients needed by microbes derived from raw materials, without any additional supply of nutrients from the outside in the form of the addition of urea or TSP. The purpose is not to the addition of these nutrients is to determine the length of microbes to degrade the waste out there in the feed biogas. [4,6].

Observation of the formation of gas volume is done every day starting from the first reactor to operate up to 17 days. The increase in water level in the gas holder showed large volume of gas produced each day.

From the results it is observed that on the ninth day with Ph 6 has seen an increase in water level as high as 0.5 cm. Having tested using the flame turns the flame color indicated yellowish orange. It can be seen that the gas is not methane.

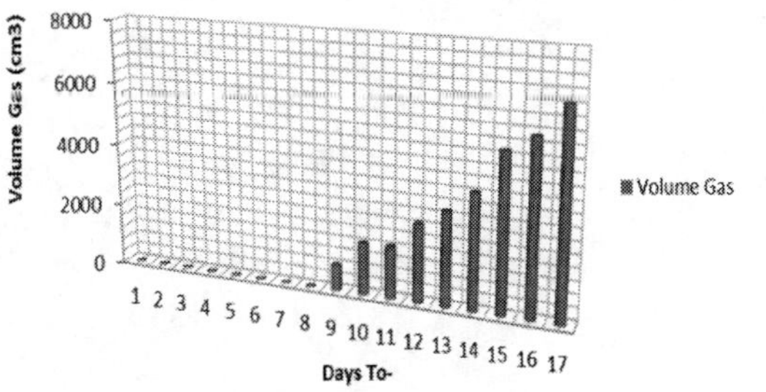

Fig. 1. Graphic Relation to the Establishment Time Gas Volume

From the results, the methane gas started coming out on the 14th with a pH of 6.5. It can be seen flames that already shows there is blue appearance, but still in small quantities. Then progressively increasing the amount of additional methane gas that is formed is also increasingly being seen from the blue color of the flames.

The formation of gas and flame testing would not be separated from how much heat is generated from the gas. Where the calorific value is only visible when the produced gas is methane gas (in CO2 gas is not obtained calorific values). Testing calorific value is done by connecting the device biodigester with a stove and used to heat 100 ml of water and observed when a temperature change in the water. Observations on the calorific value of this

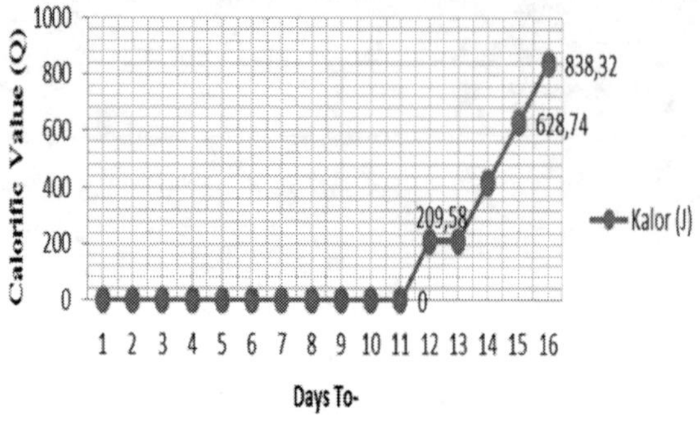

Fig.2. Relationship with Calorific Value of Time Trial Results

In figure 2 it can be seen that the calorific value obtained on the 12th day. From Table 1 it can be seen that the methane gas produced at day 14 seen from the flames. But the composition of the biogas in a small percentage have not been able to give a calorific value in the test has been done. Then on the 12th day 209.58 Joule calorific values obtained by calculation using the formula $Q = m.c.\Delta T$ and continued to rise in the following days. Until the 17th day of the calorific value generated at 1047.9 Joule. Broadly speaking, it can be concluded that the value of the heat produced will be greater and greater as the amount of methane produced.

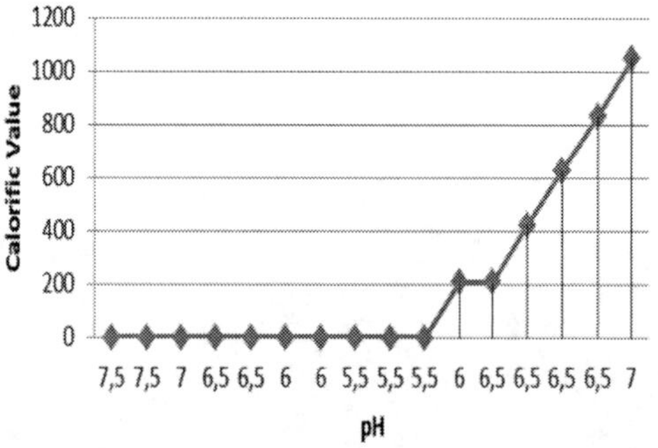

Fig.3. Relationship Graphic Calorific Values of pH

In figure 3 the most visible pH optimum in the formation calorific value that is at pH 7 wherein the calorific value generated at 1047.9 with the color blue flames and a time for 40 seconds. This is caused by anaerobic bacteria require optimal pH between 6.2 to 7.6, but the good thing is from 6.6 to 7.5 [6,7]. Proven by controlling pH 7 can generate the calorific value enough time biogas formation 17 days. The outline of the larger formation of the calorific value should be neutral pH conditions. [6,8,9,10]

4 Conclusions

In making biogas from waste carried out using a batch-type fixed dome biogas reactor can be concluded as follows: gas begins to form on day 9 with flame orange color yellow (not a methane gas). The methane gas begins to form on day 14, the color of a blue flame. The heating value of methane gas is as much as 419.16 joules. Even with the addition of EM4, formation of biogas with 17 days yet to obtain maximum results. Temperatures in the digester is not optimum because the range between 270C to 290C. Fixed dome biogas production with biogas reactor is very effective because the optimum gas volume increases.

Acknowledgements

Our gratitude is dedicated to Diponegoro University supporting financial for this research.

References

1. Ditjen Migas. *Statistik Minyak Bumi*. 2010. Available from: www.esdm.go.id/batubara/doc.../1256-statistik-minyak-bumi-2012.html diakses 21/06/14
2. De M. *Methanogenic archaeon as biogas producer in psychrophilic conditions,* 76:190–5, (2014)
3. MA. Usman, OO. Olanipekun, OM. Kareem, International Journal of Research in Chemistry and Environment Biogas Generation from Domestic Solid Wastes in Mesophilic Anaerobic Digestion. 2012;2(1).
4. P. Laopaiboon, A. Thani, V. Leelavatcharamas, L. Laopaiboon, Bioresource Technology Acid hydrolysis of sugarcane bagasse for lactic acid production. (2009)
5. M. De, Optimizing low-temperature biogas production from biomass by anaerobic digestion. 2014;69
6. MM. Kabir, C, Niklasson, MJ. Taherzadeh, I. Sárvári, Bioresour Technol.; 2014;161:446–50.
7. W. Li, L. Zheng, Y. Wang, J. Zhang, Z. Yu, Y. Zhang, et al. Bioresour Technol 2015 Jun [cited 2015 Jul 16];194:276–82.
8. S. Cheng, Z. Li, H. Mang, E. Huba, Renew Sustain Energy Rev Elsevier; 2014;34:387–400.
9. E. Nutiu, Procedia Technol. Elsevier B.V.; 2014;12:632–6.
10. IM. Alfa, SO. Dahunsi, OT. Iorhemen, CC. Okafor, SA. Ajayi,. Bioresour Technol. Elsevier Ltd; 2014;157:270–7.

A Study of Sedimentation at the River Estuary on the Change of Reservoir Storage

Iskahar[1,*], *Suripin*[2], and *Isdiyana*[3]

[1]Departement of Civil Engineering, Engineering Faculty, University Muhammadiyah Purwokerto, Purwokerto - Indonesia
[1] Doctoral Student of Civil Engineering, Enginering Faculty, Diponegoro University, Semarang - Indonesia
[2] Department of Civil Engineering, Diponegoro University, Semarang - Indonesia
[3] Center River, Center of Research and Development Water Resources, Surakarta - Indonesia

Abstract. Estuary of the river that leads to the reservoir has characteristics include: relatively flat, there is a change in the increase of wet cross-sectional area and backwater. The backwater will cause the flow velocity to be reduced, so that the grains of sediment with a certain diameter carried by the flow will settle in the estuary of the river. The purpose of this research is to know the distribution and sedimentation pattern at the river estuary that leads to the reservoir with the change of water level in the reservoir storage, so the solution can be found to remove / reduce sediment before entering the reservoir. The method used is the experimental, by making the physical model of the river estuary leading to the reservoir. This study expects a solution to reduce sedimentation, so that sedimentation can be removed / minimized before entering the reservoir. This research tries to apply bypass channel to reduce the sedimentation at the river estuary. Bypass channels can be applied to overcome sedimentation at the river estuary, but in order for the sediment to be removed optimally, it is necessary to modify the mouth of bypass channel and channel angle.

1 Introduction

The sedimentation process and pattern of the river mouth leading to the reservoir is influenced by the basic sediment and river currents. The estuary of the river that leads to the reservoir has characteristics such as: relative ramps, there is a change in the increase of the wetland area, the backwater due to the water level of the reservoir storage, causing the flow velocity to decrease. This sediment reduces the channel's inflow capacity, which will finally affect the distribution and sedimentation pattern of the reservoir. Sedimentation in reservoirs will be result in decrease in reservoir storage capacity and functional life of reservoir and decrease of water quality of reservoir that will affect the performance and operation of reservoir. [1]

The purpose and objective of this research is to know the distribution and sedimentation pattern at the river estuary to the reservoir with the change of water level in the reservoir storage, so that solution or alternative can be found to reduce / remove sediment from the estuary of the river. In the experiment try to apply bypass channel to reduce sedimentation at the river estuary.

2 Literature Review

2.1 Sediment Dynamics on the Reservoir delta

Sediments in the delta area at the highest water level of the reservoir can cause gradation (deposition) upstream of the reservoir. This sedimentation reduces the channel's inflow capacity. Delta deposits are also the most visible part of the reservoir sedimentation. Reservoir deltas have the following basic characteristics [2]:

a) There is an abrupt change between the slope of the topset and foreset deposits.
b) Sediment particles on the topset bed are coarser than on the foreset beds, and there is an abrupt change in the particle diameter between the topset and foreset deposits.
c) The elevation of the transition zone from the topset to the foresets bed depends on the reservoir operating rule and pool elevation.

In deep reservoirs which have been operated at different levels, distinct deltas may be formed at different water levels. Reservoir deltas always grow in the downstream direction, and in some cases their vertical and upstream growth may also be significant. Because the upstream area of the reservoir is shallow and has little storage

* Corresponding author: iskaharoke@gmail.com

https://doi.org/10.1051/e3sconf/20183103001

capacity the longitudinal growth of the delta may initially be very rapid. As the delta progresses downstream into deeper and wider portions of the reservoir, its rate of longitudinal growth slows.

As the delta advances downstream, the topset and foreset deposits prograde over previously deposited bottomset beds of finer sediments. This process can create alternating lenses of coarse and fine material as large events deliver coarse material deep into the impoundment, which may be subsequently overlaid by finer material. In general, fully penetrating cores in an extensively sedimented reservoir will reveal the finest sediment at the bottom, coarsening until the surface level is reached.

Sedimentation patterns reflect transport processes in reservoirs and as information on sediment delivery and distribution processes, so that predictable types of service disruptions and possible remedial strategies can be predicted. Sedimentation patterns in each reservoir will vary, depending on hydrological conditions, reservoir pond geometry, inflow discharge and sediment grain size characteristics and reservoir operations. Bed load and coarse fraction of the suspended load will soon settle to form the delta deposit at the end of the reservoir, while the fine sediment with a lower settling velocity will be transported deeper into the reservoir (see Figure 1).

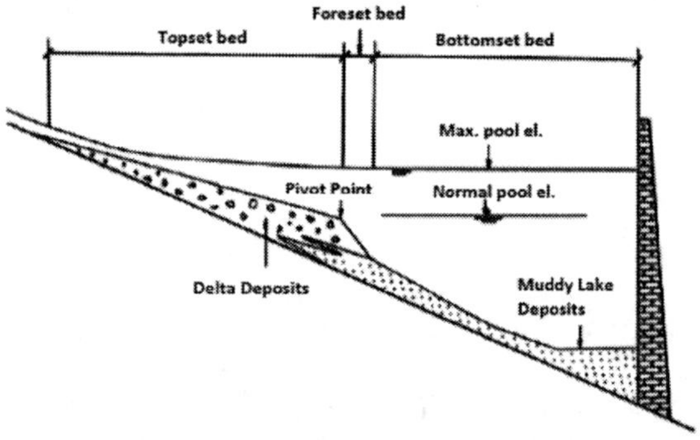

Fig. 1. Generalized depositional zones in a reservoir [2].

The longitudinal sedimentation zone in the reservoir can be divided into three main zones as follows [2]:

a. Topset bed; Composed of sediment material with larger grain size (coarse). The sediment is formed upstream to the end of the backwater, which forms a delta downstream to the end of the transport of the base material in the reservoir.
b. Foreset bed; Is a transitional part from the base of the top to the bottom of the base, starting from the forward delta into the reservoir. The difference is with the topset bed on increasing the slope and decreasing grain size.
c. Bottomset bed; Composed of fine sediment, formed from the base of the transitional portion toward the dam. This precipitate is formed because the sedimentary material is carried by turbidity or nonstratified flow (see Figure 1).

The presence of inflow from tributary, Decrease of reservoir water level, base landslides and extreme floods will cause the rough material to be transported to the normally deposited fine material zone, so that sedimentation or local variations will occur on the grain size of the sediment.

2.2 Initial sediment motion

The beginning of the Movement is important in the study of sediment transport, channel degradation, and stable channel design. Due to the stochastic nature of the sediment movement along the alluvial basis, it is difficult to define exactly whether the sediment particle flow conditions will begin to move. On channels with uniform grained sand material, which flows with certain discharges, the basic particles of the channel will surely experience a hydrodynamic force.

As the velocity and flow rates increase, the hydrodynamic forces that arise will increase in size, so that the particles can not survive and will begin to drift, where these are the so-called critical conditions or the beginning of movement. The force acting on a round sediment particle at the bottom of the open channel, which should be considered is the drag force F_D, the lift force F_L, the submerged weight of Ws and the resistance force FR. [1] (See Figure 2)

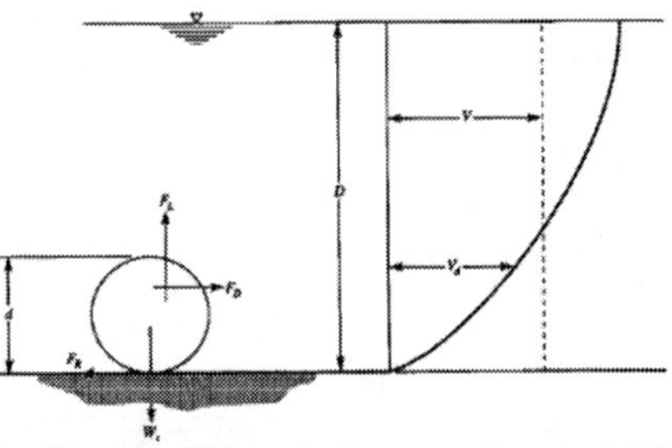

Fig. 2. The diagram of a force acting on sediment particles in an open channel [1].

A sediment particle is at a state of incipient motion when one of the following conditions is satisfied:

$$F_L = W_s \ or \ F_D = F_R \ or \ M_o = M_R \qquad (1)$$

where :

F_D = drag force,	F_R = resistence force
F_L = lift force,	M_o = overturning moment due to F_D and F_L,
W_s = submerged weight	M_R = resisting moment due to F_L and W_s

Most incipient motion criteria are derived from either a shear stress or a velocity approach.

2.3 Critical Shear Stress

The results of research conducted by Shield on a flat base were the most widely used to estimate the critical condition of the initial movement of a particle. In the study the material is not uniform used size D_{50}. Shield in his study found a relationship between critical drag force (ψ_c), critical shear stress (τ_{oc}), and critical shear velocity (U^*c), as follows [3]:

$$\psi_c = \frac{\tau_{oc}}{(\gamma_s - \gamma).d} = \frac{(U_c^*)^2}{\left(\frac{\rho_s}{\rho} - 1\right).g.d} = f\left(\frac{U_c^*.D}{\nu}\right) = f(R_g^*) \quad (2)$$

In the above formula it is shown that the factors affecting the grain motion are shear stress, shear velocity, water mass densities of fluid and sediment, particle diameter, kinematic viscosity number and gravity. The Shield has depicted a graph showing a silent particle and begins to move (see Figure 3).

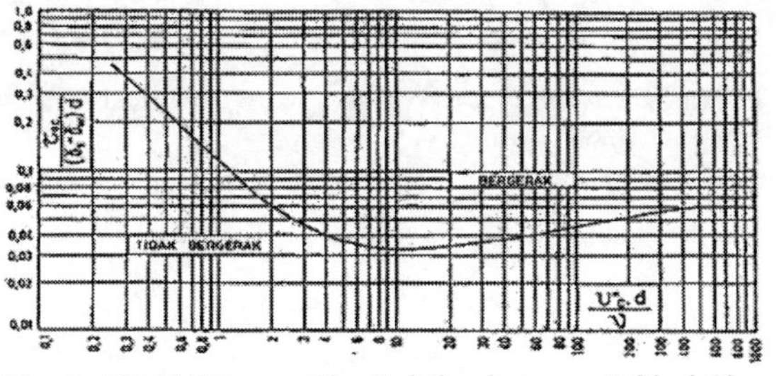

Fig. 3. Shield Diagram The Relation between Critical Shear Stress to Reynolds Numbers [3]

2.4 Sediment Control Strategy

Some opinions on strategies that can be done to control sediment according to some experts are as follows:

- According to Auel and Boes, some measures that can be taken to control sedimentation are by reducing sediment yields, sediment routing, and sediment removal, as shown in Figure 4. [4]

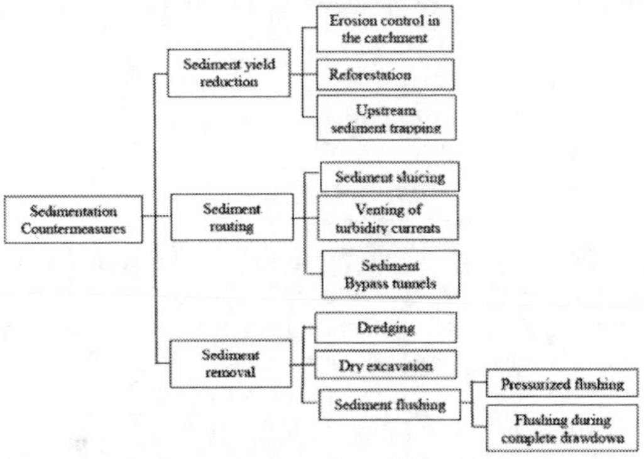

Fig. 4. Sedimentation control measures [4]

- According to Morris and Fan, to control sedimentation of the reservoir, there are several methods: [2]:
  a. Minimize the entry of sediments into reservoirs by means of watershed conservation, constructing sediments (check dam) or with vegetation screen, bypass channels to degrade flow with high sediment concentrations in order not to enter the reservoir.
  b. Minimize the occurrence of sediment deposition in the reservoir, by removing the sediment from the reservoir before sediment settles.
  c. Removing sediment that has been settled in the reservoir, with flushing method and dredging
  d. Replacing lost storage capacity by raising a dam or constructing a new reservoir.
- According to Palmieri, the methods that can be used for handling sedimentation reservoirs are: [5]
  a. Reduce sediment inflow with watershed management, upstream control structures, bypass Reservoirs and off channel storage channels.
  b. Regulates sediment in reservoirs with operating rules and tactical dredging.
  c. Removing the sediment from the reservoir with flushing, sluicing, density current venting and mechanical dredging.
  d. Restore lost storage capacity by raising high dams and constructing new dams.
  e. Dam decommissioning
- According to Yang, the capacity of the lost reservoir can be restored by flushing, dredging, sluicing and siphoning. Flushing is more effective when emptying, or with lower outlet having sufficient capacity. Dredging is an expensive method for restoring reservoir capacity, done when flushing is ineffective, and bypass sediments are not possible. Siphoning is a fairly effective method for removing sediment from small reservoirs. [6]
- According to Liu et al. to control reservoir sedimentation used different methods such as bypassing, dredging, flushing, sluicing and retaining sediment upstream. To obtain maximum results are usually used a combination of the above methods. [7]

2 Research Methods

The method used in this research is the experimental method, with the physical model of the river estuary that goes to the reservoir, conducted in the field laboratory of "Balai Sungai Surakarta".

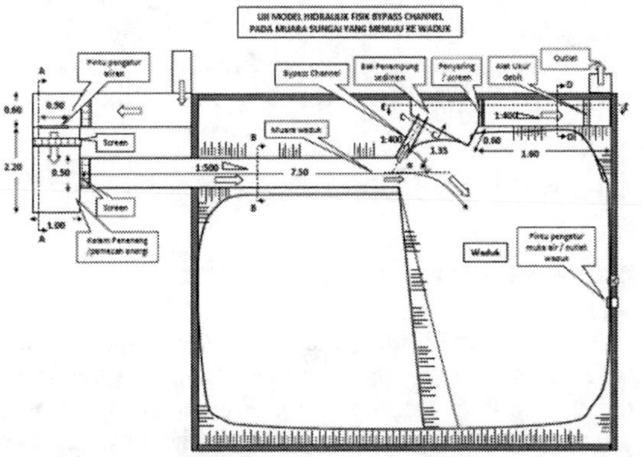

Fig. 5. Hydraulic model layout estuary leading to the reservoir with a bypass channel

Fig. 6. Photo of Hydraulic model layout estuary leading to the reservoir with a bypass channel

In this study, it consists of two scenarios / series of tests, that is :

- Series 0 (Original)

To determine the natural behavior of sediments on estuary river that to reservoir, at an elevation of Low Water Level (LWL), Medium Water Level (MWL) and High Water Level (HWL), with variations of river flow discharge of 8 lt/s, 13 lt/s and 18 Lt/s.

- Series 1 (with bypass channel)

To determine the movement of sediments on estuary river that to reservoir and the weight of the sediments coming out through the bypass channel, at an elevation of Low Water Level (LWL), Medium Water Level (MWL) and High Water Level (HWL), with variations of river flow discharge of 8 lt/s, 13 lt/s and 18 Lt/s.
Observations or measurements are: flow velocity, flow depth, river flow discharge, outflow bypass channel, flow pattern, movement pattern and sediment contour after sediment flushing at the river estuary leading to the reservoir and the weight of sediment entering the reservoir and sediment out through the bypass channel.

Storage elevation on the medium water level (MWL) is the height of the reservoir water level parallel to the base estuary of the river mouth. Whereas the elevation of the reservoir at low water level (LWL) is set 1cm below the MWL and the elevation of the reservoir on the high water level (HWL) is set 1cm above the MWL.

4 Results and Discussion

4.1 Flow Patterns

The flow pattern in the series 0 and series 1 experiment is relatively similar, ie when the flow enters the stream the flow pattern forms a straight line along the river to the mouth of the river. Furthermore the flow enters to reservoir on a large area, the flow begins to twist to form the number eight and eventually breaks down after touching the wall of the reservoir boundary. As for series 1 when entering the branching flow will be divided, a small part of which goes into the bypass channel and mostly goes into the reservoir. The stream entering the bypass channel as it enters crashing the right wall then moves again crashing the left channel wall, so that the flow pattern line is tilted to the right and tilted to the left against the channel wall to the end of the channel bypass channel. (See Figure 7).

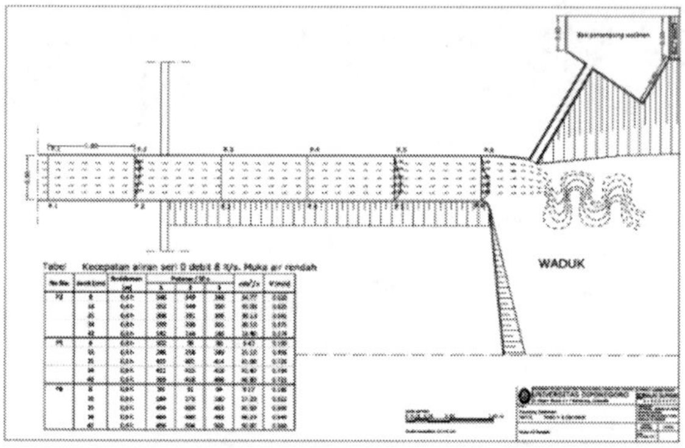

Figure 7. Experimental flow pattern series 0 (original / without bypass channel)

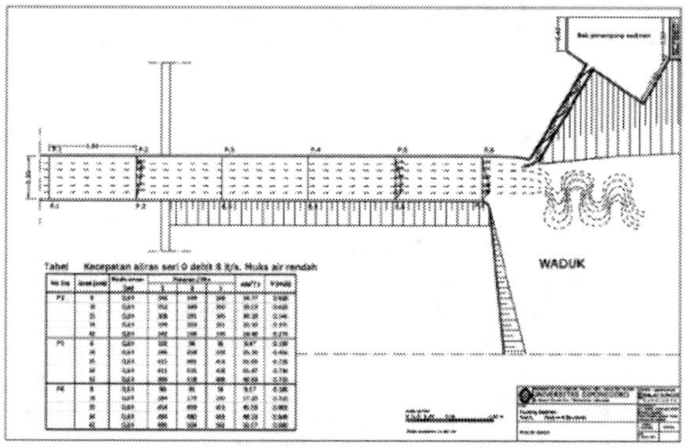

Figure 8. Experiment flow pattern series 1 (with bypass channel)

4.2 Sediment Flushing

In the sediment flushing experiments recorded data on the relationship between river flow discharge and outflow bypass channel (see table 1.). From the table it can be seen that the higher elevation of the reservoir storage, the outflow through the bypass channel will be greater. The greater river flow discharge, of course, the outflow through the bypass channel will be increase. So it can be estimated that with the higher elevation of the reservoir storage and the greater the river flow discharge, then the amount of sediment out through the bypass channel will be increase.
The sediment flushing experiment was carried out for 60 minutes, with three variations of discharge (8 l/s, 13 l/s and 18 l/s), three variations of sediment thickness (5 cm, 7 cm and 9 cm) and three variations of reservoir storage

 https://doi.org/10.1051/e3sconf/20183103001

(LWL, MWL and HWL). For the experimental results of sediment flushing can be seen in Table 2.

Table 1. The relationship between river flow discharge and outflow bypass channel

River Flow Discharge (l/s)	Outflow bypass channel (l/s)		
	LWL (l/s)	MWL (l/s)	HWL (l/s)
8	0.438	0.465	0.802
13	0.620	0.726	1.012
18	1.012	1.105	1.526

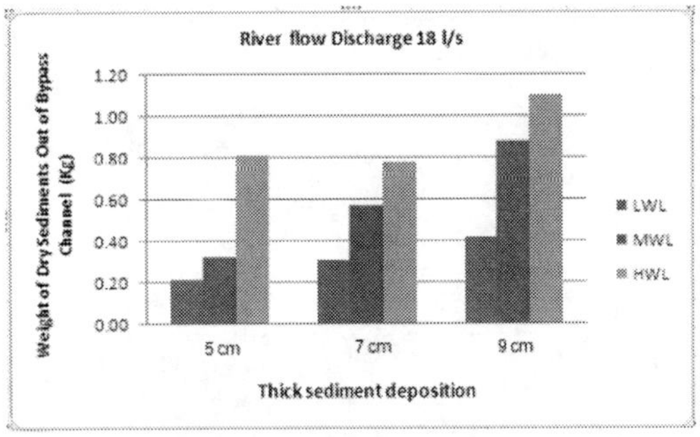

Fig. 11. Weight of Sediment Out bypass channel at Q= 18 l/s

Table 2. Recapitulation of sediment flushing experiment in series 1

River flow Discharge	Thick sediment deposition	Total weight of sediment	Weight of dry sediments left in the river			Weight of dry sediments out of bypass channel			Weight of dry sediments left in the reservoir		
			LWL	MWL	HWL	LWL	MWL	HWL	LWL	MWL	HWL
(l/s)	(cm)	(kg)	(kg)	(kg)	(kg)	(kg)	(kg)	(kg)	(kg)	(kg)	(kg)
8	5	63.936	3.45	8.00	15.05	0.13	0.28	0.31	60.36	55.65	48.57
8	7	85.248	3.30	7.30	17.50	0.26	0.51	0.72	81.69	77.44	67.02
8	9	106.560	4.10	10.00	7.00	0.37	0.66	0.87	102.09	95.90	98.69
13	5	63.936	0.18	1.67	11.80	0.15	0.30	0.50	63.61	61.97	51.64
13	7	85.248	1.20	1.80	11.30	0.28	0.55	0.76	83.77	82.90	73.19
13	9	106.560	2.20	3.10	10.50	0.38	0.76	0.97	103.99	102.70	95.09
18	5	63.936	0.00	0.71	0.00	0.21	0.32	0.81	63.73	62.91	63.13
18	7	85.248	0.00	0.20	2.40	0.31	0.57	0.78	84.94	84.48	82.07
18	9	106.560	0.00	0.82	9.30	0.42	0.88	1.10	106.14	104.86	96.16

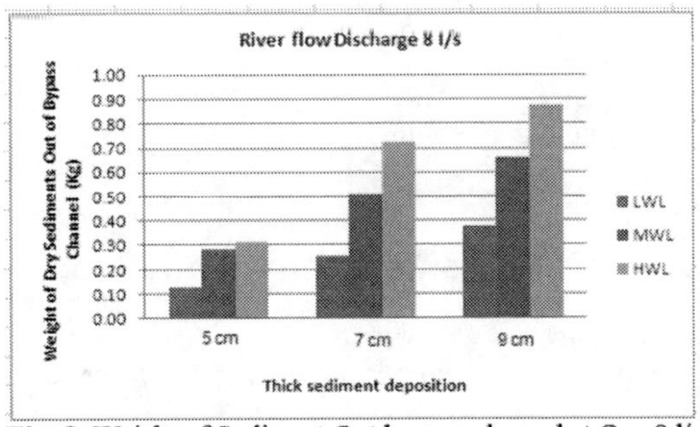

Fig. 9. Weight of Sediment Out bypass channel at Q = 8 l/s

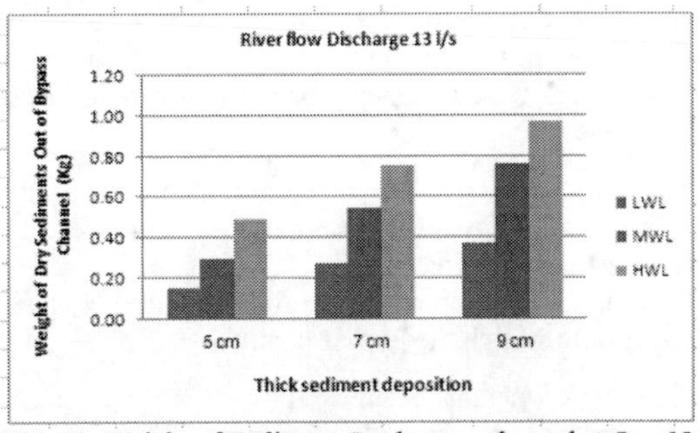

Fig. 10. Weight of Sediment Out bypass channel at Q = 13 l/s

From the table and graph shows that the higher elevation of the reservoir storage, the sediment that goes into the reservoir decreases and the sediment out through the bypass channel increases. The greater the flow discharge in the river, the sediments entering the reservoirs and sediments out through the bypass channels are increasing. Likewise, with increasing sediment depth thickness, sediments entering reservoirs and sediments through the bypass channels are increasing. So to get more bypass channel outlet sediments, the elevation of the reservoir storage in HWL and with large river flow discharge.

For example the image of sediment contours after flushing for 60 minutes can be seen in Figure 11 (for series 0) and Figure 12 (for series 1). In this case the examples described are selected for flushing at 18 l/s flow discharge, and 9 cm of thick sediment deposition, as both are the largest discharge and largest thick deposits of the variations used in this study, in which the sediment contour image obtained, it is clear the difference between the reservoir storage LWL, MWL and HWL.

https://doi.org/10.1051/e3sconf/20183103001

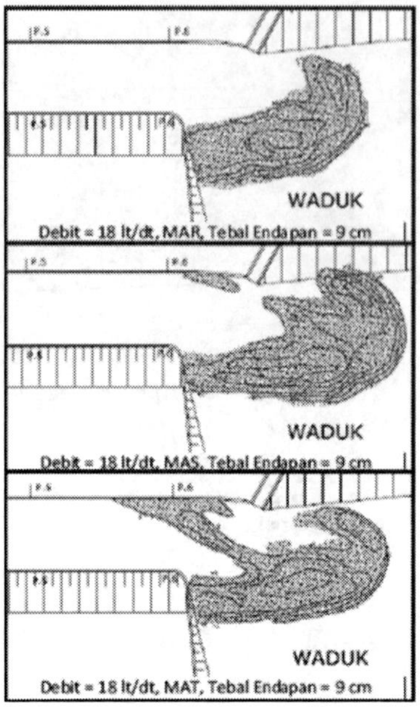

Fig. 12. Examples of contour images of sediments in the experimental series 0

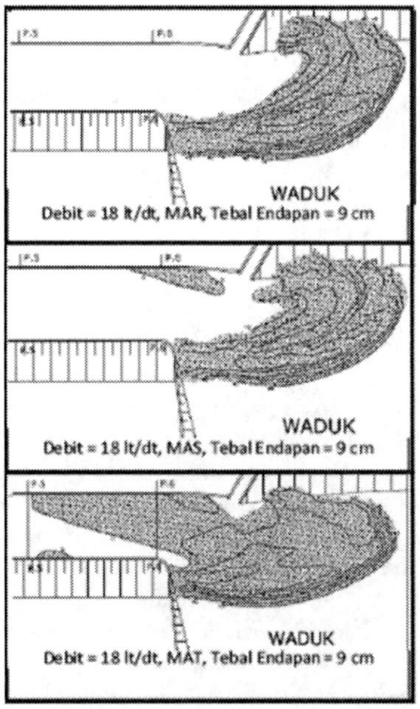

Fig. 13. Examples of contour images of sediments in the experimental series 1

In the series 0 experiment, at a flow discharge of 18 l/s with a sediment thickness of 9 cm, the LWL showed that sediment deposits entirely entered the reservoir (nothing left in the river). In MWL it appears that most sediment deposits enter the reservoir and only a small amount of sediment is left in the river. Here it appears the effect of backwater has appeared, but not so strong. While the HWL seen sediment deposit that are still left in the river, this is because the influence of backwater reservoirs storage on HWL is strong enough.

In the series 1 experiment, at a flow discharge of 18 l/s with a 9 cm thick deposit, it appears that the sediment LWL entirely enters the reservoir (nothing left in the river) and only a few enter the bypass channel. In MWL it appears that most sediment deposits enter the reservoir, only a small amount of sediment is left in the river, and only a few enter the bypass channel. This shows that the effect of backwater reservoirs storage in MWL has emerged, but not yet so strong. In the HWL it appears that most sediment deposits enter the reservoir, the sediments left in the river are slightly higher, and those that enter the bypass channel are slightly increased. This shows that the effect of backwater reservoir storage on HWL is strong enough.

Thus, in order for the sediment that comes out more bypass channels, it is necessary to have a guidance structure in the form of submerge dike, to collect and direct the sediment into the bypass channel and the need to change the branching angle from $\alpha = 60°$ to $\alpha = 45°$ in order for more sediment to enter bypass channel. This is also supported by research conducted [8] which states that the maximum secondary flow stress occurs in the bend section with a transfer angle of 45 °. In addition, the transfer from the stream to the bypass channel angle made stream line, so sediment will be easier to enter the bypass channel.

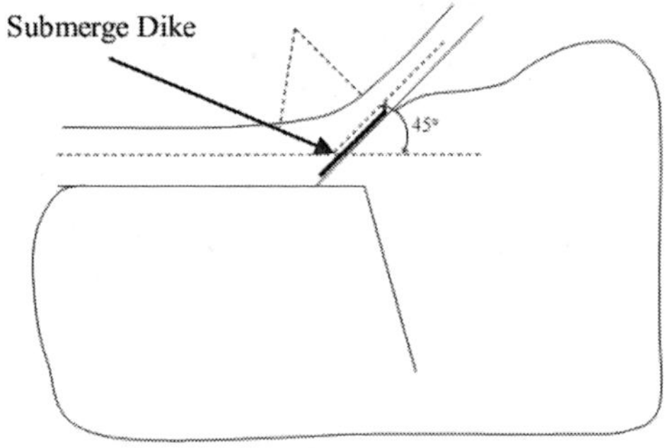

Fig. 14. Sketch of hydraulic model modification

5 Conclusions and recommendations

Some conclusions can be mentioned here related to sedimentation at the mouth of the river that leads to the reservoir, namely:

- In the sediment flushing experiment at the river estuary leading to the reservoir, the effect of reservoir backwater began to appear at the elevation of the MWL and the increasing backwater effect at the HWL elevation.
- The elevation of the reservoir storage strongly influences the sedimentation pattern occurring at the river estuary leading to the reservoir and also affects the amount of sediment in the river mouth that can be come out through the bypass channel.
- The experimental results show that the higher elevation of the reservoir storage, the sediment entering the reservoir decreases and the sediment that

goes through the bypass channels is increasing. The greater the flow discharge in the river, the sediments entering the reservoirs and sediments come out through the bypass channels are increasing. Likewise, with increasing sediment depth thickness, sediments entering reservoirs and sediments come out through the bypass channels are increasing. So to get more bypass channel outlet sediments, the elevation of the reservoir storage in HWL and with large river flow discharge.

To get the sediment deposition out through the bypass channel more optimally, the author's suggestion is that additional series modifications are needed, that is by adding a submerge dike, as a structure for collecting and directing the sediment into the bypass channel and the need for change the branching angle from $\alpha = 60^o$ to $\alpha = 45^o$ to allow more sediment into the bypass channel. In addition to the channel diversion from the stream to the bypass channel angle is made smooth or stream line, so sediment will be easier to enter the bypass channel.

References

1. 1 Anonymous, *Reclamation Managing Water in the West – Erosion and Sedimentation Manual*, , Bureau of Reclamation - US Department of the Interior, Denver, Colorado (2006).
2. 2. G.L. Morris, and J. Fan, *Reservoir Sedimentation Handbook : Design and Management of Dams, Reservoirs, and Watersheds for Sustainable use,* Electronic version ver 1.04, McGraw-Hill Company (2010)
3. 3 Yiniarti, *Book sediment transport courses*, Bandung (1999).
4. 4 Auel dan R. Boes, *Sedimentation Reservoir of Alpine Reservoirs and Possible Counter Measures*, Vienna (2011).
5. 5. E. H.Coker, R.H. Hotchkiss, and D.A. John, Journal of The American Water Resources Association. Vol **45**, No. 4, pp. 815-827 (2009).
6. 6 C.T. Yang, *Sediment Transport Theory and Practice*, McGraw-Hill Company (1996).
7. 7 J. Liu, , S. Minami, H. Otsuki, , B. Liu and K. Ashida, J. Hydraulic Eng., Vol. **130** Number. 11, pp.1089- 1096 (2004).
8. 8 H. Montaseri, M. Ghodsian , M. Shafieefar, S.A.A. Salehi Neyshaboouri, and A.A. Dehghani, J. Agriculture Science .Nature Resources, Vol.**15**, Number 2 (2008).

https://doi.org/10.1051/e3sconf/20183103002

Ozone application for tofu waste water treatment and its utilisation for medium growth of microalgae *Spirulina sp*

Hadiyanto Hadiyanto[1,2]

[1]Chemical Engineering Department, Faculty of Engineering, Diponegoro University, Indonesia
[2]Master Program of Environnemental Sciences, School of Postgraduate Studies, Diponegoro University, Indonesia

Abstract. Tofu industries produce waste water containing high organic contents and suspendid solid which is harmful if directly discharged to the environment. This waste can lead to disruption of water quality and lowering the environmental carrying capacity of waters around the tofu industries. Besides, the tofu waste water still contains high nitrogen contents which can be used for microalgae growth. This study was aimed to reduce the pollution load (chemical oxygen demand-COD) of tofue wastewater by using ozone treatments and to utilize nutrients in treated tofu waste water as medium growth of microalgae. The result showed that the reduction of COD by implementation of ozone treatment followed first order kinetic. Under variation of waste concentrations between 10-40%, the degradation rate constant was in the range of 0.00237-0.0149 min^{-1}. The microalgae was able to grow in the tofue waste medium by the growth rate constants of 0.15-0.29 day^{-1}. This study concluded that tofu waste was highly potent for microalgae growth.

Introduction

Tofu is local food in Indoensia, and its consumption reaches 7.4 kg / person / year [1]. The production process of tofu produces liquid waste in a large quantity with the average amount of 17 Litres per kg of processed soybeans[2]. The waste still contains organic content such as protein and lipids of about 40-60% and carbohydrates about 25-50% [1]. Besides, tofu waste water still contains total nitrogen contents of 188.34 mg/L, and total phosphorus of 1.69 mg/L. These nutrient contents can be a potent as a nutrient for microalgae [3].

Microalgae is microorganism that capable to use nutrient of nitrogen and phosphorous for photosynthesis to produce biomass and oxygen[4]. The nitrogen is required for the formation of protein, the cell walls, and nucleic acid. Microalgae also requires a source of carbon and sunlight as energy sources.

On the other hand, the tofu waste has a very high COD content of 8640 ppm and BOD content of 6586 ppm[3]. High levels of COD/BOD will interfere the growth of microalgae, therefore a pre-treatment is required to reduce COD/BOD level in the waste. There are many kind of pretreatment for wastewater treatments such as physical, biological and chemical treatments. These methods require longer processing time. Oxidation using ozone is one methods to reduce the COD level in the waste. Ozone is a very strong oxidizing agent (E = 2.08V) compared to other oxidizing agents such as hydrogen peroxide (E = 1.78V) [5]. Ozone has been extensively used in industrial processes to eliminate the presence of pollutants in wastewater. Ozone is also able to decompose organic components into simpler and more biodegradable compounds. Moreover, ozone is also commonly used as a disinfectant in drinking water, preserving food and reducing odor from bacteria in waste.

The efficiency of ozone higly depends on pH as an indication of hydrogen ion activity. A high pH will facilitate the formation of OH radicals because the presence of OH ions can initiate ozone decomposition as a first step in the production of OH radicals [6], whereas at low pH the ozone molecule remains as its initial oxidant.

The objective of this study was evaluate the use of ozone for oxidation of tofu waste such that COD will be degraded. The treated waste will be also used medium growth of microalgae *Spirulina sp*.

Material and Methods

Material

Tofu wastewater was obtained from local indsutry in Gunung Pati, Semarang. The microalgae *Spirulina* was prodived by CV Neoalgae.

Sample Preparation and Ozone Treatment

The tofu waste was varied in the range of 10 -40% (v/v) in 1 L erlenmeyer. The sample was placed in the container and

*Corresponding author: Hadiyanto@live.undip.ac.id

https://doi.org/10.1051/e3sconf/20183103002

continously imposed by ozone. The sample was taken every 10 min to measure its COD concentration.

Cultivation of Microalgae

The treated waste was then used as medium growth of microalgae. The control used fresh water with external nutrients of 0.05 gr of urea fertilizer, 1.5 gr of fine salt, 0.1 ppm vitamin B12, and 0.05 g of NPK fertilizer. The sample was measured for its opticald ensity at 620 nm. Cultivation was carried out for 7 days and the concentration was monitored by daily measuring its optical density .

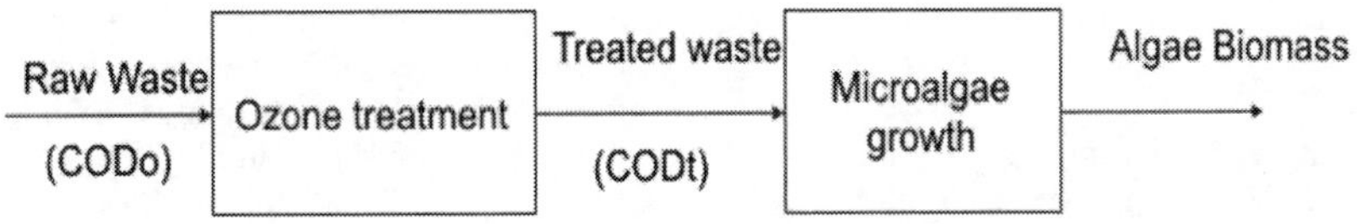

Figure 1. Research step of tofu treatment and utilization of its nutrient for microalgae growth

Cell Number Analysis

Take 5 ml of microalgae sample and diluted up to 10 dilutions. Take 1 ml of diluted sample and inserted it into the Sedgwick-Rafter plate. Count the number of cells with a 10x10 magnification microscope with a calculated 7x7 plate area.

Result and Discussion

Effect of Waste Concentration on COD Degradation

The research was conducted on 4 variations of waste concentration, i.e. 10%, 20%, 30% and 40% of tofu waste. The COD sample values were analyzed every 15 minutes, during the 2 hour ozone treatment. The effect of the concentration of tofu waste on the decrease of COD concentration is depicted in Figure 2. Figure 2 shows that the degradation of COD at high waste concentration (>40%) goes in slow decrease. It might due to high amount of total solids in waste which difficult to oxidize by ozone treatment. The degradation rate was determined by fitting the first order reation kinetic to the experimental data. The result is shown by Table 1.

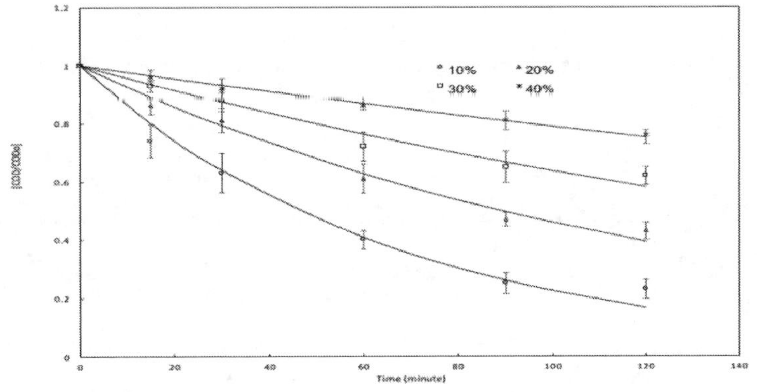

Figure 2. First-order kinetics of the ozone decomposition

Table 1. First-order kinetics of the ozone decomposition of waste water

Concentration	Kinetic rate constant (min^{-1})
10%	0.0149
20%	0.00779
30%	0.00453
40%	0.00237

Table 1 also proves that ozone is a powerful oxidizing agent capable of oxidizing various compounds. Ozone oxidizes compounds in water in two ways: direct reaction and indirect reactions [7]. For direct method, the oxidation is carried out by ozone itself that dissolved in water. While indirect method, the oxidation occurs by producing OH radicals that would oxidize other compounds [6]. The longer the ozonation time, the more ozone will be formed that will oxidize the compound, resulting in lower COD values due to the simpler compounds.

The utilization of nutrient for microalgae growth

Spirulina platensis requires nutrients C, H, O, N and P in their growth to perform photosystasis. The growth of microalgae in treated waste at 10% waste concentration has higher growth rate constant than 20-40%. Moreover, this finding also support the hypothesis that the nutrients in the waste water can be used to replaced the external nnutrients added to the medium. Although the growth of microalgae grown in waste medium was lower than the control (Figure 3), but this findings showed that the waste has high potent to replace the external nutrients. The growth rate constant is shown by Table 2. The growth rate constant at high waste consentration is slower than algae growth in lower concentration. This is due to high organic content whcih will disturb the growth and inhibit the light penetration to the culture.

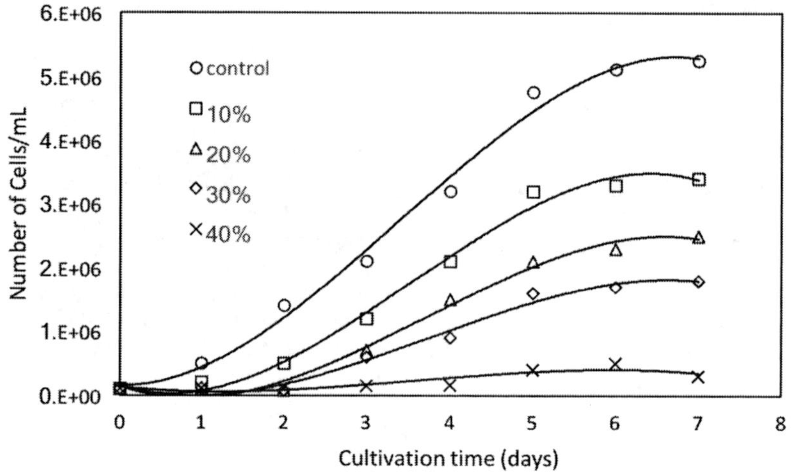

Figure 3. Growth of microalgae in variation of tofu waste volume

Table 2. Growth rate of *Spirulina platensis* in cultivation medium

Medium (% tofu wastewater)	μ/day
10%	0.2848
20%	0.2339
30%	0.2103
40%	0.1497

Conclusion

This study showed the potency of tofu waste water as nutrient source of microalgae growth. The degradation of COD by using ozone oxidator followed the first orde kinetic with rate constant of 0.014 for 10% tofu concentration. The treated tofu waste could be used as nutrient source of *Spirulina* growth with growth rate of 0.28 day^{-1} at 10% concentration. However, at concentration higher than 40%, the microalgae has lower growth rate.

References

1. Dianursanti, Rizkyata, B.T., Gumelar, M.T. and Abdullah, T.H. *Energy Procedia*, **47**, 56 – 61 (2014)

2. Hadiyanto and Nur, M.M.A. *World Applied Sciences Journal,* **31(5)**,959-967 (2014) .
3. Hadiyanto, Christwardana, M., Soetrisnanto, D. *Journal of Environmental Science and Technology* **6(2)**, 79-90 (2013)
4. Wen, Z., Gross, M., Ninno, M. D., & Michael, C. *Bioresource Technology,* **179**, 473-482 (2015).
5. Rajeswari, R., Kanmani, S. *Desalination,* **242**, 277e285 (2009).
6. Nur, M.M.A and Hadiyanto. *Journal of Engineering and Technological Sciences* **47(5),** 487-497 (2015).
7. Balasubramaniam N., Preethi V., Parama Kalyani, K.S., Iyappan, K., Srinivaskannan, C. *Journal of Hazardous Materials,* **166**, 150–154 (2009).

Effect of light Intensity and photoperiod on growth of *Chlorella pyrenoidosa* and CO_2 Biofixation

Teuku Johar Gunawan[1,*], *Yusni* Ikhwan[2], *Fajar* Restuhadi[3], and *Usman* Pato[3]

[1]Post graduate program of Environmental Science, Universitas Riau,Pekanbaru, Riau, Indonesia
[2]Post graduate program of Environmental Science, Universitas Riau,Pekanbaru, Riau, Indonesia
[3]Faculty of Agriculture, Universitas Riau, Pekanbaru, Riau, Indonesia

Abstract. Microalgae have been viewed as one of potential solution for CO_2 biofixation or CO_2 sequestration. However, many factors need to be evaluated to support development of CO_2 biofixation. One important environmental factor for the growth of micro algae is related with light requirement. The aim of this study was to evaluate the effect of light intensity and photoperiod on growth of *Chlorellapyrenoidosa* (*C.pyrenoidosa*) and CO_2 biofixation. Experiments were carried out in 1000 mL semi batch photo bioreactors, purged continuously with air (0.034% CO_2). An Experiment of Factorial Design was employed in which the light intensity was evaluated 4 level at 2000, 4000, 6000 and 8000 lux with 3 level of photo period at L/D (light /dark) 8 hours/16 hours; L/D 12 hours/12 hours and L/D 16 hours/8 hours. The result indicated that both light intensity and photo period had significant effect ($p< 0.05$) on growth of *C. pyrenoidosa*. However, the photo period showed stronger effect relative to light intensity on growth of *C.pyrenoidosa* within the range reviewed. The interaction between the two factors was indicative but statistically not significant. Best growth profile sustained at combination of L/D 16 hours/8 hours of photoperiod and light intensity of 8000 lux with the highest average biomass observed at 0.516 ± 0.069gr/L. An increase in CO_2biofixation rate of around 2 times was also observed between highest setting (8000 lux; L/D 16/12 hours) relative to that of lowest setting (2000 lux; L/D 8/12 hours).

1 Introduction

The global warming has become an attention to the international community and an important challenge[1].This global warming is due to an increase of several gases called green house gas. One of the contributing gases is carbon dioxide released by human activities such as the gas coming out from industrial and transportation sectors. The result of global warming is an increase in earth temperature.

The negative implication of global warming to human has been acknowledged such as a negative effect to human health [2] as well as to global environment in term of unusual weather and storm, and impact to the plant and animals [3]. The environmental impact of global warming has triggered efforts to reduce emission from the green house gas[4], [5], [6].

Carbon dioxide (CO_2) is one of important component of green house gas. Simulation on global and regional scenarios estimated that the carbon resulted from usage of fossil fuel (this includes gas and crude oil) will still be main source of green house gas emission in the next century. It also has estimated that in year 2100 carbon dioxide may reach to the level of 70 to 80% of total emission of green house gas [7]. Nowadays, around 40% of emission of CO_2 is resulted from power generation fuelled by fossil fuel including gas, crude oil and diesel [4].

Many non biological methods which use physical and chemical process have been proposed to sequester carbon dioxide (carbon fixation or carbon capture) by researchers. Example of those methods are chemical absorption using Amine called Amine Absorption, adsorption using solid in example Zeolites or using Activated Carbon , physical process using membrane polymer and the use of cryogenic or very low temperature process to capture carbon [4],[8],[9]. Part of the method is also the idea to capture carbon and then to store it under (sub surface storage),[10], however another researcher reported the environmental impact of the sub surface storage mentioned above [11].

Biological method is another way to sequester carbon dioxide. Microalgae are viewed as one of potential solution for CO_2 sequestration or CO_2 biofixation beside major potential for biodiesel development which has better fuel characteristic [12], and has high lipid content [13]. Microalgae can also be cultivated along the year, have better sun light utilization and higher yield relative to terrestrial plants and advantage to be used as one method to sequester CO_2 [14].

However many factors need to be evaluated to support development of CO_2 Biofixation. One of

* Corresponding author: tjg@consultant.com

important environmental factor for the growth of microalgae is related with light requirements. Various reports were available on effect of different light intensity but limited report on effect of photoperiod to different strain of algae. One good paper reported the effect photoperiod on *Apanoteche microscopic Nageli* but at a single fixed light intensity of 150 μmol/m^2.s through-out the experiment[15].

There was opportunity in term of factors evaluation to different strains especially on *Chlorella pyrenoidosa* as data available evaluating both factors was limited. Therefore the aim of our study was to evaluate the effect of light intensity and photoperiod on growth of *Chlorella pyrenoidosa* (*C.pyrenoidosa*) and CO_2 biofixation.

Our approach is to evaluate both photoperiod and light intensity using the factorial design to see the main effects of both factorsand their significance on growth of *C.pyrenoidosa* and to see if there is a factor interaction. This approach is difference from traditional approach of one factor at a time evaluation.

The result of this study has contribution to provide a baseline for *Chlorella pyrenoidosa* in term of light regime requirements of both photoperiod and light intensity when purged by air with CO_2 0.034%. The knowledge has important implication for further cultivation application using flue gas as well as optimization schemes and economic valuation later on.

2 Materials and Methods

2.1. Microalgae Strain and Medium

The culture of *Chlorella pyrenoidosa* used in this study was from Jepara region, Java, Indonesia. Initial incubation was conducted in room temperature between 27 – 31 °C, exposed to continuous lighting for 24 hours with light intensity of 2000 lux. Initial culture was grown in several photo bioreactor of 25 litres volume to provide sufficient volume for experimental runs.

Medium for growth used in the experiment was analternate novel medium fertilizer complex (G15Deptan) with composition of 5% NO_3, 18% K_2O, 3% MgO, 8% S, 0.35% Fe, 0.02% Mn, 0.02% Zn, and 0.015% Boron.

2.2. Experimental set up, design and analysis

The photo bioreactor used for the experiment were made of glass (Schot, German) with volume of 1000 mL, diameter of 9.5 cm. Total of 24 reactors were used for all experimental runs as per design.

Source of light was LED (light emitting diode) with total of 24 unit of 10 Watt LED (Hannoc, Indonesia) for each of reactor run. LED was position at adjustable distance to provide light intensity of 2000, 4000, 6000 and 8000 lux at the surface of the reactors measured with lux meter (LX-101A, Taiwan).Cycle of photoperiod was adjusted for 3 levels at L/D 8 hours of lighting and 16 hours of dark period, L/D 12 hours of lighting and 12 hours of dark period, and L/D 16 hours of lighting and 8 hours of dark period. Time was set for beginning and end of lighting accordingly for those 3 levels. A partition was put surrounding each reactor and an aluminum foil were used to cover top portion of the partition to isolate light from specific LED source to its associated single reactor.

Small air compressors were used to feed air continuously into each reactor. The objective is to supply CO_2 (contains CO_2 approximately 0,034%) to be used for microalgae but also to help fluid mixing in the reactor. Overall configuration is shown in Figure 1.

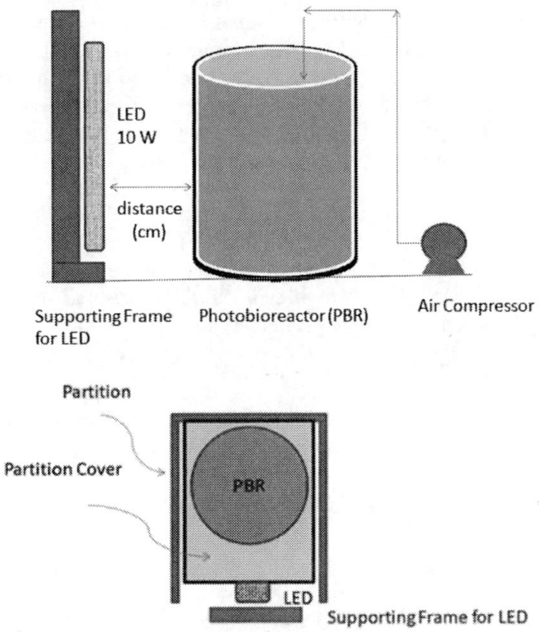

Fig. 1. Experimental set up and the top view of it.

Experiment was conducted semi batch. Microalgae fed into the reactor at the beginning of experiment and air was purged into the reactor continuously.

Design of experiment (DOE) used factorial design with 4 levels of light intensity (2000, 4000, 6000 and 8000 lux) and 3 levels of photoperiod (8, 12 and 16 hours of lighting and associated dark period accordingly). There were 24 runs out of 12 light intensity and photoperiod combinations including replicate.

Analysis of variance (ANOVA) was employed to see the significance of the main effect and interaction of factors involved. Tukey Post hoc Method wa sused for pair wise comparisons of means at 95% confidence level. Statistical package (Minitab-16) was used to help statistical analysis.

2.3. Growth measurement

Growth of the algae was determined daily by measuring optical density at wavelength of 680 nm (nano meter) using a spectrophotometer (Genesys 20, Thermo Scientific, USA). Prior to experimental run, curve of absorption spectrum of *Chlorella pyrenoidosa* was plotted at various wavelengths to determine the wavelength that will be used in the experiments. The wavelength of 680 nm was used as base for optical density measurement during the experiment.

2.4. Dry weight determination

Proxy method was used to get correlation between optical density and dry weight of biomass is used to determine dry weight of biomass in gram per litre (gr/L) when measured optical density data is available.

An amount of known volume of culture was taken into several Erlenmeyer flasks and then diluted to different ratio to provide range of optical density readings. Liquid was then transferred into test tubes (10 mL), spinned for 60 minutes using centrifuge (Gemmy, Taiwan), filtered with filter paper (Whatman, German) and naturally dried in room temperature for 50 hours. Dry weight of biomass was determined gravimetrically using analytical balance (Kenko, Japan).

The resulted correlation given in Equation (1) is then used as a calibration curve to calculate the biomass dry weight in gram/litre based on the measured optical density data from experimental run.

$$y = 1.227.x + 0.126(R^2 = 0.982) \quad (1)$$

Where y is the dry weight in gr/L, x is the measured optical density or absorbance from the experiment.

2.4. Biomass productivity and growth kinetic

Biomass productivity is an increase in biomass per unit of time and calculated as per Equation (2) and the biomass productivity maximum was calculated by Equation (2).

$$P = (X_i - X_{i-1})/(t_i - t_{i-1}) \quad (2)$$

Where $P$is daily productivity, $X_i$is the biomass at time t_i, $X_{i-1}$is the biomass at time t_{i-1}.

Specific growth rate of microalgae was determined using the equation (3).

$$\mu = [\log_e (X_t / X_{t-1})] / \Delta t \quad (3)$$

Where μ is the specific growth rate (unit of 1/day), Xi is the biomass at time ti and Xi is the biomass at initial and Δt is elapse time in reviewing μ. The specific growth rate was calculated on the growth phase of the *C.pyrenoidosa*.

2.5. Rate of CO_2 biofixation

The rate of CO2 biofixation was calculated as follows:

$$RCO2 = C.\ P\ .\ (MCO_2 / MC) \quad (4)$$

$R\ CO_2$ is the rate of CO_2 biofixation. C is the carbon content of biomass. P is the productivity. M CO_2 is the molar mass of carbon dioxide and MC is the molar mass of carbon. Typical carbon content of biomass ranges from 41 – 51 % of weight of biomass[16]. In our analysis the approach is to use 50% average of carbon content of biomass in Equation (4) following the average carbon content (% weight) resulted from elemental analysis across several CO_2 concentration on *C.pyrenoidosa*[17].

3 Results and discussion

Light is required for photosynthesis process of microalgae. Light intensity and photoperiod is an important part that not only can influence the growth of microalgae but in turn influence also the rate of CO_2 biofixation.

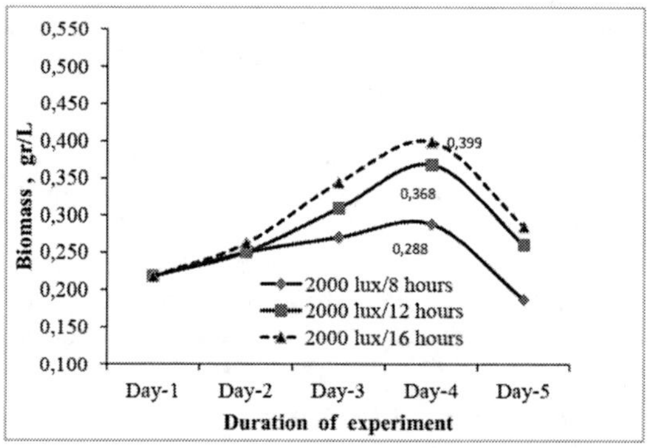

Fig. 2. Growth profile of *C.pyrenoidosa* at 2000 lux

3.1. Growth profile and kinetics

The growth profiles of *Chlorella pyrenoidosa* are presented as a plot of biomass versus time (Fig 2 to Fig 4). The growth profiles follow a general pattern of growth curve for a typical batch cultures phases from initial, logarithmic, declining growth, stationary and death phase, although clear distinction between phases differs from each condition in the experiment.

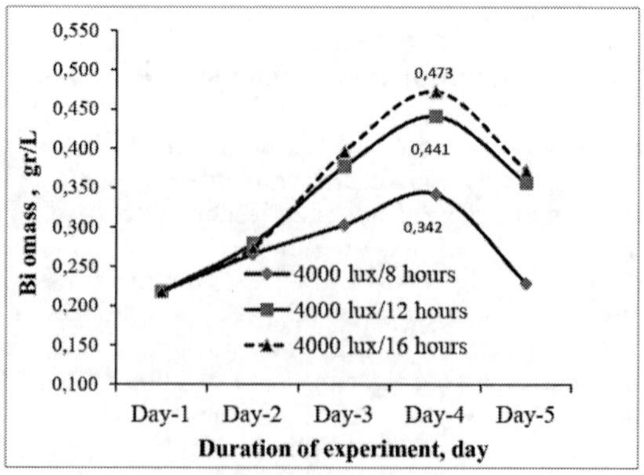

Fig. 3. Growth profile of *C.pyrenoidosa* at 4000 lux

The initial or lag phase were almost unnoticeable indicating high adaptability to the new growth medium

https://doi.org/10.1051/e3sconf/20183103003

after inoculums transfer to the reactors. The batch cultures reached the highest point on day 4 of cultivation and then entered the death phase for all condition as nutrient available for growth diminished, except the culture with light intensity of 8000 lux and 16 hours photoperiod that still sustained until day 6 of cultivation with highest biomass at 0.516 ± 0.069 gr/L before entering death phase afterward.

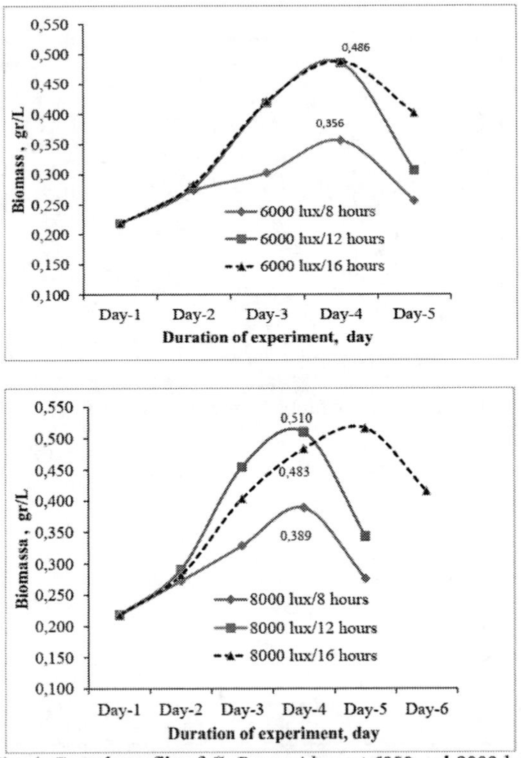

Fig. 4. Growth profile of *C. Pyrenoidosa* at 6000 and 8000 lux

The kinetic growth was reviewed at the logarithmic phase. The highest specific growth, μ (d^{-1}) of *C. Pyrenoidosa* was 0.447 at 8000 lux and 12 hours of photoperiod which was 5.7 times higher than the specific growth rate at 2000 lux and 8 hours of photoperiod and 1.2 times higher than specific growth at 8000 lux and 16 hours of photoperiod (Fig.5).

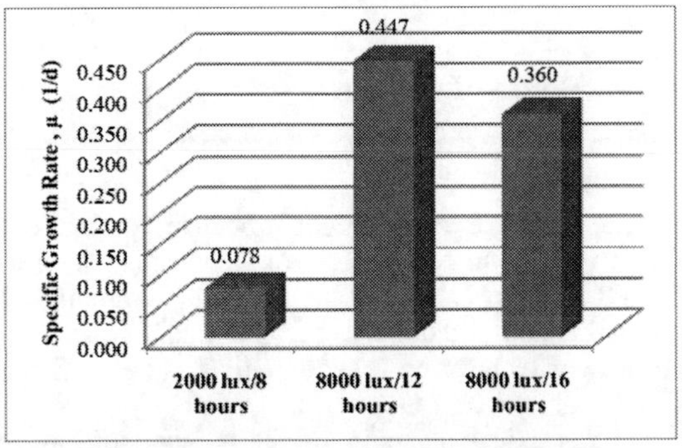

Fig. 5. Specific growth rate of *C.pyrenoidosa* between 2000 lux and 8000 lux at 12 and 16 hours of photoperiod.

Other researcher reported specific growth rate of several Chlorella strains at various conditions. It was reported that specific growth of *Chlorella sorokiniana* was 0.4 d^{-1}; *Chlorellavulgaris* under 4% CO_2 and 8% CO_2 were at the 0.41 d^{-1} and 0.51 d^{-1}respectively. *Nanochloropsisgaditana,* another strain was reported also at 4% of CO_2 and 8% of CO_2 with specific growth value of 0.44 d^{-1} and 0.45 d^{-1} [16].

All of those reported values are in the same order with the result of this study for *Chlorella pyrenoidosa* with specific growth rate of 0.45 d^{-1} (when subjected to air with around 0.034 % CO_2, light intensity of 8000 lux and photoperiod of 12 hour) and growth rate of 0.36 $d^{-1}$when subjected to 8000 lux but with photoperiod of 16 hours. Other reports had used CO_2 while *C.pyrenoidosa* in our study used air nevertheless the specific growth is comparable showing the potential of *C.pyrenoidosa* to be cultivated to biofixed CO_2.

3.2. Factors main affect and interaction

We observed that the biomass increases with the increase of light intensity. The similar trend was also observed with the increase of photoperiod. However the increase of biomass with increase of light intensity appeared to be more gradual than an increase of biomass due to an increase of photoperiod. The value of biomass reached0.399 gr/L at the 16 hours of photoperiod with light intensity of 2000 lux (Fig.2), while the value reached 0.473 gr/L with the same photoperiod but at 4000 lux light intensity (Fig.3). That was an increase of 18.5%.

There was an increase of 27.8 % and 38.5% of biomass if we move from photoperiod of 8 hours to 12 hours and from 8 hours to 16 hours respectively within the same light intensity of 2000 lux.

Similar trend was observed at 4000 lux intensity with the increase of biomass of 28.9 % and 38.3%, if we move from photoperiod of 8 hours to 12 hours and from 8 hours to 16 hours respectively.

The biomass differences at light intensity of 6000 lux and 8000 lux (Fig.4) also follow the same fashion when comparing value of biomass response for photoperiod 8 hours to 12 hours and 8 hours to 16 hours respectively. However, the percent increase of biomass was lower at 8000 lux compared to 6000 lux for the same duration of photoperiod 8 hours to 16 hours (32.6% and 36.5% respectively).

In addition, the same percent increase was observed between 8 hours to 12 hours and 8 hours to 16 hours photoperiod (36.5%).

The analysis of main effect of factors of light intensity and photoperiod revealed actual influence of factors to mean of response. The response of biomass had steeper increase from light intensity of 2000 lux to 4000 lux and from 8 hours to 12 hours photoperiod (Fig.7).

Mean of biomass gradually increased from light intensity of 4000 lux to 8000 lux and from 12 hours to 16 hours of photoperiod.

Interaction between light intensity and photoperiod was indicative from the result above. Analysis of interaction between the factors shows that although increasing photoperiod will certainly increase the mean of response as indicated in the main effect; however the response was dependent on the level of light intensity to some degree. The interaction was less obvious from 8 hours to 12 hours of photoperiod relatively to 12 hours to 16 hours of photoperiod across the range of light intensity (Fig.8).

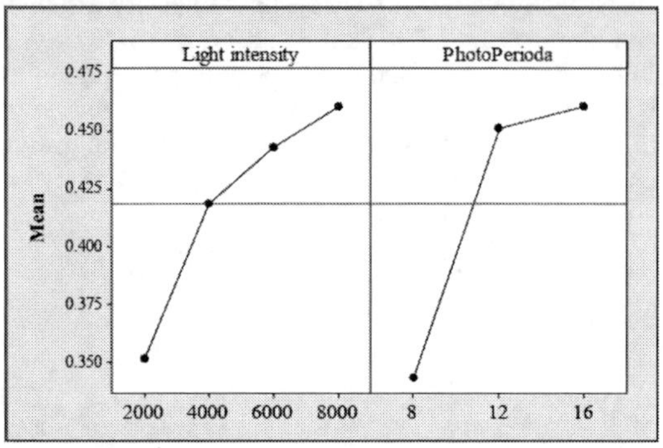

Fig. 7. Main effect plot for biomass (light intensity in lux and photoperiod in hours) of *C.pyrenoidosa*

Results above indicated that the photoperiod had greater affect to biomass relatively than the changes in light intensity within the range reviewed in this experiment. One good report on the effect of photoperiod, although evaluated to different microalgae strain - *Apanoteche microscopic Nageli* and used only a single fixed light intensity of 150 μmol/m^2.s through-out the experiment -, had observed that the photoperiod was a determinant factor and linear reduction in response was also observed with reduction of photoperiod.[15]. Our study confirmed the observation as well.

In order to examine the statistical significance of factor affects, the analysis of variance (ANOVA) was employed.

Analysis of variance result (Table.1) confirmed that the two factors of light intensity and photoperiod were statistically significant ($p< 0.05$), with p value of light intensity factor greater than photoperiod although still less than 0.05, confirming the visual observation of stronger effect of photoperiod relative to light intensity.

Interaction term between light intensity and photoperiod although indicative from the plot was statistically not significant ($p = 0.885$ which is > 0.05). In order to see which of combination of light intensity and photoperiod were significant out of 12 experimental runs, we employed Tukey pairwise test at 95% confidence.

TukeyTest provided difference of means of biomass response among level of light intensity and photoperiod. The difference response of biomass between light intensity of 2000 lux to all other light intensity from 4000 lux, 6000 lux and 8000 lux were statistically significant (all value of $p< 0.05$). The p value for difference biomass response between 4000 lux to 6000 lux and 8000 lux ligh intensity were not statistically significant ($p = 0.618$ and 0.269 respectively). The difference biomass reponse between 6000 lux and 8000 lux was also not significant ($p =0.8545$).

The grouping information using the same method at 95% confidence showed that 16 hours and 12 hours of photoperiod were within the same group and 8 hours of photoperiod were in another difference group in term of differencesin responses (Table.2)

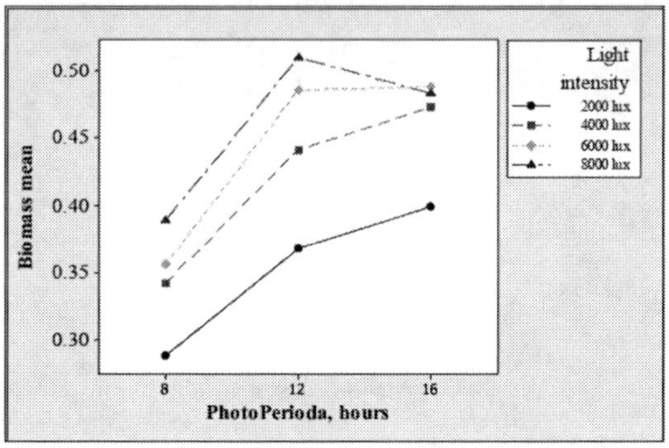

Fig. 8. Interaction plot for biomass of *C.pyrenoidosa*

Table 1.ANOVA from Minitab output.

Analysis of Variance for Biomass, gr/L, using Adjusted SS for Tests

Source	DF	Seq SS	Adj SS	Adj MS	F	P
Light Intensity, lux	3	0.040989	0.040989	0.013663	9.57	0.002
Photoperiod, hours	2	0.067667	0.067667	0.033833	23.71	0.000
Light Intensity, lux* Photoperiod, hours	6	0.003157	0.003157	0.000526	0.37	0.885
Error	12	0.017124	0.017124	0.001427		
Total	23	0.128936				

S = 0.0377755 R-Sq = 86.72% R-Sq(adj) = 74.54%

Table 2. Tukey pairwise comparison of means of biomass to photoperiod (L/D) - Minitab output.

Grouping Information Using Tukey Method and 95.0% Confidence

Photoperiod, hours	N	Mean	Grouping
16	8	0.5	A
12	8	0.5	A
8	8	0.3	B

Means that do not share a letter are significantly different.

The Tukey Test at 95% confidence confirmed that the difference in response were statistically significant from 8 hours of L/D to 12 hours and 16 hours of L/D ($p = 0.0003$ and $p = 0.0001$ respectively, all p were < 0.05), but L/D of 12 hours to L/D 16 hours was not statistically significant ($p = 0.8749 > 0.05$).

The practical meaning of the comparison above is that there is a threshold value of light intensity and photoperiod setting for cultivation of *C.pyrenoidosa*. Light intensity of 2000 lux is too low for cultivation as it did not provide sufficient energy to achieve best growth. The 8 hours of photoperiod or L/D of 8 hours of light exposure and 16 hours associated dark hours was not sufficient in term of timing to accumulate light energy necessary for the growth.

On the other hand, 12 hours of photoperiod with L/D 12 hours of light and 12 hours dark was sufficient for growth provided the appropriate light intensity is available. The additional 4 hours of light exposure at L/D 16 hours light /8 hours dark did not significantly different than the 12 hours of light exposure in term of biomass response; however. It differed in term of sustainability prior to entering the death phase.

The comparison indicated also that there was more flexibility in term of light intensity choices beyond the 4000 lux to 8000 lux provided the photoperiod is within the range of 12 hours of light exposure to 16 hours of it, although results from this study suggested that best region is close to 8000 lux with 12 hours or 16 hours of photoperiod if we neglect the possibility of curvature (where maximum point may be located) between those two photoperiods.

3.3. Rate of CO_2 biofixation of *C.pyrenoidosa*

CO_2 biofixation rate was determined based on calculated biomass daily productivity using Equation (2). The average rate of CO_2 biofixation of *C.pyrenoidosa* at 12 factorial combination of photoperiod and light intensity ranges from 0.082 ± 0.03 $gr/L^{-1}.d^{-1}$ at 2000 lux and 8 hours of photoperiod to 0.267 ± 0.02 $gr/L^{-1}.d^{-1}$.That was an increased of about 2 times or the ratio of almost 3 folds value of the rate from the low setting of light intensity and photoperiod to the higher setting of them (Table. 3).

Table 3. Rate of CO_2 biofixation by *C.pyrenoidosa* at 12 experimental conditions of light intensity and photoperiod.

Experimental condition (light intensity & photoperiod)	CO2 Biofixation, ($gr/L^{-1}.d^{-1}$) means ± sd*
2000 lux/8 hours	0.082 ± 0.03
2000 lux/12 hours	0.132 ± 0.04
2000 lux/16 hours	0.165 ± 0.03
4000 lux/8 hours	0.133 ± 0.02
4000 lux/12 hours	0.211 ± 0.01
4000 lux/16 hours	0.223 ± 0.01
6000 lux/8 hours	0.152 ± 0.01
6000 lux/12 hours	0.237 ± 0.00
6000 lux/16 hours	0.245 ± 0.03
8000 lux/8 hours	0.170 ± 0.01
8000 lux/12 hours	0.267 ± 0.02
8000 lux/16 hours	0.239 ± 0.06

* sd - standard deviation

of light intensity and photoperiod for cultivation from this study could provide comparable result of CO_2 biofixation rate to other cultivation condition reported.

Our study indicated the important of length of exposure to light (photoperiod) and light intensity level selection to achieve desired result of CO_2 biofixation out of microalgae cultivation. In general the result showed an increase of CO_2 biofixation rate at the increasing combination of factors, however the highest average rate of CO_2 biofixation was achieved at high setting of light intensity (8000 lux) and 12 hours of photoperiod not at the 16 hours photoperiod. Interaction plot of biomass also reveal the same phenomenon (Fig.8). This was due to higher interaction occurring at 8000 lux and 16 hours relative to lower light intensity and lower photoperiod. In addition, the phenomenon also may indicate the existence of curvature on the response which may happen between photoperiod of 12 hours to 16 hours. It means that the maximum response could be located between 12 hours to 16 hours of photoperiod.

Our study also indicated that to achieve the comparable rate of CO_2 biofixation (purged with air), relatively lower light intensity range used in this experiment (2000 lux to 8000 lux) could suffice the light requirement for cultivation. Economic wise, this is a desirable aspect for mass cultivation.

The result provided practical implication in term of region of light intensity and length of light exposure to achieve best CO_2 fixation rate within the range of review. Light intensity needs to be between 6000 lux and 8000 lux and photoperiod to be between 12 hours to 16 hours to be in the region of highest rate. The best route for higher rate of CO_2 biofixation is diagonal upward to the right of contour plot (Fig.9).

However, we need to note that the best region in the contour plot CO_2 biofixation rate mentioned above does not necessarily pointed out the actual maximum point that can be achieved by *C.pyrenoidosa* but it indicated regions of combination of photoperiod and light intensity that we can expect to be the best route within the range reviewed in this experiment.

The actual maximum point location could be located within the region but to pin point the actual maximum location, the optimization study will need to be performed later on due to the possibility of curvature between light intensity of 6000 and 8000 lux and photoperiod of 12 hours and 16 hours based on contour plot review.

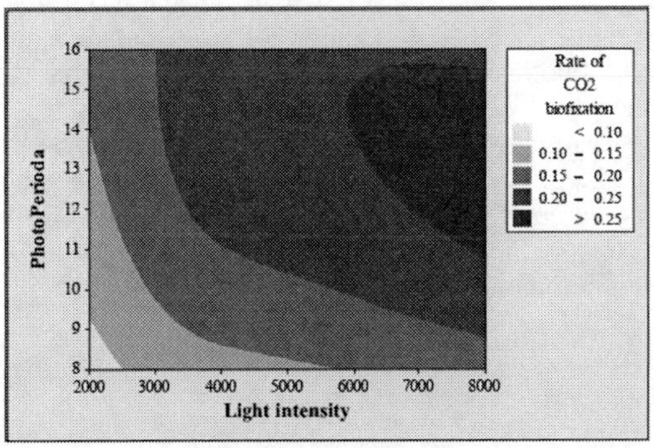

Fig. 9. Contour plot showing region of light intensity and photoperiod for CO_2 biofixation of *C.pyrenoidosa* (Air with CO_2 0.034%).

Therefore the result of this study is important baseline for further study on the optimization of *C.pyrenoidosa*.

Report(s) available on rate of CO_2 biofixation of *Chlorella pyrenoidosa* is limited. The previous reported value of CO_2 biofixation rate on the strain was available only with one level of light intensity used at 150 μmol/(s.m^2) and one photoperiod with 10% CO_2 and air [17], therefore to some extend could obscure the important of both effects of photoperiod and light intensity from those reports.

The rate of CO_2 biofixation of *C.pyrenoidosa* from our study was comparable than the value previously reported by other researcher under different culture condition with 10 % CO_2, whilethe result of this study showed around 1.7 to 2 times higher than the value reported by other research under the same condition with air. (Table.4).

Table 4. Several reported CO_2 biofixation rate.

Mikroalgae	CO2 %	Rate of CO2 Biofixation (gr/l/d)	Condition / References
C.pyrenoidosa (Tang et al, 2011)	CO2, 10%	0,26	180 μmol/(s.m^2) ~ 10,000 lux, BG11 / [17]
C.pyrenoidosa (Tang et al, 2011)	CO2, 0.03 %	0,134	180 μmol/(s.m2) ~ 10,000 lux, BG11 / [17]
C.pyrenoidosa (this study, 2017)	Air, 0.034% CO2	0,239	average value at condition of 8000 lux, 16 hours of photoperioda, novel fertizer complex medium
C.pyrenoidosa (this study, 2017)	Air, 0.034% CO2	0,267	average value of 8000 lux, 12 hours of photoperioda, novel fertilizer complex medium

4 Conclusions

The light intensity and photoperiod as part of the light requirement were both statistically significant to the growth as well as the ability of *C.pyrenoidosa* to fix CO_2. Photoperiod showed higher influence than light intensity to the growth and rate of CO_2 biofixation.

It is important and economically desirable feature to be able to set suitable light regime for microalgae cultivation. The outcome of this study provided the best route of biomasss, specific growth rate and CO_2 biofixation rate of *C. Pyrenoidosa* to be within region of 6000 lux to 8000 lux and between 12 hours and 16 hours of photoperiod (length of exposure to light).

The result can be used as a baseline for further evaluation of *C.pyrenoidosa* with other source of CO_2 (i.e. flue gas), as well as optimization and economic valuation later on.

References

1. M. K. Lam, K. T. Lee, and A. R. Mohamed, Int. J. Greenh. Gas Control, **10**, 456–469 (2012)
2. M. H. R. Anstey, Global. Health, vol. **9**, no. 1, 4 (2013)
3. A. Yakub, P. Shrivastava, K. Krishnamurthi, S. N. Mudliar, S. Saravana, G. S. Kanade, S. K. Lokhande, and T. Chakrabarti, Bioresour. Technol., **138**, 382–386 (2013)
4. Hadiyanto, Azimatun Nur, M.M.World Applied Sciences Journal, **31 (5)**, 959-967) (2014)
5. M. Meinshausen, N. Meinshausen, W. Hare, S. C. B. Raper, K. Frieler, R. Knutti, D. J. Frame, and M. R. Allen, Nature, **458**, no.7242, 1158–1162 (2009)
6. J. Ball, Wall Str. J. East. Ed., **252**, 70, A13 (2008)
7. T. Kram, T. Morita, K. Riahi, R. A. Roehrl, S. Van Rooijen, A. Sankovski, and B. De Vries, Technol. Forecast. Soc. Change, **63**, no.2–3, 335–371 (2000)
8. A. Hussain and M.-B. Hägg, J. Memb. Sci., **359**, no. 1–2, 140–148 (2010)
9. M. J. Tuinier, M. van Sint Annaland, and J. a. M. Kuipers, Int. J. Greenh. Gas Control, **5**, no. 4, 694–701 (2011)
10. J.-P. Nicot, S. Solano, J. Lu, P. Mickler, K. Romanak, C. Yang, and X. Zhang, Energy Procedia, **37**, 4552–4559 (2013)
11. J. Koornneef, A. Ramírez, W. Turkenburg, and A. Faaij, Prog. Energy Combust. Sci., **38**, no. 1, 62–86 (2012)
12. Y. C. Sharma, B. Singh, and S. N. Upadhyay, Fuel, **87**, no. 12, 2355–2373 (2008).
13. E. Suali and R. Sarbatly, Energy Rev, **16**, no. 6, 4316–4342 (2012)
14. B. Wang, Y. Q. Li, N. Wu, and C. Q. Lan, Appl. Microbiol. Biotechnol., **79**, 707–718 (2008)
15. E. Jacob-lopes, C. Henrique, G. Scoparo, L. Mara, and C. Ferreira, Chem. Eng. Process. Process Intensif., **48**, 306–310 (2009)
16. M. Adamczyk and J. Lasek, App Biochem Biotech., **179**, 1248–1261 (2016)
17. D. Tang, W. Han, P. Li, X. Miao, and J. Zhong, Bioresour. Technol., **102**, no. 3, 3071–6 (2011)

Pollution Impact and Alternative Treatment for Produced Water

Yusran Hedar[1,*] and *Budiyono*[2]

[1]Master Program of Environmental Science, School of Postgraduate Studies, Diponegoro University, Semarang - Indonesia
[2] Department of Chemical Engineering, Faculty of Engineering, Diponegoro University, Semarang - Indonesia

Abstract. Oil and gas exploration and production are two of the activities that potentially cause pollution and environmental damage. The largest waste generated from this activity is produced water. Produced water contains hazardous pollutants of both organic and inorganic materials, so that the produced water of oil and gas production cannot be discharged directly to the environment. Uncontrolled discharge can lead to the environmental damage, killing the life of water and plants. The produced water needs to be handled and fulfill the quality standards before being discharged to the environment. Several studies to reduce the contaminants in the produced water were conducted by researchers. Among them were gravity based separation - flotation, separation technique based on filtration, and biological process treatment. Therefore, some of these methods can be used as an alternative waste handling of produced water.

1 Introduction

The significance of oil and natural gas in modern civilization is well known. Nevertheless, oil and gas exploration and production activities produce solid, liquid, and gas waste with 80% liquid waste composition, even it reaches 95% in the old oil fields. Produced water is the largest liquid waste generated by the activity [1]. Oilfield wastewater or produced water contains various organic and inorganic components. Discharging produced water can pollute surface and underground water and soil [2].

The produced water can be from the saline water and formation water. There are three sources of saline water: Flow from above or below the hydrocarbon zone, Flow from within the hydrocarbon zone, Flow from injected fluids and additives resulting from production activities. The last category is called "connote water" or "formation water" and becomes produced water when saline water mixed with hydro-carbons comes to the surface [3]. Global produced water production is estimated at around 250 million barrels per day compared with around 80 million barrels per day of oil. As a result, water to oil ratio is around 3:1 that is to say water cut is 70%. Oil fields are responsible for more than 60 % of produced water generated worldwide [2,4]. The large number of the produced water can be used for various purposes, such as underground injection to increase oil production, irrigation, livestock or wildlife watering and habitats, and various industrial uses (e.g., dust control, vehicle washing, power plant makeup water, and fire control) [2,3].

Continuation of large produced water discharges is expected because the water cut (the relative amount of water to oil) increases with the age of the production wells [2,14]. Thus, the question of biological effects of produced water discharges is a matter of continued relevance. Usually, produced water is treated to remove dispersed oil by either flotation or gravity in separation tank batteries, and then discharged to the environment. Alternatively, it may be injected into a nearby non-producing well [15].

Before being disposed to the environment, the produced water need treatment to meet the quality requirement of waste water set by the Government. Production quality standard in Indonesia is regulated by the Regulation of the Minister of Environment Number 19 Year of 2010 regarding the quality standard of waste water for businesses and/or activities of oil and gas and geothermal. The regulation determines that the maximum permissible concentration for COD is 200-300 mg / L, oil and fat are 25-50 mg / L, dissolved sulfide is 0.5 mg / L, ammonia is 5 mg / L and phenol is 2 mg / L [5]. The permitted oil and grease (O&G) limits for treated produced water discharge offshore in Australia are 30 mg/L (milligram per liter) daily average and 50 mg/L instantaneous [6]. Based on United States Environmental Protection Agency (USEPA) regulations, the daily maximum limit for O&G is 42mg/L and the monthly average limit is 29 mg/L [7]. As regards the significant matter of environmental concern, many countries have implemented more stringent regulatory standards for discharging produced water. The monthly average limits of O&G discharge and chemical oxygen demand (COD) prescribed by the Peoples Republic of China are 10 and 100 mg/L, respectively [8]. Based on the Convention for the Protection of the Marine Environment of the North-East Atlantic (OSPAR Convention), the annual average limit for discharge of

* Corresponding author: yusranhedar@gmail.com

dispersed oil for produced water into the sea is 40mg/L [2].

The utilization of the produced water and the disposal of untreated produced water containing many hazardous materials can interfere with the environmental sustainability. The characteristics of the produced water must be known to provide an appropriate way to reduce the content of hazardous substances in the water prior to the disposal. Appropriate management and handling are needed, so the produced water can be disposed to the environment safely and can be utilized properly in order not to interfere with environmental sustainability. In this review will be discusses about the introduction of produced water, the impact of produced water discharge on the environment, and technology that has been used in handling of produced water.

2 Produced Water

Produced water (also called formation water, brine or saltwater) is water from underground formations that is brought to the surface during oil or gas production. The water is discharged in an oily form after its separation from the real oil. During oil and gas exploration other forms of wastewater are produced, these include injected water, little quantity of water that is condensed and traces of some chemicals used among which produced water is the highest generated by-product [24]. The physical and chemical properties of produced water vary considerably depending on the geographic location of the field, the geo-logic formation, and the type of hydrocarbon product being produced [24,25]. Dispersed oil, aromatic hydrocarbons and alkylphenols (AP), heavy metals, and naturally occurring radioactive material (NORM) are of particular environmental concern [25,26]. Because the water has been in contact with hydrocarbon-bearing formations, it contains some of the chemical characteristics of the formations and the hydrocarbons. It may include water from the reservoir, water previously injected into the formation, and any chemicals added during the production processes. The composition and characteristics of naturally occurring chemical substances in produced water are closely associated to the geological characteristics of each reservoir [25]. Produced water is mostly discharged to the immediate aquatic environment; the organic and inorganic compounds in produced water have higher toxicity to the environment than crude oil.

2.1. Characteristic of produced water

Produced waters characteristics depend on the nature of the producing/storage formation from which they are withdrawn, the operational conditions, and chemicals used in process facilities. The composition of produced water from different sources can vary by order of magnitude. However, produced water composition is qualitatively similar to oil and/or gas production [2,9,10].

Produced water is a mixture of organic and inorganic materials. Some factors such as geological location of the field, its geological formation, lifetime of its reservoirs, and type of hydrocarbon product being produced affect the physical and chemical properties of produced water [3]. The major compounds of produced water include:

a. Dissolved and dispersed oil compounds
b. Dissolved formation minerals,
c. Production chemical compounds,
d. Production solids (including formation solids, corrosion and scale products, bacteria, waxes, and asphaltenes),
e. Dissolved gases [2,11,12]

Although produced water varies significantly among wells and fields, several groups of constituents are present in most types of produced water. The major constituents of concern in produced water are salt content (expressed as salinity, total dissolved solids, or electrical conductivity), oil and grease, and various natural inorganic and organic compounds [16]. Sulphate reducing bacteria may also be present in produced water [24].

2.2. Produced water from oilfield

The produced water from the oil processing may contain groundwater or seawater injected to maintain the reservoir pressure and also the content of solid particles and bacteria. Most produced water contains more saline water than sea water. Besides, the produced water contains chemical compounds used in the drilling process and the production process as well as the process of separation of water and oil. The chemical compounds for processing are usually the complex mixture of many molecular compounds. The mixture may contain:

a. Corrosion inhibitor and oxygen remover to reduce the risk of corrosion of equipment
b. Crust inhibitor that limits deposits of the mineral crust
c. Biocide compounds to reduce bacterial fouling
d. Emulsion destroyer and emulsion purifier to destroy water emulsion in oil and vice versa
e. Coagulants, flocculants, and purifiers to remove solids
f. Solvent to reduce paraffin deposit

In the produced water, these chemical compounds can affect the coefficients of partition, toxicity, bioavailability, and biodegradability. In line eith the development of the oil fields, these compounds are needed in large quantities to ensure flow in the underwater piping system [16]. Table 1 summarizes a range of produced water characteristics in different oilfields in the world. The data show ranges of pollutants and constituents that are present in produced water.

Table 1. Summary of oilfield-produced water parameters in world [2, 17].

Parameter	Values	Heavy metal	Values (mg/L)
Density (kg/m^3)	1014–1140	Calcium	13–25800
Surface Tension (dynes/cm)	43–78	Sodium	132–97000
TOC (mg/L)	0–1500	Potassium	24–4300
COD (mg/L)	1220	Magnesium	8–6000
TSS (mg/L)	1.2–1000	Iron	<0.1–100
pH	4.3–10	Aluminum	310–410
Total oil (IR; mg/L)	2–565	Boron	5–95
Volatile (BTX; mg/L)	0.39–35	Barium	1.3–650
Base/neutrals (mg/L)	<140	Cadmium[a]	<0.005–0.2
(Total non-volatile oil and grease by GLC/MS) base	275	Chromium	0.02–1.1
Chloride (mg/L)	80–200,000	Copper	<0.002–1.5
Bicarbonate (mg/L)	77–3990	Lithium	3–50
Sulfate (mg/L)	<2–1650	Manganese	<0.004–175
Ammonia nitrogen (mg/L)	10–300	Lead[a]	0.002–8.8
Sulfite (mg/L)	10	Strontium	0.02–1000
Total polar (mg/L)	9.7–600	Titanium	<0.01 –0.7
Higher acids (mg/L)	<1–63	Zinc[a]	0.01–35
Phenols (mg/L)	0.009–23	Arsenic[a]	<0.005–0.3
VFA's (volatile fatty acids) (mg/L)	2–4900	Mercury	<0.001–0.002
		Silver[a,b]	<0.001–0.15
		Beryllium	<0.001–0.004

a Analyzed by atomic absorption.
b Value should be regarded as a minimum due to poor solubilities.

2.3. Produced water from gas fields

In gas fields, water injection is not utilized; therefore, the produced waters are mixture of formation water and condensed water. Their chloride content varies from almost those of fresh water to salty formation water with chloride concentration about 14 times that of seawater. Its acidity is greater than that of produced water from oilfields [2,18].

The volume of produced water from gas field is less than in oilfields. A wide range of gas treatment chemicals is used in gas fields including methanol, ethylene glycol, and triethylene glycol. About one-third of these chemicals are discharged in produced water [2,6]. Volatile components concentrations in produced water from gas fields are higher than those in produced water from oilfields [2,19]. Table 2 shows concentrations of constituents in produced water from gas fields.

Table 2. Constituents (mg/L) in natural gas produced waters (pH is presented in standard units) [2,20].

Parameter	Minimum	Maximum	Parameter	Minimum value	Maximum value
pH[a]	4.4	7.0	Iron[a]	ND	1100
pH[b]	3.1	6.47	Iron[b]	39	680
Conductivity[a] (umhos/cm)	4200	180,000	Lead[b]	<0.2	10.2
Conductivity[b] (umhos/cm)	136,000	586,000	Lithium[b]	18.6	235
Alkalinity[b]	0	285	Magnesium[a]	0.9	4300
TDS[a]	2600	310,000	Magnesium[b]	1300	3900
TDS[b]	139,000	360,000	Manganese[a]	0.045	6.5
TSS[a]	14	800	Manganese[b]	3.59	63
TSS[b]	8	5484	Nickel[a]	ND	0.02
BOD5[a]	75	2870	Nickel[b]	<0.08	9.2
COD[a]	2600	120,000	Potassium[b]	149	3870
Aluminum[a]	ND	0.4	Silver[b]	0.047	7
Aluminum[b]	<0.50	83	Sodium[a]	520	45,000
Arsenic[a]	0.004	1	Sodium[b]	37,500	120,000
Arsenic[b]	<0.005	151	Strontium[a]	–	6200
Barium[a]	ND	26	Sulfate[a]	<0.1	47
Barium[b]	9.65	1740	Sulfate[b]	ND	19
Boron[a]	ND	56	Tin[a]	ND	1.1
Bromide[b]	150	1149	Zinc[a]	ND	0.022
Cadmium[a]	ND	0.015	Zinc[b]	<0.02	5
Cadmium[b]	<0.02	1.21	TOC[a]	67	38,000
Calcium[a]	ND	25,000	Surfactants[b]	0.08	1200
Calcium[b]	9400	51,300	Benzene[a]	1.8	6.9
Chloride[a]	1400	190,000	Benzene[c]	<0.010	10.3
Chloride[b]	81,500	167,448	Toluene[a]	0.857	3.37
Chromium[a]	ND	0.03	Toluene[c]	<0.010	18
Copper[a]	ND	0.02	Oil/grease[a]	6	60
Copper[b]	<0.02	5	Oil/grease[b]	2.3	38.8

a [21].
b [22].
c [23].

3 Environmental Effects of Produced Water

The effect of produced water in a certain environment depends on the physical, chemical and biological composition of such environment. Findings indicate that in spite of all the level of toxicity of produced water effluent, there is paucity of information on their real impact on the exposed ecology [24,27]. Produced water from oil and gas industries often is permitted to be discharged to the environment [2,28]. In offshore drilling, produced water are mostly discharged to the immediate aquatic environment. The organic and inorganic compounds in produced water have higher toxicity compared to that of crude oil. The discharge of these toxic constituents and contaminants to the aquatic environment pose threat to aquatic life and agricultural resources by altering the natural state of the aquatic environment [24,29]. Water's toxicity and organic loading can generally characterize the impact of discharging produced water into the sea [2,11].

Produced water is a mixture of inorganik and organic compounds. Salinity is a general attribute of produced water. Salinity or salt concentration, describe as TDS, can vary in produced waters from 1.000-400.000 mg/L [33]. Environmental effect of produced water salts can occur in all regions where oil and gas have been produced [2,30]. Sodium is a major dissolved constituent in most produced waters and it causes substantial degradation of soils through altering of clays and soil textures and subsequent erosion. High sodium levels compete with calcium, magnesium, and potasium for uptake by plant roots, therefore, excess sodium can prompt deficiencies of other cations. Elevated levels of sodium also can cause poor soil structure and inhibit water infiltration in soil [33]. Produced water salts seem to have the most wide-ranging effects on soils, water quality, and ecosystems [30]. It is as a major contributor of toxicity [6]. Salinity is higher in produce water than some sea water which could result to aquatic destruction in fresh water [24,31]. Inorganic ions (e.g., sodium, potasium, calcium, and chloride) are not concern in produced water discharges to the ocean [36] but are of environmental concern when the treated water dischargesd to land or surface fresh or brackish water [26].

The produced water content that has the potential to cause damage to the environment is the organic material. Organic matter in produce water exist in two form: dispersed oil and non-hydrocarbon organic material. Dispersed oil is small, discrete droplets suspended in the water. Nonhydrocarbon organic material is dossolved in the water [33,34]. Dispersed oil and droplets do not precipitate at the bottom of sea but rise to the surface of water. Volatile and/or toxic compounds evaporate. These materials increase the biochemical oxygen demand (BOD) of the affected water [2,32]. Worldwide research has proven that produced water effluents are as-sociated with high level of biological oxygen demand (BOD) and chemical oxygen demand (COD) which are generated from compounds of fatty acids [24]. Nonpolar organics from different sources of produced water are consistently toxic [2,36]. Produced water toxicity can be expressed as acute or chronic toxicity. Acute toxicity can be measured by the LC50 test, but long-term effects or chronic toxicity are more difficult to quantify [2,11].

Some produced waters contain chemicals that are highly toxic to sensitive marine species, even at low concentrations. When discharge is to shallow, enclosed coastal waters, or when discharge is of a low-density produced water in an area with low water turbulence and current speeds, concentrations of produced water chemicals may remain high for long enough to cause ecological harm [6]. The chemicals of greatest environmental concern in produced water, because their concentrations may be high enough to cause bioaccumulation and toxicity include aromatic hydrocarbons, some alkylphenols, and a few metals. Highly alkylated phenols (octyl- and nonyl-phenols) are well-known endocrine disruptors, but rarely are detected in produced water at high enough concentrations to cause harm to water column animals following initial dilution [26,37].

Most metals and naturally-occurring radionuclides are present in produced water in chemically reactive dissolved forms at concentrations similar to or only slightly higher than concentrations in seawater and, therefore, are unlikely to cause adverse effects in the receiving water environment [6]. Heavy metal toxicity is less than nonpolar organics in produced water [36]. If produced water is discharged to shallow estuarine and marine waters, some metals and higher molecular weight aromatic and saturated hydrocarbons may accumulate in sediments near the produced water discharge, possibly harming bottom living biological communities [26].

Schifter et al. [16] studied the effect of produced water effluents in Sonda de Campeche located in Gulf Mexico, This study is based on direct field sampling of effluent released into the ocean in the years 2003 and 2013. The result showed that Nicel, Cr, Pb, and As has increases values in 2013. The sum of average metal concentration increased from 2,72 to 11,04 μg/L in the decade. Otherwise, it was found that TPHs increased by two in the water at diffuser area during the decade. Fish tissues at the discharge zone and reference site are contaminated with PAHs, dominated by 2- and 3-rings; 4-ring accounted for less than 1% oh total PAHs in 2003, but increased to 7% in 2013.

Some production treatment chemicals are toxic and, if they are discharged at high concentration in produced water, could cause localized harm [26]. Water-soluble production chemicals do not have toxicity effects in the aqueous phase compared to oil soluble production chemicals at the same concentration. However, some production chemicals can increase partitioning of oil compounds into the aqueous phase at high concentrations [29]. Treatment chemicals can precipitate and accumulate in marine sediments [39].

Produced water from gas production tend to have higher higher content of low molecular weight aromatic hydrocarbon such as benzene, toluene, ethylbenzene, and xylene (BTEX) than produced water from oil production. Studies indicate that the produced water discharge from gas/condensate platform are about 10 times more toxic than produced water discharged from oil platform [18,33].

4. Produced Water Treatment

Produced water is considered an oilfield waste. Whether waste or commodity, produced water management has a cost. For managing produced water, a three-tiered pollution prevention hierarchy is followed:

- Employing technologies to minimize produced water production,
- Reuse and recycling,
- If neither of these tiers is practical, disposal is the final option [2,40].

Some of the options available to oil and gas operators for produced water management proposed by Arthur et al. [41] are as follows:

- Injection: injection of produced water into the same formation from which the oil is produced or handle to another formation.
- Discharge: treatment of produced water to meet onshore or offshore discharge regulations.
- Reuse in oil and gas operation: treat the produced water to meet the quality required to use it for usual oil and gas fields operations.
- Consume in beneficial use: produced water treatment to meet to quality required for beneficial uses such as irrigation [42], rangeland restoration, cattle and animal consumption, and drinking water.

Because produced water may contain many different types of contaminants and the concentration of contaminants varies significantly, numerous types of treatment technologies have been proposed to treat produced water. Treatment of produced water is an effective option for produced water handling. Treatment of produced water has the potential to be a harmless and valuable product rather than a waste. The general objectives for operators for treating produced water are as follows[41] :

a) De-oiling: removing dispersed oil and grease,
b) Soluble organics removal,
c) Disinfection,
d) SS removal: removing of suspended particles and sand,
e) Dissolved gas removal: removing of light hydrocarbon gases, carbon dioxide, and hydrogen sulfide,
f) Desalination: removing dissolved salts,
g) Softening: removing excess water hardness,
h) Miscellaneous: removing NORM.

Most often, an effective produced water treatment system will consist of many different types of individual unit processes used in series to remove a wide suite of contaminants that may not be removed with a single process [33]. In general, some of the technologies used for treating the produced water are gravity based separation - flotation, separation technique based on filtration, and biological process treatment.

4.1. Gravity Separation

Most commonly used water treating equipment depends on the forces of gravity to separate the oil droplets from the water continuous phase. The oil droplets, being lighter than the volume of water they displace, have a buoyant force exerted on them. This is resisted by a drag force caused by the vertical movement through the water [43]. Oil and grease removal methods depend on the end usage of treated water and composition of oil in the produced water [41]. Table 3 shows typical performance for oil removal treatment as expressed by oil particle size.

Table 3. Oil and grease removal technologies based on size of removable particles [41,44].

Oil Removal Technology	Minimum size of particles removed (microns)
API gravity separator	150
Corrugated plate separator	40
Induced gas floatation (no flocculants)	25
Induced gas floatation (with flocculants)	3 – 5
Hydroclone	10 – 15
Mesh coalescer	5
Media filter	5
Centrifuge	2
Membrane filter	0.01

4.1.1 Corrugated Plate Interceptors (CPI)

The use of flow through parallel plates to help gravity separation in skim tanks was pioneered in the late 1950s as a method of modifying existing refinery horizontal, rectangular cross-section separators to treat oil droplets less than 150 μm in diameter. Various configurations of plate coalescers have been devised. These are commonly called parallel plate interceptors (PPI), corrugated plate interceptors (CPI), or cross-flow separators. All of these depend on gravity separation to allow the oil droplets to rise to a plate surface where coalescence and capture occur [43].

The first form of a plate coalescer was the PPI. This involved installing a series of plates parallel to the longitudinal axis of an API separator (a horizontal, rectangular cross-section skimmer). The plates form a "V" when viewed along the axis of flow so that the oil sheet migrates up the underside of the coalescing plate and to the sides. Sediments migrate toward the middle and down to the bottom of the separator, where they are removed [43].

The most common form of PPI used in production operations is the CPI. This is a refinement of the PPI in that it takes up less plan area for the same particle size removal, it makes sediment handling easier, and it has the added benefit of being cheaper than a PPI [43]. Corrugated plates are packed to enhance the performance of gravity separation tanks (Figure 1). The oil droplets coalesce and form larger oil droplets as the corrugated plates provide a longer path for the oil droplets to travel to the top of the tank. It is a simple operation that allows the compact design of the API separation tank; however, the efficient oil removal limits the oil droplet size of 40 microns and larger. Removal of smaller oil droplets is difficult with corrugated plate separator [41].

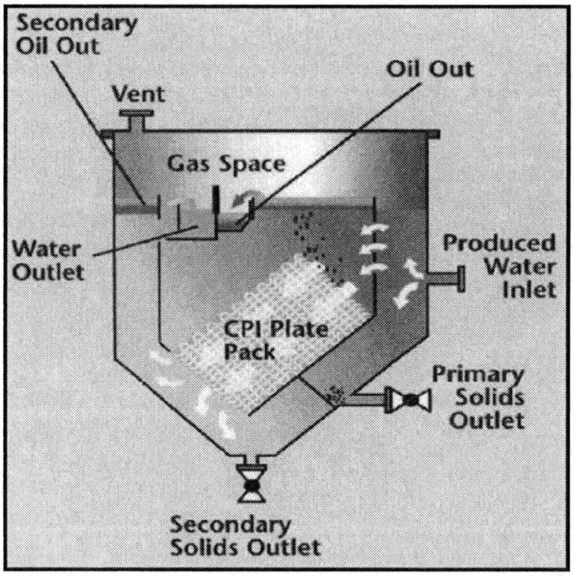

Fig. 1. Corrugated Plate (CPI) packing separates oil and solids from produced water [41,43].

Deng et al. [2,45] proposed crossflow oil–water separator for polymer flooding. The system consisted of different coalescence and separator sections. Sludge was removed from the bottom and oil and gas was separated from the top. Oil concentration in treated produced water fell to less than 100 ppm. Fig. 2 shows a schematic of the system. Van den Broek and Van der Zande [2,46] compared oil removal efficiencies of different de-oiling systems. Their rankings with respect to performance of three physical separators were: centrifuges, hydrocyclones, and plate separators. Besides the low removal efficiency of these systems, dissolved and hazardous components could not be removed.

https://doi.org/10.1051/e3sconf/20183103004

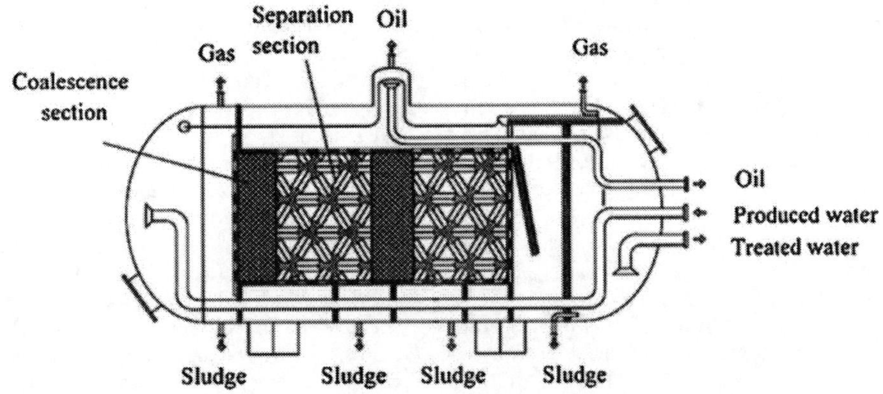

Fig. 2. Schematic illustration of crossflow oil–water separator [2,45]

4.1.2 Flotation

Flotation is a process in which fine gas bubbles are used to separate small, suspended particles that are difficult to separate by settling or sedimentation. Gas is injected into the water to be treated, and particulates and oil droplets suspended in the water are attached to the air bubbles, and they both rise to the surface. As a result, foam develops on the surface, which is commonly removed by skimming. The dissolved gas can be air, nitrogen, or another type of inert gas. Dissolved air/gas flotation also can be used to remove volatile organics and oil and grease. Dissolved air flotation units have been widely used for treatment of produced water [33,47,48].

Gas flotation technology is subdivided into dissolved gas flotation (DGF) and induced gas flotation (IGF). The two technologies differ by the method used to generate gas bubbles and the resultant bubble sizes. In DGF units, gas (usually air) is fed into the flotation chamber, which is filled with a fully saturated solution. Inside the chamber, the gas is released by applying a vacuum or by creating a rapid pressure drop. IGF technology uses mechanical shear or propellers to create bubbles that are introduced into the bottom of the flotation chamber, shows at figure 3. Coagulation can be used as a pretreatment to flotation [33,41]. Efficient performance is limited to oil droplet size of greater than 25 microns. To achieve higher efficiency if smaller droplets are present. Produced water treatment systems based on micro-bubble floatation system have been developed which use 5-50 micrometer bubbles through the reactor [41,49]. Smaller bubbles more effectively separate oil from the produced water which results in low skim volume [41]. Flotation works well in cold temperatures and can be used for waters with both high and low TOC concentrations. It is excellent for removing natural organic matter (NOM) and can be used to treat water containing TOC, oil and grease, and particulates < 7% solids [33]. In one reported study, flotation achieved an oil removal of 93% [50]

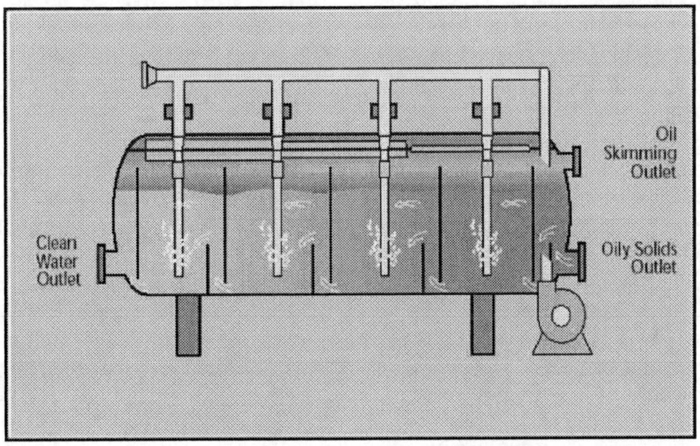

Fig. 3. Induced gas floatation cell [41]

4.1.3 Hydrocyclones

Hydrocyclones have been used extensively to treat produced water and are marketed by numerous companies for produced water treatment [33,53]. In offshore installations because of space constraints, compact systems with small and light characteristics are favorable. Seureau et al. [52] designed a three-phase cyclonic separator to remove solids and oil from offshore-produced water. The system combines the functional characteristics of both desanding and de-oiling hydrocyclones.

Hydrocyclones can be used to separate liquids and solids or liquids of different densities. Hydrocyclones can be used to remove particulates and oil from produced water. Depending on the model of hydrocyclone employed, they can remove particles in the range of 5-15 micrometers (μm) [33,51]. Hydrocyclones will not remove soluble oil and grease components [33,47].

Hydrocyclones generate spinning motion of the fluid that creates centrifugal force to push heavier water outward and lighter oil into the middle core of the cones. The water continues down and exits out the tapered end. The improved gravity effect provides a higher rate of separation [41]. Hydrocyclones can be used to treat water with high solids and organic chemical concentrations and can reduce oil and grease concentrations to 10 ppm. High product water recovery is possible with this technology [33].

4.2. Adsorption

Adsorption is a widely accepted technology for the removal of soluble hydrocarbons from the produced water. Adsorption columns are packed with porous solid material known as adsorbent. The hydrocarbons present in the produced water adhere onto the surface of adsorbent and are eventually retained within the porous structure. The effluent from the adsorption column contains little or no hydrocarbons. Highly porous adsorbents with higher surface area offer better performance [41].

Adsorption can be accomplished using a variety of materials, including zeolites, organoclays, activated alumina, and activated carbon. Chemicals are not required for normal operation of adsorptive processes. Chemicals may be used to regenerate media when all active sites are occupied. Periodically, the media is backwashed to remove large particulates trapped between the voids in the media. Typically, these processes can be gravity fed and do not require an energy supply, except during backwash [33].

Adsorbents are capable of removing iron, manganese, total organic carbon, BTEX compounds, heavy metals, and oil from produced water. Adsorption is generally utilized as a unit process in a treatment train rather than as a stand-alone process. The adsorbent can be easily overloaded with large concentrations of organics, so this process is best used as a polishing step rather than as a primary treatment process. Adsorption is capable of removing over 80% of heavy metals and can accomplish nearly 100% product water recovery[33].

Activated carbon can remove soluble BTEX but organoclay can remove insoluble free hydrocarbons that contribute to total petroleum hydrocarbons (TPH) and O&G measurement [2]. Copolymer beads are prepared based on methylmethacrylate (MMA) and divinylbenzene (DVB) by suspension polymerization. These copolymers can reduce oil content of produced water to around 85% [2,54]. Zeolites are often used as ion-exchange resins. Use of hydrophobic zeolite pellets in a fixed bed to adsorb dissolved organic compounds in produced water is proposed [2,11]. Janks and Cadena [55] used "tailored" zeolites, to remove BTEX from saline produced water. They were made by adsorbing neutralized amines onto natural zeolites. The projected process removal efficiency was 70–85% [2].

Doyle et al. tested the combination of ET Ventures' ET #1 (a modified polymer or bentonite or an organoclay) and granular activated carbon in packed bed adsorption column. The treatment system could remove hydrocarbons consistently and effectively. Results showed that the system reduced total petroleum hydrocarbon and O&G to non-detectable levels, and reduced soluble hydrocarbons: benzene, ethylbenzene, toluene, and xylene too to barely detectable levels. The main drawback of adsorption systems is the need for frequent regeneration of materials and the generation of waste [2,56].

4.3. Membrane Separation Technologies

Membranes are thin films of synthetic organic or inorganic materials, which selectively separate a fluid from its components. Membrane separation processes such as Microfiltration (MF), Ultrafiltration (UF), Nano filtration (NF), and Reverse Osmosis (RO) can be used to separate different sized materials. MF is separation of suspended particles, UF is the separation of macromolecules, and RO is the separation of dissolved and ionic components [2]. NF membranes are generally designed to be selective for multivalent ions rather than for univalent ions. RO membranes are designed to reject all species other than water. They are unable to offer a significant barrier to dissolved gases and certain low-molecular weight organic molecules [2]. Various applications of the pressure driven membrane technologies are listed in Table 4. Molecular Weight Cutoff (MWCO) is the ability of a membrane to reject the species of certain molecular weight measured as Daltons [41].

MF is a process of sieving particulates based on the pore size of the membrane. The pore size of MF is about 0,1 μm or more, therefore its application in produced water treatment should be considered as a pre-treatment in order to enhance the effectiveness of UF, NF or RO processes [59]. Abbasi et al [60] studied usage of mullite and mullite–alumina ceramic MF membranes for the treatment of produced water. Result obtained are the TOC rejection of 94% for the synthetic oily wastewater but only 84% of the TOC rejection for the real produced water using the same membrane.
Zhu et al. [61] reported the use of hydrophilic and oleophobic *polyvinylidene fluoride* (PVDF) hollow fiber MF membranes for the oil-water separation. The modified polymers were added as additives to the

PVDF dope solution in order to make hydrophilic and oleophobic PVDF hollow fiber MF membranes. Three different types of oils namely 'H-oil' (Hexane in water), 'C-oil' (crude oil in water) and 'P-oil' (palm oil in water) were studied for the oil-water separation using the new hydrophilic and oleophobic PVDF hollow fiber membrane. Result of their experiment are The 'H-oil' removal efficiency of >98% was achieved and the 'Coil' removal was impressive, >99%. However the 'P-oil' showed only about 70% separation efficiency due to the smaller size of oil droplet in the feed water [59].

Table 4. Applications of advanced membrane filtration technologies [41].

Membrane Filtration	Separation Specifications	Applications/Removal
Microfiltration (MF)	>100,000 Daltons 10 - 0.1 μm	bacteria, viruses, suspended solids etc
Ultrafiltration (UF)	10,000 to 100,000 Daltons 0.05 - 5 e^{-3} μm	proteins, starch, viruses, colloid silica, organics, dyes, fats, paint solids etc
Nanofiltration (NF)	1,000 to 100,000 Daltons 5 e^{-3} - 5 e^{-4} μm	starch, sugar, pesticides, herbicides, divalent ions, organics, BOD, COD, detergents etc
Reverse Osmosis (RO)	salts and lower MWCO 1 e^{-4} - 1 e^{-5} μm	metal ions, acids, sugars, aqueous salts, dyes, natural resins, monovalent salts, BOD, COD, ions etc

UF is one of the most effective methods for oily wastewater treatment, especially for produced water, in comparison with the traditional separation methods because of its high oil removal efficiency, there is no necessity for chemical additives, energy costs are low, and space requirements small [2,57]. In a study, Safitri et al. [58] studied Ultrafiltration process using membrane polyethersulfone (PES) from NADIR, Germany, with pore size of 10 kDa to treat oilfield produced water. Result of their experiment showed that 90% of COD rejection, 99,5% of O&G rejection, 83% of toluene rejection and 82% of xylene rejection.

Bilstad and Espedal compared MF and UF membranes in pilot trial to treat the North Sea oilfield-produced water. Results showed that UF, but not MF, could meet effluent standards for total hydrocarbons, SS, and dissolved constituents. By UF membrane treatment with molecular weight cut-off (MWCO) was between 100,000 and 200,000 Da, total hydrocarbon concentration could be reduced to 2 mg/L from 50 mg/L (96% removal). Benzene, toluene, and xylene (BTX) were reduced by 54%, and some heavy metals like Cu, and Zn were removed to the extent of 95% [2].

Both NF and RO membranes are operated at relatively high pressure compared to MF and UF membranes. In RO membrane, the water permeation flux is directly proportional to the operating pressure whereas the salt permeation is independent of pressure. Therefore, membrane is more selective when operated in high pressure [62]. Both NF and RO membranes are effective in removing inorganic minerals. The major difference between NF and RO membrane is the selectivity. The RO membrane rejects all the ionic species including monovalent ions whereas the NF is more selective for the divalent ions and partially allows the monovalent ions like Na^+ and Cl−. Most of the NF and nearly all of the modern RO membranes are made as thin film composite (TFC) membrane on an asymmetric UF membrane support. Since these membranes are highly selective at the high pressure operation, these membranes are easily prone to fouling and therefore require relatively clean feed water in terms of suspended solids, clay, organic foulants etc [59].

4.3. Biological Treatment

Aerobic and anaerobic microorganisms were used in studies of biological treatment of produced water. In aerobic treatment, researchers used activated sludge, trickling filters, sequencing batch reactors (SBRs), chemostate reactors, biological aerated filters (BAF), and lagoons [2].

Activated sludge is the usual method for treating wastewater. In a continuous-flow pilot plant, an oil skimmer was used to remove oil before treatment in an activated sludge system. Naturally occurring microbial growth was used in an aeration tank. The activated sludge treatment unit could maintain a total petroleum hydrocarbon (TPH) removal efficiency of 98–99% at a solids retention time (SRT) of 20 days [2]. Freire et al. [63] studied COD removal efficiency of acclimated sewage sludge in SBR with different percentages of produced water and sewage. In 45% and 35% (v/v) mixtures of wastewater, COD removal efficiencies varied from 30% to 50% [2].

BAF is a well established technology and has been used for produced water treatment for many years. BAF can remove oil, suspended solids, ammonia, and nitrogen, chemical oxygen demand, biological oxygen demand, iron, manganese, heavy metals, soluble organics, trace organics, and hydrogen sulfide. Iron and manganese removal in BAFs is mainly due to chemical oxidation rather than a biological process. BAF is most effective on waters with chloride levels below 6,600 mg/L, oil concentrations less than 60 mg/L; COD less than 400 mg/L, and BOD less than 50 mg/L [33]. Approximate removal capabilities of BAFs technology are 60 to 90% nitrification; 50 to 70% total nitrogen; 70 to 80% oil;

30 to 60% COD ; 85 to 95% BOD; 75 to 85% suspended solids [33].

Hidayat and Kardena [64] studied the removal of COD as an organic compound in produced water using indogenous bacteria originating from Total E & P waste water treatment site. The results showed that indogenous bacteria can be used in organic compound removal with removal efficiencies of 88.21%. However, nitrogen and phosphorus bind to indogenous bacteria in degrading organic compounds

In biological oxidation, harmless bacteria, algae, fungi, and protozoa convert dissolved organics and ammonia compounds into water and CO2, nitrates/nitrites, respectively, but have no effect on TDS [2]. The dominant mechanism of hydrocarbon removal by microorganisms in biological treatment is biodegradation and occluding particles by microorganisms similar to bio-flocculation. Activated sludge has the property of adsorbing and occluding not only soluble but also insoluble materials. Bacteria produce surface-active compounds such as surfactants (biosurfactants) and emulsifiers (bioemulsifiers) that enhance the local pseudo-solubility of hydrocarbons and thus improve mass transfer to biodegrading bacteria. Biodegradation of less complex oil components, e.g., normal alkanes is easier than of complex and large molecules. Less biodegradable oil molecules attached to microorganisms will remain in the aeration tank. These components are removed along with sludge in excess-sludge removal processes. The mixture of hydrocarbons and microorganisms are a source of hazardous material which has to be disposed [2].

4. Conclusion

The produced water is the largest liquid waste generated from oil and gas production activities. It contains hazardous materials such as, (1) organic matters which include dispersed, dissolved and emulsified oil, and grease, (2) inorganic matters which is commonly referenced as TDS, and (3) Heavy metals. The materials can contaminate the environment. The produced water produced prior to the disposal to the environment must already meet the required quality standards. Several handling technologies can be done to reduce and eliminate the contaminants in the produced water, namely gravity separation, floatation, adsorption, membrane separation and biological treatment. The combination use of several technologies is needed to achieve the optimal results. Therefore, the produced water that has been processed can be recovered and disposed safely to the environment.

References

1. S. Hasiany, E. Noor, and M. Yani, *Journal of Natural Resources and Environment Management*, **5**(1), pp.25–32, (2015).
2. F. Ahmadun, A. Pendashteh, and L. Chuah, *Journal of Hazardous Materials*, **170**, pp.530–551, (2009).
3. P. Ekins, R. Vanner, J. Firebrace, *J. Clean. Prod.*, **15**, 1302–1315, (2007).
4. Z. Khatib, P. Verbeek, Water to value-produced water management for sustainable field development of mature and green fields, in: SPE International Conference on Health, Safety and Environment in Oil and Gas Exploration and Production, Kuala Lumpur, Malaysia, 20–22 March, (2002).
5. Ministry of Environment, Regulation of the Minister of Environment Number 19 Year of 2010 regarding the quality standard of waste water for businesses and/or activities of oil and gas and geothermal, (2010).
6. J.M. Neff, Bioaccumulation in Marine Organisms: Effect of Contaminants from Oil Well Produced Water, Elsevier, The Netherlands, (2002).
7. USEPA, http://www.epa.gov
8. G.T. Tellez, N. Nirmalakhandan, J.L. Gardea-Torresdey, *Advances in Environmental Research*, **6**, 455–470, (2002).
9. B.R. Hansen, S.H. Davies, *Chemical Engineering Research & Design*, **72**, 176–188, (1994).
10. J.P. Fillo, S.M. Koraido, J.M. Evans, *Environmental Science Research*, **46**, 151-162, (1992).
11. D.T. Bostick, H. Luo, H. Hindmarsh, Characterization of Soluble in Produced Water, Oak Ridge National Laboratory, (2002). https://digital.library.unt.edu/ark:/67531/metadc740840/m2/1/high_res_d/814231.pdf, Online:03-07-2017
12. About produced water (Produced water 101) Available: http://aqwatec.mines.edu/produced_water/intro/pw/ Online:03-07-2017
13. A white paper describing produced water from production of crude oil, natural gas, and coal bed methane. Available: www.ipd.anl.gov/anlpubs/2004/02/49109.pdf Online:04-07-2017.
14. Z. Chen, L. Zhao, K. Lee, and C. Hannath, *Water Quality Research Journal Canada*, **42**, 303–310, (2007a).
15. J. F. Payne, C. Andrews, L. Fancey, B. French, and Lee, K, Risks to fish associated with barium in drilling fluids and produced water: A chronic toxicity study with Cunner (Tautogolabrus adspersus). In K. Lee, & J. Neff (Eds.), (pp. 401–417). New York: Produced Water, Springer, (2011).
16. I. Schifter, et al.,. *Environ Monit Assess*, **187**, p.723, (2015)
17. P.J.C. Tibbetts, I.T. Buchanan, L.J. Gawel, R. Large, *Environmental Science Research*, **46**, pp. 97–113, (1992).
18. R.P.M.W. Jacobs, R.O.H. Grant, J. Kwant, J.M. Marquenie, E. Mentzer, *Environmental Science Research*, **46**, pp. 13–22, (1992).
19. T.I.R. Utvik, Composition, Characteristics of produced water in the North Sea, in: *Produced Water Workshop*, Aberdeen, Scotland, 26–27 March, (2003)
20. B. M. Johnson, L. E. Kanagy, J. H. Rodgers Jr., J.W. Castle, *Water Air Soil Pollut*, **191**, pp. 33–54, (2008).
21. J.P. Fillo, J.M. Evans, *American Gas Association Operation Section Proceedings*, pp.448–459, (1990).

22. United States Environmental Protection Agency (USEPA). EPA Office of Compliance Sector Notebook Project: Profile of the Oil and Gas Extraction Industry. EPA/310-R-99-006, (2000).
23. M.C. Shepherd, F.L. Shore, S.K. Mertens, J.S. Gibson, *Environmental Science Research*, **46**, pp. 163–174, (1992).
24. Gazali, A.K. et al., *Journal of Engineering (IOSRJEN)*, **7**(6), pp.22–29, (2017).
25. T. Bakke, J. Klungsøyr, and S. Sanni, *Marine Environmental Research*, **92**, pp.154–169, (2013).
26. J. Neff, K. Lee, E.M. DeBlois, Produced water: overview of composition, fates, and effects. In: Lee, K., Neff, J. (Eds.), Produced Water. Springer, New York (Chapter 1), (2011)., Available: https://id.scribd.com/document/267721195/Produced-Water-Overview-of-Composition-Fates-and-Effects Online : 05-07-2017
27. P. Odeigah, O. Nurudeen, and O. O. Amund, Hereditas **126** (2), 161-167, (1997)
28. J.M. Neff, T.C. Sauer, N. Maciolek, *Environmental Science Research*, **46**, pp.371–386, (1992).
29. O. Obire, and F. Amusan, *Journal of Applied Sciences and Environmental Management*, **7** (1), 61-66, (2003).
30. DJ.K. Otton, Environmental Aspects of Produced-water Salt Releases in Onshore and Coastal Petroleum-producing Areas of the Conterminous, (2006). https://pubs.er.usgs.gov/publication/ofr20061154, Online : 05-07-2017
31. S. Isehunwa, and S. Onovae, *Journal of Engineering & Applied Sciences*, **6** (8), 66-71, (2011).
32. T.I.R. Utvik, *Chemosphere*, **39**, 2593–2606, (1999).
33. Oil and gas produced water management and beneficial use in the western united states, https://www.usbr.gov/research/dwpr/reportpdfs/report157.pdf Online : 05-07-2017
34. M.T. Stephenson, *Environmental Science Research*, **46**, pp. 1–11, (1992).
35. D.A. Pillard, J.E. Tietge, and J.M. Evans. 1996b. Estimating the acute toxicity of produced waters to marine organisms using predictive toxicity models. Pages 49-60 In: M. Reed and S. Johnsen (Eds.). 1996. Produced Water 2. Environmental Issues and Mitigation Technologies. Plenum Press, New York.
36. N. Elias-Samlalsingh, J. B. R. Agard, *Environ. Toxicol. Chem.* **23**, 1194–1203, (2003).
37. S. Boitsov, S.A. Mjøs, and S. Meier. *Mar. Environ. Res.* **64**, 651-665, (2007).
38. S.B. Henderson, S.J.W. Grigson, P. Johnson, B.D. Roddie, *Mar. Pollut. Bull.* **38**, 1141–1151, (1999).
39. S.J.W. Grigson, A. Wilkinson, P. Johnson, C.F. Moffat, A.D. McIntosh, *Rapid Commun. Mass Spectrom.* **14**, 2210–2219, (2000).
40. J.V. Veil, Research to Improve Water-use Efficiency and Conservation: Technologies and Practice, (2007). https://www.gpo.gov/fdsys/pkg/CHRG-110hhrg38533/pdf/CHRG-110hhrg38533.pdf
41. J.D. Arthur, B.G. Langhus, C. Patel, Technical Summary of Oil and Gas Produced Water Treatment Technologies, (2005). http://www.all-llc.com/publicdownloads/ALLConsulting-WaterTreatmentOptionsReport.pdf
42. D.B. Boysen, J.E. Boysen, J.A. Boysen, Creative Strategies for Produced Water Disposal in the Rocky Mountain Region, (2002). http://ocdimage.emnrd.state.nm.us/Imaging/FileStore/SantaFeAdmin/CF/61750/14015_638_CF.pdf,
43. M. Stewart, K. Arnold, Emulsion and Oil Treating Equipment, Elsivier Inc., Oxford, (2009).
44. J.A. Veil, M.G. Puder, D. Elcock and R. Jr. J.Redweik, A White Paper Describing Produced Water from Production of Crude Oil, Natural Gas, and Coal Bed Methane, *US DOE* (January 2004) W-31-109-Eng-38
45. S. Deng, R. Bai, J.P. Chen, Z. Jiang, G. Yu, F. Zhou, Z. Chen, *Sep. Purif. Technol*. **29**, 207–216, (2002).
46. W.M.G.T. Van den Broek, M.J. Van der Zande, Comparison of plate separator, centrifuge and hydrocyclone, in: SPE International Oil and Gas Conference and Exhibition, Beijing, China, 2–6 November, (1998)
47. T. Hayes, and D. Arthur, Overview of Emerging Produced Water Treatment Technologies. 11th Annual International Petroleum Conference, Albuquerque, New Mexico, (2004).
48. M. N. Cakmakce, Kayaalp, et al. *Desalination,* **222**(1-3), 176-186, (2008).
49. Broussard P. "Skip" Jr, Floatation Pump Proves Successful for The Treatment of Produced Water, paper presented at the 10th IPEC, (2003).
50. ALL Consulting. *Handbook on Coal Bed Methane Produced Water: Management and Beneficial Use Alternatives*. Prepared by ALL, Consulting for the Ground Water Protection Research Foundation, U.S. Department of Energy, and U.S. Bureau of Land Management, (2003). http://www.all-llc.com/CBM/BU/index.htm.
51. NETL. "Produced Water Management Technology Descriptions Fact Sheet.", http://www.netl.doe.gov/technologies/pwmis/techdesc/.
52. J.J. Seureau, Y. Aurelle, M.E. Hoyack, A three-phase separator for the removal of oil and solids from produced water, in: SPE Annual Technical Conference and Exhibition, New Orleans, Louisiana, USA, 25–28 September, (1994)
53. Sinker, A. *Produced Water Treatment Using Hydrocyclones: Theory and Practical Applications.* 14th Annual International Petroleum Environmental Conference, Houston, Texas, (2007).
54. M.S. Carvalho, M.D. Clarisse, E.F. Lucas, C.C.R. Barbosa, Evaluation of the polymeric materials (DVB copolymers) for produced water treatment, in: SPE International Petroleum Exhibition and Conference, Abu Dhabi, UAE, 13–16 October, (2002)
55. J.S.Janks, F. Cadena, *Environmental Science Research*, **46**, pp. 473–488, (1992).
56. D.H. Doyle, F. Daniel, A.B. Brown, Field test of

produced water treatment with polymer modified bentonite, in: SPE Rocky Mountain Regional meeting held in Casper, Wyoming, USA, 18–21 May, (1997).

57. Y. He, Z.W.Jiang, Treating oilfield wastewater: technology review, Filtr. Sep. June (2008)14–16
58. Safitri, H.I. et al., *Journal of chemical and industrial technology*, **2**(4), pp.205–211, (2013).
59. S. Munirasu, M. Abu, and F. Banat, *Process Safety and Environmental Protection*, **100**, pp.183–202, (2016).
60. M. Abbasi, M. Mirfendereski, M. Nikbakht, M. Golshenas, T. Mohammadi, *Desalination*, **259**, 169–178, (2010).
61. Zhu, X., Tu, W., Wee, K.-H., Bai, R., *J. Membr. Sci.* **466**, 36–44, (2014).
62. Baker, R.W. (Ed.), Membrane Technolgoy and Applicatoins., second ed. John Wiley & Sons, West Sussex, England, (2004).
63. D.D.C. Freire, M.C. Cammarota, G.L. Sant'Anna, *Environ. Technol.* **22**, 1125–1135, (2001).
64. S. Hidayat, and E. Kardena, Removal Organic Compounds with Batch Reactor using Indogenous Bacteria and Additional Nutrition, (2007). Available at: http://www.academia.edu/31854780/25310008-Syarif-Hidayat.pdf

Integrating Electrokinetic and Bioremediation Process for Treating Oil Contaminated Low Permeability Soil

Bimastyaji Surya Ramadan[1*], *Agus* Jatnika Effendi[2], and *Qomarudin* Helmy[2]

[1]Master Program of Environmental Engineering, Environmental Engineering Department, Bandung Institute of Technology, Indonesia
[2]Environmental Engineering Department, Bandung Institute of Technology, Indonesia

Abstract. Traditional oil mining activities always ignores environmental regulation which may cause contamination in soil and environment. Crude oil contamination in low-permeability soil complicates recovery process because it requires substantial energy for excavating and crushing the soil. Electrokinetic technology can be used as an alternative technology to treat contaminated soil and improve bioremediation process (biostimulation) through transfer of ions and nutrient that support microorganism growth. This study was conducted using a combination of electrokinetic and bioremediation processes. Result shows that the application of electrokinetic and bioremediation in low permeability soils can provide hydrocarbon removal efficiency up to 46,3% in 7 days operation. The highest amount of microorganism can be found in 3-days operation, which is $2x10^8$ CFU/ml using surfactant as flushing fluid for solubilizing hydrocarbon molecules. Enhancing bioremediation using electrokinetic process is very potential to recover oil contaminated low permeability soil in the future.

1 Introduction

Traditional oil mining activities in Indonesia is increasing every year especially in places with high crude oil reserves. Oil resources are non-renewable energy resource that is being a primary fuel for society and industries. The development of this industry certainly has an impact, either directly or indirectly. Economic, environmental, and socio-cultural impacts can be either positive or negative. Counter and recovery measures from the negative impacts that will occur should be done so it will not harm the environment [1].

Crude oil is a combination of various hydrocarbon chemical compounds from short to long chains that are difficult to mineralize. Various activities of oil mining industry such as exploration, exploitation, extraction, processing, transportation and marketing of processed oil products potentially produce waste (byproduct) that can pollute the environment. Oil contains aliphatic and aromatic compound and some hazardous materials such as benzene, ethyl benzene, toluene, and xylene (BTEX) that can accumulate in human bodies and living things, destroy the ecosystems, and pollute the soil and groundwater environment [2]. Indonesian government has been already concerned about soil contaminated with petroleum hydrocarbon as stated in Ministerial Decree of Environment Number 128/2003 about Procedures and Technical Requirements for Processing Waste Oil and Contaminated Soil. In that regulation, petrochemical compound in contaminated soil is measured as Total Petroleum Hydrocarbon (TPH).

Various kinds of remediation strategies due to crude oil pollution have been widely studied from physics, chemistry, and biology methods. Remediation of contaminated soil using chemical-physics methods such as solidification, stabilization, thermal, soil washing, gas and liquid extraction or a combination of various technologies is usually expensive, and can produce hazardous compound [3]. Therefore, bioremediation then used as a alternative technology that attract great attention from practitioner because it can eliminate pollutants easily through biodegradation process at low invesment cost. Nevertheless, the process of bioremediation requires long operational time so that there is a need for finding a right strategy to improve the process efficiency [4].

Electrokinetic technology is known to be used as a alternative for treating contaminated soil using constant low voltage gradient [5], [6]. The electrochemical process is used to improve bioavailability and biomixing on contaminated soil so that improving the efficiency and effectiveness of the bioremediation process. The weakness of this technology is the availability of electrical energy resources to supply the needs of electric current. In addition, removal efficiency of contaminant is strongly influenced by the complexity of the soil environment both in terms of chemical, physical and biological. Drastic changes in pH and high transport velocity of soil compounds or ions due to electrochemical process can negatively affect the soil environment to be toxic to microorganisms. This technology has a good removal efficiency on clay soil that have low hydraulic permeability [7].

* Corresponding author: bimastyajisurya@gmail.com

Implementation of electrical current can directly affect the soil condition chemically and biologically so that the need for research that discusses the interaction between the application of electrokinetic remediation technology by indigenous microorganism. The combination of electrokinetic and bioremediation technologies is also known to have greater efficiency when compared to the application of individual technologies [8]. Therefore, a laboratory scale study is needed as a guide for the application of this technology in Indonesia.

In this area of research, there are 4 variations that we compared: bioremediation (bio), electrokinetic bioremediation without any development (EK-Bio), electrokinetic bioremediation with polarity reversal (PR-EK-Bio) and electrokinetic bioremediation with addition of tween 80 surfactant (EKSF-Bio). The study was conducted on a laboratory scale by taking the contaminated soil from around traditional oil mining in Grobogan District, Central Java, Indonesia.

2 Methods

2.1 Lab Scale Set-up

The reactor set up that has been used in this study can be seen in Figure 1. The reactor is made of acrylic glass which is divided into 3 main chamber: overflow / collector chamber, electrolyte chamber, and soil chamber. The dimension of soil chamber is 18x12x10 cm^3. Contaminated soil is located in the center of the reactor and flanked by an electrolyte chamber containing carbon electrodes with dimensions of 9x9x0.5 cm^3. The carbon electrode is connected to a power supply device (Sanfix, DC Power supply) with a fixed voltage of 0.5 V / cm. Both electrolyte chambers are connected with an collector chamber to accommodate overflow of water due to electroosmosis and electrolysis process. Electrolyte and soil chamber is separated by whatmann filter paper and porous acrylic glass with a pore diameter of 0.3 cm and 0.5 cm in space for each pore.

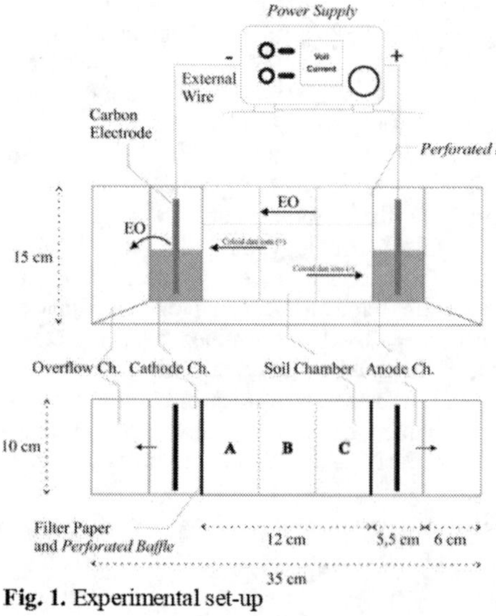

Fig. 1. Experimental set-up

2.2 Materials

2.2.1 Soil

Oil-contaminated low permeability soil is derived from Bandarharjo village, Grobogan, Central Java around tradtional oil drilling well which managed independently by the local community. Physical tests of soil parameter were conducted in Soil Mechanics Laboratory of Civil Engineering ITB while soil chemical properties were carried out at the Center for Research and Development of Mineral and Coal Technology, Bandung. Microbiology testing was conducted in Environmental Microbiology Laboratory of ITB. From the XRD test results, it is known that the soil contains quartz, montmorillonite, kaolinite and calcite mineral. Detailed soil characteristics can be seen in table 1.

Prior to use in the study, soils that had been taken from the field were prepared first. The soil is air-dried and pounded using mortar. Then, soil is sieved in 2 mm pore sieve to remove rack, twigs and grass. In order to increase homogeneity, soil is mixed using cone and quartering methods. An electrokinetic reactor requires 1,500 grams of dry soil.

Table 1. Soil properties

Parameter		Parameter	
Gravel (%)	5,2	Cation Exchange	30,48
Sand (%)	15,81	Capacity (meq/100g)	
Silt (%)	45,46	Organic Carbon (%)	5,07
Clay (%)	33,53	N Total (%)	0,137
Hydraulic	0,014	P Total (mg/kg)	329
Conductivity	$9,6x10^{-7}$	C/N Ratio	37,08
Permeability		Electrical	243
Plasticity Index	111	Conductivity	
pH	5,5-6	(µS/cm)	

2.2.3 Micro-organisms

The microorganisms consortium was obtained from the original soil around traditional oil well and propagated in SBS solution with medium composition per 500 ml of aquadest was 0.45 g KH_2PO_4; 0.135 g K_2HPO_4; 0.135 g (NH_4) $2SO_4$; 0.06 $MgSO_4$; And 0.135 yeast extract. The microorganism propagation was then continued using aquadest with dextrose substrate for 2 months until TSS reached ± 2.000 mg / l of concentration.

The ideal water content for bioremediation process is 30-60%. In this study, aquadest was added to dry soil of 750 ml and 1% of microorganism suspension. The soil, water and microbial suspension are then mixed manually and placed in an electrochemical reactor.

2.2.2 Hydrocarbons

In this study, crude oil is analyzed quantitatively and qualitatively. The concentration of oil was analyzed quantitatively in Total Petroleum Hydrocarbon units and qualitative using Gas Chromatography Mass Spectroscopy (GC-MS) method. The soxhlet extraction method is used to extract oil from the soil in accordance with EPA 9071B on n-hexane extractable material (hem) for sludge, sediment, and solid samples. The carbon compound contained in the oil extraction can be seen in Table 2. The concentration of TPH at the beginning of the study was 3.3509%.

Table 2. Hydrocarbon compound from soil extraction

No	*Compound Name*	*MW*	*Formula*
1	Alpha-Muurolane $$ Naphthalene, decahydro-1,6-dimethyl-4-(1-methylethyl)-	208	$C_{15}H_{28}$
2	4,5-alpha,alpha-Eudesmane $$ 4a,8-dimethyl-2-isopropyl perhydronaphthalene	208	$C_{15}H_{28}$
3	Phytane $$ Hexadecane, 2,6,10,14-tetramethyl-	282	$C_{20}H_{42}$
4	Hexadecane	226	$C_{16}H_{34}$
5	Cadalene $$ Naphthalene, 1,6-dimethyl-4-(1-methylethyl)	198	$C_{15}H_{18}$
6	Norphytane $$ Pentadecane, 2,6,10,14-tetramethyl	268	$C_{19}H_{40}$
7	Eicosane	282	$C_{20}H_{42}$
8	Heneicosane	296	$C_{21}H_{44}$

2.3 Experimental Procedure

This study was conducted by varying four batch studies conducted at room temperature which applying a constant voltage of 18 V in three reactor and without applying an electric field in the last reactor. The first test uses bioattenuation placed on the reactor in the absence of an electrokinetic treatment (Bio). The second test is an electrokinetic bioremediation test without reversing the polarity during running time by using aquadest as its electrolyte solution (EK-Bio). The third test is in the same principle as the second test but uses tween 80 as a flushing fluid as well as an electrolyte (EKSF-Bio). The fourth test uses aquadest as electrolyte and polarity reversal (PR-EK-Bio). The surfactant used in this study is a biodegradable tween 80 surfactant which has a low CMC value. The concentration of surfactant used was 1 g/L. The research duration is 7 days which is intended to see the efficiency of TPH removal in a short period of time. In summary, the conclusions of the four tests that have been done can be seen in Table 3.

Table 3. Summary of EK Test

Test	*Explanation*	*Electrolyte*
Bio	Bioremediation	No
EK-Bio	Electrokinetic Bioremediation	Aquadest
EKSF-Bio	Electrokinetic Soil Flushing Bioremediation	Tween 80
PR-EK-Bio	Polarity Reversal Electrokinetic Bioremediation	Aquadest

2.2 Sampling and Analysis

The main parameters to be analyzed are the number of microorganisms represented in total plate count (TPC) and total petroleum hydrocarbon (TPH). Both parameters are taken on a daily basis of 0, 1, 2, 3 and 7th day of the study. Other parameters that may affect processes such as temperature, pH, and moisture are tested daily. Multimeters are used to ascertain whether the voltage generated by each power supply is the same and constant every day. Detailed sampling location can be seen in Figure 2. The entire sample was tested in triplicate.

Temperature and pH were tested using a portable sampler available in the laboratory. Water content was analyzed by gravimetry by taking 30-40 g samples and air-dried it then weighed again. The difference between wet and dry weight is the moisture content of the sample. The TPH analysis is performed by soxhlet extraction as described in point 2.2.2. TPC was analyzed using pour plate method on NA medium with composition: 1 g beef extract, 2 g yeast extract, 5 g pepton, 5 g NaCl, and 15 g agar and mixed it in 1 liter of distilled water.

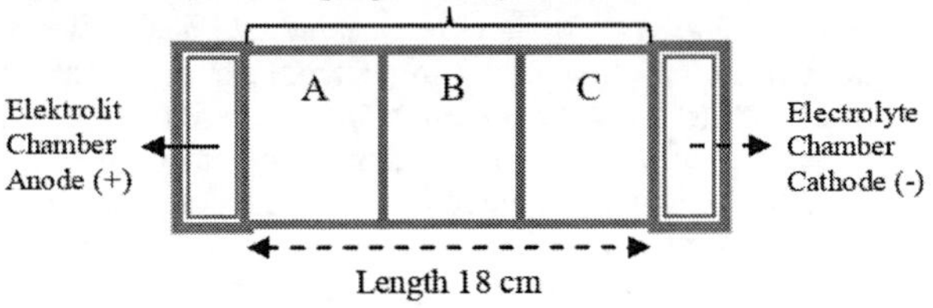

Fig. 2. Sampling point

3 Results and Discussion

3.1 Effect on Soil Condition

Water content is an important factor for microorganisms activity in the soil. Water can stimulate microorganisms to perform metabolism. Figure 3 shows that in the Bio test, no electroosmosis process occurs so that the water in the soil evaporates over time. The existence of electroosmosis process allows the movement of water from electrolyte chamber into the pores of the soil. In the EK-Bio and EKSF-Bio tests, the water in both electrolyte chambers is significantly reduced in the anode chamber. The process of electrolysis of water at the anode ($2H_2O \rightarrow 4e^- + 4H^+ + O_2$) and in the cathode ($2H_2O + 2e^- \rightarrow 2OH^- + H_2$) and electroosmosis in the electrolyte chamber causes water volume reduction [4]. In contrast to PR-EK-Bio, in this test the water in both electrolyte chambers tends to be stable and does not disappear too much during electrokinetic process. This phenomena occurs because most of the electrolyzed ions re-form water (neutralization) when the polarity reversal occurs [4], [9].

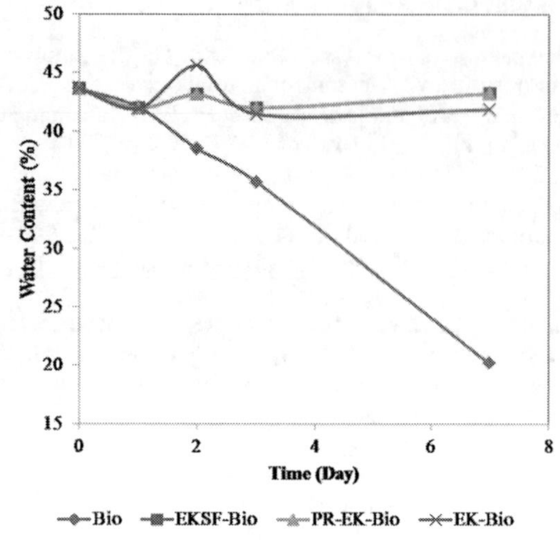

Fig. 3. Water content in soil profile

The distilled water may be replaced with mineral salt as both buffers and surfactants. Mineral salts may also serve as nutrients for microorganisms. Surfactant tween 80 used in this study serves to reduce the surface tension of oil droplet so it can solubilize in water and transported through electroosmosis mechanism. Surfactants are also capable of making micelle oil that can increase bioavailability and make oil suspended in soil-water pore and move through the electrophoresis process. PR-EK-Bio and EKSF-Bio have stable levels of moisture content when compared to EK-Bio so combination of surfactants with polarity reversals technique may provide favorable environmental conditions for electrokinetic bioremediation processes [5], [10].

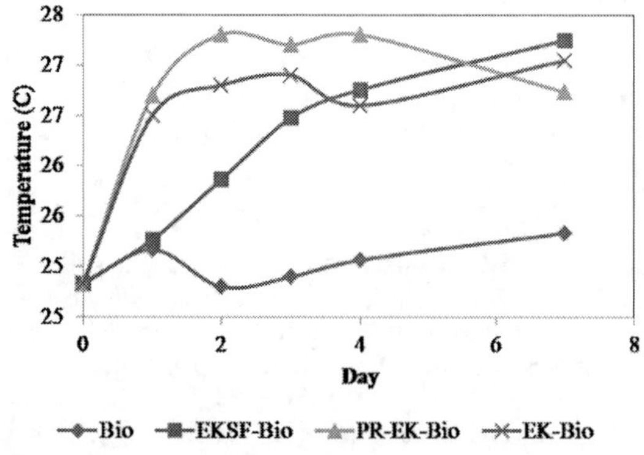

Fig. 4. Soil temperature profile

Bearing pH values to protect microorganisms from death is an important factor in this study. Acidity / alkalinity of the system is highly dependent to water hydrolysis and ion migration due to electro migration. The anode chamber will be filled with negative ions from the soil and produce an acidic condition whereas the cathode chamber is filled with positive ions and produces an alkaline condition. In this study, a very rapid pH change occurs where on the third day, the average pH drop to the range of 4-5. Overall, the average pH decreases toward acidic ambience. This limitation can be maintained by using electrolyte solution recirculation or polarity reversal technique. Changes in pH may affect microorganism consortium as it may alter some contaminant properties, catabolic pressure and also toxicity. However, the increasing of acidity will make soil bonds molecules release of hydrocarbon compounds easier [8], [13], [14]. Figure 5 shows the pH change over time.

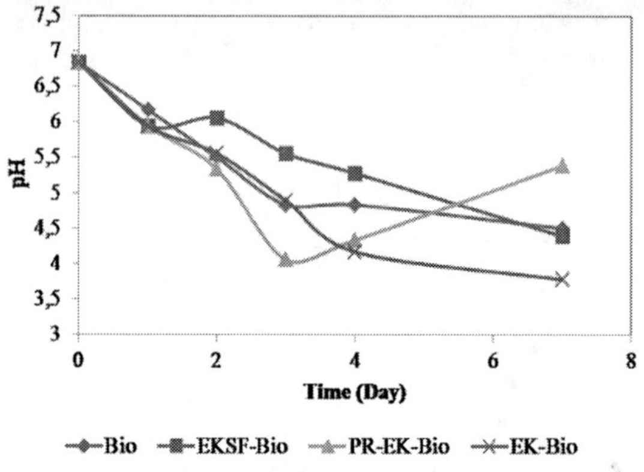

Fig. 5. Soil pH profile

3.2 Microbial Distribution

Electrokinetic process can improve bioremediation process because in addition to delivering nutrients, acceptor and electron donors and contaminants, this process is able to move the microorganisms through the pores of the low permeability soil. The movement velocity of microorganisms ranges from 0.05 - 0.15 cm^2 / Vh and hydrophobic contaminants of 0.001 - 0.26 cm^2 / Vh depending on their specific nature / substance. In the process of electrokinetic bioremediation, electroosmosis is the primary transport mechanism that carries microorganisms and contaminants through the soil pores. The major obstacles in the bioremediation process in clay soil can be overcome by using this transport mechanism [7].

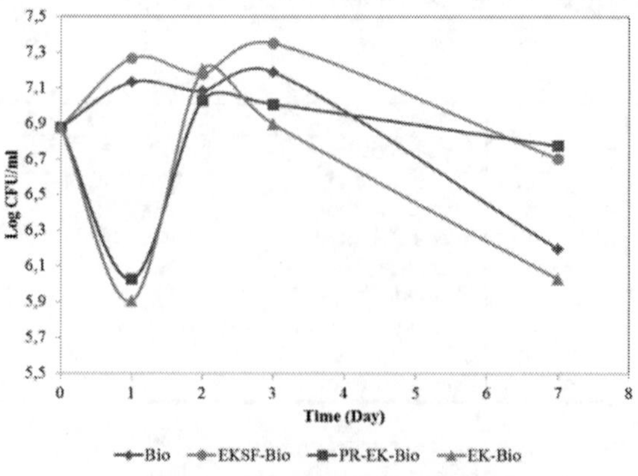

Fig. 6. Total microorganism in soil (log cfu/ml)

Figure 6 shows that EKSF-Bio has the greatest number of microorganisms when compared with other tests. This relates to the surfactant's ability to make the oil dissolved in water. In addition, the surfactant may serve as a useful co-substrate for microorganisms. The microorganisms at the beginning of the study amounted to $8x10^6$ CFU / ml and on average increased on the third day before finally decreasing until the seventh day. This decrease occurs due to changes in drastic environmental conditions such as decreased pH and moisture content.

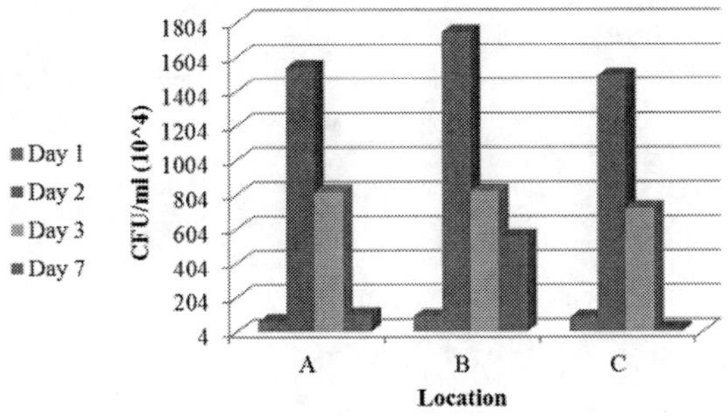

(a)

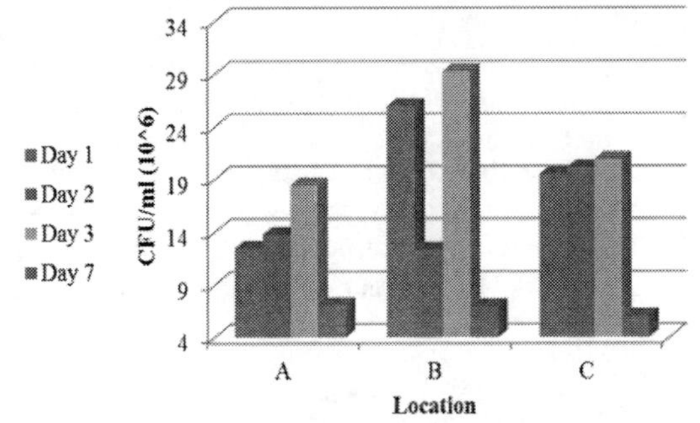

(b)

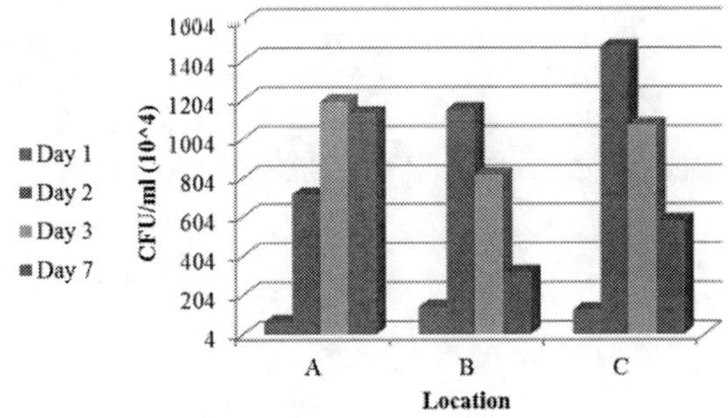

(c)

Fig. 7. Microorganism distribution under different test: a. EK-Bio; b. EKSF-Bio; c. PR-EK-Bio

The lowest number of microorganisms was found in the Bio test of $2x10^6$ CFU / ml. PR-EK-Bio can maintain the stability of the number of microorganisms in the range of $7x10^6$ - $1x10^7$ CFU / ml. The number of these microorganisms is greater when compared with other tests. This suggests the potential for the development of a combination of PR-EK-Bio and surfactant in subsequent studies. EK-Bio has the smallest number of microorganisms in the last day due to the high pH gap in the test reactor. In the anode, the pH value can reach 2 and 12 at the cathode. In contrast to EKSF-Bio is still able to maintain the pH between 3-11 and PR-EK-Bio of 4-10 at the end of the study.

The microorganisms near the cathode and anode chambers are lower than those located in the center of the reactor. In accordance with the results of the study shown in Fig. 7, microorganisms in the anode have smaller concentrations than cathodes. This relates to the acidic conditions near the anode chamber and the neutral conditions in the cathode chamber. This neutrality is due to much larger acid strength and almost affects the entire reactor. It also indicates the high negative ions contained in the soil. The distribution of microorganisms in polarity reversal tests is slightly different from the others due to transient effects on the system [12], [15].

3.3 TPH Concentration

Figure 8 shows the percentage of TPH in all four treatments conducted during the 7 days of the study. Table 2 shows that the volatile compounds are not left in the contaminated soil. The remaining hydrocarbon compounds are compounds with long carbon chains ranging from 15 to 21. TPH removal in this study is assumed to be done by biological process and electrochemical phenomena since there is no further significant treatment of the process. The concentration of TPH fluctuates, this may occur as a result of contaminant transport factors at a particular point in the tested soil. Nevertheless, the downward trend is shown by all three electrokinetic tests.

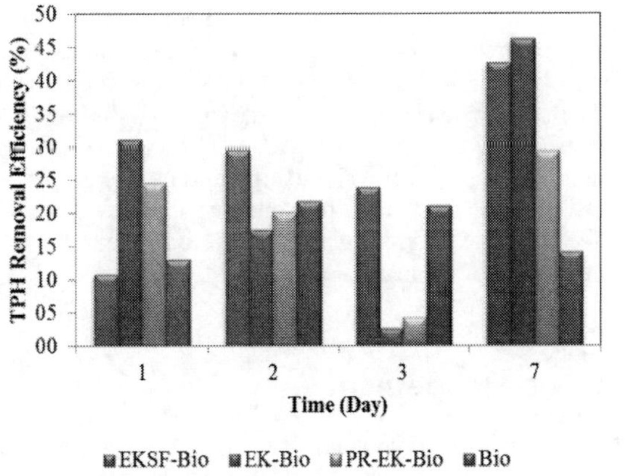

Fig. 8. TPH removal efficiency after 7 days *run*

The largest percentage of removal was in EK-Bio that was 46,3%, followed by EKSF-Bio equal to 42,7% and PR-EK-Bio equal to 29,2%. This condition is not in accordance with the original hypothesis which states that the reversal of polarity will be more effective when compared with ordinary electrokinetic processes. This can be happen because EKSF-Bio and EK-Bio have a lower pH value when compared to PR-EK-Bio. The pH value accelerates the process of solubilizing the hydrocarbon compounds into water so that the hydrocarbon compounds accumulate near the electrolyte chamber. In short, the electrolytes in both tests are rapidly depleted by evaporation and electrolysis processes. In bioremediation test (Bio), there is a decrease of TPH concentration although not as drastic using electrochemical process. This may be due to the mixing process and transport of microorganisms, soil and water at the beginning of the test causing contact between the active microorganisms and the contaminants so that the degradation process can occur [10], [14].

The percentage of degradation obtained is also not too large because of the very low in situ biodegradation. The TPH degradation is caused by two things, the transportation of contaminants due to the application of electrochemical and biological processes occurring in the soil. In EKSF-Bio, surfactants are able to mobilize TPH to microorganisms attached to the pores of the soil. At the end of the study, the surfactant was spread evenly on the soil and obtained an average degradation that was evenly distributed at each point. The electroosmosis process allows the presence of petroleum in the anode chamber. In this study, there is no research to know the TPH concentration in the electrolyte chamber. However, a larger TPH removal is likely to occur when prolonging the research process. Development carried out in this research can be an option in supporting the effectiveness of electrokinetic bioremediation process [16], [17].

4 Conclusion

The results show that in situ bioremediation process can be integrated either by electrokinetic process or by soil flushing. Polarity reversal technique is known to maintain the environmental conditions that are compatible with the growth of microorganisms. But there needs to be some improvement strategies to increase the degradation efficiency. Surfactants are known to improve the electrochemical process. A greater percentage of degradation is still possible if appropriate design process is found or extending the running time of the study. Electrokinetic bioremediation can be used to treat petroleum contaminated soil.

Acknowledgement

The authors acknowledge the financial support for this study from Environmental Engineering Department, ITB.

References

1. Q. Helmy, R. Laksmono, and E. Kardena, *2nd Humbold Kolleg in conjunction with International Conference on Natural Sciences*, vol. **14**, pp. 326–333, (2015)
2. T. Li *et al.*, *Electrochim. Acta*, vol. **187**, pp. 567–575, (2016).
3. D. Huang, Q. Xu, and J. Cheng, *Int. J. Electrochem. Sci.*, vol. 7, pp. 4528–4544, (2012).
4. I. Hassan, E. Mohamedelhassan, E. K. Yanful, and Z. Yuan, *Adv. Microbiol.*, vol. **6**, no. January, pp. 57–72, (2016).
5. E. M. Ramírez, C. S. Jiménez, J. V. Camacho, M. A. R. Rodrigo, and P. Cañizares, *Electrochim. Acta*, vol. **181**, pp. 192–199, (2015).
6. E. M. Ramirez, J. V. Camacho, M. A. Rodrigo, and P. Canizares, *Sci. Total Environ.*, vol. **533**, pp. 307–316, (2015).
7. R. T. Gill, M. J. Harbottle, J. W. N. Smith, and S. F. Thornton, *Chemosphere*, vol. **107**, pp. 31–42, (2014).
8. L. Y. Wick *et al.*, *Sci. Total Environ.*, vol. **408**, no. 20, pp. 4886–4893,(2010).
9. I. Hassan, E. Yanful, E. Mohamedelhassan, and M. W. Bo, "Enhanced electrokinetic bioremediation by pH stabilisation," (2016).
10. E. Mena, J. Villaseñor, P. Cañizares, and M. A. Rodrigo, *Electrochim. Acta*, vol. **190**, pp. 294–304, (2016).
11. T. Li, S. Guo, L. Zhang, and F. Li, *2010 4th Int. Conf. Bioinforma. Biomed. Eng. iCBBE 2010*, pp. 1–4, (2010).
12. Y. Yuan, S. Guo, F. Li, B. Wu, X. Yang, and X. Li, *J. Hazard. Mater.*, vol. **320**, pp. 591–601, (2016).
13. L. Y. Wick, P. A. Mattle, P. Wattiau, and H. Harms, *Environ. Sci. Technol.*, vol. **38**, no. 17, pp. 4596–4602, (2004).
14. O. H. P. and G. N. P. Adrián J. Acuña, *New Technol. Oil Gas Ind.*, pp. 29–48, (2012).
15. F. Li, S. Guo, and N. Hartog, *Electrochim. Acta*, vol. **85**, pp. 228–234, (2012).
16. T. Li, S. Guo, B. Wu, L. Zhang, and Y. Gao, *J. Chem. Technol. Biotechnol.*, vol. **90**, no. 3, pp. 441–448, (2015).
17. T. Li, S. Guo, B. Wu, F. Li, and Z. Niu, *J. Environ. Sci.*, vol. **22**, no. 9, pp. 1381–1386, (2010).

Numerical Solution of a 3-D Advection-Dispersion Model for Dissolved Oxygen Distribution in Facultative Ponds

Sunarsih[1,2,*], Dwi P. Sasongko[3], Sutrisno[1]

[1] Department of Mathematics Faculty of Science and Mathematics, Diponegoro University,Semarang – Indonesia
[2] School of Postgraduate Studies – Doctorate Program of Environmental Science, Diponegoro University,Semarang – Indonesia
[3] Department of Physics Faculty of Science and Mathematics, Diponegoro University, Semarang –Indonesia

Abstract. This paper describes a mathematical model for the dissolved oxygen distribution in the plane of a facultative pond with a certain depth. The purpose of this paper is to determine the variation of dissolved oxygen concentration in facultative ponds. The 3-dimensional advection-diffusion equation is solved using the finite difference method Forward Time Central Space (FTCS). Numerical results show that the aerator greatly affects the occurrence of oxygen concentration variations in the facultative pond in the certain depth. The concentration of dissolved oxygen decreases as the depth of the pond increases.

1 Introduction

Dissolved oxygen (DO) is one of the indicators of water quality. This is because dissolved oxygen plays a role in the oxidation process and reduction of organic and inorganic materials. In aerobic conditions, the role of oxygen is to oxidize organic and inorganic materials. Under anaerobic conditions, the resulting oxygen reduces the chemical compounds to be simpler in the form of nutrients and gases. This oxidation and reduction process that makes dissolved oxygen have an important role in reducing pollution load in waters either naturally or artificially [1].

The main source of oxygen in waters comes from a diffusion process of free air and products photosynthetic of organisms living in these waters. In the surface layer, high oxygen levels due to the diffusion process between water with free air and the process of photosynthesis. Then the oxygen level will decrease along with the increase of water depth. This is because the process of photosynthesis is reduced and oxygen levels are widely used for respiration and oxidation of organic and inorganic materials. Hence there is variation in the dissolved oxygen distribution in waters with different depths [2].

The process of reducing pollution loads in waters commonly used in Indonesia is the facultative pond system. The pond system is a process of wastewater treatment in a shallow pond with long residence time and using aeration process. Aeration is the giving of air into the water to increase the oxygen content by emitting water or passing air bubbles into the water so that the dissolved oxygen in the water is higher [3].

The aeration process in the facultative pond system, of course, will increase the variation of dissolved oxygen dissemination occurring in waters. Context of movement in facultative ponds can be divided into 2, namely advection and dispersion. Advection is a mechanism that moves a substance or matter from one position to another in space, or in other words refers to mass transfer by fluid speed [4]. Dispersions relate to fine insoluble or only slightly soluble particles distributed throughout a continuous medium [5]. Given the horizontal and vertical gradient of the pollutants in the pond room, the transport pollutant phenomenon with this advection and dispersion mechanism can be illustrated in 3-D model. This 3-D advection-diffusion mathematical model is represented by using partial differential equations.

Various studies have been conducted relate to advection-dispersion mechanisms with analytical and numerical solutions, that is [6] comparing several finite difference methods as FTCS, Upstream, Dufort Frankel and Crank Nicolson to solve one dimensional advection-dispersion equations. [7] gives 3 numerical methods for solving 2 advection-diffusion problems with initial conditions and specific boundary conditions. [8] provides examples of numerical solutions of advection-dispersion equations 3-dimensional using finite difference methods FTCS (Forward Time Central Space) with different initial values and boundary conditions. [9] develops and compares several different numerical techniques to solve a three dimensional advection-diffusion equation.

Research has also been carried out by authors in the waste stabilization ponds, that is [10-11] solving the Chemical Oxygen Demand (COD) concentration

* Corresponding author: narsih_pdil@yahoo.com

deployment model on one side of the waste stabilization pool (1-D) and on a plane (2-D) with based on advection-diffusion mechanisms using finite difference Dufort Frankel and [12] solving advection-diffusion model of Biological Oxygen Demand (BOD) concentration of 2-dimension using finite difference method.

Based on the description, this article provides a numerical solution of the 3-Dimensional advection-dispersion model of the variation distribution dissolved oxygen (DO) using finite difference methods FTCS scheme.

2 Materials and Methods

2.1 Distribution Model of Dissolved Oxygen (DO) of Advection-Dispersion

The development of the dissolved oxygen mathematical equation with 1-way flow considering the depth can be described as follows [8]:

$$\frac{\partial C}{\partial t} = -\frac{\partial (uC)}{\partial x} + D_{mx}\frac{\partial^2 C}{\partial x^2} + D_{my}\frac{\partial^2 C}{\partial y^2} + D_{mz}\frac{\partial^2 C}{\partial z^2} \qquad (1)$$

2.2 Finite Difference Method Forward Time Central Space (FTCS)

In general, the basis of the finite difference method is to obtain the value of the variable as a function of space in the time step t + Δt with the spatial distribution in the time step to t is known. The Forward Time Central Space Scheme in question is derivative approximation of time by using forward finite difference and space by using central finite difference. The derivative approximation by using finite difference FTCS is [7]:

$$\frac{\partial C}{\partial t}(i\Delta x, j\Delta y, k\Delta z, t\Delta t) = \frac{C_{i,j,k}^{t+1} - C_{i,j,k}^{t}}{\Delta t} - O(\Delta t) \qquad (2)$$

$$\frac{\partial C}{\partial x}(i\Delta x, j\Delta y, k\Delta z, t\Delta t) = \frac{C_{i+1,j,k}^{t} - C_{i-1,j,k}^{t}}{2\Delta x} - O(\Delta x^2) \qquad (3)$$

$$\frac{\partial^2 C}{\partial x^2}(i\Delta x, j\Delta y, k\Delta z, t\Delta t) = \frac{C_{i+1,j,k}^{t} - 2C_{i,j,k}^{t} + C_{i-1,j,k}^{t}}{\Delta x^2} + O(\Delta x^2) \qquad (4)$$

$$\frac{\partial^2 C}{\partial y^2}(i\Delta x, j\Delta y, k\Delta z, t\Delta t) = \frac{C_{i,j+1,k}^{t} - 2C_{i,j,k}^{t} + C_{i,j-1,k}^{t}}{\Delta y^2} + O(\Delta y^2) \qquad (5)$$

$$\frac{\partial^2 C}{\partial z^2}(i\Delta x, j\Delta y, k\Delta z, t\Delta t) = \frac{C_{i,j,k+1}^{t} - 2C_{i,j,k}^{t} + C_{i,j,k-1}^{t}}{\Delta z^2} + O(\Delta z^2) \qquad (6)$$

This research is a case study by assuming dissolved oxygen is present in a plane with a certain depth. The points on a plane are given in the same horizontal and vertical direction, but with 3 different depths. Thus will be obtained 3 fields with 9 different points. The pond discretization regarding this case study is shown in Figure 1.

3 Results and Discussions

Equation (1) would be discrete form using a finite difference method FTCS scheme by substituting equation (2) - (6) to equation (1), so the following result are obtained:

$$\frac{C_{i,j,k}^{t+1} - C_{i,j,k}^{t-1}}{\Delta t} = -u\frac{C_{i+1,j,k}^{t} - C_{i-1,j,k}^{t}}{2\Delta x} + D_{mx}\frac{C_{i+1,j,k}^{t} - C_{i,j,k}^{t} + C_{i-1,j,k}^{t}}{\Delta x^2}$$
$$+ D_{my}\frac{C_{i,j+1,k}^{t} - C_{i,j,k}^{t} + C_{i,j-1,k}^{t}}{\Delta y^2} + D_{mz}\frac{C_{i,j,k+1}^{t} - C_{i,j,k}^{t} + C_{i,j,k-1}^{t}}{\Delta z^2} \qquad (7)$$

With algebraic manipulation, equation (7) could be written as follows:

$$C_{i,j,k}^{t+1} = C_{i-1,j,k}^{t}\left[\frac{u\Delta t}{2\Delta x} + \frac{D_{mx}\Delta t}{\Delta x^2}\right] + C_{i,j,k}^{t}\left[1 - 2\frac{D_{mx}\Delta t}{\Delta x^2} - 2\frac{D_{my}\Delta t}{\Delta y^2} - 2\frac{D_{mz}\Delta t}{\Delta z^2}\right]$$
$$+ C_{i,j+1,k}^{t}\left[\frac{D_{my}\Delta t}{\Delta y^2}\right] + C_{i,j-1,k}^{t}\left[\frac{D_{my}\Delta t}{\Delta y^2}\right] + C_{i+1,j,k}^{t}\left[-\frac{u\Delta t}{2\Delta x} + \frac{D_{mx}\Delta t}{\Delta x^2}\right]$$
$$+ C_{i,j,k+1}^{t}\left[\frac{D_{mz}\Delta t}{\Delta z^2}\right] + + C_{i,j,k-1}^{t}\left[\frac{D_{mz}\Delta t}{\Delta z^2}\right] \qquad (8)$$

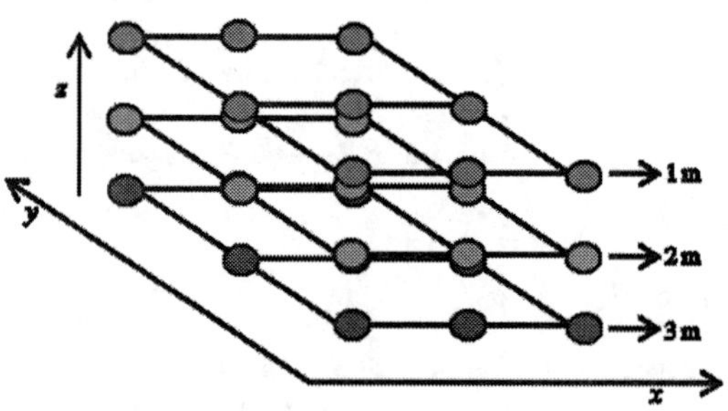

Fig. 1. Pond Discretization Based on Depth Level

Using this equation (8) we would calculate the value of $c_{i,j}^{t+1}$ using the initial values and bounds in each plane up to the desired iteration. Because the calculation process was very long, it would be solved with the help of MATLAB.

To solve the equation, given the values of the diffusion coefficient parameter $D_{mx} = 8.23 \times 10^{-5}\, m^2/s$, $D_{my} = 8.23 \times 10^{-5}\, m^2/s$, $D_{mz} = 8.23 \times 10^{-5}\, m^2/s$, flow velocity $u = 1.09\, m/s$, flow length $t = 24\, hour$, depth $h = 4m$. Then given grids with sizes $\Delta x1 = 11, \Delta x2 = 27.5, \Delta y1 = 10$, $\Delta y2 = 25, \Delta h = 1$ and $\Delta t = 1$. Determination of the grid is adjusted to the size of the pond, which is the length of the pond = 77 meters and the width of the pond 70 meters.

The simulated results of dissolved oxygen distribution in the pond plane with a depth of 1 meter at time $t+1$ can be shown in Figure 2.

Figure 2 shows that the high dissolved oxygen concentration in the middle of the pond is slowly spreading and decreasing at the edges of the pond. This is thought to be due to an aerator in the middle of the pond.

https://doi.org/10.1051/e3sconf/20183103006

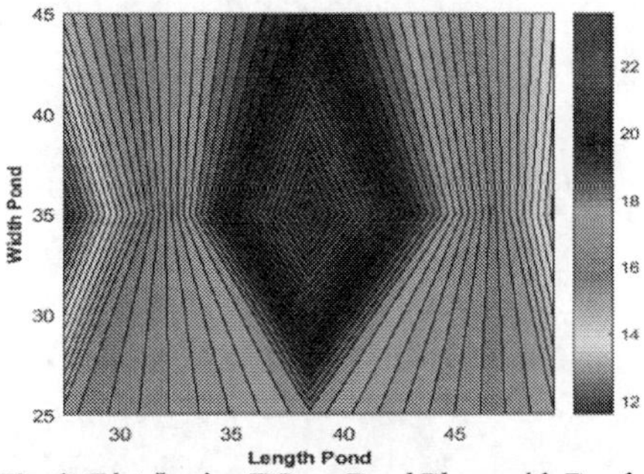

Fig. 2. Distribution DO on Pond Plane with Depth 1 m

The simulation results of dissolved oxygen distribution in the plane of pond with a depth of 2 meters at time t + 1 can be shown in Figure 3.

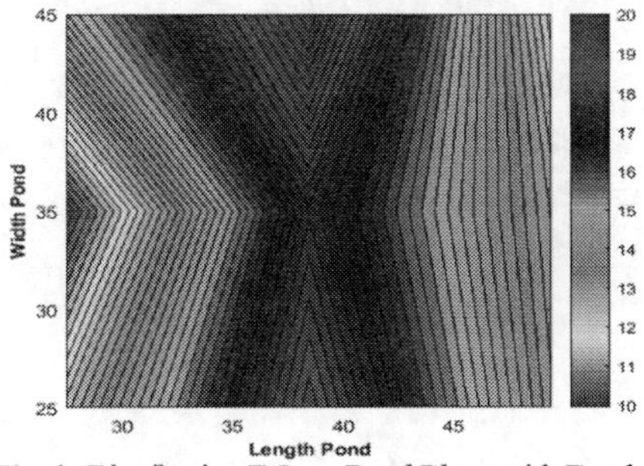

Fig. 3. Distribution DO on Pond Plane with Depth 2 m

Figure 3 states that the initial high concentration of dissolved oxygen on the upper edge of the center of the pond slowly decreases until it reaches the center of the pond. But it increases back to the bottom edge of the pond.

Based on the observations of Figures 2 and 3, it is seen that dissolved oxygen distribution tends to differ in the middle of the pond. The dominant spread which occurs in the middle of the pond as if moving towards the top and bottom edge of the pond. The dissolved oxygen distribution that occurs on the right and left edge of the pool tends to look the same. This indicates that the aerator in the center of the pond does not affect the dissolved oxygen distribution at a depth of 2 meters.

The simulated results of the dissolved oxygen distribution in the pond plane with a depth of 3 meters at time t + 1 can be shown in Figure 4.

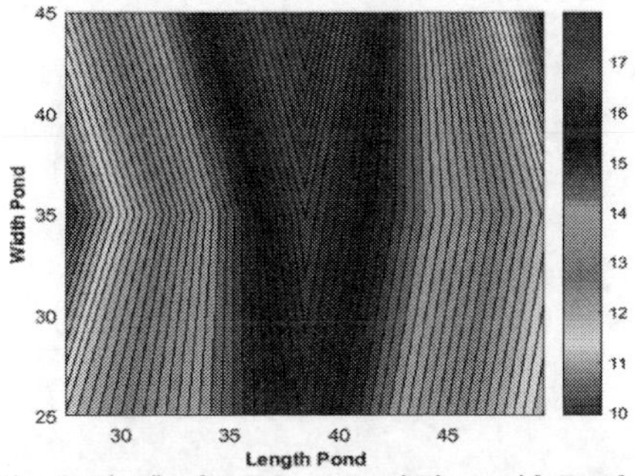

Fig. 4. Distribution DO on Pond Plane with Depth 3 m

Figure 4 states that the previously high concentration of dissolved oxygen at the upper edge of the center of the pond slowly decreases to the bottom edge of the pond. However, the dissolved oxygen concentration at the upper edge of the center of the pond at a depth of 3 meters is lower than at a depth of 2 meters.

Based on the observations of Figures 2 through 4, it appears that the dissolved oxygen concentration decreases with increasing depth of the pond. In addition, dissolved oxygen distribution also looks different. This indicates that there is a variation in the concentration of dissolved oxygen in the facultative pond, especially at a certain depth.

The variation in the dissolved oxygen distribution is more clearly seen in the 3-dimensional image as shown in Figure 5.

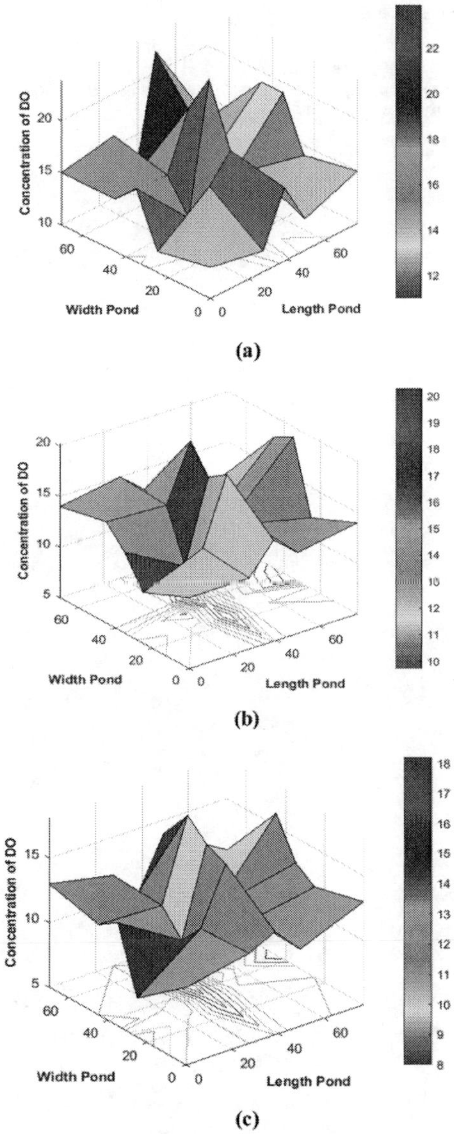

Fig. 5. Variation of Dissolved Oxygen Distribution (a) Depth of 1 meter (b) 2 meters (c) 3 meters

4 Conclusions

Based on the results and discussion, it can be concluded that there is a variation in the concentration of dissolved oxygen in the facultative pond, especially at a certain depth. The dissolved oxygen concentration decreases based on the depth of the pond and its distribution differs only in the middle of the pond. This shows that the aerator greatly affects the occurrence of variations in oxygen concentration. However, this aerator can function optimally on the surface of the pond to a depth of 1 meter. At a depth of more than 1 meter, the process of photosynthesis occurs less than the maximum so that the aerator also can not help increase the concentration of oxygen with the maximum as well.

References

1. D. A. Fuchs, *An Institutional Basis for Environmental Stewardship* (Kluwer Academic Publishers, London, 2003)
2. F. R. Spellman, *Handbook of Water and Wastewater Treatment Plant Operations* (CRC Press Taylor & Francis Group, 2014)
3. A. Boyce, *Introduction to Environmental Technology* (John Wiley & Sons Inc, New York, 1997)
4. A. James, *An Introduction to Water Quality Modelling* (John Wiley & Sons Inc, England)
5. E. Kissa, *Dispersions Characterization, Testing, and Measurement* (Marcel Dekker Inc, New York, 1999)
6. H. S. Najafi, H. Hajinezhad, Appl. Math. Sci. **2**, 2611-2618 (2008)
7. A. R. Appadu, J. K. Djoko, H. H. Gidey, Appl. Math. Comp. **272**, 629-647 (2016)
8. M. Thongmoon, R. McKibbin, R. Tangmanee, Thai J. Math. **5**, 91-108 (2007)
9. M. Dehghan, Appl. Math. Comp. **150**, 5-19 (2004)
10. Sunarsih, D. P. Sasongko, Sutrisno, Adv. Sci. Lett. **23**, 2383-2385 (2017)
11. Sunarsih, Farikhin, H. Rya, Anies, Adv. Sci. Lett. **23**, 2273-2276 (2017)

Effectiveness of Nitrification Inhibition on Various Species of *Brachiaria* Grass Rhizosphere

Purwanto[1,*], *Supriyadi*[1], and *Aniek* Hindrayani[2]

[1]Department of Soil Science, Sebelas Maret University, Surakarta - Indonesia
[2]Department of Economic Education, Sebelas Maret University, Surakarta - Indonesia

Abstract. Nitrification has the potential to decrease the efficiency of nitrogen utilization by plants. The use of nitrifying inhibitory chemicals proved to be effective in controlling nitrification, but also affects beneficial soil microbes. Another attempt to inhibit the more environmentally-friendly nitrification is to use plants that have allelochemical nitrification inhibiting compounds such as the grasses of *Brachiaria*. The aim of this research is to know the effectivity of *B.mutica, B.decumbens,* and *B.humidicola* as inhibitors of nitrification rate in soil. The experiment was carried out by pot experimental method based on nondestructive sampling and Complete Randomized Design, consisting of Brachiaria plant types and various doses of N fertilizer, 100 kg/ha, 150 kg/ha, 200 kg/ha. The results of this study show that 1) *B.mutica, B.decumbens, and B.humidicola*, highly significant to the soil potential nitrification, but the treatment of various doses of N fertilizer is not significant to the soil potential nitrification. 2) the highest soil potential nitrification in *B.mutica* rhizosphere was 5.160 mg NO_2^-/g of soil/5h, while the lowest soil potential nitrification in the rhizosphere of *B.humidicola* plant was 0.414 mg NO_2^-/g/5h. 3) From the four treatment of *Brachiaria* plants can be concluded *B.humidicola* plant more effective in inhibition of nitrification.

1 Introduction

One of the main problems in agricultural cultivation is low nitrogen (N) used efficiency due to loss of N through nitrification, erosion, leaching and volatilization [1]. Approximately 67% of N fertilizers in the world's cereal crops (equivalent to US $ 15.9 billion year-1) are leached in NO_3^- form, volatilized as NH_3, N_2O and N_2 gases and lead to complex environmental problems [2]. Chemical compounds of nitrification inhibitors that have been developed include nitrapyrin, dicyandiamide (DCD) and 3,4-dimethyl-pryrazole-phosphate (DMPP) [1]. Although these synthetic compounds effectively reduce soil N losses, they have negative impacts on non-targeting microbes such as N_2 fixing (diazotroph) bacteria and mycorrhiza fungi [3]. Several types of grasses Brachiaria (*B. decumben*, *B. humidicola* and *B. brizantha*) have been shown to be effective in inhibiting nitrification in the rhizosphere plant [1], [4], [5].

Researchers wanted to get models to improve nitrogen use efficiency and nitrate leaching by utilizing *B. mutica*, *B. decumbens* and *B. humidicola* as nitrification inhibitor plants. The purpose of this study was to determine the effectiveness of the rhizosphere of *B. mutica*, *B. decumbens* and *B. humidicola* in inhibiting nitrification in various doses of nitrogen fertilizer, whether the root activity of *B. mutica*, *B. decumbens* and *B. humidicola* plants were able to inhibit the nitrification of Alfisols soil. Of the three plants tested, researchers wanted to know which species of Brachiaria was most effective in inhibiting nitrification in Alfisols soil.

This research method is experimental pots in green house. The results of this study showed that *B. mutica*, *B. decumbens*, and *B. humidicola*, capable of inhibiting nitrification are shown with the result of all three having a very significant effect on the potential of nitrification; *B. humidicola* with a fertilization dose of N 200 kg/ha is most effective in inhibition of nitrification, indicated by the lowest nitrification potential value of 0.414 mg NO_2^-/g soil/5 hours.

2 Methods

The detail of the experiment is as follows:

2.1 Experimental Design

2.1.1 Material

Materials used for this research were *B. mutica*, *B. decumbens* and *B. humidicola*, and chemical for laboratory analysis. The soil used in this study is Alfisols.

2.1.2 Research Design

This research is a pot experiment with functional relationship with variable approach based on nondestructive sampling and using factorial Completely Randomized Design (CRD). The treatment factor of this research is the combination of planting of three species of Brachiaria plant, namely *B. mutica*, *B. decumben* and *B. humidicola*, with the use of three different N fertilizer

* Corresponding author: purwahadi@yahoo.com

doses of 100 kg/ha, 150 kg/ha and 200 kg/Ha, so there are 9 (nine) different treatments with the control without plants and without fertilizer. Each treatment was repeated three times to obtain 30 treatments.

Table 1. Combination Treatment Factor of Plant Type and Urea Fertilizer Dose

Types of plants	Dose of Urea Fertilizer as a source of N			
	N0 (0 kg/ha)	N1 (100kg/ha)	N2 (150kg/ha)	N3 (200kg/ ha)
B1 (*B. mutica)*	B1N0	B1N1	B1N2	B1N3
B2 (*B. mecumbens)*	B2N0	B2N1	B2N2	B2N3
B3 (*B. humidicola)*	B3N0	B3N1	B3N2	B3N3

2.1.3 Data Analysis

Table 2. Methods and Units for Measuring Treatment Variables

Parameter	Units	Method
pH H_2O	pH	1 : 2.5 (soil : H_2O)
Organic Matter	%	Wet oxidation
CEC	cmol kg^{-1}	Elektrometric
Base Saturation	%	Titration
Organic-C	g kg^{-1}	Wet oxidation
Total-N	g kg^{-1}	
Nitrification Potential	mg NO_2^-/g soil/ 5 hours	Schinner *et al.*, .

The data of the research were analyzed with F 5% test to know the difference between the treatments. Calculation and comparison between treatments done with Duncan Multiple Range Test (DMRT) 5%. Data analysis used Minitab14 and Excel software.

2.2 Experiment Implementation

The experiments were conducted with the following stages:

(1) Initial soil sampling, to find out Organic-C, total-N, available-N (nitrate and ammonium), CEC, C/N ratio, pH H_2O, temperature, and soil moisture.

(2) Preparation of planting media, the soil Alfisols that pass 2 mm sieve.

(3) Nitrogen fertilizer, ie urea fertilizer mixed into the soil before planting *B. mutica*, *B. decumben* and *B. humidicola* into the pot. Determination of nitrogen fertilizer is to meet the nutrient needs of Brachiaria plants as well as nitrification substrate. The doses administered were adjusted for each treatment.

(4) Planting of Brachiaria plants with plant propagation, then transferred into pots intact with the rhizosphere of the roots.

(5) Maintenance and irrigation by cleaning weeds and watering.

(6) Measurement of variables and soil sampling. Soil samples for potential nitrification measurements were taken aseptically at a depth of 0-20 cm. Potential nitrification was measured from the amount of NO_2^- formed from soil samples after added $(NH_4)_2SO_4$ and incubated at 25° C for 5 hours[6].

3 Results and Discussion

3.1 Characteristics of Alfisols Soil

Table 3. Result of Soil Analysis Before Treatment

No	Soil Properties	Result	Valuation
1.	pH H_2O	5.2	Acid *)
2.	OM	4.1 %	Low *)
3.	CEC	22.28 cmol kg^{-1}	Moderate *)
4.	Base Saturation	36 %	Moderate *)
5.	Organic C	3.89 g kg^{-1}	Moderate *)
6.	Total N	0.28 g kg^{-1}	Moderate *)

Source: Lab. Analysis
Note: *) Valuation according to the Soil Research Institute, 2009.

3.2 The Influence of Various Brachiaria Plants to Soil Nitrification Potential

The amount of NO_2^- formed by a union of time, due to the biological NH_4^+ oxidation process is called nitrification potential [6].

Table 4. Results of Nitrification Potential Diversity Analysis

Source of Variations	F test	P
Plants	18,27	0,000**
Fertilizer	0,15	$0{,}826^{ns}$
Plants*Fertilizer	1,00	$0{,}416^{ns}$

**: significant; ns: not significant

Table 4. shows that the Brachiaria plant species has a very significant effect on the value of soil nitrification potential (P <0.01). The treatment of various doses of nitrogen fertilizer as nitrification substrate proved to have no significant effect on the value of soil nitrification potential (P> 0,05).

Table 5. Results of Diversity Analysis Influence of Planting of Some Species of Brachiaria Plants to Soil Nitrification Potential

Source of Variations	F tetst	P
B. mutica	20,14	0.000**
B. decunbens	98,98	0.005**
B. humidocola	21,75	0.009**

**: significant; ns: not significant.

Table 6. Results of Potential Diversity Analysis of Nitrification Influence of Planting of Some Species of Brachiaria Plant on Each Incubation

Source of Variations	Nitrification Potential (mg NO_2/g soil/5 hours) per incubation (Weeks)					
	1	2	3	4	5	6
B. mutica	0.029*	0.010*	0.006**	0.012*	0.012*	0.017*
B. decunben	0.013*	0.007**	0.000**	0.006**	0.031*	0.061*
B. humidicola	0.018*	0.032*	0.073*	0.110^{ns}	0.100^{ns}	0.137^{ns}

Note:*) significant;**)very significant; ns:not significant.

Based on result of analysis of variance Table 5 it is known that the planting of various Brachiaria have a very significant effect (P value <0.01) to the potential of nitrification. The results of the analysis of variance of table 6 of the treatment of various Brachiaria species showed different effects on the potential value of nitrification at each incubation time. *B. mutica* plants at each incubation period showed a significant effect on the potential value of nitrification, and very significant during the third incubation period. Plant *B. decumben* at the second incubation until the fourth incubation showed a very significant effect (P value <0.01) on the nitrification potential, whereas for *B. humidicola* plant the fourth incubation until the end of incubation did not significantly affect the nitrification potential.

Differences in the effect of planting of various Brachiaria plants during the incubation period can be indicated that each of the Brachiaria plant rhizosphere has different inhibitory activities.

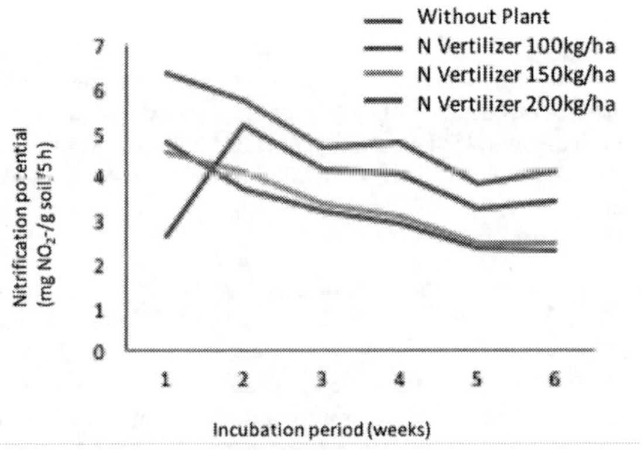

Fig. 1. Potential Nitrification (NO_2^-) of soil in *Brachiaria mutica* plant per incubation time

The potential value of nitrification of *B. mutica* plants has a similar pattern at each N fertilization dose, except in the N fertilizer treatment of 200 kg/ha at incubation time of the first week showed the lowest nitrification potential value of 2.582 mg NO_2^-/g/5 hours later Increased to the highest value of 5.160 mg NO_2^-/g of soil/5 hours at the second incubation period, and subsequently began to decrease. The decrease in the potential value of nitrification after the third incubation period indicates the onset of inhibition of nitrification by the *B. mutica* root activity. Treatment of *B. mutica* cultivation with N 100 kg/ha fertilization has the lowest nitrification potential value (2,245 mg NO_2^-/g of soil/5 hours in fifth incubation) compared with N 150 kg/ha fertilizer and N 200 kg/ha fertilizer.

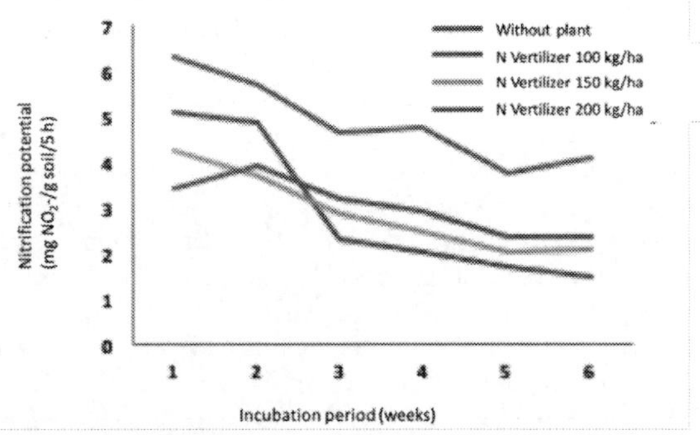

Fig. 2. Potential Nitrification (NO_2^-) soil in *Brachiaria decumben* plant per incubation time

Figure 2 shows that the potential value of nitrification at planting *B. decumben* has a large value difference compared to the potential value of nitrification on soils without planting *B. decumben*. The highest value of 5.114 mg of NO_2^-/g of soil/5 hours was found in the first N 100 kg/ha incubation fertilization treatment, but then it will substantially decrease from the second incubation to the sixth week. In the third incubation period until the last incubation period the measured potential value of nitrification was 1.493 mg NO_2^-/g of soil/5 hours. In general, treatment of addition of N 100 kg/ha fertilizer has the lowest nitrification potential value from the addition of N 150 kg/ha and N 200 kg/ha fertilizer.

Based on figure 2 above it can also be seen that in each treatment, the decrease of potential nitrification value occurs in the third incubation which then shows a stable nitrification potential value until the end of incubation.

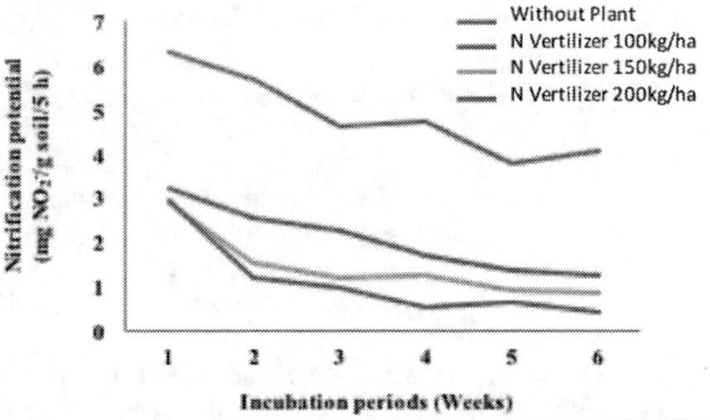

Fig. 3. Potential Nitrification (NO_2^-) soil in *Brachiaria humidicola* plant per incubation time

From the observation of potential value of soil nitrification on *B. humidicola* plant has a measured value that is much lower than the potential value of soil nitrification without planting Brachiaria species. The potential value of nitrification on the N 200 kg/ha fertilizer treatment was the lowest compared to the treatment of N 100 kg/ha and 150 kg/ha, ie 0.414 mg NO_2^-/g/5 hours. In the treatment of *B. humidicola* plant it can be seen that the decrease of nitrification potential indicates a value that is not too different from each incubation.

Treatment with planting of three types of Brachiaria plants showed a lower nitrification potential (NO_2^-) value than a plantless and non-fertilized treatment. This indicates that all three species of Brachiaria plants are capable of inhibiting nitrification. Soil without Brachiaria plants assumed to be a common condition of Alfisols soil proved to show a high potential value of nitrification. The activity of root exudate of *B. humidicola* plant is able to decrease NO_2^- quantities in soil during incubation period, compared with control[16]. This suggests that the hydrolysis of N-fertilizers in soils at *B. humidicola* planting treatment is still largely in NH_4^+ and not oxidized to NO_3^-

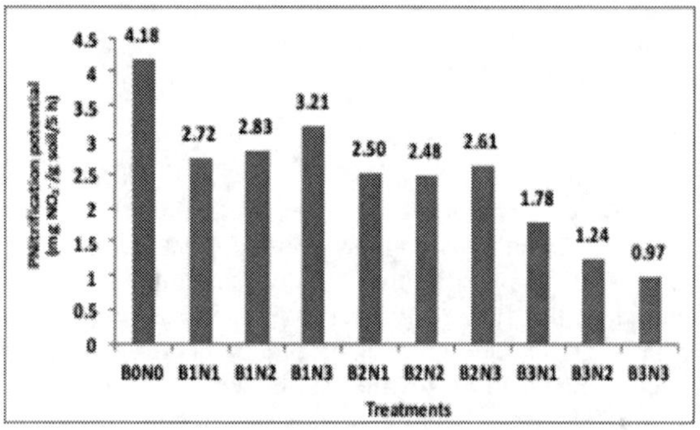

Fig. 4. Potential Nitrification at Various Treatments

From the average range of treatments compared with control (without crops and without fertilizers) showed that treatment with the Brachiaria plant had a lower nitrification potential value than that of non-plant treatment. This is because most of NH_4^+ of urea fertilizer will be utilized as nitrification substrate so the nitrification potential is higher than treatment with Brachiaria plant. In the three treatments of the Brachiaria plant, (Figure 4), the treatment of *B. mutica* plants has a higher nitrification potential value, followed by the treatment of *B. decumben* and *B. humidicola* plants.

The average number of potential nitrification (3.21 mg NO_2/g soil/5 hour) on a variety of plant species Brachiaria of the entire incubation period is the planting of *B. mutica* treatment with dosages of 200 kg/ha, while the lowest value (0.97 mg NO_2^-/g of soil/5 hours) was found in the treatment of *B. humidicola* with fertilizer N 200 kg/ha.

The treatment plant *B. mutica* and *B. decumben* with fertilizer dose of 200 kg N/ha shows nitrification potential value that is greater than the treatment plant *B. mutica* and *B. decumben* with fertilizer dose of 100 kg N/ha and a dose of fertilizer N 150 kg/Ha. This is thought to be due to a larger N source so that inhibition of nitrification is less effective than the treatment of smaller N fertilizer doses. Based on comparative test (DMRT 5%) between interaction treatment of various species of Brachiaria and N fertilization dose to nitrification potential showed no significant difference.

Figure 4 shows that *B. humidicola* plant yields the lowest potential nitrification potential of the two other Brachiaria species. The rhizosphere activity of the three *B. mutica*, *B. decumben* and *B. humidicola* plants was shown to have a lower nitrification potential value than the non-plant and without fertilizer treatment. *B. humidicola* plant looks most effective in inhibiting nitrification with nitrification potential value of the lowest (0.414 mg NO_2^-/g soil/5 hour) at a dose of fertilizer N 200 kg/ha than *B. mutica* and *B. decumben*.

Some of these nitrifying inhibitors have been isolated and identified from plant root exudates using bioassay purification tests. Biological nitrification inhibitor compound at several plants Brachiaria, have been identified in the form of free fatty acids are unsaturated[7],[8],[9],[10]. In a further study, it was found to be more specific that *B. humidicola* plants contain linoleic and alpha-linoleic acid compounds in appropriate amounts as a nitrification inhibitor. Nitrification inhibitors compound inhibit the chain allows the enzyme ammonia monooxygenase (AMO) and the enzyme hydroxylamin oxidoreductase (HAO) on nitrification [7], [11].

By proving the nitrification inhibition of various plant species Brachiaria in this study is expected to be developed model of the nitrification inhibition by utilizing Brachiaria plant as a crop mix (multiple cropping) together with other types of forage crops or other crops so as to increase the effectiveness of the use of fertilizer N.

4 Conclusion

Nitrification is an adverse process in relation to the nitrogen use efficiency of plant and raises complex environmental problems. This study aims to find a way of controlling biological nitrification that is environmentally friendly by utilizing plants that produce alelochemical nitrification inhibitors compounds. The study was conducted by pots experiments in a greenhouse.

The research result of a variety Brachiaria planting with various doses of fertilizer N sources indicate that: (1) *B. mutica*, *B. decumben*, and *B. humidicola* plant, capable of inhibiting nitrification is indicated by the results of three very significant effect on nitrification potential; (2) *B. humidicola* with a dose of fertilizer N 200kg/ha of the most effective in the inhibition of nitrification, nitrification potential value indicated by the low at 0.414 mg NO_2^-/g soil /5 hours. *B. humidicola* planting as mixed crops along with other types of grasses or other cultivation plants can be used to increase the efficiency of N nutrient utilization in the soil. To ensure the effectiveness of nitrification inhibition in the rhizosphere of Brachiaria plants need to do further research on the influence of the rhizosphere of various Brachiaria species on the ratio of NH_4^+ and NO_3 content in soil (actual Nitrification in soil).

References

1. T. Watanabe, I. Osamu, *World Journal of Agricultural Science*, **4**, 1, 106-113 (2008)
2. W.R. Raun and G.V. Johnson, *Agronomy Journal*, **91**, 357 – 363, (1999)
3. M. Rao, T. Ishikawa, O. Ito, M.P. Hurtado, E. Amezquita, E. Barrios, and C. Lascano, *JIRCAS Working Report* 71, 107–112 (2009)
4. T. Ishikawa. *JIRCAS Research Highlight News Letter*, **6** (3), 3 (1999)
5. V. Damin, H.C.J. Franco, M.F. Moraes, A. Franco, P.C.O. Trivelin. *Agriculture Sci.*, **65**, 4, 402-407 (2008)
6. F. Schinner, R. Ohlinger, E. Kandeler, and R. Mergesin, (eds.) *Methods in Soil Biology*, Springer-Verlag Berlin Heidelberg, 146-149 (1995)
7. G.V. Subbarao, M. Kishii, K. Nakahara, T. Ishikawa, T. Ban, H. Tsujimoto, T. S. George, W. L. Berry, CT Hash, and O. Ito, *Breeding Science* **59**, 529–545 (2009)
8. G.V. Subbarao, M. Rondon, O. Ito, T. Ishikawa, M. Rao, K. Nakahara, C. Lascano, and W.L. Berry. *Plant Soil*, 294, 5–18 (2007)
9. G.V. Subbarao, T. Ishikawa, K. Nakahara, O. Ito, M. Rondon, I.M. Rao, and C. Lascano, *JIRCAS Working Report*, **51**, 99–106 (2007)
10. S.T. Gopalakrishnan, S.J. Watanabe, Pearse, O. Ito, Z.A.K.M. Hossain, and G.V. Subbarao. *Soil Science. Plant Nutr*, **55**, 725–733 (2009)
11. B. Tomohiro, K. Masahiro, O. Ito, H. Samejima, H.Y. Wang, S.J. Pearse, S. Gopalakrishnan, K. Nakahara, A.K.M. Zakir Hossain, H. Tsujimoto, and W.L. Berry, *Plant Soil*, **310**, 55–64 (2008)
12. A.J. Erickson, R.S. Ramsewak, A.J. Smucker, and M.G. Nair, 2000. *Journal of Agricultural and Food Chemistry*. **48**, 12 6174-6177 (2000)
13. A.M. Fernandez, Souza, EFC. Gabriel, A, Marcelo, and A. Ciro, (Rosolem World Congress of Soil Science, Soil Solutions for a Changing World. Brisbane, Australia, 2010)
14. D.I. Dinnes, D.L. Karlen, D.B. Jaynes, T.C. Kaspar, J.L. Hatfield, T.S. Colvin, and C.A. Cambardella, *Agronomy Journal*, 94, 153–171 (2002)
15. G.V. Subbarao, K Nakahara, T. Ishikawa, T. Yoshihashi, O. Ito, H. Ono, M. Ohnishi-Kameyama, M. Yoshida, N. Kawano, and W.L. Berry, Plant Soil, **313**, 89–99 (2008)
16. I.R.P. Fillery, *Plant Soil*, **318**, 1–4 (2009)
17. J.A. Delgado, A.R. Mosier, *Journal on Environmental Quality*, **25** (5). 1105 – 1111 (1996)
18. S.J. Dennis, K.C. Cameron, H.J. Di, J.L. Moir, V. Staples, P. Sills, and K.G. Richards, *Biology and The Environment*, **112B**, 79–89 (2012)
19. S. Warneke, L.A. Schipper, D.A. Bruesewitz, I. McDonald, S. Cameron, *Ecol Eng*, **37**, 511–522 (2011)
20. Soil Research Institute, Technical Guidelines for Soil Chemical Analysis, Plants, Water and Fertilizers. Soil Research Institute. Agency for Agricultural Research and Development. Ministry of Agriculture, Bogor (2009)

 https://doi.org/10.1051/e3sconf/20183103008

Performance of Anaerobic Baffled Reactor with Three Compartments in Removal of COD of Wastewater of Chilly Sauce

Indro Sumantri[1*], B. Budiyono[2], P. Purwanto[2]

[1] Doctoral Program of Chemical Engineering, Faculty of Engineering, Diponegoro University, Semarang - Indonesia
[2] Department of Chemical Engineering, Faculty of Engineering, Diponegoro University, Semarang - Indonesia

Abstract. The objective in this study is to examine the performance of each compartment of the number of compartments of anaerobic baffled reactor (ABR) to the COD removal of the chilly sauce wastewater. Three-compartments of ABR were conducted in this experiment with the total volume of 60 l. ABR is very suitable for processing waste water with high content of COD. Wastewater conducted in this research is a degradable chilly sauce synthetic and high content of organic compounds. While the COD parameter is the main parameter to indicate the achievement of wastewater treatment plant. Stepwise in the research starting with the preparation of raw materials such as sample preparation of synthetic wastewater and preparation of activated sludge. Variable used is the time digestion in the ABR, sludge volume (50% and 70%), and initial COD concentrations (6000 – 14000 mg/L). The response is observed up to 7 days process. For a load of organic compounds, the first compartment has high degree of decomposition of organic compounds than 2nd and 3rd, it is shown that the COD removal the second and third compartment increase insignificantly compare the first compartment. As for the different height of the activated sludge indicated that for organic load of of 6170 mg/L up to 14265 mg/L, the first compartment has removal efficiency 79-73%, in the second compartment is 81-75%, 81-77% and third compartment.

1 Background

Chilly sauce is a complement food with basic ingredients of chilly, it has long been known as an evocative and appetite enhancer. Recent progress of the times, the chilly sauce is now not only produce in traditional process with a simple tool, but it has also been available in the form of ready-made chilly factory output. In Indonesia, there are about 24 companies that produce chilly sauce. The number of company increased when compared in 2003 with number of 16 companies. The number of these companies indicated growing production of chilly sauce in Indonesia which means it also increases the amount of waste generated. Wastewater from the production process resulting from the leaching chilly feedstock, washing equipment, washing of the bottle or packaging. The content of the wastewater is generated mostly organic compound with COD of 14,000 mg/L. Before the wastewater released into the environment, it must be treated through wastewater processing in order to fulfill the threshold value in accordance with the standard quality according to the regulation of Central Java Province (number 5 in 2012) for the chilli sauce that is of 150 mg/L.

Wastewater treatment can be done in physics, chemistry, and biology. Biological wastewater treatment can take place in two main conditions, namely aerobic and anaerobic. In the treatment of organic wastewater on anaerobic condition can utilize activated sludge i.e. suspension biological or microbial mass which is a very active to degrade dissolved organic materials. According to [1], activated sludge technology able to remove COD (chemical oxygen demand) of 76% with retention time for 6 days. The wastewater with COD content is greater than 4000 mg/L usually use anaerobic process. On the anaerobic process, micro organisms do not require or less oxygen to treat wastewater. If in terms of energy consumption, the aerobic process will require a lot more energy electricity while the anaerobic process is assessed more effectively and efficiently.

The effort to obtain intensive contact wastewater-sludge active is intended so that the waste water residence time became shorter so that will save land for anaerobic reactor [2]. In addition to the effort to get intensive contacts, another attempt was with the optimization of the active sludge whose performance can be improved with the addition of micro-nutrients [3].A small number of compounds that are included in the categories of micro nutrient elements such as nickel, iron, molybdenum, cobalt, copper and zinc that are added into the anaerobic treatment system gives a positive performance against a decline in the levels of organic compounds wastewater loads and resistant to change [4].

* Corresponding author: indrotekim@yahoo.com

2 Research Method

2.1 Raw Material Preparation

Raw material wastewater is synthetic wastewater and is set up by dissolved of certain quantity of chilly sauce to the distilled water and this synthetic wastewater is analyzed to obtain the quantity of COD content. While activated sludge obtained from small-scale industrial. This sludge is seeding for two weeks with addition of nutrient of micro organisms with proportion of COD : N : P of 300 : 5 : 1 and also addition of Calcium ion of 40 mg/L to prevent the alkalinity.

2.2 Experiment equipment

A lab-scale ABR bioreactor with total volume of 60 L was fabricated using transparent plexi glass sheets and designed has 3-pairs down and upflow, the dimension of the ABR is 60 cm of length, 25 cm of wide, 40 cm of height, 10 cm of distance between baffles, 1,5 cm of distance of baffle and bottom reactor. The ABR consisted of three equal compartments and each compartment had a pair of baffles of downflow and upflow (hanging and standing baffles) and sample ports.

The temperature difference of inside reactor, due to the changes in the temperature during the day and night was less than 3°C. Therefore, the reactor did not require cooling or heating system to maintain the temperature. A peristaltic pump was used to feed the ABR with synthetic wastewater.

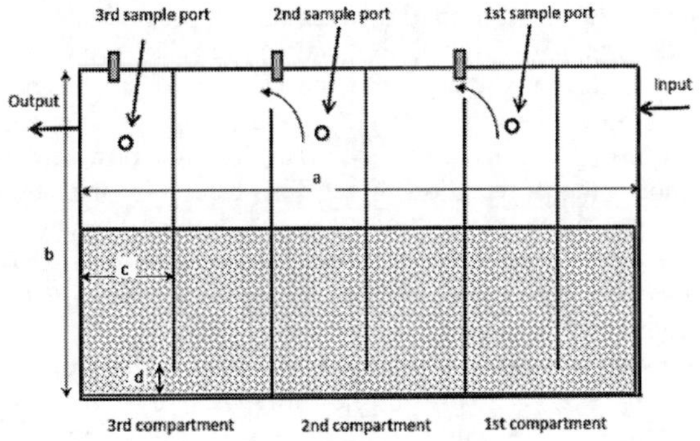

Fig. 1. Diagram of the anaerobic baffled reactor (ABR) and its baffles with dimensions (cm) a = 60 cm; b = 40 cm; c = 105 cm; d = 1.5 cm; wide: 25 cm[5]

2.3 Experiment sets up

Experiments conducted with volumetric feed rate of 20 L/day by using peristaltic pump, initial pH is 7-7,5. Addition of trace elements per gram COD (Co : 500 µg, Cu : 300 µg, Fe : 4000 µg, Mo : 150 µg, Ni : 6000 µg, and Zn : 600 µg). Room temperature, atmospheric pressure, calsium ion (40 mg/L). The research variables are the height of activated sludge (50 and 70 %) and initial COD concentration (6000, 10000, and 14000 mg/L)[3].

2.4 Parameter response

Parameter observed for this research is COD. The selection of COD parameter is based on the quantity of the organic material could be quantity as COD[6].

3 Results and discussion

3.1 Performance of ABR with 50 % sludge height and various COD contents

The content of the COD conducted this research is: 6170 mg/L, 10170 mg/L, and mg/L 14265. Observations of COD content done for each day until the seventh day with the consideration that the wastewater is easily degraded hence, the retention time is short. Performance of ABR with three-pair compartments for various contents of wastewater were observed for each compartment to show the COD removal: for first compartment (71.83 – 67.29-62.90%), second compartment (73.31 – 72.20-65.70%), and third compartment (74.55 – 74.88-68.57%).

For the condition of 50 % height of activated sludge, the higher the concentration of wastewater will be the lower removal of COD in this anaerobic process. Increasing the amount of wastewater will increase the load of activated sludge to degrade the organic substances, actually the activity of the micro organisms indicated advance in degradation of the organic substances but in the term of percentage of COD removal, it showed a decrease with an increase in the load of wastewater. The efficiency of the COD removal for first to second and third compartments indicated less significant improvement, with total similar amount of activated sludge in each compartment the COD removal of first to second and third compartments are very small (less than 5%). This indicated that with the increasing load of wastewater is required sufficient amount of microorganisms to enhance degradation of organic compounds. The presence of a preservative substance in chilly sauce affect the performance of ABR because preservatives will influence the performance of microorganisms. The addition of the preservatives substance into the food and drink is to maintain freshness and prevent food damage or food ingredients by suppress the growth of micro-organisms (Maier et al., 2010; Porsby et al., 2008).The first compartment is a compartment with a heavy load because the first contact of wastewater and activated sludge occurred in this compartment. The COD removal is caused by anaerobic bacteria activity on the activated sludge [7]. These bacteria growth optimum at pH neutral. These bacteria work with degradation of glucose. The value of COD removal indicated that organic load which degraded by microorganisms in the activated sludge. Increasing the amount of bacteria will affect to the greater organic load that can be degraded by microorganisms. This is due to amount of substrate contacted with the microorganisms.

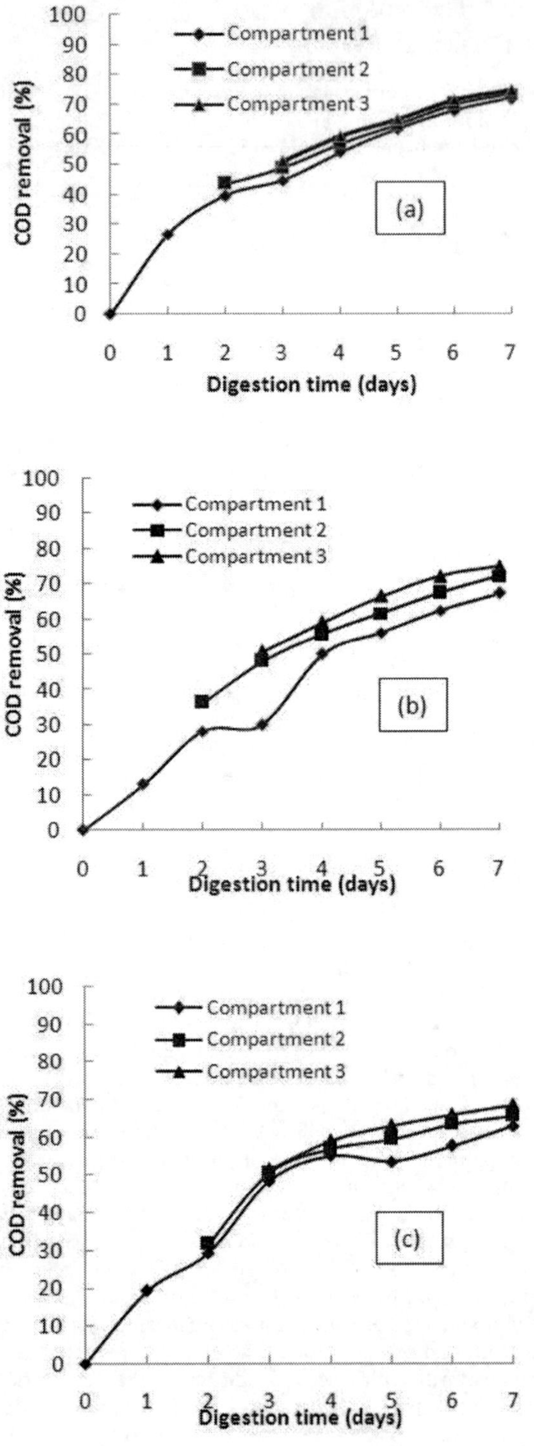

Fig. 2. Performance of ABR at 50 % sludge height with various COD content (a) 6170 mg/L, (b). 10.127 mg/L, and (c). 14265 mg/L

While, even at the shortest retention time of3 days the system can achieve high performance as above77.3% COD removal [8], also an retention time of 4.25 d, the total COD removal efficiencies reached 88.7% [9]. Attributed to the incomplete anaerobic digestion. The main reason for the relatively poor efficiency was the accumulation of suspended solids in the sludge blanket resulting from suspended solids overload in this condition. Consequently, it caused sludge washout, followed by a decrease in COD removal from 95.13 to 78.01%. It was noted that after the occurrence of sludge washout, the reactor efficiency was observed [9].

3.2 Performance of ABR with 70 % sludge height and various COD contents

Figure 3 (a)-(c) indicated the performance of ABR with 70% height of activated sludge. It is generally shown that COD removal until the third day shows a sharp removal in rates, 60% of COD can be removed. Compared to 50% height activated sludge then the COD removal for 70 % height of activated sludge is much larger. This is because the more sludge is utilized then the more microorganisms obtained, so, that will degrade on the organic compounds in the wastewater [10]. The amount microorganisms availability will be increased degradation of the organic compounds, hence, the COD content in wastewater will be getting smaller. The availability of food sources of microorganisms will enhance the growth and amount the number of microorganisms, hence, decomposition of organic compounds in wastewater is also faster, it causes a decrease in the value of COD content sharply[11].

Increased efficiency of the first compartment to the second and third compartments is small and insignificant. The small increment of improvement in the efficiency of the COD removal in the second and third compartment is due to the short retention time, hence, this three compartments are nearly have the same composite on. It can be said that the high flow rate or short liquid retention time will result in the three compartments have similar composition or the same as the type of Upflow Anaerobic Sludge Blanket (UASB) reactors. Figure 3 shows that for all three types of COD content of wastewater have shown that the stationary condition has almost achieved, this can be seen from the small enhancement of the efficiency of the COD removal. The correlation between the number of activated sludge with the efficiency of the COD removal is very clearly indicated that the higher number of activated sludge then it will be more and more organic compounds are degraded. Total COD removal is between 77,53-82,06%. The value of the COD removal is lower if compared to that done by previous researchers [8,9]. The presence of a preservative in chilly sauce played an important role so that the efficiency of decreasing COD becomes smaller compared to previous research.

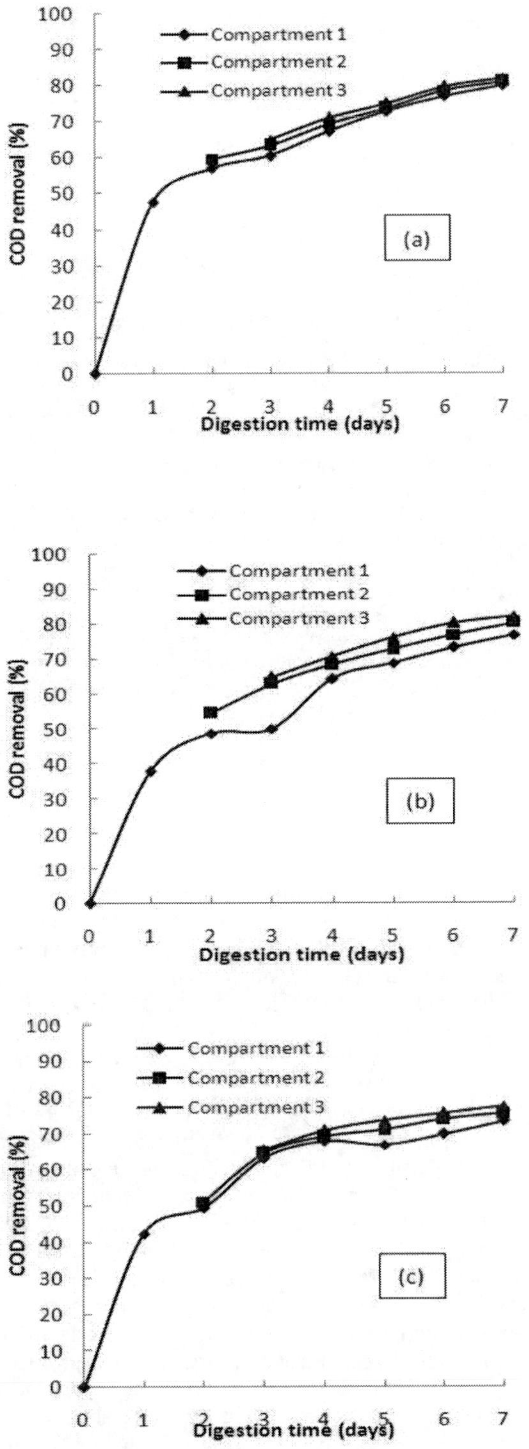

Fig 3. Performance of ABR at 70 % sludge height with various COD content (a) 6170 mg/L, (b). 10.127 mg/L, and (c). 14265 mg/L

4 Conclusions

The amount of activated sludge plays an important role in the COD removal, the higher of activated sludge the higher the efficiency of degradation. Three compartments observed for the COD parameter has concentration is closed to each compartment, it showed that the retention time of wastewater in the ABR is a main factor to be controlled.

5 Acknowledgement

We would like to thank the UNDIP financial scheme of *APBN DPA SUKPA LPPM Universitas Diponegoro Year 2017*

References

1. Setiyani. . "*Bioreaktor Pengolahan Limbah Cair*". Pusat Antan, Universitas Bioteknologi, Institut Teknologi Bandung (2010).
2. G.F Zhu, Li, J. Z., Wu, P., Jin, H. Z., & Wang, Z. Bioresource Technology, ***99***(17), 8027–8033.(2008).
3. J. Ma, L.J. Mungoni, W. Willy Verstraete, M. Carballa, Bioresource Technology **100** (2009) 3477–3482, (2009).
4. K. Izumi, Y. Okishio, N. Nagao, C. Niwa, S. Yamamoto, &T. Toda, *International Biodeterioration & Biodegradation*, **64**(7), 601–608,(2010).
5. I. Sumantri, Purwanto, and Budiyono.,AIP Conference Proceedings 1699, 050017 (2015).
6. *APHA. Standard methods for the examination of water and wastewater*. 21 st edition, 2005. Washington, DC
7. K. Dhamodaran& S.K. Ajay, J.Env Chem Eng.Sci. vol.**2**, pp.1821-1830, (2014).
8. M. Faisal &H. Unno, Biochemical Engineering Journal, **9**(1), 25–31,(2001).
9. M. Pirsaheb, M. Rostamifar, AM. Mansouri, a.a.L. Zinatizadeh, &K. Sharafi, Journal of the Taiwan Institute of Chemical Engineers, 47, 137–148.(2015).
10. Suyasa, I Wayan Budiarsa dan I Made Arsa.2013. *Penurunan Kadar COD Air Limbah Industry Jamu menggunakan Lumpur Aktif*.
11. J.P. Guyer . *Introduction to Secondary Wastewater Treatment (*2011)..

 https://doi.org/10.1051/e3sconf/20183103009

Heavy Metal Content in Terraced Rice Fields at Sruwen Tengaran Semarang - Indonesia

Yulis Hindarwati[1,*], *Tri* Retnaningsih Soeprobowati[2], *Sudarno*[3]

[1]Master Program of Environmental Science, School of Postgraduate Studies, Diponegoro University, Semarang - Indonesia
[2]Departement of Enviromental Engineering, Faculty of Engineering, Diponegoro University, Semarang - Indonesia
[3]Departement of Biology, Faculty Science and Mathematics, Diponegoro University, Semarang - Indonesia

Abstract. The presence of heavy metal on agricultural soils can be caused not only natural factors but also due to human intervention. Differences in management and lack of understanding of farmers in the production input of fertilizers and pesticides ensued in land ravaged. Periodic testing of paddy fields is necessary to minimize the contaminants from being absorbed by plants that will have an impact on health decline. The purpose of the assessment was to identify the heavy metal content in the terraced rice field in Sruwen Village, Tengaran District, Semarang Regency. Survey was conducted in February 2017. Sampling on terraced rice fields of different heights consisted of upper, middle, and upper down. Taken as many as eight single points and composed at a depth of 0-20 cm and 20-40 cm. The identification results showed that heavy metal content of Pb, Cd, and Cu were present at all altitudes. Heavy Metals Pb and Cd at a depth of 0-20 cm were higher from 20-40 cm in the upper and lower rice fields but lower in the middle rice field. Cu heavy metal at a depth of 0-20 cm was higher than 20-40 cm in all altitude land. The heavy metal content of Pb, Cd, and Cu was still below the heavy metal standard set by the European Union and India.

1. Introduction

The demand for increased productivity as an effort to fulfill food needs is very much affecting the farmers. To increase production with excess input in the form of fertilizers and chemical pesticides basically accelerate the process of environmental degradation. The impact caused by environmental pollution is the chemical residue in the soil of heavy metal. The presence of natural content of heavy metals in agricultural soils can derived from soil parent material and when it accumulates causing impaired soil function and affect plant growth [1].

The lack of understanding of farmers on environmental sustainability and inherited practices is a difficult social factor to tackle. The experience of farmers shows that the success of agricultural production still depends on the use of excess fertilizers and pesticides. This is triggered by the difficulty of receiving technological innovation. Differences in wetland management have different effects from both nutrient content and heavy metal contamination. Other factors can also be caused by natural factors, proximity to roads, and water sources. Excessive and fertilized irrigation water, especially the direct use of fertilizers and pesticides in agricultural activities will seep into the soil and contaminate groundwater [2].

Utilization of agricultural waste is actually more efficient and economically profitable than the use of chemical inputs both fertilizers and pesticides. Besides the price is more expensive, scarcity and government subsidies are still lacking. Nevertheless, farmers still use inorganic fertilizers and pesticides in the hope that productivity can be higher than organic. But in fact, precisely with the use of inorganic fertilizers soil quality and quantity of results is not necessarily as expected even may decrease.

The heavy metal content is not only found in rice fields close to the industry, but there are also in paddy fields that are turning to organic development. Results of research using external inputs of site-specific chemical fertilizers still contained heavy metals in the soil (0.54-1.32 ppm) [3]. Similarly, results of research in Belgrade Serbia reported that heavy metal contaminants from the Power Plant industry had detected Pb (13.4-27.4 mg / kg), Cd (0.11-0.27 mg / kg), Cu (20, 5-34.4 mg / kg), Zn (75.3-142 mg / kg) of land on agricultural land is still below established national and international standards [4]. Based on the above consideration, it is necessary to conduct an assessment aimed at identifying heavy metal content in terraced rice field in Sruwen Village, Tengaran District, Semarang Regency.

2. Materials And Methods

Soil sampling was conducted in February 2017 in Sruwen Village, Tengaran District, Semarang Regency. Examples of soil are taken from each different rice field

* Corresponding author: yulis_hindarwati@yahoo.co.id.

height or terrace in one spreadsheet. The rice field consists of the upper rice field at an altitude of 705.9 m above sea level, the center at a height of 702.2 m asl, and the bottom at an altitude of 696.0 m asl. The soil is taken ten points then composited by using a soil drill (auger). *Zig-zag* intake at soil depth 0-20 cm and 20-40 cm. Examples of water are taken from the main irrigation door and each canal into the compressed upper, middle, and bottom rice field. Examples of composted soil were tested to the Laboratory of Environmental Research Institute of Pati to determine the content of heavy metals Pb, Cd, and Cu. Using the method of wet destruction in in the mixture of HNO3: HClO4 [5], then measured with Variance Atomic Absorption Spectrophotometer (AAS) AA240 FS Variants [6].

3. Results And Discussion

3.1 The Characteristics of Location

Semarang regency consists of 18 districts of which is District Tengaran which is located at 729 m above sea level. In 2015 the amount of rainfall of 3,699 mm with average rainfall of 1,683 mm in a year. Year 2014, district of Tengaran has a land area of 883.43 ha, irrigated rice field area of 652.85 ha and rain-fed rice field 230.59 ha. In 2015 the use of wetland farming area of 883.44 hectares and not rice fields 1,883.93 ha. Area of rice harvest 1.261.00 ha with production 6,909.02 ton and productivity 5.48 ton/ ha [7]. Research sites with different altitudes are listed in Table 1.

The location of paddy fields from upper, middle, and lower paddy fields has different soil properties either from nutrient content or chemical residue. The presence of heavy metals in the soil is influenced by soil physical and chemical properties, such as pH, C-organic, and Cation Exchange Capacity (CEC). Assessment of soil content on land from each section is shown in table 2.

Types of rice fields located in the village of Sruwen is grumusol with soil texture structure Silt loam. The content of soil pH of upper and lower rice field is neutral and somewhat acidic in middle soil. Neutral pH in the soil up and down is because the fields tend to be flooded continuously (flooded soil) in rice cultivation and near the water source irrigation.

Table 1. Location, coordinates, and height of rice fields in Sruwen Village

Code	Coordinate	Altitude (m dpl)
Upper rice field	S.07°25'18.35" E. 110°32'25.41"	696.0
Middle rice field	S.07°25'17.41" E. 110°32'31.35"	690.0
Bottom rice field	S.07°25'15.75" E. 110°32'30.54"	688.0

Table 2. Properties of rice fields at Sruwen Village, Tengaran District, Semarang Regency

Parameter	Upper Soil	Middle Soil	Bottom Soil
Texture			
- Sand (%)	23.30	19.13	22.94
- Silt (%)	58.40	61.58	53.86
- Clay (%)	18.30	19.30	23.20
pH H_2O	6.64	6.21	6.95
pH KCl	5.28	4.94	5.52
C-Organic (%)	3.51	3.02	2.71
N-Kjeldahl (%)	0.43	0.33	0.32
P_2O_5 provided Olsen (ppm)	42.66	43.73	36.24
P_2O_5 HCl 25 % (mg/100g)	128.59	109.45	105.44
K_2O HCl 25 % (mg/100g)	45.05	38.65	39.41
CEC (cmol(+)kg^{-1})	16.00	19.77	16.38
Exchangable Cation (cmol (+) kg^{-1})			
K	0.04	0.04	0.04
Na	0.05	0.04	0.04
Ca	1.17	0.85	1.16
Mg	0.29	0.20	0.33

The content of C-organic as an indicator of organic material in the lower rice field is medium. The content of optimal soil organic matter is closely related to clay content and soil pH. The content of N of the three parts of paddy soil has medium value. While the content of Phosfor (P2O5) is very high. The CEC value in upper and lower rice field is low but middle rice field is medium.

3.2 Rice Field Management

Farmers as the main actors in the management of rice fields and local experience until today is still the basis of farming so less attention is given to environmental sustainability. The important role of government agencies is indispensable in the management of natural resources, especially agricultural and technological innovation. The use of fertilizer and pesticides in wetland management is strongly influenced also from the economic level. Utilization of water resources that have always been available to encourage farmers to make rice fields always produce. This is as one of the triggers of farmers using inorganic fertilizers in addition to organic fertilizer.

Differences in management, land ownership, source of irrigation and the same goal that the achievement of maximum productivity impact on differences in heavy metal content on paddy fields from each farmer. Management differences are also strongly influenced by the socio-economic conditions of the farmers themselves. Economically among them is the arrangement of cropping pattern within a year. The high cost of wetland management especially in the purchase of fertilizers and pesticides has an effect on work in a year even though most farmers are not hired. The rice field located on the land refers to the cropping pattern of paddy-*palawija*-paddy in a year. The most commonly used fertilizer in the terraced rice terraces are inorganic fertilizers in the form of urea, phonska, and manure for rice plants. Crops are often cultivated intercropping between sweet corn and mustard, some are just corn. The fertilizer used is urea fertilizer, TSP and organic fertilizer in the form of manure. While organic fertilizer comes from cattle dung waste. Therefore, the amount of use of fertilizers and pesticides both organic and inorganic may be influential on the amount of heavy metal content on paddy fields.

Based on the ownership, the farmers of the upper and lower rice fields are owner farmers, while those of the middle rice field are working farmers. Differences in ownership is very influential on the success rate of farming. One of the most influencing social factors is the lack of care for the impact of land damage. Environmental sustainability factors are often unthinkable due to lack of awareness and understanding of farmers. The availability of existing water sources does not significantly affect the level of land management as it is available throughout the growing season. The presence of heavy metals in paddy fields an be derived from irrigation water flowing from the upper rice field to the rice fields below it. Other factors may also be possible from water seepage after fertilization from the upper rice field. Accumulation also can be due to the flow of water, especially the soil layer (top soil). In addition to human activities that manage the agricultural environment, the climate factor is very influential including rainfall, wind direction, humidity, sunlight. The higher the rainfall, the higher the eroded soil layer.

Location of the altitude aligned with the direction of water flow, this is one of the triggers of the increase of heavy metal content. Conservation technology which is still needed on terraced land is irrigation network improvement because it is still uneven even though the water requirement is fulfilled. Another technology that still needs to be extracted is by the exploitation of phytoremediation plants that grow a lot on any watery rice field that is still regarded as a weed by farmers. The plant is azolla that can act as a heavy metal absorber plant. Azolla is able to absorb several types of heavy metals so that it has potential aSa waste phitoabsorber containing heavy metals [8]

3.3 The Metal Content in Soil of Rice Field

Soil is a receptor of most of the pollutants of human activity. Degradation of soil quality can be caused by the presence of inorganic pollutants such as metals. The problem that actually needs to be explored is the monitoring of the level of environmental damage, especially in the rice fields that are quantitatively through laboratory testing. The use of excessive inorganic fertilizers in the long term as substitution of fertilizer on the soil that is used up in plant growth will accumulate. Consequently, pollution, although still below the threshold according to the standards of different heavy metal thresholds in each country. The standard and limits of heavy metal content are found in Table 3.

Table 3. Standards and Limits of Heavy Metal Content of Pb, Cd, dan Cu in Soil

Location	Heavy Metal (mg/kg)		
	Pb	**Cd**	**Cu**
EU Standard	300	3	140
Indian Standard	250-500	3-6	135-270
Normal Limit *	2-300	0,01-2,0	2-250
Critical Limit *	100-400	3-8	60-125

*Source: Alloway, 1995

Heavy metal is a substance that is so persistent that it can accumulate in the food chain and cause accumulated fects on humans. According to the direction of the slope or height, the upper rice field contains the highest heavy metals of both Pb, Cd, and Cu compared to lower paddy soils with different depths. The bigger the depth of soil, the less the content of the heavy metal.

3.4 The Content of Heavy Metal Pb

The heavy metal of Pb is found at all altitude of paddy field both at 0-20 cm and 20-40 cm depth which shows different data. The heavy metal content of Pb is still below the standard provisions of India (250-500 mg / kg), Europe (300 mg / kg), and the critical limits in the soil are 100-400 mg / kg, but fall within the normal range of 2-300 mg / kg [9,10,11]. The heavy metal content of Pb is possible from pesticide spraying by accumulated farmers and vehicle fumes close to the village road. The distribution of Pb in the soil is correlated with the depth that is decreased according to the depth caused by the formation of the Pb bond with the organic soil commonly found in the topsoil [2]. The contamination content of heavy metal Pb on the ground based on the height is shown in Figure 1.

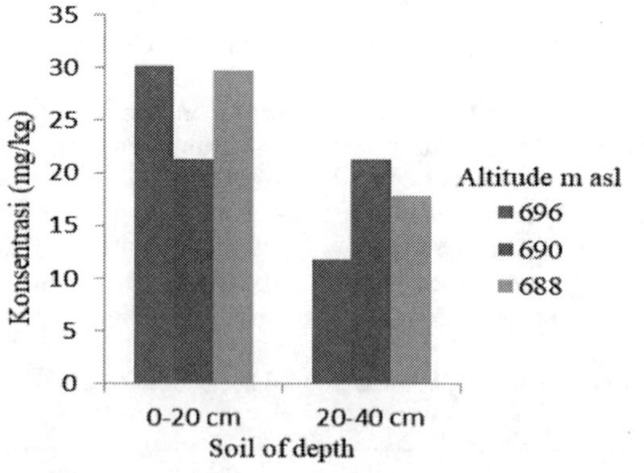

Fig 1. The content of heavy metal Pb at Semarang Regency

Based on the results of testing on paddy soil with an altitude of 696,0 m above sea level, heavy metal content of Pb is higher than the height of 688,0 m above sea level. The content also decreases when the height decreases. This is due to the location of the upper rice fields adjacent to the river as the first irrigation source and water recipient in irrigation containing Pb heavy metals although similar in ownership and management of the land. The decrease in heavy metal content is correlated with the depth of soil that allows heavy metals to be absorbed in plants growing on it. While in the middle rice field is increasingly rising. The increase is possible because of differences in management such as the use of fertilizers and pesticides so that heavy metals which are generally common in the layer if they are finally dissolved in the deeper soil.

3.1. The Content of Heavy Metal Cd

The heavy metal Cds are present in all of the upper, middle and lower paddy fields at 0-20 cm and 20-40 cm depths which are all still below the standard of India (3-6 mg / kg), Europe (3 mg/kg) And the critical limit in soil is 3-8 mg / kg, but above the normal limit in soil is 0.01-2.00 mg / kg [9,10,11]. Manure that is commonly used is fertilizer derived from cow dung. The presence of heavy metal content of Cd in the soil at all altitudes because during this time farmers still use phosphate-based chemical fertilizers is TSP and phonska fertilizers.

Fertilizer TSP that is supplied excessively in rice fields in the long term will be accumulated and cause pollution because TSP fertilizer contains heavy metals. Heavy metals are also present in natural phosphate rocks used as raw material for the manufacture of fertilizer P [12]. The addition of fertilizer will increase the concentration of P in the soil [2]. The presence of heavy metal content of Cd is also possible from the rest of the farm although the amount is not much because the phosphate compound is bound by the soil and cannot be carried away by the irrigation water flow. Another factor is possible because the application of some pesticides in the form of fungicides containing heavy metals Cd and Cu. The content of contamination in the form of heavy metal Cd on the ground based on altitude is shown in figure 2.

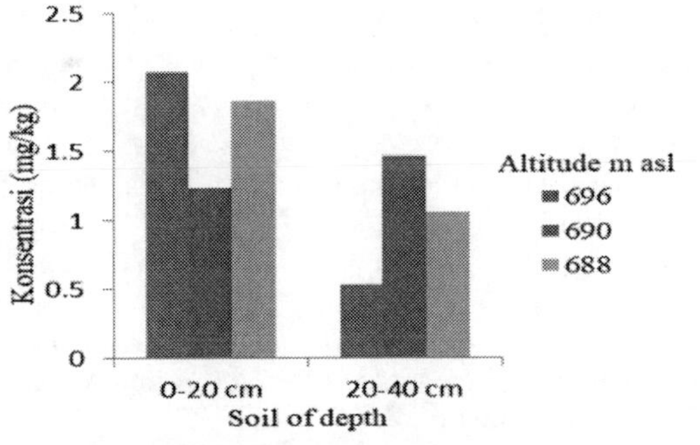

Fig 2. The Content of Heavy Metal Cd at, Semarang Regency

Variations of heavy metal content Cd decrease based on height and depth of soil depth in upper and lower rice field. While in the middle rice field content of Cd tends to increase. The high metal weight is not only from the use of chemical fertilizers and pesticides but can also be from manure containing heavy metals. Feed sources containing heavy metals will produce fertilizers containing heavy metals as well.

3.3. The Content of Heavy Metal Cu

Provision of chemical fertilizers supply certain nutrients in the form of high concentration of inorganic compounds and easily soluble so it can cause contamination of waterways, especially ground water if done repeatedly [13]. Higher metal content, other than the natural factor of other factors as a supplier of heavy metals in these fields is the input of fertilizer used before and during the accumulated planting. Similarly, the use of some insecticides contains heavy metals Pb.

Cu heavy metals are present at all altitudes of 0-20 cm and 20-40 cm deep, all still below the standard provisions of India (135-270 mg / kg), Europe (140 mg / kg), and the critical limits in soil are 60 -125 mg / kg, but still within the normal range in the soil ie 2-250 mg / kg [9,10,11]. The heavy metal content of Cu on the soil is shown in Table 3.

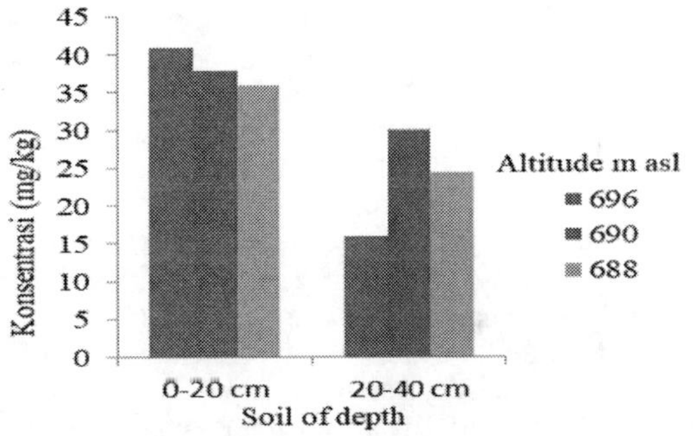

Fig 3. The Content of Heavy Metal Cu at Semarang Regency

Cu elements are present only in soils whose existence is not supported by irrigation water sources because they are undetectable based on the metal test results. One factor contributing Cu heavy metals to wetland is not only from pesticide applications but can be derived from natural mineral erosion events. The heavy metal content in the soil is still below the set threshold. Cu heavy metal content based on both height and depth is correlated. The lower the height and depth, the lower the heavy metal Cu content is. This can be because the metal content does not dissolve in the lower altitudes and is only found in the soil though it accumulates at 40 cm of soil depth. Sources of heavy metals found at the site not only from the supply of fertilizers and natural factors, but water sources that have been used in irrigation. The content of heavy metals in water is shown in table 3.

Table 4. The results of irrigating water testing

Example Code	Heavy Metal (ppm)		
	Pb	**Cd**	**Cu**
Water irrigation	0,0148	nd	nd
Value of LoD	0,0024	0,0007	0,0009
Value of LoQ	0,0080	0,0023	0,0030

Where:
LoQ : Limit of Quantification: The lowest measurable heavy metals concentration by the HClO4 + HNO3, AAS method).
Nd : Not detected or lower than the value of LoQ
LoD : Limit of Detection: The lowest measurable heavy metal concentrations measured by the instrument (AAS).

Based on the results of the irrigation water sample analysis, only the heavy metal content of Pb was detected. Although the water looks clear but household waste is still washed away. This is possible as one source of pollutants due to human activities in land management, especially the use of fertilizers and chemical pesticides. In addition to these effects, natural factors in the water source can be derived from a scientific weathering process [14]. The result of excessive fertilization is pollution carried by runoff into rivers [15].

Water condition is an indicator of spatial area both from the problem of the intensity of the causes of floods and droughts or the quality of which contains pollutants or sediments [16]. Use of land that is less concerned with the sustainability and health of river water can increase the pollutant content into river water bodies [17]. Related to this effort of water quality monitoring needs to be done so that the accumulative can be prevented. Management of water quality and control of water pollution to conserve water function is contained in Government Regulation of Republic of Indonesia No. 82 Year 2001. While water that can be used for agriculture purposes contained in Government Regulation No. 20 of 1990 on Water Pollution Control implemented in 1990-2001. The function of water sources that flow along the rice fields can be classified into class two, three, and four because in addition to irrigation of crops used can also in the cultivation of fisheries. Although the water source can serve to irrigate crops but in terms of water quality still contains inorganic chemical pulp that is Pb contamination. Heavy metal of 1 mg / l still under the inorganic chemical parameters of water quality criterion based on class IV as water irrigation [18]

Not all heavy metal content is present in irrigation water sources found in irrigation canals and inlet channels of each rice field. These pollutants can come from fertilizer and pesticide applications that flow from the paddy fields of water drainage. Innovative technological efforts are required to prepare irrigation systems and water resources management that meet the future economic and agricultural development needs [19]. Reducing the accumulative heavy metal residues such as the utilization of bio accumulator plants such as water hyacinth and *kiambang* which are easily available on certain irrigation channels.

4. Conclusion

The content of heavy metal of Pb, Cd, and Cu on all land with different depth and altitude is still below the established threshold. The presence of heavy metal content is also because the differences in the management of rice fields. Lack of understanding of farmers makes they still apply conventional cultivation with chemical input which is not appropriate with local specific recommendation.

5. suggestion

Environmental friendly technology needs to be implemented to minimize the contamination by utilizing organic resources such as agricultural and livestock waste as well as human resources for the development and empowerment of the community for acceptable technology, and the utilization of azzola.

References

1. Y-B. Guo, H. Feng, C. Chen, C-J. Jia, F. Xiong, Y. Lu. 2013. *Heavy metal concentrations in soil and agricultural products near an industrial district*. Pol. J. Environ. Stud. Vol. **22**, No. 5.1357-1362 (2013)
2. S. Notodarmojo. *Pencemaran Tanah dan Air Tanah*. Institut Teknologi Bandung. Bandung (2005)
3. Y. Hindarwati, F.D. Arianti, and H. Anwar. *Identifikasi logam berat Cd dalam tanah dan beras di lokasi pengembangan padi organik di Kabupaten Semarang*. Prosiding seminar nasional. Faperta UGM. Yogyakarta (2014)
4. Markovic, M., S. Cupac, R. Durovic, J. Milinovic, and Kljajic. *Assesment of heavy metal and pesticide levels in soil and plant products from agricultural area of Belgrade*, Serbia. Arch Environ Contam Toxicol 58:341-351 (2010)
5. Eviati and Sulaeman. Petunjuk Teknis Edisi 2. *Analisis kimia tanah, tanaman, air, dan pupuk*. Badan Penelitian dan Pengembangan Pertanian. Kementerian Pertanian (2012)
6. Varian. Analytical Methods: Flame Atomic Absorption Spectrometry. Varian Australia. Victoria (1989)
7. Badan Pusat Statistik. Kabupaten Semarang dalam angka (2016)
8. Z. Stêpniewska, R.P. Bennicelli, T.I. Balakhnina, K. Szajnocha, A. Banach, and A. Wolińska. *Potential of Azolla caroliniana for the removal of Pb and Cd from wastewaters*. Int. Agrophysics, **19**, 251-255 (2005)
9. S.K. Awashthi. Prevention of food adulteration act no.37 of 1954. Central and state rules as amended

for 1999. Ashoka Law House. New Delhi. India (2000)

10. European Union. *Heavy metals in wastes in wastes*. European commission on environment.http://ec.europa.eu/environment/waste/studies/pdf/heavy metalsreport.pdf (2002)
11. B.J. Alloway. *Heavy Metal in Soils*. Blackie Academic and Profesional (1995)
12. U. Kurnia, H.Suganda, R. Saraswati, and Nurjaya. *Teknologi pengendalian lahan sawah tercemar*. Balai Penelitian Tanah. http://balittanah.litbang.pertanian.go.id/eng/dokumentasi/buku/tanahsawah/tanahsawah9.pdf. Accessed on 5 September 2016
13. R. Dharmawan Margolang, Jamilah, M. Sembiring. Karakteristik beberapa sifat fisik, kimia, dan biologi tanah pada sistem pertanian organik. Jurnal online agroekoteknologi. ISSN No. 2337- 6597 Vol.**3**, No.2 : 717 – 723 (2015)
14. R.M. Mulia. *Kesehatan lingkungan*. Graha Science. Jakarta (2005)
15. H.R. Mulyanto. *Ilmu lingkungan*. Graha Science. Jakarta (2006)
16. A. T. Sastrawijaya. *Pencemaran lingkungan*. Rineka Cipta. Jakarta (2009)
17. A.B. Supangat. *Pengaruh berbagai penggunaan lahan terhadap kualitas air sungai di kawasan hutan pinus di Gombong, Kebumen, Jawa Tengah*. Jurnal penelitian hutan dan konservasi alam.Vol. V. No. **3** : 267-276, (2008)
18. Peraturan pemerintah No. 82/2001 tentang pengelolaan kualitas air dan pengendalian pencemaran air (2001)
19. E. Pasandaran. Reformasi irigasi dalam rangka pengelolaan terpadu sumberdaya air. Analisis kebijakan pertanian. E-jurnal Litbang Pertanian. Vol. **3** No. 3, September 2005 : 217-235 (2005)

Integration of Cleaner Production and Waste Water Treatment on Tofu Small Industry for Biogas Production using AnSBR Reactor

Suparni Setyowati Rahayu[1, 2,*], *Budiyono* Budiyono[2, 3], *Purwanto* Purwanto[2, 3]

1Mechanical Engineering Department, State Polytechnic of Semarang, Indonesia
[2]Doctoral Program of Environmental Science, School of Postgraduate Studies, Diponegoro University, Semarang
[3]Department of Chemical Engineering, Faculty of Engineering, Diponegoro University, Semarang

Abstract. A research on developing a system that integrates clean production and waste water treatment for biogas production in tofu small industry has been conducted. In this research, tofu waste water was turned into biogas using an AnSBR reactor. Mud from the sewage system serves as the inoculums. This research involved: (1) workshop; (2) supervising; (3) technical meeting; (4) network meeting, and (5) technical application. Implementation of clean production integrated with waste water treatment reduced the amount of waste water to be treated in a treatment plant. This means less cost for construction and operation of waste water treatment plants, as inherent limitations associated with such plants like lack of fund, limited area, and technological issues are inevitable. Implementation of clean production prior to waste water treatment reduces pollution figures down to certain levels that limitations in waste water treatment plants can be covered. Results show that biogas in 16 days HRT in an AnSBR reactor contains CH_4 (78.26 %) and CO_2 (20.16 %). Meanwhile, treatments using a conventional bio-digester result in biogas with 72.16 % CH_4 and 18.12 % CO_2. Hence, biogas efficiency for the AnSBR system is 2.14 times greater than that of a conventional bio-digester.

1 Introduction

Environmental management can be carried out by prevention and control of pollutants. Environmental management in a cluster of tofu small industry should integrate both aspects. Measures to control pollution usually come with significant cost for equipment and operation that they are often neglected, but these measures become a must when pollutant sources have been identified. On the other hand, measures to prevent pollution are generally both effective and beneficial whenever they are possible to be conducted. Nonetheless, possible effects from pollution are still likely to occur. Therefore, environmental management in a cluster of tofu small industry must be initiated with prevention measures that are integrated with measures to control pollution [1].

A tofu small industry cluster can prevent environmental pollution by, among other things; re-use its waste water as to avoid polluting the water ways. Even though it is not a panacea of all water pollution problems from tofu industry, it helps minimizes those problems. Therefore, the cost and burden for waste water treatment will be reduced.

Environmental management must be carried out sustainably and this is only possible when related measures are easy, cheap, and even beneficial. Hence, measures for environmental management such as the clean production approach is very much applicable in a small industry cluster. A clean production approach that is integrated with pollution control attempts like waste water treatment should be the best approach for environmental management in small industries.

Environmental management has always been identical with being costly and not directly beneficial. This leads to the need for an alternative measure in the form of clean production, which has some benefits [2]. Clean production that is integrated with waste water treatment will reduce the load of waste water to be treated at a treatment plant. In turn, this load reduction reduces the budget for construction and operation of a waste water treatment plant [3]. Some limitations are associated with the ability of a plant to treat waste water, for instance, fund, land acreage, and technology. The practice of clean production prior to waste water treatment should reduce pollution load down to certain levels that those aforementioned limitations can be minimized [4].

* Corresponding author: suparnirahayu@yahoo.co.id

Both physical and chemical treatments are meant to take care of the solids suspended. Meanwhile, anaerobic biological treatment (AnSBR) is an interesting alternative to treat organic compounds in tofu waste industry. Almost all developments in the technology of anaerobic treatment are aimed at containing microorganism in the reactor other than improving the efficiency of biogas formation [5].

Biogas production technology is currently less developed due to slow rate and less efficient production, especially in terms of clean production and waste water treatment in small tofu industry. The problem to be solved by this research is lack of environmental management strategy, and the lagging implementation of clean production in small tofu industry [6].

Based on those considerations, observations and analyses on the effective integration of clean production and waste water treatment in tofu small industry for biogas production were carried out. Further analyses on eco-efficiency of water and energy were also conducted. These are necessary, as achievement and effectivity of waste water treatment in small tofu industry for biogas, and the implementation of clean production, as well as eco-efficiency of energy and water is important information for tofu business people and stakeholders, and especially the policy makers [7].

2 Experimental Details

Measures of environmental management involving the small industry can be carried out individually by all business entity or together by a cluster [8]. Individual efforts by a business entity mean that both cost and resources required are from that business entity itself. No communication and cooperation among different business entities are required to do environmental management [9]. This type of effort is usually laden with challenges especially concerning resources, both fund and human resource [10].

On the other hand, a cooperative effort in by a cluster is conducted by either all or some business entities in a cluster. Therefore, both cost and resources required are shouldered by all parties involved. This cluster approach has some advantages compared to the individual one [11].

Integrated approach of clean production and waste water treatment is in essence: a synergy between attempts to improve economic efficiency whilst ensuring environmental well-being (more efficient use of natural resources with less waste and pollution), and efforts for final waste water treatment. These efforts are intertwined and support each other in a cycle of planning, monitoring, and evaluation in a sustainable way [12].

3 Materials and Methods

Research activities conducted include five (5) main works: (1) workshop; (2) supervising; (3) technical meeting; (4) network meeting; and (5) technical application.

The Workshop discussed materials on the basic principles of clean production and eco-efficiency, non-product output (NPO), inefficient production process, identification of sources for NPO, and ways to calculate NPO cost.

The Supervising is meant to train small industries to formulate their own NPO flow chart, calculate NPO cost, and identify NPO priority based on predicted NPO cost.

The Technical Meeting was aimed at informing the activities of the working team in each cluster and taking input or suggestion from the other clusters, for the improvement and success of activities in each cluster.

The Networking Meeting was directed at informing and synergizing activities that have been and will be done by stakeholders and local government units.

The Technical Application dealt with applying processes in the AnSBR system that consist of five stages; filling, reacting, decanting/drawing, and stabilization/idling. Cycle time for the AnSBR system is around 3 - 24 hours, depending on the characteristics and purposes of treatment. AnSBR system can be modified to accommodate carbon oxidation, nitrification, de-nitrification, and phosphor elimination. Prior to operation, AnSBR system underwent flushing using 95.95% nitrogen (Industrial *grade*) for 30 minutes to get rid of oxygen [13].

Experiment was carried out using AnSBR reactor that is filled with 20 % ditch mud, while the conventional digester contains yoghurt plastic bottles and bamboo plaid for bacteria proliferation. HRT treatment was done for 10, 20, 30, and 40 days. AnSBR operated in five stages of filling (25%), reacting (35%), precipitation (20 %), decanting (15 %), and stabilizing (5%), of cycle operation. These processes are conducted every day for 12 hours each [14].

4 Results and Discussion

An integrated approach is possible when the whole process of planning for clean production and waste water treatment are carried out, and each is taken into proper consideration, and each process also provide feedback for each other [15]. Implementation of clean production that was integrated with waste water treatment was capable of reducing the load of waste to be treated in the plant. In turn, this load reduction reduced the cost of construction and operation of a waste water treatment plant. Some limitations are associated with the ability of a plant to treat waste water, for instance, fund, land, and technology. The practice of Clean Production prior to waste water treatment could reduce pollution load down to certain levels that those aforementioned limitations can be minimized [16].

During the stage of technical application, research was run for 24 days to produce biogas that was then stored in a storing tank. Biogas was measured for each HRT, starting at HRT 2 (day 2), up to HRT 24 (day 24). On the 10th day, 5815 mL of biogas was obtained, and this figure kept on growing through day 24. This research tried to figure out and compare the amount of biogas produced using AnSBR system, and the conventional bio-digester.

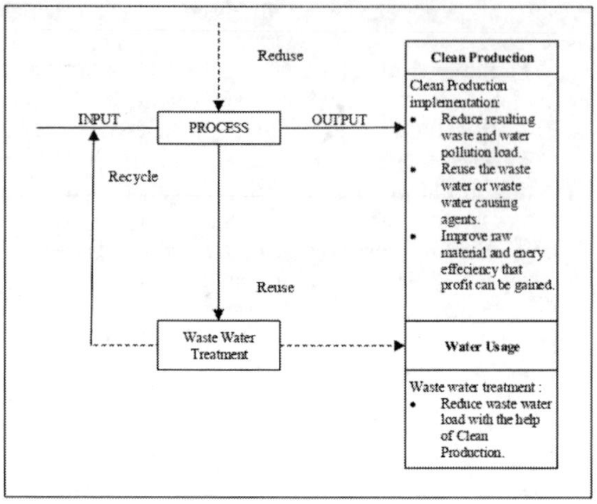

Fig. 1. Clean production that is integrated with waste water treatment.

It is also targeted at finding out the effect of HRT holding time on AnSBR system operated by batches to produce biogas. AnSBR that is operated using tofu waste water and ditch mud as input involving stages of filling, reacting, precipitation, decanting, and stabilization is known as an energy-independent AnSBR, whereas AnSBR that only operates using waste water is called a bio-digester. In this system, microbes are proliferated using bamboo plaid and yoghurt plastic bottles inside it. The effect of HRT on biogas production biogas production was analyzed by setting HRT for the 10th, 20th, 30th, and 40th day. The ratio of input between tofu waste water and ditch much is 5: 1. The AnSBR was operated at a temperature and a maximum mud concentration of 30 °C and 20 %, respectively. Results of analyzes on the effect of HRT on biogas production are given in Table 1.

Table 1. The effect of HRT on the production of biogas.

HRT (day)	Biogas Production (mL)	
	AnSBR	Conventional Bio-digester
10	5720	1845
20	12435	5374
30	12972	6577
40	13625	7116

Figure 2 shows that high HRT yields high biogas as well. On HRT 10, 20, 30, and 40, AnSBR system had biogas output of 5720, 12435, 12972, and 13625 mL, respectively. Meanwhile, the conventional digester produced biogas output of 1845, 5374, 6577, and 7116 mL, on the same HRT respectively. On average, the ratio of biogas production between AnSBR reactor and conventional bio-digester is 44752 mL against 7116 mL. Hence, AnSBR reactor produces 2.14 times more biogas than conventional bio-digester.

During the running period, characteristics of biogas were monitored to the 16th day. This examination included contents of CH_4, CO_2, and its heat value. Results show that CH_4 content in the biogas was 78.26 %, while the CO_2 content was 20.16 %.

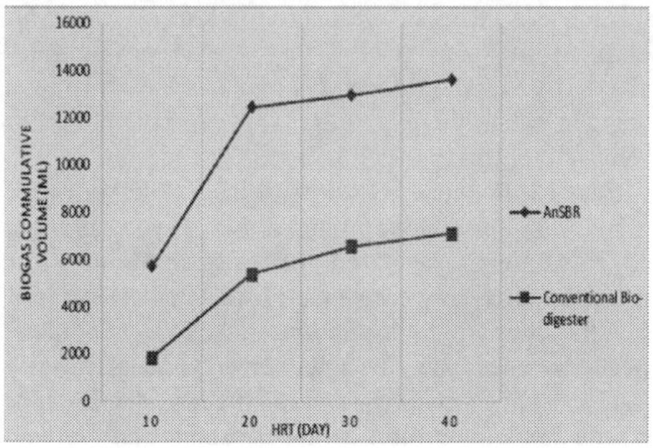

Fig. 2. The effect of HRT on biogas production.

5 Conclusions

Implementation of clean production that is integrated with waste water treatment does reduce the waste load to be treated in a treatment plant. This reduction in waste load reduces the cost for construction and operation of a treatment plan. At times, a waste water treatment plant's capacity to treat waste water is compromised by, among others: lack of fund, land acreage, and technology. Implementation of clean production prior to treating waste water should be able to reduce the waste load to certain points that the inherent lacks of a treatment plant can be minimized.

Biogas on HRT 16 has a CH_4 and CO_2 content of 78.26 % and 20.16 %, respectively. They were produced by an AnSBR reactor with a batch operation that includes filling, reaction, precipitation, decanting, and stabilization. On the other hand, the conventional bio-digester yielded biogas with a composition of 72.16 % CH_4 and 18.12 % CO_2. Therefore, biogas production efficiency for AnSBR reactor is 2.14 times greater than that of the conventional bio-digester.

Acknowledgement

The writers would like to thank all parties contributing to this research, especially DRPM and the tofu industry in Pesalakan, Adiwerna sub-district, Tegal regency.

References

1. Azimi, A. A. and Zamanzadeh, M... Int. J. Environ. Sci. Tech. **I**, 1, pp. 51-57 (2014)

2. Balsam, J., Burke, D.A.2011. *Dairy Waste Anaerobic Digestion Handbook*, Environmental Energy Company, Hill Street Olympia, pp. 17-27 (2016)
3. Borja, Caudia ET, Magali C Carmarota, Bio. Tech. **81**(1), pp. 61 – 69 (2008)
4. Carucci,G.,F.Carrasco,K.Trifoni,M.Majone, and M.Beccari. J. Environ. Eng. **131** 7, pp. 1037-1045 (2015)
5. Demirer, G..N., and S. Chen, Jou. Chem. Tech. and Bio. **79**, pp. 1381-1387 (2014)
6. El-Gohary, Fatma A., Hala M. El-Kamah, and Sabah A. Badr., Australian Jou. Basic and App. Sci., **2** (4), pp. 1345-1355 (2008)
7. El-Sheikh, Mahmoud A., Hazem I. Saleh, Joeseph R. Flora, Mahmoud R. AbdEl-Ghany., Desalination. **276**, pp. 253 -259 (2011)
8. G. Nakla, A. Lugowski, J. Patel, Bio. Tech., **97**, 1, pp. 1-14 (2011)
9. Henze, M. and Ledin A. *Types, Characteristics and Quanities of Classic Combined Wastewaters* ,in P. Lens. G. Zeeman, G Lettinga. *Decentralized Sanitation and Reuse,* IWA Publishing, UK., pp. 57-72 (2012)
10. M. Merzouki, N Bernet, JP Delgenes, Bio. Tech. **96**, 12, pp. 1317 – 1322 (2005)
11. Parawira,W., J.S.Read,B. Mattiasson, and L. Bjornsson, Bio.and Bio., **32,** pp. 44—50 (2013)
12. Purwanto, *Penerapan Teknologi Produksi Bersih untuk Meningkatkan Efisiensi dan Mencegah Pencemaran Industri*, Universitas Diponegoro pub. (2009)
13. Rajeshwari, K.V., M.Balakrishnan, A. Kansal, K. Lata, and V.V.N.Kishore, Ren. and Sus. En. Rev., 2, pp. 135-156 (2013)
14. Sarapatka,B., Bio. and Bio., **5,** 5, pp.387-393 (2013)
15. Syafila, Mindriany, A. H. Djajadiningrat, and M. Handajani, PROC. ITB Sains & Tek., **35** A, 1, pp. 19-31 (2003)
16. Yasar, Abdullah, and Tabinda, A. Bari, Polish J. of Env. Study **19**, 5, pp.1051-1061 (2010)

Preparation and Characterization of NiMo/Al_2O_3Catalyst for Hydrocracking Processing

Aditya Widiyadi[2], *Gema* Adil Guspiani[2], *Jeffry* Riady[2], *Rikky* Andreanto[2], *Safina* Dea Chaiunnisa[2], and *Widayat* Widayat[1,2*]

[1]Laboratory of Advance Material, Central Laboratory for Research and Service, Diponegoro University, Semarang - Indonesia
[2]Departement of Chemical Engineering, Faculty of Engineering, Diponegoro University, Semarang - Indonesia.

Abstract. Hydrocracking is a chemical process used in petroleum refineries for converting high boiling hydrocarbons in petroleum crude oils to more valuable lower boiling products such as gasoline, kerosene, and diesel oil that operate at high temperature and pressure. Catalyst was used in hydrocracking to reduce temperature and pressure. Hydrocracking catalyst are composed of active components and support. Alumina is widely used in hydrocracking process as catalyst support due to its high surface area, high thermal stability, and low prices. The objective of this research was preparated NiMo/Al_2O_3 catalyst that used as hydrocracking catalyst. Catalyst was synthesized by wetness impregnation method and simple heating method with various kind of Al_2O_3. The physicochemical properties of catalyst were investigated by X-ray diffraction (XRD) to determine type of crystal and scanning electron microscopy (SEM) to determine morphology of the catalyst. The NiMo/Al_2O_3 catalyst prepared by aluminium potassium sulfate dodecahydrate exhibited the highest crystallinity of 90.23% and it is clear that MoO_3 and NiO crystallites are highly dispersed on the NiMo/Al_2O_3 catalyst which indicates as the best catalyst. The catalytic activity in hydrocracking process was successfully examined to convert fatty acid into hydrocarbon.

1 Introduction

Energy needs in Indonesia keeps increase from year to year. This occurred in all sectors which include industrial, transportation, commercial, and household sector. The transportation sector is one of the largest sector in consuming energy compared to other sector that were still reliant on fossil fuel [1] Gasoline, diesel, and kerosene was the most widely used in this sector. The growing demand of gasoline, diesel, and kerosene continues to increase progressively in the recent years. Many efforts have been conducted to overcome the crisis energy in the transportation sector. Biofuels as a predominant source of energy from biomass may offer a solution to this problem [2]. Biofuel as alternative energy source have many benefits with respect to environmental concerns [3].

Plant oils cannot be used directly as fuel due to several drawbacks. This problem can be resolve by using some processes such as pyrolysis and hydrotreatment. Although both processes produce hydrocarbon as main product, there exits significant differences between both process. Pyrolysis is less complicated process. However, pyrolysis produces organic acids short chain cracking and aromatic products rather than pure hydrocarbon. Hydroprocessing is much different process from pyrolysis, it can remove all oxygen content from the plant oils and produce pure low boiling point hydrocarbon [4]. Hydroprocessing consist hydrotreating and hydrocracking processes that needs hydrogen gas to convert fatty acid directly to hydrocarbons and water. Hydrotreatment offers a simpler and better option because it can be used in the existing refinery and produce hydrocarbons that similar to hydrocarbons produced from petroleum [5].

Catalyst is needed to reduce the temperature and pressure of this process. Many studies have been conducted on catalyst used in hydroprocessing of vegetable oil. Catalyst support like alumina, zeolite, and activated carbon were used for hydrocracking reactions. All this support were impregnated with active metals like Pd, Pt, NiMo, and CoMo for hydrogenation/dehydrogenation [6-9]. Catalyst which has highly porous and thermostable is not only able to disperse the metal but also can increase its thermal stability. The use of support can enhance the activity of catalyst due to its higher surface area and thermal stability [10].

Simple heating method can increase the surface area of the catalyst. In simple heating method, high molecular weight polymer is needed as a continous media [11]. Mixed solution of metal and continous

*Corresponding author: yayat_99@yahoo.com

media were stirred and heated to evaporate water in the mixed solution. This continuous media must be remained until the end of process because it can avoid agglomeration of catalyst particle. Finally, the continuous media will be vaporized get nano particle catalyst by heating the solution above the polymer decomposition temperature [12]. Support like alumina is widely used in hydro processing due to its high surface area, high thermal stability, and low prices [13]. However, the pore size distributions are not efficient for catalytic reaction.

In this study, NiMo/Al_2O_3catalyst was prepared with various kind of alumina by simple heating and wetness impregnation methods. In order to know the physicochemical properties of the catalyst with various kinds of alumina, the NiMo/Al_2O_3 catalyst were characterized by several techniques including scanning electron microscopy (SEM) to determine morphology and composition of catalyst and x-ray diffraction (XRD) to determine crystal size and type of crystal.

2 Experimental

2.1 Materials

Aluminium potassium sulfate dodecahydrate ($KAl(SO_4)_2.12H_2O$), aluminium nitrate nonahydrate ($Al(NO_3)_3.9H_2O$, Merck), aluminium hydroxide ($Al(OH)_3.xH_2O$, Merck), alumina (Al_2O_3, PT. Inalum), commercial gamma-alumina (γ-Al_2O_3), nickel(II) nitrate hexahydrate ($Ni(NO_3)_2.6H_2O$, Merck), ammonium heptamolybdatetetrahydrate ($(NH_4)_6Mo_7O_{24}.4H_2O$, Merck), and poly ethylene glycol (PEG-4000).

2.2 Catalyst Preparation

The Al_2O_3 support for NiMo/Al2O3 catalyst were prepared by various kind of Al_2O_3 (A-E) using (A) aluminium potassium sulfate dodecahydrate ($KAl(SO_4)_2.12H_2O$), (B) aluminium nitrate nonahydrate ($Al(NO_3)_3.9H_2O$), (C) aluminium hydroxide ($Al(OH)_3.xH_2O$), (D) alumina (Al_2O_3), and (E) gamma-alumina (γ-Al_2O_3). The NiMo/Al_2O_3 catalyst (A, B, C) were prepared by simple heating methods and (D,E) were prepared by wetness impregnation method. These are summarized in Table.1. The metal loading of nickel and molybdenum were assigned as 5% wt of NiMo/Al_2O_3 for all catalyst. For simple heating method, ammonium heptamolybdatetetrahydrate ($(NH_4)_6Mo_7O_{24}.4H_2O$) and nickel(II) nitrate hexahydrate ($Ni(NO_3)_2.6H_2O$) were dissolved in aquadest. Then, various kind of Al_2O_3 (A-C) and required amounts of polyethylene glycol was added to the solution with contstant stirring and heating at 100°C for 1 hours to evaporate the water in the solution. The solution was dried at 110°C for 1 hours and calcined at 600°C for 30 minutes. For wetness impregnation method, ammonium heptamolybdatetetrahydrate ($(NH_4)_6Mo_7O_{24}.4H_2O$) and nickel(II) nitrate hexahydrate ($Ni(NO_3)_2.6H_2O$) were dissolved in aquadest. Then, various kind of Al_2O_3 (D,E) was added to the solution with constant stirring for 12 hours. The solution were dried at 110°C for 24 hours and calcined at 600°C for 5 hours.

Table 1. NiMo/Al_2O_3 catalyst prepared from various kind of aluminium compounds

Catalyst	Methods	Al Compound
A	Simple Heating	$KAl(SO_4)_2.12H_2O$
B	Simple Heating	$Al(NO_3)_3.9H_2O$
C	Simple Heating	$Al(OH)_3.xH_2O$
D	Wetness Impregnation	Al_2O_3
E	Wetness Impregnation	γ-Al_2O_3

2.3 Catalyst Characterization

The physicochemical properties of NiMo/Al_2O_3 catalyst were characterized by X-ray diffraction (XRD) to determine type of crystal and crystallinity of the catalyst and scanning electron microscopy (SEM) to determine morphology and composition of the catalyst.

Table 2. NiMo/Al_2O_3 catalyst crystallinity

Catalyst	Crystallinity (%)
A	90.23
B	62.99
C	61.84
D	59.69
E	53.19

2.4 Hydrocracking of Dirty Palm Oil

The hydrocracking process was carried out to check the activity of NiMo/Al_2O_3 catalyst. Dirty palm oil was used as feed for hydrocracking process. Hydrocracking of dirty palm oil was carried out in custom made batch autoclave reactor with an internal diameter of 90 mm and length of 160 mm. 100 gr of dirty palm oil and 1 gr of NiMo/Al_2O_3 catalyst was loaded into the reactor and the H_2 was injected with initial pressure of 3 atm. The reactor was heated to 550°C and the reaction was performed for 1.5 hours. The reactor was cooled and the liquid product was collected for GCMS analysis.

3 Results and Discussion

3.1 XRD Characterization

Fig.1 shows the XRD patterns of NiMo/Al_2O_3 catalyst with various kind of alumina. The characteristic peak position were observed at 2θ. Three strongest peak of NiMo/Al_2O_3 catalyst A (2θ = 22.40°, 31.4°, 33.33°), B (2θ = 23.53°, 26.40°, 27.50°), C (2θ = 23.37°, 25.47°, 28.85°), D (2θ= 23.33°, 26.74°, 66.85°), E (2θ = 22.09°, 23.31°, 26.20°) were obtained.

It can be seen from XRD diffractogram that there is no nickel and molybdenum diffraction peaks found in XRD patterns of catalyst (B,C,D,E) which indicate that nickel and molybdenum species are well dispersed on the support. The peak of catalyst (A) were comparatively weaker than the supports because the

https://doi.org/10.1051/e3sconf/20183103011

material was covered by the active metal component after impregnation and diffraction peaks of NiO and MoO_3 can be clearly seen which indicate the growth of MoO_3 and NiO crystals over the catalyst.

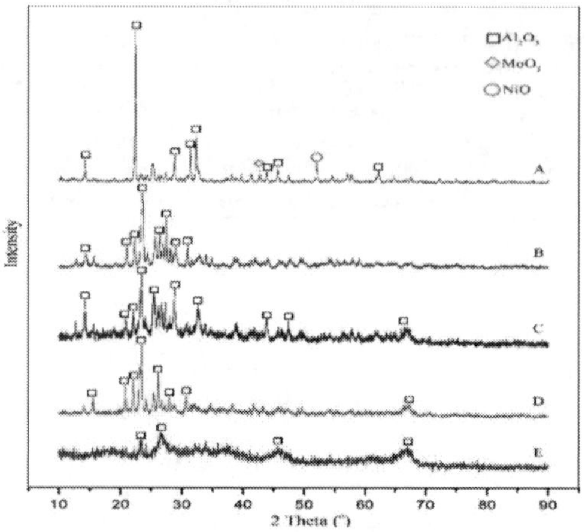

Fig. 1.XRD diffractogram of the $NiMo/Al_2O_3$ catalyst prepared from various kind of aluminium compounds

The intensity of the peak represents the crystallinity of the catalyst [10].Sharper peak means particles tend to get higher crystallinity High crystallinity will affect catalytic activity and stability at higher temperature [12]. Crystallinity of of$NiMo/Al_2O_3$ catalyst with various kind of alumina were summarized at Table. 2. It can be observed that simple heating methods has higher crystallinity than wetness impregnation methods due to the addition of PEG as continuous media avoid the agglomeration of impurities in solution that make the impregnation in simple heating methods free from impurities. Catalyst (A) have the highest crystallinity than the other catalyst because it have the highest acidity in alumina support so the impregnation of catalyst (A) produced the best crystal of $NiMo/Al_2O_3$.

3.2 SEM Characterization

The morphological properties of the catalyst were determined using scanning electron microscopy(SEM). Catalyst $NiMo/Al_2O_3$ (A) have the highest crystallinity

Fig. 2.SEM photographs of the $NiMo/Al_2O_3$ catalyst (A) at (a)3000x, (b)10000x, (c)20000x, and (d)30000x magnification

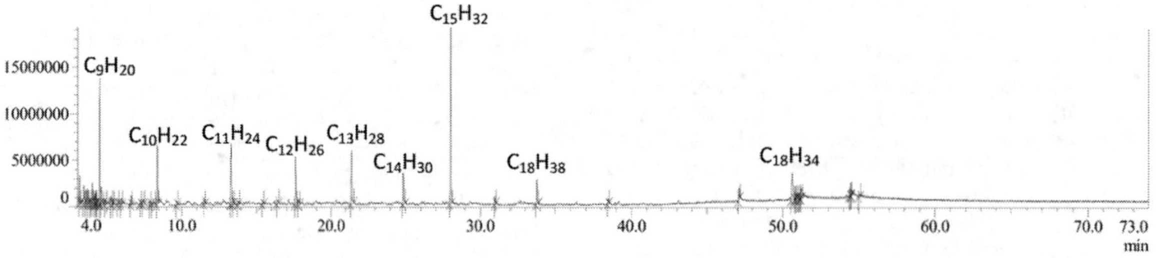

Fig. 3. Gas Chromatograms of hydrocracking product

Table 3. Hydrocracking product

Fraction	Percentage(%)
$<C_5$	7.67
C_6-C_8	8.28
C_9-C_{15}	70.79
C_{16}-C_{19}	8.28
$>C_{19}$	4.97

than the other catalyst (B-E). The SEM micrographs of catalyst $NiMo/Al_2O_3$ (A) from 3000x,10000x, 20000x,and 30000x magnification are displayed in Fig. 2 respectively. Morphology plays an important role in activity and selectivity of the catalyst [10]. The presence porous agglomerates was revealed that this agglomerates are composed of small spherical particles show irregular shapes of alumina which were coated by MoO_3 and NiO fine particle. Coating tended to occur on cracked alumina [14]. In other words, small size alumina particles coated with MoO_3 and NiO havea higher thermal stability and it is clear that MoO_3 and NiO crystallites are highly dispersed on the $NiMo/Al_2O_3$ catalyst.

3.3 Hydrocracking Process

The GCMS analysis of the hydrocracking product shows that the $NiMo/Al_2O_3$catalyst successfully convert fatty acid in dirty palm oil into hydrocarbon. The hydrocarbon yield was ranged from light hydrocarbon ($<C_5$), naphtha (C_6-C_8), kerosene (C_9-C_{15}), diesel (C_{16}-C_{19}), and heavy hydrocarbon ($>C_{19}$).

The dirty palm oil used in this study was composed of mainly C_{16}-C_{18} fatty acids. The hydrocarbon product consists7.67% light hydrocarbon,8.28% naphtha, 70.79% kerosene, 8.28% diesel, and 4.97% heavy hydrocarbon.

4 Conclusion

In this study, $NiMo/Al_2O_3$ catalyst with five kinds alumina supports were prepared by simple heating method using $KAl(SO_4)_2.12H_2O$, $Al(NO_3)_3.9H_2O$, $Al(OH)_3.xH_2O$ and wetness impregnation methods using Al_2O_3 and γ-Al_2O_3 to be alumina resources. The XRD characterization of catalyst have shown the nickel and molybdenum species are well dispersed on the support. Simple heating methods has higher crystallinity than wetness impregnation methods due to the addition of PEG as continuous media avoid the agglomeration of impurities in solution that make the impregnation in simple heating methods free from impurities. $NiMo/Al_2O_3$ catalyst (A) have the highest crystallinity than the other catalyst because it have the highest acidity in alumina support so its produced the best crystal of $NiMo/Al_2O_3$.

The SEM characterization of catalyst (A) shows that alumina particles were coated with MoO_3 and NiO and it is clear that MoO_3 and NiO crystallites are highly dispersed on the $NiMo/Al_2O_3$ catalyst. This two characterizations of catalyst indicates the simple heating method with $KAl(SO_4)_2.12H_2O$ as alumina resources was the best catalyst due to its highest crystallinity and high metal loading dispersion in the catalyst. The catalytic activity of the $NiMo/Al_2O_3$ catalyst in hydrocracking process was successfully examined to convert fatty acid in dirty palm oil into hydrocarbon.

References

1. ESDM,*Outlook Energi Indonesia 2014*, 66-84 (2014)
2. N.K. Patel, S.N. Shah, and S. Ahuja (Ed.), *Food, Energy, and Water: The Chemistry Connection*, 278-279 (2015)
3. D. Verma, B.S.Rana, R. Kumar, M.G. Sibi, A.K. Sinha, J. Appl. Cat. A**490**, 108-116 (2015)
4. M. Aslam, L.J. Konwar, A.K. Sarma, N.C. Kothiyal, J. Anal. Appl. Pyrol., **2** (2015)
5. M. Anand, S.A. Farooqui, R. Kumar, R. Joshi, R. Kumar, M.G. Sibi, H. Singh, A.K. Sinha, Fuel Proc. Tech.**151**, 1-2 (2016)
6. S. Liu, Q. Zhu, Q. Guan, L. He, W. Li, Biosour. Tech.**183**, 95-99 (2015)
7. M. Anand, A.K. Sinha, Biosour. Tech.**126**, 149-155 (2012)
8. A. Srifa, K. Faungnawakij, V. Itthibenchapong, N.V. Empikul, T. Charinpanitkul, S. Assabumrungrat, Biosour. Tech.**158**, 81-89 (2014)
9. H. Liu, C. Liu, C. Yin, B. Liu, X. Li, Y. Li, Y. Chai, Y. Liu, Cat. Today, **8** (2016)
10. B.H. Susanto, M.B. Prakasa, M.H. Shahab, INSIST **1No.2**, 43-44 (2017)
11. Liherlinah, M. Abdullah, Khairurrijal, J. Nano Saintek. Ed. Khusus, 91-93 (2009)
12. M. Abdullah, Khairurrijal, A.R. Maruly, Liherlinah, M. Sunny, J. NanoSaintek.**1**, 1-6 (2008)
13. H. Liu, Y. Li, C. Yin, Y. Wu, Y. Chai, D. Dong, C. Li, C. Liu, J. Appl. Cat. B, 4-5 (2016)
14. M. Lu, N. Fatah, A.Y. Khodakov, J. En. Chem. **25**, 1003-1004 (2016)

Optimization of Chitosan Drying Temperature on The Quality and Quantity of Edible Film

Endah Sri Wahyuni[1] and *Fahmi* Arifan[1,*]

[1]Chemical Engineering, Diponegoro University, Semarang, Indonesia

Abstract. Edible film is a thin layer (biodegradable) used to coat food and can be eaten. In addition edible film serves as a vapor transfer inhibitor, inhibits gas exchange, prevents aroma loss, prevents fat transfer, improves physical characteristics, and as an additive carrier. Edible film made of cassava starch, glycerol and chitosan. Cassava starch is used as raw material because it contains 80% starch. Glycerol serves as a plasticizer and chitosan serves to form films and membranes well. The purpose of this research is to know the characteristic test of edible film by using ANOVA analysis, where the variable of drying of the oven is temperature (70°C, 80°C, 90°C) and time for 3 hours and variables change chitosan (2 gr, 3 gr, 4 gr). The result of this research was obtained the most optimum for water content and water resistance in temperature variable 80 ^{o}C and chitosan 4 gr. The best edible films and bubbles on temperature variables are 80 ^{o}C and chitosan 4 gr.

1 Introduction

Foodstuffs are generally very sensitive and susceptible to quality degradation due to environmental, chemical, biochemical and microbiological factors. Decrease in quality can be accelerated in the presence of oxygen, water, light, and temperature. One way to prevent or slow the phenomenon is with proper packaging [1].

Plastic packaging materials are widely used with economic consideration and provide good protection in preservation. About 60% of polyethylene and 27% of polyester are produced to make packaging materials used in food products. However, the use of synthetic materials has an impact on environmental pollution [2]. Therefore, at this time research is needed on packaging materials that can be described (biodegradable) [2,3].

The development of edible films on food in addition to providing better product quality and extended durability, can also be an environmentally friendly packaging material. Edible film provides an alternative packing material that has no impact on environmental pollution because it uses renewable materials and is cheap [4-6]. The application of edible films on food products is not a new concept and has been extensively studied extensively. The application of edible films can extend the shelf life and maintain the quality of various food products [5,6].

Polysaccharides such as starch can be used as raw material for edible film making. Starches are often used in the food industry as biodegradable films to replace plastic polymers because they are economical, renewable, and provide good physical characteristics [4]. Cassava starch is often used as a supplementary ingredient in the food industry and starch-based industry due to its high stone content [7,8]. High starch production, easy planting, and easy to find in Indonesia make cassava potential to be used as the basic material of edible film [9].

This study aims to produce quality edible film, both physically and chemically, this research tries to use temperature and chitosan variation on edible film drying of cassava starch and to study the influence of each factor on the measured parameter.

2 Methods

The experiment were conducted on March 2017 in Chemical Lab in Chemical Engineering Diponegoro University.

2.1 Raw Material

2.1.1 Starch solution

Edible film obtained from the main ingredient of starch flour. 6 gr starch flour plus 100 ml aquadest in 1000 ml beaker glass. Mix starch and water flour on magnetic stirrer for 10 minutes at room temperature to form a starch suspension. The resulting suspension is stirred and heated at 60°C for 15 minutes with a stirring scale of 6 to form a starch gel.

2.1.2 Chitosan solution

Chitosan variable 2, 3, 4 gr plus 100 ml of 1% acetic acid concentration in beaker glass 500 ml. Mix both materials on magnetic stirrer with temperature 50°C for 20 minutes with stirring scale 6. Formed yellow light chitosan gel.

* Corresponding author: fahmiarifan80@gmail.com

2.2 Edible film Preparation

Edible film, starch gel mixed with chitosan gel. Then stirred and reheated at 60^{0}C for 20 minutes. At 10 minutes added 4 ml of glycerol. Stirring is resumed until the solution is completely mixed (± 15 minutes). Edible Solution The resulting film, removed its impurities and bubbles, then settled momentarily at room temperature. Set up a 20x5cm glass plate that has been cleaned and taped. Edible film solution is poured in glass plate slowly evenly. The glass plate along with the edible film is dried in an oven with variable temperature (70^{0}C, 80^{0}C, 90^{0}C) for 3 hours. The dried film is released from the glass plate. Edible film formation of cassava starch and shrimp chitosan with the addition of glycerol plasticizer

2.3 Water Content Analysis

Determination of water content in food can be done by drying method (with regular oven), where calculation of moisture content based on dry matter (dry base). Dry base is the ratio between the weight of the water in the material and its dry weight. Dry matter is the weight of the original material after being reduced by the weight of the water [10].

$$\%\,Water\ content\ = \frac{(initial\ initial{-}weight\ weight)}{(final\ weight)} x\ 100\%$$

2.4 Water Resistance Analysis

The water resistance test procedure is by weighing the initial weight of the sample to be tested (wo), then inserted into a container containing the distilled water for 10 seconds. The sample is removed from the container containing the distilled water and the water contained on the plastic surface is removed with a paper tissue, after which the weighing is done. Samples are put back into the container containing the distilled water for 10 seconds. Then the sample is removed from the container and recalculated. The weighing is repeated until the final sample weight is obtained [11]. Furthermore the water absorbed by the sample is calculated through the equation:

$$water(\%) = \frac{W - Wo}{Wo} x100\%$$

2.5. Organoleptic Analysis

Organoleptic analysis or sense-test is a way of testing using the human senses as the primary means of measuring the reception power of the product. Organoleptic testing has an important role in the application of material quality. Organoleptic test ie product color analysis and bubble analysis

3 Results and Discussion

3.1 Water Content Analysis

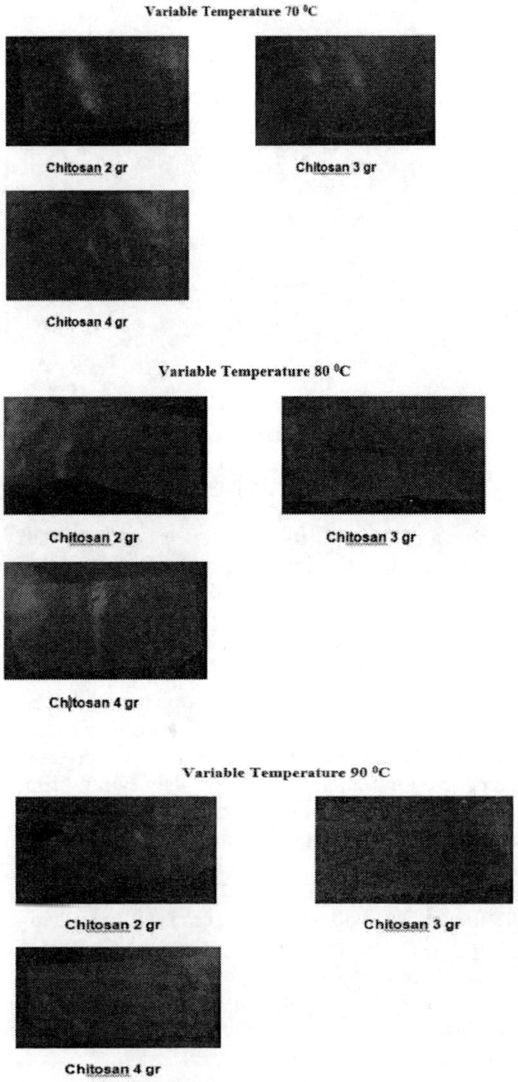

Fig. 1. Edible film in the temperature variable 700C, 800C, 900C.

Water content based on dry matter (dry basis is the ratio of the weight of water in the material to dry weight Dry material is the weight of the original material after it is reduced by the weight of the water.

This water content analysis was performed by using 3 gram samples from each of the oven variables with temperature of 1050C for 15 minutes.

Table 1. Water content analysis edible film

Variable	Temperature (°C)	Time (hours)	Water content(%)		
			A	B	C
1	70	3	34,3 %	25,3 %	21,7 %
2	80	3	19 %	14 %	13,3 %
3	90	3	10,7 %	9 %	7,3 %

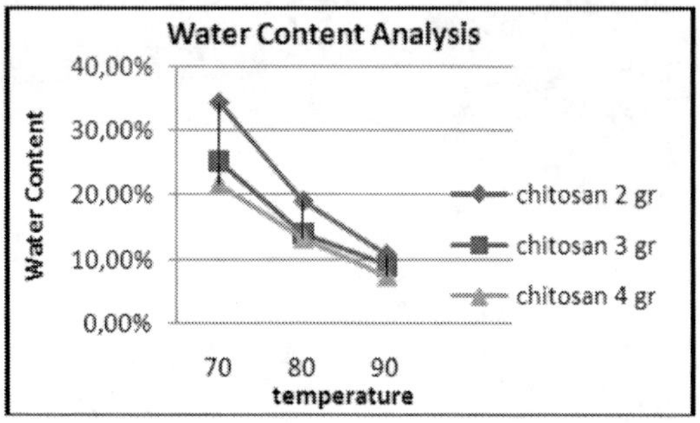

Fig 2. Water Content Analysis

Based on fig 2 of the water content analysis tendency to decrease the edible water content of the film along with the increase of added chitosan concentration. This is due to the hydrophobic nature of chitosan. Hydrophobic is the inability of a compound to bind water, so edible film with the addition of higher chitosan causes the water content in the material to decrease and the resulting water content becomes low [11,12]. In addition, the use of starch as a base material in the manufacture of edible film causes the resulting film has high water content so that the edible film added khitosan with low concentration causes the water content increases, as in the variables with the addition of chitosan 2 gr and 3 gr. This happens because the starch has the ability to absorb water because it has a hydroxyl group. The starch molecule contains a very large hydroxyl group so the ability to absorb water is also very large. This supports that with the addition of chitosan with high concentration will produce edible film with a small water content [13].

Based on the theory above the most appropriate water content is a temperature variable of 800C with the addition of chitosan 4 gr is obtained water content of 13.3%. With the water content is produced edible films that are not dry only so not fragile and not moist, other than that high water content susceptible to microbial growth.

3.2 Water resistance analysis

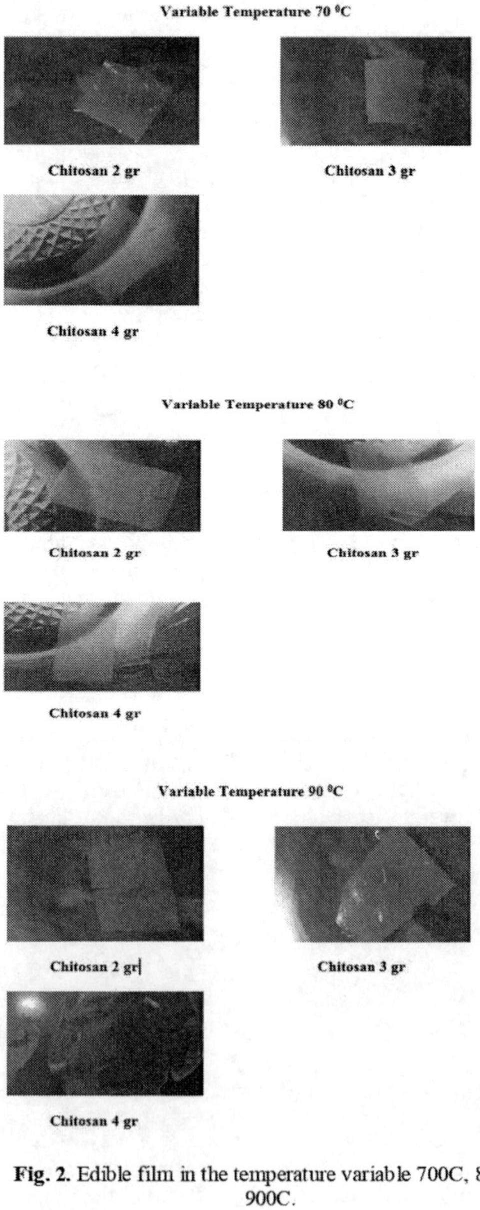

Fig. 2. Edible film in the temperature variable 700C, 800C, 900C.

Table 2. Results of Resilience to Water edible film

Variable	Temperature (^{0}C)	Time (hours)	Water resistance		
			A	B	C
1	70	3	17,4 %	18,2 %	16,9 %
2	80	3	17,86 %	17,6 %	16,4 %
3	90	3	17,72 %	15,49 %	15,28 %

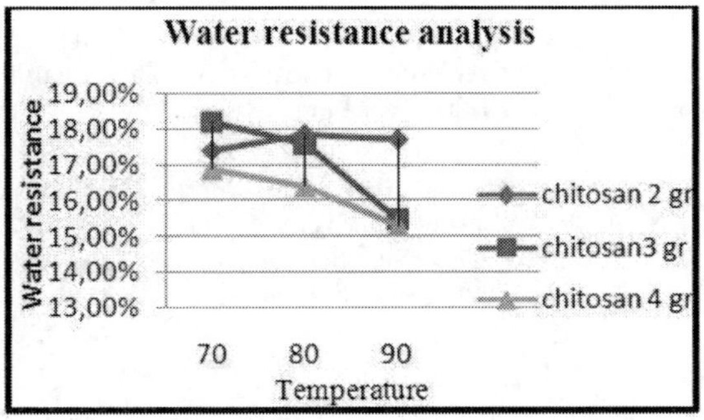

Graph 2. Water resistance analysis

Edible resistance of film to water can be done using swelling test. Swelling is the ability of a polymer to absorb the solvent and experience a certain volume swelling. The water resistance test in this study was conducted with edible sample size of 1 × 2 cm where weighing the initial weight of the sample to be tested (wo), then inserted into the container containing the distilled water for 10 seconds. The immersion and weighing procedure is repeated until the final sample weight is obtained constantly.

The value of water resistance in the film is said to be good if the percent swelling is low. So the lower percent swelling that is obtained then the resulting film more water resistant. In general there is a tendency to decrease the sweeling value of the edible film as the added concentration of chitosan is added. The higher the concentration of added chitosan, the smaller the swelling percentage [4]. This is because the nature of chitosan is hydrophobic (do not like water). Chitosan has insoluble properties in water and organic solvents because chitosan has a crystal structure composed of intramolecular and intermolucular hydrogen bonds [14,15].According to Muzarelli et al, [15] stated that chitosan has a negatively charged hydroxyl group (-OH) and an amin (-NH2) group that is positively charged so that the chitosan is able to bind strongly ionic. The presence of negatively charged hydroxyl groups on chitosan causes chitosan to be hydrophobic and has an effect in increasing the percentage of water resistance of edible film produced.

Based on the water resistance test, the best result is the temperature variable of 800C with the addition of chitosan 4 gr that is obtained% water resistance of 0,164 With% water resistance is produced edible film that absorbs water with small quantities and not destroyed when the swelling test.

3.3. Organoleptic Analysis

Table 3. Organoleptic Analysis Edible Film

Variable	Color			Bubble		
	A	B	C	A	B	C
1	Brown	Brown	Brown	Many	Little	Little
2	Transparent white	Transparent white	Transparent white	Little	Little	-
3	Transparent white	Transparent white	Transparent white	Little	-	-

The color of the edible film is determined by the visual (subjective) test. Factors affecting the color of the edible film is the drying temperature. The most optimum color is the temperature variable 80^{0}C and 3 hours of drying time is transparent white. While at the temperature of 70^{0}C and the duration of drying 3 hours of brownish white edible film this may be due to the temperature is less hot so that the film is still thick and the color can not be transparent.

The number of air bubbles means the worse the quality of edible film because the more air bubbles the worse the water resistance of edible film. The number of air bubbles caused by the process of formation of cassava starch gelatin is not perfect.

4 Conclusion

The most optimum variable on water content analysis and water resistance is temperature of 800C with the addition of chitosan 4 gr that is got water content equal to 13,3%. And water resistance of 16.4%. Organoleptic analysis on edible film obtained results include color and bubbles, where the most optimal result is on the temperature variable 800C and 3 hours of drying time is transparent white and no bubbles. The number of air bubbles means the worse the quality of edible film because the more air bubbles the worse the water resistance of edible film.

Acknowledgement

The authors acknowledge with gratitude the support given by Diponegoro University for our research in this area.

References

1. Y. H. Hui, Handbook of Food Science, Technology, and Engineering Volume II CRC Press, USA, (2006)
2. Henrique, C. M., R. F. Teofilo, L. Sabino, M. M. C. Ferreira, and M. P. Cereda, Journal of Food Science. 74: E184-E189, (2007)
3. S. Kittur, K.R. Kumar and R.N. Tharanathan, Functional packaging properties of chitosan film. Z. Lebesm Unters Forsch A. 206, (1998)
4. T. Bourtoom, 2007. Effect of Some Process Parameters on The Properties of Edible Movie Prepared From Starch. Department of Material Product Technology, Songkhala. (On line) Avaliable at: http://vishnu.sut.ac.
5. M. C. Chen, G. H. C. Yeh, B. H. Chiang, J. Food Processing and Preservation 20: 379-390, (1996)
6. W.S Subowo, and S. Pujiastuti, Proceedings of National Symposium of Polymers IV, Bandung, Center for Informatics Research-LIPI, pp. 203-208, (2003)
7. Susilawati, Biodegradable Plastics From A Mixture Of low density Polyethylene.http: //download.portalgaruda.org, (2011)
8. Vendy, Utilization of Cassava Starch As Raw Material.http: //.uns.ac.id, (2010)
9. Winarno, Biofermentase and Protein Biosynthesis. PT. Space. Bandung, (1992)
10. Wulan, Determination of Water Content Method Oven.Https://wulaniriky.Wordpress.com, (2011)
11. Hargono, et al. Making Chitosan From Shrimp Shell Waste And Its Application In Reducing Cholesterol Fat Goat.http: //staff.uny.ac.id.
12. Fehragucci, Helmi, Practical Addition of Plasticizer And Chitosan To The Charible Edible Film Ca-Algina. Https://dglib.uns.ac.id, (2012)
13. Puspita, Tyas, Pengaruh Addition Of Khitosan And Plasticizer Glycerol On Biodegradable Plastic Characteristics Of Cassava Skin Waste Starch: http://digilib.its.ac.id, (2012)
14. M. Manullang, Bul. Technology and Food Industry. Vol **IX** no. 1: 34-43, (1998)
15. Muzarelli, R. A. A. Farsi, R. Filippini, O. Giovanetti, E. Biagini, G. Varaldo, P. E, Antimicrobial Agents Chemonth 34: 2019-2023, (1990)
16. Chen, C. S. Liau, W. Y. and Tsai, G. J, J. Food Prot 61: 1124-1128 .: 44-47, (1998)

Potential application of biodrying to treat solid waste

Badrus Zaman[1], *Wiharyanto* Oktiawan[1], *Mochtar* Hadiwidodo[1], *Endro* Sutrisno[1], *Purwono*[1,*], *Irawan* Wisnu Wardana[1]

Department of Environmental Engineering, Faculty of Engineering, Diponegoro University, Semarang - Indonesia

Abstract. The generation of solid waste around the world creates problems if not properly managed. The method of processing solid waste by burning or landfill is currently not optimal. The availability of land where the final processing (TPA) is critical, looking for a new TPA alternative will be difficult and expensive, especially in big cities. The processing of solid waste using bio drying technology has the potential to produce renewable energy and prevention of climate change. Solid waste processing products can serve as Refuse Derived Fuel (RDF), reduce water content of solid waste, meningkatkan kualitas lindi and increase the amount of recycled solid waste that is not completely separated from home. Biodrying technology is capable of enhancing the partial disintegration and hydrolysis of macromolecule organic compounds (such as C-Organic, cellulose, hemicellulose, lignin, total nitrogen). The application of biodrying has the potential to reduce greenhouse gas emissions such as carbon dioxide (CO_2), methane (CH_4), and dinitrooksida (N_2O). These gases cause global warming.

1 Introduction

Solid waste generation worldwide can cause problems if not properly managed. Solid waste management with a decentralized system in residential areas is able to reduce solid waste well. Intensity and roving activity produce many biodegradable solid wastes that must be managed to avoid negative impacts on the environment [1]. The interesting technologies of recent years are Mechanical Biological Treatment (MBT) and Biological Mechanical Treatment (BMT). Biodrying (Biological Drying) where solid wastes are readily bio compatible biological-mechanical [2].

Biodrying rework process (biological access). In practice, the biodiesel reactor entrusts high-level solid waste that has been chopped up and produces the output of biodrified waste that will undergo further mechanical processes. The heat generated from the aerobic decomposition process by using excess air serves to dry out solid waste [3].

The results of processing using biodiesel can serve as fuel substitute fuel (RDF) which is fuel produced from various types of waste such as waste, industrial waste or commercial waste. RDF can be used as a coal substitute [2]. The most obvious uses of the bio drying process are the decrease in odor, volume, and weight of waste, which can improve dental, transport and solid waste disposal at the time of large solid waste volumes. Furthermore, biodiesel has also become an effective way to eliminate some pathogenic microorganisms.

The bio-drying process has the potential to generate air emissions derived from aeration and aerobic decomposition processes such as Volatile Organic Compounds (VOCs), carbon dioxide (CO_2), methane (CH_4), and dinitroxide (N_2O) [4]. The gas contributes to global warming or global warming which is a phenomenon of increasing global temperature from year to year due to the occurrence of the greenhouse effect (greenhouse effect). According to [5], the global average earth temperature increases over the last 100 years with an increase temperature of 0.74 ± 0.18 ° C caused by increased emissions of gases such as carbon dioxide (CO_2), methane (CH_4), dinitroxide (N_2O) And chlorofluorocarbons (CFCs) or can be called GHGs so that solar energy is trapped in the Earth's atmosphere.

The following discussion is a potential application of bio-drying to treat solid waste, especially waste water content and Municipal Solid Waste (MSW).

2 Refuse Derived Fuel (RDF)

In the bio-drying process, the heat generated from the aerobic decomposition process of organic compounds combined with excess air serves to drain waste [3]. The result of processing using bio-drying can serve as Refuse Derived Fuel (RDF) which is fuel produced from various types of solid waste such as urban waste, industrial waste or commercial waste. RDF can be used as a substitute for coal [6-9]. According to [10], the criteria of a material can be used as fuel can be seen from chemical properties (eg elemental analysis, trace analysis, ash analysis), mechanics (eg bulk density, particle size / particle size distribution), calories (eg LHV, adiabatic Combustion temperature, minimum air consumption) and the reaction (Volatile Organic Compounds, emission gas). In practice, inert solid waste must be separated first, enriching the organic material and increasing the

* Corresponding author: purwono.ga@gmail.com

lower heating value (LHV). The final LHV value of the bio-drying product is higher than the initial MSW LHV [1]. For example comparison of LHV values in some fuels such as Coke 29.5 MJ/kg, Wood chips 12.0 MJ / kg, RDF 11.7 MJ / kg. According to [11] RDF requirements are (1) Calorific value on average at 15 GJ/ton, (2) Moisture content <20 % mass, (3) Chlorine content <1%.

3 Water Content and Leachate Quality

The bio-drying process has an impact on the decrease of water content due to high temperature and adequate ventilation. According to [12-13], proper aeration control (eg, debit and direction of flow) and temperature can lower water levels efficiently (66.7% of the initial moisture content). Solid waste can be burned properly when the water content is about 20%. Aerobic bio-drying technology principle is to encourage evaporation using the energy generated as a result of the degradation of organic matter. When the water content ratio of the solid waste of the biodegradable organic material is too high, then the heat arising is not enough to evaporate the water. On the other hand, solid waste will decay due to high water content. The amount of water caused by spoilage (leachate) is limited by a cell wall or a crushed membrane. Their aeration increasing disintegration and partial hydrolysis of organic compounds macromolecules (such as C-Organic, cellulose, hemicellulose, lignin, total nitrogen [14]. The process bio-drying solid waste with high moisture content they produce leachate. According to [15], [16] the parameters of pH, Chemical Oxygen Demand (COD), Biological Oxygen Demand (BOD), ammonium (NH_4-N), nitrate (NO_3-N), and the volume of leachate is much lower than without bio-drying.

4 Sorting

The main principle for every country related to solid waste management is prevention. The second principle is recycling (recycle) and energy production (energy production). The final principle is the disposal of solid waste without recycling and generating energy. Biodrying technology supports sustainable environment related to solid waste processing. The amount of solid waste that will enter the final processing place (TPA) will substantially decrease. Significantly increase the amount of recycled solid waste by separating materials (glass, inert, iron and non ferrous metals) that are not completely separated at home [12], [17], [18].

The concept of MBT consists of three stages. Pre-treatment stage is shredders. The second stage is bio-drying with the aim of drying solid waste. Solid waste is placed in the pile, covered with a membrane and aerated for approximately 3 to 4 weeks. The purpose of this step is to drain the water content of <20% mass%. Post-treatment stage of separation of ferrous metals, sieves, density separators, inert screens, manual sorting, re-shredders and storage. The purpose of this step is to sort out recyclables that can be treated efficiently and RDF (particle size ≤ 50 mm, moisture content ≤ 20%) of waste and to reduce the amount of waste residue for disposal (amount for disposal should be ≤ 20%) [11].

5 Gas Emission

In the bio-drying process, volatile organic compounds (VOCs) such as foul-smelling (sulfuric), aromatic, chlorine, terpene and ketones will be released into the air. This VOC became one of the main problems with air pollution [4]. Many previous studies have suggested that VOCs and/or odors will be released into the air derived from solid waste aerobic composting (MSW, bio waste, fruits, food scraps, mixed paper, and yard waste) at various scales [19], [20], [21], [22]. Non-methane organic compounds during the decomposition of solid waste under aerobic and anaerobic conditions have been compared by [23]. Whereas [18] states that foul-smelling sulfur compounds and terpene compounds are the most widely released compounds into the air in the early phases of bio-drying, whereas aromatic and ketone compounds are released later during the bio-drying process. The ammonia emissions resulting from the conversion of organic matter biologically can be absorbed using the acidic compound to form ammonium sulphate solution which serves as a substitute for artificial fertilizer [24]. Gases causing the greenhouse effect have not been much discussed by researchers. The gases are carbon dioxide (CO_2), methane (CH_4), and dinitrooksida (N_2O). However, according to [25], methane gas probably will not be formed because in the process of bio-drying is the aerobic process.

6 Conclusion

The processing of solid waste using bio drying technology has the potential to produce renewable energy as Refuse Derived Fuel (RDF), reduce water content of solid waste, Improve leachate quality and increase the amount of recycled solid waste that is not completely separated from home. Biodrying technology is capable of enhancing the partial disintegration and hydrolysis of macromolecule organic compounds. The application of biodrying has the potential to reduce greenhouse gas emissions such as carbon dioxide (CO_2), no methane (CH_4), and dinitrooksida (N_2O). Gases causing the greenhouse effect have not been much discussed by researchers.

7 The acknowledgments

Thank you to DRPM DIKTI for funding this research through PUPT grant with contract number 007 / SP2H / DRPM / V / 2017 financing year 2017

References

1. E. C. Rada, M. Ragazzi, V. Panaitescu, Sci. Bull., Series D, Vol. **71**, Iss. 4, (2009).
2. E. Rada, and M. Ragazzi, Biological Treatment of Solid Waste: Enhancing Sustainability, **219**, (2015).
3. C.A. Velis, P.J. Longhurst, G.H. Drew, R. Smith, S.J.T. Pollard. 2009. UK: Bioresource Technology **100**, 2747–2761, (2009)
4. He, P J, Shao L M, Qu X, Li G J, Lee D J, Chemosphere, **59**(6): 837–844, (2005)
5. Muhi, A. Hanapiah, *Pemanasan Global (Global Warming)* (Skripsi). Institut Pemerintahan Dalam Negeri (IPDN). Jatinangor, (2011)
6. E. C. Rada, I. A. Istrate, M. Ragazzi, Environmental Technology, Vol. **30**, No. 7, 651–661, (2009).
7. E. C. Rada, M. Ragazzi, V. Panaitescu, T. Apostol, Stintific Buletin, seria C: Electrical Engineering, vol **67**, nr. 2, 69-76, (2005).
8. Gleis Markus, Technische Lösungen zur Abfallentsorgung in Europa –Strategien und Erfahrungen aus 20 Jahren, Proceedings of Tagung:Gesellschaft, Umweltund Gesundheit, Ecocenter Bolzano, (2009).
9. Elena Cristina Rada, PhD, Biological Treatment of Solid Waste Enhancing Sustainability. Canada: Apple Academic Press, Inc, (2016).
10. M. Beckmann, M. Pohl, D. Bernhard, K. Gebauer, Waste Management & Research. vol. **30**, S. 354 – 369, April (2012)
11. Virginie Herbst, Dr. Ludwig Streff, Andrea Humeniuk, New ways to make MBT sustainable Case Study – Bio-drying for RDF production in Sulaimaniyah, Iraq, 2013. Germany: Ingenieurgesellschaft Prof. Czurda und Partner mbH (ICP), (201
12. F. Adani, D. Baido, E. Calcaterra, and P. Genevini, Bioresource technology, **83**(3), 173-179, (2002).
13. M. Sugni, E. Calcaterra, F. Adani, Bioresource Technology, **96**: 1331–1337, (2005)
14. P. H. L. Nguyen, P, Kuruparan, C. Visanathan, Bioresour Technol, **98**(2): 380–387, (2007)
15. W. Oktiawan, Purwono, H. Mochtar, R. Arya, Penerapan teknologi biodrying dalam pengolahan sampah hight water content menuju zero leachate. Prosiding Seminar Nasional Lahan Basah Tahun 2016 Jilid 2: 572-576, (2006)
16. H.N. Rahmawati, Sudarno, Purwono, Skripsi. Departemen Teknik Lingkungan, Fakultas Teknik, Universitas Diponegoro, (2016)
17. R. M. Negoi, M. Ragazzi, T. Apostol, E. C. Rada, and C. Marculescu, UPB Sci. Bull. Serie C, 71, 193-204, (2009).
18. H. L. Choi, T. L. Richard, and H. K. Ahn, Compost Science & Utilization, **9**(4), 303-311, (2001).
19. K. S. Das, Odor related issues in commercial composting. Workshop Presented at the Y2K Composting in the Southeast Conference & Expo. October 9, 2000, Charlottesville, VA, (2000)
20. B. D. Eitzer, Environmental Science & Technology, **29**: 896–902. Germany, (1995)
21. P. Pierucci, E. Porazzi, M. P. Martinez, F. Adani, C. Carati, F. M. Rubino et al., Chemosphere, **59**: 423–430, (2005)
22. X. M. Wu .T. Wang, Environmental Science & Technology, **42**: 3265–3270, (2008)
23. B .F. Staley, Xu F X, S. I. Cowie, M. A. Barlaz, G. R. Hater, Environmental Science & Technology, **40**: 5984–5991, (2006)
24. Winkler MK1, MH. Bennenbroek, FH. Horstink, van Loosdrecht MC, van de Pol GJ. 2013. Bioresour Technol, Nov;**147**:124-9, (2013)
25. G. Tchobanoglous, and F. Kreith, Handbook of Solid Waste Management, 2nd Ed., McGraw-Hill, (2000)

Oil Palm Empty Fruit Bunches (OPEFB): Existing Utilization and Current Trends Bio Refinery in Indonesia

Rame[1,*]

[1]Center of Industrial Pollution Prevention Technology, Semarang - Indonesia

Abstract. In a future carbon-constrained global economy, the use of fossil fuels will be restricted. Biomass resources will be increased demand for renewable products. Oil Palm Empty Fruit Bunches (OPEFB) can be used as lignocellulose feedstock. The production of biofuels from lignocellulose feedstock can be achieved through biochemical or thermo-chemical routes. OPEFB contain chemical blocks of cellulose, hemicellulose and lignocellulose. Due to these substances, OPEFB can be converted into bio-products and chemical. Special attention to biorefinery approach that is present at relatively high potential in bio-products such as polymers, nutraceuticals, chemical building blocks, biofuels, and bioenergy. Different utilization types were considered and reviewed, and the most common and efficient process were discussed. In general, there is no single product which could be considered a solution to the utilization of managing OPEFB – in this review a number of product are more economic, effective and environmentally friendly.

1 Introduction

Oil Palm Empty Fruit Bunches (OPEFB) are the solid residue from fresh fruit bunch (FFB) of oil palm tree that are mainly cultivated in several countries, including Indonesia, Malaysia, Thailand and other. The OPEFB contain chemical blocks of cellulose, hemicellulose and lignocellulose, while its present utilization as a biofuels, fertilizer, and compost production [1] [2], however its utilization can lead to emission resulting in organ damages and impact on the environmental systems.
Besides, OPEFB been reported to possess many distinct advantages, some of them are: pulp [3], feedstuff [4], carbon briquette [5], activated carbon [6], and fillers [7]. The OPEFB are also used as foods [8], oleo chemicals [9], and soap [10].

Indonesia is one of the largest palm oil producing countries in the world. Crude palm oil (CPO) is a mainstay industry for Indonesia's national income and foreign exchange, which can be seen from the export value of CPO commodities. In 2014, total CPO exports reached US $ 17.46 billion with volume reaching 22.89 million tons. The growth of oil palm plantation area in Indonesia has also continued to increase, by 2014 the area of oil palm plantation 10.75 million hectares with an estimated increase of 11.67 million hectares in 2016 with production 33.50 million tons CPO [11].

The development of biofuel industry in Indonesia represents the emerging phase, because of the impetus of government commitment to reduce greenhouse gas emission regulations which are decanted in President Decree No. 61 the year 2011. Indonesia has committed to reducing greenhouse gasses by 26% below 2005 levels, equivalent to 767 million tons of greenhouse gasses on their own. National Energy Policy set to blending bioethanol in premium 10% and 20% in 2020 and 2050 respectively [12]. Accumulation biomass from agroindustry in large quantity causes big problem. Biomass processing can help reducing agricultural waste also can added high value of this feedstock [13].

The characteristics of OPEFB is that it act as a renewable energy sources and environmentally friendly. OPEFB may be converted to useful compounds, some of them are: Micro fibrillated cellulose [14], bio vanillin [15], Sugar [16], Ethanol[4], p-hydroxybenzoate [17], Lignin-containing cellulosic Nano fibril [18].

Since OPEFB is a cellulose chain, it can be used as a particle board [1], furniture [10], and fibre boards [19]. Recently, many biorefinery approach have been reported [20], [10]. In order to understand OPEFB and its existing utilization and current trends biorefinery in Indonesia, this paper will provide the complete review of information on Utilization of OPEFB; its current trends biorefinery; production of bio-products such as polymers, nutraceuticals, chemical building blocks, biofuels, and bioenergy.

2 Oil Palm Empty Fruit Bunches (OPEFB)

The availability of OPEFB will continue to increase along with the increase in the production of fresh fruit bunches of palm oil in Indonesia. OPEFB has enormous potential to be developed into high value-added products. Chemically the OPEFB contains lignin> 20%, hemicellulose> 25% and lignocellulose> 35%. All three

* Corresponding author: aryanagasasra@gmail.com

components can be converted into various chemicals, materials and valuable products.

The development of a highly prospective palm oil market in the global market has problems mainly from other vegetable oil producing countries. The black campaign on environmental, sustainability and health issues poses the greatest obstacle to the export of world oil and vegetable oil markets. In addition, the weak development of downstream industries in the country caused Indonesia's competitiveness and added value to be low in obtaining optimal benefits from the global palm oil market. Another major problem is the low productivity compared to its potential. The problem demands a huge research role in supporting strategic palm oil commodities into a formidable industry. Aspect of OPEFB utilization as one of palm oil industry waste is needed to increase productivity, added value, efficiency, and environmentally friendly.

As mentioned earlier, it is widely believed that lignocellulose is the major component that is responsible for the difficulty of OPEFB utilization. However, when decomposed in delignification using microbe, lignocellulose becomes toxic and causes mortality of microbe in the bioprocessing. Apart from lignocellulose, cellulose and hemicellulose is also available in OPEFB [21]. Figure 1 depicts the method pre-treatment and structure of OPEFB.

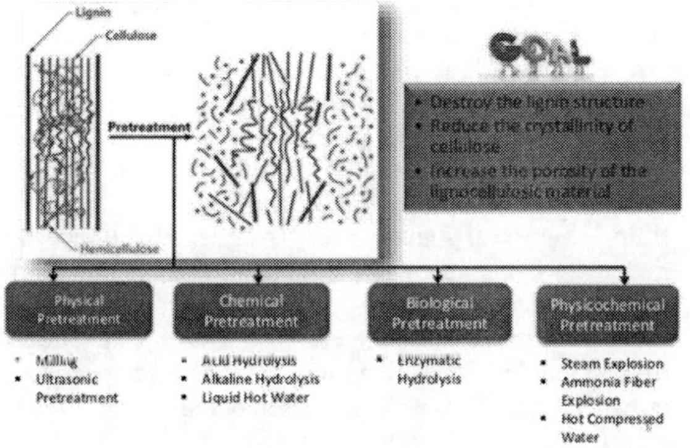

Fig. 1. Chemical blocks and structure of OPEFB [22].

2.1. Existing Utilization of OPEFB

OPEFB were extensively largely dumped and traditionally been burnt in incinerator of palm oil mill and ash recycled into the plantation as fertilizer. Utilization of soil improver and co-composting has been widely applied in Indonesia. However, due to the environmental pollution, the incineration of OPEFB has been avoided. In order to protect the environment and to ensure the sustainability of the oil palm industry, the OPEFB must be fully converted.

The building block of native OPEFB fibre is made up from a complex matrix of three main polymers which are cellulose, hemicellulose and lignocellulosic. The lignocellulosic material from OPEFB has been considered as a very good source of fermentable sugar for conversion into value added products. Cellulose and hemicellulose can be hydrolysed chemically by acid or enzymatically into glucose and a variety of pentose and hexose sugars, which can then be fermented to produce bioethanol.

Biorefinery approach that is present at relatively high potential in bio-products such as polymers, nutraceuticals, chemical building blocks, biofuels, and bioenergy are under development. The possible cause for the late biorefinery approach because the price of bio product is still not economical due to the high cost of production. Figure 2 represent conversion route of OPEFB via biological and thermo-chemical routes.

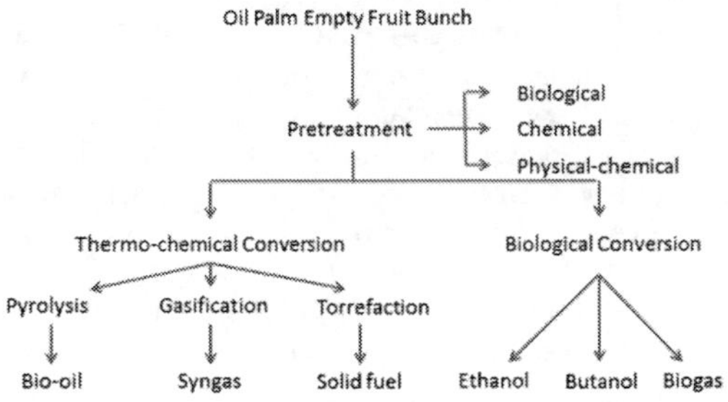

Fig.2. Biofuel production from OPEFB [23].

2.2 Biorefinery Approach

The development of the national palm oil industry in addition to positively impacting the economic sector also has considerable challenges with environmental issues, especially related to greenhouse gas emissions due to the impact of oil palm industry activity. An increased awareness of the importance of environmental protection, and the possible impacts associated with the product. In terms of production has increased interest in the development of methods to overcome these impacts. One of the techniques developed is an environmentally friendly biorefinery.

In the biorefinery, through different processes (physical, chemical and/or biological), OPEFB is transformed into a wide range of products with high added value. The whole process includes turning carbohydrates, lignin and other products into fuels, high value-added chemicals and other materials, with an approach to the prospect of not generating waste and with processes that ideally involve minimal energy and environmental impact.

2.3 Production of bio-products

The production of biofuels from lignocellulose feedstock can be achieved through biochemical or thermo-chemical routes. Majority of OPEFB bio-products were via the thermo-chemical route, in some minor cases via. OPEFB contain chemical blocks of cellulose, hemicellulose and lignocellulose. Due to these substances, OPEFB can be converted into bio-products and chemical. Special attention to biorefinery approach that is present at relatively high potential in bio-products such as polymers, nutraceuticals, chemical building blocks, biofuels, and bioenergy. The most significant biorefinery with highly prospective involves the bioenergy and polymers.

Approach production of bioethanol using methods that have been done on similar lignocellulosic biomass is very relevant to do, such as: from coconut fibre using alkaline pre-treatment and acid hydrolysis method [44], rice straw biomass by direct saccharification and fermentation [45], microwave-NaOH pre-treatment and hydrolysis enzyme using Trichoderma reesei - Aspergillus niger [46], chemical pre-treatment for cocoa pod [47].

OPEFB as waste from palm-oil processing are abundantly potential offering ligno-cellulosic fibre stuffs, which have not been utilized effectively hence suggesting their potential use in those paperboard industries. Center for Forest Products Research Development Center Bogor using semi-chemical hot soda process in the semi-pilot scale closed (pressurized) digester for manufacturing fancy paperboard from the mixture of OPEFB [24] [25]. The manufacture of composite boards from OPEFB to produce environmentally friendly furniture products [26], material for pulp and paper manufacture [27]. Research by creating composite from OPEFB to be particle board have fulfilled standard of SNI 03-2105-1996 [28]. Cellulose is a natural polymer having good mechanical properties and is widely used as a polymer matrix. Compared with synthesis polymers, the use of cellulose as a polymer gets a lot of attention because it does not contain health hazardous materials, is biodegradable and does not cause environmental emissions. One product that can be developed from empty palm bunch cellulose is cellulose ester as a bioplastic raw material.

OPEFB used to increase the quality of briquettes from cassava skin waste [29]. The pyrolysis method was used to convert the palm empty fruit bunch from biomass to bio-oil. Slow pyrolysis at a temperature of 450-600 °C which is fed by nitrogen gas. In the bio-oil product identified the presence of ethanol, benzene, toluene and xylene [30]. Pre-treatment on OPEFB with NAOH and used for fibre concrete [31] [32]. The spent alkaline (black liquor, BL) that obtained from the pilot plant scale pre-treatment of OPEFB is used as a solvent in next OPEFB pre-treatment [33].

OPEFB production potential as a source of bioethanol in Indonesia is huge, but it needs more complicated and longer process technology than the production of starch and sugar base bioethanol [34]. Pre-treatment at 220 °C in 30 minutes and saccharification using accretion of cellulase & BG optimash and fermentation with Saccharomyces cerevisiae IR-2 was 0.12 ml / g bioethanol [35]. Relatively low biofuels cause carbon burning that can damage the environment, but the production process can have a major impact on environmental destruction because it raises 10 times more carbon burning than fossil energy production [36].

Processing OPEFB into compost [37] [38]. Treatment OPEFB compost 50 g and coco peat 50 g produce in the growth palm oil seedlings better than the other treatments on medium subsoil ultisol in pre nursery [39]. The compost fertilizer product that has been obtained already fulfill the standard of fertilizer cultivation issued by SNI and fertilizer quality standard issued by PT. PUSRI. The C / N standard can be achieved at 3 months composting age for 25 grams of mushroom addition, whereas without the addition of mushroom fungus, the condition is only reached at 4 months composting age [40]. Composition of OPEFB compost + sub soil + sand 3:1:1 + 4 g urea/pot is the best composition as a substitution of top soil [41]. The quality of the compost produced in laboratory-scale inoculum testing in principle meets the standard, SNI 19-7030-2004 [42].

Utilizing the isolated lignin from OPEFB as a sodium lignosulfonate surfactant. Highest purity analysis results 78%. While the lowest purity value of 66% [42]. Conversion OPEFB into base catalyst has been reported [43].

2.4 Managing OPEFB

The concept of managing OPEFB is attracting increasing attention in all aspects, especially in industrial conversion. An example of managing OPEFB is the environmentally friendly utilization of OPEFB as natural resources, taking place through the recovery of by-products and/or waste generated by utilization or conversion activity. One of the main current challenges of the industrial utilization is the development of products that can fulfill new market requirements with an ecological commitment. Nowadays, Indonesia is awakening and it is convinced of the needs for consuming products that have been produced through environmentally friendly processes.

The biorefinery of OPEFB is becoming increasingly important. This process uses OPEFB of low cost with little commercial significance. Using OPEFB biomass as raw material is intended to take advantage of all the fractions derived from it, obtaining fuels (such as bioethanol or biodiesel) and various chemical products of commercial interest (products for food, pharmacy, polymers etc.), thus producing a maximum plant OPEFB valorization and a minimization of residues.

Many researchers are developing the OPEFB industry to adapt the alternative conversion process to the above-mentioned managing OPEFB concept. Not only the improvement of the conversion involved in the OPEFB industry through new approach process, but also their modification in order to integrate the OPEFB industry into a biorefinery, are being undertaken. Thus, the need for the use of each one of the OPEFB cell components is imperative and under this concept, it can be achieved. On the other hand, the reality is that in Indonesia, residues of OPEFB from palm oil mill have been largely dumped. Considerable efforts are dedicated to searching new biorefinery process and studying their feasibility to be used in OPEFB industry.

2.5 Trends Biorefinery

The presence of biorefinery approach suggests the Presence of a number of product are more economic, effective and environmentally friendly. The production of bio product from OPEFB feedstock can be achieved through biochemical or thermo-chemical routes. Delignification process are very necessary because of the presence of lignocellulose component (i.e., decomposed to lignin and cellulose after process) by the pre-treatment facility. There are no scaling up data to guide the successfully treatment of lignin neutralization or separation. The combination of thermo-chemical and biochemical routes may be required to treat OPEFB more economic and effective. Co-Pyrolysis should be used with caution due to the alternating output of gas and heat during OPEFB pyrolysis [48]. Residues of OPEFB from palm oil mill have been largely dumped. Utilization of soil improver and co-composting has been widely applied in Indonesia. Processing into biogas is under development. While biorefinery becomes a highly prospective product through pyrolysis, bioprocess, and combination.

OPEFB consists of cellulose, hemicellulose and lignocelluloses that have great potential for reuse after being converted to bioethanol and other industrial materials. But the price of bioethanol is still not economical due to the high cost of production. The production of bioethanol through an on-site chemical-enzymatic integration process carried out through a three-way integrated approach is potentially enormous. The first bioethanol unit that processes OPEFB into bioethanol, by hydrolysis with ammonia, alkanol amine, and other auxiliary materials. Followed by the saccharification process using cellulase enzyme obtained on-site from microbes such as *T. viride* and *N. crassa* using OPEFB substrate. The critical points on the production of cellulase enzymes on-site include moisture, substrate size, and nutrients. Both IPAL units that process liquid waste from bioethanol units become biogas. The three generator units that convert biogas into steam and electricity can be used in bioethanol units, IPAL units and palm oil operations. The OPEFB Biorefinery becomes an integrated and commercially potential bioethanol where the biogas / electricity energy generated can be used for other production operations.

The palm oil industry, which converts OEFB into bio-ethanol as a fuel and produces biogas, will save costs in its energy supply and overcome environmental problems. The use of bioethanol and biogas is also very influential on the reduction of greenhouse gas and global warming effects, so it is environmentally friendly[49]. Gaining Added Value and Palm oil revenues will increase with the acquisition of bioethanol and biogas as well as additional revenue derived from CER (Certified Emission Reduction). The utilization of treated clean water and fertilizer from the IPAL unit will lower the operational costs for the palm oil industry.

3 Conclusions

In general, Oil Palm Empty Fruit Bunches (OPEFB) can be converted into bio-products and chemical. To present, no single product which could be considered a solution to the utilization of managing OPEFB. Thus, a number of product are more economic, effective and environmentally friendly.

References

1. M. Vakili et al., Env. Dev Sustain, doi:10.1007/s10668-014-9581-2, (2014)
2. Direktorat Jenderal Perkebunan, Statistik Perkebunan Indonesia 2014-2016 (Jakarta, 2016).
3. H. Harsono, A. Samodra, P. Roni, J. Wood Sci. (2015), doi:10.1007/s10086-015-1526-3.
4. A. Christia et al., Int. J. Green Energy ISSN. **13**, 566–572 (2016).
5. M. Afif, A. Wan, M. Faizal, W. Mahmood, Z. Harun, J. Mater. Cycles Waste Manag. (2016), doi:10.1007/s10163-016-0518-8.
6. O. Nb, N. Shamsuddin, Y. Uemura, AProcedia Eng. **148**, 758–764 (2016).
7. J. W. Kiama, V. Raman, T. H. H. Patrick, J. Phys. Conf. Ser. **622** (2015), doi:10.1088/1742-6596/622/1/012025.
8. Z. Zhang et al., Cellulose (2016), doi:10.1007/s10570-016-1043-6.
9. F. B. Ahmad, Z. Zhang, W. O. S. Doherty, I. M. O. Hara, T. Crops, Biofuels, Bioprod. Bioref. (2016), doi:10.1002/bbb.
10. J. C. Kurnia, S. V Jangam, S. Akhtar, A. P. Sasmito, A. S. Mujumdar, Biofuel Res. J. **9**, 332–346 (2016).
11. Badan Pusat Statistik, Statistik Kelapa Sawit Indonesia (Jakarta, 2016).
12. S. Hasibuan, N. Nazir, Int. J. Adv. Sci. Eng. Inf. Technol. **7**, 276–283 (2017).
13. A. C. Djohan, U. Perwitasari, Int. J. Adv. Sci. Eng. Inf. Technol. **6**, 40–44 (2016).
14. K. Yin, G. Yern, C. Ching, Cellulose. **23**, 379–390 (2016).
15. A. Zulkarnain, E. Kamal, B. Norhayati, R. Lai, Y. Phang, Waste and Biomass Valorization (2016), doi:10.1007/s12649-016-9745-4.
16. N. Zulkiple, M. Yusuf, O. Hassan, Pretreatment of Oil Palm Empty Fruit Fiber (OPEFB) with Aquaeous Ammonia for High Production of Sugar. **18**, 155–161 (2016).
17. F. Lu et al., Naturally p - Bioenerg. Res. (2015), doi:10.1007/s12155-015-9583-4.
18. M. Ago, A. Ferrer, O. J. Rojas, ACS Sustain. Chem. Eng (2016), doi:10.1021/acssuschemeng.6b01279.
19. J. Paltseva, S. Searle, C. Malins, "Potential For Advanced Biofuel Production From Palm Residues In Indonesia" (Washington, 2016).
20. Y. Zhu, F. Xin, Y. Zhao, Bioprocess Biosyst Eng (2014), doi:10.1007/s00449-014-1209-2.
21. Departemen Kehutanan, Pedoman Pengelolaan Limbah Industri Kelapa Sawit (Jakarta, 2006).
22. M. D. B. M. Ali, "Conversion of OPEFB to Fermentable Sugar" (2014).
23. A. Geng, in Liquid, Gaseous and Solid Biofuels - Conversion Techniques (2013).

24. H. Roliadi, "Manufacturing and Qualities of Paperboard from the Mixture of Empty Oil-Palm Fruit Bunches and Pulp-Mill Sludge" (2009).
25. D. Anggraini, H. Roliadi, Penelit. Has. Hutan. **29**, 211–225 (2011).
26. R. Lusiani, Sunardi, Y. Ardiansah, J. Tek. Mesin Untirta. **I**, 46–54 (2015).
27. K. Syamsu, L. Haditjaroko, G. I. Pradikta, H. Roliadi, J. Ilmu Pertan. Indones. **19**, 14–21 (2014).
28. U. Fathanah, Sofyana, J. Rekayasa Kim. dan Lingkung. **9**, 137–143 (2013).
29. K. Abdullah et al., J. Din. Penelit. Ind. Vol. **27**, 49–58 (2016).
30. E. Saputra, S. Bahri, E. Hs, J. Rekayasa Kim. dan Lingkung. **6**, 45–49 (2007).
31. E.-A. Program, Buku Panduan Pabrik Kelapa Sawit Skala Kecil Untuk Produksi Bahan Baku Bahan Bakar Nabati (BBN) (2009).
32. N. Gurning, TELAAH J. Ilmu Pengetah. dan Teknol., 13–20 (2013).
33. A. Cahyono, W. A. Nugroho, D. Kadarisman, J. Keteknikan Pertan. Trop. dan Biosist. **4**, 33–40 (2016).
34. M. Ulya, T. I. Pertanian, U. T. Madura, in Prosiding Konferensi Nasional "Inovasi dalam Desain dan Teknologi" (2011), pp. 349–356.
35. Suyanto, in Seminar Rekayasa Kimia Dan Proses (2010), pp. 1–4.
36. J. Supraniningsih, Widya. **29** (2012).
37. A. A. L. Tbk, "Laporan Tahunan 2013 Annual Report Contents" (2013).
38. Kartika, L. N. Sulistyaningsih, Z. P. Negara, in Prosiding Seminar Nasional Lahan Suboptimal 2015, vol. **2012**, pp. 1–8, (2015)
39. S. Andri, S. I. Saputra, Di pre nursery. J. Agroteknologi. **7**, 1–6 (2016).
40. Nasrul, T. Maimun, J. Rekayasa Kim. dan Lingkung. **7** (2009).
41. Y. Sitio, gede wijana, i. G. N. Raka, Periode Pre Nursery. E-Jurnal Agroekoteknologi Trop. **4**, 264–273 (2015).
42. A. Suryani, E. Hambali, J Tek. Ind. Pert. **18**, 127–137 (1999).
43. I. Tahir, J. Kimia, F. Matematika, P. Alam, Mns. dan Lingkung. **14** (2007).
44. A. M. Jannah, F. Asip, Bioethanol Production from Coconut Fiber using Alkaline Pretreatment and Acid Hydrolysis Method. **5**, 320–322 (2015).
45. A. B. Juanssilfero, A. C. Djohan, A. Purnawan, International J. Adv. Sci. Eng. Inf. Technol. **5**, 1–5 (2015).
46. B. D. Argo, Y. Hendrawan, D. M. Maharani, A. W. Putranto, International J. Adv. Sci. Eng. Inf. Technol. **6**, 20–26 (2016).
47. N. Nazir, E. Juita, C. Amelia, R. Fatli, International J. Adv. Sci. Eng. Inf. Technol. **6**, 403–409 (2016).
48. S. Zullaikah, T. S. Zigmawiko, S. Wafa, in Prosiding Seminar Nasional Teknik Kimia "Kejuangan" Pengembangan Teknologi Kimia untuk Pengolahan Sumber Daya Alam Indonesia Yogyakarta, pp. 1–8, (2015)
49. Hadiyanto, D.Ariyanti, A.P.Aini, D.S. Pinundi, Energy Procedia **47**,108-112 (2014)

 https://doi.org/10.1051/e3sconf/20183103015

Biodegradation of COD in Household Wastewater with Aerobic Biofilm Technology by Adding Sediment Drainage Sewerage

Sri Sumiyati[1*], *P.* Purwanto[2], *Endro* Sutrisno[3],*S.* Sudarno[3], *Jalu* Arthawidya[3]and *Humam* Izzudin[3]

[1]Doctoral Program of Environmental Sciences, School of Post Graduate, Diponegoro University, Semarang, Indonesia
[2]Department of Chemical Engineering, Faculty of Engineering, Diponegoro University, Semarang, Indonesia
[2]Department of Environmental Engineering, Faculty of Engineering, Diponegoro University, Semarang, Indonesia

Abstract. Household wastewater contains contaminants that harm the environment. One of the pollutants found COD. If being discharged into the environment directly, COD concentrations exceeding the existing quality standard will disrupt the ecosystem in the receiving water body. One of the technologies that can degrade COD is biofilm technology with honeycomb tube media. This research aims to analyze the decrease of COD concentration present in wastewater household with biofilm technology of honeycomb media. The reactor used in was made of glass with a thickness of 4 mm, a volume of 18 litres and operated continuously. The media used is made of a PVC pipe, cut to 3 cm in size and then glued to one another, forming a honey comb. The results showed that there has been a decrease in COD concentration of household wastewater.

1Introduction

Household wastewater is a very important issue recently. If it is not well managed and in right ways, it will be harmful to the environment. Household wastewater contains numerous organic pollutants, such as Biological Oxygen Demand (BOD) and Chemical Oxygen Demand (COD). Besides, it exists other pollutants, i.e. TSS, Ammoniac, Phosphate, pH, temperature, and fat. The existence of pollutant's parameter in the household wastewater is if the concentration is over the specified standard, it will bring negative impact to the balance of ecosystem of nature. That is why the household wastewater needs to be proceeded first before being drained into the environment. One of the household wastewater treatments is by biological wastewater treatment. Biological wastewater treatment is divided into two, i.e. anaerobic and aerobic. One kind of biological wastewater treatments is biofilter and biofilm technology.

Biofilter and biofilm technology are wastewater treatment technology by using the microorganism activities, where those microorganisms attach on the surface of a medium and form biofilm layer. According to [1] biofilm layer was alive bacterial community and stuck on the surface of abiotic. Biofilm technology had numerous advantages than other types of treatments. According to [2] biofilm technology had numerous advantages compared to the suspension system and active mud. According to [3] biofilm technology had a high tolerance to the shock loading.

In this research, we chose biofilm technology due to the consideration that it was efficient and effective technology to decrease the pollutant parameter of household wastewater. Median that was used in this research was honey comb tube. Consideration using honey comb tube as a median was because it was resistant to clogging and had large specific surface area. Some researchers who had used bio-filter technology and honey comb tube as median were [4] reported that bio-filter technology of honey comb tube could reduce the wastewater COD of chicken slaughterhouse up to 78-87%. The aim of this study was to analyze the efficiency of COD decreasing on the household wastewater by using biofilm technology in aerobic condition.

2 Material and Methods

2.1 Synthetic Wastewater

The wastewater that was used in this research was synthetic wastewater, with the material compositions as follow: skim powder, K2PO4, NH4CI and fresh water. Those materials include the materials of Pro Analysis (PA) standard that were bought on Laboratory supply store. This synthetic household wastewater COD was set on concentration 543.12 mg/I.

2.2 Biofilter Media

Biofilter media that was used in this research was created from PVC pipes with the diameter ½ inch, and those PVC pipes were cut 5 cm long. PVC pipes that had been cut off were glued together and formed becoming media that resembled honey comb tube. The biofilter

* Corresponding author: srisumiyati71@gmail.com

media in the form of honey comb tube was shown on Figure 1.

Fig. 1. Biofilter Media in a shape of Honey Comb Tube

2.3 Research Instruments

Table 1. The instruments in this research were shown

No.	Description	Explanation
1.	Reactors: • Materials • Length • Wide • Height	 Glass, with the thickness of 4 mm 30 cm 20 cm 18 cm
2	Biofilter media • Materials • Length of the piece • Glue • Diameter	 PVC pipes 5 cm Pipe glue ½ inch
3	Inlet and outlet pipes	PVC with the diameter of ½ inch
4	Submersible pump AA – 106	Voltage: 220-240 V, Frequency: 50/60 Hz, Watt: 85 W, F Max: 4000 L/H, Hmax: 4 meters
5	Air Supply Aerator	Brand: Amara, 1 hole, 2.5 Watt
6	First Basin	Plastic material, Volume: 200 liter
7	Equalization basin	Material: Plastic, Volume: 200 liter
8	Discharge controller	Discharge regulator
9	Last Basin	Material: Glass, with the thickness 4 mm

2.4 Operational of the Reactor

The reactor was made of glass and in shape of blocks. The reactor operation was by draining the wastewater to the first basin and then it was drained to the equalization basin. From the equalization basin, the wastewater was drained into the aerob reactor and at last, it was accommodated in the effluent basin. The design of aerob biofilm reactor used Honey Comb Tube media was shown on Figure 2.

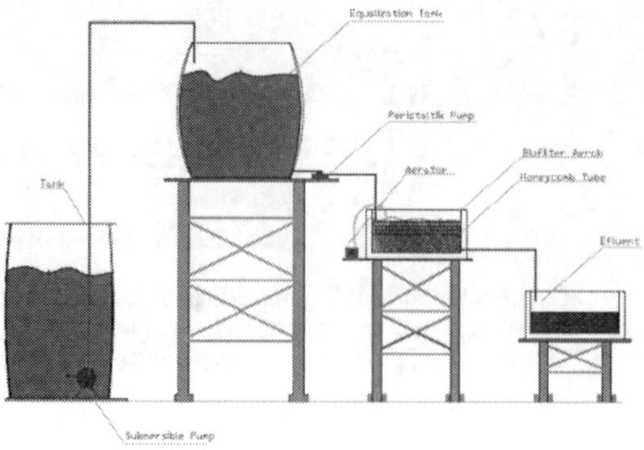

Fig. 2. Design of Aerobic Biofilm Reactor Used Honey Comb Tube

2.5 Sample Collection

The sample collection was done at certain points in the reactor. The collected samples were taken to the laboratory to be tested the COD concentration in the wastewater.

3 Result and Discussion

3.1 Cultivation of Microorganism (Seeding Step)

The process of microorganism cultivation was done naturally, by adding the mud in the drainage channel. Consideration to use mud addition in the drainage channel was due to the mud contained numerous microorganism that had been adapted to the condition of household wastewater. The time selected on acclimatization step was 15th days. In those 15th days, it was expected that the microorganism, which created biofilm layers, had begun to be mature and steady state to proceed the household wastewater continuously. The research of [5], used 10 days of the acclimatization time. [2] also reported that biofilm to reach maturity needed some days. The process in the reactor of aerob biofilm was said stable if the process run well, with the characterized by the covering of the entire bio-filter media surface by biofilm layers and happened the decreasing of COD concentration constantly. According to the physical observation inside the aerob biofilm reactor, it could be seen that on the first step of acclimatization, the biofilm layers were not formed yet, and the honey comb tube media seemed clear. After the 3rd day to 7th day, the biofilm layer had started to be formed although not the entire surface of biofilter media had been covered. After the 7th day, which was the 8th day to 15th day, visually it could be seen the biofilm layers that stuck on the surface of the honey comb tube media almost covered the entire surface of the media. The honey comb tube media was almost unseen, the wastewater that was drained from the aerob reactor was also clear and clean compared to the un-proceeded household wastewater. The efficiency of COD decreasing on the last step of acclimatization had reached stable number, which was on between 60-75 75%.

3.2 Content of the Research

After the step of microorganism cultivation, the next step was the reactor running step. pH of the wastewater on this research was set up on a range of 7.0 – 8.5 and on temperature of 26 - 27°C. After the biofilm layer was on steady state condition, the wastewater was then drained continuously on the discharge of 8.5 ml/ minute inside the aerob bio-filter reactor. Time to stay for wastewater in this research was set up to 12 hours. The chart of COD decreasing efficiency on the aerob reactor was shown on Figure 3.

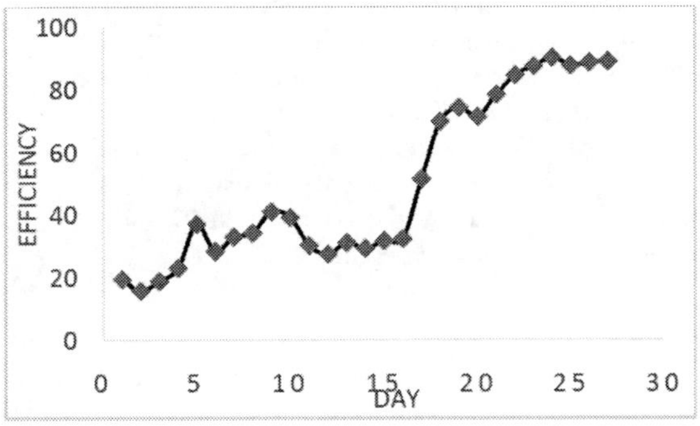

Fig. 3. Chart of COD Concentration Decreasing Efficiency

The chart of COD Concentration Decreasing Efficiency was shown on Fig 3. From that chart; it could be seen that at the first time running the reactor until the 3rd day, the value of decreasing efficiency was still low, which was less than 20%. Besides, visually the wastewater was still stink and muddy. It was because of the change of the steps from acclimatization to running step, so the microorganisms that formed new biofilm layers adjusted the set-up condition on the running step, which was the incoming discharge was set continuously, including pH and temperature of the wastewater. As each passing day of reactor running, it seems that after the 3rd day, the efficiency value started increasing, the wastewater stink was lessen and the color had been closer to the clear. On the 15th day (about more than 2 weeks), the value of COD concentration decreasing efficiency reached the value of about 20-30%. After two weeks, the efficiency value was continue rising, the chart trend started rising, the efficiency value reached above 50%. On Picture 3, it could be seen that after the 20thday, the efficiency value reached stable values, which were between 80-90%. This condition showed that microorganism that formed biofilm layers had worked well. The microorganism activities that stacked on the surface of biofilter media, so the process of suspense solidity extraction and decomposition of organic compound in the wastewater run well, it was shown by the value of COD decreasing efficiency from time to time increased up to 90%. Besides, the condition of wastewater that had been proceeded was not muddy and the stink was started to go.

4 Conclusion

Result on this research showed that the aerob biofilm technology, using honey comb tube by adding mud on the drainage channel, was able to decrease the COD concentration up to 90%.

References

1. M.D. Macia, E. Rojo-Molinero and A. Oliver, Clin Microbiol Infect,**20**,981–990, (2014)
2. S. Anderson. Characterization of Bacterial Biofilms for Wastewater Treatment. Printed by Universitetsservice US-AB Drottning Kristinas väg 53B SE-100 44 Stockholm Sweden ISBN 978-91-7415-255-5 TRITA-BIO Report 2009:3 ISSN 1654-2312 (2009)
3. S. Schlegel and H. Koeser,Water Sci.Technol. **55**(8–9), 83–89. (2007)
4. N.I. Said and Firly. JAI, **1**, 3, Teknologi Lingkungan, BPPT Indonesia, (2005)
5. M.C. Lee, Yen-Hui Lin, Huang-Wei Yu. 2014. Biodegradation **25**,849–865. DOI 10.1007/s10532-014-9705-2, (2014)

Decreasing of BOD Concentration on Artificial Domestic Wastewater Using Anaerob Biofilter Reactor Technology

Sri Sumiyati[1,*], *P.* Purwanto[2],and *S.* Sudarno[3]

[1]Doctoral Program of Environmental Science, School of Postgraduate Studies, Diponegoro University, Semarang - Indonesia
[2]Department of Chemical Engineering, Faculty of Engineering, Diponegoro University, Semarang - Indonesia
[3]Department of Environmental Engineering, Faculty of Engineering, Diponegoro University, Semarang - Indonesia

Abstract. Pollution of domestic wastewater becomes an urban problem. Domestic wastewater contains a variety of pollutants. One of the pollutant parameters in domestic wastewater is BOD. Domestic wastewater which BOD concentrations exceeding the quality standard will be harmful to the environment, particularly the receiving water body. Therefore, before being discharged into the environment, domestic wastewater needs to be processed first. One of the processing that has high efficiency, low cost and easy operation is biofilter technology. The purpose of this research was to analyze the efficiency of BOD concentration reduction in domestic wastewater with anaerobic reactor biofilter using volcanic gravel media. The type of reactor used is an anaerobic biofilter made of glass which volume of 30 liters while the biofilter media is volcanic gravel. In this research the established HRT were 24, 12, 6 and 3 hours. The results showed that the efficiency of BOD concentration reduction in artificial domestic wastewater reached 80%.

1 Introduction

Domestic wastewater contains numerous pollutants that are harmful to the environment. Pollutant parameters that exist in domestic wastewater, i.e.: BOD, COD, TSS, Ammoniac, fat, pH and temperature. If the pollutant concentration reaches over the quality standard and directly thrown away to the environment; then it will negatively affect to the human being and other living creatures. That is the reason why before throwing it to the environment, the domestic wastewater needs to be processed first. Today, the technology of domestic wastewater processing has been applied in the fields. One of the technologies, which have many advantages, is bio-filter/ bio-film technology. The advantages of bio-filter technology, i.e.: cheaper, easy to use, and effective to decrease the pollutant parameter. As stated by [1], the bio-filter/ bio-film technology had very high efficiency on wastewater processing. [2] also stated that on applying the bio-film technology, it is rarely found the "bulking" matters so the processing becomes easy. According to [3], the used bio-filter media can be used for long term period without regeneration. Usage of bio-filter media needs cleaning to remove the dirt on the surface. Besides, the bio-filter technology also does not produce much mud, so it does not need mud handling on its processing.

Some researchers used bio-film technologies had been done by some researchers, i.e.: [4] proceeded simulation wastewater using the decreasing of nitrogen and phosphor by bio-film algae. The result showed that the decreasing efficiency for each, Total Phosphor (TP) 98.17%; Total Ammoniac (TN) 86.58%; Ammonia Nitrogen (NH3-N) 91.88% and COD 97.11%. Besides, [5] did the domestic wastewater processing using bio-filter of Moroccans sand, it was got the decreasing result as NH4-N and COD for 65% and 85%.

That was the reason why the technology chosen for this research was a technology of anaerobic bio-filter. The choosing of bio-filter technology was due to cheap, easy to use and effective according to the high-efficiency value of the pollutant concentration decreasing. While the consideration to use anaerobic system was due to generally in any field, the application used was mostly anaerobic. Besides, the anaerobic system was not expensive on its operational. Different from aerobic system, it needed clower or aerator to add oxygen in the reactor, so it needed electricity that would increase the operational cost.

Media used in this research was volcanic gravel. Consideration to take volcanic gravel was due to it was available abundance in nature and the characteristic of volcanic gravel had the rough and uneven surface so it was effective as a place to stick the microorganism that would create the bio-film layer. The aim of the research was to analyze the concentration decreasing of BOD in the domestic wastewater by using anaerobic bio-filter technology. Pollutant parameter analyzed was BOD concentration on artificial domestic waste.

2 Material and Methods

Domestic wastewater that was processed on this research used artificial wastewater. Materials to create artificial

* Corresponding author: srisumiyati71@gmail.com

https://doi.org/10.1051/e3sconf/20183103016

domestic wastewater on this research were: skim milk powder (Source C), KH_2PO4 (Source P), NH_4Cl (Source N) and water [6]. All materials used to create artificial domestic wastewater were included pro analyst (p.a) material. Bio-filter media used, was volcanic gravel taken from one of the rivers in Yogyakarta, Indonesia. The size of volcanic gravel used in this research was about 2-5 cm, the shape of gravel was heterogenic, the chosen gravels were the ones which had the rough surface and not fragile. The used equipment on this research were: glass with thickness of 4 mm, rectangle shaped with dimension of 40 cm x 25 cm x 30 cm, submersible pump of AA 106 brand with the specification of voltage: 220 – 240 V, frequency: 50/ 60 Hz, Watt: 85 W, F Max: 4000 L/H, Hmax: 4 meters. Other pieces of equipment used in this research were: peristaltic pump with the debit of 28 ml/ minute, plastic barrel with the volume of 200 liters, PVC pipe with the diameter of ½ inch, covered bucket with the volume of 20 liters. Equipment to measure the BOD concentration in a laboratory were a spectrophotometer, BOD bottle, pH meter, thermometer, DO meter, volumetric flask, measuring cup, measuring pipette and BOD incubator. The sample taking was done in some sites. After sample taking in some sites, then the sample was tested for the BOD concentration in Environmental Laboratory, Department of Environmental Engineering, Faculty of Technique, Diponegoro University, Indonesia.

Design of anaerobic bio-filter reactor with the bio-filter media of volcanic gravel was shown on Figure 1.

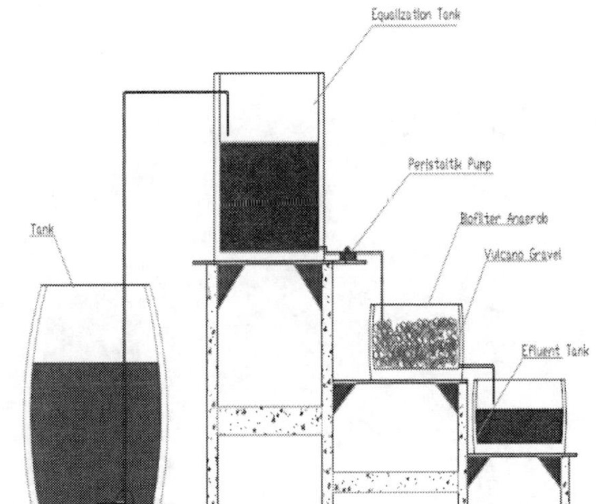

Fig. 1. Design of Anaerobic Bio-filter Reactor with Media of Volcanic Gravel.

3 Results and Discussion

3.1. Research's Procedure

This research was divided into three steps, i.e.: Seeding, acclimatization and running. The three steps were related each others. Seeding step was stepped to breeding micro-organism. The purpose of this step was to grow the microorganism. Seeding on this step was done naturally. The waste was flown into the biofilm reactor with the anaerobic condition. A reactor that had been fulfilled by media and wastewater was then made the condition so that the micro-organism could grow on the median surface of the volcanic gravel. Furthermore, the acclimatization steps or called as a step of micro-organism adaptation where in this step the micro-organism had started growing and adapting inside the anaerobic reactor. The microorganism was breeding and sticking on the gravel media surface creating the bio-film layer.

This bio-film layer contained various micro-organism types, such as protozoa, bacteria, fungus, etc. As stated by [7], the bio-film layer contained some types of microorganism, such as bacteria, diatoms, fungus, algae, protozoa and non-cellular materials. According to [8], the bio-film layer was a micro-organism aggregate that adhered on the solid surface and the existence of Extracellular Polymeric Substance (EPS). The acclimatization step was ended up after the condition of bio-film layer reached steady state. Next, it was done the step of reactor operating (running). Running was done by flow the wastewater continuously with the debit set up according to the time to stay that had been set. The setup time-to-stay in this research were 24 hours, 12 hours, 6 hours and 3 hours.

3.2 Analysis of BOD Concentration Decreasing Efficiency on Domestic Wastewater According to Residence Time

On this research, the analyzed pollutant parameter was BOD concentration. The BOD concentration of domestic wastewater was set on the concentration of 492.35 mg/ l, pH 7.1 – 8.1 while the temperature was set on 27.6 – 28.3°C. The chart of efficiency of BOD concentration decreasing on wastewater according to time to stay was shown on Figure 2.

On Figure 2, it was shown that the chart of efficiency of BOD concentration decreasing generally experienced assertion. In the 24 hours HRT time, the efficiency of BOD concentration decreasing started from the lowest of 10.81% until reached the highest efficiency on the 8th day that was 66.39%. While on HRT 12 hours, the efficiency had been increased on the first day 69.62% and reached the highest value on the second day, which was 72.85%. On HRT 6 hours, the lowest efficiency was 34.68% and the highest was 73.82%. On HRT 3 hours, the lowest efficiency was 57.34% while the highest was 88.36%.

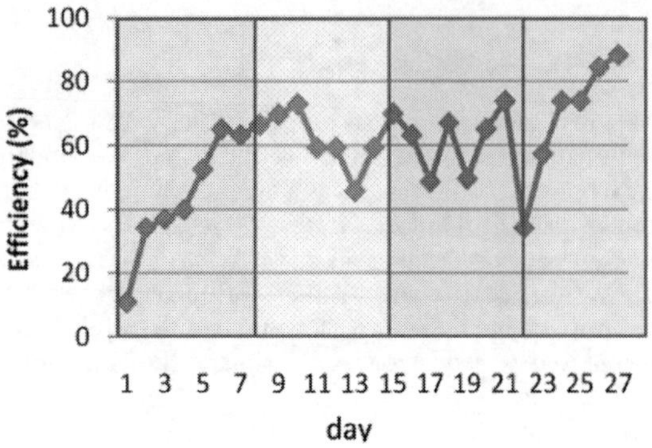

Fig.2. Chart of efficiency of BOD concentration decreasing on Anaerobic condition.

On the result of this research, it could be seen that on all HRT happened the degradation of organic compounds in the domestic wastewater that was marked by BOD concentration decreasing compared to the earlier concentration. From the Figure 2, it could be seen that for HRT 24 hours happened the increasing of efficiency on almost all time observations. Whereas on HRT 12 hours and 6 hours the charts trend tended to be stationary with the efficiency that was relatively stable. On HRT 3 hours, the efficiency also experienced the ascension in all time observations until reached the highest efficiency, which was 88.36%.

3 Conclusion

The wastewater processing used anaerobic biofilter technology with volcanic gravel medium was effective to decrease the BOD concentration, which the highest efficiency of 88.36%.

References

1. N. Houda, C. Abdelwaheb, B.R. Asma, M. Ines, L. Ahmed and H.Abdennaceur,*CurrMicrobiol*. **70**:602–609.(2015)
2. N.I. Said,*JAI*.**1**, 1 (2005)
3. D.S. Chaudhary, S. Vigneswaran, H.H. Ngo, H.S.D. Moon, *Korean J. Chem. Eng*. **20**, 6, 1054-1065. (2003)
4. Q. Wei andG. Zhiguan, *Front. Environ. Sci.Engin*. **2,** 4, p:446-451 (2008)
5. K.I. Yettefti, S. Etahiri, D. Malamis, O. Assobhei, *Carpathian Journal of Earth and Environmental Sciences* **8,** 3 p: 207-218 (2013)
6. I. Nopen, C. Capalozza and P.A. Vanrolleghem, *Technical Report*. Department of Applied Mathematics, Biometrics and Process Control. Universiteit Gent. Belgium. July,p : 4 (2001)
7. S.G. Wesley, S. Satheesh, *Journal of Experimental Marine Biology and Ecology*, **379**, p: 1-7. (2009)
8. Donian, *Biofilms: Microbial Life on Surfaces. Emerging Infectious Diseases* **8**, p: 881- 890 (2002)

 https://doi.org/10.1051/e3sconf/20183103017

Optimization of Wastewater of Batik Buaran Pekalongan by Using Photocatalytic Membrane Bioreactor

Fahmi Arifan[1], *FS* Nugraheni[1], and *Niken* Elsa Lianandaya[1]

[1]Chemical Engineering Department, Vocational School, Diponegoro University, Semarang, Indonesia

Abstract. The purpose of this study is to determine the final COD concentration reduction by changing COD and MLSS concentration on the performance of submerged membrane bioreactor (MBRs) as a waste treatment of Batik in Buaran Pekalongan. The method is covers the process of seeding, the acclimatization process and the main process. Description of the process that we take an active mud from IPLT Buaran Pekalongan, then we analyze the sludge MLSS, MLVSS, COD, BOD, and TSS. After that we enter the active sludge in the bath nursery that has been given aerator (a tool for aeration) and made provision in the form of NPK nutrients and glucose at a ratio of 1:10. Activated sludge from the acclimatization process is inserted into the MBRs (membrane bioreactor submerged) that is equipped with an aerator. Then prepare influent(waste to be lowered concentration of COD). How, liquid waste of Batik Pekalongan Buaran COD diluted concentration of 10,000 mg / l and 15,000 mg / l, and then inserted in influent tub. After that liquid waste of Batik Buaran Pekalongan influent flowed into Photocatalytic Membrane Bioreactor, of MPB effluent flowed into the tub (result).

1. Introduction

Nowadays, the development of technology in all fields to bring a considerable impact on human life. One negative aspect of these new technologies is the environmental degradation caused by the waste. Types of waste can be, among others, solids, liquids, and gas. Waste is very dangerous because it contains a high concentration of chemical substances, therefore it must be lowered levels of COD (Chemical Oxygen Demand) [1,2]. Process wastewater treatment with submerged membrane bioreactor, have many advantages over waste treatment with other membrane processes. It is seen from the standpoint of effectiveness and efficiency of the process itself is simple and automatic operating costs are also much smaller with better results.

Theoretically, the combination of activated sludge processing and submerged membrane will produce better processing performance than conventional activated sludge (sludge and clarifier) [3]. This is because the ability of the membrane filtration is not influenced by the characteristics biofloc formed. Membranes can be generally defined as a selective barrier between the two phases. Said selective here describe something that is typical of the membrane and the process by using a membrane, but did not explain matters related to the structure and function of the membrane itself. The membranes have different thicknesses. No structure or homogeneous. Transport of molecules across the membrane can be active or passive. Passive transport occurs because of differences in concentration, pressure, and temperature. Because of very varied shape and characteristics of the membrane, the membrane is carried classification. One classification, membrane divides into two parts, namely biological membranes and synthetic. Biological membrane is in the body of living creatures, as well as cell-cell life.

These membranes can be divided into the membrane life (living membrane), and the membrane inanimate (non-living membrane). Synthetic membranes can be divided into organic membrane (polymer) and inorganic Separation techniques contained in membrane technology there are two kinds of cross-flow system (crossflow) and passing flow (dead end). Cross flow (crossflow) is the flow of the bulk solution including the parallel flow on the membrane surface and only a few solutions left on the surface of the membrane. Named "crossflow filtration" for bulk flow immediately follow the permeate flow. While applied in this study is a "dead-end". The flow of dead-end, including the bulk solution flow across the membrane surface and bulk solution including the solids trapped on the membrane surface. As a consequence, the collected solids filtered continuously and form clogging the membrane surface. In a membrane bioreactor, SS concentration of the bulk solution, such as activated sludge, which has a high concentration on a dead-end where the separation is

* Corresponding author: fahmiarifan80@gmail.com

generally not possible due to the active sludge will form quickly clog the membrane [4,5].

There are two configurations of membrane bioreactor (MBR) that: membrane of bioreactor located outside and submerged membrane bioreactor. In the configuration of external MBR, mixed liquor is pumped from the aeration tank to the pressurized membrane system located outside the bioreactor where suspended solids are retained and fed back into the bioreactor, while effluent penetrating the membrane. Membranes need in backwash periodically to remove suspended solids formed and accumulates and is chemically washed when the pressure becomes too high operational. So that high-quality effluent that can be achieved without depending on MLSS concentrations and deposition characteristics biofloc. Although there is still the possibility of the outbreak of microbial floc (biofloc) due to high shear stress caused by high crossflow velocity [6]. High energy recirculation pump which can be an obstacle. Submerged membrane separation activated sludge process is important in the treatment of wastewater, because it can overcome the problem of microbial lyses shear because high crossflow velocity. In a submerged MBR configuration, low-pressure membranes immersed in aeration tanks operated by vacuum pressure. The membrane is agitated by air bubble aeration system that can help prevent the accumulation of suspended solids on the membrane surface. Periodically submerged membrane and washed chemically in backwash when the operating pressure becomes too high. But does not require a circulating pump.

The use of a membrane allows the MBR operate without depending on the quality of the deposition of silt and eliminate the need clarifier. Consequently submerged MBR can typically be operated at MLSS concentration between 5000-40000 mg / L. MBR submerged in operation at the age of sludge exceeds forty days. Operational parameters submerged MBR is different than conventional biological wastewater systems. This study seeks to develop a modified activated sludge treatment that is more compact in terms of construction of the building through a particular configuration. That is, with the removal of the building after the building activated sludge clarifier. Clarifier used as a substitute for submerged membrane bioreactor installed activated sludge. With the modifications, this research tried to explore the factors that influence the combination of activated sludge and membrane submerged [5, 7, 8]. By trying some of the variables will be known whether the combination of activated sludge and submerged membrane is quite reliable and capable of overcoming the problems which arise in the ordinary activated sludge treatment. Research to be carried out include the initial experiments to test the efficiency of the membrane-based process and ultrafiltration process separately. The main experiment is done by combining both instruments. The material to be processed is wastewater of Batik Buaran Pekalongan. Membranes used an ultrafiltration membrane. By observing the performance of membrane reactors will be determined ultra filration operating parameters that affect the water purification process and will obtain optimum operating conditions. In this way an attempt to reduce the waste problem that happened lately.

1.1 Aim

To determine the final COD concentration reduction by changing the concentration of COD and MLSS on the performance of submerged membrane bioreactor (BMRt) as a waste treatment of Batik in Buaran Pekalongan

1.2 Benefits of Research

The benefits of this research are as follows:

a. Developed an alternative solution to conventional biological treatment Lumpur active, so that problems at the operational activated sludge process can be reduced as well as enhance the performance of activated sludge in biological wastewater treatment.
b. Studying the effect of variation of organic load (COD), a load of solids in the form of microorganisms concentration (MLSS) on the performance of the combination of activated sludge and submerged membrane.

Preliminary analysis of the wastewater Batik Buaran Pekalongan conducted to determine the BOD and COD waste or liquid. With this analysis it can be seen BOD-COD concentration, temperature, pH, and total solids. This data is used later to calculate the type and amount of nutrients need to be added and the conditioning phase of acclimatization microbes. For purposes of analysis, COD concentration is done with the titration method and MLSS, MLVSS determined by Standard Methods For Examination of waste and wastewater.

2. Methods

2.1 Seeding Phase

Seeding is done by taking an active Lumpur obtained from domestic unit in aerobic wastewater. Activated sludge obtained from a waste management unit Stool (IPLT) Sukolilo Surabaya. To increase the MLSS, the active Lumpur obtained aerated and added glucose substrate and nutrient required. The amount of glucose addition estimated to be sufficient for microbial energy needs for growth and maintenance. At this stage, MLSS analysis carried up to25000mg/l

2.2 Acclimatization stage

Microorganisms are the adjustment phase to the surrounding environment that can degrade organic material from the waste Batik Buaran Pekalongan.

 https://doi.org/10.1051/e3sconf/20183103017

2.3 Characterization Membrane

To characterize the membrane to determine the permeability of the membrane. The feed solution that is used to characterize the membrane obtained from the mud that has acclimatized. Characterization of membrane also be obtained from the critical flux membranes used. To obtain the critical flux carried combinations using the technique of back flushing.

2.4. Top Trial Phase

Variables that are used to perform the main experiment are:
- Concentration of Microorganisms (MLSS): 15,000; 20,000 and 25,000 mg / l
- Organic Load (Concentration COD): 10,000 and 15,000 mg / l

3. Results And Discussion

The process of seeding is done to increase the number of microorganisms present in activated sludge. In this case, the mud used is derived from the waste treatment plant sludge Buaran Pekalongan. Mud obtained came from the OD (Oxidation Dict) and clarifier. In the initial condition is known that the amount contained in the sludge MLSS of 11880 mg / l, MLVSS of 6040 mg / l, COD of 710 mg / l, BOD of 285 mg / l, TSS of 5270 mg / l, having a pH of 7.8 and the temperature is 270C.

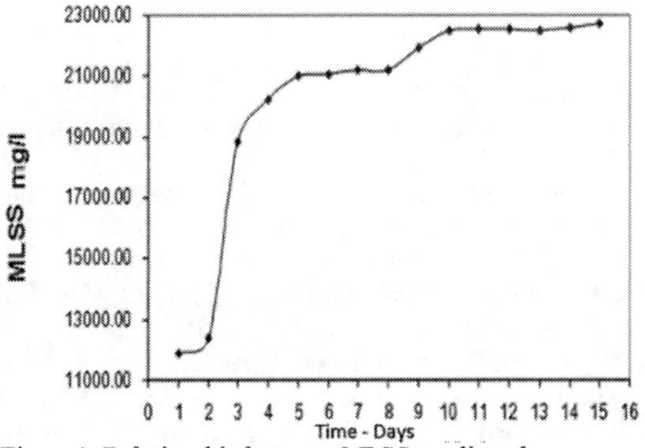

Figure 1. Relationship between MLSS seeding phase

The process of seeding sludge is done by mixing the activated sludge from the clarifier plus of OD, then after that added nutrients in sludge with aeration process. nutrients
given consists of two kinds of substances, namely NPK and Glucose with a ratio of 1:10. Giving nutrient done to increase the number of MLSS in the activated sludge. Seeding process is stopped when the amount of MLSS has been fulfilled. Here's a picture chart of the process of seeding.

4. Conclusion

This study can conclude:
- On the MLSS concentration of 15,000 mg / l for initial COD concentration of 10,000 and 15,000 mg / l, the concentration of COD can be derived is 98.36%.

- In the MLSS concentration of 20,000 mg / l for initial COD concentration of 10,000 and 15,000 mg / l, the concentration of COD can be derived is 98.41%.

- In the MLSS concentration of 25,000 mg / l for initial COD concentration of 10,000 and 15,000 mg / l, the concentration of COD can be derived is 98.42%.

From the research results in decreased concentrations of COD by using active Lumpur with MLSS concentration of 15,000 mg / l, 20,000 mg / l, 25,000 mg / l for waste of batik in Buaran Pekalongan with COD concentration10,000 mg / l and 15,000 mg / l obtained results inaccordance with the standard quality standard, so this study was successful, and can be applied to industries that treat their own wastes. Especially for waste Batik BuaranPekalongan.

References

1. Sugiharto. *Dasar–dasarPengelolaanAir Limba.* UI – Press,Jakarta (1987)
2. Surizal, Ariy. ,Whin Andreanne, *Pengaruh COD Terhadap Kinerja Kombinasi Lumpur Aktif dan Membran terendam dalam mengolah Limbah Cair.* ITATS Surabaya.
3. Cote, Piere., Herve Buisson, Charles Pound, Greg Arakaki. Desalination. **113**.189-196. (1997)
4. Cicek,N., J. P. Franco, M. T. Suidan, V. Urbain, J. Manem. Wat Environ. Res.71,64-70. (1999)
5. Ghyoot. W, W. Verstraete,.Wat. Res. **34**, 205-215. (1999)
6. Chan A.H,. Water Science and Technology, **44**, 187-195. (20010
7. M. Noguchi, *Membrane-Cleaning Method for Membrane Bioreactor Process*, Patents No. 7,422,689 B2, 2008
8. Metcalf & Eddy., Watewater Enginering. Mc.Graw – Hill International Editor. (1991)

Dose of Biocoagulant-Mixing Rate Combinations for Optimum Reduction of COD in Wastewater

Maria Faustina Patricia[1, *], *Purwono*[1], *Mochamad* Arief Budihardjo[1]

[1]Department of Environmental Engineering, Faculty of Engineering, Diponegoro University, Semarang - Indonesia

Abstract. Chemical oxygen demand (COD) in domestic wastewater can be treated using flocculation-coagulation process with addition of Oyster mushroom (*Pleurotus ostreatus*) in powder form as biocoagulant. The fungal cell wall of Oyster mushroom comprises of chitin that is high polyelectrolyte and can be function as an absorbent of heavy metals in wastewater. The effectiveness of flocculation-coagulation process in treating wastewater depends on dose of coagulant and mixing rate. Therefore, this study aims to determine the best combination of three variation of dose of biocoagulant which are 600 mg/l, 1000 mg/l, and 2000 mg/l and mixing rate which are 100 rpm, 125 rpm, and 150 rpm that give the most reduction of COD in the wastewater. The result indicates that the combination of 1000 mg/l of biocoagulant and 100 rpm of mixing rate were found to be the most optimum combination to treat COD in the wastewater with COD reduction of 47.7%.

1 Introduction

Wastewater of canteen Dean of Faculty of Engineering Diponegoro University and canteen of Electrical Engineering Diponegoro University is one of the waste water that contributes to pollution if not processed first. Contaminant sources contained in waste water canteen of Dean of Faculty of Engineering Diponegoro University and canteen Electrical Engineering Diponegoro University, among others, is pathogenic bacteria and organic and inorganic solids. So that the result of pollution caused by canteen waste water make the channel become clogged, and odor and waste of processed product still not fulfill the required quality standard.

In domestic liquid waste canteen requires the main ingredients that can be used to wash the existing dishes of detergent or soap. Detergent is a derivate organic substance in which its accumulation will cause increased organic content in the waste water. Waste with organic content that continuously contaminates water bodies will cause negative impacts, such as foul odor, source of aquatic diseases, and water quality bodies that are not in accordance with its designation [1]. The detergent or soap used in the own washing industry can cause the resulting liquid waste to contain some contaminants which, if excessive, can be harmful to the environment is COD.

One of the treatment alternatives that can be done to degrade the pollutants is by using the coagulation and flocculation method in the jartest method. Coagulation-flocculating water purification generally uses aluminum salt coagulant [2]. However, due to environmental reasons, many doubted the use of aluminum coagulant. Other alternatives such as iron salts and synthetic polymers actually began to popular, but the application is still tripping over the price factor and environmental problems that may arise. In recent years, interest in natural coagulants from plants has increased especially in the field of water and wastewater treatment in many developing countries [3].

Biocoagulant is said to have a bright future and attracts many researchers because of its abundance, low price, eco-friendly, multifunctional and biodegradable nature [4]. The use of extracts from plants has a function of coagulation and anti microbials that are safe for human health [5]. This research utilizes biocoagulant from an oyster mushroom (Pleurotusostreatus). White oyster mushroom is a type of oyster mushrooms that many farmers cultivated in Indonesia because of its adaptive to environmental changes and have high productivity [6]. And oyster mushrooms have environmentally friendly properties. The main constituent component of Basidiomycetes classroom cell wall is Kitin [7]. Chitin is known to possess properties such as bioactivity, biodegradability and has the properties to bind metallic ions such as Fe, Cu, Cd, Hg, and has adsorption properties [8]. and has high polyelectrolyte properties that can act as adsorbents against heavy metals Wastewater [9].

* Corresponding author: mariafaustinaps@gmail.com

2 Materials and Methods

An oyster mushroom biocoagulant is first made by crushing oyster mushrooms by using a blender tool mixing 0.25 kg oyster mushrooms with 500 L of water. Once in the blender, then squeezed the mushroom in the napkin to remove the water. After that biocoagulant can be used.

The method used in this research is Jar test method. Jar test is a laboratory-scale experiment to determine optimum operating conditions for water and wastewater treatment processes. This method can determine the variation in the addition of coagulant dose and polymer, rotational speed, coagulant or polymer type variation, on a laboratory scale to predict actual water treatment requirements. The jar test method simulates the flocculation coagulation process to remove suspended solids and suspended organic matter [10]. Jar test is a procedure to study the effect of adding water coagulant; Used to determine the necessary dosage and operating conditions for effective coagulation and flocculation [11].

Table 1. Variation of mixing rate and bio coagulant dose

The dose of biocoagulant (mg/l)	Mixing rate (RPM)
600	100
1000	125
2000	150

At the stage of implementation, first conducted preliminary test that is the initial characteristic of the domestic liquid waste both domestic canteens. The initial characteristics of the liquid wastes being examined are the values of COD and the pH of both wastewater. The value of COD from both domestic wastewater far exceeds the quality standard, so both require processing.

Table 2. Tools used in this research

No	Tools	Information
1.	Measuring cup	Waste water sample container before screening
2.	Desiccators Chemistry	Coolant after heating
3.	Volumetric flask	place of dilution samples to be tested
4.	Test Tube	To react liquid waste with some reagent compounds
5.	COD reactor	To analyze the oxygen content in water
6.	Pipet	To take the solution used for the COD and Turbidity test
7.	Spektrofotometer	Measures the results of COD absorbance in wastewater samples.
8.	Jartest	As a tool for coagulation and flocculation
9.	Stopwatch	To see the stirring time

Then after knowing the initial characteristics then done jar test stage by using some variation to the speed of jar test quick stirring and variation of dosage of oyster mushroom biocoagulant used. The selected variations are as show in Table 1.

Of all the variations obtained, this research was done duplo. In addition to the variations in the study, there are also fixed variables used in this study: jar test time is 12 minutes for rapid mixing rate stirring, 3 minutes for slow mixing rate stirring, and 1 hour for settling time. And for the slow mixing rate stirring speed used is 45 RPM . Tools and materials used in this research show in Table 2 and 3.

Table 3. Materials used in this research

No	Materials
1	Domestic liquid waste from Dean of Engineering Faculty of Diponegoro University and Canton of Electrical Engineering of Diponegoro University.
2	Oyster Mushroom (*Pleurotusostreatus*)
3	High Digestion
4	H_2SO_4
5	Aquadest

3 Results and Discussion

In the preliminary test the initial characteristics of COD value of domestic waste received a value of 399.4 mg / L for the canteen Dean of the Faculty of Engineering Faculty of Diponegoro University and 636.9 mg / L for the Diponegoro University Electrical Engineering canteen.Where both COD values exceed the existing quality standard.

The results obtained from this research are as follows:

A. Canteen of Dean of Faculty of Engineering Diponegoro University

For fast mixing rate stirring speed 100 RPM, 125 RPM, 150 RPM, and control obtained COD value as show in Table 4.

Table 4. COD concentration of wastewater from Canteen of Dean, Faculty of Engineering, Diponegoro University.

Fast speed of mixing rate	Dose of Biocoagulan (mg/l)	COD (mg/l)
100 RPM	600	272.26
	1000	208.90
	2000	248.28
	Control	320.20
125 RPM	600	303.08
	1000	226.02
	2000	280.82
	Control	327.05
150 RPM	600	256.84
	1000	217.46
	2000	229.45
	Control	323.63

And get the total value of COD canteen of Dean of Faculty Engineering, Diponegoro University as Fig.1

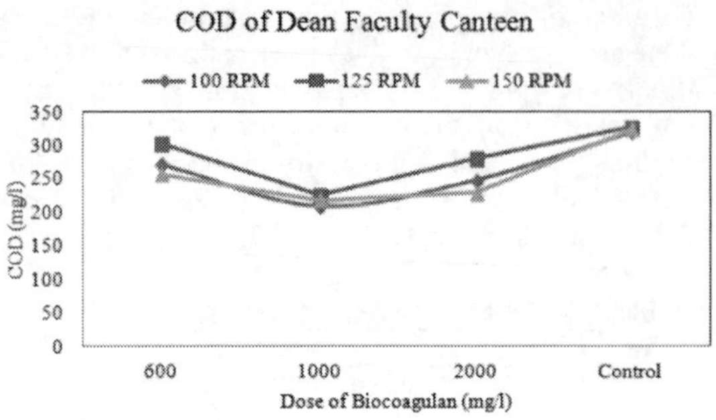

Fig.1 The COD concentration of wastewater from Canteen of Dean, Faculty of Engineering, Diponegoro University

B. Canteen of Electrical Engineering, Diponegoro University

The COD concentration of wastewater from the electrical canteen with mixing speed of 100 RPM, 125 RPM, 150 RPM, and control is shown in Table 5.

Table 5. COD concentration of wastewater from Canteen of Electrical Engineering, Faculty of Engineering, Diponegoro University (100 rpm)

Fast speed of mixing rate	Dose of Biocoagulan (mg/l)	COD (mg/l)
100 RPM	600	385.27
	1000	333.21
	2000	436.64
	Control	594.17
125 RPM	600	452.05
	1000	426.36
	2000	464.04
	Control	607.87
150 RPM	600	424.65
	1000	301.36
	2000	457.19
	Control	597.60

And get the total value of COD canteen of Electrical Engineering Diponegoro University as show in Fig. 2.

From the above results can be seen that the value of COD in the canteen Electrical Engineering is higher than the value of COD in the canteen Faculty of Engineering Diponegoro University due to the composition of domestic wastewater electro canteen more has more organic compounds such as vegetables. So the value of COD in the canteen Electrical Engineering is higher than the value of COD in the Faculty Engineering Faculty canteen canteens.

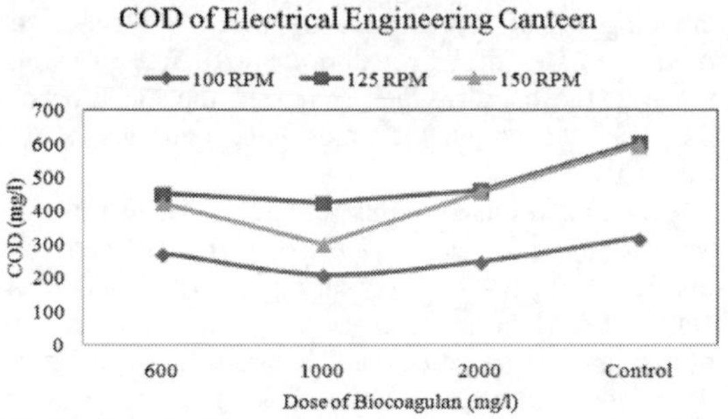

Fig.2 The COD concentration of wastewater from Canteen of electrical Engineering, Faculty of Engineering, Diponegoro University

And for the declining result of COD value on both domestic wastewater both have the efficiency of decreasing COD content equal to 47,7%. And both of them can get the biggest decrease when using variation dose of fungus biokoagulan of 100 mg / L and with fast speed of stirring at jar test is 1000 RPM.

4 Conclusions

Oyster mushroom proven to decrease the levels of pollutants present in the liquid waste is COD. With coagulation and flocculation processing using oyster mushroom, it was proven to decrease COD content by 47.7% for domestic wastewater at variation for mixing rate at 1000 RPM and 100 mg / L biocoagulant dose.

References

1. Adillah, K. W., *Perencanaan Instalasi Pengolahan Air Limbah Rumah Sakit Pupuk Kaltim Prima Sangatta. Surabaya*: Fakultas Sains dan Teknologi, Universitas Airlangga. (2012).
2. Bratby J.R.,*Coagulation and Flocculation, With an Emphasis on Water and Wastewater Treatment*, Uplands Press Ltd, Croydon.1980
3. Jahn S.A.A., *Traditional Water Purification in Tropical Developing Countries: Existing Methods and Potential Application*. Eschborn, GTZ, Germany. 1981
4. Madhavi, T. P., and R. Rajkumar. International Journal of ChemTech Research. Vol. **5** No.3, (2013)
5. Muyibi,S.A.&Okuofu,C.A. International Journal of environmental studies,**46**,263-273 (1995)
6. Cahyana,Y.A., Muchrodji, M., Bakrun, *Jamur Tiram*: Pembibitan, Pembudidayaan, Analisis Usaha. Cetakan VI. Jakarta: Penebar Swadaya. (2001).
7. Griffin, D. H., *Fungal Physiology*. John Wiley and Sons, New York, (1981).

https://doi.org/10.1051/e3sconf/20183103018

8. Hirano, S., *Chitin and Chitosan. In Ullman's Encyclopedia of Industrial Chemistry. Completely Revised Edition.*Weinheim. New York, (1986)
9. Widyastuti, *Khitin Ranjungan Sebagai Biokoagulan Dalam Penurunan Turbidity, BOD, COD, dan TSS Dalam Pengolahan Air LimbahFarmasi PT. Phapros, Tbk* Semarang. Semarang: Diponegoro University., (2014)
10. Nugraheni, D. T. (2014). *Cangkang Udang Sebagai Biokoagulan Untuk Penyisihan Turbidity, TSS, BOD dan COD pada Pengolahan Air Limbah Farmasi PT. Phapros Tbk. Semarang*: Environment Engineering, Diponegoro University.
11. Crittenden,J.C., *MWH's Water Treatment Principles and Design* 3rd Edition. New Jersey: John Wiley & Sons, Inc, (2012).

Application of Pesticide Phytoremediation in Irrigated Rice Fields System Using Eceng Gondok (*Eichhornia crassipes*) Plants

Ika Kartika Febriani[1,*] and *Hadiyanto*[1,2]

[1]Master Program of Environmental Science, School of Postgraduate Study Diponegoro University
[2]Chemical Engineering Dept, Faculty of Engineering, Diponegoro University

Abstract - The problem of environmental pollution especially urban water pollution becomes major issue in Indonesia. The cause of water pollution is not only from industrial factory waste disposal but also other causes which become pollution factor. One cause of water pollution is the existence of agricultural activities with the use of the amount of pesticides that exceed the threshold. As regulated in Government Regulation No. 82/2001 on Water Quality Management and Water Pollution Control, it is necessary to manage water quality and control water pollution wisely by taking into account the interests of current and future generations as well as the ecological balance. To overcome the problem of water pollution due to agricultural activities, it is necessary to conduct research on phytoremediation technique by utilizing eceng gondok plant. It is excepted that using this phytoremediation technique can reduce the problem of water pollution due to the use of pesticides on agricultural activities.

1 Introduction

Problems of environmental pollution, especially the problem of water pollution in cities in Indonesia is quite serious. The cause of water pollution is not only from industrial factory waste disposal but also other causes which become pollution factor. One cause of water pollution is the existence of agricultural activities that exceed the threshold. Namely excessive use of pesticides. The use of pesticides in agriculture has increased since agricultural intensification programs were proclaimed by the government in 1970. Even most farmers already depend on pesticides. They assume that pesticides are the saviors of pest attacks. Farmers use pesticides with doses that are sometimes incompatible with the procedure. They do not know what the impact will be on the environment if pesticides are used excessively.

Pesticides that are not absorbed by plants will return to nature. So it becomes a residue that is quite detrimental to the environment. The residue settles on the ground and is carried by water to the river. Then the river water polluted by the residue was reused by farmers to irrigate their fields. So the residue was reabsorbed by plants, especially rice plants that need water that pooled during its growth. To overcome the problem of water pollution in rice fields due to pesticide residues, farmers can use eceng gondok as a heavy metal absorber carried by water. Eceng Gondok is one of the floating aquatic plants that have the Latin name *Eichhornia crassipes*. This plant is often regarded as a weed that destroys the aquatic ecosystem because of its fast growing nature and easily spread from one body of water to another body of water.

Eceng gondok is one type of aquatic plant that has the ability to absorb and accumulate heavy metals. This plant has the potential to absorb heavy metals because it is a plant with high tolerance that can grow well in waste, rapid growth and absorb and accumulate metal well in a short time. Eceng gondok can also reduce the value of Biological Oxygen Demand (BOD), Total Suspended Solid (TSS) and Chemical Oxygen Demand (COD) of liquid waste [1]. Eceng gondok absorbs heavy metals using Phytoremediation Technique. Phytoremediation is a process involving chlorophyll plants to reduce the pollutant content present in soil, air and water [2]. Phytoremediation is a technique that uses plants to reduce or decrease the levels of pollutants in the environment so that it becomes harmless [3]. Phytoremediation is gaining attention because it is one of the alternative technologies to clean up the environment from pollutants, especially for soil and water, because it is considered effective, efficient, more economical and sustainable [4].

There are several categories in phytoremediation: phytoextraction, phytofiltration, phytostabilization,

*Corresponding author: ikakf@yahoo.com

phytovolatization, and phytodegradatio depending on their remediation mechanism [5].

Phytoextraction involves the use of plants to remove contaminants in the soil. Pytofiltration is the process of removing metals from water contaminated by plant roots or plants. While phytostabilization involves the roots to absorb pollutants from the soil and store them in the rhizosphere, and reduce the spread of pollutants. Phytovolatization involves the use of plants to remove pollutants through the evaporation process in leaf foliage, as in Se and Hg pollutants. Pyhtodegradation is the use of plants to associate with microorganisms in reducing levels of pollutants [6].

1.1 Eceng gondok plant

Eceng gondok (*Eichornia crassipes*) is a water plant that grows in swamps, lakes, reservoirs and rivers that flow calmly. Historically, water hyacinth in Indonesia was brought by an American botanist to the Bogor Botanical Garden. Due to rapid growth (3% per day), water hyacinth is able to cover the entire surface of a pond. According to Hadiyanto et al [7] explains that water hyacinth is a weed plant in aquatic area that lives floating in deep water or develops rooting in mud in shallow water. Hyacinth multiply very quickly, both vegetatively and generatively. Proliferation by vegetative means doubled twice within 7-10 days.

The proliferation of eceng gondok occurs vegetative and generative, vegetative propagation occurs when new shoots grow from the leaf's armpits, then enlarge and eventually become new plants. Every 10 hyacinth plants are capable of breeding to 600,000 new plants within 8 months, this is what makes water hyacinth used for wastewater treatment.

The ability of eceng gondok is much so that it is used to treat waste water, because human activity is capable of processing domestic waste water with a high level of efficiency. Eceng gondok can reduce BOD, suspended particles biochemically (lasts a bit slowly) and is able to absorb heavy metals such as Cr, Pb, Hg, Cd, Cu, Fe, Mn, Zn well, the ability of absorbing heavy metal union of eceng gondok at a young age than in old age.

1.2 The Benefits Of EcengGondok

Water hyacinth causes a lot of problems of river and reservoir pollution, but has the following benefits [8] : It has a biological nature as a water filter contaminated by various industrial-made chemicals; As a cover material, compost in agriculture and plantation activities; As a gas source, among others, ammonium sulphate gas, hydrogen gas, nitrogen and methane obtained by fermentation; Raw material of plant fertilizer containing NPK elements which are the three main elements needed by plants; As an industrial material of artificial board paper and activated carbon materials.

1.3 Loss of Eceng Gondok

Adverse conditions that arise as a result of uncontrolled growth of eceng gondok are: The decrease of the amount of light entering into the water causes the decreasing of oxygen solubility in water (DO: Dissolved Oxygen); Disrupt traffic (transport) of water, especially for people whose lives are still dependent from the river; Increased habitat for human disease vector and decrease aesthetic value of aquatic environments.

1.4 Absorption by Eceng Gondok

This plant has a rapid regeneration power because the pieces of vegetative pieces that carry will continue to develop into adult eceng gondok. Eceng gondok is very sensitive to the conditions in the water that the element is insufficient, but its response to high levels of nutrients is also large, causing eceng gondok can be used as a controller of environmental pollution. Plant root cells commonly contain ions with a higher concentration than the normally negatively charged surrounding medium. This absorption involves energy, as a consequence and its existence, the cation shows the passive entry into the cell passively into the electrochemical gradient, while the anion must be transported actively into the root cell of the plant according to the gradient state of concentration against the electrochemical gradient.

The root part, the ordinary plant changes the pH and forms a chelate called fitosiderofor. It is this substance that then binds the metal and is then brought into the root cell. In order for the absorption of metals to increase, then this plant to form the redottase molecule in the root membrane. While the transport model in the plant body is the metal brought into the root cell then to the transport network of xylem and phloem, go to other plants, while the localization of metal on the network aims to prevent metal poisoning against the cell, the plant will do detoxification, such as stockpiling of metals into certain organs such as roots. There are two ways of ion absorption into plant roots: The mass flow of ions in the water moves rapidly toward the potential gradient roots caused by transpiration; concentration gradient diffusion is generated by ion removal at the root surface.

In the absorption of these ions there are two important things that must be considered, namely first, the metabolic energy needed in the absorption of nutrients actually a little nutrients, resulting in respiration. Second, the collection process is selective, the plant has the ability to select the absorption of certain ions in wide environmental conditions.

2 Heavy Metal

Heavy metals are dangerous water pollutants, including Pb, Cu, Hg and Cd. Various cases of pollution by Pb, Cd, Hg and Cu have been widely reported, because heavy metal ions can cause poisoning in humans.

Severe damage to the kidneys, liver, brain and central nerves and inhibits Hb synthesis. There are 80 species of 109 chemical elements on earth that have been identified as heavy metals. From a toxicological point of view, these heavy metals can be divided into two types. First, the essential heavy metal, where its presence in a certain amount is needed by living organisms, but excessive amounts can have toxic effects. Examples of these heavy metals are Zn, Cu, Fe, Co, Mn and others, while the second type is non-essential heavy metals or toxic, where its presence in the body is still unknown or even toxic, such as Hg, Cd, Pb, Cr and others.

2.1 Heavy Metal Pollution

Water contamination occurs when the substance of the condition (including heat) lowers the quality of the water body so that water can't meet the standard quality or can't be used for a particular purpose [8]. Pollution of heavy metals present in the soil, water and air with a mechanism of entry into living things. Heavy metals that enter into the human body will interact with other enzymes, proteins, DNA, and other metabolites. The presence of heavy metals in the body will obviously affect the body. When the amount is excessive, it will be harmful to the body. Pollution of heavy metals into the environment can be through three ways, namely:

2.1.1 Water

Water pollution occurs in water sources of lakes, rivers, oceans and ground water caused by human activities. Water is said to be contaminated if it can't be used according to its function. This pollution can be caused by industrial waste, housing, agriculture, household, industry and fishing by using poison. Industrial pollutants include organic pollutants (wastewater), inorganic pollutants (solids, heavy metals), fuel residues and kerosene spills are sources of water pollution, especially ground water. Water pollutants include chemical elements, pathogens and changes in the physical and chemical properties of water. Many chemical elements are toxins that contaminate water.

Water pollution is a regional and global environmental problem and is closely related to air pollution and land or land use. Water pollution consists of various types, and their effects on the environment and living things. One of the causes of water pollution is the presence of inorganic chemicals such as Pb, Cd and Hg. The presence of metal toxic substances such as Pb, Cd, Hg in high levels cause the water is not good for consumption.

2.1.2 Soil

Pollution is mostly caused by waste, both organic and inorganic. Substances of waste that seep into the soil can't be lost for long periods of time. The waste substances that enter the soil are absorbed by the plant and remain in the plant's body, because the plant can't decipher it. Industrial waste that pollutes the soil is usually excessive fertilizer and the use of herbicides and pesticides.

The content of heavy metals in the soil greatly affect the metal content of plants growing on it, unless there is interaction between the metal so that the metal absorption occurs by the plant. The accumulation of metals in plants depends not only on the content of metals in the soil, but also depends on soil chemical elements, metal types, soil pH, and plant species.

2.1.3 Air

Air pollution is caused by exhaust fumes such as CO2, SO, SO2, CFC, CO and cigarette smoke. Pollutants can come from factories, machines that use fossil fuels and from burning wood.

2.2 Heavy Metal Pollution in the Waters

Heavy metal pollution in many waters is sourced from mining, metal smelting, and other types of industries [9]. Heavy metals are usually found very little in water naturally less than 1 µg / l. To determine the water quality to the concentration of metal in dissolved therein. Concentrations of toxic metals such as Cd, Pb, Hg and As in waters are naturally very small. The presence of heavy metals in the water, dangerous directly to the life of the organism, as well as its effects indirectly to human health. Heavy metals are chemical elements of a type greater than 5 gr / cm^3, located in the lower right corner of the periodic system, have a high affinity for the element S and are usually atomic number 22 to 92 from the period of 4 to 7. Some heavy metals such as Lead (Pb), cadmium (Cd), and mercury (Hg) are harmful pollutants. The high affinity of the S element causes this metal to attack the sulfur bond in the enzyme, so that the enzyme becomes inactive.

Many heavy metals, both toxic and essential, dissolve in water and contaminate both fresh and marine water. The source of this pollution comes largely from mining, smelting metal, and other types of industries, and can also come from agricultural land using fertilizer containing metals. In water the metal is usually bonded in a chemical compound or in the form of an ionic metal, depending on the compartment on which the metal is located. The degree of metal content in each compartment varies considerably, depending on location, type of compartment and level of contamination. It has been widely reported on the concentrations of metals in water and living biota therein. Usually the concentration levels of heavy metals in water are differentiated according to the level of pollution, heavy pollution, medium pollution, and non polar. A waters with heavy pollution levels usually contain heavy metals in water, and the organisms that live in them are quite high. At a moderate level of pollution, the heavy metal content of water and biota

https://doi.org/10.1051/e3sconf/20183103019

living within it are within marginal limits. While at the non polar level, the heavy metal content in water and organisms that live in it is very low, even undetected.

3 Phytoremediation

The term phytoremediation comes from English phytoremediation; The word itself is composed of two parts of the word, the phyto derived from the Greek phyton ("plant") and remediation derived from the Latin remedium ("heal", in this case also means "solving the problem by fixing the error or deficiency"). Thus phytoremediation can be defined as the use of plants to remove, move, stabilize, or destroy the contaminants be it organic or inorganic compounds Alternative simple wastewater treatment is by phytoremediation using eceng gondok plant. Phytoremediation is the use of plants and their parts for waste decontamination and environmental pollution issues either ex-situ using artificial or reactor ponds or in-situ (direct field) on soil or contaminated areas of waste.

The method used is experimental method (experiment). While the experimental design used in this study was Completely Randomized Design (RAL) with 4 treatments and 3 replications.
Treatment design:
A: Control (Without Phytoremediation)
B: 5 eceng gondok / bucket
C: 8 eceng gondok / bucket
D: 10 eceng gondok / bucket

Then taken a sample of polluted water from several fields. Samples taken as many as 4 random stamps on several fields. Sample 1 is a water seal that has not yet entered the rice fields and water that has not been contaminated with pesticides. Sample 2 is water from a rice field that is close to irrigation. Then sample 3 is water from the rice fields in the middle. And the 4th is the water sample from the rice field plot leading to irrigation. For example illustration can be seen in the picture :

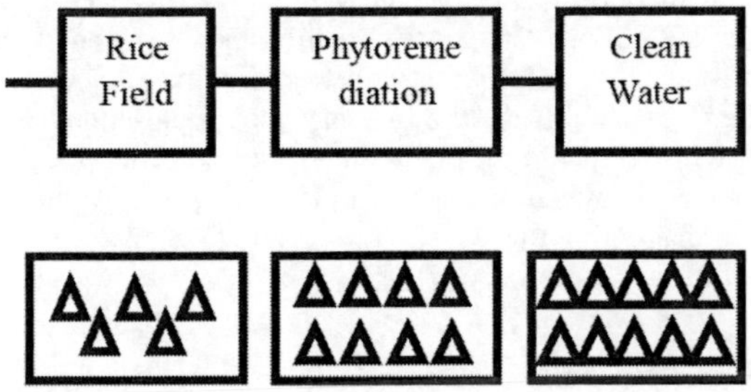

Fig. 1. Phytoremediation process

4 Conclusions

From the above reviews it can be concluded that agricultural activity is one of the pollution factors that occur in the environment, especially in the waters. To overcome this problem it must be phytoremediation so that the waters are not polluted by agricultural residues. Eceng gondok is selected as a plant for phytoremediation in waters contaminated by pesticides.

References

1. S.Kris, and Wariningsih. Jurnal Teknologi Technoscientia, **Vol. 4** No 1. Hal : 17 – 22, (2011).
2. Zayed. Journal Environmental Quality,**Vol. 27**: 715-721 , (1998).
3. Lahenda, Shelga Sapta. Ellyke. Khoiron,E-Jurnal Pustaka Kesehatan, **Vol. 3** No. 2, (2015).
4. Chaney, R.L., Malik, M., Li, Y.M., Brown, S.L., Brewer, E.P., Scott Angle, J., Baker, A.J.M.. Curr. Opin. Biotechnol, **Vol. 8**(3):279-284, (1997).
5. D.E. Salt, R.D. Smith, L. Raskin. Plant Mol. Biol". **Vol. 49** (1):643-668, (1998).
6. M.I. Lone, Z. He, P.J. Stoffella, X. Yang, Journal Of Zhejiang University SCIENCE B, **Vol. 9**(3):210-220, (2008).
7. H. Hadiyanto, M. Christwardana, D. Soetrisnanto, Journal of Environmental Science and Technology,**Vol. 6**(2):79-90, (2013).

Analysis of Heavy Metal Content (Pb) on Waters and Fish at The Floating Cages BPPP Ambon

Rachel L. Wattimena[1], *Debby* A.J. Selanno[2,*], *Semuel* F.Tuhumury[2], and *Juliana* W. Tuahatu[2]

[1]Magister of Marine Science, Postgraduate Pattimura University, Ambon-Indonesia
[2]Fisheries and Marine Sciences Faculty, Pattimura University, Ambon-Indonesia

Abstract. Coastal waters play important roles due to highly in natural resources and developing of environmental services. However, there are highly intensity of natural resources utilization, environment and settlement. Consequently, environment and natural resources would be degraded such as in the Ambon Bay. One of the potency at the Ambon Bay is mariculture area namely the floating cages (KJA) which belongs to Fisheries education and training (BPPP) Ambon. The research aimed to analyze physical-chemical of waters (temperature, pH, salinity and current speed), to analyze heavy metal concentration (Pb) on water and fish from floating cages (KJA) and to analyze waters pollution status at KJA BPPP Ambon. The average salinity of each floating cage ranged from 30.09 - 30.34°C, pH ranged from 8.03 – 8.44, salinity ranged from 31.36 – 33.34 PSU, and current speed at spring tide ranged from 0.5 – 55.8 Cm/sec while neap tide ranged from 0.1 – 9.8 Cm/sec. Heavy metal concentration (Pb) on waters was below the standard for waters quality and the average concentration was 0.002 mg/l. Whilst, the heavy metal concentration (Pb) on fishes was below standard for floating cages (floating cages 2-6) which was 0.05 and 0.17mg/l. Otherwise, floating cage 1 had been above maximum standard for fish food and its processing following SNI 7387:2009 (0.3mg/l) which was 0.31 mg/l. The status of waters pollution at KJA BPPP Ambon belonged to C class and could be categorized as moderate based on standard for waters quality issued by State Ministerial Decree for the Environment No. 51 Year 2004.

1 Introduction

Coastal waters play important roles due to highly in natural resources and developing of environmental services. However, there are highly intensity of natural resources utilization, environment and settlement. In line with highly intensity of natural resources, development and communities necessities, caused waste in coastal area and sea becomes varied with highly volume. This highly waste volume will have an impact on quality of waters environment [1]. The Inner Ambon Bay and its surrounding are coastal waters and have highly intensity of natural resources utilization, and environmental services, so it causes changed in natural resources and environment. One of the potency of the Inner Ambon Bay is mariculture area where there is floating cages (KJA) which belongs to Balai Pendidikan dan Pelatihan Perikanan (BPPP) Ambon. In addition, the Inner Ambon Bay has been used for transportation such as ferry boat, speed boat and ship repair.

The successful of fish mariculture in floating cages is determined by waters quality. The highly intensity of spatial utilization in the Inner Ambon Bay is possible to change its waters quality especially for the entry of pollutant materials into the ecosystems. The pollutant materials could be toxic. One of the pollutants which harmful for the mariculture biota are heavy metal [2]. According to Palar [3,4] one of the toxic and dangerous heavy metal as pollutant and tend to affect the waters organisms is Pb metal.

The previous study indicated that heavy metal (Pb) in the waters of Ambon Bay ranged from 0.004 mg/l-0.008 mg/l [5]. Otherwise, the heavy metal content (Pb) in the body of rabbit fish (*Siganus canaliculatus)* and caranx (*(caranx sexfasciatus)* ranged from 0,007-0254 ppm [6]. Based on the some facts as explain above, it could be assumed that the concentration of heavy metal (Pb) has been increasing due to highly intensity of space in the Inner Ambon Bay. This would be threatened the sustainability of floating cages mariculture. Therefore, the research aimed were to analyze physical-chemical of waters (temperature, pH, salinity and current speed), to analyze heavy metal concentration (Pb) on water and fish from floating cages (KJA) and to analyze waters pollution status at KJA BPPP Ambon.

2 Methods

2.1 Study sites

The research was conducted from April to June 2014 and consisted into two stages: fieldwork and laboratory analysis. The fieldwork was done in floating cages (KJA) BPPP Ambon, located in Tanjung Martha Fons. The sampling station of six floating cages was situated at

*Correponding author: debby_st@yahoo.co.id

128°13'30" E - 128°13'36" E and 03°37'47,12" - 03°37'50,12" S (Figure 1).

The analysis of waters quality including physical-chemical such as waters current, temperature, pH and salinity were done as in situ by using CTD and current meters in 6 floating cages at 0.5-10m depth. Whereas, the analysis of heavy metal content (Pb) both in waters and fish were conducted in the testing laboratory of Oceanography Research Centre LIPI Ancol – Jakarta. The equipment and materials for testing heavy metal (Pb) in the laboratory follow SNI 2354.5:2011 to determine heavy metal content (Pb) of fisheries product.

2.2 Methods of sampling water and fish

Water sample were collected in 6 floating cages (KJA) with the size of $3x3x3m^3$ at 1 meter depth by using *Van Dorn bottle sampler*. Then, the water was placed into polyethylene bottle and preserved by adding HNO3 to pH < 2. Finally, the water samples were put into the cool box. In addition fish sample were collected at each floating cages (6 units) especially *Caranx* approximately at age 6 months with average weight 250 grams and length 21 cm. Fish were put into plastic PE then it's are frozen. The analysis of heavy metal content (Pb) both in waters and fish were conducted in the testing laboratory of Oceanography Research Centre LIPI Ancol – Jakarta.

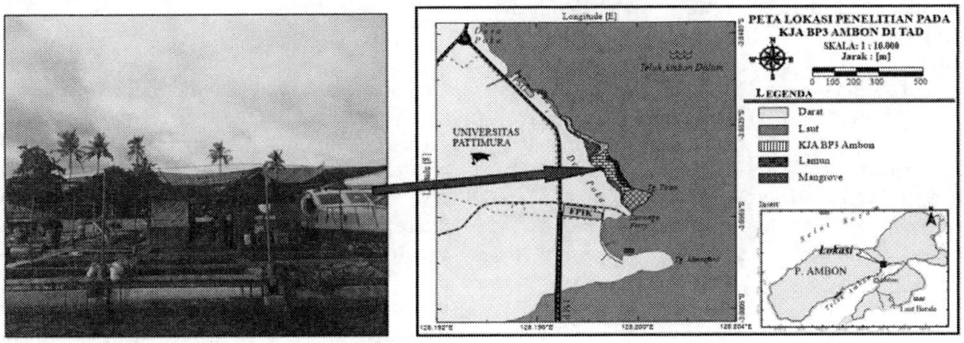

Fig.1. Study site

2.3 Data analysis

The results of the analysis correspond to standard for seawater quality issued by State Ministerial Decree for the Environment No. 51/200 [7,8]. The results of heavy metal (Pb) analysis both in water and fish were also corresponded to standard for heavy metal especially for water and biota (fish) and could be categorized as polluted or unpolluted.

To determine the water status, used scoring system US-EPA (Environmental Protection Agency), is classified into four classes namely:

1. A: very good , score = 0 = fulfilled the standard
2. B : good, score = -1 to -10 = lightly polluted
3. C : moderate, score = -11 s/d -30 = moderate polluted
4. D : bad,score ≥ -31= heavily polluted

The procedure to determine status of water quality by using STORET Method as follows: Data (physical, chemical and biology) obtained should be calculated the minimum, maximum and average values then compare to the standard for seawater quality with some criteria:

- If water quality data fulfilled measurement of the standard with ≤ standard quality, score 0,
- If water quality data unfulfilled measurement of the standard with ≥ standard quality, scoring as follows Table 1.

Table 1. Determination of value system for water quality status

Parameters	Value	Parameters		
		Physic	Chem	Bio
< 10	Maximum	-1	-2	-3
	Minimum	-1	-2	-3
	Average	-3	-6	-9
≥ 10	Maximum	-2	-4	-6
	Minimum	-2	-4	-6
	Average	-6	-12	-18

(Appendix State Ministerial Decree for the Environment No. 115 / 2003) [9].

3 Results and Discussion

3.1 Physical-Chemical Waters Quality and Its Correlation with Pb

3.1.1 Temperature

Temperature distribution during the research at each floating cages ranged from 27.47 – 30.51°C with the average 29.52°C. The average temperature of each floating cage was represented in figure 2. The results indicated that the temperature at water surface (0.5 m depth) of each floating cage was higher (30.09 - 30.34°C) than that of the bottom (10 m depth) 28.55 –

28.70°C. According to Nontji water surface is warm layer due to getting sun radiation in the day [10,11],.

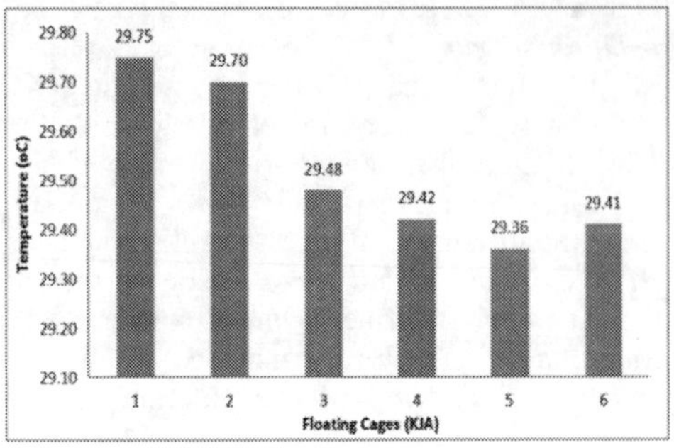

Fig. 2. Temperature at floating cages KJA BPPP Ambon

On the other hand, the increasing of water depth in the water column, decreasing temperature due to less of sun radiation. The water depth has influenced temperature of water body [12, 13]. Based on the standard for seawater quality (State Ministerial Decree for the Environment No. 51 / 2004), recommended temperature ranged from 28-30°C and temperature in Indonesian waters ranged from 28-31°C.

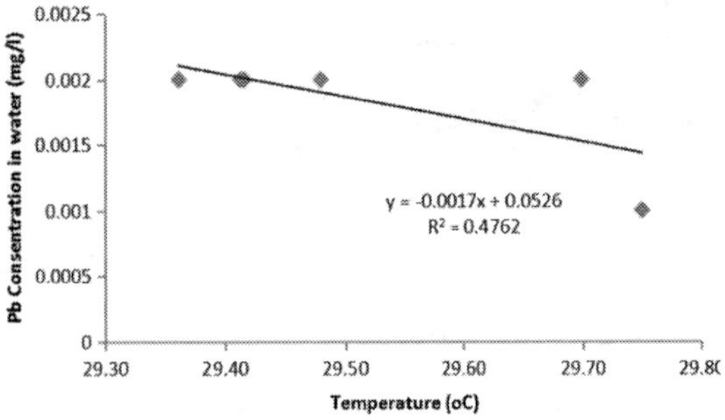

Fig. 3. The correlation of Pb concentration and water temperature

Therefore, the temperature at 10m depth in the KJA BPPP Ambon was normal and could be tolerated by waters organism. Temperature distribution at each floating cage with different depth showed that the temperature was suitable for mariculture and the temperature could decrease negative effect of metal on the marine organism (mariculture fish). The presentation of Pb in the water colum could be pushed by the reaction of physical-chemical parametres. The simple correlation was represented in Figure 3, which showed a weakness correlation of 47.62 % of water temperature and Pb concentration.

3.1.2 pH

pH is a parameter which explains dissolved hydrogen content in water. It could influence element content and chemical compound found in the waters. Palar stated that toxicity of heavy metal are also influenced by pH changed which toxicity of heavy metal will increase, if pH decrease. The research conducted by Ambon city government and Unpatti showed that pH at water surface in the Inner Ambon Bay ranged from 8.3-8.6 whereas in the depth 30-40m ranged from 7.3-8.6. This depth the pH varied compared to other depth. Selanno stated that pH at each monsoon ranged from 7.99-8.28 [14]. The results indicated that pH ranged from 8.03-8.44 with the average 8.17. The average pH at each floating cages could be seen in Figure 4.

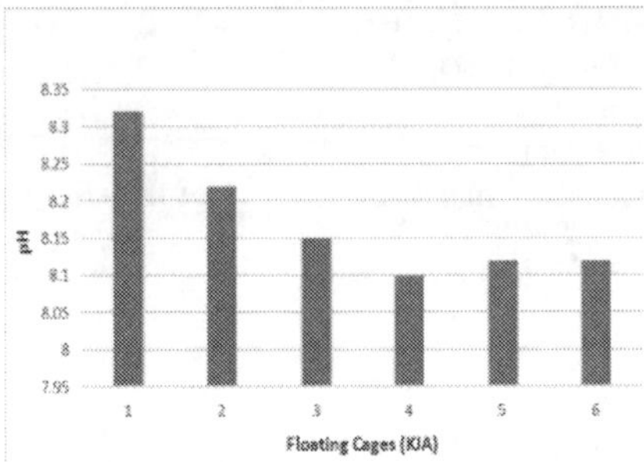

Fig. 4. Waters pH at KJA BPPP Ambon

pH of the waters could influence accumulation of heavy metal in fish because the lower water pH and sediment, the higher heavy metal dissolved (ion). So, it was easily to penetrate in the fish through gill, food or diffusion [15]. According to the standard for seawater quality issued by State Ministerial Decree for the Environment No. 51/2004, for marine organism pH 7 – 8,5. Therefore, the water pH at KJA BPPP Ambon was in recommended, could be tolerated by organisms.The dissolved of heavy metals in the water could be controlled by water pH. The increased of water pH would be decreased the dissolved of metal in the water.

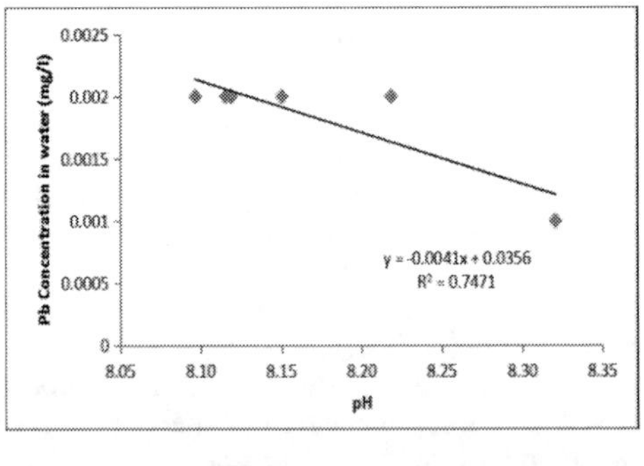

Fig.5. The Correlation of Pb concentration and water pH

Therefore, the increased of water pH will be changed the stability of carbonate shape to hydrocide that formed a bond with some particles in the water column and might be sediment as a mud (Effendi, 2003). Figure 5, showed a significant correlation of 74.71% of Pb concentration in the water and water pH.

3.1.3 Salinity

There are some factors which influenced waters salinity such as evaporation, water circulation, rainfall and river flows. The changes of the salinity could be varied both temporal and spatial due to changes of the sea condition, evaporation, and distance from the sun. The research conducted by Ambon city government and Unpatti

(2002) showed that salinity the Inner Ambon Bay was 27.00 – 32.00 PSU. Selanno also found salinity at each monsoon ranged from 29.45 – 35.27 PSU. The results indicated that salinity at KJA BPPP Ambon ranged from 31.36 – 33.34 PSU with the average 32.87 PSU. The average salinity at each floating cages could be seen in Figure 6 and their correlation could be seen in Figure 7. There was a significant correlation of 55.56% of Pb concentration and water salinity.

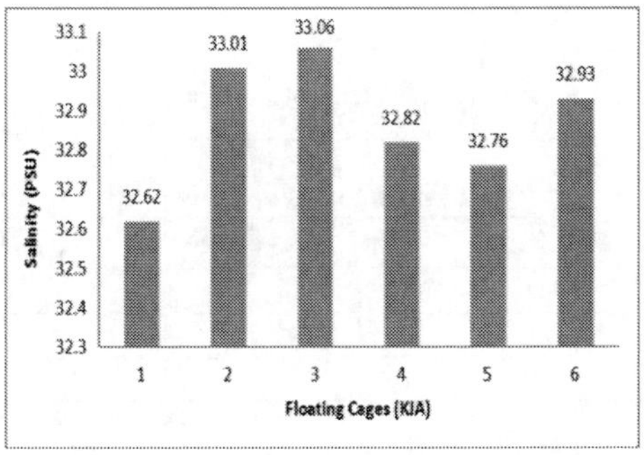

Fig. 6. Waters salinity at KJA BPPP Ambon

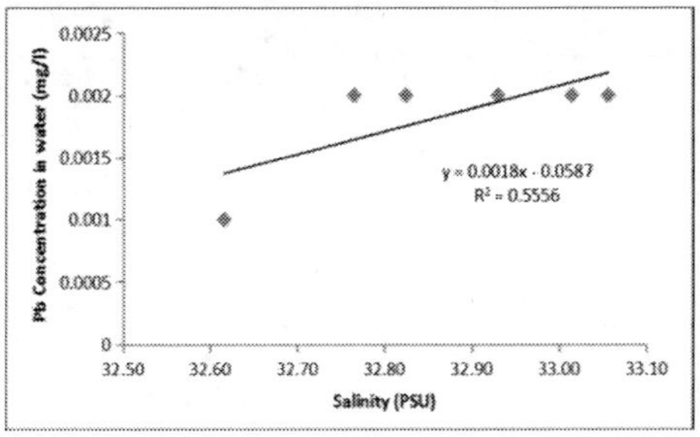

Fig. 7. The correlation of Pb concentration and water Salinity

3.1.4 Current

Waters current will influence the existence of heavy metals in the waters. Hoshika stated that the occurrence of heavy metals in the waters was influenced by current pattern [11]. Current could distribute dissolved heavy metal in the sea. The Inner Ambon Bay is semi-enclosed waters and tends to have limited physical characteristics, such as slow current speed, being protected from wave with limited water circulation and followed tidal pattern [13]. The tide in the Ambon Bay is mixed semidiurnal so it would also affect waters condition. The current when high tide was 0.5 – 55.8 Cm/sec whilst during low tide was 0.1 – 9.8 Cm/sec. The tidal current plays important role in nutrient and plankton transport. In addition, it is also diluted and controlled waste which going through to the sea [12].

3.1.5 Heavy metal concentration (Pb) in water at KJA BPPP Ambon

The analysis of heavy metal concentration (Pb) in water which collected from each floating cages could be seen in Figure 8.

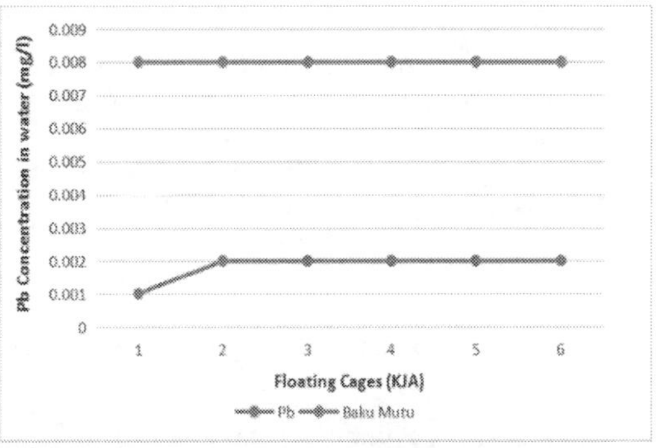

Figure 8. Pb concentration in water at KJA BPPP Ambon

It could be seen from Figure 8, that heavy metal concentration (Pb) at KJA BPPP Ambon station (floating cages 2, 3, 4, 5, 6) has similar concentration of Pb which was 0,002 mg/l except for floating cage 1 0.001 mg/l. So, it meant that the heavy metal concentration (Pb) in waters at KJA BPPP Ambon was less than the determined standard namely 0.008 mg/l. Nevertheless, the concentration of Pb could increase due to time and increasing of human activities both in land and sea.

3.1.6 Heavy metal concentration (Pb) in fish at KJA BPPP Ambon

Fish is bioindicator of environmental pollution including chemical pollution. This is showed that fish reacts on pollution in the waters in certain concentrations such as activity changes, abnormal growth to death [15]. The figure 9 indicated that the heavy metal concentration (Pb) in fish of floating cages 2, 3 and 4 was 0.05 mg/kg is less than the standard of Pb concentration for food and fish product. On the other hand, the heavy metal concentrations (Pb) in fish at floating cages 5 and 6 was 0.17 mg/kg reaching the maximum standard whereas at floating cage 1 was 0.31 mg/kg, higher than the standard of heavy metal pollution for food and fish products based on SNI 7387 : 2009, that is 0.3 mg/kg.

Heavy metal concentration in fish at KJA BPPP Ambon was possible occurred due to accumulation of heavy metal (Pb) in fish bodies earlier. It also probably the accumulation began when fish reached juvenile stage from previous places transported to the KJA BPPP Ambon.

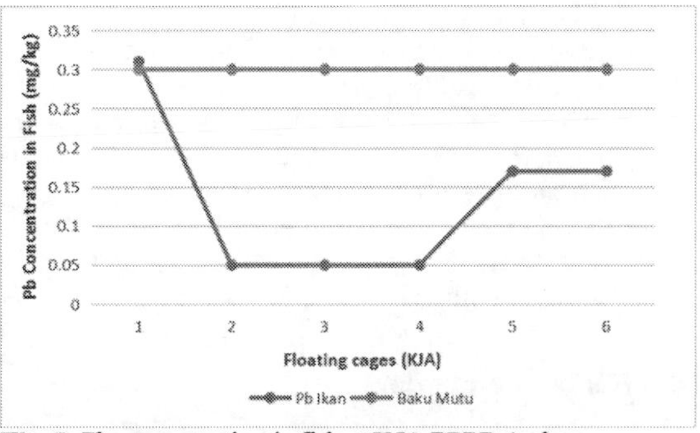

Fig. 9. Pb concentration in fish at KJA BPPP Ambon

Besides that there were also some reasons namely the location of KJA in coastal area where highly community activity, next to jetty, transportation area for ferry and other ships. The accumulation of heavy metal (Pb) occurred in fish due to continually of bioaccumulation process and biomagnifications through food chain of aquatic animals [15]. However, some of the floating cages had lower heavy metal concentration than the standard and also higher than the standard based on, SNI 7387 : 2009 for food and fish products, the residue of this heavy metal must take into account. The impact of this residue is dangerous for human life.

3.1.7 Water Pollution Status at KJA BPPP Ambon

The STORET method is used to determine the status of pollution. The analysis indicated that the total score was -13. So, it could be categorized that water quality at KJA BPPP Ambon belonged to class C moderate or moderate polluted (Table 2).

Table 2. Water Pollution Status at KJA BPPP Ambon According to STORET Value System

No.	Parameter	Unit	Standard Quality	Floating cages			Total Score
				Max	Min	Average	
	Physical						
1	Water temperature	°C	28 – 30	30.51	27.47	29.52	-2
2	Current	M	>5	19.8	0.1	7.8	-1
	Chemical						
1	pH		7 - 8.5	8.44	8.03	8.17	0
2	Salinity	PSU	18 – 32	33.34	31.3	32.84	-8
	Heavy metal						
1	Lead (Pb) water	mg/l	0.008	0.002	0.001	0.002	0
2	Lead (Pb) fish	mg/l	0.3	0.31	0.005	0.083	-2
	Total						**-13**

The status of moderate water polluted explained that waters surround KJA BPPP Ambon would be dangerous and lead to intolerant condition for growth of marine organisms especially for KJA activities. This status would be a warning for stakeholder KJA management namely BPPP Ambon especially for fish consumption for today and future. Therefore, the community surround the Ambon Bay should be careful in doing activities and utilizing marine resources. It was needed to BPPP Ambon to continually monitor water quality.

4 Conclusions and Recommendations

4.1 Conclusions

1. The waters quality of physical and chemical are still in determined standard.
2. Heavy metal concentration (Pb) in water is less than determined standard.
3. Heavy metal concentration (Pb) in the fish is less and higher than determined standard for food and fish products as SNI 7387 : 2009.
4. The status of water pollution at KJA BPPP Ambon using STORET method belonged to C class, moderate polluted.

4.2 Recommendations

Based on the results, it is needed to analysis heavy metal at waters, sediment and marine organisms at different monsoon so the heavy metal pollution (Pb) in KJA BPPP Ambon and its surrounding will be complete. It could be used to determine the source of pollutant. BPPP Ambon should continually monitor water quality in order to know the status of water pollution at the KJA.

References

1. D.A.J. Selanno, *Beban Pencemaran Pada Ekosistem Teluk : Prespektif Pengelolaan Kualitas Lingkungan Perairan*. IPB Press. (2011)
2. H. Ghufran, K. Kordi. A.B. Tancung, *Pengelolaan Kualitas Air Dalam Budidaya Perairan*. Penerbit Rineka Cipta. (2005)
3. H. Palar , *Pencemaran dan Toksikologi Logam Berat*. Penerbit Rineka Cipta. Jakarta, (1994)
4. Mukhtasor. *Pencemaran Pesisir dan Laut*. Jakarta : Penerbit PT. Pradnya. (2007)
5. BTKLPPM. *Kandungan Logam Berat Hg, Cd, Cr, Cu, Pb, dan Zn di Perairan Teluk Ambon*. Diakses melalui http://digilib.its.ac.id/public/ITS-Master-11532-4107205807-Chapter. pdf (12 Desember 2013; Pkl. 22.00 WIT) (2008)
6. J.M.F. Sahetapy, and N. Tuhumury, Chr. *Akumulasi dan Kandungan Logam Berat Timbal (Pb) dan Cadmium (Cd) dalam Tubuh Beberapa jenis Ikan Laut di Teluk Ambon Dalam*. Prosiding KONAS Pesisir dan Pulau-Pulau Kecil. DKP. Jakarta (2008)

https://doi.org/10.1051/e3sconf/20183104001

7. Menteri Negara Lingkunga Hidup. Keputusan Menteri Negara Lingkungan Hidup RI Nomor 51/Tahun 2004 tentang Baku Mutu Air Laut. Jakarta : KLH (2004).
8. S.E. Manahan, *Environmental Chemistri*, Seventh Edition, Lewis Publisher, New York. (2002)
9. Menteri Negara Lingkunga Hidup.. Keputusan Menteri Negara Lingkungan Hidup RI Nomor 115/Tahun 2003 tentang Index Storet. Jakarta : KLH (2003)
10. A. Nontji, *Laut Nusantara*. Penerbit Djambatan. Jakarta . (2007)
11. A. T. Hoshika, K. Shiozawa, Kawana and T. Tanimoto,. *Heavy Metal Pollution in Sediment from The Seto Island, Japan.* Marine Pollution. Bull (1991)
12. H. Effendi, *Telaah Kualitas Air Bagi Pengelolaan Sumberdaya dan Lingkungan Perairan*. Penerbit Kanisius. Yogyakarta. (2003)
13. D.W. Connell, and G.J. Miller. *Kimia dan Ekotoksikologi Pencemaran*. Diterjemahkan oleh Yanti Koestoer. Universitas Indonesia Press. Jakarta. (1995)
14. D.A.J. Selanno, *Analisis Hubungan Beban Pencemaran dan Konsentrasi Limbah Sebagai Dasar Pengelolaan Kualitas Lingkungan Perairan Teluk Ambon Dalam [disertasi]*. Bogor : Institut Pertanian Bogor.p.373. (2009)
15. I.S. Chahaya, *Ikan Sebagai Alat Monitoring Pencemaran*. IPB-Press. (2007)

https://doi.org/10.1051/e3sconf/20183104002

The Benefit Impact of Air Pollution Reduction Through ATCS Implementation at Intersections

Mochamad Arief Budihardjo[1*], *Haryono* Setiyo Huboyo[1], *Budi* Prasetyo Samadikun[1]

[1] Department of Environmental Engineering, Faculty of Engineering, Diponegoro University, Semarang - Indonesia

Abstract. The field study in five intersections that had ATCS, such as Krapyak, Tugu Muda, Polda, Bangkong and Fatmawati were done to investigate the effectivity of air pollution reduction. The study was done by estimating the differences between the vehicle speed due to the cycle duration of green light by field observation and video recording in each intersection. In five intersections that had been observed, the percentage of fuel consumption savings for two-wheeled vehicles were between 15 – 18%, meanwhile for four-wheel vehicles were between 30 – 46%. Based on the calculation that adopt the emission factor from CORINAIR and USEPA emission, the emission reduction based on pollutant types were TSP (12-17%), NOx (22-36%), CO (15-25%), HC (16-28%) and SO2 (22-35%). The result to the vehicles' speed that passed the intersections through ATCS has also indicated that the ATCS could increase vehicle speed and consequently reduce the emission.

1 Introduction

Nowadays, transportation has become priority for people, mostly for those who live in municipalities. There are various types of transportation, however the most preferred one is road transportation [1]. Similar to other big cities in Indonesia, the demand for road transportation in Semarang has also increased drastically due to economic progression. The number of private cars and motorcycles has grown enormously over public transportations and most of them are powered by combustion engine [2].

As a result of combustion process, the combustion engine will discharge emission. Large amounts of this pollutant will degrade air quality. Highway is one of the largest sources of air pollution emitted from transportation, especially in such complex road systems like road system with the bridge and intersections [3]. Emissions in the highway will disperse to the surrounding areas. This will affect the surrounding environment and increasing the risk of exposure to humans living around the highway. There are about 200 thousand deaths each year due to outdoor pollution that affect urban areas, and about 93% of cases occur in developing countries [4].

The air pollution generated from transportation mainly consist of 60% of carbon monoxide and about 15% consists of hydrocarbons. The most harmful pollutant to health are particles, followed by nitrogen oxides (NOx), sulphur oxides (SOx), hydrocarbons (HC), and the least toxicity is carbon monoxides (CO_2) [5].

In most municipalities, main roads and highways are equipped with Area Traffic Control System (ATCS). ATCS is a control system of highway traffic intersection using traffic light (traffic light) where traffic light settings at each intersection coordinated with each other, so that road users get a minimum of delay [3]. With the implementation of an ATCS, there will be efficiency of movement, increasing capacity of the intersection to serve traffic, cutting travel time, decreasing the level of risk of injury to the rider, creating opportunity to provide higher safety for pedestrians and providing better comfort for road users. ATCS very well be applied to the junction point with busy traffic movement and high volume of traffic [6].

Even the implementation of ATCS has some advantages on the traffic, however, the benefit of ATCS on pollution reduction has not been reported yet. Thus, this study aims to investigate the contribution of ATCS on pollution reduction by performing filed investigation at intersections in Semarang City, Indonesia.

2 Methodologies

This research was conducted at five intersections in Semarang city i.e Krapyak, Tugu Muda, Polda, Bangkong and Fatmawati from October 2016 to November 2016. Location of research is depicted in Fig. 1. Recorded videos of ATCS (Area Traffic Control System) center were used for observing number and speed of vehicles pass through the intersection. We intensively observed the video during peak hours i. e in the morning from 06.00 – 10.00 AM and 03.00 – 06.00 PM each day. Field validation of speed was conducted for several samples in the surrounding of intersections.

* Corresponding author: m.budihardjo@ft.undip.ac.id

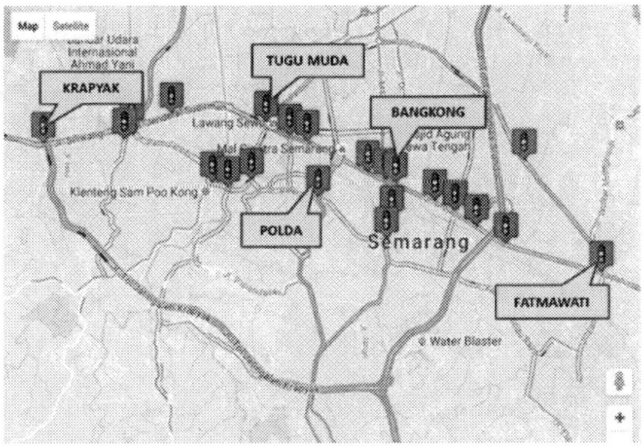

Fig. 1. Location of the Study

At the ATCS center, we record the phase modification i.e changing the duration of green light to accommodate peak vehicles. Thus we were able to know the changing vehicle speed due to changing phase of traffic light.

From each observation we record the vehicle speed, we estimate the emission reduction due to increasing speed. We did not estimate any increasing emission on other intersection lane due to increasing red light. However we assume the increasing on other lane at intersections are much lower than the lane we observed as we our observing lane is the main road (higher road class than other roads in the intersection). For estimating air pollution emission, we adopt SITRAMP (2004) study combining with emission factor derived from IPCC, CORINAIR (2009) and USEPA (1995).

3 Results and Discussion

3.1. Vehicle Characteristics

The vehicle number during observation at each lane are shown in Table.1. It can be seen that the number of vehicles were varied during observation for the same lane with high vehicles volume in Tugu Muda, Bangkong and Fatmawati. The number of vehicles during green light are much higher than during red light except for Krapyak intersection.

Table 1. Average vehicle volume on observed intersection

Intersections	Vehicle number during green light		Vehicle number during red light	
	2 wheels	4 wheels	Motor	Mobil
Krapyak	67*	32 ± 9	74 ± 51	30 ± 16
Tugu Muda	132 ± 42	45 ± 18	93 ± 54	52 ± 20
Polda	54 ± 18	33 ± 9	32 ± 17	23 ± 12
Bangkong	132 ± 50	36 ± 9	85 ± 52	32 ± 13
Fatmawati	139 ± 45	38 ± 7	119 ± 48	25 ± 14

* only once observation

Changing phase of operational traffic light by extending duration of green light may increase the vehicle speed as depicted in Fig 2, both for 4 wheels (Fig.2A) and 2 wheels (Fig.2B). This figure illustrate vehicle speed increase as the function of green light duration at Tugu Muda intersection.

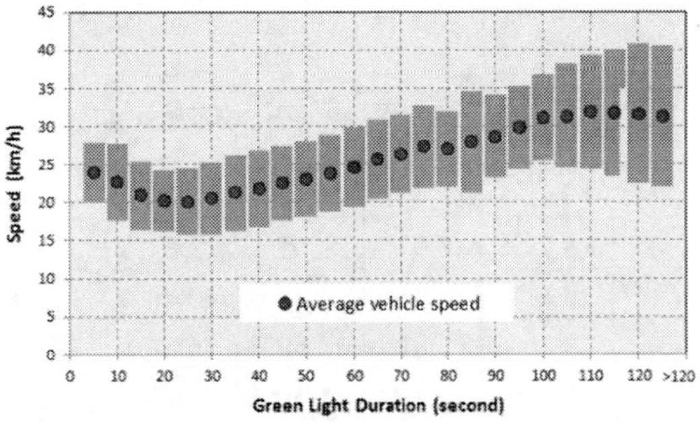

(A)

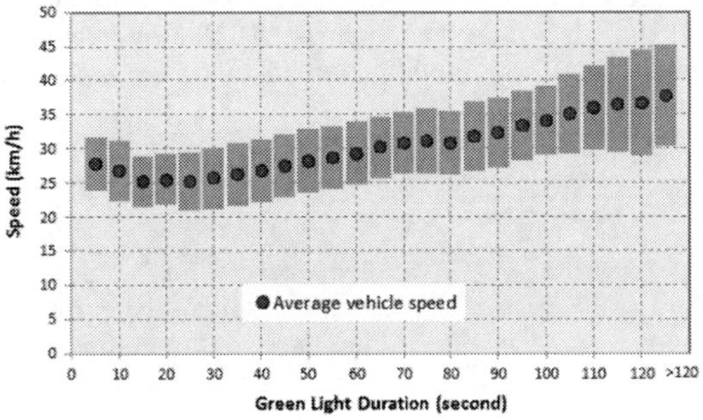

(B)

Fig.2 Increasing vehicle speed, 4 wheels (A) and 2 wheels (B) due to elongated green light duration

3.2 Emission Reduction

As the speed of vehicles at the ATCS implementation tend to increase at certain level then it could reduce the emission based on formula of fuel consumption of SITRAMP study [7]. The vehicle speed increase before and after implementation of ATCS could be seen in Figure 3 as follows:

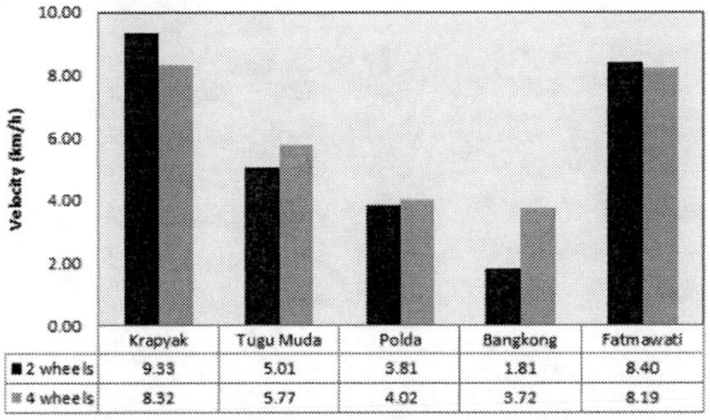

Fig.3 Speed increase due to ATCS implementation

It is obviously, both two wheels and four wheels vehicles had increasing speed because of modification of traffic light operation.

We estimate the emission reduction after ATCS operation in Table 2. Highest emission reduction percentage was performed for NOx parameter, while the lowest emission reduction was shown by TSP parameter.

This emission reduction is the aggregation of emission reduction of both two wheels and four wheels.

Table 2. Emission Reduction After ATCS Operation

	Krapyak	Tugu Muda	Polda	Bangkong	Fatmawati
TSP	17.47%	15.42%	15.05%	12.69%	15.10%
NOx	36.78%	33.95%	34.07%	22.81%	23.87%
CO	25.11%	20.98%	21.56%	15.40%	17.07%
HC	28.25%	23.81%	24.59%	16.88%	18.24%
SO2	35.58%	32.24%	32.60%	21.73%	22.72%

It should be noted that the emission reduction we calculate based on operation of ATCS during peak hour where the phase modification in this period was very often.

4 Conclusion

The effect of ATCS at five intersections in Semarang City ie. Krapyak, Tugu Muda, Polda, Bangkong and Fatmawati has been studied to investigate the effectiveness of ATCS towards reduction of air pollution. The result showed that the implementation of ATCS has increased the vehicle's speed from 2 to 9 km/h and reduced the pollutants emitted by vehicle. The reduction of emission based on pollutant types were TSP (12-17%), NOx (22-36%), CO (15-25%), HC (16-28%) and SO2 (22-35%).

This research was funded by DIPA Faculty of Engineering-Diponegoro University No. 172/SK/UN7.3.3/V/2017.

References

1. W. Wen, Expert Systems with Applications **34**, 4 (2008)
2. S.J. Santosa, T. Okuda, and S. Tanaka, CLEAN–Soil, Air, Water **36**, 5-6, (2008)
3. F. Costabile and I. Allegrini, Environmental Modelling & Software **23,** 3 (2008)
4. P.K. Hopke, et al., Science of the Total Environment, **404** 1 (2008)
5. A.F. Both, Science of the Total Environment **443** (2013).
6. B. Zhou, Vehicular technology conference fall (VTC 2010-Fall) IEEE 72nd (2010)
7. JICA-BAPPENAS, Pacific Consultants International and ALMEC Corporation (2003).

Acid Rain Contribution from Pesticide Distribution to Rice Farmers in Pati Regency

Ahmad Qosim[1,*], *Anies*[1,2], and *Henna* Rya Sunoko[1,3]

[1] Doctoral Program of Environmental Science, School of Postgraduates Studies Diponegoro University, Semarang - Indonesia
[2] Department of Environmental Science Medicine Faculty of Diponegoro University, Semarang – Indonesia
[3] Department of Environment Science Graduate School of Diponegoro University, Semarang - Indonesia

Abstract. Productivity rate of rice fields in Regency has been in a surplus condition annually. The fields have produced 7 to 8 tons per hectare, making the total annual rate of 600 tons. The regency, therefore, is considered to be capable of fulfilling its own need for rice and to contribute significantly to the rice needs in Central Java Province. Agriculture coexists with the presence of pesticides. While helping the farmers to combat the plant diseases, pesticides have still been greatly necessary by the local farmers. Distribution by means of transportation devices plays an important role for the dissemination of the pesticides from the producers to their end users. Problem arises due to emission produced during the transportation activities. Transportation emits SO_2 as the major contributor to acid rain. To make worse, application in practice by the farmers also emit the similar substance. Annual use of pesticides in Pati Regency has reached 605 tons with SO_2 emission of 13,697 kg. It is recommended that distribution management and selection of pesticides are performed by applying an integrated pest control in order to reduce the pesticide emission.

1 Introduction

Pati Regency is one of agricultural centers in Central Java Provence with a total agricultural area of 150,368 hectares, comprising 58,448-hectare rice fields. Local farmers in the regency performed three harvesting times annually. Effort of pest control depends on the use of pesticides. These substances are believed to be the key to the harvest success. Whether the pests truly exist, the farmers have been get used to spraying the pesticides on their fields. The use of pesticides is usually performed by mixing 3 to 7 different types of pesticide in 5 to 9 times during the season for rice and 12 to 17 times during the season for vegetables [1]. Such condition has become the productive marketplace for the pesticide producers.

The use of pesticides often causes multi-facet impacts on ecosystem. Birkved develops a model of estimating the pesticide emission during its application towards air, surface water, and ground water according to the pesticide types and time of application, vegetation species and stage of development, geological and meteorological condition, local application, characteristics of active materials contained by the pesticide for LCA-based agricultural estimation of pesticide emission [2]. Life Cycle Impact Approach (LCIA) is a method usually applied to evaluating toxic effect of the pesticides on human health and ecosystem [3]. Juraske et al. adds that toxic effects can be found by fate, exposition, effect, and damage analysis for assessing cumulative and potential impacts [4]. Panichelli et al. assume that pesticides are totally emitted into soil compartment [5]. Margni et al. find that the distribution rate of the pesticide emission on soil, plant, and air are 85%, 5%, and 10%, respectively[3]. In addition to emission, the pesticide application also causes contamination before and during the formulation, filling sprayer tank, and after the application where the users treat the residual mixture in the sprayer, including the formulation washing [6].

Effects of pesticides on environment can be directly found from production to final cycles [7]. The LCA applies to assessing the environmental aspect and potential effects of the products by calculating eco-efficiency rate. Eco-efficiency is believed to be important towards sustainable development. Pesticide use and treatment relate to transformation process from raw materials, chemical substances, and spraying treatment by the end users. In the first stage, pesticide production provides economic benefit to the producers. However, it also pollutes in the forms of wastewater, harmful solid waste, etc. The following stages have revealed that the use of pesticides contributes to the improvement of agricultural volume. However, the negative impacts also come across in the form of ecosystem pollution by residuals and wastes the pesticides produce.

LCA applies to finding out the environmental effects, including potential course of global warming, greenhouse effect, etc. Pesticide distribution by transportation emits gases, including SO_2. The pesticide application process causes vaporation and air movement, producing gas phase pesticide in the atmosphere and

* Corresponding author: ahmadqosim75@yahoo.com

reacting with water molecules. As it occurs, acid rain develops.

2 Research Methodology

2.1 Research Location and Time

This study was performed in Pati Regency in 2016.

2.2 Research Design

This study applied a quantitative methodology using primary and secondary data on the pesticide distribution lifecycle of the Life Cycle Assessment (LCA) approach as recommended by ISO 14044 on environmental management system.

2.3 Research Scope

The study focused on measurement of energy, materials, and wastes during the pesticide application. The pesticides selected were those popularly applied by the rice farmers in Pati Regency in one unit per ton.

2.4 Population and Samples

The research populations were rice fields with pesticide use of one ton per unit by 87 respondents.

2.5 Research Variables

Dependent variable of this study consisted of environmental effect of the pesticide distribution and use related to acid rain issue, whereas the independent variables consisted of energy needed, pesticide material distribution and use, and waste volumes from the pesticide distribution at the scale equal to SO_2 as the major substance that contributes acid rain.

2.6. Data Processing

Data collected were processed by an Eco-Indicator95 method using the LCA approach, beginning with pesticide distribution by the producer, to pesticide storage at whole sellers, to transportation and storage by retailers, and, finally, to the use and application by the farmers. The calculation also used SimaPro 7.1. application software.

3 Results

3.1. Research Location Profile

Pati is one of 35 regencies under the administrative territory of Central Java Province. It has a strategic position as it is situated between major cities of the North coast of the Java Island, from Jakarta to Semarang to Surabaya. Pati has a geographic feature of 100°15'-111°15 East and 6°25'-7°00' South with the total area of 158,368 hectares. Of these total territories, 58,448 hectares are cultivated for rice fields and 91,036 for other needs. Pati Regency is situated at the elevation of 0-1,000 meters above the sea level, experiences two seasons, dry and rainy, with the more wet than dry months. The average rainfall of the regency, as recorded in 2014, was 2,734 mm with 132 rainy days per year. The lowest and the highest temperature recorded are 23 °C and 39 °C, respectively.

Pati is the place of quite large rivers, where 93 rivers flow across the region. In general, the rivers have motives and trees, with the river mouth at Java Sea. The popular use of the rivers is for agricultural irigation. Unfortunately, during the dry season most of the rivers suffer from drought, whereas during the rainy days, they cause floods. Some of the rivers store the water source, both many of them do not. Most of the rivers have their upstreams at Mount Muria, in particular those situated in the Northern part of the regency.

3.2 Effects of pesticide distribution on acid rain

Life Cycle Assessment (LCA) (EPA, 2006) is a "cradle to grave" assessment, an approach to evaluating particular industrial system.[9] The "cradle to grave" system begins with raw material collection from nature to be converted as particular product(s), and ends at the point where all of the materials are driven back to the nature. The LCA allows the estimation of the cumulative effect on the environment caused by all stages or phases of the product life cycle.

Previous studies have revealed that LCA is a form of technological development to find out the effects of pesticides by a consistent manner to enable the comparison of impacts of other practices in agriculture. The LCA method is useful for determining the pesticide effects on human health and ecosystem. In case of human toxicity, the estimation of the pesticide residuals reveals that foods consumed have the highest toxicity exposition, approximately 103-105 higher than that of consumed by drinking or inhaled from the air [3]. To perform the LCA, one needs adequate inventories of all resources necessary for chemical substance emission calculation, including agricultural input and output inventory, such as pesticides and plant fertilizers [10].

Pesticide distribution consists of several phases, as follows: distribution from producers to distributors; unloading/loading at the distributors' yard; distribution to agricultural whole sellers and kiosks; management at the whole sellers and kiosks; distribution from kiosks to farmers; and pesticide application by end users.

The distribution begins from the producers. In this study, the national producers were located in Gresik Regency, East Java Province, as the 28% of the total respondents reported.

The pesticides were transported by trucks with the trip distance of 212 kilometers. The pesticides were then distributed to the agricultural whole sellers and kiosks using MPVs with the average trip distance of 15 kilometers. The farmers bought the products using motorcycle transportation with the trip distance of 4 kilometers from the shops to the rice fields.

The pesticide distribution causes damaging emission during the transportation from the producers to the consumers. The pesticide emits substances, energy, and components due to over-application, causing damage of particular system. Gas emission is produced by engine combustion, causing greenhouse effect and global warming. The composition of the emission produced during the pesticide distribution is carbon compounds. These compounds are found in gasoline and may cause health problems and disorders, such as eye irritation, headache, memory disorder, reflect disorder, cough, skin irritation, asthma and even lung cancer. The monoxide gas emission causes headache, and cardiac disorder, respiratory tract infection, and lung cancer. Furthermore, SO_2 (sulphuric oxide) emission causes blood formation system toxicity, leading to red blood cell development disorder, anemia, high blood pressure, kidney dysfunction, and reproductive disorder.

The SO_2 emission assessment of the pesticide distribution using eco-indicator 95 method and SimaPro 7.1 revealed that in Pati Regency, at every tone of pesticide, the emission produced was 22.638887 kg SO_2/year. Annually, the pesticide use in Pati Regency was 605 tons. Therefore, the total SO_2 emission was 13,697 kg SO_2/year.

The largest SO_2 emission was found in the pesticide application and from carbofuran pesticide, as Birkved has reported.[2] This result also illustrated the effect of the pesticide emission, assuming that pesticides are totally emitted within the soil compartment.[12, 5]. The emission was divided into 85% to the soil, 5% to the plants, and 10% to the air.[3] The LCA approach can be used for finding out the emission effects on the environment, whereas in this study the LCA revealed the SO_2 emission rate that affected the incidence of acid rain.

The pesticide distribution using transportation devices is believed to cause gas emission, including that of SO_2. The process of the pesticide application deals with evaporation and air movement, causing gas phase pesticide in the atmosphere and reaction with water molecules, where the acid rain develops.

Table 1. Assessment of SO_2 emission during pesticide distribution

No	Component	Impact category (Kg SO_2)
1	Motorbike I	0.10974424
2	Trailer diesel FAL	0.39311314
3	Transport, van <3.5t/CH S	0.20529393
4	Carbofuran, at regional storehouse/CH S	5.7990893
5	Glyphosate, at regional storehouse/CH S	3.626833
6	Tap water, at user/CH S	0.66123444
7	ABS 30% glass fibre I	0.19149345
8	Modified starch, at plant/RER S	0.20860855
9	Pyridazine-compounds, at regional storehouse/RER S	3.6666452
10	Phenoxy-compounds, at regional storehouse/RER S	2.5156454
11	Nitro-compounds, at regional storehouse/RER S	1.9707039
12	Dithiocarbamate-compounds, at regional storehouse/RER S	1.1643351
13	2,4-D, at regional storehouse/RER S	2.1021127
	Total Kg SO_2	**22.638887**

Source: Simapro 7.1

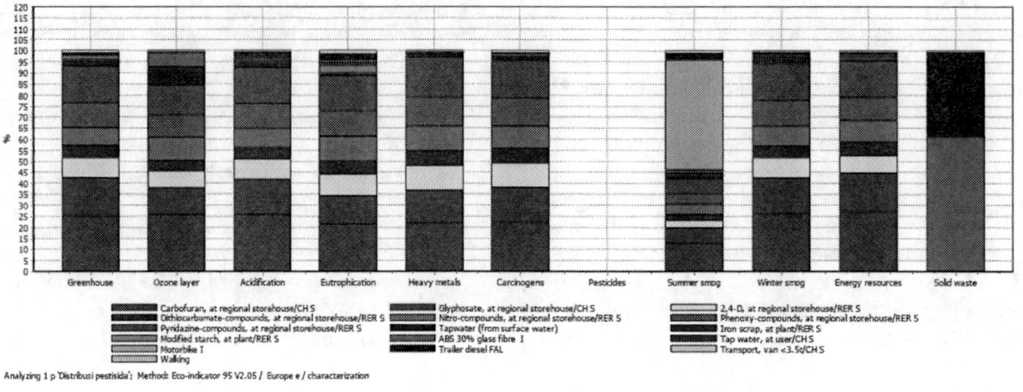

Fig. 1. Emission during pesticide distribution Source: SimaPro 7.1

The above figure presents the varied effects of the pesticide distribution, including SO_2 emission. Acid rains develop during such process. Furthermore, the below figure illustrates a flow diagram of such emission.

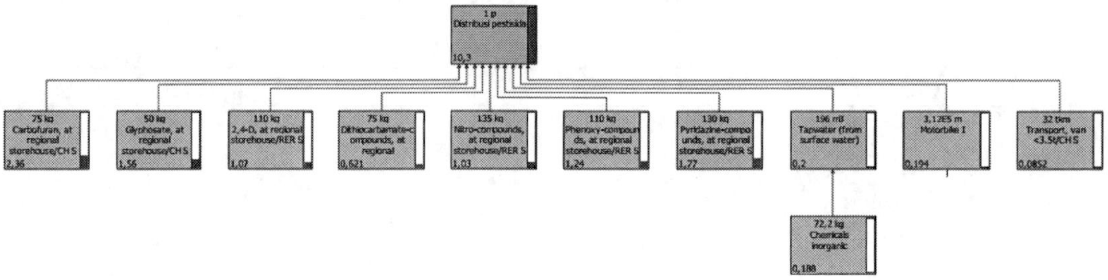

Fig. 2. Flow diagram of pesticide distribution

The study observed that of 100% composition of the pesticide effects it revealed as follows: 5% affected the target plants (the rice fields); 85% affected the nearby environment (land, air, and water contamination); and 10% vaporized

4 Conclusion

The pesticide distribution began from the producers in Gresik Regency, East Java Province and ended at the end users. Using eco-indicator 95 and SimaPro 7.1 the study found that in every tone of pesticide and 16 litres of the pesticide application, it resulted in 22.638887 kg SO_2/ton of pesticide applied by the farmers. Annually, the pesticide use in Pati.

References

1. BLH Pati-PPLH UGM. (2003).
2. M. Birkved, and M.Z. Hauschild, Journal Ecologcal Modelling, **198**, Pp.433–451, (2006)
3. M. Margni, et al., Journal Of Agriculture Ecosystems & Environment, **93**, Pp.379–392, (2002)
4. R. Juraske, et al., Journal Of Chemosphere, **77**(7), Pp.939–945, (2009)
5. Z. Zhu, K. Wang, and B. Zhang, Journal Of Cleaner Production, **69**, Pp.67–73, (2014)
6. R. Zelm, P. Lassalle, and P. Roux, Journal Of Chemosphere, **100**, Pp.175–181, (2014)

Modeling Nitrogen Decrease in Water Lettuce Ponds from Waste Stabilization Ponds

Gitta Agnes Putri[1,*], and *Sunarsih*[1,2]

[1]Magister Program of Mathematics, Faculty of Science and Mathematics, Diponegoro University, Semarang – Indonesia
[2]Doctoral Program of Environmental Science, School of Postgraduate Studies, Diponegoro University, Semarang-Indonesia

Abstract. This paper presents about the dynamic modeling of the Water Lettuce ponds as a form of improvement from the Water Hyacinth ponds. The purpose of this paper is to predict nitrogen decrease and nitrogen transformation in Water Lettuce ponds integrated with Waste Stabilization Ponds. The model consists of 4 mass balances, namely Dissolved Organic Nitrogen (DON), Particulate Organic Nitrogen (PON), ammonium (NH_4^+), Nitrate and Nitrite (NO_x). The process of nitrogen transformation which considered in a Water Lettuce ponds, namely hydrolysis, mineralization, nitrification, denitrification, plant and bacterial uptake processes. Numerical simulations are performed by giving the values of parameters and the initial values of nitrogen compounds based on a review of previous studies. Numerical results show that the rate of change in the concentration of nitrogen compounds in the integration ponds of waste stabilization and water lettuce decreases and reaches stable at different times.

1 Introduction

Waters is one environment that has an important function for human life. But now, the supply of source life is increasingly limited and increasingly polluted by human activities. The increase of development activities in various fields, either directly or indirectly will affect the environmental damage, primarily pollution of waters environment originating from domestic and non domestic waste, such as factories and industries.

Waste is a mixture of household activities, offices and industries. The strength of the waste water is measured by BOD or COD. One of the important ingredients / compounds contained in the waste, especially liquid waste is the nitrogen compound [1].

Nitrogen is an important nutrient in biological systems. Nitrogen accounts for about 12 percent of bacterial protoplasm and 5 to 6 percent of mold protoplasm [2]. In water, the main source of nitrogen is not only in the form of gases, but rather inorganic and organic nitrogen. Inorganic nitrogen consists of ammonia (NH_3), ammonium (NH_4), nitrites (NO_2), nitrates (NO_3) and nitrogen molecules (N_2) in gaseous form. While organic nitrogen in the form of proteins, amino acids and urea. These nitrogen forms undergo a transformation, as part of the nitrogen cycle [3].

The high of nitrogen compounds in waters (wastewater) giving an impact, such as being able to stimulate unlimited algae growth and decrease dissolved oxygen content causing fish death. In addition, waste containing high of nitrate concentrations can also cause poisoning in infants and toddlers [4]. Therefore before re-use, processing is necessary until it reaches the permissible quality requirements.

In wastewater treatment processes, especially those containing organic compound pollutants, the most commonly used technology is the activity of microorganisms that can decompose organic pollutants, commonly called biological processes. One of the most widely used biological treatment processes is a waste stabilization pond system. This is due to the low cost and simplicity of construction, operation, and maintenance [5]. However, the main problem facing the operation of this pool is the minimal reduction in the amount of excessive nitrogen compounds in waste caused by algae activity especially during the summer.

Research on the transformation and decrease of nitrogen compounds has previously been widely, one of them is by using water hyacinth pond which is integrated with waste stabilization pond. However, the use of water hyacinth as a biofilter still has some deficiencies, namely the number of roots owned by water hyacinth, the growth of root buds and leaves which are relatively long, and requires a relatively wide place as a growing medium [6]. Therefore, this article provides an effective method for separating algae and other particles from waste in waste stabilization ponds by considering the shortcomings of water hyacinth, ie by the use of a water lettuce pond.

*Corresponding author: narsih_pdil@yahoo.com

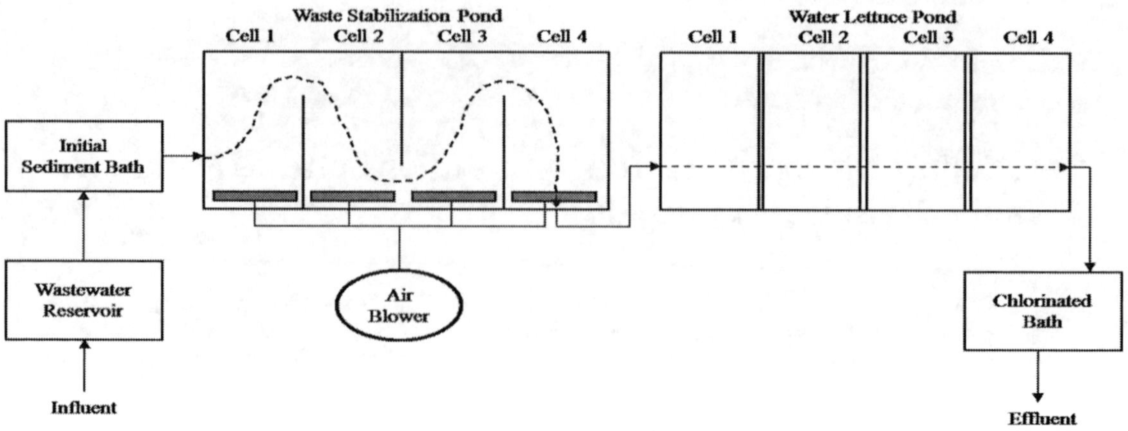

Fig. 1. Pilot System Layout

The Water Lettuce pond is the name for a pond filled with floating aquatic plants, similar to lettuce, the apu woods that have the latin name *Pistia Stratiotes*. The apu wood plants have long, dangling roots such as the hair that hangs in the water [7]. At the root of this plant occurs the process of sedimentation and filtration that can reduce the number of particles of nitrogen and other compounds. In fact some people use this plant with soap to wash clothes, bowls and other household utensils [8]. The apu wood plants do not require such a vast growth medium and can multiply more rapidly because buds of apu wooden shoots emerge from the armpit of the parent's leaf, which can grow apart by having roots.

Previous research on the reduction of organic and inorganic compounds in wastewater using apu wood has also been done [6, 9]. However, the weaknesses in the research are in the design and operation of ponds systems, due to lack of design and simulation methods as appropriate. Several studies related to the design and simulation methods which have been done, only on the facultative primary stabilization pool, maturation pond and facultative stabilization ponds planted with macrophytes, ie research of [10-12].

Therefore, this article provides a dynamic model that can be used to simulate the operation of a water lettuce pool. This model can be used to predict nitrogen decrease as well as nitrogen transformation processes in water lettuce pools integrated with waste stabilization ponds.

2 Materials and Methods

The pilot system integrating the waste stabilization pool with the apu pond is treating secondary waste in sewage treatment plants using an aerobically suspended culture (with air / dissolved oxygen). The suspended culture used is the most common and widely used, ie the process of processing with the standard activated sludge process. Wastewater treatment through active sludge process is chosen because it has the advantage, that is to treat wastewater with a large BOD (Biological Oxygen Demand) load, so it does not require a large place or pond. The wastewater treatment generally consists of an initial sediment bath, an aeration bath and a final settling bath, and a chlorination bath for killing pathogenic bacteria [13].

In this article, wastewater treatment consists of a wastewater reservoir, an initial sediment bath, an aeration bath (stabilization pool), a final sedimentary bath (a pumice pond) and a chlorinated bath. The waste stabilization pond and water lettuce pond are divided into four cells as shown in **Figure 1**.

3 Development of Mathematical Models

Biological waste processing is generally only a transformation process, not a purge. In all cases, it means the change of dissolved organic matter into particulate matter (bacterial colonies) which can then be removed by subsequent process steps, usually by sedimentation or filtration [14]. To construct the model, the following assumptions are given:

1. The process of nitrogen transformation is only in the water lettuce pond, as a refinement of the stabilization pond.
2. The process of nitrogen transformation is assumed to be a filtration process that produces sedimentation in which there are also several processes, namely hydrolysis, mineralization or amonification, nitrification, denitrification, bacterial uptake and plant uptake.
3. The process of hydrolysis, nitrification, denitrification occurs only at the root.
4. The apu wood and bacteria absorb ammonium and nitrate as a source of nitrogen. Nitrates and ammonium are the main sources of nitrogen that are key nutrients for plant growth [4].
5. Bacterial growth occurs in sedimentation. This is because bacteria tend to stick to sediments and other solid materials [4].
6. The waste flow rate in both the waste stabilization pool and the apu pond is assumed to be constant over time.

For the model system components shown in **Table 1** and the conceptual model of the nitrogen transformation in the apu wooden pool can be shown in **Figure 2**.

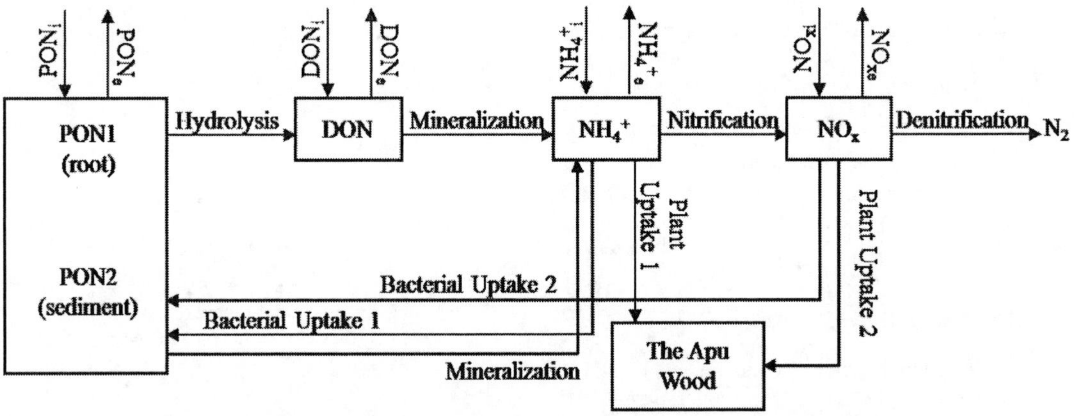

Fig. 2. Conceptual Model of Nitrogen Transformation in the Water Lettuce Pond

Table 1. System model components

Symbol	Definition
PON	Particulate Organic Nitrogen
DON	Dissolved Organic Nitrogen
NH_4^+	Concentration of Ammonium
NOx	Concentration of Nitrate and Nitrite

Consider from the actual natural, the phenomenon that affects the concentration is very complicated. The modeling requires some simplification to form a mathematical equation. That equation is a differential equation which is the rate of change or growth of a substance or species in the stabilization pond. Simplification is intended to limit the factors that affect the degradation process in ponds beginning with chemical reaction rate, biochemical reaction rate and growth rate of bacteria and plants. Based on **Figure 2** and the above reaction rate equations, the model for predicting the components given in **Table 1** are equation (1) to (4):

$$\frac{dDON}{dt} = \frac{Q}{V}(DON_i - DON_e) + r_h - r_{m1} \quad (1)$$

where the influent and effluent of the DON is described as DON_i and DON_e. r_h is the rate of the process of PON hydrolysis and r_{m1} is the rate of the process of DON mineralization.

Eq. (2) is used to predict the ammonium:

$$\frac{dNH_4^+}{dt} = \frac{Q}{V}(NH_4^+{}_i - NH_4^+{}_e) + r_{m1} + r_{m2} - r_n - r_p - r_b \quad (2)$$

where r_{m2} is the rate of PON mineralization, r_n is the rate of nitrification and r_p, r_b are the rate of the plant and bacteria absorption.

The mass balance of nitrate and nitrite (NO_x) is given by the following equation:

$$\frac{dNO_x}{dt} = \frac{Q}{V}(NO_{xi} - NO_{xe}) + r_n - r_d - r_p - r_b \quad (3)$$

where r_d is the NO_x denitrification rate.

$$\frac{dPON}{dt} = \frac{Q}{V}(PON_i - PON_e) + r_b - r_h - r_{m2} \quad (4)$$

The rate of process of hydrolysis and mineralization based on the biochemical transformation rate equation (Yi et al, 2009) are:

$$r_h = k_h.PON \quad (5)$$

$$r_{m1} = k_{m1}.DON \quad (6)$$

$$r_{m2} = k_{m2}.PON \quad (7)$$

where k_h, k_{m1}, k_{m2} are the constant rate of hydrolysis of PON, mineralization of DON and PON, respectively.

The nitrification process rate equation is built from the growth of nitrifying bacteria dependent on the pH, temperature, and concentration of ammonia and dissolved oxygen [10-11], as follows:

$$r_n = \frac{\mu_n}{Y_n}\left(\frac{NH_4^+}{K_1 + NH_4^+}\right)\left(\frac{DO}{K_2 + DO}\right)C_T C_{pH} \quad (8)$$

where μ_n is the maximum growth rate of nitrifying bacteria, Y_n is the yield coefficient/production rate of nitrification bacteria, K_1, K_2, C_T and C_{pH} are the saturation constant of bacteria that depends on temperature and oxygen, a nitrifying rate that depends on temperature and pH, respectively.

The denitrification process rate equations are modeled using the first-order Arrhenius kinetic equation [10-11] as follows:

$$r_d = R2_{20}\,\theta^{(T-20)} NO_x \quad (9)$$

where $R2_{20}$ is denitrification constant, θ is Arrhenius constant and T is temperature.

The equation of bacteria and plant absorption rate, modified from [5, 10-11] based on the monod equation, as follows:

$$r_b = \mu_b (i_{NB} Y_A + i_{NB} Y_H) \quad (10)$$

$$i_{NB} = \left[\frac{NH_4^+}{K + NH_4^+}\right]PON \text{ atau } i_{NB} = \left[\frac{NO_x}{K + NO_x}\right]PON \quad (11)$$

$$r_p = i_{NP}\, g_p \quad (12)$$

$$i_{NP} = \frac{NH_4^+}{K + NH_4^+} \text{ atau } i_{NP} = \frac{NO_x}{K + NO_x} \quad (13)$$

where μ_b is the maximum growth rate of bacteria, g_p is the maximum growth rate of plant, i_{NB}, i_{NP} are nitrogen content absorbed by bacteria and plant, Y_A, Y_H are level of production of autotrophic and heterotrophic cells and K is the saturation constant for ammonium or nitrite/nitrate.

4 Results and Discussions

To find out the system behavior of equations (1), (2), (3) and (4) we need the values of the model parameters which can be seen in **Table 2**.

Table 2. Model Parameters

Parameters		Value	Source
	Y_A	0.24	[5]
	Y_H	0.67	[5]
First-order rate	k_H	0.01	[15]
	k_{m1}	0.1	[15]
	k_{m2}	0.1	[15]
Bacteria	μ_b	0.18	[10]
	K	1.8[a], 2.0[b]	[10]
Plant	g_p	0.5	[5]
	K	1.8[a], 2.0[b]	[10]
Nitrification	μ_n	0.008	[10]
	Y_n	0.13	[10]
	K_1	$1^{(0.051(T-1.58))c}$	[10]
	K_2	1.3	[10]
	C_T	$e^{0.098(T-15)c}$	[10]
	C_{pH}	1.0[d]	[10]
		1 - 0.833(7.2 - pH)[e]	[10]
Denitrification	$R2_{20}$	0.1	[10]
	θ	1.02	[10]

[a] ammonium
[b] nitrite/nitrate
[c] depend on the temperature
[d] If pH ≥ 7.2
[e] If pH < 7.2

The solution graph of the nitrogen reduction system (DON, NH_4^+, NO_x and PON) is given in **Figures 3** to **7**.

Based on **Figure 3**, it can be seen that the rate of change in the concentration of nitrogen compounds (DON, PON, NH_4^+, and NO_x) in ponds that integrate the waste stabilization pool and water lettuce ponds decreases. **Figure 4** showed that the dissolved organic nitrogen (DON) compound decreases until it reaches stable around the 80th day. However, for particulate organic nitrogen (PON) seen on **Figure 5**, the first 10 days increased up to 35 mg / l and then decreased until it reached stability on the 80th day as well.

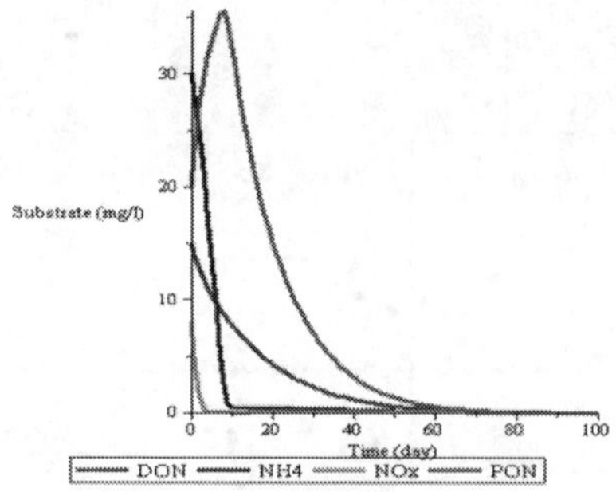

Fig. 3. Behavior of Substrate Concentration

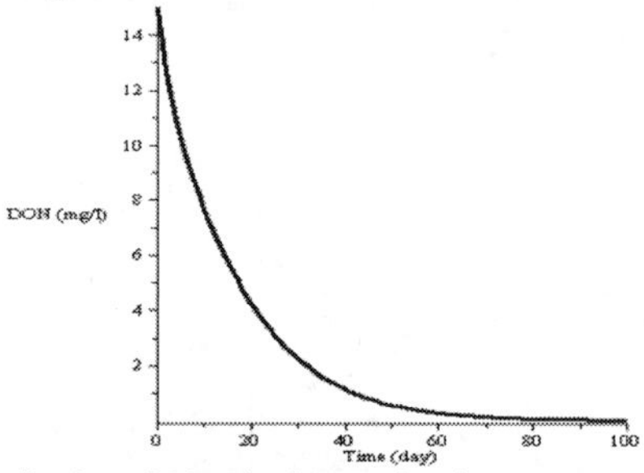

Fig. 4. Behavior of Dissolved Organic Nitrogen Concentration

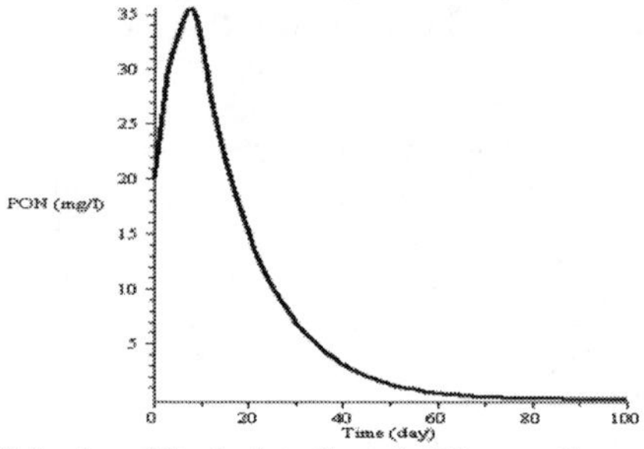

Fig. 5. Behavior of Particulate Organic Nitrogen Concentration

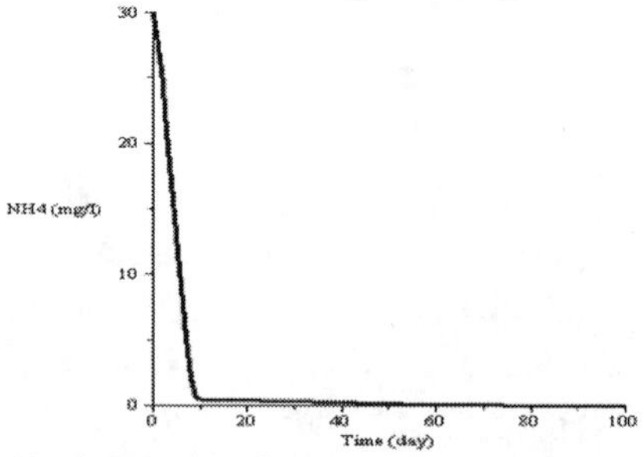

Fig. 6. Behavior of Ammonium Concentration

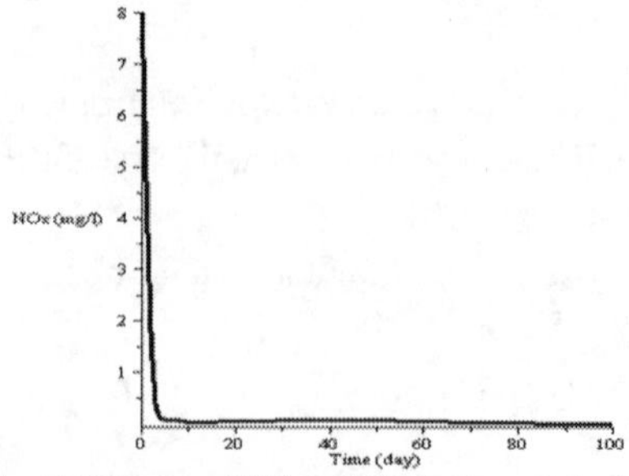

Fig. 7. Behavior of Nitrite/Nitrate Concentration

This is thought to be caused by bacteria. Bacteria decomposed dead animals or vegetation into several organic compounds, including organic nitrogen compounds, causing organic nitrogen to increase rapidly. However, with the process of sewage treatment designed in such a way, for example by aerator, conventional active sludge system, the bacteria switch function to convert organic compounds into inorganic and can be utilized by other living things, including himself. Thus, the inorganic nitrogen that was originally increased, over time becomes decreased and stable, so it can be channeled into the vast waters. The question arises, why not so with dissolved organic nitrogen? This is because bacteria are usually inside solids / particulates or in sedimentation (sediment).

For inorganic nitrogen compounds, both ammonium (NH4 +) and nitric-nitrite (NOx) as shown on **Figure 6** and **Figure 7** decreased until reach stability at a faster time than organic compounds, ie at about day 10 and day 5. Based on these results, the wastewater treatment by integrating the waste stabilization pond and the water lettuce pond can be said to be capable of assisting in the case of the reduction of nitrogen compounds.

5 Conclusions

Based on the results and discussion, it can be seen that the rate of change in the concentration of nitrogen compounds in the integration pond of waste stabilization and water lettuce decreases and reaches stable at different times. This suggests that a pond of waste stabilization integrated with a water lettuce pond provides a good effect in the reduction and transformation of nitrogen.

References

1. P. G. Smith, J. G. Scott, *Dictionary of Water and Waste Management* (IWA publishing, London, 2002)
2. O. A. M. Lewis, *Plants and Nitrogen* (Edward Arnold Ltd, London, 1986)
3. M. Negulescu, *Municipal Waste Water Treatment* (Elsevier Science, New York, 1985)
4. H. Effendi, *Telaah Kualitas Air Bagi Pengelolaan Sumber Daya dan Lingkungan Perairan* (Kanisius, Yogyakarta, 2003)
5. Q. Yi, H. C. Hur, Y. Kim, Eco. Eng. **35**, 75-84 (2009)
6. S. A. Mustaniroh, Wignyanto, E. S. Bernardus, J. Tekn. Pert. **10**, 10-18 (2009)
7. W. S. Don, E. Threes, H. Cherry, *Tanaman Air* (Gramedia Pustaka Utama, Jakarta, 2000)
8. O. C. Hean, *Tanaman Hiasan Khasiat Makanan dan Ubatan* (Yeohprinco SDN. BHD, Kuala Lumpur, 2006)
9. H. I. Owamah, M. A. Enaboifo, O. C. Izinyon, Agri. Wat. Manage. **146**, 262-269 (2014)
10. M. A. Senzia, A. W. Mayo, T. S. A. Mbwette, J. H. Y. Katima, S. E. Jørgensen, Eco. Model. **154**, 207-215 (2002)
11. W. A. Mayo, M. Abbas, Phy. and Chem. Earth. **72-75**, 77-82 (2014)
12. S. M. Grau, A. G. Sfinchez, J. M. Clavel, J. S. Aniorte, M. D. M. Grau, Eco. Model. **91**, 77-103 (1996)
13. N. S. Idaman, *Pengolahan Air Limbah Domestik di DKI Jakarta* (Badan Pengkajian dan Penerapan Teknologi, Jakarta, 2008)
14. A. S. Siregar, *Instalasi Pengolahan Air Limbah* (Kanisius, Yogyakarta, 2005)
15. B. Beran, F. Kargi, Eco. Model. **181**, 39-57 (2005)

FLASH Technology: Full-Scale Hospital Waste Water Treatments Adopted in Aceh

Rame[1,*], *Adeodata* Tridecima[2], *Hadi* Pranoto[2], Moesliem[3], *Miftahuddin*[3]

[1] Center of Industrial Pollution Prevention Technology, Semarang - Indonesia
[2] Bartec Utama Mandiri, Semarang - Indonesia
[3] Public Health Office, Aceh - Indonesia

Abstract. A Hospital waste water contains a complex mixture of hazardous chemicals and harmful microbes, which can pose a threat to the environment and public health. Some efforts have been carried out in Nangroe Aceh Darussalam (Aceh), Indonesia with the objective of treating hospital waste water effluents on-site before its discharge. Flash technology uses physical and biological pre-treatment, followed by advanced oxidation process based on catalytic ozonation and followed by GAC and PAC filtration. Flash Full-Scale Hospital waste water Treatments in Aceh from different district have been adopted and investigated. Referring to the removal efficiency of macro-pollutants, the collected data demonstrate good removal efficiency of macro-pollutants using Flash technologies. In general, Flash technologies could be considered a solution to the problem of managing hospital waste water.

1 Introduction

Hospital waste water pollution is a major environmental problem in the world, and usually, it is associated with hazardous chemicals and harmful microbes [1][2][3]. Hazardous chemicals sources could be come from pharmaceutical active compounds (PhACs), heavy metals, detergents, X-ray contrast media, and disinfecting agents [4] while harmful microbes sources come from pathogenic microorganisms such as viruses, bacteria, fungi, protozoan, and helminthes [5]. Hospital waste water pollution in the environmental is one of the major quality issues in many countries because maintenance of water quality and sanitation infrastructure did not increase along with population and urbanization growth [4].

Hazardous chemicals contributed threat to public health and environment due to its potential toxicity as well from highly infectious potential that releases contamination [6]. Harmful microbes in hospital waste water pollution have also been associated with pathogen dissemination and in antibiotic resistance spread into the environment. As hazardous chemicals are non-biodegradable and accumulative in environment, the elevated hazardous chemicals and their deposition over time can lead to the contamination of the surface environment [7][8].

Nangroe Aceh Darussalam (Aceh) is chosen as a focus in this study with the objective of treating hospital waste water effluents on-site before its discharge. Flash Full-Scale Hospital waste water Treatments in Aceh from different district have been adopted and investigated [9]. Flash technology uses physical and biological pre-treatment, followed by advanced oxidation process based on catalytic ozonation and followed by GAC and PAC filtration [9]. This efforts have been carried out in Aceh considered as community health centre (Puskesmas) which government-mandated community health clinics, whereas provide healthcare for population on sub-district level. Puskesmas becomes a mini hospital which provided maternal and child health care, general outpatient curative and preventative health care services, pre- and postnatal care, immunization, and communicable disease control. With an increase of Puskesmas growth, hospital waste water around Aceh area will produce more hazardous chemicals, harmful microbes and other types of pollution [10].

It is significant to note that macro-pollutants level are indicate of pollution levels [5]. Some macro-pollutants parameter were identified from the smell, color, taste, pH, turbidity, COD, BOD_5, and heavy metal content, as well as total coliform as be measured for hospital waste water treatments [11]. However, the macro-pollutants level is sensitive to technology which used for hospital waste water treatments [12]. The objective of the study is to determine the removal efficiency of macro-pollutants in Flash Full-Scale Hospital waste water Treatments in Aceh from different district. The results can also be used to determine whether Flash technologies could be considered a solution to the problem of managing hospital waste water.

2 Materials and Method

Hospital waste water pollution including both the hazardous chemicals and harmful microbes have a

* Corresponding author: aryanagasasra@gmail.com

particular significant in contamination, since they are highly persistent and have all the potential to be toxic to living organisms. The macro-pollutants that are generally determined from effluent of flash full-scale hospital waste water treatments in Aceh include physical parameters (temperature, dissolved solids, suspended solids), chemical parameters (pH, COD, BOD, oils and fats, MBAS, ammonia nitrogen, heavy metals), and biological parameters (total coliform). Six sampling Puskesmas were chosen from 21 (twenty-one) districts in Aceh as shown in Fig. 1. Three sampling Puskesmas (Puskesmas 1, 3, 4) were located at the east area of Aceh, Puskesmas 2 was in the south of Aceh and Puskesmas 5 was in the north of Aceh and Puskesmas 6 was located at the west of Aceh.

The water samples were processed for analysis in the laboratory of Baristand Industri Aceh. Heavy metals were determined using Atomic absorption spectroscopy (AAS) Shimadzu. Other parameters were analysed using Ultraviolet-visible spectroscopy (UV / Vis) and standard method according to Indonesian National Standard (SNI). Evaluation removal efficiency of macro-pollutants can be known from the analysis of physical, chemical, and microbiological parameters. The results of this analysis are then compared to water quality standards in accordance with their designation stipulated by regulation of the Minister of Environment Republic of Indonesia number 5:2014 about quality of waste water quality.

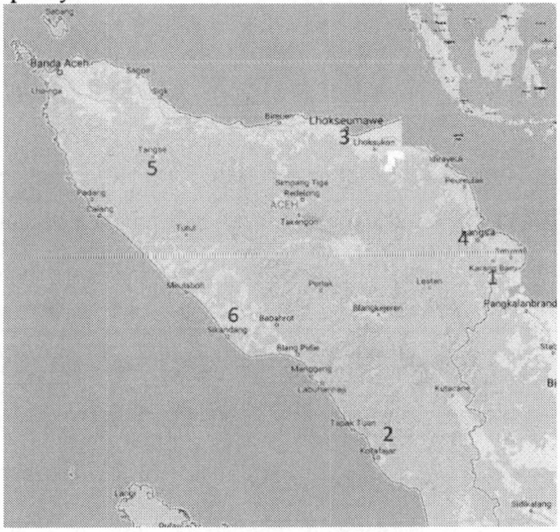

Fig. 1. Location of seven effluent FLASH sampling in Aceh.

3 Results and Discussion

Thirty-one of parameters were analysed from the water samples which were collected from the seven sampling Puskesmas (Puskesmas 1 in aceh tamiang = P1, Puskesmas 2 in aceh selatan = P2, Puskesmas 3 in Kota Lhokseumawe = P3, Puskesmas 4 in Kota Langsa = P4, Puskesmas 5 in Pidie = P5, Puskesmas 6 in Nagan Raya = P6) in the Aceh such as physical parameters (pH, temperature, total dissolved solids = TDS, total suspended solids = TSS), chemical parameters (COD, BOD, oils and fats, MBAS, Ammonia, iron, Manganese, barium, copper, zinc, Hexavalent chromium, Chromium, cadmium, Mercury, lead, Arsenic, Selenium, nickel, Cobalt, cyanide, sulphide, Fluoride, Chlorine free, nitrate, Phenol, Nitrite), and biological parameters (total coliform). Each of them will be described as follows.

3.1 Physical parameters

The range of pH at six sampling puskesmas was from 7.04 to 8.24, which the highest (8.24) was recorded at P3 and the lowest (7.04) was recorded at P4 as shown in (Fig. 2a). The value of pH from six sampling puskesmas in the study area was still below the maximum threshold value of 9 (P0) and above the minimum threshold value of 6.

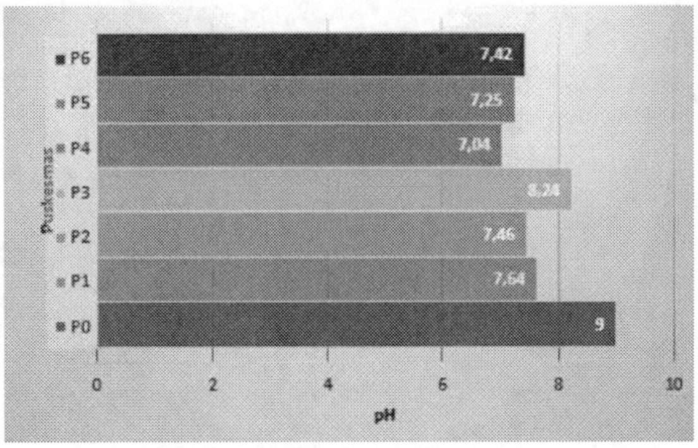

Fig. 2a pH values at six sampling puskesmas

The range of temperature at six sampling puskesmas was from 27.8 to 30.3, which the highest (30.3) was recorded at P3 and the lowest (27.8) was recorded at P1 as shown in (Fig. 2b). The value of pH from six sampling puskesmas in the study area was still below the maximum threshold value of 38 (P0).

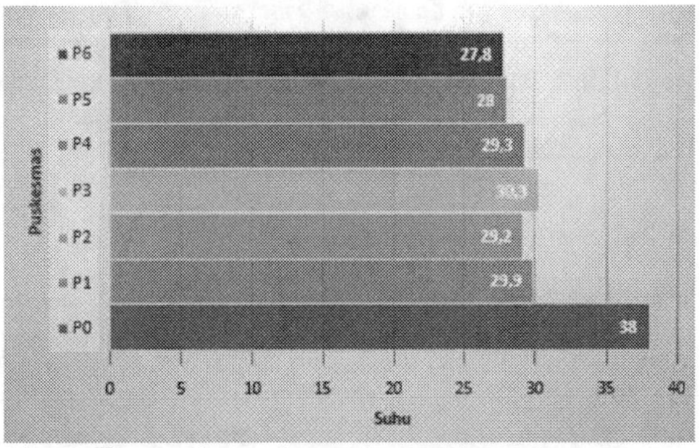

Fig. 2b Temperature values at six sampling puskesmas

The range of TSS at six sampling puskesmas was from 8.57 to 130, which the highest (130) was recorded at P5 and the lowest (8.57) was recorded at P4 as shown in (Fig. 2c). The value of pH from six sampling puskesmas in the study area was still below the maximum threshold value of 200 mg/L (P0).

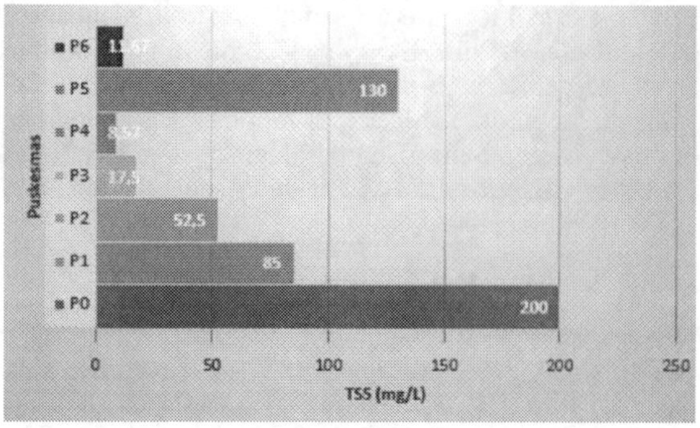

Fig. 2c TSS values at six sampling puskesmas

The range of TDS at six sampling puskesmas was from 121.2 to 389, which the highest (389) was recorded at P5 and the lowest (121.2) was recorded at P6 as shown in (Fig. 2d). The value of pH from six sampling puskesmas in the study area was still below the maximum threshold value of 2000 mg/L (P0).

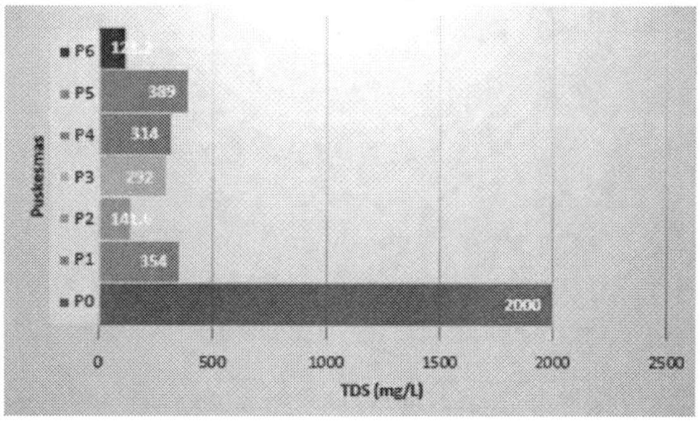

Fig. 2d TDS values at six sampling puskesmas

Physical parameters showed that there were no significant differences between six sampling Puskesmas. Based on value of pH, temperature, TSS, TDS, the level of parameters is classified as safety. This value contributed to the environment through water conditions is not threatening the lives of living things, especially microorganisms that play a role in the balance of environmental ecosystems.

3.2 Chemical parameters

The range of BOD_5 at six sampling puskesmas was from 3.84 to 9.49, which the highest (9.49) was recorded at P2 and the lowest (3.84) was recorded at P5 as shown in (Fig. 3a). The value of BOD_5 from six sampling puskesmas in the study area was still below the maximum threshold value of 50 mg/L (P0).

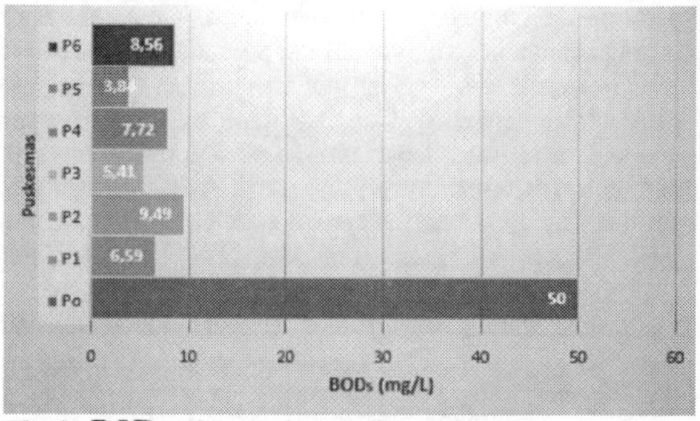

Fig. 3a BOD_5 values at six sampling puskesmas

The range of COD at six sampling puskesmas was from 16.09 to 56.34, which the highest (56.34) was recorded at P2 and the lowest (16.09) was recorded at P3 as shown in (Fig. 3b). The value of COD from six sampling puskesmas in the study area was still below the maximum threshold value of 80 mg/L (P0).

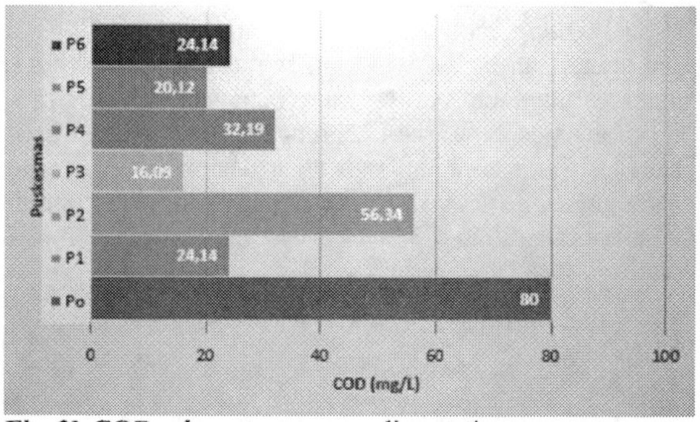

Fig. 3b COD values at seven sampling stations

The collected data of other chemical parameters such as oils and fats, MBAS, ammonia, iron, manganese, barium, copper, zinc, hexavalent chromium, chromium, cadmium, mercury, lead, arsenic, selenium, nickel, cobalt, cyanide, sulphide, fluoride, chlorine free, nitrate, phenol, nitrite show all still below the maximum threshold value. Chemical parameters showed that there were no significant differences between six sampling Puskesmas. Based on the value of chemical parameters in the study area is classified as safety to environment.

3.3 Biological parameters

The range of Total coliform at six sampling puskesmas was from 50 to 1600, which the highest (1600) was recorded at P2 also P3 and the lowest (50) was recorded at P6 as shown in (Fig. 4). The value of total coliform from six sampling puskesmas in the study area was still below the maximum threshold value of 5000 APM/100 mL (P0).

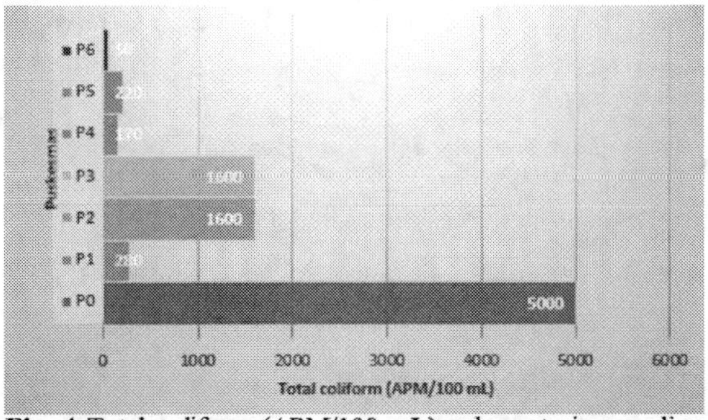

Fig. 4 Total coliform (APM/100 mL) values at six sampling puskesmas

Biological parameters as total coliform analysis showed that there were no significant differences between six sampling puskesmas. Based on the value of total coliform, the level of total coliform is classified as low. With respect to the value of total coliform, the study reveals that there are expected additions of quantities of coliform effluents in P2 and P3 that are particularly associated with the air quality in the area of waste water treatments. Air is another pathway by which coliform enters the water in open space via the dust surface and the coliform becomes bound to the surfaces of dust or other particles [13].

Fig. 5 Flash technology

Flash technology uses physical and biological pre-treatment, followed by advanced oxidation process based on catalytic ozonation and followed by GAC and PAC filtration as shown in (Fig. 5). The Catalyst System ozonation system as one of the advanced oxidation process (AOP) methods is based on ozone decomposition and the formation of stronger and reactive hydroxyl radicals than ozone. The AOP process can be carried out with the use of ozone together with some chemicals or combination with UV radiation and some chemicals [O_3 / H_2O_2, O_3 / UV, O_3 / H_2O_2 / UV] as well as catalytic ozonation [14][7][11]. Ozone that does not react with the catalyst will decompose the contaminants directly or decompose back into the oxygen gas in the waste water as shown in (Fig. 6).

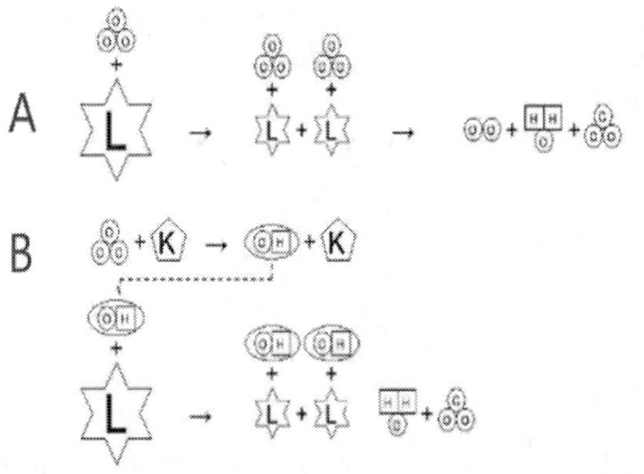

Fig. 6 Mechanism of ozonation reactions: (A) directly, (B) catalytic [*15*]

Data analysis showed that there were significant difference between values of water from output of six sampling Puskesmas adopted Flash Full-Scale Hospital waste water treatments and the maximum threshold value. Referring to the removal efficiency of macro-pollutants, the collected data demonstrate good removal efficiency of macro-pollutants using Flash technologies from six sampling Puskesmas.

The result of Flash technologies is in accordance with some study data, ie O_3 + PAC + sand filtration can process micropollutants (including PhACs), and in the study carried out in a HWWTP in the Netherlands, none of the PhACs (32 different compounds) were detected in the Effluent after tertiary treatment based on GAC + ozone [16]. Biological treatment CAS and MBR is the most widely used worldwide especially in indonesia. However, biological treatment of hospital wastewater does not provide some of the compounds such as pharmaceuticals / PhACs and some pathogenic microbes. Only additional advanced steps like ozonation, activated carbon, or AOPs will enable a better elimination of these compounds [16]. Flash technologies should be adopted for better removal of macro pollutants, harmful bacteria and PhACs. In general, Flash technologies could be considered a solution to the problem of managing hospital waste water around Aceh.

4 Conclusions

The results of the analysis of the thirty-one at the six sampling Puskesmas from Aceh could be concluded that the value of all parameters still below the maximum threshold value. Based on the analyses show that the removal efficiency of macro-pollutants, the collected data demonstrate good removal efficiency of macro-pollutants using Flash technologies. Flash Full-Scale Hospital waste water treatments in Aceh from different district have been adopted and investigated. Referring to in general, Flash technology could be considered a solution to the problem of managing hospital waste water. Monitoring should be carried out from time to time of the hospital waste water treatments to control and prevent contamination as well to identify macro-pollutants, especially to those that can have a direct impact on environmental pollution.

5 Acknowledgment

The authors would also like to thank to Baristand Industry Aceh for the use of their research laboratory facilities. The authors would like to thank RR Kartika for the willingness to read and edit this manuscript.

References

1. J. Ferre-aracil, Y. Valcárcel, N. Negreira, M. L. De Alda, D. Barceló. Sci. Total Environ. **556**, 70–79 (2016).
2. A. Arslan, S. Veli, D. Bingöl, Sep. Purif. Technol. **132,** 561–567 (2014).
3. T. Azuma et al., Sci. Total Environ. 548–549, 189–197 (2016).
4. T. Chonova, F. Keck, J. Labanowski, B. Montuelle, Sci. Total Environ. **542**, 965–975 (2016).
5. P. Verlicchi, M. Al Aukidy, E. Zambello, Sci. Total Environ. **514,** 467–491 (2015).
6. M. Umar, F. Roddick, L. Fan, H. A. Aziz, Chemosphere. **90,** 2197–2207 (2013).
7. L. Blaney, Ozone Treatment of Antibiotics in Water. Water Reclam. Sustain. (2014), doi:10.1016/B978-0-12-411645-0.00012-2.
8. M. Escolà et al., Sci. Total Environ. 530–531, 383–392 (2015).
9. Bartec Utama Mandiri, Annual report 2015 (2015).
10. A. Tridecima, "FLASH: WWT Catalyst System" (Bartec Utama Mandiri, 2017).
11. R. Kıdak, Ş. Doğan, Ultrason. Sonochem. (2017), doi:10.1016/j.ultsonch.2017.01.033.
12. Y. Lee, L. Kovalova, C. S. Mcardell, Water Res. **64**, 134–148 (2014).
13. N. F. Gray, Ozone Disinfection (Elsevier, Second Edition., 2014; http://dx.doi.org/10.1016/B978-0-12-415846-7.00033-0).
14. K. M. S. Hansen et al., Chem. Eng. J. **290**, 507–514 (2016).
15. Rame, A. Purwanto, A. Budiarto, Res. J. Ind. Pollut. Prev. Technol. In Press, (2017).
16. E. cooperation Project, "Pharmaceutical residues in the aquatic system: a challenge for the future" (Gelsenkirchen, 2012), (available at www.pills-project.eu).

Tidal Influence on Water Quality of Kapuas Kecil River Downstream

Rizki Purnaini[1*], *Sudarmadji*[2] and *Suryo* Purwono[3]

[1]Doctoral Program of Environmental Science, School of Postgraduate Studies, Gadjah Mada University, Yogyakarta - Indonesia
[2]Department of Environmental Geography, Faculty of Geography, Gadjah Mada University, Yogyakarta - Indonesia
[3]Department of Chemical Engineering, Faculty of Engineering, Gadjah Mada University, Yogyakarta - Indonesia

Abstract. The Kapuas Kecil River is strongly influenced by tidal, in the dry season the intrusion of surface water is often a problem for the WTP because it causes the change of raw water quality to be processed. The purpose of this study was to examine the effect of sea tides on water quality of the Kapuas Kecil River. The study was conducted in Kapuas River downstream along ± 30 km from the upper boundary to the estuary. Water sampling is carried out during the dry and rainy season, when the tidal conditions at 7 (seven) locations of the monitoring station. Descriptive analysis methods and regression-correlation statistics are used to determine the effect of tides on water quality in Kapuas River downstream. In general, the water quality of the Kapuas Kecil River has exceeded the criteria of first class water quality, ie water that can be used for drinking water. The status of water quality of the Kapuas Kecil River based on the pollution index calculation shows the condition of the river is "mild to medium pollutants". The result of multiple linear regression analysis got the value of coefficient of determination (adjusted R square) = 0,760, which in whole show that independent variable (tidal and distance) influence to dependent variable (value of TDS) equal to 76%.

1 Introduction

River is one of the water resources that has many functions and benefits for humans, but inevitably the river also serves as a place to dump the remains of human activities both solid waste and liquid waste. The entry of various kinds of discharges into the river either directly or indirectly can lead to a decrease in river water quality.

The ability of self-purification of the river will decrease with the influx of pollutants from various pollutant sources with the increase of population so that the river becomes polluted and can not be utilized again as intended. As one source of raw water, surface water must be maintained in order to maintain its quality to be utilized in accordance with its allocation, therefore water quality monitoring and policy are necessary in water quality management and water pollution control.

The Kapuas Kecil River as part of the Kapuas River Basin that flows from upstream to downstream through Kubu Raya District, Pontianak City and Mempawah District. Kapuas Kecil River divides and passes through Pontianak City before heading to estuary to Karimata Strait. This river passes through the city with densely populated settlements and various community activities such as trade, services, and large-small industries. Pontianak City has many small rivers that form naturally or artificially that serves as a drainage and domestic sewage. The main sources of pollutants entering Kapuas Kecil River downstream come from domestic waste settlements, trade and service activities, large and small industrial waste, fish cages and urban runoff.

Kapuas Kecil River is the source of raw water for Pontianak City WTP and Sei Raya City WTP. Some people in the city still use river water directly for daily utility. The Kapuas Kecil River is strongly influenced by tidal, in the dry season the intrusion of surface water is often a problem for the WTP because it causes the change of raw water quality to be processed.

River is dynamic so the water quality is also dynamic. Fluctuations in water quality are caused by the type and amount of pollutants from various sources of pollutants and diverse areas and the process of physics, chemistry and biology in the waters. In addition, the hydrological conditions of the river and the effect of fluctuating tides will also affect water quality.

River flows and seasonal changes are one of the important parameters in circulation in the estuary. River flow depends on the characteristics of hydrology and the watershed. In the rainy season the flow of large flow, while in the dry season is small. In general, river discharge is much smaller than the discharge caused by the influence of the tides. The effect of flow discharge is

*Corresponding author: rizkipurnaini@yahoo.co.id

more dominant in the upstream estuary compared with downstream. At the time of flooding, river discharge pushes pollutants into the sea, while at a small discharge the pollutant moves upstream. Tidal currents affect the shifting of pollutants along the estuary, moving upstream at high tide and downstream at low tide [1].

Water quality can be highly dependent on river flow conditions. River flow affects water quality in several ways, namely: 1). Dilution; Large flow volumes can dilute the concentration of pollutants being dumped into the river. 2). Residence time; High flow rates can reduce the residence time and affect the amount of material that can be produced or degraded in the river. 3). Mixing; High flow velocity leads to increased mixing in rivers, enhanced river assimilation capabilities, and reduced gradient of pollutant concentrations. 4). erosion; High flow rates can erode the material along the flow [2].

This research was conducted with the aim to study the influence of tides on water quality in the lower Kapuas River downstream so as to get description of water quality characteristics of the river and the ability of the river to accommodate incoming pollutants, so that the sustainability of river utilization can be maintained.

2 Materials and Methods

The study was conducted in August 2016 to represent the dry season and January 2017 to represent the rainy season in the Kapuas Kecil River downstream of ± 30 km from the upstream boundary of the research site to the estuary. Upper boundary of research location is intake location of Sei Raya WTP, while the downstream boundary is mouth of river. Sampling using purposive sampling method at 7 (seven) location of monitoring station during tidal condition (Fig. 1).

The data used in this study are primary data including water quality parameters :temperature, pH, TSS, TDS, BOD, COD, DO, N, P, and salinity; hydrometry; and bathymetry. Secondary data include tidal forecast and land use and drainage maps.

Sampling of river water is done in accordance with SNI 6989.57: 2008, using boat and water sampler tool. Samples were collected at each monitoring station divided into 3 dots across the river and each at three depth layers to be combined (composite). Subsequently the water sample for the BOD parameter was directly transferred into a 500 ml BOD bottle while for other parameters into the 2.5 liter plastic jewel that has been labeled. Water quality were analyzed insitu and exsitu. Temperature, TDS and salinity parameters using Extech instruments, ExStick EC 400; pH with pH meter type ATC (pH-009 (I) A; DO with LUTRON DO meter DO-5509, while other parameters were analyzed exsitu in Sucofindo Laboratory Pontianak.

Measurement of vertical distribution of salinity at each monitoring station. Samples were taken at three layers of depth ie surface layer (0.2d), middle (0.6d) and bottom of river (0.8d) with water sampler and examined its salinity by insitu with Extech instruments, ExStick EC 400.Cross-sectional measurements of river and river depth at each monitoring station are done by boat, GPS and echosounder. Measurement of water level elevation using HOBO onset water level logger S / N: 10949770 installed in intake of Pontianak City WTP.

Water quality data will be analyzed by statistical regression-correlation method to determine the effect of tides on water quality and its relation to river discharge and hydrological condition.

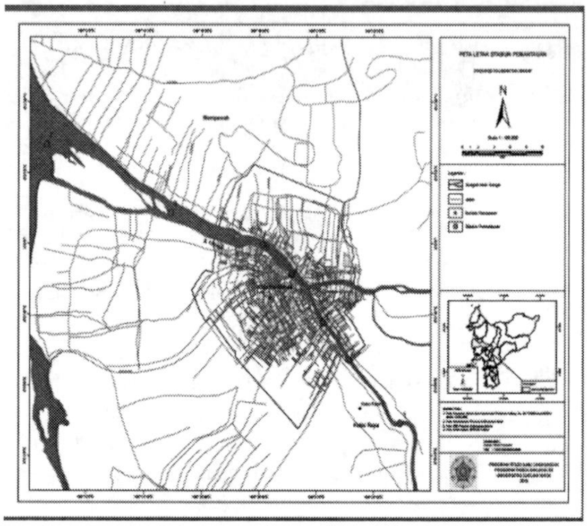

Fig. 1. The map of sampling Location

3 Results

The results of tidal observations at the study sites in August 2016 and January 2017, the tidal type in the Kapuas Kecil River is mixed tide prevailing diurnal, in one day there is one tide but sometimes temporarily occurs twice tide with very different heights and periods (Fig. 2 &3). Maximum water level elevation of Kapuas Kecil River in August 2016: 1.6 m at high tide and 0.5 m at low tide, while January 2017: 1.7 m at high tide and 0.4 m at low tide [3].

3.1 Salinity distribution in Kapuas Kecil River downstream

The horizontal salinity distribution in Kapuas Kecil river from upstream to downstream tends to increase at high tide (Fig. 4). The effect of tides on salinity increases significantly with the increasing distance of observation stations progressively toward the downstream / river estuary. Distance of sea water is relatively far, approximately 10 km upstream (monitoring station 6), salinity value at this location is about 6 ppt. In the longitudinal direction, the salinity will increase or increase along with the distance from the upstream to the downstream or vice versa due to the influence of tidal, estuary characteristics, and river discharge [1].

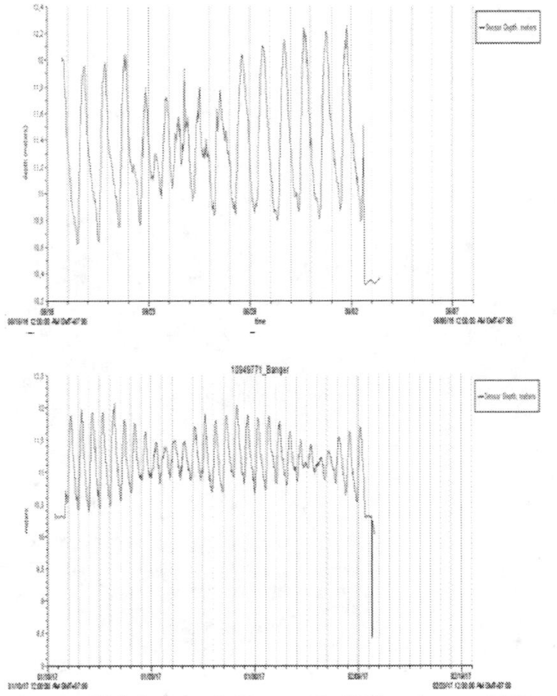

Fig. 3. Tidal elevation in Kapuas Kecil River January 2017

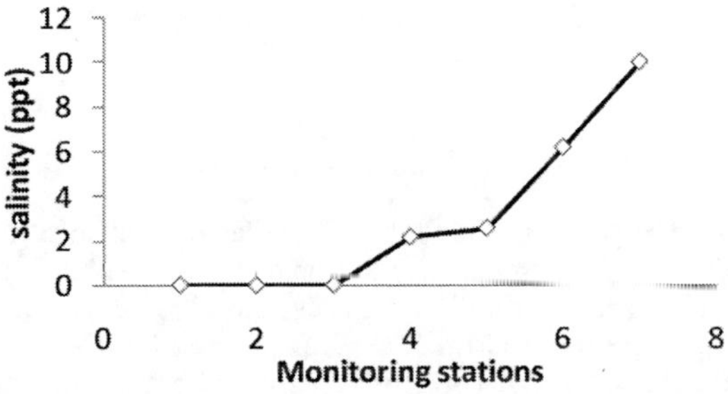

Fig. 4. The horizontal salinity distribution in Kapuas Kecil river

The vertical salinity distribution in the three depth layers of each monitoring station at the study sites is shown in Fig. 5 & 6. Layer depth of the river affects the salinity distribution vertically. Salinity increases with depth, due to the greater sea-water density than freshwater, so that the salinity value will increase on the bottom of the river.

Based on the salinity structure, Kapuas Kecil River estuary has a "partially mixed" estuary type. This is based on the characteristics of this estuary type described in Triatmodjo, salinity varies in longitudinal and vertical directions. In the longitudinal direction, the salinity is reduced from the mouth of the river to the upstream, while in the vertical direction decreases from the base to the surface.

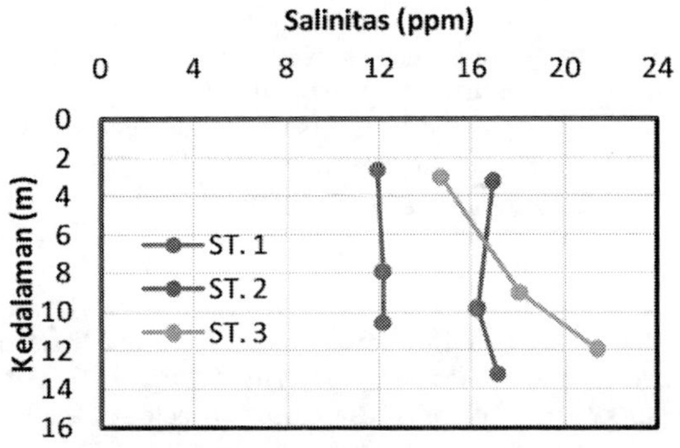

Fig. 5. The vertical salinity distribution in uptream site

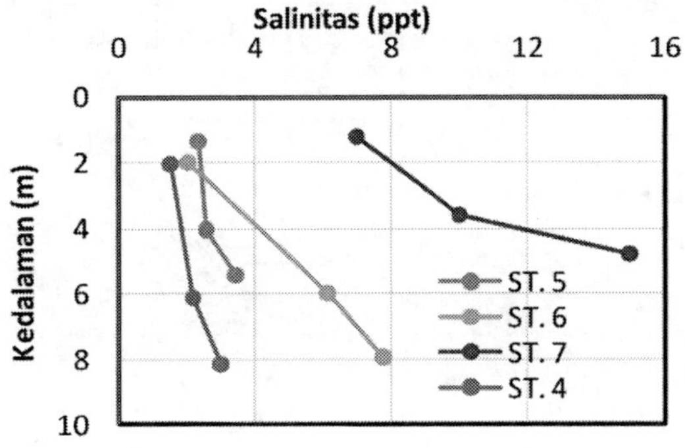

Fig. 6. The vertical salinity distribution in downstream site

3.2 Water Quality of Kapuas Kecil River downstream

Water quality of Kapuas Kecil River must conform with the criteria of water quality class I PP RI No. 82/2001, that is, water which can be used for drinking water, and / or for other uses which require the same water quality as the activity [4].

Insitu data measurement showed that temperature 29 – 34°C and pH value 6,0 - 7,8.

BOD5 and COD levels during the rainy and dry season at high and low tides have exceeded the water quality criteria established for Class I, ie BOD 2 mg / l and COD 10 mg / l, with upstream to downstream trend increasingly (Fig.7&8). High levels of BOD and COD in the study sites indicate that river water is contaminated by organic material derived from discharges entering the river along the stream, while self purification of the river is exceeded. Most of the discharges coming into the river come from urban runoff, domestic waste settlements, waste of trade and services activities (markets, hotels, restaurants, malls), large industrial wastes (factories) along streams and small industrial waste (home industry).

The results of DO insitu measurements at the study sites from upstream to downstream during the rainy and dry seasons at high and low tide generally fluctuate and still meet the water quality criteria set for Class I of minimum 6 mg / l. Dissolved oxygen levels fluctuate on a daily and seasonal basis, depending on the mixing and movement of water masses, photosynthesis, respiration and effluent activity into water bodies [5].

DO values should be inversely proportional to BOD and COD values, since BOD and COD represent the total amount of oxygen required to oxidize organic materials either by microorganisms or chemically. The higher the value of BOD and COD then the DO value will be lower. The value of DO measurements obtained in the field is in fact incompatible with the logic of existing theory. This can be caused by human error factors at the time of measurement or factor of the tool [6].

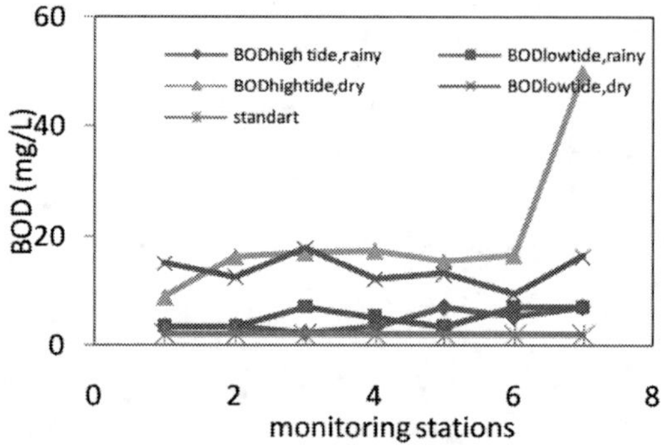

Fig. 7. BOD content in Kapuas Kecil River

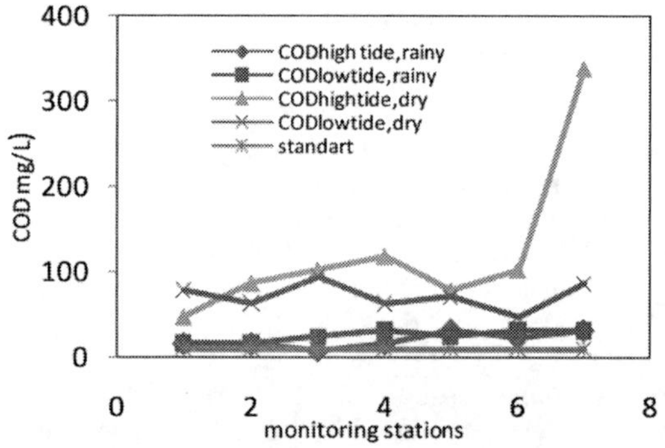

Fig. 8. COD content in Kapuas Kecil River

Nitrate value at research location during rainy season and dry season at high and low tide generally conform with water quality criteria for Class I that is 10 mg / l. Nitrate levels greater than 5 mg / l illustrate the occurrence of anthropogenic contamination derived from human activities and animal feces [5].

The total value of phosphorus at the study sites during the rainy and dry seasons at high and low tide has exceeded the Class I water quality criteria of 0.2 mg / l. Anthropogenic sources of phosphorus are industrial and domestic waste, ie phosphorus derived from detergents. Runoff from agricultural areas using fertilizers also contributes substantially to the presence of phosphorus [5].

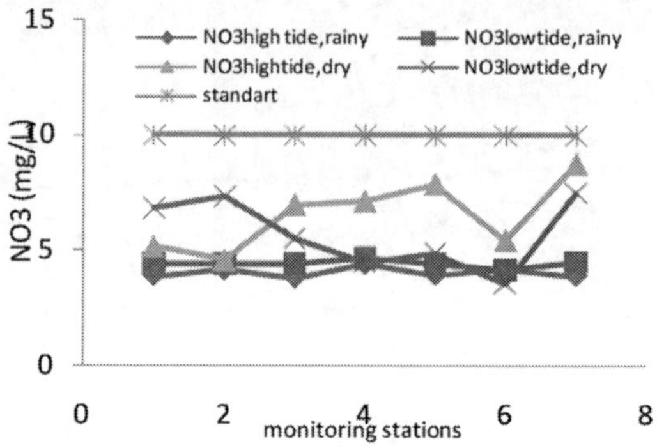

Fig. 9NO3 content in Kapuas Kecil River

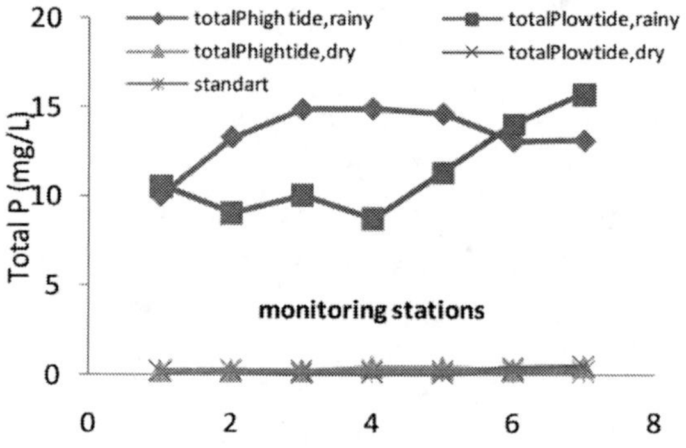

Fig. 10. Total P content in Kapuas Kecil River

The value of TSS in the study sites from upstream to downstream during the rainy and dry seasons at high and low tide generally fluctuates, and exceeds the water quality criteria set for Class I of 50 mg / l. TSS (Total Suspended solid) consists of mud and fine sands and microorganisms, mainly due to soil erosion or soil erosion brought to water bodies [5].

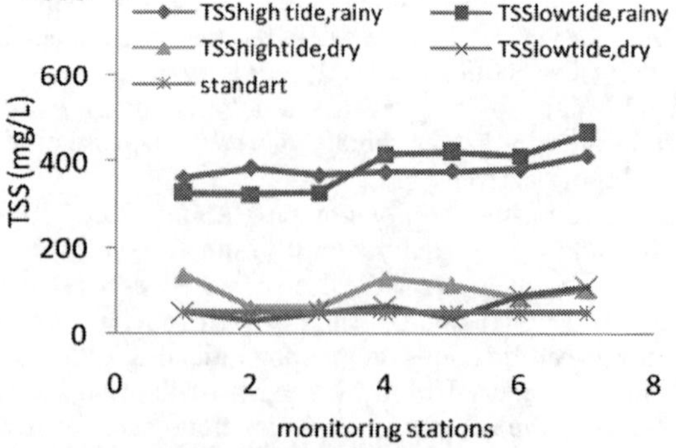

Fig. 11. TSS content in Kapuas Kecil River

TDS (Total Dissolved Solid) describes dissolved ions in water. The value of TDS at the study sites during the rainy and dry seasons at high and low tide is generally increasing downstream and exceeds the Class I water quality criteria of 1,000 mg / l, but at the low tide, conform with the criteria of Class I water quality. Increasing the value of TDS at downstream indicates the effect of tides that enter the river, high TDS value also shows a high salinity value [5].

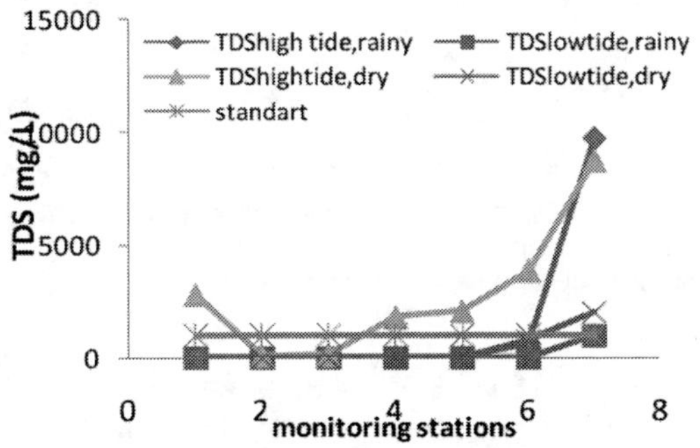

Fig. 12. TDS content in Kapuas Kecil River

The result of pollution index calculation to determine the status of water quality of Kapuas Kecil River based on KepMenLH 115/2003 found that river condition in general "mild to medium pollutant" [7]. This indicates that there has been a decrease in water quality of Kapuas Kecil River so that it is not in accordance with its intended to be utilized as raw water for drinking water, so it is necessary efforts to water quality management and water pollution control.

3.3. Tidal Relationship to Distance Salinity

The results of analysis using multiple linear regression found that there is a relationship between tidal elevation and distance to water quality of 91%, seen from the correlation coefficient (R) as shown in Fig. Next got the coefficient of determination (adjusted R square) = 0.760; This value as a whole indicates that the independent variable (tidal and distance elevation) influence to water quality (TDS value) equal to 76%, while the rest influenced by other variable not examined in this research (Table 1).

Table 1.Multiple regression analysis results

Model	R	R Square	Adjusted R Square	Std. Error of the Estimate	Durbin-Watson
1	.916[a]	.840	.760	1783.42781	2.447

a. Predictors: (Constant), Tidal Rain level (X2ph), Distance (X1)

b. Dependent Variable: TDS Tidal Rain (Yph)

4 Conclusion

- Based on its salinity structure, the Kapuas Kecil estuary has a "partially mixed" estuary type, salinity varies in longitudinal and vertical directions.
- In general. the water quality of the Kapuas Kecil River has exceeded the criteria of first class water quality, ie water that can be used for drinking water.
- Status of water quality of Kapuas Kecil River based on calculation of pollution index shows the condition of the river "mild contamination" to "medium pollutant".
- The result of multiple linear regression analysis got the value of coefficient of determination (adjusted R square) = 0,760, which overall showed that tidal and distance influence to water quality (TDS value) equal to 76%.

5 Acknowledgements

This research was supported by Doctor Dissertation grant from Directorate of Research and Community Service, Directorate General of Research and Development Reinforcement, Ministry of Research, Technology and Higher Education, In accordance with the Contract of Research Number: 107 / SP2H / LT / DRPM / IV / 2017 April 3, 2017.

References

1. B. Triatmodjo, Teknik Pantai, (1999)
2. Z. G. Ji.*River Fate and Transport*, chapter in J.S. Gulliver (ed.), *Transport and Fate of Chemicals in the Environment*, (2012)
3. Dishidros TNI AL Indonesia, *Daftar Pasang Surut Kepulauan Indonesia*. (2016).
4. Peraturan Pemerintah No. 82 tentang *Pengelolaan Kualitas Air dan Pengendalian Pencemaran Air*(2001)
5. H. Effendi, *Telaah Kualitas Air; Bagi Pengelolaan Sumberdaya dan Lingkungan Perairan*(2003)
6. Salmin, *Oksigen Terlarut (DO) dan Kebutuhan Oksigen Biologi (BOD) Sebagai Salah Satu Indikator Untuk Menentukan Kualitas Perairan*. Oseana, Volume XXX, No. 3, 21-26(2005)
7. Keputusan Menteri Negara Lingkungan Hidup No.115 Tahun tentang *Pedoman Penentuan Status Mutu Air,* (2003)

https://doi.org/10.1051/e3sconf/20183104007

Strategy of Water Pollution Control Base On Social Economic Activitiy, in Karang Mumus River, Samarinda East Kalimantan, Indonesia

Vita Pramaningsih[1*], *Slamet* Suprayogi[2], and *Setyawan* Purnama[2]

[1]Doctoral Program of Environmental Science, School of Postgraduate Studies, Gadjah Mada University, Yogyakarta – Indonesia (Department of Environmental Health, Universitas Muhammadiyah Kalimantan Timur, East Kalimantan – Indonesia)
[2]Department of Environmental Geography, Faculty of Geography, Gadjah Mada University, Yogyakarta - Indonesia

Abstract. Water Pollution in Karang Mumus River caused society behavior along the river. Daily activity such as bath, washing and defecate at the river. Garbage, sediment, domestic waste and flood are river problems should be solved. Purpose this research is make strategy of water pollution control in the Karang Mumus River. Method used observation in the field, interview to the society, industry, public activity along the river and government of environment department. Further create data using tool of Analysis Hierarchy Process (AHP) to get the strategy to control water pollution in the river. Actors have contribute pollution control are government, industry and society. Criteria to pollution control are society participation, low, human resources and sustainable. Alternative of pollution control are unit garbage storage; license loyalty for industry and waste; communal waste water installation; monitoring of water quality. Result for actor priority are government (0.4); Industry (0.4); Society (0.2). Result for priority criteria are society participation (0.338), low (0.288), human resources (0.205) and sustainable (0.169). Result for priority alternative are unit garbage storage (0.433); license loyalty for industry and waste (0.238); communal waste water installation (0.169); monitoring of water quality (0.161).

1 Introduction

Water pollution in Karang Mumus River is worse caused social economic activity along the river. There are communities in flood plain area had done daily activity such as bath, washing and defecate in the river. Garbage, domestic waste and flood are problems in Karang Mumus River should be solved.

Water Quality in Karang Mumus River has polluted base on Government of Environment Department at Samarinda around 2009 until 2012 [1]. Report from Government of Environment Department during 2010 – 2015 found Biological Oxygen Demand (BOD), Chemical Oxygen Demand (COD) and *Fecal Colliform* along the Karang Mumus River already out of standard base on Government Regulation in East Kalimantan No. 2 / 2011 about Management of Water Pollution. Upstream to downstream land use are residence; city, market, a lot of residences in flood plain area have not relocated; market, a lot of residences in flood plain area have been relocated and port in the estuary [2]. Land use a long the Karang Mumus River shown in Figure 1.

Society has not awareness to keep the river. Society behaviour trough waste and defecate in the river make polluted. It's happen because of there are not enough garbage storage in the community. Then nothing communal waste water installation to manage domestic waste. Community in the flood plain area know that water quality of the river is poor but still use it for daily activity such as bath, washing and defecate [3].

Fig. 1. (a). Head water with vegetation; (b) centre segment with a lot of community in the flood plain; (c) down stream with relocation of flood plain.

Measurement 2016 of Biological Oxygen Demand (BOD) in Karang Mumus has been polluted, shown in Table 1. The result compared with East Kalimantan Provice Regulation No. 2 / 2011 about Water Quality Management and Water Pollution Control.

* Corresponding author: vitapramaningsih21@gmail.com

Table 1. Pollution of BOD along Karang Mumus River

No.	Location	Code	BOD (mg/L)	BOD Standard (mg/L)
1	Waduk Benanga Bridge	A1	17,1	3
2	Lempake Jaya River	B2	17,1	3
3	Bengkuring River	B3	38	3
4	Tepian Lempake Bridge	A2	20,9	3
5	Mugirejo-Gn. Lingai River	B4	7,6	3
6	Gunung Lingai (Jl. P.M. Noor) Bridge	A3	13,95	3
7	Sempaja River	B5	16,2	3
8	Drainage A.Yani (Gelatik-Pemuda)	B6	15,2	3
9	Pramuka-UNMUL River	B7	9,5	3
10	Gelatik Bridge	A4	9,5	3
11	Drainage Lembuswana-Vorvoo	B8	23,4	3
12	S. Parman Bridge	A5	12,55	3
13	Perniagaan Bridge	A6	10,8	3
14	Drainage Jl. Gatot Subroto	B9	22,8	3
15	Drainage Jl. Lambung Mangkurat	B10	39,9	3
16	Drainage P. Hidayatullah	B11	12,6	3
17	Sei Dama Bridge	A7	9	3

Base on those problems needed strategy to control water pollution in the river. It shown society behaviour, government low and instrument to waste management. Analysis Hierarchy Process (AHP) used for choose the alternative as strategy to control water pollution in the river. AHP is a theory of measurement through pair wise comparisons and relies on the judgment of experts to derive priority scales [4]. AHP as comprehensive safety evaluation method combining qualitative and quantitative analysis, has been used in many fields of safety and environmental science [5]. Main idea of AHP is to judge the important level between pair wise indicators by the separation of a complex problem into several hierarchies and several factor, that is, to construct a judgment matrix and determine the maximum eigen value of this matrix and its eigenvector by calculation, which could yield the weights of important level of different plans and provide the basis for the best plan [6].

The fundamental scale of absolute number to construct pair wise comparison matrices use the priorities obtained, shown in Table 2. Expert respondents (actor) are government, industry and communities along the Karang Mumus River. Respondents are choose the people lives and known the environment, condition, human activity along Karang Mumus River. Criteria in the hierarchy matrix to make pollution control strategy are sustainable development, communities participation, human resources and low regulation. Criteria is point of think to chose alternative. Then Alternative in the hierarchy matrix had chosen are industry and waste management license, water quality monitoring, waste water treatment monitoring and garbage storage unit. Priority alternative of pollution control in Cibanten River, Banten, Indonesia was water quality monitoring, city spatial layout review, low regulation, stakeholder, socialization and waste water treatment communal [7].

Table 2. Scale of Absolute Number in AHP

Intensity of Importance	Definition	Explanation
1	Equal importance	Two activities contribute equal to the objective
3	Moderate importance	Judgement slightly favour one activity over another
5	Strong importance	Judgement strongly favour one activity over another
7	Very strong importance	An activity is favoured very strongly over another
9	Extreme importance	An activity with the highest possible order of affirmation

2 Materials and Methods

The map of sampling location along Karang Mumus River, shown in Figure 2.

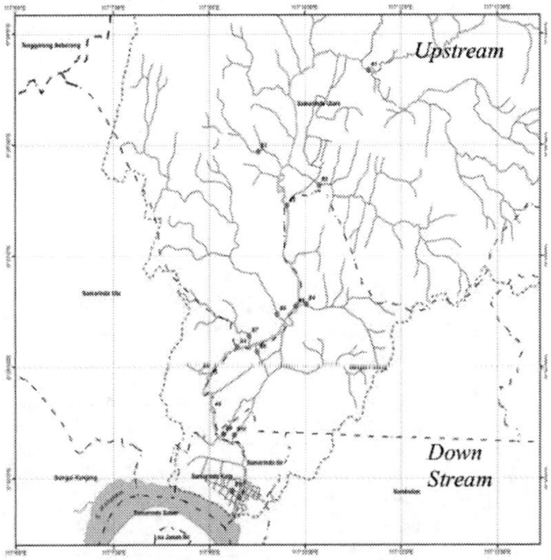

Fig. 2. Sampling Location Spot

Each respondent provided judgement to estimate the dominance factor to get strategy water pollution control follow as the hierarchy matrix. Judgement base on pairwise comparison using scale. Analysis strategy to control water pollution in the Karang Mumus River, used AHP to establish a hierarchical model/matrix with three layers included, shown in Figure 3. Actors judgement have analysed by expert choice 11 tools to get priority value.

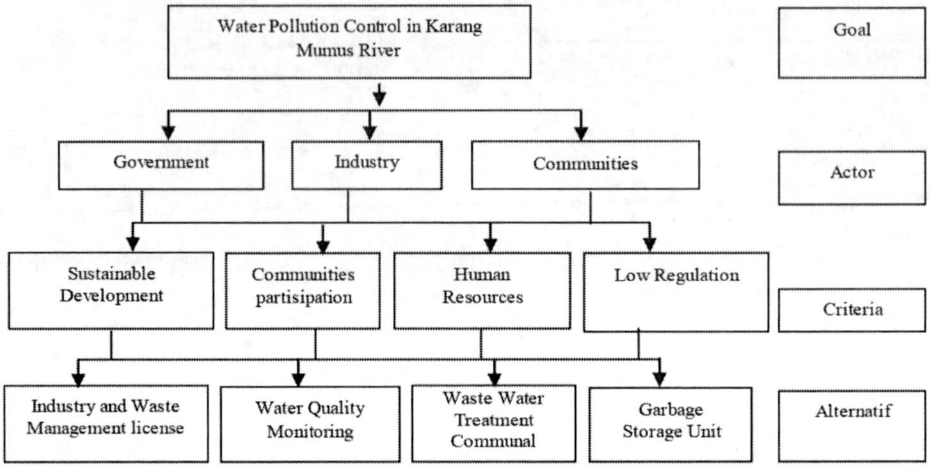

Fig. 3. Hierarchical Matrix

3 Results and Discussion

Result of expert choice to analysed alternative of pollution control strategy, shown in Figure 4.

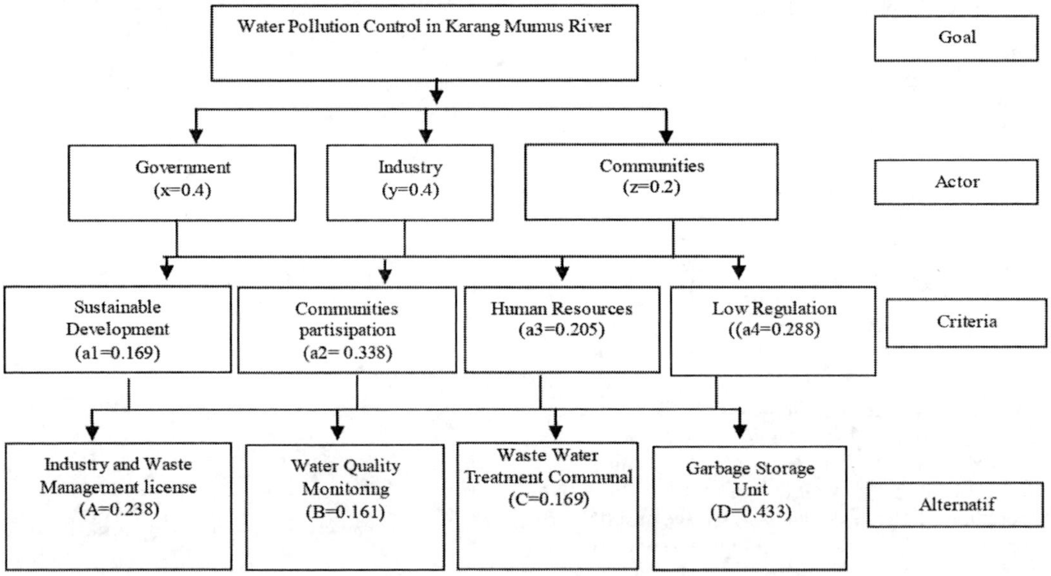

Fig. 4. Hierarchical Matrix Base on Priority

Graphic of priorities for actor contribution to pollution control, shown in Figure 5. Government and industry have same priority value. It's mean they are should be manage environment along the river. Specifically for waste water management and water quality monitoring. Cooperation inter discipline knowledge, stakeholder and government side are important things in sustainable development [7]. Three pillars sustainability development are social, economic and environmental [8].

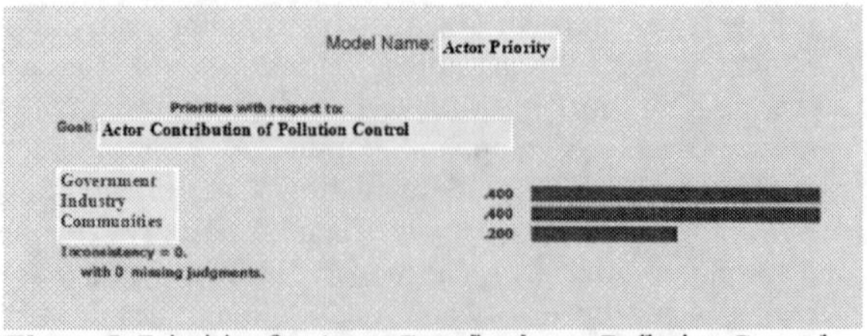

Figure 5. Priorities for Actor Contribution to Pollution Control

Graphic priorities for criteria contribution to pollution control, shown in Figure 6. Communities participant has highest value then low regulation. It's mean communities participation is key to solve the river water pollution control problems. Water quality in surface or ground have been deteriorated because of some importance factor as increasing population, industrialization, urbanization, etc. Then water quality in the river is depend on influence various activity along the river [9]. Urbanization along the river banks to be cause river water pollution [10]. Low regulation have powerful to manage environment pollution.

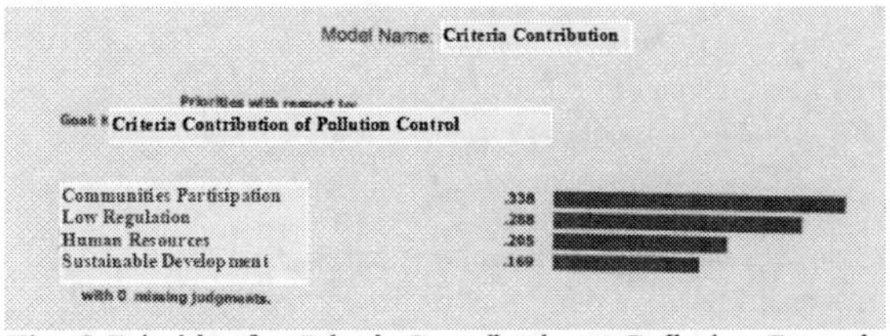

Fig. 6. Priorities for Criteria Contribution to Pollution Control

Graphic priority for alternative to river water pollution control, shown in Figure 7. Garbage storage unit have highest value than others. Human activity are produced garbage as waste. Quantity of garbage storage especially along the river banks will be impact to the garbage decrease in the river. It's depend on communities participant to keep environment. Communities along the Karang Mumus river banks said that they through garbage to the river because of nothing garbage storage near their house. Therefore needed cooperation between government, industry and communities to prepare garbage storage unit especially along the river banks.

Priority alternative of pollution control in Cibanten River, Banten, Indonesia was water quality monitoring, city spatial layout review, low regulation, stakeholder, socialisation and waste water treatment communal [7].

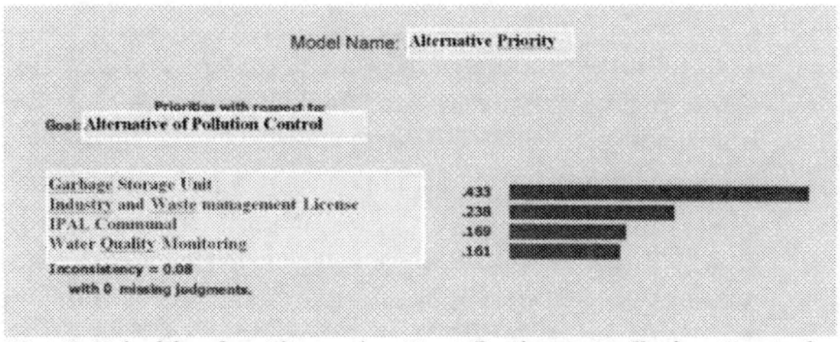

Fig. 7. Priorities for Alternative Contribution to Pollution Control

4 Conclusion

Government have highest priority than industry and communities to arrange river water pollution control strategy. Then communities participation is important factor to help government reach the goal. Finally alternative priority are provide garbage storage unit, industry and waste management license, waste water treatment communal and water quality monitoring.

References

1. http://e-journal.fhunmul.ac.id/index.php/beraja (in Indonesia)
2. PERDA KALTIM No. 2 Tahun 2011 tentang Pengelolaan Kualitas Air dan pengendalian Pencemaran Air (in Indonesia).
3. I. Yuliana, eJournal Sosiatri-Sosiologi, **1** (1): 20-30 (2013). http://e-journal.sos.fisip-unmul.org. (in Indonesia).
4. T.L. Saaty, Int. J. Services Sciences, Vol. **1**, No. 1 (2008)
5. J. Guo, Z. Zhang, Q. Sun, China Saf Sci J **18**(5):148-153(2008)
6. O. Tong, S. Shuai, Z. Yun, C. Yu, L. Su, S.Z. Shu, Clean Techn. Environ. Pollicy **14**: 857-868 (2012).
7. Baherem, Suprihatin, NS. Indrasti, Jurnal Pengelolaan Sumberdaya Alam dan Lingkungan Vol. **4** No.1, 60-69, (Juli 2014) (in Indonesia).
8. R. Rossi, M. Gastaldi, G. Gecchele, Eur. Transp. Res. Rev, **5**:11-26, (2013)
9. Effendi Hefni, 2015, Elsevier B. V, Procedia Environmental Sciences 33, 562-567, (2016)
10. S. Kalavaty, T. R. Sharma, P. Sureshkumar, 2011, Arch. Environ. Sci, **5**, 55-61, (2011).

Preliminary Design of Industrial Symbiosis of Smes Using Material Flow Cost Accounting (MFCA) Method

Rahayu Siwi Dwi Astuti[1,*], *Arieyanti* Dwi Astuti[1], *Hadiyanto*[1,2]

[1]Master Program of Environmental Studies, School of Postgraduate Studies Diponegoro University, Semarang - Indonesia
[2]Chemical Engineering Department, Faculty of Engineering Diponegoro University, Semarang - Indonesia

Abstract. Industrial symbiosis is a collaboration of several industries to share their necessities such material, energy, technology as well as waste management. As a part of industrial ecology, in principle, this system attempts to emulate ecosystem where waste of an organism is being used by another organism, therefore there is no waste in the nature. This system becomes an effort to optimize resources (material and energy) as well as minimize waste. Considerable, in a symbiosis incure material and energy flows among industries. Material and energy in an industry are known as cost carriers, thus flow analysis in this system can be conducted in perspective of material, energy and cost, or called as material flow cost accounting (MFCA) that is an economic and ecological appraisal approach. Previous researches shown that MFCA implementation could be used to evaluate an industry's environmental-related efficiency as well as in planning, business control and decision making. Moreover, the MFCA has been extended to assess environmental performance of SMEs Cluster or industrial symbiosis in SMEs Cluster, even to make preliminary design of an industrial symbiosis base on a major industry. This paper describes the use of MFCA to asses performance of SMEs industrial symbiosis and to improve the performance.

1 Introduction

Nowadays most of industries are facing demands not only to increase production and economic benefits but also responsibility for environmental sustainability and health at the same time. However the production process will always be related to the environment; raw materials extracted from the environment, energy used and emissions released into the environment. This relationship encourages the development of industrial ecological concepts.

The concept of industrial ecology relies on attempt to emulate ecosystem principles, where there is a mutual relationship between living organisms. In the nature, the relationships have a consequence of zero waste system. The waste generated by the organism will be utilized by other such that nothing is wasted. This concept is used in industrial ecology as a base of the concept of resource optimization and waste minimization.

An important part of industrial ecology is industrial symbiosis, where inter-industrial relations are realized by the exchange of resources, knowledge and technology or waste co-management. As the flow of material or energy plays an important role in this symbiosis, the performance of this system can be measured by assessing the efficiency of this material/energy flow. In the industrial world, the efficiency of material/energy flow will be more meaningful when expressed in economic value. One of assessment tools that can be used to measure is Material Flow Cost Accounting (MFCA).

The MFCA assesses the performance of a production process by analyze the material/energy flow from the initial process to final, step by step. It will show the efficiency of material/energy use throughout the production process. This method has been clearly stated in ISO 14051. This method is helpful for entrepreneurs to improve the efficiency of their business while improving their environmental performance. The MFCA is described in section 3.

MFCA concepts have also been implemented in supply chains [1][2]as explained in ISO 14052 (2017). This method assesses the economic and environmental performances of a supply chain by calculating materials/energy flows among involved industries. Ecological and economic value of materials/energy of each industry, output and material loss can be calculated and used to design optimization and efficiency of raw materials/energy. MFCA implementation in a supply chain is illustrated in section 4.

*Corresponding author: siwidwiastuti@gmail.com

The concept of MFCA implemented to supply chain is mostly applied to industrial symbiosis, that can be regarded as expansion of supply chain. In a supply chain system involved upstream and downstream industries of a product system – i.e system of canned fish products. The supply chains of the canned fish industry may consist of fishing industry, fish freezing, canning fish, tomato sauce industry, and tincan industry. In industrial symbiosis, the system involved not only industries supplies raw materials for the main products, but also involve industries that utilize by-products even provide/share waste management, resulting in various products. For example in fish canning industry, industries that can be involved are fishing industry, fish freezing, canning fish, tomato sauce industry and canned industry, fish meal industry, fish protein concentrate industry, fish collagen industry, fish paste, silage, waste treatment, water treatment and fertilizer. In this system, the main raw material is not only fish, but also by-product such as stomach, bone, head, even waste water. An overview of the implementation of MFCA in industrial symbiosis is illustrated in section 5.

Assessment of material/energy flow in an area also can be used to determine whether, in a region, industrial symbiosis can also be developed[3]. In the area, input-output materials/energies are identified, then materials that have not been optimally utilized are identified and appropriate technologies are selected. The congruence of available material/energy flow and selected technology are base of the initial design of industrial symbiosis development. This initial design is illustrated in section 6. Scientists believe the development of appropriate industrial symbiosis will help to improve the economic and environmental performance of an industrial estate..

2 Industrial Symbioses

Industrial symbiosis is part of the industrial ecology where several industries establish mutually beneficial cooperation relationships that enhance competitive advantage by involving material, energy, water and/or by-product exchange[4]. Examples of excellent and inspiring applications are those built in Kalunborg, Denmark. The definition of industrial symbiosis has been so far developed. According to Chertow [4] the key to industrial symbiosis is the possibility of collaboration and synergism in the presence of geographical proximity. Maltin [5] defines industrial symbiosis as a long-term symbiotic relationship between firms involving physical exchange and human or technical exchange as an effort to improve environmental performance and enhance collective competitive advantage, so that this collaboration includes the transformation of economic outlook, environmental and social performance. The exchange involved material, energy, water, by-product, even infrastructure, experience and knowledge[5]. In line with Maltin[5], Lombardi & Laybourn [6], explained the synergistic relationship in industrial symbiosis can include creating and sharing knowledge about new input resources, value-added output products and improved processes and techniques. But they did not consider the role of geographical proximity to be an important factor[6]. The repeated implementation of industrial symbiosis produces beneficial industrial ecosystems reducing the use of raw materials and pollution, improves systemic energy efficiency, reduces waste volumes, increases the number and type of outputs of economic value [7].

The design of industrial symbiosis should pay attentions: 1). The corresponding input output of the industries involved; 2). The technological process efficiency used by each industry that determines the volume of waste emissions and the number of primary inputs [8]. Murat et al. [8] describes three conditions of industrial symbiosis: 1). Perfect symbiosis; 2). Excess waste; 3) lack of waste. To approach the condition of the perfect symbiosis and overcome the second and third conditions, Murat et al. [8] suggested two stages of symbiotic design as: 1). Physical and technological conditions must be determined; 2). Related to economic comfort, contract, and willingness to persist in the implementation of industrial symbiosis.

3 Material Flow Cost Accounting (MFCA)

MFCA is one of the tools of Environmental Management Accounting (EMA) that develops material flow models, traces and quantifies the flow and supply of materials within an organization in physical and monetary units [9]. In this method, the costs in the production system are distributed on lost products and materials to enable companies to identify the added value they do not get as a result of the missing material [10]. Furthermore, Tachikawa [9], added that by identifying each material and its cost (including material costs, waste treatment and handling), we can look at each source of waste and identify its reduction opportunities.

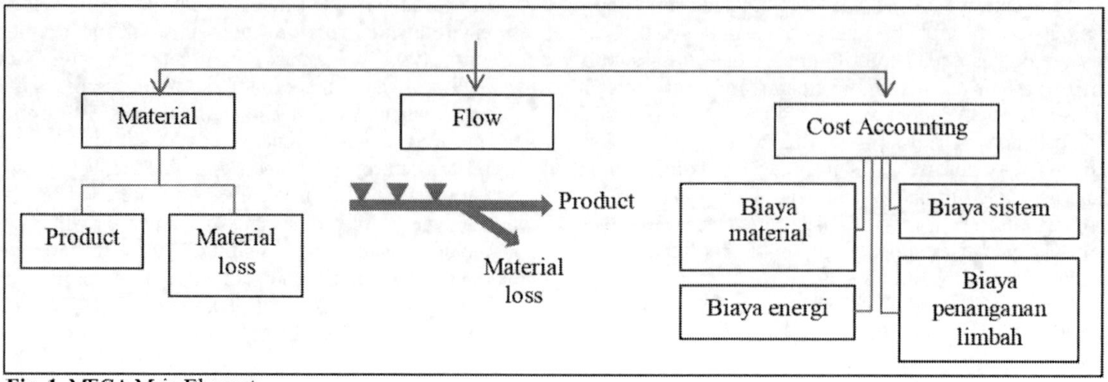

Fig. 1. MFCA Main Element
Source: adapted from Tachikawa [9]

ISO 14051 (2011) explain MFCA guidance. It describes the fundamentals and principles of MFCA. The principles of MFCA include understanding material flow and energy use, linking physical and monetary data, ensuring the accuracy, completeness and compatibility of physical data, and estimating and linking costs with material losses. MFCA Principles consist of:

1. Quantity center
   The quantity center is selected part or part of the process that measured in physical and monetary units. Quantity centers are basis for data collection. The data collected includes: 1). Material flow and energy use; 2) material costs, energy costs, system costs and waste management costs.
2. Material balance
   The material entering and exiting a quantity center must be balanced. Comparisons between inputs and outputs and inventory changes are made to identify significant missing materials or other data gaps.

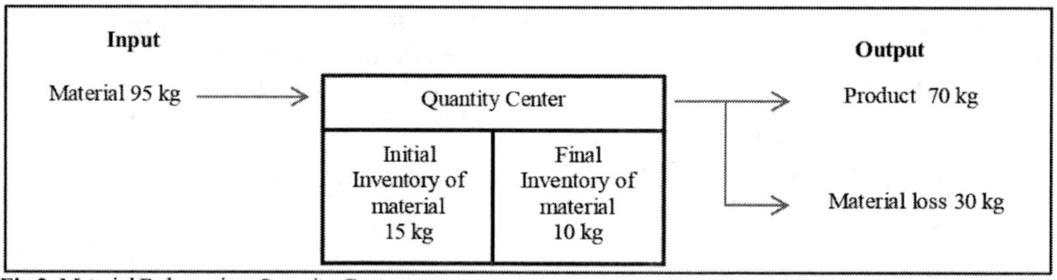

Fig 2. Material Balance in a Quantity Centre
Source: ISO, 2011

3. Cost calculation
   Calculation is done on three cost, namely: material cost, energy cost and system cost.

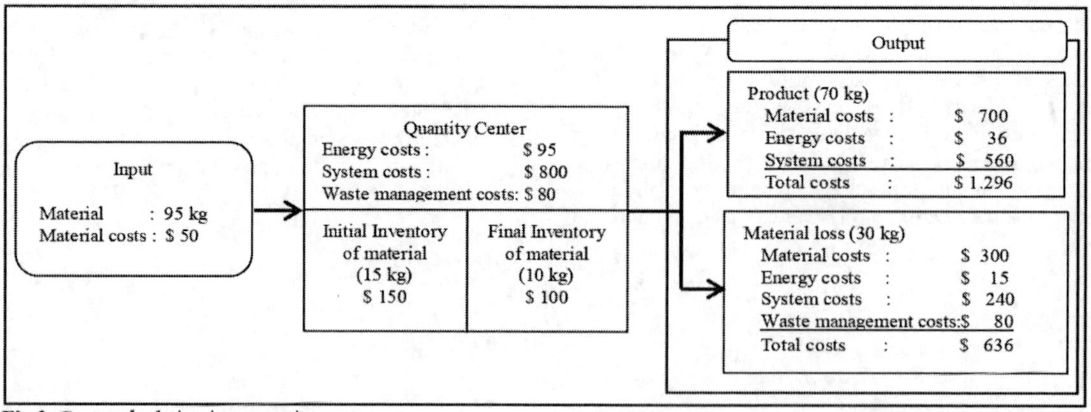

Fig 3. Costs calculation in a quantity centre
Source: ISO, 2011

4. Material flow model
   Production, recycling, and other systems are illustrated by the visual models of some quantity centers in which the material is distributed, used or modified and the movement of materials among the quantity centers.

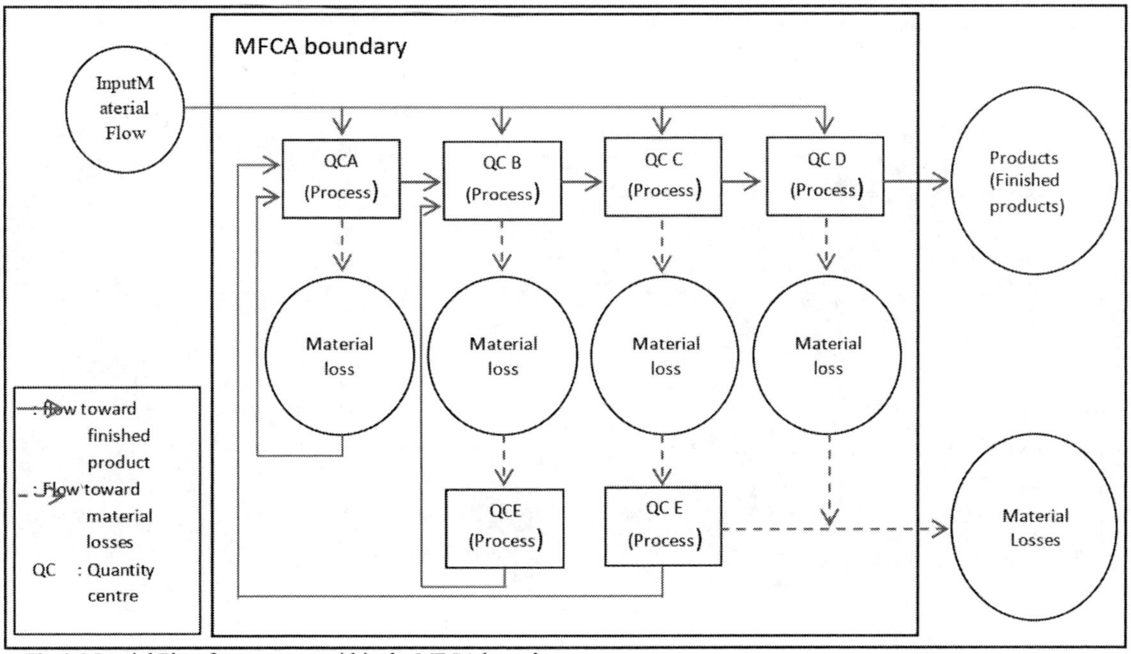

Fig 4. Material Flow for a process within the MFCA boundary
Source: ISO, 2011

4 Extension of MFCA to a supply chain

Prox [1] illustrated well MFCA implementation in supply chain. It is mentioned that the internal or environmental losses identified by the MFCA are influenced by the supply of suppliers or the final product characteristics requested by the customer. The solution to this problem is the collaboration between suppliers and customers as well as an understanding of the overall production system. Prox [1] illustrated every company's inefficiencies are potentially improved. According to him, inefficiencies in small percentages experienced by each stage of the process, eg 1-2% can be accepted, but when accumulated from the whole system can reach 20-60%.

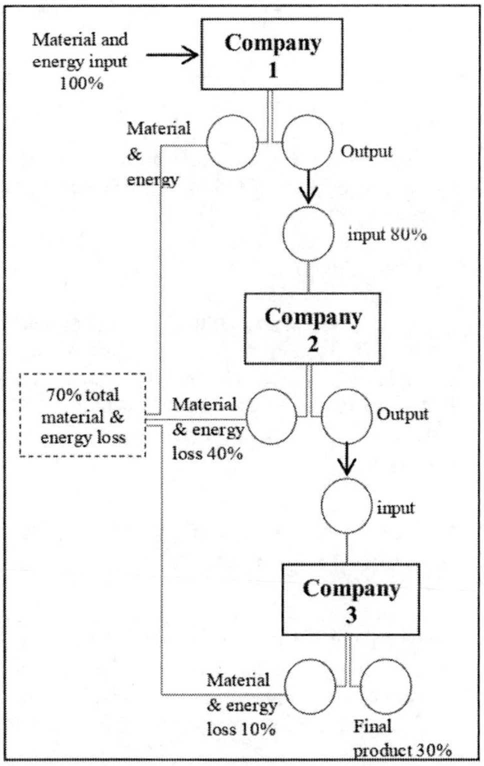

Fig 5.Material loss in supply chain
Source: adapted from Prox [1]

5 Implementation of MFCA to SMEs Industrial Symbiosis

An industrial symbiosis can be illustrated such as supply chain system. But industrial symbiosis involved some different flows of material/energy among different industries. In other word, a industrial symbiosis might has several supply chains. Material flows throughout the system can be either by-product or waste. As effort of optimization of resource utilization (materials/energy) and waste minimization, the researchers found that industrial symbiosis can provide significant advantages both in terms of environment and economy.

A simple example of industrial symbiosis can be described as follows.

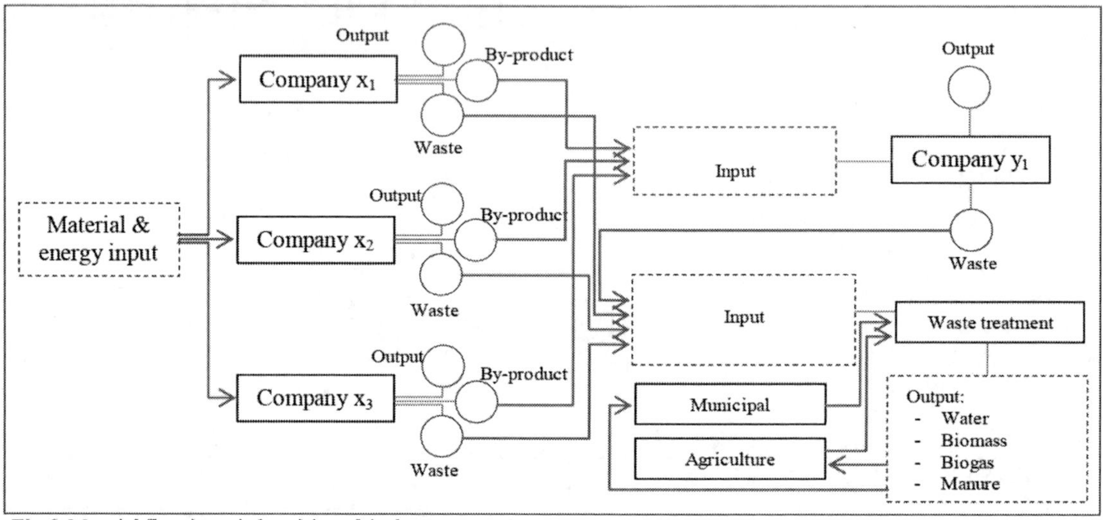

Fig 6. Material flow in an industrial symbiosis

Cimren, et al. [11] used MFCA to optimize the by-product synergy (BPS) network in Kansas City, Missouri. The study found that cooperation to improve the profitability of Kansas City's network system could save $ 15 million annually with a total cost reduction of 29%, reduction 28.5% of the company's average cost, reduction in 30%carbon dioxide emissions (CO2), and reduction of 37%waste to final landfill[11]. The MFCA application was used by Gu, et al. [12] to model the mathematical optimization control of his research in Le Havre France. This control model aimed to help industries reduce environmental costs and reduce environmental impacts simultaneously[12][13].

6 Preliminary Design of Industrial Symbiosis Using MFCA

Ulhasanah & Goto [14] provided an example of how MFCA is used to see the possibilities of industrial symbiosis can be built and provide benefits to the cement industry in Indonesia. The MFCA provides an overview of how the input-output efficiency of a group of industries and how the utilization has been done. Furthermore, taking into account the technology to be applied we can estimate the inputs and outputs of the industry to be developed, and also estimate opportunity of implementation[8].

For example, Ulhasanah and Goto[14] proposed two industrial symbiotic scenarios (scenarios 2 and 3) that were compared to existing conditions (scenario 1). Scenario 2 involved recycling cement kiln dust (CKD) and utilized municipal waste and surrounding industries as alternative energy sources. Scenario 3 involved scenarios 1 and 2 coupled with the utilization of waste from other cities around it as an alternative energy sources. MFCA was used to calculate the total product cost, system cost and material loss cost of the three scenarios and compared it, and concluded that scenarios 2 and 3 can save raw materials by 37 billion IDR/year, surplus energy costs in scenario 2 of 52.028 billion IDR/year , while in scenario 3 the surplus energy cost is 61.390 billion IDR/year[14].

 https://doi.org/10.1051/e3sconf/20183104008

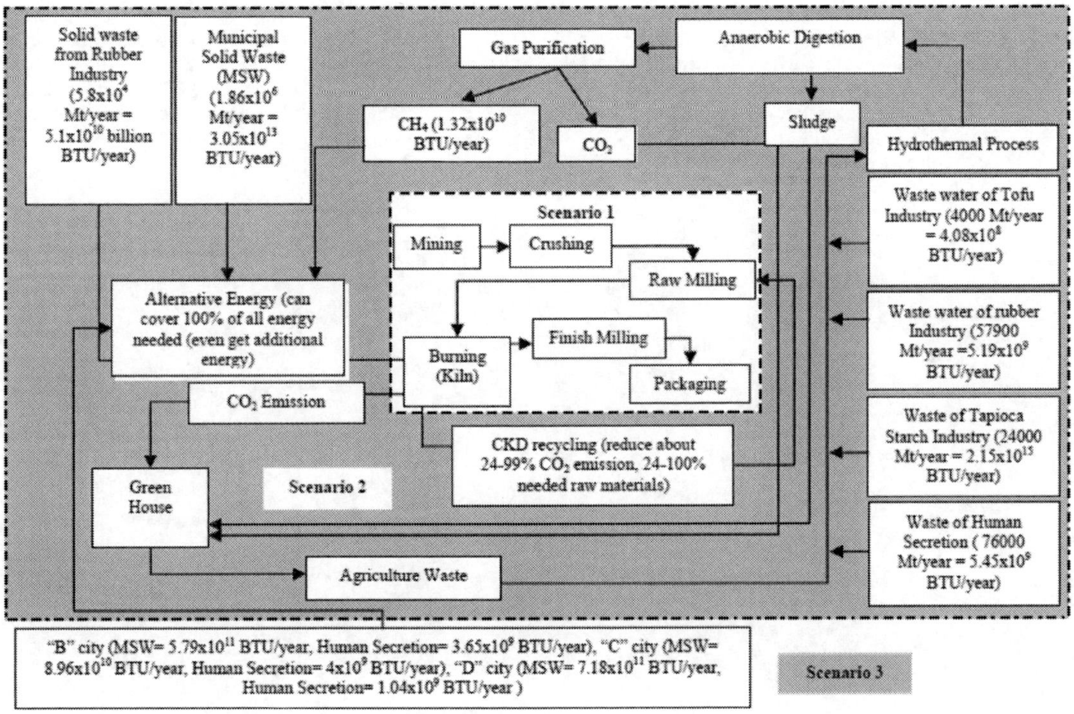

Fig. 7. Scenarios of design of alternative material exchange
Source: Ulhasanah & Goto[14].

We often find small and medium industries built in urban areas, spread among settlements and other buildings. Its presence in society is often followed by a simple symbiotic relationship between the industry and the urban community. For example small food processing industries such as tofu, tempeh or fish processor. These industries produce waste and/or by-products that can be utilized by the community, such as tofu dregs or fish guts that can be used as animal feed, even some types of by-products still have nutritional value and taste that can be sold as food, for example tofu dregs, fish head, and others. Although this relationship seems to be beneficial and can reduce the waste, its benefits, effectiveness and efficiency need to be tested. Alternatively, many technological options can be used to construct several energy symbiotic scenarios, depending on the suitability of the input-output of the production processes. By calculating the total cost of product, system and material loss and comparing it, we may propose conclusion and recommendation scenarios as the initial design of industrial symbiosis as did Ulhasanah and Goto [14].

7 Conclusions

The conclusions that can be drawn from the above description are:

1. MFCA can be used to assess the environmental and economic performance of an industry, supply chain and industrial symbiosis,
2. By calculating the total cost of product, system and material loss of an industrial or industrial estate and inventory of waste/by-product utilization technology, it can be designed the initial scenario of industrial symbiosis.

Reference:

1. M. Prox, *Procedia CIRP 29*, **vol. 29**, pp. 486–491, (2015).
2. M. Nakajima, A. Kimura, and B. Wagner, *J. Clean. Prod.*, **vol. 108**, pp. 1302–1309, (2015).
3. M. R. Chertow, "' Uncovering ' Industrial Symbiosis," **vol. 11**, no. 1, (2007).
4. M. R. Chertow, "Industrial Symbiosis : Literature and Taxonomy," (2000).
5. M. Maltin, *IIIEE Reports*, no. October,(2004).
6. D. R. Lombardi and P. Laybourn, **vol. 16**, no. 1, (2012).
7. N. Gertler, *Industrial Ecosystems: Developing Sustainable Industrial Structures. Thesis*. Massachussets: Massachussets Institute of Technology, (1995).
8. D. Murat, V. Alessio, and V. Albino, *J. Clean. Prod.*, **vol. 129**, pp. 537–547, (2016).
9. H. Tachikawa, *Manual on Material Flow Cost Accounting ISO 14051*. Tokyo: Asian Productivity Organization,(2014).
10. M. Schmidt, *J. Clean. Prod.*, pp. 1–10, (2014).
11. E. Cimren, J. Fiksel, M. E. Posner, and K.

Sikdar, **vol. 15**, no. 2, (2011).

12. C. Gu, S. Leveneur, L. Estel, and A. Yassine, “Industrial Symbiosis Optimization Control Model for the exchanges of the material / energy flows in an industrial production park,” pp. 1015–1020, (2013).
13. C. Gu, S. Leveneur, L. Estel, and A. Yassine, *Energy Procedia*, **vol. 36**, pp. 243–252, (2013).
14. N. Ulhasanah and N. Goto, *Int. J. Environ. Sci. Dev.*, **vol. 3,** no. 6, (2012).
15. ISO 14051 : 2011(E). Environmental Management - Material Flow Cost Accounting - General Framework

Cultivation of Microalgae *Chlorella sp* on Fresh Water and Waste Water of Tofu Industry

Widayat[1,2,*], *John* Philia[2] and *Jessica* Wibisono[2]

[1]Center of Biomass and Renewable Energy (C-BIORE) Center of Research and Service Unit Diponegoro University (CORES DU)
[2]Department of Chemical Engineering, Faculty of Engineering, Diponegoro University

Abstract. Chlorella sp. is a microalgae that potential for food supplement, pharmaceuticals, animal feed, aqua culture and cosmetics. Chlorella sp. commonly growth in sea water. Indonesia as a producer of tofu generated more liquid waste. Nutrient that contained in the tofu wastewater are very useful for the production of microalgae. Cultivation carried out for 7 days at different percent volume of tofu liquid waste showed that the more volume of tofu liquid waste make them longer process decipherment of polymer compounds in the waste, that's make the growth rate of Chlorella sp. are slowness. Variable of10%V has the fastest growth rate. While, 90% v/v variable has the highest concentration of algae. It shows that Chlorella sp. better to grows in tofu wastewater than seawater.

1. Introduction

Indonesia is one of the tofu producers that quite productive either home or industrial scale with a number of industry reached 84.000 business units. Data obtained from BPPT (Agency for the Assessment and Application of Technology) showed that for every 80 kg tofu produced, 2610 kg waste produced [1]. It means every year, for about 240 million people in Indonesia, 5,7942 x 10^{10} kg industrial tofu waste produced annually.

Gases commonly found in tofu waste are nitrogen (N_2), Oxygen (O_2), hydrogen sulfide (H_2S), ammonia (NH_3), carbon dioxide (CO_2) and methane (CH_4).The gases are derived from the decomposition of organic materials contained in tofu wastewater [2]. If the tofu wastewater is being discharged directly into the environment, there would be precipitation of organic substances in the water body, the process of decay, and growth of pathogenic microorganisms. To resolve the problem, there are methods that can be solved tofu wastewater as medium cultivation microalgae.

Tofu wastewater contains some organic materials. The protein in tofu wastewater occupies about 60% of all solid fractions. Through amino acids fermentation in anaerobic treatment process, protein can be degraded into various organic compounds [3], which some of them can be used either for macronutrient or micronutrient for the autotrophic organism, such as microalgae.

Nowadays, micro algae are consumed as food supplement, pharmaceuticals, animal feed, aquaculture and cosmetics[4]. Some microalgae are able to remove the pollutant and able to produce lipids [5-7]. A Microalgae *Chlorella pyrenoidosa* including, *Scenedes musobliquus, Chlorella* and *Spirulina platensis, Minutissima* [6,8-10]. That microalgae are able and effective to remove the pollutant in the waste, but have a low productivity, low lipid content and intolerant in extreme environment condition [8,11].

Chlorella sp. is the most cultivated eukaryotic green micro alga, as it is widely used as a health food and feed supplement. It contains proteins, carotenoids, lipids, immune stimulator compounds, polysaccharides, vitamins, antioxidants and minerals [12]. The culture medium has a direct influence on cellular growth, as well as on concentration of microalgae [13]. *Chlorella* sp. is usually cultured on the nutrient solution that simulates sea water composition, with the addition of B1 and B12 vitamins. Air is injected into this culture medium to furnish carbon dioxide as the only carbon source [14].

In previous studies on *Chlorella* sp. various methods have been adopted to increase growth rate using sea water (nutrient composition, pH, light effect, temperature), but only a few studies on culture medium have been published. Because of this, then conducted research to know the growth rate and algae concentration on cultivation microalgae types of *Chlorella*sp. in medium tofu wastewater.

2. Fundamental Theory

Microalgae cultivation can be carried out with freshwater, seawater, water from estuaries, wastewater from agricultural, industrial operations and domestic wastewater [15]. Algae media refers to the solution or culture in which algae grow. All the media haveseveral

*Corresponding author: widayat@live.undip.ac.id

components in common: sources of nitrogen (in the form of nitrate, nitrite and ammonia), phosphorus, vitamins and trace metals. However the specific types of these nutrients, their concentrations and ratios vary between the media. It is important to know which media is best for optimum growth of the algae [16]. Some research suggests that tofu wastewater contains some nutrients which are very useful for the production of microalgae. Tofu wastewater contains some organic materials which mostly consist of proteins and lipids about 40-60% (226.06 mg/L to 434.78 mg/L) and other compound which are carbohydrate (25%-50%), and fat (10%) [17,18]. The wastewater also contains nitrate and phosphate [17]. Nutrient that contained in the tofu wastewater are very useful for the production of *Chlorella sp*.

3. Material and Method

The material used in this research is tofu wastewater and are obtained from tofu home industry n Mrican village, inoculums of microalgae *Chlorella*sp. are obtained from BPAP Jepara that cultivated using seawater and fertilizer NPK, $NaHCO_3$technical grade, B1 and B12 vitamin. Tools are used include flask, aerator, neon lights, spectrophotometer Optima SP 300, salinity meter and glassware.

Research begins with the preparation of materials of tofu wastewater. Waste is filtered to reduce the total solid and then without sterilization, directly mixed with freshwater according variable percent volume (10%; 30%; 50%, 70%, 90%) with added nutrition 30 ppm. Also add nutrition 30 ppm to seawater and freshwater. Microalgae *Chlorellasp.*10% V added to the flask containing media with lighting and aeration for 24 hours. Measurement of the optical density is done every day by the spectrophotometry method using a spectrophotometer Optima SP 300 at a wavelength 680 nm.

4. Result and Discussion

4.1. Effect of Tofu Liquid Waste Concentration to Optical Density

Based on Figure 4.1 shows that optical density values of all the variables tend to rise at the beginning of cultivation, and then decline in the mid-and rise in later cultivation. Increase in Optical Density (OD) at the beginning of the cultivation is the effect of the process of homogenization between sea water and total suspended solids of liquid waste. Dilution of total suspended solids leads to increased concentrations of tofu liquid waste. So, the value of OD in media at the beginning for cultivation 1-4 days tends to rise. While the decline of OD in cultivation in the middle occur because chemical compounds of polymer components in the waste degradation into simple compounds. Tofu liquid waste containing polymer compounds of carbohydrate such as oligosaccharides, monosaccharide as well as containing other compounds such as proteins, amino acids, isoflavone, saponin, P, Ca, Fe and other nutrients [19].

Optical density is measurement of the level of turbidity or density of a solution, more concentrated a solution be values of optical density will be increased. Waste polymer compounds degradation proses resulted in reduced concentration.

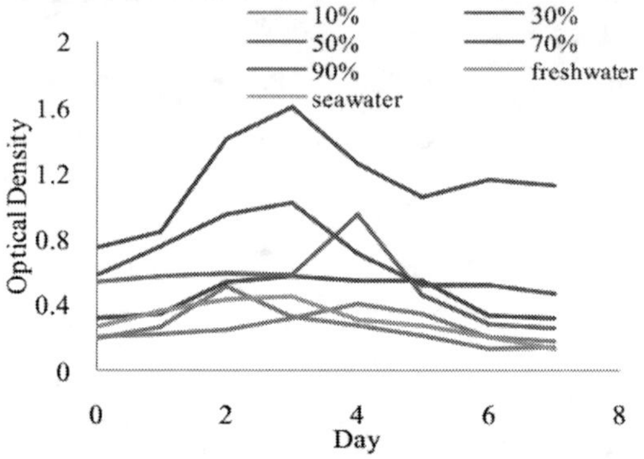

Fig.1. Relationship of optical density and% volume of tofu wastewater

Level decomposition of polymer compound on wastewater largely determines the growth rate of microalgae. Faster the polymer breaks down; it will be faster growth rate of *Chlorella* sp. anyway. Based on the Figure 4.1, fastest growth rates occurred at 10% volume that Optical Density had decrease on the third day. Variable more than 10% volume, polymer compounds decomposition takes a longer time so that less effective for the purpose of processing waste and biomass production. An addition of liquid waste is greater than 10% volume has a higher concentration of the polymer medium so that microalgae takes much longer for decomposition, that result slowness of the process of growth.

During the first four days of cultivation, the cells had better growth in tofu wastewater than in the freshwater or seawater. After four days, almost all variables reached stationary phase. Based on the figure 4.1, the highest algae concentration occurred at 90% volume that had the highest Optical Density. The larger of percent volume tofu wastewater will be the highest concentration of algae production. Rapid growth in 10% volume and high concentration of alga in 90% volume shows that *Chlorella* sp. usually grows in seawater, better to grows in tofu wastewater.

4.2. Effect of Carbon Metabolism to Medium Salinity

Based on figure 4.2, the salinity for almost of all variables is increased. Figure 4.3 shows that the biggest salinity increase occurred at 10% volume. High rates of growth and respiration need lot of energy, that obtained from carbon metabolism. Energy produced impact to evaporation of the culture medium, so, the salinity of the medium increases.

https://doi.org/10.1051/e3sconf/20183104009

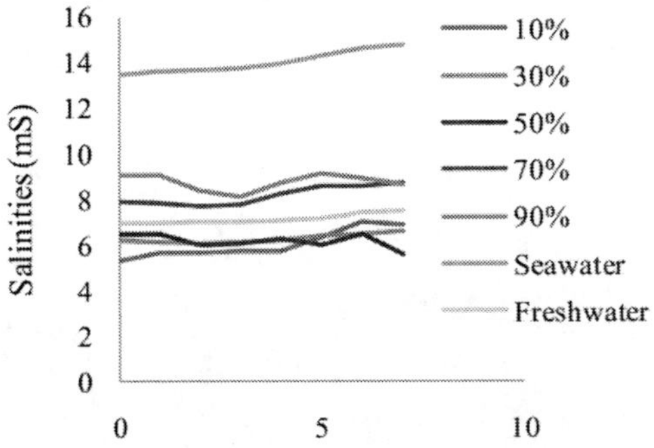

Fig.2. Time behaviour of salinity along *Chlorella sp.* cultivation

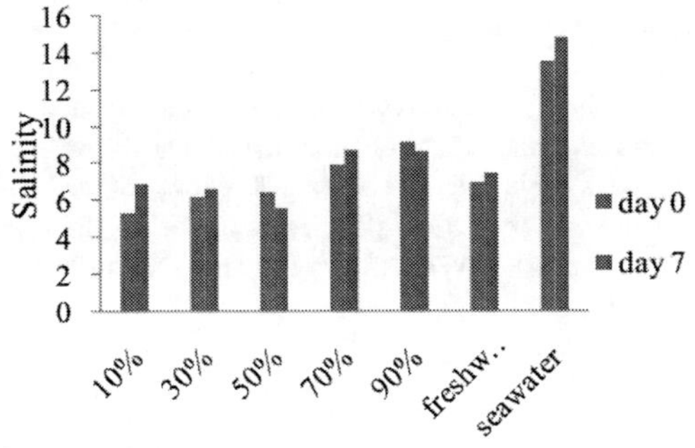

Fig.3. Ratio of salinity

5. Conclusion

Percent volume of tofu liquid waste effect on optical density. The larger of percent volume liquid waste will be the longer process decipherment of polymer compounds in effluent which resulted in the slowness of the process of growth of *Chlorella* sp. But, be increased in percent volume of liquid waste cause higher concentration of algae.

Rapid growth in 10% volume and high concentration of alga in 90% volume shows that *Chlorella* sp. uusually grows in seawater, able to grows in tofu wastewater. Algae growth and respiration need energy that obtained from carbon metabolism, it cause the salinity of the medium increases because of medium evaporation.

References

1. Herlambang A. Teknologi Pengolahan Limbah Cair Industri Tahu-Tempe. *Teknologi Pengolahan Limbah Cair*. Badan Pengkajian dan Penerapan Teknologi, Jakarta Pusat. 2002.
2. Kaswinarni, Fibria, *Kajian Teknis Pengolahan Limbah Padat dan Cair Industri Tahu,* Thesis (2007)
3. Ramsay IR and Pullammanappallil PC. Biodegradation. 2001;**12**:247-257.
4. Safi C, Areview. Renewable and Sustainable Energy Reviews. 2014;**35**,:265-278.
5. Garcia OP, Escalante FME, de-Bashan LE and Bashan Y. Water Research. 2011;**45**:11-36.
6. Bhatnagar A, Bhatnagar M, Chinnasamy S and Das KC. Applied Biochemistry and Biotechnology. 2010;**161**:523-536.
7. Li Y, Horsman M, Wu N, Lan CQ and Calero ND. Biotechnology Progress. 2008;**24**(4):815-820.
8. Hong Yang S and others. Bioresource Technology. 2011;**102**:9884–9890.
9. Martinez ME, Sànchez S, Jimènez JM, El Tousfi F and Munoz L. Scenedesmus obliquu. Bioresource Technology. 2000;**73**(3):263-272.
10. Yang CF, Ding ZY and Zhang KC. Microbiology Biotechnology. 2008;**24**:2919–2925.
11. Chisti Y.Biodiesel from microalgae. Biotechnology Advances. 2007; **25** (3):294–306.
12. Sharma R, Singh GP and Sharman VK. *Effect of culture condition on growth and biochemical profile of Chlorella vulgaris. Plant Pathology and Microbiology*. 2015;**3**(5).
13. Honrard AA and others. African Journal of Microbiology Research. 2015;**9**(21):1431-1439.
14. Frumento D and others. Biotechnology Progress. 2016;**32**(2):279-284.
15. Hadiyanto, Azimatun Nur, M.M. World Applied Sciences Journal. 2014., **31 (5):** 959-967.
16. Idenyi JN, Ebenyi LN, Ogah O, Nwali BU and Ogbanshi ME. IOSR-JPBS. 2016;**11**(3):24-28.
17. Nurhasan and Pramudyanto. *Penanganan Air Limbah Tahu*. Yayasan BinaKarya, Jakarta Selatan. 1991.
18. Singh J and Gu S.Renewable and Sustainable Energy Reviews. 2010;**14**(9):2596-2610.
19. Barbosa ACL, Lajolo FM and Genovese MI. Food *Chemistry*. 2006;**98**(4):757-766.

Effect of Batik Waste Water on Kali Wangan Water Quality in Different Seasons

S. Lestari [1,3*], *Sudarmadji* [1], *S.D.* Tandjung [1] and *S.J.* Santoso [2]

[1] Environmental Science Study Program at Gadjah Mada University
[2] Chemistry Study Program at Gadjah Mada University
[3] Biology Faculty at Jenderal Soedirman University

Abstract. Sokaraja Batik Center is one of batik industrial centers in Banyumas Regency. The craftsmen in Sokaraja Batik Center dispose of their waste water directly to a river named Kali Wangan. This study aims at figuring out the quality of Kali Wangan in dry and rainy seasons. The research is conducted along the Wangan River in January – November 2015. The research method used is survey with Purposive Random Sampling. The Kali Wangan water is sampled in four observation stations. The obtained data are analyzed descriptively and compared against the environmental quality standards. The research results show that the quality of Kali River water is found contaminated by the batik waste water, all parameters are below the class III standards quality based on Government Regulation Number 82 Year 2001 during dry and rainy season

1 Introduction

Sokaraja Batik Center to contaminate the environment since their batik waste water are disposed of to Kali Wangan. Kali Wangan is used by the people living along it for irrigation, fish cultivation, and even for bathing and washing clothes. The batik waste water disposed of to Kali Wangan without being treated in advance could contaminate and harm water biota life. The TSS, BOD and COD levels in Wangan River have exceeded the environmental quality standards at 540.13 mg/L; 540.42 mg/L and 672.78 mg/L respectively, and the heavy metal parameter exceeding the quality standards are Cd and Cr at 0.018 mg/L and 0.231 mg/L [1].The batik waste water contains coloring substance rhodamine Breaching to 0.344ppm,methylene blue 0.179ppmand methyl orange 0.779ppm [2].The batik waste water contains such metals as Zn, Crand Cuat 2ppm,2ppmand 0.2ppm respectively. Naphthol and indigosol colorants contain large amount of heavy metals such as zinc (Zn),chrome (Cr) and copper (Cu) [3]. Addition to colorants and heavy metals, the batik liquid waste also contains high organic materials.

Visually, the batik waste water has color. The existence organic materials in water could increase turbidity, thus prevent light penetration and oxygen diffusion to the water. The inorganic materials and heavy metals are hard to degrade in water [4]. Their existence in aquatic ecosystem will harm the metabolism process of water organisms. Any organism exposed to metals will accumulate metals in their body and experience biomagnification, thus the metal level gets increasingly higher according to the organism position in the food chain.

Seasons have some influence on the contaminant concentration in the environment. Dry season with low rainfall causes the waste concentration to be higher, while during rainy season the waste concentration is low since rainwater dilutes it. Study the quality of river water contaminated by batik liquid waste [5,6].The studies are conducted in large rivers. This research, on the other hand, study the quality of river water contaminated by batik liquid waste in a small river (primary irrigation channel). The downstream of Kali Wangan is utilized directly to irrigate the rice fields, leading any contamination by batik liquid waste to have direct impact on the plants. This research aims at figuring out the quality of Kali Wangan water that contaminated batik waste water during dry and rainy seasons.

2 Research Method

The research is conducted in January - November 2015. It is performed using survey as its method. The Wangan River water sample is taken using Purposive Sampling in 4stations (Figure 1). The first station is a place where no waste has been disposed of, the second station is a place right where the waste is disposed of, the third station is a place where the waste has been disposed of for some time and the fourth station is a place where another waste is disposed of in addition to the batik liquid waste. The sampling is performed during dry and rainy seasons each 3 times at an interval of one mouth. The parameters measured are based on the class III Water Quality Standards for agricultural irrigation, fresh water fish

* Corresponding author: djalaltandjung@yahoo.com

 https://doi.org/10.1051/e3sconf/20183104010

cultivation and water recreation infrastructures / facilities under the Government Regulation Number 82 year 2001 which are limited to the parameters for textile waste (Provincial Government Regulation of Central Java Number 5 year 2012),i.e. TSS, pH, BOD_5, COD, NH_3, total chrome, sulfide, oil-grease and phenol.

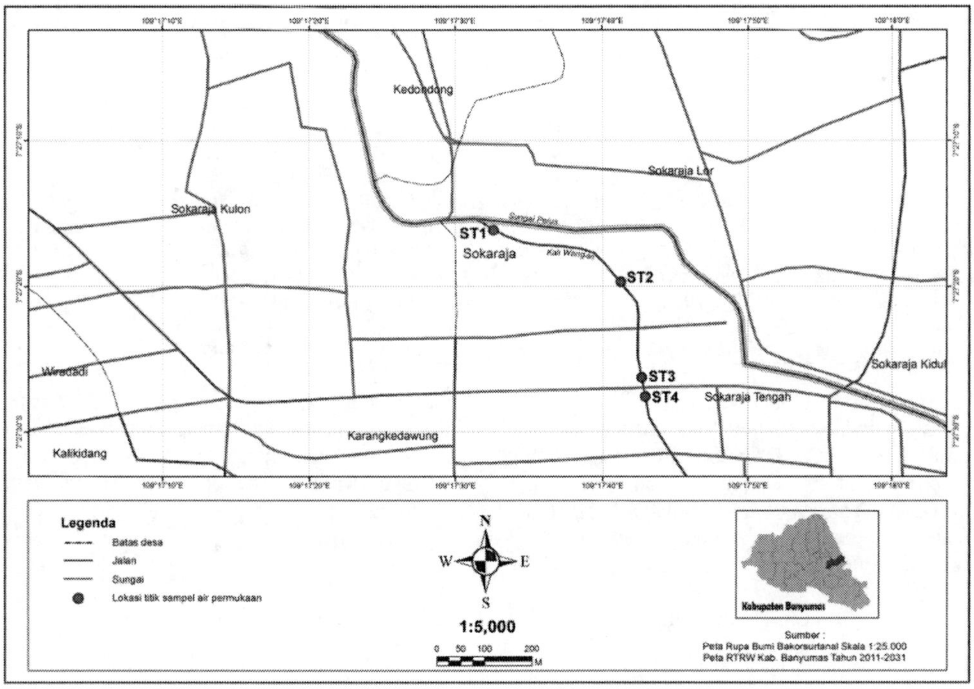

Fig. 1. Sampling location map

3 Results and Discussion

The research results indicate that the quality of Wangan River water during dry and rainy seasons is still below the quality standards. The complete data on the physical and chemical of Kali Wangan water during dry and rainy seasons are presented in detail in Tables 1 and 2.

Based on Tables 1 and 2,it is found that all parameters, both the physical and chemical ones, have the same pattern i.e. they are low in station 1 then increase in station 2 and 3 and decrease in station 4. The station 1 observation is performed in upstream part, close to the sluice of Pelus River, thus it is relatively uncontaminated by the batik waste. Stations 2 and 3 are the river segments receiving the batik waste, resulting in an increased concentration of contaminant in these segments. Meanwhile, station 4 is the river segment after receiving the disposed batik liquid waste. The contaminant concentration decreases since a dilution by the river water occurs. The BOD and COD parameters increase once again in station 4, since this is the site where other wastes of Sokaraja Soto and Gethuk are disposed.

Table 1. Physical-chemical parameters of Kali Wangan water during dry season

No	Parameter	Unit	Mean per station				Quality Standart (Class III)
			I S $07^0$25'36.0" E $109^0$14'12.3"	II S $07^0$27'19.5" E $109^0$17'41.3"	III S $07^0$27'25.4" E $109^0$17'42.4"	IV S $07^0$27'34.4" E $109^0$17'46.1"	
1.	Temperature	^0C	26.33±0.58	26.33±0.578	27±1	28.33±0.57	deviasi 3
2.	pH		6.92±0.02	6.64±0.18	6.77±0.14	6.20±0.025	6-9
3.	TSS	mg/L	4.0±0.01	9±1	8±1	10±1	400
4.	BOD	mg/L	1.43±0.153	2.1±0.1	1.8±0.2	1.8±0.1	6
5.	COD	mg/L	13.1±0.1	17.1±0.1	15.93±0.152	24.16±0.25	50
6.	Ammonia	mg/L	0.45±0.01	0.66±0.02	0.64±0.025	0.56±0.005	negative
7.	Total Chrome	mg/L	0.0014±0	0.0014±0	0.0014±0	0.0014±0	0005
8.	Sulfide	mg/L	undetected	undetected	undetected	undetected	0.002
9.	Oil and grease	mg/L	4.01±0.01	12.09±0.03	26.02±0.01	16.07±0.02	1000
10.	Phenol	mg/L	0.05±0.0002	0.116±0.0007	0.021±0.0003	0.066±0.0002	1

Table 2. Physical-chemical parameters of Kali River water during rainy season

No	Parameter	Unit	Mean per Station				Quality Standart (Class III)
			I S 07⁰25'36.0" E 109⁰14'12.3"	II S 07⁰27'19.5" E 109⁰17'41.3"	III S 07⁰27'25.4" E 109⁰17'42.4"	IV S 07⁰27'34.4" E 109⁰17'46.1"	
1.	Temperature	⁰C	28.53±0.416	29.27± 0.20	30.2±0.2	30.2±0.2	deviasi 3
2.	Ph		7.2±0.1	7.1±0.1	6.7±0.2	6.67±0.15	6-9
3.	TSS	mg/L	1.8±0.6	9.033±0.15	8.2±0.264	5.567±0.378	400
4.	BOD	mg/L	2.97±0.208	4.47±0.35	5.1±0.36	4.067±0.152	6
5.	COD	mg/L	0.54±0.052	18.03± 1.040	18.433±0.503	25.867±0.568	50
6.	Ammonia	mg/L	0.54± 0.052	0.823±0.073	0.676± 0.092	0.663±0.005	Negative
7.	Total Chrome	mg/L	0.0014±0	0.0014±0	0.0014±0	0.0014±0	0.05
8.	Sulfida	mg/L	undetected	undetected	undetected	undetected	0.002
9.	Oil and Grease	mg/L	3.89±0.03	8.05±0.01	8.02±0.01	6.04±0.04	1000
10.	Phenol	mg/L	0.021±0.001	0.0719±0.003	0.054±0.004	0.046±0.004	1

The heavy metal in Wangan River is still below the environmental quality standards both during dry and rainy seasons. Even its Cr(VI) is found to be below the device detection limit. Based on Tables 1 and 2, the chrome both during rainy and dry seasons is found to be tiny, far below the environmental quality standards. It shows that despite the fact that Wangan River is where the batik liquid waste is disposed of its water quality is found to be uncontaminated. A preliminary study by measuring the concentration of several metals in Wangan River1. The concentration of several metals resulting from this preliminary study is presented in Table 3 below.

Table 3. The concentration of heavy metal on Kali Wangan water in preliminary study

No	Parameter	unit	Station			Standart Quality (Class II)
			I S 07⁰25'46.0" E 109⁰15'11.6"	II S 07⁰27'19.5" E 109⁰17'41.3"	III S 07⁰27'25.4" E 109⁰17'42.4"	
1.	Cd	mg/L	0.007	0.018	0.014	0.01
2.	Cr^{6+}	mg/L	0.015	0.231	0.191	0.05
3.	Cu	mg/L	0.010	0.025	0.015	0.02
4	Pb	mg/L	0.016	0.033	0.023	0.03
5.	Hg	mg/L	<0.001	<0.001	<0.001	0.002
6.	Zn	mg/L	0.028	0.044	0.034	0.05
7.	Cd	mg/L	0.007	0.018	0.014	0.01

Note: Environmental quality standards is based on Government Regulation No 81 year 2001 concerning Management of Water Quality and Class II Water Contamination Control.

Based on Table3.the Cr, Cu and Pb parameters in station II have been above the quality standards, while in station III only Cr is above the quality standards. In station I, all metals are still below the environmental quality standards. Station I is the station close to the Pelus River sluice, hence it is logical that the contaminant there is still few, while station II is the station right where the batik liquid waste is disposed of, and station III is the one after station II which receives the waste flow. The preliminary research prove that Wangan River is contaminated by chrome resulting from batik liquid waste. Batik waste water comes from basic or advanced coloring substances which use toxic, hazardous chemical mixture during the coloring process. During the coloring process, only 5%of the colorants are absorbed and the remaining 95%will be disposed of as waste[7].

The batik waste water disposed of to Kali Wangan has not deteriorate the quality of Kali Wangan water. Both physically and chemically speaking, Kali Wangan water can still serves its purpose for irrigation both during dry and rainy seasons. Even though all parameters are still below the environmental quality standards, the contaminant concentration in Kali Wangan during dry season is higher than during rainy season in each observation station (Table 1 and 2). The content of contaminants in a body of water during rainy season is lower since dilution occurs7.During dry season, however, it is higher for it is concentrated. Rainfall has some influence on the dilution of these contaminants. In addition to rainfall, the river water discharge also has some influence on the fluctuation of contaminant level in the body of water.

The fact that some batik waste water gets into Kali Wangan has not caused its water quality to decrease. This is due to the relatively stable river water discharge both during rainy and dry seasons. The Kali Wangan water discharge during dry and rainy seasons per station is presented in Table 4.

Table 4. WanganRiver water discharge during dry and rainy seasons

Season	**River Water Discharge (m^3/second)**			
	Station 1 S 07°25'36.0" E 109°14'12.3"	**Station 2** S 07°27'19.5" E 109°17'41.3"	**Station 3** S 07°27'25.4" E 109°17'42.4"	**Station 4** S 07°27'34.4" E 109°17'46.1"
Dry	0.309	0.267	0.276	0.049
Rainy	0.310	0.220	0.204	0.124

Based on Table4.Kali Wangan water discharge during dry and rainy seasons is relatively the same. During rainy season, the river water discharge in stations 1 and 4 is higher than during dry season, while in stations 2 and 3 the dry season has lower water discharge than the rainy season. It is because the Kali Wangan water flow depends on the activity of opening and closing the sluice of Pelus River by the Water Resources and Mineral Office of Banyumas Regency. The relatively stable river water discharge does not significantly influence the quality of Kali Wangan water. Nevertheless, if the incoming contaminants have high quantity, then it may increase the water contamination load. The high quantity may result from several things, such as increased production capacity and increased number of batik craftsmen. An increased production capacity could rise the number of liquid waste volume the industry generates. The same applies to the increased number of batik craftsmen. However, an improved knowledge of batik craftsmen on waste management could minimize the contamination load in Kali Wangan.

4 Conclusion and Suggestions

Based on the research results, it can be concluded that the quality of Kali Wangan is contaminated of batik waste water is disposed of during rainy and dry seasons still suits its utilization under the Government Regulation Number 82 year 2001 on Class III.

Continuous monitoring to the Kali Wanganis necessary as an early warning system for any environmental contamination resulting from batik waste water.

References

1. S. Lestari, Sudarmadji, S. D. Tandjung, and S. J. Santosa, in *Seminar Nasional Pengelolaan Sumber Daya Alam dan Lingkungan* (2015).
2. K. Riyani, T. Setyaningtyas, and R. Andreas, *Laporan Penelitian Riset Unggulan Daerah.* (2007).
3. P. Andarani, and D. Roosmini, *EM*7, 1–12 (2009).
4. D. Connel, and G. Miller, *Chemistry and Ecotoxicology of Pollution.* (1985).
5. Zaenuri. Int. J. Educ. Res.**2,** 309–318 (2014).
6. P. Supenah, E. Widyastuti, and R. E.Priyono, Biosfera **32,** 110–118 (2015).
7. D. P. Sasongko, and W. P. Tresna, *J. Ilmu Pengetah. dan Teknol. TELAAH* **27,** 22–27 (2010).

Application of Life Cycle Assessment (LCA) in Sugar Industries

Arieyanti Dwi Astuti[1], *Rahayu* Siwi Dwi Astuti[1], *Hadiyanto* Hadiyanto[1,2]

[1]Master Program of Environmental Science, School of Postgraduate Studies Diponegoro University, Semarang - Indonesia
[2]Chemical Engineering Department, Faculty of Engineering Diponegoro University, Semarang - Indonesia

Abstract. Sugar is one of the main commodities that are needed for human life. The demand of sugar is very high with the trend increase from year to year. This condition makes the sugar industry become a leading industry that must be maintained sustainability. The sustainability of the sugar industry is influenced by the use of energy and natural resources and the resulting environmental impacts. Therefore, an effort is needed to analyze the environmental aspects and potential environmental impacts resulting from a product (sugar), by using Life Cycle Assessment (LCA). LCA is a very important tool for the analysis of a process/system from its cradle to grave. This technique is very useful in the estimation of energy usage and environmental load of a product/system. This paper aims to describe the main elements of sugar industries using Life Cycle Assessment.

1 Introduction

The sugar industry is one of the vital industries whose existence is needed to meet the basic needs of the community for sugar, making this industry must have a sustainability (sustainability industry). In fact, it is found that the large volume of sugar entering Indonesia through imported faucets as a concrete step of Government policy indicates that the domestic sugar industry has not been able to meet the national sugar demand. In fact, Indonesia's sugar industry is actually able to meet the national sugar demand from the potential of agro-ecosystem owned. The sugar industry is one of the industries considered as economic assets as well as social assets. This is based on the consideration that the prospect of sugar market in Indonesia is quite promising that the consumption of sugar is 4.2-4.7 million tonnes/year [1]. From the social aspect, the sugar industry is able to provide a source of income for 900 thousand sugar cane farmers [2].

The rush of imported sugar is exacerbated by the high cost of producing sugar industry when compared to production cost of other basic products [3], making sugar industry sustainability highly vulnerable. The sustainability of the sugar industry in general is also strongly influenced by the use of energy and natural resources and the resulting environmental impacts in its production process.

Therefore, an effort must be made by the sugar industry in order to achieve efficiency in the production process (sugar production) in order to support the achievement of sustainability industry. The efficiency of a production can be achieved by managing the main elements in the process of producing sugar, which includes the reduction of electricity consumption, energy and natural resources in a single product life cycle. There are various techniques that can be done one of them Life Cycle Assessment (LCA). This technique include maximizing output with the same inputs, reducing raw material and transportation use, the implementation of pollution control, emission reduction in the environment and waste utilization [4]. The aim of this study is to describe the main elements in a sugar industries system using Life Cycle Assessment (LCA).

2 Life Cycle Assessment (LCA)

LCA is a tool used to analyze the interaction between human activity and the environment [5]. LCA is an important and comprehensive method for analyzing the environmental impact of a single product or service. One of the things that determine whether or not an efficient production process, can be seen from the using of energy in the production system. Efficient system optimization can generate significant energy and cost savings while minimizing CO_2 emissions. LCA scan be carried out with life cycle approaches and can be used for carbon accounting and to evaluate other important environmental impacts such as acidification, ozone depletion, eutrophication, smoke, and health effects on humans [6].

There are five key stages in the life cycle of a product (Figure 1):

- Raw Materials, sourcing the materials required for the product
- Production, converting raw material the products
- Distribution, getting the product to the end user
- Use, where the user derives the direct value from the product

* Corresponding author: ariey_antik@yahoo.com

- End of life, what happens when the end user has finished the product.

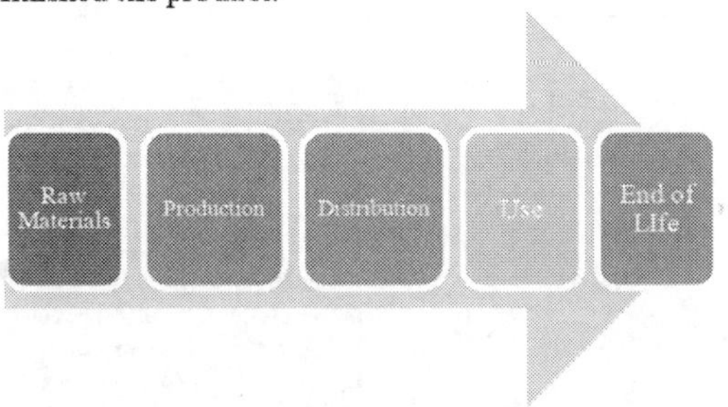

Fig. 1. Stages in The Life Cycle of a Product [7]

The methodological framework for LCA consists of four phases are definition of goal and scope, inventory analysis, impact assessment and interpretation of results as shown in Figure 2 [8].

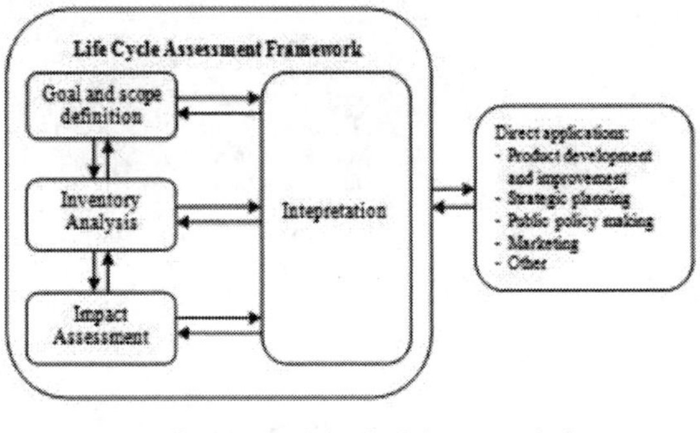

Fig. 2. The Methodological framework for LCA

3 Sugar Producing Materials

Sugar is one of the basic needs in the fulfillment of human life. Generally, sugars are produced from sugar cane crops found in tropical climates. In addition to sugarcane, sugar can also be produced from corn, beet, cassava and so on [9]. However, in some countries, sugar cane becomes the mainstay raw material in the production of sugar due to several reasons, such as several aspects as considered as over all cost, environmental effects, rate of sugar production and percentage of sucrose present in different materials [10].

4 Sugar Industries

In earlier time, the sugar industry only produced the main products of white sugar. But nowadays, besides producing sugar, sugar industries also produce other products such as cogeneration energy of bagasse, and ethanol. In general, the process of production in the sugar industry includes several main process, including the process of milling, clarification, crystallization and centrifugation [11].

4.1 Sugar milling

Initially, sugarcane is washed to remove excessive quantity of soil, rocks, and trash. After cleaning and fibrising of sugarcane, it passes through several types of mill to get extract raw juice of sugar cane. This sugar cane contains sucrose as the main ingredient of sugar making. Before milling, sugar cane will be crushed first which aim to simplify the process of sugar cane extraction. This process will produce two types of products namely bagasse and sugar cane extract. Both products are separated in different places, baggase used for energy cogeneration raw materials, while sugar cane extract is transferred to the next process, juice clarification and treatment.

4.2 Clarification

In this process, raw juice will be treated with the addition of chemicals such as sulfur (SO_2) and lime. The addition of this chemical aims to coagulate the impurities contained in the juice so easily deposited to be separated [11]. The chemical reactions of these chemicals will separate the juice into 2 parts, in liquid form (syrup), and in solid form (mud or filter cake) [4].

4.3 Crystallization

The syrup from the clarification process is then evaporate using heat. The purpose of this evaporation is to remove the water content so that it will be obtained thick syrup and ultimately produce *massecuite* that is the solution mixture of crystals of sugar and molasses.

4.4 Centrifugal separation

After crystallization, massecuite is passed through centrifugal tanks. These tanks are used to separate sugar and molasses. Molasse is the remaining massecuite that no longer contains sugar crystals. Sugar that comes out from these tanks are passed though vibrating device that addition serve as a means of transport also serves as a dryer. After drying, sugar is ready to be packed.

5 LCA of Sugar Industries

LCA in sugar industry has been done in many sugar producing countries. The LCA has different results depending on the system boundary, goal, inventory result and selected impact analysis (Figure 1). Research related to LCA aims to improve the balance of energy in sugar industry. This can be done by reducing the amount of water use, using steam in heating raw juice, using electricity from the use of baggase and heating raw juice by condensation. Analysis LCA in the sugar industry in Australia is able to contribute significantly to the environmental impact [10]. LCA study of sugar industry for different countries is shown in Table 1.

Table 1. Result of LCA Study in Sugar Industries

No	Author	System Boundary	Functional Unit	Input	Output
1	Toolseram Ramjeawon(12)	Cane production and harvest, cane burning, transportation, fertilizer and herbicide manufacture, and sugar processing, electricity generation	1 tonne of raw cane sugar	- Land, 0.12 ha - Water, 553 m^3 - Fossil fuel, 1995 MJ - Total energy (fossil fuel and baggase), 14235 MJ	- Emissions to air: 1.7 kg TSP, 1.21 kg SO_2, 1.26 kg NO_x, 1.26 kg CO, 0.065 kg VOC, 0.002 kg CH_4, 0.26 kg N_2O, 160 kg CO_2 from fossil fuel - Emissions to water: 1.7 kg N, 0.002 kg herbicides, 19.1 kg COD, 6.3 kg BOD_5, 13.1 kg TSS, 0.37 kg PO_4^{3-}, 0.1 kg oil and grease - By-product: 0.27 kg molasses, 0.3 tonnes baggase (65 KWh) electricity
2	Livision Mashoko, Charles Mbohwa, and Valerie M. Thomas(13)	The growing and harvesting of sugar cane all the way to the production of sugar and co-generation of electricity from baggase	1 tonne of raw cane sugar	- 846 tonnes of cane - Total fossil energy 1893 MJ/tonne of raw sugar produce - Coal, 70.8 kg (1397 MJ/tonne of raw sugar produced) - Water, 17000 m^3/ha - Land, 0.15 ha	- 35 KWh of electricity from 1 tonne of cane crushed - Emissions from cane burning : 280 kg of tops and dry leavesat 50% moisture per metric tonne of cane harvested - Ashes and slags, 368 kg/tonne sugar - Emissions to air : 7.5 kg CH_4, 196 kg CO_2, 0.5 kg N_2O, 2.18 kg SO_2, 7.5 kg NO_2, 0.85 kg suspended particulate matter - Emissions to water : 6.6 kg BOD_5, 19 kg COD, 12 kg NO_3^-, 0.15 kg PO_4^{3-}, 0.05 kg suspended solids, 0.00126 kg Fe - By-Products : 0.56 tonne filter cake, 0.38 tonne molasses
3	Wichita Witayapairot, and Chatpet Yossapol(3)	Cane cultivation, sugar production process, electricity production, ethanol production	1 tonne of cane	Cane cultivation (1 tonne): - Cane seed, 136.74 kg - Water 138.1 m^3 - Fertilizer 11.39 kg - Herbicide 0.18 kg - Diesel 3.26 L Sugar Production : - Cane, 1 tonne - Lime 1.5 kg - Flocculants 6.54 kg - NaOH 0.43 kg - Water 0.046 m^3 Electricity Production : - Baggase, 280 kg - Water, 112.71 kg - Diesel, 4.85 L Ethanol Production : - Molasses, 55 kg - Water, 0.137 m^3 - Yeast, 5.79 x 10^{-3} kg - Electricity 3.994 KWh - Steam 1.484 kg	Cane cultivation: - Cane, 1 tonne - Emission to air : 9.7 kg CO_2, 3.8 kg CO, 0.9 kg NO_x, 0.4 kg SO_2, 0.114 kg SPM - Emission to soil : 0.11 kg N_2O, 0.38 kg nitrogen, 0.042 kg BOD, 0.0073 kg Phosphorus Sugar Production : - Raw sugar 60.02 kg - Refin esugar 31.71 kg - Super refine sugar 16.81 kg - Waste water : 7.65 x 10^{-3} mg/l BOD, 3.44 x 10^{-4} mg/l nitrate, 6.92 x 10^{-3}phospate - By-Products : 280 kg baggase, 40.25 kg filter cake, 55 kg molasses Electricity Production : - Electricity, 69.4 KWh Ethanol Production : - Ethanol 99.5% → 14.47 L - Vinasse 6.51 x 10^{-2} kg

6 LCA of by-product Sugar Industries

Besides producing sugar, the sugar industry also produces several "by-product" such as baggase and molasses which are the raw material for other plants. Baggase is used to source raw materials from power plants, while molasses are used as a source of raw materials for ethanol production. The utilization of these two materials aims to improve the environmental profile of the industry as well as to increase the economic value of the waste. These by products from sugar industries can produce electricity, fuel, paper and organic chemicals[14]. The production of electricity from baggase and trash cane has an important role in reducing CO_2 emissions.

6.1. Baggase

As explained earlier that the sugar industries are now using the waste generated, such as baggase used as a raw material for cogeneration energy. Every 100 tons of sugar cane is capable of producing 11.8 tonnes of sugar and 28-30 tonnes of baggase which contain about 50% water[15]. In South Africa, 1 tonnes of raw sugar produced by 8.46 tonnes of sugarcane, 17.000 m^3 of water, 0.15 ha of land and 71 kg of coal. However, the production process also produces byproducts of 0.56 tonnes of filter cake and 0.38 tonnes of molasses, 2.4 tonnes of baggase and 368 kg of ashes and slugs. Production of 1 ton of sugar requires 23.800 MJ of renewable and non renewable energy which 18.400 MJ renewable energy is provided from baggase[13]. While the sugar industry in Mauritius, 1 tonnes of sugar can be produced using resources of 0.12 land, 553 m^3 of water,1995 MJ fossil fuel, 14.235 MJ total energy consumption (fossil fuel and baggase)[12] In Thailand, 1 tonnes of sugar cane with 4.6 x 10^{-2} m^3 water is able to produce 60.02 kg raw sugar, 31.71 kg refine sugar, 16.81 kg super refined sugar and 55 kg molasses. By product that produced are 280 kg baggase and 40.25 kg of filter cake [3]. From another study, 1 tonnes of sugarcane production requires 0.017 ha of land and produced 103.6 kg of sugar, 45.2 kg of molasses, 247 kg baggase, and 16.94 kWh of electricity[16].

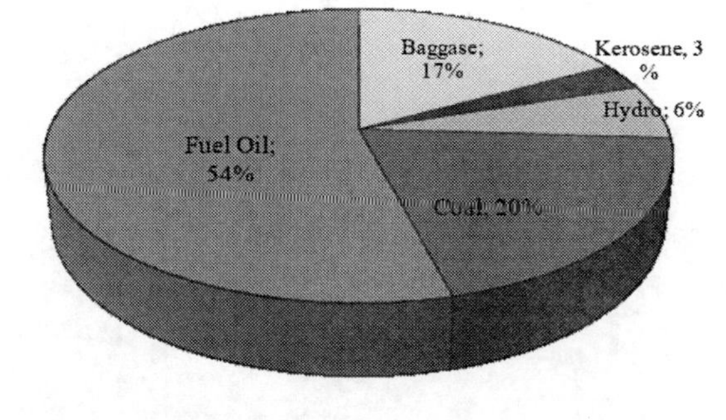

Fig. 3.Resource required of electricity in Mauritius [17]

Baggase is a by-product of the sugar industry used as inputs for power generation sources and capable of meeting industrial electricity needs. Even in some countries are able to supply electricity from the use of baggase for use outside the sugar industry. The fact that the sugarcane plant provides its own source of energy from sugar production in the form of baggase has long been a special feature of the sugar industry. In Mauritius, in traditional approach, sugar industries co-generate just enough steam and electricity to meet their on-site needs. With the availability of advanced co-generation technologies, sugar industries today can produce surplus electricity for sale to the national grid or directly to other electricity users. Such industries burn baggase during the crop season and use coal as a complementary fuel, during the intercrop. Around 44% (750 GWh) of electricity in Mauritius comes from the sugar industries, of which 21% (360 GWh) is from baggase. Electricity generated from baggase combustion provides an environmental advantage over electricity generated from fossil fuels (oil and coal)[17]. The use of baggase as a source of raw material for electricity is able to meet 17% of electricity demand in Mauritius compared to other sources of power, as shown in Figure 3.

6.2 Molasses

Molasses is also one of the most economically important by-products of sugar industries which is used as raw material of ethanol generation. Ethanol analysis has been done either directly from sugar cane or indirectly from by product sugar industry, molasses [13]. In a day, molasses produced in the sugar industry X in Indonesia is 60.000 kg. To produce ethanol, molasses are mixed with 12.000 kg water/day, 3.000 kg of H_2SO_4 solution per day, *Saccahro mycescerevicesiae* 1.500 kg/day, 600 kg urea per day. Fermentation and distillation process produces 13.413 kg/day of bioethanol, water vapor 17.020 kg/day and vinasse 46,667 kg/day[18].

An average of 1 ton sugarcane produces 23 L of molasses. Molasses have many industrial uses i.e generation of alcohol, preparation of animal feeds, and food stuffs. Molasses containing large fractions of fermentable sugars which is diluted with water and allowed to ferment in the presence of yeast culture either by batch or continuous process of fermentation [19].

7 Conclusions

In the past, sugar industries produced only sugar crystal but nowadays sugar industries are involved in the production of sugar, electricity and ethanol. So, sugar industry is now called as the cane industry. From several studies LCA of sugar industries, it can be concluded that the main elements of sugar industries are cane cultivation and harvesting, transportation, fertilizer and herbicide use, sugar production, co-generation of electricity, and ethanol production. Electricity and ethanol are products that generate from by-products of sugar industries, baggase and molasses.

Reference

1. Direktorat Jenderal Industri Agro dan Kimia. Roadmap Industri Gula. (2009).
2. Togi Y, Marpaung F, Hutagaol P, Limbong WH, Kusnadi N, Belakang L. Indones J Agric Econ.**2**, 1–14 (2011).
3. Witayapairot W, Yossapol C. Life Cycle Assessment of Sugar Production in Northeastern Thailand. In: International Conference on Green and Sustainable Innovation. (2009).
4. Chauhan MK, Varun, Chaudhary S, Kumar S, Samar. Renew Sustain Energy Rev.;**15(7)**,3445–3453 (2011).
5. Kautzar GZ, Sumantri Y, Yuniarti R.. J Manajemen, dan Sist Ind.**3(1)**, 200–211 (2014).
6. Aziz R, Chevakidagarn P, Danteravanich S..

Walailak Journal of Science and Technology (WJST) **13(3)**, 221–33 (2016).
7. H. Lehtinen, A. Saarentaus, J. Rouhiainen, M. Pitts, A. Azapagic. A Review of LCA Methods and Tools and their Sustainability for SMEs.Europe Innova, (2011).
8. ISO 14040. International Standard ISO 14040 - Environmental Management - Life Cycle Assessment - Principles and Framework. 1997.
9. Eggleston G, Lima I. Sustain. **77(9)**, 12209–12235 (2015).
10. Renouf MA, Wegener M. 29th Conf Aust Soc Sugar Cane Technol.**29**:385–400 (2007).
11. Purwaningsih IW. Penilaian Daur Hidup (Life Cycle Assessment) Gula Tebu di PG Subang, Jawa Barat. Universitas Diponegoro, Semarang; (2016).
12. Ramjeawon T. Int J Life Cycle Assess. **9(4)**, 254–60 (2004)
13. Mashoko L, Mbohwa C, Thomas VM. Journal of Environmental Planning and Management **53(6)**, 793–807 (2010).
14. J. M. Paturau. Alternative Uses of Sugarcane and Its By Products in Agroindustries. Available from: http://www.fao.org/livestock/agap/frg/AHPP72/72-24.pdf
15. Crop W. South African Sugar Industry. 2012.
16. Nguyen TLT, Gheewala SH. Int J Life Cycle Assess. **13(4)**,01–11 (2008);.
17. Balaji A, Karthikeyan B, Sundar Raj C. J Clean Prod.**16(1)**,1727–34(2008).
18. Putra, A. S . Molekul.**8(2)**,:123–30(2013).
19. Dotaniya ML, Datta SC, Biswas DR, Dotaniya CK, Meena BL, Rajendiran S,. Int J Recycl Org Waste Agric**5(3),**185–94 (2016)

Distribution of heavy metals (Cu and Fe) in sea water of Gresik coastal area

Ayu Nindyapuspa[1,*], *Achmad* Chusnun Ni'am[2]

[1,2] Department of Environmental Engineering, Faculty of Civil and Planning, Institut Teknologi Adhi Tama, Surabaya - Indonesia

Abstract. The improvement of industrial activities at Gresik Regency will increase the heavy metals concentration on the seawater at Gresik Regency. Therefore, the research of Fe and Cr distribution on the seawater at Gresik Regency has been conducted. Methods that were used is sampling by Nansen water sampler at three sampling points (housing in northern coastal Gresik Regency, Maspion V Industrial Estate, and Petrokimia Port). Samples were analyzed by Atomic Absorption Spectrophotometry (AAS) to determine the concentration of heavy metals. The results showed that the highest Fe and Cr concentration are located at Maspion V Industrial Estate (0.452 mg/L and 0.081 mg/L respectively). Meanwhile, Fe and Cr concentrations at the housing in northern coastal are (0.408 mg/L and 0.081 mg/L respectively). The concentration of Fe and Cr at Petrokimia Port are 0.174 mg/L and 0.021 mg/L respectively.

1 Background

Coastal area is the fishery field that is very important. Therefore, it need a special focus because amount of hazardous waste such as heavy metals will end up at the sea. Heavy metals can be derived from industrial, transportation, mining, and farming. Most of the Gresik Regency is the coastal area with length coastline 69 km, start from Kebomas, Gresik, Manyar, Bungah, Sidayu, Ujungpangkah, and Panceng sub-regency. Moreover, Gresik Regency has a coastline 71 km at Sangkapura sub-regency and Bawean Island. The most economic activity that was conducted in Gresik is industrial sector, with percentage 46,53%. There are 494 middle and large industries in Gresik Regency. As an industrial city, the improvement of industrial activity will increase the generated pollutant. The pollutant is disposed to the sea caused by industrial activity, agricultural activity, and household activity. One of the pollutants that are disposed of by industries is hazardous waste. The coastal area more susceptible with hazardous waste disposal. The toxicity of heavy metals spread widely contaminating the environment at the coastal area in Indonesia. Heavy metals contamination is located at the northern beach on Java Island and eastern beach on Sumatra Island. Seawater contamination caused a problem globally, especially for the marine ecosystem. [1-2]. Heavy metal is the natural element of the marine ecosystem. Heavy metals such as Fe and Cr are pollutants in the marine environment. Heavy metals that are penetrated by the body water such as lake, river, and groundwater can cause healthy damaged [3]. To protect the seawater quality and human health at the Gresik coastal area, heavy metals measurement is needed. The purpose of this research is to determine the distribution of heavy metals pollutant, especially Fe and Cr. Data of Fe and Cr distribution can be used as a database for advanced environment monitoring.

2 Literature Study

Heavy metals that penetrate into the aquatic environment will be sediment, diluted and dispersed, then absorbed by aquatic organism [4]. Based on the toxicology, heavy metal was divided into two types. The first type is essential heavy metal whereas at the certain concentration are needed by organisms, but at the large concentration, it will be toxic, such as Fe concentration at the seawater. The excess of Fe can result vomiting, intestinal damage, premature aging, and death [5]. The second type is non-essential heavy metal, whereas the benefit for human health still not found and toxic such as Pb [6]. Cr is toxic heavy metal. The toxicity depend on the valence of its ion. The toxicity of Cr^{6+} is hundred times higher than the toxicity of Cr^{6+} [7]. The Cr^{6+} is corrosive and carcinogenic.

3 Methods

Sampling was conducted at two areas with a dense industrial activity, seawater at the industrial area on Manyar sub-regency and seawater at around PT Petrokimia Gresik Port. There are three stations for sampling location. Sampling was conducted in the morning at 07.00 AM - 10.00 AM. The wind blew to the north west. The method that was used for this research is purposive sampling. Seawater was taken by Nansen water sampler. Fe and Cr were measured by Atomic Absorption Spectrophotometer (AAS) in the laboratory.

* Corresponding author: ayunindyapuspa@itats.ac.id

The results of Fe and Cr analysis will be compared with Ministry of Environment Regulation Number 51 The year 2004 about Seawater Quality Standard. Climatological data (rainfall, wind direction, wind velocity) while sampling was conducted were taken from Meteorology station Perak I Surabaya. The amount of industries in Gresik was taken by Industrial Agency of Gresik Regency.

4 Results and discussion

Sampling of seawater were conducted at the coastal area at Gresik Regency. There was coastal area at Gresik sub-regency and Manyar sub-regency. These sub-regencies were chosen because there are many industries such as PT Petrokimia Gresik and PT Maspion. Based on the Gresik Regional Development Planning Agency 2016, there are 45 industries at Gresik sub-regency and 22 industries at Manyar sub-regency. The seawater sampling location shown in Figure 1.

Fig. 1. Seawater Sampling Location

The station 1 is located at PT. Petrokimia Port, station 2 is located at marine area near Maspion V Industrial Area, and Station 3 is located at downstream at Manyar sub-regency. The coordinates of sampling location were described in Table 1.

Table 1. Coordinates of Sampling Location

Location	Coordinates
Station 1	7° 8'36.40"S and 112°39'25.80"E
Station 2	7° 7'17.50"S and 112°38'50.10"E
Station 3	7° 6'8.40"S and 112°37'58.80"E

Samples were analyzed at laboratory using AAS. Data were compared with Ministry of Environment Regulation Number 51 The year 2004 about Seawater Quality Standard. The seawater quality standard of Fe were unregulated. The results of Fe and Cr analysis shown in Table 2.

Table 2. Result of Fe and Cr Analysis

Location	Fe (mg/L)	Cr (mg/L)	Cr Quality Standard (mg/L)*
Station 1	0.408	0.078	0.002
Station 2	0.452	0.081	0.002
Station 3	0.174	0.021	0.002

* Based on the Ministry of Environment Regulation Number 51 The year 2004 about Seawater Quality Standard for Marine Tourism

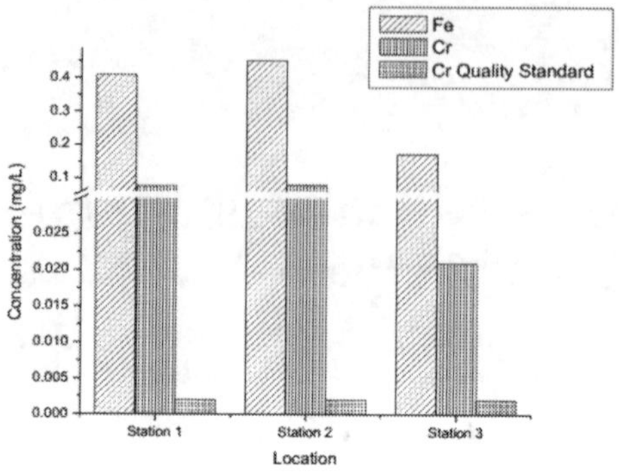

Fig. 2. Graphic of Fe and Cr Distribution in Sea Water of Gresik Regency

The results showed that the highest Fe and Cr concentration is located at Station 2 (Figure 2). The lowest Fe and Cr concentration are located at Station 3. Based on the Ministry of Environment Regulation Number 51 the year 2004 about Seawater Quality Standard, the Cr concentration at 3 stations did not meet the standard. Meanwhile, the quality standard of Fe concentration is unregulated. Because Fe unregulated, comparing with the previous research is needed. Data were compared with the previous research, but similar with the sampling location. Data were compared with [6] and [9]. These two researches were taken at Paranggi Village's Seaport [6] and Taipa's Ferry Harbor [9]. These location are similar with Gresik Regency. The Fe concentration near with Paranggi Village's Seaport was 0.307 mg/L. Meanwhile, the Fe concentration near with Taipa's Ferry Harbor was 0.546 mg/L. Dissolved iron can exist in seawater in two oxidation states, Fe(II) and Fe(III), free or complexes with inorganic and organic ligands [10].

The Cr species most frequently found in marine media are: (i) chromate $(Cr^{VI}04^{2-})$; (ii) Cr^{III} hydroxo complexes, $Cr(OH)_3$, $Cr(OH)_2^+$ and $Cr(OH)_2^+$. Chromium(VI) is considered to be highly toxic mainly on account of its ability to interact favourably with biological substances. According to thermodynamic calculations, only the hexavalent form of Cr should be present oxygenated sea-waters at pH 8 [11].

Coastal ecosystems are affected by industrial and anthropogenic activities such as wastes from iron and steel factories, shipyards, shipbuilding/breaking. Heavy metals concentration depends on seawater condition. When human activities are getting higher at the land or the beach, the concentration of heavy metals will be enlarged. The amount of Fe concentration caused by daily activity at household, such as domestic waste disposal containing Fe and effect of pipe corrosion so that Fe concentration is getting higher [8]. The increasing of Fe concentration caused by mineral rock erosion because of sea waves. Ship and pile port corrosion may contribute to sea water pollution by Fe [6]. In this case, industrial activity at Maspion V Industrial Area may contribute Fe and Cr to the marine pollution.

5 Conclusion

The Fe and Cr pollutant at Station 2 are the highest of all with concentration 0.452 mg/L and 0.081 mg/L respectively. The Fe and Cr pollutants at Station 3 is the lowest of all with concentration 0.174 mg/L and 0.021 mg/L. The concentrations of Cr and Fe in the sea water at Gresik coastal area was exceeded quality standard based on the Ministry of Environment Regulation Number 51 The year 2004 about Seawater Quality Standard Based on the quality standard, the seawater at Gresik Coastal Area is polluted by Cr. Cr pollution caused by industrial, port, and household activities.

6 References

1. P. Censi, S.E. Spoto, F. Saioano, M. Sprovieri, S. Mazzola, Chemosphere. Vol.**64**: 1167-1176 (2006)
2. J. Pote, L. Haller, J.L. Loizeau, A.G. Bravo, V. Sastre, and W. Wildi, Bioresource Technol.**99**: 7122–7131 (2008)
3. R. Verma, and D. Pratiwi, Recent Research in Science and Technology. Vol. **5**(5): 98-99 (2013)
4. J. Murtini, and Rosmawaty. Jurnal Perikanan, 8(2), 177-184 (2006)
5. Tahril, P. Taba, N. L. Nafie, and A. Noor, Jurnal Natur Indonesia, **13**(2), 105-111. [2011]
6. Fiskanita, B. Hamzah, and Supriadi. J. Akad. Kim. **4**(4):175-180 (2015)
7. Perdana, M., et al. Fotoelektrokatalisis kromium
8. elektroda timbal dioksida (PbO2). Chem Info Journal, **1**(1), 11–17. (2013)
9. Ika. Tahril. and Said, Irwan. J. Akad. Kim, **1**(4): 181-186 (2012)
10. X. Liu, and F.J. Millero, Marine Chemistry (77): 43-54 (2002)
11. A. Boughriet, et al. Journal of Analytical Atomic Spectrometry **(99)**: 1135-1142 (1994)

 https://doi.org/10.1051/e3sconf/20183105001

The Acoustical Properties of the Polyurethane Concrete Made of Oyster Shell Waste Comparing Other Concretes as Architectural Design Components

Erni Setyowati[1,*], *Gagoek* Hardiman[1], and *Purwanto*[2]

[1] Department of Architecture, Faculty of Engineering, Diponegoro University, Semarang - Indonesia
[2] Department of Civil Engineering, Faculty of Engineering, Diponegoro University, Semarang - Indonesia

Abstract. This research aims to determine the acoustical properties of concrete material made of polyurethane and oyster shell waste as both fine aggregate and coarse aggregate comparing to other concrete mortar. Architecture needs aesthetics materials, so the innovation in architectural material should be driven through the efforts of research on materials for building designs. The DOE methods was used by mixing cement, oyster shell, sands, and polyurethane by composition of 160 ml:40 ml:100 ml: 120 ml respectively. Refer to the results of previous research, then cement consumption is reduced up to 20% to keep the concept of green material. This study compared three different compositions of mortars, namely portland cement concrete with gravel (PCG), polyurethane concrete of oyster shell (PCO) and concrete with plastics aggregate (PCP). The methods of acoustical tests were conducted refer to the ASTM E413-04 standard. The research results showed that polyurethane concrete with oyster shell waste aggregate has absorption coefficient 0.52 and STL 63 dB and has a more beautiful appearance when it was pressed into moulding. It can be concluded that polyurethane concrete with oyster shell aggregate (PCO) is well implemented in architectural acoustics-components.

1 Introduction

Innovation in concrete materials has been developed significantly. The use of materials and filler additive for designation are examined. One of the efforts to reduce cement and lighten the concrete material is the use of a foam additive such as polyurethane [1], [2] and polystyrene [3]. The use of foam additive in concrete material is also beneficial for the invention of the acoustic-based cement composite material. While other studies used polyesther in concrete for the purpose of increasing harshness on building materials[4], [5].

This research is focused on the acoustic performance of polyurethane concrete material made of oyster shell waste compared to other concrete materials.

2 Theoretical Review

The theoretical review will discuss the oyster shell concrete and polyurethane concrete as basic material in this study. Previous research stated that the oyster shell concrete made of anadara granosa lynn species has the strongest compressive strength comparing to the other shell concrete[6], [7]. This study focused on the use of the polyurethane in concrete to lighten it and improve the capability of acoustics.

2.1. Oyster Shell Concrete

Previous research has compared three different types of concrete material made of three species of shells. Three shell concrete specimens made of Anadara granosa Lynn, Perna viridis and Placuna placenta Lynn have tested both the ability of compressive strength as well as acoustic abilities. The result is that shell concrete of Anadara granosa Lynn has highest compressive strength up to 200 kg/cm^2 [6]. Meanwhile, shell concrete made of Perna viridis Lynn has the best acoustic capabilities, with 7.28 dB on it's Sound Transmission Loss [7].
Because this study observing the concrete material based on the previous research, then the selected species is Anadara granosa Lynn either as fine aggregate or coarse aggregate, due to it's performance either in compressive strength or acoustical performances. While for other selected materials are polymeric concrete (PCP) and normal concrete with gravel aggregates (PCG).

2.2 Acoustical Performance of Shell Material

Research on the shell concrete material for noisy urban region has been already conducted. The result is that material of oyster shell with porous additive substances can be used to absorb sound in housing near the airports[8]. Oyster shell material without additive substance will not have a good acoustic performance,

* Corresponding author: ernisyahdu@gmail.com; ernisetyowati@arsitektur.undip.ac.id

due to the low absorption coefficient between 0.03 – 0.04[7]. Even, the fly ash mixed with oyster shell will reduce pores and reduce the ability of sound absorber [9].

2.3 Previous Research of Concrete

The normal concrete with coarse aggregate (PCG) and Polymeric Concrete (PCP) have been already examined in the previous researches. The method used was a mix design with DOE Methods. On polymeric concrete (PCP), the aggregate of polymeric waste used to replace gravel aggregate and a part of cement was substituted by baggase ash [10], [11]. Meanwhile, the normal concrete is portland cement concrete with gravel (PCG). PCG has a normal compressive strength that is equivalent > Q-250. The harshness of normal concrete tends to have capabilities as a reflector.

The normal concrete (PCG) and polymeric concrete (PCP) have the composition ratio of cement (plus baggase ash), sand and gravel or polymeric waste as 2,600; 2,220; 0,635 respectively, while the Polyurethane concrete with shell aggregate has composition of cement, oyster shell, sand and polyurethane as 1,600; 0,400;1,000; 1,200 respectively. The moulding of material used are cubical mouldings as seen in fig. 1. The figure below shows the mixture of materials.

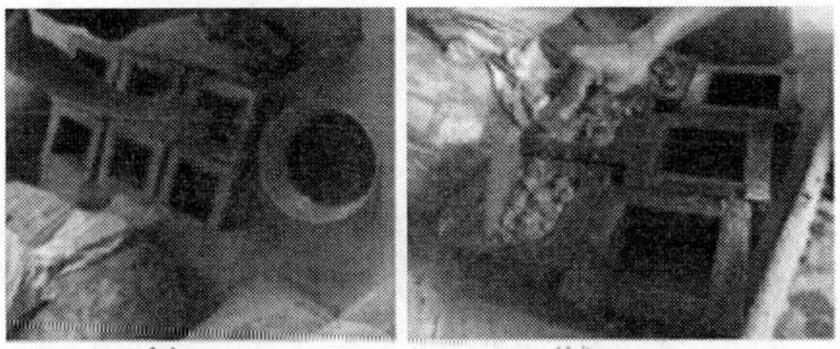

(a) (b)

Fig. 1. (a) cubical moulding (b) mix design process of materials

3 Research Methods

Acoustic characteristics in this study consist of absorption coefficient (α) and Sound Transmission Loss (STL).

3.1. Absorption Coefficient

The impedance tube with 1/3 octave frequency filter was used in the sound absorption test. The data were connected into Personal Computer (PC) by software refer to ASTM E413-04[12]. The sound absorption coefficient (α_0) was calculated by measuring sound pressure that come on a surface of material and was reflected by it. A good absorber must have an absorption coefficient at least 0. 2[7], [13–15].

$$\alpha_0 = \frac{4}{n + \left(\frac{1}{n}\right) + 2} \quad (1)$$

Where α_0 is sound absorption coefficient and n is standing wave ratio. The standing wave ratio (n) is measured by substituting resistance with attenuation, determining the ratio of the standing waves from the difference of sound pressure (L) dB using equation as follows:

$$n = 10^{\left(\frac{L}{20}\right)} \quad (2)$$

With n is a standing wave ratio and L is the difference of sound pressure in *deci* Bell. In the process of the absorption coefficient test (α). The impedance tube was equipped by microphones to transmit sound at low to high frequencies. The Frequency scale that was used in acoustics laboratory is between 200-6000 Hz.

3.2. Sound Transmission Loss (STL)

The Sound Transmission Loss (STL) Test used impedance tube equipped with 4 microphones (see fig.2) that have sensitivity to high frequency sound.

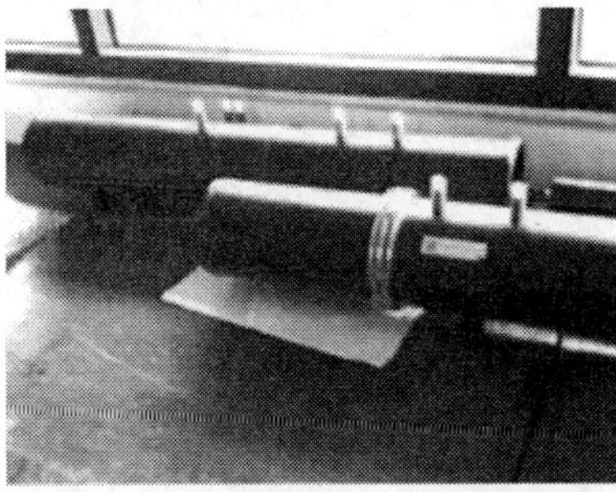

Fig.2. The Impedance Tube with microphones

Sound Transmission Loss/STL of a partition is defined as the ratio of logarithmic between the transmitted sound powers (Wt) and the sound power of partition material which comes to the surface (Wi). In general it can be formulated as:

$$TL = 10 \log \frac{W_t}{W_i} \quad (3)$$

$$TL = 10 \log \frac{1}{r} \quad (4)$$

With r is the sound transmission coefficient of such material, i.e. the ratio between the transmitted sound power through partition of material against the sound felt through the material's surface. The STC value is a parameter to know how the material can reduce sound energy[7]. The STC is determined based on the raw quality ASTM E 413 issued by the American Society for Testing and Materials[12].

4 Results and Discussion

4.1. Absorption Coefficient

The absorption coefficient test results of materials, PCG, PCO and PCP are as follows:

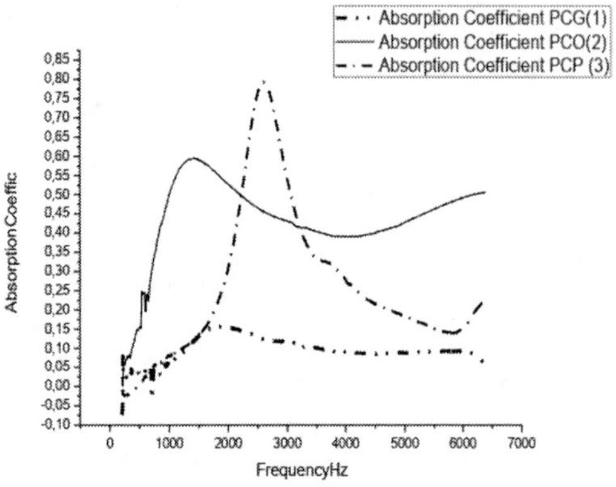

Fig. 3. Absorption Coefficient Graph of materials

As shown in the graph (see fig. 3), PCG and the PCP have characters on absorption coefficient of octave frequencies which are almost similar. PCG has a coefficient of absorption peaks in 0.155 at frequency 1968 Hz, while the PCP has the highest coefficient of 0.793 on frequency 2, 6 k (see Table 1).

Table 1. Absorption Coefficient on octave band frekuensi

Frequency (Hz)	α PCG	α PCO	α PCP
1000	0,0626	0,4989	0,0783
1432	0,1211	0,5951	0,1266
1968	0,1550	0,5315	0,2892
2600	0,1238	0,4559	0,7929
3000	0,1193	0,4305	0,5381
4000	0,0912	0,3906	0,2770
5000	0,0867	0,4271	0,1813

Unlike the PCG and PCP, hence the absorption coefficient ranges of PCO is better, because the value of the coefficient is at a wider broadband frequency. Maximal absorption coefficients that can be achieved is 0.59508 in 1432 Hz, but it is able to absorb sounds at lower frequencies than the other materials. The variations in the ability of absorbency at low to high-frequency sound are required in building materials, especially for the concert hall building design.

The polyurethane concrete material with shell aggregates (PCO) has a pretty good absorption coefficient because it has a capability to absorb sound energy in a wide broadband of octave frequency. But, in the future, this material need to be improved on acoustical performance by modification of resonator efforts as several researches conducted. Some modification efforts to improve absorption coefficient of resonators for examples with the use of flexible panels [16], coupled- Helmholtz resonator[17], [18], recycling material from industry[19], [20], sustainable absorber from biomass[21].

4.2. Sound Transmission Loss (STL)

Unlike normal concrete, polyurethane concrete with oyster shell (PCO) and polymeric concrete (PCP) have STL characters which are almost the same. Both materials have range of STL between 54-56 dB starting from around 2000 Hz. At the frequency of <500 Hz, PCP has range of STL lower than PCO is, due to the polyurethane compound in PCO that absorbs and reduces sound energy more than polymeric concrete (PCP).

Sound Transmission Loss test results of all materials are as follows:

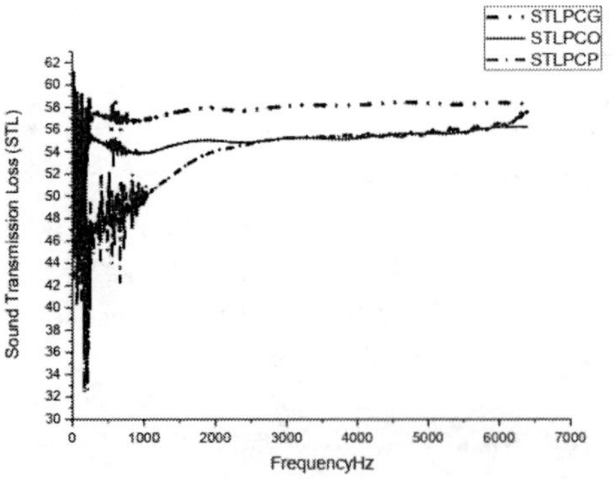

Fig. 4. Sound Transmission Loss (STL) of PCG, PCO and PCP

The graph shows that the normal concrete (PCG) with aggregate gravel has the highest STL highest among other materials. The value of the STL of PCG ranged between 57-58 dB (see fig.4 and table 2).

The role of polyurethane here is as sound traps, so that the sound of falling on the surface are not forwarded to penetrate the material. The increased efforts of the STL

https://doi.org/10.1051/e3sconf/20183105001

has been done by a number of researchers, such as by adding material on acoustic resonator system[22].

Table 2. Absorption Coefficient in octave band frequencies

Frequency (Hz)	STL PCG (dB)	STL PCO (dB)	STL PCP (dB)
1000	56,9132	53,9376	49,456
2000	57,8971	55,0005	54,0774
2400	57,6903	54,8373	54,6301
3000	58,1237	55,2227	55,2424
4000	58,1805	55,2048	55,3544
5000	58,364	55,6168	55,7178

4.3. The Use of Polyurethane Concrete in Architectural Design

The use of Polyurethane concrete made of oyster shell waste in buildings have wide variations. Polyurethane concrete can be printed in color and texture on the surface. Here is the usage of polyurethane concrete on the application component of building materials (see fig. 5 and 6).

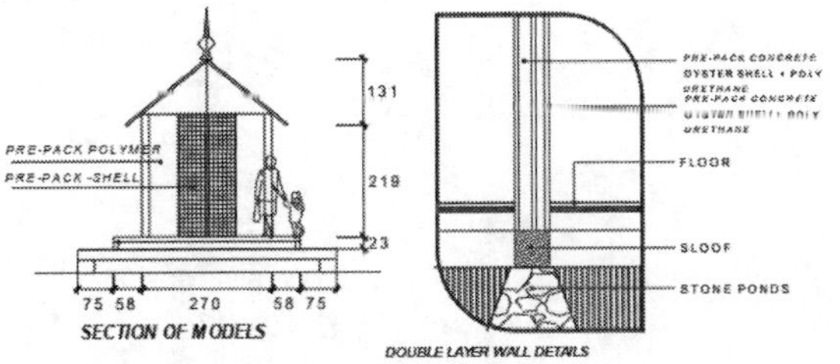

Fig. 5. Textured Polyurethane concrete in exterior double wall panel

Fig. 6. Polyurethane concrete and colour innovation in Japanese industry.

Green architecture uses green concept materials, therefore discharging the innovation material that is capable of utilizing waste and reduces the use of cement is indispensable in the development of a sustainable green construction in architecture world.

5 Conclusions

The research was concluded a few things as follows:

1. Polyurethane concrete with oyster shell (PCO) are able to absorb sound in a wider frequency range than other materials.
2. Polyurethane concrete (PCO) can be used as absorber material on buildings as it can absorb sound energy at low to high frequency.
3. Normal concrete has the highest STL compared to other materials. While PCO has a better STL than PCP (polymeric concrete), because the content of polyurethane which is capable of trapping the sound.
4. Polyurethane concrete usage with shell aggregate (PCO) are excellent for building materials in the noisy urban region, because it was able to absorb the energy of the sound properly.
5. Green architecture is the architecture that is capable of utilizing waste based materials and is able to reduce the use of cement. By Polyurethane concrete material, the green construction can be realized because the material is lightweight and environmentally friendly.

This research is financed by the budget of the Research Grant from Ministry of Research, Technology and Higher Education, fiscal year 2017 with the scheme of Competency Grant and University of Diponegoro Grant Number: 276-47/UN7.5.1/PG/2017. The funders do not have any intervention on the data and the content of the manuscript. Authors thank all private sectors who have contributed in this research and gave the valuable opinion, advices and discourses.

References

1. I. Golpazir, A. Ghalandarzadeh, M. K. Jafari, and M. Mahdavi, Constr. Build. Mater., **118**,104–115 (2016).
2. G. Sung, J. W. Kim, and J. H. Kim, J. Ind. Eng. Chem., **44**, 99–104 (2016).
3. B. a. Erickson, *Experimental study on the dynamic stress-strain behavior of expanded polystyrene Geofoam using cyclic triaxial tests*, no. December 2011, p 45, (2011)
4. G. Martínez-Barrera, C. Menchaca-Campos, and O. Gencel, Constr. Build. Mater., **41**, 204–208(2013).
5. I. J. Chiou, C. H. Chen, and Y. H. Li, Constr. Build. Mater., **64**, 480–487 (2014).
6. E. Setyowati, G. Hardiman, and P. Purwanto, J. Teknologi , **78 (5)**, 203-207 (2016).
7. E. Setyowati and G. Hardiman, J. Eng.

Technol., **3(3)**, 1–6 (2015)

8. E. Setyowati and H. Trilistyo, Procedia - Soc. Behav. Sci., **227**, 294–299 (2016).
9. H. Wang, W. Kuo, C. Lin, and C. Po-yo, Constr. Build. Mater., **41**, 532–537 (2013).
10. E. Setyowati and Purwanto, *The Acoustical Characteristics of Polymer Mortar as a Green Concrete Material,* in *The 14th International Conference on Quality in Research*, 2015, pp. 983–987.
11. E. Setyowati and Purwanto, *The Polymer Brick as Nano-Technology based Material to Support Green Building Construction*, in *The 3rd Annual International Conference Proceedings on Architecture and Civil Engineering*, 2015, pp. 224–228.
12. ASTM, *ASTM E413-04 Classification for Rating Sound Insulation*. .
13. E. Setyowati, G. Hardiman, and E. E. Pandelaki, Structural , *Acoustic and Aesthetic Performances of Double Wall Layer Made Of Oyster Shell And Polymer As Green Material In Green Construction*," proceeding of ICENIS, December (2016).
14. E. Setyowati, A. Satyapratama, S. T. Atmadja, and G. Hardiman, J. Teknologi , **78(5)**, 289-293 (2016).
15. E. Setyowati, G. Hardiman, and S. T. Atmaja, J. of Applied Mechanics and Material, **747**, 221–225 (2015).
16. A. Sanada and N. Tanaka, Appl. Acoust., vol. **74(4)**, 509–516 (2013).
17. S. K. Tang, J.of Sound and Vibration, **279**, 1085–1096 (2005).
18. D. Li and L. C. Ã, J. of Sound and Vibration, **305**, 272–288, (2007).
19. F. Asdrubali, A. L. Pisello, F. D. Alessandro, F. Bianchi, M. Cornicchia, and C. Fabiani, Energy Procedia, **78**, 321–326 (2015).
20. C. Buratti, E. Belloni, E. Lascaro, G. Anna, and P. Ricciardi, Energy Procedia, **101**, 972–979 (2016).
21. D. J. Oldham, C. A. Egan, and R. D. Cookson, ble acoustic absorbers from the biomass, Appl. Acoust., **72(6)**, 350–363 (2011).
22. K. T. Chen, Y. H. Chen, K. Y. Lin, and C. C. Weng, Appl. Acoust,. **54(1)**, 71-82 (1998).

Utilization of Natural Zeolite from Ponorogo and Purworejo for Naphthol Substance Adsorption

Sundus Imandiani[1,*], *Christine* Indira[1], *Anthony* Johan[1] *and Budiyono*[1]

[1]Department of Chemical Engineering, Faculty of Engineering, Diponegoro University, Semarang – Indonesia

Abstract. Indonesia has many zeolite producing areas yet untapped. Researchers developed the utilization of natural zeolites useful for the adsorption of naphthol dyes commonly found in batik waste. In this study researchers used natural zeolites from Purworejo and Ponorogo that are activated using hydrochloric acid that is used for adsorption. The purpose of this research is to know the effect of natural zeolite activation from Ponorogo and Purworejo on the effectiveness of adsorption of naphthol dyes widely used in batik industry. Natural zeolite was activated using HCl concentration of 1.3N; 1.8N; 3.2N; and 3.9N for 60 minutes. The methods are preparation of natural zeolite from Purworejo and Ponorogo, dealumination using hydrochloric acid, adsorption process of naphthol dyes using activated zeolite, and test of adsorption result with uv-vis spectrophotometry. The test results showed that the higher HCl concentration will increase adsorption capacity. This can be known from the concentration of naphthol dye which decreased both using natural zeolite Ponorogo and Purworejo. While the effectiveness of adsorption shows natural zeolite Purworejo has a greater adsorption capacity than Ponorogo with optimum conditions of dealumination using concentration HCl 3,9N.

1 Introduction

One of the most developed industries in Indonesia is batik industry. Batik industry in 2006 amounted to 48,287 business units spread over 17 provinces, and employed 792,300 people [1]. The main source of waste in the batik industry is dyes, which are contain naphthol dyes. If the waste is continuously discharged directly into the waters and exceeds the ability of the river to clean themselves (self-purification), there will be pollution of the waters that negatively affect the life of aquatic biota and public health utilizing the river water.

Natural zeolite is a group of minerals produced from hydrothermal processes in alkaline igneous rocks, and is a precipitate of volcanic activity which contains many silica elements [2]. Zeolite crystals are very regular with interconnected cavities in all directions and make the zeolite surface area so large that it is very well used as an adsorbent [3]. The composition of natural zeolite varies greatly, depending on the species. Zeolites occur naturally on the soil surface. Today many types of natural zeolites have been found and grouped by their structural similarity. Although synthetic zeolites have also been produced, natural zeolites have an important role because of their abundant availability in nature, especially in Indonesia.

In contrast to synthetic zeolites whose predictable structure of their constituent compounds, natural zeolites have structures that are not always the same, depending on the conditions of their formation in nature. Therefore, in the use of natural zeolite as an adsorbent, the activation process is required. This activation process is needed to improve the specific properties of zeolites and remove impurities. In this research used dealumination method with HCl.

The purpose of this study is to examine the effect of HCl concentration in the zeolite adsorbent activation process on the effectiveness of naphthol adsorption dyes

2 Methods

2.1 Preparation and Activation

Natural zeolites that have been obtained from Ponorogo and Purworejo are smoothed to a size of 100 mesh, then zeolite washed with demineralized water and dried using oven to constant weight.

The process of activating the natural zeolite was carried out using HCl solution. Zeolite as much as 100 and 10 gram mixed for 60 minutes with HCl solution with concentration of 1.3 N; 1,8 N; 3,2N; 3,9 N as much as 100 ml, then filtered. The zeolite precipitate is then dried with oven to constant weight.

2.2 Adsorption Process

In the adsorption process, 10 grams of natural zeolite was mixed with naphthol dye with a concentration of 400 ppm, 600 ppm, 800 ppm, and 1000 ppm of 100 ml

* Corresponding author: christineindirarinai@gmail.com

https://doi.org/10.1051/e3sconf/20183105002

with constant stirring. After that the solution was filtered to separate the filtrate, then the filtrate was measured its absorbance by UV-VIS Spectrophotometer, then calculated the final concentration of the naphthol dye after adsorption process.

3 Results And Discussion

3.1 Naphthol Dye Adsorption Test

The naphthol absorption analysis was carried out based on absorbed concentration, and absorption percentage. From the dealumination process that has been done on zeolite from Ponorogo (ZAP) and Purworejo (ZAW) using HCl solution with concentration of 1.3 N; 1.8 N; 3.2 N; and 3.9 N. Natural zeolite then used as a naphthol dye adsorbent at various concentrations of 400 ppm, 600 ppm, 800 ppm and 1000 ppm. The final concentration of the solution after the adsorption process presented in Figure 1 and Figure 2.

Based on Figures 1 and 2 showing the zeolite dealumination process with a concentration of 3.9 N HCl solution gives the best possible adsorption results with the final concentration of the least naphthol dye with an average percentage of naphthol dye absorbed up to 98.61% in the natural zeolite Ponorogo And 98.92% in natural zeolite Purworejo.

The amount of adsorbed naphthol dye stuff indicates an increase in surface area in activated natural zeolites. The completion of the dealumination process will cause the rise of zeolite surface area due to the reduction of impurity metal covering the pores of the zeolite with the increase of surface area will result in the absorption process happening the greater [4].

While activation of natural zeolite with HCl concentration solution 3, 2 N; 1.8 N; and 1.3 N continue to decrease with the smaller concentration of HCl solution used, the average percentage of the least absorbed naphthol dyestuff is by the activation treatment at HCl 1.3 N solution concentration of 97.8% for Ponorogo zeolite and 98 % For Purworejo zeolite. This is because the natural zeolite still contains many impurities that cover the zeolite surface and also because the natural zeolite has not formed silanol group due to activation with acid so that adsorption is less effective.

SEM Characteristic show the difference of natural zeolite particle size before and after activation, using HCl 3, 9 N. The particle size of natural zeolite before activation is 2,41 μm, while after activated using HCL the particle size changes to smaller that is 0,5 – 0,8 mm as presented in Figure 3. The smaller particle size will increased the surface area of zeolite and make the adsorption process better.

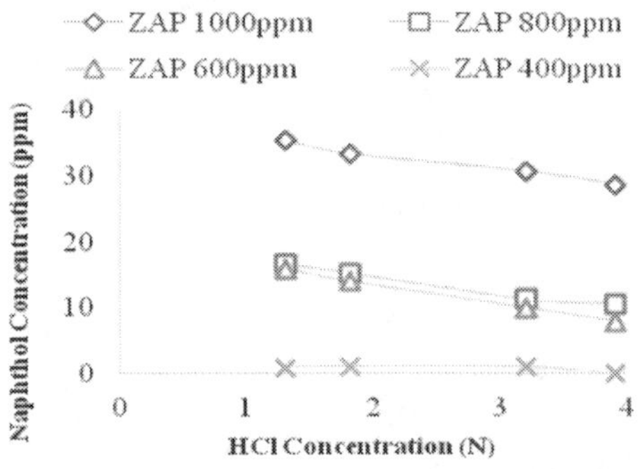

Fig. 1. Naphthol Concentration after Adsorption (ZAP)

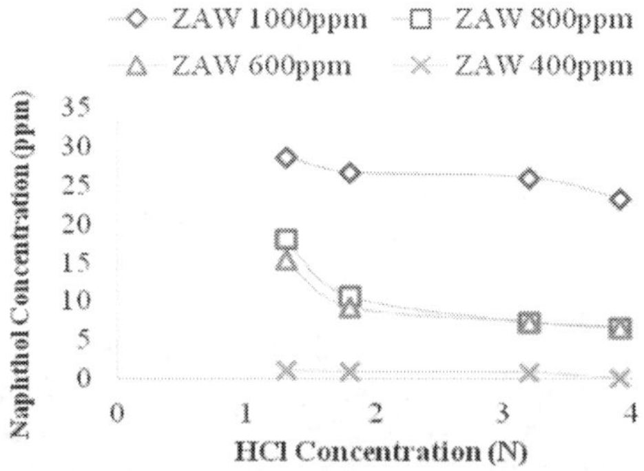

Fig. 2. Naphthol Concentration after Adsorption (ZAW)

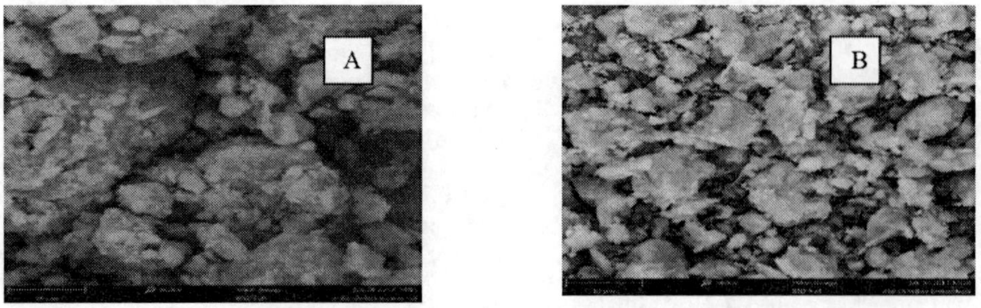

Fig. 3. SEM Characteristic on ZAP before Activation (A); ZAP after Activation (B)

 https://doi.org/10.1051/e3sconf/20183105002

3.2 Zeolite Dealumination

EDX test results show the composition of natural zeolite from Ponorogo and Purworejo that are activated using hydrochloric acid solution 3.9 N.

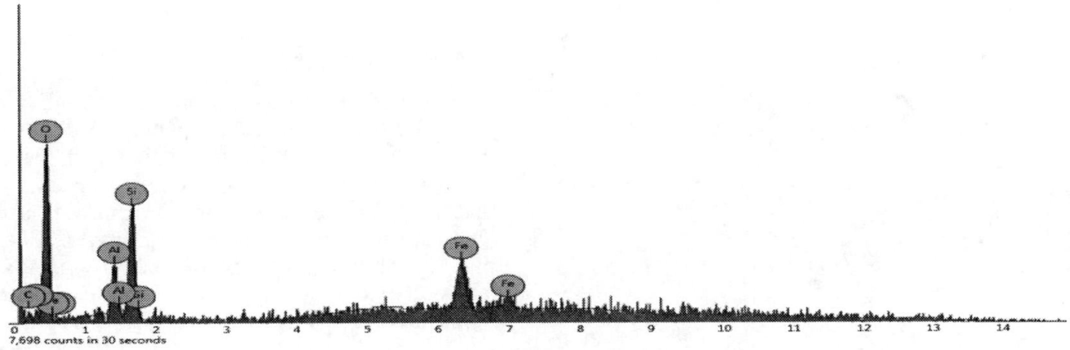

Fig. 4. EDX Analysis on Activated Natural Zeolite from Ponorogo

Table 1. Ponorogo Natural Zeolite (ZAP) Composition

Element Number	Element Symbol	Weight Concentration
8	O	42.8
14	Si	10.6
26	Fe	26.2
13	Al	5.4
7	N	7.0
6	C	8.0

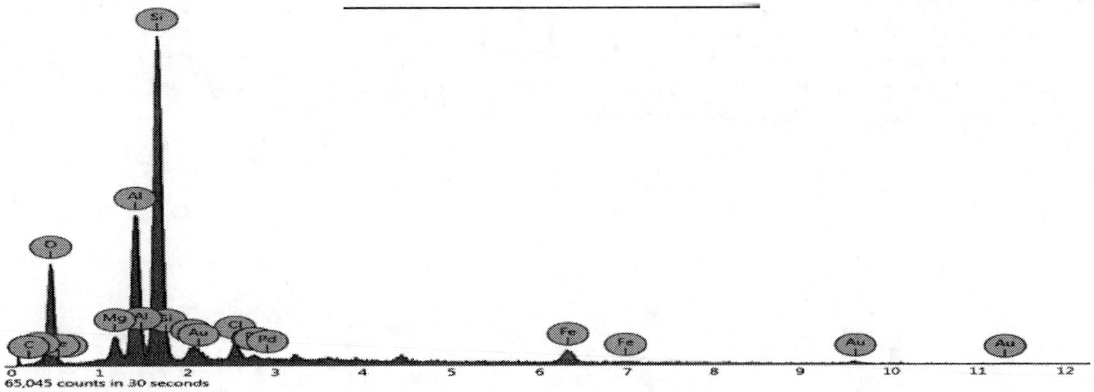

Fig. 5. EDX Analysis on Activated Natural Zeolite from Purwoerjo

Table 2. Purworejo Natural Zeolite (ZAW) Composition

Element Number	Element Symbol	Weight Concentration
14	Si	5.3
13	Al	2.6
8	O	7.1
12	Mg	0.5
17	Cl	0.5
26	Fe	1.0
46	Pd	0.4
7	N	0.6
79	Au	81.8
6	C	0.2

From the analysis it can be seen the ratio of Si/Al on ponorogo zeolite is 2 and zeolite Purworejo is 2.04. So the result of adsorption between natural zeolite Ponorogo and Purworejo are not much different.

The influence of Si / Al ratio on a material will affect the properties of the material. The higher the Si / Al ratio of a material the material is hydrophobic. Dealumination is the most important commercial method of getting the desired amount of Al. An increase in Si / Al ratio will cause changes in the electrostatic magnetic field in the zeolite, thus affecting the interaction of zeolite adsorption [1].

The accumulation of natural zeolite as aluminosilicate with HCl can be seen as a process of replacing aluminum with hydrogen which involves altering the structure of Si-O-Al aluminosilicate into a silica structure in the form of silanol -Si-OH group [1]. The greater the concentration of HCl used to activate the zeolite causes the crystallinity of the zeolite to be relatively decreased. This is due to the increase in HCl concentration which is getting bigger and causing the dealumination (Al out of the zeolite framework) more and more. With the escalation of Al atoms more and more zeolites cause decrystallization which will lead to increased irregular crystal structure resulting in less [5].

4 Conclusions

The activation process performed on both zeolites derived from ponorogo and purworejo showed that with the increased HCl concentration used, the ability of adsorption in natural zeolites increased. It is indicated by the percentage of adsorbed naphthol concentration. The activation process using HCl causes the Si / Al ratio on the zeolite to increase. And also causes the pore surface area of zeolite to increase due to the loss of impurity components in the zeolite

References

1. Ranjani V.Siriwardane, Ming-Shin Shen, and Edward P. Fisher, *Adsorption Of CO_2, N_2, and O_2 on Natural Zeolit, West Virginia* 26507-0880. (2002).
2. Wang, Shaobin and Peng, Yuelian. *Natural Zeolites as Effective Adsorbents in Water and Wastewater Treatment.* Department of Engineering, Curtin University of Technology, (2009)
3. 2 A. Ertan, and Ozkan. Adsorption, Vol **11**, 151-156, (2005)
4. Worch. Eckhard, Adsorption Technology in Water Treatment, Dresden Germany, (2012).
5. Eunjoo Kim et al, *Mono-dispersed DDR Zeolit Particles by seeded Growth and their CO_2, N_2, and H_2O Adsorption Properties* (2016).
6. Benefield. *Process Chemistry for Water and Waste Water Treatment.* Prentice Hall Inc: New Jersey, 1982

https://doi.org/10.1051/e3sconf/20183105003

Sustainable Development Strategy Of Domestic Waste Infrastructure In The City Of Surakarta

Arya Rezagama[1,*], *Purwono*[1] and *Verika* Damayanti[1]

[1]Department of Environmental Engineering, Faculty of Engineering, Diponegoro University, Semarang - Indonesia

Abstract. Shifting from traditional system to large, centralised infrastructure domestic waste is widely complex challenge. Most of the sanitary system on household in Surakarta use on site septictank, 17% sewerage system reached and16,0% stll open defecations. Sanitation development sustained aims to develop policy and strategies waste management domestic Surakarta in the long term (20 years). The projection use quantitative method and institutional condition approach by SWOT analysis. Surakarta City get priority sanitation urban planning from Indonesian government in Presiden Joko Widodo era. The domestic waste management systems that is Surakarta divided into system on-site and system off site. Waste Water Treatment Plant (WWTP) mojosongo, WWTP pucangsawit and WWTP Semanggi will be developed to treat 30% domestic waste of Surakarta Residence. While on-site system will are served 70% residence by service programs Regular Cleaning Septictank. The toughest challenge is how to increase community participation in waste management and improve the company"s financial condition. Sanitation sustainable development is going to happen if supported by facility development also good, institutional development, the arrangement that oversees, and the public participation.

1. Introduction

Sustainable Development Goals (SDGs) has placed availability and sustainable management of water and sanitation as the sixth goal. Open defecation were practiced by 892 million people (12 per cent of the global population), continues to pose a serious health risk[1]. Large cities in the world obtain 78 ± 3% of their water from surface sources, some of which are far away: cumulatively, large cities moved 504 billion liters a day (184 $km^3 yr^{-1}$) a distance of 27,000 ± 3800 km [2]. Urban growth like in Indonesia Cities is increasing the demand for freshwater resources, hampering efforts to assess the sewerage can causes of urban water stress.

The high level of domestic pollution will impact significantly to the quality of the health of people living along flood plains such in Surakarta area [3]. The development of sanitation has an important function as it pertains to health, lifestyle and environmental conditions that can provide comfort in daily life. This problem is the responsibility of the Indonesia government and society. Evaluation of wastewater management in developing countries is important, where sanitation conditions are usually poor[4]. Indonesia is developing countries with a backlog in achieving targets for the implementation of wastewater collection, treatment and recovery systems [5]. The Challenge of development waste water treatment in developing country is very complex. Infrastructure still weak and human resources is limited. In other side, people still get bad habits throw rubbish into the river. We will found easily trash so much along river. Sometimes trash can block the water flow and cause flood. Hence, waste water treatment need to be addressed precisely and sustainable, so the water out safe into a body of water for public health and the environment.

A number experiment in qualitative or case study design research support implementation sustainable development infrastructure in domestic waste is still rare in Indonesia. However there are few studies using case study in Indonesia city that have complex challenge, social rejection, and developing country problem. There has been limited use of studies such consistency of regulation government, weak of implementation and complex relation of government department. The present study will focus describe detail strategy to implement sustainable infrastructure in domestic waste. The current study aims to determine total amount domestic waste produced, describe strategy how to treat waste by system on site and off site and strategy planning to implement.

This paper empirically examines the governance needs of strategy to develop domestic waste infrastructure, using Indonesian water sector and the challenges it faces as a rich source of insight. Derived from extensive qualitative research drawing on the projection waste water produce, the paper tests the current governance approaches considered suitable for sustainable infrastructure. Finally key questions and challenges arising from this research are identified

2. Methods

Surakarta is one of the cities in Central Java with total area of 44.4 km2, with a population of 512,226 inhabitants, with a density of 11,631 people/km^2[6]. The research methodology carried out in several phases includes preparation, data collection, processing and data

* Corresponding author: arya_tl@ft.undip.ac.id

analysis, conclusion. Primary data collection is done by direct observation in the field, calculation of water source pipeline system and WWTP. While the secondary data obtained through interviews, documents from relevant respondents.

First, calculate population growth projection with comparative study of regression analysis of linear, exponential and logarithmic statistical models. Data input uses the population for the previous ten years. The method projection base on Regression modeling strategies: with applications to linear models, logistic and ordinal regression, and survival analysis book. Projection Demand Wastewater used to determine the amount of water demand for the next 20 years (2016-2030)

Second, analysis domestic waste water existing condition use Strengths Weaknesses Opportunities Threats Matrix. SWOT analysis aims to identify the key internal and external factors seen as important to achieving an objective. SWOT Analysis is instrumental in strategy formulation and selection. It is a strong tool, but it involves a great subjective element. It is best when used as a guide, and not as a prescription. The method analysis base on SWOT Analysis. Idea, Methodology And A Practical Approach [7]. Limited resources and investment requires strategy to make prior planning. The priority use scoring method base on aspect of health, infrastructure, pollutant load, and environmental condition.

On site system is taken into account with the projections by 2020 will be served by 70% using the method of Fecal Sludge Services scheduled and declined to 40% in 2035. Waste water is waste water derived from the households are feces human waste of neighborhood (real estate), restaurants, office, commercial, the apartment and dormitory. The domestic wastewater needs to treat before discharge to environment. Total wastewater load is calculated using a formula to use a formula as follows 80% of water supply consumption. The total demand water supply meet water Work Surakarta City. When urban village serviced by the pipe waste water service means the area included in the recommendations priority off-site system, while areas have not passed by a pipe service included in on-site service system.

3. Result And Discussion

In the UN in 2017 report said 600 million people still used a "limited" service an improved facility shared with other households. 2.9 billion people used a "safely managed" sanitation service, defined as a basic facility that safely disposes of human waste and another 2.1 billion people used a "basic" service—an improved facility that is not shared [1]. Household waste water or dirty water from community activities Surakarta ranks highest in the discharge of about 89%, and then 11% of industry and hospitals [3].

Waste water infrastructure is one of the most important socio-technical systems underpinning the development and functioning of modern cities like in developing countries [8]. Quality of river has socioeconomic relationship of water-life quality. It can be measured by several indicators like communities and population access to water sources, water influence on the health of the population, water use, water quality, public perception on the supply of water[9].

Base on SDGs target, Indonesian government have target to reduce lack sanitation access until 0% in 2019. The urban poor are also highly affected by the environmental problems such as inadequate urban services (water, sanitation) and poor housing conditions. Surakarta City get priority sanitation urban planning from Indonesian government in Presiden Joko Widodo era. The city of Surakarta in the province of Central Java has emerged as a very prominent and celebrated case of policy innovation within and beyond Indonesia[10]. Several program to reduce open defecation to river had been doing. "Bedah Kampung" Program was Relocated people in Solo began in 2011 and built upon the existing interest of the municipality in conserving and revitalizing a traditional waterfront market. Program has contributed to sustainable housing development of the area [11]. The relocation of squatters notably from the banks of the Bengawan Solo River to self-built permanent new homes partly subsidized by the city government and complete with property rights was not, a prelude to lucrative private sector development but instead oriented towards the creation of new civic spaces within the city [10].

Waste water services in the city of Surakarta, should be developed in order to be able to handle the waste water load caused by everyday household activities, commercial (commercial) activities, and social events. Appropriate time span of planning in this master plan, the calculation of the load of waste water services, consisting of the total waste water discharge and the organic load, will be performed for 20 years into the future (2016 – 2035). Base on city environment report, A half river in Surakarta City was polluted by domestic waste in middle risk. The value of COD rise until 80 mg/l and even more than 100 mg/l in several places. The colour of water look dark, smelly, and high turbidity. The impacts of discharges waste water directly on aquatic ecosystems include eutrophication, deoxygenation, fish kills and accumulation of toxic metals and compounds in sediments, plants, fish and other animals[8].

The Surakarta waste water management plan is prioritized in the future using offsite systems. Service levels for offsite systems are designed to increase from year to year. As for the system onsite as much as possible maintained at the maximum sludge waste as 250 m^3/day on the target service 70% population by 2020. Besides the offsite system is a system that has advantages in controlling the quality of waste water. So that waste water discharged into the environment / river body will be more maintained quality.

On site system uses the septic tank that located in every house. Onsite systems in Surakarta city has grown to include Septic tank, Communal households WWTP, industry WWTP. Future planning, all onsite household will become a customer of sludge service regularly

https://doi.org/10.1051/e3sconf/20183105003

program. Every 3 year, costumer of water work will automatically maintain to be waste water service. Sludge in septic tank will be drained by truck and deliver to domestic sludge handling unit. The location is on side of final disposal Putri Cempo. Unit process sludge handling consist of imhoff tank, anaerobic pond, facultative pond, maturasi pond and sludge drying bed.

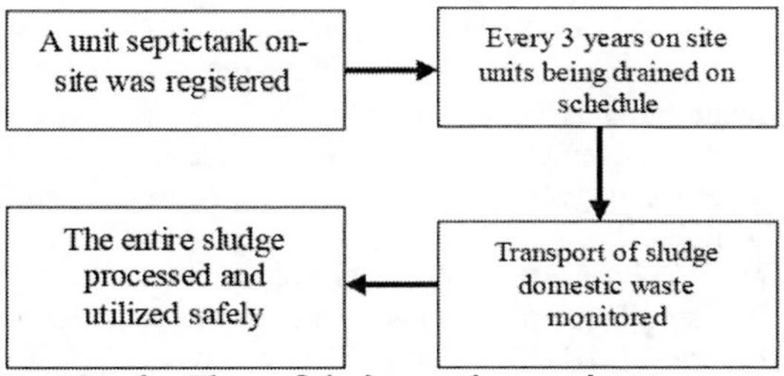

Fig. 1 Service Flow of sludge service regular program

Communal system, area system, and the system of the region need waste water treatment facilities in order to lower organic compounds and solids (suspended solids) that is contained in waste water until it reaches the target processed (effluent) desired. Installation of some more sophisticated designed also capable of lowering the polluter other content, as the nutrient (a compound of nitrogen and phosphate). Installation of water for domestic waste almost always apply biological process to lower the womb of organic compounds, both in the aerobic, anaerobic, and facultative. The cost estimates for the assessment of required wastewater investment and operational budgets in the planning process depend on land prices and revenues from recovered resources that were varied. It showed that the effect of high land prices on land intensive WWT systems have a major impact[5]

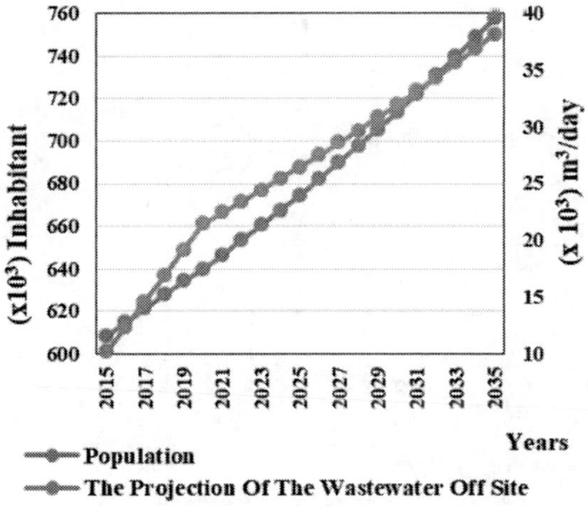

Fig. 2 The Projection Of The Wastewater Off Site

Centralized system or off site system is a system where the waste water from the whole area of service collected in collectors riol, which then streamed into the city riol to processing facility and thrown into a body of water recipients. Now, All WWTP was serving 14,000 connection, or less than 15 % from Surakarta Population. The plan targeted offsite system service levels can serve 30% in total population in the year 2019. It will then be increased 41% in 2035. Domestic waste water services in the city of Surakarta was divided into 3 zones. In the Southern area there are WWTP Semanggi with capacity of 60 l/s serving about 8928 connection. In the central there are WWTP Pucangsawit 40 l / s serving about 576 connection. And in northern area there are WWTP mojosongo with capacity of 50 l / dt and WWTP mojosongo with capacity 60 l / dt serving 5420 connection. Offsite system development plan starting from Enhancing the capacity of waste treatment Installations. Perform optimization of the WWTP mojosongo from 50 l/sec to 70 l/sec. build capacity with 2 WWTP Semanggi 100 l/sec. at this stage it also needs revamping and infrastructure, institutional, regulatory, community participation and dissemination Governments against the system.

The second phase target wastewater services city of Surakarta is 34% off site system of 297 l/sec with 38,216 Household. At this stage it is expected Surakarta sanitation can be increased with reduced waste in rivers, as well as the absence of households dumping waste water without being processed.

The third phase wastewater services target the city of Surakarta is 37% off site system of 342 l/sec with 43,994 KK. At this stage are targeted over the function of water quality to be well marked with the development of fish endemic to the river that is an indicator of good water. In addition to that obtained in ground water quality are not contaminated by the bacteria e-coli. The fourth stage waste water services target city of Surakarta is 41% off site system of 403 l/sec with 51,763 KK. In addition to improving service levels, at this stage it is also targeting the improvement quality of service by using the latest technology. The number household connection of sewer will improve the quality of water river [12].

In other side, projection waste water on site system of waste water are on the site in 2015 – 2019 increase very rapidly. But in the year 2020 - 2035 the resulting fecal discharge tends to be reduced. However by the year 2020 - 2035 discharge the black water which are produced tend to be lacking. This is due to the increase of the population of the system on the site will be maintained at influence of 250 m^3/day starting from the year 2020 – 2035 and the addition of residents will be on point with the system of the site. In future, capacity of domestic sludge handling unit reaches 250 m^3/day are still far from enough because the condition woke up now only 90 m^3/day.

There are two main reasons the need for the development of the sewerage system, such as: first meanwhile residential environment, and the degree of human health and to achieve the target the millennium development goals sector sanitation [13]. In Aprvous research, Gajah Putih river be a place emissions waste from activity for settlement and some plants, causing a decrease in the quality of water in the river [14]. In addition to the Gajah Putih river, the domestic waste water into the Pepe River still up of quality standard according to local regulations central java no. 5 2012, so that required planning the sewer and wastewater treatment domestic to prevent water pollution river by using land in border the river [12].

Onsite systems development plans divided into five stages. The first phase is targeting an improvement in healthy 86% of the toilets worked 100% in areas all over the city. Target customer sludge service regularly program reached 70% of the population of the city of Surakarta or 447,809 the soul of domestic sludge handling unit needs 250 m^3/day. At this stage of the need for improving the infrastructure, institutional, regulatory, community participation and dissemination Governments against the system. The second stage – fourth, retained on site system remains on the design discharge 250 m^3/day.

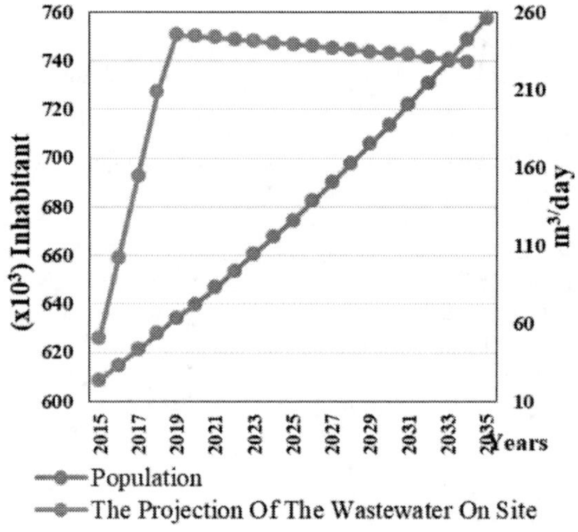

Fig. 3 The Projection Of The Wastewater On Site

Service system determined from a map service areas of waste water city of Surakarta. When urban village serviced by the pipe waste water service means the area included in the recommendations priority off-site system, while areas have not passed by a pipe service included in on-site service system. Zone priority on site and off site can be seen in Table 1. Scope of priority zone off site is presented in Figure 3.

The shift from traditional, centralized, large sanitation infrastructure to alternative and distributed technologies is widely accepted as essential sustainable water management concept. The challenges facing urban water management, such as wastewater treatment technologies, models and assessment tools, the shift from traditional to more sustainable approaches remains slow significant needs advances in developing technical solutions. [8], [15]. Surakarta City will develop domestic sludge handling system onsite in every district. The proper location will be chosen to decrease distance transportation truck. Despite the land acquisition problem because the price and social conflict.

Opportunities manifest in areas where high resource recovery potentials from municipal wastewater and solid waste match their demand, such as the urban areas on Java. Projection in 2035, the resource recovery from this can supply significant fractions of the Indonesian demand for phosphorus (15%), compost (35%), duckweed (7%), plastic (66%), and paper (18%) [16].
Development done by means of stages, despite the limited resources and the other needs also urged. Hence, there needs to be a priority areas services will be built. Several criteria that important but not included as the conditions for the city Surakarta relatively ramps and be mostly have had the system of waste off site. The overall health have data similar in every sub district and less visible to scoring.

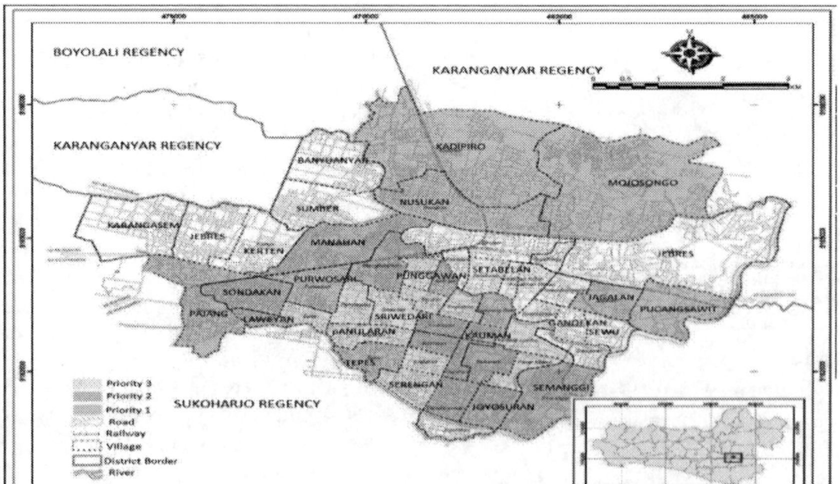

Fig. 4 Priority Zones Off Site

For stability in water work company that handle waste program, serious effort should be made by the top management to clearly identify organization visions and mission, objectives, past records. They should consider major strength and weakness and how handled it. What opportunities they have and how they utilized it and finally the threat and how they handled it [7].

Solo water work company have a job to handle sewerage treatment in there. One of the toughest challenges from outside is the lack public awareness of treating domestic waste. So, it makes low willingness to pay contribution. The financial of operating treatment plant and organization is deficit. The challenges from within are bad condition of infrastructure of sewer pipeline and lack of competence of employees.

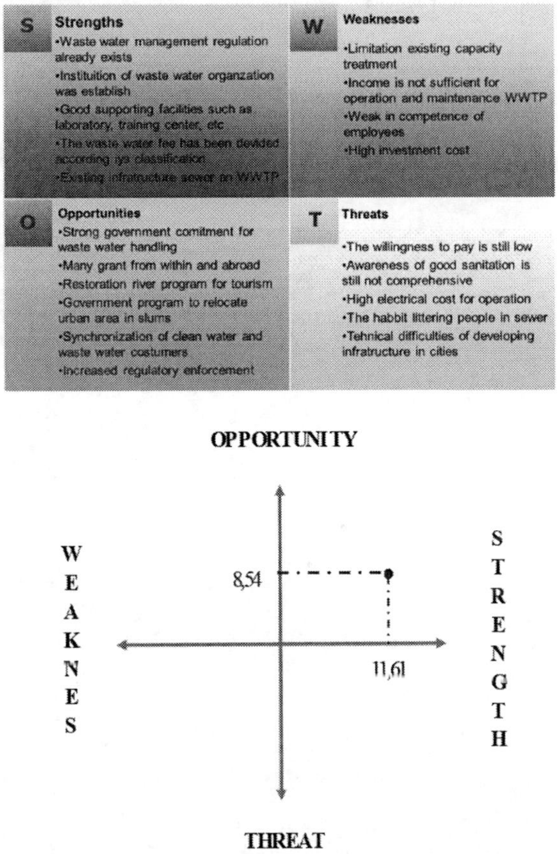

Fig 5. SWOT Analysis Domestic Waste Water Organization on Surakarta City

Water Work of Surakarta City is currently in position III (Growth / Aggressive), aggressive, meaning the organization is in good condition and steady so it is possible to continue to expand, increase growth and achieve maximum progress. Thus the Water Work can harness the power it has to maximize available opportunities. It should be noted that in the internal aspect, the contributors of positive values are mostly because Water Work have excellent value in terms of corporate management, engineering and non-technical or financial systems and the company's operating system.

On the external side it is known that the main cause of more open opportunities than threats is due to several things. The good economic condition of the community certainly encourages high interest to subscribe to Water Work are the things that make the opportunity of Surakarta City Water Work to expand its business quite open. Development urban water utilities are required to consider broad environmental impacts and water cycle interdependencies of water supply investments and yet quantitative assessments of broad impacts [17].

4. Conclusion

The establishment, domestic waste management systems that is Surakarta divided into two types of system on-site and system off site. Service system off site divided into 3 service areas which is stored on WWTP mojosongo, WWTP pucangsawit and WWTP Semanggi. While on-site system will are served by service programs Regular Cleaning Septictank. The target of waste water service the city of surakarta in the year 2035 is 41% system off site of 403 L/s with 51.763 household and 59% system on site amounting to 250 m^3/day. Sanitation sustainable development is going to happen if supported by facility development also good, institutional development, the arrangement that oversees, and the public participation. Water Work of Surakarta City is currently in position rapid Growth that mean the organization is in good condition and steady so it is possible to continue to expand, increase growth and achieve maximum progress. The toughest challenge is how to increase community participation in waste management and improve the company's financial condition.

Reference

1. UN, "The Sustainable Development Goals Report," pp. 1–56, 2016.
2. R. I. Mcdonald *et al.*, "Water on an urban planet : Urbanization and the reach of urban water infrastructure," vol. **27**, pp. 96–105, 2014.
3. R. P. Lestari, *"water sampling quality in Mojosongo water treatment plant,"* Surakarta, 2011.
4. L. Lam, K. Kurisu, and K. Hanaki, *J. Clean. Prod.*, vol. **104**, pp. 185–198, 2015.
5. S. M. Kerstens, I. Leusbrock, and G. Zeeman, Sci. Total Environ., vol. **530–531**, pp. 53–65, 2015.
6. S. G. Board, *Surakarta in numbers 2016*. Surakarta: Statistic Goverment Board, 2017.
7. I. O. R. and N. J. Ifediora Christian Osita, "of Innovative and," *Organ. Stab. Product. role SWOT Anal. an Acron. strength, weakness, Oppor. Threat*, vol. 2, pp. 26–33, 2014.
8. S. Bell, "Renegotiating urban water," *Prog. Plann.*, pp. 1–28, 2014.
9. S. Iacob, "The Wastewater - a Problem of Integrated Urban Water Management," *Procedia Econ. an Financ.*, vol. 6, no. 13, pp. 436–443, 2013.
10. N. A. Phelps, T. Bunnell, M. A. Miller, and J. Taylor, Cities, vol. **39**, pp. 37–49, 2014.
11. W. Astuti and D. A. Prasetyo, Procedia Environ.

https://doi.org/10.1051/e3sconf/20183105003

Sci., vol. **20**, pp. 593–601, 2014.
12. R. Januarita, G. Samudro, S. Pepe, and A. L. Belakang, J. Tek. Lingkung., vol. **5**, no. 4, 2016.
13. P. W. Ministry, *Regulation of Public Work Ministry No: 16/PRT/M/2008 about strategy development domestic treatment system*. Jakarta, 2008.
14. Y. K. Panjaitan, *"Structure population Guppy fish (Poecilia Reticulata Peters) in Gajah Putih River* Surakarta," Salatiga, 2016.
15. S. J. Van De Meene, R. R. Brown, and M. A. Farrelly. Glob. Environ. Chang., vol. **21**, no. 3, pp. 1117–1127, 2011.
16. S. M. Kerstens, A. Priyanka, K. C. Van Dijk, F. J. De Ruijter, I. Leusbrock, and G. Zeeman. Resour. Conserv. Recycl., vol. **110**, pp. 16–29, 2016.
17. J. M. Kandulu, J. D. Connor, and D. Hatton. J. Environ. Manage., vol. **145**, pp. 43–53, 2014.

Waste Bank Revitalization in Palabuhanratu West Java

Budi Prasetyo Samadikun[1], *Dwi* Siwi Handayani[1], and *Muhamad* Permana Laksana[1*]

[1] Department of Environmental Engineering, Faculty of Engineering, Diponegoro University, Semarang - Indonesia

Abstract. Palabuhanratu Village has three waste banks, one of them was established since 2010, the others built in 2016. However, waste processing from the source is still not optimal, it's only reduced waste about 5% of the total waste generated to the final waste disposal site. The performance of waste banks is still minimal, because one waste bank can not serve the entire area of the village. Furthermore, organic waste processed by some communities of Palabuhanratu Village to be compost can not be a mass movement, due to the lack of public knowledge. The purpose of this research is to know the existing condition of waste management in Palabuhanratu Village and to formulate the revitalization of existing waste bank. The research used survey research method by using questionnaire, in depth interview, and observation. Analytical technique using quantitative and qualitative analysis. The findings of the research indicate that the residents of Palabuhanratu Village who often do waste sorting from the source only from the residents of RT 01 / RW 33. The number of existing temporary waste disposal site in Palabuhanratu Village is still lacking, so it requires addition up to 5 units that integrated with waste bank in this village.

1 Introduction

Waste management that are not going well , would cause the decreasing quality of the environment. In SNI (Indonesia national Standard) 19-2454-2002 it was stated that solid waste consist of an organic and inorganic substance , regarded as useless and must be managed so that could not harm the environment and protect development investment [1].

According to interviews with Environmental Agency (BLH) Sukabumi Regency [2] said that there are three waste bank which is in Palabuhanratu Village, one of them have been constructed since 2010 and two others began to stand up since the beginning of 2016 .Waste volume in 2015 is 134,89 m3 /day that includes 89 % Palabuhanratu Villages area. Waste produced is from domestic activities, activities in fish auction venue, public facilities, with the varying quantity and composition. However, waste management from the source is not optimal because waste reduced still about 5 percent of the total waste that sent into final disposal site (FDS) Cimenteng which is 70 km away from Palabuhanratu Village.

Far mileage between Palabuhanratu Village and final disposal site (FDS) Cimenteng cause high operational cost and a little ritation, so the service that conducted by the government of Sukabumi Regency not going optimally. Besides, the lack performance of the existing waste bank in Palabuhanratu Village, one waste bank only able to serve one Rukun Warga (RW). Apparently, three temporary disposal site is not able to serve all areas of Palabuhanratu Village.

In Indonesia, the *Rukun Warga* (RW) is the lowest hierarchy of community organizational system which can implement creative and innovative arrangements to support solid waste management activities with less financial requirement [3]. But, a review of the institutional aspects shows that the existing waste bank Palabuhanratu Village still do not have a clear structure. In daily operation, there has been no party responsible for the sustainability. In addition, organic waste processed by some communities of Palabuhanratu Village to be used as compost, cannot be applied by all people in society, because of limited knowledge of the community to process organic waste become compost.

Seeing the phenomenon that occurred in Palabuhanratu Village, it is necessary to revitalize the waste bank in order to better function in waste management in Palabuhanratu Village. Attention and public awareness has become an important factor in supporting the success of waste management, so the environment quality in this village is maintained. Indrianti stated that high participation from the community supported by harmonious relationship among community members enables the waste bank to run effectively[4].

This paper purposes to know the existing condition of waste management in Palabuhanratu Village Sukabumi Regency and also to formulate existing waste bank revitalization efforts as a support party for waste management in final disposal site (FDS) Sukabumi Regency.

2 Research method

The research used survey research method by using questionnaire, in depth interview, and observation.

* Corresponding author: budisamadikun@gmail.com

Analytical technique using quantitative and qualitative analysis.

Research site located in Palabuhanratu Village Sukabumi Regency. The existing temporary disposal site located in three RW, namely RW 24, RW 30, RW 32 Palabuhanratu Village.

The number of respondents as many as 100 people spread over 35 RW, determined by purposive sampling technique. Criteria of respondents are the Head of Family (KK) who live in Kelurahan Palabuhanratu with the highest population density and found waste scattered in the environment.

3 Waste bank revitalization

3.1 Waste bank

Based on the Regulation of the Ministry of Environment Number 13 of 2012 [5], Waste Bank is a place for sorting and collecting of waste that can be recycled and / or reused which has economic value. Waste Bank can serve as implementing reduce, reuse, and recycle. The scope of Waste Bank includes:

1. Waste bank requirements
   Based on the Regulation of the Ministry of Environment Number 13 of 2012 [5], the requirement to establish a waste bank should have a building and management system. The management system required in the waste bank is the name of the waste bank, the address of waste bank, the waste bank depositor, the waste bank administrator, the waste collector / buyer / recycling industry, waste management in the waste bank, and the role of the waste bank administrator.
2. Waste bank working mechanism
   Waste bank working mechanism includes waste sorting , waste delivery to waste bank, weighing of waste, recording, waste result of sale submitted into saving book, and sharing of waste sale between depositor and executor.
3. Implementation of waste bank
   Implementation of the waste bank includes the determination of working hours, withdrawal of savings accounts, borrowing money, savings books, waste collection services, savings types, waste type, pricing, waste conditions, minimum weight, and waste containers.
4. Waste bank administrator
   Waste Bank administrator have a role as facilitator in construction and implementation of waste bank, providing waste collector / buyer data for Waste Bank, providing recycling industry data, providing reward for Waste Bank

3.2 Revitalization

Gouillart and Elly in Santoso [6] argue that the essence of revitalization is to revive or reenact a place or organization that has potential assets. Organizational revitalization as a planned change takes place through a long-term process that is divided into systematic and detailed planning phases. It is not only limited to physical revivals such as completion of infrastructure, utility support or other development, but also the planning of new creative and innovative activities that have been prepared along with its management mechanisms. Organizational revitalization can be achieved through 3 approaches, ie :

a. Achieve Market Focus
b. Invent New Business
c. Changing The Rules trough Information Technology.

3.3 Lesson learned from application of waste bank

3.3.1 Waste bank in Depok and Surabaya Indonesia

Despite the lack of success of government in solving waste problem, some communities in Depok established what so called '*Bank Sampah*' (waste bank). Waste bank is a bank that uses inorganic waste as deposit. It emerges as a community-based program driven by waste bank administrators who work on voluntary basis with little incentive in terms of money. Established in 2011, now there are more than 500 waste banks in Depok Municipality. During June 2013 – December 2013 alone, the number of waste bank increased from 60 to almost 400 locations (7).

Waste bank program as community initiative supports government program by changing people's behavior to sort their waste. Both inorganic and organic waste creates economic value after the sorting: the inorganic waste will be recycled and the organic waste will be composted. The organic waste is composted either in UPS (waste treatment facility for organic waste) or through homemade compost. Depok Municipality targeted that in 2014 Depok will have 2,000 locations to reduce 40 percent of the waste generated in Depok.

Waste bank as a business is owned by people who consider waste as a valuable economic commodity and savings, has instruments that involving community in waste management. In Surabaya, waste bank grows rapidly and has supported community's livelihood and encourage people's self-reliance in environmental management [8].

3.3.2 Waste bank in Malaka Sari Village Jakarta and Kepanjen District Malang Regency Indonesia

Waste Bank Malaka Sari, is one of the waste bank in Jakarta. Currently, the waste bank located at *Jalan* Delima III Number. 190 Malaka Sari Village, Duren Sawit Sub-district, East Jakarta, has been a pilot in waste management both from within and from abroad [9].

Since 2009, residents of Malaka Sari Village in East Jakarta do community-based waste management by applying the principles of 3R (Reduce, Reuse, and Recycle) as a social engineering. As a social movement, the waste bank is built upon public awareness of the importance of waste. The main goal of waste bank is reducing solid waste at community level & increasing effectiveness of integrated waste.

Citizen gets money from the sale of waste. The price of the waste varies; for instance, cardboard sold for USD 0.10-0.15 per box, soft drink cans sold for USD 0.44 per can, and so on. Then, this waste is processed by the caretakers. The organic waste is processed into fertilizer, while other waste is recycled into craft items such as handbags, wallets, tissue boxes, umbrellas, and more. All processed products are sold back. In 2012, business profits reached $ 698.21. Through this method, all parties gain mutual benefits. Customers are increasingly eager to collect as much waste. The more waste collected, the more money earned.

Waste reduction through waste bank has been initiated in Kepanjen District Malang Regency [10] because of the landfill in this area is approaching its maximum capacity. Kepanjen District has three kind of participation rate which are low, medium, and high participation. Consequently, the scenarios proposed are pessimist scenario, fair scenario, and optimist scenario for low, medium, and high participation respectively. Pessimist scenario focuses on avoiding the community to do open burning causing air pollution and health injury. Meanwhile, the fair scenario starts to introduce the institution to the community to conduct the capacity building of the community in waste separation. The optimist scenario proposes the implementation of waste bank involving the community.

3.3.3 Waste bank in Ban Nong Fab School, Map Ta Phut, Rayong province, Thailand

The success of the first Waste Recycle Bank project in Ban Nong Fab School, Map Ta Phut, Rayong province has encouraged Indorama, the world's leading manufacturer of integrated polyester chain, to expand the project to cover two more schools in Rayong ne [11]. Indorama started the project in November, 2010, in Ban Nong Fab School. Following the success of the maiden project, the company now plans to run waste recycle banks in two more schools in Rayong province next year.

Richard Jones, head of investor relations and corporate communications, recently said that as a manufacturer of polyethylene terephthalate (PET), a raw material for manufacturing plastic drinking bottles, Indorama can support the recycling to reduce waste. Indorama has supported the school by inviting experts in waste recycling to educate the students and teachers on how to select and screen garbage. Meanwhile, teachers in the project have coordinated with recycling companies in the province to buy waste from the bank.

The garbage accepted by the bank includes paper, plastic drinking bottles, and drinking cans. The purchasing price for each waste is different. Pieces of paper, for instance, get Bt2 per kilogram, general plastic garbage can be sold for Bt7 per kilogram, while PET drinking bottles fetch as high as Bt14 per kilogram.

The school is confident that this problem can be managed with the well-planned project. It foresaw that most of workers who move to work here are poor. Thus, if they can generate revenue from litter, they can increase their income while saving the environment. The project was born out of this idea [11].

4 Result and discussion

4.1 Existing condition of waste management in Palabuhanratu Village Sukabumi Regency

Based on the results of the questionnaire, the general condition of waste management in Palabuhanratu Village is 42% good and 58% enough. Public perceptions on the condition of waste management vary depending on the standards of hygiene and public awareness in waste management

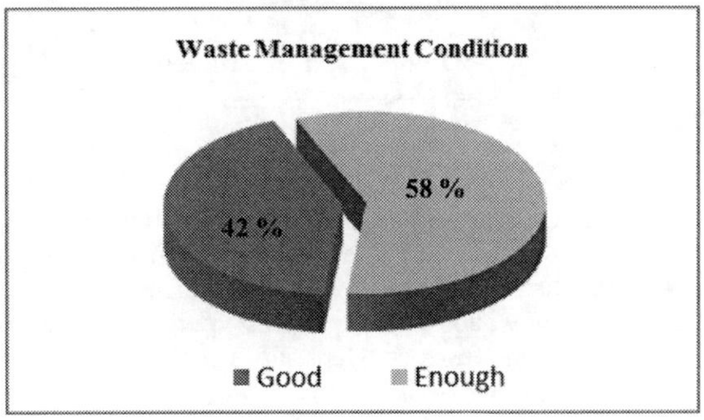

Fig. 1. Waste management condition percentage

The implementation of waste management is influenced by several factors, one of which comes from the aspect of community participation that is influenced by the community's knowledge of 3R concept (Reduce, Reuse, and Recycle) as well as the willingness and capability in the implementation of the 3R. The community's experience in waste sorting is 20% of respondents always do waste sorting, 79% of respondents do not do waste sorting and 1% of respondents sometimes do waste sorting, as seen in Figure 2. From these results can be stated that most people are still less experienced in sorting waste. One thing that is needed in waste management is the willingness of people to go directly in the waste management. This can be done like throwing waste in place, sorting from source, utilizing waste that can still be used to reduce waste generation, etc.

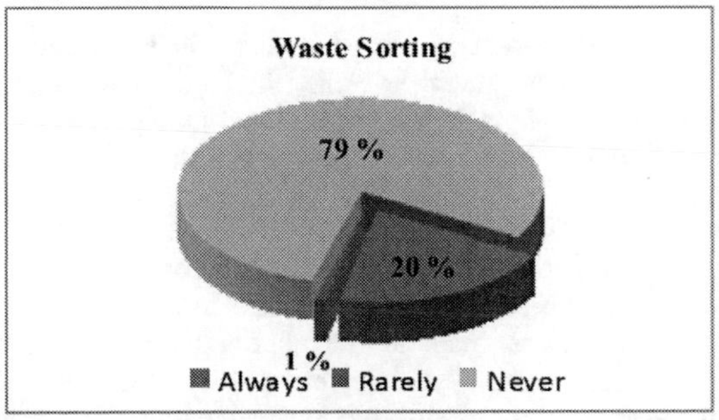

Fig. 2. Waste sorting condition percentage

Most of the residents of Palabuhanratu Village have understood in the procedure of composting, that is as much as 69% of respondents. There are 31% of respondents who do not understand the composting process is supported by field findings that are still found by residents who throw organic waste directly into the trash, as shown in Figure 3.

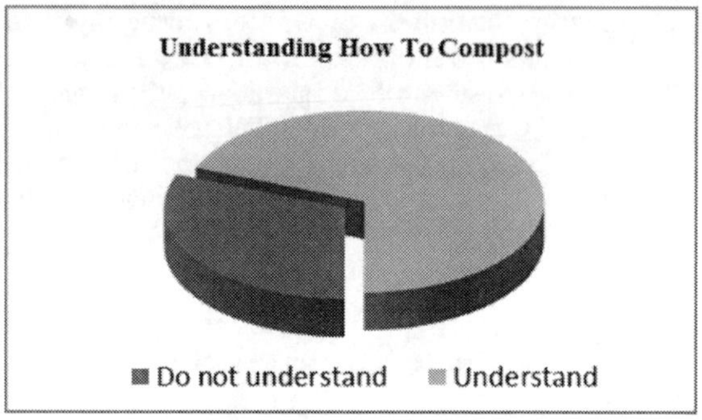

Fig. 3. Societys understanding of composting percentage

Palabuhanratu residents have a good understanding of the waste bank. This is evidenced from the results of the questionnaire, that 77% of Palabuhanratu Village residents know about waste banks, and only 23% of respondents do not know. Departure from this condition, it is expected that in the future for inorganic waste sold to waste bank can be separated from its source to reduce waste generation.

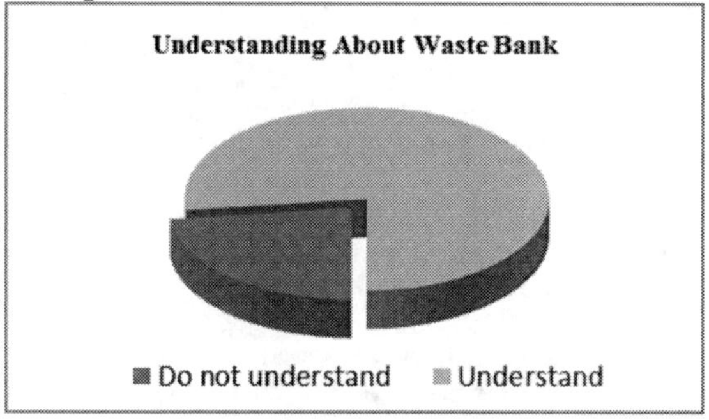

Fig. 4. Societies understanding of waste bank percentage

4.2 Ratu Indah Waste Bank

There are 3 units of waste bank in Palabuhanratu Village. Two garbage bank units are no longer operational due to lack of capital and human resources, while those that operate today are Ratu Indah waste bank.

Waste Bank Ratu Indah is located in RT 01 / RW 33 residing in the Housing of Bank Tabungan Negara (BTN) Pantai Ratu Indah. The waste bank manager has three members with a member structure, namely chairman, treasurer, and secretary. Ratu Indah's waste bank service area is still only one RT with service level of 67% that is 60 Head of Family from total 90 Head of Family. Residents in BTN housing that is not a customer of Ratu Indah garbage bank is a very busy activity so they do not have time to sort out waste. Waste Bank Ratu Indah per month can produce 200 Kg of garbage consisting of iron, plastic, cardboard, and aluminum cans. The results of the purchase of waste from residents of BTN housing serve as raw materials for making handicrafts and sold to collectors. The profit earned per month is 500 thousand. Some of the work of Bank Sampah Ratu Indah is shown in the following figure:

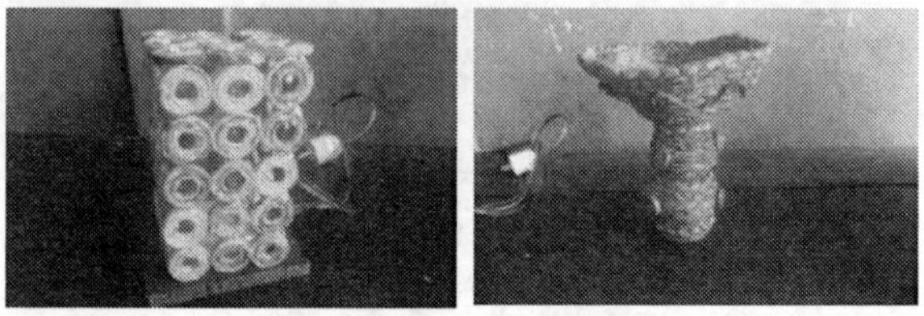

Fig. 5. Product of Ratu Indah waste bank

Based on interviews with Mrs. Maryatun as the Chairman of Waste Bank Ratu Indah stated that the purpose of the waste bank was formed is to introduce the meaning of waste and its impact to the community and also to socialize indirectly about the sorting of waste. Since the establishment of the waste bank, 60-70% of the people around the waste bank have become aware of the functions and benefits of the waste bank. The obstacles faced in establishing the waste bank is the lack of human resources to manage the waste bank so that management has not been able to work optimally.

4.3 Waste bank revitalization in Palabuhanratu

4.3.1 Establishment of NGO for Waste Management Regency

In order to support waste management in Palabuhanratu Village, an institutional structure at the village level is arranged by the community through community meetings and consists of various components of society, in the form of Non Governmental Organization (NGO). This NGO which will carry out and oversee the way of waste management accompanied by relevant agencies. The organizational structure of NGO in PalabuhanRatu Village is as follows:

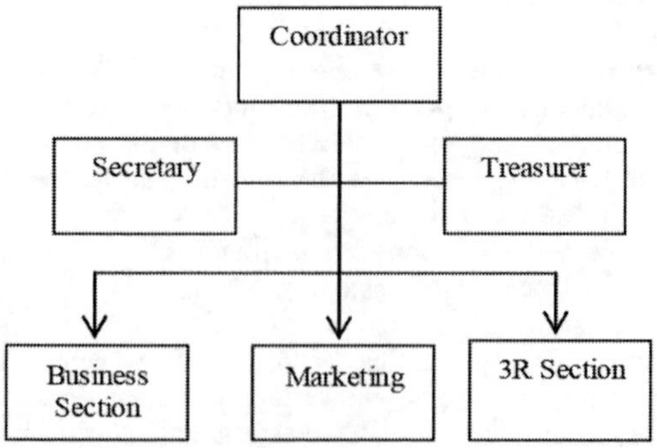

Fig. 6. Structure of non-governmnetal organization in Kelurahan Palabuhanratu

 https://doi.org/10.1051/e3sconf/20183105004

In line with that stated by Gouillart and Elly in Santoso [6] related to the three approaches in organizational revitalization, the organizational structure that plays an important role in the development or activation of Waste Bank is the Business Section, Marketing Sector, and 3 R (Reduce, Recycle, Reuse) Sector. The business field is responsible for finding customers for the garbage bank in temporary disposal site (TDS) 3R and also in charge of technical gathering. The marketing field coordinates with garbage buyers (industry, collectors) to conduct trading and also coordinate with local government and other KSM. Field 3R is tasked and responsible for doing 3R in TDS 3R, as has been done in Malaka Sari Village.

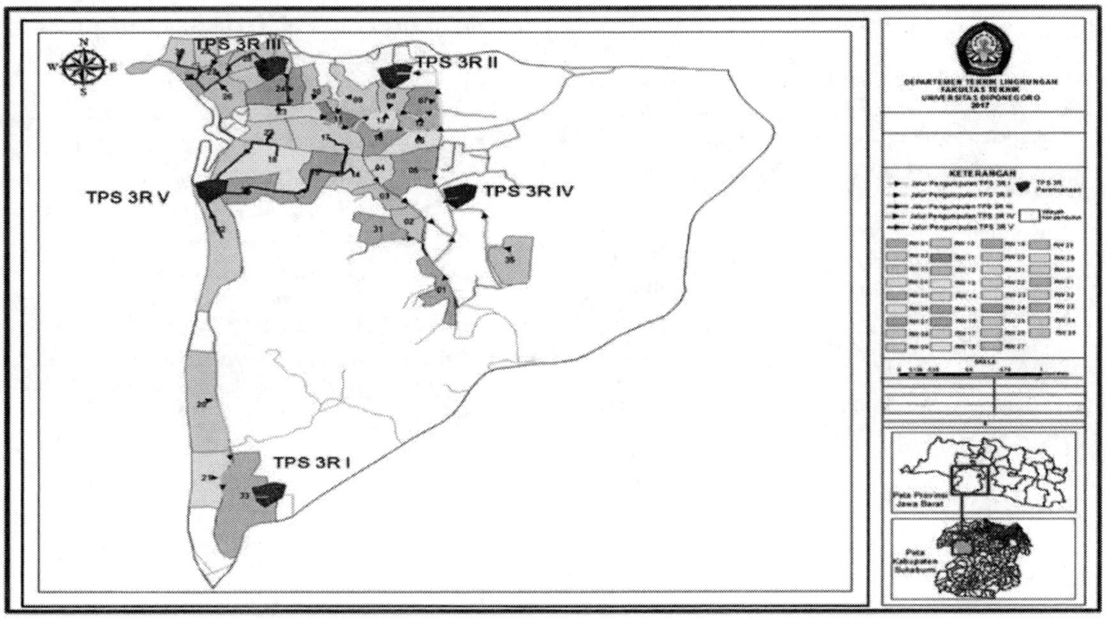

Fig. 7. Service area temporary disposal site (TDS) reduce-recycle-reuse (3 R)/waste bank

4.3.2 Increasing the number of waste bank in temporary waste disposal

Existing temporary disposal site (TDS) in Palabuhanratu Village have 3 RW, namely RW 24, RW 30, and RW 32, all of which have not functioned 3R (Reduce, Recycle, Reuse) and have less strategic location to serve waste transportation in Palabuhanratu Village.

Table 1. Service Area TPS 3R

TDS 3R	Total Service Area (RW)	Service Area (RW)	Waste Volume (m^3/day)
I	3	20,21,33	11,116
II	9	6,7,8,9,10,11,12,13,16	21,364
III	9	23,24,25,26, 27,28,29,30,34	24,390
IV	7	1,2,3,4,5,31,35	22,366
V	7	14,15,17, 18,19,22,32	20,34
Total	35		99,576

The condition and number of existing TDS inadequate must be added in number and function, that is 5 TDS which function to do 3R. Based on Figure 7 there are 5 TDS 3R integrated with Waste Bank. Each TDS 3R has its own service area, TDS 3R I serves RW 20, RW 21, and RW 33, because the location of RW is very far apart from other RW so that waste management is made separately. While for TDS 3R II-V serving more than 3 RW because of its adjacent RW location. The service area of each waste bank can be seen in the following table:

5 Conclusions

Waste landfill in temporary disposal site (TDS)and final disposal site (FDS) is not a wise solution because every day waste volume always be increased. The shelter continues to decrease, and the adverse impacts around the shelter locations are growing. Waste Bank is one of concrete efforts to minimize the amount of waste in society that has not been well managed.

Through this paper, the authors present a real example of the implementation of waste banks that have not been optimal in Kelurahan Palabuhanratu. There are two things that can be done in revitalizing a garbage bank, including:

1. to create a solid waste management structure in village coordinated by Non Governmental Organization (NGO). NGO will serve as a driver and waste bank activist in the Palabuhanratu Village

2. to add number of temporary disposal site (TDS) and functioning as TDS 3R in which integrated with waste bank

The success of waste bank revitalization will require government support to make special regulation related to waste bank, so that the managers and customers of the garbage bank have more certainty and also the spirit to continue to play an active role in this realm.

References

1. Badan Standar Nasional SK SNI 19-2454-2002 *Tentang Tata Cara Teknik Operasional Pengolahan Sampah Perkotaan*. Jakarta: Balitbang DPU. (2002).
2. Badan Lingkungan Hidup Kabupaten Sukabumi (2017).
3. Towolioel, S., Permana, A.S., Aziz, N.A., Ho, C.S., Pampanga, D.G. (2016). The Rukun Warga-Based 3Rs and Waste Bank as Sustainable Solid Waste Management Strategy. *PLANNING MALAYSIA: Journal of the Malaysian Institute of Planners SPECIAL ISSUE IV (2016)*, 181 – 196
4. N Indrianti,. *Procedia - Social and Behavioral Sciences* 224, 158 – 166(2016)
5. Ministry of Environment. *Peraturan Menteri Lingkungan Hidup No. 13 Tahun 2012 tentang Pedoman Pelaksanaan Reduce, Reuse, dan Recycle Melalui Bank Sampah*. Jakarta. (2012).
6. W.Santoso, *Revitalisasi Dewan Pendidikan dalam Peningkatan Mutu Pendidikan di DKI Jakarta*, Tesis Program Pascasarjana Ilmu Administrasi FISIP UI. (2009).
7. Halimatussadiah http://www.lpem.org/the-economic-and-environmental-benefit-from-thewaste-bank-bank-sampah-in-depok-municipalitywest-java/ Accessed on July, 12th 2017 (2014).
8. D.R., Wijayanti, S Suryani,. *Procedia - Social and Behavioral Sciences* 184, 171 – 179 (2015),
9. http://www.republika.co.id/berita/inpicture/jabotabek-inpicture/16/08/10/obp6wj314-bank-sampahmalaka-sari. Accessed on July, 12th 2017
10. D.H Purba, C.Meidiana, D.W Adrianto, *International Journal of Environmental Science and Development,* Vol. 5(2), (2014)
11. Viboonchart,Nalin.http://www.nationmultimedia.com/business/Waste- banks-in-schools-30172942.html, (2012). Accessed on July, 13th 2017

 https://doi.org/10.1051/e3sconf/20183105005

Effectiveness of Reducing P Fertilizer and Adding Fish Pond Mud Waste on Growth and Yield of Soybean in Peatland

Erina Riak Asie[1], *Nyahu* Rumbang[1,*], *Sih* Winarti[1], *Soaloon* Sinaga[1]

Faculty of Agriculture, University of Palangka Raya, Indonesia

Abstract. The objective of the study was to assess the effectiveness of P fertilizer reduction and the addition of fish pond sludge waste on the growth and yield of soybean crop in peatland. Research used Complete Randomized Design factorial with two factors. The first factor was the reduction of P fertilizer from the dose of 150 $kg.ha^{-1}$ consisting of 4 levels, namely P0: 100% (2.944 g/polybag), P1: 75% (2.208 g/polybag), P2: 50% (1.472 g/polybag), and P3: 25% (0.736 g/polybag). The second factor was the addition of fish pond mud waste (L) from the dose of 15 $ton.ha^{-1}$ consisting of 4 levels, namely L0: 25% (73.595 g/polybag), L1: 50% (147.19 g/polybag), L2: 75% (220.78 g/polybag), and L3: 100% (294.38 g/polybag). Each treatment combination was replicated 3 times to obtain 48 experimental units. The results showed that (1) fish pond mud waste was effective to reduce the use of P fertilizer, (2) the reduction of P fertilizer up to 50% from recommendation dosage by addition of fish pond sludge waste at 75% dose of 15 ton/ha was the best combination due to providing the best plant growth and the highest P concentration of plant tissue. The highest number of pods and weight of seed obtained in the combination were 60.33 pods/plant and 7.30 g/plant, respectively.

1. Introduction

The soybean crop (*Glycine max* (L.) is a high-quality vegetable-producing plant of high-quality vegetable protein, but until now, the domestic soybean requirement is highly dependent on imports. Efforts to reduce the volume of soybean imports are the extensive cultivation system and yield improvement through the application of appropriate cultivation technology.

The conversion of peatlands into agricultural land has been increasingly difficult to avoid as a result of the food availability fulfillment. Peat swamp forest conversion to other land uses comes at a great cost in terms of large quantities of greenhouse gas emissions [1]. Utilization of peatlands as agricultural land has threatened the ecological function of peat as a carbon store, because it releases CO_2 into the atmosphere, which contributes substantially to greenhouse gases.

Peatland utilization as agricultural land is constrained by peat fertility low levels so that application of high amount of lime and inorganic fertilizer input should be conducted. Substantial organic matter decomposition occurs in both oxic and anoxic conditions in tropical peat [2]. Rewetting of organic soils results in a decrease in CO_2 and N_2O emissions, DOC losses and GHG emissions based on the global warming potential but also leads to an increase in CH_4 emissions [3].

Another alternative used as ameliorant as well as substitution of the plant nutrient, especially P element is fish pond mud waste. Aquaculture activity produced significantly N and P affecting not only the quality of peatland water, but also the environment as a whole. The levels of N and P on fish food and the efficiency used by fish affected the amount of these elements excreted into the environment [4].The other research showed that the nutrient content of P in fish pond sludge was very high, ie 309.68 ppm and 6 times higher than P in peat soil about 40.93 ppm. Furthermore, the combination of fish pond mud sediment and peat resulted in a sharp growth of red dragon fruit plant in line with the increasing volume of sediment pond sludge [5]. Combination of mud and fish pond water before planting sweet potato and corn may increase the C-organic, N-total, P-Bray, CEC, and base content in the soil [6]

Mud of fish pond is one of aquacultural waste that has not been widely used for agricultural activities, because farmers have not known the potential nutrient content in fish pond mud. However, fish pond mud was rich in humid content and feed residues that contained lots of nutrients N, P, and K [7]. Therefore, mud of fish ponds was potential to be utilized as ameliorant and to

* Corresponding author: nyahurumbang@yahoo.com

substitute inorganic fertilizers, especially P fertilizers that may reduce the use of inorganic fertilizers which in turn may reduce production costs buying fertilizer and lime. In addition, the utilization of fish pond sludge waste for crop cultivation was not only proposed at improving peatland chemical properties, but also as waste management efforts.

2. Materials and Methods

The study was conducted from May - September 2016 in Palangka Raya, Central Kalimantan Province. The materials used were varieties of Wilis soybean seeds, fish pond mud waste, SP-36 fertilizer, Urea, KCl, and Furadan. The equipments used were pH meter, meter, ruler, paper label, analytical scale, polybag size 50 cm x 50 cm. Complete Randomized Design (CRD) factorial with two factors was applied. The first factor was the reduction of the dosage of SP-36 (P) from the recommended dose of 150 kg/ha, consisting of 4 levels, P0: 100% (2.944 g/polybag), P1: 75% (2.208 g/polybag), P2: 50% (1.472g/polybag), and P3: 25% (0.736 g/polybag). The second factor was the addition of fish pond mud (L) from the dose of 15 ton/ha, consisting of 4 levels, namely L0: 25% (73.595 g/polybag), L1: 50% (147.19 g/polybag), L2 : 75% (220.78 g/polybag), and L3: 100% (294.38 g/polybag), so there were 16 treatment combinations with three replications. The data were analyzed using variance analysis (F test) at $\alpha = 5\%$ and followed by LSD test. In all Figures, the same small letters upper each column were not significantly different according to the LSD test $\alpha = 0.05$.

3. Results and Discussion

3.1. Root Volume

The effect of the interaction between the reduction of P fertilizer and the addition of fish pond sludge was significantly shown. The volume of soybean roots at the lowest dosage of fish pond sludge (25%) at various given P fertilizer doses was quite different from that of the volume of plant roots given fish pond sludge at higher doses (Figure 1). This was closely related to the fish pond mud ability to improve the fertility of peat chemicals such as increasing pH and increasing the amount of nutrients in the planting media. Accumulation of fish pond sludge was rich in nutrients and organic matter and potentially as ameliorant. Porous texture of peat and fine texture and clay from fish pond mud may be complement each other in improving the physical properties of planting media, so that aeration and absorption capacity of water and nutrients may be better, which in turn the roots of plants may grow well [7]. The combination of P-fertilizer application at moderate dose (50%) and fish pond mud at 75% was able to give the best root volume, ie 19.66 mL/plant (Figure 1)

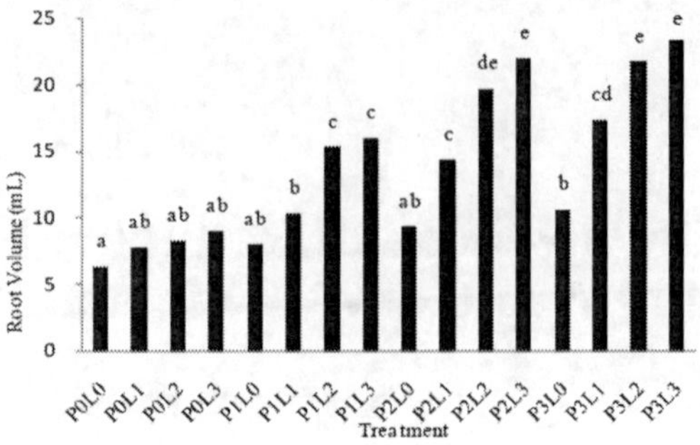

Fig. 1. Root volume

3.2. Shoot Dry Weight

The effect of the interaction between the reduction of P fertilizer and the addition of fish pond sludge to the dry weight of the shoot was seen (Figure 2).An increase in the dry weight of the shoot due to the increase in fish pond sludge was caused by the addition of fish pond sludge and P fertilizer that caused an increase in supply and availability of N and P nutrients. In addition, fish pond mud may improve soil physical properties that may cause a good impact on the growth and development of plant root systems, so that nutrient uptake may also increase. Increased nutrient uptake, especially N, increased the formation of chlorophyll so that the rate of photosynthesis increased and in turn the more biomass was produced. The P element was needed to activate plant metabolism processes such as photosynthesis due to its presence in the ATP used to reduce CO_2 to carbohydrates, so that increased the formation of plant biomass, which was seen with the increase in dry weight of the plant shoot. The provision of marine sludge was able to increase the growth and yield of soybean crops on peat soils [8]

Plant P concentration was higher with higher doses of fish pond sludge given up to 75% of the 15 ton/ha doses and the application of P fertilizer at a dose of 50% of the recommended dose (Figure 3).Application of fish pond sludge up to 75% dosage was expected to increase the soil P content. This was due to the high content of P in fish pond mud. The P content of fish pond sludge was 309.68 ppm [5]. In addition to the supply of P derived from fish pond sludge, P was also added through SP-36 fertilization. Therefore, the combination of P at 50% dosage and fish pond sludge at 75% dosage was able to increase the availability of P soil so that the P concentration in soybean plant tissue grown on the combination of treatments showed the highest value of 2402.97 ppm or 0.2402%, classified as adequate [9] . Phospor absorption by plants was an interconnected simultaneous process between P-soil, P diffusion flow and plant metabolism [10]. The increase of P uptake by soybean plants in combination of 50% P treatment and 75% of fish pond sludge was related to good root system development. The root of the soybean crop with a large root volume on the combination of treatments was obtained (Figure 1). The root system development, especially hair root growth, was influenced by the distribution of nutrients, especially P

https://doi.org/10.1051/e3sconf/20183105005

element. With a good root system development, the root ability to reach or obtain nutrients was greater, so the absorption and concentration of P on the soybean plant tissue were to be better [11]

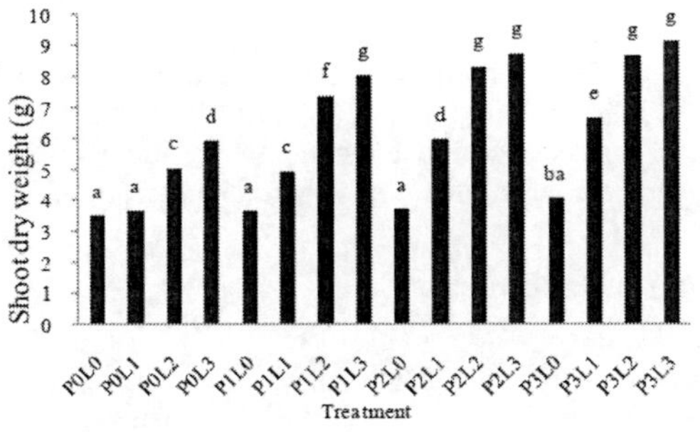

Fig. 2. Shoot dry weight

3.3. Plant Tissue P Concentration

Plant P concentration was higher with higher doses of fish pond sludge given up to 75% of the 15 ton/ha doses and the application of P fertilizer at a dose of 50% of the recommended dose (Figure 3). Application of fish pond sludge up to 75% dosage was expected to increase the soil P content. This was due to the high content of P in fish pond mud. The P content of fish pond sludge was 309.68 ppm [5]. In addition to the supply of P derived from fish pond sludge, P was also added through SP-36 fertilization. Therefore, the combination of P at 50% dosage and fish pond sludge at 75% dosage was able to increase the availability of P soil so that the P concentration in soybean plant tissue grown on the combination of treatments showed the highest value of 2402.97 ppm or 0.2402%, classified as adequate [9].

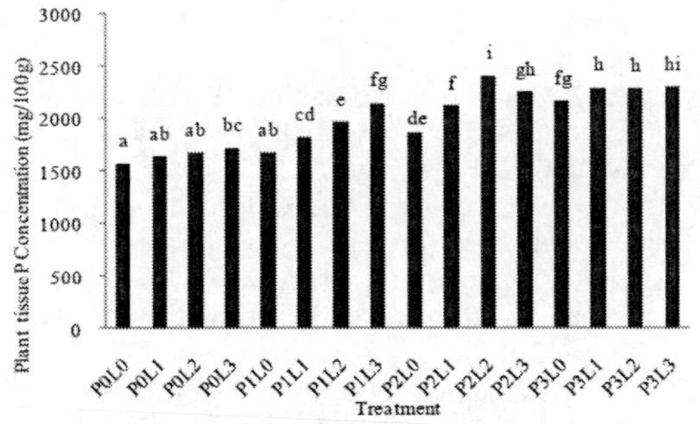

Fig 3. Plant tissue P Concentration

Phospor absorption by plants was an interconnected simultaneous process between P-soil, P diffusion flow and plant metabolism [6]. The increase of P uptake by soybean plants in combination of 50% P treatment and 75% of fish pond sludge was related to good root system development. The root of the soybean crop with a large root volume on the combination of treatments was obtained (Figure 1). Root system development, especially hair root growth, was influenced by the distribution of nutrients, especially P

element. With a good root system development, the root ability to reach or obtain nutrients was greater, so the absorption and concentration of P on the soybean plant tissue were to be better [11].

3.4. Number of pods per plant

The number of pods per plant was more in line with higher doses of fish pond sludge given from 25% to 75% dosage on P fertilizer as much as 50% of recommended dosage (Figure 4).Root volume, shoot dry weight and P concentration of soybean tissue given fish pond sludge increased up to 75% dosage along with medium-dose (50% recommended dose) of P fertilizer (Figure 1 and 2). It is seemed that this caused the growth of soybean crops better and then may produce more pods of content. Supply P derived from fish pond mud and P fertilizer may increase the availability of P-soil. In addition to P element, fish pond mud also contained elements of N and K [7]. P element may play a role in the formation of plant generative organs such as pods. However, the K element acted as an activator of various essential enzymes in photosynthetic and respiratory reactions and functions in photosynthesis translocation. K mud of fish pond was relatively high about 1.43 me/100g [5].

The photosynthesis process may run well with the fulfillment P and K element, and other nutrients so that the photosynthesis result also increased. The photosynthesis result was partially transferred to the plant generative parts for the pod feeding process so that the number of pods per plant increased. The highest number of pods was obtained in a combination of 75% fish pond mud and 50% of P fertilizer, ie 60.33 pods.

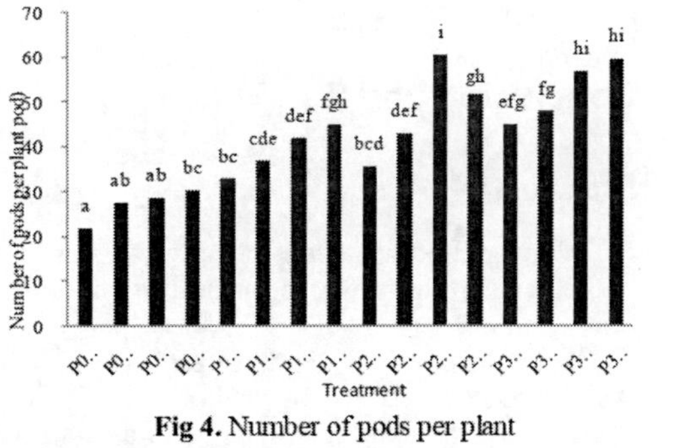

Fig 4. Number of pods per plant

3.5. Dry Seed Weight per Plant

The response of dry seeds of soybean crops to P fertilizer application was increase along with the increase of fish pond sludge added from 25% dosage up to 75% dose (Figure 5). This was due to the improvement of physical, chemical and biological properties of peat soil coming from fish pond mud. The presence of basic cations such as Ca in fish pond mud potentially increased the saturation of the peat soil base, so that soil pH increased. An increase in soil pH may accelerate the development of microbes in the soil that will improve the process of peatland decomposition. Use of sludge from sea increased the pH of the peat so that it may spur the development of bacteria in peat soil causing a decrease C / N ratio of peat soil [12].

An increase in soil pH may also affect the availability of other nutrients such as N, P, K, Ca, and Mg [13]. The increase of N, P, and K elements due to the application of P fertilizer and fish pond sludge caused the photosynthesis process to occur smoothly so that the photosynthesis may meet the needs of plants for growth and development balanced by translocation of most photosynthesis result to the generative, so that the crop may be improved. Proper supply of nutrients may greatly assist plant growth to achieve high yields. Soybean plants applied fish pond sludge at 75% dose with P fertilizer at moderate dose (50%) showed better growth. Plants with good growth may provide a high yield, because the result was a manifestation of growth. The highest yield of dry seeds per plant in this study was obtained on medium dosage of P fertilizer (50%) and pond sludge at 75%, ie 7.30 g/plant.

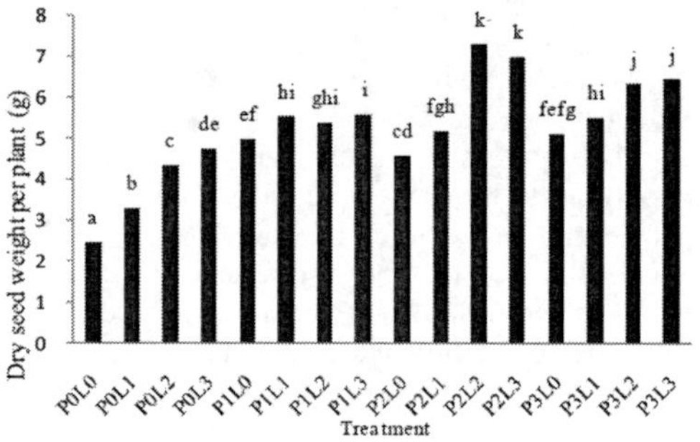

Fig 5. Dry seed weight per plant

4 Conclusion and Suggestion

4.1Conclusion

1. Fish pond mud waste was effective to reduce the use of P fertilizer on soybean cultivation in inland peatlands.
2. The reduction of P fertilizer to 50% of recommended dosage by addition of fish pond sludge with a dose of 75% was the best combination due to producing the best growth: the largest root volume and the heaviest dry weight of shoot, of 19.66 mL/plant and 8.33 g/plant, respectively.
3. The highest amount of plant tissue P concentration, number of pods and dry seed yield per plant was obtained in combination of P-fertilizer application at moderate dosage (50% recommended dosage) and fish pond mud at 75%, namely 2402.97 ppm/plant, 60.33 pods/plant and 7.30 g/plant, respectively.

4.2. Suggestion

1. Fish pond mud waste may be used not only as ameliorant, but also as alternative material to add P nutrient.
2. For the soybean crop cultivation in the peat soil it is advisable to provide P fertilizer with a dosage of 50% recommendation dosage along with the fish pond mud with 75% of the dose 15 ton/ha,
3. Further research to study the residual effect of fish pond sludge waste should be performed.

Acknowledgments

Authors thank to (1) Rudi Ansyah for assisting in field research, and (2) Palangka Raya University for providing financial support through DIPA PNBP Universitas Palangka Raya Fiscal Year 2016, No: 2899/UN24 /PL/2016, dated August 19, 2016.

References

1. I. Basuki, J. B. Kauffman, , D. Murdiyarso, G. Anshari. Proc 15th International Peat Congress 2016: *Peatlands in Harmovy, Agriculture, Industry and Nature*. Kuching, Sarawak, Malaysia. 258-263 (2016)
2. J. Jauhiainen, H. Silvennoinen, M. Könönen, S. Limin, H. Vasander. Proc 15th International Peat Congress 2016: *Peatlands in Harmovy, Agriculture, Industry and Nature*. Kuching, Sarawak, Malaysia. 127 -131 (2016)
3. J. O. Rieley, D. Wilson, D. Blain, J. Couwenberg, C. D. Evans, D. Murdiyarso, S. E. Page, F. Renou-Wilson, A. Sirin, M. Strack, E-S. Tuittila. Proc 15th International Peat Congress 2016: *Peatlands in Harmovy, Agriculture, Industry and Nature*. Kuching, Sarawak, Malaysia. 301-306 (2016)
4. R. Lazzari, B. Baldisseroto. B.Inst. Pesca, Sao Paulo, **34**(4) 591-600 (2008)
5. Nyahu , S. Sinaga, E. R. Asie. *Aplikasi Kombinasi Endapan Lumpur Kolam Ikan dan Gambut Sebagai Amelioran dan Media Tanam Buah Naga Merah* (*Hylocereus sp.*). Lembaga Penelitian Universitas Palangka Raya: Palangka Raya (2014)
6. S. H. Wijayanti. *Perubahan sifat kimia tanah yang diberi perlakuan lumpur dan air kolam ikan dalam 2 kali penanaman di desa Petir, Dermaga*. Bogor. http://repository.ipb.ac.id (verified 18 Feb.2016) (2015)
7. R. Sutanto R. *Penerapan Pertanian Organik (Pemasyarakatan dan Pengembangannya)*. Kanisius. Yogyakarta. (2012)
8. T. Abdurrahman, Radian, M. Safwan, M. Pertumbuhan dan hasil kedelai dengan pernggunaan lumpur laut cair pada beberapa tingkat kematangan gambut. *Pembangunan pertanian berkelanjutan berbasis kedaulan*

pangan dan energi untuk meningkatkan perekonomian nasional. Fakultas Pertanian Universitas Palangka Raya. 137-142 (2015)

9. B. Lakitan. *Dasar-dasar Fisiologi Tumbuhan*. Rajagrafindo Persada. Jakarta (1996)
10. S. A. Barber, S.A. *Soil Nutrient Bioavailability*. 2nd Edition. Jhon Wiley and Sons, Inc. New York (1995)
11. P. J. Gregory. Root growth and activity. *Physiology and Determination of Crop Yield*. ASA,CSSA,SSSA, Madison, WI. 65-96 (1994)
12. T. Abdurrahman. Indonesian J. of Applied Science. 3(**3**) : 78-83 (2013)
13. K. A. Hanafiah. *Dasar-dasar Ilmu Tanah*. Rajagrafindo Persada. Jakarta (2004)

Preliminary Evaluation of Method to Monitor Landfills Resilience against Methane Emission

Noor Amalia Chusna[1,*], *Maryono* Maryono[2]

[1]Master Program of Environmental Studies, School of Postgraduate Studies Diponegoro University, Semarang-Indonesia
[2]Departemen of Urban and Regional Planning, Faculty Engineering, Diponegoro University, Semarang-Indonesia

Abstract. Methane emission from landfill sites contribute to global warming and un-proper methane treatment can pose an explosion hazard. Stakeholder and government in the cities in Indonesia been found significant difficulties to monitor the resilience of landfill from methane emission. Moreover, the management of methane gas has always been a challenging issue for long waste management service and operations. Landfills are a significant contributor to anthropogenic methane emissions. This study conducted preliminary evaluation of method to manage methane gas emission by assessing LandGem and IPCC method. From the preliminary evaluation, this study found that the IPCC method is based on the availability of current and historical country specific data regarding the waste disposed of in landfills while from the LandGEM method is an automated tool for estimating emission rates for total landfill gas this method account total gas of methane, carbon dioxide and other. The method can be used either with specific data to estimate emissions in the site or default parameters if no site-specific data are available. Both of method could be utilize to monitor the methane emission from landfill site in cities of Central Java.

1 Introduction

Global Climate Change is one of the un-expected and then to be dangerous and irreversible magnitude of climate impact. Several aspects of these climate impacts are already noticed: global warming, sea level rise, intense rain, as well as more frequent and severe heat waves. Many nations around the world have been developing innovative policies and business approaches to build low-carbon economies and adapt to changing climate. In particular, control of greenhouse gas emission from various sources. Several global effort have been made to understand, quantify, and manage greenhouse gas emissions. Major actions have been taken to stabilize the green house gas concentrations in the atmosphere to prevent the dangerous climate change impact [1].

Methane (CH_4) is the second most important anthropogenic greenhouse gas after carbon dioxide (CO2). Per mass of the compound has been estimated to be more than 28 times of that of carbon dioxide [2]. Globally, over 60% of total methane emissions come from human activities [3,4]. Mostly, methane is emitted from industry, agriculture, and waste management activities. Methane emissions from waste management are dominated by the decomposition of organic matter.

Landfill sites produced landfill gas (LFG) which containing decomposable organic waste. The major components of LFG are methane and carbon dioxide, as a product of the biological decomposition of organic material. Despite the impact, methane is very potential to explore as an energy alternative resource [5]. Landfills are one of the main anthropogenic biogenic emissions of CH_4 [6], and still one of the most common waste management option globally [7]. In developing countries CH_4 emissions from landfills rapidly increase because of rapid population growth and accelerated urbanization process [8].

The characterization of landfill emissions is a complicated task, primarily because emissions are the result of a complex matrix of biological, physical, and engineering factors (i.e. CH_4 generation, oxidation, migration, storage, and recovery) [9]. These factors depend on parameters such as organic content, age and distribution of the waste [10,11], climate [12], and soil cover properties (e.g. water content, nutrient availability, pH, texture, porosity, fissures, and cracks) [13,14,15]. Given the number and variability of these factors, CH_4 emissions can vary greatly spatially and temporally.

Landfills gas could be generated by the biologically breaking down processes of waste components [16]. The production of landfill gas depends on the quantity and composition of the disposed waste [12], i.e., the content of organic waste and its degradability. In general, landfill gas is mainly

* Corresponding author: chusna.amalia95@gmail.com

composed by CH_4 and CO2 (50%–60% and 40%–45%, respectively), although N_2, O_2, H_2S, NH_2 and more than two hundred organic compounds are also present [17]. Landfill gas production also depends on waste age and other parameters, such as temperature, water content, nutrients, and inhibiting compounds [18]. Young landfills (5–30 years) produce gas with higher methane content (30%–60%) than older landfills [19].

Various researchers have reported the spatial variability of methane emissions. Several researchers illustrated the high spatial variation in emissions, caused by the high heterogeneity of waste. Emissions at one spot can be 1000 fold of emissions from a spot a located a few meters away [20]. There is no correlation between emission at a spot at the landfill and the emission six meters away [21]. Moreover, They estimate that 50% of emissions are released at 5% of the landfill surface. Other research found and estimate that 70% of methane emissions are released through short cuts [22]. The typical distribution of methane emissions at a landfill is similar patterns [23], but there are some variation emissions in the square meter at a small scale emissions [20,24,25].

Refers to the condition, this study is proposed to comparing and assessing of method to estimate methane missions in a landfill. This study evaluated two method such as LandGEM [26], and using gas prediction of the Guidelines for National Greenhouse Gas Inventories [8].

2 Method to monitor landfills resilience against methane emission

2.1 IPCC Method

The IPCC developed two methods for estimating methane emissions from MSW landfills in each country: the default method and the FOD method [27,8]. The default method is based on a mass balance approach, which estimates total national emissions based on a number of constant empirical parameters. The FOD method estimates the change of methane emissions with time. This method assumes that the degradable organic carbon (DOC) in waste decays slowly over a few decades when CH_4 is formed. The most appropriate method is based on the availability of current and historical country specific data regarding the waste disposed of in landfills. The methane emissions for each region can be calculated as follows :

$$ECH_4 = \sum_T((e^{-(T-1)k} - e^{-Tk}) \text{ x MSW x MCF x DOC x DOCf x F x } 16/12 - R) \text{ x } (1\text{-OX})$$

$$K = (\ln 2)/t^1/_2$$

where ECH_4 represents the methane emissions from a MSW landfill, T is the inventory year for which emissions are calculated, k is the reaction rate constant, t1/2 is the half-life of methane, MSW is the total MSW disposed of in the landfill, MCF is the methane correction factor based on the method of disposal and depth available at the landfill, DOC is the fraction of degradable organic carbon in the MSW, DOCf is the fraction of total DOC that actually degrades, F is the fraction of methane in the landfill gas, 16/12 is the conversion ratio (CH_4/C), R is the methane recovered value, and OX is the oxidation factor. The values of the parameters used here were derived from the IPCC [8], and the latest research results and local emissions factors.

On the basis of the evaluation of the methane formation potential in landfill site are estimated. The gas prediction is carried out via a 1st order approach of the IPCC Guidelines for National Greenhouse Gas Inventories [8]. The results from the gas modeling show that the primary gas production already takes place during the filling phase and the first years of the closure phase. As, particularly in this phase, only minor methane oxidation in the upper waste layers can be assumed, the highest emissions occur during this time. They decrease significantly with the application of a surface cover or a surface sealing system designed for methane oxidation.

2.2 LandGEM Method

LandGem is an automated tool for estimating emission rates for total landfill gas, methane, carbon dioxide and can be used either with site specific data to estimate emissions or default parameters if no site-specific data are available. LandGem contains two sets of default parameters. The CAA defaults are based on requirements for MSW landfills laid out by the Clean Air Act (CAA), including the inventory defaults based on emission factors in the U.S. Environmental Protection Agency's (EPA's) Compilation of Air Pollutant Emission Factors (AP-42). This set of defaults yields average emissions can be used to generate emission estimates for use in emission inventories and air permits in the absence of site specific test data. LandGem is based on a first-order decomposition rate equation for quantifying emissions from the decomposition of landfilled waste in MSW landfills and to estimate annual emissions over a time period based on user specification. The equation which is carried out is as follows [11]:

$$QCH_4 = \sum_{i=1}^{n} \sum_{j=0.1}^{1} \left(\frac{Mi}{10}\right) e^{-k\, tq}$$

Where QCH_4= annual methane generation in the year of the calculation (m3/year); i = 1 year time increment; n= (year of the calculation) - (initial year of waste acceptance); j= 0.1 year time increment; k = methane generation rate ($year^{-1}$); L0 = potential methane generation capacity, meter cube per metric ton ($m3/t^{-1}$).This factor depends on rate of decay and composition of waste ; Mi = mass of waste accepted in year (t); tij = age of the section of waste mass Mi accepted in the year (decimal years, e.g. 3.2 years) [11]. L0 factors use Clean Air Act (CAA) default values for the measurement of the parameters [11]. Methane generation rate (k) has default value 0.04 yr-

1for areas receiving 635 mm or more annual rainfall precipitation and 0.02 yr-1forareas receiving less than 635 mm and 0.30 yr-1for wet landfills [11,28]. Potential methane generating capacity (L0), depend on the land fill site, waste composition and ultimate methane yield of each component have default values ranging from 6 to 270 m3/t-1[11].

Comparative studies showed that emission models may give different results, even when employed input-data are the same. Some comparison methane emission models with methane emission measurements has been conducted by J. Jacobs et, all [29]. They showed that emission models resulted into high divergent estimations. They also stated that in the near future further development of methane emission measurement techniques may provide more reliable tools than modeling The methane output score is calculated by comparing the predicted methane output rate to the actual methane output rate for a given site. The methane output is predicted using the LandGEM model for the ideal methane output. The score is expressed as the percentage deviation from the predicted value:

$$Mx,t = (Ba - Bi)/Bi$$

Where M is the methane output score for site 'x' at time 't', 'BA' the actual methane output (m3/yr) and 'BI' the ideal value for methane output (m3/yr). Therefore, a score of 0% represents the actual methane output being equivalent to predicted output. The methane output score is given a red, yellow or green light to highlight good, average or poor methane output rate. A green light indicates a score higher than 30% which is determined by defining the error margin of the LandGEM model to be 30% either side of the actual score [20]. A yellow light represents a score of -30% to 30% whilst a red light represents a score below -30%. The boundaries over which red, yellow and green lights are given can be changed in further versions of the model.

3 Methodology

This study is quantitative evaluation research which uses two possibility method to monitor the landfill resilience against methane emission. Landfill site sample is selected according to the availability of data requirement from those parameter of two method methane emission calculation in Indonesia. This study selected at least 10 landfills in Indonesia's cities.

4 Result and discussion

This study comparing the IPPC method and LandGem Method to monitor methane emission. The perspective of resilience here associated to engineering and ecological resilience at which the landfills site methane emission could be assess and account by the method [30]. In practical to achieved the landfill resilience, the stakeholder intention must be assess. There some factor concerning the stakeholder knowledge [31]. Moreover in the context of disaster event a huge disaster waste will be discharge to the landfill site

The engineering resilience refers to the amount of the methane emission that could be account, while ecological resilience associated to the degree of environmental degradation such as climate change impact due to increasing of methane.

This study shown that both of the method could be use to calculate and assess the amount of methane emission. Such as shown in table 1, both of method methane emission have a positive correlation with capacity of methane, at which more bigger capacity of landfill also more bigger methane emission. According to the calculation, the bigger of methane calculation is Bantargebang landfill with total emission of CO2 are estimated $79.2E^{+06}$ m3 per year by using IPPC method and estimated $49.68E^{+06}$ by using landGem method. Table 1 shown that the emission number by using IPPC method is more bigger comparing to the LandGem Method. The varian of the differences from each landfill sample is vary from 30% until 90%. According to the result the two method could not be use in same time due to the high of deviation.

Table 1. Comparison calculation of methane emission by using IPPC and LandGEM Method in several landfill in Indonesia.

No	Landfill	Landfill Capacity x1000 ton	Method o	
			IPCC m3/year	LandGEM m3/year
1	TPA Supit Urang, Malang	4,800	$6.6E^{+06}$	$4.14E^{+06}$
2	TPA Galuga Cibung bulang, Bogor	5,000	$10.42E^{+06}$	$6.7E^{+06}$
3	TPST Bantargebang, Bekasi	28,800	$79.2E^{+06}$	$49.68E^{+06}$
4	TPA Jatibarang, Semarang	3,360	$8.66E^{+06}$	$5.47E^{+06}$
5	TPA Banyuurip, Magelang	8900	$2.06E^{+06}$	$1.33E^{+06}$
6	TPA Jeruklegi, Cilacap	4,220	$8.25E^{+06}$	$5.17E^{+06}$
7	TPA Bestari, Probolinggo	5,250	$10.48E^{+06}$	$6.57E^{+06}$
8	TPA MuaraFajar, Pekanbaru	1,400	$6E^{+06}$	$3.6E^{+06}$
9	TPA Basirih, Banjarmasin	1,200	$4.8E^{+06}$	$3.1E^{+06}$
10	TPA Benowo, Kota Gresik	13,800	$38E^{+06}$	$23.88E^{+06}$

According to the calculation, each of method have a different characteristic. In practical the two method have some specific function and parameter that could not be comparing due to their characteristic of data and parameter that utilize for assessment. Moreover, the characteristic of engineering resilience or ecological resilience in practical monitoring of methane emission could not be comparison. Both of method have a difference characteristic and parameter concerning to engineering resilience and ecological resilience.

5 Conclusions

1. Method monitoring the estimate methane emissions can be using predicted using the LandGem. The calculation is based on user input of age, waste acceptance and potential methane generation capacity (L0). The methane output is predicted using the LandGEM model for the ideal methane output.
2. Method monitoring the estimate methane emissions can be using gas prediction IPCC. The results from the gas modelling show that the primary gas production already takes place during the filling phase and the first years of the closure phase.

6 Acknowledgment

This research is funded by the Directorate of Research and Community Services, Ministry of Research, Technology and Higher Education for year 2017 with contract number 344-42/UN7.5.1/PP/2017

Reference

1 Kormi, Tarek., Mhadhebi, Safa., Aji, NizarBel Haj., Abichou, Tarek., Green, Roger. Waste Management, (2016)

2 IPCC, Climate Change 2013: The Physical Science Basis, Contribution of Working Group I to the Fifth Assessment Report of the Intergovernmental Panel on Climate Change, 1535, (2013)

3 EPA, Methane and Nitrous Oxide Emissions From Natural Sources. U.S. Environmental Protection Agency, Washington, DC, USA, (2010)

4 E. Nisbet, E. Dlugokencky, P. Bousquet, Methane on the rise—again. Science **343** (6170), 493–495, (2014)

5 Conestoga-Rovers and Associates, Landfill Gas Management Facilities Design Guidelines, British Columbia, Ministry of Environment, (2010)

6 M. Yang, Y. Xue, Y. Jiang, K. Wang, Y. Zhao, J. China Coal Soc. **34**, 1349e1353, (2009)

7 L. Ziyang, W. Luochun, Z. Nanwen, Z. Youcai, Martial 651 recycling from renewable landfill and associated risks: a 652 review. Chemosphere **131**:91–103, (2015) http://dx.doi.org/10.1016/j. 653 chemosphere.2015.02.036.654

8 EPA, Global Mitigation of Non-CO2 Greenhouse Gases: 2010–2030. U.S. Environmental Protection Agency, Washington, DC, USA, (2014)

9 K. Spokas, J. Bogner, J. Chanton, J. Geophys. Res.: Biogeosci. 116 (G4).n/a-n/a, (2011)

10 I. Georgaki, P. Soupios, N. Sakkas, F. Ververidis, E. Trantas, F. Vallianatos, T. Manios, Sci. Total Environ. **389** (2–3), 522–531, (2008)

11 USEPA (Unites States Environmental Protection Agency), Landfill Gas Emissions Model (LandGEM) Version 3.02 User's Guide. (PDF) (56 pp., 1.4 Mb) Publication No. 600/R-05/047,(2005) <http://www.epa.gov/ttncatc1/dir1/landgem-v302-guide.pdf> (accessed 29.07.2015).

12 J. Chanton, T. Abichou, C. Langford, G. Hater, R. Green, D. Goldsmith, N. Swan, Sci. Technol. **45** (1), (2011)

13 J. Bogner, K. Spokas, Landfills. In: Reay, D., Smith, P., Van Amstel, A. (Eds.), Methane and Climate Change. Routledge, London & Washington, D.C., pp. 175–200, (2010)

14 J. Gebert, I.U. Roewer, H. Scharff, C.D.L. Roncato, A.R. Cabral, Waste Manage. 31 (5), 987–994, (2011)

15 L. Giani, J. Bredenkamp, I. Eden, J. Plant Nutr. Soil Sci. **165** (2), 205–210, (2002)

16 D. Widory, E. Proust, G. Bellenfant, O. Bour, Waste Manag. **32**:1685–1692. (2012). http://dx.doi.org/10.1016/j.wasman.2012.04.008

17 A. Beylot, J. Villeneuve, G. Bellenfantat, Waste Manag. **33**: 401–411. (2013). http://dx.doi.org/10.1016/j.wasman.2012.08.017.54.

18 C. Scheutz, A.M. Fredenslund, J. Nedenskov, J. Samuelsson, P. Kjeldsen, Waste Manag. **31**:946–955, (2011) http://dx.doi.org/10.1016/j.wasman.2010.10.021.

19 O.P. Karthikeyan, S.Y.M. Jane, R. Technol. **217**:205–209, (2016). http://dx.doi.org/10.1016/j.biortech.2016.01.059

20 H. Oonk, Innovat. Environ. Technol. 75, (2010)

21 P.M. Czepiel, J.H. Shorter, B. Mosher, E. Allwine, J.B. McManus, R.C. Harriss, C.E. Kolb, B.K. Lamb, Waste Manage. **23**, 593–598, (2003)

22 P. Bergamaschi, C. Lubina, R. Konigstedt, H. Fischer, A.C. Veltkamp, O. Zwaagstra, J. Geophys. Res.-Atmosph. **103**, 8251–8265, (1998)

23 A.N. Nozhevnikova, A.F. Lifshitz, V.S. Lebedev, G.A. Zavarin, Chemosphere **26** (1–4), 401–417, (1993)

24 K.R. Mackie, C.D. Cooper, Environ. Model.Softw. **24**, 1223–1232, (2009)

25 I. Rachor, J. Streese-Kleeberg, J. Gebert, 2009. Proceedings Sardinia 2009, Twelfth International Waste Management and Landfill Symposium, S. Margeritha di Pula, Cagliari, Italy; 5–9 October 2009

26 Emkes, Harriet., Coulon, Frederic., Wagland, Stuart., Waste Management, (2015)

27 S. Kumar, S. Gaikwad, A. Shekdar, P. Kshirsagar, R. Singh, Atmos Environ, **38**:3481–7, (2004)

28 H.R. Amini, D.R. Reinhart, Waste Manage. **31**, 2020–2026, (2011)

29 J. Jacobs, H. Scharff, Comparison of Methane Emission Models and Methane Emission Measurements. Workshop on Inventories and Projections of Greenhouse Gas Emissions from Waste, EEA, Copenhagen, Denmark, (2005)

30 CS Holling - Engineering within ecological constraints, 1996

31 Maryono, H. Nakayama, T. Shimaoka, Memoir of faculty Engineering, Kyushu Univeristy, Vol.**74** no 3, pp.79-98

 https://doi.org/10.1051/e3sconf/20183105007

The Removal of Turbidity and TSS of the Domestic Wastewater by Coagulation-Flocculation Process Involving Oyster Mushroom as Biocoagulant

Astrid Pardede[1,*], *Mochamad* Arief Budihardjo[1], and *Purwono*[1]

[1]Department of Environmental Engineering, Faculty of Engineering, Diponegoro University, Semarang - Indonesia

Abstract. Oyster mushroom (*Pleurotus ostreatus*) can be utilized as biocoagulant since it has chitin cell wall. Chitin has characteristics of bioactivity, biodegradability, absorption and could bind the metal ions. In this study, Oyster Mushroom is micronized and mixed with wastewater to treat turbidity and Total Suspended Solid (TSS) using coagulation-flocculation process employed jartest method. Various doses of Oyster mushroom, 600 mg/l, 1000 mg/l, and 2000 mg/l were tested in several rapid mixing rates which were 100 rpm, 125 rpm, and 150 rpm for 3 minutes followed by 12 minutes of slow mixing at 45 rpm. The mixture then was settled for 60 minutes with pH level maintained at 6-8. The result showed that the Oyster mushroom biocoagulant was able to remove 84% of turbidity and 90% of TSS. These reductions were achieved with biocoagulant dose of 600 mg/ L at 150 rpm mixing rate.

1 Introduction

Coagulation flocculation as a step in water treatment processes is applied for removal of turbidity in raw water that comes from suspended particles and colloidal material [1]. Amongst the available methods of water treatment, coagulation and flocculation (CF) is a low cost, simple, reliable, and low energy consuming process that is commonly practiced. This is because it requires no exclusive or complex machineries; also no energy consumption is required for the operation, once an effective coagulant is obtained. It is an established process that effectively removes soluble, colloidal, and suspended particles through induced aggregation of both micro and macro particulates into larger-sized ones followed by sedimentation. The Coagulation Flocculation operation is not limited to the Aqua removal of particulates alone but also applicable in the removal of other types of pollutants e.g., color, organic compounds, micro-pollutants, and oils and fat [2].

Materials that are used in this stage of water treatment can be inorganic coagulants, synthetic organic polymers or coagulants from natural sources [1]. Some of the most commonly used coagulants are aluminum salts, but they have serious disadvantages, such as, production of harmful sludge and indications that their residues in treated water can be harmful for out health [3,4,5] In spite of its undoubtful effusiveness in turbidity removal, aluminum increases concerns towards ecotoxicological impact when introduced into the environment as post-treatment sludge having large volumes [1]. They are still lacking in terms of green chemistry. In the 1960s, detrimental effects of chemical coagulants on human health were published [6] Further, many researchers have related Alzheimer's disease to the residual aluminum ions in the treated waters [7]. For the application of synthetic organic polymers, concerns about the presence of residual monomers which are neurotoxicity and carcinogenic have impeded its use in water treatment process [8]. One possible solution to these problems may be natural coagulants which are preferable as an alternative excellent method for removing turbidity from water. These coagulants must be safe for health and biodegradable.

In recent years, numerous studies on a variety of plant materials which can be used as source of natural coagulants have been reported. Natural coagulants are widely investigated because of their efficiency, abundant source, low price and biodegradability [5,8,9]. Many materials of plant origin were studied as a source of natural coagulants [10,11,12]. The potential of *C. obtusifolia* seed gum, as a natural coagulant, in the treatment of raw and undiluted pulp and paper mill effluent (PPME) was investigated [13] *Ocimum basilicum* as a natural coagulant for textile wastewater [14] active coagulant agent from Jatropha Curcas seeds for using in turbidity removal [15].

This research has used oyster mushroom as a natural coagulant to reduce turbidity and TSS in canteen's domestic wastewater. Oyster mushroom has chitin as its cell wall [16,17]. In recent years, chitin has been applied as flocculants in water and other liquid treatment [18,19] The chitin, polysaccharide composed of β-(1-4)-N-acetyl-D-glucosamine units (>50%) and glucosamine units, is a linear semi-crystalline biopolymer which displays α, β, and γ polymorphic forms. This biopolymer is regarded as biocompatible and biodegradable, among many other qualities [20].

* Corresponding author: astrid_pardede@ymail.com

This research varied the biocoagulant dose to 600 mg / l, 1000 mg / l and 2000 mg / l to obtain the optimum dose. The given rapid mixing velocity was 150 rpm for 3 minutes followed by slow mixing at 45 rpm for 12 minutes and then precipitated for 1 hour.

2 Methods

2.1 Biocoagulant

- ¼ kg of oyster mushrooms were blended with 500 ml of water
- Then the obtained result was squeezed with water until the solids are separated from the liquid
- The solids were ready to serve as biocoagulant

2.2 Jartest Implementation

- The biocoagulant according to the existing variations of 600 mg / l, 1000 mg / l and 2000 mg / l was inserted into the 1-liter glass beaker.
- Then, the canteen's domestic wastewater was mixed up to the volume of 1 liter.
- Do the rapid mixing at 150 rpm for 3 minutes
- Then, slow mixing with 45 rpm of speed for 12 minutes
- Then, the jartest result was precipitated for 1 hour.

2.3 Turbidity Test (SNI 06-6989.25-2005)

- Put the 2/3 of sample volume of wastewater into the cuvette, cover and shake.
- Dispose the sample of wastewater and repeat twice
- Clean the cuvette with a tissue
- Press "ON / OFF" button to turn on the tool
- Enter the cuvette in the tool
- Press "READ"
- Read the sample's turbidity

2.4 TSS Test (SNI 06-6989.25-2005)

- Heat a clean cup at 105°C for 1 hour, then put into desiccators for 15 minutes.
- Weighing the mass of the cup that has been cooled to constant. Record the weight gained (A)
- Heat an ash-free filter paper that has been dampened with aquades at 105°C for 1 hour, then put into the desiccators for 15 minutes.
- Weigh the mass of filter paper that has been cooled to constant. Record the weight gained (B)
- Fold the filter paper that previously has been weighed, according to the shape of the used cone. Put the cone with the folded paper on Erlenmeyer.
- Shake the sample up to 100 ml to become homogen.
- Filter the sample using the cone with the filter paper.
- Heat the used filter paper at 105°C for 1 hour, then cool it in desiccators for 15 minutes.
- Weigh the filter paper that has been cooled to constant and record the weight gained (C)
- Counting the TSS value with the formula:

$$TSS\left(\frac{mg}{l}\right) = \frac{1000}{100} x \{C - (A + B)\} x\, 1000 \quad (1)$$

3 Results and Discussion

3.1 Initial Characteristics of Wastewater

Table 3.1 Initial Characteristics of Canteen's Domestic Wastewater

No	Contaminants	Result	Regulation of Environment Minister No. 68/2016	Information
1.	Turbidity	267 NTU	-	-
2.	TSS	591 mg/l	30	Do not comply
3.	pH	7,41	6-9	Comply

It can be seen in the table that the TSS parameter does not meet the quality standard required by Regulation of the Minister of Environment and Forestry no. 68 of 2016 on the Quality Standards of Domestic Wastewater. Characteristics of canteen's wastewater are influenced by several factors such as the types of food served every day, hours of activity, and the number of buyers who visit. Pollutants that exceed the quality standard can pollute the environment if it is discharged directly into the water body. Therefore a further treatment is required for the effluent before being discharged into the channel.

3.2 Turbidity Test Results

The results of turbidity content test in canteen's wastewater after the coagulation flocculation processing by utilizing oyster mushroom biocoagulant are presented in the following table:

Table 3.2 Turbidity Test Results after Processing

No	Dose	Initial Value	Final Value	Efficiency
	mg	mg/l	mg/l	%
1.	600	267	42,44	84,1
2.	1000	267	43,2	83,82
3.	2000	267	49,2	81,57

The chart of turbidity value after coagulation flocculation processing based on oyster mushroom variation is presented in the following graphic image:

 https://doi.org/10.1051/e3sconf/20183105007

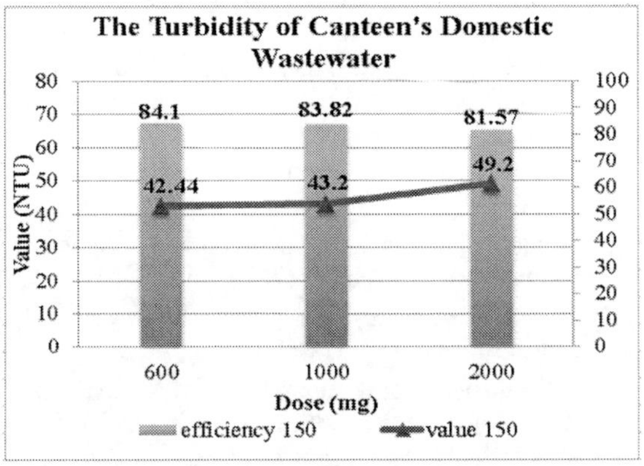

Fig. 1. The Chart of Turbidity Test Result from Canteen's Domestic Wastewater

Table 3.3 Wastewater pH after processing

No	Dose	Initial Value	Final Value
	mg/l		-
1	600		6.19
2	1000	7,41	6.21
3	2000		6.31

The optimum dose of biocoagulant in the canteen's wastewater was 600 mg/l that could lower the concentration of turbidity to 42.44 NTU from the initial condition of 267 NTU with 84.1% efficiency. This phenomenon could be explained by the high content of amine groups in chitin that provide cationic charge at acidic pH and can destabilize colloidal suspension to promote the growth of large rapid-settling flocs that can then flocculate. As chitin is a long-chain polymer with positive charges at natural water pH, it can effectively flocculate natural particulate and colloidal materials which are negatively charged through mechanism of adsorption, charge neutralization, interparticle bridging as well as hydrophobic flocculation. Chitin gave the highest total solids and total suspended solids reduction at both laboratory and industrial scale [21] Chitin has been widely used as effective coagulant–flocculant for a broad range of suspended solids in various food and fish processing industries and suspension containing mineral colloids in water [13, 21] This technique results in the destabilization of colloidal particles and subsequently, the increment in particle size for the ease of sedimentation [22].

3.3 TSS Test Results

The results of TSS content test in canteen's wastewater after the coagulation flocculation processing by utilizing oyster mushroom biocoagulant are presented in the following table:

Table 3.4 TSS Test Results after Processing

No	Dose	Initial Value	Final Value	Efficiency
	mg	mg/l	mg/l	%
1.	600		55	90,69
2.	1000	591	89,5	84,86
3.	2000		113	80,87

The TSS value chart after coagulation flocculation processing based on oyster mushroom variation is presented in the following chart figure:

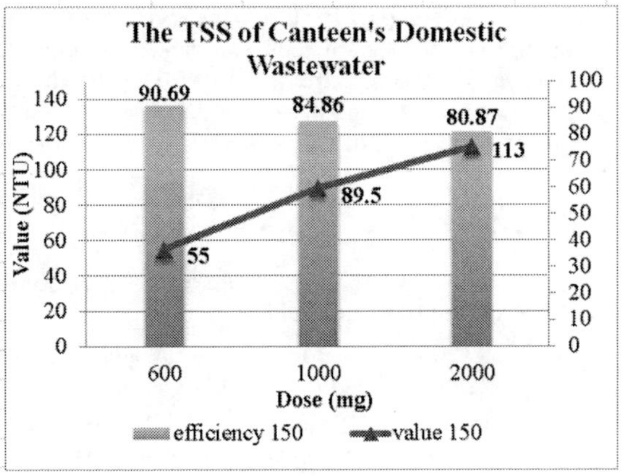

Fig. 2. The Chart of TSS Test Result from Canteen's Domestic Wastewater

Table 3.5 Wastewater pH after processing

No	Dose	Initial Value	Final Value
	mg/l		-
1	600		6.19
2	1000	7,41	6.21
3	2000		6.31

The optimum dose of biocoagulant in the canteen's wastewater was 600 mg/l that could lower the concentration of TSS to 55 NTU from the initial condition of 591 NTU with 90.69% efficiency. The reactivity of chitin during flocculation of suspended particles results from several mechanisms including electrostatic attraction, sorption and bridging. The contribution of each mechanism depends on the pH of the suspension. This phenomenon could be explained by the high content of amine groups in chitin that provide cationic charge at acidic pH and can destabilize colloidal suspension to promote the growth of large rapid-settling flocs that can then flocculate. As chitin is a long-chain polymer with positive charges at natural water pH, it can effectively flocculate natural particulate and colloidal materials which are negatively charged through mechanism of adsorption, charge neutralization, interparticle bridging as well as hydrophobic flocculation. Chitin gave the highest total solids and total suspended solids reduction at both laboratory and industrial scale [21]. Chitin has been widely used as effective coagulant–flocculants for a broad range of suspended solids in various food and fish processing industries and suspension containing mineral colloids in water [13, 21] This technique results in the destabilization of colloidal particles and subsequently, the increment in particle size for the ease of sedimentation [22].

4 Conclusion

This research aims to see the ability of oyster mushrooms as biokoagulan to reduce levels of turbidity contaminants and TSS in domestic canteen's domestic wastewater. The optimum dose of oyster mushroom biocoagulant is 600 mg / l with an efficiency of 84.1% and turbidity decrease of 90.69%.

References

1. M. Antov, M. Sciban, N. Petrovic. Bioresource Technology **101**, 2167–2172 (2010).
2. N.A. Oladoja. v Journal Of Water Process Engineering **6**, 174–192 (2015).
3. M.L. Jodi, U.A. Birnin-Yauri, Y. Yahaya, M.A. Sokoto. Global Adv. Res. J. Chem. Mater. Sci. 1 (4), 71–75 (2012).
4. S.C. Bondy. Neurotoxicology 31, 575–581 (2010).
5. H. Betatache, A. Aouabed, N. Drouiche, H. Lounici. Ecol. Eng. **70**, 465–469 (2014).
6. G.S. Simate, S.E. Simate, S. Ndlovu, M. Heydenrych, L.F. Walubita. Environ. Int. 39 (1), 38–49 (2012).
7. H. Aylin Devrimci, A. Mete Yuksel, F. Dilek Sanin, Desalination **299**, 16-21 (2012)
8. M. Šćiban, M. Klašnja, M. Antov, B. Škrbić, Bioresour. Technol. **100**, 6639–6643 (2009).
9. M. Antov, M. ˇS´ciban, J. Prodanovi´c. Ecol. Eng. **49**, 48–52 (2012).
10. Ghebremichael, K., Abaliwano, J., Amy, G., J. Water Supply: Res.Technol.– AQUA **58**, 267–276, (2009)
11. Teh, C.Y., Wu, T.Y., Juan, J.C., Ecol. Eng. **71**, 509–519, (2014)
12. Fedala, N., Lounici, H., Drouiche, N., Mameri, N., Drouiche, M., Ecol.Eng. **77**, 33–36, (2015)
13 W. Subramonian, T.Y. Wu, S.P. Chai, Ind. CropsProd. **61**, 317–324, (2014)
14. S. Shamsnejati, N. Chaibakhsh, A.R. Pendashteh, S. Hayeripour, Ind. Crops Prod. **69**, 40–47, (2015)
15. Z. Abidin, N. Shamsudin, N. Madehi, S. Sobri. *Optimisation of a method to extract the active coagulant agent from Jatropha Curcas seeds for use in turbidity removal*
16. RA. Muzzarelli, J. Boudrant, D. Meyer, N. Manno, M. DeMarchis, MG. Paoletti, Carbohydrate Polymers **87**: 995e1012, (2012)
17. P. Maftoun, H. Johari, M. Soltani, R. Malik, N. Z. Othman, and H. A. El Enshasy, International Journal of Biotechnology for Wellness Industries, **4**, 67-83, (2015).
18. M.C. Menkiti, O.D Onukwuli. AIChE J. **57**, (2011)
19. F. Gassara, C. Antzak, C.M. Ajila, S.J. Sarma, S.K. Brar, M. Verma. Journal of Food Engineering **166** (2015) 80–85
20. P. Kandra, M.M. Challa, H. Kalangi Padma Jyothi, Applied Microbiology & Biotechnology, **93**(1), 17-29. (2012).
21. F. Gassara, C. Antzak, C.M. Ajila, S.J. Sarma, S.K. Brar, M. Verma. 2015. Journal of Food Engineering **166**, 80–85, (2015)
22. S.Y. Choy, K.M.N. Prasad, T.Y. Wu, M. E. Raghunandan, R. N. Ramanan. Journalo of environmental Sciences **26** 2178-2189, (2014)

Mathematical Analysis for the Optimization of Wastewater Treatment Systems in Facultative Pond Indicator Organic Matter

Sunarsih[1,2,*], *Widowati*[1], *Kartono*[1] and *Sutrisno*[1]

[1] Department of Mathematics, Faculty of Science and Mathematics, Diponegoro University, Semarang - Indonesia
[2]Doctoral Program of Environmental Science, School of Postgraduate Studies, Diponegoro University, Semarang-Indonesia

Abstract. Stabilization ponds are easy to operate and their maintenance is simple. Treatment is carried out naturally and they are recommended in developing countries. The main disadvantage of these systems is large land area they occupy. The aim of this study was to perform an optimization of the wastewater treatment systems in a facultative pond, considering a mathematical analysis of the methodology to determine the model constrains organic matter. Matlab optimization toolbox was used for non linear programming. A facultative pond with the method was designed and then the optimization system was applied. The analyse meet the treated water quality requirements for the discharge to the water bodies. The results show a reduction of hydraulic retention time by 4.83 days, and the efficiency of of wastewater treatment of 84.16 percent.

1 Introduction

In today's life, settlements are more centered on a residential area. This poses a new problem especially in sewerage systems. The precise way of disposing of wastewater that is flowing into the river was considered successful. At first, it was not cause problems because the amount of liquid waste is small compared to the river flow. But with increasing waste disposal, the pollution is increasing to pollute the environment. Centralized settlements make the collection of domestic waste water in sewerage very high. If the waste is throw without first processing, it can reduce the quality of river water. To solve the problem, wastewater needs the processing process first before it is streamed into the water system. One of the processes is using a centralized Wastewater Treatment Plant unit (WWTP).

Waste Stabilization Ponds are used to improve the quality of waste water by relying on natural processes that treat wastewater by utilizing the presence of bacteria, algae and zooplankton to reduce the organic pollutants contained in wastewater [1-4].

Commonly used waste water disposal systems are waste water coming from toilets flowed into septic tanks and runoff from septic tanks impregnated into the soil or discharged into waterways, whereas non-toilets that come from bathing, washing and kitchen waste is discharged directly into waterways.

Improper waste water management causes problems, for example, health risks due to contamination by pathogens, odors, and loss of biodiversity. The purpose of a pond system is to keep and treat wastewater with the specified retention time [5-6]. Stabilization ponds may function independently or in combination with other processing systems. Stabilization ponds are usually designed with a system which consist of 2 types of pond that is facultative and maturation, because it is easy to operate and easy to maintain [7]. The purpose of this research is to know the optimum waste load and the level of degradation of organic matter. Mathematical analysis is used by applying linear program, so the results of wastewater treatment fulfill the quality standard.

2 Materials and Methods

2.1. Linear Program Application

Linear program is applied to wastewater treatment system at Wastewater Treatment Plant (WWTP), Sewon, Bantul, Yogyakarta. Optimization is reviewed from the treated waste load by analyzing the efficiency of wastewater treatment so as to meet the quality standard.

The formulation of the model begins by determining the objective function to maximize wastewater treatment in facultative stabilization ponds. To determine the constraint function, the model is limited by the concentration of Biochemical Oxygen Demand (BOD) with the efficiency of wastewater treatment in each pond to meet the quality standard. The wastewater treatment system scheme in the stabilization pond for model formulation assuming no consider to the maturation pond can be seen in **Figure 1**.

* Corresponding author: narsih_pdil@yahoo.com

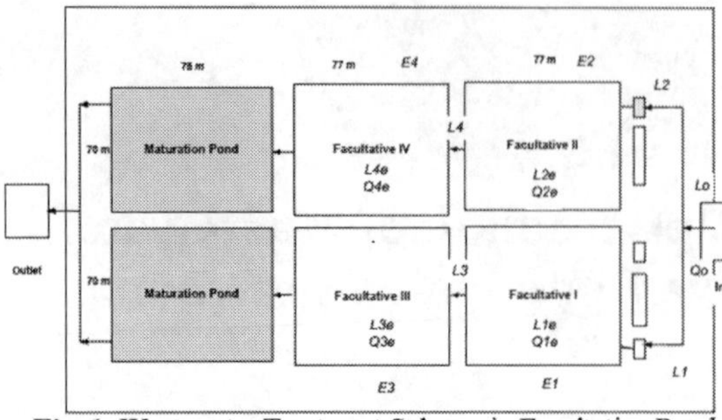

Fig. 1. Wastewater Treatment Scheme in Facultative Pond

with:

L_0 : The initial waste load that goes into the inlet (kg/day)
Q_0 : Debit of waste water in inlet (m^3/day)
Q_{ie} : Debit of waste water in pond *i* (m^3/day)
L_i : Wastewater load before entering pond *i* (kg/day)
L_{ie} : The amount of wastewater treatment in pond i (kg/day)
C_i : Concentration of BOD of pond i (mg/L)
L : Wastewater load (kg/day)
E_i : Efficiency of wastewater treatment in pond i
BM : Wastewater quality standards

From **Figure 1** can be formulated into a model with a linear program on a water treatment system in WWTP Sewon, Bantul as follows.

Objective Function:
Maximize $Z = L_{1e} + L_{2e} + L_{3e} + L_{4e}$

Constraint Function:

- $E_1C_1 \le BM$
- $E_2C_2 \le BM$
- $E_3C_3 \le BM$
- $E_4C_4 \le BM$
- $L_{1e} \le L_1$
- $L_{2e} \le L_2$
- $L_{3e} \le L_3$
- $L_{4e} \le L_4$

The amount of incoming waste load on the wastewater treatment system in the facultative stabilization pond is the debit of waste water along with the organic material Biochemical Oxygen Demand (BOD) formulated as $L_{ie} = \frac{(Q_{ie}.C_i)}{1000}$ and $C_i = \frac{(1000\ L_i)}{Q_{ie}}$, the linear program model becomes as follows.

Objective Function:
Maximize $Z = L_{1e} + L_{2e} + L_{3e} + L_{4e}$

Constraint Function:

- $E_1\frac{(1000\ L_1)}{Q_{1e}} \le BM$
- $E_2 e\frac{(1000\ L_2)}{Q_{2e}} \le BM$
- $E_3\frac{(1000\ L_3)}{Q_{3e}} \le BM$
- $E_4\frac{(1000\ L_4)}{Q_{4e}} \le BM$
- $L_{1e} \le L_1$
- $L_{3e} \le L_1 - L_{1e}$
- $L_{2e} \le L_2$
- $L_{4e} \le L_2 - L_{2e}$

To solve the equation, we assume the model with the facultative stabilization pond conditions as follows.

1. The amount of waste water debit in inlet Q_0 with period of time for 1 year (January - December 2016) is 11,238.38 m^3/day.
2. The waste retention time in the facultative pond is an average of 4.83 days and the retention time of waste in each month as in **Table 1**.

Table 1. Waste Retention Time the Period from January to December 2016

No	Month	Debit (m^3/day)	Retention Time (day)
1	January	10.997,22	3,97
2	February	11.853,80	3,68
3	March	15.951,28	3,94
4	April	13.937,84	4,13
5	May	13.187,30	4,31
6	June	12.071,40	4,82
7	July	10.616,44	4,11
8	August	8.243,10	5,6
9	September	7.534,32	5,8
10	October	8.044,08	5,43
11	November	10.378,02	6,3
12	December	12.045,77	5,83
Average		11.238,38	4,83

3. The waste load on the Facultative I (L_1) and Facultative II Pond (L_2) are half of the inlet initial load (L_0). The organic waste load is known from the number of residents connected with the Sewon WWTP system which includes the number of households as many as 54,440 people, social places (worship places and others) as many as 8,550 people, commercial places as many as 49,350 people and hotels as many as 7,650 people, so wastewater comes from 119,990 people. The standard clean water requirement in developing countries is 120 L/person/day and each person produces 40 grams of BOD per day. So the BOD concentration of Sewon is equal to 0.41667 gr/L or 416,67 mg/L and total BOD load or inlet load (L_0) is 119,990 x 40 gr/person/day which is 4,799,600 gr/day Or 4,799.6 kg/day. Thus the load of BOD (L_1) and (L_2) is 2,399.8 kg/day.
4. The coefficient of degradation of BOD (k) for the effective volume of 86.240 m^3 in 1 year is 1.1, with C_{in} is the rate of change of BOD concentration at

the inlet and c_{ef} is the rate of change of BOD at the outlet.

Thus, the formulation of the level of degradation of organic material which capable doing in the facultative pond system is as follows.

$$E = \frac{k.t}{1 + k.t}$$

with:

k : Rate of degradation BOD

t : Waste retention time (days)

From the calculation result obtained that effectivity of wastewater treatment system in facultative pond able to degrade waste equal to 84,16%.

3 Results and Discussions

The system of linear program equations is done by simplifying the equation, i.e. by entering the value of known variables. Known variables are debit of waste water in each pond (Q_n), waste load in each pond (L_n), efficiency of pond (E_i) and predetermined quality standard (BM). The quality standards of waste water based on the Decision Letter of the Governor of Yogyakarta Special Region No. 214/KPTS/1991 on the domestic waste water quality standard is 50 mg/L. By entering these values of variables, so the objective function of the constraint function is obtained as follows.

Objective Function:
Maximize $Z = L_{1e} + L_{2e} + L_{3e} + L_{4e}$

Constraint Function:

- $0.84 \times \frac{(1000\ L_1)}{13.475} \leq 50$
- $0.84 \times \frac{(1000\ L_2)}{13.475} \leq 50$
- $0.84 \times \frac{(1000\ L_3)}{13.475} \leq 50$
- $0.84 \times \frac{(1000\ L_4)}{13.475} \leq 50$
- $L_{1e} \leq L_1$
- $L_{3e} + L_{1e} \leq L_1$
- $L_{2e} \leq L_2$
- $L_{4e} + L_{2e} \leq L_2$

Equation systems solved by using software POM-QM version 3.0. This software is used to help math calculations by using the simplex method.

To know the suitability of the model with conditions in the field, so the model is verified with observation data in Sewon's IPAL, Bantul. The observational data were organic BOD concentrations in inlets range from 68.3 mg/L - 177.2 mg/L and at outlets range from 13 mg/L-18.8 mg/L.

4 Conclusions

Wastewater treatment systems in facultative ponds using mathematical analysis (linear program) are able to degrade organic matter. The intended organic matter is BOD with degradation rate of 84.16% and can meet the domestic waste water quality standard with waste retention time 4.83 days.

References

1. B. Beran, F. Kargi, Eco. Model. **181**, 39-57 (2005)
2. S. Kayombo, T. S. A. Mbwette, A. W. Mayo, J. H. Y. Katima, S. E. Jorgensen, Eco. Eng. **18**, 287-291 (2002)
3. L. Puspita, E. Ratnawati, I. Suryadiputra, *Lahan Basah Buatan di Indonesia* (Wetlands International Indonesia Programme, Ditjen PHKA, Bogor, 2005)
4. Sunarsih, Purwanto, W. S. Budi, J. Urb. And Env. Eng. **7 (2)**, 293-301 (2013)
5. K. Naddafi, M. S. Hassanvand, E. Dehghanifard, D F. Razi, S. Mostofi, N. Kasaee, R. Nabizadeh, M. Heidari, Iran. J. Env. Health. Sci. Eng. **6 (1)**, 41-46 (2009)
6. H. Abbas, R. Nasr, H. Seif, Eco. Eng. **28**, 25-34 (2006)
7. A. N. Shilton, D. D. Mara, Wat. Sci. Tech. **51**, 103-106 (2005)

Facultative Stabilization Pond: Measuring Biological Oxygen Demand using Mathematical Approaches

Ihsan Wira S[1,*] and *Sunarsih* Sunarsih[1,2]

[1]Master Program of Environmental Science, School of Postgraduate Studies, Diponegoro University, Semarang - Indonesia
[2]Faculty of Science and Mathematics, Diponegoro University, Semarang - Indonesia

Abstract. Pollution is a man-made phenomenon. Some pollutants which discharged directly to the environment could create serious pollution problems. Untreated wastewater will cause contamination and even pollution on the water body. Biological Oxygen Demand (BOD) is the amount of oxygen required for the oxidation by bacteria. The higher the BOD concentration, the greater the organic matter would be. The purpose of this study was to predict the value of BOD contained in wastewater. Mathematical modeling methods were chosen in this study to depict and predict the BOD values contained in facultative wastewater stabilization ponds. Measurements of sampling data were carried out to validate the model. The results of this study indicated that a mathematical approach can be applied to predict the BOD contained in the facultative wastewater stabilization ponds. The model was validated using Absolute Means Error with 10% tolerance limit, and AME for model was 7.38% (< 10%), so the model is valid. Furthermore, a mathematical approach can also be applied to illustrate and predict the contents of wastewater.

1 Introduction

The capital cities in Indonesia have some problems about environment quality, especially at settlements, so the greater population the more pollution increase [1, 2]. The waste of settlements are always domestic wastes. Untreated domestic wastes from settlements will cause water pollution at the river. So, to decrease pollutant concentration, we need to treat those domestic wastewater before disposed to environment [3, 4]. Wastewater Stabilization Pond (WSP) is considered as the most suitable system to treat urban wastewater in tropical and subtropical regions and used for improving quality of water [2, 4, 5, 6, 7, 8]. Stabilization pond also relies on natural processes of wastewater treatment by using the presence of microorganisms on it [2, 3, 9]. Those microorganisms could reduce organic matter of pollutant contained in pond [2]. One part of stabilization pond system is facultative pond such as Waste Water Treatment Plants (WWTP) Sewon in Bantul, Indonesia. WWTP Sewon was established to treat domestic wastewater from Yogyakarta City, Sleman Regency, and Bantul Regency.

Biological Oxygen Demand (BOD)is the amount of oxygen required by bacteria to oxidize organic matter in a wastewater [6]. The presence of BOD can be used as pollutant indicators. The greater oxygen in a wastewater, the smaller the contamination will be [2].The change of BOD in wastewater also can describe the quality of that wastewater [2]. To know a quality of water, wastewater need to meet a requirement value of Biological Oxygen Demand (BOD). We need water sample data to know the amount of BOD of wastewater. However, the data only describe the existing condition of quality of wastewater in facultative ponds. So, we need a method to know condition/ prediction value of (BOD) by using previous data. It is worth it to use mathematical approach to predict BOD values next few days or months [3].

The purpose of this research is to depict degradation process by using mathematical model of BOD in facultative pond and to validate the model.

2 Materials and Methods

2.1.Research Area Description

Domestic wastewater from Yogyakarta City, Sleman Regency, and Bantul Regency is treated through WWTP Sewon in Bantul. The WWTP Sewon is located in Bantul regency and services domestic wastewater from almost 63.015 people [10].

WWTP Sewon has four facultative ponds and two maturation ponds. **Fig.1** schematically shows the pond system. This research collected data from facultative II , as we see on the **Fig.1** as well. So this research was using facultative II as taken sampling point to validate a model we have.

*Corresponding author: matematison@gmail.com

https://doi.org/10.1051/e3sconf/20183105009

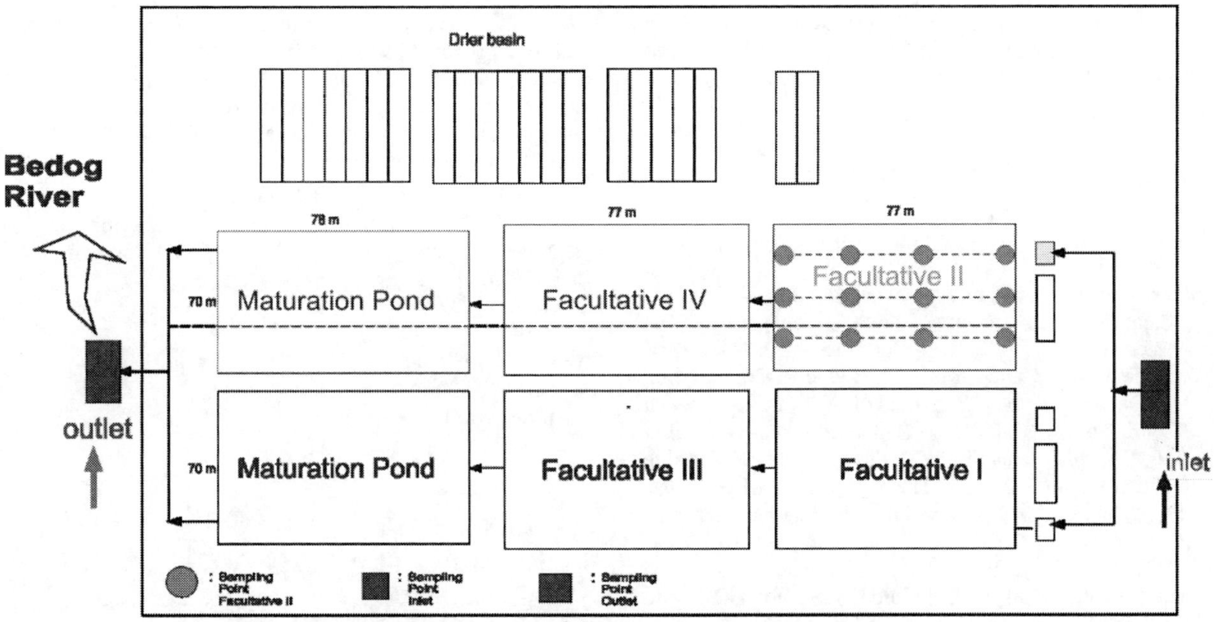

Fig. 1. Map of IPAL Sewon [4]

2.2. Mathematical Model

The model that could depict system process and environment as quantitative is mathematic model. The mathematical model is defined as the formulation of engineering problems into forms of mathematical equations based on the basic laws of physics, chemistry, and biology [2, 11]. The following models are presented on an organic waste degradation Biological Oxygen Demand (BOD) process system in a stabilizing pond:

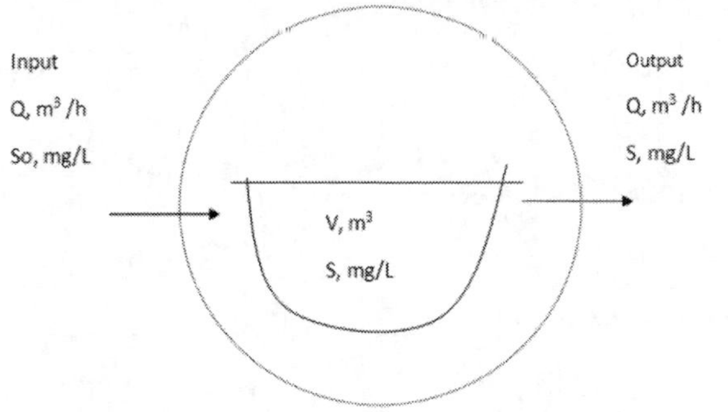

Fig.2. Diagram of stabilization pond [11]

And the following equation describes a degradation of BOD process in waste stabilization pond: [11].

$$S = \left(\frac{QS_o}{Q+kV}\right) + \left[S_o - \left(\frac{QS_o}{Q+kV}\right)\right] e^{-\left(\frac{Q+kV}{V}\right)t} \quad (1)$$

Where :

S = BOD outfrom stabilization pond, mg/ L

S_o = BOD in stabilization pond, mg/ L

k = degradation rate constant,

V = volume of pond, m^3

Q = debit, m^3/ time

t = time period,

a valid model can be applied to simulate the system we have made [11].

2.3. Validation Model

A good model is a model that represents the real condition, physic process for specific condition, enough data, and requires the test of validation [9, 12]. There are two validation of model, structure validation and performance validation. Structure validation is to get conviction about the construction of valid model scientifically or based on theory, while performance validation is to know how far model is appropriate/ compatible with performance real system or empiric data [1].

The model is validated to know the appropriateness of the formulated model, to know whether the model represented field condition or not with difference deviation between calculated data and observed data [9]. If the deviation is less than (10% tolerance limit), so the model is valid [9, 13].

Absolute Means Error (AME) is deviation between the average simulation value and actual value [13]. The following equation for AME :

$$AME = \left(\frac{|D_a - D_s|}{D_a}\right) \times 100\% \quad (2)$$

where : D_s = Simulation data

D_a = Actual data

AME = Absolute Means Error

The model is valid if AME is less than 10%, and if AME is more than 10% then the model must be evaluated.

2.4. Sample and Data Analysis

The water samples were measured on the facultative II in different time to get repeated measurement for concentration of BOD. The results of measurement were used to validate the model. The model that had been validated by comparing simulated model and observed model will be analyzed by AME.

3 Results and Discussion

3.1BOD of Facultative Stabilization Pond

Wastewater discharge at Sewon Bantul WWTP has indicated approximately 10,697.53 $m^3/time$. The wastewater that had been treated through WWTP Sewo will be disposed to environment, Bedog River. Based on the Decree of the Governor of Yogyakarta Special Region No.214/KPTS/1991, Bedog River is registered as Class II with Standard Local Environment BOD value (effluent) below 50 mg/L. The total volume of pond was 43,680 m^3 and effective volume for facultative II was 8,085 m^3 with 1.5 m depth.

The BOD inlet for the stabilization pond is approximately 180 mg/L as initial value for BOD with the coefficient of degradation was 1.85 (k)[2].

Table.1. BOD Observation of Facultative II

Times	Biological Oxygen Demand (BOD) (mg/L)			
	06:00 am	12:00 pm	10:00 pm	Average of certain sample
Sample 1	93.72	72.34	64.32	76.79
Sample 2	63.22	51.42	74.52	63.05
Sample 3	71.32	58.48	75.16	68.32
Sample 4	51.32	75.88	133.84	87.01
Sample 5	48.32	75.64	112.60	78.85
Sample 6	88.54	80.42	64.08	77.68
Sample 7	49.52	66.36	103.56	73.15
Sample 8	82.30	77.24	52.20	70.58
Sample 9	67.58	91.57	64.52	74.56
Sample 10	67.20	63.32	73.96	68.16
Sample 11	48.32	58.32	72.96	59.87
Sample 12	41.89	62.84	63.96	56.23
Average of certain time	60.55	69.5	79.64	71.18

The **Table.1.** indicated that the amount of BOD observation per day on facultative II was 71.18 mg/L. That value was still more than 50 mg/L as a limit for Quality Standard as Class II to dispose to Bedog River. It happened because that amount is only on facultative II, there are still more to go to treat wastewater on the next ponds in WWTP Sewon.

The results of BOD simulation by equation (1),was run by *Maple 2016* software describes the degradation process, where S = BOD out from stabilization pond, mg/ L
S_o= 250 mg/ L
k =0.6
V = 8,085 m^3
Q = 10,697.53 $m^3/time$
t = time period,

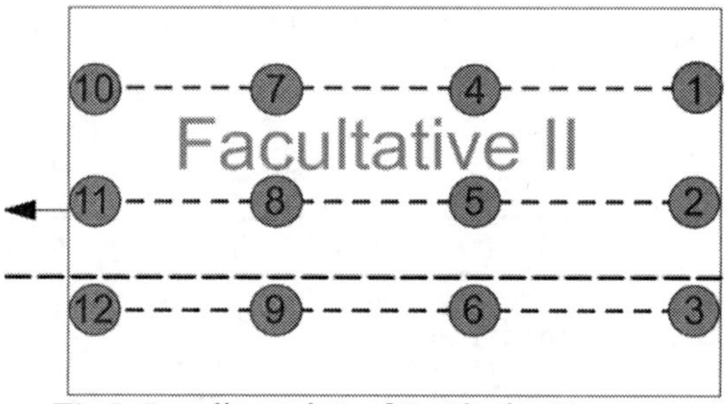

Fig.3. Sampling points of Facultative II

We have exponential equation

$$S = 75.05639885 + 104.9436012e^{-3.173132962\,t}$$

the graphic showed that the BOD decreased exponentially from 180 mg/L.

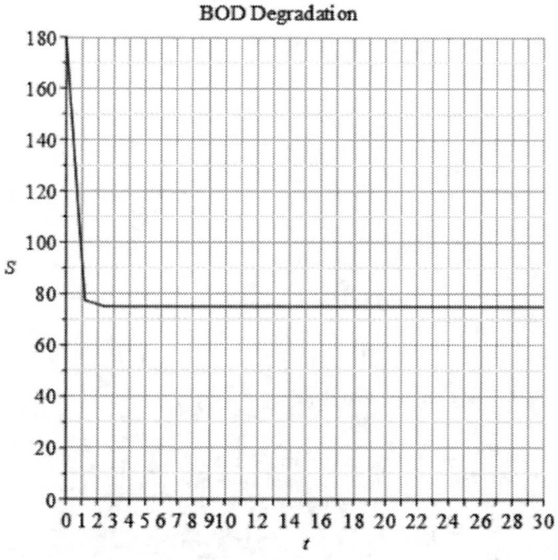

Fig.4.BOD Degradation

3.2.Validation Testing

The Absolute Means Error was calculated to validate the model. By using equation (2), the results of AME between BOD observation and BOD simulation is presented on **Table.2.**below.

 https://doi.org/10.1051/e3sconf/20183105009

Table.2. AME between observation data and simulation data

Time	Biological Oxygen Demand (BOD)	
	BOD observation	**BOD simulation**
06:00 am	60.55	75.056
12:00 pm	69.5	75.056
10:00 am	79.64	75.056
Average	69.897	75.056
AME	**7.38%**< 10% (tolerance limit)	
Conclusion	**Valid**	

4 Conclusion

The mathematical model can be applied to know and predict the value of BOD on the facultative pond. The model was validated by Absolute Means Error and the result was 7.38%. It indicated the model is valid.

The amount of BOD is still greater than Quality Standard class II River. It happened because the measurement and model was only on the facultative II. Actually the BOD will continue be treated by next facultative and maturation ponds.

The authors really appreciate to the leaders and staff WWTP Sewon, Public Works Department, Housing and ESDM Local Government Yogyakarta who have given permission for the research and data collection of wastewater quality control at Sewon Bantul WWTP

References

1. Sunarsih. *Permodelan Lingkungan Kualitas Air Limbah Pada Kolam Stabilisasi Fakultatif.* Disertasi Semarang: Sekolah Pascasarjana, Universitas Diponegoro (2013)
2. Sunarsih.,Purwanto., W. S. Budi. IJTech **4** 689-698 .(2015)
3. Sunarsih, D.P. Sasongko,Sutrisno.AJER**5**,01-05. (2016)
4. V.D.Leite ,G.B.A. Junior, J.T. Sousa, W.S.Lopes, I.N. JUEE **3** 58-62 (2009)
5. M. Farzadki, M.H.Ehrampoush, S. Sadeghi, M. Kermani, M.T. Ghaneian, E.A. Mehrizi. EHEMJ **1** 7-12 (2014)
6. Mara, Duncan. 2004. *Domestic Waswater Treatment in Developing Countries*. EarthScan: UK and USA.
7. WHO-World Health Organization *Health guidelines for the use of water in agriculture and aquaclture. Geneva: World Health Organization.* (Technical Report Series), 77 (1989)
8. S. Kayombo., T.S.A.Mbwette, A.W.Mayo, J.H.Y.Katima, S.E Jorgensen. EM **127**21-31. (2000)
9. S. Sunarsih, P. Purwanto, W. S. Budi.JUEE **7** 293-301. (2013)
10. Sumarlan, N. Heksa,*Physical Study to Determine Wastewater Quality of Was Water Treatment Plant (WWTP) Sewon, Bantul, Yogyakarta at The Opened Drain Canal.* Skripsi Yogyakarta : Universitas Gadjah Mada.
11. Purwanto. 2005. *Permodelan Rekayasa Proses dan Lingkungan*. BadanPenerbitUniversitasDiponegoro: Semarang.
12. Muhammadi.,Aminullah,E., Soesilo, Budhi *Analisis Sistem Dinamis*. UMJ Press. Jakarta. 2001.
13. C. Musi, S. Anggoro, Sunarsih.JEE **18** 25-34.

Potential Of Microalgae *Chlorella vulgaris* As Bioremediation Agents of Heavy Metal Pb (Lead) On Culture Media

Endah Rita Sulistya *Dewi*[1*], *Riza* Nuravivah[2]

[1,2] Department of Mathematics, Sciences and Information Technology, PGRI Semarang University, Indonesia

Abstract. The purpose of this study to determine the ability of *Chlorella vulgaris* in absorbing Pb (lead) and the effect of the variation of Pb metal concentration on the growth of *Chlorella vulgaris*.This study using an experimental study with complete random design with 4 treatments, namely control (without the addition of metal), Pb1 (addition of metal 1 mg / l), Pb3 (3 mg / l) and Pb5 (5 mg / l), respectively 3 replications. Exposure Pb ion in *Chlorella vulgaris* for 7 days. Analysis of the metal content of Pb concentration performed on culture media after exposure it at 3 hours after dispersion *Chlorella vulgaris* and on day 7 of culture using the AAS method. Do also counting the growth of cells each day. The results of the analysis of the average metal content of Pb in the culture medium at the end of the study was the control (0.1980), Pb1 (0.1453), Pb3 (0.4144) and Pb5 (0.5305). While the average growth of *Chlorella vulgaris* at the end of the study were control (630.1116 x 10^4), Pb1 (829.0012 x 10^4), Pb3 (1069.9446 x 10^4) and Pb 5 (808.94450 x 10^4). The results of the analysis of the content of Pb in the F test shown that the difference in concentration of water Pb given real influence on the ability of *Chlorella vulgaris* in absorbing Pb and growth. The conclusion of this study was *Chlorella vulgaris* has the ability to absorb metals in the waters, and the provision of various concentrations of Pb can affect the growth of *Chlorella vulgaris*.

Introduction

Control of the environment from pollution by industrial waste is one of the issues that need to be addressed, because it is potentially causing changes in water quality due to the influx of waste from these activities, particularly industrial waste containing heavy metals, which will lead to increasing pollutants carried by streams towards the estuary and will accumulate in the ocean.

These problems can affect the marine life that is therein to impact the lives of human dependence on marine environment is very large. One of the sewage treatment process to reduce pollution by heavy metals problem is to bioremediation, which is the process of converting waste contamination in an attempt to repair the quality of the environment by utilizing living things. Living creature that has great potential as a bioremediation agent is microalgae. Biological wastewater treatment to reduce heavy metals from polluted water into an alternative technology that has the potential to be developed. One of them utilizing ion exchange capacity and absorption ability of microorganisms to absorb heavy metals [3]; [10], Microalgae *Chlorella vulgaris* is an example that has the ability to absorb metal ions. Where the use of *Chlorella vulgaris* has not been maximized, only limited as a dietary supplement and ingredient manufacture of cosmetics.

Metallic lead most of its compounds are toxic. Lead and compounds can be in the water body naturally and as a result of human activity. Naturally, Pb can get into the water body through crystallization Pb in the air with the help of rain water. In addition, the process corrofication of mineral rocks due to the pounding waves and the wind. While Pb that goes into the water body as a result of the activities of human life there are various forms. Among them is the waste water (sewage) from Pb-related industries, the waste water from the black tin ore and industrial waste discharge the battery. Wastes will fall in pathways such as the waters of the tributaries to then be carried over to the ocean. Generally the discharge line of industrial waste material which uses Pb will damage the system enters aquatic environments (make the river and polluted plot).

The purpose of this research activity, namely: Knowing decreased levels of Pb (lead) to the culture medium *Chlorella vulgaris*, knowing the effect of various concentrations of Pb (lead) on the growth of *Chlorella vulgaris*, knowing the results of research on concentration reduction of Pb (lead) and the growth of *Chlorella vulgaris* in the culture medium

2 Research Methods

Obtained from microalgae *Chlorella vulgaris* BBPBAP Jepara, and cultured in the laboratory of natural feed BBPBAP Jepara. For the analysis of metal test Lead (Pb) is performed at the Laboratory of Chemical UNDIP

* Corresponding author: endahrita@yahoo.co.id

Semarang. This research method using the experimental method completely randomized design (CRD) with 4 treatments three replications.

This research is *Chlorella vulgaris* cultured 7 days old alone for 7 days, the object of research is seawater samples by administering varying concentrations of Pb.

Tools and materials used are 12 glass jars volume of 3 litre, rack culture, haemocytometer, microscopes, pH sticks, aerators, hose aeration, straws, measuring cup, beaker glass, refractometer, fluorescent lights long, pipettes, micro pipettes, filter paper, mineral water bottles, erlenmeyer, stove, *Chlorella vulgaris*, sea water (as an example), fertilizer walne, vitamin B12, heavy metals $Pb(NO_3)_2$, NaOH, HCl, 70% solvent, soap and distilled water.

Working procedures implementing this study begins with the sterilization of tools that will be used. The sea water is also sterilized by boiling until boiling and let cool then filtered using filter paper. To get the salinity of sea water, in this study using a salinity of 25 ‰, using a dilution formula as follows:

VI x S1 = V2 x S2
Information:
V1 = Volume of sea water is sought (L)
S1 = initial seawater salinity (‰)
V2 = volume of sea water necessary to the culture medium (L)
S2 = desired Salinity (‰)
[2]

Mixing of compounds Pb(NO3)2 on each vessel with four variations of concentrations (0 mg / l, 1 mg / l, 3 mg / l, and 5 mg / l). *Chlorella vulgaris* cell density used is ± 1 x 106 cells / ml. The calculation of the number of cells using haemocytometer using the following formula:

$$\text{The number of cells / ml} = \frac{(\text{total number of cells in 3-4 blocks})}{(\text{number of blocks})} \times 10{,}000 \times P$$

[6].

To get cell density of ± 1 x 106 cells / ml, then the dilution calculation using the formula as follows:
V1 x V2 x N1 = N2
Information:
V1 = Volume of *Chlorella vulgaris* necessary for culture
N1 = density cell culture seedlings on stock
V2 = volume of seawater used as a culture medium
N2 = density desired cells for culture
(6).

The calculation of the rate of growth of *Chlorella vulgaris* done every day, using the formula:

$$\text{The number of cells/ml} = \frac{(\text{total number of cells in 3-4 blocks})}{(\text{number of blocks})} \times 10{,}000 \times P$$

Information:
P = Dilution (to do if experiencing difficulty in the calculation for the number of cells is too much).
[6].

Chlorella vulgaris in cultivation for 7 days in media Pb with walne as fertilizer nutrients. Control variables such as temperature, salinity and pH is controlled every day. *Chlorella vulgaris* cultures was centrifued to separate the sea water as a cell culture medium with *Chlorella vulgaris* (as solids). Pb concentration measured using AAS method after 3 hours after the stocking of seeds of *Chlorella vulgaris* and day 7 and measure the growth of *Chlorella vulgaris* every day using a haemocytometer.

3 Results and Discussion

3.1 A decrease in concentration of metals Pb (Lead) by *Chlorella vulgaris*

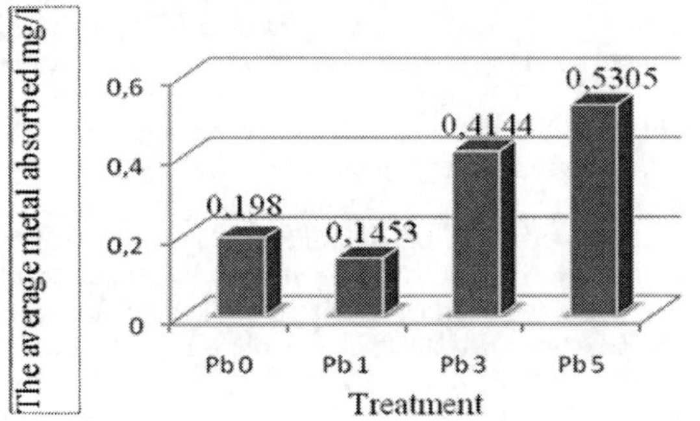

Fig. 1. The decline in the average concentration of Pb

Information:
Pb 0 = Media *Chlorella vulgaris* without the addition of Pb
Pb 1 = Media *Chlorella vulgaris* with the addition of Pb 1 mg / l
Pb 3 = Media *Chlorella vulgaris* with the addition of Pb 3 mg / l
Pb 5 = Media *Chlorella vulgaris* with the addition of Pb 5 mg / l

From the diagram of Figure 1 shows that the highest concentrations of Pb absorbed contained on the treatment Pb 5 (Pb5 = 0.5305), this means that the more the concentration of each metal were absorbed by the *Chlorella vulgaris*. This is in accordance with the results of research by [3], that an increase in metal biosorption ability is directly proportional to the increase in concentration. This absorption occurs because of the OH groups in cellulose *Chlorella vulgaris* which is polar.

There are many things that cause absorption mechanism of Pb (lead) by *Chlorella vulgaris* occurred in the treatment Pb1, Pb3 and Pb5, such as the cellulose in the cell walls of *Chlorella vulgaris*. *Chlorella vulgaris* cell wall structure of cellulose. In the chemical structure of cellulose has an OH group. Cellulose is potentially large enough to be used as a metal ion catcher because of the OH groups are bound to interact with the adsorbate, adsorbate in this study is Pb. OH group causes the polar nature of the adsorbent, the adsorbent in this case is the cellulose contained in the cell wall of *Chlorella vulgaris*. Thus the OH groups in cellulose can absorb more strongly polar substances of the substance less polar. The presence of the OH group causes metal ions Pb absorption mechanism in the cell walls of *Chlorella vulgaris*. Interactions between the cellulose in the cell walls of *Chlorella vulgaris* with Pb ion is an extracellular detoxification mechanism or mechanisms of tolerance. Detoxification is the process of conversion of heavy metals into non-toxic form.

Reduced levels of Pb in the culture medium *Chlorella vulgaris* it can also takes place through the formation of the metal-binding protein found in microalgae, among others metalotionein and fitokelatin. According [11], *Chlorella sp*, cells of phytoplankton metabolic processes can synthesize metal chelating protein fitokelatin to respond to the negative effects of heavy metals. The protein can bind to heavy metals because it has sulfhydryl groups (-SH) and will accumulate in the vacuole, through enzymatic processes. According [13], which contains sulfhydryl groups can easily bind to metal ions that enter the body. Fitokelatin will form complexes with heavy metal and serves as detoksifikan [4]. Fitokelatin synthesized from a tripeptide derivative (glutathione), if the environment is polluted by Pb, then it will form fitokelatin glutathione. Fitokelatin will further establish fitokelatin-Pb which will be forwarded to the vacuole [11]. Microalgae generally have a protective mechanism against toxic metals to keep to his own life. This mechanism involves the formation of metal complexes with proteins in the cell membrane so that the metal can accumulate in the cell without interfering with growth. If so high metal concentrations, accumulation may inhibit cell growth because the organism protection system is no longer able to offset the effects of the toxic metal, other than that it is also related to the size of the cell *Chlorella vulgaris* [12] ; [7].

Associated with the role, Pb is in the cell work in assisting the action of the enzyme for its metabolism. According [13], in carrying out its duties enzymes often require metal or a vitamin or a combination of both as a cofactor and activator. Generally, the active center of the enzyme is a group of metal ions. However, enzymes that have metal ions as active centers tend to be volatile. This is due to metal ions contained in a group of enzymes can often be replaced by other metals that come inside. The circumstances shift metal ions contained in a group of enzymes will be very easy to happen when there is deficiency. Zn and Fe deficiency often causes influx of Pb to replace the function of the metal ions from the group of enzymes. It turned out that the metal ions are signed to replace metal ions that should play a role (in this case Pb), has made the cause of the obstruction of the work ability of the related enzyme. Pb ion thus herein referred to as inhibitors (barrier) if the concentration exceeds the threshold. If there is one type of enzyme that disturbed her, certainly can affect the work of other enzymes. If this is allowed to continuously, it will interfere with the metabolism of the cell body. Simply put, the reaction can be illustrated as follows:

Enzyme + Substrate ⟶ enzyme (enzyme active),
in the presence of metal ions into the reaction,
Enzymes + Metal ⟶ enzyme (enzyme active).

However, if the optimal absorption of Pb concentrations found in treatment Pb5 associated with optimal growth of *Chlorella vulgaris* cells in treatment Pb3. This means the treatment Pb5 (Pb 5mg / l), the more metals are absorbed would interfere with cell growth of *Chlorella vulgaris*. This is in accordance with the previous explanation, that could be a barrier Pb action of an enzyme that would result in disruption of cell metabolism. While on treatment Pb3 optimal cell growth occurred *Chlorella vulgaris*, meaning the concentration of Pb 3 mg/l can still be tolerated by *Chlorella vulgaris*. This is consistent with the statement of [8]; [1], that the metal disc at the microalgae will become an obstacle in the metabolism of cells at high concentrations, but will be useful in the process of metabolism in moderate concentrations.

In addition, the mechanism of absorption of Pb by *Chlorella vulgaris* can also be due to the alginate to the wall sel *Chlorella vulgaris*. According to the research [3[, the alginate is a polysaccharide constituent contained in *Chlorella vulgaris* cell wall can bind with metals Pb, so the presence of metal ions Pb in the water levels are less toxic to the organism are others nearby. The mechanism of ionic bonds between the alginate with Pb are as follows:

3NaAlg + PB3 + Pb $(Alg)_3$ + 3NA.

Based on the chemical reaction, Pb ion binds to alginate so that Pb does not pollute the water because it forms ions are already bonded.

3.2 The growth of *Chlorella vulgaris* cells in Media Metals Pb (Lead)

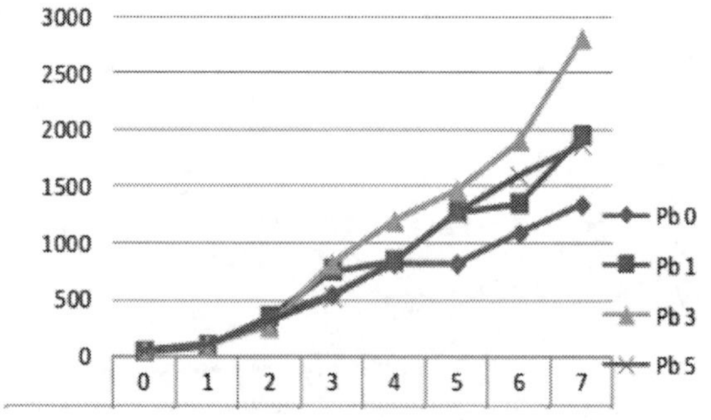

Fig. 2. Growth of *Chlorella vulgaris* cells for 7 days

Information:

Pb 0 = Media *Chlorella vulgaris* without the addition of Pb

Pb 1 = Media *Chlorella vulgaris* with the addition of Pb 1 mg / l

Pb 3 = Media *Chlorella vulgaris* with the addition of Pb 3 mg / l

Pb 5 = Media *Chlorella vulgaris* with the addition of Pb 5 mg / l

Based on the growth data, all treatments are in a phase of adaptation in the day during a 3 hour stocking of seedlings *Chlorella vulgaris*, and day 1 to day 7 an exponential phase. Growth is highest in the treatment of

Pb 3, the lowest in the control treatment was followed treatment Pb 5. This means that at certain concentrations of Pb is required, but at high concentrations it will hamper growth. According to [7], microalgae generally have a protective mechanism against toxic metals to keep to his own life. If so high metal concentrations, accumulation may inhibit cell growth because the organism protection system is no longer able to offset the effects of toxic metals. Decreasing the concentration of Pb is also affected due to nutritional factors and increase the number of cells in the culture medium.

Another factor influencing the growth of *Chlorella vulgaris* cells is salinity. Initial salinity used in this study was 25 ‰ and continues to increase each day until it reaches the highest salinity on the 7th day that is 50 ‰. The increase in salinity during the culture can occur due to the metabolism of the cell or precipitation of salts and nutrients in the medium. The salt concentration in the medium increases due to the evaporation of sea water used as a culture medium for the heat lamp TL which is close to the jar for culture. This is evidenced by the discovery of salt deposits on the surface of the white wall of the jar mouth and culture during the study. According [5] ; [9], environmental influences such as temperature, salinity, pH and dissolved oxygen levels in the water can also affect the toxicity of metals in water to living things in it. With the decline in value of salinity will increase the toxicity of the metal.

4 Conclusion and Recommendations

1. The microalgae *Chlorella vulgaris* giving effect to the decrease of various concentrations of Pb in the culture medium, with the highest absorption is Pb metal on the treatment Pb5, and absorption of Pb smallest Pb1 treatment.
2. Related to the provision of various concentrations of Pb influence on the growth of *Chlorella vulgaris*. At concentrations of Pb 1 mg / l and 3 mg / l can be tolerated by the cell *Chlorella vulgaris*, and optimal growth contained in the treatment Pb3, while at a concentration of 5mg / l can no longer be tolerated by cells of *Chlorella vulgaris*, it is characterized by decreasing the number of cells *Chlorella vulgaris* in the culture medium.

References:

1. Andersen. *Algal Culturing Techniques*. Oxford: Elsevier Academic Press, (2012)
2. Arrokhman. Survival Rate Ikan Bawal Bintang *(Trachinotus blochii)* dalam Media Pemeliharaan Menggunakan Rekayasa Salinitas Jurnal Sains Dan Seni Its Vol. 1, No. 1, (Sept. 2012) ISSN: 2301 – 928x, (2012)
3. Aunurohim. Jurnal Sains dan Seni Pomits Vol. **3**, No. 1, 2337 – 3520, (2014)
4. Cobbet, CS. *Phytocelatin Biosynthetis and Function in Heavy Metal Detoxification*. Curr. Opin. Plant. Biology.3, (2000)
5. Darmono. *Lingkungan Hidup dan Pencemaran Hubungannya dengan Toksikologi Senyawa Logam*. Jakarta: UI Press, (2001)
6. Fardias, S. *Teknik Dan Prosedur Dasar Laboratorium*. Jakarta: PT Gramedia, (1985)
7. Hala,Y., Paulina Taba, and Febrianti Rut Langan. Potensi Mikroalga Laut *Nannochloropsis salina* Sebagai Biosorben Ni^{2+}, Cu^{2+}, dan Zn^{2+} dalam Medium Conwy Terkontaminasi Campuran Tiga Logam Chem. Prog. 5 (2),86, (2012)
8. Held, Paul and Keri Raymond. *Determination of Algal Cell Lipids Using Nile Red Using Micro Plates to Monitor Neutral Lipids in Chlorella vulgaris* AN071211_08, Rev. 07/12/11. bioTek Application Note, (2011)
9. Hermanto, Perancangan Bioreaktor Untuk Pembudidayaan Mikroalga. Jurnal Teknologi Pertanian Vol. 12 No. 3 (Desember 2011) 153-162, (2011)
10. Hlihor, RM., Diaconu, M. Chelaru, C. Sandu, I., Tavares, T. and Gavrilescu J. Environ. Res., 7(3), 581-594, ISSN: 1735-6865, (2013)
11. Lehniger, A L., Nelson D.L., and Cox M.M. *Principles Of Biochemistry*, Second Edition. New York: Worth Publishers (1993)
12. Manuela D.Machado, Eduardo V. Soares, Helena. MVM. Soares. Journal of Hazardous Materials. **180**: 347-353, (2010)
13. Palar, Heryando. *Pencemaran dan Toksikologi Logam Berat*. Jakarta: Rineka Cipta, (2008)

Good Housekeeping Implementation for Improving Efficiency in Cassava Starch Industry (Case Study : Margoyoso District, Pati Regency)

Wijayanto Setyo Aji[1,*], *Purwanto* P[1,2], and *Suherman* S[1,2]

[1]Master Program of Environmental Science, School of Postgraduate Studies, Diponegoro University, Semarang - Indonesia
[2]Department of Chemical Engineering, Faculty of Engineering, Diponegoro University, Semarang - Indonesia

Abstract. Cassava starch industry is one of the leading small-medium enterprises (SMEs) in Pati Regency. Cassava starch industry released waste that reduces the quantity of final product and potentially contamined the environment. This study was conducted to observe the feasibility of good housekeeping implementation to reduce waste and at the same time improve efficiency of production process. Good housekeeping opportunities are consideration by three aspect, technical, economy and environmental. Good housekeeping opportunities involved water conservation and waste reduction. These included reuse of water in washing process, improving workers awareness in drying section and packaging section. Implementation of these opportunities can reduce water consumption, reduce wastewater and solid waste generation also increased quantity of final product.

1 Introduction

Tapioca starch industries are one of the leading Small and Medium Enterprises (SMEs) that played important role of Pati Regency economy. Tapioca starch industries in Pati can produced 146.491 ton tapioca starch per year and have production value approximately 200 million dollars per year [1]. Tapioca starch is produced from cassava root and used for food industries especially cracker industries. Currently, the number of tapioca starch industry in Pati Regency were 600 units in 2014 and increasing every years. Tapioca starch manufacture spread in five districts with the highest number of manufactures located in Margoyoso district. Generally, production process of tapioca starch running for nine months, depending on the harvest season and raw material availability.

Production process of tapioca starch consist of 7 major step, cutting rods cassava, washing and peeling cassava, rasping cassava, pulp extraction, starch pulp separation, drying, milling and packaging.

Currently, tapioca starch industry are often causes environmental problems because of the waste, either from inefficient production process or from by-product of production process. According to the study of Prayitno (2009), the generation of wastewater at the tapioca starch industries in Margoyoso was 7.11 m^3[2]. Chavalparit & Ongwandee reported the approximate generations of wastewater and solid waste are 12 m^3 and 3 kg per ton of tapioca starch [3]. In addition, on 2016 farmers fish-ponds in Langgenharjo wasn't completed harvest because of water ponds contaminated with waste from tapioca starch activities. Therefore, it needs action to reduce the negative impact of tapioca starch to environment.

An approach which can be applied to improve efficiency and reduce waste generation is cleaner production. Cleaner production is the continuous application of an integrated, preventive environment strategy towards processes, products, and services in order to increase overall efficiency and reduce damage and risks for humans and the environment [4]. Cleaner production as a proven effective and economically viable approach for waste prevention and management for SMEs in developing countries [5].

One of the tools that used in the application of cleaner production is good housekeeping (GHK). Good housekeeping associated with changes the procedural action, administrative and institutional with purpose to preventing the generation of waste, enhance organizational and workplace safety. Good housekeeping activities such as, human resources development, prevention of loss material and resources, repair procedures for handling and controlling the supply of materials or resources, improvement of production schedules and separation of waste by type.

The implementation of good housekeeping is relatively easy and can be accomplished with low cost even no investment cost [6]. As we know, the reasons why SMEs do not engage in environmentally friendly practices or make environmental improvements are their perceptions that these represent a cost and most of them are skeptical about potential cost savings [7]. Thus, this tool are particularly convenient for Small and Medium Enterprises especially for tapioka starch industry. Based on study of Kumalasari et al, stated that GHK action on

* Corresponding author: setyoajiwijayanto@gmail.com

small scale cracker industry by increasing the awareness of employees at production process and clean impurities on production tools before washing can reduce water consumption to 22.04% [8]. Changes in standard operational procedure for process of shipping wine it has been proven to reduce environmental impact to 13 percent [9].

The objective of this paper is to develop good housekeeping opportunities for tapioca starch industry in order to improve production efficiency and reduce waste generation.

2 Method

This study was located on UD Sinar Cerah in Margoyoso District. In this study, a systematic methodology to improve efficiency and reduce waste generation consists of five step as follows :

1. Observed of production process and collecting information related consumption of resources and wastes generation.
2. Detailed evaluation and measurements of consumption of resources and wastes generation. During evaluation, production process were analyzed in detailed and translated into a flowsheet diagram that showing input and output of resources nad waste generation at each step of production process. This flowsheet shows the type of resources consumed, the type of waste generation and the sources of loss in the production process. These results served as basic to develop good housekeeping opportunities.
3. Proposing good housekeeping opportunities based on evaluation and measurement results.
4. Evaluation the feasibility of good housekeeping opportunities in term of technical, economical benefits and environmental impacts. Feasibility studies were performed to prioritize the most feasible and beneficial options. However, since small and medium scale enterprises have very limited financial resources and capabilities, the alternatives good housekeeping must therefore be economically reasonable and affordable [10].
5. Recommendation of alternatives good housekeeping opportunities to improve efficiency and reduce waste generation.

3 Results and Discussion

3.1 Company Profile

Production capacity of UD Sinar Cerah per day approximately 10 ton of tapioca starch and using cassava roots about 50 ton. The numbers of workers are varies because of their status only daily labor dan hired labor, but the numbers on average is 40 workers per day. Production process run daily from 07.00 to 15.00 with break time on 11.30 until 12.30. Sales area for UD Sinar Cerah is cover up Bandung and Surabaya areas to provided cracker industry.

3.2 Production Process

Production process of tapioca starch at the study location shown in Fig. 1. The flowsheet diagram are based on 50 ton of cassava as raw material.

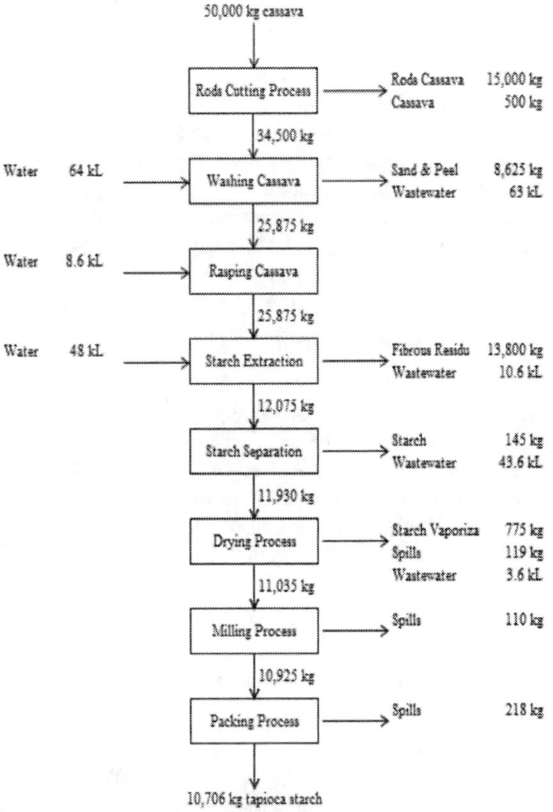

Fig. 1. Flowsheet diagram of tapioca starch production process.

Cassava roots are firstly delivered to a cutting rods process. This process held to remove stick that attached on the roots during harvesting, because the rods would affecting in the grinding process of cassava. This process is realized manually with special knife. After cutting, cassava are sent to washing process to sand removal and peeling process. In the end of peeling process, cassava roots sprayed again with clean water to ensure cassava roots was clean and ready for the rasping process. The rasping process aims to minimize size of cassava so that starch content can be extracted. Pulp cassava the results of rasping process then mixed with water to facilitate the process and then pumped to extraction process.

Extraction process has purpose to separate starch slurry and fibrous residue. During extraction process, water sprayed constantly from above to take starch from pulp. Fibrous residue (cellulose) will restrained above filter to be passed to shelter fiber and water contains tapioca slurry will pass through a filter and pumping into deposition tank to separate starch from others substances. Deposition process aims to separate pure tapioca starch by the other part such as water or contaminant substances. Deposition process lasted for approximately 4 hours, and resulted starch cake with

https://doi.org/10.1051/e3sconf/20183105011

water content about 40%. Water and others contaminant substances are dispose to environment without any treatment.

Drying process have objective to reduce water content of tapioca cake to achieve tapioca starch with water content less than 12%. Water content above 15% can cause tapioca starch be moist and quickly broken. Drying process are undertaken by conventional way by using sunlight because of costumer requirement. Drying process take time approximately 4-6 hours depend on weather and thickness of tapioca layer in drying floor. The result of drying process are coarse tapioca starch. Coarse tapioca starch then forwarded to milling process in order to size reduction and finally going to packaging process.

3.3 Non Product Output (NPO)

Non product output (NPO) is raw materials, supporting material, water and energy that used for production process but not contained in final product. By that definition, then waste and loss of production is part of NPO. According to the analysis of production process and cost production calculation, shows that NPO of tapioca starch production process reached 64.19%. NPO breakdown per step of production process shows in fig. 2 as below.

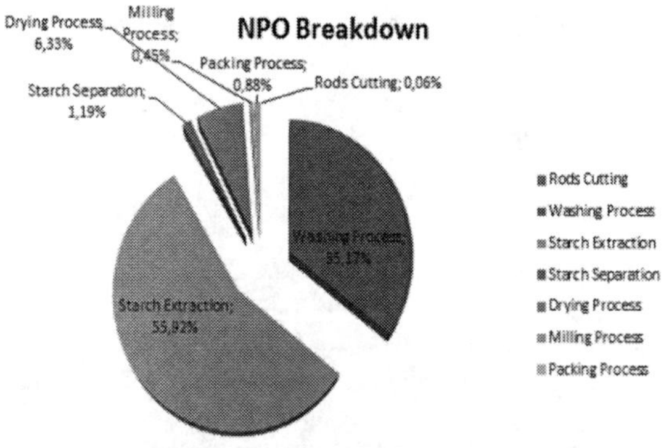

Fig. 2. NPO breakdown per step of production process

Fig. 3. Workers was doing cutting rods cassava.

3.4 Good Housekeeping Alternatives

Based on analysis of tapioca starch production process, several alternatives of good housekeeping were suggested and have been potentially implemented to improve production efficiency and reduce the waste generation. The good housekeeping alternatives were:

3.4.1 Utilize pads to rods cutting process

Currently, the process of rods cutting running with one hand held cassava and the other hand cut the rods without using pads. This certainly risk of occupational safety and potentially reduce the quantity of raw materials if workers careless and cutting the cassava come together rods. Alternative that can suggested is utilize pads while rods cutting process. The utilize of pads expected the cutting process more precisely and can minimize potential loss of cassava so the quantity of raw materials will rise.

3.4.2 Re-use water of washing process stage 2.

Washing process stage 2 is spraying process to cleaning cassava after peeling process. At this time, water from this process directly disposed to drain. Actually, the water from this process can be reused to washing process stage 1. Change of this process will able to reduce water consumption in the washing process.

3.4.3 Increasing employee's awareness at drying process.

Fig. 4. Condition of the floor after drying process.

After drying process finished and tapioca starch moved to milling and packaging room, there were still founded tapioca starch left on the drying floor. This thing happened because of lack of employee's awareness about production efficiency and waste prevention. Tapioca starch that left behind should be can still continue to next process and increasing the number of final product. Good housekeeping alternative that can be suggested is to increasing employee's knowledge so can raise awareness to not leaving material that can still continue to next process.

https://doi.org/10.1051/e3sconf/20183105011

3.4.4 Collecting Spills of Tapioca Starch at Milling and Packaging process.

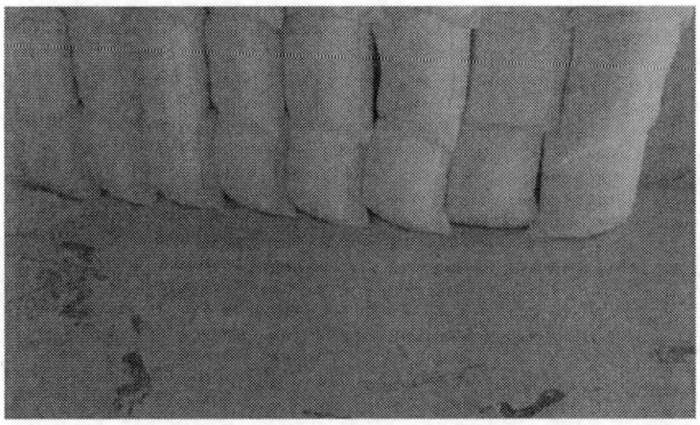

Fig. 4. Floor condition of milling and packaging room.

At milling and packaging process, spills of tapioca starch happened because of size of starch is small so easy to down. In addition, spills of tapioca starch also happens when packing process that still running manually with put the first starch on the flour before moved to sack. Good housekeeping opportunities can be applied is recollecting spills of tapioca starch using broom. Actually, broom starch has been provided, but it was rarely used by workers to collect the spills. By giving knowledge and counseling to workers that the spills of tapioca starch has actually product that can be sold but they have wasted is expected to reduce amount od spills and increase quantity of final product.

4 Conclusions

Tapioca starch manufacture is important in Pati regency, but often causes environmental problems because of their inefficient process and by-product generated. One of approaches that can be implemented to prevent the problems is good housekeeping. Based on analysis, alternatives good housekeeping opportunities that can be suggested such as, utilize pads on rods cutting process, increasing employee's awareness at drying process, collecting spills of tapioca starch at milling and packaging process.

References

1. S. Suroso, Department of Research and Development Kabupaten Pati (2016) (Unpublished).
2. H.T. Prayitno, Thesis Master of Environmental Science School of Postgraduate Studies Diponegoro University (2009)
3. O. Chavalparit, M. Ongwandee, J. Clepro **17,** 105-110 (2009).
4. United Nation Industrial Development Organization (UNIDO), *Manual on the Development of Cleaner Production Policies – Approaches and Instruments*, UNIDO CP Programme (2002)
5. R. Rahim, A.A.A. Raman, J. Clepro **101,** 215-221 (2015).
6. Sustainable Bussiness Associates (SBA), P3U Working Paper No. 9e (1998).
7. F. J. Saez-Martinez, C. Diaz-Garcia, A. Gonzalez-Moreno. J. Sustainability **8,** 898 (2016).
8. N. Khuriyati, W. Wagiman, D. Kumalasari, Journal of Agriculture and Agricultural Scienci Procedia **3,** 102-107 (2015).
9. D. Amienyo, C. Camilleri, A. Azapagic, J. Clepro **72,** 110-119 (2014).
10. H. Schollenberger, M. Treitz, J. Geldermann, J. Clepro **16**, 1856-1864 (2008).

 https://doi.org/10.1051/e3sconf/20183105012

Use of a germination bioassay to test compost maturity in Tekelan Village

Wiharyanto Oktiawan[1,*], *Badrus* Zaman[1], and *Purwono*[1]

[1]Department of Environmental Engineering, Faculty of Engineering, Diponegoro University, Semarang - Indonesia

Abstract. Livestock waste from cattle farms in Tekelan village, Getasan Subdistrict, Semarang Regency can be grouped into three types, namely solid waste, slurry and waste water. Solid waste (cow dung) was processed into compost, while slurry and waste water were used to make liquid fertilizer. This compost was used as a component of planting media in horticultural crops and potted plants production. We evaluated the toxicity (phytochemical and ecotoxicological) test of compost by using germination index (GI). *Vigna radiata* seeds are sown on filter paper dampened with compost extract for different times. GI was calculated by relative germination (G) and relative radical length (L). The germination index (GI) = G / G0 x L / L0 x 100, where G0 and L0 are values obtained by distilled water as a control. The results showed that germination bioassay and radical length using aquades and groundwater in Tekelan village did not affect the radical length of *Vigna radiata* . Technically, groundwater in Tekelan village can be used as a germination bioassay control. The cow dung compost substrate appears to have a major influence on compost toxicity. Mature compost was produced on day 14 with a GI of 104.03.

1 Introduction

Tekelan Village Community, Getasan Sub-district, Semarang Regency mostly work as vegetable farmers and cattle farm. Livestock waste from cattle farms can be grouped into three types, namely solid waste, slurry and waste water. Communities utilize farm and animal waste to make. This is an appropriate step for treating solid waste. This compost was used as a component of planting media in horticultural crops and potted plants production. These activities are carried out in order to realize organic farming in Tekelan Village.

According to [1], [2] the cultivation of plants in the soil can cause problems due to changes in salinity, soil pathogens, and changes in some of the soil physical and chemical properties that are difficult to solve. This change is due to microbiological activity between compost and biological activity in the compost. Biological processes will continue in a reversible manner until the rob nut nutrients of soil [3].

In order to guarantee the safety of the compost during its use in the agriculture, or as an amendment in degraded soils, certain quality criteria should be filled. These are related to the content of pathogens, heavy metals, organic matter, nutrient content, stability and maturity [4-7]. Good compost quality for plant growth is based on maturity [8] which shows the relationship between compost quality and plant growth. The aerobic biological activity is based on the compost stability of the relationship between compost quality and biological activity in compost [3], [8]. Inadequate and stable compost will inhibit plant growth and negatively impact on soil quality [9]; [10]; [11].

The maturity level of compost can be tested in the laboratory by means of physical, chemical and biological parameter analysis, including microbiological activity [12]. Physical parameters include odor, color, temperature, particle size and inert material. Chemical parameters include carbon and nitrogen analyses (C / N ratio in solid and water extract), cation exchange capacity (CEC, total organic-C ratio), water-soluble extract (pH, EC, organic-C, ions) Nitrogen minerals (NH_4-N content, NH_4-N/NO_3-N ratio), pollutants (heavy metals and organics), Organic matter quality (organic composition: lignin, complex carbohydrates, lipids, sugars), and humification (humification indices and Humic-like substances characterization: elemental and functional group analyses, molecular). Biological parameters include microbial activity indicators, phytotoxicity (germination and plant growth tests), pathogen and ecotoxicity tests [13].

Based on [14] phytotoxicity tests such as germination rate (GR) and RE tests have many advantages. The test is simple, inexpensive and requires only a relatively small sample. Moreover, the seeds can still be used for a long time. Simple and highly reproducible methods, applicable in situ and in vitro, there is no requirement for main equipment, minimal maintenance costs, few required samples (eg water, effluent, soil, sediment), no season, seeds can be easily purchased everywhere, the seeds remain long, and the germination is fast.

In this study, we evaluated the use of bioassay germination as an indicator of compost maturity made by

* Corresponding author: purwono.ga@gmail.com

 https://doi.org/10.1051/e3sconf/20183105012

Tekelan Village Community, Getasan Subdistrict, Semarang Regency, Indonesia.

2 Method

The compost samples used for the germination index test were from tekelan village, Getasan Subdistrict, Semarang regency, Indonesia with coordinates 7 ° 25'50.03 "S and 110 ° 26'17.22" T. The sample was taken from the composting process pile on days 0, 7, 14, and 21. Samples were extracted using aquades with solids ratio: water 1:10 gr / l. The compost mixture is shaken and played with jartest for 30 minutes. Then the sample is filtered with filter paper. Germination test using petri dish (diameter 10 cm) which has been included media planting (tissue paper). Add 10 *Vigna radiata* to the medium, then add 10 ml of compost extract. Petri dishes were incubated in the dark for 72 hours. The control is carried out in the same way by replacing 10 ml of the compost extract replaced by 10 ml of aquades. After 72 hours incubation, observe the number of seeds affected by negative effects, normal germinated seeds and *radicle lengths* on tissue paper, both in the addition of compost extract and control.

Relative seed germination and relative radicle length were calculated using equations (1) and (2). The germination index (GI) is calculated using the formula GI = (G / G0) x (L / L0) x 100, where G is the number of normal germinated seeds on the sample substrate, G0 is the number of normal germinated seeds in the control. Where L is the average of plant roots on the sample substrate and L0 is the average length of plant roots on the control substrate.

3 Results and Discussion

3.1 Water control

Aquadest and groundwater are used as bioassay germination control substrate. Table 1 shows germination data and *radicle length* of green beans (*Vigna radiata*) which are fed to different control substrates. The test is done on green beans *Vigna radiata* which is equal to 10 unit. The germination of green beans (Vigna radiata) is said to be normal when the seeds show the potential to continue growing into complete plants including roots, hypocotyls, cotyledons, epicotes, and leaves [15]. The results showed that seed germination and *radicle length* did not show significant differences between aquades and groundwater control substrates. An average number of seeds averaged aquades substrate and ground water substrate respectively as much as 9.70 and 9.80 pieces. While the average teaching length of 2.15 and 2.16 mm.

Based on field studies, real conditions in Tekelan Village, ground water is derived from the spring under the slopes of Mount Merbabu. This indicates that the groundwater can be used instead of aquadest. For practical reasons, the control water selected for mature compost testing is ground water.

Table 1. Germination data and radicle length of green beans Vigna radiata on the control substrate of aquadest and groundwater.

substrate	Normal seed (unit)	radicle length (mm)
Aquadest	10	2,4
	10	2,5
	10	2,0
	10	2,5
	9	2,5
	10	2,0
	10	2,0
	10	1,8
	9	1,8
	9	2,0
Average	**9,70**	**2,15**
Groundwater	10	2,5
	9	2,2
	10	2,0
	10	1,9
	10	2,5
	10	2,0
	10	2,5
	9	2,0
	10	2,0
	10	2,0
Average	**9,80**	**2,16**

3.2 Germination seed

The compost sample was taken from the composting stack of cattle dung on days 0, 7, 14, and 21. This selection was based on the influence of composting time on the quality of the compost identified from the temperature parameters (mesophilic, thermophilic, cooling). In other words, we will know the mesophilic, thermophilic, and cooling effects on the quality of the compost. The germination data and the length of roots of green beans (Vigna radiata) which were incorporated into the different substrate of compost substrate are shown in

Table 2. The test was performed on Vigna radiata which was equal to 10 pieces.

Substrate on day	Average normal seed (unit)	Average radicle length (mm)
0	3,0	2,0
7	6,5	2,3
14	8,0	2,8
21	9,5	3,5

The results showed that the average number of green beans (Vigna radiata) most commonly produced from 21 days compost substrate as much as 9.5 pieces. On day 0, 7th, and 14th has a value below 9.5 pieces. This difference indicates the presence of germination inhibitors on each substrate. Phytotoxic compounds such as phenol, ethylene and ammonia compounds, excessive salt accumulation, and organic acids can inhibit seed germination [16]. The same thing happens to the roots of the roots of green beans. The average *radicle length* of 3.5 mm produced from the compost substrate on the 21st day. At the end of the test of germination, the seeds that do not sprout are said to be dead, dead seeds and hardened seedlings [15].

3.3 Germination Index (GI)

The response of green bean plants *Vigna radiata* to the toxicity of the compost extract during the cow dung composting period in relative seed germination and the relative root extension percentage is shown in Table 3.

Table 3. Germination Index (GI) value of cow dung compost extract at composting time of 0, 7, 14, and 21 days

Substrate on day	GI
0	29,13
7	58,31
14	104,03
21	173,35

At the beginning of composting, green bean GI values *Vigna radiata* were very low with an initial value of 29.13 compared with groundwater controls, probably due to the phytotoxic effects of ammonia and low molecular weight organic acids [16], but when composting took place, GI increases. According to [16], compost is considered mature if the GI value is higher than 60%, compared with control. By the time the composting lasted for 14 days, the value increased to 104.03 and on the 21st day to 173.35. This suggests that the phytotoxicity of the compost pile has been lost. Loss of phytotoxicity is a measure of the maturity level of compost [16, 17].

Based on GI value, cow dung compost in Tekelan village matured on the 14th day.

Conclusion

This study showed that germination test and root elongation test using aquades and groundwater in tekelan village did not affect the length of roots of green beans (*Vigna radiata*). Technically, groundwater in tekelan village can be used as a germination test control. The cow dung compost substrate appears to have a major influence on compost toxicity. Mature compost was produced on day 14 with a GI of 104.03. Further research will be more interesting by ensuring the tekelan villagers can apply this test to other plant fertilizers.

The acknowledgments

Thank you to DRPM DIKTI for funding this research through IbM grant with contract number 007/SP2H /PPM/DRPM /V/2017 financing year 2017.

References

1. N. F. Gariglio, A. Pilatti, D. E. Gonzalezrossia, M.R. Acosta, New Zealand Journal of Crop and Horticultural Science, Vol. **30**: 135-139, (2002).
2. E. Martinez Caldevilla, Garcia Lozano, M. Cultivos sin suelo. Hortalizas en clima mediterraneo. Compendia de Horticultura **3**, 123 p, (1993).
3. S. Mangkoedihardjo, J. Appl. Sci. Environ. Mgt. September, Vol. **10** (3) 83 – 85, (2006)
4. M. Soumaré, F.M.G. Tack, M.G. Verloo, Waste Manage. **23** (6), 517–522, (2003)
5. C. Tognetti, M.J. Mazzarino, F. Laos, Bioresour. Technol. **98** (5), 1067–1076, (2007)
6. M.T. Barral, A. Moldes, Y. Cendón, F. Diaz-Fierros, Waste Manage. Res. **25** (2), 99–108, (2007)
7. P. Wang, C. M.Changa, M. E. Watson, W. A. Dick, Y. Chen, H. A. J. Hoitink Soil Biol Biochem **36**: 767-776, (2004).
8. R. H. Llewelyn, *Development of standard laboratory based test to measure compost stability. The Waste & Resources Action Programme* The Old Academy, 21 Horse Fair, Banbury, Oxon OX16 0AH, (2005).
9. R. Riffaldi, R. Levi-Minzi, A. Pera, M. De Bertoldi, Waste Manage. Res. **4** (4), 387–396, (1986)
10. L. Wu, L.Q. Ma, G.A. Martinez, J. Environ. Qual. **29** (2), 424–429, (2000)
11. J.C. Tang, Y. Inoue, T. Yasuta, S. Yoshida, A. Katayama, Soil Sci. Plant Nutr. **49** (2),273–280, (2003)
12. A. Cesaro, V. Belgiorno, M. Guida, Resour. Conserv. Recycl. **94**, 72–79, (2015)
13. M.P. Bernal, J.A. Alburquerque, R. Moral, A review. Bioresource Technology **100**, 5444–5453, (2009)
14. A. Priac, B. Pierre-Marie, G.C. Gre, C. R. Biologies **340**, 188–194, (2017)
15. *International Seed Testing Association, International rules for seed testing*, Seed Sci. Technol., 27: 333, (1999)
16. S. M. Selim, M. S. Zayed, H. M. Atta, Nature and Science **10**(2), (2012)
17. F. Zucconi, and M. de Bertoldi, Elsevier Applied Science. pp, 30-50, (1987).

A Review on Landfill Management in the Utilization of Plastic Waste as an Alternative Fuel

Nurul Hidayah[1,*] and *Syafrudin*[1,2]

[1]Master Program of Environmental Science, School of Post Graduate Studies, Diponegoro University - Indonesia
[2]Environmental Engineering Dept, Faculty of Engineering, Diponegoro University - Indonesia

Abstract. Wastes from landfills originate from many spheres of life. These are produces as a result of human activities either domestically or industrially. The global plastic production increased over years due to the vast applications of plastics in many sectors. The continuous demand of plastics caused the plastic wastes accumulation in the landfill consumed a lot of spaces that contributed to the environmental. In addition, economic growth and development also increased our demand and dependency on plastics which leads to its accumulation in landfills imposing risk on human health, animals and cause environmental pollution problems such as ground water contamination, sanitary related issues, etc. The management and disposal of plastic waste have become a major concern, especially in developing cities. The idea of waste to energy recovery is one of the promising techniques used for managing the waste of plastic. Hence, this paper aims review at utilizing of plastic as an alternative fuel.

1. Introduction

The rising environmental problems, involving greenhouse gases and air and water pollution together with the energy crisis and resource scarcity will become more imminent in the coming decades, dictating that we need to take actions towards a more sustainable society. Energy are mainly produced by using fossil resources which causes the release of CO_2 in the atmosphere together with other toxic compounds (such as volatile organic compounds and nitrogen oxides). It is estimated that 90% of the global emission in CO_2 (which reached approximately 34billion tons of greenhouse gases in 2011) was derived from the combustion of fossil fuels [1]. Important challenge is the rising accumulation of wastes in the environment. The waste accumulation in the environment has raised the public awareness due to the problems the amount disposed into the environment causes. On the other hand, the huge volumes that are produced globally alongside with the diversity that these waste present makes them ideal candidates to be used for high value applications [2].

Conventional methods that are widely used for the treatment and management of municipal solid waste (MSW) include, landfilling, incineration, composting, etc., present some negative characteristics. Waste disposal by landfilling cause severe environmental issues such as uncontrolled release of methane in the atmosphere, a gas that has 20 to 23 times higher GHG (greenhouse gas) potential than CO_2, production of leachate which contaminates the soil and the ground water, unpleasant odors and spread of pathogenic microorganisms [3]. For example, more than 95% of food waste (FW - estimated to be between 25% and 70% of MSW) are ending up in landfills, which has a catastrophic impact on the climate due to the release of methane and other GHGs (which is calculated to be 125 m^3 of gas per ton of land filled FW) [4]. Nevertheless, the potential of wastes such as a plastic to produce valuable fuels is an attractive alternative solution that has gain the interest of both scientific and public opinion [5].

Waste management, i.e., the collection, transport, recovery and neutralization of waste, as well as the supervision of these activities and the locations where the waste is neutralized, are among the most important issues in terms of environmental protection. The importance of waste management continues to grow due to the progress of civilization and the increase in waste quantity [6]. Rapid and unplanned urbanization leaves municipalities largely overwhelmed when it comes to the collection and disposal of increasing amounts of waste. Municipal solid waste management continues to remain one of the most neglected areas of urban development and this has called for sustainable means of management. It is characterized by a number of waste types and of these are plastic solid wastes (PSW) [7]. Currently, various forms of plastics are used around the world. Large amounts of plastic are used in packing films, wrapping materials, shopping and garbage bags, fluid containers, toys, household industrial products and building material. However, the benefits of plastic use are suppressed by its harmful impacts on the

*Corresponding author: nurul.hida1@gmail.com

environment. Plastics are a small but significant component of waste streams [8].

The invention of plastics is a major milestone that led to improvement in the quality of the lives of human beings. Since its first synthesis in early 1900s, plastics have substituted many types of materials such as wood, metals and ceramics in production of consumer products, as they are light, durable, resistant to corrosion by most chemicals, diversity of applications, ease to processing and low cost. Other than the mentioned advantages of plastics, studies have shown that plastic-based products are responsible for reduction in cost of production in different fields of human endeavor [9].

Plastics are produced in large quantities due to the high demand of their use in agriculture, households, automobiles, packing materials, toys, electronics, and a variety of other applications [10]. Plastics are basically long hydrocarbon chained organic compounds synthesized from petroleum products.

They have become highly popular in short time because of their unmatched usability, wide range of application, non-degradable nature and low cost. As per the estimate given by Plastic Europe, the global production of plastic has crossed 280 million tones in 2011 and it is increasing exponentially [11]. The ever increasing demand for plastic has also produced a huge amount of plastic waste and they pose a serious threat to the environment because of their disposal problems[12]. A survey report of the Central Pollution Control Board, India in 2012 shows nearly 5.6 million tones of plastic waste is generated annually in India. Out of this only up to 60% is recycled that means an astounding 6500 tones of plastic waste is going as landfills per day.

In order to minimize the adverse environmental impacts, Plastic Waste Management was implemented for controlling and reducing the plastic waste generated [13].The increase in plastics use correspondingly increases the amount of plastic wastes being produced. Plastic waste can be classified into at least two categories: municipal and industrial [14]. Industrial plastics are generally more homogeneous and contamination free, making them useful for down cycling into lower-grade plastic products.

Municipal plastics tend to be more heterogeneous and contain extraneous materials. In general, about 10.6 ±5.1wt% of municipal wastes consists of plastics [15]. Municipal plastic waste consists primarily of low-density polyethylene (LDPE), high-density polyethylene (HDPE), poly(ethylene terephthalate) (PET), polypropylene (PP), poly-styrene (PS), and poly(vinylchloride) (PVC) [14].

Overall, about 50 - 70% of the total plastic waste is packaging materials derived from polyethylene, polypropylene, polystyrene, and polyvinylchloride [16]. On average, polyethylene makes up the greatest fraction of all plastic wastes (69%), especially plastic bags [15], and polyethylene comprises 63%of the total packaging waste [16].

2. Environment Global Issues

Plastic solid waste is a major contributing factor towards the waste generated on a global level. Disposal of polymer is becoming a global issue due to high production and consumption of polymer materials [17]. Many countries have their different waste generation level based on their income level become a serious issue for disposal and management of PSW. Management of waste is a complex process because of the requirement of various information from different sources such as influencing factors in waste generation, forecasts of vast quantities and reliable data [18]. In Eastern and Central Asia, the waste generated per year is at least 93 million tones. The per capita waste generation ranges from 0.29 to 2.1 kg per person per day, with an average of 1.1 kg/capita/day [19].

Solid waste management contributes less than 5% of global greenhouse gas (GHG) emission[20]. In response to growing concerns about the threat of climate change, international action aimed at reducing greenhouse gas (GHG) emissions is accelerating and the solid waste management sector is expected to contribute [21]. Land filling leads to contamination of earth's surface. Further, land filling leads to emission of carbon dioxide gas. Plastic also contributes towards the health related issues like Skin corrosion/irritation, Aspiration hazard, serious eye damage/eye irritation etc. It takes much time for polymerization reactions to complete and, therefore, unreacted residual monomers often found in the polymeric material, many of which are hazardous to human health and the environment [22].

Plastic disposal is a major concern in many countries, including the United State. After its initial use, over 60% of the total plastic solid waste (PSW) produced is discarded in landfills throughout the world [15]. Less than 10% of plastic waste is recycled. This is problematic, as plastic waste is a major environmental threat due to its non-degradability[23], its potential health risks to aquatic and terrestrial animals [24], and its impact on environmental pollution[22]. The waste plastics ending up in ocean makes a huge plastic soup and garbage patch like the great Pacific garbage patch risking the health of aquatic animals. In the last 40 years the plastic waste in the North Pacific ocean has increased by 100-fold. The plastic foot- print is consider more dangerous than carbon footprint[10]. Some of the proposed solutions for PSW management are incineration or mechanical recycling. However, incineration can contribute to pollution by causing harmful and toxic emissions [25]. Additionally, both of these processes are costly and may or may not be economically viable in different situations[26].

Aside from the challenge of plastic waste disposal, another global issue is the energy crisis. Transportation consumes one third of the world's energy. The main energy sources for transportation are fossil fuels, coal, oil, and natural gas, all of which are non-renewable sources of energy. Fossil fuels are also major sources of environmental pollution, greenhouse gases, and ocean acidification [17,27].

Today, these fuels are being consumed at an unsustainably high rate throughout the world [26]. Even though more than a hundred billion tons of oil and gas have been discovered in the last 40 years [27], the rate at which it is consumed has also increased. The US alone consumes one quarter of the global oil supply, while only having 1.6% of the total oil reserve. At the current rate of consumption, the global supply of fossil fuels will be depleted within 40–70 years. Among 195 countries in the world, only 40 can produce fossil fuel, and for some countries, their independent oil reserve still do not satisfy all of their energy needs [17]. Many countries spend a significant portion of their gross national income (GNI) to purchase oil and gas. A number of studies have been conducted to investigate alternative ways of producing energy. Some of the alternatives for fossil fuel energy are biomass energy, wind energy, hydroelectric energy, and nuclear energy [12]. Bio-based oils such as palm oil, soybean oil, corn oil, cottonseed oil and their derivatives are used as lubricants and fuel additives to replace petrochemicals [28]. However, even these additional energy sources may not completely solve the problem of increasing energy demand.

3. Plastic Waste

The increase in plastics use correspondingly increases the amount of plastic wastes being produced. Plastic waste can be classified into at least two categories: municipal and industrial [14]. Municipal plastic wastes (MSW) collected as household wastes, although normally remain a part of municipal solid wastes. The various sources of MSW plastics includes agricultural (like mulch films, feed bags, fertilizer bags, and in temporary tarpaulin-like uses such as covers for hay, silage, etc.), wire and cable, automobile wrecking, etc domestic items (like food containers, packaging foam, disposable cups, plates, cutlery, CD and cassette boxes. fridge liners, vending cups electronic equipment cases, drainage pipe, carbonated drinks bottles, plumbing pipes and guttering, flooring. cushioning foams, thermal insulation foams, surface coatings, etc.), [26].

Separation of plastics from household wastes is required to recycle municipal plastic wastes. Mechanical separation equipment is currently available for separation mixed plastics [29]. For example, separating using a wet process mixed plastics can be separated into two groups: those with a density greater than water such as polystyrene and polyvinyl chloride, and those with a density lower than that of water such as polyethylene, polypropylene, and expanded polystyrene. The larger group is the latter group. Consequently, recycling of municipal plastic wastes should deal with plastic mixtures of polyethylene, polypropylene and polystyrene, proved that the above separation procedures are practiced. Although MSW separation technologies have been studied extensively, it is still not possible to obtain marketable fractions and classify MSW mechanically. So waste separation at the household would be a better option, and separately disposed into three parts: (i) combustibles such as paper, kitchen waste, textiles, and wood, (ii) incombustibles such as metals, glass, ceramics, and (iii) plastics [40].

Industrial plastic wastes (called primary Waste) are those arising from the large plastics manufacturing, processing and packaging industry. The industrial waste plastic mainly comes from construction and demolition companies (e.g. polyvinyl chloride pipes and fittings, tiles and sheets) electrical and electronics industries (e.g. switch boxes, cable sheaths, cassette boxes, TV screens, etc.) and the automotive industries spare-parts for cars, such as fan blades, seat coverings, battery containers and front grills). Most of the industrial plastic waste have relatively good physical characteristics i.e. they are clean enough in the production and there is no contamination and are available in sufficient quantities. During the manufacturing process it's use high temperatures which may have decreased its characteristics, but it has never been used to product applications [26]

Municipal plastic wastes are heterogeneous and industrial plastics wastes are homogeneous in nature. To recycling homogeneous plastic wastes can use repelletization and remolding, it's seem to be a simple and effective. But if plastic wastes are heterogeneous or consist of mixed resins, they are unsuitable for reclamation. In this case thermal cracking into hydrocarbons may provide a suitable means of recycling, which is termed chemical recycling [40].

The last 40 years have witnessed an explosive growth of the plastic industry. The production of synthetic polymers signified by polyethylene (PE), polypropylene (PP), polystyrene (PS), and polyvinyl chloride (PVC) worldwide has increased more than 100 times in the last four decades. These plastics are widely used in very important day to day applications such as fashion, household appliances and in automotive products and aerospace [23]. As there are various plastic based materials are available in the market or as waste. Two basic categories of plastic are designated as thermosetting (long strands) and thermoplastic (short link) materials. Thermosetting plastic materials are those which cannot be recycled again and thermoplastic are those which have certain recycling capacity [30].

The treatment of waste plastics becomes an unavoidable circumstances and imminent issue. In this regard, it can be safely stated that urgent effective ways to recycle waste plastics. Recently new ways of environmentally-friendly waste plastic recycling have been of interest, and among them, the use of waste plastics as a supplemental fuel with coal in the steel making industry has attracted interest Attention is also focused on using oil derived from waste plastics in diesel engines. Diesel engines are the most preferred power plants due to their excellent drivability and higher thermal efficiency [23].

Plastics are non-biodegradable polymers mostly containing carbon, hydrogen and few other elements. According to a nationwide survey conducted in India in the year 2000 approximately 6000 t of plastic wastes were generated every day, and only 60% of it was recycled. In India alone, the demand for plastics is about 8 million tons per year. More than 10,000 metric tons/ day are produced in India and balance is imported from other countries. Most of the plastics are recycled and sometimes it is not done so due to lack of sufficient market value. Of the waste plastics not recycled about 43% is polyethylene and most of them is in the form of containers and packaging materials. The extent of conversion of plastics or plastic derived waxes into light engine fuels can be increased by the application of stable hydro cracking catalysts. Similar to petroleum derived cracking products, the fractions from plastics processing contain appreciable quantities of aromatics and unsaturated hydrocarbons [29].

4. Production of Fuel from Plastic

The challenges of plastic waste management and increasing energy demand can simultaneously be addressed by the production of fuel from plastics, and a number of research groups are currently developing this capacity. The fuels produced from plastics can be clean and have fuel properties similar to fossil fuels [31]. HDPE, LDPE, and PP are polymers containing only carbon and hydrogen. Unlike bio-fuels, the absence of oxygen and higher carbon and hydrogen content in plastic fuels avoids the need of further upgrading. The absence of water in plastic fuels makes the caloric value very high, and the absence of oxygen content makes the fuel non-acidic and non-corrosive, unlike bi-ofuel[31–33]. Therefore, conversion of these plastic wastes to usable oil is a growing and important field of study that can potentially mitigate the energy crisis. However, the techno-economic evaluation plays an important role in the commercial success of the plastics to fuel conversion.

The production of fuel from plastics is one of the most demanding research topics throughout the world. Thermo chemical, and catalytic conversion are the most widely used methods for fuel production from plastics [26]. Among them, thermo chemical conversion/pyrolysis treatment seems to be the dominant mode used.

4.1 Thermo chemical

Thermo chemical treatment breaks large polymers into smaller hydrocarbons of various carbon numbers and boiling points in an inert, air-free, or controlled environment at elevated temperature [13]. The hydrocarbons between the boiling points of 35 and 185 °C can be used as motor gasoline, between 185 and 290 °C as diesel #1, between 290 and 350 °C as diesel #2, between 350 and 538 °C as vacuum gas oil, and 4538 °C as the residue [31]. These fractions can be obtained in different proportions by the appropriate choice of thermal treatment/degradation temperature, type of plastics, degradation time, and catalyst type [33-35].

The product yields of liquid, gas, and residue vary depending on the type of polymer, degradation temperature, and atmosphere. Some data on product yields obtained for thermal treatments at 400–800 °C. Although the results vary somewhat because they come from different studies, thermo-chemical treatment seems to be a promising method of producing hydrocarbons from polymers.A subset of the thermal treatment is the gasification process. In this case the plastic is decomposed to combustible gaseous products at high temperatures, usually with the help of an oxidizing or a gasification agent. Both of these techniques to convert plastics to fuel do not have negative impacts on the environment [36,37].

4.2 Catalytic conversion

In catalytic conversion, catalysts are added to pyrolysis reactions to improve conversion, improve fuel quality, increase selectivity, and lower the pyrolysis temperature and residence time [13]. Pyrolysis is generally defined as the controlled burning or heating of a material in the absence of oxygen. In plastics pyrolysis,
the macromolecular structures of polymers are broken down into smaller molecules or oligomers and sometimes monomeric units. Further degradation of these subsequent molecules depends on a number of different conditions including (and not limited to) temperature, residence time, and the presence of catalysts and other process conditions. The pyrolysis reaction can be carried out with or without the presence of catalyst. Accordingly, the reaction will be thermal and catalytic pyrolysis. Since majority of plastic used are polyolefins, so extensive research has been done on this polymer which is summarised as below [26].

4.1.1 Thermal pyrolysis

The non-catalytic or thermal pyrolysis of polyolefins is a high energy, endothermic process requiring temperatures of at least 350–500 8C [42,43]. In some studies, temperatures as high as 700– 900 8C are essential in achieving decent product yields [44,45]. Thermal pyrolysis of both virgin and waste plastics as well as other hydro-carbonaceous sources has been studied extensively in the past. A good number of these thermal cracking studies are on polyethylene, polystyrene, and polypropylene[42,45]. On the other hand, only a few have worked on the thermal decomposition of other common plastics. Generally, thermal cracking results in liquids with low octane value and higher residue contents at moderate temperatures, thus an inefficient process for producing gasoline range fuels [42].

The gaseous products obtained by thermal pyrolysis are not suitable for use as fuel products, requiring further refining to be upgraded to useable fuel products [46,47]. A few researchers have sought to improve thermal pyrolysis of waste polyolefins without employing the use of catalysts; however these changes either yielded insignificant improvements or added another level of complexity and costs to the system [48].

https://doi.org/10.1051/e3sconf/20183105013

4.2. Catalytic cracking

Addition of catalyst enhances the conversion and fuel quality. As compared to the purely thermal pyrolysis, the addition of catalyst in polyolefin pyrolysis [26].

1. Significantly lowers pyrolysis temperatures and time. A significant reduction in the degradation temperature and reaction time under catalytic conditions results in an increase in the conversion rates for a wide range of polymers at much lower temperatures than with thermal pyrolysis.
2. Narrows and provides better control over the hydrocarbon products distribution in Low density polyethylene (LDPE), High density polyethylene (HDPE), polypropylene and polystyrene pyrolysis. Again, oils obtained by catalytic pyrolysis contain less olefins and more branched hydrocarbon and aromatic content.
3. Increases the gaseous product yields. Under similar temperatures and reaction times, a much higher gaseous product yield is observed in the presence of a catalyst for polyethylene.

Catalytic degradation of plastics is found to have greatest potential to be developed into a commercialized process [49]. One of the most successful example being the Alka Zadgaonkar's Unique Waste Plastic Management & Research Company plant in India which could produce fuel oil from waste plastics at par with the regular gasoline. From the research, it is clear that the oil produced in this process resembles the regular petrol in all respects (from color until reactivity with heavy metal) and reported to give better mileage as compared to commercial petrol. In addition, the cost of production is also reported to be very less.

5. Conclusion

The utilization of plastic waste cannot reduced landfill waste significantly, but most of the plastic waste can definitely produced may be treated with suitably process method to produce fossil fuel substitutes. The method should be superior in all respects (ecological and economical). Fuel production from plastics is attractive because it simultaneously addresses the issues of waste management and alternative energy generation. The optimization of conversion parameters such as the choice of catalysts, pyrolysis temperature, and plastic-to-catalyst ratio plays a very important role in the efficient processing of gasoline and diesel grade fuel. The use of a catalyst for thermal conversion lowers the energy required for conversion, and catalyst choice is important for efficient fuel production.

So, a suitable process which can convert waste plastic to hydrocarbon fuel if designed and implemented then that would be a cheaper partial substitute of the petroleum without emitting any pollutants

Reference

1. RAD. Arancon, CSK. Lin, KM. Chan, TH.Kwan, R.Luque. *Energy Sci*Eng**1**:53-71. (2013)
2. CSK. Lin, LA. Pfaltzgraff, L.Herrero-Davila, EB. Mubofu, S. Abderrahim, JH. Clark, et al. *Energy Environ Sc*;**6**:426-64 (2013)
3. L. Matsakas, U. Rova, P. Christakopoulos. *BioResources*;**11**:5482-99. (2016)
4. TPT. Pham, R. Kaushik, GK. Parshetti, R. Mahmood, R. Balasubramanian, *Waste Manage*; **38**:399-408. (2015)
5. E. Svensson Myrin, PE. Persson, S. Jansson. *Fuel* ;**132**:165- 9. (2014)
6. Y. Qu, O. Zhu, J. Sarkis, Y. Geng, Y. Zhong. *J Clean Prod*.;**17** :1175-82. (2013)
7. Bupe G Mwanzaa, Charles Mbohwab. *Procedia Manufacturing* **8**.649 – 656. (2017)
8. M.P. Subramanian, *Conservation and Recycling* **28**, 253–263. (2000)
9. D. Hoornweg, P. Bhada-Tata, *The World Bank*; (2012).
10. F. Pinto, P. Costa, I. Gulyurtlu, CabritaI. *J Anal Appl Pyrolysis*; 51(1- 2):39–55. (1999)
11. Sriningsih, Wiwin, Saerodji, Monica Garby, Trisunaryanti, Wega, Triyono, Armunanto, Ria, Falah, Iip Izul, *Procedia Environ. Sci.* **20**, 215–224. (2014)
12. Antony, Raja, Advaith, Murali,. *J. Mater. Sci. Eng.* B1, 86–89. (2011)
13. AK. Panda, RK. Singh, DK. Mishra. *Renew Sustain Energy Rev*;**14** (1):233–48.(2010)
14. C. Zhou, W. Fang, W. Xu, A. Cao, R. Wang. Characteristics and the recovery potential of plastic wastes obtained from landfill mining; **80**:80–6.(2014)
15. P. Lettieri, S. Al-Salem,Thermochemical treatment of plastic solid waste. In: Letcher TM,ValleroD, editors.Waste :A handbook for management;,p. 233–42.(2011)
16. DS. Scott, SR. Czernik, J. Piskorz, DSAG. Radlein. *Energy Fuels*. ;4(4):407–11.(1990)
17. M. Sarker, M. M. Rashid, S. Rahman, M. Molla, Conversion of Low Density Polyethylene (LDPE) and Polypropylene (PP) Waste Plastics into Liquid Fuel Using Thermal Cracking Process 2(1), 1–11. (2012)
18. [18] D. Grazhdani, *Waste Management* **48**, 3–13.(2016)
19. OECD, Global forum on environment focusing on sustainable materials. A material case study. (2010)
20. J. Bogner, M. Abdelrafie Ahmed, C. Diaz, A. Faaij, Q. Gao, S. Hashimoto, et al., 2007. Waste management. In: Metz, B., Davidson, O.R., Boesch, P.R., Dave, R., Meyer, L.A. (Eds.), Climate Change 2007: Mitigation Contribution of Working Group III to the Fourth Assessment Report of the International Panel on Climate Change. Cambridge University Press, Cambridge, UK . (2007)

21. D. A. Turner, I. D. Williams, S. Kemp, *Resources, Conservation and Recycling*, **105**, 186–197. (2015)
22. Matlack AS. Introduction to green chemistry. New York: Marcel Decker Inc; (2001).
23. PT. Williams, E. Slaney. *Resour Conserv Recycl*;**51** (4):754–69.(2007)
24. RC. Thompson, SH.Swan, CJ.Moore, FSV. Saal. Our plastic age. Philos Trans R Soc B;364(1526):1973–6.(2009)
25. J. Aguado, DP. Serrano, JM Escola, Ind Eng Chem Res;**47**(21):7982–92.(2008)
26. FJ. Passamonti, U. Sedran. Appl Catal B: Environ;**125**:499–506. (2012)
27. NA. Owen, OR. Inderwildi, DA. King, Energy Policy; **38**(8):4743–9.(2010)
28. SZ. Erhan, BK. Sharma, JM. Perez. Ind Crops Prod; **24**(3):292–9.(2006)
29. Plastic Wastes: Resource Recovery and Recycling in Japan. Tokyo: Plastic Waste Management Institute; 1985
30. L. Gu, T. Ozbakkaloglu, 2016. Use of recycled plastics in concrete: A critical review. Waste Management. Vu P, Nishida O, Fujita H, Harano W. Reduction of NOx and PM from diesel engines by WPD emulsified fuel. SAE Tech Pap;1:0152. (2001)
31. BK. Sharma, BR. Moser, KE. Vermillion, KM. Doll, N. Rajagopalan. Fuel Process Technol; **122**:79–90.(2014)
32. N. Miskolczi, A. Angyal, L. Bartha, I. Valkai, Fuel ProcessTechnol; **90**(7–8):1032–40. (2009)
33. G. Elude, M. Olazar, G. Lopez, M. Amutio, M. Artetxe, R. Aguado, J. Bilbao. J Anal Appl Pyrolysis; **85**(1–2):345–51.(2009)
34. M. Syamsiro, H. Saptoadi, T. Norsujianto, P. Noviasri, S. Cheng, Z. Alimuddin, K. Yoshikawa. *Energy Procedia*;**47**:180–8.(2014)
35. S. Kumar, R. Prakash, S. Murugan, RK. Singh, Energy Convers Manag;**74**:323–31. (2013)
36. G. BelgiornoV De Feo, C. Della Rocca, RMA. Napoli, *Waste Manag*; **23**(1):1–15.(2003)
37. RP. Singh, VV. Tyagi, T. Allen, MH. Ibrahim, R. Kothari, Renew Sustain Energy Rev; **15**(9):4797–808.(2011)
38. K. Murata, K. Sato, Y. Sakata, Journal of Analytical and Applied Pyrolysis; **71**(2): 569–89 (2004)
39. L. Sorum, MG. Gronli, JE. Hustad. Fuel; **80**(9):1217–27. (2001)
40. A. Demirbas, Journal of Analytical and Applied Pyrolysis ;**72**:97–102. (2004)
41. FJ. Mastral, E. Esperanza, P. Garcı´a, M. Juste. Journal of Analytical and Applied Pyrolysis; **63**(1):1–15. (2002)
42. SS. Kim, S. Kim. Chemical Engineering Journal; **98**(1–2):53–60. (2004)
43. OS. Woo, N. Ayala, LJ. Broadbelt. Catalysis Today ; **55**(1–2):161–71. (2000).
44. OS. Woo, LJ. Broadbelt, Catalysis Today; **40**:121–40. (1998).
45. JH. Chan, ST. Balke. Polymer Degradation and Stability; **57**(2):135–49. (1997).
46. AR. Songip, Applied Catalysis B Environmental; **2**(2–3):153–64. (1993).
47. HS. Joo, JA. Guin. Continuous upgrading of a plastics pyrolysis liquid to an environmentally favorable
48. DP. Serrano, J. Aguado, M. Escola, E. Garagorri, Applied Catalysis B Environmental;**44**(2):95–105. (2003).
49. Y. Sakata, MA. Uddin, A. Muto. Journal of Analytical and Applied Pyrolysis;**51**(1–2):135–55. (1999)
50. Alka Zadgaonkar. Conversion of Waste Plastic into Liquid Hydrocarbons/Energy—A major breakthrough in the arena of non-conventional sources of energy, Information Brochure and Technical Write-Up.

The Effectivity of Green Coconut Water To Reduce Mercury Level In The Blood And To Improve Blood Profiles And Liver Cells Appearance (*Study In Sprague Dawley Rats*)

Abdulrzag Ehmeeda M [1,*], *Tri* Nur Kristina[2], *Ari* Suwondo[3], *Henna* Rya Sunoko[2]

[1]Doctoral Program of Medical Sciences Diponegoro University, Semarang, Indonesia
[2]Faculty of Medicine, Diponegoro University, Semarang, Indonesia
[3]Faculty of Public Health, Diponegoro University, Semarang, Indonesia

Abstract. When people are exposed to mercury chloride, it can produce a variety of health effects in the blood and liver. Coconut water contains Zn, Fe, Vit. C, Vit B11, Vit. B6, and Se to reduce mercury chloride level in the blood and improve blood profile and liver cells. Aim of this study was to analysis the effect of green coconut water supplementation in overcoming the toxic effect of Hg chlorid in the blood and liver of *Sprague dawley* rats exposed to Hg chloride. Samples were randomly about 36 animals rats exposed to $HgCl_2$ through forced feeding by 20 mg/kgBW sondage per day for 14 days, which divided into control group, and intervention groups were given fresh green coconut water in each by 6, 8, and 10 mL/kgBW for intervention 7 and 17 days. The result of this study showed that there is a significant effect and the decrease in mercury levels in the blood. There is no significant affect on the hemoglobin level, hematocrit level and platelet count with the treatment of green coconut water in the mice with exposure Hg. There is no significant effect between treatments using green coconut water with SGPT levels; there is a decrease in SGPT levels at the increasing number of doses of green coconut water and the length of treatment.

1 Introduction

Heavy metals such as arsenic, cadmium and mercury are very dangerous for human health. There are two main sources of heavy metals contamination, i.e., natural (e.g. volcanic eruption) and anthropogenic activities (e.g. mining). Gold mining is one of the world's top source of contamination by Hg, especially in developing countries [1]. Blood is being an important component in the body, an exposure to contaminants will matter to its constituents of hemoglobin, platelets, erythrocytes, and leukocytes. The pollutants will bind to the active groups of enzymes Aminolaevulinic acid dehydratase (ALAD) in the body. Bond is formed between the metal Hg with clusters of Aminolaevulinic acid dehydratase (ALAD), which results in the formation of intermediates porphobilinogen. Poisoning due to air pollutant Hg can result in disruption of the blood components (blood profiles) that elevated levels of levulinic Amino acid (ALA) in the blood and urine, improving protoporphyrin levels in red blood cells, lowering the amount of red blood cells, lowering the number of erythrocytes causing hematopoietic and increase hematocrit levels in the blood. It can be seen from the value of MCV (mean corpuscular volume/volume of blood cells), MCH (Mean Corpuscular Hemoglobin/Weight Hemoglobin average in 1 erythrocytes), and MCHC (Mean Corpuscular Hemoglobin Concentration / Hemoglobin concentration erythrocytes average) [2].

Coconut water (coconut liquid endosperm), with its many applications, is one of the world's most versatile natural product. This refreshing beverage is consumed worldwide as it is nutrition's and beneficial for health. There is increasing scientific evidence that support the role of coconut water in health and medicinal application. Coconut water is traditionally used as a growth supplement in plant tissue culture or micro propagation. The wide applications of coconut water can be justified by its unique chemical composition of sugars, vitamins, minerals, amino acids and phytohormones [3].

Most of the detoxifying methods of Hg are problematic and have to be adjusted to the patient individually. So, Hg can be absorbed through the intake of selenium, but then it is often deposited in the body and cannot be excreted. Folic acid may also absorb Hg; however, in this way organic Hg chloride can emerge which is even a hundredfold more toxic. So-called chelating agents like DMPS (dysmyelopoietic syndrome) absorb Hg very well, but at the same time absorb all other metals in the body as well, so that a long-term treatment is not possible: only single use for a one time "remove" of heavy metals. Single homeopathic detoxifying methods have to start with low potencies which, however, can lead to a re-contamination with Hg because the oxidation [3]. Coconut water contains fatty acids which can extract Hg stored in fatty tissue [4]. Based on this background, this research will examine the

* Corresponding author: libya6566@yahoo.com

https://doi.org/10.1051/e3sconf/20183106001

effects of green coconut water supplementation to overcome the toxic effect of Hg in the blood and liver of Sprague Dawley rat exposed to Hg chloride.

The benefit of this study is to contribute ideas for the implementation and development of the substance in the medical health faculty, especially about the effectiveness of green coconut water to reduce Hg level in blood and improve blood profile.

2 Methods

The populations in this study were Sprague dawley rat originated from Animal Care Laboratory Pharmacology and Therapeutics Faculty of Medicine Gadjah Mada Unversity total of 36 Rats. The sampling technique have been done by Simple random sampling is the selection of a group of Sprague dawley rat based on equality chances every individual of Sprague dawley rat to be a member of the experimental Group I, II, or III.

Experimental Design: All animals were acclimated and put in place with good environment, food, and at libitum drinks for 7 days. The experimental rats were exposed to $HgCl_2$ through forced feeding by 20 mg/kgBW sondage per day for 14 days. Aminals are randomly about 36 animal, it divides as a control group, and intervention. Intervention groups give fresh green coconut water in each by 6, 8, and 10 mL/kgBW for intervention for 7 and 17 days. The amount of fresh green coconut water was served according to each rat's body weight by the doses of 6 mL/kgBW, 8 mL/kgBW and 10 mL/kgBW for group I, II, and III respectively for 7 and 17 days. Measurement of Hg level in that blood sample was done at "BPTLPP YOGYAKARTA Laboratory" using Mercury analyzer. Researcher has observed any changes of blood mercury level, blood profile, and liver enzyme, and liver histopathology on the experimental animal.

Data recording, processing and analyzing were done according to the latest methodology available and assisting by computer. The first data was analized normality to find the suitable of statistics analize. Secondly, statistical analysis of univariate data presented (mean, SD), bivariate (Wilcoxon, Pairs t-test), bivariate (Kruskal Wallis) unless the data is no normality, and bivariate (Pairs t-test, ANOVA / Post Hoc Duncan's Multiple Range Test) by significance level 0.05.

3 Result and discussion

3.1. Results

The results (Table 1) of this study revealed that there the mean values of blood Hg, hemoglobin, hematocryt, WBC, platelet, and SGPT with the mean values of the control group..

Table 1. Result Analysis Before and After Treatment with Green Coconut Water

	Day		
	0	7	17
Dose I (6ml/kg BW) (Group A)			
Blood Hg (μg/l)	9.80	8.10	8.13
Hemoglobin (g/dl)	17.32	17.07	16.65
Hematocryt (%)	51.15	51.30	49.77
WBC (μl)	11550	9350	11866
Platelet (Mel)	1050	1235.25	1272.75
SGPT (U/l)	216.75	264.25	232.75
Dose II (8ml/kg BW) (Group B)			
Blood Hg (μg/l)	8.74	2.21	3.48
Hemoglobin (g/dl)	16.95	16.60	16.75
Hematocryt (%)	48.02	46.52	49.52
WBC (μl)	11050	9825	9000
Platelet (Mel)	929.25	1207.75	1118.75
SGPT (U/l)	101.75	87.75	87.25
Dose III (10ml/kg BW) (Group C)			
Blood Hg (μg/l)	5.96	3.00	1.85
Hemoglobin (g/dl)	15.52	15.62	16.90
Hematocryt (%)	49.05	46.55	48.77
WBC (μl)	10775	9400	10125
Platelet (Mel)	1304.25	1289	1418.50
SGPT (U/l)	113	60.5	71.75

3.1.1 Blood Hg level

The distribution of Hg levels after 14 days was exposed by 20 mg/KgBW before they were treated by fresh green coconut water in animal experiments can be seen in table 1. The highest Hg level was group A, and the lowest was group B. The distribution of mercury levels after 7 days treatment of fresh green coconut water in animal experiments, the highest mercury level at day 7 was group A, and the lowest was group B. The distribution of Hg levels after 17 days treatment of fresh green coconut water in animal experiments, the highest Hg level at day 17 group was group A, and the lowest was group C.

In second group day 0 and day 17 there are significant differences in blood Hg level before and after 17 days intervention with dose I of fresh green coconut water. There is significant difference in blood Hg level before and after 7 and 17 days of intervention with dose I of fresh green coconut water. There is significant difference in blood Hg level before and after 7 and 17 days of intervention with dose III of fresh green coconut water.

3.1.2 Blood profile

Hemoglobin level

The distribution of hemoglobin levels after 14 days was exposed by 20 mg/KgBW before they were treated by fresh green coconut in animal experiments can be seen in table 1. The highest hemoglobin level was group A, and the lowest was group C. The distribution of hemoglobin levels after 7 days treatment of fresh green coconut water in animal experiments, the highest hemoglobin level at day 17 was group A, and the lowest was group B. Data distribution of hemoglobin levels after 17 days treatment of fresh green coconut water in animal experiments, the highest hemoglobin level at day 7 was group C, and the lowest was group A.

Hematocrit level

The distribution of hematocrit levels after 14 days was exposed by 20 mg/KgBW before they were treated by fresh green coconut in animal experiments can be seen in table 1. above. The highest hematocrit levels was group B, and the lowest was group C. The distribution of hematocrit levels after 7 days treatment of fresh green coconut water in animal experiments, the highest hematocrit levels at day 7 was group C, and the lowest was group B. Data distribution of hematocrit levels after 17 days treatment of fresh green coconut water in animal experiments, the highest hematocrit level at day 17 was group A, and the lowest was group B.

White Blood Cell (WBC) count

The distribution of WBC count after 14 days was exposed by 20 mg/KgBW before they were treated by fresh green coconut in animal experiments and after they were treated by fresh green coconut in animal experiments can be seen in table 1 above. The highest WBC counts was group C, and the lowest was group B. The distribution of WBC count after 7 days treatment of fresh green coconut water in animal experiments, the highest WBC count at day 7 was group A, and the lowest was group A. The distribution of WBC count after 17 days treatment of fresh green coconut water in animal experiments, the highest WBC count at day 17 was group A, and the lowest was group B.

Platelet count

Data distribution of platelet count after 14 days were exposed by 20 mg/KgBW before and after they were treated by fresh green coconut in animal experiments can be seen in table 1 above. The highest platelet count was group C, and the lowest was group A. Data distribution of platelet count after 7 days treatment of fresh green coconut water in animal experiments, the highest platelet count at day 7 was group B, and the lowest was group A. Data distribution of platelet count after 17 days treatment of fresh green coconut water in animal experiments, the highest platelet counts at day 17 was group C, and the lowest was group B.

3.1.3. Liver enzymes (SGPT)

Data distribution of SGPT levels after 14 days were exposed by 20 mg/KgBW before and after they were treated by fresh green coconut in animal experiments can be seen in table 1 above. Data distribution of SGPT levels 0 days treatment of fresh green coconut water, the highest SGPT levels before treatment was group B, and the lowest was group A. Data distribution of SGPT levels after 7 days treatment of fresh green coconut water in animal experiments, the highest SGPT levels at day 7 was group A, and the lowest was group C. Data distribution of SGPT levels after 17 days treatment of fresh green coconut water in animal experiments, the highest SGPT levels at day 17 was group A, and the lowest was group B.

3.1.4 Liver Histopathology

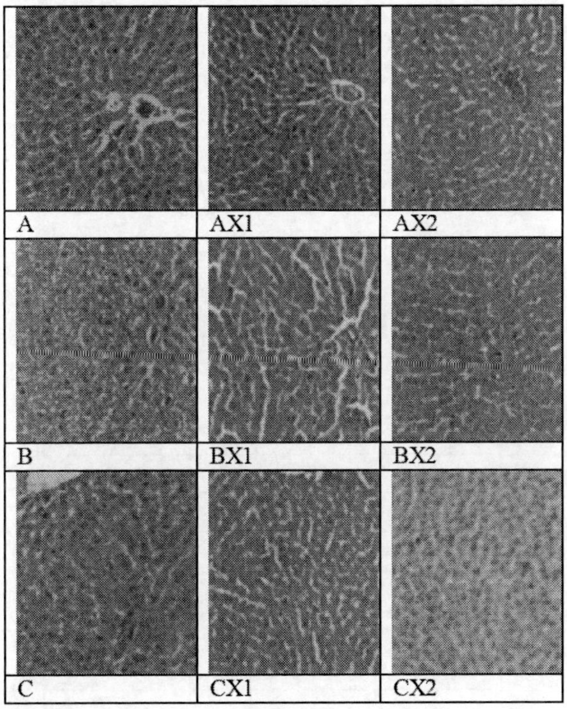

Fig. 1. Coloring Result of Sprague dawley tissues zoomed 400x.

Notes : A, B, C = Sprague dawley Hepar tissues exposed by HGCl2 20mg/kg of weight; AX (1,2) = Sprague dawley Hepar tissues treated by fresh coconut water with the dosage of 6 mL/kg of weight (7,14 days); BX (1,2) = Sprague dawley Hepar tissues treated by using fresh coconut water with the dosage of 8 mL/kg of weight (7,14 days); CX (1,2) = Sprague dawley Hepar tissues treated by using fresh coconut water with the dosage 10mL/kg of weight (7,17 days).

The figure shown that the damage was reversible (fat and hydropic degeneration). Fresh green coconut water treatment proven can reduce number damage of the cells. Even fat degeneration still happens after 7 days treatment of fresh green coconut water, but in 17 days the damage reduce significantly.

Hematoxylin-Eosin coloration on mouse liver tissue showed the architecture and general pattern of the liver. In the normal liver, hepatocytes are supported by a network of reticular fiber. The red dotted line indicates a portion of the classic lobule. The diamond-shaped hepatic acinus is located between two central veins and two portal triads. After exposured by mercury for 14 days at a dose of 20 mg/KgBW, an increase in the number of degenerated cells and necrosis accompanied by sinusoid dilation, but no fibrosis and no sirosis too. This result shows that cell experience hydropic degeneration. The level of cell damage is shown by the changes of some cytoplasm organelles, such as reticulum endoplasm that become pockets full of water, so that in microscopic examination, the cytoplasm will be seen as having vacuole, however the cell nucleus is still in the center. Mean of degenerated liver cell before fresh green coconut water (A,B,C) is 33 cells per 100 cells. After 7 and 17 days tratment of fresh green coconut water degeneration of liver cells reduced become 4 cells per 100 cells and 2 cells per 100 cells respectively. The detail showed in the table 2. There is also a fat degeneration depicted from hepatocytes that are pushed to one side. But also the nucleus of hepatic cells experiencing pycnosis, namely the existence of solid and shrunken dark blue in the cell nucleus.

Table 2. Wide of the liver damage

Name of Portal area	Num of damage cell per 100 cells	Num of normal cell per 100 cells
Cx 8	2	98
Cx7	1	99
C1	15	85
Bx5	2	98
Bx3	4	96
B2	53	47
Ax8	2	98
Ax1	2	98
A4	69	31
A3	23	77
A2	19	81
A1	22	78

3.2. Discussions

3.2.1 Blood Hg level

There is a significant effect and the decrease in mercury levels in the blood. This is because coconut water contains fatty acids which can extract mercury stored in fatty tissue. At the same time the sulfur-containing amino acids of coconut water bind the mercury and prevent redistribution in the body. Mercury levels in the blood reduce by treatment fresh green coconut water, because for a definite detoxifying, sulfuric amino acids in the green coconut were required to absorb mercury and to allow excretion through urine and purgation [5].

Coconut water was rich in free amino acids, the source of L-arginine and vitamin C, which can prevent heart disease and peroxidation [6]. A high blood mercury level indicated that the level of free radicals in the body was high. The main target of free radicals is a protein, unsaturated fatty acids, as well as elements of DNA, including carbohydrates. The three molecular targets, unsaturated fatty acids were those most susceptible to free radical attack. A high level of free radicals in the body can be demonstrated by the low activity of antioxidant enzymes and high levels of malondialdehyde (MDA) in the blood plasma [7]. Coconut water contains antioxidants like vitamin C, vitamin B1, vitamin B6, amino acids such as methionine, L-Arginine, selenium, cytokines and minerals that were helpful to protect the body's cells against the risk of free radical attack and prevent the effects resulting from exposure to mercury [6]. A component such as the amino acid methionine found in tender coconut water is the source for cysteine sulfhydryl. Methionine acts as a precursor for the formation of cysteine was the main compound in the synthesis of glutathione (GSH). Methionine was being synthesized into S-adenosyl methionine (SAM) is assisted by the enzyme methionine catalyst adenosiltransfrease. Then, S-adenosilmetionin (SAM) will be converted into the S-adenosilhomosistein (SAH). SAH is converted into homocysteine by the enzyme adenosylhomocysteine and by methionine synthase (MS) homocysteine can back into methionine. In the synthesis of cysteine, homocysteine is converted into cystathionine with the help of the enzyme cystathionine β-synthase (CBS) and vitamin B6, and then converted to cysteine with the help of cystathionine lyase enzyme and vitamin B6, cysteine eventually be synthesized into glutathione (GSH) [8].

GSH was a substrate for the antioxidant glutathione peroxidase (GPx) to decompose hydrogen peroxide. If glutathione synthesis would be disturbed, it will cause a decrease in GSH. There will eventually dump hydrogen peroxide and hydroxyl radicals, so as result it will be no increase is more dangerous to the human body. GSH also serves as a substrate in the reaction radical regeneration of vitamin C, so that if a drop of GSH occurs, it will interfere with the reaction and will cause lipid peroxidation. Methionine works in synergy with vitamin B6. Vitamin B6 acts as a cofactor of the enzyme of cystathionine β-synthase and the lyase cystathionine cysteine and GSH synthesis process.

Coconut water as a source of methionine and vitamin B6 [9]. The high content of L-arginine in tender coconut water can be used to reduce the generation of free radicals, increase antioxidant activity and inhibit the process of lipid peroxidation [6]. L-arginine plays an important role in the detoxification of mercury and could potentially serve as a preventive and therapeutic strategy against occupational or environmental exposure to mercury [10]. NO can inhibit xanthine oxidase (XO), increases levels of SOD, the number of thiol (T-SH), vitamin C, the number of antioxidants (TAC) [11]. If the XO activity decreases, the amount of superoxide also is decreased so that the rate of increase SOD, because of the need to decipher superoxide to hydrogen peroxide is also lacking. Superoxide can also be formed due to the reduction of O2 by NO synthase, when deprived of arginine and tetrahydrobiopterin (BH4) occurs, so that L-arginine may be used as an alternative to reduce the production of superoxide.

Thiol levels after administration of L-arginine can improve ascorbic acid radical conversion into ascorbic acid [11]. Treatment with L-arginine can increase the activity of GPx in rats exposed to Pb [12]. The high content vitamin C in tender coconut water makes coconut water effective to raise the level of antioxidants such as vitamin C can reduce superoxide radicals, hydrogen peroxide and other reactive oxygen derived from activated neutrophils and monocytes. As an antioxidant, ascorbic acid was not directly give birth membrane-bonding antioxidants, such as α-tocopherol, with increasing peroxyl radicals and singlet oxygen. Vitamin C works synergistically with vitamin E. Vitamin E is oxidized by free radicals can react with vitamin C, after obtaining the hydrogen ion of vitamin C and then be converted into tocopherols [7]. Tender coconut water contains selenium, GPx activity was strongly influenced by the presence of selenium, selenium deficiency in the body can reduce the activity of GPx [12]. Mercury has a strong affinity for binding selenium [13], and the two antagonists [14]. Provide gentle coconut water after exposure to mercury can be used as a source of selenium to increase the activity of GPx.

High nutrient content in tender coconut water can be used as a source of minerals that are important for the body, one of which is as natural cofactor antioxidant SOD contained in the body. Mineral deficiency Cu, Zn and Mn can lower the activity of Cu-Zn SOD and Mn-SOD. Mn-SOD function as a catalyst for the reaction of superoxide anion to hydrogen peroxide (H_2O_2) and oxygen (O2) in the mitochondria, while the Cu-Zn SOD function as a catalyst for the reaction of the anion superoxide dismutase into hydrogen peroxide (H2O2) and oxygen (O_2) in the cytosol. In the Cu-Zn SOD is, Cu minerals required for the proper functioning of the catalytic enzyme, whereas Zn is important for the structural function. Mineral Cu, Zn and Mn work synergistically [7].

3.2.2 Blood profile

Hemoglobin level

After 7 days of treatment have no significant affect on hemoglobin levels and also on the increasing doses of green coconut water given after 17 days of treatment no significant affect on the hemoglobin level with the treatment of green coconut water in the mice with exposure Hg.

The process of formation of red blood cells (erythropoiesis) and of hemoglobin synthesis occurs in red bone marrow. Iron contained in coconut water is thought to have a role as a main constituent of hemoglobin chemical bonds that became one component of red blood cell formation. Heme iron by binding to globulin protein to form hemoglobin [7]. The absorption of iron from foods that will be used in the formation of hemoglobin allegedly influenced by other substances, namely vitamin C. Vitamin C is contained in coconut water has been implicated in increasing the absorption of iron in the body to speed up the synthesis of hemoglobin. Vitamin C serves to reduce the ferric iron (Fe 3+) to ferrous (Fe 2+) in the small intestine so easily absorbed [15]. Vitamin C and iron to form iron ascorbate complex compounds soluble and easily absorbed to the synthesis of hemoglobin in erythroblast [16]. Vitamin B6 is also contained in coconut water has been implicated as a coenzyme in the reaction of the formation of hemoglobin. Other substances that alleged role in the formation of red blood cells are also found in coconut water is folic acid has a role in the synthesis of DNA to control the maturation of the core erythroblast (cells found in red bone marrow, which will develop into erythrocytes) [17]. However, in this research the iron and the vitamin C contain in coconut water could not help to rise the hemoglobin level, probably due to the iron contain was not enough to repair hemoglobin level.

Hematocrit level

Based on the results based on the increasing doses of green coconut water given on control group, after 7 days of treatment, increasing doses of green coconut water given after 17 days of treatment have no significant affect on hematocrit level with the treatment of green coconut water in the mice with exposure Hg. Based on the iron content and Vitamin C contained in the green coconut water, green coconut water should be able to increase or improve blood profile. However, in this study there is a decrease in blood profile in mice with exposure by Hg. This is because the Hg content in the drinking water of mice had damaged tissue cells in mice that treatment using green coconut water does not have a significant effect on the blood profile of rats. Hg directly damage the cells of tissues of living beings with their deployment to air, water and food contamination [18].

Hg inhibits the synthesis of heme and shorten the life of erythrocytes. This is due to the activity of various enzymes necessary for heme synthesis is inhibited. One example of the enzyme glutathione peroxidase, this enzyme has sulfhydryl groups where the presence of Hg in the body of the sulfhydryl groups are bound with Hg so that the enzyme becomes inactive and enzyme concentration decreases. The decline in the activity of glutathione peroxidase will also interfere with the process of glycolysis resulting in energy reduction of erythrocytes is reduced so as to cause a short erythrocyte lifespan [18].

Hg has a high affinity against erythrocytes, approximately 95% bound in blood erythrocytes. Hg has a half-life in the blood very slowly about 25 days, at 40 days and the soft tissue on the bones of 25 years. Given the nature of this very slow excretion of Hg easily accumulate in the body. Hg deposited in the red blood cells and cause damage to red blood cells into the blood Hg of blood, 95% attaches to red blood cells, 5% in the blood plasma. Damage to red blood cells that contain Hg i.e. before the rupture of red blood cells mature red blood cells, resulting in a decrease in the number of red blood cells in the blood. Eritrosit regenerated cell, so Hg accumulated momentarily in erythrocytes, will be lessened follow released of red blood cells, but if exposure to Hg continuously at levels above the normal threshold will together of red blood cells to the heart and other organs, namely bone marrow and accumulate settles [9].

White blood cell count

The increasing doses of green coconut water given to control group have a significant effect on the WBC level. After 7 days of treatment and also on the increasing doses of green coconut water given after 17 days of treatment no significant affect on the WBC level with the treatment of green coconut water in the mice with exposure Hg. Coconut water also contains folate which is also known as vitamin B9. These were identified in the late 1930s as nutrients needed in reducing anemia in pregnancy which also helped prevent mitochondrial toxicity caused by methanol metabolism [12]. This contrasts with the results of research that states the number of leukocytes decreases with increasing doses of green coconut water. High mercury content can inhibit heme red blood cell synthesis after poisoning [7]. Anemia is caused not only by increased erythrocyte impairment [15] but also by decreasing the synthesis and release of erythrocytes into the blood circulation [16]. Connects anemia with erythropoiesis disorders caused by the direct effects of metals on hematopoietic (kidney / spleen) centers, accelerated erythematic due to increased membrane permeability and / or increased mechanical fragility [17].

Platelet count

No significant affect on the platelet count with the treatment of green coconut water in the mice with exposure Hg. A disturbance in the bone marrow lead to failure or suppression of stem cells that result in disruption of the process of formation of blood cells (hematopoiesis). Hematopoiesis disorders can occur in the process of formation of red blood cells, which cause a reduction in the number of red blood cells. All blood cells originate from stem cells pluripotential which then differentiate into: stem cell lymphoid forming lymphocytes and plasma cells, stem cells are multi potential myeloid (non-lymphoid) which then develop into different types of hematopoietic cell to another, such as platelets, erythrocytes, granulocytes and monocytes [17].

3.2.3 Liver Enzyme (SGPT)

There is no significant effect between treatments using green coconut water with SGPT levels, there is a decrease in SGPT levels at the increasing number of doses of green coconut water and the length of treatment. The decline in SGPT levels mean there were improvements in liver enzymes of the liver damage caused by Hg.

The high level of SGPT in the blood occurs due to free radicals Hg would interfere with the synthesis and metabolism of proteins in the liver that cause hepatocellular damage. Abnormalities hepatocytes as a result of Hg characterized by increased SGOT and SGPT also can trigger the disorder caused due to secretion of bile flow is impaired, resulting in increased ALP [19]. In the presence of Hg exposure, the levels of AST and ALT rise in male rats [20]. A significant increase in serum SGOT and SGPT mice with exposure $HgCl_2$ [21], an increase in SGOT and SGPT of mice because of Hg [22]. The increase of chronic exposure to $HgCl_2$, resulting in an increase of SGPT and SGOT in acute or chronic exposure to $HgCl_2$ in teleost fish [23].

3.2.4 Histopathology

To determine the damage level and tissue repair of Sprague dawley exposed by Hg compared to rats treated by fresh coconut water was conducted by using Hematoxylin-Eosin coloration, as shown in Figure 1. Pathophysiological disorder on hepatic tissues is determined by the changes of the smallest hepatic functional unit that consists of parenchyma cells around arterioles, venule, ductus biliaris terminal and between two central vena.

In the control group exposed by chloride Hg with the dosage of 20mg/kg of weight, there is an increase of cell that experience degeneration and necrosis and sinusoid widening. The accumulation of triglyceride in cytoplasm, especially in hepatocyte caused by triglyceride metabolism disorder in hepatic will cause fat degeneration. In the process, fat drops into cytoplasm and form globules that will get bigger and push hepatocyte nucleus to one side. However, in the process of forming microscopic preparations, fat accumulation in hepatocyte cytoplasm is dissolved by the alcohol so that it will be seen as a cavity or vacuole around hepatocyte nucleus.

The frequency of chloride Hg exposure in this research that was conducted for 17 days is determined to cause pycnosis in hepatic nucleus. Pycnosis occurs when there is a solid mass that wrinkles and becomes dark blue around cell nucleus area. It is initiated by clouding swelling that is resulted from water hoarding inside the cell. The increase of free radicals caused by chloride Hg exposure results in energy metabolism disorder inside the cell and cell membrane damage so that the activity of sodium ion pumping outside cell will be disturbed.

The increase of sodium ion concentration inside the cell will stimulate the process of abnormal water osmosis inside the cell. Then, hoarding and swelling will occur in the cytoplasm and other organelles, such as mitochondria, reticulum endoplasm. Cell can experience hydropic degeneration if there is a continued increase of water influx. The level of cell damage is shown by the changes of some cytoplasm organelles, such as reticulum endoplasm that become pockets full of water, so that in microscopic examination, the cytoplasm will be seen as having vacuole, however the cell nucleus is still in the center.

The condition of *Rattus norvegicus* Hepar tissues after treated by fresh coconut water with the dosages of 6mL/kg of weight, and 10mL/kg of weight shows an improvement on the hepatocyte cells. Based on the research results, there is an increase of improvement on the hepatocyte cells that maximally occurs after treated by fresh coconut water with the dosage of 10mL/kg of weight. The antioxidant content in fresh coconut water, such as sugar, vitamin B, vitamin C, mineral, and free amino acid, has a contribution on the improvement of hepatocyte cells. High Potassium level can reduce blood pressure. L-arginine amino acid also has important antioxidant activity that can reduce free radical development so that there will be an increase of Glutathione level (GSH). Glutathione level (GSH) is an enzyme that organizes immune system. Besides, L-ascorbate acid holds up fat peroxidase significantly.

Fresh coconut water therapy on rats has a positive control in increasing the level of L-arginine that can manage NO signaling and fix the function of endothelial vascular cell. L-arginine is one of NO physiology precursor that determines the speed of NO radical biosynthesis. NO radical has a connection with microvascular regulation, permeability, and epithelium cell function, such as immune system, mucus regulation, and epithelium cell fluid production. L-arginine synthesis occurs in rats' livers when oxidative stress occurs. L-arginine addition is determined to reduce MDA level, myeloperoxidase and xanthine oxidase that cause lipid peroxidase. Lipid peroxidase retardation mechanism is possible to be conducted due to stimulating activation of NF-kB transcription factor that is followed by the increase of glutathione level. Thus, L-arginine is one of approaches to restore NO signaling dysregulation and dysfunction of endothelium cell of hepar tissue damage.

4 Conclusions

Green coconut water can decrease Hg concentrations in the blood after supplementation with coconut water given to Sprague dewly rats exposed to mercuric chloride. There were no significant differences in blood profile (hemoglobin, hematocrit, WBC and platelet count) between before and after supplementation green coconut water. There were no significant differences between liver enzymes (SGPT) before and after supplementation gree coconut water. Decreased levels of Hg in the blood that is effective in treatment after a day 7 at a dose of 8 ml of green coconut water. Increased hematocrit effective in treatment after day 17 with a dose of 6 ml and 8 ml of green coconut water, increasing WBC effective in the treatment after 7 days with a dose of 8 ml and treatment after day 17 with a dose of 10 ml of green coconut water, whereas the increase in platelet count that is effective on the 17th day of treatment using green coconut water to Sprague dewly rats exposed to mercuric chloride. SGPT levels increased in treatment after day 17 at a dose of 6 ml despite the downturn in the treatment after 7 days using a green coconut water. Green coconut water can decrease liver cell damage after supplementation green coconut water.

References

1. United Nations Environment Programme. *Global Mercury Assessment. In. Geneva, Switzerland: UNEP (United Nations Environment Programme) Chemicals.* 2002. Website: http://www.chem.unep.ch;
2. J. Woods, Can J Physiol Pharmacol, **(74)**:210–215, (1996).
3. Hoffbrand, A. Victor, Daniel Catovsky, Edward g.d. Tuddenham, Anthony R. Green. *Postgraduate Haematology Sixth Edition*. Willey Blackwell, (2011).
4. C. Broker, *Ensiklopedia Keperawatan*. Alih bahasa: Andry Hartono, Brahm U. Pendit, Dwi Widiarti, Jakarta: ECG, (2008).
5. AM. Fonseca, FJQ. Monte, da Conceic M, de Oliveiraaão F., Journal of Molecular Catalysis B: Enzymatic. **57**: 78-82, (2009).
6. D. Bhagya, L. Prema, T. Rajamohan, Asian Pasific Journal of Tropical Medicine. **270**-6, (2012).
7. H. Winarsi, *Antioksidan Alami & Radikal Bebas. 3 ed*. Yogyakarta: Kanisius, (2007).
8. SM. Baker, *New York: A Clinician's Journal New York*. 88-95, (2007).
9. S. Ethel, *Anatomi dan fisiologi untuk pemula*. Jakarta: EGC, (2004).
10. M. Bracci, M. Tomasetti, M. Malavolta, V. Bonacucina, E. Mocchegiani, L. Santarella, Industrial Health. **46**: 567-74, (2008).
11. P. Tripathi, M. Chandra, Misra MK. Oxid Med Cell Longev., **2**(4): 231-7, (2009).
12. H. Tkachenko, N. Kurhalyuk, Pol J Environ Stud., **20**(5): 1319-25, (2011).
13. SamZiff. *The Toxic Time Bomb Can The Mercury in Your Dental Fillings Poison You. 4 ed*. USA: Aurora Press, (2002).
14. JWH. Yong, Ge L, Ng YF, Tan SN. Molecules., **14**: 5144-64, (2009).
15. A. Sembiring, M. Tanjung, and E. Sabri, *Pengaruh Ekstrak Segar Daun Rosela (Hibiscus sabdariffa L.) terhadap Jumlah Eritrosit dan Kadar Hemoglobin Mencit Jantan (Mus musculus L.) Anemia Strain Ddw melalui Induksi Natrium Nitrit (NaNO2)*. Medan: Departemen Biologi FMIPA Universitas Sumatra Utara, (2012).

16. G. Argana, Kusharisupeni. and D. M. Utari, Jurnal Kedokteran Trisakti. **23** (1) : 6-14, (2004).
17. A. V. Hoffbrand, and P. A. H. Moss, *Kapita Selekta Hematologi Edisi 6*. Jakarta: Penerbit Buku, (2013). Kedokteran EGC.
18. EPA. *Mercury*. Human Health, (2008).
19. SP. Daniel, MK. Marshall, USA; JB Lippincott publications; 205-239, (1999).
20. Mohammad A.M. Wadaan. Journal of Pharmacology and Toxicology, **4**: 126-131, (2009).
21. G. Jagadeesan, and S.S. Pillai. J. Environ. Biol., **28**: 753-756, (2007).
22. V. Singh, D. Joshi, S. Shrivastava and S. Shukla, Indian J. Exp. Biol., **45**: 1037-1044, (2007).
23. K.V. Sastry, and K. Sharma. Toxicol. Lett., **5**: 245-249, (1980).

Prevalence of Hookworm infection and Strongyloidiasis in Cats and Potential Risk Factor of Human Diseases

Blego Sedionoto[1,2,*], *Witthaya* Anamnart[1]

[1]Doctoral Program of Biomedical Sciences, School of Allied Health Sciences and Public Health, Walailak University, Thasala, Thailand
[2]Department of Environmental Health, Faculty of Public Health, Mulawarman University, Samarinda- Indonesia

Abstract. Hookworm infection and Stronyloidiasis are public health problem in the worldwide which both of them could infective in human by penetrated on skin and they have potential risk from Gastrointestinal zoonotic helminths of pets, including cats. We investigated the prevalence soil transmitted helminths infection in human and cats used modified Formal-Ether Concentration and agar plate culture. Fecal samples of 23 cats and human from Naitung and Subua Villages (area study 1), and fecal samples of 15 cats and 17 humans from Thasala Beach villages (area study 2) were collected. Result of study in area study 1 showed prevalence of infection in human was not hookworm and strongyloidiasis but 10% humans have infected Ascaris and Tricuris, and in cats have infected by hookworm 75.2% and S. strercoralis 8.5%, toxocara 13%, spirometra 13% and overall prevalence 82.5%. In area study 2 showed in human has infected by Trichuris 100% and S. stercoralis 29.4% and in cats have infected by hookworm 100% and S. strercoralis 40%, toxocora 20%, and spirometra 20%. Helminth infection found in both humans in two areas study are S. strercoralis. Hookworms were the most common helminth in cats but did not connection with infection in human, while S. strercoralis was helminth infection in cats which has potential zoonotic disease to human.

1 Introduction

Dogs and cats play a significant role as reservoir hosts for gastrointestinal zoonotic parasites including protozoa, trematode, cestode and nematode [1, 2, 3]. Humans can be infected via contact with a dog or cat or via contamination of infective stages in food or water [4, 5].

Worldwide, there is a significant variation in the prevalence of gastrointestinal zoonotic helminths in dogs and cats [6, 3]. High infection rates of zoonotic parasites including hookworms, *Trichuris spp.*, *Spirometra spp.*, *Taenia spp.*, *Toxocara spp.* and *Opisthorchis spp.* have been reported [7,8,6,3]. Infection of zoonotic helminths has previously been researched in Thailand.

In the central area, a high prevalence of hookworm Ancylostoma ceylanicum was reported among dogs in temple communities in Bangkok [9]. The infections of zoonotic helminths, hookworms, *Trichuris spp.*, *Toxocara spp.* and *Spirometra spp.* were found in dogs and cats in animal refuges [10].

In the Northeastern area, a high infection rate of liver fluke, Opisthorchis viverrini (O. viverrini) in dogs and cats, was found in communities where O. viverrini infection in human was high [3]. In Thailand, infections of hookworms and O. viverrini are the major public health problems [11, 12, 13, 14, 9]. Infections of zoonotic hookworms, A. ceylanicum and A. caninum, have been reported in many areas [13, 9]. Molecular analysis showed A. ceylanicum is prevalent in humans and dogs in the Central and the Northeastern areas of Thailand [13, 9].

Another STH, *Strongyloides stercoralis*, is often neglected in helminth surveys [15, 9], yet previous studies show high *S. stercoralis* infection rates in Cambodia [16]. School-aged children in the developing world are at highest risk of morbidity due to STHs and intestinal protozoan infections [17].

However, mass treatment only focuses on three major STHs (*Ascaris*/hookworm/*Trichuris*). Other nematodes like *S. stercoralis*, trematodes and protozoan infections are not addressed. In rural Southeast Asia,little is known about the zoonotic potential of IPIs in humans and animals. Therefore of domestic animals, such as cats, dogs and pigs, as contributors to human STHs and as reservoir hosts for zoonotic parasites remains unexplored and/or the data are inaccessible.

Although surveys of zoonotic gastrointestinal helminths in dogs and cats had been done in Thailand, most of the studies have focused on the Central or Northeastern region [18, 19, 10, 20]. This study to investigate prevalence of zoonotic helminth infection in cats that potential risk factors to human.

* Corresponding author: blego_kesling@yahoo.com

2 Materials and methods

2.1 Ethical considerations

The study protocol was approved by the Ethics Committee of Walailak University. All participants and relevant parties were informed of the purpose of the study. Written informed consent was obtained from all individuals prior to enrolment. All infections diagnosed in humans and animals were treated at the end of the study according to the Thai treatment guidelines.

2.2. Study design and area

The study was carried out in 2016 at village No 4, Thasala district, Nakhon Si Thammarat province, Southern Thailand). The climate is tropical, with warm and hot temperatures all year round and alternating dry and wet seasons. Households from Thasala District were randomly. All household members and animals (cats) were assessed for STHs using a single diagnostic test approach on one stool sample for each human and one sample for each animal. Only animals owned by the household were included in the survey. Risk factors for infection of humans and animals were assessed based on information collected through interviews and observations.

2.3. Field procedures and sample collection

On the day of the first visit, informed consent was obtained from all household members and interviews were conducted with enrolled participants. Interviews with young children were conducted with the help of a parent or legal guardian.

All enrolled participants received a prelabelled stool container. Participants were asked to fill the container with feces passed the following morning. Upon collecting the first sample, a second stool container was given to participants for filling. The collected stool samples were transported within two hours following defecation to a laboratory in Medical Technology laboratory Walailak University.

Stool samples from each human and cat present at the time of the visit and belonging to the household were obtained. For each animal, approximately five grams of feces were collected from around houses, placed into a sterile plastic fecal container and chilled immediately in a box containing ice. For each human, one stool sample given on consecutive day was analysed and for each, animal, one sample was analysed.

2.4. Laboratory procedure

For each human stool sample, the following tests were performed: Koga Agar culture [21], formalin-ether concentration technique (FECT) [22] analysis as they arrived in the laboratory, human samples were processed as follows:

First, MFECT, Samples that were found to be negative by direct smear examination were re-examined using formol-ether concentration technique [22]. In this, about 1 g of the feces was collected with a stick, mixed with physiological saline and put in a screw-cap bottle containing 4 ml of 10% formol water. The bottle was capped and mixed by shaking for about 20 seconds.

Thereafter, the feces were sieved and the suspension collected in a beaker. The suspension was poured into a tube and 3 ml of ether was added. The tube was closed and mixed by shaking for 1 minute, after which, the stopper was removed and the setup was centrifuged immediately at 3000 r.p.m. for 1 minute. An applicator stick was used to loosen the layer of faecal debris from the side of the tube after centrifugation. The layers of ether, debris and formal water that formed were decanted off, while the sediment was mixed, transferred to a slide and covered with a cover slip. The slide was examined under the microscope using 10x followed by 40x objective to identify the hookworm eggs.

Second, a Koga Agar test was prepared by placing a piece of stool (3–5 g) on a freshly produced Agar plate. The plates were then incubated for 48 hours at 28 °C. Larvae were washed from the plate into a tube, the liquid was centrifuged and the entire sediment was read at 40x magnification for hookworm and *Strongyloides stercoralis* larvae.

3 Results

3.1. Humans and cats samples

In this study, fecal samples from 15 cats at three houses Village No 4 Thasala District with 17 samples from human were collected from households Thasala Beach Village, Thasala district, Nakhon Si Thammarat province, Thailand.

3.2. Prevalence in animals

Result of study showed in two area studies such as at area study 1 hookworm 18 (78.2%,Strongyloidiasis 2(8.5%), Tococara spp 3 (13%), spirometra 3 (13%) and Overal Prevalence 19(82.5%) , and at area sudy 2 :hookworm infections and strongyloidiasis found in cats including hookworms 15 (100%), Strongyloides stercoralis 6(40%), Toxocara spp Toxocara (20%), and Spirometra 3 (20%)

Table 1. Prevalence Parasite in Animals

Parasite	Area Study 1	Area Study 2
Hookworm	78.2%(18/23)	100%(15/15)
Strongyloides stercoralis	8.5%(2/23)	40%(6/15)
Toxocara spp	13%(3/23)	20%(3/15)
Spirometra	13%)3/23)	20%(3/15)
Overall Prevalence	82.5%(19/23)	-

Dogs and cats are important reservoir hosts of various zoonotic helminths [1, 23, 3], many of which cause serious public health problems. Result studied at Norhern Thailand reported the prevalence of zoonotic intestinal helminths as 40.1 % (79/197) in dogs and 33.9 % (61/180) in cats, respectively.Zoonotic helminths found included hookworms, *Spirometra* spp., *Toxocara*, *O. viverrini*, *Taenia* spp. *Strongyloides* and *Trichuris* [24,3].

The occurrence of macroparasites was studied from 543 stray cats in four urban cities from the west (Kuala Lumpur), east (Kuantan), north (Georgetown) and south (Malacca) of Peninsular Malaysia from May 2007 to August 2010 showed up to nine species of helminths were recovered with overall high prevalences of infection of 83% in Kuantan, followed by 75.1% in Kuala Lumpur, 71.6% in Georgetown and 68% in Malacca. The helminth species comprised five nematodes, *Toxocara malaysiensis*, *Toxocara cati*, *Ancylostoma braziliensis*,*Ancylostoma ceylanicum*, *Physaloptera praeputialis*, two cestodes *Taenia taeniaeformis*, *Dipylidium caninum* and one trematode, *Playtnosomum fastosum*. [25].
Zoonotic hookworm, *A. caninum*, was found to have low infection rates in both dogs and cats. Similar to other areas, prevalence of *A. caninum* was lower than that of *A. ceylanicum* [21]. Although its infection rate was low, this hookworm can result in eosinophilic enteritis and chronic abdominal pain in human [17,9]. Other zoonotic hookworm, such as *A. braziliensis*, was not found in this area.

Among zoonotic helminth infections in dogs in the lower Northern area of Thailand, hookworms were the most prevalent helminth, and Spirometra was the second most prevalent. Our results confirmed that hookworm infection in dogs is common in Thailand [10-12]. The high prevalence of hookworm infections in dogs can contribute to the occurrence of zoonotic ancylostomiasis in human [26].

Zoonotic helminth infections in cats were different from dogs. *Spirometra* was the most prevalent, while hookworms were the second most prevalent helminth. High rates of *Spirometra* infection might be a reflection of the fact that most cats roam freely and had access to small prey as a food source. High infection rates of *Spirometra* spp. in cats might indicate a high infection rate of plocercoid and plerocercoid in intermediate hosts in the area. The infection of *Spirometra* spp. in cats and dogs can lead to a high risk of sparganosis in humans who have the habit of eating undercooked meat [27, 28]. However, human sparganosis in Thailand is rare. In the period 1943-2010, only 53 cases had been reported [29,27].

Dogs are associated with more than 60 zoonotic parasites worldwide, many of which pose serious public health concerns [30]. Compared with some other studies in South Asian countries, the overall prevalence of IPIs in dogs in Cambodia was higher (81.9%) than previously reported for dogs in rural India, for example [30].

This research in contrast with Combodian research that dogs in rural Cambodian villages such as Dong village are largely kept as guard dogs and allowed to roam freely, especially during the day. The dogs are also allowed inside the house and around rice and vegetable fields and ponds. At night-time, the dogs then often stay in or around the house.

Dogs, therefore, pose a serious zoonotic risk as they have the potential to transmit zoonotic parasites through their close association with household members as well as through heavy contamination of the environment, including soil, fresh produce and waterways, with parasite eggs and oocysts, in our observation sew behavior of cats almost all day and night stayed around houses and rare contact with ponds and did not stay around rice and vegetable fields.

3.3 Prevalence in humans

Prevalence parasite diseases in humans ain this research such as : area study 1 ; T. trichiura 10% (2/23) and in area study 2 :, *T. Trichiura* (100%), *Strongyloides strercoralis* (29.4%).

Table 2. Prevalence Parasite Infection in Human

Parasite	Area Study 1	Area Study 2
T. Trichiura	10% (2/23)	100% (17/17)
Strongyloides stercoralis	-	29.4% (5/17)
Ascaris	10% (2/23)	

The present study showed similar patterns of parasite infections in humans compared to previous surveys conducted in Cambodia [31]. The major parasite infections found in humans were hookworms (63.3%), Entamoeba spp. (27.1%), *S. stercoralis* (24.3%), *G. duodenalis* (22.0%) and *Blastocystis* (18.4%) . In total, 14 different parasite species were diagnosed, including eight helminthic and six protozoan parasites. Of the 218 participants, 27 (12.8%) were negative in all examinations. More than a quarter of the human participants (64, 29.4%) were infected with one parasite and a third (72, 33.0%) with two or more parasites.

Three (1.4%) and one (0.5%) participant(s) harboured five and six parasites, respectively, the prevalences of parasites (those with the highest infection rates) are given for the different age-groups. For hookworm, the prevalence increases from less than 50.0% in children up to the age of ten to more than 60.0% in adolescents and then remains above 60.0% in all subsequent age-groups. For *S. stercoralis*, the prevalence also increases over age, reaching its peak in age-groups 30 years and older. That research shows that the average number of helminthic co-infections increases over age, whereas the average number of protozoan co-infections is highest in children and lowest in adults older than 51 years. However, in this study, no cases of human *Ascaris spp*. infection were detected by microscopy. This coincides with the findings of Park and colleagues [30]. We demonstrated that in all age-groups, the average number of co-infections is about the same, yet helminthic co-infections accumulated over time, with a peak in 30–50 year old individuals. The trend for protozoan co-infections is reversed, with the highest number of protozoan co-infections occurring in children.

This pattern might reflect higher exposure of children. Alternatively, it could be because of higher infection intensities rather than prevalence in children, as microscopy can miss low-intensity protozoan infections [14], although this applies also for helminthic infections.

3.4 DNA analysis

Identify species of hookworm is *N. americanus*, but cats usually could be infectived by *Ancylostoma caninum*, condition of data this research not similar with research in Northern Thailand that it has significant zoonotic hookworms include A. ceylanicum, A. braziliensis and *A. caninum* [14, 28, 32].

Molecular analysis revealed that the most prevalent hookworm (over 80%) found in dogs and cats in the lower Northern area was *A. ceylanicum*. *A. ceylanicum* is highly prevalent in many areas in Asian countries [28, 29, 34] and is known to produce potent infections in humans. *A. ceylanicum* is the second most common hookworm infection in humans that can lead to anemia [33, 27].

PCR and sequencing were used for detection and identification of parasites in various specimens with high sensitivity and specificity [20, 14]. In our survey, molecular analysis was applied for two significant helminths infection, hookworms and *O. viverrini*. Morphological identification of hookworm larvae or eggs to species is difficult, and molecular identification provides great results in this regard [27].

This research has deference with research in Cambodia which it showed that in humans about half of the infections (51.6%) were *Ancylostoma ceylanicum* and the remaining *Necator americanus* infections. In dogs over 90% were *A. ceylanicum* indicating that most probably dogs are the source of infection. We hypothesize that regular deworming in communities lead to a replacement of *N. americanus* by *A. ceylanicum*. Parallel deworming of the dog population is likely to reduce the incidence in humans [14].

3.5 Zoonotic risk factors

Behavior of defecation cats and environmental factors also personal hygiene and sanitation have contribution became spread STHs infected from animal to human, defecation of cats observated had not potential risk infection from cats to human because cats usually closed with soil after defecated and cat anly defecated round houses and the fecal dry by solar contact then parasite was killed, deferent with studied in Nortern Thailand that Zoonotic hookworm, *A. caninum*, was found to have low infection rates in both dogs and cats. Similar to other areas, prevalence of *A. caninum* was lower than that of *A. ceylanicum* [20, 28]. Although its infection rate was low, this hookworm can result in eosinophilic enteritis and chronic abdominal pain in human [27, 28].

Environmental factors have potential zoonotic determined of hookworm infection and strongylidiasis where this area researched location have poor sanitation, without wastewater drainage that made wet surrounding houses and sandy soil round houses could make easy to spread hookworm and *Strongyloides stercoralis* by directed penetration from cats to human, defecated cats could not spread hookworm infection and strongyloidiasis because cats behavior on defecation made larvae of soil transmitted helminth not completed to infective filariae form larvae especially hookworm and *Strongyloidisis stercoralis*.

4 Conclusions

This research given statement that cats was not equal with human hookworm infections which have not zoonotic potential also *S. stercoralis, Ascaris lumbricoides*. and *T. trichiura* because cats behavior on defecation made larvae of soil transmitted helminth not completed to infective filariae form larvae especially hookworm, but *S. stercoralis* became risk factors zoonotic disease by directed contact cats to human. Helminth infection found in both humans in two areas study are *S. stercoralis*. Hookworms were the most common helminth in cats but did not connection with infection in human, while *S. strercoralis* was helminth infection in cats which has potential zoonotic disease to human.

References

1. F. Fang, J. Li, T. Huang, J. Guillot, and W. Huang,. BMC Vet. Res. **11**: 211, (2015)
2. A.P. Oliveira-Arbex, E.B. David, T.C. Oliveira-Sequeira, S. Katagiri, S.T. Coradi, and S. Guimarães, J. Helminthol. **11**: 1-6, (2016)
3. Overgaauw, P.A., van Zutphen, L., Hoek, D., Yaya, F.O., Roelfsema, J., Pinelli, E., van Knapen, F. and Kortbeek, L.M. Vet. Parasitol. **163**: 115-122, (2009)
4. R. Nijsse, L. Mughini-Gras, J.A. Wagenaar, F. Franssen, and H.W. Ploeger, Parasit. Vectors **8**: 397, (2015)
5. V. Wiwanitkit, and W. Waenlor, Rev. Inst. Med. Trop. Sao Paulo **46**:113–114, (2004)
6. D. Ng-Nguyen, S.F. Hii, V. A. Nguyen, T. Van Nguyen, D. Van Nguyen, and R.J. Traub, Parasit. Vectors **8**: 40, (2015)
7. K. Ash, T. Orihel, *Atlas of human parasitology, 4th edition*. American Society of Clinical Pathologists, U.S.A.; (1997)
8. T. Inpankaew, F. Schär, A. Dalsgaard, V. Khieu, W. Chimnoi, C. Chhoun, D. Sok, H. Marti, S. Muth, P. Odermatt, and R. J. Traub. Emerg. Infect. Dis. **20**: 976–982.(2014)
9. Q. Liu, M.W. Li, Z.D. Wang, G.H. Zhao, and X.Q. Zhu, The Lancet Infect. Dis. **15**: 1226–1235, (2015)

10. RJ. Traub, T. Inpankaew, C. Sutthikornchai, Y. Sukthana, RC. Thompson. Vet Parasitol, **155**:67–73, (2008)
11. W. Rojekittikhun, K., Chaisiri, A. Mahittikorn, S. Pubampen, S. Sa-Nguankiat, T. Kusolsuk, W. Maipanich, R. Udonsom, and H. Mori, Southeast Asian J. Trop. Med. Public Health **45**: 31−39, (2014)
12. A.R. Jex, Y.A. Lim, J.M. Bethony, P.J. Hotez, N.D. Young, and R.B. Gasser, Adv. Parasitol. **74**:231-265, (2011)
13. P. Jongsuksuntigul, and T. Imsomboon, Acta Trop. **88**: 229-232, (2003)
14. I. Phosuk, P.M. Intapan, T. Thanchomnang, O. Sanpool, P. Janwan, P. Laummaunwai, W. Anamnart, N. Morakote, and N. Maleewong, Korean J.Parasitol. **51**:747-749, (2013)
15. P. Sithithaworn, R.H. Andrews, V.D. Nguyen, T. Wongsaroj, M. Sinuon, P. Odermatt, Y. Nawa, S. Liang, P.J. Brindley, and B. Sripa, Parasitol. Int. **61**:10-16, (2012)
16. F. Schär, U. Trostdorf, F. Giardina, V. Khieu, S. Muth, H. Marti, et al. PLoS Negl Trop Dis,**7**:e2288, (2013)
17. V. Khieu, F. Schär, H. Marti, S. Sayasone, S. Duong, S. Muth, et al. PLoS Negl Trop Dis, **7**:e2035, (2013)
18. B. Speich, S. Knopp, KA. Mohammed, IS. Khamis, L. Rinaldi, G. Cringoli, et al. Parasites Vectors, **3**:71, (2010)
19. S. Aunpromma, P. Tangkawattana, P. Papirom, P. Kanjampa, S. Tesana, B. Sripa, and S. Tangkawattana, Parasitol. Int. **61**: 60−64, (2012.)
20. T. Inpankaew, R. Traub, R.C. Thompson, and Y. Sukthana, Southeast Asian J. Trop. Med. Public Health **38**: 247−255, (2007)
21. R.J. Traub, T. Inpankaew, C. Sutthikornchai, Y. Sukthana, and R.C. Thompson, Vet. Parasitol. **155**: 67−73, (2008)
22. I. A. Nnanchi. O. O. Odikamnoro, A. U. Nnachi, O. C. Ani, A. C. Uhuo, C. C. Egwuatu, Nigeria International Journal of Tropical Disease & Health **10**(4). Pages: 1-10, (2015)
23. A.P. Oliveira-Arbex, E.B. David, T.C. Oliveira-Sequeira, S. Katagiri, S.T. Coradi, and S. Guimarães, J. Helminthol. **11**: 1-6, (2016.)
24. P.A. Overgaauw, L. van Zutphen, D. Hoek, F.O. Yaya, J. Roelfsema, E. Pinelli, F. van Knapen, and L.M. Kortbeek, Vet. Parasitol. **163**: 115-122, (2009)
25. Siti Nursheena Mohd Zain, Norhidayu Sahimin, Paul Pal, John W. Lewis. Veterinary Parasitology.**196**, Pages 469-477, (2013).
26. P. Prociv, and J. Croese, Acta Trop. **62**: 23−44, (1996.)
27. Fabian Schär, Tawin Inpankaew, Rebecca J. Traub, Virak Khieuf, Anders Dalsgaard, Wissanuwat Chimnoi, Chamnan Chhoun, Daream Sok, Hanspeter Marti, Sinuon Muth, (2014)
28. P Odermatt. Parasitology International **63**, 597–603, (2014)
29. SK. Park, DH. Kim, YK. Deung, HJ. Kim, EJ. Yang, SJ. Lim, et al. Korean J Parasitol, **42**:201–3, (2004)
30. RJ. Traub, T. Inpankaew, SA. Reid, C. Sutthikornchai, Y. Sukthana, ID. Robertson, et al. Acta Trop, **111**:125–32, (2009)
31. Y.C. Hsu, and J.T. Lin, N. Engl. J. Med. **366**: e20, (2012)
32. R.J. Traub, T. Inpankaew, C. Sutthikornchai, Y. Sukthana, and R.C. Thompson, Vet. Parasitol. **155**: 67−73, (2008)
33. T. Inpankaew, F. Schär, A. Forrer, V. Khieu, W. Chimnoi, C. Chhoun, et al. 2012. Emerg Infect Dis 2014 [June].
34. D.D. Bowman, S.P. Montgomery, A.M. Zajac, M.L. Eberhard, and K.R. Kazacos, Hookworms of dogs and cats as agents of cutaneous larva migrans. Trends Parasitol. **26**: 162−167, (2010)

Mapping Of Leptospirosis Environmental Risk Factors and Determining the Level of Leptospirosis Vulnerable Zone In Demak District Using Remote Sensing Image

Siti Rahayu[1], *Mateus* Sakundarno Adi[2], and *Lintang* Dian Saraswati[1*]

[1]Department Epidemiology and Tropical Diseases, Public Health Faculty, Diponegoro University, Semarang, Indonesia, 50275
[2]Master Program of Epidemiology, School of Postgraduate Studies Diponegoro University, Semarang, Indonesia

Abstract. Leptospirosis, a zoonotic disease, transmitted to human trough contact with contaminated animal urine and contaminated environment. Demak District is an endemic area where cases increased in the past 2 years. The aim of the study was to map environmental risk factor of Leptospirosis and to determine Leptospirosis vulnerable zone using cross-sectional study design. There were 42 cases mapped by GPS and overlaid using remote sensing (Quickbird image) by using ArcView program then interpreted by Spatial Feature and Spatial Analyses. Leptospirosis cases were spread out and grouped in Demak Sub District area. More cases were males (61.9%), 21-50 years old age group (59.3%) and farmers (40.4%). Spatial analyses showed that all the leptospirosis cases took place in the area with low plain <47 msl, rainfall ≥220 mm/month (64.7%), clay soil (100%), buffer river <50 m (71.4%), presence of rat (100%), wastewater disposal (100%), waste disposal facilities (97.7%), flood's profile (28.6%), tidal inundation's profile (7.1%), vegetation (59.5%). Leptospirosis high-risk zone was in 37,801.8 ha (41.32%), moderate risk zone was 43,570.23 ha (48.55%), and low-risk zone was 9,090.96 ha (10.13%). Densely populated housing, bad environment condition, and the presence of rat and puddles that were contaminated by rat's urine were risk factors of Leptospirosis cases in Demak District.

1 Introduction

Leptospirosis is a worldwide public health problem, especially in tropical and subtropical countries with high rainfall, and is also a major problem in temperate regions [1]. Leptospirosis, caused by pathogenic spirochetes belonging to the genus Leptospira, is a zoonosis that has important impacts on human and animal health worldwide [2]. Transmission to mammals occurs via direct contact with leptospira-infected urine or tissues or indirectly through contact with contaminated soil or water. Although infection may take place through unbroken skin after prolonged immersion, *Leptospira sp* usually gain entry to the host via abrasions or cuts in the skin or through exposed mucosae (eyes, nose, *etc.*). Incidence is seasonal, peaking in summer and fall in temperate climates and during the rainy season in tropical areas, mirroring the ability of the bacteria to survive in the external environment. Soil [3], [4], mud [5], and surface waters contaminated with urine from chronically-infected reservoir hosts remain important sources of human leptospirosis transmission worldwide [6]–[8].

The incidence of Leptospirosis in subtropical countries is in the range of 0.1-1 incidence per 100,000 population per year, whereas in tropical countries it ranges from 10-100 incidence per 100,000 population per year [1]. Indonesia as a tropical country becomes one of them. The *International* Leptospirosis *Society* states that Indonesia is one of the tropical countries with relatively high leptospirosis death cases, ranging from 2.5%-16.45% or an average of 7.1% and is ranked third in the world. Leptospirosis cases in Indonesia in 2012 among 239 patients, 29 people died (CFR 12.13%), in 2013 there are an increasing number of Leptospirosis cases with 640 patients, 60 people died (CFR 9.38%). Leptospirosis cases in 2014 slightly decreased by 519 patients, 61 people died (CFR 11.75%) [1], [8].

Demak District is one of the endemic areas of Leptospirosis in Central Java Province with a mortality rate of Leptospirosis exceeding the national average (7.1%). According to Health Office of Demak District in 2015, there are 13 patients of Leptospirosis cases and 2 people died in 2012 (CFR 15.3%). In 2013, there are 17 patients of Leptospirosis cases and 2 people died (CFR 11.7%), in 2014 it's increased by 30 patients and 5 people died (CFR 16.6%), and in 2015 there are 12 patients and 1 person died (8.3%).

* Corresponding author: lintang.saraswati@live.undip.ac.id

The occurrence of infectious diseases is influenced by three factors: Agent, Host, and Environment [9]–[11]. The agent of Leptospirosis disease is *Leptospira* bacteria [12], [13]. The main hosts are rats and pets such as dogs, cats, and cows [2], [12], [13]. The accidental host is human [14], [15]. In the tropical climatic zone, where environmental conditions are most favorable for survival of Leptospira and the highest morbidity is noted, extreme weather events such as cyclones and floods occurring in recent years with increasing frequency and greater intensity, may potentially result in an upsurge in the disease incidence, as well as the magnitude of leptospirosis outbreaks [16]–[18].

The high prevalence rate of Leptospirosis in tropical and subtropical climates can be attributed to adverse environmental conditions that allow the environment to be a suitable place to live and develop of *Leptospira* bacteria [5], [19]. Our previous study showed mapping and analysis of environmental conditions are needed to determine the picture of environmental factors existed and have a risk for the incidence of Leptospirosis [20].

The mapping can be done by utilizing Geographic Information System (GIS). Geographic Information System is a special case where database consist of observations on spatially distributed features, activities, or events, which are definable in space as point, lines, or areas [21], [22]. Remotely sensed imagery is an effective data source for urban environment analysis that is inherently suited to provide information on urban land cover characteristics and their changes over time at various spatial and temporal scales without being in contact with it [23]–[25].

Both technologies are very useful in the management of spatial information of earth surface condition [26], [27]. Recently, the use of GIS and spatial representation of various health issues make professionals arrive at conclusions in a faster and better way in the field of both public health and decision-making. The use of these systems has a wide impact on the public health and lots of studies are based on them [17,18].

2 Methods

An observational study with the cross-sectional design was used for mapping environmental risk factors and determining the level of Leptospirosis vulnerable zone using remote sensing image in Demak District. Primary data were an environmental risk factor and coordinate location point using GPS. Secondary data were a number of leptospirosis cases and digital SHP map of Demak District in 1:25,000 scale (Administration Border, Land Use, River, Road, Contour, Altitude, Soil Texture, Flood Area, and Rainfall). There were 42 leptospirosis cases.

The data obtained was mapped with the GPS Mobile Topographer with the ArcView program built with Quickbird image. After visual interpretation and screen digitalization to visual identification of environmental risk factors. Spatial modeling analysis of environmental factors and distribution of Leptospirosis cases with overlay process to determine the level of Leptospirosis vulnerable zone.

3 Results and discussions

The result of this study indicates that more than half of respondents are male by 61.9% or as many as 26 cases. The result is different from the condition of the population in Demak District, where the female population is larger than male. Both men and women are equally at risk to be infected by leptospirosis, but men are at greater risk for Leptospirosis infection due to several things, such as activity and occupation.

Table.1 Characteristic of Respondents by Sex, Education Level, and Occupation (n=42)

Characteristics of Respondents	Frequency	Percentage (%)
1. Sex		
a. Male	26	61.9
b. Female	16	38.1
2. Occupation		
a. Entrepreneur	4	9.5
b. Labor	3	7.1
c. Civil Servant	1	2.3
d. Housewife	7	16.6
e. Farmer	17	40.4
f. Scavenger	2	4.7
g. Merchants	4	9.5
h. Pedicab Driver	1	2.3
i. Fisherman	2	4.7
j. Student	1	2.3

Table.2 Frequency Distribution of Leptospirosis Cases in Leptospirosis Case Area in Demak District in 2014-2015

Variable	Frequency	Percetg. (%)
1.Topography		
a. <47 masl	42	100
b. ≥47 masl	0	0
2.The distance between River and House		
a. <50 m = at risk	30	71.4
b. 50-300 m = moderate	12	28.6
c. >300 m = good	0	0
3.Rainfall		
a. ≥220 mm/month	27	64.3
b. <220 mm/month	15	36.7
4.The presence of Rats		
a. Present	42	100
b. None	0	0
5.Wastewater Disposal Facility		
a. Good	42	100
b. PooR	0	0
6.Waste Disposal Facility		
a. Good	41	97.7
b. Poor	1	2.3

Variable	Frequency	Percetg. (%)
7.The presence of Pet		
a. Present	29	69
b. None	13	31
8.Flood History		
a. Present	12	28.6
b. None	30	71.4
9.Tidal Flood History		
a. Present	6	14.3
b. None	36	85.7
10.Vegetation		
a. ≥3 types	25	59.5
b. <3 types	17	40.5

As of the case in Demak District, the activity of men who will directly clean the environment after flood causes the possibility of exposure to the contaminant environment even larger.

Occupation of the respondents is also at risk to exposure to contaminants or animals infected with Leptospirosis. It can be seen from the result of this research that the characteristic of the occupation of respondents is labor farmer, and scavenger. Farmers are at risk to the rice fields contaminated with Leptospirosis from rat droppings. Construction laborers are at risk to the dirty water contaminated by *Leptospira* when working, i.e. during stirring of cement. Scavengers are at risk of direct contact with the rat droppings while searching for rubbish in the trash, as rats often roam for food in the trash. Respondents' location to look for rubbish is in Semarang City. Semarang City is one of the endemic areas of Leptospirosis, so there is a high possibility of transmission occurred when looking for rubbish.

Demak District is an area with high rainfall that is 534-3.331 mm/year. More than half of cases occur in areas with high rainfall ≥220 mm/month (64.7%).

The condition of Demak District which is mostly lowland so more than half of its area is used for irrigation rice field. The existence of large river or river irrigation becomes a strong factor of flood during rainy season. A flood occurs due to the overflow of river water. The habit of throwing rubbish in the river make the volume of river shrinks so water will be overflowed when the water debit start to increase. In addition, the result of this study also indicates that the respondents are living in housing with poor sewerage conditions and poor waste disposal facilities.

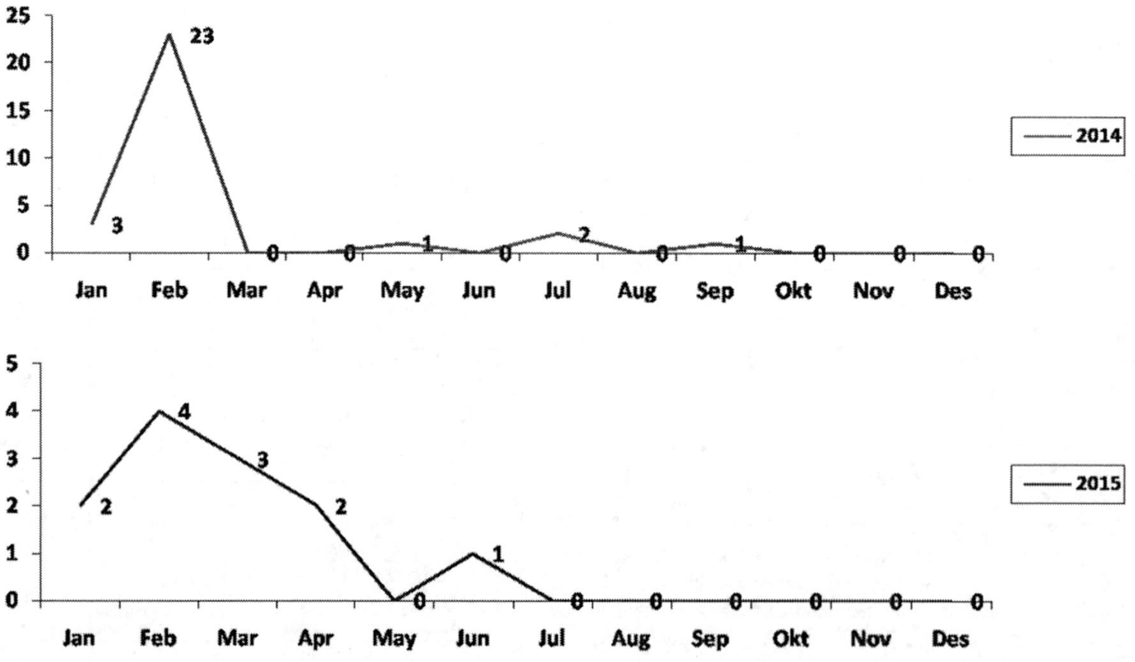

Fig.1 The Incidence of Leptospirosis Cases in Demak District Every Month in 2014-2015

https://doi.org/10.1051/e3sconf/20183106003

The incidence of Leptospirosis in Demak District in 2014-2015 is increasing during high rainfall (January-February) and still occurred during low rainfall. High rainfall is the cause of flood and resulted in the formation of puddles. The presence of puddles is supported by clay soil type in Demak District, so the water can stagnate longer [3], [29].

The altitude from sea level is an important variable for Leptospirosis cases spreading. Altitude condition is strongly associated with flood area and the location of permanent water logging [30], [31]. Mostly (98.9%) Demak District area is lowland, except for half of Mranggen Sub-district and Karangawen Sub-district that has ≥47-100 masl

Flooding and heavy rainfall have been associated with an increase of leptospirosis in specific in Albania. The study showed that Leptospirosis is spread in 47% teritory of the country with increase of incidence during the last two years. The most affected areas are the flooded areas of Shkoder, Fier and Lezha and also Berat and Tirana with single cases in other areas. The level of incidence has been increased from 0.1 cases/10 000 in 2005 toward 1.8 -2.5 cases/ 10 000 in 2010 and 2011. Climate change and floods may potentially result in an upsurge in the disease incidence over the last years [20].

Distribution of tidal flood history in Leptospirosis incidence in Demak District shows that flood history minimum once in 2 years occurred in 3 respondents houses (7.1%). Tidal flood and flood are the same events that cause the widespread formation of water puddles that last for a long time. However, from the distribution of cases, there appears to be a difference between the distribution of cases in the area with flood history and the distribution of cases in the area with tidal flood history. In the distribution of cases in the area with flood history, there are a lot of cases occurred in Mangunjiwan Village area, that occurred 7 cases after the flood. As for the area with tidal flood history in Monosari Village, Sayung Subdistrict there are only 2 cases. This is possible because the pH of tidal flood water ranges from 8.3-8.5, whereas pathogenic *Leptospira sp* survive in moist soil and fresh water for long periods of time, especially when the pH is slightly alkaline [13], [32]. Rats are the primary reservoir animals in Leptospirosis cases in Demak District. This can be seen from the results of this research showing the presence of rats either inside or outside the house of all respondents. Respondents claimed to have seen rats either inside or outside the house, to have seen rat's traces in the house or rat droppings inside the house.

The renal carrier state of rodents still remains a basic factor for the possible survival and spread of *Leptospira sp*. The primary source is the excretor rodent, from whose proximal renal tubules *Leptospira sp* are excreted into the environment with urine. Although they are susceptible to environmental factors, in particular drying, they can survive for long periods in water and wet soil [18]. Respondents with rats in their house have 22.6 times greater risk of Leptospirosis compared to those who do not have rats in their house. And respondents who use vegetables ate by rodents have 4.6 times greater than who don't use vegetables ate by rodents [29].

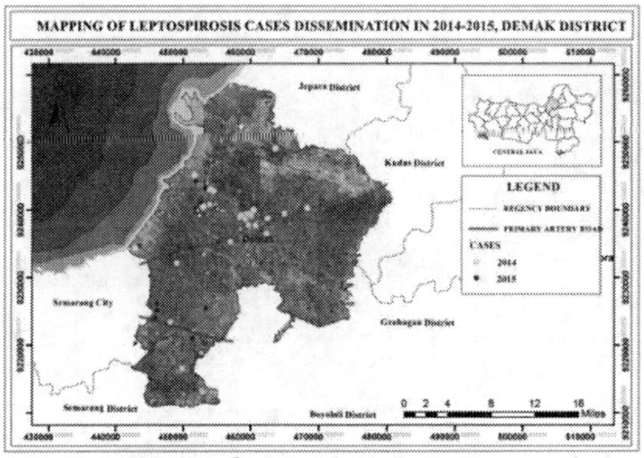

Fig.2. Dissemination of Leptospirosis Cases

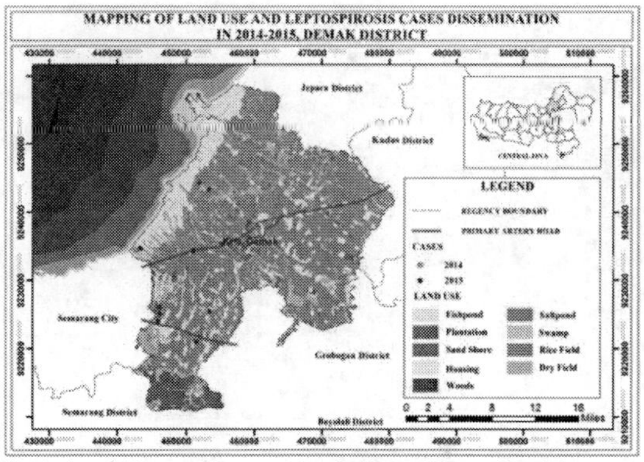

Fig.3. Land Use and Leptospirosis Cases Dissemination

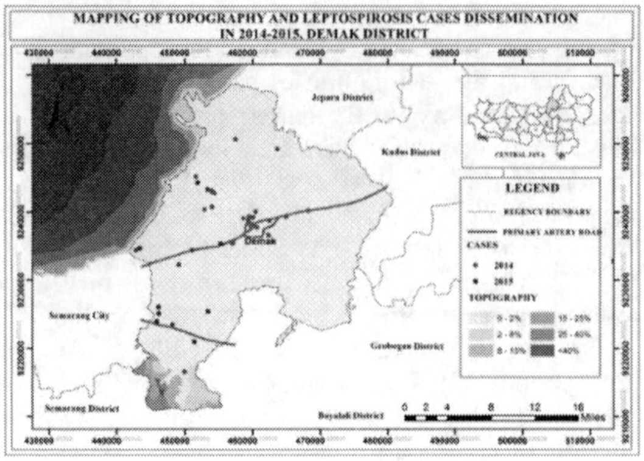

Fig.4. Topography and Leptospirosis Cases Dissemination

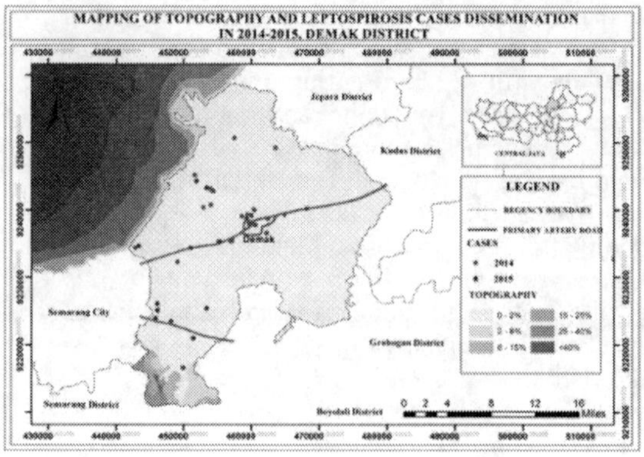

Fig.5. Type of Soil and Leptospirosis Cases Dissemination

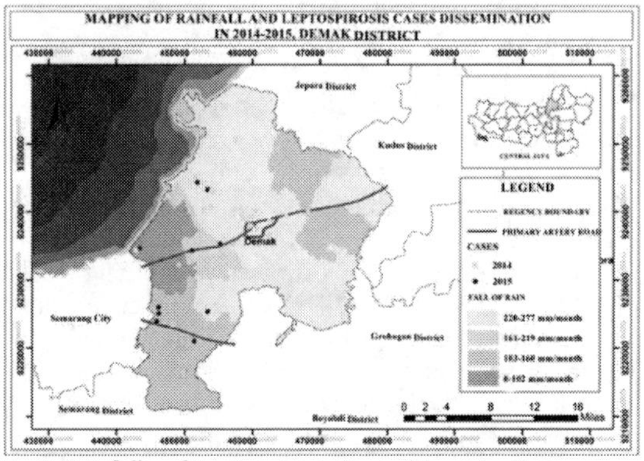

Fig.6. Rainfall and Leptospirosis Cases Dissemination

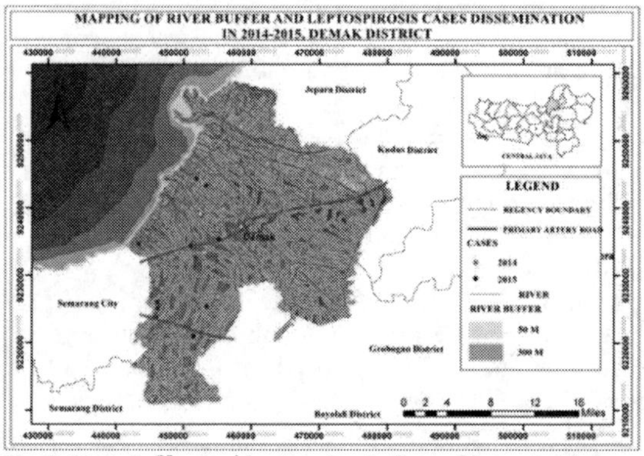

Gig.7. River Buffer and Leptospirosis Cases Dissemination

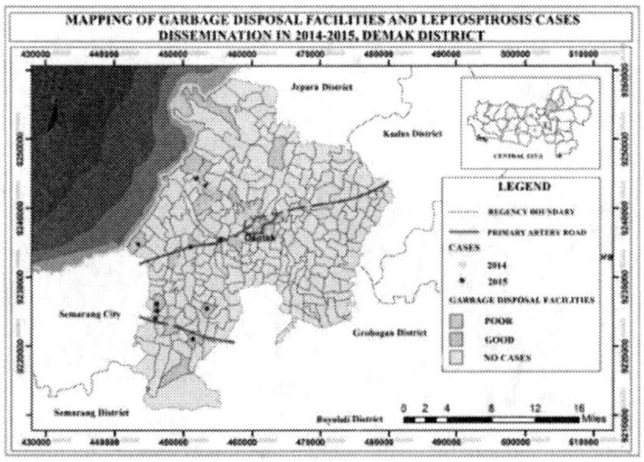

Fig.8. Garbage Disposal Facilities

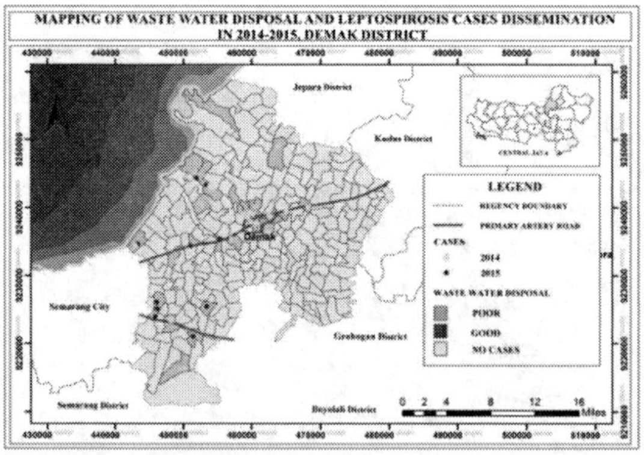

Fig.9. Waste Water Disposal Facilities

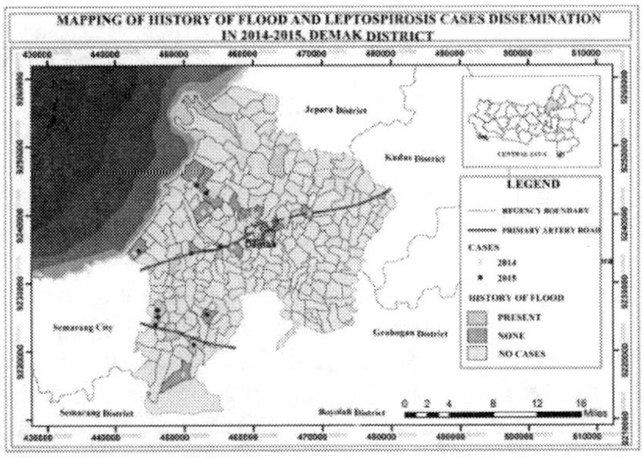

Fig.10. History of Flood

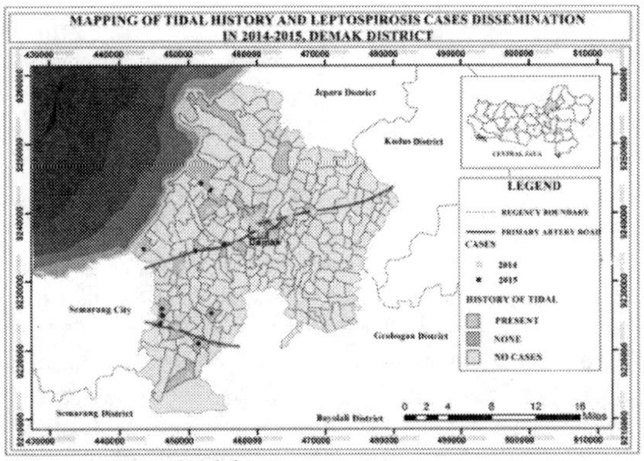

Fig.11. History of Tidal

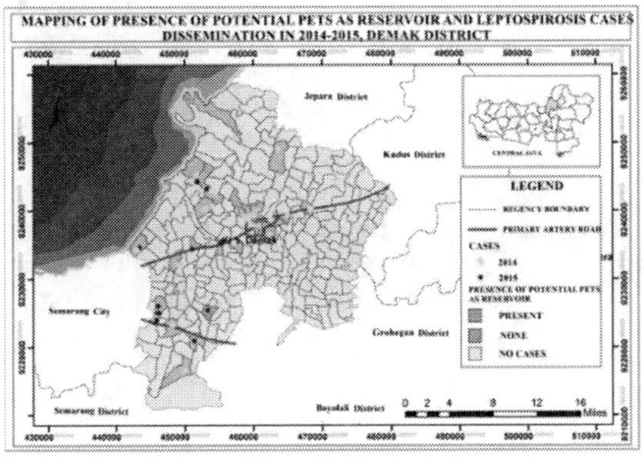

Fig.12. The presence of Potential Pests As Reservoir

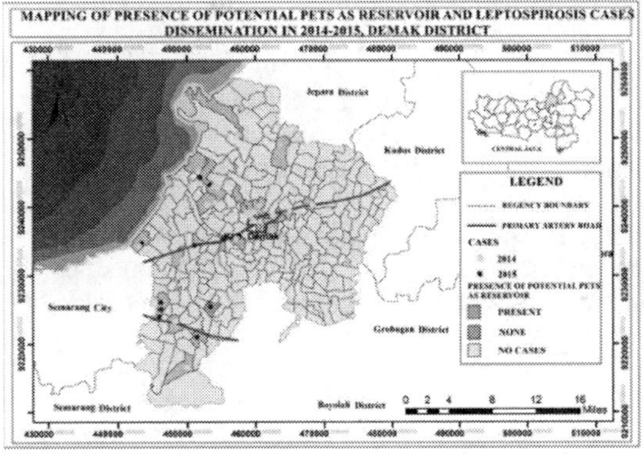

Fig.13. Vegetation Number

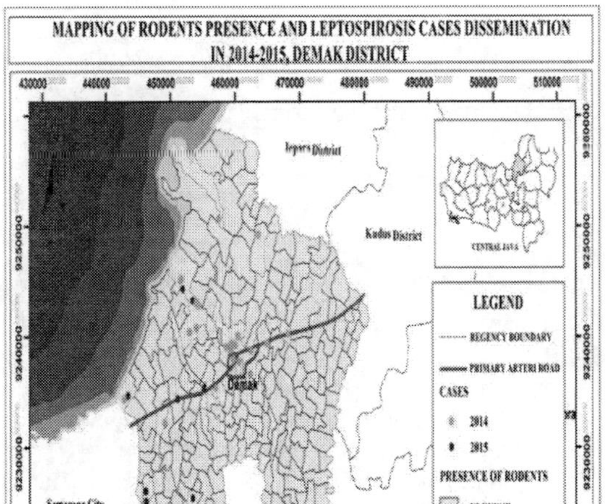

Fig.14. Presence of Rodents

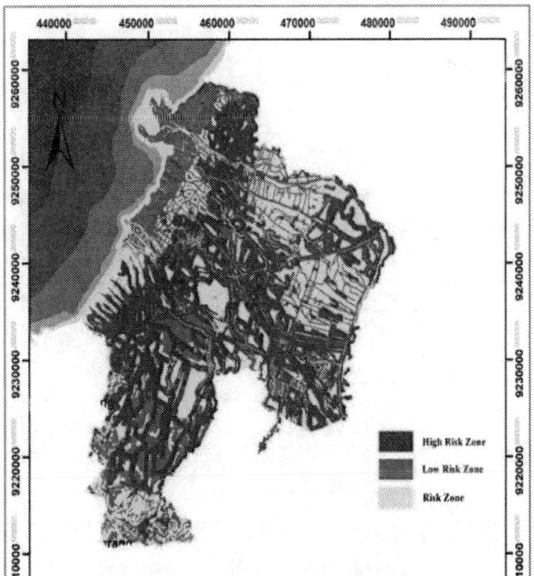

Fig.15. Mapping of Leptospirosis Vulnerable Zone in Demak District

This can be seen from the results of this research showing the presence of rats either inside or outside the house of all respondents. Respondents claimed to have seen rats either inside or outside the house, to have seen rat's traces in the house or rat droppings inside the house.

The renal carrier state of rodents still remains a basic factor for the possible survival and spread of *Leptospira sp*. The primary source is the excretor rodent, from whose proximal renal tubules *Leptospira sp* are excreted into the environment with urine. Although they are susceptible to environmental factors, in particular drying, they can survive for long periods in water and wet soil [18]. Respondents with rats in their house have 22.6 times greater risk of Leptospirosis compared to those who do not have rats in their house. And respondents who use vegetables ate by rodents have 4.6 times greater than who don't use vegetables ate by rodents [29].

Demak District is 89,743 ha which has 14 subdistricts and 243 villages. All of it is not leptospirosis risk zone. Handling determining and leptospirosis intervention is prioritized at a specific zone that is risk zone.

Making of leptospirosis vulnerable zone be based on leptospirosis vulnerable zone which is determined by environmental parameters that are land use, type of soil, river buffer, rainfall, topography, and vegetation.
Scoring and marking of all variables were based on variable influence great or environmental variable toward leptospirosis incident. Basic of scoring and marking are the result of first research and the result of the modification.

Table.2 Score, Mark, and Value at Environmental Parameters of Determining of Leptospirosis vulnerable zone in Demak District.

Environmental Parameters	Score	Mark	Value
1. Land Use	3: Housing	6	18
	2: Rice field, Plantation, woods, Swamp		12
	1: Fishpond, Tegalan, Dry field		6
2. River Buffer	3: <50 m	5	15
	2: 50-300 m		10
	1. >300 m		5
3. Vegetation	2: ≥3 Types	4	8
	1: <3 Types		4
4. Rainfall	2: ≥220 mm/month	3	6
	1: <220 mm/month		3
5. Type of soil	2: *Clay, Loam, Silt*	2	4
	1: *Sand*		2
6. Topography	2: <47 masl and 0-2%	1	2
	1: >47 masl and>2%		1

Final scoring of all environmental parameters procured from a score of environmental parameter mark. Overlay executed after summing the value from each environmental parameters. Dissolve at the end of overlay executed with the aim to agglomerate the final result that has the same value at each environmental parameters so class-wide of leptospirosis vulnerable zone can be calculated.

After scoring, marking and value each environmental parameters, determination of classification leptospirosis vulnerable zone is determined. Spatial classification of leptospirosis vulnerable rate is using Sturgess method as reference. It divides the result of the number of maximal value decrement with the number of minimal value be based on the number of the class wanted. The determination result of scoring, marking, and value can be known that the number f maximum value is 53 and the number of minimal value is 21. Classification of leptospirosis vulnerable zone is divided 3 class which are high-risk zone, risk zone, and low-risk zone.

Classification = (The number of maximal value – The number of minimal value) /The number of class

Classification = (53-21)/3

Classification = 10

Table.3 Class of Leptospirosis Vulnerable Zone

Class	Interval	Zone
1	43-53	High-Risk Zone
2	32-42	Risk Zone
3	21-31	Low-risk Zone

Leptospirosis high risk zone in Demak district is 37,801.8 ha (41.32%), risk zone is 43,570,23 ha (48.55%), and low risk zone is 9,090.96 ha (10.13%).

Leptospirosis risk zone in Demak is spread evenly in all of Demak area. It is caused by the area of Demak district which has low plain exploited as rice field. the leptospirosis high-risk zone spreads in all of Demak district area with forming a pattern on the river current because Demak district is the area which has many big rivers, a small river, and irrigation that can cause flood or puddle when the river water overflowing at the moment that raining. Leptospirosis high-risk zone is crowd and dirt housing which has near a location with river and rice field.

4 Conclusions

1. More than half of the respondents are male with average age of 44 years old. The respondents' risk occupations are a farmer, construction laborers, and scavengers.
2. Leptospirosis cases in 2014 and 2015 have the same pattern of cases that are occurring during high rainfall (January and February) and during low rainfall or dry season (June and July) although the rain intensity is lower.
3. The condition of Demak District which has lowland area with clay soil type and high rainfall attributed to the occurrence of flood and puddles. The condition is exacerbated by poor sewerage and poor waste disposal facility.
4. Rats are the main reservoir of Leptospirosis cases in Demak District.

5 Suggestions

1. Cooperate with sectoral agencies regarding the prevention of flood by rejuvenating the functions of the river and digging the shallowing rivers so the water will flow smoothly during the rainy season.
2. Cooperate with relevant cross-sectoral agencies in rejuvenating the function of sewer and constructing a healthy sewer in residential areas.
3. Conducting efforts to eradicate rats, given the fact that rats are present in all areas of Leptospirosis cases.

6 References

1. World Health Organization, *Human leptospirosis : guidance for diagnosis, surveillance and control*. Geneva: World Health Organization, (2003).
2. A. R. Bharti *et al.*, *Lancet Infect. Dis.*, vol. **3**, no. 12, pp. 757–771, Dec. (2003).
3. D. Benacer, P. Y. Woh, S. N. Mohd Zain, F. Amran, and K. L. Thong, *Microbes Environ.*, vol. **28**, no. 1, pp. 135–40,(2013).
4. D. Benacer *et al.*, *Asia. Pac. J. Public Health*, vol. **28**, no. 4, pp. 290–302, May (2016).
5. M. A. S. Johnson *et al.*, *Emerg. Infect. Dis.*, vol. **10**, no. 6, pp. 1016–22, Jun. (2004).
6. P. R. Hunter, *J. Appl. Microbiol.*, vol. **94** Suppl, p. 37S–46S, (2003).
7. A. M. Monahan, I. S. Miller, and J. E. Nally, *J. Appl. Microbiol.*, vol. **107**, no. 3, pp. 707–716, Sep. (2009).
8. M. Sakundarno, D. Bertolatti, B. Maycock, J. Spickett, and S. Dhaliwal, *Asia. Pac. J. Public Health*, vol. **26**, no. 1, pp. 15–32, Jan. (2014).
9. A. J. Mcbride, D. Athanazio, M. Reis, and A. I. Ko, *Leptospirosis Curr. Opin. Infect. Dis.*, vol. **18**, no. 5, pp. 376–386, (2005).
10. T. Fraga, E. Carvalho, L. Isaac, and A. Barbosa, in *Molecular medical microbiology*, 2nd Ed., Y.-W. Tang, M. Sussman, D. Liu, I. Poxton, and J. Schwartzman, Eds. Elsevier, pp. 1973–1990, (2014).
11. S. Craig *et al.*, *Ann Trop Med Parasitol*, vol. **103**, no. 7, pp. 647–51, (2009).
12. B. Adler and A. de la Peña Moctezuma, *Vet. Microbiol.*, vol. **140**, no. 3–4, pp. 287–96, Jan. (2010).
13. P. N. Levett, *Clin. Microbiol. Rev.*, vol. **14**, no. 2, pp. 296–326, Apr. (2001).
14. G. André-Fontaine, *Vet. Microbiol.*, vol. **117**, no. 1, pp. 19–24, Oct. (2006).
15. J. F. Prescott, B. McEwen, J. Taylor, J. P. Woods, A. Abrams-Ogg, and B. Wilcock, *Can. Vet. J. = La Rev. Vet. Can.*, vol. **43**, no. 12, pp. 955–61, Dec. (2002).
16. C. L. Lau, L. D. Smythe, S. B. Craig, and P. Weinstein, *Trans. R. Soc. Trop. Med. Hyg.*, vol. **104**, no. 10, pp. 631–638, Oct. (2010).
17. S. C. Sehgal, A. P. Sugunan, and P. Vijayachari, *Natl. Med. J. India*, vol. **15**, no. 1, pp. 22–3, (1995).
18. P. Vijayachari, A. P. Sugunan, and A. N. Shriram, *J. Biosci.*, vol. **33**, no. 4, pp. 557–69, Nov. (2008).
19. R. B. Reis *et al.*, *PLoS Negl. Trop. Dis.*, vol. **2**, no. 4, p. e228, (2008).
20. S. N. Fajriyah, A. Udiyono, and L. D. Saraswati, *IOP Conference Series: Earth and Environmental Science*, vol. **55**, no. 1, (2017).
21. A. Butt, R. Shabbir, S. S. Ahmad, and N. Aziz, *Egypt. J. Remote Sens. Sp. Sci.*, vol. **18**, no. 2, pp. 251–259, Dec. (2015).
22. K. J. Dueker, Ntherlands: Elsevier Scientific Pub. Co, (1979).
23. V. Mesev, *Comput. Environ. Urban Syst.*, vol. **21**, no. 3–4, pp. 175–187, May (1997).
24. M. Haq, M. Akhtar, S. Muhammad, S. Paras, and J. Rahmatullah, *Egypt. J. Remote Sens. Sp. Sci.*, vol. **15**, no. 2, pp. 135–141, Dec. (2012).
25. M. K. Jat, P. K. Garg, and D. Khare, *Int. J. Appl. Earth Obs. Geoinf.*, vol. **10**, no. 1, pp. 26–43, Feb. (2008).
26. O. Berke, *Prev. Vet. Med.*, vol. **71**, no. 3–4, pp. 173–182, Oct. (2005).
27. R. K. Raghavan, K. M. Brenner, J. A. Harrington, J. J. Higgins, and K. R. Harkin, *Geospat. Health*, vol. **7**, no. 2, pp. 169–82, May (2013).
28. W. Poeppl *et al.*, *Eurosurveillance*, vol. **18**, no. 25, p. 20509, Jun. (2013).
29. K. Swain, S. Pattanshetty, Ns. Nair, and R. Kamath, *J. Glob. Infect. Dis.*, vol. **6**, no. 1, p.

3, Jan. (2014).
30. A. M. Dechet *et al.*, *PLoS One*, vol. **7**, no. 7, p. e39672, Jul. (2012).
31. C. Barcellos and P. C. Sabroza, *Cad. saúde pública*, vol. **17** Suppl, pp. 59–67, (2001).
32. E. Kupek, M. C. de Sousa Santos Faversani, and J. M. de Souza Philippi, *Braz. J. Infect. Dis.*, vol. **4**, no. 3, pp. 131–4, Jun. (2000)

Plankton And Heavy Metal Correlation From Commercial Vessels In Port Of Tanjung Emas Semarang

Agus Tjahjono[1*], *Aziz* Nur Bambang[2], *Sutrisno* Anggoro[2]

[1]Engineering Department, Semarang Merchant Marine Polytechnic, Semarang, Indonesia
[2]Master Program of Environmental Science, School of Postgraduate Studies, Diponegoro University, Semarang-Indonesia

Abstract. The commercial vessels activity have a big role to increase the flow of number of cargoes from a port to another port. However , the impact of these activities are the disposal of ballast water from port area to the destination port. The purpose of this research was to analyze the correlation of phytoplankton, zooplankton, and heavy metal which were contained inside the ballast water of commercial vessel towards in waters of the port of Tanjung Emas Semarang. The concentration of heavy metal either from commercial vessels or the waters in port area analyzed by Atomic Absorption Spectrophotometer (AAS). The result showed that the correlation of zooplankton and phytoplankton in the water ballast at commercial vessels have a medium correlation to zooplankton and phytoplankton in waters of Port of Tanjung Emas Semarang (PTES) were 48.9% and 58.3%. Correlation of heavy metal Cd, Zn, Cu, Zn and Pb in ballast water of commercial vessel toward each metal in waters of PTES area has a strong correlation in contribution were 76.7%, 75.6%, 71.4% and 73.8%. It showed us that the loading activity of commercial vessels in port are contributed towards the pollution in waters.

1 Introduction

Commercial vessels play an important role in cargoes distribution around the world that more than 80% of the global trade volumes are carried out by sea transportation [1]. On the other hand, developing countries are positioned as the global trade booster [2]. In two decades, the centre of global economy and industry gradually has been moved from Northern Atlantic to Asia. Thus, the change definitely influences the role of Asia in the world trade [3].

The rise of world trade increases the demands of means of transportation one of which is commercial vessel. On one side, the use of commercial vessels can be use as the useful transportation. However, on the other hand, negative impacts are also resulted one of which is ballast water disposal. Ballast water is carried in vessel's ballast tanks to improve stability, balance, and trim which will be discharged when the cargoes are loaded. In Asia, as the effect of ballast water disposal, predatory fish *Scieanops occellatus* emerges [4]. Blenny and Goby fish are also found around the waters of Port Baltimore, Norfolk, and Maryland caused by the commercial vessels activities. These predatory gobies threaten the existance of endemic fish [5, 6].

Port's activities are also considered as the source of metal pollutants [7]. Moreover, this heavy metal pollutants can also be induced by industry and domestic activities which correlate positively to the marine organisms [8,9]. These pollutants increase the Pb content in scallop shells *Amusium pleuronectes* and red fish *Lutjanus erythropterus*, Chromium content in blood clams *Anadara granosa* [10,11,12,13]. The pollutants also affect tape seagrass, *Enhalus acoroides* which live in the waters which has been contaminated by Pb and Cd [14].

Port of Tanjung Emas Semarang (PTES) has been used as a location where both domestic and foreign commercial vessels berth and execute their port activities. The activities more or less give some effects to the port besides their domestic activities and dockyard. Moreover, within the port, we can also find Kali Baru estuary which is the accumulation of domestic activities and industry. This research was aimed to analyze the correlation of ballast water brought by commercial vessels toward the PTES waters.

2 Experimental details

The research was conducted in the PTES, Cetral Java, Indonesia. It was located on the northern coastline $6^0 53'$ South and $110^0 24'$ East. The sample was collected from the commercial vessels in PTES waters from December 18, 2014 until October 21, 2015. The sample was taken twice from a tank of ballast water on a vessel with DWT (Dead Weight Ton) over 400 tons. Purposive random

* Corresponding author: a_agus_tjahjono70@yahoo.co.id

sampling was applied to port's six piers : Pusri (*Pupuk Sriwijaya*) and oil, passengers, domestic, wheat, container and LPG (Liquid Petroleum Gas).

The sample of ballast water was collected by putting a hose into the sounding pipe. A portable pump and suction hose were needed to take the sample [15]. The researcher used portable pump Sanyo, P-WH137C, with voltage source of 220 V~50 Hz, output power of 125 W, and maximum water flow capacity of 30 litre/minutes. The diameter of the suction hose was 0,019 m and it was 10 m length. A valve was installed at the end of the hose. The sample was taken from both surface and base of the ballast tank. The surface water sample was obtained by raising the end of the suction hose while the bottom water sample was obtained by lowering the end of the suction hose.

The sample water of the surface of the tank can also be obtained by opening the manhole on the ballast tank. The sample water of the bottom of tank was taken by using a ten litre bucket.

The sample water was then filtered with filter paper Whatman 40 (0,40 μm) and washed with HNO_3 with pH < 2. The water was then put into water sampler with 5 litres of volume and brought into the laboratory. The water (100 ml) was blended and put into a beaker. After that, 5 ml of citric acid was poured into the mixture and heated up. Then, 50ml of the distilled water was poured. The mixture was then put into a 100 ml graduated flask [16, 17,18,19]. The content of Pb, Cd, Cu, Cd in the ballast water depended on the AAS (Atomic Absorption Spectroscopy) of Shimadzu AA-6300 by using the flame-mixed of air acetylene.

The water sample for the plankton was collected through filtration by using a plankton net with a diameter of 30 cm and mesh of 20 μm. In order to take the water sample from the base of the tank, a bucket was used. The water was then filtered into the plankton net [15,20]. Next, the water sample was put into a 50 ml bottle and a drop of lugol was added. Then, the water sample was filtered into a counting chamber (1 ml). It was important to make sure that there was no air bubbles inside. The water was then observed using a microscope with low magnification in order to obtain the general description. The plankton was counted using 5 field of view (5 columns). Last, genuses on the field of view was identified and counted.

Statistical analysis using SPSS for Windows (Versi 22.0) was applied. Correlation analysis was used to compare the phytoplankton abundance in PTES waters and within the ballast water of commercial vessels in PTES in October 2015.

3 Result and discussion

During the High Tide Level (HTL), the concentration of Cd reached 0.0432-0.0513 mg/litre, the highest concentration in Kali Baru estuary which was from the domestic disposal. On the other hand, the lowest concentration was found in the petroleum pier. The concentration of Cu was about 0.0555-0.066 mg/l, the highest concentration was found in Kali Baru estuary and the lowest was found in the outer area of port waters. Based on the finding, the farther the water were from the estuary, the lower the concentration of Cu would be. The concentration of Pb reached 0.627-0.8417 mg/l.The highest concentration was found in Kali Baru estuary and the lowest was found in the passenger pier. The concentration of Pb in all station had exceeded quality standard. It showed that PTES has been polluted by oil. The concentration of Zn reached 0.0483-0.0783 mg/litre in which the highest concentration was found in Kali Baru estuary and the lowest was found in the cargo vessel's pier. Concentration of heavy metals are in order Cd, Cu, Pb, and Zn were found during the HTL (Figure 1).

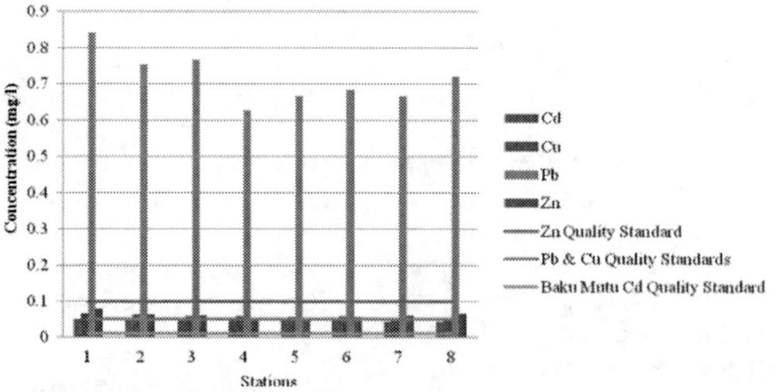

Fig. 1. Concentration of heavy metal during the HTL in October until December 2015 in PTES waters.

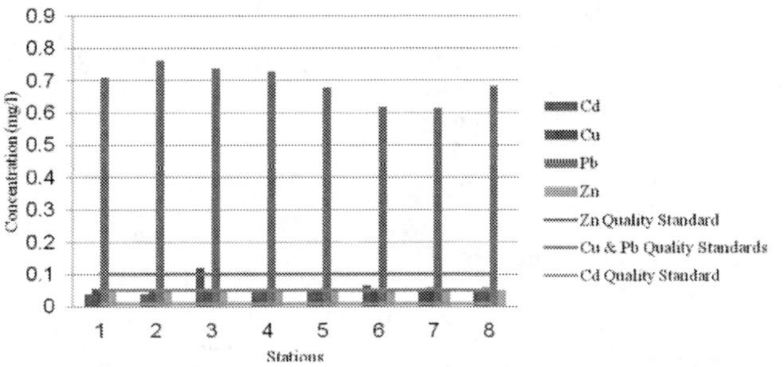

Fig. 2. Concentration of heavy metal during the LTL from October until December 2015 in PTES waters.

Based on the research, it was found that tidal range affects to the concentration of heavy metal. During the Low Tide Level (LTL), water flows from the river into the sea. Thus, the concentration of Cd ranged from 0.039 up to 0.119 mg/l. The maximum concentration occurred in bulk grain pier and the lowest concentration occurred in Kali Baru estuary. The concentration of copper ranged from 0.048-0.059 mg/l, the highest concentration was found in the middle position between container and petroleum piers whereas the lowest was found in bulk grain pier. The concentration of lead (Pb) ranged from 0.615 and 0.763 mg/l, the highest concentration was found in the fertilizer pier and the lowest was found in the middle between container and LPG piers. The concentration of zinc ranged from 0.045-0.057 mg/l, the highest concentration was found in the middle between container and LPG piers and the lowest was found in the bulk grain pier. Those metal concentrations was found in the estuary during the LTL. Based on the data, highest concentration metal was Pb followed by Cu, Zn, and Cd (Figure 2).

The ballast water within the commercial vessels in PTES contained heavy metal pollutants are Pb 0.37192 mg/l, Cd 0.001-0.46 mg/l and Zn reaching 0.001-2.464 mg/l [21,22]. Based on the identification of phytoplankton the HTL of October, it was found that there were 27 genuses of class *Baccilariophyceae, Dinophyceae, Chrysophyceae* and *Cyanophyceae*. During the LTL, there were 30 genuses which included the classes of *Bacillariophyceae, Dinophyceae, Chrysophyceae* and *Cyanophyceae*. The occurance of *Skeletonema* was the highest of all during the HTL followed by *Thalassionema, Thalassiothrix dan Asteroinella*. On the other hand, *Skeletonema, Thalassiothrix, Chaetoceros* and *Asteroinella* was those whose occurance were high during the LTL (Figure 3).

Based on its genus, Skeletonema was identified as phytoplankton with the highest level of occurance during both HTL and LTL in the eight stations [23]. The most dominant genus in PTES waters belongs to class *Baccilariophyceae*. It was also found that there was genus *Chaetoceros* which reached 3,535 individual/l during the HTL and 4,873 individual/l during the LTL. Moreover, genus *Ceratium* was also identified. *Algae* from genus *Chaetoceros* and *Ceratium* can possibly lead to blooming which is definitely dangerous for the fish [24].

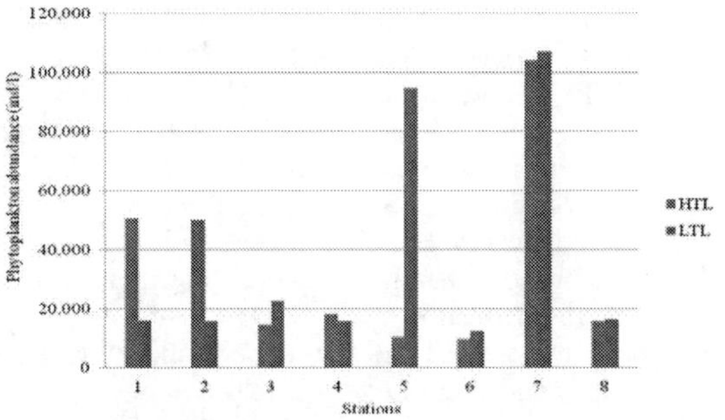

Fig. 3. The Histogram of Phytoplankton abundance in PTES waters during HTL and LTL in October 2015

Phytoplankton abundance during the HTL reached 9,655-104,319 individual/l and 12,318-10,7229 individual/l during the low tidal level, phytoplankton and genus abundance found during the LTL were higher than the HTL. It was because of the fact that the LTL occurs during daytime that the sunlight can easily shine through the waters which leads to the high numbers of phytoplanktons. In the eight research locations, the highest phytoplankton abundance was found in the middle of port waterways which is located between container and petroleum piers. It was due the fact that within this area waters, pollutants were less than some other area of the port. On the other hand, the least numbers of phytoplankton can be found in the container pier (Figure 3).

During the LTL, genus *Gyrosigma* of class *Bacillariophyceae* and *Ornithocercus* of class *Dinophyceae* were found. On the other hand, they were not found during the HTL. Moreover, *Pyrocytis* of class *Chrysophyceae* was also identified during the HTL.

Based on the identification within the ballast water tank of commercial vessels, eight classes were found *Bacillariophyceae, Dinophyceae, Euglenoidea, Conjugatophyceae, Cyanophyceae, Charophyceae, Chrysophyceae, Conjugatophyceae* and *Chlorophyceae*. *Skeletonema* occurance dominated within the commercial vessels followed by *Chaetoceros* and *Nizschia*. Phytoplankton abundance within the commercial vessels in PTES was about 0.32 -412 individual/l. The highest phytoplankton abundance was found in GW vessel in petroleum pier which was from Situbondo in the ballast tank no 4 port and starboard side. Sample was also taken from drain pipe of Ballast

Pump in JM vessel from Palembang waters. The lowest phytoplankton abundance were found in the A, DF2, MP, and PI vessels. *Skletonema* was frequently found within the ballast tank of GW reaching 412 individual/l and J reaching 206 individual/l (Table 1).

Within commercial vessels, some different genus/species were found in PTES waters including *Gyrosigma* (class of *Bacillariophyceae*) within the SR 68 vessel, *Diatoma vulgare* (class of *Bacillariophyceae*) within the HS vessel, *Pinnularia tabellaria* within C 8 vessel, *Euglena acus ehenberg* (class of *Euglenoidea*) and *Spirotaenia condensata* (class of *Conjugatophyceae*) within the L vessel, *Lyngbya* (class of *Cyanophyceae*) within S, A dan SB vessels; *Oscillatoria* (class of *Cyanophyceae*) within C 8, BC S, GD, GNA and OK vessels; *Gonatozygon* (class of *Charophyceae*) within A, DF,C 8, BC S and GNA vessels; *Ankistrodesmus* (class of Chrysophyceae) within BC S vessel, *Tetmemorus laevis* (class of *Conjugatophyceae*) within GD. In a series, the SR 68, HS, C8, L, S, A, SB, BC S, GD, GNA, and OK vessels were from Jakarta, Tuban, Dumai, Jakarta, Cilacap, Jakarta, Singapore, Singapore, Sorong, Jakarta and Palembang.

The genus/species and the vessel's origin, in which *Oscillatoria* and *Euglena* were found, can possibly cause eutrophication to the new coastal waters. These genus were frequently used as an indicator of organic pollutant in ballast water within commercial vessels (Table 2). The toxic substance within these genus may threaten the organisms as they are consumed. Then, within a food chain, a bigger bioaccumulation may take place. *Oscillatoria* which belongs to genus Cyanobacterium is able to carry out a fermentation (producing energy in an anaerobic condition) in a dark environment and potentially produce toxin [25].

Gyrosigma is a unicelulular diatom which belongs to *Chromophyta* division of class Bacillariophyceae and ordo *Bacillariales* (pennate Diatom), family of *Naviculaceae* [26, 27]. These pennate diatoms of ordo *Bacillariales* has upper valve epitheca and lower valve hypotheca. Some of diatom pennates have raphe and some others do not. Some have *pseudoraphe*, a small axial area on the valve. Some species of diatom pennates have apical pore fiel on their valve but do not possess any raphe.

Most of diatoms are in form of diatom pennate. In both freshwater and seawater, diatoms form colonies into substrate. Diatoms and bacteria form biofilms with some organisms and build a colony. Diatoms also grow in a living substract such as plants and algae. Diatom's growth can possibly produce dark brown liquid [28]. Diatoms which grow in a polluted waters show that these organisms are pollutant tolerant which are mostly dominated by *Navicula, Nitzchia, Pinnularia* and *Eunotia*. These genuses can be used in an experiment of ecosystem's improvement [29]. Euglena shows some animal's characteristics including having no sellulose cell wall. Its outer membrane is flexible. Some species have a clear eyespot. Inside its cell, it has contractile vacuole and fibrils/microtubules. Yet, Euglena is able to make its own food by photosynthesis since it has chloroplast and is an facultative autotrophic [30].

Table 2. Genus/species, vessel's origin and port of origin

No	Genus/Spesies	Vessels	Port of origin
1	*Gyrosigma*	SR 68	Jakarta
2	*Diatomae vulgare*	HS	Tuban
3	*Pinnularia tabellaria*	C 8	Dumai
4	*Euglena acus ehrenberg*	L	Jakarta
5	*Spirotaenia condensata*	L	Cilacap
6	*Lyngbya*	S, SA, SB	Cilacap, Jakarta, Singapore
7	*Oscillatoria*	C 8, BC S, GD, GNA, OK	Dumai, Singapore, Sorong, Jakarta, Palembang
8	*Gonatozygon*	A, DF, C 8, BC S, GNA	Jakarta, Dumai, Singapore, Jakarta
9	*Ankistrodesmus*	BC S	Singapore
10	*Tetmemorus laevis*	GD	Sorong

Genus *Okeania, Lyngbya, Simploa, Phormidium, Oscillatoria* and *Sprirulina* belong to cynobacteria. This cynobacteria causes bloom that have occured in Old Providence Island, 300 km from shore of Mosquito, Nicaraguay. It depresses coral reef community in the island. It was caused by overbalanced nutrient and global warming [31]. In Roebuck bay in which nutrients was overbalanced, biodiversity has been significantly changed. Therefore, it increases cynobacteria abundance such as *Lynbia* which causes anoxia and hypoxia. It also impacts to the decrease of benthic invertebrate [32].

Some pollution-tolerant genuses were also found including *Navicula, Nitzschia, Scenedesmus, Coscinodiscus, Ankistrodesmus, Chhlorella, Oscillatoria, Phormidium*. Those genuses dominated four research stations of seafood factory waste. Those genuses caused eutrophication [33].

Some microalgae including *Euglena, Navicula, Nitzschia, Ankitrodesmus* and *Scenedesmus* were identified as tolerant toward organic tolerant [34]. It showed that algae found in ballast tank of commercial vessels were pollution tolerant organisms.

Chlorophyll of Genus *Ankistrodesmus* can be significantly produced by adding CO_2 and controlling its value on 10.7 mg/l, 10.57 mg/l and 7.84 mg/l [35]. It shows that genus *Ankitrodesmus* is tolerant to enclosed space inside the ballast tank of the commercial vessels

Table 1. Composition and phytoplankton abundance within ballast water in commercial vessels

No.	Class/Genus	L	S B	C 1	JM	GW	HS	A	MB2	BM79	DF2	M	C8	MP	GD	OK	P I
	BACILLARIOPHYCEAE																
1	*Nizschia vermicularis*	+++	-	-	-	-	-	-	-	-	-	-	-	-	-	-	-
2	*Hemialus sp.*	+	+	-	-	-	-	-	-	-	-	-	-	-	-	-	-
3	*Nizschia curvula*	+	-	-	-	-	-	-	-	-	-	-	-	-	-	-	-
4	*Skeletonema sp.*	-	+	-	++++	++++	-	-	-	-	-	-	-	-	-	-	-
5	*Rhizosolenia sp.*	-	+	+	+	-	-	-	+	-	-	+	-	+	-	+	+
6	*Coscinodiscus sp.*	-	+	-	+	-	-	-	-	-	-	-	-	-	-	-	-
7	*Pleurosygma sp.*	-	+	+	+	-	-	-	-	-	-	-	-	-	-	-	+
8	*Ceratium sp.*	-	-	+	-	-	-	-	-	-	-	-	-	-	-	+	-
9	*Chaetoceros sp.*	-	-	-	+	++++	-	-	-	-	-	-	-	-	-	-	-
10	*Thallaxiothrix sp.*	-	-	-	-	+	-	-	-	-	-	-	-	-	-	-	-
11	*Navicula sp.*	-	-	-	-	+	-	-	-	-	-	-	-	-	-	-	-
12	*Gyrosygma sp*	-	-	-	-	-	-	+	-	-	-	+	-	-	-	-	-
13	*Nitzschia sp.*	-	-	-	-	-	-	-	-	+	+	+	+	-	+	-	-
14	*Diatoma vulgare*	-	-	-	-	-	+	-	-	-	-	-	-	-	-	-	-
15	*Synedra sp.*	-	-	-	-	-	-	-	-	-	-	+	-	-	-	+	-
16	*Pinnularia tabellaria*	-	-	-	-	-	-	-	-	-	-	-	+	-	-	-	-
	DINOPHYCEAE																
17	*Peridinium sp.*	-	-	-	-	-	-	-	-	-	-	-	-	-	-	+	-
18	EUGLENOIDA																
19	*Euglena acus Ehrbg.*	+	-	-	-	-	-	-	-	-	-	-	-	-	-	-	-
	CONJUGATOPHYCEAE																
20	*Spirotaenia condensata*	+	-	-	-	-	-	-	-	-	-	-	-	-	-	-	-
21	*Tetmemorus leavis*	-	-	-	-	-	-	-	-	-	-	-	-	-	+	-	-
	CYANOPHYCEAE																
22	*Lyngbya*	-	-	-	-	-	-	-	-	-	-	-	-	-	-	-	+
23	*Oscillatoria sp.*	-	-	-	-	-	-	-	-	-	-	-	+	-	+	+	-
	CHAROPHYCEAE																
24	*Gonatozygon sp.*	-	-	-	-	-	-	+	-	-	+	-	+	-	-	-	-
	Total	157	11	8	219	703	0.3	3	0.3	2	1	8	4	1	6	6	1

Note : + 0-50, ++ 51-100, +++ 101-150, ++++ > 151 ind/l.

Based on the research, planktons which cause *red tide* in Indonesia were found. Within SB, JM and GW vessels, *Skletonema sp.* was found up to 2.55 sel/l, 206.4 sel/l and 412 sel/l. Moreover, plankton *Chaetoceros sp.*, which also causes red tide, was also found inside the ballast water of JM and GM vessels reaching 3.82 sel/l and 282.8 sel/l. *Red tide* is defined as a phenomenon in which plankton population (plankton bloom) increases significantly [36]. This phenomenon causes the rising of mass fish and biota die-off due to the sudden decrease of DO content. Thus, red tide is also known as Paralytic Shellfish Poisoning (PSP), a kind of disease brought by poisonous seafood which was contaminated by microorganisms in which red tide takes place. This phenomena once occured in Ancol in May 2004 as *Skletonema sp., Thalassiosira sp.* and *Chaetoceros sp.* were found [37].

In June and September 2011, four genuses which cause *red tide* were found in the estuary of Cisadane and Untung Jawa Island. The genuses were *Ceratium, Pseudonitzshia, Dinophysis dan Chaetoceros.* Plankton which causes *red tide* was also found in the C1 vessel which was from Pontianak whose numbers up to 1.275 sel/l [38].

Red tide occured in Kao Bay in 1994 was resulted by the decrease of DO or exotic species such as *Phyrodinium* (class of *Dinofragellate)* [39]. There was also found in the ballast tank of a commercial vessel from Palembang, OK vessel, reaching 0.955 sel/l. Genus *Nitzschia spp.* which causes red tide was also found in the L vessel. The species were *Nitzshia vermicularis* (124 cell/l) and *Nitzshia curvula* (29.3 cell/l). Genus *Nitzshia sp.* was found in the RB, S, BM 79, DF, M, C, BC S, GD and GNA vessels whose numbers reached 1.275 cell/l, 0.32 cell/l, 1.6 cell/l, 0.32 cell/l, 3.505 cell/l, 1.595 cell/l, 0.32 cell/l, 3.505 cell/l and 8,6 cell/l. Species that can cause Diarrhetic Shellfish Poisoning (DSP) are *Nitzschia pungens* which can cause the effects of Amnesic Shellfish Poisoning (ASP) [40].

Zooplankton found within the ballast water in the commercial vessels have a moderate level of correlation to zooplankton found in PTES waters. The correlation reaches 0.489 or 48.9%. Moreover, phytoplankton found within the ballast water in the commercial vessels also have a moderate level of correlation to phytoplankton found in PTES waters reaching 0.583 or 58.3% (Table 3).

Table 3. the Correlation between phytoplankton and zooplankton in ptes waters and phytoplankton and zooplankton within ballast water of commercial vessels

Biota	Phytoplankton (coastal waters)	Zooplankton (coastal waters)
$Phytoplankton_{vessel}$	0.489*	
$Zooplankton_{vessel}$		0.583*

* Significant correlation $p < 0.01$ (2-tailed)

The hypothesis of heavy metal in PTES waters and heavy metal within ballast water of commercial vessels stated that H_1 : there is a correlation between heavy meatal in PTES waters and heavy metal found in the ballast water of the commercial vessels. The correlation of Cd, Zn, Cu, Zn dan Pb found in PTES waters and those which were found in the ballast water was high up to 0.776 (76.6%), 0.756 (75.6), 0.714 (71.4%) and 0.738 (73.8%) (Table 4).

Table 4. The correlation between heavy metal in PTES waters and in ballast water of commercial vessels

Element	$Cd_{coastal\ waters}$	$Cu_{coastal\ waters}$	$Zn_{coastal\ waters}$	$Pb_{coastal\ waters}$
Cd_{vessel}	0.776*			
Cu_{vessel}		0.756*		
Zn_{vessel}			0.714*	
Pb_{vessel}				0.738*

* Significant correlation p=0.05 (2-tailed)

Lead (Pb) can possibly give bad impacts to the mangrove *Avicennia marina.* The highest content of lead was found within mangrove's stems (5.89 ppm). Lead also damaged mangrove's cells within leaf and fruit [41]. The effect of copper (Cu) exposure to *Moerella iridescens* in Maluan Bay has correlated positively to the dissolved particle of the sediment. It also contributed to the metal content within the shells [42]. Metal exposure on gastropod has greater effect than the exposure on crustacean and fish due to its lower tropical nature. On the other hand, metal content within fish liver is greater than what has been found in its gills and muscles tissue [43].

4 Conclusion

Within the commercial vessels, some different genus/species were found including Gyrosigma (class of Bacillariophyceae), Diatom vulgare (class of Bacillariophyceae), Pinnularia tabellaria, Euglena *acus ehenberg* (class of *Euglenoidea*) and *Spirotaenia condensata* (class of *Conjugatophyceae*), *Lyngbya* (class of *Cyanophyceae*), *Oscillatoria* (class of *Cyanophyceae*), *Gonatozygon* (class of *Charophyceae*), *Ankistrodesmus* (class of Chrysophyceae), *Tetmemorus laevis* (class of *Conjugatophyceae*). These genus/spesies were brought from various ports that they can possibly influence the organisms of PTES waters.

Zooplankton within the ballast water of commercial vessels has a moderate level of correlation with zooplankton found in PTES waters which number reaches 0.489 or 48.9%. Phytoplankton within ballast water of commercial vessels has a moderate level of correlation to phytoplankton found in PTES waters which number reaches 0.583 or 58.3%. There is a moderate level of correlation between phytoplankton in PTES waters and those found in the commercial vessels.

Metal Cd, Zn, Cu, Zn and Pb within ballast water of commercial vessels have a high level of correlation to each metal found in PTES waters. The correlation of Cd, Zn, Cu and Pb to the heavy metal in PTES waters reach 0.776 (76.6%), 0.756 (75.6), 0.714 (71.4%) and 0.738 (73.8%).

It is concluded that it is better for the commercial vessels to discharge their ballast water offshore before they reach PTES waters in order to prevent and minimize the emerge of some alien species in the port waters. A useful ballast water management is definitely needed.

5 Appreciation

We would like to thank the following people for their support, without whose help this work would never been possible: Port Authority and Harbormaster Office Head of Tanjung Emas Semarang, officers of commercial vessels for permitting us to take research sample of ballast water. Faculty of Fisheries and Marine Science staffs at Diponegoro University; Mr. Andreas, Sigit, and Vian who helped to identify phytoplankton of both PTES waters and ballast water of commercial vessels.

References

1. Rivaldigia. Perdagangan internasional. http://rivaldigia.wordpress.com/2011/07/06/perdangangan internasional accessed on August 9, 2016 (2011).
2. UN (United Nations). World economic situation and prospects, (2012). http://www.un.org/er/developmentdesa accessed on April 18, 2015 (2012).
3. Asian Development Bank (ADB). 2011. Asia's changing role in world trade : prospects for south-south trade growth to 2030. http://www.adb.org/sites/default/files/publication/29010/economics, accessed on 9 Agustus 2016 (2011).
4. Y-C. Liao, L-S Chen, K-T. Shao. Biological Invasions, **12**:1961-1965 (2010).
5. M.J.Wonham, J.T. Carlton, G.M. Ruiz, L.D. Smith. Marine Biology **136** : 1111-1121 (2000).
6. P.M. Kocovsky, J.A. Tallman., D.J. Jude, D.M. Murphy, J.E. Brown, C.A. Stepien. Biological Invasions, **13**:2775-2784. DOI 10.1007/s10530-011-9962-5 (2011).
7. M. Kara, Y. Dumanoglu, H. Altiok, T. Elbir, M. Odabasi, A. Bayram. Environ Earth Sci, 73 : 405-421 (2015)
8. K. Srikanth, I. Ahmad, J.V. Rao. Water Air Soil Pollution **225**, (269) 1-10 (2014)
9. L. Maslukah. Buletin Oseanografi Marina, **2** : 55-62 (2013)
10. S. Usman, N.L. Nafie, M. Ramang. Marina Chimica Acta **14**, (2) 49-55 (2013)
11. H. Azhar, I. Widowati, J. Suprijanto. Journal of Marine Research **1**, (2) 35-44 (2012)
12. N.A. Suprapti. BIOMA **10**, (2) 36-40 (2008)
13. A.W. Wahab, Mutmainah. Marina Chimica Acta, **6**, 2, 21-24. (2005)
14. A.R. Pratiwi, A. Pratomo, N. William. Jurnal Zarah, **2**,1 (2014)
15. M.J. Garret, J.L. Wolny, B.J. Williams, M.D. Dirks, J.A. Brame, R.W. Richardson. Algae, **26**(2) : 181-192. DOI : 10.4490/algae (2011)
16. Badan Standardinasi Nasional, Cara Uji Timbal dengan Spektrofometri Serapan (AAS)-nyala, SNI 06-6989.51 (2009)
17. Badan Standardinasi Nasional, Cara Uji Kadmium dengan Spektrofometri Serapan (AAS)-nyala, SNI 06-6989.16 (2009)
18. Badan Standardinasi Nasional, Cara Uji Seng dengan Spektrofometri Serapan (AAS)-nyala, SNI 06-6989.7(2009)
19. Badan Standardinasi Nasional, Cara Uji Tembaga dengan Spektrofometri Serapan (AAS)-nyala, SNI 06-6989.16 (2009)
20. T. Dodgshun, S. Handley. Cawthron Report 417 (1997)
21. A. Tjahjono, A.N. Bambang, S. Anggoro. Journal of Ecological Engineering, **18**, 2, 7-11 (2017)
22. A. Tjahjono, A.N. Bambang, S. Anggoro. IOP Conf. Series : Earth and Environment Science. **55**,012024 (2017)
23. A. Hartoko. Oceanographic character and plankton resources of Indonesia. Yogyakarta : Graha Ilmu (2013)
24. R.M. Rompas. Toksikologi Kelautan. Sekretariat Dewan Kelautan Indonesia. Jakarta (2010)
25. E.A Pastich., S.M. Barbosa, L . Florencio, S. Gavazza, M.T. Kato. Interciencia, Vol. **41**, (5) 330-333 (2016)
26. I.Yamaji. Ilustrations of marine plankton of Japan. Osaka : Hikusha Publishing Co., Ltd. (1986)
27. Widianingsih, R. Hartati, H. Endrawati, R. Azizah. Mikroalga laut. Semarang : Badan Penerbit Universitas Diponegoro Press (2009)
28. P. Sze. A biology of the algae thrid edition. Massachusetts : WCB McGraw-Hill (1998)
29. A.S Stenina. Contemporary problems of Ecology, Vol.1, No. 4, pp. 449-453 (2008)
30. H. Budiati. Biologi : untuk SMA dan MA kelas X. Jakarta : Pusat Perbukuan, Departemen Pendidikan Nasional (2009)
31. M. Puyana, A. Acosta A, K. Bernal-Sotelo, T. Velasques-Rodrigues, F. Ramos. Universitas Scientiarum **20** (1):83-105 doi: 10.11144/Javeriana.SC20-1.sscb (2015)
32. S.M. Estrella, A.W. Storey, G. Pearson, T. Piersma. Journal of the Royal Society of Western Australia, **94**:171-179 (2011)
33. V. Vidya, K. Sumathy, G. Prasad. Nature Environment and Pollution Technology Journal, Vol. **13**, no.2, pp.234-244 (2014)
34. B.S.Giriyappanavar, P.B.Shivalli. Nature Environment and Technology Vol.**14** , No. 4, pp. 943-946 (2015)
35. M.A. Salim. Journal of Asian Scientific Research, **3** (1):75-84 (2013)

36. S.S. Bates. Ecophysiology and metabolism of ASP toxin production- in : Anderson D.M., Cembella A.D., Hallegragraef G.M. (eds), Physiological ecology of harmfull algal blooms, NATO ASI series , Series G, Ecological sciences , vol.**41**, Springer-Verlag, Berlin, pp. 405-426 (1998)
37. .I. Sachoemar, H.D. Wahyono. JA.I Vol.**3** , No.1, 1-14 (2007)
38. D.D. Rengganis, T. Aunorohim, Hikmah. Fitoplankton penyebab algae blooms (HABs) di perairan Teluk Jakarta. (2011)
39. N.N. Wiadnyana. Jurnal Oseanologi dan Limnologi di Indonesia No. 29 : 15-28 (1996)
40. , R. Widiarti, W. Wardhana. Jurnal Akuatika Vol. **III.** No. 1, 28-39 (2012)
41. K.R. Arisandy, E.Y. Herawati, E. Suprayitno. Jurnal Penelitian Perikanan **1**, (1) 15-25 (2012)
42. Z. Wang, C. Yan, Q. Pan, Y. Yan. Environ Monit Assess, (**175**) 239-249 (2011)
43. A.A. Pereira, B. van Hattum, J. de Boer, P.M. van Bodegom, C.E. Resende, W. Salomons. Arch Environ Contam Toxicol, (**59**) 464-477 (2010).

Physicochemical Characteristics of Artificial Rice from Composite Flour: Modified Cassava Starch, *Canavalia ensiformis* and *Dioscorea esculenta*

Siswo Sumardiono[1], *Isti* Pudjihastuti[2], *Noer* Abyor Handayani[1], *Heny* Kusumayanti[2,*]

[1]Chemical Engineering Department, Faculty of Engineering, Diponegoro University, Semarang, Central Java, Indonesia
[2]Vocational Program Study of Chemical Engineering Diponegoro University, Semarang, Central Java, Indonesia

Abstract. Indonesia is the third largest country on the global paddy rice production and also considered as a rice importer. Even, Indonesia has the biggest per capita consumption of paddy rice (140 kg of paddy rice per person per year). Product diversification using local commodities. Artificial rice is potential to be developed as a new value product using different types of grains. It is one of appropriate solutions for reducing imported rice rate. Artificial rice was produced using high nutrition composite flours (modified cassava starch, corn, *Canavalian ensiformis, and Dioscorea esculenta*). This study consists of three main stages, preparation of composite flour, formulation, and artificial rice production using hot extruder capacity 10 kg/day. The objectives of this studies were to investigate some formulation in compare with commercial paddy rice. Artificial rice has been successfully conducted using prototype of hot extruder with the temperature 95°C. Physical analyses (color and water absorption) were carried out to artificial rice product and commercial paddy rice. Chemical analyses (nutrition and amylose content) of product will be also presented in this study. The best formulation of artificial rice was achieved in 80% modified cassava starch, 10% *Canavalian ensiformis*, and 10% *Dioscorea esculenta*, respectively.

1 Introduction

Artificial rice could be developed as a new value product using different types of grains with added nutrients and functionalities. The terms of artificial rice, enriched rice, or reformed rice were introduced by several authors [1-5]. The developed products could avoid deficiency in nutrients as compared to the natural rice without major changes in consumers' dietary habits. The main advantages could be the easy process of developing instant rice and the possibility of incorporating different nutrients with desired textural characteristics. In addition, it could be developed using wastes or byproducts with low cost and utilizing energy efficient process, such as extrusion [6].

Artificial rice can be obtained from carbohydrates sources, such as potatoes, tubers and cereals [7-9]. Modified cassava flour and *Dioscorea esculenta* have a high carbohydrate content of up to 90% thus could be alternative raw materials for artificial rice [10]. They are also easy to obtain, affordable price and have a capability to be used with another component for improving artificial rice physicochemical properties. Protein is a kind of nutrition which is very important for human growth. *Canavalia ensiformis* and corn can be developed into an preference food product because of its high protein content (23-27.6%) [11,12]. Yellow corn from corn indicates the existences of carotenoid 6.4-11.3 μg/g [11]. Utilizing composite flours from modified cassava, corn, *Dioscorea esculenta*, and *Canavalia ensiformis* flour will meet the human nutrition needs (Figure 1).

In general, artificial rice production can be conducted using granulation method and extrusion method [8,9]. There are differences of gelatinization techniques and molding stages from these methods. Artificial rice has the shape of granules and oval using granulation and extrusion method, respectively [13]. A number of studies are reported on the development of artificial rice [1-4,6]. Based on the shortcomings from previous studies, this study focused on artificial rice production from composite flour from modified cassava starch, *Canavalia ensiformis* and *Dioscorea esculenta* using hot extrusion process. Innovation processes of artificial rice might be an appropriate alternative solution for Indonesia to produce high nutritional value and good physicochemical properties of imitative rice in compare with commercial paddy rice. The design and fabrication of hot extruder capacity 10 kg/day were conducted to produce artificial rice. Some formulations were investigated on proximate analysis, water absorption, amylose content, and color brightness properties in compare with commercial paddy rice (Rojolele).

* Corresponding author: henykusumayanti@live.undip.ac.id

2 Material and Methods

2.1 Materials

The raw materials used in this study were composite flour from modified cassava starch, corn, *Canavalia ensiformis*, and *Dioscorea esculenta*. Skim milk, Gliserol Monostearat (GMS), salt, and oil were also involved in this study. Artificial rice was produced using hot extrusion as represented on Figure 2. The formulation of artificial rice was shown in Table 1.

2.2 Methods

The composite flours, skim milk, GMS, water, salt, and oil were mixed at 80-90°C for 20 minutes. This blend components were proceed on extruder to produce artificial rice. Products were immersed into boiled salt solution for 2 minutes, washed by water, and then dried at 60°C for 4 hours.

2.3 Analysis

Artificial rice products were analyzed for color brightness, water content, proximate properties (carbohydrate, protein, and fat content), water absorption, and amylose content. Color brightness and water content were analyzed using Minolta Chromameter CR-300 and oven. Protein, fat, and amylose content were tested using Kjedahl, extraction, and iodo-iodi calorimetric method, respectively. Carbohydrate content followed by the difference method.

3 Result And Discussion

3.1 Proximate analysis

Table 2 represented the proximate analysis of artificial rice in compare with commercial paddy rice "Rojolele". Water content of artificial rice ranged from 10.23 to 13.44%, which approached the moisture content of rice "Rojolele". They also have met the water content standard by SNI 01-6128-2008 (max 14% wt). Rice moisture content of less than 14% (wt) could prevent the growth of mold thus could extend its shelf life [10]. Protein analysis of various formulation of artificial rice ranged from 4.77% to 7.75% (wt). This protein content was slightly below "Rojolele". In general, fat content of artificial rice were range between 0.7%-2.05% (wt) and lower than "Rojolele". Some case, formula 4 and 10, have a greater fat content in compare with commercial paddy rice "Rojolele" (Figure 2). Carbohydrate status in various compositions was 65-80% (wt) and similar with the content of "Rojolele" (77.11% wt).

3.2 Amylose content

Amylose content is one of the key parameters to determine the physical properties of rice. The unbranched polymer of amylose has the very strong bond and difficult to digest. Thus, they have a strongly correlation with low glycemic indexed. Accordance on amylose content, rice can be classified into five categories, waxy (1-2%), very low (2-12%), low (12-20%), intermediate (20-25%), and high amylose content (25-33%) [14].

Figure 3 showed the amylose content of artificial rice ranged from 13.88 to 17.56%. This artificial rice included in low amylose group (12-20%), thus it will become sticky and shiny when it cooked [14]. However, the amylose content should not influence the level of people preferences due to their different appetite.

Fig. 1. Morphology of Cassava (a), Dioscorea esculenta (b), Canavalia ensiformis (c), and corn (d)

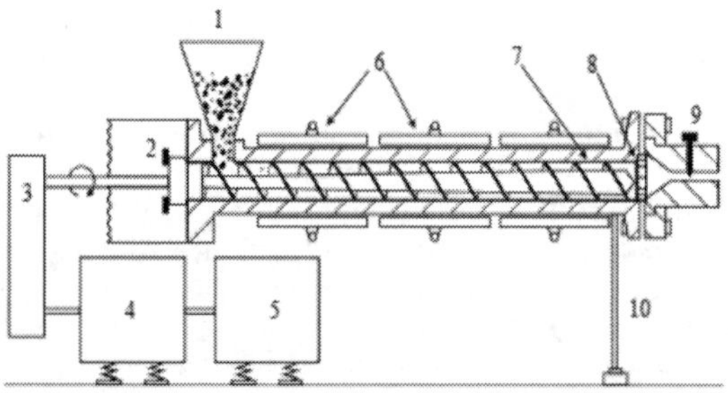

Fig. 2. Hot extrusion

3.3 Water absorption

Water absorption properties are one of the key parameters which influence the level of cooked rice. Table 4 represented the water content of artificial rice. Water absorption of artificial rice was 0.8 to 1.84 (Table 4). The average water absorption of rice in Indonesia was 2.5. Artificial rice had lower water absorption rate than ordinary rice, thus the fewer water will be needed to cook rice.

Table 1. The formulation of artificial rice from composite flour: modified cassava, *Canavalia ensiformis,* Corn and *Dioscorea esculenta* flour

Formulae	Modified starch	*Canavalia ensiformis*	Corn flour	*Dioscorea esculenta*
1	60	30	10	-
2	60	30	-	10
3	60	20	10	10
4	60	10	20	10
5	60	10	10	20
6	60	30	10	-
7	60	30	-	10
8	60	20	10	10
9	60	10	20	10
10	60	10	10	20
11	60	30	10	-
12	60	30	-	10

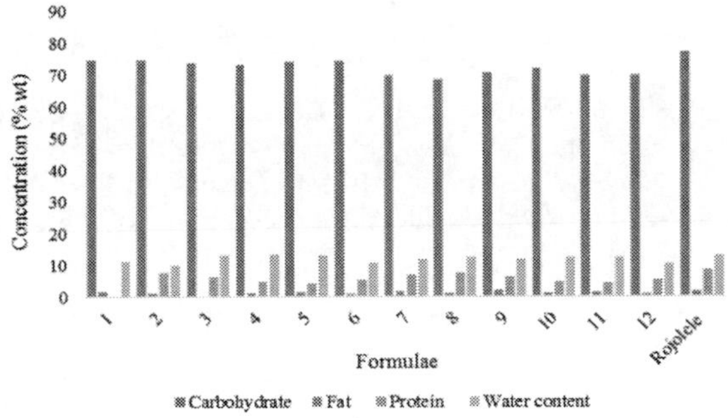

Fig. 2. Proximate analysis of artificial rice

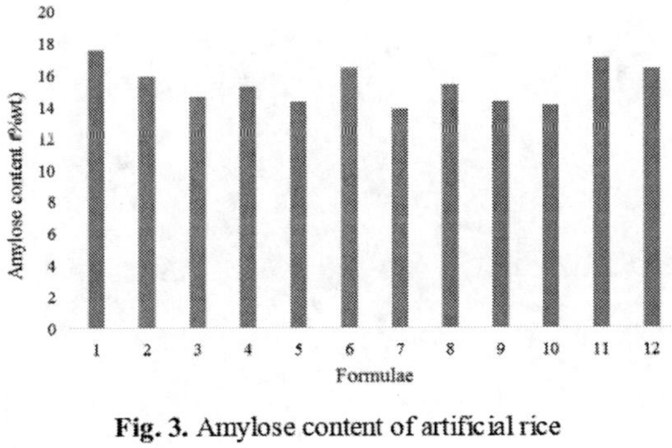

Fig. 3. Amylose content of artificial rice

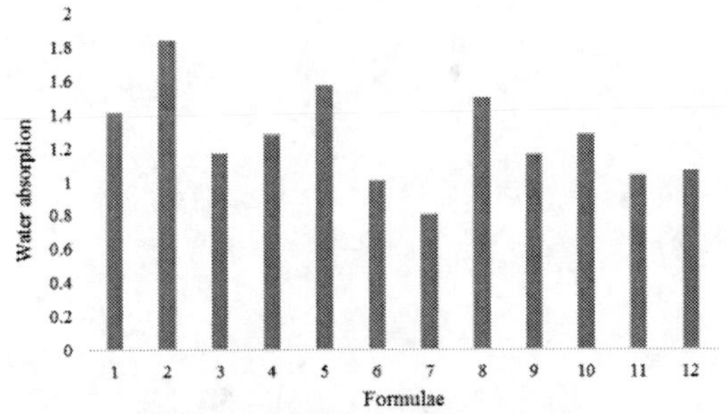

Fig. 4. Water absorption of artificial rice

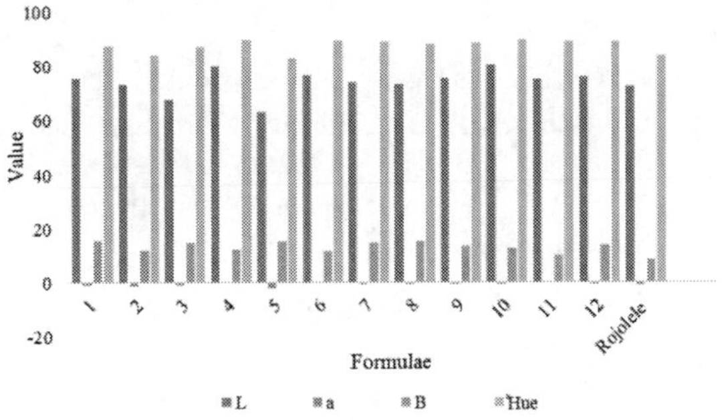

Fig. 4. Color analysis of artificial rice

3.4 Color analysis

Color measurement has three parameters to determine the quantity of color (L, a, and b). "L" is indicate lightness, "a" is the red or green coordinate, and "b" is the yellow or blue coordinate. Figure 4 and 5 showed the color analyzes and products of artificial rice. It indicates that all samples have a degree of yellow hue and the brightness range 63.29 to 80.45. When compared with Rojolele rice, rice analog portion has a degree of hue and brightness level that approximates the color of rice Rojolele. The brightness levels approaching rice is expected to increase consumer acceptance of analog rice.

Fig. 5. Artificial rice from modified cassava starch, *Canavalian ensiformis*, and *Dioscorea esculent*

4 Conclusion

The best formula of artificial rice contains 12.51% water, carbohydrates 71.94%, 4.86% protein and 1.1% fat. It also had 14.09% amylose content and belongs to low amylose groups. Starch modification process should be optimized to improve the amylose content therefore the rice becomes fluffier. Optimization of operating conditions is also needed at extrusion and cutting stage. A sharp knife is required so that the shape of the product will be more uniform. In fact, the artificial rice is expected to be alternative products non paddy rice which could reduce people's reliance on rice and might be used as a staple food in Indonesia

References

1. T. Yoshida, T. Ojima, US Patent No. 3628966. (1971)
2. D.A. Harrow, J.W. Martin, US Patent No. 4325976 (1982)
3. H. Kurachi, US Patent No. 5403606 A (1995)
4. S. Budijanto, D. Yulianti, Jurnal Teknologi Pertanian **13**(3), 177–186 (2012)
5. N. A. Handayani, H. Santosa, and H. Kusumayanti, Reaktor, **15**(2), 111-116 (2014)
6. J.P. Wang, H.Z. An, Z.Y. Jin, Z.J. Xie, H.N. Zhuang, J.M. Kim, J. Food Sci. Technol. **50** (4), 655–666 (2013)
7. B. Slamet, and Yuliyanti, Jurnal Teknologi Pertanian, **13**,177- 186 (2013)
8. F. Hulliandini, *Formulation and Characterization of Analogue Rice Made of Sweet Potato (Ipomoea batatas), Sago Starch (Metroxylon sp.), and Maize Flour (Zea mays)* (2014)
9. S. Sumardiono, I. Pudjihastuti, A.R. Poerwoprajitno, and S.M. Suswandi, World Applied Sciences Journal **32** (6), 1140-1146 (2014)
10. X. Wang, L. Chang, Z. Tong, D. Wang, Q. Yin, D. Wang, and Q. Huang, Scientific report, **6**, 19643 (2016)
11. O. Moreno, C. Pastor, J. Muller, L. Atarez, C. Gonzalez, and A. Chiralt, Journal of food engineering, **141**, 27-36, (2014)
12. G. Sasipriya, and P. Siddhuraju, Tropical animal health and production, **45**(3):811-820, (2013)
13. S. S. Widara, and S. Budijanto, *Study of Rice Analogue Production from Various Carbohydrate Sources Using Hot Extrusion Technology* (2012)
14. Z. Lu, T. Sasaki and Y. Li, Food Hydrocolloids, **23**: 1712-1719 (2009)

 https://doi.org/10.1051/e3sconf/20183106006

The Effect of Cloud Ear Fungus (*Auricularia polytricha*) on Serum Total Cholesterol, LDL And HDL Levels on Wistar Rats Induced by Reused Cooking Oil

Ratih Budinastiti [1], *Henna* Rya Sunoko [1,2,*], *Nyoman* Suci Widiastiti [1]

[1] Faculty of Medicine. Diponegoro University, Semarang - Indonesia
[2] Doctoral Program of Environmental Science, School of Postgraduate Studies, Diponegoro University, Semarang - Indonesia

Abstract. The usage of reused cooking oil affects the increase of serum total cholesterol (TC) and LDL, also the decrease of serum HDL. This condition escalates the risk of atherosclerosis, which could lead to the incidence of cardiovascular disease. Cloud ear fungus is a natural antioxidant that contains polysaccharides, flavonoids, niacin, and vitamin C, which can improve the lipid profiles. Objective of this research is to analyze the impact of water from boiled cloud ear fungus on total cholesterol, LDL, and HDL level of Wistar rats that have been given reused cooking oil. This study is a true experimental research with post test only control group design, using 12 weeks-aged male Wistar rats (n = 24) that were randomly divided into 4 groups. K1 as the negative control, K2 was given reused cooking oil and standard diet, K3 was given water from boiled cloud ear fungus and standard diet, and K4 was given reused cooking oil, water from boiled cloud ear fungus and standard diet. Serum total cholesterol, LDL, and HDL levels were measured by the CHOD-PAP method after 28 days treatment. The study showed that TC mean value of K1 (80.2217 ± 3.61 mg / dL), K2 (195.8483 ± 5.47 mg / dL), K3 (75.5800 ± 4.02 mg / dL), and K4 (110.8683 ± 5.82 mg / dL); p = 0.000. LDL mean value of K1 (29.9200 ± 1.53 mg / dL), K2 (78.4167 ± 1.77 mg / dL), K3 (24.3167 ± 1.77 mg / dL), and K4 (40, 1617 ± 2.84 mg / dL); p = 0.000. HDL mean value of K1 (65.8950 ± 1.99 mg / dL), K2 (24.3233 ± 1.44 mg / dL), K3 (73.2300 ± 1.92 mg / dL), and K4 (54 , 9550 ± 2.04 mg / dL); p= 0.000. **Conclusion:** Water from boiled cloud ear fungus decreases the serum total cholesterol and LDL, and increases serum HDL levels of Wistar rats that has been given reused cooking oil.

1 Introduction

The practice of using heated cooking oil repeatedly or usually called *jelantah* oil to cook was done by people frequently to save the expense. Heating the cooking oil repeatedly caused composition alteration such as increased saturated fatty acids and trans fatty acids content and reactive oxygen compound appeared. It caused the total amount of cholesterol, LDL, TG, free fatty acids, phospholipids and cerebrosides content increased and HDL content decreased [1,2, 3]

The increased total amount of cholesterol, trigliserida and LDL content and the decreased amount of HDL content indicated a state called dyslipidemia. Dyslipidemia increased the risk of complex pathology occurred which was called atherosclerosis. If the cholesterol content in the blood managed to be controlled, the development of atherosclerosis would be resisted [4]. Preventing and taking medical treatment were something than can be done. Preventing actions included improving nutrition, controlling blood pressure and doing physical exercises. On the other hand, taking medical treatment could be done by having synthetic and traditional medicine, and one of them was cloud ear fungus [6].

Edible mushrooms were widespread used in order to prevent degenerative diseases popped out. It was because edible mushroom, mainly cloud ear fungus had been proven contained antioxidants. [6]Cloud ear fungus were one of foodstuffs containing polysaccharides which could be used to lower cholesterol and prevent platelet aggregation [7, 8]. Besides, it also contained natural antioxidant which could be used as an alternative in resisting oxidation process. A new research showed that the use of synthetic antioxidant was limited as it showed potential danger [9, 10].

2 Method

This study was an experimental study with post-test only group design. Branches of science covered in this study were pharmacy and pharmacology, biochemistry, and

* Corresponding author: hennarsunoko@gmail.com

also environmental health. This study was done for 35 days, starting from March 22nd until April 27th 2016 in Inter-University Study Centre Laboratory for Food and Nutrition (*PAU*) Gadjah Mada University. The subjects of this study was 12 week male galur Wistar white mouses weighing 150-220 grams and had no anatomical disorder. The free variable of this study was the supply of cloud ear fungus broth dosed 3.6 gram/200gramBB and *jelantah* oil dosed 3ml/200gramBB each day for 28 days, while the dependent variable used was total cholesterol, LDL, and HDL serum content.

The number of samples referred to the WHO standard which were minimum 5 mouses. This study used 6 mouses for each group. The total number of mouses used in this study was 4 mouses which were divided randomly into 4 groups. All mouses were adapted by individual denned and given standard diet for 7 days. After adaptation process, they were divided into 4 groups randomly. Group K1 was given standard diet and aquades. Group K2 was given standard diet, aquades and *jelantah* oil per sonde 3ml/200gramBB. Group K3 was given standard diet and cloud ear fungus broth dosed 3.6gram/200gramBB per sonde as much as 2 ml/200gramBB. Group K4 was given standard diet, *jelantah* oil per sonde 3ml/200gramBB. After 28 day treatment, the researcher took their blood sample as much as 3 ml using hematocrit pipet from the mouse's plexus retroorbita and measured mouse's total cholesterol, LDL, and HDL serum content using CHOD-PAP approach in spectrophotometry manner and stated in mg/dL.

The process of making *jelantah* oil used 5 time heated bulk oil which had been used to fry sweet potatoes. The process started by taking the cooking oil to the pan as much as ± 2500mL, and it was heated until 180°C. A 1 kilogram sweet potato which was bought in Kopeng market was fried for 10 minutes. Then, the oil was put aside until it got cooler and then the researcher continued to re-heat the oil until the 5th heating. The oil used was the same (there was no replacing or adding new oil volume). The procedure of making cloud ear fungus broth required 85 gram dried cloud ear fungus which had been cleaned, cut and boiled in 1800ml water until the volume was 600ml left. After that, it was strained using flanel material and steamed at 90° until the volume was 50 ml left.

The data showed in this study was a primary data measured in laboratories in a form of: total cholesterol, LDL, and HDL content. It was analyzed in descriptive manner which was presented into tables after its normality was measured using *Shapiro-Wilk* test. In order to differentiate the content of each experimental group, it was analyzed using *One WayAnova* and then proceed to *Post Hoc Bonferoni* statistic test using *SPSS 21.0 for Windows.*

3 Result and discussion

3.1 Result

3.1.1. Sample Analysis

Based on the research, out of 24 mouses from adaptation stage to after experimental stage, there was no mouses which were found dead. So, the total amount of mouses whose blood were taken to be measured for its total cholesterol, LDL, and HDL serum content was still 6 mouses for each group.

Weight measurement was done each week to determine both *jelantah* oil and cloud ear fungus broth doses. The mouses' weight measurement result was shown in Table 1.

Weight	Experimental Group			
	K1	K2	K3	K4
Before Experiment	193,83 ± 4,53	196,33 ± 2,66	193,33 ± 3,72	194,00 ± 5,02
After Experiment	218,17 ± 4,58	233,00 ± 2,37	217,50 ± 3,51	226,33 ± 5,28

Table 1. Mouses' Weight Description (g)

The result of statistic analysis indicated that there was no significant differences from the weight average between experimental groups before experiment showed by *p=0,597* (p>0,05) through *One Way Anova* test. It could be concluded that each experimental group had no significant different variations or characteristics.

3.1.2. Total Cholesterol Content

The average of mouses' total cholesterol serum content examination from each group was presented in Table 2.

Table 2. Average of Total Cholesterol Serum Content

Group	Average (mg/dL) ± SB
K1 (n=6)	80,22±3,61
K2 (n=6)	95,85 ±5,48
K3 (n=6)	75,58±4,02
K4 (n=6)	110,87 ±5,82

The result of statistics test showed that the data was in normal distribution and had homogen variation and then proceeded to *One Way Anova* parametric test. *One Way Anova* test of total cholesterol serum content showed that at least there was a significant difference of total cholesterol content from 2 groups (p=0,000). In order to know which groups showing a significant difference of total cholesterol serum content, the process was continued by *Post-hoc Bonferoni* analysis. It was found that there were significant differences between group K1 and group K2 (p=0,000), group K2 and group K4 (p=0,000), and group K3 and group K4 (p=0,000). However, there was no significant difference between group K1 and group K3 (p=0,667).

3.1.3. LDL Cholesterol Content

The average of mouses' LDL cholesterol serum content examination from each group was presented in Table 3.

Table 3. Average of LDL Serum Content

Group	Average (mg/dL) ± SB
K1 (n=6)	29,92 ± 1,53
K2 (n=6)	78,42± 1,77
K3 (n=6)	24,32 ± 1,77
K4 (n=6)	40,16± 2,84

The result of statistic test showed that the data was in normal distribution and had homogen variation and then proceeded to *One Way Anova* parametric test. On the LDL serum cholesterol content data, the *One Way Anova* test indicated that at least there was a significant difference of LDL cholesterol content between two groups (p=0,000). In order to know which groups showing a significant difference of LDL cholesterol content, the process was continued by *Post-hoc Bonferoni* analysis. It was found that there were significant differences between group K1 and group K2 (p=0,000), group K1 and group K3 (p=0,001), group K2 and group K4 (p=0,000), and group K3 and group K4 (p=0,000).

3.1.4. HDL Cholesterol Content

The average of mouses' HDL cholesterol serum content examination from each group was presented in Table 4.

Table 4. Average of HDL Serum Content

Group	Average (mg/dL) ± SB
K1 (n=6)	65,89 ± 1,99
K2 (n=6)	24,32 ± 1,44
K3 (n=6)	73,23 ± 1,92
K4 (n=6)	54,95 ± 2,04

The result of statistic test showed that the data was in normal distribution and had homogen variation and then proceeded to *One Way Anova* parametric test. The HDL serum cholesterol content data was analyzed using *One Way Anova* test and showed a result where p= 0,000. Therefore, it could be concluded that at least there was a significant difference of HDL cholesterol content between two groups (p<0,05). Then, *Post-hoc Bonferoni* analysis test was done in order to know which groups showing a significant difference of HDL cholesterol content. It was found that there were significant differences between group K1 and group K2 (p=0,000), group K1 and group K3 (p=0,000), group K2 and group K4 (p=0,000), and group K3 and group K4 (p=0,000).

3.2 Discussion

This study reported that giving *jelantah* oil as much as 3 ml/200gram BB for 28 days causing the total cholesterol content decreased. This result supported the previous studies about giving *jelantah* oil. Based on several studies, giving *jelantah* oil caused the total cholesterol content on the examined animals decreased. Experimental studies on examined animals which were given repeatedly heated palm oil diet stated that there were total cholesterol, trigliserida, LDL cholesterol content addition and HDL cholesterol reduction on examined animals which had been given repeatedly heated palm oil diet (*jelantah* oil). [2] The other result indicated that on a group of animals which was given refined palm oil, there was also total cholesterol, trigilserida, and LDL cholesterol content addition and HDL cholesterol content reduction [11].

Giving cloud ear fungus broth could lower the toal cholesterol content as much as 43.39% and LDL cholesterol as much as 49.78% and increased the HDL cholesterol content as much as 55.73% on mouses which had been induced by *jelantah* oil. However, compared to the group which was given *jelantah* ol without cloud ear fungus broth, the reduction was statistically significant (p<0,05). The result of this present study supported the previous studies where cloud ear fungus extract had antihypercholesterolemia effect. This was in line with the previous study in 2014 which reported that there was a significant reduction on hipokolesterolemia mouses' total cholesterol content which were given straw mushroom compared to controlled group [12].

By having significant differences between a group of mouses which had been given *jelantah* oil and cloud ear fungus broth (K4) and a group of mouses which had only been given *jelantah* oil and aquades (K2), so it could be concluded that cloud ear fungus broth dosing 3.6 gram/200gramBB was potential to help preventing oxidative stress as the result of giving *jelantah* oil that it could lower the total cholesterol, LDL cholesterol content and increase the HDL cholesterol content of mouses' serum.

The reduction of total cholesterol, LDL cholesterol content and the addition of HDL cholesterol was happened because of several cloud ear fungus's nutrient compound components which influenced the changing of blood lipid profile. Some of the compounds were polysaccharides, arkobat acid, niasin, dan flavonoid.[7, 8, 13, 14] The previous studies mentioned that cloud ear fungus (*Auricularia polytricha*) contained soluble polysaccharides compound which had antihypercholesterol effect. [8] Polisakarida was used as the fiber source which could be used to decrease cholesterol content. β-glucan polysaccharide was one of the fiber in cloud ear fungus which considered as a strong inhibitor to impede gatrointestinal lipase enzyme so that it could lower blood cholesterol content. [14] One of the roles of fiber toward fat metabolism was to impede gastrointestinal lipase enzyme reaction such as pancreatic lipase enzyme. Besides, the food fiber would block enterohepatic cycle (gall reabsorbance in the intestines to liver) with its ability and viscosity trapping micelles containing gall acid in the intestines and freed them from the bundle by luminal membrane of the intestinal epithelium transporter. The process reduced fat absorbance and reabsorbance including cholesterol and fat acid that it would increase the feces excretion.

As compensation after losing cholesterol and gall acid stock in the liver, there would be some addition of cholesterol which converted into gall acid by liver. It would result in free cholesterol reduction in the liver due to endogenous cholesterol synthesis and caused *7α-hydroxilase* dan *HMG CoA reductase* activitiy increased. This condition would trigger hephatic LDL cholesterol receptor regulation to store LDL cholesterol serum spare. Fiber also caused longer feeling of satiety which would result in controlled calorie and cholesterol consumption [15].

Another possibility that would influence blood cholesterol, LDL and HDL content was the amount of antioxidant in cloud ear fungus [10]. Flavonoid was one of the antioxidants. It reduced LDL serum content through VLDL blocking mechanism. LDL was formed from VLDL which had been gone through lipid separation cascade so that the LDL content would be determined by VLDL content. The VLDL's component composers were just the same as LDL's which were trigliserida, ester cholesterol and apolipoprotein B, and also some other lipid components. VLDL content reduction by flavonoid was done through MTP protein transfer and ACAT enzyme blocking. MTP (*microsomal triglyceride transfer protein*) was a transfer protein which was responsible in combining trigliserida, ester cholesterol and Apo0-B process, while ACAT enzyme (*Acyl Co-A Cholesterol Acyl transferase*) was an intraselular enzyme whose role was to catalyze ester cholesterol from cholesterol and facilitate Apo B translocation acrossing endoplasmic reticulum membrane, from cytoplasm heading to lumen. MTP and ACAT blocking would reduce VLDL content and lower LDL serum and total cholesterol content [16, 17].

The reduction of bloods' LDL cholesterol content influenced bloods' HDL cholesterol content to decrease. The previous studies mentioned that fiber helped to bind gall acid which was then excreted along with the feces. To synthesize gall acid, liver needed cholesterol, and if the supply was empty, the liver would send a message to the brain and it would respon by sending signal to HDL to take cholesterol lied in the tissue to be brought back to the liver.

Other component which was presumed to be influential was ascorbic acid (vitamin C). The result of experimental study on animals which tried to see the effect of giving vitamin C to lipid profile reported that there was a significant reduction on total cholesterol content, and LDL cholesterol in white mouses after being given vitamin C. Besides, there was also trigliserida content reduction and HDL cholesterol which was increased though it was not statistically significant. [20] An explanation which might seem to be quite logic in this case mentioned that it was the result of *7α-hydroxylase* enzyme activation by ascorbic acid which increased the conversion of plasma cholesterol into gall acid and further caused the total cholesterol serum content decreased. Another mechanism to decrease total cholesterol serum content was bby hydroxylate steroid hormone synthesis which needed cholesterol as its precursors [20, 21].

Niasin or nicotinic acid was a part of vitamin B complex which also known as vitamin B3. Niasin could affect lipoprotein containing apo-B such as VLDL and LDL, and it could also increase lipoprotein containing Apo-A 1 such as HDL. The newest invention showed that niasin inhibited *Diacylglycerol Acyltransferase-2 (DGAT-2)* enzyme activity directyly whereas it was the most important enzyme in triglycerides synthesis. Inhibiting TG synthesis by niasin would result in apo-B intraseluler degradation and decrease VLDL and LDL secretion. Another mechanism was that niasin controlled TG content by inhibitting lipolysis on adipocytes so that it would decrease TG content in plasma. To influence HDL cholesterol content, niasin played a role as the inhibitor of HDL cholesterol and Apo-A 1 absorbance and transference [22, 23].

Cloud ear fungus broth had a way to prevent oxidative stress by being exogen antioxidant which would decrease the amount of free radicals formed after consuming *jelantah* oil. Therefore, the more antioxidants, the more active antioxidants which could neutralize free radicals formed after being given *jelantah* oil.

4. Conclusion and Suggestion

Based on the result, we could conclude that cloud ear fungus broth played a role to prevent total cholesterol and LDL content increased and HDL cholesterol serum content decreased. It was necessary to do further study by using various doses of cloud ear fungus broth to see the most effective doses to improve blood lipid profile. Besides, it was also necessary to do further analysis on active compound components on cloud ear fungus broth.

References

1. N. Chun-Yi, Yusof Kamisah, Othman Faizah, Zakiah Jubri, HjMohd Saad Qodriyah, Kamsiah Jaarin. International Journal of Vascular Medicine (2012).
2. SK. Adam, Ima Nirwana Soelaiman, Nor Aini Umar, Norhayati Mokhtar, Norazlina Mohamed, Kamsiah Jaarin. McGill Journal of Medicine, **11**,146-151. (2008)

3. N. Chun-Yi , Xin-Fang Leong, Norliana Masbah, Siti Khadijah Adam, Yusof Kamisah, Kamsiah Jaarin. Vascular Pharmacology **61**:1-9 (2014)
4. H. Istiadi, Endang Sri S. Media Muda Medika **4**:1-10 (2010)
5. E, Braunwald, *Approach to the patient with cardiovascular disease*, **16th ed**. In: Kasper D, et al., editors. *Harrison's principles internal medicine*. New York: McGraw-Hill Medical Publishing Division, 1301-4 (2005)
6. Hung PV, Nhi NNY. Research Journal **19**:611-615 (2012)
7. H. Puspitasari, Sri Peni Fitrianingsih, Lanny Mulqie. *Pengaruh Pemberian Ekstrak Jamur Kuping Hitam terhadap Penurunan Kadar Kolesterol Mencit Swiss Webster Jantan*. FMIPA. Bandung: Universitas Islam Bandung, (2015).
8. S, Zhao, Chengbo Ronga, Yu Liua, *et al.* Carbohydrate Polymers **122**:39-45 (2015)
9. Septiyana. *Uji Toksisitas Akut Air Rebusan Jamur Kuping Hitam (Auricularia polytricha (Mout) Sacc.) Serta Gambaran Histopatologi Organ Hepar dan Bobot Limpa pada Mencit Putih (Mus musculus) Jantan Galur BALB/C. Farmasi.* Semarang: Sekolah Tinggi Ilmu Farmasi Yayasan Pharmasi, (2010).
10. Sun Y-X, Ji-Cheng Liua, John F. Kennedy. Carbohydrate Polymers **82**:299-304 (2010).
11. SL. Wansi, *Et al*. Internationl Journal of Pharmacrutical, Chemical, and Biological Sciences (IJPCBS) **3**:627-634 (2013).
12. AE. Damayanty, *Pengaruh Pemberian Ekstrak Jamur Merang (Volvariella volvacea) Terhadap Kadar Kolesterol Total, Enzim LpPLA2 dan MDA Darah*. Magister Ilmu Gizi. Semarang: Universitas Diponegoro, (2015).
13. S. Falakh, *Aktivitas Antioksidan Ekstrak Jamur Kuping Hitam (Auricularia polytricha).* Fakultas Matematika dan Ilmu Pengetahuan Alam. Bogor: Institut Pertanian Bogor, (2008).
14. Smith, Rowan, Sullivan. *Medicinal Mushrooms: Their therapeutic properties and current medical usage with special amphasis on cancer treatments*, (2002).
15. E. Theuwissen, P. Ronald, Mensink. Physiology & Behavior **94**:285-292 (2008).
16. A. Pramono, UK. Solikah, HT. Nurul, AY. Rahma Mutiara Medika **11**:130-143 (2011).
17. PA. Mayes, Botham KM. *Metabolisme asilgliserol dan sfingolipid*. Biokimia Harper. Jakarta: EGC, (2009).
18. I, Zaki, *Pengaruh Pemberian Jus Mangga Terhadap Profil Lipid dan Malondialdehyde pada Tikus yang Diberi Minyak Jelantah*. Program Studi Magister Ilmu Gizi. Semarang: Universitas Diponegoro, (2014).
19. B. Hartoyo, I. Irawan, N. Iriyanti, Animal Production **7**:27-33 (2005).
20. MU. Eteng, HA. Ibekwe, TE. Amatey, BJ. Bassey, FU. Uboh, DU. Owu. Journal Of Physiological Sciences **21**:15-19 (2006).
21. Z. Prakoso. *Pengaruh Pemberian Vitamin C terhadap Kadar Kolesterol LDL dan HDL Serum Tikus Wistar Jantan Hiperlidemia Setelah Perlakuan Jus Lidah Buaya (Aloe vera Linn).* Fakultas Kedokteran. Semarang: Universitas Diponegoro, (2006).
22. VS. Kamanna, ML. Kashyap. American Journal of Cardiology **101**:20B-26B (2008)
23. J. Haseeb, HK. Richard, TK. Jeffrey. Clinical Lipidology **4**:565-57 (2009).

https://doi.org/10.1051/e3sconf/20183106007

Work Environment Factors and Their Influence on Urinary Chromium Levels in Informal Electroplating Workers

Yuliani Setyaningsih[1,*], *Adi Heru Husodo*[2], and *Indwiani Astuti*[2]

[1]Public Health Faculty , Diponegoro University, Semarang - Indonesia
[2]Medical Faculty, Universitas Gadjah Mada, Yogyakarta - Indonesia

Abstract. One of the informal sector which absorbs labor was electroplating business. This sector uses chromium as coating material because it was strong, corrosion resistant and strong. Nonetheless hexavalent chromium is highly toxic if inhaled, swallowed and contact with skin. Poor hygiene, the lack of work environment factors and sanitation conditions can increase the levels of chromium in the body. This aimed of this study was to analyze the association between work environment factors and levels of urinary chromium in informal electroplating worker. A Purposive study was conducted in Tegal Central Java. The research subjects were 66 male workers. Chi Square analysis was used to establish an association between work environment factors and level of urinary chromium. There is a relationship between heat stress and wind direction to the chromium levels in urine ($p < 0.05$), but there is no relationship between humidity and levels of chromium in the urine ($p > 0.05$). This explains that work environment factors can increase chromium levels in the urine of informal electroplating workers.

1 Introduction

Chromium is commonly used in various industrial processes such as stainless steel production, welding, tannery and metal coating. Chromium is one of the heavy metals that are often used as a metal coating because of its anti-corrosive properties and can beautify the surface into a shiny white [1-2]. Clinical and laboratory evidence suggests that chromium (VI) is responsible for most poisoning events. Chromium (VI) is highly toxic by inhalation, in contact with skin and ingested through consumption of foods and beverages containing chromium (VI). It can cause lung cancer, nasal irritation, nasal ulcers, hypersensitivity reactions such as contact dermatitis and asthma [3-4].

Inhalation routes are the main entrance and exposure of workers. Workers can inhale chromium in the form of gas, dust or particles while working around the electroplating tub. Particles with a diameter size less than 5 μ can reach the alveoli and dissolve into the bloodstream. Chromium is reduced by macrophages in the bronchial branches and pulmonary parenchymal cells [5-6]. The respiratory tract is the main target of Cr (VI) toxicity and is associated with chronic and acute exposure [4,7]. Exposure to Cr (VI) in the body mainly through inhalation of aerosols can cause health problems in the respiratory tract, carcinogenic, liver, kidney and immune system disruption. Several in-vitro studies indicate that Cr (III) concentrations in cells can cause DNA damage [8].

The work environment can be the determinant variable of air pollution dynamics such as wind direction and speed, humidity and working climate. An increased temperature can increase the reaction rate of a chemical [9]. High levels of chromium levels in the urine can be used as an indicator of high chromium exposure indoors [10].

From the preliminary survey, it is known that some workers work in a work area that integrates with residence. Limited ventilation and poor sanitation hygiene can increase air pollutant levels and harm workers' health conditions. Therefore the researchers wanted to examine the relationship between environmental factors with chromium levels in the urine of informal metal coating workers.

2 Research method

This research is observational with cross sectional design. The study was conducted on an informal electroplating worker in Tegal, Central Java. The subjects consisted of 66 male workers with criteria already working for 1 year and exposed by chromium. Chromium in the urine of the worker can be used as an indicator of the amount of chromium that enters the body. A total of 10 mL of urine was collected from the respondents after working for 4 hours. Level or urinary chromium was measured by flameless atomic absorption spectrophotometer (AAS) with a graphite furnace (GF-3000) and auto-sampler (PAL-3000) [10]. Working climate and humidity are measured using the heat stress area of the Questemp 34 brand monitors with reference to the standard SNI No. 16-7061-2004. Wind speed is measured by an Anemometer. Data were analyzed by

* Corresponding author: joeliani_kesja_undip@yahoo.com

using bivariate analysis covering environmental and chromium levels in urine by using Chi-Square

3 Result and analysis

A total of 66 respondents were male. All of them do not use personal protective equipment when working. Age ranges from 17-56 years and has a working period of 1-40 years.

3.1 Work Environment Factor

Work environment factors are assessed by measuring the working climate, humidity and wind direction in the metal coating production site. The heat stress measured at 12 workplaces ranged from 27.8 o C - 29.5 °C. According to Ministry of Manpower & Transmigration Regulation No. Per.13 / MEN / X / 2011, the ISBB threshold is permitted for 75% -100% working hours arrangement with moderate work, a permissible working climate is 28 0 C. Humidity measured In 12 workplaces ranged from 73% - 87%. The humidity allowed in the work space is between 65-85%. As for the wind direction is divided into work exposed to the wind direction parallel and opposite to the wind direction. Workers exposed by environmental factors can be seen in table 1.

Table 1. Frequency Distribution of Metal Coating Workers exposed to Environmental Factors in Talang Sub-district, Tegal Regency, 2014

No.	Respondent exposed	N	%
1.	Heat stress		
	a. Non - appropriate ISBB	40	60.6
	b. Appropriate ISBB (28 ^{0} C)	26	39.4
2.	Humidity		
	a. Non - appropriate	32	48.5
	b. Appropriate (65-85 %)	34	51.5
3	Wind Direction		
	a. Opposites	46	69.7
	b. Linear	20	30.3

Metal coating workers work in an environment with an unsuitable heat stress ISBB of 60.6%. In addition, there are 48.5% of workers in a production room with unsuitable humidity for the workplace. A total of 69.7% of workers were exposed to chromium vapor directly because while in the production room working opposite to the wind direction. The longer workers work, the risk of exposure to sharp-smelling chemicals is greater, given that workers do not use personal protective equipment such as masks and gloves [11]. The Environmental Protection Agency (EPA) and the International Agency for Research on Cancer (IARC) indicate that Cr (VI) is inhaled otherwise carcinogens in humans and latent danger occurs for 20 years and the shortest duration of exposure is 2 years [12].

3.2 Urinary examination in electroplating workers

Worker's urine was taken after work for 4 hours and before lunch break of + 5 ml. The urine sample was then collected in a polypropylene container, and urine chromium (5 ml) was examined. The results of urine examination of metal coating workers can be seen in Table 2.

Table 2. Frequency distribution of Chromium content, MDA levels and 8-OHdG levels of Metal Coating Workers in Talang Sub-district, Tegal Regency, 2014

Variable	N	%
Chromium Levels		
a. ≥ 11.77 µg/L	32	48.5
b. < 11.77 µg/L	34	51.5

As many as 51.5% of workers with chromium levels were less than median (11.7 µg / L). Chromium is actually a natural heavy metal and is an essential micronutrient to support insulin work. However, chromium is highly toxic by inhalation, in contact with skin and swallowed [3-4]. Chromium levels in the worker urine in this study had a minimum value of 2.811 µg / L and a maximum value of 145.340 µg / L. This value is higher than the normal range for chromium urine in humans is between 0.1 µg / L -0.5 µg / L while the threshold value of chromium in urine 0.05 µg / L [13]. Chromium in urine shows absorption for 1-2 Days before the examination [4].

3.3 Complaints pain

Table 3. Types of complaints Pain suffered by Metal Coating Workers in Talang Sub-district, Tegal Regency, 2014.

No	Complaints Pain	Total	%
1.	Itchy	8	12.12
2.	Cough	2	3.03
3.	Shortness of breath	3	4.55
4.	Musculoskeletal disorder	7	10.60
5.	Itchy and Cough	3	4.55
6.	Itchy, Cough and shortness of breath	2	3.03
7.	Itchy, Cough and Musculoskeletal disorder	2	3.03
8.	No complaints	39	59.09

Table 3 shows how many complaints of workers such as itching, coughing, and shortness of breath and musculoskeletal complaints. There are 10.58% of metal coating workers who experience 2 or more complaints of illness due to their work while 59.09% admitted to having no complaints during their work.

Of workers who feel sick 45.5% of workers choose to buy medicine in stall when feeling pain, 25.8% choose to Community Health Center - puskesmas and the rest choose not to cure. Informal sector workers have a higher risk in relation to health problems suffered by the work [11]. Zeins says that high chromium content can affect respiratory symptoms and increased bronchial asthma [14].

All workers do not use PPE. To prevent inhaled chemicals and skin contacts the worker should improve personal hygiene such as regular hand washing and showering after work hours. General ventilation can be used to drain the air into the room so that the levels of harmful chemicals become lower in the air. In addition, it can also use local exhauster, the installation that sucks air in a certain place through the canopy for dangerous chemicals to be flown out of the workplace space [15]. Workers should use PPE as the last preventive measure of exposure to chemicals. PPE is not a substitute for hazard control either engineering or administrative engineering but if hazard control is not possible then APD can be used wherever danger can be absorbed, swallowed or inhaled [16].

3.4 Statistical test

The variables of working environment factors measured include working climate, humidity and wind direction. Table 4 shows the test results of the relationship between environmental factor variables and chromium levels in the urine of metal coating workers.

Table 4. Test Result Relation between work capacity variable with chromium content in urine Metal Coating Worker, Talang Sub-district, Tegal Regency, 2014

No.	Variable of environment factors	*p*	OR	95 % CI
1.	Heat stress with chromium level (ISBB)	**0.038**	**3.375**	**1.186 – 9.604**
2.	Humidity with chromium level	0.617	0.691	0.262 – 1.824
3.	Wind direction with chromium level	**0.005**	**6.222**	**1.791-21.618**

The results of statistical tests show that the working climate and wind direction are related to chromium levels in the worker's urine ($p < 0.05$). Only humidity is not associated with chromium levels in the worker's urine ($p > 0.05$). Workers in climatic working environments not in accordance with ISBB (> 28 0 C) have a chance of 3.37 times having chromium levels greater than 11.77 μg / L in their urine compared with workers working in a heat stress-conditioned room in accordance with ISBB (OR = 3,375). In addition, workers working in the opposite wind direction of the electroplating tub have a chance of 6.22 times having chromium levels of more than 11.77 μg / L in their urine compared to workers working on the parallel wind direction of the electroplating bath (OR = 6.222).

The results of statistical tests show that the working climate and wind direction are related to chromium levels in the worker's urine ($p < 0.05$). Only humidity is not associated with chromium levels in the worker's urine ($p > 0.05$). Workers in climatic working environments not in accordance with ISBB (> 28 0 C) have a chance of 3.37 times having chromium levels greater than 11.77 μg / L in their urine compared with workers working in a heat stress-conditioned room in accordance with ISBB (OR = 3,375). In addition, workers working in the opposite wind direction of the electroplating tub have a chance of 6.22 times having chromium levels of more than 11.77 μg / L in their urine compared to workers working on the parallel wind direction of the electroplating bath (OR = 6,222). The temperature and chemical concentrations of the electrochemical tub are regulated during the metal coating process. When the tub starts to be heated, smog will appear and exposure to the worker will increase [13]. Guillemin and Berode in Greenberg [13] also stated that the concentration of chromium content in the air is higher in the hard plating process and the chromium levels found in urine are also higher. Hard plating uses higher temperatures and more chemicals in the electroplating tub than with bright plating.

The worker as a dye is closer to the electroplating tub than the non-dyeer. Metal coating workers are exposed by particle and hexavalent Cr fog resulting from the electrolysis process occurring in the electroplating basin. These particles and fogs occur as a result of bursts on the liquid surfaces in the electroplating basins coming from oxygen bubbles and hydrogen out of the electrodes during the coating process [5].

In addition, the multivariate test using smart PLS proved that there is a relationship between environmental factors to chromium levels in urine. Environmental factor variables are formed by wind direction indicators and heat stress (ISBB). The overheated heat stress will increase the process of vaporization of Cr in the tub. It is still exacerbated by a working position opposite to the wind direction; consequently, workers will be exposed to greater chromium during inhalation [4]. Mukono [17] states that increased temperatures plus the presence of sunlight, including the occurrence of reactions of 2 or more secondary pollutants, may cause increased effects of toxic substances in the air. Secondary pollutants have unstable physical and chemical properties so it is very easy to react and change their chemical makeup in the presence of other chemicals in the air as well as in the body. Husaini [18] states that a hot climate or working temperature will greatly affect the rate of chemical reactions in the air especially in the worker's body where each work is always exposed.

4 Conclusions

Heat stress and wind direction are related to chromium levels in the urine of informal electroplating workers while the moisture is not related to chromium levels in the urine of informal electroplating workers. The presence of chromium in urine can be used as an indicator of high chromium exposure in the workplace.

Acknowledgements

The authors would like to thank to the Ministry of Research and Technology and Higher Education of the

Republic of Indonesia, which has provided the doctoral scholarship program; Biochemistry and Parasitological Laboratory of Medical Faculty, Universitas Gadjah Mada in Yogyakarta; and BBTKLPP for analyzing urinary chromium.

References

1. A. Gambelunghe, R. Piccinini R , M. Ambrogi , M. Villarini , M. Moretti , C. Marchetti , G. Abbritti, G. Muzi. Toxicology ***188***, 187-195. (2003)
2. Caglieri, Andrea, G. Matteo, A. Olga, A. Roberta, V. Maria, C. Massimo, A. Pietro, M. Antonio, Environmental Health Perpectives. **Vol 114**. No 4. (2004)
3. R. Shrivastava, R.K. Upreti, P.K. Seth, U.C. Chaturvedi , *FEMS* Immunology and Medical Microbiology, **34** , 1-7, (2002.)
4. Agency for Toxic Substances and Disease Registry (ATSDR). *Toxicological Profile for Chromium.* (2012)
5. Y. Hedberg, Johanna Gustafsson, Hanna L Karlsson, Lennart Möller, Inger Odnevall Wallinder. Particle and Fibre Toxicol.7:23. (2010)
6. A.P. Das, S. Singh, J. Occup Environ Med . Jan-Apr; **15**(1): 6–13 (2011)
7. De Flora S, Carcinogenesis **21**:533–541 (2000)
8. D.A. Esmond, J.T. Macgregor, R.S. Slesinski, Trivalent, Crit Rev. Toxicol, **38** :173-190 (2008)
9. UF. Achmadi, *Manajemen Penyakit Berbasis Wilayah*. Edisi revisi. (2014)
10. Y. Setyaningsih. Adi Heru Husodo. *Indriani Astuti. Detection Urinary8-Hidroxydeoxyguanosine (8-OHdG) Levels as a Biomarker of Axydative DNA Damage Among Home Industry Workers Exposed to Chromium.* **Vol 23**. 290-296. (2015)
11. Anies. *Penyakit Akibat Kerja.* (2005)
12. V. Kapol, J. Keogh, *Case Studies in Environmental Medicine.* (1990)
13. Greenberg, I. Michael, J.H. Richard, D.P. Scott, J.M. Gayla, *Occupational Industrial and Environmental Toxicology.* E **2**, (2003)
14. M. Zeins, C.I. Rivard, JL. Malo, D. Gautrin, *Is Metal Fume Fever a Determinant of Welding Related Respiratory Symptomps and/or Increased Broncial Responsiveness?* **62**. 688-694. (2005)
15. Soedirman. *Higiene Perusahaan.* (2012)
16. Damanik, HL. Adi Heru Husodo. Totok Gunawan. Pramono Adi. *Model Pengendalian Kesehatan Tenaga Kerja pada Kegiatan Pengecoran Logam Tradisional Studi Kasus di Kawasan Industri Batur Klaten-Jawa Tengah*. **Vol 4**. 101-198. (2015)
17. HJ. Mukono, *Toksikologi Lingkungan.* (2005)
18. H. Husaini, N E S., H H. Marsetyawan Adi, *Studi Kasus : Koreksi Terhadap Pengukuran Polutan di Udara Unit Perajin Logam dan Dampaknya Terhadap Kesehatan*, *Buletin Penelitian Kesehatan*, **Vol 44**, No 2 (2016).

The Presence Of Rat And House Sanitation Associated With *Leptospira sp.* Bacterial Infection In Rats (A Cross Sectional Study In Semarang, Central Java Province, Indonesia)

Endang Setiyani[1], *Martini* Martini[2], and *Lintang* Dian Saraswati[2,*]

[1]Animal Borne Disease Control Unit Banjarnegara, National Institute of Health Research and Development, Ministry of Health of Republic of Indonesia, Jl. Selamanik 16A, Kutabanjarnegara, Banjarnegara, Indonesia, 53415
[2]Department Epidemiology and Tropical Diseases, Public Health Faculty, Diponegoro University, Semarang, 50275

Abstract. The *Gajah Mungkur* sub-district in Semarang, Indonesia had highest leptospirosis cases (reported in human with seven infected and one dead) in 2015. The purpose of this study was to analyze the association between house sanitation and density of rats with *Leptospira sp.* infection in rats. The study design was cross sectional observational analytic. The number of 308 trapswere placed in study sites over three consecutive nights afterwards. Every houses were placed with four traps, inside and outside. Trapped rats were anesthetized with atropine dose from 0.02 to 0.05 mg/kg body weight of rats continued with Ketamine HCL dose of 50-100 mg/kg body weight of rats by injecting in the thick thigh muscle of it. After that, identification of rats by species and gender then continues with surgery in which a kidney sample was taken to confirm the presence of bacteria *Leptospirasp* using PCR techniques. The trap installed in 77 houses which later had further observation on house sanitation which includes the existence of a pile of used goods, food storage, garbage can, and the presence of the ceiling, windows and other ventilation. Data was analyzed using distribution frequency and bivariate chi-square test. We had 100 rats captured with live traps as the samples.The proportion of *Rattusnorvegicus*was 27% (14.8% positive *Leptospira sp.*infection) and *Rattustanezumi* 73% (11%positive *Leptospira sp.*infection). The proportion of male and female rats were almost equal. The statistic test result was significant between the density of rats (p = 0.0001, OR 12.833, 95%CI: 1.565-105.261) and sex of rats (p = 0.019, OR 0.095, 95%CI: 0.012–0.769) with *Leptospira sp.* infection in rats. The number of rats may increase the infection of *Leptospirasp.*, especially female rats and poor condition of house sanitation. It is recommended to improve house sanitation and regularly trapping rats.

1 Introduction

Leptospirosis is neglected infectious disease caused by *Leptospira* bacterium infection, in group of *Spirochaeta* [1]–[3]. The advanced molecular techniques allowed the identification of 14 pathogenic and intermediately pathogenic species, and 7 non-pathogenic species, distinguishing three clades in Leptospira genus. These clades comprise more than 260 serovars. Among these, 7 species: L. interrogans, L. borgpetersenii, L. santarosai, L. noguchii, L. weilli, L. kirschneri and L. alexanderi are considered as the main agents of leptospirosis [4]–[9]. Portals of entry include cuts and abrasions or mucous membranes such as the conjunctival, oral, or genital surfaces. Exposure may occur through either direct contact with an infected animal or through indirect contact via soil or water contaminated with urine from an infected animal [8], [10]–[12].

In 2015, Gajah Mungkur district was one of highest case in Semarang within leptospirosis case. The data acquired in 2015 reported 7 patients and 1 died (CFR = 16.67%). Higest case found in Petompon and Sampangan Village. Both area was passed by river streams of West Flood Canal which made it possible as a vehicle of transmission of leptospirosis.

Leptospirosis is a good example of the interaction between humans and their environment, and particularly human interface with domestic animal. In the tropical climatic zone, where environmental conditions are most favourable for survival of leptospires and the highest morbidity is noted, extreme weather events such as cyclones and floods occurring in recent years with increasing frequency and greater intensity, may potentially result in an upsurge in the disease incidence, as well as the magnitude of leptospirosis outbreaks, such as our previous study [11]–[15]. Exposure to animals could be a potential risk factor for the disease. In the present study, rearing domestic animals at home was identified as a risk factor with an unadjusted OR of 1.21 (95% CI: 1.02-1.43) [9]. The presence of rodents and

* Corresponding author: lintang.saraswati@live.undip.ac.id

seeing more than five rats per day at home or work place were found to be risk [16], [17]. Practices such as use of vegetables ate by rats, contact with contaminated soil with rat's urine surrounding home were found to be highly associated as an independent factors [9], [13], [18]. Determination of *Leptospira* bacteria presence in mice are tested by using PCR (Polymerase Chain Reaction) with high sensitivity which reaches 90-100% approximately [4].

This study aims to determine the relationship of house sanitation and density of rats with bacterial infection by *Leptopsira sp* in the subvillage Petompon and Sampangan in Gajah Mungkur District of Semarang.

2 Methods

2.1. Study Design

An analytic observational cross sectional study was conducted to estimate bacterial infection in rats and other related factor. The trap installed in 77 houses which later undertake further observations on house sanitation, which includes the existence of used goods piles, the existence of food storage, the presence of trash, the presence of the ceiling, windows condition and the presence of ventilation. The data is processed and statistically analyzed using the chi-square test to determine the relationship of home sanitation and density of mice with bacteria *Leptospira* sp infection in mice.

2.2 Study Sites

The study site were in *Petompon* village and *Sampangan* village located in Gajah Mungkur Sub-District in Semarang City.

2.3 Population and Sample

Population of the study were all rats lived at or nearby *Petompon* village and *Sampangan* village. Samples were captured at research sites using live traps.

We installed 308 trap in 77 houses on location over three consecutive nights afterwards. Every house installed with four traps, inside and outside of the house. The traps which caught rats or other animals like a shrew, mongoose, squirrels and other were taken and then replaced with empty traps which has been washed and dried. The caught rats put into a cloth bag, and then brought into field laboratories for the identification and PCR test. Other animals that caught were released, and the trap was immediately washed.

2.4 Laboratory Testing

Rats were anesthetized with atropine dose from 0.02 to 0.05 mg / kg body weight of rats continued with Ketamine HCL dose of 50-100 mg/kg. Body weight of rats by injecting in the thick thigh muscle of it The anaesthized rats were identified by species and gender then undergoes a surgery. Kidney sample was taken to confirm the presence of bacteria *Leptospira* with using PCR techniques.

3 Results and Discussion

3.1. The existence of Rats.

From the two variables was studied, there is one variable that showed a significant relationship with the bacteria *Leptospirasp* infection in mice, the sex of mice (Table 1). Gender mice with $p \leq 0.05$ and value OR = 0.095, it means that the sexes of mice not a risk factor for the infection of bacteria *Leptospira sp*. in mice. Based on the results of catching rats, 100 rats caught split by mostly 56 a female, so the female rats had higher activity than male rats. This is in line with previous studies, female rats were more common than the male rats [19]–[21]. Female rats have the nature of post-partum oestrus (estrus ability mother after giving birth). One to two days after giving birth, the mother ready to marry male rats [22]. This further increased the *Leptospira* bacteria in mice.

The mean life time of in days of the males rats 628±17.3 and the female rats 707±15.9[22] It can be said the life span of female mice longer than male rats, so the possibility of female rats exposed to the bacteria *Leptospira* has longer [23], [24].

The genders of mice that were caught in this study, can be used to determine the mobility of the population of rats in a residential area. Activities of mice is usually aim for eating, drinking, sex, and orientation of the range region. During the orientation area, these animals will recognize the environmental situation especially preferred feeding areas, water, and shelter to save themselves [15], [24]–[26].

Table 1. Identification of captured rats and *Leptospira sp* Bacterial Infections in Rats at Petompon and Sampangan Village 2016

Variable and Category	*Leptospira sp* bacterial infection in mouse				Total		Value	OR (95% CI)
	Positive (n=12)		Negative (n=88)					
	f	%	f	%	f	%	p	
Species								
- Rattus norvegicus	4	14.8	23	85.2	27	100	0.750	1.413
- *Rattus tanezumi*	8	11.0	65	89.0	73	100		(0.389–5.38)
Gender								
- Male	1	2.3	43	93.7	44	100	0.019	0.095
- Female	11	19.6	45	80.4	56	100	*	(0.012–0.769)

https://doi.org/10.1051/e3sconf/20183106008

3.2. House Sanitation

Of the seven variables studied, only one variable that showed a significant relationship with the bacterium *Leptospira* sp infection in mice that the density of the mice in the house. The density of rats with p≤ value of 0.05 and value OR = 12.8, means that the house is solid rat has a 12.8 times greater risk of occurrence of Leptospira positive bacterial infections in mice compared to a house without mouse. With the discovery of mice in residential areas, may increase the occurrence of bacterial infections *Leptospira sp*. in mice. A recent study indicated that young non reproductive rats, which had just left the nest, showed a prevalence of up to 30% of kidney colonization [4], [27].

Tabel 2. House Sanitation and *Leptospira sp* Bacterial Infections in Rats at Petompon and Sampangan Village 2016.

VariabledanCategory	*Leptospira sp* bacterial infection in mouse				Total		Nilai	OR (95% CI)
	Positif (n=12)		Negatif (n=65)					
	f	%	f	%	f	%	P	
Piles of used goods								
a. Yes	10	14.5	59	85.5	69	100	0.603	0.508
b. No	2	25.0	6	75.0	8	100		(0.090-2.883)
Food storage								
a. Open	8	14.0	49	86.0	57	100	0.498	0.653
b. Covered	4	20.0	16	80.0	20	100		(0.173-2.460)
Garbage can								
a. Open	10	16.1	52	83.9	62	100	0.9	1.250
b. Covered	2	13.3	13	86.7	15	100		(0.244-6.414)
Ceiling								
a. No	3	27.3	8	72.7	11	100	0.363	2.375
b. Yes	9	13.6	57	86.4	66	100		(0.529-10.661)
Window								
a. Not tightly fitted	1	11.1	8	88.9	9	100	0.9	0.648
b. Tightly fitted	11	16.2	57	83.8	68	100		(0.073-5.711)
Ventilation								
a. Open	11	16.4	56	83.6	67	100	0.9	1.768
b. Covered	1	10.0	9	90.0	10	100		(0.203-15.402)
Population density of mouse/rats								
a. High (> 7%)	12	29.3	29	70.0	41	100	0.0001	12.833
b. Low (< 7%)	0	0	36	100	36	100	*	(1.565-105.261)

Infection at a very young age (>2 months) suggests that transmission may occur also in utero or to neonates, possibly through infectious milk, or by other routes within the nest including close contact with an infected mother via fomites or uro-genital cleansing of neonates by the dam [22], [21], [28], [24].

In this study, the success of catching rats (Trap success) is as high as 11%. The success of catching mice at home habitat is when more than 7% [4], [22], [21]. With the success of these arrests rate, can illustrate that the research sites have high density rats population as a potential transmission of leptospirosis. *Leptospira* bacteria infection occurred because a lot of rats were found in residential areas so in this case, contamination by rat urine, will be easily affected leptospirosis case in human .

The high density of the mice in this study, it is possible for a distance of one another house huddled together with poor sanitary conditions so that the house can live rats and mice in the house proliferate [13], [15], [25]. The existence of mice in the house is very dangerous to humans because rats was the main reservoir of leptospirosis disease. Leptospirosis can be debilitating, particularly for the young and old, and severe cases can lead to liver and kidney failure and internal bleeding (Weil's syndrome) [1], [3], [29]. Indeed, there are many diseases transmitted by rodents either through rodent faeces (salmonella), urine (leptospirosis) airborne viruses (haemorraghic fever) and vector insects (plague and murine typhus). Considering the close proximity rodents have with humans in these environments, many rodent diseases are likely to be having debilitating effects which reduce productivity and cause mortality [9], [13], [29].The presence of rats in a residential area needs serious attention, as a form of awareness of disease transmission, especially leptospirosis.

4. Conclusion

Based on the results of the analysis can be seen that there is a significant correlation between the density of rats (p = 0.0001), Gender of rats (p = 0.019) with infection of bacteria *Leptospira sp*. in rats and there was no significant correlation between rat species, piles of junk, used to store food, trash, where the ceiling, where the windows and the presence of bacterial infection ventilation with *Leptospira sp*. in mice. Advice to the Community to keep the house in order clean for not to become a breeding ground for rats and mice catching efforts regularly and continuously to reduce the rat population, especially in Subvillage Petompon and Sampangan.

References

1. T. Fraga, E. Carvalho, L. Isaac, and A. Barbosa, "Leptospira and Leptospirosis," in *Molecular medical microbiology*, 2nd Ed., Y.-W. Tang, M. Sussman, D. Liu, I. Poxton, and J. Schwartzman, Eds. Elsevier, pp. 1973–1990, (2014).
2. D. A. Haake and P. N. Levett, "Leptospirosis in Humans," *Curr. Top. Microbiol. Immunol.*, vol. **387**, pp. 65–97, (2015).
3. A. J. Mcbride, D. Athanazio, M. Reis, and A. I. Ko, "Leptospirosis," *Leptospirosis Curr. Opin. Infect. Dis.*, vol. **18**, no. 5, pp. 376–386, (2005).
4. U. Kositanont, P. Naigowit, A. Imvithaya, C. Singchai, and P. Puthavathana, *J. Med. Assoc. Thai.*, vol. **86**, no. 2, pp. 136–42, Feb. (2003).
5. D. Benacer *et al.*, *Parasit. Vectors*, vol. **9**, no. 1, p. 117, Dec. (2016).
6. J. P. dos Santos *et al.*, *Trop. Anim. Health Prod.*, vol. **44,** no. 1, pp. 101–6, Jan. (2012).
7. S. M. de Carvalho, A. L. B. B. Mineiro, V. Castro, M. E. Genovez, S. S. Azevedo, and F. A. L. Costa, *Trop. Anim. Health Prod.*, vol. **46**, no. 2, pp. 491–4, Feb. (2014).
8. A. Dreyfus *et al.*, *PLoS Negl. Trop. Dis.*, vol. **10**, no. 8, p. e0004858, Aug. (2016).
9. M. Lelu, C. Muñoz-Zanzi, B. Higgins, and R. Galloway, *BMC Vet. Res.*, vol. **11,** p. 31, (2015).
10. M. P. Ward, L. F. Guptill, and C. C. Wu, *J. Am. Vet. Med. Assoc.*, vol. **225**, no. 1, pp. 72–77, Jul. (2004).
11. A. P. Sugunan *et al.*, *Indian J. Med. Res.*, vol. **130**, no. 1, pp. 67–73, Jul. (2009).
12. A. M. Dechet *et al.*, *PLoS One*, vol. **7**, no. 7, p. e39672, Jul. (2012).
13. R. K. Raghavan, K. M. Brenner, J. J. Higgins, J. M. Shawn Hutchinson, and K. R. Harkin, *Prev. Vet. Med.*, vol. **106**, no. 3–4, pp. 324–31, Oct. (2012).
14. S. C. Sehgal, A. P. Sugunan, and P. Vijayachari, *Natl. Med. J. India*, vol. **15**, no. 1, pp. 22–3, (1995).
15. S. N. Fajriyah, A. Udiyono, and L. D. Saraswati, in *IOP Conference Series: Earth and Environmental Science*, 2017, vol. **55**, no. 1, (2017).
16. J. G. Songer, C. J. Chilelli, R. E. Reed, and R. J. Trautman, *Am. J. Vet. Res.*, vol. **44**, no. 10, pp. 1973–6, Oct. (1983).
17. B. Wasiński, *Przegląd Epidemiol*, vol. **65**, no. 3, pp. 471–6, (2011).
18. P. Vijayachari, A. P. Sugunan, and A. N. Shriram, *J. Biosci.*, vol. **33**, no. 4, pp. 557–69, Nov. (2008).
19. F. Costa *et al.*, *Vector-Borne Zoonotic Dis.*, vol. **14**, no. 1, pp. 33–40, Jan. (2014).
20. F. Costa *et al.*, *PLoS Negl. Trop. Dis.*, vol. **9**, no. 6, p. e0003819, Jun. (2015).
21. L. H. Krøjgaard, S. Villumsen, M. D. K. Markussen, J. S. Jensen, H. Leirs, and A.-C. Heiberg, *Epidemiol. Infect.*, vol. **137**, no. 11, pp. 1586–92, Nov. (2009).
22. C. G. Himsworth, K. L. Parsons, C. Jardine, and D. M. Patrick, *Vector-Borne Zoonotic Dis.*, vol. **13**, no. 6, pp. 349–359, Jun. (2013).
23. C. G. Himsworth, K. L. Parsons, C. Jardine, and D. M. Patrick, *Vector-Borne Zoonotic Dis.*, vol. **13**, no. 6, pp. 349–359, Jun.(2013).
24. C. G. Himsworth *et al.*, *PLoS Negl. Trop. Dis.*, vol. **7**, no. 6, p. e2270, Jun. (2013).
25. F. Costa *et al.*, *PLoS Negl. Trop. Dis.*, vol. **8**, no. 12, p. e3338, Dec. (2014).
26. M. A. S. Johnson *et al.*, *Emerg. Infect. Dis.*, vol. **10**, no. 6, pp. 1016–22, Jun. (2004).
27. C. A. Ganoza *et al.*, *PLoS Med.*, vol. **3**, no. 8, p. e308, Aug. (2006).
28. M. Runge *et al.*, *Pest Manag. Sci.*, vol. **69**, no. 3, pp. 403–408, Mar. (2013).
29. P. N. Levett, "Systematics of Leptospiraceae," in *Leptospira and Leptospirosis*, 1st Ed., B. Adler, Ed. Springer-Verlag Berlin Heidelberg, pp. 11–20, (2015).

https://doi.org/10.1051/e3sconf/20183106009

Analysis of Work Design in Rubber Processing Plant

Dini Wahyuni[1,*], *Harmein* Nasution[1], *Irwan* Budiman[2] and *Khairini* Wijaya[1]

[1]Department of Industrial Engineering, Faculty of Engineering, Universitas Sumatera Utara, Medan - Indonesia
[2] Master Program of Industrial Engineering, Faculty of Engineering, Universitas Sumatera Utara, Medan - Indonesia

Abstract. The work design illustrates how structured jobs, tasks, and roles are defined and modified and their impact on individuals, groups, and organizations. If the work is not designed well, the company must pay greater costs for workers' health, longer production processes or even penalties for not being able to meet the delivery schedule. This is visible to the condition in a rubber processing factory in North Sumatra. Work design aspects such as layouts, machinery and equipment, worker's physical working environment, work methods, and organizational policies have not been well-organized. Coagulum grinding machines into sheets are often damaged, resulting in 4 times the delay of product delivery in 2016, the presence of complaints of heat exposure submitted by workers, and workstation that has not been properly arranged is an indication of the need for work design. The research data will be collected through field observation, and distribution of questionnaires related aspects of work design. The result of the analysis depends on the respondent's answer from the distributed questionnaire regarding the 6 aspects studied

1 Introduction

North Sumatra is one of the provinces in Indonesia which has a pretty extensive rubber plantations, totaling 395,017 hectares (sumutprov.go.id, 2014), or approximately 10.95% of all rubber plantations in Indonesia amounted to 3,606,245 Ha [1]. Production of processed rubber in Indonesia mostly RSS (Ribbed Smoked Sheet) with commodity SIR I to III (Standard Indonesian Rubber). On the initial observation in a rubber processing plant in North Sumatra, the design aspects of work such as layout, machinery and equipment, physical work environment, workers, work methods, and policy organizations have not well-ordered. Coagulum grinder machine into a rubber sheet often suffered damage, so it happened 4 times the delay in delivery of the products in the year 2016, heat exposure complaints submitted workers, and workspaces that have not been well groomed is an indication of the need for an analysis of the work design done.

Work design discusses the technical aspects and the social aspect of the job. The technical aspects of the process, in the form of a collection of tasks that must be done and have the means in the form of technology, procedures, and equipment. Social aspects referring to the workers, that is related to the ability and attitude as well as the office that determines their roles and responsibilities [2,3].

Work design illustrates how occupations, tasks, and roles are structured are set and modified as well as its impact on individuals, groups, and organizations. When the design work is not good, the company must pay health workers a larger, longer production process or company can't fill orders in a timely manner. [4,5,6,7]

Design improvement of work covering several aspects that should be improved as methods of work relating to machinery and equipment, the condition of the physical work environment, layouts, methods of work, organization, and policy workers, which need to be analyzed in advance in order to make known the shortcomings as well as its limitations.

2 Method

Figures and tables, as originals of good quality and well contrasted, are to be in their final form, ready for reproduction, pasted in the appropriate place in the text. Try to ensure that the size of the text in your figures is approximately the same size as the main text (10 point). Try to ensure that lines are no thinner than 0.25 point.

Design aspects of work that are to be analyzed are the layout, machinery and equipment, physical work environment, workers, work methods, and policy organizations. To get the actual condition of the object of research, using the 5 steps beginning from 10 steps Macroergonomics Analysis and Design (MEAD) [2]:

1) Examination of the environmental conditions and the organizational subsystem
2) Definition of the type of production system and setting performance expectations
3) Definition of unit operations and processes of work
4) Identification of variance

* Corresponding author: diniwahyuni2015@gmail.com

5) The making of the matrix variance

To identify the problems of work design used questionnaire. In this study, the use of questionnaires is a staple in data collection. The purpose of the subject matter of making the questionnaire is to obtain information relevant to achieving the objectives of the research by way of asking a question to respondents are selected.

The closed questionnaire was compiled by following the format of the scale Likert. Likert type format is designed to allow respondents to answer in many levels of any granule questions or statements filed. As for the Likert, in the scale of importance respondents to a question in the question form is classified as follows:

1) Strongly agree, abbreviated with the symbol (SS).
2) Agree, abbreviated with the symbol (S)
3) Neutral, abbreviated with the symbol (N)
4) Disagree, abbreviated with the symbol (TS)
5) Strongly Disagree, abbreviated with the symbol (STS)

Testing the validity of the questionnaire done by using correlation Product Moment with the formula:

$$r_{xy} = \frac{N\sum XY - (\sum X)(\sum Y)}{\sqrt{[N\sum X^2 - (\sum X)^2][N\sum Y^2 - (\sum Y)^2]}} \quad (1)$$

Where: r_{xy} is a coefficient of product moment, X is the number of answers the entire respondents per question, Y is the number of answers all the questions per respondent, and N is the number of respondents.

As for testing the validity of measures is as follows:

1. Establish a research hypothesis (H_o: the questionnaire is an instrument that is valid and can be used in data collection; H_1: the questionnaire is not a valid instrument)
2. Assign the significance level $\alpha = 0.05$
3. Determine the critical area of product moment with $\alpha = 0.05$ and N = 25, namely $r_{tabel} = 0.381$
4. Calculate the product moment with the formula (1)
5. In conclusion, set accept H_o if $r_{hitung} > r_{tabel}$.

Questionnaire reliability testing done using Cronbach Alpha Coefficient in questions that have a score in the range of 1 to 5. The formula used in calculating Cronbach Alpha coefficient: [8,9]

$$r_{11} = \left[\frac{k}{(k-1)}\right]\left(\frac{\sum \sigma_b^2}{\sigma_t^2}\right) \quad (2)$$

Where: r_{11} is the reliability of the instruments, k is the number of questions in the grain instrument, $\sum \sigma_b^2$ is the sum of the variance points questions, and σ_t^2 is the variance in total.

Collecting data instrument reliability is said to have indicated high reliability or if test Alpha Cronbach a coefficient giving bigger than r_{kritis}

3 Result and Discussion

3.1.The Actual Conditions of the Working System

The spread of the open questionnaire aims to find out the condition of the working system in one of the rubber processing plant. Conditions that are known to want to include some aspects of compliance with research variables are used, namely: machinery and equipment, the condition of the physical work environment, layouts, methods of work, organization, and policy workers.

Details of the open questionnaire on question formulated based on aspects of the work design and include 3 initial steps on the methods which have been presented.

The entire answers of respondents against the grain of the open questionnaire questions are then summarized in a table and presented in the form of recapitulation answers accompanied by the percentage of each answer. Recap of the open questionnaire answers can be seen in Table 1.

Table 1. Recap Of The Open Questionnaire

No.	Questions	Answers of the Respondents	
		Description	Percentage
1.	The difficulties workers in operating the machine.	Care	52%
		Repair	32%
		Damage	8%
		Others	8%
2.	The difficulties of workers in the use of the equipment.	Care	40%
		Storage	28%
		Others	32%
3.	Environmental factors that impede the work.	Hot	40%
		The Smell of	28%
		Dark	20%
		Others	12%
4.	The suitability of the layout work.	Not set	24%
		Space limited	20%
		Traffic materials	16%
		Others	40%
5.	The difficulties experienced by workers in implementing the SOP or work instruction.	Field conditions	40%
		Difficult to understand	28%
		Experience	28%

 https://doi.org/10.1051/e3sconf/20183106009

No.	Questions	Answers of the Respondents	
		Description	Percentage
		Others	4%
6.	The company's efforts to maintain the safety and occupational health.	APD Warning Briefing	52% 28% 20%
7.	The company's efforts to control the work.	Foreman Asis. Analyst Lab test	64% 28% 8%
8.	The difficulties in increasing workers' job skills.	Experience Job Difficult to understand	52% 28% 20%

3.2. Identification of Variance

The identification data of variance were conducted to classify variables existing problems to facilitate subsequent analysis process. Data variance is the problems that occur in the production process and obtained based on the recapitulation of the open questionnaire. Data is grouped in the tree variance factor of the problem can be seen in Figure 1.

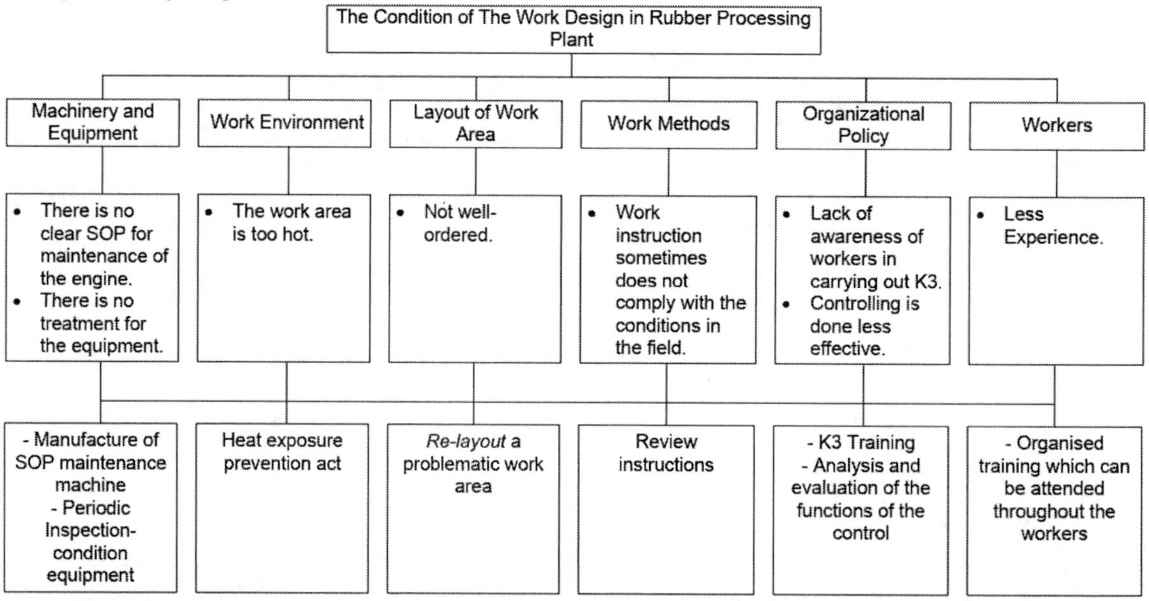

Fig. 1. Factor Tree Problems

3.3. Preparation of Matrix Variances

Data collection instruments used in the step of making matrix variance is a closed questionnaire. Matrix variances are arranged to illustrate the relationship between the variance. The process of preparing the matrix the variance using a Likert-scale measurement to look at the severity (severity) of a variance (Hendrick, 2002). The values in the matrix the variance is a score given to the respondent on a closed questionnaire which is then categorized according to their respective operations unit.

Dissemination of the questionnaire covered aims to know the perception of workers and assessment of the current system working conditions (current condition) as well as a comparison to achieve the desired system working conditions (desired condition). The closed questionnaire consists of aspects that correspond to variables such as research on the questionnaire. The closed questionnaire is composed of 8 rounds each of the questions using the Likert scale. The answer to "strongly agree" interpreted with nominal 5, "Agrees" with nominal 4, "neutral" with nominal 3, "did not agree" with nominal 2, and the last "strongly disagree" with nominal 1.

Validity and reliability testing was done on the closed questionnaire and the questionnaire results obtained valid and reliability with the value of each 0.8279 and 0.9516 so it can be used as an instrument of research. Matrix variance presented in tables frequency corresponds to the number of answers that appear in the recapitulation of the questionnaire. Matrix variance can be seen in Table 2. The variance is the variance key-variance that significantly impact on the performance of the process of providing and/or interact with other variances so that the effect is affecting more than one operating unit.

Table 2. Matrix Variances

Variances	Frequency of Answer				
	Strongly Disagree	Do Not Agree	Neutral	Agree	Strongly Agree
There is no fixed schedule maintenance machine	2	3	3	11	6
There is no maintenance for equipment	2	5	6	7	4
Work area is too hot	0	2	4	9	10
The layout is not set properly	2	7	10	5	1
Instructions do not match field conditions	1	6	8	7	3
Less awareness of maintaining work safety and health	0	5	10	9	1
Controlling is done less effective	4	5	7	9	0
Less experience	8	3	4	7	3

From the table above, the specified key variance with Pareto approach (just a little bit that becomes the key variance), which has a total answer to agree and strongly agree the greatest. Analysis of work design has 2 variance key that will be completed first, namely a work area to overheat, and the absence of engine maintenance schedule periodically.

References

1. Indonesian Directorate of General Plantation, *Statistics of General Plantation* (2015)
2. H. W. Hendrik, *Macro-ergonomics. Santa Monica: HFES*. (2001)
3. N. Stanton, *Handbook of Human Factor and Ergonomic Method. USA: CRC Press*. (2005)
4. A. Rositaningrum. *Analisa Implementasi Ergonomi Makro Terhadap Keuntungan Perusahaan (Studi Kasus : Merpati Maintenance Facility Juanda - Surabaya). ITS: Surabaya* (2014)
5. A.M. Grant, *Giving Time, Time After Time: Work Design AndSustained Employee Participation In Corporate Volunteering. The Wharton School: University Of Pennsylvania*. (2012)
6. S. Mayang,. *Evaluasi Tracer Study Untuk Pembelajaran dengan Pendekatan Ergonomi Makro. Serang: Universitas Sultan Ageng Tirtayasa.* (2013)
7. I. F. Mukti, *Desain Perbaikan Lingkungan Kerja Guna Mereduksi Paparan Panas Kerja Operator di PT. XY. Medan: Universitas Sumatera Utara*. (2013)
8. K. Parsons, *Human Thermal Environments: The Effect of Hot, Moderate and Cold Environment on Human Health, Comfort and Performance. Second Edition. London: Taylor & Francis*. (2003)
9. A. Auliciems, S. V. Szokolay. *Thermal Comfort.* Brisband.

https://doi.org/10.1051/e3sconf/20183106010

Identification of Soil Properties and Organophosphate Residues From Agricultural Land in Wanasari Sub-District, Brebes, Indonesia

Tri Joko[1,2,*], *Sutrisno* Anggoro[1,3], *Henna* Rya Sunoko[1,4], and *Savitri* Rachmawati[2]

[1]Doctorate Program of Environmental Science, School of Postgraduate Studies, Diponegoro University, Semarang –Indonesia
[2]Majoring of Environmental Health, Public Health Faculty, Diponegoro University, Semarang – Indonesia
[3]Faculty of Fisheries and Marine Science, Diponegoro University, Semarang – Indonesia
[4]Faculty of Medicine, Diponegoro University, Semarang – Indonesia

Abstract. Organophosphates have been used to eradicate pests and prevent losses from harvest failures caused by pest attack. It is undeniable that the organophosphate persist in soil. This study aims to identify the organophosphate residue and soil properties include pH, soil texture, and permeability. The soil samples were taken from cropland in 10 villages, Wanasari sub-district, Brebes, Indonesia. Organophosphate residue determined by gas chromatography using Flame Photometric Detector. Soil texture was determined by soil texture triangle from NRCS USDA, and the permeability value was determined by falling head method. The mean value of chlorpyrifos, profenofos, diazinon were 0.0078; 0.0388; 0.2271 mg/l respectively. The soil texture varies from clay, silt clay, loam, silt loam, and silt clay loam with permeability value at 10^{-7} with the soil pH value between 6.4 – 8.1. The results showed that organophosphate residues found in the soil and its potential affect the soil fertility decline. We recommend to conduct routine soil quality analysis to prevent soil damage in the agricultural environment.

1 Introduction

Agriculture is a potential sector in Indonesia. Total agricultural land in Indonesia reaches 8,114,829 ha in 2014. Central Java Province is the region with the second largest agricultural area after East Java province, with an area of 966,647 Ha. The area of agricultural land is also proportional to the production of food crops and horticulture in Indonesia. As a tropical country various types of plants can grow fertile, one of them is onion. Brebes Regency located in Central Java Province is a central area of onion. Onion production in Brebes Regency in 2014 reached 3,759,742 Kw, an average production of 121.46 Kw / Ha. Wanasari District as the region with the highest onion production in Brebes Regency is able to produce 1,025,680 Kw, the average production is 144.97 Kw / ha [1]. According to previous research conducted by Joko T, *et al* (2017) that onion farmers in Kecamatan Wanasari use large amounts of pesticides, and average to apply pesticides every three or four days. The use of pesticides in large numbers is due to onion plants are very vulnerable to pests, so farmers must apply insecticides to prevent damage to the plant. In his research, Joko T, *et al* (2017) showed that found farmers who use insecticides class of organophosphates with chlorpyrifos active ingredients [2].

Large and massive applications of organophosphate pesticides can cause adverse effects such as environmental pollution. The results of previous studies conducted Gafar *et al.* (2012) also showed that there was a decrease in soil parameters tested during addition of organophosphate pesticides. This research was conducted to find out the residue of organophosphate pesticide in soil and to know soil properties assessed through pH parameter, permeability, and soil texture [3].

2 Material and methods

2.1 Organophosphorus residue test method

The principle of testing the residual organophosphate is to use extraction with acetone. The extract was evaporated until it was almost dry and the residue was dissolved in n-Hexane. Generally no cleaning is required, and set by gas chromatography using an FPD detector. Reactors used are Acetone, Sodium Sulphate, n-Hexan, and Celiite 545. Equipment used is shaker, evaporator, gohch cup, round flask, gourd flask, erlenmayer, and gas chromatograph equipped with FPD detector. The extraction procedure is carried out by weighing 50 grams of soil samples, inserting into the Erlenmeyer covered sharpening, then adding 100 ml of acetone. The extracted results were filtered using a goblet plate that had been given celitte 545 pulp, and the sieve was then dried using a evaporator (evaporated). The residue present in the circular flask is diluted using n-Hexane as much as 8 ml. To remove the water content

* Corresponding author: trijokoundip@gmail.com

https://doi.org/10.1051/e3sconf/20183106010

in the sample, insert ± 2 g of anhydrous Sodium Sulphate into a round flask. Next n- Hexane is poured into test tube and implements up to 10 ml volume. Determination of residue is done by injecting 1-2 µl of extract into gas chromatograph. The calculation is performed by comparing the mooring time and the height or peak area of the chromatogram obtained from the sample solution and the reference standard solution.

2.2 Permeability determination method

Standart Procedure / Method used measurements with changing voltages (Variable Falling Head). The equipment used is glass tube pipette, silent 1.47 cm with an area of 1.696 cm2, brass tube samples, weight = 525 gr, diameter = 6, 36 cm and height 9, 60 cm, glass bottom tube weighing 175 gr, 5 , 9 cm and 9.85 cm. A brief description of the implementation of the sample is printed with a tube ring and weighed, tube rings containing soil samples soaked for 24 hours, tube rings mounted on permeability devices (top and bottom of samples given sample paper), after the sample is assembled. And it is observed whether it has entered the entire sample and is closed by the bottom, and the reading begins after positioning the water level in a glass guide glass zero (0) and the tap is opened. The results of the investigation were analyzed using laboratory seepage (k) values in cm/second.

2.3 Soil texture determination method and pH

The method used is hydrometer. Equipment used is measuring cup 1000 ml diameter 6.5 cm, container hydrometer, thermometer, mixer, a set filter, scales, oven, stopwatch. The test procedure was carried out by wasting soil samples with dry weight ± 50 grams with 100 mg of aquades and 25 mm waterglass. Stirred until smooth and soaked ± 24 hours, after 24 hours the mixture is dimixed ± 15 minutes then transferred in a measuring cup and added aquades until the mixture becomes 1000 ml. The mouth of the tube is tightly closed with the palm of the hand, then shake in the horizontal direction ± 1 minute after being shaken the tube carefully and stopwatch pressed. The scale number is read on ½, 1, and 2 then carefully lifted the float, washed with distilled water and inserted in a tube containing distilled water. The hydrometer buoy is inserted, the hydrometer is read at 5, 15, 20 minutes, 1.4 and 24 hours after each reading is washed and returned the hydrometer buoy into distilled water, the mixed temperature is measured once in the first 15 minutes then each subsequent reading, after reading Last transferred to filter no. 200 was washed until the washing water was clear and left the water flowing out. Items left on the filter no. 200 in dry, filtered and weighed. The results obtained in the form of graphs in the logarithmic scale show the percentage of fineness of the grains on the vertical axis with the grain of the soil on the horizontal axis.

3 Result and discussion

Organophosphate insecticides are hydrocarbon compounds containing one or more phosphate atoms and these phosphates have relatively short residence times in biological systems. Many organophosphate pesticides disappear from the soil within weeks after application [4]. If in this study found several types of organophosphate residues, then this shows the truth that farmers sprayed pesticides organophosphate continuously in amounts that exceed the recommended limits.

Organophosphate residues are found almost in all sample sites, except in the Pebatan village. Chlorpyrifos as a group of organophosphates is an insecticide which is found almost in all sampling sites. This is consistent with the results of this study showed that chlorpyrifos are the most widely used pesticides by farmers. Chlorpyrifos residue with the highest level of 0.0630 ppm in Tanjungsari village, while the lowest residual content of 0.0014 ppm was found in Klampok village (Figure 1, Table 1).

Other organophosphate group pesticides found were diazinon with levels of 0.0388 ppm in Sigentong village. Organophosphate residues contain active metidation with levels of 0.0140 ppm was found in the village of Sisalam. The organophosphate residue of profenofos was found in the village of Sigentong with levels of 0.1829 ppm and in the village of Klampok of 0.2714 ppm. The residual organophosphate type of malathion was found in four villages, namely Tanjungsari village with residual content of 0.1370, Sisalam village with content of 0.1450 ppm, Dukuhwringin with residual malathion 0,3630 ppm, and Wanasari with residual content of 0.1490 ppm (Table 1).

Based on the result of organophosphate residue test on soil sample can be known pattern of insecticide usage that farmer tends to use 1-3 kinds of certain insecticide. Farmers in Tanjungsari village, Dukuhwringin, Wanasari apply insecticide type of malation and chlorpyrifos, Sisalam village apply insecticide type metidation, malation, and chlorpyrifos. Farmers in the village of Sigentong use kind of diazinon and chlorpyrifos. While the village of Jagalempeni, Pesantunan, Klampok, Siasem, Siwungkuk, and Tegalgandu applying the type clorpyrifos on their onion farm (Table 1).

Clorpyrifos and Diazinon are phosphorothioates compounds that have greater hydrolylic stability under aqueous conditions. Diazinon is a more active insecticide. Malation is a compound of phosphorodithioates [4].

In this study also measured the value of soil pH, soil permeability, and soil texture at the sampling site. Laboratory test results showed that soil pH values ranged from 6.4 to 8.1 with the lowest pH value of 6.4 in soil samples in Sigentong and Pesantunan villages, then the highest pH value was 8.1 in the soil samples in Tanjungsari village. The soil permeability across the sample ranged from 3.848 x 10^{-7} to 9.240 x 10^{-7}. The soil permeability value <10-6 is clay type with practically impervious properties (impermeable) (Table 1).

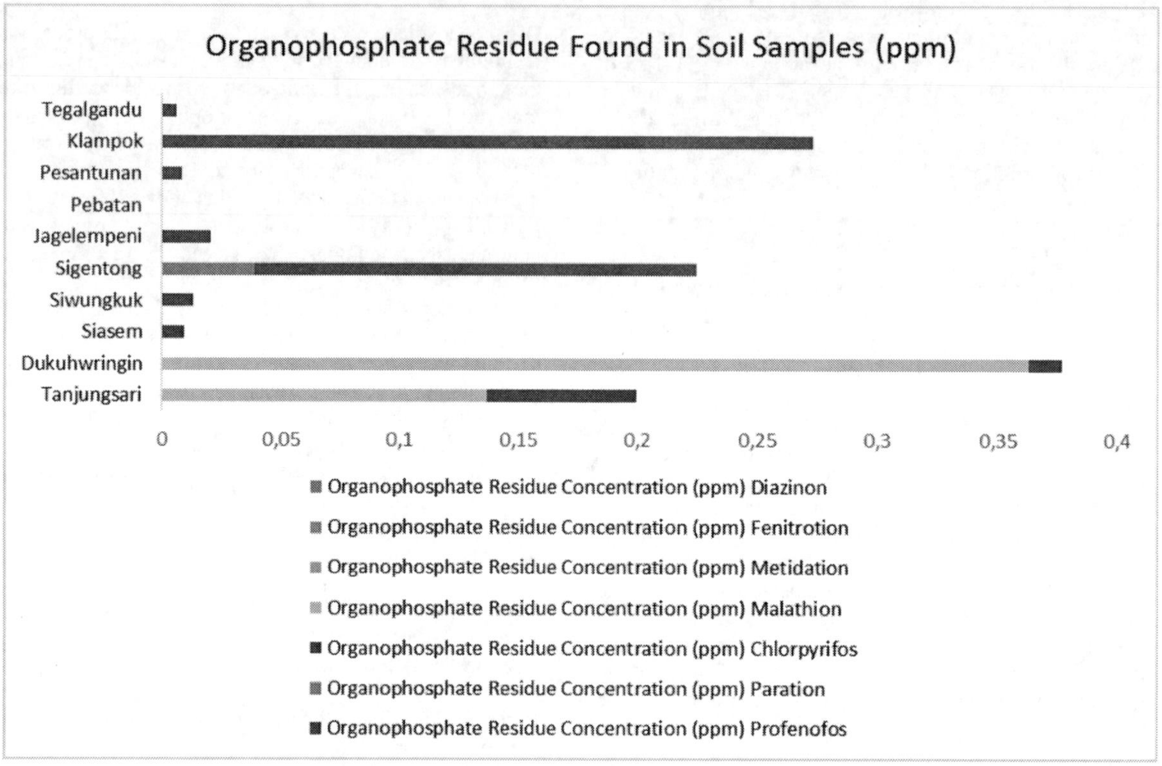

Fig. 1. Graphic of Organophosphate Residue in Distributed Sampling

Table 1 . Soil Properties and Organophosphate Residue Level

No	Sampling Location	Organophosphate Residue Concentration (ppm)							pH H_2O	Soil Permeability (cm/s)	Soil Texture
		Diazinon	Fenitrotion	Metidation	Malathion	Chlorpyri-fos	Paration	Profenofos			
1.	Tanjungsari	<LOD	<LOD	<LOD	0,1370	0,0630	<LOD	<LOD	8,1	7,708 x 10^{-7}	Clay
2.	Dukuhwringin	<LOD	<LOD	<LOD	0,3630	0,0140	<LOD	<LOD	6,7	7,920 x 10^{-7}	Silt loam
3.	Siasem	<LOD	<LOD	<LOD	<LOD	0,0091	<LOD	<LOD	7,3	3,848 x 10^{-7}	Silt clay
4.	Siwungkuk	<LOD	<LOD	<LOD	<LOD	0,0130	<LOD	<LOD	7,6	3,848 x 10^{-7}	Silt loam
5.	Sigentong	0,0388	<LOD	<LOD	<LOD	0,0033	<LOD	0,1829	6,4	4,613 x 10^{-7}	Sandy clay loam
6.	Jagelempeni	<LOD	<LOD	<LOD	<LOD	0,0202	<LOD	<LOD	7,0	3,954 x 10^{-7}	Silt clay
7.	Pebatan	<LOD	<LOD	<LOD	<LOD	<LOD	<LOD	<LOD	7,5	3,848 x 10^{-7}	Silt clay loam
8.	Pesantunan	<LOD	<LOD	<LOD	<LOD	0,0084	<LOD	<LOD	6,4	3,954 x 10^{-7}	Loam
9.	Klampok	<LOD	<LOD	<LOD	<LOD	0,0014	<LOD	0,2714	7,8	9,240 x 10^{-7}	Dust
10.	Tegalgandu	<LOD	<LOD	<LOD	<LOD	0,0061	<LOD	<LOD	7,2	7,708 x 10^{-7}	Silt clay loam

The result showed that residue with the highest level of Malathion (0, 3630 ppm) in Dukuhwringin village was found in soil pH 6.7 and silt loam soil type (Table 1). J H. Huddleston (1996) has mentioned that there are several factors causing pesticide contamination on soil properties of the soil, properties of the pesticides, hydraulic loading on the soil, and crop management practise [5]. Land that has a property to transmit pesticides is quickly referred to as sensitive soils. Factors that affect the sensitivity of soil properties are permeability, water table conditions, organic matter content, and clay content. In this case, soil permeability can control leaching potential. Lands that have high leaching potential tend to be more sensitive than those with low leaching potential. While the organic matter and clay content can control the potential sorption of the soil. Soil that has low sorption potential will be more sensitive to contamination compared to soils that have high sorption potential.

Soils that have a very soft texture such as clay have very small pores and have very low permeability. While the soil with loam texture, silt loams, and clay loam have intermediate rates of soil permeability. Clay content has very small and flat particles that have a large surface area. This condition makes the surface of the soil is chemical reactive. The soil that has clay cotent is higher then the soil has a number of binding to the pesticide [5]. The soil sample in Tanjungsari village is clay type, and found two types of pesticide namely malathion and chlorpyrifos. Clay soil type is a low sensitivities soil because this soil is a group of slow or very slor permeabilities (7,708 x 10^{-7} cm/s in this sampling location) and high sorption potentials. The soil sample in Dukuhwringin village also found two types of pesticide malathion and chlorpyrifos on the silt loam type of soil, this type of soil is a group of moderate sensitivities, because it has slow permeabilities and relatively high sorption potentials. So also with the location of sampling which type of soil silt clay like in the village of Siasem and Jagalempeni.

Research conducted by Fosu-Mensah *et al* also found residuals of chlorpyrifos and profenofos in soil samples on cultivated cocoa farms in Ghana [6]. Supriyadi *et al* also conducted a study to determine the presence of organophosphate residues in agricultural land located in mojogedang, karanganyar central java province. The results show the residue of profenofos found on horticultural farms. In his research, also found soil texture and pH is the nature of soil that correlation is close to the residue of organophosphate in the soil. PH H_2O in horticultural soil sample is 5.7 and sandy clay type of soil. In this study, the soil pH of the ten test samples ranged from 6.4 to 8.1 [7].

In this study found residual chlorpyrifos found almost in all sampling sites, unlike other types of pesticides are only found in one or two sampling sites only. The use of chlorpyrifos pesticides can affect the soil quality. It has been found that the use of chlorpyrifos can reduce the bacteria reinforcing aerob dinitrogen and dinitrogen fixation. Chlorpyrifos also significantly decreased the activity of acids and alkaline phosphatase and dehydrogenase, but the condition recovered after 14 days without the use of chlorpyrifos [8]. Farmers in Kecamatan Wanasari use intensive pesticides in large quantities, especially chlorpyrifos. It is also evident in almost all sampling sites found residual chlorpyrifos.

Based on E. P. Lichtenstein and Schulz (2014) that most insecticides are in dry soil, whereas insecticides are not present in high water soils [9]. The results of laboratory studies showing the degradation of chlorpyrifos insecticides have been done by Yucel *et al.* The results showed that for a 50% reduction in chlorpyrifos insecticide levels it took 10 days. This reason also shows that after application of insecticide chlorpyrifos as well as other types of organophosphates make residues still found in the soil as in the results of this study [10].

3 Conclusion

Soil texture was determined by soil texture triangle from NRCS USDA, and the permeability value was determined by falling head method. The mean value of chlorpyrifos, profenofos, diazinon were 0.0078; 0.0388; 0.2271 mg/l respectively. The soil texture varies from clay, silt clay, loam, silt loam, and silt clay loam with permeability value at 10^{-7} with the soil pH value between 6.4–8.1. The results showed that organophosphate residues found in the soil and its potential affect the soil fertility decline. We recommend to conduct routine soil quality analysis to prevent soil damage in the agricultural environment.

References

1 BPS. 2014. *Luas Lahan Sawah Menurut Provinsi (ha)*, 2003–2014. [online] https://www.bps.go.id/linkTableDinamis/view/id/895.Accessed on July, 18 2017

2. T. Joko, S. Anggoro, H. R. Sunoko, and S. Rachmawati, Applied and Environmental Soil Science, (2017). Article ID 5896191

3. M.O. Gafar, Journal of Agricultural Science and Soil Science (ISSN.2251- 0044) vol. **2**(2) pp058-062-2012, (2012).

4. S. U. Khan, *Pesticides in the soil environment.* Elsevier, (2016)

5. J.H. Huddleston. 1996. *How soil properties affect groundwater vulnerability to pesticide contamination. EM 8599.* Oregon State University.

6. B. Y. Fosu-Mensah, E. D. Okoffo, G. Darko, and C. Gordon, Environmental Systems Research, **5** (1), 10, (2016)

7. S. Supriyadi, A. D. Utami, H. Widijanto, and S. Sumani, Modern Applied Science, **9** (6), 87, (2015)

8. C. Pozo, M. V. Martinez, Toledo, V. Salmeron, B. Rodelas, and J. Gonzalez Lopez, Environmental Toxicology and Chemistry, **14** (2), 187-192, (1995)

9. E. P. Lichtenstein, and K. R. Schulz, Journal of Economic Entomology, 57 (5), 618-627, (2014)
10. U. Yucel, M. Ylim, K. Gozek, C. S. Helling, and Y. Sarýkaya, Journal of Environmental Science & Health Part B, 34(1), 75-95, (1999)

Hazard Identification and Risk Assessment in Water Treatment Plant considering Environmental Health and Safety Practice

Fajrul Falakh[1*], *Onny* Setiani[2]

[1]Master Program of Environmental Science, School of Postgraduate Studies, Diponegoro University, Semarang-Indonesia
[2] Departement of Environmental Health, Faculty of Public Health, Diponegoro University, Semarang - Indonesia

Abstract. Water Treatment Plant (WTP) is an important infrastructure to ensure human health and the environment. In its development, aspects of environmental safety and health are of concern. This paper case study was conducted at the Water Treatment Plant Company in Semarang, Central Java, Indonesia. Hazard identification and risk assessment is one part of the occupational safety and health program at the risk management stage. The purpose of this study was to identify potential hazards using hazard identification methods and risk assessment methods. Risk assessment is done using criteria of severity and probability of accident. The results obtained from this risk assessment are 22 potential hazards present in the water purification process. Extreme categories that exist in the risk assessment are leakage of chlorine and industrial fires. Chlorine and fire leakage gets the highest value because its impact threatens many things, such as industrial disasters that could endanger human life and the environment. Control measures undertaken to avoid potential hazards are to apply the use of personal protective equipment, but management will also be better managed in accordance with hazard control hazards, occupational safety and health programs such as issuing work permits, emergency response training is required, Very useful in overcoming potential hazards that have been determined.

1 Introduction

Occupational safety and health is heavily influenced by occupational hazards identified and managed in a competent risk assessment process. Hazards in the workplace can be physical, chemical or psychological and can lead to workplace incidents and work-related injuries, which have an impact on organizational productivity and profitability. [1]

Hazard Identification Risk Assessment (HIRA) is a method for determining and providing hazards based on their probability, frequency and severity and evaluating adverse consequences, including potential loss and injury. The work process in the industry must pay attention to aspects of environmental health and safety in order to support the effectiveness of the industry. The industry must identify hazards, assess the associated risks to tolerate continuous levels, risk assessments have been made using risk guidelines and standards. [2][3]

Water Treatment Plant (WTP) is an important infrastructure to guarantee human and environment health. As water supply, they have a key role in giving healthy clean water access to the society. In the progress of this, environmental and health safety aspect becomes things to be concerned about. [4] The high hazard impacts in workplace often become the cause of work accident and occupational disease.

Working in the field of water treatment is considered dangerous, especially as it can lead to frequent deaths in confined spaces. Occupational safety and health is not particularly noticed in this area, many decision makers consider it to be somewhat less dangerous at the moment, but processing workers are still experiencing the possibility of health problems and deaths, especially exposure to chemicals as materials for water purification. [4][5].

Water treatment companies use machines and equipment that are likely to cause injury to workers. Accidents that occur can be caused by negligence of workers when operating machinery and equipment or unsafe working environment conditions. Potential hazards that often occur are defects in operation, exposure to chemicals and work fatigue.

The study was conducted using the HIRA (Hazard Identification and Risk Assessment) method to identify potential hazards found in the workplace. By identifying potential hazards and work risks, it is expected to facilitate the company in the management and control of safety in the workplace and minimize the possibility of accidents.

2. Literature review

Hazard Identification is a proactive process to identify hazards and eliminate or minimize/reduce the risk of injury/illness to workers and damage to property, equipment and the environment. It also allows us to show our commitment and due diligence to a healthy and

* Corresponding author: fajrulfalakh.hse@gmail.com

safe workplace. We must identify hazards and potential hazards in the workplace in order to be able to take action to eliminate or control them. [8]

Table 1. Description of Likelihood Level

Level	Likelihood	Expected or actual frequency experienced
1	Rare	May only occur in exceptional circumstances; simple process; no previous incidence of non-compliance
2	Unlikely	Could occur at some time; less than 25% chance of occurring; non-complex process &/or existence of checks and balances
3	Possible	Might occur at some time; 25 – 50% chance of occurring; previous audits/reports indicate non-compliance; complex process with extensive checks & balances; impacting factors outside control of organisation
4	Likely	Will probably occur in most circumstances; 50-75% chance of occurring; complex process with some checks & balances; impacting factors outside control of organisation
5	Almost certain	Can be expected to occur in most circumstances; more than 75% chance of occurring; complex process with minimal checks & balances; impacting factors outside control of organisation

This is a step by step process to guide responsible persons to an effective hazard identification, assessment and controls system. The steps include:

- Hazard Assessment: identifying the hazards and potential hazards, determining the risks and the risk designation (rating) associated to the hazard based on: Likelihood and severity
- Hazard control - controlling the hazards and the risks associated with the hazard
- Providing information, education, training and supervision on the hazards, risks and controls for employees affected by the hazards
- Review of the hazard assessment and control process

Table 2. Risk Assessment Matrix

Likelihood	5	5	10	15	20	25
	4	4	8	12	16	20
	3	3	6	9	12	15
	2	2	4	6	8	10
	1	1	2	3	4	5
		1	2	3	4	5
		Severity				

Extreme	: 15-25
High Risk	: 8-12
Medium Risk	: 4-6
Low Risk	: 1-3

Table 3. Description of Severity Level

Level	Description of Severity
1	• Minor onsite injuries (first aid and nondisabling, reportable injuries). • Property damage less than base level amount * • Minor environmental impact (no remediation). • Loss of production less than base level amount * • No offsite impact or damage. No public concern or media interest.
2	• Serious onsite injuries (temporary disabling worker injuries). • damage from 1 to 20 times base level. • Moderate environmental impact (cleanup or remediation in less than 1 week and no lasting impact on food chain, terrestrial or aquatic life). • Loss of production from 1 to 20 times base level • Minor offsite impact (public nuisance—noise, smoke, odor, traffic). • Potential adverse public reaction. Some media awareness.
3	• Permanent disabling onsite injuries or possible fatality. • Property damage from 20 to 50 times base level. • Significant environmental impact (cleanup or remediation less than 1 month and minor impact on food chain, terrestrial or aquatic life). • Loss of production from 20 to 50 times base level. • Moderate offsite impact limited to property damage, minor health effects to the public or first aid injuries. • Adverse public reaction. Local media concern.
4	• Onsite fatality or less than four permanent disabling worker injuries. • Property damage from 50 to 200 times base level. • Serious environmental impact (cleanup or remediation requires 3–6 months and moderate impact on food chain, terrestrial and/or aquatic life). • Loss of production from 50 to 200 times base level. • Significant offsite impact property damage, short-term health effects to the public or temporary disabling injuries. • Significant public concern or reaction. National media concern.
5	• Multiple onsite fatalities or four or more permanent disabling onsite injuries. • Property damage greater than 200 times base level.

	• Extensive environmental impact (cleanup or remediation exceeding 6 months, significant loss of terestrial and aquatic life or damage to food chain uncertain). • Loss of production greater than 200 times base level • Severe offsite impact property damage, offsite fatality, long-term health effect, or disabling injuries. • Severe adverse public reaction threatening facility continued operations. International media concern

During the interview process, worker and safety officer is given a table containing the scale / category of likelihood and severity, so that the worker and the safety contractor can know and determine for themselves the category level of likelihood and severity. The value scale for likelihood is 1-5 ranging from an unlikely to almost certain level of probability.

While the value scale for severity is 1-5 ranging from insignificant to catastrophic severity. After the value of relative risk obtained then analyzed using Risk Assessment Matrix table. [7]

Risk is a measurement to analyze and evaluate the hazard. The measurement is made by identification on how severe and when likely of the hazard. In other words, the risk assessment is an in-depth look to specify situations, process and other harmful activities or hazard at workplace. [8]

Rating the hazard is one way to help determine which hazard is the most serious and thus which hazard to control first. Priority is usually established by taking into account the severity and Likelihood. By assigning a priority to the hazard, you are creating a rating or an action list.

The following factors play an important role:

- Severity of exposure - impact when exposed to the hazard.
- Likelihood - that an incident will occur when exposed to the hazard.

When the hazard is identified, determine the controls which are already in place to ensure this information is taken into account when assigning a risk designation.

Risk is presented in variety of ways to communicate the distribution of the risk throughout a plant and area in a workplace. The results of risk assessment that presented in a risk matrix are essential to make decision on risk control. Risk can be calculated using the following formula:

Risk (R) = Likelihood (L) x Severity (S). [7]

The phase of risk identification is essential, because it puts the bases of the risk analysis. Indeed, the data of risk identification will be the input of the evaluation, Therefore it is necessary to make an identification phase in an exhaustive way, to obtain the best results. [9]

4 Methodologies

The type and design of this research based on time research is cross sectional because the process of collecting data and observation of the variables done at once or at one particular time. While in terms of place, this research includes field research, because the research conducted and the way researchers in getting the data is directly plunged into the field by conducting interviews and observation

When viewed from the way of data collection, this study is observational because researchers obtain data through observations and interviews to workers and related parties in the company. In addition, the objects in this study were not treated during the course of the observational / observational study. Based on the nature of the problem and its data analysis, this research is included in descriptive research because this research does not make comparisons or connections between variables. This study describes a situation objectively.

Variables to be studied in this research are hazard identification, risk assessment, and risk level determination on water installation process. The data collected in this study there are two types of primary data and secondary data. Primary data obtained through observation and interview. These observations and interviews are used to determine the potential water hazard clearance process, the magnitude of occupational risks and the working environment.

While the secondary data collected is a general description of the company, the work procedure, the number of workers, tools and hazardous materials in the water treatment process and accident control efforts that have been done.

Processing techniques and data analysis conducted based on observation and interview data. Based on the results of observation and interviews are known potential hazard and value. Identification of potential risk hazards in the water treatment plant will be very effective if done on the basis of the factual conditions of the workplace and existing work processes, this is an effort that can be done so that industrial health and environmental health programs can be done well in accordance with policies and Regulations that have been set.

4. Result

Risk Assessment is performed using the Risk Matrix as described in the literature study, the results obtained from this risk assessment are the 22 potential hazards present in the water purification process, these findings are based on assessments of workshops and processing units at subsequent water treatment plants Described in detail in table 4. According the existing categories of extreme risk, high risk, medium risk and low risk then the findings are grouped into each risk category.

Extreme categories that exist in the risk assessment are chlorine leak and industrial fires. Leakage of chlorine and fires get the highest value because their impact threatens many things, such as industrial disasters that can harm human life and the environment.

Hazards Chlorine can be absorbed through the skin and cause burns ranging from mild to severe depending on the length of the contact In addition chlorine can also be absorbed through the eye, causing burning or discomfort, irregular blinking, unconscious closure of

eyelids, redness, and tearing. Large amounts of chlorine in the air can cause severe burns, pain, and blurred eyesight.
Therefore workers in storage must conduct a well-scheduled inspection so that the presence of chlorine can be safely maintained, besides that it is also necessary to have an emergency management control system that refers to leakage of chlorine and industrial fires, Prevention efforts from known potential hazards.

Table. 4 Hazard Identification Risk Assessment in Water Treatment Plant

No	Workshop or Treatment unit	Hazard / Hazardous Situation	Potential Risk	Consequence	Risk Assessment Matrix		
					L	S	Risk Value
1	Flow Meter Chamber	Possible entrance and entering the flow meter chamber	Chlorine inhalation by operator working inside the chamber	Lost-time accident up to fatality due to chlorine inhalation	3	2	6 (medium Risk)
2	Demolition in Chemical Building	Work at height (8m) for demolishing walls and floors	Falling from a height of approximate 8 meter to the ground	Permanent injury up to fatality	2	4	8 (High Risk)
3	Control Room	Electrical Hazard	Electric short circuit	fractures, Fatality, Disaster	3	5	15 (Extreme Risk)
4	Process of Treatment	Cleaning accumulation sludge in channel raw water inlet of accerator 1,2 at once a month	harmful atmosphere, difficulty of entry/exit access	Fatality accident more than 1 person	2	5	10 (High Risk)
5	Process of treatment	Cleaning once a month sludge extraction	All confined space risks, including fall, electrical shock	Fatality accident	2	4	8 (High Risk)
6	Process of treatment	Working at height for routine operating main drain valve, noise, smell	Falling, slippery	Fatality accident	1	4	4 (Medium Risk)
7	Process of treatment	Working/cleaning over compartment	Falling, drowning	Fatality accident	1	4	4 (Medium Risk)
8	Water treatment line	Working at height around sand filter	Drowning, Falling	Concussion, fracture	3	2	6 (Medium Risk)
9	Backwash pump room	Rotating part, noise, slippery at backwash pump	Injury at arm or hand in Projection of loose bolt, noise, fall	Arm amputation, fracture, hearing disorders, concussion	3	3	9 (High Risk)
10	Gear box of the turbine	Exposed rotating parts	Injury at arm or hand in Projection of loose bolt	Arm amputation, shut down machine, fracture	2	3	6 (Medium Risk)
11	Chlorine Facilities	Crash inside the site	Personnel crashed by chlorine vehicle	Fatality accident	3	4	12 (High Risk)
12	Chlorine Facilities	Chlorine Leakage	Inhalation of chlorine gas	Fatality accident, Disaster	3	5	12 (Extreme Risk)
13	Chlorine Facilities	Falling container when loading/unloading	Impacted by falling heavy objects (weight: +1.8 ton, height: 1.5 m)	Fracture/irreversible harm	2	2	4 (Medium Risk)
14	Purchasing	Wrong/miss specification when purchased devices, equipments, tools or materials and services	Use of improper devices, equipments, tools, materials	Stop production, fatality	5	1	5 (Medium Risk)
15	Filter gallery	Working at height	Falling at a height (6 M)	Fatality accident	2	4	8 (High Risk)

16	Laboratory	Exposure to UV radiation (The device has a UV germicidal lamp with a wavelength of 257.7 nm, including UV-C type)	Erythema (damage / disorder) on the skin and damage the cornea of the eye	Skin disorders or skin and cataracts in the eyes for the long term	4	2	8 (High Risk)
17	Reservoir	Confined space hazard	Inadequate O2 content, high toxic gas content, leakage current on electrical	Fainting, poisoning, death	2	4	8 (High Risk)
18	Acid material	Strong acid vapor in the laboratory	Inhalation, skin contact	Respiratory disorders, Skin sores	3	1	3 (Low Risk)
19	Storrage & dosing system H2SO4	Exposure to H2SO4	Steam H2SO4 strong acid operation of H2SO4 dosing pump for water neutralization	Blind, burns, irritation	2	4	8 (High Risk)
20	Submersible Pump	Electrical	Electric caution, Fire	Disaster, Injury	1	5	5 (Medium Risk)
21	Storage of hazardous chemical waste	Waste spills	Hazardous chemical waste spills at the time of pouring, the occurrence of waste leakage	Environmental pollution, poisoning, fainting	3	2	6 (Medium Risk)
22	Storage Poly Aluminum Chloride	Poly Aluminum Chloride	Leaks on pipe connections and ball valves	Irritation, impaired vision, indigestion	3	2	6 (Medium Risk)

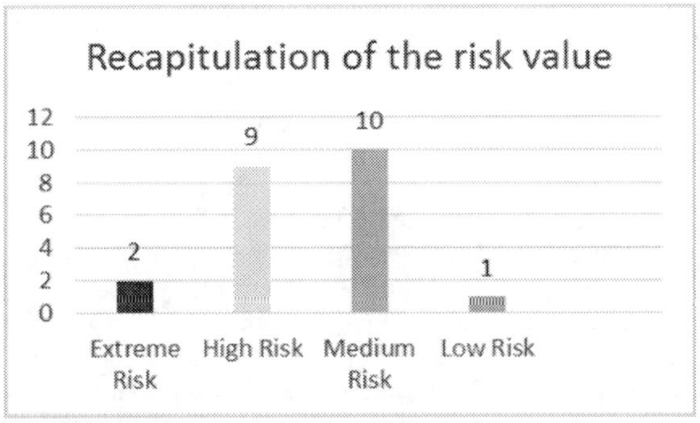

Fig 1. Recapitulation of the risk value

4. Conclusion

Potential occupational hazards in the water treatment industry are exposed to chemicals to workers, the potential danger of leaking chlorine gas can also greatly affect the safety and health of the industrial environment. Control measures undertaken to avoid potential hazards are to apply the use of personal protective equipment, but management will also be better managed in accordance with hazard control hazards, occupational safety and health programs such as issuing work permits, emergency response training is required, Very useful in overcoming potential hazards that have been determined.

References

1. S. T. Bahn, (pp. 1-9). Gold Coast, Queensland. Griffith University. (2012)
2. Islam, Tanveer; Ryan, Jeffrey R. *Butterworth-Heinemann is an imprint of Elsevier* (2016)
3. R. Ramesh, M. Prabu, S. Magibalan, P. Senthilkumar, *International Journal of ChemTech Research*. (2017)
4. N. J. Brown, Ithaca, NY: *Cornell University, Chemical Hazard Information Program.* (2016)
5. A. Vantarakis, S. Paparrodopoulos, P. Kokkinos, G. Vantarakis, K. Fragou, and I. Detorakis. Journal of Environmental and Public Health Volume **2016** Article **ID 8467023** (2016)
6. AY. Ambarani, *The Indonesian Journal of Occupational Safety and Health* (2017)
7. S. Ruchi, P. Praveen, International Journal of Engineering Research & Technology. Vol **3** e-ISSN: 2278-0181 (2014)
8. Capital Health. *Hazard Assessment Control Form.* www.cdha.nshealth.ca/.../hazard-assessment-program (Accesssed 2 July 2017)
9. European Agency for Safety and Health at Work, *Safety and health at work is everyone's concern*, Numberg, Germany, (2007).
10. SJ. Moja, CS. Van Zuydam, Mphephu. J Geogr Nat Disast **S6**: 006 (2016)

Indoor Air Pollution in Non Ac Passenger Bus

Iksiroh El Husna[1,2*], *Yan El* Rizal Unzilatirrizqi D[3], *Yudi* Karyanto[3] , and *Henna* R Sunoko[1,4]

[1]Doctoral Program of Environmental Science, School of Postgraduate Studies, Diponegoro University, Semarang-Indonesia
[2]Nautical Departement, Merchant Marine Polytechnic Semarang, Jingosari 2A, Semarang, Indonesia
[3]Automotive Engineering, Polytecnic of Road Safety Transport , Jl. Semeru No. 3, 50125,Tegal, Indonesia
[4]Faculty of Medicine, Diponegoro University, Jl. Prof Sudharto SH, 5027, Semarang, Indonesia

Abstract. Passenger buses have been one of favorite means of transportation in Indonesia due to its affordability and flexibility. Intensity of human activities during the trip in the buses have a potential of causing indoor air pollution (*polusi udara dalam ruang*; PUDR). The indoor air pollution has an impact of 1000-time bigger than outdoor air pollution (polusi udara luar ruang; PULR) on lung. This study aimed to find out indoor air pollution rate of non air conditioned buses using an approach to biological agent pollutant source. The study applied an analysis restricted to microorganisms persistence as one of the sources of the indoor air pollution. The media were placed in different parts of the non AC buses. This study revealed that fungs were found in the non AC buses. They became contaminants and developed pathogenic bacteria that caused air pollution.

1 Introduction

Passenger buses have been one of favorite means of transportation in Indonesia due to its affordability and flexibility. Buses are capable of transporting large number of passengers and ranging varied distances during the trip. However, few studies have been undergone concerning bus-borne indoor air pollution. [1] Indoor air pollution is derived from many sources, including building materials, biological agent, and even humans who occupy the room. Pollution cause by externalities brought in the room may take the forms of soil-borne gases. These pollution sources are derived from burning, evaporation, biological agent, and radon. Burned materials come from gasoline and wooden stoves, cigarettes, chimneys, as well as room heater. Biological agents can be produced by organisms in domesticated animals and human. [2] Such criteria are relevant to the incidence of the indoor air pollution in the buses that transport humans.

1.1 Indoor Air Pollution

Indoor air quality in a particular room is a scientific aspect that focuses on the air quality in particular room, which is to be transferred to the room or building occupied by humans. Indoor air quality can be defined as a result of interaction between place, temperature, building system, construction technique, contaminant source, building equipment, humidity, process, and activity within the room as well as other sources of externalities and workers. [3].

Studies in the United States and Europe proved that individuals spent most times indoor; children, patients and urban dwellers even stay longer indoor, spending 90% of the total daily activities. [4] Indoor air pollution does not only occur in factories and urban settlements, but also in rural areas, which produce burning materials from wood, charcoal, husk, and gasoline stove. In developing countries, more than a million population burn wood and other biomass fuels without installing adequate chimneys in their houses. [5]

There have been several studies on the indoor air pollution, such as that in India, concerning water quality inside the buses using CO2 parameter. [1] Another study in 2015 was performed to find out the indoor air quality using a biological parameter approach, but in other locus than bus. The study revealed that indoor air pollution rate was beyond the quality thresholds, resulting in 508 CFU/m3 germs. The result suggested the need for a special effort in preventing the indoor air pollution, while maintaining human health. [3]

1.2 Biological agent

Heating process, ventilation, and air conditioner (AC) affect indoor humidity and air quality. Vapor accumulation in a building construction causes humidity and microbial growth. Symptoms for such phenomenon ranges from change in color, material surface leakage, wet spots, and attachment, and fungal odor. Sources of humidity are rain water, surface water, groundwater, and local water with poor drainage systems, which are condensated. [4] Bacteria and fungi can also be found in foods. They grow abundantly in water tract, air conditioner, moisturizer, air cleaning filter, carpet and poor ventilated room in the basement. Viruses and germs are transported into the room by human, domesticated

* Corresponding author: iksirmahfudz@yahoo.co.id

animals or pets, and insects. Insects may grow abundantly in sofa, chair, carpet, and bedroom. [6]

Biological agents exposure are often correlated to non-specific symptoms of upper and lower respiratory tract disorders; some episodes may include sick building syndrome, a situation that causes acute symptoms, such as skin irritation, mucosal membrane, and other symptoms developed by workers/inhabitants of modern buildings. However, the syndrome does not identify specific diseases and the causes are unidentified. These complaints can occur locally in a specific room of the building, but may also spread over the building. [7]

2 Research methodologies

This study was performed to analyze air quality restricted to biological agent. The purpose was to find out the indoor air quality in the buses. The study applied a viable air sample method and used blood agar and nutrient agar media to grow microorganisms trapped within buses in Tegal, Central Java. The media were placed in different points of the buses justified as an intense rooms occupied by human. This method applied a growth medium of agar in petri dish or agar strips to grow the trapped microorganisms. The culture method had an advantage of detecting bacteria (in addition to spore), providing a broad illustration of the number of the microorganisms living in the air, and determining microorganism species because it comprised a single colony for further culture. In this study biological parameter used was germs and fungi persistence as an indicator for the indoor air pollution. The parameter was then compared to the quality threshold pursuant to the Decree of the Minister of Health Affairs of the Republic Indonesia 1204/Men Kes/SK/X/2004 on the Requirements for Environmental Health.

3 Results and discussion

This study was performed to find out the indoor air quality by focusing on microorganism indicators, i.e. germs and fungi. Results obtained were then compared to the quality threshold pursuant to the aforementioned Ministerial Decree on the Requirements for Environmental Health (indoor: 200 CFU/m3). The results found that such germs as *Bacillus subtilis*, *Staphylococcus aureus*, *Staph epidermidis, Klebsiella sp., Streptococcus,* and *Escherechia coli*, all being pathogenic germs, had caused indoor air pollution. In addition, the study also found some fungi species, i.e. *Aspergillus fumigatus, Penicillum sp*., mod, and *Aspergillus niger* as the contaminants that caused the air pollution and pathogenic germs, as well. Details of the results are presented in the following table.

Table 1. Germs Identification

NO	Room	Germ Population (CFU/M^3)	Germs
1	1.1	17	*Bacillus Subtilis, Staphylococus aureus, staph epidermidis*
2	1.2	24	*Bacillus Subtilis, Staphylococus aureus, staph epidermidis, Klebsiella sp*
3	4.1	22	*Bacillus Subtilis, Staphylococus aureus, staph epidermidis*
4	4.2	49	*Bacillus Subtilis, Staphylococus aureus, staph epidermidis*
5	5.1	52	*Bacillus Subtilis, Staphylococus aureus, staph epidermidis, Streptococcus, and Escherichia coli*
6	5.2	38	*Bacillus Subtilis, Staphylococus aureus, staph epidermidis, Streptococcus, and Escherichia coli*
Max Threshold		200	

Bacillus subtilis were found in the buses. They usually grew in soil and human digestive system. This block-shaped bacillus is capable of developing hard protecting endospora, which allows it to tolerate extreme environmental conditions. These bacteria take part into the enzymatic production secreted by industrial activities, such as biotechnological factories. Staphylococcus aureus bacteria are the most dangerous of all staphylococcus genus. These bacteria spread by having a direct contact with infected individuals, mediated by contaminated materials, or inhaling the infected drops dispersed by sneezing and cough. The Staphylococcus aureus may have access to all body organs, from outer skin to cardiac system, lung, bone and produce toxins. Staphylococcus epidermis is one of bacterial species from the genus of Staphylococcus, which potential causes opportunistic infection (infecting individuals with low immunity). These bacteria are found in human mouth and colon and are pathogenic in nature. The pathogenic Streptococcus may cause several diseases, such as pneumonia, meningitis, necrotizing fasciitis, erisipelas, throat inflammation, and endokarditis. [8]

Furthermore, the table also proved the persistence of Escherichia coli. These bacteria are usually found in human colon. Most of them are harmless, but some species, such as E. coli type O157:H7, produce toxins that contaminated human foods, causing blooded diarrhea due to exotoxin derived from verotoxin. Klebsiella cause pneumonia, urinary tract infection, septikemia, meningitis, diarrhea, and soft tissue infection. This species also contributes to pathogenesis of ankylosing spondylitis and other spondyloarthropaties. Most of Klebsiella infections in human are caused by Klebsiella pneumoniae and Klebsiella oxytoca. The Klebsiella infection often occurs to very young as well as very old individuals, indicating the development of cancer. This species usually contaminated invasive medical equipments. [8]

Table 2. Fungi Identification

NO	Room	Fungi Population (CFU/M^3)	Fungi
1	1.1	12	*Aspergilus fumigatus, Penicilium sp,* Mold
2	1.2	9	*Aspergilus fumigatus*
3	4.1	12	*Aspergilus fumigatus, Penicilium sp, Aspergilus Niger*
4	4.2	3	*Aspergilus fumigatus, Penicilium sp*
5	5.1	6	*Aspergilus fumigatus, Aspergilus Niger*
6	5.2	2	*Aspergilus Niger*
Max Threshold		0	

Infections due to Aspergillus cause significant morbidity and mortality. Most of the infections are correlated to Aspergillus fumigatus, Aspergillus flavus and Aspergillus terreus. Whereas, Aspergillus niger was rarely reported as the cause of invasive diseases, more likely related to automikosis [9], cutaneous infection [10], penumonia, fatal lung infection, and long-term steroid for COPD (chronic obstructive pulmonary disease). [11] An examination of the COPD patients with invasive lung aspergillosis revealed that 3.6% of the total cases were caused by Aspergillus niger. [2] Other case indicated that Aspergillus niger related to 75% of the mortality rates in the infected patients. [12] The mold may grow indoor and outdoor and they are found in abundance in warm and humid environment. Whereas, Penicillium species usually found indoor was Penicillium chrysogenum. This species produces toxins with a moderate toxicity and causes allergy and infection, in particular in the patients without immediate intervention and medication.

4 Conclusions

The study on the non AC buses using the biological agent parameter as the indicator for the indoor air pollution revealed that the non AC buses contained bacteria, i.e. Bacillus subtilis, Staphylococcus aureus, Staph epidermis, Klebsiella sp., Streptocossus, and Eschericia coli. The rate was still considered below the maximum threshold. However, they produced pathogenic germs harmful for human health. The study also found fungi, i.e. Aspergilus fumigatus, Penicillium sp., mold, and Aspergillus niger, all pathogenic and harmful for human health.

References

1. A. Gajewski, *Indoor air quality in a bus. Safety and Security Engineering V 749* Department of Heat Engineering, Faculty of Civil Engineering and Environmental Engineering, Bialystok University of Technology, Poland, (2016)
2. JR. Balmes, IB. Tager, MD. Eisner, Air pollutant. In: JR. Manson, JF. Murray, VC. Broaddus, JA. Nadel, eds. 4th ed. Philadelphia:Elsevier Saunders, pp.1800-14, (2005)
3. WHO. *Indoor air pollutan and household energy.* [cited 2005 Des 2006]. Available from: http://www.who.int/heli/risks/indoor.air/en/index.html
4. U.S Enviromental protection agency. *Indoor air pullutan: An introduction for health professionals.* [cited 2005 Des 4]. Available from: http://www.cpsc.gov/cpscpub/455. Html
5. L. Fianchi, M. Picardi, L. Cudillo, L. Corvatta, L. Mele, G. Trape, C. Girmenia, and L. Pagano. Mycoses **47**, 163–167. [PubMed], (2004).
6. J. Wiggins, T. J. Clark, and B. Corrin, Thorax **44**, 440–441. [PMC free article] [PubMed], (1989).
7. K. W. Loudon, A. P. Coke, J. P. Burnie, A. J. Shaw, B. A. Oppenheim, and C. Q. Morris, J Hosp Infect **32**, 191–198. [PubMed] , (1996).
8. P. Bulpa, A. Dive, and Y. Sibille, Eur Respir J **30**, 782–800. [PubMed], (2007).
9. KJ. Ryan; CG. Ray (editors). *Sherris Medical Microbiology (4th ed. ed.).* McGraw Hill. ISBN 0-8385-8529-9 (2004)
10. Husna, Iksiroh El & Yan El Rizal. *Analisis Mikroorganisme Untuk Mengetahui Tingkat Pencemaran Udara Dalam Ruang di Gedung PKB PKTJ Tegal* (National Conference On Conservation For Better Life 2014
11. JM. Samet, MJ. Utell, Pulmonary Diseases and Disorders. New York: McGraw-Hill Co, pp. 941-62, (1998)

Comparison of Sellers's Awareness to Environmental Hygiene of Market Bulak, Market Klender and Market Rawamangun, East Jakarta

Maulidya Hasana P.[1,*]

[1]Master Program of Environmental Science, School of Postgraduate Studies, Diponegoro University, Semarang-Indonesia

Abstract. This research is motivated by the difference of market hygiene condition, where the market hygiene level is influenced by the environment around the market. In general, markets located near densely populated housing tend to be overlooked, while markets near elite housing tend to be clean. This condition is also influenced by marketers' awareness of market hygiene. If the market is near the elite neighbourhood, the level of awareness of sellers on cleanliness will be high. If the market is located in a densely populated area, sellers generally do not pay attention to cleanliness. The purpose of this research is to know the sellers's awareness of environmental cleanliness of Market Bulak, Klender Market and Rawamangun Market. Respondents in this study are sellers and buyers who make transactions in these 3 markets. This type of research is descriptive analysis with the method of observation and interview to 10 sellers in each market. Seller hygiene awareness are poor.

1 Introduction

Healthy, clean and comfortable environment is everyone`s dream in life. To achieve this is a very complex in making humans as the main actors who are truly aware of the meaning of clean, healthy life so as to create a clean, healthy and comfortable environment. It is a very complex to achieve in making someone as the main actors who are truly aware of the meaning of clean, healthy life so as to create a clean, healthy and comfortable environment. It needs to be instilled the importance of hygiene to members of the community so that they have an awareness to maintain the environment [1].

Healthy, clean and comfortable environment also expected to exist a market, as well as traditional market. The goods in traditional market usually household daily needs. Traditional markets are usually associated with a negative impression, namely dirty and less comfortable. If traditional market conditions are not change, it is not impossible later it will be abandoned by consumers. Later this possibility will be more clear considering the rapid growth of modern market especially in Jakarta. However, traditional markets have a natural competitive advantage that is not directly owned by modern markets. Strategic location, large sales area, complete diversity of goods, low prices, a bargaining system that demonstrates intimacy between sellers and buyers is an advantage possessed by traditional markets [2].

Creating a clean and healthy environment is a shared responsibility. Especially the people around the neighborhood. They have an important role in protecting the environment and creating a clean and healthy environment .Creating a clean and healthy environment in the market is a shared responsibility of all stakeholder. Each one of them have an important role in protecting the environment and creating a clean and healthy environment. The awareness and consciousness level of most of seller in traditional market on the maintenance of the environment is still lacking.

In Yulianto study said education determines how seller think and understand in the sorting of waste. Low educated seller does not understand how to sort waste properly because do not understand the benefits it generated after. The higher the education level of a person then the more high the way of thinking and understanding in sorting waste [3].

The availability of a dumpster and trash can prepared by the market management can trigger a person to dispose waste init place. In other case a lot of this trash is missing this is very burdensome to the market management. A good trash must be strong, not easy to leak or crack, easy to open and closed again, so that the smell of garbage is not smelled and invisible from the outside.

Researcher interested to compare the awareness and consciousness of seller to the cleanliness of the environment in three traditional markets, namely Bulak Market, Klender Market and Rawamangun Market in East Jakarta.

Bulak Market is located in Klender village, Duren Sawitsubdistrict. Owned by Privat Company. In front and back side of Bulak Market there is crowded densely populated housing and in the right sight are well-resettled settlements from middle-class society. At the left side of Bulak Market lies Citra Mall Klender with four stories building, and at forth story there is super market but less visitor come there. People prefer buy vegetables in Bulak Market which cheaper. The Bulak Market building is one-

* Corresponding author: lydia_maulidya91@yahoo.com

floor ward-shaped building and partitioned by the sellers themselves according to the area they rented. In this ward there are vegetables mix with fruit stall.

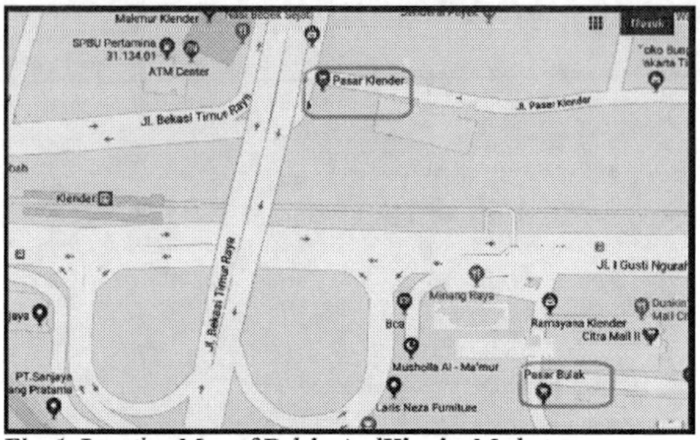

Fig. 1. Location Map of Bulak AndKlender Market

No live Poultry, but in separate building for fish sellers and dairy slices. Buying and selling activities in this market is different from most other traditional market, here last from afternoon until morning. There are no activities from eight o'clock in the morning until around four in the afternoon. Sellers in Bulak Market mostly sell their goods in larger quantity, because many of their costumer are vegetables retailer. Some of them sell vegetables, spices, herbs and fruits by their cart go around village. Catering entrepreneurs, and even sellers from Market Rawamangun who buy their goods here to sell in their stall.

Fig.

Fig. 2. Bulak Market Condition

Klender Market is located in Jatinegara Kaum Pulo village Gadung district Klender Market is by the side of a railway, and surrounded by a densely populated settlement that is generally inhabited by the weak economic community. This market is three-storey building and vegetable department is on the basement no fruit stall in this basement. There is meat stalls, dairy slices in addition to fish sellers, all located in the back part on one same flor. Living poultry sellers are on the back of the market building. Outside around market building there so many vegetables, fruit and also fish and meat seller as well as cooked food and cookies seller Consumers are generally housewives and cooked food maker. Quantity goods bought by customer are small. Fortunately this market easy to reach by customer who lives far away, because many public transportation passes and also near railway station. Many fruit seller of any kind of fruit especially banana sell their goods on footpath side by car way mostly in the afternoon until evening.

Fig. 3. Klender Market Condition

Rawamangun Market lies in a well-tended settlement of middle-class society. A three-story building with a vegetable department is on the ground floor. Vegetable area mix with fruit stall. The poultry sellers, meat and fish slices is located on the back of the basement stalls. Outside there are also life poultry sellers. Consumer Market Rawamangun generally housewives.

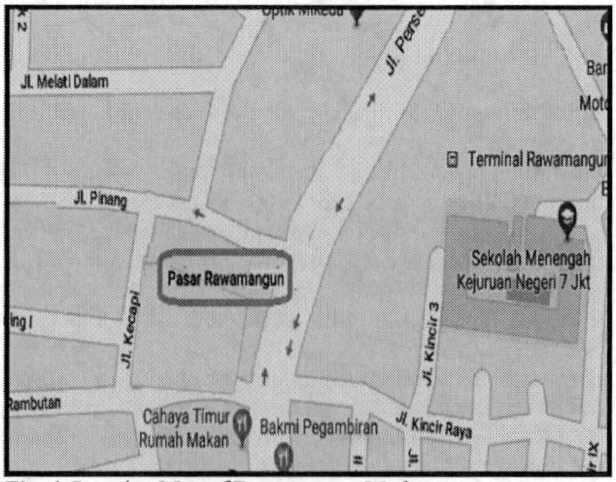

Fig. 4. Location Map of Rawamangun Market

The part of the market observed is the vegetable part, because in this part the amount of waste is relatively large plus the behavior of marketers who are less concerned about the cleanliness of thc environment. Waste of vegetables easily to decay. According to Anam [2] quickly decay or decomposition of garbage will more quickly create an unpleasant odor that leads to the impression of unhealthy, dirty and uncomfortable for the environment. Waste decay can lead to the appearance of pathogenic microorganisms and invite animal transporters or disease spreaders (vectors). The more vegetable waste the faster the environmental problems arise, the more uncomfortable the environment, this can not be separated from the seller's awareness of environmental cleanliness. Therefore, the waste must be managed properly, making the environment healthy, clean and add aesthetic value. Faridha (2001 in Hermawan Y) found a significant relationship between knowledge and seller behavior in managing environmental hygiene.

Fig. 5. Rawamangun Market Condition

The people live around the market affects the reason of consumers to the market, some consumers visit the market for reasons of near so that more economical for them, some because of the lower price, quality and completeness of vegetables, there is also a visit for convenience reasons. Middle class people are more selective in choosing quality (freshness) and convenience shopping, for them the price is the second consideration. This community is willing to come to a market that is located relatively far provided that there is a quality product and shopping convenience. For people who are classified as a weak economy are more concerned with product prices and ease of reaching the market, for example by walking.

2 Research methode

This type of research belongs to qualitative research. Qualitative research is research that is intended to understand the phenomenon of what is experienced by the subject of the researcher. The research was conducted in Bulak Market, Klender Market, and Rawamangun Market, East Jakarta. In order to obtain data and information relevant to the research problem, the data collection is done by determining the research informant. Selection of research informants conducted by purposive sampling, because by using purposive sampling technique the researcher can determine their own research informants based on research objectives, and also expected sample criteria obtained really in accordance with the research to be done. Informants in this study are 10 sellers and 10 buyers of each market. Sellers who are made in informants are people who are selling vegetables. Types of data in this study is primary data. Primary data is data that is directly collected in research. Sources of data in this study is the primary data obtained from interviews and observation,

3. Research result and discussion

Most of vegetables seller of Market Bulak are male which age around thirty to fifties. The goods quantity of each seller in Market Bulak is larger than Market Klender and Market Rawamangun seller. Vegetables are more varied in Market Bulak. Vegetable seller in Market Klender and Rawamangun seen older, male and female almost balance.

Based on researcher observation the hygiene of the three market environment is still not in accordance with expectations, even in the middle class environment. There are still many piles to the results of market activity in the market aisles. In market Bulak this is allegedly due to the low awareness, knowledge, education, habits, attitudes and experience of sellers about the importance of maintaining the cleanliness of the market environment. Lack of caring attitude of sellers to market hygiene problem will cause big impact to public health, and also negative impact to surrounding environment, hence need of cooperation among sellers in maintaining cleanliness of market environment.

The alley seen not too clean, in market Klender and market Rawamangun, there are some garbage heap. In market Bulak the alley clean lyness is worse, there are some alley covered by vegetables waste. Seller habits maintaining cleanliness still bad because of the many of them do not throw garbage in its place. This is due to the absence of responsibility the market management service in providing garbage space.

Writer interview result with seller show that awareness of them in maintaining hygiene of the market is still low due most mof them only collect waste of his wares, and put the garbage nearby their place, and some of them provide temporary dump like a sack.

Most of seller know that decaying garbage not good for health and aesthetic, according to most of them that cleanliness of alley not their responsibly, market management should pay attention to the hygiene of the market. According to them only few market visitor or buyer complaining dirty environment. Cleaning market alley is done twice a day, then the garbage put in certain place until dump truck from local government cleanliness agency.

Table 1. Characteristic of Markets

Market	Location	Market hygiene	Seller hygiene awareness
Bulak	Between decent and dense settlements	Bad	Poor
Klender	Dense settlements	Worse	Poor
Rawamangun	Decent settlements	Not too bad	Poor

Most seller tend to underestimate the issue of environmental hygiene conditions in which to trade and to attitudes toward health. The low level of awareness and consciousness of seller on environmental conditions can be seen from the attitude of most merchants that do not reflect the culture of clean and healthy living. The problem that has been the obstacle is caused by the seller in the market, because they do not show cooperative attitude and participation to help manage waste. Just as there are merchants who rarely sweep the merchandise so that it seems dirty. Although market management have tried to provide briefing and guidance about awareness and consciousness of the environment around their stall.

In general, most often buy vegetables in the market is housewives, the researchers interviewed the housewives who visited the market under study. They were asked why they chose the market they visited, how often they visited, what they opinion about the market condition, how it was compared to other traditional markets. The result that although they are not satisfied with market environment condition but they can tolerate. Some of the reason they choose the destination market is because it is close to their residence, the vegetable varied and more fresh also cheaper. According to them other traditional market condition almost same with destination market, except for newly renovated markets. Most of buyer are loyal buyer, it mean they always go to the market which they usually visit.

5 Conclusion And Suggestion

The environments cleanliness of the three market, almost same, even in middle-class neighborhoods. There are still many piles of market activity results in the market aisle. The seller awareness in the lack of caring attitude of sellers to the problem of market hygiene it will impact on public health, and also have negative impact on the surrounding environment, it is necessary cooperation between sellers in maintaining the hygiene of the market environment.

References

1. Y. Hermawan, H.O. Roesman, Jurnal Bumi Lestari Vol **8** No. 2 hal 186-192, 2008
2. C. Anam, Jurnal Sain danTeknologi Vol **7** No. 1, 2014
3. B. Yulianto, Jurnal Kesehatan Komunitas, Vol. **3** No, 2, 2016
4. S.Lestari, *Perilaku Pedagang Dalam Membuang Sampah*, FISIP UNILA, Skripsi, 2016
5. Sadino, J.A. Syahbana, Biro Penerbit Planologi Undip Volume **10** (2): 205-217 Juni 201
6. R. Sihotang, S. Afiddudin, Rahmanta, Jurnal Ekonom, Vol **17**, No 4, Oktober 2014. 182

Implementation of Geographical Information System for Bacteriological Contamination Analysis on Refill Drinking Water Depot (Study in Tembalang District)

Amelia Rahmitha[1,*], *Endang* Sri Utami[1], *Marya* Yenita Sitohang[1]

[1]Bachelor Program of Public Health, Diponegoro University, Semarang - Indonesia

Abstract: People used refilled-drinking-water for household and food stall because its efficient and low cost. Based on Indonesian Health Ministry regulation, it should not have any coliform bacteria. This study aimed to describe the bacteriological contamination of refilled drinking water using geographical information system (GIS). In this research, it was used an analytic observational method. The samples were from all available (37) depots in Tembalang district, one form each depot took used a sterile bottle. Contamination of bacteria was identified by Most Probable Number (MPN) method lactose broth media, Mac Conkey media, and IMVIC media. The depot samples were then plotted on (GIS). This study showed 95% samples were not feasible to consume since they contamined coliform. All sub-district had one that contaminated by coliform, 75% sub-districts had depots that contaminated Escherichia coli, while 55% sub-districts had depots that contaminated with other bacteria. The internal risk factors of the contamination were the absence of hygiene-sanitation worthy certificate (95%), depots location near to pollution sources (5%), and the misused of UV light. The external risk factor was lack of quality control that was not as the sterilization from office health Semarang city. Policy reinforcement should be done to all of the depots.

1 Background

Drinking water is water that goes through the processing or without processing that required of health and can be directly drunk. Safe supply of drinking water is essential for human life and should not pose a significant risk to humans[1].The requirement quality of drinking water have to be followed and obeyed by all producers so that drinking water is safe for consumption. These requirements include physical, microbiological, chemical, and radioactive aspects[2]. Biological parameters are directly related to health. The regulation on Minister of Health Republic of Indonesia number 492/Menkes/Per/IV/2010 are *Escherichia coli* and total bacteria coliform maximum allowable is 0 per 100 ml sample[2].

Escherichia coli that such pathogens can be transmitted through water and food contamination, and contact with people[3]. *Escherichia coli* that contaminate water has a risk to consumers, such as diarrhea[4]. According to the Health Profile of Semarang City, Incidence Rate of diarrhea is 25 per 1,000 population in 2014. Meanwhile, worldwide according to UNICEF Indonesia, 88% of children die since diarrhea[5]. *Coliform* bacteria are organisms in the environment and the feces of warm-blooded animal and human. The presence of coliform in drinking water indicates that disease-causing organisms (pathogens) can occur through the water system [6].

Nowadays, refilled drinking water has become a common choice by middle-income people[7]. The water is produced by the Drinking Water Refilled Depot: a business entity that manages drinking water for public purposes in bulk and not packaged[8]. The refilled drinking water business is growing well due to the higher price of bottled mineral water. However, the process of refill drinking water uses the same technology as bottled mineral water. However, refilled drinking water does not use packaging and distribution costs.[7].

Tembalang Sub-district of Semarang City has 156.865 people in 2015[9]. This area is a densely populated area due to the development of several campuses such as Diponegoro University, Poltekkes, Pandanaran University, University of Muhammadiyah Semarang, Husada School of Health Sciences and AsihHusada Health Academy. This condition increased demand for drinking water, both in terms of quality and quantity so that the risk of decreased feasibility and safety of drinking water due to contamination of microorganisms. Based on a survey conducted by the author, the number refilled drinking water depot in Tembalang sub-district is 37 depots.

The more use for the vital needs of the community and the indication of the lack of safety refilled drinking water depot, it is necessary to supervise or monitor and

* Corresponding author: rahmithahelmi@gmail.com

adequate testing. To determine whether or not the condition is contaminated, clinical research or testing in the laboratory is needed [10]. To ensure adequate water to consume, there are several point to note: regulatory compliance, the ability of a water laboratory to conduct complete and accurate water testing and analysis, proper operation and maintenance of water purification equipment, local health service skills for sanitation surveying and monitoring water quality, water source protection programs, and strong information campaigns to keep water safe at home [11]. In 2016, microbiological quality test was conducted by Semarang City Health Office 14 by sampling, the result is 100% sample containing total coliform and 14,3% sample containing *E. coli*[12]. This condition because not all depots to do the processing properly and correctly [13]. In addition, only 2 of 37 DAMIUs have sanitation sanitation-worthy certificates [14].

Based on the background, it is necessary to test bacteriological contamination on refilled drinking water in Tembalang Sub-district. The laboratorium result will be implemented on Geograpghical Information System (GIS) to describe the contamination of Escherichia coli and coliform contamination. Literally, GIS is a component consisting of hardware, software, geographic data and human resources that effectively to capture, store, repair, update, manage, manipulate, integrate, analyze, and display data in any information geographically based. [15]. In this study, GIS is used to visualize microbiological analysis on depots products in Tembalang Sub-district in map form. It is very efficient for storing and communicating geographic information[16].

2 Method

This study used observational analytic method in Tembalang Sub-district of Semarang City from April to June 2017. Samples were taken in each depot using sterile bottles with total of 37 depots in Tembalang Sub-district. To test the consistency of the results, this study conducted two laboratory tests at 2-month intervals. The first study used 37 samples and the second study used 29 samples. This is because as many as 8 depot owners are unwilling to do second phase testing. Unstructured interviews were also conducted on depot owners and observations around depots location. *Coliform* contamination was identified using the Most Probable Number (MPN) method on the lactose broth medium. Further positive results were tested with Mac Conkey media to identify *Eschericha sp*. To know Eschericha coli species used IMVIC method. Samples of drinking water depots were subsequently plotted using Geographical Information System (GIS).

3 Results

The test was conducted twice to the product samples in Tembalang sub-district in April and June 2017. The first test was successful in examining 37 samples but in the second test only 29 samples were tested because 8 depots owners refused to return to sample.

The table above shows that the presence of coliform bacteria in the drinking water refill sample has decreased from the first test to the second test. As for the presence of Escherichia coli showed improvement. Based on the result of paired T-test, it was found that there was no difference of first and second sample test result (p> 0,05). In water samples also showed contamination of *Escherichiavalneris*, *Escherichiafergusonii*, *Staphylococcus*, *Enterobacter* and *Pseudomonas* bacteria which consisted in the same sample both at the first and second test. Therefore it can be concluded that there is consistency in the test results so that in this study used 37 samples from the first test results.

Table 1. The results of Coliform and Escihierichia coli on the first and second tests on refilled drinking water in Tembalang Sub-district

Test	Status of Bacteriology			
	Coliform		*Escherichia coli*	
	Not Qualify	Qualify	Positive	Negative
1st	23 (79,3%)	6 (20,7%)	9 (31%)	20 (69%)
2nd	21 (72,4%	8 (27,6%)	13 (44,8%)	16 (55,2%)

3.1 Contamination of bacteriology at refilled drinking water depot

Laboratory test on microbiological quality parameters of refilled drinking water in Tembalang Sub-district. These microbiological parameters are total *coliform* and *Esherichiacoli*.

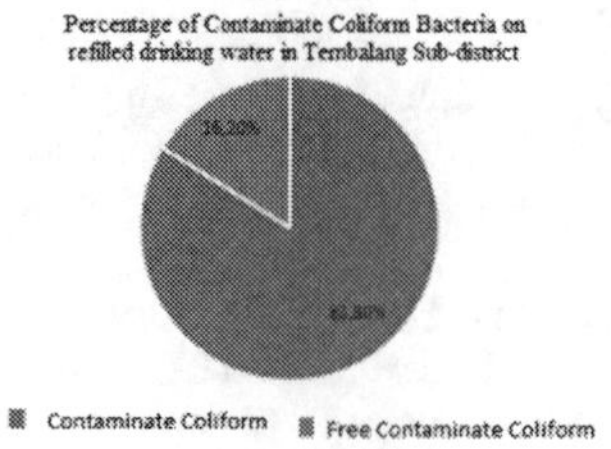

a

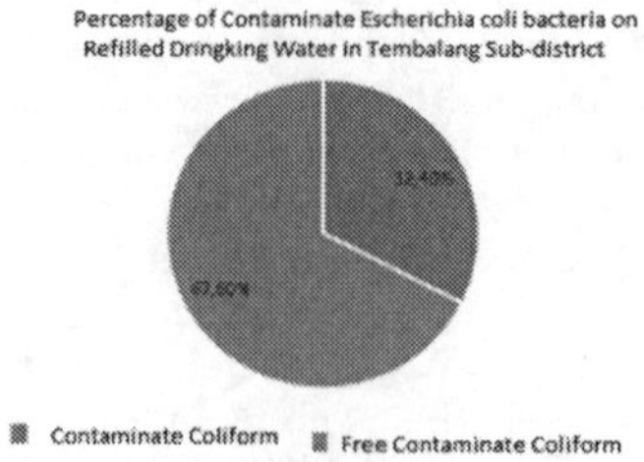

b

Fig.1. Contamination of coliform (a) and Esherichia coli (b) at refilled drinking water in TembalangSubdistrict

Based on Figure 1 shows that most of refilled drinking water contaminated *coliform* exceeding the maximum limit as recommended in Permenkes RI Number 492/MENKES/PER/IV/2010 is 0/100 ml water sample. While less than some refilled drinking water proven contaminated *Escherichiacoli* bacteria.

3.2 Map of bacteriology contamination at the refilled drinking water depot

The distribution of bacteriological contamination at at refilled drinking water from depots in Tembalang Sub-district is described using Geographical Information System (GIS). The distribution map is shown in the following figure.

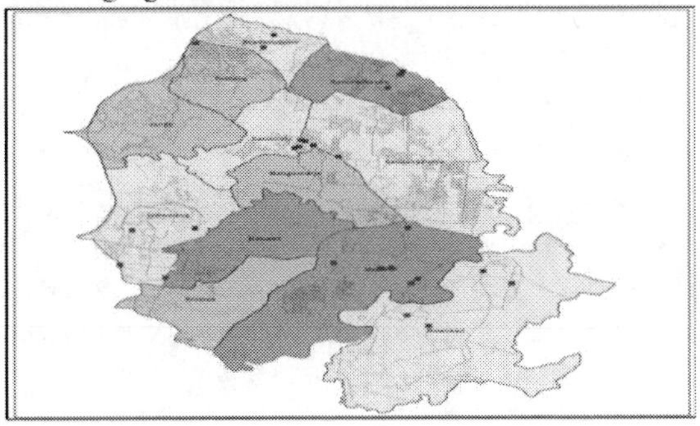

Fig. 2. Map Distribution of Coliform Bacteria at AMIU in Tembalang Sub-district

The analysis was performed by calculating the percentage of depots containing positive bacteria on all depots examined in the villages. After that the villages is categorized in the following categories:

0 : Free of Pollution
1-33% : Light Pollution
34-66%: Medium Pollution
67-100%: Heavy contamination

Based on figure 2 shows that from 37 depots, 3 villages namely Meteseh, Bulusan and Kedungmundu are in medium contamination category, while Tembalang, Rowosari, Sendangmulyo, Sendangguwo and Sambiroto villages are included in medium contamination.

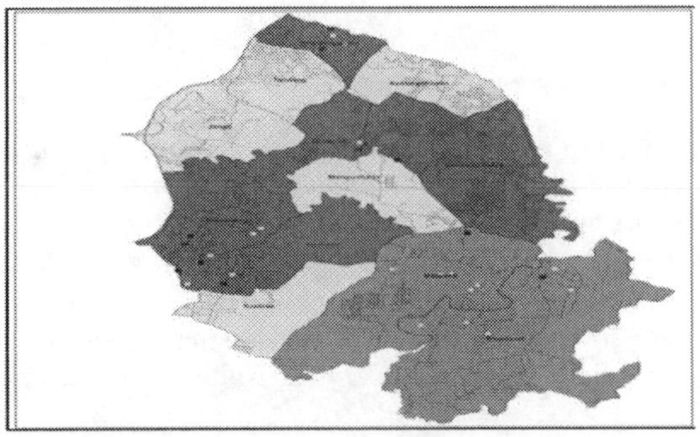

Fig. 3. Map of E. coli Bacteria Distribution at AMIU in Tembalang District

Villages based on the risk of contamination are classified into high risk (MPN> 100), medium (MPN 10-100) and low (MPN 1-10). In Figure 3 it is known that 2 villages has drinking water contains mild *E.coli* bacteria ie in Rowosari and Meteseh. While for the medium category is Sendangguwo, Sendangmulyo, Sambiroto, and Tembalang. Based on Figure 4, bacteriologic effects of drinking water can also identify other bacteria *E. fergusoni, E.valneris, Enterobacter, Pseudomonas* and*Staphylococus*. The villages covering the light category is in Rowosari, Sambiroto and Sendangguwo. While for the medium category is Sendangmulyo and Tembalang.

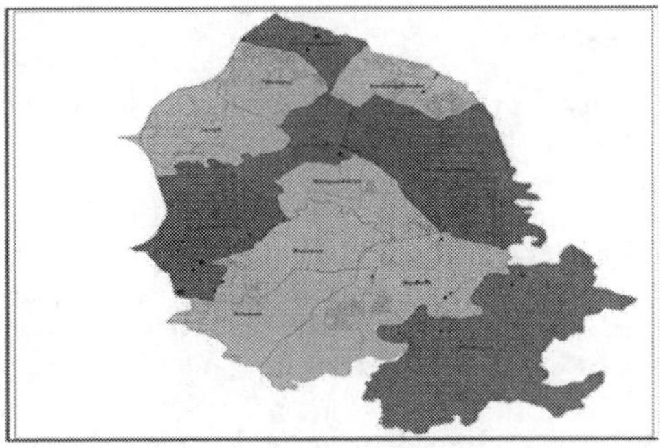

Fig. 4. Map of Bacteria Distribution at depots in Tembalang Sub-district

4 Discussion

The existence of the bacteria *Coliform* and *Escherichiacoli* at refilled drinking water in Tembalang District can be caused by various factors. Here's an explanation of some of the risk factors for the contamination:

4.1 External Factors

External factor at risk for bacteriological contamination at refilled drinking water is the quality control of drinking water produced by depot less than optimal. This observation is conducted by District Health Office which includes sanitation inspection, drinking water sampling at least 1 month 1 time[19].

Based on the condition of the field shows that the District / Municipal Health Office performs inspection and sampling so it is not comprehensive in supervising the quality of depot circulating in the community. This shows that the implementation of the program has not been in accordance with the regulation of Minister of Health RI Number 43 Year 2014about Hygiene Sanitation Drinking Water Depot. Things that affect the condition in accordance with the Health Department of Semarang City Data is related to Human Resources (HR) and funds. The results show that by 2016, the number of depots under surveillance is 13 of the 37 in Tembalang Sub-district.

4. 2 Internal factors

Internal factors at risk for bacteriological contamination in refilled drinking water can be derived from itself. Internal factors include sanitation hygiene. Depot sanitation hygiene is an effort to control the risk factor of contamination from place, equipment and handler to drinking water for safe consumption[19]. Based on the data of the health office of Semarang City 2016 shows that the depot that has certificate of sanitary hygiene is only 2. Aspects of hygiene sanitation observed by researchers include:

4.2.1 Location

Locations used for depots should be in areas that are free from contamination, ie away from pollution areas such as puddles and swamps, sewerage and garbage, accumulation of junk or hazardous and toxic materials (B3). In addition, the site should also be away from polluting areas such as paint workshops, welding, limestone, asbestos and the like as well as public sewerage sites, bus terminals or other pollution-intensive areas [17].

Based on the Regulation of the Minister of Health No. 1096/MENKES/PERVI/2011 on hygiene and sanitation of catering services, the location of food processing from the source of pollution is at least 500 meters [18]. Field observations indicate that there are 2 depots (15%) located near motor and car repair shops.

4.2.2 Use of ultra violet lamp

Ultra Violet (UV) lamps is used for the disinfection process. Disinfection is an attempt to shut down the remaining microorganisms in the process, especially addressed to the pathogen. Disinfection is intended to kill pathogic germs contained in water so that drinking water produced is safe to drink.Observations show that most depots do not light UV light in drinking water treatment processes.

4.2.3 Behavior of handkerchief

The handler is the person who directly handles the drinking water management process at depot to serve the consumers. A handler must have good personal hygiene such as not smoking or scratching body parts while serving customers, washing hands before serving customers and using clean and decent clothing. The observation results show that only 10% of owner depots have a handyman with good personal hygiene. This is due to lack of training or notice so that the handlers do not have sufficient knowledge. In addition, the lack of facilities also affects the application of personal hygiene by the handlers.

5 Conclusion

This research uses depots31 samples from Tembalang sub-districts which is a densely populated area and campus development area. Through laboratory test results on 37 samples, it was found that 83.8% of samples contained coliform bacteria and 32.4% of depot products were contaminated with *Escherichia coli* bacteria. Based on GIS analysis found contamination of coliform bacteria in depots in 3 villages classified as heavy (100%). While the contamination of bacteria *Escherichiacoli* exist in depots in 3 villages with different categories. Contamination of *Escherichiasp* bacteria other species only not found in depot in Bulusan village and contamination of gram positive bacteria (+) only not found in depot in Kramas village. Other types of Escherichia bacteria encountered were *Escherichia valnerisand Escherichia fergusonii.* However, the type of gram-positive bacteria can not be determined because of laboratory tests that do not further identify the type of bacteria.

6 Suggestion

- For the Government: it is expected to provide regulation for depot to maintain the quality of refilled drinking water products through maintaining and improving the quality of sanitation as well as conducting inspection of drinking water in the laboratory on a regular basis that is at least 6 months. In addition, the government must also conduct equity in order to check on depots done in all places and not only in certain places.
- For the owner of depots: it is expected to maintain the quality of refill drinking water products through maintaining and improving the quality of sanitation as well as conducting periodic inspection of drinking water in the laboratory
- For the Community: more vigilant in drinking water consumption,and wise to choose producer refilled drinking water.
- For Researchers: expected to be able to conduct further research to develop science and dedication to improve the quality of life of the community, especially the health of the local community.

References

1. WHO (World Health Organization). *Fourth edition drinking water guidelines.* World Health Organization, Geneva. (2011)
2. Depkes. *Regulation of the Minister of Health of the Republic of Indonesia Number* 492/Menkes/Per/IV/2010 Concerning the Quality of Drinking Water. Ministry of Health of the Republic of Indonesia. (2010)
3. CDC. *Escheria coli.* Https://Www.Cdc.Gov/Ecoli/General/Index.Html Accessed on 16 June 2017. (2015).
4. BPOM.*Dangers of Microbes On Packaging Water.* http://Ik.Pom.Go.Id/V2014/Artikel/Bahaya-

Mikroba-Pada-Air-Kemasanfinal.Pdf Accessed on 16 June 2017. (2014).

5. UNICEF Indonesia. *Water Supply, Sanitation and Sanitation*. Jakarta: UNICEF Indonesia. (2012).
6. The Washington Health Ministry.*Coliform Bacteria In Drinking Water*. Http://Www.Doh.Wa.Gov/Communityandenvironment/Drinkingwater/Contaminants/ColiformAccessed on 16 June 2017. (2017).
7. Keman, Soedjajadi. *Folia MedicaIndonesiana,* **Vol. 41 No. 1** .(2005).
8. Rb. Tombeng, B. Polii, S. Sinolungan. *Journal of Public Health of Sam Ratulangi University*. Agustus 2013;**1(7)** (2005)
9. BPS Semarang City. City of Semarang in Figures 2016 Semarang: BPS Kota Semarang (2016)
10. Rumondor, P Perisai, JohnPorotu'o, and Olivia Waworuntu, Jurnal*E-Biomedik (Ebm)*. **Volume 2, No 2**(2014)
11. Magtibay. Water refilling station: an alternative source of drinking water supply in the Philippines. Acessed 8 July 2017.http://wedc.lboro.ac.uk/resources/conference/30/Magtibay.pdf(2004).
12. Semarang City Health Office. Semarang City Rechargeable Water Quality Inspection Results 2016. Not published. (2016)
13. Nuria, Maulita Cut dkk. *Test of Escherichia Coli Bacteria In Drinking Water Refill From Recharge Depot In Kabupaten Rembang. Yogyakarta :Universitas Gadjah Mada*.**Vol 5. No 1. Page** 27 – 35. (2009)
14. Semarang City Health Office. List of Certificate of Sanitation Hygiene Worthy of Drinking Water Refills in Semarang City 2016. Not published. (2016)
15. Puntodewo, Atie. Geographic Information System for Natural Resource Management. *Center For International Forestry* (2003).
16. Doktafia. Geographic Information System(GIS). http://doktafia.staff.gunadarma.ac.id/Downloads/files/30525/SISTEM+INFORMASI+GEOGRAFIS+-+1.pdfAccessed on 4 May 2017. (2017)
17. Sabariah. *Influential Factors Against Water Contamination Refill by Escherichia Coli in Denpasar City*. Bali: Udayana University. (2016)
18. Ministry of Health RI. Regulation of the Minister of Health No. 1096/MENKES/ PERVI/2011 concerning Food Catering Hygiene and Sanitation. Jakarta; (2011)
19. Ministry of Health RI. Ministry of Health Regulation Number 43 Year 2014 About Sanitation Hygiene Drinking Water Depot. Jakarta, Indonesia: Ministry of Health; (2014)

Measuring Carbon Footprint of Flexible Pavement Construction Project in Indonesia

Jati Utomo Dwi Hatmoko[1,*], *Arif* Hidayat[1], *Apsari* Setiawati[1], and *Stefanus* Catur Adi Prasetyo[1]

[1]Civil Engineering Department, Diponegoro University, Semarang - Indonesia

Abstract. Road infrastructure in Indonesia is mainly dominated by flexible pavement type. Its construction process, however, has raised concerns in terms of its environment impacts. This study aims to track and measure the carbon footprint of flexible pavement. The objectives are to map the construction process in relation to greenhouse gas (GHG) emissions, to quantify them in terms of carbon dioxide equivalents (CO_2e) as generated by the process of production and transportation of raw materials, and the operation of plant off-site and on-site project. Data collection was done by having site observations and interviews with project stakeholders. The results show a total emissions of 70.888 tonnes CO_2e, consisting of 34.248 tonnes CO_2e (48.31%) off-site activities and 36.640 tonnes CO_2e (51.687%) on-site activities. The two highest CO_2e emissions were generated by the use of plant for asphalt concrete laying activities accounted 34.827 tonnes CO_2e (49.130%), and material transportation accounted 24.921 (35.155%). These findings provide a new perspective of the carbon footprint in flexible pavement and suggest the urgent need for the use of more efficient and environmentally friendly plant in construction process as it shows the most significant contribution on the CO_2e. This study provides valuable understanding on the environmental impact of typical flexible pavement projects in Indonesia, and further can be used for developing green road framework.

1 Introduction

The construction industry is one of important sectors in national infrastructure development. While it supports national economy, a series of construction activities is potentially harmful to the environment. In 2010, the construction sector contributed 18% directly or indirectly to greenhouse gas emissions [1]. In 2007, China's construction sector accounted for nearly 50% of total energy use in which major energy–producing contributors were; materials, heating, fuels, and electricity supply [2]. As the largest CO_2 emitter since 2013 China's emissions even showed an upward trend as a quadratic polynomial [3].

One of the construction sectors that produce significant greenhouse gas emissions is road construction, which will continue to increase along with the growth of the road length built by the government. Road infrastructure in Indonesia is dominated by flexible pavement which allegedly produces high greenhouse gas emissions due to changes in land conditions and functions, and the consumption and exploitation of resources. This is in line with a study in the Netherland which compared emissions from different types of pavement and that found that flexible pavement emits carbon emissions higher than rigid pavement and brick roads mostly from production and construction activities [4]. Life cycle assessment on flexible pavement construction activities showed that it generated twice higher carbon emissions than that of rigid pavement construction activities [5].

The carbon emissions have been closely linked to global warming and will leave carbon footprint. Although it has now become one of main environmental indicators [6], however it has been defined differently, e.g. [7-11]. One of the most cited common definitions of Carbon footprint is; a measure of the whole sum of CO_2 emissions resulted directly or indirectly from activities of individual, organization, process, industry sectors over the life cycle of a product (goods and services) [12]. The importance of relating the environmental aspects, in terms of energy consumption and carbon emissions, with the decision of pavement design has been promoted as opposed to typically cost consideration [13].

This study aims to track and measure the carbon footprint of flexible pavement construction. The objectives are to map the construction process in relation to greenhouse gas (GHG) emissions, to quantify the GHG emissions in terms of carbon dioxide equivalents (CO_2e) as generated by the process of production and distribution of raw materials, and the operation of plant in the project.

2 Highway Pavements

The highway pavement is a section of highway which is hardened by a certain construction layer. It has a certain

* Corresponding author: jati.hatmoko@ft.undip.ac.id

thickness, strength, and stiffness, and stability in order to be able to channel the traffic load on it to the ground safely. Other factors, such as environmental condition, temperature, seepage and properties of materials also influence the behaviour of the pavement [14].

Flexible pavement is one type of road pavement that uses asphalt as a binder. Typical layers of flexible pavement include sub-grade, sub-base, and base courses, binder course, and surface course and have the function to carry and transfer traffic loads to the sub-grade (Fig. 1). Based on temperature during mixing and compacting, asphalt concrete is divided into hot mix, warm mix, and cold mix. This study focuses on hot mix asphalt concrete as it is the most commonly works method and is considered to meet pavement requirements. A typical cross section of the flexible pavement is shown in Fig. 2.

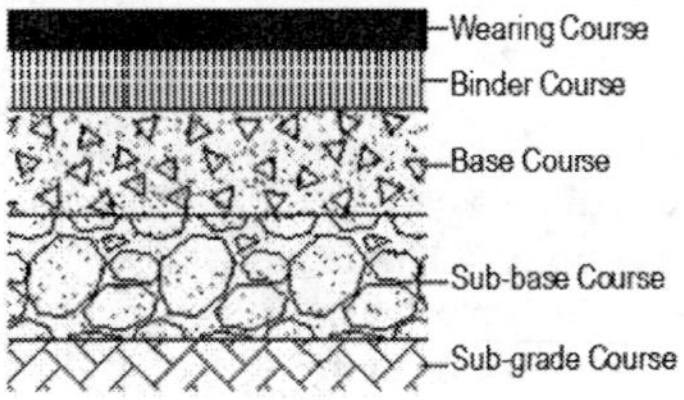

Fig. 1. Typical layers of flexible pavement

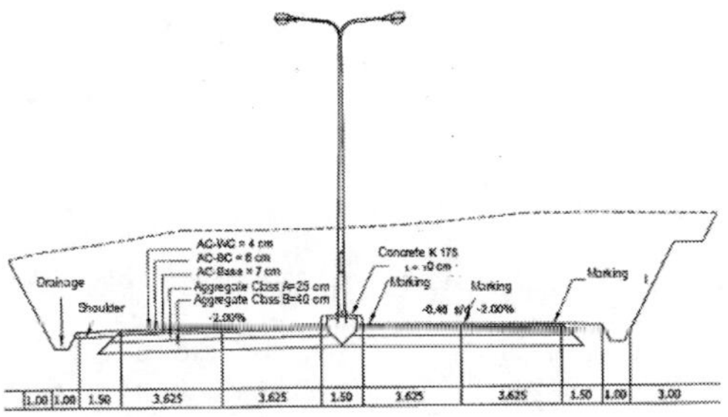

Fig. 2. Cross-section of the flexible pavement

3 Construction Method of Flexible pavement

The stages of flexible pavement work include preparatory work, sub base and base course work, and surface course work. Preparatory work related to the survey, and preparation of body of the road. Sub-base and base work include spreading work using grader, and compaction work using roller and worked after the sub-grade meets the elevation and density requirements.

Surface course work includes prime coat work, and asphalt concrete work using asphalt finisher, dozer, and pneumatic tired roller. Table 1 shows the sequence of works of a flexible pavement project as observed from Salatiga Ring Road project from STA. 9 + 285 to STA. 9 + 385.

Table 1. Sequence of work of typical flexible pavement

No	Sequence of Work	Description
1	Sub-base Course	Mixing and loading the aggregate in a Dump Truck using Wheel Loader in Base Camp. Transporting aggregate to the job site by Dump Truck and overlaid it using Motor Grader. Overlay aggregate wetted with Water Tank Truck before compacted using Tandem Roller and pneumatic tandem roller. During compacting, a group of workers will tidy up the edge of the expanse and surface level using a tools.
2	Base Course	Wheel Loader mixes and loads aggregate into Dump Truck in Base Camp. Dump truck transports aggregate to the job site and overlaid it with Motor Grader. Overlay aggregate wetted with Water Tank Truck before compacted using Tandem Roller and Pneumatic Tandem Roller. During compacting,, a group of workers will tidy up the edge of the expanse and surface level using a tools.
3	*Asphalt Concrete-Base*	Wheel Loader loads aggregate into asphalt mixing plant's (AMP) Cold Bin. Aggregates and asphalts are mixed and heated by AMP and to be loaded directly into the Dump Truck and transported to the job site. The AC-Base hot mix is overlaid by Finisher and compacted by Tandem Roller and Pneumatic Tandem Roller. During compacting,, a group of workers will tidy up the edge of the expanse and surface level using a tools.
4	*Asphalt Concrete - Base Course*	Wheel Loader loads aggregate into AMP's Cold Bin. Aggregates and asphalts are mixed and heated by AMP and to be loaded directly into the Dump Truck and transported to the job site. The AC-BC hot mix is overlaid by Finisher and compacted by Tandem Roller and Pneumatic Tandem Roller. During compacting,, a group of workers will tidy up the edge of the expanse and surface level using a tools.
5	*Asphalt Concrete – Wearing Course*	Wheel Loader loads aggregate into AMP's Cold Bin. Aggregates and asphalts are mixed and heated by AMP and to be loaded directly into the Dump Truck and transported to the job site. The AC-Wearing Course hot mix is overlaid by Finisher and compacted by Tandem Roller and Pneumatic Tandem Roller. During compacting,, a group of workers will tidy up the edge of the expanse and surface level using a tools.

4 Carbon Emissions of Construction Resources

In the construction industry the CO_2 emission varies along the project life cycle, however the density of the CO_2 emissions is the highest in construction phase compared to operation and maintenance phases [15, 16]. Construction materials and the use of plant during construction phase have the potential to generate greenhouse gas emissions through the exploitation of materials in nature, and the use of fossil fuels during the production process. Plant refers to all heavy machinery and equipment utilised for construction works.

According to the Kyoto Protocol 2007 [17], there are six main greenhouse gases, namely carbon dioxide (CO_2), methane (CH4), nitrous oxide (N2O), hydrofluorocarbons (HFCs), perfluorocarbons (PHCs) and sulphur hexafluoride (SF6). Internationally, the effect size of a process on the environment is gas emissions equalized with CO_2 equivalent (CO_2e). CO_2e emission conversion factors generated by material production and the use of fossil fuels are presented in Table 2.

Table 2. CO_2 equivalent emission conversion factors [18, 19]

No	Resources	Emission Factor	Source
Materials			
1	Coarse aggregate	1.067 kg CO_2/ ton	US EPA, 2004
2	Fine aggregate	1.067 kg CO_2/ ton	US EPA, 2004
3	Filler	1.067 kg CO_2/ ton	US EPA, 2004
4	Asphalt	11.91 kg CO_2/ gal	Climate Registry Default Emission Factor, 2016
Fossil fuels			
1	Motor gasoline	2.32 kg CO_2/ liter	US EPA, 2004
2	Diesel fuel	2.66 kg CO_2/ liter	US EPA, 2004
3	LPG (HD–5)	1.52 kg CO_2/ liter	US EPA, 2004

Carbon footprint can be classified as embodied carbon (EC) and operational carbon (OC) [20]. EC is emissions resulted from the production, transportation and construction stages. OC is carbon emissions resulted from the utilization of built facilities, which is normally believed to be higher than the EC. However, estimating the EC values are challenging due to variability of project characteristics, e.g. construction type, design, site condition, etc.

The carbon footprint measurement in this study covers off-site and on-site project activities (Fig. 3). Off-site activities include material production and transportation, while onsite activities examine the operation of plant use for asphalt concrete production and laying for road sub-base, base, and surface courses. The material production includes the processing of natural materials, such as coarse aggregate and fine aggregate, fillers and asphalt. Material transportation generates carbon emissions by bringing the natural materials from the sources to project site, using dump trucks for aggregate and filler, and tank truck for asphalt. The production of asphalt concrete operates AMP and wheel loader. While for asphalt concrete laying more plant were used, such as dump truck, motor grader, vibrator roller, water tanker, etc.

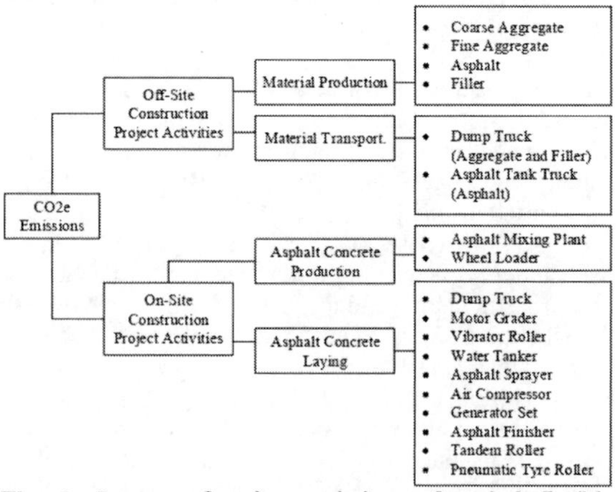

Fig. 3. Sources of carbon emissions of typical flexible pavement project

5 Research method

The object of this study is Salatiga Ring Road project from STA. 9 + 285 up to STA. 9 + 385 with four 3.625-meter-width lanes and flexible pavement type (Fig. 2) Data collection was done by having regular site observations to flexible road projects, interviews with project stakeholders, as well as document analysis. Site observation was carried out to understand the construction methods (see Table 1), procedure, including the materials and equipment used in the project. Interview with project staffs enhanced the understanding the common practice in the typical projects.

The method of this study is a case study method carried out by quantifying the CO_2e emissions generated during the construction of flexible pavement including material production and transportation, and the plant operation. Material production stage includes the amount of CO_2e emissions associated with the mining, processing, and production of materials construction. Material transportation stage includes the amount of CO_2e emissions associated transportation of construction materials from the source or manufacture to the site. Plant use stage includes the amount of CO_2e emissions associated with the fuels consumption during the construction of sub base and base course, and surface course. Data analysis techniques of this study is quantitative analysis by multiplying each factor causing CO_2 emissions in construction with emission factor as presented in equations 1, 2, and 3.

$$CO_2e_{MP} = V \times EF_{MP} \quad (1)$$

$$CO_2e_{MT} = TFC_{MT} \times EF_{MT} \quad (2)$$

$$CO_2e_{PU} = TFC_{PU} \times EF_{PU} \quad (3)$$

where CO_2e_{MP}, CO_2e_{MT}, CO_2e_{PU} denote the carbon emissions equivalent for material production, material transportation, and plant use, respectively (ton CO_2e). V denotes the volume of materials. EF_{MP}, EF_{MT}, and EF_{PU} denote the emission factor for material production (kg CO_2/ton), material transportation (kg CO_2/litre), and plant use (kg CO_2/ litre), respectively. TFC_{MT} and TFC_{PU} denote Total Fuel Consumption (litre) for material transportation and plant use, respectively.

6 Results and analysis

6.1 CO_2e emissions from off-site construction project activities

6.1.1 CO_2e Emissions of Material Production

For material production, the CO_2e emission is calculated by multiplying the volume of each material with emissions factor as presented in Table 3. The total CO_2e emission in material production for all road layers is 9.327 tonnes CO_2e. Asphalt contributes the highest CO_2e of 7.564 tonnes CO_2e, followed by coarse aggregate contributes of 1.247 tonnes CO_2e, fine aggregate contributes of 0.515 tonnes CO_2e, and filler contributes 0 tonnes CO_2e (Fig. 4).

Table 3 The CO_2e emissions in material production

No	Components	Materials	Volume (Kg)	Emissions Factor kg CO_2 / ton	Total Emissions (Tonnes CO_2e)
(1)	(2)	(3)	(4)	(5)	(6) = (4) x (5) / 1000
1	Subbase course (t=15 cm)	Coarse agg.	609	1.067	0.650
		Fine agg.	203	1.067	0.217
2	Base course (t=15 cm)	Coarse agg.	380.625	1.067	0.406
		Fine agg.	126.875	1.067	0.135
3	Prime coat	Asphalt	1.951	11.91	5.924
4	Tack coat	Asphalt	0.525	11.91	1.595
5	AC – Base (t=6 cm)	Coarse agg.	73.892	1.067	0.079
		Fine agg.	53.998	1.067	0.058
		Filler	0.004	1.067	0.000
		Asphalt	0.006	11.91	0.019
6	AC – Base course (t=6 cm)	Coarse agg.	60.9	1.067	0.065
		Fine agg.	70.644	1.067	0.075
		Filler	0.005	1.067	0.000
		Asphalt	0.006	11.91	0.017
7	AC– Wearing course (t=4 cm)	Coarse agg.	44.66	1.067	0.048
		Fine agg.	28.42	1.067	0.030
		Filler	0.002	1.067	0.000
		Asphalt	0.003	11.91	0.010
		Total			9.327

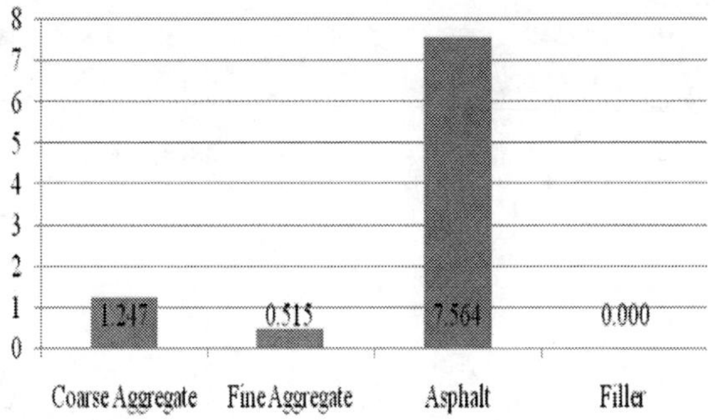

Fig. 4. CO_2e emissions of material production

The main constituent materials of flexible pavement are asphalt and coarse aggregate. Asphalt production has a long production process of petroleum distillation residue and has an emission factor of 11.91 kg CO_2 / gal [19]. Whereas, the aggregate production process is generated by the volcanic working mechanism so that it has a smaller emissions factor of 1.067 kg CO_2/ ton [18]. The differences in production process and in the value of the emission factor affect the total emissions generated by each of the flexible pavement constituent materials.

6.1.2 CO_2e Emissions of Material Transportation

For material transportation, the CO_2e emission is calculated by multiplying the total of fuel consumption associated material transportation with emissions factor as presented in Table 4. With a total CO_2e emission of 24.921 tonnes CO_2e, fine aggregate contributes the highest CO_2e emission (18.876 tonnes CO_2e), followed by coarse aggregate (5.712 tonnes CO_2e), asphalt (0.306 tonnes CO_2e), and filler (0.027 tonnes CO_2e). The distance between the fine aggregate source and the AMP is very far (i.e. 201 km), and the large demand of fine aggregate affects total fuel consumption and the release of CO_2e emissions.

Table 4 CO_2e emissions in material transportation

No	Mat	Vol	CoP	DfS (km)	TD (km)	TFC (liter)	TE (Tonnes CO_2e)
1	Coarse agg.	835.055 m^3	4 m^3	18	7515	2147.284	5.712
2	Fine agg.	344.955 m^3	4 m^3	144	24837	7096.217	18.876
3	Asphalt	2392.5 liter	13200 liter	201	402	114.857	0.306
4	Filler	0.008 m^3	4 m^3	18	36	10.286	0.027
			Total				24.921

Note : CoP = Capacity of plant, DfS= Distance from source (km),
TD = Total Distance [DfS*Vol/CoP*2], TFC=Total Fuel Consumption = TD*0.126 [Diesel = 0.286 l/km], TE= Total Emission = TFC*2.66/1000 [Diesel = 2.66 kgCO_2/litre]

6.2 CO_2e emissions from on-site construction project activities (asphalt concrete production and laying)

The CO_2e emissions in terms of plant operation are released during the process of compaction and asphalting. The CO_2e emissions is calculated by multiplying the total of fuel consumption associated plant use in construction site and AMP with emissions factor as presented in Table 5. It shows the total CO_2e emission related to the plant operation is 36.64 tonnes CO_2e.

Tabel 5 CO_2e emissions for plant operation on on-site

No	Components	Vol. (m^3)	Units	Plants	Plant Coef. (hour/m^3)	Operati-onal Time (hour)	Fuel Consump / hour (gal/hour)	Total Fuel Consump (litre)	Total Emissions (Tonnes CO_2e)
(1)	(2)	(3)	(4)	(5)	(6)	(7) = (3) x (6)	(8)	(9) = (7) x (8) x 3.79	(10) = (9) x 2.66 / 1000
1	Sub base course (t = 15 cm)	580	m^3	Wheel Loader	0.061	35.380	2.08	278.908	0.742
				Dump Truck	0.292	169.418	5.50	3531.518	9.394
				Motor Grader	0.011	6.612	6.00	150.357	0.400
				Vibro	0.017	10.092	10.50	401.611	1.068
				Water Tanker	0.021	11.948	5.00	226.415	0.602
2	Base course (t = 15 cm)	362.5	m^3	Wheel Loader	0.061	22.112	2.08	174.317	0.464
				Dump Truck	0.292	105.886	5.50	2207.199	5.871
				Motor Grader	0.011	4.132	6.00	93.973	0.250
				Vibro	0.017	6.308	10.50	251.007	0.668
				Water Tanker	0.021	7.468	5.00	141.509	0.376
3	Prime coat	1885	liter	Asph. Sprayer	0.335	0.631	0.26	0.631	0.002
				Air Compress	0.000	5.613	1.50	31.910	0.085
4	Tack coat	507.5	liter	Asph. Sprayer	0.335	0.170	0.26	0.170	0.000
				Air Compress	0.000	5.613	1.50	31.910	0.085
5	AC – Base (t = 6 cm)	101.5	m^3	Wheel Loader	0.048	4.842	2.08	38.167	0.102
				AMP	0.054	5.491	4.00	83.246	0.221
				Genset	0.054	5.491			
				Dump Truck	1.329	134.934	5.50	2812.701	7.482
				Asph. Finish	0.068	6.861	8.52	221.623	0.590
				Tandem Roll	0.028	2.842	4.00	43.085	0.115
				Pneumatic Tyre Roller	0.021	2.121	4.00	32.160	0.086
6	AC – Base Course (t = 6 cm)	87	m^3	Wheel Loader	0.048	4.150	2.08	32.714	0.087
				AMP	0.054	4.707	4.00	71.354	0.190
				Genset	0.054	4.707			
				Dump Truck	1.392	121.139	5.50	2525.138	6.717
				Asph. Finish	0.068	5.881	8.52	189.963	0.505
				Tandem Roll	0.028	2.436	4.00	36.930	0.098
				Pneumatic Tyre Roller	0.021	1.818	4.00	27.565	0.073
7	AC – Wearing Course (t = 4 cm)	58	m^3	Wheel Loader	0.002	0.110	2.08	0.869	0.002
				AMP	0.002	0.122	4.00	1.846	0.005
				Genset	0.002	0.122			
				Dump Truck	0.105	6.113	5.50	127.430	0.339
				Asph. Finish	0.003	0.151	8.52	4.871	0.013
				Tandem Roll	0.002	0.093	4.00	1.407	0.004
				Pneumatic Tyre Roller	0.002	0.128	4.00	1.934	0.005
				Total					36.64

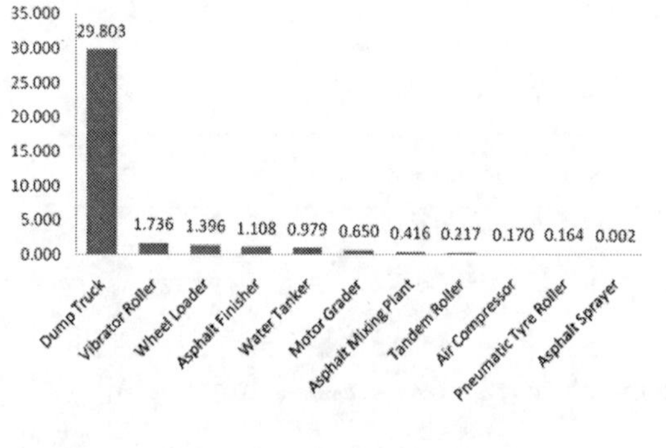

Fig. 5. CO_2e emissions for each type of plant

Figure 5 shows that operation of dump trucks contributes the highest CO_2e emissions of 29.803 tonnes CO_2e, which leaves a significant gap compared with other types of plant.

The great volume of asphalt as well as great distance from AMP to site (approximately 80 km) have great impact on the fuel consumption, which linked to significant CO_2e emissions emitted.

7 Discussions

Table 6 summarizes the results of the research that the total emissions of this project is 70.888 tonnes CO_2e. It can be seen that the highest CO_2e emissions were generated by the use of plant for on-site activities 36.640 ton CO_2e (51.687%), particularly for asphalt concrete laying activities accounted 34.827 (49.130%). The CO_2e emission of material transportation is the second highest accounted 24.921 (35.155%). Both asphalt concrete laying activities and material transportation indeed operate plant, particularly dump trucks, intensively during construction works, which obviously will generate significant amount of CO_2e. This condition is worsened by the great distance to transport materials (fine aggregate) from the quarry to project site, and asphalt from the AMP location to project site.

Table 6. Summary of CO_2e emission of Flexible Pavement

CO_2e emission off-site project (ton CO_2e)		CO_2e emission on-site project (ton CO_2e)		Total CO_2e emissions (ton CO_2e)
Material production	Material Transport	Asphalt Concrete production	Asphalt Concrete Laying	
9.327 (13.157 %)	24.921 (35.155 %)	1.813 (2.557 %)	34.827 (49.130 %)	70.888
34.248 (48.313 %)		36.640 (51.687 %)		

In this context, where great distance of transporting materials contributes significantly to the amount of CO_2e emissions, it is very important to manage plant and resource allocation to get optimum results. The AMP should be located as near as possible to the site to reduce time travel as well as reducing the carbon emissions. Closer distance will also minimize risks of travelling, such as traffic jam, accident, while at the same time maintaining the quality of hot asphalt from the AMP. In addition, with the significant contribution of plant operation of carbon emissions, there is also an urgent need for the use of more efficient and environmentally friendly plant in the construction process.

As coarse and fine aggregates are dominant materials in the flexible pavement, the use of recycled materials is one option to reduce the amount of energy and carbon emissions. Recycling will minimize the need for material extraction from nature, processing and production which eventually will contribute to reducing carbon emissions. This has been addressed by several studies, e.g. [21-23].

The results of this research provide a new perspective in comparison to the results of a study in China [24]. Using similar method to estimate carbon emissions from construction steps, i.e. production of materials, transportation of materials, and on-site construction activities, they found that most CO_2e emissions (over 80%) were contributed by production of raw materials. On-site activities and transporting materials generated only 10% and 3% of the total CO_2 emissions. While Wirahadikusuma et al [25] found that aggregate drying process at the AMP contributed most on energy consumption (68%) and carbon emissions (70-75%).

As each project is unique, the difference can be caused by distance of sources of materials, types of plant used, construction method, site condition, etc. These factors may explain why various results may be obtained from various researches. Nonetheless, the findings of this research provides a new perspective of the carbon footprint in flexible pavement in Indonesia, and suggest opportunities for developing research areas related to road pavement technology which is more environmentally friendly. On top of that, the use of more efficient plant, recycled materials, and better management will help reducing carbon emission from the construction sector to tackle global warming.

8 Conclusions

This research has successfully tracked and measured the carbon footprint of flexible pavement, using Salatiga Outer Ring Road STA. 9 + 285 to STA. 9 + 385 with four 3.625-meter-width lanes as the case study. The total emissions of this project is 70.888 tonnes CO_2e, consisting of 34.248 tonnes CO_2e (48.31%) off-site activities and 36.640 tonnes CO_2e (51.687%) on-site activities. The two highest CO_2e emissions were generated by the use of plant for on-site activities for asphalt concrete laying activities accounted 34.827 tonnes CO_2e (49.130%), and material transportation accounted 24.921 (35.155%).

The findings of this research provide new information and portray pictures of carbon footprint sources and quantification model of carbon emissions in flexible pavement project. It has also provided a case study which may help stakeholders including government in formulating policies for national green road development. To gain more comprehensive understanding on the carbon footprint, it is recommended for future research to include also emissions at the operational and maintenance stages, as well as sub grade compaction work. In addition, as one case study may have limitation for generalization, more number of flexible pavement projects may provide more evidence and more comprehensive results.

References

1. IPCC, *Climate Change 2014–Impacts, Adaptation and Vulnerability: Regional Aspects* (Cambridge University Press, Cambridge, (2014))
2. Y. Chang, R.J. Ries, Y. Wang. Energy Policy, **38**(11), 6597-6603. (2010)
3. Y. Lu, P. Cui, D. Li, Building and Environment **95** (2016)
4. E.E. Keijzer, G.A. Leegwater, S.E. de Vos-Effting, M.S. de Wit, Environmental Science & Policy **54** (2015)
5. Y. Liu, Y. Wang, D. Li, Journal of Cleaner Production **144** (2017)
6. A. Galli, T. Wiedmann, E. Ercin, D. Knoblauch, B. Ewing, S. Giljum, Journal of Ecological Indicators **16** (2012)
7. M. Bishop, *Home to Reduce Carbon Footprint* (Crabtree Publishing Company, Canada, 2008)
8. ETAP, *The Carbon Trust Helps UK Business Reduce their Environmental Impact* (Press Release, Europe, 2007)
9. GFN, *Ecological Footprint Glossary* (Global Footprint Network, Oakland, 2007)
10. M. Kumar, L. Sharma, P. Vashista, International Journal of Emerging Technology and Advanced Engineering **4**, 1 (2014)
11. S. Kar, A. Behl, P.K. Jain, A. Shukla, Indian Highways **43**, 12 (2015)
12. T. Wiedmann, J. Minx, Ecological Economics Research Trends **1** (2008)

13. D. Chong, Y. Wang, The International Journal of Life Cycle Assessment **22**, 6 (2017)
14. P.G. Chandak, A.B. Tapase, S.S. Sayyed, A.C. Attar, *A state – of – the – art review of different conditions influencing the behavioral aspects of flexible pavement* (*International Congress and Exhibition, Sustainable Civil Infrastructures : Innovative Infrastructure Geotechnology*, (2017))
15. D. Huisingh, Z. Zhang, J.C. Moore, Q. Qiao, Q. Li, Journal of Cleaner Production **103** (2015)
16. J.M. Barandica, G. Fernández-Sánchez, Á. Berzosa, J.A. Delgado, F.J. Acosta, Journal of Cleaner Production **57** (2013)
17. IPCC, *IPCC Fourth Assessment Report : Climate Change 2007* (Cambridge University Press, Cambridge, 2007)
18. EPA, *Climate Change Indicators in The United States 2016 Fourth Edition* (2016), available at www.epa.gov
19. TCR, The Climate Registry, Climate Registry Default Emission Factor 2016, available at http://www.theclimateregistry.org/ (2016)
20. G. Kang, T. Kim, Y.W. Kim, H. Cho, K.I. Kang, Energy and Buildings **105** (2015)
21. V.J. Ferreira, A.S.D.G. Vilaplana, T. García-Armingol, A. Aranda-Usón, C. Lausín-González, A. M. López-Sabirón, G. Ferreira. Journal of Cleaner Production **130**, 175-186 (2016)
22. R. B. Mallick, M.J. Radzicki, M. Zaumanis, R. Frank. Resources, Conservation and Recycling **86**, 61-73 (2014)
23. M. Hoy, S. Horpibulsuk, A. Arulrajah. Construction and Building Materials, 117, 209-219 (2016).
24. X. Wang, Z. Duan, L. Wu, D. Yang, Journal of Cleaner Production **103** (2015)
25. D.R. Wirahadikusuma, P.H. Sahana, Jurnal Teoritis dan Terapan Bidang Rekayasa Sipil **19** (2012)

The Influence Of Highway Transportation Infrastructure Condition Toward Commodity Production Generation for The Resilience Needs at Regional Internal Zone

Juang Akbardin [1,3,*], *Danang* Parikesit [2], *Bambang* Riyanto [3], *Agus* Taufik Mulyono [2]

[1]Department of Civil Engineering, Universitas Pendidikan Indonesia, Bandung, Indonesia
[2]Department of Civil Engineering,Universitas Gadjah Mada, Yogyakarta, Indonesia
[3]Department of Civil Engineering,Universitas Diponegoro, Semarang, Indonesia

Abstract. The poultry commodity consumption and requirement is one of the main commodities that must be fulfilled in a region to maintain the availability of meat from poultry. Poultry commodity production is one of the production sectors that have a clean environment resistance. An increasing of poultry commodity generation production requires a smooth distribution to arrive at the processing. The livestock location as a commodity production is placed at a considerable far distance from residential and market locations. Zones that have poultry commodity production have an excess potential to supply other zones that are lacking in production to the consumption of these commodities. The condition of highway transportation infrastructure that is very diverse with the damage level availability in a zone has an influence in the supply and demand of poultry commodity requirement in the regional internal of Central Java province. In order to know the effect of highway transportation infrastructure condition toward the poultry commodity movement, demography factor and availability of freight vehicles will be reviewed to estimate the amount of poultry commodity movement generation production. Thus the poultry commodity consumption requirement that located in the internal - regional zone of central java province can be adequated from the zone. So it can be minimized the negative impacts that affect the environment at the zone in terms of comparison of the movement attraction and generation production at poultry commodity in Central Java.

1. Introduction

Increased demand for poultry commodities is increasing in line with the increasing demand for food consumption of the population in a developing region. Poultry commodities have an important role in food security in a region to increase protein nutritional needs and human resource quality development in the region. Production of poultry commodities in a region is increasing with the economic level per capita region and increasing demographic conditions. Increased poultry commodity production will directly increase the growth of traffic movement in distributing poultry commodities. Production generating of the poultry commodities movement requires the smooth distribution that is affected by the condition of the road infrastructure in each zone as well as the regional internal zone as a whole for a stability of the consumption tenacity needs of sustainable poultry commodity [1-3].

This research has intend to know the influence of road transportation infrastructure condition to the production generating of poultry commodity movement for the resilience of regional internal zone in Central Java Province with special purpose:

1. Identify and estimate the generating and attraction movement model of the poultry goods commodity sector based on the zones in the internal Regional Central Java Province
2. Analyze the generating and attraction movement model of the poultry commodity sector in the Regional Internal Central Java Province
3. Analyze the distribution model and the effective distribution distance in the movement for inter-zone production resilience in the internal region [1,3,4,5].

2. Literature Review

2.1 Model of Generation –Attraction Movement

One approach to transport planning in the four-stage transportation planning model is traffic generation (Trip Generation). This traffic generation depends on land use aspect, transportation and traffic flow can also be used approach quantitatively. The Trip Generation model generally estimates the number of trips for each travel destination based on the characteristics of land use and the socio-economic characteristics of each zone. The goal of Trip Generation planning is to estimate as accurate as possible the current traffic generation, which

* Corresponding author: akbardien@upi.edu

will be used for future predictions. Trip Generation modeling is used to predict the number of traffic generated for a given zone characteristic condition. The rise of movement is the number of moves generated by a zone of origin (Oi) and the number of moves that are interested in each destination zone (Dd) contained in the study area. The generating and attraction movement are strongly influenced by two aspects:[5,6]

2.1.1 Type of land use.

The amount of the land use activity (and intensity). The land use type has the characteristics of different movement generations, namely:

a. Different types of land use result in different movements
b. Different types of land use result in different types of movement
c. Different types of land use produce movement at different times.
d. Factors that influence the modeling of movement generations for people and goods are as follows::

2.1.2 Trip Production for human and goods travel determined :

a. Population income in a region and GDP of the area
b. Number of vehicle ownership as one of transportation services
c. Structure and size of households within a population of communities in residential areas that determine density
d. The value of land within the area determining the residential or commercial location
e. The accessibility of movement will encourage the rate of rise of movement

2.1.3 Trip Attraction for human movement and goods travel determined:

a. Industrial centers, commercial, shops and public services will bring workers and service users to a system of activities in a region
b. Number of employment, number of marketing places, wide of each industry.

The modeling used in movement generation is divided into two parts: Analytical model (Regression); Categories Model. In this study modeling the production and attraction movements using multiple regression analysis models [5,6].

2.2 Spatial Interaction Model of Goods Transportation

Black model (1972) is one of the gravity models that gives the formulation of the goods transportation total volume according to the transported commodity type from one place to another, as follows (Equation 1)

$$T_{idk} = \frac{S_{ik} . D_{dk} f_{id}}{\sum D_{dk} . f_{id}} \quad ....(1)$$

where :
T_{idk} = the number of k commodities that produced in the i region and sent to the d region d
S_{ik} = the total number of shipments of k commodities from i region
D_{dk} = the total number of demand of k commodity in d region
f_{id} = Friction/ barriers Factor (= 1 / $d_{id}\lambda$)
λ = parameter

2.3. Model Gravity (GR)

The most famous and often used method of synthetic (spatial interaction) is the Gravity Model (GR) because it is so simple that it is easy to understand and use. This model uses the concept of gravity introduced by Newton in 1686 developed from the analogy of the law of gravity.[6]

$$F_{id} = G . \frac{m_i . m_d}{d_{id}^2} \quad ...(2)$$

where G is the gravity constant
In geography, force can be considered as a movement between two regions; While mass can be replaced with variables such as population or movement generating and attraction, as well as distance, time, or cost as a measure of accessibility. Thus, for transportation purposes, the GR model is expressed as: (Equation 3)

$$T_{id} = k . \frac{O_i . O_d}{d_{id}^2} \quad ...(3)$$

where k is a constant

Thus, in mathematical form, the model GR can be expressed as:

$$T_{id} = O_i . D_d . A_i . B_d . f(C_{id}) \quad ...(4)$$

Both Barrier equations are satisfied if the constants A_i and B_d are used, which are associated with each zone of generating and attraction. The constants are called balancing factors ;[6]

$$A_i = \frac{1}{\sum_d (B_d D_d f_{id})} \quad ...(5)$$

$$B_d = \frac{1}{\sum_i (A_i O_i f_{id})} \quad ....(6)$$

2.4. The Model of Attraction Limitations (ACGR)

The ACGR method defines that the movement attraction is required. By using the basic formula Gravity model used the restriction provisions that by the formula :[6]

$$B_i = \frac{1}{\sum_{i=1}^{N} (A_d . O_i . f_{id})}$$ For all d and Ai = 1 for all i

2.5. Model Testing

2.5.1. Correlation Test

The correlation coefficient is a measure of how close the relationship between variables in the model. Correlation Test:[6,7]

$$r = \frac{N\sum_{i=1}^{N}(X_iY_i) - \sum_{i=1}^{N}(X_i).\sum_{i=1}^{N}(Y_i)}{\sqrt{\left[\sum_{i=1}^{N}(X_i)^2 - \left(\sum_{i=1}^{N}(X_i)\right)^2\right].\left[N\sum_{i=1}^{N}(Y_i)^2 - \left(\sum_{i=1}^{N}(Y_i)\right)^2\right]}} \quad ...(7)$$

2.5.2. Determination Test

$$R^2 = 1 - \frac{\sum_{i=1}^{N}\sum_{d=1}^{N}(\hat{T}_{id} - T_{id})^2}{\sum_{i=1}^{N}\sum_{d=1}^{N}(\hat{T}_{id} - T_1)^2} \quad ...(8)$$

$i \neq d$

2.6. Gravity Model Calibration

Gravity Model Calibration of Linear regression method with negative exponential barrier function. Calibration method by linear regression analysis to find the model parameters is done through the steps in the following equation :[3,6]

$$\exp(-\beta C_{id}) = \frac{T_{id}}{A_i.B_d.O_i.D_d}$$

$$log_e(\exp(-\beta C_{id})) = \log_e\left[\frac{T_{id}}{A_i.B_d.O_i.D_d}\right] \quad ....(9)$$

$$-\beta C_{id} = \log_e T_{id} - \log e(A_i.B_d.O_i.D_d)$$

$$\log_e T_{id} = \log_e(A_i.B_d.O_i.D_d) - \beta C_{id}$$

By linier transformation where : $\log^e$ Tid = Yi dan Cid = Xi

$$-\beta = B = \frac{N\sum_{i=1}^{N}(X_i.Y_i) - \sum_{i=1}^{N}(X_i).\sum_{i=1}^{N}(Y_i)}{N\sum_{i=1}^{N}(X_i^2) - \left(\sum_{i=1}^{N}(X_i)\right)^2} \quad (10)$$

$$A = \overline{Y} - B\overline{X}$$

3. Methodology

In this research, the modeling study of movement distribution use secondary data so for simplify it is used the development of zones or cities within the Central Java province with its development. Approach to the research of the movement distribution model of a very wide area that is much influenced by factors of land use in each zone requires a lot of data and information that is very complex for the accuracy Determination of Variable Model of Generation and Attraction [1-7]

3.1 Variable Dependent

Variables that affecting in movement generation (Oi) are:

Dependent variables
Y1 = Oi = Generated Movement of poultry Trip Transportation in the zone area.

Independent variables
X1 = Variable of population,
X2 = Variable of GRDP (Gross Regional Domestic Product),
X3= independent variable of IO Commodities poultry - zone area,
X4 = independent variable length of national roads in the district or city in Central Java,
X5 = independent variable length of provincial roads in the county and the city in Central Java,
X6 = independent variable length of local roads in the District and City in Central Java,
X7 = independent variable road conditions in both the District and the City in Central Java,
X8 = independent variable road conditions were in the District and City in Central Java,
X9 = independent variable condition of roads damaged in the District and City in Central Java,
X10 = independent variable road conditions, severely damaged in the District and City in Central Java,
X11 = independent variable number of goods vehicles in the county and the City of Central Java ,
X12 = independent variable number of goods vehicles with individual ownership status,
X13 = independent variable number of goods vehicles to the ownership status of the company[1-7]

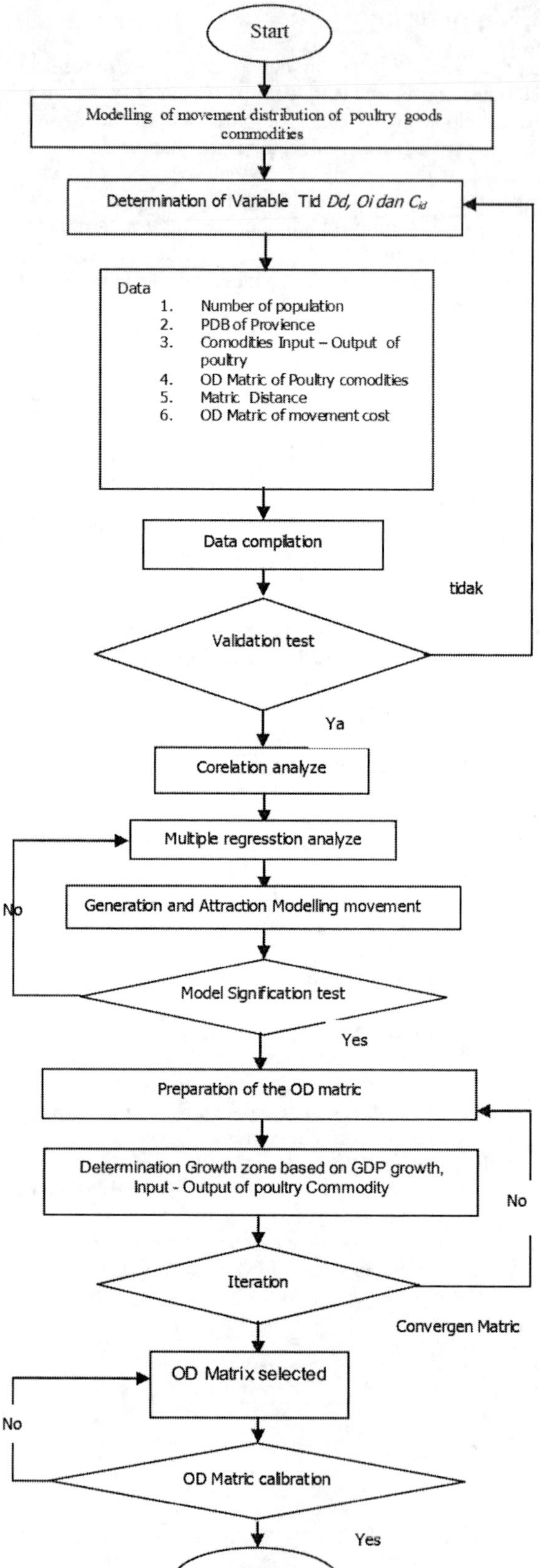

Fig 2. Modeling Flow Chart

4. Result and Discussion

4.1 Modelling of Generation and Attraction Movement of Poultry Goods Commodities

The socio-economic variables, the infrastructure condition (infrastructure) and the condition of facilities (moda) used in this study are: Projection based on trends (trend), that is based on the historical trend of development of independent variable parameters defined. Projection based on the pattern to be targeted, that is based on the direction of development to be achieved, generally this prediction is associated with spatial planning and economic development strategy in RTRW and MP3EI design. Future travel distributions in the form of MAT are obtained by including estimation of the received variables into the model. The future generating /attraction of travel can be predicted using multiple linear regression methods based on several variables of the coastal territorial condition. The study area is divided into zones, in accordance with the assumption in transport modeling that the movement starts and ends from / to a point in the zone commonly referred to as the zone centroid. While the determination of the zone system (including its limits) is based on the administrative boundary system (county and city). The zone is defined from the number of districts and cities in Central Java. [3,5,6,7]

Table 1. Results of Double Regression Analysis Phase II

Yi = 2,98 + 0,864 X1 + 0,111 X2 + 0,0637 X3 + 0,0343 X4 + 0,0063 X5 - 0,038 X6 + 0,0090 X7 - 0,0568 X8 + 0,0195 X9 - 0,0145 X10 - 0,0924 X13

Predictor	Coef	SE Coef	T	P	VIF
Constant	2,977	2,010	1,48	0,152	
X1	0,8642	0,1413	6,12	0,000	6,4
X2	0,11121	0,08889	1,25	0,223	1,7
X3	0,06368	0,04862	1,31	0,203	2,1
X4	0,03430	0,03196	1,07	0,294	1,7
X5	0,00629	0,05292	0,12	0,906	3,2
X6	-0,0379	0,1197	-0,32	0,755	5,4
X7	0,00901	0,07012	0,13	0,899	1,9
X8	-0,05677	0,08185	-0,69	0,495	3,3
X9	0,01948	0,08339	0,23	0,817	4,9
X10	-0,01453	0,02687	-0,54	0,594	1,5
X13	-0,09243	0,06446	-1,43	0,165	1,8

S = 0,197069 R-Sq = 90,8% R-Sq(adj) = 86,4%

Analysis of Variance

Source	DF	SS	MS	F	P
Regression	11	8,79493	0,79954	20,59	0,000
Residual Error	23	0,89323	0,03884		
Total	34	9,68815			

Source : Data Analysis, 2016

4.1.1. Normality test

Normality test used in this research is One Sample Kolmogorov-Smirnov Test with complete observation through scatterplot chart. In the One Sample

Kolmogorov-Smirnov test, it is known that the normal distribution of data if the residual has Asymp. Sig (2-tailed) above 15% and vice versa if residual has Asymp. Sig (2-tailed) below 15% then the data distribution in regression analysis is not normal. The normality test is shown in Fig. 3.[7]

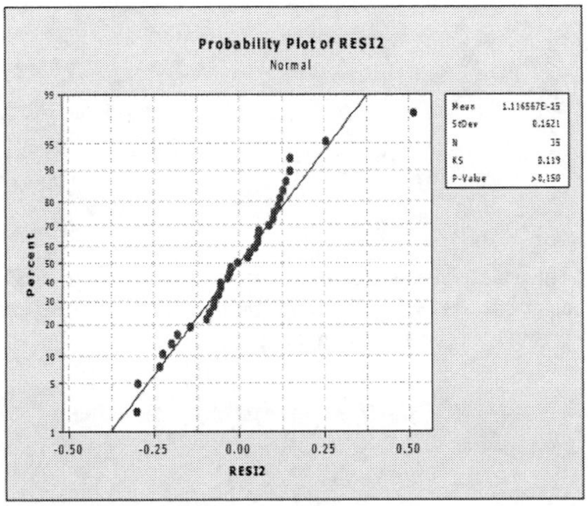

Fig. 3. Graph of Normality Test of Poultry Commodities

4.1.2. Anova Test (Analisis Of Variance)

Anova test shown in Table 1 yielded multiple regression model of commodity movement based on F test value resulting F value counted 20.59 bigger than F table (F value table with dfl = 11 and df2 = 23 ie 2,24) with significance level 0,000. The probability of significance is much smaller than 0.05, then the regression model can be used to predict Y as the Awakening (Oi). Assessment of model parameters from the regression results using the least square estimation method shown in Table output. 1. the result of analysis of computation program Minitab 16 is formulated in the formation of equation with Logarithm Natural (Ln) shown in equation. [4]

$$\text{Ln } Y_i = 2{,}98 + 0{,}864 \text{ Ln } X1 + 0{,}111 \text{ Ln } X2 + 0{,}0637 \text{ Ln } X3 + 0{,}0343 \text{ Ln } X4 + 0{,}0063 \text{ Ln } X5 - 0{,}038 \text{ Ln } X6 + 0{,}0090 \text{ Ln } X7 - 0{,}0568 \text{ Ln } X8 + 0{,}0195 \text{ Ln } X9 - 0{,}0145 \text{ Ln } X10 - 0{,}0924 \text{ Ln } X13 \quad \ldots (11)$$

4.2 Estimated results of the Attraction Modelling of Poultry Commodity Movement on Zones in Central Java

Furthermore, based on the result of modelling that use movement attraction data. Estimated results from attraction modelling of poultry commodity movement is estimated based on data condition from each zone in Central Java. By using Matlab computation method, the simulation model of movement generation in each zone according to the magnitude of movement is shown in Figure 4.

The rise of poultry production movement in each zone in Central Java has characteristic dominated by condition of middle passage path and connecting lane from region internal of Central Java province region.[4,-,7]

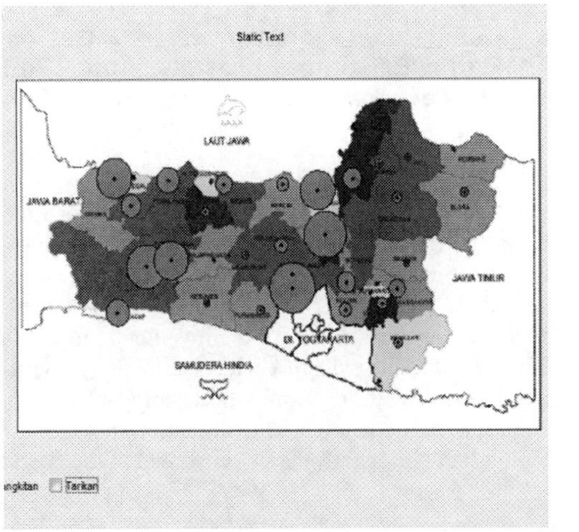

Fig. 4. Estimation of the Poultry Commodities Movement and Its Results by Zone In Central Java

4.3 Estimated results of Poultry Movement Modelling Model in Zones in Central Java

Furthermore, based on the model results using tensile motion data, the result estimation of the poultry commodities movement modelling simulated in each zone in Central Java by using Matlab computation method shown in Figure 5.

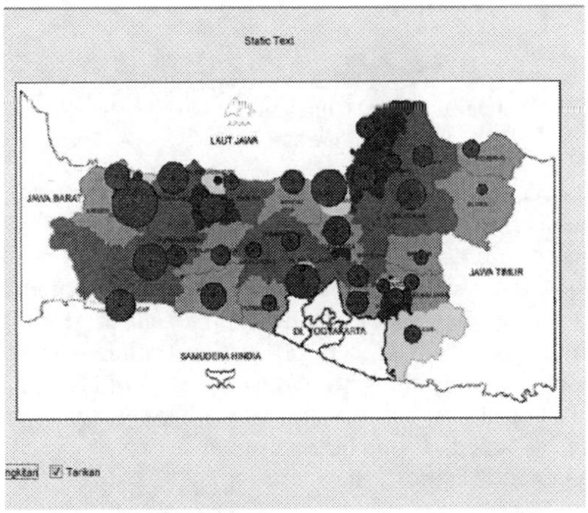

Fig. 4. Estimated Attraction Movement of Poultry Commodities and Results by zone In Central Java

The Estimation Result of the Movement modeling in each zone in Central Java shows that most of the pull of poultry commodity movement has almost uniform characteristics with higher quantity in cities with demographic condition and higher GRDP level. Especially in big cities in Central Java have the pull of movement of poultry commodity is greater. This shows that the consumption of poultry commodities and derivatives is quite large within the regional internal of Central Java Province. [4-7]

4.4 Comparison of Movement Generaton Modelling Estimated results of Poultry Commodities Movement at Zones in Central Java

By knowing the magnitude of the generation and attraction of poultry commodities movement in each zone in the region of Central Java Province, Supply and demand conditions of poultry commodities can be obtained by knowing the comparison between the movement generation and the movement attraction of poultry commodities from each zone in the province of Central Java. Results Simulation comparison of the production of generating and attraction movement by using matlab computation method shown in the figure. 5

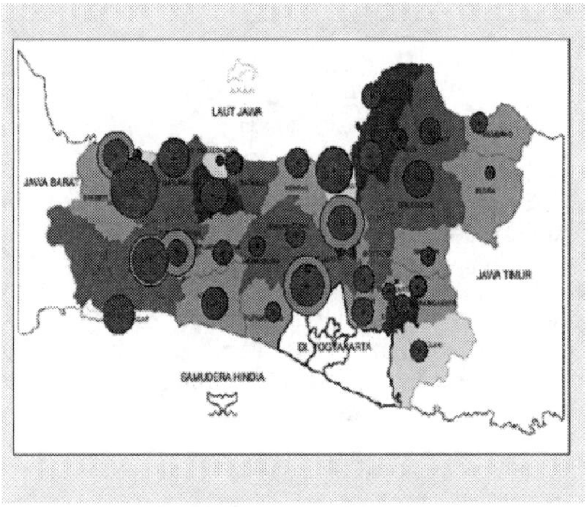

Fig. 5. Comparison of Generation and Atraction Movement Estimation of Poultry Commodities and Results by zone In Central Java

Based on the estimation of simulation results, the comparison of poultry commodity movements generation and attraction in each zone in Central Java shows that the pull of movement of poultry commodities is greater than the production of movement generation. In zones with higher production volumes indicate contribution to the surrounding zones. So the resistance of a zone to the consumption of poultry commodities can be obtained from the surrounding zone based on production capacity in a more surplus zone.[4-7]

4.5 Model Distribution Movement of Poultry Commodities with Limit of Attraction Movement on Zone in Central Java

By using the exponential negative resistance function ACGR stages are as follows: The calculation process on distribution model estimation is done by the iteration process shown in equation 6

$$B_1 = \frac{1}{[A_1.O_1.\exp(-\beta C_{1,1}) + A_2.O_2.\exp(-\beta C_{2,1}) + \ldots + A_{35}.O_{35}.\exp(-\beta C_{35,1})]}$$

$$B_2 = \frac{1}{[A_1.O_1.\exp(-\beta C_{1,2}) + A_2.O_2.\exp(-\beta C_{2,2}) + \ldots + A_{35}.O_{35}.\exp(-\beta C_{35,2})]}$$

$$B_3 = \frac{1}{[A_1.O_1.\exp(-\beta C_{1,3}) + A_2.O_2.\exp(-\beta C_{2,3}) + \ldots + A_{35}.O_{35}.\exp(-\beta C_{35,3})]}$$

$$\ldots \qquad . \quad (12)$$

$$B_{35} = \frac{1}{[A_1.O_1.\exp(-\beta C_{1,35}) + A_2.O_2.\exp(-\beta C_{2,35}) + \ldots + A_{35}.O_{35}.\exp(-\beta C_{35,35})]}$$

The value of Bd subsequently used obtained is used to calculate the value of each matrix cell using the basic formula of the gravity model in equation 7

$$\begin{aligned}
T_{1.1} &= O_1.A_1.B_1.D_1.\exp(-\beta(C_{1.1}) \\
T_{1.2} &= O_1.A_1.B_2.D_2.\exp(-\beta(C_{1.2}) \\
T_{1.3} &= O_1.A_1.B_3.D_3.\exp(-\beta(C_{1.3}) \\
&\ldots \qquad ..(13) \\
T_{2.1} &= O_2.A_2.B_1.D_1.\exp(-\beta(C_{2.1}) \\
&\ldots \\
&. \\
&. \\
T_{35.35} &= O_{35}.A_{35}.B_{35}.D_{35}.\exp(-\beta(C_{35.35})
\end{aligned}$$

By using the basic equation of the gravity model, the matrix iteration calculation is done by computing programming with MATLAB 2012 so as to get the value of β in the convergent matrix condition.[6,7]

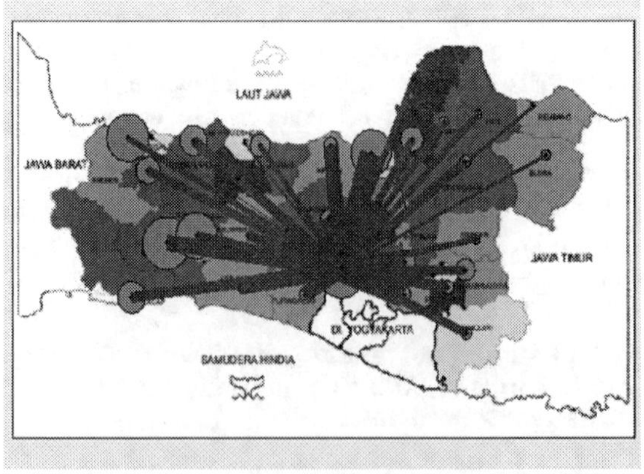

Fig. 6. Simulation Sample of Distribution from production zone to zones in Central Java Province.

Figure 6 shows that the production of a surplus poultry commodity will supply the less-productive zone of poultry production.

https://doi.org/10.1051/e3sconf/20183107002

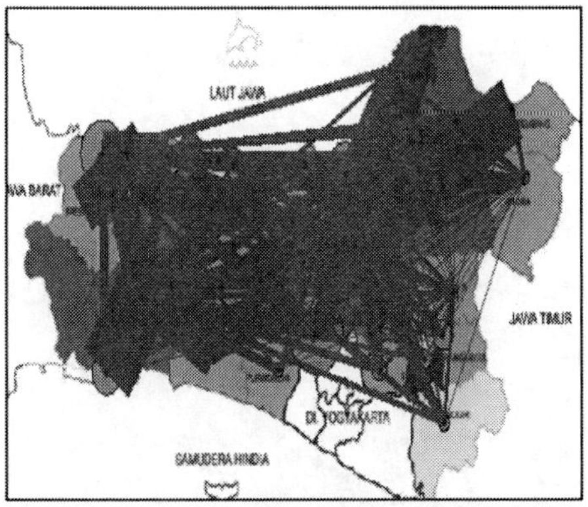

Fig. 7. Simulation of Distribution between production zones to zones throughout the Central Java Province.

Figure 7 shows that the magnitude of the *dissire line* from and to the production zone has large magnitudes, indicating that the volume of distribution movements between production zones of poultry commodity movements generation and attraction has an effect on the distance of movement between zones. The interaction of cultivation. The resilience of poultry commodities is determined from the level of production and consumption at zones in accordance with the growing needs and market potential of poultry commodities within the internal - regional of Central Java province.[5-7]

4.6 Distribution Distances of Poultry Commodities Movement with the Atraction Movement Limit at Zone in Central Java

Based on the modeling results of the distribution of poultry commodity movements in each zone in Central Java province shows that the generation and attraction movement of poultry commodities has a very large distance distribution volume at the distance between zones 100-200 km shown in Figure 8

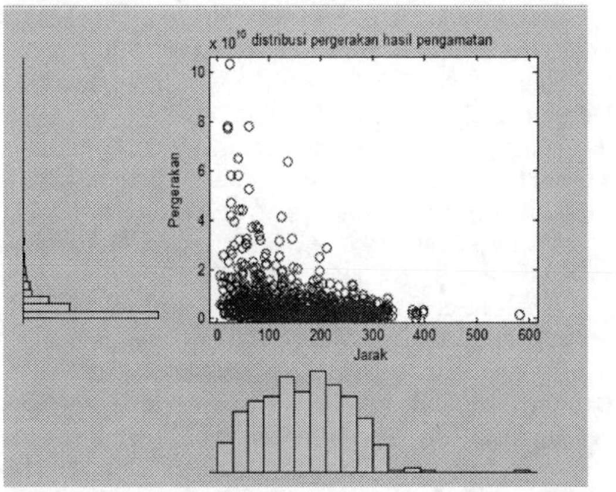

Fig .8.. Distribution Distance of movement generation and attraction between poultry commodities zone.

It shows that the interaction between zones with the distance has a very strong relationship on the system of inter-zone dependence on supply and demand needs required. The resilience of poultry production needs has a great contribution to the price stability of poultry consumption in a region independently. Resilience based on the stability of goods movement distribution requires the support of good road infrastructure condition and smoothly to maintain a smooth flow of distribution. So that will reduce the impact of environmental decline due to the movement of shipping poultry production delivery traffic in the distribution path channel that resulted in environmental pollution due to odor generated due to delivery of poultry from the production zone to the consumption zone. [5-7]

5 Conclusion

Based on data analysis and modeling from the generation and attraction of poultry commodity production in the Central Java province region can be concluded that the condition of road infrastructure has a significant influence on the amount of generation and attraction movement production of poultry commodities. A smooth cross-zone distribution system with the support of road infrastructure and adequate transportation conditions will strengthen the resilience of the production and consumption needs of stable and independent poultry commodities within a regional internal area. Effective and efficient distribution distances will reduce the impact of odor pollution generated in the movement of poultry commodities between zones. So that further requires the better arrangement of supply chain system from poultry commodity movement system in distribution system.

References

1. G. Bergeron, S. Morris, and J. Manuel Medina Banegas. Food Consumption and Nutrition Division Discussion Paper Number 43. Washington, D.C.: International Food Policy Resarch Institute, (1998).
2. Coates, Jennifer, Patrick Webb and Robert Houser. Washington, D.C.: Food and Nutrition Technical Assistance Project, FHI 360, (2003)
3. J. Holguin-Veras, and E. Thorson, Transportation Research Record 1707, pp37-48(2000)
4. Friesz, T.L.,J. Gottfried and E.K. Morlok. Transportation Science, 20 (1), pp. 80-91(1986)
5. C. Winston, Transportation Research Part A, 17.pp.419-427. (1983)
6. Tamin Z. Ofyar *Perencanaan dan Permodelan Transportasi,* Edisi kedua, ITB Bandung. (2000),
7. Ghozali, Imam *Aplikasi Analisis Multivariate dengan IBM SPSS 19*, Badan Penerbit Universitas Diponegoro , Semarang (2001)

The Effectiveness of Building Permit Regulation for Green Open Space at Housing Estates: Case Study of Kendal Regency, Central Java, Indonesia

Wiwik Yulianti[1] and *Sudharto* P. Hadi[1]

[1]Master Program of Environmental Studies, School of Postgraduate Studies Diponegoro University, Semarang – Indonesia

Abstract. Increasing demand for settlements steamed by population growth declines the quality of the environment specifically at urban area. The existing spatial planning could not able to prevent the change of land use for settlement and other infrastructures. The Act no. 26 of 2007 on spatial planning stipulates that green open space must reach 30% of the total area, consisting of 20% public open space and 10% private open space. The existing condition of urban area at Kendal Regency reach 245,6 million m^2 with 88.145,5 m^2 green open space or 0,036% out of total area. An effort to increase green open space in urban areas taken by the Government of Kendal Regency is by promulgating a local regulation stipulating that each housing developer request a building permit is obliged to provide a green open space at least 10 percent of the total housing area. This paper reviews the effectiveness of building permit regulation, the problems encountered and the concept proposed to make the local regulation work. The area of sample taken is three urban districts out of five urban districts, the resource persons chosen are those from relevant offices (Dinas) involved at the implementation of the local regulation. The data collection techniques employed are the Analytical Hierarchy Process (AHP), Geographic Information System (GIS) technology, social observation and informal interview. The data gathered will be analyzed quantitatively and qualitatively.

1 Introduction

Physical development as an implication of population grow this inevitable because of the need for the development growth of residential areas, industrial estates, trade and service areas, and supporting facilities such as roads, terminals, ports, airports, etc [1]. The population's need for space for shelter, activities, and activity support is following the rate of population growth in a region.

The change of land use shows an increase in human needs, especially the primary need to build houses. Currently, the provision of housing has become a potential business project [2]. The increase in housing demand is driving the emergence of housing provided by housing developers. Land use change can not be denied occurs in every inch of the city, changing the original area of the natural area into the built area. Green areas are displaced into residential areas.

Therefore, the direction of development is not only focused on the acceleration of economic growth but also on improving the quality of human life through balancing between development and the environment. Development in a region must always consider the social conditions of the population and environmental aspects known as sustainable development. In the world that more urbanized, the need for sustainable urban and residential development is important [3].

The implementation of the concept of sustainable urban development began in the early 1990s, preceded by the UN Conference on Environment and Development (UNCED) in 1992, and the 15th UNCHS in Jakarta 1995, which identifies key sustainable development measures for human settlements. The Commission demonstrates that sustainable development is not only a new way of environmental protection, but a 'new concept of economic growth that ensures justice and opportunity for everyone in the world without destroying natural resources and without reducing the world's carrying capacity'. In 1996, UNCHS or UN-Habitat expanded the concept of sustainable development for urban planning. That 'Settlement planning plays an important role to ensure management and urban development achieve sustainable development goals'[4].

The concept of sustainable cities is closely linked to economic development, protection of resources and the environment, which in turn leads to achievement of

* Corresponding author: anti.juli@gmail.com

https://doi.org/10.1051/e3sconf/20183107003

acceptable minimum quality of life [5]. And there needs to be an ongoing effort to address the issue of air pollution, congestion, human populations and the availability of green open spaces. A strong, healthy and habitable city depends on a healthy environment, a strong economy and adequate employment opportunities for its citizens [6].

Housing policies should be directed towards achieving sustainable housing development goals. Tolba's statement as quoted by Choguill [5], sustainable housing can be achieved with 4 (four) criteria, that is economically feasible, socially acceptable, technically feasible and environmentally friendly.

Government efforts to accommodate sustainable settlement fulfillment are held through the development of Green Open Space. Urban green open space is one of the important elements of the habitable city [7]. In principle, green open space is intended to suppress the negative effects of urban built environments, such as decreasing the rate of water absorption, increasing air temperature and humidity, pollution, etc. [8].

Increasing demand for settlements steamed by population growth declines the quality of the environment specifically at urban area. The existing spatial planning could not able to prevent the change of land use for settlement and other infrastructures. The Act no. 26 of 2007 on spatial planning stipulates that green open space must reach 30% of the total area, consisting of 20% public open space and 10% private open space. The existing condition of urban area at Kendal Regency reach 245,6 million m2 with 88.145,5 m2 green open space or 0,036% out of total area. An effort to increase green open space in urban areas taken by the Government of Kendal Regency is by promulgating a local regulation stipulating that each housing developer request a building permit is obliged to provide a green open space at least 10 percent of the total housing area.

Through the mechanism of the technical requirements of the building permit expected growth of urban green space area following the growth of housing in urban areas, Kendal. The results of Kristianova's research[9] indicated that the supply of green open space in residential areas is often defeated by social and economic interests, such as the development of housing itself (densification) and the development of facilities both commercial services (parking) and community (places of religious activity) Additional facilities previously unplanned during the housing development planning.

This article applies the combination of visualization tools with a specialized analysis method to evaluate the effectiveness of building permit regulation for green open space at housing estates in Kendal Regency, Central Java, Indonesia. This paper reviews the effectiveness of building permit regulation, the problems encountered and the concept proposed to make the local regulation work. The area of sample taken is three urban districts out of five urban districts, the resource persons chosen are those from relevant offices involved at the implementation of the local regulation. The data collection techniques employed are the analytic hierarchy process (AHP), Geographic Information System (GIS) technology, social observation and informal interview. The data gathered will be analyzed quantitatively and qualitatively.

2 Study Case

Figures and tables, as originals of good quality and well The area of sample taken is three urban districts out of five urban districts in Kendal Regency, Central Java. The locus of research include: Kendal District, Kaliwungu District, and Weleri District

Three districts in the research samples selected with the justification that the three districts in the corridor of the north coast of Java that in fact the urban level higher than the urban areas are in the southern region of northern coastal road.

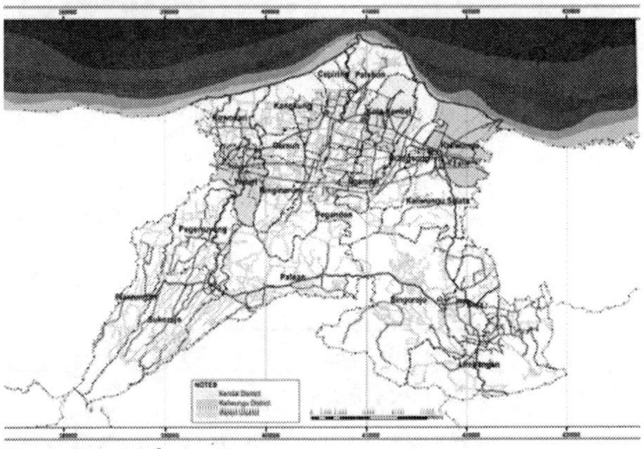

Fig. 1. Case Study Area

3 Effectiveness Analysis

The Geographical Information System (GIS) application is used to clarify the visualization of the research object. GIS applications are used to identify housing estates areas and green open spaces in urban areas through delineation green areas with remote sensing data of Upright High Resolution Satellite Imagery of Kendal scale 1: 5000. The results of the calculation of the existing green open spaces of the effectiveness analysis stage.

Measurement of effectiveness is done by adapting the Satries et al. [10] method with the appropriate adjustment of research materials, then the formula used as follows:

$$\textit{Percentage Open Space Area} = \frac{\textit{The area of green open space in housing estates}}{\textit{Housing Estates Area}} x\ 100\%$$

 https://doi.org/10.1051/e3sconf/20183107003

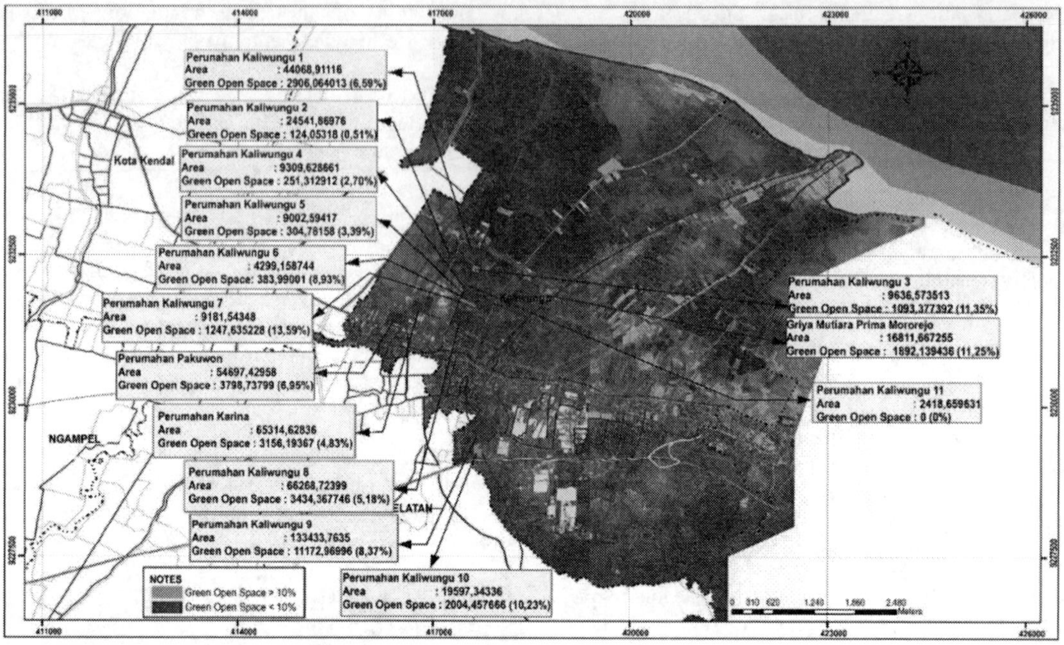

Fig. 2. Housing Area in Kaliwungu District

Table1. Green Open Space Percentage of Housing Estates in Kaliwungu District

No	Housing Estates	Housing Estates Area (m2)	Green Open Space Area (m2)	% Open Space	Note
1	*Perumahan Kaliwungu 11*	2.418,66	0,00	0,00	<10%
2	*Perumahan Kaliwungu 2*	24.541,87	124,05	0,51	<10%
3	*Perumahan Kaliwungu 4*	9.309,63	251,31	2,70	<10%
4	*Perumahan Kaliwungu 5*	9.002,59	304,78	3,39	<10%
5	*Karina*	65.314,63	3.156,19	4,83	<10%
6	*Perumahan Kaliwungu 8*	66.268,72	3.434,37	5,18	<10%
7	*Perumahan Kaliwungu 1*	44.068,91	2.906,06	6,59	<10%
8	*Pakuwon*	54.697,43	3.798,74	6,95	<10%
9	*Perumahan Kaliwungu 9*	133.433,76	11.172,97	8,37	<10%
10	*Perumahan Kaliwungu 6*	4.299,16	383,99	8,93	<10%
11	*Perumahan Kaliwungu 10*	19.597,34	2.004,46	10,23	>10%
12	*Griya Mutiara Prima Mororejo*	16.811,67	1.892,14	11,25	>10%
13	*Perumahan Kaliwungu 3*	9.636,57	1.093,38	11,35	>10%
14	*Perumahan Kaliwungu 7*	9.181,54	1.247,64	13,59	>10%
Average				**6,70**	

Source: Spatial Analysis, 2017

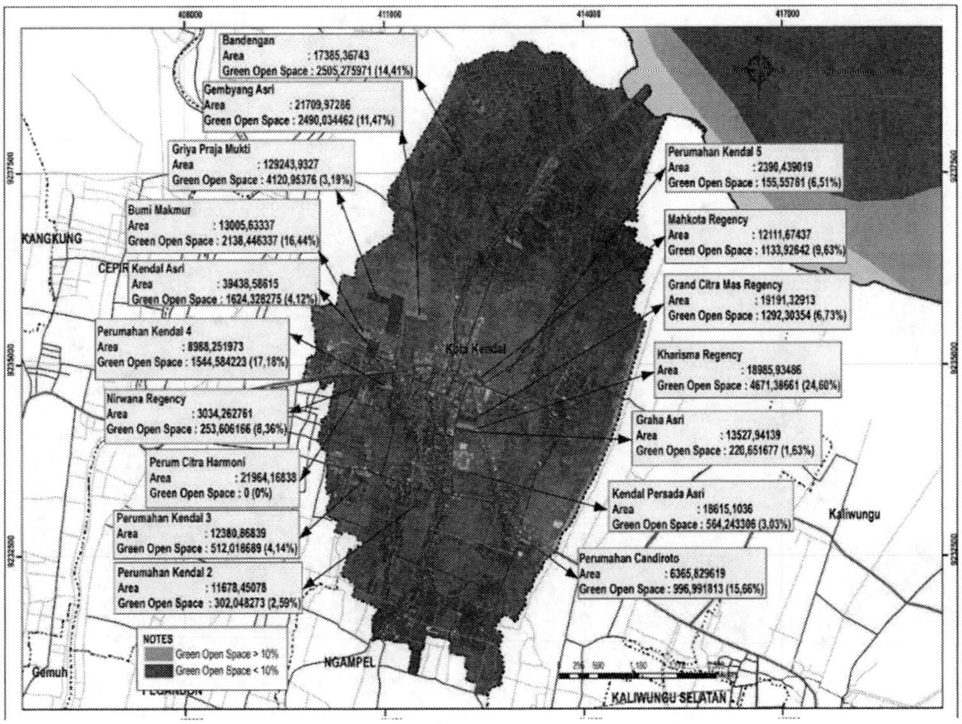

Fig.3. Housing Area in Kendal Kota District

Table2. Green Open Space Percentage of Housing Estates in Kendal Kota District

No	Housing Estates	Housing Estates Area (m2)	Green Open Space Area (m2)	% Open Space	Note
1.	*Perum Citra Harmoni*	21964,16838	0	0,00	<10%
2.	*Graha Asri*	13527,94139	220,651677	1,63	<10%
3.	*Perumahan Kendal 2*	11678,45078	302,048273	2,59	<10%
4.	*Kendal Persada Asri*	18615,1036	564,243306	3,03	<10%
5.	*Griya Praja Mukti*	129243,9327	4120,95376	3,19	<10%
6.	*Kendal Asri*	39438,58615	1624,328275	4,12	<10%
7.	*Perumahan Kendal 3*	12380,86839	512,018689	4,14	<10%
8.	*Perumahan Kendal 5*	2390,439019	155,55761	6,51	<10%
9.	*Grand Citra Mas Regency*	19191,32913	1292,30354	6,73	<10%
10.	*Nirwana Regency*	3034,262761	253,606166	8,36	<10%
11.	*Mahkota Regency*	12111,67437	1133,92642	9,36	<10%
12.	*Gembyang Asri*	21709,97286	2490,034462	11,47	>10%
13.	*Bandengan*	17385,36743	2505,275971	14,41	>10%
14.	*Candiroto*	6365,829619	996,991813	15,66	>10%
15.	*Bumi Makmur*	13005,63337	2138,446337	16,44	>10%
16.	*Perumahan Kendal 4*	8988,251973	1544,584223	17,18	>10%
17.	*Kharisma Regency*	18985,93486	4671,38661	24,60	>10%
Average				**8,79**	

Source: Spatial Analysis, 2017

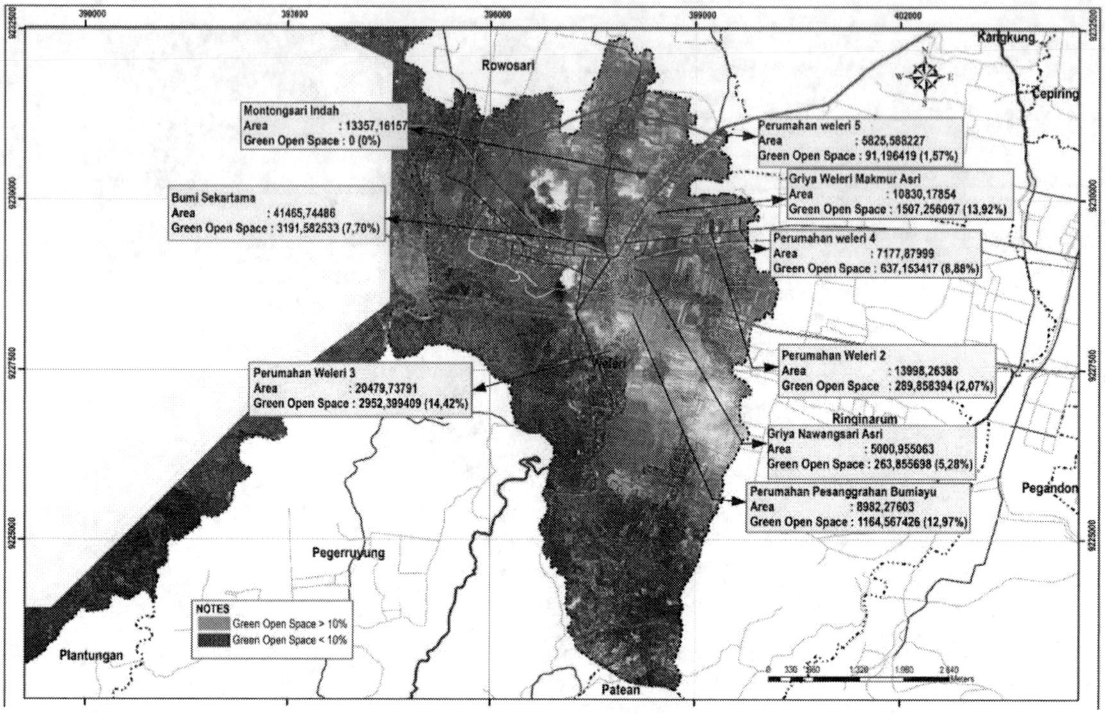

Fig.4. Housing Area in Weleri District

Table3. Green Open Space Percentage of Housing Estates in Weleri District

No	Housing Estates	Housing Estates Area (m2)	Green Open Space Area (m2)	% Open Space	Note
1.	*Montongsari Indah*	13357,16157	0	0,00	<10%
2.	*Perumahan Weleri 5*	5825,588227	91,196419	1,57	<10%
3.	*Perumahan Weleri 2*	13998,26388	289,858394	2,07	<10%
4.	*Griya Nawangsari Asri*	5000,955063	263,855698	5,28	<10%
5.	*Bumi Sekartama*	41465,74486	3191,582533	7,70	<10%
6.	*Perumahan Weleri 4*	7177,87999	637,153417	8,88	<10%
7.	*Perumahan Pesanggrahan Bumiayu*	8982,27603	1164,567426	12,97	>10%
8.	*Griya Weleri Makmur Asri*	10830,17854	1507,256097	13,92	>10%
9.	*Perumahan Weleri 3*	20479,73791	2952,399409	14,42	>10%
Average				**7,42**	

Source: Spatial Analysis, 2017

The calculation results of the effectiveness of each housing then recapitulated and calculated the average value. Based on Arikuntoat Satries et al., Interpretation of the result of the percentage of the average value refers to the following table .

Table 4. Effectiveness Value Interpretation Standard

Number of Green Open Space Area <10%	Value Interpretation
80% - 100%	Very Low
60% - 79,9%	Low
40% - 59,9%	Rather Low
20% - 39,9%	Enough
0% - 19,9%	High

Table 5. The Result of Effectiveness Value Interpretation

No	District	Green Open Space >10%	Green Open Space <10%	Percentage	Effectiveness Value Interpretation
1	Kaliwungu	4	10	71,43	Low
2	Kota Kendal	6	11	64,71	Low
3	Weleri	3	6	66,67	Low

Based on the interpretation of the effectiveness value, the value of building permit effectiveness in the fulfillment of green open spaces of housing estates area can be determined.

Based on spatial analysis it is seen that the three districts have low effectiveness value in enforcement of building permit related to the fulfillment of green open space standard in housing estates area.

4 Regulation Implementation Effectiveness Strategies

Strategy formulation combines the corresponding theories such as the effectiveness of law enforcement theory, public policy, and environmental management. The formulation of indicators, criteria, and variables for effectiveness of regulation implementation are as follows.

Table 6. The Formulation of Indicators, Criteria, and Variables for Regulation Implementation Strategies

NO	INDICATORS	CRITERIA [11]	VARIABLES
1.	INPUT ([12],[13],[14])	A. Organization Characteristics[15]	1. Institutional Climate([11],[16],[17])
			2. Institutional Goals ([12],[18],[19],[20])
			3. Institutional Regulation ([17], [19], [21], [22], [23])
			4. Standard Operating Procedures (SOP) [24]
			5. Violation Detection System [25]
			6. Source of Budget Funds ([15], [22], [24], [26])
			7. Facilities and Infrastructure ([11], [12],[15],[24])
			8. Incentives dan Disincentives, ([16],[22], [26])
			9. Sanctions [24]
		B. Actors Characteristics ([19],[21])	1. Leadership ([18], [19], [22], [27])
			2. Rational Actors[26]
			3. Competence ([3], [22], [26])
			4. The divison of Roles ([3], [11])
			5. Commitment([11],[15],[18], [19],[26])
2.	PROCESS ([12], [13], [14], [17])	A. Management Characteristics	1. Implementation Strategy[12]
			2. Instrument Implementation[11]
			3. Duty and Authority([3],[23])
			4. Communication ([11],[19],[26])
			5. Coordination ([22], [26])
			6. Responsiveness[24]
			7. Accountabilty[28]
			8. Monitoring [12]
			9. Evaluation [15]
		B. Environment Characteristics (Internal and External) ([12], [19], [21], [27])	1. Task alignment[12]
			2. Alignment of Goals[26]

NO	INDICATORS	CRITERIA [11]	VARIABLES
			3. Policy Alignment[23]
			4. Participation ([11], [15], [20], [28])
			5. Obedience ([16], [26],[28])
			6. Rule of Law ([15],[20],[22], [23], [27])

Expert Choice (EC)

Expert Choice (EC) v11 issued as one of the tools to help decision makers in making decisions. Based on priority assessment by Kendal District Licensing Team, the priority of indicator, criterion, and evaluation variable of the maturity of the regulation implementation.

Hierarchy of priority assessment of indicators, criteria, and variable assessment of the maturity of regulatory implementation as following fig.

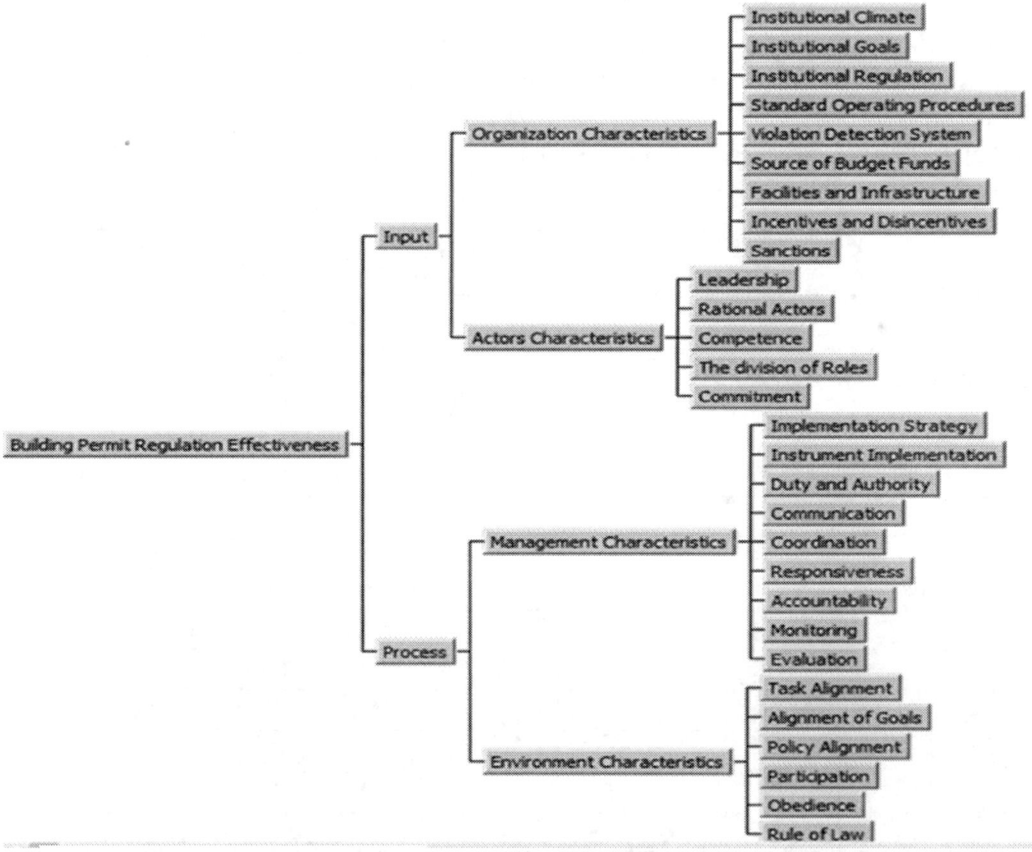

Fig. 5. Hierarchy of Priority Assessment

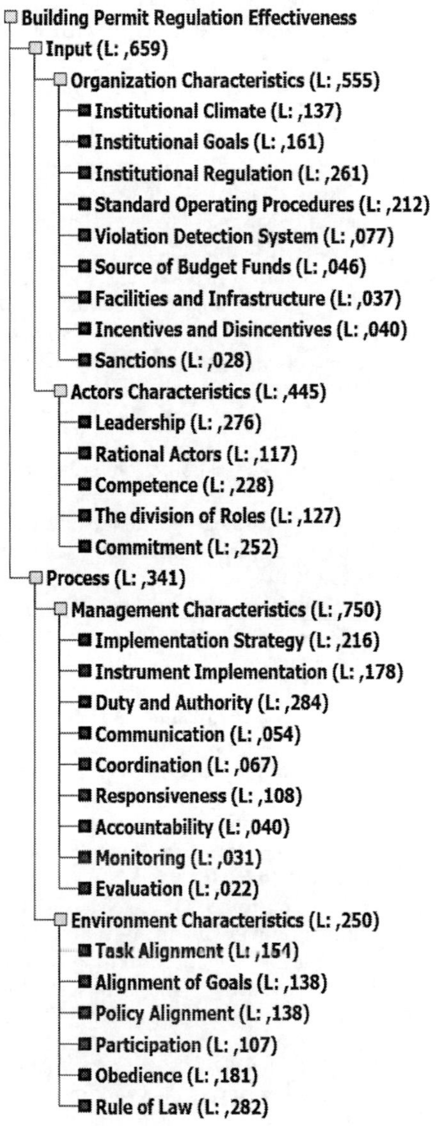

Fig. 6 The Result of AHP

To realize the effectiveness of the implementation of the regulation, the implementation of activities in accordance with the above priority values.

5 Conclusion

The conclusions that can be drawn from the above description are:

1. By calculating the priority of indicators, criteria, and variables of maturity regulation implementation above, it can be designed the initial scenario of regulation implementation strategies.
2. In accordance with the results of the study, the effectiveness of building permit regulations in the fulfillment of private green open space standards in residential areas is influenced by input implementation, especially organizational internal characteristics (institutional goals, institutional climate, violation detection system, source of budget funds, incentives and disincentives, facilities and infrastructure, and sanctions). The second factor affecting effectiveness of regulatory implementation is actors characteristics (leadership, commitment, competence, the division of roles, and rational actors).
3. The third and fourth factors affecting the effectiveness of building permit regulations in the fulfillment of private green open space standards in residential areas are part of the implementation process, ie. management characteristics (duty and authority, implementation strategy, instrument implementation, responsiveness, coordination, communication, accountability, monitoring and evaluation). The last factor affecting the effectiveness of regulatory implementation is the factor deriving from the object of the regulation itself, namely Environment Characteristics (ie. rule of law, obedience, task Alignment, alignment of goals, policy alignment, and participation)

References

1. S. P. Hadi, *Manusia dan Lingkungan*, Cetakan II. Semarang: Badan Penerbit Universitas Diponegoro (2013)
2. B. Widodo, R. Lupyanto, B. Sulistiono, D. A. Harjito, J. Hamidin, E. Hapsari, M. Yasin, and C. Ellinda, "Analysis of environmental carrying capacity for the development of sustainable settlement in Yogyakarta urban area," *Procedia Environ. Sci.*, vol. **28**, no. SustaiN 2014, pp. **519–527** (2015)
3. L. Smeddle-Thompson, "Implementing Sustainable Human Settlements," Stellenbosch University (2012)
4. UN-Habitat, "Planning Sustainable Cities: Global Report on Human Settlements 2009," London (2009)
5. C. L. Choguill, "The search for policies to support sustainable housing," *Habitat Int.*, vol. **31**, no. **1**, pp. **143–149**(2007)
6. C. V Katju, "Sustainable Cities," *Vol. **10** No. 7 July* (2000). [Online]. Available: http://www.devalt.org/newsletter/jul00/lead.htm. [Accessed: 07-May-2017].
7. C. Y. Jim and S. S. Chen, "Comprehensive greenspace planning based on landscape ecology principles in compact Nanjing city, China," *Landsc. Urban Plan.*, vol. **65**, no. **3**, pp. **95–116**(2003)
8. Hasni, *Hukum Penataan Ruang dan Penatagunaan Tanah dalam Konteks UUPA-UUPR-UUPPLH*, Ed. **3**, Cet. Jakarta: Rajawali

Pers (2016)

9. K. Kristianova, "Post-Socialist Transformations of Green Open Spaces in Large Scale Socialist Housing Estates in Slovakia," *Procedia Eng.*, vol. **161**, pp. **1863–1867** (2016)
10. W. I. Satries, "Efektivitas Program Pemberdayaan Pemuda pada Organisasi Kepemudaan Al Fatih Ibadurrohman Kota Bekasi," Universitas Indonesia (2011)
11. T. W. Steers, Richard M and Lee, "Facilitating Effective Performance Appraisals: The Role of Employee Commitment and Organizational Climate," in *Performance Measurement and Theory*, New Jersey: Lawrence Erlbaum Associates, Inc(1983), pp. **75–87**.
12. S. Oberthür and L. Groen, "The Effectiveness Dimension of the EU's Performance in International Institutions: Toward a More Comprehensive Assessment Framework," *J. Common Mark. Stud.*, vol. **53**, no. **6**, pp. **1319–1335** (2015)
13. P. Sparrow and C. Cooper, "Organizational effectiveness, people and performance: new challenges, new research agendas," *J. Organ. Eff. People Perform.*, vol. **1**, no. **1**, pp. **2–13** (2014)
14. S. A. Hejazi and K. R. Jackson, "A neural network approach to efficient valuation of large portfolios of variable annuities," *Insur. Math. Econ.*, vol. **70**, pp. **169–181** (2016)
15. J. L. Harrison, "Bureaucrats' Tacit Understandings and Social Movement Policy Implementation: Unpacking the Deviation of Agency Environmental Justice Programs from EJ Movement Priorities," *Soc. Probl.*, vol. **63**, no. **4**, pp. **534–553** (2016)
16. J. Wallner, "Legitimacy and Public Policy: Seeing Beyond Effctiveness, Efficiancy and Performance," *Policy Stud. J.*, vol. **36**, no. **3**, pp. **421–443**(2008)
17. R. L. Glicksman, "Effectiveness of Government Interventions at Inducing Better Environmental Performance: Does Effectiveness Depend on Facility or Firm Features?," *Coll. Environ. Aff. Law Rev.*, vol. **35**, no. **3**, pp. **479–512** (2008)
18. P. Sabatier and D. Mazmanian, "THE IMPLEMENTATION OF PUBLIC POLICY :," (1979)
19. P. Hill, Michael and Hupe, *Implementing Public Policy*. London: SAGE Publications Ltd(2002)
20. M. Watiah, "Efektivitas Pembinaan Dinas Pengelolaan Pasar Terhadap Pedagang Kaki Lima di Kota Bandar Lampung (Studi pada Pasar Bambu Kuning Kota Bandar Lampung," Universitas Lampung (2011).
21. S. Maryati, "Conceptual Model of Green Infrastructure Implementation," in *Proceeding International Seminar Innovations in Accelerating Infrastructure Competitiveness and Sustainability*, pp. **51–58**(2014)
22. C. Li, J. Li, M. Liu, Y. Wang, and Z. Wu, "Anti-misconduct policies, corporate governance and capital market responses: International evidence," *J. Int. Financ. Mark. Institutions Money*(2016)
23. UKP4, "Strategi Pemberantasan Kejahatan Kebakaran Hutan dan Lahan," no. November, pp. 1–23 (2014)
24. A. Syaprillah, "Penegakan hukum administrasi lingkungan melalui instrumen pengawasan," *Bina Huk. Lingkung.*, vol. Volume **1**, no. **32**, pp. **99–113** (2016)
25. T. Badjuri, Abdulkahar dan Yuwono, *Kebijakan Publik Konsep dan Strategi*. Semarang: Universitas Diponegoro (2002)
26. M. Juntti, D. Russel, and J. Turnpenny, "Evidence, politics and power in public policy for the environment," *Environ. Sci. Policy*, vol. **12**, no. **3**, pp. **207–215**(2009)
27. D. N. Bengston, J. O. Fletcher, and K. C. Nelson, "Public policies for managing urban growth and protecting open space: Policy instruments and lessons learned in the United States," *Landsc. Urban Plan.*, vol. **69**, no. **2–3**, pp. **271–286** (2004)
28. Tachjan, *Implementasi Kebijakan Publik*. Bandung: AIPI Bandung (2006)

Total Economic Value of Applied Used Green Line Street Model For Tamarindus Indica in Rembang District

Abdillah Rangga Fajar[1,*], *Bambang* Azis Nur[2]

[1]Master Program of Environmental Science, School of Postgraduate Studies, Diponegoro University, Semarang-Indonesia
[2]Faculty of Fisheries and Marine Science, Diponegoro University, Semarang - Indonesia

Abstract. One of the green line street functions is to providing land of carbon sink crops. Javanese tamarind (*Tamarindus indica*) is a type of productive crop that has a carbon absorbent function. The type of Javanese tamarind crops production is wood products and tamarind fruit harvest products. The purpose of this study to determine the total economic value (NET) a calculation of the model of the applied use of green line street for Javanese tamarind plants .The economic value that can be calculated from the modeling are Direct Value and Indirect Value. Direct benefit value consist of value of wood and value of crop. While indirect benefits value consist of value of carbon stock and biodiversity value. This research was conducted in May 2017 in Rembang District, Rembang Regency, Central Java. The methods used are allometric modeling method and economic valuation method. Allometric modeling method is used to determine the potential of carbon and wood reserves along the green line street. While the economic valuation method is used to calculate total economic value of the application of this modeling result. Applying of this model has result in total economic value Rp 1,942,788,721 per year. It can be concluded that the green liner street has great potential utilization and should be optimally utilized.

1 Introducing

The existence of Green Open Space is required by a city or region. One of the green open space forms in a city is urban forest. The urban forest is divided into the following classifications; green line, garden city, gardens and courtyards, botanical gardens, forest and zoos, protected forests, cemetery and heroes parks [1].

At this time Green Open Space public areas of Rembang City area provide 73.10 Ha or only 2.3% is calculated from the total area of Rembang City (31183.76 ha), the number still far from the minimum requirements. According to Minister of Public Works Regulation no. 05 of 2008 requires that the green open space public urban area must be fulfilled at least 20% of the total area of the city [2]. Increased green open space is necessary, one of them is to optimize the potential of the green line street.

The green line street that constitutes the urban forest certainly has a very important role for city. Long-lived plants or trees grown in forests as well as in mixed gardens are places where carbon stocking or storage is much larger than in annual crops. Therefore, long-lived tree species are the highest C storage warehouse capable of reducing the greenhouse gas effect [3].

How important the green line street in a city can be measured its economic value, either from environmental services or from direct use of resources on the green line street. To find out the economic value is required a model to calculating the utilization of green line street with productive plants, like Javanese Tamarind. Javanese Tamarind or *Tamarindus Indica* in Latin name was chosen because the existing location of research has been found. In addition Javanese Tamarind have the potential of ecology as a carbon sink and has the economic potential of wood and fruit produced. Besides this Javanese Tamarind does not require special treatment throughout the season so it does not require treatment costs.

2 Research Method

2.1 Time and Place of Research

This research was conducted by modeling the calculation of district roads in Rembang, Rembang, Central Java, which was implemented in June 2017.

2.2 Variable Measurement Concept

Variable measured in this research are

a. The green open space area of green line street can be utilized as planting land of Javanese Tamarind.
b. The direct benefits value from resulting the direct utilization of green open space green roads such as: timber potential and fruit potential of Java Acid (Rp / year)

* Corresponding author: rangga.abdillah1989@gmail.com

c. The indirect benefits value from resulting indirect utilization of green open space green roads such as; Economic value of carbon exchange (Rp / year).
d. Cost. Cost incurred by the beneficiaries of the green open space on the green line street to get the commodity from the green road area (Rp / year)

2.3 Collecting Data Method

2.3.1 The Length of road

The length of the road calculated in this study is the entire length of the road belonging to the district in the District of Rembang. This data is obtained through the Decree of the Rembang Regent No. 900/893/2017 on the Stipulation of Roads of District and Poros of Village in Rembang Regency.

2.3.2 Javanese Tamarind Crop

This study is a simulation of calculation of a planting model, then data about tree height, tree circumference, and fruit harvest quantity using previous research results to obtain uniform data. So it is easier in the process of calculating the simulation results.

2.4 Processing Method and Data Analysis

2.4.1 Green Line Street Analysis

The area of the green line street is obtained from the area of each road, then the total number of all existing road segments.

2.4.2 Standing Volume Wood Analysis

Analysis of standing volume to find out the volume of existing mangrove wood. To obtain the volume of wood, it must first be known the value of the height and also the circumference of the circle as high as the chest (1.3 m) tree planned in the modeling. The data obtained are included in Table 1.

Table. 1. Diameter and Height of Wood Planned

No	Tree Type	Diameter (cm)	Height (m)	Volume (m^3)
1	Javanese Tamarind			

source : researcher, 2017

The volume of this wood is obtained by using equation [4]:

$$V = (Lbd \times t) \quad (1)$$

That:
V = Volume
Lbd = Broad base area $\{[(diameter/100) \times 0.5]^2\} \times \Pi$
T = height (m)

The analysis of standing volume will illustrate the condition of the green line street of the road on each hectare. It can also be used as the initial calculation of the potential economic value of Javanese Tamarind wood.

The stand value can be determined by calculating the resulting cubic timber multiplied by the selling price per m^3 multiplied by the green line street area.

2.4.3 Biomass and Carbon Stock Analysis

In calculating the amount of carbon it uses allometric equations for trees because Java Acid is included in the category of hard trees. Allometri used to measure the hard tree as follows [5]

$$B = a\,DBH^b \quad (2)$$

That :
B = Biomass (ton)
a and b = Constanta
DBH = Diameter at Breast Height

Constant values for hard trees are as follows:

a = 0,133 and b = 1,164.

The amount of Biomass is the amount of the biomass of the trees. To know the total amount of biomass then it must be multiplied by the number of trees that exist. Modeling the number of trees on the green line street of the road can be seen in the following table:

Table. 2. Tree Planting Distance Planned

No	Total of Length Road (m)	Distance Planting (m)	Amount of Side Planting	Amount of tress
1				

Carbon stocks are estimated by multiplying the value of tree biomass with a conversion value of 50% that roughly 50% of the tropical wood biomass content is composed of carbon [6]

$$Carbon\ (C) = Y \times 0.5 \quad (3)$$

That C = Carbon (kg), and Y= Biomass (kg).

2.5 Green Line Street Economical Valuation

The assessment of green line street economical valuation done by using two stages of approach: the stage of identification of benefits and functions of existing resources, then proceed with the phase of calculation all benefits and functions into the value of money.

2.5.1. Identify the benefits and functions of the green line street resource

The economic value of a greenline street resource is divided into two, namely direct and indirect values

a. *direct use value*
The value of direct benefits is the value that results from the direct utilization of a resource. Direct benefits can mean the benefits that can be consumed. The direct benefit value of this green line is calculated by the following equation [7]:

$$DUV = \Sigma\, DUVi \quad (4)$$

That:
DUV = Direct use value
DUV 1 = wood benefit
DUV 2 = fruit benefit

b. *Indirect use value*
Indirect benefits are the value of the benefits of an indirectly utilized green line resource. The indirect benefits of a green line street are carbon sinks. This benefit can be calculated through a carbon exchange mechanism.

2.5.2. Quantify all benefits and functions into the value of money.

Once all the benefits are identified, the next step is to quantify all the benefits into the value of money with several values:

a. Market value
This market value approach is used to calculate the economic value of commodities that can be directly utilized from the pathway green resources.
b. Indirect Price
This approach is used to assess the indirect benefits of planned road green paths, such as the value of carbon exchanges
c. Total Economic Benefit Value The total useful value of a green open space area in the form of a green pathway is the sum of all economic values of the green line benefits identified and quantified into the value of money. The total benefit value is calculated using the following equation [7] with the modification of the researcher:

$$TEV = DUV + IUV \quad (4)$$

TEV = Total Economic Value
DUV = Direct Use Value
IUV = Indirect Use Value

3 Result And Discussion

3.1 Regional General Condition

Rembang Regency is located in Rembang District, Central Java Province, Indonesia. Border area of Kec. Rembang as follows:

- North side = Java Sea
- East side = Lasem sub-district
- South side = Sulang sub-district
- West side = Kaliori sub-district

Based on data from the Regional Government of Rembang Regency, the area of Rembang District 5,881 Ha with District road length of 77.9 Km.

Fig. 1. Rembang District Map

3.2 Modeling Description

This modeling is a simulation of the potential calculation of the planting of Java Acid on the green path of the road in the District of Rembang. The focus of this research is only on the simulation result of the calculation only. So that the existing plant conditions on the path of this green road is not included. This effort is done to make it easier to get the calculation of potential planting. The simulation step is as follows:

a. Determining the total area of road in the district of Rembang. This total area is considered a broad green pathway. Through direct observation at the location where there is a Javanese Acid plant, this Java Acidic canopy cover covers almost the entire area under it. As shown in the picture below.

b. Determine the desired dimension of tree dimension (tree age, tree trunk circumference and number of fruit harvest) using previous research results.
c. Calculate the potential benefits of applying the planting model.

Fig. 2. The Javanese Tamarind plant head covering the area of the road

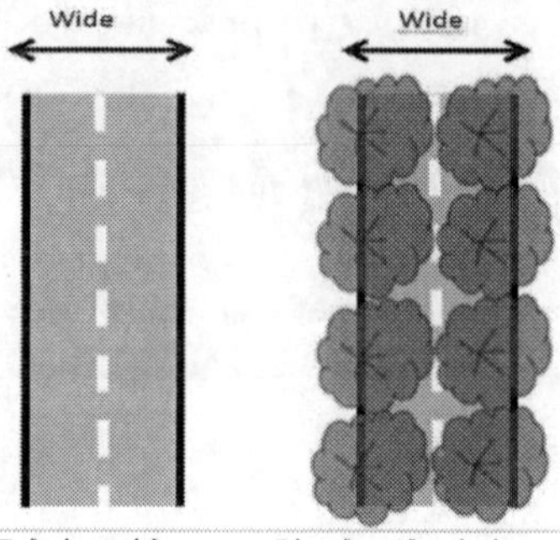

Fig. 3. Planting Apply Ilustration

3.3 Total of Green Line Street Land

Table 3. Area Roads District in District Rembang

No	Road Size Name	Lenght (m)	Width (m)
1.	Sumberjo - Seren	6100	4
2.	Besi - Kebonagung	2700	3.5
3.	Besi - Pranti	2000	3
4.	Ngadem - Nganguk	2400	4
5.	Ngadem - Nganguk	2500	3
6.	Bagel	1600	3
7.	Pasar Pentungan - Pulo	2700	7
8.	Pasar Pentungan - Pulo	3400	7
9.	Sumberjo - Pulo	2100	4
10.	Jl. KS. Tubun	900	4
11.	Jl. Sumberejo	1000	4
12.	Jl. Sidodadi	300	4
13.	Jl. M. Yamin	400	4
14.	Jl. Yos Sudarso (ex Jl. Kp. Baru)	1000	7
15.	Jl. Achmad Yani (ex Jl. Sawahan)	1300	4
16.	Jl. Achmad Yani (ex Jl. Grajen)	500	3
17.	Jl. Palen	400	3
18.	Jl. Kulit	400	3
19.	Jl. Erlangga	500	3
20.	Jl MH. Tamrin	500	3
21.	Jl Dampo Awang	300	4
22.	Jl. R. Shaleh	300	4
23.	Jl. Dr. Wahidin	1100	8
24.	Jl. Lumba- lumba	1400	4
25	Jl. Dorang	900	4
26.	Jl. Dorang	400	3
27.	Jl. Dr, Sutomo	800	4
28.	Jl. Veteran	600	6
29.	Jl. Setia Budi	300	4
30.	Jl. WR. Supratman	500	3
31.	Jl. Bisri Mustofa	1200	5
32.	Jl. Mojopahit	1200	4
33.	Jl. Asnawi	300	5
34.	Jl. K.H.A. Chafidz	300	4
35.	Jl. K.H. Mansyur	300	10
36.	Jl.P. Sedolaut	300	10
37.	Jl. Gatot Subroto	300	4
38.	Jl. HOS Cokroaminoto	1200	12
39.	Jl. Notoprojo	600	4
40.	Jl. Pahlawan	1400	6
41.	Jl. Adipati Honggojoyo	900	4
42.	Jl. Sukoharjo	300	3
43.	Jl. Pesantren	1000	4
44.	Jl. Piere Tendean	600	4
45.	Ngotet - Tireman	4800	8
46.	Kedungrejo - Ps. Banggi	9200	3
47.	Tireman - Japerejo	9700	7
48.	Tireman - Padaran	1100	3.5
49.	Kasreman - Punnjulharjo	3900	6
	Total Length Of Road (m)		**77900**
	Total Of Green Line Width (m2)		**391400**

By using the data length and road width, the total area of green path is 391400 m^2 or 39.1 Ha .

3.4 Direct Use Potential Analysis

3.4.1 Economical of Javanese Tamarind Wood Potential

Analysis of standing volume is used to determine the volume that Java Acid wood can produce in each hectare. This data is used to find the value of the direct benefits of the green road path of potential timber that can be produced.

At adult tree age about 50 years diameter 1.6 m logs and also high tree acid 25 m Java [8], obtained the potential volume of Java acid volume Of 70.84 m3 per ha as shown in the Table 4.

Table 4. Calculation of Wood Volume Per Hectare

No	Tree Type	Diameter (cm)	Height (m)	Volume (m^3)
1	Javanese Tamarind	160	25	70,84

Source : Processed data, 2017

The total area of the green road is 39.1 Ha. So the volume of Java Acid wood as a whole is 2773 m3. Based on data from *Perum Perhutani* in 2016, Java Asam wood entered in the second type of wood forest usage, that is for non-structural purposes. The price of Java Acid wood in this class is Rp 2,225,000 per m3 [9], so the potential economic value of Java Asam wood on the green line is Rp 6.169.925.000, -. If it is assumed to be a cutting cycle for 50 years, then the benefits of production earned amount to Rp 123,398,500, - per year.

3.4.2 Economic Potential of Javanese Tamarind Fruit

The potential of Javanese fruit harvest is as big as 0.82 tons per hectare per year [10], so the number of potential harvests on the road green path with an area of 39.1 ha is 32 tons or 32,000 kg per year. While the current market price of this research is done in the year 2017 is worth Rp 10.000, - per kilo. Thus, the economic value of Javanese Acid fruit in planting model on the green road is Rp 320,000,000 per year

3.5 Indirect Use Potential Analysis

3.5.1 Economic Potential of Carbon Stock

To calculate the value of carbon stock, it is necessary to find the amount of carbon uptake or biomass at that location. Using allometric equation for plant species in tropical forest, the biomass value as shown in the following Table shows.

Table 5. Result of Measuring Biomassa value/ tree

No	Tree Type	constanta (a)	constanta (b)	DBH (cm)	Biomassa (ton)
1	Javanes e Tamari nd	0,133	1,164	160	48,9

Source: Processed data, 2017

After the amount of biomass in each tree is found then multiplied by the number of trees that can be planted in this model. The number of trees can be seen in the following

Table 6. Result of Measuring Amount Plan Trees

No	Total Length Road (m)	Distance Planting (m)	Amount of Side Planting	Amount of tree
1	77900	10	2	15.580

Source : Processed data, 2017

In this calculation obtained biomass value of 761,862 Ton. The value of carbon stocks stored in the biomass is 380,931 Ton, or 50% of the existing nico biomass [6].

The economic value of stored carbon stocks can be calculated by carbon trading schemes. The economic value of carbon stock or carbon price varies according to various sources. According to Saloh and Clough states that carbon prices in carbon trading are US $ 15 - 20 US $ per ton of carbon absorbed [10]. In this calculation take the lowest price of US $ 15 per ton of carbon absorbed. So the economic value of carbon stocks on the green road is US $ 5,713,965 or Rp 75,904,311,060, - (The rupiah exchange rate at the time of this research is Rp 13,284). This benefit is obtained within 50 years of planting time, so the economic value of carbon stock is Rp 1,518,086,221 per year.

3.6 Price Proceed Analysis

In the cultivation of Javanese Acid this is actually no significant management costs because it does not require watering and hold during the dry season [8]. So the cost that appears in this modeling is the cost of procurement of seeds of Java Acid for this green path of 15,580 seeds. Based on the market price at the time of the research, the price of seeds of Java Acid Rp 50.000, - and the cost of planting is assumed Rp 10.000, - per seedlings. So the total cost for this road green road modeling is Rp 934,800,000, -. The fee is for the value of the investment utilization for 50 years. So the average cost incurred in the application of this model is Rp 18,696,000 per year.

3.7 Total Economic Value

The total economic value of the utilization of the green road path can be seen in the following table

Table 7. Green Line Street Total Economic Value

No	Benefit Type	Type	Value (Rp/year)
1	Direct use value	wood	123.398.500
		fruit	320.000.000
2	Indirect use value	Carbon Stock	1.518.086.221
3	Price		- 18.696.000
	Total economic value		1.942.788.721

Source : Processed data, 2017

4 Conclusions and Suggestion

4.1 Conclusion

Based on the result, the total economic value of green road utilization is Rp 1,942,788,721 per year, calculated from direct benefit (Rp 323,398,500 per year), indirect benefit (Rp 1,518,086,221 per year) and reduced by cost (Rp 18,696,000 per year)

4.2 Suggestion

Seeing the huge economic value of the results of Java Acid planting modeling on the green road in Rembang District, should the government and the community cooperate in realizing the modeling of this planting. Besides beneficial from the ecological side of the application of Java Asam planting is also beneficial from the economic side.

References

1. Dahlan, N.E,.*Hutan Kota untuk Pengelolaan dan Peningkatan Kualitas Lingkungan Hidup*. Asosiasi Pengusahan Hutan Indonesia. Jakarta. (1992)
2. Mashuri et al. Jurnal Teknik Sipil dan Perencanaan,Vol **14**, Semarang (2012)
3. K. Hairiah, and S. Rahayu, *Pengukuran Karbon Tersimpan di Berbagai Macam Penggunaan Lahan*. ICRAF. Bogor, (2007)
4. Santoso. Valuasi Ekonomi Ekosistem Hutan Mangrove di Kawasan Pondok Bali, Desa Legonwetan, Kecamatan Legonkulon, Kabupaten Subang, Jawa Barat. Bogor. Institut Pertanian Bogor, (2005)
5. N. S. Dudley and J. H. Fownes *Journal of Tropical Forest Science* Vol. **5**, No. 1 (September 1992), pp. 68-73
6. S. Brown, *Estimasi Biomass and Biomass Change Of Tropical Forest*. FAO forestry Paper. Roma, (1997)
7. Bishop, J.T. (ed.), Valuing Forests: A Review of Methods and Applications in Developing Countries. International Institute for Environment and Development: London, (1999)
8. R. Rukmana, "Budidaya Asam Jawa." *Kanisius, Yogyakarta* (2005).
9. Surat Direksi Perum Perhutani Nomor : , Tahun 2016
10. Tree Corp Estate of Indonesia, Spices and Beverage Crops. Ministry of Agriculture Repiblic of Indonesia Directorate General o Estate Crops. (2014)

A Resilience Pattern in Village level: The Case Babalan Village, Pati, Central Java Indonesia

Ragil Nurwahyudi[1,*], and *Maryono* Maryono[2]

[1]Master Program of Environmental Science, School of Postgraduate Studies, Diponegoro University, Semarang-Indonesia
[2]Department of Urban and Regional Planning, Faculty Engineering,, Diponegoro University, Semarang - Indonesia

Abstract. Based on the Indonesia Disaster Prone Index 2013, Pati Regency is a high risk area of disaster and is ranked 11th level Central Java province while nationally ranked 156. Babalan Village located on the edge of Juwana River has disaster history from 2006-2014 shows flood disaster Giving the greatest probability and impact followed by rat pest, tornado, drought, fire. The public recognizes the signs of a continuous flooding of heavy rains accompanied by clouds all over the edge, the continuous rise of the Juwana River surface to overflow, ants, isoptera, and animals out of its nest, "Yuyu Bule", earthworms out, clear water for "Rowo floods ", Brownish water for the flash floods. Most residents have boats and can make rafts from makeshift materials (jerry cans, bamboo, banana stems). Make "Ranggon" at home for those who do not evacuate for a place to stay during the flood. Citizens elevate the kitchen (to evacuate people and goods / household furniture). Breeding *Tyto Alba* owl for rats pest control post-flood and controllers in the fields. Develop vegetable crops in the yard with viticulture pattern (upstairs) if flood can be moved and can eat vegetables during flood. Have food reserves for stock before outside help comes. Citizens initiate "Water Bath honesty" to meet the water needs during the dry season.

1 Introduction

According to The Indonesian National Agency for Disaster Management (BNPB), there were 12,494 disasters in the period of 2002–2012 with 190,087 numbers of casualties [1]. The first Pati regency is one of the 35 districts / cities in central java province that are vulnerable to disaster, both caused by natural, non-natural and human factors. Based on the Indonesia Disaster Risk Index (IRBI) 2013, Pati District is a high risk area of disaster and is ranked 11th level Central Java province while nationally ranked 156.

Central Java Province is one of the strategic province in Indonesia, that hold in significant role the perform of energy resilience. From the demand view side, Central Java was dynamic place. Central Java province is also a prone province since many natural hazards hit, such as flood, earthquake, and volcano eruption. One conceptual indicator that can be used to assess energy resilience is the stakeholer's effort to understand energy needs in the event of a disaster [2].

There are several disasters that can happen in Pati Regency. In the rainy season Pati district is an area prone to floods, flash floods, landslides or hurricanes. This is influenced by the geographical condition of Pati Regency which has 93 rivers that are spread evenly throughout the region. In addition Pati Regency also has an area with a slope of more than 40% and areas prone to landslides are Cluwak District, Gembong, Tlogowungu, Gunungwungkal, Sukolilo, Kayen, Tambakromo, Winong and Pucakwangi [3]. While in the dry season in some areas of Pati Regency also occurs drought disaster.

Disasters have destroyed the results of hard-earned development. Funds used for emergency response and post-disaster recovery have also reduced the budgets that should be utilized for national development and poverty eradication programs. In the event of a disaster, the poor and marginalized who live in vulnerable areas will be the most disadvantaged, since the greatest number of victims usually comes from this group and the impoverishment caused by the disaster will mostly befall them.

Considering the significant impact of natural disaster, it is important to determine the level of disaster risk in a country's area. A deep understanding of the matter will help government to develop a comprehensive framework or policy to minimize the negative impact of disasters. In addition, understanding risk level should also be followed up by the assessment of resilience level toward the disasters. As mentioned by Mayunga [4].

Disaster resilience is the capacity or ability of a community to anticipate, prepare for, respond to, and recover quickly from impacts of disaster. Furthermore, disaster resilience along with economic vitality, environmental quality, social and inter-generational

* Corresponding author: ragilenvironment@gmail.com

equity, quality of life, and participatory process are the six principles of sustainability [5].

The behaviour alternation to the less preparedness community to cope disaster may be anticipated by internal mechanism. The lesson learned in Indonesia of disaster countermeasures is dominantly done through external mechanism, i.e., the disaster countermeasures is done by actors coming from outside of the community. So far, the disaster countermeasures in Indonesia were done by extensive intervention by externals actors, while the local communities were only becoming an object of the programs. That external mechanism may create powerless community in the reduction of disaster vulnerability. The external factor did not involve the local community and considering the local potential resources to reduce the disaster risk. Moreover, the community response, it was also expressed by the local leader, of *'nantang perkoro'* which mean expecting somewhat bad in the future. According to the community, 'We are now living in safe and peace; we don't have to prepare ourselves for the uncertain coming disaster [6]

The United Nations International Strategy for Disaster Reduction (UNISDR) defined resilience as "the ability of a system, community or society exposed to hazards to resist, absorb, accommodate to and recover from the effects of a hazard in a timely and efficient manner, including through the preservation and restoration of its essential basic structures and functions [7].

In addition, Razafindrabe et al. [8] proposed that there were five dimensions of resilience towards climate disaster, namely physical, social, economic, institutional, and natural (hazard frequency and hazard density). Meanwhile, Cutter et al. [9] proposed indicators of community resilience which included several dimensions, such as ecological, social, economic, institutional, infrastructure and community capacity.

A disaster is a serious disruption in the functioning of a community, which exceeds its capacity to cope within its own resource [10]

In accordance with the disaster management paradigm according to Law number 24 of 2007 that the disaster management paradigm must be integrated and comprehensive preventive with the participation of all parties, both the government of the business community and society.

Babalan Village is part of Gabus Subdistrict Pati Regency with geographical location as follows:

North	:	Juwana River, Gajah Mati Village and Banjarsari Village
East	:	Koripandriyo Village
South	:	Plumbungan Village
West	:	Tanjang Village

Administratively divided into 3 (three) *dukuh,* 3 (three) *Rukun Warga* and 15 (fifteen) *Rukun Tetangga* (RT). Total area of 235.20 Ha consisting of rice field 161,20 Ha; Not 7.96 Ha of rice fields and non-agricultural land of 66.36 Ha with the type of land of Babalan village is alluvial (clay). Babalan Village through 3 (three) rivers are 1 (one) big river (Juwana river) and 2 medium river. Babalan village area belongs to a group of areas with a height of 0-8 meters above sea level [11]. This condition causes brackish conditions during the dry season, as sea water rises through the Juwana River.

History of disaster events in Babalan village from the last 9 (nine) years 2006 to 2014, the most frequent occurrence of natural disasters due to floods that occur almost every year, droughts, high winds and catastrophic crop failures due to rat pests [12]. Disaster management is a systematic and comprehensive effort to tackle all disaster events quickly, accurately and accurately to reduce victims and losses [3].

Article 3 of Law No. 24 of 2007 on Disaster Management states that disaster management is based on the principles of: humanity, justice, equality of positions in law and government, balance and harmony, law and order of law, togetherness, environmental sustainability, science and technology. In addition, disaster management should also be based on the following practical principles: fast and precise, prioritized, coordinated and integrated, effective and effective, transparency and accountability, partnership, empowerment, non-discrimination and non-prosecution [13].

Sustainable development of the flood plains is one of the crucial development concerns around the world. Theoretically, sustainable development can be seen as a process of positive biophysical and socio-economic change or creating wealth (capital) that meets the needs of present generation and can be continued indefinitely without diminishing the natural systems upon which it depends or excluding the range of opportunities available to future generations. This is a continuous adaptation process to evolving environmental, economic, and social systems [14,15].

Community in the context of disaster management is a group that may share one or more things in common, such as living in the same environment, having similar disaster risk exposure, or having been affected by a disaster [16].

A similar definition, though more closely related to the built environment, is proposed by Ginige [17-11] and Amaratunga [18]. They define community as individuals and groups sharing a natural and built environment that is vulnerable to hazards. In other words, community is the general public; the users and occupants of the built environment and the beneficiaries of post-disaster reconstruction.

Boughton explained that community, environment, and hazard are element of risk. The risk itself is a result of interaction between hazard that may create damage and lost, community that may be affected, and the specific situation of environment which may reduce the level of hazard [19].

Risk assessment is based on a review of both technical features of hazards such as location, intensity, frequency and probability; And also an analysis of the physical, social, economic and environmental dimensions of the vulnerability and vulnerability of Exposure, while taking account of the handling capabilities associated with the Vulnerability Risk Scenario measures the physical, social, economic, and

 https://doi.org/10.1051/e3sconf/20183107005

environmental conditions or processes that increase the vulnerability of a community For hazard impact [20].

Disasters can erase the benefits of development investments, and poorly planned development interventions may become a source of hazard. Therefore, disaster planning is a necessary step and is needed to realize the MDGs and sustainable development. These are elements of a consensus that was reaffirmed at the World Conference on Disaster Reduction (WCDR) in Kobe, Japan, on 18–22 January 2005. Also at that meeting climate change was recognized as posing an immediate and long-term threat to the achievement of the MDGs and sustainable human development, and, as such, should be an integral part of disaster planning [21].

2. Method

Qualitative method is a research procedure that will produce qualitative data in the form of written or oral words of the people and an observed process. This type of research is a problem-solving procedure that is investigated by describing the state of the subject / object of research (community, a process and others) at the present moment based on facts that appear or as is [22].

3. Result and Discussions

3.1 Risk Identification

Risk identification and Assessment are usually performed by key actors at the national, provincial, district, districts and village levels [23].

The focus of disaster management is to reduce the risk posed by actual and potential hazards [24]. The capacity of a system, community or society to resist or to change in order that it may obtain an acceptable level in functioning and structure. This is determined by the degree to which the social system is capable of organizing itself and the ability to increase its capacity for learning and adaptation, including the capacity to recover from a disaster [25].

What needs to be done before estimating flood warning standards is to understand the risks of a flood disaster. The following things need to be done to determine the level of Risk of a region to flash flood disaster.

1. Identification of disaster areas based on flood prone maps;
2. Conduct a survey of infrastructure in vulnerable areas including the following:
   - Residential location;
   - Flood retaining buildings;
   - Flood detection infrastructure.
3. Conduct a survey on the social conditions of the community including:
   - Population by gender, age, illness sickness, occupation, education level, wealth, family members;
   - Knowledge of the causes and processes of disasters;
   - disaster preparedness.
4. Conduct analysis through the survey results data above to determine the level of risk of a region;
5. Conduct standard warning forecasts

Data The last nine years from 2006 to 2014 disaster occurrence in Babalan village, natural factor is Flood almost happened every year, Drought and Wind fast while non natural factor is Failed Harvest due to Rat Attack. Floods occur in 2006, 2008, 2009, 2010, 2011 and 2014. Pest attacks in 2012 and 2013. Strong winds occur in 2014 and drought in 2006.

From the above data it can be seen that the disaster that ever happened in Babalan Village is Flood, Drought and Wind, then it is estimated that probability is scored by category:

1. Score 5 Definitely (almost certainly 80 - 99%).
2. Score 4 Most likely (60-80% occur next year, or once in the next 10 years)
3. Score 3 Possible occurrence (40-60% occur next year, or once in 100 years)
4. Score 2 Small Possibilities (20-40% in 100 years)
5. Score 1 Possibly small (up to 20%)

If the above probability is supplemented by the estimated impact if the disaster does occur with consideration of the impact factors such as:

1. the number of victims;
2. Loss of property;
3. Damage to infrastructure and facilities;
4. The wide range of affected areas; and
5. Socio-economic impacts.

Even this impact is given weight as follows:

1. Score 5 Very Severe (80% - 99% devastated and total paralyzed area)
2. Score 4 Severe (60 - 80% of the area destroyed)
3. Score 3 Medium (40 - 60% of affected area is damaged)
4. Score 2 Lightweight (20 - 40% of damaged area)
5. Score 1 Very Light (less than 20% broken area)

It is estimated that the scores obtained are as follows (figure 1)

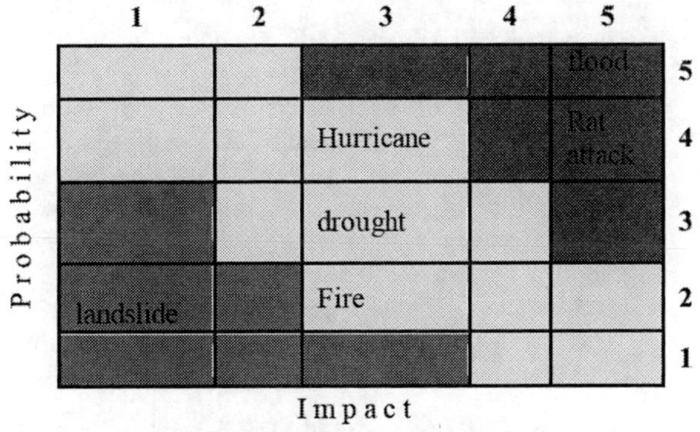

Fig. 1. Disaster Risk Assessment

Table 1. Results of the Focus Discussion Group (FGD) together with community representatives

Character	Character Details	Threat Profile	Risk	
			Individual / Family	**Community / Village**
Cause (origin)	Heavy rains, the Juwana River Flood and the Sani river	Starting from continuous cloudy rain, while the river Juwana experiencing a continuous increase, eventually overflowing, water inundated rice fields, roads, settlements with high puddle 0.5 s / d 3 m long puddle of one week to 4 months resulting in physical damage and environment and loss of property: Population displaced, disease epidemic: Diarrhea, Skin Disease, Coughs Cold, Fever and Muntaber. Rice fields failed to harvest, damaged roads and people stopped their activities in earning a living		
Destructive Power	Long standing puddles for rowo floods, mud and strong currents for flash floods		Food shortages, Decreased health conditions, outbreaks of disease, can not work, crop failure, Livestock loss / loss, damage / loss of furniture, damage to homes, displacement	Cessation of village government services, road damage, cessation of community processes, public facilities can not be used
Warning & Signs	Continuous heavy rain accompanied by clouds all over the edge, the rise of the surface of the river Juwana continuously to overflow, Ants, terminet, and animals out of the nest, crab *(Parathelphusa convexa)* called "Yuyu Bule", Earthworms out, clear water for the old flood "rowo flood", brownish water for flash flood			
Interrupted Time	Fast and Slow			
Speed is present	Fast and slow			
Frequency	Flood often almost every year (dukuh Gilis area) big flood 3-5 years (whole village)			
Genesis Period	December to March			
Duration	Mountain or flash flood from Sani river 1 week, "rowo" flood from Juwana river 3-4 months			
Intensity	Flooding is in Gilis Hamlet area of the entire flood of villages			
Position of the source of the threat	Juwana River and Sani River in the village			

Types of hazard threats that need to be addressed. Threat assessed danger level by scale (3-1):

1. High danger / threat value 3 (red) is flood and mouse attack.
2. Danger / threat is the value of 2 (yellow) is a hurricane, drought and fire
3. Danger / low threat value 1 (green) is landslide

From the results of the Focus Discussion Group (FGD) together with representatives of the community elements of the disaster picture of Babalan Village in Table 1.

The pattern of resilience of the Babalan villagers is:

1. Recognizing the signs of the coming of continuous heavy floods of rain accompanied by cloudy throughout the tip, the rising surface of the river Juwana continuously to overflow, Ants, larons, and animals out of the nest, *Parathelphusa convexa* called "Yuyu bule", Earthworms out, clear water to "flood rowo", Brownish water for Flash floods.
2. There is a model of giving information to the residents about the flood (SMS, Mosque Speaker, etc.)
3. Can make a raft of makeshift materials (jerry cans, bamboo, banana stems, etc.)
4. Make "ranggon" at home for those who do not evacuate. Ranggon is a place of survival when flooding in a house made of wood or bamboo.
5. The average citizen has a boat (each house / head family).
6. Average citizens have raised the kitchen (to evacuate people and goods / household furniture).
7. Breeding *Tyto Alba* owl for rat pest control (Fig. 2)
8. Develop vegetable crops in yard with verticulture pattern (upstairs) if flood can be moved and can eat vegetables during flood.
9. Have food reserves for stock before outside help comes.
10. Initiate "Water Bath of Honesty" in managing the provision of clean water.
11. Make a flood map of the village in a simple way (Fig. 3).

Fig. 2. Captive breeding of *Tyto Alba* in Babalan Village, Pati, Central Java

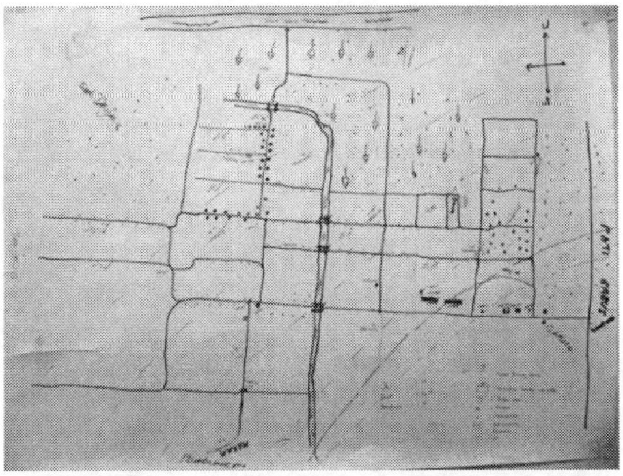

Fig. 3. Map of Flood Directions by Babalan village people

3.2 Early warning system

One must also juxtapose 'top down' disaster management in both MDCs and LDCs with the self-protection efforts made by households and communities them selves actions based on local knowledge and the activities of the institutions of civil society that work on natural and technological hazards from 'the bottom up' [26]. Moreover this study also showing the stakeholder intention in preparedness, at which have been discuss for disaster waste management [27]

In addition to EWS tools, early warning systems implemented by the Babalan villagers are (Fig.4) :

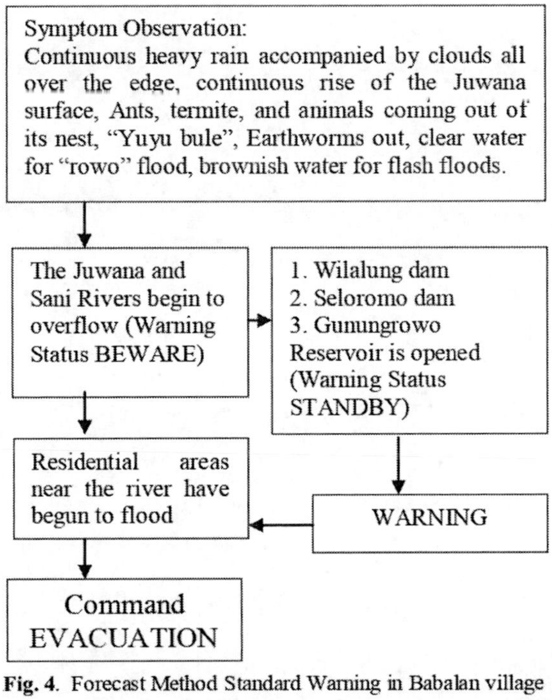

Fig. 4. Forecast Method Standard Warning in Babalan village

4 Conclusion

The robustness pattern of a resilient village in Babalan Village can prepare people living in disaster prone areas. Babalan Village Community is now able to identify safe areas and create evacuation routes. It is also shown that stakeholder have been intention to

References

1. BNPB, Data dan informasi bencana Indonesia (data and information of disaster in Indonesia), Retrieved from: 〈www.dibi.bnpb.go.id〉. 2013
2. Maryono et al. Procedia - Social and Behavioral Sciences 227, 146 – 151(2016)
3. Peraturan daerah kabupaten pati nomor 5 tahun 2011 tentang Rencana Tata Ruang Wilayah Kabupaten Pati tahun 2010-2030
4. J.S. Mayunga, Understanding and applying the concept of community disaster resilience: a capital-based approach, A draft working paper prepared for the summer academy for social vulnerability and resilience building, Munich, Germany, 22–28 July (2007).
5. [5-16]Natural Hazards Center (University of Colorado) *Holistic Disaster Recovery: Ideas for Building Local Sustainability After a Natural Disaster* Natural Hazards Center, Boulder, Colorado (2006)
6. S. Maarif, et.all. Indonesian Journal of Geography, Vol. **44**, No. 2, December 2012 : 173 – 182
7. UNISDR. Terminology, Retrieved from: 〈http://www.unisdr.org/we/inform/terminology〉, 2012 (accessed on 30.06.12).
8. B.H.N. Razafindrabe, G.A. Parvin, A. Surjan, Y. Takeuchi, R. Shaw, J. Environ. Disaster Manag., pp. 101-116.1 (2009).
9. S.L. Cutter, L. Barnes, M. Berry, C. Burton, E. Evans, E. Tate, J. Webb. Glob. Environ. Change, pp. 598-606 18 (2008),
10. Bradley, D.T., MacFarland, M., Clarke, M., The effectiveness of disaster risk communication: A systematic review of intervention studies. http://www.ncbi.nlm.nih.gov/pubmed/25642365 (2014)
11. Yayasan Sheep Indonesia. *Benang Kusust Daerah Aliran Sungai Juwana*. Yayasan Sheep Indonesia: Yogyakarta. (2010)
12. BPBD. *Rencana-rencana Penanggulangan Bencana*. 2014.
13. Nurjanah,dkk.. *Manajemen Bencana*. Bandung: ALFABETA. (2012)
14. [14-10] B. Dalal-Clayton, B. Sadler. *Sustainability appraisal: a sourcebook and referenceguide to international experience.* Routledge. (2014)
15. [15-20].B. Sadler, *On evaluating the success of EIA and SEA. Assessing impact:Handbook of EIA and SEA follow-up* 248-285.(2004)
16. [16-1].I. Abarquez, and Z. Murshed, *Community-based Disaster Risk Management: A Field*

Practioners' Handbook, ADPC, Pathum Thani. (2004)
17. [17-11]K. Ginige, and D. Amaratunga, "Capacity development for post-disaster reconstruction of the built environment", in D. Amaratunga, and Haigh, R.P. (Eds), Post-disaster Reconstruction of the Built Environment: Rebuilding for Resilience, Wiley, Chichester, pp. 13-29. (2011)
18. [18-3]. D. Amaratunga, and R.P. Haigh, "Introduction", in D. Amaratunga, and R.P. Haigh, (Eds), Post-disaster Reconstruction of the Built Environment: Rebuilding for Resilience, Wiley, Chichester, pp. 1-12. (2011)
19. [19-6] Boughton, G. the community: central to emergency risk management in The Australian Journal of Emergency Management, 13(2): 2-5 (1998),
20. [20] Ramli, Soehatman. *Pedoman Praktis Manajemen Bencana*, Dian Rakyat; Jakarta. (2010)
21. [21-24]. UN/ISDR *World Conference on Disaster Reduction Kobe, 18–22 January 2005.* http://www.unisdr. org/. (2005)
22. [22-15] Moleong, Lexy. *Metodologi Penelitian Kualitatif.* Bandung: PT. remaja Rosdakarya. 2002
23. [23-14].Moe, Tun Lin and Pairote Pathranarakul. An International Journal, Vol. **15** Iss 3 pp. 396 – 413(2006).
24. [24-2]. D. Alexander, *Principles of Emergency Planning and Management*. Terra Publishing, Harpenden. (2002b)
25. [25-23] [25-23]..UN/ISDR, Preliminary version prepared as an interagency effort coordinated by the ISDR Secretariat, Geneva. (2002)
26. [26-9]. T. Cannon, Vulnerability Analysis and Disasters'. In D. Parker (ed.) Floods. Routledge, London. pp. 43–55. (2000)
27. Maryono, H. Nakayama, & T. Shimaoka, Memoirs of the Faculty of Engineering, Kyushu University, 74(3), PP. 79-98 (2015)

Preliminary Identification of Urban Park Infrastructure Resilience in Semarang Central Java

Aji Uhfatun Muzdalifah[1], *Maryono* Maryono[2]

[1]Master Program of Environmental Science, School of Postgraduate Studies, Diponegoro University, Semarang – Indonesia
[2]Department of urban and Regional Planning, Faculty Engineering, Diponegoro University, Semarang-Indonesia

Abstract: Park is one of the spot green infrastructure. There are two major characteristic of park, first Active parks and second passive park. Those of two open spaces have been significant on the fulfillment of urban environment. To maintenance the urban park, it is very importance to identify the characteristic of active and passive park. The identification also needs to fostering stakeholder effort to increase quality of urban park infrastructure. This study aims to explore and assess the characteristic of urban park infrastructure in Semarang City, Central Java. Data collection methods conduct by review formal document, field observation and interview with key government officer. The study founded that urban active parks infrastructure resilience could be defined by; Park Location, Garden Shape, Vegetation, Support Element, Park Function, and Expected Benefit from Park Existence. Moreover, the vegetation aspect and the supporting elements are the most importance urban park infrastructure in Semarang.

1 Introduction

Currently, environmentally sound in urban planning the facing urban climate hazard which its impact such as urban infrastructure resilience. Fostering the urban infrastructure resilience for sustainable cities life becoming major requirement. One of the main issues in urban spatial planning to provide comfort, beauty and health city is the resilience of urban green open space. Increasing empirical evidence of climate hazard, such as flood, windstorm, drought, indicates that the presence of natural disturbance should be manage properly [1].

For the institutional resilience perspective, Indonesia government has issued guidelines concerning to the provision and utilization of green open space trough the Regulation of the Minister of Public Works No. 05 / PRT / M / 2008 [2]. Refers to the guidelines provision and utilization of green open space in urban area, some parameter should be indentify such as open land which belongs to a unity of non-natural green open space, social aspect in green open space, aesthetic functions. recreative, educational or other activities that may use the green open space. The urban green open space resilience consist of socio-economy and engineering resilience [3]. Moreover, in the perspective of institution resilience, urban park infrastructure resilience in developing cities also facing issues associated to the flood hazard and disaster waste, for example handling urban green waste in disaster event [4]

Resilience of parks, open space and other forms of green space also provide essential services that are critical to both urban ecological functioning and integrity [5]. The resilience of green open space such as park have a positive correlation with the ecological impact on the city environment, for example, contribute to the balance of physical urban development and human development in the context of city amenity, comfort and convenience [5,6]. However, characteristic of physical urban open space and green urban open space mostly tend to more influence for the degree of resilience [6]

2 Literature review

2.1 Green Infrastructure

Regional and Urban Development need green infrastructure for ecological balancing. Sufficient Green infrastructure will perform human life by saving water, air from any pollution and disturbance. Therefore green infrastructure resilience is significant during rapid development, irregular and dangerous environment hazard. Green infrastructure consist of two major part, the first one is space and vegetation and the second is physical form of space at which could be define as natural and man made material [6,7].

Green infrastructure consists of space, vegetation and its element. Green infrastructure could be defined as open spaces, natural areas, urban woodland and parks; green streets, squares and public realm; sustainable drainage systems and healthy waterways, cycleways and pedestrian routes within our city environments; and smaller scale green roofs, walls and facades [8]. Green Infrastructure is the network of green places and

* Corresponding author: maryono@pwk.undip.com

water systems that delivers multiple environmental, social and economic values and services to urban communities[9].

2.2 Urban park infrastructure resilience

Urban parks are defined as delineated open space areas, mostly dominated by vegetation and water, and generally reserved for public use[10]. The scope and speed of current changes demands that urbanists define compelling visions and integrated design measures for shaping resilient cities [11]. While Urban resilience could be define as the ability of urban asset, location and/ or system to provide predictable performance – benefits and utility and associated rents and other cash flows – under a wide range of circumstances [12,13]. According to the definition urban park infrastructure resilience could be defined as ability of an urban park infrastructure system and all its consitutent socio-ecological and socio-technical networks across temporal and spatial scales-to mantain or rapidly return to desire functions in theface of disturbance, to adapt to change, and to quickly transform systems that limir currentor future adaptive capacity [12,13,14]

Assessing the urban park resilience should be viewed as an important part of urban and community development rather than just as settings for recreation and leisure. Urban infrastructure and form encompass the built environment such as buildings, transportation networks, energy, and water grids (utilities), along with urban green space and parks [12,13]. Moreover, assessment of urban park resilience include should assess the domestic gardens which it is important component in many people's lives. However stakeholder intention and attitudes towards this resilience issues is very poor. They not yet understand that the importance function of domestic gardens such as maintenance the of the thermal comfort in their residents, reduce domestic energy consumption and minimize storm run-off. Moreover, stakeholder not yet realized that domestic garden provide environment from greenhouse gas emissions rather than off setting them.

Resilience of urban park infrastructure is need to against climate change impact. From the perspective of infrastructure system, resilience of urban park will importance for ecological aspect such as providing enough water in long term impact[15]. Moreover, garden cities have been use as one of the principal theme in cities development[16]. maintenance urban park resilience will be arise other social benefit such as economics perspective and ecosystem benefit such as biodiversity, moreover, degree urban environment also will be be increase within the availability of parks and other open spaces. There are three major component of urban park resilience; (1) urban parks in economic development strategies; (2) parks asrecreaton spaces to improve citzens health outcomes; and (3) function of green infrastructure parks associated with the improvement stormwater [17]. Other attributes associated with urban park resilience are; the utility of park, physical activity that usually occur in park, its features, conditions, aethetics, safety and social environments [3].

3 Methodlogy

3.1 Study area and issues

As shown in figure 1, Semarang is one of
the Unique Cities not only for his citizen, but also for Central Java Province and Indonesia. This city is one of 100 Resilience Cities network at which getting more intention in climate change adaptation [18,19,].

In recent development, Semarang faces various complex environmental problems, such as flood and inundated area [20, 21]. The inundated area, climate hazard and other environmental issues in Semarang is tend to increase have an impact on the deterioration of the quality of the environment. This study aimed to addressed effort especially for fostering in urban park resilience.

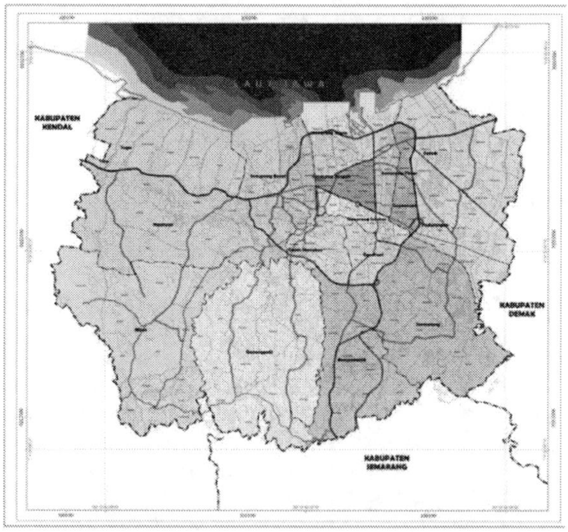

Fig. 1. Administrations Map of Semarang

The government of Semarang City have been realize concerning to their environmental issues. They state formally in the local regulation No. 6 of 2010 concerning to Long Term Development Plan Area (RPJPD) Semarang City periode 2005-2025 [22]. Refers to the document, the main environmental issue in Semarang are the degradation of environmental quality and global warming. Moreover, in their Regional Spatial Plan of Semarang City Year 2010-2030, explained that green open space is less then 20%, so they must increase the number of their urban green open space. one of the importance component is city park [23]. A city park is a park intended to serve residents of one city or part of the city. The park serves a minimum of 480,000 residents with a minimum standard of 0.3 m2 per city dweller, with a minimum park area of 144,000 m2 [2]. The park can be shaped as green open space (green field), equipped with recreational facilities and sports, and sports complex with a minimum of 80% - 90% green open space. All these facilities are open to the public. The vegetation species selected in the form of annual trees, and shrubs are planted in clusters or spread as microclimate creatures or as a barrier between activities.

Urban parks and green open space besides serve as a function of beauty, also serves as a space of community interaction, sports facilities, and the lungs of the city. Awareness of Semarang City Government in the procurement and management of green open space is indicated by the allocation of land of green space utilization in regulation (Spatial Plan of Area) which is quite dominant. In Semarang City Center, the presence of green open space is in: private gardens, 90 units parks currently managed by the Housing and Settlement Service and from the greening / tree by the roadside.

3.2 Data collection and analysis

The focus of this study is to identify the infrastructure component of urban park that influence the degree of it resilience. The study will classify the urban park infrastructure base on two major characteristic of urban park namely active park and passive park. Moreover this study will explore the condition, status of utilization in active and passive park.

This research is a qualitative and quantitative approach. Primary data conducted by field observation, and interview with key stakeholder, while Secondary data in this study were obtained from literature and related institutions. Primary data getting fromsurvey and interview result. Moreover, data collection is also conducted by examining the various documents related to the recent study of urban parks, stakeholder and government.

The statistics descriptive used to analyze and utilize the data that has been collected. Moreover, the description content of factors urban park resilience such as features, conditions, aethetics, safety and social environments.

4 Result and discussion

Base on the field survey, this study have classify the characteristic of urban park in Semarang as urban active park and urban passive park [24]. The urban active park is that the characteristic of park which community use urban infrastructure for refreshing, play or use for physical exercise and sport. While the passive urban park is the there is no urban park infrastructure for refreshing, play or use for physical exercise and sport [24, 25]. In passive urban park, the urban park infrastructure is design only for aesthetics, security and safety, and amenities. Urban active park in semarang mostly use for jogging, sport and other physical exercise, playing, and several urban park there economic activities such as urban street vendors.

Table 1 shown observation result concerning to the characteristic of urban active park and urban passive park. As shown in table, there are 76 urban active park and 167 urban passive park in Semarang. In accordance Mostly urban park found in Semarang Tengah District with totally 47 urban park. While the smallest availability of urban park founded in Tembalang District.

Table 1. Amount of Urban Parks Infrastructure in Semarang

Districts	Type Of Park		Amount	Total Area (m^2)
	Active	Passive		
Banyumanik	6	8	14	18,595.1
Candisari	1	10	11	14,206.5
Gajahmungkur	3	15	18	24,102.5
Gayamsari	2	2	4	2,016.0
Mijen	2	1	3	36,000.0
Ngaliyan	1	1	2	600.0
Pedurungan	30	5	35	68,396.0
Semarang Barat	2	33	35	32,865.3
Semarang Selatan	9	13	22	55,473.8
Semarang Tengah	1	46	47	30,793.6
Semarang Timur	9	25	34	26,383.0
Semarang Utara	9	6	15	7.276.4
Tembalang	0	1	1	2,578.0
Tugu	1	1	2	17,645.0
Jumlah	**76**	**167**	**243**	**336,931.2**

This study founded that the characteristic of urban park follow the direction of the minister of interior number 14 of 1988, that the greening density of urban parks in Semarang is between 80% - 90% of the land area. Totally there are 336,931.2 m^2. As classify in district level, Pedurungan district have totally 68,396 m^2, it is the largest area comparing to the other of district. Mijen district have a smallest area of park compare to other district. The characteristics of urban park in Semarang mostly near the central of cit for passive urban park, and near the public activity such as housing and office for active urban park.

Table 2. Typology of Urban Park Infrastructure Resilience in Semarang central Java

Type of Urban Park	Typologi Urban Park Infrastructure Resilience
Active	The probability of resilience issues (sorted from the highest disturbance) 1. Social resilience 2. Amenities resilience 3. Economic resilience 4. Spatial boundary of plan resilience 5. Education resilience 6. Ecological resilience
Passive	The probability of resilience issues (sorted from the highest disturbance) 1. Ecological resilience 2. Spatial boundary of plan resilience 3. Education resilience 4. Economic resilience 5. Social resilience 6. Amenities resilience

Many literature have been discussion related to resilience such as, ecology, engineering, psychology, complex adaptive system, economy, and social perspective [26,27,28]. This study identify probability of urban park infrastructure resilience refers to the characteristic, function [6], at which the resilience can be identify as shown in table 2. This study founded the six resilience of urban park infrastructure in Semarang. The resilience may vary depending on the characteristic of park.

The highest disturbance from the urban park infrastructure is concerning to social resilience. While for passive urban park infrastructure the highest disturbance is related to ecological resilience. The social resilience refers to the main function of active urban park related to to human activity such as for recreation, sport activity and other physical activity. Ecological resilience refers to the main function of the passive urban park infrastructure at which dominant for ecological perspective. Both for active and passive urban park infrastructure, economic resilience may be disturb at the middle level, since for the economic perspective there is not yet found the importance of economic both in the perspective of activity in park or economic value of park.

Amenities of urban park infrastructure resilience is that associated to the physical form of urban park. For example when the disaster attack such as flood the form of urban park will destroyed and huge of green waste will be generated in urban area [3]. For the Economic resilience of urban park infrastructure, this study found that main issues of resilience mostly concerning to the basic understanding of economic value of activity in urban park or value of infrastructure of park. There is no study or formal document concerning to the economic valuation of availability park in Semarang. Educational resilience of urban park infrastructure associated to the flora and fauna that may life in urban park. urban park could be design as a place for increase biodiversity

5 Conclusion

Base on the field observation, interview and cross check with the secondary data, this study found characteristic of urban park infrastructure resilience as follow:

- Urban park infrastructure resilience could be defined best on the its physical condition, form and function for city life
- In Semarang estimated found 243 urban infrastructure park. Consist of 76 active urban park infrastructure and 167 passive urban park.
- The total area of urban park infrastructure are 336,931.2 m^2. Refers to the number, Semarang Tengah district found the highest of urban park with 47 urban parks. Refers to the area, Pedurungan district have the largest urban park with 68,396.0 m^2
- The major classification of urban park infrastructure resilience could be define as resilience of active urban park infrastructure and passive urban park infrastructure.
- In active urban park infrastructure resilience, the disturbance that may reduce are Social resilience, Amenities resilience, Economic resilience, Spatial boundary of plan resilience , Education resilience and Ecological resilience
- In passive urban park infrastructure resilience, the disturbance that may reduce are Ecological resilience, Spatial boundary of plan resilience, Education resilience, Economic resilience , Social resilience and Amenities resilience

6 Acknowledgment

This research is funded by the Directorate of Research and Community Services, Ministry of Research, Technology and Higher Education for year 2017 with contract number 344-42/UN7.5.1/PP/2017

References

1 A. Chiesura, Landsc. Urban Plan., vol. **68**, pp. 129–138, (2004).

2 Indonesia, Ministry of Public Work No. 5 Year 2008 Guidelines of Providing and Utilization of Urban Green Open Space. (2008)

3 B. Stephan, P.John, E.Henrik. Urban studies, **52**.7, pp 1321-1338. (2015)

4 H. Nakayama, T. Shimaoka, K. Omine, Maryono, P. Patsaraporn, O. Siriratpiriya, (2013). Journal of Disaster Research, **8** (3), pp. 456-464, (2013)

5 J. R. Wolch, J. Byrne, and J. P. Newell, "Landsc. Urban Plan., vol. **125**, pp. 234–244,(2014).

6 K.Tzoulas, K.Korpela, S.Venn, V. Yli-Pelkonen, A.Kaźmierczak, Jari Niemela, P. James, andsc. Urban Plan., vol. **81**-3, pp. 167–178,(2007).

7 M.A. Benedict, E.T. McMahon, Green infrastructure: linking landscapes and communities, Monograph S Eries (2012)

8 C. Luebkeman and J. Hargrave, *Arup*, (2014).

9 M. Herold, J. S. Latham, a. Di Gregorio, and C. C. Schmullius, J. Land Use Sci., vol. 1, no. 2–4, pp. 157–168, Dec. (2006).

10 GR McCormack, M Rock, AM Toohey, D Hignell, Health & place, vol. **16**, pp. 712-726, (2010)

11 C. C. Konijnendijk, M. Annerstedt, a. B. Nielsen, and S. Maruthaveeran, *Int. Fed. Park. Recreat. Adm.*, no. January, pp. 1–68, (2013).

12 S.Meerow, J.P, Newell, M. Stults, Land and urban planning, vol **147**, pp. 38-49. (2016)

13 L Chelleri - Documents d'Anàlisi Geogràfica, 2012 - dag.revista.uab.es

14 D. Levitz, "The Role of Parks in Shaping Successful Cities," no. May, (2014).

[15 G. R. Mc Cormack, M. Rock, A. M. Toohey, andD. Hignell, *Heal. Place*, vol. **16**, no. 4, pp. 712–726, (2010).

16 Cameron, Ross W.F, Urban Forestery & Urban Greening 11 (2012) 129-137. Elsevier, (2012)

17 McCormack, G. R., Rock, M., Toohey, A. M., & Hignell, D. (2010). **16**(4), 712–726.

18 http://www.100resilientcities.org/cities/semarang/

19 D.Harwitasari, J.A Van Ast , J.flood risk management, 2011

20 I.Buchori, A.Sugiri, S.P. Hadi, American Journal of Environmental Sciences, vol **11**.2, pp. 62-75. 2015)

21 P. Khadiyanta, S. Dewantari - Procedia-Social and Behavioral Sciences, (2016)

22 Semarang City Government, Long Term Planning of Semarang Development 2005-2025, bappeda.semarangkota.go.id, (2010)

23 The government of Semarang, Local State Regulation No. 14 year 2011 concerning Spatial Planning of Semarang City 2011-2031

24 K.M. Roche, J. Gittelsohn, A.F. Yan, N.M, Astone, J of Andolescent Health, Vol. **45,** pp. S64-s70 (2009)

25 D.A. Cohen, T.Marsh, S. Williamson, K.P. Derose, H.Martinez, C.Setodji, T.M Mckenzie, Preventive Madicine, vol. **50,** pp. S9-S12, (2010).

26 P.M-Breen, J.M. Anderies, Resilience: A literature review, Bellagio Initiative partners: Institute of Development Studies (IDS), the Resource Alliance and the Rockefeller Foundation. (2011).

27 WN Adger, Progress in human geography, vol 3, (2000)

28 C Folke - Global environmental change, vol 16, pp. 253-267. (2006)

Analysis of Urban Forest Needs as Anthropogenic (CO_2) Gas Absorbent in Semarang City

Anisa Putri Febriani[1], *Tri* Retnaningsih Soeprobowati[2], and *Maryono* Maryono[3]

[1]Master Program of Environmental Science, School of Postgraduate Student Diponegoro University, Semarang-Indonesia
[2]Departemen Biology, Faculty of Science and Mathematic, Diponegoro University, Semarang-Indonesia
[3]Department Urban and Regional Planning , Faculty Engineering, Diponegoro University, Semarang-Indonesia

Abstract. Green open space in cities in significant needs to maintenance environment quality. On of the critical function is to absorb increasing number of gas CO_2. Therefore, developing urban forest in cities is very importance. The objective of the study is to determine the area of urban forest as CO_2 gas anthropogenic absorb which is formed from fuel, diesel fuel, liquid petroleum gas. The study consists of (1) Analyzing the number of CO_2 gas emission by calculating the needs of petroleum and gas based on the number of population, (2) Analyzing the power of gas absorption, (3) Measuring the air concentration of CO_2 gas ambient based on daily traffic activities. This study shown that from year 2013 to year 2017, the increasing of urban forest is not so significant. For year 2013 the green open space in Semarang City are 373.67 hectares (7.5 percent from Semarang City area), consists of 239 parks, 11 public cemeteries, production forests, community forests, and urban forests, however the area of urban forest is not increase. The study assess that Antidesmabunius is one of the green species which high absorb capacity planted for Semarang. This trees produce 31,31 ton annually. This study proposed to fostering Antidesmabunius as one principle threes in Semarang urban forest.

1 Introduction

Semarang is the principal city in central Java. Semarang has a function of of central Java government, centre of service socio-economic, centre of trade and industrial. Which high growth on population, economic, and physical development, Semarang needs sufficient green infrastructure to support urban life [1]. Recently, public green open space in Semarang only 7,5% of the area. While according to the spatial planning regulation, a city requires at least 20% of the area. One of the problem to increase the green open space is land constraints [2]. It is known that the limitation of green open space will be increase air pollution, since CO_2 emission is not converted in urban environment.

Nowadays the concentration of CO_2 in tend to increase highly due to cities activities. Major of contributor is burning fossil fuels, electricity consumption, industries activities. Those activity estimated 65% of total emissions. In addition, about 14% comes from agricultural activities, 18% deforestation, domestic activity and waste decomposition [3]. One effort to maintain and control the concentration of CO_2 is to increase the area of green open space. Green open space is a part of urban forest [4].

The real impact of the rapid development of physical facilities and infrastructure are the reduction of green open space and the increasing consumption of fossil energy. the increasing consumption of fossil energy makes the urban environment become polluted. The increasing of CO_2 will make the urban environment unhealthy and reduce human health, therefore the concentration of CO_2 in the air should be maintenance. One way to reduce in urban areas is to reduce carbon emissions and build urban forests [5]. Urban forests are the most effective carbon sinks that can reduce carbon emissions in the atmosphere. Photosynthesis is an important process in the role of carbon cycle and maintaining in the atmosphere at the same time also play the oxygen cycle. Shifting cultivation, logging and forest fires in absorbing gas decreases. To help overcome the decline in forest function, it is necessary to build urban forest [6].

The balance of population growth with the availability of green open space will maintain health and comfort for the community. Therefore, the greening movement in Semarang City to neutralize the impact of air pollution should be further intensified. In order to achieve the air quality of Semarang City in the present and future period, it is necessary to research concerning

* Corresponding author: anisaputrifebriani43@gmail.com

to urban forest as an anthropogenic gas absorber in order to maintain the air quality in Semarang City [7,8].

The green of the city not only makes the city beautiful and cool but the aspect of sustainability, harmony and balance of natural resources will make comfort, freshness, freedom of the city from pollution and noise. it makes the citizen of the city healthy and intelligent [8].

2. Literature Review

Urban forest is an ecosystem. It is not like forest definition in general. Urban forest is a community of plants in the form of trees and its association which grows in urban land or around the city, path shaped, spreading or clustering with the structure that imitate natural forests, forming a habitat which allows life for animals and leads to a healthy, comfortable and aesthetic environment [9]. Urban Forest according to Fakuara [10] is defined as a woody vegetation plant in urban areas that provide enormous environmental advantages in the use of protection, recreation, and environmental aesthetics. In addition, there are two approaches used in building urban forest; The first approach, all areas in a city are basically the area for urban forest. In this approach, the components of a city such as settlements, offices and industries are viewed as a part of a forest in the city [5]. The second approach, urban forest is built in certain locations. In this approach urban forest is part of a city.

Urban forest is considered to have an advantage in absorbing CO_2 compared to parks, since it occupies a wider expanse of the park and consists of several height stratums

2.1 Carbon Dioxide gas

Normal air contains a gas consisting of 78% nitrogen; 20% oxygen; 0.93% argon; 0.03% (300 ppm) of carbon dioxide, and the remainder consists of heon, helium, methane and hydrogen. This composition can support human life. Carbon dioxide (CO_2), methane (CH 4), and nitrogen oxide (N_2O) are some examples of greenhouse gas (GHG) which cause greenhouse effect (GHE).Greenhouse Effect (GHE).

Greenhouse effect is useful for living things on earth. If there is no greenhouse gas, the average temperature on Earth is only 18°C. This temperature is too low for most living things, including humans. However, due to the greenhouse effect, the average temperature on earth becomes 33^{0}C higher, compare to the normal condition which estimated 15°C. This temperature is suitable for the life of living beings [11]. Carbon dioxide is the most dominant greenhouse gas that occurs naturally and had significant role in the biological system of the world. Carbon dioxide and water are raw materials for photosynthesis. The carbon flow from atmosphere to vegetation is a two-way flow, which is the binding of CO_2 to the atmosphere through the process of decomposition and combustion and the absorption of CO_2 by plants.

2.2 Fuel Oil and Gas

Humans need oil fuel obtained from petroleum. Petroleum is a complex mixture composed mostly of hydrocarbons. Hydrocarbons contained in petroleum are mainlyC_nH_{2n+2}, and cyclotane (C_nH_2). Crude oil contains about 500 types of hydrocarbons with C - 1 to 50 C atoms. In organic chemistry, hydrocarbon compounds, especially paraffinic and aromatic ones, have their own boiling points, of which the hydrocarbon chain rods are proportional to their boiling point. Meanwhile, the length of the hydrocarbon chain is proportional with greater boiling point and density. Therefore, petroleum processing (purification) is carried out by multilevel distillation, in which crude oil is separated into clusters [12]. After being distillated, the boiling point of petroleum fraction can be differentiated into oil and gas fuels as seen in Table 1:

Table 1 Fossil Fuel and CO_2 Emission

No	Type of Fuel	Range Carbon Chain	Boiling point	Allotment
1	Gas	C1 – C5	0^0 - 0^0 C	Gas tube.
2	Gasoline (Petrol)	C6 – C11	50^0 - 85^0C	Fuel Motor
3	Kerosene	C12 – C20	885^0 - 105^0C	Domestic and motor fuel, industrial fuel, petrochemical process feeds
4	Solar	C21-C30	105^0 - 135^0C	Motor fuel, industrial fuel.

Source : [13]

2.3 Photosynthesis

Photosynthesis is the process of metabolism in plants with the help of chlorophyll and light, producing carbon dioxide and water into carbohydrates and oxygen molecules. The process of photosynthes is takes place on mesophyll tissue, because the tissues there is chloroplast where there is also chlorophyll. Chloroplast consists of two parts: (1) Tilakoid composed of that allows the conversion of light energy into chemical energy. (2) Lamela the liquid part (less solid) which is where CO_2reduction occurs in the dark reaction. Carbon dioxide gas as the main ingredient of photosynthesis enters through stomata. Plant productivity can be appropriately estimated by measuring both oxygen and carbon dioxide used in photosynthesis because the amount of C in $CO_2$is directly proportional to the amount of C bound in sugar during photosynthesis, productivity can be expected by the disappearance ofCO_2 in the environment [14].

2.4 Plant As a Gas Absorber CO_2

Plants that are in and around the city can be directed to overcome the greenhouse effect. The greenhouse effect is a symptom of warmer air temperatures, especially in the city center. Forest and city parks can absorb CO_2 gas. City Forest can create a cool and comfortable micro climate Therefore, the greenhouse effect can be overcome well by the vast city forest. Plants can absorb CO_2 gas through the process of photosynthesis by the formula:

6 mol CO_2+ 12 mol $H_2O \rightarrow$ 1 mol $C_6H_{12}O_6$ + 6 mol O_2 + 6 mol H_2O.

The ability of plants to absorb CO_2 bgas varies. According to Prasetyo et al, forests that have different types of vegetation have a different ability or absorption to CO_2 [17]. Absorption of various types of vegetation to CO_2 can be seen in table 2:

Table 2. Characteristic of absorption power of CO_2

Number	Closure Type	Gas absorption powerCO_2 (kg/ ha/ hour)	CapacityCO_2 (ton/ha/year)
1	Tree	129, 92	569, 07
2	Shrubs	12,56	55,00
3	Meadow	2,74	12,00
4	Rice fields	2,74	12,00

Source: [15]

3 Methodology

This method refers to the scientific research conducted by Philipi de Rozari and Suwari [19]. The research uses quantitative descriptive method, ie problem-solving procedure that is investigated by describing or describing the state of the current research object based on facts that appear or as it is in quantitative way. The research stages include (1) determining the condition of existing CO_2 emissions and estimation of CO_2 concentration in the future,(2) identification of forest constituent tree species, (3) analysis of CO_2 (*carbon sink* absorption by various forms of green open space (RTH) of urban forest, and (4) determining the amount of urban forest area requirement as CO_2absorber. Research conducted in the city of Semarang is in Forest City Tinjomoyo.

3.1 Materials and tools

The materials used in this study were leaf samples, aquadest, HCl, NaOH, $Ba(OH)_2$, alcohol, reagent Cu, red phenols, Nelson reagents, ZnSO4, filter paper, cutex, RTRW and green open space Semarang city.

The equipment used in the study includes: a set of glassware for chemical analysis, analytical balance, water bath, mortar, porcelain cup, oven, and Cecil spectrophotometer.

3.2 Data collection

The data used in this study consists of primary data and secondary data. Primary data collection was conducted through field surveys and laboratory analysis. Secondary data are obtained from various related institutions and literature tracking, such as research results and other relevant scientific documentation. Secondary data collected include population, electricity consumption, fuel oil and gas consumption in Semarang City.

3.3 Data analysis

The levels of anthropogenic CO_2 emissions sourced from electricity consumption and the use of fuel oil and gas were analyzed using CO_2 emission factor values established by the Energy Information Administration (EIA) 2000, DEFRA (2001) and The National Energy. Foundation (NEF) in 2005, while CO_2 absorption by various plants was analyzed by carbohydrate measurement method [16]. Detail of analysis method as follow:

1) CO_2 emissions from electricity consumption. The approach used to determine CO_2 emissions from electricity consumption is based on the amount of electricity consumption in Semarang City and emission factor CO_2. Total emissionsCO_2 is calculated from electricity consumption / year of observation multiplied by emission factor(gram CO_2/kWh). The value of the emission factor is the value of CO_2 emissions from Electricity consumption set by the Energy Information Administration (2000), which is gram CO_2/kWh. The total value of emissions is the total CO_2 emissions resulting from the use of electrical energy.
2) CO2 emissions from Premium Consumption
The calculation of CO_2 emissions from premium / gasoline is done through the approach of the amount of consumption and the value of emission factor.
3) CO2 emissions from solar consumption
The amount of CO_2 emissions from diesel consumption is calculated based on total diesel fuel consumption multiplied by CO_2 emission factor from diesel consumption set by The National Energy Foundation (2005),
4) CO2 Emissions from Kerosene Consumption
The determination ofCO_2 emissions from the use of kerosene in Semarang City is based on kerosene consumption multiplied by kerosene emission factor.
5) CO2 Emissions from Fuel Gas Consumption
The amount ofCO_2 emissions from gas fuel consumption is calculated based on the total consumption of gas fuel (BBG) multiplied by CO_2 emission factor from diesel consumption set by The National Energy Foundation (2005),
6) Need for BBMG and Prediction of Future Needs
The data used is the data of the use of fuel oil and gas (BBMG) taken from PT. Pertamina. With regard to the current population of Semarang City and its development in the coming year, it can be calculated the use of BBMG (premium, diesel, kerosene, diesel oil and LPG)
7) Analysis of Sink Gas Power CO_2 Different Types
Plants with Carbohydrate Method For the purpose of measuring the absorption of CO_2 gas, carbohydrate levels were determined in the sample of plant leaves

measured at 5:00 am and 10:00 am. Three examples of leaves from two trees of each species The plant is taken with scissors then fixed by soaking it in 70% alcohol for 15 minutes. Leaf samples are then dried in the hot sun, after dry oven at 70°C for 2 days. Examples are then ground until smooth. A total of 20 grams of leaf powder was put in a glass container then added with 20 mL of 0.7 N HCl and hydrolyzed for 2.5 hours in a water bath. Strain in a 100 mL measuring flask and add a few drops of red phenol and 1 N NaOH solution until a solution changes from blue to pink. Then add 5 mL $ZnSO_4$ 5% and 5 mL Ba (OH) 2 0.3 N, then dilute with distilled water up to 100 mL volume. After filtering, a clear solution of supernatant is taken with a 1 mL pipette.

Then stored in a beaker. A series of standard solutions containing 0, 5, 10, 15, 20 and 25 mg carbohydrates were prepared and then added 2 mL Cu reagents and heated in a water bath for 10 minutes and then cooled. After that, add Nelson's reagents and 20 mL of water on

Each standard solution, then shaken and left for 20 minutes. The sample and standard solutions were then measured by a CeCil spectrophotometer at the optimum wavelength. Selected a standard solution which gives the absorbance average closest to absorbance [17]

Carbohydrate levels account with the formula as follow:

$$\text{Carbohydrate levels} = \left(\frac{\frac{A}{S} x \frac{100}{0,2} x \frac{20}{1} x 100\%}{1000}\right)$$

Where : A = the average value of uptake of sample solution

S = the average value of absorption of standard solution

The equations used to determine the absorption of CO_2 by various types of plants are as follows:

CO_2 sinks per leaf area (D) =CO_2mass / leaf area (of 20 gram samples)

CO_2 sink per unit area per hour (Dt) = D/t

CO_2 sink per leaf = Dt x area per leaf sheet

CO_2sinks per tree = L. P / JD

CO_2 sinks per tree per hour (Dn) = Dt x d x area per leaf sheet

CO_2sinks per tree per year (Dy) = [{Dnx t} + {Dn x (A-t) x 0.46}] x 365

With:

L = average leaf area per 20 g wet weight of leaf

P = number of leaves per tree

JD = number of leaves per 20 g wet weight of leaf

□t = 4 hours

A = average maximum duration of irradiance per day (hour / day)

T = actual average irradiance value per day (hour / day)

0.46 = average comparison rate per day of photosynthesis rate on cloudy and sunny days

365 = number of days in 1 year

8). Analysis of Urban Forest Needs

To analyze the needs of the forest area City as an anthropogenic CO2 absorber Used the following equation [18]:

$$L = \frac{aV + bW + cX + dY + eZ}{K}$$

with,

L: The area of urban forest

A: The amount of CO_2 emissions produced by a human (g / hour)

B: The amount of CO_2 emissions generated from combustion Premium (g / liter)

C: The amount of CO_2 emissions generated from combustion Diesel (g / liter)

D: The amount of CO_2 emissions generated from combustion Kerosene (g / liter)

E: The amount of CO_2 emissions generated from combustion LPG (g / g)

V: Population (soul)

W: Total consumption of gasoline (liter / hour)

X: Total consumption of diesel (liter / hour)

Y: The amount of kerosene consumption (liters / hour)

Z: The amount of oxygen consumption (g / hour)

K: The ability of forest to absorb CO_2 (75 kg / hour / ha).

4 Discussion

Urban forest is need to absorb the anthropogenic CO_2 gas emissions. This study proposed the formula to account and analyze the needs of the forest area City as an anthropogenic CO_2 absorber Used. This formula is use to assess the parameter oil gas such as gasoline, diesel, kerosene and LPG. The urban forest formula not yet account for the oxygen need for human life in cities. This formula could not be use directly to design the area of urban forest since the parameter of urban forest need should accommodate many function, such as, social, economic, ecology, psychology, amenity and leisure.

5 Conclusions

Anthropogenic CO_2 gas emissions increase every year. This is due to the increasing demand for fuel oil and gas as the population increases. Each tree has a different absorption of carbon dioxide gas depending on the leaf area per tree and the absorption of CO_2 per cm^2. Trees that have CO_2 gas absorption. The addition of urban forest area to minimize the increase of anthropogenic CO_2 gas in Semarang City

6 Acknowledgments

This research is funded by the Directorate of Research and Community Services, Ministry of Research, Technology and Higher Education for year 2017 with contract number 344-42/UN7.5.1/PP/2017

Reference

1 Gratimah, Guti. *Analysis of Urban Forest Needs as Anthropogenic CO_2 Anthropogenic* Gas Absorbent in Medan City Center. Thesis. University of Northern Sumatra, Sumatra

2 Ministry of Indonesia Secretary, 2007, Undang-Undang No 26 Tahun 2007 about *Spatial Planning*

3 Anonimous. 2008. Forest Sinks. Article. *http://www.climatechange.gov.au/land/forest-sinks.html*. accessed July14, 2017.

4 EN. Dahlan, *Membangun Kota Kebun Bernuansa Hutan Kota*. IPB Press, Bogor, (2004)

5 EN. Dahlan, *Forest City for Improving the Quality of the Environment*. Jakarta: APHI, (1992)

6 JR. Etherington, Environment and Plant Ecology. New Dehli: Wiley Eastorn Limited, (1976)

7 Y. Fakuara, Urban Development Concept. Department of Forest Management, Faculty of Forestry, Bogor Agricultural University. Bogor, (1987)

8 R.B. Palmquist, J.Urban Econ. **11**, 333 – 347, (1982)

9 I. Heriansyah, N. Mindawati, Journal of Forest Research and Nature Conservation: Vol **II** No.2; Halaman 105 – 111, (2005)

10 Y. Fakuara, Conception of Forest Development City. Department of Forest Management, Faculty of Forestry, Bogor Agricultural University. Bogor, (1987)

11 O. Soemarwoto, Ecology, Environment and Development Jakarta: Djambatan, (1991)

12 D. Altrell, Bangladesh Foresh Department, Ministry of Environment and Forest, p, 118, (2007)

13 CF. Zuhra, Oil Refining, Processing and Use. Faculty of Mathematics and Science. University of Northern Sumatra

14 Harjadi MM S. Introduction to Agronomy. Jakarta: Gramedia, (1979)

15 R. S. Tinambunan, Management of Natural Resources and Environment. Graduate School of Bogor Agricultural University, (2006)

16 L. Popoola, O. Ajewole, Arboricultural journal **25**.1-22, (2001)

17 W. De Jong, Biotropica **42**, 3-9, (2010)

18 A. Bernatzky, *Tree Ecology and Preservation*. Elsivier Science Co., Amsterdam, (1978)

Comparing Value of Urban Green Space Using Contingent Valuation and Travel Cost Methods

Dea Chintantya[1,*], *Maryono* Maryono[2]

[1]Master Program of Environmental Science, School of Postgraduate Studies, Diponegoro University, Semarang - Indonesia
[2]Department of Urban dan Regional Planning, Faculty of Engineering, Diponegoro University, Semarang - Indonesia

Abstract. Green urban open space are an important element of the city. They gives multiple benefits for social life, human health, biodiversity, air quality, carbon sequestration, and water management. Travel Cost Method (TCM) and Contingent Valuation Method (CVM) are the most frequently used method in various studies that assess environmental good and services in monetary term for valuing urban green space. Both of those method are determined the value of urban green space through willingness to pay (WTP) for ecosystem benefit and collected data through direct interview and questionnaire. Findings of this study showed the weaknesses and strengths of both methods for valuing urban green space and provided factors influencing the probability of user's willingness to pay in each method.

1 Introduction

Urbanization causes urban population increase so demanding development in various aspects, especially the physical development of cities and infrastructure. Unfortunately, the city has limited land and space, so in the end urban green space that tends to dormant and has a low economic value become into alternative land development. Urban green space is a natural city resource and has an important role in maintaining the quality of urban environment. Urban green space may include urban forests, city parks, green fields (sports and golf fields), farmland, riverbanks, and un-built land [1]. Urban green space provides very diverse benefits. It can be used as a conservation land to preserve various types of flora and fauna. The urban green space that has enough vegetation can also balance the urban temperatures that tend to be high, absorb air and sound pollution, absorb carbon, reduce rainwater runoff, and prevent floods [2].

It also plays an important role for social life and urban communities, especially public spaces such as urban parks because they are able to provide access to all circles of society without distinction of age, gender, and occupation. Park city has been widely used by the community to gather and conduct socialization, it's also used by the community for various outdoor activities that can improve physical and mental health. For children, city parks are a space that gives them a chance to play safely [3]. It is also able to influence the price of houses and surrounding buildings [4]. In some cities of the world, the provision of urban green spaces is done as part of the city's regeneration plan because the increase in the value of houses and buildings will increase the taxes earned. it also has the potential to attract tourists [5] and increase sales of local products.

Although urban green space has many benefits but it is difficult to get the economic value because most benefits are non use value, public goods, and do not have market prices. For example, city parks, city park users are never charged for accessing them, but that does not mean the city park has no economic value because if its existence is lost then there will be consequences or losses that require other costs. An assessment of the urban green space economic value has been widely used to compare the costs of provision and benefits gained in the scale of land use planning and management [6]. Most commonly used method to estimate economic value from urban green space are contingent valuation (CV) and travel cost method (TCM). Both of these methods basically measure the perceived benefits through the willingness to pay (WTP) by visitors and measure value of non-market good [7]. However, there are fundamental differences where TCM is only able to estimate the economic value of urban green space based on recreational benefits while the CV is able to estimate the overall economic value.

The purpose of this study was to compare the use of CV and TCM methods in estimating the economic value of urban green space. Weaknesses and strengths of both methods will be reviewed based on the results of previous studies. There are at least 30 national and international journals with economic value of urban green space including the value of urban forest, City Park, riverbanks, and agricultural land using CV and TCM method. This paper has the structure and framework as follows, section 2 provides the overview the use of CV and TCM methods in estimating the urban green space economic value. Section 3 provides

* Corresponding author: chintantya@gmail.com

comparison between CV and TCM and section 4 provides conclusions.

2. Contingent Valuation and Travel Cost Method Overview

2.1 Contingent Valuation Overview

The main concept of contingent valuation is to model individuals' responses in specific hypothetical situation. CV method is used to estimate the total economic value of various types of benefits or services produced by an ecosystem. The economic value is estimate through a question directly to the respondent who is a urban green space user about willingness to pay to provide a desired benefit or compensation of loss of a benefit. The WTP question modes have several variants such as dichotomous choice (DC), iterative open-ended bidding (OE) and payment cards. The receptor used in the CV is a user or resident around a location that has an age of 15-75 years or is in a productive age. This is because at that age, they already have income so as to provide real value of the WTP. Although CV may be used to estimate economic value based on use and non use benefits but CV is used more to estimate economic value related to non use benefits. This is why most of the research related to the economic value of urban green space using CV has object urban forest. However, the estimates of economic value obtained different from each other [8], in accordance with those proposed by other research hat estimates of economic value can be different from one to another due to differences in variables, time, and location [9]. Some research results on the economic value of green space also shows that the highest economic value of urban parks has the highest economic value compared with urban forest, agricultural land, riverbanks, or un-built land. This is because the estimated economic value of the urban green space depends on the number and type of its usefulness and the availability of substitute.

In addition to estimating the economic value, research conducted in general describes the motivation of respondents in providing the value of WTP. The motivation of respondents becomes something very interesting to discuss because the benefits of urban green space are public goods whereby anyone can access and without pay or the cost of management is charged to the government. Even from some sample research found not a few respondents who are not willing to pay for the benefits they feel or commonly called "protest" responses [10]. Based on the set of various studies and journals some of the most common motivations are (1) the respondents consider that urban green space will make the urban environment better; (2) as an inheritance for future generations; and (3) to protect and conserve resources especially in urban areas. The magnitude of the WTP given by respondents is also related to the socio-economic characteristics of the respondents. The results of 8 out of 10 studies show that income is the main factor affecting the value of one's WTP. Furthermore, the level of education and number of visits also have a positive effect on the value of the WTP. Other findings on some studies that involve not only visitors but also local residents are value for WTP influenced by visitor and residence status, residents will be willing to pay more for surrounding urban green space compared to visitors, as they can feel the impact directly related to prices of houses and buildings.

Although the use of CV techniques tends to be easier and more flexible to use but there are some weaknesses that have the potential to cause consistent errors in estimating a value or called "bias". Some of them are:

1. CV rely on the hypothesis and assumptions of the respondents so that the value of WTP does not have a direct impact on the respondent, it is not uncommon to cause excessive assessment by the respondents
2. CV only estimates value based on compensation and profit equivalent to benefits obtained but can not estimate value identically or specifically
3. The implementation of the CV is often constrained by the problems of the survey design and the given questionnaire, this may affect the value of the WTP
4. Unresponsive responders' responses such as refusing to pay or providing less than average WTP value due to lack of understanding of indirect benefits of urban green space such as rainwater absorber and flood prevention
5. Urban green space has a very difficult number of visitors to determine because of the unavailability of data on the number of visitors, this causes the samples used do not represent actual populations especially during certain seasons or vacation times.

2.2 Travel Cost Method Overview

The travel cost method is used to estimate the value of ecosystem use or the location used for recreation. The estimated cost or value is related to:

- Changes in access fees for a recreational location
- Removal of the existence of a recreational location
- Addition of a new recreation site
- Changes in environmental quality at a recreational location [11].

The basic assumption of the travel cost method is that the travel time and expenses people spend traveling to a location present a "price" for accessing a location [12]. Thus the willingness to pay when visiting the site can be estimated based on the number of trips with different travel expenses. The use of TCM methods to estimate the economic value of green open space is widely used in urban forests and urban parks, as TCM is only able to estimate the economic value based on the benefits of recreation. In TCM there are two approaches, namely zonal travel cost method and individual travel cost method. ZTM estimates the economic value by gathering information about the number of visits at a site from several different locations that have been grouped previously, while ITCM estimates the economic value of recreation based on the individual cost of travel approach [13]. In some studies TCM requires data on the number of annual visitors to estimate the number of requests and the economic value generated during the year. This led to most of the research being undertaken starting with pilot survey activities and using secondary data on the number of visitors [14].

Estimation of economic value using TCM is usually followed by an analysis of access or road network, modes of transportation, frequency of visits and socio-economic conditions of its respondents. Some things that have a positive relationship with the value of WTP is the type of activity, mode of transportation, and income. This is because the three factors are very determine the cost of travel responder. Some of the obstacles seen in estimating the economic value of green open space through TCM are:

1. One of the requirements of TCM is that the respondent must have 1 travel destination (not multiple trips), this can minimize the number of samples because many visitors encounter more than one travel destination [15].
2. Most research difficulties in estimating the value or cost of time sacrificed to visit a site, so that the use of time is negligible [16].
3. The use of TCM in urban areas has many limitations, this is related to the number of replacement locations, various types of transportation, and the relatively close distance, resulting in very low travel costs [17].
4. Specifically on ZTCM the use of zip codes and zone determination usually ignores road routes and uses only linear lines, it also has the potential to cause error measurement [18].

3. Comparation of CV and TCM

CV and TCM is a method to estimate economic value based on ecosystem benefits obtained by willingness to pay (WTP). However there are several fundamental differences between TCM and CV studies. A first difference is that CV analyzes stated behaviour and preferences in hypothetical situations, while TCM studies analyze actual behavior and preference based on what they purchase. Secondly, CV can estimate value for both use and non-use value as a whole, but TCM can only estimate use value for recreational benefits, so when compared to CV application will be more flexible and easy compared to TCM. In addition there are also some differences in the results of economic value obtained. In some studies that have one object and analyzed through both these methods show the economic value through CV method has a lower value than TCM. This is because the value by using CV is obtained based on the assumption of respondents who have no consequences on what they get while the TCM value is obtained based on real market price. In addition there is a tendency that residents will provide higher WTP values in CV compared to visitors because of the benefits they feel directly, but instead with TCM, the residents will provide a zero value because they do not incur transportation and time costs to access that place. In general, the use of both methods is also followed by an analysis of the related socio-economic factors of respondents, and most of them, whether the value of WTP generated from CV or TCM has a positive relationship to income, number of visits, and facilities at a particular location. This positive relationship means that people with higher incomes will have the opportunity to visit a location more often, especially those with tourist attraction. While urban green space with various types of supporting facilities will attract more visitors (19).

4. Conclusion

CV and TCM are the most commonly used methods in various studies to estimate the economic value of urban green spaces. Each method has certain weaknesses and strengths in estimating economic value. For example, CV can be estimate economic value for both use and non-use, where TCM is only able to estimate the value of use value on recreational benefits. Several studies on CV and TCM also analyzed the relationship between the WTP value provided by the respondent and the socio-economic condition. The result is the factor of income, the type of facility, and the number of visits is the most common factor.

References

1. L. M. Brander, M. J. Koetse, Journal of Environmental Management, **92**, 2763-2773 (2011).
2. S. Orr, J. Paskins, S. Chaytor, *Valuing Urban Green Space : Challenges and Opportunities*, UCL Policy Briefing (2014).
3. H. Woolley, S. Rose, *The Value of Public Space*, CABE Space (2014).
4. X. Song, X. Lv, C. Li, Acta Ecologica Sinica **35**, 89-94 (2015).
5. S. Majumdar, J. Deng, Y. Zeng, C. Pierskalla, Urban Forestry and Urban Greening **10**, 275-280 (2011).
6. L. Tyrvainen, Journal of Environment Management **62**, 75-92 (2011).
7. Jala, L. Nandagiri, Aquatic Procedia **4**, 1315-1321 (2015)
8. C. Y. Jim, W. Y. Chen, Cities **26**, 187-194 (2009).
9. M. Verbic, R. S. Erker, M. Klun, Land Use Policy **56**, 58-67 (2016).
10. A. Y. Lo, C. Y. Jim, Ecological Economics **114**, 58-66 (2015)
11. S. T. Djajadiningrat, Y. Hendriani, M. Famiola, *Ekonomi Hijau*, Rekayasa Sains (2011)
12. J. T. Bishop, *Valuing Forest : A Review of Methods and Applications in Developing Countries*, Environmental Economic Programme, Internatioanl Institude for Environment and Development (IIED).
13. T. A. Kassaye, World Scientific News **62**, 124-135 (2017).

 https://doi.org/10.1051/e3sconf/20183107008

14. R. Mokhtari, S. M. Hosseinifar, European Online Journal of Natural and Social Sciences **2**, 549-555 (2013).
15. C. Betram, N. Larondelle, Ecological Economics **132**, 255-263 (2017)
16. C. M. Fleming, A. Cook, The Recreational Value of Lake McKenzie, Flaser Island : An Application of the Travel Cost Method, Tourism Management 29, 1197-1205 (2008)
17. P. Iamtrakul, K. Hokao, K. Teknomo, Procceddings of the Eastern Asia Society for Transportation Studies **5**, 1249-1264 (2005).
18. M. M. Hanauer, J. Reid, Journal of Environmental Management **198**, 50-65 (2017)
19. T.A Kassaye, World Scientific News, **62**, pp.124-135, 2017

Assessing Method to Identifying Water Resilience Against Natural and Climate Change Hazards.

Rofi Amril [1] and *Maryono* Maryono[2]

[1]Master Program of Environmental Science, School of Postgraduate Studies, Diponegoro University, Semarang-Indonesia
[2]Department of urban and Regional Planning, Faculty Engineering, Diponegoro University, Semarang-Indonesia

Abstract. A geographic region may become vulnerable toward water resources in a variety of ways. Common issues arise when man-made infrastructure such as housing, industrial, agriculture and other spatial land use policy implementation exceeds more than desired level. Vulnerability of a region due to water resources could be interpreted as the inability of the region to sustaining economic and social activity associated to socio-economic water availability. This study assess four aspects of water resilience: water quantity, water distribution, water quality, and water requirements. Literature review then followed by interview with academic expert used as method of study. This study found that four aspect of water vulnerability mostly have been applied to asses water resource vulnerability. Each aspect have a specific characteristic and could be define more specific and detail indicator according to the local content.

1 Introduction

Currently have been more understandable that climate hazard impact influence the balancing of hydrological cycle as shown on the duration of dry and wet season. Water quality and quantity, for example, in Indonesian's river tend to polluted [1]. Moreover, the link between economic development and water resources tend to negative at which cost for water purification becoming bigger day by day [2]. It is a clear sign that water resource vulnerability clearly arise. Assessing the water contribution for development and the vice versa the impact of development toward increasing water have been conducted, however the increasing of water vulnerability should be evaluate since the issues will influence more significant the relationship. Moreover increasing the living standard also influence to the water competition among users.

The demand for water already exceeds the available supply in many regions and river basins, especially in parts of Asia, North Africa and the Middle East, and sub-Saharan Africa – where the majority of poor people live and where the population is growing most rapidly. Global fresh water withdrawals have been growing at a rate of about 1% annually since the 1980s, almost entirely in developing countries. Global water demand will be increase for 20 – 130 % for post 1990, and it is significant change compare to the increasing of water global demand pre 1990 [3]. The rapid increasing of water cause by dominant sector such as industrial and domestic use, energy production, and agriculture [3,4].

Over the past several decades, food production as a result of the green revolution has improved dramatically, concerning to access to drinking water that increase for 2.4 billion people, while access to sanitation facilities improved for 600 million since 1980 [4]. Moreover, concerning to the data trend, a systemic water crisis due unsustainable management of water resources, a poor social, environmental, economic policies is then clearly understandable. Therefore at the same time, there is a need to increase water supplies and services to fulfill population growth, in an environmentally sound manner. It is also clearly note that the traditional problems concerning the provision of water supply and sanitation services in urban and rural areas will require more specific efforts to solve, since more than a billion people do not have access to potable water supply and 3 billion do not have adequate sanitation [4]. If there is no policy transformation, the rapid population growth, food security and growing energy and electricity, domestic supply, and industrial development will face a huge problem concerning to the water availability. Moreover the climate hazard will impact seriously to the water resilience.

Concerning to the availability and vulnerability of water, this study will assess the possibility method to identifying water resilience.

2 Literature review: water resilience

Many study concerning to water resilience have been conducted worldwide. One of the principle water resilience is associated to the Institutional and ecological engineering, for example, concerning to parameter of policy of land use, spatial waters management [5]. Water resilience also have a correlation with the adaptive

* Corresponding author: maryono@pwk.undip.ac.id

government especially for the ability of government to design law enforcement to cover the spatial planning issues [6]. Water resilience could be defined associated with the influence to the food production [7], and sustainable cities and regional economic development [8,9]. Other importance issue of water resilience is the water supply associated with the water demand especially in drought season [10]. The characteristic of the water resilience in dry season usually suffer in rural and mountain area. However, in specific condition also suffer in coastal zone.

This study assess four basic aspects of water resilience: water quantity, temporal distribution, water quality, and water requirements. Water quantity associated to the availability to support development both in local or regional region is decrease in recent decade [3,4]. Therefore the development agenda such as SGs 2030 need specific and applicable indicator for water issues. In the context of water distribution water issues also need specific spatial management to solve the un-balance situation [3,8]. It is known that the Water quality is the most importance parameter in water services [11]. Increasing the quality of life and degree of economic development recently affect to the requirement of water, in vice versa climate hazard arise from the un proper rapid development.

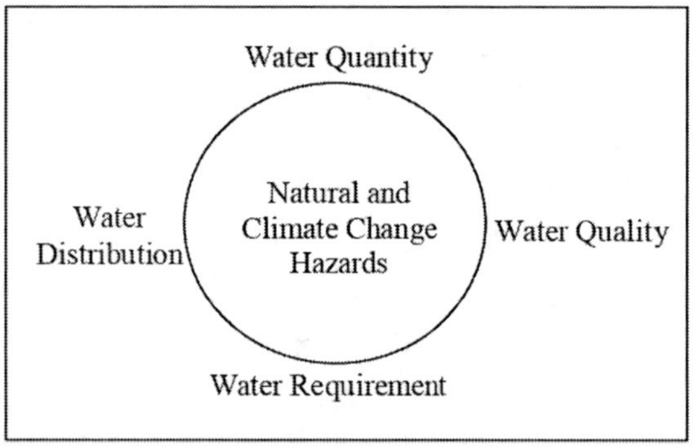

Fig. 1. Parameter and aspect of water resilience against natural and climate change hazards.

3 Methodology

3.1 Study area and issues

As shown in figure 2, Agam Regency is selected as study area. The Agam Regency is part of West Sumatra Province. The west Sumatra , including Agam Regency is one of the vulnerable region in Indonesia. It was attacked huge earthquake in 2009. The victim such as 1,117 people death, 1214 people hard injure, 1168 medium injure. 135.448 housing was hard damage, 65.380 housing was medium damage, and 78.604 housing was low damage [14]. While in 2010 Tsunami disaster attacked Mentawai Island near to the study area also with huge victim, damage and losses[15].

In 2010 the number of population are 455.484 and the growth rate of population is 0.93% , The population is distributed within 16 district, 82 Nagari and 467 Jorong [13]. Since this regency in one of the vulnerable region due to the earthquake, tsunami and climate hazard, the study of water resilience is very importance to fostering the community resilience.

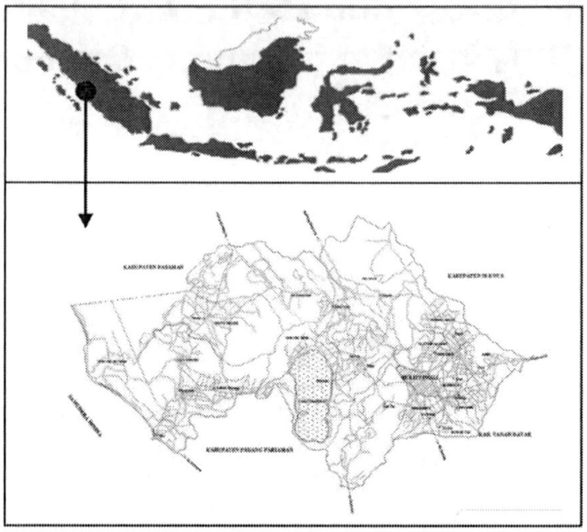

Fig. 2. Location and Map of Agam Regency [13]

3.2 Scope, data collection and analysis

This study assessed four aspect of water resilience; water requirement, water quality, water quantity and water distribution. Assessing of water requirement and water distribution by using quantitative model while water quantity and water quality by using qualitative descriptive refers to the formal regulation. Data concerning to the water resilience will be collected from the secondary data. Quantitative model of water resilience describe as follow.

3.2.1 Water requirement

Water requirement assessed by using the formula and model of the water requirement as follow :

$$Wt = (Po(1 + r)^n) * 60 \frac{L}{capita}$$

Where

W_t : water requirement for year t
P_o : Population in base year assessment
r : rate of population growth
n : the different of year (year prediction and base assessment)
L : unit of water in Liter ($0.001m^3$)
Capita : unit of people.

3.2.2 water distribution

Water distribution resilience assessed by using the model of total water requirement divided by total requirement in each district. Water distribution calculated by model as follow

https://doi.org/10.1051/e3sconf/20183107009

$$Wi = ((Po(1+r)^n) * 60\frac{L}{capita})/WT$$

Where

w_i : Water distribution for region i for year t
P_o : Population in base year assessment
r : rate of population growth
n : the different of year (year prediction and base assessment)
L : unit of water in Liter ($0.001m^3$)
Capita : unit of people.
WT : Total Water requirement for year t for

4 Result and discussion

One of the aspect of the water resilience is concerning to the requirement. As shown in table 1, water requirement in Agam for year 2040 totally 33.077 m^3. The number is calculated by using the assumption that the population growth is 0,93% and it is fix for 30 years of calculation. In the district level, the highest of requirement is Lubuk Basung regency with estimated 5.402 m^3. While the lowest is Malalak regency estimated 737 m^3. When the government supply the number of the water requirement the ideal resilience of water will be achieved. Moreover, the degree of resilience is depending of the level of water services.

Table 1. Water Requirement in Agam Regency for year 2040

District in Agam Regency	Year 2010	Year 2040	Water Requirement 2040 (m3)
Tanjung Mutiara	28.311	37.373	2.242
Lubuk Basung	68.198	90.028	5.402
Ampek Nagari	22.570	29.795	1.788
Tanjung Raya	33.307	43.969	2.638
Matur	16.944	22.368	1.342
IV Koto	23.036	30.410	1.825
Malalak	9.299	12.276	737
Banuhampu	36.059	47.602	2.856
Sungai Pua	23.042	30.418	1.825
Ampek Angkek	43.347	57.223	3.433
Canduang	21.886	28.892	1.734
Baso	33.016	43.585	2.615
Tilatang Kamang	34.027	44.919	2.695
Kamang Magek	19.972	26.365	1.582
Palembayan	29.426	38.845	2.331
Palupuh	13.044	17.219	1.033
Total of Agam Regency	455.484	601.286	36.077

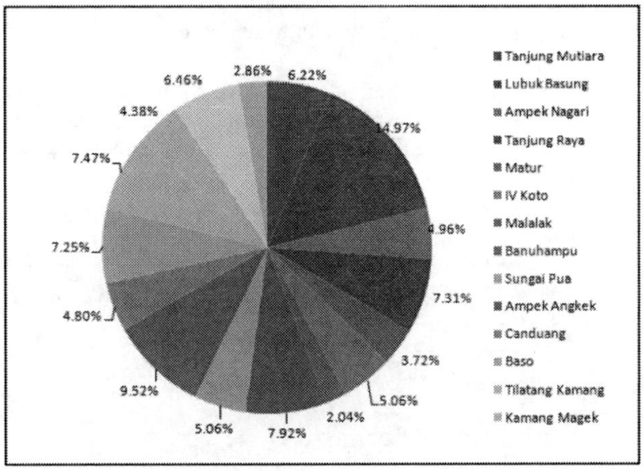

Fig. 2. Water Distribution at Agam Regency for year 2040 base on district level.

The distribution of water requirement in Agam for year 2040 depicted in figure 2. According the calculation the highest propeortion of distribution is lubuk basung district with estimated 14,97% while the lowest proportion is malalak district with estimated 2,04% comparing to the total of water requirement in Agam. The calculation of the distribution is conducted by using the formula of total requirement in district divided by the total number of regency. The assumption of this proportion is that the number of population growth is fix for 0.93% per year for 30 years and the number of percapita of water requirement also fix 60 l/capita.

Water quality for year 2040 in Agam regency may be change depending on the policy and law performance. Control of water quality need stakeholder effort and detail understanding of whole resilience aspect. WHO have been issued guidelines for water drinking quality [16]. Moreover stakeholder intention to prepare and mitigate the water pollution is very needed. Moreover to prepare and fostering resilience in water and sanitation there were some factors that may be considered [17].

Water quantity is different with water requirement. The water quantity content terminology of logistic. It is concerning to the number of water that storage in region in the condition and high level of readiness. The quantity of water could be defined as the capability of government to provide water not only the current situation, but also for future years. Water quantity means conducting sustainable water provision [17]. The parameter of assessment also associated with the number of water for all purpose not only for domestic sector.

5 Conclusion

- This study assessed four method to assess and identify water resilience namely water quantity, water distribution, water quality, and water requirements.
- Method to identifying waster requirement resilience could be developed by using simple mathematic model until complex model. This

study proposes the model by combining population growth and standard of daily water need percapita.

- The assessment of water distribution could be asses by development mathematical model such as base on district level. This study propose model according to the requirement of water in district level.
- This study proposed that method to identifying the water resilience concerning to the water quality and quantity is by assessing the stakeholder intention. Since this aspect in associated to the understanding and behaviour.

6 Acknowledgment

This research is funded by the Directorate of Research and Community Services, Ministry of Research, Technology and Higher Education for year 2017 with contract number 344-42/UN7.5.1/PP/2017

References

1. M.A. Shannon, P.W. Bohn, M. Elimelech, J.G. Georgiadis, B.J. Marinas, A.M., Mayes, Nature, Vol **452**, pp. 301-310, (2008).
2. B.P. Resosudarmo, Int.J. Environ and Sus Dev, Vol **2**, no.1. pp. 62-77. (2003).
3. U.A. Amarasinghe, V smakhtin, Global Water Demand Projections: Past, Present and Future Global water demand projections: past, present and future. (IWMI Research Report 156). doi: 10.5337/2014.212, (2014)
4. J. Alcamo, P Doll, . Hendrichs, F. Kaspar, B. Lehner, T. Rosch, S. Siebert, Hydrological Sci. J. vol **48**.no3. pp. 339-348, (2003)
5. M Falkenmark, J Rockström , J. Water Resour. Plann. Manage. **136**(6): 606-610. (2010)
6. M. H. Clarvis, A. Alan, D.M. Hannah, Envi Science & Pol, **43**, pp 98-110 (2014)
7. J Rockström, M Falkenmark, L Karlberg, H.Hoff, S. Rost, D. Gerten, Water Resources, **45** (2009)
8. S Barthel, C Isendahl - Ecological Economics, **86**.pp. 224-234 (2013)
9. A Rose, S.Y Liao – J. Reg Sci, **45**.1. pp. 75-112. (2005)
10. J Rockström, Physics and Chemistry of the Earth, **28**. 20 pp. 869-877 (2003)
11. A. Milman, A short, Glob Envi Change,**18**, pp. 758-767 (2008)
12. J.Rockström, M. Falkenmark, C. Folke, M. Lannerstad, J. Barron, E. Enfors, L. Gordon, J. Heinke, H. Hoff, C. Pahl-Wost, Water resilience for human prosperity. Cambridge University Press, (2014).
13. The government of Agam Regency, Peta Kabupaten Agam, 2017,available on http://www.agamkab.go.id/?agam=profil&se=peta, access January, 2017
14. The BNPB, Bappenas, Provincial and District/City Governments of West Sumatra and Jambi and international partners Damage, Loss and Preliminary Needs Assessment, Joint report (2009)
15. A. V. Newman, G. Hayes, Y. Wei, & J. Convers, Geophysical Research Letters, **38**.5. (2011)
16. WHO, Guidelines for drinking-water quality, 2004
17. Maryono, H. Nakayama, T. Shimaoka, Memoir of faculty Engineering, Kyushu Univeristy, Vol.**74** no 3, pp.79-98
18. P. Chaves, T Kojiri, Y Yamashiki , Hydrological Processes, **17**(14), 2769-2793. (2003)

 https://doi.org/10.1051/e3sconf/20183107010

The Population Growth and Carrying Capacity in Semarang City

Hariyanto[1,*], *Sudharto* P Hadi [1], and *Imam* Buchori[2]

[1]Doctoral Program at Environment Science of Post Graduate School Diponegoro University Indonesia
[2]Department of Urban Plannung, Engineering Faculty Diponegoro University Indonesia

Abstract. Population growth and development of city activities take some lands to carry them. As a result, land use competition happens among persons, society or sector. Land necessity for settlement, industry, or sector has taken over farm land, therefore farm land has been converted intensively and massively. Chronologically, population growth will cause land necessity increase. Unproductive land, especially farm land will be converted. Furthermore, farm land conversion will cause carrying capacity change. Carrying capacity has certain bio capacity. With the population growth, it will increase resource consumption; on the other side, farm land conversion will decrease carrying capacity. The objective of the study is to know about the influence of population growth towards carrying capacity (bio capacity) in Semarang city. Land consumption per capita is indeed influenced by city population, the higher the population is, the lower the land consumption per capita. With the population growth, it will influence carrying capacity. Carrying capacity here is the ratio of area to population. Analytical descriptive method is applied in the study with all sub-districts in Semarang city as the analysis unit. Population here is sub-district area and population per sub-district in Semarang city. Population growth data period is from 2000 until 2015. Main variables of the study are area per sub-district, population, population growth, carrying capacity. Result of the study shows significant influence of carrying capacity decrease, especially some outskirts in Semarang city. This condition happens because the outskirts in Semarang city tend to have dense population growth. Range of carrying capacity in Semarang city is from 0,007 to 0,117 of 0 to 1. Almost all sub-districts in Semarang city show miserable condition, except Mijen and Tugu. The conclusion of the study is that population will decrease carrying capacity. Therefore, the government should control population growth by paying attention to its distribution.

1 Background of the study

Land is a kind of important natural recourse for human life as it is needed in every human activity such as agriculture, industry, settlement, and so on. In other word, it is said that land has direct and indirect multiple purpose use. Land is useful to fulfil material need, health, psychological need, and it is also important to protect other natural resources like vegetation and water [1]. Land supply which is relatively constant will make it as a scare commodity, whereas population in Indonesia is growing.

Based on census in 2010, population in Indonesia was 237,556,363, and it became 245,862,034 in 2014 with 58 percent population distribution that was still concentrated in Java Island, followed by 21 percent in Sumatera Island. The growing population mostly happens in some big cities such as Jakarta, Surabaya, Semarang, etc. Semarang has an average population growth of 2.04 percent per year. Data from BPS shows that population in Semarang city has reached 1,763,370 (2015). It is bigger than population in 2001 that was 1,350,005.

In order to fulfil some needs of the growing population, land demand is also increased for many kinds of activities. This case has changed the function of the land (land conversion) that gives some disadvantages for environment conservation.

The function change often happens to the land which has important function for human life, like fertile farm land that is changed into settlement, infrastructure, and so on. The function change of farm land has changed rapidly and massively.

Kusumawati stated that industrial sector takes over agriculture sector, the higher the function change of farm land, the higher the miserable land will be [2]. Settlement that is extended to some rural areas causes the fertile farm land does not produce any more food material to fulfil people's need. This condition shows that farm land will keep on decreasing due to population problems and growing development in Indonesia. But unfortunately, the decrease of farm land is not followed by extensive farm (creating rice field out of Java island), therefore growing population affects the decrease of land capacity.

Farm land conversion is not merely caused by growing population, but also caused by growing

* Corresponding author: hariyantogeo@yahoo.com

economy of an area in form of growing urban sprawl. The growing urban sprawl can be infrastructure, settlement, industry, activity centre, etc. that will converse the farm land. The phenomena of farm land conversion are absolute impact for an urban growth [3].

2 Related literatures

Regulation of Indonesia Republic No.41 year 2009 about Sustainable Protection for Food Farm Land indicates that the growing population and economy and also industry will cause degradation, function change, and fragmentation for farm land. This phenomenon threaten nationally to have self-sufficiency in food. Through this case the government of Indonesia cannot deny that growing population can threaten farm land, or even worse than that, it can threaten self-sufficiency in food. The government of Indonesia used to appeal self-sufficiency in food, but actually Indonesia has imported some food materials instead [4].

3 Land capacity

Land capacity is carrying capacity to support life of the human beings and other living things that live around. It is measured by land demand which means maximum population that can be carried by available natural resources stated by absolute number in form of α [4].

4 Method of the analysis

Land capacity is calculated based on Regulation of Environment Minister Number 17 year 2009 about Guidance of Carrying Capacity Determination in an Area (Region) Spatial Planning. According to the regulation, in order to get to know land capacity of an area (region), the following steps should be conducted:

1. Land supply calculation with the following formula:

$$SL = \frac{\Sigma\,(Pi \times Hi)}{Hb} \times \frac{1}{Ptvb} \quad (1)$$

Note:

SL : Land supply (ha).
Pi : Actual production for each commodity (its unit depends on type of commodity). Calculated commodity includes agriculture, plantation, forestry, husbandry, and fishery.
Hi : Price of each commodity per unit (Rp/unit) on producer level.
Hb : Price of rice per unit (Rp/kg) on producer level.
Ptvb : Productivity of rice (kg/ha).

2. Land demand calculation with the following formula:

$$DL = N \times KHLL \quad (2)$$

Where : DL : Total of land demandequals to rice (ha)
N : Population (person/people)
KHLL : Land area that is needed for well-lived per individual [5]

Land area that is needed for well-lived per individual (KHLL) is a demand of welfare life for every individual divided by local rice productivity. The demand of welfare life is as worth as 1 ton of rice per capita per year. Data from national rice productivity on average, which is 2400 kgs/ha/year can be used
For some regions that have no data about local rice productivity

Table 1. Population in Semarang in 2001 – 2015

No	Sub-district	Population in 2001	Population in 2010	Population in 2015	Population in 2001-2015
1.	Mijen	38.843	48.923	69.382	78,6
2.	Gunungpati	58.130	65.465	91.075	56,7
3.	Banyumanik	106.834	121.855	147.449	38,0
4.	Gajahmungkur	58.482	61.668	66.499	13,7
5.	Semarang Selatan	84.103	85.591	84.496	0,5
6.	Semarang Utara	122.929	126.765	143.469	16,7
7.	Semarang Tengah	106.090	74.228	69.896	-34,1
8.	Semarang Barat	148.753	159.425	176.575	18,7
9.	Semarang Timur	84.044	81.747	82.542	-1,8
10.	Candisari	78.336	77.937	89.983	14,9
11.	Tembalang	106.090	127.008	177.333	67,2
12.	Pedurungan	141.695	163.562	200.726	41,7
13.	Genuk	63.904	80.600	108.252	69,4
14.	Gayam sari	64.104	70.782	78.395	22,3
15.	Tugu	24.400	26.976	34.622	41,9
16.	Ngalian	92.548	109.108	140.676	52,0
	Total	**1.350.005**	**1.481.640**	**1.763.370**	30,6

Source: Statistics Central Bureau (BPS) of Semarang city in 2016 [6]

3. Determination of land capacity status

Land capacity status is obtained from the ratio between land supply (SL) and land demand (DL). If SL > DL, it means surplus land capacity, but on the contrary, if SL < DL means deficit land capacity.

5 Discussions

The population growth will increase resource consumption, not only in quantity but also in quality. This causes the decrease of carrying capacity, for sure. Something noticeable from the table above is that outskirts of the city like Mijen, Gunungpati, Genuk, Tembalang, and Ngaliyan Sub-districts have high population growth. While subdistrics in the downtown like *Semarang Tengah* (Central Semarang) and *Semarang Timur* (East Semarang) have lower population, due to some big projects that have taken over the settlements. The growing population in the outskirts is because of urbanization, or urban people who just want to have a comfortable place to live.

5.1 Land capacity formula

Land capacity can be determined by capacity, and it is calculated based on land function area divided by population. So, the concept of land capacity formula is on the contrary to population density concept.

The formula is:

$$A = L/P$$

Where, A = Land capacity
L = Land area (ha)
P = Population.

Land consumption depends on population of the city. The more population in a city, the land consumption per capita will be smaller. Look at the following table.

Table 2. Land Consumption per Capita

No	Population (person)	Land Consumption per Capita (ha/capita)
1	10.000	0,100
2	25.000	0,091
3	50.000	0,086
4	100.000	0,076
5	250.000	0,070
6	500.000	0,066
7	1.000.000	0,061
8	2.000.000	0,057

Based on data of Semarang city population, the most populous sub-district is Pedurungan with the population of 200,726 persons, therefore the land consumption is as follows:

Population 250,000 – 100,000 , land consumption 0.076 – 0.070

150,000 = 0.06

25,000 = 0.01

If the population in Pedurungan 200,726 people, it is between 100,000 and 250,000.

Yeates scale limit = ((200,726 – 100,000) x 0.01) – 0.076 = 0.072.

With the same way, the Yeates scale limit can be applied for all sub-districts

Table.3 Carrying Capacity in Semarang City

No (1)	Sub-district (2)	Population (3)	Land area (ha) (4)	Carrying capacity (5)=(4)/(3)	Yeates scale limit (6)	Carrying Capacity Status (7)= (5)-(6)
1	Mijen	69.382	5.755	0,083	0,087	Threshold
2	Gunungpati	91.075	5.411	0,059	0,082	Overshot
3	Banyumanik	147.449	2.569	0,017	0,073	Overshot
4	Gajahmungkur	66.499	907	0,014	0,084	Overshot
5	Semarang Selatan	84.496	593	0,007	0,081	Overshot
6	Semarang Utara	143.469	1.097	0,008	0,075	Overshot
7	Semarang Tengah	69.896	605	0,008	0,081	Overshot
8	Semarang Barat	176.575	2.174	0,012	0,074	Overshot
9	Semarang Timur	82.542	770	0,009	0,080	Overshot
10	Candisari	89.983	654	0,007	0,079	Overshot
11	tembalang	177.333	4.420	0,024	0,074	Overshot
12	Pedurungan	200.726	2.072	0,010	0,072	Overshot
13	Genuk	108.252	2.739	0,025	0,079	Overshot
14	Gayamsari	78.395	618	0,008	0,081	Overshot
15	Tugu	34.622	3.178	0,117	0,090	Quite safe
16	Ngaliyan	140.676	3.799	0,035	0,077	Overshot
	SEMARANG	**1.763.370**	37.370	0,021	0,063	Overshot

Source: Data from Analysis Statistics

Based on yeates scale, the carrying capacity status of 14 sub-districts out of 16 sub-districts (87.5%) is overshot. The other two are quite safe (Tugu) and threshold (Mijen). The highest range of carrying capacity is Tugu (0.117) and the lowest is Candisari (0.007). Mijen and Tugu are predicted to be overshot in the near future due to high population growth in the outskirt areas.

5.2 Conclusion and suggestion

Rapid population growth in Semarang city tends to happen in the outskirt areas such as Mijen, Gunungpati, Genuk, Tembalang, and Ngaliyan. This indicates that outskirt areas are more attractive than downtown areas, especially for immigrants, even downtown dwellers have moved to outskirt areas.

Among 16 sub-districts in Semarang, only Tugu sub-district is quite safe and Mijen sub-district is in threshold. It indicates that 87.5% areas in Semarang city has overshot carrying capacity, especially downtown areas like Candisari and South Semarang sub-districts.

The government of Semarang city should control population growth by restricting Building Permit (IMB) or controlling farm land conversion.

References

1. Rahardjo, Mugi. *Pengantar Ekonomi Sumber daya alam.* UNS Press, Surakarta. 1997.
2. Kusumawati, Agni. *Rantai Nilai (Value Chain) Agribisnis Labu Di Kecamatan Getasan Kabupaten Semarang.* Skripsi. Fakultas Ekonomika Dan Bisnis Universitas Diponegoro Semarang. Semarang. 2013.
3. Hadi. Sudharto, P. *Bunga Rampai Manajemen Lingkungan.* Dua Satria Offset. Yogyakarta, (2014)
4. Muta'ali, Lutfi. *Daya Dukung Lingkungan untuk Perenanaan Pembangunan*, Badan Penerbit Fakultas Geografi UGM, Yogyakarta, (2012)
5. Soemarwoto. *Ekologi, Lingkungan Hidup dan Pembangunan.* Yogyakarta : Djambatani.1983.

The role of health and safety experts in the management of hazardous and toxic wastes in Indonesia

Supriyadi[1*], *Hadiyanto*[2]

[1]Postgraduate Program of Environmental Science, School of Postgraduate Studies, Diponegoro University, Semarang - Indonesia
[2] Master Program of Environmental Science, School of Postgraduate Studies, Diponegoro University

Abstract. Occupational Safety and Health Experts in Indonesia have an important role in integrating environmental health and safety factors, including in this regard as human resources assigned to undertake hazardous waste management. Comprehensive knowledge and competence skills need to be carried out responsibly, as an inherent professional occupational safety and health profession. Management leaders should continue to provide training in external agencies responsible for science in the management of toxic waste to enable occupational safety and health experts to improve their performance in the hierarchy of control over the presence of hazardous materials. This paper provides an overview of what strategies and competencies the Occupational Safety and Health expert needs to have in embracing hazardous waste management practices.

1 Introduction

Development as a conscious effort in processing and utilizing resources natural resources to increase the prosperity of the people, both to achieve prosperity and to achieve the inner satisfaction of the increased development activities as an effort to improve the welfare of life always contain the risk of pollution and destruction of the environment, so that the basic structure and function of ecosystems that become life support can be damaged. Therefore, the use of natural resources must be harmonious, harmonious, and balanced with environmental functions.

Hazardous waste is considered to be highly toxic and therefore waste disposal requires the proper management and human resource support in order to reduce the possibility of environmental hazards. The diversity of this type of waste will depend on industrial activity and other waste generators [1].

Starting from the use of raw materials, the selection of production processes and so on will affect the character of waste that can't be separated from the industrial process itself. However, not all industrial waste is a hazardous waste, but only partially. And in fact, as large as hazardous waste does come from industrial activities and should be handled specifically [1-2].

Pollution of the environment itself can be interpreted as an undesirable environmental event, where the occurrence causes disruption or environmental damage and even can lead to health threats until death. Unwanted things that can be called pollution, such as odorless air, cloudy water, land dumped garbage. It can evolve from simply unwanted to interference [2].

The polluted air either by dust, gas or other chemical elements can be painful to the respiratory tract, the eyes become spicy or red and watery. If the contaminant contains hazardous and toxic substances, it may be fatal.

Regulation of the Minister of Manpower of Indonesia no. 2 of 1992 has set out the procedures for the appointment of Occupational Safety and Health Experts. Any company with 100 or more employees, or at high risk of employment, is required to have a safety and health supervisory committee and also a minimum Occupational Safety and Health Expert. Occupational Safety and Health Expert is an extension of the government in overseeing the work of its workplace, in accordance with the requirements of legislation set by the government, thereby reducing the risks and incidents, whether accidents or occupational diseases, including in the case of waste management [3].

Basically every company and labor anywhere does not want accidents, occupational diseases, and environmental pollution. A potential risk of accidents, fires, environmental pollution and occupational diseases may arise due to errors in the use of equipment, lack of information on the work area, and the ability and skills of the less competent workforce [4].

Efforts to enforce occupational safety and health both institutional and work attitude is one way to create a good work area so as to keep the workforce healthy, comfortable, safe and prosperous both during work and after completion of work so that ultimately the level of productivity At the company can reach the highest level. One of the series on examination aspect safety and

* Corresponding author: Supriyadi2219@yahoo.com

https://doi.org/10.1051/e3sconf/20183107011

health is about Environment and Hazardous and Toxic Waste. Field checks that focus on OSH Environment and Hazardous and Toxic Waste need to be done because it is closely related to the level of awareness of a company on the health of the work area environment [4-5].

Occupational safety and health experts in this case as supervisors recommended by the government, can have a good knowledge of the management of materials waste. So that the impact of pollution and safety can occur well.

The purpose of writing this paper is to provide knowledge to the Occupational Safety and Health experts regarding safety and environmental factors with real practices in the application of requirements and guidance of occupational safety and health in the workplace that includes the conditions and labor facilities, the attitude of the workforce in operating the work equipment, The handling of hazardous chemicals, the nature of the work and the working environment.

2. Methods

The principle of hazardous waste management that has been widely discussed in various findings, will be combined with the possibility of optimizing the internal human resources of the company, in this case the role of occupational safety and health experts. Occupational safety and health experts in Indonesia generally have the duty and authority to integrate health, safety and environmental factors. Therefore this important perceived role will be developed more practically for the management of hazardous toxic wastes.

3. Principles of hazardous waste management in Indonesia

Understanding of toxic waste according to Law Indonesia no. 32 of 2009 on the Protection and Management of the Environment [5-6]

Article 1 paragraph 21, "Hazardous and toxic material is a substance, energy, and / or other components due to the nature, concentration, and / or amount, either directly or indirectly, can pollute and / or damage the environment, and / or Endangering the environment, health, and the survival of human beings and other living things " [6].

Article 1 verse 22 "Hazardous and toxic waste, hereinafter referred to as B3 Waste, is the remainder of a business and / or activity containing hazardous chemicals"

Article 1, paragraph 23, "Management of toxic waste is an activity that involves the reduction, collection, storage, transport, use, processing, and / or landfill

3.1 Why Need Toxic Waste Treatment?

Increased use of hazardous chemicals in various activities, including on industrial activities, mining, energy, oil and gas, health, household and other activities; Increased air pollution control and water pollution control, which will result in dangerous and toxic sludge or dust; significant impacts or pollution caused by toxic waste disposal to the environment;

Indonesia is one of the destination countries for B3 waste disposal; as the implementation of the legislation both the local, national and international (ministerial regulations, local regulations, peraruran government, the Law and the International Convention / Basel Convention).

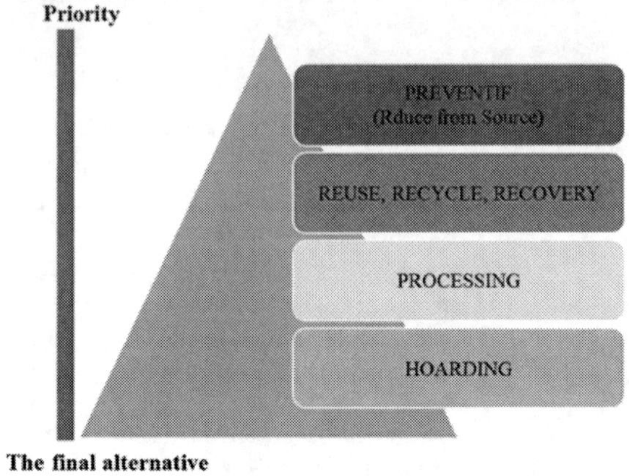

Fig. 1. Hierarchy of hazardous waste management

3.2 Principles of Hazardous Waste Management

It is necessary to establish a special type of waste, the criteria of Hazardous Waste, and its management arrangements;

Minimization of Hazardous Waste is a priority;

To minimize risk, the processing should be as close as possible to where the waste will be produced (proximity);

"Polluter pays principle" applies, meaning that anyone who produces Hazardous Waste shall be responsible for managing it;

The principle of monitoring of waste management of Hazardous Waste is "from cradle to grave"

3.3 Minimization of Hazardous Waste

Reduction at source;

Substitution of materials, is replacing the use of materials which have the potential to generate Harmful Waste in large quantities and are highly toxic with substances which have the potential to cause Dangerous Waste lower and less toxic and not even toxic;

Arrangement of operation of activities, namely arranging the process of production process in a systematic and planned by considering the selection of production process that can reduce the occurrence of pollution;

Application of clean technology; Waste Exchange [7].

Hazardous waste is identified as a chemical with one or more characteristics. According to its nature and character, Hazardous waste is distinguished into: (1) explosive; (2) flammable; (3) reactive; (4) toxic; (5) causes of infection; And (6) are corrosive. While viewed from the source, hazardous wastes are categorized into 3 (three) ie hazardous waste of specific sources, non-

https://doi.org/10.1051/e3sconf/20183107011

specific sources, and expired chemicals; spill; Leftover packaging; [7].

Product discharge that does not meet specifications.

1. Explosive waste is defined as waste that through chemical reactions can produce gas with high temperature and pressure that can quickly damage the environment.
2. Flammable waste is waste that when adjacent to a fire, sparks, friction or other ignition sources will be easily ignited or burned and when it has been burning will continue to burn great for a long time.
3. Reactive waste is a waste that causes a fire because it releases or receives oxygen or peroxide organic wastes that are unstable in high temperatures.
4. Toxic waste is a toxic waste that is harmful to humans and the environment. Hazardous waste can cause death or pain when entering the body through respiration, skin or mouth.
5. Waste that causes infection is laboratory waste infected with disease or waste containing germs, such as human body parts amputated and human body fluids infected.
6. Corrosive waste is a waste that causes skin irritation or corrosion of steel, which has a pH equal to or less than 2.0 for acidic waste and greater than 12.5 for a bias

3.4 The Nature of hazardous Chemicals in Industries

Hazardous waste sites pose a multitude of health and safety concerns, any one of which could result in serious injury or death. These hazards are a function of the nature of the site as well as a consequence of the work being performed. They include:

- Chemical exposure.
- Fire and explosion.
- Oxygen deficiency.
- Ionizing radiation.
- Biologic hazards.
- Safety hazards.
- Electrical hazards.
- Heat stress.
- Cold exposure.
- Noise.

Several factors distinguish the hazardous waste site environment from other occupational situations involving hazardous substances. One important factor is the uncontrolled condition of the site. Even extremely hazardous substances do not endanger human health or safety if they are properly handled. However, improper control of these substances can result in a severe threat to site workers and to the general public. Another factor is the large variety and number of substances that may be present at a site. Any individual location may contain hundreds or even thousands of chemicals.

Frequently, an accurate assessment of all chemical hazards is impossible due to the large number of substances and the potential interactions among the substances. In addition, the identity of the substances on site is frequently unknown, particularly in the initial stages of an investigation.

Occupational Safety and Health experts will often be forced to select protective measures based on little or no information. Finally, workers are subject not only to the hazards of direct exposure, but also to dangers posed by the disorderly physical environment of hazardous waste sites and the stress of working in protective clothing [7-8].

3.5 Legal Basis of Hazardous Waste Management to be known by Occupational Safety and Health Expert in Indonesia

Given the importance of processing and utilization of B3 waste, the government considers it necessary to enact legislation to regulate this B3 waste. These rules include:

1. Law Number 32 Year 2009 on Environmental Protection and Management.
2. PP RI Number 18 Year 1999 Jo. PP No. 85/1999 on the Management of Hazardous and Toxic Waste as a revision of PP RI No. 19 of 1994 Jo. PP RI Number 12 of 1995 on Waste Management B3.
3. Kepdal 01 / BAPEDAL / 09/1995 on Technical Procedures and Technical Requirements for Storage and Collection of Waste B3.
4. Kepdal 02 / BAPEDAL / 09/1995 on the B3 Waste Document.
5. Kepdal 03 / BAPEDAL / 09/1995 on Technical Requirements for B3 Waste Treatment.
6. Kepdal 04 / BAPEDAL / 09/1995 on Procedure of Stockpiling of Processing Result, Location Requirement of Former Treatment and Location of B3 Waste Landfill.
7. Kepdal 05 / BAPEDAL / 09/1995 on Symbols and Labels.
8. Kepdal 68 / BAPEDAL / 05/1994 on Procedures to Obtain the B3 Waste Management Permit.
9. Kepdal 02 / BAPEDAL / 01/1998 on the Management of Waste Management of B3.
10. Kepdal 03 / BAPEDAL / 01/1998 on B3 Control Program.
11. Kepdal 255 / BAPEDAL / 08/1996 on Procedures for Storage and Collection of Used Lubricant Oil.

3.6 Orientation/Education Occupational Safety and Health Expert to Manage Waste Hazardous

Department Heads are responsible for ensuring that their employees are trained and competent regarding the safe use of the hazardous materials with which they may be working. For employees who have direct contact with hazardous materials and waste this includes precautions for selection, handling, storage, use, and disposal.

They are also instructed about emergency procedures for hazardous material releases and exposures including the health hazards of mishandling hazardous materials and emergency reporting procedures. All new employees attend New Employee Orientation, which includes information regarding hazardous materials and hazardous waste safety issues, along with an overview of available resources. In addition, employees' role in maintaining a safe patient care environment is explained.

Every employee is expected to read each Environment of Care section of the Staff Information Handbook and take the posttest, annually. A perfect score is considered passing. Wrong answers are corrected and reviewed with the employee at that time [8].

The Safety routinely conducts Chemical Safety, Hazard Communication and Material Safety Data Sheet training for appropriate departments. In addition, large, open-invitation, hazardous materials seminars are also available as needed. Every employee is given department-specific safety training annually which may include a hazardous materials component, if appropriate. In addition, specific job or area-specific training may be given more frequently to address potential safety hazards associated with the introduction of hazardous materials.

- Training related to Hazardous Chemicals:
- Hazard Communication,
- MSDS Program (HCP)
- Hazardous (Chemical)
- Waste Disposal
- Liquid Nitrogen
- Chemical Spill or Release
- External Hazardous Material
- Release (Toxic Cloud
- Mercury Spill Response Procedures
- Compressed Gas Cylinder
- Transport
- Storage and Use Management of Contaminated Patients
- Cytotoxic Agents

3.7 Performance Improvement for Occupational Safety and Health Expert to Manage Waste Hazardous

Performance improvement initiatives are generated from risk assessments, analyses of performance monitoring activities and staff and manager observations and feedback. The primary goal is to provide a safer environment for staff, patients and visitors. 21 Performance improvement projects' selection criteria include their ability to be easily measured, their relevance and statistical validity.

They are also based on risk analysis, quality and effectiveness issues, customer satisfaction, productivity and financial considerations, and regulatory compliance. The process for obtaining measurement data must be reproducible. Reproducible measurement techniques result in advances in quality, productivity, customer satisfaction, cost containment and safety. A threshold level is attached to every performance improvement project. The threshold is the acceptable minimum level of frequency or occurrence. Should an indicator fall below that threshold, the following actions occur:

An investigation is conducted to determine possible reasons for failure to meet or sustain threshold. Based on the findings, a plan for corrective action is developed and implemented.

The standard continues to be monitored to determine if corrective actions were effective in re-establishing threshold. If not, the process is repeated. Each PI indicator is measured on a routine and ongoing basis. If the indicator is an event, then the indicator is measured at each occurrence. If the indicator is monitored consistently over time, then the measurement takes the form of a trend analysis [9].

Performance Monitors are as follows:

- Environmental Rounds results (twice annually)
- Staff Interview results (twice annually)
- Hazardous materials spills and releases (as occur)
- Hazardous materials exposures (as occur)
- Relevant regulatory agency activities (as occur) Respiratory Protection program improvements including medical clearance, initial and annual fit testing. (quarterly)

4 Conclusions

Occupational Safety and Health experts as an important element in a company's management play a very strategic role to assist the government in managing hazardous waste. Comprehensive knowledge and competence skills need to be done responsibly, as an inherent professional occupational safety and health profession. Management leaders must continue to provide training in external agencies in charge of science in the management of toxic waste in order to provide opportunities for occupational safety and health experts to improve their performance in the hierarchy of control over the presence of hazardous materials.

References

1. S. Salhofer, N. Isaac, Environmental Management **30** (1), 68–76, (1999)
2. Gianpaolo Ghiani, Waste Management, Volume **32**, Issue 7, Pages 1291-1296, July (2012)
3. Ana Pires,. Journal of Environmental Management, Volume **92**, Issue 4, Pages 1033-1050, April (2011)
4. P. Agamuthu, Waste Management and Research **21** (6), 487–497, (2003)
5. Santa Monica, UCLA Medical Center. (2013)
6. S. Salhofer, N. Isaac, Environmental Management **30** (1), 68–76, (1999)
7. *Law Number 32 Year 2009 on Environmental Protection and Management*. Republic Indonesia (2009)
8. Government Regulation Indonesia Republic Number 18 Year 1999 Jo. PP No. 85/1999 *on the Management of Hazardous and Toxic Waste 3.* (1999)
9. Mehmet Salim Oncel, Sustainable Environment Research **27** (2017) 188e194. (2017)

The Application Potential of Eco-Efficiency for Greening Company

Lukman Eka Prasaja[1*], *Hadiyanto*[2]

[1]Master Program of Environmental Science, School of Postgraduate Studies, Diponegoro University, Semarang-Indonesia
[2] Departement of Chemical Engineering, Faculty of Engineering, Diponegoro University, Semarang - Indonesia

Abstract. Eco-Efficiency emerged in the 1990s as a measure of "the efficiency that ecological sources use to meet human needs." As a tool in economic and environmental integration, Eco-efficiency needs to be promoted further so that regulation in government and industrial management can include it as an important instrument. This paper provides several approaches that can help various industries to develop effective eco-efficiency principles. The approach used is to maximize the role of the Steering Committee of the company's internal environment. Utilization of natural resources such as water, forests, mines and energy needs need to be balanced with Eco-Efficiency so that natural exploitation can be well controlled so that sustainable development aspired by the world can be realized.

1 Introduction

The growing industrial development has a positive impact on growth Economy in Indonesia. However, the development of industry is also not free from negative impacts Caused to environmental sustainability. Waste and also pollutants produced by various industrial activities are problems that can lead to degradation environment. Such conditions are certainly more worrying.

Natural Resources exist with various forms and distribution. Some can be renewed, others are non-renewable. There are also areas that are rich in natural resources, otherwise there are areas that are poor in resources. It all seems to form a balance that should be maintained. Abundant areas of certain natural resources can meet the needs of the under-developed areas. Non-renewable resources are sought to balance with the principles-based management of eco-efficiency and sustainability. So it is with other natural resources. In essence, the sustainability of natural resources can be achieved with the use of the eco-efficient, manage it with sustainable and insightful guidelines [1].

Public awareness of Efforts to conserve the environment itself is also still low, while government policy in Dealing with environmental degradation due to the activities of the industrial sector is still in question. One effort Overcoming the problems of industrial environmental degradation is by applying the concept Sustainable development as a paradigm in industrial sector development, be it industry Large, medium and SMEs [2].

The world's financial and economic problems caused a crisis that has caused economic stagnation in many countries. One way to overcome the crisis is to re-adjust the growth of green economy.

The concept of eco-efficiency could have been much adopted by the world's companies in making improvements and prevention of industrial pollution to the environment [3].

There are many concepts related to the environment, such as environmental management, environmental conservation, sustainable development, natural resource conservation, eco-efficiency and much more. When read at a glance, the concepts have almost no significant difference, but in reality each term has a different emphasis. In order to know the meanings and also the interrelationship between the concepts, it is necessary to study theoretically related to the concepts and also the relationship [4].

Eco-efficiency improves business and environmental performance, and helps companies get out in front of market and regulatory trends, to reduce costs, to gain competitive advantage and to ensure long-term profitability and sustainability. Being efficient is always a high priority for every company. Eco-efficiency is suitable to be applied on green industry conceptual, because it can synergize economic and environmental aspects. Application of eco-efficiency will lead to competitiveness, environmental sustainability, and quality of life better

In this essay will be presented notions in every concept that is often used in the discussion of environmental problems. In addition, it will also disclose the differences and continuity between concepts until it can be drawn a conclusion that will be accompanied by case examples. Hope to be achieved by the author of this essay is the reader can know the position of each concept

* Corresponding author: bungekaprasaja@gmail.com

in environmental problems, so that can be drawn a conclusion that is easy to understand.

2. Methods

The proven principle of eco-efficiency in some business approaches, to reduce losses from a variety of environmental damage effects and financial losses, becomes an opportunity that can be used in a variety of industry sectors. In this paper the method used is to provide various kinds of approaches that allow for some types of industries to be performed

2.1 Principle of eco-efficiency

Eco-efficiency is management philosophy that aims at minimizing ecological damage while maximizing efficiency of the firm's production processes, such as through the lesser use of energy, material, and water, more recycling, and elimination of hazardous emissions or by-products [5].

Since the 1960s, corporations have witnessed an era of compliance marked by the promulgation of numerous environmental laws, followed by an era of voluntary environmental management to help prevent pollution, emergencies and environmental disasters such as the Exxon Valdez spill and the Bhopal catastrophe. In the developed world, many corporations are now implementing voluntary environmental management systems (EMS) to prevent and minimize environmental impacts. However, site location, local environmental regulations, and the level of legal enforcement will often dictate the stage of a company's environmental management or compliance [6].

An EMS can be developed according to various published guidelines but the ISO 14000 series of standards developed by the International Standardization Organization (ISO) are the most internationally recognized. A few thousand companies across the world have already obtained their ISO 14001 EMS registration and many more are in the process of becoming registered. Although the ISO standard is not prescriptive, it requires companies to have a formal EMS that includes: [6]

- Documented environmental policy honored by senior management;
- Documented procedures to prevent or minimize all significant environmental impacts;
- Employee environmental training program to ensure environmental awareness and respect of environmental regulations and documented EMS procedures;
- Regular internal and external audits of the EMS; and,
- an annual management review of the system.

The implementation process of eco-efficiency adopting from cleaner production usually consists of 5 basic steps:

(1) Planning and organizing the resources;
(2) Analyzing the research object to identify the wastes and its characteristics;
(3) Proposing the possible solutions/methods;
(4) Studying the feasibility of the solution;
(5) Implementing the chosen solution.

However, cleaner production or eco-efficiency is not a one-time-project. In other to achieve the sustainable development and long-term benefits, the factory needs a system to maintain and improve continuously [5,7].

2.2 Steering committee eco-efficiency for green company

This study was conducted on a management practice for green company as a case study. This formula is adapted from what has been done by rice companies in Vietnam. The commitment from the leader is very important, which could support establishing the cleaner production team. This team usually has the scale depending on the organization size. However, it should consist of members who have knowledge and skills as presented in Table 1. [7].

This organizational management approach can be used in several sectors related to the possible implementation of eco-efficiency.

Table. 1. Eco-Efficiency Team Formulation

Member	Required knowledge and skills	People in charge (suggestive)
Team leader	Decision Making Techniques. Leadership. Project Management	Vice Director
Member 1	Environmental and energy use requirements in the processing product. Programs and applications related to environmental and energy consumption management. The concept of environmental management systems	General Manager
Member 2	The process of financial decision making. Risk management. Cost-benefit analysis of the selected solution	Finance and accounting staff
Member 3	Production processes. The environmental measurement, control and assessment process. Industrial and systems engineering	Technical staff

3. Approach for implementation of the principle of eco-efficiency for green industry

This paper will explain the approach that can be used by various types of industries in support of eco-efficiency, it is important to realize the successful implementation in the green economy. Various natural resource assets in Indonesia encourage people to use as much as possible. This is very natural because people always feel not enough for the needs of life, so that human exploitation of natural resources. In order for the sustainability or potential of natural resources to exist, the following ways are required:

- Inventory and evaluation of natural resources need to be improved to better understand the utilization of natural resources
- In the exploitation and exploration of natural resources should be used appropriate technology appropriate, so that the potential of natural resources can be maintained.
- In the implementation of development should always be held a careful assessment of its impact on the environment.
- Rehabilitation of natural resources is done through an integrated approach.
- Utilization of natural resources of coastal areas, sea and air areas needs to be improved without undermining the quality and environmental sustainability.

Each development activity always produces a by product to the environment, that is industrial pollution as part of the construction machinery discharges waste into the environment. The process can not be avoided and the size depends on the nature of the goods produced. Although pollution always occurs, it can be reduced to a minimum.

3.1 Eco-efficiency principle in forest utilization

Forest utilization should use the principle of eco-efficiency. Illegal logging should be stopped. Deforestation should be carried out under the principle of eco-efficiency by breeding, selective logging, and reforestation. The objectives of reforestation and forest rehabilitation are as follows:

a. Increase the sustainability of forests, soil, and water
b. Expand the supply of valuable raw materials for the community.
c. Rescue the results of development efforts in the field of irrigation.

Forest conservation efforts include the following:
Felling trees is selective and replaces trees with trees that have an important role for the environment and the economy

- Balance between logging and reforestation should be sought
- Deforestation of forest functions as a preservative of water sources, soil, and recreation areas should be encouraged.

3.2 Eco-efficiency principle in mining utilization

Mining goods are non-renewable natural resources. The principle of mine utilization eco-efficiency can be done with the following steps:

a. Savings in usage by always remembering the next generation
b. Exports mining materials as finished or semi-finished goods.
c. Conducting investigations and research to find new mining areas
d. Cultivated substitute material. For example the use of fuel is replaced with solar, gas or alcohol

3.3 Eco-efficiency principle in water utilization

Water conservation efforts are carried out in the following manner:

a. To maintain the existence of forests so that springs are not dry, especially forests in the upper river area
b. Keep the river water from being polluted. The disposal of industrial waste into the river must be neutralized first
c. Keep the well water clean. For example, keep the distance between wells and landfills at least 10 meters
d. Prevent the disposal of nuclear waste or industrial wastewater directly into the sea. Avoid leaks on oil tanker ships. Avoids carelessness, such as collisions between tankers and does not dispose of waste and waste into the sea

3.4 Eco-efficiency principle in manufacture industry

Unused materials and energy in a production process system will be wasted into waste and lead to increased social costs for further processing. Therefore, it is necessary step-step implementation of eco-efficiency principles in the industry as follows:

a. Minimize the use of raw materials and energy
b. Minimize the release of toxic waste into the environment
c. Produce recyclable products
d. Utilization of Renewable Resources
e. Able to produce durable products

3.5 Eco-efficiency principle in meeting energy source needs

Various human activities require energy such as coal, petroleum, geothermal, hydrocarbon, water, sunlight, wind, and waves. The current global problem is the diminishing of world petroleum reserves, while the need for fuel should always be available. The application of the principle of eco-efficiency in meeting the needs of fuel, among others: the use of alternative renewable energy sources such as wind power, sunlight, water, biomass, and organic materials.

a. Wind power
   Wind power is generated by windmills to pump water with 5 liters of discharge used for daily necessities and agricultural needs. In the principle of eco-efficiency, wind power is useful to replace the energy of fuel pumping windmills.
b. Solar energy
   Indonesia is actually a country that is very rich in solar energy because of Indonesia's position on the equator. Solar energy is a very potential alternative energy. This energy can be used for power generation to replace the water energy to generate electricity in megawatt level because the solar heat is able to supply more than 219,000 thousand billion kWh of energy every year. In addition to being utilized as a powerhouse, solar energy can also be used to heat the green house and dryer energy of agricultural products.
c. Water
   Utilizing water as a source of power generation has been implemented by the Indonesian government on a large scale. But the actual use of water as a power plant can also be done alone by remote communities where there are small water sources, so that people dependence on electricity supply from electrical supply industry can be reduced by utilizing natural resources in their own.
d. Biomass and Organic Waste Materials
   Agricultural biomass, especially agro-industry waste is an alternative source of energy, in lieu of oil and gas fuels. Thus the utilization will have a positive impact, both for business and for the welfare of the environment as a whole. In Indonesia, agro-industry wastes can be used as alternative energy sources such as charcoal briquettes, biogas, biodiesel and acetone, butanol, and ethanol.

4. Conclusion

Eco-Efficiency is very necessary to be done by various sector of Industry, it is important as effort to keep exploitation of natural resources. Utilization of forests, water and energy needs to be well managed, eco-efficiency can be an appropriate approach to be consistent with the principle of sustainable development. The results of this paper are expected to provide a stimulus of initial views by policy makers, so that it can be an action that gives a positive impact in environmental management seems to be government, especially on policy and finance. It needs radical action: one action, one large funding and strong institutional collaboration.

References

1. Environmental Protection Agency (EPA), United States, *"Green Servicizing: Building a More Sustainable Economy"*, Internal working draft, EPA, Washington, D.C, (2009)
2. FORA, *Green Business Models in the Nordic Region: A key to promote sustainable growth"*, Green paper for the Nordic Council of Ministers, Copenhagen, (2010)
3. *Measuring Eco-efficiency: A Guide to Reporting Company Performance.* (2000).
4. World Business Council for Sustainable Development. Mickwitz, Per et al. "Regional Eco-efficiency Indicators E A Participatory Approach". Journal of Cleaner Production, Volume **14**: 1603-1611, (2006)
5. L. Kenneth, Mu'lholland, *Identification of Cleaner Production Improvement Opportunities.* A joint publication of the American Institute of Chemical Engineers and John Wiley & Sons, Inc, (2006)
6. Phil Ribiel. Eco-Efficiency: *The Next Step Toward Sustainable Development,* (2015)
7. Tran Cuoc. *Feasibility of cleaner production for Vietnam rice processing industry.* 13th Global Conference on Sustainable Manufacturing - Decoupling Growth from Resource Use, (2016)

Management to Insulate Ecosystem Services from the Effects of Catchment Development

Peter Gell[1,*]

[1]Wetlands Research Network, Federation University Australia, Mt Helen, Vic. – Australia

Abstract. Natural ecosystems provide amenity to human populations in the form of ecosystem services. These services are grouped into four broad categories: provisioning – food and water production; regulating – control of climate and disease; supporting – crop pollination; and cultural – spiritual and recreational benefits. Aquatic systems provide considerable service through the provision of potable water, fisheries and aquaculture production, nutrient mitigation and the psychological benefits that accrue from the aesthetic amenity provided from lakes, rivers and other wetlands. Further, littoral and riparian ecosystems, and aquifers, protect human communities from sea level encroachment, and tidal and river flooding. Catchment and water development provides critical resources for human consumption. Where these provisioning services are prioritized over others, the level and quality of production may be impacted. Further, the benefits from these provisioning services comes with the opportunity cost of diminishing regulating, supporting and cultural services. This imbalance flags concerns for humanity as it exceeds recognised safe operating spaces. These concepts are explored by reference to long term records of change in some of the world's largest river catchments and lessons are drawn that may enable other communities to consider the balance of ecosystems services in natural resource management.

1 Introduction

Human societies have reaped food, water and materials from river catchments. While climate variability at a range of time scales has mediated the supply of these resources at regional scales, the sedentarisation of human communities through the Holocene, and the attendant increases in population and technology, has increased the intensity of resource exploitation. The Millennium Ecosystem Assessment reveals the further amplification of impacts of human resource exploitation from the mid-20th century identifying the Great Acceleration, which has prompted calls for the demarcation of a new geological epoch, The Anthropocene [1,2].

While ethical arguments can be mounted that natural systems warrant conservation for intrinsic reasons, the Ecosystem Services they provide humans is increasingly being used to justify investment in wise management [3]. It is recognised that the demand for consumptive resources such as food, water, energy, timber and minerals for the construction of shelter and fibre for clothing is impacting negatively on the other services provided humanity by the natural environment. In market based economies there remain opportunities for the price of consumption to reflect merely the cost of production, with little requirement for it to reflect the trade-off in the loss of assets and services, that are valuable, but represent a challenge to quantify economically. Without full cost accounting of the trade-offs between services society risks undermining the support afforded by the less quantifiable phenomena and, ultimately, the ongoing supply of provisioning services.

The most readily identifiable services provided by natural ecosystems are usually those that provide directly for human needs. These Provisioning Services comprise potable water and food, including those harvested directly such as fish and native fruit, as well as those sown by people such as crops and stock raised for milk and meat. As a resource timber was used by early hominids as an energy source and then for shelter as technology became more sophisticated. Extracted minerals have replaced timber as a provider of shelter and this fibre is now directed in large volumes to the creation of paper. Most of humanity's energy is now provided by extracted fossil fuels that were largely unavailable before the industrial revolution.

The natural environment also affords considerable benefit to humanity by means that are not defined as provisioning. Natural systems regulate the habitat used by people by moderating microclimatic extremes (e.g. shade, shelter) and by controlling irruptions of pests, predators and disease carrying organisms that may impact negatively on people. It may also mitigate the risk of environmental hazards – coastal and riparian vegetation play's a clear role in protecting human settlements from floods and, as witnessed in 2004, tsunamis. Natural ecosystems also provide support to society that underpins the provision of food and water through the pollination of flowers that beget seed and fruit and the purification of water to mitigate the

* Corresponding author: p.gell@federation.edu.au

personal health of consumers. Lastly, nature provides considerable cultural benefits and there is an increasing body of evidence that reveals the wellbeing benefits that accrue from close association with natural systems such as wetlands [4].

From even before the Brudtland Report *Our Common Future* global ecosystem assessments have revealed, with increasing sophistication and accuracy, the degree to which provisioning services have been exploited at the expense of regulating, supporting and cultural services [5]. The rapid increase in the rate of human impact from the mid 19th century is identified as the Great Acceleration with global scale surveillance revealing exponential increases in human population; and in rates of overfishing, deforestation, water catchment regulation, desertification etc. Variations in socio-political circumstances has yielded spatial variations in these patterns but, for example, China has developed very rapidly and its ecosystems have suffered accordingly, joining the developed world in threatening, and in some circumstances, exceeding safe operating spaces for humanity [6,7].

Long term evidence, however, reveals that humanity has a long history of exploiting its water catchments and impacting on aquatic ecosystems [8]. The development of agriculture and mineral extraction industries is associated with nutrient, metal and sediment pollution as well as salinisation from the mid- to late Holocene. The mere increase in populations in settlements, in the absence of waste management infrastructure, in the 19th century was sufficient to drive the eutrophication of receiving waters and widespread and unprecedented anoxia in the world's deep lakes [9]. So, the trade-off in ecosystem services by human populations has a considerable history [10]. To understand the challenge of restoring the regulating, supporting and cultural services afforded by ecosystems it is important to understand the threats emerging from new technology but also the momentum of ecosystem degradation that is the legacy of human exploitation of provisioning services over millennia.

2 The Murray Darling Basin

Many of the world's water resource catchments have come under considerable pressure from land and water development for human consumption. In Australia the Murray Darling Basin (MDB), the continent's largest catchment, represents a case study that reflects clearly the trade-off between services over time [11]. Colonisation of the continent by non-indigenous peoples was relatively late and so the technology brought by the colonists was relatively sophisticated. Further the catchment lies in a spatially and temporally variable climate with low catchment efficiency exposing it to the risks of water resource development. Further, the floodplain soils carry high loads of native phosphorus and are highly erodible, and in many parts, overly saline water tables. In combination these boundary conditions conspired to lead to a rapid degradation of natural resources to provide for an expanding, consumption intensive society. The management of the MDB today represents Australia's greatest environmental restoration challenge with considerable funds directly towards the MDB Plan [12].

Humans have lived in the MDB for at least 46,000 years [13] over which time hydroclimates have changed greatly [13,14]. The hydrogeomorphology of the rivers stabilised from ~ 12,000 years ago and many wetlands evolved from ~ 5,000 years ago as their meandering habit became entrenched. Human populations likely increased from this time as climate, and so resources, became more reliable. Over the millennia preceding European colonisation traditional resource use comprised the harvesting of wildlife and aquatic plants. While people buried foods to render them safe to consume, and sowed tubers to harvest at a later time, agriculture was limited, although plants collected typically represented the largest part of the economy. Ecosystems provided timber for fuel, weapons and water craft and clean drinking water. Floodplain vegetation provided shade and shelter and communities had strong cultural associations with freshwater systems and their biota.

The MDB was a focus for exploration soon after the colonisation of the continent by Europeans and settlements were established within the catchment from the early-to-mid-1800s. Very soon after settlement the plains supported large numbers of stock run for wool and food. Intensive gold mining in the watershed from 1848 and the extraction of water for irrigation agriculture, most from 1880, added to the destabilisation of the landscape. The state parliament in Adelaide, near the base of the Basin, first discussed the threat of water use in the upper reaches as early as 1887 and the Interstate Royal Commission into the basin testified to eutrophic river waters impacted by mining sludge [15]. Soldier settlement schemes following the two world wars promoted land clearance and accelerated river regulation and water consumption which ultimately drew saline water tables to the surface driving widespread dryland and wetland salinisation, particularly after the wet La Nina phases in the 1950s and 1970s [16]. While cyanobacterial blooms were recognised from 1878 [17] the Darling River suffered an unprecedented bloom of over 1000 km in length in 1991; a risk that remains for most surface waters in the southern basin. To turn this Basin in the nation's food bowl, provided in excess of 40% of its agricultural gross domestic product, most land is now cleared, the native fishery is greatly diminished, resilience to climate variability has been compromised and natural systems have suffered from the increasing flux of salts, sediments and nutrients [18] to the point where the basin is regarded as one of ten Australian ecosystems most at risk of exceeding a tipping point [19].

While there has long been concern on the state of the MDB waterways, the first comprehensive assessments came in the 21st century, mostly associated with the call to provide river flows to restore ecosystems. The Sustainable Rivers Audit [20] and other commissioned reports revealed the Basin to be widely degraded and the waters over-exploited for the provision of irrigation agriculture to produce food and fibre. This drew contest from those with a water allocation and this escalated into a politicised debate over water volume. The emerging Water Act (2007) provided for the Murray Darling Basin Plan which released $13B AUD to recover 3200 GL (~ 25%) of mean annual flow to restore the aquatic ecosystems of the Basin. This modern debate over water volume overlooked the long recognised impact of catchment disturbance on the services afforded both people and ecosystems.

3 Long Term Change

In 1917 Australia embarked on a program of river regulation and commissioned a suite of weirs along the River Murray and other MDB tributaries from 1922 to 1936. Consistent with the accelerated development of water resources in the northern hemisphere the construction of large dams from the 1950s lead to significant, additional river regulation. In the absence of water quality monitoring or biological assessments water diversions and reduced flows became identified as the principal driver of ecosystem decline. In particular, the terminal lakes, deemed to be predominantly fresh [21] but estuarine elsewhere [15] were considered to be at risk of salinisation through inundation by seawater owing to the diminished freshwater flows [22]. This dissonance of view highlights the opacity of evidence derived from historical sources. Clear evidence of past condition is highly instructive in terms of scoping options for restoration, but also in understanding the trade-offs between various ecosystems services that have occurred under past management eras.

While archaeological sources document the presence of people, and by inference the effect they may have had on the river and its floodplain, artefact are often poorly constrained in time and sites rarely report on the response of the ecosystems. Sediment-based paleolimnological records on the other hand, while not recording directly the presence of people, are usually well dated, are continuous in time and provide detailed inferences as the condition of lentic waterways. Over 50 records of wetland change have been documented across the southern Basin. Several provide evidence of pre-European baseline conditions and point to long periods of stasis mediated by regional hydroclimatic variation [23]. While the impact of indigenous people on wetlands has not been demonstrated, but is not discounted, substantial and often unprecedented changes occurred from the period when new people, their technologies and stock developed the catchment to provide for a rapidly increasing population. These changes included widespread increase in waterway turbidity, rapid sedimentation of shallow wetlands, eutrophication and salinisation [24], and loss of aquatic plant communities. While most wetlands changed after river regulation, in many instances on account of a regime of more frequent flooding, many impacts are evident from soon after European settlement and are attributable to direct impacts of erosion and the release of pollutants.

A shift from wetlands rich in aquatic plants to systems dominated by phytoplankton was first documented by Ogden [25] and then more widely by Reid *et al.* [26]. Gell and Reid generated a typology of wetlands vulnerable to this switch attributable to the attenuation of light in the water column with elevated water turbidity [27]. While Ogden showed associated shifts towards pelagic zooplankton; Kattel *et al.* documented a decline in faunal diversity [28]. The structure of the decline documented from the sediment sequences is consistent with the model of regime shifts or the critical transitions documented for Lake Erhai in Yunnan province by Wang *et al.*, [29]. Given the challenge in demonstrating changes to ecological feedback effects from sediment records, and the unlikelihood of removing the likely stressor of change, it is challenging to demonstrate a regime shift based on the criteria identified by Capon *et al.* [30]. As such the change may well represent a press response to ongoing sediment and nutrient flux [31]. Critically, the replication of these changes in Murray floodplain wetlands, and the identification of the river channel as a source of sediments [32] leads to the suggestion that the critical transitions are occurring at a sub-basin scale rather than wetland-by-wetland. Irrespective, these records attest to the diminishing capacity to regulate water quality and cultivate the supporting services that provide ecosystem resilience. The heightened contest over water resources reveals that society is recognising the decline in cultural services once afforded by the biodiversity rich wetlands.

4 Recovery – The Basin Plan

The Australian and relevant State Governments implemented The Murray Darling Basin Plan that is to invest in the purchase of water allocation licenses and water delivery infrastructure to recover 2750 GL of irrigation allocation for use in environmental flow programs across the Basin. The intent is to allocate a supplementary 450 GL if it can be demonstrated that this will have little negative impact on regional communities. This Plan reveals that Australian society has accepted that the provisioning services made possible through the regulation and diversion of the Basin's water for agriculture have been at the expense of other services, and the intrinsic value of the natural aquatic ecosystems. It does, however, identify water volume as the principal cause of this decline, possible as water over allocation can be readily rectified using economic mechanisms whereas water quality issues cannot. A Long Term Intervention Monitoring program has been implemented to assess the ecological recovery from the release of environmental water. Yet, as with any monitoring program, the perceived success or otherwise can be influenced by the selected baseline against which recovery is assessed. Clearly, assessing performance against decade old monitoring, rather than paleoecological records that can report on pre-regulation, and even pre-industrial conditions, will paint the restoration success in a brighter light.

Meanwhile, the decision to implement the Plan brought an abrupt impact to food and fibre producers who were encouraged to forego water allocation. Having been encouraged by past governments to develop irrigation plots through settlement schemes this generation of growers have been required to adapt abruptly to a withdrawal of water in a rapid switch to a management regime that values other services provided by the same water. The implementation has been abrupt and, likely owing to poor socio-economic assessment and modelling, measures to moderate the impacts are being implemented retrospectively. The reaction to the Plan has been acute having pitted growers against the authorities and rural interests against urban. This prompted a Federal Senate Select Inquiry into the Basin Plan in 2015 where communities were pitted against each other in making their case for access to water. Now, a Royal Commission has been called by the South Australian Government to investigate impediments to the provision of what they perceive to be their right to water.

5 Balancing Ecosystem Services

The degraded state of the Murray Darling Basin and the fraught nature of the political contest over critical water resources are clear symptoms of the prioritisation of provisioning services. While the South Australian Parliament were debating the impact of upstream water use on the water that flowed into their jurisdiction as early as 1887 there has been a state sanctioned drive to use Basin water to provide food and fibre for domestic and overseas markets for more than a century since. Over this time there has been little attention paid to the regulating, supporting or cultural services provided by the same water. This has left the landscape and communities more vulnerable to climate change which is forecast to lead to lower rainfall and runoff. The legacy of the narrow focus on provision is the substantial direct cost of the Plan - $13B and the inestimable indirect costs to communities, so this demands much economic readjustment to repair these other services. These include reinstating the heavily impacted community cohesion; the risk to wellbeing of rural communities and the endangered natural biodiversity. Even so, given the compromised quality of the water to be allocated to the environment, it appears unlikely even this investment will address issues of turbidity, sedimentation and eutrophication to allow for complete recovery. There remains a risk that thresholds exist whereupon regulating services become permanently diminished or the cost of recovering them is prohibitive.

Clearly, had those advocating so enthusiastically for catchment development over a century ago had the foresight to value other services as much as they did provisioning, then the system resilience would not have become so compromised. Clearly it would have been better to have adapted through time and empowered the community to produce food and fibre, but also protect the natural systems and the benefits that accrue from them and to be prepared for unforeseen circumstances. While the time has passed to avoid many of the consequences of overexploitation, it is essential to implement adaptation now, as the Plan intends, to avoid the need for transformational change [33]. A just approach may be to distribute the cost of the implementation of an adaptation cycle across multiple generations in an adaptation pathways approach [34] rather than levelling the full cost on today's providers. The degraded state of the Murray Darling Basin serves as a lesson of overexploiting the provisioning services of natural systems. Davis *et al.* (2016) observe that Australia would do well to apply the lessons of the MDB crisis should it seek to develop the landscape in the tropical North.

Globally, developing countries would do well to apply the same lessons. The modern condition of the Semarang River reflects a similar enthusiasm for the development of provisioning services in the form of agriculture. This is likely an outcome of the necessity for providing for an increasing regional population and the primacy of food provision as a socio-economic necessity. Here, steep landscapes have been cleared of rainforest and terraced, lakes impounded and water flow regulated to service market gardens. To maximise production nutrients have been added and these are carried in return waters to river tributaries. The opportunity cost of this is the degradation of the waterways and the loss of this resource at a potable water supply. In Melbourne early planners protected the water catchments and still today, the supply of water to Melbourne is drawn from forested catchments ensuring that the surface waters remain the principal source of potable. The compromised quality of the potable water of Semarang leaves authorities with the choice of importing water or extracting groundwater. The modern focus on groundwater is an expensive solution which fails to address the driver of the problem – waterway management in the upper catchment. Often groundwater resources are not recharged at the rate water is abstracted and so groundwater based solutions to water provision are often unsustainable in the long term. Further, groundwater plays a critical role in landscape stability and unsustainable extraction of groundwater can lead to land subsidence, coastal inundation and diminished regulating services in coastal settings.

In both the MDB and the Semarang River the adoption of an holistic ecosystem services approach at the outset may have avoided the trade-offs between services and the level of degradation experienced today. Now realising that degradation and the diminished amenity the challenge to recover ground and adopt a more sustainable path is expensive and challenging. While short political cycles and discount rates act to encourage governments to differ expensive measures that may mitigate overexploitation of natural resources, intergenerational equity demands that society mitigates the decline in services at the same time as it reaps the benefits of harnessing the provisioning services of its natural ecosystems.

References

1. *Millennium Ecosystem Assessment*. Millennium Ecosystem Assessment Findings. Millennium Ecosystem Assessment. (2005)
2. Steffen, W., Crutzen, P.J. & McNeill, J.R. Ambio, **36** : 614-621. (2007)
3. Costanza, R., d'Arge, R., de Groot, R., Farber, S., Grasso, M., Hannon, B., Limburg, K., Naeem, S., O'Neill, R.V., Paruelo, J., Raskin, R.G., Sutton, P. & van den Belt, M. Nature, 387 : 253-260. (1997)
4. Finlayson, C.M., Horwitz, P. & Weinstein, P. Wetlands: Ecology, Conservation and Management, volume **5**. Springer, Dordrecht, 263 pp. (2015)
5. WCED. *Our Common Future*. OUP. 383pp. (1987)
6. Rockström, J., Steffen, W. ., Noone, K., Persson, A., Chapin III, F.S., Lambin, E., Lenton, T.M., Scheffer, M., Folke, C., Schellnhuber, H., Nykvist, B., De Wit, C.A., Hughes, T., van der Leeuw, S., Rodhe, H., Sörlin, S., Snyder, P.K., Costanza, R., Svedin, U., Falkenmark, M., Karlberg, L., Corell, R.W., Fabry, V.J., Hansen, J., Walker, B., Liverman, D., Richardson, K., Crutzen, P., Foley, J. Nature, **46**: 472-475. (2009)
7. Dearing, J.A., Wang, R., Zhang, K., Dyke, J.G., Haberl, H., Sarwar Hossain, Md., Langson, P.G., Lenton, T.M., Raworth, K., Brown, S., Cartstensen, J., Cole, M.J., Cornell, S.E., Dawson, T.P., Doncaster, C.P., Eigenbrod, F., Flörke, M., Jeffers, E., Mackay, A.W., Nykvist, B. & Poppy, G.M. Global Environmental Change, **28** : 227-238. (2014)
8. Dubois, N., Saulnier-Talbot, E., Mills, K., Gell, P., Battarbee, R., Bennion, H., Chawchai, S., Dong, X., Francus, P., Flower, R., Gomes, D.F., Gregory-Eaves, I, Humane, S., Kattel, G., Jenny, J-P., Langdon, P., Massaferro, J., McGowan, S., Mikomägi, A., Ngoc, N.T.M., Ratnayake, A.S., Reid, M., Rose, N., Saros, J., Schillereff, D., Tolotti, M. and Valero-Garcés, B. *First human impacts and responses of aquatic systems: A review of palaeolimnological records from around the world*. The Anthropocene Review. (2017)
9. Jenny, J-P., Francus, P., Normadeau, A., Lapointe, F., Perga, M-E., Ojala, A., Schemmelmann, A. and Zolitschka, B. *Global spread of hypoxia in freshwater ecosystems during the last three centuries caused by rising local human pressure*. *Global Change Biology*, doi: 10.1111/gcb.13193. (2016)
10. Dearing, J.A., Yang, X., Dong, X., Zhang, E., Chen, X., Langdon, P.G., Zhang, K., Zhang, W. and Dawson, T.P. *Extending the timescale and range of ecosystem services through paleoenvironmental analyses, exemplified in the lower Yangtze basin*. Proceedings of the National *Academy of Sciences*, US 109: E1111-1120. (2012)
11. Davis, J.A., O'Grady, A.P., Dale, A., Arthington, A.H., Gell, P., Driver, P., Bond, N., Casanova, M., Finlayson, M., Watts, R., Capon, S., Nagelkerken, I., Tingley, R., Fry, B., Page, T.J., Specht, A. *When trends collide: the challenge of protecting freshwater ecosystems under multiple land use and hydrological intensification scenarios*. Science of the Total Environment. doi: 10.1016/j.scitotenc.2015.03.127 (2016)
12. Murray Darling Basin Authority. *MDBA Basin Plan*. MDBA, Canberra. (2013)
13. Bowler, J.M., Johnstone, H., Olley, J.M., Prescott, J.R., Roberts, R.G., Shawcross, W.F., & Spooner, N.A. Nature, **421** : 837-840. (2003)
14. Mills, K., Gell, P., Hesse, P.P., Jones, R., Kershaw, P., Drysdale, R. & McDonald, J. Australian Journal of Earth Sciences, **60**, 547-560. (2013)
15. Davis, J., Murray, S. and Burchell, F.N. *Interstate Royal Commission on the River Murray, representing the states of New South Wales, Victoria, and South Australia: report of the Commissioners with Minutes of evidence, appendices, and plans*. Parliament of Victoria, Melbourne. (1902)
16. Macumber P. *Interaction between groundwater and surface systems in northern Victoria*. Department of Conservation and Environment,Victoria, Melbourne. (1991)
17. Francis, G. Nature, **2**: 11-12. (1878)
18. Gell, P., Fluin, J., Tibby, J., Hancock, G., Harrison, J., Zawadzki, A., Haynes, D., Khanum, S., Little, F. & Walsh, B. Geomorphology, **108**: 122-126. (2009)
19. Laurance, W.F., Dell, B., Turton, S.M., Lawes, M.J., Hutley, L.B., McCallum, H., Dale, P., Bird, M., Hardy, G., Prideaux, G., Gawne, B., McMahon, C.R., Yu, R., Hero, J-M., Schwarzkopf, L., Krockenburger, A., Douglas, M., Silvester, E., Mahony, M., Vella, K., Saikia, U., Wahren, C-H., Xu, Z., Smith, B. & Cocklin, C. Biological Conservation, **144**: 1472-1480. (2011)
20. Davies, P.E., Harris, J.H., Hillman, T.J. & Walker, K.F. Marine and Freshwater Research, **61** : 764-777. (2010)
21. Sim, T. and Muller, K. *A Fresh History of the Lakes: Wellington to Murray Mouth, 1800s to 1935*. River Murray Catchment Water Management Board, Strathalbyn: 75pp. (2004)
22. Gell, P.A. *Watching the Tide Roll Away – Advocacy and the Obfuscation of Evidence*. Pacific Conservation Biology. (2018)
23. Gell, P., Bulpin, S., Wallbrink, P., Bickford, S. & Hancock, G. Marine and Freshwater Research **56**: 441-456. (2005)
24. Gell, P. & Reid, M. The Anthropocene, **8**: 39-45. (2014)
25. Ogden, R.W. *Regulated Rivers: Research & Management* **16**: 497–512. (2000)
26. Reid, M.A., Sayer, C.D., Kershaw, A.P. & Heijnis, H. Journal of Paleolimnology **38**: 191-208. (2007)
27. Gell, P. and Reid, M. Frontiers in Ecology and Evolution, **4**, 16. (2016)
28. Kattel, G., Gell, P., Perga, M., Jeppesen, E., Grundell, R., Weller, S., Zawadzki, A. & Barry, L. Freshwater Biology **60**(4): 711-723. (2014)
29. Wang, R., Dearing, J. A., Langdon, P. G., Zhang,

E., Yang, X., Dakos, V. and Scheffer, M. Nature, **492**, 419–422. (2012)
30. Capon, S.J., Lynch, J.J., Bond, N., Chessman, B.C., Davis, J., Davison, N., Finlayson, C.M., Gell, P.A., Hohnberg, D., Humphrey, C., Kingsford, R.T., Nielsen, D., Thomson, J.R., Ward, K. and MacNally, R. Science of the Total Environment, **534**, 122-130. (2015)
31. Kattel, G., Gell, P., Zawadzki, A. & Barry, L. Hydrobiologia **787**(1): 269-290. (2017)
32. Grundell, R., Gell, P., Zawadzki, A. & Mills, K. Journal of Paleolimnology. **47**: 205-219. (2012)
33. Walker, B. & Salt, D. *Resilience Thinking: Sustaining Ecosystems and People in a Changing World*. Island Press, Washington, DC. (2006)
34. Butler, J.R.A., Suadnya, W., Yanuartati, Y., Meharg, S., Wise, R.M., Sutaryone, Y., and Duggan, K. Priming adaptation pathways through adaptive –co-management: Design and evaluation for developing countries. Climate Risk Managememnt, **12**: 1-16. (2016)

Find the future from the past: Paleolimnology in Indonesia

Tri Retnaningsih Soeprobowati[1, 2], *Sri* Widodo Agung Suedy[12], *Hadiyanto*[1, 3]

[1]School of Postgraduate Studies, Diponegoro University, Semarang Indonesia
[2]Department of Biology, Faculty of Science and Mathematics, Diponegoro University, Semarang Indonesia
[3]Department of Chemical Engineering, Faculty of Engineering, Diponegoro University, Semarang Indonesia

Abstract. Agriculture, hydro electricity power, fisheries, source of drinking water, recreation, and industrial use has led to wide-spread impairment of Indonesia surface waters. Indonesia is rich with various freshwaters, including 840 unique lakes with different size and types, 735 ponds, and 162 dams that covering about 21,000 km^2 of area, contain about 500 km^3 of water functions as both economically and ecologically purposes. Indonesia is also one of the biggest listed freshwater endemic species in Asia. However, an increase of population and human activities affect a decrease in quantity and quality of several lake ecosystems that able to induce the destruction up to extinction of the lake ecosystem. An increase of lake utilization, with partial management system without considering the preservation and sustainability of ecosystem had reduced lake functions. Paleolimnological approach is one of science and technological approach to support the successfull of grand design of Indonesian lakes conservation and rehabilitation, specifically on the program of lakes' development ecosystem monitoring, evaluation, and information system. The studied on (paleo) limnology in Indonesia had been collected and compared. The vertical distribution of trace elements of Ni, Mg, and B were analyzed by Inductively Coupled Plasma (ICP) with mass spectrophotometry for 44 cm sediment core from Telaga Pengilon Dieng, Java. The paleolimnological studies in Indonesia have been limited, particularly for Indonesian as the first author. Lakes'sediment records the changes condition of catchment area, like a diary book. Reconstruction of the past environmental condition is required to predict the future condition. This can be done using the organism preserved in the sediment, such as diatoms and pollen. Diatoms are unicellular algae that are used extensively in paleoecology. The diatom assemblages in sedimentary records can make direct and indirect inferences about past environmental conditions. The palynological data on the terrestrial and aquatic ecosystem are used as an indicator of recent and paleoecological changes, particularly on the climate, history and floral diversity. Paleolimnological analysis increasingly attention is being given to developing the technique to model human activities.

1 Introduction

Indonesia own more than 840 lakes by which 521 are having more than 10 hectares wide that spread out in Sumatra, Kalimantan, Papua, Java and Bali, Sulawesi, NTT and NTB, and Maluku. Indonesia also has 735 ponds, 162 dams that totally covering about 21,000 km^2 of area and resettling about 500 km^3 water volume. In Indonesia, lakes and reservoirs are important water resources, mainly for water supply and irrigation, while the remaining serves as storage basin for agricultural usage, hydroelectricity power, flood control, recreational activities, fisheries, uniqueness of tradition, culture and heritage, habitat for high diversity of potentially endemic, rare, and endangered species of flora and fauna [1].

At present, lake degradation has progressed at an accelerating rate due to changes in land use pattern, resulting in the deterioration lakes' function and services. The development of lake management strategies have to consider long term changes and dynamics and its variability. Sediment core from the lake is a diary book that tells the story about what happen in the catchment area. Indonesia rich with the type and variety lakes that mostly utilized only for water resources with sectorial management, based on their own interests that in turn induce conflict of interest.

As endorses at the National Conference of Indonesian Lake II commenced on 13-14 October 2011 in Semarang, the Ministry of Environment restated that the 15 lakes are nationally treasured as the priority of environmental crisis to be resolved following Bali Agreement on Sustainable Management of Lake on 13 August 2009. At the first phase (2010-2014) 15 lakes priority were Lake Toba, Maninjau, Singkarak, Kerinci, Tondano, Limboto, Poso, Tempe, Matano, Kaskade

* Corresponding author: trsoeprobowati@live.undip.ac.id

Mahakam Semayang-Melintang-Jempang, Sentarum, Sentani, Rawa Danau, Batur, and Rawapening. The second phases (2015-2019) there are consideration for another 15 lakes, i.e Lake Diatas, Dibawah, Ranau, Dendam Tak Sudah, Lindu, Towuti, Mahalona, Bratan, Paniai, Laut Tawar, Aneuk Laot, Kelimutu, Rinjani, Taliwang, and Tasik.

It is certainly a challange task to do paleolimnological research as a basic for developing suitable lake management. Save Indonesia Lake References book were developed for each lake, basically with 3: approaches science and technology application for remediate water and catchment area, governance, and community involvement for lake conservation. One of super prirority program that have to be considered in Save Indonesia Lakes was paleolimnological studies [2]. One of recommendation from 16th World Lake Conference, November 2016 was that paleolimnology can play valuable roles in addressing many lake issues including risk management and disaster preparedness.

Paleolimnology was freshwater studies focused on the intepretation of sequent sediments and diagenetic process that influenced the sedimentary record. The aims of paleolimnology were to figure out the past condition, productivities, and changes of parameters than induce changes of lakes productivity [3]. Paleolimnology is the study on the interrelation watershed geology, watershed climate, mode of Lake formation/evolution, inflow/outflow hydrology, internal circulation, and organic productivity, which were ultimately linked to regional climate, geology, and human activities [4]. Paleolimnology is multidiscipline approach of using physical, chemical, and biological informations from sediment to reconstruct past environmental condition, particularly for freshwater ecosystem. Paleolimnology is study about long-term changes on geomorphological lakes [5]. Paleolimnology has been used for decades to reconstruct past environments and more recently help the scientist to understanding and observed changes that occurred in the past [6]. Paleolimnology also provide knowledge about environmental process in ancient time and useful to determine geochemistry mechanism in lake [7]. Therefore, paleolimnology is basically learn about stratigraphy sediment of the lakes where fossils well preserved in the sediment and provides information about past condition of the catchment area. However, paleolimnological studies in Indonesia was very limited [8].

2 Methods

This research is combination qualitative and quantitative methods. Qualitatively, data collection for research on paleolimnology in Indonesia was done through searching in Research Center for Limnology Indonesia, communicate with Society of Indonesian Limnology, and searching in internet. The result was listed in the table and several stratigraphy of pollen and diatoms were selected and compare. Quantitatively, this paper provides stratigraphy of heavy metals Nickel (Ni), Magnesium (Mg), and Boron (B) from Telaga Pengilon Dieng. Sediment cores were collected from 3 sites of Telaga Pengilon using modified core barrel. After retrieval the cores were maintain in vertical orientation. Slicing sediment core was every 1 cm using plastic spatula and ruler. Sediment were placed in plastic zip lock and store at 4°C. In this study, only 1 core has been analyzed down to 44 cm to provide Ni, Mg, and B from Telaga Pengilon. A manuscript incorporating long term history of catchment area of these lakes based on diatoms were being produced and will be published elsewhere. Analysis of heavy metals was conducted in Western Carolina University US using Inductively Coupled Plasma (ICP) with mass spectrophotometry. The data of Ni, Mg and B were log10 transformed and graphed using the software package C2 version 1.7.7 [9].

3 Result and Discussion

Based on the tracer study, at least 30 research on limnology in Indonesia had been recorded since Sunda Limnological Expedition organized by Ruttner in between September 1928 and July 1929 (Table 1). This expedition emphesize to study the biology of numerous lakes, ponds, swamps, paddies fields and other aquatic ecosystem in Sumatra, Java, and Bali [10]. The diary, stereoscopic photographs, and all scientific data of Ruttner and Thienemann provided the basis for tropical limnology. Limnology has shown its capacity to provide the very basis for a cautious and environment oriented development. For the commemoration of the 65 anniversary of the Ruttner-Thienemann Sunda Limnological Expedition, the International Conference on Tropical Limnology was held on 4-8 July 1994 at the Christian University of Satya Wacana Salatiga, Central Java, Indonesia that was attended by experts from various countries and wide spectrum of topic. The proceeding entitle Tropical Limnology was the valuable document that consist of 3 volume. Volume I Present Status and Challanges, volume II Tropical Lakes and Reservoirs, and volume III Tropical Rivers, Wetlands and Specials Topics [10].

The other important note of Expedition Indodanau was initiated in 1991 by the Research Institute for Water Resource Development, affiliated to the Agency for Research and Development of the Indonesian Ministry of Public Works, in cooperation with Finnish counterpart. Expedition Indodanau dealed with physical, chemical, and biological characteristics of 38 lakes and reservoirs situated in Sumatra, Java, Bali, Lombok, Flores, Sulawesi, and Irian Jaya. Expedition Indodanau provide information and database on physical, chemical, phytoplanton variables for contemporary evaluation and comparison, and as basis for longterm monitoring programs [11]. Now, the data from those expedition became the baseline for paleolimnological analysis. The newest expediton was in Towuti Lake Sulawesi, 1 out of 3 lake known as Malili Lakes (Matano, Mahalona and Towuti Lakes) the highest publication as one of ancient lake in the world.

Table 1. Research on paleolimnology of Indonesian lake

No	Island	Lake	Source
1	Sulawesi	Towuti	[11],[12],[41], [13],[14],[15]
2	Papua	Kutubu	[39],[40]
3	Sumatra	Singkarak Toba	[26], [43],[30]
4	Sumatra	Aceh	[27]
5	Java	Logung	[22]
6	Java	Rawapening	[23]
7	Java	Pekalongan, Brebes, Semarang, Demak, Rembang	[24]
8	Kalimantan, Papua	Berau, Raja Ampat	[42]
9	Sulawesi	Matano	[16][17][18][19]
10	Sulawesi	Wakatobi	[20]
11	Kalimantan	Batulicin	[33]
12	Sulawesi	Tondano	[21]
13	Java	Borobudur	[25]
14	Java	Dieng	[32]
15	Sumatra	Banda	[29]
16	Sumatra	Laut Tawar,Toba, Maninjau, Diatas, Dibawah, Singkarak, Kerinci, Ranau,	[11]
17	Java	Saguling (R), Cirata (R), Jatiluhur (R), Darma (R), Sempor (R), Mrica (R), Rawapening, Gadjah Mungkur (R), Kedung Ombo (R), Sengguruh (R), Lahor (R), Sutami (R), Wlingi (R), Selorejo (R)	[11]
18	Flores	Kelimutu	[44]
19	Halmahera	Island	[28]
20	Sumatra	Padang	[31]

At least 12 publications on the lakes in Sulawesi, such as by Goudge et al,[11]; Morlock et al, [12]; Costa et al, [13], Russell et al, [14]; Russell & Bijaksana, [15]; Jones et al, [16]; Vailant et al, [17]; Tamuntuan et al, [18]; Crowe et al, [19]; Engelhart et al, [20], Dama et al, [21], and Lehmusluoto et al, [11]. At least 6 publications related to the paleolimnology in Java ([22], [23]; [24]; [25], [11], and [10]). In Sumatra there were at least 8 researchs on paleolimnology, which were by Setyobudi et al, [26]; Grand et al, [27]; van der Kaars et al, [28]; van der Kaars et al, [29]; Lehmusluoto et al, [11]; Nishimura et al, [30]; Morley, [31]; and Ruttner, [10]. There were also research on paleolimnology in Papua, Kalimantan, Flores, and Halmahera (Table 1). Most of those publications were the output of project with the international funding. There is still limited publication on the paleolimnlogy with the first author is Indonesian. Pudjoarinto [32] studied about pollen stratigraphic as an human activity evidence in Dieng. But that was his last research before pass away. Yulianto [33] followed research on holocene pollen. Soeprobowati et al ([23],[24]) initiated to study on diatoms for paleolimnlogical tool for reconstruction past environmental changes. Setyobudi et al [26] focused on the sedimentology and limnology. Tamuntuan [18] studied the magnetic properties of tectonic lake sediment. Bijaksana was Indonesian geologist that have many publications related to the Towuti project.

The research that purely studied on paleolimnology in Indonesia that were used biological tools of pollen and diatoms were in Telaga Balekambang Dieng [32], Lake Tondano [21], and Lake Rawapening [23]. Pudjoarinto & Cushing [32] found that pollen preserved in Telaga Balekambang Dieng indicated pattern of vegetation in the catchment area. There were 5 zones that indicated human settlement and cultivation. The dominance of *Plantago major* indicated that Hindu-Javanese religious was centered in Dieng since 700 BP. Land changes from montane forest into agriculture was indicated by a decrease of trees and an increase of cultivated plants that was started in Dieng since 1920 (Figure 1).

Based on the pollen preserved in the sediment, in 33,000 until 30,840 yr BP indicated low land - sub montane forest around Lake Tondano. A high diverse of non-diatoms indicated lacustrine deep water. The dominance of epiphytic diatoms species of *Cocconeis placentula, Epithemia adnata*, and planktonic species of *Aulacoseira granulata* indicated that Lake Tondano was deep eutrophic lake that have abundant of submerge plants. The dominance of small diatoms of *Fragillaria* spp and *A. granulata* indicated that in 33,000 yr BP was full lacustrine, open shallower water. An absent of *A. granulata* in 30,840 yr BP indicated shallowing the lake. The lack of diatoms and the appearance of Cyperaceae and Pteridophyte indicated the expose of lake shore and swampy dryland in 30,840 – 11,930 yr BP. The appearance diatoms again with the dominance species of *F. pinnata* and *Cymbella cistula* in the upper CP-1, indicated a swamp to shallow lake with abundant of aquatic vegetation. The succession from *A. granulata* dominant into *F. tenera* reflected a decrease of water depth about 1-2 m. In the late Holocene (1,500 yr BP until present), a succession from *C. cistula* to aerophylous diatoms of *Navicula converacea* and *F. pinnata* indicated the transition perennially flooded and swampy alluvial plain (Figure 2). The fluctuation of diatom abundance associated with water depth fluctuations during early Holocene [21].

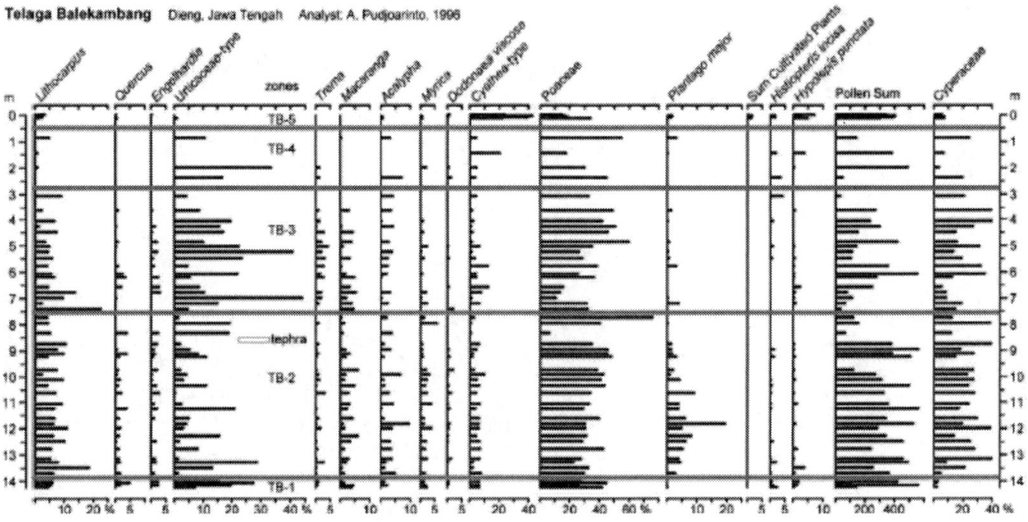

Fig. 1. Pollen stratigraphy of Telaga Balekambang, Dieng [32]

The reconstruction of past environmental changes in vicinity of Rawapening Lake based on diatoms indicated that eutrophic condition of Rawapening Lake was occurred since 1967, as the longest sediment cores collected in 2008 was 63 cm (Figure 3). The eutrophic condition was indicated by the dominant of *Gomphonema parvulum* (Kützing) Kützing, *Luticola goeppertiana* (Bleisch) Mann, *Luticola mutica* (Kützing) Mann, *Mayamaea* (syn:Navicula) *atomus* (Kützing) Lange-Bertalot, *Nitzschia palea* (Kützing) W.Smith and *Nitzschia umbonata* (Ehrenberg) Lange-Bertalot. By the time, there were a shift dominant species into *A. ambigua* (Grunow) Simonsen, *A.granulata* (Ehrenberg) Simonsen and *U. acus* (Kützing) M. Aboel that indicated high pH and hypereutrofic condition [34]. Based on monitoring data on pH, indeed the pH was tend to increase [34, 35].

In this paper, the heavy metals of Ni, Mg and B from one coring of Telaga Pengilon showed that stratigraphically, there were trend of increasing Ni, Mg and B form the bottom core (44 cm) to the top soil (0 cm) with the similar trend. The highest concentration of Ni was in 4-6 cm, Mg was in 14-16 cm, and B was in 22-24 cm. Seems that the layer of 22-24 cm recorded a sharply increase concentration compare to the layers before (Figure 4). The trace element composition and distribution are the net result of processes in the catchment area (allochthonous influences) and in the lake itself (autochthonous influences) can be observed using traced elements. A concentration of B was reported from July 1994 and July 1999 by van Bergen et al, [36]. Ni, Mg, and B in the southeast part of Telaga Pengilon which mainly derived from rock weathering and other geochemical processes. Ni together with Cu and Cr of Aibi Lake, Northwest China were derived from geological process, whereas Cd, Pb, Hg, and Zn in the sediment were mainly from pollution [37]. The concentration of Pb, Cd, Cr, and Cu in Telaga Warna Dieng were exceeding the Indonesian water quality standard [38]. Telaga Warna and Telaga Pengilon are connected but the pH were very differ along with different depth.

https://doi.org/10.1051/e3sconf/20183108002

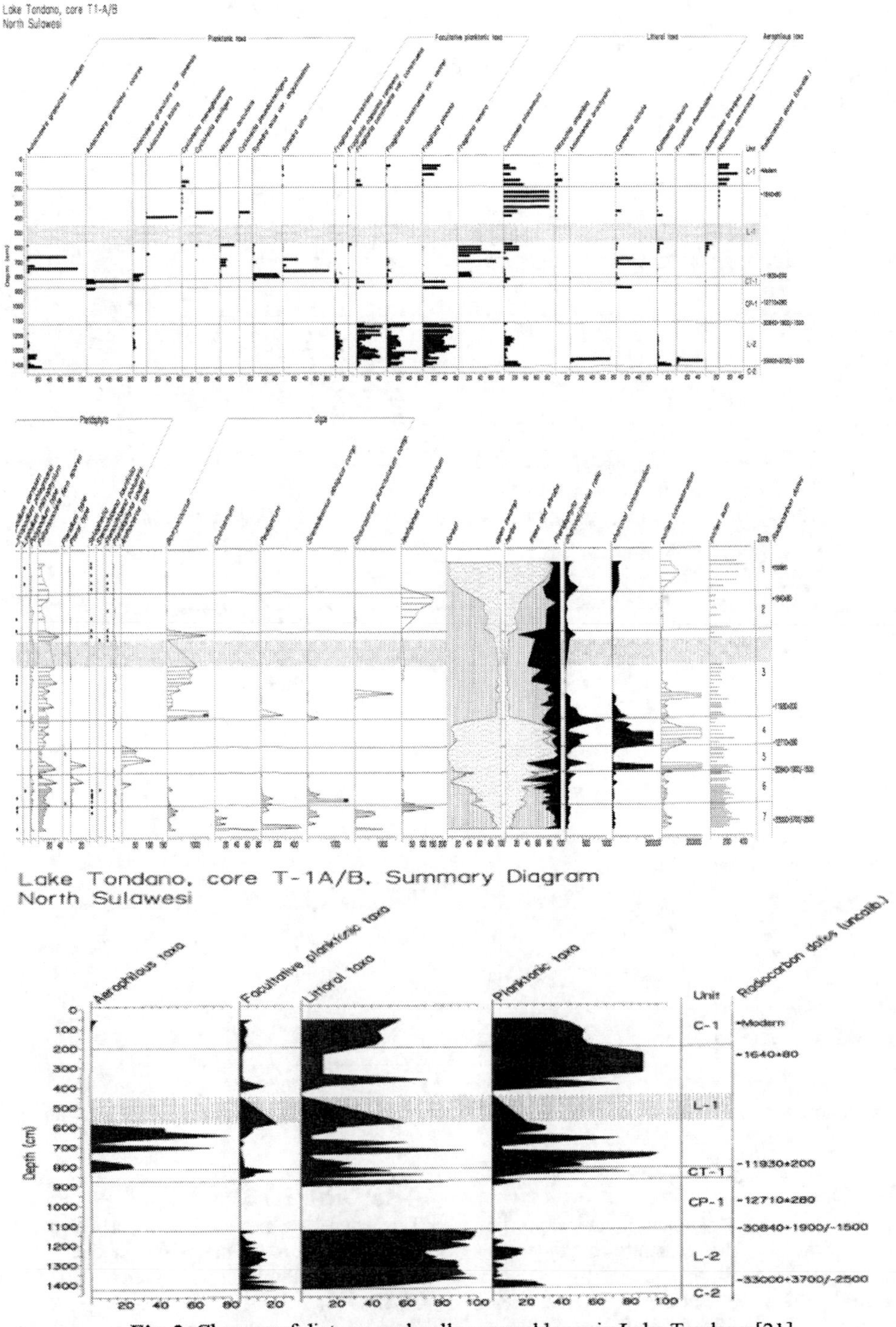

Fig. 2. Changes of diatoms and pollen assemblages in Lake Tondano [21]

Based on the sediment core profile of tracer elements from Lake Kutubu Papua New Guinea, shown that oil/gas extraction activites are significantly increase the concentration of tracer elements in the lakes as well as intensification of human inhabitants in the catchment area [39]. The occurence of tephras in Lake Kutubu marked the timing of climate change, the rate of vegetation change and development of agricultural in Lake Kutubu, that all related to human impact on Lake Kutubu [40].

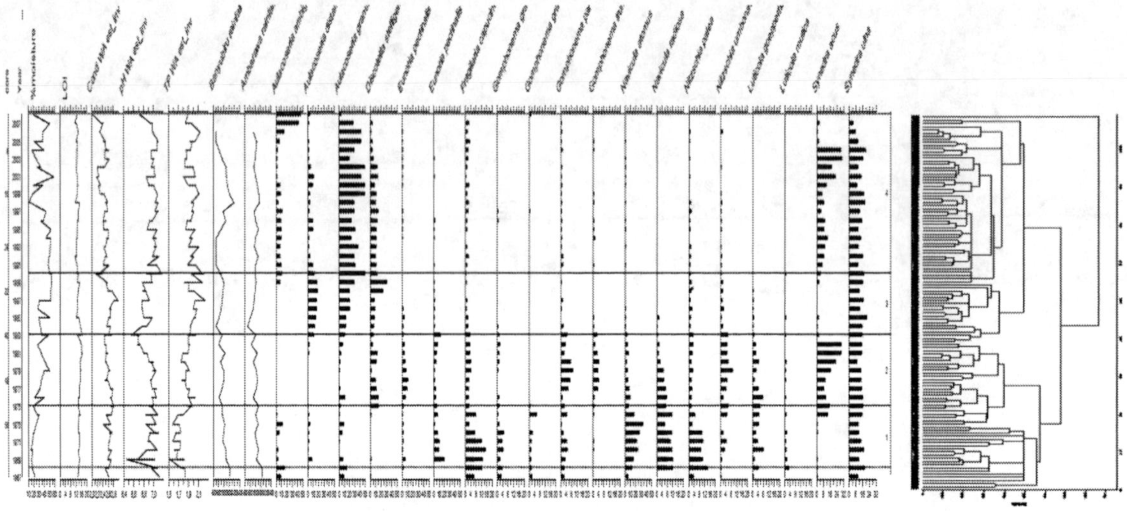

Fig. 3. Diatom stratigraphy of Asinan site from Rawapening Lake, Java and 4 zones [23]

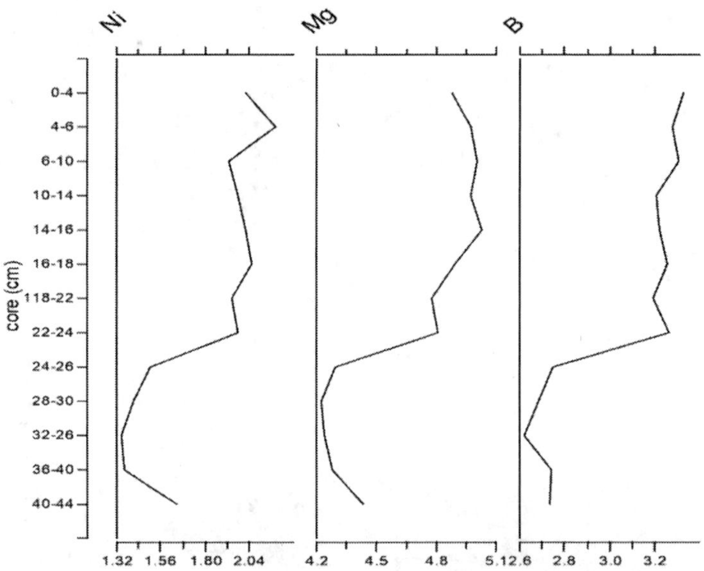

Fig. 4. The stratigraphy of heavy metals of Ni, Mg, and B from 44 cm sediment core from Telaga Pengilon and Dieng

CONCLUSION

Indonesia rich with unique and specific characters of tropical lakes that currently degradate at the accelerating rate due to changes in land use pattern, that affected lakes function and services. Paleolimnology provide significantly information about past environmental change both naturally or human activites in the lakes or cathment area, therefore, provide information to develop suitable environmental management. It is certainly a challanging task to develop paleolimnology in Indonesia to reach the sustainable lakes in the future.

AKNOWLEDGMENT

This article was produced as a part of project supported by Diponegoro University Indonesia through an International Publication Research Grant Number: 276-39/UN7.5.1/PG/2017, 30 March 2017. A heartfelt thanks go to Kenanga Sari, Geyga, Alam Lazuardi, and M Hadi El Amin for field work and diatom analysis. Thanks also go Global Innovation Initiative Project that provides opportunity for Kenanga Sari to analyze heavy metals in Western Carolina University USA supervised by Jerry Miller.

References

1. MEF (Ministry of Environment and Forestry). The grand design of Indonesian lakes conservation and rehabititation. Ministry of Environment and Forestry, p.94(2016).
2. ME (Ministry of Environment). Gerakan Penyelamatan Danau (Germadan) Rawapening. Ministry of Enviroment. p.67 (2011).
3. Wetzel RG.. Limn ology: lakes and river ecosystems. Academic Press. 1006 (2001)
4. A.S. Cohen. Paleolimnology: the history and evolution of lake systems. Oxford University Press, NY. P. 525 (2003)
5. J.P. Smol. Pollution of Lakes and Rivers A Paleoenvironmental Perspective. 2nd Ed. Blackwell Publishing, USA(2008)
6. I.R.Tobler. Using Paleolimnology for Lake Restoration and Management. Frontier in Ecology and Evolution (2016)
7. E.S.Talbot Paleolimnology as a Tool to Achieve Environmental Sustainability in the Anthropocene: An Overview. Geoscience (2016).
8. P.E. Hehanusa and G.S. Haryani. Klasifikasi morfogenesis danau di Indonesia untuk mitigasi dampak perubahan iklim. Makalah disampaikan dalam Konferensi Nasional Danau Indonesia I, Sanur-Denpasar-Bali, 13-15 Agustus 2009
9. S. Juggins. The European Diatom Database User Guide Version 1.7.7. University of New Castle, New Castle, UK (2016).
10. F.Ruttner. Arch. Hydobiol. Suppl. 8, 197-454 (1931)..
11. P. Lehmusluoto, B. Machbub, N. Terangna, S. Rusmiputro, F. Achmad, L. Boer, S.S. Brahamana, B. Priadi, B. Setiadji, O. Sayuman and A. Margana. National inventory of the major lakes and reservoirs in Indonesia. Edita Oy Helsinki 72 pp(1992)
12. M. Morlock, H.Vogel, V. Nigg, A. Hasberg, M. Melles, J.M. Russel, S. Bijaskana, and TDP science team. 2016. Geochemical characterization of Lake Towuti, Indonesia: setting the stage for climate reconstructions over several glacial-interglacial cycles. 14th Swiss Geoscience Meeting, Geneva 2016.
13. K.M. Costa, J.M.Russell, H. Vogel, and S. Bijaksana. Palaeogeography, Palaeoclimatology, Palaeoecology 417 467–475(2015).
14. J.M. Russell, H. Vogel, B.L. Konecky, S. Bijaksana, Y. Huang, M. Melles, N. Wattrus, K. Costa, and J.W.King. B.P. 111(14), 5100–5105(2014).
15. J.Russell and S. Bijaksana. The Towuti Drilling Project: Paleoenvironments, Biological Evolution, and Geomorphology of a Tropical Pacific Lake. Scientific Drilling, No. 14, September 2012.
16. C. Jones, S.A. Crowe, A. Sturn, K.L. Leslie, LCW Maclean, S. Katsev, C. Henny, D.A. Fowle, and D.E.Canfield. Biogeosciences, 8, 2977–2991(2011)
17. J.J. Vaillant, G.D.Haffner, and M.E.Cristescu. Integrative and Comparative Biology 51(4), 634–643(2011)
18. G. Tamuntuan, S. Bijaksana, E.Gaffar, J. Russell, L.O.Safiuddin, and E. Huliselan E. ITB J. Sci. 42 A (1), 31-48(2010)
19. SA Crowe, AH O'Neill, S Katsev, P Hehanusa, GD Haffner, B Sundby, A Mucci, and DA Fowle. Limnol. Oceanogr. 53(1): 319–331(2008)
20. SE Engelhart, BP Horton, DH Roberts, CL Bryant, and DR Corbett. Marine Geology, 242(1–3): 65–81(2007).
21. RAC Dama, J Fluin, P Suparan, and S van der Kaars. Palaeogeography, Palaeoclimatology, Palaeoecology 171: 147-183(2001)
22. JR Rodysill, JM Russell, S Bijaksana, ET Brown, LO Safiuddin, and H Eggermont. Journal of Plaeolimnlogy 47(1), 125-139(2012).
23. T.R Soeprobowati. S Hadisusanto, P Gell and A. Zawadski. American Journal of Environmental Science (3), 334-344(2012).
24. T.R Soeprobowati, S.W.A Suedy and P Gell. Journal of Coastal Development 15(2), 197-208(2012).
25. H Murwanto, Y Gunnell, S Suharsono, S Sutikno, and F Lavigne. The Holocene 12(3), 459-463(2004).
26. PT Setyobudi, PA Suandhi, ZL Tarigan, A Bachtiar, AGR Jayanti, and L Budin. Sedimentology and limnology of Singkarak and Toba Lakes, Sumatra, Indonesia: depositional and petroleum system model for tropical Fluvio-lacustrine and volcanic related rift basins in Southeast Asia. Proceedings Indonesian Petroleum Association, Forth Annual Convention & Exhibition May 2016, IPA16-660-G
27. C.A Grand, B.P Horton, H.M Kelsey, C.M Rubin, A.D Hawkes, M.R Daryono, G Rosenberg, S.J Culver. *Quaternary Science Reviews* 54, 142-151(2012).
28. W.A van der Kaars. *Palaeogeography, Palaeoclimatology, Palaeoecology. 85*: 239–302 (1991).
29. S van der Kaars, X Wang, P Kershaw, F Guichard, and DA Setiabudi. Palaeogeography, Palaeoclimatology, Palaeoecology 155(1–2):135–153 (2000).
30. S Nishimura, E Abe, J Nishida, T Yokoyama, A Dharma, P Hehanusa, and F Hehuwat. Techtonophysiscs 109(3-4), 253-261 (1984).
31. RJ.Morley J. Biogeogr. 9, 151–190 (1982)
32. A Pudjoarinto, and EJ Cushing. 2001. Palaeogeography, Palaeoclimatology, Palaeoecology, 171(3–4), 329–340 (2001).
33. E Yulianto, AT Rahardjo, D Noeradi, DA Siregar, and K Hirakawa. 2005. A Holocene Pollen Record of Vegetation and Coastal Environmental

Changes in the Coastal Swamp Forest at Batulicin, South Kalimantan, Indonesia.

34. T.R Soeprobowati.; S.D Tandjung.; Sutikno; S Hadisusanto.; and P Gell. Biotropia 23(2): 96-104 (2016).
35. T.R Soeprobowati, S.D Tandjung, Sutikno, S Hadisusanto, P Gell. H Hadiyanto and S. W. A Suedy. Biodiversitas 17(2), 657-664(2016).
36. van Bergen MJ, Bernard A, Sumarti S, Sriwana T, and Sitorus K. 2000. Crater lakes of Java: Dieng, Kelutd, and Ijen. Excursion Guidebook, IAVCEI General Assembly, Bali 2000.
37. J Abuduwaili, ZY Zhang, FQ Jiang. PLoS ONE 10(3): e0120001(2015).
38. T.R Soeprobowati.; S.W.A Suedy.; and H. Hadiyanto. IOP Conf. Series: Earth and Environmental Science 55 (2017): 012051 (2017).
39. L Schneider, SG Haberle, WA Maher, F Krikowa, A Zawatdzki, and H Heijnis. Chemosphere 148, 369e379 (2016)
40. L Schneider, B Alloway, , JR Blong, GS Hope, SJ Fallon, FC Pain, WA Maher, and SG Haberle. Journal of Quetanary Science 1-13(2017)
41. TA Goudge, JM Russell, JF Mustard, JW Head, and S Bijaksana. GSA Bulletin 129 (7-8) (2017).
42. LE Becking. 2012. Marine lakes of Indonesia. Naturalis Biodiversity Center, Faculty of Science, Leiden University
43. S van der Kaars, MAJ Williams, F Bassinot, F Guichard, E Moreno, F Dewilde, and EJ Cook. *Quaternary International*, *258*, 45–53(2012)
44. GB Pasternack and JC Varekamp. Geochemical journal 28, 243-262(1994)

https://doi.org/10.1051/e3sconf/20183108003

Analysis Of Landslide Materials Spreading In Bendan Dhuwur Village Gajahmungkur Subdistrict Semarang City

Devina Trisnawati[1,*], Najib[1], *Istiqomah Ari* Kusuma[1] and *Anissa Fitratul* Husna[1]

[1] Geological Engineering Department, Diponegoro University, Semarang - Indonesia

Abstract. Bendan Dhuwur is one of area in Semarang city, which continuously has landslide problem. This problem resulted in damage of some buildings and main road. Landslide materials/coluvial have been estimated lays on under those infrastructures and tend to move during rainy season. Therefore, it needs to understand the spread of coluvial to minimize the effect of landslide. Remote sensing method has been used to analyze multi temporal image for mapping landslide materials from different years recorded direction of creep and spread of coluvials. This method has been combined with surface and subsurface data from mapping and resistivity data. The analysis result on map which show that the coluvial material spreads on the south side, beneath the University of Tujuh Belas Agustus construction and Pawiyatan Luhur road. Its move to east leads to the Kaligarang river.

1 Preliminary

Landslide is one process that works in changing the face of the earth. This process can work quickly or slowly. Generally, the process of land slide has either work quickly or slowly a negative effect on society. One of the negative effect of the land slide slowly felt by the people in Bendan Dhuwur Village, Gajahmungkur Sub-district, Semarang City, especially in Pawiyatan Luhur road.

The process of land slide slowly resulted in Pawiyatan Luhur road which is the nearest access from Sampangan to Jatingaleh experiencing subsidence. Road repair periodically done by the Public Works Department does not have much impact on the road, because subsidence still occurs especially in the rainy season. According to [1] colluvial material on the slope bend causes the material always move especially in the rainy season. Based on this, the research is aimed to find out the distribution of colluvial material which is estimated to be under the Pawiyatan Luhur road. Material distribution can be interpreted by remote sensing method.

Remote sensing method makes it easier to observe the research object in quite wide locations research. In addition, the use of remote sensing for the compilation of landslide databases has been done a lot nowadays because it can load past information data up to now [2,3,4,5]. The use of remote sensing methods combined with surface geology and resistivity data to compare remote sensing interpretation results with actual geological conditions (ground checking). From the interpretation result of the distribution of colluvial material then directed the effort of controlling the movement of creeping during the rainy season.

2 Study Area

The location of research belongs to administrative area of Semarang City, Gajahmungkur sub district Bendhan Dhuwur Village (Fig.1). In general, the study area has a morphological condition that characterizes the result of structural formation [6].

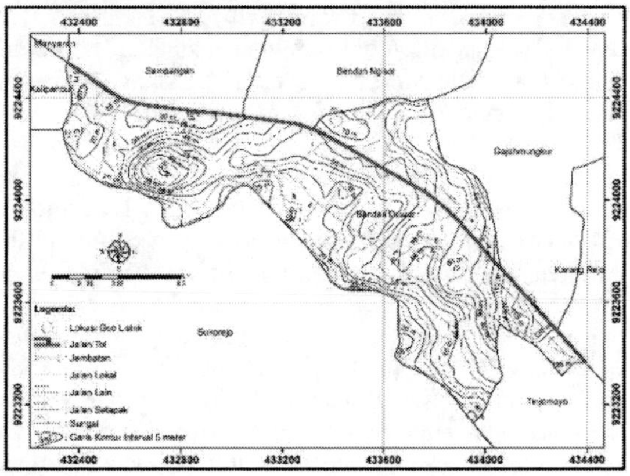

Fig. 1. Research Area.

* Corresponding author: tdvina812@gmail.com

3 Research Method

3.1 Image Processing

Image processing is intended to interpret changes in the form of land in the location research. Imagery used in the form of Landsat 7 ETM+ image from different year (2014 and 2016) got from USGS. Image processing is done with Er Mapper 7.1 and Arc Map 10.1 software.

Information about landslide from remote sensing is closely related to morphology, especially the slopes. The interpretation of the landslide is based on the introduction of changes occurring in the landform which is associated with the slope [2]. Changes in landforms especially on slope buckling may be indicated as avalanche material.

Image processing begins with geometric correction aim to correcting recording errors, resulting in uniformity of coordinate system and scale in image. The corrected research image is subsequently cut in accordance with them. The next process is image masking to get the desired display. The Red Green Difference method is used in analyzing changes in the landform. The contrast of green and red colors from different year images is an indication of a change in the landform. Digitization is done to record the changes of existing land form so that the limitation of the spread of landslide material is obtained.

3.2 Interpretation of Surface and Subsurface Data

In addition to image data that has been processed, still required surface data conditions and subsurface research area. Surface data was obtained by mapping the soil / rock type. Mapping is needed to ensure the difference of landslide material results with residual soil or in situ rocks in the research area. The landslide material consists of earth and debris, where the earth has a material composition with grain size smaller than 2mm as much as 80%, while debris has a material composition with grain size of more than 2 mm as much as 20 - 80% [7]

The subsurface data used in this research is resistivity data. The resistivity value reflects soil / rock conditions [8] which is obtained from 6 measurement points in the research location. The measurement point is placed in the area with the expected lithology of a landslide material. Measurements were performed with a total line of 100m by the Schlumberger method. Furthermore, the analysis is done with Progress 3.0 software to determine the resistivity value of soil / rock. According to[9] the resistivity value from another research presented in Table 1.

Table 1. Parameter of resistivity value [9].

Resistivity Value	Interpretation of Rock
101-1.000 Ωm	Breccia
51-100 Ωm	Tuffaceous Sandstone / Tuffaceous Sand
20-50 Ωm	Tuff
1-50 Ωm	Claystone

4 Results and Discussion

4.1 Image Interpretation

Visual interpretation of Landsat ETM 7+ images with different years, which is 2014 and 2016 has done. The interpretation process using red green difference method, this method using two layers with different raster data sets. The green layer contains image of year 2014 and red layer contains image of year 2016, by using band 4 of each layer. The changes of the landform show on red and green colors contrast on the raster dataset (Fig. 2).Area without any significant changes will show yellow colors.

4.2 Surface and Subsurface Condition

4.2.1 Lithology Mapping

In general, the research area is composed of volcanic rocks [10]. Morphology of the hills with the lowest point 17 msl and the highest 100 msl. Based on the mapping results (Fig.3) the lithology of research area consists of:

a. Alluvium deposit
   It is a precipitation or deposition of river activity, so it is commonly found around the river. This deposit is encountered in a loose, light gray with a variation in the size of clay until boulder (Fig.4a).
b. Colluvial Deposition
   Choluvial deposition is a precipitate of soil movement results. On the field is characterized by a light brown color - dark brown. It is a loose deposit with variations in clay grain size (Fig.4b).
c. Tuffaceous Sandstone
   This type of lithology is found in almost 50% of the research area. The appearance of the field (Fig.4c) can be distinguished by the lithology of the brownish-yellow, with grain size 2 to 32mm. In general, lithology in a fresh state - decayed with a thickness varying between 50 cm – 1meters.
d. Volcanic Breccia
   This lithology is under the tuffaceous sandstone layer. Characterized by a light brown color, the composition of the grain-sized matrix varies from clay - sand and gravel - sized fragments. Found in massive circumstances with a slight decay (Fig 4d).

https://doi.org/10.1051/e3sconf/20183108003

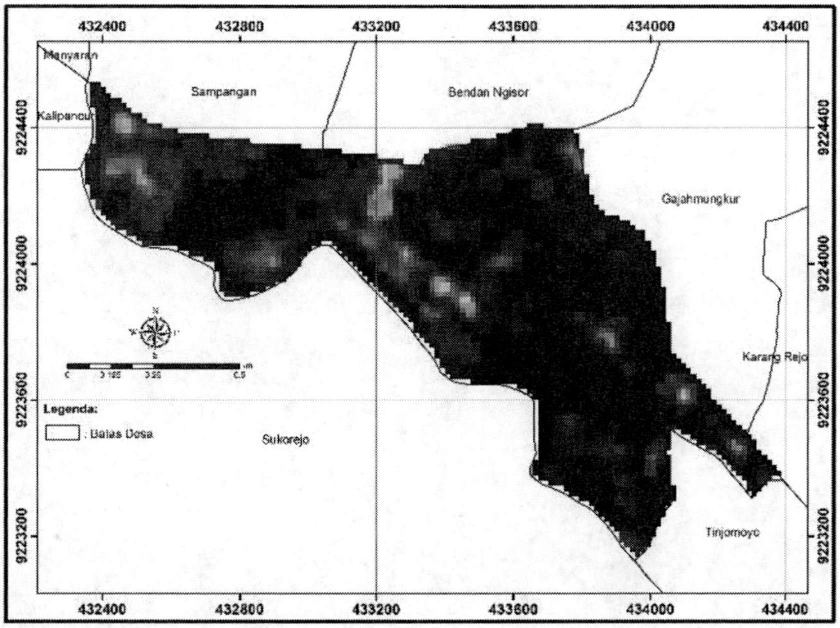

Fig. 2. Image of Red Green Difference Interpretation Result

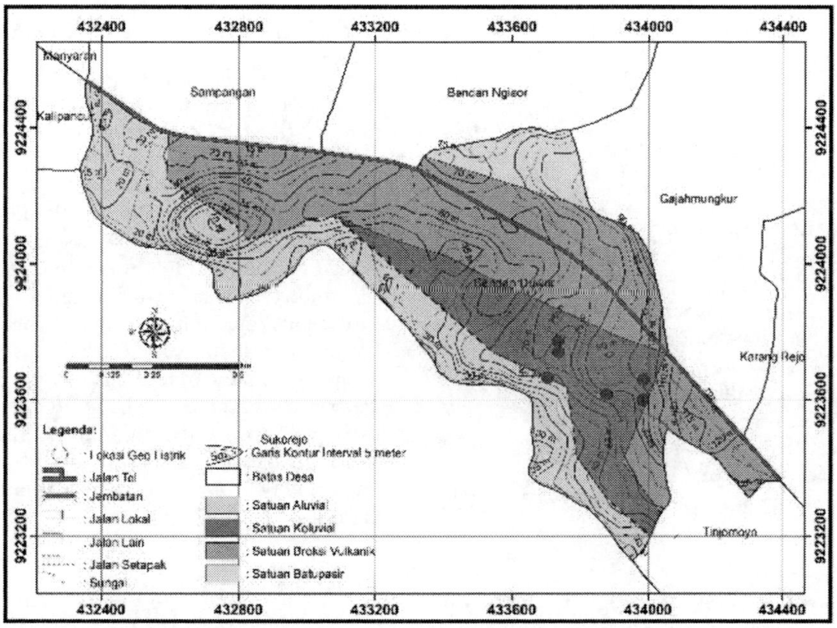

Fig. 3. Lithology Distribution Map in Research Area

Fig. 4. Lithology in research area a. Alluvial Deposit, b. Colluvial Deposit, c. Tuffaceous Sandstone, d. Volcanic Breccia.

4.2.2 Resistivity

Resistivity methods are performed at 6 point measurements (Fig.3). The result of the interpretation using matching curve on the false resistivity data obtained at the time of resistivity measurement and software analysis of Progress 3.0, known that in the research area found lithology variation from surfaceare:

a. Colluvial deposits with resistivity value from 7,12 Ωm to 27,1 Ωm, with a thickness ranging from 0 to 2 m.
b. Clay layer below colluvial deposits with resistivity value range from 0.18 Ωm to 10.58 Ωm.
c. Tuff with resistivity value range 11,72 Ωm - 36,08 Ωm, found at depth above 3 m.
d. Tuffaceoes sand with resistivity range 57,24 - 77,5 Ωm.

4.3 Analysis of Distribution of Material Landslide

Based on the results of the interpretation from image shows changes occur almost all areas of research. From overlay Landsat 7 ETM+ and DEM shows clearly that the changes on the eastern side of the study area with the dominant land use of shrubs are caused by the movement of the soil. This is indicated by a change which parallel to the lineament, where the lineament has a pattern parallel to the river. Changes that occur this can be an indication of changes in landform . In contrast to the south side of the study area, the changes that occur massive (almost in all locations) with relatively equal elevation. The change is in residential land use, so it can be indicated that changes occur in the land cover.

Based on the results of lithology mapping and resistivity measurements shows the correlation of the type of landslide that occurred at the study area. Landslide type is creep and material type of spread flow. Tuffaceous sandstone that lies above the volcanic breccia are more susceptible to weathering due to exposure on the surface. The tuff properties that absorb water accelerate the weathering process, resulting in the tuff being converted into clay. The correlations of resisitvity measurements at points 4.5 and 6 (Fig.7) show that the clay layer is above the tuff indicate that the clay is the result of tuff weathering.

5 Conclusion

Based on the results of research concluded that, landslide that occurred in the research area is creep then crown of landslide is in the southeast of the research area. A landslide leading to south, leads to the Kaligarang river, proving that the pawiyatan road is above the landslide material. Colluvial material is the result of rupting tuffaceous sandstone above the volcanic breccia.

https://doi.org/10.1051/e3sconf/20183108003

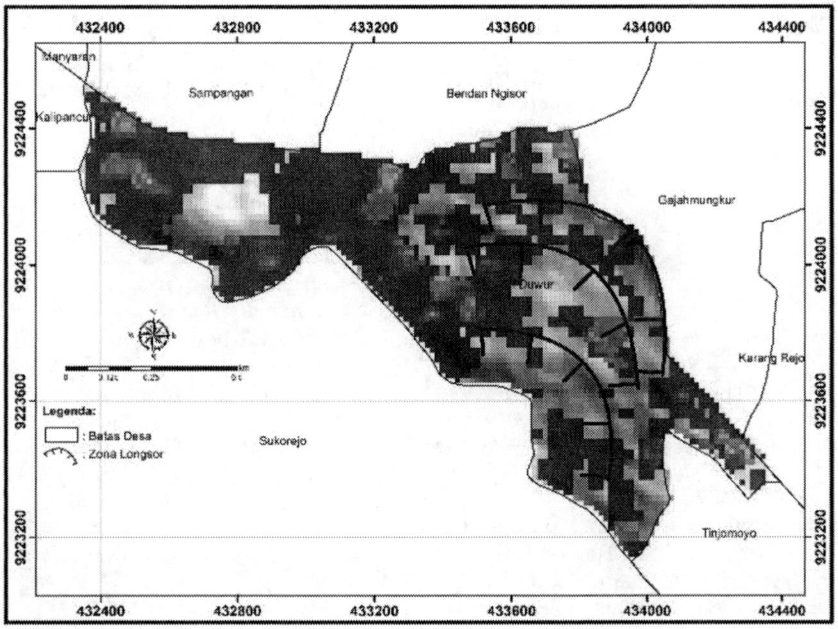

Fig. 6. Overlay of landsat image processing and DEM

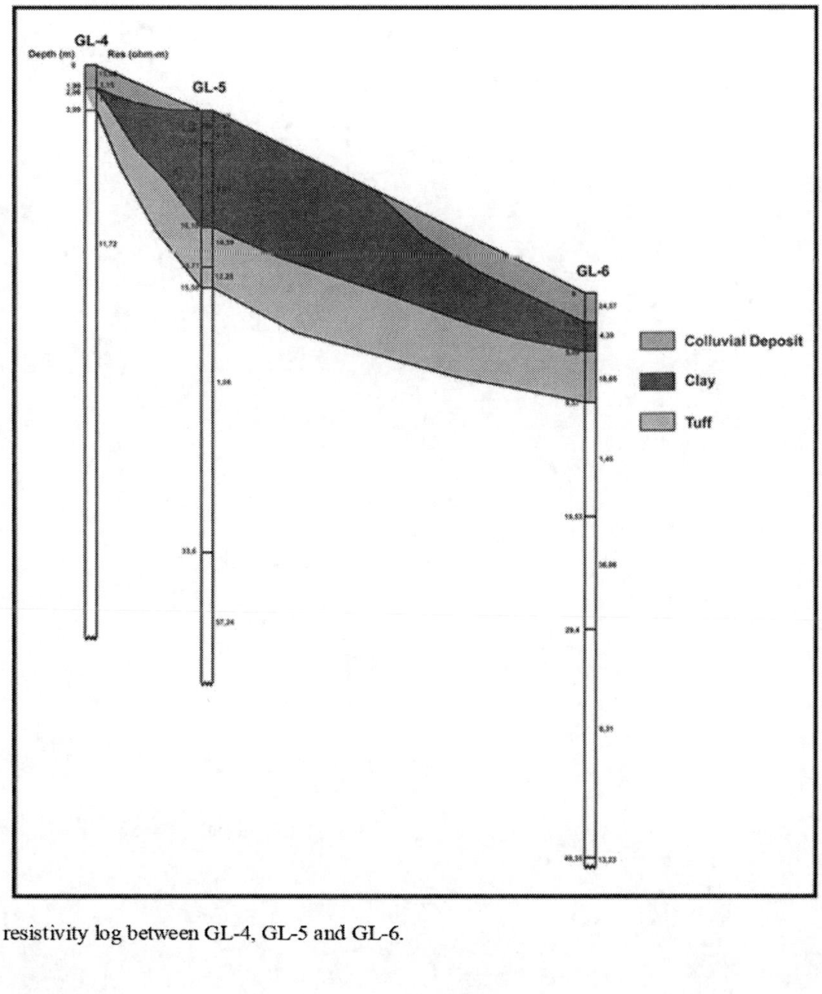

Fig. 7. Correlation resistivity log between GL-4, GL-5 and GL-6.

Acknoledgment

The author thanks to Geological Department of Diponegoro University that supply support to do research and research funding. The thanks also go to parties who have been willing to provide corrections, criticisms, suggestions and input, so that the research and writing of this paper can be resolved.

References

1. D. Karnawati. *Bencana Alam Gerakan Massa Tanah di Indonesia dan Upaya Penanggulangannya* (2005)
2. *R. Soeters, and C.J. van Western. Landslides: Investigation and Mitigation, Transportation Research Board, National Research Council, Special Report 247, 129-177 (1996)*

a. Ciampalini, F. Raspini, S. Bianchini , W. Frodella, F. Bardi, D. Lagomarsino, F. Di Traglia, S.Moretti, C. Proietti, P. Pagliara, R. Onori, A.Corazza, A. Duro, G. Basile, N.Casagli.J. Geomorphology **249**, 103–118 (2014)

3. N. Casagli, F. Cigna, S. Bianchini, D. Hölbling, P. Füreder, G. Righini, S. DelConte, B. Friedl, S. Schneiderbauer, C. Iasio, J. Vlcko, V. Greif, H. Proske, K. Granica S. Falco, S. Lozzi, O. Mora, A. Arnaud, F. Novali, M. Bianchi. J.RSASE **4**, 92–108 (2016)
4. C. H. Hsu, T. C. Tsao, C. M. Huang, C. F. Lee, Y. T. Lee.*Procedia Engineering 143*, 708–716 (2016).
5. P. Rachwibowo. *Proceedings PIT IAGI Ke 39* (2010)
6. D.M. Cruden and D. J. Varnes. *Landslides: Investigation and Mitigation, Transportation Research Board, National Research Council, Special Report 247*, 36-75(1996)
7. W. M. Telford, L. P. Geldart and R. E. Sheriff. *Applied Geophysics* (1990)
8. T. T. Putranto, N. Susanto, D. J. Suprapto, N. Anatoly, A. Rifqi. *Proceeding Seminar Kebumian Ke-8*, 87–97 (2015)
9. R. E. Thanden, H. Sumadirdja, P. W. Richards, K. Sutisna and T.C. Aim. *Peta Geologi Lembar Magelang dan Semarang, Jawa Skala 1:100.000* (1996)

Vulnerability Assessment of Mangrove Habitat to the Variables of the Oceanography Using CVI Method (Coastal Vulnerability Index) in Trimulyo Mangrove Area, Genuk District, Semarang

Rifandi Raditya Ahmad[1,*], *Muhammad* Fuad[2]

[1]Master Program of Environmental Science, School of Postgraduate Studies, Diponegoro University, Semarang - Indonesia
[2]Faculty of Mathematics and Science, Diponegoro University, Semarang - Indonesia

Abstract. Some functions of mangrove areas in coastal ecosystems as a green belt, because mangrove serves as a protector of the beach from the sea waves, as a good habitat for coastal biota and for nutrition supply. Decreased condition or degradation of mangrove habitat caused by several oceanographic factors. Mangrove habitats have some specific characteristics such as salinity, tides, and muddy substrates. Considering the role of mangrove area is very important, it is necessary to study about the potential of mangrove habitat so that the habitat level of mangrove habitat in the east coast of Semarang city is known. The purpose of this research is to obtain an index and condition of habitat of mangrove habitat at location of research based on tidal, salinity, substrate type, coastline change. Observation by using purposive method and calculation of habitat index value of mangrove habitat using CVI (Coastal Vulnerability Index) method with scores divided into 3 groups namely low, medium and high. The results showed that there is a zone of research belonging to the medium vulnerability category with the most influential variables is because there is abrasion that sweeps the mangrove substrate. Trimulyo mangrove habitat has high vulnerable variable of tidal frequency, then based on value variable Salinity is categorized as low vulnerability, whereas for mangrove habitat vulnerability based on variable type of substrate belong to low and medium vulnerability category. The CVI values of mangrove habitats divided into zones 1; 2; and 3 were found to varying values of 1.54; 3.79; 1.09, it indicates that there is a zone with the vulnerability of mangrove habitat at the study site belonging to low and medium vulnerability category.

1. Introduction

In general, coastal areas of Semarang are coastal areas that have considerable pressure on coastal vulnerability. It is proved by Tidal flood phenomena currently befalling in 7 districts out of 16 districts, the result is an area of 3,915.16 Ha affected by tidal flood [1]. One of the increases of tidal flood areas caused by sea level rise during 1985-2008 recorded up to 5,536 cm per year, while land subsidence at the port area reaches 5-7.5 cm per year [2]. And according to Marfai, [3] Semarang is the most significant erosion process.

This leads to the occurrence of Coastal vulnerability, and coastal vulnerability has occurred in almost all coastal areas around the world including occurring in Indonesia [4], More than 20 villages along the Semarang City coastline suffered from the coastal inundation which produces tidal flood and enhanced the land Subsidence phenomena. Trimulyo Village is one of the tidal and sea surge, affecting the sea level rise above the critical height above the coastal land and flooding the residential area [5].

Nevertheless in Trimulyo there is a good condition of mangrove ecosystem located east of Trimulyo. That mangrove ecosystem is formed by the results of rehabilitation activities conducted over the last 10 years in the former ponds. Now that has a mangrove area of more than 30 Ha.

Mangrove ecosystem has the function to minimize the potential of vulnerability in coastal areas. Actually mangroves are the dominant ecosystems that line the coasts of tropical coastlines around the world. The mangrove ecosystem has very important benefits. One of them is mangroves can be absorb the force of strong waves that have been built up on the open seas, especially during times of tropical storms. Additionally mangroves resist erosion which waves cause over time. But the mangrove ecosystem becomes vulnerable if the erosion intensity is too large so the wave will sweep the mangrove substrate [6].

In addition to shoreline changes characterized by erosion and abrasion, there are several other parameters that affect the vulnerability of mangrove habitats, including salinity, height and duration of tidal and substrate types.

* Corresponding author: radityarifandi@gmail.com

CVI (Coastal Vulnerability Index) is one of the methods developed for assessing the vulnerability of coastal areas caused by climate change. CVI is an index-based method and has been commonly used in assessing coastal vulnerability to shoreline changes impacts due to erosion and inundation and sea level rise impacts [7].

This paper discusses coastal vulnerability in Trimulyo Village, Genuk Sub-district, Semarang City based on Coastal Vulnerability Index (CVI) value. The CVI parameters used include coastal coastline changes, tidal inundation height and length of tidal inundation time, salinity, substrate type. The purpose of this study is to determine the grade value of each coastal vulnerability parameter and to determine the CVI value of the mangrove ecosystem of Trimulyo, Genuk Sub-district, Semarang City so it can be an urban village reference in the rehabilitation of mangrove ecosystem such as mangrove replant as well as installation of supporting tools of rehabilitation activities.

2. Method

This research used CVI method (Coastal Vulnerability Index). CVI refers to Borruf research, 2008 [8] which is used to assess the magnitude of the effect of sea level rise on coastal conditions in the Americas. The CVI method uses some data that is used as a tickling variable to measure the vulnerability conditions of a coastal region. This study uses mangrove habitat object to be studied vulnerability level by using several oceanographic variables used as supporting data to measure vulnerability level of mangrove habitat.

2.1. Vulnerability Index Variables of Mangrove Habitat

Variables used in the Mangrove Habitat Vulnerability Index analysis include four variables consisting of primary data directly put in location are coastal shoreline phenomenon, salinity and type of substrate, while the secondary data is tidal data of Semarang City July 2016 - June 2017 obtained from the station Meteorology uploaded through the www.pasanglaut.com site, then processed to obtain the average tidal inundation time of monthly and maximum height of tidal inundation.

2.2 Measurement Data Variable Index of Mangrove Habitat

The CVI data measurements at the study sites were conducted on 8 observation points divided into 3 zones, with the distribution of observation points 1; 2; and 3 located on the western side of the mangrove habitat are zone 1, point 4 and 5 included in the zone 2 located on the north side of the mangrove area or directly adjacent to the sea, and observation points 6; 7; and 8 are zone 3 located on the eastern side of mangrove habitat. The determination of the research point is based on the area considered to represent the zone and location of the study.

2.2.1 Coastal Shoreline

The method of selecting the location of observation was done by purposive sampling method. The Coastal Shoreline assessment is based on visual observations at eight points of observation by skirting the area to find out firsthand the signs of the coastal shoreline change phenomenon. With previous we know the signs of coastal shoreline change and make direct observations it will be known whether the area has accretion or sedimentation.

2.2.2 Salinity

Measurement of salinity using a refractometer and using seawater in research point as a sample. Measurement of susceptibility index of mangrove habitat of salinity variable is done by purposive sampling method that is measurement of variable which done in sampling location based on area of representation of object to be studied so that result obtained can represent from area - object area of studied mangrove habitat.

2.2.3 Tidal Inundation Time and Height Maximum of Tidal Pool

The variable of time duration and height of puddle (h) the maximum inundation is obtained from secondary data obtained from Semarang Maritime Meteorology Station by calculating the average length and height of maximum inundation each month for a year, July 2016 - June 2017, The average duration of the average tidal inundation time per month and also the maximum of tidal pool (h) at the observation site.

2.2.4 Substrate Type

Substrate sampling method is done by purposive sampling method. Measurements of substrate type variables were performed by taking the test samples at five points at the observation sites scattered in the study area and were considered to represent the location of the observation zone, point 1 and 2 of observation were in zone 1, the observation points 3 were in zone 2, then the observation points 4 and 5 are in zone 3. Sampling method in the observation location for the 5 samples of substrate that have been obtained will be done by the grand size analysis test at soil mechanic Laboratory, Faculty of Engineering, Diponegoro University, so that the content of substrate type dominates in the observation location.

2.3 CVI Analysis

Vulnerability assessment in the CVI method has a simple concept. The potential vulnerability (rank) of each variable is assessed based on its value range

 https://doi.org/10.1051/e3sconf/20183108004

according to the ranking system defined by the USGS. [9] The variable values are 5 classes (1 = very low, 2 = low, 3 = moderate, 4 = high, 5 = Very high), but in this study only use 3 class that is class 1 = low, class 2 = moderate, class 3 = height of next potential of simultaneous vulnerability (index of vulnerability) assessed by result of calculation each value of each variable through equation, as follows:

$$\mathrm{CVI} = \sqrt{\frac{a * b * c * d * e}{5}}$$

CVI = value (score) of Mangrove Habitat Frame Index, a = value of coastline change variable b = value of tidal variable, c = maximum value of maximum tidal pool variable, d = value of salinity variable, e = value of substrate type variable. Based on the scores of the equations, the distribution of vulnerability categories (indexes) along the mangrove habitats dominated by the *Avicennia marina* species assessed using the Boruff, [8] has been modified.

Through this approach, the vulnerability index of mangrove habitat is grouped according to their respective values against predetermined category classes. Based on the rule, the index category of vulnerability of mangrove habitat based on CVI scores are divided into scores; Low (0.45-2.31), medium (2.32-4.62), high (4.63-6.93).

3. Results and Discussion

Based on the calculation of the Coastal Vulnerability Index (CVI) of each parameter, so susceptibility of coast vulnerability class for each parameter is obtained, as shown in Table 1. The results of calculating the vulnerability values in mangrove habitat are shown in Figure 1. In Figure 1 it is found that a high susceptibility index is on the shoreline, then vulnerability is present in commensurate to the Babon River, whereas areas with low vulnerability exist on the eastern side of the mangrove habitat in the former pond site.

Table 1. Modification of Habitat Vulnerability Index of Mangrove Trimulyo

Variable	Bobot		
	Low (1)	**Medium (2)**	**High (3)**
Coastal shoreline	Accretion	Stable	Abrasi
Tidal Pooling Time (total/month)	10 - 19	20	<10 or >20
Tidal Maximum Pooling (m)	< 0,5	0,5 - 1	> 1
Salinity (‰)	15-30	10 – 15 ; 30-33	< 10 ; > 33
Substrate type	Silt and Clay	Sand	Rock and Gravel

Low to High vulnerable class in Trimulyo Mangrove Habitat with area of ± 30 Ha as shown in Fig 1. Based on the calculation of CVI value, it is found that the parameters that highly affect the value of coastal vulnerability in Trimulyo mangrove habitat is the parameter of shoreline change and tidal factor.

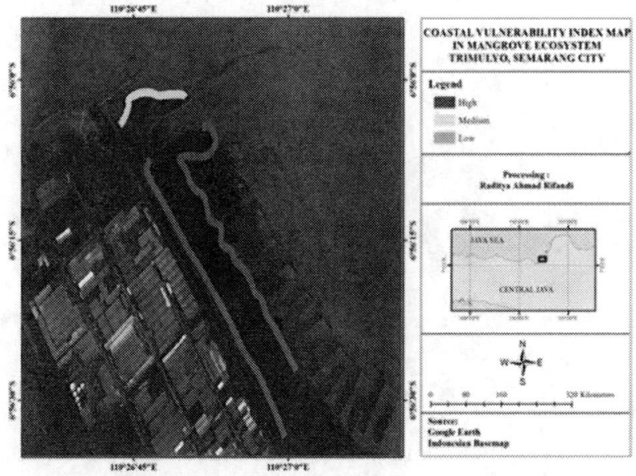

Fig. 1. Vulnerability Index Map In Mangrove Habitat Trimulyo, Semarang.

3.1 Coastal shoreline

Coastal shorelines are always subjected to changes due to coastal processes, which are controlled by wave characteristics and the resultant near-shore circulation, sediment characteristics, beach form, etc.

Fig. 2. Damage to mangrove plants due to abrasion

An abrasion occurring in the zone 2 is marked by the erosion of the substrate and mass mortality of mangrove plants along the coastline of Trimulyo (fig 2). Then for zone 1 has a static tendency characterized by the condition of mangrove stands and sediments that are still standing firm and the absence of marked substrate is covered. While in zone 3 happened accretion marked with silting area of former pond and several time this often used as location of planting program (fig 3). Coastal areas that has additional land due to sedimentation are considered to have a tendency to be less vulnerable areas) [10], this is due to the growing substrate that became the medium of growing seedling mangrove.

Mangroves have effective function to maintaining the sediment-carrying carried by the river, the overall coastal wetlands should be preserved [11]. This function is very important in estuaries adjacent to areas with high sediment loads such as the Babon River flow to the mangrove ecosystem is able to withstand sediment and accumulate in zone 3.

Fig. 3. Sedimentation on the side of mangrove habitat

3.2 Tidal Inundation Time and Maximum Height of Tidal Inundation

Based on observations of tidal data in July 2016 - June 2017 can be obtained from the Meteorological Station which is downloaded in website www.pasanglaut.com. So get the type of tidal Semarang City included in the tidal type of mixed inclines to a single daily, meaning that the tidal events in the city of Semarang on average in one day happened one time the tide and one time receded, but sometimes also happened two times sea tides up and two times of tides down, with tidal's high and very different periods recorded every month. Tidal type of Semarang City in July 2016 - June 2017 are presented in Table 2.

Table 2. Type of Tidal in Semarang City

Month	Type of Tidal	Value
July	Mixed semidiurnal tend to diurnal	3
Agustus	Mixed semidiurnal tend to diurnal	3
Sepetember	Mixed semidiurnal tend to diurnal	3
October	Mixed semidiurnal tend to diurnal	3
November	Mixed semidiurnal tend to diurnal	3
December	Mixed semidiurnal tend to diurnal	3
January	Mixed semidiurnal tend to diurnal	3
February	Mixed semidiurnal tend to diurnal	3
March	Mixed semidiurnal tend to diurnal	3
April	Mixed semidiurnal tend to diurnal	3
May	Mixed semidiurnal tend to diurnal	3
June	Mixed semidiurnal tend to diurnal	3
Mean	Mixed semidiurnal tend to diurnal	3

Source: www.pasanglaut.com

Average of maximum high tide of Semarang in 2016 - 2017 is 0,87 m. The data is obtained after calculating high tidal average in Semarang City every month in July 2016 - June 2017. The monthly maximum high tide in Semarang City can be seen in Table 3.

Table 3. Tidal level in Semarang City

Month	High of Tide (m)	Value
July	0,9	2
Agustus	0,9	2
Sepetember	0,9	2
Oktober	0,8	2
November	0,9	2
December	0,9	2
January	0,9	2
February	0,9	2
March	0,8	2
April	0,8	2
May	0,9	2
June	0,9	2
Mean	0,87	2

Source : www.pasanglaut.com

Tidal type Semarang City is tend to diurnal, that mean to tidal in Semarang included in the observation location there are one up tide and one down tide in a day. That's mean to the tidal frequency vulnerable variables in the mangrove habitat in Trimulyo include in high, because the optimum condition of mangrove habitat is generally inundated of tidal about 10 – 19 days a month.

Effect of high time frequency of inundated of mangrove growth due to tidal regulate the availability of brackish water so that influence salinity level at mangrove habitat, root system Long time inundation also affect mangrove rooting system as at pairs and roots completely submerged then reduce the supply of oxygen which will be absorbed by root, Then when the inundation time will make the mangrove seedling difficulty obtaining oxygen and disturbing the transfer of nutrients into the mangrove habitat [12] [13].

Furthermore, the maximum tide height variables affect the survival of mangrove because it will impact on the role of mangrove root for nutrient absorbent on the substrate then high water at soaking tidal should also be considered [12]. In mangrove habitat Trimulyo, the maximum water level at tidal is 0,87 m, value of this indicates that maximum tidal height variable of CVI in Trimulyo mangrove habitat is two, that mean vulnerable level is medium.

3.3 Salinity

Salinity is one of variables which affected the condition of mangrove plants. Many factors affected the value of salinity in a water area. Affecting of the salinity value in a watershed area such as the openness of the land or area, the area, and the tidal activity and the effect of sea water intrusion in the area.

The salinity values obtained at direct measurements at 8 observation points using a refractometer showed that the range of salinity values obtained at the observation point was the difference between several observation points. In zone 2 has a salinity value of 30 ‰ and 31 ‰. While the results obtained at observation points located in zones 1 and 3 have salinity value 28 ‰ - 30 ‰. With the result as follows so that salinity value in zone 2 in vulnerability index of mangrove habitat is included in moderate category with value of salinity variable is two, whereas at zone 1 and 3 have low susceptibility with weight of vulnerability are one.

Specifically mangrove is special plant of halophyta which has a high water-retaining capacity of the salty substrate. This plant is able to maintain water balance well due to the mechanism of regulation of salt in the form of morphology adaptation, such as stomata behavior, osmotic settlement, level of salinity, and salt removal [14]. However, although mangroves have high salinity tolerance, the less optimal salinity is either too high or too low, it tends to disrupt the growth of mangrove plants. With the disruption of mangrove growth resulting in vulnerability in mangrove habitat.

3.4 Substrate

Observation of substrate type in the location of the research conducted by the substrate content analysis on mangrove habitat through grand size analysis The results obtained are presented in Table 4.

Table 4. Substrate Type in Observation Location

Station	Substrate (%)				Value
	Gravel	Sand	Silt	Clay	
1	0,00	2,80	36,49	60,71	1
2	0,00	3,20	41,94	54,86	1
3	19,40	51,75	26,14	2,71	2
4	0,00	10,40	41,97	47,63	1
5	0,00	9,86	50,8	39,34	1

Source : Data Research (2017)

Type of substrate obtained at the research site from the laboratory test results was obtained in five test samples contained various types of substrate, the sample taken in zone 2 contain sand up to 51.75%, the value indicates the sand content is more dominant than silt and clay, so vulnerability index in this zone included in medium with the value on the variable type of substrate is two. While the mean for four observation points in zone 1 and 3 more dominated by silt and clay with mean value at zone 1 silt 48.6% and clay 57.78% and mean in zone 3 silt 46.38% and 43.48%, So that the Coastal Vulnerability Index belongs to a low vulnerability category with a value on substrate type variables at several observation points that zone 1 and 3 are one.

In the Amin *et al*, study, [15] mangroves averages were fewer in a habitat with harder substrates and more gravel textures than mud. Several types of mangroves can grow well on muddy soil substrate to tolerate muddy-sandy substrate but which is more dominated by silt [16].

Therefore, the type of substrate dominated by silt, that's mean to vulnerability index value of this variable is low and the vulnerability index value will be higher if the substrate has more rough and hard characteristic.

3.5 Coastal Vulnerability Index Value

Determination of Coastal Vulnerability Index in this study is based on the calculation of variables affecting vulnerability of mangrove habitat such as tidal frequency, maximum inundation height, salinity value, and substrate type. Based on the measurements and observations on coastal shoreline (CS) variables, tidal frequency (TF), maximum height of tidal inundation (h), salinity (S), and substrate type (ST), the values of Coastal Vulnerability Index (CVI) Table 5.

Table 5. CVI Value in Observation Location

Zone	CS	TF	h	S	ST	CVI
1	2	3	2	1	1	1,54
2	3	3	2	2	2	3,79
3	1	3	2	1	1	1,09

Fig 4 shows that the difference of vulnerable rates of some Trimulyo mangrove habitat zones, in zone 1 and 3 are in the low vulnerability category because the CVI values are within the range 0.45 – 2.31, while in zone 2 the vulnerability category being in the range of 2.32 – 4.62, the value is differentiated based on the vulnerability index classification of mangrove habitat in Table 6.

Table 6. Classification Index of Coastal Vulnerability Index

CVI Value	0,45 – 2,31	2,32 – 4,62	4,63 – 6,95
Vulnerable Rate	Low	Medium	High

Vulnerability index values for each variable considered to affect vulnerability in mangrove habitat can be used as reference of management of efforts to improve and rehabilitate mangrove habitat such as for tidal variable in Trimulyo mangrove habitat including into high category, so that tidal incident in that area can make mangrove habitat condition at the research area has a risk of damage especially for young mangrove plants, in addition to other variable tides such as salinity and substrate types also affect vulnerability although the range of vulnerability is still in the low and medium category.

Zone with vulnerability index value that has more susceptibility in zone 2, one of the most influential variable is the abrasion resulting mangrove ecosystem damaged on the side to the coastline because the substrate to which the mangrove swept by wave. In addition in this zone also has a higher salinity value because when the tide occurs the of sea water is more dominant.

4. Conclusion

The conclusion obtained from this research is there is different vulnerability index value of mangrove habitat value in Trimulyo, in zone 1 and 3 with value 1.73 and 1.22 located at west side of mangrove habitat in Babon river side and east side of mangrove habitat which is an area of the pond in is belonging to the low vulnerability category. While in zone 2 located on the north side of mangrove habitat which is the coastline has a value of 3.79, then coastal vulnerability index category in this zone belongs to the moderate vulnerability category.

References

1. Hadi, S.P. *In Search for Sustainable Coastal Management: A Case Study of Semarang, Indonesia.* 2nd International Conference on Tropical and Coastal Region Eco Development (2016)
2. Miladan N. Communities' *Contribution to Urban Resilience Process: a Case Study of Semarang City Toward Coastal Hydrological Risk.* Unpublished PhD Dissertation. Universite Paris-Est, French (2016)
3. M.A. Marfai, H. Almohammad, S. Dey, B. Susanto, L. King, Environ Monit Assess **142**:297–308 (2008)
4. M.A. Marfai, Malaysian Journal (2011)
5. M.A. Marfai, and D.R. Hizbaron, 2011. *Analele UniversităŃii din Oradea – Seria Geografie.* **21**:209–221 (2011)
6. B.M. Wolf, NRES **323**. University of Wisconsin-Stevens Point (2012)
7. E. Ramieri, A. Hartley, A. Barbanti, F. D. Santos, P. Laihonen, N. Marinova, dan M. Santini, *Methods for Assessing Coastal Vulnerability to Climate Change. ETC CCA Background Paper.* European Environment Agency, Copenhagen. Denmark (2011)
8. B.J. Boruff, C. Emrich, S.L. Cutter, Journal of Coastal Research. **21** (5). 932-942 (2005)
9. USGS. 1999. *National Assessment of Coastal Vulnerability to Sea-Level Rise: Preliminary Results for the U.S. Atlantic Coast.* Unites States Geological Survey (USGS). http://pubs.usgs.gov/of/1999/of99-593/pages/data.html. 9 June 2017.
10. T.S. Kumar, R.S. Mahendra, S. Nayak, K. Radhakrishnan, K.C. Sahu, Journal of Coastal Research, **26**(3), 523–534. West Palm Beach (Florida), ISSN 0749-0208 (2010)
11. M.F. Adame, D. Neil, S.F. Wright, C.E. Lovelock, Elsevier Ltd .Estuarine, Coastal and Shelf Science **86**. 21–30 (2010)
12. A. Wahyudi, B. Hendrarto, A. Hartoko, Diponegoro Journal of Maquares, **3**(1), 98-98 (2014)
13. A.R. Zaki, C.A. Suryono, R. Pribadi, Journal of Marine Research **1** (2). 88–97. Diponegoro University. Semarang Indonesia (2012)
14. R.A. Balsamo, and W.W. Thomson, American J. of Bot., **82** (4) : 435-440 (1995)
15. D.N. Amin, H. Irawan, A. Zulfikar, Density of *Rhizopora* sp. Vegetation in Mangrove Forest Nyirih River Tanjungpinang Sub-district, Tanjungpinang City (2015)
16. M. Suriani, *Land Quality and Growth of Rhizophora mucronata in Mangrove Rehabilitation Area in Aceh Besar dan Banda Aceh.* Marine Science, Fisheries and Marine Science Faculty, Syiah Kuala Darussalam Darussalam University-Banda Aceh. Indonesia (2013)

Eco-efficiency Analysis of Furniture Product Using Life Cycle Assessment

Dyah Ika Rinawati [1], *Sriyanto*[2], *Diana* Puspita Sari [3], and *Andana* Cantya Prayodha [4]

[1,2,3,4]Department of Industrial Engineering, Faculty of Engineering, Diponegoro University, Semarang - Indonesia

Abstract. Furniture is one of Indonesia's main commodities strategically role in economic growth and employment in Indonesia. In their production process there many wastes resulted, such as such as sawdust, cuttings - pieces of wood, components that do not conform to specifications and the edges of wood from a log. Contrast with requirement of timber for furniture industries, availability of raw material sources decrease because of limited forest areas. Beside that, using electricity and chemical material in furniture production process have impact to environment. This study aim to assess the eco-cost and eco-efficiency ratio of the product so strategic recommendations to improve the eco-efficiency of products can be designed. The results of data processing showed the environmental costs of the furniture production process amount Rp 30.887.84. Eco-efficiency index of furniture products studied was 4,79 with the eco-efficiency ratio of 79,12%. This result means that the measured furniture products already profitable and sustainable, as well as its production process is already fairly efficient. However, improved performance of the production process can still be done to improve the eco-efficiency by minimizing the use of raw materials.

1 Background

Furniture industry has strategic role and significant contribution to economic growth for Indonesia. Based on data from the Ministry of Industry Indonesian, wood furniture export value continued to increase in recent years. In 2012 the export value of wooden furniture Indonesia reached US $ 1.6 billion, later increased to US $ 1.7 billion in 2013 and then in 2014 increased to USD 1.9 billion. However the industry strategically and contribute to the economic development, it also contributed to the pollution of the environment and ecosystems. Based on a report from the United Nations Development Program (UNDP) in 2008 stated that in the forestry sector, carbon emissions released as a result of deforestation reached 80 percent, while 20 percent are caused by degradation. This triggered destruction of forests and environmental damage to the ecosystem [1].

The furniture industry that used wood raw material produces several types of Non-Product Output (NPO) or waste. Waste generated from each stage in the form of solid waste such as sawdust, wood chips, small wood pieces, and waste water which causes water pollution such as residual solvent finishing, resin particles and adhesive residual. Wastes in form of sawdust, wood chips and small wood pieces almost reached 30% in Jepara SME's [2]. This waste will have a serious impact on the environment, health and safety in the absence of a strong commitment to make change for sustainable development, clean production and waste reduction [3]. Measuring of sustainability has been developed by [4] known as the eco-efficiency index (EEI) that finding the ratio between profit and production costs added by environmental impact. Eco-efficiency is a strategy that combines the concepts of economic and ecological efficiency in the use of natural resources. Simply put eco-efficiency can be understood as a strategy to produce a product with better performance, using less energy and natural resources. Eco-efficiency can be regarded as a business strategy that has more value because of less use of natural resources and reducing the amount of waste and environmental pollution [5]

Appraisal of product eco-efficiency based on production and environmental costs (eco-cost). Production cost is obtained from the cost of goods manufactured data, while the environmental costs previously obtained by measuring the environmental impact of the production process. Impact measurements of the use of wooden material can be done by using the Life Cycle Assessment (LCA). LCA measurement is intended to calculate the cost of the environment (eco-cost) of the use of material making up the product and then calculated the eco-efficiency level of products and propose recommendations to improve the eco-efficiency of products [5].

In order to reach sustainable production in the furniture industry, it is needed a measurement of the eco-efficiency index as a baseline for making improvement. This study aim to measure the environmental impact and value of environmental costs or eco-costs of the furniture production process and analyze eco-efficiency index and eco-efficiency ratio of furniture products.

* Corresponding author: dyah.ika@ft.undip.ac.id

2 Research methodology

Research methodology of this study consist of Life Cycle Assessment (LCA) and eco-efficiency measurement.

1.1 Stage of Life Cycle Assessment (LCA)

Life Cycle Assessment (LCA) evaluates holistically the environmental consequences of a product system or activity, by quantifying the energy and materials used, the wastes released to the environment, and assessing the environmental impacts of those energy, materials and wastes [6]. Stages of LCA consist of four phases, first, it is goal and scope phase, then followed by life cycle inventory, the third is life cycle impact assessment and interpretation phase is the final stage.

Life Cycle Inventory (LCI) is second stage in LCA and consists of activities related to the search, collection and interpretation of the data requirements for the measurement of environmental impacts. Gate-to-gate LCI information being used in this study. Chair product has been observed because this product are the highest volume of export product. This product exported in form of unfinishing condition. Detailed of LCI given in Fig. 1. It can be seen that process production of the observed product conducted in three areas, there were Rough Mill, Smooth Mill and Sanding.

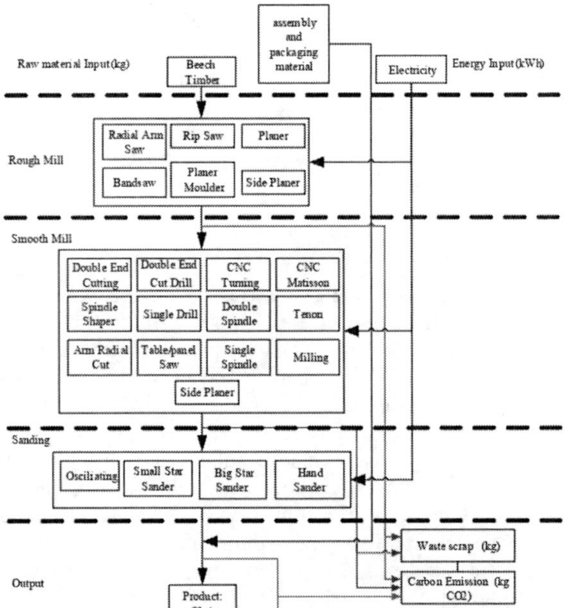

Fig.1. Life cycle inventory of chair product

Assembly, disassembly and packing were assumed include in Sanding Department because no physically product changing. Upholstery and finishing processes were not conducted in the observed product. Input of chair product was Beech timber, assembly and material packaging and electricity. Output of the process consists of product output and non product output (NPO). NPO consist of sawdust, wood chips and carbon emission. Measurement of energy consumption is done by logging all the machines used in producing bye product in the boundary of production system under study. Then the power on each machine (kw) is identified and multiplied by the time of the product (hour) by the machine so that the machine energy consumption is obtained to produce the product in kwh.

In Life Cycle Impact Assessment (LCIA), types and amounts of environmental impact calculated using software Simapro Ecoinvent database v 7.1 and 2.0 as well as the impact assessment method Eco costs 2012 v 3.03. In the Eco costs method, LCIA calculation consists of several stages, among others, characterization, normalization and a single score from the resulting environmental impacts. The results of this indicator category is obtained by assigning the results of the Life Cycle Inventory for emissions to soil, water and air of the substances that are used to group representing environmental issues so-called impact category. Normalization step in the Eco costs 2012 method is calculation in the preventive marginal cost of each impact category obtained by multiplying the result by a normalization factor indicator category which is the value of eco cost of each unit of emissions. For single stage score, based on the marginal cost of preventive obtained from normalization stage then summed from each category to a total eco-impact cost. Eco-cost given in Euro that the exchange rate assumed as Rp 14.036,93/euro based on rate in 22 December 2016.

Life cycle interpretation is the last phase of the LCA process. Life cycle interpretation is systematic techniques to identify, measure, examine, and evaluate information from LCI and LCIA results.

1.2 Eco-efficiency measurement

Eco-efficiency measurement consist of four phases. There are cost-benefit analysis, eco-efficiency Index (EEI), eco-cost value Ratio (EVR) and eco-efficiency ratio (EER) Rate calculation.
Cost-benefit analysis was conducted to determine the value of the net value of the product. Net value is obtained by reducing the benefits the company obtains in the form of product selling price with production cost represented by cost of goods manufactured. Costs calculated in the determination of cost of goods manufactured include material costs that include raw materials in the form of wood and supporting materials in the form of material assembly and packaging, labor costs either directly or indirectly, the cost of electrical energy, and the cost of overhead companies. Price of the product were US$40 or Rp 538.319 (based on exchange rate in 22 December 2016, Rp 13.457,99/US$). Cost of good manufacture was Rp 393.929, so net value of the product were Rp 144.390.
The calculation of the eco-efficiency index (EEI) is conducted to assess the feasibility of a product, whether

the product is economically efficient (profitable) and ecologically efficient (sustainable). Formula for measuring EEI is net value divided by eco-cost [7]. If EEI more than 1, so it called profitable and sustainable. If EEI are in range 0-1, it called profitable but unsustainable. While EEI less than 0, it is unprofitable and not sustainable.

The calculation of eco-cost per value ratio is obtained by comparing the amount of eco-cost incurred by the value of the furniture product. The output of the EVR calculation is the ratio of the two values to the inputs in this calculation.

EER Rate is the final calculation of the eco-efficiency measurement of the production process of furniture products. The EER Rate calculation is obtained by reducing the product's net value with the eco-cost of the production process and then the reduction result is subdivided by the net value of the product.

3 Results

Based on data processing, the results are described in tables 1., table 2 and table 3. Table 1. give information about input and output of life cycle inventory for the product being studied.

Table 1. Input & Output of Product Life Cycle Inventory

Type	Classification	Volume		Unit
Input	Material	37.225	Kg	Timber
	Energy	10.20	kWh	electricity
Output	Product Output	21.036	Kg	Chair
	Non Product Output	16.189	Kg	Waste
		Output of LCIA		Emission

Table 2. Characterization of Output of Chair Product

Impact category	Unit	Rough Mill	Smooth Mill	Sanding
climate change	kg CO2 eq	5.007862	3,146243	0.,859937
Acidification	kg SO2 eq	0.027109	0.,009499	0.,000981
Eutrophication	kg PO4-- eq	0.004970	0.001863	0.,000239
photochemical oxidant formation	kg C2H4 eq	0.001534	0.,000455	0.000015
fine dust	kg PM2.5 eq	0.005243	0.,001531	0.000041
Human toxicity	CTUh	0.000000263	0.000000077	0.000000002
Ecotoxicity	CTUe	3.955764	1.,155359	0.030606
Metals Depletion	Euro	0	0	0
Oil&Gas Depletion excl energy	kg oil equ	0	0	0
Waste	MJ	0	0	0
land-use	Euro	0	0	0
Water Stress Ind	WSI factor	0.,036176	0.,010566	0.000280

Table 3. Eco-cost of Chair Product (in Euro)

Impact category	Rough Mill	Smooth Mill	Sanding	Total
climate change	0,676061	0,424743	0,116092	1,216896
Acidification	0,223650	0,078365	0,008093	0,310108
Eutrophication	0,019383	0,007264	0,000932	0,027579
photochemical oxidant formation	0,014877	0,004416	0,000150	0,019443
fine dust	0,178268	0,052067	0,001379	0,231714
Human toxicity	0,270554	0,079021	0,002093	0,351669
Ecotoxicity	0,005637	0,001646	0,000044	0,007327
Metals Depletion	0	0	0	0,000000
Oil-Gas Depletion excl energy	0	0	0	0,000000
Waste	0	0	0	0,000000
land-use	0	0	0	0,000000
Water Stress Indicator	0,027493	0,008030	0,000213	0,035736
Sum				**2,200472**

Table 2. show characterization output of chair product and Table 3. Give information that the total of eco-cost are 2,200472 euro or equal to Rp 30.888. So, calculated EEI are 4,67. Because the score more than 1, the product are profitable and sustainable.

$$EEI = \frac{Rp\ 538.319 - Rp\ 393.929}{Rp\ 30.888} = 4{,}67 \quad (1)$$

Table 4. Comparison between cost, EVR and EER of The Product

Benefit (Rupiah)	Prod Cost (Rupiah)	Net value (Rupiah)	Eco-cost (Rupiah)	EVR	EER
538.319	393.929	144.390	30.888	0,21	78,6 %

Table 4. Show comparison between cost, EVR and EER. Eco-cost per value ratio (EVR) or eco-indicator is a comparative indicator that takes into account the economic and ecological aspects of the seat product. Ecological aspect factor is calculated as net value while ecological aspect is expressed in eco-cost. EVR value is obtained from eco-cost division as ecological value, while net value of product as economic value. The larger the net value of a product, the smaller the EVR value of the product where it indicates that the environmental costs that need to be removed from product production are only a small part of the net value of the product. The smaller EVR value of the product means better and worthy for product to be produced. The EER value represents the value obtained from net product value reduction with eco-cost of the product and then result of the deduction is divided by net value of the product itself. So far the company has not considered eco-cost as part of product cost. Through this measurement, net-value value reduced by eco-cost indicates that eco-cost is also a cost factor calculated in product value.

4 Discussion

Table 1 and Table 2 illustrate that highest waste and carbon emission resulted by Smooth Mill department. In Smooth Mill were processed formations of chair components into more detail so that the physical changes of product components are also getting bigger. The chair production process using Beech wood as raw materials and electricity generated from burning coal or crude oil that has an impact on the environment. Environmental measurement endpoints measured using Simapro Software are shown in eco-cost. Eco-cost is cost that is required to address the emerging environmental impacts, but the environmental costs are not mandatory for the company. Since there is no subjective factor in eco-cost measurement, the level of significance and urgency of the impact is seen from the value of the eco-cost impact.

Climate change measured in the production process is 9.014 Kg CO2 (carbon dioxide) equivalent, which means use of resources to produce 1 unit of seats in the form of 0,0517 m3 or 37,225 kg of wood and the use of electric power of 13.252 kWh contributes to the generation of 9.014 kg of carbon dioxide that affects climate change. Carbon dioxide in this study arises from the felling of trees as raw materials for production because the tree plays a role in converting carbon dioxide gas into oxygen through photosynthesis.

Acidification measured in the seat production process is 0.0376 Kg SO2 (sulfur dioxide) equivalent, which means the use of resources to produce 1 unit of seats in the form of 0.0517 m3 or 37.225 kg of wood and the use of electric power of 13.252 kWh contributes to the generation of 0, 0376 Kg of sulfur dioxide gas. This compound is a toxic gas with a pungent odor released by some industrial processing such as coal combustion and petroleum also contains sulfur compounds. Sulfur dioxide is produced from burning coal and petroleum in electricity production [8].

Eutrophication measured in a chair production process of 0.0071 Kg PO4 (phosphate) equivalent, which means the use of resources to produce 1 unit of Brant Chair chairs in the form of 0.0517 m3 or 37.225 kg of wood contributes to 0.0071 kg of phosphate. Phosphate is produced by the use of wood because the tree plays a role in binding phosphate substances.

Photochemical Oxidant Formation measured in the process of production of 0.002 Kg C2H4 (ethylene) equivalent, which means the use of resources to produce one unit of erupa seat 0.0517 m3 or 37.225 kg of wood and electric power consumption of 13.252 kWh contribute to the generation of 0.002 Kg of ethylene substances. Photochemical Oxidant Formation is a forming phenomenon through chemical reactions due to exposure to sunlight. Photochemical oxidants or more defined as smog are the result of reactions that occur due to volatile organic compounds (VOCs) exposed to UV radiation. One type of volatile organic compound is ethylene gas. The substance formed here is the ozone layer which at the right altitude of the atmosphere is the protector of the Earth from the sun's heat but ozone in this case is formed at low atmospheric levels making it harmful to living things [9]. The contribution of the production process to this environmental impact is due to the combustion process in the production of electricity and the use of wood as fuel because the tree plays a role in airborne toxic substances in the air.

Fine Dust in the process of production of wooden chair reaches 0.0068 Kg PM 2.5 equivalent, which means waste in the form of wood pulp contribute to 0.0088 kg of measured dust particles. Wood dust particles are very commonly released into the air because of their very micro size so often unhandled. These dust particles are air pollution because it is very disturbing breathing and can result in various respiratory problems.

Human Toxicity measured in a chair production process of 3.41 x 10-7 Kg Benzo (a) pyrene equivalent which means the use of resources to produce 1 unit of chair in the form of electricity power consumption of 13.252 kWh contributes to 3.41 x 10-7 Kg Benzo (a) pyrene. Benzo(a)pyrene is a compound resulting from incomplete combustion of coal and organic material. In this study the Benzo compounds (a) pyrene arise from the combustion process for electricity production as production energy [10]. Benzo(a)pyrene is a toxic compound that if exposed to living things for a long time can cause damage to the nervous system, decreased immune system, damage to the reproductive system and even potentially cause cancer cells [11].

Ecotoxicity measured by eco-cost method in a chair production process of 5.14 Kg Zn (Zinc) equivalent which means the use of resources to produce 1 unit of ch in the form of electric power consumption of 13.252 kWh gives an impact equivalent to 9.014 kg of zinc. Ecotoxicity is an environmental impact that refers to the potential of substances in polluting the aquatic ecosystem. The pollutant substance measured by Simapro Software with this eco-cost method is a zinc substance. The process of extracting natural oils generally leads to zinc content which is then released into the aquatic ecosystem. Zinc is a type of heavy metal that is very dangerous if it goes into the human metabolism system for example through the processing of water containing zinc into drinking water. Excessive zinc content in human metabolic system can lead to malfunction and maintenance of skin, pancreas and male reproductive organs. In addition, the zinc content can also lead to deficiency of other minerals in the body such as iron is very important in the process of blood formation [12].

Water-stress indicator measured in the production process Chair of 0.047 WSI factor which means the use of resources to produce 1 unit Chair of 0.0517 m3 or 37.225 kg of wood that comes from trees and the use of electric power of 13.252 kWh contribute to the increase in water saturation .

Climate Changed has the greatest environmental impact value that caused by the use of wood raw materials. In order to reduce the impact of Climate Change, it is recommended to decrease the use of wood as raw material for production. The difference in length of the coarse size and average final size for each component is 20 mm. One of the components of the arm post even has a gap size difference and the final size of 60 mm.

Reduction of wood material usage can be done by cutting the length of the crude component by 10 mm. If this is done then it can save the use of wood materials by 0.000421 m3. At a glance, the aforementioned savings value is small but this will be significant if it takes into account large-scale production. The size of the production lot of wooden chairs used as the measurement object of this research is 700 units. If the size reduction is done on this production lot, the saving of wood material can reach 0.2946 m3 where the savings value is equal to the wood material needed to produce 10 units of chair.

5 Conclusion

Based on the result of gate-to-gate life cycle inventory of wooden chair product, it can be conclude that the product was profitable and sustainable because the EEI is more than 1. However, the product has climate change impact that can be reduced by decrease the tolerance of crude component. This scope of the study can be enlarged to be gate-to-cradle or cradle-to-cradle approach and also for local market product, so it can describe the eco-efficiency of furniture industry more clearly.

References

1. C. P. P, Purba, S. G. Nanggara, M. Ratriyono, I. Apriani, L. Rosalina, N. A. Sari, A. H. Meridian, Potret Keadaan Hutan Indonesia. Bogor: Forest Watch Indonesia. (2014)
2. M. Campaka, S. Hartini, D. I. Rinawati, *Perancangan Pemanfaatan Limbah pada Industri Mebel untuk Menuju Produksi Berkelanjutan.* eprints.undip.ac.id/32981/1/L2H_006_046.pdf. (2010)
3. ----. *An environmental management guide for wood-based furniture industry*. Philippines: The Development Bank of the Philippines. (2004)
4. J. G. Vogtlander. *International Journal of Ecodynamics*, **1** (2), 136-148, (2006)
5. D. P. Sari, S. Hartini, D. I. Rinawati, T. S. Wicaksono, *Jurnal Teknik Industri*. **14** (2), 137-144
6. C. J. Gonzalez, S. Kim, M. R. *The International Journal of Life Cycle Assessment*, **5** (3) 153–159 (2000)
7. T. Hur, S. T. Lim dan H. J. Lee . *A Study on The Eco-efficiencies for Recycling Methods of Plastics Wastes*. Seoul: Departement of Material Che-mistry and Engineering Konkuk University. (2003)
8. H. Müller, in *Ullmann's Encyclopedia of Industrial Chemistry*. Weinheim: Wiley-VCH. (2005)
9. G. T. Miller, *Living in the Environment: Principles, Connections, and Solutions* (12th Edition). Belmont: The Thomson Corporation. (2002)
10. B. M. Lee, G. A. Shim, J Toxicol Environ Health A. **70** (15–16), 1391–1394 (2007)
11. M. M. McCallister, M. Maguire, A. Ramesh, Q. Aimin, S. Liu, H. Khoshbouei, M. Aschner, F. F. Ebner, and D. B. Hood, *NeuroToxicology*. **29** (5), 846–854 (2008)
12. A. S. Wernersson, Aquatic Ecosystem Health & Management. **7** (1), 127–136. (2004)

Evaluation Of Water Quality At River Bian In Merauke Papua

Irba Djaja [1,*], *P.* Purwanto[2] and *H. R.* Sunoko[3]

[1]Doctoral Program of Environmental Science, School of Postgraduate Studies, Diponegoro University, Semarang Indonesia
[2]Department of Chemical Engineering, Faculty of Engineering, Diponegoro University, Semarang Indonesia
[3]Department of pharmacy, Faculty of Medical, Diponegoro University, Semarang Indonesia

Abstract. River Bian in Merauke Regency has been utilized by local people in Papua (the Marind) who live along the river for fulfilling their daily needs, such as shower, cloth and dish washing, and even defecation, waste disposal, including domestic waste, as well as for ceremonial activities related to the locally traditional culture. Change in land use for other necessities and domestic activities of the local people have mounted pressures on the status of the River Bian, thus decreasing the quality of the river. This study had objectives to find out and to analyze river water quality and water quality status of the River Bian, and its compliance with water quality standards for ideal use. The study determined sample point by a purposive sampling method, taking the water samples with a grab method. The analysis of the water quality was performed by standard and pollution index methods. The study revealed that the water quality of River Bian, concerning BOD, at the station 3 had exceeded quality threshold. COD parameter for all stations had exceeded the quality threshold for class III. At three stations, there was a decreasing value due to increasing PI, as found at the stations 1, 2, and 3. In other words, River Bian had been lightly contaminated.

1 Introduction

Use River Bian is one of the rivers situated in Merauke Regency, Papua Province, which has been included within the Bikuma River Stream (DAS Bikuma). The River Stream consists of Bian, Kumbe and Maro, covering a total area of approximately 9,493.80 km^2 [1]. The change in land use by a deforestation nearby DAS Bian for investment activities, such as plantation and industrial forest, has mounted a pressure on the River Bian. This change in land use has led to the decrease of the river water quality. The use of the river water for vary necessities should have been done sustainably. River is a very important ecosystem for human. It also provides water for many activities such as agriculture, industry, and domestic needs. Therefore, river resources must be protected in order to be used with a good practice by the human as well as other living creatures. River Bian plays a significant role for the Papuans, in particular the Marind, who live along the river. It has become a remarkable source for food materials (in forms of fish, shrimp, crab, etc., transportation media connecting between villages, and a place of traditional and cultural ceremonies for the local people [1].

Increasing domestic, agricultural, and industrial activities may affect the water quality of the river. In particular, domestic activities contribute to the largest BOD concentration to the river body (Minister of Environment Decree 115/2003) [2]. A river is considered having a decrease in water quality when it is no longer used according to a normal water quality status. The water quality status is the degree of water quality that proves a contaminated or non-contaminated conditions in a particular water source in a particular time comparable to the predetermined water quality threshold.

The determination of the water quality status is done, among others, by applying a Pollution Index Method. The Pollution Index is used for determining a pollution rate relative to a permissible water quality parameter. The Pollution Index (PI) is determined for a particular purpose, and then developed for some other purposes of the whole parts of the water body or parts of a river [2]. This study had purposes to analyze the river water quality and to find out water quality status of the River Bian before the change in land use from forestry to industrial (HTI) and its compliance with the water quality threshold as required.

2 Research method

This study applied a case study design by collecting primary data from the field. Information about water quality was obtained by collecting samples of water quality directly from the field at the observation points determined in each period from 2007 to 2015. The water samples at the River Bian (DAS Bikuma) in Merauke, Papua, were taken from three observation points namely Bian upstream, Bian middle stream, and Bian

* Corresponding author: plhdjaja@gmail.com

 https://doi.org/10.1051/e3sconf/20183108006

downstream, all situated in the area of the Industrial Forest (HTI).

The study used five broad categories of instruments, as follows: 1) pH meter and thermometer to calculate the pH and the temperature on locations; 2) media for samples; 3) icebox to preserve the samples; 4) Global Positioning System (GPS) to determine coordinate points of the sampling location (latitude, longitude, and elevation) on the map; and 5) writing devices to record the observation results.

The sampling points were determined by a purposive sampling technique, considering accessibility, cost, and time consumed during the research. The three sampling points of River Bian consisted of as the followings (Figure 1):

Station 1 : Bian Downstream (07054'45.6'' S – 140007'23,9" N)

Station 2 : Bian Middle stream (07055'43.1''S – 140007'50" N)

Station 3 : Bian Upstream (07054'45.6''S – 140003'42,4" N)

Sampling at the river water was performed by a grab sample technique. It was a direct sampling technique from the water body under the observation and only informed the water characteristics at the time the samples were taken [3]. Once the water sampling had been determined at each station, the samples collected were then subject to a laboratory test according to the predetermined standard. Treatments of the water samples consisted of labeling on each samples media), samples preservation (cooling or addition of chemical substances), and samples transportation (from the sampling stations to the laboratory). The samples preservation was intended to prevent them from any physical and chemical changes.

The river water quality test in this study was intended for pH and temperature parameters. The test was performed directly on the field using pH meter and thermometer. In case of TSS, DO, COD, BOD, phosphate, and nitrate parameters, the study performed a test at the Water Quality Laboratory Bandung. The water quality test was performed using a method pursuant to the ongoing standards. Table 1 contains the water quality testing method used in the current study.

The data collected from the physical and chemical parameters test of the river water, either on the field or in the laboratory, were then subject to a water quality analysis for the River Bian. The analysis was compared the test results to the class III water quality threshold pursuant to the Government Act 82/2001 on Water Quality Management and Water Pollution Control.

The water quality status was determined by using a Storet model with the following steps:

1. Water quality data collection in a periodical manner as such that it forms time series data.
2. Calculated data from each water parameter comparable to quality threshold by the water class.

The determination of the river water quality status in the current study applied a Pollution Index method [4] and used the following Pollution Index equation:

$$PI_j = \sqrt{\frac{(C_i/L_{ij})^2_M + (C_i/L_{ij})^2_R}{2}}$$

where :

Ij : Concentration of water quality parameter recorded in the quality threshold for water (j)

Ci : Concentration of water quality parameter (i)

PI : Pollution Index for purpose (j)

(Ci/Lij)M : Maximum value of Ci/Lij

(Ci/Lij)M : Average value of Ci/Lij

Results of the Pollution Index informed the degree of the pollution suffered by River Bian by comparing it to the quality threshold by the water class as required by the Government Act 82/2001. By doing so, valuable information is expected to determine whether the water in the river can be used for purposes according to its water class.

3 Results and discussion

3.1 Analysis of water quality of River Bian

The water quality analysis was performed to find out the use of the water for particular purposes by comparing it to the water quality by the water class. According to the purpose, River Bian was included in the class III. The observation results of the physical parameters (temperature and TSS) and the chemical parameters (pH, DO, COD, BOD, phosphate, and nitrate) at each station were then compared to the class III water quality threshold [5]. The followings are the analytical results of each parameter:

3.1.1 Temperature

Figure 2 informs no significant difference in (or relatively stable of) the observation results of the water temperature parameters of the River Bian from 2007 to 2015 at each observation station.

The changing patterns of the temperature of the River Bian at the three observation points were generally stable, approximately 3 as compared to first conditions before the change in land use nearby the river. The average temperature at stations 1, 2, and 3, were 20.60 °C, 28.26 °C, and 27.13 °C, respectively. Compared to the water quality threshold of class III, which requires a deviation (+3) of the natural condition6, the temperature parameter proved that the water quality of River Bian was still within the water quality threshold for its purpose.

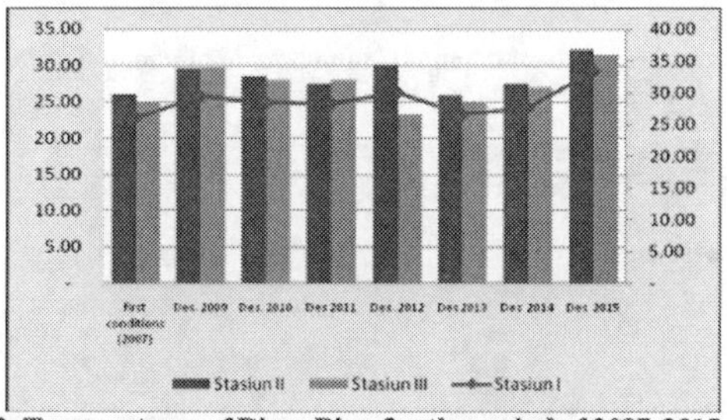

Fig 2. Temperatures of River Bian for the period of 2007-2015

Figure 2 summarizes that the temperatures of all stations fluctuated, as compared to the early 2007, ranging from 25 to 26 °C. An increase then occurred in the following year with the deviation of + 3, with the highest deviation taking place in 20015 (4-5 degrees). The increasing temperature of the river in 2015 was triggered by the seasonal condition since 2015 had the longest summer.

3.1.2 TSS (Total Suspended Solid)

The TSS parameter at each station revealed an increase from the station 1 to the station 3. The TSS value at the station was 222.13 mg/l, decreasing at the station 2 to 78.60 mg/l, and increasing at the station 3 to 165.18 mg/l. Therefore, compared to the class III water quality threshold under the Government Act 82/2001 (400 mg/l), the water quality condition of the River Bian was still in the water quality threshold for its purpose by the TSS.

The TSS was composed of muds and smooth sands as well as microorganisms, in particularly because of erosion carried to the water body [3]. In the river water, the source of the suspended solids is a run-off from the terrain nearby the river and the microorganisms weathering and animal corps carried by surface current to the water body. Writes that the suspended solids can decrease light intensity that absorbs into the water column, therefore disturbing growth and photosynthetic process of the vegetations living in this water area [5].

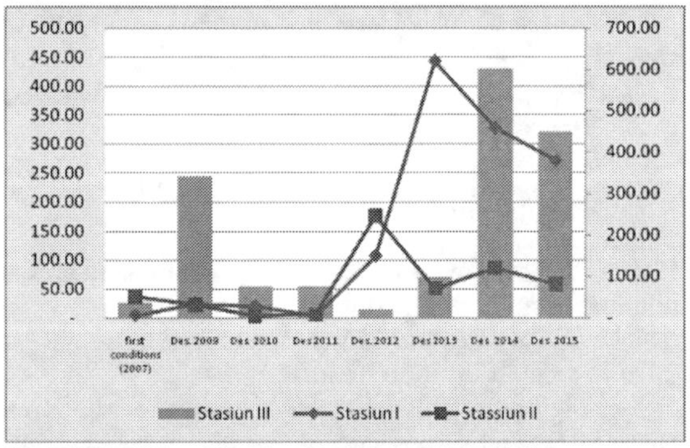

Fig 3. TSS of River Bian for the period of 2007-2015

The water TSS of River Bian at the stations 1 and 3 was high (Figure 3) because these stations were subject to deforestation for settlements, offices, power plant site, and industrial forest nearby the river flow. Suspended soils then inevitably formed by the run-off process. Furthermore, Table 2 contains the TSS for the fishery culture at the River Bian [6], 78.60-222.13 mg/l. These figures had caused the river not appropriate to the fishery activities [6].

Table 2. Appropriateness of the water for fishery purpose by the TSS

Grade TSS (mg/liter)	effect on interests of the fisheries
< 25	no effect
25 – 80	little effect
81 – 400	unfavorable to interests of the fishery
>400	not good for interests of the fishery

Source: Alabaster dan Lloyd, 1982

3.1.3 pH

The pH observation at each station revealed a fluctuative results from the station 1 to the station 3. The pH at the stations 1, 2, and 3 were 7.39, 7.42, and 7.26, respectively. Compared to the pH required by the class III water quality threshold under the Government Act 82/2001 (6-9), the water quality of the River Bian from the pH parameter was still within the water quality threshold for its purpose (Figure 4).

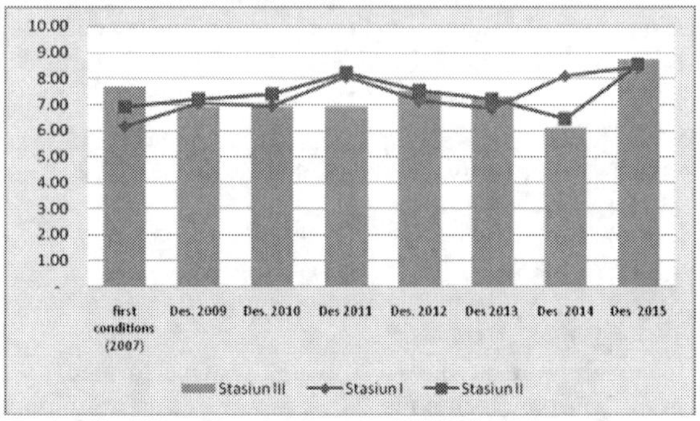

Fig 4. pH of River Bian for the period of 2007-2015

Figure 4 demonstrates that pH of the River Bian in the initial condition (2007) at the three stations ranged from 6.16 to 7.71. These fluctuative pH values were due to the disposal of organic waste from domestic activities, vegetation and animal weathering, and waste as the by-product of the industrial forestation nearby the River Bian. Besides, inorganic waste disposal also contributed significantly. However, the pH fluctuations did not significantly affect the river water quality. The normal water that fulfilled the requirements of a life system had a pH ranging from 6.5 to 7.5 [7]. The pH for the non-polluted water traditionally is close to neutral (pH 7) and fulfills the needs of most aquatic organisms [8]. In conclusion, the River Bian was still an ideal habitat for the aquatic organisms because it contained pH of 7.26-7.43.

3.1.4 Biochemical Oxygen Demand (BOD)

BOD is the amount of dissolved oxygen required by decomposing bacteria to decompose organic pollutants in the water. The higher the BOD concentration of an aquatic area, the higher organic materials concentration in the river water [9]. The higher BOD means the more severe pollution. Results of the BOD parameter observation revealed that a fluctuation occurred from the station 1 to the station 3. The average BODs at the station 1, 2, and 3 were 3.001 mg/l, 2.84 mg/l, and 4.13 mg/l, respectively. Pursuant to the Government Act 82/2001, the class III water quality threshold for the BOD parameter is 6 mg/l. Compared to the observation results of the River Bian water quality, the water quality condition at the station 1 was still within the quality threshold, but at the stations 2 and 3 it had exceeded for its purpose.

https://doi.org/10.1051/e3sconf/20183108006

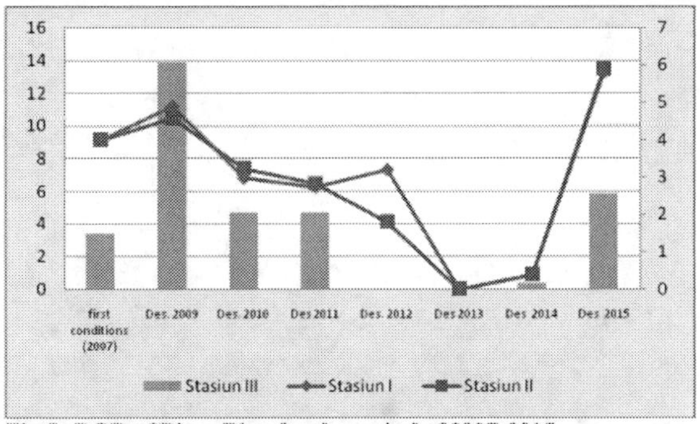

Fig 5. BOD of River Bian for the period of 2007-2015

Figure 5 shows an increase in the BOD content in 2009 at the station III, exceeding the class III quality threshold (quality threshold 6). It was because in 2009 there was a deforestation for development area. Contend that the increase in the BOD can be due to organic materials from the domestic and other wastes [10]. The high BOD is believed to be due to waste disposal from the settlements and from the agricultural fields to the river [6]. The Figure 5 also exposes that the water quality of the River Bian was still within the threshold but the higher BOD from the upstream to the downstream signed a polluted water area due to the domestic and agricultural waste disposal.

3.1.5 Chemical Oxygen Demand (COD)

COD is the amount of the oxygen needed to oxidize organic materials chemically [9]. The observation results of the BOD content at each station proved a decrease from the station 1 to the station 3. The average COD at the stations 1, 2, and 3 were 20.34 mg/l, 25.82 mg/l, and 42.43 mg/l, respectively. Compared to the class III water quality threshold for the COD parameter pursuant to the Government Act 82/2001 (50 mg/l) the water quality condition of the River Bian for the stations 1, 2, and 3 was still within the water quality threshold for its purpose.

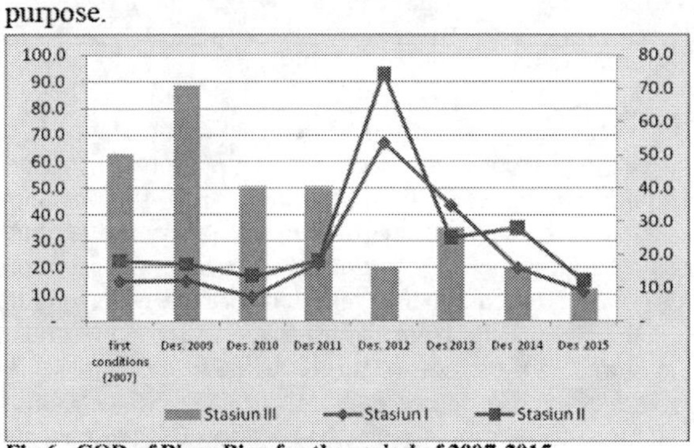

Fig 6. COD of River Bian for the period of 2007-2015

Figure 6 summarizes an exceeding rate of the initial COD (as of 2007) at the station and a further increase by 2009. The high COD indicated the more pollutants found in the water body [9]. The increase in COD at the River Bian was caused by the change in land use/deforestation for investment, where the by-products and the waste were found nearby the river. Aquatic area with high COD is no longer suitable for fishery or agricultural purposes. The COD in the non-polluted water is generally less than 20 mg/l, whereas in the polluted water the rate can rise more than 200 mg/l [3]. The COD tends to decrease over the years after the deforestation. Therefore, the water quality of the River Bian for the COD parameter was still supportive for the fishery and agricultural purposes.

3.1.6 DO

Dissolved oxygen (DO) can be used as an indicator of the water freshness [8]. The oxygen plays an important role as a quality indicator in the water because DO helps oxidation and reduction processes of both organic and inorganic materials. Because of oxidation and reduction processes the DO becomes significant to help ease the pollution burden in the water naturally [11].

The results of the DO parameter observation at each station proved fluctuative rates from the station 1 to the station 3, as it might be seen from the DO at the station 1 (4.65 mg/l), station 2 (4.63 mg/l), and station 3 (4.67 mg/l). Compared to the class III water quality threshold for the DO parameter pursuant to the Government Act 82/2001 (3 mg/l), the water quality condition of the River Bian at all stations had been exceeding the minimum requirement for its purpose.

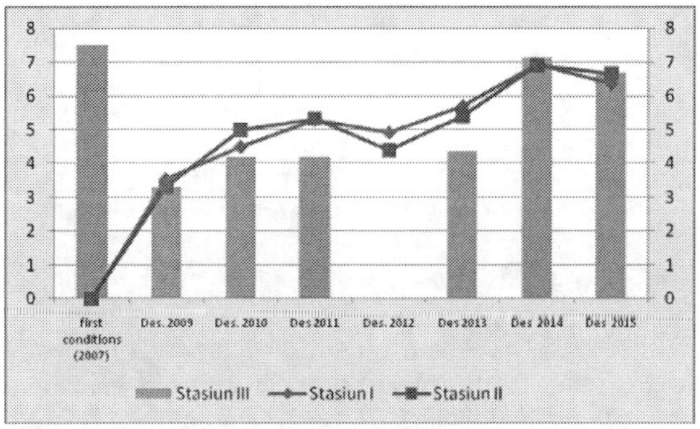

Fig 7. DO of River Bian for the period of 2007-2015

Generally, the study only found a small proportion of the polluted water. The more organic disposal waste causes the lower content of the dissolved oxygen in the water [12]. Human activities such as agricultural and waste disposal have caused the decreasae in the dissolved oxygen [13]. The decrease in the BOD of River Bian at each station was an accumulation of the waste disposal from the domestic and agricultural activities nearby the river. Water can be categorized good with a low pollution if the dissolved oxygen content is >5 mg/l [11]. The study indicated that the water quality of the River Bian for the DO parameter had a low pollution rate.

3.1.7 Phosphate

The observation of the phosphate parameter at each station showed a decrease from the station 1 to the station 3, as it might be proven by the phosphate content of the stations 1, 2, 3, which steadily decreased from

0.22 mg/l to 0.20 mg/l to 0.19 mg/l. compared to the phosphate content of the class III water quality threshold pursuant to the Government Act 82/2001 (1 mg/l), the water quality condition of the River Bian from the phosphate parameter was still within the water quality threshold for its purpose.

Phosphate is a phosphorus variant, which can be valuable for vegetations. Its characteristics are very different from the other substances as it can help form biosphere and it is not available in the atmosphere [3]. In agricultural area phosphate is produced from fertilizers absorbed in the river through drainage and rainfall waterflow [14]. Even though whole part of the River Bian was subject to the agricultural activities, the phosphate rate at each station was proven low (Figure 8). It was probably due to a natural dissolution along the River Bian so that the concentration of the pollutants decreased. This natural recovery process, both totally or partially, helped the river to reclaim itself and to perform a self purification [12].

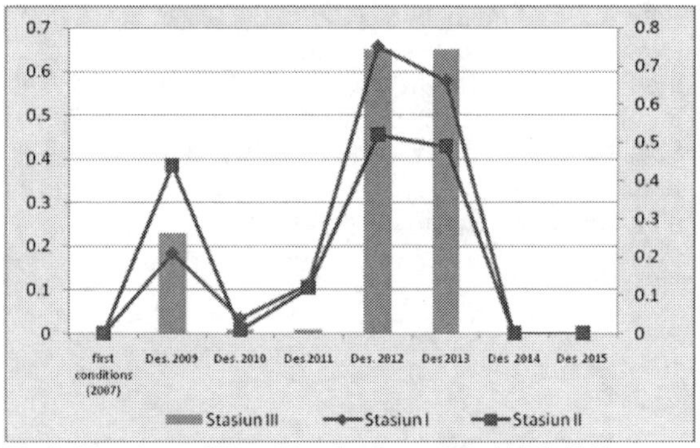

Fig 8. Phosphate of River Bian for the period of 2007-2015

3.1.8 Nitrate

Nitrate (NO3) is the main form of nitrogen in natural waters and is the major nutrient for the growth of vegetations and algae. The nitrogen nitrate is water dissolved and stable. This compound is derived from a complete oxidation process of the nitrogen in the water [15]. Nitrate is a stable compound and is derived from the disposal of the agricultural activities, fertilizers, and composes [14]. The study observation resulted in an increase in the nitrate content from the station 1 to the station 3, as it was proven by the average nitrate at the stations 1, 2, and 3 of 2.21 mg/l, 2.57 mg/l, and 3.97mg/l, respectively. Figure 9 shows a fluctuative rate of the nitrate, where a drastical increase occurred in 2012 and 2013. However, the increase was still below the quality threshold. The class III water quality threshold pursuant to the Government Act 82/2001, the maximum rate of the nitrate parameter required is 20 mg/l. therefore, compared to the water quality condition, the River Bian was still within the water quality threshold for its purpose albeit the factual condition proved vary activities found along the river, including defecation and waste disposal. It was possibly due to a dissolution process along the river.

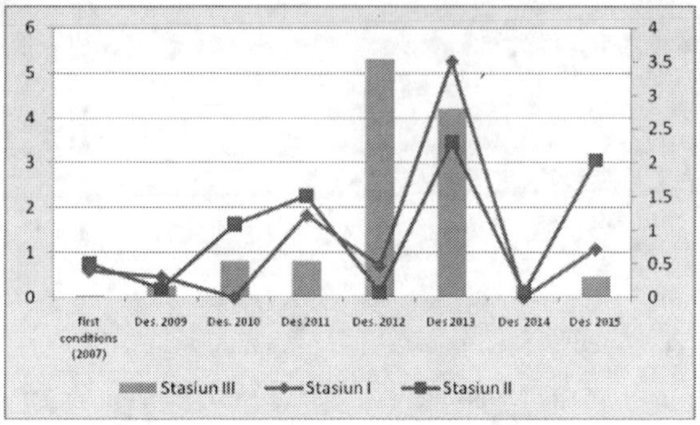

Fig 9. Nitrate of River Bian for the period of 2007-2015

3.2 Analysis of water quality status of River Bian

Water quality status is a degree to which the water quality condition indicates either pollution or non-pollution in particular water source and particular time as comparable to the predetermined water quality threshold. In this study the parameters used for the analysis of the water quality status were TSS, pH, CO, COD, BOD, phosphate and nitrate. The legal parameter used was class III water quality threshold pursuant to the Government Act 82/2001.

Table 3. Correlation of IP to Water Quality Status

Pollution Index	Water Quality
$0 \leq$ PIj $\leq 1,0$	Good condition (fulfilling quality threshold)
$1,0 <$ PIj $\leq 5,0$	Light pollution
$5,0 <$ PIj $\leq 10,0$	Moderate pollution
PIj $> 10,0$	Severe pollution

Source : Government Act 82/2001

The analysis of the water quality status was performed according to a case determination of the water quality status as required [16], using a Pollution Index method. The calculation of the water quality status of the River Bian using the PI method is summarized in Table 3 below. The calculation revealed that the water quality of the River Bian from the upstream to the downstream at three observation stations tended to increase as marked by the decrease in the PI. The water quality status at the station 1 (PI=1.61), station 2 (PI=1.63), and station 3 (PI=1.82) proved that the river was in a "moderate pollution status". Therefore, the water quality of the River Bian at the stations 1, 2, and 3 could still be utilized according to the class III water purpose (freshwater fishery culture, farm, irrigation, and other related purposes). These results proved that the investment activities did not mount any pressure on the water quality of the River Bian. The local people who live nearby the river would still be able to make use of the water for their daily needs.

4 Conclusion

1. Water quality of River Bian according to BOD and COD parameters at three stations had exceeded the quality threshold for the class III.
2. Water quality status of River Bian at the stations 1 and 2 proved a "moderate pollution status".

References

1. *Laporan Status Lingkungan Hidup Kabupaten Merauke* tahun 2007. BLH Kabupaten Merauke. (2007)
2. Kementerian Lingkungan Hidup. 2003. *Keputusan Menteri Negara Lingkungan Hidup Nomor 115 Tahun 2003 tentang Pedoman Penentuan Status Mutu Air.* Kementrian Lingkungan Hidup Republik Indonesia, Jakarta. (2003)
3. H. Effendi, *Telaah Kualitas Air bagi Pengelolaan Sumberdaya dan Lingkungan Perairan.* Penerbit Kanisius. Yogyakarta, (2003)
4. Kementerian Lingkungan Hidup. *Peraturan Pemerintah Republik Indonesia Nomor 82 Tahun 2001 tentang Pengelolaan Kualitas Air dan Pendendalian Pencemaran Air.* Kementrian Lingkungan Hidup Republik Indonesia, Jakarta. (2001)
5. S. M. Haslam, *River Pollution: An Ecological Perspective.* John Wiley and Sons. Chichester. 253 h, (1990)
6. Alabaster, JS and R. Lloyd. *Water Quality Criteria for Freshwater Fish. Second Edition.* Food and Agriculture Organization of United Nations by Butterworths. London, (1982)
7. W.A. Wardhana, *Dampak Pencemaran Lingkungan.* Penerbit Andi, Yogyakarta, (2004)
8. A. Sutriati, Jurnal Sumber Daya Air, **7**. 61-76, (2011)
9. S. Yudo, *Kondisi Kualitas Air Sungai Ciliwung di Wilayah DKI Jakarta Ditinjau Dari Parameter Organik, Amoniak, Fosfat, Deterjen dan Bakteri Coli.* Pusat Teknologi Lingkungan, Badan Pengkajian dan Penerapan Teknologi (BPPT), Jakarta. Hal: 1-9, (2010)
10. S. Rahayu, and Tontowi. "Penelitian Kualitas Air Bengawan Solo Pada Saat Musim Kemarau". Jurnal Sumber Daya Air, **5**. 127-136, (2009)
11. Salmin, Jurnal Oseana, **30**. 21-26, (2005)
12. R. Vagnetti, P. Miana, M. Fabris and B. Pavoni, Jurnal of Chemosphere, **52**. 1781-1795, (2003)
13. B.A. Anhwange, E. B. Agbaji, and E.C. Gimba,: International Journal of Science and Technology, **2**(5):248-254, (2012)
14. I.N.A. Winata, A. Siswoyo, and T. Mulyono, Jurnal Ilmu Dasar, **1**. 24-28, (2000)
15. K.K. Blume, J.C. Macedo, A. Meneguzzi, L.B. Silva, D.M. Quevedo, and M.A.S. Rodrigues. Journal of Biology, **70**. 1185-1193, (2010)
16. Kementerian Lingkungan Hidup. *Keputusan Menteri Negara Lingkungan Hidup Nomor 115 Tahun 2003 tentang Pedoman Penentuan Status Mutu Air.* Kementrian Lingkungan Hidup Republik Indonesia, Jakarta. (2003)

The Environmental Impact Study Of Micro Hydro Power In Pekalongan Indonesia

Suwarto[1*], *Sudharto* P Hadi[2] and Hermawan[3]

[1]Doctoral Program of Environmental Science, School of Postgraduate Studies, Diponegoro University, Semarang - Indonesia
[2]Department of Environmental Science, Diponegoro University, Semarang – Indonesia
[3]Faculty of Engineering of Diponegoro University, Semarang – Indonesia

Abstract. Curugmuncar II micro hydro power (MHP) located in Petungkriyono sub district is one of three MHPs installed in Pekalongan district. This study aims to analyze the MHP operation environmental impact. The study used qualitative method, with interviews, observations, and material testing. The data used are primary and secondary data. This research was conducted in Curugmuncar Village, Petungkriyono Subdistrict, Pekalongan Regency, Indonesia. MHP has power capacity of 100 KW with power usage of 50 KW. MHP used by 155 users with load capacity 2 A 220 volt AC. The community more used of lights as the houses and street lighting. The MHP operation had several environmental factors such as: sociology, technically feasible, hydrology, physical and chemical water quality, ergonomics, economically feasible, irrigation, clean water supply, government policy, and others. The supporting factors sustainability of MHP were sociology, irrigation, ergonomics, clean water supply, physical and chemical water quality, hydrology, and government policy. The inhibiting factors of MHP operation were technically feasible, economically feasible, and government policy. The results showed that the MHP environment requires a professional management system to achieve the MHP sustainability

1 Introduction

Energy plays an important role in a country's economy. Renewable energy sources are potential solutions to meet energy needs, especially in rural areas. The use of renewable energy is considered to contribute to reducing the impacts of climate change and environmental degradation [1].

Data from the statistics of Indonesia's EBTKE related to renewable energy and energy conservation (2016) showed that the new renewable energy mix in Indonesia has increased by an average of 0.36% annually. The new renewable energy mix as primary energy source still occupies the lowest position with a percentage of 6.2% compared to 43.0% for petroleum, 28.7% for coal and 22.0% for natural gas by 2015. The achievement of new renewable energy has increased since 6 Last year, In 2010, the renewable energy (RE) achievement of electricity amounted to 48,184,711 barrel of oil equivalent (BOE) and continued to increase until reaching 73,497,500 BOE by 2015.

New renewable energy potentials in Indonesia are abundant as water, wind or wind, solar and bioenergy. Most of Indonesia is waters and mountains. Many areas of Indonesia have the potential for forests and mountains that are the source of water. The potential of hydro power in Indonesia is 75,091 MW, with the potential of hydro power in Central Java province at 813 MW. The micro hydro potential in Central Java Province is 1,044 MW with total micro hydro potential in Indonesia of 19,385 MW. In Indonesia there are about 41 micro hydro power built in various regions with total energy capacity of 134.21 MW [2].

The electricity aspect in Indonesia is managed by a State-Owned Enterprise, namely *Perusahaan Listrik Negara* PT. PLN. Indonesia's electrification ratio in 2015 was 88.30%. That is, there are still 11.7% of the area in Indonesia that is still not yet powered by electricity. The government targets electrification ratio in 2019 of 97.35% [3]. The number of villages in Indonesia is 74,093 villages with 39,086 of them including lagging villages, and 17,268 villages are underdeveloped. According to data from the Central Bureau of Statistics in 2015 as many as 2519 villages have not electricity, and as many as 20,493 villages using electricity non PLN, and amounted to 69,531 have access to PLN electricity [4].

Electricity is a key strategic in the equity of social justice that has an effect on improving people's lives. One of the obstacles to the provision of rural electricity access to rural areas is quite difficult, especially in mountainous areas, such as in Curugmuncar Village, Pekalongan Regency, Indonesia. Specifically, through the program of

* Corresponding author: suwarto.dil@gmail.com

the Ministry of Energy and Mineral Resources in Indonesia has provided funding for the construction of micro hydro power in rural areas that have not yet electricity.

The benefits of electrical energy are enormous in human life. The goal of rural electrification programs in developing countries is to realize socio-economic development of households [5]. Electrical energy has a very important role in various industrial sectors such as education, health, food. So that electric energy can have an impact on economic development in a region. A study conducted by Uduak Akpan et al to find out the impact of rural electrification on the development of rural SMEs in the Niger Delta, Nigeria, that small rural industries connected to power grids are 16.2% more profitable than small rural industries not connected to power lines. The study also shows that micro-business owners are fully aware that access to electricity is critical and affects the profitability of their businesses [6].

Rural electrification is also assessed to serve socio-economic development goals. Increased access to electricity can encourage rural industry, increase agricultural productivity, and provide productive time for children to study at night, and not close the possibility for people to work at night [7].

Many international organizations are calling for increased use of renewable energy in an effort to reduce CO_2 emissions and tackle climate change. The results of previous studies conducted by Brantley Liddle and Perry Sadorsky (2017) that a one percent increase of the share of non-fossil fuel power plants reduces CO_2 emissions per capita from power plants by about 0.82% [8]

Indonesia is targeting to build micro hydro power (MHP) from 2010-2014 with an average capacity of 85 MW / year and 2015-2019 with an average capacity of 127 MW / year. Curugmuncar II MHP installed in 2010-2014 with an average capacity of 100 MW / year. The development of MHP in Indonesia in the period 2010-2014 exceeded the target by 15%. Thus, the development of MHP in Indonesia has the potential to become a sustainable source of renewable energy sources. This study aims to make an environmental agreement about operating MHP in Curugmuncar pekalongan.

2 Materials and Methods

The method used is qualitative with interview and observation. The data used are primary data and secondary data. This research conducted in Curugmuncar Village, Petungkriyono Sub district, Pekalongan Regency, Indonesia in May 2017. The object of research is Curugmuncar II MHP which started to operate in 2010. Aspects analyzed in this research include Sociology aspect, technical feasibility, policy Government hydrology, river water quality, ergonomics, clean water supply and irrigation, and other environmental aspects. This research use descriptive qualitative approach. The data used in this study are primary data and secondary data. Primary data were obtained from interviews, observation, and laboratory testing. While the secondary data obtained from the results of literature review obtained from scientific journals, the study of Meteorology and Geophysics Agency, as well as studies in the field of Mineral Resource Energy.

Interviews conducted to obtain the information such as current state of MHPs, informants of this research were managers of the MHPs and villages heads. At the time of observation, the researcher made an observation on the environmental aspects of MHP in Curugmuncar Village and collected data on everything related to the operation of MHP, electricity distribution, interaction between the community and the management, the behavior of the local people's electricity usage. Data analysis used qualitative descriptive method to describe phenomenon that occurred in Curugmuncar II MHP based on variables that have been studied. In addition, observation data are presented in graphs and percentages.

In this study also conducted analysis through economic and technical approach with the following calculation formula:

Total profit of MHP without decrease of investment value
Profit = total income - operational costs
Total profit of MHP with decrease of investment value
= Total income - (operating costs + decreased investment)

Furthermore, we also conducted a mathematical analysis to determine the reliability of a system, reliability is defined as "the ability of an item to perform a required function, under given environmental and operational conditions and for a stated period of time". Reliability of a system can be known by calculation (*Mean Time to Failure*, number of failures in some time interval, and others) [9, 10].

Mean Down Time (MDT)

"Mean Downtime (MDT) is the average Total Downtime required to restore an asset to its full operational capabilities. MDT includes the time from reporting of an asset being down to the asset being given back to operations / production to operate"[11].

Mean Time Between Failure (MTBF)

"MTBF is the mean operating time (up time) between failures of a specified item of equipment or a system. MTBF (years, hours) is most often used to express the overall reliability of equipment" [12].

Following is the formula used to calculate MTBF ,

$$MTBF = \frac{\text{Total operating time}}{\text{Number of failures}}$$

$$\lambda = \frac{\text{Number of failures}}{\text{Total operating time}}$$

$$MTBF = \frac{1}{\lambda}$$

" λ (hr-1, pmh, FITs) – is convenient to use for components, and it is easy to calculate the MTBF of an item of equipment from the sum of the component λs."

Mean Time To Repair (MTTR)

"Mean Time To Repair (MTTR) is the time needed to repair a failed hardware module. In an operational system, repair generally means replacing a failed hardware part (S. Stanley, 2011)."
Following is the formula used to calculate MTTR :

$$MTTR = \frac{Total\ Maintenance\ Time}{Numbers\ of\ Repairs}$$

Operational Availability

Following is the formula used to calculate Operational Availability,

$$Operational\ availability = \frac{\text{MTBM}}{MTBM + MDT}$$

Where " MTBM is mean time between corrective and preventive maintenance actions" [13]

Forced Autage Rate (FOR)

Forced outages is " when unit (s) is out of service due to failure called also unscheduled (unplanned) outage" [14]
The formula used to calculate FOR is,

$$FOR = \frac{\text{Time unit is on forced outages}}{\text{Time unit on service} + \text{Time on outages}}$$

3 Results and Discussions

Curugmuncar II MHP was built in Curugmuncar Village, Petungkriyono Subdistrict, Pekalongan Regency which is 48 km distance from the district center, and 7 km from the center of the district. Curugmuncar II MHP is located at an altitude of 1300 MDPL with an ambient air temperature ranging from 18-30^{0}C. Curugmuncar II MHP has a capacity of 100 KW with power used 50 KW. Curugmuncar II MHP is used by 155 KK with load capacity 2 A 220 volt AC. Curugmuncar II MHP is built in forest and plantation areas. Community forest area of 13,000 Ha, state forest 341.142 Ha, state or private plantation 1.107 Ha, and garden as much as 56.110 Ha.

Curugmuncar II MHP was built in 2008 and started to operate in 2010. The length of the water flow is 600 meters with the length of the river built 400 meters and the length of the pipe 200 meters with a diameter of 60 cm. Initially, the water of Curugmuncar II was used as agricultural irrigation, but after 2012 Curugmuncar I MHP was built. This proves that the water of MHP can still be used to generate more energy. Residents utilize water from dams for daily uses such as cooking, drinking, washing, bathing, etc. While spillway is used for agricultural irrigation channel and put into steam Curugmuncar MHP. Penstock passes the agricultural path of citizen and slightly disrupt agricultural activity.

(a) (b)

Fig. 1. (a) Power House of Curugmuncar II MHP (b) Generator

Sociology Aspects

As many as 16% of users feel very satisfied with the existence of MHP; 74% of MHP electricity users were satisfied. As many as 90% of MHP's were happy and proud of the MHP existence in their village. In addition, they also revealed that that became one aspect of this community acceptance due to the process of cheap electricity payment and no problem if they pay late. This means no specific sanctions from the manager for users who have not paid the electricity charge at the time specified. In addition, the community is very happy with the existence of MHP because it can build the economy and education in their village. Some users revealed that after the existence of electricity MHP then they can open a sewing business, furniture, electric welding, and also increase educational and religious activities.

As many as 6% feel quite satisfied; 4% not satisfied with MHP power. As many as 10% of users who were less satisfied with this MHP stated that they want electricity from PLN for various reasons. According to them, by using PLN electricity, electricity problems die during floods or landslides will not interfere with their daily activities. Usually, if MHP's electricity dies due to flood or landslide, it can take days to make improvements. Surely this is quite disturbing community activities.

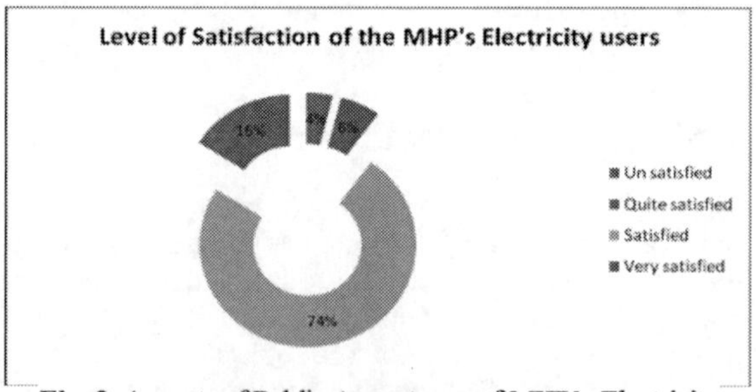

Fig. 2. Aspects of Public Acceptance of MHP's Electricity Users

The existence of Curugmuncar II MHP since 2010 has changed the behavior of community in Curugmuncar village. Until now people's lives have grown in the presence of electricity. Society has known technology such as television, mobile phone, computer, and others. Therefore, information and communication technology that enter into this area is a positive impact of the

 https://doi.org/10.1051/e3sconf/20183108007

existence of MHP electricity. Today's society can get information from things they need more easily. The community uses more lights as the lighting of houses and roads. The types of lights used by the community are fluorescent lights, LEDs, and tungsten bulb. Type of TV used is plasma TV, LCD TV. Types of communication tools used are smartphone, parabola, mobile phone, laptop, modem. Types of work equipment used are electrical, electrical, electrical sewing machines. Household appliances used are refrigerator, fan, blender, mixer, magic com.

Economics Feasibility Aspects

The total income of MHP was 14979.30 USD with operating costs was 4821.23 USD. Based on the calculation, the total profit of MHP without decreasing investment value was 10158.07 USD.

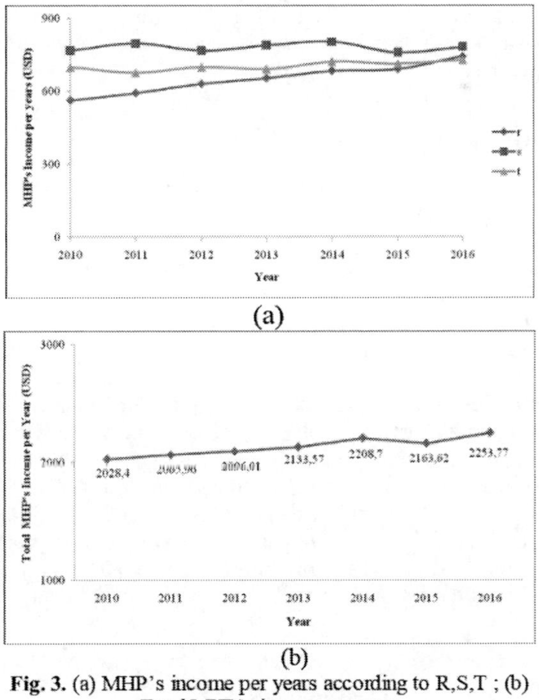

Fig. 3. (a) MHP's income per years according to R,S,T ; (b) Total MHP's income per year

The investments decrease were 9995.83 USD. From the calculation results obtained the total MHP profit with the decreased of investment value were 162.25 USD. This value indicates that MHP Curugmuncar II is not feasible for profit aspect but for sustainability aspect is still feasible.

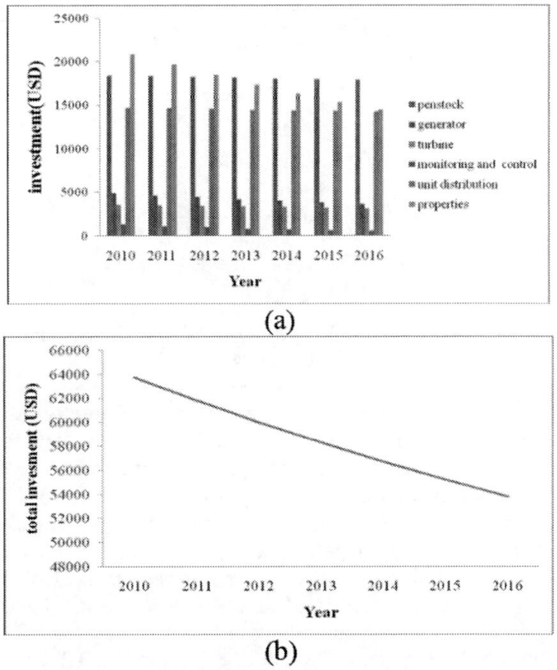

Fig. 4. (a) Decrease of MHP Investment Value (b) Decrease of Total MHP Investment Value

Technical Feasibility and Ergonomics Aspects

Curugmuncar II MHP is managed by an organization formed by the village community. The village head is responsible as a guardian. Next there is the structure of chairman, treasurer secretary, and two operators. The ability to master the maintenance of waterways consists of understanding the maintenance of diversion weir and intake, settling basin / sand trap, headrace, headtank, spillway, penstock, turbine, control valve, trailrace. The capability of controlling the power plant unit consists of spare part of the generating unit, installation of generating unit, maintenance and repair of electric generator, AVR (Automatic Voltage Regulator), panel / instrumentation, mechanical transmission, safety and environment of power plant unit. The ability to master electrical transmission system maintenance consists of wiring system, electrical transmission system spare parts, installation electrical transmission system, and safety and environment electrical transmission system. The management capability of MHP's management consists of safety and environment, financial, technical, logistics, administrative, quality, human resources, communication and public relations. Assessment of each component aspect is presented in the following figure:

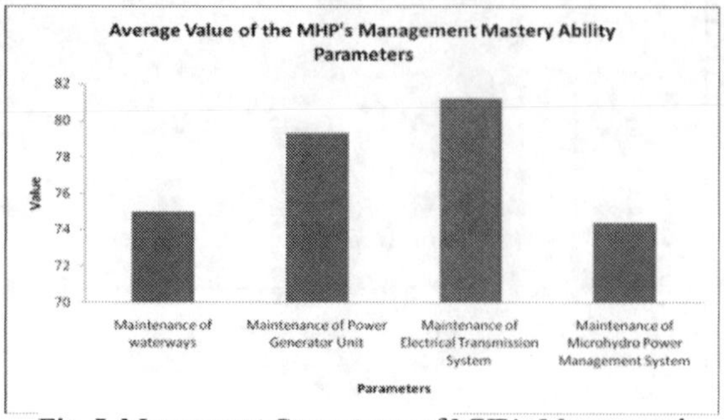

Fig. 5. Management Competency of MHP's Manager and Operators

Based on the assessment results, data obtained the value below the number 85, it was imply that the professionalism management aspect still very less. Though it takes a number of at least 85 or more to know the professionalism in the management of MHP. Factors that influence the unprofessionalism in management was the lack of salaries, because of the management of MHP were based on the basis of family and non-profit. Based on the calculation results show that Curugmuncar II MHP has MDT 70.4 hours, MTBF 3780 hours, MTTR 21.25 hours, and Availability 98.1%. In a study conducted by Sapkota, D et al in 2014 showed at Sunkoshi Hydro Power Station, Nepal conducted for four years showed availability of hydropower is 89% [15]. Furthermore, at Balimela Hydro Electric Power Station known availability 92% tested for 7 years [16].

Furthermore, the value of FOR Curugmuncar II MHP is known to be 1.9% so that the unavailability of electricity or the opportunity for electricity to die is 7 days / year, this is not in accordance with the Minister of Energy and Mineral Resources Decree of the Republic of Indonesia Number 5899 2016 which explains that the maximum electricity availability 1 day /year. The reliability criteria used are Loss of Load Probability (LOLP) smaller than 0.274% or equivalent with probalility outage 1 day of the year. In developed countries requires high reliability, many electrical systems are designed with LOLP criteria of 0.15 days or about 4 hours per year.

Government Policy Aspects

The implementation of micro hydro in Indonesia has not been widely regulated in national, provinces, or local regulations. This can be one of the weaknesses of micro hydro development built in the interior. Until now there is no special standard or special regulation that regulates the micro hydro institutional and micro hydro electricity tariffs enjoyed directly by the community. Therefore, the management of micro hydro in the village is implemented based on the agreement of the village apparatus and the manager.

On the other hand, the Law of the Republic of Indonesia Number 30 Year 2007 on Energy in Article 4 that New energy resources and renewable energy resources are regulated by the state and utilized for the greatest of the people's sincerity. In Article 19 that everyone has the right to gain energy and society, both individually and in groups, can play a role in the development of energy for the public interest. Furthermore, in the Law of the Republic of Indonesia Number 30 Year 2009 on Electricity, mentioned in Article 5 that the authority of the municipal government in the electricity sector involves the stipulation of regulation of regency/ municipality in the field of electricity.

Seeing the development of micro hydro technology that is widely applied in Indonesia, it is appropriate that local governments specifically regulate the management standards of MHP and tariffs established by considering the aspects of socio-economic in the region. It aims to maintain the sustainability of MHP and establish a good management pattern. However, not only local governments can establish a regulation for MHP, the village level government is also entitled to make village regulations that specifically regulate the management and tariff of the MHP in their village. So with the existence of clear rules, then hope management can be done more professional.

Hydrology Aspects

Based on data from the government of Pekalongan Regency, that rainfall in 2013 is on average per year 2,992 mm, higher than in 2012 which experienced 2,243 mm rainfall. The average rainy day of 2013 is 143 days, higher than the average rainy day of 2012 of 120 days [17]. The highest rainfall occurred in Kecamatan Lebakbarang on average per year 5,321 mm. Petungkriyono District is located in a mountainous area with Lebakbarang sub-district so it can be assumed that the rainfall that is sampled in Lebakbarang sub-district is almost similar to rainfall in Petungkriyono sub-district. Based on the analysis, it is estimated that the water cycle in Petungkriyono sub-district within 1 year is 39,151 M^3. The volume of water required by MHP for one year is about 3765 M^3. The volume of water required is about 10% of the water volume of one sub-district within a year.

Water Quality of River Water, Clean Water Supply and Irrigation Aspects

The river water used to drive turbines at Curugmuncar II MHP is sourced from waterfalls in the slopes of Mount Ragajambangan. The river water is utilized by the community as irrigating rice fields and also for daily needs such as washing clothes, bathing, and others. According to the information from the local community that the operation of the MHP does not interfere with their activities in irrigating rice fields and daily water use. The operation of MHP is known not to have a negative impact on the quality of river water. The measurement results show mean TDS of river water input 34 ppm and output 31 ppm with average pH of water output of 7.07 and pH of water output of 7.02. The output water flow from the powerhouse to the trailrace is reused as the next generation micro hydro source. MHP that does not cause residue is very environmentally friendly and supports forest conservation. It educates the community indirectly to preserve nature in the forest area as a water storage, so that the availability of water is always abundant throughout the year.

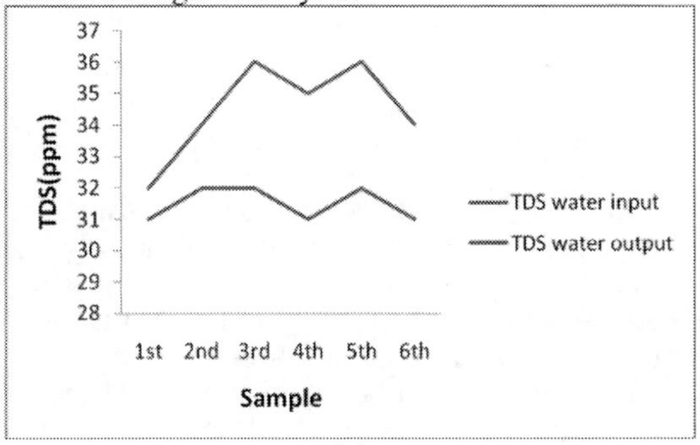

(a)

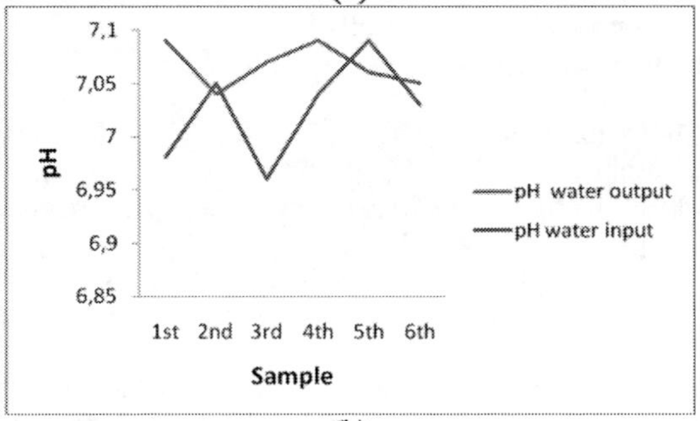

(b)

Fig. 6. Graph of River Water TDS and pH Condition from Input and Output the Power House

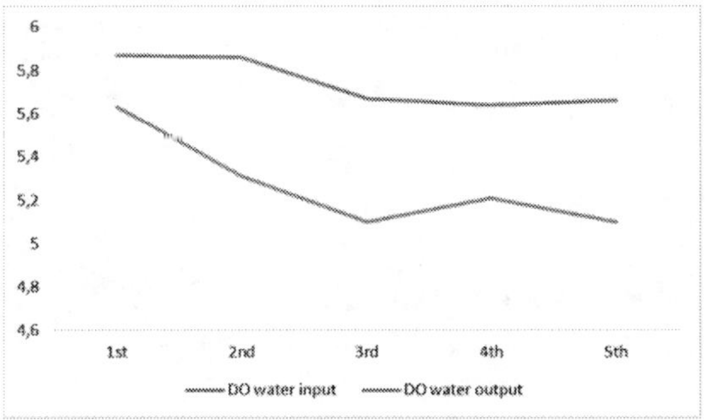

(a)

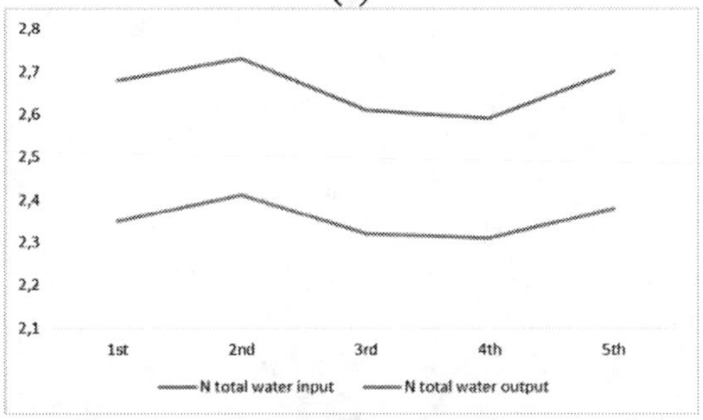

(b)

Fig 7. Dissolved Oxygen and Nitrogen Total Water from Input and Output the Power House

The condition of dissolved oxygen and total nitrogen content in water did not change significantly. Water samples are taken from the river stream at the input and output of the power house. The highest DO level of 5.87 on the input and the highest DO level of output was 5.63. The sample water also has the highest total nitrogen content of 2.73 (input water) and 2.41 (output water).

Other Environmental Aspects

The measurement results showed that the operation of Curugmuncar II MHP has no impact on the surrounding environment. Noise generated due to the operation of the system inside the power house averaged 74 db and noise outside the power house averaged 56 db. The average temperature inside the power house is 38°C and the average temperature is outside 29 °C. Activities around the power house is destined for agricultural land and quite far from the settlement so as not to interfere with community activities. The following figure shows a comparison graph between the noise and operating temperature of the Curugmuncar II MHP inside and outside the power house from a radius of 50 m.

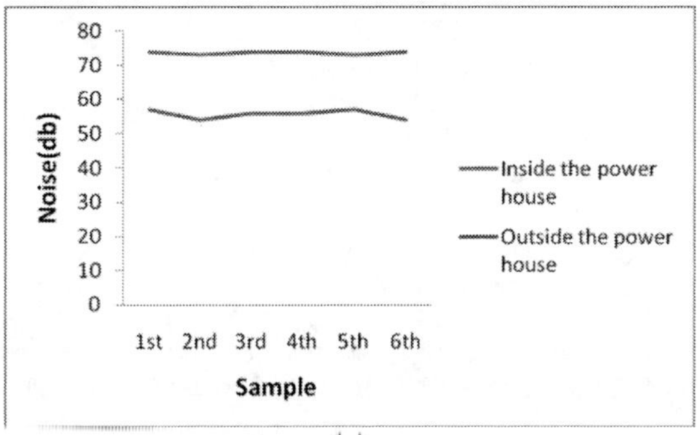

(a)

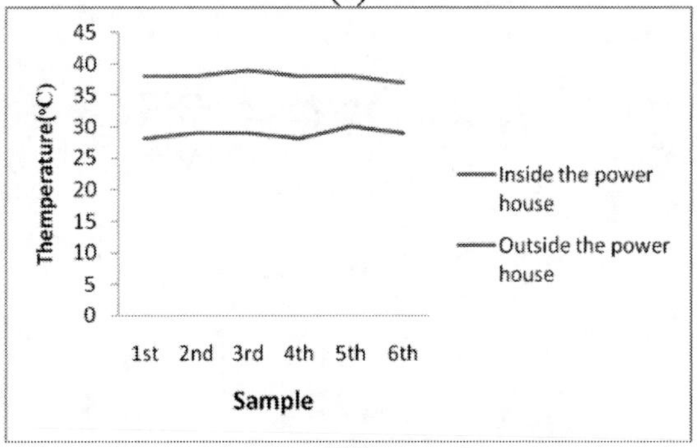

(b)

Fig. 8. Noise and Temperature Conditions from Input and Output the Power House

4 Conclusions

Curugmuncar II MHP with a power capacity of 100 KW with the use of 50 KW power. MHP serves 155 users with load capacity 2 A 220 volt AC. The community uses more lights as the lighting of houses and roads. MHP's operations has several environmental factors such as sociology, technically feasible, hydrology, physical and

 https://doi.org/10.1051/e3sconf/20183108007

chemical water quality, ergonomics, economically feasible, irrigation, clean water supply, government policy, and environmental impact other. The factors support the MHP sustainability were: sociology, irrigation, ergonomics, clean water supply, physical and chemical water quality, hydrology, and government policy. The factors hamper MHP sustainability were: technically feasible, economically feasible, and government policy. From the results of the study shows that the environment aspects of MHP requires a professional management system to ensure the MHP sustainability.

References

1. Sazib Mollik, M.M. Rashid, M. Hasanuzzaman, M.E. Karim, M. Hosenuzzaman, Renewable and Sustainable Energy Reviews. Volume **65**, November 2016, Pages 553–567, (2016)
2. Directorate General of Renewable Energy and Energy Conservation. 2016. EBTKE Statistics 2016. [online] http://ebtke.esdm.go.id/post/2017/03/07/1583/statistik.ebtke.2016 Accessed on May, 21 2017.
3. Directorate General of Renewable Energy and Energy Conservation. 2016. Buletin of Electricity Edition 46th. 2016. ISSN: 1979-6579[12].
4. Central Bureau of Statistics. 2015. Number of Village According to the Presence of Family of Electricity Users and Source of Lighting of Village Main Road. [online] https://www.bps.go.id/linkTabelStatis/view/id/1759 Accessed on 21 Mei 2017.
5. C. L Azimoh, P.Klintenberg, F.Wallin, B. Karlsson, C.Mbohwa, Energy Conversion and Management. **[110]** Pp. 268–277, (2016)
6. U. Akpan, M. Essien, S.Isihak, Energy for Sustainable Development **[17]** 5. Pp 504–509, (2013)
7. T. Slough, J. Urpelainen , J. Yang. 2015. Light for all? Evaluating Brazil's rural electrification progress, 2000–2010. Energy Policy. **[86]**. Pp. 315–327
8. B. Liddle and P. Sadorsky, How much does increasing non-fossil fuels in electricity generation reduce carbon dioxide emissions?. Applied Energy. **[197]** pp. 212–221, (2017)
9. Rohani, H., & Roosta, A. K. Calculating Total System Availability. [online] https://pdfs.semanticscholar.org. Accessed on June, 5, 2017
10. Stanley, S. 2011. MTBF, MTTR, MTTF & FIT Explanation of Terms. *IMC Network*, pp16.
11. Office of the Assistant Secretary of Defense for Logistics & Materiel Readiness. [online] http://www.acq.osd.mil/log/mr/mean_down_time.html . Accessed on June, 5, 2017
12. Crouse-Hinds, C. Availability, Reliability, SIL. *Eaton, TX, US.*
13. Humble, T. X. (1997). Availability, Reliability, Maintainability, and Capability.
14. Billinton, R. (1970). *Power system reliability evaluation*. Taylor & Francis.
15. D. Sapkota, T. R. Bajracharya, & M. C. Luintel, In *Proceedings of IOE Graduate Conference* (pp. 197-202), (2014).
16. S. Dash, & D. Das, (2014). Availability Assessment of Generating Units of Balimela Hydro Electric Power Station (510 MW)–A Markovian Approach. Volume **3** *Issue 2–February 2014*, 44.
17. Pekalongan Regency Goverment. 2017. Iklim. [online] http://www.pekalongankab.go.id. Accessed on June, 5, 2017

Analysis of a Fishery Model with two competing prey species in the presence of a predator species for Optimal Harvesting

Sutimin[1,*], *Siti* Khabibah[1], and *Dita* Anis Munawwaroh[1]

[1] Department of Mathematics, Faculty of Sciences and Mathematics, Diponegoro University, Semarang - Indonesia
[2] Department of Environmental Engineering, Faculty of Engineering, Diponegoro University, Semarang - Indonesia

Abstract. A harvesting fishery model is proposed to analyze the effects of the presence of red devil fish population, as a predator in an ecosystem. In this paper, we consider an ecological model of three species by taking into account two competing species and presence of a predator (red devil), the third species, which incorporates the harvesting efforts of each fish species. The stability of the dynamical system is discussed and the existence of biological and bionomic equilibrium is examined. The optimal harvest policy is studied and the solution is derived in the equilibrium case applying Pontryagin's maximal principle. The simulation results is presented to simulate the dynamical behavior of the model and show that the optimal equilibrium solution is globally asymptotically stable. The results show that the optimal harvesting effort is obtained regarding to bionomic and biological equilibrium.

1 Introduction

Introducing red devil fish species in freshwater ecosystems has threatened the presence of commercial fish (such as, tilapia and goldfish) in the habitats. In the freshwater ecosystems, red devil (*Amphilophus Labiatus*) growth highly and abundantly. Red devil is as one of the major threats to global biodiversity [1, 2]. The red devil species introduced in the ecosystems through spreading the seeds of the commercial fish. The red devil species is from Nicaraguan lakes. The presence of a red devil fish in the ecosystems has become the leading threat of the survival of other fish in the habitats. Red devil has preyed on other commercial fish, such as carp and tilapia. Red devil fish is a type of predator that interferes the survival of other fish in which as the predator, red devil fish has low commercial value.

A number of the investigations regarding to fisheries resources have been conducted. The fishery model of two competing species was addressed by Clark *et al.* to discuss the aspect bionomics and the policy of optimal harvesting policy [2]. The combined harvesting of two competing species from the aspect of bionomic was also studied and discussed for the optimal equilibrium policy [3-7].

Das *et al.* [8] and Chaudhuri *et al.* [9] proposed the model one prey – one predator with combined harvesting. They analysed the dynamical behaviour of the model and the optimal harvesting policy. They assumed that the growth of both prey and predator are formulated in the logistic terms. Das *et al.* [8] used the functional response of predator that it is limited by density of prey. Rao *et al.* [10] investigated analytically the dynamical behaviour of the model for three species with one prey and two competing predators. They also discussed the policy of optimal harvesting. .

Kar *et al.* [11] analyzed a harvesting model for two competing species taking into account the presence of a predator species which feeds on the two competing species, as the third species which was not harvested. They also analyzed bionomic equilibrium solution and the policy of optimal harvesting.

In this paper, we propose and analyze an ecological model of multispecies harvesting of two competing prey species and a predator species. We consider a combined harvest effort that imposes the exploitation of each species. The growth model of each species is modeled by logistics terms and we consider that the functional response of both competitor and predator of other species is assumed in bilinear term. The existence of the possible equilibrium states of the model and the stability of nontrivial equilibrium state by using a Lyapunov function is discussed. Next, we discuss the possibilities of the existence of a bionomic equilibrium. The optimal harvesting policy is studied and the optimal solution is derived in the nontrivial equilibrium case by using Pontryagin's maximum principle. Finally, some numerical examples are discussed.

2 Model formulation

We consider an aquatic ecosystem, there are two prey species of fish that competes each other for using a common resource. In the presence of a red devil fish, an invasive species (as predator) threat the survival of these prey species. Besides competing for the use of same resources, red devil fish also eats the small other fish. All

* Corresponding author: sutimin@undip.ac.id

these species are imposed continuously harvesting. We propose the logistic growth function for both two prey species and the predator (that is, the population of each species compete for the same resource), in addition to the third species increases due to in the presence of prey populations. The model equations for three species in which two competing prey species and a predator species with harvesting on all species is given by the following,

$$\frac{dx_1}{dt} = r_1x_1\left(1-\frac{x_1}{K_1}\right) - \alpha_{12}x_1x_2 - \alpha_{13}x_1x_3 - q_1Ex_1$$

$$\frac{dx_2}{dt} = r_2x_2\left(1-\frac{x_2}{K_2}\right) - \alpha_{21}x_1x_2 - \alpha_{23}x_2x_3 - q_2Ex_2 \quad (1)$$

$$\frac{dx_3}{dt} = r_3x_3\left(1-\frac{x_3}{K_3}\right) + \alpha_{31}x_1x_2 + \alpha_{32}x_2x_3 - q_3Ex_3$$

where $x_1(t), x_2(t)$ and $x_3(t)$ are the biomass densities of the first prey, the second prey and predator populations with the natural growth rates r_1, r_2 and r_3 respectively. The parameters K_1, K_2 and K_3 are the carrying capacities of the first prey, the second prey and predator populations, respectively. Parameter α_{12} is rate of decrease of the first prey due to the competition with the second prey, α_{13} is rate of decrease of the first prey due to inhibition by the predator, α_{21} is rate of decrease of the second prey due to the competition with the first prey, α_{23} is rate of decrease of the second prey due to inhibition by the predator, α_{31} is rate of increase of the predator due to successful attacks on the first prey, α_{32} is rate of increase of the predator due to successful attacks on the second prey, q_i for $i = 1,2,3$ are catch ability coefficient of the first prey, the second prey and predator species respectively. E is the harvesting effort and $q_iEx_i, i = 1,2,3$ are the catch-rate functions based on the catch-per-unit-effort hypothesis. In the analysis of the system, we assume that $r_i - q_iE > 0$ for $i = 1,2,3$.

3 Model analysis

In the section, we analyze the existence and the stability of equilibria.

3.1 Existence of equilibrium

We determine the conditions for the existence of the equilibrium points of the system (1). By equating the right hand side of the system (1) to zero, we obtain the equilibrium states. The possible equilibrium states are given as follows,

P_0 : The state of all washed out $x_1^* = 0; x_2^* = 0; x_3^* = 0$.

P_1 :The state in which only the predator survives, two prey species are washed out. The equilibrium state is $x_1^* = 0; x_2^* = 0; x_3^* = \frac{K_3(r_3 - q_3E)}{r_3}$.

P_2 : The state in which the first prey and predator species survive and the second prey species extinct out, is given by

$$x_1^* = \frac{\alpha_{13}(-r_3 + q_3E) + \frac{r_3}{K_3}(r_1 - q_1E)}{\alpha_{31}\alpha_{13} + \frac{r_3r_1}{K_1K_3}}; \quad x_2^* = 0;$$

$$x_3^* = \frac{\alpha_{31}(r_1 - q_1E) + \frac{r_1}{K_1}(r_3 - q_3E)}{\alpha_{31}\alpha_{13} + \frac{r_3r_1}{K_1K_3}}.$$

It exists when $\alpha_{13}(r_3 - q_3E) < \frac{r_3}{K_3}(r_1 - q_1E)$.

P_3 : The state in which the second prey and predator exist, while the first prey extinct out. It is given by

$$x_1^* = 0; \quad x_2^* = \frac{\alpha_{23}(-r_3 + q_3E) + \frac{r_3}{K_3}(r_2 - q_2E)}{\alpha_{32}\alpha_{23} + \frac{r_2r_3}{K_2K_3}};$$

$$x_3^* = \frac{\alpha_{32}(r_2 - q_2E) + \frac{r_2}{K_2}(r_3 - q_3E)}{\alpha_{32}\alpha_{23} + \frac{r_2r_3}{K_2K_3}}.$$

It exists, when $\alpha_{23}(r_3 - q_3E) < \frac{r_3}{K_3}(r_2 - q_2E)$.

P_4 : The state in which both the first and the second prey exist, while the predator extinct, It is given by

$$x_2^* = \frac{\alpha_{21}(r_1 - q_1E) - \frac{r_1}{K_1}(r_2 - q_2E)}{\alpha_{21}\alpha_{12} - \frac{r_1r_2}{K_1K_2}}; \quad x_3^* = 0.$$

This state exists, when

$$\alpha_{12}(r_2 - q_2E) > \frac{r_2}{K_2}(r_1 - q_1E), \alpha_{21}(r_1 - q_1E) > \frac{r_1}{K_1}(r_2 - q_2E),$$

$$\alpha_{21}\alpha_{12} > \frac{r_1r_2}{K_1K_2}.$$

P_5 : The co-existence state, the state in which two prey and predator species exists. The equilibrium state is given by,

$$x_1^* = \frac{1}{\bar{X}}\left[\left(-r_1 + q_1 E\right)\left(\alpha_{32}\alpha_{23} + \frac{r_2 r_3}{K_2 K_3}\right)\right]$$

$$+\frac{1}{\bar{X}}\left[\left(r_2 - q_2 E\right)\left(\alpha_{32}\alpha_{13} + \frac{r_3}{K_3}\alpha_{12}\right)\right]$$

$$+\frac{1}{\bar{X}}\left[\left(r_3 - q_3 E\right)\left(\alpha_{12}\alpha_{23} - \frac{r_2}{K_2}\alpha_{13}\right)\right]$$

$$x_2^* = \frac{1}{\bar{X}}\left[\left(r_1 - q_1 E\right)\left(\alpha_{31}\alpha_{23} + \frac{r_3}{K_3}\alpha_{21}\right)\right]$$

$$+\frac{1}{\bar{X}}\left[\left(r_3 - q_3 E\right)\left(\frac{r_1}{K_1}\alpha_{23} - \alpha_{21}\alpha_{13} + \right)\right]$$

$$+\frac{1}{\bar{X}}\left[-\left(r_2 - q_2 E\right)\left(\alpha_{31}\alpha_{13} + \frac{r_1 r_3}{K_1 K_3}\right)\right]$$

$$x_3^* = \frac{1}{\bar{X}}\left[\left(r_1 - q_1 E\right)\left(\alpha_{21}\alpha_{32} - \frac{r_2}{K_2}\alpha_{31}\right)\right]$$

$$+\frac{1}{\bar{X}}\left[\left(r_2 - q_2 E\right)\left(\alpha_{12}\alpha_{31} - \frac{r_1}{K_1}\alpha_{32}\right)\right]$$

$$+\frac{1}{\bar{X}}\left[\left(r_3 - q_3 E\right)\left(\alpha_{21}\alpha_{12} - \frac{r_1 r_2}{K_1 K_2}\right)\right],$$

where,

$$\bar{X} = \left(\alpha_{12}\alpha_{23} - \frac{r_2}{K_2}\alpha_{13}\right)\alpha_{31} + \left(\alpha_{21}\alpha_{13} - \frac{r_1}{K_1}\alpha_{23}\right)\alpha_{32}$$

$$+\frac{r_3}{K_3}\left(\alpha_{21}\alpha_{12} - \frac{r_2 r_1}{K_1 K_2}\right)$$

3.2 The stability of the equilibrium state

We analyze the global stability of the system (1) by constructing a suitable Lyapunov function. The global stability of the system (1) for the co-existence equilibrium state is given in the following theorem.

Theorem 3.1: The equilibrium state $P_5(x_1^*, x_2^*, x_3^*)$ is globally asymptotically stable if,

(i) $\alpha_{13} = \alpha_{31}, \alpha_{23} = \alpha_{32}$,

(ii) $4\frac{r_1 r_2}{K_1 K_2} > \left(\alpha_{12} + \alpha_{21}\right)^2$.

Proof: To examine the globally stability of the system (1), we define a Lyapunov function

$$V(x_1, x_2, x_3) = x_1 - x_1^* - x_1^* \ln\left(\frac{x_1}{x_1^*}\right) + x_2 - x_2^* - x_2^* \ln\left(\frac{x_2}{x_2^*}\right)$$

$$+ x_3 - x_3^* - x_3^* \ln\left(\frac{x_3}{x_3^*}\right)$$

We see that V is definite positive. The time derivative of V along the solutions of the system (1), is given by,

$$\frac{dV}{dt} = \frac{x_1 - x_1^*}{x_1}\frac{dx_1}{dt} + \frac{x_2 - x_2^*}{x_2}\frac{dx_2}{dt} + \frac{x_3 - x_3^*}{x_3}\frac{dx_3}{dt}.$$

After simplifying, $\frac{dV}{dt}$ can be rewritten as,

$$\frac{dV}{dt} = -\mathbf{X}^T \mathbf{A} \mathbf{X}, \quad (2)$$

where,

$\mathbf{X}^T = \left[x_1 - x_1^*, x_2 - x_2^*, x_3 - x_3^*\right]$, and matrix $\mathbf{A}$ is

$$\begin{bmatrix} \frac{r_1}{K_1} & -\frac{1}{2}\left(\alpha_{12} + \alpha_{21}\right) & \frac{1}{2}\left(\alpha_{31} - \alpha_{13}\right) \\ -\frac{1}{2}\left(\alpha_{12} + \alpha_{21}\right) & \frac{r_2}{K_2} & \frac{1}{2}\left(\alpha_{32} - \alpha_{23}\right) \\ \frac{1}{2}\left(\alpha_{31} - \alpha_{13}\right) & \frac{1}{2}\left(\alpha_{32} - \alpha_{23}\right) & \frac{r_3}{K_3} \end{bmatrix}$$

We observe that $\frac{dV}{dt} < 0$ if the matrix $\mathbf{A}$ is definite positive. The matrix A is definite positive if, $\frac{1}{2}\left(\alpha_{31} - \alpha_{13}\right) = 0, \frac{1}{2}\left(\alpha_{32} - \alpha_{23}\right) = 0$, and $4\frac{r_1 r_2}{K_1 K_2} - \left(\alpha_{12} + \alpha_{21}\right)^2 > 0$. So, we conclude that the co-existence equilibrium is globally asymptotically stable, if $\alpha_{13} = \alpha_{31}$, $\alpha_{23} = \alpha_{32}$ and $4\frac{r_1 r_2}{K_1 K_2} > \left(\alpha_{12} + \alpha_{21}\right)^2$.

This completes the proof.

3.3 Bionomic equilibrium

The concept of bionomic equilibrium is the concepts that are related to the biological and economic equilibrium. The biological equilibrium is derived from the right hand sides of the system (1) equal zero. The economic equilibrium is obtained by equaling the total revenue that is obtained by selling the harvested biomass with the total needed efforts for harvesting.

Let c is the fishing cost per unit effort, p_i for $i = 1, 2, 3$ are the price per unit biomass of the first, the second and the third species respectively. The net economic revenue at any time t is given by

$$R = \left(p_1 q_1 x_1 + p_2 q_2 x_2 + p_3 q_3 x_3 - c\right)E. \quad (3)$$

We can obtain the bionomic equilibrium $P\left(x_{1b}, x_{2b}, x_{3b}, E\right)$ that are given in the following equations,

$$x_1 = 0,\ E = \frac{r_1}{q_1}\left(1 - \frac{x_1}{K_1}\right) - \frac{\alpha_{12}}{q_1}x_2 - \frac{\alpha_{13}}{q_1}x_3$$

$$x_2 = 0,\ E = \frac{r_2}{q_2}\left(1 - \frac{x_2}{K_2}\right) - \frac{\alpha_{21}}{q_2}x_1 - \frac{\alpha_{23}}{q_2}x_3 \qquad (4)$$

$$x_3 = 0,\ E = \frac{r_3}{q_3}\left(1 - \frac{x_3}{K_3}\right) + \frac{\alpha_{31}}{q_3}x_1 + \frac{\alpha_{32}}{q_3}x_2$$

Eliminating E from (4), we obtain the nontrivial biological equilibrium. The possibility of the bionomic equilibrium is determined in the following cases.

Case I. When $p_1q_1x_1 + p_2q_2x_2 + p_3q_3x_3 < c$, the fishing cost per unit effort exceeds the total revenue, so the biomass is not harvested.

Case II. When $p_1q_1x_1 + p_2q_2x_2 + p_3q_3x_3 > c$, the fishing cost per unit effort is less than the total revenue, so the biomass can be harvested. The economic equilibrium is given by

$$R = p_1q_1x_1 + p_2q_2x_2 + p_3q_3x_3 - c = 0 \qquad (5)$$

We refer to (4) as the economic equilibrium point. The bionomic solution (x_{1b}, x_{2b}, x_{3b}) is obtained by solving both the equations (4) and (5) simultaneously.

3.4 Optimal harvesting policy

Our objective is to maximize the present revenue value J of a continuous time that is given by

$$J = \int_0^{\infty} e^{-\delta t}\left(p_1q_1x_1 + p_2q_2x_2 + p_3q_3x_3 - c\right)E(t)\,dt \qquad (6)$$

where δ represents the instantaneous annual rate of discount. In this problem, we maximize J subject to the state equations (1) by applying Pontryagin's maximal principle [6, 11]. The value of the control variable $E(t)$ is between 0 and E_{max}, so that the control function $V(t) \in [0, E_{max}]$. The Hamiltonian of the problem is

$$\begin{aligned} H &= e^{-\delta t}\left(p_1q_1x_1 + p_2q_2x_2 + p_3q_3x_3 - c\right)E \\ &+ \lambda_1\left(r_1x_1\left(1 - \frac{x_1}{K_1}\right) - \alpha_{12}x_1x_2 - \alpha_{13}x_1x_3 - q_1Ex_1\right) \\ &+ \lambda_2\left(r_2x_2\left(1 - \frac{x_2}{K_2}\right) - \alpha_{21}x_1x_2 - \alpha_{23}x_2x_3 - q_2Ex_2\right) \qquad (6) \\ &+ \lambda_3\left(r_3x_3\left(1 - \frac{x_3}{K_3}\right) + \alpha_{31}x_1x_3 + \alpha_{32}x_2x_3 - q_3Ex_3\right) \end{aligned}$$

where $\lambda_i, i = 1,2,3$ are the adjoint variables.

By using Pontryagin's maximal principle, we write the adjoint equations regarding to optimal equilibrium solution as $\frac{d\lambda_i}{dt} = -\frac{\partial H}{\partial x_i}$ for $i = 1,2,3$. We obtain adjoint equations as follows,

$$\frac{d\lambda_1}{dt} = -e^{-\delta t}p_1q_1E + \frac{\lambda_1r_1x_1}{K_1} + \lambda_2\alpha_{21}x_2 - \lambda_3\alpha_{31}x_3 \qquad (8)$$

$$\frac{d\lambda_2}{dt} = -e^{-\delta t}p_2q_2E + \lambda_1\alpha_{12}x_1 + \frac{\lambda_2r_2x_2}{K_2} - \lambda_3\alpha_{32}x_3 \qquad (9)$$

$$\frac{d\lambda_3}{dt} = -e^{-\delta t}p_3q_3E + \lambda_1\alpha_{13}x_1 + \lambda_2\alpha_{23}x_2 + \frac{\lambda_3r_3x_3}{K_3} \qquad (10)$$

Next, we consider the state variables, x_1, x_2, x_3 as constants. Eliminating λ_1 and λ_3 from (8) and (10), we find the differential equation in λ_2

$$\left(D^3 + a_2D^2 + a_1D + a_0\right)\lambda_2 = A_2e^{-\delta t} \qquad (11)$$

where,

$$a_2 = -\left(\frac{r_1}{K_1}x_1 + \frac{r_2}{K_2}x_2 + \frac{r_3}{K_3}x_3\right)$$

$$a_1 = \left(\frac{r_1r_2}{K_1K_2} - \alpha_{21}\alpha_{12}\right)x_1x_2 + \left(\frac{r_1r_3}{K_1K_3} + \alpha_{31}\alpha_{13}\right)x_1x_3 + \left(\frac{r_2r_3}{K_2K_3} + \alpha_{32}\alpha_{23}\right)x_2x_3$$

$$a_0 = \left(\alpha_{32}\alpha_{13}\alpha_{21} - \frac{r_1\alpha_{32}}{K_1}\alpha_{23} + \alpha_{31}\alpha_{13}\alpha_{23}\right)x_1x_2x_3 + \left(\frac{r_3}{K_3}\alpha_{12}\alpha_{21} - \frac{r_2}{K_2}\alpha_{31}\alpha_{13} - \frac{r_1r_2r_3}{K_1K_2K_3}\right)x_1x_2x_3$$

$$\begin{aligned} A_2 &= \left[\left(\alpha_{32}\alpha_{13} + \frac{r_3}{K_3}\alpha_{12}\right)x_1x_3 + \alpha_{12}\delta x_1\right]p_1q_1E + \\ &\left[\left(\alpha_{31}\alpha_{12} - \frac{r_1}{K_1}\alpha_{32}\right)x_1x_3 - \alpha_{32}\delta x_3\right]p_3q_3E \\ &-\left(\frac{r_1r_3}{K_1K_3} + \alpha_{31}\alpha_{13}\right)x_1x_3p_2q_2E \\ &-\left(\frac{r_1}{K_1}\delta x_1 + \frac{r_3}{K_3}\delta x_3 + \delta^2\right)p_2q_2E \end{aligned}$$

The solution of the equation (10) is

$$\lambda_2(t) = C_1e^{m_1t} + C_2e^{m_2t} + C_3e^{m_3t} + \frac{A_2}{N}e^{-\delta t}, \qquad (12)$$

where $C_i, i = 1,2,3$ are arbitrary constants, and $m_i, i = 1,2,3$ are the roots of the equations $a_3m^3 + a_2m^2 + a_1m + a_0 = 0$, and $N = a_0 - a_1\delta + a_2\delta^2 - a_3\delta^3 \neq 0$. It is clear from the equation (11) that $\lambda_2(t)$ is bounded if and only if $m_i < 0, i = 1,2,3$ or $C_i, i = 1,2,3$ identically equal zero. It is difficult to identify whether $m_i < 0, i = 1,2,3$. For this, we assume that $C_i = 0, i = 1,2,3$, then $\lambda_2(t) = \frac{A_2}{N}e^{-\delta t}$. By a similar

 https://doi.org/10.1051/e3sconf/20183108008

process, we get $\lambda_1(t) = \frac{A_1}{N} e^{-\delta t}$ and $\lambda_3(t) = \frac{A_3}{N} e^{-\delta t}$ where

$$A_1 = \left[\left(\alpha_{23}\alpha_{31} + \frac{r_3}{K_3}\alpha_{21}\right)x_2 x_3 + \left(\alpha_{21}\delta - \frac{r_2}{K_2}\delta\alpha_{31}\right)x_2\right]q_2 p_2 E$$

$$-\left[\left(\alpha_{32}\alpha_{23} + \frac{r_2 r_3}{K_2 K_3}\right)x_2 x_3 - \frac{r_2}{K_2}\delta x_2 - \frac{r_3}{K_3}\delta x_3\right]q_1 p_1 E$$

$$+\left[\left(\alpha_{32}\alpha_{21} - \frac{r_2}{K_2}\alpha_{31}\right)x_2 x_3 + \alpha_{31}\delta x_3\right]q_3 p_3 E$$

$+2\delta^2\alpha_{31}q_2 p_2 E - \delta^2 q_1 p_1 E$, and

$$A_3 = \left[\left(\frac{r_2}{K_2}\alpha_{13} - \alpha_{12}\alpha_{23}\right)x_1 x_2 - \delta\alpha_{13}x_1\right]q_1 p_1 E +$$

,

$$\left[\left(\frac{r_1}{K_1}\alpha_{23} - \alpha_{21}\alpha_{13}\right)x_1 x_2 - \delta\alpha_{23}x_2\right]q_2 p_2 E +$$

$$\left[\left(\alpha_{12}\alpha_{21} - \frac{r_1 r_2}{K_1 K_2}\right)x_1 x_2 + \frac{r_2}{K_2}\delta x_2 + \frac{r_1}{K_1}\delta x_1 - \delta^2\right]q_3 p_3 E$$

We observe that the shadow prices $\lambda_i(t)e^{\delta t}, i = 1,2,3$ of each fish species is bounded when $t \to \infty$ and we get $\lambda_i(t)e^{\delta t} = \frac{A_i}{N} = \text{constant, for } i = 1,2,3.$

We maximize the Hamiltonian in the equation (7) for $0 \le E \le E_{max}$. We consider that the optimal equilibrium does not occur either at $E = 0$ or $E = E_{max}$. There is a singular control that satisfies the equation,

$$\frac{\partial H}{\partial E} = e^{-\delta t}\left(p_1 q_1 x_1 + p_2 q_2 x_2 + p_3 q_3 x_3 - c\right) -$$

$$\lambda_1 q_1 x_1 - \lambda_2 q_2 x_2 - \lambda_3 q_3 x_3 = 0, \qquad (13)$$

We conclude that the total harvesting cost per unit effort equals to the discounted value of the future profit at the steady state [5, 11]. By eliminating parameters λ_1, λ_2 and λ_3 into equation (11), we obtain

$$q_1 x_1\left(p_1 - \frac{A_1}{N}\right) + q_2 x_2\left(p_2 - \frac{A_2}{N}\right) + q_3 x_3\left(p_3 - \frac{A_3}{N}\right) = c \qquad (14)$$

The optimal equilibrium solution $x_{1\delta}$, $x_{2\delta}$, , and $x_{3\delta}$ can be solved simultaneously from (4) and (14) for a given value δ. We observe that if $\delta \to \infty$, we have $\frac{A_i}{N} \to 0$. It implies that $p_1 q_1 x_{1\infty} + p_2 q_2 x_{2\infty} + p_3 q_3 x_{3\infty} = c$. Thus, we have $R(x_{1\infty}, x_{2\infty}, x_{3\infty}), E) = 0$. It shows that the net economic revenue is fully lost when the discount rate is infinite. This conclusion is also supported by Kar *et al.* [11] for the combined harvesting of two competing species, while there is no harvesting for predator.

4 Numerical simulations

In this section, we present the simulation results to illustrate the dynamical of the model. We pick up for some parameters value from with $r_1 = 2.09$, $r_2 = 2.07$, $r_3 = 2.1$, $\alpha_{12} = 0.001$, $\alpha_{21} = 0.001$, $\alpha_{13} = 0.01$, $\alpha_{23} = 0.01$, $\alpha_{31} = 0.01$, $\alpha_{32} = 0.01$, $q_1 = 0.03$, $q_2 = 0.03$, $q_3 = 0.03$, $p_1 = 5$, $p_2 = 5$, $p_3 = 2$, $c = 50$ and $\delta = 0{,}05$. In our optimal problem, we have an equilibrium solution that satisfies the necessary conditions of the maximum principle. For the parameter values as is given, we find that both the biological equilibrium $P_5(33,32,116)$ and the bionomic equilibrium $R(39,38,137)$ exist and achieve to the optimal equilibrium solution. We find that the optimal harvesting effort is $E = 5.6$ units, regarding to the optimal equilibrium $(36,35,126)$. The simulations are presented in the following figures.

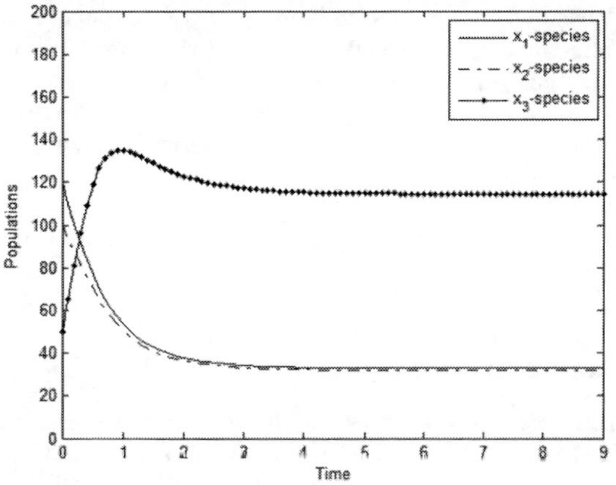

Fig. 1. Graph of the populations against time with initial conditions $x_1 = 120$, $x_2 = 100$, $x_3 = 50$

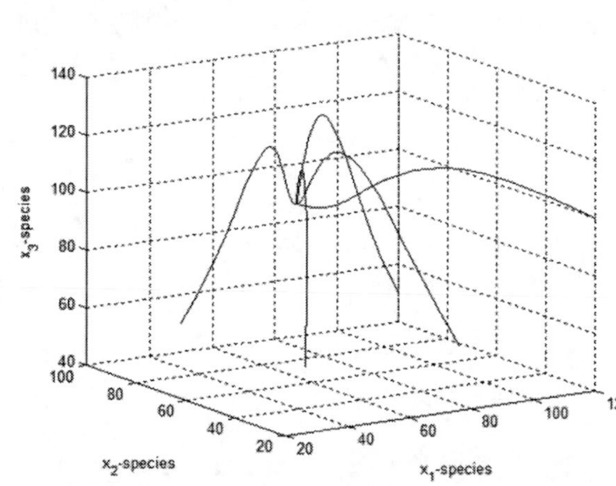

Fig. 2 Graph of the populations corresponding to the optimal harvesting effort $E = 5.6$ units, with different initial conditions.

In figure 1 and figure 2 shows that the trajectories of the solutions close to the optimal equilibrium $P(36,35,128)$. It indicates that the optimal equilibrium solution is globally asymptotically stable.

5 Conclusion

An ecological model was proposed and analysed to study the dynamics for two competing prey species and one invasive species (as predator) in an aquatic habitat. In this paper, we have attempted to study the effects of harvesting of two competing prey species and one predator species, in which predator species eats other two prey species and also the predator competes for using the same resources among this species. The existence of the possible steady states and bionomic equilibrium of the system have been studied. The globally stability of the system in co-existent state is examined by using Lyapunov function.

The optimal harvesting problem was addressed to obtain the optimal equilibrium solution. It is shown that the shadow prices of each biomass remain constant over time, so the boundedness and positivity condition are satisfied. An ecological model is developed from one prey-one predator model, in which the growth of each population is formulated in logistics term for among the same species. For interaction in other species is proposed in bilinear term. In the next research, will be developed for interaction of predator and prey species based on the limited density of prey population, in which it is more reasonable in an ecological system.

Acknowledgements

This work is partly founded by DIPA Faculty of Sciences and Mathematics, Diponegoro University, Semarang, Indonesia.

References

1. A.K. Sakai, F.W Allendorf, J.S Holt, Lodge D.M, Molofsky J, Kimberly A, S. Baughman, R. J. Cabin, J.E. Cohen, N.C. Ellstrand, D.E. McCauley, P. O'Neil, I.M. Parker, J.N. Thompson, S.G. Weller, Annu. Rev. Ecol. Syst., **32** (2001).
2. M. Clavero, Garcý´a-Berthou E. *Trends Ecol Evol* **20** (2005).
3. C.W. Clark, *John Wiley and Sons*, New York,NY, USA, (1990).
4. T. Das, R.N. Mukherjee, K.S. Chaudhuri, App. Math. Modell. **33**:228-2292 (2009).
5. T.K. Kar, U.K. Pahari, K.S. Chaudhuri, J. Appl. Math. & Computing, **19**(1-2), 311-326 (2005).
6. R. Zhang, J. Sun, H. Yang, Appl. Math. Sci. **1** : 2481-2492 (2007).
7. B.R. Reddy, Int. J. Math. Trends and Tech. **4**(2), (2013).
8. T. Das, R.N. Mukherjee, K.S. Chaudhur, J. Biol. Dyn. **3**(5):447-462 (2009).
9. K.S. Chaudhuri, S.S. Ray, J. Biol. Syst. **4**: 373–389, (1996).
10. A.V.P. Rao, K.L. Narayan, S. Bathul, BSOMASS, **1**(1), (2012).
11. T.K. Kar, K.S. Chaudhuri, ANZIAM J. **45 :** 443-456 (2004).

 https://doi.org/10.1051/e3sconf/20183108009

Conservation Management of Agriculture Land using Geospatial Approach (A Case Study in the Bone Watershed, Gorontalo Province, Indonesia)

Sri Maryati[1]*, *Sunarty* Eraku[1], and *Muh* Kasim[2]

[1]Geography Education Study Program, Department of Earth Science and Engineering, Faculty of Mathematics and Natural Sciences, Universitas Negeri Gorontalo, 96128 Indonesia
[2]Engineering Geology Study Program, Department of Earth Science and Engineering, Faculty of Mathematics and Natural Sciences, Universitas Negeri Gorontalo, 96128 Indonesia

Abstract. Bone Watershed is one of the major watersheds in Gorontalo Province. Bone watershed has a very important role for the people of Gorontalo Province. The role of Bone Watershed is mainly related to the providing clean water, producing oxygen, controlling flood, providing habitat for endemic flora fauna and other environmental functions. The role of Bone Watershed for the community's economic sector is also very important, the Bone watershed provides livelihood for surrounding communities includes fertile land resources for agriculture and plantations, forest products, and livestock feed. This research is important considering the Bone watershed has limited availability of land for agriculture and the high risk of natural disasters such as floods and landslides. Geospatial data includes topography map, landform map, soil map, integrated with field survey results and soil properties were analized to determine conservation management of agriculture land in the Bone Watershed, Gorontalo Province, Indonesia. The result of this study shows that based on soil properties and physical land characteristics, land use for agriculture should consider appropriate conservation techniques, land capability and respect to local wisdom.

1 Introduction

Bone Watershed is one of the major watersheds in Gorontalo Province. Bone watershed has a very important role for the people of Gorontalo Province. The role of Bone watershed is mainly related to the providing clean water, producing oxygen, controlling flood, providing habitat for endemic flora fauna and other environmental functions. The role of Bone Watershed for the community's economic sector is also very important, the Bone watershed provides livelihood for surrounding communities includes fertile land resources for agriculture and plantations, timber forest products, rattan, resin and livestock feed.

Bone Watershed also has its own characteristics with the existence of National Park Bogani Nani Wartabone in Bone Watershed which is the habitat of endemic flora and fauna such as Anoa, Maleo birds, and forest orchids. Area of National Park Bogani Nani Wartabone which belongs to Bone Watershed area is 73,477.12 Ha or 70.65% of total watershed area. Bone Watershed sustainability must be maintained in an integrated manner considering the watershed is a complex system. Environmental damage from one part of the watershed will affect other parts of the watershed and lead to the possibility of natural disasters and other economic losses.

This study focuses on the management of conservation for agriculture in the Bone Watershed, Gorontalo Province. Environmental conservation should be implemented for environmental sustainability preserved and supports the concept of sustainable development. This becomes very important because the Gorontalo province in December 2016 was appointed as a conservation province by the Indonesia Minister of Forestry and Environment.

Similar to other regions in Gorontalo Province, Bone Watershed community choose corn commodities because corn is easy to be cultivated in a wide range of physical conditions of the land. Besides, corn has short growing period and is one of the staple foods other than rice for the people of Gorontalo. Agricultural management that considers environmental sustainability can improve agricultural productivity, increase community income and minimize opening of agricultural land in forest areas, including protected areas.

Human intervention in watershed management is important as there is interaction between human activities and natural process in watershed [1]. This is due to the increased human pressure on nature due to the increasing population which causes various development activities in watershed [2]. A good watershed management will have an impact on

* Corresponding author: sri.maryati@ung.ac.id

increasing people's incomes, increasing employment opportunities, reducing migration to cities, as well as reducing poverty [3].

The benefits of watershed management for the agricultural sector including the improvement of cropping patterns and provide better agricultural products [4]. The upland of watershed provide forestry products, livestock and fruits while the lowland of the watershed provide food crop [5]. The benefits of watershed management on agricultural land include increased agricultural productivity by increasing irrigated areas and soil conservation, crop diversification, high value crop planting and fodder availability. This is ultimately improves people livelihood [6].

This research aims to determine appropriate conservation management of agricultural land using geospatial approach in Bone Watershed, Gorontalo Province, Indonesia. This research utilizes various geospatial data, field survey, laboratory analysis, and determination of appropriate conservation management based on the consideration of land physical characteristics.

2 Study Area

The study area is Bone Watershed, as can be seen in Figure 1 located in Bone Bolango Regency, Gorontalo Province, Indonesia which covers an area of 1017.55 km^2. Geographic coordinate of the study area is 0° 23' 48.26" N to 0° 42' 18.354" N and 123° 3'45.522" E to 123°32'56.235" E.

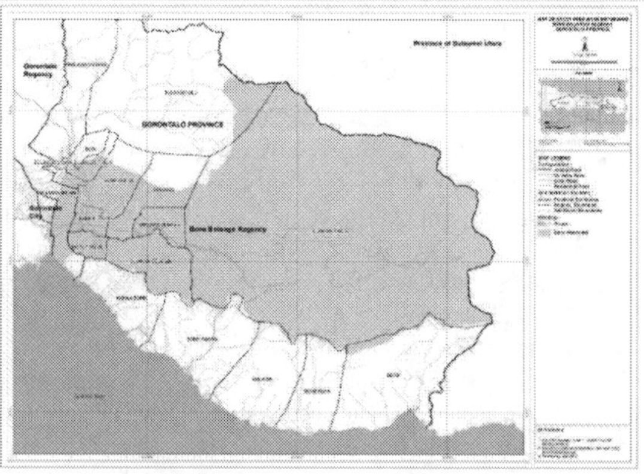

Fig. 1. Map of Study Area

Based on administrative boundaries, Bone Watershed is located in Bone Bolango Regency and Gorontalo City. There are 17 districts of Bone Bolango Regency and 6 Subdistricts of Gorontalo City which are included in Bone Watershed area. The largest subdistrict is Pinogu with an area of 47.398.25 Ha and Suwawa Timur which is 33,084.67 Ha. Based on Map Function Area, there are 5 types of area functions in Bone Watershed including national park, protected forest, production forest, limited production forest and non forest area.

3 Methodologies

This study used various spatial data include topographic map, landform map, soil map, climate map, geological map, and area function map. The unit of analysis used in this research is land unit. Land unit map was generated by overlaying landform map, slope map, and land map. Land unit map is presented in Figure 2.

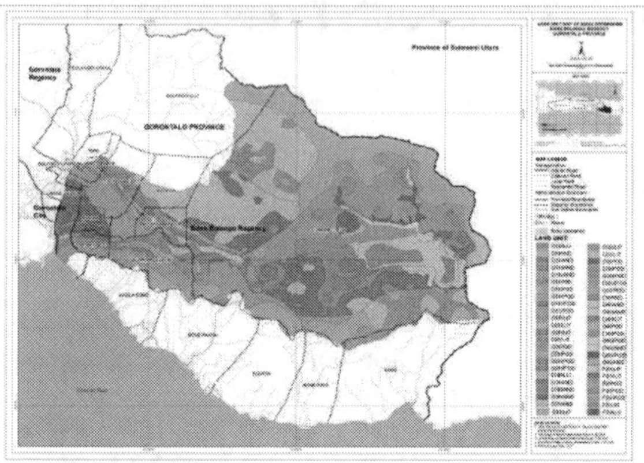

Fig. 2. Land Unit Map

Field survey was conducted to identify physical characteristics of land and soil sampling. Field survey results, soil properties, land capability, area function map, local wisdom characteristics were analyzed to determine conservation management of agriculture land in the Bone watershed, Gorontalo Province, Indonesia. This study is limited to areas that have status non protected forest and non the national park area.

4 Research Finding and Discussion

4.1 Topography

Topographic factors influence stability of land and determine velocity and volume of runoff. The main topographic factors are angle of slope, length of slope, and elevation. The longer the slope length is, the greater the volume of accumulated water is. The accumulated water flows in the land surface increases the erosive power. The greater slope angle is, the greater runoff velocity and also the greater ability of runoff to transport the soil particles are. Agriculture land located at higher elevation is more susceptible to erosion and landslides compared to the land located in the lowland area. Based on the slope map presented in Figure 3, the topography of the study area varies widely from flat, undulating, rolling, steep, and very steep. Based on the percentage, Bone Watershed is dominated by steep topography (34.66%) and very steep topography (34.61%).

 https://doi.org/10.1051/e3sconf/20183108009

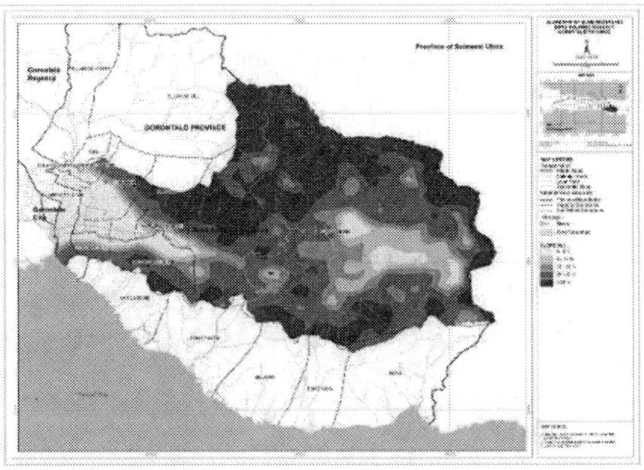

Fig. 3. Slope Map

4.2 Landform

Landform in the study area, as can be seen in Figure 4 Consist of alluvial plain, intermountain basin, the slope of denudasional hills, denudasional mountain, and denudasional hills. The landform became one of the important data inputs in this study because the landform is closely related to topography, parent material, and soil type hence it plays an important role in agricultural planning.

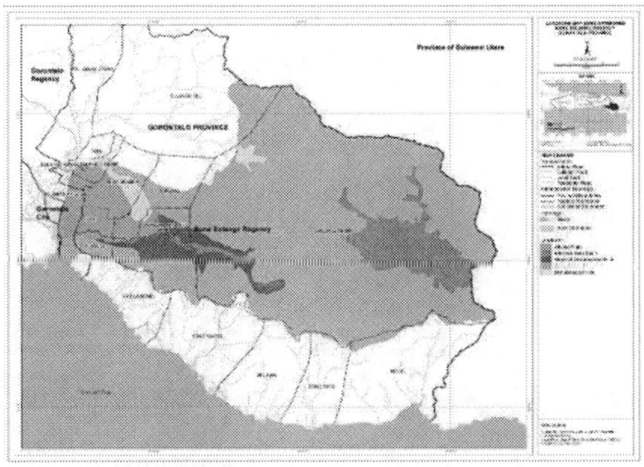

Fig. 4. Landform Map

4.3 Soil Properties

Chemical and physical properties of soil influence soil fertility and soil productivity. Thus soil properties are very important to determine suitable crop, appropriate soil treatment, proper soil improvement, and proper amount of fertilizer. Analysis of soil properties provided important information of soil condition both chemical and physical properties. The soil properties can be improved and maintained to increase the quality of soil by adding organic matter, adding appropriate amount and type of fertilizer, stabilizing topsoil against soil erosion and water detachment, and ripping to increase soil porosity.

Based on soil map, soil types in the study area include alluvial, litosol, podsolic and andosol. Field survey results showed soil depth in the study area varies from 25 -40 cm, 40-60cm, and> 60 cm; soil texture of top soil include clay, sandy clay, sandy clay loam; percent of coarse material <5% and 5 - 15%. Related to risk hazard, generally no risk and low for flood and erosion. Some areas which has steep slope has severe erosion.

4.4. Land Capability

Land capability analysis is conducted to evaluate the potency of land for sustainable long term use. Result of land capability analysis can be used to support decision making for sustainable development in agriculture sector. Land capability assessment is used to determine land management to prevent land degradation both longterm conservation and type of land utilization[7]. Land capability analysis utilize topography data, soil properties data, and risk hazard data. Result of land capability in this study is presented in Figure 5.

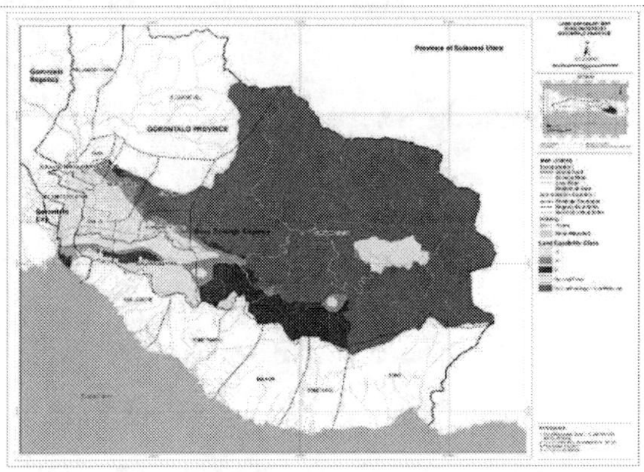

Fig. 5. Land Capability Map

Based on Figure 5, class of land capability in the study area consist of 3 class : Class III, Class IV, and Class VI. The study area is dominated by land capability Class III (51.39%), followed by Class VI (33,44%), and Class IV (15.17%).

4.6. Conservation Management

Referring to result of land capability analysis, the study area has 3 class of land capability : Class III, Class IV, and Class VI. Land classified in Class III and Class IV can be utilized for agriculture with moderate and high conservation effort. Whereas land categorized in Class VI could not be used for cultivation, this land may be utilized for grazing or plantation. Type of limiting factor for class VI is topography condition (steep slope). Topographic conditions in steep slope land are natural factors that are difficult to improve in conservation, because it will change the landscape that may cause negative impacts to the environment such as erosion, landslides and land degradation triggering. Hence, class VI should be a forest or grazing land.

The limiting factors for class III and class IV are soil drainage, soil depth, soil texture, slope, and % of coarse material. These limiting factors can be treated by soil improvement and soil conservation. Soil improvement according to the limiting factors will improve the capability class, the use for agricultural land will be productive and sustain. Topographic factors as a limiting factor in classes III and IV can be overcome by the selection of appropriate conservation techniques namely vegetative techniques, mechanical techniques, or combined both techniques.

5 Conclusions

The results of this study are expected to be the basis for decision making and direction in the land utilization for agriculture in the Bone Watershed. Utilization of land for agricultural areas can be carried out on land that has the capability class III and class IV with appropriate soil conservation efforts and respect to local wisdom. The selection of agricultural commodities can consider the suitability of land for optimal productivity and land can be used for sustainable agriculture.

Acknowledgements

The authors' wish to express their deepest gratitude to Directorat General Higher Education, Ministry of Research, Technology and Higher Education Indonesia for research funding. The authors also acknowledge Rector of Universitas Negeri Gorontalo, Head of Institute for Research and Community Services Universitas Negeri Gorontalo, and Dean of Faculty of Mathematics and Sciences for permit and support to participate and to attend The 2nd International Conference on Energy, Environment and Information System (ICENIS).

References

1. X. Cai, R.Vogel, R. Ranjithan, Journal of Water Resources Planning and Management **139 (5)** (2013)
2. A.Mirchi, D.Watkins Jr, K.Madani, *Modeling for Watershed Planning, Management, and Decision Making* (Nova Science Publishers, Inc, 2009)
3. P.Pathak, A.K. Chourasia, S.P.Wani, R.Sudi, Journal of Water Resource and Protection **5** (2013).
4. S.S.Nerkar, A.Pathak, C.S.Lundborg, A.J. Tamhankar, International Journal of Environmental Research and Public Health **12** (2015)
5. G.B.Thapa. *Integrated Watershed Management : Basic Concepts and Issues. In Human Settlement Development* (Eolss Publishers Co.Ltd, 2009)
6. S.P.Wani, KH.Anantha, TK. Sreedevi, R.Sudi, SN. Singh, and M. D'Souza, Journal of Sustainable Watershed Science and Management **1(1)** (2011)
7. T. Notohadiprawiro, *Kemampuan dan Kesesuaian Lahan: Pengertian dan Penetapannya, Lokakarya Neraca Sumberdaya Alam Nasional,* DRN Kelp. II- BAKOSURTANAL, (1991).

AUTHOR INDEX

AUTHOR INDEX

AUTHOR INDEX

AUTHOR INDEX

AUTHOR INDEX

2nd International Conference on Energy, Environmental and Information System (ICENIS 2017)

E3S Web of Conferences Volume 31 (2018)

Semarang, Indonesia
15 – 16 August 2017

Part 2 of 2

Editors:

Hadiyanto
Sudarno
Maryono

ISBN: 978-1-5108-5877-0

Printed from e-media with permission by:

Curran Associates, Inc.
57 Morehouse Lane
Red Hook, NY 12571

Some format issues inherent in the e-media version may also appear in this print version.

Printed by Curran Associates, Inc. (2018)

For additional information, please contact EDP Sciences – Web of Conferences at the address below.

EDP Sciences – Web of Conferences
17, Avenue du Hoggar
Parc d'Activité de Courtabœuf
BP 112
F-91944 Les Ulis Cedex A
France

Phone: +33 (0) 1 69 18 75 75
Fax: +33 (0) 1 69 28 84 91

contact-edps@webofconferences.org

Additional copies of this publication are available from:

Curran Associates, Inc.
57 Morehouse Lane
Red Hook, NY 12571 USA
Phone: 845-758-0400
Fax: 845-758-2633
Email: curran@proceedings.com
Web: www.proceedings.com

TABLE OF CONTENTS

PART 1

PART 2

Carbon Value Analysis of Batang Gadis National Park, Mandailing Natal Regency, North Sumatera Province, Indonesia

Dini Novalanty Ohara Daulay[1], and *Jafron* Wasiq Hidayat[2]

[1]Master Program of Environmental Science, School of Postgraduate Studies, Diponegoro University, Semarang-Indonesia
[2]Department of Biology, Faculty of Science and Mathematics, Diponegoro University, Semarang-Indonesia

Abstract. Global warming is an important issue in the world which it gives a negative effect on human life. One indicator of global warming is increasing greenhouse gas i.e. carbondioxide from human activities. Deforestation and forest degradation are the second largest contributor of carbon into the atmosphere, after the use of fossil fuels by industry and transportation. As lungs of the world, forest is enable to produce renewable energy sources i.e. biomass. Forest carbon stock in above ground biomass (AGB) is the greatest effect source on deforestation and forest degradation. Therefore, it is necessary to perform a study the potential of carbon in forest. The purpose of this research is to determine carbon stock value in Batang Gadis National Park, Mandailing Natal Regency, North Sumatera Province, Indonesia. The carbon potential stored in this forest vegetation is calculated using AGB allometric equation by using data in diameter at breast height (dbh = 1.3 m), height, and density of the wood for trees. Data obtained from secondary data is Asset Assessment Report which State Controlled Forest Natural Resources Batang Gadis National Park, 2016. Study locations were Pagar Gunung and Sopo Tinjak Villages. Carbon stock values were calculated and analyzed with assumption that a half of biomass part is carbon stock which using Australian carbon price about AUD $ 11.82 Australia (Australian dollars) and EU € 5 (US $ 6). The results showed that the total biomass in Pagar Gunung and Sopo Tinjak Villages amounted to 259.83 tonnes and 160.89 tonnes. From the results of the total biomass, the total carbon stocks (C) and CO_2 stocks in both villages are 210.36 tonnes (129.92 tonnes in Pagar Gunung Village and 80.45 tonnes in Sopo Tinjak Village) and 772.03 tonnes (476.79 tonnes in Pagar Gunung Village and 295.24 tonnes in Sopo Tinjak Village). By using the carbon price prevailing in the market place Australia Emission Trading System (ETS) and the EU ETS (AUD $ 11.82/t CO_2e and € 5 (US $ 6)/t CO_2e), the value of carbon stock that can be produced from Batang Gadis National Park (Pagar Gunung and Sopo Tinjak Villages) is about Rp. 92,499,921.72 (in AUD $) or Rp. 61,654,433.67 (in US $).

1. Introduction

Forest has an important role in maintaining the survival of human life as well as other living creatures. Based on function, forest has three functions, namely conservation, protection, and the production functions [1]. As lungs of the world, forest produces oxygen necessary for human and the ability of absorbing carbon. Forest vegetation that has chlorophyll that capable of absorbing CO_2 from the atmosphere through photosynthesis. The results of this photosynthesis is stored as biomass that makes vegetation grow bigger or higher. This growth will continue until the vegetation is physiologically stopped growing or harvested [2].

Forest can reduce excessive carbon emissions in the atmosphere and store it in a variety of separate parts such as vegetation, litter and soil organic matter. Measurement the amount of carbon stored in the body plant life or biomass in an area can describe the amount of carbon dioxide in the atmosphere. Plants or long-lived tree that grows in the forest or in the garden is a mixture of carbon storage dumps or much larger than annual crops. Like-wise woody vegetation is the highest carbon sink [3].

Indonesia is one country that has the largest forest in the world. According to the Ministry of Environment and Forestry data [4], forest area in Indonesia is 124 million hectares nowadays. However, the rate of deforestation is very high every year reached 684.000 hectares. This condition places Indonesia as the second

* Corresponding author: dinidaulay29@gmail.com

highest country to lose its forests after Brazil. Forest damage can threaten the extinction of animals and plants in it, so it has an impact on biodiversity degradation. In addition, forest degradation and deforestation also cause an increase in greenhouse gases resulting from CO_2 into the atmosphere, causing global warming.

Batang Gadis National Park (BGNP) as a nature conservation area in Indonesia has considerable benefits, tangible and intangible, one of which is the storage and/or sequestration. BGNP comprising several mountainous landscape of the save function as water catchment areas and water sources for some watersheds. This area has a high diversity of flora and fauna are rare flora and fauna (protected). It also has interesting natural beauty, with the cool air, and there are various tourisms potential that can be developed for nature tourism. Unfortunately, this area is also facing of the activities of encroachment caused by forest conversion. These conditions should be solved immediately. Therefore, the calculation of above ground biomass and carbon stock should be undertaken as a first step in simulation the emission modeling and absorption of greenhouse gases in an area. Data or information regarding the carbon value is important as a consideration in the conversion. With the availability of accurate data on carbon stock in some forest types and how much carbon emissions released when forests are converted, is expected to help the government in taking appropriate policy related efforts to reduce carbon emissions from forest deforestation and degradation.

2. Method

Batang Gadis National Park (BGNP) is located in Mandailing Natal (Madina) Regency, North Sumatra Province, precisely in Bukit Barisan Mountains of Northern Sumatra. Geographically, BGNP is located at 99 ° 12 '45 "up to 99 ° 47' 10" east longitude and 0 ° 27 '15 "up to 1 ° 01' 57" north latitude. Administratively, this area is in Mandailing Natal (Madina) Regency, North Sumatra Province which includes thirteen districts and sixty-eight villages, where one village of them, the Batahan Village is an *enclave* in BGNP. The name of this national park is named the main river that flows and splits the Madina Regency i.e. Batang Gadis River. Location BGNP is ± 519.1 km from the capital of North Sumatra Province, Medan.

This area is a stretch of tropical rain forest in the plain topography, the hills to the mountains with an elevation between 300-2145 meters above sea level (top of Sorik Marapi Mountain). Based on the distribution of vegetation, BGNP area are divided into two sub-types of forest formations. First, sub-type Air Bangis-Singkil formation which is part of the Bukit Barisan Western-type formations hills of medium elevation (300-1000 m asl). Second, sub-type of the Forest of Montana (1000-1800 m asl), which is part of the Bukit Barisan type formations above 1000 meters above sea level [5].

The secondary data was obtained from the survey i.e. tree list data complete with diameter that comes from Asset Assessment Report which State Controlled Forest Natural Resources Batang Gadis National Park, 2016. In the secondary data, the locations were used as place of research are Pagar Gunung and Sopo Tinjak Villages. Widely used research areas are 0.6 hectares (Pagar Gunung Village) and 0.76 hectares (Sopo Tinjak Village). Estimation of above ground biomass tree is calculated by allometric approach that is based on dbh *(diameter breast height* /dbh) 1.3 m above ground level. Allometric equations to estimate the biomass of vegetation, namely:

$$W = 0.11 * \rho * D^{2+c} \quad (1)$$

Where, W = biomass per tree (kg),
D = DBH (cm),
P = density of the wood or the density of the wood (g / cm^3 or t / m^3),
c = coefficient that describes the relationship between the diameter and tall tree (used c = 0.62 [6].

The carbon content is assumed to use the value of the biomass allometric equation where 50% of the biomass of a forest vegetation is composed of the elements carbon [7]. To convert carbon (C) into CO_2, then eq. 2 is used [8]:

$$CO_2 = C*3.67 \quad (2)$$

The assessment poles and trees are used to determine the value of carbon in BGNP area. The determination of the carbon is using a carbon price basis ranged Australia AUD $ 11.82 (Australian dollars) [9] and the EU € 5 (US $ 6) [10].

3. Results and Discussion

Based on the results obtained from secondary data, it can be seen that the species found 418 individual of trees in Pagar Gunung Village and 391 individual of trees in Sopo Tinjak Village. The distribution of tree species is presented in Table 1.

Table 1. The distribution of tree species in Pagar Gunung and Sopo Tinjak Villages

No.	Species	Pagar Gunung Village		Sopo Tinjak Village	
		Total plot attendance	Total individual trees	Total plot attendance	Total individual trees
1	*Anona spec* (tree)	0	0	1	1
	Anona spec (pole)	0	0	1	2
2	*Aquilaria mallaccenis* (tree)	2	2	1	1
	Aquilaria mallaccenis (pole)	1	2	0	0
3	*Alseodaphne peduculans* (tree)	2	3	1	1

No.	Species	Pagar Gunung Village		Sopo Tinjak Village	
		Total plot attendance	Total individual trees	Total plot attendance	Total individual trees
4	*Cassia fistula* (tree)	3	6	0	0
	Cassia fistula (pole)	2	4	0	0
5	*Cinnamomum subavenum* (tree)	2	4	0	0
	Cinnamomum subavenum (pole)	1	1	0	0
6	*Cratoxylon arborescens* (Tree)	0	0	3	3
	Cratoxylon arborescens (pole)	0	0	3	16
7	*Cryptocarya tomentosa* (tree)	0	0	2	8
	Cryptocarya tomentosa (pole)	0	0	1	1
8	*Dipterocarpus gracilis* (tree)	1	5	4	10
	Dipterocarpus gracilis (pole)	1	1	2	3
9	*Dipterocarpus kunstleri* (tree)	4	9	2	6
	Dipterocarpus kunstleri (pole)	6	10	4	6
10	*Ficus benjamina* (tree)	0	0	1	1
11	*Ficus drupacea* (tree)	1	3	1	3
	Ficus drupacea (pole)	1	1	1	1
12	*Ficus gibbosa* (tree)	1	1	1	1
13	*Ficus glandulifera* (tree)	1	3	0	0
14	*Ficus toxicana* (tree)	4	7	0	0
	Ficus toxicana (pole)	3	3	0	0
15	*Garcinia dioica* (tree)	4	6	2	9
	Garcinia dioica (pole)	3	4	2	2
16	*Gluta renghas* (tree)	0	0	1	2
17	*Grewia acuminate* (tree)	3	21	0	0
	Grewia acuminate (pole)	1	2	0	0
18	*Hopea beccariana* (tree)	1	1	0	0
19	*Ilex pleiobrachiata* (tree)	0	0	1	1
20	*Knema conferta* (tree)	3	4	1	2
21	*Laportea stimulant* (tree)	1	1	1	1
	Laportea stimulant (pole)	0	0	2	2
22	*Liex pleiobrachiata* (tree)	6	11	0	0
23	*Litsea brachystachys* (tree)	11	31	5	8
	Litsea brachystachys (pole)	5	6	7	10
24	*Litsea cubeba* (tree)	0	0	1	2
25	*Litsea firma* (tree)	0	0	2	2
26	*Litsea odorifera* (tree)	2	2	0	0
	Litsea odorifera (pole)	1	1	2	2
27	*Litsea resinosa* (tree)	5	24	4	48
	Litsea resinosa (pole)	1	3	2	4
28	*Litsea sp.* (tree)	0	0	5	5
	Litsea sp. (pole)	0	0	3	3
29	*Macaranga lowii* (tree)	1	15	2	16
	Macaranga lowii (pole)	1	1	1	1
30	*Myristica lowiana* (tree)	1	1	1	1
31	*Myrtales ningra* (tree)	3	3	1	1
	Myrtales ningra (pole)	0	0	1	1
32	*Muraya paniculata* (tree)	1	1	0	0
33	*Neesia sp.* (tree)	7	19	0	0
34	*Neonauclea calycina* (pole)	1	1	0	0
35	*Palagium gutta* (tree)	2	2	0	0
36	*Palaquim obovatum* (tree)	1	1	0	0
37	*Palaquim rostratum* (tree)	1	1	0	0
38	*Palaquim spp.* (tree)	2	2	7	9
	Palaquim spp. (pole)	0	0	2	2
39	*Palaquim sumatranum* (tree)	4	7	0	0
40	*Parkia speciosa* (tree)	1	1	2	2
	Parkia speciosa (pole)	1	1	1	1
41	*Phyllanthus indicus* (tree)	1	1	0	0
	Phyllanthus indicus (pole)	1	1	0	0
42	*Podocarpus imbricatus* (tree)	0	0	1	1
	Podocarpus imbricatus (pole)	0	0	1	1
43	*Podocarpus menifolius* (tree)	2	2	0	0
44	*Pternandra cordara* (tree)	0	0	1	1
45	*Quercus gemelliflora* (tree)	3	17	15	32

No.	Species	Pagar Gunung Village		Sopo Tinjak Village	
		Total plot attendance	Total individual trees	Total plot attendance	Total individual trees
	Quercus gemelliflora (pole)	1	2	6	8
46	*Shorea hopeifolia* (tree)	0	0	1	1
47	*Shorea iliginosa* (tree)	1	2	0	0
48	*Shorea maxwelliana* (tree)	0	0	3	6
	Dipterocarpus gracilis (pole)	1	1	2	3
49	*Styrax paralleloneurus* (tree)	5	6	2	2
	Styrax paralleloneurus (pole)	0	0	1	1
50	*Syzygium sp.* (tree)	15	131	12	105
	Syzygium sp. (pole)	7	13	8	14
51	*Tabernaemontana macrocarpa* (tree)	0	0	1	1
	Tabernaemontana macrocarpa (pole)	0	0	1	1
52	*Urandra scorpiodes* (tree)	2	2	0	0
53	*Vitaca micrantha* (tree)	0	0	1	1
54	*Zanthoxylum limonelia* (tree)	1	1	0	0
55	*Other species*	0	3	0	15
	Total		418		391

Table 1 indicates different species of trees found in both locations. The majority of species of tree in the Pagar Gunung and Sopo Tinjak Villages is *Syzygium sp.* (Fam: *Myrtaceace*), with a number of individuals as much as 144 trees in Pagar Gunung Village, and 119 trees in Sopo Tinjak Village. Then, *Litsea brachystachys* (36 trees), *Litsea resinosa* (27 trees) in Pagar Gunung Village and *Litsea resinosa* (52 trees), *Quercus gemelliflora* (40 trees) in Sopo Tinjak Village. The composition of species diversity in these two locations is very diverse. There are 39 species composition in Pagar Gunung Village and 35 species in Sopo Tinjak Village. From the forest inventories conducted at both locations, approximately 18 trees (3 trees on Pagar Gunung Village and 15 trees in Sopo Tinjak Village) recorded are unknown species. The unknown species consists of 3 species in Pagar Gunung Village and 6 species in Sopo Tinjak Village. Distribution of the unknown species in Pagar Gunung and Sopo Tinjak Villages occupy 3 plots and 8 plots both levels of the tree or pole. They are classified as other species and totally of them is very few, only about 2.22% of the population.

Among the variety of species in both locations, there is one dominant plant species, with the number of plots the presence of 15 (tree) and 7 (pole) from 16 plots observed were *Syzygium sp.* in Pagar Gunung Village by 31.34% relative density (tree) and 3.11% (pole), the relative dominance of 15.05% (tree) and 0.69% (pole), and Important Value Index (IVI) for 56.46% (tree) and 8.50% (pole). In the Sopo Tinjak Village, which has a relative abundance of species and the highest IVI is *Syzygium sp.* with a value of 26.85% relative density (tree) and 3.58% (pole), as well as the value of IVI of 53.82% (tree) and 10.05% (pole). As for the relative frequency and relative dominance, species that have the highest ratings are *Quercus gemelliflora*, the relative frequency of 10.56% (tree) and 4.23% (pole), and the relative dominance of 18.93% (tree) and 0.60% (pole). This is probably because *Quercus gemelliflora* most commonly found in plots measured. Total plot the presence of this species as many as 15 (trees) and 6 (poles) of the total of 19 plots were observed. In addition, total basal area *Quercus gemelliflora* is higher than the species *Syzygium sp.* namely 3.35 and 0.11. The dominance of these species due to the kind found in large amounts (high density) and is spread almost evenly throughout the area. The types of dominant indicate that these species are more adaptive to environmental conditions, in other words, these species are able to adapt to the environment in its life. In the forest, as a result of competition, certain types of more powerful (dominant) than other types.

3. 1. Distribution of tree diameter class

Distribution of tree diameter class are shown in Table 2. Approximately 88.04% and 92.07% of the trees in Pagar Gunung and Sopo Tinjak Villages are under 40 cm in diameter, belongs to the class with a diameter ≥ 5-10 cm followed by a tree in diameter class > 40 cm (Pagar Gunung Village), 10.1 to 15 cm (Sopo Tinjak Village). In Pagar Gunung Village, the lowest diameter class (5-10 cm) is dominated by tree species *Syzygium sp.* While the class with the highest diameter (> 40 cm) is dominated by tree species *Neesia sp.*

Table 2. Distribution of tree diameter class

No.	Tree diameter class (cm)	Total individual trees	
		Pagar Gunung Village	Sopo Tinjak Village
1.	≥ 5-10	176	194
2.	10,1-15	28	38
3.	15,1-20	42	31
4.	20,1-25	34	30
5.	25,1-30	32	29
6.	30,1-35	33	25
7.	35,1-40	23	13
8.	>40	50	31
	Total	418	391

Meanwhile, in Sopo Tinjak Village, the lowest diameter class (≥ 5-10 cm) is dominated by tree species

Syzygium sp. While the highest diameter class (> 40 cm) is dominated by tree species *Syzygium sp*.

The composition of the BGNP forest area in Pagar Gunung and Sopo Tinjak Villages is overgrown vegetation with a variety size of tree diameter. However, in this area is dominated by vegetation with a small diameter. These conditions ensure the survival of the stand to reach the levels of the tree. The larger the diameter of a vegetation, then the numbers will decline. Similar results were shown by research Onrizal, et al. [11] at different locations namely that the tree population in Batang Toru Forest (Aek Game-game forest and Aek Silemes forest) consists of a mixture of a whole class of diameter which is dominated by trees that have a small diameter and density of tree will continue decreases with increasing diameter of the tree so that as the curve "L" which follows a negative exponential function. This is consistent with the model developed by Meyer [12] which described the distribution of diameter (dbh) usually skewed to the right. This classical model showing the condition of the forests are balanced (balanced forest), so that continuity is assured naturally able to stand in the future. However, at this exponential model of individual growth and death of trees are independent to size of the tree, such as disturbance [13].

Diameter of the tree is one of several parameters that affect the biomass in an ecosystem other than the individual density, diversity of tree species, and soil types. Indirectly all the parameters that affect the biomass will also affect carbon stock on a stand. Stem diameter is proportional to the value of biomass. The larger the diameter at breast height (dbh) indicates the old aged trees and old trees store more carbon than the young trees [14].

3. 2. Distribution of tree height class

Distribution of tree height class on Pagar Gunung and Sopo Tinjak Villages are quite diverse. At level of the tree, the tree density is dominated by height class < 10 m. Then, continue to decrease until the maximum height class (> 50 m) although at tree height class from 20.1 to 30 m in Pagar Gunung Village has increased the number of individual tree after falling on the tree height class 10-20 cm. This is as shown in Table 3.

Table 3. Distribution of tree height class

No.	Tree height class (m)	Number of individual trees	
		Pagar Gunung Village	Sopo Tinjak Village
1.	<10	188	222
2.	10-20	89	127
3.	20,1-30	103	41
4.	30,1-40	36	1
5.	40,1-50	1	0
6.	>50	1	0
	Total	418	391

Table 3 shows that the forest area in Pagar Gunung Village composed by a mixture of high-class whole tree. Density of tree height class in low (< 10 m) in Sopo Tinjak Village forest denser than Pagar Gunung Village forest. At Pagar Gunung plot, there is only one tree with a height which included high grade > 50 m, i.e *Ficus gibbosa*. While, in the Sopo Tinjak Village plot is not found a tree that reaches such a height. The maximum tree height in a Sopo Tinjak Village forest encountered only reached 31 m, which is belong to *Syzygium sp*. (Fam: *Myrtaceae*). Pagar Gunung and Sopo Tinjak Villages have complete canopy stratum i.e. stratum A, B, and C although when seen from the existing deployment tree height, Pagar Gunung Village more unevenly distributed than Sopo Tinjak Village. Canopy stratum on Pagar Gunung and Sopo Tinjak Villages generally dominated by the stratum C. This is as stated by Soerianegara and Indrawan [15] that tree with a height of 4-20 meters are categorized in stratum C with continuous canopy characteristics, low tree, and many branches. While the trees with a total height of over 30 meters with a regular layer, diconstinue canopy, tall and straight trunk with a high branch-free trunk categorized as stratum A, and trees with a total height of 20-30 meters with a continuous canopy generally categorized as stratum B. The existence of the canopy layer is very important in supporting the diversity of wildlife, and reducing the kinetic energy of rainwater, which in turn can reduce the risk of erosion.

3. 3. Biomass and Carbon Forest

Total above ground biomass (AGB) was calculated mainly from trees and poles (i.e, vegetation with height value of more than 1.30 m). These classes are defined based on the diameter (dbh) and height (h). Calculation of AGB Pagar Gunung Village forest and Sopo Tinjak Village forest are 405.99 tonnes/ha and 211.70 tonnes/ha. In total the highest average AGB in Pagar Gunung Village for all stands of trees found in the class > 40 cm (291.78 tonnes/ha) and the average of the lowest AGB found at diameter class ≥ 5-10 cm diameter class with total about 2.33 tonnes/ha. While, in Sopo Tinjak Village found the highest average total AGB for all stands at the trees with diameter class > 40 cm (139.96 tonnes/ha) and the average of the lowest AGB found in diameter class ≥ 5-10 cm with a total average 2.36 tonnes/ha. From these results show that more than 94% of the total AGB consists of a tree. This is as stated in Table 4.

Table 4. Biomass tree with tree diameter class

No.	Tree diameter class (cm)	Biomass (tonnes/ha)	
		Pagar Gunung Village	Sopo Tinjak Village
1.	≥ 5-10	2.33	2.36
2.	10,1-15	2.69	2.73
3.	15,1-20	9.16	5.67
4.	20,1-25	14.27	9.55

5.	25,1-30	20.07	15.74
6.	30,1-35	32.16	19.71
7.	35,1-40	33.54	16.00
8.	>40	291.78	139.96
	Total	405.99	211.70

Above ground biomass (AGB) on the Pagar Gunung and Sopo Tinjak Villages are 405.99 tonnes/ha and 211.70 tonnes/ha. When multiplied by the area of each area is 0.64 ha for Pagar Gunung Village and 0.76 ha for Sopo Tinjak Village, the total biomass of both locations are 259.83 tonnes and 160.89 tonnes. In Table 5, it shows that the diameter of the tree is very affected by biomass. The bigger the diameter of the tree, the bigger the biomass. Diameter growth was positively correlated with increasing biomass of trees as well as the amount of carbon stock in vegetation. This condition is consistent with the Windardi's statement [16] which stated that the vegetation is strongly influenced by amount of biomass that is owned. Biomass in every part of the tree increases proportionally with increasing magnitude of the tree diameter so that the tree biomass in each section have a relationship with the tree diameter [17].

Forest biomass is particularly relevant to the issue of climate change. Forest biomass has an important role in biogeochemical cycles, especially in the carbon cycle. Of the total forest carbon, about 50% of them are stored in forest vegetation [18]. The carbon content can be estimated by using biomass values obtained from allometric equation where biomass is carbon stored. From the previously known biomass value, the total carbon storage (C) in Pagar Gunung and Sopo Tinjak Villages are 129.92 tonnes and 80.45 tonnes (Figure 1). Potential carbon stocks (C) is equivalent to 476.79 tonnes of CO_2 for Pagar Gunung Village and 295.24 tonnes of CO_2 for Sopo Tinjak Village. These results suggest that carbon stock in both locations are quite high. This is in line with the Windardi's statement [16] which stated that if the value of vegetation biomass is large then the resulting carbon value conversion is also large.

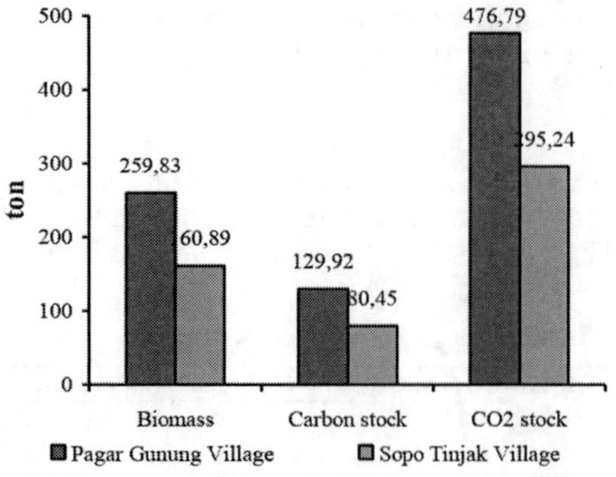

Fig. 1. Biomass, carbon stock, CO_2 stock in the BGNP forest (Pagar Gunung and Sopo Tinjak Villages)

Carbon stock in an area affected by type of vegetation. A trunk diameter of vegetation would indirectly also affect carbon storage on a stand [14]. When compared with the results of the two locations of Batang Toru primary forests i.e. Aek Silemes forest and Aek Game-game forest, AGB and carbon stock in forests BGNP (Pagar Gunung and Sopo Tinjak Villages) are lower. At Batang Toru primary forests (Aek Silemes forest and Aek Game-game forest), AGB are ranged from 544.4 to 583.0 tonnes/ha on Aek Silemes forest and 604.5 to 613.6 tonnes/ha on Aek Game-garne forest. While, carbon stocks are 272.2 to 291.5 tonnes C/ha on Aek Silemes forest and 302.2-306.8 tonnes C/ha on Aek Game-game forest. Location altitudes in Batang Toru primary forests are an altitude of 850-975 m above sea level (Aek Silemes forest) and 625-800 m above sea level (Aek Game-game forest) [11]. While, location altitudes of BGNP forests (Pagar Gunung and Sopo Tinjak Villages) are an altitude of ± 1000 to 1150 m asl ± 1200 to 1500 m asl. From these results indicate that the higher a site that would lower carbon stock. Conversely, if the lower one location then carbon stock will be higher. Altitude affects the species composition and diameter class [19.11]. It is also associated with rainfall and temperature. The higher somewhere, then the higher the rainfall intensity and the temperature will be lower. This condition is in line with Onrizal, et al. statement [11]; Zilikifli, et al. [20] which stated that forest biomass is influenced by climatic factors (rainfall and temperature) associated with the rate of increase of forest biomass and climatic differences (differences in the rate of production of organic material).

Carbon emissions trading is now a critical need given the increasing climate of the world up to 5 degrees Celsius. If this condition is allowed, it can bring damage to the environment. To achieve the target of reducing carbon dioxide emissions, then Protoko Kyoto (1997), which discussed the framework of the climate change convention (Framework Convention on Climate Change, FCCC) has agreed that industrialized countries will reduce emission levels averaged 5.2% below the level 1990 in 2008 to 2012 including emissions trading mechanisms. Carbon trading is a way relieve the burden on the industrialized countries to reduce their emissions. Usually, the trade is done between the developed and developing countries [21].

Determination of the carbon value in the global carbon trading market can be done through several mechanisms i.e. joint implementation (JI) and clean development mechanism (CDM). At joint implementation mechanism, an industrialized country invests emission reduction projects in other industrialized countries. Emission reduction credits gained from the implementation of the project will be given to countries that make investments. At CDM mechanism, industrial and developing countries working together to undertake clean development. Through CDM, industrialized countries can reduce their emissions by undertaking emission reduction projects while developing countries will receive financial and technological compensation from such cooperation. Emissions trading may be mandatory (required by government) or voluntary. Nowadays, the countries that have been trading carbon in their exchanges such as The European Union, The United States, Australia, New Zealand, and Japan i.e. European Climate Exchange, NASDAQ OMX European Commodities, PowerNext, Commodity Exchange Bratislava, European Energy Exchange, Carbon Trade Exchange, and Chicago Climate Exchange. Determination of carbon values in this study are using carbon price in the Australian Emission Trading System (ETS) AUD \$ 11.82/t CO2e, and European Union ETS € 5 (US \$ 6)/t CO2e. The amount of carbon dioxide stock (CO2) in Pagar Gunung and Sopo Tinjak Villages is 772.03 tonnes, so the total carbon value is Rp. 92,499,921.72 (in AUD \$) or Rp. 61,654,433.67 (in US \$). From these results indicate that carbon value in the forest area of BGNP, especially Pagar Gunung and Sopo Tinjak Villages is relatively high. This indicates that forest condition in both locations is still good. Therefore, various conservation efforts need to be done continuously in the forest area in Pagar Gunung and Sopo Tinjak Villages in order to maintain the balance of environment ecosystem.

4. Conclusion

Above ground biomass (AGB) found in BGNP forest area i.e. Pagar Gunung and Sopo Tinjak Villages are 405.99 tonnes/ha and 211.70 tonnes/ha. If multiplied by the area of each area of 0.64 ha for Pagar Gunung Village and 0.76 ha for Sopo Tinjak Village, the total biomass of both sites are 259.83 tonnes and 160.89 tonnes. At the highest mean total AGB in Pagar Gunung Village for all stands found in trees with diameter class > 40 cm (291.78 tonnes/ha) and the lowest average AGB is found in diameter class ≥ 5-10 cm with average total 2.33 tonnes/ha. While in Sopo Tinjak village found the highest total average AGB for all stands in trees with diameter class > 40 cm (139.96 tonnes/ha) and the lowest average AGB was found in ≥ 5-10 cm diameter class with average total 2.36 tonnes/ha. From the total biomass result, the total carbon stock for both locations are 210.36 tonnes (129.92 tonnes in Pagar Gunung Village and 80.45 tonnes in Sopo Tinjak Village) or equivalent with a total of CO_2 is 772.03 tonnes (476.79 tonnes in Pagar Gunung Village and 295.24 tonnes in Sopo Tinjak Village). By using the applicable carbon price in the Australian Emission Trading System (ETS) AUD \$ 11.82/t CO2e, and European Union ETS € 5 (US \$ 6)/t CO2e, the carbon value that can be generated from BGNP forest areas (Pagar Gunung and Sopo Tinjak Villages) is Rp. 92,499,921.72 (in AUD \$) or Rp. 61,654,433.67 (in US \$). From these results indicate that carbon value in the forest area of BGNP, especially Pagar Gunung and Sopo Tinjak Villages is relatively high. This indicates that forest condition in both locations is still good. Therefore, various conservation efforts need to be done continuously in the forest area in Pagar Gunung and Sopo Tinjak Villages in order to maintain the balance of environment ecosystem.

References

1. Undang-Undang Nomor 41, *Kehutanan* (1999).
2. K. Borje, *Unasylva*, **41**, 4 (1990).
3. H. Kurniatun, R. Subekti., *Pengukuran karbon tersimpan di berbagai macam penggunaan lahan* (2007).
4. C. Hendra, Setiap tahun, hutan Indonesia hilang 684.000 hektar, *http://regional.kompas.com/read/2016/08/30/15362721/setiap.tahun.hutan.indonesia.hilang.684.000.hektar* (Accesed on July 11, 2017).
5. Balai Konservasi Sumber Daya Alam Sumatera Utara II, *Rencana Pengelolaan Taman Nasional Batang Gadis (RP-TNBG)* (2005).
6. K. Quirine, R. Coe, M. V. Noordwijk, Y. Ambagau, C. A. Palm, *Forest Ecology and Management*, **146**, 199-209 (2001).
7. B. Sandra, *FAO Forestry Paper*, **134** (1997).
8. M. V. Martin, *Carbon budget accounting at the forest management unit level: an overview of issues and methods* (Canada's model forest program, 2000).
9. Australian Government Clean Energy Regulator, *Auction April 2017*, *http://www.cleanenergyregulator.gov.au/ERF/Auctions-results/april-2017* (Accesed on July 11, 2017).
10. World bank group climate change, Ecofys, *Carbon Pricing Watch* (2016).
11. Onrizal, Ismail, E. A. Perbatakusuma, H. Sudjito, J. Supriatna, I. H. Wijayanto, *Jurnal Biologi Indonesia*, **5**, 2, 187–199 (2008).
12. Meyer, H. A, *Journal of Forestry*. **50**, 2, 85 92 (1952).
13. L. De Renato Augusto Ferreira, J. L. F. Batista, P. I. Prado, *Forest Science*, **61**, 2, 320–327 (2015).
14. R. Subekti, B. Lusiana, M. V. Noordwijk, Pendugaan cadangan karbon di atas permukaan tanah pada berbagai sistem penggunaan lahan di Kabupaten Nunukan, Kalimantan Timur, *ICRAF* (2007).
15. S. Ishemat, A. Indrawan, *Ekologi Hutan Indonesia* (2005).
16. W. Aditya C, Struktur komunitas hutan mangrove, estimasi karbon tersimpan dan perilaku masyarakat sekitar kawasan Segara Anakan Cilacap, *Thesis* (2014).
17. A. W. Catur, K. Sidiyasa, *Jurnal Penelitian Hutan dan Konservasi Alam*, **3**, 1, 103-117 (2006).
18. S. Dandun, *Penghitungan biomassa sebuah pengantar untuk studi karbon dan perdagangan karbon* (2009).
19. K. Wanda, B. S. Antoko, *Jurnal Penelitian Hutan dan Konservasi Alam*, **5**, 4, 337–354 (2008).

20. Z. Hilda, Y. Windusari, I. Yustian, D. Heriinawati, *Prosiding seminar nasional dan rapat tahunan dekan bidang ilmu-ilmu pertanian badan kerja sama perguruan tinggi negeri (BKS-PTN) wilayah barat*, **3** (2011).
21. S. Barliana S, Indonesia produsen emisi karbon dunia, *https://www.bappebti.go.id/id/edu/articles/detail/2997.html* (Accesed on July 11, 2017).

The Effectiveness of Hybrid Structure in Overcoming Coastal Abration in Trimulyo, Genuk Subdistrict Semarang City

Domas Kurnia [1,*], *Denny* Nugroho [2,3]

[1] Master Program of Environmental Science, School of Postgraduate Studies, Diponegoro University, Semarang, Indonesia
[2] Department of Oceanography, Faculty of Fisheries dan Marine Science, Diponegoro University, Semarang, Indonesia
[3]Center fo Coastal Disaster Mitigation and Rehabilititation Studies, Diponegoro University, Semarang, Indonesia

Abstract. Trimulyo is one of coastal village in Genuk Subdistrict, Semarang City which now facing serious coastal abrasion. Such a thing has been causing loss of ponds and settlements. One of solution which currently carried is hybrid structure which combining permeable structure to break up the waves and trap sediment. The hybrid structure is designed as agitation dredging, which increase suspended sediment in sea water. The goals of this research were to studying the effectiveness of hybrid structure in handling coastal abration and to counting the volume of sedimentation during 20 months as well as rate of sedimentation. To reach the goals, high resolution satellite imagery year 2015 and 2016, scaled stick and sediment trap were applied to the study. Image processing was conducted by using Arc GIS 10.3 software. The effectiveness of hybrid structured was determined by series of field survey of existing condition. Rate of sedimentation measured during before and after hybrid structure built (20 months). The results showed that hybrid structure was effective to reduce coastal abrasion, it proven by a large amount of sediment was trapped behind the structure and coastline was upward along 170 meter since it was built. The volume of sediment during 20 months is 81.500 m^3. If it assumed that the rate of sedimentation constantly, monthly rate of sedimentation is 4.075 m^3/month or daily rate is 135,8 m^3/day. The sediment that has formed highly recommended to use as mangrove conservation area in Semarang City.

1 Introduction

According to Dinas Kelautan dan Perikanan Kota Semarang, Semarang city is one of coastal regency in northern Java Sea that seriously facing coastal degradation called abrasion, land subsidence and water inundation. Due to coastal degradation, there are many impacts material and immaterial damage, for instance loss of ponds, settlements and infrastructures. Trimulyo village is one of the most coastal area that facing such degradations [1].

There were many efforts have done by many institutions, communities and universities. The efforts include the built of hard structure and soft structure that called hybrid structure. Hybrid structure was developed to reduce wave energy, this system was adopt from mangrove ecosystem. Permeable structure was designed to trap the sediments in the structure. In late of 2015, hybrid structure was built in Trimulyo coastal, but the effectiveness of this structure has not been studied.

This research was done to give information to the stakeholder about the effectiveness of hybrid structure to reduce coastal abrasion, counting volume and rate of sedimentation.

2 Materials and methods

2.1 Time and location

The research was done on January-February 2017 located in Trimulyo village, Sayung Sub District, Semarang city. (Figure 1).

Fig. 1. Map of Research Location

* Corresponding author: queendomas@gmail.com

The material used in this research were primary and secondary data. The primary data were high resolution satellite imagery and sediment data in hybrid structure. The secondary data were bathymetric map of scale 1:25.000 published by Bakosurtanal and tides data.

2.2 Methods, processing and analyses

2.2.1 Image processing, coastline delineation and sediment spreading in hybrid structure

High resolution satellite imagery was applied to detect the change of sediment before and after hybrid structure built (2015 and 2017). Those high resolution satellite imagery was download from google earth and digitally processing to georeferenced into real coordinate to the earth.

Satellite image composite was done by using visible and invisible bands to mapping objects using fusion techniques in 0.5 m colour spatial resolution, filtering also applied to colour enhancement. This applied to simplify objects identification for visual interpretation. The methods applied was linear contrast enhancement for reduce spectral and visual distortion of satellite imagery.

2.2.2 Costline delineation and Sediment spreading

Coastline delineation anf sediment spreading was done using on screen digitations on Arc GIS 10.3 software. Those objects was clearly can be determined by visual interpretation on high resolution satellite imagery.

2.2.3 Accuration test and reinterpretation

Accuration test performed to validate result of interpteration and ground check. Accuration test done using the matching of interpretation and real condition by ground checking (Short in Sutanto,1992).

$$\text{Accuration} = \frac{\text{True interpretation}}{\text{Total sampel}} \times 100\% \quad \ldots\ldots\ldots(1)$$

2.2.4 Sediment spreading counting, volume and rate of sedimentation

Spread of sediment was calculated using Arc GIS 10.3. This methods was developed due to matching of visual interpretation and ground checking. The volume of sedimentation was calculated by data of length, width and the depth of sediment trapped. Systematically the formulas as follow:

$$\text{Volume} = \text{length} \times \text{width} \times \text{depth} \ldots..(2)$$

Depth of sedimentation was calculated by bathimetric data and scaled stick that has been placed in trapped sediment. Meanwhile the rate of sedimentation calculated by the accumulation of sediment during 20 months. It was assumed that the rate of sedimentation is constant during all months. Monthly rate of sedimentation formulas are as follows:

$$v = \text{Volume} / 20 \text{ month} \ \ldots\ldots..(iii)$$

v= monthly rate of sedimentation (m^3/month)

2.2.5 Sampling sediment methods

Sediment sample was done once in three stations. Sample was taken using eckman grab (500 gram), stored in labelled plastic and analysed in laboratorium. Sediment analyses was using sediment fraction and wet gravel, using pipet texture analyses Rifardi, stated [2]

Dyer, Davis subscribed that Based on Wenworth scale, sediment s classify by fraction size, clay, silt, gravel, pebble, cobble dang boulder. That scale refers to standart size of fraction in micron to millimeter with continuous spectrum [3] [4].

2.2.6 Hidro oseanography

The oceanographic factors take a very important role in the sedimentation in the coastal, namely tide and sea current. Those two parameters were very close related to the dynamic of coastal area. Those factors were analyses by secondary data.

3 Results and discussion

3.1 Image processing, digitations and accuration test

High resolution satellite imagery was downloaded in JPG Format. Series of processing was done using Arc GIS 10.3 with the regional spatial plannnig (*Rencana Tata Ruang Wilayah- RTRW)* Semarang City as basemap. Geometric correction was done using georeferenced tool in Arc Map 10.3

Fig. 2. Geometric image processing

Based on digital processing, trapped sediment in hybrid structure was clearly seen by visual interpretation as figure 3 below:

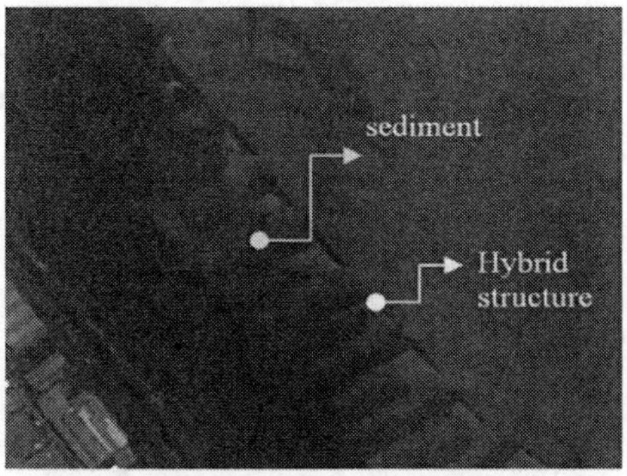

Fig.3. Trapped sediment in hybrid structure identified using high resolution satellite imagery.

Daruati subscribed that accuration test was done to validate visual interpretation using satellite imagery and real condition by field survey. Accuration test was done in 10 location within the hybrid structure. Based on accuration test, the accuracy of interpretation was 90 percent, because there was one missing objects between interpretation and field survey. Based on USGS criteria, this accuration can be use because more than 85% [5].

3.2 The Effectiveness of hybrid stucture

The effectiveness of hybrid structure was analysed descriptively by series of field survey. Based on existing condition, hybrid structure is effective in redusing coastal abration , It is proof by the trapped sediment in the structure. Field documentation of the effectiveness of this structure as figure 4 below.

Fig.4. Existing condition in hybrid structure

Based on series of field survey, hybrid structure is very effective to trapped sediment and recover coastal abration. Since it was built 20 months ago there is much sediment trapped within the structure. The sedimentation occurred as the impact of the design of permeable hybrid structure that allows sediment to entering the structure and trapped in it.

Hybrid structure can trap the sediment because permeable structure that not reflecting the wave but deflect the wave energy. The energy of long shore current and wave will and all sediment taken will be trapped. This structure is very effective to recover losses abration area. The design of hybrid structure has been considering the wave and long shore current, so sediment easily trapped in the structure.

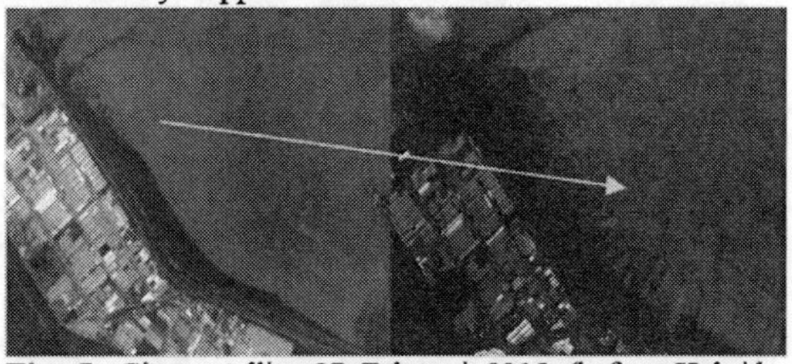

Fig. 5. Citra satellite 27 Februari 2015 (before Hybrid structure) Citra satellite 13 September 2016 (before Hybrid structure)

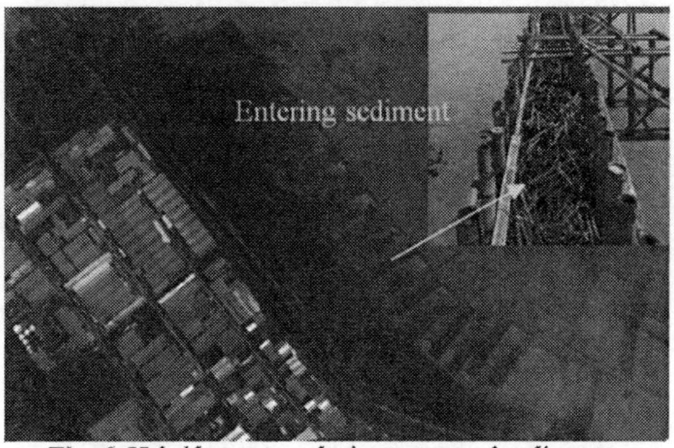

Fig. 6. Hybrid structure design to trapped sediment.

3.3 Calculation of volume and rate of sedimentation

Based on delineation and field survey, there were 81.500 m^3 sediment was trapped in the structure within 20 months. If it assumed that the rate of sedimentation is constant, the rate of sedimentation as 4.075 m^3/month or 135.8 m^3/day. The map of sediment within the hybrid structure as follows.

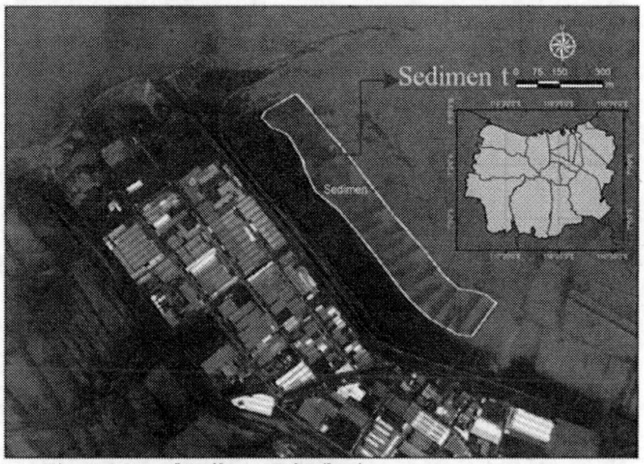

Fig. 7. Map of Sediment Distribution

As seen on figure 7, sediment distribution trapped all in hybrid structure. It proved that hybrid structure was very effective to recover coastal abration.

3.4 Sediment analyses

Trapped sediment in the hybrid structure visually is mud. For more information, the sediment was analysed in laboratory. Based on laboratory analyses (figure 7), sediment composition were 90,9% silt, mud 3%, fine sand 5,6% dan coarse sand 0,5%. Those result showed that silt is the most material formed the sediment in hybrid structure.

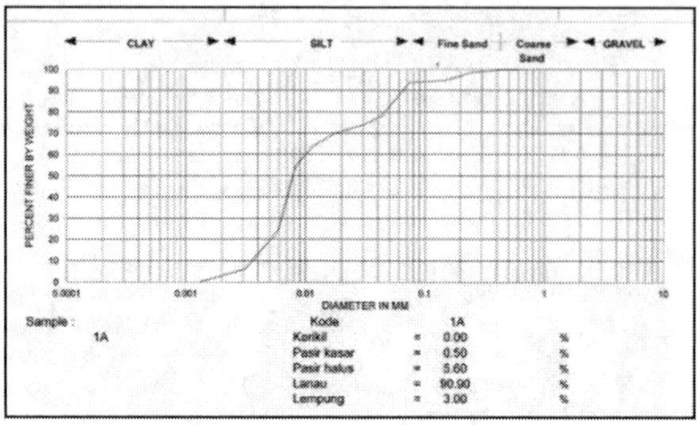

Fig. 7. Laboratory analyses of sediment

Bowles described that lanau is fine-grained soils that are smaller than 0.074 mm (No. 2000. Silt and clays mostly found in the coastal and estuary [6].

3.5 Oceanographic parameters

3.5.1 Sea current

Ocean currents is a horizontal mass of seawater caused by the wind , sea waves and tides. Based on Fisheries Services of Semarang City, the current pattern in May 2015 – April 2016 the maximum speed of long shore current was 0.064 m/s and the minimum speed was 0.008 m/s that moving toward the west of Semarang. The highest current occurred in December with 0.19 m/s in speed and the minimum speed occurred in August 2016 with speed 0.008 m/s. The average of long song current speed was 0.1 m/s [1].

3.5.2 Tides

Tides are the rise and fall of sea levels caused by the combined effects of the gravitational forces exerted by the Moon and the Sun and the rotation of Earth. The times and amplitude of tides at any given locale are influenced by the alignment of the Sun and Moon, by the pattern of tides in the deep ocean, by the amphidromic systems of the oceans, and the shape of the coastline and near-shore bathymetry. While tides are usually the largest source of short-term sea-level fluctuations, sea levels are also subject to forces such as wind and barometric pressure changes, resulting in storm surges, especially in shallow seas and near coasts. Hakim, et al. described that tidal type in Semarang City was tend to diurnal [7].

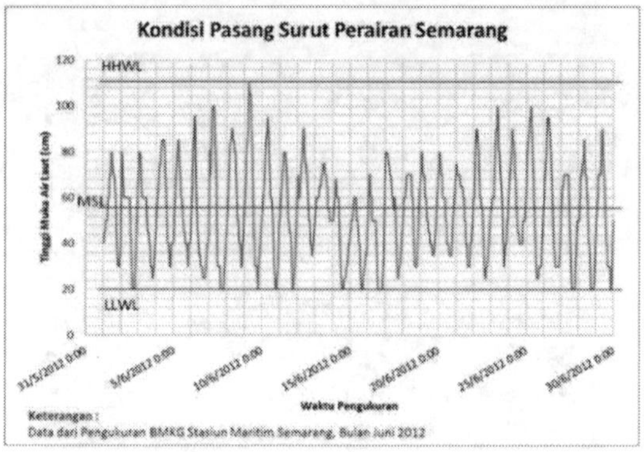

Fig. 8. Tidal pattern in Semarang City between January to August 2017

3.5.3 Wave

Danial subscibed that waves are the forward movement of the ocean's water due to the oscillation of water particles by the frictional drag of wind over the water's surface.[8]

Waves is one of the parameters of oceanography very affect the coast .Triatmodjo stated that can give rise to a wave of energy to form the coast , give rise to the current , well as result in energy which works on the the coast .Three factors that determine the characteristics of a wave that is fueled by the wind i.e. (1) blowing duration of the wind, (2) wind speed and (3) fetch (the distance traveled by the wind from the direction of stirring up a wave or the region of the generation of waves. [9]

According to Hakim et. Al. the height of waves in Semarang relatively small, ranged from 0,35m up to 0,89 m with the period between 3.5 seconds to 5.8 seconds. This indicates that a wave that formed in waters Semarang can be described is strongly influenced by a wind occurring [7].

Oceanographic factors has been considered in the process of designing hybrid structure. There were many inlets in hybrid structure to trapped sediment. Based on field observation, those structure was very effective to trapped the sediment along hybrid structure. It proven by the sediment formed and coastline forward 170 m since it was built.

4 Conclusion

Based on this research, the conclusions are:

a. The hybrid structure in Trimulyo village, Genuk Sub district, Semarang City is very effective in overcoming coastal abration, it proved by the sediment formed in the structure and the coastline forward 170m since it was built.
b. The volume sedimentation during 20 month was 81.500 m^3. If it assumed that the rate of sedimentation were constant during all months, the monthly rate of sedimentation is 4.075 m^3/month or 135,8 m^3/day.

c. Formed sediment in hybrid structure highly recommended for mangrove conservation in Semarang City.

References

1. Dinas Kelautan dan Perikanan Kota Semarang, *Kajian Penanganan Pesisir Kota Semarang*. (DKP Kota Semarang, 2016)
2. Rifardi, *Ukuran Butir Sedimen Perairan Pantai Dumai Selat Rupat Bagian Timur Sumatera*. (Jurnal Ilmu Lingkungan Universitas Riau. ISSN 1978-5283, 2008)
3. K.R. Dyer, *Coastal and Estuarine Sedimen Dynamic*. (New York: John Wiley and Sons Ltd, 1986)
4. Davis, R.A.Jr, *An Introduction to Sedimentology and Stratigraphy Depositional System*. (New York: Precite Hall-Englewood Cliff, 1993)
5. D. Daruati, Penggunaan Citra Satellite Landsat 7ETM+ Untuk Kajian Penggunaan Lahan DAS Cimanuk.LIMNOTEK, **XV**,40-50, (2008)
6. Bowles, E. Joseph, *Sifat-sifat Fisis dan Geoteknis Tanah* (Erlangga,Jakarta, 1991)
7. Hakim, A. Buddin, Suharyanto, Wahyu Krisna Hidayat, *Buletin Oseanografi Marina* (Juli 2013.2, 81 – 93, 2013)
8. M. Danial, *Rekayasa Pantai*. (Alfabeta, Bandung, 320, 2008)
9. Triatmodjo, Bambang, *Hidrologi Terapan*. (Beta Offset, Yogyakarta, 2008)
10. W. Mahardi, *Kajian Efektivitas Metode Struktur Alat Pemecah Ombak (APO) dan Hybrid Engineering (HE) sebagai studi pengembangan konsep rehabilitasi pantai di Kabupaten Demak*. (Thesis: Universitas Indonesia, 2014)
11. I Alifdini, D N Sugianto, Y O Andrawina and A B Widodo, Bali. Indonesia. IOP Conference Series: Earth and Environmental Science. **55** (2017)

Fish Marketing of Ribbon Fish (*Trichiurus* sp.) in Nusantara Fishing Port (NFP)at Palabuhanratu, West Java

Azis Nur Bambang[1,*]

Faculty of Fisheries and Marine Science, Diponegoro University,

Abstract. The objective of this research is to study the marketing process of ribbon fish (*Trichiurus sp.*), including the marketing margin, marketing agencies, traders and marketing channels The research was carried out for 3 mo in Nusantara Fishing Port (NFP), Palabuhanratu, Sukabumi, West Java. A case study was used in this research. A purposive sampling method was used to collect data from 55 respondents of fish marketing, consisting of fishermen, agents, traders, and retailers, who were involved in the marketing of ribbon fish in NFP Palabuhanratu. The result of the research showed that ribbon fish production in Palabuhanratu fluctuated from year to year. There are two types of ribbon fish marketing, i.e. type one is from fishermen to retailers, and type two is indirect marketing from fisherman to consumers through intermediate traders (exporters). The greatest marketing margin was obtained from the first type, while the smallest marketing margin was obtained from type two. The form of the market was considered to be oligopsony market. Fisherman's share is greatest in the collectors and the smallest share is on retailers. Marketing process in traders is efficient due to its lowest margin and highest fisherman's share.

1 Introduction

Sukabumi, West Java, is an area where most of it is directly adjacent to the Indian Ocean. Sukabumi regency has a coastline of 117 km, and it has the Nusantara Fishery Port (NFP) located in Palabuhanratu district. The marine waters of Palabuhanratu is a potential and strategic area for fishing because of the great economic value from the captured fish in Nusantara Fishing Port [1]. One of the main commodities of fisheries is ribbon fish (*Trichiurus* sp.), tuna (*Thunnus* sp.), small tuna (*Euthynnus* sp.), snapper (*Lutjanus* sp.), and mackerel (*Scomberromorus* sp.).

Ribbon fish (*Trichiurus* sp.) is one of the major fish commodities in NFP Palabuhanratu. The amount of ribbon fish captured in NFP Palabuhanratu in 2014 was 61 560 kg and production value of IDR 1.546299 $\times$ 10^9 [2]. Ribbon fish in NFP Palabuhanratu was captured using stretching and longline fishing method. The total production was abundant, thus, requiring a good marketing to distribute the fish to consumers in the form of fresh and processed fish.

2 Materials and Methods

Respondents used in this study were the subjects of ribbon fish marketing, consisting of fishermen, agents, traders, wholesaler and retailers. Descriptive analysis method was used in this research, and followed by the implementation using survey techniques [3]. The data were collected to describe the marketing of ribbon fish in NFP Palabuhanratu succinctly and objectively.

Purposive sampling and snowball sampling method used in this research. The primary data and secondary data were used. Samples of respondents consist fishermen, wholesalers, agencies, and ribbon fish retailers (Table 1).

Table 1. Respondents sample

No.	Respondents	Sample
1.	Fishermen	30
2.	Wholesaler	5
3	Agent	10
4.	Retailer	10
Total		55

* Corresponding author azis_undip2013@yahoo.com

Technical data analysis methods of the fishing by a descriptive analysis, such as value and production of ribbon fish and ribbon fish fishing areas [4]. Marketing data processed by marketing margin analysis, including costs and benefits as well as fisherman share were analyzed. Analysis of the advantages of each marketing agency is based on the following formula [5]:

$$\pi = TR - TC \quad (1)$$

π = Profit (IDR), TR = Total Revenue (IDR) and TC = Total cost (IDR).

Marketing margin analysis was used to determine the distribution of the costs and benefits of any intermediaries as well as parts of prices received by producers. Marketing margin analysis was also used to determine the level of competence of the subjects involved in the marketing of fish distribution. Marketing margin can be calculated as follows [6] :

$$Mm = Pe - Pfi \quad (2)$$

Mm = Marketing Margin (IDR · kg^{-1}), Pe = Price on seller (IDR · kg^{-1}) or selling price,
Pf = Price on producer (IDR · kg^{-1}) or purchasing price.

The average net profit per day from marketing agencies can be obtained by the following formula [7]:

$$M = B + \pi \quad (3)$$

M = Marketing Margin (IDR · kg^{-1}), B = Cost (IDR · kg^{-1}) and

π = Profit market seller (IDR · kg^{-1}).

To determine the efficiency of marketing in any involved marketing agencies, the following formula was used [7]:

$$Ep = \frac{Marketing \quad cos\ t}{Value \quad of \quad marketing \quad product} \quad (4)$$

The marketing cost is in IDR · kg^{-1}, and the value of marketing product is in IDR · kg^{-1}. If the Ep value > 1, the marketing process is not efficient, whereas if Ep value < 1, the process is efficient.

Fishermen's share analysis can be formulated as follows:

$$F = \frac{Hn}{Hk} \quad 100\ \% \quad (5)$$

in which F = shares received by fishermen (%), Hn = price of ribbon fish at fishermen level (IDR · kg^{-1}) and Hk = price of ribbon fish at consumer level (IDR · kg^{-1})

Table 2. Value and production of ribbon fish, periode 2010 to 2014

No	Year	Production (kg)	Value (IDR)
1	2010	36730	556 035 500
2	2011	147 864	2 114 744 500
3	2012	185 914	3 420 045 200
4	2013	83 936	1 750 018 000
5	2014	61 560	1 546 299 000

Source : NFP, Palabuhanratu [2]

According to Maisaroh fishermen's share is the share of the fishermen (producer), in which the greater the fishermen's share, the smaller the marketing margin, thus, marketing distribution runs efficiently [8].

3 Results and Discussion

3.1 Overview

The Nusantara Fishery Port (NFP) Palabuhanratu is one of the five regional fishery ports converted into industrialization of fisheries. NFP Palabuhanratu is selected as a base or center for fisheries management for economic development, *i.e.* creation of jobs and public revenues through the reform of fisheries business service centers that functions to serve, encourage, and stimulate regional development of fisheries in West Java. Ribbon fish (*Trichiurus* sp.). as a dominant fish commodities at NFP o.
Trichiurus sp. is one of the leading commodity fish in NFP Palabuhanratu. Ribbon fish production in Palabuhanratu fluctuated from year to year. Table 2 shows the production of ribbon fish in Palabuhanratu NFP from 2010 to 2014. According to the Table 2, from 2010 to 2014, ribbon fish production continued to fluctuate because of the uncertain weather and unstable fishing season. Some factors that influence the success of ribbon fish fishing are season, weather and oceanographic condition. Production and price of ribbon fish in NFP Palabuhanratu is fluctuative each month depending on fishing season (Table 3).

Production of ribbon fish in NFPPalabuhanratu fluctuated every month. The supply fluctuation does not particularly affect the price. Price of ribbon fish depends on the agreement between sellers and buyers (retailers and collectors), not on season and supply.

3.2 Fishermen / Producer Ribbon Fish

Ribbon fish production in NFP Palabuhanratu using fiber boats was given a booster in the form of wood. Ribbon fish fishing boat was operated by two or three crews within 1 d of fishing. The average trip of ribbon fish fishing boat is 175 trip each year. The boat uses an engine of 5.5 hp (horsepower) [1 hp = 746 W]. Ribbon fish fishing gear consists of a spindle, main rope, swivel, rope branches, rods, weights, wireleader, and hook.Ribbon fish fishing is usually done during the day or night, at 4:00 a.m. to 15:00 noon, or at 4:00 p.m. to 7:00 a.m. Fishing ground area is located around Palabuhanratu gulf which is determined based on the experience of fishermen in the previous fishing activities.

Table 3. Production, value, and price of ribbon fish at Palabuhanratu in 2014

Month	Production (kg)	Value (IDR)	Price (IDR)
January	3 982	109 675 000	27 543
February	3 746	92 747 000	24 759
March	28 108	702 700 000	25 000
April	11 917	296 040 000	24 842
May	4 237	105 925 000	25 000
June	904	22 600 000	25 000
July	716	17 900 000	25 000
August	624	15 562 000	24 939
September	1 293	32 325 000	25 000
October	1 428	35 700 000	25 000
November	2 485	62 125 000	25 000
December	2 120	53 000 000	25 000

Source: Annual Report Palabuhanratu NFP, 2014 [2]

USD 1 = IDR 12 287

Table 4. Revenues, expenses and profits of fishermen per trip

No	Items	Total (IDR per day)
1	Revenues	325 000
2	Expenses	
	- depreciation cost	12 500
	- maintenance cost	6 600
	- operational cost	120 000
	Total cost	139 100
3	Profits	185 900

USD 1 = IDR 12 287

The total expense of ribbon fish fishing which consists of fixed costs and variable costs is quite low because it takes only one day for fishing and the number of crew is only two or three people. Average catch of ribbon fish is 10 kg in each trip. Revenues, expenses and profits of fishermen per trip are presented in Table 4.

3.3 Agent/ Seller

An agent is a person who mediates fishermen by middlemen. After fishing operation is completed, fishermen land their catch in NFP Palabuhanratu. The ribbon fish are not sold in auction, but directly sold by agent to collectors. Due to cooperation between fishermen and seller agents, the ribbon fish are not sold directly to collectors, but it has to be through seller agent.

Contact between fisherman and seller agent occurs because not all fishermen are able to finance operational costs which make them borrow fishing supplies such as fuel, ice blocks, oil, *etc.* from agents. Thus, in return, fishermen have to sell their catch through the agency.

Ribbon fish agents can earn 10 % of total sales revenue and they can sell 47.5 kg of ribbon fish each day. The determination of selling price of ribbon fish is based on agreement between the agent and traders. In this study, average selling price of ribbon fish is IDR 32 500 · kg^{-1}. An average commission sales for agents is about IDR 155 000 per day.

3.4 Wholesaler/ Collector

Traders require substantial capital investment. They have to purchase ribbon fish through an agent. Ribbon fish in traders are subsequently distributed to some industrial fisheries in the region of Palabuhanratu, *i.e.* PT. AGB Palabuhanratu, Duta Ratu, and Jaya Mitra, to be sent to Jakarta and surrounding areas, then exported. Besides industries, collectors also sell the fish to retailers or small traders. The size of the fish are divided into small (100 g to 250 g), and large (over 250 g). Usually, large fish are sold to industries, while the small size fish are sold to retailers.

The price of fish from the agent is between IDR 30 000 · kg^{-1} to IDR 35 000 ·kg^{-1}. The average price of fish purchased from sellers is IDR 32 500 and then resold for IDR 42 500. Below is a table of revenues, expenses and profits earned by wholesalers each day.

Table 5. Revenues, expenses and profits earned by wholesaler per day

No	Items	Total (IDR)
1	Revenue	2 720 000
2	Cost	
	– operational cost	60 000
	– material cost	2 080 000
	– Total cost	2 130 000
3	Profit	580 000

USD 1 = IDR 12 287

3.5 Retailer

Retaillers in PPN Palabuhanratu buy ribbon fish by skipping the service from agents. Profit obtained by retailer is quite big, because no transportation expense is needed. Consumers come directly to retailer sale outlet around the Fish Auction Place.

Table 6. Revenues, expenses and profits earned by retailer

No	Item	Total (IDR)
1	Revenues	618 750
2	Expenses	
	- depresiation cost	30 000
	- maintenance cost	406 250
	- Total Cost	436 250
3	Profit	182 500

USD 1 = IDR 12 287

https://doi.org/10.1051/e3sconf/20183108012

3.6 Consumer

Ribbon fish marketing stops at final consumer level. Ribbon fish is not only marketed to local merchants but also for export commodities. Ribbon fish is quite attractive for consumers overseas.

Export of fish ribbon fish is done by a fish processing company in Indonesia. Consumers expect ribbon fish with low price and good quality. The average price of local ribbon fish at the consumer level is IDR 45 000 · kg^{-1}.

3.7 Marketing Distribution

First type of ribbon fish marketing channels is from fishermen to retailers and then to consumers, the distribution of fishery products to the distribution of semi-direct [9]. The size of ribbon fish marketed in type 1 is smaller, *i.e.* below 250 g, but there are also some fish weighing above 250 g. Sales percentage of ribbon fish at retailer level is lower than in traders, because retailers do not have much capital, thus the volume of buying and selling is also low. Retailers in this type of marketing channels sell their fish at the stalls of NFP Palabuhanratu.

Meanwhile, for marketing type 2, distribution of ribbon fish is from fisherman to agents, continued to traders, then to PT. AGB, exporters and consumers as the final chain. Fish marketing distribution channel is called the distribution of fishery products to the indirect distribution [9]. Ribbon fish that are sent to PT. AGB are stored in cold storage. The fish commodity is then distributed to exporters in Jakarta and surrounding areas (Figure 1) after fulfilling exporter's demand.

3.8 Marketing Margin and Profit

Margin is the difference between selling price and purchasing price at the marketing agency. The marketing margin when ribbon fish in which the fish are distributed from the fishermen to retailers in marketing type 1 is IDR 12 500 · kg^{-1}.
Highest marketing margin is IDR 12 500 · kg^{-1} and obtained during distribution from fisherman to retailers in marketing type 1

The high margin value is influenced by the operational costs spent by retailers only to purchase ice cubes and plastic, buyers will come to their own kiosk retailers in NFP Palabuhanratu. Besides retailers market ribbon fish directly to consumers.

Ribbon fish price at consumer level ranges between IDR 40 000· kg^{-1} to IDR 5 0000 · kg^{-1}. In the marketing channel type 2, the marketing margin (IDR 10 000 · kg^{-1}) is lower than the marketing channel type 1. PT. AGB has determined the price between IDR 40 000 · kg^{-1} to IDR 450 000 · kg^{-1}, on the other hand there is also a transportation cost for mobilization to PT. The greatest marketing profit is obtained by retailers in distribution type 1. Retailers have high marketing profit because retailers have greater marketing margin than collector or wholesaler.

The high margin value is influenced by how much profit can be obtained by retailers and marketing costs, such as exposure margin above, retailers do not require transportation cost and the selling price to the consumer of fish is between IDR 50 000· kg^{-1} to IDR 40 000· kg^{-1}

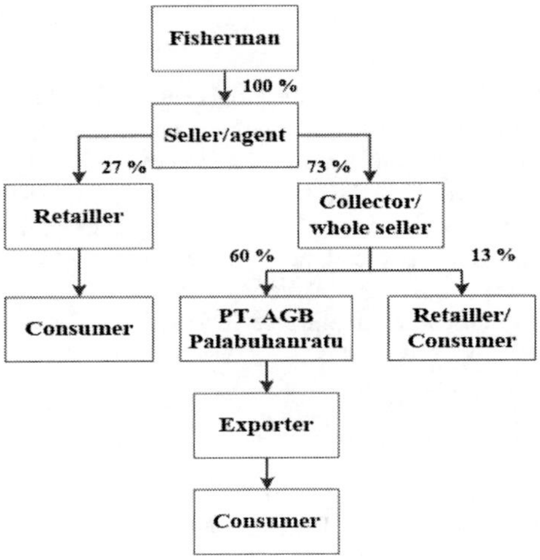

Fig. 1. Marketing distribution of ribbon fish at NFP Palabuhanratu

3.9 Marketing Efficiency and Fisherman's Share

Analysis of ribbon fish marketing efficiency was performed by counting marketing costs (IDR ·kg^{-1}) compared to the price (IDR · kg^{-1}) of ribbon fish in the marketing chain. Marketing efficiency in the distribution of type 1 or type 2 value is <1, therefore, both types are efficient [7].

From the two distribution types, the most efficient distribution is obtained by traders with efficiency values of 0.02, which is smaller than the value of efficiency at retailers (0.05). In addition to using efficiency value, marketing efficiency can also be seen from the number of fisherman's share. The biggest fisherman's share is on trader, i.e., 76.5 %. This value is the share that is obtained by the fisherman. This value also represents the amount of fisherman influence towards the price. Moreover, fisherman's share in the retailer is 72.2 % (Table 8). According to Maisaroh *et al.*, smaller marketing margin means that distribution to the market runs efficiently [8].

Table 7. Marketing margin (IDR · kg^{-1}) of ribbon fish at NFP Palabuhanratu

Distribution type	Marketing institution	Paying price (IDR · kg^{-1})	Selling price (IDR · kg^{-1})	Margin (IDR · kg^{-1})	Marketing Cost (IDR · kg^{-1})	Marketing profit (IDR)
1	Agent		32 500			
	Seller (Retailler)	32 500	45 000	12 500	2 400	10 100
2	Agent		32 500			
	Seller (Wholesaler)	32 500	42 500	10 000	937	9 063

USD 1 = IDR 12 287

While according to Dewayanti, the high margin is usually used as an indicator of the inefficiency of a marketing system, however, this is not always right [10]. By comparing fisherman's share, efficiency of a marketing system for fishery commodities can be determined. Therefore, the most efficient marketing agencies are traders because the margin are low and the fisherman's share is the greatest.

Conclusion

There are two types of ribbon fish marketing, *i.e.* type one is from fishermen to retailers, and type 2 is indirect marketing from fisherman to the consumer through intermediate traders (exporters). The greatest marketing margin is obtained from type 1, while the smallest marketing margin is obtained in type 2. The structure of the market is oligopsony market.

Fisherman's appearance has a market share which is greatest in the collectors and Fisherman's smallest share is on retailers. Marketing process in traders is efficient due to its lowest margin and highest fisherman's share. The market structure is oligopsony. To avoid domination from buyers. (oligopsony), fish distribution from fisherman directly to fish auction place is recommended.

References

1. Sudrajat, S.M. Nurul, A. Rosyid and A. N. Bambang, Journal of Fisheries Resources Utilization Management and Technology **3** (III). 141-19, (2014) [in Bahasa Indonesia].
2. Pelabuhan Perikanan Nusantara (PPN) Palabuhanratu, Laporan Tahunan Pelabuhan Perikanan Nusantara Palabuhanratu Tahun 2013. Direktorat Jenderal Perikanan Tangkap Kementerian Kelautan dan Perikanan.[*Annual Report of Nusantara Fishing Port Palabuhanratu 2013*. Jakarta: Ministry of Marine Affairs and Fisheries, (2014).
3. S.F. Shalichaty, A. K. Mudzakir, and A. Rosyid, Journal of Fisheries Resources Utilization Management and Technology, **3**(3), 37–43, (2014). [in Bahasa Indonesia].
4. I.M. Wirartha, *Economic social research methodologies*. Yogyakarta: Andi Offset, (2006).
5. T. Nurasa, dan V. Darwis, Jurnal Akta Agrosia, **10**(10), 40–48, (2007). [in Bahasa Indonesia].
6. A. Supriatna, (Study case in Cirebon district, West Java)].Agrin, **14**(2), 97–113, (2010). [in Bahasa Indonesia].
7. N. Rasuli, M. A. Saade, and K. Ekasari, Agrisistem Journal, **3**(1), 36–43, (2007). [in Bahasa Indonesia].
8. N. Maisaroh, Ismail, and H. Boesono, Journal of Fisheries Resources Utilization Management and Technology, **3** (3), 131–140, (2014). [in Bahasa Indonesia].
9. N. Rahardi, and K. Regina, *Agribisnis perikanan* [Fisheries agribusiness]. Jakarta: Penebar Swadaya Press, (2001). [in Bahasa Indonesia].
10. N.C. Dewayanti, Analisis Pemasaran Ikan Laut Segar di Kabupaten Cilacap. [Marketing Analysis of Marine Fish in Cilacap District] Universitas Sebelas Maret , Surakarta, (2003).

Community Structure Of Coral Reefs In Saebus Island, Sumenep District, East Java

Mada Rizmaadi[1,2,*], *Johannes* Riter[1,2], *Siti* Fatimah[1,2], *Riyan* Rifaldi[1,3], *Arditho* Yoga[1,2], *Fikri* Ramadhan[1,2], and *Ambariyanto* Ambariyanto[2,4]

[1]UKSA-387 Diving Club Diponegoro University, 50275 Jl. Prof. Soedarto SH, Tembalang Campus, Semarang, Indonesia
[2]Faculty of Fisheries and Marine Science, Diponegoro University, 50275 Jl. Prof. Soedarto SH, Tembalang Campus, Semarang, Indonesia
[3]Faculty of Engineering, Diponegoro University, 50275 Jl. Prof. Soedarto SH, Tembalang Campus, Semarang, Indonesia
[4]Coastal Disaster Rehabilitation and Mitigation Center, Integrated Laboratory, Diponegoro University, 50275 Jl. Prof. H. Soedarto, SH, Tembalang, Semarang, Indonesia

Abstract. Increasing degradation coral reefs ecosystem has created many concerns. Reduction of this damage can only be done with good and proper management of coral reef ecosystem based on existing condition. The condition of coral reef ecosystem can be determined by assessing its community structure. This study investigates community structure of coral reef ecosystems around Saebus Island, Sumenep District, East Java, by using satellite imagery analysis and field observations. Satellite imagery analysis by Lyzenga methods was used to determine the observation stations and substrate distribution. Field observations were done by using Line Intercept Transect method at 4 stations, at the depth of 3 and 10 meters. The results showed that the percentage of coral reef coverage at the depth of 3 and 10 meters were 64.36% and 59.29%, respectively, and included in fine coverage category. This study found in total 25 genera from 13 families of corals at all stations. The most common species found were *Acropora*, *Porites*, and *Pocillopora*, while the least common species were *Favites* and *Montastrea*. Average value of Diversity, Uniformity and Dominancy indices were 2.94, 0.8 and 0.18 which include as medium, high, and low category, respectively. These results suggest that coral reef ecosystems around Saebus Island is in a good condition.

1 Introduction

Saebus Island is one of the islands that located in Kangean Islands, East Java, Indonesia. Administratively, this island is located at Sapeken Sub-district, Sumenep District of East Java Province. Coastal areas of this archipelago have high potential of natural resources, one of the largest marine resources is coral reef ecosystem [1]. According to [2], Sumenep District waters has areas of coral reefs covering 47,760.9 Ha, is the largest coral reefs area in East Java Province.

Coral reefs are known as tropical marine ecosystems with high productivity, which has important role not only on biodiversity including genetic but also on fisheries productivity [3,4]. As one of habitat, food source, spawning and growth spots for fish communities [5], coral reefs become very important in Kangean Archipelago region where the majority of coastal communities have livelihoods in the fisheries sector [6].

However, on the past few years, coral reef ecosystems on the region experienced drastic degradation due to destructive fishing activities using explosives and potassium [6]. Destructive fishing has been reported in Indonesia marine areas [7,8], and has caused reducing biodiversity and increasing extinction of marine organisms [9,10]. Generally, almost 45% of coral reefs in Indonesia are under high threat due to human exploitation activities as well as threat of temperature change and coral bleaching [11].

In order to reduce the rate of coral reef degradation, better management has to be implemented based on existing condition both threats and coral ecosystem. Therefore, ecological research on coral reef ecosystem is needed to base line condition of coral reef in the region [12]. This research aims to determine the structure of coral reef community in Saebus Island waters that includes measurement of cover percentage, diversity, uniformity, and dominance of coral reef communities.

* Corresponding author: madarizmaadi95@gmail.com

2 Methods

The research is conducted on September to November 2016 at Saebus Island, Sapeken Sub-district, Sumenep District, East Java. The research stations determined by purposive sampling method, based on Satellite imaging by Lyzenga methods. The observations were conducted at 4 stations, based on fairly good living coral coverage area from satellite imaging, also represent 4 different cardinal points. See Figure 1.

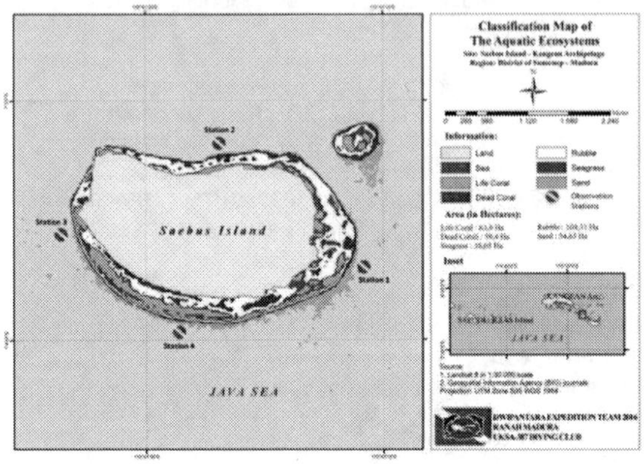

Figure 1. Map of research location

2.1. Field observation

Coral reef condition analyzed using Line Intercept Transect (LIT) observation [13] at 4 different stations at the depth of 3 and 10 m. coral reef data were recorded by measuring the diameter and record coral genus found along the transect.

2.2. Coral percent cover

Percentage of coral reefs community cover was obtained using [14] formula. Coral reefs conditions were categorized based on coral cover percentage with reference to Environmental Ministry Decree No.4/2001 (Table 1) [15] about fixed criteria of coral reefs damage.

Table 1. Category of coral reefs condition based on percent of live coral cover [15]

Category		Percentage
Bad	Poor	0 – 24.9
	Fair	25 – 49.9
Good	Good	50 – 74.9
	Excellent	75 – 100

2.3 Diversity, uniformly and dominancy index

Diversity index is measurement of community abundance, observed from the number of species/genus in one area, calculated with Shannon-Wiener Formula on [16].

Uniformity index represents the number of individual among species in one community, calculated with formula of [17]. The more balanced distribution of individual among species, the balance of the ecosystem will be increased.

Low Uniformity and diversity index represents dominancy of one species towards other species, and vice versa. Dominancy index calculated with Equation of [18].

3 Results

3.1 Coral reef conditions

From satellite imaging by Lyzenga algorithm displayed that Saebus Island has 227.61 Ha of coral reef area, with coverage of 63.9 ha of living coral, 59.4 ha of dead coral and 104.31 of abiotic substrate (Figure 1). Coral reefs on Saebus Island are categorized as Fringing reef type. Almost all of stations have similar characteristic. On the depth of 1 to 3 m form coastline to slope forms flat coral reef overlay, while on the depth of more than 3 m is reef slope. But, 4th station have quite different characteristic, on the depth of 1-10 m have low sloping degree and tends to form reef flat overlay from coast line to open ocean.

Table 2. Percent coral cover(%), coral reef condition, and number of genus in Saebus Island

Stations	Depth (m)	Coral cover (%)	Conditions	Number of genus
St.1	3	57.46	Good	8
	10	77.46	Excellent	10
St.2	3	62.62	Good	14
	10	61.85	Good	12
St.3.	3	74.58	Good	10
	10	44.83	Fair	14
St.4	3	62.89	Good	15
	10	52.48	Good	14

Based on coral reef coverage percentage classification (Table 2), coral reef at Saebus Island categorized on "Fair" to "Excelent". On 1st Station (10 m of depth), Coral reef condition categorized on "Excellent" condition with coverage percentage is 77.46%. While on the 3rd station (10m of depth), coral reef condition categorized on "Fair" condition with 44.83% of area covered by coral. While the rest categorized on good condition with 52.48% - 74.56% area covered. Generally coral reef condition is good, there's no station categorized as "Poor" condition.

3.2 Diversity of corals

Based on genera diversity, station 4 (3 m) has the highest number of corals genus i.e. 15 genera were found. Then in a row 2nd station (3 m), 4th station (10 m) and 3rd station (10 m) with 14 genera found. While 1st station (3m) have least genera found with only 8 genera (Tabel 2).

Generally there are 25 genera of coral found on Saebus Island. Those 25 genera belongs to 13 famili, such as Acroporidae (5 genera), Astrocoeniidae (1 genus), Faviidae (5 genera), Fungiidae (1 genus), Helioporidae (1 genus), Merulinidae (1 genus), Milleporidae (1 genus), Mussidae (2 genera), Oculinidae (1 genus), Pectiniidae (1 genus), Pocilloporidae (3 genera), Poritidae (2 genera), and Siderastreidae (1 genus).

The most common coral found belongs to Acropora with high percentage at every station, ranging from 20.75% to 42.43%. Followed by Porites and Pocillopora with coverage percentage of 0.63%-10.75% found on every observation station. While the least found coral is Favites and Montastrea with less than 1% of coverage, and found only in 1 station. All of the data presented on Table 3.

Table 3. Kind of genus list and hard coral coverage (%) in Saebus Island

Family	Genus	Station 1		Station 2		Station 3		Station 4	
		3	10	3	10	3	10	3	10
Acroporidae	Acropora	29.40	42.43	27.59	26.13	39.11	20.75	28.90	23.53
	Anacropora	-	0.95	3.88	12.63	-	0.25	-	-
	Astreopora	-	-	-	-	-	0.69	0.25	0.25
	Montipora	8.40	-	-	-	-	-	2.00	1.19
	Palaclavarina	-	7.89	4.40	-	-	-	-	-
Astrocoeniidae	Palauastrea	-	3.47	-	-	-	-	-	-
Faviidae	Cypastrea	-	-	4.13	1.25	-	-	1.13	0.88
	Favia	6.60	1.58	-	1.00	6.73	-	-	-
	Favites	-	-	-	0.63	-	-	-	-
	Goniastrea	-	-	0.63	2.94	9.09	0.20	0.50	0.50
	Montastrea	-	-	0.50	-	-	-	-	-
Fungiidae	Fungia	-	0.63	0.13	3.63	1.50	0.13	-	
helioporidae	Heliopora	-	-	-	0.50	-	0.13	2.25	0.38
Merulinidae	Hydnopora	-	12.78	-	-	-	-	0.50	0.38
Milleporidae	Millepora	-	-	5.25	4.25	4.50	9.68	-	-
Mussidae	Lobophyllia	0.10	-	1.88	-	-	0.13	0.75	0.63
	Symphyllia	-	-	-	-	-	1.25	0.25	-
Oculinidae	Galaxea	-	-	0.50	-	-	-	0.50	0.50
Pectiniidae	Oxypora	-	-	1.50	2.13	-	-	0.50	0.50
Pocilloporidae	Pocillopora	3.10	4.10	1.75	0.63	2.08	0.56	10.75	9.50
	Seriathopora	3.10	-	-	-	2.25	0.13	2.75	1.13
	Stylophora	-	-	-	-	1.13	-	-	-
Poritidae	Goniopora	3.00	2.21	0.38	-	-	3.05	5.05	7.30
	Porites	3.76	1.42	10.10	6.13	3.13	7.50	6.81	5.81
Siderastreidae	Psammocora	-	-	-	-	5.06	0.38	-	-

3.3 Diversity, uniformly and dominancy of coral reefs community

As Known in Table 4, coral reef diversity rates on Saebus Island Categorized as fair to high category with indices ranging from 2.52 to 3.24. Refers on [16], the diversity value of 1 > H' > 3 is categorized as fair diversity and fair community, while H'> 3 is a high diversity and high community category. 1st and 3rd stations have fair diversity levels with indices of 2.81,

2.65 and 2.52. While other stations have a high level of diversity. Uniformity index at all of the station categorized as high with index ranging from 0.66 to 0.99. Refers on [17], uniformity index ranging 0.6 < E< 1 categorized as high uniformity and stable community. While dominancy index represents low dominancy rate from all of the station, ranging from 0.14 to 0.24. According to [18], dominancy index ranging from more than 0 to less than 0.5 categorized as low dominancy can be concluded that there's no dominant coral on the area.

Table 4. Diversity, Uniformly and Dominancy indices

Stasiun	H'		E		C	
	3m	10m	3m	10m	3m	10m
St.1	2.81	2.65	0.74	0.70	0.19	0.24
St.2	3.03	3.08	0.80	0.81	0.18	0.16
St.3	3.05	2.52	0.99	0.66	0.19	0.23
St.4	3.24	3.16	0.85	0.83	0.15	0.14

4 Discussions

4.1 Coral diversity and composition

Overall, 25 corals genera were found at the survey sites which consist of 13 families. The most common genera found are Acropora, Porites and Pocillopora. While the least genera found are Favites dan Montastrea genera. [4] Reported that, Saebus Island has a coral reef that is still beautiful with the potential of coral reefs on the type of Acropora and Porites. The statement is similar to the results of this research which shows that the genus Acropora is the dominant coral species in each observation station with percent cover reach 20.75% - 42.43%, followed by Porites with percentage of 1.42% - 10.10%, then genus Pocillopora with percentage 0.63% - 10.75%. Dominancy of Acropora was also reported in other islands such as Gangga island [19].

Generally, Acropora branching corals are more likely to dominate at a location with supporting water conditions because they have higher growth rates than other species [12]. Acropora branching is able to grow very rapidly with a growth rate of 5-20 cm per year, inversely proportional to the Porites massive lifeform whose radial growth rate is only 1-2 cm per year [12]. The existence of resistant species against environmental stresses such as massive Porites lifeform is directly proportional to the level of diversity in an area, the higher its existence the higher the diversity in the area. Otherwise if a community is dominated by susceptible species such as Acropora and Pocillopora then the species diversity will be lower [3].

The results showed that the level of diversity in the research area with an average value of 2.94 index included in the category of fair diversity and fair community condition. This situation is thought to be due to variations in the presence of resistant species and susceptible coral species [3]. However, the dominance level in the area belongs to the low category with the index of 0.14 - 0.24, indicating that there is not actually a genus that dominates excessively in the area. This is supported by a high level of uniformity with an index of 0.66 to 0.99. The higher uniformity index will indicate the higher uniformity of the population in the community, meaning that the spread of individuals of each species around Saebus island waters is fairly quite and the condition of the community in the location tends to be stable [17,20].

4.2. Coral reefs condition

Sapeken Sub-district has the largest coral reef area compared to 21 other Sub-districts in Sumenep district with an area of 21,790.70 Ha [2]. The result of satellite image analysis shows that Saebus Island has coral reef area of 227.61 Ha which approximately 0.01% of coral reefs area in Sapeken Sub-district waters. The results of this study also show that coral reefs in Saebus Island can be categorized into "fair", "good", to "excellent" with live coral cover percentage of 44.83% - 77.46%. This condition is better if it is compared with other islands around Kangean Island, such as Mamburit Island with corals live cover between 35% - 75% [6], and also Kangean Island with coral reefs area that classified as "poor" condition with coral live cover about 9.40% - 18.36% [21].

The condition of coral reefs on Saebus Island has potential in the fisheries and tourism sectors. Acording to [22], fisheries and coastal resources in Sapeken waters are the greatest and realistic potential for sustainable empowerment with more than 30 fish families identified in this area. But based on this study, at several observation stations were found unnatural coral reefs damage, indicated by the impact of explosives or potassium. It was characterized by a rubble area that forms a crater among other coral reefs communities in that area. Based on the observation result, the percentage of rubble substrate covered 0.25% - 25.50%. High percentage of rubble can indicate that was damaged by human activities that affect the environment [21]. Damaged coral reefs in Kangean Archipelago are impacted of illegal fishing activities and overfishing use explosives and potassiums that often occur in the region [6,4]. Many fishermen are aware of their fishing activities that can damage the environment. However, with the perspective that protection / conservation policies will reduce the chance of catching fishes, then subsequent damage to aquatic resources is no longer cared [23].

By looking at the potential of coastal resources and coral reef damage issues in this area, there needs to be a protective action against the aquatic environment especially coral reefs by the authorities, as well as rehabilitation action on the damaged water environment [24]. However, very little information can be accessed to determine the condition of coral reefs in this region. Therefore, further research is needed to support relevant parties to develop appropriate and beneficial policies for the environment and civil society. This research has provided general information about the condition of coral reefs community on Saebus Island, so it is expected to be a useful reference for other research that will be conducted around the area.

5 Conclusions

The conditions of coral reefs ecosystem at Saebus Island was categorized as "good" with average coral cover respectively 64.36 and 59.29 at depths of 3 and 10 meters. There were 25 genera of corals belongs to 13 families from all stations. The most common genera found were *Acropora*, *Porites*, and *Pocillopora*, while the least common genera were *Favites* and *Montastrea*. The Diversity levels of corals was included medium category with average Index of 2.94. Uniformity levels are included in high category with the average index of 0.80, while the dominance levels are included in the low category with the average index of 0.18.

Acknowledgements

This research was held by UKSA-387 Diving Club Diponegoro University on Dwipantara V Expedition: Kangean Archipelago "Jelajah Pesona Bahari Tampuk Timur Madura". We thanks to UKSA-387 Diving Club families who have fully supported and big appreciation to the Dwipantara V Expedition Team who has worked hard in this project. This project depends on logistical and field support from a number of contributed peoples. We thanks to the Department of Fisheries and Marine of East Java Province, Indonesia. Mr. H. Muhammad Ali, Mr. Usnan, Saur-Saebus village and Sapeken Sub-district governments, as well as the entire community of Saebus and Sapeken Islands for supports this research. We also give an apreciation to Prof. Ambariyanto, MSc, Ph.D., Diponegoro University, as our Supervisor who has directed and gave helpful suggestions.

References

1. M.A. Sutanto, R. Bedjo, and C.Y. Elishabeth, Petra Christian University. Surabaya (2013). (in Bahasa).
2. F.F. Muhsoni, M. Syarief, and Effendi, M. Jurnal Kelautan, **4**,1: 96-101, (2011). (in Bahasa).
3. Munasik and R.M. Siringoringo, Ilmu Kelautan (Indonesian Journal of Marine Sciene), **16**,1: 49-58 (2011). (in Bahasa).
4. A.S. Ackiss, S. Pardede, E.D. Crandall, M. Ablan-Lagman, A. Carmen, P.H. Barber, and K.E. Carpenter, Marine Ecology Progress Series, **480**: 185–197. DOI: 10.3354/meps10199 (2013).
5. J.W. Nybakken, Marine Biology An Ecological Approach. 457 pp (PT. Gramedia, Jakarta, 1988). (in Bahasa).
6. B. Tamam, A. Apri, and M. Saleh, Jurnal Kelautan, **6**,2: 120-127. DOI: http://dx.doi.org/10.21107/jk.v6i2.785 (2013). (in Bahasa).
7. L. Pet-Soede, and M. Erdmann, SPC Live Reef Fish Information Bulletin, **4**: 28-36 (1998).
8. P. Mous, L. Pet-Soede, M. Erdmann, H. Cesar, Y. Sadovy, and Pet, J. Collected essays on the economics of coral reefs: 69-76 (2000). Kalmar, Sweden: CORDIO, Kalmar University.
9. H. Cesar, L. Burke, and L. Pet-Soede, (Cesar Environmental Economics Consulting (CEEC), 2003).
10. Ambariyanto. IOP Conference Series: Earth and Environmental Science, **55**,1: 012002 (2017). IOP Publishing.
11. C.L. Huffard, MV. Edmann, and T. Gunawan, (Ministry of Maritime Affairs and Fisheries & Marine Protect Areas Govermence, Jakarta, 2012).
12. T. Toda, T. Okashita, T. Maekawa, B.A.A. Alfian, M.K. Rajuddin, R. Nakajima, W. Chen, K.T. Takahashi, B.H. Othman, and M. Terazaki, Journal of Oceanography, **63**: 113-123 (2007).
13. S. English, C. Wilkinson, and V. Baker, *Survey Manual for Tropical Marine Resources*. 368 pp (Australian Institute of Marine Science, Townsville, 1994).
14. E. D. Gomez, and H. T. Yap, *Monitoring Reef Condition. P:187-195 in R. A. Kenchington and B.E.T. Hudson (eds.), Coral Reef Management Hand book.* (UNESCO Regional Office for Science and Technology for South East Asia, Jakarta, 1988).
15. State Minister for the Environment of Republic Indonesia. Minister of Environment Decree No. 04 in 2001. About Fixed Criteria of Coral Reef Quality Standard. Jakarta (2001). (in Bahasa).
16. J. A. Ludwig, and J.F. Reynolds, *Statistical Ecology, A Primer on Methods and Computing*. 337 pp (Jhon Wiley & Sons, Inc. Toronto-Canada, 1988).
17. C. J. Krebs, *Ecologycal Methodology*. 620 pp (Wm. C. Brown Publisher, Dubuque, 1989).
18. E. P. Odum, *Fundamental of Ecology, 3 rd Edition*. 564 pp (W.B. Sounders Co., Philadelphia and London, 1971).
19. Souhoka, J. Ilmu Kelautan, **18**,4: 213-224. DOI: https://doi.org/10.14710/ik.ijms. 18.4.213-224 (2013). (in Bahasa).
20. V.H. Anwar, I.J. Zakaria, and S. Afrizal Jurnal Biologi Universitas Andalas, **3**,1: 20-26 (2014). (in Bahasa).
21. Insafitri. Jurnal Kelautan, **3**,2: 112-116 (2010). (in Bahasa).
22. A. Romadhon, F. Yulianda, D.G. Bengen, and L. Adrianto, Tata Loka, **15**,3: 218-234. DOI: 10.14710/tataloka.15.3.218-234 (2013). (in Bahasa).
23. B. Riegl, A. Bruckner, L. Steve , Coles, P. Renaud, and R.E. Dodgea, The Year in Ecology and Conservation Biology : Ann. N.Y. Acad. Sci.

1162 : 136–186. DOI : 10.1111/j.174 9-6632.2009.04493.x (2009).

24. L.S. Hoek, and E.K . Bayoumi,. IOSR Journal of Pharmacy and Biological Sciences (IOSR-JPBS), **12**,II: 59-63. DOI: 10.9790/3008-1202025963 (2017).
25. E.F. Ardiansyah, Hartoni, and L. Litasari, Maspari Journal, **5**,2: 111-118 (2013). (in Bahasa).
26. Z. Hidayah, National Seminar of Marine VIII, **B3** : 88-96 (2012). (in Bahasa).
27. K. Kleemann, *Tropical Marine Biology II : Classification of Scleractinian (Stony) Corals*. (University of Vienna, Vienna, 2002).
28. McClanahan, T.R. *In Coral Reef Conservation. I.M.* Cote et al., Eds.: 147–182. (Cambridge University Press. Cambridge, UK, 2006).
29. A. Muqsit, P. Dewi, and T. Zamdial, Jurnal Enggano, **1**,1: 75-87 (2016). (in Bahasa).
30. M.R. Syahrir, A. Jayadi, Adnan, Yasser M., and Hanjoko, T. Internationl Journal of Science and Engineering, **8**,1: 60-64. DOI: https://doi.org/10.12777/ijse.8.1.60-64 (2015).

Implementation Of Conservation Policy Through The Protection Of Life Support System In The Karimunjawa National Park

Nur Anisa Eka Ariyani[1], *Kismartini*[2]

[1]Master Program of Environmental Science, School of Postgraduate Studies, Diponegoro University, Semarang –Indonesia
[2]Department of Doctoral Public Administration Program, School of Postgraduate Studies, Diponegoro University, Semarang - Indonesia

Abstract. The Karimunjawa National Park as the only one marine protected area in Central Java, managed by zonation system has decreased natural resources in the form of decreasing mangrove forest area, coral cover, sea biota population such as clams and sea cucumbers. Conservation has been done by Karimunjawa National Park Authority through protection of life support system activities in order to protect the area from degradation. The objective of the research is to know the implementation of protection and security activities of Karimunjawa National Park Authority for the period of 2012 - 2016. The research was conducted by qualitative method, processing secondary data from Karimunjawa National Park Authority and interview with key informants. The results showed that protection and security activities in The Karimunjawa National Park were held with three activities: pre-emptive activities, preventive activities and repressive activities. Implementation of conservation policy through protection of life support system is influenced by factors of policy characteristic, resource factor and environmental policy factor. Implementation of conservation policy need support from various parties, not only Karimunjawa National Park Authority as the manager of the area, but also need participation of Jepara Regency, Central Java Provinces, communities, NGOs, researchers, developers and tourism actors to maintain and preserve existing biodiversity. Improving the quality of implementors through education and training activities, the availability of the state budget annually and the support of stakeholders is essential for conservation.

1 Introduction

Extinction is a natural process, but the extinction rate that reaches 1000-10,000 times from its natural processes [1] is quite alarming. More than 77,300 species have been included in the IUCN Red List in 2015. This extinction occurs due to habitat degradation, over exploitation, pollution, disease and climate change. The threat of extinction of flora and fauna is mainly due to the high rate of population growth in Asia and Africa including Indonesia [2].

Indonesia ranks is 4th out of 20 potential countries have extinction of biodiversity, where there are 1126 endangered species [3]. Efforts to save biodiversity both in situ and ex situ have not been satisfactory. According to the Ministry of Environment and Forests [4], Indonesia has allocated no less than 27 million hectares of land as a protected area. 51 areas designated as national parks have not been able to be a shelter because of threats of damage including The Karimunjawa National Park.

The threat of ecosystem damage occurring in the Karimunjawa Islands has prompted the government to conserve. Determination of Karimunjawa Marine Nature Reserve on April 9, 1986 through Decree of Minister of Forestry no. 123 / Kpts-II / 1986 was the first step, then on February 22, 1999 Karimunjawa was designated as a national park through Decree of Minister of Forestry number 78 / Kpts-II / 1999. The area of 111.625 Ha consists of 22 islands with 5 The major ecosystems of coral reef ecosystems, sea grass beds, mangrove forests, coastal forests, and lowland tropical rainforests [5,6].

Karimunjawa Island and Pulau Kemujan as the largest island in The Karimunjawa National Park threatened damaged mangrove ecosystem. According to research conducted by Kamal et al [7], the area of mangrove forest from 2009 to 2012 on both islands is degraded by 23.8 Ha, due to changes in land use. Similarly, what happens to coral reef ecosystems, technical reports from Wildlife Conservation Society Indonesia Program [8], shows that there is a decrease in hard coral cover from 57.86% in 2013 to 49.89% in 2016 in The Karimunjawa National Park.

Coral reef ecosystem is a habitat for various marine biota including clams and sea cucumbers. clams (*Kima*) as a protected animal according to PP. 7 Year 1999 About Preservation of Plant and Animal Species has decreased population in 3 islands in The Karimunjawa National Park that is Cemara Besar Island, Geleang Island and Katang Island [9]. Besides clams, sea cucumber population also decreased. The results of sea cucumber monitoring conducted by BTNKJ in 2016 found 166 sea cucumbers decreased when compared with the population of sea cucumbers monitoring results

* Corresponding author: nur_anisa0429@yahoo.com

in 2012 as many as 315. Population decline occurs due to habitat destruction and the over exploitation of clams and sea cucumbers.

Mangrove degradation, decreased hard coral cover and decline in marine biota population are an indication that the Ministry of Environment and Forestry through the Karimunjawa National Park Agency (BTNKJ) needs to improve its performance in the context of conservation. Efforts to manage the area must be increased, conservation through protection of life support system activities as mandated by Law No. 5 of 1990 on Conservation of Biological Natural Resources and Ecosystems have various challenges in its implementation. TNKJ still have many problems like illegal logging, illegal fishing, exploitation of protected plants and animals, destruction of infrastructure and protection of zone functions. Its problems are indicated of the lack of optimum implementation of conservation policies in The Karimunjawa National Park.

Based on the phenomenon, research on the implementation of conservation policy through the protection of life support system in The Karimunjawa National Park is important to do. The purpose of this research is to know the implementation of protection and security activities in The Karimunjawa National Park for the period of 2012 - 2016.

2 Literature review

Public policy according to some experts [10,11] is the government's decision to do or not to do something about public problems that occurred. The public policy process consists of several stages: problem identification, agenda formulation, policy formulation, policy approval, policy implementation, and policy evaluation.

Implementation of policy as one of the stages of the public policy process is very important to note. The process at this stage can be said to be crucial, because no matter how good a policy when implementation is bad then the policy will not get to the target of policy well.

The success of an implementation is influenced by several things, with reference to previous research (George C. Edward III; Donald S. Van Meter and Carl E. Van Horn; Merilee S. Grindle; Daniel A. Mazmanian and Paul A Sabatier; David L Weimer and Aidan R Vining; Maier, C., & Winkel, G; Kalaba, FK; Muhumuza, M. , & Balkwill, K) [4,11,12] in this research will be seen the influence of policy characteristics, resources and policy environment to the successful implementation of conservation policy.

Conservation policies in Indonesia are governed by Law No. 5 of 1990 on the Conservation of Biological Natural Resources and their Ecosystems. This regulation explains that one of the conservation activities is the protection of life support system. The protection of the life support system is a natural process of biological and non-biological elements that ensures the survival of ecological creatures to improve the welfare of society and the quality of human life. The protection of the life-support system is carried out through prevention, prevention and restriction of damage caused by human, livestock, nature, invasive species, pests and diseases, and effectively guarding the area. Protection and security of conservation area is an activity that must be done by the institution appointed to manage the conservation area one of them is the national park.

National park is a natural conservation area that has native ecosystem, managed by zoning system which is utilized for research, science, education, cultivation, tourism and recreation purposes. Criteria of a region can be designated and designated as a national park area is to have unique natural resources and unique and unique ecosystems that are still intact and natural and unique natural phenomena; Has one or more intact ecosystems; Has sufficient breadth to ensure the sustainability of natural ecological processes; And is an area that can be divided into core zones, utilization zones, jungle zones, and / or other zones as required.

3 Research methods

Research conducted in The Karimunjawa National Park as the only one Marine Protected Area in Central Java with high biodiversity. The research method used qualitative approach, processing secondary data from Karimunjawa National Park Agency and interview with key informants.

Secondary data itended of Strategic Plan of Karimunjawa National Park 2015-2019, Performance Report 2012 - 2016 and report on the results of protection activities conducted in the last 5 years (2012-2016). The data will be collected and grouped and then analyzed qualitatively. The required secondary data is obtained from Karimunjawa National Park Agency as the technical implementation unit of the Ministry of Environment and Forestry in implementing the conservation policy.

The implementation of the conservation policy described in this study is limited to the protection of life support system activities which is conducted by BTNKJ in the last 5 years. Interview with key informant is done by researcher to be able to describe the secondary data obtained and know the factors that influence the implementation and to know whether there is a decrease of the disturbance to the area after the protection of the area. Key informants are BTNKJ employees of forest police elements and forest ecosystem controllers who doing protection and security activities of the area.

4 Results and discussion

The Karimunjawa National Park is geographically located at 5 ° 40'39 "- 5 ° 55'00" LS and 110 ° 05'57 "- 110 ° 31 '15" BT which is administratively belong to the sub-district of Karimunjawa Regency of Jepara Central Java Province can be saw in figure 1.

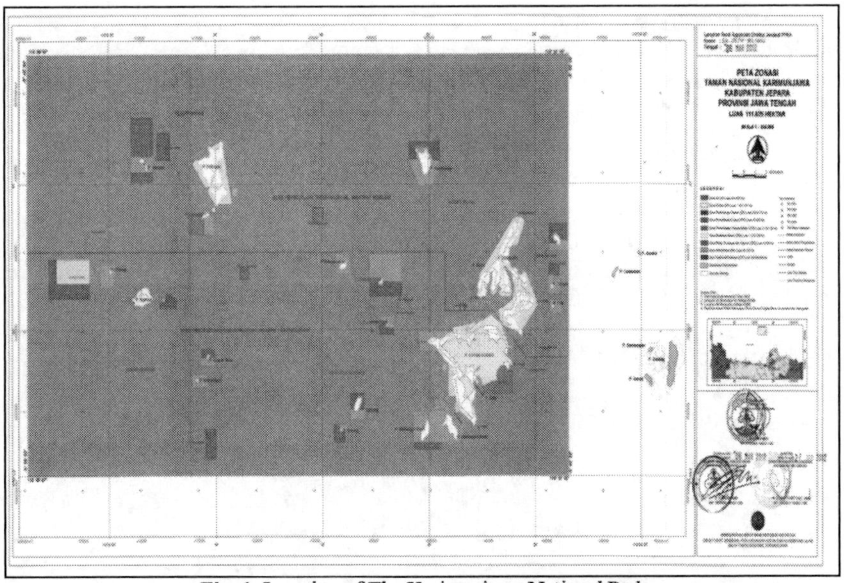

Fig. 1. Location of The Karimunjawa National Park

The total area is 111.625 Ha with details of 1.285.50 Ha is the mainland area of Karimunjawa Island in the form of lowland tropical rain forest ecosystem, 222.20 Ha is the mainland area in Kemujan Island in the form of mangrove forest ecosystem and 110,117.30 Ha is the water area. TNKJ has two National Park Management Section (SPTN) namely SPTN I Kemujan and SPTN II Karimunjawa.

As a national park, TNKJ is managed by a zoning system. Based on the Decision Letter of the Director General of PHKA. SK 28 / IV-SET / 2012 on Zoning of Karimunjawa National Park, there are 9 zones within this area: core zone, jungle zone, marine protection zone, land use zone, marine tourism zone, marine culture zone, religious zone, culture and History, rehabilitation zone, traditional fishery zone.

4.1 Implementation of the protection of life support system

The protection of life support systems is aimed at the maintenance of ecological processes that support the viability of life to improve people's welfare and the quality of human life. To achieve these objectives, the protection of the area in The Karimunjawa National Park is implemented with 3 activities, namely Preemptive, preventive and repressive activities.

4.1.1 Preemptive activities

Preemptive activities are activities aimed at preventing, eliminating, reducing, closing the intentions of a person or group to commit forestry crimes. The Karimunjawa National Park Office conducts pre-emptive activities through the manufacture and maintenance of fire breaks, maintenance of the boundary limit, maintenance of reference points, light beacon maintenance, patrol tracking, zonation boundary marking, establishment and guidance of fire-concerned community (MPA), borderline maintenance, , Facilitation, guidance, assistance of community polhut partners (MMP), socialization of regulation / zoning, and fire control.

Preeminative activities through the manufacture and maintenance of fire breaks are intended to prevent the spread of flames by separating flammable materials and also to facilitate the movement of equipment and personnel in handling fires. In the Year 2012 there are 2 activities of manufacture and maintenance of fire breaks, 1 activity in 2013, 2015, 2016, and in 2014 there is no implementation of fuelbreak activities.

The next preemptive activity is the maintenance of the boundary. Borders are fixed boundary marks of a certain size made of concrete with iron frame or of durable I / II grade wood or other boundary marks installed along boundary routes. Boundary maintenance activities are very important in the work area of the Karimunjawa National Park Authority, especially those directly related to the community's landrights so that the location, position and condition of the boundary are still good to ensure legal certainty and integrity of conservation forest areas. In the last 5 years there are only 2 border maintenance activities ie 1 activity in 2013 and 1 activity in 2016.

From the results of maintenance activities on 2016 limits at Resort Legon Lele Regional Park Management Section of Karimunjawa Region II of 292 border frontier that limits the jungle zone area with community owned land, Pal TN 1 to Pal TN 60 is located in the mangrove forest stretching from Nyamplungan Until Kemloko, TN 61 to TN 72 at Legon Moto, while TN 73 to TN 292 stretched from Kemloko to Legon Janten. 20 of them damaged in Kemloko area up to Cikmas and have been repaired through this activity.

In pre-emptive activities, BTNKJ involves communities around the region by establishing fire-caring community (MPA) and the community of polhut partners (MMP). MPA is a community that voluntarily cares for the control of forest fires and land that have been trained / provided and can be empowered to assist forest fire control activities while MMP is a community group around the forest that helps polhut in the implementation of forest protection under coordination, coaching and Supervision of BTNKJ.

Preemptive activities can be done well if it gets enough budget from the government. The highest preemptive activity budget realization was in 2013 and the lowest was in 2015. Budget support and BTNKJ preemptive activities in the last 5 years can be seen in figure 2.

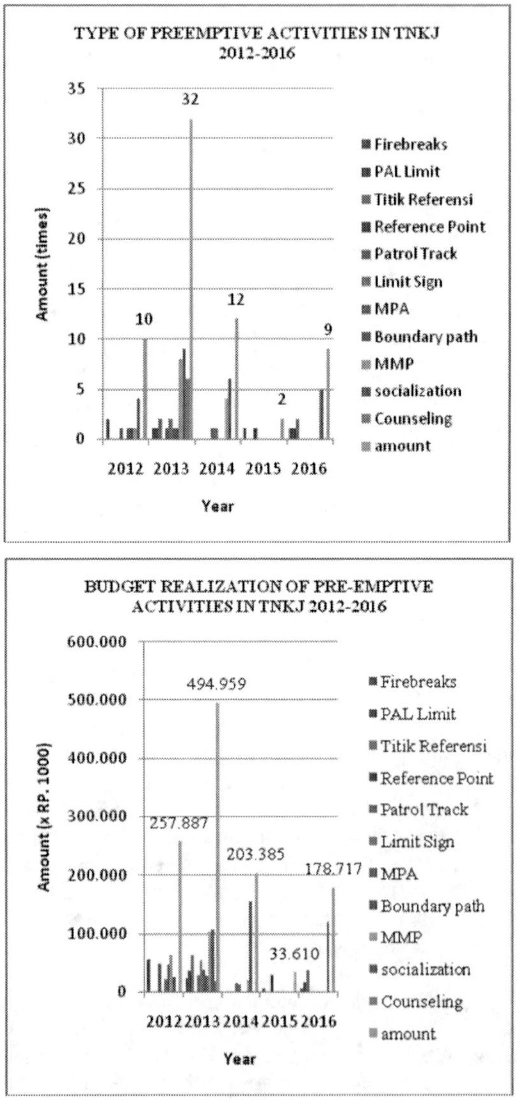

Fig. 2. Types of activities and budgets of 2012-2016 preemptive activities

4.1.2 Preventive activities

Preventive activities are activities aimed at preventing, eliminating, reducing, closing a person's or group's opportunity to commit forestry crimes. Preventive activities in The Karimunjawa National Park are carried out with forest fire prevention patrols, MMP joint patrols, water patrols, ground patrols, monitoring of zoning fishermen compliance, data analysis of regional security disturbances, forest fire standby apples. Patrol activity is a guarding activity to narrow the space for violation. This activity is carried out at guarded posts that have been determined based on the point of prone to the disturbance.

The frequent preventive activities in The Karimunjawa National Park are fire patrol and water patrol. Of the annual performance report data, 12 fire patrol activities were conducted in 2013, 12 water patrol activities in 2015 and 16 water patrols in 2016.

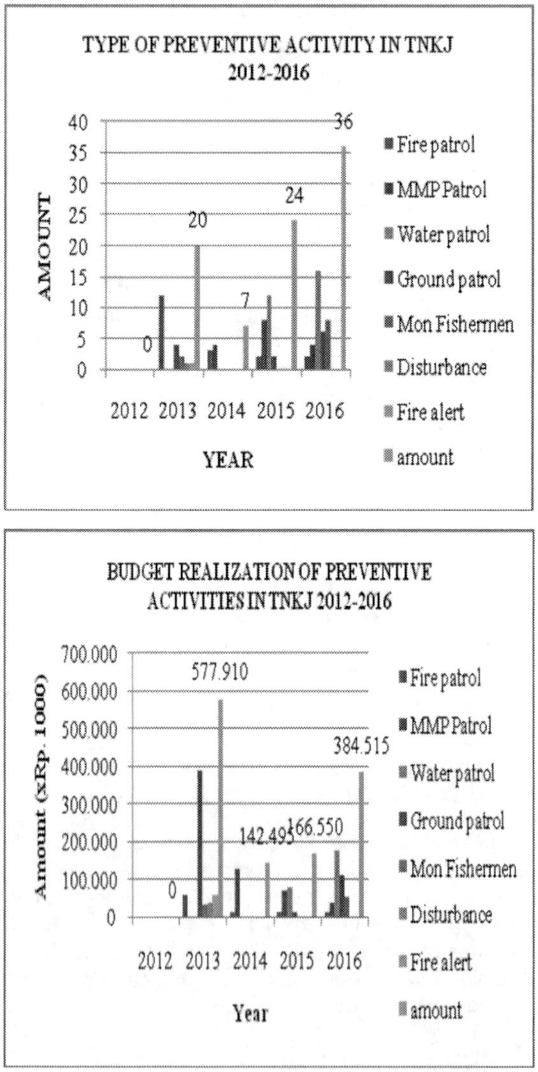

Fig. 3. Type of activities and budget of preventive activities in 2012-2016

4.1.3 Repressive activities

Repressive activities are non-judicial law enforcement activities to reduce, suppress or stop forestry crimes committed by a person or group. This repressive activity is carried out if the situation and condition of the disturbance of the security of the region occurs and tends to continue to take place even increased so that the need to immediately take action against the perpetrators. Repressive activities in The Karimunjawa National Park are conducted through investigation activities, evidence handling, justice operations, intelligence operations, aquaculture operations, joint security operations, controlling operations of TSL, ground security operations and destruction of evidence.

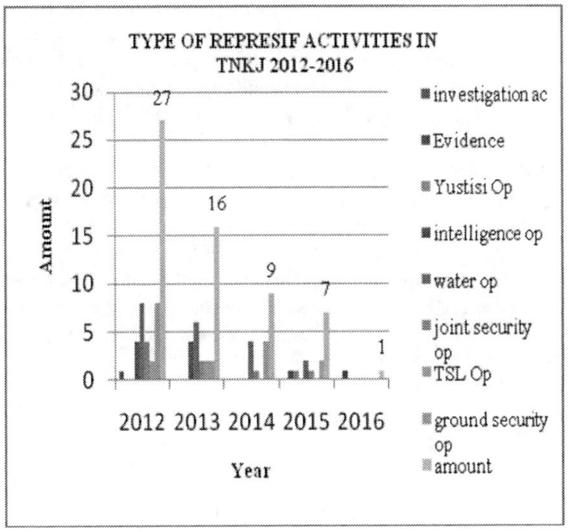

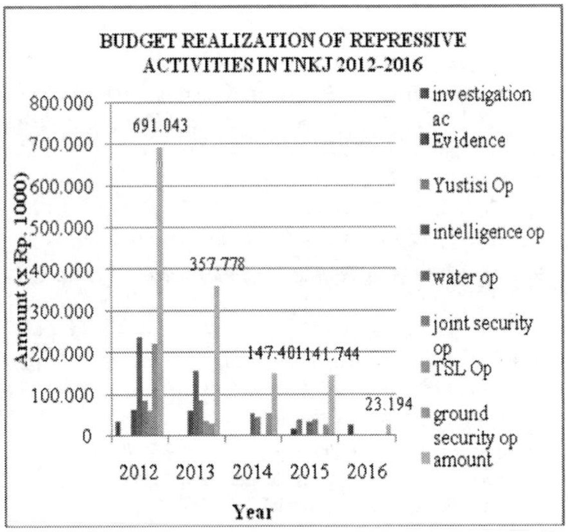

Fig. 4. Type of activities and budget of repressive activities in 2012-2016

Repressive activities frequently conducted in The Karimunjawa National Park are marine security operations and ground security operations while intelligence operations and joint security operations are sometimes performed. From the data analysis there are 20 watershed operations, 16 ground security operations, 8 activities for each intelligence operation and joint security operation in the last 5 years.

There is only one repressive activity in 2016, namely the destruction of evidence. This is because the repressive activities that were previously the authority of the directorate general (director general) Conservation of natural resources and ecosystem (KSDAE) turned into the authority of the director general of law enforcement. All forms of repressive activities are handled by this division. Type of activities and budget realization of repressive activities 2012 - 2016 can be seen in figure 4. Number of activities and budget realization tends to decrease every year. This is a good signal for the decrease in disturbance to the region.

4.2 Factors affecting the implementation of conservation policies

4.2.1 Factors of policy characteristics

The results of interviews with key informants, conservation policies in life protection system protection activities have been in accordance with the objectives to be achieved is to ensure the survival of ecological creatures to improve the welfare of society and the quality of human life. The interests of the target group in this community around The Karimunjawa National Park have been accommodated as well as the conservation interests.

The protection activities carried out are in accordance with the master plan and zone standards so as to reduce disturbance to the area. The clarity of the program also felt that the implementor was good enough so that the activities could be implemented. There is consistency of the program, it is seen from the existence of life protection system protection activities in the last five years.

4.2.2 Resource factor

The Karimunjawa National Park Agency has 92 employees consisting of 75 civil servants and 17 daily employees who work in The Karimunjawa National Park Ageny, National Park Management Section I (Kemujan), National Park Management Section II (Karimunjawa) as shown in table 1.

Table. 1 State of Employment Hall of Karimunjawa National Park Based on Group and Sex of 2016

No	Organizational unit	civil servants Class				daily employees	gender		total number
		IV	III	II	sum		Male	female	
1	The Karimunjawa National Park	2	27	4	33	7	27	13	40
2	National Park Management Section I (Kemujan).	0	15	2	17	6	21	2	23
3	National Park Management Section II (Karimunjawa)	0	18	7	25	4	27	2	29
	Total number	2	60	13	75	17	75	17	92

From the interviews with key informants, employees in TNKJ have been sufficient in number and function. Every employee is given equal opportunity to attend education and training as an effort to improve conservation skill. The availability of APBN budgets through DIPA (List of Budget Usage Fulfillment) has been felt sufficiently since 2012-2016. Funding from outside parties whether NGOs or CSR companies support the better conservation in The Karimunjawa National Park.

4.2.3 Environmental factors policy

The policy environment includes the community surrounding the park, the support of relevant agencies and outside support is indispensable in the context of conservation. Given the limited capacity that is still owned by Karimunjawa National Park as a manager, BTNKJ has 4 (four) partners in realizing the conservation in TNKJ namely Wildlife Conservation Society-Indonesia Program (WCS-IP), Jakarta Animal Aid Network (JAAN), RARE and Marine and Fisheries Agency of Central Java Province. In addition to those parties, support from scientists, tourism actors and developers is also needed.

From the interviews with key informants, there is an increase in public awareness about the importance of conservation. TNKJ has also been included as a conservation area within the RTRW of Jepara District and has been proposed to be included as a conservation area within RZWP3K of Central Java Province.

5 Conclusions and suggestions

Implementation of conservation policy through protection of life support system activities within the period of 2012 - 2016 implemented with 3 activities of preemptive, preventive and repressive activities. Karimunjawa National Park Agency conducts pre-emptive activities through the manufacture and maintenance of fire breaks, maintenance of the boundary limit, maintenance of reference points, light beacon maintenance, patrol tracking, zonation boundary marking, establishment and guidance of fire-concerned community (MPA), borderline maintenance, facilitation, guidance, assistance of community *polhut* partners (MMP), socialization of regulation / zoning, and fire control.

Furthermore, preventive activities in Karimunjawa National Park are conducted by patrolling forest fire prevention, MMP joint patrolling, water patrolling, land patrolling, monitoring of zoning fishermen compliance, data analysis of regional security disturbances, forest fire standby apples.

For repressive activities in Karimunjawa National Park conducted through investigation activities, handling of evidence, justice operations, intelligence operations, marine security operations, joint security operations, controlling operations of TSL, ground security operations and destruction of evidence.

Implementation of conservation policies through life protection system protection activities in the period 2012 - 2016 is influenced by factors of policy characteristics, resource factors and environmental policy factors. The policy characteristics of BTNKJ are appropriate and workable to achieve the goal of conserving natural resources. Similarly, with the resources owned, the number and function of the implementor in maintaining the area of degradation has been considered sufficient. In terms of funding, the APBN has been able to support conservation activities in TNKJ. Environmental policy factors include related parties such as communities around the region, Jepara regency, Central Java Provinces, NGOs, scientists, tourism actors and developers have provided support in policy implementation in The TNKJ.

Conservation policy conservation through life protection system protection activities needs to be maintained or where possible enhanced for the preservation of biological natural resources in the TNKJ. Improving the quality of implementors through education and training activities, the availability of the state budget annually and the support of stakeholders is essential for conservation.

References

1. *Laporan Tahunan Balai Taman Nasional Karimunjawa 2012-2016* [in Indonesian]
2. Joko Widodo, Bayumedia Publishing, (2006). [in Indonesian]
3. IUCN (2015) https://cmsdocs.s3.amazonaws.com/keydocuments/IUCN_Red_List_Brochure_2015_LOW.pdf diakses tanggal 10 Juli 2017
4. Kalaba, F. K. (2016). Journal of Forest Policy and Economics **69**, 40–44. https://doi.org/10.1016/j.forpol.2016.04.004
5. *Statistik Balai Taman Nasional Karimunjawa, 2012-2016* [in Indonesian]
6. Wildlife Consevation Society Indonesia Program, *Laporan Teknis Monitoring Ekosistem Terumbu Karang Taman Nasional Karimunjawa,* Bogor (2016). [In Indonesian]
7. http://lipi.go.id/berita/ancaman-kepunahan-spesies-indonesia-tertinggi-di-dunia/1669 diakses tanggal 10 Juli 2017 [in Indonesian]
8. E. H. W. Chan, and J. Hou, **46**, 35–43, (2015). https://doi.org/10.1016/j.habitatint.2014.10.018
9. *Balai Taman Nasional Karimunjawa,* (2016) [in Indonesian]
10. Kamal, M., Hartono, H., Wicaksono, P., Adi, N. S., & Arjasakusuma, S. Journal of Geomatics and Planning, **3**(2), 107, (2016). https://doi.org/10.14710/geoplanning.3.2.107-116
11. C. Maier, and G. Winkel, Journal of Forest Policy and Economics, (2016). https://doi.org/10.1016/j.forpol.2016.12.015

12. M. Muhumuza, and K. Balkwill, International Journal of Biodiversity (2013).

The Application Of Liquid Fertilizer Made Of Traditional Market Organic Wastes On Growth Of Setaria Grass (*Setaria splendida* Stapf)

Eko Hendarto[1,*] and *Suwarno*[1]

[1] Faculty Of Animal Husbandry, Jenderal Soedirman University, Purwokerto – Indonesia

Abstract. There are hugh amount of traditional market organic wastes that may polute the environment. In general, the wastes are utilized for compost making and liquid fertilizer as well for plant. The use of liquid fertilizer from organic wastes of traditional markets opens up opportunities for misplaced cultivation of Setaria grass (Setaria splendida Stapf), which is required by ruminant farms. This research was conducted to evaluate the best mixture of water to the fertilizer in term of its effectiveness on the variables and experimental method using Completely Randomized Design. The treatments were: 6 doses of mixtures namely 0, 10, 20, 30, 40 and 50 liters of water, each of which was mixed with 10 liters of liquid fertilizer. The variables measured were the height, the numbers of tillers, the numbers of leaves, and canopy. The results of the study showed that the doses of water in the fertilizer did not indicate any significant differences ($P > 0.05$) on all variables being studied, however, the linear equation showed that greater concentrations of water in the fertilizer tended to decrease the growth of Setaria grass. Suggested use of water on the liquid fertilizer mixture should be not greater than 30 l – 10 l fertilizer.

1 Introduction

In all of the cities, traditional markets produce large amount of wastes that may pollute the environment, one of which is organic wastes. However, the waste can be used to support human life by converting it to become composts. Via the technology of processing, the wastes produce liquid fertilizer that can be used to support the growth of plants. One of the advantages of using liquid organic fertilizers into plants is the easily absorbed nutrients in the fertilizers by the soil that can readily be utilized by the plants.

The product of forage plants is forage that can be used to feed the ruminants, therefore, it functions to support the development of ruminant farmings [1]. As a feed, the forage consists of leaves and twights, which is the most nutritious parts of the forage plants [2]. Setaria grass is famous to the livestock farmers as the most preferred feedstuff by the ruminants.

Liquid organic fertilizers generally have low concentration of chemical compounds with fairly good nutrient content to fulfill the plant requirement. If there is an excessive amount of the fertilizer in the soil, the plants can easily control its absorption [3]. The mentioned characteristics of liquid organic fertilizers is also owned by the traditional market-waste organic fertilizer, in the other side, the making or processing the wastes into the fertilizer can reduce the pollution and prolong the utilization of WEP (Waste End Processing) site, due to the reduced amount of wastes being dumped into the site. The waste processing also opens the opportunity to decrease unemployment. A traditional market can produce the daily wastes up to a value greater than 100 m^3 [4], therefore, the potent to produce the liquid fertilizer is huge in order to guarantee the long-term availability of the liquid organic fertilizers made of traditional market wastes.

In common, liquid fertilizer is only applied for ornamental plants such as orchids, palms, roses, jasmines, and so on. However, it can also be used for grasses to widen it utilization in the livestock farmings, in an effort to support the environmental awareness activities. The utilization of organic fertilizers can improve the soil quality for plant growth, therefore, increase its productivity. The productivity aspects that can be on served include plant height, the numbers of tillers and leaves, and the width of canopy.

The increase of forage plant productivity one of which is via fertilization management of Setaria finally increases the animal growth and products [5]. In relation to the ideas mentioned above, a study concerning the effects of the utilization of traditional-market-waste organic fertilizer on the growth of Setaria *(Setaria splendida)* was conducted to support the development of ruminant farming in order to facilitate the human life.

* Corresponding author: ekohendarto17@gmail.com

2 Materials and Method

A study of the fertilizer utilization management was conducted involving the traditional-market-waste liquid organic fertilizer and Setaria grass. The study used Completely Randomized Design, consisted of 6 (six) treatments and 4 (four) replicates in each treatment. The treatments were the applications of traditional market waste liquid organic fertilizer that was mixed with water of various concentrations (Table 1).

Table 1. The doses of liquid organic fertilizer of the treatments.

No	Teatment Code	Liquid Organic Fertilizer	Water	Total	Width of land being fertilized	Volume of Fertilizer/ m^2
		Liter			m^2	ml
1	S-0	0	0	0	100	0
2	S-1	10	10	20	100	200
3	S-2	10	20	30	100	300
4	S-3	10	30	40	100	400
5	S-4	10	40	50	100	500
6	S-5	10	50	60	100	600

The results of laboratory analyses showed that the fertilizer contained nitrogen as much as 0.038 percent, P_2O_5 81.748 mg/l, K_2O 0.134 percent and C/N = 0.92. The size of each plot was 3 m^2. The variables that were observed at the plant age of 50 days were the phenomena of plant growth included the height, the numbers of tillers and leaves, and the width of canopy. Chronologically, work procedures included land processing, cultivation, fertilization, plant management, and harvest. When the age of Setaria was 2 weeks, the plants were fertilized using liquid organic fertilizer with the doses as written in table 1, by spraying the fertilizer onto the soil surface. The data were analyses using analysis of variances, continued by using the Orthogonal Polynomial, and regression analysis for the doses of fertilizer.

3 Results and Discussion

3.1 The conditions of experimental site

The study was conducted at a field without tree canopy. The results of soil laboratory analysis showed that the soil content of Nitrogen was 49.35 ppm (very high), available P_2O_5 was 0.198 ppm (very low) and available K_2O was 0.496 percent (fair). The texture class was sandy loam, containing sand fraction as much as 24.63 %, dust fraction of 9.15 % and clay of 66.22 %, with a soil pH of 5.11 (The Laboratory of Soil Science, Faculty of Agriculture, Jenderal Soedirman University, Indonesia).

On the basis of pH, the soil was acidic [6], however, on the basis of its mineral contents, especially its high N content, the soil fairly support the growth of grasses. According to Huang et al., nitrogen is a macro mineral that highly required by plants [7]. Related to the above condition, according to Thomas and Schultze-Kraft the land in this study is capable of supporting the growth of certain plants, including grasses [8]. Aminudin and Hendarto report, Setaria (*Setaria splendida*) grass is able to grow under heavy condition of soils, therefore, better soil condition gave a more beneficial information of the study, especially in relation with the fertilization management [9].

3.2 The heights of Setaria grass

The results of the applied traditional market waste liquid organic fertilizer onto the Setaria (*Setaria splendida*) was showed in Table 2. The phenomenon of growth in term of height, is an aspect that can be used to know the quality of growth of plants. The mean value of height of Setaria due to the fertilization treatments was 50.17 cm, with a range of 45.93 cm – 52.05 cm. The results were in accordance with the genetic value of the grass, which indicated that the genetic aspect cannot be erased due to the environmental factors [9].

The control grass without fertilization showed the tallest growth, might be due to the very high soil content of N. Rahman et al., states, nitrogen functions to stimulate the whole of plant growth, including height [10].

Tabel 2. Effect of traditional-market-waste liquid organic fertilizer on the growth phenomena of Setaria grass

Treatment	Plant Height (cm)	Number of tillers	Number of Leaves (sheet)	Canopy (cm^2)
S-0	52.05[a)]	11.98[b)]	72.00[c)]	40.85[d)]
S-1	52.58[a)]	11.75[b)]	64.00[c)]	40.68[d)]
S-2	52.00[a)]	10.98[b)]	70.40[c)]	40.05[d)]
S-3	51.65[a)]	10.40[b)]	64.15[c)]	40.18[d)]
S-4	48.75[a)]	10.73[b)]	63.58[c)]	38.10[d)]
S-5	45.93[a)]	10.38[b)]	56.85[c)]	37.83[d)]
Mean Value	50.16	11.04	65.18	39.58

Note : Similar superscripts in th samme collumns showed similar value ($P>0.05$)

The usage of liquid fertilizer according to Budiraharjo et al., in an effort to utilize organic wastes has a potent to be applied in the fields [11]. According to Marpaung et al., the differences between traditional-market-waste organic compost to that of liquid is, the minerals from the compost need time space in order the minerals to be available to plants, on the contrary, those of liquid fertilizer, the minerals are available as soon as the fertilizer is applied to the soil [12]. It is reported by Hanolo that the decomposition process of compost causes the late availability of its minerals [13]. It is added by Hendarto at the time the minerals in the compost is available to the plants, the plants have arrived to their generative stage of growth that does not need minerals in excessive amount [14]. It is possible that minerals from the decomposition process of compost will be used for the next growth, namely after harvesting of the current grass. The results of this study were in line with those of Marpaung et al., (2014), that at the soil with high N content, the addition of N into the soil gave no significant effect on the height of plants. Wachjar and Kadarisman also report that the applications of various levels of liquid organic fertilizer gave no significant effect on the height of cocoa plants [15].

The results of analysis of variances showed tha there was no significant effect (P > 0.05) of liquid organic fertilizer on the heights of Setaria grass, which indicated that the heights of the grass that received different concentrations of the fertilizer were relatively similar. However, the doses of the water showed a linear regression, Y = 55.71 – 0.2218 X (Figure 1), which indicated that higher concentrations of water in the fertilizer resulted in lower heights of plants, might be to the lower content of N in the mixtures.

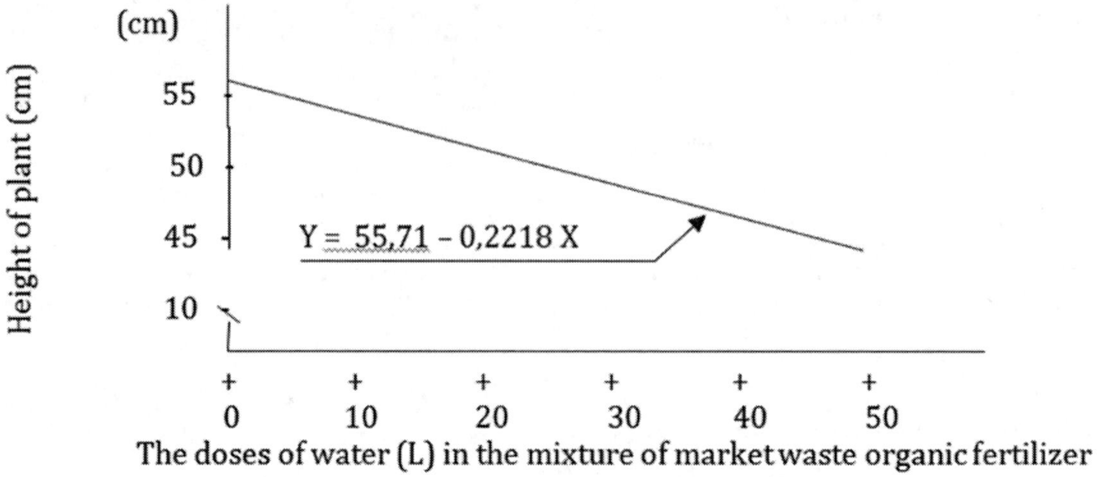

Fig. 1. Relationship between water mixture in the market waste organic fertilizer with the mean heights of Setaria Grass

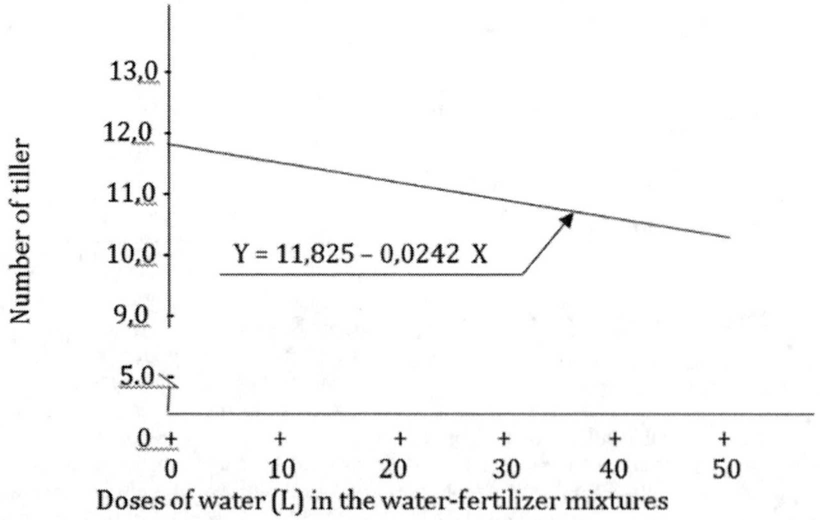

Fig. 2. Relationshi between water dosen in the market-waste lizuid organic fertilizer mixtures with the number of tillers of Setaria grass

Figure 1 showed, the higher the dilution of the market-waste liquid organic fertilizer with water, the lower the height of Setaria grass. Although the anova showed a uniform results, the doses of water showed a linear response, therefore, the conclusion is, the less dilution of the fertilizer with water resulted in a higher height of Setaria (*Setaria splendida*). It is possible that the ratio of water to fertilizer of 3 to 1 still resulted in a good condition.

3.3 The number of tillers of Setaria

The productions of forage plants are greatly affected by the number of tillers. In general, the higher the number of tillers, the higher the production and the number of animal units that can be fed by the forage. The market-waste liquid organic fertilizer is a natural resources as a product of human activity, represented an added value of the disposed wastes. Basically, all kinds of wastes have utility and economic values for human being, including the utility of wastes as a liquid fertilizer, finally, as a stimulant for livestock performances. The forage is a whole, accumulative products of all parts of forage plants, the number of tillers in the management of grass agronomy has a pivot role viewed from the production function. The higher the number, the higher the production [16].

The mean value of the number of tillers as a phenomenon of growth of Setaria grass was 11.04/ bunch of plant (Table 2) with a range of 10.38 – 11.98 tillers / bunch. Higher doses of water in the mixture of water-fertilizer decreased the number of tillers. The fact was in line with the statement of Setiyowati et al., that the concentrations of minerals in the fertilizer determine the growth aspects of plants [17].

The analysis of variances and differences test showed, there was no significant differences ($P > 0.05$) among the treatments on the numbers of tillers. However, the doses of water in the water-fertilizer mixtures resulted in a significant, linear relationship ($P < 0.05$), $Y = 11.825 - 0.0242\ X$ in term of the numbers of tiollers of Setaria grass (figure 2). Table 1 showed that Setaria garss (*Setaria splendida*) without fertilization had the greatest number of tillers compared to those fertilized with market-waste liquid organic ferilizer. Greater volumes of water in the mixtures resulted in vewer numbers of tillers.

Figure 2 showed, the all 5 doses of water resulted in the decrease of the number of tillers of Setaria (*Setaria splendida*). This was in line with the repost, the decrease in mineral concentrations in the mixtures of water-fertilizer due to dilutions resulted in the decrease of plant growth, including the numbers of tillers.

These results were in accordance with the height of plant, the greater the doses of water in the water-fertilizer mixtures, the fewer the height and the numbers of tillers. Although the anova showed no significant effects of treatments on the number of tillers, however, the linear response of the doses of water on the number of tillers concludes that higher concentrations of fertilizer in the water-fertilizer mixtures resulted in the greater number of tillers of Setaria grass (*Setaria splendida*). Table 1 indicated, the water doses until up to 3 (three) times as those of the fertilizer still resulted in a fairly good growth condition.

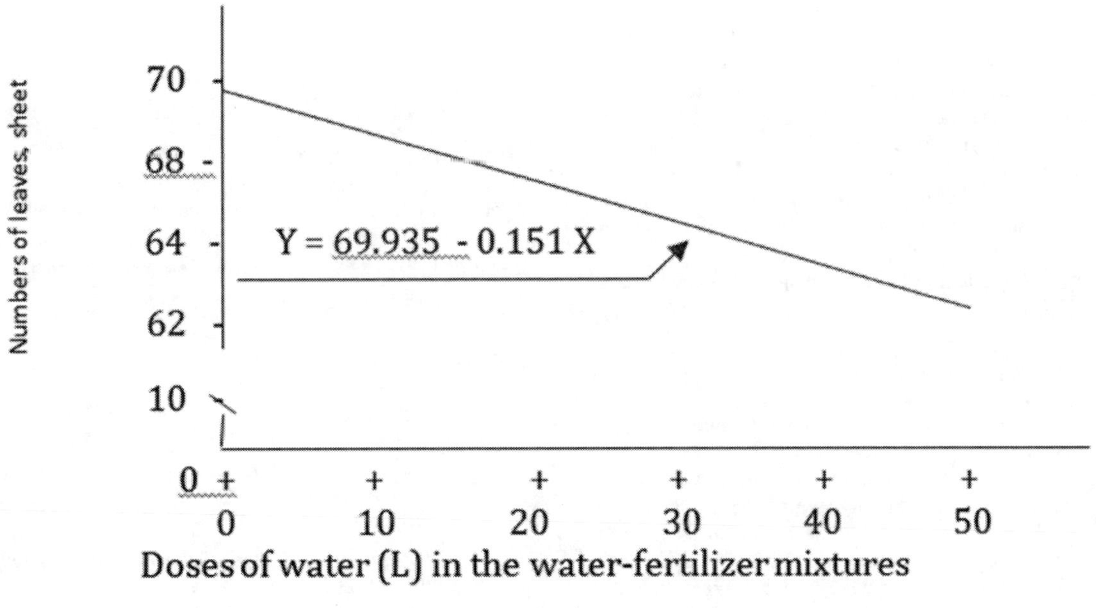

Fig. 3. Relationship between the doses of water in the water-fertilizer mixtures with the numbers of leaves of Setaria grass

3.4 Numbers of leaves of Setaria

The performances of production and quality of forage are greatly affected by leaf condition [18]. The productions of forage plants are expressed after harvest, one of which is represented by leaf as one of the production components [19]. Each addition of a leaf increases fresh matter weight. According to Hendarto farmers prefer leaf parts of forages plants rather than stems, because they are preferred by livestocks more than stems in order to enhance the livestock performances [4].

The mean of the numbers of leaves per bunch of Setaria (*Setaria splendida*) at the first defoliation were 65.16 sheets with a range of 56.85 – 72.00 sheets (Table 2). The bunches without fertilization had the greatest numbers of leaves, with a mean of 72.00 sheets of leaves. Whereas for the fertilized bunches of plants, there were various mean values of the numbers of leaves, with a tendency that the greater volumes of water in the water-fertilizer mixtures resulted in a lower numbers of leaves. This was in line with the results of Abdullah et al., study that liquid organic fertilizer that was applied to indigofera showed a decreasing numbers of leaves due to the lower concentrations of minerals in the mixtures of water-fertilizer [20]. It was added that the leaves represent an important biomass in the productions of forage plants, because leaves are the vegetative parts of plants that have the highest quality of plant products. Taufika added that the taller plants generally have a greater number of leaves [3].

The results of anova showed that the treatments had no significant effect ($P > 0.05$), however, the doses of water affected significantly ($P < 0.05$) on the number of leaves of Setaria (*Setaria splendida*) as expressed by a linear equation, $Y = 69.935 - 0.151 X$ (Figure 3). The figure showed that each dose of water addition resulted in a decrease of the number of leaves up to the 5 times as much water relative to the volume of the liquid organic fertilizer.

Other observations also showed that greater volumes of water in the water-fertilizer mixtures resulted in lower numbers of leaves. The response of figure 3 showed that higher concentrations of liquid organic fertilizer increase the number of leaves of Setaria grass (*Setaria splendida*). According to Setiyowati et al., and Rizqiani etal., liquid organic fertilizer can be used to fertilize all species of plants because it contains 13 kinds of macro and micro minerals that are absolutely required for the growths of plants; at the proper concentration, it is able to improve the quality of plants, including the number of leaves [17, 21].

3.5 Canopy of Setaria

Canopy may represents the vigority of plants to grow, that is needed to optimize the growth of plnats in order to achieve high level of production. The results of this study (Table 2) showed tha the mean value of canopy of Setaria (*Setaria splendida*) that was fertilized with market-waste liquid organic fertilizer was 39.58 cm^2 at a range of 37.83 – 40.85 cm^2 which indicated that the plant was uniformly and vigorously grew vertically high in all the treatments. The area of field that was exposed to sunlight supported the fact. Huang, et al., reported similar facts that the addition of N from the market-waste liquid organic fertilizer resulted in a higher chlorophyl concentrations in leaves which in turn, promote photosynthesis and canopy formation [7]. It was added by Marpaung et al., that liquid organic fertilizer is bale to increase the vigority of plants, as ar esult, the plants become strong and the leaf canopy becomes wider [12].

In relation with the use of liquid fertilizer on the plants, it is possible that the use of fertilization techniques via leaves according to Rahmi and Jumiati, will reap better growth to increase the production of plants through the functions of stomata of plants. The canopy as a result, grows wider due to the application of the liquid organic fertilizer [22].

The anova showed that the treatments and the doses of water in the water-fertilizer mixtures did not give any effect ($P > 0.05$) on the width of canopy of Setaria (*Setaria splendida*), which indicated that the widths of canopies of the treated and untreated plants were similar. This result was in accordance with that of Parman study, the uses of several concentrations of liquid organic fertilizer on potato plants did not give any differences on the parameters of growth [23]. However, the additions of water doses up to 5 times the volumes of fertilizer tended to produce shorter canopy. Table 2, indicated that the additions of water volumes up to 3 times the volume of fertilizer showed fairly good growths.

The traditional-market-waste liquid organic fertilizer is abundant natural resources. Although the waste is one of the urban pollutants as the results of human activities, however, it is basically very useful for human life through its support to plant growth. The enter of liquid fertilizer into the agronomy management of plant will increase the diversification of businesses that have economy values and develop socio-cultural aspects in the life of human race.

4 Conclusions

On the basis of the discussion, the conclusion is, traditional-market-waste liquid organic fertilizer plus water as much as 3 times the volume of the fertilizer can be used to support the growth of Setaria (*Setaria splendida*) in order to increase the performances of ruminant animals.

References

1. E. Hendarto, *Dimensi Lingkungan Tata Ruang Peternakan Sapi Perah Rakyat Di Kabupaten Banyumas, Provinsi Jawa Tengah. Disertasi.* Program Pascasarjana. Universtas Diponegoro. Semarang. (2011)
2. H. Greatheat, Proceedings of the nutrition Society. **62**: 279-290 (2003)
3. R. Taufika, Jurnal Tanaman Hortikultura. **8** (2): 1-10 (2011)
4. E. Hendarto, E. Widyastuti, D.R. Puspita, A. Iqbal, E. Hilmi, Muslihudin, E. Dewanto, A. Anwar, E.B. Riyanto and S.A. Amir, *Analisis Permasalahan Lingkungan Di Daerah dan Solusi yang Dapat Dikembangkan. Pusat Penelitian Lingkungan Hidup, Lembaga Penelitian dan Pengabdian*

Kepada Masyarakat, Universitas Jenderal Soedirman. Purwokerto. (2011)

5. E. Hendarto, *Pengaruh Kombinasi Pupuk Organik dan Taraf Urea terhadap Kualitas Visual dan Produksi Rumput Raja* (*Pennisetum purphoides*). *Jurnal Pembangunan Pedesaan.* Lembaga Penelitian Universitas Jenderal Soedirman. Purwokerto. (2005)
6. J.F.S. Ferreira, M.V. Cornacchione, X. Liu and D.L. Suarez, Nutritient Composition, Forage Parameters, Journal Agriculture. **5**: 577-597 (2015)
7. Z.A. Huang, D.A. Jiang, Y. Yang, Y.W. Sun and S.H. Jin. Photosynthetica. **42** (3): 357-364 (2004)
8. D. Thomas, and R. Schultze-Kraft, Tropical Grassland. **24**:87-92 (1990)
9. S Aminudin, dan E. Hendarto. *Buku Ajar Agrostologi.* Fakultas Peternakan. Universitas Jenderal Soedirman. Purwokerto (2000)
10. Rahman, D. Lasamadi, S.S. Malalantang and Rustandi, Jurnal Zootek (2012)
11. K. Budiraharjo, M. Handayani and G. Sanyoto, Jurnal Ilmu-ilmu Pertanian, Mediagro. **7** (1): 1-9 (2011)
12. A. E. Marpaung, B. Karo and R. Tarigan, Journal Horticultura. **24** (1): 49 – 55 (2014)
13. W. Hanolo, Jurnal Agrotropika **1** (1): 25 – 29 (1997)
14. E. Hendarto, Penampilan Aspek Agronomi Rumput Gajah (*Penntsetum Purpureum*) Di Bawah Pengaruh Berbagai Jenis Dan Dosis Pupuk Organik Dan Anorganik. *Prosiding Seminar Nasional Peran Pertanian Dalam Menunjang Ketahanan Pangan dan Energi Untuk Memperkuat Ekonomi Nasional Berbasis Sumberdaya Lokal.* Fakultas Pertanian. Universitas Jenderal Soedirman. Purwokerto. (2012)
15. A. Wachjar, and L. Kadarisman. Buletin Agronomi. **35** (3): 212 -216 (2007)
16. E. Hendarto, dan Suwarno, Bionatura. Journal of Life and Physical Sciences. **15** (2): 86 -91 (2013)
17. Setiyowati, S. Haryanti and R.B. Hastuti, Bioma. **12** (2): 44-48 (2010)
18. S. Tudsri, S.T. Jorgensen, P. Riddach and A. Pookpakdi, Tropical Grasslands. **36**: 248 – 252 (2002)
19. I. Herdiawan, L. Abdullah, D. Sopandi, P.D.M.H. Karti and N. Hidayati. Journal Animal and Veteriner Science. **17** (2): 276 -283 (2012)
20. Abdullah, L. D.D.S. Budhie and A.D. Lubis, Pastura. **1**(1): 5 – 8 (2011)
21. N.F. Rizqiani, E. Ambarwati, and N. W. Yuwono, Jurnal Ilmu Tanah dan Lingkungan. **7** (1): 43 – 53 (2007)
22. A. Rahmi, and Jumiati, Agritrop. **26** (3): 105 -109 (2007)
23. S. Parman, Buletin Anatomi dan Fisiologi. **15** (2):21-31 (2007)

Vegetation Structure of Ebony Leaf Monkey (*Trachypithecus auratus*) Habitat in Kecubung Ulolanang Nature Preservation Central Java-Indonesia

Rahmawati Ervina[1,*], *Hidayat* Jafron Wasiq[2]

[1]Master Program of Environmental Science, School of Postgraduate Studies, Diponegoro University, Semarang - Indonesia
[2]Department of Biology, Faculty of Mathematics and Science, Diponegoro University, Semarang – Indonesia

Abstract. Kecubung Ulolanang Nature Preservation is ebony leaf monkey's habitats in Central Java Indonesia. Continuously degradation of their population is caused by illegal hunting and habitat degradation that made this species being vulnerable. Habitat conservation is one of important aspects to prevent them from extinction. The purpose of this research was to analyze the vegetation's structure and composition, which was potentially, becomes habitat and food source for the monkeys. Data collected using *purposive sampling* with line transect method of four different level of vegetation. Data analysis used Important Value Index and Diversity Index. There were 43 species of vegetation at seedling stage, 18 species at sapling stage, 8 species at poles stage and 27 species at trees stage. Species that had the highest important value index at seedling was *Stenochlaena palustri* , at the sapling was G*netum gnemon*, at pole was *Swietenia mahagoni* and at tree was *Tectona grandis* . Species of trees those were potentially to become habitat (food source) for ebony leaf monkey were *T. grandis*, *Dipterocarpus gracilis*, *Quercus sundaica* and *Ficus superba*. The highest diversity index was at seedling gwoth stage.

1 Introduction

Extinction threat of wild plants and animals which was occurred naturally or caused by over exploitation becomes a necessary issue about environment. Human population growth and development in many sectors have been cause a narrowing animal space-range, decreasing quality and fragmentation of animal habitat.

Ebony leaf monkey (*Trachypitecus auratus*) is an endemic species of Java, Bali and Lombok Island [1]. Nowadays, ebony leaf monkeys have an conservation status vulnerable from IUCN [1,2], because of their decreased as an effect to illegal hunting and habitat degradation [3], so they need conservation action as soon as possible [4]. Such decrease and population change were not only in java, but also in Bali (Leca, et al., 2013) [5].

According to Gunawan et al. [6], Central Java Province had lost its dry land natural forest about 446.561,09 hectares or 88% for duration of 16 years (1990-2006). The remains of dry land natural forest usually left on mountain peaks which are hardly touched by human activities. Further, according to [6], natural forest fragmentation in Central Java which were occurred among 1990 - 2000 made a total edge (TE) increasing from 42,43 km to 133,88 km, usually caused by forest conversion to farmland, man-made forest, plantation, settlement area and infrastructure development (artery road, highway and high tension transmission lines (SUTET)).

To conserve habitat and the species of ebony leaf monkey, it has to do in-situ and ex-situ conservation actions. Habitat conservation is important to prevent the extinction of this species. Kecubung Ulolanang Nature Preservation is a lowland rainforest type with high biodiversity and becomes an in-situ habitat of ebony leaf monkeys [7]. To understand about habitat condition, it is need data about its vegetation structure because ebony leaf monkeys use the trees for providing their food and their nest. Ebony leaf monkeys choose their own trees for food; vegetation structure of ebony leaf monkeys' habitat can be identified from the vegetation data i.e. seedling, sapling, pole and tree of four growth stage.

The purpose of this research was to analyze the vegetation's structure and composition, which was potentially, becomes habitat and food source for the ebony leaf monkey.

2 Method

2.1. Study Area

The study was conducted in Kecubung Ulolanang Nature Preservation Batang (6°51'46" - 7°11'43" N and 109°40'19" - 110°03'06" E) Central Java Indonesia.
Kecubung Ulolanang Nature Preservation as a nature preserve based on Ministerial Decree of Forestry and Plantation no. 435 / Kpts-II / 1999 dated June 15, 1999. The topography of the area was hillary with long slopes to the south, with a river that is use as a source of water for the surrounding community and a natural boundary

* Corresponding author: ervina.bksdajateng@gmail.com

for this region. The area is located at 165 meters above sea level with temperature minimum 24.4 °C and maximum 29°C, and annual precipitation of 277.7 mm. In addition, the soil classified as latosol developed from felsic igneous rock and intermediary.

2.2. Material and Method

In this study, data was collected in April to May 2017 using survey and observation method. Vegetation sample were accumulated through purposive sampling with line transect method on the habitat of ebony leaf monkey. In this method the plot were taken at four growth stage, i.e. 2 x 2 m for seedling, 5 x 5 m for sapling, 10 x 10 m for poles and 20 x 20 m for trees. Total plots area for seedling level was 96 m^2, at sapling level was 600 m^2, then 2,400 m^2 at poles level and at tress level was 9,600 m^2.

Seedling are categorized from sprout to young of 1.5 m including herbs and shrubs. Furthermore, sapling are plants with DBH of less 10 cm and higher than 1.5 m, then Poles are plants with DBH between 10 - 20 cm, and trees are plant with DBH more than 20 cm [8]. Each plot (20 m x 20m) was repeated 3 times, located in eight different places.

2.3. Data Analysis

In this study focused on the vegetation areas that became the habitat of Ebony Leaf Monkey. Community struct including diversity and important value of vegetations.

It was quantified relative density, relative frequency, relative dominance, and importance value index (IVI) for all species following Dombois, et al. (1974) [9]. To determine species diversity, we use the Shannon-Wiener index. The formula for diversity index and important value index are presented below.

$$H' = -\Sigma(ni/Ni) \ln (ni/Ni) \quad (1)$$

$$RDi = \frac{\text{Density of a species}}{\text{Total density of all species}} \text{ X } 100 \quad (2)$$

$$RF = \frac{\text{Frequency of a species}}{\text{Total frequency of all species}} \text{ X } 100 \quad (3)$$

$$RDo = \frac{\text{Dominance of a species}}{\text{Total dominance of all species}} \text{ X } 100 \quad (4)$$

$$Di = \frac{\text{Number of a species}}{\text{Total area sampled}} \text{ X } 100 \quad (5)$$

$$F = \frac{\text{Area of a plot in which a species accours}}{\text{Total area sampled}} \quad (6)$$

$$Do = \frac{\text{Total basal area of a species}}{\text{Total area sampled}} \quad (7)$$

3 Result

3.1. Species Abundance and Diversity

Observation result revealed total 978 individuals belonging to 68 species and 33 families of growth stage was reported in the 24 plots. There were 43 species at seedling stage, then 18 species at sapling stage, 8 species at poles stage and 27 species at trees stage. Species abundance at growth stage present at Table.1 below.

Table.1. Species abundance at growth stage.

Growth Stage	Species	Family	Ni	Species	Family	Ni
Seedling	*Achasma coccineum*	Zingiberaceae	16	*Grewia panicuta*	Malvaceae	27
	Ageratum conyzoides	Asteraceae	69	*Imperata cylindrica*	Poaceae	34
	Aglaia odoratissima	Meliaceae	10	*Lantana camara*	Verbenaceae	32
	Caesalpinia sappan	Fabaceae	2	*Leea indica*	Leeaceae	11
	Calamus sp	Arecaceae	5	*Leersia hexandra*	Poaceas	7
	Chloranthus officinale	Chloranthaceae	4	*Mallotus paniculatus*	Euphorbiaceae	12
	Coffea sp	Rubiaceae	2	*Mimusops elengi*	Sapotaceae	4
	Commelina nudiflora	Commelinaceae	30	*Nephelium lappaceum*	Sapindaceae	3
	Corypa utan	Arecaceae	1	*Pandanus sp*	Pandanaceae	20
	Costus speciosus	Costaceae	24	*Piper baccatum*	Piperaceae	12
	Cyclea barbata	Menispermaceae	13	*Piper cubeba*	Piperaceae	22
	Dalbergia latifolia	Fabaceae	4	*Prema integrifolia*	Verbenaceae	27
	Dioscorea hispida	Dioscoreaceae	8	*Quercus sundaica*	Fagaceae	4
	Dipterocarpus geacilis	Dipterocarpaceae	22	*Stachytarpheta jamaicensis*	Verbenaceae	18
	Donax canniformis	Marantaceae	17	*Stenochlaena polustris*	Blechnaceae	116
	Erioglossum rubiginosum	Sapindaceae	1	*Streblus asper*	Moraceae	5
	Eugenia polyantha	Myrtaceae	2	*Tetracera scandens*	Dilleniaceae	41
	Euphatorium odoratum	Asteraceae	19	*Tridax procumbens*	Asteraceae	6
	Ficus hispida	Moraceae	1	*Urena lobata*	Malvaceae	66
	Flagellaria indica	Flagellariaceae	9	*Xanthophyllus excelsum*	Polygalaceae	8
	Garcinia sp	Clusiaceae	2	Krotak		10
	Globba marantina	Zingiberaceae	26			
Growth Stage	**Species**	**Family**	**Ni**	**Species**	**Family**	**Ni**
Sapling	*Artocarpus elastica*	Moraceae	1	*Leea indica*	Leeaceae	5
	Coffea sp 4	Rubiaceae	1	*Mallotus paniculatus*	Euphorbiaceae	1

Growth Stage	Species	Family	Ni	Species	Family	Ni
	Dipterocarpus gracilis	Dipterocarpaceae	3	*Nephelium lappaceum*	Sapindaceae	1
	Erioglossum rubiginosum	Sapindaceae	1	*Pithecolobium lobatum*	Fabaceae	1
	Garcinia sp	Clusiaceae	1	*Quercus sundaica*	Fagaceae	1
	Gluta renghas	Anacardiacea	1	*Streblus asper*	Moraceae	1
	Gnetum gnemon	Gnetaceae	3	*Swietenia mahagony*	Meliaceae	1
	Grewia paniculata	Malvaceae	5	*Syzygium densiflora*	Myrtaceae	1
	Hibiscus tiliaceus	Malvaceae		*Tectona grandis*	Verbenaceae	1
Poles	*Albizzia procera*	Fabaceae	1	*Nephelium lappaceum*	Sapindaceae	1
	Erioglossum rubiginosum	Sapindaceae	1	*Pithecolobium lobatum*	Fabaceae	1
	Gnetum gnemon	Gnetaceae	2	*Swietenia mahagony*	Meliaceae	2
	Mallotus paniculatus	Euphorbiaceae	1	*Tectona grandis*	Verbenaceae	1
Trees	*Albizzia chinensis*	Fabaceae	1	*Ficus superba*	Moraceae	3
	Albizzia procera	Fabaceae	10	*Gluta renghas*	Anacardiacea	1
	Anacardium occidentale	Anacardiacea	1	*Gnetum gnemon*	Gnetaceae	1
	Artocarpus elastica	Moraceae	1	*Mallotus paniculatus*	Euphorbiaceae	1
	Artocarpus integra	Moraceae	1	*Nephelium mutabile*	Sapindaceae	1
	Bambusa sp.	Poaceae	21	*Otophora spectabilis*	Sapindaceae	1
	Barringtonia racemosa	Lecythidaceae	1	*Pangium edule*	Achariaceae	1
	Calamus sp.	Arecaceae	26	*Planchonella nitida*	Sapotaceae	1
	Cocos nucifera	Arecaceae	1	*Quercus sundaica*	Fagaceae	16
	Corypa utan	Arecaceae	1	*Salacca edulis*	Arecaceae	13
	Dipterocarpus gracilis	Dipterocarpaceae	14	*Swietenia mahagony*	Meliaceae	13
	Dracontomelon dao	Anacardiacea	1	*Tectona grandis*	Verbenaceae	18
	Enterolobium cyclocarpum	Fabaceae	3	*Toona sureni*	Meliaceae	1
	Eugenia polyantha	Myrtaceae	1			

Ni is the number of individuals.

The abundance and Diversity Index were present in Table.1 below. Diversity Index (H') at seedling was highest than other growth stage. Accordingly to the diversity index value, the species abundance of seedling stage was highest than sapling, poles and trees stage [5]. It's because many species at seedling stage include herbs and shrubs besides seedling trees could be found in the plot areas.

The highest diversity index value at seedling level i.e. 3.24, it's means that the diversity was abundant, so do of tree level of 3. 2. Furthermore, at sapling and pole stage the diversity index were lower at 2.64 and 2.02 respectively, indicated that the diversity was average. For details can be seen at Table. 2.

Table.2. The abundance and Species Diversity at Renegeration Level in Kecubung Ulolanang Nature Preservation Central Java Indonesia.

	Seedling	Sapling	Poles	Trees
N	772	32	10	155
S	43	18	8	27
H'	3.24	2.64	2.02	3.2
D (ind ha^{-1})	80	533	42	161

N is the number individuals registered. S is the total number of species censuses. H' is the Shannon- Wiener diversity index,. D is the number of density (ind ha^{-1}).

3.1.1. Seedling

Based on inventory vegetation's result, there were 43 species of 772 individual including trees and shrubs at seedling stage. The highest species density was *Stenochlaena polustris* at 12 ind ha^{-1}, whereas *Corypha utan, Erioglossum rubiginosum* and *Ficus hispida* had the lowest species density at 104.16 in ha^{-1}. Likewise, the species with the highest relative density and relative frequency was *Stenochlaena polustris*. It was the important species at seedling level with the highest IVI at 22.85%. The Important Value Index at seedling and sapling level were present in Table.3 below.

Table.3. Important Value Index (IVI) at Seedling and Sapling Level.

(A) Species	Relative Density (%)	Relative Frequency (%)	IVI (%)
Stenochlaena polustris	15.02	7.82	22.85
Urena lobata	8.94	7.82	16.76
Ageratum conyzoides	8.55	5.22	13.77
(B) Species	Relative Density (%)	Relative Frequency (%)	IVI (%)
Gnetum gnemon	12.50	14.81	27.31
Grewia paniculata	15.62	11.11	26.73
Leea indica	15.62	11.11	26.73

(A) is at seedling level, (B) is at sapling level. IVI = relative density + relative frequency.

Herbs and shrubs at this growth stage were found abundant rather tree seedlings. Even they s on the bottom rank i.e. *E. rubiginosum* and *F. hispida*.

3.1.2. Sapling

There were 18 species of 32 individuals at sapling level. The species which the highest relative density and relative frequency and also the most important species with the highest IVI was *Gnetum gnemon* at 27.31%

(presented in Table.2). It's had the species density at 67 ind ha^{-1}, less than *Grewia paniculata* and *Leea indica*, but it's had relative frequency higher than both at 16.67%. Not all species could be found at seedling stage also found at sapling stage. It was because spatial competition.

3.1.3. Poles

The inventory's result there were 8 species of 10 individuals at poles level. Total individuals at this growth stage were enumerated less than sapling. This was due to the cover of tree stand canopy limited of sunlight. Especially the shade-intolerant species. *Swietenia mahagony* was the highest density at 8 ind ha^{-1}, and also it's had the highest frequency and basal area. Consequently, it was the important species at poles level with the highest IVI at 58.47%.

Species were found at seedling and sapling stages were rare or not found at all at pole stage. Total individuals were registered also more less. *G. gnemon* was the species that also could be found at sapling stage. It had the second highest IVI at 54.82%.

3.1.4. Trees

here were 27 species determinate of 155 individuals at trees stage. *Tectona grandis* was the species with the highest density value at this stage there are 19 ind ha^{-1}, followed by *Dipterocarpus gracilis* (15 ind ha^{-1}) and *Quercus sundaica* (17 ind ha^{-1}). *T. grandis* was also the important species with the highest IVI at 46.50% (Fig.1).

According to the sunlight tolerance, *T. grandis, Q. sundaica* and *D. gracilis* were shade-tolerant species at adult phase and shade-intolerant species at seedling phase, so they could grow optimally and become the dominant vegetation at there. At seedling and sapling stage, these three species were rare or not found in this study, but they found at the pole stage, even they could be died or languish before at tree stage.

Fig.1. IVI species were potentially to become the habitat (food source) for ebony leaf monkey.

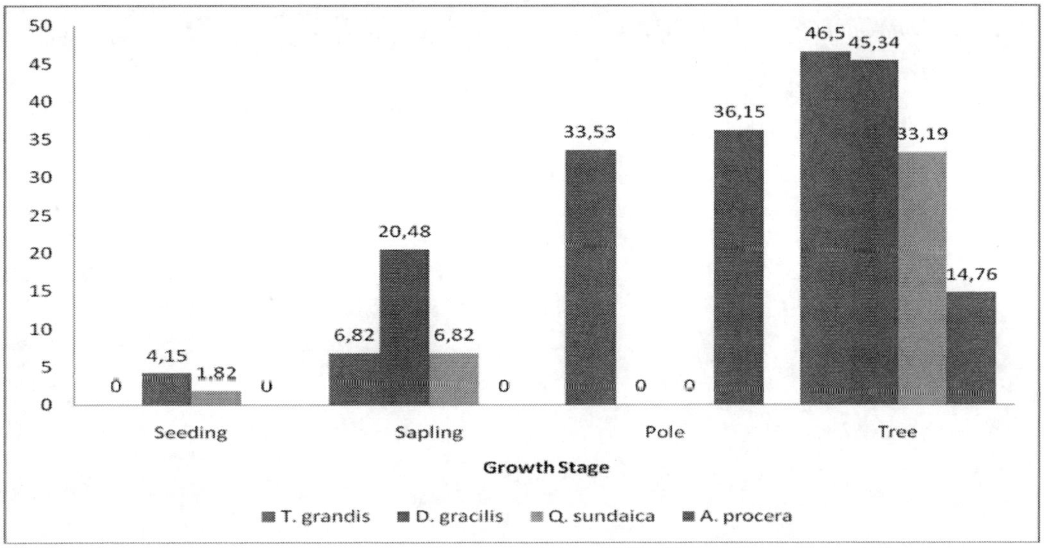

3.2. Vegetation Composer Ebony Leaf Monkey's Habitat

According to the vegetation analyze, the species of trees stage were potentially to become the habitat and food source for ebony leaf monkey were *T. grandis*, *D. gracilis*, *Q. sundaica*, *Albizzia procera* and *Ficus superba*. These species were found with high abundant at trees stage and they also had the high IVI, except *F. superba*. These value were 46.50%, 45.35%, 33.20% and 14.76% respectively. Besides those species, *G. gnemon* and *S. mahagony* were also potentially to become the habitat and food source for ebony leaf monkey. *G. gnemon* was dominant species at pole and sapling stage. The result of the IVI analysis were presented in figure.1.

These trees as ebony leaf monkey's habitat shows that they grow steadily in this ecosystem, but has not been able to maintain its existence (regeneration) since some of them were least successfully grown in their growth stage. At the tree level, *T. Grandis* had the highest relative frequency and relative diversity value, means it had existence (frequency and abundance) and also spreads more evenly than other species. However, this species was not found at seedling and sapling stages (consociation), it was a competition to grab and fight the same habitat. Due to the spatial and light competition.

In Kecubung Ulolanang Nature Preservation, ebony leaf monkey had many kinds of tree to be their habitat, although there was a little concern about the regeneration of these feed trees.

4 Conclusion

The result were identify of 978 individuals belonging to 68 species of growth stage was registered in the 24 plots. There were 43 species of 772 individual at seedling stage including trees and shrubs, 18 species of 32 individuals at sapling stage, 8 species of 10 individuals at poles stage and 27 species at of 155 individuals at trees stage. The highest diversity index was found at seedling level (3.24), at tree level (3. 2). Furthermore, at sapling and pole level the diversity index were 2.64 and 2.02 respectively.

Moreover, species that had the highest important value index at seedling level was *Stenochlaena palustri* , at the sapling level was G*netum gnemon*, at pole level was *Swietenia mahagoni* and at tree level was *Tectona grandis*. The tree species of potentially to become habitat and food source for ebony leaf monkey were *T. grandis*, *D. gracilis*, *Q. sundaica* and *F. superba*.

References

1. V. Nijman, *Contribution to Zoology*.**69**(3), 157-177, (2000)
2. C. Roos, R. Boonratana, J. Supriatna, J.R. Fellowes, C.P. Groves, & D. Stephen. *Asian Primate Journal*. **4 (1)**: 2-38, (2014)
3. E. Sulistyadi, PA. Kartono, I. Maryanto, Pergerakan Lutung Jawa Trachypithecus auratus (E. Geoffroy 1812) Pada Fragmen Habitat Terisolasi Di Taman Wisata Alam Gunung Pancar (Twagp) Bogor. *Berita Biologi*.**12**(3),383-395, (2013)
4. Q. Ayunin, S. Pudyatmoko, AM. Imron, *Jurnal Penelitian Hutan dan Konservasi Alam*. **11** No. 3 : 261-279, (2014)
5. J.B. Leca, N. Gunst, A. Rompis, G. Soma, I.G.A. Putra, I.N. Wandia, *Primate Conservation Journal*. **26**: 133-144, (2013).
6. H. Gunawan, BL. Prasetyo, A. Mardiastuti, PA. Kaartono, *Jurnal Penelitian Hutan dan Konservasi Alam*. **VII** No.1 : 75-91, (2010)
7. BKSDA, Rencana Pengelolaan Cagar Alam di Kabupaten Batang. Not published, (2015)
8. C. Kusmana, Metode Survey Vegetasi. PT. Institut Pertanian Bogor. Bogor, (1997)
9. Mueller-Dombois, L.D. Ellenberg, H.1974. Aims and Methods of Vegetations Ecology. John Wiley & Sons, Inc. New York.

Indigenous Peoples Involment At The Environmental Impact Assessment (EIA) Process in Tabi Mamta Area Of Papua Province

Ferdinand Saras Dhiksawan[1*], *Sudharto* P. Hadi[2], *Adji* Samekto[3], *Dwi* P. Sasongko[4]

[1]Doctoral Program of Environmental Science, School of Postgraduate Studies, Universitas Diponegoro
[2]Doctoral Program of Environmental Science, School of Postgraduate Studies, Universitas Diponegoro
[3]Law Faculty, Universitas Diponegoro
[4]science and Mathematic Faculty, Universitas Diponegoro

Abstract. The purpose of this study is to find a picture of the involvement of Indigenous Peoples of Tabi Mamta in the process of environmental impact assessment (EIA) in Tabi Mamta customary territory. The method and type of research used is non-ethnographic qualitative research with data collection techniques using limited observation techniques. Data and information in the field will be analyzed using constructivism paradigm. The paradigm of constructivism is based on an interpretive understanding called hermeneutics (hermeneuien) in the sense of interpreting, giving understanding, translating data and information obtained in the research location as a result of social reality. The results of this study indicate that the customary community of Tabi Mamta is a unit of customary community that still has territorial customary territory, has a customary leadership structure, still visible relationship of kinship, cultural values as well as customary norms and sanctions, and has environmental wisdom in maintaining existence Natural resources. In the socio-cultural system of customary communities there are components such as customary stratification, permissiveness, communication, reciprocity, past history, cultural values, customary norms and sanctions, religious and customary leadership. Components in the socio-cultural system of indigenous and tribal peoples play a role in the EIA process in the Tabi Mamta customary area especially in the environmental feasibility decision making process. The components of custom stratification, cultural values and customary norms play a role in the EIA process. In customary stratification there is uncustomary structure in the ondoafi, Iram and Tribal Leadership. Components in a socical cultural system is a unity resulting from interaction between individuals and groups to prevent environmental damage and disturbance of natural resources. Natural resources are considered as ancestral symbols passed down by ancestors from generation to generation

1 Introduction

Papua is located in the eastern part of Indonesia. It has two provincial administrative areas, namely West Papua Province and Papua Province, where there are 7 (seven) customary areas, namely Tabi Mamta, La-Pago, Ha-Anim, Mi- Pago, Saireri, Bomberai, and Domberai. In the Amdal process, indigenous and tribal peoples feel that there is injustice, inequality, disagreement based on the existence of customary values and norms within the customary law community's living order. The purpose of this study is to find a picture of the involvement of Indigenous Peoples of Tabi Mamta in the process of analyzing environmental impacts in the Tabi Mamta customary territory.

The type of research used is non-ethnographic qualitative research with data collection techniques using limited observation techniques. Data and information in the field will be analyzed using constructivism paradigm. The paradigm of constructivism is based on an interpretive understanding called hermeneutics (hermeneuien) in the sense of interpreting, giving understanding, translating data and information obtained in the research location as a result of social reality. The results describe the components in the social cultural system of customary and tribal peoples in the Environmental Impact Analysis process in Tabi Mamta Customary Area. Furthermore, in this research, the existence of Tabi Mamta customary Community is a community which still has territorial of customary territory, has a clear and respected leadership structure (Ondoafi, Iram, and Chief of Tribe), the relationship of kinship, cultural values and customary norms and sanctions, and have environmental wisdom in maintaining the existence of Natural Resources.

The values and norms in the life of indigenous peoples, as well as the existence of natural resources as property rights of Indigenous People of ancestral lands

* Corresponding author: dhiksawan@yahoo.co.id

are recognized and protected by the Regional Government through legislation.

2 Material and Research Methods

The type of research used is non-ethnographic qualitative research with data collection techniques using limited observation techniques. Data and information in the field were obtained through in-depth interviews using unstructured structured question materials (using recording instruments) on key figures (key informants). Determination of key figures as a source of data and information obtained through snowball technique (snowball sampling).

Further data and information obtained will be analyzed using constructivism paradigm. The paradigm of constructivism is based on an interpretive understanding called hermeneutics (hermeneuien) in the sense of interpreting, giving understanding, translating data and information obtained in the research location as a result of social reality

3 Research Result

Tabi Mamta customary territory is located in the northern part of Jayapura (in the east by Papua New Guinea) up to Mamberamo (western) and South bordering Jayawijaya Mountains. Administratively, the territory of Tabi Mamta Indigenous Peoples covers Keerom Regency, Jayapura City, Jayapura Regency, Sarmi Regency and Mamberamo Raya Regency.

The Tabi Mamta People is called TABI MAMTA because it comes from the Sunrise (TABI) which then resides in the Mamberamo River Basin and the Tami River Basin (MAMTA). The history of the journey of the Tabi Mamta People begins with the existence of a Sun and Moon god who bears the Sun and Moon. The God of the Sun and the Moon began the journey from where the sun began. Furthermore, the first stopover of the Sun and Moon gods from the east is called Tabatji village (first village). After a night of exhaustion, then the journey continues back towards the West. Arriving in one place, the Sun and Moon gods no longer continue their journey. Pikulan in the form of a Sun and the Moon then placed somewhere that causes the occurrence of a large hole.

The place is now called Bhuyakha (Water Place) or called Lake Sentani. It is from this place that later generations to the genaration develop and inhabit the Mamberamo River Basin and the Tami River Basin. For that reason, the Tabi Mamta Indigenous Peoplehighly appreciates their leaders because the people still consider that their leader is the incarnation of the sun god and the moon. Their leader is called Ondofolo or ondoafi, and for certain customary territories is called Iram. The view of Indigenous Peoplewhich still holds the leadership of Ondofolo said that Tabi Mamta People comes from Tabi meaning sunrise, up to Mamberamo River Basin Region and Tami River Basin. While the view of the Indigenous Peopleof Iram argues that the origin of the Tabi Mamta Customary Law spread from the upper Mamberamo River across the mountains in the Mamberamo region to the Sarmi region and spread to Fak Fak area (found the same clan name that is the "Wou" clan) which is administratively Is in West Papua Province. According to the story of the ancestors this incident stems from the separation of brother and sister due to tribal wars that resulted in scattering toward the West

4 Indigenous Peoples Leadership System Keondoafian

The system of keondoafian is a system of leadership of the Tabi Mamta Customary Law Community, led by a hereditary patent (patrinial) who has birthright, is appointed and respected in indigenous peoples. Indigenous PeopleTabi Mamta ondofolo leadership considers that ondofolo is the incarnation of the sun and the moon god. The territory of the system of keondoafian is in the Sentani valley, the Grime valley to the north coast. Leaders in the Sentani valley and Grime valley are called ondofolo, while on the coast is called ondoafi. The word Ondoafi or Ondofolo comes from the language of Tepera namely Ondewafi which means Head of Customary Country or "Take care of the People", "Master who care, give life for the people".

The language of Tobaji calls Harsori, the Nimboran language of Iram or Dekening, the language of Kemtuk and Gresi mentioning Duguea and Dugeno, Jouwawary Language mentioning Matawun Pan or Marar Matawun, Nachaiba Language (Big and Small Ormu) called Tubwe and Moi Language mentioning Done word which means taking care of goods Pet. In the customary region Tabi Mamta is familiar with custom symbols. Customary symbols are highly respected, respected and have a high religious value. These custom symbols are communal lands, forest resources, and other natural resources. The symbols of customary have been traditionally adhered from the ancestors of the Tabi Mamta Customary Law Community. Indigenous symbols have always been under traditional rule. The highest customary authority governing custom symbols is in the hands of Ondofolo or Ondoafi. The customary structure of the cynosure system begins with a leader called ondofolo. Ondofolo in his leadership was assisted by the clans awarded Abhu Affa who was assigned as a substitute ondofolo when ondofolo was absent at an event, as well as the title of Abhu Akho who served as an intermediary or spokesman ondofolo, and Koselo (Ondofolo Yo) Stairs inside Yo (kampung). The structure under Koselo (ondofolo Yo) is a clan or Akhona, which is authorized to keep the customary treasures, and then the lowest structure is the people or the customary community called Akha Beakhe. The following figure shows the traditional structure of keondofoloan system.

5 Indigenous Peoples Leadership System Iram

The Tabi Mamta Indigenous People which has the leadership pattern of the Iram system is in the Grime valley customary territory. According to the beliefs and beliefs of Indigenous Nambloung people that 3 (three) traditional leaders are symbols of the Sun. The customary leader or Iram is assisted by Takai who serves as uncustomary adviser or judge, as well as Duneskingwou who is a customary spokesman or daily executive. Iram's duties within the customary village are to protect the people from all threats and to promote the welfare of the people. Meanwhile, when out of the village customary Iram's duty is to establish relationships with Iram who was in another village. In traditional talks the spokesperson or daily executive (Duneskingwou) can not express his opinions before the advisor (Takai) expresses his opinion in the form of advice to Iram. So the decision issued by Iram as the supreme leader is considered to be the highest decision and is of a scral nature, and can not be violated. If the decision is violated then there will be custom sanctions that are believed to bring disaster to those who violate it.

In carrying out his duties then Iram assisted Duneskingwou as spokesperson or daily executive and Neskingwouw as head of general government. The Public Administration Board is divided into 4 (four) sections namely Yrung Neskingwoow as head of general government as well as Indigenous Council, Iram Tang as head of general government in charge of external affairs and in indigenous kampong, Takay Tang Neskingwoow as head of the affairs of the judiciary and community protector , And Takay Tang Sa Neskingwoow as head of general affairs in the field of customary justice. The meaning of Tang is clan (Clan) or clan that has line line of descentIram, Takay or Duneskingwoow.In addition there is also Undo Hamong You Hlu / Hamung bu lelou who served as a financial supervisor or Examination Agency treasures treasures and customary finance. While Undo Hamong Bu Hlu as financial executor or Head of Kampung Finance Agency duty to receive, store, maintain and issue treasures customary treasures or money (dwei). In the leadership structure of Iram there are legal advisors called Hlu Waji or Ki Waji, as well as Wue Leng's duties as a messenger, and Flomewu Hlu as Commander of War.

6 Indigenous Peoples of Collective Leadership System

This pattern is located in the northern coastal area of Jayapura Regency, especially in Ambora, Tablanusu, Yepase, Dormena, Ormu, Tarvi and some areas of Keerom up to the border area of Papua New Guinea State (PNG). In addition, collective leadership patterns were also found in the villages of Meikari, Skori, Demetim, Sabeyab Kecil, Mamda and Aimbe. The Customary Law Society This Collective Leadership was formed due to the acculturation of culture, due to mixed marriage among tribes, and the inclusion of other tribes outside the majority tribe forming customary village. For example the occurrence of cultural mixing that can be seen from the names of the clan. For example the occurrence of cultural mixing that can be seen from the names of the clan. Like the clan name "Pangkatana" in custom area of doyo, which is a mixing culture Sentani and Genyem. Then clan "Marweri" which is a mix between Sentani culture, especially in Sentani lake area with Genyem. Similarly the "Wasangge" clan in the Kanda Boroway region is a mixture of Kemtuk culture with Genyem.

There is also a cultural mix between cultures of indigenous and tribal peoples who live in villages west of Lake Sentani with customary law communities residing in the Ormu region to the west of the Dormena region, Waib to the Kaptiaw region.Another example in the Tarvi region is the Subu tribe of Ronggowai tribe in Sarmi. So if there are customary events ondoafi will come to the Sarmi region to follow the traditional ceremony, which should follow the traditional ceremony in the region Tarond keondoafian. Tarvi region actually entered the administrative area of Sarmi Regency, while muaib entered the administrative area of Jayapura Regency.

The process of territorial capability begins with a system of exchange of natural resources (burter). Beginning with indigenous peoples of Genyem (Jayapura District), they have their fields swapped for the marine products brought by the Tarvi indigenous people. Then because it has been for a long time then the Genyem people build settlements in the coastal area of the Tarvi. The place is now called the village of Muaib.

Thus Tarvi and Muaib have different languages than other villages in the territory of Tarvi's terondoafian. The pattern of collective leadership is usually directed to the odowapi or the head of the Tribe. Ondowapi or Chief Tribe has the full right (Veto) to decide ancustomary decision. Leadership ondowapi or chief of the tribe is selected based on the offspring of the eldest son in the largest tribe that forms the customary village. For example the Tabi Mamta Customary Law of the tribe in the village of Kriku and Skrofo. The customary structure of this village begins with the highest stratum is the Head of the Tribe and then oversees the clan leaders (keret). The tribal chief holds the highest authority, followed by the clan chiefs.

7 Social and Cultural Aspects of Tabi Mamta Indigenous Peoples In Amdal Process

Development in Papua Province when not yet involve all indigenous and tribal peoples. Indigenous and tribal peoples involved in development are intended to provide consideration in accordance with local wisdom applicable in the area. But in reality the current society is not involved and only as a spectator especially in Papua.
In addition, development in Papua is currently aimed only at infrastructure development but not human development. So there is no change in the mental attitude

of the Papuans. This is evident with the consumptive lifestyle of Papuans. Especially with the policy of providing assistance to the villages, do not give an impact on the mental development of Papuans.

There was the sale of customary lands, while the value of land for the Tabi Mamta Indigenous People was sacred. Customary land rights and powers are an integral part of customary structure, so that when customary lands are sold, there is a fundamental change to the authority and power of the Tabi Mamta Customary Law Community.

8 The Indigenous Peoples Stratification

The customary structure of indigenous people's life in Tabi Mamta is seen in the customary stratification that still exists today. The stratification has been customary and structured by the ancestors of indigenous peoples. The form of social stratification in Tabi Mamta indigenous people is known by the customary stratification of Ondofolo as well as in the section of Tabi Mamta custom territory called Iram system. The indigenous structures formed in customary stratification have been established since ancestors.

Ondoafi or Iram is the highest level in traditional structures in Tabi Mamta customary territory. For example in the Sentani customary area there are 33 Ondoafi that protect tribes and clans, while in other parts there are several Iram and collective leadership systems (tribal chief). Relation to development especially in the field of Amdal, the stratification of customary is currently not running in accordance with its function. This is because indigenous peoples are not involved and given the authority according to customary stratification in the development process in Amdal field. There should be a strengthening of indigenous peoples at the village level by the government, by providing assistance to customary components in traditional stratification, as well as providing opportunities for customary leaders such as Ondoafi, Iram or Chief to play a role according to their functions, including in the Amdal decision-making process.

The mechanism of development planning including environmental planning is through the EIA process has been running until now, but the implementation is not maximized considering the government's intervention is still high, by not giving full authority to the Indigenous People to regulate themselves in the development planner in the field of environment. Indigenous and tribal peoples have been able to plan the development of the environment from generation to generation. Customary planning mechanisms have been established before the government exists.

For example in environmental planning, especially in Amdal field, the stages in the customary structure are carried out by firstly giving Ondoafi information to the entire community that there will be an activity plan of a company, then the second mechanism of Ondoafi invites the traditional chiefs and ulayat rights owners to the meeting customary discusses the intended activity plan.

The place of the customary meeting took place in Obhe. All responses and aspirations of customary chiefs and owners of customary rights will be discussed in Obhe, such as the aspiration to refuse activities because there are sacred places that the ancestors should not be touched by strangers or animals. This process is a process of collecting bottom-up indigenous peoples' aspirations carried out by Tabi Mamta indigenous peoples. This mechanism does not work. This is due to the role of the customary Mechanism has been replaced by the role of Musrembang which is more normative and guided by village officials.

In addition, the mechanism of crafting the development aspirations from the lower levels is not working because the Musrembang apparatus, which is a medium to mediate this process, does not understand customary knowledge, traditional rules and local culture. Therefore the government should strengthen the indigenous peoples by providing a portion of authority to ondoafi and supporting customs with funds and facilities. Ondoafi and Iram and tribal chiefs hold the highest authority in safeguarding and exploiting customary symbols in the form of customary lands, forests, waters and others contained within the customary territory. Utilization of custom symbols begins with talks with ondoafi, because ondoafi is the representative of the people as well as leaders and leaders.

For example, in Nimbugresi District of Jayapura there is a Levite tribe in which there is a Wou clan who was appointed osebagai ondoafi by the clans of Sanuel, Klemen, Traper, Mebri, Mes, Nian, Tegai, Srom, Rumei and Tabisu. The decision-making process in the customary stratification begins with the customary. All clans gathered to agree on the problems faced by the customary party. Furthermore, from the agreement and the consent of the community is endorsed by ondoafi as a decision. As in Kampung Yetti and Kriko there are Manam tribes. Manam tribe appointed Ondoafi Manam which is customary structure is recognized as the elder of the five villages namely Yetti Village, Wembi Village, Kampung Atibai, Kampung Ujung Karang and Kampung Skuare.

In these five villages there are several clans of the clan Tungkir, Marga Less, Marga Bewangkir and Marga Sakor. In the customary territory of Tabi Mamta Manam Tribe there is Manama Customary Council (DAM). The duties and functions of the Manama Tribal Council (DAM) are to deal with issues of violations that occur within the community.

Beginning with customary land use issues for development purposes up to issues of promises of unapplied companies to indigenous peoples Tabi Mamta Suku Manam. Tabi Mamta Customary Law Community, especially Manam tribe in the village of Kriko wants if there is a problem between the company and society customary law then the settlement is done together.

Starting from the process of meeting up to the decision-making process. Customary parties want a joint meeting to avoid arising suspicion or distrust of indigenous peoples against the company. In addition, the customary community does not feel disadvantaged by the decisions issued by the company. Indigenous people want openness among various parties. Coordination among the troubled parties continues to be undertaken by customary parties, companies and governments. Continuous coordination will provide customary benefits, such as solving non-discontinuous problems. Indigenous peoples do not want any unilateral or unilateral settlement

In addition, in solving the problem must be attended by parties who really have the right to take decisions and can not be represented. Indigenous peoples have a habit in customary meetings that all thoughts put forth in the meeting will be a fixed and immediate decision to be implemented. Indigenous and tribal peoples do not want a mechanism of stewardship of issues that do not take sides with honesty and balance. Indigenous Peoples Tabi Mamta Manam tribe in solving problems using the principle of honesty and balance in indigenous peoples.

The party is responsible for all the decisions made, because the decision also represents the spirit of their ancestors. For that the Indigenous People of Tabi Mamta Manam Tribe does not want the customs problems resolved by companies and government. Governments and corporations can only provide material impulses to indigenous and tribal peoples, but do not interfere to the affairs of decision making. By carrying out such a function, then indigenous peoples themselves feel that they have been recognized by the government and the company.

The existence to manage their own natural resources and create their own welfare. The Indigenous Peopleof Tabi Mamta Tribe Manam believes that programs produced by the government and companies will be accepted if the programs benefit the indigenous people. Despite the views of indigenous peoples, it is beneficial but if it can not be realized the Indigenous Peopleof Tabi Mamta of Manam tribe will not accept the program and potentially create conflict among indigenous people of Tabi Mamta tribe of Manam tribe with company or government, Tabi Mamta customary law Manam in decision-making is not based solely by individuals, but decisions collectively among indigenous and tribal peoples themselves.

The collective decision will then be voiced by Ondoafi or Chief Tribe. Chiefs are very active role in the life of indigenous people of Manam tribe. Before voicing the results of indigenous peoples' decisions, the chieftains coordinate with ondoafi and clan heads. Communication between heads of tribes, ondoafi and clan heads (keret) is done continuously. This communication is not interrupted, if interrupted it will lead to conflicts within indigenous peoples.
In addition, conflicts can also occur if the company or government to communicate continuously to certain people by excluding indigenous peoples Tabi Mamta Tribe Manam collectively. Government or companies need to know the custom structures that exist in the life of indigenous peoples Tabi Mamta Manam tribe.

For example in the tribe of Manam there are clans (rail). For example, in Suskun village there are Marga Bogofier, Konondoi clan, Giryar clan, Mesas clan, number clan, Abar clan, Krunggir clan and Kumryagir clan. While in the village there is Marga Kera, Putui, Itungkir, Nyumbori, while in the village of Kriku there Menarga Marga and Marga Bewangkir. These clans recognize the principle of decision-making that is the birthright of a son (Patrenial).

Also in decision-making is not only based on the decision of the Tribe Manam, but they will communicate with other tribes around them like the tribe Ewir. In Senggi region there are several tribes, including Knap Tribe, Mela Tribe, Nanai Tribe, Woi Tribe, Mangu Tribe, and Tribe Wetlai. In this region the chief of the tribe oversees the Chairs of Keret.

The decision-making process on indigenous land use begins with the customary village (Fin) meeting, with the approval of the community and clan leaders (keret) of the customary land owners. In Kwansuvillage there are 5 (five) tribes of Waru, Bemai, Yaram, Bano, and Kwano. Each has a tribal chief and ondoafi.Dalam traditional stratification Tribal heads play an active role to control the life of his tribe, while ondoafi role validate and protect the results of existing decisions.
On the other hand ondoafi can not make its own decisions. If ondoafi takes his own decision then the consequences will be death. The stratification of is also seen in the life of the Balli tribe who inhabited the Nimboran area. The clans in the Balli tribe are Demonggrang, Demotekai and Yaung. The customary structure is seen with the selection of the Balli genus as Iram which is determined by descent. All decisions are taken jointly by the existing clan.

Iram's function is to give seriousness or confidence to all existing decisions. The decision-making process begins by listening to the aspirations of the clans within the Iram. The task of unifying the clan aspirations is the task of a Tekai. Task Tekai as well as advisor will be assisted Duneskingwou as spokesman. Furthermore, in the event -meeting as a regulator of the event conducted by a clan Rum status. After all the decision of the meeting is produced, Iram can not directly issue the decree statement, but must hear from the clan with status Bemai who is in charge of carefully assessing all decisions that will be issued by Iram.

In this region, each tribe has a different Iram. If there is one decision in one Iram, the other Iram can not interfere with Iram's customary business. However, if the decision concerns the territory under Iram's other authority, coordination will be made between Iram before making a decision. The clues about the clan that became Iram in a region were conducted using a patristial hereditary approach.

In Beniekvillage there are Sawa tribes. The Sawa tribe was led by the chief. Chiefs have played an important role since their ancestors existed. The concept of ondoafi had just been accepted by this tribe in the Dutch era. Thus with the influence of the outside of the customary structure began to include ondoafi in the custom stratification in this tribe. In addition there is a tribe in Gerusa Village Jack. In this ethnic life the role of ondoafi is very dominant in decision making.

Decision making is done following the starlification of the cohesion. Ondoafi will consult with the chiefs and clan heads before making a decision. In addition there is a stratification in Village Santosa where there is a general ondoafi and ondoafi clan. The decision-making process takes place within the customary meeting, involving the clans (keret) within the tribe.

The role of the general odoafi is very active in the decision-making process. In Tarvi village there are 5 (five) tribes namely Ondi (Taudufu), Bernepu, Kaisiu, Fetowin and Taurui. Indigenous stratification shows that in the village of Tarvi the role of ondoafi is very active. Ondoafi will consult with the customary land owner before making a decision. Ondoafi has more power to make decisions because people believe that a direct, inner connection occurs between ondoafi and the spirits of the ancestors.

For example when someone breaks a ban ondoafi to go somewhere, and if violated it will appear storms at sea and on land, even though it was still the shady season. In Sawesuma village there are 4 (four) tribes of Tekbo, Digan, Tribe Jasa and Tribe Bungu. Each tribe has a tribal chief, and above the tribal chief there are ondoafi who play an active role in indigenous peoples. The decision-making process is in the hands of ondoafi after asking for consideration from each tribal chief.

9 Permissive

Permissive attitude is an open attitude in giving permission to the utilization of natural resources. Development to be held in the territory of Tabi Mamta Indigenous Peopleis required to obtain permission from customary parties. The development plan to be held in Tabi Mamta's customary territory of Ondoafi's leadership system is done in advance through notification to the local government.

Furthermore, the government convey the purpose of development to ondoafi, and ondoafi will convey to the tribes and clans that have ulayat right on land yuang will be used as media of development.

In indigenous communities Tabi Mamta the process of licensing of the party starts from the talks conducted by families in a clan or called a rail. After the clans or clans in a family speak for themselves in their extended family and a decision is obtained, then the result of the decision is brought by the chief to be discussed in the . Indigenous peoples will gather in the (Obhe) to discuss the purpose of the development plan.

Ondoafi will invite heads of tribes and trolleys to attend and provide advice, feedback and opinions. Suggestions, inputs and opinions are given by first considering the customs and culture of the local community. While in the Iram system, the decision-making arrangement is carried out through a custom process involving the house's eyes (family in one clan), tribal chiefs, and clans appointed as spokespersons (Duneskingwou), advisors (Takai), organizers (Rum) As well as the decision maker (Bemai). Such a decision-making process is called "The Process of Sincerity".

In Benyom village there are Kekri (Yantewo), Tecuari Tribe, Bano Tribe and Kasmando Tribe. Each tribe has Iram and other devices respectively. The licensing process begun with the holding of the meeting at the family or clan level then continued to the village level attended by indigenous peoples and Iram devices in each tribe. The highest decision is at customary meeting in village. Furthermore, one of the Tabi Mamta custom territory Mix System between ondoafi and chiefs of tribal power occurs in the tribal region of Manam.

In this region there are companies that operate. The company when entering in the territory of the Manam tribe did not approach the chief or ondoafi residing in the village. Approach done by the company through local government. The Tabi Mamta Customary Law Society of Manam felt that the presence of the company had bypassed the customary authority of the region.

There are 2 (two) conditions of permissive attitude shown by indigenous people in Amdal process. First, when indigenous and tribal peoples no longer have bargaining power over their natural resources. Indigenous and tribal peoples will be permissive of investors' requests to exploit natural resources because previously investors have licenses from the government. As a result of not having the bargaining position, the Indigenous People is resigned in giving permission openly. Second, when indigenous and tribal peoples feel fully involved in making decisions at the level.

The result of the decision to grant permission will be shown openly through the mechanisms occurring within the customary structure. The permissive attitude often encountered within the Tabi Mamta Indigenous Peopleis the first permissive attitude, where indigenous and tribal peoples are willingly and openly allowing investors to exploit their customary symbols because investors have received prior permission from the government.

This situation will lead to conflicts in the future among indigenous peoples, among indigenous and tribal peoples with the government and investors. Therefore, before the customary community gives permission to use customary land, it needs to be mediated by the government. So far, the mediation established by the government through Musyawarah Pembangunan (Musrenbang) at the village level does not take into consideration the results of decisions made by indigenous and tribal peoples in the (Obhe). Development Deliberation (Musrembang) in villages is only normative, oriented towards securing government programs.

Thus the needs of indigenous peoples from the cultural side are often not accommodated. For example the need to preserve sacred customary lands, the need to invest, the need to become masters on their own land, are the things that are often ignored by the government. The government is more oriented towards material development, infrastructure than development of cultural aspects of Tabi Mamta indigenous people. Tabi Mamta Indigenous People has a habit since hereditary in giving permission of customary land use starting from para-para adat (Obhe).

Para-para adat (Obhe) is a place where people give suggestions, opinions and responses to development plans that will be held in Tabi Mamta customary territory. Suggestions, opinions and responses include permitting, rejecting development, until making any decision on behalf of indigenous and tribal peoples.

10 Continuity of Communication

Tabi Mamta Indigenous People is a Indigenous People that is open in terms of communicating with outsiders. The Tabi Mamta Indigenous Peopleholds 2 (two) principles, namely the principle of balance and the principle of relationship when meeting with outsiders who will utilize Natural Resources owned by customary law community.

The outsiders are migrants from outside the customary law of Tabi Mamta. The principle of equilibrium in Tabi Mamta's Indigenous People implies that unbalanced use of natural resources will bring disaster to Tabi Mamta's customary law community, whereas the relation is meant to be the brotherly relationship between the Tabi Mamta law community and the outsider is not a momentary relationship when one The parties have felt benefited then left the customary party.

For that Communication between the Indigenous Peopleof Tabi Mamta with the immigrants done continuously, believes in the sense that there is a learning process on the customary side and seriousness. If these principles are not exercised by outsiders, what will happen in Tabi Mamta's Indigenous Peopleis suspicion and mistrust of outsiders.

The migrants enter the Tabi Mamta customary territory in groups such as transmigrants from Java, Bali Island and Nusa Tenggara Island. In addition there are also migrants imported by the company such as PT. Rimba Matoa Lestari, PT. Tandan Sawita Papua engaged in Oil Palm, PT. Semarak Timber is swarming in the field of Forestry, PT. Berlin Krida Jaya engaged in the processing of Sago cultivation.

Some companies enter the Tabi Mamta customary law area through licensing from both the District and City governments. Communication between companies and customary law communities Tabi Mamta has little intensity and continuity, when compared to the intensity and continuity between the company and Provincial, District and City Governments. This is evidenced by the loss of information on the presence of companies in the territory of customary law of Tabi Mamta.

The Tabi Mamta Customary Law Society did not get any information about the presence of the company in the customary law of the Tabi Mamta community. For example, the Tabi Mamta Indigenous People of the Manam tribe, who wishes that everything related to the problems between companies and indigenous peoples should be continuously communicated. If the communication is interrupted it will potentially conflict.

One form of conflict that occurs is the result of the company's decision which then led to indigenous people's discontent. This dissatisfaction will then lead to adversity leading to anarchist destructive activities. In addition, the form of communication that is expected by the Indigenous People of Tabi Mamta Manam tribe in the form of a proportional communication. This communication will occur if the company contacts people based on existing custom structures.

For that the company or government must understand the people who are in the customary structure and have the authority to voice the voice of indigenous peoples. Such communication will result in decisions or agreements that can solve problems among the troubled parties. Misinformation in communication between indigenous peoples and companies or governments is usually due to some information that does not reach the parties in communication.

For that the initial process of communication must be maintained sustainability. All information to be communicated is acceptable to all parties. For example, often the Indigenous People of Tabi Mamta Manam tribe does not know with whom to communicate, who is authorized to make decisions in the organizational structure of the company.

So often when the Indigenous People of Tabi Mamta Manam tribe invites the Company and Government present together to discuss the problem, the presence is only represented by people in the company that can not take decisions at that time. At the end of the meeting there will be dissatisfaction with the resulting decisions. To avoid a sense of suspicion and a high sense of respect for the Indigenous People Tabi Mamta Manam tribe, then the implementation of communication is not implemented in large hotels

The Costumary Law Community Tabi Mamta Manam tribe feel highly respected existence when the communication is done in the village customs. The kampong is a place considered sacred by the indigenous people of Manam tribe because of that place their ancestors until the present generation held a meeting to discuss the existing customary issues. Often in the Amdal process there is an unsustainable communication between the Tabi Mamta Indigenous Peopleand the investor who will manage customary symbols belonging to the customary law community. Unsustainable communication will have an impact on the distrust of the Tabi Mamta Indigenous People against other investors who will enter to utilize custom symbols in the Tabi Mamta customary territory.

11 Reciprocity

The concept of repayment (reciprocity) does not apply in the life of the Tabi Mamta Customary Law Community. The concept of repayment in the form of assistance from the company is opposed by the Tabi Mamta Customary Law Community. The provision of assistance made by the company that will invest in the Tabi Mamta Customary Law Area is the company's effort to master the customary rights of the Community.

In the Amdal Process in the Tabi Mamta region, the concept of repayment will only bring suspicion among indigenous and tribal peoples, especially in the clans of ulayat owners. The suspicion of fellow clans will bring conflict in the Tabi Mamta customary law community. The customary rights to land, water and all the plants and animals that live therein are hereditary rights that can not be bought or replaced in the form of compensation or material assistance.

The right is not a sole right but a right of communal or customary right inherited from generation to generation. In the ulayat right there is order which has been regularly inherited by the ancestors. The order has a Magical-Religius value especially related to the values of life's morality. For example, humans have a strong relationship with the land due to human reasons derived from the soil.

There is also a view that the clan in certain tribes derived from the names of plants and animals. In addition there is also the view of the Indigenous People about the importance of respecting the lake (the story of the occurrence of Lake Sentani), rivers and springs (story of the occurrence of rivers coming from the Cycloops mountains). Therefore the Tabi Mamta Indigenous Peopleconsiders that the compensation system or material aid as compensation for the use of the right to land, water and plants and animals which is above it, is a capitalist system.

Tabi Mamta Indigenous People from generation to generation does not recognize the system of repayment with migrants, foreigners and communities outside the Tabi Mamta Customary Law Community. The reciprocity system applies only between families and clans in the Tabi Mamta customary law. This is evidenced by the existence of repayment system in the ceremony of marriage, grief ceremony, and traditional ceremonies associated with the ritual of honor to the ancestors. Tabi Mamta Indigenous People viewed reciprocity among society with company or better known as compensation rights ulayat really not known in life of custom law community Tabi Mamta.

Usually the compensation made by an outsider begins with a tap door compensation. The act of compensation is an economic act that is very much against the customs and customs of the people. Such as tap door compensation, road and bridge compensation, scholarship compensation, traditional market compensation, village development compensation are all since the ancestors were not known in the Tabi Mamta customary law community. Indigenous and tribal peoples in meeting their economic needs are carried out with a local cultural approach. Such as the exchange of livestock products with the results of the garden in meeting the needs of the wedding ceremony, grief and other ritual ceremonies. All this is done to maintain the principle of balance and principle of relations within the Tabi Mamta customary law community.

The Indigenous People of Tabi Mamta holds that the compensation given by outsiders to indigenous and tribal peoples is a form of outsourced tribute to the ancestors and customs of Tabi Mamta customary law community. However, what happens today is compensation is a substitute for customary lands. When the compensation took place for the Tabi Mamta Indigenous People was the beginning of the loss of rights to the lands given by their ancestors. Companies present in Tabi Mamta customary areas often approach with indigenous and tribal peoples by delivering promises to the community As an example that occurs in the territory of Tabi Mamta Manama Tribe.

The company promises clean water network, customs like customary children, house building, every clan gets the help of four-wheeled vehicles, and livestock assistance. All company's promises are submitted to the Indigenous People of Manam Tabi Mamta. These promises if not fulfilled will potentially lead to a sense of distrust and conflicts that prolonged. Therefore, in response to the promises demanded by indigenous and tribal peoples, in a meeting between the company and the Indigenous People of Tabi Mamta of the Manam tribe, the meeting must be attended by the chairman of the company who can take the decision at the time.

Thus indigenous peoples feel highly appreciated, the level of trust of indigenous peoples can be maintained, and the level of suspicion will decline further. In Beniek village there are Sawa tribes. In this region often occurs communication between Sawa tribe with existing companies. Assistance from companies is often given to the Sawa. Suku Sawa do not recognize the process of buying and selling land, which there is to provide land for use by companies, not to own. Often the company promises promises that attract the attention of the Customary Law Community. The promises are intended to enable the company to have customary land for investment purposes.

For example, in the village of Gerusa the company promises to the Terss, Manoa, Jasa, Jack, Yess, Digam and Tetbo tribes, to be given 4 (four) hectares (Ha) of land to be owned by every household head. The fields will be equipped with certificates. In Santosa village there are Targualam, Yandu, Sasbe, Meigar, Kabri, Worto, Wakan and Tuku Owe. These tribes agreed before the exploitation of indigenous lands were required to do "gifts of door money" to those tribes.

In Sawesuma village there are also company promises of honorariums to be given to ondoafi, educational scholarships for village children. These promises are given by the company to the ondoafi with the aim that customary lands can be utilized by the company.

12 History of the Past

The history of the past about the failure of relations among firms investing in the territory of Indigenous Peoples had an impact on the existence of the Tabi Mamta Customary Law Community. The failure of this relationship greatly affects mental, character, competitiveness even psychology of society. In the Amdal Process there are promises set forth in the Amdal document as a form of community development program from the company to indigenous and tribal peoples. These promises are often incompatible with the needs of the Tabi Mamta customary law community.

Tabi Mamta Indigenous People has a life philosophy Promise to keep, and Promise to be executed.

 https://doi.org/10.1051/e3sconf/20183108017

Past failures between the company and the Tabi Mamta Customary Law have the potential to cultivate suspicion, distrust, pretense, and communication gaps. Forms of past historical failures include companies only communicating with the government, many companies making promises to the public, using the apparatus (militarism) to intimidate some societies that do not support the presence of companies in the Tabi Mamta customary law.

This is felt by the Indigenous People Tabi Mamta Tribe Manam where the promise to complete the Ropes of Love has not been implemented by the company. Ropes of Love which will be completed gradually for 4 (four) Years with the value of each location of 10 Ha will be given compensation of 20 Million. This decision comes from the company itself not from indigenous peoples. Company promises that are often delivered and unsigned will increase indigenous peoples' distrust of the company.

In addition, the vagueness of the party responsible for the promise is often experienced by the indigenous peoples of Manam Tribe. The unclear responsibilities were strongly influenced by the political and economic situation that existed at the time. For example, the company's promises as outlined in 9 (nine) forms are not through the process that has been practiced by indigenous peoples since their ancestors existed to this day. In this case indigenous peoples are involved when it is required by governments and companies to form cooperatives.

13 Culture value

The Tabi Mamta Indigenous Peoplehas cultural values that have been maintained to this day. Cultural values embedded in the life of the Tabi Mamta Indigenous People are manifested in respect for the existence of customary land, protection of forests with everything living in them, the rivers of the sea and the coast. These values provide prosperity and comfort in the life of the Tabi Mamta Customary Law Community. The cultural values of the Tabi Mamta Indigenous People are filters of new cultural values (diffusion, assimilation and acculturation) in the life of the Tabi Mamta Customary Law community.

Indigenous peoples of Tabi Mamta are strengthening their cultural values through strengthening institutional relationships, either in customary (Obhe) customary meetings, as well as family gatherings in clans (tribes) of each tribe.

One of the cultural values within the Tabi Mamta Indigenous Peopleis the Land Value of Indigenous People. Tabi Mamta Indigenous People does not recognize the system of buying and selling custom land. Customary land is an integral part of the customary structure of Tabi Mamta Customary Law Community. So when customary lands are sold then there will be a change of power and authority within the customary structure. In the EIA process in the Tabi Mamta customary area, there are often violations of cultural values of indigenous and tribal peoples. The cultural values of indigenous and tribal peoples are the value of land.

Often companies assess low land by trying to calculate the economy against the value of land. Potential conflicts arise when investors make land compensation based on investor size. On the one hand, indigenous and tribal peoples are not given the opportunity to give value to the land in the EIA process based on the view of the Tabi Mamta customary law community.

For example the Value of Land in the Indigenous Demutru region. Tabi Mamta Indigenous People in Demutru customary area believe that Land is Mama or Mother or in Nambloung language called Me. According to the teachings of (Woi Iram), the land should be kept, protected and not trafficked to foreigners. Whoever sells Me, to a stranger then he will be burned with the heat of the sun, and his body will be eaten by worms and result in prolonged civil war.

The decision-making authority over Tanah is on Ondoafi, after receiving various inputs from the clans of each tribe within the village. Ondofolo has supra natural power, because ondofolo is considered by the Tabi Mamta Indigenous People as the incarnation of god. The introduction of new values in the form of the entry of religion, science and technology into the cultural values of the Tabi Mamta customary law cause the weakness of power and authority of ondofolo. Customary order will shift, and experience the decline of values when it is brought into the political and economic aspects. Even not only the decline of values, but will lead to poverty.

The Costumary needs to be restored to pure, simple values, and uphold customs and appreciate customary leadership. Indigenous values that are still embedded in people's lives are maintained and preserved. Such values are the religious values of a place (shrine), animals, trees, water sources. These values still exist and are maintained by the Tabi Mamta customary law community.

Evidence of customary values in environmental management (Environmental Wisdom), among others, restrict the collection of certain types of trees, the collection of water at certain water sources that are only allowed for Ondoafi, Iram or Chief Tribe, hunting animals such as Bird Cenderawasih Which is considered a bird that comes from a god. Therefore, violations of customary values are the responsibility of the government. The command is responsible because the government as the ruler has a development policy that is not owned by the Tabi Mamta customary law community. Therefore, the Government is responsible for shifting customary values.

One form of government responsibility is to recognize and restore the role of customary existence in the customary order as the owner of natural resources. The government conducts guidance, empowerment and strengthening of the customary. Efforts to restore customary values and customs need to be sustainable. Customary values will be a force for development. Traditional values existed before the Government and Religion.

As an example before the existence of religious values entered into the customary order, then the values of trust have existed before. For example, the Metu Debi Declaration contains custom values that are understood as a philosophy of "We Can Hold The Wheel, But Can Not Determine The Direction of Wind and Current".

The Tabi Mamta Indigenous People still holds the basic principle of sovereignty over custom symbols in the form of Natural Resources and everything contained within it. The values of beliefs are closely related to customary symbols. Utilization of symbols will greatly affect the life order of indigenous peoples. For example, the conflict between companies and indigenous peoples is due to the utilization that is done in places that are considered sacred by indigenous peoples. In addition, in the customary order there are values of mutual cooperation. The value of gotong royong can be seen in the loyalty of indigenous peoples to traditional leaders in indigenous villages. These values need to be reappointed by the government and investors.

With the inclusion of new cultural values into the life of the Tabi-Mata customary law community, there is a change of attitude and behavior of people in their lives. The inclusion of village governance systems, Agrarian Law, Forestry Law, information-technology, communications and migrants brought in by the government and companies have brought significant changes in the life of the Tabi Mamta customary law community.

Changed values such as the loss of customary stratification of a centralized village government on one leader (Ondoafi, Iram or Tribal Chief) are displaced by the dualism of leadership, the headman and village head, the displacement of the ulayat rights or the right of collectiveness to the land to individual ownership The existence of Agrarian law, the displacement of cultural utilization zone by Tabi Mamta Indigenous People such as sacred forest and forest that can be utilized is changed by the area of cultivation and protected area regulated by the forestry law. The government has not yet screened new cultural values that affect the cultural values of the Tabi Mamta customary law community. The community has not felt the strengthening of the government's base.

The strengthening of the base needed by the Tabi Mamta Indigenous People is strengthening the knowledge of the Tabi Mamta Indigenous Peoplethrough formal and informal education level, besides that the community wishes to strengthen the values of the land, strengthening the change of cultural acculturation.

14 Norms and Customary Sanctions

The Tabi Mamta Indigenous People still maintains customary norms in its life. The customary norms that are still maintained among others relate to customary land ownership. Form of customary norms that still apply the rules of use ulayat land rights land, rivers, lakes, coastal and sea. The ownership of customary rights applies only to the clans, rags and tribes belonging to the structure.

In the Amdal process it often does not accommodate all customary norms prevailing in the Tabi Mamta customary area where investors will manage the symbols of custom symbols. One example of customary norms still prevailing in the life of the Tabi Mamta Indigenous Peopleis currently in the Demutru customary territory. The customary norms relating to the management of territories and natural resources are listed in the Waidemuo Law (the Dematru Customary Law) in Articles 4 and 5 stating "Do not take the goods of others without the permission of the concerned persons and do not rob people of other lands or gardens or families other".

Customary norms or customary rules prevailing in the Indigenous Demutru region consist of the Laws of Waidemuo, Demutru and Mako. Waidemo's Law consists of 6 (six) articles. Article 1 reads; Fear and honor Wai Iram; Article 2, Fear and respect for parents; Article 3, Do not covet the wives of other persons or women; Article 4, Do not take the goods of others without the consent of the person concerned; Article 5, Do not rob land and Article 6, Do not kill anyone else if it is not for violating Waidemuo's Law and other customary laws called Demutru.

Furthermore, Demutru Law consists of 15 articles. Article 1 reads; When people are anywhere in the world, Wai-Iram always sees and cares for all his behavior and deeds; Article 2, Grays or elders are the representatives of Wai Iram in this world; Article 3, It may see a woman's tuning or feet but is not allowed to see and notice the upper thigh calves of a woman because it is not good; Article 4, It is possible to kill another person if he uses his wife or seizes the land of a family's right; Article 5, Must not undermine Iram's authority because it can be killed; Article 6, People who are many mistakes and do no good, if they do not want to repent will be attacked by disease; Article 7 Each boy, starting at the age of twelve or less, must enter the dormitory at the initiation house (Injo Yamo), which is nurtured by the government (in this case Iram, Bemai and Tekai as learners). For adolescent males who do not go to education can be killed by the Spirit or the Magical Creatures (Injo);

As for the things taught in the initiation house (Injo Yamo) in the form of fighting skills, hunting, gathering, farming, planting and other skills related to survival. Article 8, Women are educated by mother and father at home until adulthood; Article 9, carving and songs belonging to one family (Tang) should not be followed by another family (Tang); Article 10, War only occurred on the announcement of Iram who gave the mandate to Plo Iram (Commander of War); Article 11, All men shall be educated as members of the armed forces; Article 12 Education is nurtured by a container by the name of Injo, and those who run Injo are called Injo-Hlu; Article 13, the crystal bracelet (Hamong) is held and stored by Iram.

Bracelet has the highest value and the most expensive; Article 14, Everyone is not allowed to organize a major dance party event (Kebi) on a large scale because it is Iram's right; And Article 15 states that if any person continuously violates the provisions and has committed a crime, if not repentant, although never punished or given advice then Duneskingwouw that is the head of Tang or clan is entitled to kill him. Other customary norms are Law Makko. Makko's Law consists of 10 Articles.

Article 1 states that every person has the right to devote to Wai-Iram; Article 2, If there is a gray or an elderly person sitting, you pass by in front of him or defend him by squatting or sitting with him; Article 3, An unmarried young man is forbidden to eat or drink and or to gather with married adults; Article 4 If a person kills another person for taking possession of his land, then the plot of land shall be left to the damaged family (this is specific to the adjacent lands); Article 5, Iram is prohibited from eating and drinking with commoners or commoners; Article 6, Iram's orders must be upheld whenever continuous resistance must be punished; Article 7, any adolescent or youth who misbehave beforehand, when entering Injo Yamo's boarding school then must be killed; Article 8, therefore the role of parents is required to foster or teach their children before entering boarding Injo Yamo; Article 9 A man may marry if he has been able to perform the usual work of an adult such as making a garden, making a house, or shooting a pig. Likewise a woman if she can already plant in the garden and others; Article 10 states that marriage must be accompanied by a dowry which is an entertainment for the discharge of a woman from the family (Tang), her family and relatives.

In addition, also found customary norms that are still valid in the Community of Law Tabi Mamta Moy tribe. Such norms are marriage laws called Kampug daray; The law of protection of women called Kampugyanggok; Forest protection law called Basyo nembut; The law of hunting is called maney nembut and law of Mutual cooperation called Krat. Further to this day, the Tabi Mamta Indigenous People does not recognize private ownership of mainland land, rivers, coastal lakes and seas. A form of violation of customary norms is the transfer of land ownership of land, rivers, lakes, coastal and sea from indigenous communities to private ownership. For example, land ownership is owned by individuals and companies.

This violation will be followed by custom sanction which is granted by the customary party in this case the owner of customary rights against the party who surrenders the customary land to the company or individual. The sanctions are in the form of alienation from the Tabi Mamta Indigenous People by excluding all customary processes that are taking place in the Tabi Mamta Customary Law Community. Tabi Mamta Indigenous People does not recognize custom sanctions in the form of compensation. This is because the compensation provided by the company or individual to the ownership of customary lands is contrary to the customary values that the land is a common property and can not be traded.

Business values can not enter in traditional values due to different color spectra. The Indigenous People of Tabi Mamta recognizes sanctions derived from ancestors and ancestral spirits, and sanctions derived from nature. The concrete form of the sanctions can not be seen but the consequences of the application of sanctions can be seen with the sacrifice of life. The stronger the belief of the Tabi Mamta Indigenous People is the higher the value of the fine or sanction given. For example, the displacement of some sacred places in the customary territory of Tabi Mamta Manama Tribe gives high customary sanction in the form of a fine of hundreds of millions to companies doing eviction.

The place is considered sacred because according to the belief of the tribe of Manam there is the place of the first human existence of their ancestors, and they assume that the spirits of their ancestors still exist today. In addition, custom sanctions are granted to companies that do not fulfill their promises to the Tribal Peoples Tabi Mamta Manuku Community. For example the completion of the Ropes where the value of the Rope of Love that should be determined by the Tribal Law Community of Tabi Mamta Manama Tribe based on customary values applicable, but the value is determined by the company. Furthermore, in the region of the Balli Tribe who know the customary structure of Iram, there are sanctions for those who violate customary rules. In the event of a violation of customary regulations, the applicable sanctions are murder.

While in the life of the Waru tribe of Kwansu village, in case of violation of, the sanction process will be handed over to the council. Furthermore, the council led by the head of the tribe will take decisions after talking with the clans that exist. In Beniek village there are Sawa tribes. This tribe does not recognize the system of buying and selling custom land. Suku Sawa never sells customary land, but allows their customary land to be used for development. Conflict often occurs when companies insist on buying customary land, especially lands that are considered sacred.

While in the region of Senggi there are customary norms that regulate the sacred place's that should not be violated. In the event of a violation it will be subject to custom sanctions. Customary sanctions are death. Customary norms that exist in Santosavillage still valid with the ceremonies of marriage customs. The marriage treasures are Tomakobatu, beads and axes. Customary sanctions are given to those who violate marriage customs such as in the right and left thighs or other physical laws. While in Sawesuma village there is a traditional ceremony of marriage with marriage property in this region in the form of Bracelet, Stone ax and Masete (a kind of bone which both ends are flattened).

15 The Customary Leadership

The environmental feasibility decision-making process often ignores processes that should be done through mechanisms that apply to Tabi Mamta's customary law community. The focus of environmental feasibility decision-making is more focused on the mechanisms occurring within Amdal Commission Meetings. On the other hand the process of taking through customary mechanisms is neglected. The decision-making process through customary mechanisms follows the customary structures within which there is a customary stratification which is highly respected by indigenous

https://doi.org/10.1051/e3sconf/20183108017

and tribal peoples. All customary decision-making processes are conducted through customary consultation. Indigenous deliberations are conducted through the mechanisms (paths) of the path of the cohesion. The mechanism takes place in the customary (Obhe) with the involvement of clans, rags and tribes within the traditional structure.

Customary symbols are customary land, forest, waters and other natural resources under the power of ondoafi. Ondoafi is a leader in charge of serving indigenous and tribal peoples. In the Tarvi region, the role of ondoafi is very active. Legal issues that occur in the village must be known and the permission of ondoafi. Ondoafi is considered to have magical powers due to having a relationship with the ancestors. Ondoafi is considered to have supernatural powers that can bring death to a person. In addition, ondoafi is considered to recover people from illness through communication with the ancestors. All custom meetings or traditional ceremonies are held in the ondoafi house. Above ondoafi there is Tribal Council (DAS) which serves to coordinate the ondoafi in every village.

The Chairman of the Watershed Council is considered to be the general Ondoafi who oversees the village's ondoafi. Furthermore, the village's ondoafi oversees the tribal chiefs in the village of Tarvi, Muaib, Yogapsa, Ambora, Little Muris, Great Muris (Yokari) and Demta.

16 Religion

The Tabi Mamta Indigenous People is very respectful of their ancestors. Community religious life is reflected in the life of each family. Families still believe in places that are considered sacred. Belief in the existence of ancestors is felt by the Tabi Mamta customary law community. Especially when the natural resources of land, water, forest and life are disturbed by the incoming investment and change the order of the close relationship between the life of the people of Tabi Mamta customary hulum with nature and ancestral spirits. The beliefs of the people who have been to the sacred places in the Tabi Mamta customary area still exist and are planted in every generation. Problems often occur when places considered sacred are used by outsiders who do not understand the cultural background. This is often the case with the Amdal process, where the places considered sacred by indigenous and tribal peoples are neglected due to the lack of investor's understanding of the importance of sacred places.

On the other hand indigenous people believe the sacred places are considered to have high religious values. These places are considered to be the place where the first human existed, the place is considered that the ancestors still live up to the present time. These places are considered to be the place where the ancestral spirits reside, where the demons dwell even ordinary place is where the former tribal war occurred. In this place also Chief Tribe can call the spirits of ancestors who died when the tribal war occurred. In the area of Nimbugresi village there is a tribe of Levi. This tribe still holds the belief that certain places in the indigenous territory lies with ancestral spirits.

The guidance of the ancestral spirits is the existence of a very sacred old village. Anyone who enters the area if they do not get permission from ondoafi will get custom sanctions. The customary sanction of a fine is determined by custom. As a result of a fine that is not resolved, then the concerned can not speak anymore (mute). The forms of fines are Tomakobatu (Uda), and Beads (Ngoi). In the Genyem village region there is the Balli Tribe. The tribe of Balli gives the sanction of murder or the person will lose his memory forever in the event of a violation of the sacred places. The sacred places were determined by the Balli tribe based on the belief that the place used to form the crossroads and even the settling places of the ancestors. Ancestral spirits are believed to have been transformed into trees, stones, wood, and spirits. Due to the ancestral spirits dwelling in the place then the place is highly valued for generations. In this region only Iram can communicate with the ancestral spirits.

The form of communication is done by speaking directly with the ancestral spirits. The Indigenous Peopleassumes that their ancestral spirits heard when the communication took place. Communication and spiritual contact can not occur between the owner of the communal right elsewhere with the local ancestral spirits. In Kwansu village a strong relationship exists between nature and indigenous peoples. Chief or ondoafi can communicate with nature using mother language (Soil Language). The characteristics of the land language are usually seen from naming the names of places, villages, rivers or water sources, and usually those names are associated with the origin of humans in the region. Not everyone can communicate like this. The ability to communicate with nature occurs from generation to generation according to the right of inheritance (patrinial).

Inheritance is hereditary, if the head of the tribe or ondoafi not memiilki boy, then the right of inheritance fell on the younger brother and so on. In addition in the Tabi Mamta customary territory there is a village of Gerusa. In this area there are places that are considered sacred places. Sacred places are usually dintandai with the forest protected by indigenous peoples. Customary sanctions that occur when there are people who enter the region will be dipanah thigh. In the village of Tarvi there are sacred places that are determined by indigenous peoples. A place is used as a sacred temple because the place is believed to be where their ancestors rest when traveling long. For that the places are highly appreciated by the local community. In case of violation in sacred places it will be sanctioned by custom sanction of custom size. In the village of Sawesuma there are sacred places.

The sacred places are shaped hills covered with trees. Trees that have been marked with clear signs such as red paint. The sacred places are highly respected by indigenous peoples. In the event of violation of sacred places, it will be subject to physical sanctions for those who violate it. Contextual of each of the above social and cultural variables in the AMDAL process can be clearly seen in the description of the dynamics of environmental feasibility decision making in the EIA process that occurred in the Tabi Mamta Indigenous People in Papua Province.

17 Conclusion

1. Indigenous Peoples Tabi Mamta is a community of Customary Law that still has territorial custom territory, genealogically has kinship, has customary values and norms and sanctions, possesses cultural objects, and has environmental wisdom in maintaining the existence of Symbols Custom in the form of Natural Resources.
2. In the socio-cultural system of customary law communities there are components such as stratification, permissiveness, communication, reciprocity, past history, cultural values, customary norms and sanctions, religious and leadership.
3. Components in a socio-cultural system constitute a unity resulting from the interaction between individuals and groups in order to prevent the occurrence of environmental damage and disruption to the existence of environmental and natural resource functions.
4. Components in the socio-cultural system of indigenous and tribal peoples play a role in the EIA process in the Tabi Mamta customary area especially in the environmental feasibility decision making process.
5. Natural resources are considered as ancestral symbols inherited from ancestors from generation to generation.

Bibliograph

1. Afrizal. *Metode Penelitian Kualitatif. Sebuah upaya mendukung penggunaan penelitian kualitatif dalam berbagai disiplin ilmu.* Jakarta: Raja Grafindo Persada, (2014) [in Bahasa]
2. Alan Bond., Juan Palerm., Paul Haigh. Environmental Impact Assessment Review: **24**:617–641, (2004)
3. Anne N. Glucker., Peter P.J. Driessen., Arend Kolhoff, Hens A.C. Runhaar. Environmental Impact Assessment Review: **43**: 104–111, (2013)
4. Badan Pusat Statistik. *Papua Dalam Angka* (Papua in Figures), (2015) [in Bahasa]
5. Bambang Rudito., Melia Famiola. *CSR (Corporate Sosial Responsibility)*, Bandung: Rekayasa Sains, (2013) [in Bahasa]
6. Bambang Rudito., Melia Famiola. *Sosial Mapping Metode Pemetaan Sosial Teknik Memahami Suatu Masyarakat atau Komuniti, edisi revisi*, Bandung: Rekayasa Sains, (2013) [in Bahasa]
7. Bao Cun-kuana., Lu Yong-sena., Shang Jin-cheng. Environmental Impact Assessment Review; **24**: 27–46, (2004)
8. Berger R. Thomas. Northern Frontiers, Northern Homeland: The Report of the Mackenzie Valley Pipeline Inquiry – Report Volume **1**. Otawa. Canada, (1977)
9. Bo-sin Tang., Siu-wai Wong., Milton Chi-hong Lau. Environmental Impact Assessment Review; **28** : 57–72, (2008)
10. BP Tangguh LNG. *Dokumen Analisis Mengenai Dampak Lingkungan (Amdal), Dokumen Final Andal lampiran 8.3.1,* (2001) [in Bahasa]
11. Canter L.W. *Environmental Impact Assessment.* Second Edition. New York: Inc Mc Graw-Hill, (1996)
12. Chaunjit Chanchitpricha, Alan Bond. 2013. Conceptualising the effectiveness of impact assessment processes, Environmental Impact Assessment Review **43**, 65–72, (2013)
13. Ciaran O'Faircheallaigh. Environmental Impact Assessment Review; **27**: 319–342, (2007)
14. Ciaran O'Faircheallaigh. Environmental Impact Assessment Review: **30**:19–27, (2010)
15. Death Carl. *Mengelola Pembangunan yang Berkelanjutan (Kemitraan, Protes, dan Kekuasaan Di KTT Dunia)*, Yogyakarta; Pustaka Pelajar, (2014) [in Bahasa]
16. Dewan Suku Tepera Yewena Yosu. Profil Kampung Yepase. *Tim Kerja DAS Tepera Yewena Yosu. Distrik Depapre-Refenirara* Kabupaten Jayapura, (2015) [in Bahasa]
17. Dewan Suku Moy. *Proposal Pengusulan Kampung Dosay Dosay Kabupaten Jayapura*, (2015) [in Bahasa]
18. Dinas Kebudayaan Provinsi Papua. *Peta Suku Bangsa Di Tanah Papua,* Jurusan Antropologi Universitas Cenderawasih (Uncen), Summer Institute of Linguistic (SIL), Dewan Papua (DAP) dan Badan Pusat Statistik (BPS) Provinsi Papua, (2008) [in Bahasa]
19. Dwi Narwoko J., Bagong Suyanto. Sosiologi: *Teks Pengantar dan Terapan,* Jakarta: Kencana Prenada Media Group, (2004) [in Bahasa]
20. Dwi Susilo K. Rachmad. *Sosiologi Lingkungan dan Sumber Daya Alam Perspektif Teori dan Isu-isu Mutakhir.* Yogyakarta: Ar-Ruzz Media, (2012) [in Bahasa]
21. Einar Leknes, Environmental Impact Assessment Review : **21**: 309–334, (2001)
22. Endraswara Suwardi. *Filsafat Ilmu.* Yogyakarta: CAPS (Center for Academic Publishing Service), (2012) [in Bahasa]
23. Erari K. Phill. *Tanah Kita, Hidup Kita (Hubungan Manusia dan Tanah di Irian Jaya sebagai PersoalanTeologis)*, Jakarta, Pustaka Sinar Harapan, (1999) [in Bahasa]

24. French P. David., Stephen Sutton., Susie J. Hennings J O Mitchell., Nicholas J. Wareham., Simon Griffin.,Wendy HArdeman., Ann Louise Kinmont. Journal of Applied Sosial Psychology: **35** (9):1824-1848, (2005)
25. Fuady Munir. *Teori-teori Dalam Sosiologi Hukum.* Jakarta: Kencana Prenada Media Group, (2011) [in Bahasa]
26. George Ritzer. *Eksplorasi Dalam Teori Sosial Dari Metateori sampai Rasionalisasi.* Yogyakarta: Pustaka Pelajar, (2013) [in Bahasa]
27. Gordon Walker. Environmental justice, impact assessment and the politics of knowledge: Theimplications of assessing the sosial distribution of environmental outcomes Environmental Impact Assessment Review: **30**: 312–318, (2010)
28. Hadi P. Sudharto. *Aspek Sosial AMDAL Sejarah, Teori dan Metode.* Yogyakarta: Gadjah Mada University Press, (2009) [in Bahasa]
29. Hadi P. Sudharto. *Resolusi Konflik Lingkungan. Semarang*: Badan Penerbit Universitas Diponegoro, (2010) [in Bahasa]
30. Hadi P. Sudharto. *Dimensi Lingkungan Perencanaan Pembangunan.* Yogyakarta: Gadjah Mada University Press, (2012) [in Bahasa]
31. Jan. J.J.M. Wuisman. *Teori dan Praktek Memperoleh kembali kenyataan supaya memperoleh masa depan.* Jakarta: Yayasan Pustaka Obor Indonesia, (2013) [in Bahasa]
32. Javier Toro., Oscar Duarte., Ignacio Requena., Montserrat Zamorano. Environmental Impact Assessment Review:**32**:107–117, (2012)
33. Javier Toro., Ignacio Requena., Oscar Duarte., Montserrat Zamorano. Environmental Impact Assessment Review: **43**: 9–20, (2013)
34. Jennifer L. Walker., Bruce Mitchell., S. Wismer, Environmental Impact Assessment Review: **20** : 513–535, (2000)
35. Jesper Persson. Environmental Impact Assessment Review: **26**: 605–613, (2006)
36. John A Sinclair., Laura Sims., Harry Spaling. Environmental Impact Assessment Review: **29**, (2009)
37. John scott. *Teori Sosial Masalah-masalah Pokok Dalam Sosiologi.* Yogyakarta: Pustaka Pelajar, (2012) [in Bahasa]
38. Juan R.Palerm. Environ Impact Assess rev; **19**:201–220, (1999)
39. Karl Polyani, The Great Transformation, Societies and economic systems, Boston: Beacon Press; Chapter IV, pp 43-44, (1944)
40. Lawrence W Neuman, *Metode Penelitian Sosial: Pendekatan Kualitatif dan Kuantitatif.* Jakarta: PT. Indeks, (2013) [in Bahasa]
41. Linda Stalker Prokopy, World Development Vol. **33**, No. 11: pp. 1801–1819, (2005)
42. Mansoben Johszua Robert. 1995. Sistem Politik Tradisional Di Irian Jaya, Jakarta, Lembaga Ilmu Pengetahuan Indonesia (Indonesian Institute of Sciences), (Seri LIPI-RUL: no.5) ISBN 979-8258-06-1.
43. Marion Hourdequin., Peter Landres., Mark J. Hanson., David R. Craig, Environmental Impact Assessment Review; **35**: 37–44, (2012)
44. Mathius Awoitauw, *Pemimpin Transformatif Berbasis Kearifan Lokal.* Jakarta. Indomedia Global Mandiri, (2015) [in Bahasa]
45. Nazir. Moh, *Metode Penelitian.* Bogor: Ghalia Indonesia, (2011) [in Bahasa]
46. Nicola Hartley., Christopher Wood., Environmental Impact Assessment Review: **25**: 319–340, (2005)
47. Obaidullah Nadeem., Thomas B. Fischer., Environmental Impact Assessment Review :**31**: 36–47, (2011)
48. Reed Isaac., Alexander C. Jeffrey, Sosiologi Kebudayaan dalam Bryan S.Tuner (ed): Teori Sosial Dari Klasik Sampai Postmodern. USA: Blackwell Publishing. Ltd. p.642-661, (2012)
49. Masyarakat Kementrian Lingkungan Hidup. *Pedoman Tatacara Inventarisasi Pengakuan Keberadaan Masyarakat Hukum , Kearifan Lokal, dan Hak Masyarakat Hukum yang Terkait Dengan Perlindungan dan Pengelolaan Lingkungan Hidup*, Jakarta. [in Bahasa]
50. Sairinen R, Environmental Impact Assessment Review; **30**: 289–292, (2010)
51. Salim Agus, *Teori dan Paradigma Penelitian Sosial (Dari Denzin Guba dan Penerapannya).* Yogyakarta: PT. Tiara Wacana Yogya, (2001) [in Bahasa]
52. Salim Momtaz, Environmental Impact Assessment Review: **25**: 33–45, (2005)
53. Schaefer T. Richard, *Sociology.* 12th ed. New York: McGraw-Hill, (2012)
54. Sidanius Jim., Pratto Felicia, *Sosial Dominance; An Intergroup Theory of Sosial Hierarchy and Oppression,* United Kingdom: Cambridge University Press, (1999)
55. Soemarwoto Otto, *Analisis Mengenai Dampak Lingkungan.* Yogyakarta: Gadjah Mada University Press, (2001) [in Bahasa]
56. Suranto Aw, *Komunikasi Sosial Budaya.* Yogyakarta: Graha Ilmu, (2010) [in Bahasa]
57. Thomas F. King, Environmental Impact Assessment Review: **20**: 5–30, (2000)
58. Watloly Aholiab, *Sosio Epistemologi Membangun Pengetahuan Berwatak Sosial.* Yogyakarta: Penerbit Kanisius, (2013) [in Bahasa]
59. York Richard., Dunlop E. Riley, *Sosiologi Lingkungan (Hidup) dalam George Ritzer* (ed): The Wiley-Blackwell Companion Sosiologi, UK:Blackwell Publishing Ltd, John Wiley & Sons Ltd. p. 851-879, (2013)

Water Footprint Assessment in the Agro-industry: A Case Study of Soy Sauce Production

Alfiana Aulia Firda[1,*], *Purwanto*[2]

[1]Master Program of Environmental Science, School of Postgraduate Studies, Diponegoro University, Semarang - Indonesia
[2]Department of Chemical Engineering, Faculty of Engineering, Diponegoro University, Semarang - Indonesia

Abstract. In terms of global water scarcity, the water footprint is an indicator of the use of water resources that given knowledge about the environmental impact of consuming a product. The sustainable use of water resources nowadays bring challenges related to the production and consumption phase of water intensive related goods such as in the agro-industry. The objective of the study was to assessment the total water footprint from soy sauce production in Grobogan Regency. The total water footprint is equal to the sum of the supply chain water footprint and the operational water footprint. The assessment is based on the production chain diagram of soy sauce production which presenting the relevant process stages from the source to the final product. The result of this research is the total water footprint of soy sauce production is 1.986,35 L/kg with fraction of green water 78,43%, blue water 21,4% and gray water 0,17%.

1. Introduction

Human activities are related with consume and contaminate a lot of water. On a global scale, most of the water use occurs in agricultural production, but there are also a volume of water consumed and polluted in the industrial and domestic. Water consumption and pollution can be associated with specific activities, such as irrigation, bathing, washing, cleaning, cooling and processing. The total water consumption and pollution are generally regarded as the amount of water requires a lot of activities and pollution evenly. A little attention is paid to the fact that, ultimately, total water consumption and pollution are relate to what and how much communities consume and to the structure of the global economy that supplies the various consumer goods and services. The issue of water consumption and pollution of the entire production and supply chain has been mentioned as being very important for the water management practices [1]. The virtual water content of a product is useful as an instrument to achieve water security and efficient water use, thus, tells about the environmental impact of consuming that product [2]. This concept evidences a complex net of relationship concerning to water and provides significant information for policy and economic actor, user and manager in order to planning and taking decision on this resources [3]

The water footprint has been introduces by Hoekstra (2002). This is a tool that assesses the amount volume of freshwater by considering both direct and indirect uses to produce the goods and services. It can also tell how much water is being consumed by a particular country or globally in a specific river basion or from an aquifer. The water footprint is expressed in water volume per unit of mass (m^3/ton or L/kg), unit of time (m^3/month or m^3/year). The water footprint is an aggregate and multidimensional indicator of water use, showing different sorts of water consumption as a function of space and time and differs from the traditional concept of water balance [1]. Water balance describes the flow of water in and out of a system considering only consumptive water use. Therefore, the water footprint is a way to describe virtual water volume and include the water volume offered by rainfall (green water), the water volumes polluted (gray water) and excludes water volume consumption (blue water) insofar as water is returned to where it comes from. The blue water refers to the consumption of fresh surface water or groundwater that evaporates. It is incorporated in the product, it does not return to the same catchment area, or indeed return, or indeed return but in the different time. The green water footprint is a volume of rainwater consumed during the production process. The different between the blue and green water footprint is important because the hydrological, environment and social impact, as well as the economic opportunity cost of surface and groundwater use for production. The gray water footprint does not represent of the polluted water volume, but it is indicating the level of pollution of the water, expressed in terms of the freshwater volume required to assimilate the existing load of pollutant below the threshold value or exo-toxicological end point [4].

The water footprint assessment is a helpful tool for the calculation of water use throughout the supply chain, providing valuable insights of the largest components and location of water consumption and the potential effects on local watersheds and future water availability

* Corresponding author: falfianaulia@gmail.com

to serving the need of communities, nature, companies, producers and suppliers [5]. The water footprint assessment framework aims to illustrate the full impact of water consumption during the life cycle of a product, from direct water extraction to water pollution [6]. Water footprint assessment is a developing method that is being increasingly applied to quantify water use, priorities reduction, assess sustainability and provide information to achieve sustainable efficient and equitable water use [7]. The water footprint consist of three components, there are blue, green and gray water footprint. The blue water footprint refers to surface and groundwater in a catchment area that evaporates during crop growth. The green water footprint refers to the rainwater that evaporated during crop growth. The gray water footprint is the volume of freshwater for assimilating waste water base on ambient water quality standards [1].

Water footprint can be relevant to environmental sustainability metric by addition of green water consumption in the analysis [8]. Nearly 80% of green water footprint are associated with global agriculture production [9], which includes wheat, animal products, cotton and bioenergy [10].

Soy sauce is a fermented product of vegetable or animal high protein in salt solution. The raw material of soy sauce are soybean, brown sugar and salt. There are 3 main stage in the process of soy sauce production is boiling and drying soybean, fermentation process and mixing and cooking soy sauce.

2. Materials and methods

The objective of the study was to assessment of the total water footprint from soy sauce production plant. Firstly, the water footprint from crop land of soy bean has been obtained from Research Report Series: The Water Footprint of Indonesian Provinces Related to The Consumption of Crop Products. The next step was to calculate of the water footprint for the operational and supply chain stages. The research method followed the systematic guidelines of the Water Footprint Assessment Manual by [1].

2.1. Materials

Research material and all data that obtain through field investigation. If domestic data is not available, a combination of available literature and provincial or national data was used international literature is referred. Field investigation was conduct in the soy sauce industry in Grobogan Regency.
The water footprint of the soybean crop in Central Java can be seen in table 1.

Table 1. Water footprint from the soybean crop in Central Java [11].

Green water footprint (L/kg)	1.558
Blue water footprint (L/kg)	418
Gray water footprint (L/kg)	-

From the data that obtained from field investigation, this industry uses 3300 L water to produce soy sauce of 360 kg. The water used for the production process is obtained from the PDAM (is a regional water supply company) Grobogan Regency. The gray water footprint estimate the volume of water required to dilute the pollutant that reaches the water body.

2.2.Methodology

Based on the Water footprint Assessment Manual: Setting the Global Standard [1], The water footprint was calculated, as follows:

$$WF_{total} = WF_{green} + WF_{blue} + WF_{gray} \quad (1)$$

The blue water footprint (WF_{blue}) is the volume of surface and consumed by the process. The green water footprint (WF_{green}) is the volume of rainwater evapotranspirated and incorporated by the process and the gray water footprint (WF_{gray}) is an estimate of the freshwater needed to dilute the load of pollutans. Real (blue and green) and virtual (gray) water volumes are summed to compute the total of water footprint. The gray water footprint of a product is an indicator of freshwater pollution that can be associated with the production of the product over its full suplly chain.
The water footprint of product is equal to the sum of the relevant process water footprints divided by production quantity of the product:

$$WF\ prod = \frac{\sum_{s=1}^{k} WF\ proc\ [s]}{P[p]} \quad (2)$$

in which $WF_{proc}[s]$ is the process water footprint of process step s (volume/time) and P[p] the production quantity of product p (mass/time).
The water footprint assessment manual provides a clear guideline for accounting approach of the assessment.

3. Result

Based on the equation 2, can be obtained the water footprint from soy sauce production process as seen in the table below.

Table 2. Water footprint of soy sauce process production in Grobogan Regency.

Green water footprint (L/kg)	-
Blue water footprint (L/kg)	6,94
Gray water footprint (L/kg)	3,41

The water footprint of soy sauce production consist 2 part, there are soybean crop stage and soy sauce process production. Using equation 1, obtained the total water footprint of soybean crop stage 1.976 L/kg is divided into green and blue water footprint. While the total water footprint on soy sauce production process amounted to 10,35 L/kg is divided into blue and gray water footprint. The total water footprint of soy sauce production is 1.986,35 L/kg with fraction of green water (78,43%), blue water (21,4%) and gray water (0,17%). Green water component account for more than 50% of the total water footprint for soy sauce production in Grobogan Regency, indicating that these crops are more reliant on natural rain water, making them more suitable for selection based on input criteria.

Table 3. Water footprint of soy sauce production.

Soybean crop stage				Soy sauce production process			
Green water footprint (L/kg)	Blue water footprint (L/kg)	Gray water footprint (L/kg)	Total water footprint of soybean crop stage (L/kg)	Green water footprint (L/kg)	Blue water footprint (L/kg)	Gray water footprint (L/kg)	Total water footprint of soy sauce production process (L/kg)
1558	418	-	1976	-	6,94	3,41	10,35

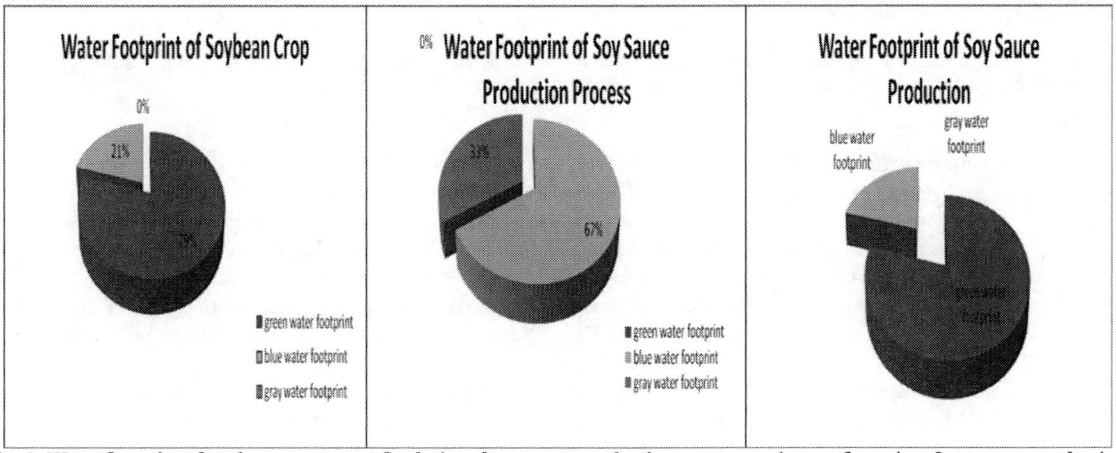

Fig. 1. Water footprint of soybean crop water foodprint of soy sauce production process, and water footprint of soy sauce production

4. Conclusion

The study shows the water footprint of soy sauce production in Grobogan Regency. It has show the volume of freshwater use per average soy sauce product yield. By calculating the water use over the whole production process including green water, blue water and gray water, so that awareness is given to the soy sauce industry to highlighting the water intensive stages that serving as a basic for formulating strategies in order to reduce water use. Water footprint assessment identified where the water was used in the soy sauce production and what type of water used.

The water footprint of soy sauce production reached 1.986,35 L/kg with fraction of green water (78,43%), blue water (21,4%) and gray water (0,17%). Green water component account for more than 50% of the total water footprint for soy sauce production in Grobogan Regency, indicating that these crops are more reliant on natural rain water

References

1. 1 A.Y. Hoekstra, A.K. Chapagain, M.M. Aldaya and M.M. Mekonnen. Water Footprint Network (2011)
2. 2 A. El-Sadek. Virtual water: an effective mechanism for integrated water resources management. *Agric. Sci*. E. **2** 248-261
3. 3 A.K. Chapagain, S. Orr. *J. Enviro. Management*. E. **90** 1219-1228 (2009)
4. 4 L. Lamastra. N.A. Suciu, E. Novelli, M. Trevisan. *Journal Science of Total Environment*. E **490**, 748-756 (2014)
5. 5 TCCC and TNC. Product Water Footprint Assessment: Practical Application in Corporate Water Stewardship. *The Coca-Cola Company* (2010)
6. 6 I. Herath, S. Green, R. Singh, D. Horne, S. van der Zijpp, B. Clothier. *J. Cleaner Production*. E **41**, 232-243 (2013)
7. 7 S.A. Munro, G.C.G. Fraser, J.D Snowball, M. Pahlow. J. Cleaner Production. E **135**, 668-678 (2016)
8. 8 M. Mekonnen and A.Y. Hoekstra. The green, blue and grey water footprint of farm, animals and animal product.
9. 9 J. Liu, A.J. Zehnder, H. Yang. 2009. *Water Resour. Res. 45* (2009)
10. 10 W. Garbens-Leenes, A.Y. Hoekstra, T.H. van der Meer. *Proc. Natl. Acad. Sci. U.S.A.* **106** 10219-10233 (2009)
11. 11 F. Bulsink, A.Y. Hoekstra, M.J. Booij, *Research Report Series* No. **37**. UNESCO-IHE Institute for Water Education (2009)

The existence of High Conservation Value Forest (HCVF) in *Perum Perhutani* KPH Kendal to support Implementation of FSC Certification

Sri Sulistyowati[1,*], *Sudharto* P. Hadi[2]

[1] Master Program of Environmental Science , School of Postgraduate Studies, Diponegoro University 50241, Semarang - Indonesia
[2] Faculty of Social and Political Science, Diponegoro University 50241, Semarang - Indonesia

Abstract. High Conservation Value Forest (HCVF) is the identification of High Conservation Values that are important and need to be protected. Under FSC certification mechanism, HCVF becomes one of Principles and Criteria to attain certification. In this study, we identify the existence of HCVF in Perum Perhutani KPH Kendal to support implementation process of FSC certification. Qualitative method was conducted through observation and secondary data from Perum Perhutani KPH Kendal. Data analysis showed through ecolabel certification, Perum Perhutani KPH Kendal has been identified HCVF area covering 2,715.5 hectares consists of HCV 1 until 6. Secondary Natural Forest (HAS) Subah and Kaliwungu for Ulolanang and Pagerwunung Nature Reserve buffer zone include as HCV 1.1, conservation area of leopard (*Panthera pardus melas*) and Pangolin (*Manis javanica*).for HCV 1.2, conservation area of lutung (*Trachypiyhecus auratus*) as endemic species for CITES App I and Critically Endangered species include as HCV 1.3, Goa kiskendo for bats species habitat include as HCV 1.4, regions of interest species for Deer (*Cervus timorensis*) and Kepodang (*Oriolus chinensis*) as HCV 2.3, Germplasm Protection Region/ KPPN area with high biodiversity include as HCV 3, river border area and water springs for HCV 4. While, utilization of firewood, grass for cattle fodder include as HCV 5 and 14 cultural sites include as HCV 6. From monitoring and evaluation of HCVF data, showed that in 2011-2015 the level of diversity for flora and fauna were increased.

1 Background

Indonesia's forests are the third largest tropical forest in the world after Brazil and Congo covering forest area of 1,860,359.67 km2 and second ranks for biodiversity after Brazil. Sustainable forest management in tropical countries using mandatory command and control approaches are seen as unsuccessful project by green consumers, proving that the environment of tropical forests is getting decrease, including forest in Indonesia. Forest degradation in Indonesia has occurred with a very large scale area. According to the analysis of Forest Watch Indonesia in 2011, rate of deforestation in Indonesia during these three periods decreased due to the depletion of Indonesia's forest area of 1.8 million ha / year within the period of 1985-1997, about 2.84 million ha / year in the period of 1997-2000 and approximately 1.51 million ha / year during 2000-2009 [1].

Perum Perhutani as a state owned company that manage forest products area in Java and Madura Island, take a responsibility to create a sustainable forest. Ecolabel certification is a forest management instrument with the aim to maintain the sustainability of forest resources and their functions. The implementation of ecolabel certification in Perum Perhutani KPH Kendal uses Forest Stewardship Council (FSC) scheme. This study aims to determine the existence of High Conservation Value Forest (HCVF) in Perum Perhutani KPH Kendal and management forms undertaken to maintain environmental protection. HCVF is an area that has one or more High Conservation Values (HCVs). HCV is something that has high conservation value at the local, regional or global level that includes ecological values, environmental, social and cultural values [2].

2 Research method

This research was conducted at Perum Perhutani KPH Kendal, Regional Division of Central Java. Overall research method used qualitative method. Data collection was obtained through field observation and literature review analysis using Perum Perhutani KPH Kendal documents. Data analysis was conducted using descriptive qualitative method.

* Corresponding author: iik.pht@gmail.com

3 Result and discussion

3.1 HCVF (High Conservation Value Forest) Identification result

HCVF was introduced by FSC in 1999 based on FSC principle number 9. The concept of HCVF is intended to identify the existence of high conservation values existing in an area (forest) and the establishment of it's management and monitoring plan to maintain and / or enhance the conservation value.

HCVF is an area that has one or more High Conservation Values (HCV). HCV is something that has high conservation value at the local, regional or global level that includes ecological values, environmental, social and cultural services. In the HCVF concept, high conservation values (HCV) classified into six HCV values. Forest areas could be classified to be of high conservation value if they possess one or more of the following characteristics [2]:

HCV1: Forest areas that has a globally, regionally and locally important concentrations of biodiversity values (e.g. endemic species, endangered species, refugia).

HCV2: Forest areas that has locally important in global, regional and local landscape area, within or with a management unit, in which most species populations, or all species naturally present in the region, in distribution patterns and natural abundance.

HCV3: Forest areas within or rare, threatened or endangered ecosystems.

HCV4: Forest areas that act as natural regulators in critical situations (e.g. watershed protection, erosion control).

HCV5: Forest areas that is importance to meet the basic needs of local communities (e.g. fulfillment of basic needs, health).

HCV6: Forest areas that are critical to the traditional cultural identity of local communities (important cultural, ecological, economic, religious areas identified with local communities).

According to the HCV identification in Kendal forest area of 2,715.5 Ha consist of HCV 1, HCV 2, HCV 3, HCV 4, HCV 5 and HCV 6.

3.1.1 HCV 1

HCV 1.1 In the form of Secondary Natural Forest (HAS) Subah (Bufferzone CA Ulolanang) of 362.9 ha and Kaliwungu HAS (bufferzone CA Pagerwunung) covering 254.4 Ha.

HCV 1.2 is leopard (*Phantera pardus melas*) and Pangolin (*Manis javanica*).

HCV 1.3 in the form of Germ Plasm Protected Area compartment 9 RPH Besokor area of 70.6 ha which is the habitat of leopard (*Panthera pardus melas*) habitat and HAS Subah area of 362.9 Ha as Lutung habitat (*Trachypithecus Auratus*)

HCV 1.4 in the form of Kiskendo Cave of 6.5 Ha as the habitat of temporary animals (bats).

3.1.2 HCV 2

HCV 2 identified in KPH Kendal known as HAS Kaliwungu area of 254.4 Ha, HAS Curam area of 405.2 Ha and compartment 9 RPH Besokor (KPPN) of 70.6 Ha. This area is habitat for interest species like Leopard, Kepodang and Deer.

3.13 HCV 3

HCV 3 is a high diversity area located in Kendal KPH, compartment 9 (KPPN) of 70.6 Ha, compartment 44 of 49.5 Ha, compartment 61a of 57,2 Ha and compartment 39 f of 12 Ha at BKPH Sojomerto.

3.14 HCV 4

HCV 4.1 in KPH Kendal identified as springs, located in 13 sites covering 6.1 Ha and River border area of 1.443, 4 Ha. Another HCV 4.2 is identified as Steep Secondary Natural Forest area of 405.2 Ha scattered in the forest area of KPH Kendal.

Table 1. HCVF areas in Perum Perhutani KPH Kendal

No	HCV Target	HCV Category	Information
1	HAS (Secondary Natural Forest) Subah	1.1 ; 1.3	Bufferzone Cagar Alam Ulolanang; Lutung (*Trachypithecus Auratus*) Habitat
2	HAS Kaliwungu	1.1 ; 2.3	Bufferzone Cagar Alam Pagerwunung; Deer Habitat
3	Compartment 98 (KPPN)	1.3; 2.3; 3	Leopard (*Panthera pardus melas* habitat; RTE Ecosystem
4	Copartment 44a	3; 4.2	RTE Ecosystem; steep area
5	HAS Curam (Steep)	2.3; 4.2	Kepodang habitat; erosion control area,
6	Compartment 61a (Curug sewu)	1.3; 3; 4.2	Lutung & Kepodang habitat; RTE Ecosystem; Steep area
7	Compartment 39	2.3; 3; 4.2	Kepodang habitat; RTE Ecosystem; Steep area
8	Compartment 67j (Kiskendo cave)	1.4; 2.3; 4.2	Bats & Kepodang habitat, HAS Curam (Steep)
9	Spring (13 location)	4.1; 5	Clean water resources
10	Riparian zone	4.2	Water reservoir
11	Agro forestry location, Livestock forage, Tree branch (rencek)	5	Food resources
12	Cultural sites (14 location)	6	Tomb cultural sites

Source: Monitoring and Evaluation of HCVF Document in KPH Kendal 2015

3.1.5 HCV 5

HCV 5 is related with the fulfillment of basic human needs. Some of these activities, contributed by Perum Perhutani KPH Kendal to fulfill the basic needs such as intercropping system, utilization of firewood by the community from the forest, the fulfillment of forage needs and the source of clean water from the spring.

3.1.6 HCV 6

Perum Perhutani KPH Kendal identified 14 cultural sites which used by the community for the fulfillment of religious and cultural needs. Cultural sites located in KPH Kendal such as cultural tombs. These cultural sites usually used by local communities and outside areas to perform rituals such as alms earth, religious tourism, certain rituals (ascetic).

Detail identification of HCVF areas can be seen in the Table 1.

3.2 Management of HCVF areas

In order to maintain and enhance the existence and function of HCV Areas, Perum Perhutani KPH Kendal undertakes some management such as:

- Maintenance of HCVF boundary.

-Counseling /socialization of illegal activities on HCVF areas

Counseling of illegal activities on HCVF areas in KPH Kendal such as illegal cutting prohibition, illegal forest fire prohibiton, wildlife hunting prohibition and wild cultivation and HCVF areas management to the stakeholders and communities who lives near forest areas.

Fig. 1. Socialization activities on HCV areas

-Collaborative management (joint patrol) with BKSDA Central Java especially in HCVF area near Nature Reserve location.

-Prohibition and information boards establishment on HCVF protection at strategic HCVF locations.

Fig. 2. Information boards near HCVF areas

-Maintenance of cultural sites

Perum Perhutani KPH Kendal has conducted maintenance in some cultural sites such as road improvement to site, information board establishment.

-Enrichment program conducted in protection areas in order to increase the diversity of vegetation by using local species. Enrichment is carried out in the area of Local Protection (KPS) and Secondary Natural Forest (HAS) which is in open area condition (open plek).

Table 2. Enrichment program in protected area

No	Year	Total Area	Species
1.	2011	124.9 ha	Rbc
2.	2012	69.8 ha	Kepuh, Trengguli, Salam, Beringin, Johar, Duwet, Bendo, Wungu
3.	2013	360.8 ha	Kepuh, Trengguli, Salam, Beringin, Johar, Bambu, Duwet, Bendo, Wungu
4.	2014	105.8 ha	Kepuh, Trengguli, Salam, Beringin, Johar, Bambu, Duwet, Bendo, Wungu
5.	2015	130.6 ha	Kepuh, Trengguli, Salam, Beringin, Johar, Bambu, Duwet, Bendo, Wungu

Source: Annually planning Document in KPH Kendal 2011-2015

3.3 HCV conditions

3.3.1 Diversity of Flora in HAS Subah, HAS Kaliwungu and HAS Curam, Local Protection (KPS) (HCV 1.1 ; 1.3 ; 2.3 ; 4.2)

Table 3. Diversity of vegetation

No	Parameter	HAS (Secondary Natural Forest) Subah				
		2011	2012	2013	2014	2015
1	Aboveground vegetation	1,945	1,920	2,031	2,366	2,121
2	Seedling	1,358	1,589	1,593	2,221	2,598
3	Sapling	0,941	0,958	0,982	1,433	2,700
4	Pole	0,340	0,854	1,026	1,424	1,065
5	Tree	0,216	0,941	0,998	1,195	1,337

No	Parameter	HAS (Secondary Natural Forest) Kaliwungu				
		2011	2012	2013	2014	2015
1	Aboveground vegetation	3,003	2,351	2,688	2,664	2,818
2	Seedling	1,826	2,011	2,183	2,480	2,578
3	Sapling	2,389	2,385	2,402	2,373	2,461
4	Pole	0,000	0,566	0,637	1,303	1,695
5	Tree	0,142	0,189	0,247	0,266	0,337

No	Parameter	steep > 40% (HAS Curam, Pt 9, Pt 61, pt 67j, pt.39f, pt 44)				
		2011	2012	2013	2014	2015
1	Aboveground vegetation	2,206	2,596	2,662	2,288	2,288
2	Seedling	3,584	3,215	3,429	3,446	3,492
3	Sapling	3,213	3,244	3,245	3,129	3,052
4	Pole	2,197	2,238	2,319	2,441	2,466
5	Tree	2,396	2,432	2,436	2,467	2,472

No	Parameter	Riparian zone				
		2011	2012	2013	2014	2015
1	Aboveground vegetation	2,796	3,023	2,570	2,788	2,788
2	Seedling	2,795	2,820	2,068	2,456	2,490
3	Sapling	2,384	2,539	2,560	2,533	2,084
4	Pole	2,341	2,447	1,647	2,374	2,383
5	Tree	1,663	1,735	1,737	1,450	1,460

Source : Monitoring and Evaluation of Biodiversity Document in KPH Kendal 2016

3.3.2 *Biodiversity of fauna in HAS Subah, HAS Kaliwungu and HAS Curam, Local Protection (KPS) (HCV 1.1 ; 1.3 ; 1.4 ;2.3 ; 4.2)*

Table 4. Diversity of fauna

No	Tahun	HAS Subah			HAS Kaliwungu		
		aves	mamalia	herpet	aves	mamalia	herpet
1	2011	2.300	0.000	0.598	3.150	1.369	1.630
2	2012	2.490	0.000	1.471	3.118	2.480	1.427
3	2013	2.604	0.000	1.101	3.122	2.533	1.810
4	2014	2.920	0.474	1.600	3.507	1.593	1.992
5	2015	2.675	1.265	1.838	3.002	1.110	2.041

No	Tahun	steep > 40% (HAS Curam, Pt 9, Pt 61, pt 67j, pt.39f, pt 44)			riparian zone		
		aves	mamalia	herpet	aves	mamalia	herpet
1	2011	3,142	1,504	2,332	3,093	1,310	1,346
2	2012	3,143	1,721	2,572	3,126	1,740	1,555
3	2013	3,452	1,811	2,575	2,922	1,659	1,824
4	2014	3,060	2,335	2,579	2,813	1,659	1,763
5	2015	2,864	2,074	2,530	2,642	1,647	1,195

Source : Monitoring and Evaluation of Biodiversity Document in KPH Kendal 2016

In areas which identified as HCV 1 to 4 above, HCV with good conditions are characterized by the increasing levels of different types of flora and fauna in some of these areas.

3.3.3 *Springs debit (HCV 4.1 ; 5)*

Table 5. Springs debit data

No	Spring name	debit (m3/second)				
		2011	2012	2013	2014	2015
1	Cokro tulung	0,00025	0,00009	0,0001	0,0005	0,0003
2	Kali alang	0,00046	0,0005	0,003	0,09	0,003
3	Kuthung mano	0,00022	0,0002	0,004	0,02	0,01
4	Tuk cabe	0,00002	0,00004	0,0004	0,0004	0,0001
5	Tuk kalibathang	0,00012	0,08	0,07	0,006	0,0002
6	Tuk pancuran	0,0016	0,003	0,008	0,0002	0,0001
7	Ngasman	0,004	0,005	0,0005	0,002	0,004
8	Petak (compartment) 27 a	0,00058	0,004	0,003	0,009	0,0009
9	petak 56 a	0,00023	0,0003	0,001	0,0001	0,0003
10	Sekenyes	0,00023	0,0002	0,001	0,004	0,0004
11	Tuk dandang	0,00014	4,9	0,007	0,0005	0,004
12	Tuk tumpang	0,0000033	0,00002	0,0006	0,001	0,002
13	Si rendeng	0,000018	0,0001	0,005	0,004	0,02

Source: Monitoring and Evaluation of HCVF Document in KPH Kendal 2015

3.3.4 *Tumpangsari (agroforestry), forage animal feed (HMT) and Rencek/ branches (HCV 5)*

From monitoring and evaluation result of HCVF KPH Kendal in 2015, showed that the community near forest can fulfill their life needs from intercropping, forage animal feed (HMT) and rencek. Annual data shows an increasing income, which can be seen in the following table:

Table 6. Data of Income from Tumpangsari (agroforestry)

Year	Comodity					
	Rice		Corn		Others	
	Total (TON)	Price (X1000)	Total (TON)	Price (X1000)	Total (TON)	Price (X1000)
2011	90,3	270.900	3.475	6.950.256	152	5.390.543
2012	132	396.000	3.357	7.384.647	569	3.368.059
2013	44	202.200	2.434	6.298.138	169	7.384.866
2014	81	328.850	784	7.329.509	584	785.255
2015	-	304.500	2.926	8.341.475	1.155	8.687.820
2016	53	157.500	2.788	8.687.820	645	2.597.091

Source : Annual report of social management in 2011-2016

Table 7. Data of Income from branches

No	Year	Volume (bundle)	Total (IDR)
1	2011	244.477	1.222.385
2	2012	279.922	1.804.977
3	2013	87.189	828.295.500
4	2014	84.237	768.170.000
5	2015	129.681	1.695.075.000

Table 8. Data of Income from forage animal feed (HMT)

No	Year	Forage animal feed	
		Volume (bundle)	Total (IDR)
1	2011	44.671	53.604.720
2	2012	489.663	587.645.000
3	2013	60.528	565.659.875
4	2014	82.234	764.182.500
5	2015	72.848	828.311.500

Source : Annual report of social management in 2011-2016

3.3.5 *Cultural sites*

The cultural sites identified in the KPH Kendal are still exists and used by the community for several activities such as alms, hermits and traditions of local communities known as khoul activities at the Mbah Demang Sulang site compartment 105e, Ds. Kutosari Kec. Gringsing, Batang regency, the path to the onion bone site has been done. This shows that the cultural sites managed by KPH Kendal are still used by the community.

4 Conclusions and suggestions

4.1 Conclusions

1. HCVF areas were identified in Perum Perhutani KPH Kendal are HCV 1.1 for HAS Subah and HAS Kaliwungu as bufferzone of Ulolanang and Pagerwunung Nature Reserves, HCV 1.2 is leopard (*Phantera pardus melas*) and Pangolin (*Manis javanica*). HCV 1.3 for HAS Subah, pt 9 KPPN RPH Besokor, Pt 61a as lutung (*Trachypithecus auratus*) habitat and Pt 9 KPPN RPH Besokor as Macan tutul (*Panthera pardus*) habitat and HCV 1.4 for Kiskendo cave as temporary habitat for some animals. HCV 2 for HAS Kaliwungu, HAS Curam (steep) Pt 9 RPH Besokor (KPPN) as spesies intereset habitat (Macan Tutul, Kepodang and Rusa), HCV 3 for Pt 9 (KPPN), Pt 44, Pt 61a and Pt 39 f with total area of 12 Ha BKPH Sojomerto, HCV 4 for springs in 13 locations, Local protected area (KPS) and HAS Curam (steep) which spread in the forest area of KPH Kendal, HCV 5 for tumpangsari (agroforestry), forage animal feed (HMT) and rencek/branches for firewood and HCV 6 for cultural sites which identified about 14 sites.
2. HCV area management activities conducted by Perum Perhutani KPH Kendal showed that the management could maintain and even improve the function and condition of HCVF area.

4.2 SUGGESTIONS

1. Maintaining and controlling the existence of HCVF area of *Perum Perhutani* KPH Kendal.
2. In order to protect HCVF areas, it is necessary needed to involve the participation from community around forest area.

References

1. Forest Watch Indonesia. *Potret Keadaan Hutan Indonesia Periode Tahun 2000-2009* (2001)
2. Konsorsium Revisi HCV Toolkit Indonesia. *Panduan Identifikasi Kawasan Bernilai Konservasi Tinggi di Indonesia.* Tropenbos International Indonesia Programme. Balikpapan, (2008)
3. Perum Perhutani KPH Kendal. *Laporan Monitoring dan Evaluasi Kawasan Bernilai Konservasi Tinggi (KBKT) tahun 2015* (2015)
4. Perum Perhutani KPH Kendal. *Laporan Monitoring dan Evaluasi Biodiversity KPH Kendal Tahun 2016* (2016)
5. Perum Perhutani KPH Kendal. *Laporan Kelola Sosial tahun 2011-2016* (2016)

Suitability Analysis For Scuba Diving To Develop Marine Tourism At Saebus Island, East Java, Indonesia

Putranto Wijaya[1,2,*], *Tri* Putra[1,2], *Fatra* Hidayat[1,3], *Chandra* Levraeni[1,2], *Mada* Rizmaadi[1,2], and *Ambariyanto* Ambariyanto[2,3]

[1]UKSA-387 Diving Club Diponegoro University, 50275 Jl. Prof. Soedarto SH, Tembalang Campus, Semarang, Indonesia
[2]Faculty of Fisheries and Marine Science, Diponegoro University, 50275 Jl. Prof. Soedarto SH, Tembalang Campus, Semarang, Indonesia
[3]Coastal Disaster Rehabilitation and Mitigation Center, Integrated Laboratory, Diponegoro University, 50275 Jl. Prof. H. Soedarto, SH, Tembalang, Semarang, Indonesia

Abstract. Indonesian government currently has policies to improve the performance of the tourism sector, including marine tourism. One of the attractions of marine tourism is the coral ecosystem especially through scuba diving activities. The purpose of this study was to determine the suitability of the coral ecosystem on Saebus Island, East Java, to find appropriate locations for scuba diving activities. Purposive samplings were done around the island to determine four stations which will be assessed through suitability analysis. Tourism Suitability Index was used to assess all stations for scuba diving activities. The result showed that all four stations were categorized as very suitable with the score: 85%, 85%, 85% and 83%, respectively. Several aspects that need to be improved and anticipated for diving at all stations are coral coverage and water current. These results suggest that there are several spots around Saebus Island that are suitable for diving site, and can be promoted as marine tourism destination.

1 Introduction

Coral reef ecosystem is known to have important role and high biological diversity. However, this ecosystem is also one of the most vulnerable marine ecosystems damaged by human activities. Several anthropogenic activities can caused marine organisms become endangered including those live in coral reef ecosystem [1].

Conservation action is one of the ways to prevent damage on coral reef condition. Non extractive activities such as properly manage marine tourism can also reduce the impact of damage caused by human exploitationon coral reefs [2]. At the same time, it provides solutions for conservation activities in the area. Therefore, with its high biological diversity and its uniqueness, coral reef ecosystem can be used as a marine tourism [3,4,5].

Saebus Island is known to have pristine coral reefs around the island. It is located in District Sapeken, East Java, which has 126 islands with an area of 2093.45 km^2. The utilization of marine resources by the community in the area is mainly on capture fisheries [6]. High pressure on marine resources through fishing activities will have an impact on the sustainability of marine resources [7]. Destructive fishing methods such as using bombs and cyanide poison at coral reef ecosystem are still common in the area [2]. These activities will damage the ecosystems and losing its diversity. Whilst as for other activities such as marine aquaculture and marine tourism is still not maximally utilized.

Coastal Saebus Island has a stretch of white sand beaches and clear waters. In addition, diversity and abundance of marine organisms are also high [6]. Clear waters and natural beauty under the sea with an abundance of coral reefs and very divers marine organisms on the island of Saebus is potential to be developed into a dive site. However, there is no scientific study on the possibility of developing marine tourism in the area. Several reports show suitability for marine tourism [8,9]. This study determines the suitability of Saebus Island coral reef ecosystem for diving activities by looking at biological and no biological parameters, as part of the development of marine tourism.

2 Methods

This research was conducted in November 2016 at Saebus Island, Sapeken, East Java. Determination of four survey locations was done by purposive sampling method [10]. Considerations taken was the existence of coral ecosystems by using Aster satellite image results. See Figure 1.

* Corresponding author: putrantokondang232@gmail.com

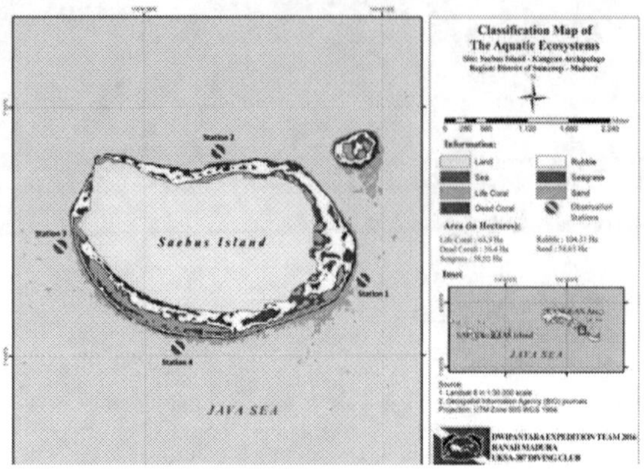

Figure 1. Map of research location

2.1. Field observation

2.1.1 Coral coverage and species

Coral coverage was measured with Line Intercept Transect (LIT) method by using 100 meters roll meter with precision in centimetre (cm). Percentage of coral cover was measured based on English et al. formula [11]. Coral reef measurement procedure begins with the installation of a 50-meter-wide line transect or line without intervals with the accuracy of the line transect scale in centimetres (cm) [12].

2.1.2 Fish species

Coral fish observation was done by using visual census method as the line with coral coverage observation. The measurement of reef fish data was done by recording the number of species of reef fish up to 2.5 meters to the left and 2.5 meters to the right of the transect line [13]. Then identify the fish using illustrated instruction manuals [14].

2.1.3 Water parameters

Water transparency was measured by using secchi disk and calculated by Effendi's formula [15]. Water current was measured using a floating ball with rope [16]. The depth of the water was measured by utilizing the depth gauge, part of diving console regulator tools [17]. Depth measurements were made when it reaches the bottom the waters.

2.2. Data analysis and suitability analysis

2.2.1. Scoring and assessment

Scoring and assessment were done by using all parameter data. Scoring was based on the value of the parameter, while for the assessment was given on each parameter with weight value of 5, 3, and 1 showing the effect of parameters towards suitability of diving tourism. The greater the influence of these parameters the greater the assessment given. The description given for the parameter is Very Appropriate (S1), Quiet Appropriate (S2), Appropriate Conditionally (S3) and not Appropriate (N). Scoring and assessment of the dive tourist suitability index are outlined in Table 2.

2.2.1. Suitability tourism index analysis

The index can be measured from the parameter value divided by max value of from tourism category [18].The value of Suitability Tourism Index (IKW) were divided into four suitability classes (see Table 1).

Table 1. Suitability matrix for diving tourism [18]

Parameter	Value	Category S1	Score	Category S2	Score	Category S3	Score	Category N	Score
Visibility (%)	5	>80	3	50–80	2	20–<50	1	<20	0
Coral coverage (%)	5	>75	3	>50–75	2	25–50	1	<25	0
Fish species (n)	3	>100	3	50–100	2	20–<50	1	<20	0
Coral lifeform (n)	3	>12	3	>7–12	2	4–7	1	<4	0
Water current (cm/s)	1	0 – 15	3	>15–30	2	>30–50	1	>50	0
Coral depth (m)	1	6 – 15	3	3–<6	2	>20–30	1	<3	0

Table 2. Suitability diving tourism categories [18]

Category	Index Value
Very Appropriate (S1)	>83-100%
Quite Appropriate (S2)	>50-83%
Appropriate Conditional (S3)	>17-50%
Not Appropriate (N)	17%

3 Results

3.1 Field observation

Suitability study for scuba diving sites, consist of several parameters. Parameters that measured in field observations were percentage of coral cover, coral diversity, diversity of fish species, visibility, current velocity, coral reef depth and coral reef condition. The result of parameters measurement obtained from field observation is presented in table below:

Table 3. Result of parameters measurement on Saebus Island waters

Parameters	Stations			
	1	2	3	4
Visibility (m)	10	10	10	10
Coral coverage (%)	68	63	58	60
Fish species (n)	36	32	30	49
Coral lifeform/genus(n)	13	17	16	15
Sea current (cm/s)	0,05	0,05	0,06	0,17
Coral depth (m)	10	10	10	10

Generally, the condition of Saebus Island waters still looks natural. Based on the results, the condition of the waters of Saebus Island has good clarity with visibility reaches up to ⊥ 10 meters in each station, and belong to S1 category. Very good for scuba diving activities that will make the diver comfortable. Current speed at station 1 - station 3 is classified S1 with speed 0.05-0.06 cm/s. While at station 4 pertained S2 because current velocity 0,17cm/s. Coral cover is quite good with the range 58-63% included in the category S2. The coral genus was found on average 15 genera of corals in each station classified as S1 category. The diversity of fish in the study sites was classified into S3 category in each station with the findings of 30-49 fish species in each station. The station that has the highest diversity of fish is the 4th station. The depth of coral at saebus island is very suitable (S1) for scuba diving activity with depth up to ±10 meters.

3.2 Suitability analysis

Suitability analysis for scuba diving tourism sites was analyzed based on parameter measurement results (Table 3) [18]. The results of the assessment of the suitability of scuba diving sites are presented in the table 4.

The suitability of the location obtained for diving in Saebus Island waters shows that for the four stations observed included in the category of 'Very Appropriate' (S1) with the value of 85%, 85%, 85%, and 83%. These value meets the S1 category because it is within the value range> 83-100% [18].

Table 4. Dive tourism index value

Parameters	Stations			
	1	2	3	4
Visibility (%)	15	15	15	15
Coral coverage (%)	10	10	10	10
Fish species (n)	2	2	2	2
Coral lifeform/genus (n)	9	9	9	9
Sea current (cm/s)	3	3	3	2
Coral depth (m)	3	3	3	3
Total	49	49	49	48
Suitability Tourism Index (%)	85	85	85	83
Category	S1	S1	S1	S1

4 Discussions

4.1 Waters current

Sea currents in Saebus Island ranged from 0.05 - 0.17 m/s. This current condition affects the level of safety and comfort in tourism activities. Based on these results, it is included to very appropriate category. The current belonging to a suitable state for diving activity ranges from 0-15 cm/s [19]. Flow is a limiting factor in the suitability of dive tourism that affects the movement of divers to regulate its balance. According to [18], the current with relatively weak flow velocity is ideal for diving tourism.

4.2 Visibility

Transparency is one parameter for diving activity which affects the visibility of the water. The result shows that all stations have water visibility value of 10 and can be categorized as good aspect. The category for visibility that suitable for diving area is good if the value between 80–100% [18]. Therefore, all stations were scored 3 on the four stations. The high scores can be affected by shallow water depths that have been penetrated by sunlight into the water column.

4.3 Coral Reefs Areas

Based on the field observation, it was found that coral cover value from the four stations ranged from 58% - 68% so it was rated 2 for each observation station. The scores are categorized in appropriate [18] said, that coral coverage between 50 – 70% categorized as appropriate for diving activity. The species that dominate on Saebus island are *Acropora*sp.,*Porites* sp. and *Pocillopora* sp.The depth of coral reefs in Saebus Island were categorized into shallow waters with the fringing reefs characteristic. Generally, this type is located in the edge of the island [21].The results showed that the depth of coral reefs at all four stations were classified into two grades. Stations 1 and 3 have a score of 0 that fall into the inappropriate category, while stations 2 and 4 scored 3 which can be classified as very suitable. A very suitable depth for diving is located on stations 2 and 4 with depths of 9 and 10 meters. The depth of coral reefs suitable for diving activities ranged from 6 to 15 m [18]. Diving at a location with a strong current is not recommended but, if forced to dive then the divers involved must be experienced [22].

4.4 Coral fish

The number of species of reef fish found in the waters of Saebus Island is categorized as Quite Appropriate (S2). This is because the number of species of fish found more than 20 species. The types of reef fish found in each research station were Pomacentridae and Abudefduf. Report shows that Pomacentridae fish is very common in coral reefs [23]. Four Chaetodon fishes were found at station 3. These fish are indicator of coral reef health. This is in accordance with the value of coral cover at station 3 i.e. 60% which is included in the good category [18]. Fish from Chaetodontidae family are coral health indicator fish with habit of eating coral polyp as food source [23].

4.5 Determination for tourism suitability

Parameters measured in field observations are taken into consideration to assess how appropriate the location to scuba diving sites refers to conformity assessment [18]. The results show that the waters of all stations at Saebus Island can be included in the suitability class S1 (Very Appropriate) with the Suitability Diving Tourism Index at 85%, 85%, 85%, and 83%, respectively.

Determination of these values can not be separated from the relationship between the parameters of the suitability of diving tourism. Some reports show that coral coverage affects the number and biodiversity of coral fish [24, 25]. It is evident that each station has an appropriate coral cover level proportional to the number of reef fish available. Although each parameter affects each other in the suitability of the dive tour, not every parameter will be well correlated. Stations 1, 2, and 3 have a rating of the suitability of diving tourism index of 91% and classified as very appropriate (S1), but these locations have coral cover score categorized as appropriate (S2). This value indicates that the coral cover in the area is quite good. As for station 4, has a value of 89%. The conformity index value at station 4 is lack of coral cover parameters and current velocity. Each value belongs to a category quite appropriate (S2). It should be noted that for diving activities with existing current conditions there is a need of assistance in order to improve safety diving [26].

5 Conclusions

Saebus Island waters is suitable for scuba diving area and can be develop for marine tourism. The value of suitability index for diving tourism in every station are 85%, 85%, 85%, and 83%. So categorized in very Appropriate Category (S1).

Acknowledgements

This research was held by UKSA-387 Diving Club Diponegoro University on Dwipantara V Expedition: Kangean Archipelago "Jelajah Pesona Bahari Tampuk Timur Madura". We thanks to UKSA-387 Diving Club families who have fully supported and big appreciation to the Dwipantara V Expedition Team who has worked hard in this project. This project depends on logistical and field support from a number of contributed peoples. We thanks to the Department of Fisheries and Marine East Java Province, Indonesia. Mr. H. Muhammad Ali, Mr. Osnan, Saur-Saebus village and Sapeken Sub-district governments, as well as the entire community of Saebus and Sapeken Islands for supports this research.

References

1. Ambariyanto. In IOP Conference Series: Earth and Environmental Science. **55**,1: 012002. doi: doi:10.1088/1755-1315/55/1/012002 (2017).
2. B. Tamam, A. Apri, and Saleh. J. Kelautan, **6**,2: 120-127. DOI: http://dx.doi.org/10.21107/jk.v6i2.785 (2013).
3. N. Riyan, Thesis. Fakultas Perikanan dan Ilmu Kelautan Institut Pertanian Bogor (2007).
4. H. Apriyanto, A. Wahyu and Umroh. Akuatik, **10**,2: 30-38 (2016)
5. M.H. Natha, A. Tuwo, and F. Samawi, J. Sains & Teknologi, **14**,3: 259 – 268 (2014).
6. Sutanto, M.A., Bedjo, R., and Elisabeth, C.Y. Petra University, Surabaya, (2013)
7. V. Siwat, A. Ambariyanto, I. Widowati, AACL Bioflux, **9**,4:915-922 (2016).
8. P. Subardjo, and N.I. Hidayat, Ilmu Kelautan: Indonesian Journal of Marine Sciences, **12**,4:211-218. DOI: https://doi.org/10.14710/ik.ijms.12.4.211-218. (2007)
9. A. Fahrudin, and Y. Wardiatno, *Research Report.* Intitut Pertanian Bogor, Bogor. (2010)
10. Sugiyono. Alfabeta, Bandung, 330 p (2011)
11. English, S., Wilkinson, C., and Baker, V. Australian Institute of Marine Science, Townsville, 368 p (1994)
12. S. Adibrata, Jurnal Kelautan, **6**,1: 19-28. DOI: http://dx.doi.org/10.21107/jk.v6i1.829 (2013)
13. Dedi. Thesis. Departemen Pengelolaan Sumber Daya Perairan, Universitas Bangka Belitung, Bangka (2012).
14. G.R. Allen, Periplus Perth, Australia, (1997)
15. H. Effendi, Living Aquatic Resources Management, **292**. DOI:http://repository.ipb.ac.id/handle/123456789/79927 (2003)
16. H. Huttagalung, and A. Rozak. Puslitbang Oseanologi-LIPI, Jakarta, 35 pp. (1997)
17. B.A. Adi, A. Mustafa, and R. Ketjulan. Jurnal Mina Laut Indonesia, **01**,01: 1-13 (2013)
18. F. Yulianda, Makalah Seminar Sains, **21** (2007).

19. Supriharyono. Pustaka Pelajar, Yogyakarta, 428 p (2007).
20. L.S.H. Yulius, Ramdhan, M., Arifin, T., and Purbani, T. Globe, **15**,2 : 129-136. DOI: http://dx.doi.org/10.24895/MIG.2013.15-2.%25 x (2013).
21. Suharsono. LIPI, Jakarta, 372 pp. (2008)
22. M. Philips, University of Queensland, Australia, 126 p. 2006
23. F. Kurniawan, R.F.D. Destilawaty, S.G.A. Dedi, and Tabunan, J.M. Prosiding Seminar Nasional Perikanan dan Ilmu Kelautan,V:432-442 (2015).
24. J.D. Bell, and R. Galzin, Marine Ecology Progress Series, **15**,3: 265-274 (1984).
25. G.P. Jones, M.I. McCormick, Srinivasan, M. and J.V. Eagle, Proceedings of the National Academy of Sciences of the United States of America, **101**,21: 8251-8253 (2004).
26. Abidin, S.Z., and Mohamed, B. A.SHS Web of Conferences, **12**, 01093. DOI: 10.1051/ shsconf/20141201093 (2014).

Diversity of Coral Fish At Saebus Island, East Java, Indonesia

Siti Fatimah*[1,2], *Tri* Widya Laksana Putra[1,2], *Putranto* Kondang[1,2], *Suratman*[1,2], *Larossa* Gamelia[1,2], *Hendry* Syahputra[1,2], *Rahmadayanti*[1,3], *Mada* Rizmaaadi[1,2], *Ambariyanto* Ambariyanto[2,4]

1UKSA-387 Diving Club Diponegoro University, 50275 Jl. Prof. Soedarto SH, Tembalang Campus, Semarang, Indonesia
[2]Faculty of Fisheries and Marine Science, Diponegoro University, 50275 Jl. Prof. Soedarto SH, Tembalang Campus, Semarang, Indonesia
[3]Faculty of Public Health, Diponegoro University, 50275 Jl. Prof. Soedarto SH, Tembalang Campus, Semarang, Indonesia
[4]Coastal Disaster Rehabilitation and Mitigation Center, Integrated Laboratory, Diponegoro University, 50275 Jl. Prof. H. Soedarto, SH, Tembalang, Semarang, Indonesia

Abstract. Coral reef ecosystem is known as an important place to live various types of fish, where coral conditions will affect the diversity and abundance of the fish. In healthy coral reef ecosystems generally can be found many types of fish with high density. This research aims to investigate the diversity and abundance of coral fishes at Saebus Island, East Java. The observation conducted at 4 stations, according to cardinal point by UVS (underwater visual census) methods with belt transect with the visibility of 2,5 m horizontally, and 5 m vertically. The length of transect was 100 m parallel with coastline, with the area of observation is 500 m2. The censuses were conducted at 2 different depths (3 and 10 m). This study found 70 kinds of coral fish originated form 20 family at all stations. These fishes were from 3 different fish categories i.e. 7 target fishes, 13 indicator fishes, and 50 major fishes. Three different fishes that dominated target fish, indicator fish and major fish were *Epinephelus fasciatus*, *Chaetodon baronessa* and *Aulostomus chinensis*, respectively. There was similar value of fish diversity index at two different depths which were 3.635 and 3,623. While uniformity index at the depth of 3m was 0.153 and at 10m was 0.217, and domination index at the depth of 3m was 0.11 and at 10m was 0.167. These values suggest that diversity of coral fish at Saebus island can be categorized as high diversity.

1 Introduction

Coral reef ecosystem has important and diverse benefits, ecologically and economically. The benefits are include as living area for fishes, as the object of marine tourism, and for research as well as utilization of aquatic organisms existing on coral reefs [1]. Disturbances of this ecosystem, such as destructive fishing, boat anchor, pollution, diseases, and climate change, will gives detrimental effects on the ecosystem [2,3,4,5,6]. In the condition of worst scenario it will bring about changes in biodiversity within the ecosystem and leading to extinction [7]

The existence of coral reef as a coastal area ecologically functioned as a habitat for many organisms, including reef fishes. The function of coral reef for these fishes can be as their shelter [8], spawning, nursery grounds [9] and food sources [10]. Therefore, the existence of reef fishes community become important indicator for health level coral reefs ecosystems [11].

Saebus Islands administratively located in Sumenep Regency. Geographically located between $6^0 50'$ SL-$115^0 25'$ WL. Kangean Islands consisted of several islands, among of them are Sadulang Besar Island, Sadulang Kecil Island, Pagerungan Besar Island, Pagerungan Kecil Island, Sapeken Island, Sepanjang Island, Saubi Island, Paliat Island, Sepapan Island, Sasiil Island, Sepangkur Island, Sabuntan Island, Saebus Island, Saor Island, and Mamburit Island [12].

Saebus Island based on Bakorsurtanal (2009) survey result have the most large area of coral reef among other 7 district on East Java, with the area of 40.083,71 Ha from the total of 49.257,98 Ha. Saebus Island have a good coastal ecosystems as pictured by the growth of sea grass ecosystems, mangrove ecosystems and large area of coral reefs. The livelihoods of Saebus people are fishermen and farmers. Fishermen in Saebus Island take advantage of natural resources, especially fish as a source of income. Many fisherman catch fish by destructive fishing methods and destroying coral reef ecosystems [13].

Saebus has largest area of coral reefs, with high rate of fishing from local residents, but ecological study for reef fish information still hard to get. Ecological study for reef fishes such as reef fish community, including abundance, diversity, uniformity and dominancy of coral reef is needed to justify that coral reef ecosystem conditions is

* Corresponding author: sitifatimahh@hotmail.com

optimally functioned. This study investigates the diversity, uniformity and dominancy from reef fish abundance at Saebus island.

2 Methods

This research was conducted at Saebus Island, Sapeken, Sumenep, Madura, East Java (Figure 1). The fish surveys were done at four stations from 4 cardinal point of the Island.

2.1 Data collection

Coral reef condition was determined by using Line Intercept Transect (LIT) observation [14] at 4 different stations at the depth of 3m and 10m by UVS (Underwater Visual Sensus) method [15] Surveys were done by scuba diving for 100 m long with the radius of belt transect 2,5 m both to the left and right sides and 5m upright. Therefore, total area of survey was 500 m^2 for each station and each water depth.

Coral reef data were collected by measuring the diameter and record coral genus found along transect lines. Percentage of coral reefs community cover was obtained using Gomez and Yap formula [16]. All fish found were grouped as target, indicator and major fishes.

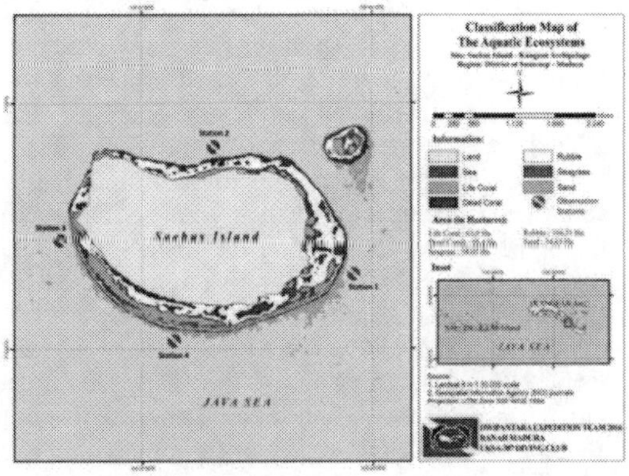

Fig. 1. Four stations for survey reef fishes at coral reef ecosystems around Saebus island waters

2.2 Data analysis

The number of fishes was estimated by equation [17]. Correlation value between coral coverage and fish abundance was calculated according to Pratisto [18]. Description of a population can be determined by the number of individuals of each species in the community [19]. The structure of the population can be seen by value of diversity index, which is categorized based on EPA [20]. Community balance measurement was calculated by uniformity index (E) ranging from 0 to 1 [21]. Dominancy index represent dominancy rate of one area and the value ranging from 0 to 1 [19].

3 Results

3.1 Fish abundance

Total number of reef fish found was 1633 individuals within total area of 4000 m^2 at all 4 stations at depth of 3 and 10m, dominated by major fish (1500 indv.) compared with indicator fish (73 indv.) and target fish (60 indv.) see Table 1. There were 7 species of target fish from 3 families i.e. Serranidae, Haemulidae and Lutjanidae. Eleven species of indicator fish that categorized as Chaetodontidae. While for major fish, there were 50 species of fish from 15 families.

There are differences on the number of fish found in each station. More than half of target fish were found at station 4, and only one fish at station 1. Similarly, no indicator fish was found at station 3, while at other station the number was quite similar. On the other hand, major fishes were dominated at all stations and the number was much higher compared with other fishes

Table 1. Total of individual of target fish, indicator and major at each research stations on Saebus Island

Station	Target	Indicator	Major
1	1	26	352
2	15	26	438
3	10	0	232
4	34	21	478
Total	**60**	**73**	**1500**

3.2 Correlations of fish abundance and coral coverage

There was different correlation of reef fish abundance and coral coverage percentage at two water depth. See Table 2.

Table 2. Abundance Value (N) of Coral Fish and Coral Coverage

Station	Depth (m)	N	Coral Cover (%)
1	3	0.510	57.38
	10	0.448	78.08
2	3	0.404	62.59
	10	0.354	61.81
3	3	0.196	74.56
	10	0.288	44.80
4	3	0.434	62.89
	10	0.632	52.45

Based on the correlation analysis between fish abundance value and coral reef coverage percentage on the depth of 3 m have negative relationship. This result shown by the negative regression value which is -0,9875 (Figure 2). The least fish abundance value is on station 3 on the depth of 3 m. At this station and at the same depth shown high coral reef coverage which is 74.56%.

Correlation value between fish abundance and coral reef coverage percentage on the depth of 10 m at each station have value of R= 0.0164 (Figure 3). This correlation value is very low, which mean the correlation rate of 2 variables only shown on the 0.0164 interval. While the positive value indicates that the higher fish abundance the higher percentage of coral coverage [22].

This result is not in accordance with other studies which generally found that the higher the percent cover of corals, the higher the abundance of the fish [23]. There are two possibilities in explaining this result. First, the number of samples probably too small (only four points). Secondly, abundance of fish is influenced by many aspects including the complexity of the habitat [24].

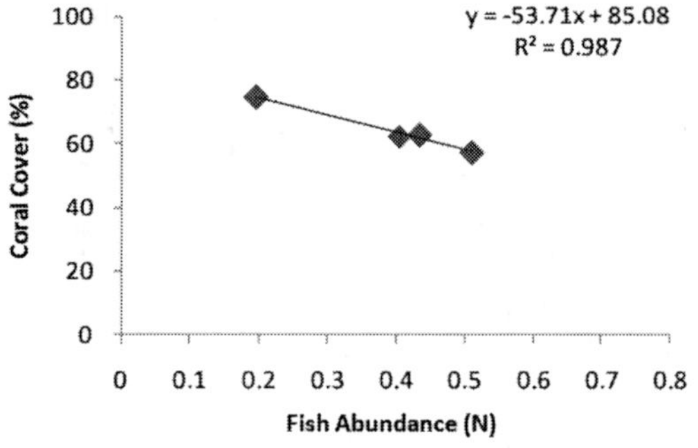

Fig. 2. Correlation of coral reef fish abundance and coral coverage at 3 meters depth

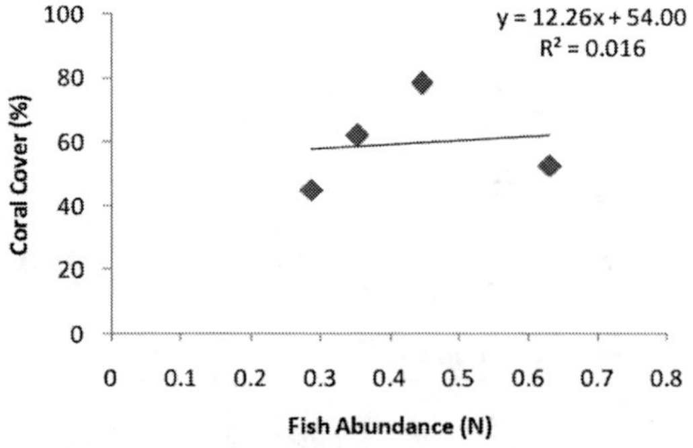

Fig. 3. Correlation of coral reef fish abundance with coral coverage at 10 meters depth

3.3 Ecological index of reef fish

The value of coral reef fish diversity indices (H') from four stations, although they are not similar but all lay in medium category.

Table 3.Saebus Island's Ecological Indices

Station	H'	E	D
1	3.054	0.273	0.224
2	2.727	0.166	0.112
3	2.782	0.174	0.118
4	3.202	0.126	0.101

The values of uniformity indices were quite similar and they are in low category since the values are all below 0.4. While based on the result of the dominant indices where no value is close to 1, which means that no individual dominates in the population. See Table 3.

4 Discussion

4.1 Reef fish abundance

The smallest number of individuals of target fish was found at station 1. It is suspected that the coral reef fishing activities in Saebus Island is relatively intensive especially in station 1. The abundance of target fishes, a group of small fish predator such as Serranidae, Lutjanidae, Lethrinidae and Carangidae, became low due to intensive fishing activities [25]. Structural changes of target fish communities show variations in environmental factors because the target fish have properties that are integrated with the coral reef condition [26, 27,28].

The very low appearance of indicator fish, in station 3, can be caused by the hydro-oceanography factor of the waters causing the instability of the water balance. The indicator fish found was from Chaetodontidae family which is known as important and widely studied fish family in coral reef[29].Fish belonging to the Chaetodontidae family group have strong associations with coral and can be used as an indicator of coral health [30, 31]. Total number of Chaetodontidae species have significant correlation to hard coral coverage [32]. However, not at all Chaetodontidae fish is hard coral eaters (scleractinian coral), there are also octocoral (soft corals eaters) such as *Chaetodon melannotus* [33]. This observation found 19 indv. which is dominated by the genus Chaetodon.

A total of 50 major fish species were found during the observation comes from 15 families. The families considered to be significant in the group are the Pomacentridae, Labridae and Pomachantidae groups. Most of fish have relatively small body size and in nature play an important role in the food chain, especially as food supply for carnivorous fish. The study also found some of types of fish have beautiful body colour so potentially exploited by ornamental fish enthusiasts or for diving attractions.

4.2 Fish abundance value and coral coverage correlation

The correlation between the complexity of coral reef types and the diversity of fish communities indicates that the structure of the reef fish community can be affected by the physical complexity of the substrate [34]. Based on correlation analysis of reef fish abundance value with coral cover percentage at depth of 3 meter in all station have correlation relationship. The phenomenon of low coral reef abundance in station 3 of 3 m depth can be caused by hydro-oceanographic condition of waters causing the instability of water balance [22]. On the other hand, this depth is an area that can be accessed directly by fishermen, but the behaviour of local communities in fishing is still not good, such as using potassium. The factors of behaviour to fulfil economic needs are exploitation without considering sustainability of the fish, resulting in abundance of fish that is inversely proportional to coral cover. The activities of local communities gave negative impact to the balance of coral reef ecosystems including fish in it [35]. The less complex habitat in coral waters will provide smaller ecological niches for associated marine organisms [36]. At this location the ecological niche areas are relatively small so that fish abundance has a negative correlation on coral cover. Abundance of fish is influenced by many aspects including the complexity of the habitat [24].

A positive correlation between the abundance of reef fish and the percentage of coral coverage is shown in the depths of 10 m. There is a positive linear relationship between the percentage of live coral coverage and the number of reef fish species [37,38]. Reef fish have narrow ecological niches so more species can inhabit (accommodate) coral reefs. Consequently, certain types of reef fish are limited and localized only in certain areas of coral reefs [39]. Luckhurst and Luckhurst [38] examined relationship between reef fish community parameters and substrate diversity, and abundance of coral species indicated that significant relations were found.

4.3 Diversity, uniformity and dominancy indices

The diversity of reef fish in the Saebus Island are medium diversity category. These results suggest that the surrounding environmental conditions are in stable condition, both the water quality and the coral reef ecosystem. Diversity and abundance of reef fish is increasing with increasing complexity of coral habitat [41]. Reef fish have various roles, ie, as plant-eaters, predators, and also play a role in calcium erosion to produce sediment [42]. However, diversity will significantly decline if the ecosystems have to face destructive events and activities [7].

The result of uniformity index is E <0,4 which mean low level of uniformity. This illustrates that ecosystem equilibrium occurs in the environment because of the high diversity of fish, there is no high level of fish dominance or abundance of one species of reef fish.

5 Conclusion

It has been found the number of individual reef fish 1633 individual which comes from 7 species of target fish of 3 families. Butterfly fish from Chaetodontidae found 13 species and major fish as many as 50 species from 15 families. The largest abundance of reef fish was the major fish group.

The correlation between abundance of reef fish and the percentage of coral coverage shows a negative correlation at 3 m and positive correlation at 10 m depth. This is probably caused by hydro-oceanographic conditions causing instability of aquatic conditions. The diversity of reef fish can be categorized as medium with the value of 2,272 - 3,202. The Uniformity Index (E) value of 0.273-0.126 categorized as low category. The value of the dominance index (D) 0.224-0.112 indicates there is no dominance between fish species in Saebus Island.

Acknowledgements

This research was held by UKSA-387 Diving Club Diponegoro University on Dwipantara V Expedition: Kangean Archipelago "Jelajah Pesona Bahari Tampuk Timur Madura". We thanks to UKSA-387 families who fully supported and big appreciation to the the Team who has worked hard in this project. This project depends on logistical and field support from a number of contributed peoples. We thanks to the Department of Fisheries and Marine East Java Province, Indonesia. Mr. H. Muhammad Ali, Mr. Osnan, Saur-Saebus village and Sapeken Sub-district governments, as well as the entire community of Saebus and Sapeken Islands for supporting this research.

References

1. Sukandarrumidi. Yayasan Pustaka Nusantara, Yogyakarta. (2009)
2. *E.N. Edinger, J. Jompa, G.V. Limmon, W. Widjatmoko, and Risk, M.J. Marine Pollution Bulletin,* ***36****, 8: 617-630 (1998)*
3. J.A. Maynard, K. R. N. Anthony, S. Afatta, L. F. Anggraini and D. Haryanti, Ambariyanto,

A. Pacific Conservation Biology, **14**, 4: 242-243 (2009).

4. Hoegh-Guldberg, O., Hoegh-Guldberg, H., Veron, J.E.N., Green, A., Gomez, E.D., Lough, J., King, M., Ambariyanto, A., Hansen, L., Cinner, J., Dews, G., Russ, G.,Schuttenberg, H.Z., Peñaflor, E.L., Eakin, C.M., Christensen, T.R.L., Abbey, M., Areki, F., Kosaka, R.A., Tewfik, A., Oliver, J. WWF Australia. (2009)
5. Ambariyanto. Ilmu Kelautan, **16**, 1: 30-34 (2011)
6. A. Sabdono, O. K. Radjasa, A. Ambariyanto, A. Trianto, D.P. Wijayanti, D. Pringgenies, Munasik. International Journal of Zoological Research, **10**, 2: 1-10 (2014)
7. Ambariyanto. IOP Conference Series: Earth and Environmental Science. 55, 1: 012002 (2017)
8. A. Yanuar, and Aunurohim. J. Science and Art, **4**,1: E19-E24 (2015).
9. F. Setiawan, T.B. Razak, E. Idris, J. Trop. Mar. Sci. Technol.**5**,2: 377-390 (2013)
10. I Manembu, L. Adrianto, D.G. Bangen, and F. Yulianda. J. Fisheries Mar. Trop, **8**,1:28-32 (2012)
11. T.F. Hourigan, C.T. Timothy, and E.S. Reese, *Marine organisms as indicators*. pp: 107-135. Springer New York. (1988)
12. Djojoprajitno, S. Bul. Kangean Nyiur Melambai. (2005)
13. *B. Tamam, J. Kel.* ***6****,2:1907-9931 (2013)*
14. S. English, C. Wilkinson, and V. Baker, Australian Institute of Marine Science, Townsville,368 pp. (1994).
15. I. Hill, and C. Wilkinson. Australian Institute of Marine Science, Townsville, 117 pp (2004)
16. E. D. Gomez, and H. T. Yap, UNESCO Regional Office for Science and Technology for South East Asia, Jakarta, (1988)
17. Suharsono. Oceanology Research and Development Center - LIPI, Jakarta, (1995)
18. A. Pratisto, Elex Media Komputindo,Jakarta, 282 pp. (2005)
19. Odum, E.P. W. B. Sounders Company Ltd, Philadelphia, (1971)
20. EPA. US Environmental Protection Agency, Wangsinton, 595 pp.(1973)
21. J.E. Brower, and Zar, J.H.WM Brown Comp. Pub., Dubuque-Lowa, 194 pp. (1977)
22. Mujiyanto, B.I. Purnawati, Hartati, S.T. Proc. Nat. Forum of Fish Res.**II**:KR-13 (2009)
23. J.D. Bell, and R. Galzin, *Influence of live coral cover on coral-reef fish communities. Marine Ecology Progress Series,* pp.265-274, (1984)
24. C.M. Roberts, and R.F. Ormond, 1987. Habitat complexity and coral reef fish diversity and abundance on Red Sea fringing reefs. Marine Ecology Progress Series, pp.1-8.Russ, G.R.and A.C. Alcala. Mar. Ecol. Prog. Ser, **10**: 13-27. (1989)
25. J.H. Choat, D.R. Bellwood, Department of Zoology University of New Hamshire Durham. p. 39-47 (1991).
26. A.J. Dartnall, M. Jones, Australian Institute of Marine Science,Townsville, 166 pp. (1986)
27. R.H. Kuiter, Gramedia Pustaka Utama, Jakarta, (1992)
28. D. McB. Williams, and A.I. Hatcher. Mar. Ecol. Prog. Ser,**10**: 239-250 (1983)
29. Reese, E.S. Bull. Mar. Sci, **31**: 594-604. (1981).
30. T.F. Hourigan, T.C. Tricas, E.S. Reese, New York: Springer – Verlag 107 – 135 p (1988)
31. White, A.T. Proc. 6th Int. Coral Reef Symp.**2** (1988)
32. P.M. Alino, P.W. Sammarco, J.C. Coll, 6th Int. Coral Reef Symp. **3:**31-36. (1988)
33. J.D. Bell, and R. Galzin, Mar. Ecol. Prog. Ser., **15**: 265-274 (1984)
34. M. Adrim, S.A. Harahap, K. Wibowo, Ilmu Kelautan**,** **17**,3:154-163. doi: https://doi.org/10.14710 /ik.ijms.17.3.154-163 (2012).
35. Hartoni, A. Damar, and Y. Wardiatno, MASPARI Journal Institut Pertanian Bogor, **4**,1: 46-57 (2012).
36. Mujiyanto. *Prosiding Seminar Nasional Perikanan Indonesia*: 317-325 (2012).
37. H. Malikusworo, Training Course in Coral Reef Research Methods and Management. March 25 – April 30. Bogor. Indonesia. Seameo-Biotrop. (1986).
38. Ilham. Thesis. Faculty of Marine Science and Fisheries, UNHAS Makassar, (2007).
39. B. E. Luckhurst, and L. Luckhurst. Mar. Biol. **49**: 317-323. (1978).
40. Risk, M. J. Atoll Research Bulletin **153**. Washington UNEP. (1993).
41. Lowe-McConnell, R. H.Cambridge University Press, Cambridge, (1987).
42. T. A. Yadi, A. Pratomo, F. Yandri, *J. Mar. Sci. Prog*. (2013)

Motivation and Perception of Tourists as Push and Pull Factors to Visit National Park

Jumrin Said[1*], *Maryono*Maryono[2]

[1]Master Program of Environmental Science, School of Postgraduate Studies, Diponegoro University, Semarang - Indonesia
[2]Department of Urban and Regional Planning, Faculty of Engineering, Diponegoro University, Semarang - Indonesia

Abstract. Push-pull theoretical framework is a popular theory to explain the reason why the tourists decide to visit the destination rather than other place, the kind of experience they want to get and the type of activity they want to do. In this paper, it is explained the motivation as push factors and the perception as pull factors of the tourist in deciding the destination based on previous literature and research using descriptive method. The framework asumed that tourists are motivated to fulfill their needs, including to reduce the psychological imbalance and to gain recognition of social status. National Park is one of destination based on nature or commonly knowns as ecotourism. In choosing the destination, the tourists tend to classify their alternative choice based on several criteria, such as the domination perception of tourist from one destination (pull factor), self motivation (push factor) and the available time and money (situational constraints).

1Introduction

The International Union for Conservation of Nature and Natural Resources (IUCN) has defined a six-category system of protected areas to clarify the differences between various objectives for protected areas. The protected areas are strict nature reserve/wilderness area, national park, nature monument or feature, habitat/species management area, protected landscape/seascape and protected area with sustainable use of natural resources. The category system has adopted in Indonesia by defining Nature reserve area, national park, world heritage site, forest park, nature tourism park, and hunting park.

The main objectives of management National Parks are preservation of species and genetic diversity; maintenance of environmental services; and tourism and recreation [1]. Based on Law no.5 year 1999 about conservation of biological natural resource and its ecosystem , national park has three functions, they are protection on life support system, preservation of the diversity of plants and animals, and sustainable use of biological natural resources and its ecosystems. Tourism and recreation are some activities that can be done in this park.

Sustainable Biodiversity and natural resources in national park area is one of the important factors support in tourism activities in this park. The amazing beauty of nature and its unique also benefits of the environment offered makes the visitors interesting to visit national park [2,3].

Besides cultural and customs potential, Indonesia located around equator is well known for mega biodiversity that may attract the tourists [4,5]. The Increasing of tourism activities in Indonesia can be seen from the number of foreign tourist who visited in 2016 as much as 12 million, or grew 15,54 % compared to the number of foreign tourist previous year. In the previous year was only 5,5 million tourists visited protected area [6]. It makes tourism sector as the 4th biggest national foreign exchange contributor following oil and gas, coal, and palm oil commodities as much as 12, 23 million USD [7].

Visitors' behavior doing tourism activities in national park may influenced by perception and motivation. It has been explained by Rossi et al. [8] in a conceptual model that the factors mediate visitors behavior visiting national park area are motivation and perception itself. One of the theory explains about it is push-pull framework.
This study aims to determine motivation as a push factor and perception as a pull factor for tourists in visiting a tourist destination based on the results of research that has been done.

2 Method

This Paper uses secondary data from research journal. Analysis method that has been used in this paper is qualitative method, by describing the result of secondary data review that has been shown to explain push-pull theoretical framework, motivation as push factor and

*Corresponding author: jumrin46@gmail.com

perception as pull factor based on the previous researches.

3 Results

3.1 The push-pull framework

Motivation bases Visitors' behavior can be explained using push-pull framework approach [9, 10]. Push-pull theoretical framework is a popular theory to explain the reason why the tourists decide to visit the destination rather than other place, the kind of experience they want to get and the type of activity they want to do [11]. When we discuss about tourism behavior, the approachment with framework is easy to be used and very effective [12].

The psychological encouragement is one of push factors such as social interaction, the desire for escape, adventure, relaxation, and self-exploration1[12]. Push factor is the things underlying and directing someone's behavior to do a travelling [11].

Some previous Researches have been done to find out Push-pull factors when they do travelling and tour(Table 1). Every researcher tries to identify Push-pull factor in tourism however they are different from the focus of identification. There is a researcher who identifies both of them (Baloglu and Uysal 1996; Crompton 1979; Oh, Uysal, and Weaver 1995; Turnbull and Uysal 1995; Uysal and Jurowski 1994; Yuan and McDonald 1990), only push factor (Cha, McCleary, and Uysal 1995; Dann 1977; Fodness 1994), or only pull factor (Sirakaya and McLelland 1997). The approachment that have been used to identify Push-pull factor are also different, one researcher uses qualitative approaches like personal interview (Crompton 1979), another uses scale development approaches (Dann 1977; Fodness 1994), and the other uses multivariate analyses of existing survey data (Baloglu and Uysal 1996; Cha, McCleary, and Uysal 1995; Oh, Uysal, and Weaver 1995; Sirakaya and McLellan 1997; Turnbull and Uysal 1995; Uysal and Jurowski 1994; Yuan and McDonald 1990) [10].

Table 1. Previous Studies Examining Push And Pull Factors

Researcher(s)	Research Approach Used	Push Factors Identified	Pull Factors Identified
Dann (1977)	Scale/survey development and analysis	Anomie, ego enhancement	
Crompton (1979)	Unstructured in-depth interviews	Escape, self-exploration and evaluation, relaxation, prestige, regression, enhancement of kinship relationships, social interaction	Novelty, education
Yuan and McDonald (1990)	Factor analyses of 29 motivational/push items and 53 destination/pull items	Escape, novelty, prestige, enhancement of kinship relationships, relaxation/ hobbies	Budget, culture and history, wilderness, ease of travel, cosmopolitan environment, facilities, hunting
Fodness (1994)	Scale development	Ego-defense, knowledge, reward maximization, punishment avoidance, value expression, social adjustive	
Uysal and Jurowski (1994)	Factor analyses of 26 motivational/push items and 29 destination/pull items	Re-experiencing family and togetherness, sports, cultural experience, escape	Entertainment/resort, outdoors/nature, heritage/culture, rural/inexpensive
Turnbull and Uysal (1995)	Factor analysis of 30 motivational/push items and 53 destination/pull items	Cultural experiences, escape, re-experiencing family, sports, prestige	Heritage/culture, city enclave, comfort/ relaxation, beach resort, outdoor resources, rural and inexpensive
Oh, Uysal, and Weaver (1995)	Canonical correlation analysis of30 motivational/ push items and 52 destination/pull items	Knowledge/intellectual, kinship/social interaction, novelty/adventure, entertainment/prestige, sports, escape/rest	Historical/cultural, sports/activity, safety/upscale, nature/outdoor, inexpensive/budget
Cha, McCleary, and Uysal (1995)	Factor analysis of 30 motivational/push items	Relaxation, knowledge, adventure, travel bragging, family, sports	

Researcher(s)	Research Approach Used	Push Factors Identified	Pull Factors Identified
Baloglu and Uysal (1996)	Canonical correlation analysis of 30 motivational/ push items and 53 destination/pull items	Four canonical variate pairs of push and pull items were identified but were not labeled. These variates were used to identify four market segments labeled sports/activity seekers, novelty seekers, urban-life seekers, beach/resort seekers	
Sirakaya and McLellan (1997)	Factor analysis of 56 destination/pull items		Local hospitality and services, trip cost and convenience, perceptions of a safe/secure environment, change in daily life environment, recreation and sporting activities, entertainment and drinking opportunities, personal and historical link, cultural and shop ping services, unusual and distant vacation spot

Source:[10]

In the latest research that has been done using analysis factor has reduced 56 (fifty six) attributes into 9 (nine) factors as a pull factor [13]. These 9 factors are local hospitality and services, trip cost and convenience, perceptions of a safe/secure environment, recreation and sporting activities, entertainment and drinking opportunities, personal and historical link, cultural and shopping services, and unusual and distant vacation spot. The research respondent towards the first factor " local hospitality and services" as the most important factor attracts their interest to visit [10, 13].

3.2 Push factors : motivations

Motivation is understood as the underlying power of rise and directly affect behavior [14]. Motivation appears when a person wants to fulfill the needs [15].

Motivation commonly related to Push-pull factors that affects visitors' behavior. Push factors is a socio-psychological visitor contract that affects on their motivation to visit an attraction and destination (peacefull, loneliness, to feel a new experience, etc). While The pull factors are The quality of the setting that attracts them to come to an attraction or spesific destination[14].

Perception and motivation can influence visitors' behavior to do tourism activities in national park.It is explained in a hierarchy conceptual model to the factors that mediates visitors' bahavior when they visit national park (Fig.1) [8].

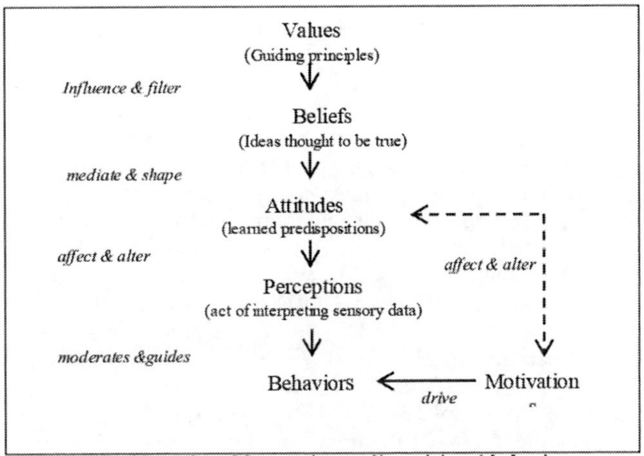

Fig. 1. The hierarchy of factors that mediate visitors' behavior to the national park

Source : [8]

In the beggining of the exploration research about tourism motivation for vacation, identified some basic motivation dimension, for instance socio-pshycology, prestige, cultural, social, education, and benefit. The following research simplifies the number of motivation into four main domains that are climate or environment, relaxation or escape, adventure, dan personal [14,16].

The motivation is briefly explained [16] as follow :

- Environment or climate is a term in a common meaning. Temporary desire, move to dry and warm climate. The destination where the physical environment differs from visitors normal environment.

- Relaxation is spend the time to do interesting activity, including sport activity such as golf and scuba diving. This category is including a motive to get out from daily routine activity, enjoy the time and romantic experience.
- Adventure is socio-culture motive to looking for the new thing and curiosity, including to see different culture in society and a certain view. Among all the researcher, adventure is a culture motive that gives attraction, opposite of the temporary change from environment that causing climate push factor.
- Personal, personal reason including prestige, nostalgia, strengthen relationship, exploration, and social interaction facility. This category is included visiting family and relative.

4 Pull factors: perceptions

Visitors' perception as a whole perception to the tourist site is the image of destination [17]. The image of destination is a key factor and stimulus for the tourists when they choose destination [11,17].

Perception of someone's experience related to his feel involves between experience from surrounding environment stimulant and followed by activity as a responses or answer from the stimulant [18]. Perception refers to how human feels, processes mentally and acts to the information that received from surrounding environment [8].

Positive image perception from a destination become the most favorite choice among the other at the same alternative. So that negative image perception will lessen the number of visiting [11,12]. Factors that influence tourists to visit a destination are the attitude towards the destination, the opinion from relatives and friends, experience from previous travelling and the limited of time and financial. Based on tourism prespective, Darnell and Johnson (2001) found that the level of satisfaction causes the intention to come back to the destination [19].

Positive perception or the expression of joy gained by the visitors from a visiting destination is a kind of visitors' satisfaction from the tourism destination [20]. It can be seen from satisfaction main criteria consist of common characteristics [21]:

- Personnel . The criteria refers to services provided by national park's staff in following sub criteria : knowledge, services(accuracy, fast service), communication visitors, and courtesy.
- Nature characteristics. This dimension related to physical characteristic of national park including flora and fauna and panorama
- Infrastructure. This Main satisfaction criteria refers to the most part for infrastructure that developed by national park. It is consist of path, hostel, informatin center, and bird watching site. Besides that, there is support access (the availbility of road network on the way to national park), and also road condition in national park.
- Recreation facility. This criteria related to all recreation facility available in national park area, such as seat, watching site, stall, picnic, toilet, and trash can.
- Information-communication. Satisfaction dimension related to information that given to visitors through available information boards, line marker, and map, including materia that could be buy by visitors (suc as pictures, CD/ DVD, handicraft, etc).

Component of visitors satisfaction are expectation, performance (performance of all sector: manager, service staff, visitors not only individual but also group, and local community); disconfirmation wishes (distinction between wish and performance), attribute (locus consideration), security(stability), management by manager, emotion (such as afraid, shocked), and justice (the feeling of justice) [22].

5 Conclusions

In choosing the destination, the tourists tend to classify their alternative choice based on several criteria, such as self motivation (push factor), the domination perception of tourist from one destination (pull factor), and the available time and money (situational constraints). motivation as push factor could be seen from four main groups that is environment or climate, relaxation, adventure and personal. On the other hand, perception as pull factor could be seen from visitor satisfaction components. So that protected area manager especially national park makes main function balance in preserving the sustainability of national park and satisfying visitors experience.

References

1. J.Suh, and S. Harrison, Management Objectives and Economic Value of National Parks: Preservation , Conservation and Development, (2005).
2. C.Fandeli, and M. Nurdin, *Pengembangan Ekowisata Berbasis Konservasi di Taman nasional*. Yogyakarta: Fakultas Kehutanan UGM, Pusat Studi Pariwisata UGM dan Kantor Kementerian Lingkungan Hidup, (2005).
3. T.Kamri, and A. Radam, Procedia - Social and Behavioral Sciences, **101**, 495–505, (2013).
4. I.Dunggio, andH. Gunawan, H. Jurnal Analisis Kebijakan Kehutanan, **6**(1), 43–56, (2009).
5. Y. T.Latupapua, S.Pudyatmoko, C.Fandeli, and M. Baiquni, Kawistara, **5**(3), 250–259, (2015).
6. Ministry of tourism. Jumlah Kunjungan Wisatawan Mancanegara Menurut Pintu Masuk Dan Kebangsaan Bulan Januari - Desember 2016. http://kemenpar.go.id/userfiles/desember(1).pdf. Diakses tanggal 9 April 2017.
7. Ministry of Environment and Forestry. Statistik Kementerian Lingkungan Hidup dan Kehutanan 2015. Jakarta (2016).

8. S. D.Rossi, J. A.Byrne, C. M.Pickering, and J. Reser, Geoforum, **66**, 41–52, (2015).

9. G. M. Dann, Anomie, Annals of Tourism Research, **4**(4), 184–194, (1977).

10. D. B. Klenosky, Journal of Travel Research, **40**, 385–395, (2002).

11. G.Prayag, and S. Hosany, Tourism Management, **40**, 35–45, (2014).

12. L. J.Chen, and W. P. Chen, Tourism Management, **48**, 416–425, (2015).

13. E.Sirakaya, and R. W. Mclellan, Anatolia: An International Journal of Tourism and Hospitality Research, **8**(3), 31–44, (1997).

14. A.Beh, and B. L. Bruyere, Tourism Management, **28**(6), 1464–1471, (2007).

15. V.Gundersen, M.Mehmetoglu, O.Inge Vistad, and O. Andersen, Journal of Outdoor Recreation and Tourism, **9**, 77–86, (2015).

16. H.Bansal, and H. A. Eiselt, ETourism Management, **25**(3), 387–396, (2004).

17. T. H. Lee, Leisure Sciences, **31**(3), 215–236, (2009).

18. K. Cherry, (2013). Perception and the Perceptual Process. Retrieved April 11, 2017, from http://psychology.about.com/od/sensationandperception/ss/perceptproc.htm

19. C.-N.Lai, Yu, T.-K., and J.-K. Kuo, Social Behavior and Personality: An International Journal, **38**(4), 509–514, (2010).

20. E. Pouta, International Journal of Sociology, **40**(3), 50–69, (2010).

21. G., Arabatzis, and E. Grigoroudis, V Forest Policy and Economics, **12** (3), 163-172, (2010)

22. D. Bowen and J. Clarke, Journal of Vacation Marketing, **8** (4), 297-308, (2002).

Community Concern on Environmental Conservation

Fajar Adie Nugraha[1,*], and *Maryono* Maryono[2]

[1]Master Program of Environmental Studies, School of Postgraduate Studies Diponegoro University, Semarang - Indonesia
[2]Urban and Regional Planning Department, Faculty of Engineering Diponegoro University, Semarang - Indonesia

Abstract. Research on the relationship between humans and the environment is always very interesting to be studied. This paper is one of the studies of sustainable development in its implementation with the empowerment of society that comes from the community itself. So far, community studies related to development occasionally target the growth of the economic side only. Community study on the environment becomes an alternative choice, compared with human relationships with humans themselves, or humans with human needs themselves. The study of community development by looking at the environment can be a wise choice, where all activities of fulfilling human needs are always inseparable from the element of interaction with the environment. Community development that is based on the environment itself, will give a better impact, just solely. Various methods of learning human relationships. A community-based environmental assessment study can be an alternative choice to support a sustainable development mission, which is development that has a positive impact on the present and the future.

1 Introduction

Development, people and the environment are interrelated. Self-development is one of human effort in its main activity as the fulfillment of basic life needs. Without a development, Human can not develop, and extinction becomes an answer to that problem. The concept of the fulfillment of human needs has actually started since human civilization itself existed, the initial concept of development appears in the use of human survival methods that started from hunting and gathering, then developed into cultivation, and until now developed with various methods of fulfilling human needs. Self-development can not be separated from the environment, because why? The sustainability of a development itself is part of environment, which makes the various developments inevitably have a direct impact on the environment. Lopez and Luigi (2008) Said that The consequences of this worldview tends to be focused on exploitation of nature, resource extraction, efficiency management, and a belief that technology can solve all problems [8]. So far we have confidence, that when the development is growing rapidly, and technology is growing, so as if every environmental problem will be solved easily. However, the situation in the field finds that with the development that does not put forward the environmental assessment, no matter how sophisticated the technology used, the impact of damage to the environment tends to increase.

Hawkins (2010) points out that To be relevant in the contemporary world, social work must move beyond our traditional focus on social and economic justice. We must actively advocate for environmental justice and pursue sustainable development so that all people can live in a clean, safe, and healthy environment [6]. Our use of economic and social benchmarks to a successful development is worth reviewing. So far we tend to ignore the environmental impact directly. As a result both directly and indirectly we encounter a lot of environmental degradation occurs around the development area. Yet we as human beings, are the main perpetrators of development. So far we tend to ignore the environmental impact directly. As a result both directly and indirectly we encounter a lot of environmental degradation occurs around the development area. Yet we as human beings, are the main perpetrators of development. Denpaiboon and Amatasawatde (2012) even say that People in this generation will become an important group to run community further [3]. So that we as individual human beings can establish unity with other human individu by forming a community or society that has a role for the interests that will come.

When development has been successfully aligned with the environmental interest by us as a society. Harmony in development will be created, so the human image as a contributor to environmental damage can be

*Corresponding Author :geografifajar@gmail.com

minimized as possible. Although the data found by Howe where the current reality, studies quantifying success of conservation and sustainable development projects are also limited. Without quantitative data on either of these aspects it is difficult to translate research into action, which is vital if conservation and sustainable development strategies are to succeed [7]. Can be used as our reference to awaken themselves to the need for study of sustainable development projects, the impact on society, environment and achievement as a means of fulfillment of basic human needs. Humans are the highest link in the environment, all natural phenomena more or less affected human activities. With that understanding we should change the paradigm as the most powerful, be the most responsible, and should be the most aware of the changes that occur in the environment. Coates (2003) even discusses the need for a new ecological paradigm in social work, which emphasizes a core theme that environmental exploitation results from the same pressures that create social injustice [2].So the real problem of development can not be separated from our social life, as well as environmental problems is actually our social problems as well.

2Community participation in Development

Rapley argues that development was considered largely synonymous with industrialization. Its ultimate goal was fairly clear: to raise incomes and in the process give poor people access to the range of goods and services then widespread in developed societies [13]. It can be said that development is a means that can not be inevitable to the concept of fulfilling the needs of human life. Now as if development is a duty in a human development. Many land use changes are in the name of development. When reflecting on the existing development in the world today, no wonder impacts arise due to development, such as social and environmental issues. Social problems struck in the development process has actually been seen since the industrial revolution in Britain several centuries ago, which when the replacement of human labor into machinery resulted in the number of production of goods and services increased messages, and also the cost of production can be reduced to a minimum. The phenomenon was a remarkable development progress at that time, as if humans at that time entered into a new age of an industrial age. But behind the success of development in the field of industry, there are social problems that are quite inconvenient for the UK at that time the early symptoms of population blasting. Workers who quit their jobs automatically do not have busy in their daily lives, so they spend more time with their wives at home, and household activities contribute to increased birth rates in the UK, and indirectly increase the number of residents in the UK. The example is one form of the impact of uncontrolled development, and did not rule out, when the current development is not controlled, problems of similar problems will begin to appear to the surface.

In Reser's study he found that the condition and state of the world's natural environment is one of unprecedented fragility, escalating threats and human impacts, and massive and in many cases irreversible natural environment degradation and habitat and species loss [14].This should be a warning for us, where a lot of research related to the negative impact of uncontrolled development raises various problems. Humans unconsciously or in the realize already contribute significant changes in the environment. Myller and Spoolman also examine the topic of development, according to him, economic development is devoted to making aneconomy better by improving human well being meeting fundamental human needs for items such food, shelter, physical and economic security, and good health [12].The result of the study is actually a trigger for an advanced study, about the significant impact of a development on the environment, both quantitatively and qualitatively. The more studies on the impact of development, the more theories will be obtained as a reference in the management of a development, so that in a planning and implementation of a development can promote more complex studies, and not just economic and social.

3 Community concern to environment

As the subject of all the issues, humans is always in every situation called the problem. When as an individual sometimes the consequences of a problem just seem simple because it only concerns the individual, but when humans interact and form a system of society, the problems will be more complex. Society in this case is a group of people who do a development, is one of the main objects of a study of scholarship. As much as any study of a physical environmental problem, but when ignoring the study of social and human behavior, then the study just seemed to evaporate into a theory without the application. Society must be equipped with the knowledge and knowledge of environmentally conscious behavior, where when people are aware of environmental conditions, they will use that knowledge consciously in the application of their life activities, including the development context even. Babylon (2003) says there needs to be an approach to society, especially in relation to the environment, although in implementing a community-based approach to conservation it is often incompatible with established goals because too few organizations are aware of the ins and outs of the communities they are trying to mobilize. Rather than regard the communities as a small (in both population and geographic area) unified whole, more attention should be given to the processes within communities and the varying influences of different players, whether citizens or public officials [1]. The problems in the community approach are complex, the system of society, either from the simplest form of the neighborhood or village, to the government and the stakeholders is not as easy as it is planned. Several existing case studies show that there is a need for in-depth study, about the approach to society, especially in relation to development and environmental mindset. The non-formal approach may be one of the efforts of community empowerment. Lynch (2001) finds an alternative solution to environmental approaches and environmental empowerment strategies , one of the most popular strategies that has emerged is environmental education (EE). Conservation organizations and governmental entities alike have embraced environmental education as an integral part of their agendas. Yet, there is often a lack of critical analysis regarding how EE programs are contributing to conservation goals [9]. The community must be given a real solution, and need an intensive assistance, it is necessary, because one of the best ways to instill environmental values to the community is to familiarize them in their daily activities. Without such habituation, we can be sure that various efforts of empowerment and facilitation of environmental insight will be less than the maximum.

Mallen and Corbera in his research argued that decision making around natural resource management based on traditional ecological knowledge was identified as a key factor contributing to successful conservation in both types of community-based conservation initiatives [10]. Based on the research study, it can be taken several things, among others, raises the intention of community initiative, and the provision of knowledge to the community. Both of these can be alternatives to other solutions, in addition to various methods that can be used, but to find the right formula must be tailored to the conditions where the community is located, so that the study and results in accordance with what is wanted.

4 Environment pollution and degradation

As a place where all the problems caused by humans are, so far as if the environment is still a priority to so many by humans. The environment is more often ignored in the name of other interests, not even rare environmental issues are seen as a stumbling block a development progress. Reser suggests environmental degradation is in large part caused by human behaviours and directly affects human health and well being [14]. It can be said that when a human ignores the environment, he is unknowingly trying to dig his own grave hole. Development should use the guidance of an environmental perspective as a basic reference in its planning, and as a human being subject to all problems, it must be used in all decision-making, especially related to development and the environment. Lynch (2001) argues in a particular case that due to the undisputed need for improved education in most tropical areas, environmental education proposals are rarely questioned or critically evaluated [9]. This is a form of abandonment of the concept of environmental education, which in fact in terms of a development and public relations environment education is an obligation in the implementation.

In its development, when development, society and environment have tried to be synchronized into one unity that can not be separated. It still needs the initiative of the community as the main actor of all policies, theories, and its application. Mallen and Corbera added that a community-based conservation initiatives, which combine traditional and official regulations as well as informal and formal sanctions established by cross-scale institutions, traditional ecological knowledge was considered the basis for sound decision making around natural resource management, alongside scientific knowledge [10]. Public awareness of environmental issues is a fundamental requirement today. As mentioned before, sometimes the local culture of society has shown the form of environmental science, but its shape is not yet packed with a more scientific concept. But it does not mean that the local culture of society is not necessarily far from the word environment rather than scientifically own scholarship. If the line drawn back, the study of the environment is already a lot, both physically environment, as well as the concept of the environment. But it should be underlined, although with all these studies the implementation and application of science is still limited and not maximized. So the study of scientific applications in the environment that is applied to the community can be a discourse of the development of human interaction studies with the environment.

5 Sustainable Development

The study of interaction between the environment, human and development now began to become a serious discourse and study by environmentalists. This is evident in the changing direction of world development goals that begin to lead to the sustainability of a development. Where sustainable development itself is a form of development that put forward the concept of fulfilling human needs today, but also still pay attention to the concept of needs and availability of resources for the future. DEE and QCA put forward the view that education for sustainable development, through developing pupils' skills in, and commitment to effective participation in the democratic and other decision making processes that affect the quality, structure and health of environments and society and exploring values that determine people's actions within society, the economy and the environment [4]. It should be underlined that in the implementation of sustainable development needs a commitment to active participation, in this case not necessarily charged to the stakeholders, or government, but also need to include all layers of human life order, so that in the execution of policy making, can supervise each other to the value The value of the environment in a development. When comprehensive monitoring, the implementation of sustainable development can be comprehensive in all lines of society, whether from individuals, to the sphere of government or related institutions. This is reasonable considering that in environmental education, there needs to be a habituation of the concept of environmental knowledge in everyday life, so that in the application of sustainable development which is the development of the concept of development and environment so also become an thing absolutely necessary. Therefore it is necessary to study more deeply about the application of environmental education in society, in relation to the achievement of sustainable development.

Dobson added that Sustainable development is at least as much about values as about techniques and technologies. For example, science might be able to tell us what the threshold tolerances of nitrogen in the atmosphere are for any given species, but it cannot tell us which species we should be concerned about. The key questions, then, are not technical – they are normative. Bearing this in mind, we have to say that we will be short-changed by any ecological citizenship curriculum that does not confront normative questions of this sort [5]. Application of such an analogy dobson proposed can indeed be a reference in determining the concept of study on environmental education, as well as how to link it to sustainable development. Because without a concept adapted to environmental conditions, or simply placing a theory on a particular study and without being based on a scientific understanding of the object of the study, the steps in decision-making or method may be less than the maximum, so the results can be far Of the unity itself. Matti, S responds that sustainable development is applied in the community, he insists that environmental citizenship must have a presence in the social, political and economic life of societies aiming at sustainable development. It is tempting to say that fiscal incentives and environmental citizenship can be pursued simultaneously, but recent research from Sweden suggests that self-interest based policy initiatives can 'crowd out' citizenly behaviour, making it less likely that such behaviour is either followed or fomented [11]. Studies of community development by looking at the environment can be a wise choice, where all activities of fulfilling human needs are always inseparable from the elements of interaction with the environment. Community development based on the environment itself, will have a better impact.

6 Conclusions

Human beings, and the environment, are a major component of development, the interaction between the components can lead to how development works. During this study of human interaction with the environment is still limited to the environmental impact discourse will be a human activity. Still less popularly studied to discuss the prevention because of the incompatibility of human interaction with the environment occurs again, mostly focusing on improving the consequences of the incompatibility of interaction between humans and the environment. Community studies on the environment can be an alternative choice, compared to human relationships with humans themselves, or humans with human needs themselves. The study can also support the sustainable development mission, which is development that has a positive impact on the present and the future.

References

1. Babylon, G. Ruth,*The Use of Community-Based Conservation in Natural Resource Management: Case Studies from The Nature Conservancy of Virginia*. Thesis. Virginia Polytechnic Institute and State University, (2003)
2. J. Coates, *Ecology and social work: Toward a new paradigm*. Halifax, NS: Fernwood, (2003)
3. Denpaiboon, Chaweewan and Amatasawatde Kornchakorn, *Japanese Studies Journal Special Issue : Regional Cooperation for Sustainable Future in Asia*. Thammasat University, (2012)
4. Department for Education and Employment (DEE) and the Qualifications and Curriculum Authority (QCA). The National Curriculum for England: Citizenship. Department for Education and Employment and the *Qualifications and Curriculum Authority*: London, (1999)
5. Dobson, Abdrew, Environmental Citizenship: Towards Sustainable Development.*Sustainable. Development*. 15, 276285. DOI: 10.1002/sd, (2007)
6. A. Hawkins, Catherin, Sustainability, *Critical Social Work*, Vol. **11**, No.3, (2010).
7. Howe, Caroline,*The Role of Education as a Tool for Environmental Conservation and Sustainable Development*. Disertation. Imperial College London, (2009)
8. V. L. Lopez, andV. L. Luiggi, Participatory sustainability: Building sustainability for complexity and change. In Osborne, R. & Kriese, P. (Eds.). *Global community, global security*. New York, NY: Rodopi. pp. 217-227, (2008).
9. Lynch, A. Kathryn,*Environmental Education and Conservation in Southern Ecuador: Constructing an Engaged Political Ecology Approach*. Disertation. University Of Florida, (2001)
10. Mallen-Ruiz, Isabel and Corbera Esteve, *Ecology and Society* .18(4):12, (2013)
11. Matti S. C. Berglund,*Environmental Politics*. 15(4): 550–571, (2006)
12. Myller, G. Tyler and Spoolman, E. Scott,*Environmental Science*. National Geographic Learning, (2015)
13. Rapley, Jhon. 2007. *Understanding Development: Theory and Practice in the Third World THIRD EDITION*.Copyright © 2007.
14. Reser, P. Josep, Psychology and The Natural Environment. A Position Statement Prepared for The Australian Psychological Society. *The Australian Pshycology Society*. Ltd, (2007)

Economic Valuation as an Instrument to Determine The Management Strategy of Baros Mangrove Forest, Bantul, Yogyakarta, Indonesia

Irawan Waluyo Jati[1,*], *Rudhi* Pribadi[2]

[1]Master Program of Environmental Science, School of Postgraduate Studies, Diponegoro University, Semarang - Indonesia
[2]Department of Marine Science, Faculty of Fisheries and Marine Science, Diponegoro University, Semarang - Indonesia

Abstract. The Baros mangrove forest in Bantul Regency is now beginning to develop. Many government and private sectors programs are rolled out to support its development. The development of the Baros mangrove forest must be in accordance with the rules of conservation so that it will not damage the mangrove ecosystem. Mangrove forest has high economical and ecological value but is very vulnerable if lack of wisdom in maintaining, preserving and managing them. The involvement of government and other stakeholders are essential in determining management policies. Unawareness of society and the government to the importance of mangrove ecosystem can cause development of it becomes uncontrolled, consequently can destroy it. Mangrove forests are an important natural resource in coastal environments, and have three main functions: physical, biological, and economic functions. To quantify the functions of mangrove forests as the basis to determine the policy is required a research instrument called economic valuation. The approach of this study is the literature review from various studies before to perceive the influence of economic valuation in determining the management strategy of Baros mangrove forest in Bantul Regency, Yogyakarta, Indonesia.

1 Introduction

Mangrove is one of the rare ecosystems, because the area is only 2% of the earth's surface. Indonesia is the largest mangrove ecosystem in the world [1]. Mangroves as an ecosystem have unique properties and forms and have diverse functions and benefits for humans and other living things [2].

The large potential of mangrove forests encourages excessive resource exploitation that can threaten its sustainability. Pramudji mentions that mangrove damage is not only caused by natural processes, but also due to human activities [3]. The existence of mangrove forest exploitation for the fulfilment of excessive human needs and ignoring the rules of conservation will be able to damage the mangrove itself. The ignorance of the community and the government of the importance of mangrove forest ecosystem causes development in the framework of mangrove forest area development becomes uncontrolled so that it will become damaged [4].

The existence of mangrove forest spread in Indonesia, one of them is in Baros Hamlet, Village of Tirtohargo,Kretek sub district, Bantul Regency. The mangrove ecosystem is precisely in the estuary / encounter between the southern sea and the mouth of the Opak River. The Baros mangrove forest is a mangrove conservation area that has the appeal of biological natural resources, geological formations, and / or natural phenomena that can be developed for the benefit of the development of science, research, education and awareness raising of the conservation of biological natural resources, marine tourism and recreation. Baros mangrove forest became one of the coastal and marine ecosystems that are potential for improving the welfare of the people of Baros Hamlet in terms of economic, social and environmental.

Based on interviews with the manager that the current Baros mangrove conditions as a special interest attraction has grown. The existence of programs from the government both central and regional as well as the private sector in order to develop the Baros mangrove area would need new spaces. Programs Development of mangrove areas without regard to ecological function will actually damage the existence of mangrove and this will cause problems in the future. [5] mentions that the destruction of mangrove forest one of them caused by the mistake of government policy.

Mangrove forests have high economic and ecological value but are very vulnerable to damage if not prudent in maintaining, preserving and managing them [6]. The damage usually experienced by mangrove is caused by human activities directly in the utilization of mangrove forest ecosystem. To avoid these problems, it is necessary to assess the utilization of mangrove forests, not only economic benefits but also the consideration of

* Corresponding author: irawan.bantul@gmail.com

https://doi.org/10.1051/e3sconf/20183108024

ecological benefits that can become a reference in the mangrove forest management strategy.

This study aims to determine the total economic value of mangrove forests and the implementation of sustainable mangrove management strategies in the mangrove area of Baros, Tirtohargo, Kretek, Bantul.

2 Methods

2.1 Data collection technique

This research is done by using data collection method which divided into secondary data and literature review. Secondary data were collected through document review and documentation of activities related to Baros mangrove forests. While the literature review comes from similar studies and research related to Baros mangrove forest. The analytical methods used are quantitative and qualitative methods, which are then described to explain the total economic value and strategy of Baros mangrove forest management.

2.2 Analysis technique

The economic assessment of all the benefits of natural resources of mangrove forest refers to the concept of natural or environmental value according to [7] [8]. The economic values contained in a natural resource, both useful and functional, must be taken into account in formulating management policies so that their allocations and alternative uses can be properly determined and targeted [9]. The formulation is as follows:

TEV = TUV + NUV = (TDUV + TIUV + OV)
Where :

TEV : Total Economic Value
TUV : Use Value
NUV : Non Use Value
TDUV : Total Direct Use Value
TIUV :Indirect Use Value
OV : Option Value

3 Results and discussion

This research was conducted in Baros mangrove forest area, Tirtohargo village, Kretek subdistrict, Bantul district located at 110 ° 17 '9199 "BT and 8 ° 0' 48.066" LS. The location of this study was chosen because it is the only mangrove forest in Bantul district and is one of the icons of tourism and conservation. The location of the research can be seen in Figure 1.

Fig. 1. Map of Mangrove Baros Forest Area

3.1 Mangrove forest ecosystem condition

The mangrove forest ecosystem in Baros Village is managed by Pemuda Pemudi Baros Family (KP2B) since 2003. Mangrove planting is done in Sultan Ground land in Opak River estuary of 25 Ha. Until now the well-grown Mangrove is 3 - 4 Ha. The type includes *Avicennia sp.* 60%, *Rhizhopora sp.* 20%, *Brugueira sp.*10%, and *Nipah* 10% [10]. Composition of mangrove species in Baros mangrove forest can be seen in Figure 2. Based on inventory conducted by [11] in Baros mangrove area recorded four (4) mangrove species: *Avicennia sp., Rhizophora sp., Bruguiera sp.,* And *Nypa frutican*.

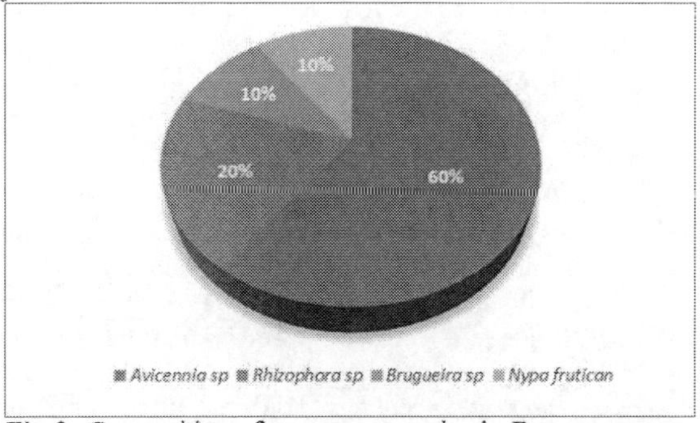

Fig 2. Composition of mangrove species in Baros mangrove forest

In an effort to protect and conserve biodiversity in the Baros mangrove ecosystem, the Local Government through the Regent of Bantul Regency on April 28, 2014 establishes Baros mangrove area as a Mangrove Conservation Area through the Decree of Bupati Bantul Number 284 Year 2014 on Coastal Reserve Conservation Area. Zoning in the Baros mangrove conservation area consists of three zones, namely core zone (10 ha), other zones (94 ha) and limited utilization zones (28 ha) [12]. Map of reserve area of conservation of mangrove coastal park of Bantul Regency can be seen in Figure 3.

 https://doi.org/10.1051/e3sconf/20183108024

Fig. 3. Map of reserve area of conservation of mangrove coastal park Bantul Regency

3.2 The value of economic benefits of baros mangrove forest

3.2.1 Direct use value

a. Fisheries

Vegetation Mangrove is a convenient habitat for crabs, is a place to live plankton as a source of food for crabs, shrimp and fish. So many fishermen who get sea products around the Mangrove forest. Communities around the Baros mangrove forest get a lot of benefits, one of them is fishery. Based on research conducted by [13] that the fishery products obtained in the area of Baros mangrove forest such as *Scylla sp., Mugil sp., Macrones sp.* And *Oreochromis*. The cost of fishing is Rp. 20.467.542, - / year with total revenue Rp. 100.716.000, - / year. The profit earned by fishermen is Rp. 80.248.458, - / year.

b. Mangrove seeds

Making mangrove seedlings ready for planting (Avicenia) is able to create jobs for group members. Currently, the types of mangrove seeds made by KP2B are sold to those in need. [10] states that the mangrove seedlings per year reaches 5,200 *Avicenia sp* mangrove seedlings sold at Rp 2.000,00 / seedlings, the total value of *Avicenia sp* mangrove seedlings is Rp 10.400.000, - / year.

3.2.2 Indirect use value

Coastal Protection (Breakwater)

Mangrove is one of the unique and unique ecosystems that can protect the coast from abrasion [14]. The indirect benefits of mangrove forests as abrasion holder are diminized through replacement cost with the construction of breakwater buildings. Based on the research [6] The cost of breakwater construction for a distance of 1 m with endurance for 10 years is Rp. 21.986.700, - and the length of Baros mangrove forest is 500 m, the function value is Rp 10,993,350,000, - / 10 years or Rp 1,099,335,000 / year.

3.2.3 Option value

The benefit of choice is a value that indicates a person's willingness to pay in order to preserve the mangrove ecosystem for future use. This value is approached with reference to the biodiversity value of mangrove forests in Indonesia. As a reference value of biodiversity in Indonesia used the results of research from [7] that is US \$ 1500 / km^2/ year. The area of Baros mangrove land used is referring to the Decree of the Regent of Bantul Number 284 of 2014 on the Reserve of Coastal Conservation Area. Zoning in the Baros mangrove conservation area is the core zone (10 ha). Thus, the value of the choice of mangrove ecosystem is: Area of land x US \$ x Rupiah = 10 Km^2 x US \$ 1500 x Rp 13,350.00 = Rp 200,250,000.00 per year.

3.3 Assessment of economic benefits of total mangrove forest

The total economic value of the mangrove ecosystem is the sum of the value of direct benefits, indirect benefits, and benefits of choice. Quantification of the benefits of mangrove forests as a whole shows that, the total benefit value of Rp.1.390.233.458,00 / year, where indirect benefits have more value High compared to other types of benefits with a percentage of 79.1%. The total value of benefits can be seen in Table 1.

Table 1. Total Economic Benefit Value of Mangrove Forest

No.	Type of Benefits	Use Value (Rp/years)	(%)
1.	Direct Use Value	90,648,258	6.5
2.	Indirect Use Value	1,099,335,000	79.1
3.	Option Value	200,250,000	14.4
	Total Economic Value	1,390,233,258	100.0

4 Conclusion

Based on the results of research that has been done then it can be concluded as follows:

1. Utilization of mangrove resources in Baros hamlet that can be directly felt by the community, namely: fishery utilization *(Scylla sp., Mugil sp., Macrones sp., Oreochromis)* and utilization of mangrove seedlings.
2. Based on the research results obtained the total economic value of mangrove forests in Baros hamlet of Rp. 1,390,233,258 per year, calculated from direct benefits Rp. 90,648,258 per year, indirect benefit Rp. 1,099,335,000 per year and a choice of Rp.200.250.000 per year.

Acknowledgements

We would like to thanks the Master Program of Environmental Science, School of Postgraduate Studies, Diponegoro University, Semarang for the knowledges and experiences. Also thanks to Dwi Ratmanto and KP2B for valuable information and supports for this paper.

References

1. AD. Setyawan, K. Winarno, Biodiversity, **7** (April): 159-63, (2006)
2. Gunarto. *Conservation of Mangrove as Supporting Biological Resources of Coastal Fishery.* (2004)
3. Pramudji. Oseana, **XXV (2)**: 13-20, (2000)
4. H. Purnobasuki, Bul PSL, **25**: 3-6, (2011)
5. Suthawan Sathirathai. *Economic Valuation of Mangroves and the Roles of Local Communities in the Conservation of Natural Resources : Case Study of Surat Thani , South of Thailand.* Research Reports. (1995)
6. WD. Kalitouw, *Economic Valuation of Mangrove Forest In Tiwoho Village, Wori Subdistrict, North Minahasa District.* J Univ Sam Ratulangi. (2015)
7. HJ. Ruitenbeek, *Mangrove Management: An Economic Analysis of Management Options with a Focus on Bintuni Bay, Irian Jaya. Jakarta and Halifax;* (1992)
8. B. Barbier, Valuing Environmental Functions: Tropical Wetlands Edward, **70** (2):155–73, (1994)
9. Center for Natural Resource Surveys. *Guidelines for Preparation of Balance Sheets and Values of Coastal and Marine Natural Resources Economies. Cibinong:* Bakosurtanal; (2005)
10. KP2B. *Activity Profile of Community Based Mangrove Forest Development.* Bantul; (2014)
11. Trialfhianty TI. *Kondisi Ekosistem Mangrove Dusun Baros.* (2013)
12. *Decision of Bupati of Bantul Number 284 Year 2014 concerning Coastal Reserve Conservation Area, No. 284,* (2014)
13. Ismi T. *Economic Valuation of Mangrove Resource In Baros Coast Tirtohargo Village Sub-District of Kretek.* Kawistara. (2014)
14. M.G.H.Kordi. *Mangrove Ecosystem "Potential, Function and Management."* Jakarta: PT Rineka Cipta; (2012)

The Profile Quality of Pond In Kendal Regency to Diversification Aquaculture

Ligar Novi Ayuniar[1] and *Jafron* Wasiq Hidayat[2]

[1]Master Program of Environmental Science, School of Postgraduate Studies, Diponegoro University
[2]Biology Dept, Faculty of Science and Mathematic, Diponegoro University

Abstract. Water quality, particularly coastal areas, is systematically tropogenic. The decline in water quality is caused by industrial waste pollution, soil erosion carried by the river, and the depletion of mangrove areas. The decrease of water quality can affect the fishery cultivation activities that exist in the region. It also affects the quality of the cultivated fish. Fish cultivated in ponds with poor water quality can be harmful to the health of the people who consume the fish. One effort to manage the feasibility of pond waters is by identifying the quality. The purpose of this research is to know the profile of pond water quality and to know the diversity potential of aquaculture. Based on the nature of the problem this research is a field research, while the purpose of this study is descriptive and explanatory research. The method used in this research is research by using survey method. Aquatic profile results are essential to improve the quality and quantity of Fisheries, especially in diversifying fisheries.

1. Introduction

Pond is one type of habitat that is used as a place for brackish water cultivation activities located in coastal areas. Continuous aquaculture activities lead to environmental degradation, which is characterized by a decrease in water quality. Environmental constraints encountered in the cultivation activities include the arrangement of the area or the arrangement of cultivation development space that does not pay attention to the carrying capacity of the environment due to improper management, resulting in environmental problems with all aspects of complications in a long period [1]. Good water quality in fish ponds benefited fish for prospering plankton as foods, buffering harmful matters, etc., and poor water quality easily resulted in lessening fish yields and cheapening production quality. How to balance water qualities was a technical skill in pond farming, and monitoring parameters of water quality might be the first step. On the other hand, aquaculture pond effluents can potentially harm the environment, although in some cases, the effluents have been environmentally negligible. Governments and organizations have developed limitation guidelines and standards for aquaculture effluents. A few variables have been used to evaluate aquaculture water qualities, for example, total suspended solids (TSS), turbidity (Tur), total particulate organic matter (POM), total nitrogen (TN), total phosphorus (TP), 5-day biochemical oxygen demand (BOD), and total settleable solids (Set). Aquaculture producers ought to measure variables to know the status of fish pond waters and comply with the standards for aquaculture effluents, and the more variables that were measured, the higher the cost [2].

Kendal Regency consists of 20 districts, has a fairly complete area, consisting of beaches, lowlands, highlands and mountains. With a coastline of approximately 42.2 km, it has enough potential resources to be cultivated and developed, both for freshwater / pond, brackish water and marine aquaculture. Based on data from the Livestock, Marine and Fisheries Service Office (DPKP) of Kendal Regency, for brackish water aquaculture (tambak) is located in 7 sub districts of Kendal namely Kaliwungu, Brangsong, Kendal, patebon, Cepiring, Kangkung and Rowosari. Area of ponds in the region of Kendal reached 3,292.22 ha, with the number of fish cultivators as many as 1,084 people. Leading commodities consist mostly of milkfish, shrimp vename, and tiger shrimp [3]. Seeing the state of Kendal Regency considerable potential for the development of aquaculture, it is possible this has a negative impact on the environment, for example, the declining quality of the water. So there needs to be further research on pond water quality and diversification of aquaculture in ponds of Kendal Regency.

2. Study Literature

Water quality is the nature of water and the content of living organisms, substances, energies, or other components in the water expressed by several parameters such as physical parameters (temperature,

* Corresponding author: ligar.novi@gmail.com

 https://doi.org/10.1051/e3sconf/20183108025

turbidity, suspended solids, etc.), chemical parameters (pH, DO, BOD, and Etc.), and biological parameters (plankton, bacteria, and Etc.)[4]. Water quality has a very important role for cultivation life. For example, water temperatures can affect many processes in both biology, physics and water chemistry. The increase in temperature that can still be tolerated by the organism will be followed by an increase in the degree of metabolism and the activity of photosynthesis of natural food (phytoplankton). Similarly, water temperature will affect survival, morphological growth, reproduction, behavior, rate of skin turnover (for shrimp) and shrimp metabolism. The magnitude of the influence of water temperature on life is determined by the tolerance and the speed of changes in water temperature. The degree of acidity (pH) of water shows the levels of hydrogen ions or protons contained in water. pH affects the process and speed of chemical reactions in water media as well as biochemical reactions in the shrimp body, affecting the toxicity of a compound, the survival and growth of shrimp. Each type of organism has a different tolerance to changes in pH [5].

Brightness is a measure of water transparency, which is determined visually by using a secchi disk. Brightness is measured directly in the field in a way until it can be distinguished between black and white on the secchi disk [6]. Water depth is also measured using secchi disk, by reversing the secchi disk with black and white disk position above it after measuring the depth of the pond by inserting the disk secchi and record the result.

The amount of dissolved oxygen in the water is very important for aquatic organisms. This is because dissolved oxygen affects the growth, survival, distribution, behavior, and physiology of shrimp and other aquatic organisms. The oxygen distribution also strongly influences the solubility of organic nutrients. The presence of oxygen is also used to determine whether the environment is in aerobic or anaerobic conditions [7]. Salinity can affect the oxygen content in the water, the higher the salinity level the dissolved oxygen will also be lower. Plankton also has an optimum growth range at high salinity. Nitrites are also affected by dissolved oxygen levels so that nitrite is an unstable gas. In naturally occurring waters nitrite is typically found in small amounts and displaces the transitional forms between ammonia and nitrate (nitrification), and between nitrate and nitrogen (denitrification). The source of nitrate can be industrial waste or domestic waste [8]. Ammonia is a substance that causes a very sharp smell and piercing the nose. Ammonia content (NH3) is a parameter of water quality that must be considered, because the ammonia found in waters when the levels exceed the limit will be able to harm the aquatic organisms [6].

Uncontrolled water quality can have a negative impact on the environment around the pond, one example is the occurrence of eutrophication. Of all the pollution sources, aquaculture is one of the main reasons for the high level of nutrients. Aquaculture accelerates eutrophication in lakes and reservoirs due to its adverse effects on water quality and the ecosystem [9]. The discharge of aquaculture pond effluents could be an important contributor to the problem of water pollution and eutrophication in the receiving waters of the coastal zones in China. Effective treatment of aquaculture pond effluents before discharge will become an important challenge in the future in alleviating the pressures of eutrophication in the coastal zone [10].

Water conditions must be strictly controlled to maintain water quality conditions in order to be able to support the life and health of aquaculture organisms [11], but there is also a need to limit the number of ponds so as not to have a negative impact on the environment. Whereas in Indonesia, most of the fishery cultivation waste is feed that is not consumed by fish around the pond. Clearly there is a need to limit the number of ponds in a particular area so that the ecological carrying capacity of the water body in which the pond is installed is not exceeded, with undesirable biota and water quality changes [12]. In addition to these two things, the need for consideration for the location of ponds that will be used. For example for brackish water ponds. This as most brackish water ponds are set in river deltas where fresh water historically has been lost to the ocean anyways. Redirecting the flow of water into shrimp ponds could, however, affect natural estuarine ecosystems as these have been shaped by natural fluctuations in salinity. Future research should therefore identify at which scales redirecting river flows into brackish water ponds start to have negative impacts on local ecosystems, and whether zoning and area based management could reduce these negative effects[13].

Water Quality Parameters

The range of these parameters is based on the Decree of the Minister of Marine Affairs and Fisheries Number: Kep. 28 / Men / 2004 About the General Guidelines for Shrimp Farming in Ponds [14].

Table 1. Sources Water Quality Parameters

Parameter	Range
Salinity (ppt)	5-35
pH	7,0 – 9,0
Alkalinity (ppm)	>50
H_2S (mg/l)	0,001
Organic matter (ppm)	<55
Total phosphate (ppm)	0,05-0,50
BOD (ppm)	<25
COD (ppm)	<40
TSS (ppm)	25-500

Information:
BOD = Biochemycal Oxygen Demand
COD = Chemycal Oxygen demand
TSS = Total Suspended Solid

Table 2. Shrimp Pond Quality Effluent Standards

Parameter	Unit	Value
TSS	mg/l	≤200
pH		6-9
BOD	mg/l	<45
H2S	mg/l	<0,03
NO3	mg/l	<75
NO2	mg/l	<2,5
NH3	mg/l	<0,1

Based on Government Regulation No. 20 of 1990 that water used for fishery and livestock activities is found in class C [15].

Table 3. Water quality data that can be used for fisheries

Parameter	Unit	Maximum levels [15]	Optimum range [16]
NH_3	Mg/l	0,02	
NO_2	Mg/l	0,06	
DO	Mg/l	Required ≥3	>5
pH	-	6-9	6,5-9
temperature	°C		28 -32
brightness	cm		30-40
Salinity	ppm		5-25

In this study, the presence of plankton also play a role in the life of aquaculture cultivation. Plankton calculation is done by collecting plankton using plankton net. Since plankton is the basis of the food web, there is a close relationship between plankton abundance and fish production [17].

Once we know the quality standards of water quality for aquaculture, then we can know the potential of diversification of any fishery business that can be done in the area. Suppose the cultivation of shrimp-milkfish, milkfish-seaweed, or seaweed-shrimp. The purpose of this diversification is to minimize the risk of failure of fishery cultivation business, is expected to increase the income of coastal communities, especially fish farmers, take advantage of spare time or take advantage of assets owned, and preserve nature.

3. Method

This method refers to scientific research conducted by Anggoro Prihutomo [18]:

Materials and Research Tools

Materials used in this study include area maps or maps of pond farm use, GPS, sample bottles, label paper, secchi disk, pH meter, DO meter, refractometer, thermometer, plankton net, stationery, observation sheet, questionnaire, and camera.

Types and Research Methods

Based on the nature of the problem this research is a field research, while the purpose of this study is descriptive and explanatory research [19]. Judging from the analytical approach, this research is a kind of research with qualitative and quantitative approach, whereas from the depth of the analysis this research is a kind of inferential research which relies the conclusion of research result on a probability of rejection of null hypothesis. The method used in this research is research by using survey method, that is method that does not make changes (no special treatment) to the variables studied [20].

Assessed Parameters

1. Physical parameters include: temperature, brightness, suspended particle charge, and salinity.
2. Chemical parameters include: dissolved oxygen (DO), pH, BOD, NO_2.
3. Biological parameters include: Phytoplankton and Zooplankton.

Research Assessment

The scope of the study includes the description of the existing condition of the area and the environmental quality study. Description includes the general condition of the area, the activities developed and the environmental management undertaken. The study of pond quality quality is water quality description in a water quality index value in ponds with various aquaculture system.

Sample Research

The sample in this research is pond water from various aquaculture system. Selection of pond plot taken as sample is done non-random or non probability sampling with purposive sampling method. The pond samples were determined based on criteria based on the extensive representation of ponds and ponds with semi-intensive or intensive systems, the representation of the location of pond blocks, and the cultivation lifespan.

Data collection technique

Primary data obtained from observation and interview. Observations were conducted in both insitu and exsitu through laboratory analysis and interviews with actors of aquaculture activities in the study sites. Fill interviews through the aquaculture system undertaken, technical production processes from input to output. Secondary data is obtained from the literature study, activity report, previous production data archive related to this research. Secondary data is needed to support the completeness of primary data to obtain more complete information.

Determination of Pond Sample

Determination of ponds to be sampled is done through the results of initial observation visually to map the aquaculture system used as well as aquaculture commodities that are cultivated. In the aquaculture / pond technicians interviewed the maintenance system and maintenance age. After that the ponds with the technical condition of maintenance and maintenance age are almost the same calculated to be samples. Total ponds that develop intensive and traditional brackish water commodities are the study population, and from that population it is determined 10% to be the research sample. The distribution of pond samples is attempted to represent every area block in the research area and the aquaculture system.

Pond Water Sampling Technique

Water sampling is done on the mornings once a week for one month. Special parameters DO and temperature are measured in the morning and afternoon in situ, assuming the highest oxygen solubility (afternoon) and lowest (morning). Samples of water analyzed exituately in the laboratory were taken using bottles of 600 ml. Sampling is done at the midpoint of the water pool. Water quality parameters to be observed include physical parameters, chemical parameters, and biological parameters. Physical parameters include temperature, brightness, water depth, light intensity, and turbidity. Water chemistry parameters include DO, pH, BOD, NO_3, and PO_4. Biological parameters include density and plankton diversity.

Sample Analysis

The temperature parameters determined in situ using a mercury thermometer, dissolved oxygen were analyzed in situ using a DO meter, pH was measured using pH meters, brightness and depth by insitu using a secchi disk, and salinity using a refractometer. Total ammonia was tested using indofenol reaction, while nitrite with sulfanilamide staining method [21]. Plankton samples were taken by filtering 20 L of water using a net plankton with a hole diameter of 25 μm. Plankton observations include the abundance and diversity index (H ') of plankton present in pond water.

Plankton abundance

Calculations of plankton abundance were performed using the Haemocytometer aids seen using a microscope tool. The abundance of plankton species was calculated based on the following equation principle [22]:

$$D = (l/p)\ q\ l/v$$

Information:
D = number of unity volume plankton
Q = number of plankton in subsample
P = subsample volume
L = sample volume
V = volume of filtered water

Plankton Diversity

Plankton diversity expressed in a Shannon-Wiener Index is used to calculate the diversity index according to Odum by the following formula:

$$H' = -\sum_{i=1}^{S}\left(\frac{ni}{N}\right)\ln\left(\frac{ni}{N}\right)$$

Information:
H '= Shannon-Wiener diversity index
Ni = the number of individual types of i
N = total number of individuals throughout the genera
(Where S = number of types)

Data Analysis Research

Analysis of the existing condition of the aquaculture area is the data obtained is analyzed descriptively, so it can describe the aquaculture area. Determination of quality status of pond environment is done by analyzing descriptively water quality data of measurement result into a quantitative value in the form of Water Quality Index (WQI).

$$WQI = \Sigma(Si\ x\ Wi)$$

Information:
Si = Scores of water quality indicator values
Wi = Weight index of each water quality

The value of Si and Wi is standardized from 0 to 1, and therefore the WQI value is also calculated by the range of values 0 to 1. The scoring method or weighting of each parameter is adjusted to the importance of the function of each parameter in supporting the life of aquaculture organism.

Weighting is given subjectively. Physical, chemical, and biological properties are given almost equal weight, since the three indicators have an equally important function (weight value 1). Each indicator is assigned a factor called sub-weight (weighted value 2) whose weight assessment may differ for each individual depending on the individual's role of the water parameter.

The determination of the effect of the aquaculture system on the environmental status of the pond was done statistically by analyzing the variety (ANOVA) and if there was a significant effect followed by the mean test (compare means) using the Tukey HSD (Honestly Significant Different) test. Analysis of the influence relationship between parameters of water quality was analyzed using multiple linear regression statistic analysis.

4. Conclusion

Water quality plays an important role to the success of aquaculture, by knowing the water quality profile in a region then we can identify the potential of diversification of aquaculture what can be done in the pond. As is known, the purpose of this diversification is to minimize the risk of failure of fishery cultivation business, is expected to increase the income of coastal communities, especially fish farmers, take advantage of spare time or take advantage of assets owned, and preserve nature.

Reference

1. [S. M. Niti. J Saintek Fisheries **Vol. 4**, No. 1, 50 – 55(2008).
2. Zhongneng Xu , Claude E. Boyd. *Reducing the monitoring parameters of fish pond water quality. Aquaculture.* **465,** 359–366 (2016).
3. Ristiyani Dwi. *Evaluation of Land Suitability for Fish Pond Fisheries in Kendal Coast.* Geo Image **1,** 12-18 (2012).
4. H. Effendi, *Water Quality Study for Resource Management and Water Environment.* Yogyakarta: Kanisius. (2003).
5. Suwarsih, Marsoedi, Nuddin Harahab, Mohammad Mahmudi. Proceedings of the National Marine Seminar (2016)
6. Putra Reza Rammiko, Dr. Dedi Hermon, Mp, Farida S.Si, M.Sc. *Brackish Water Quality Study For Aquaculture In Coastal Area Linggo Sari Baganti Sub-district South Pesisir Regency.*
7. AA. Ekubo, JFN. Abowei, Applied Sciences, Engineering and Technology **3**. No 12 (2011) 1342-1357.
8. Makmur, Rachmansyah, and Mat Fahrur. *The Relationship Between Water Quality and Plankton in Ponds Tanjung Jabung Barat Regency Jambi Province.* The Center for Brackishwater Aquaculture Research. South Sulawesi.
9. Huai-dong ZHOU, Cui-ling JIANG, Li-qin ZHU, Xin-wei WANG, Xiao-qin HU, Jun-yu CHENG, Ming-hua XIE. Water Science And Engineering, (2011), **4**(1): 92-100.
10. Ping Yang, Derrick Y.F. Lai, Baoshi Jin, David Bastviken, Lishan Tan, Chuan Tong. Science of the Total Environment 603–604 (2017) 256–26.
11. Kilawati Yuni, Yunita Maimunah. Research Journal Of Life Science E April- (2015) **Vol 02** No. 01: 50-59.
12. [Edwards Peter. Aquaculture **447** (2015) 2–14.
13. Henriksson Patrik John Gustav, Nhuong Tran, Chadag Vishnumurthy Mohan, Chin Yee Chan, U-Primo Rodriguez, Sharon Suri A, Lara Dominguez Mateos, Nur Bambang Priyo Utomo, Stephen Hall, Michael John Phillips. Journal Of Cleaner Production **162** (2017) 1482-1490.
14. Decree of the Minister of Marine Affairs and Fisheries Number: Kep. 28 / Men / 2004 About the General Guidelines for Shrimp Farming in Ponds.
15. Government Regulation number 20 of 1990 on Water Container Control.
16. M. Kordi, H. Ghufran, Andi Baso Tancung. (2005). *Water Quality Management in Aquaculture.* Publisher Rineka Cipta.
17. A. Bhatnagar and P. Devi, International Journal of Environmental Sciences, **Vol. 3**, No. 6, (2013) 1980-2009.
18. Prihutomo Anggoro. *Quality Status of Biological Physics and Chemistry (BIOFISKIM) and Supporting Power of Pond Environment in BLUPPB Karawang Aquaculture Area of West Java.* Diponegoro University. (2015).
19. Suryana, (2010) *Research Methods Practical Model of Quantitative and Qualitative Research*, Bandung: UPI.
20. Dharminto, (2007), *Research Methods and Sample Research.* http://eprints.undip.ac.id/5613/1/METODE_PENELITIAN_-_dharminto.pdf.
21. S. Nimrat, S. Suksawat, P. Maleeweach & V. Vuthiphandchai. (2008). Aquaculture. **285**: 123-129
22. W. Wardhana, (2003). *Sampling Techniques Preservation and Plankton Analysis.* [Journal]. Presented in Sampling Technique Training and Plankton Identification. Fishery Quality Development and Testing Center. Jakarta.

Study of Carrying Capacity Assesment for Natural Fisheries in Jatibarang Reservoir In Semarang City

Bambang Sujono[1], and *Sutrisno* Anggoro[2]

[1]Master Program of Environmental Science, School of Postgraduate Studies, Diponegoro University, Semarang - Indonesia.
[2]Fisheries and Marine Sciences, Diponegoro University, Semarang - Indonesia

Abstract. Jatibarang reservoir serves as water supply in dry season and controlling flood in Semarang City. This reservoir is stem Kreo River which cathment areas of 54 km^2, pool of area 110 ha and volume is 20 billion m3. This reservoir is potential to develop as natural fisheries area. The goals of this research were to explore existing condition of physical, biological as well as chemical parameter; carrying capacity assessment for natural fisheries; determining appropriate fish species to be developed in Jatibarang reservoir. This research was done in descriptive explorative scheme. Field survey and laboratory analyses were conducted to identify physical, chemical and biological parameters of the water. Physical parameters measured were temperature and water brightness. Chemical parameters measured were pH, DO, phosphate, Ammonia, nitrites and nitrate, while biological parameter measured were chlorophyll-a concentration. Carrying capacity analyses was done referred to the Government Regulation Number 82, 2001 that regulate the management of water quality and water pollution control. Based on the research, it showed that the existing condition of physical, chemical and biological parameters were still good to be used for natural fisheries. Based on TSI index, it classified as eutrofic water. Furthermore, tilapia fish (*Oreochromis mossambicus*), nile tilapia (*Oreochromis niloticus*) tawes (*Barbonymus gonionotus*) and carper fish (*Cyprinus carpio*) were considered as best species for natural fisheries in Jatibarang Reservoir.

1 Introduction

Jatibarang reservoir that is built in Kandri Village, Gunung Pati sub-district, Semarang city is strategic area of environmental support, with the major function as flood controller and can be develop as tourism. Reservoir of Jatibarang serves as water storage in the dry season and controlling flood during rainy season. This reservoir was stemming Kreo River which catchment areas is 54 km^2, broad puddle 189 ha and reservoir volume is 20,4 millionm^3 [1]. The such condition has a chance to develop as natural fisheries.

2 Material and methods

This Research location was Jatibarang Reservoir, Semarang City, Central Jawa. The research was divide into two activities, field survey and laboratory analyses. Primary and secondary data was collected during field survey and water quality analyses was conducted in Laboratory of Fisheries Department, Diponegoro University, Semarang City. Water sampling was done during February, 2017 in 4 location (Figure 1). Physical parameters measured were water temperature, water brightness; Chemical parameters measured were pH, DO, Phosphate, Ammonia, Nitrite, and nitrate. Furthermore biological parameters measured were chlorophyll-a concentration.

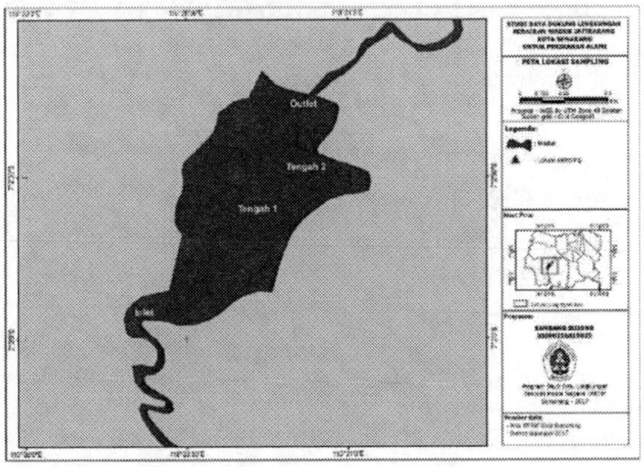

Fig. 1. Sampling Location in Jatibarang Reservoir

*Corresponding author: abee4833@gmail.com

Water sampling were done in the surface water in inlet, two location in reservoir water body dan one location in outlet.

Physical, chemical and biological analyses were conducted in laboratory. (Table 1).

Table 1. Water Quality Analyses Methods

Parameter	unit	Methods/tool	Description
Physical			
Temperature		Thermo meter	*In-situ*
Brightness	cm	Visual, *Secchi disk*	*In-situ*
Chemical			
pH	-	pH meter	*In-situ*
DO	mg/L	DOmeter	*In-situ*
Ortofosfat	mg/L	Spektrofotometer	*Ex-situ*
Total Fospat	mg/L	*Molybdate ascorbic Acid*, Spektrofotometer	*Ex-situ*
Amonia	mg/L	*Phenate*, Spektrofotometer	*Ex-situ*
Nitrit	mg/L	*Sulfanilamide*, Spektrofotometer	*Ex-situ*
Nitrat	mg/L	*Brucine*, Spektrofotometer	*Ex-situ*
Biologycal			
Chlorophyll-a	mg/m^3	Spektofotometer	*Ex-situ*

Series of analyses was done including physical, chemical and biological parameters to describes water fertilities and the status of carrying capacity of natural fisheries[2] Tropic State Index (TSI) was applied to determine water fertilities [3], which mainly using three parameters (water brightness, phosphate total and chlorophyll-a).

TSI-P = 14,42 x Ln[TP] + 4,15 (µg/l).....(i)
TSI-Cla = 30,6 + 9,81 x Ln[Chlor-a] (µg/l)..(ii)
TSI-SD = 60 – 14,41 x Ln[Secchi] (meter) ..(iii)
TS I Average = (TSI-P + TSI-Cla + TSI-SD)/3..(iv)

Which:
TSI-P is *Trophic State Indek* for phospat
TSI-Cla is *Trophic State Indek* for chlorophyll-a
TSI- SD is *Trophic State Indek* for secchi disk depth

Based on TSI index, level of water fertilities were*ultra oligotrophic (<30), oligotrophic (30-40), mesotrophic (40-50), eutrophic (50-60), weight eutrophic (60-70), hipereutrohic (70-80) and alga scum (>80)*(Carlson,1977).

Fish production and carrying capacity for natural fisheries fully depends on plankton production. Therefore natural fisheries carrying capacity was analysed using plankton primary productivity.

Primary production productivity was common method to predict fish production and ranching. Primary productivity (∑PP)calculated by maximum chlorophyll a concentration [4].

Based on TSI and chlorophyll concentration, stated that chlorophyll concentration in eutrofic condition is 20 mg/m^3[3]. Chlorophyll a concentration used to predict primary productivity based on [5].

$$\sum PP = \frac{483*Chla^{1,33}}{9+1,15*Chla^{1,33}}$$

Primary productivity was conversed to calculated annual fish production based on conversion table [6] below:

Table 2. Annual Primary Productivity Conversion based on Beveridge,1984

PP	Annual fish production
(gC/m^2/year)	(g fish C/m^2/year)
<1000	1,0 -1,2
1000 -1500	1,2 -1,5
2000 -2500	2,1 -3,2
2500 -3000	3,2 -2,1
3000 -3500	2,1 -1,5
3500 -4000	1,5 -1,2

Furthermore, fish species analyses was done using catched fish during field survey in Jatibarang reservoir.

3 Results and discussion

3.1 The condition of Jatibarang reservoir

Geographically, Jatibarang reservoir was located in coordinate 110°20'49.71"E and 7° 2'24.67"S. Based on administrative located in Kandri Village, Gunung Pati Sub district, Semarang City (Figure 2). This reservoir was stemming Kreo River which catchment area 54 km^2, broad puddle 184 ha and reservoir volume is 20 m^3, and 10-15 meter in depth. This reservoir was supply water for West Semarang District which debit is 1.050 lt/second, maximum water level is +155.30m, level above foundation 74 m and peak elevation +157m, 200 m peak length.

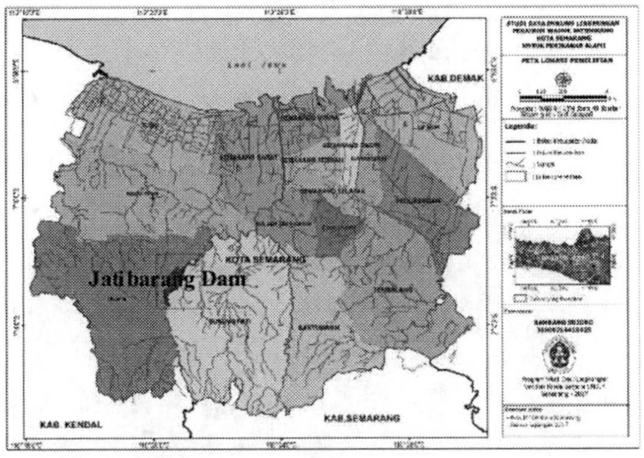

Fig. 2. Map Of Jatibarang Reservoir

3.2 Physical, Chemical and Biological Parameters

The results of in situ and ex situ parameters were as shown in Table 3 below.

Table3. Physical, Chemical and Biological Parameters of Jatibarang Reservoir.

Parameter	Unit	St 1	St 2	St 3	St 4
Temperature	^0C	29.5	30,3	30.8	30,9
pH	-	6.9	7.1	7.2	7.1
DO	mg/L	4,90	5,00	5,10	5,10
Brightness	M	0,80	1.45	1.6	1,20
Nitrat	mg/L	1,21	0,91	1,19	0,79
Nitrit	mg/L	0,06	0,05	0,06	0,04
Amonia	mg/L	0,08	0,05	0,04	0,20
Ortofosfat	mg/L	0,06	0,07	0,07	0,06
Total Fosfat	mg/L	0,53	0,31	0,15	0,16
Chlorophyll-a	mg/m^3	55,15	70,45	60,31	35,12

The physical, chemical and biological parameters of Jatibarang reservoir was various during the research (Table 3). Water temperature range was 29,5-30,9°C which average30,37°C. Optimum temperature for trophical fish is 25 - 32°C [7].Water brightness range 0,8 - 1,6 m which average 1,26 m. Lowest water brightness located in inlet area in correlation with concentration of total dissolved suspend. The level of water brightness describe euphotic zone indirectly, it correspond to nutrient concentration in water body [2]. pH range was netral 6.9-7.1 which average 7.05. Mostly aquatic species sensitive with pH dan prefer in range 7 – 8,5 [8]. Dissolved oksigen range 4.9-5.1 mg/l which average 5,06 mg/l. Commonly DO concentration 5 ppm which water temperature 20-30°C suitable for fish [9].

Dissolved oxygen parameter is vital parameter for the water organism. DO concentration tends to the weather. The oxygen levels have an real to water living [10]. Oxygen plays an important role as an indicator of the quality of waters, because oxygen dissolved participate in the process oxidation and reduction of organic matter and inorganic. In addition, oxygen also determines biological done by an organism aerobic or anaerobic .The aerobics, the role of oxygen is to oxidize organic matter and inorganic with the end result is nutrients which in turn would give fertility [11] .The main source of oxygen dissolved my water is diffusion from the air and results photosynthesis organisms have chlorophyll living in waters. Diffusion of oxygen from the air into the water was very slow, therefore phytoplankton is the main source of dissolved oxygen in the water (Moriber,1974) [10].

Fishes use oxygen to their activities, growth , reproduction etc. Hence oxygen for fish determine a circle of fish activities, feed conversion. Growth rete similarly depend on oxygen with the provisions of another factors is optimum [12].

The nitrites concentration between 0.04 -- 0,06 mg / l with average of 0.05 mg /l. Concentration of ammonia 0,02-0,8 with an average 0,092 mg/l, is still in range within the limits of tolerance for life fish. Pescod (1973) and samuel et. al. described it on tropical waters no more than 1.0 mg/l for fisheries. The concentration of orthophosphate 0,06-0,07 mg/l with an average 0,067 mg/l, meanwhile the total phosphorus range 0,15- 0,63 mg/l with an average 0,28 mg/l.

3.3 Carrying capacity for natural fisheries

Carrying Capacity is sum or quantity maximum a fish that can supported by a body of water in the long time, influenced by time laundering, volume of water body, and load waste in to the waters. Natural fisheries in reservoir are herbivorous and fully utilizing phytoplankton in waters as a source of nutrients.

The basic principle of natural fisheries is based on the capacity of waters which includes the condition of physics, chemical and biological waters. This is because natural fisheries fully rely on natural food chain and to avoid any chemicals derived from the rest of artificial feed.

Based on field survey and laboratory analyses, the physical, chemical and biological parameters in Jatibarang reservoir still in good condition for natural fisheries activities based on level 3 standart on Government Regulation Number 82, 2001 concerning in water quality dan water pollution control [13].

Meanwhile TSI index range between 60-65, therefore Jatibarang Reservoir indicated medium eutrophic. It's indicates that productivity of waters high with the low diversity. Based on it, it is highly recommended that natural fisheries to be developed.

The most fish catched during the research were tilapia fish (*Oreochromis mossambicus*) and nile tilapia (*Oreochromis niloticus*). This two species were mostly recommend as fish species for natural fisheries in Jatibarang reservoir in Semarang City.

4 Conclusion

Based on this research, can be concludes :

a. Existing condition of chemical and biological parameters of Jatibarang reservoir is still good for natural fisheries based on level 3 Government Regulation Number 82, 2001 that regulate the management of water quality and water pollution control.
b. Carrying capacity of Jatibarang Reservoir based on TSI Index was 60-65, therefore called medium eutrophic. It showed that water productivities is high and low in diversity.
c. Tilapia fish (*Oreochromis mossambicus*), nile tilapia (*Oreochromis niloticus*), tawes (*Barbonymus gonionotus*) and carper fish (*Cyprinus carpio*) were recommended species for natural fisheries in Jatibarang reservoir.

References

1. Wening Indriani, Sahala Hutabarat, Churun A'in. Diponegoro Journal of Maquares **5**(4): 258-264.(2016)
2. Pratiwi, T.MNiken,S, Hariyadi, Inna Puspa Ayu, Novita MZ, Tri Apriadi, Jurnal Biologi Indonesia **11**(2):267-274. (2015)
3. R. E. Carlson, A trophic State Index for Lakes. Limnology and Oceanography **22** (2): 361-369.(1977)
4. De Silve SS, Amarasinghe US, Nguyen TTT. *Better practice approaches for culture based fisheries development in Asia*. ACIAR Monograph 120 (2006)
5. Smith VH, Journal of Plankton Research **29** (1): 1-6. (2007)
6. Beveridge MCM. Cage and pen fish farming: carrying capacity model and environmental impact. FAO. Fisheries Technology Paper **(255)**: 131 p. (1984)
7. Mulyanto. *Lingkungan Hidup Untuk Ikan*. Departemen Pendidikan dan Kebudayaan, Jakarta. (1992)
8. V. Novotny, and H. Olem, *Water Quality: Prevention, Identification and Management of Diffuse Pollution.* Van Nostrand Reinhold. New York. 1054 p. (1994)
9. Rifa'I RS, Pertagunawan K. *Biologi Perikanan 1*. Penerbit CV Kayago, Jakarta (ID). (1983)
10. Edward and MSTarigan, Makara Sains **7**,(2),82-89(2003)
11. Salmin, Oksigen Terlarut (DO) Dan Kebutuhan Oksigen Biologi (BOD) Sebagai Salah Satu Indikator Untuk Menentukan Kualitas Perairan. Oseana, Volume **XXX**, Nomor 3, p. 21 – 26 (2005)
12. Sutimin, *Model Matematika Konsentrasi Oksigen Terlarut Pada Ekosistem Perairan Danau.* Jurusan Matematika, FMIPA, Universitas Diponegoro, Semarang (2006)
13. Peraturan Pemerintah No. 82. *Pengelolaan Kualitas Air dan Pengendalian Pencemaran Air* (2001)

Sustainability Activities In The Mining Sector: Current Status And Challenges Ahead Limestone Mining In Nusakambangan

Theresia Vika Ayuningrum[1*] and *Hartuti* Purnaweni[2]

[1]Master Program of Environmental Science, School of Postgraduate Study Diponegoro University
[2]Public Administration Dept, Faculty of Social and Political Science, Diponegoro University

Abstract. Potential Karst area in Nusakambangan has an important role in maintaining the balance of nature. But with the existence of mining activities, will automatically change the environmental conditions there. In order for the utilization of resources to meet the rules of optimization between the interests of mining and sustainability of the environment so in every mining sector activities required a variety of environmental studies. The purpose of this study is to find out how the analysis of environmental management due to limestone mining activities in Nusakambangan so that it can be known the management of mining areas are optimal, wise based on ecological principles, and sustainability. In qualitative research methods, data analysis using description percentage, with the type of data collected in the form of primary data and secondary data.

1 Introduction

The mining industry has a marked influence on the economy of several countries, mostly through job creation and cascading effects on other sectors. However, mining activities often have major environmental impacts that incur long-term costs for governments. In response to mounting criticism, the mining industry has paid increasing attention to the environmental and social impacts of its activities, notably by embracing the concept of sustainable development [1]. However, the capacity of the mining industry to be sustainable is often contested [2], mostly because non-renewable resources are exploited [3]. Moreover, while civil society considers sustainable development a necessity, industries such as mining still largely see it as a compromise [4].

Adopting sustainable development principles can be a major challenge for the mining industry [4,5]. Important efforts have been made to better take into account environmental and social issues across the sector [6,7]. These issues are increasingly mentioned in the sustainable development reports published by mining companies[8,9,10]. However, the efforts of mining companies to ensure that their activities are respectful of the environment and society are often below the expectations of the population [11,12,13].

Limestone mining on Nusakambangan Island has been done by PT Holcim Indonesia Tbk (formerly PT Semen Nusantara) since 1977 ago. This activity is done to get limestone as raw material of cement making. This study aims to determine the physical environment changes due to limestone mining and to know how the activities of sustainable environmental management post mining. This research uses descriptive qualitative method based on interview result and direct observation in field and supported by secondary data. The results of research conducted in several regions indicate that limestone mining activities have a negative impact, such as the formation of very precipitous slopes that are very dangerous, air pollution, the number of open land, the decrease of the flow of water springs. Nevertheless, mining activities are still running because mining activities also have a positive impact on the original revenue (PAD) Cilacap regency.

2 Sustainable Indicators

The minerals industry is often divided into four subsectors: Energy minerals (e.g. coal, oil); Metallic minerals (e.g. iron, copper, zinc); Construction minerals (e.g. natural stone, aggregates, sand, gravel, gypsum), and Industrial minerals (e.g. borates, calcium carbonates, kaolin, plastic clays, talc). Although these four sub-sectors share many common characteristics with regard to sustainability issues, they are also quite different. For example, a key issue associated with the use of energy minerals is depletion of nonrenewable resources. Other minerals on the other hand are (strictly speaking) not depleted, but dissipated in use and can (at least

*Corresponding author: vikaayuningrum89@gmail.com

theoretically) be recovered and reused. These and other differences require a particular care in the development of an appropriate set of sustainability indicators that, in turn, is contingent upon identification of the key sustainability issues that are relevant for an industrial activity. In the development of the indicators proposed here, the emphasis has been on the key issues relevant to three mineral sub-sectors: metallic, construction and industrial minerals.

2.1 Economic indicators

Economic indicators measure economic impact of a company on its internal and external stakeholders and on economic systems at the local, national and global levels. Therefore, the economic indicators need to include the usual measures of financial performance such as profits and shareholder returns, but also to go beyond the traditional fiscal indicators to reflect the wider contexts in which companies operate. The economic indicators within this framework are based on the five GRI categories of performance each comprising a number of specific indicators such as net sales, return on investments and taxes paid to the public sector.

In addition to these, as indicated, it has been necessary to add two more types of indicator, related to products (e.g. value added) and local communities (e.g. revenue distribution). This makes the information provided under this category more transparent and compatible with the way the environmental and social indicators are presented. The categories of economic indicators proposed within this framework are summarized, which shows the type of information provided by each indicator category as well as the economic issues and stakeholders they affect. The full list of economic indicators. All together, 24 economic indicators have been proposed, of which six are additional to those proposed by the GRI. The additional indicators include generally-applicable measures of economic performance such as earnings before interest and tax (EBIT) as well as the sector-specific indicators, such as total investments for mine closure and rehabilitation.

2.2 Environmental indicators

Of the three types of sustainability indicators, environmental measures of performance are most developed and have achieved the highest degree of consensus among experts. They measure a company's impacts on natural systems, including humans, ecosystems, land, air and water. These impacts can be local, regional or global thus affecting a wide range of stakeholders. For example, depletion of minerals resources has both local and global implications. Locally, it will affect the employees and their job prospects as well as the local communities' prospects to benefit in the longer term from mining activities. Globally, depletion of minerals reserves affects both the current and future generations, as they can no longer rely on these resources to satisfy their needs.

It is therefore important that the environmental indicators reflect not only the key environmental issues but also the scale of magnitude of the impacts. This can be achieved by using both the absolute measures of environmental performance (e.g. tonnes of mineral resources extracted per year) and normalized units (e.g. tonnes of a mineral resource depleted per year relative to the world reserves of that resource). Normalized units also enable comparisons between companies of different sizes, particularly if they are expressed per tonne of product.

As discussed earlier, another important aspect in developing the environmental indicators is life cycle thinking which provides a holistic approach required in addressing sustainable development. The life cycle approach is particularly useful for the identification of 'hot spots' in the supply chain as it shows the most important stages and impacts in the life cycle so that they can be targeted for maximum improvements. Life cycle assessment (LCA) is the usual tool used for quantification of environmental impacts of products and processes along the supply chain.

However, it is appropriate to make a distinction between the scope of the analysis for mineral products and mine facilities. Given a myriad of uses that most minerals are put to, in most cases it would be impossible for mineral companies to follow all their products from 'cradle to grave'; instead, a reduced, 'cradle-to-customer' scope may be more suitable but also more practical. This includes all activities along the supply chain, from extraction to delivery of the mineral product to the customer. However, in cases where companies are also involved in the recovery of waste products, a full 'cradle-to-grave' approach should be taken. For the mine and processing facilities, however, the full 'cradle-tograve' approach must be taken, encompassing design, operation, decommissioning and rehabilitation.

2.3 Social indicators

While the economic and environmental performance are relatively easy to measure and the indicators are generally well developed and agreed upon, measuring the level of social sustainability of a business or a sector is not an easy task. One of the reasons is that social indicators must take into account the many interests of both employees and those of the wider communities to reveal company's social impacts at the local, national and global levels. Furthermore, in social and ethical dimensions of a company's activity, many of the variables such as protection of human rights or cultural values are hardly quantifiable, and cannot even be defined in physical terms [14]. Yet, without addressing these issues, the assessment of sustainability would not be complete.

Therefore, the set of social indicators proposed within this framework has been designed to include both qualitative and quantitative indicators, as appropriate. Following the GRI framework, they have been grouped into five categories, related to labour practices, human rights and broader issues affecting consumers, local communities and other stakeholders..The GRI recommends the use of 53 different social indicators. This number has been reduced to 45 in the framework proposed in this paper as some of the GRI indicators are more appropriate for companies in the consumer products sector rather than to the mining and minerals industry.

3 Overview Of Nusakambangan Island

Nusakambangan Island is located in the south of Java island and is the outermost small island bordering Australia. Geographically located at 7°30'-7°35 'LS and 108°53'-109°03'BT. To the north of the island there is a strait known as the seal of tillers. This strait separates the island of Nusakambangan with mainland Java island, especially with city of cilacap. To the south of the island nusakambangan is an Indian ocean famous for its great waves. Nusakambangan island is about 210 km^2 or about 21,000 ha. It extends from west to east. Administratively Nusakambangan island is included in the area of Tambakerja subdistrict Cilacap South, Cilacap, Central Java.

Nusaakambangan has an area of 21,000 Ha. In it there are four small conservation areas, namely Nusakambangan Barat Nature Reserve (928 ha), East Nusakambangan CA (277 ha), CA Wijayakusuma (one ha), and CA Karangbolong (0.5 ha) Its status since the Dutch era.To the north of Nusakambangan island lies Segara Anakan with a series of small islands along with swamps and mangrove forests on the southern coast of Cilacap. Segara Anakan waters is part of the Indonesian Ocean with river estuaries and the brackish forest of Cilacap and Ciamis. Together with Nusakambangan forest,SegaraAnakan and its surrounding areas constitute an ecosystem structure composed of small island ecosystem units with lowland forests in Nusakambangan, coastal and coastal, lagoon ecosystems in Segaraa nakan, mangrove swamps, both in the North coast of Nusakambangan, south coast of Cilacap and east coast of Ciamis. This environmental arrangement is influenced by the ocean waters of the Indonesian Ocean and freshwater from several watersheds such as Citanduy from the west, Cibeureum and others from the north, and the Donan River near Cilacap.

Natural panorama such as natural caves that number more than 25 and still growing and growing in terms of growth staklamit and stalatitnya, with an average depth of 3-4 meters below sea level and consists of old volcanic rock. White sand, wet tropical rain forests that are rare in Central Java, and historic buildings such as prison houses built by the Dutch, Portugal fortress, is a natural and historical potential that is very possible to be a different tour.

Nusakambangan tropical natural forests, now partially preserved as Nature Reserve (forest cover), for the conservation of germplasm, maintaining endemic species, as well as wildlife habitats such as panthers, javan hawks, langurs, pangolins, and some rare bird species.

Topography is generally hilly and bumpy, as well as slightly flat areas on the north and west ends. Pucak-peak is quite high ranging from 150-200 MASL (metres above sea level) with the highest peak reaches 190 m in the eastern part of the island. The area of ridge-ridge that runs along the northern part of the island consists of limestone hills, while along the southern part it consists of old volcanic rocks.

Nusakambangan until the 1980s, is a closed area, but now closely related to the utilization of natural resources. Mining of cement raw materials appears to be one of the limestone limestone factors, although rehabilitation measures have been designed and implemented. The current form of nusakambangan management consists of nature reserves, protection forests, mining areas, agricultural lands and penitentiary complexe.

4 Limestone mining in Nusakambangan

Limestone mining on the island of Nusakambangan has been done by PT Holcim Indonesia Tbk (formerly PT Semen Nusantara) since 1977. Holcim's mining permit area reaches 1,000 ha, but only mined about 350 ha or 0.45% of the area of Nusakambangan Island. The remaining 650 ha will be used as conservation forest.

Limestone mining on Nusakambangan Island and clay in Jeruklegi by Holcim, has legal and licensing grounds, among which, Adendum No.017 / - LCA.DIR / I / 2012 Letter of Cooperation Agreement between PTSemen Cibinong Tbk (PT Holcim Indonesia Tbk) With the Ministry of Justice and Human Rights No: E.PL.03.06-629 0034 / Dir / XI / 2001 on Limestone Mining on Nusakambangan Island with a period of 30 years, valid from November 27, 2001 to November 26, 2031. In addition, the Decree of the Governor of Central Java Number 540/32 / - Year 2000 dated 18 September 2000 on the granting of Mining Business License (SIPD) Exploitation of Lime Stone Material to PT Semen Cibinong Tbk valid until 19 September 2023. Conservation Forest Decision Java Governor Central Number 540/11/2006 on Amendment to Decree of Central Java Governor Number 540/32/2000 concerning Provision of Mining Permit (SIPD) Exploitation of Lime Stone Material to PT Semen CibinongTbk is valid until September 19, 2023. Then Decree of Regent of Cilacap no. 545/421/19/2011 dated September 12, 2011 concerning the granting of Mining Operation Permit (IUP) of Non-Metal Mineral Production Operation to PT Holcim Indonesia Tbk valid until September 19, 2023. All mining land has been reclaimed according to the document of analysis Environmental impact (AMDAL), RPL and RKL, by closing the ex-mine area with soil, and then greening by planting Nusakambangan native plants.

Overall, limestone reserves in the SIPD area of 1,000 hectares is estimated at 290 million tons. In the former area of PT Holcim Indonesia mine is re-planted into a forest with Nusakambangan native plants. The A and B zones that are being cultivated are subdivided into quarry quarries. The limestone mining in Nusakambangan area begins by blowing limestone hills with dynamite that once blast is installed in about seven dots. Based on the company's warning board, the preparation siren sounded at 11.45-11.55, blasting sirens at 12.00 and 16.00, the safe siren at 12:15 and 16:15.

5 Issues related to the impact of limestone mining in Nusakambangan

Exploitation of limestone in Nusakambangan has been carried out continuously with a very large amount of production. This condition leads to significant change in the topography of the mining area. Mining often substantially alters the landscape in ways that affect groundwater, surface water and enviromental resources. Streams, springs, and wetlands are often substantially altered or oven removed from the landscape. Surface mines, by their nature, results in the removal of the landscape within the footprint of the open pit. The removal of key hydrologic landscape elements, such as sinkholes, streams and springs, may result in substansial alteration of ground water flow patterns, quantity and quality.

Karst is basically an area vulnerable to environmental damage. Mining of lime in the karst area that currently becomes uncontrolled leads to environmental damage. The damage originated from the cutting down of cover vegetation to support limestone mining activities. The deforestation that occurred in the karst hills caused the area to be unable to catch the rain. Impact, the area will be dry. Reduced vegetation around the karst hills also causes a decrease in the oxygen content in the air so that the temperature in the region becomes higher. The least vegetation in the karst area also causes the soil to become increasingly unstable. The absence of a buffer / buffer in the form of trees makes the area very vulnerable to the movement of land masses, especially landslide hazards. In addition, other environmental problems that may arise are decrease in soil productivity, erosion and sedimentation, and disturbance to flora and fauna that have habitat in karst area.

Here are some of the features of the mining area on the island nusakambangan photographed from google earth.

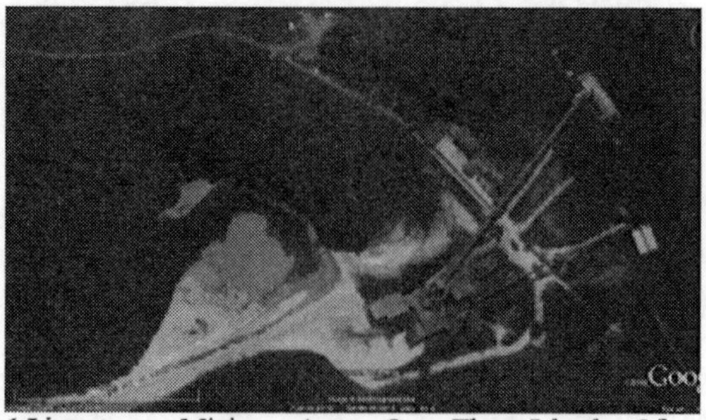

Fig 1. Limestone Mining Area On The Island Of Nusakambangan

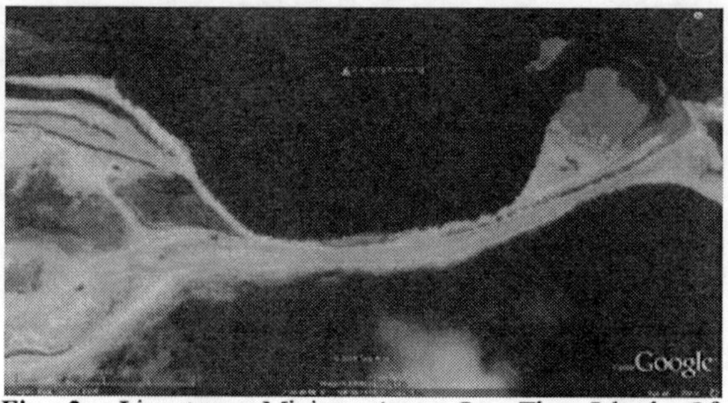

Fig 2. Limestone Mining Area On The Island Of Nusakambangan

Reference

1. Royaume du Maroc, Projet de loi no 33-13 relative aux mines. Royaume du Maroc, Ministère de l'Énergie, des Mines, de l'Eauet de l'Environnement, Département de l'Énergie et des Mines. Sater, J.N., 2010. Morocco: Challenges to Traditi, (2015)
2. K. Babi, Perceptions du développementminier durable par les acteurslocaux, gouvernementauxetindustriels au Maroc. Masters thesis, Université du Québec en Abitibi-Témiscamingue,Rouyn-Noranda.
3. J.E.Tilton, Resour. Policy **22**, 91–97, (1996)
4. Akcil, A., Koldas, S, J. Cleaner Prod. **14**, 1139–1145, (2006)
5. Andrews-Speed, P., Ma, G., Shao, B., Liao, C., *Economic responses to the closure of small-scale coal mines in Chongqing*, China, (2005)
6. Azapagic, A., Resour. Policy **30**, 39–54. (2004)
7. J. Cleaner Prod. **12**, 639–662.
8. Burneo, M.L., Warnaars, X., Dev. Change **39**, 887–914, (2008)
9. J.Bhattacharya, Mineral Resour. Eng. **9**, 451–464, (2000)
10. B.Bossé, B.Bussière, R.Hakkou, A.Maqsoud, M.Benzaazoua, Mine Water Environ. **32**, 152–167, (2013)
11. F.Vanclay, Conceptualizing social impacts. Environ. Impact Assess. **22,** 183–211, (2002)
12. Van Dam, R.A., Humphrey, C.L., Martin, P., Toxicology **181/182**, 505–515, (2002)
13. D.J.Vogel, Manag. Rev. **47**, 19–45.
14. Wang, Y., Shi, J., Wang, H., Lin, Q., Chen, X., Chen, Y., 2007. Ecotoxicol. Environ. Saf. **67**, 75–81, (2005)
15. A.Whitmore, J. Cleaner Prod. **14**, 309–314.
16. Woertz, E., 2014. *Mining strategies in the Middle East and North Africa.* Third World Q. 35, 939–957.

17. Worrall, R., Neil, D., Brereton, D., Mulligan, D., 2009. J. Cleaner Prod. **17**, 1426–1434, (2006)
18. Younger, P.L., Total Environ. **265**, 309–326, (2001)
19. *Dokumen Status Lingkungan Hidup Kabupaten Cilacap* Tahun 2007.
20. Partomiharjo, Tukirin & Ubaidillah, Rosichon. *Daftar Jenis Flora dan Fauna Pulau Nusakambangan, Cilacap Jawa Tengah* (2004)

Mangrove Cultivation For Dealing With Coastal Abrasion Case Study Of Karangsong

Feti Fatimatuzzahroh[1,*], *Sudharto* P. Hadi[2], *Hartuti* Purnaweni[2]

[1]Doctorate Program of Environmental Studies, Diponegoro University, Semarang, Indonesia
[2]Study Program of Environmental Studies, Diponegoro University, Semarang, Indonesia

Abstract. Coastal abrasion is consequence from destructive waves and sea current. One of cause is human intervention. The effort to solve of abrasion is by mangrove cultivation. Mangroves are halophyte plant that can restrain the sea wave. Mangrove cultivation required participation community that give awareness the importance of mangrove in coastal sustainability. Mangroves in coastal Karangsong, Indramayu west java, in 2007 was through abrasion approximately 127.30 ha. Mangrove cultivation in Karangsong has been replanting since 1998 to 2003, but there was no maintenance and management. In 2007 until 2015 Karangsong replanting mangroves and has been succeed. Karangsong became the center of mangrove study for west java area in 2015. This achievement is result of cooperation between community, NGO, and local government. In addition, this effort made not only overcome the abrasion problem but also give community awareness about the importance of mangrove cultivation in preventing coastal abrasion throughout community development. This paper reviews abrasion in Karangsong and the impact for local community and empowerment in mangrove cultivation. To achieve the success mangrove cultivation required community development approach from planning process, planting, maintenance and management.

1 Introduction

Coastal abrasion is a common problem facing coastal Indonesia. At least recorded from the 99,093 km long coastline of Indonesia, The ministry of Marine Affairs and Fishery said that rate of abrasion in Indonesia is 1.950 ha per years with average coastline 420 km/ years. Abrasion is caused by a wave of seawater that hit the shoreline with no wave retention so as to erode the erosion of the beach. The destruction of coastal areas has had a social and economic impact on the surrounding communities. Most coastal damage is an abrasion caused by the loss of mangrove forests as a natural protector in coastal areas. Mangrove cultivation is required to restore the affected coastal areas. Community development approach is needed in successful mangrove cultivation to overcome coastal abrasion.

Karangsong is a coastal area located in Indramayu, West Java, has been started mangrove cultivation activities to overcome coastal abrasion. Since 2007 the government with the local community has been replanting mangrove but because there is no good management, mangrove cultivation did not succeed and produce mangrove forest along the coastline of Karangsong. In 2015 Karangsong set as the center of mangrove studies for the western part of Java. The factor that moves society and government Karangsong to do mangrove cultivation is the number of ponds that have been affected by abrasion so that people Karangsong both ponds and fish yield from ponds also decreased. This journal reviews abrasion in Karangsong and its impact on local residents. The questions addressed are (1) How cultivation of mangroves in Karangsong? (2) What is the impact of mangrove cultivation on abrasion in coastal Karangsong? (3) And what are the barriers to mangrove cultivation in Karangsong related to community empowerment?

2 Literature review

Community development in coastal area is to build and maintain local biological diversity and to solve their problem with their local potential. Community refers to village-level or single goods-level activities [1]. Indonesia has a rich biological diversity that Foucat and Jones argued rationale for developing community-based management is to build upon and maintain local, ethnic and biological diversity[2,3]. Community involvement in management and conservation activities encourages to a local conservation commitment and can improve a

*Corresponding author: feti.faza@gmail.com

community's ability to sustainably manage natural resources through generating appropriate norms and rules and enhancing trust and reciprocity.

Community development has a variety of strategies that are less effective than others [4]. Furthermore, Pomeroy (1994) mentions that community management approach represents an essential practice to reach sustainability. And community management can have the effect of intensify long state control (where some may desire while others resist) and denying people the space to change, to privatize property, and to invent alternate resource management regimes in their own interest [5].

3 Research methodology

This research used descriptive analysis method that depict mangrove cultivation and its impact in overcoming abrasion. The study included (a) mangrove cultivation (b) impact of mangrove cultivation on abrasion (c) barriers to mangrove cultivation. Data collection used primary data which is information from impact of mangrove planting to abrasion and secondary data which is abrasion data and mangrove cultivation. The case study was conducted in Karangsong, Indramayu, West Java.

4 The result and discussion

Mangrove cultivation in Karangsong has been done since 1998. At that time there was an offer of coastal reforestation cooperation with Wetland International and was an effort of environmental conservation group that focuses on pond program with environmentally based. Care and maintenance of mangrove is done in pond owned by pond farmers. Furthermore, the mangrove cultivation effort was replanting in 2004 with the government program from the Ministry of Forestry, but failed due to the internal problems of the group. In addition, there was no management and maintenance in mangrove cultivation became the main factor of such failure so that mangrove cultivation were abandoned. In 2008 mangrove cultivation was carried out by Pantai Lestari group with focus of coastal greening and nurturing mangrove seedlings, procurement of buffer zone with planting, and maintenance and mangrove embankment. The mangrove cultivation is in cooperation with Ministry of Public Works, Regency Government, Ministry of Marine Affairs and Fisheries, Ministry of Environment, CSR PT PertaminaBalongan RU VI, PT Traktor Nusantara Jakarta, KEHATI Foundation and MfF Indonesia. Here is the data of mangrove planting in Karangsong.

Table 1 Mangrove Planting in Karangsong, Indramayu

Year	Data of Mangroves Planting	Amount (Trees)
2008	BBWS Cimanuk-Cisanggarung (Ministry of Public Works and Public Housing)	25.000
2009	TP – APBN DISKANLA Regency of Indramayu	50.000
2010	RU VI – PT. PertaminaBalongan – CSR	5.000
2011	KEHATI Foundation, Jakarta	11.000
	BPHM Region 1 Bali- Ministry of Environment and Forestry	50.000
2012	RU VI – PT. Pertamina Balongan – CSR	10.000
	Dekon Ministry of Marine Affairs and Fisheries	25.000
	PT. Traktor Nusantara Jakarta - CSR	6.000
2013	Dekon Ministry of Marine Affairs and Fisheries	20.300
2014	Dekon Ministry of Marine Affairs and Fisheries	21.000
	MFF Indonesia (PBB – UNDP)	50.000
2015	Dekon Ministry of Marine Affairs and Fisheries	21.000
	Ministry of Marine Affairs and Fisheries	306.500

Source: Pantai Lestari Group (2015)

The impact of mangrove cultivation is the return of shifting beach into green area by dividing area for buffer zone and pond area in Karangsong. Currently, the mangrove area in Karangsong reaches 20 hectares of mangrove land and set as mangroves eco-tourism, 5 ha for arboretum, and 49 ha for mangroves reforestation with land belonging to a group of fishermen purchased and stipulated. To avoid mutual claims of the region in mangroves forest, in 2009 set Perdes (Village Regulation) that ruled that mangrove forest is a conservation and protection areas.

Barriers to mangrove cultivation related with community empowerment that community participation is still oriented to direct economic benefits. People who invited to participate in mangroves cultivation used as planting labor for later given wages. This is because people still do not want to be involved if there is no direct economic benefits associated with the necessities of life.

5 The problem

The efforts of community involvement are still planting as laborers. That's because people are still reluctant to participate in planting personal consciousness without providing direct economic benefits. So that the awareness of the importance of mangrove is still limited from the economic side. Maintenance is also still done by Lestari Beach Group which is a local group in Karangsong which initiated Karangsong coastal reforestation program.

6 Conclusion and recommendation

Mangrove cultivation required community development approach from planning process, planting, maintenance and management. In addition, long-term assistance is needed in community engagement efforts for mangrove cultivation in an effort to cope with abrasion.

References

1. R. S. Pomeroy and M. B. Carlos, "Community-based coastal resource management in the Philippines : a review and evaluation of programs and projects , 1984-1994," vol. 21, no. 5, pp. 445–464, 1997.
2. V. S. A. Foucat, "Community-based ecotourism management moving towards sustainability , in Ventanilla , Oaxaca , Mexico," *cean Coast. Manag.*, vol. 45, pp. 511–529, 2002.
3. Jones, "COMMUNITY-BASED ECOTOURISM The Significance of Social Capital," vol. 32, no. 2, pp. 303–324, 2005
4. T. K. Bradshaw, "Theories of Poverty and Anti-Poverty Programs in Community Development Theories of Poverty and Anti-Poverty Programs in Community Development," *J. Community Dev. Soc.*, vol. 38, no. November 2014, pp. 37–41, 2007.
5. [5] G. P. Manatee and J. M. Belsky, "Misrepresenting Communities : The Politics of Community-Based Rural Ecotourism i n," vol. 64, no. 4, pp. 641–666, 1999.

The Effect of Mangrove on Fish Catch Using *Belat* at Teluk Pemedas and Sanipah of Kutai Kartanegara Regency, East Kalimantan Province, Indonesia

Bustam Sulaiman[1,5*], *Azis* Nur Bambang[1,2,3], *Hartuti* Purnaweni[1,3,4], *Mohammad* Lutfi[6]

[1]Doctoral Program of Environmental Science, School of Postgraduate Studies, Diponegoro University, Semarang-Indonesia
[2] Faculty of Fishery and Ocean, Diponegoro University, Semarang, 50275, Indonesia
[3]Master Program of Environmental Science, School of Post Graduate Studies, Diponegoro University, Semarang, 50275,Indonesia
[4]Faculty of Social Science and Political Science, Diponegoro University, Semarang, 50275, Indonesia
[5]Department of Oil and Gas Processing Engineering, STT MIGAS, Balikpapan, 76127, Indonesia
[6]Department of Petroleum Engineering, STT MIGAS, Balikpapan, 76127, Indonesia

Abstract. The increasing population at Teluk Pemedas and Sanipah coastal areas of Kutai Kartanegara Regency, East Kalimantan Province, has led to coastal degradation. The problem was mainly caused by conversion of mangrove into aquaculture, agriculture, plantation industrial, and residential areas. This research was a study case, used descriptive analysis to describe the local fishermen activity using *belat* traditional device to catch fish. The results revealed that there were many fishermen still did not know the benefits of mangrove for economy growth, ecological, ecotourism etc. This condition affects to the decline of the mangrove function. Fisherman who used *belat* often move from one location to another location in order to find places that have many catches. The *belat* which are placed in good environment mangrove areas produces lots of catches that greatly benefit the fisherman, therefore mangrove ecosystem must be well preserved for the sustainability of fishermen's efforts to earn a living.

1 Introduction

Coastal area is an area with a very dynamic environment condition compared to other locations in the world [1]. Coastal area development in Indonesia has been challenged by various problems such as high rate of local poverty, low and minimum infrastructure quality, and high rate of coastal degradation including degradation of its mangrove area [2]. This case also occurred in Teluk Pemedes and Sanipah, coastal areas of Kutai Kertanegara Regency, East Kalimantan Province.

Mangrove areas at Teluk Pemedas and Sanipah have been experience degradation due to the natural phenomenom and human activities such as the conversion to aquaculture and settlements. Mangrove forest contributes large spectra of ecosystem services. Its environmental functions include nutrition supply and regeneration, pollutant recycle, water cycle and quality preservation [3]. In three decades Indonesia has lost 40% of its mangroves, mainly as a result of aquaculture development. Annual mangrove deforestation in Indonesia is only 6% of its total forest loss. The lost of mangrove forest also contributes to 42% of gas emission which cause a greenhouse effect due to the destruction of coastal ecosystem. The prevention of mangrove deforestation in Indonesia is believed to fulfill a quarter of 26% target of the emission reduction by 2020 [4].

Mangrove forest plays an important role to reduce wave energy and mangroves height can potentially reduce associated damage. All evidence suggests that mangroves can reduce the height of wind and swell waves over relatively short distances. wave height can be reduced by between 13 and 66% over 100 m of mangroves [5]. Early warning system in coastal areas can be built by considering numerical modeling of hydrodynamics [6] and mangrove condition based on the satellite imagery. Mangrove can also reduce disaster due to hydrometeorological activity resulted from climate change, such as greater maximum tides and rising sea level, so that the cost for building the flood control can be reduced [7].

The socio-economic values of mangrove are varied. It can be useful for many purposes such as for building material, fire wood, fishing scheme, port, railway, furniture to handicraft and colouring materials [8].

Mangrove has a great potential for social and economic development in coastal area and especially for the fishermen, therefore, mangrove should be well planned for its sustainability. Ecological function of mangrove forest is beneficial for among others provide nutrition to aquatic biota and nursery ground for various biota, shelter for abrasion, typhoon attack, and tsunami, ocean water prevention, and waste absorber [9].

* Corresponding author: bustams@gmail.com

https://doi.org/10.1051/e3sconf/20183108029

Mangrove ecosystem ecologically acts as nutrition provider, nursery ground and feeding ground, coastal and erosion protection, and nursery and feeding for fish species and crustacea, sea water intrusion and natural disasters, such as tsunami and typhoon. It also functions as mangrove seeding area [10-13].

The other benefit of the mangrove system is also to promote the effort and plan for ecotourism because mangrove is very important for ecotourism sustainable development and improvement of local community life [14].

The actual benefits of the mangrove system also deals with social and cultural functions. Mangrove forest supports religious and spiritual values, in addition to aesthetic and recreation for the ecotourism, therefore, community participation is expected in the ecotourism planning because it is important towards the ecotourism sustainable development and in fulfilling the social and economic equilllibrium of ecotourism [15, 16].

Mangrove ecosystem provide sound area for catching fish among the fishermen, especially for fishermen in Teluk Pemedas and Sanipah who make use of belat, a traditional fishing net popular among the local fishermen.

This research aims to describe the benefit of *belat* for the fishermen of Teluk Pemedas and Sanipah, who usually place the *belat* in the mangrove area.

2 Method

This study was conducted at Teluk Pemedas and Sanipah, Samboja District, Kutai Kartanegara Regency, East Kalimantan Province, Indonesia. The study applied a case study, an approach to learning, explaining, and interpreting a case in its natural context without any external intervention. The case study used in this study was intrinsic due to the fact that the phenomenon of mangrove forest in a replicated manner that has caused degradation of the coastal area.

The research used an analytical-descriptive manner for comprehending descriptive effort of phenomena found during the study. The case study research model can also meant a report of particular event, situation, or development constructed in a comprehensive and detailed manner. The phenomena found in this study were fishermen activities who used *belat,* a local, unique fishermen's device for catching fish. The selection of this case study was based on the questions of how and why of a phenomenon as well as of intensive, in-depth, detailed, and comprehensive case assessment.

This study restricted and focused on the effects of mangrove on the catches of the fishermen who used *belat* in the research area. This study applied a descriptive method using interviews as data collection instrument. The samples (event, human, situation) were expected to provide information.

The interviews were directed to local people that could provide representative data. The local community consist of fishermen and community leaders from which reliable information about the functions of mangrove forest and *belat* could be obtained.

The recapitulation data of population at Teluk Pemedas and Sanipah are presented in Table 1.

Table 1. Number of Population 2017

No.	Village	Household	M	F	Total
1	Teluk Pemedas	791	1447	1376	2823
2	Sanipah	2311	2931	2152	5083

This study used a purposive sampling to get particular objective of the research with desirable research outputs. The purposive sampling technique requires research subjects to represent the population being examined.

Sampling was determined by Slovin [18] equation as follow:

$$n = \frac{N}{1 + N e^2}$$

n = sample size
N = population size
e = percent of tolerated or preferable errors due to sample miscollection, e.g. (= 0.1)

The population size (Table 1) resulted in 97 samples from Teluk Pemedas and 98 samples from Sanipah.

Table 2. Number of respondents at Teluk Pemedas

Occupation	Total	%	Respondent
Civil servants	71	11,95	11
Private employees	321	54,04	52
Entrepreneurs	84	14,14	14
Farmers	37	6,23	6
Fishermen	34	5,72	6
Service providers	47	7,91	8

Source: Monograph of Teluk Pemedas Village 2017

Tabel 3. Number of respondents at Sanipah

Occupation	Total	%	Respondent
Civil servants	106	4,76	5
Private employees	755	33,93	33
Entrepreneurs	266	11,96	12
Farmers	595	26,74	26
Fishermen	266	11,96	12
Service providers	237	10,65	10

Source: Monograph of Sanipah Village 2017

3 Results and discussions

Belat (Barrier Trap) is a kind of trap for catching fish, usually placed at tidal areas. It consist of two nets as walls and a kind of "bag" between the two nets. The fishermen usually place these nets in a V or U shape in the inner mangrove areas during maximum high tide and draw the nets during low tidal wave in which the fish were trapped in the bag. The picture of *belat* is as follows :

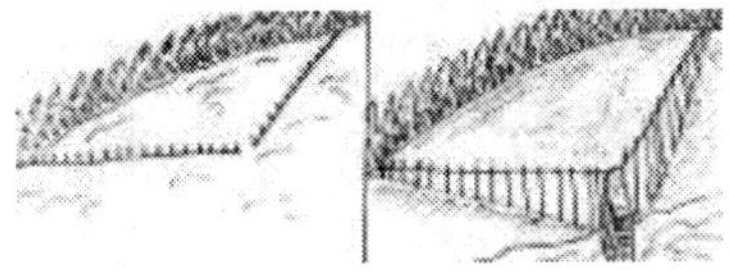

Fig 1. Belat (Barrier Trap) [17]

The respondents were the local fishermen of Teluk Pemedas and Sanipah. The observation revealed that fishermen who used *belat* had the larger number of catches around the mangrove area than that of away from the mangrove area. Mangrove ecosystem significantly proved that it is home of many fish, enabling the fishermen able to find good catch using traditional fish catch device of *belat*.

The *belat* fishermen proved that their economic life was adequate, proved by their belongings of motorcycles and/or cars, television, and good furniture in their house, to mention some.

In additional to being *belat* fishermen, these local people also had side-jobs such as working in the plantation (cocoa, rubber, etc.) in their leisure times. The average income of the *belat* fishermen provided that they earned income from vary occupations, are more than earned by civil servants, private employees, farmers, entrepreneurs, and service provider. The following graphic illustrations and tables compare the proportion of the income.

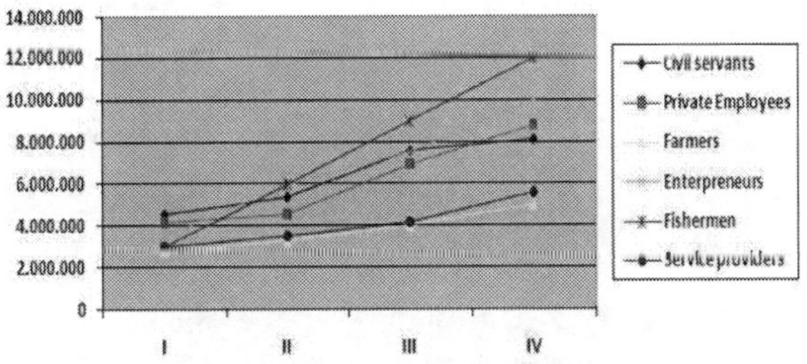

Fig. 2. Income comparison in Indonesian Rupiah (IDR) of *belat* fishermen by occupation at Teluk Pemedas.

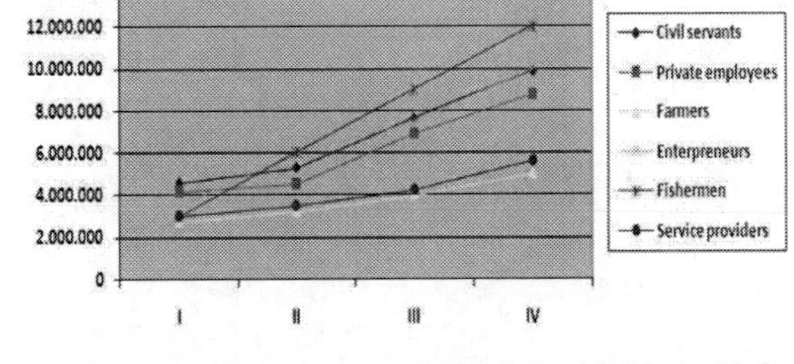

Fig. 3. Income comparison in Indonesian Rupiah (IDR) of *belat* fishermen by occupation at Sanipah.

The fisherman often moved their *belat* from one location to another in order to find places that have many catches. The average income of *belat* fishermen was Rp. 3.000.000/*belat*. Whereas *belat* ownership varied from one to four units per fisherman. The fishermen informed that *belat* was one of the most practical techniques because they do not need to wait too long to catch the fishes, it needs only one-day process. Their income had been adequate to meet their family's day-to-day need. Furthermore, there is no negative effect of the existence of the *belat* to the mangrove ecosystem because they used their *belat* at tidal areas adjacent to the mangroves. Their knowledge about the benefits of mangrove forest was satisfactory, even though more efforts should be taken to get them more informed towards the sustainability of the mangrove forest.

4 Conclusions

The *belat* which are placed at good environment of mangrove areas produces lots of catches. The fisherman able to routinely catch good amount of fish, enabling them have the average income around Rp. 3.000.000 per *belat*/day. The fisherman used *belat* very effectively and efficiently just by simply utilizing tidal zone in shallow water around their settlements to earn a good living. Good mangrove proved to be very beneficial for the fishermen

According to the results, there are several recommendations to preserve the mangrove forest in this research location, namely:

1. Coastal community should manage and preserve mangrove for its sustainability. It must become a mutual responsibility.
2. Mangrove ecosystem had been well preserved. *Belat* did not have adverse impact on the ecosystem and provided larg number of catches to the fishermen
3. Concerted evaluation of the damaged mangrove site is necessary to restore and flourish its existence.
4. The government is expected to make a continued socialization of the role and benefit of the mangrove forest for its sustainability.
5. There must be a synergy between the government and the citizens towards sustainable mangrove.

References

1. M. A. Marfa'i, Quaestiones Geographicae **31** (3), 2012, pp 47-55.
2. H. Purnaweni, Kismartini, S.P. Hadi, T. R. Soeprobowati. Advance Science Letter. **23** (3), March 2017, pp. 2582-2583 (2). https://doi.org/10.1166/asl.2017.8711
3. H. J. Ruitenbeek. Ecological economics, **10**, 3 (1994), pp. 233-247
4. D. Murdiyarso, J. Purbopuspito, J. Kauffman, J.B. Warren, M.W. Sasmito, S.D. Donato, D.C. Manuri, S. Krisnawati, H. Taberina, S. Kurnianto. Nature Climate Change **5**, pp. 1089-1092 (2015). Doi: 10.1038/nclimate2734
5. A. L. McIvor, L. Möller, T. Spencer, M. Spalding. Natural Coastal Protection Series: Report 1.

Cambridge Coastal Research Unit Working Paper **40** (2012) ISSN 2050-7941.

6. M. Lutfi, R. S. Afifah, B. Sulaiman, Risna. Indian J. of Sci. and Tech, **10,** 29 (2017). Doi: 10.17485/ijst/2017/v10i29/116502
7. T. James, V. J. Savin, P. M. Claire, R. Md. Masumur, T. Arnel, M. Lutfi. J. of IDEC, **19**, 3 (2013) pp. 131-145. Doi: http://doi.org/10.15027/35045
8. L. S. Hamilton, S. C. Snedaker. *Handbook for mangrove area management. Honolulu: East-West Environment and Policy Institute.* (1984)
9. R. Dahuri, J. Rais, S. P. Ginting, M. J. Sitepu, *Pengelolaan sumberdaya wilayah pesisir dan lautan secara terpadu.* Pradnya Paramita, Jakarta, (1996).
10. C. Giri, E. Ochieng, L. L. Tieszen, Z. Zhu, A. Singh, T. Loveland, N. Duke. (2011). Global Ecology and Biogeography, **20**, 1, pp. 154-159.
11. C. Kuenzer, M. Ottinger, M. Wegmann, H. Guo, C. Wang, J. Zhang, M. Wikelski. International Journal of Remote Sensing, **35**, 18 (2014) 6599-6647.
12. C. Kuenzer, M. Ottinger, M. Wegmann, H. Guo, C. Wang, J. Zhang, M. Wikelski. International Journal of Remote Sensing, **35**, 18 (2014) 6599-6647.
13. M. Boer, N. T. Pratiwi. J. Ilmu dan Teknologi Kelautan Tropis, **7,** 2 (2016).
14. H. Masood, S. Afsar, U. B. Zamir, J. H. Kazmi. *Application of Comparative Remote Sensing Techniques for Monitoring Mangroves in Indus Delta, Sindh, Pakistan.* In Biological Forum **7**, 1 (2015), p. 783.
15. B. C. Adeleke, T. Nzama, *Assessment of community participation in ecotourism and conservation at Hluhuwe-Umfolozi Park, South Africa.* Journal of Environment and Earth Science. (2013)
16. J. A. Robinson, K. P. Lulla, M. Kashiwagi, M. Suzuki, M. D. Nellis, C. E. Bussing, L. J. McKenzie. Conservation Biology, **15**, 4 (2001), pp. 876-884.
17. C. J. León, J. de León, J. E. Araña. M. M. González. Ecological Economics, **118** (2015) pp. 21-29.
18. Rudi Sanjaya,
    https://www.academia.edu/3304700/Jenis_-_*Jenis_Alat_Tangkap_Ikan_dan_Pengoperasiannya*
19. C. G. Sevilla, *Research Methods.* Rex Bookstore, Inc. (1992)

Impacts of Agricultural Practices and Tourism Activities on the Sustainability of Telaga Warna and Telaga Pengilon Lakes, Dieng Plateau, Central Java

Sudarmadji[1*] and *Hermin* Pudjiastuti[2]

[1] Department of Environmental Geography, Faculty of Geography, Gadjah Mada University,
[2] Environmental Study Program, Graduate School, Gadjah Mada University

Abstract. Telaga Warna and Telaga Pengilon are two volcanic lakes in the Dieng Plateau offer some unique phenomena which are interested for tourists to visit. Telaga Warna and Telaga Pengilon are located side by side in the Dieng Palteau. Those two lakes also have specific ecosystem which differ to other lakes. However as land use in the surrounding area is now gradually changing, the lake is now facing to environmental degradation. The land use in the surrounding area is for intensive agricultural which main crops are vegetable, especially potatoes. Meanwhile, the number of tourist visiting those two lakes is increasing; it may also give some impact to the lake environment. This research aims to study the impacts of agricultural practices and tourism activities to the lake environmental which lead to the environment sustainability of the lakes. The field survey was conducted to collect some data on lakes characteristics, agricultural and tourism activities. Some interviews to local people and tourists were also conducted. Some water and sediment samples were collected followed by laboratory analyses. Some secondary data from previous study was also collected. Data analysis was conducted based on qualitative and quantitative techniques. The study found that agricultural practices of potatoes plantation uses water from the Telaga Pengilon to irrigate the plant by pumping out the water using water pump and distributes the water over the plantation area. Agricultural practices lead to soil erosion, which contribute sediment to the lake carried by surface runoff. Therefore, the volume of lakes is gradually decreasing. The use of fertilizer in the agricultural practice contribute nutrient into the lake carried by surface runoff, leading to the eutrophication, due to the excess used of fertilizer. The study concludes that agricultural practices and tourism activities have some positive economic impacts to the local community, however it also give some adverse affects on the lakes, both quantitatively and qualitatively. The volume of lakes is gradually decreasing due to sedimentation. In the long term periods the lake may not be interesting object for tourism.

1 Introduction

Around the world a lot of lakes have some environment problem such as sedimentation and water pollution [1-3]. Telaga Warna and Telaga Pengilon are two lakes which are located within one of the most strategic complexes in the Dieng Plateau. Both of these lakes are well known among tourists, especially local tourists. However, many foreign tourists also come to visit the lakes. Foreign tourists do not merely visit the lake, because actually the uniqueness of the Telaga Warna and Telaga Pengilon are not separated by the uniqueness of the Dieng Plateau region as a tourist objects. Telaga Warna has colorful water caused by the minerals and the organisms contained, combined with the sunlight that strikes the lake and then reflected, thus giving a colorful effect. Telaga Pengilon has a clear water, make the lake looks like looks like a mirror. As volcanic lake, these two lakes have some uniqueness like other volcanic lakes [4].

The exotic panoramic of Telaga Warna and Telaga Pengilon make many tourists to visit, therefore, some opportunities for local governments and local community to develop them as natural attractions, which can increase local revenue, as well as expected to improve the welfare of the surrounding community. However, on the other hand, the increasing number of tourists visiting the Talaga Warna and Telaga Pengilon has also some negative impacts to occur, due to the behavior of tourists who are not environmentally friendly. The tourist's activity may create some wastes that may pollute the environment [5].

On the other side the catchment area of Telaga Warna and Telaga Pengilon has been gradually

* Corresponding author: sudarmadji@ugm.ac.id

changed and developed for potato cultivation. The changes of land use of the surrounding area into agricultural area (with potato cultivation) bring economic benefits to the community [6]. The changes of land use of the surrounding area into agricultural land produce sediment, and contribute part of the residual fertilizer and pesticide into the lake [7]. Therefore, the lakes environment will gradually threaten that make the lakes no longer attract tourists to visit. Based on the above background, this research aims to: 1) identify the potential of Telaga Warna and Telaga Pengilon as a natural tourism object either at present or their development in the future. 2) Analyze the activities of the surrounding community on the possibility of decreasing the environmental quality of the Telaga Warna and Telaga Pengilon,

2 Methods

2.1. Research Location

The research was conducted in Telaga Warna and Telaga Pengilon, Dieng Plateau; administratively the lakes are located in Wonosobo District (Figure 1.)

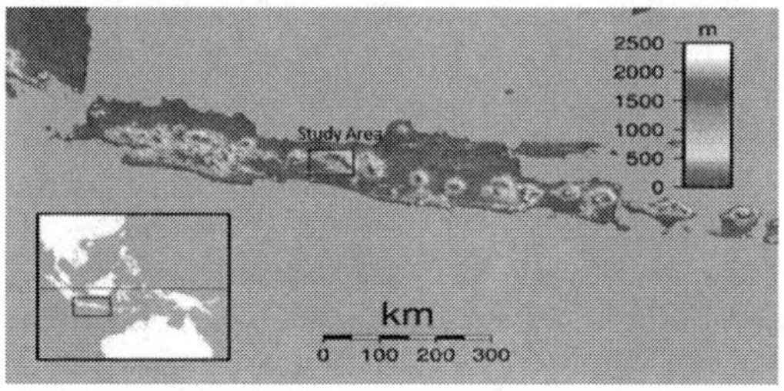

Fig. 1. Research Location

2.2. Materials and Tools

a. Map of Wonosobo District of 1:50.000;
b. Photographic Map scale 1: 100.00 Wonosobo and Banjarnegara regencies related to the location of the lake;
c. Geomorphology map scale 1: 100,000 Wonosobo and Banjarnegara regencies related to the location of the lake and its catchment area;
d. Geological Map scale 1: 100,000 Wonosobo and Banjarnegara regencies related to the lake location;
e. Maps the location of tourism objects of Wonosobo and Banjarnegara regencies is obtained from previous research, including from the Tourism Office, the local Public Works Department;
f. GPS and Satellite Imagery;
g. A set of instruments for water sampling (pH meter, EC-meter) water sample bottle;
h. Questionnaires for interviews with local residents as well as local community leaders who use water from the lake and tourists who visit the lake, both local and foreign tourists;
I. A set of computer units to perform data analysis.

2.3. Field Work and Data Collecting

a. Determine the location of the studied lake using Satellite Imagery.
b. Conduct observations of lake conditions and measurements and taking necessary water samples
c. Conducting interviews with residents and community leaders to find out how people use the lake, manage including how to conserve lakes;
d. Conducting interviews with local tourists and foreign tourists
e. Conducting a laboratory analysis of the samples taken to determine the water quality

2.4. Data Analysis

a. Analyzing the spatial presence of natural lakes is associated with environmental conditions as well as other activities around it.
b. Processing and analyzing the results of interviews to obtain the way of the community to manage the lake, especially related to the culture of the local community and associated with tourism activities.

3 Result And Discussion

3.1. Potency of Telaga Warna and Telaga Pengilon

Telaga Warna and Telaga Penggilon both located in Dieng Plateau, Wonosobo District. Many of the attractions associated with the two lakes, however, this research focused on the lakes as volcanic lakes with all of the characteristics and uniqueness. Nevertheless, the surrounding areas where the community conducting activities that likely to affect the condition of the two lakes are also studied. Telaga Warna and Telaga Color potential in general can be seen from Satellite Imagery (Figure 2). The two lakes appear to differ, in which the Color Lake shows a brighter color whereas Pengilon's lake shows a darker appearance. Both lakes are surrounded by a densely vegetal cover, while in a distance from the lakes the area is already covered by agricultural land and settlements

 https://doi.org/10.1051/e3sconf/20183108030

Fig. 2. Telaga Warna and Telaga Pengilon from Satellite Imagery (*Google earth, 6th Dec. 2015)*

The Telaga Warna when viewed closely shows the calm and cool atmosphere provided by Telaga Warna and Telaga Pengilon due to the charm of its own natural beauty.

The tourist objects close to Telaga Warna and Telaga Pengilon are as follows: 1). Bukit Sidengkeng, 2). Sikendang Crater, 3). Semar Cave, 4). Well Cave, 5). Pond in Sumur Cave, 6). Jaran Cave, 7). Batu tulis. 8). Mliwis Hill, 9). Wlingi Grass.10). Lake Pengilon, 11). Mandar Batu, 12). Rest Area of Bumi Pertolo. These objects are mostly unique volcanic formations that become the main attraction (caves, craters, even flora and fauna). Very interesting color of the lake water is partially whitish, greenish, bluish, and yellowish due to the sulfur and algae and content and because of the reflection of sunlight (Figure 3, right). The silver colors because of the calm looks like a big mirror. There is also a crater that emits sulfuric gas (Figure 3, left). The potency of Telaga Warna has been supported by facilities such as information boards, offices, parking areas, tour guides, flying fox, and spots to rest and walking tracks around the lake (Figure 4). However, the facility has not been supported with adequate maintenance.

Fig. 3. Yellowish Color of Telaga Warna (Left) and Sikendang Crater (Right) as Unique Tourism Objects (Doc. 2015)

Fig. 4. Supporting Facility and *Flying Fox (Doc. 2015)*

3.2. Community Activities around Telaga and Tourists

The local people mostly work as farmers with main vegetable crops, mainly potatoes (Figure 5). The figure shows a preparation for planting potatoes that make the soil in such a way to create soil erosion. In irrigating the potatoes plants the water of the lake is pumped up using water pump and distributes the water to their farming area with pipes (Figure 6).

The development of Telaga Warna as natural tourist attractions, make some local people to have some new jobs to support tourism activities. Culinary business and selling souvenirs are open, and utilized by the local community. Culinary shops and stand of souvenirs are found around Telaga Warna. Similarly, the profession becomes parking operator, tour guide is also open (Figure 7). This means that the development of Telaga Warna provides a positive impact for the surrounding community, in the form of opening of business fields and increasing revenue, not only to the local government, but also for the surrounding community.

Fig. 5. Preparing Land for Potatoes Plantation *(Doc. 2015)*

Fig. 6. Irrigation Facilities (water pumps and pipes*) (Doc. 2015)*

Fig. 7. Souvenir shops (left) and Parking Area (right) *(Doc, 2015)*

Fig. 8. Tourists activities in the Jogging Track *(Doc. 2015)*

Tourists who visit Telaga Warna can perform various activities, from just relaxing to enjoy the beautiful scenery and atmosphere of Telaga Warna and Telaga Pengilon, enjoy the uniqueness of the natural phenomenon around Telaga Warna, walking around the lake through the jogging track, or trying the flying fox (Figure 8).

3.3. Environmental Impacts

The development of Telaga Warna as a natural tourist attraction has provided many benefits and poses a positive impact for the local government and the surrounding community, especially from the economic point of view i.e. the increasing of income and the opening of business opportunities [5, 8]. However, tourism activities can also cause negative impacts to the environment, although it should be noted that the negative impacts experienced by Telaga Warna not only comes from tourism activities.

Local activities, such as agricultural activities and other daily activities, also generating waste, disposing into the environment, and can cause environmental impacts [9,10]. Agricultural activities of potato cultivation in the surrounding area of Telaga Warna increase soil erosion which further affects sedimentation in Telaga Warna. Although there is no erosion measurement yet, but the effect of sedimentation in Telaga Warna is already visible (Figure 9). The use of fertilizers and pesticides has also affected water quality in Telaga Warna and Telaga Pengilon; it appears that in some places water plants began to grow with "fertile", which in turn cause the shrinking volume of Lake Color and Telaga Pengilon as also occur in other lakes [9, 10, 11].

Nitrate concentration in Telaga Pengilon is higher than those in Telaga Warna, indicates that some impact of agicultural activities have occured.

Water quality data from both Telaga appear to be much different, especially from the pH and TSS parameters. The pH at Telaga Warna indicates that the water of the lake is acidic with high TSS, while Telaga Pengilon exhibits a lower of TSS (Table 1 and 2).

Fig. 9. Jogging Tracks as Sediment Source (left) and Sediments in Telaga Warna (right)

Table 1. Water Quality Analysis of Telaga Warna

Parameters	Unit	Number of Water Sample		
		TW A 2	TW A 3	TW A 5
Temperature	°C	25.9	27.6	27.3
EC	µS/cm	1497	552	1384
pH	-	2.90	4.66	2.87
Nitrate	mg/l	0.275	0.402	0.466
BOD	mg/l	0.37	0.39	-
COD	mg/l	1.46	0.74	6.67
Phosphate	mg/l	≤0.02	≤0.02	≤0.02
Coliform	MPN/100ml	7	11	7

Sources: Laboratory Analysis

Table 2. Water Quality Analysis of Telaga Pengilon

Parameters	Unit	Number of Water Sample		
		TPA 1	TPA 2	TPA 3
Temperature	°C	25.7	26.0	25.6
EC	µS/cm	202	176	176
pH	-	8.36	8.30	8.88
Nitrate	mg/l	1.616	1.041	0.339
BOD	mg/l	0.79	0.49	0.65
COD	mg/l	6.42	12.59	10.85
Phosphate	mg/l	≤0.24	≤0.11	≤0.11
Coliform	MPN/100ml	7	7	11

Sources: Laboratory Analysis

In Telaga Warna volcanic activity is still occurring intensively. This water is not good for irrigation uses. Tourist activities can also have impacts on Telaga Warna and Telaga Pengilon. Jogging track land is vulnerable to soil erosion, and it contributes to the sedimentation of the lake.

Tourist's activities can contribute wastes, which is often discarded in the improper place. This may also occur in Telaga Warna and Telaga Pengilon, although warning signs have been installed and garbage disposal bean has been provided (Figure 10). It is also important to note that some vandalism is observed in some places.

Based on the results of observation it is very clear that Telaga Warna and Pengilon have some potencies to be developed in term of tourist attraction. The development can include the addition of tourism supporting facilities, such as the road to the lake Telaga Warna, facilities in the Telaga Warna such as places or spots to enjoy the view, camping ground, and adequate office. Toilets should also be provided adequately with good maintenance. Native plants in this area may be offered as a tourist attraction that if it is managed properly. If in this area a botanical park with a typical plant is managed properly, it is something can attract for tourists to visit. In this case, the involvement of local community is very important [8, 12, 13].

Fig. 10. Warning Sign in Telaga Warna *(Doc, 2015)*

The negative impacts of tourism activities and local agricultural activities, especially originated from potatoes farming. It will threaten the preservation of Telaga Warna and Telaga Pengilon, mainly due to high sedimentation rate. The sedimentation that has occurred, decreased the volume of the lake, is apparent during the dry season (Figure 11). The result of sediment grain analysis obtained shows that the sediment contained in Telaga Warna and Telaga Pengilon has the similar characteristics with the soil located on either right or around the lake. This means that the land around the lake has create high erosion followed by transporting sediment material into the lake. Some solid wastes is also resulted from tourism activities, all of which must be controlled (if not preventable) to maintain the sustainability of natural attractions of Telaga Warna and Telaga Pengilon.

Fig. 11. Sediment to Decrease the Volume of Telaga Warna *(Doc. 2015)*

4 Conclusions

Telaga Warna and Pengilon is a volcanic lake that becomes main attraction in Dieng area. Mineral content and algae that vary along with the reflection of sunlight produces unique and natural colors for tourists to enjoy. The famous Telaga Pengilon has a very clear. Other uniqueness of the two lakes is that, even though they located side by side, the water color is significantly difference. The uniqueness of the lakes is a potential for tourism attraction, therefore it should be developed to maintain the sustainability of Lake Color and Pengilon as natural tourism attractions.

Community activities around Telaga Warna and Pengilon are very diverse. The existence of Telaga Warna and Pengilon as a tourist attraction has some positive impacts, because it opens the opportunities to find new job that can increase people's income, such as traders, tour guides or parking operator. Telaga Warna and Telaga Pengilon is the source of water used by the community for the irrigation of its plants.

However, there are some community activities that have negative impact to Telaga Warna and Pengilon. Potato farm activity around the steep slopes contributes considerably to the sedimentation. Potato cultivation results in soil erosion, fertilizers and pesticides dissolved into the lake thus degrading the water quality of the lake. Dissolved nutrient in water originated from the rest of the fertilizer may cause blooming of water plants in some places [6,9,14]. It affects the light penetration and dissolved oxygen levels in the waters, so that organisms in the water such as fish are difficult to survive. The farmers take water from the lake for irrigation using water pumps. The management prohibits the surrounding community from using water for irrigation sources. Although there is a limitation from using water, but the farmers do not stop from pumping up the water.

As a leading tourist attraction for Wonosobo Regency, Telaga Warna and Telaga Pengilon cannot be developed independently without linking with other tourism objects. This means that the development of this tourist attraction must be combined with other attractions in an integrated manner, and made a package with other attractions. Community involvement needs to be improved so that the negative impact of the development of this tourism object can be minimized. The development should also consider the risk disasters, as Telaga Warna and Telaga Pengilon are located in an active volcanic area. Natural disasters may occur unpredictably, therefore mitigation is needed. Management of tourist area of Dieng as a whole should also consider, the existence of home stay, restaurant and other facilities such as road connecting tourist objects around the area.

REFERENCES

1. International Lake Environment Comitee (ILEC) Foundation. *How can We Stop Degradation of the World's Lake Environment.* ILEC, Shiga Japan.
2. Kowenje, C. O. and Agunya, E.O. Chemical Polution of Lake Victoria A Case Review of Winam Gulf. *Proceeding Seminar on Five Years of Exceed* (2014)
3. Ministry of Environment of the Republic of Indonesia. *Indonesian Lake Management*. The Indonesian Movement for lake Ecosystem Conservation and Rehabilitation. Ministtry of Env. of Republic of Indonesia, Jakarta. (2014)
4. Setiawan, S., Wibowo, H., Santosa, A. B., Nomosatryo, S., dan Yuniarti, I. Karakteristik Danau Asal Vulkanik, Studi Kasus: Danau Tolire, Pulau Ternate. *Limnotek,* Vol. 21. No. 2. pp. 103-114. (2014)
5. Sudarmadji, Eko Haryono, Darmakusuma Darmanto dan Widyastuti. Pengembangan Wisata Alam: Implikasinya terhadap Lingkungan Hidup dan Risiko Bencana yang Dihadapi di Daerah Istimewa Yogyakarta. *Research Report*. Sekolah Pascasarjana UGM, *Yogyakarta*. (2013)
6. Rusiah, M. , Satya, N and Wahyudin, A. Dampak Aktivitas Pertanian Kentang terhadap Kerusakan Lingkungan Obyek Wisata Dataran Tinggi Dieng. PELITA, Vol. **1**, No. 1, August. (2005)
7. Sudarmadji. Fungsi Waduk Dalam Ekosistem Daerah Aliran Sungai dan Masalah yang Dihadapi. *Proceeding Seminar on Limnology*. Fakultas Biologi UGM. (2004)
8. Sudarmadji, Suprayogi, S., Thianton, T.A., dan Werdiningsih. Survey Hidrologi untuk Pengelolaan Sumberdaya Air Berbasis Masyarakat: Kasus di Telaga Towet Desa Girisekar, Kecamatan Panggang, Kabupaten Gunungkidul. *Laporan Penelitian,* Fakultas Geografi UGM Yogyakarta.
9. Piranti, A.S. *Kajian Kriteria Nutrien sebagai Dasar dalam Penilaian Status Tropik Perairan Waduk Mrica Banjarnegara, Jawa Tengah*. Disertation. Universitas Gadjah Mada, Yogyakarta. (2012)
10. Wantasen, S. Spasial Ekologi Nitrogen di Danau Tondano, Provinsi Sulawesi Utara. *Disertation.* Sekolah pascasarjana UGM. Yogyakarta. (2009)
11. Widyastuti, M., Fadilah, L.N., and Rasyadi, F.A. Kajian Potensi Sumberdaya Air Danau untuk Pertanian, Studi di Danau Merdada Kecamatan Dieng, Kabupaten Banjarnegara. *Research Report*., Fak. Geografi UGM. (2014)
12. Partomo, Mangkuprawiro, S., Hupeis, A. V.S., dan Adrianto, L. Pengelolaan Danau Berbasis *Co-Management* : Kasus Rawa Pening. Jurnal Pengelolaan Sumberdaya Alam dan Lingkungan. Vol. **1** No.2 pp. 114-119. (2011)
13. Sudarmadji, Slamet Suprayogi dan Setiadi. Konservasi Mataair Berbasis Masyarakat di Kabupaten Gunungkidul untuk Mengantisipasi Perubahan Iklim. *Research Report*. Sekolah Pascasarjana UGM, Yogyakarta. (2010)
14. Nomosatryo, S dan Lukman. Klasifikasi Trofik Danau Toba, Sumatera Utara. Jurnal Limnotek, Perairan Darat Tropis di Indonesia, Vol. **19**. No. 1. pp. 13-21. (2012)

Does Green Investment Increase Financial Performance? Empirical Evidence from Indonesian Companies

Anis Chariri [1,*], *Gretta* Ratna Sari Br Bukit[1], *Octrine* Bethary Eklesia[1], *Bourinta* Uly Christi[1] and *Daisy* Meirisa Tarigan[1]

[1]Faculty of Economics & Business, Diponegoro University, Semarang - Indonesia

Abstract. The negative effects of globalization and rapid growth of industries on environment have changed the business paradigm from profit issues to profit, people and planet (triple bottom line). Consequently, a number of companies have invested their money in environmental issues (called as green investment). This study aims to investigate the effect of firm characteristics on green investment and how green investment influences financial performance. Using annual reports of companies receiving the Program for Pollution Control, Evaluation and Rating (PROPER) award and listed on the Indonesia Stock Exchanges in the year of 2009-2014 as research data, the findings showed that firm size, foreign ownership, industry profile, and frequency of audit committee meeting significantly influenced green investment whereas ISO14001 management certification had no effect on it. Interestingly, green investment positively determined an increase in firm financial performance. This reveals that the better the green investment, the higher the financial performance of the companies. The findings contribute to the importance of adopting green investment as a company's strategy to increase profit without destroying the environment. Secondly, this finding can be used by government as a reference for formulating any regulations concerning business and environment. Finally, the finding contributes to the importance of including environmental issues in business education.

1 Introduction

Green investment is seen as interesting issues in business and accounting areas since there is a shift in the business paradigm from single P (Profit) to Triple P (Profit, People and Planet). A number of companies listed in capital markets have implemented green investment to attract potential shareholders. Green investment can be defined as company's efforts in managing environmental issues by reducing the negative impact of business activities on environment [1]–[3]. It is believed that green investment can increase competitive advantage, firm reputation and value [4], [5]. Therefore, a number of studies have been carried out to investigate the determinants and consequences of green investment.

Unfortunately, in the business context, previous studies have been focused on environmental disclosure [6]–[9]) and environmental performance [10]–[12]. The studies tend to ignore the determinants of green investment from the perspectives of firm characteristics and corporate governance. In addition, most of the studies were conducted in developed countries and it is not easy to find similar studies on green investment in emerging countries, including Indonesia. Thus this study is intended to investigate the effect of firm characteristics (industry profile, firm size, ISO14001 management certification) and corporate governance (foreign ownership and audit committee meeting) on green investment. The findings may contribute to the importance of adopting green investment as a company's strategy to increase profit without destroying the environment. Secondly, this finding can be used by government as a reference for formulating any regulations concerning business and environment. Finally, the finding may contribute to the importance of including environmental issues in business educations.

2 Hypotheses development

2.1 Firms size and green investment

Firm size reflects the number of assets and resources owned by companies to achieve their business objectives. As large companies have more opportunities to invest than that of smaller companies, stakeholders are more concerned with large companies as such investment may influence their interests [13]–[15]. In fact, as claimed by legitimacy theory, large companies are committed to environmental issues as such commitment can be used to gain and maintain legitimacy. Previous studies showed that big companies are more transparent in implementing and reporting their environmental and social policies [16]–[19]. Moreover, government puts more attention on larger companies than smaller ones in regard to any company policies concerning social and environmental issues. Hence, the larger the company, the greater the ability of the

* Corresponding author: anis_chariri@live.undip.ac.

company to implement environmental-related policies [7], [20]–[23], including environmental investment and green management [17], [20], [23]–[25]. This study proposes the following hypothesis:

H1. Firm size positively affects green investment

2.2 Foreign onwership and green investment

Stakeholder theory claims that shareholders are seen as powerful stakeholders in influencing company's policies, including green investment. Such influence depends on how powerful are the stakeholders compared to other ones. Foreign investors—mostly institutional investors—are seen as powerful stakeholders who may put pressures on environmental issues. Borrowing Reference [26] arguments, companies with large foreign ownership are seen as powerful in monitoring management, especially environmental issues [11], [27]–[32]. Some studies on carbon emission policies also support this claims[28], [33]. Thus, this study proposes the following hypothesis

H2: Foreign ownership positively affects green investment

2.3 Industry profile and green investment

Industry types play important roles in determining green investment. In line with its sensitivity to the environmental issues, the types of industry can be divided into two main clusters, namely high-profile and low-profile industry. Borrowing Reference [9], high-profile industry refers to an industry on which their consumer visibilities, political risks, competitions are high (for example, petroleum, chemical, forest and paper, automobiles, aircraft, extractive, agricultural, liquor and tobacco, and media and communications industries). In the Indonesian setting, the Government (in the PROPER assessment) commonly monitors the palm oil industry, oil and gas industry, and textile industry (PROPER Assessment Report, 2011), which are similar to high-profile industry proposed by Reference [9]. Hence, it is claimed that this type of industry may affect green investment as described by the legitimacy theory. The theory insists that the company seeks to get legitimacy from all stakeholders [34] by implementing policies (including green investment) that are consistent with their interests and values. Indeed, green investment help companies to gain legitimacy. The more sensitive the types of industry on environmental issues, the more serious the companies in environmental issues [17], [35]–[39]. Therefore, the hypothesis is proposed as a follow:

H3. Companies in high-profile industry positively influence green investment

2.4 Audit committee meeting and green investment

Audit committee is part of governance mechanism, which is responsible for overseeing the implementation of financial and accounting policies [40]–[43], including green investment policies. Therefore, the green investment policy cannot be separated from the role of audit committees. Borrowing previous studies on the role of audit committees in various corporate policies such as compliance with regulations [44], [45], and firm performance [46], it is believed that the effectiveness of the audit committee determines the reason why companies implement green investment. Studies on the relationship of audit committees and environmental issues can also be traced to Reference [47]. Thus, it is argued that the more effective the audit committees in monitoring company's policies, the higher the green investment of the company. The effectiveness of audit committees can be related to the frequency of the audit committee members in holding formal meeting. Based on the argument, this study proposes the following hypothesis

H4. Frequency of Audit Committees Meeting positively affects green investment

2.5 ISO14001 management certification and green investment

The ISO14001 standard can be used as a framework for improved corporate environmental management in a multinational perspective [48]. Companies implementing ISO14001 standards because such implementation can increase companies performance. Previous study revealed that there is a positive relationship between environmental management systems and financial performance [49]. Moreover, other studies suggested that ISO14001-certified facilities positively influenced environmental performance because they decreased their pollution emissions faster compared to non participants [50], [51]. The finding implies that companies with ISO14001 certificates are committed to higher green investment than those without the certificate. When the companies believe that ISO14001 management certification contributes to the companies' performance, then they will increase their funding in green investment. Consequently, this study proposed the following hypothesis.

H5: ISO14001 management certification positively influences green investment

2.6 Green investment and financial performance

Legitimacy theory points out that companies actively look for and maintain their legitimacy [34] by aligning company values, policies and strategies to the community values. Green investment can be seen as a company's strategy to gain and maintain legitimacy. The reason is that green investment enables companies to manage the impacts of their business on environment by minimizing the use of energy and decreasing carbon emissions and other negative impacts [1]–[3]. Green investment can increase company reputation, and competitive advantages of companies [4], [5]. Success in managing environmental issues can eventually increase

firm values[12], [37], [52]–[58]. Therefore, the proposed hypothesis is as a follow.
H6: Green investment positively affects firm financial performance.

3 Research method

This study employed five variables: green investment, financial performance, foreign ownership, industry profile, firm size, ISO14001 certification and frequency of audit committee meetings. Green investment refers to the total investment incurred by the company to reduce the negative impact of company activities [54]. Unfortunately, it is not easy to find monetary expenditure of green investment on annual reports. Hence, green investment (GI) is measured by the rank of PROPER Award received by companies with measurement scales as follows: five (5) for Gold (excellent), four (4) for Green, three (3) for Blue, two (2) for Red, and one (1) for Black (very poor). The reason is that, companies with higher ranks of PROPER Awards actually show that the companies are serious in investing their money to manage the negative environmental impacts of their business.

Foreign ownership (FO) is measured by proportion of shares owned by foreign shareholders to total outstanding shares. Firm Size (FS) is measured by Ln total assets of the company[17], [19], [54]. Industry profiles (IP) are considered as a dummy variable which refers to low profile or high profile industry (high profile industry is scored by one, otherwise zero). ISO14001 management certification (IS) is also dummy variable, which is measured by one (1) if the company has ISO14001 management certification, otherwise zero. Meanwhile, frequency of audit committee meetings (FM) is measured by the number of audit committee meetings within one year. Finally, financial performance (FP) is measured by ROA (Earnings After Tax divided by Total Assets).

Population of this study consists of all companies listed on the Indonesia Stock Exchanges in the year 2009-2014. Samples are chosen based on the following criteria: the company published annual reports in the year 2009-2014 and won PROPER awards in the observation year. Data were then analyzed using ordinal logistic regression and linear regression based on the following models.

$$GI = \alpha + \beta_1 FS + \beta_2 FO + \beta_3 IP + \beta_4 ISO + \beta_5 FM + e \quad (1)$$

$$FP = \alpha + \beta_1 GI + e \quad (2)$$

Where GI represents green investment; FP is financial performance; FS showa firm size; FO is foreign ownership; IP represents Industry Profile; IS shows ISO14001 management certification; and FM is Frequency of Audit Committee Meetings.

4 Findings and discussion

Based on the availability of data, 172 companies have received PROPER Awards from 2009 to 2014, but only 135 companies met sample criteria (27 companies receiving the PROPER Award were not those listed on the Indonesia Stock Exchanges). Table 1 indicated descriptive statistics of empirical data.

Table 1 demonstrated that most companies (54.80%) received Blue category of PROPER award. This reveals that the companies have arranged their green investment at the minimum level. In other words, they invested their money in green activities because of mandatory reasons, which is just to qualify the minimum requirements specified by regulators. Hence, the initiatives of the companies to voluntarily invest in environmental issues exceeding the minimum requirements were not indicated in the sample companies. In fact, only 9% and 26 % of the total sample received Gold and Green category of PROPER awards respectively.

Table. 1 Descriptive Statistics

Variables	Indicators	Number	Percentage
EI	Gold	12	8.90%
	Green	40	29.60%
	Blue	74	54.80%
	Red	8	5.90%
	Black	1	0.70%
IP	low profile	30	34.80%
	high profile	105	65.20%
IS	With ISO14001	130	96.30%
	Without ISO14001	5	3.70%
	Valid	135	100.00%
Variables	**Min**	**Max**	**Mean (St.D)**
FS (Ln)	21.97	32.08	28.86(1.64)
FO	0.00	100.00	36.60(35.37)
FM	1.00	57.00	8.90 (9.90)
FP	-19.57	54.40	9.20(11.40)

In terms of firm size, the data showed that on average, the firm size of the samples (LnFS) was 28.86 (equivalent to 3.3 trillion Rupiahs). The description of foreign ownership (FO) also reveals that foreign shareholders only hold 36.60 outstanding shares. It is interesting to note that there is a sample company by which 100% of its shares are owned by foreign shareholders. Concerning industry profiles (IP), the samples (65.20%) are dominated by high profile companies. Furthermore, the audit committees on average held eight formal meetings (FM) a year. The formal meetings were seen as quite high for publicly listed companies in Indonesia. Finally, the majority of samples were dominated by companies with ISO14001 management certification (96.30%).

Table. 2 Ordinal Logistic Regression: Model 1 (Dependent= Green Investment)

Variables	Estimation	Wald	Sig.	Results (Ha)
FS	0,122	3,670	0,049*	Supported
FO	0,012	14,671	0,000*	Supported
IP	0.361	10.684	0.001*	Supported
FM	0,024	4,601	0,032*	Supported
IS	-0,910	1,514	0,218	Not supported
-2 log Likelihood (Chi-Square) = 257.006 (Sig. = 0,000) Coefisien of Determination (Nagelkerke) = = 0,302				

Note: *Significant at 5%

While descriptive statistics described the characteristics of research sample, hypothesis testing were run by using ordinal logistic regression (Model 1) and simple regression (Model 2) to confirm whether empirical data support the hypothesis. Table 2 demonstrated the empirical findings. Table 2 indicated that the Chi-Square value is equal to 257.006 (Sig. = 0.000), which implies that the model can be employed to explain the determinants of green investment. Based on the results of Table 2, it can be inferred that the predicted variables, which significantly affected the green investment were firm size (FS), foreign ownership (FO), industry profiles (IP), and audit committee meetings (FM)) However, ISO14001 management certification (IS) did not significantly influence green investment. Finally, Nagelkerke coefficient has a value of 0.32, which implied that the degree to which firm size (FS), foreign ownership (FO), industry profiles (IP), and audit committee meetings (FM)) influenced green investments was only 32%.

Table. 3 Results of Regression: Model 2 (Dependent= Financial Performance)

Variable	Stand. Coeff	t	Sig.	Results (Ha)
GI	0.322	3.926	0.000*	Supported
Coeffisien of Determination (Adj R^2) = 0.104				

Note: *Significant at 5%

The second model was intended to investigate how green investment (GI) affects firm financial performance (FP). Table 3 showed the results of statistical test. It can be inferred from Table 3 that the empirical data supported the hypothesis (Sig. 0.000). This reveals that green investment significantly and positively influences financial performance. Thus, green investment can be used as a strategy to boost financial performance (profitability). The findings of hypothesis testing are further discussed below.

The first hypothesis claimed that firm size positively influences green investment. The finding supported the hypothesis. Companies with larger assets tend to have better green investment policies. This finding is consistent with legitimacy theory, which argues that larger companies tend to pay more attention on social and environmental issues than smaller ones because they want to align their interests to the community and environment ones [13]–[15] for legitimacy reasons. Furthermore, the finding is in line with previous studies concerning firm size and social and environmental issues [7], [16], [19]–[23], [59] including pollution and green management [17], [20], [23]–[25]

The second findings of this study indicated that foreign ownership significantly affects green investment. This means that companies implemented green investment because of pressures from foreign shareholders. The finding also reveals that foreign shareholders have significant power to influence management in the implementation of green investment policies as claimed by stakeholder theory. The finding is also consistent with findings by previous studies claiming that the greater the ownership, the greater the pressure on the company to manage all policies relating to the environment issues [8], [11], [27]–[32] and carbon emission policies [24], [28], [33]

The third hypothesis proposes argument that industry profiles positively influence green investment. The finding indicated that the empirical data supported this hypothesis. This reveals that environmentally high sensitive-companies (high profile industries) tend to have better green investment policies than low profile ones. It is, therefore, not surprising that industry profiles in Indonesia, which are intensively overseen by the Ministry of Environment are those in palm oil industry, oil and gas industry, and textile industry, which are similar to high profile industries. As claimed by legitimacy theory, to gain and maintain legitimacy and get public supports, companies must identify any activities, which are consistent with the public expectations [20], [23], [60]–[64], including activities related to green investments. In addition, the finding supported previous studies arguing that company concerns on environmental issues will increase when their business activities have significant impacts on the environment [17], [35]–[39]

The fourth hypothesis argues that the frequency of audit committees meetings—representing the effectiveness of the audit committees—positively determines the green investment. The hypothesis was supported by empirical data, which means that audit committees play an important role in overseeing the green investment policy of the companies. The average meeting held each year reached nine times, and these meetings might discuss any policies related to green investment issues. Based on the similar studies on the role of audit committees in various corporate policies such as compliance with regulations [44], [45], [65], and firm performance [46], this study supported the finding of the previous studies [47]. Indeed, audit committees play important roles in monitoring company policies on environmental issues [47]. This reveals that the effectiveness of the audit committee determines the reason why companies implement green investment.

The fifth hypothesis examines the effect of ISO14001 management certification on green investment. The empirical data did not support the hypothesis, which infers the ISO14001 management certification did not have effect on green environment of the companies. This empirical data was interesting as the number of companies used as samples were mostly dominated by those with ISO14001 management certification. The reason for this is that companies get ISO14001 management certification because it provides them with business advantages in terms of "improved corporate image" or for being seen as "responsible citizens" [50] [50]. Indeed, previous studies found that there is great variability in the implementation of ISO14001 standards [66], [67] and consequently, the standard is not effective in terms of environmental performance improvements [68].

Finally, it is believed that green investment positively influences firm financial performance. The empirical data of this study supported the hypothesis. This means that companies with better green investment tend to have higher financial performance. The research finding supported legitimacy theory [34] insisting that environmental issues (including green investments) can be used to build corporate images to gain and maintain their legitimacy. This will lead to the increase in company reputation and competitive advantages [4], [5] and eventually increase financial performance. This study supported other studies that firm value or financial performance will improve when the company implements an adequate environmental investment policies [12], [37], [52]–[58].

5 Concluding remarks

Studies on environmental issues and financial performance, including their antecedents have been conducted in many countries, but it is not easy to find the similar studies in Indonesia. This study aimed to investigate the effect of firm size, foreign ownership, industry profile, and frequency of audit committee meeting, and ISO 14001 management certification on green investment of companies listed on the Indonesia Stock Exchanges (IDX) and receiving PROPER award. Secondly, we also examined how green investment increases firm financial performance.

The findings showed that the level of companies' green investment is moderate. This can be seen from the PROPER award received by the companies (as a proxy of green investment) which are mostly on the Blue category. The research findings inferred that companies were committed to green investment because of regulation reasons, namely to qualify requirement set by the Minister of Environment, Republic of Indonesia. Moreover, this study found that firm size, foreign ownership, industry profile and frequency of audit committee meetings determined the companies' green investment. However, ISO 14001 management certification did not affect green investment. Meanwhile, it is confirmed that green investment positively influenced firm financial performance.

The findings of this study provide us with useful contributions. Firstly, firm size, foreign ownership, industry profile and frequency of audit committee meetings are important determinants of green investment. Thus, this study enriches prior findings, which mostly related firm size, foreign ownership, industry profile and frequency of audit committee meetings to social and environmental disclosures and tend to ignore the effect of these predictors on green investment. Secondly, the government or regulators may utilize the findings as reference in making regulations concerning environmental issues on business, especially for large-scale companies and those which are sensitive to the environmental issues. Thirdly, the positive correlation of green investment and financial performance provides companies with ideas of how to earn profits without destroying the environment. Finally, the results of this study can be utilized by accounting academicians on the importance of including environmental issues into business education.

Despite its contributions, this study suffers from limitations. Firstly, this study only used limited samples from companies receiving PROPER awards and were listed on the Jakarta Stock Exchanges. Thus, the findings cannot be generalized to all companies listed on the Indonesia Stock Exchanges and other markets. Secondly, this study only covered five main variables as predictors of green investment. We suggest that the future studies should include more companies listed on the Indonesia Stock Exchanges or Asian emerging markets and should consider more variables such as the activity of independent board of directors, audit committee expertise/skills, and institutional ownership as predictors of green investment.

Reference

1. D. Berliner and A. Prakash, Law Soc. Rev., vol. **47**, no. 2, pp. 345–373, (2013).
2. D. D. Minatti Ferreira, J. A. Borba, S. Rover, and F. Dal-Ri Murcia, Environ. Qual. Manag., vol. **23**, no. 4, pp. 71–86, (2014).
3. F. Testa, N. M. Gusmerottia, F. Corsini, E. Passetti, and F. Iraldo, Corp. Soc. Responsib. Environ. Manag., p. n/a--n/a, (2015).
4. L. A. Bagur-Femenías, J. A. Perramon, and O. B. Amat, Total Qual. Manag. Bus. Excell., vol. **26**, no. 7–8, pp. 840–853, (2015).
5. B. C. Bonifant, M. B. Arnold, and F. J. Long, Bus. Horiz., vol. **38**, no. 4, pp. 37–47, (1995).
6. E. Banasik, M. Barut, and L. Kloot, Aust. Account. Rev., vol. **20**, no. 4, pp. 387–399, (2010).
7. E. M. Barbu, P. Dumontier, N. Feleagă, and L. Feleagă, Int. J. Account., vol. **49**, no. 2, pp. 231–247, Jun. (2014).

8. G. E. Iatridis, Emerg. Mark. Rev., vol. **14**, pp. 55–75, Mar. (2013).
9. D. Hackston and M. J. Milne, Accounting, Audit. Account. J., vol. **9**, no. 1, pp. 77–108, (1996).
10. J. H. Sun, J. Hu, J. M. Yan, Z. Liu, and Y. R. Shi, Energy Procedia, vol. 16, Part A, pp. 377–382, (2012).
11. H. Wahba, Bus. Strateg. Environ., vol. **19**, no. 8, pp. 495–511, (2010).
12. A. Rokhmawati, M. Sathye, and S. Sathye, Procedia - Soc. Behav. Sci., vol. **211**, pp. 461–470, Nov. (2015).
13. M. E. Nawaiseh, Am. J. Appl. Sci., vol. **12**, no. 12, pp. 967–981, (2015).
14. G. J. A. Yu, K. M. A. Kwon, J. B. Lee, and H. A. Jung, Sustain., vol. **8**, no. 3, (2016).
15. H. Youn, N. Hua, and S. Lee, Int. J. Hosp. Manag., vol. **51**, pp. 127–134, (2015).
16. Y. Tan, Z. Zhu, C. Zeng, and M. Gao, Int. Rev. Financ. Anal., vol. **36**, pp. 212–222, Dec. (2014).
17. C. H. Cho, M. Freedman, and D. M. Patten, Accounting, Audit. Account. J., vol. **25**, no. 3, pp. 486–507, (2012).
18. J. V. . Frias-Aceituno, L. . Rodríguez-Ariza, and I. M. . Garcia-Sánchez, Bus. Strateg. Environ., vol. **23**, no. 1, pp. 56–72, (2014).
19. L. A. Chang, W. B. Li, and X. C. Lu, Bus. Strateg. Environ., vol. **24**, no. 1, pp. 1–19, (2015).
20. Z. Borghei-Ghomi and P. Leung, Account. Financ. Res., vol. **2**, no. 1, pp. 110–127, (2013).
21. I. Gallego-Álvarez and I. A. Quina-Custodio, Online Inf. Rev., vol. **40**, no. 2, pp. 218–238, (2016).
22. M. A. Hou, H. B. Liu, P. C. Fan, and Z. A. Wei, Asia Pacific J. Manag., vol. **33**, no. 1, pp. 195–228, (2016).
23. S. Yunus, E. Elijido-Ten, and S. Abhayawansa, Manag. Audit. J., vol. **31**, no. 2, pp. 156–179, (2016).
24. N. Hrovatin, N. Dolšak, and J. Zorić, J. Clean. Prod., p. , (2016).
25. P. A. Nath and R. B. Ramanathan, Int. J. Prod. Econ., vol. **171**, pp. 427–437, (2016).
26. R. Chakroun and H. Matoussi, J. Account. Manag. Inf. Syst., vol. **11**, no. 3, pp. 335–370, (2012).
27. F. Calza, G. Profumo, and I. Tutore, Bus. Strateg. Environ., p. n/a--n/a, (2014).
28. Y. M. Nulla, Corp. Ownersh. Control, vol. **13**, no. 1CONT9, pp. 844–854, (2015).
29. H. H. A. Hsiung, J. L. B. Wang, and W. T. B. Ku, Int. Res. J. Financ. Econ., vol. **88**, pp. 132–145, (2012).
30. N. Ortiz-de-Mandojana, J. Alberto Aragón Correa, and J. D. Ceballosj, Cuad. Econ. y Dir. la Empres., vol. **14**, no. 4, pp. 222–230, (2011).
31. B. B. Lahouel, J.-M. Peretti, and D. Autissier, Corp. Gov., vol. **14**, no. 3, pp. 363–381, (2014).
32. M. Hadani, J. Bus. Res., vol. **65**, no. 7, pp. 944–950, (2012).
33. G. Marsden and S. Groer, J. Transp. Geogr., vol. **51**, pp. 170–179, (2016).
34. J. Dowling and J. Pfeffer, Pac. Sociol. Rev., vol. **18**, no. 1, pp. 122–136, (1975).
35. J. H. A. Chen and S. I. B. Wu, Total Qual. Manag. Bus. Excell., vol. **26**, no. 7–8, pp. 778–792, (2015).
36. S. K. Fuisz-Kehrbach, Resour. Policy, vol. **46**, pp. 101–115, (2015).
37. X. A. Xie, J. A. Huo, G. B. Qi, and K. X. C. Zhu, IEEE Trans. Eng. Manag., vol. **63**, no. 1, pp. 101–112, (2016).
38. N. A. Sariannidis, G. B. Konteos, and G. B. Giannarakis, Corp. Ownersh. Control, vol. **12**, no. 2, pp. 92–106, (2015).
39. G. Giannarakis, G. Konteos, and N. Sariannidis, "Financial, governance and environmental determinants of corporate social responsible disclosure," Manag. Decis., vol. **52**, no. 10, pp. 1928–1951, 2014.
40. L. Spira, Bus. Ethics A Eur. Rev., vol. **8**, no. 4, pp. 262–273, (1999).
41. R. M. Hayes, J. Account. Econ., vol. **58**, no. 2–3, pp. 231–239, (2014).
42. C. Ghafran and N. O'Sullivan, Int. J. Manag. Rev., vol. **15**, no. 4, pp. 381–407, (2013).
43. F. T. Dezoort, Accounting, Organ. Soc., vol. **23**, no. 1, pp. 1–21, (1998).
44. M. K. Bepari and A. T. Mollik, J. Appl. Account. Res., vol. **16**, no. 2, pp. 196–220, (2015).
45. M. Bryce, M. J. Ali, and P. R. Mather, Pacific-Basin Financ. J., vol. **35**, pp. 163–181, Dec. (2014).
46. B. S. . Kallamu and N. A. M. . Saat, Asian Rev. Account., vol. **23**, no. 3, pp. 206–231, (2015).
47. A. J. . Trotman and K. T. . Trotman, Auditing, vol. **34**, no. 1, pp. 199–230, (2015).
48. M. Epstein and M.-J. Roy, Eur. Manag. J., vol. **16**, no. 3, pp. 284–296, (1998).
49. T. Feng, D. Cai, D. Wang, and X. Zhang, J. Clean. Prod., vol. **113**, pp. 781–791, Nov. (2015).
50. D. Rondinelli and G. Vastag, Eur. Manag. J., vol. **18**, no. 5, pp. 499–510, (2000).
51. M. Potoski and A. Prakash, J. Policy Anal. Manag., vol. **24**, no. 4, pp. 745–769, (2005).
52. L. A. Jackson and D. Singh, Tour. Manag. Perspect., vol. **14**, pp. 25–33, Apr. (2015).
53. M. J. A. Teng, S. Y. B. Wu, and S. J. H. B. Chou, Environ. Policy Gov., vol. **24**, no. 1, pp. 16–27, (2014).
54. E. Nakamura, Eurasian Bus. Rev., vol. **1**, no. 2, pp. 91–111, (2014).
55. S. L. Hart and G. Ahuja, Bus. Strateg. Environ., vol. **5**, no. 1, pp. 30–37, (1996).

56. E. Claver, M. D. López, J. F. Molina, and J. J. Tarí, " J. Environ. Manage., vol. **84**, no. 4, pp. 606–619, (2007).
57. V. I. F. Orellano and S. Quiota, RAE Rev. Adm. Empres., vol. **51**, no. 5, pp. 471–484, (2011).
58. W. Q. Judge and T. J. Douglas, J. Manag. Stud., vol. **35**, no. 2, pp. 241–262, (1998).
59. J. V Frias-Aceituno, L. Rodríguez-Ariza, and I. M. Garcia-Sánchez, Bus. Strateg. Environ., vol. **23**, no. 1, pp. 56–72, (2014).
60. G. J. M. M. Braam, L. Uit de Weerd, M. Hauck, M. A. J. J. Huijbregts, L. U. de Weerd, M. Hauck, and M. A. J. J. Huijbregts, J. Clean. Prod., vol. **129**, pp. 724–734, Apr. (2016).
61. D. Liang and T. Liu, J. Clean. Prod., vol. **142**, Part 4, pp. 2985–2998, (2017).
62. D. Li, M. Zheng, C. Cao, X. Chen, S. Ren, and M. Huang, J. Clean. Prod., vol. **141**, pp. 41–49, (2017).
63. G. O'Donovan, Accounting, Audit. Account. J., vol. **15**, no. 3, pp. 344–371, (2002).
64. C. Pellegrino and S. Lodhia, J. Clean. Prod., vol. **36**, pp. 68–82, (2012).
65. Y. J. Kang, A. J. Trotman, and K. T. Trotman, Accounting, Organ. Soc., vol. **46**, pp. 59–76, (2015).
66. H. Yin and P. J. Schmeidler, Bus. Strateg. Environ., vol. **18**, no. 7, pp. 469–486, (2009).
67. A. Prakash and M. Potoski, Int. Stud. Q., vol. **51**, no. 3, pp. 723–744, (2007).
68. B. Fei-baffoe, G. K. Botwe-koomson, I. F. Mensa-bonsu, and E. A. Agyapong, J. Waste Manag., vol. **2013**, no. Article ID 935843, (2013).

Environmental Policy of Mangroves Management in Rembang Regency

Ali Roziqin[1,*]

[1]Magister of Public Administration, Faculty of Social Science and Politic Science, University of Gadjah Mada, Yogyakarta – Indonesia

Abstract. Mangrove area is an area overgrown mangrove in a natural or artificial, to maintain the environmental sustainability of coastal areas. In addition to maintaining the ecosystem of biodiversity, the mangrove area also has a role to social-economic, and socio-cultural. Rembang regency is one of the districts on the north coast of Java which has a large mangrove area. However, due to the high economic activity in the region of Rembang Regency, the mangrove area becomes less and damaged. This research to describe how environmental policy to manage mangrove area in Rembang regency with qualitative descriptive approach. The result is the role of government and society gradually able to restore mangrove ecosystem. Moreover the district government through Environmental Agency has made a masterplan for the development of mangrove ecotourism in Pasarbanggi Village. The existence of sustainable mangrove conservation has a positive impact on the environment and society.

1 Introductions

1.1 Background

Coastal ecosystems are natural resources that have dynamic, unique and complex characteristics [1]. The coastal area is also home to a variety of environmental, economic and social activities. A lot of human activity is highly dependent on this place. As a result, if not well controlled, there will be imbalances between fields that will threaten the ecosystem of coastal environment. Coastal areas are basically composed of various ecosystems of each other interconnected [2]. One of the coastal ecosystems we know so far is the mangrove area. Mangrove area is a region that is formed either natural or artificial that serves to maintain the ecosystem of coastal environment. Mangrove area can be called "rainforest of the seas.

Referred to as the "rainforest of the seas", mangrove zones functions as the backbone of coastal and offshore food webs. Not only are these forested wetlands a superb nursery habitat for marine life, but they also provide shoreline protection and are accordingly beneficial for local economies in approximately 123 countries/territories. Diminishing of mangroves area is as a result of shrimp farming (aquaculture), coastal development, and clear cutting for timber production. from 1980 to 2005, nearly 35,600 square kilometers of mangroves were destroyed. There is no estimate of original global mangrove cover, but according a general consensus that it was once over 200,000 square kilometers. And now it appears that more than 50.000 square kilometers, approximately one-quarter of the original mangrove area, have been lost [3]. With human populations increasing at an astonishing rate in coastal areas, the next future it is crucial situations which need a policy and program solutions are implemented in the all mangrove-inhabited nations [4].

In the international, many organizations are have detailed guidelines to help their members develop more effective mangrove management plans. These organizations include but are not limited to, the United Nations (UN) Food and Agriculture Organization (FAO), International Tropical Timber Organization (ITTO), the Ramsar Convention, the International Society for Mangrove Ecosystems (ISME) [3]. Good mangrove management involves all stakeholders such as state, regional, private and non-profit organizations. Because the success of mangrove management policy is determined by good cooperation across sectors.

Mangroves plants have good adaptation characteristics. Like living in extreme environments with stagnant soil conditions, soils that have high salinity or unstable mud substrate conditions. In Indonesia there are about 202 species of mangrove plants, covering 89 species of trees, 5 species of palm, 19 species of climbers, 44 species herbs, 44 types of epiphytes and 1 type of nails. Some species are true mangroves and others are mangroves [5]. The Several mangrove plants are many things that can be utilized by the community ecologically, socio-economic and socio-cultural. As a result, in recent decades the extent of damage to the mangrove area tends to be very fast.

Rembang regency is one of the areas located on the north coast of Central Java Province. As a coastal area, Rembang regency has some mangrove areas, such as those located in Pasarbanggi village, Rembang District. Past mangrove conditions in the north coast of Java,

* Corresponding author: ali_roziqin@ymail.com

especially in the Rembang Regency is very abundant. But because of the high activity of the community, of course have a negative impact on the ecosystem of mangrove areas. Central java north beach recorded the worst damage compared with java west pantura or east java pantura reach 96.5% [6]. The easy accessibility has contributed to the destruction of the mangrove area. In addition, factors that contribute to mangrove damage in Rembang include aquaculture, mangrove logging, reclamation and sedimentation and environmental pollution. According to Kismartini [7] reduced mangrove area in Rembang originated from the conversion of mangrove land into milkfish ponds in the 1960s and embankment of windu in the 1990s.

Mangrove species present in Rembang regency are 27 species, consisting of 12 major species, 2 minor mangrove species and 13 species of mangrove asosiation [8]. While the mangrove area in Pasarbanggi village, Rembang District is dominated by the type of rhizophora with sonneratia standing to the sea and avicenia standing to mainland. The condition of the mangroves area that suffered damage will certainly have a bad impact for the community and the sustainability of the surrounding environment. On that basis the district government through several related agencies began to pay attention and try to improve the mangrove area, especially in the village Pasarbanggi Rembang District. Therefore in this research will explain how environmental policy related to mangrove management in Rembang Regency.

1.2 Theory

1.2.1 Public policy

Public policy is an applied discipline that is multidisciplinary. The initial definition of public policy proposed by Harold Laswell and Abraham Kaplan [9] is public policy as a projected of goals, values, and practices. While George C Edwards III and Ira Sharkansky [10] define public policy as a government action in the form of government programs to achieve goals or objectives.

Meanwhile, according to David Easton [11] defines public policy as the allocation of values by force (legitimate) to all members of society. Public policy in its speakers is not in the void of values, it is in a complex organization and invites much value from its environment. Based on the opinion of some experts, the researcher concludes that public policy is something effort taken by institution or organization in this case sovereign government to solve the problem in order to be able to realize social welfare.

1.2.2 Environmental policy

As an applied discipline that can accommodate from other sciences. Policy studies can not stand alone. Therefore in this study researchers focused on environment-based policies in more specific policies in mangrove management. According to RI (Indonesia Republic) law number 32 of 2009 on Environmental Protection and Management [12]. The environment is the unity of space with all things, power, circumstances, and living things, including human beings and their behavior that affect nature itself, the survival of life, and the welfare of human beings and other living beings.

Based on this environmental policy according to the researcher is a effort taken by the institution or organization in this case the government, to maintain the continuity of human environment and other living things. The environmental policy referred to in this research is the policy in the management of mangrove area in Rembang regency. In addition mangrove management policy in Rembang Regency purposes to restore and preserve the environment, also beneficial for economic, social and cultural and tourism sector for the surrounding community.

1.3 Research method

In this research, researcher used a qualitative descriptive approach aims to explain how environmental policy in mangrove management in Rembang regency. Qualitative descriptive approach [13] is a study that aims to describe social phenomena that occur in society. Then the data obtained from the primary and secondary data from the government, literature review and other documents related to mangrove area management.

In addition, to obtain the data the researchers conducted interviews to several informants supported by the results of observations in the field. Then the results are analyzed using techniques described by Miles and Huberman [14] consisting of three stages of data reduction, data presentation and conclusion or verification.

2 Results and discussions

The damage occurring to the mangrove ecosystem in Rembang Regency over the past 1 decade resulted in increased stakeholder awareness that the importance of mangrove ecosystems in coastal areas. Therefore, in this study will be presented environmental policy by the Government of Rembang regency in mangrove area management in the village Pasarbanggi, District Rembang. Before explaining the mangrove management policy, researchers will explain the problems that become obstacles in mangrove management.

2.1 The problems of mangrove management

The decrease of mangrove area in Rembang Regency that occurred in the period of 2000-2010 is due to the lack of understanding of the importance of mangrove functions and benefits ecologically and economically. Conversion of mangrove land into aquaculture areas, salt ponds, and population settlements will certainly have a negative impact on capture fisheries production and damage to the environment. Because the mangrove ecosystem is where fish and other plants breed. For mangrove land, itself is around 6.2 Ha only or quarter of pond area. To more details will be explained in the table below:

Table 2.1 Coastal land use in Rembang Regency

No	Land Use	Are (ha)
1	Settlement	2.69
2	Rice Fields	4.71
3	Embankment	24.42
4	Mangrove	6.2
5	Warehouse	1.35
	Totally	39.37

Source: Environmental Agency of Rembang Regency, 2015.

According to Mr. Sahal as a head of mangrove farmers [15], mangrove areas close to economic activities such as fishponds are often converted by the community for fishponds. This is because people still consider the mangrove area is a less functioning area so they are free to do all the activities. The lack of awareness of the community towards the environmental protection has a negative impact on the mangrove area. Like doing wild hunting and mangrove logging in the area of mangrove.

Then another problem is accessibility. Accessibility issues that exist in the region are divided into two, namely access to residential areas and access to the mangrove area. The second, access area is very important to support all economic activities or tours made by residents. If the accessibility to the mangrove area is good then the people will be helped in the mobility of daily activities. Besides that is the problem of facilities and infrastructure: in the form of roads, garbage facilities, drainage, sanitation needs. In addition, after the mangrove area opened for public / tourism, there are only visitors who do not heed the manager's appeal by dumping the garbage not on the place that has been provided in the mangrove area. Therefore, In the future, the government and other stakeholders must solve these problems to make the mangrove area of Pasarbanggi village be a better.

2.2 The environmental policy of mangrove management

As a coastal area that has a diversity of ecosystems environment. District really need policies to protect the environment. Local law No. 8 of 2007 on the management of coastal areas, marine and small islands [16], one of the purposes is to protect the mangrove ecosystem in Rembang. It is as set out in several articles such as article 12, which reads as follows:

1. Protection of coastal areas forested mangrove (Mangrove) is done to preserve the mangrove forests as a shaper of mangroves and breeding grounds for a variety of marine life as well as coastal protection from erosion of the sea and protective cultivation ponds behind him including the designation.
2. Criteria mangrove forested coastal area is a minimum of 130 (one hundred thirty) times the value of the average difference in the highest and lowest tides measured annual low tide line towards the west.

It is also supported with the support of government policy making local village (local Policy) is Perdes number 03 of 2014 [17] which specifically regulate and aims to protect the mangrove area in the village of Pasarbanggi, Rembang. This Perdes be agreement and commitment pasarbanggi villagers not to do illegal logging or other destructive interests of the mangrove areas. As outlined in article 20, paragraph 1.

"Each person / group / organization banned cutting down trees in the forest damaging mangrove Pasarbanggi Village area. Penalties for offenders who violate the mangrove planting of 100 (one hundred) rods with maintenance ".

With the adoption of these policies is gradually people began to be aware and proactive in supporting environmental conservation efforts mangrove areas. Since it, Management of region becomes more manageable because of their youth are active in promoting the mangrove areas. Pasarbanggi mangrove ecosystem is one of the best on the north coast of Central Java. Mangrove in Rembang is a blend of natural mangrove and rehabilitation outcomes. With the support of the role of government, private and society. Effort to management through rehabilitation and conservation of mangrove ecosystems were able to restore gradually. For more details, following one of the image spot Pasarbanggi mangrove area.

Fig. 2.1 Condition of Mangrove Area in the Village of Pasarbanggi

2.3 Master plan of environmental agency in mangrove management

Rembang district government through issuing BLH Masterplan development Pasarbanggi mangrove area in the village. Management of mangrove areas in the village Pasarbanggi is a demonstration of the Government of Rembang through Environmental Agency in cooperation with the Department of Marine and Fisheries as well as private parties. This is done to protect the ecosystem of the mangrove areas negative actions of people who are not responsible. Other than that expected by the master plan of the management of mangrove areas, in the future this area will benefit the community both in ecological, economic and socio-

https://doi.org/10.1051/e3sconf/20183109002

cultural. The business initially had difficulties and obstacles. Many peoples around who feel that the plan of conservation land is hers. But aided by community leaders and youth in village areas Pasarbanggi, the public finally became aware and willing to work together to be active in the conservation of the mangrove area.

In the Master Plan for the management of mangrove areas will be developed as the follows:

2.3.1 Ecotourism of Mangrove Area

Ecotourism is meant is combining ecology (Environment) to the development of tourism. Ecotourism is a form of travel to natural areas that is done with the goal conserve the environment and preserve the life and welfare of the local population [18]. In the utilization of natural areas for ecotourism use approach to the conservation and utilization. Both approaches are implemented with emphasis on "preservation" than the utilization. Ecotourism development in the region can ensure the integrity and preservation of mangrove ecosystems. Ecotourism is more use of conservation approaches.

Fig. 2.2 Mangrove Conservation Area By Environmental Agency of Rembang Regency in the Village of Pasarbanggi

Mangrove conservation efforts began well managed in 2014 with the construction of facilities to enjoy the mangroves area in the form of form of a bridge and gazebo which surrounds its area. Construction of the gate. Construction of access roads more adequate provision of facilities such as bins, toilets, area of the pavement.. Construction of bird observation post. During the observation conducted by researchers identified only the construction of bird observation post that had not been realized.

Fig. 2.3 Red Bridge of Mangrove Area in The Pasarbanggi Village.

In the Mangrove conservation of Pasarbanggi contain 15 mangroves in mangrove plant species consisting of five true mangrove species and 10 types of associations. *Rhizophora mucronata* is the dominant species. In addition, the area mangroves as a habitat for 22 species of birds from 14 families were 7 of which are protected species [19].

Fig. 2.4 Type of Mangrove in Mangrove Conservation Area

Development of ecotourism in mangrove areas, the government is also building a board walk that we know as the red bridge. Opening of mangrove area as a tourist a positive impact on the environmental, social and economic communities. From the opening environment of mangrove areas for ecotourism in accordance with the carrying capacity of the area [8]. So that this will not damage the environment or the mangrove ecosystem.

Fig. 2.5 Management Activity by Community of Pasarbanggi Village

In the past this region has a fairly extensive mangrove areas are scattered from the District to the District Kaliori nest. However since that time people do not notice their mangroves around their environment, then long mangrove ecosystem begins to decline because of the high economic activity from coastal communities. But over time with the management and improvement are done together. This region often become a pilot and be a tree planting mangroves. So that the existing ecosystem is becoming increasingly widespread and well ordered. In addition to being a place of environmental conservation and animal biodiversity. Pasarbanggi mangrove is also a tourist destination based environment.

2.3.2 Village tourism

Village tourism is a form of integration between the interaction, accommodation and support facilities are presented in a structure of a society that blends with the applicable rules and traditions [20]. In mangrove areas, the population still has a cultural tradition and a relatively original. There was also a special food, agricultural systems and social systems that foreshadowed tourism villages. Development of rural tourism will certainly accentuate on the side of life, social and cultural rights of coastal communities. So expect the tourists will get to know how everyday life of coastal communities survive.

2.3.3 Water travel area (the area of concept waterfront and water park)

In this mangrove area in the future will be developed tourist areas, namely water Waterfront and Waterpark. Waterfront is area that may include building or activity that is not directly above the water, but it is bound to visual / historical / physically associated with water as part of a broader scheme [21]. With the waterfront is expected to revive community life activities, such as trade and also liaison between the banks. While the construction of the waterpark will increase tourist traffic in the local, so the impact on the growing economy of local people and income for the local government.

3 Conclusions

In this paper described environmental policy in Rembang, particularly in relation to mangrove management. Mangrove management policy is a step taken by the government to preserve the environment. Besides the mangrove areas are also having an impact on other sectors of public life around like economic, social, cultural and tourism, so we need a policy that guarantees the existence of mangrove areas for the life to come.

Mangroves region is an area that is very important in people's lives and vast coastal areas. Besides as an ecosystem for the environment is also beneficial for all aspects of human life. Therefore we need a commitment to the triangle actors ranging from governments, private sector, and the community itself to support environmental policy (Master Plan) management of mangroves in Rembang. With the master plan, the policy of mangrove management to be more structured and clear. So that conservation efforts mangroves region will go well, as well as plans to build ecotourism mangroves area in Rembang sub-district village Pasarbanggi can be realized.

Commitments undertaken by Environmental Agency of Rembang in developing the next mangroves area should be more adopted by other agencies. Synergy between actors also must be intensified in the future. It is quite important to avoid sectoral ego or the construction of a single area. Because the policy-making related to coastal resources, especially the development of mangrove areas is quite complex. Construction of Master Plan for development of mangrove areas is evidence of local governments pay attention to the environmental sector as a vital sector in society.

References

1. Kismartini. Jurnal Ilmu Sosial **Vol.13** No.1 Universitas Diponegoro, Semarang (2014).
2. Sutrisno. *Kantor Penelitian dan Pengembangan Kabupaten Pati* (2015).
3. Haille N. Carter, Steffen W. Schmidt, and Amy C. Hirons. (2015).
4. *Mangrove and Coastal Wetlands Protection Draft Polcy and Regulation*. USAID (1996)
5. Y.R. Noor, M. Khazali, I.N.N. Suryadiputra. *PKA-Wetland International Indonesia Program*, Bogor (1999).
6. Puryono. *Desertasi Program Doktor Manajemen Sumberdaya Pantai* Pascasarjana Universitas Diponegoro, Semarang. (2009).
7. Kismartini. *Disertasi. Program Doktor Ilmu Administrasi, Minat Administrasi Publik. Fakultas*

Ilmu Administrasi. Universitas Brawijaya. Malang. (2012)

8. Setyawan, Eko. *Tesis Magister Ilmu Lingkungan*. Universitas Diponegoro. (2015).
9. Howlett, Michael and M. Ramesh. Oxford University Press, Oxford. (1995)
10. Suwitri, Sri. Badan Penerbit Universitas Diponegoro. Semarang. [2008].
11. Islamy, Irfan. *Prinsip-Prinsip Perumusan Kebijaksanaan Negara*, PT. Bina Aksara: Jakarta (1986).
12. Undang-Undang nomor 32 tahun 2009 tentang Perlindungan dan Pegelolaan Lingkungan Hidup
13. Singarimbun, and S. Efendi, *Metode Penelitian Survei*. Jakarta: LP3ES. (1989)
14. Miles, B. Metthew, A. Michael Huberman and Johnny Saldana. (3th. Ed.). Sage Publications, Inc, (2014)
15. Roziqin, Ali. *Skripsi, Jurusan Administrasi Publik*. Universitas Diponegoro, Semarang. (2016).
16. Perda Kabupeten Rembang Nomor 8 tahun 2007 *tentang Pengelolaan Wilayah Pesisir, Laut dan Pulau-Pulau Kecil*.
17. *Perarturan Desa Pasar Banggi Nomor 3 Tahun 2014*, Kabupaten Rembang.
18. *The Ecoutorism Society*. (1990).
19. *Master Plan BLH Kabupaten Rembang*. (2015)
20. Nuryanti, Wiendu. Concept, Perspective and Challenges. Paper from of the International Conference Report about Pariwisata dan Culture, (1993)
21. City Planning Directorate of Urban and Rural, Cipta Karya DPU, (1998).

The Process of People Gold Mining in Paningkaban Village Banyumas Indonesia

Muslihudin Muslihudin [1,2,*], *Azis* Nur Bambang [3], *Eko* Hendarto [4], and *Thomas* Triadi Putranto[5]

[1] Faculty of Social and Political Sciences, Jenderal Soedirman University, Purwokerto-Indonesia
[2] Doctoral Program of Environmental Sciences, School of Postgraduate Studies, Diponegoro University, Semarang-Indonesia
[3] Faculty of Fisheries and Marine Sciences, Diponegoro University, Semarang-Indonesia
[4] Faculty of Animal, Jenderal Soedirman University, Purwokerto-Indonesia
[5] Geology Department of Technic Faculty, Diponegoro University, Semarang-Indonesia

Abstract. Gold mining in Paningkaban Banyumas conducted by the community is called the People gold mining. At the beginning, many miners from outside the region have involved and transferred of method, technic and knowledge about gold mining to local people. The aim of the study is to identify the existing process of public gold mining. The method of the study is qualitative by using observation and interview. The result showed that the mining process are: 1. Determining the location of mining well; in this determination there are two references; rational and intuition 2. Mining; at this stage, a deep well is drawn about 50-100 meters that leads vertically and horizontally. It is the most high-risk stage because of work accidents that occurred and potentially environment destruction. 3. Pulverization; this stage is classified as the lowest level of difficulty and risk, therefore in this work many woman included. 4. Rolling; in this stage involves enough technology, electrical mechanic and energy with the dynamo and using mercury that potentially contaminate environment. 5. Filtering; this stage is a quite risky because the workers contact directly with mercury. 6. Burning; is the shortest process to separate mercury with gold grains. 7. Sales to local buyer guided by the international gold market in every Thursday.

1 Introduction

Mining of natural resources is quite a lot done in Indonesia. This is very reasonable, because the potential that exists in Indonesia is very rich in natural resources. One form that is developing in many areas is gold. The history of gold mining at this time is a new moment coming on the earth archipelago that always exploit the spices, also glance at the mineral wealth. Traces of mining activities undertaken by the Dutch during valid in Indonesia can still be found starting from Sumatra, Java, Kalimantan, and Sulawesi. Mineral mining [1] describes long before the Dutch come, the Indonesia is already well known to be rich in gold.

Gold mining has many variants which include mining that belongs to the traditional category, illegal, small scale, and usually on behalf of the people. Mineral Tambang [1], illustrates that public mining is a mining whose method is applied to underground mining under Gophering method, which is an unsystematic mining method, no need for development works and excavation only following the direction of the ore crops. Therefore the size of the holes (stope) is also indefinite, depending on the size of the ore deposit in that place and generally without good buffering. This mining method is generally without adequate buffers and excavation is generally done without mechanical tools. This method of gold mining is commonly applied in various areas of Indonesian mining operations, such as in Ciguha, Pongkor-Bogor; Gunung Peti, Cisolok-Sukabumi, Mount Subang, Tanggeung-Cianjur; Cikajang-Garut; Cikidang, Cikotok-Lebak; Cineam-Tasikmalaya; Kokap-Kulonprogo; Selogiri-Wonogiri, and including those in Banyumas.

The simple method in the mining process makes gold mining of this kind very vulnerable to the risk of work accident. Almost in every public gold mining have an accident that killed miners, including in Banyumas district. Jennifer et al [2], illustrates that public gold mining is often accompanied by degradation of environmental quality. Hg amalgamation not only has a serious impact on miners but also on the environment, especially with regard to the food chain. Work accidents and mine failures such as landslides are also common.

In addition, there are also legal and illegal variants. Illegal mining is unlicensed mining, often abbreviated as PETI [3]. One of the problems that faced by the Department of Energy and Mineral Resources (DESDM) is the rampant illegal mining activities (PETI). The term PETI was originally used for unlicensed gold mining, but in subsequent developments PETI issues not only on gold mining commodities but also applied to unlicensed mining for other minerals either class A, B or C (PP No. 27 Year 1980 About Classification Excavation Materials) which are usually included in small-scale mining.

Factors that encourage increased PETI activities are:

* Corresponding author: muslihudin1963@yahoo.com

https://doi.org/10.1051/e3sconf/20183109003

1. Mining business in general provides relatively high material benefits, because some excavated materials without going through direct processing can be sold.
2. Relatively high demand for the quarry market, especially in areas with high physical development growth.
3. Easy level of exploitation, intended for class C excavation materials.
4. Very short mining periods, some of which are shorter than the length of the licensing process, since most are small-scale mining [4].

This gold mining seems to have tremendous appeal than mining in other fields. This is understandable because the price of gold is very high compared to the price of other mining materials. Gold is still regarded as a reliable investment instrument, especially in the midst of the current unstable global economic situation. Gold could be a safe-haven investment, unlike other instruments that fluctuate quite a lot.

The purpose of this research is to know about the public gold mining process in Banyumas Regency. How this gold mining process is important to know because it can be used as a basis for consideration management or further development so as not to harm miners, communities and the environment.

2 Research methods

This study was designed as a qualitative descriptive study, using case studies. Data collection methods are in-depth interviews and documentation. The analytical method used is qualitative analysis [5]. The process of qualitative analysis begins from the collected data is condensed in the form of research findings that are relevant to the writing material, and then presented in a narrative. Condensation and data presentation are two components of analysis performed simultaneously with the data collection process. The next process is the conclusion, which is done after the data collection process, presented, described and then given a logical interpretation.

The final goal of the analysis is to gain some understanding of the focus of this research. The activity of the two components interacts until the correct conclusion is obtained. If the conclusion is inadequate, it is necessary retesting activity that is by searching data again in the field and try to interpret it, thus the activity of analysis with data collection is the process of the cycle until the research is completed.

3 Result and discussion

The gold mining in Paningkaban village of Gumelar sub-district of Banyumas Regency is considered as the public mining. Public mining in the sense of doing mining is a non-government or official institution recognized by the government. In other words this kind of activity is illegal. Illegal activities do not have operational standards that can be justified either jurisdictionally or technically and environmentally.

This gold mining process of the people includes the following important stages.

1. Determination of the location of the mine site. In this study, there are two references used by miners, namely rational logic and intuition or on the basis of guidance of spiritual leaders.
2. Mining. In this stage, site-digging is done within 50-100 meters, vertically and horizontally. This mining process is considered to be the most high-risk stage because of work accidents at this stage. Every site always has a kind of standard operating procedure taught by miners from outside of the area who are the majority of Tasikmalaya people. They are famous as miners who know much about the technical knowledge of gold mining. Security standards carried out in mining are:
   2.1 Each hole should be made of wooden steger to prevent from landslides and as well as stairs in and out of miners.

Fig. 1 Mining Site

Figure 1 above is a vertical hole with a relatively sturdy wooden stager installed, as well as a hole that is horizontal. However long it has never been avalanched because the soil type is hard so it is not easy to slide.

   2.2 Oxygen filling, this is very important because conditions in deep sites are often filled with acid gases that are very harmful to the miners. Therefore, before the miner enters the well, a blower is inserted into the well by using a compressor through the pipe and fed by electricity. This blower is turned on before the miners enter the hole about 1-2 hours (figure 2). This blower function is very important to prevent accident because of the acid gas. Because of the acid gas, miners are banned to do the excavation at night. In addition, the oxygen above the hole is also slightly absorbed by the respiration of plants.
   2.3 Lighting, the conditions in the sites are dark, thus the miners need lighting to do their activities. The lighting used is powered by electricity. Therefore if the power is off, then the mine activities automatically stop.

https://doi.org/10.1051/e3sconf/20183109003

Electricity energy is quite dominant in the mining process. Electricity is not only used as energy for lighting but also as a dynamo generator in the next process.

Fig. 2 Electricity and air lines to the site

3. Mining product mashing. This stage is classified as the lowest level of difficulty and risk. That is why women workers are often involved in this work. Ababio et al [6] describes that women are involved in traditional gold mining in Ghana as well. Even in Ghana there are certain signs for women involved in traditional gold mining.

Fig. 3 Women participation in the gold mining

4. Rolling, in this stage is a sufficient work related to technology. The technology used in the mechanical roll that turns the logs using the electrical energy with the dynamo which then rotate the log. This stage also uses chemical liquid called mercury. Mercury is used to tie the existing gold in the log. Rolling for about 3 hours. After the first 2 hours, a bottle of mercury is poured for each drum. Then, the mercury is rolled again for about 1 hour. After that the drums are opened and spilled which then flowed into the tub / pond shelter and some also flow into the river. Then, the content of the bucket which is processed further filtering. On the other hand [7,8] concluded that the gold mining in Gumelar in general has polluted the environment that Hg levels spread over the Tajum River with values exceeding the environmental quality standard that is 0.002 mg / liter.

Fig. 4 Rolling

5. Filtering, this stage is a phase that is quite risky because workers directly in contact with mercury-containing fluid. Filtering is done by bare hands using a soft cloth. Although miners are daily in direct contact with mercury, they do not feel threatened by their health.

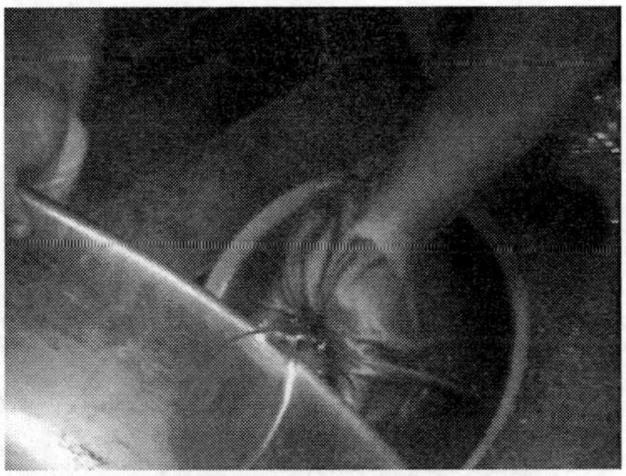

Fig. 5 Filter of fluids containing gold

6. Combustion. Combustion is a relatively short stage to separate the mercury with gold grains. Combustion is using gasoline with a welding tool which is then directed to a lump of gold that is still mixed with mercury. The workers were not using any safeguards. Thus, they are very likely to be contaminated with air from combustion containing mercury. The result of this burning so a golden blob whose color is still whitish and has hardened. In this combustion phase, there is shrinkage of about 50 percent of the previous weight. The gold content of this combustion is about 70 percent, which is usually used as gold jewelry.

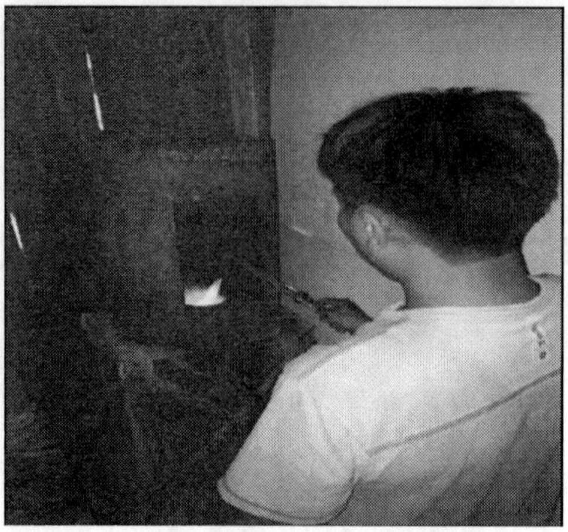

Fig. 6 Separation of mercury with gold grains

7. Selling, this stage is the stage of gaining the economic value of the gold itself. The gold price follows the international price. Although gold mining areas including rural areas even somewhat remote, but the price of gold always follow the real price in international markets that are known through the internet. The price of gold when this research is done is Rp380.000, - / gram. Sales are done every Thursday, as a weekend day because Friday is a holiday. On Thursday was also the result of a mine within a week known. Acceptance of results by miners is accepted. As an illustration of mining products for the present condition that is classified as being low. One log of 4/5 drums a week can be one gold item as shown in the following figure.

Fig. 7 Sample of ready-to-sell mining products

The weight of gold in the image is about 14.5 grams. The gold content of the time was 70% (percent), sold at Rp 380.000, - per/gram. Thus, the money earned is Rp 5.510.000, - (five million five hundred and ten thousand rupiah). Alfonsus et al [9] describes that traditional gold mining has a positive impact on the people's economy so that they also have a positive perception of the mining activity, despite the fact that there is a negative impact on the environment.

3 Conclusions

The public gold mining in Gumelar Banyumas Sub-district is a gold mine that involves a lot of energy. People's gold mining is sustained by the labor of the local population. At first this mining is mostly done by personnel from outside the region. As mining proceeds, there is a process of knowledge transfer from migrants to local residents so that many of the local people are good already at doing their own mining. The gold mining process is quite long and very risky for the safety and health of workers. The most risk-bearing job is excavation into a deep site that can reach hundreds of meters deep. At this stage of the excavation, there are a lot of work accidents. The filtering and combustion phase are also dangerous in which both workers are in direct contact with mercury. Contamination of mercury to the environment is also indicated to have occurred. On that basis the central government through BPPT (Agency for the Assessment and Application of Technology) in collaboration with ESDM (Institution of Energy and Mineral Resources) and the Department of Environment and the Health Office is trying to find the alternative technology that can minimize mercury pollution to the environment.

4 Suggestions

The security efforts of miners should be improved so as to minimize / eliminate the number of victims of work accident. The government is still needed to monitor the gold mining conditions of the people even though their status is illegal. However, these activities are very helpful to the community's economy, not only in the villages where there are mining activities but also to the wider community.

5 Acknowledgements

Thank to the head of Paningkaban Village, Mr. Sukarmo who has given permission and much information about various matters relating to gold mining in his region. Thank you also to the gold miners of the people who have provided information related to their activities.

References

1. *Mineral of Mining,* Jakarta, (2013) http: //www.tambang-penambang-emas, [accessed on April 15, 2015].
2. H. Jennifer, M. Marcello, Tadeu C. Veiga, *International Journal of Cleaner Production,* Vancouver, Canada, (2003)

3. Refles, *Mining Activities of the People's Peoples and Its Implications on Socio-Economic Condition of the Community In Kenagarian Mundam Sakti Sub-district Nagari, Sijunjung Regency,* Andalas University, Padang, (2012)
4. F. Irna, Ismadi and H.S. Dewi, *Public Health,* Andalas University, Padang, (2010)
5. B.M. Miles, M. Huberman, and J. Saldana, Third Edition Copyright Sage Publications, Inc, Washington, (2014)
6. Ababio E, OFOSU M, 2010, Nordic Journal of African Studies **19**(2): 124–147, University of Ghana, Ghana, (2010)
7. Fahmi, Budianta and Idrus, *Impact of Mercury Pollution on Geological Media at the People's Mining in Banyumas, Central Java.* Proceedings of the 7 th National Geological Engineering Seminar, Faculty of Engineering, Gadjah Mada University, Yogyakarta, (2014)
8. S. Widiadi, Postgraduate Program, Jenderal Soedirman University, Purwokerto, (2013)
9. Alfonsus, H. Harianja and S. Asep, *Forest Research Institute,* Aek Nauli, Medan, (2014)

Implementation of State Obligations and Responsibility Ensuring the Availability of Clean Water in Karimunjawa Islands

Rahayu[1,2]*, *Tri Retnaningsih Soeprobowati*[2,3]

[1]Faculty of Law, Diponegoro University
[2]School of Postgraduate Studies, Diponegoro University
[3]Faculty of Science and Mathematics, Diponegoro University

Abstract. This article aims to analyze the implementation of state obligations and responsibility ensuring the availability of clean water as part of human rights in Karimunjawa islands. The analysis based on principle of the State obligations and responsibility to fulfill their citizen right. Water sources in Karimunjawa Islands is very limited. It depend on forest conservation. Around 9.600 peoples live in Karimunjawa Islands, but Karimunjawa is non groundwater basin region. It means, Karimunjawa doesn't have groundwater potential. The quantity of water depends on the season. The solution to maintain the sustainability of clean water is piping from water reservoir to residential areas. The problem is there are so many hotels in Karimunjawa islands, it disrupted the fulfillment of clean water. Besides utilizing water from reservoir, many hotels drilled the ground to get water. It had impact to the availibity of water in dry season and affected to fulfillment of water supply for Karimunjawa people. There is no specific regulation and policy to solve this problem. Clean water management is doing by Karimunjawa's people. Meanwhile, based on Mahkamah Konstitusi Decree number 85/PUU-XI/2013, state is a rights holder to dominate the water in accordance with the Articles 33 paragraph (2) and (3) UUD NRI 1945, so the government has an obligation to make a policy, regulations, management, and supervision.

1 Introduction

1.1 Background

Water is the source of life. For human, water possesses social, economic, and religious values and functions. Water resources have always become actual issues because the right of water is fundamental for human life. In practice, however, the need for clean water for most people in the world, including in Indonesia, has never been fulfilled.

Global water problems also indicate the similar problems with that of Indonesia. Naturally, Indonesia is the fifth richest country in the world, having an annual rainfall rate of 2,799 mm, with a great deal of water resource potential. However, such abundance raises some problems because of inequal distribution through the year. Due to global climate change, the climate pattern continues to change significantly. Rainy season in particular country will affect the climate of the other country. Therefore, water climate-related water problems are considered global issues, no longer local or national ones [1].

International community concern of the water-related problems has been indicated by an agreement of global development by the issuance of a document namely 'Sustainable Development Goals' (SDGs). This document is signed by 193 country members of the United Nations at the General Assembly on September 25, 2015 in New York. The SDGs document promotes 17 goals and 169 development objectives and its sixth goal, 'clean water and sanitation', requires the sustainable clean water and sanitation for everyone.

Karimunjawa Islands, one of districts in Jepara Regency, are regions with limited clean water, in particular when it comes a dry season. In the dry season, many local people use brackish water to fulfill their daily necessities. The islands lack water supply and have a forest zone conservation area. A policy is necessary to regulate the water supply and use to ensure the sustainable right of water for the local people of the Karimunjawa Islands. The obligation to fulfill the need for clean water is among the primary responsibilities of the government.

Indonesia is a country where mutual welfare has become the ultimate objective. This concept is not only becoming a legal image (*rechtsidee*), but also a state image (*staatsidee*). From the concept of welfare state, the state does not only act as security and tranquility watch for its people, but also has a responsibility for putting in practice social justice, mutual welfare and people prosperity [2-3].

* Corresponding author: rahayu_undip@yahoo.com

1.2 Theory

1.2.1 Policy on Water Resource Management

Water resource is understood as a capacity and potential of water to be exploited by human for social-economic activities. There are different water resources that can be used by the community, such as marine/sea water, surface water, rain water, and ground water. The use and control of water resources deal with water quality and quantity. Water quality is an important point in the water exploitation because it does not only depend on how much water that can be obtained to fulfill the needs, but also how much the quality of the water in line with the human needs [4].

In terms of water resource management as part of the natural resources, the government has an authority to issue related policies, laws and licenses. Since the Decree of Constitution Court 85/PUU-XI/2013, which negated Law 7/2004 on Water Resources, has been taken in effect, the water management has once again referred to the Law 11/1974 on Water Management [5]. According to the Law 11/1974 water resources are water places and media, either over or under the land surface (Article 1, point 4). Water and the water sources, including their natural richness, have social functions by which the people can exploit and use for their prosperity. The Law also requires that the water and the water resources are possessed by the state, pursuant to the Constitution of the Republic of Indonesia 1945, Article 33(3).

According to the Decree of the Constitution Supreme Court, the state expands its authority as it has responsibility to make policy (*beleid*), regulation (*regelandaad*), provision (*bestuursdaad*), management (*beheersdaad*), and monitoring (*toezicthoudensdaad*) [5]. These five authorizing functions become integrated instruments for the state towards the ultimate goal, i.e. prosperity for the all citizens. Therefore, the phrase 'to be possessed by the State' in the Article 33(3) of the 1945 Constitution must be understood within the concept of public law related to the principles of people sovereignty within the 1945 Constitution, either those concerning political democracy or economic democracy. It does necessarily mean that 'to be possessed' is 'to be owned' [6].

1.2.2 Citizen's right on water under legal standard and norm and State's obligations to fulfill

In principle, water is the essence of life because its existence is indispensable. Like oxygen, human cannot live without water. Even, every living creature on this planet will not survive without water. Therefore, water is the most fundamental right for mankind.

In addition to object, right is the object of law that can be controlled or owned by legal subjects. In general, rights can be distinguished into absolute rights and relative rights. Human rights are included in absolute rights.

Vary legal instruments promoted at international level indicate that water is not only the right, but also an aspect in which the access must be enabled for everyone. In other words, water has also become the part of human rights. Such defense has been promoted by the UDHR (Preambule and Article 3); Decision 1999/108 the United Nations on Human Rights; Resolution of Sub Commission on Prevention of Discrimination and Protection of Minorities 1998/7; Resolution Number 2000/8 Sub-Commission on Human Rights; "Promotion of the Realization of the Rights of Drinking and Sanitation"; Resolution of Commission on Human Rights 2003/71; "Human Rights and the Environment as part of Sustainable Development"; Article 11 and 12 International Covenant on Economic, Social and Cultural Right (ICESCR); etc. In the General Comment No. 14/2008 CSECR, in particular in the paragraph 12, the right of access to water comprises four elements, i.e. availability, accessibility, acceptability, and quality. Accessibility relates to respect for non-discriminatory principle, physical and economic access, and information.

In the context of the national law of Indonesia, the right of water has been constitutionally given under the 1945 Constitution, Articles 28A and 28B, 33(2) and 33(3). Furthermore, its implementation also has already been regulated, such as by Law 39/1999, in which Article 77 requires that "the government has an obligation to and responsibility for the respect, protection, enforcement, and honor the human rights under this Law, other regulations, and any international law concerning the human rights...". It means that the state has the obligation in three different matters, i.e. (1) to respect by self-control of any possible human rights violation, such as polluting rivers by which the people use for daily activities; (2) to protect by taking measures to prevent third parties' vilation of the important elements of the right of water, and (3) to fulfill by taking legislative, administrative, budgetary, and judiciary measures to fulfill the need for and the right of water. These obligations point at the involvement level of the state and government [7].

1.3 Research Method

In this research social reality is defined as a reality that becomes part of the awareness, knowledge of society within a State. The social reality that underlines this research is the empirical reality, the reality that seems visible and is believed to be the real reality [8].

1.3.1 Research approach

This research applied a socio legal study approach. Such approach requires that legal aspect cannot only be conceptualized as a norm, but also a means of behavior. Therefore, the investigation of realties is expected to give the light whether the positive law or normative law within the patterns between subjects in the community have been fairly applied.

The use of socio legal study approach is necessary because a phenomenon cannot be solved only by providing information about the legal regulations. Instead, it also has to integrate approaches from other perspectives based on reality towards the ultimate goal of the law, i.e. creating regularity, justice, and prosperity.

1.3.2 Research type and materials

This research applied a qualitative-descriptive method with the aim at providing description of the right of water fulfillment for the local people of Karimunjawa Islands amidst the lacking clean water supply by exploration and clarification of the object under observation.

Data used for the research were both primary and secondary. The primary data were obtained by in-depth interview and focus group discussion with key informants. These data implied legal attitudes and behaviors of the local people of Karimunjawa Islands to fulfill their right of the clean water. Whereas, the secondary data were collected from literature study and documentary study, in the forms of primary, secondary, and tertiary legal materials.

1.3.3Research Output Analysis

The data obtained were subject to qualitative analysis, which went on the following steps:

1.3.3.1 Data reduction, irrelevant data elimination, data unit abstraction;
1.3.3.2 Categorization, data were categorized;
1.3.3.3 Category correlation and correlated interperpretation;
1.3.3.4 Interpretation and conclusion.

2 Results and Discussion

2.1 Access to clean water for local people of Karimunjawa Islands

The Karimunjawa Islands are under the administrative area of Jepara Regency, Central Java Province. This area becomes one of districts, which consists of four villages and 12 neighborhood area (*dukuh*). The islands spread 125 kilometer square on the geographical coordinate of 110°05'57-110°31'15 East and 5°40'39-5°55'00 South and are parts of the Java Sea group of islands. They are situated Northwest of the capital city of Jepara surrounded by the Java Sea. Karimunjawa Islands are composed of 27 islets and atolls, only four of them are inhabited with the total population of 9,802 [9].

Karimunjawa Islands do not have big rivers, but they possess five water springs, i.e. Kapuran (Pancuran Belakang), Legon Goprak, Legon Lele, Cikmas and Nyamplungan. These water springs are used by the locals for daily needs. According to *Peta Cekungan Air Tanah* established during *RAKEPRES Tentang Cekungan Air Tanah* dated September 17, 2007, the Karimunjawa Islands do not have ground water potenial and are included in the *Non-CAT* category. Water resources in the islands are merely dependent on surface water and season [10-11].

Water quantity in the islands fluctuated seasonally so that it was unpredictable. Effort had to be made towards sustainable supply for the local people. The government had provided pipelines by connecting points of water springs to settlements. Such effort was coordinated by a community union for clean water work unit namely "TIRTA KENCANA KARIMUNJAWA" [9]. The unit was established under the Letter of Decree of Karimunjawa District Head and was composed of local people to organize and to manage water distribution for mutual needs, in particular households in the islands. In addition, in some islets, such as Kemujan, Parang, and Nyamuk, where the water springs was not found, the people digged 8-20 meter deep wells to get the ground water for their daily needs [9].

Karimunjawa Islands were the home for 9,802 population with the domestic water need of 60-100 litres/individual/day. Whereas, non-domestic water needs consisted of commercial use, institutional need, and industrial needs, which could only apply to an area with less than 100,000 population. There was ony 25% of the total water needs for households and domestic needs. According to water availability compared to the population size, the water need in Karimunjawa Islands had been fulfilled, although the population kept growing annually and the water supply was constant. Problems arose when Karimunjawa Islands began to develop resort nd hotels for tourism activities.

Of five hotels and two resorts available in the Karimunjawa Islands, one needed to drill the ground water and the rests still depended on pipeline systems from the Karimunjawa National Park. The drilling activity affected the clean water supply in the dry season, in which water springs began to dry. It was predicted that the local people will no longer have access to the water if the drilling continues and the resorts and hotels grow in numbers.

2.2 State obligation and responsibility for clean water availability in Karimunjawa Islands

Dynamic development of and change in aspects of life, nationwide or worldwide, have contributed to the shortage of water resources due to severe crises. The change in land function and environmental destruction have worsened the situation and damaged the water quantity and quality. The population boom and economic growth need more supplies of water than ever before.

In terms of Indonesia, the use of water must be monitored and supervised by the government because water is a *res commune*. It is what the 1994 Constitution requires in Article 33(2) and 33(3), along with Article 28H, in which water is part of human rights with non-derogable nature. Furthermore, Indonesia has also ratified ICESCR in the Law 11/2005. At international level, in addition to the ICESCR, the right of water has been also pointed by the General Comment No. 15 (2002) adopted by the UN Committee on ECSCR entitled 'the right to water'. Some of these legal norms, either at national or international scales, have exemplied the legal stance of the right to water as part of the human rights. In consequence, the state has obligation to respect, to fulfill, and to protect the right to water of everyone indiscriminately [7].

Concerning the water management in the Karimunjawa Islands to fulfill the needs for water, it is important to notice that the islands have difficulties to find the water sources. The limited supply of water at the conservation area must be taken account seriously and wisely. The government has a responsibility for performing five duties as follows:

Firstly, the government, i.e. Local Government of Central Java and Local Government of Jepara Regency, have responsibilities for determining a policy line (*beleid*) on the waer use in the Karimunjawa Islands. The policy made by the Local Governments should not be apart from legal politics already required by the 1945 Constitution, in particular Article 33(2) and 33(3). It means that the policy on the water use in Karimunjawa Islands must consider the aspect of right fulfillment while consistently protecting the forest sustainability as the main sources of the water. The local governments must firmly decide what can be done or want cannot be done by the local people towards the access to water. The geographical condition where the water supply in the Karimunjawa Islands is situated makes the islands depend on the forest as the major water supplier. A clear and rigid regulation is necessary to fill technological gaps applied by the local people to access the water.

Secondly, the state as the possessing party must establish and decide regulation (*regelandaad*). In case of the water use in Karimunjawa, no regulation has been made by ether provincial or regency governments. Therefore, daily and business activities related to the water use in the islands have not been under control. The regulation is greatly needed to manage the access to the water towards it sustainability.

Thirdly, the state has a mandatory to take a provision (*bestuursdaad*), which means that it must fulfill the water need in Karimunjawa Islands, in partiular the local governments. Even though in the context of the water management the responsibility is within the hand of the government, but the Leter of Decree on Karimunjawa Islands also requires a concerted work between the local government and the local communities at the district level by establishing a work unit an license from the Karimunjawa National Park Office. They represent the government at the local level and facilitate the need for water. However, the differentiation between the household need and business need (for resort and hotel) has not been explained and regulated.

Fourthly, the state has a responsibility for management (*beheersdaad*) as required by the previous paragraphs that the water management in Karimunjawa Islands have not been directly performed by the government. Instead, it is managed by a self-sufficient local dwellers.

Fifthly, the state, in performing the rights of water possession, is responsible for monitoring (*toezichtthoudensdaad*). Hence, because the water sources are situated in the Karimunjwa National Park, the monitoring is performed by District Government and the Karimunjawa National Park Office, in particular those concerning license prolongation.

3 Conclusion

Karimunjawa Islands do not have plenty of water sources. The limited supply requires a wise management by considering fair distribution while paying attention to environmental sustainability. No policy has been made on the water resources and its sustainable use in the Karimunjawa Islands.

Water management in the Karimunjawa Islands, in particular those related to distribution for the local people need, was still managed locally by the local people under the permission of the District Government and the Karimunjawa National Park Office. No regulation has been established for water use and water access. The differentiation between the household need and business need (for resort and hotel) has not been explained and regulated.

References

1. Irianto, *Air dan Perubahan Iklim Global*, Kompas, 10 Juni 2003.
2. Bagir Manan, *Politik Perundang-undangan Dalam Rangka Mengantisipasi Liberalisasi Perekonomian*, Makalah FH UNILA, Lampung, 1996.
3. Nanik Tri Hastuti, *Community Development* di Sektor Pertambangan Ditinjau dari Kosep Negara Kesejahteraan, *Masalah-Masalah Hukum FH UNDIP, Vol. XXXIII No. 1 Januari-Maret 2004.*
4. Raja Siregar, *Privatisasi Air*, dalam *WWW.Wahli.or.id*.
5. Decree of Constitution Court Number : 85/PUU-XI/2013.
6. Yance Arizona, *Konstitusionalisme Agraria,* (Yogyakarta : STPN Press, 2014).
7. Rahayu, *Hukum Hak Asasi Manusia,* (Semarang : Badan Penerbit Universitas Diponegoro, Edisi Revisi 2015).
8. Adji Samekto, *"Perkembangan Ranah Kajian Hukum",* Orasi Ilmiah yangDisampaikan pada Dies Natalis Ke-48 Fakultas Hukum UNDIP di Semarang pada 9 Januari 2005.
9. Laporan Akhir *"Kajian Nilai Ekonomi Pemanfaatan Sumber Daya Air Taman Nasional Karimunjawa"* tahun 2010.
10. Robert J. Kodoatie and Roestam Sjarief, *Tata Ruang Air,* (Yogyakarta : Penerbit Andi, 2010).
11. Kodoatie, Robert dan M. Basoeki, *Kajian Undang Undang Sumber Daya Air,* (Yogyakarta : Andi Offset, 2005).

Spatial Distribution of Conventional Air Pollutant and GHGs from Land Transportation in Two Developing Cities and Main Co-Benefit Actions For Reducing It.

Haryono S Huboyo[1,2*], *Endro Sutrisno*[1], *Ana M Sutrisno*[1], *Velida L Tiarani*[1]

[1] Department of Environmental Engineering, Faculty of Engineering, Diponegoro University, Semarang – Indonesia
[2]Center for Environmental Studies, Diponegoro University, Semarang - Indonesia

Abstract. Surakarta and Yogyakarta are the emerging cities which now struggle to manage its pollution from transport sector. This study aims to calculate the emission, to describe spatial distribution and to analyze existing co-benefit actions related to land transportation in Surakarta and Yogyakarta in 2015. The main method used for this analysis comes from Ministry of Environment and Forestry. The VKT values were aggregated for Surakarta city and Yogyakarta city and it showed 27.36 km/day, 37.52 km/day and 27.71 km/day for motorcycle, car and truck respectively. At Surakarta city, the emission load from transport sector in 2014 were 449.95 tons/ year (TSP), 5134 ton/ year(NOx), 243 ton/year (SO_2), 50,605 ton /year (CO) and 421,594 tons/year (CO_2e). Villages of Kemlayan, Timuran and Keprabon showed the highest emission. While in Yogyakarta city in 2014, the burden of TSP was 58,409 tons/year, NOx was 8,058 tons/year, SO_2 was 285.37 ton/year, CO was 75,008 tons/year and CO_2e by 601,068 tons/year. The village of Pringgokusuman and Sosromeduran showed the highest emission. Several programs were adopted in Yogyakarta city and Surakarta city for mitigating air pollution i.e ITS-ATCS, BRT system, car free day.

1 Introduction

Vehicle emission is considered to be major emission of conventional air pollutants as well as greenhouse gases (GHGs) particularly in urban cities [1]. Carbon dioxide (CO_2), methane (CH_4) and nitrous oxide (N_2O) are the major of GHGs. Conventional air pollutants such as particulate matters ($PM_{2.5}$, PM_{10}), SO_2, NO_2, VOCs, ground level O_3 are blamed for numerous health impacts if we related to the human exposure. Moreover fine particulate matter ($PM_{2.5}$) and ozone are the main contributor of human health problem from transport sector [2]. Health effects related to traffic sources emission has been broadly studied and summarized by HEI [3].

Emission from transport sector could be governed by engine combustion technology, sort-quality of consumed fuel, exhaust device control, sort of road surface-terrain and vehicle operation. These factors could play simultaneously for worsening the emission.

Air pollution emission inventory in urban area needs a robust vehicle population estimation [4]. To find the real number of operational vehicles in the road is a big challenge as it requires intensive and broad sampling in the field. Two methods are generally used for conducting emission inventory i.e top-down and bottom-up method. Several studies adopted these methods for estimating air pollution load in emission inventory [5-7].

Both of these methods have advantages and disadvantages so that selecting the method use depend on the availability resources and time allocation.

Big urban cities in Indonesia now are struggling with traffic jam, congestion and huge emitted air pollution related to transportation sector. Private vehicles growth is much higher than public transport means. Therefore, several policies were adopted to lessen this emission. These policies-related to reduction of air pollution are Intelligent Transportation System (ITS), implementing Traffic Impact Control, introducing Bus Rapid Transport (BRT) system, developing non-motorized transport, renewing paratransit public transportation and introducing smart driving training. These measures usually are not well prepared for reduction calculation methodology. This study is addressed to know the emission load through emission inventory from transport sector at two cities in Indonesia.

2 Methodologies

This research was conducted at two cities i.e Surakarta city at Central Java Province and Yogyakarta city in Special Province of Yogyakarta (See Fig.1). We collected data on June to September 2015. For calculating the emission we collected secondary data (registered vehicle number, main road length, road network) from local government agency related to transport sector and state-owned oil and gas company

* Corresponding author: huboyo@gmail.com, huboyo@undip.ac.id

(Pertamina) for knowing the oil consumption. Based on these secondary data, we derived the vehicle kilometer travelled (VKT) either for line source and area source. Actual data on odometer for each vehicle on year basis were also gathered from several vehicle repairing workshops.

For estimating the spatial distribution, we gridded the city every 500 m length. The calculation for each grid follows this formula:

$$VKT_{b,c,k} = (K_1^k + K_2^k) \times VKT_{b,c} \quad (1)$$

where:

$VKT_{b,c,k}$ = VKT for vehicle b category using c fuel at grid k

$VKT_{b,c}$ = VKT for vehicle b category using c fuel

$K_1^k = \frac{\alpha pk}{pt}$ = weighing for population density

$K_2^k = \frac{(1-\alpha) Lk}{Lt}$ = weighing for road area network

pk = population number at grid k, pt = total population, Lk = road area at grid k, Lt= total area of road,

α = facktor (in fraction 0 - 1) which define population and road area affecting VKT. For Indonesian case, the number is 0.35.

In order to know the vehicle emissions, we adopted several secondary emission factor [8-10]. Overall the method we used for estimating the emission at city basis is derived from Ministry of Environment and Forestry [11].

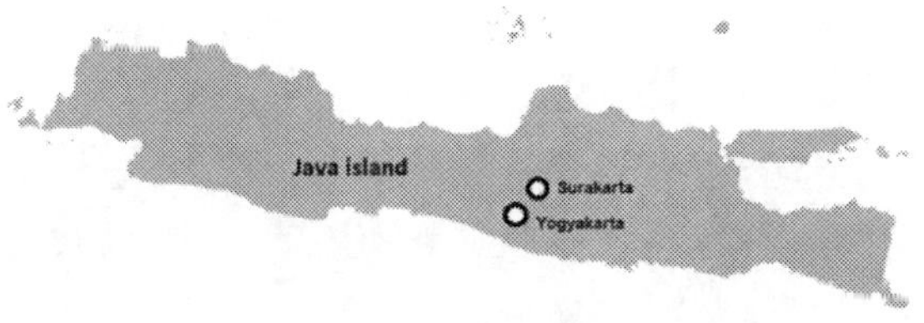

Fig. 1. Location of the Study

3 Results and Discussion

3.1. Vehicle Number and VKT

The first step to quantify the emission related to transport sector is determining the VKT. This method corresponds to bottom-up method. The VKT could be derived from volume number of vehicles passing through the main road combined road length. The VKT at main road (VKT Line Sources) and VKT for road network (VKT Area Sources) are depicted at Fig.2. From the Fig.2 it is clearly seen that the VKT between Surakarta and Yogyakarta is comparable for area sources. However it differs for VKT line sources. It could be caused by method for selecting the road. For motorcycle and bus, it showed comparable, while for car and it was quite different. The registered number of car for two cities were comparable at 2014, while for motorcycle the registered number at Yogya city is little bit higher than those at Surakarta city. The number of bus at Surakarta city was three times than those at at Yogyakarta city in 2014.

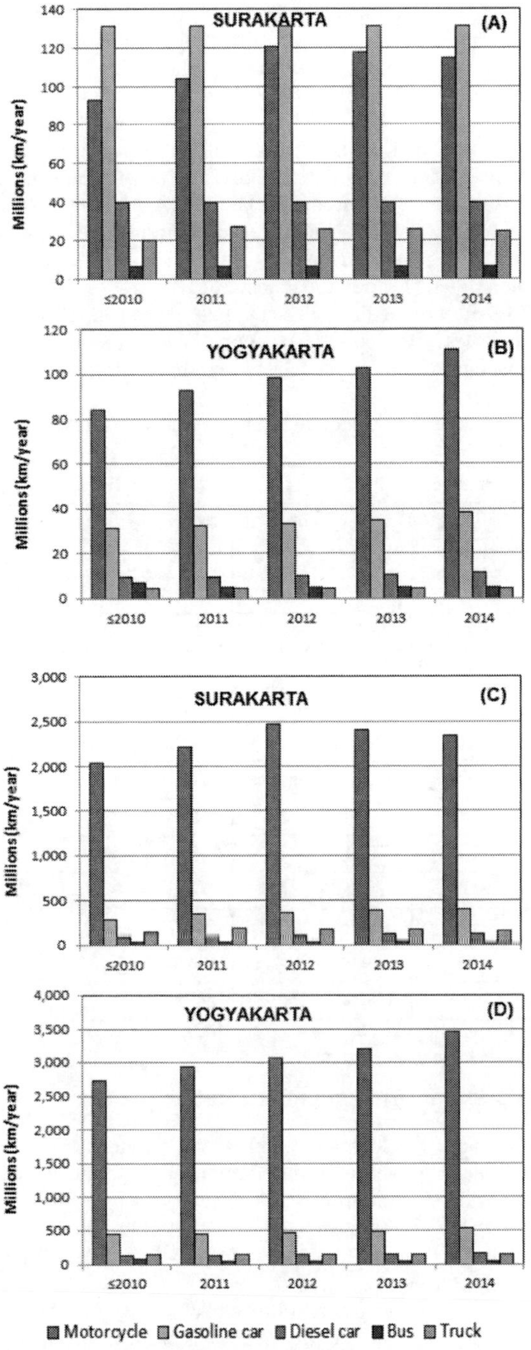

A. VKT Line Sources (Surakarta)
B. VKT Line Sources (Yogyakarta)
C. VKT Area Sources (Surakarta)
D. VKT Area Sources (Yogyakarta)

Fig. 2. VKT for each vehicle mode

3.2 Emission Load

The VKT total (line and area sources) was used to quantify the emission load. The result of emission load (ton/year) is listed in Table 1. Based on the calculation results, the biggest contributor for air pollution for both cities was motorcycle. At Surakarta city, motorcycle contributes 62%, 88%, 65%, 48% for TSP, HC, CO and CO2e respectively. While at Yogyakarta city motorcycle contributes 73%, 89%, 67%, 51% for TSP, HC, CO and CO2e respectively. The biggest contributor of NOx and SO2 was truck both in Surakarta and Yogyakarta city. It showed more than 50% in Surakarta city and around 30 – 40% in Yogyakarta city. The least contributor for air pollution was bus. The biggest emission load of bus was for SO2 parameter. It contributes 12 – 16% of SO2 emission. The other parameter (TSP, NOx, HC, CO, CO2e) were contributed by bus less than 8%. This indicates that public transport is the preferable for reducing air pollutants related to transport sector.

Table.2 Emission load of air pollutant (ton/year)

Parameter	MC	GC	DC	Bus	Truck
Surakarta city					
TSP	281.15	1.97	31.2	21.14	114.49
NOX	679.45	788.44	412.14	359.42	2895
SO2	18.74	10.25	51.81	28.09	134.11
HC	13823	1577	23.55	39.26	294.39
CO	32801	15769	329.71	332.23	1374
CO2 E	202889	110969	37872	19485	50379
Yogyakarta city					
TSP	42804	285	4514	334	10472
NOX	1034	1141	2667	568	2648
SO2	2854	1483	7495	4437	12268
HC	21046	2281	34.07	62.03	269.29
CO	49939	22811	476.96	524.83	1257
CO2 E	308891	160527	54786	30781	46083

MC: motorcycle, GC: gasoline car, DC: diesel car

3.3 Spatial Emission Distribution

To know the location which has high emission of air pollutant, we show the distribution of air pollutant based on gridded location for each city. In this paper we only show two parameters i.e TSP and CO2e. Overall, the spatial distribution for two cities are depicted in Fig.3 - Fig.6.

From this figure, for Surakarta city Villages of Kemlayan, Timuran and Keprabon showed the highest emission. While in Yogyakarta city, the village of Pringgokusuman and Sosromeduran showed the highest emission. It should be noted that the emission we calculate here was derived from quantifying vehicle activity based on grid. Thus the location or grid which has high volume number of vehicle in the road will have high emission.

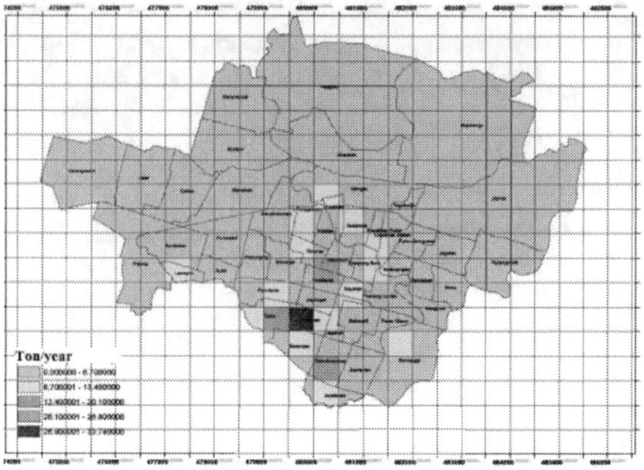

Fig.3. Spatial TSP distribution at Surakarta city

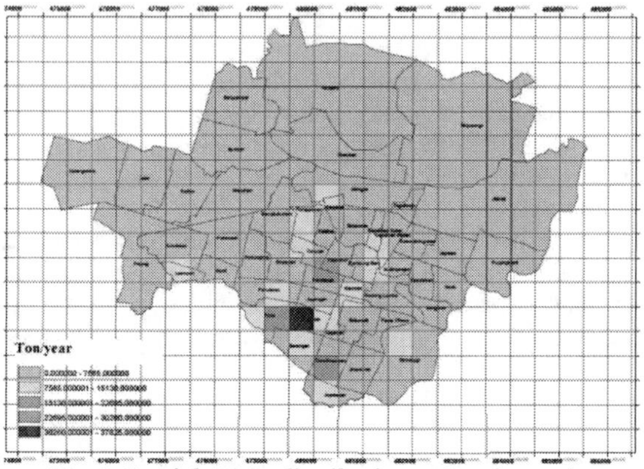

Fig.4. Spatial CO2e distribution at Surakarta city

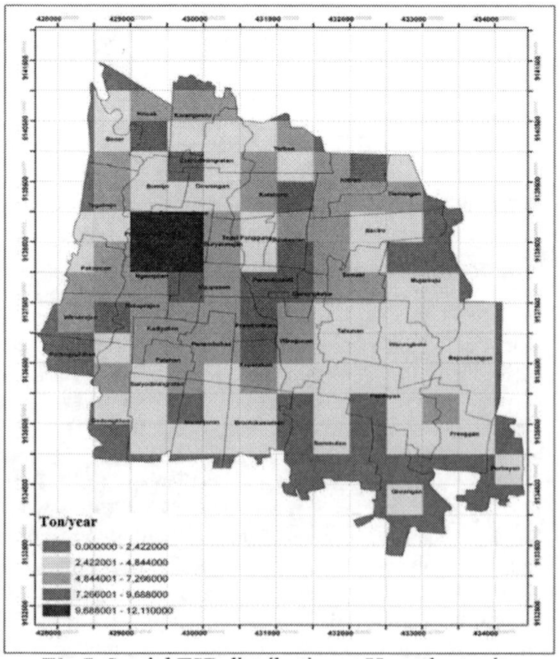

Fig.5. Spatial TSP distribution at Yogyakarta city

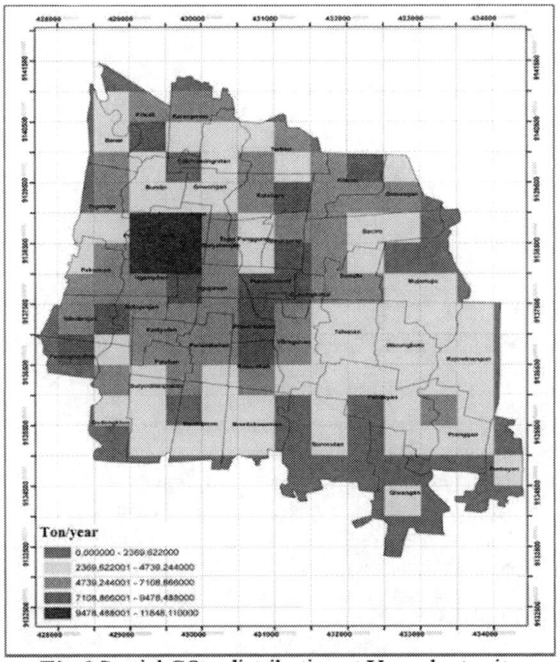

Fig.6 Spatial CO_2e distribution at Yogyakarta city

3.4 Emission Reduction Measures

Using formula generated by Ministry of Transportation, we estimate the emission reduction of CO_2e as a results of several measures i.e implementation of ATCS (*area traffic control system),* operational of bus rapid transit (BRT) and car free day (CFD). Based on our calculation the results of emission reduction of those measures are listed in Table.

Table 2. Emission Reduction After Measures Operation (ton/year)

Measures	Surakarta	Yogyakarta
ATCS	3042 (0.62%)	4964 (0.83%)
BRT	12795 (2.6%)	17596.6 (2.93%)
CFD	1427 (2.01%)	440.97% (1.56%)

In Yogyakarta, other measures also be adopted such as park management, smart driving which could reduce GHGs emission 6262.28 ton/year (1.04%) and 1229.5 (0.2%) respectively.

4 Conclusions

Emission of air pollutants in urban area were estimated using bottom approach with VKT parameter. The VKT values were aggregated for Surakarta city and Yogyakarta city and it showed 27.36 km/day, 37.52 km/day and 27.71 km/day for motorcycle, car and truck respectively. At Surakarta city, the emission load from transport sector in 2014 were 449.95 tons/ year (TSP), 5134 ton/ year (NOx), 243 ton/year (SO_2), 50,605 ton / year (CO) and 421,594 tons/year (CO_2e). Villages of Kemlayan, Timuran and Keprabon showed the highest emission. While in Yogyakarta city in 2014, the burden of TSP was 58,409 tons/year, NOx was 8,058 tons/year, SO_2 was 285.37 ton/year, CO was 75,008 tons/year and CO_2e by 601,068 tons/year. The village of Pringgokusuman and Sosromeduran showed the highest emission. Several programs were adopted in Yogyakarta city and Surakarta city for mitigating air pollution i.e ITS-ATCS, BRT system, car free day. However the reduction achievement was quite small compare to overall emission.

Acknowledgment

This study was supported by Cluster Research Group on Air Pollution Management in Environmental Engineering Department Diponegoro University.

References

1. K. Papagiannaki, D. Diakoulaki, Energy Policy; **37**:3259–67 (2009)
2. C. Guerreiro, F. de Leeuw, F. & V Foltescu. *Air Quality in Europe: 2013 report. EEA report No. 9/2013*. Luxembourg: Publications Office (2013).
3. Health Effects Institute. *Traffic-Related Air Pollution: A critical review of the literature on emissions, exposure, and health effects*. HEI special report 17. Boston, Mass.: Health Effects Institute (2010)
4. S. Soylu. Energy Policy.**35**:4088–94 (2007)
5. AG. Progiou, IC. Ziomas. Sci. Total Environ **410**, 1–7 (2011)
6. HK. Wang, CH. Chen, C. Huang, LX Fu. Sci. Total Environ **398**, 60–67 (2008)
7. R. Borge, I. de Miguel, D. de la Paz, J. Lumbreras, J. Perez, E Rodriguez. Atmos. Environ **62**, 461–471 (2012)
8. IPCC. *Guidelines for National Greenhouse Gas Inventories*. Volume **2** : Energy (2006)
9. JICA. *Study of Integrated Transportation Master Plan for Jabodetabek* (SITRAM Phase II). (2014)
10. CORINAIR. *Atmospheric Emission Inventory Guidebook 3th Edition*. European Environment Agency (2006).
11. Kementerian Lingkungan Hidup dan Kehutanan. *Pedoman Teknis Penyusunan Inventarisasi Emisi Pencemar Udara di Perkotaan*. Jakarta (in Indonesian) (2013).

Critical Review of Spatial Planning of CAT Watuputih, Rembang Zone, Central Java, Indonesia

Sudharto P. Hadi[1,*]

[1]Study Program of Environmental Studies and Faculty of Social and Political Sciences, Diponegoro University, Semarang, Indonesia

Abstract. The Act 26 of 2007 on spatial planning stipulates that spatial planning at national, provincial and local level must be based on environmental carrying capacity and environmental carrying capability. Provincial governments generally finished its spatial planning in 2010 and the city and regency's governments in 2011.This paper reviews the spatial planning of Central Java and Rembang Regency regarding the use of CAT (ground water basin) Watuputih, Rembang Zone. Both spatial planning determined that CAT Watuputih was allocated for conservation and for mining. The mixed use zoning stimulates conflict between private sector and government on one side and local people along with academician and NGOs on another side. The SEA (Strategic Environmental Assessment) studies initiated by central government found that CAT Watuputih has strong indication to be KBAK (natural landscape area of karst) need to be conserved while at the moment there have been 21 mining permit holders operating since 1998. The lesson learned from the review is that formulation of spatial planning must be conducted participatory by involving relevant stakeholder, objective and accountable.

1 Introduction

The issues regarding spatial planning revolve around inconsistency and the quality of document. The inconsistency is related to the use of space which is not suitable with allocation determined by spatial planning. This was steamed by the orientation of development heavily toward economic growth as noted by Triana [1]. The phenomena have been occurring at most cities in Indonesia and it is popularly called market driven development.

Many green open space, parks, agriculture land, forest have been easily changed for housings, malls, industries, infrastructures and other commercial facilities. Historical sites have been turn down and changed to be commercial facilities. Quality of spatial planning is related to the spatial planning which are based on environmental carrying capacity and environmental carrying capability as mandated by the Act 26 of 2007 on spatial planning. This paper reviews the policy in determining spatial planning of CAT (ground water basin) Watuputih, the environmental and social implications of that policy and proposed recommendation to deal with these problems.

2 Literature review

Based on the Act 26 of 2007, spatial planning must incorporates economic, social and environmental aspects. However, according to Budihardjo [2], spatial planning is oriented more on economic and physical aspects without taking into account social and cultural aspects. Act 26 of 2007 initiated by the Ministry of Environment stipulates that the spatial planning must be based on environmental carrying capacity and environmental carrying capability. However, even, most spatial planning have been accompanied by Strategic Environmental Assessment (SEA) as required by the Laws, the environmental carrying capacity and environmental capability were not fully incorporated by other Ministries, provincial and local governments in formulating their spatial planning. Zhou et al [3] tried to establish a coordinated spatial planning system by taking a case study of Yulin City in Shaanxi Province in China and found that conflict among relevant department revolves around "integration of multiplanning"

The problems must be dealt are how to position and integrate key departmental planning. Grêt-Regamey et al [4] on their studies are aimed to integrate ecosystem services into spatial planning and identified conflict of interests between developers and farmers aiming at securing their fertile land. Lerouge et al [5] evaluated ecosystem services to explore scenarios for adaptive spatial planning and noted that the ecosystem service concept shows a great potential to contribute to an adaptive spatial planning paradigm, however, it is not yet a mainstream practice in spatial decision making. Natarajan [6] found a possible role for 'local knowledge' in producing planning knowledge. The 'spatiality' of local knowledge is a particular interest in constructing spatial planning,

*Corresponding author: sudhartophadi@yahoo.co.id and sudhartophadi@undip.ac.id

3 Research methods

The type of research is analytical descriptive outlining the driving factors of determining spatial planning and identifying environmental and social impacts. A case study taken is CAT Watuputih, Rembang Zone, Central Java, which is a conflicting area between the government and cement factory on one side and local people, NGOs and academician on other side. The data collection techniques include document (literature) review, content analysis from mass media, physical and social observation and informal interview with relevant interests consisting of government, local people and mining license holders. The data obtained is analyzed qualitatively.

4 Results and discussions

Spatial planning of Rembang Regency of 2011-2031 was based on National Spatial planning and provincial spatial planning. Based on National Spatial Planning of 2008, CAT Watuputih is allocated for other uses meaning that this area could be allocated for any use considered suitable with the type of land. Provincial regulation (Perda) number 6 of 2010 on spatial planning states that CAT Watuputih is determined as protection area in which a recharge area categorized as geological conservation area. This Perda recommends that spring area such as Watuputih must be allocated for conservation area.

This Perda also determines that Kendeng Utara mountains including CAT Watuputih is used as cultivation area such as mineral metal and non metal mining, stone and coal, even though without being accompanied with map of spatial pattern. While spatial planning of Rembang regency stipulated at Perda (Regency Regulation) number 14 of 2011 in which CAT Watuputih which has function as recharge area is allocated for conservation. However, the attachment of this regulation stipulates that Watuputih area is allocated for mining use.

This inconsistency of regulation is also caused by inadequate national regulation which does not elaborate in detail the use of CAT. The consequences are that there have been now 21 mining permits issued and operated at CAT Watuputih since the first permit issued in 1998. Both spatial planning at provincial and local (Regency) levels have been accompanied with Strategic Environmental Assessment (SEA) incorporating environmental carrying capacity and environmental carrying capability. However, the involvement of relevant stakeholders including those concerns with environment was considered late where the draft of SEA almost finished. Consequently, the need, interest and aspiration of stakeholder specifically local people were not taken account well. Local people are worried about water resources being degraded which is currently utilized for potable water and irrigation of their rice field. Another reason for inadequacy of environmental consideration are the lack of guidance for measuring environmental carrying capacity and environmental carrying capability. It is the fact that based on SEA studies initiated by the Ministry of Environment and Forestry and the Office Secretariat of President, CAT Watuputih has a strong indication to be KBAK (natural landscape of Karst need to be conserved. CAT Watuputih is considered fulfilling the indicators of KBAK include (a) as a scientific function (b) as water recharge (c) as water conservation (d) as permanent spring water (e) has caves forming underground rivers

5 Conclusions and recommendation

Act 26 of 2007 mandates to improve the quality of spatial planning document by incorporating environmental carrying capacity and environmental carrying capability. This stipulation has been followed formally and not substantially by the government at all levels in formulating its spatial planning. Consequently, spatial planning do not express the need, interest and aspiration of local people and stimulates conflict between private sector and the government orienting toward economic growth on one side and local people and their allies who are worried about water resources degradation.

It is recommended that (1) the existing minings at CAT Watuputih must be environmentally audited to improve their performance and to identify the termination of permit (2) the mining permit holder which has not have operated recommended to operate at other places within Rembang zone parallel with the revised spatial planning (3) the technical guidance of measuring environmental carrying capacity and environmental carrying capability is urgently issued by the Ministry of Environment and Forestry (4) the existing spatial planning at local level should be revised by involving all stakeholders including local people.

References

1. N. Triana. *Kebijakan Penataan Ruang Bopunjur. Disertasi PDIH UNDIP. (Policy of Spatial Planning of Bogor, Puncak and Cianjur, West Java. Dissertation of Doctorate Program of Law, Diponegoro Universit.* (2013)
2. E. Budihardjo. *Prepared for EIA Course. Environmental Study Centre, Diponegoro University).* (2003)
3. X. Zhou, X. Lu , H. Lian [a], Y. Chen [c], Y. Wu. *Habitat International*. **65.** 32-48. (2017)
4. A. Grêt-Regamey, J. Altwegg, E.A. Sirén, M.J. van Strien, B. Weibel. *Landscape and Urban Planning*. **165.** 206-219. (2016)
5. F. Lerouge, H. Gulinck, L. Vranken. *Journal homepage.www.elsevier.com.locate/ecolink*. (2017)

https://doi.org/10.1051/e3sconf/20183109006

6. L. Natarajan. *Progress in Planning*. **111**. 1-22. (2015)
7. Departemen Pekerjaan Umum. *Undang-Undang no 26 tahun 2007 tentang Penataan Ruang (Ministry of Public Works. Act 26 of 2007 on Spatial Planning)*. (2007)
8. Kementerian Lingkungan Hidup. *Undang-Undang no 32 tahun 2009 tentang Perlindungan dan Pengelolaan Lingkungan Hidup.(Ministry of Environment. Act 32 of 2009 on Environmental Protection and Management)*. (2009)

Comparative Study on the Characteristics of Community-Based Tourism between Pentingsari and Nglanggeran Tourism Village, Special Region Yogyakarta

Novia Purbasari[1], *Asnawi* Manaf[2]

[1] Research Assistant in Urban Development Laboratory, Urban and Regional Planning Department, Diponegoro University- Semarang-Indonesia.
[2] Senior Researcher in Urban Development Laboratory, Urban and Regional Planning Department, Diponegoro University- Semarang-Indonesia,

Abstract. Community-based tourism is one of the tourism development models that effectively used as a tool to alleviate poverty through empowerment strategy of the local community. Nevertheless, many people do not have adequate understanding on the characteristics of community-based tourism, which are used as a determinant in the tourism development. This article describes the comparison on characteristics of community-based tourism between Pentingsari and Nglanggeran. These villages were chosen because Pentingsari was a tourism village that able to apply the principles ethical codes of world tourism, shown by an award from the World Committee on Tourism Ethics Code and Nglanggeran was awarded as Best Tourism Village award in Indonesia from ASEAN Community Based Tourism Award 2017.The objectives of this study is to explore the characteristics of community-based tourism applied in the Pentingsari and Nglanggeran, and to identify any indicators that could be used to indicate those characteristics. The research achieves through in-depth interviews, observation, and review of documents. There were 17 persons as informants. Further, the observation was reached by directly observing in the both study cases. In addition, the data obtained through the review of secondary data from the local manager of tourism village. Generally, Pentingsari has characteristics as a *community-based rural tourism*, while Nglanggeran has characteristics as *community-based ecotourism*.

1 Introduction

Development of rural tourism in Indonesia is able to move other sectors and give impacts on rural economies in general and specifically to involve the communities. It is related to the role of rural tourism capable to creating jobs, improving the living conditions of rural communities and moving the creative economic activities of rural communities. "Tourism may be highly relevant to slow down the loss of biodiversity through the planning and management of protected areas for improvements as foster local participation" [7]. "Rural tourism has a value of exploiting social environment, preservation of culture, and the spirit local empowerment" [6]. One development of tourism activities performed in the form of community-based tourism, which can be managed by community with simple management. "Community Based Tourism is an opportunity to move all the potential and dynamics of society, in order to balance the role of large-scale tourism businesses. This is because the community empowerment in community based tourism is able to give community an opportunity to build social capital which is defined to sharing profits as a result of tourism activities which are ran together. Community based tourism does not mean only small and local efforts, but should be placed in the context of global public partnership" [5]. Nevertheless, many people do not have the adequate understanding of the characteristics of community-based tourism, due to community-based tourism is depended on different geographical conditions around the world. "This created a different way to look for alternative traditional mass tourism, and it created a destination creations possible providing welfare for local communities and creation of jobs" [8]. Many characteristics of community-based tourism can't be appropriately applied to the Tourism Village, considering the community-based tourism is one approach that emphasizes empowerment tourism.

This article described the results of qualitative study using two cases study about the characteristics of community based tourism. It was taken at Nglanggeran and Pentingsari Tourism Villages. Both Tourism Village were located in Special Region of Yogyakarta. "Special Region of Yogyakarta was a province that second ranked after Bali in terms of tourism potential. This was because

* Corresponding author: veapurbasari@gmail.com

the Special Region of Yogyakarta had diversity of object factors and the object with a variety of specifications steady and unique character" [9]. Both tourism villages were used as cases study because through community based tourism Nglanggeran Tourism Village been successful in conservation efforts, while Pentingsari Tourism Village has succeeded in the conservation of cultural and empowerment of local communities. Some awards have also been achieved Nglanggeran Tourism Village one of them is "Cipta Award 2011 in the management of Natural Tourism Attractions insightful Environment National Level [10]". While Pentingsari Tourism Village was a village that be successful as a tourism village that able to apply the principles ethical codes of world tourism, including the application of empowering local communities. This was shown by an award from the World Committee on Tourism Ethics Code. Both Tourism Village have different characteristics concept. Therefore, through this study, later with both of these case studies, will be compared to understand both community based tourism's characteristics. The uniqueness of this study compared to studies that have been done are indicators that can be used to identify the characteristics of community based tourism.

2 Literatures

2.1 Tourism and tourism system

"Tourism is the activities of the people who travelling and living in places outside of their usual environment for not more than one consecutive year for leisure, business, and other purposes. Environmental usual term meant to exclude the trip in ordinary residential areas, frequency and regular trips between domicile and place of work, and other community trips routine [4]. "Tourism is also a process, activities and results that arise from the relationship and interaction among tourists, tourism suppliers, local authorities, host communities, and the environments are involved in attracting and hosting visitors [4]. Activities occurring in tourism [4] as follows: Tourism is an activity trip or tourist movements from place tourist stay to tourist destinations. Tourism is also always involves the fulfillment of goods and tourist services which far from their place. Tourism is an activity enjoyed attractiveness provided by the attractions and also involves the interaction between tourists as visitors with communities living on tourism.

Definition of tourist is every person who traveling to a place from their daily environment in a period of less than one year (12 months) to perform an activity and contributing to the places visited. Tourist also can be defined everyone traveling between two or more countries or between two or more areas within / usual residence country. All types of travelers in tourism are described as a visitor, a term that is the basic concept from entire system of tourism statistics. International visitors are people who travel for a period of not more from twelve months to another country from which in which they often live and whose main purpose is other than the implementation of income in the place visited. Internal visitors are people who travel to a destination in their own country, which is outside their usual environment, for a period of not more from twelve months [4]. In addition, the tourism system also consists from the elements of tourism [15], namely: Main, which appeal being a trigger tourism activities and the population as tourism actors and the host tourism; Prerequisite, which transportation for ongoing process of tourism; Supporting, which information and promotions to build and encourage interest traveled; Support services that tourism element which makes the process of tourism activity becomes easier, comfortable, secure and enjoyable such as hotels, guesthouses and others.

2.2 Community based tourism

Community based tourism generally is a small scale tourist activity and involves the interaction between visitors and the local community, particularly suited for rural areas. CBT is understood to be managed and owned by the community, for the community. It is a form of 'local' tourism, supporting local service providers and suppliers and focused on interpreting and communicating the local culture and environment [1]. The definition of community based tourism is tourism that prioritizes environmental sustainability, social and cultural rights in the account. It is managed and owned by the community, for the community, with the purpose of enabling visitors to increase their awareness and learn about the community and the local way of life [14]. While different opinion that "community based tourism is tourism which realize continuity of cultural, social and environmental. This form of tourism is managed and owned by the community in order to help tourists raise their awareness and learn about the way of life of local communities. It has a different concept of mass tourism. It is a model of tourism development which assumes that tourism must start from the awareness of values the needs of society in an effort to build tourism which is more beneficial to the needs, initiatives and opportunities for local communities [9]. It based on the active participation from local communities this makes the creation of a community can support this type of tourism. While at the same time it can also create relationship between the local community and visitors [8].

The principles that can be used as a tool of community development for local communities [9], namely: recognize, support and promote community-owned tourism; involve community members from the beginning in every aspect; promotes community pride; Improving the quality of life; ensure environmental sustainability; maintaining the character and the unique local culture; helping to develop cross-cultural learning; Respecting cultural differences and human dignity; distribute benefits equitably among community members; and to contribute a percentage determined for income community projects.

In addition the above, other matters relating to Community Based Tourism is a measure of the success criteria for Community Based Tourism obtained through evaluation research in several countries in Asia [11]: involving mass of people; benefits can be distributed equitably to all communities; good tourism management; strong partnerships both inside and outside; unique attractions; environmental conservation has not gone unheeded. Related to that, Prabawati has identified the success factors Community Based Tourism [11], among others: the uniqueness of the location; facilitation of funding for embryos existing activities; extensive community involvement as tourism actors; figures mover; and links.

Community Based Ecotourism is a form of ecotourism which emphasizes the development local communities; allowing for local residents to have great control over; involvement in development and management; and a large proportion of the benefits remain within the community. Community Based Ecotourism should encourage sustainable use and collective responsibility, but also include individual initiatives within the community [3]. Furthermore, Community Based Ecotourism implies that people take care of the natural resources in the context of income earning through the operation of tourism enterprises and use it income to improve the lives of its members. Therefore, CBE involves conservation, businesses enterprise, and community development [13].

3 Research methods

This study used qualitative approach and case study strategy with multiple case studies type [2] because researchers explored the differences of both case and the results depicted in the comparison of characteristics. Research of both case studies resulted in two community sustainability lessons learned from the two characteristics of the different community based tourism and some indicators that can be used to indicate the characteristics.

The unit of analysis was a case study of the characteristics of typology community based tourism. In this case the informant acted as a resource. Determination of informants in this study used purposive and snowball strategy. Purposive method was performed by interviews of government agencies while the first informant acted as keeper gathering. Besides, the snowball method was performed by the determination of community informants. The role of researchers in this study was as a key instrument, where researchers collected their own data through documentation, behavioral observation and interviews with participants

Analyses were performed by using the analysis techniques of qualitative descriptive. The analysis was began by identifying characteristics of both tourisms and comparing characteristics of each typology that resulted in two different characteristics and indicators that could be used as a tool indicate it.

4 Result and discussion

This article aimed to explore the understanding of the characteristics of community based tourism that was applied in the tourism village, as well as identifying any indicators that could be used to indicate those characteristics. This analysis performed at Nglanggeran and Pentingsari Tourism Village. The findings of indicators characteristic community based-tourism reference to success criteria of community-based tourism [11] and the success factors of community-based tourism [11]. Criteria and these factors were taken with naturalistic in the field conditions just for structuring in order to obtain indicators of community-based tourism characteristics. Below are the findings of research related to community-based tourism in Nglanggeran and Pentingsari Tourism Village.

4.1 Coverage area

Nglanggeran Tourism Village has covered by one administration area, which involving the residents from five hamlets to be a tourism community. Meanwhile, Pentingsari Tourism Village covered one hamlet area which involving the whole residents of Hamlet. The differences between those covered areas, affect the form of intervention by other parties, the government in particular, because of direct relevance to the territory authorities. The village government in Nglanggeran contributes to tourism activities, due to the management of tourism managed by *Pokdarwis* (organization under BUMDes/village-owned enterprises). Different from Nglanggeran, Pentingsari Tourism Village just has the one hamlet of coverage area, can stand independently without involves village government in the management of the tourism.

4.2 Concept and form of attraction

Tourism attractions in Nglanggeran Tourism Village are the attraction of the initial appearance of the appeal (the natural attractions) in the form of "Gunung Api Purba" [12] and evolve over time raising other attractions. Other nature attractions are "Kedung Kandang" waterfalls located around the area of Nglanggeran Village Tourism. Besides nature, there is also a tourist attraction that utilizes artificial natural potential. One of the most interesting is "embung" of fruit made reservoirs to irrigate the orchards, but presents a beautiful view. In addition, there are also points of outbound around "Gunung Api Purba", whether it's flying fox, raffling, and climbing. Everything was based nature "Gunung Api Purba" that serves a wonderful view. There is also spa tourism, the relaxation sensation supported by a village atmosphere that makes tourist become more relaxed. Other attractions are the educational and cultural travel that is presented in the form of tour packages such as travel packages agriculture, live in, local arts, culture, love of nature, and culinary. It is like what has been described by the facilitator Nglanggeran Village Tourism:

"... firstly "Gunung Api Purba", but now there are three destinations of the first "Gunung Api Purba", "Embung Buah" Nglanggeran; and "Kedung Kandang" Waterfall... we are also a village tour that sells educational packages, So we have agriculture, animal husbandry, there is also culture and art, then the third is the adventure attraction, we provide many packages of adventure activities option as well, there are raffling there is rock climbing, then outbound..." (SH,W1A,28-04-2016, 4-13)

Attractions in Pentingsari Tourism Village are combine of creativities and local wisdom that supported by the nature potential in valley of Merapi Mountain, which rarely found in other places [12]. Community still upholds the local wisdom, and this is attract for visitors, especially for live-in activity. The attractions that presented in here, can not be enjoyed only in one part. However, it must be enjoyed as travel package. This make the length of the visitors staying become longer. Tourist attractions are packed in many packages, in addition to the natural attractions and local wisdom life is also an educational and cultural. Tourists can learn about art, culture, farm life, the arts, mushroom cultivation, processing food, travel bulbs (about variety of bulbs and other dairy products), etc. In addition, there are some historical relics which can be used as historical knowledge. For visitors who love to camping, Pentingsari Tourism Village also provides camping ground that can be used as place for camping. Tourist attractions which will serve as the main magnet to attract tourists to come and visit the tourist village. Overall tourist attraction developed by the communities themselves through the tourist village board, whose members come from the community itself. Here are excerpts of interviews related to it:

"Attractions are combination of physical village with creativity, not art but the creativity of residents in performing art, in farming, in social activities. It is the attraction with the physical village which is the environment is the village environment, well that there will be homestay accommodation, that's the attraction, Nah, if people stay that hotel is a standard, if staying in the homestay is an attraction that combines the physical village is interesting, tracking people are also interesting, it's also an attraction. "(DY,W2B,10-05-2016,3-10)

The concept of attraction on both the Tourism Village affects to the length of the visitors staying. The attractions of conservation in Nglanggeran Tourism Village give choice to visit in a short time (less than one day) or visit in a long time (live in that take more than one day). While the concept of the attractions presented Pentingsari Tourism Village, visitors must make a visit in a long time (more than one day) to enjoy the attractions offered by the package.

4.3 Tourism actors and community involvement

Nglanggeran Tourism Village involved the community as subjects and objects of tourism activities. They entered the groups of homestay providers, providers of culinary, tour guides, art activists, agricultural activists, and organizer of tourism activities. Besides being the actors, community also becomes the direct beneficiaries of tourism activities expressed through the interview excerpts below:

"Our communities makes the first and main beneficiaries, because our area is ecotourism, community-based environmental insight, so the community will be the perpetrators"(SH,W1A,28-04-2016,75-78).

Fig. 1. Involvement Farmers Group And Culinary Groups In Nglanggeran Tourism Village

Source: Documentation Pokdarwis Nglanggeran, 2014.

The community also involved in monitoring and evaluation activities of the Tourism Village to interact with the tourists directly. It is also directly related to the decision-making in the management and development of tourism village. It through the forum meetings are held every "Malam Selasa Kliwon" (Tuesday Kliwon Night in Javanesse calendar) involving Podarwis, community and local village government. Here are excerpts of interviews with the Chairman of the Youth:

"The role, function and community involvement more to decision-making, monitoring and evaluation, providing input and so on. We have a forum called "Malam Selasa Kliwon ...." (AB,W8A,9-05-2016,98-112).

Community involvement applied in Pentingsari Tourism Village is the involvement of the community directly as a subject in every aspect of tourism activities as well as objects that attract tourists. Communities participate in the management, planning, development undertaken by the village board tourism, while regarded as an object, as a tourist attraction from all over the community activities and community life. Community also serves as direct beneficiaries, as tourism activities in Pentingsari Tourism Village is an activity undertaken of the people, by the people and for the people. Forms of community involvement in tourism activities are to be homestay provider, the provider of the attractions done by farmer groups and arts, then there is the PKK, which acts as a provider of culinary and souvenirs, youth also had a role as tour guides. The involvement contributes greatly to the development of rural tourism. Pentingsari Tourism Village able to get an appreciation of the

Committee on Tourism Ethics Code of the World (World Committee on Tourism Ethics - WCTE) as a tourist village that is able to apply the principles of the Code of Ethics for Tourism of the World, including the implementation of efforts to empower local communities or community-based tourism. Here are excerpts of interviews related to it:

> "... They can contribute with their own abilities. I have a homestay, it means they will be involved as a homestay owner of 60 families, then their involvement in culinary and consumption, yes they have women's groups, PKK. Continue, involvement in attractions, agriculture-based, cultural-based, nature-based attractions, it's got people too. So they donate and contribute according to their own character. "(DY,W2B,10-05-2016,70-79)

Fig.2. Involving PKK as Culinary Group In Pentingsari Tourism Village

Source: Documentation Authors, 2016.

4.4 Initiator community-based tourism

The initiation of Nglanggeran village as one of tourism destinations came from activities of communities to conserve nature through conservation efforts due to the environmental degradation associated with the availability of water. However, as the natural potential has been welcoming visitors solely for camping, people were initiated to develop it into a tourism activity particularly when "Karang Taruna" (youth organization) began to follow the race in the field of conservation. Achievement gained increasingly spur communities to develop tourism activities. Here are excerpts of interviews related to it:

> " We set out from youth activities, our background is more to conservation activities .... "(AB,W8A,9-05-2016,127-128)

Pentingsari Tourism Village emerged from the local people without the intervention of any party. It started from discussion conducted by several prominent people. They saw the rural potential and local knowledge which were rare found in this modern era. Seeing the needs of urban communities about the sense of comfort village, the idea to developing Pentingsari came as a tourist destination that sells life and village atmosphere. This initiation came up with preparing the the initial program and the next steps. After consideration, the results of these community leaders conveyed to the Pentingsari's residents through deliberation as well. According to the agreement of these deliberations, Pentingsari prepared and opened as tourism activities. Sleman District Government through the Department of Tourism provided the guidance when it was opened. Here are excerpts of interviews related to it:

> "... In the beginning it was from Mr. Sumardi, Mr. Ajun, Mr. Toni Sukoyo, all the time since it was the pioneer or the initiator of the Tourism Village, they continued to my place. Automatically even though it was initiated like that, if there is no public support, it certainly is not there ... "(R, W7B, 13-05-2016,39-43)

Fig. 3. Initiation of Tourism Activity Starts From Conservation Activity in Nglanggeran Tourism Village

Source: Documentation Pokdarwis Nglanggeran Tourism Village , 2016.

4.5 Tourism asset ownership

Tourism assets are important for the ongoing tourism activities. Ownership status of tourism assets affects the tourism activities related to land ownership, ownership of tourist facilities and tourist activities. Land ownership in tourism activities at Nglanggeran Tourism Village belongs to the people at the homestay and some land for tourist activity, and local government. Besides, the land of 48 ha around "Gunung Api Purba" is "Sultan Ground" owned by the Sultan lent to the local government via "Right to Wear" through "PANITIKESMO", this right was submitted to Pokdarwis for managed, maintained and used towards public welfare. Facilities in tourism activities belong to the peoples who through Pokdarwis managed and maintained for tourist activities. They are also the owner of tourism activities. The peoples are the subject and object in travel are the key passage of a tourist activities in the Tourism Village Nglanggeran. The peoples as the subject and object in tourism are the key passage of tourism activities in the Nglanggeran Tourism Village. Some of the villages owned assets and the government-owned assets (Sultan) then Pokdarwis formed are under *BUMDes* (Village-owned enterprises) Nglanggeran. Here are excerpts of interviews related to it:

> "... if for the ownership of the land itself mostly especially for this "Gunung Api Purba" area belonging to Sultan's

Ground, but some also belong to the citizens, and also the residents themselves make some facilities for our activities, so We are working with the community to provide facilities ... "

The land of Pentingsari was almost owned by community, it used as homestay and other tourism activities. The land used as a camping ground is an area of land the village treasury is deliberately hired by Pentingsari Community Tourism at the lower cost than the rent cost elsewhere. Here are excerpts of interviews related to it:

"...mostly belonging to the community, but the camping belongs to the village treasury ..." ...."(S,W1B,10-05-2016,67-68)

Ownership of tourism activities also one part of the tourism assets that affect tourism activities in Pentingsari Tourism Village. Tourism Activities in Pentingsari Tourism Village are all owned by the community, run by the community and its benefits to the community.

4.6 Tourism management and decision making

Management of both Tourism Villages have brought success in community empowerment. Here are some form of managements found at the both Tourism Villages:

4.6.1 Management of tourism

Management of tourism activities in Nglanggeran Tourism Village is semi-modern management of "social enterprise management." The management of tourism activities was monitored, reported to the community. Public also control of tourism activities. The social principle is also becoming one of the basic as well as its governance over to the local knowledge grew in society. However, to increase the capacity of human resource people do enterprise systems (Semi modern) in which each party involved in tourism activities, using a promissory note agreement, so that if the party violates the agreement, there will be a warning in accordance with a written agreement. Besides the use of technology is also used in the management of tourism, one of them with their e-ticketing are connected directly to the intranet manager. Here are excerpts of interviews related to it:

"Our management is more to the management can be called the management of a company, but you can call it as a social company." (AB,W8A,9-05-2016,161-164).

Tourism management in Pentingsari Tourism Village managed based on the benefits. All forms and management activities carried out by the community with the principle of high familial and social capital. The management was done from the community, by the community, and for the community. Local wisdom in Pentingsari still very thick felt, it seen from the social capital invested community towards tourism activities itself. Moreover, the manager of Pentingsari Tourism Village also does the bookkeeping and management in a transparent manner with a known and through community meetings decision. Here are excerpts of interviews related to it:

" We managing tourism activities based on benefit. So we managing it together, we counting it together and we enjoying it together. So who contribute will get the benefit, who have not contributed, we encourage them to contribute ... "(DY, W2B, 10-05-2016,93-97)

4.6.2 Organization of Tourism Management

Management of Nglanggeran Tourism Village is managed by the community, through organizations formed under BUMDes (Village-owned enterprises) Nglanggeran, namely Pokdarwis. Pokdarwis (Travel Awareness Group) of Nglanggeran Village Tourism Village is a management organization established by the Government Nglanggeran as a business community for management forms of Pokdarwis are shown as follows:

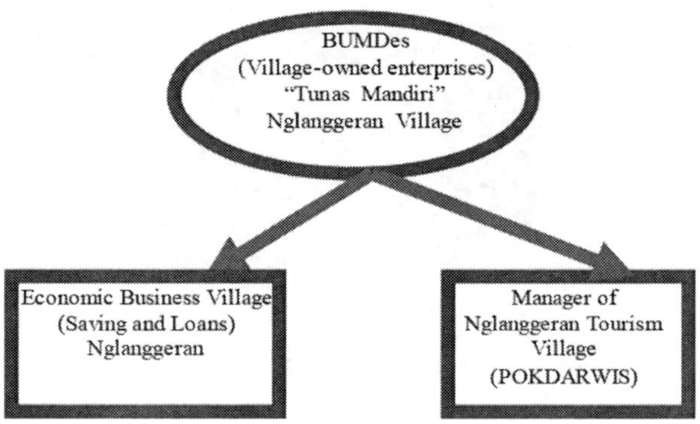

Fig. 4. *Pokdarwis* Position In The Nglanggeran Tourism Village

Source: Pokdarwis Nglanggeran, 2014.

Pentingsari Tourism Village also has management organization. The management is not managed by the local government but managed by the village board tourism. This Tourism Village board still traditionally have come from communities that are social. In this management are the division of tasks according to predefined positions, but their position remains the same as the board equally from the public and social nature. Here are excerpts of interviews related to it:

"... The important thing is we have a board, the Tourism Village organization is everything were set up." (N,W5B,11-05-2016,62-63).

Tourism Village Board is to manage based on fields that have been distributed according to their respective capabilities. All involved were local residents , it is because this management still a social group in society.

4.6.3 legality of management

Legality manager is one of the important issues related to the legal status and certainty in the management of tourism activities. Legality manager in Nglanggeran Tourism Village can be seen from the Decree of the Head of Village Nglanggeran given to Pokdarwis Nglanggeran to manage and develop the Tourism Village. The status of Pokdarwis is under BUMDES Nglanggeran, where there is accountability reports and the distribution portion of the proceeds from tourism village for public welfare, as expressed in the interview excerpts below:

> ".. we as village Government establish BUMDes that have a legal entity, it overshadow tourism activities, including Pokdarwis, village level micro-finance activities.....we make Decree related to POKDARWIS, include in relation to cooperation with the implementation of contribution from the local government ... "(S, W4A, 2-05-2016,52-58)

Management status of Pentingsari is registered by the Culture and Tourism Office of Sleman. This Tourism Village do not have legality as Pokdarwis, because it is given at the Tourism Village coverage administrative territory of the village, while the Pentingsari Tourism Village is just one hamlet. With the existence of this registered status, indicating that Pentingsari Tourism Village is one that is accompanied by the Culture and Tourism Office of Sleman. There is not a full government intervention in its management. For institutional and managers all remain in the hands of the population, while the government is only acting as a company. Here are excerpts of interviews related to it:

> "We only registered, so it is recorded in the Tourism Office...." (DY,W2B,10-05-2016,106-108)

4.6.4 Sharing profit system and levy

Related to the legal status manager Tourism Village Nglanggeran under BUMDES, it affects to their profit-sharing system. This revenue sharing profit system do Pokdarwis to the village, so that approximately 10-15% of cash coming into the village. Although basically the village government is not demanding and requires a system for these results, but for the welfare of the whole society Nglanggeran, then the sharing system is held, for cash village itself is used for the construction of public facilities for the community. Here are excerpts of interviews related to it:

> "... if for our local village government there is income of 10-15% every month.... "(AB,W8A,9-05-2016,423-428)

Nglanggeran Tourism Village is one that implements the withdrawal of levies such as those contained in Gunung Kidul District Regulation No.17 Year 2013. It is intended as an umbrella law at the Tourism Village Tourism activities Nglanggeran. Withdrawal fees are due at the Tourism Village Nglanggeran there are a few attractions. However, for the withdrawal of these levies through the door with entrance tickets are serviced by Pokdarwis. Amount of levy to be paid visitors amounted to IDR 2000.00 per month and the levy paid to the local government as the PAD village. Will be refunded by 35% and from village PAD will return to Pokdarwis by 60% as operating costs. Here are excerpts of interviews related to it:

> "Meanwhile, if the local government there is a share of 2000 rupiah per tourist in the form of retribution every month to City. Then, there is a return to us 35%. So the nominal 2000 rupiah goes to the PAD of the Village, and from the PAD Village, 60% of it will return to us for operational costs. "(AB, W8A, 9-05-2016,428-432).

Profit-sharing system and the withdrawal of the levy do not apply in Pentingsari Tourism Village due to the influence of local government policy. This shows that the community plays its full course of tourism activities, while government intervention does not dominate.

4.6.5 Monitoring, evaluation and networking aspirations

Tourism management in Nglanggeran has their whole local people as the Pokdarwis workers. In order to monitor and evaluate the performance of Pokdarwis, every "Malam Selasa Kliwon" (Tuesday Kliwon Night in Javanesse calendar) they have a meeting in which discusses all the problems, obstacles, and being a vessel performance of public aspirations. The meeting is attended by all tourism stakeholders, community leaders and local government (village government). This activity is very positive, as it can quickly find the solutions of the problems and monitor the performance of all actors in Nglanggeran. Here are excerpts of interviews related to it:

> "We have a forum called "Malam Selasa Kliwon", so every Tuesday Kliwon is our meeting from Pokdarwis, the village government, community leaders submit the work program, then accept the aspirations of the community in the form of criticism and constructive suggestions. Well later in a forum that arose agreement. Well that agreement is what we will use. "(AB, W8A, 9-05-2016,100-105)

Fig. 5 "Malam Selasa Kliwon" Meeting as Monitoring and Evaluation Activities In Nglanggeran Tourism Village
Source: Documentation Pokdarwis Nglanggeran Tourism Village, 2016.

To perform monitoring and evaluation of tourism activity, it is done in consultation with the meeting. This meeting was held at fixed times, even if it is necessary to discuss something, then the meeting is held. All decisions are in the hands of the community. Therefore, in the monitoring and evaluation, community plays a fairly big in decision-making. Here are excerpts of interviews related to it:

> "... all rights belong to the community. We only carry out the tasks of the people's decisions. So if you want to make a chairman ... "(S, W1B, 10-05-2016,70-73)

4.7 Concept of tourism

The concept of tourism departs from the vision and mission of the establishment of the Tourism Village. Tourism Village Nglanggeran vision is: "Tourism Village being featured with environmentally friendly ecotourism and community based." To realize this vision, Pokdarwis Nglanggeran develop a mission as follows:

- Protecting the environment in the area of " Gunung Api Purba" Ecotourism, good culture, flora, fauna and also the uniqueness of rock.
- Engage communities as actors and managers of Village Tourism and "Gunung Api Purba" Ecotourism Area Nglanggeran.
- Improve long-stay travelers (length of stay) in the district of Gunung Kidul and DIY
- Improve public welfare

Concept of tourism development in the area of Nglanggeran Village Tourism is an ecotourism concept that manages the potential for the welfare of society while still promoting the principle of conservation of nature, culture, flora and fauna as well as other potential. If classified in the concept of travel, the Tourism Village Tourism Village Nglanggeran is implementing the concept of community based eco-tourism. See the main purpose as conservation activities, but also did not rule out the needs of the economy and the welfare of society. In addition the development is done through empowerment of local communities, so that local communities grow as an independent community. As disclosed in the interview excerpts below:

> "..., finally they also read the potential there, to save the existing conservation and then packed into attractions." (M, W5A, 2-05-2016, 82-85)

Fig. 6. Concept of Community Based Eco-Tourism In Nglanggeran Tourism Village
Source: Documentation Pokdarwis Village Nglanggeran, 2014.

Pentingsari Tourism Village has concept lift local wisdom as a reference for the presentation and development of tourism activities in the Pentingsari Tourism Village. Took away result of community creations combined with the rural atmosphere and local wisdom of rural communities, directing the Tourism Village Pentingsari grown into a tourist destination to apply the concept of community-based rural tourism. This is evident from the efforts of local people serving life atmosphere of the village with the principles of community empowerment. As we can see in the following interview excerpt:

> "... combining the physical village with the creativity, not the art but the creativity of community in performing art, in farming, in social activities, it's the attraction with the physical village which is the environment is the village environment." (DY, W2B, 10-05- 2016,2-7)

Fig. 7 Concept Of Community-Based Rural Tourism In Pentingsari Tourism Village
Source: Documentation Authors, 2016

4.8 Comparison of characteristics both community based tourism

From the description of the characteristics of both the Tourism Village above, in outline, comparison of characteristics between both Community-Based Tourism are as follows:

https://doi.org/10.1051/e3sconf/20183109007

Table 1 Comparison Characteristics Of Nglanggeran And Pentingsari Tourism Village

INDICATORS CHARACTERISTICS	PENTINGSARI TOURISM VILLAGE	NGLANGGERAN TOURISM VILLAGE
Coverage Area	One hamlet territory authoritiy	One village territory authority
Attraction Concept	Tourism attraction formed from a blend of creativity and wisdom of local communities that are supported by the potential of nature	Attraction that the initial appearance of the appeal (the natural attractions) in the form of “Gunung Api Purba” that based on conservation effort.
Attraction Form	Package Of Live In, Camping, Tracking, Objects Of Heritage. “Jathilan” Traditional Dance , “Punokawan”, “Karawitan” (Traditional Of Javanesse Music), Creations Of Leaf, “Kenduri”, Tubers Attraction, Traditional Foods Processing, Outbound	“Gunung Api Purba” Object, “Fruit Embung” Object, “Kedung Kandang” Waterfalls Object, Outbound, Package Of Live In, Musical Art, Calung Art, Batik Masks, Spa Tourism, Shopping Tourism, Culinary Tours, Traditional Foods Processing.
Tourism Actors	Local Communities	Local Communities
Community Involve	Direct Community Involvement	Direct Community Involvement
Forms / Types Of Community Involvement In Tourism Activities	Tourism Performers (Homestay Provider, Local Art, Guide), A Provider Of Of Culinary And Souveniers, Management, Development Planning, And Supervision Of Tourism Activities Through The Attraction Village Board	Performers Tourism (guide, Art, Provider Homestay, Provider Souvenir and distinctive culinary), parking attendants, security, marketing, food seller around the object of “Gunung Api Purba”, management, development planning, and supervision of tourism activities through Pokdarwis and local community leaders.
Initiator	Village leaders are holding discussions to develop tourism activities in Dukuh Pentingsari.	Youth Organization that starts from conservation activities
Benefits Of The Tourism Village For The Local Community	With the existence of tourism village Pentingsari, communities has become more open to external. more confident for exploration potential of the area and the ability themselves.	Community, especially the youth obtain jobs through tourism activities.
Tourism Management	Managed by the community through Attraction village board, stood by indepent, using **traditional management**. The division of tasks based on the organizational structure of the management of tourism village. **Share results to all communities involved, the advantages for the construction of public facilities.**	Managed by the community through Pokdarwis whose position is under BUMDes Nglanggeran **(a semi modern management by using IT**). The division of tasks based on the organizational structure and working groups Pokdarwis. **A system of profit sharing to communities (tourism actors), BUMDes (Kas village) and also withdrawal of the levy as PAD village.**
Decision Making	Community has full authority to control the development of tourism (There is a mechanism of communication and decision-making)	Community has full authority to control the development of tourism (There is a mechanism of communication and decision-making is by their association as a forum “Malam Selasa Kliwon” evaluation and discussion of the work program)

INDICATORS CHARACTERISTICS	PENTINGSARI TOURISM VILLAGE	NGLANGGERAN TOURISM VILLAGE
Tourism Assets Owned	All homestays and tourism activities are communal land and resident's own, but a camping ground is the village treasury land leased by the board for tourism	All homestays and tourism activities are communal land and the resident's own, but the land on 48 ha attraction is Sultan's Ground administered using the Right to Use, partly also the land of the village treasury
Manager Legality	Without legacy manager from Decree of Local Government, but registered status by Tourism Departement of Sleman	Decree of Pokdawis from the the village government is known by local government
Conservation Effort	Nature conservation and the conservation of local wisdom that are still maintained, for example, people do not develop excessive homestay. This is done to avoid going over the land as a homestay	Giving priority to nature conservation, so that conservation is the core concept from the Nglanggeran Tourism Village. Some of the actions carried out by the Nglanggeran villagers is the presence of selective logging, slash fertilizer, replanting and planting trees as a backup source of water. These conservation efforts can not be separated from the role of collaboration partners who are concerned with the environment. It also minimizing their capital investment from the outside investors.
Concept of Tourism	*Community Based -Rural Tourism*	*Community based-Eco tourism*

Source: Analysis of Authors, 2016.

Table above shows several similarities and differences of the characteristics in both Tourism Villages. Similarities characteristics are shown by tourism actors, community involvement and the role of the community. Besides, the differences characteristics are shown by covered area; Nglanggeran covers one village and Pentingsari covers one hamlet. This is certainly an impact on collaborations and role of local government. Additionally there are other differences in the form and the main attractions are presented. For Nglanggeran Tourism Village, in addition to the lives of people who become attractions, the existence of three objects that are part from the Gunungsewu geopark be one part of the attraction. For the management of tourism, Pentingsari Tourism Village more on traditional management with communities as a manager through the Tourism Village board whose status is registered by the Tourism Office of Sleman and profit used for the construction of public facilities, some are also distributed to the community. While the Nglanggeran Tourism Village management is the management of social entrepreneurs conducted by Pokdarwis as a manager with legal status in accordance with the Decree of the Headman Nglanggeran Village. It's management is semi-modern because already using the technology, there is also a system of profit sharing to the village and District Locally-generated revenue via Village Locally-generated revenue obtained from retribution withdrawal. In outline, Nglanggeran Tourism Village implementing the concept of community based eco-tourism, while the Pentingsari Tourism Village implementing the concept of community-based rural tourism.

4.9 Iimplications findings against theories

From the comparison above, it can be identified the indicators for identifying the characteristics of community based tourism.

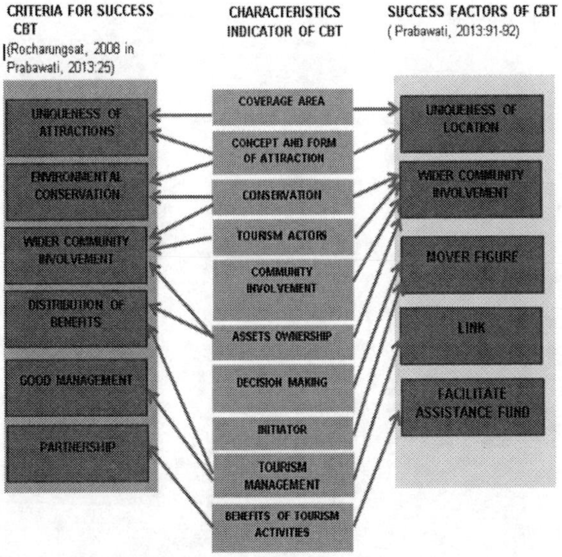

Fig. 8. Success Factors of CBT

These indicators include: Coverage Area; Concepts and Forms of Attraction; Tourism Actors; Involvement and Community Involvement; Benefits of Tourism Activity; Tourism Management; Decision Making; Tourism Asset Ownership; Initiator; Legality of Manager; and Conservation.

5 Conclusion

Identifying the characteristics of community based tourism require several indicators. In this study, it is obtained the indicator characteristics that can be used, among other things: covered area; the concept of attraction; forms attractions; tourism actors; community involvement; initiator; forms of community involvement; the benefits of tourism activities; tourism management; decision-making; tourism asset ownership; legality manager; and forms of conservation.

From the comparison using these indicators, it was obtained two characteristics of community based tourism which was found in both tourism village. These characteristics include:

1. The first characteristic is a community-based rural tourism which appeared on Pentingsari Tourism Village. This characteristic covered one hamlet of case study, which will be managed independently by local communities. Government interventions in tourism activities do not dominate. It can be seen from the Tourism Village Board, who belong to the social communities. As the covered area is not wide, local people are able to manage the area more independently. These characteristics prioritize the principle of local wisdom which relates to the main attractions formed by the creativity and habits of people.
2. The second characteristic is a community based eco-tourism which appeared on Nglanggeran Tourism Village. This characteristic covered one village of case study where the influence of governmental in this village is potentially high. It can be seen from the manager of Tourism Village through Pokdarwis formed by BUMDes. Government influence can be seen also from the profit-sharing system and the withdrawal of levy into local government policy. With wide regional coverage (one village), then the people involved even more opportunities for involvement of other parties. These characteristics promote conservation, it relates to the main attractions in the form of natural potential that needs to be preserved.
3. Basic differences between two characteristics of community based tourism at the Nglanggeran and Pentingsari Tourism Village are related to the covered area, the main attractions and the concept from the tourism activities. It will affect the policies and the form of government intervention. It also affects the mindset of local people as tourism actors managing the tourism activities. In addition, the concept presented by the attraction will affect the

duration of lenght visit to visitors and relate to the concept of tourism.

6 Acknowledgements

The authors are very grateful to DRPM DIKTI Kemenristekdikti RI, Indonesia, that has been generously funded the research of this article. The authors are also sincerely thank for all crews of Pentingsari Tourism Board and Pokdarwis Nglanggeran for sharing information with us during the course of this research and the facilitation from the Department of Urban and Regional Planning, Faculty of Engineering, Diponegoro University, Indonesia, which has made the research possible and well completed.

Reference

1. S. Asker, L. Boronyak, N. Carrard, and M.Paddon, *Effective Community Based Tourism*. (HengMui Keng Terrace, Singapore: Sustainable Tourism Cooperative Research Centre.2010).
2. Creswell, W. John. *Pendekatan Kualitatif, Kuantitatif dan Mixed*. Terjemahan Achmad Fawaid. (2010).
3. R. Denman, Guidelines for community-based ecotourism development. *Report*, (July), 1–24. (2001).
4. C. R. Goeldner, and J. R. B. Ritchie, *Tourism Principles, Practice, Philosophies* (Eleventh). 5-8.(New Jersey: John Wiley & Sons, Inc.2009).
5. S. S. Gumelar, *Konsep Pemberdayaan Masyarakat Berbasis Pariwisata (Community Based Tourism)*, 1–17. (not publish)
6. S. S. Hadiwijoyo, *Perencanaan Pariwisata Perdesaan Berbasis Masyarakat (Sebuah Pendekatan Konsep)*.(2012).
7. H. Job, and F. Paesler, *Journal of Outdoor Recreation and Tourism*. **1–2.** 18–28.(2013).
8. T. López-guzmán, and S. Sánchez-cañizares, *Spring*, **6(1),** 69–84. (2011).
9. I. Muallisin, *Jurnal Penelitian Bappeda Kota Yogyakarta*, **2,** 15–23. (2007
10. Pokdarwis Nglanggeran. *Laporan Akhir PNPM Mandiri Pariwisata 2013 di Kawasan Ekowisata Gunung Api Purba dan Desa Wisata Nglanggeran*. (not publish).
11. H. J. P. Prabawati, *Faktor-Faktor Keberhasilan Community Based Tourism Dalam Pengembangan Desa Wisata (Studi Kasus:PNPM Mandiri Pariwisata di Dataran Tinggi Dieng)*. (not publish)
12. N. Purbasari, and A. Manaf, Jurnal Magister Pembangunan Wilayah Dan Kota. **13**.100-113.(2017)
13. K. W. Sproule, *Yale F&ES Bulletin*, ***99***, 233–250. (1996)
14. P. Suansri, *Community Based Tourism : Principles and Meaning*. 14. (Bangkok: Responsible Ecological Social Tours Project.2003).
15. S. Warpani, and I. Warpani, *Pariwisata Dalam Tata Ruang Wilayah*. 23.(Bandung: Institut Teknologi Bandung.2007)

Environmental Analysis of The Impacts of Batik Waste Water Polution on The Quality of Dug Well Water in The Batik Industrial Center of Jenggot Pekalongan City

Slamet Budiyanto[1)], *Anies*[2)], *Hartuti* Purnaweni[3)], *Henna* Rya Sunoko[2)]

[1]Doctoral Program of Environmental Science, School of Postgraduate Studies, Diponegoro University, Semarang-Indonesia
[2)] Faculty of Medical Diponegoro University
[3)] Faculty of Social Science and Political Science Diponegoro University

Abstract. The city of Pekalongan known as "Kota Batik" is one of Batik Industrial Centers in Indonesia with 917 batik industry. There are 203 batik industries located in Jenggot Village, which is the biggest batik industrial center in Pekalongan City. The process of making batik requires a dye derived from synthetic dyes containing heavy metals. Most of the waste is directly discharged into the environment without going through the processing first. This is due to the lack of optimal management of existing WWTP as well as lack of public awareness of environmental conservation. This condition has a negative impact on the surrounding community, especially in terms of environmental health. The result of measurement of 5 (five) batik industrial waste outlets and 5 point of batik waste water in residential sewer shows almost equal number for 3 (three) parameters of heavy metals Cd, Cr and Pb with average number: Cd 0.07 Mg / L, Cr 0.76 mg / L and Pb 0.78 mg / L. These three parameters exceed the maximum level of quality standard established by Government Regulation No.82 of 2001 on Water Quality Management and Water Pollution Control. The average result of the water quality measurement of the well digging population to the heavy metal content are: Cd 0,001 mg / L, Cr 0,002 mg / L and Pb 0.04 mg / L. Of the three parameters of heavy metals, heavy metals of Pb are on average higher than the maximum level of quality standards established by Decree of the Minister of Health Number. 492 / Menkes / Per / IV / 2010 regarding Water Quality Requirements. Potential occurrence of dug well water contamination due to infiltration of batik waste water is big enough. Survey results of 15 dug wells show that the construction of dug wells is not sufficient. There is a dug well with a damaged outer wall of 16.1%, damaged inner wall of 17.9% and a damaged well floor of 19.7%. Improper well construction impacts on the infiltration of batik waste water into the well. Survey results of physical well water quality showed 3.6% snoring, 16.1% smelly, 9.8% color and 10.7% taste. Pollution to dug well water can affect public health because it is used for daily use to drink, cook, bathe and wash because> 65% do not have other sources of clean water. The recommendation of this research is the need for waste water treatment batik starting from source through WWTP both on the household and communal scale with the role of local government as a facilitator. Society and batik industry players need to be involved in managing environmentally friendly batik industrial center.

1. Introduction

Water is one of the most important natural resources for the fulfillment of the needs of human life and other living things. One of clean water sources used by residents is ground water from shallow wells. Therefore, the quality of shallow well water is needed by people who use it for their daily needs such as for bathing, washing, drinking and cooking.

Shallow well water quality is strongly influenced by the quality of the surrounding environment. The existence of environmental pollution around the well has an impact on water quality. The occurrence of pollution on the environment caused by several things such as industrial activities in the area. Industrial activities will not affect the environment if they are managed properly. if the industrial activities are not managed properly, it will have an adverse impact on the environment.

One kind of industries that has the potential to pollute an environment is batik industry. The process of the making of batik in batik industry includes stages of painting, wax covering, coloring, wax cleansing, washing, and drying uses raw material gray cloth, synthetic dyes, wax, energy and water.

Pollution caused by batik industry is dyestuff waste. As it is known that many coloring in batik dye synthetic dyes because the price is relatively cheaper when it is compared with natural dyes. The use of synthetic dyes in

* Corresponding author: ibudks@yahoo.com

the batik industry produces dyestuff waste that degrade water quality. Synthetic dyestuff waste which discharged in waters is non-biodegradable, and cannot be decomposed by microorganisms [1].

The coloring for batik clothes that it is now done by crafts men uses synthetic dye. Synthetic dyes are widely used by artisans as base dyes and Diazo direct. Synthetic dyes contain heavy metals such as Cd, Cr, Pb, Co, Cu, Hg, Ni, Mg, Fe and Mn. Heavy metals are used to increase the bonding strength between dyes and fabrics [2].

Suprihatin [3] states that one of the most popular examples of dyestuffs used in the textile industry is the black, red and golden yellow remazol. In staining, this compound is only used about 5% while the rest of 95% will be disposed of as waste. This means that the dyestuff which effectively binds the fabric is only 5% while the rest are wasted.

The city of Pekalongan known as "Kota Batik" is one of the cities located in the coastal region which has batik industries as the most dominant activities of its population. According to the document of Environment Agency of Pekalongan City, in the year of 2014, Pekalongan City has 917 batik industries spread in 4 (four) subdistricts namely 247 batik industries in West Pekalongan District, 124 batik industries in East Pekalongan District, 104 batik industries in North Pekalongan District, and 442 batik industries or about 48.2% batik industries are located in South Pekalongan District.

Of the 442 batik industries located in South Pekalongan District, about 46.3% of them or about 203 batik industries are located in Jenggot Village, which is the biggest batik industrial center in Pekalongan City, occupying 35.382 m2 area, and having production per month 25,095 *codies* or about 501,900 batik products per month, consisting of batik cap and batik combination, using clean water 15,505 m3 per month and disposing waste water as much as 12,404 m3 per month mostly directly to the environment without prior processing and with potency of a negative impact on the surrounding community, especially for the health of the environment health.

The activities of Batik industries in Pekalongan City can be the cause of environmental damage such as the occurrence of water pollution and soil contamination due to waste water disposal by batik industry which reached 73,878 m3 per month to the environment without going through a prior waste treatment. If the industry disposes liquid waste, then the waste course will pass through the waters around the settlement. Thus the quality of the neighborhood where the population lives is decreasing. The waste can increase organic content such as COD, BOD, TSS and pH. If this exceeds the permissible threshold, then the most easily known symptom is the death of aquatic organisms [4].

Some batik dyes containing Cd, Pb, Cu, and Cr such as $Na_2Cr_2O_7$ or $Na_2Cr_3O_7$ are very dangerous and other compounds included in the category of hazardous and toxic substances (B3) are highly reactive H_2O_2 and toxic HClO. Water pollution by heavy metals has long been a serious problem that needs to be addressed, given the ever-increasing volume of waste, the toxic properties of heavy metals, and the introduction of heavy metals into water bodies can affect water quality [5] Dyestuffs in batik waste are disposed of without prior water treatment into waters, which indirectly can endanger the health. [6]. The purpose of this study is to determine the quality of batik waste water discharged directly to the residential channel and its impact on the quality of well water dug in the center of batik industry Jenggot Pekalongan City.

2. Methodology

2.1. Place and time of research

This research was conducted at Industrial Centre Batik Jenggot Pekalongan City area by assigning places of waste water source where the batik industry exists. and sewage waste is in the sewer in residential area and at well dug of its population. The measurement was conducted in August 2016, while the interviews of respondents were conducted in January 2017.

2.2. Research design

The design of this research is analytical descriptive by measuring the quality of industrial waste water batik as a source of pollution, especially on the parameters of heavy metals Cd, Cr and Pb, then measuring the quality of puddles of waste water in the settlement sewer. The result of the measurement is compared by using the quality standard according to Government Regulation No.82 of 2001 [7]. If the value is greater than the value of the quality standard, it means it has been polluted. Next, measuring the quality of dug well water of residents which is located around the puddle waste on the parameters of heavy metals Cd, Cr and Pb. The result is compared by using the quality standard according to Kepmenkes RI Number. 492 / Menkes / Per / IV / 2010 [8] If the value is greater than the value of the quality standard, it means the well water has been polluted.

2.3. Population and sample

The population of this research is the environment of Batik Industry Centre Jenggot Pekalongan City. The sample taken as a purposive sampling is at the source of waste water in the batik industry centre of Jenggot, the waste water puddle in the residential sewers, in which there are 5 sample points for each, and some well water in the population around the puddle of sewage in the residential, 3 sample points for each, so that the number of samples of dug wells are 15 samples.

2.4. Variables of research

The variables of this study are the quality of dug well water with heavy metal parameters Cd, Cr and Pb as dependent variables, and its independent variables are the values of heavy metal parameters Cd, Cr and Pb in batik industry waste water and waste water puddle in the sewer population.

2.5. Data processing

Waste water samples at industrial source of batik as well as in puddles in residential sewer were taken then the value of the heavy metal consisted of Cd, Cr and Pb measured at Water Laboratory and then analyzed. The value of heavy metals parameters is then compared with the river water quality standard in accordance with Government Regulation no. 82 Year 2001. The condition is polluted if the value of Cd, Cr and Pb result of measurement in Laboratory is greater than the value on the quality standard. Likewise, for the samples of dug well water around the puddles of waste water are measured for their values of heavy metals consisted of Cd, Cr and Pb at the same laboratory. The result is compared with the quality standard according to Kepmenkes RI Number. 492 / Menkes / Per / IV / 2010 regarding Water Quality Requirements. If the value is greater than the value of the quality standard, it means the well water has been polluted is not qualified for drinking water.

3. Result and discussion

3.1 Description of area

The study was conducted at Industrial Batik Centre of Pekalongan City, Central Java Province. Batik Jenggot Industrial Centre is chosen because it is the largest Batik Industrial Centre in Pekalongan City and the batik industry activity has been running long enough as a hereditary activity. The spatial scope of the research location of Batik Industry Centre of Pekalongan City covers its area and the area around it.

a. Area of ± 35,382 m2, as a place to agglomerate various types of batik, stamp and printing industries that are heterogeneous and are thought to be the source of impact for the environment within the batik industry centre of Jenggot as well as in the surrounding environment. In batik industrial centre there are 203 batik industries with total production per month of 501,900 pieces of batik products and most of the waste water disposal without going through waste management of 12.404 m3.

b. The area around the batik industry centre of Jenggot is identified as directly affected by the activities in the industrial center. Batik Industrial Centre in Pekalongan City is a coastal ecosystem consisting of land sub-systems (industry, settlements, offices and commerce). The landscape is located in South Pekalongan District shown in Figure 1. As follows:

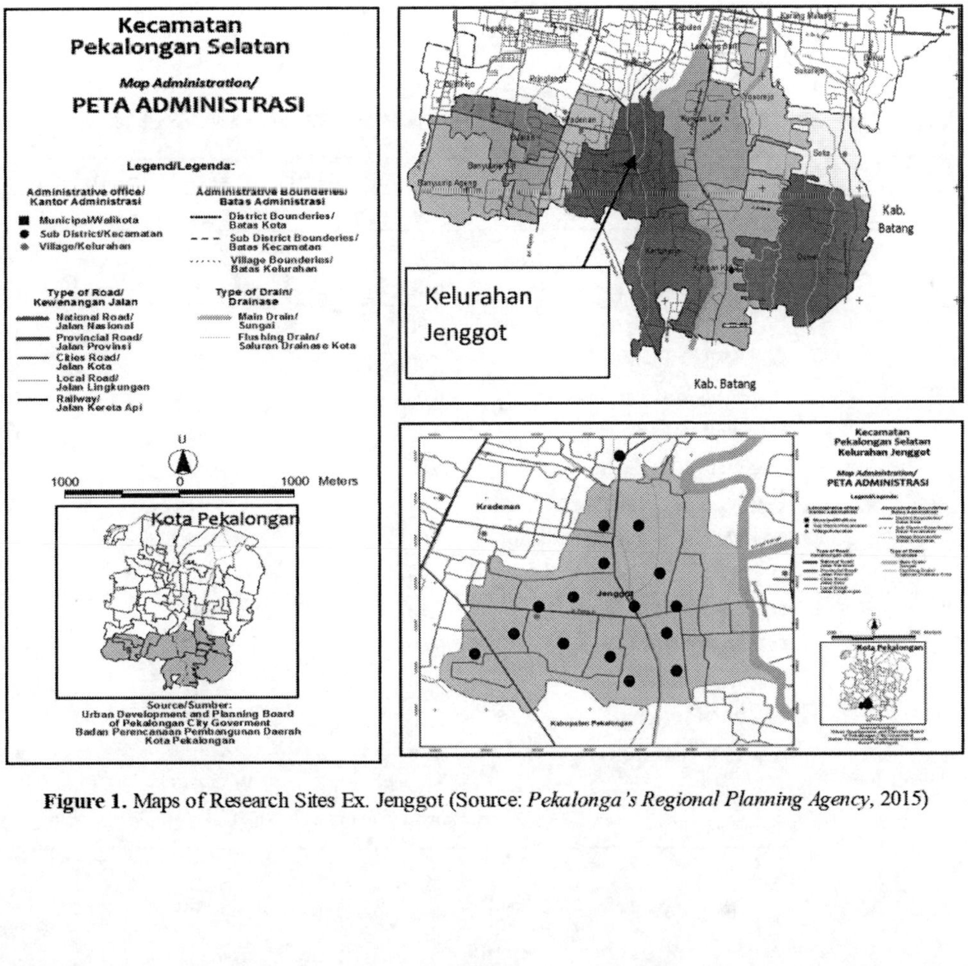

Figure 1. Maps of Research Sites Ex. Jenggot (Source: *Pekalonga's Regional Planning Agency*, 2015)

3.2 Waste water quality from the sources

Table 1. Quality of Waste water at Batik Industrial Outlet Baku heavy metal Parameter Quality Content Waste water

Parameters of heavy metals	Quality Standard parameters Max	Method	Waste water Source				
			Outlet 1	Outlet 2	Outlet 3	Outlet 4	Outlet 5
Cadmiun (Cd)	0,003 mg/L	SNI 6989 16 2009	0,07	0,07	0,07	0,07	0,07
Total Crom (Cr)	0,05 mg/L	SNI 6989 17 2009	0,76	0,76	0,76	0,76	0,76
Lead (Pb)	0,01 mg/L	SNI 6989 18 2009	0,79	0,79	0,77	0,79	0,78

Source: Field data (2016).

The result of the measurement of heavy metal content in waste water at the source of 5 points of batik industrial outlet is as above. Based on table 1 above, the scoring value is almost the same in each outlet. This indicates that the materials, composition and dosage used in the batik process are almost identical. When compared with the Standards of Quality, then all the numbers for the parameters of heavy metals Cd, Cr and Pb has exceeded the maximum levels set. The highest figure contained in the parameter measurements Heavy metal Pb which are much higher when compared with the maximum levels set as a quality standard.

3.3 The quality of waste water of sewer puddle of population

The result of the measurement of heavy metal content in the waste water puddle in residential sewer that at 5 points of puddles is as follows:

Table 2 Waste water Quality in the sewer puddle of Population

Parameters of heavy metals	Quality standards Max	Method	Waste Water Puddle				
			Sewer 1	Sewer 2	Sewer 3	Sewer 4	Sewer 5
Cadmiun (Cd)	0,003 mg/L	SNI 6989 16 2009	0,07	0,07	0,07	0,07	0,07
Total Chrom (Cr)	0,05 mg/L	SNI 6989 17 2009	0,76	0,78	0,76	0,76	0,76
Lead (Pb)	0,01 mg/L	SNI 6989 18 2009	0,76	0,79	0,79	0,77	0,75

Source: Field data (2016).

Table 3 Quality of Well Water Dug around Wading Batik Batik Waste on Residential Sewage

Location		Parameters of heavy metals		
		Cadmiun (Cd)	Total Chrom (Cr)	Lead (Pb)
Puddle 1	Dug Wells < 10 meters	0,001	0,02	0,04
	Dug Wells 10-15 meters	0,001	0,02	0,04
	Dug Wells > 15 meters	0,001	0,02	0,04
Puddle 2	Dug Wells < 10 meters	0,001	0,02	0,04
	Dug Wells 10-15 meters	0,001	0,02	0,04
	Dug Wells > 15 meters	0,002	0,02	0,04
Puddle 3	Dug Wells < 10 meters	0,001	0,02	0,04
	Dug Wells 10-15 meters	0,002	0,01	0,04
	Dug Wells > 15 meters	0,001	0,003	0,04
Puddle 4	Dug Wells < 10 meters	0,001	0,003	0,04
	Dug Wells 10-15 meters	0,001	0,003	0,04
	Dug Wells > 15 meters	0,001	0,003	0,04
Puddle 5	Dug Wells < 10 meters	0,001	0,003	0,04
	Dug Wells 10-15 meters	0,001	0,003	0,04
	Dug Wells > 15 meters	0,001	0,003	0,04
Quality Standards Max		0,003 mg/L	0,05 mg/L	0,01 mg/L
Method		SNI 6989 16 2009	SNI 6989 17 2009	SNI 6989 18 2009

Based on table 2 above, the scoring value is almost the same in every sewer. Comparing to the numbers from the measurement result at outlet, the scoring value shows an almost the same result. This indicates, the waste water flowing from outlet to the settlement area is not changing. If it is compared with the standard quality, all numbers for heavy metal parameters Cd, Cr and Pb have exceeded the specified maximum level. The highest measurement figures are in the parameters of Pb heavy metals whose numbers are much higher when compared to the maximum levels set as the quality standard.

3.4 Dug well water quality

The results of the measurement of heavy metal content in the dug well water of residents, 3 points for each puddle with a distance of <10 meters, 10-15 meters and> 15 meters are tabulated in Table 3.

Based on Table 3, the number of measurements of heavy metals in dug wells for parameters Cd and Cr is below the maximum level on the quality standard, while the parameter of heavy metal Pb measurement figures are above the maximum level on the quality standard. This indicates that there has been contaminations on dug wells which exist in the population settlement around the pool of batik waste water. Azumi [9] analyzed the impact of industrial waste water on and Barthi [10] surface water and found several types of heavy metals present in such surface waters as zinc, copper, chromium, lead, iron, cadmium, and mangan. The well water is a shallow groundwater which starts from infiltration of rain water into the soil. Dyestuffs in aquatic environments can actually decompose naturally by the presence of sun light, but this reaction is relatively slow, due to the intensity of UV light that reaches the surface of earth is relatively low so that the accumulation of dye to the bottom of the water or soil is faster than the photodegradation [4]. If the industry disposes liquid waste, then the waste stream will pass through the waters around the settlement. Thus the quality of the neighborhood where the population lives down.

3.5 Potential of impact pollution to environment and population health

3.5.1 Construction of Dug Well

Dug wells which exist in Jenggot Village have potentials to be contaminated by batik waste water. From the survey of 112 dug wells in Batik Jenggot Industrial Center, the construction aspect is described in Table 4 below:

Table 4 Construction Condition Dug Well Population

Indicator Observed	Construction Condition						Total	
	Good	%	Medium Damage	%	Heavy Damage	%	Amount	%
Outer Wall	94	83,9	18	16,1	0	0	112	100,00
Inner wall	92	82,1	14	12,5	6	5,4	112	100,00
Wells Floor	90	80,4	17	15,2	5	4,5	112	100,00

Source: Field Data (2017)

Based on the data in Table 4, there are still more pollution potentials to dug wells because there are still 16.1% of damage of outer wall, 17.9% of damage of inner wall and 19.7% well with damage of well floor. Improper well construction will potentially infiltrate batik waste water into the well.

3.5.2 Physical Quality of Dug Well Water

The pollution to dug well water can be known earlier based on its physical quality. From the survey results of 112 dug wells in the Batik Jenggot Industrial Center, and the aspect of physical quality of dug well water is described in the table 5.

Table 5 Physical Quality of Well Water Dug Population

Water Physical Indicator Observed	Conditions of Well Water Dug				Total	
	Yes	%	No	%	Amount	%
Clear	108	96,4	4	3,6	112	100,00
No Smell	94	83,9	18	16,1	112	100,00
Colorless	101	90,2	11	9,8	112	100,00
No Taste	100	89,3	12	10,7	112	100,00

Source: Field Data (2017)

Based on the data in table 5, there are water wells that physically are not qualified. There are 3.6% turbid well water, 16.1% smelly water, 9.8% colored water and 10.7% taste.

3.5.3 Utilization of Dug Well Water

The existence of dug wells is one of the important things in Jenggot Urban Village.
From the approximately 1,200 heads of households, there are approximately 75% of people still relying on dug wells as a source of clean water. From the survey of 112 digging wells in Batik Jenggot Industrial Center, the utilization of dug well water is described in Table 6.
Based on the data in Table 3.6, dug well water is used daily drinking, cooking, bathing and washing. If the water used is polluted by heavy metals due to batik waste, it will cause the entry of heavy metals into the human body, especially through ingestion process (digestion). As many as 65.2% of families say they do not have other sources of clean water, the potential for bioaccumulation of heavy metals in the body due to daily use can occur. Heavy metals of Pb or lead is one element of heavy metals that are toxic to living things including humans. The influence of Pb [12] on human health can damage the nervous system, kidney and reproductive system. Hasan [13] states that the implementation of sustainable waste water management including proper treatment of waste water and industrial waste is essential to protect ecosystems and also to prevent the further damage of the surface water quality in this area.

Table 6 Utilization of Well Water Dug Population

Utilization	Condition				Total	
	Yes	**%**	**No**	**%**	**Amount**	**%**
Drinking water	85	75,9	27	24,1	112	100,00
Cook	93	83,0	19	17,0	112	100,00
Bath	111	99,1	1	0,9	112	100,00
Wash	111	99,1	1	0,9	112	100,00
Don't have another Source of Clean Water	73	65,2	39	34,8	112	100,00

Source: Field Data (2017)

4. Conclusion and Recommendation

4.1. Conclusion

Batik waste water from batik industry outlets contains heavy metals Cd Cr and Pb which exceed the required standard of environmental quality. This happens also on batik waste water pools in residential sewers and on dug well water around the gutter. On dug well water around the gutter only Pb which levels exceeds the required environmental quality standard. There were 16.1% of damaged wells, 17.9% of damaged internal walls and 19.7% of wells with damaged well floors. The potential for infiltration of batik waste water into the well. The physical condition of dug wells has the potential to contaminate batik waste water, there are 3.6% turbid well water, 16.1% smelly water, 9.8% colored water and 10.7% taste. Pollution to dug well water can have an impact on public health because it is used for daily use to drink, cook, bathe and wash because> 65% has no other sources of clean water.

4.2. Recommendation

The recommendation of this research is the need for waste water treatment batik starting from source through WWTP both on the household and communal scale with the role of local government as a facilitator. Society and batik industry players need to be involved in managing environmentally friendly batik industrial center.

References:

1. S. Widyaningsih, D. W Dwiasi., and D. Hidayati., *9*(2), 166–174. (2014).
2. Yuniarto, D. R dan Iqbal Achmad, Jurnal Teknologi Pengelolaan Limbah (Journal of Waste Management Technology), ISSN 1410-9565 Volume 16 Edisi Suplemen 2013 (Volume **16**, Supplement Edition), Pusat Teknologi Limbah Radioaktif (Radioactive Waste Technology Center), (2013)
3. Suprihatin, *Kandungan Organik Limbah Cair Industri Batik Jetis Sidoarjo Dan Alternatif Pengolahannya*, Pusat Penelitian Lingkungan Hidup Universitas Riau.(2014)
4. Al-Kdasi, A., Idris, A., Saed, K. and C.T. Guan. Global Nest The Int. J. **6**: 222-230. (2004)
5. D. Bashyal, P.L. Homagai, K.N. Ghimire. Journal of Nepal Chemical Society. Vol 26:53-60. (2010)
6. Grezechulska and Morawskin, App. Catal B: Environmental, 36, 45.(2002)
7. Peraturan Pemerintah No.82/2001, *Pengelolaan Kualitas Air Dan Pengendalian Pencemaran Air.*
8. Mepmenkes RI Nomer. 492/Menkes/Per/IV/2010 tentang *Persyaratan Kualitas Air Minum*
9. S. Azumi, M. Bichi. International Journal of Engineering & Technology, **10**(1), pp.101–110.(2006).
10. P.K Bharti; P Kumar; and V. Singh. Journal of Applied and Natural Sceience, **5**(1), pp.132–136.(2013)
11. Supriyono, Ety susilowati, and Suci Dwi Suryani. *Analisa Kadar Logam Timbal dan Seng di Pantai Slamaran Pekalongan secara Spektrofotometer Serapan Atom.* Media Farmasi Indonesia Vol **3** No 1.(2008)
12. K. Hasan, and M. Miah. Impacts of Textile Dyeing

Industries Effluents on Surface Water Quality: A Study on Araihazar Thana in Narayanganj District of Bangladesh. Journal of Environment and Human, (**3**), pp.8–22.(2014)

Coastal Community Group for Coastal Resilient in Timbulsloko Village, Sayung, Demak Regency, Indonesia

Hartuti Purnaweni[1,2], *Kismartini,*[1,2] , *Sudharto* P. Hadi[1,2,3] , *Ike* Soraya[4]

[1] Master and Doctorate Program of Environmental Science, School of Post Graduate Studies, Diponegoro University, Semarang, 50241, Indonesia
[2] Public Administration Department, Faculty of Social Science and Political Science, Diponegoro University, Semarang, 50275, Indonesia
[3]Business Administration Department, Faculty of Social Science and Political Science, Diponegoro University, Semarang, 50275, Indonesia
[4] Alumni Master Program of Public Administration, Faculty of Social Science and Political Science, Diponegoro University, Semarang, 50275, Indonesia

Abstract. Coastal areas are very dynamic and fragile environment, demanding for policies to preserve these areas as materialized in the Resilient Coastal Development Program (PKPT) by the Indonesian government. Amongst the targeted area was Timbulsloko Village in Sayung District, Demak Regency, which coastal areas is severely damaged by erosion. This article analyzed the development of the Coastal Community Group (CCG) related to the PKPT program in Timbulsloko village, especially in how the group is empowered to increase the community's resilient in facing the disaster. This study, applied an analytical descriptive method, used the development of the CCG as phenomenon. Primary data was collected through observation and in-depth interviews with stakeholders, accompanying the secondary data. The result shows that the PKPT funding was mostly spent on infrastructure development and used for project management, not for optimizing local economic empowerment. After the completion of the PKPT, there are no actions or following programs to keep the physical results constructed by the CCG. Accordingly, the orientation towards the CCG building capacity for Timbulsloko community's ecological resilience had not been optimally implemented. This study recommended a "putting the last first" policy approach to preparing the local community. The government must play a stronger role in encouraging a self-help local group for strong human development

1 Introduction

Coastal areas are one amongst the most dynamic natural environments on earth, providing a range of goods and services that are very important to human social and economic existence. Coastal zones constitute the transitional area between the world's land and oceans, characterized by highly diverse ecosystems such as coral reefs, mangroves, and beaches. Since the beginning the mode of transportation was highly relied on sea transportation, therefore, coastal areas have been favorable for many people to live in, to take its benefit for food production, transportation, recreation and other human activities.

A large part of the global human population now lives in coastal areas: estimates range from 20.6% within 30 km of the sea to 37% in the nearest 100 km to the coast [1]. In 2010 a total of more than a billion people worldwide, most of them in Asia, live in low-lying coastal regions, within 20 meters of mean sea level on land measuring about 8 million square kilometers. It is roughly estimated that more than 200 million people worldwide live along coastlines, which is less than 5 meters above sea level. By the end of the 21st century this figure is estimated to increase to 400 to 500 million [2].

Indonesia is a country that has huge potential in many sectors including fisheries and marine. This is an implication of Indonesia's characteristic as archipelago country. The potential is so large that should be utilized optimally for the welfare of its people, especially for communities in coastal areas. But in reality, these potencies can only be utilized about 20% alone that it brings a bad impact on coastal communities. This is seen from various problems hampering the people's welfare such as poverty, environmental degradation of coastal ecosystems, lack of organizational village self-sufficiency, and constraints of infrastructures [3]. The increasing problems in the coastal zones contributed to the high vulnerability to natural disasters and climate change in Indonesia, including in Java. Climate change is a crucial problem, both in impact and policy. Climate change does not appear on the "empty space," but it exists in socio-political context also [4]. This coastal degradation problem will more severe when it is doubled with climate change problem.

* Corresponding author: hartutipurnaweni@gmail.com

 https://doi.org/10.1051/e3sconf/20183109009

One forms of environmental degradation is coastal erosion, including in Java, the mostly populated island in Indonesia. Central Java is the most degraded province in this case, figuring to 261.27 kilometers of impacted coastal area in 2014, followed by East Java (245.99 km2), West Java, Banten, and Jakarta [5] as can be seen in Figure 1.

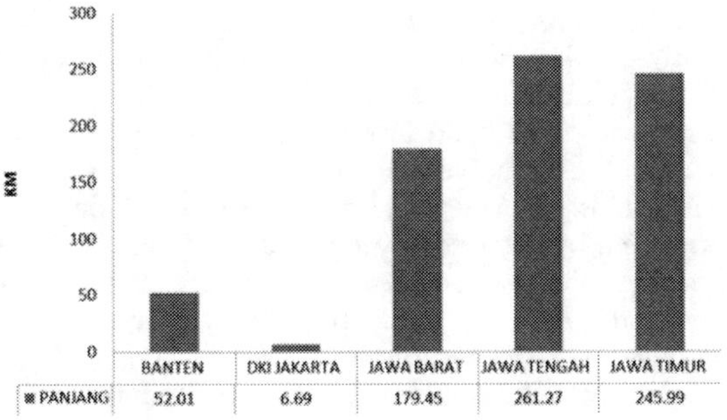

Fig. 1. Coastal abrasion in Java, 2014

One of the most vulnerable areas in the coastal area of Central Java is Demak Regency, especially in Sayung District [6], in which Timbulsloko village is one among the most severely affected by coastal erosion.

To address the high vulnerability of coastal villages to disasters, the government in this case the Ministry of Marine and Fisheries issued a policy in the form of *Program Kawasan Pesisir Tangguh* (PKPT) or the Resilient Coastal Development Program, which was stipulated in the Regulation of Minister of Marine Affairs and Fisheries Number 2/2013 on the Guidelines of the Implementation of the Self-Help National Program for Community Empowerment (PNPM Mandiri) of Marine and Fisheries. This program includes efforts to improve the resilience of coastal communities not only in terms of economy but also resilience to disaster. One of the villages appointed to implement the PKPT program is Timbulsloko village. Through the PKPT program, it is expected that the resilience and responsiveness of the community can be increased. Due to this, one among the program activities was the establishment of Coastal Community Group. This article analyzed the development of the Coastal Community Group related to the PKPT program in Timbulsloko village, especially in how the group is empowered to increase the community's resilient in facing the disaster.

2 Methodology

Location of this research was at Timbulsloko Village, Sayung District, Demak Regency which was severely affected by the erosion, therefore badly need disaster resilience program. The study applied an analytical descriptive method. The phenomena were the development of the Coastal Community Group. Primary data was collected through observation and in-depth interviews with stakeholders as most of them were mentioned by Kismartini et al [7], consisting of Head and staffs of Demak Regencial Office of Marine and Fisheries, Demak Regencial Government, Demak Regencial Office of Public Works, Head and staff of Timbulsloko Village Office, local community, members of the Coastal Community Group (CCG), and facilitators of the CCG. Meanwhile the secondary data was collected from various research results, GIS data, reports, maps, and files from various sources especially from Demak Regencial Office of Marine and Fisheries.

3 Results and Discussion

In the PKPT program, Human Development is one among its intended results, among which was facilitated with the establishment of Coastal Community Group, which is defined as organized group whose members live in coastal areas and practice commercial and non-commercial marine and fisheries activities related to environmental conservation [8]. This group was supported by a facilitator, which is a person which has educational background and competence in marine and fisheries, live in the benefited community, and responsible in accompanying the local community group during the program period [8] for increasing their resilient to disaster.

The term resilience is often used in the same manner as the notion of "bouncing back" that reflects its Latin root *resiliere* which means "to jump back" [9]. The concept of resilience originates three decades ago, in which Holling [19] is frequently cited as maybe the first to both use and define the concept of resilience in the field of ecology in his article entitled *Resilience and Stability of the Ecological Systems*. Holling defined the term resilience for an ecosystem as the measure of the ability of an ecosystem to absorb changes and still persist. He compared the concept of resilience with stability which he defined as the ability of a system to return to its equilibrium after a temporary disturbance. According to him, the more rapidly the system returns to its equilibrium, the more stable it is.

Timbulsloko Village

Timbulsloko village has 462.50 hectares of land area, with 3.5 kilometers coastal line since it is geographically located at coastal zone and directly bordered with Java Sea. Administratively this village is divided into 4 *dukuh* (sub-village) consisting of Wonorejo, Karanggeneng, Bogorame and Timbulsloko. Timbulsloko is 8 kilometers away from Sayung District office, and 17 kilometers from the capital of Demak Regency

Most of the villagers (49%) were farmers, 33% of which were labor-farmers. The farmers are fishpond-farmers, not those who work in the rice fields. In the second group were blue collar workers (17% for industry, 13% for construction), followed by traders (7%) and fishermen (5%) [11].

Coastal Community Group for Ecological Resilience

Abrasion which then causing inundated area constitute as a disaster in coastal areas such as Timbulsloko encouraged Ministry of Marine and Fisheries stipulated Resilient Coastal Development Program by among others establishing Coastal Community Group (CCG) to mediate development capacity of ecological resilient in the affected areas, including in Timbulsloko. This coastal group is a group of people who gather to help one another and cooperate in developing financial service source to encourage and to develop productive activities and enhancing standard of living among the members and their families. The community need to be gathered in group because it s difficult for the needy /the poor if they have to rely on themselves. Cooperation is hoped to be a better way to develop their economic condition.

The CCG is responsible for optimizing any resources provided in the Timbulsloko Village to minimize the disastrous impact of the abrasion, i.e. coastal inundation (local term: *rob*), which has caused economic and health losses as well as environmental damages. The PKPT in Timbulsloko has allowed the village to obtain financial assistance from the government towards further development efforts for human, environment, infrastructure, administration, and resilience [8]. Aided by facilitators, the CCG plays a significant role in (1) performing identification, selection, and verification of the CCG candidacy; (2) proposing the CCG candidacy verification results; (3) helping develop the RPKP plan; (4) prioritizing activities according to the existing RPDP documents; (5) facilitating the CCG in the RKK development; and (6) reporting.

Table 1. Community Involvement in CCG Samudera Jaya [12 and interview]

No	Activity	Target Group	Total	Forms of Involve-ment
1	Socializa tion	Men	4	Participant
		Women	3	
2	CCG Identifi – cation	Men	10	Participant, Executor, Decision maker
		Women	3	
3	Location survey and measure-ment	Men	3	Participant, Executor
		Women	1	

During the CCG socialization the facilitators informed activity plan of the PKPT to all stakeholders at a village meeting. The meeting featured discussion of plan and evaluation of the rural development through vary funding schemes, CCG candidacies, terms and conditions and size of the group members. The committees and members of the CCG had to be natives as proven by their ID Cards. Furthermore, they could not be members if they were village government officers or staff, government apparatus or police/army members. There are 13 people in the group, 60% of which were, according to the standard, economically near-poor. CCG in Timbulsloko was then named "Samudra Jaya" (Wealthy Ocean)

However, in practice, the funding was more focused and prioritized on infrastructure development and used for project management. The fund is not optimized for economic empowerment. In Timbulsloko, most of the fundings were allocated to spending dumping materials of the local roads, in order to prevent these roads from flood during inundation. The group also bought a motor vehicle for collecting waste from households. The member group also joined field trips organized by the Demak Regencial government to visit Rembang Regency and Bali Province for learning about salt and mangrove cultivation. They also spent the money for joining carnival at Demak Regency [12]. Indeed, empowerment of the coastal communities should include comparative study and guided tour as suggested by the Ministry of Marine and Fisheries; however other programs such as economic trainings in terms of mangrove seedlings, processing of local product, or silvofishery should be given priority along with institutional building involving local government, village apparatuses and community groups [5].

After the completion of the projects in Timbulsloko, there are no actions or following programs to keep the physical results achieved by the Coastal Community Group. Therefore, when infrastructure development results deteriorated, there are no efforts to rebuild or rehabilitate. The waste motor vehicle was not operated anymore due to flood, lack of fund for gasoline, and miss management.

Even more, the CCG, which used to participate in the development, could no longer take action because it did not have any activity to offer. In the word, the group did not have any sustainability of actions and programs. There was not any economic activity supported through the PKPT scheme. The informants said that this was due the absence of any economic potential among the local communities that was worth of being developed.

Accordingly, the orientation towards the CCG building capacity for the ecological resilience for the local villagers of Timbulsloko had not been optimally implemented. The implication of such lackluster was high risk and disastrous impact of abrasion, i.e. coastal flooding. The local community resilience against the disaster was very low. Therefore, there is a need for alternative policy on improving the ecological resilience of the local villagers. Community's response and adaptive capacity is a one of the key roles on the disaster management and risk reduction program [13], therefore this has to be strengthened.

Assessing community resilience is a complex process because of the dynamic interactions of people, community, societies, and the environment. There are currently many conceptual frameworks proposed to measure this concept [9]. Generally, most of these frameworks conceptualize disaster resilience in the same way, in which they all focus on similar factors that could reduce vulnerability and increase community resilience.

Such factors include economic resources, assets and skills, information and knowledge, support and supportive networks, access to services, and shared community values. These frameworks can be categorized as "*community capital*". The optimization of these communities' capital needs to be balanced. Coastal management policies must facilitate harmonic coexistence between human activities and the coastal system, between hard and/or soft components [13], relationship between natural resources and human resources [14]. The end results anticipated from the program is environmental sustainability and high responsiveness to disasters and climate change [15]. All stakeholders, consisting of coastal managers, ecologists, engineers, decision makers and society in general are jointly responsible of the future of the fragile and dynamic coasts [16], not only the fragile coastal community.

4 Conclusions

This study recommended a "putting the last first" policy approach to preparing the local community. The local government apparatus of Demak Regency and the rural government of Timbulsloko Village must play their role in performing a self-help institution promotion by taking measures from assistance, facilitation, to the promotion of the ecological resilience for the local villagers of Timbulsloko, especially in the establishment of the Coastal Community Resilience Group.

In term of institutional aspect this study recommended a community training towards the establishment of a sustainable disaster prevention group. To obtain the funding the group must be directed towards developing Village Income and Spending Budget (APBDes) to increase the funding allocation. The group is also expected to co-operate with private and other sectors.

References

1. RJT Klein, PhD Thesis for Mathematisch-Naturwissenschaftliche Fakultat of the Christian Albrechts Universitat, Kiel University, Germany (2002), https://www.oceandocs.org/bitstream/handle/1834/789/Klein.pdf?sequence=1
2. World Ocean Review 1, *Living with the oceans, A report on the state of the world's oceans,* (2010), http://worldoceanreview.com/en/wor-1/coasts/living-in-coastal-areas/
3. Y. Anggraeni et al, J Administrasi Publik, Vol **3**:11, pp. 1862-1867 (2010)
4. Rusmadi, Sudharto P. Hadi, H.Purnaweni, Advance Science Letter, Vol **23**:3 March (2017), pp. 2556-2558, https://doi.org/10.1166/asl.2017.8697
5. M.E Rudianto, *Rencana Penanganan Kerusakan Pantura Jawa* (Rehabilitation Plan for Java's Northern Coastal Area), Direktorat Pesisir dan Lautan, Kementerian Kelautan dan Perikanan (2015)
6. A.M. Marfai, Quaestiones Geographicae **31** (3), 2012, pp 47-55.
7. Kismartini, P. Hartuti, Sudharto P. Hadi, *Proceeding ICM-MBT*, pp 81-87, (2016) http://pksplipb.or.id
8. Direktorat Pesisir dan Lautan, KKP, *Pedoman Teknis Pengembangan Kawasan Pesisir Tangguh* (2015)
9. J. Mayunga, *Understanding and applying the concept of community disaster resilience*, https://www.ucursos.cl/usuario/3b514b53bcb4025aaf9a6781047e4a66/
10. C.S. Holling, Annual Review of Ecology and Systematics, **4**, 1–23 (1973)
11. *Monografi Desa Timbulsloko* (2015).
12. *Pusat Penyuluhan Kelautan dan Perikanan, Badan Pengembangan SDM Kelautan Perikanan, KKP, Laporan Kegiatan PKPT Desa Timbulsloko Kecamatan Sayung Kabupaten Demak,* (2015)
13. A.M. Marfai, and H. Diah, 2011, Seria Geografie Year XXI, no. 2/2011 (December), pp. 209-221 ISSN 1454-2749, E-ISSN 2065-1619 Article no. 212106-536.
14. M. A.Delgadillo Calzadilla, E.Mendoza, R. Silva, J.A. Gonzalez-Vázquez, and D. Infante-Mata, J of Coastal Research Special Issue **71**- Coastal Erosion and Management along Developing Coasts, (2014), pp. 114 – 121 , https://doi.org/10.2112/SI71-014.1.
15. Y.A. Supianto, *BPBD Kabupaten Garut*, http://www.garutkab.go.id/download_files/article/Desa-Tangguh-%20Bencana.pdf
16. P. Hartuti, Kismartini, Sudharto P. Hadi, Tri R. Soeprobowati, Advance Science Letter, Vol **23** (3) March 2017, pp 2582-2583 (2) https://doi.org/10.1166/asl.2017.8711.
17. M.L Martínez, R. Silva, D. Lithgow, E. Mendoza, P. Flores, R. Martínez, and C. Cruz, C., J of Coastal Research, Special Issue **77** pp. 143–153 (2017)

Urban Environment Development based on Universal Design Principles

Bangun IR Harsritanto[1,*]

[1]Architecture Department of Engineering Faculty, Diponegoro University, Semarang - Indonesia

Abstract. Universal Design is a design which facilitated full range of human diversity. By applying Universal design principles, urban environment can be more functional and more user-friendly for everyone. This study examined five urban streets of South Korea as a country experienced on developing various urban street designs based on universal design. This study aimed to examine and compare the South Korea cases using seven principles of universal design. The research methods of this study are literature study, case study, and site observation. The results of this study are: South Korea cases are good practices, urgency of implementing the direction into local regulations; and change of urban development paradigm.

1 Introduction

Urban streetscape which designed and developed in detail with a special theme or purpose of urban development. However, not all people can access the specialized street easily because of their different ability (as the elderly, children and people with special needs).

Universal design (also called inclusive design or accessible design) refers to facility designs that accommodate the widest range of potential users, including people with mobility and visual disabilities and other special needs [11]. Universal design should be comprehensive from origin to destination for the greatest possible range of potential users. Universal design should consider all possible obstacles that may exist in buildings, transportation terminals, sidewalks, paths, roads and vehicles.

South Korea is one of the countries that applied universal design principles in public spaces and public buildings. South Korea applied it to ease the access for the elderly and disabled. The condition was happened because the number of this last category in South Korea is steadily increasing due to congenital disabilities, aging and the increase of physical impairments from accidents. This study examined five urban streets of South Korea as a country experienced on developing various urban street designs based on universal design.

2 Method

This study was performed with three research methods, such as:

1. Literatures study was compiled and analyzed several related literature to construct evaluation sheet of this study.
2. Site observation was used to inspect South Korea urban streets with evaluation sheet which constructed before, and measured by interval scales of five points.
3. Case study was used to explain several phenomena happening on the site observation and later on can be generalized into the bigger scope of research area.

3 Discussions

3.1 Literature summary

Summarized from the 7 existing specialized street guidelines (see table 1), there were substantial issues that need to be fixed and completed by this dissertation study, such as:

1. the previous studies were purposed only for tourism street facilities [1];[2];[3] which resulted an artificial disneyfication a of specialized street without concerning on daily activities [4]. This artificial disneyfication of tourism oftenly force the natural expression of the urban society to develop their neighborhood.
2. the previous guidelines were oriented to design local district [5];[6] which resulted the unsustainable specialized street fabric from and to the site. This orientation of design built exclusive area on the city.
3. the previous guidelines were a lack of universal design appliance especially: [7]; [8] which promote distinct disable people as a different kind of person.

[a] A process of how real world is becoming more like a theme park (eg: Disney world) ; standardisation; popular culture; superficial and globalization

* Corresponding author: bangunirh@arsitektur.undip.ac.id

The facilities were designed specifically for them who concerned on the referenced literature [9].The designer should learn about the barriers, difficulties and especially the various needs of people and [10] reduce dependency to other people helps.

3.2 Site observation scale and objects

Present condition analysis by universal design principles which derived from literature study and resumed as an evaluation sheet to compare with cases study and the site survey result. The numerical results were generated from the interval scale criteria which consist of 5 scales valued from the worst is 1 and the best is 5.

From three major cities, five famous specialized streets were chosen to be studied on this dissertation :

1. Insadong and Gwanghwamun area in Seoul (the biggest city in South Korea)
2. Nampodong and Seomyeon area in Busan (second biggest city)
3. Yangyeonsi, Banwoldang in Daegu (third biggest city

The observation objects were [12] :

1. Parking lots
2. Bus stop/subway gates
3. Linkage design
4. Street furnitures
5. Street utilities
6. Signages
7. Street materials
8. Pedestrian spaces

4. Results

The statistical result showed that there are two dominant streets to be good examples which are Gwanghwamun and Insadong (see figure 1). The Banwoldang Yangnyeonsi at Daegu has the worst result on 6 sections with scores 3.19 on bus stop/subway, 2.77 on street furniture, 2.79 on street utilities, 2.79 on signages, 2.57 on pedestrian space, and 3.08 on linkage design. The 2nd worst result is Seomyeon which score 3.13 in parking lot section and 3.11 in street material

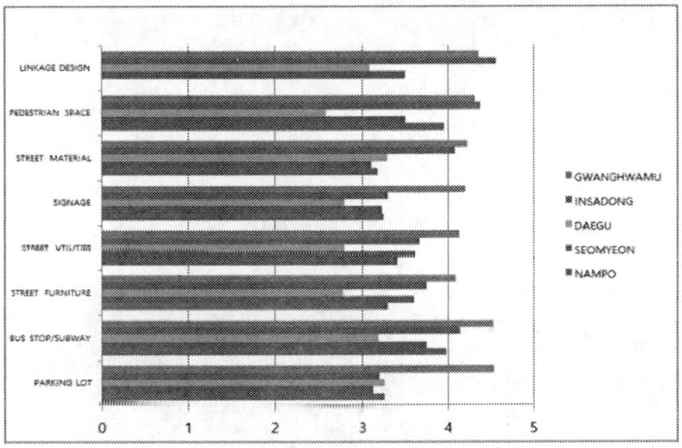

Fig. 1. South Korea's Urban Street Section's Statistical Analysis

In highlight score of universal design application, Gwanghwamun got all highest score, then followed by Insadong, Nampo and Seomyeon. In the opposite Daegu, Yangnyeonsi street got all lowest score in universal design applications. More detail on the findings will be explained in next sub-chapter along with the universal design related value.

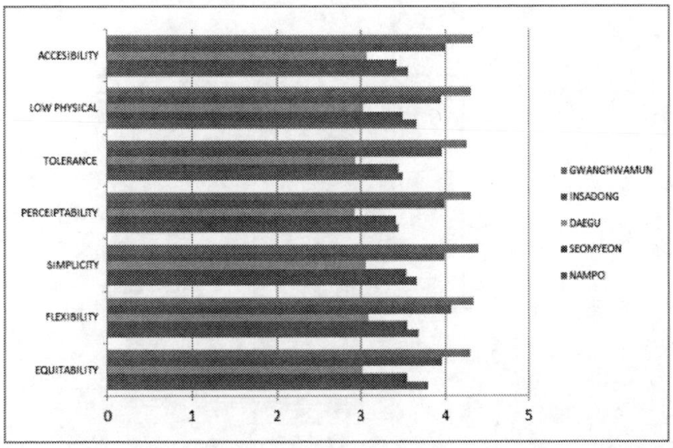

Fig. 2. Universal Design Application Statistical Anaysis of South Korea's Specialized Streets

Table 1. Literature Review Summary

no	item	Inclusive Urban Design	Islington Streetbook A borough-wide streetscape manual	Design principles for blind and partially sighted people	Abu Dhabi Urban Street Design Manual	Street Design Guideline for Landcome	city of Scottsdale Streetdesign guidelines	Torbay Streetdesign guidelines	this study
	Purpose	Creating environments that maximize people's quality of life-based on dementia people preference research	Built an attractive, safe and accessible streetscape by setting out guidelines and specifications for materials, street furniture and layout design for the public highway and footways	Accommodate the differing needs and expectations of all those who use the street space especially for blind and partially sight people	To design streets that create a safe environment for all users, recognizing varying levels of driver education and cultural differences	Built environment providing routes for vehicles and public transport and accommodate utility services and drainage system	Establishing a strong identity and distinctive visual character for scottsdale road as a "signature" corridor along its entire length	Create attractive,pedestrian environment we will create pleasant spaces that encourage greater pedestrian activity - with improved safety and health benefits to the community, and environmental benefits through car use reduction	To build streetscape that can be used by all user regardless sexes, ability and ages based on principle of universal design
1	Method	Interview	Literature study	Literature study	Literature study	Literature study	Literature study	Literature study	Literature study
2		Literature study	Community-participatory research	Case study	Participatory action research	Action plan research	Participatory action research	Participatory action research	Site survey
3		Case study		Participatory action research					Study case
	Result	Dementia and elderly friendly street	Tourism streetbook	Blind and partially blind sighted street	Abudhabi design guidelines by estidama's principle	Australia's streetbook guideline	Tourism streetbook	Tourism streetbook	Universal design based urban streetbook
	Cons	Concern on dementia-elderly	Concern on local tourism	Concern on blind-dog walker	Desert climate response	Concern on residential area	Concern on local tourism	Concern on local tourism	Concern on urban streetscape
	Pros	Dementia and elderly friendly, strong memory/history aspect	Promote tourism, minimal material use	Blind people dan dog walker friendly	Strong traffic management, estidama local content	Legibility force, design community-based	Signage concern, transit oriented	Local conservation item, sustainability	Resumed, add urban identity concept, universal design based

 https://doi.org/10.1051/e3sconf/20183109010

According to the evaluation, the first major problems in the accessibilities were the facilities of the parking lot and street utilities. Even though the location of parking lots was near (less than 200m from the specialized street), the amenities for disabled persons in special parking lots and the signage were not well supported. In Nampodong area, the parking lots were located on several buildings in a small alley that provided difficulties for car on wayfinding and access. More added problems were the locator signs which were in the Korean alphabet without guidance any picture for people who didn't study Korean. In other facts at Insadong, the location of parking lots was more accessible horizontally but some raised levels and small bombs burden the wheelchair accessibility. Most of the signage attached is bilingual, which helps international people to notice it. Since the designed parking spaces should be provided for all, normal and disabled passengers, every passenger shall be able to open the door fully to get in or out the vehicles and manoeuvre from the street to the lots in a low physical effort. The locations of parking lots on the main street with a minimum width of 3m for single car manoeuvres. The parking lot also must be equipped with firm, suitable ramps, slip resistant and tactile pavements for the special needs lots. In addition, the locator signages shall be written in international languages and better to put a braille sign at the decision point.

The second point that can be marked is the bus stop and subway facilities. According to the evaluation, the first point of local character that give notice to the passengers about the specialized street was unnoticeable. The implication of uniformity on subway exits and bus stop was also provided by the visual surrounding adaptation to give assistance to the passenger. The effect of unnoticeable spot also may attract blockade and visual barrier which made difficulties to wayfinding in this area. The accessibility and low effort of elevators and escalators facility were not installed properly, especially on Insadong but these items were provided on Nampodong gate of Gwangbok street which can indicate that instalment of subway station related to specialized street design. Those situations also appear on bus stop instalment in both specialized streets. The bus stops of Insadongs were located disparsely on around the areas. At Nampodong, the facilities were located on start, end, and priority points, but both areas were covered 500m radius of bus stops and subway exits. In the specific enclosure of the point, the Insadong bus stop was not equipped with good enclosure and in the oversized. On three spots of Nampo bus shelters were erected in 2,1m height roof linked with several special design bus for a wheelchair. The signage of this spot was installed with international signs, symbols and words, even though they weren't provided by a good lighting.

The design of specialized streets were planned to support special theme which indicated on material differences compared to the surrounding areas. The connectivity over the specialized street smoothly planned with hierarchy on the material and facilities differences [13]. The small block was injected in both specialized streets to give easy access, nevertheless the Insadong's street wasn't designed with a proper windbreaker in the street which results in multi-use access. In other hands, the passengers will get trouble walking on this area during windy seasons. The strict reinstatement about the specialized street impacts the traffic calming protection over pedestrian way, maintenance, and installation of public art. According to the evaluation of Nampodong, mix use access of pedestrians, cars and motorbikes in the specialized street gives low protection for passengers access. The installment of the median island also able to give wind breaker effect. The rest area facilities were provided even though the distance between each area was more than standard of 20m.

Street furniture's major problems were maps, hand railing, public amenities that only few on the specialized street. That conditions may impact on difficulties to get information on the site and difficulties on access for the disabled persons. In addition to this, Insadong has permanent and more memorable street furniture with simple and local character design rather than Nampo that use uniformity on street furniture.

The street utilities are almost similar to the street furniture section [14]. In a further evaluation, drainage control design on several spots at Insadong may give difficulty with open type and unsafe manhole design.

The signage section had been a vital part in the tourism industry of South Korea since this country using a different alphabet, but it may cover with international form / drawings of signage. Non visual signage such sound and touchable sign were not provided on both country's specialized streets may burden the disabled person to find the information on this site. Most of building information was designed well to give the first impression to the passengers with local art instalments and contrast adjusts the designs. In Nampodong, the map area was not perceptible enough to locate where we are.And the reduplication of building signage makes disorientation of the spot thus problem may solve with a portable map but not solve in environmental design point. The permanent public art installation gives landmark point on the area to locate the spot. In some area, dark spot was found on the pedestrian way that give visual and psychological barrier for the passengers.

The street materials on both specialized streets are well planned designs that give identity and non-slippery, non-glare, safe access to the passenger. Furthermore, the permanent ramp installment makes competition to the accessibility, but the improper tactile location makes some problems to the users especially for distance connection. Soft material such rubberized tile perform better for pedestrian access and signage but was not provided in urban streets and prefer to hard material like cobble stone, marble, and low grate concrete pavements which cause pedestrian fatigue faster than the soft materials.

5. Conclusion

Based on analysis and comparisons of specialized streets which examined by the evaluation sheets, this study concluded with some points as follows : South Korea cases are good practices in universal design principles application, urgency of implementing the universal design based direction into local regulations so some problems can be solved; and change of urban development paradigm from partial based design into universal design based.

References

1. Islington Council, *Islington Street book A borough-wide streetscape manual, Islington* Council London (2005)
2. Otak Team, *Scottsdale Road Streetscape Design Guidelines, city of Scottsdale*: Arizona (2008)
3. Torbay council , *Torbay Streetscape Design Guidelines, Torbay*: United Kingdom (2004)
4. Putra, Budi A., *Cultural Representation of Vernacular Housing in Jambi*, Doctoral's thesis , Seoul National University, South Korea
5. Abu Dhabi Urban Planning Council , *Abu Dhabi Utility Corridor Design Manual ver 1*, Emirates of Abu Dhabi Urban Planning Council (2017)
6. Landcome, *Street Project guideline, Landcome* : New South Wales, Australia (2008)
7. Burton, Elizabeth; Lynee, Mitchell, *Inclusive Urban Design: Street for Life, Elsevier* Linacre House, Jordan Hill, Oxford (2006)
8. Guide Dogs on Behalf of Action for Blind People, Inclusive Streets: *Design Principles for Blind and Partially Sighted People, Association of Visually Impaired Office Staff*, Deafblind UK (2010) http://www.guidedogs.org.uk/media/1497778/Inclusive_Streets_Design_Principles_booklet_Guide_Dogs_2010.pdf)
9. Shehu, Vjosa, *Universal Design Strategies On Developing Country, Case Study: Tirana*, Master's thesis, Eposka University Tirana, Albania (2011)
10. Guimares, Marcello, *An Assessment Of Understanding Universal Design Through Online Visual Resources and Role-Playing Simulation Exercises,* North Carolina State University (2005)
11. Aslaksen , *Universal design: Planning and Design for All,* Cornel University (1997) accessed at http://digitalcommons.ilr.cornell.edu/gladnetcollect
12. Department of Transport, *Manual for Street, Thomas Telford,* London (2007)
13. Indrakusumo R.H., *Bangun, Design Characteristics of Specialized Streets for Indonesia through Cases of South Korea* Departement of Architectural Engineering, The Graduate School, Pukyong National University (2016)
14. Mehta, Vikas, *Lively Streets: Exploring The Relationship Between Built Environment And Social Behavior,* Doctoral's thesis, University of Maryland, College Park (2006)

Tourism Village Model Based on Local Indigenous: Case Study of Nongkosawit Tourism Village, Gunungpati, Semarang

Kurniasih[1], *Dyah* Maya Nihayah[2], *Syafitri* Amalia Sudibyo[3], and *Fajri* Nur Winda[4]

[1] Student of Economics Development , Faculty of Economy, Universitas Negeri Semarang, Semarang - Indonesia
[2] Department of Economics Development , Faculty of Economy, Universitas Negeri Semarang, Semarang – Indonesia
[3] Student of Economics Development , Faculty of Economy, Universitas Negeri Semarang, Semarang – Indonesia
[4] Student of Economics Development , Faculty of Economy, Universitas Negeri Semarang, Semarang - Indonesia

Abstract. Officially, Nongkosawit Village has become a tourism village since 2012. However, the economic impact has not been received by the society yet because of inappropriate tourism village model. Therefore, this study aims to find out the best model for the development of Nongkosawit Tourism Village. This research used Analytical Hierarchy Process method. The results of this research shows that the model of tourism village which was suitable to the local indigenous of Nongkosawit Tourism Village was the cultural based tourism village with the percentage of 58%. Therefore, it is necessary to do re-orientation from the natural-based village model into the cultural-based village model by raising and exploring the existing culture through unique and different tourism products.

1 Introduction

Tourism is a sector that provides great foreign exchange for many countries. According to the World Travel and Tourism Council (WTTC), Indonesia's travel and tourism sectors have great contributtion to the economy. It was 8.4 percent in 2013. The, in 2014, the GDP generated from the tourism sector reached 394.52 trillion rupiah, and in 2015 it reached 461,36 trillion rupiah [1]. This condition indicates that the tourism sector from year to year always contributes greatly to the economy of Gross Domestic Bruto (GDB).

The tourism development that has been done by both government and private sector can increase the number of tourist arrivals. One of Central Java government strategies in improving the tourism sector is creating a tourism village. The chosen village should have rich natural resources and uniqueness. A tourism village or rural tourism should have good integration aspects such as various attractions, good accommodation and completed facilities. Those aspects should be presented in the structure of society life include rules and traditions applied at the society [2]. The tourism village is a way to apply a sustainable community-based tourism development. By developing the tourism village, it is expected that equal distribution with the concept of sustainable tourism development can occur [3]. The products in Working Farm, Passive Contact Agrotourism enable farmers to increase their income without disrupting the agriculture [4]. Besides getting the benefit from the harvest, the farmers can also earn additional income through agrotourism.

One of the tourism villages in Semarang is Nongkosawit Village which has been established by the Mayor of Semarang. It is supprted by the establishment of the Decree of Semarang Mayor, Number 556/407. Nongkosawit Village is launched as Semarang City Tourism Village which has been valid since December 21, 2012. It is one of the efforts in realizing the program "Ayo Wisata ke Semarang" which is proclaimed by the Semarang Mayor, Hendrar Prihadi. The purpose of having this program is to increase the attractiveness of Semarang tourism, to improve the economy of Semarang City, especially in tourism sector, and to optimize the local wisdom of Nongkosawit Village.

Local wisdom-based tourism is an activity where local people directly involves in providing and controlling the culture as the essence of an attraction [5]. The tourism object which is designed with the local wisdom approach becomes an alternative tourism object [6]. It is to for offering an alternative tourism object to the tourists who feel bored because of the conventional tourism object. The alternative tourism object can decrease unemployment. Thus, the income of people living in the area will increase [7].

As a matter of fact, Nongkosawit as a tourism village has many potential tourism products to be developed. However, since it was officially launched as a tourism village in 2012, the economic impact for the surrounding region has not been increased significantly. Based on some observations done, it is find that only few people directly involve in the management and development of this tourist village. Next, they are still passive in developing the village, so that the coordination has not been optimally conducted.

Based on the data from Bappeda, the number of Nongkosawit visitors is less than 100 visitors per month

* Corresponding author: kurniasihpeto@gmail.com

[8]. In fact, the attractions and available facilities in the Nongkosawit are various. According to Bapak Suwarsono as the Chairman of Nongkosawit Tourism Village Management, several attractions and facilities owned by Nongkosawit Tourism Village can be seen in table 1.

Table 1. The Available Attraction and Facilities in Nongkosawit

The Attraction Offered	The Facilities support the Tourism Village
1. Farms for cows, sheeps and hens	1. *Homestay*
2. Fruits and vegetables plantation	2. Tour cart
3. Traditional games and *gamelan*	3. Wali mosques
4. Historical buildings and spring waters or *sendang*	4. Public transportation
5. Tape and local snacks factories	5. Restaurants
6. Painting batik	6. shops
7. Scarce bird breeding	
8. Kirab Kyai Bendhe	
9. Bamboo craft	

Sources: Processed Primary Data, 2016

Unfortunately, based on Figure 1, the tourism potential is not optimally utilized. Many problems come out such as lack of sign posts, bad management tourism object, bad road, and unclean water spring with many destructions. Moreover, many tourism objects there actually have interesting places if they are managed well, and have historical and educational values.

Fig. 1. The Tourism Objects of Nongkosawit

Sources: personal documentation

Based on the existing problems, the researchers initiate to conduct a research related to the relevant model of tourism village with its potential and local indigenous. It aims at improving the village through the development of tourism. This village model is based on sustainable tourism development. Then, it is expected to be able to support the realization of the Vision of Central Java tourism. The vision is making Central Java as the center of culture and the main tourism destination with good personality and prosperity. The local Indigenous based tourism village is intensified to realize the sustainable development goals in 2030.

2 Method

This research used descriptive-statistical qualitative method to determine the model of indigenous local based tourist village. The data used in this study was the primary data obtained from interviews and questionnaires to some key persons. The selected Key persons in this study were the manager of village tours, the Village and District staffs, and business owners. They were determined by using purposive sampling technique.

The data analysis method used in this research was Analytical Hierarchy Process (AHP). Then, the variable used was arranged in a hierarchy or priority. This analysis was based on pairwise comparison matrix. It was the elements that functioned as the judgment of the decision maker. Further, it was compared to a pair of top row elements with left column level.

To develop a tourism village that has a high selling value and to attract visitors to come, several things such as human resources capabilities as subjects and objects in the context of development of tourist villages should be improved.

The human resources become very important because they have uniqueness, creativity and innovation. Their capabilities are influenced by the quality level of human resources. The higher the qualities level of human resources, then the larger the tourism village growth. The human resources capabilities that can be developed to support the indegenous local based tourist village are their genuine and natural life with its uniqueness and moral values.

Besides having good human resources, facilities and play important rules to support the development of village. The facilities and infrastructure need to be built, so that the Nongkosawit Village local wisdom can be explored. The examples are ancestral relics or old buildings, tombs of *wali* and *sendang* or spring waters. Nongkosawit Tourism Village provides facilities such as rice planting training, pottery painting, gamelan training, fruit picking, and water games are applied with iwak flute.

After improving the human resources capabilities, facilities and infrastructure, another important thing that should be done is marketing. It is to inform the village potentials and attract the tourists to come. The marketing done by Nongkosawit Tourism Village managers is tourism product planning based on local customs such as community meetings conducted by *kelompok sadar wisata*. The management of a tourism village run by *kelompok sadar wisata* is as a participation and mutual cooperation based on the principle of kinship. It has reflected good institutionality.

Based on figure 2, the four main elements used as the criteria in determining the model of indegenous local based tourism village were Human Resource (HR) capability, facilities and infrastructure, marketing and institution. The determination of these criteria was based on the urgency of each criteria explained at the previous discussion. The criteria are then elaborated into sub-criteria elements to explain the criteria more clearly.

To know the consistency of opinions, then the next step was testing the validity of the data by calculating the Consistency Ratio. This calculation was done for having the valid elements. Consistency Ratio (CR) is obtained from the comparison between Consistency Index (CI) and Ratio Index (RI). By implementing *Analytical Hierarchy Process* (AHP), some relevants models were yielded. Then, the most appropriate model was choosen to be implemented at Nongkosawit to increase the society income from agrotourism sector.

3 Results and discussion

Based on the field studies done, there are still many residents in Nongkosawit Village who do not know about the Decree of the Semarang Mayor regarding to the establishment of Nongkosawit Village as a tourism village. This causes the roles of the community in the development of tourism village has not been maximized. Next, inadequate community participation can not create maximum benefits. Several factors that hampered the development of Nongkosawit Tourism Village are human resources, limited funds, less cooperation between citizens, inadequate facilities and infrastructure, and limited information related to the marketing and development of tourist villages.

From the aspect of human resource capability view shown in table 2, most of the villagers are elementary school leavers, so that the people who have insight and skills related to tourism village development are still few.

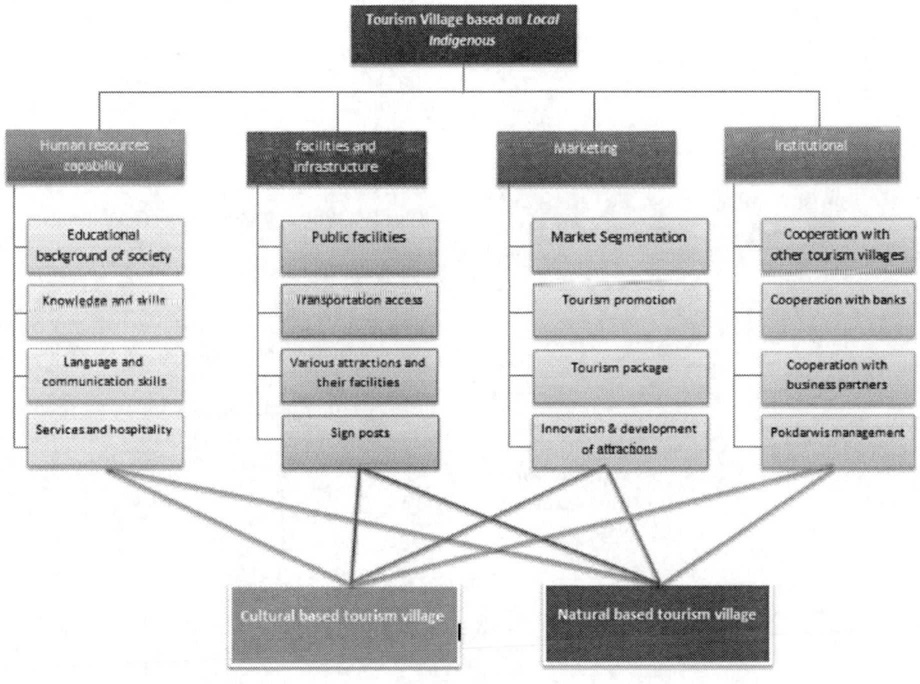

Fig. 2. The Hierarchy of Indigenous Local Based Tourism Village Model

Further, the low capabilities of the human resources there are because of the less workshops related to the tourism village. Unfortunately, the training is for the managers and some related parties only, so that the people who are the subject in the development of tourism village does not have much understanding about the concept of tourism village.

Table 2. The number of Nongkosawit Villagers based on their Educational background in 2016

No.	Graduates	Total
1.	Infants and children under the age of four years	483
2.	Uncompleted Elementary School	1039
3.	Elementary School	1274
4.	Junior High School	943
5.	Senior High School	933
6.	Bachelor	51
7.	Undergraduate	268
8.	Illiterate	0

Sources: Monograph of Kelurahan Nongkosawit, 2016

Table 2, depicts that the majority of Nongkosawit Villagers are only elementary school leavers. In 2016, the monograph data of Nongkosawit recorded as many as 1274 people are elementary school graduates / equal, then as many as 1039 people did not complete elementary school. It can be concluded that the capability of Human Resources of Nongkosawit is low. In fact, the capabilities of Human Resources play great role to develop the tourim village.

Nongkosawit Tourism Village had a variety of unique, and rare attractions which were different from other tourism villages, but it did not have enough facilities and infrastructures . It could be seen from incomplete and broken directions signs. Although there were complete public facilities such as worship places, stalls and kiosks but public transportation did not support the tourism. It was only private vehicles available at Nongkosawit Tourism Village.

There were two attractions at Nongkosawit Tourism Village; they were natural attractions and cultural tourism. The natural tourism was in the form of educational tours such as training of planting rice, planting fruit, nourishing the fruit until harvesting the fruit.

Besides; there were also a milking training conducted at cattle ranching, rare bird tourism, and natural attractions such as waterfalls, and natural scenery. Then; the cultural tourism at Nongkosawit Tourism Village which was known Kirab Kyai Bendhe conducted once a year on the 1st Rajab. There were also some annual cultural tourism events which could be seen by the tourists; such as, shadow puppets, traditional plays, *gamelan* and others.

For developing the tourism village, the management of Nongkosawit Tourism Village which was incorporated in *Kelompok Sadar Wisata* (Pokdarwis) Kandhang Gunung as a group which initiated the village tourism already mapped some attractions into the tour packages. The tour packages were made not only using the attractions at Nongkosawit Tourism Village, but also the attractions at Kandri Tourism Village. The cooperation between Nongkosawit and Kandri Tourism Village managements was a mutual support and complementary efforts between tourism villages. The available tour packages were one day tour, soft tracking, nature exploration, *onthel* bike tour and camping ground.

Then, the management already promoted the tourism via internet; such as creating website and instagram. It made the market segmentation in determining the tourists' target clear and directional. The market target of Nongkosawit Tourism Village were academicians, especially for educational tourism places. However, the development through travel packages and promotions through websites and instagram did not give any significant impact in increasing the number of tourists yet. It was due to the lack of innovation in developing the attractions; there were only simple and modest attractions.

If it was seen from the terms of institution, cooperation was only with other tourism villages. Cooperating with other business partners and banking was limited. The lack of the cooperation made Nongkosawit Tourism Village experience the various obstacles, one of them was financial or funds difficulty. The fund problems which was experienced by Nongkosawit Village Tourism happened because the lack of cooperation with companies or business partners; such as getting CSR to Nongkosawit Village. The ease of banking credit also were needed as a village fund injection to develop the tourism.

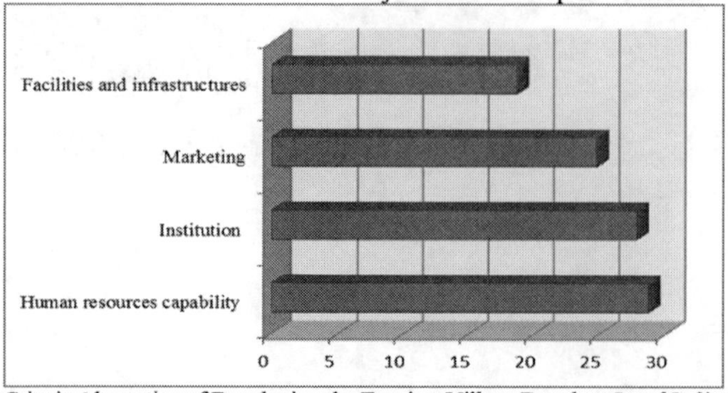

Fig. 3. The Criteria Alternative of Developing the Tourism Village Based on Local Indigenous

The calculation results of Analytical Hierarchy Process (AHP) as described at Figure 3 shows that human resources capability was the main priority to develop the tourism village for 28.7%. It is the most influential factor in rural development because all ideas and innovations come from human and community as the subject of the tourism village in general and the rural tourism management in particular. The higher the capability of human resources, the larger opportunities for the development of tourism villages.

The second priority criterion of developing the tourism village based on local indigenous is institutional. Institutional, in this case, is an effort made by the tourism village management to cooperate with the internal management and other tourism villages, banking and business partners. The cooperation between the tourism village manager and the members could solve the problems which hamper the development of tourism villages. The next two criteria are marketing and infrastructure. Marketing becomes the important criterion to increase the number of tourists. The best innovations and attractions which are provided at the tourism village could not attract the tourists if the marketing is done traditionally and less conceptualized way. Marketing could be done by giving the general information related to tourism villages and offering the alternative tour packages.

The inconsistency average value calculated from the overall value is 0.14. It means that the respondents' answers in formulating the alternative decision is consistent. The respondents' consistency showed that the information was valid and suitable with the circumstances of Nongkosawit Tourism Village.

A more detailed policy alternative was done by looking at the value of each sub-criteria generated from the calculation process. Based on the table 3 below, services and hospitality attitudes of the community for the visitors or tourists were the most important priorities. One of the attractions of the tourism village was the down to the earth, friendly, and warm daily life of the village community. Community became the considered factor since they were the tourism subject and object at the tourism village.

The second sub-criterion which had the high priority value of 11.1% was insight and skill. Insight and skills were the perspective of Nongkosawit Tourism Villagers; they were the village administration, the village tourism management and the general public toward the development of the tourism village. Skills were needed to transform an untapped potential into the unique, exciting, different and rare attractions.

To create the better condition of the tourism village, it could not be separated from the managerial ability of the tourism management (Pokdarwis). In this study, the managerial ability of Pokdarwis was the third rank of the priority scales. The Pokdarwis management was the way of the chairman and its members in planning, managing and evaluating all performances by upholding the cultural values which were existed in the community, such as; *musyarawah mufakat* (discussion) and *gotong royong* (mutual cooperation).

Table 3. The Sub-Criteria Alternatives in Developing the Tourism Village based on *Local Indigenous*

No.	Sub-Criteria	Priority Values
1.	Services and hospitality	11.3 %
2.	Knowledge and skills	11.1 %
3.	Pokdarwis management	11.0 %
4.	Innovation and development of attractions	9.8 %
5.	Language and communication skills	8.6 %
6.	Various attractions and their facilities	7.4 %
7.	Tourism promotion	6.3 %
10.	Tourism package	6.0 %
11.	Sign posts	4.9 %
12.	Transportation Access	4.8 %
13.	Public facilities	4.7 %
14.	Educational background of society	4.3 %
15.	Market Segmentation	3.5 %
16.	Cooperation with business partners	2.4 %
17.	Cooperation with banks	2.2 %
18.	Cooperation with other tourism villages	1.6 %

The stakeholders or the policy makers should make the priority levels and prepared the strategic plan for developing Nongkosawit Tourism Village generated from the valid calculations. The attraction innovation and development should be done to attract the tourists; the attractions should not be boring and monotonous. Some attractions should be considered with the current trend in today's society. It was expected that the attractions could increase the number of tourists.

Some sub-criteria which should be considered were the local residents' language and communication abilities to interact with tourists, the completeness and diversity of attractions offered as the tourists' choices, the tourism promotions to introduce the tourism villages and to attract tourists to visit Nongkosawit Tourism Village. Next, cooperation with business partners, banking and other tourism villages was also the priority to develop the tourism villages. The inconsistent value derived from the calculation of sub-criteria for the development of a tourism village was 0.10; it means that the answers for data processing were consistent.

Nongkosawit tourism village was a tourism village which had a lot of potential attractions, but the number of these attractions did not have any significant influence on the number of visitors. It happened because the less focused or less conceptual model of tourism village which was suitable with the local indigenous of Nongkosawit Village. Some attractions offered actually resembled to the attractions offered by the near tourism village, Kandri Tourism Village. But, if it was well-explored, there were many uniqueness owned by Nongkosawit Tourism Village which was not owned by other tourism villages, such as; annual carnival event which attracted all Nongkosawit Villagers at one time; it was Kirab Kyai Bendhe.

Based on the results of AHP calculations in Figure 4 which was done by processing the data taken from some key persons; it shows that the model of tourism village which was suitable to the local indigenous of Nongkosawit Tourism Village was the cultural based tourism village with the percentage of 58%. The culture referred to in this model was a habit, the work of creation, such as; dance, *wayang*, historical heritage, the habits of the village community which became the daily habit which was different from other tourism villages or other places.

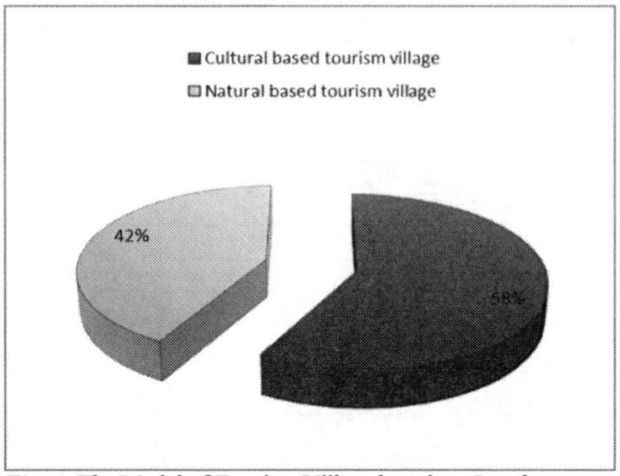

Fig. 4. The Model of Tourism Village based on *Local Indigenous*

One of the cultures which became the icon of Nongkosawit Tourism Village was Kirab Kyai Bendhe on Thursday Wage in Rajab (Islamic Month before Ramadhan). This annual event was a series of events conducted from the early morning until the late afternoon. The first event began with the activity of cleaning the village elder's tomb commonly called *Nyadran*. After doing *Nyadran*, the heirloom of Kyai Bendhe was a relic of the Guardian of Shaykh Hasan Munadi, it was cleaned by using seven flowers and water from the seven springs called *Jamasan*. After the heirloom was cleansed, it was brought to go around the village. Furthermore; there was also the earth alms collected from the community; it consisted of two parts called *Tumpeng Lanang* and *Tumpeng Wedhok*. *Tumpeng Lanang* was created from nutmegs, such as; corn, beans and tubers, whereas; *Tumpeng Wedhok* was created from vegetables and fruits. Then; the march consisted of *Domasan, Kuda Lumping, Rebana* and others.

Moreover; there were also other cultures at Nongkosawit Tourism Village, they were *Kuda Lumping*, Traditional Plays, Masjid Wali, Bedug, the Relic of Kyai Shaykh Hasan Munadi, *Sendang, Karawitan, Seni Lukis, Campursari Drama*, and others.

4 Conclusions

Nongkosawit tourism village was a tourism village which had many different attractions which were potential to be developed, but the number of these attractions did not give any influence on the number of visitors. It happened because the less conceptualized model of tourism village based on the local indigenous Nongkosawit Village. Some key elements or criteria which had the important influence to the development of tourism villages were the human resources capabilities, institutions, facilities and infrastructure, and marketing.

These four key elements were the indicator to assess the tourism condition at Nongkosawit village. Indicators at each tourism village were different because each village had its certain characteristics. Each indicator has sub-criteria which were tailored to the potential of the existing area. Nongkosawit village itself had characteristics and uniqueness in the field of culture and tradition, so the indicators and sub-criteria were made in accordance with the potential related to the social and historic society, so it would produce a research with a coherent historical storyline.

The study results showed that the chosen alternative model to develop Nongkosawit Tourism Village in accordance with the local indigenous was a cultural-based village model. It was supported by the fact that the various culture owned by Nongkosawit Tourism Village had an opportunity to attract tourists. One of the annual events which became the icon of tourism at Nongkosawit was Kirab Kyai Bendhe which was done on Thursday Wage in Rajab month. Besides, there were also *Kuda Lumping, Traditional Dolanan, Masjid Wali, Bedhug, The Relic Of Kyai Shaykh Hasan Munadi, Sendang, Karawitan, Painting, Campursari* and others.

Therefore; it is necessary to do re-orientation from the natural-based village model into the cultural-based village model by raising and exploring the existing culture through unique and different tourism products.

References

1. World Travel and Tourism Council (WTTC). 2014. Economic Impact Report. http://www.wttc.org
2. Chusmeru, Analisis Pariwisata. Vol. **10**: 1, hlm. 16-23. Denpasar: Fakultas Pariwisata Universitas Udayana, (2010)
3. Susyani, Dewi Winarni, Jurnal Ekonomi dan Bisnis. Vol. **12**: 1, hlm. 33-36, (2013)
4. Phillip, Sharon, Colin Hunter and Kirsty Blackstock, Tourism Management Journal. DOI:10.1016/j.tourman.2009.08.001, (2010)
5. Zeppel, Heather, *Indigenous Ecotourism: Sustainable Development and Management. Australia*: Cromwell Press, (2005)
6. N. M. Ariani, and I. W. Suardana, *Seminar Penelitian*. Program Studi Pariwisata. Universitas Udayana. Bali, (2009)
7. Kurleto, Malgorzata, Slovak Scientific Journal. Vol. **3** No. 2, page. 51-57, (2014)
8. Badan Perencana Pembangunan Daerah (Bappeda) Kota Semarang dan Badan Pusat Statistik Kota Semarang. 2012. *Kecamatan Gunungpati dalam Angka 2011*. Semarang: Bappeda Kota Semaran (2012)

Towards Water Sensitive City: Lesson Learned From Bogor Flood Hazard in 2017

Muhammad Ramdhan[1,*], *Hadi* Susilo Arifin[2], *Yuli* Suharnoto[3], *Suria* Darma Tarigan[4]

1Doctoral Program of Environmental Management, School of Postgraduate Studies, Bogor Agricultural University, Bogor – Indonesia
[2] Department of Landscape Architecture, Bogor Agricultural University, Bogor – Indonesia
[3] Department of Civil and Environmental, Bogor Agricultural University, Bogor – Indonesia
[4] Department of Soil And Land Resources, Bogor Agricultural University, Bogor – Indonesia

Abstract. Bogor known as rain city and it's located at an altitude range of 190-330 meters above sea level. In February 2017 Bogor experienced a series of natural disasters related to heavy rainfall that fell during that time. The hazard in the form of flash floods that cause casualties was shocked, due to the location of Bogor city that located in the foothills with a fairly steep slope. There is a problem with the drainage system in the city of Bogor. Australia Indonesia Center in cooperation with Bogor city government held a focus group discussion to seek a permanent solution for the problems and so that similar incidents do not occur in the future.

1 Introduction

Bogor city is one of Indonesia's important and historical cities. Geographically Bogor City lies in the coordinate range of 106° 43 '59.2 "E - 106° 50' 54.8" E; 6° 30 '38.6 "S - 6° 40' 48.5" S, with an altitude range of 190-330 meters above sea level. It is important, because in Bogor city lay a presidential palace that is often used for visits of state guests. Historic because it has stood since the 15th century as the center of the Pajajaran Kingdom and became a city of rest for the officials in a Colonial era [1]. Bogor in Sundanese language means palm tree root [2], it illustrates that in the past the landscape of Bogor City is naturally overgrown with palm trees. The climate condition of the city of Bogor that known well as the rain city, with the topography of steep terrain, is perfect for the growth of water absorbing trees such as palm trees [3,4]. Palm trees will store water in their stems in the wet season and will release them through roots in the dry season, a natural mechanism that benefits the sustainability of water resources.

Development of Bogor City requires a change of land cover. Natural vegetation land is increasingly reduced in number, converted to built up area [5,6] as shown on figure 1. The coefficient of rainwater runoff on built land is greater than open or vegetated so the ability of the land to absorb water becomes lower [7]. This causes the city of Bogor experiencing problems water flooding in the rainy season, and the difficulty of clean water in the dry season.

In February 2017 there has been an event that shocked the citizen of Bogor. Flash floods occurred in Bogor, with two deaths [8]. This raised a question for the academics, what is the cause of that disaster. This paper will try to provide an overview of the causes of flooding, then looking for a solution for handling it from the perspective of urban water management.

This paper will explores the problem that happen in Bogor City and its surrounding. Regarding the flood event that happened on February 2017.

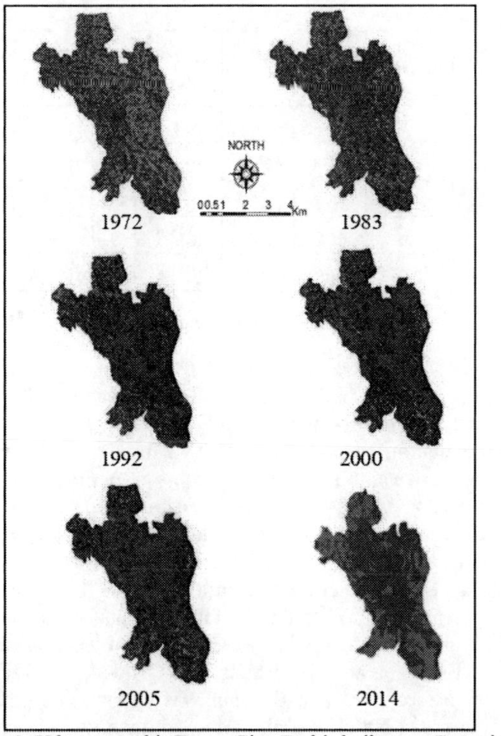

Fig. 1. Urban sprawl in Bogor City; Red is built area, Green is vegetation or open space.

* Corresponding author: ramdhan_98@apps.ipb.ac.id

https://doi.org/10.1051/e3sconf/20183109012

2 Methods

Urban flooding is a complex system, with a rainfall as one of the component on it [9]. We try to find a rainfall event on the date that the flooding event occurred in Bogor city. Then we will examine about the infrastructure for the drainage system in Bogor city. After that we search the literature on the Google scholar about the state the art of urban water management for looking a solution that can be implemented to reduce a flooding event in Bogor city.

The rainfall data recorded by Agency of Meteorological Climatology and Geophysics (BMKG) in Darmaga weather station. Its provide a daily records of rainfall event in Bogor city. We also use the Tropical Rainfall Measuring Mission (TRMM) data that recorded by JAXA-NASA using satellite imagery.

The flood was occurred on February 27th, 2017. at 15.00 West Part Indonesian time zone. It happened after more than 2 hour rain event at Bogor area.

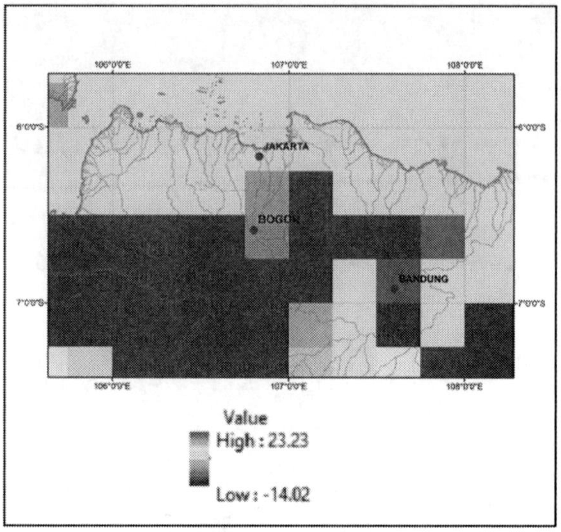

Fig. 2. TRMM data for Bogor region at February 27th, 2017.

3 Result

3.1 Rainfall event

From the BMKG weather station in Darmaga-Bogor, we find the hourly data that shows maximum rainfall that occurred in Bogor with in a period of every 2, 5, 10, 20 and 50 years. As shown by Table 1. Bogor city in every 2 year have a maximum rainfall of 97,6 mm with the 2 hours duration of the rain .

Table 1. Bogor rainfall event

Re-period (year)	1-hrs	2-hrs	3-hrs	6-hrs	12-hrs
2	80.6	97.6	108	115	118.8
5	89.4	115	129	139	141.1
10	95.2	126	142	154	155.9
20	101	137	155	169	170
50	108	151	172	188	188.3

Source : BMKG Darmaga-Bogor

It was confirmed by TRMM data that on February 27th, 2017 in southern part of Bogor region heavy rain was happened for more than 3 hours. It shown by the precipitation number value of 23,3 mm/hour. It is not a maximum value as shown by 2-hrs re-period in 2 year for Bogor area as reported by BMKG. But still it will send very huge amount of water to Bogor city. Due to the area that the rain was very large as shown on figure 2. In a simple equation we could calculate the amount of water per Km^2. With the rainfall of 1 mm, it will give a volume of $1 \text{ mm} \times 1 \text{ km}^2 = 10^{-3} \times 10^6 \text{ m}^3 = 1000 \text{ m}^3$ of water. Bogor area is 118.50 Km^2. And in the TRMM data we can see that the rain was happen not only in Bogor area but almost all part of southern west java had the rain event at that time.

3.2 Irrigation system versus drainage system

Bogor was an agricultural area which has many paddy fields and the centre of rubber plant in Indonesia. In the Dutch colonial era, the government built an irrigation system to distribute the water to the planting area. Such as in figure 3, The water gate was built at 1872 BC in Empang-Bogor to irrigate more than 4000 Ha paddy fields. But right now the paddy fields has been converted to settlement area.

The irrigation system is used by the developer of the settlement area as a drainage system. There is a different purpose of irrigation system and drainage system. Irrigation purpose is to distribute water, so it has structure from large canals to small canal. In drainage system, the purpose is to throw the water to the larger canals so it will prevent the settlement to get flooded.

Fig. 3. Water gate for irrigation system in Empang, Bogor

3.3 State the art of urban water management

To find a state of art in the urban water management, we conduct a literature review by using Google Scholar search engine search facility. Google Scholar provides an easy way to search the broad academic literature [10]. Search results can be: peer-reviewed papers, theses, books, abstracts, and articles, from academic publishers, professional communities, universities, and other academic organizations.

The analysis results on the distribution of literature show that there are 675 scientific publications on "urban water" in the last 20 years (1997 - 2017). The paper has been cited as 6403 times, with 337 citation / year. As for the h-Index of the search topic is 35, this means there are 35 papers that minimize at least 35 times.

There is an increase in the number of literature from the first five years (1997-2002) of 58 papers to 181 papers in the second five years (212%). This shows that in the second period (2003-2007) many new innovations in the topic "urban water". For the next period it appears to be stabilizing with an average increase of 21 papers per fifth year (Figure 4.).

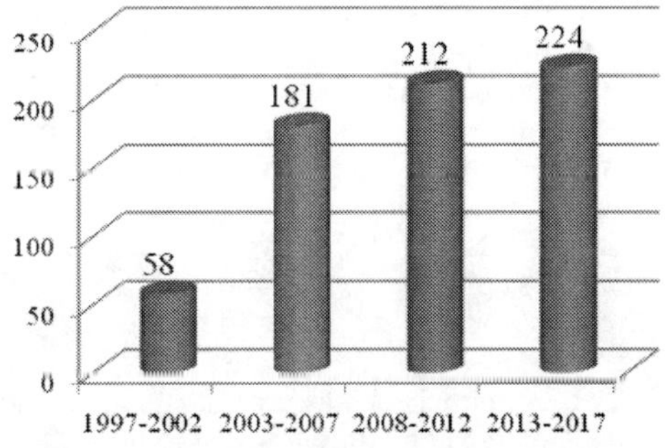

Fig. 4. Number of published literature with the topic of urban water management according to Google scholar

From the 675 publications there are literature that explicitly indicate the location of the study. Australia became the most discussed country as research locus which is found in 90 literatures. While Indonesia is only mentioned in 2 literatures. This shows that research on urban water still needs to be done in Indonesia. And we need to learn from other countries, especially Australia who has discussed many things related to urban water and water sensitive city.

To view more specifically on the themes discussed in urban water research, search results literatures are divided into 10 thematic themes: Management, Engineering, Modeling, Waste, Social, Hydrology, Mitigation, Economics, Climate Change and Conservation. The paper on management topics occupies the most portion of 197 literatures. Followed by a 183 Technique literatures and which is the most rarely discussed topic is the issue of conservation (5 literatures).

3.4 Water sensitive city

Water sensitive cities concept arise from the mainstreaming of sustainable development issues. In the early development of a city, population growth will trigger an increase in the need for water, so that the management is to meet the needs of water for its residents. This stage is called "water supply city". The next stage, once the population is met the need for water, it appears the demand for water that is available and that have been used does not have an impact on their health. Then came the "Sewered City" with the management response of providing separate sewerage channels with clean water channels. Furthermore, population growth and development continues to grow causing a city vulnerable to flood disasters. "Drained city" will provide management response for good drainage channels and flood defence.

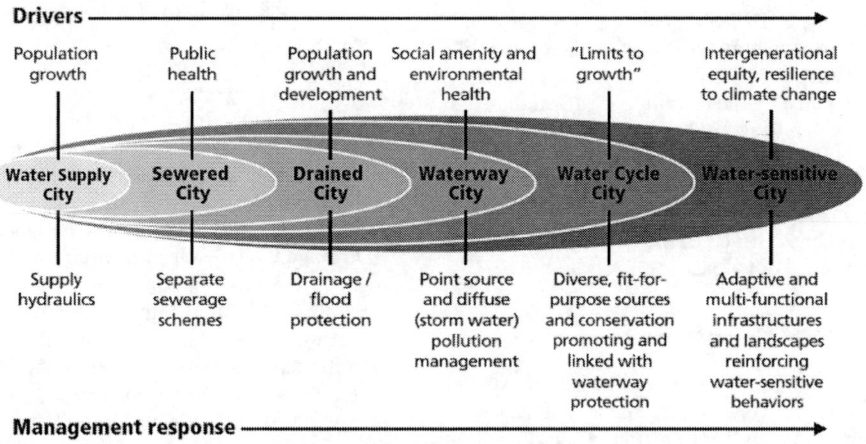

Fig. 5. Urban water management stage [11]

People who already have a good standard of living are increasingly demanding for the comfort and health of the environment, "waterway city" comes with response to waste handling and the use of more water resources such as the utilization of rainwater runoff. Furthermore, urban communities will have an awareness that the city has a capacity limit and environmental carrying capacity, forming a "watercycle city" that handles urban water resources with new regulations on the use of suitable for-source resources, for example the use of waste water Rinse for flush toilets, and also promote the conservation of protection against the water cycle. Finally, water sensitive city, driven by the spirit of sustainable development that demands intergenerational equality in water resources utilization, is also encouraged by the desire to better adapt to climate change. Water sensitive cities implement adaptive water resource management policies, provision of multifunctional infrastructure and urban landscape arrangement that takes into account the cycles and properties of water (Figure 5).

Centre for Water Sensitive Cities defines the water-sensitive city as "a flexible philosophical foundation in the provision and use of water resources to meet all user needs related to the collection and movement of water, may be a technology to facilitate the physical movement of water by a design that recognizes and appreciates the visual appearance of water ".

4 Discussions

4.1 What really happened during the hazard

High rainfall level in the southern region of Bogor city triggers the occurrence of flash flood. This can be seen from the results of TRMM data analysis that provides an overview of precipitation condition when incident is happened, February 27, 2017. Although the rainfall that occurs is not the maximum value of the annual cycle as it is recorded from BMKG station. However, high rainfall in the upstream area will increase the probability of flash flood events [12].

It is getting worse by the condition of urban drainage infrastructure in the Bogor city. Because it using the former irrigation channels as drainage channels. Interesting facts found in the field that on irrigation channels that flow into the location of flash flood, found many channels that have been closed and switch functions to become built area.

4.2 Who should be responsible?

Government as the first actor who make the spatial planning and flood risk management policies and practices (including their coordination capacity) got to take a responsibility to protect it territory from such a disaster [13]. Academia as the intellectual actor should be involved to give the best solution according the latest science and technology. The research about urban water in Bogor city must be prioritize.

Businessman or private sector, as the one who get economical profit for their business in Bogor city. They should care to the water issue. Because they are the actor who spend most of the water supply in many place. Community or the people who live in Bogor area should be involve also, they are the actor and a victim of the unwell planed urban water system in Bogor city. Community action will be more effectively give a direct impact in the field, the media, they are the one whose responsible to give a right information to the people. They can be a leading actor to make an issues become interesting and mainstream for the daily live of the people.

4.3 Towards water sensitive city in Bogor

Bogor City need to be change and do a transformation in management the water in its territory. To overcome the same case will be happened again in the future. The success story of Melbourne city [14] can be implemented here. The outcome is not from a blueprint plan and a top-down management process, but of a strategic steering process of transition navigation over a long period of time. It seems that as long as there is a small group of normatively aligned, loosely connected lead actors from a range of sectors across the niche and regime levels, who are committed to long-term change and are supported by strategic, responsive bridging organizations, which collectively embrace a collaborative, learning-by-doing approach, then the shape, direction and speed of a transition may be influenced.

The role of frontrunners was critical in encouraging and guiding the transition through disrupting old and creating new institutional routines, it was the dynamic interplay between the frontrunners and the creation (and modification) of institutional structuring features and processes that has provided the vehicle for shaping the transition.

Overall, the actor-dynamics in Melbourne suggest that in order to enable and influence a sustainability transition, key actors need to be involved in designing, generating and connecting (shadow) actor networks and relevant bridging organizations, which reflect the stage and sophistication of niche development, while collectively promoting a collaborative learning environment. To exert the necessary influence on the regime, bridging organizations should first focus on generating understanding, collating evidence and nurturing relationships, then build confidence amongst practitioners, up-skill and train a broad range of actors within the sector, then focus pressure towards generating policy shifts. Based on the polycentric institutional design and the patterns of institutional work used by key individual and organizational actors, and the emergence and purpose of bridging organizations, policy makers and urban water strategists may be able to identify critical leadership deficits and possible disrupting and creating strategies so they can target their work to shape a more enabling platform for change. Ultimately, no one component will be responsible for influencing the pathway of a transition.

5 Conclusion

The flooding hazard that happen in Bogor city at February 2017 was a combination of flash floods that come from intensive rainfall on the southern part of Bogor, and also the poorness of local urban water management condition due to loss of water retention area /water parking lot and the unfit function of using irrigation system as a drainage system.

Bogor city must adjust the existing channel condition from the original as irrigation infrastructure become the current requirement for Bogor city become urban drainage system. Urban drainage channels need to be built according to existing urban infrastructure conditions to accommodate an abundance of rainwater that exceeds river flow capacity, which will address flood disasters as a whole, integrated, and environmentally friendly. This adjustment could provides an opportunity for the availability of raw water sources for urban clean water from previously provided for agricultural irrigation of 4000 liters /s.

To achieve a water sensitive city level. Bogor city must have a long term planning to get all stake holder sit together and collectively embrace a collaborative, learning-by-doing approach. Bidging organizations should first focus on generating understanding, collating evidence and nurturing relationships, then build confidence amongst practitioners, up-skill and train a broad range of actors within the sector, then focus pressure towards generating policy shifts.

6 Acknowledgement

The writer want to thank you to *Australia Indonesia Centre* who give an opportunity for Bogor city as one of case study for the development of water sensitive city in south east Asia. Also for the rector of Bogor Agriculture University who give all facility during study literature and discuss. For the ministry of marine and fisheries affairs who giving a scholarship program.

References

1. Jatmiko, P. *Sejarah Kota Bogor.* Karta Nagari. Bogor : Karya Cipta Anak Negeri. (2015)
2. S. Danasasmita, *Sejarah Bogor - Volume **I**.* Bogor: Pemerintah Daerah Kotamadya DT II Bogor. (1983)
3. S. N. Ichwani, *Skripsi. Departemen Konservasi Sumberdaya Hutan Dan Ekowisata. Fakultas Kehutanan.* Institut Pertanian Bogor. Bogor. (2013)
4. D. Saputra, *Diploma thesis*, Universitas Andalas. Padang. (2013)
5. Arkham, H. S. Arifin, R. L. Kaswanto, N. H. S. Arifin, *Prosiding Lokakarya Nasional dan Seminar Forum Komunikasi Perguruan Tinggi Pertanian Indonesia.* Bogor: Institut Pertanian Bogor. (2013)
6. E. F. Irianti, *Skripsi. Program Studi Arsitektur Lanskap. Fakultas Pertanian.* Institut Pertanian Bogor. Bogor. (2008)
7. K. W. F. Howard, R. G. Israfilov. Canada: Kluwer Academic Publishers. Springer-Science Business Media. (2002)
8. http://megapolitan.kompas.com/read/2017/03/01/20583781/banjir.di.bogor.terjadi.di.kawasan.rentan. accesed on June 12th 2017.
9. Chang L. Huang S. Landscape and Urban Planning J. E **143**, 13 (2015)
10. N. K. Susrini, *Google: Mesin Pencari yang Ditakuti Raksasa Microsoft*, Book. Mizan, B-first, 264 page. (2009)
11. T. H. F. Wong, R. R. Brown, Water Science & Technology. IWA Publishing. Melbourne-VIC. (2009)
12. D. Audrey, R. Hélène, G. Pierre-André, L. Kévin, L. David, D. Denis, Journal of Hydrology **541** (2016)
13. F. Maria, D. Marcin, T. Yuting, C. Faith, S. Dominic, Progress in Planning **114** (2017)
14. R.R. Brown et al., Global Environmental Change **23** (2013)

https://doi.org/10.1051/e3sconf/20183109013

Implementing Cognitive Intervention to Educate and Improve Resident's Preparedness in Landslide Areas

Novie Susanto[1,*], *Thomas Triadi* Putranto[2], *Heru* Prastawa[1] and *Ema* Amalia Ulfa[1]

[1]Industrial Engineering Department, Faculty of Engineering, Diponegoro University, Semarang – Indonesia
[2]Geological Engineering Department, Faculty of Engineering, Diponegoro University, Semarang - Indonesia

Abstract. Semarang city has the potential landslides were quite high in almost all regions. The previous research shows that the Manyaran and Kembang Arum is classified as "not ready" with vulnerability-prone of landslide areas. Therefore, design and implementation of cognitive interventions in human are needed to educate and improve the preparedness of the residents against landslide. This study aims to implement the various cognitive interventions to the residents in landslide areas and analysis of the different interventions toward the preparedness index. The study is conducted on 40 respondents from Kembang Arum and 40 respondents from Manyaran. They are aged $\geq$ 17 years, illiteracy and a RT/RW/PKK cadres. The independent variables in this study are Knowledge and Attitude, Emergency Planning, Warning System, and Resources Mobilization. The dependent variable is the preparedness index. The design of cognitive interventions is generated according to the demographic characteristics of the respondent and the result of Fault Tree Analysis. The preparedness index of the residents against landslides in Kembang Arum increases about 71.71% and in Manyaran up to 90.06%. Implementation of cognitive interventions with module, video and discussion in the Manyaran is more effective than using posters, videos and discussions in Kembang Arum.

1 Introduction

The city of Semarang has high landslide potential in almost all areas [1]. Recently, some studies were focused only on technical aspects of disaster management such as logistic system [2-5] and geological aspect [6-7]. Public awareness of landslide disaster needs to be improved through community behaviour towards disaster prevention, understanding of equipment factors and the formation of community commitment to security plans made by Semarang City Government together with the community. Policies and programs undertaken by Semarang City Government are still focused on government institutions as those who serve emergency and post-disaster response even though the general population, especially the people in disaster prone areas, needs a proper education about the condition of the region. Based on data of geoelectric measurement result in 19 sites of soil prone area in Semarang City conducted by Putranto et al. [1], it was found that threesitesare very vulnerable, eightsitesare vulnerable, sixsitesare quite vulnerable, and two sites are slightly vulnerable. Manyaran and Kembang Arum villages are vulnerable. Based on previous research [8], analysis of residents preparedness to the landslide disaster in Semarang City based on the calculation of the index of preparedness in 19 areas of Semarang City categorisesManyaran and Kembang Arum Subdistrict in "not ready" level.

Based on the data of Manyaran and Kembang Arum sub-districts in 2016, the occurrences of landslide disaster in Manyaran and Kembang Arum sub-districts is quite high. The high potential of these hazards leads to the necessity of community empowerment to improve the success of disaster management and prevent the occurrence of material losses and casualties in the event of landslide disaster. Analysis of the causes of low residents preparedness index has been undertaken as an effort to empower communities by using the Fault Tree Analysis (FTA) approach [9-10].A cognitive intervention is a form of psychological intervention, technique and therapy in counselling exercises [11]. Acognitive intervention in disaster management has been considered as the key mental aspect of successful disaster management as explained in previous research. As the next stage of the research, this study is focusing on the implementation of cognitive intervention design as proposed by Susanto et al. [9] in two sub-districts (Manyaran and Kembang Arum) to improve the individual preparedness index against landslide disaster. This study develops in detail, the phase of cognitive intervention implementation as well as analyses each phase to obtain a valid and reliable design implementation to be adapted by each research area.

* Corresponding author: novie.susanto@ft.undip.ac.id

2 Literature Reviews

Disaster management is an organizational and resource management and responsibility for addressing all aspects of humanitarian emergencies, special preparedness, response and recovery to mitigate disaster impacts [12]. Disaster management aims to reduce or avoid potential harm, ensure prompt and appropriate assistance to disaster victims and achieve rapid and effective recovery. In cycles, disaster management figures out the ongoing process planned by governments, businesses and civil society plans to mitigate impact, react immediately in the event of a disaster as well as disaster recovery [13]. Preparedness of the resident is a part from the disaster management that should be considered.

Regarding LIPI-UNESCO/ISDR policy [14], preparedness index was categorised into five parameters: knowledge and attitude, emergency planning, warning system, resource mobilization capacity and policy statement. The critical factors can be explained as follows:

1. Knowledge and Attitude) is a basic knowledge and attitude about disasters such as disaster types and factors, disasters, and procedures, locations and disaster evacuation routes. There are three variables in this parameter: natural and disaster events (type, source, quantity, and location), understanding of environmental vulnerability, and attitude during a landslide disaster.
2. Emergency Planning is the plan/action required to handle emergency situations in the event of disaster preparedness such as map making, temporary shelter, information hotline number, post, training/simulation, risk analysis and contingency planning. It includes disaster preparedness, evacuation plan and identification of emergencies.
3. Warning System is a series of systems to inform the occurrence of natural events, can be a disaster or other natural signs. In this case, the system relates to information systems, information delivery, and development of early warning systems, training and simulation.
4. Resource Mobilization Capacity is a training program (resources), community action, funds, social capital.

Measurements were made using a Likert scale questionnaire which is then weighted for each parameter with the weight of each parameter of 25%. After weighting, calculation of Preparedness Index (PI) can be taken into account by multiplying total real score of parameter with 10 and dividing the result by the maximum score of parameter.

$$PI = \frac{Total\ Riil\ Score\ of\ Parameter}{Maximum\,Score\ of\ Parameter} x10 \quad (1)$$

From the calculation of the index, it will be obtained awareness index which then can be categorised based on the detail in Table 1.

Table1.Category of awareness index [11]

Nr.	Index score	Category
1	80-100	Very ready
2	65-79	Ready
3	55-64	Almost ready
4	40-54	Less ready
5	0-39	Not ready

The determination of cognitive intervention was varied according to the characteristic of the residents in the each of district that obtained from the result of the survey [9].

3 Methods

This research was conducted in Manyaran and Kembang Arum district of Semarang city. The sample in this study were 40 people for each area aged ≥ 17 years, illiteracy and the board of neighbourhood (RT) or Hamlet (RW), and Family Welfare Programmed (PKK) cadres thereby contributing to disseminate information to the other resident. The sampling technique in this research is purposive sampling or judgmental sampling.

This study has four independent variables; they were Knowledge and Attitude (KA), Emergency Planning (EP), Warning System (WS), and Resources Mobilization (MR) from LIPI-UNESCO/ISDR [14]. The dependent variable in this study is the index of awareness of residents to landslides. Detail of variables used in the questionnaire of landslide preparedness index can be seen in Table 2.

The cognitive intervention implemented in this study was designed using the result of previous study [9]. Manyaran district was identifiedas an area with insufficient knowledge and information of landslide occurrences and by doing so, module, video and discussion were utilised to improve the awareness index. Kembang Arum has different characteristic of residents. It described as a prone area to landslides disaster and the area is often used as a place of research on landslide disaster. Therefore, the comprehension level of the residents regarding the landslide was highly expected. However, based on the previous result, it is found that the preparedness index of the residents is in the low level [8]. It requires different model of cognitive intervention such as poster rather than a module. Besides, video and discussion were utilised to strengthen the residents comprehension as well. The model can be seen in Figure 1.

The analysis of cognitive intervention is conducted by considering the pretest and posttest results so that the most effective cognitive intervention can be seen based on the difference in the increase of residents' preparedness.

https://doi.org/10.1051/e3sconf/20183109013

Table2. Variables of preparedness index in landslide occurrences

PARAMETER	VARIABLE	INDICATOR
Knowledge and attitudes	Natural events and disasters (type, source, quantity, location)	Explain the condition of disaster prone areas
		Understanding of the disaster
		The extent of the disaster
		Explaining the impact
		Explaining the scale of the disaster
		Explaining the signs of the disaster
		Explaining the source of the landslide
	Physical vulnerability	Explains the vulnerability of communities to natural disasters
	Attitude when the landslide occurred	Describes the self-rescue measures in time of disaster
		Describes rescue action during a disaster
Plan for emergencies	Disaster preparedness	Preparedness against disasters
		Preparation for disaster occurrences
	Evacuation plans	The availability of escape destination
		Availability of hazard maps
		Availability of evacuation maps, route/lane
		Availability of signs alarm
		Understanding of self-preservation procedures
	Identification of emergencies	Weather information
		Availability of basic observation facilities disasters
Disaster warning system	Disaster warning systems that traditionally evolved/ heredity applied and/or local agreements	Warning system agreed
		the level of understanding
		the effectiveness of the warning system
Resource Mobilization	Resource Mobilization	Understanding resource mobilization
		Resource mobilization activities
		the effectiveness of conducted resource mobilization done

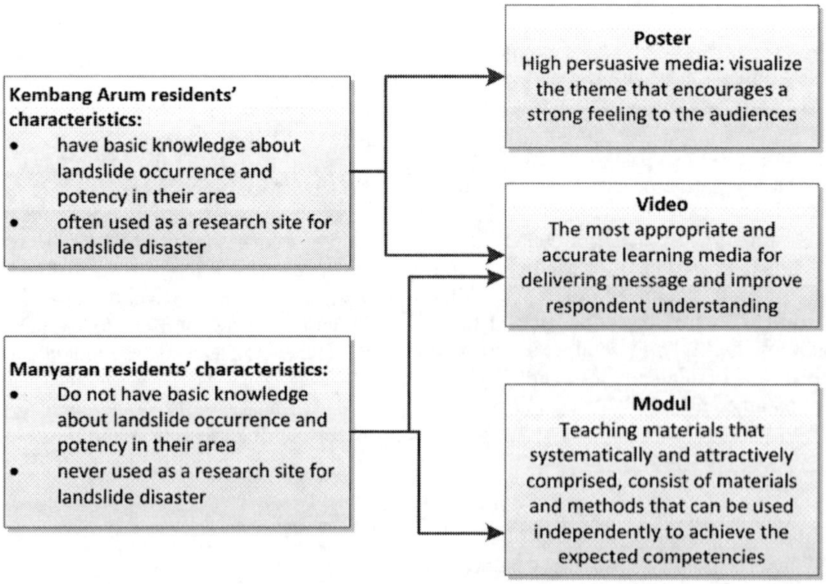

Fig 1. Design of cognitive intervention

https://doi.org/10.1051/e3sconf/20183109013

4 Result and Discussion

Respondents in Kembang Arum sub-district are 40 people who live in RW IX, RW X, and RW XI. Selection of RW and RT was based on Arum Kembang Arum 2016 data and recommendation from Head of Kembang Arum sub-district. District Disaster Awareness (KSB) formed by BPBD Semarang City in Kembang Arum sub-district is less well run and less active, so the KSB should not be able to disseminate information about landslide disaster that exists in its territory to other residents. Lack of resource mobilization undertaken and activities aimed at improving landslide disaster preparedness are rare. In addition, Kembang Arum is often used as a research site for landslide disaster, so residents already know that the area is prone to landslide disaster. Respondents in Manyaransub-district are 40 people who live in RW V, RW VI, RW IX. Selection of RW and RT was based on Manyaran 2016 data and recommendation from Head of Manyaransub-district. District Disaster Awareness (KSB) formed by BPBD Semarang City in Manyaransub-district is less well run and less active, so the KSB should not be able to disseminate information about landslide disaster that exists in its territory to other citizens. The residents in Manyaran have a low understanding and comprehension regarding the condition of the prone area.

The pretest stage was carried out to find out the index of preparedness of Kembang Arum and Manyaran before implementing the cognitive intervention. The average pretest result regarding landslide disaster in Kembang Arum is 36.20 which is categorized as "Not Ready". The average pretest result of residents preparedness index against the landslide disaster in Manyaran is 32.20 which is categorized as "Not Ready". The detail can be seen in Figures 2 and 3.

The result of qualitative analysis from FTA showed that the causes of low score of plan for emergencies in Manyaran district were of these events:

- Lack of the preparedness against landslides
- The absence of a landslide hazard map
- The absence of agreement about the escape destination
- The absence of the signs of landslides prone pins
- The absence of evacuation route
- The absence of observation facilities for landslides
- Lack of awareness from the district official about the landslide
- Less active of District Disaster Awareness (KSB).

Undevelopment Event consisting of the absence of landslide disaster observation facility and less active District Disaster Awareness (KSB). Designing the cognitive intervention is conducted by emphasizing the Basic Event because Undevelopment Event is a certain failure event that can not be found its cause and the incident is not quite related to the research undertaken. The result of qualitative analysis from FTA that explain the causes of low score of plan for emergencies in Manyaran was studied by Susanto et al. [9].

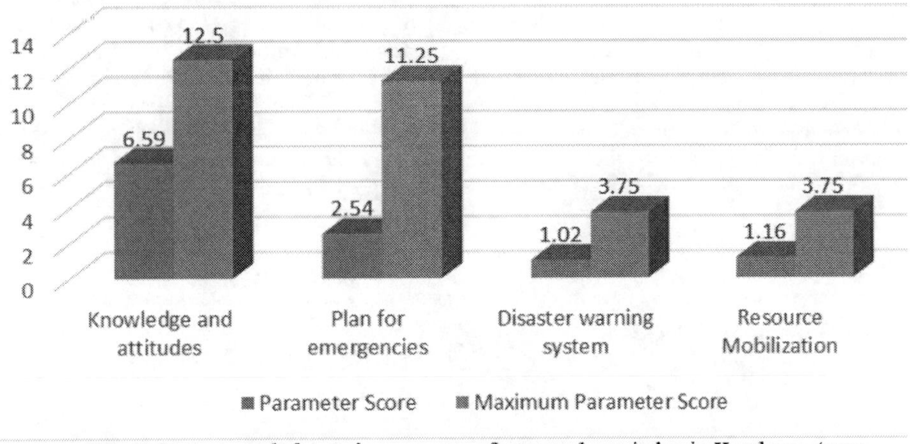

Fig. 2. Pretest result for each parameter of preparedness index in Kembang Arum

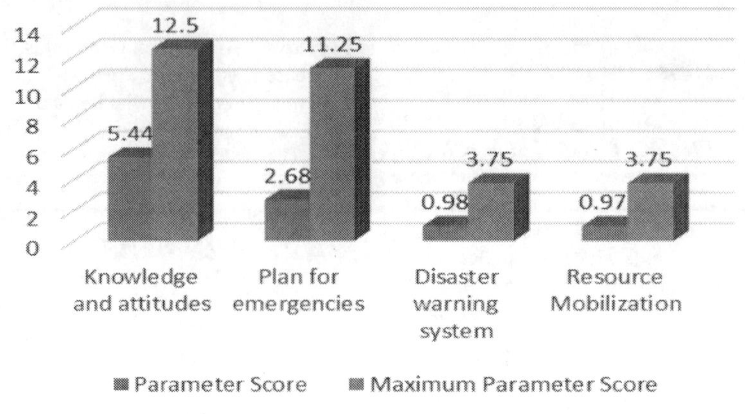

Fig. 3. Pretest result for each parameter of preparedness index in Manyaran

The poster used for cognitive intervention consists of three posters of different materials. The material was decided based on Fault Tree Analysis analysis of the low plan for emergency in Kembang Arum. The posterwas expected can provide more understanding to the residents of Kembang Arum than the module, because the residents of Kembang Arum have a basic knowledge about landslide disaster from the counselling held from BPBD and research conducted in the sub-district. The poster provided includes four parameters of the preparedness index but more emphasized on the plan parameters for emergencies.

The module provided includes basic information covering the material that became the Basic Event of Fault Tree Analysis and the index parameters of landslide disaster preparedness. It also includes short comics about the importance of residents' awareness of the landslide. Manyaransub-district has residents with the lack understanding of the prone area. It related to the lack of information about landslide from sub-district government and KSB. The module is applied to Manyaran with the aim to obtain basic information about landslide disaster and optimize the delivered information.

The selection of cognitive intervention media using video is based on are commendation from Regional Disaster Management Agency (BPBD) Semarang City which suggest making a video about landslide disaster as media of socialization and delivery of information about landslide disaster to the residents in the prone area. The discussion that was held on cognitive intervention to the residents in Kembang Arum was started by giving discussion material with slide show media from apower point. The use of slide show media aims to provide more information and maximize the implementation of the cognitive intervention.

Posttest is conducted by using the same questionnaire that is a Likert scale questionnaire. Manyaran and Kembang Arum were given different treatment according to the characteristic of the residents. The result of posttest average in the form of preparedness index against landslide disaster in Kembang Arum is 62.16 which is included in the category of "Almost Ready". The detail can be found in Figure 4. The posttest result in the form of citizens' preparedness index against the landslide disaster in Manyaran is equal to 61.20 which is included in the category "Almost Ready". The detail can be found in Figure 5.

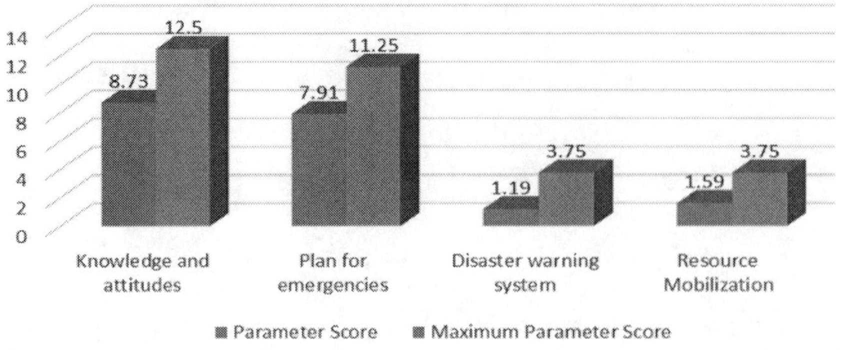

Fig. 4. Posttest result for each parameter of preparedness index in Kembang Arum

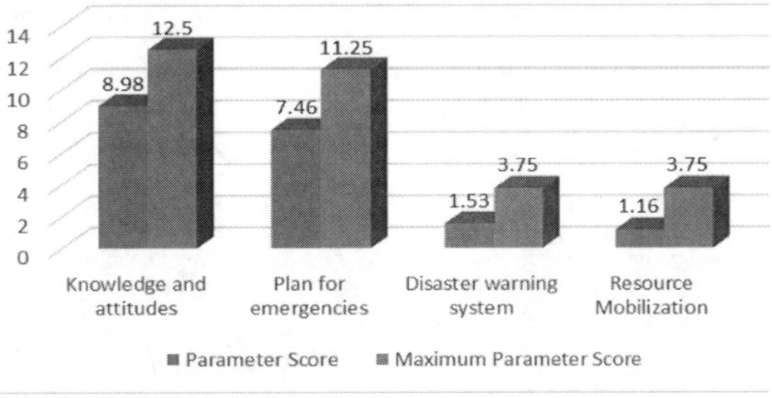

Fig. 5. Posttest result for each parameter of preparedness index in Manyaran

The increases in the index of preparedness in Kembang Arum and Manyaran after the implementation of cognitive intervention are expected to increase the awareness and preparation to face landslide disaster and reduce losses caused by landslide disaster.

The result of a comparison test of pretest and posttest of Kembang Arum and Manyaran by using Paired T-Test is <0,05, so there is an influence of cognitive intervention in preparedness index to landslide disaster.

The provision of modules and posters were distinguished between Manyaran and Kembang Arum to evaluate the effectiveness level between cognitive interventions. Differences in the results of cognitive intervention implementation in Kembang Arum and Manyaran can be seen from the improvement of resident preparedness index to landslide disaster before (pretest) and after the intervention (posttest). Kembang Arum with the media of cognitive intervention in the form of posters, videos and discussions has been increased the index of preparedness to the landslide disaster of 25.96points from pretest index 36.20 to 62.16 at posttest value. Manyaran with the media of cognitive intervention in the form of module, video and discussion have an increase of preparedness index to landslide disaster by 29 points from pretest index value 32.20 to 61.20 at posttest value.

The difference in the increase of citizens' preparedness index to the landslide disaster shows the effectiveness of different cognitive interventions in both sub-districts. Implementation of cognitive interventions with media modules, videos and discussions are more effective than cognitive interventions using poster media, video and discussion. This can be an input for the BPBD Semarang to carry out socialization about of landslide disaster in the city of Semarang. Table 3 describes the general analysis of results from this study.

5 Conclusion and further studies

The design of cognitive intervention in the landslide areas applied in each sub-district has a different approach depending on the characteristics of the residents. Residents in Kembang Arum already know the area is prone to landslide disaster and often used as a place of research on landslide disaster, so the cognitive intervention was given in the form of posters, video, and discussion. Manyaran has characteristics of the people who do not know the area is prone to landslide disaster. Therefore, the intervention was given in the form of module, video, and discussion.

Kembang Arum with the implementation of cognitive intervention media in the form of posters, videos and discussions has increased the index of preparedness to landslide disaster by 25.96 (71.71%) from the pretest value of preparedness index to landslide disaster 36.20 to 62.16 at posttest value.

Table 3. Intervention cognitive for residents in landslide areas.

	Study Areas	
	Kembang Arum	**Manyaran**
Characteristic	-Prone area of landslide -Residents know the area is prone to landslides -Often used as a place of research on landslides	-Prone area of landslide -Residents did not know the area is prone to landslides -There is no extension activity about landslides
Intervention media	Poster, video, discussion	Modul, video, discussion
Pre-test	36.20 (Not ready)	32.20 (Not ready)
Pos-test	62.16 (Almost ready)	61.20 (Almost ready)
Increasing	71.71 %	90.06 %

Manyaran with the implementation of cognitive intervention media in the form of modules, videos and discussions have increased the index of preparedness to landslide disaster by 29 (90.06%) from pretest grade of preparedness index to landslide disaster is 32.20 to 61,20 at posttest. The difference in the increase of preparedness index to the landslide disaster shows the effectiveness of different cognitive interventions in both areas. So it can be concluded that the implementation of cognitive intervention with module media, video and discussion in Manyaran is more effective than cognitive intervention using poster media, video and discussion at Kembang Arum.

Based on the conclusions, the following suggestions are given: further research should be focused on other stakeholders in disaster management such as the government and government institutions (BPBD Semarang City). This follow-up study aims to measure the level of preparedness of government and agency BPBD because both directly contribute to the index of preparedness. Further research on collaborative design among stakeholders is needed to improve the overall preparedness index for landslide disaster.

References

1. T.T Putranto, N. Susanto, J.S Dwiyanto, N. Anatoly, A. Rifqi. *Proceeding Seminar Nasional Kebumian Ke-8*. Graha Sabha Pramana. Jakarta. (2015).
2. T. Hale T and C.R. Moberg. *International Journal of Physical Distribution and Logistic Management.* 35(3) 195-207. (2005).
3. A. Haganiand A.M. Afshar. Department of Civil and Environmental Engineering, University of Maryland. (2009).
4. A. Cozzolino.*Humanitarian Logistic.*Springer Briefs in Businness. (2012).
5. J. Hong, K. Jeong, and Y. Xie, *International Journal of Industrial Engineering: Theory, Applications and Practice* 22(4) 4-12. (2015).
6. J.O. Purba. *Bachelor report in Geodetic Engineering Department Diponegoro University*http://eprints.undip.ac.id/42838/ [Online accessed Dec 20th, 2015]. (2014).
7. A. Sriyono.*Bachelor report inGeographic Department*, Semarang State University. (2012)
8. N. Susanto and T.T. Putranto. *Jurnal Teknik*..Vol 37 No 2. (2016).
9. N. Susanto, T. T. Putranto, E. A. Ulfa. *IOP Conference Series: Earth and Environmental Science*, Volume 55, conference 1. (2016).
10. W.E. Vesely., F.F. Goldberg, N.H. Robert, D.F. Hasl.*Fault Tree Handbook*. Washington D.C. : U.S. Nuclear Regulatory Commision. (1981).
11. B. Mary. *Psychological interventions: a guide to strategies*, Greenwood Publishing Group. (1995).
12. International Federation of Red Cross and Red Crescent Societies (IFRC). *About Disaster Management*. Retrieved from https://www.ifrc.org/en/what-we-do/disaster-management/about-disaster-management/ [online accesed October 20th, 2015] (2015).
13. Warfield, C.*The Disaster Management Cycle*. Retrieved from http://www.gdrc.org/uem/disasters/1-dm_cycle.html [online accesed: October 20th, 2015]. (2015).
14. J. Sopaheluwakan. *Kajian Kesiapsiagaan Masyarakat Dalam Mengantisipasi Bencana Gempa Bumi & Tsunami.Jakarta* : LIPI-Unesco/ISDR. (2006)

Utilization of Information Technology for Non Domestic Waste Management in Semarang City

Muhammad Ali[1,*], *Sudharto* P Hadi[1], and *Maman* Soemantri[2]

[1]Doctoral Program of Environmental Science, School of Postgraduate Studies, Diponegoro University, Semarang - Indonesia
[2]Department of Electricall Engineering, Faculty of Engineering, Diponegoro University, Semarang - Indonesia

Abstract. Garbage problem is often very complex in urban areas. The handling pattern of collecting, transporting and disposing that has been applied up to this day has not yet produced an appropriate solution. This is evident from the data of statistic centre institution in 2015 that 76.31% of the existing waste in the community has not been sorted, while 10.28% sorted to be used and 13.41% sorted to be discarded, showing the community amount of unsorted garbage large enough to necessitate managerial efforts at the waste sources. In designing a systematic and structured waste management system, the generations, compositions, and characteristics of the waste are indispensable. Therefore, a research is conducted on these three dimensions to the non-domestic waste in Semarang City, which involves commercial waste (from the markets, restaurants, and hotels), institutional waste (from the offices and schools).From the research result the average of 0,24kgs/person/day in weight unit of the City's non-domestical waste generation is derived. The waste composition is dominated by organic waste of around 61.95%, while the rest percentage is inorganic. The management policy is directed with the application of Management Information System model based on Information Technology because of the system's abilities to effectuate the waste management.

1 Introduction

An increase in a population will increase the amount of the waste generations. Factors causing waste generations such as lifestyle changes and the increasing varieties of consumption patterns in the community increase the volume of waste [1]. The problem of garbage in urban areas is often a very complex problem, in the manner that it requires a systematic and structured handling mechanism. The program of management at its sources which has been applied in Indonesia has not yet made a successful solution. The evidence from the Central of Statistics (BPS) data in 2015 showed that 76.31% of the existing waste in the community has not been sorted for disposal, 10.28% is sorted to be used and 13.41% is sorted to be discarded [2].

Based on its sources, garbage can be broadly grouped into domestic and non-domestic waste. The domestic waste is the waste generated from household activities or environments whereas the non-domestic waste is the waste derived commercially, industrially, or institutionally from the buildings' waste, municipal service waste, mud of sewage treatment plants and other remnants, and agricultural waste [3]. Waste that is not managed properly will cause various problems such as dirtiness, odor, disease vectors, and groundwater contamination. As a preventive measurement, it is necessary to design a waste management system.

The weaknesses in the current waste management can not be separated from the improper handlings of waste planning, implementation and supervision. One of the causes of this weakness is that the garbage management system has not been able to provide accurate information starting from handling waste at its sources, temporary shelters, and eventual final processing sites. The condition will pose difficulties to the authorities who handle the waste in formulating the policy.

The city of Semarang is currently experiencing a very rapid development, marked among others by more and more standing hotels and shopping centers as well as other infrastructural developments. Referring to Semarang's data in number of 2014, by then Semarang City already had 10 department stores, 52 supermarkets, and 47 traditional markets. The city of Semarang also had 48 hotels, 1616 schools and 126 offices. A development of a city must certainly bring benefits to the community, but behind it there also lie impacts, and one of the most visible is garbage coming from these non-domestic sources.

The method of determining and setting the number of samples of generations and compositions of cities waste in Indonesia has been arranged based on SNI-19-3964-

* Corresponding author: ilamuha@yahoo.com

https://doi.org/10.1051/e3sconf/20183109014

1994. According to the SNI method, the determination of waste generations and compositions in the city is carried out on all sources of waste, namely domestic and non-domestic, which include commercial, institutional, municipal services (from road sweeping) and industrial waste. By knowing the generations, compositions and characteristics of waste, especially those from more representative sources, problems in waste management can be prevented and anticipated as early as possible. These data can also be used to design waste management (especially at source) and appropriate waste processing technology, so that final waste disposal to the landfill can be reduced, in accordance with the patterns the developed countries have been applying in the recent years.

This research will start from a survey to collect the data of non-domestic waste generations followed with a study to the mechanism of the waste handling system that has been applied by the city government, investigation to the weaknesses of the existing system, and consequent formulation of a proper policy direction for the waste management.

2 Research methods

The stage in this research includes literature studies, secondary data collections and primary field analysis, as well as data processing. Secondary data collection is available from the existing garbage data, garbage infrastructure and supporting facilities of Semarang City from BPS of Semarang city, as well as the garbage management of Semarang City Sanitation and Gardening Department, and identification of a number of facilities that produce non domestic waste.

This research is descriptive quantitative in nature, using the data from field observation and inputs from some reliable sources. Given the research basis of focusing on an object, the data collection methods therefore will be executed more with selected key informants.[4]

Primary data collection is accomplished with observation and non-domestic waste sampling. Observations made during field research are useful as additional information to support primary data in the field. Field research in the form of garbage sampling from each source of non-domestic waste to obtain data of waste generations in both weight and volume units, the compositions of the waste based on the types of the organic waste (food waste, paper, plastics, textile, rubber, yard waste, wood) and the inorganic (glass, cans, metals, etc.), compaction factors, and the specific weights of the non-domestic waste. The samplings and their measurements are based on the method SNI-19-3964-1994. Obtained from SNI calculation are 43 non-domestic garbage samples, which can be grouped into 8 categories of sources including 7 markets, 2 shops (grocery and photocopy shops), 5 restaurants, 5 hotels, 6 offices, 10 schools (kindergartens, elementary, junior, and senior high schools, as well as universities), 2 hospitals, 3 roads and 3 industries.

3 Results and discussion

3.1 Existing condition of non-domestic waste management in Semarang city

Handlings of the non-domestically sourced garbage, including sweeping or cleaning service in the market environments up to temporary shelters, are the responsibility of the Trade Service Office (as the market management). Waste from the TPS (temporary shelters) is transported and disposed of to the landfills (TPA/final disposal) by using trucks from the Sanitation Service. SubDistrict offices nearby markets also have temporary shelters.

The handlings of the garbage along the protocol roads, especially sweeping and collecting are carried out by the KSM (Society's Self-Help Groups) coordinated by the Head of Villages. The garbage collected along the protocol roads is then transported by the Third / Private Parties using trucks and directly disposed of to the Final Disposal. The garbage on the roads that are not cooperated with private parties are then transported by the Subdistrict's.

Semarang City Government has tried to overcome the waste. The government has optimized all existing facilities and infrastructure, making local regulations on waste. The government has also established cooperation with third parties (industrial parties) in waste processing. However, these efforts still can not handle the waste problem optimally. There is still a large amount of waste that has not been handled.

3.2 Non-domestic waste generations

The determination of non-domestic waste generations has considered factors of correction, compaction and recycling. The determination of waste generations is expressed in the units of weight (kgs) and volume (liters).

Table 1. Non-Domestic Waste Generations Of Semarang City

No	Name Of Source	Amount Of Generation	Unit
	Institutional		
1	Schools	0,03	Kg / p / d
2	Offices	0,24	Kg/ p / d
3	Hospitals	0,49	Kg/ p / d
	Average	**0,253**	**Kg / p / d**
	Commercials		
4	Shops	0,23	Kg/ p / d
5	Foodstalls	0,14	Kg/ p / d
6	Markets	0,42	Kg/ p / d
	Average	**0,263**	**Kg / p / d**
7	Hotels	0,27	Kg / p / d
8	Industries	0.17	Kg / p / d

From the primary data processing, the average of non-domestic waste generation in Semarang is 0.24 kg / person / day in weight unit. The biggest contributing source of waste came from hospitals waste (0.49 kg/ p /

d), followed by markets waste (0.42 kg / p / d). Table 1 shows the non-domestic waste generation in Semarang in kg / person / day.

3.3 Non-domestic garbage generation per day

Based on SNI-19-3964-1994 the field research was conducted eight days in a row, and from the research the non-domestic waste generation per day is obtained. Figure 3 shows the daily, non-domestic waste generation in Semarang City for each source's category.

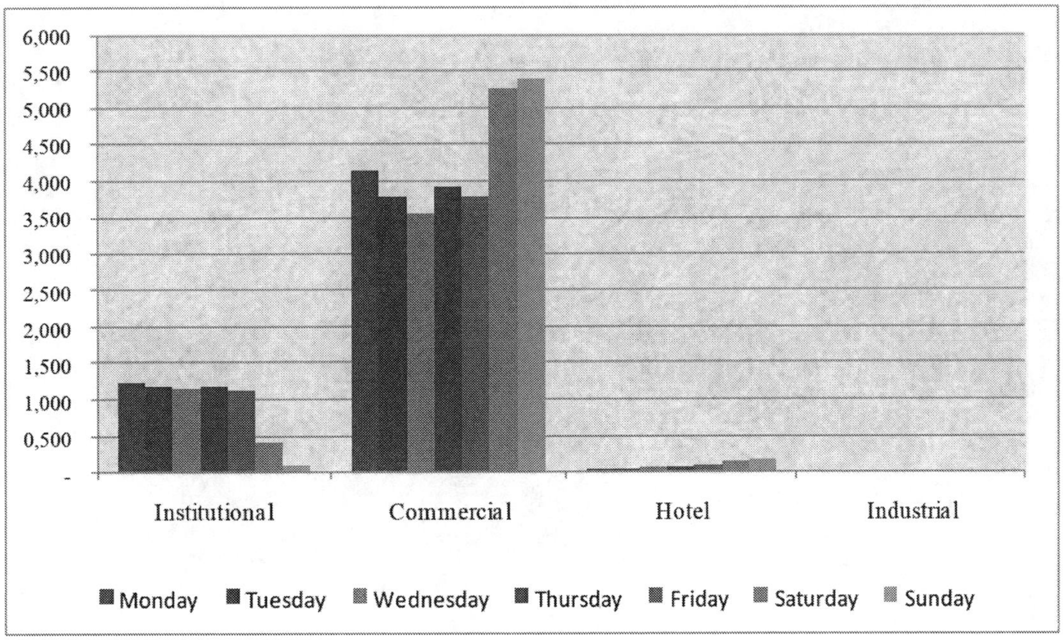

Fig 1. Non-Domestic, Daily Waste Generations Of Semarang City

The most generation of commercial waste occurs on Sundays because they are holidays, in average of about 4.6% of the total generations. In the meantime, the institutional waste generations of Semarang City increased on working days, reaching 1.2% of the total generations with a declining tendency at weekends due to the decrease of working activities during holidays. The average of the hotel-waste generations also takes increases on Sundays because not only that they are holidays but the roads facilities in front of the hotels are also used by the non inhabitants who visit Semarang City in addition to the local passers by, which produce of 0,169% of the total waste generations. In the picture above it is seen that the generation of industrial waste is almost constant due to the daily operations of productions, which is an average of 0.02% of the total existing waste generations.

3.4 composition of non-domestic waste generations

Waste composition is the ratio of the wet weight of a component or a type of waste to the whole waste. The result of research on the composition of the non-domestic waste in Semarang City shows that the most waste is the organic with 62% in percentage consisting of food and leaf waste, whereas the inorganic consists of 12.26% paper, 1,727% glass, 13,39% plastics and 1.8% metals, as well as 1.55% textile, 0.50% rubber, and 6.8% other waste. The composition of non-domestic waste in Semarang City can be seen in more detail in Figure 5.

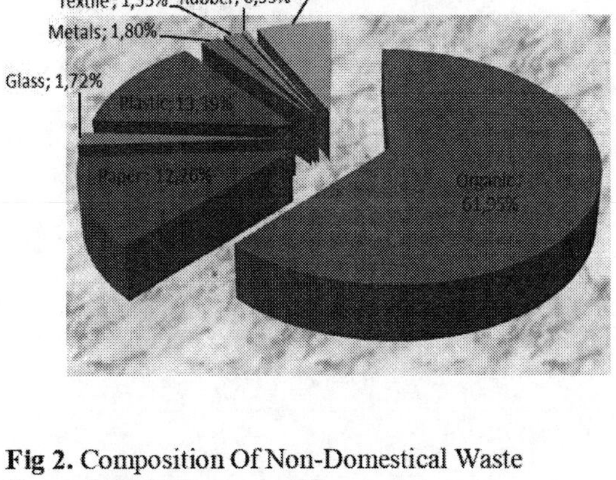

Fig 2. Composition Of Non-Domestical Waste Generations In Semarang City

https://doi.org/10.1051/e3sconf/20183109014

3.5 policy direction of non-domestic waste management through the utilization of information technology

Technology is created by men to simplify natural processing, activities, and everything that is bound to human needs. Technology is a way and an effort to improve quality of human life and to welfare it (budget funds 2013). Utilization of technology has become an important part in environmental management. Not only that technology become a tool to fulfill people's needs, but it also maintains and preserves the quality of the environment. Technology also serves to manage it, in order to remain compatible and hormonius with human life. Technological advancements have brought about shifts in human behaviors in their daily activities. Many of the human activities utilize technology either mechanically to electronically.

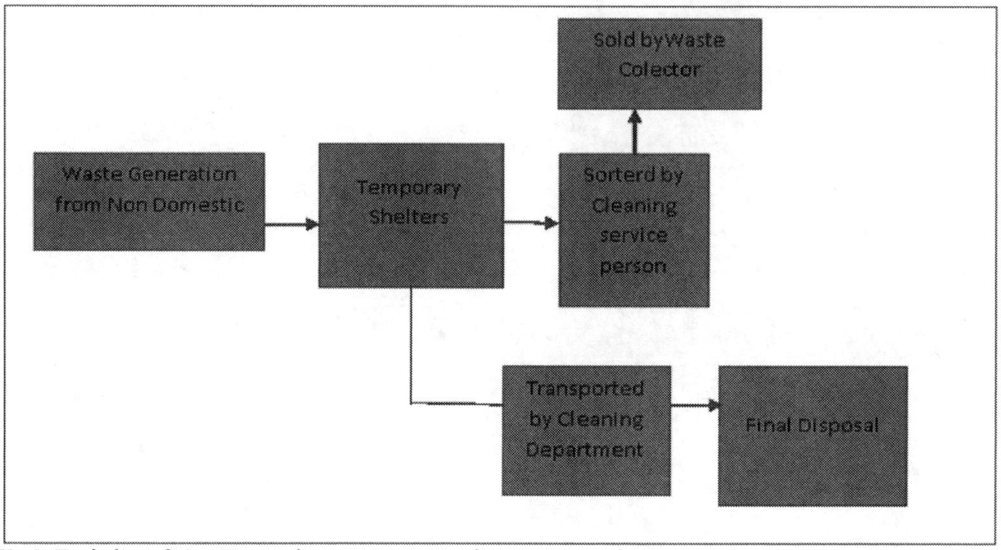

Fig 3. Exsisting of Non Domestic waste managent in Semarang City

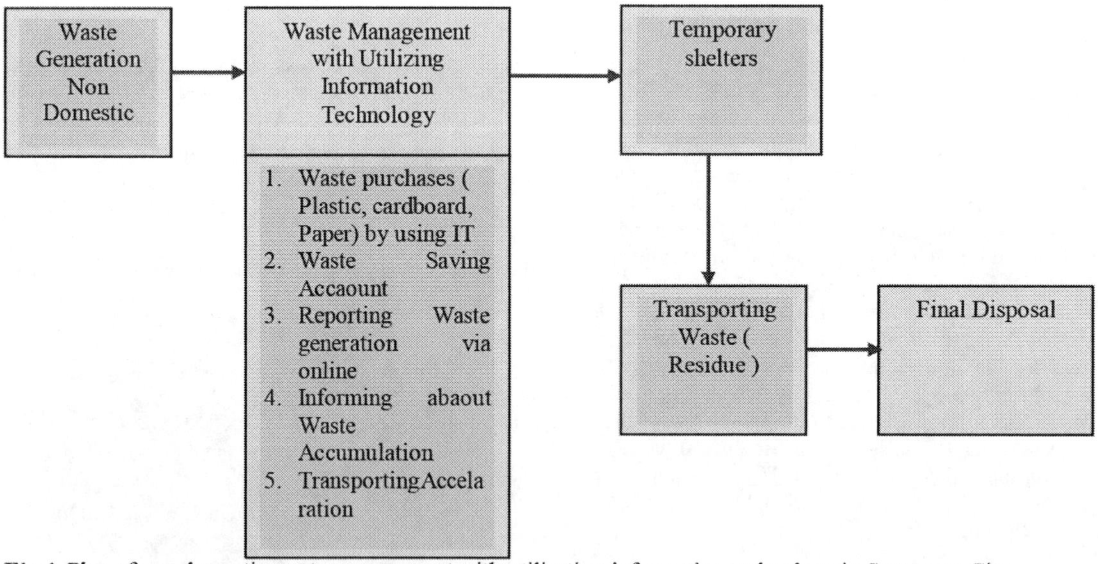

Fig 4. Plan of non domestic waste management with utilisation information technology in Semarang City

https://doi.org/10.1051/e3sconf/20183109014

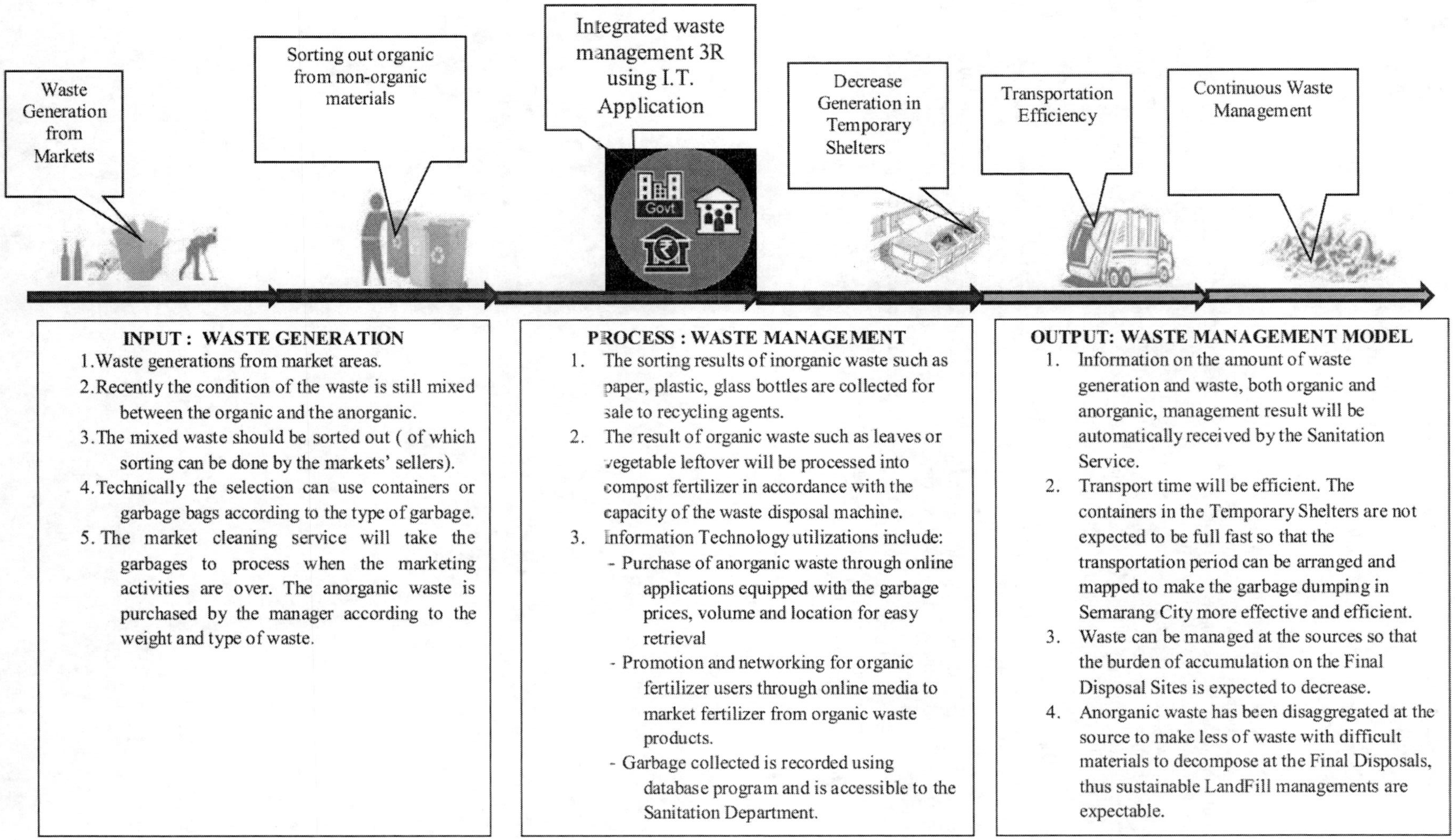

Fig.5 Input and output Waste management With Utilitazion information technology

4 Conclusions

1. The non-domestic waste generation of Semarang City is an average of 0.24 kg / person / day in weight unit. The largest contributing source of this waste comes from the hospital waste (0.49 kg / p / d), followed by the market waste (0.42 kg / p / d).
2. The daily generation of the non-domestic waste occurs mostly in the commercial areas, which is of 4.9% of the total waste generations on Sundays.
3. Non-domestic waste composition of Semarang City shows that the most waste is organic garbage with 62% percentage consisting of food waste and leaves. The rest of 38% is inorganic waste in the form of paper, plastic and so on.
4. Information System Management model based on Information Technology can simplify and streamline waste management. This is in the manner that the output system can generate information about the state of waste and handling mechanism based on certain criteria without having to do repetitive data processing.
5. A system model based on information technology can avoid data inconsistencies and can reduce files and time of recording garbage handling.

5 Suggestion

The implementation of the waste management policy direction can be an early model to be tested in Semarang City for further research that can be developed at the national level.

Reference

1. T. Karak, R. M. Bhagat, and P. Bhattacharyya, *Crit. Rev. Environ. Sci. Technol.*, vol. **42**, no. 15, pp. 1509–1630, (2012).
2. Central Of Statistic (BPS), *"Percentage of Households by Province and Treatment of Sorting Waste Easy to Decompose and Not Easy to Decompose."* 2015.
3. S. E. Vergara and G. Tchobanoglous, *Municipal Solid Waste and the Environment: A Global Perspective*, vol. **37**, no. 1. (2012).
4. Sudharto P Hadi, *Quantitative Social Research Methods, Quantitative and Action Studies*. (1997).
5. W. Wardiana, *Univ. Komput. Indones.*, (2002).

Policy Implementation Study on Spatial Planning for Environmental Conflict (Study Location: Rembang Regency)

Indraya Kusyuniadi[1,*]

[1]Master Program of Environmental Science, School of Postgraduate Studies, Diponegoro University, Semarang, Indonesia

Abstract. This study aims to see the impact and benefits as an outcome of a policy, where this policy is in the form of spatial planning (Regional Planning). As known RTRW is a product that can be regarded as "the book of development" in every region both in the provincial and regional levels. One of them is as a decision tool for investors (investors) in increasing local development investment, spatial planning (RTRW) is also expected to maintain the environment, in order to support the sustainability of regional development. In reality, there are still many conflicts of interest in the implementation process of regional development, especially between economic and environmental interests. Often the interests of regional sustainability are placed at a lower level (less priority) than investment / economy. Land conversion that is inconsistent with district / city spatial planning RTRW is relatively still occurring, especially for economic purposes. Lack of policy called spatial plan in this case RTRW Province and Regency in responding to existing condition in field. How can a product that is said to be "Scripture" a regional planning is powerless in fulfilling the space for investment in the form of industry, commercial, housing and so forth. There are several results that can be concluded in this study. Basically, the importance of the environment at least can be used as the basis or priority of the main decision makers above economic interests and other politic interests. The current Spatial Plan / RTRW document still holds a big question whether at the time of compilation it follows the norms and rules in a plan (data accuracy, through input process from the community).

1 Preliminary

1.1 Background

Based on the contents of Law No. 32/2009 (UUPPLH) Chapter I on General Provisions, Article 1 point 1 Definition of the environment: "The environment is the unity of space with all objects, power, situations, and living things, including humans and their behavior, Nature itself, the survival of life, and the welfare of human beings and other living beings. "The need for a document in the form of Strategic Environmental Assessment (KLHS) becomes very important because it is a" Tool "instrument for the prevention of pollution / environmental damage.

While spatial planning is a multi-sectoral activity, in accordance with the mandate of Law no. 26 of 2007 on Spatial Planning, and Government Regulation No. 15 of 2010 on Spatial Planning mentioned that, in the implementation of spatial planning there are three main processes that are interconnected with one another, namely spatial planning, space utilization, and spatial use control. Spatial planning material is contained in Law Number 26 Year 2007 on Spatial Planning, in one of its articles (35). Space utilization control is carried out through the establishment of zoning regulations, licensing, incentives and disincentives, and the imposition of sanctions.

Increasing population and the growing influence of private sector investment, triggering more rapid land use changes that can trigger a variety of environmental conflicts is a conflict between exploitation and destruction. The solution is to assume that natural resources are public goods. A category of goods that are considered to belong together, the numbers are fixed, and free.

In fact, spatial planning (RTRW) is a legal product as outlined in the Regional Regulation (*Perda*) which is expected to become a tool in sustainable regional development still encountered obstacles so as to create a conflict of space. Non-compliance with the existing condition is marked by the shift of conservation function land for industry, housing, agriculture etc.

1.2 Problems

Based on the above description, the occurrence of sectoral interest is the economic sector (investment) with the environmental sector as a limiting factor in the perspective of regional development. These conditions need to be viewed and studied in depth in policies and rules. The question of why there should be a conflict and a shift in the function of space, if every region already has a "Holy Book" called RTRW, which is a tool for

* Corresponding author: indraya_kus@yahoo.co.id

distributing space legally, let alone this document is equipped with Strategic Environmental Assessment (KLHS).

Some of the existing problems that occurred in the District of Rembang, Kebumen Regency associated with the conflict of space between the industry and protected areas. This conflict has long dragged on to much energy consuming both socially, economically and politically. Another problem is the shifting of other protected functions into cultivation areas. In addition, the current policy is the time to evaluate the RTRW document with the adjustment of the pattern of space.

To address the above problems this study will attempt to describe chronologically the policies and events, identifying them as part of the analysis process. In addition, in this study also the data of land change especially on protected land (preservation), where in its presentation can also be analyzed will deviation of land at district RTRW with existing condition through "Time Series" data.

2 Design and location research

To examine the implementation of policy in the District RTRW in this study was conducted two (2) the scale of the research location, namely: 1) the scale of the area is the plan of the cement industry area and 2) the regional scale that changes the pattern of land use in the period 2009-2015.

a. Research design
   1) Stages of data inventory in the form of:
      - Inventory of existing policies and regulations at national, provincial and local levels.
      - Inventory of numerical and spatial data (area and map statistics) in time series (Year 2009-2015).
      - Excavation of data through observation and interviews measurable with several sources of information.
   2) Reviewing and assessment:
      - Policy or rules, through chronological description and analysis.
      - The pattern of land change through time series data (Year 2009-2015), by looking at future trend and projection.
      - The pattern of deviation of existing land with RTRW Disctrict, by looking at future trends and projections.

b. Research sites Location

   In this study will take the location on the scale of the area of Gunem sub District Rembang sub District and regional scale in Rembang regency.

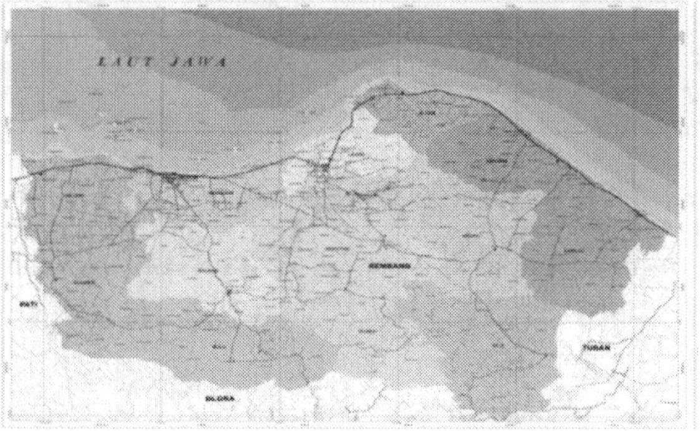

Fig. 1. Map of administration of Rembang Regency

3 Results

3.1 Policy and regulatory data

1) Decision of Rembang Regent Number 545/68/2010 About Provision of Mining Business License (WIUP) for Tras Exploration to PT Semen Gresik (Persero) Tbk - from 20 December 2012 to PT Semen Indonesia (Persero) Tbk
2) Decision of Rembang Regent Number 545/4/2011 About Mining Business License (IUP) Exploration on PT Semen Gresik (Persero) Tbk Name - since December 20, 2012 to PT Semen Indonesia (Persero) Tbk;
3) Decision of Rembang Regent Number 591/040/year 2011 On the Granting of Location Permit to PT Semen Gresik (Persero) Tbk, from 20 December 2012 to PT Semen Indonesia (Persero) Tbk for the Construction of Cement Plant, Raw Material Mine and Other Supported Facilities;
4) Decree of the Governor of Central Java Number 660.1/10 Tahun 2012 On the Environmental Feasibility of Mining Plans and Construction of Cement Plant by PT Semen Gresik (Persero) Tbk, in Rembang Regency, Central Java Province dated June 7, 2012;
5) Decree of the Governor of Central Java Number 660.1/17 year 2012 About Concerning Environmental Permit for Mining Activities by PT Semen Gresik (Persero) Tbk, in Rembang Regency, Central Java Province dated June 7, 2012;
6) Decision of Rembang Regent Number 545/0230/2013 About the Granting of Mining Business License (IUP) of Clay Rock Production Operation at PT Semen Gresik (Persero) Tbk;
7) Civil Court of Justice (PTUN) Semarang, Central Java On Thursday (16/04/2015) again held a trial lawsuit residents of Rembang with the agenda of reading a decision related to the granting of environmental permits by the Governor of Central Java for mining and the establishment of PT Semen Indonesia factory in Rembang. Decided not to accept the lawsuit of citizens.

8) Re-review (PK) proposed by Rembang farmers and NGO *Wahana Lingkungan Indonesia* (Walhi), related to the environmental permit for PT Semen Indonesia's factory construction in Rembang. The Supreme Court (MA) granted the Decree, making the environmental permit issued by Central Java Governor for PT Semen Indonesia to be canceled (Based on decision number 99 PK / TUN / 2016, 5 October 2016).

Regulation and other supporting rules, including:

1) The Constitution of the Republic of Indonesia Law Number 26 of 2007 concerning Spatial Planning;
2) The Constitution of the Republic of Indonesia Number 32 of 2009 on Environmental Protection and Management;
3) Minister of the Environment Regulation No. 27 of 2009 On Guidelines for the Implementation of Strategic Environmental Assessment
4) Regional Regulation province of Central Java Number 6 of 2010 Concerning Regional Spatial Plan of Central Java Province of 2009 - 2029;
5) Regional Regulation Regency of Rembang Number 14 of 2011 About Spatial Plan of Rembang Regency Year 2011-2031;

3.2 The potential of nature in the Kendeng Mountains

The location of a quarry that has the potential to build a Cement Plant is located in Gunem District of Rembang Regency where this area / region is part of Kendeng Mountains. The North Kendeng mountains range from Kudus Regency to Tuban Regency, including Grobogan, Blora, Pati and Rembang districts. The North Kendeng mountains are batuk Hills. Hills that have undergone natural processes within geological space and time limits. The phenomenon of the Karst North Kendeng Landscape is reflected through the numerous limestone hills of the cones, the emergence of springs in rock fractures, the flowing of underground rivers with cave corridors as their corridors [1].

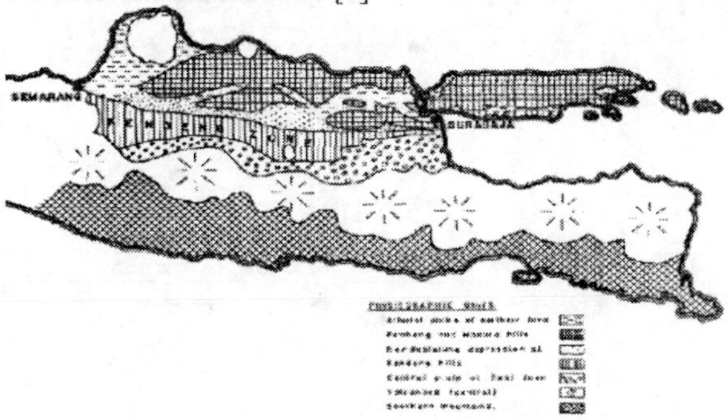

Fig. 2. Photographic Sketch of East Java Island [2]

The Kendeng zone is also often called Kendeng Mountains and those who call it Kendeng Deep, is a west-east anticlinorium. In the north bordering Depression Randublatung, while the southern part of the ranks of volcanoes (Zone Solo). The Kendeng Zone is a continuation of the North Serayu Mountain Zone that develops in Central Java. Mandala Kendeng stretches from Salatiga to the east to Mojokerto and plunges beneath the Brantas River alluvial, the continuation of these mountains can still be followed to the bottom of the Madura Strait.

3.3 Data of Land Area Year 2009 - 2015

Based on data from the Regional Public Work Water Resources and Spatial Planning (DCKTR) Province of Central Java Year 2016 shows there is a shift in the area of land between the period of 2009-2015, with details in the table below [3].

Table 1. Land area based on land use classification Year 2009

Land Use Classification 2009	Area (Ha)
Industry and Tourism	113.65
Kebun / plantation	32,407.40
Lahan Berhutan / Forested Land	844.93
Lahan Terbuka / Open field	173.75
Perairan / Waters	2,650.90
Perkebunan / Plantation	20,965.80
Settlement	5,616.11
Dryland Agriculture	5,108.73
Rice fields	33,462.50
Semak Belukar / Shrubs	2,090.86
Total Area	103,434.63

Table 2. Land area based on land use classification Year 2015

Land Use Classification 2015	Area (Ha)
Industry and Tourism	144.84
Kebun / plantation	29,873.20
Lahan Berhutan / Forested Land	838.33
Lahan Terbuka / Open field	162.53
Perairan / Waters	2,607.06
Perkebunan / Plantation	20,606.30
Settlement	10,962.50
Dryland Agriculture	149.88
Rice fields	133.04
Semak Belukar / Shrubs	35,949.60
Total Area	2,007.49
Industry and Tourism	103,434.78

3.4 Data deviation of existing land with RTRW

Existing land area of the data Year 2015 with map RTRW Province Year 2009-2029. For 6 years the data on land deviation with the RTRW document obtained data as in table 3.

Table 3. Area of land deviation Existing with spatial planning

Land Classification RTRW	The Land Classification of Existing Year 2015	The area of land deviation (Ha)
Physiographic Protected Areas	Industry and Tourism	625.50
Wetland Farming Area	Industry and Tourism	3,184.47
Dryland Agricultural Area	Industry and Tourism	958.41
Area of Hard Crop Agriculture	Industry and Tourism	106.21
Rural	Industry and Tourism	2,719.09
Nature preserve	plantation	77.68
Protected forest	plantation	9,619.94
Rural	plantation	28,973.40
Urban	plantation	2,668.47
Rural	Forested Land	3,542.57
Protected forest	Perkebunan	11,946.30
Protected forest	Settlement	2,206.35
Production forest	Settlement	11,469.90
Limited production forest	Settlement	6,848.70
Wetland Farming Area	Settlement	185,401.00
Dryland Agricultural Area	Settlement	87,481.90
Production forest	Open Mining	40.04
Limited production forest	Open Mining	45.14
Dryland Agricultural Area	Open Mining	322.49
Protected forest	Dryland Agriculture	2,106.61
Protected forest	Rice fields	2,286.17
Wetland Farming Area	Settlement	185,401.00
Dryland Agricultural Area	Settlement	87,481.90

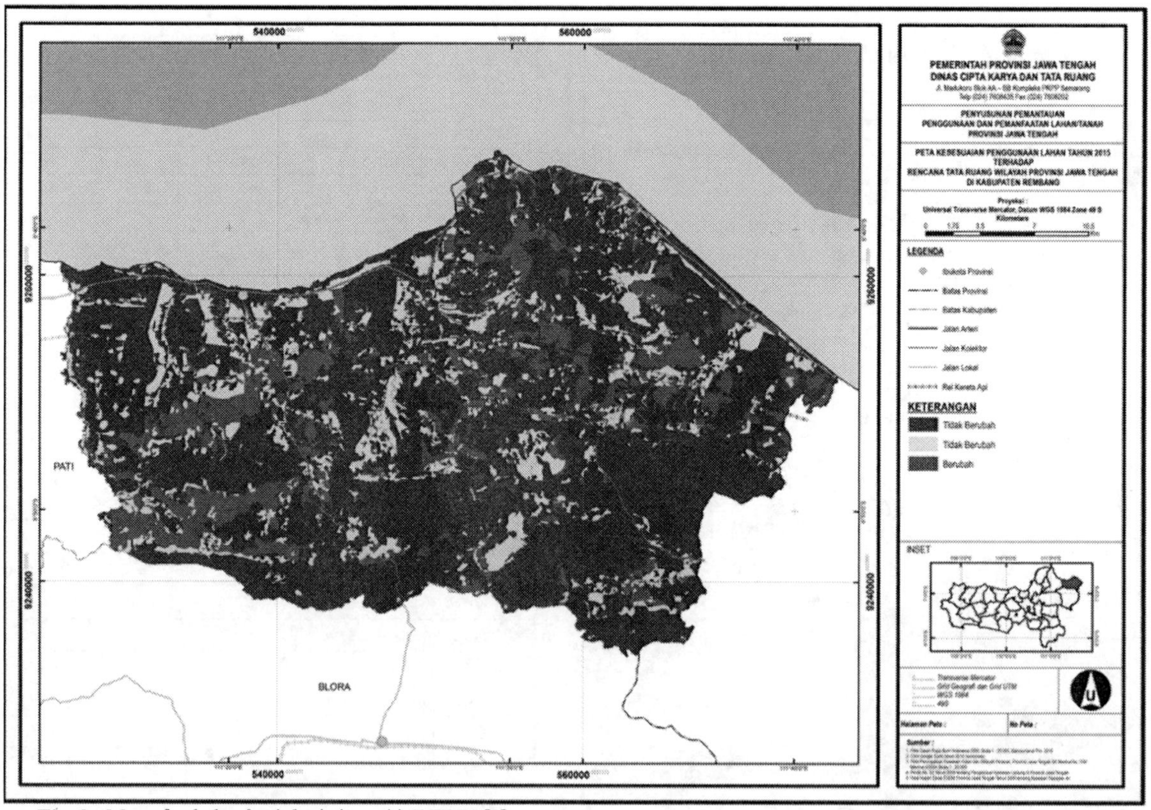

Fig. 2. Map of existing land deviation with RTRW [3].

4 Review of the study

4.1 Policy Review

Keep in understanding the problems arising digging today, that is by describing will chronologically be the cement plant (decision number 99 PK/TUN/2016), including:

https://doi.org/10.1051/e3sconf/20183109015

Table 4. Chronological review of policy in the conflict of Rembang cement plant space

Chronological Policy	Analysis / regulation review
October 14, 2010 Decree of Rembang Regent Number 545/68/2010 concerning Provision of Mining Business License (WIUP)	1. Regional Regulation of Central Java Province RTRW Applies, July 21, 2010 2. Regional Regulation on RTRW of Rembang Regency, August 25, 2011 3. Presidential Decree number 26/2011 as one of the basins of groundwater (CAT) that should be protected. Analysis: 1. Reference issuance of Decree of the Regent has been referring to RTRW Central Java but not based on spatial plan more detailed (from map scale) that is RTRW Kabupaten Rembang 2. Conditions allow for different map corrections between the detailed Provincial RTRW (1: 250,000) maps with District RTRW (1: 25,000) 3. References to the Presidential Decree No. 26/2011, also not yet available when the Provision of Mining Business License (WIUP)
January 18, 2011 Rembang Regent issued Decree No. 545/4/2011 concerning Exploration Business License (IUP)	Under the provision of WIUP, an exploration IUP can be established In this stage there is no problem, because in accordance with existing requirements
The Regent of Rembang issued Decision Number 591/040 / Year 2011 on the Granting of Location Permit to PT Semen Gresik Tbk - from 20 December 2012 to PT Semen Indonesia (Persero) Tbk - For Construction of Cement Plant, Mining Land of Raw Materials and Other Supporting Facilities.	Under the provision of WIUP, and IUP of exploration, the location permit bias is made. In this stage there is no problem, because in accordance with existing requirements
Decision No. 660.1 / 17 Year 2012 on Environmental Permit for Mining Activities and Construction of Cement Plant	Enviromental Impact Assessment (AMDAL) has been conducted The Minister, the governor, or the regent / mayor in accordance with his / her authority shall cancel the environmental permit if: a. The requirements submitted in the permit application contain legal defects, errors, misuse, and / or falsification and / or falsification of data, documents and / or information; b. The party responsible for the business / activity does not implement the requirements and obligations in the environmental permit so that there is pollution and destruction of the environment
- The environmental permit granted by the Governor of Central Java is a condition of obtaining a business license - This process begins with an AMDAL as a condition of environmental permit - The AMDAL document should prioritize the appropriate process and no interest - This document is intended as a guide to facilitate the preparation of AMDAL integrated activities of Cement and Mining industry. This guide is expected to): 1. Control the ways that are less prudent to the environment, so as to preserve the ecological function. 2. Supporting efforts that take account of ecological processes among ecosystems, as life support systems are of significant significance for the sustainability of development and improvement of the welfare of the population around the development area. 3. Provide guidance and understanding to AMDAL compilers of integrated activities of Cement and Mining industry. In this case the AMDAL should be, made participatory by involving the entire community, while the AMDAL compilation team is the Company's choice (Sponsor)	

Chronological Policy	Analysis / regulation review
In this case AMDAL is questionable?	
Decree of Rembang Regent Number 545/0230/2013 About Granting of Mining Business License (IUP) of Clay Rock Production Operation to PT Semen Gresik (Persero) Tbk	Administrative review is in accordance with the procedure because it has done AMDAL
Governor Decree No. 660.1 / 4 Year 2017 dated January 16, 2017 on the revocation of Governor's Decree No. 660.1 / 30 of 2016 on Environmental Permit for Raw Material and Construction Activities and Operation of Cement Plant of PT Semen Indonesia (Persero) Tbk	Supreme Court ruling (MA), Environmental permit issued by Governor of Central Java for PT Semen Indonesia must be canceled (Based on decision number 99 PK / TUN / 2016, 5 October 2016).

4.2 Identify patterns of land change

To see how big the land use change is in all of Rembang regency administration, it can be seen that the biggest change in the use of settlement land is 5,169% (increase from 2009-2015) then the loss of plantation land is 2,450% (shrink from 2009-2015). The following is the result of identification of land change pattern in Rembang regency.

Table 5. Identification of Land Use Change Pattern Year 2009 - Year 2015

Classification of Land Use	Land change (Ha)	%
Industry and Tourism	31.19	0.030%
Garden	-2,534.20	-2.450%
Forested Land	-6.59	-0.006%
Open field	-11.22	-0.011%
Waters	-43.84	-0.042%
Plantation	-359.50	-0.348%
Settlement	5,346.39	5.169%
Dryland Agriculture	-0.03	0.000%
Rice fields	2,487.10	2.405%
Shrubs	-83.37	-0.081%

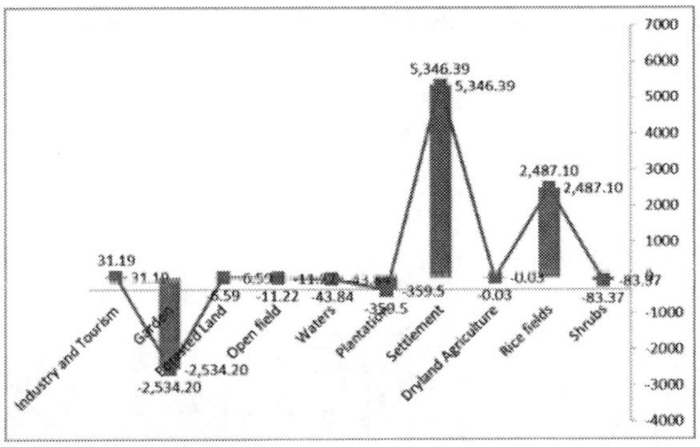

Fig. 3. Graph for the extent of land change Year 2009-2015 in Rembang Regency

4.3 Identify deviations of land to RTRW

Need to classify spatial pattern based on RTRW that is the function of protection and cultivation, as shown in the picture below.

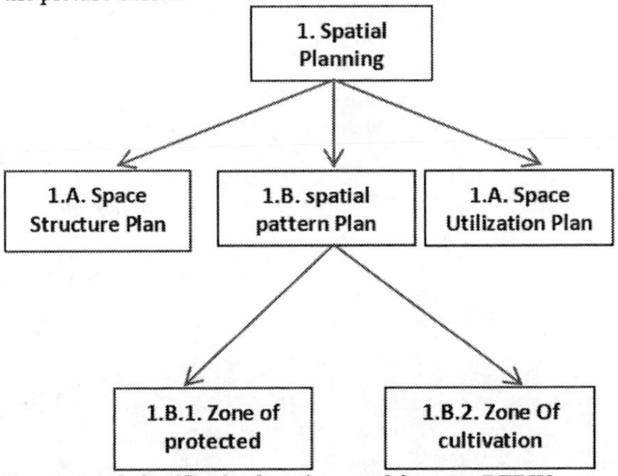

Fig. 4. Land classification based on spatial pattern RTRW

By integrating the land area of 2015 with the extent of the spatial pattern plan (land classification RTRW) the biggest change is in the function of wetland farming area of 58.88% of the total land change. Changes of concern are on protected functions such as 1) physiographic protected areas (0.1%), 2) Protected forests (4.43%), due to this function is to maintain environmental balance. Where this function is expected to be preserved / preservation.

Table 6. The area of deviation of land suitability with RTRW

Classification of Land Use RTRW	Land area not in accordance with the RTRW	%
Physiographic Protected Areas	625.50	0.10%
Wetland Farming Area	373,986.47	58.88%
Dryland Agricultural Area	175,922.21	27.70%
Area of Hard Crop Agriculture	106.21	0.02%
Rural areas	35,235.06	5.55%
Nature preserve	77.68	0.01%
Protected forest	28,165.37	4.43%
Urban	2,668.47	0.42%
Production forest	11,509.94	1.81%
Limited production forest	6,893.84	1.09%
Total	635,190.74	100.00%

5 Conclusion

Implementation of spatial planning (RTRW) on cases of economic and environmental space conflicts:

1) Overlap / decisions that precede existing policies or rules, such as granting WIUP permits by the Mayor. Where not yet based on RTRW Regional Regulation Rembang, and only based on RTRW Province with a scale of low level of accuracy. Similarly, there is no regulation of Presidential Decree number 26/2011.RTRW has not been able to respond explicitly to space for investment and space that must be protected environmentally
2) RTRW is prepared as if only to fulfill the requirements of Law no. 26 Year 2007, with the depth of the weak technical content due to the lack of data power and the lack of public participation in socialization.

Implementation of RTRW to change of spatial pattern between existing condition with spatial pattern plan

1) There is a shift in the pattern of space, especially in the function of protection, making the function of RTRW as a controlling tool was only at the level of documents and maps.
2) Weak supervision of the use of space, especially the function of protection causes RTRW not optimal.
3) RTRW should be a planning document according to the author here is a document that can be used in a long time or unlimited, so if the RTRW document is not necessarily in the continuous revision, unless there is a "force major" event. Especially for the

function of protected areas must be properly preserved and strictly defined in the RTRW.

4) Given the complexity of the content of the RTRW document, it is appropriate that the local government in this case the provinces and municipalities should prepare good sectoral documents. Given the sectoral document, it can be used as the basis for preparing the RTRW.
5) Enhancement of supporting documents in the form of Strategic Environmental Assessment (SEA) as an integral part of RTRW document must be made with accuracy and applicable rules. The SEA document preparation should be separated from the part of the RTRW making activity to be more "independent" in providing input.

References

1. Wacana et al., 2008
2. De Genevraye and Samuel, 1972
3. Data base on: *Regional Public Work Water Resources And Spatial Planning* (DCKTR) Province of Central Java Year 2016
4. Rifa'I RS, Pertagunawan K. *Biologi Perikanan 1*. Penerbit CV Kayago, Jakarta (ID). (1983)
5. Edward and MSTarigan, Makara Sains **7**,(2),82-89(2003)
6. Salmin, Oksigen Terlarut (DO) Dan Kebutuhan Oksigen Biologi (BOD) Sebagai Salah Satu Indikator Untuk Menentukan Kualitas Perairan. Oseana, Volume **XXX**, Nomor 3, p. 21 – 26 (2005)
7. Sutimin, *Model Matematika Konsentrasi Oksigen Terlarut Pada Ekosistem Perairan Danau*. Jurusan Matematika, FMIPA, Universitas Diponegoro, Semarang (2006)
8. Peraturan Pemerintah No. 82. *Pengelolaan Kualitas Air dan Pengendalian Pencemaran Air* (2001)

Climate Change Anticipation on Supporting Capacity of Fishing Environment in the Coastal Area of Tanjungmas Semarang City

Indah Kurniasih Wahyu Sari[1,*], *Sudharto* P.Hadi [2]

[1]Master Program of Environmental Science, School of Postgraduate Studies Diponegoro University, Semarang-Indonesia
[2]Doctoral Program of Environmental Science, School of Postgraduate Studies Diponegoro University, Semarang-Indonesia

Abstract. Climate change is no longer a debate about its existence but already a problem shared between communities, between agencies, between countries even global for handling serious because so many aspects of life and the environment is affected, especially for communities in coastal environments This climate change is a threat to the Earth, because it can affect all aspects of life and will damage the balance of life of Earth Climate change happens slowly in a fairly long period of time and it is a change that is hard to avoid. These Phenomena will give effect to the various facets of life. Semarang as areas located to Java and bordering the Java Sea are at high risk exposed to the impacts of climate change Also not a few residents of the city of Semarang who settled in the northern part of the city of Semarang and also have a livelihood as farmers/peasants and fishermen Many industrial centers or attractions that are prone to impacted by climate change. Thus, the anticipation of climate change on resources support neighborhood of fishermen in the coastal area of Tanjungmas Semarang interesting for further review. This study aims to find out more the influence of climate change on the environment of fishing identify potential danger due to the impacts of climate change on coastal areas of Tanjungmas Semarang The research was conducted through surveys, interviews and field observation without a list of questions to obtain primary and secondary data As for the analysis undertaken, namely the analysis of climate change on the coastal environment, the analysis of productivity of fishermen as well as the analysis of the likelihood of disaster risk at the coast due to climate change. From the results of the study the occurrence of sea rise as one of the indicators of climate change in the coastal City of Semarang to reach 0.8 mm/year and average soil degradation that ranged between 5 – 12 cm/year cause most coastal communities as well as the social life of the agricultural areas of its economy relies on the resources becoming increasingly erratic.

1. Introduction

Climate change is one of the natural phenomena where there is a change in the value of climate elements either naturally or accelerated by human activity on earth. Since the industrial revolution began, it has caused increase to global temperature. Currently, climate change is no longer a matter of debate about its existence but has become a common problem between communities, between agencies, between countries and even global to get handled because so many aspect of life get affected, especially fishery sector. Productivity and progressivity of fishery sector is affected by many factors, especially climate change and anomaly.

Government concerns related to adaptation and mitigation to climate change are regulated in the Minister of Home Affairs Regulation Number 30 Year 2010 About Guidelines for Resource Management in Marine Areas, in section four article 18 affirmed that the Regional Government in planning the management of resources in the marine areas, shall include materials containing adaptation effort and mitigation of impact from climate change. Furthermore, referred in article 19 that adaptation and mitigation of climate change is done by involving the government, regional government and/or society.

According to [1] climate change occurs around the world and has a significant impact on the fisheries sector in Java. Demand for fishery products increase in line with population growth and nutritional needs but productivity is relatively inelastic to meet the demand. The impact of climate change has been detected since 1980 to 2008 either ecologically or oceanographically. Sea surface temperature are positively detected experiencing anomalies, extreme season change have caused change to sea breeze and wave height, and so forth.

As an archipelagic country, Indonesia is very vulnerable to the impacts of climate change. With the condition as a developing country, our ability to adapt to climate change has not been as good as developed countries [2]. The City of Semarang as an area located in the north of Java Island and directly adjacent to Java Sea has a high risk to get affected by climate change. Not a few residents of Semarang City who live in the northern part of Semarang City that works as a farmer and fisherman. Many industrial center or tourist attractions are prone to climate change impact [3].

* Corresponding author: indahkurniasihw@gmail.com

Climate change phenomena has actually happened and currently still going on today until the upcoming times. In principle, climate change occurs because some climate elements intensity deviates from the normal condition towards a certain direction.

One of climate change impacts is the high rainfall occurring over a period of time. The presence of high rainfall in the city of Semarang can be a threat to coastal areas like Tanjungmas. This is because rain water flows from the higher topography such as Semarang hills to the lower topography like coastal areas through the river flow [4]. If the river condition can't contain the water volume optimally then the potential for the river to overflow to the land will increase. For Tanjungmas area is also flowed by the river flow which can potentially overflow the surrounding area at any time. On the other side of Tanjungmas located in the Semarang city coast is also prone to the threat of rising sea levels. Some locations become routine areas that are constantly inundated by flood and every year the height of the flood is increasing. Based on the results of the Indonesian fishermen condition and the existence of climate change that occurred, as a concrete solution in providing support to the people and the environment, it is necessary to have a proper study for the people in adapting to the change. Proper strategies implementation for dealing with climate change needs to be prepared so that people are not vulnerable to the rate of change [5]. Therefore the fundamental question of this research is, How to anticipate fishermen people in the coastal areas in the face of climate change?

Based on the above background, the problems to be discussed in this research are:

1. What is the effect of climate change to the supporting power of the coastal areas environment?
2. What is the adaptation strategy of the fishermen in the coastal area of Tanjungmas in facing the climate change impacts?

Based on the problems above, the purpose of this study are: (1). Analyzing the effect of climate change to the socio-economic condition of Tanjungmas fishermen, (2). Identify adaptation strategies undertaken by Tanjungmas fishermen in facing the climate change impacts.

This research is expected to provide benefits for various parties [6] : (1). For academics, this research can be a further reference on the adaptation of fishermen to climate change in the coastal area of Tanjungmas in particular, (2). For the people, the results of these study can be models of adaptation patterns that can be useful for the development of adaptation and mitigation of climate change in coastal areas, (3). For the government. This research can be used as information that can influence decision making that can determines development policy, especially development in fishery sector.

2. Theoretical Approach

On climate change which states " Climate change means a change of climate which is attributed directly or indirectly to human activities that alters the composition of the global atmosphere and which is in addition to natural climate variability observed over comparable time periods. "[6] explains that climate change is a change in elements over a long period of time (50-100 years) that is influenced by human activity that produce greenhouse gas (GHG) emission. From Diposaptono explanation, it is seen that climate anomaly that occured in a short time is not referred to as climate change. When interpreting climate change with both definitions, climate change is a change of climatic elements that occurred over a long period of time and can be compared. For example, [7] by observing the temperature data and seeing that there is a tendency to rise over time and the fluctuation is getting bigger, or climate anomaly has become more frequent than the previous time period, it can be said that climate change have occurred.

Explained that climate change originating from a global level where global warming is caused by the increase of carbon emission (CO_2) that can melt ice at the poles and increase the sea levels. Satria also added, global warming occurred due to the increase of global temperature because of greenhouse effect caused by the increasing emission of gases such as carbon dioxide (CO_2), Methane (CH_4), and CFC, trapping the solar energy in the earth's atmosphere. Thus, global climate change is the accumulation of economic activity derived from fossil energy and worsening deforestation.

3. Methods

This study is an empirical study on climate change anticipation to supporting power of the fishermen environment in Tanjungmas coastal area of Semarang City. The method used to analyze the data in this study is a qualitative approach descriptively. Qualitative data analysis is done by interview to the fishermen subjects and key-person.

3.1 Impacts of Climate Change on Fisherman's Environment

Given the impact of climate change in Indonesia, the most vulnerable category of people are fishermen, especially coastal people. Basically, fishermen in everyday activity are very dependent on nature. Drastic change to nature that are difficult to anticipate, will further add panic to natural changes that will eliminate their catch. Fishermen live in uncertainty of their work, as they depend on nature (season and weather). With the impact of climate change that directly affects the environment, the uncertainty is increasing to the livelihood aspects of fishermen.

Climate change also affects the social, cultural, and economic conditions of fishermen. Indicators and impacts from climate change to the fishermen environment are as follows [8] :

https://doi.org/10.1051/e3sconf/20183109016

1. Indicators of seasonal patterns change and wind pattern adopted by fishermen make it difficult for the fishermen to predict sea condition, disrupt fishing activity and safety of fishermen.
2. The disappearance of some animal that mark the change of the season, the shifting of the season start and end time, *untrackable* wind and waves that have been believed as the marker of the beginning and the end of a season, the changing speed of the wind in certain seasons as well as the period of wind and waves in a certain seasons have an impact on the uncertainty to sail.

3.2 Environmental Condition of Coastal Fishermen

Explains that coastal society is groups of people who together inhabit coastal areas and have a distinctive culture associated with the dependence on the utilization of coastal resources. Furthermore, it is also explained that coastal people are not only fishermen, but also fish farmers, fish processors and even fish traders. Coastal people have economic system that are not much different with the fishermen in general [9]. Coastal people solidarity is very strong due to the geographical condition that make them have to struggle together to survive in all aspect of life associated with the sea. Visible from the transportation, logistic entry, and work source as well as largest natural resources comes from the sea.

Fishermen who work in the sea are often confronted with uncertainty and ecological conditions that are hard to control. Fishermen are different from farmers, where they are faced with a difficult ecological situation to control their products considering capture fisheries are an open access so that fishermen have to move around and the risks that must be faced are greater than that faced by farmers. It is also explained that, fishermen also have to deal with the harsh marine life that makes them generally tough, firm, and open, which distinguishes them with farmers [10].

Characteristic of the small fishermen is that they are still using traditional technology, because of economical condition, they can not afford it. Usually the small fishermen are called *gurem* fishermen, where they live subsistent only to fulfill basic daily needs. However in their development, small fishermen also commercialize their catch for sale, they usually submitted their catch to the bosses. The relationship between fishermen with their bosses has a patron-client bond to avoid the threat such as reducing emissions or reducing flood impact by building embankment.

Adaptation concept declared by [9] cited by [11] is one part of cultural evolution process, a process that includes a series of people efforts to adapt or giving respond to the change of physical and social environment that happened temporally. Such responsive behavior allow them to organize certain systems for the action or behavior.

Here is the adaptation strategies undertaken by the fishermen to face climate change, namely:

1. In the economic aspect, emerges a strategy of livelihood shifting and diversification.
2. In the technical and technological aspect of capture, there are strategies of diversifying fish catching gear and changing the fishing areas.
3. In the socio-cultural aspect, there is a strategy to utilize social relations and mobilizing family members.

4. Result and Discussion

This research is conducted in the coastal area of Tanjungmas, North Semarang sub-district, to see how the anticipation strategy caused by climate change to the environmental supporting capacity of fishermen in Semarang city. Fishermen who become respondents in this research are those who still use traditional fishing gear and small boat. According to the majority of respondents, the condition of fishermen today is only sufficient to eat and must have side jobs. While the previous conditions are more sufficient for fishermen compared to the current lacking conditions for the fishermen who are mostly small fishermen[9].

Geographical and Nature Condition

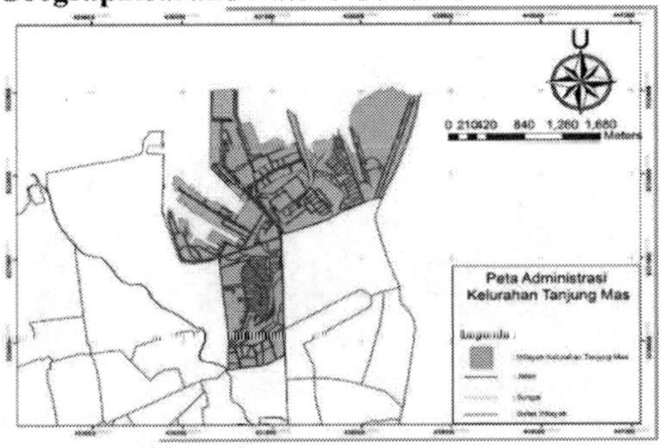

Fig. 1. Semarang Utara Sub-Distric Map

Tanjungmas urban village located in the northern sub-district of Semarang, covering an area of 323,782 Ha consists of 271,782 Ha dried lands (yard/building/emplacement) and 52 Ha wetlands (embankment). Tanjungmas urban village includes two neighborhood areas, namely Tambak Lorok village in the north and Sidodadi village in the south. Tanjungmas urban village is one of many urban villages which is included in the north semarang sub-district. The role of this area with high activity is because of its high access value, strategic location, close to the activity center, the city center and the transportation hub. One of the areas in Tanjungmas is Tambak Lorok. The graph above shows the total population based on the number of men and women in Tambak Lorok. The majority of Tambak Lorok residents are immigrants who moved because of their work in Tambak Lorok. For more detailed data based on per hamlet (RT) and average income per neighbourhood (RW).

Tambak Lorok area is located in Tanjungmas urban village, North Semarang Sub-district. This area is located on the coast of Java Sea which is crossed by Kali Banger. This area is famous for fishing settlements since 1950. Tambak Lorok area consists of 5 neighborhood that is RW 12,13,14,15,16. With an area of 46.8 Ha. Height 0,5 Above Mean Sea Level (AMSL). The geographical location of Tambak Lorok Estate is viewed from Semarang City, BWK III of Semarang City, where most of the land used as settlements. Physically, the area of Tambak Lorok is adjacent to:

North : Java Sea
East : Kali Banger
South : North Artery Road
West : Steam and Gas Powered Power Plant

Table 1. Total Population by Gender and Age Group

Age(Years)	Male	Female	Total
0-4	1.250	1.223	2.473
5-9	1.238	1.183	2.421
10-14	1.194	1.270	2.464
15-19	1.320	1.394	2.714
20-24	1.301	1.719	3.020
25-29	1.585	1.829	3.414
30-34	1.340	1.494	2.834
35-39	1.095	1.251	2.346
40-44	1.012	1.103	2.115
45-49	816	1.106	1.922
50-59	785	963	1.748
60-64	376	384	760
>65	119	195	314
Total	13.431	15.114	28.545

Tanjungmas coastal area, North Semarang Sub-district, has a fairly high number of poor people compared to other areas. From the data obtained shows that the level of poor people in North Semarang has the second highest number after the sub-district of West Semarang. For North Semarang sub-district, the majority of the poor people are in the Tanjungmas urban village. Whereas, if viewed from the potential, Northern Semarang is one part of Semarang city that has an economic activities that are developing due to the industrial and trading area. By looking at the condition of residency in Tambak Lorok, it can be categorized into the poor people in Semarang city. This condition is quite alarming because many poor people who must face the climate change disaster that requires them to make various adaptation efforts[12].

4.1 Population and Livelihood

The people of Tambak Lorok village, as much as 80% is people with income as fishermen, the rest are factory workers and construction workers. Men as heads of families are at sea assisted by their teenage and adult boys. Most fishermen here use private boats with a group system or use a rented vessel with a payment model through revenue share for the catch obtained. Most of the fishermen's catch is sold directly to the broker who will sell the products outside of Tambak Lorok area especially to Semarang City and its surroundings as well as to the restaurants that wanted certain quality and type. A small portion of the catch is sold directly to the market in Tambak Lorok, at the restaurant.

Fig. 2. Fishermen's catch

So most of the fish and other catch in TPI is the catch from fishermen with bigger boats owned by fishermen outside of Tambak Lorok. This is because most fishermen in Tambak Lorok use a relatively small boat with a smaller assets of course, so that the effect on the range or distance of the ocean that can be sailed and ultimately affect the quality and size of the catch. The people that works as fishermen, laborers, and fish traders have an average income of Rp.500.000 - Rp.1.500.000 / month. Every day the fishermen get a gross income from the sale of the catch of Rp. 500.000 then it was cut to buy diesel fuel for the ship ammounting Rp.300.000.

Fig. 3. Environmental Condition of Tanjungmas Fishermen

4.2 Economic Condition of Tanjungmas Fishermen Environment

The existing condition of existing settlements in Tambak Lorok village strongly does not support the economic growth around. Bad road conditions, poor drainage systems and littering problems are the problems in this village. This is due to the problems that exist in Kampung Tambak Lorok, such as flood which always becomes an integral part of this area caused the condition of settlement in Tambak Lorok getting worse and slum. The condition of this slum occurs due to the deterioration of environmental quality.

In addition, there is no availability of adequate facilities and infrastructure that will affect the existing economic activities in Tambak Lorok village. Furthermore, the problems that exist in Tambak Lorok village studies in Tambak Lorok village are as follow.

Tabele 2. Number and Percentage of Respondents by Age

Age	Respondent	
	n	%
Young (18-30 Years)	6	13,3
Adult (31-50 Years)	20	73,3
Old (>50 Years)	4	13,3
Total	30	100,0

Natural resources potential in the form of sea catch, including squid, shrimp, shellfish, red snapper, lotung. Infrastructure, many environment access are bad due to frequent rob, the boats are still traditional, communal washroom, electricity is available, the source of clean water is still limited. Geographic conditions with flat topography (0-2%), while the mean maximum tidal height is 0.40-0.60m. Based on the zoning of the coastal area of Semarang City, Tambak Lorok belongs to the Tanjungmas village area located in the subsident Zone of> 0.20 m per year. This area is strongly influenced by tidal sea water. Almost the entire residential area was inundated by the time the sea was in high tide[14]. This makes residents periodically raise peil (ground floor height points) of their home building.

Table 3. Number and Percentage of Respondents by Level of Education

Education level	Respondent	
	n	%
Low (Uneducated or Elementary School Graduate equivalent)	20	80,0
Medium (Junior High School Graduate / equivalent)	10	20,0
High (Senior High School Graduate / equivalent)	0	00,0
Total	30	100,0

Table 4. Number and Percentage of Respondents Based on Duration of Work as Fishermen

Duration as Fishermen	Respondent	
	n	%
Low (0-10 years)	2	6,7
Medium (11-20 years)	10	33,3
High (> 21 years)	18	60,0
Total	30	100,0

The table above shows the respondents of Tanjungmas People types of work. The majority of people have odd jobs, which are temporary or alternating jobs. Then followed by traders (majority of fish traders), unemployed, services, labor (for the industry around), and fishermen. Although the smallest number are fishermen, Tanjungmas is also called village of fishermen. The researcher assumes this is because the odd jobs, services, laborers, unemployed, and the traders will become fishermen (in the sense of fishing) during their free time to supplement their income so that there are "impromptu fishermen" at certain times[13].

Respondent Characteristic

Age of Respondents

Age is the range of years the respondent was born until the year at the time of the study. The number and percentage of respondents by age can be seen in table. 5. The age of respondents varied, the lowest ranging from 25 years and the highest 54 years. The data in table. 3. shows that the average age of respondents is 39 years out of a total of 30 respondents. Young respondents are 4 people or about 13.3 percent, adults are 22 people or about 73.3 percent, and elderly are 4 people or about 13.3 percent. In conclusion, most of the fishermen respondents in Tanjungmas urban areas have more adults than the young and old age category.

Education of Respondents

Respondent education is the highest level of education ever pursued by respondents. The number and percentage of education level of respondents can be seen in table. 6. In the table, it is seen that the distribution is not evenly distributed because it is the condition of fishermen in Tanjungmas village with maximum education only up to junior high school and no senior high school or university level.

Respondents with low education category were 24 people or 80 percent, medium category was 6 people or 20 percent, and none were highly educated. The conclusion that can be taken is the average of respondent of fishermen in Tanjungmas Urban area many people categorized as low level of education and nil for the category of respondents with high education [15].

Working Duration as Fishermen

Working duration as fishermen is the span of time in the early years of respondents working as fishermen until research is started. Respondents who are categorized as low 2 people or 6.7 percent, while 10 people or 33.3

percent and 18 people or 60 percent of the total of 30 respondents. The conclusion that can be taken is that most of the respondents are categorized have been working as fishermen for more than 20 years.

4.3 Climate change and its impact on the environmental carrying capacity of fishermen

Climate change also affects the social, cultural and economic conditions of fishermen. Indicators and impacts of climate change on social, cultural and economic fishing are:

1. Changes in seasonal patterns (the shifting of starting time and the end of the season) and the pattern of the wind adopted by fishermen make it difficult for fishermen to predict the sea condition, disrupt fishing activities and safety of fishermen. To understand climate change historically both the rate and impact of climate change occurring in Tanjungmas Urban Village on its socioeconomic condition.
2. Influence the fishermen's income level from fishing. Climate change that is happening up to now is considered by society affecting their income in fishery sector. People feel that their income since climate change that started from 2000 until now, decrease their income.
3. Fisherman's income, if only sourced from fishing alone will tend to show low per capita income. The fishermen responding to the impact of climate change that occurs and fish conditions that are uncertain, they work around it by doing additional work outside the fishery sector.

4.4 Climate Change Indicators on Ecological Aspects

Global climate change has resulted in enormous losses at the local level. Climate change impacts on all sectors, especially coastal areas.

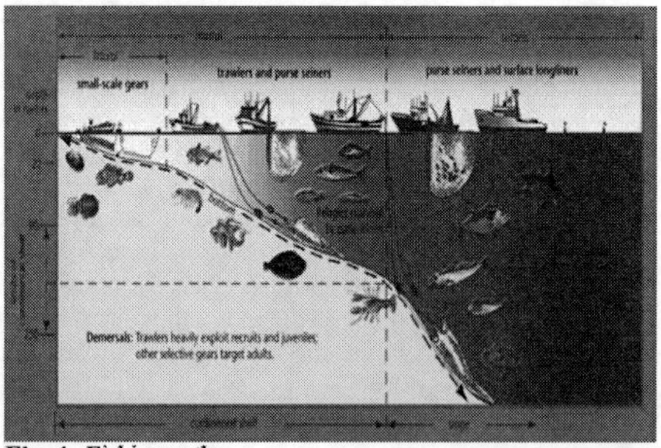

Fig. 4. Fishing paths

(1) Fishing Path I as referred to in Article 3 letter a. shall consist of :
Fishing lane IA, covering coastal waters up to 2 (two) nautical miles measured from sea level at lowest tide; and
Fishing lane IB, covering coastal waters beyond 2 (two) nautical miles up to 4 (four) nautical miles.

(2) Fishing lane II as referred to in Article 3 letter b, covering waters outside Fishing Lane I up to 12 (twelve) nautical miles measured from sea level at lowest tide.

(3) Fishing lane III as referred to in Article 3 letter c, covers ZEEI and waters off fishing lane II.

4.5 Anticipating Climate Change on Climate and Weather Conditions

Semarang City has a tropical climate and there are two seasons, the dry season and the rainy season which has a turnover cycle every six (6) months. The year-round rainfall with average annual rainfall varies about 902 mm each month with maximum monthly rainfall occurring in November with rainfall of 1,971 mm. Semarang City on the coastal area has an air temperature ranging from 25.80 c to 29.30 c. Semarang city has air humidity varies between 62 - 84%. While the wind direction mostly moves from the southeast to the northwest with an average speed of 5.7 km / [1].

Table 5. Rainfall in Semarang City 2010-2015 (mm / year)

No	Year	mm/year
1	2010	1.845
2	2011	14.349
3	2012	10.822
4	2013	10.822
5	2014	10.544

The rainfall condition in Semarang in 2010-2015 has fluctuated with the highest rainfall in 2011 with an average rainfall of 14,349 mm / year. The lowest rainfall in 2010 with an average rainfall of 1,845 mm / year. Factors that affect the average value of rainfall per year is:

1. Latitude factor causes the difference in the quantity of rainfall, the lower the latitude the higher the rainfall potential is received, because in the lower latitudes the temperature is greater than the temperature at high latitudes, this high temperature will cause high evaporation, Then it will rain by condensation first.
2. Altitude factor, The lower the altitude where the more potential rainfall received, because in general the lower the altitude, the higher the temperature in the area.
3. The distance from the water source (evaporation), the closer the place will increase the rainfall potential.
4. Wind direction, the wind that passes through the evaporating source will bring water vapor, the farther the area from the water source the rainfall potential will decrease.

5. Conclusion

The anticipation that fishermen have in coping with climate change is to mobilize family members. Fishermen in Tanjungmas usually mobilize or include family members in the work. Many fishermen in Tanjungmas take their sons who are on holiday or relatives to work together to catch fish. This is done for the efficiency of paid labor and unpaid family labor. In fishing with net there are some fishermen who invite family members to sail to the sea, besides fishermen's wife also have side jobs to help husband in fulfilling requirement such as selling processed food from fish, or helping husband take beach sand for building, make cracker, Shrimp paste and there are also looking for pangsik (a kind of shell attached to the coral sea).

Processed foods from fish that are usually sold are fish meatballs, fish crackers, and shrimp paste. Per day usually fishermen's wife can rack up money ranging from Rp. 30,000, to Rp.60.000, from the sale. Fisherman's wife can usually pocket the income Rp.70,000, - per kilogram from the manufacture of shrimp paste and fish crackers. The last job of the fisherman's wife is to search for pangsik which predominantly found in coral, the selling price per kilogram of which is usually Rp.70,000, - which will be used for processed food.

References

1. Badan Pusat Statistik Jawa Tengah. *Jawa tengah Dalam Angka* (2016).
2. Dasanto, Bambang Dwi. Jurnal Hidrosfir Indonesia.Vol 5 No 2 Hal 43-53
3. Fauzi, Akhmad. *Ekonomi Sumbe Daya Alam dan Lingkungan*. Jakarta : Gramedia Pustaka.(2006)
4. P. Manurung, J. Ananto, A. Restu, R. Marni, dan S. Barlianto, *Seminar Nasional Pengaruh Global Warming terhadap Pesisir dan Pulau-Pulau Kecil Ditinjau dari Kenaikan Permukaan Air Laut dan Banjir*, Jakarta.(2002)
5. Menteri permukiman dan Prasanrana Wilayah. Tinjauan Aspek Penataan Ruang Dalam Pengelolaan Wilayah Laut dan Pesisir, *Seminar Umum Dies Natalies ITS ke-43*, (2003)
6. P.H, Nikijuluw Victor. *Politik Ekonomi Perikanan*. PT Fery Agung Corporation (FERACO): Jakarta.(2005)
7. V.P.H, Nikijuluw. *Rezim Pengelolaan Sumberdaya Perikanan*. Jakarta : P3R.(2002)
8. Pomeroy, Robert, Len Garces, Micahel Pido, Geronimo Silvestre. Journal of Elsevier: Marine Policy Vol **34**: 298-308.(2009)
9. J. G., Pope, D. S., MacDonald, N., Daan, J. D. Reynolds, and S. Jennings. ICES Journal of Marine Science, **57**, 689-696 (2000).
10. R., Rositasari, W.B. Setiawan, I.H. Supriadi, Hasanuddin, and B. Prayuda. Jurnal Ilmu dan Teknologi Kelautan Tropis, Vol. **3**, No. 1, Hal. 52-64 (2011)
11. Satuan Tugas Walikota Untuk Perubahan Iklim, Risiko Bencana & Masyarakat Miskin Perkotaan, *Laporan Perubahan Iklim DKI Jakarta bekerjasama dengan World Bank*, Jakarta (2011)
12. Statistik Perikanan Kota Semarang, Pemkot Semarang, Dinas Perikanan.(2016)
13. Suparmoko. *Ekonomi Sumberdaya Alam dan Lingkungan: Suatu Pendekatan Teoritis*. Yogyakarta : BPFE.(2008)
14. Susandi, Armi; Herlianti, Indriani, *Dampak Perubahan Iklim terhadap Ketinggian Muka Laut di Wilayah Banjarmasin*, Institut Teknologi Bandung, Bandung. (2008)
15. A.D., Wirakusumah,. dan S. Lubis, *Seminar Nasional Pengaruh Global Warming terhadap Pesisir dan Pulau-Pulau Kecil Ditinjau dari Kenaikan Permukaan Air Laut dan Banjir*, Jakarta (2002).

 https://doi.org/10.1051/e3sconf/20183109017

Climate Change In Indonesia (Case Study : Medan, Palembang, Semarang)

Yadi Suryadi[1]*, *Denny* Nugroho Sugianto[1,2] and Hadiyanto[1,3]

1 Master Program of Environmental Studies, School of Postgraduate Studies, Diponegoro University Semarang
2 Oceanography Department, Faculty of Fisheries and Marine Science, Diponegoro University Semarang.
3 Chemical Engineering Department, Faculty of Engineering, Diponegoro University Semarang.

Abstract. Indonesia's maritime continent is one of the most vulnerable regions regarding to climate change impacts. One of the vulnerable areas affected are the urban areas, because they are home to almost half of Indonesia's population where they live and earn a living, so that environmental management efforts need to be done. To support such efforts, climate change analysis is required. The analysis was carried out in several big cities in Indonesia. The method used in the research was trend analysis of temperature, rainfall, shifts in rainfall patterns, and extreme climatic trend. The data of rainfall and temperature were obtained from Meteorology and Geophysics Agency (BMKG). The result shows that the air temperature and rainfall have a positive trend, except in Semarang City which having a negative rainfall trend. The result also shows heavy rainfall trends. These indicate that climate is changing in these three cities.

1. Introduction

Indonesia's maritime continent is one of the vulnerable areas to the impacts of climate change. Temperature changes, rainfall changes, sea level rise, increased hydro-meteorological disasters, robust, infrastructure damage, increased fire hazard, ecosystem damage, health problems and increased discomfort for residents[1,2,3,4,5] and directly or indirectly affect the economy.

One of the most vulnerable areas affected is city. Cities are vulnerable to climate change both because extremes weather events can be distrutive to complex urban systems and because so much of the world's urban population live in low-lying coastal areas, particularly in Asia[6], so environmental management efforts need to be done. To support these efforts, it is necessary to identify climate change. Climate change refers to changes in climatic conditions that can be identified (eg by using statistical tests) by changes in the mean and / or variability of their properties, over a long period, usually decades or longer either due to natural variability or as a result of human activities[7]

Climatic factors that become the main indicators of climate change in Indonesia are changes in patterns and intensity of temperature and rainfall parameters.[8] Indications of climate change can be done with time series analysis that can provide information on trends, cycle analysis or shifts around the average in the long term.

The study aims to identify climate change in the cities of Semarang, Medan and Palembang based on trend patterns, inter-time analysis (decadal) and analysis of rainfall tendencies at the height of the rainy season and at the peak of the dry season.

2. Material and Method

2.1 Location

The study sites were Semarang, Medan and Palembang.

2.2 Data

This research was done by processing the rainfall data and the temperature of the land. The data were secondary data obtained from BMKG observation / meteorology station. The data were downloaded from http //: www.dataonline.bmkg.go.id. Semarang (Period 1985-2016), Medan City (period 1982-2016) and Palembang (Period 1980-2016).

2.3 Analysis Stages

This research performed the following stages:

2.3.1 Normality Test

This test aimed to know the distribution of data distribution compared with the normal data distribution using Kolgomolov-smirnov method

* Corresponding author: yadsur80@gmail.com

2.3.2 Trend Analysis

The trend analysis used a simple regression using MS Office Excel 2013 to the annual, monthly and seasonal data.

2.3.3 Analyze the 10-year (decadal) data

Analyze the 10-year (decadal) data by dividing the period of data into Period I (1987-2006); Period II (1997-2006) Period III (2007-2016) then performing the average difference test (t test) temperature and rainfall per data period. This test was to determine whether or not there was an average change as an indication of climate change

3 Results and Discussion

3.1 Temperature

3.1.1 Normality Test of Data Distribution

Table 1 Test of Normality of Annual Temperature

	Kolmogorov-Smirnov[a]			Shapiro-Wilk		
	Statistic	df	Sig.	Statistic	df	Sig.
Semarang	,092	32	,200*	,963	32	,332
Medan	,123	32	,200*	,952	32	,168
Palembang	,115	32	,200*	,969	32	,485

*. This is a lower bound of the true significance.
a. Lilliefors Significance Correction

Normality test results for the annual temperature data using Kolmogorov-Smirnov and Shapiro-Wilk for the three cities showed a significance value > 0.05. This value indicates that the residual regression model is normally distributed. So it can be done for further parametric test.

3.1.2 Temperature Trends

The trend of temperature change in the three cities is calculated using trend line on Microsoft excel, the result is a positive trend in the three cities (Figure 1). Semarang City showed an upward trend of 0.0279 ° C / year (1985-2016), Medan City positive trend of 0.00331 ° C / year (1982-2016) and Kota Palembang positive trend of 0.0212 ° C / year (1980-2016). This is consistent with the average rise in surface temperatures of Indonesia's land area of 0.016 ° C / year[5], while the Earth's surface temperature increased by 0.85 °C (0.65-1.06 ° C) during the period of 1880-2012 or 0.002879/ year[9].

3.1.3 Temperature Patterns

Temperature patterns obtained by summing the temperature data per month during the data period then obtained the results as shown in Figure 2. Temperature patterns in Semarang and Palembang City have two peak temperatures, namely in May and October and peak minimum temperature at the peak of dry season and rainy season. While in Medan only has one peak of maximum temperature in May and a peak of minimum temperature in December-January

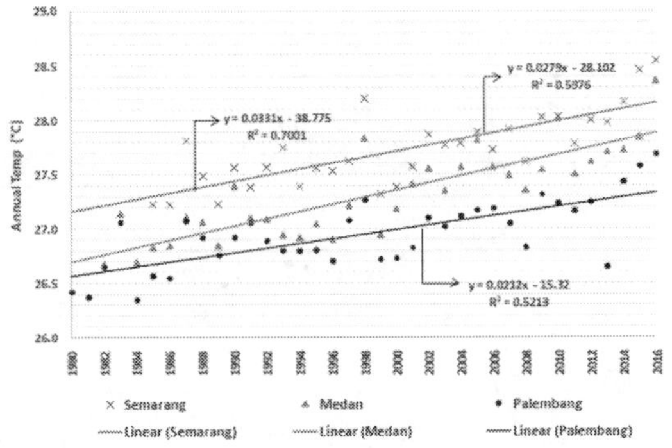

Fig. 1. Temperature Trends

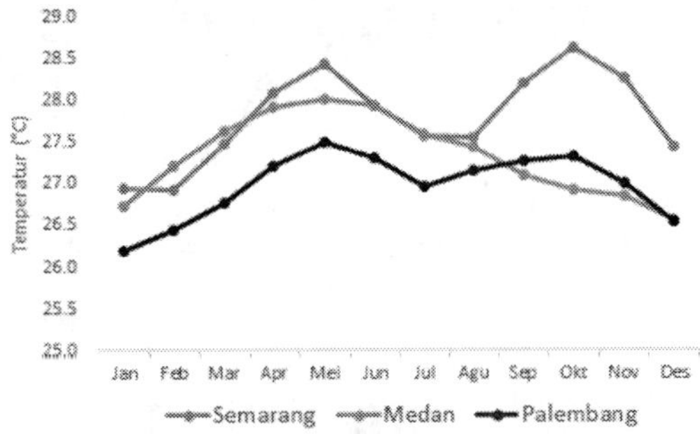

Fig. 2 Seasonal Temperature Pattern

3.1.4 Different Test of 10 Year Annual Temperature

In order to obtain temperature change information for a shorter period of 10 years, the data are grouped into three periods, and then the temperature difference test is done between periods as the result in Table 2.

Table 2 Test of Differential Temperatures of 10-Year Average

City/ Period	Test statistic	df	t Table	Sig. (2 Tailed)	Conclusion
Semarang					
III-II	8.573	11	2.201	0.000	Reject H_0
II-I	5,960	11	2.201	0.000	Reject H_0
III-I	23,833	11	2.201	0.000	Reject H_0
Medan					
III-II	8,729	11	2.201	0,000	Reject H_0
II-I	9.519	11	2.201	0,000	Reject H_0
III-I	32,729	11	2.201	0,000	Reject H_0
Palembang					
III-II	9,546	11	2.201	0,000	Reject H_0
II-I	8,317	11	2.201	0,000	Reject H_0
III-I	22,905	11	2.201	0,000	Reject H_0

The above difference test obtained t value greater than the critical value (t table) with significance <0.05, this means that the decision can be taken that H0 hypothesis is rejected for all periods in each city, so it can be concluded that there is a significant difference between Period III-II, II-I and III-I. Because the value of t-statistic positive indicates that the average temperature of Period III is higher than period II and Period I. Similarly period II has a higher average temperature than period I. Thus the temperature in the three cities consistently increased from the period previous.

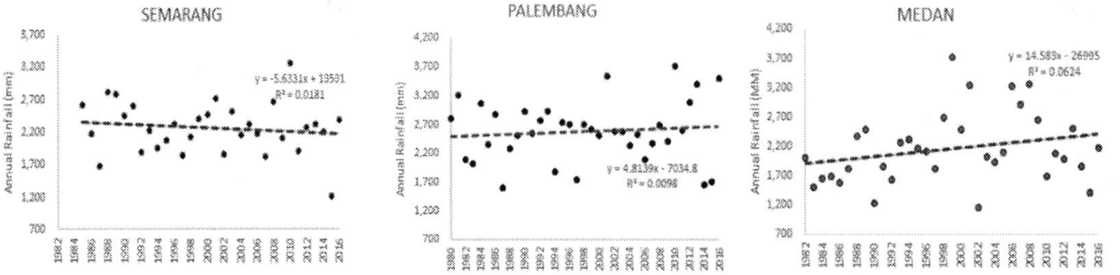

Fig. 1. Annual Raifall Trends

However, rising urban temperatures can be attributed to the effects of urban heat island (UHI) effects on temperature measurement data (*Bappenas, 2010*). Some factors affect UHI including urbanization, land cover change rate. Nevertheless, the impact is mutually reinforcing with climate change, as global climate change will exacerbate UHI's effects in urban areas[7].

3.2 Rainfall

3.2.1 Normality Test of Annual Rainfall Data

Table 3 Test of Normality of Annual Rainfall Data

Tests of Normality

	Kolmogorov-Smirnov[a]			Shapiro-Wilk		
	Statistic	df	Sig.	Statistic	df	Sig.
Semarang	,076	32	,200*	,984	32	,900
Medan	,112	32	,200*	,963	32	,336
Palembang	,109	32	,200*	,957	32	,220

*. This is a lower bound of the true significance.
a. Lilliefors Significance Correction

Normality test (Table 3) showed significance value> 0.05 either Kolmogorof-Smirnov or Shapiro-Wilk test showed that the rainfall data in the three cities followed the normal distribution.

3.2.2 Annual Rainfall Trends

The annual rainfall trends in the three cities (Figure 3) show a negative trend in Semarang city of -5.6331 mm / year, while in the other two cities are positive trend (Medan 14,583 mm/year and Palembang 4,814 mm/year)

3.2. 3 Monthly Rainfall

Normality test (Table 4) showed significance value> 0.05 either Kolmogorof-Smirnov or Shapiro-Wilk test showed that the rainfall data in the three cities followed the normal distribution.

Further analysis were conducted on monthly rainfall trends (Table 4). The negative rainfall trend in Semarang city from 1985-2016 periods is also explained by the monthly rainfall which is dominated by negative trend. There are 7 months with negative trends and 5 months with positive trends. In general, the city of Semarang has decreased both in the rainy season and in the dry season.

Table 4 Monthly Rainfall

Month	Tren					
	Smg[1]	R^2	Mdn[2]	R^2	Plb[3]	R_2
Jan	-2.785	0.016	-0.828	0.0063	**2.432**	0.086
Feb	-0.588	0.001	-0.4038	0 .002	-0.284	0.001
Mar	-2.803	0.089	**1.5984**	0.0223	**0.758**	0.003
Apr	**2.192**	0.080	**0.5717**	0.0062	**3.007**	0.146
Mei	**0.813**	0.010	**1.6226**	0.025	-1.254	0.029
Jun	-1.636	0.024	**0.9958**	0.014	-1.021	0.015
Jul	**0.423**	0.005	**2.7155**	0.0776	-0.907	0.017
Aug	-0.651	0.010	**3.2577**	0.0878	**0.038**	0.000
Sep	**1.391**	0.022	**1.4558**	0.0219	-1.477	0.019
Oct	-1.477	0.018	**3.7513**	0.201	-0.049	0.000
Nov	**1.167**	0.012	-1.0876	0.0116	**1.712**	0.017
Des	-2.937	0.054	**0.9337**	0.005	**1.859**	0.034

Medan city showed positive rainfall trend in period 1982-2016 with 9 month positive trend and 3 month negative trend. The high rainfall increase occurred in August (slope 3.2577) and October (slope 3.37523).

While Palembang has rainfall with a rising trend with relatively small increase of the city of Medan with the number of months with the positive trend equal to the number of months with negative trends. The increase in rainfall occurs in the rainy season and decreases rainfall during the dry season. The highest increase in rainfall in January (2,432) and April (3,007).

3.2.4 Monthly Decadal Rainfall Patterns

In Indonesia, rainfall patterns are influenced by several factors, including monsoon, Inter-Tropical Convergence Zone (ITCZ), IODM, ENSO, and other regional circulation found in the Pacific Ocean and Indian Ocean[8]. According to Aldrian and Susanto, [10] Indonesia's rainfall pattern is divided into three main regions as follows: a) Region A/Monsoonal characterized with a one peak rainfall season between November-March; b) Region B/ Equatorial characterized with two peak rainfall season in October-November and March-Mei; c) Region C/ Local climate area, it is characterized with one peak in June-July and one trough in November-February.

Monthly decadal rainfall analysis used to find out rainfall pattern changes. From Fig. 4 shows that Semarang is included in region A, characterized with a clear rainy and drought pattern with a rainfall peak in December-January-February (DJF) and peak dry season in June-July August (JJA). While the climate pattern of Palembang and Medan is an equatorial climate pattern, characterized with two rainfall peaks (October and March). In this climate pattern most of the year is the rainy season.

From the 10-year trend of monthly rainfall in Semarang city, there was a consistent decrease of rainfall during rainy season (DJF) period 1996-2006 and 2006-2006 to rainfall period 1986-1996 (baseline), decreasing rainfall in DJF period Especially in December-January while in February experienced an increase in the period 1996-2006 and then decreased in the period 2006-2016 although the value is still above the baseline period.

Climatic patterns in the city of Medan are included in an equatorial climate type characterized by two peak rainy season in July-December with two peaks of the nearby rainy season. The Medan rainy season observed in the period 1986-1996 occurred in August and November, during the period 1996-2006 peak rainy season shifted in october and December, and currently peak rainfall is still in the same month with the tendency of decline in month October and consistent increases in December and the highest rainfall in the period 2006-2016.

From these conditions can be searched the conclusion whether there is a change or not, it is indicated by the difference in the average of each period. To take conclusion then t test is performed to climate data per period at 95% confidence interval (table 5).

In general there is no real difference between the period III to II and Period II against Period I in the city of Semarang and Palembang. While in the city of Medan there is a significant change in rainfall between periods

The quarterly rainfall shows the current condition in Semarang city decreasing rainfall trend in rainy season (-14.54) and the increase of rainfall in dry season / JJA (7.5905), The same thing happened in the city of Medan experienced a decreasing trend of rainfall in the dry season / DJF (-7.6673)

Table 5 Test of Different Rainfall 10 Year

Periode	Semarang			Medan			Palembang		
	t Statistic	T-table	Conclusion	t-Statistic	T calc	Conclusion	t calc	t table	Conclusion
III-II	-0,131	2.201	Accept H_0	-2,203	2.201	Reject H_0	1,859	2.201	Accept H_0
II-I	-0,790	2.201	Accept H_0	2,933	2.201	Reject H_0	0.344	2.201	Accept H_0
III-I	-0,539	2.201	Accept H_0	1.474	2.201	Accept H_0	1,390	2.201	Accept H_0

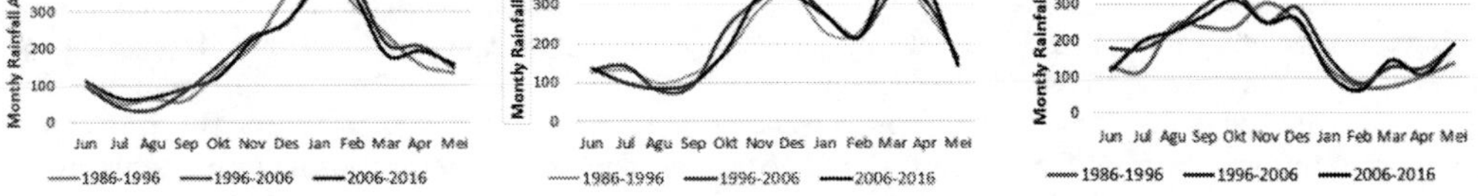

Fig. 4 Decadal Rainfall Patterns

3.2.5 Test of Different Seasonal Rainfall

To ensure the significance of the changes that occur, the statistic test of annual rainfall between decades was done. Table 6 presents the results of different test between decades showing that only the city of Medan that has an absolute value t arithmetic greater than t table so that it can be drawn conclusion H0 In rejection, it means that there are significant differences in period I - Period II (rainfall increase) and Period II - Period III (rainfall decrease)

3.2.6 Seasonal Trends

In general, the seasons are derived to rainy season (December - January - February / DJF), intermediate seasons (March - July - May / MAM), dry season (June - July - August / JJA) and In between (September - October - November / SON).

Table 6 Seasonal Trends

Season	Semarang[1] Decade			Medan[2] Decade			Palembang[3] Decade		
	I	II	III	I	II	III	I	II	III
DJF	-8.2648	-2.5043	-14.54	**0.8505**	-11.105	-7.6673	**14.448**	**2.698**	**1.0574**
MAM	-3.6121	**5.957**	-1.8097	**3.6768**	**2.0747**	-16.211	-3.9879	**2.0723**	-3.5325
JJA	-3.8057	-1.992	**7.5905**	**2.6121**	**5.2343**	-6.1673	**8.8707**	**2.7788**	**1.0394**
SON	**7.2978**	**1.0695**	-1.9034	**0.7313**	**2.0747**	**16.211**	**6.2194**	-5.2071	**3.9503**

[1]Stasiun Meteorology Ahmad Yani; [2] Station Climatology Belawan; [3]Stasisun Climatology Kenten;

Table 7 Test of Different Rainfall in Rainy Season

Decade	Semarang t calc	Semarang t table	Semarang decision	Medan Asymp. Sig (2 tailed)*	Medan decision	Palembang t calc	Palembang table	Palembang decision
III-II	-0,466	2.262	Accept H_0	0,959	Accept H_0	-0,029	2.262	Accept H_0
II-1	-0,082	2.262	Accept H_0	0.508	Accept H_0	-0,659	2.262	Accept H_0
III-1	-0,429	2.262	Accept H_0	0,959	Accept H_0	-0,483	2.262	Accept H_0

* *Wilcoxon Sihned Ranks Test* (Not Normal distribution)

Table 8 Test of Different Rainfall in the Dry Season

Decade	Semarang t calc.	Semarang t table	Semarang decision	Medan T calc	Medan t table	Medan decision	Palembang t calc	Palembang t table	Palembang decision
III-II	-0,656	2.262	Accept H_0	-0,485	2.262	Accept H_0	-0,198	2.262	Accept H_0
II-I	-0,693	2.262	Accept H_0	1,336	2.262	Accept H_0	-0,490	2.262	Accept H_0
III-1	-0,205	2.262	Accept H_0	0.966	2.262	Accept H_0	-0,951	2.262	Accept H_0

Table 9 Test of Different Rainfall in SON and MAM

Dekade	Semarang t calc	Semarang t table	Semarang decision	Medan T calc	Medan t table	Medan decision	Palembang t calc	Palembang t table	Palembang decision
MAM									
III-II	-0,537	2.262	Accept H_0	0,16	2.262	Accept H_0	0,584	2.262	Accept H_0
II-I	0,705	2.262	Accept H_0	1,466	2.262	Accept H_0	0,719	2.262	Accept H_0
III-1	0,98	2.262	Accept H_0	1,577	2.262	Accept H_0	1,649	2.262	Accept H_0
SON									
III-II	-0,300	2.262	Accept H_0	-0,548	2.262	Accept H_0	0,687	2.262	Accept H_0
II-I	0,781	2.262	Accept H_0	1,479	2.262	Accept H_0	0,750	2.262	Accept H_0
III-1	0,423	2.262	Accept H_0	0,754	2.262	Accept H_0	1,903	2.262	Accept H_0

The results of analysis (table 6) note that the rainfall trend during peak rainy season / DJF in Semarang city of rain has decreased. While in Medan the peak of the rainy season occurred in the SON months, with a trend that continues to increase from the period I, II and III. In the city of Palembang, seasonal trends tend to rise in all seasons in each period, except in the MAM month which has a negative trend in period I and III, as well as period II in SON.

While the dry season / JJA rainfall in Semarang City decreased in the period I and II, while the third period of rainfall trend is positive. In the city of Medan JJA months is the beginning of the rainy season with a positive trend in the period I and II while experiencing increasing in the period III.

Table 7.8 & 9 is the result of the test of the seasonal rainfall data (DJF / MAM / JJA / SON) with the value of t-statistic below the critical point (t table) so the decision taken is to accept H0, it is concluded that the average of bulk Seasonal rain between periods did not differ significantly.

3.2.7 Rainfall Extreme Events Identification

The identification of extreme climate is part of the effort to identify climate change in a region. Identified rainfall is heavy rainfall category (> 50mm / day) and very heavy category (> 100mm / day)..

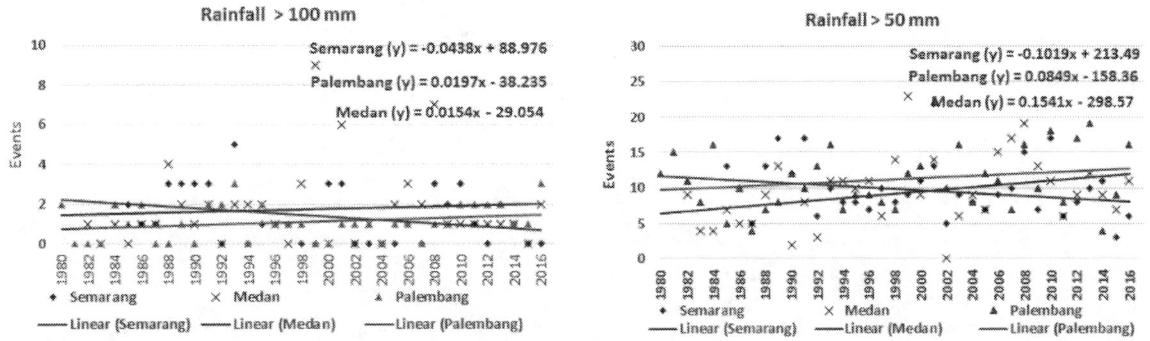

Fig. 5. Rainfall Trends over 50 mm / day and 100 mm / day

Figure 5 shows an increase in extreme climate of rainfall above 50 mm / day and above 100m / day in Medan City and Palembang City, while in heavy rainfall trend and very heavy drop in Semarang City.

4 Conclusions

- Trends in air temperature in Semarang City, Medan City and Palembang increased by 0.0279 ° C / yr , 0.00331 ° C / yr (1982-2016) 0.0323 ° C / year (1980-2016).
- Trend of rainfall in the three cities showed no real difference between periods, but has a tendency to fall in the city of Semarang and the tendency to ride in the city of Medan and the city of Palembang.
- Trend of heavy and very heavy rainfall in the city of Medan and Palembang, while the downward trend in the city of Semarang.
- Changes in temperature trends become a strong indicator of climate change occurring in all three cities.

References

1. Mercy Corps, ACCCRN, URDI, ISET. *Kajian kerentanan dan adaptasi terhadap perubahan iklim di kota semarang* 2010. 2010.
2. Arifin B. *Pemanasan Global dan Ketahanan Pangan Nasional.* J Pangan. 2009;Edisi No 5.
3. Yananto A, Dewi S. *Analisis Kejadian El Nino Tahun 2015 dan Pengaruhnya Terhadap Peningkatan Titi Api di Wilayah Sumatera dan Kalimantan, Indonesia*. 2016;17(1):11-19.
4. *Kebencanaan I, Teraktual B. Info bencana.* 2015;(September):1-4.
5. *Kementerian Perencanaan Pembangunan Nasional/Badan Perencanaan Pebangunan Nasional (Bappenas). Rencana Aksi Nasional Adaptasi Perubahan Iklim.* 2014.
6. OECD. *Cities and climate change: national governments enabling local action. Change.* 2014:276. doi:10.1787/9789264091375-en.
7. *IPCC. Climate Change 2007 The Physical Science Basis* The. Vol **53**.; 2007. doi:10.1017/CBO9781107415324.004.
8. Aldrian E, *Budiman, Mimin Karmini. Adaptasi dan Mitigasi Perubahan Iklim di Indonesia.* 2011;(November).
9. IPCC. *Summary for Policymakers.*; 2014. doi:10.1017/CBO9781107415324.
10. Aldrian E, Dwi Susanto R. Int J Climatol. 2003;**23**(12):1435-1452.

 https://doi.org/10.1051/e3sconf/20183109018

Community Capacity in The Face Of Landslide Hazards in the Southern Of Semarang City

Heri Tjahjono [1,*], *Suripin* [2], and *Kismartini* [1]

[1]Doctoral Program of Environmental Science, School of Postgraduate Studies, Diponegoro University, Semarang-Indonesia

Abstract. The study was done at Semarang, Central Java. The aims of the study are: (a) to know the variation in the level of community capacity in dealing with landslide hazards in the southern of Semarang city; (B) to know the factors that affect the capacity of communities in facing the hazards of landslides. This research was conducted by the sample method with a sample of 198 people, taken by purposive sampling. Samples taken are people living in areas that have experienced landslide or in areas that are expected to be vulnerable to landslides. The variables used in this research are (1) regulatory and institutional capacity in the prevention of landslide disaster, (2) early warning system in community, (3) education of disaster skill training, (4) mitigation to reduce basic risk factor, and (5) Preparedness on all fronts. Data were collected with questioner and interviews. Data analysis was performed by percentage descriptions, and map overlay analysis using ArcGIS release 10.3 technology. The result of the research shows that there are 5 variations of society's capacity level in facing the landslide hazard in southern Semarang city, that is the very high capacity of society as much as 4,35 % of the people that researched, the high community capacity is 7,25 % of the people that researched, the medium community capacity is 30.43 %. of the people that researched, low community capacity as much as 36.23 % of the people that researched and very low community capacity as much as 21.74% of the people that researched. Based on the result of overlay map of landslide threat in southern Semarang City with map about variation of community capacity level in facing landslide hazard indicate that community capacity with very high criterion and high occupancy area of threat of landslide with high and medium criterion which have been experienced landslide. While the capacity of the community with the criteria of medium, low and very low occupies the threat of landslide areas with high, medium, low and very low criteria that have never experienced landslide. The existence of the experience of a landslide disaster is one of the factors that encourage the community to increase the community capacity in facing the landslide.

1 Introduction

Based on the data of several types of natural disasters occurring in Indonesia, landslide are ranked third compared to other disasters seen from the number of fatalities form 2010 to 2014. More than 15% or as many as 3.074 people died from landslide. These year to year incidents in Indonesia show a significant number. The presence of a relatively high landslide incident followed by great losses requires serious attention [1].

Central Java is one of the provinces in Indonesia that is vulnerable to natural disasters, such as earthquake, volcanic eruption, and landslide. The location is traversed by the path of young mountains resulting in its several places are still moving and prone to landslide. Moreover, the global climate changes has caused an increase in rainfall. However, the land quality due to deforestation is declining. These result in high landslide occurrences. Landslide incidents in Central Java from 2011-2015 has the highest frequency compared to other disasters, i.e. 568 incidents [1]. Primarily, for landslide, the data shows that the incidents in the city of Semarang has a high frequency. In 2012 itself, there are 39 incidents, 44 times in 2013, and 123 times in 2014[2]. There is a significant increase from 39 times of incidents in 2012 to 123 times in 2014. The escalation of landslides is always followed by increased loss of

* Corresponding author: heriridlo@yahoo.com

both casualties and properties. This is due to the lack of disaster risk management.

The example of a landslide incident in the city of Semarang was the one happened in Terangkil Sejahtera and Terangkil Baru Residencies on January 23rd, 2014. The disaster had destroyed 23 houses, meanwhile others are severely damaged and unlivable.

Semarang is one of the cities in Central Java that often experiences landslides. Geologically, areas in Semarang consist of some formations. They are Damar, Marine/Kalibiuk, Kaligetas/Notopuro, Central Ungaran, Jongkong, and Alluvium formations. Each of them composes of different types of rock and soil. The existing rocks varies from marine layers, damar formation sedimentary rock, white claystone, volcanic sedimentary breccia, volcanic sedimentary lava of Central Ungaran Mount, and volcanic rocks of Old Ungaran Mount. These variations may lead to various landslide disasters [3]

The impact of disaster varies. It depends on the condition of environmental vulnerability and the capacity of society. If the society is unprepared (having low capacity), disasters will cause panic, prolonged suffering or sadness, such as; injuries, deaths and economic pressure resulting from loss of jobs and property, loss of family members and damaged to infrastructure, also environment. This means, the society management towards disasters needs to be reorganized [4].

Human or society in the context of a disaster is the object as well as subject of the disaster itself. They not only deal with pre-disaster threat, but also take the risk of losing lives and belongings due to the disaster, and they still have to face the post-disaster condition; recovering themselves physically and mentally. Therefore, there needs to be an effort to improve the ability or capacity of the society in facing disaster [5]

The objectives of this research is (a) finding out varies level of society capacity in facing the danger of landslide, in southern Semarang city. (b) finding out factors influencing the society capacity in coping with landslide.

The Concept of Landslide

Landslide is a process of transfer of soil mass or rocks with a sloping direction from its original position, so that it is separated from a stable mass due to the influence of gravity; rotational and translational movements [6]. Landslide is one type of soil mass or rock motions, or a mixture of both, moving out or down falling the slope due to the disturbance of soil stability or the constituent rocks of the slope [7]. Landslide is a potential mechanism for mobilizing and spreading pollutant [8]. Landslide and rock falls may occur in the city, and the government pours a considerable spending on investigation, design and mitigation implementation, and preventive measures to reduce the liability of life loss and economic losses [9].

According to a guideline of the spatial planning on areas proning to landslide, the process triggering the occurrence of landslides is the infiltration of water into the soil and a very high level of the slopes. The present of water seeping into the soil will increase its weight due to high rainfall. If the water penetrates to the impermeable soil that acts as the slip-plane, it will become very slippery and the obsolete soil above it will move along the slopes. Fakhruddin [10] stated that landslide is a transfer of slopes forming materials in the form of rocks, or mixed material downfalling or out of the slope.

Suripin [11] defines, landslide is a form of erosion where the transportation or movement of the soil mass occurs at the same point in and in a relatively large volume. In terms of movement, there are some erosion caused by. They are creep, rock fall, and mud flow. The moving mass in the landslide is usually massive and resulting in fatalities; environmental, agricultural land, residencies, possession and infrastructural damages, not to mention loss of lives. According to Directorate of Geology and Environmental Governance, soil movement or landslide is a product of the balance disturbance process of the slope causing the soil mass and rocks to down move.

Based on the definitions above, thus, landslide in this study is defined as mass movement type which move/slide or rotate, and the flowing object are mixed/modified material, soil and rocks due to gravity, that differs with others in terms of movement and less water content.

The Capacity of the Society

Capacity is elaborated as a combination of all strengths existing in a community, society and organization that may reduce the impact of a risk or a disaster [12]. Some elements influencing the capacity are policy, readiness, and social participation. Increasing the capacity is aimed to develop a safety culture, where community members are aware of the present danger, knowing how to self-protect, and supporting the effort to protect others and society in general.

Institutional strengthening, whether governmental, public, or even private is the key factor in the effort of disaster management. The community has an important role to prevent calamity. Disaster prevention is a series of activities undertaken to reduce the risk of disaster, through the reduction of threats or those vulnerable to disaster threat [13]. Capacity is a set of ability which enable the society to have higher endurance against threatening danger, and increasing both a defense as well as capability of the community to overcome the impact of harmful circumstance. This may relate to sources, skills, knowledge, organizational competence and attitude to act and response toward a crisis [14].

The risk of an occurred disaster relates to human capacity in dealing with it. Meaning, the action directing to a catastrophe management will be able to minimize the occurrence, so that the massive loss is avoidable. In relation to that, the social capacity in the area prone to disaster needs to be investigated for developing. The next is building the institutional capacity in response to dangerous occurrence to effectively overcome it [10].

There are many researches relevant to threats/dangers and landslide vulnerability. The prior studies mainly focuses on the factors causing landslides and the efforts to cope with landslides physically, socially, and policy [15,16]. However, only few researchers focus on the capacity of the communities in the management of landslide. Whereas, the community is the most disadvantaged party from the disaster. Hence, as a researcher, the writer would like to review more about the management of disaster risk reduction by increasing the capacity of the community as an effort to minimizing loss and casualties of landslides [15,17].

Capacity is a set of ability which enable the society to have higher endurance against threatening/disastrous danger, and increasing both a defense as well as capability of the community to overcome the impact of harmful circumstance. This may relate to sources, skills, knowledge, organizational competence and attitude to act and response toward a crisis [14,18].

There are several kinds of capacity in dealing with disaster. (1) Physical capacity, ability to obtain materials/stuffs needed to rebuild infrastructures in the community; (2) Social Economic capacity, when the demand for various goods is available, the need for organized special skillful human resources also emerges, in order to rebuild their community; (3) Organizational/institutional capacity, the presence of family and community based institution with its leader and systemized decision making; (4) Economic capacity, the presence of ability in business sector to repair and restore the community economy; (5) Motivational capacity, positive and strong motivational people, with determination to survive, love, and care for others, courage and willingness to help each other.

Vulnerability and calamity threats make it an absolute capacity to develop. The bigger the capacity and community ability to manage disaster, the smaller impact on loss and casualties will be. This is what is pioneered in disaster risk reduction.

2 Methods

This research is conducted in the southern of Semarang city, the Province of Central Java. Considering the frequent occurrence of landslides in Semarang causing loss, both belongings and human lives. The research on 'social capacity in dealing with landslide' is an observational one, using quantitative descriptive analysis. It makes a description about on-field condition with systematical, factual, and accurate illustration of the facts, characteristics, and reviews the correlation between investigated phenomena [15].

The population of this research is the field physical condition and the residences/people living in a developed area in Semarang. The sample is taken to field unit and conducted purposively, based on certain considerations and objectives. The consideration stated is that people living in areas having experienced landslides or are prone to landslide. The sample community is 196. While the physical field to sample a total of 69 terrain units.

Material used covers soil and rocks, which will be analyzed in soil laboratory. The devices used in this research are GPS, Geology Compass, Clinometer, Abney level, Indonesia Earth Map, satellite imagery, foothills Map, Geology Map, Population Distribution and Density Map, Soil Map, Plastics for sample, Soil Knife, Analysis Scale, A set of soil texture test device, and a set of soil consistency test device.

The researched variables are (1) **Variable of Threat**, threat is a potential disaster in area scale, time and inhabitants, this includes some indicators: (a) Geomorphology (declivity of slopes), (b) soil texture, (c) rocks type (geology), (d) history of landslide occurance, (e) structure of rocks layering, (f) Water plan of foothills/seepages, (g) Vegetation density, (h) excavating/cutting off slopes, (i) rainfalls; (2) **Variable of Social Capacity.** Scoring for Social capacity indicator by using *Framework for Actions Hyogo* basically consists of five components, (a) regulation and institution of disaster management, (b) early warning and disaster risk assessment, (c) disaster education, (d) basic risk factor lessening, and (e) all-lines readiness development.

The first data taken from the field is on landslide threats. This covers (a) Geomorphology (declivity of slopes), (b) soil texture, (c) rocks type (geology), (d) history of landslide occurrence, (e) structure of rocks layering, (f) Water plan of foothills/seepages, (g) Vegetation density, (h) excavating/cutting off slopes, (i) rainfalls. The second data is about the Social capacity in coping with landslide disaster. It covers (a) regulation and institution of disaster management, (b) early warning and disaster risk assessment, (c) disaster education, (d) reduction of basic risk factor, and (e) all-lines alertness development. The later data is obtained by doing questionnaires and interviews with the sample residents. It is then analysed using enhancement scoring, AHP analysis, and *overlay* analysis assisted by ArcGIS technology.

https://doi.org/10.1051/e3sconf/20183109018

3 Result and Discussion

3.1. Description of Researched Area

This research is conducted in the city of Semarang which is the capital of Central Java. Geographycally, it is positioned in between 109^0 35' – 110^0 50 ' East Longitude and 6^0 50' – 7^0 10' South Latitud and occupying an area of 38.449,94 Ha. Semarang consists of 16 districts with administered borders as follows: in the North with Java Sea, in the South with Semarang regency, in the East with Demak and Grobogan Regencies, and in the West with Kendal Regency.

The researched area has a quite high rainfall. The maximum rainfall is 2265 mm/year, can be seen in Kandri Station, and the minimum is 1483 mm/year can be seen in Candi Rain station. According to Schmidt Ferguson the researched area has the same type of climate, type C (slightly wet), grade Q (comparison between the average of dry months and wet months) which is slightly different.

Geological condition, both structural and rock formation will affect to the existence of the main rock and soil development., so that the soil physical and geotechnical characteristics are unable to separate from the main rock character. The researched area consists of some rock formations. (1) Alluvium sedimentary, (2) Kaligetas formation, in Van Bemmelen (1941) is called Notopuro formation, (3) Damar formation, (4) Kalibeng formation (5) Kerek formation, by Van Bemmelen (1941) Kalibeng dan Kerek formations are called Kalibiuk/Marine formations, (6) volcanic sedimentary lava of Central Ungaran Mount, and (7) Jongkong formation or volcanic rocks of Old Ungaran Mount.

3.2 Result of the Study

3.2.1 Social Capacity in Generally Coping with Landslide Disaster

Based on tabulation result and research data analysis, the social capacity in generally coping with landslide disaster is describes in Table 1.

Table 1. Social Capacity in Coping with Landslide

No	Value Interval	Criteria	Frequency	Percentage (%)
1	1 - < 1.8	Very Low	15	21.74
2	1.8-< 2.6	Low	25	36.23
3	2.6 - < 3.4	Medium	21	30.43
4	3.4 - < 4.2	High	5	7.25
5	4.2 - 5	Very High	3	4.35
		Total	69	100

Table 1. explains that the social capacity in generally dealing with landslide is 21.74% of the investigated respondence belongs to very low criteria, 36.23 % belongs to social capacity with low criteria, and the medium criteria is 30.43 %. Next is 7,25 % for those belong to high criteria, and 4,35 % for Social capacity with very high criteria.

In detail, the social capacity for each sub-variable is described as follows.

A. Regulation and institution of disaster management

Table 2. The Social Capacity in Regulation and Institution

No	Value Interval	Criteria	Frequency	Percentage (%)
1	0.2 - < 0.36	Very Low	10	14.49
2	0.36-< 0.52	Low	13	18.84
3	0.52 - < 0.68	Medium	28	40.58
4	0.68 - < 0.84	High	13	18.84
5	0.84 - 1.00	Very High	5	7.25
		Total	69	100

Table 2. shows that the social capacity in regulation and institution in dealing with landslide is 14,49 % of the investigated respondence belongs to very low criteria, 40.58 % belongs to medium criteria, 18.84 % is high criteria, and 7.25 % of the respondence includes in very high criteria rest in lower case.

B. Early Warning System

Table 3. The Social Capacity in Early Warning System

No	Value Interval	Criteria	Frequency	Percentage (%)
1	0.2 - < 0.36	Very Low	35	50.72
2	0.36-< 0.52	Low	26	37.68
3	0.52 - < 0.68	Medium	8	11.59
4	0.68 - < 0.84	High	0	0
5	0.84 - 1.00	Very High	0	0
		Total	69	100

Table 3 explains that 50,72 % of the investigated respondence of the social capacity on early warning in coping with landslide belong to very low criteria, 37.68 % included in low criteria, 11.59 % of the respondence belong to medium, while for high and very high criteria the percentages are not found.

C. Education and Training in Disaster

Table 4. The Social Capacity in Disaster

No	Value Interval	Criteria	Frequency	Percentage (%)
1	0.2 - < 0.36	Very Low	18	26.09
2	0.36-< 0.52	Low	28	40.58
3	0.52 - < 0.68	Medium	13	18.84
4	0.68 - < 0.84	High	9	13.04
5	0.84 - 1.00	Very High	1	1.45
		Total	69	100

Table 4 shows that the social capacity on disaster education in facing landslide is 26,09 % of the investigated respondence belong to very low criteria , 40.58 % is included in low criteria, 18.84 % belong to medium, 13.04 % is high criteria, and 1.45 % of them belong to very high.

D. Reduction of Basic Risk Factors

Table 5. The Social Capacity in the Reduction of Basic Risk Factors in Disaster

No	Value Interval	Criteria	Frequency	Percentage (%)
1	0.2 - < 0.36	Very Low	9	13.04
2	0.36-< 0.52	Low	21	30.43
3	0.52 - < 0.68	Medium	20	28.99
4	0.68 - < 0.84	High	14	20.29
5	0.84 - 1.00	Very High	5	7.25
		Total	69	100

Table 5 explains that 13,04 % of the investigated respondence in this criteria belong to very low, 30.43 % belong to low, 28.99 % is medium criteria, 20.29 % is included in high, and 7.25 % is very high.

E. All-Lines Alertness Planning.

Table 6. The Social Capacity in All-lines Alertness Development

No	Value Interval	Criteria	Frequency	Percentage (%)
1	0.2 - < 0.36	Very Low	13	18.84
2	0.36-< 0.52	Low	28	40.58
3	0.52 - < 0.68	Medium	12	17.39
4	0.68 - < 0.84	High	9	13.04
5	0.84 - 1.00	Very High	7	10.14
		Total	69	100

Based on Table 6 above, it is shown that 18.84 % of the investigated respondence belongs to very low criteria, 40.58 % of them belong to low, 17.39 % is medium, 13.04 % is included in high, and 10.14 % of the respondence belong to very high criteria.

3.2.2 Factors Influencing The Social Capacity in Facing the Danger of Landslide

The result of this study shows that, there are some factors influencing the high-low or strength-weakness of the community capacity in dealing with landslide disaster, such as; (a) regulation and institution of disaster management, (b) early warning system, (c) education, training and skills on landslide disaster, (d) mitigation to reduce basic risk factors, (e) all-lines alertness planning, (f) experience in disaster, (g) concern attitude, (h) economy capacity, i.e. the presence of economy-business ability to repair and restore the community economy, (i) information network, affluence of access and management for useful information will increase the community capacity in facing disaster, (j) building organizational cooperation network

3.2.3 Discussion

This study on the community capacity in coping with landslide risk shows a result dominated by very low criteria, low, up to medium. Owning the very low, low, to medium criteria are generally the areas with the same characters of landslide threat, i.e. very low, low, and medium. While the community capacity with high criteria, generally occurs in areas with the characters of medium, high, and very high landslide threat. For those belong to high and very high criteria, generally exist in areas that had experienced landslide before.

The community capacity relating with regulation and institution for managing landslide disaster, is further explained that most of the community have not noticed about regulations or local government law which regulate about disaster in the city of Semarang. This is due to the lack of regulation socialization to the community. Based on the interview with BPBD, the government has socialized the regulation, starting from district level to the head of village (Lurah). However, this does not continue to the level of Neibourhood/Community associations (RT/RW), causing part of the community not to know about disaster law or regulation.

Community capacity which relates to early warning system has very low and low criteria. The result of this study shows that most areas do not have early warning system toward landslide disaster, nor the information shared about it. Moreover, there have not been cooperation between the community with other institution about landslide disaster early warning, nor any evacuation paths relating to a disaster in the vicinity areas.

Social capacity in education, training and skills on landslide disaster, has very low, low, and medium criteria. The study shows that in most area, the community has not received training on landslide risk management, disaster simulation events in the vicinity of residential area. Moreover, there are still lacking of facility provided by the community or government to facilitate them accessing information about landslide disaster.

Community capacity in mitigation to reduce basic risk factors, on most areas is still dominated by very low, low, and medium criteria. The result of this research shows that in most of the areas, there are absence of, (a) social affair, in its effort to decrease the threats and the residents vulnerability toward landslide, (b) requirements to fulfill, in constructing buildings related to landslide risk, (c) information, stated that the particular area is a prone to landslide, (d) special groundwork, done by the community to face landslide threat that is likely to happen in the future.

Community capacity of all-lines alertness in most areas is still dominated by very low, low, and medium capacity. Result of study shows that in most areas, there are absence of (a) mechanism of emergency management disaster to lower the risk, (b) policy regulating emergency management disaster and its risk-decline, (c) appointed location as a shelter in case a landslide happens, (d) fund availability that is specially allocated for managing disaster in sub-district level, (e) logistic assistant availability in its effort to effectively control the crisis during landslide, (f) special officer in the level of

Neighborhood Organization (RT) who is responsible for decision maker when landslide occurs, (g) special mechanism in the process of post-disaster recovery.
Experience about disaster may support to induce the community capacity. A person who had experienced a disaster, survived from it, even only had seen will indirectly motivate him/herself to survive another disaster. This is called inducing self-capacity to reduce disaster risk when happens.

Concern/caring attitude may support the increase of community capacity in coping with landslide disaster. A person's self-concern to survive from disaster and care for helping other will strengthen human capacity to deal with disaster. Caring attitude towards others may invite the community in a landslide mitigation.

4. Conclusion and Recommendation

The research and discussion conclude that: (1) There are 5 various levels of the community capacity in dealing with landslide disaster; very low, low, medium, high, and very high. On most areas the criteria are dominated by very low, low, up to medium. Owning the very low, low, to medium criteria are generally the areas with the same characters of landslide threat, i.e. very low, low, and medium. While the community capacity with high criteria, generally occurs in areas with the characters of medium, high, and very high landslide threat. For those belong to high and very high criteria, generally exist in areas that had experienced landslide before. (2) Factors influencing the ups and down or strength and weakness of community capacity in dealing with landslide, are; (a) regulation and institution of disaster management, (b) early warning system, (c) education, training and skills on landslide disaster, (d) mitigation for reducing basic risk factors, (e) all-lines alertness planning, (f) experience in disaster, (g) concern attitude, (h) economy capacity, i.e. the presence of economy-business ability to repair and restore the community economy, (i) information network, affluence of access and management for useful information will increase the community capacity in facing disaster, (j) building organizational cooperation network. Based on the results of this research, it is recommended that the government (BPBD), along with civil state apparatus, stuffs of villages, community leaders, and the society must be cooperating, assisting each others in landslide risk management, and trying to increase the community capacity in order to reduce or minimize disaster risk due to landslides.

References

1. BNPB, *Panduan Perencanaan Kontijensi Menghadapi Bencana*. Edisi II Jakarta, (2015).
2. BPBD. *Data Bencana Tahunan di Kota Semarang,* Badan Penanggulangan Bencana Daerah Kota Semarang, (2014).
3. Thanden, Sutisna, *Peta Geologi Lembar Semarang-Magelang*, Direktorat Jendral Geologi Dan Sumber Daya Mineral, Bandung, (1996).
4. Hidayati, I. Y., & Setyono, J. S, *Jurnal Teknik PWK*, 4(4), 591-603, (2005).
5. Lassa, J., Pujiono, P., Pristiyanto, D., Paripurno, E. T., Magatani, A., & Purwati, H. *Pengelolaan Risiko Bencana Berbasis Komunitas (PRBBK).* Jakarta: PT. Gramedia Widiasarana Indonesia.
6. Peraturan Menteri Pekerjaan Umum No. 22/PRT/M/2007, Pedoman Penataan Ruang Kawasan Rawan Bencana Longsor, *Departemen Pekerjaan Umum*, (2007).
7. Imanda, Amy. *Jurnal perencanaan Wilayah dan Kota* vol 24 no 2. *TML Energy*, (2013).
8. Göransson, G., Norrman, J., Larson, M., Alén, C., & Rosén, L, *Journal Science of the Total Environment,* 472, 481-495, (2014).
9. Kwong, A. K. L., Wang, M., Lee, C. F., & Law, K. T, *Journal Engineering Geology*, 76(1), 27-39, (2004).
10. Fakhruddin, S. H. M., & Chivakidakarn, Y, *International Journal of Disaster Risk Reduction*, 9,159-180, (2014).
11. Suripin, *Pelestarian Sumberdaya Tanah dan Air. Andi*, Yogyakarta, (2002).
12. UN-ISDR (United Nation secretariat of the International Strategy for Disaster Reduction), *Living with risk : A global review of disaster reduction initiatives,* Geneva: UN Publications, (2004).
13. Undang-undang Republik Indonesia Nomor 24 tahun 2007 tentang Penanggulangan Bencana, (2007).
14. Andersson-Sköld, Y., Bergman, R., Johansson, M., Persson, E., & Nyberg, L, *International Journal of Disaster Risk Reduction,* 3, 44-61, (2013).
15. Karnawati, Dwikorita, "Geology for Regional Development," *Modul Pelatihan Jabatan Fungsional Perencana Madya,* Teknik Geologi Universitas Gajah Mada, Yogyakarta, (2005).
16. Yilmaz, I, *Journal Geosciences*, 35(6), 1125-1138, (2009).
17. Paripurno, Eko Teguh, *Penanggulangan Bencana oleh Komunita,* Yogyakarta: Pusat Studi Manajemen Bencana UPN Veteran Yogyakarta, (2006).
18. Paripurno, E.T. Manajemen Berbasis Komunitas : Seperti apa?. *Bahan Diskusi pada Lokalatih Bencana Kulonprogo*. Kulonprogo, 30-31 Januari 2001, (2001).
19. Nazir, M, *Research Metod*, Ghalia Indonesia. Jakarta, (2011).

The Land Use Change From Agricultural to Non-Agricultural in Bungo Regency, Jambi Province, Indonesia

Fajar Ifan Dolly[1], *Kismartini* Kismartini[1,2], *Hartuti* Purnaweni[1,2*]

[1] Master Program of Public Administration, Faculty of Social Science and Political Science, Diponegoro University
[2]EDP [2] Master and Doctorate Program of Environmental Science, School of Postgraduate Studies, Diponegoro University, Semarang 50241

Abstract. This study aimed at observing the development of agricultural land use in Bungo Regency, Jambi Province, for other purposes, such as plantation, mining, and other commercial buildings. According to the sustainable agriculture supposed by the government, a change in land use has become an important issue to be taken into account as such that the change does not tend to damage the environment. The research findings from Bungo Regency demonstrated the change in agricultural land into copra and rubber plantation areas. Local people had changed their mindset towards reluctance to become farmers, which caused the loss of farmer regeneration and weakened the farmer exchange rate towards the agricultural commodities.

1 Introduction

Although many forest area have been converted into agricultural land [1], but the opposite in which change in agricultural land use is a practice frequently occurred in many parts of Indonesia, such as in which agricultural land was turned into industrial area [2]. This is despite the fact that agricultural land is crucial for food security in this highly populated country. Therefore, one of the most important issues in today's development is sustainable agriculture. To this end, priority must be put into agricultural land, in particular crop agricultural land for fulfilling the food need. The protection of the sustainable agricultural land closely relates to environmental sustainability [3].

Change in land use, or land conversion, from agricultural to non-agricultural purpose refers to a shift of the land or area from its original use, for agricultural purpose, to other uses, such as mining, building, commercial facility, or settlement. The conversion of agricultural land to non-agriculture is a change of land function or part of what was originally converted into land such as mines, buildings, and shops that resulted in the function of the land to switch the proper function.[4]

Current government under President Joko Widodo has established *Nawa Cita* (Eight Plans) for development. One of the plan is maintaining sustainable agricultural land. Towards sustainability, the following efforts must be taken: irrigation reconstruction, farmer exchange value improvement, establishment of bank for farmers and small-to-medium enterprise, and acceleration control for the change in agricultural land use. In essence, there must be efforts of maintaining the agricultural land and performing improvement at distribution sectors as well as farmer empowerment programs, and policy improvement. Such efforts are necessary for controlling practices of changing the land use and preserving the food supply at domestic level. According to the data provided by Indonesia Statistic Office (BPS) in 2017 the food needs, such as rice, at the domestic level from January to February reached 14,473 tonnes. The rice were imported from Pakistan, India, Thailand, China, and Vietnam [5].

The government has been making an important effort by establishing a policy on preventing the change in agricultural land use to other purposes through Decree 41/2009 on Protection of Sustainable Agricultural Land. The central and local governments are working in concert to perform the sustainable agricultural practices in simple, self-help, and strong manner towards food resilience and security. The local government is responsible for preventing the conversion of the agricultural land into other purposes, which tend to damage the environmental sustainability.

This study aimed at identifying and analyzing the change in agricultural land use in Bungo Regency, Jambi Province. The study would find out whether the policy on preventing the change in agricultural land use into non-agricultural purposes had been properly implemented.

2 Methodologies

This study applied a qualitative methodology in a descriptive analysis. Research informants were selected by a purposive sampling technique. The informants consisted of individuals with adequate and reliable

* Corresponding author: hartutipurnaweni@gmail.com

knowledge about agriculture and were of the Government Agencies of Bungo Regency, such as Agricultural Affairs, Local Development Plan, Commission II of Local People's Assembly, Environmental Affairs, District Agricultural Extension Staff, and Law Enforcers. In addition, a snowball technique was also applied to gather informants from the local community. This group of informants was represented by a farmer union namely *Kelompok Tani Sri Bulan* from Tanah Periuk. To analyze the data the study applied a Miles and Huberman model [5].

3 Result and discussion

3.1 Agricultural land in Bungo Regency

Bungo Regency consists of a terrestrial area with large portions of lowland and marsh. Its soil composition was dominated by andosol, which spreads over areas in the districts of the regency. Andosol is a volcanic soil with fertile quality and contains rich nutrient, suitable for agriculture. In 2010 Bungo Regency covered 15,381 hectares of agricultural area. However, in 2013-2015 period, the Bungo Regency Office of Agriculture and Horticulture reported that the ricefield area had declined as much as 1.03% (1,105 hectares). Furthermore, the area for vegetables and holticultures also declined as much as 0,8% (380 hectares). These figures indicated a decrease in the total agricultural area in Bungo Regency [6].

Land use in Bungo Regency was divided into forestry (35,2%), rubber, copra, and mix plantation (34.7%), rice field agriculture (13.2%), and settlement (1.5%) [7]. The change in agricultural use had become the third greatest change after the change in forestry area into other purposes. The change in agricultural use into other purposes had raised the existence of cultivated plantation for rubber, copra, and mix plants. The distribution of the agricultural land was fairly equal for each district in the regency. Table below describes the agricultural land use for the period of 2010-2015 in Bungo Regency.

Table 1.
Change in Agricultural Land Use in Bungo Regency 2010-2015

No	**Year**	**Description**			**Total (Ha)**
		Rice field and plantation (Ha)	**Secondary crops (*Palawija*) (Ha)**	**Horticulture (Ha)**	
1	*2*	*3*	*4*	*5*	*6*
1	2010	12,553	1,756	1,009	15,381
2	2011	12,666	1,432	988	15,086
3	2012	11,807	1,201	1,045	12,972
4	2013	8,123	887	1,097	10,107
5	2014	10,689	1,459	1,869	13,917
6	2015	8,830	1,967	1,068	11,865

Source: Bungo Regency Office of Agriculture, Food Crops, and Horticulture (Bungo.co.id)

The above table explains the diminishing rice field area over time. The dramatic decline occurred in the span of three years from 2010 to 2013, reaching 6.5 per cent and affecting 8,569-hectare area. Such phenomenon was caused by much lower rice field economic value than that of for other purposes. However, in 2014 there was an increase of agricultural area up to 1.5 per cent, in particular that of horticulture crops (2.2 per cent). The increase, unfortunately, was only temporary because in 2015 the agricultural area continued to decline. In other words, the rate ratio between agricultural use and other uses was dynamic and dependent upon economic and commodity values.

a. Policy

At national level, the National Land Office (*Badan Pertanahan Nasional*) has recommended a policy on controlling the rice field conversion under three zoning criteria, i.e.: (1) rice field eligible for conversion; (2) rice field eligible for restricted conversion; and (3) rice field ineligible for conversion (protected zone) [8]. To control the agricultural land conversion practice the government has issued Law 41/2009 on Protection for Sustainable Food Crops Agricultural Land. This law is expected to enhance the availability of the agricultural land for food self-fulfillment, resilience, and security. The Law 41/2009 has objectives as follows: (a) to perform a sustainable protection of the agricultural area and land; (b) to secure a sustainable food crop agricultural land; (c) to provide food self-fulfillment, resilience, and security; (d) to protect farmers' agricultural land tenure; (e) to improve farmers and community welfare and prosperity; (f) to improve protection and empowerment of the farmers; (g) to increase more job opportunities for happier life; (h) to maintain ecological equilibrium; and (i) to create agricultural revitalization.

Because of its fertile agricultural land Jambi Province has a potential to be a center for food supply in 2025. It is going to be a disaster if the changes in agricultural use into plantation, industrial sites, and mining continue to happen. The Provincial Government of Jambi has therefore ratified the Law 41/2009 by establishing a policy on rice field conversion control through Governoral Decree 14/2009 on Prevention of Food Crop and Horticulture Land Conversion for Other Purposes in Jambi Province. This decree has major objectives as the followings: (a) to implement sophisticated, self-dependent, and resilient agriculture to support food and horticulture security using sustainable food crop and horticulture lands; and (b) to prevent the change in food crop and horticulture land use into other purposes to maintain sustainable agriculture. However, the conversion practices have been continuing to grow.

In order to prevent the change in agricultural land use in Bungo Regency the local government, following the Law 41/2009 and the Governoral Decree 14/2009, has established the Regency Decree 7/2014 on Protection of Sustainable Food Crop Agricultural Land. Unfortunately, this decree has not been supported by technical guidance for its implementation by the Bungo Regency Office of Agriculture. Accordingly, the prevention of the change in agricultural land use into other purposes in Bungo Regency is finding a deadlock.

A series of complementary regulations have also been put in practice from the central to the local levels. However, the practice of changing the agricultural land use has not been immediately prevented, making it difficult to obtain the dream of being a center for food supply. In practice, the change in agricultural land use closely relates to the economic needs. Most of local people tend to pursue more promising land uses than merely agriculture [9].

A series of activities related to the change in agricultural land use have caused the local people to have more motivation to cultivate copra and rubber. The increasing plantation area in some regencies in Jambi are motivated by the higher price of copra and rubber commodities. Compared to other agricultural commodities, copra and rubber are considered to give more benefits in terms of cash income, either periodically or immediately [8]. To the date, 10,000-hectare agricultural land has been changed into copra and rubber plantation, settlement, and rural gold mining, all without license. When this trend continues, rice supply from the rice field for Jambi is going to decrease over time [10].

Studies have shown that the change in agricultural land use into copra plantation may cause misery in the long term because it affects multiple aspects, i.e. ecology, social and culture, land and resource conflicts, environmental pollution, and food scarcity [11].

Local people of Bungo Regency has been practicing the change in agricultural land use for generations. Such practice is complicated because it involves various actors in interest. In the long term, the change in agricultural land use into other purposes will threaten the food security. Indeed, another problem arises for the low farmer exchange value for production so that most of the farmers tend to be hopeless in cultivating their land. Lacking infrastructure maintenance is also problematic because it relates to maintenance cost. The irrigation system is inadequate so that it is too vulnerable towards natural climate and climate change.

The Bungo Regency Government has not been responsive to opinions, discussions, and inputs from the rural people concerning problems they face in their agricultural activities. Agricultural extension is inadequate so that knowledge and technological exchange do not take place, causing the delay of the agricultural production process.

4. Conclusion

Summaries : (1) Change in agricultural land use into copra and rubber plantation in Bungo Regency still frequently occurred; (2) Change in agricultural land use has threatened long-term food security along with inevitable demographic bonus; (3) Change in agricultural land use into other purposes has caused losses. Many negative impacts have been experienced by either community or environment; (4) Change in agricultural land use involves many actors of interest, such as local people, civil society, business community/persons, and the farmers.

Recommendation: (1) Change in agricultural land use into other purposes in Bungo Regency can be prevented by intensive co-operation and coordination between stakeholders; (2) To minimize the further opening of the rubber and copra plantation a special team is necessary to be established to secure the sustainable agricultural land by determining limitation of the area eligible for the plantation; (3) The local government of Bungo Regency must develop a co-operation with rural officials, traditional organs, agricultural academics, non-governmental organizations, environmentalists, and local leaders to prevent the change in agricultural land use; (4) The local government of Bungo Regency must give legal protection to all villages and rural officials who implement regulations and punishment according to local tradition to those who perform the illegal change in agricultural land use.

References

1. A.Mufubi, Y.Setiawan, H.Effendi, Procedia Environment Sciences **33** (2016), 654-660.
2. P.A. Permatasari, A.Fatikhunnada, Liyantono, Y.Setiawan, Syartinilia, A.Nurdiana, Procedia Environment Sciences **33** (2016) 27-35.
3. MF. Anita.W.Handari, AN. Bambang, H. Purnaweni, Ekosains, (2012), Vol **4** (3), pp 19-26

4. Rustiadi, Ernan dkk, 2012, *Penyelamatan, Tanah, Air dan Lingkungan*, Crespent Press, Jakarta.
5. Detik Finance, Kamis 16 Mar 2017. *https://finance.detik.com/berita-ekonomi-bisnis/3448448/ri-impor-beras-di-awal-tahun-2017-paling-banyak-dari-pakistan*
6. *Bungo Dalam Angka, 2015*. Badan Pusat Statistik Kabupaten Bungo, Jambi (2015)
7. Suryani, Lili, *Analisis perubahan penggunaan lahan, komoditas unggulan perkebunan dan arahan pengembangannya di Kabupaten Bungo, Provinsi Jambi* *http://repository.ipb.ac.id/handle/123456789/78796*. (2015)
8. Minsyah, Endrizal, dkk, 2012, *http://www.litbang.pertanian.go.id/buku/konversi-fragmentasi-lahan/BAB-IV-3.pdf*
9. Puspasari, Anneke, *Faktor-faktor yang mempengaruhi alih fungsi lahan pertanian dan dampaknya terhadap pendapatan petani (studi kasus Desa Kondangjaya, Kecamatan Karawang Timur, Kabupaten Karawang)*, Tesis IPB, Bogor. (2012)
10. Suara Pembaruan, 26 Februari 2015. *http://sp.beritasatu.com/ekonomidanbisnis/jambi-beri-sanksi-pada-pelaku-alih-fungsi-lahan-tanaman-pangan/79382*.
11. Herman, Fahmuddin Agus, dan Irsal Las, *Analisis finansial dan keuntungan yang hilang dari pengurangan emisi karbon dioksida pada perkebunan kelapa sawit*, http://*pustaka.litbang.pertanian.go.id/publikasi/p3284092.pdf* (2009)

Public Participation Planning of Environmental Impact Assessment (EIA) and Regulations: Analysis of Inconsistency for Some Cases in Indonesia

Aniek Hindrayani[1,*], and *Purwanto*[2]

[1] Department of Economic Education, Sebelas Maret University, Surakarta - Indonesia
[2] Department of Soil Science, Sebelas Maret University, Surakarta - Indonesia

Abstract. The failure in community involvement during the environmental documents planning may result in the failure of the planned project implementation. This study aims to determine the gap between practices and regulations that apply to the process of community involvement in the environmental documents planning, and find out inconsistency of implementation on each stakeholder in the planning of the Environmental Impact Assessment (EIA) and the environmental permit. The method used was qualitative through interview and literature study which is analyzed using triangulation model and presented in the form of concept map. The results of the study indicate that 1) the determination of community representatives based on the criteria of the impacted communities is not clearly described, 3) suggestions, opinions, and responses to the environmental impact management are not well implemented by the project proponent, 3) implementation of the environmental management of other licensed activities affecting the behavior (4) stakeholders (project proponent, EIA consultants, and EIA appraisal committee) do not play their role as mandated in applicable legislation.

1 Introduction

Public consultation in the EIA study of a proposed project is a preliminary step that must be undertaken and largely determines the sustainability of the proposed project. The results show that only 40% of Indonesians participate in public consultations, and such engagement is limited to the subjective norm [1]. There are still many unclear public participation in development country caused by negative attitudes of government officials, weak public environmental awareness, weak supervision and penalty mechanism, bribery and corruption issues, political intervention, and poor economic conditions [2]. The low level of community involvement in the EIA process also occurs in Kenya's development plan which is largely due to lower policies, laws and regulations [3].

This research aims to determine the gap between practices and regulations that apply to the process of community involvement in the environmental documents planning, and find out inconsistency of implementation on each stakeholder in the planning of the Environmental Impact Assessment (EIA) and the environmental permit.

This research method is descriptive qualitative, through interview and literature study. Data analysis was done with triangulation. The results showed that the determination of community representatives based on the criteria of the impacted communities is not clearly described. Suggestions, opinions, and responses to the environmental impact management are not well implemented by the project proponent. Implementation of the environmental management of other licensed activities affecting the behavior. Stakeholders (project proponent, EIA consultants, and EIA appraisal committee) do not play their role as mandated in applicable legislation.

Furthermore, this paper discusses about literature review, research methods, results and discussion, and concludes.

2 Literature review

2.1 Review of Legislation

The regulations on public involvement are generally described in Law No.32 of 2012 and Govt. Reg. No.27 of 2012. While regulations and guidelines on the implementation of public consultation are specifically listed in the Regulation of the State Minister of the Environment No.17 of 2012. But in practice, it is often the proponent/EIA consultant ignores the objectives, principles and procedures in accordance with the law so that there is turmoil, rejection and social conflict in the pre-construction and or construction stage of the implementation of its development. As stated in the Regulation of Environment No.17 of 2012, the objectives of the Community Involvement in EIA Process are:

* Corresponding author: aniekhindrayani@gmail.com

a. In order to obtain information about proposed projects that have an important impact on the environment for the communities in the vicinity
b. The public may submit their suggestions, opinions and responses related to the proposed project,
c. Communities may be involved in the decision-making process regarding the feasibility or inappropriateness of the proposed project
d. Communities may submit their suggestions, opinions and responses on the application and issuance of environmental permit from the proposed project

Community involvement in the EIA process should be implemented on the following basic principles:

a. Providing transparent and complete information,
b. Equality of positions among the parties involved,
c. Problem solving that is fair and wise,
d. Coordination, communication and cooperation among the parties concerned.

The content in public consultations to be submitted includes:

- Name and address of the proponent;
- Location and extent of proposed project with information on administration limits from project site location and project site map;
- A detailed description of the proposed project;
- Potential environmental impacts (potential impact of initial judgment of the initiator) in the presence of a proposed project;
- Important environmental components (cultural, ecological, socioeconomic, defense, etc.) to be affected;

EIA proponents/consultants may use public methods that are considered effective and efficient to encompass public tax returns such as public meetings, workshops, seminars, focus group discussions, and other methods that can be used for two-way communication. In modern context, communication is more effective using ICT as gadget, email, or short message [4]. Communities involved in the public consultation include (1) the affected, (2) environmentalists, and (3) who are affected by all decisions in the EIA process.

2.2 Previous research

Community involvement is more consultative than participation. Consultations simply meet the regulations, where the activity is seeking advice from the public, while public participation more to the democratic process of discussing, thinking, deciding, planning, and playing an active part in the development and operation of services that affect their lives according to Wathern [3]. Previous research found that community involvement in EIA process specially in developing country was not yet optimum [1], [2], [3].

2.3 Case of crude palm oil, hotel & apartment, and cement industry

In the case of oil palm plantations and factories, regulated in Minister of Agriculture Regulation No.98 2013 and updated with No.29 2016. The regulatory reforms eliminate some articles that generally harm the surrounding community.

In the case of hotel & apartment, there are no specific regulations, so it refers to the more general rules of building and construction. Not far with in the cement industry, the regulations available more to the industry itself, have not specifically pointed to the affected communities.

3 Research Methods

This research is a qualitative descriptive research. Researchers do triangulation, i.e method, data source, and theory to analyze data. Triangulation method is done by interview and observation. Triangulation of data source is primary data of interview result and secondary data from literature study, while triangulation theory is based on literature study and previous research.

Interviews are conducted to stakeholders, in this case community representatives, appraisal commissions, and initiators. Observations were made on the EIA documents of hotel & apartment, oil palm plantation, and cement plant. We also doing literature study of applicable regulations related to public consultation and description of activities to control policy and implementation in the field. The results of previous relevant studies are a reference for this study to strengthen the results of the analysis.

We selected several cases in Indonesia represented by palm oil plant & industry, housing & apartment, and cement industry cases in three different regions, which is in East Kalimantan, Special Region of Yogyakarta, and Central Java.

4 Results and discussion

Generally, in the process of community involvement, stakeholders mapping really should represent, not the amount, but represent the community that approximately support and not support the proposed project. When the consultant appoints community representatives, it is not really oriented to involve the community in exploring the suggestions, opinions, and responses to provide an opportunity for decision-making, but mostly just to meet formal requirements, the administrative has conducted public consultation.

At the time of appointment are also not described tasks and functions as a representative in capturing the aspirations of society that has not been stated/achieved during the Focus Group Discussion, so that accumulated at the end of the process of drafting environmental permit documents.

This is related to the Regulation of the Minister of Manpower No. 122 of 2016 concerning the Stipulation of Indonesian National Work Competency Standards Professional Activity, Scientific and Technical Category Other Activities Professional, Scientific and Technical Activities at the Working Position of the EIA. The process of community involvement in some cases is described below.

4.1 Case of crude palm oil (cpo) in east Kalimantan

In the case of oil palm plantation, the plasma plantation gets a 20% share that should be managed by the community in the vicinity of the activity site, but in fact it is sold. According to the regulation, the distribution of smallholdings is intended to empower the community, so there is no social gap. As a result, local people look for more land in the forest, leading to deforestation. If studied from the production chart, up to the 5th year, the garden is immature, in the next 15-20 years, peak production, then decreases to 25-30 years. In the period of production decline, the population should start replanting, but in the reality, they do not want to invest for replanting, should be some replanting, some up to 30 years, so there is sustainability.

Involving the community in the process of drafting the documents to the extent they do not have difficulty to meet basic needs and be involved in the operation of the plants and factory only. Other issues related to long-term sustainability have not been the demands of affected communities, whether they are sustainable for economic welfare or environmental conservation.

4.2 Case of hotel & apartment in Yogyakarta

In the case of hotel & apartment development plans in Special region of Yogyakarta, the main issue that arises is the occurrence of water crisis. Hotel & apartment should not use groundwater, while there is no regulation, the consideration is only carrying capacity and capacity. While many hotels are already in operation, it becomes a cumulative impact that has not been studied in depth. Utilization of water by government department is required, while the use of ground water is only for reserves, but then the supervisor has not been well laid out.

People began to reject most of the hotel & apartment development plans for that reason, as well as other reasons such as the use of public roads, funerals and the use of public facilities. This is caused by the lack of fulfillment of community representatives in the process of preparing EIA documents and communication between stakeholders.

4.4 Case of cement industry in Central Java

In the case of the cement industry, the community is not involved in the process of composing the environmental documents in its entirety, the process of involving the community is not participative, not the essence, and the socialization is not clear.

Citizens are not given the opportunity to voice violations that have been committed during the preparation of development projects

Citizens are not given the opportunity to voice violations that have been committed during the preparation of the plant construction project. Citizens do not know clear information about plant plans. There is no socialization involving the villagers in general, but only the village apparatus. Environmental documents are not give in to the residents so there is no clear explanation of the negative impacts

4.5 Community Involvement Patterns

The process of preparing the EIA, the application of environmental permits and the involvement of communities based on regulation is illustrated in the following concept maps:

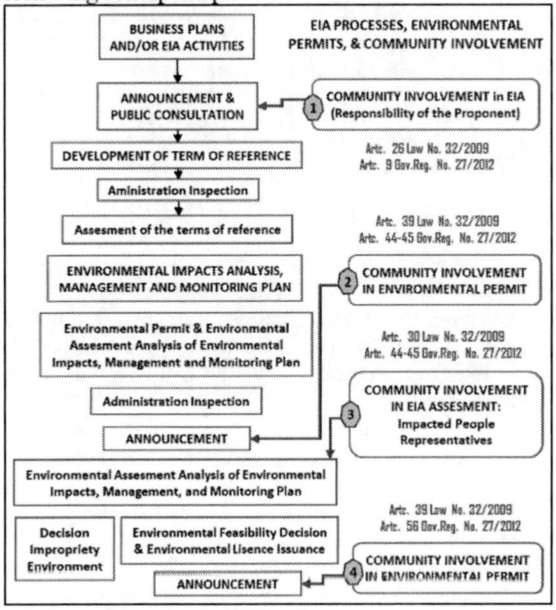

Fig 1. EIA Processes, Environmental Permits, and Community Involvement

From the figure above, we can say that in the selection of suggestions, opinions, and responses from the community, among others, determined the important hypothetical impact that needs to be studied more deeply. Communities are involved in the announcement of the environmental permit request from the proponent to the public prior to the Technical Team meeting or before the EIA Assessor Commission session. The public acts as a member of the EIA's appraisal commission in assessing documents to engage in the determination of eligibility or inadequacy of the environment. The community also plays a role to provide suggestions, opinions, and responses prior to the issuance of environmental permits.

While the mechanism of public consultation in the field should go along the following pattern as Figure 2

The consultation mechanism begins by understanding the objectives of public consultation (collecting suggestion, opinions, and response from the community regarding the planned activities, what are the expectations, concerns and concerns of the community), then planning and preparing public consultations in the form of facilities and infrastructure, who will be invited / stakeholders mapping to be true representing a wide range of affected communities, stakeholder consultation mapping and public consultation exercises to gain legitimacy from local formal leaders).

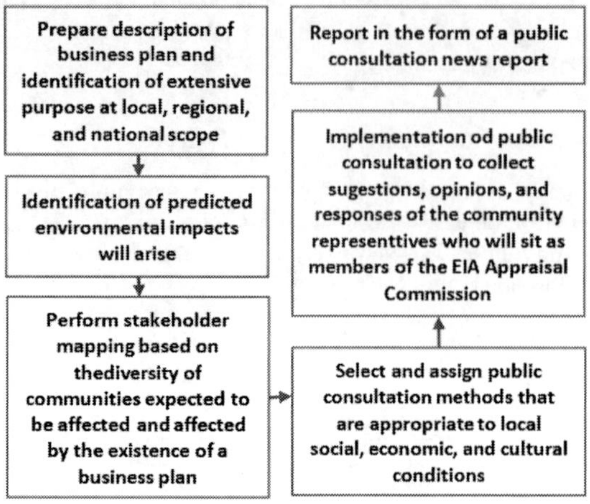

Fig 2. Consultation Mechanism

The consultant/proponent predicts significant impacts that will arise (as part of the socialization material), identify areas of consultation (field orientation and limited interviews with key figures), determine the objectives of the consultation (who will be invited to the public consultation and who is expected to become community representatives as members of EIA assessment commission). Next is to prepare an information exchange plan, by preparing the most effective media, forms, and methods of public consultation according to the level of education and local cultural conditions, selecting consultation techniques, establishing places, dates, techniques and methods of public consultation after obtaining input from local community leaders, and implementation of public consultations. Last is to summarize suggestions, opinions and community responses to be reported in terms of reference to the EIA commission.

Anticipating the implications of the ineffectiveness of community involvement processes can be done in the following way (Table 1):

Tabel 1. Problems in EIA Process and Solution Alternatives

No	Problems	Solution Alternatives
1	The project plan will be rejected	• Reminds the scope of the problem • Learn the groups that resist and object to activity plan
2	The project does not meet the expectations of the community	• Expressed to the public, that suggestions will be considered & discussed with the Regional Government • Describe project limitations
3	Your project will be stunted	• Community consultation preparation of social survey • Careful scheduling
4	The project plan confronts itself with the community	• Ask the local government for help to organize a meeting forum • Make sure the apparatus understands the purpose of the consultation
5	Project budget enlarged	• Good budget planning

5 Conclusions

Community involvement in the preparation of environmental documents is essential to EIA. This research aims to find the inconsistency between practices and regulations that apply to the process of community involvement in the environmental documents planning, and find out the implementation on each stakeholder in the process of the Environmental Impact Assessment (EIA) and the environmental permit. We used triangulation in qualitative methods analyzed the data.

The result of this study showed that in the preparation of environmental documents, communities were not included from the applicable regulations, such as participatory, non-essential, and no socialization. Stakeholder mapping has not really representation of the community that is predicted to support and reject the proposed project. When appointing community representatives, consultants, and technical teams, it does not really show the principal task and its function as a "representative" that shows the legitimacy of society. The appointment of community representatives is still carried out by consultants and technical teams, who should be chosen by the community itself.

On community empowerment, which should be provided to empower communities for sustainable prosperity, but not fully implemented, leading to deterioration in economic conditions and social inequalities, and environmental degradation. There is a gap between carrying capacity and implementation of policies and regulations. For the stakeholders, especially the EIA assessment commission, there are no legal consequences. Further research need to accomplished with more cases to enrich the discussion.

References

1. R. Nadlifatin, M. Razif, S. Lin, SF. Persada, P.F. Belgiawa. PES. **28**. 3-10. (2015)
2. J. Chen. *Public Participation Provision in EIA Legal System*. Master Thesis. Universiteit Utrecht. (2013)
3. D. N. Ngonge. *Evaluation of Public Participation in EIA of The Southren Bypass Road in Nairobi, Kenya*. A Thesis. (2015)
4. S.F. Persada, S. Lin, R. Nadlifatin, M. Razif. Open Access JIS. (2015)
5. A. Marzuki. TERUM. **3**(12). 126-136. (2009)
6. C. Wasseman, *Application of Public Participation in Environmental Management in Indonesia*. (2012)
7. C. Wood. *New Directions in Impact Assessment for Development*. (2003)
8. D. Annadale. Impact Assessment and Project Appraisal. **19** (3). 187–193. (2001)
9. D. Y. Khalajabadi. *A Thesis*. (1994)
10. G. O'Faircheallaigh. EIAR. **30**. 19-27. (2010)
11. J.O. Kakonge. Impact Assessment. **14** (3). 309-320. (2012)
12. S. Bastidas. *Canadian Trade Policy*. (2004)

13. S. Momtaz. *The Intl. Journal of Environmental, Cultural, Economic, & Social Sustainability*. **2**(4). (2006)
14. T. Webler, H. Kastenholz, O. Rena. MEP. 443-463. (1995)
15. Y. Rahmatwati. Kompasiana, April 4, 2017.

Environmental Kuznets Curve Hypothesis: A Perspective of Sustainable Development in Indonesia

Citrasmara Galuh Nuansa[1,*], and *Wahyu* Widodo[2]

[1]Master Program of Environmental Science, School of Postgraduate Studies, Diponegoro University, Semarang-Indonesia
[2]Faculty of Economics and Business, Diponegoro University, Semarang - Indonesia

Abstract. Sustainable development with three main pillars, namely environmental, economic, and social, is the concept of country's development to achieve inclusive economic growth, good environmental quality, and improvement of people's welfare. However, the dominance of economic factors cause various environmental problem. This phenomenon occurs in most of developing countries, including in Indonesia. The relationship between economic activity and environmental quality has been widely discussed and empirically tested by scholars. This descriptive research analysed the hypothesis called Environmental Kuznets Curve (EKC) from a perspective of sustainable development in Indonesia. EKC hypothesis illustrates the relationship between economic growth and environmental degradation forming an inverted U-curve, indicating that at the beginning of development, environmental quality will decrease along with increasing economic growth, and then reached a certain point the environmental quality will gradually improve. In this paper will be discussed how the relationship between environmental quality and economic growth in Indonesia was investigated. The preliminary results show that most of the empirical studies use the conventional approach, in which the CO_2 emission used as the proxy of environmental degradation. The existence of inverted U-curve is also inconclusive. Therefore, the extension research on the relationship between economic growth and environmental quality in Indonesia using the EKC hypothesis is required.

1 Sustainable Development

1.1 Definition of sustainable development

The term sustainable development became a familiar thing to be discussed. Sustainable development is defined as development that aims to meet the needs of the present without compromising the ability of future generations to meet their own needs. The Brundtland Report described the definition of sustainable development into two key concepts [1], namely: 1) the concept of needs that give priority to the needs of the world's poor; and 2) the concept of limitations on the ability of the environment to meet the needs of present and future. From these definitions, inter-generational and intra-generational equity are the keys to sustainable development [2].

Sustainable development has three main pillars: economic, environmental, and social that presented as three interconnected rings as shown in Figure 1 [3]. This figure shows that these three aspects should be applied in a balanced state policy to achieve sustainable development.

1.2 The dominance of economic aspect

In practice, the successful development of a country tends to be measured in purely economic indicators, such as economic growth. Per capita income is used as a measure of the country's economic growth, as it has been used by the World Bank to classify the countries level of development.

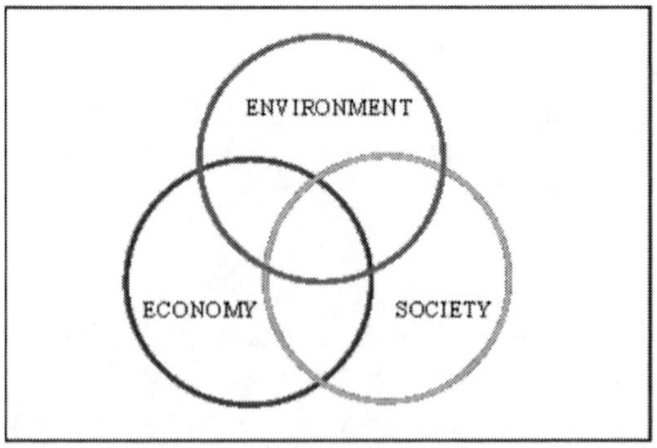

Fig. 1. Three main pillars of sustainable development [3]

Therefore, countries with low income will continue to increase economic growth in their development objectives, so that environmental and social issues to be neglected. This indicates that the economic side still dominates the practice of development compared to the social and environmental aspects. In other words, the balance of the three aspects of development has not been

*Corresponding author: galuh.citra@gmail.com

fully implemented as it was intended from the concept of sustainable development.

In addition, the existence of political factor provides the flexibility of economic aspects to continue exist [3]. Political factor and economic interest are two inseparable things. Political and economic actors need each other to maintain their existence. As a result, the environment, including the natural resources contained in it is used as capital for exploitation. Environment and society becomes victim because it used as a place to dispose of problems such as waste or disease.

In fact, economic growth requires two basic requirements, that is physical needs and social needs. Physical needs including raw materials, fuel and other natural resources. Social needs becomes very important to support the economy, in case of employment as well as developments in science and technology [4].

With only concerned with the economic aspect in the development of a country, it is no wonder if there is a wide range of environmental issues, including in Indonesia.

2 Economic and environmental conditions in Indonesia

Based on per capita income, economic growth in Indonesia tends to increase from 1970 to 2015 (Figure 2). In 1970, Indonesia's per capita income was US$ 84.09. Per capita income has increased from year to year up to 2015 to US$ 3346,487. Based on the data from the Central Bureau of Statistics (Badan Pusat Statistik) [5], manufacturing; agriculture, forestry, and fishing; mining and quarrying dominate the contribution of Indonesia's gross domestic product in 2016 (Figure 3). Therefore, Indonesia still relies on industrial sector and utilization of natural resources as capital development.

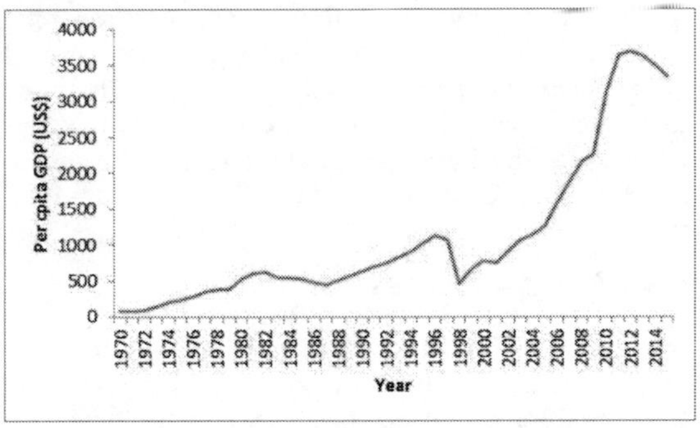

Fig. 2. Indonesia's per capita income [6]

Indonesia's economic growth was not in line with the maintenance of environmental quality. Various environmental problems in Indonesia, perhaps, produced from the development activities that are only pursuing economic growth. One is the increasing CO_2 emissions. It is the major cause global warming. According to a report from the World Resources Institute [7] Indonesia has become a top 10 contributors to global greenhouse gas emissions.

In addition, deforestation also occur in Indonesia (see Figure 4) caused by the conversion of forest lands [7]. The existence of the rate of deforestation in Indonesia is high, i.e. more than 1% per year [8] will worsen air quality and will further increase global warming.

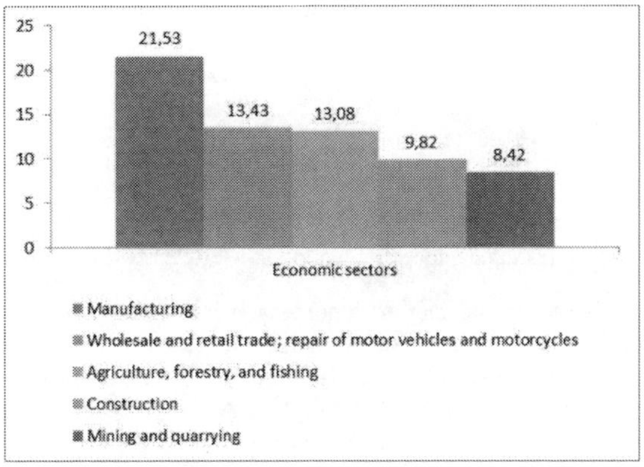

Fig. 3. Top five economic sectors contribution of Indonesia's GDP in 2016 (percent) [5]

Ecological Footprint can also be used to look at the quality of the environment. Ecological footprint illustrates the amount of land required by the human population in its activities to meet the needs [9]. In Indonesia, the ecological footprint is showed in Figure 5, in which over the years the ecological footprint in Indonesia has increased. When compared to its natural capacity (bio capacity), Indonesia had been an ecological deficit since 2002. This means that per capita consumption of Indonesian society has exceeded the carrying capacity of the natural environment as seen in Figure 5.

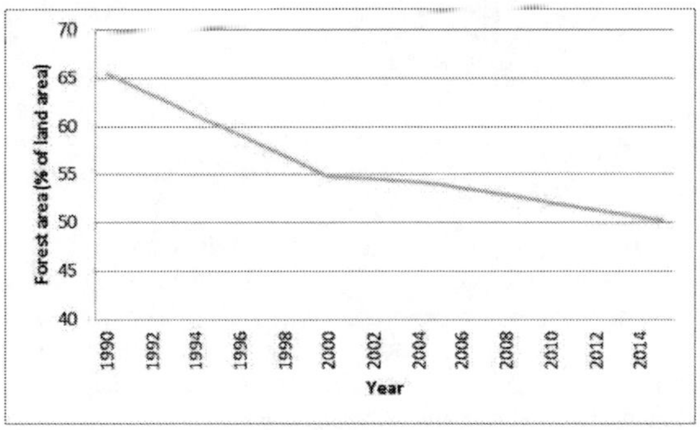

Fig. 4. Forest area in Indonesia [10]

The low quality of Indonesian environment can also be seen from the *Environmental Performance Index* (EPI). In 2016, there were 180 countries assessed in terms of its efforts in addressing environmental issues with high priority are divided into two categories, that is the protection of human health and the ecosystem [11]. Indonesia was only able to rank 107. Meanwhile, other ASEAN countries like Singapore, Malaysia, Thailand, and Philippines are ranked 14, 63, 91, and 66 respectively.

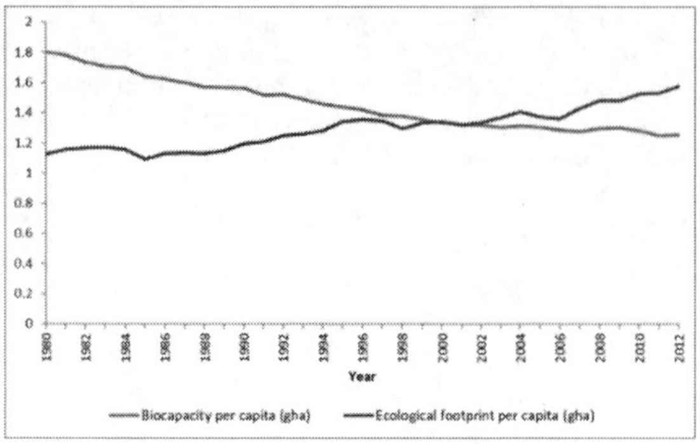

Fig. 5. Ecological footprint Indonesia 1980-2012 [12]

In terms of energy, Indonesia is still relying on non-renewable energy. Based on the data from the National Energy Council [13], nearly 50% of Indonesia's primary energy is supplied from oil, followed by coal at 31%, while renewable energy is only used less than 5% (see Figure 6). Though Indonesia has the potential of renewable energy is large enough, such as hydropower that is equal to 75 GW, but unfortunately only 8,111 MW installed. There are also geothermal energy which reached 29,475 MW and biomass by 32 GW. However, the utilization of renewable energy is far under the existing potential.

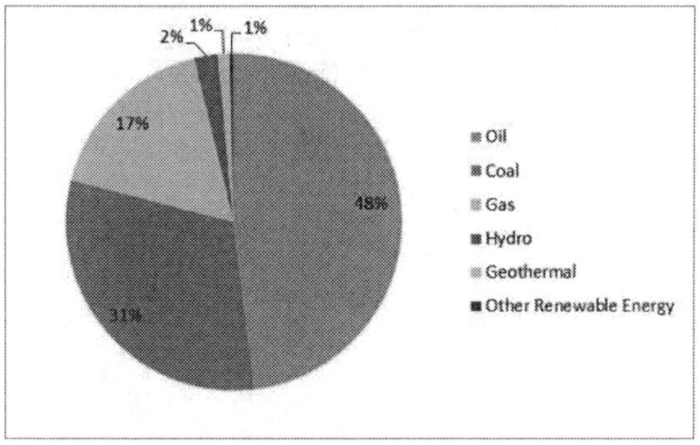

Fig. 6. Indonesia's primary energy mix 2014[13]

Based on the Figure 6, Indonesia may be experiencing a more severe environmental degradation and natural resource crisis, when the development is only concerned in pursuing the income per capita, without realizing the importance of maintaining environmental quality.

From the phenomena figured out above, there are some important questions to be addressed, whether the economic growth will always be detrimental to the quality of the environment? How is their relationship? Will it continue to contradict?

3 The relationship between economic growth and environmental quality

Economic growth and the environment have a close relationship. Economic activity consists of the production and consumption activities that can't be separated from the environment, where they are located. It is because all economic activities depend on the resources contained in the environment [14, 15]. This dependency will also have environmental impacts as a result of economic growth.

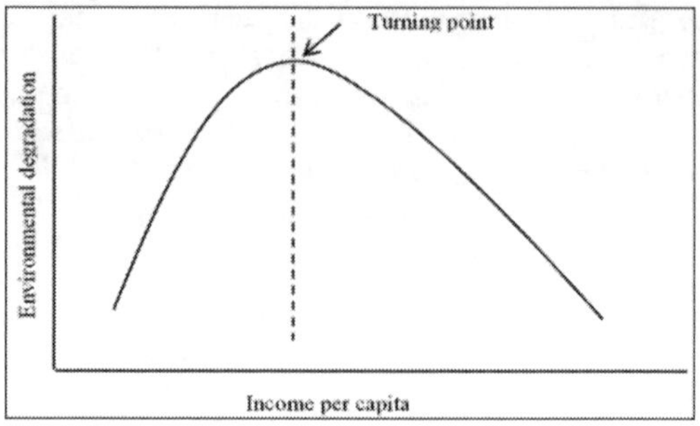

Fig. 7. Environmental Kuznets Curve [16]

The relationship between economic growth and environmental quality, becomes a concern by experts. Various studies have been conducted to determine the relationship between them. One of these studies resulted in a hypothesis called hypothesis Environmental Kuznets Curve (EKC) [16, 17].

EKC hypothesis illustrates the relationship between economic growth indicated by per capita income and environmental degradation, forming an inverted U-shape curve (Figure 7). This curve shows that at the beginning of development, per capita income will increase with increasing the environmental degradation to a certain point, and then the environmental quality will be improved along with the increase of income per capita. This is because at the beginning of development, the concern for environmental issues is still low and environmentally friendly energy is not yet available. At the higher level of development, there will be awareness about the importance of maintaining the environmental quality as a result of previous economic growth, the enactment of environmental protection regulations by the government, and the rise of science and technology so as to create an environmentally friendly technology. The existence of this development will lead to environmental degradation gradually decreases [18].

4 EKC hypotheses in Indonesia

In general, to investigating EKC hypothesis using two approaches, viz. using cross-country panel data analysis (see [19-23]) and using a single country time series analysis (see [24-28]). By using cross-country panel data analysis will only describe the general EKC hypothesis, in which suggestions and conclusions can't be applied in each country. Therefore, in order to provide advice and input to environmental policy in certain countries, the approach to single-country time series analysis is more appropriate.

The research investigation of EKC hypothesis in Indonesia itself is still fairly limited. Based on the search of scientific articles on the Science Direct, it was found three relevant articles that discuss the investigation of EKC hypothesis using single country time series analysis. These studies were conducted by Saboori, et. al. [29], Jafari, et. al. [30], and Sugiawan & Managi [31], which briefly presented in Table 1.

These studies use CO_2 emissions as an indicator of environmental quality, and per capita income is used as an indicator of economic growth. Per capita income is used as a measure of economic growth due to, in fact, the per capita income is an fundamental indicator shows the growth or development of a country [16]. Meanwhile, CO_2 emission is used as variable that indicate the environmental degradation. This is associated with the issue of global warming that become the world's attention, and CO_2 emissions is one of the main causes.

By choosing CO_2 emissions as an environment variable, it will be able to provide input on policies relating to the reduction of CO_2 emissions in order to keep in line with the increase of economic growth. In Indonesia, CO_2 emissions reduction issue becomes government's particular concern. However, the environmental issues and problems faced by Indonesia are also increasingly complex.

Table 1. Empirical evidences of EKC hypothesis in Indonesia

Author	Saboori, et al. [29]	Jafari, et. al. [30]	Sugiawan & Managi [31]
Study period	1971-2007	1971-2007	1971-2010
Method	ARDL	Toda-Yamamoto (TY) procedure	ARDL
Variables	Per capita CO_2 emissions, per capita GDP, commercial energy consumption per capita, trade openness ratio	CO_2 emissions, GDP, energy consumption, urban population, capital stock	Per capita CO_2 emissions, per capita GDP, per capita electricity production from renewable sources, energy consumption, total factor productivity
Result	Do not support the EKC hypothesis	Do not support the EKC hypothesis	Support the EKC hypothesis

Need to put more deeply concern on the use of variables that able to represent the environmental quality. Choosing the proper environment variables in the EKC hypothesis will describe the relationship between economic activities and the environment better. Some aspects need to be considered in the selection of environmental variables. The main issue of the development and environment quality is an important side in the EKC hypothesis research. The research should describe the condition of the country as it is. In addition, the selection of environmental variable also needs to consider the availability of the data in a long period of time as required in the analysis of EKC.

In the study using single country time series analysis, the method of autoregressive distributed lag (ARDL) is widely used by researchers. This is because the method of ARDL has several advantages [32-34], among others, that:

1) This model uses only single equation
2) ARDL can be used on data that are integrated in different orders
3) The ARDL approach more suitable for small samples when compared with using other approaches
4) ARDL model is sufficient to simultaneously correct for residual serial correlation and problem of endogenous variables

Research conducted by Saboori, et. al. [29], Jafari, et. al. [30] does not prove EKC hypothesis in Indonesia. However, a different result was found by Sugiawan & Managi [31] that prove EKC hypothesis in Indonesia in the long-run.

EKC hypothesis verification research in Indonesia provides different results. Although all of them use CO_2 emissions and per capita income, but the explanatory variables are used differently. Saboori, et. al. [29] use commercial energy consumption per capita and trade openness ratio. Jafari, et. al. [30] use energy consumption, urban population, and capital stock. Meanwhile Sugiawan & Managi [31] use per capita electricity production from renewable sources, energy consumption, and total factor productivity. The use of different variables will produce different models, so the results will be different too. Therefore, in the research to proof EKC hypothesis, it's very important to choose the appropriate variables.

5 Conclusions

The relationship between economic activity and environment in the development of a country is an issue that has been discussed for a long time. Research to determine the relationship of the two produces the Environmental Kuznets Curve hypothesis. This hypothesis illustrates the relationship between economic growth indicated by per capita income, and environmental degradation that forms an inverted U-shape curve.

Investigation of EKC hypothesis in a country can be used to determine the extent of economic development undertaken affect the environmental quality. Environmental quality is one of the pillars of sustainable development, so that the EKC hypothesis shows a picture of the sustainability of development in a country. From the proof of EKC hypothesis, can be obtained input for policy development of a country seen from the various variables used and the resulting correlation.

From various studies to investigating the hypothesis EKC in Indonesia, need additional studies with specific updates. The variable is one of the important things in the EKC hypothesis, notably variables describing environmental degradation. Investigation of EKC hypothesis is carried out over a long period of time.

Therefore, the study should conducted in a period of time were able to describe the past historical conditions up to current conditions.

Reference

1. S. El Serafi, *Environmentally Sustainable Economic Development: Building on Brundtland*, (1992)
2. I. Moffat, Ecol Econ, **32**:359-62 (2000)
3. B. Hopwood, G. O'brien, B. Giddings, *Sustain Dev*, **10**:187-196 (2002)
4. D. H. Meadows, D. L. Meadows, J. Randers, et al., *The Limits to Growth*, Universe Books, New York (1972)
5. Badan Pusat Statistik, [Seri 2010] *PDB Triwulanan Atas Dasar Harga Konstan 2010 Menurut Lapangan Usaha (Miliar Rupiah), 2014-2017,* (2017)
6. World Bank, GDP per Capita, (2016)
7. World Resources Institute, *Infographic: What Do Your Country's Emissions Look Like?* (2015)
8. Kementerian Lingkungan Hidup, *Status Lingkungan Hidup Indonesia 2013*, Jakarta, (2014)
9. W. E. Rees, Environ Urban, **4(2)**:121–30 (1992)
10. World Bank, *Forest Area*,(2016)
11. Yale Center for Environmental Law and Policy,*Global Metrics for the Environment*, (2016)
12. Global Footprint Network, *National Footprint Accounts 2016,* (2016)
13. National Energy Council, *Executive Reference Data National Energy Management*, Jakarta (2015)
14. K. Arrow, B. Bolin, R. Costanza, P. Dasgupta, C. Folke, C. S. Holling, et al., *Science*, **268**:520-521 (1995)
15. A.Kahuthu, Environ Dev Sustain, **8(1)**:55-68 (2006)
16. T. Panayotou, *ILO Working Papers*, (1993)
17. S.Kuznets, Am Econ Rev, **45(1)**:1-28 (1995)
18. S.Dinda, Ecol Econ, **49(4)**:431-55 (2004)
19. G.Koop, L. Tole, J Dev Econ, **58(1)**:231-44 (1999)
20. M. Bagliani, G. Bravo, S. Dalmazzone, Ecol Econ, **65(3)**:650-661 (2008)
21. M. Hervieux, O. Darn, *HAL – Arch*, (2013)
22. U. Al-Mulali, C. Weng-Wai, L. Sheau-Ting, A. H. Mohammed, Ecol Indic, **48**:315-323 (2014)
23. M. M. Alam, M. W. Murad, A. H. M. Noman, I. Ozturk, Ecol Indic, **70**:466-479 (2016)
24. J. B. Ang, Energy Policy, **35(10):**4772-4778 (2007)
25. A. Jalil, S. F. Mahmud, Energy Policy, **37(12)**:5167-5172 (2009)
26. U. F. Akpan, A. Chuku, Munich Pers RePEc Arch Pap, **31241** (2011)
27. M. Shahbaz, H. H. Lean, M. S. Shabbir, Renew Sustain Energy Rev, **16(5)**:2947-2953 (2012)
28. B.Saboori, J. Sulaiman, S. Mohd, Energy Policy, **51**:184-191 (2012)
29. B. Saboori, J. Sulaiman, S. Mohd, Int J Econ Financ, **4(2)**:243-251 (2012)
30. Y. Jafari, J. Othman, AHSM Nor, J Policy Model, **34(6)**:879-889 (2012)
31. Y.Sugiawan, S. Managi, Energy Policy, **98**:187-198 (2016)
32. M. H. Pesaran, Y. Shin, *Econom Econ Theory 20th Century Ragnar Frisch Centen Symp*, 371-413 (1999)
33. U.Al-mulali, B. Saboori, I. Ozturk, Energy Policy, **76**:123-131 (2015)
34. U. Al-Mulali, S. A. Solarin, I. Ozturk, Nat Hazards, **80(3)**:1729-1747 (2016)

Planning Of Beef Cattle Development in District Blora, Central Java, Indonesia

Budi Santoso[1*], and *Bambang* Waluyo Hadi Eko Prasetiyono[2]

[1]Master Program of Environmental Science, School of Postgraduate Student Diponegoro University
[2]Animal Sciences Dept, Faculty of Animal and Agriculture Sciences, Diponegoro University

Abstract. Continuity of meat supply availability is generally related to the number and production of livestock in a region. Therefore, a framework of sustainable livestock development is needed to increase the production and productivity of livestock. Blora Regency is one of the areas in the Province of Central Java with the largest number of large livestock, primarily beef cattle. Blora Regency has a population of 199.584 beef cattle. Agricultural waste results in Blora Regency can be used as supporting the availability of feed for livestock sector. This is supported by the availability of forage feed which is very abundant.Based on these potentials, it is necessary to assess the characteristics of natural land for the development of beef cattle farms. Therefore, the objectives of this study are (1) to assess the environmental suitability of the environment for the development of cattle ranching that is grazed and stacked; (2) to analyze the potential of forage source of fodder and bearing capacity for beef cattle farming; (3) to analyze the centers of activity of development of beef cattle; (4) to prepare direction and strategy of beef cattle development in Blora Regency.

1 Introduction

Continuous and targeted development of the livestock sector has contributed to regional development, directly in the formation of Gross Regional Domestic Product (PDRB), employment absorption, income generation, and indirectly through the creation of conditions conducive to the implementation of development and synergistic relationships with Other sectors [1]. Along with the increasing population, there is an increasing demand for food. Therefore, the availability of food supplies is needed (especially the continuity of supply availability) both from plants and animals. Increased public awareness of nutrient fulfillment encourages increased demand for food derived from animal protein such as meat derived from ruminants (beef, buffalo, sheep and goats).

Availability of meat supply is generally related to the number and production of livestock in a region. However, the number and production of livestock is not only indicated by the number of livestock population, but also related to its productivity. This needs to be considered to keep the demand for livestock products (meat) can be fulfilled. Therefore, a sustainable livestock development is needed to increase livestock production and productivity [2]. The main objective of livestock development is to improve the supply of safe, healthy, whole, halal (ASUH) animal products and farmers' welfare through sustainable competitive livestock policies and programs, optimizing the utilization of local resources.

Livestock ruminannsia that have high economic value in Indonesia is beef cattle. Nevertheless, the number of beef cattle available in Indonesia is still far from meeting the demand for meat for the community. The availability of national beef to date is still experiencing a deficit of 220 thousand tons, so it must be overcome with imports [3]. Rouf *et al* [4]stated that beef imports are relatively larger compared to other types of meat imports. This is reinforced by data that beef imports accounted for 21.44% of the total value of livestock imports while the value of livestock imports constituted 18.29% of the total value of agricultural products imported nationally.

Based on the data of average population of beef cattle in 2011-2015 seen that 10 provinces contribute up to 78.49% of the total beef population in Indonesia. Center population of beef cattle in Indonesia is found in three provinces, namely: East Java, Central Java and South Sulawesi. Blora Regency is one of the areas in the Province of Central Java with the largest number of large livestock, primarily beef cattle. Blora Regency has a population of 199.584 beef cattle. Agricultural waste results in Blora Regency can be used as supporting the availability of feed for livestock sector. This is supported by the availability of abundant feeding feed or carrying capacity of 236,242 AU consisting of 152,485 AU from agricultural waste, 83,029 AU of field grass and 729 AU of superior grass [3]. The carrying capacity of the area for livestock

* Corresponding author: budibudisan8@gmail.com

development is demonstrated by the ability of the region to produce feed, especially forage that can accommodate and meet the needs of a number of livestock populations [5].

Refers to the Minister of Agriculture No. 41 / Permentan / OT.140 / 9/ 2009 concerning Technical Criteria Appropriation Region Agriculture, explained that the area of the farm must have the following characteristics: (1) the location refers to the RTRW provincial and district / city and refers to the suitability of land, (2) was built and developed by the government, local government, private and / or public in accordance with the biophysical, social, economic and environmental, (3) based on livestock commodities featured national and local and / or strategic commodities, (4) development of farmers groups and business groups, (5) can be integrated on other cultivated areas, (6) supported by the availability of water, food, technology, institutional, and market. The regulation refers to the development of animal husbandry in Blora need to consider several aspects such as the environmental aspects as a vehicle for raising livestock and carrying in it as well as socio-economic aspects as a driver of growth regions.

Based on data from BPS Blora Regency in 2016, the population of beef cattle in Blora Regency has been increasing for the last three years. In 2013 the number of beef cattle by 198,806 tails, in 2014 of 199.584 tails and in 2015 to 211,559 tails. The population of beef cattle is spread in 16 sub-districts with the largest concentration in the territory of Todanan Sub-district [6].

Livestock development is actually influenced by several factors namely physical factors, socio-cultural and economic. Physical factors are influenced by two things: genetic factors and environmental factors, while socio-cultural and economic factors are more influenced by human resources. Genetic and environmental factors affect the performance of the livestock itself. Livestock usually shows optimal physical condition if it has superior genetic properties and is supported by the suitability of growing cattle. Environmental factors also affect the growth of cattle between temperature, air humidity and solar radiation [7]. Kadarsih [8] added that the difference in altitude of the place affects the performance of cows. French [9]argues that the richness and composition of grassland species provide forage quality and potential benefits to livestock so as to meet the nutritional requirements of beef.

Socio-cultural factors that influence the development of livestock is the behavior of farmers in raising livestock. The farmer's response in raising beef cattle is influenced by business scale, farmer's condition, and maintenance objectives [10]. An enterprise is considered profitable if the business is economically feasible, meaning it can provide short-term profits and can be a long-term investment. Socio-cultural and economic factors are also related to the physical factors of the region in determining the success of beef cattle farming. The placement of livestock in a region should consider the suitability of land, the availability of forage feed, agricultural waste, and skilled human resources [11]. Physical (environmental), social, and economic factors are interrelated systems of one another.

Based on the potential of the region in Blora Regency, it is necessary to assess the natural characteristics of the land for the development of beef cattle farming. Evaluation of the physical (environmental), social and economic factors is closely related to the development planning of land-based cattle breeding cattle. This evaluation can be a development direction that is able to encourage the acceleration of regional development while maintaining environmental sustainability.

2 Materials And Methods

The research will be conducted in Blora Regency, Central Java Province. The data used in this study include primary data and secondary data. Primary data was obtained through questionnaires and interviews to respondents. Secondary data used consist of tabular data and spatial data obtained from related government agencies. Table 1 shows the relationship between objectives, data types, analysis techniques and outputs.

2.1 Land Suitability for Cattle Development

Land as an ecological base of feed support and cultivation environment should be optimized utilization. Optimal land for the development of beef cattle is a land that is ecologically capable of producing sufficient forage feed, quality and continuous. There is a close relationship between land, livestock and fodder. Therefore, the development of agribusiness areas need to consider the suitability of land and agro climate that support the advantages of the location concerned.

Land plays an important role in agricultural systems. Land serves as a place of production activity and is a factor in the production of livestock, especially as a source of animal feed. The interaction of livestock with land consists of three aspects: (a) biological adaptation of livestock, (b) the ability of land to produce livestock feed, and (c) pattern of maintenance and capacity of existing area [12]. The livestock industry in Asia relies heavily on land resources as a key input to the raw crude fiber forage feed that is needed and can be supplied from local sources [13].

Table 1. Relationship between objectives, data types, analysis techniques and outputs.

No	Purpose	Data Type	Data Source	Technique Analysis	Output
1.	Assessing the suitability of the ecological environment for the development of beef cattle farms in Blora Regency	- Land use map - Land resources map of Java Island - Climate data and rainfall	- Blora regional planning agency - *Puslittanak* - *Dinas PDSAP* district Blora	- Overlay SIG - Matching	- Map of ecological conformity of beef cattle development
2.	Analyze the potential of forage source of forage and the carrying capacity of forage for beef cattle in Blora Regency	- Land use map - Data in agricultural crop production - Livestock population data	- Bappeda - BPS - Animal husbandry department	- Overlay SIG - Matching - Analysis availability of forage	- Map conformity of forage livestock feed - Map forage support capability
3.	Analyzing the centers of activity of the development of beef cattle in Blora Regency	- Livestock population data - Commodity price data of livestock	- Animal husbandry department - BPS	- Regional centralization analysis	- The value convergence of beef cattle farming activities
4.	Develop strategies and direction of development of beef cattle in Blora Regency	- Data from interview / questionnaire - Maps of overlay results of ecological, forage, and carrying suitability	- Interview	- Analysis Hierarchy process - SWOT	- Directive of beef cattle development in District Blora

Land suitability is the state or level of suitability of a plot of land for a particular use. Land suitability at the level of the order is divided into two namely Suitable (S) for suitable and (Non Suitable / N) land for unsuitable land. Land suitability for livestock, especially ruminants should be determined in an effort to increase productivity for large scale businesses, livestock industry, or small businesses (community farms) [14]. Based on the land requirement for livestock, it is known as land base agriculture. Land-based livestock business is a cattle business with its feed components consisting mostly of forage crops (grass and leguminous). Therefore, land is an important factor as the environment and feed support.

Farming cultivation area at this time feels increasingly difficult to find, because of the increasing competition of land use. On the other hand, the need and consumption of meat is increasing as the population grows and the welfare of the population is getting better. Associated with the spatial arrangement, the area intended for the livestock business should refer to the Regional Spatial and Regional Plans (RTRW) in accordance with Permentan. 41 of 2009 on Technical Criteria Area of Agriculture Allotment.

2.2 Environmental factors (ecology) that affect the development of beef cattle

The development of ruminant livestock is influenced by genetic and environmental factors. Genetic factors affect the quality of livestock seeds (offspring) produced while the appearance (performance) of production and productivity is more influenced by environmental factors. The genetic factor of livestock determines the ability of a livestock, while the environmental factor provides an opportunity for the livestock to display its ability [15].

Environmental factors affecting livestock can be direct or indirect. Direct environmental influences include production levels (through basal metabolism), food consumption, food rate movement, maintenance needs, growth reproduction, and milk production, whereas indirect environmental influences are related to the quality and availability of food. Environmental factors are factors that have a considerable effect on the level of production.

Environmental factors that affect the physiology of livestock, among others, climate factors. Temperature and air humidity are two climate factors that influence dairy cattle production, because this factor can cause balance changes in the balance of heat in the body of livestock, water balance, energy balance, and balance of livestock behavior. The ambient temperature corresponds to the altitude of a place. The higher the location of the area from the sea surface will be the lower daily air temperature.

Optimal environmental temperature for livestock growth varies between species of livestock. The environmentally comfortable thermo netral / temperature zone for European cattle ranges from 13-18 ° C (McDowell 1974) and 4-25 ° C (Yousef). Frisian Holstein (FH) cows exhibit the best production performance when placed at ambient temperature of 18.3 ° C with 55% moisture [16]. The relation between temperature and humidity is called Temperature Humidity Index (THI) and THI application can show the effect of temperature and humidity on stress level in livestock. Stressed cattle will usually make physiological adjustments and behavior. Longer heat stress on cattle will usually have an impact on increased drinking water consumption, decreased milk production, increased urine volume, and decreased feed intake.

In addition to temperature and humidity, rainfall, and long drought also affect the livestock life. The number and pattern of rainfall per year affect the productivity of livestock. The length of the dry season shows the level of water availability as a medium that supports life with the ground material from the soil. Topographic factors that affect the growth of livestock is the height of the place and the slope (slope). The percentage of slope turned out to be a limiting factor because the higher the percentage further narrowed the chances of livestock to flourish [17].

2.3 Suitability of Forage Livestock Feed

Beef cattle are one of ruminants and are herbivores. Herbivorous cattle require forage as food. Beef cattle require forage for more than 60% of all food consumed, both fresh and dried. Therefore, the provision of forage feed that is sufficient and has a high quality needs to get the main attention. Thus the location for the development of ruminant livestock should also be supported by the availability of quality feed.

Feed is one of the production inputs that determine the success of livestock business because it directly affects the productivity and efficiency. In ruminants, forage for feed is still a major component of the feed system and is a relatively inexpensive source of feed in most agro-ecosystems in Indonesia. Forage in the tropics like Indonesia tends to have lower nutritional quality than forage in temperate regions, due to the high proportion of fiber, low protein content and the potential deficiency of some mineral elements. In addition, its availability fluctuates considerably, primarily due to the effects of rainfall. The quantity and quality of forage will decrease during the dry season and cause the production of livestock to decrease drastically. The availability of feed crops that have high nutritional quality and are able to grow throughout the year are expected to overcome these fluctuations [18]. On the other hand, ruminant livestock can utilize the remaining agricultural produce and its follow-up products in the form of forage from the annual crop to meet the needs of feed. With the management of a good farming system, then some of the feed can be met from farmland. Based on the needs of livestock and forage production, every one hectare of dry land can support the needs of 2-6 cows or 16-30 sheep / goats, depending on the applied cropping pattern.

The ownership of livestock (especially ruminants) in each farming family is generally limited to only 3 - 5 sheep / goats or 1 - 2 cows / buffaloes and this is related to limited land and capital ownership. Even if sufficient capital, the availability of domestic workers will limit the increase in the amount of livestock raising, in which case there is competition between the forage seekers with energy for cultivation activities. In this regard, it is necessary to consider an effort that can streamline the system and crop production and livestock cycles, for example by storing and treating agricultural waste as the main feed source [19].

Agricultural waste is a by-product of food crop processing whose production is highly dependent on the type and number of planting areas or cropping patterns of food crops in a region [20]. Further explained some types of agricultural waste that can be used as animal feed ingredients include: rice straw, corn straw, peanut stover, and sugar cane. The yield of palawija plant waste is generally higher in nutritional value than rice straw or corn straw. Utilization of agricultural waste for livestock will support the integration of livestock business with the agricultural business of both food crops, horticulture and plantation.

Lack of forage feed in the dry season, is a routine problem every year for farmers who raise ruminant livestock. In ruminant livestock rations, grass is more widely used because in addition to cheaper is also more easily obtained. In addition, grasses have higher production and are more resistant to defoliation pressure (cutting and grasping). To increase the production and productivity of livestock, the availability and continuity of forage of fodder is necessary. For that we need to realize the existence of land used as forage gardens fodder and grazing.

The technical standards for extending grazing areas according to the Directorate General of PSP (2013) are as follows: (a) the land for developing grazing land is on relatively fertile soil; (B) the slope / topography of the pasture land up to 15%; (C) the slope / topography of the land for expansion of forage forage area maximum of 40% (d) grazing area at least 10 ha per expanse; (E) planted grasses are in grass and 20 to 25% of the grazing area covered by cut grass [21].

2.4 Land carrying capacity

Carrying capacity indicates the enormous environmental capacity to support animal life, expressed in the number of tails per unit area of land. The number of animals that life can support depends on the biomass (plant organic material) available to animals. Carrying capacity is determined by the abundance of plant organic material formed in the process of photosynthesis per unit area and time, called primary productivity.

The determination of environmental carrying capacity shall be contained in the Regulation of the State Minister of the Environment Number 17 Year 2009 on Guidelines for the Determination of the

Capacity of the Environment in Spatial Planning, which is done by knowing the capacity of the natural environment and resources to support human activities / residents who use space for their survival. The magnitude of environmental carrying capacity in a place is influenced by the condition and characteristics of the resources contained in the expanse of the space in question. The carrying capacity of the land is related to the availability and need of land that can meet the needs of biological production for the people living in a region [22].

Land carrying capacity is determined by many factors. Biophysical, social, economic, and cultural factors influence each other. Carrying capacity depends on the percentage of land that can be used for certain sustainable and sustainable allocations. The percentage of land is determined by the suitability of land for a certain designation. According Ardhani [23], the carrying capacity of the region to livestock is the ability of the region to accommodate a number of livestock population optimally, which is very specific between agro-ecosystems. With this understanding then:

a. The carrying capacity of the territory against traditional livestock is the ability of the region to produce feed, especially in the form of forages that can be sufficient for the needs of a number of livestock population both in the form of fresh and dry without going through special processing and addition. The carrying capacity is derived from the total forage of indigestible feed available to a number of livestock populations in the region by considering the value of other benefits optimally.
b. Real carrying capacity is the ability of land in the area concerned to produce forage feed available and commonly used and affordable utilization.
c. The potential carrying capacity is the ability of the land to produce forages in the form of opportunities for development, cultivation, and processing (save pruning).

2.5 Geographic Information system

Geographic information systems (GIS) became one of the most widely used tools for spatial analysis. Geographic information systems are literally defined as a component consisting of hardware, software, geographic data and human resources that work together effectively to capture, store, repair, update, manage, manipulate, integrate, analyze and display data in a Geographic-based information. SIG is a system consisting of various components that can not stand alone [24].

The main components of GIS are computer systems, geospatial data, and users. The computer system for GIS consists of hardware, software, and procedures for the preparation of data entry, processing, analysis, modeling, and viewing of geospatial data. Geospatial data sources are digital maps, aerial photographs, satellite images, statistical tables, and other documents relating to the data to be processed. The user as a target to select the required information, create standards, create an efficient updating schedule, analyze the results issued for the intended use and plan the application.

The use of GIS in the livestock sector has been developed by academics and overseas scientists, among others: information on variations of grazing and excretion behavior on ruminant species, land use type, lawn gardening, and grazing management with agropastura systems in West Nigeria [25]. Impact on the adequacy of water in cattle grazing in South Georgia [26], parameter estimates of several factors in livestock integration systems in West African savannas [27], and analyzes the application of technology and land use on dairy farms in Kenya [28].

3 Conclusion

Based on the potential of the region in Blora Regency, it is necessary to assess the natural characteristics of the land for the development of beef cattle farming. Evaluation of the physical (environmental), social and economic factors is closely related to the development planning of land-based cattle breeding cattle.

References

1. Department of Animal Husbandry and Animal Health of Central Java Province, *Masterplan of Livestock Area Development in Central Java Province*. Semarang: Department of Animal Husbandry and Animal Health of Central Java Province, 2016.
2. E. Suhaema, "Area Analysis For The Development Of Beef Cattle In Cianjur District," Institut Pertanian Bogor, 2014.
3. Kementan, *Beef Outlook*. Jakarta: Ministry of Agriculture RI, 2016.
4. A. A. Rouf, Daryanto, and Fariyanti, "Business Competitiveness of Beef Cattle in Indonesia: Domestic Approach Resources Cost," *Wartazoa*, vol. Vol.**24**, no. 2, pp. 97–107, 2014.
5. Sumanto, E. Juarini, Wibowo, and Ashari, "Potential Areas For The Dissemination And Development Of Animal Husbandry In Yogyakarta Special Region," *Semin. Nas. Peternak. dan Veieriner*, 2000.
6. BPS, *Blora In Figures 2016*. Blora: Central Bureau of Statistics of Blora Regency, 2016.
7. N. Silanikove, "Effects of Heat Stress on the Welfare of Extensively Managed Domestic Ruminants," *Livest. Prod. Sci.*, vol. **67**, no. October, pp. 1–18, 2000.
8. S. Kadarsih, "Performance of Bali Cattle Based on Elevation Place in Bengkulu Transmigration Area : II. Reproductive performance," *J. Penelit. UNIB*, vol. **10**, no. 2, pp. 119–126, 2004.
9. K. E. French, "Species Composition Determines Forage Quality and Medicinal Value of High Diversity Grasslands in Lowland England," *Agric. Ecosyst. Environ.*, vol. **241**, pp. 193–204, 2017.

10. Rusdin, "Several Influential Factors Against Community Responses Cattle Breeding In Parigi Moutong District," *J. Agrol.*, vol. **16**, no. 4, pp. 301–308, 2009.
11. Atmiyati, "Carrying Support Forage Support to Livestock Development in Sambas District," *Temu Tek. Nas. Tenaga Fungsional Pertan.*, pp. 96–100, 2006.
12. Suharyanto, "Development Strategy of Buffalo-Based Plantation Oil Palm Plantation in Bengkulu Province," Bengkulu, 2006.
13. N. Lasmawati, "Swot Analysis of Ruminant Animal Husbandry Development Based on the Potency of Forage Feed Using Visual Basic Programming Language 6.o," Bogor, 2006.
14. S. Hardjowigeno and Widiatmaka, *Evaluation of Land Suitability and Land Use Planning*, Printed 2. Yogyakarta: Gadjah Mada University Press, 2011.
15. G. A, K. Jamal, and C. Sumantri, "Estimation of Body Weight Through Morphometric analysis with Best Regression Approach Best - Subset on Garut Sheep Type Pedaging, Agile and Cross," pp. 1–6, 2008.
16. A. Yani, H. Suhardiyanto, R. Hasbullah, and B. P. Purwanto, "Analysis and Simulation of Air Temperature Distribution on Dairy Cage Using Computational Fluid Dynamics (CFD)," *Media Breeders.*, vol. **30,** no. 3, pp. 218–228, 2007.
17. N. Rusmana, Atmiyati, and Ridwan, "Making Ecological Conformity Map For Ruminant Livestock On The Scale of Review," *Temu Tek. Nas. Tenaga Fungsional Pertan.*, 2006.
18. Haryono, *Indigofera As Animal Feed*. Jakarta: Ministry of Agriculture, 2013.
19. K. Diwyanto, B. Prawiradwiputra, and D. Lubis, "Integration of Crops in the Development of Agribusiness Competitiveness, Sustainable and Sourced," *Pros. Semin. Nas. Teknol. Peternak. dan Vet.*, vol. **1**, no. 1, pp. 17–26, 2002.
20. H. Makkar, "Applications Of The In Vitro Gas Method In The Evaluation Of Feed Resources, And Enchancement Of Nutritional Value Of Tannin-Rich Tree/Browse Leaves and Agro-Industrial By-Products," *Dev. F. Eval. Anim. Feed Suppplementation Packag. Prooceeding Final Rev. Meet. An IAEA Tech. Coop. Reg. AFRA Proj. Organ. By Jt. FAO/IAEA*, pp. 23–40, 2002.
21. Ditjen PSP, *Technical Guidelines for Extension of Farming Areas*. Jakarta, 2014.
22. KLH, *Guidance on Determination of Environmental Support Capacity in Spatial Planning*. Jakarta: Ministry of Environment, 2009.
23. F. Ardhani, "Potential Areas for Livestock Development in Bulungan District, East Kalimantan Province," *EPP. Vol.5. No.1. 2008*, vol. **5**, no. No.1, pp. 36–43, 2008.
24. J. T. Puntodewo.A, S.Dewi, "Geographic Information System for Natural Resource Management," *Cent. Int. For. Res.*, no. 2003, 2003.
25. E. Schlecht, P. Hiernaux, I. Kadaouré, C. Hülsebusch, and F. Mahler, "A spatio-temporal analysis of forage availability and grazing and excretion behaviour of herded and free grazing cattle, sheep and goats in Western Niger," *Agric. Ecosyst. Environ.*, vol. **113**, no. 1–4, pp. 226–242, 2006.
26. M. M. Colby and Y. J. Johnson, "Potential uses for geographic information system-based planning and decision support technology in intensive food animal production," *Anim Heal. Res Rev*, vol. **3,** no. 1, pp. 31–42, 2002.
27. V. M. Manyong, I. Okike, and T. O. Williams, "Effective dimensionality and factors affecting crop-livestock integration in West African savannas: A combination of principal component analysis and Tobit approaches," *Agric. Econ.*, vol. **35**, no. 2, pp. 145–155, 2006.
28. S. J. Staal, I. Baltenweck, M. M. Waithaka, T. deWolff, and L. Njoroge, "Location and uptake: integrated household and GIS analysis of\ntechnology adoption and land use, with application to smallholder\ndairy farms in Kenya," *Agric. Econ.*, vol. **27**, pp. 295–315, 2002.

 https://doi.org/10.1051/e3sconf/20183109023

Students Environmental Awareness of Ar Ridho Nature School Semarang

Agustina Prihatiningsih*

Doctoral Program of Environmental Science, School of Postgraduate Studies, Diponegoro University, Semarang - Indonesia

Abstract. In order to meet the goal of environmental education, creating generation who are environmentally concerned and responsible, Ar Ridho Nature School Semarang (SAA), put the message of such education on their gardening subject. The subject has designed for grade 1 to 6 as the school commitment to build a civilization who are able to live in harmony with nature. The purpose of gardening is not only giving students environmental knowledge but also building environmental awareness and skill to preserve nature. Thus, this study aims at presenting environmental awareness among students of SAA and finding out whether there is a relationship between environmental knowledge and environmental awareness. The third grade of students were selected as participants since they were in the middle level of elementary education which has not applied 2013 Curricula (K13). A questionnaire survey was applied to 62 students face to face with considering the level of students for understanding the given questions. Descriptive statistics, ANOVA, and Pearson coefficient of correlation are the techniques used to analyze the data.

1 Introduction

Environment, one of the sustainable development pillars becoming ubiquitous topic discuss worldwide today. People has realized that the development of both economy and society should be in accordance with the environmental condition, thus the environmental protection is considered an integral part of the development process and cannot be isolated from it. In an attempt to conservation and preservation, Environmental Education has been applied to school curricula in every country, including Indonesia.

Environmental education according to UNESCO in Tbilisi Declaration 1978 is "a learning process that increases people's knowledge and awareness about the environment and associated challenges, develops the necessary skills and expertise to address the challenges, and fosters attitudes, motivations, and commitments to make informed decisions and take responsible action" [1].

In Indonesia, Environmental education has applied since 1996, after the two department ministries (environment department ministry and education and culture department ministry), signed Memorandum of Understanding (MoU) stated that the content of environment education should be integrated in school curricula.

Since then, many programs rolled out to schools' environment education. From the application of school subject *Pendidikan Lingkungan Hidup* to newest one, *Adiwiyata* School Program. In accordance with the spirit of sustainability, *Sekolah Alam ArRidho* (SAA) put gardening as the teaching method of environmental education. Gardening, from UNESCO (2012) is defined an alternative method in delivering Environmental Education since its activities provide students experience in real-life context.

Gardening is actually an amalgamation of the philosophies behind experiential education, ecological literacy and environmental awareness and agricultural literacy [2]. It involves teaching children through personal discovery in natural settings, where they learn ecological principles that govern all life, as well as develop a sense of connection with the land. Moreover, according gardening facilitate the children to open their minds to their natural existence, develop their sense of responsibility and of self-dependence, train them to respect the resources of the earth, teach them the obligations of citizenship, interest them sympathetically in the occupations of men, touch their relation to human life in general, and touch their imagination with the spiritual forces of the world [3]. Furthermore, gardening is expected to influence students environmental awareness. Environmental awareness (EA) is all about being conscious of the environment around us. EA is classified into two aspects: perception of environmental problems and behavioral inclination to protect the environment. The perception is the peoples should have knowledge of environment and their issues. The behavior inclination is to protect the environment includes two major aspects; firstly, the value of environmental protection in peoples mind which is indicated by the balance between environmental protection and economic development and also the willingness to pay for the protection of the environment;

* Corresponding author: agustinaprihatiningsih@gmail.com

secondly, attitudes about participating in environmental protection [4].

In delivering EE for the elementary level, Maria Montessori , the founder of the Montessori method of child education stated that a garden could help children in developing their moral and appreciation in nature because the first education for children is the senses, then the education of the intellect [5]. She stated that nature can be a great basis for an introduction to the pleasures and later on the necessity for nurturing our living things [6]. Gardens can provide an environment in which students can learn to work with peers, teachers, parents and volunteers while growing plants and wild life [7]. Thus, supported gardening to become teaching method of EE. Teacher todays cannot get a child early enough to learn how to protect life and how to support the weak, but in gardening teachers teach how that damaged plants get a bandage to heal and weak ones a support [8]. Gardening then, expected to improve students' environmental awareness, as

The gardening effects has studied by many researchers. Waliczek and Zajicek studied the effects of an environmental curriculum, Project GREEN, on elementary and junior high students in Kansas and Texas. They found that participants' environmental attitudes improved after participating in the program [9]. Another study by Dirks and Orvis (2005) found that 3rd grade participants in the Junior Master Gardener Program increased in environmental knowledge and attitudes compared to non-participants [10]. In brief, schools with a garden demonstrated more positive gains than those without [11].

Thus, the purpose of the study is to examine the influence of students environmental knowledge from gardening activities with environmental awareness.

2 Material and Methodology

This paper investigated the extent of students' environmental awareness after taking gardening activities.

2.1. Sample

The study sample included 63 students of grade III Sekolah Alam Ar Ridho Semarang in the first semester of academic year 2016-2017.

2.2 Measuring Tools

In this study, the five-point Likert-type consists five possible responses: Disagree (1), Strongly disagree (2), Don't Know/ Don't Care (3), Agree (4) and Strongly Agree (5) Environmental Awareness towards Plants Questionnaire adapted and modified from Childs (2011) was used to assess students environmental awareness. The questionnaire consists of 21 questions. The questionnaire's reliability was a Cronbach alpha of .949. the data obtained from the questionnaire were analyzed by using the descriptive statistics, ANOVA, and Pearson coefficient of correlation in SPSS 21 for windows statistical software.

3 Research Finding

3.1 Descriptive Statistics, ANOVA and Pearson coefficient of correlation

The study included total of 62 students grade III in three parallel classes. Of the 24 students were female (38,7%) and 38 students were male (61,3%).

In the Table 1 below, Cronbach Alpha produced in each variable was <0.60. Whereas, *Corrected Item – Total Correlation* showed in each variable was < 0,30. It means that the variables used were reliable.

Table 1. Item-Total Statistics

	Scale Mean if Item Deleted	Scale Variance if Item Deleted	Corrected Item-Total Correla-tion	Cronbach's Alpha if Item Deleted
q1	67,81	197,700	,847	,944
q2	67,84	197,580	,717	,946
q3	67,84	197,580	,717	,946
q4	68,06	196,553	,699	,946
q5	67,35	206,921	,483	,949
q6	66,85	208,585	,565	,948
q7	67,00	214,656	,395	,950
q8	67,76	203,465	,610	,947
q9	67,35	204,397	,696	,946
q10	67,40	204,835	,634	,947
q11	67,79	201,283	,659	,947
q12	68,10	202,056	,653	,947
q13	68,16	200,859	,697	,946
q14	68,19	198,027	,698	,946
q15	67,81	197,700	,847	,944
q16	67,69	200,905	,726	,946
q17	67,69	200,905	,726	,946
q18	68,06	196,553	,699	,946
q19	68,39	205,290	,501	,949
q20	67,71	199,718	,732	,946
q21	67,71	199,718	,732	,946

Table 2. Model Summary[b]

Model	R	R Square	Adjusted R Square
1	,726[a]	,527	,519

a. Predictors: (Constant), Environment Knowledge (x1)

b. Dependent Variable: Environment Awareness (x2)

Determinant coefficients value showed in Table 2, on Adjusted R square column wass0,519 means 51.9% the change of Environmental Awareness (EA) variable determined by the change of Environmental Knowledge

(EK) variable, while the rest of 43,5% determined by other variables which did not explained in this study.

3.2 Observation

The study has conducted for six months (from July to December 2016. The researcher observed the gardening activies in grade III in three classes. Teachers in every class were asked to observed students' environmental awareness during the six months gardening activities. The result from the observation showed students who were not interested in gardening performed less awareness toward plants than the students who enjoyed gardening.

3.3 Interview

In the study, the findings about the interviews with two students representatives in every class confirmed the observation findings. During the interviews, they were asked about the gardening activities they experienced. Some examples of the interview findings were presented below:

Interviewer: What did you learn from gardening?

All of the 6 students answered the questions similarly. The explained what they get for gardening.

"I learnt how plant vegetables, from seeding to taking care of them and finally how to identify vegetables which are ready to harvest and the teacher used to remind us the gardening rules, what we shoul do and should not do in the garden"

Interviewer: Are you enjoy gardening activities? Why or why not?

The responds of this questions were varied. Four students answered positively, while two of them considered gardening as an boring activities.

"I did enjoy gardening, I love taking care of plants and the most expecting moment for us was harvesting."

While another students answered:

"I didn't like gardening, the activities were so monotonous and I had dirty." (Students 1)

"I didn't like gardening, I felt bored with the activities, I do hope that the teacher would create the interesting ones." (Students 2)

Interviewer: Do you have any plan to caring for plants in your spare time?

As the four students responsed positively for the previous question, the answer of the four students were the same.

"Of course, I would. In fact, I really love taking care of plants. they gave us oxygen, food and aesthetic." (Students 3)

Interviewer: Do you think gardening effects your awareness towards plants?

Four of students answered "yes".

" Of course, the gardening activities triggered our sense of belonging to plants and I would feel sad if we didn't take care of them". (Students 4)

While another students added the answer, " I would feel so mad if I saw someone step on the plants and destroyed them."

The interview conducted not only with sudents but also teachers. Teachers were asked about the gardening curriculum.

Interviewer: How do you find this gardening curriculum?

" I think we do need a revision, the exist curriculum have not covered the complete materials and activities variation."(Teacher 1)

" Ya... the activities has not been designed specifically, as the result, each teacher count on their knowledge and ability of gardening." (Teacher 2)

4. Conclusion

The study found that Environment Knowledge (EK), that the students got from gardening activities influenced 51,9% of students Environmental Awareness (EA). The result of both observation and interview confirmed this result. Most of the students enjoyed the gardening activities, which the rest didn't. From the interview with both students and teacher determined that gardening activities had effected on students environmental awareness. Furthermore, in order to develop the effects of gardening in Ar Ridho Nature School the newest revision of gardening curriculum is needed.

References

1. Tbilisi Declaration (Final report of the International Conference of EE:Tbilisi,(1978)
2. Subramaniam, GBL in Basic Ed, (2002).
3. L.H. Bailey, THSI ., (1990)
4. X. Xietal, X., Lihong, and D. Xueming, PEA in China, (1998)
5. M. Montessori, The absorbent mind, (1912)
6. M. Montessori, The Montessori method, (1964)
7. C.D. Klemmer, T.M. Waliczek, J.M. Zajicek.. GM:. HorT Tech, **15**(3):448-452, (2005)
8. K. Green, Encouraging nurturing behavior of two to seven year olds by introducing plants and flowers, (1994)
9. S.M. Skelly, and J.M. Zajicek.. HortTech. **8**(4):579-583, (1998)
10. A.E. Dirks and K. Orvis,. HortTech, **15**(3):443-447, (2005)
11. A.W. Fleener,, Masters Thesis, (2008)

The Control of Environment Management Through Administrative Court

Aju Putrijanti[1,*]

[1]Faculty of Law, Diponegoro University,Semarang – Indonesia

Abstract. Environment is important in human life. Conflict of interest comes between development of economy sector, citizenship needs and Governance, as it becomes completely difficult to analyze. The environment's lawsuit is increase from the beginning of the Court established. The duty of Administrative Court are to investigate, decide and settle administrative disputes. The Governance has to pay attention before issuing the Government's decree by put principle of good governance as priority. The issue in this paper is strengthening the role of Administrative Court to maintain the environment reuse by settle environment disputes based on the importance of environment. The administrative decisions in environment field may cause a loss or damage for the people. When the public officer did not put the appreciation to the reuse of environment and principle of good governance, it will become problems. The decision should be environmentally friendly. There should be certified judge to settle the dispute. The method of this research by examines the Judge's verdict in environment disputes, and its relation with regulations and the newest issues. The conclusion is increase the role of the Administrative Court to maintain the environment by law enforcement through settle environment disputes.

1 Introduction

Environment is very important for human life, we depends on the sustainability of it. It is important to control the environment management by the Government. The Government has a power based on Constitution to supervise it. As stated in Article 28 section (1) that everyone has rights to live with prosperity, residence, get a well and healthy environment, health services. The development in every sector should regulate quite well because there will be conflict of interests. This appears between private sector, government, citizenship as stakeholder of environment. The issue in this article is how to strengthening the role of Administrative Court to maintain the environment reuse by settle environment disputes based on the importance of environment.

2 Method

This research is use judge's verdict as secondary data, besides regulations. In this judge's verdict, we can know the legal consideration. Legal consideration is important, because it shows the judge's thinking of the case and maybe there is an important opinion for the development of law.

3 Discussions

3.1 Administrative court

The idea of Administrative Court came from F.J. Stahl in Rechtsstaat, as one of the elements of State, in Civil Law System, and to protect citizenship from abuse of power by government. On the beginning, the Administrative Court started in France, followed by Dutch, Germany and some European country.

Concept of Administrative Court in Indonesia are to protect the human rights, and independence of judiciary system. Administrative Court has to control government's act in public sector, by cancel the decree which against the law.

Administrative Court is one of the judiciary body under The Supreme Court based on Article No. 24 section (2) of Constitution. The competence is to investigate, decide and settle administrative disputes. The disputes are between the individual or private company as Defendant and the Government as Plaintiff, with administrative decree as an object to the Court. Due to the fast changing in every sector as implication of development, the environment's lawsuit increase. The lawsuit put in the Court with claim to cancel the government's decree as an object.

The main part in Judge's verdict is law consideration by Judge. Judge will give explanation based on law and knowledge. Judge has freedom to explain his consideration as one of the independence of judiciary, without intervention.

* Corresponding author: aputriyanti@yahoo.com

3.2 Government

In the welfare state, the government has a wide range to supervise the management of every sector to develop the nations. Government has to put principles of good governance along with the law. Principles of good governance as a norm, unwritten law, guidance is to govern the government. The British and Irish Ombudsman Association put ten guide principles, i.e.: independence, effectiveness, openness and transparency, accountability, integrity and clarity of purpose [1].

The different characteristic by UN Economic and Social Commission for Asia and the Pacific, i.e: participatory, consensus oriented, accountable, transparent, responsive, affective and efficient, equitable and inclusive and follows the rule of law. In Law No 30 Year 2014 of Governance Administration, Article 10 verse 1, there are 6 (six) principles of good governance i.e : legal certainty, expediency, impartiality, accuracy, not abuse the authority, openness, public interest, good service.

In a modern state, government has to cooperate with private sector in order to develop the country. Private sector has to ask permission from government, and government's decree issue for the certain purpose. It becomes an object to Administrative Court when it's infringe the law.

3.3 Environment

Environment is actually global in nature, it is multidisciplinary subject including physics, geology, geography, history, economic, physiology, biotechnology, remote sensing, geophysics, soil science and hydrology, etc. Environment is sum total of water, air and land interrelationship among themselves and also with the human being other living organism and property [2].

3.4 The control of administrative court

To build a plant, private sector must get a permission decree from relevant government's office. In 2009 there was administrative dispute with Decree of Chief of Integrated Licencing System in Pati No. 540/040/2008 dated 5 November 2008 re Mining Permission Exploration Region of Karst In Section C. on behalf Ir. Muhammad Helmi Yusron as an object. Defendant is Yayasan Wahana Lingkungan Hidup Indonesia (WALHI).The aim is to cancel the decree because it against the law and principle of good governance.

Administrative Court has its special principle, that Judge must be active (dominus litis, actieve rechter) so he can decide who has to prove, what has to prove, what should be proven by the party, what should be proven by the Judge, tools of evidence and the strengthen of the evidence . The Court granted the Plaintiff's claim and stated that first, formally procedural, the decree has followed the law. AMDAL's document based on Article 15 verse (1) Law No 23 Year 1997 should be completed, and did not respect to indigenous people. This was against the principle of good governance, i,e., principle of openness, principle of wisdom, principle of protecting the way of life.

Judges not only based on legal positivism, but also on socio cultural, because environmental has a multidisciplinary approach along with progressive law which has a dynamic evolutive character. As stated in Article 28 verse (1) Law No 4 Year 2004 of Judiciary Power, Judge has to dig, follow and respect value of law and sense of justice in civilization.

In the appeal of Administrative Court No 138/B/2009/PTTUN. Srby dated 30 November 2009, the Judges cancel the judge's verdict in the first level. The Supreme Court cancelled the judge's verdict of Pengadilan Tinggi Tata Usaha Negara (PTTUN) Surabaya. In this case, the Supreme Court granted the Plaintiff's claims and accepted their last legal effort.

Based on article by De Hoon and Verberk , that the new judge involved in the needs of litigants and its decides, together with (and not for) the litigants, which dispute resolution is the best chances for viable and sustainable decision [3]. This is a new approach to settle dispute, when it is necessary to implement. The use of the new approach should be wisely and properly.

Judges use the independence by put the indigenous principle of Sedulur Sikep in a proper way, and to maintain the reuse of environment with their customary. The role of Judge is important, especially in the modern governance, as a part of judicial governance, it has to follow and change the way of thinking, modernization of the court, also management of the court.

Judicial independence requires that judge impartial and unbiased when they judging, in order to keep trust on the case, as stated by Gar Yein Ng. [4]. It is important to look in a broader view the meaning of judicial independence, and it must relevant with the case, local norms and values The respect of principle of protecting the way of life shows that there is a philosophy thinking for reuse of environment. Judges collaborate between the principle of protecting the way of life, reuse of environment, Constitutional rights in appropriate legal reasoning, this could be a new way to control the environment management. It is necessary to use sociolegal approach to settle dispute it is means that we can use positive law, unwritten law included valuable principles in society, habits, traditions or in another way, we have tried to combine the use of doctrinal and non doctrinal approach [5].

In the environment dispute as above, there should be consider wisely because it is also involve the indigenous people with their belief. On the other side the law has to be obey and enforced. This approach is needed to understand between the law and the case, and might be appropriate in the development country [6].

In another case of administrative dispute, between Plaintiff and Defendant, with object the Governor's Decree of Environment Permit of Mining and Developing Plant Factory of Cement of PT Semen Indonesia in Rembang District Central Java Province No. 660.1/17 Year 2012 dated 7 June 2012. In the first Administrative Court, the Judge's verdict was refuse the suit, because it was over the period scheduled. In the Article 55 of Law No 5 Year 1986 of Administrative

Act, stated that the lawsuit should be submitted in 90 days after announced or informed to the applicant.

In the appeals court, has stated that they affirmed the first Administrative Court's verdict. Judges's have the same opinion and legal reason with the first Court. In the judicial review by The Supreme Court, it has withdrawn the second Court's verdict, and has its own legal reason.

One of the important legal reasoning of the Supreme Court was that they put principle of good governance as guidance to government to took prior of refuse the damage rather than the benefit. The principle of prudent and punctilious as the principle of good governance which should be used in this case.

Judges said that in processed of AMDAL's document, did not followed the regulations neither the principle of continuity, principle of prudent and punctilious. The supporting documents of AMDAL is very important, that is why it should be based on principle of good governance as well as regulations.

The different legal reasoning of the Judges, was important. It showed the independence of judiciary, and judge has to explicate their opinion based on law. To build a good governance, this is strongly needed in judiciary system.

There are some interested points in the two judge's verdict. First, both are administrative disputes of mining permission. This mining's project related with Sedulur Sikep an indigenous community, which had lived for century in the mining's area project, and they have their own way to reuse the environment.

Second is the different legal reasoning from judge when they consider and decide the dispute. One is using sociolegal approach to analyzed, and the other using normative approach, although in the last effort of the Supreme Court, it seems that judge using sociolegal approach but did not mention clearly.

Third this is another way to control the use of environment through Administrative Court. The judge's verdict which used sociolegal approach to settle administrative dispute in mining permission, is a good idea and possible to develop in forthcoming dispute.

The cancellation of government's decree is a control of environment. With consideration to respect the Sedulur Sikep way of life in reuse of environment, not only based on normative. Culture is important role and in a system of law as stated by Friedman, besides structure and substance of law.

Culture in each state and society has an important role to build the nations. In Indonesia, it is plural and heterogenous, and its has Adat Law for each region.

The decree of government is object to Administrative Court and it gives permission to applicant to use the environment. In the concept of welfare state, there are govenrment, society and private sector which collaborate to develop the state. This concept is vary in every state, depends on the ideology and aims of the state [7].

Government has an authority to control the environment, but on the other side there is an economic interest for development, and the decree is issued by the government. The permission to use the environment, actually as an instrument to protect it, but it had changed as caused of the environment contamination. Hikmahanto Juwana also said that government is able to become Defendant [8].

Administrative Court has to control the government's duty, because it is related with the well being for everyone, and to fulfill the constitution's rights to civil and political right. There are regulations which should be followed before issued decree.

The principle of good governance is important and should be as a legal basic in issued the decree. This is a universal principle, it means that this principle is use in every state to run the state. Since every state has its own philosophy, ideology and aims, so the principles will be adjust. Principle of good governance as an ethic and guidance for government to perform the State [9]. This is now become as a positive law as stated in Law No 30 Year 2014 of Administrative Governance, so it has more strengthen as legal basis.

Furthermore the control by Administrative Court as implementation to get access of justice in environmental disputes. This sociolegal approach can be use as a new method to get access of justice, since the disputes is multidisciplinary. As a comparative study from Brazil, in the article "Access to environmental justice in Brazil ", the legal environmental system is the possibility of inserting legal – environmental conflicts in the legal-institutionalsphere, which influencing the elaboration of final decision, which must be guided in the principle of environmental justice [10]. This means that to settle disputes of environment and to get access to environment justice, there should be comprehensive approach, and fair justice in court.

4 Conclusion

Control of environment management can be done by Administrative Court by cancel the government's decree which against the principle of good governance and law. Before issue the decree, government has to consider the principle of good governance as legal basis correctly and wisely.

Principle of good governance has to be used as a legal basis to control the environment reuse. Related to the dispute by put respect on principle of protecting the way of life, principle of wisdom, principle of openess are part of principle of good governance which is vary in every State.

Sociolegal approach can be used as a method to control the reuse of environment, because its combine the law and social value of society. This approach might be developed by Judge with mature legal basis and appropriate dispute.

References

1. https://www.bioa.org.uk
2. www.unescap.org
3. De Hoon & Verberk. http://www.utrechtlawreview.org Volume **10** Issue 4 (November 2014).

4. Gar Yein Ng. http://www.utrechtlawreview.org Volume **7**, Issue 1, (January 2011).
5. Sarah Blandy. Pp 166 – 175.(2014). http://eprints.whiterose.ac.uk/
6. Sulistyowati Irianto. " Memperkenalkan Studi Sosiolegal dan Implikasi Metodologisnya" dalam *Metode Penelitian Hukum Konstelasi dan Refleksi.*Jakarta : Yayasan Pustaka Obor Indonesia.pp. 173-175.(2011).
7. Mohammad Tavip. *"The Dynamic Concept of Welfare State in Indonesian Constitution". Doctoral Program of Legal Studies, Faculty of Law, Brawijaya University, Malang*, pp. 1-17., (2013). Download.portalgaruda.org.artikel/php
8. A'an Effendi. *Hukum Penyelesaian Sengketa Lingkungan*. (Jakarta : Sinar Grafika. Pp. 10 – 11, 2016)
9. Hotma P. Sibuea. *" Asas-asas Umum Pemerintahan yang Layak Sebagai Sarana Penguji Peraturan Kebijakan dalam Perspektif Negara Hukum Kesejahteraan"* dalam *Asas Negara Hukum, Peraturan Kebijakan dan Asas-asas Umum Pemerintahan yang Baik.* Jakarta: Erlangga. pp. 152. (2010)
10. Mariana Almeida Passoss de Freitas. International Journal of Court Administration.Special Issue Vol.**8**, No. 3. (July 2017). URL http://www.iacajournal.org

Undergraduate Students' Pro-Environmental Behavior in Daily Practice

Widiaswati Dewi[1,*], *Sawitri* Dian R[2]

[1]Master Program of Environmental Science, School of Postgraduate Studies, Diponegoro University, Semarang - Indonesia
[2]Faculty of Psychology, Diponegoro University, Semarang - Indonesia

Abstract. Pro-environmental behavior is an individual action as a manifestation of one's responsibility to create a sustainable environment. University students as one of the agent of change can adopt pro-environmental behaviors concept, even through simple things to do on daily activities such as ride a bicycle or walk for short distance, reuse the shopping bags, separate waste, learn about environmental issues etc. Many studies have examined pro-environmental behavior from various approaches. However, the study about university students' pro-environmental behavior is lacking. The aim of this paper is to examine the undergraduate students' pro-environmental behaviors level. We surveyed 364 first year undergraduate students from a state university in Semarang. The survey included six aspects of pro-environmental behavior in daily practice which include energy conservation, mobility and transportation, waste avoidance, recycling, consumerism, and vicarious behaviors toward conservation. Findings of this study showed the level of pro-environmental behavior of first year undergraduate students is medium. Recommendations for undergraduate students and future researchers are discussed.

1 Introduction

Pro-environmental behavior (PEB) is an individual action as a manifestation of one's responsibility to create a sustainable environment. It will be based on the presence of a wide range of environmental problems that threaten environmental sustainability [1] for example, IPCC reported that climate change had affected the nature and human life [2].

Therefore to ensure environmental sustainability, young people need to be led to behave which environmental friendly [3], because those will be exposed to the effects of environmental problems in the future [4]. In addition, young people are critical stakeholders [5], including in it is the university students.

The university students can adopt pro-environmental behavior in everyday life, e.g., recycling, carpooling or driving hybrid vehicles, using energy-efficient appliances, making ecologically conscious purchases [6]. In previous studies, pro-environmental behavior at an individual level are influenced by demographic factors [7], whereas in general literature makes it clear that PEB are influenced by social-psychological and social-demographic factors [3].

First year for the university (undergraduate) students is an important phase to adapt to life at the university, where the literature consistently mentions that first year was an important stage in shaping student attitudes and affects the persistence [8]. In this situation, pro-environmental behaviour adopting the process is very possible through the influence at the university current conditions.

The study about pro-environmental behavior has increased along with the popularity of sustainable concepts on academics, practitioners, government etc.. However, the study about freshmen' pro-environmental behavior is lacking, even though it is one of the important points to predict the students' potential. The aim of this paper is to examine freshmen' pro-environmental behaviors level.

2 Method

2.1 Participants

Participants were 364 first year undergraduate students (freshmen) from a state university in Semarang, which include two major programs (social & humanity and science & technology). The participants' profiles are shown in Table 1.

2.2 Survey materials

All responses were made using a 5-point Likert-like scale (1= never, 2= seldom, 3= occasionally, 4= often, 5= always).

The 36-item Pro-Environmental Behavior Scale was modified from General Ecological Behavior Scale [9]. It measures adolescent six aspects of pro-environmental

* Corresponding author: dewiwidiaswati@gmail.com

https://doi.org/10.1051/e3sconf/20183109025

behavior in daily practice which include energy conservation, mobility and transportation, waste avoidance, recycling, consumerism, and vicarious behaviors toward conservation. A sample item, "I ride a bicycle, take public transportation or walk to campus". Cronbach's alpha of this 36-item scale has been reported as .669.

Table 1. Profile of survey participants (n = 364)

	Frequency	Percentage
Major		
Social & Humanity	119	32.7
Science & Technology	245	67.3
Gender		
Female	231	63.5
Male	133	36.5

2.3 Data collection procedure

Research assistants administered the questionnaires to students in classroom. Participation in this study was completely voluntary. Ethics approval was obtained.

3 Results and discussion

3.1 Freshmen' pro-Environmental behavior in daily practice

Participants indicated that their pro-environmental behavior higher while outside the campus, except for waste avoidance (see Fig. 1). It is influenced by the type of activity and several infrastructures which is around an individual location when carrying out the activities.

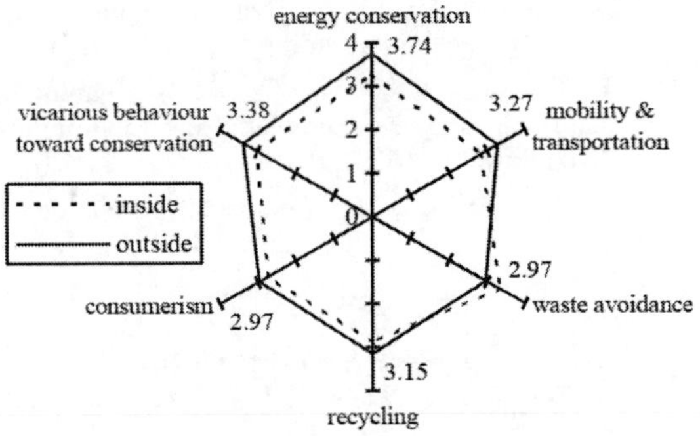

Fig. 1. Freshmen' Pro-environmental behavior in daily practices (inside & outside campus)

In general, the students show that they are most concerns with energy conservation. It was similar with case in Ohio Wesleyan University (OWU) which in general the students are concerned with energy conservation and recycling than another aspect of pro-environmental behavior [10]. This condition is a little different from Lithuanian students. The Lithunanian students are most concerns in recycling, while the other aspect of pro-environmental behavior such as saving energy or water and using more sustainable transportation are lacked of interest [11]. The difference is shown that demographic factors play an important role to shape pro-environmental behavior, because pro-environmental behavior itself is a function of the attitude/environmental ethics and various social demographic factors [12]. In addition, pro-environmental behavior has other antecedent variables that influence in a complex way (see Table 2).

Table 2. Antecedents of university students' pro-environmental behavior

Antecedent	Main Source
Demographics	[15], [10]
Education	[10], [16]
Environmental beliefs	[17]
Gender	[4]
Major /academic backgrounds	[14], [17]
Motivation	[4]
Self-efficacy	[15]
Value orientations	[11]

Pro-environmental behavior basically aims to reduce environmental problems [13]. In daily practice, an individual face some barriers to engage in PEB, these barriers include personal cost, inconvenience, and lack of time to conduct conservation [14]. Thus, to engage in pro-environmental behavior the students can do various behavior that suitable for their conditions. It is very important because the students as one of the agent of change will become a professional, as exemplars for the general public to emulate [14].

3.2 Six aspects of pro-environmental behavior

Table 3 presented that freshmen' pro-environmental behavior in general are at medium category. These result are influenced by the condition in which first year at the university is undergoing a phase of transition from high-school students to university students. A wide range of dispositions, conceptions, motivational tendencies and prior knowledge color the way an individual students sees their new learning environment [8], so that the behavior demonstrated by everyone is the response of the overall condition.

The increase of age and year at university will be effected on pro-environmental behavior, even though it does not have a constant effect [10]. This is in line are assumed by Vicente-Molina that people with more education are more likely to develop environmental performance [4], but keep in mind that pro-environmental behaviors are shaped by complex combination factors [14].

Pro-environmental behavior of everyone has various forms, among which are:

a. **Energy conservation.** Related activities include to save energy or purchase energy captured in a cleaner way.
b. **Mobility and transportation.** Related efforts to reduce carbon emissions through the choice in the

type of environment-friendly transport (e.g., using public transport, ride bikes or walk to reach a minimum distance).

c. **Waste avoidance.** Related things that aim to avoid or manage waste that arises (e.g., minimize the use of plastic, buy refill product).
d. **Recycling.** Related activities that aim to recycle the materials used been unused.
e. **Consumerism.** Related on the use or consumption of eco-friendly product like organic products.
f. **Vicarious behaviors toward conservation.** Related activities that generally do not give a negative effect to the environment.

Table 3. Distribution of Freshmen' Pro-environmental behavior

Aspects	Description	Percentage
energy conservation	low	2.2
	medium	74.2
	high	23.6
mobility and transportation	low	11.0
	medium	77.2
	high	11.8
waste avoidance	low	4.7
	medium	87.2
	high	8.2
recycling	low	6.9
	medium	89.3
	high	3.8
consumerism	low	6
	medium	92.9
	high	1.1
vicarious behavior toward conservation	low	0.8
	medium	93.4
	high	5.4

As it has been mentioned before that the participant has the greatest attention on energy conservation. It becomes a point that Indonesia does have an intensive promotion of energy conservation. It's seen through the rank of Indonesia Environmental Performance Index (EPI) which at position 41 for indicator energy and climate in 2016 [18].

Based on the results of measurements on samples retrieved that students have a diverse response magnitude on each aspect of the pro-environmental behavior, so it can be said that each student has a different 'sustainability lifestyles'. The students' sustainability lifestyles is significant to be improved, because it can play a role in the process of promoting desirable behavioral changes. It is becoming increasingly attractive given that university students as one of the agents of change in the future have a significant role, and if future talent can take decisions that are beneficial to the environment, society is more likely to make progress along the path towards sustainability [4].

However, the limitation of the current study must be noted before these results are generalized. Perhaps the main limitation of the current study is the proportion of the sample is based on the total number on each faculty/ major program (not specific in distribution of all undergraduate programs).

4 Conclusions

The aim of this study was to examine the undergraduate students' pro environmental behaviors level. It was discovered that freshmen' pro-environmental behaviors levels in general are medium. Therefore, necessary efforts to improve undergraduate students' pro-environmental behavior, because they are the leaders on the future, and thus capable of making a difference in the next decades [19]. In terms that university as a second home to the students has an important role for creating sustainable environment, examples through stir institutional changes over and above interventions aimed at changing individual behavior [20].

Acknowledgments

We would like to thank all the participants and the research assistants for their willingness and time to participate in this study.

References

1. L. Steg, C. Vlek, Journal of Environmental Psychology **29**, 309 (2009)
2. E. Margetts, Y. Kashima, Journal of Environmental Psychology **49**, 30 (2017)
3. D. Hooi, C. Fang, C. Cheng. Journal of Hospitality, Leisure, Sport & Tourism Education **20**, 16 (2017)
4. M. A. Vicente-molina, A. Fernández-sáinz, J. Izagirre-olaizola, Journal of Cleaner Production **61**, 130 (2013)
5. A. Leeuw, P. Valois, I. Ajzen, P. Schmidt, Journal of Environmental Psychology **42**, 128 (2015)
6. A. Hergatt, B. R. Werff, J. B. Henning, K. Watrous-rodriguez, Journal of Environmental Psychology **38**, 262 (2010)
7. X. Yu, Journal of Environmental Psychology **38**, 39 (2014)
8. E. E. Ketonen, A. Haarala-muhonen, L. Hirsto, J. J. Hänninen, K. Wähälä, K. Lonka, Learning and Individual Differences **51**, 141 (2016)
9. F. G. Kaiser, B. Oerke, F. X. Bogner, Journal of Environmental Psychology **27**, 242 (2007)
10. A. Meyer, Journal of Cleaner Production **112**, 3451 (2016)
11. M. S. Poskus, *Curr Psychol*, (2016)
12. G. D. Lee, D. Hong, J. Liu, Environmental Conservation **38**, 45 (2011)
13. A. Homburg, A. Stolberg, Journal of Environmental Psychology **26**, 1 (2006)
14. J. Easton, M. Koro-ljungberg, J. C. Cheng, Applied Environmental Education and Communication **8**, 126 (2009)

https://doi.org/10.1051/e3sconf/20183109025

15. S. Kim, S. Jeong, Y. Hwang, Science Communication **35**, 2 (2012)
16. S. Ajaps, R. Mclellan, Cogent Education **28**, 1 (2015)
17. M. Cordano, S. Welcomer, R. Scherer, L. Pradenas, V. Parada, The Journal of Environmental Education **41**, 4 (2010)
18. A. Hsu, 2016 Environmental Performance Index, New Haven (2016)
19. M. Yazdanpanah, M. Forouzani, Journal of Cleaner Production **107**, 342 (2015)
20. P. Mtutu, G. Thondhlana, Habitat International **53**, 142 (2016)

Characteristics and Generation of Household Hazardous Waste (HHW) in Semarang City Indonesia

Elanda Fikri[1,2*], Purwanto Purwanto[2,3] and Henna Rya Sunoko[2,4]

[1]Department of Environmental Health, Bandung Health Polytechnic, Cimahi Utara, Bandung, Indonesia
[2]Doctoral Program of Environmental Science, School of Postgraduate Studies, Universitas Diponegoro, Semarang, Indonesia
[3]Department of Chemical Engineering, Faculty of Engineering, Universitas Diponegoro, Semarang, Indonesia
[4]Faculty of Medicine, Universitas Diponegoro, Semarang, Indonesia

Abstract. Most of Household Hazardous Waste (HHW) is currently mixed with domestics waste. So that, it can impact human health and environmental quality. One important aspect in the management strategy is to determine the quantity generated and characteristics of HHW. The method used to determine the characteristics HHW refers to SNI 19-2454-2002, while the HHW generation refers to the SNI 19-3694-1994 calculated based on weight and volume. Research was conducted in four districts of Semarang. The samples used in this study were 400 families calculated based on the proportion of Slovin Formula. The characteristic of HHW in Semarang City is mainly infectious (79%), then poisonous (13%), combustible (6%) and corrosive materials (2%). The quantity HHW generated is 0.01 kg/person/day equivalent with 5.1% of municipal solid waste (MSW) in Semarang (linear equations : y=1,278x+82,00 (volume), y=0,216x+13,89 (weight).

1 Introduction

Generally, the people in the city (85,3%) mix all components of household waste, included HHW. Many people do not know that household activities can produce dangerous waste and endanger health and surrounding environment [1-2]. Although the existence household hazardous waste in the waste appearance from city people is relatively small, but with the increasing population, the amount of waste generated will also increase.

The impact of HHW is accumulated in the water and soil is an annual. For example, research conducted by Tuthill *et al* [3] and Ololade *et al* [4] the sample shows result containing Pb, Ni, Cd and also the increase of physic parameter concentration and chemical such as pH, turbidity, conductivity and phosphate [3-4]. Considering the potential of the impact of HHW, it would require proper management. One aspect that is important as a first step draw up guidelines and HHW management strategy is to determine the characteristics and HHW generation.

2 Material and methods

2.1 Research sites

The study was conducted in the Semarang City (Indonesia), Semarang lies between 6°50' - 7°10' of south latitude and line 109°35 - 110°50' of east longitude. The west border is Kendal Regency, the east border is Demak Regency, the south border is Semarang Regency, and the north is Java Ocean with the length of 13,6 Km.

2.2 The sample of the research

The number of minimum sample measurement for Semarang City was 400 family, divided into 109 family in Pedurungan Sub-District (high economic level), 129 family for West Semarang Sub-District (medium economic level), 77 family for North Semarang Sub-District (low economic level), and 85 family for Banyumanik Sub-District (topography of high land zone) [5-7]. The measurement of generation HHW refers to SNI 19-3694-1994 (National Standard of Indonesia). The characteristic or composition of HTHW identification was based on SNI 19-2454-2002 attachment.

2.3 Primary data collection

2.3.1 HHW generation

HHW generation is based on Indonesian National Standard (SNI 19-3694-1994) and was conducted for 32 consecutive days for four districts in the Semarang City.

2.3.2 HHW characteristics

Sampling was done by the separation of organic waste and HHW. After the separation, followed by

* Corresponding author: elandafikri@yahoo.com

identification of characteristic/composition HHW based attachment Indonesian National Standard SNI 19-2454-2002.

3 Results and discussion

3.1 HHW generation

HHW generation in Semarang City is 0.01 kg/person/day (5.1% MSW) or 0.0591 l/person/day (3.9% MSW). Estimated HHW generation in 2014 reach 16,003 tons/day or 94,415 m^3/day. Projection result in the next 4 years (2019) is estimated to reach minimum 17.1 ton/day or 101 m^3/day. More over this trend will increase by 18.5 ton/day each year or 109.4 m^3/day in 2025 (linear equations : y=1,278x+82,00 (volume), y=0,216x+13,89 (weight) (**Figure 1**).

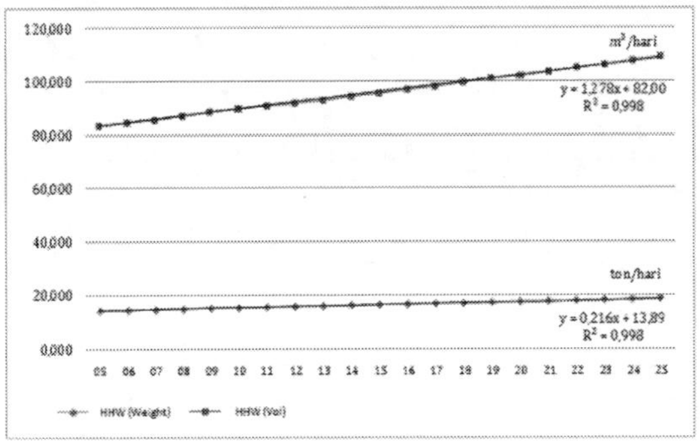

Fig. 1. HHW Generation in Semarang City

From the sampling result there should be a serious concern to manage infectious group of HHW. Until now, there is no stated regulation including in SNI 19-2454-2002. Revision is needed in that SNI. In this research, HHW infectious group was included in the analyses based on the effect on health (infection spreading) that would be caused and the degrading of environment quality. This background was supported by the categorization conducted by Delgado et al., and Jin et al., [8-9]. According to Delgado et al [8] classification HHW there are 8 categories, one of which is that it can cause an infection such as: rubber gloves, syringes, condoms, bandages and gauze bandages, while Jin et al confirm the statement "*waste segregation at the source it self is extremely important as municipal solid waste turn hazardous when mixed with hazardous waste such as pain, dyes, batteries and **human excreta***". So it is clear that the addition of one (1) characteristics/category HHW as "infectious" is one that is relevant, both from a quantitative and health impacts and environmental degradation.

Based on the sampling results of the 400 families within 32 days in the city, HHW with the characteristics of "infectious" the most dominant and commonly found types of diapers, bandages, gauze, some condoms and plabot former infusion. The diversity of the characteristics of HHW in Pedurungan found in the

3.2 HHW characteristics

Characteristic of HHW in Semarang was divided into 4 types: corrosive, combustible, poisonous and infectious. Sampling results to identify the characteristics of HHW in Semarang city indicate that HHW generation: corrosive (0.0002 kg/person/day), combustible (0.0006 kg/person/day), poisonous (0.0018 kg/person/day) and infectious (0.0074 kg/person/day).

Based on descriptive analysis in **Figure 2**, the domination characteristic of HHW based on the measurement of weight scale was infectious group with 79%, and then poisonous group 13%, combustible 6%, and the last corrosive with 2% percentage.

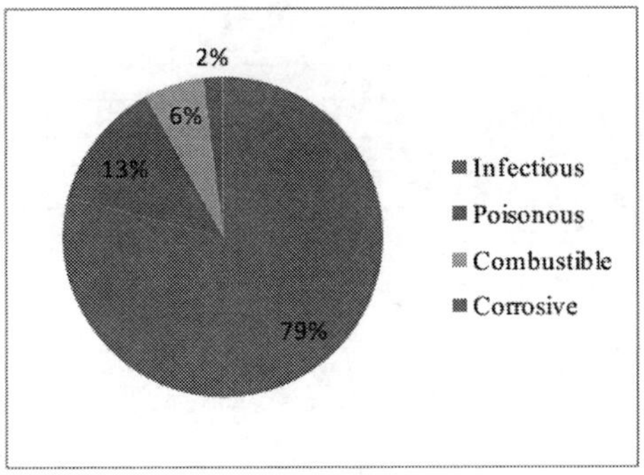

Fig. 2. Comparison of the characteristics of HHW

district which is a region on a sample group of economic strata.

Different characteristics HHW based economic strata, does not mean the same with different characteristics Based HHW topography. The statistical results using *Kruskal-Walis* test showed no differences in the characteristics HHW based topographic regions (*p-value* 0.442). So that the test results can be interpreted to have similar characteristics to HHW based topographic regions in Semarang.

3.3 Comparisons with other countries

Previous research found in seven (7) different countries, including: India, Vietnam, Switzerland, Mexico, Greece, the UK and Hungary. For example, research conducted by Lakshmikantha and Lakshminarasimaiah in the State of Karnataka (India) [10], which reported that the results of sampling HHW in the City of 0.005 kg/person/day (300 tons/day or 0.9% MSW). Results of another study, reported by Thanh et al [11] based on the results of the sampling in the Mekong (Vietnam) were reported in the city HHW generation is smaller, and only reaches 0.00053 kg/person/day (0.2% MSW) [11]. Other studies have also reported that studies are not much different, as in Switzerland which reported its HHW generation of 2 kg/person/year in 2001 with a percentage (0.5% of MSW), which means lower than generation HHW in

Semarang per year, which reached 3.65 kg/person/year with a percentage (5.1% MSW).

Research conducted by Otoniel et al in Morelia (Mexico) shows the same thing also that generation HHW City Morelia (Mexico) is estimated to 7.1 tonnes/day with a percentage of 1.6% of the total domestic waste [12]. While in Semarang in 2014 is estimated to reach 16.0 tons/day HHW with a percentage of 5.1% of the total domestic waste.

Other studies have shown different results with this research. Research conducted by Zotos et al in Greece who reported HHW generation of 7,000 tons per year [13], further research reported by Otoniel et al on HHW generation in the UK amounted to 250,000 tons HHW / year with a percentage 0.9% HHW in MSW with the largest HHW category is aerosol (26%), paint (17%), oil (15%), batteries (14%) and bleach (10%). Different also reported from Hungary in 2002, there were 17,000 tons of HHW or about 0.7% of MSW [12].

Based on the comparison HHW generation in India, Vietnam, Switzerland, Mexico, Greece, the UK and Hungary has been described previously, indicating generation HHW in Semarang higher than in the country of India, Vietnam, Switzerland and Mexico, but is smaller when compared to the generation HHW in Greek, English and Hungarian. This indicates the diversity of generation HHW level in each country. In addition to the comparison of from 11 countries, there are few studies that reported on the contribution/percentage HHW in domestic waste, including:

Table 1. Comparison of contribution/percentage HHW with domestic waste

No	Study	% of MSW	Reference
1	Tehran	10	[14]
2	***Semarang, Indonesia (Weight Scale)***	5,7	***This study***
3	Massachussets, USA	5,0	[15]
4	***Semarang Indonesia (Volume Scale)***	3,9	***This study***
5	Mexicali	3,7	[8]
6	Morelia, Meksiko	1,6	[12]
7	Greenland	1,2	[16]
8	Cuitzeo, Basin, Meksiko	1,01	[8]
9	UK	1	[17]
10	Karnataka, India	0,9	[10]
11	Mekong Delta City, Vietnam	0,2	[11]
12	Thesaloniki, Yunani	0,06	[18]
13	Hangzhou, China	0,05	[19]

Based on **Table 1** shows that the percentage HHW in Semarang is similar to research done by Tuthill et al in Massachusetts, USA with a percentage of 5.7% [3]. But the results of this study is lower to that of Ziaee et al in the city of Tehran that its HHW percentage reaches 10% of the domestic waste [14] [20].

4 Conclusions

1) Analysis by weight showed HHW generation reached 0.01 kg/person/day or 5.1% of total domestic waste (linear equations : $y=1,278x+82,00$ (volume), $y=0,216x+13,89$ (weight).
2) The characteristic of HHW in Semarang was consist of corrosive, combustible, poisonous, and infectious. The largest characteristic of HHW in Semarang is Infectious. Domination characteristic of HHW based on the measurement of weight scale was infectious group with 79%, and then poisonous group 13%, combustible 6%, and the last corrosive with 2% percentage.

Acknowledgements

This study was financially supported by Lembaga Pengelola Dana Pendidikan (LPDP), Ministry of Finance, Indonesia scheme (Contract : PRJ-598/LPDP/2014).

References

1. E. Fikri, Purwanto, R.S. Henna, Procedia Environmental Sciences. **23**, 123 – 129, (2015)
2. E. Fikri, Purwanto, R.S. Henna, Int. J. Environment and Waste Management. **17**(2), 146-157, (2016)
3. R.W. Tuthill, J.S. Edward, C. Willis, G.S. Moore, American Journal of Public Health (AJPH). **77**, 304-306, (1987)
4. I.A. Ololade, A. Adewunmi, A. Ologundudu, A. Adeleye, International Journal of Phisical Science. **4**, 22-29, (2009)
5. C.G. Sevilla, A Research Primer, Quezon City Philippines: Rex Printing Company. Inc, (1990)
6. M.M. Rivera, and R.V. Rivera, *Philippines*: Katha Publishing, (2007)
7. T.P. Ryan, *New Jersey*: John Wiley & Sons, Inc, (2013)
8. O.B. Delgado, S. Ojeda-Benitez, BL. Marquez, Journal Waste Management. **27**, 792-801, (2007)
9. J. Jin, Z. Wang, S. Ran, Journal Waste Management. **26**, 1045-1051, (2006)
10. H. Lakshmikantha, and N. Lakshminarasimaiah, *Proceeding of The International Conference on Sustainable Solid Waste Management 5-7 September 2007, Chennai, India*; 163-168, (2007)
11. N.P. Thanh, Y. Matsui, T. Fujiwara, Jounal of Environmental Management. 2307-2321, (2010)

https://doi.org/10.1051/e3sconf/20183109026

12. B.D. Otoniel, M.B. Liliana, P.G. Francelia, Waste Management. **28**, 52-56, (2008)
13. G. Zotos, A. Karagiannidis, S. Zampetoglou, A. Malamakis, I.S. Antonopoulos, S. Kontogianni, G. Tchobanoglous, Journal Waste Management; **29**,1686-1692, (2009)
14. S. Ziaee, G. Omrani, M.M.A. Agha, N. Mansouri, Advances in Environmental Biology. **6**, 676-68, (2012)
15. E.J. Stanek, R.W. Tuthill, C. Willis, G.S. Moore, Archives of Environmental Health; **42**, 84-86, (1987)
16. R. Eisted, and T.H. Christensen, Journal Waste Management. **31**, 1461-1466, (2011)
17. S.J. Burnley, Journal Waste Management. **27**, 1274-1285, (2007)
18. E. Papachristou, H. Hadjianghelou, E. Darakas, K. Alivanis, A. Belou, D. Ioannidou, E. Paraskevopoulou, K. Poulios, A. Koukourikou, N. Kosmidou, K. Sortikos, Journal Waste Management, **29**.1158-1162, (2009).
19. Y. Zhuang, S.W. Wu, Y.L. Wang, W.X. Wu, Y.X. Chen, Journal Waste Management. **28**, 2022-2030, (2008)
20. E. Fikri, Purwanto, R.S. Henna, Journal of Ecological Engineering. **18**, (5), (2017)

Understanding safety data sheets as a strategy to protect humans and the environment at the laboratory

Rizkiawalia Elza[1,*] and *Suherman* Suherman[2]

[1]Master Program of Environmental Science, School of Postgraduate Studies, Diponegoro University, Semarang - Indonesia
[2]Department of Chemical Engineering, Faculty of Engineering, Diponegoro University, Semarang - Indonesia

Abstract. Safety Data Sheet (SDS) gave important information for safe chemicals handling and widely used in communicating chemical hazards. Laboratory as one of the places associated with the consumption of a number of chemicals, then the worker in laboratory have to know the information chemicals used. The purpose of this research is to know the perception and understanding of workers in a laboratory toward SDS. The quantitative research was used and the collect data by questionnaire using Likert scale, then analyzed descriptively. The total sample of data was twenty-seven of laboratory worker people. The finding from this study showed that a great majority of people agree SDS has benefits for workers in the laboratory, agree the SDS should be available in the workplace, feel the need to know the contents of the SDS, but not so many people agree 'I access SDS while working using chemicals' and great majority of people feel 'I obtained chemical information other than SDS'.

1. Introduction

Although it involves the use of a number of chemicals that may be harmful in their activities, the laboratory is considered a safe workplace because it consumes only a small amount of chemicals [1]. However, the existence of a laboratory can have a negative impact on the environment, thus having an obligation to reduce the impact it poses as a consequence of its activities [2] Because laboratory activity contributes to waste generation in quantity but contains many types of chemicals, many of which are toxic and unknown composition [3]. Cleaner Production (CP) promotion in sustainable development becomes a holistic and integrated strategy for primary prevention due to injury, illness and casualties related to environmental pollution and unhealthy and unsafe working environment conditions [4]. CP is often used interchangeably with pollution prevention including reducing, preventing and reducing waste generation from its source [5]. Waste reduction at the source is implemented through good management when acquiring materials, replacing toxic materials with other materials at lower hazard levels and with good laboratory practice [3] In which chemical information in Safety Data Sheet (SDS) is used to inventory the chemicals used and identify the possibility of replacement of materials with lower toxicity. The study of inventory and evaluation through SDS in a cotton / polyester fabric dyeing textile mill in Turkey found 74 of 291 chemicals can be replaced with other materials that are more biodegradable and lower toxicity levels [6]. In a study on the evaluation of Cleaner Production-Pollution Prevention in the form of Toxic Use Reduction in Massachusset, using SDS from suppliers, managers and team companies work together to assess and consider chemical hazards including toxicity, cost and waste management [4]. Development of the Greenness Index, a tool to provide a thorough assessment to evaluate the impact of reagents on Safety, Health and Environment (EHS), the analysis is based on information from SDS that contains information on the nature of reagents and their impact on SHE from cradle to grave [7], Evaluation of the efficiency and safety of cosmetic products with the availability of LD50 information on SDS found 276 chemicals banned from use and 65 allowed to be used [8], Based on SDS has evaluated 166 commercial cleaning products for washing food processing appliance from various countries including detailed observations of its composition and characteristics [9].

Based on some experiences of SDS utilization above it is known that SDS is an effective system for the management of chemical safety information to protect human (workers and consumers) and the environment. The purpose of this study was to know the perception of workers in a Indonesia chemistry laboratory toward SDS by using survey research.

2. Methods

The study adopted quantitative research with descriptive analysis. Data was collected from respondents through a questionnaire with 5 points Likert scale one (1) to five (5) being : (1) strongly disagree, (2) disagree, (3) neutral, (4) agree and (5) strongly agree. The population was made up of all chemical laboratory workers with a total

* Corresponding author: elza.rizkiawalia@gmail.com

https://doi.org/10.1051/e3sconf/20183109027

of twenty-seven people. Thus a maximum sample. Their job as analysts of several environmental quality parameters (as such BOD, COD, metals, MBAS, etc.) that use chemicals daily. Questionnaire includes respondents' responses on SDS, SDS's important role for laboratory workers, access to SDS when working on chemicals, obtaining chemical information from other than SDS, SDS availability and the need to know the content of SDS. Data were analyzed by calculating the average answer based on scoring each answer from the respondent.

3. Result and Discussion

Research questions given to respondents to identify as follow :

Q1, SDS is very useful for workers in the laboratory for safe handling of chemicals.

Q2, I access SDS when working with chemicals.

Q3, I obtained chemical information other than SDS.

Q4, SDS should be available in the workplace.

Q5, I feel the need to know the contents of SDS

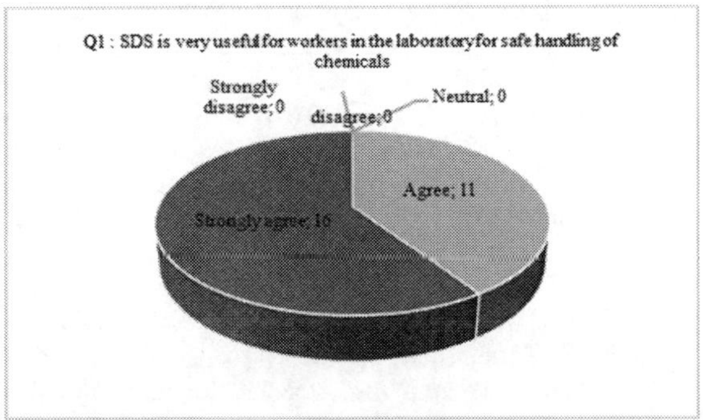

Fig. 1. Results of the first questionnaire, n=27

Score calculation results 'SDS is very useful for workers in the laboratory for safe handling of chemicals' (fig.1) It can be concluded all respondents agree SDS has benefits for laboratory workers for safety in handling chemicals with approval rating 92%.

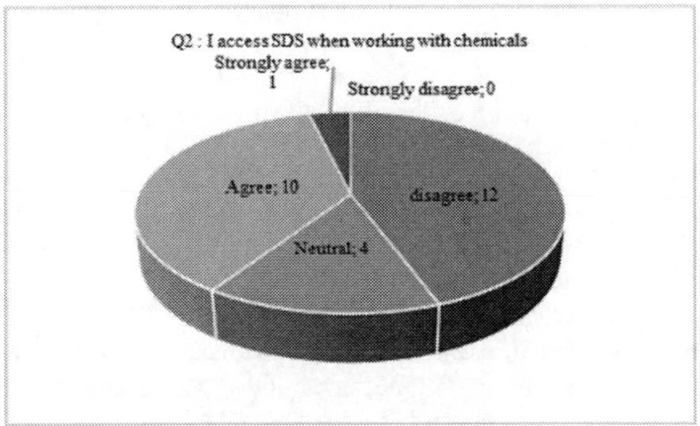

Fig. 2. Results of the second questionnaire, n=27

Dari 'I access SDS when working with chemicals' questionnaire (fig. 2) Although having only an 60% approval rating it can be concluded that laboratory workers agree to access SDS while working with chemicals. This is supported by the discovery of the third questionnaire where to 'I obtained chemical information other than SDS' (fig. 3) With an 79% approval rating, most workers agreed to feel that obtaining chemical information could be obtained other than SDS. From interviews that have been conducted, in daily work, respondents use hazardous chemicals, among others, sulfuric acid, nitric acid, chloroform, mercury (II) sulfate, silver (II) sulfate, nitrate reagent, sodium sulfate, potassium dichromate and hexane. They also feel able to obtain chemical information from other than SDS, ie from labels, TOK (transfer of knowledge) from colleagues, books, reference work instructions and internet.

The new culture of laboratory security and safety emphasizes experimental planning that includes regular attention to risk assessment and hazard considerations for yourself and others. Every worker in the laboratory should be aware of potential hazards and should reduce their numbers to a minimum. The dangers of exposure to toxic chemicals include acute toxicants, irritants, corrosive substances, allergens and sensitizers, asphyxiants, neurotoxins, reproductive toxins, developmental toxins, toxic substances, and carcinogens. A laboratory is exposed to various risks, both from within and outside the facility. Some of the risks can affect especially the laboratory itself as the dangers of flammable, explosive, and reactive chemicals pose a great risk to laboratory personnel, which if not handled properly will have an effect on the environment. In addition, almost all laboratory activities produce waste, including chemical spills. Laboratory waste includes chemical containers, filter media, aqueous solutions, and hazardous chemicals. This waste is potentially hazardous by having one or more properties of ignitability, corrosivity, reactivity, or toxicity [10]. Therefore, the knowledge and information of chemicals contained in SDS is helpful to minimize the various risks faced by workers and the environment around the laboratory.

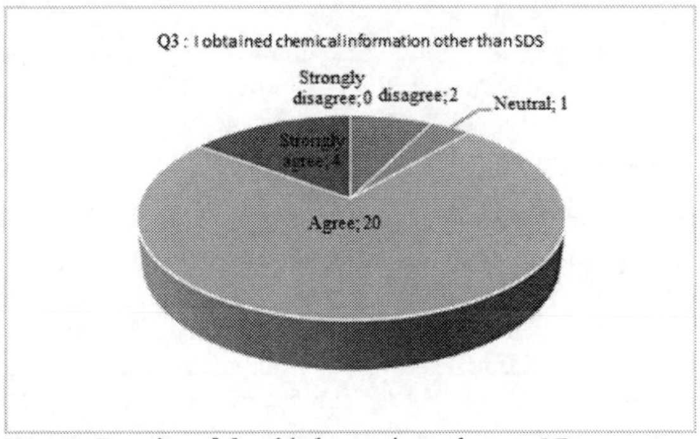

Fig. 3. Results of the third questionnaire, n=27

Among the current resources, SDS remains one of the best sources of information for evaluating hazards and assessing chemical risks [11] including the security measures that the user should follow [10].

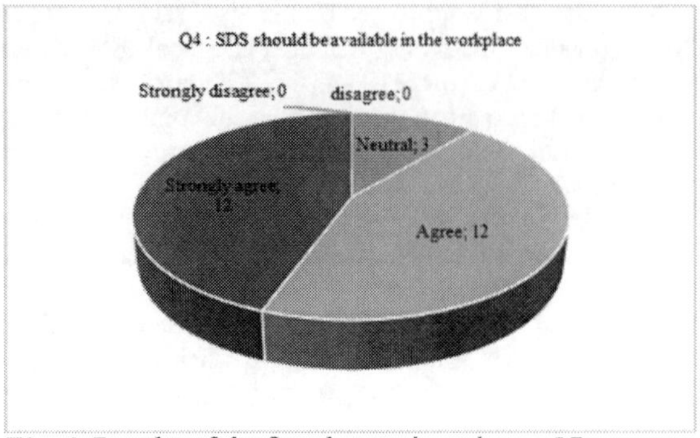

Fig. 4. Results of the fourth questionnaire, n=27

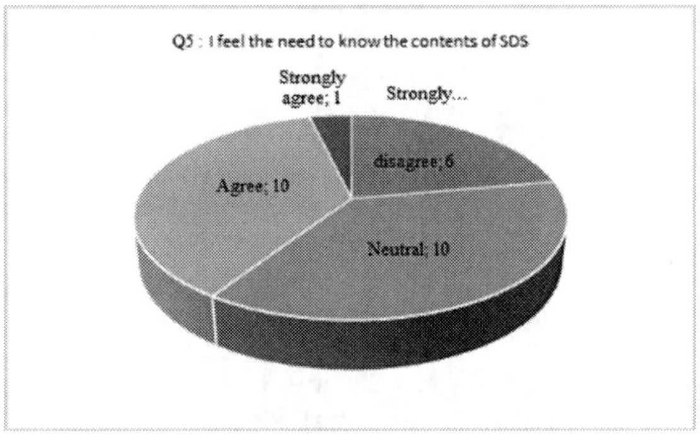

Fig. 5. Results of the fifth questionnaire, n=27

Nevertheless, with an 87% approval rating indicating the majority of workers agree that SDS should be available in the workplace (fig. 4). And feel the need to know the contents of SDS with approval rating 67% (fig. 5).

OSHA standards require manufacturers, distributors and importers of hazardous chemicals to provide SDS that explains potential hazards and other information, whereas business owners are required to provide SDS for any hazardous chemicals used including maintenance, distribution and accessibility of workers [12], Many labs now access SDS files electronically either from disk, internet or computer networks [11]. No matter what format MSDS is provided, the ability to understand its content or message is essential in protecting occupational health and safety, in fact by looking at several SDS n-Hexane from several countries: America, Korea, China and Persia interesting fact that internationally and in any language SDS has a section label that is very similar to the same 16 main sections [13] (1) Chemical product and company identification, (2) Hazard Identification, (3) Composition, information on ingredients, (4) First aid measures, (5) Fire-fighting measures, (6) Accidental release measures, (7) Handling and storage, (8) Exposure controls / personal protection, (9) Physical and chemical properties, (10) Stability and reactivity, (11) Toxicological information, (12) Ecological information, (13) Disposal consideration, (14) Information, (15) Regulatory information and (16) Other information.

4. Conclusion

The findings from this study showed that, (Q1) SDS is very useful for workers in the laboratory for safe handling of chemicals, approval rating 92%, (Q2) I access SDS when working with chemicals, approval rating 60%, (Q3) I obtained chemical information other than SDS, approval rating 79%, (Q4) SDS should be available in the workplace, 87%, (Q5) I feel the need to know the contents of SDS, approval rating 67%.

Acknowledgement

The authors would like to thank the Environment Agency of Bangka Belitung Province for their participation. The content of this paper are solely the responsibility of the authors and do not necessarily represent the organization which participated in this research.

References

1. M. Abbas, A. M. Zakaria, and M. A. Balkhyour, "Investigation of safety facilities and safe practices in chemical laboratories of a Saudi university," vol. **7**, no. 2, pp. 141–147, (2016).
2. J. B. Lopez and T. Badrick, "Proposals for the mitigation of the environmental impact of clinical laboratories," vol. **50**, no. 9, pp. 1559–1564, (2012).
3. E. D. S. Nascimento and A. T. Filho, *Brazilian J. Pharm. Sci.*, vol. **46**, no. 2, pp. 187–197, (2010).
4. K. R. Armenti, R. Moure-eraso, C. Slatin, and K. Geiser, *J. Clean. Prod.*, vol. **19**, no. 5, pp. 488–497, (2011).
5. D. A. Lopes Silva, I. Delai, M. A. S. De Castro, and A. R. Ometto, *J. Clean. Prod.*, vol. **47**, pp. 174–187, (2013).
6. E. Ozturk, H. Koseoglu, M. Karaboyaci, N. O. Yigit, U. Yetis, and M. Kitis, *J. Clean. Prod.*, vol. **130**, pp. 92–102, (2015).
7. Y. Shen, C. Lo, D. R. Nagaraj, R. Farinato, A. Essenfeld, and P. Somasundaran, *Miner. Eng.*, vol. **94**, pp. 1–9, (2016).
8. M. Uckaya, F. Uckaya, N. Demir, and Y. Demir, "Evaluation of the efficiency and safety in cosmetic products," vol. **499**, pp. 295–300, (2016).
9. M. Basso, M. Simonato, R. Furlanetto, and L. De Nardo, *J. Ind. Eng. Chem.*, vol. **53**, pp. 23–36, (2017).
10. National research council, "Chemical Laboratory Safety and Security." (2011).
11. National research council, "Prudent Practices in the Laboratory: Handling and Management of Chemical Hazards." (2011).
12. OSHA, "Laboratory Safety Guidance," (2011).
13. R. J. Willey, "2012 *International Symposium on Safety Science and Technology Understanding a safety data sheet (SDS) in regards to process safety,*" vol. **45**, no. March, pp. 857–867, (2012).

Evaluating Risk Perception based on Gender Differences for Mountaineering Activity

Novie Susanto[1,*], *Susatyo* Nugroho W.P[1] and *Ega* Rizkiyah[1]

[1]Industrial Engineering Department, Faculty of Engineering, Diponegoro University, Semarang – Indonesia

Abstract. In average 26 death events in mountaineering per year for the time span from 2003 to 2012 is reported. The number of women dying during the mountaineering is significantly smaller than males (3.5 deaths male for one female death). This study aims to analyze the differences of risk perception based on gender and provide recommendations as education basic to prevent accidents in mountaineering. This study utilizes the Kruskal-Wallis test and the Delphi Method. A total of 200 mountaineer respondents (100 males and 100 females) participated in this study. The independent variable in this study was gender. The dependent variable was risk perception including perception toward the serious accident, perception toward the probability of accident event as well as anxiety level and perception of efficacy and self-efficacy. The study result showed that the risk perception of women is higher than men with significant difference (p-value = 0.019). The recommendations from Delphi method result are by developing a positive mental attitude, showing about the risks that exist in nature, implementing Cognitive Behaviour Therapy (CBT) to raise awareness of the safety of ownself, following the climbing or mountaineer school, and using instructors to give lessons about safety in outdoor activities.

1 Introduction

Nowadays extreme sports activities in nature become the most popular activity. There are several activities that can be done in nature such as climbing the mountain. This is evidenced by the increasing number of mountaineers at all times. Increasing the number of mountain climbers can increase local revenue from the tourism sector, but it turns on some problems such as the increasing number of accident. Mountain is not a place that can be visited easily without good preparation because it can risk your life. According to Soule et al. [1] more than 30% of mountain-related deaths in France are mountain climbing, with an average death toll of 26 lives per year for the 2003 to 2012 timeframe. The possibility to die on women during the climb is significantly smaller than on men (3.5 deaths of men for one woman's death).

Some factors causing accidents that occur in the mountain according to Chamarro and Castro [2] are environmental events, equipment, medical events, errors, and time pressure. Mountain climbing is a high-risk activity. This is proven based on questionnaire results that have been distributed in the preliminary study to 258 respondents consisting of 120 male mountain climbers and 138 female mountain climbers. As many as 94% of all respondents stated that mountain climbing is an activity that has a high risk. Based on the experience of the respondents in mountain climbing, men climbed mountains on average eight times per year while women five times per year. The allegations suggest that this is because men's risk perception is lower than that of women in mountain climbing activities. As the research was done by Short, et al. [3] explains that men and women have different perceptions of risk as a whole in which men perceive lower risk than women in sports. According to Sitkin & Pablo [4], risk perceptions concerning how individuals assess the risks inherent in certain situations. Even when individuals are exposed to the same activities and circumstances, it will result in different perceptions of risk. Some studies regarding risk perception were studied [5-9]

The purpose of this study was to analyze differences in risk perceptions of mountain climbing regarding gender and provide recommendations to prevent accidents on mountain climbing.

2 Literature reviews

2.1 Perception

Sunaryo [10] explains that perception is the process of receiving stimuli through the senses preceded by attention so that the individual can know, interpret, and appreciate about the observed, both outside and within the individual. There are two kinds of perception, namely: external perception and self-perception. External perception is the perception that occurs because of the stimuli that come from outside the individual, while self-perception is the perception that occurs because of the stimuli that come from within the individual self. In this case, the object is itself.

* Corresponding author: novie.susanto@ft.undip.ac.id

Perception arises through the processes and factors that influence the perception. This is what causes each individual to have different interpretations, even though the objects seen are the same. According to Robbins and Judge [11], three factors affect one's perception are:

2.1.1 Individual concerned

Every individual sees something and tries to give an interpretation of the object being seen to be influenced by individual characteristics such as attitudes, personality, motives, interests, past experiences, and expectations.

2.1.2 The goal of perception

The characteristics of the perceptual target affect what the individual feels. Individuals who speak loudly will be more noticed than individuals who are quiet. This causes a person to tend to group people or similar events and separate them from other groups that are not similar.

2.1.3 Situation

Perception must be viewed contextually, which means the situation in which the perception arises should receive attention. The situation is a factor that contributes to the process of forming a person's perception.

2.2 Risk perception

Risk perception is a subjective assessment of the probability of a certain kind of probability of accidents occurring and to assess how much care it sees in a consequence [9]. Risk perceptions are influenced by psychological, socio-economic, and cultural processes [12].

Perceptions of risk or ability to differentiate risk associated with risk tolerance, or individual capacity to accept some risks. The official definition of risk is a measure of the probability and severity of an effect [9]. Inouye [13] explains that the inability to understand risk carefully can lead to a higher level of risk tolerance and can encourage the high-risk behavior.

2.3 Mountaineering

Mountaineering is an activity either rocks or icebergs climbing [14]. Pomfret [15] explains that mountain climbing is a tourism adventure and it is closely related to the power of emotional experience. Based on the previously mentioned description of the notion of risk perception and mountain climbing, it can be concluded that the perception of risk to mountain climbing is a subjective assessment of the probability of accident probability and to assess how much care in seeing a consequence arising on mountain climbing.

According to Chamarro & Castro [2], there are some causes of accidents that occur on mountain climbing, i.e.,

1. Environmental Events are covering difficulties, weather conditions and environmental conditions.
2. Equipment including equipment condition, the quantity of equipment, and quality of equipment used, protective equipment, and knowledge of the use of such equipment.
3. Medical Events is fatigue triggered by the activity itself.
4. Errors including errors by omissions, errors by execution, and troubleshooting.
5. Time Pressure covering speed and rush.

In a study conducted by Demirhan et al. [9] on students at the University of Turkey with male respondents totalling 1479 women and 1259 men, it showed significant results on differences in gender-based risk perceptions in extreme sports. The goal of the study is to examine the perceived risks and appeal of extreme sports by gender. Extreme sports are land sports that consist of mountain climbing, rock climbing, trekking, orienteering, cave and cycling mountain bike. Water sports consisting of rowing, canoeing, surfing, sailing, water skiing, swimming, arum rafting and scuba diving. Winter sports consisting of skiing in the Alpine, skiing in the Nordic, skiing and snowboarding. Aero Sports which consists of parachutes, hang gliding, jumping cliffs, paragliding, and bungee jumping. Motor sports are consisting of carting, off road, rally, auto track, piste motor, cross motor, and motor track. The results of the study were consistent with the hypothesis of the study that men regard extreme sports as a less risky sport than women.

In contrast to research conducted by Monasterio [14], it stated that mountain climbers are populations that have different personality characteristics. The differences are not only about the characters that can define the personality of each mountaineer, but there are other factors namely the opportunities and access to climbing activities and rock climbing, peer influence, increased popularity, media spotlight and sports commercialization. So it can be concluded that the risk perception on each mountaineer depends on the personality of each climber.

In another study conducted by Harris & Jenkins [16], it was explained that the male risk assessment is significantly seen from the severity of injuries suffered, while women assess a risk of susceptibility to any injury. This suggests that women may avoid risky situations that cause injury and men may avoid risky situations only if the injury suffered has been severe.

2.4 Gender

Gender is a set of characteristics and attributes that are socio culturally attached to men and women. The attached characteristics or attributes are created by the surrounding social or culture [17].

2.5 Delphi method

The Delphi method was used and popularized in the early 1960s by RAND, a research institute in Santa

Monica, California, USA. Delphi method is a method that is done by forming a group or group communication consisting of experts to discuss a problem. Generally the experts involved are experts with expertise in the subject matter being discussed [18].

3 Methods

The independent variable used in this research is gender. The dependent variable used in this study is the perception of risk in mountaineering. The hypothesis in this study is processed using Kruskall-Wallis test, with the aim to determine whether there is a significant difference in risk perception between men and women. The hypothesis in this research is as follows:

H_0: There is no significant difference in risk perception between men and women

H_1: There is a significant difference in risk perception between men and women

The basic decision in Kruskall-Wallis test is if the value of P Value <0.05 then it can be concluded that H_1 accepted and H_0 rejected. The questionnaires were distributed online to 200 mountaineers as respondents in this study. There were 100 respondents divided into 50 male and 50 female.

The questionnaire used in this study was designed based on adapted risk perception variables by Voetern [20] as seen in Table 1.

Table 1. Variables used in the questionnaire [19]

Variable	Sub variable	Questionnaire
Risk perception (RP)	Perceptions of the seriousness of accidents	RP1. The degree of seriousness of accidents that have occurred
		RP2. The level of seriousness of the accident in the future
	Perception of the possibility of having an accident and anxiety levels	RP3. Opportunity to experience accidents if without precaution
		RP4. Concern for accidents
	Perception of **efficacy* and ***self-efficacy*	RP5. Response to preventive measures
		RP6. Willingness to carry out preventive measures

*) *efficacy*: Ability to perform the expected action
**) *self-efficacy*: confidence that she/he can perform the expected action

In addition to the question of risk perception above, there were questions about the motivation to implement preventive measures and information needs [20]. Data processing on these questionnaires were performed descriptively. Question questionnaire and answer choice can be seen in Tables 2 and 3.

Table 2. Preventative Action Question [19]

Question	Respondent's Willingness	Answer Options
Are you willing to implement preventative advice to avoid accidents on mountain climbing activities?	Why are you willing to implement preventative suggestions? (Choose two answers)	A. Accidents that occur in ascent are very serious
		B. I feel responsible for my own safety
		C. I think I have a risk
		D. I want to prevent accidents
	Why are you unwilling to implement preventative suggestions? (Choose two answers)	A. Accidents that occur in the climb is not too serious
		B. I am not worried about self-safety
		C. I thought I had no risk
		D. I do not want to prevent accidents

Table 3. Information Needs Question [19]

Question	Answer Options
Based on the perception between men and women, what information do you want to receive right now? (Choose three answers)	A. What is the risk perception of mountain climbing that is owned by male and female climbers?
	B. Explain in detail the factors that cause accidents in the ascent!
	C. What types of accident that occur in the ascent?
	D. What actions can prevent accidents in climbing?

The Delphi method is used to make recommendations based on the opinion of experts who have expertise in mountain climbing and risk perceptions. The question posed to the expert for this recommendation was about the actions that the climber must take to be aware of the risks. Experts who advise recommendations for risk awareness on mountain climbing activities are:

1. Chairman of the Nature Lovers Student (WAPEALA) UNDIP
2. Secretary of Nature Lovers Student (WAPEALA) UNDIP
3. Lecturer of Social Psychology UNDIP

4 Result and discussion

The result of Kruskal-Wallis test for risk perception variable stated that P value (0.019) <critical area (0,05). So, it can be concluded that there is a significant difference in risk perception between men and women.

From the results of data processing, it can be concluded that the perception of men included in the level of perception of low risk because most respondents feel a serious risk of and have the possibility of an accident. While women have a high perception of risk because most of the female respondents feel a very serious risk, as well as a lot of anxiety and distress. The statement is supported by the study of Short, et al. [3] which explains that men and women have different perceptions of risk as a whole. Men perceive lower risk than women in sports.

Data showed that female climb only once a year and male climb more than five times a year. The recapitulation can be seen descriptively in Table 4.

Table 4. Climbing frequency

Gender	Climbing frequency per year	Total
Male	once	23
	Two times	23
	Three times	12
	Four times	15
	>five times	27
	Total	100
Female	once	37
	Two times	28
	Three times	18
	Four times	4
	>five times	13
	Total	100

Self-concept theory can explain other causes related to differences in risk perceptions. Individuals differ in the degree to which they understand themselves as masculine or feminine. Based on gender theory, men are referred to as masculine individuals, where highly masculine individuals believe they have many attributes, interests, preferences and skills that the community usually associates with masculinity. While women are referred to as very feminine individuals, where very feminine individuals believe they have attributes, interests, preferences, and skills associated with femininity. In addition, self-esteem can also affect men in taking risks in every action taken. Self-esteem is how people judge the dimensions of psychological masculinity. So men tend to have high self-esteem compared to women in terms of independence, assertiveness and competence [201].

In another study conducted by Harris & Jenkins [16], it was explained that the assessment of male risk was significantly seen from the severity of the injury suffered, while women assessed a risk of susceptibility to any type of injury. This suggests that women may avoid risky situations that cause injury and men may avoid risky situations only if the injury suffered has been severe. Almost everyone understands and realizes that outdoor activities such as mountain climbing have various risks. It also often heard that there was much news about accidents that occur on the climb but climber still does the activity. This fact is in line with the results of questionnaires that have been distributed to 200 respondents consisting of 100 male mountain climber respondents and 100 female mountain climbers. As many as 99% of all respondents claimed that they had known the news of the accident, and 98% of the respondents said still climbing the mountain despite an accident report. So it can be concluded that climbers still often ignore the risks that exist in mountain climbing.

According to Voeten [19], the relationship between perceptions of risk to preventive action questions can be seen from the willingness of respondents to carry out preventive measures and the belief that such precautions are effective in reducing accidents. If respondents are not willing to implement prevention and doubt the effectiveness of these preventive measures, then the respondent can have a high-risk perception. The relationship between the results of risk perception with the question of information needs can be seen from the information requirement. If the respondent does not require any more information, then it can be said the risk perception of the respondent is low, whereas if the respondent needs information about the factors - cause factors accidents, actions that can prevent accidents, it can be said respondents have a high-risk perception [19].

The results of this study indicate that the perception of men's risk is lower than that of women. The Delphi method is used to make recommendations in the opinion of experts who have expertise in mountain climbing and risk perceptions. Experts who advise recommendations for risks awareness of climbers on mountain climbing activities are the Chairman of the Nature Lovers Student (WAPEALA) UNDIP, UNDIP Nature Student Secretary (WAPEALA), and UNDIP Social Psychology Lecturer. The purpose of this recommendation is to raise the awareness of male climbers of a risk present in mountain climbing activities.

The Delphi method result suggests some recommendation to improve awareness of risks in mountain climbing activities for male climbers such as:

1. Cultivate a positive mental attitude that grows from daily climbers such as not complaining, positive thinking, not arrogant and humble. In addition, increasing the desire to live or will to live on male climbers.
2. Understand the risks that exist in nature, so climbers get a clear picture of the true risks in mountaineering activity.
3. Conduct the Cognitive Behavioral Therapy to raise awareness of personal safety.
4. Follow a mountain climbing school such as the nature lovers club (Wanadri).
5. Having instructors who can provide lessons on outdoor activities no matter how good the male climber is.

5 Conclusions

The result of Kruskal-Wallis test for risk perception variable stated that P value (0.019) <critical area (0.05). It showed that there was a significant difference of risk perception between man and woman. Based on the

analysis, the perception of men's risk is lower than that of women in mountaineering activities. Based on the implementation of the Delphi method, the recommendations to increase risk awareness of male mountain climbers are growing a positive mental attitude that grows from daily climbers such as not complaining, positive thinking, not arrogant and humble. In addition, increasing the desire to live or will to live on male climbers and showing the risks that exist in nature to get a clear picture of the true risks in mountain climbing can be conducted. Besides, performing Cognitive Behavior Therapy to awake awareness of self-salvation, following the school of mountaineers (for example Wanadri), having an instructor who can give lessons on outdoor activities can improve the awareness of the mountaineer, as well.

References

1. B. Soule, B. Lefevre, E. Boutroy, V. Reynier, F. Roux and J. Corneloup. Accidentology of mountain sports: Situation review & diagnosis. PETZL Foundation, 1-47. (2014).
2. A. Chamarro And J. F. Castro. Accident Analysis and Prevention, 197-201. (2009).
3. S. E. Short, J. Reuter, J, Brandt and M.W. Short. The Online Journal of Sport Psychology, Vol **6** Issue 3. 38-46. (2004)
4. S.B. Sitkin. and A.L. Pablo. Academy of Management Review, **Vol. 17, No. 1**. 9-38. (1992).
5. E.L. Vazquez. Environmental Management and Health, **Vol. 12 Issue: 2.** 122-133. (2001)
6. G.C. Dickson, L. Price, W.M. Maclaren and W.M. Stein. Journal of Health Organization and Management, **18(4-5):** 308-320. (2004).
7. C. Martha, X. Sanchez and M. Goma-i-Freixanet, Psychology of Sport and Exercise, **Vol 10. Issue 1.** 193-200. (2009).
8. L. Kern, A. Geneau, S. Laforest, A. Dumas, B. Tremblay, C. Goulet. Safety Science, **Vol 62.** 370-375. (2014).
9. G. Demirhan, H. Asci, M. Kangalgil. and O. Saracbasi, O. Hacettepe Journal of Sport Sciences, **25 (1).** 11-22. (2014).
10. Sunaryo. *Psikologi Untuk Keperawatan*. Jakarta: EGC. (2004).
11. S.P. Robbins, and T.A Judge. *Organizational Behavior Edition 15*. New Jersey: Pearson Education. (2013).
12. A. Marcon, G., Nguyen, M. Rava, M. Braggion, M. Grassi and M.E. Zanolin. Science of the Total Environment, **Vol. 527-528.** 270-278. (2015)
13. J. Inouye. *Risk Perception: Theories, Strategies, And Next Steps*. National Safety Council, 1-12. (2014).
14. E. Monasterio. Wilderness & Environmental Medicine, **Vol 25 Issue 2.** 1-6. (2014).
15. G. Pomfret. Tourism Management, **Vol 27 issue 1.** 113-123. (2006).
16. C.R. Harris and M. Jenkins. Judgment and Decision Making, **Vol. 1 No. 1.** 48-63. (2006).
17. H. Herdiansyah. *Gender dalam Perspektif Psikologi*. Jakarta: Salemba Humanika. (2016).
18. T.J. Gordon. *The Delphi Method*. London: Millenium. (1994).
19. H. Voeten. *Standard questionnaire on risk perception of an infectious disease outbreak.* Effective Communication In Outbreak Management for Europe, 1-10. (2015).
20. [20] S.E.Taylor, L.A Peplau and D.O Sears. *Psikologi Sosial Edisi Kedua Belas*. Jakarta: Kencana. (2009).

The development of sasi laut in the preservation of sea in Kei community

Sri Wahyu Ananingsih[1,*]

[1]Faculty of Law, Diponegoro University

Abstract. "Sasi laut" is a local wisdom of Kei society that prohibits community members to take natural resources in a particular area over a period of time. Sasi laut has an important role in the preservation of the sea. This article is based on research. The purpose of the research is to analyze causal factors of sasi laut are not practiced anymore in Kei kecil society, and the efforts to revitalize sasi laut as a local wisdom to preserve sea. This article uses socio-legal approach and the analysis is qualitative. The data consist of primary and secondary data. The results of the research show that sasi laut in several villages of Kei kecil is not practiced anymore because of the damage to sea and ecosystem, market problem, the society problem and adat institution problem. The efforts to revitalize sasi laut are improve the quality of the sea environment, legal awareness Kei society and the participation of local government. This article can be a recommendation for the local government of Southeast Maluku in order to determine the policy of sasi laut.

1 Introduction

Sasi laut is known in Indonesia as one of local wisdom Indonesian people. The regions which still recognize and practice sasi laut tradition i.a : Raja Ampat (Papua), Haruku Island, Seram Island, Aceh Island, Saparua Island, and Kei Island. Sasi laut is a traditional resource management institution. The implementation of it is dynamic through time [1]. Sasi laut can be defined as a set of rules or regulations that govern resource use. It regulates the use of poisonous plants and other chemicals destructive nets and gear, access to the sasi-governed areas, activities that are permitted and seasonal rules of entry and the harvest and other aspects of fisheries [2] [3]. It can be said as a ban on the use of the specific period [4]. The period of time of sasi laut depends on the availability of resources and the needs of the society. They usually have a prediction to decide the time to take resources in sasi laut areas. For example in Nolloth, the time to take lola (throcus Niloticus) resources as the object of sasi laut is whenever the shell has reached the minimum diameter of 6 centimeters [5]. Sasi laut is also an effort to maintain the pattern of social life of the community through the equal distribution of the benefits from the harvest of sasi [6].

Sasi laut is an important institution to preserve the sea. For the period of time, the objects of sasi are prohibited from retrieving by anyone. The aim of sasi is the natural resources will grow optimally and the sea ecosystem can be protected from damage which caused by over exploitation. These reasons have proved by Mc.Leod et al. with their research that the villages that still maintain the practice of sasi have better sea resource conditions than villages which have already disappear the practice. Besides that, sasi laut also gives profit for the community because they will get a part of the harvest. But in fact, although sasi laut has survived over approximately 400 years, it is in the process of dying out in some areas of Maluku province [7]. The same statement is also stated by Craig Thorburn:

> .....although increasing domestic and international awareness and praise of sasi, the institution is in decline in many parts of the Moluccas. This is evidenced by many islands no longer practiced at all [8].

In another place, like in Tanimbar Kei, sasi laut is still practiced, but has a change in the system from local knowledge to scientific rational [9]. Likewise in Kei community, the practice of sasi laut is in decline, especially in Kei kecil. The research was conducted to analyze the causal factors of sasi laut are not practiced anymore in Kei kecil community and the efforts to revitalize sasi laut as a local wisdom to preserve sea in Kei Island. Although many articles described sasi laut i.a Franz von Benda Beckmann et al.(1992); Kissya (1995); Craig C. Thorburn (1998); Irene Novaczek, et al.(2001); Nils Bubandt (2005) and Paulus Boli et al. (2014) but they have a different focus of study.

2 Method

This research was conducted in Kei Island, Southeast Maluku Regency, Indonesia. The framework of this research analyzes the causal factors sasi laut is not practiced anymore in Kei kecil society and the efforts to

* Corresponding author: ayu_fhundip@yahoo.com

revitalize sasi laut as a local wisdom to preserve sea in Kei Island. The data consist of primary and secondary data. Primary data were collected by observation, questionnaire and interview. Questionnaire and interview are conducted to gain information about the reason that sasi laut is not applied in Kei kecil, the prospect of sasi laut to preserve sea and the effort of local government and kei society to revitalize sasi laut. The method of questionnaire is used for 50 respondents of Kei kecil while interview for the head of ratshap danar, tubab yamlim, and rumadian (ratshap is combination of several villages (ohoi) in Kei), the head of Kei kecil sub-district, the head of maritime and fisheries office in Kei district, Ngo activist and the tetua adat in Kei. Secondary data were collected by literature study. The data were analyzed with qualitative analysis.

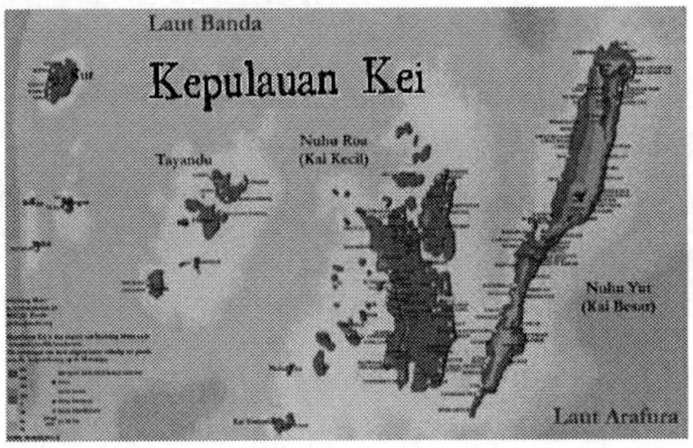

Fig 1. Kei Island

3 Result and discussion

3.1 The causal factors of sasi laut are not practiced in Kei society

Kei Island consists of kei besar (*Nuhu Yut*) and kei kecil (*Nuhu Roa*).See *Fig.1*. Kei besar is a highland area. People can reach this region by ship from Kei kecil and then continued by land transportation. Meanwhile Kei kecil can reach more easily than Kei besar. It just reached by land transportation. That's one of reasons why people in Kei kecil receive more external influence than people in Kei besar. The social life of Kei kecil community also looks more modern than Kei besar community.

Sasi laut is recognized in Kei besar as "*Yot*" and in Kei kecil as "*Yutut*". Both of Kei besar and Kei kecil are in Southeast Maluku regency. This region has a heterogeneous potential of marine resources, because it is flanked by the Arafura Sea Islands, Banda sea and the territory of Papua and Australia. Sasi laut is based on customary law and managed by adat institution. This institution has the task to plan, announce and prepare ritual of sasi laut. They also keep track of the practice of sasi laut. Sasi laut is marked by a young coconut leaf which tied up to a piece of wood (people kei named as ainum wood) in the area of sasi. People are prohibited to take resources in the area during the time of sasi. If they find a violation, they inform to the head of the village. The violator will be prosecuted to adat court. If he is proven guilty, the head of the village or the head of *ratshap* (*Kei people called it "Raja")* will determine adat sanction for him.

In the past, objects of sasi laut were all kinds of fish. But since 2000-2001, the objects are just throcus niloticus (*lola*) and sea cucumber because they are marketable in export markets. Related with the duration of sasi laut, at this time, it is longer than in the past. Sasi laut usually needs time 1 to 5 year [10].

Villages in Kei kecil no longer practicing sasi laut because 4 factors. They are:

Table 1. The factors cause villages in Kei kecil do not maintain sasi laut

No.	Factors	Problem
1	Sea and its ecosystem	Sea and it's ecosystem damage. They usually caused by the use of poisonous plants, other chemicals, destructive nets and gear. Consequently, lola and sea cucumber can not life optimally.
2	Market	A difficulty in selling the harvest of sasi laut.
3	The community	They do not support to maintain the tradition of sasi laut.
4	Adat institution	Adat institution has not authority anymore.

The four factors are related to each other, but the majority of respondents have the same opinion that *the main factor* sasi laut is not practiced in villages of Kei kecil is sea and ecosystem problem. It is *the first of problem*. It is caused by the use of poisonous plants, other chemicals, destructive nets and gear. The most fishermen who use poisonous plants and other chemicals, destructive nets and gear are outsiders, are outsiders. They want to get large amounts of fish so they use poisonous plants, other chemicals, destructive nets and gear. Surely, it is a violation of adat law. *Second,* market problem. It is a difficulty in harvest management of sasi laut. The community has difficulty to find buyers for the harvest. The buyers usually have criteria in determining the size of lola and sea cucumber. On the contrary, if the harvest of sasi are not conform the size, they are not sold out. *Third,* the attitude of Kei kecil people have change nowadays. Kei people begin do not care about the sustainability of sasi laut and customary law. The majority of them are young generation of Kei kecil. The passiveness of them are often caused by influence foreign culture. *Fourth*, adat institution problem. Although adat institution still exists in Kei, but the authority is reduced, including the control system of sasi laut.

3.2 The efforts to revitalize sasi laut in Kei

Sasi laut is an important institution and give economic benefit to community so it must be maintained its existence. Especially in Kei kecil. there are some efforts to revitalize sasi laut. The efforts of the local government are first, give socialization about the importance of the role of sasi in preserving sea resources and ecosystem to the community, especially young generation of Kei kecil. This activity can be realized through coordination with Fisheries and Marine office and education and culture. Second, increase the role of Fisheries and Marine office in the development of sustainable natural resources in Kei.

It is very important because all the time, they just behave passive role. In this case, the local government must be active. Third, the strengthening of adat institution through intensive communication with the adat leaders. It means adat institution as the main institution must exist in Kei kecil so it support sasi laut and its sustainability. Fourth, improve legal awareness of Kei society. Fifth, improve society empowering in natural resources management. Sixth, overcoming difficulties in determining buyers for selling the harvest of sasi laut. The efforts of community of Kei kecil to revitalize sasi laut are preserved sea and its ecosystem. The way to actualize that aim are prevented the use of poisonous plant and chemicals, destructive net and gear in fishing, enforce regulation of sasi laut and set adat sanction to violators.

4 Conclusion

Sasi laut is an important traditional institution, especially to preserve sea and its ecosystem. It means that sasi laut must be maintained in Kei kecil. Villages are not practice of sasi laut anymore in Kei kecil because 4 factors: sea and ecosystem problem, market problem, the community problem and adat institution problem. The efforts to revitalize sasi laut include give socialization about the importance of the role of sasi in preserving sea resources and ecosystem to the community, increase the role of Fisheries and Marine office in the development of sustainable natural resources in Kei, the strengthening of adat institution through intensive communication with the adat leaders, improve legal awareness of Kei society, improve society empowering in natural resources management and overcoming difficulties in determining buyers for selling the harvest of sasi laut.

5 References

1. C. Zerner, *Law Soc. Rev.*, vol. **28**, no. 5, p. 1079, (1994).
2. Gerd Winter (ed), *IUCN Environmental Law Centre, Germany*, 74, (2009).
3. I. Harkes, in *Proceedings of the International Workshop on Fisheries Co-management*, p. 1, (2001)
4. B. A. Benda-Beckmann Fronz, Benda-Beckmann Keebet, (1992).
5. R. V. Rugebregt, *Faculty of Law Pattimura University*. [Online]. Available: http://fhukum.unpatti.ac.id/artikel/hukum-tata-negara/45-revitalization-of-the-role-of-kewang-institute-in-traditional-government-system-of-natural-resource#_ftnref21.
6. Kissya, *Sasi Aman Haru-Ukui*. Jakarta: Sejati, (1994).
7. I. Novaczek, I. Harkes, J. Sopacua, ICLARM, (2001).
8. T. Craig, Indones. J. Coast. Mar. Resour., vol. **1**, no. 2, pp. 57–68, (1998).
9. Jayanti and Osozawa, J. Perikan. (Journal Fish. Sci., vol. **XV**, no. 1, p. 43, (2014).
10. S. W. Ananingsih, Universitas Gadjah Mada, (2015).

Minapadi Development Strategy in Supporting National Food Security

Sri Lestari[1,]*and *Azis* Nur Bambang[2]

[1]Master of Environmental Science, School of Postgraduate Studies, Diponegoro University, Semarang - Indonesia
[2]Faculty of Fisheries and Marine Science, Diponegoro University, Semarang - Indonesia

Abstract. Efforts to achieve food security, especially food self-sufficiency, face severe challenges. Intensification needs to be done in optimizing the existing land by applying integrated agriculture. One of them is by integrating agriculture and fishery aspect with implementation of minapadi. Minapadi cultivation has actually grown since a long time, but in the course of time this system Began displaced because of pesticides and chemical fertilizers. Become flagship Minapadi back in line with the increasing public awareness of the importance of sustainable environment, health, increasing demand for organic products that are free from pesticide residues which means safe for consumption. Minapadi is the answer to the fulfillment of people's need for carbohydrate and protein food. Blending the fulfillment of carbohydrates is rice as the staple food of most people with the fulfillment of animal protein from fish commodities to add nutritional value. In realizing the food security strategy is required Minapadi development. This descriptive research aims to find out various minapadi development strategies in several regions with different methods based on the literature and previous studies. The result is the formulation of minapadi development strategy in an area not necessarily can be applied in other areas. This occurs because of differences in setting criteria/ variables based on the characteristics and potential of a region, the strengths and opportunities, as well as the weaknesses and threats that the area may encounter.

1 Introduction

Law of the Republic of Indonesia Number 18 Year 2012 on Food [1] mandates that states are obliged to realize the availability, affordability, and the fulfillment of adequate, safe, quality and balanced nutritious food consumption, both at the national and local levels to individuals equally throughout the entire territory of the Republic of Indonesia over time by utilizing resources, institutional, and local culture. The definition of food according to the Elucidation of Law of the Republic of Indonesia Number 18 Year 2012, the Food is a basic human need and fulfillment is the most important part of the fundamental right of every Indonesian people. Food should always be available sufficiently, safely, quality, nutritious, and diverse with an affordable price by the community, and does not conflict with religion, faith and culture. In order to achieve this, it should be held a Food system that provides protection for both producers and consuming food.

According to Sularno, et.al, [2],The agricultural sector contributes to food availability and increases food security. Food is a primary need that all the time needed by the community. Therefore, the availability of food must be guaranteed both quantity and quality to meet the needs of the community. Based on the mandate of the Law of the Republic of Indonesia Number 18 Year 2012 concerning the Food, Food Security Agency, the Ministry of Agriculture in the Strategic Plan of Food Security 2015-2019, prioritizes increasing food sovereignty become a superior program to create economic independence by implementing common policies in RPJMN 2015- 2019, among others, aimed at strengthening food security in order to achieve food self-sufficiency through increased staple food production; improving the quality of food consumption and nutrition; and improving the welfare of food business actors [3]. This is in line with the Strategic Plan of the Ministry of Marine and Fisheries Years 2015-2019contained in sub agenda Food Sovereignty Increased Fishery Production directed at extensification and intensification of fishery business to realize the creation of food nutrition, and sub agenda of the National Economic Growth Acceleration Through Increased fishery products directed at improving quality, increasing value added and innovation of fishery technology [4].

1.1 National food security

The agricultural sector is the foundation for the growth of the national economy. The agricultural sector is

* Corresponding author: srilestari.estepe @gmail.com

developed to achieve food security, which is to meet the needs of food and proteins that society needs [2]. Efforts to achieve food security, especially food self-sufficiency face severe challenges due to the availability of land decreases caused by the conversion of agricultural land into residential and industrial. This happens because of the increasing population growth. The increase of Indonesian population from 2010-2015 is 16,942,900 inhabitants [5], while the depreciation of irrigated wetland area of 2011-2015 according to the Ministry of Agriculture [6] is presented in Table 1.

Table 1. Irrigated Wetland Area in Indonesia, 2011- 2015.

Year	Irrigated Wetland Area (Hectares)
2011	9,696,653
2012	9,768,207
2013	9,543,501
2014	9,446,141
2015	9,422,677.4

In addition to the agricultural sector, Indonesia also has a great potential in the fisheries sector. This can be seen from the number of production and market demand is increasing every year. Based on data from the Ministry of Marine Affairs and Fisheries [7], Aquaculture production in 2014 was 14,359,129 tons and increased to 15,634,093 tons in 2015. The level of fish consumption in 2014 amounted to 38.14 kg / capita / year increased to 41.11 Kg / capita / year in 2015. Figure 1 shows the level of fish consumption in Indonesia, from 2012 to 2016 as follows:

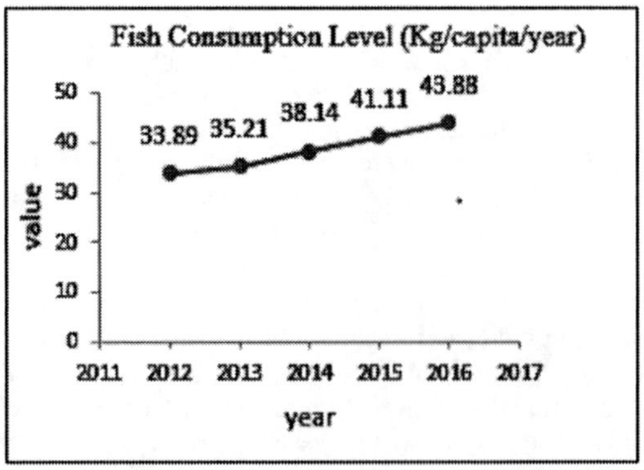

Fig. 1. Fish Consumption Level, 2012-2016

To overcome the decreasing agricultural land area that is getting lower and fulfill the needs of rice as a staple food and the increasing demand of fish consumption can be done by implementing integrated farming as proposed by Abuasir, et.al, [8], integrating agricultural aspect and fishery aspect with implementation of minapadi cultivation.

1.2 Minapadi

Minapadi cultivation has actually grown since a long time, but in the course of time this system is began displaced because of pesticides and chemical fertilizers. Minapadi cultivation is starting to become flagshipas the increasing of people awareness of the importance of sustainable environment, health, the increasing demand for organic products are free from pesticide residues means safe for consumption, and after the discovery of pest and disease resistant varieties. [9].

In tropical and sub-tropical Asia, minapadi systems constitute a unique agro-landscape. In food security and global change, minapadi is one of the important supporting factors because minapadi produce food sources of carbohydrates and proteins and in economic terms can reduce the risks that may be faced by farmers. In this integrated ecosystem, there is a symbiotic mutualism between rice and fish. As the water temperature increases, the rice provides shade for the fish, and the decaying leaves are very useful for fish feed. On the other hand, fish can increase fertility and reduce pests of disease that attack rice plants [10].The results of Nurhayati, et.al, [11], concluded that minapadi have economic and social value for the community in terms of optimizing land use. The objective of minapadi implementation is to supports the increasing of soil productivity, increasing farmer's income, improving food quality for the community. Long DN, et.al, [12] in his research mentioned that an integrated agricultural processing between fish and rice help farmers generate more income. Extra income can only be achieved if the system is integrated farming, diversification and production efficiency. The results of his research in the Mekong delta illustrates the net income of farmers who apply minapadi system is higher than the farmers who apply monocultures. In another study, Bobihoe J, et.al, [13] states that through Integrated Crop Management (ICM) is more provitable compared with the non ICM due to perform integrated crop management can increase production by 50%. Minapadi system is also very beneficial because it can increase the production of grain and straw [14].According to the Ministry of Marine Affairs and Fisheries, 2017 [15] in one hectare of farmland can earn 72 million rupiah per year in the form of rice as much as 7-8 tons per cycle, and additional fish production as much as 1.5-2 tons per cycle. This means that farmers will get a 50% added value of income compared to before applying minapadi.

Therefore, the review and analysis of minapadi in Indonesia is very important. In this paper will provide an overview of how minapadi development strategies in various regions in Indonesia in supporting national food security.

2 Methods

In this paper used the descriptive method to describe minapadi development strategies in various regions in Indonesia. There are three types of methods used to formulate minapadi development strategies, the SWOT analysis, AHP, and A'WOT.

3 Results and discussion

Research on minapadi development strategy in various regions in Indonesia have been carried out by the method and the different variables, which are summarized in Table 2.

Table 2. Minapadi Development Strategy Research in Indonesia

Author	**Sularno, et. al. (2014)**	**Nuryasri S, et. al. (2015)**	**Cahyaningrum W, et.al. (2014)**
Location	Soropadan village, district. Pringsurat, Temanggung	Village A. Widodo, district. Tugumulyo, MusiRawas	Cianjur, West Java
Method	SWOT	AHP	A'WOT
Results	a. Improve quality and production according to market demand. b. Establishing institutional market c. Applying farming greener and handling of product yield	Election implementation of species of fish in minapadi, the priority application of catfish from the tilapia or carp	a. Creating the cooperation of farmers and groups, open access to capital b. Implementing utilization policy space pattern consistent c. Conducting water quality management & use of resistant fish d. Improving the coordination of making cheap / economical feed/ providing quality seeds

According Sularno et al, [2] SWOT method used taking into account the internal and external factors that support and hinder efforts minapadi. Maximizing strength (S) and opportunities (O) and minimize weaknesses (W) and threats (T) of the business minapadi known through market surveys, interviews and observations. The strength of minapadi businesses in the village Soropadan include: irrigation, food availability, time of harvest and nutritional value of fish. Weaknesses become an obstacle, namely: the use of chemicals, sorting and grading treatment, to determine the price at the producers / market, level, and nature of perishable fishery products. Identification of external factors such as opportunities that include: high demand for rice and fish, access to juvenile fish and the high price of fish, while the factors that threaten minapadi business is rice plant pest and predator fish. In the preparation of the strategy is done by making a SWOT matrix that illustrates clearly how the opportunities and threats faced and combined with the strengths and weaknesses, so as to produce alternatives strategies as shown in Table 2.

Based on research conducted by Nuryasri S, et al, [16], the method used is AHP *(Analysis Hierarchy* Process).The data collection was done by using a structured interview using a questionnaire and literature from agencies or institutions. There are four criteria used in this study, namely production, revenue, marketing and third party support. Each of these criteria is divided into sub-criteria, which consist of: a). production: capital, labor, rice seeds and fish seed, land, pests and diseases; b). revenue: selling price, total costs and revenues minapadi; c). marketing: market segmentation and cooperation with institutions; and d). 3rd party support: Field Extension Officer, and other related agencies.

Research on minapadi development strategy by using a blend of SWOT method with AHP or A'WOT method performed by Cahyaningrum W, et.al, [17]. The data collection is done by means of a literature study and interviews. The process carried out consisted of two phases. The first step is the identification of SWOT factors that include internal factors (strengths and weaknesses) and external factors (opportunities and threats). The second stage of the weighting and rating of each of the internal and external factors. Based on the SWOT matrix, the position of Cianjur regency fisheries are in quadrant II so it is necessary to apply the strategy ST*(Strength-Threat)*as the most appropriate strategy, namely to minimize the threat faced by utilizing all the existing strength. The criteria in determining the strengths, opportunities, weaknesses and threats as follows: a). power: the potential of natural resources, fish farming techniques, manpower, availability of fish production facilities and infrastructure, national and local governments policy support, and a group of fish farmers; b). opportunities: public awareness of nutrition, demand for fish, near the area of marketing (Bandung, Jakarta, Bogor), potential food producer, wide price range, non-formal education; c). weakness: quality of mains and fish seeds, fish processing business, markets / depots fish, quality of human resources and groups, coaching and *supporting* apparatus to cultivators, and interests of the younger generation; d). threat: fish feed prices, declining water quality, land use and water conflicts, law capital capability, access to complex capital, and competition with other regions.

Based on the methods used in the study, the SWOT method is the most appropriate method in the formulation of minapadi development strategy, because this analysis can formulate the best alternative that helps to achieve the goal, and according to David FR, 2009 [18], states that the integration of the 3 stages in decision making is the most important strategy formulation technique. The stage consists of :a). input collection stage in the form of collecting information needed in determining the strategy, b). a matching stage that creates a reasonable alternative strategy with regard to internal and external factors, c). decision stage that gives priority strategy that must be done first. Therefore, this method can help identify, evaluate and select strategies. In addition, research on minapadi development strategy is limited to the internal and external environment that affect the group and Analysis of internal and external factors of groups in providing alternative minapadi development strategies. While research with AHP method has weakness among others dependence on the main-input. In the form of subjectivity from the perception of an expert so if the expert gives a wrong assessment, the model becomes meaningless. In addition, this AHP method has no confidence limits from the correctness of the model that is formed because it is a mathematical method without any statistical test [19].

4 Conclusion

There are different minapadi development strategies implemented in various regions in Indonesia. This happens because of differences in setting criteria / variables based on the characteristics and potential of a region, the strengths and opportunities, and the weaknesses and threats that the region might face. So the formulation of minapadi development strategy in an area not necessarily can be applied in other areas

References

1. *Undang-Undang Republik Indonesia Nomor 18 Tahun 2012 Tentang Pangan*. 2012.
2. Sularno, S. Jauhari,SEPA.**10**(2):268–74, (2014)
3. *Kementan RI. Rencana Strategis Badan Ketahanan Pangan Tahun 2015-2019*. 2014;
4. *KKP RI. Rencana Strategis Kementerian Kelautan dan PerikananTahun* 2015-2019. 2015;
5. BPS. *Badan Pusa tStatistik,* Jakarta. (2014) [cited 2017 Jul 1]. Available from: https://www.bps.go.id/linkTabelStatis/view/id/1274
6. Kementan RI. *Kementerian Pertanian RI*. (2016) [cited 2017 Jul 1]. Available from: http://prasarana.pertanian.go.id/lahanmy/
7. KKP RI. *Produksi Perikanan Budidaya* (2017) [cited 2017 Jul 1]. Available from: http://statistik.kkp.go.id/sidatik-dev/2.php?x=3
8. S. Abuasir, N. Hakim, Y. Sumitro. J Komun. dan Pengemb. Masy.**1**(1):30–7, (2004)
9. Sutanto R. *Penerapan Pertanian Organik: Pemasyarakatan dan Pengembangannya*. Yogyakarta: Kanisius; (2002).
10. J. Lu, X. Li, Aquaculture; **260**(1–4):106–13. (2006)
11. A. Nurhayati, W. Lili, T. Herawati, I, Riyantini. AquatProcedia. **7**:12–8, (2016)
12. Long DN et. al. *Experiment on an Integrated RicefishPolyculture System (6 Species , 1 – 2 fish / m 2) in the Mekong Delta*.140–50, (2002)
13. Bobihoe.J et. al. J Lahan Suboptimal. **4**(1):47–56, (2015)
14. M.T.H. Chowdhury, S. Dewan, M.A.Wahab, S.H. Thilsted, Bangladesh J Fish Res. **5**(2):115–22, (2001)
15. KKP RI. *Kebijakan KKP Pada Subsektor Perikanan Budidaya KementerianKelautandanPerikanan RI*. 2017
16. Nuryasri S. et al. AGRISEP.**14**(1):66–78, (2015)
17. CahyaningrumW et. a. Maj Ilm Globe.**16**(1):77–88, (2014)
18. F.R. David, *Manajemen Strategis : Konsep. Kedua belas*. Salemba Empat, Jakarta; (2009).
19. Fitria. J Inform.**13**(1):10–22, (2013)

The Importance of Pro-Environmental Behavior in Adolescent

Tyas Palupi[1,*], and *Dian R Sawitri*[2]

[1]Master Program of Environmental Science, School of Postgraduate Studies, Diponegoro University, Semarang-Indonesia
[2]Faculty of Psychology, Diponegoro University, Semarang - Indonesia

Abstract. Studies regarding pro-environmental behavior in adolescents are lacking. This study aimed to examine the importance of pro-environmental behavior in adolescents (high school and university students) by conducting literature review from previous studies on pro environmental behavior. Pro-environmental behavior is the behavior of individuals that contributes towards environmental preservation. Based on previous studies, measurement of pro-environmental behavior were investigated on several theories, namely theory of planned behavior (TPB) and value, belief, norms (VBN) by using aspects of pro environmental behavior. Young people with critical thinking, and good environmental education, are expected to behave more environmentally friendly for creating a sustainable future.

1 Introduction

The environment is one of the most important component for mankind. Interactions between humans and environment that occur continuously, will affect human behavior on the environment. Human attitudes and behavior will determine the good condition of an environment. The way humans treat their environment will have an impact on the quality of human life itself [1]. Lack of human behavior that cares about the environment, causing global environmental damage. It encourages people today to change their behavior in an effort to reduce the harmful effects of environmental damage [2]. Environmental protection and recovery are the main challenges facing our society today, therefore it is important to know and understand pro-environmental behavior in society, and what factors influence it [3].

In this case, the adolescent is part of the community, which has the potential for environmental protection [4]. Attitudes, knowledge, behaviors, concerns that the adolescent have about the environment either directly or indirectly influence future decision-making about natural resources and how their use can be sustainable [5]. Therefore, the adolescent need to be motivate, in order becomes a pioneer for pro-environmental behavior, hence in that aged the challenges and modification toward environment more easily achieved [6].

Scholars have various terms to describe behaviors that protect environment, such as; environmentally responsible behavior, environmentally concerns behavior, eco-friendly behavior, and pro-environmental behavior [7]. Pro-environmental behavior is a behavior that harms the environment as little as possible but provides enormous benefits to the environment [8]. Based on Kaiser [9], pro-environmental behavior has six (6) indicators; (1) energy conservation, (2) mobility and transportation, (3) waste avoidance, (4) recycling, (5) consumerism, and (6) vicarious behaviors toward conservation. These six indicators can be used to measure the pro environmental behavior of each individual. These six indicators can be used to explain pro environmental behavior through a widely used theory, the theory of planned behavior (TPB). Based on TPB, there are several variables used to explain pro-environmental behavior, ie attitude toward behavior, subjective norm, perception of behavioral control, and intention [10].

In order to improve pro-environmental behavior in the adolescent, the government has intervened in environmental education through "Adiwiyata Program" [11]. It was accordance revealed by Pooley & Connor [12] and Chankrajang and Muttarak [13] that education is the most effective ways to solve environmental problems. Environmental education is considered important to increase individual knowledge, positive attitude, and eco-friendly behavior. Based on this exposure, it becomes important to know the importance of pro-environmental behavior in adolescent.

2 Method

In this paper a descriptive method is used to know the importance of pro-environmental behavior in adolescent based on literature and previous studies.

3 Results and discussions

The adolescent will be affected by environmental problems starting from current actions, so they must propose with precise environmental knowledge and skills to evolve sustainable solution [14]. Pro-

* Corresponding author: michycha@yahoo..com

environmental behavior is solution that related to environmental problems [15].

Larson has been described actions that include behavior related to pro-environmental behavior through several terms, namely:

a. Ecological behavior by Kaiser , 1998
b. Environmentally responsible behaviors by De young, 2000
c. Environmentally significant behaviors by Stern, 2000
d. Responsible environmental behaviors by Cottrel, 2003
e. Pro-Environmental behaviors by Bamberg, 2007
f. Conservation behaviors by Huddart-Kennedy, 2009

Pro-environmental behavior was seen as a combination of self-interest and concern for others, subsequent generations, other living beings or the whole ecosystem [16].

Given the gap between attitudes, conservative psychologists have ceased to believe that attitudes can be traced from the record of human behavior. In contrast to this conventional understanding, Kaiser et al developed a behavior-based behavioral tool for youth. Technically, behavior measurement results based on attitudes toward the environment represent such a result of measuring human attitudes directed at conservation behavior [9].

3.1. The effect factor of pro-environmental behavior

There are several factors that affect a person, be it positive or negative in pro-environmental behavior, among others; demographic factors (gender and years of education), external factors (institutional, social, economic and cultural); and internal factors (motivation, environmental knowledge, awareness, values, attitudes, emotions, locus of control, responsibility and priority) [17]. According to Abedi Sarvestani and Shahvali (2009) in [18], human behavior (including pro-environment) influenced by beliefs, one's values and attitudes. Personal values are key to shaping attitudes towards the environment but not always followed by a pro-environmental behavior. Situational factors beyond the control of the poses in pro-environmental behavior. So, we need an intervention strategy that aims to support and removal of barriers to pro-environmental behavior [19].

3.2. Theory of pro-Environmental behavior

The existence of different types of factors that influence pro-environmental behavior encourages a variety of studies that generate a variety of theories to explain or predict which variables are significantly involved. The theory used, ie norm activation model (NAM), theory of planned behavior (TPB), Value-Beliefs-Norm (VBN) theory.

NAM describes that there were 3 variables that affected the process of the occurrence of a behavior. The three variables include personal norms, awareness of consequence and ascription of responsibility (responsibility for negative consequences if not prosocial).

One of the related theories is Ajzen's theory of planned behavior (TPB) which explains that there are three antecedent variables, namely (1) attitudes toward behavior, (2) subjective norms and (3) perceived behavior control [20]. The three antecedent variables that have been mentioned affect the formation of an intention of individual behavior in this case is pro-environmental behavior.

The value-belief-norms theory described there were several variables used to predict pro-environmental behavior, including social value orientation and pro-environmental activities; orientation value (individual and altruistic); orientation hedonist, consumer behavior (the behavior of individuals and the public) developed by [21]

Some previous studies on pro-environmental behavior, explained through TPB ie; Boldero, [22] states that the intention to recycle newspapers can be predicted directly from these recycling activities and attitudes toward recycling activities predict the intention of recycling; ; Tonglet et al. [23]suggests that attitudes can predict behavior, so attitudes show an important role in behavior. Davis et al. [24] in his research states that there is a relationship between attitudes, subjective norms of behavior. Another study on predicting pro-environment behavior Sparks and Shepherd [25], suggests that the attitude of subjective norms, and control of "Green consumerism" have a significant relationship in the intention of individuals in consuming green vegetables. Positive attitudes toward the environment can directly affect pro environmental behavior at a low cost, such as recycling. (Diekman in [17].The attitudes derived from life experience and education, markedly influence behavior [26].

Leeuw et al. [27] revealed that TPB model (attitude, subjective norms, and perceived behavioral control) made independent contribution to predict intetion. The intention was predicted behavior. Behavioral, normative, and control beliefs predicted, respectively, attitudes, subjective norms, and perceived behavioral control. Empathic concern influenced intentions and behavior indirectly by its effects on behavioral, normative, and control beliefs. Based on the data, showed that high school students have a positive intention toward pro-environmental behavior but failed in action. High school students as a young people is an important population because the initial formation of pro-environmental behavior occured because of the habit of applying environmentally freindly behavior. Examination of the effects of specific beliefs revealed important implications for designing effective behavior-change intervention.

3.3. Pro-environmental behavior in adolescent

Based on Li and Liu [28], the influence of intergenerational on the behavior of pro-environment teenager in "chines families", divided into 2 social type (influence and normative information), it was found that pro-environmental behavior in adolescents has a positive relationship between influence and normative information in the "Chinese families". Furthermore, when it was tested with variable "parental power" There are two main effects in different directions, namely the relationship between the influence of information and behavior of pro-environment in adolescents become stronger when adolescents feel the influence of parents is higher, by contrast with the normative influence.

Social influence for healthy eating in adolescents can be associated with a family member, television programs, and teacher. The government's policy to promote healthy foods do not contribute to the build subjective norms among adolescents Denmark. Young women who have a high weight of time, have the intention to eat healthy foods. Subjective norms have contributed to predict the intentions of a person to eat healthy foods. Therefore, young people in Denmark do not have a strong motivation to comply with their social norms [29].

Environmental education and pro-environmental behavior are difficult to measure reliably. Aspects that affect the behavior of school students and students are very spacious and the interrelationship of these aspects are very complex. It becomes difficult to separate the effects of the environmental education of other factors, such as the role of age in personal responsibility, the impact of various resources, customs, and norms. The survey conducted on university students and high school mainly focused on knowledge, attitudes and behavior patterns and have analyzed its relationship rather than causality. The impact of environmental education has not been measured directly in the study and identified a clear delineation between research focuses on patterns of behavior and interrelated policy document that will add to the challenges and strategies for environmental education. The main implication of this study for environmental education is that educators should be aware of the diversity of the commitment and interest of the students high level of environmental and sustainability issues when designing the curriculum. Environmental education is now likely to make a mistake with a focus on students who are less committed to the environment than students who are strongly committed to raising environmental awareness. Therefore, environmental education is expected to overcome differences in knowledge, attitudes, and behavior of students in a differentiated, targeted and effective way in the future [30].

Szagun and Pavlov [31], showed that adolescent in both countries (German and Russia) have an emotional attitude towards environmental issues and expressed willingness to engage in pro-environmental behavior, both personally and politically. Consistency in the relationship between feelings and behavior of the environment so that indifference to environmental issues associated with the lack of willingness to engage in pro-environmental behavior. Influence of gender on the environment in adolescents more obvious difference compared with culture. Pro-environmental behavior can be seen as pro-social behavior, which included a caring attitude towards nature.

Bell et al [32] showed that young people are an important population to get the energy-saving intervention for adolescent consumers of energy in the future. The intervention is done with the intention to develop the implementation of behavior change and psychological techniques are widely used. Adolescents who are already actively involved in energy saving, identified resulting in changes in behavior.

Behavioral intention have significantly affected by awareness of the consequences, attitudes toward environment, public transportation, a moral obligation to change, and easy or difficult behavior, environmentally friendly way. This will trigger a younger people to protect the environment to be sustainable. Motivation and campaigns should be routinely performed to increase awareness and understanding of students about the consequences of transportation and environmental issues [33]. However, environmental behavioral intention in disaster event is very limited. For example how intention behaviour concerning to debris cleanup, what factor have been influence [34].

Gronhoj and Thogersen [35] , states that pro-environmental of behavior in adolescents are highly influenced by the norms that dominate in the family and how it is manifested in the behavior of their parents.

Based on the above, pro-environmental behaviors among adolescents may be formed from a variety of factors, both internal and external. These factors become unity will manifest in a person's actions. Involvement of young people in environmental activities will foster positive feelings that will make a positive feeling that the realization of a sustainable pro-environmental behavior and will contribute both to the environment in the future.

5 Conclusions

Pro-environmental behavior is a individual effort to reduce the negative impact due to the destruction of nature by improving and preserving the environment. Pro-environmental behavior becomes very important, especially among adolescents. That is because adolescents are the future assets of the nation that will be the policy makers about the environment so that sustainable. Adolescence is a good representative for the change of behavior especially pro-environmental behavior.

Reference

1. Hamzah S. *Pendidikan Lingkungan: Sekelumit Wawasan Pengantar*. Bandung: Refika Aditama; (2013).
2. R. Gifford, A. Nilsson, pro-environmental concern and behaviour : A review,1–17, (2014).
3. NC. Bronfman, PC. Cisternas, López-vázquez E, Maza C De, JC. Oyanedel. Understanding Attitudes and Pro-Environmental Behaviors in a Chilean Community, 14133–52, (2015)
4. K. Lee, K. Lee. Gender differences in Hong Kong adolescent consumers ' green purchasing behavior. (2009);

5. Meinhold JL, Malkus AMYJ. ADOLESCENT ENVIRONMENTAL Can Knowledge , Attitudes , and Self-Efficacy Make a Difference? 37(4):511–32, (2005)
6. D. Hooi, C. Fang, Cheng C. Guided learning and behavioural enhancement. 2017;20(December 2016):16–26.
7. T. Hung, F. Jan, C. Yang, Conceptualizing and measuring environmentally responsible behaviors from the perspective of community-based tourists. JTMA [Internet]. Elsevier; 2013;36:454–68. Available from: http://dx.doi.org/10.1016/j.tourman.2012.09.012
8. L. Steg, C. Vlek C. Encouraging pro-environmental behaviour : An integrative review and research agenda. J Environ Psychol [Internet]. Elsevier Ltd; 2009;29(3):309–17. Available from: http://dx.doi.org/10.1016/j.jenvp.2008.10.004
9. FG. Kaiser, B. Oerke, FX. Bogner, Behavior-based environmental attitude : Development of an instrument for adolescents. 2007;27:242–51.
10. Macovei I. Determinants of Consumers ' Pro-Environmental Behavior – Toward an Integrated Model. 2015;(August).
11. RD. Iswari, SW. Utomo, Evaluasi Penerapan Program Adiwiyata Untuk Membentuk Perilaku Peduli Lingkungan di Kalangan, Siswa,15(1):35–41, (2017)
12. JA. Pooley, MO. Connor, Environment and Behavior. (2000)
13. T. Chankrajang, R. Muttarak. Green Returns to Education : Does Schooling Contribute to Pro-Environmental Behaviours ? Evidence from Thailand. Ecol Econ [Internet]. Elsevier B.V.; 2017;131:434–48. Available from: http://dx.doi.org/10.1016/j.ecolecon.2016.09.015
14. MA. Vicente-molina, A. Fernández-sáinz, J. Izagirre-olaizola. J Clean Prod [Internet]. Elsevier Ltd; 2013;61:130–8. Available from: http://dx.doi.org/10.1016/j.jclepro.2013.05.015
15. A. Homburg, A. Stolberg, Explaining pro-environmental behavior with a cognitive theory of stress $,26:1–14, (2006)
16. S. Bamberg, G. Moser Twenty years after Hines , Hungerford , and Tomera : A new meta-analysis of psycho-social determinants of pro-environmental behaviour, 27:14–25, (2007)
17. A. Kollmuss, J. Agyeman. Mind the Gap : Why do people act environmentally and what are the barriers to pro- environmental behavior ? 2002;(July 2013):37–41.
18. MS. Omran, Eur Online J Nat Soc Sci,3(3):141–51, (2014)
19. P. Mtutu, G. Thondhlana. Encouraging pro-environmental behaviour : Energy use and recycling at Rhodes University , South Africa. Habitat Int [Internet]. Elsevier Ltd; 2016;53:142–50. Available from: http://dx.doi.org/10.1016/j.habitatint.2015.11.031
20. M. Greaves, LD. Zibarras, C. Stride, Using the theory of planned behavior to explore environmental behavioral intentions in the workplace. J Environ Psychol [Internet]. Elsevier Ltd; 2013;34:109–20. Available from: http://dx.doi.org/10.1016/j.jenvp.2013.02.003
21. PC. Stern, Toward a Coherent Theory of Environmentally Significant Behavior,56(3):407–24, (2000)
22. J. Boldero, The Prediction of Household Recycling of Newspapers : The Role of Attitudes , Intentions , and Situational Factors1, 440–62, (1995)
23. M. Tonglet, PS. Phillips, AD. Read, Using the Theory of Planned Behaviour to investigate the determinants of recycling behaviour : a case study from Brixworth , UK, 41:191–214, (2004)
24. G. Davis, F. O'Callaghan, K. Knox. Sustainable attitudes and behaviours amongst a sample of non-academic staff: A case study from an Information Services Department, Griffith University, Brisbane. J Sustain High Educ. (2008)
25. P. Sparks, R. Shepherd, Self-Identity and the Theory of Planned Behavior: Assesing the Role of Identification with "Green Consumerism." Soc Psychol Q. (1992)
26. N. Newhouse. J Environ Educ, 22(1): 26–32.; (1990)
27. Leeuw A De, P. Valois, I. Ajzen, P Schmidt, J Environ Psychol [Internet]. Elsevier Ltd; 2015;42:128–38. Available from: http://dx.doi.org/10.1016/j.jenvp.2015.03.005
28. J. Li, C. Liu. Intergenerational_influence_on adolescent PEB. (2016)
29. A. Gronhoj, T. Bech-Larsen. Using theory of planned behavior to predict healthy eating among Danish adolescents. (2012)
30. Á, Zsóka, ZM. Szerényi, A. Széchy, T. Kocsis. J Clean Prod [Internet]. Elsevier Ltd; 2013;48:128–38. Available from: http://dx.doi.org/10.1016/j.jclepro.2012.11.030
31. G. Szagun, VI. Pavlov, Environmental awareness. (1995).
32. BT. Bell, N. Toth, L. Little, MA. Smith. Planning to Save the Planet : Using an Online Intervention Based on Implementation Intentions to Change Adolescent Self-Reported Energy-Saving Behavior. (2016)
33. NK. Basri, N. Sabahiah, A. Sukor, S. Asmah, An Analysis of Pro-Environmental Attributes Towards Adolescents in Penang using Extended Theory of Planned Behaviour, 802:419–24, (2015)
34 Maryono, H. Nakayama, & T. Shimaoka, Memoirs of the Faculty of Engineering, Kyushu University, 74(3), PP. 79-98 (2015)
35. A. Grønhøj, J. Thøgersen. Action speaks louder than words : The effect of personal attitudes and family norms on adolescents ' pro-environmental behaviour,33:292–302, (2012)

https://doi.org/10.1051/e3sconf/20183109032

Stakeholder Analysis in Utilizing of Environmental Services and Natural Attractions in Tuk Semuncar Utilization Zone of Gunung Merbabu National Park: A literature review

Muhammad Arif Setiawan[1,*] and *Fuad* Muhammad[2]

[1] Master Program of Environmental Science, School of Postgraduate Studies, Diponegoro University, Semarang - Indonesia
[2] Faculty of Science and Mathematics, Diponegoro University, Semarang - Indonesia

Abstract. Many stakeholders in the utilizing of environmental services and natural attraction of Tuk Semuncar Utilization Zone needs to get the attention of the Gunung Merbabu National Park Officer. The existence of natural resources has an important role for the life of the community, making the complexity of the relationship between various parties who have interests in natural resource management. The existence of stakeholders and their interests should receive attention and be considered to be accommodated as an effort to prevent the occurrence of losses to the conservation area of Gunung Merbabu National Park. Every stakeholder's interest can have a positive and negative impact on other stakeholders. Stakeholder analysis can help in understanding the conflict on the utilizing of environmental services and natural attraction that is happening, as well as input strategy in involving stakeholders for the achievement of goals.

1. Introduction

Indonesia has the potential of valuable natural resources, partly within the conservation area. The potential has an important position and role for the Indonesian nation as a development capital. The management of conservation areas and their development has the objective to promote the conservation of biological natural resources and their ecosystems so as to support efforts to improve the welfare of the community and to support the success in achieving sustainable development.

As a conservation area surrounded by densely populated settlements, the area of Gunung Merbabu National Park is inseparable from public pressure. The area of Gunung Merbabu National Park becomes an integral part of the life of the people around it. The community has strong interactions with the natural resources of the Gunung Merbabu National Park area in economic, social and cultural aspects. Communities hereditary to meet the needs of life from natural resources in the forest area of Mount Merbabu which is currently designated as Gunung Merbabu National Park area. These community activities can affect the sustainability of natural resources and the ecosystem of Gunung Merbabu National Park. This makes the area of conservation gets high pressure from the community so that it can threaten the preservation of natural resources and its ecosystem. States that all conservation areas that are public assets and managed by the government for public interest have been damaged, widespread, or contested by various parties wishing to exploit the area for other purposes [1].

Determination of Gunung Merbabu National Park area as government step for ecological protection of life buffer system, preservation of plant species diversity, wildlife and its constituent ecosystem, sustainable use of natural resources and ecosystem for science, education, research, cultivation support and tourism [2]. Ecologically, the Gunung Merbabu National Park area becomes a water catchment area for the surrounding areas and water source producers. Gunung Merbabu National Park area there is also the potential of nature tourism that can be developed as a tourist destination.

In addition to potential water sources, Tuk Semuncar Utilization Zone has a natural beauty that has the potential to become a natural tourist destination. The natural tourism opportunity of Tuk Semuncar Utilization Zone is used by some people to get the economic benefit. The number of natural tourist visitors to the Tuk Semuncar Utilization Zone caused conflict within the community. Nature tourism activities have a negative impact on community water utilizing. Conflict is allegedly a difference in interests and differences in ownership rights. As a measure to reduce conflict, the Gunung Merbabu National Park Officer decided to close all natural tourism activities to the Tuk Semuncar Utilization Zone indefinitely. Actually, the District Government sees the potential of nature tourism Tuk Semuncar Utilization Zone can be developed to improve the welfare of the surrounding community. However, until now there has been no agreement between interested parties to exploit that potential.

* Corresponding author: areefmuhammad04@gmail.com

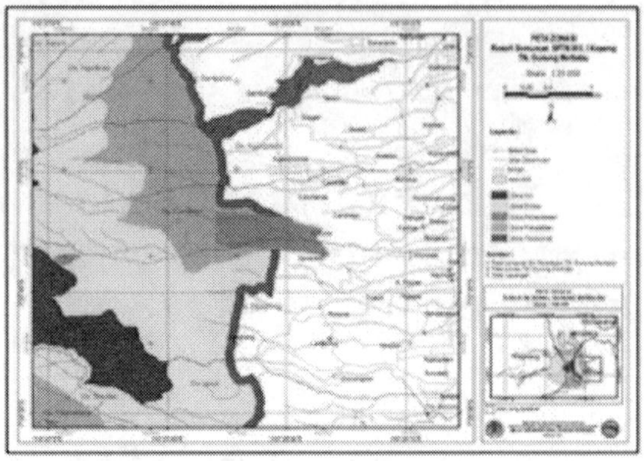

Fig. 1. Map of Tuk Semuncar Utilization Zone of Gunung Merbabu National Park

Tuk Semuncar Utilization Zone is a part of Gunung Merbabu National Park area which has abundant water resources potential and is very important for the livelihood and economy of surrounding communities [2], [3]. There are two large springs in the Tuk Semuncar Utilization Zone. The surrounding community named the two springs as Tuk Semuncar and Tuk Sipendok. Both springs flow into a river whose water is used commercially and non commercially. The people who come from the surrounding villages are very dependent on the water source.

Budimanta, who a key person, stated that the existence of natural resources has an important role for the life of the community, making the complexity of the relationship between various stakeholders in natural resources management [4]. The complexity of relations between stakeholders raises social and environmental problems that can be a constraint in the management of Gunung Merbabu National Park conservation area. Thus, an understanding of the existence and role of stakeholders is important in realizing good management and environmental, social and economic benefits [5].

The number of stakeholders in the utilizing environment services and natural attraction Tuk Semuncar Utilization Zone needs to get the attention of Gunung Merbabu National Park Officer. Utilizing of environmental services and natural attraction of Tuk Semuncar Utilization Zone by one party may cause harm to the interests of other parties. The existence of stakeholders and their interests should receive attention and be considered to be accommodated as an effort to prevent the occurrence of losses to conservation areas.

Stakeholders management is important to minimize problems in the joint use of environmental services and nature attraction of Tuk Semuncar Utilization Zone so that it can provide optimal benefits for all stakeholders. It can help improve management effectiveness and synergize the interests of all stakeholders based on the understanding and agreement in accordance with the laws and regulations. Success in managing stakeholders interest in the utilizing environmental services and nature attraction of Tuk Semuncar Utilization Zone will contribute positively to the Gunung Merbabu National Park natural resources management.

2. Who are Stakeholders?

The definition of stakeholders is individuals or groups that can be impacted or affected by the success of an organization's goals [6]. Anyone who impacts and / or is affected by development policies, programs and activities. They can be men or women, communities, socioeconomic groups, or institutions in dimensions at every level of society [7]. Each individual / group has the resources and needs of each who must be represented in the decision-making process in development activities. As parties either individually or in groups that can be influenced and / or influencing decision-making as well as achieving the objectives of an activity [8].

In the implementation of development programs, the term stakeholders are used to describe communities or organizations that are affected by activities or policies, in which a party does not always receive a fair impact. Some parties may bear the costs and some will actually benefit from an activity or policy [9].

Crosby argues that broadly the stakeholders are divided into three groups, (a) Main stakeholders, those who receive positive or negative impacts (outside the willingness) of an activity, (b) Supporting stakeholders, ie parties who mediate in assisting the process of delivering activities. They can be classified on behalf of donors, executives, supervisors, and advocacy organizations such as government organizations, NGOs, and private parties. In some activities, support stakeholders may be individuals or key groups with both formal and informal interests, (c) Key stakeholders, ie those with strong or important influences related to issues, needs, and concerns for the smoothness of activities [10].

3. Stakeholders Analysis

Stakeholder analysis is defined as a process for (1) defining aspects of social and natural phenomena that are influenced by decisions or actions; (2) identifying individuals, groups and organizations affected by or may affect parts of the phenomenon; and (3) prioritizing an individual and group for involvement in decision-making processes [11].

Stakeholder analysis is useful in identifying communities or groups of people most affected by the impact or impact of a development activity [9]. The analysis is also useful in determining priorities about the community or community groups needed in the implementation of development activities and benefits for them. Stakeholder analysis can assist in understanding the current conflict of land resource use [12].

Grimble and Chan argues that stakeholders' analysis of natural resource management has the following main objectives, (a) Improve the effectiveness of policy implementation by explicitly considering the interests and challenges of stakeholders regarding the implementation of a policy, identifying and managing stakeholder conflicts over natural resources, and providing early consideration for ways to build

https://doi.org/10.1051/e3sconf/20183109032

togetherness, complementary interests, and opportunities for cooperation and compromise, (b) Addressing the social impacts of a policy by breaking down the analysis to separately assess the interests and impacts of interventions on different stakeholders [8].

Stakeholder analysis is conducted by: (a) identifying stakeholders, (b) mapping and categorizing stakeholders, and (c) investigating relationships among stakeholders [11]. Identification of stakeholders to identify groups / organizations / individuals who have a negative / positive impact on the activities of the utilization of environmental services and natural tourism. Stakeholder mapping is done by using interest rate and influence matrix to classify stakeholders as key player, context setter, subject, and crowd [11]. The mapping of stakeholders will help the Mount Merbabu National Park Officer in the involvement of stakeholders in the achievement of objectives Investigation of relationships among stakeholders can be useful to minimize conflicts that occur so as not to cause harm to natural resources There are three types of relationships among stakeholders, cooperating, complementary, and conflict.

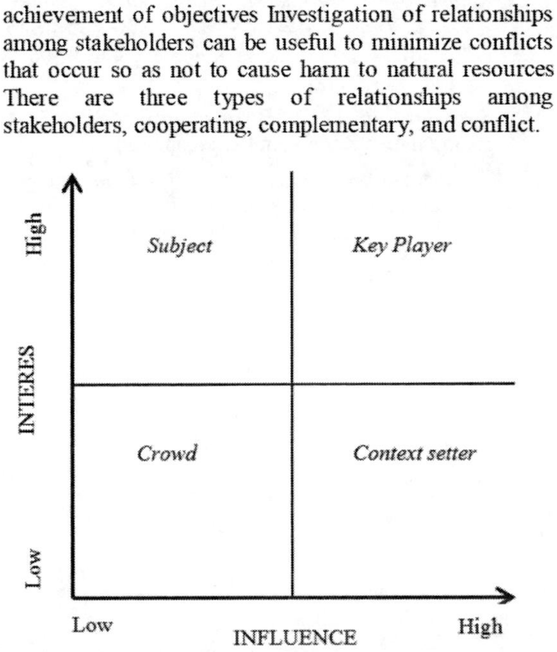

Fig. 1. Stakeholders interest and stakeholders influence matrix

4. Stakeholders Perception and Stakeholders Participation

Some important things in stakeholder analysis that also need to know is the stakeholders perception and stakeholders participation.

4.1 Stakeholders Perception

Stakeholder perceptions need attention in integrating their roles and tasks. This is important because perceptions of each stakeholder may differ from one to another, both individuals and groups [10]. Feldman argues that the perception can be a view of policies, programs, activities, and promotions made by external parties to local communities [10]. While Gabriel states that knowing the perception of each stakeholders is a process in the context of task integration and the role of stakeholders in a collective action [10]. Golder and Gawler argue that analysis of stakeholders' perceptions is important because ultimately each activity will depend on selected stakeholders with whom they will work together to achieve a goal [13].

4.2 Stakeholders Participation

Participation is a process whereby all parties can shape and engage in all development initiatives. In this case, anyone can play a role and actively have control over his own life, take a role in society, and become more involved in development [14]. Participation is a medium in achieving the objectives of the implementation of activities. Through participation, stakeholders are expected to formulate and simultaneously implement joint actions in achieving the objectives [10].

Communities are the most affected stakeholders of an activity. The success of a program is largely determined by the high level of community participation. Community participation is a process involving people to participate in local planning, implementation and management activities. The readiness of local governments and communities to accept responsibility and work together in environmental management activities is an important value of this participation. In addition, a commitment to managing resources, skills and knowledge to achieve goals, and respecting the capabilities and capacities of all partners is the most important element in the success of these community participation activities[15].

5. Conclusion

Tuk Semuncar Utilization Zone of Gunung Merbabu National Park has potential of environmental services and natural attraction. Potential is utilized by many parties. Knowledge of interested parties can minimize conflicts of interest in the utilization of environmental services and natural tourism. Stakeholder analysis can be used to mapping stakeholders by identifying, perceptions, participation, and stakeholder roles. The results of stakeholder analysis can be used as input in formulating strategies in the conservation areas management.

References

1. A. Mulyana., *"Kebijakan pengelolaan zona khusus dapatkah meretas kebuntuan dalam menata ruang Taman Nasional di Indonesia?,"* Bogor, 2010.
2. Balai TNGMb, *"Rencana Pengelolaan Taman Nasional Gunung Merbabu 2014 - 2023,"* Jawa Tengah, 2014.
3. H. Gunawan, M. Bismark, and H. Krisnawati, *"Kajian Sosial Ekonomi Masyarakat Sekitar Sebagai Dasar Penetapan Tipe Penyangga Taman*

Nasional Gunung Merbabu, Jawa Tengah," J. Penelit. Hutan dan Konserv. Alam, vol. **10**, no. 2, pp. 103–119, 2013.

4. I. Marina and H. A. Dharmawan, *"Analisis Konflik Sumberdaya Hutan di Kawasan Konservasi,"* J. Transdisiplin Sosiologi, Komunikasi, dan Ekol. Mns., vol. **5**, no. 1, pp. 90–96, 2011.
5. I. Alviya, E. Y. Suryandari, R. Maryani, and M. Z. Muttaqin, "*Meningkatkan Peran pemangku Kepentingan Dalam Pengelolaan Wilayah Hulu Daerah Aliran Sungai Ciliwung"* J. Penelit. Sos. dan Ekon. Kehutan., vol. **13,** no. 2, pp. 121–134, 2016.
6. J. M. Bryson, *"What To Do When Stakeholders Matter: A Guide to Stakeholder Identification and Analysis Techniques,"* Minneapolis, 2004.
7. J. Gonsalves, *Participatory Research and Development for Sustainable Agriculture and Natural Resource Management: A Sourcebook. Volume 1: Understanding Participatory Research and Development.* Philippines and International Development Research Centre, Ottawa, Canada, 2005.
8. A. K. Wakka, *"Analisis Stakeholders Pengelolaan kawasan Hutan Dengan Tujuan Khusus(KHDTK) Mengkendek, Kabupaten Tana Toraja, Provinsi Sulawesi Selatan"* J. Penelit. Kehutan. Wallacea, vol. **3**, no. 1, pp. 47–55, 2014.
9. D. Race and J. Millar, *Social and Community Dimensions to ACIAR Projects (ACIAR Training Manual).* Canberra: Australian Centre for International Agricultural Research, 2008.
10. M. Iqbal, *"Analisis Peran Pemangku Kepentingan dan Implementasinya Dalam Pembangunan Pertanian,"* J. Litbang Pertan., vol. **26**, no. 3, pp. 89–99, 2007.
11. M. S. Reed. J. Environ. Manage., vol. **90**, no. 5, pp. 1933–1949, 2009.
12. P. Mushove and C. Vogel, Glob. Environ. Chang., vol. **15**, pp. 184–198, 2005.
13. I. Alviya, M. Salminah, V. B. Arifanti, R. Maryani, and E. Syahadat, *"Persepsi Para Pemangku Kepentingan Terhadap Pengelolaan Lanskap Hutan di Daerah Aliran SUngai Tulang Bawang"* J. Penelit. Sos. dan Ekon. Kehutan., vol. **9**, no. 4, pp. 171–184, 2012.
14. Syahyuti, *Partisipasi*. Bogor: Pusat Analisis Sosial Ekonomi dan Kebijakan Pertanian, 2006.
15. R. Sahai, *"Community Participation in Environmental Management : Role of Women,"* Int. Conf. Recent Res. Dev. Environ. Soc. Sci. Humanit., pp. 173–180, 2015.

Literature Study on Community Participation in Community Based Rural Water Supply and Sanitation Programs

Siti Robiah Nurbaiti[1*], and *Azis* Nur Bambang[2]

[1]Master Program of Environmental Science, School of Postgraduate Studies, Diponegoro University, Semarang - Indonesia
[2]Fisheries and Marine Science Faculty of Diponegoro University, Semarang - Indonesia

Abstract. Clean water and proper sanitation are basic human needs, existing procurement in the Law of the Republic of Indonesia Number 7 of 2004 on Water Resources and Government Regulation of the Republic of Indonesia Number 16 of 2005 on Development of Water Supply System, which the state guarantees the right of everyone water for basic daily minimum needs to meet the needs of a healthy, productive, and clean life. Norms every society has the right to get clean air to meet basic daily needs. One of the points in the goal of sustainable development goals (SDGs) in the environment sector is the guarantee of the community to achieve universal access to clean water and sanitation. The SDG High Level Panel held in 2012 calls on countries around the world to do so in 2030. Fulfillment of clean air and sanitation in Indonesia is conducted through two sectoral approaches, the first through agencies, or related agencies and the second through a Society. In accordance with its community-based principles, the role itself is a key factor in the success of the program. Therefore, the purpose of this paper is to find out the forms of community participation and the factors that influence participation in community-based water supply and sanitation programs in the field of literature studies of previous research such as research journals, theses, theses, dissertations and related books This literature study topic.

1 Introduction

Water is the largest part of the planet. Water is also an important part of life on earth. The amount of water present in the earth, 97.5% is sea water (salty) and only 2.5% of fresh water which more than 2 thirds of its parts froze in the form of ice while fresh water that does not freeze only 30% can be found primarily in soil in the form of groundwater. Of the amount of clean water in the world is only 1% that can be consumed, and not all can be accessed easily by the community [1].

Clean water and basic sanitation have an important role as an indicator of poverty, especially in efforts to improve public health. The Millennium Development Goals (MDGs) place the guaranteed supply of drinking water and sanitation into the main objective to be achieved by 2015 ie the 7C target of halving the proportion of households without sustainable access to water resources is for the proportion of households with sustainable access To a viable drinking water source. The achievement of MDGs in Indonesia in 2014, for access to urban drinking water and rural areas reached 68.36% of the target set by the government of 68.87%, while the proportion of households with sustainable access to basic basic sanitation facilities, urban and rural has reached 61 , 04% of the targeted 62.41% [2].

After the MDGs end in 2015, the member country leaders agreed to continue the noble ideals of MGDs wishing to concentrate on famine and poverty in the world with SDGs (Sustainable Development Goals / SDGs) with one of the points in the goal of sustainable development in the environmental sector To ensure that communities achieve universal access to clean water and sanitation. The High Level Panels of SDGs held in 2012 called on countries around the world including Indonesia to achieve this goal by 2030. To achieve these objectives there are six targets to be achieved: to create equitable access to drinking water, to create affordable access to sanitation, to increase Water quality by reducing the level of contamination, improving water use efficiency to avoid scarcity, implementing integrated water source management and protecting as well as restoring ecosystems related to water sources including mountains, forests, wetlands, rivers, groundwater and lakes [3].

The availability of clean water and proper sanitation is a basic human need, as contained in the Law of the Republic of Indonesia Number 7 2004 regarding Water Resources and Government Regulation No. 16 of 2005 on the Development of Water Supply System, that the state guarantees the right of everyone to Get water for basic daily minimum needs to meet the needs of a healthy, productive, and clean life. This means that every community has the right to get clean water to meet basic daily needs.

Clean water in addition to drinking, is also used for most of the day-to-day activities of humans, such as washing, cooking, and others. But in reality, there are still many people who have difficulty obtaining it, causing various problems in society, especially in rural areas. One of the efforts made by the government to overcome this problem is the Community Based Drinking and Sanitation (PAMSIMAS) Program which is a continuation of the WSLIC2 program. In addition,

* Corresponding author:baityjannati14@gmail.com

many companies also contribute through Corporate Social Responsibility (CSR) as a manifestation of commitment built by the company to contribute to improving the quality of life of the community through various programs such as the provision of clean water and sanitation facilities based on community and direct water distribution to the community which requires them to be done by PT. Exxonmobile [4-5].

In accordance with its community-based principles, the active participation of the community itself is a key factor in the success of the program. Both PAMSIMAS and CSR programs are implementation of Community Development. Ife (1995) suggests that one of the principles of Community Development (CD) is participation. Participation in community development should create an active involvement of all people in the community in the process of community activity [5], where one definition of community participation is community participation in the process of identifying problems and potentials in society, Problems, implementation of problem solving efforts, and community involvement in the process of evaluating the changes

To measure the success of water supply and sanitation program using community development approach, among others by looking at the form of community participation in the program where in the implementation of course influenced by various factors both internal and external. With participation, it is expected that the program will be more sustainable because it is based on the actual basic needs of the local community.

2 Forms of Community Participation

According to Ericson in [6], the form of participation given by society in the development stage has several forms that are divided into three stages:

1) Participation in the planning stage (idea planning stage). Participation at the stage has a purpose is the involvement of a person at the stage of preparation of plans and strategies in the preparation of committees and budgets in an activity / project. The community is able to participate by providing suggestions, suggestions, criticisms through meetings;
2) Participation in the implementation stage (implementation stage). Participation at this stage means the involvement of a person at the stage of implementation of a project work. The community here can provide energy, money or material / goods, and ideas as one form of participation in the work; and
3) Participation in utilization. Participation at this stage means the involvement of a person at the stage of utilizing a project after the project is completed. The participation of the community at this stage is in the form of manpower and operating money, as well as maintaining the projects that have been built.

In general there are several forms of participation that can be provided by the community in a development program, namely money participation [7, 8], participation of property [7, 8, 9], labor participation [7,9], skills participation [7, 9] , participation of ideas [7,9], social participation [7,9], participation in decision-making processes [8], and participation of representatives [8]. The explanation can be seen in Table 1.

Table 1. Form of participation

Form of participation	Explanation
Money	A donation of money to facilitate efforts for the achievement of the needs of the people who need help.
Material	Donations of property, for example: work tools, tools
Manpower	Donations to support the success of the program
Skills	Provide encouragement through skills it has to other members of the community
Ideas	Contribution of ideas, opinions, or constructive ideas
Social	Donation Social, attention as a sign of the community or a sign of closeness
Decision Making	involved in making decisions related to common interests
Representative	Provide trust / mandate to the representative sitting in the organization or committee

At the time of program planning, the form of community participation is the contribution of the good thoughts in the form of suggestions and suggestions, nevertheless the people who come and remain silent still take part in decision making [10, 11]. In other research results, people can propose ideas but in some cases the decision is in the hands of the government [12].

The implementation phase is in the form of energy and money, and this usually has been their commitment since the beginning of planning so that the implementation process of the program can run well and not cause adverse impact to the environment and social. For those who can not contribute money, they will contribute energy by taking the time to participate in consecrated work and become a construction worker, and some provide both money and energy. Vita and chika, In addition there are people who donate property or materials in the form of work tools and building materials [11]

After the implementation of the program, the need for community participation in utilizing what has been built and it is very important. People will feel the benefits directly and can improve the quality of their lives so that the results of development into a tangible form. The form of participation after implementation to maintain that has been built is usually a donation of money in the form of contributions [10, 11, 12].

The form of community participation given in each stage of water supply and sanitation program can foster a sense of togetherness in realizing an awareness of environmental conditions in the region, a sense of ownership of existing facilities and infrastructure, so as to have a considerable effect on the smoothness and sustainability of the program.

3 Factors Affecting Community Participation

From the literature it is found that generally community participation is influenced by two factors, namely internal and external.

1) Internal factors that include individual characteristics that can affect the individual to participate in an activity, namely age, sex, family status, education level, ethnicity, religion, language, occupation, income level, the distance of the house to the location of work or activity And land ownership [13]
2) external factors are all outside parties with an interest and have influence on the program, among others Village managers, community leaders, Local Government, NGOs, third parties (NGOs, Social foundations, Universities) [14]

Individual characteristic factors can affect group activity, individual mobility and financial ability [6]. From several research results indicate that the age of the respondents significantly influence the participation of the community. This means that the high level of community participation is strongly influenced by the age society composition [10, 11, 15, 16]. In this case the number of respondents who answered based on the results of questionnaires and interviews is not a representation of the age of the community as a whole at the study site, because in the study only took a representative sample of the family as users of clean water infrastructure.

Gender influences liveliness in discussions, donations and the involvement of physical activities. Male respondents were more active in following activities than women [10, 11, 15]. According to Soedarno et. Al (1992) this can happen because within the social coating system on the basis of this sexuality, the male class has a number of privileges over the women. Thus the tendency is that men's groups will participate more [17].

Educational factors are considered important because through education, one will more easily communicate and interact with others and quickly respond to the development of knowledge and technology. The higher the education, of course, has a broad knowledge of development and the role of the given role [6]. With a high level of education a person has more knowledge in planning activities and more easily filtering outward influences [15]. The level of education affects the liveliness of the discussion. The existence of new information with simple language and practical example becomes an opportunity for people who only have elementary school graduates to find out and get involved, especially if the activities are profitable [10].
The targets of rural water supply and sanitation programs implemented in the villages are poor, thus providing opportunities for lower-level communities to take a more dominant role so that financial capability should not be an obstacle to participation. This can be seen from some people who participate in the water resources management program is a society with a range of income <Rp. 700.000, - [10, 11, 15].

For work is very diverse, but because the program is implemented in the village, it is mostly as laborers and private [10, 11]. As for ethnicity and religion it is usually related to the status of the community whether as a native or an immigrant [10, 11, 15].

Research conducted [15] by comparing community participation in two receiving villages of PAMSIMAS receiving Village *Hibah* Incentives (HIDs) and those not receiving HIDs, it was found that the factors that led to the implementation of participation in HID villages were optimal because of strong community characteristics And influential in every stages of the program so as to form a joint strength to build and target the success of the program PAMSIMAS can be more easily achieved, and the level of participation in the management phase has reached the level of full control (full control).

From the research results, the role of village officials, community leaders, local government, and accompanying teams and NGOs have an effect on the level of community participation. Approaches with attention, attending meetings, providing direction and input, and interacting directly with the community are effective in encouraging people to support the program [10].

Factors - factors above determine the shape and level of community participation that takes place in an activity. The effect that appears at each place is not the same, although it takes place in a nearby area. In addition, the ability of factors to affect each individual is also not the same.

4 Conclusion

The form of community participation in the implementation of community-based water supply and sanitation program is in the form of money, thought, energy, material or material contribution, and decision making. Factors affecting community participation are divided into internal factors derived from within the community itself, ie individual characteristics, while external factors are external factors that can encourage or hamper participation, among others the role of village officials, community leaders, local government, NGOs and accompanying teams or facilitator.

Reference

1. www.pkpu.org [Internet]. Available from: www.pkpu.org
2. Bappenas. *Laporan Pencapaian Tujuan Pembangunan Milenium di Indonesia 2014*.

Jakarta; 2015.

3. UNEP & UNDP. *Indicators and Data Mapping to Measure Sustainable Development Goals (SDGs) Targets (Case of Indonesia 2015)*. Jakarta; 2015.
4. www.exxonmobil.co.id.
5. Muhdar, Jamaludin, Irwansyah. J Bisnis dan Pembang. 2014;**1** No 1:22–8.
6. Slamet Y. Pembangunan Masyarakat Berwawasan Partisipasi. Surakarta: Sebelas Maret University Press; 1994.
7. Hamijoyo. *Pembangunan Masyarakat Berwawasan Partisipas*. Yogyakarta: UGM Press; 2007.
8. Holil Soelaiman. *Partisipasi Sosial dalam Usaha Kesejahteraan Sosial*. Bandung; 1980.
9. Pasaribu, Simanjuntak. *Rasa Percaya Diri Sendiri*. Salatiga: Pusat Bimbingan Universitas Kristen Satya Wacana; 2005.
10. Ratnasari AV. *Partisipasi Masyarakat dalam Program Penyediaan Air Minum dan Sanitasi Berbasis Masyarakat (PAMSIMAS) Di Kabupaten Pekalongan (Studi Kasus Desa Kwayangan Kecamatan Kedungwuni dan Desa Lambenggelun Kecamatan Paninggaran)*. Universitas Diponegoro. Semarang; 2014.
11. Chaerunnissa C. Jurnal Politika. 2014;**5**(2).
12. Ratnawati B. *Tingkat Partisipasi Masyarakat dalam Pengelolaan Lingkungan melalui Program Penyediaan Air Minum dan Sanitasi berbasis Masyarakat (PAMSIMAS) (Studi Kasus Desa Sodong Kecamatan Wonotunggal dan Desa Mojotengah Kecamatan Reban)*. Universitas Diponegoro. Semarang; 2012.
13. Cohen, J. and Uphoff N. *Rural Development Participation Concept and Measure for Project Design Implementation and Evaluation* [Internet]. New York: Cornell University; 1980. Available from: https://www.researchgate.net/publication/37882394_Rural_development_participation_concepts_and_measures_for_project_design_implementation_and_evaluation
14. Sunarti. Jurnal Tata Loka. Semarang; 2003;
15. Afriadi T, Wahyono H. 2012;**8**(4):341–8.
16. A. Tenriawaruwaty. *Partisipasi Masyarakat dalam Pengelolaan Sumberdaya Air Secara Berkelanjutan di Kabupaten Bulukumba*. Universitas Hasanuddin. Makasar; 2013.
17. Suciati. *Partisipasi Masyrakat dalam Penyusunan Rencana Umum Tata Ruang Kota Pati*. Universitas Diponegoro. Semarang; 2006.

Land Use and River Degradation Impact of Sand and Gravel Mining

Putra Rizal Ichsan Syah[1,*], *Purnaweni,* Hartuti[1,2,3]

[1] Master Program of Environmental Science, School of Postgraduate Studies, Diponegoro University, Semarang - Indonesia
[2] Doctorate Program of Environmental Science, School of Postgraduate Studies, Diponegoro University, Semarang – Indonesia
[3] Public Administration Department, Social Science and Political Science, Diponegoro University, Semarang – Indonesia

Abstract. Sand and gravel mining is aimed at providing materials for infrastructure development, as well as providing economical source to the miners. However, the impacts of sand and gravel mining could also cause disturbances to ecological balance, since it is closely related to land use change and river degradation, besides causing conflicts in the miners, the government, and the private relationship. Therefore the government regulation and proper supervision are needed to preserve the ecological balance and decreasing the negative impacts of this mining, and therefore guarantee sustainable development.

1. Introduction

Man and nature are in essence very much related where humans use nature for their various necessities. At the beginning of the history, the relationship between human and nature is called pan cosmism, in which nature is regarded as something sacred [1]. However, in line with the increasing number of people and much more necessities of these human need to be met, human relationship with nature changed into anthropocentrism. An important feature of anthropocentris is the belief that human beings are not part of nature since they assume themselves have a higher value than nature [2].

A thought that man is able to conquer nature and therefore can do whatever they want is among the main causes that makes the environment is degraded. This is also the case with the uncontrolled use of natural resources which cause serious environmental damage.

This condition is the impact of society's thought that nature can be maximally utilized regardless of the surrounding environment.

Among the badly needs of human beings are taken from natural resources, in the forms of renewable and non renewable resources. Their extraction must be carefully managed, so as not to disturb the existing environmental carrying capacity. Therefore, mining activities must be seriously controlled.

Mining sites are often located in areas with significant biology and environments characteristics, such as in small islands and tropical rainforests [3]. However, these sites are sometimes also located in populated areas, that mining may evoke conflict between the society and the government as a regulator. This also occurred in sand mining activities [4].

In this rapid development era, people have turned to rivers and floodplains as a major source of sand and gravel for construction [5]. Sand and gravel are components of extractable soil, underground geological sources formed from eroding mountain rocks carried by a river. The mining of sand and gravel is one of the activities that positively impact the local economy. Sand and gravel are also important sources for the economic development activities of developed and developing countries [4].

Sand is a very important construction material. UN Environmental Program of Global Environmental Alert Service noted that in concrete processing of infrastructure industry, sand is six to seven times much more consumed than other materials including cement [6]. A very significant prove that sand and gravel are very important for construction materials. However, sand and mining activities also raise various environmental issues both physical and social.

Based on several environmental impacts caused by sand mining activities, this paper will focus on mining and sustainable development, as well as environmental impacts of mining in terms of land use change and river degradation.

2. Method

This paper was a result of a library research done through reviews of books and journals tracking, focused on the environmental impacts of sand and gravel mining activities and its relation to sustainable development. This paper was only based on secondary data. The related secondary data were then reviewed to analyze the correlation between sand and gravel mining activities to environmental impacts in terms of changes in land use and river degradation that ultimately impact on the other aspects of environment. It is therefore endanger sustainable development, a paradigm that is a revision of development that was formerly focused mainly on economic benefit into focused on interrelations and interconnection of ecology, ecology and social aspects.

* Corresponding author: rizal.ichsan90@gmail.com

3 Environmental Impacts

Environmental devaluations that result from human activities such as sand/gravel mining on land disturbance ranges from poor or distorted topography, unproductive agricultural terrain, creation of puddles for pest cultivation, deforestation and general degradation of ecosystems with air, ground and land pollution [4].

One of means to meet human needs is through mining which yield various mining materials. However, excessive mining is dangerous as mining significantly change land use and left non reversible impacts, and therefore very potential in damaging the environment. It also increase natural disaster risk, in which the loss could actually exceed the benefit it creates [7].

3.1 Sustainable Development

Sustainable development become a popular term in various activities related to human activities. This term was started to be used in the second phase of the 20th century as a reaction to rapid economic growth in many parts of the world, closely related to the intensive use of natural resources. To reduce this phenomenon, in 1972 the World Commission on Environment and Development (WCED) was the first to define sustainable development as development that meets the needs of the present without compromising the ability of future generations to meet their own needs [8].

Keraf in his book entitled *Enviromental Ethics* explained that the main ideal of sustainable development is an effort to synchronize, integrate, and giving the same weight to three main aspects of development namely economic, social and culture, as well as environment [10]. This is a new paradigm compared to the old, in which development was focused on economic benefit. Development results was weighed on economic calculations such as how much the increase of income per capita, how much of the country's percentage of economic growth. The new paradigm focused not only on economic growth, as when environmental damage was also calculated in the development impacts, the percentage is surely decrease. The Three Pillars of Sustainable Development was stated by Dubinski as follows (see Figure 1)

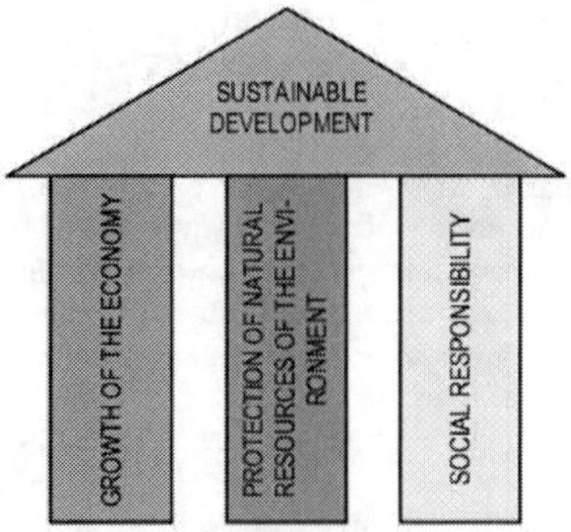

Fig 1. Three Pillars of Sustainable Development in Mining (Dubinski)

The idea behind this new paradigm is that development is a concept in which the system consist of inter related and inter connected sub systems. Each subsystem can not be separated or contradicted to each other. One of the key factors of development success depends on the balance of the three sub systems of sustainable development, as depicted by Dubinski.

The implementation of sustainable development will affect on the sustainability of activities and resources, this also implies to mining activities as shown in Figure 1.

Dubinski stated about three pillars of sustainable development in mining. He said that the intentions behind sustainable development implementation of mineral mining, is as follows : (a) Economy : technical activities that guarantee economic growth ; (b) Ecology : guarantee protection to natural resources and environment ; (c) Social Responsibility : meaning the care to the employees both in the work place and in the mining environment [8].

If sustainable development is neglected, the impact will be among others the higher risk or incidents of disaster because of the environmental damage..

In the Law No 32/2009 on Environmental Protection and Management, it is stated that environmental damage is a direct and non-direct change to the physical, chemical, and/ biological resources which exceed standard criteria of environmental damage [10].

Environmental damage is caused by two factors. Keraf stated that among disaster we experience, some of them are pure as natural disasters such as volcanic eruption and earthquake. The rest are environmental disasters as the cause are not pure natural disaster but more impacted from human activities [9]. Hadi is in the same opinion as saying that at an extreme point of the environmental disaster it threatens the continuity of human activities and therefore threatens human existence. An extreme point which must be seriously prevented if human beings are aware of its very existence [11].

Keraf in his book entitled *Krisis dan Bencana Lingkungan* (Crisis and Environmental Damage) stated that so far the environmental damage are the effects of: (a) fundamental factor or the false in view point; (b) Paradigm false and development policy ; (c) Modernization factor ; (d) Bad governance; (e)Decentralization and political liberalization ; (f) Mutual commitment weakness [12].

Mining activities are very potential to degrading the environment. Sudrajat in his book entitled *Teori dan Praktik Pertambangan* (Mining Theories and Practice) stated that mining can be regarded as *Zero Value* activities since there are many impacts of mining based on so many improper technical criteria of good mining governance [13].

3.2 Land Use Change

Land basically includes resources both inside and above the surface that can be utilized for human needs. With the increase in human needs as the population increases, land use changes will be wider. Conversion of forest land into other land uses is a long-standing phenomenon in the world and has direct impacts such as air pollution, reduced biodiversity, global warming with declining CO2 binders [14], as well as social impacts [3a/4 AHW]

A research conducted by Musah related to environmental impacts of mining activities in East Gonja District and Gunnarsholt Island, Ghana. Musah revealed that the loss of agricultural and productive land in the Eastern District of Gonja was very significant compared with conditions in Gunnarsholt, in which mining activities at Gunnarsholt are mostly done at highland and barren areas [15]. The effects of mining holes in East Gonja Distrik, Ghana is that the holes became breeding grounds for malaria-causing mosquitoes and other diseases. The results of this study was supported by Kori and Mathada's research at Nzhelele Valley, South Africa that the stagnant water in the excavation pits make a good breeding sites for disease-carrying vectors. Mosquitoes and snails pose a danger of malaria and scistisomiasis to the local communities. They mentioned that the impact of land change from mining can reduce ecological value as well as economic value [4]. The result of this research is almost the same with the research done by Dyahwati in Temanggung, Indonesia that sand mining activity caused the declining of economic value of the land [17].

3.2. River degradation

According to Boudaghpour, river gravel and sand are the best known building materials because the dissoluble and frail fragments of the gravel and sand are removed by the water flow and therefore the stable ones remain [18]. Although sand mining is expected to be regulated by law in many places, many are still illegal. It quickly became an ecological problem as demand for sand increased in industry and construction.

However, based on research by Ashraf, et al, mining in the river has a negative impact outside the mine site. Many hectares of fertile land are lost, as well as timber resources and the loss of valuable wildlife habitats. Moreover, degraded river habitats result in loss of fishery productivity, biodiversity and recreation potential [19].

Result of a study done by Devi and Rongmei mining activities in the Imphal River found that excavations along the river had a negative impact on the quality of river water. The impacts of sand mining activities on rivers include short-term turbidity at the excavation site due to sediment reduction, in addition to other environmental impacts such as increased erosion and collapse of riverbanks that increase suspended solids at the excavation sites and downstream rivers [5].

4. Conclusion

Sand mining activities carried out in watersheds and around riverbanks have ecological impacts, among others changes in land use that affect the loss of agricultural land and other productive land can cause impact on public health, such as malaria. In addition, sand mining activities in the river can cause river degradation such as sedimentation and erosion, as well as threatening ecosystems in the river water. Sand and gravel mining activities in which main objective is for infrastructure development need proper supervision and good regulation.

Therefore sand and gravel mining endanger sustainable development, which focuses on the sustainability of resources, including natural resources. Therefore, good mining practices which is based on sustainable development must be implemented.

5 References

1. R. Dhany. Semarang, (2013) http://dhanyvironment.blogspot.co.id/2013/09/manusia-kontra-lingkungan.html
2. Beckman, et al. Universidad Carlos 3 de Madrid Journal. (1997) https://portal.uc3m.es/portal/page/portal/grupos_investigacion/sociologia_cambio_climatico/Pardo%20%20Anthropocentrism%20Environmental%20Values%20(ENG).pdf
3. Devi, Bernadetta and Prayogo, Dody. Mining for Development Centre. AusAID. Australia(2013)
4. Kori, E and Mathada, H., Environment and Chemistry, IPCBEE, Vol **46**, Singapore (2012)
5. Devi, Maharabam Anjali and Rongmei, Lunghim. Asia Pacific Journal of Energy and Environment. Volume **2**. Number 2/2015. ABC Research House. (2015)
6. UNEP Global Environment Alert Service (GEAS). Sand, Rarer than one thinks. https://na.unep.net/geas/archive/pdfs/GEAS_Mar2014_Sand_Mining.pdf. (2014).
7. Purnaweni, H, Policy on Environmental Management in North Kendeng Area, J Ilmu Lingkungan, **12**, 1: 53-65, (2014)
8. Dubinsky, J, Sustainable Development of Mining Mineral Resources. J of Sustainable Mining. **12** No. 1: 1–6 (2013).
9. Keraf, S. Etika Lingkungan (Environmental Ethics), Jakarta : Kompas Press, (2002)
10. Law No 32/2009 on Environmental Protection and Management
11. Hadi, SP, Aspek Sosial AMDAL (SEA Social Aspects), Gadjah Mada University Press: Yogyakarta, (2005)
12. Keraf, S, Krisis dan Bencana Lingkungan (Environmental Crisis and Disaster), Kanisius: Yogyakarta, (2010).
13. Sudrajat, Teori dan Praktek Pertambangan (Mining Theory and Practise), Pustaka Yustisia:Yogyakarta, (2013)
14. Hidayat, et al. Journal of Regional and City Planning. 26, No.2, 130-146, Agustus 2015. IPB Repository. Bogor. (2015) (In Bahasa)
15. Musah, Jafaru Adam and Barkarson, Bjorn H. United Nations University Land Restoration

Training Programme, Agricultural University of Iceland and the Soil Conservation Service of Iceland (2009). http://www.unulrt.is/static/fellows/document/musahja.pdf
16. Wibowo, AH, The Physical and Social Impacts of Mining Policies on Environents in Bantarbolang Sub-District of Pemalang Regency, Asian Jr. of Microbiol. Biotech. Env Sc, **19**, 4:28-33, (2017)
17. Dyahwati, NI, The study of environmental degradation as impact of sand mining in Moutt Sumbing Area (Study Case at Kwadungan Gunung Village, Kledung District, Temanggung Regency), Theses at Master Program of Environmental Science, Diponegoro University, (2007).
18. Boudaghpour, S. and Monfared, SAH, J. Wseas transaction on Environment and Development. **5**, 4, May 2008. (2008). http://www.wseas.us/elibrary/transactions/environment/2008/26-102.pdf.
19. Ashraf,et al. Scientific Research and Essays, **6,** 6:. 1216-1231, (2011). Academic Journals, http://www.academicjournals.org/article/article1380635625_Ashraf%20et%20al.pdf

Expert Advisor (EA) Evaluation System Using Web-based ELECTRE Method in Foreign Exchange (Forex) Market

Satibi Satibi[13,*], *Catur* Edi Widodo[2], and *Farikhin* Farikhin[2]

[1]Master of Information Systems, School of Postgraduate Studies, Diponegoro University, Semarang - Indonesia
[2]Physics Department of Science and Mathematics Faculty, Diponegoro University, Semarang - Indonesia
[3]Department of Informatics Management, AMIK Selatpanjang, Meranti Island - Indonesia

Abstract. This research aims to optimize forex trading profit automatically using EA but its still keep considering accuracy and drawdown levels. The evaluation system will classify EA performance based on trading market sessions (Sydney, Tokyo, London and New York) to determine the right EA to be used in certain market sessions. This evaluation system is a web-based ELECTRE methods that interact in real-time with EA through web service and are able to present real-time charts performance dashboard using web socket protocol communications. Web applications are programmed using NodeJs. In the testing period, all EAs had been simulated 24 hours in all market sessions for three months, the best EA is valued by its profit, accuracy and drawdown criteria that calculated using web-based ELECTRE method. The ideas of this research are to compare the best EA on testing period with collaboration performances of each best classified EA by market sessions. This research uses three months historical data of EUR/USD as testing period and other 3 months as validation period. As a result, performance of collaboration four best EA classified by market sessions can increase profits percentage consistently in testing and validation periods and keep securing accuracy and drawdown levels.

1 Introduction

In determining the investment strategy in the foreign exchange market, decision makers have important roles and responsibilities to determine risks and profit targets. Decision makers are also required to be able to make three important decisions, namely: 1. Prices will move up or down; 2. Determine buying or selling decisions; 3. How long to hold the position that has been opened [1]. Due to the importance of decision-making role in forex trading market, it is necessary to have a system to record and evaluate the history of transactions that have been done before in real-time so that it can help to provide performance reports from previous transactions.

Nowadays it is the era of computer and information technology that has been developing quite rapidly. For sure, this evolution affecting the way of trading transactions method in the forex market to continues to grow. Utilizing current technology allows Traders to perform real-time analysis and trading over the internet. Forex trading can also be done automatically based on EA (Expert Advisor) without human intervention, with automatically analytics and transactions using EA in forex transaction negotiation is to avoid human intervention that is usually influenced by emotions, anxiety, stress and other things. Therefore, trading by using EA becomes a solution to eliminate the elements of human intervention that are volatile. Even studies in 2010 concluded 45% of negotiation system in the forex market was conducted using EA [2].

Although the percentage of the use of EA in forex trading is very large, but analysis process and transaction history evaluation usually had been done only limited to transaction history analysis that found in Metatrader software, which its not optimal when used as EA performance evaluation data. With the integration of EA supported by calculations based on real-time web-based ELECTRE method will be more optimal in providing the best choice of several EA alternatives based on trading market sessions.

In addition to being able to perform analysis and transaction in the forex market automatically, the EA also has the ability to interact with the web service, so it can be integrated with algorithms that are applied to web based applications that provide supporting information for the EA itself. Forex trading research on the forex market has been done a lot before, including forex forecasting using ANN methods [3], multi-agency financial decision support system with Fuzzy logic [4] and Web-based decision support systems with ANN for forecasting and forex trading [5]. However, this research focuses on evaluating the performance of ten open source EA that available on the official MQL4 (MetaQuotes Language 4) community website, although this evaluation system can still be used on all types of EAs either paid or created by traders who master the

* Corresponding author: satibi@batam.mobi

MQL4 programming language, by inserting a line of MQL4 code created in this evaluation system in order to interact with the web service and utilizing historical data of previous relevant currency movements to simulate the EAs.

2 Method

2.1 Data acquisition

Historical currency movement data that used to simulate trading charts movement in MetaTrader software is downloaded from Histdata.com, while the EAs trading transaction data acquired in real-time on the web-based ELECTRE application through web service which is also integrated with web socket for Displaying EA performance graphs in real-time on the dashboard. These acquired transaction data saved in database, this database will be classified based on trading market sessions to get the best performing EA in each trading market session. Classifications process of market session is recognized by trading open position that done by real-time EA transaction. The flow of data acquisition process as shown in Figure 1.

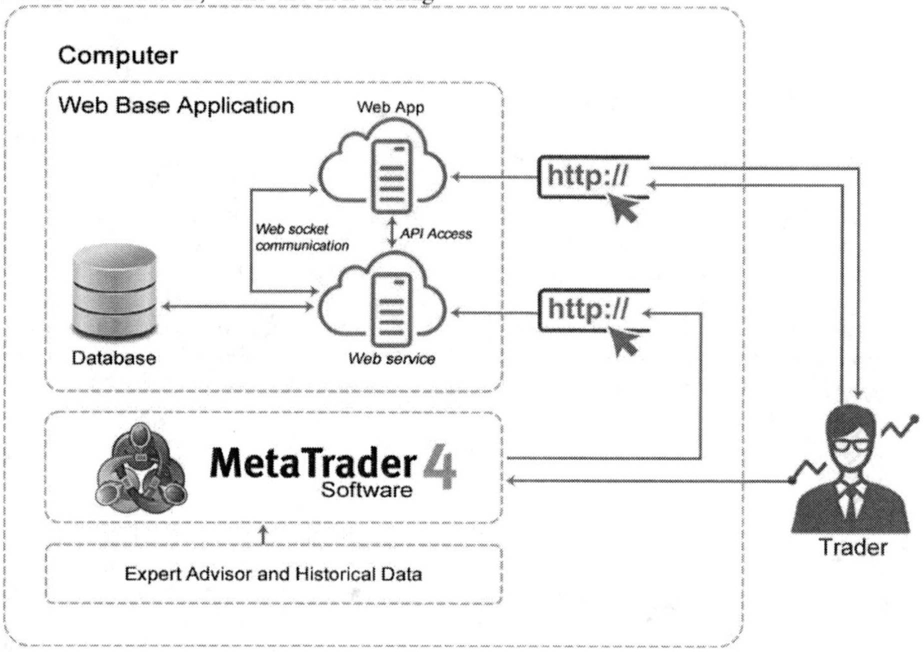

Fig. 1. Data acquisition process of EA transactions

2.2 ELECTRE method

The ELECTRE method is one of the multi criteria decision analysis methods which is an acronym of Elimination Et Choix Traduisant la Realité or in English means Elimination and Choice Expressing Reality, first introduced in 1965 by the SEMA consulting firm in Europe [6]. The modeling with the ELECTRE method includes six stages, ie :

1. Normalization of matrix
   Normalization process will produce R matrix, that calculated by :

$$\mathrm{r_{ij}} = \frac{x_{ij}}{\sqrt{\sum_{i=1}^{m} x_{ij}^2}} \quad (1)$$

2. To weight the normalized matrix, multiply that R matrix values by the weight of each criterion using following equation:

$$v_{ij} = w_i r_{ij}. \quad (2)$$

3. Determining concordance and discordance index sets.
   Concordance index calculated by :

$$C_{kl} = \{j | v_{kj} \geq v_{ij}\}; untuk\ j = 1,2, \dots, n \quad (3)$$

   Meanwhile, following equation is to calculate discordance index :

$$D_{kl} = \{j | v_{kj} < v_{ij}\}; untuk\ j = 1,2, \dots, n \quad (4)$$

4. Concordance and discordance matrix
   Concordance matrix obtained by :

$$c_{kl} = \sum_{j \in C_{kl}} W_j \quad (5)$$

   Below equation is used to obtain discordance matrix:

$$d_{kl} = \frac{\max \{|v_{kj} - v_{lj}|\}_{j \in D_{kl}}}{\max \{|v_{kj} - v_{lj}|\}_{\forall j}} \quad (6)$$

5. Dominance concordance and discordance matrix
   To determine dominance concordance matrix :

$$\underline{c} = \frac{\sum_{k=1}^{m}\sum_{l=1}^{m} c_{kl}}{m(m-1)} \quad (7)$$

Elements of F matrix specified by :

$$f_{kl} = \begin{cases} 1, jika\ c_{kl} \geq \underline{c} \\ 0, jika\ c_{kl} < \underline{c} \end{cases} \quad (8)$$

To determine dominance discordance matrix :

$$\underline{d} = \frac{\sum_{k=1}^{m}\sum_{l=1}^{m} d_{kl}}{m(m-1)} \quad (9)$$

Elements of G matrix specified by :

$$g_{kl} = \begin{cases} 1, jika\ d_{kl} \geq \underline{d} \\ 0, jika\ d_{kl} < \underline{d} \end{cases} \quad (10)$$

6. Determining the aggregate dominance matrix
   Aggregation of the dominance matrix indicates the sequence of partial preferences of the available alternatives, to obtain this aggregate dominance matrix, following equation is used :

$$e_{kl} = f_{kl}\ x\ g_{kl} \quad (11)$$

From that calculation will produce the E matrix which gives sequence of options from each available alternative, if $e_{kl} = 1$ then alternative A_k is a better alternative than A_l. So, the rows in the E matrix that having less $e_{kl} = 1$ can be eliminated, as well as other rows calculation can be eliminated until remaining first row as selected best EA alternative.

3 Results and discussion

Required data acquisition process in this research is EAs transaction data from ten simulated EA and percentage of profit, accuracy and drawdown from each EA performance.

a. Transactions data acquisition process
   Data acquisition process in web-based ELECTRE application is done in real-time when trading simulation is executed on MetaTrader software, EA will send trading transaction data to web service such as alt_id, ticket_id, open_time, close_time, type, lots, symbol, open_price, close_price and profit fields.

b. Meanwhile, the value data that consisting of profit, accuracy and drawdown were obtained by calculating the percentage of that three criteria based on transaction history data, while capital=5000, %Profit = ((profit + loss) / capital) x 100, %Accuracy = ($\sum$ profit positions / $\sum$ all positions) x 100, Drawdown = (maximum floating / capital) x 100. This calculation will be obtained value as shown in Table 1 and Table 2.

Table 1. Value of each criteria without market classification

Alt ID	Cri ID	Value	Step	Alt ID	Cri ID	Value	Step
1	1	1.89	1	1	1	-10.67	2
1	2	39.29	1	1	2	26.36	2
1	3	-0.52	1	1	3	-11.78	2
2	1	-13.69	1	2	1	3.08	2
2	2	61.11	1	2	2	80	2
2	3	-17.87	1	2	3	-1.32	2
3	1	-49.56	1	3	1	-20.76	2
3	2	51.54	1	3	2	53.62	2
3	3	-13.77	1	3	3	-23.35	2
4	1	6.76	1	4	1	4.11	2
4	2	100	1	4	2	100	2
4	3	-2.62	1	4	3	-0.17	2
5	1	-45.19	1	5	1	40.22	2
5	2	53.33	1	5	2	72.28	2
5	3	-90.29	1	5	3	-40.94	2
6	1	6.38	1	6	1	10.34	2
6	2	40.35	1	6	2	40.82	2
6	3	-1.49	1	6	3	-0.22	2
7	1	29.43	1	7	1	7.41	2
7	2	70.63	1	7	2	64.62	2
7	3	-2.64	1	7	3	-52.85	2
8	1	17.7	1	8	1	13	2
8	2	100	1	8	2	100	2
8	3	-15.21	1	8	3	-7.38	2
9	1	31.55	1	9	1	43.25	2
9	2	81.93	1	9	2	76.92	2
9	3	-42.12	1	9	3	-7.37	2
10	1	-15.72	1	10	1	1.02	2
10	2	47.2	1	10	2	49.74	2
10	3	-39.62	1	10	3	-27.18	2

Table 2. Value of each criteria by market sessions

Alt ID	Cri ID	Sydney	Tokyo	London	New york	Step
1	1	5.92	7.21	-1.53	-2.74	1
1	2	46.67	50	36.36	38.1	1
1	3	-0.8	-0.8	-1.02	-1.01	1
2	1	5	2.96	-17.65	-10.7	1
2	2	100	83.33	45.45	42.86	1
2	3	-1.5	-3.04	-6.94	-4.6	1
3	1	-6.1	-8.46	-43.79	-7.39	1
3	2	52.63	52.78	49.33	53.53	1
3	3	-2.79	-2.79	-3.97	-3.97	1
4	1	2.2	3.2	3.56	2.23	1
4	2	100	100	100	100	1
4	3	-1.43	-1.43	-4.63	-0.76	1
5	1	1.51	-6.25	-45.34	-17.43	1
5	2	71.43	50	53.85	54.55	1
5	3	-1.42	-3.88	-49.41	-21.55	1
6	1	6.63	6.35	0.37	1.96	1
6	2	53.85	60	36.36	40	1
6	3	-0.83	-0.7	-0.84	-0.84	1
7	1	18.54	10.6	16.35	5.56	1
7	2	69.64	72.09	71.64	70.18	1
7	3	-8.81	-8.81	-3.5	-6.34	1
8	1	11.86	4.11	3.23	3.81	1
8	2	100	100	100	100	1
8	3	-18.35	-21.61	-23.84	-2.98	1
9	1	6.2	6.77	21.46	15.96	1
9	2	81.25	80	84	83.72	1

https://doi.org/10.1051/e3sconf/20183110001

Alt ID	Cri ID	Sydney	Tokyo	London	New york	Step
9	3	-20.66	-20.66	-35.22	-35.22	1
10	1	1.56	5.16	-14.16	0	1
10	2	48.05	46.59	46.2	46.36	1
10	3	-7.18	-8.82	-10.76	-10.76	1
1	1	-6.68	-5.19	0.96	-5.04	2
1	2	17.31	22.81	36	28.95	2
1	3	-0.88	-0.88	-1.08	-1.08	2
2	1	4	2.46	-1.13	-2.38	2
2	2	100	83.33	75	66.67	2
2	3	-2.46	-3.04	-4.59	-4.59	2
3	1	-3.07	-6.29	-29.51	-9.13	2
3	2	38.1	44.62	53.85	59.24	2
3	3	-1.55	-2.79	-5.4	-5.4	2
4	1	0.08	1.08	2.49	2.53	2
4	2	100	100	100	100	2
4	3	-0.04	-0.12	-1.19	-4	2
5	1	4.06	5.54	35.25	25.67	2
5	2	68.42	64	75.68	70.69	2
5	3	-9.6	-19.1	-65.58	-52.86	2
6	1	2.81	2.12	3.11	4.18	2
6	2	44.44	37.5	43.48	37.5	2
6	3	-0.68	-0.74	-0.84	-0.84	2
7	1	-12.31	-28.86	51.99	31.42	2
7	2	63.64	61.11	68.75	70.59	2
7	3	-19.18	-19.18	-58.77	-85.32	2
8	1	4.3	4.19	1.78	5.69	2
8	2	100	100	100	100	2
8	3	-10.4	-8.66	-6.26	-8.14	2
9	1	8.42	10.02	25.8	15.12	2
9	2	76	69.23	77.42	77.19	2
9	3	-3.5	-20.72	-36.83	-16.12	2
10	1	-1.92	1.68	-8.22	-7.8	2
10	2	51.61	56.52	50.91	45.28	2
10	3	-7.45	-7.18	-4.78	-4.78	2

c. Determining the weight of each criteria, the criteria weight in this case obtained from online questionnaire results, the questionnaire gives criteria weight for profit = 3.83%, accuracy = 2.39% and drawdown 3.27%.

d. Normalize the matrix by specifying value for each matrix index of all alternatives per criterion, value of profit calculation example :

$$|x_1| = \sqrt{\begin{matrix}1.89^2 + (-13.69^2) + (-49.56^2) \\ + \ldots + (-15.72^2)\end{matrix}}$$

$$= 84.8390$$

To get profit value from first and second EAs :

$$r_{11} = \frac{x_{11}}{|x_1|} = \frac{1.89}{84.8390} = 0.0223$$

$$r_{21} = \frac{x_{11}}{|x_1|} = \frac{-13.69}{84.8390} = -0.1614$$

$$r \ldots = \cdots$$

e. The same way is used to determine the value of normalization on accuracy and drawdown criterias value.

f. Normalized matrix weighting is obtained by multiplying normalized matrix values by the weight of each criterion, eg. :

$$v_{11} = 0.0223\,x\,3.83 = 0.0853,$$

$$v_{21} = -0.1614\,x\,3.83 = -0.6180, \ldots$$

g. Determining concordance and discordance index sets by comparing an alternative with another alternative based on the same criterion, if selected value is greater then that index is used as a concordance index, and if the index value is smaller it will be used as a discordance index, eg. : $C_{12} = \{1,3\}, C_{13} = \{1,3\}$

h. Calculates concordance and discordance matrices. Concordance matrix is obtained by summing weights based on concordance index, eg. : $C_{12} = 3.83 + 3.27 = 7.1000$. While discordance matrix is obtained by getting value of difference between two criterion values based on discordance index then taken the highest value from it, this value will divided by the highest value of difference from all indexes comparison..

i. Dominance matrix of concordance and discordance is obtained by determining the threshold value first, then determining the dominance concordance matrix by comparing concordance and discordance matrix elements with the threshold value of each matrix, if the value of each element is greater than the threshold value will give a value of 1 And gives a value of 0 if it is smaller than the threshold value.

j. Aggregate dominance matrix is obtained by multiplying the dominance matrix element of concordance and dominance discordance matrix elements. The alternative with $e_{kl} = 1$ will be eliminated while alternative with the most of $e_{kl} = 1$ will be selected as the best alternative as shown in Figure 2.

k. These calculation steps are also used to get the best EAs in the best EA selection process based on market sessions as shown in Figure 3.

l. In order to get optimal results, if there are more than one EAs with same performances appeared in the same market session, only one EAs will be selected as best EA on that market session based on highest criteria weight priority.

3.1 Results Evaluation and validation

EAs performance evaluation and validation of ELECTRE method calculation had been tested using different test time data, October to December (2016) as testing period and January to march (2017) as validation period with different transaction data amounts, so that the results obtained as in Figure 4 and 5.

 https://doi.org/10.1051/e3sconf/20183110001

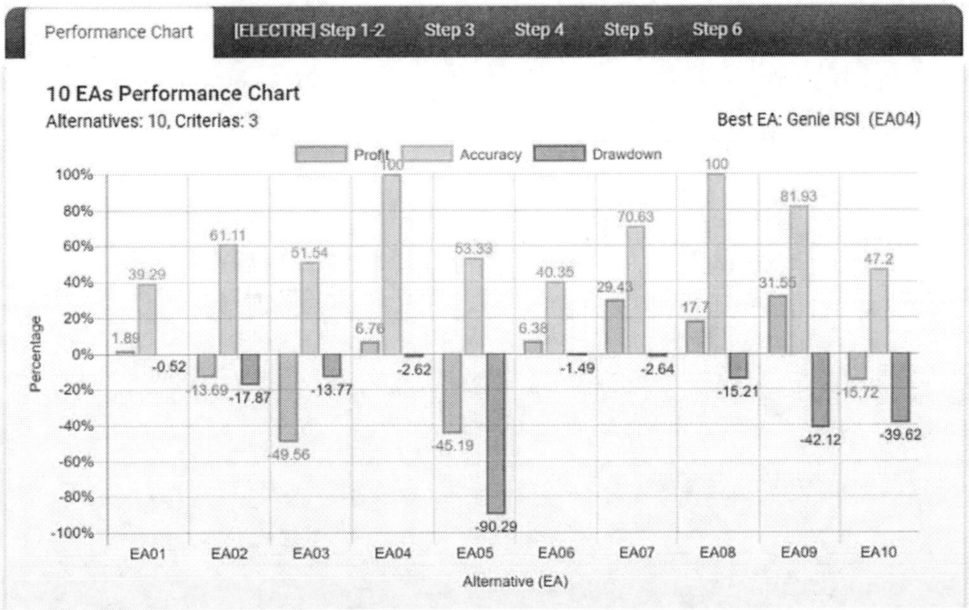

Fig. 2. Ten EAs performances in testing period

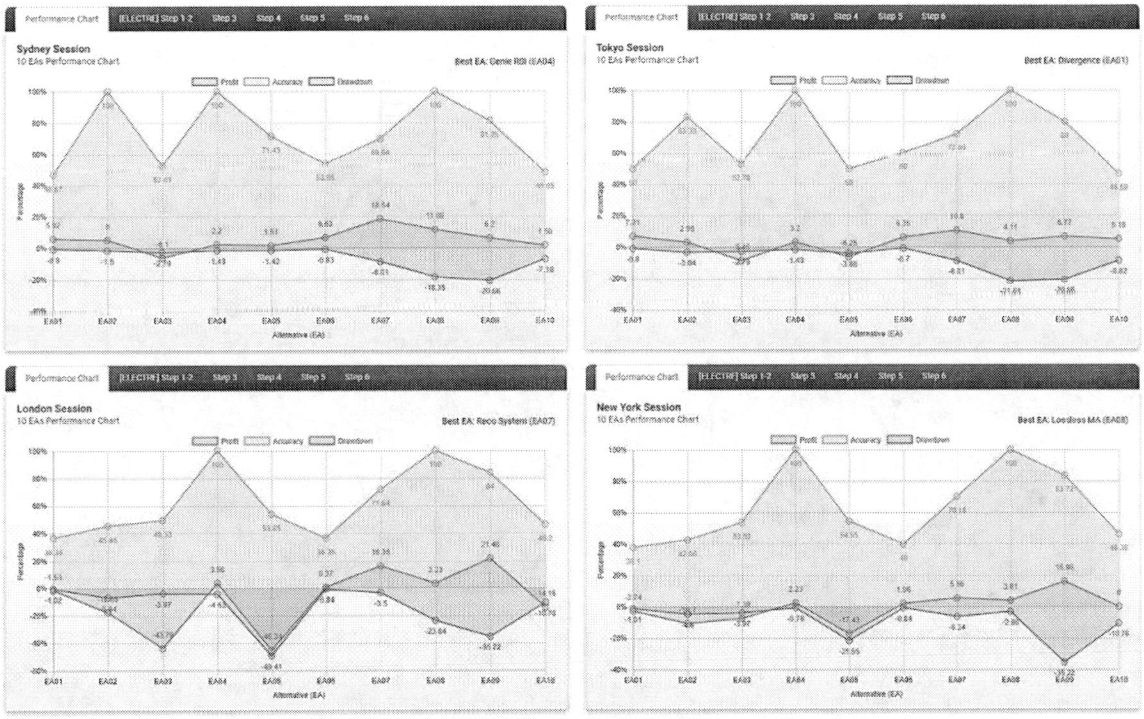

Fig. 3. Ten EAs performances by market classification

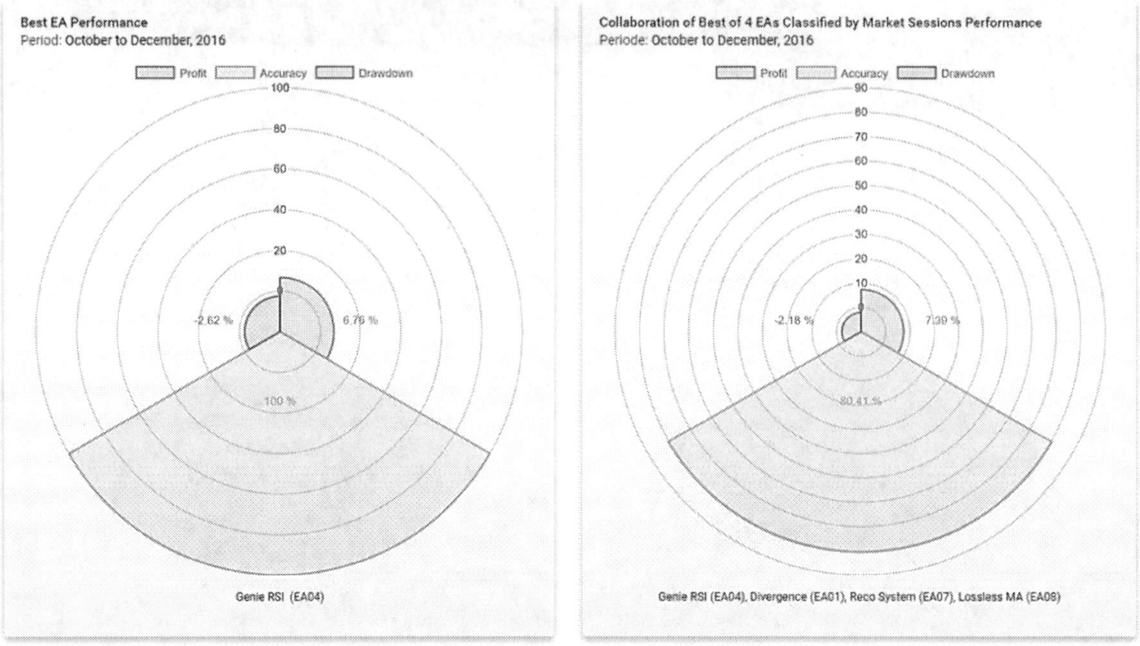

Fig. 4. EAs performance comparation

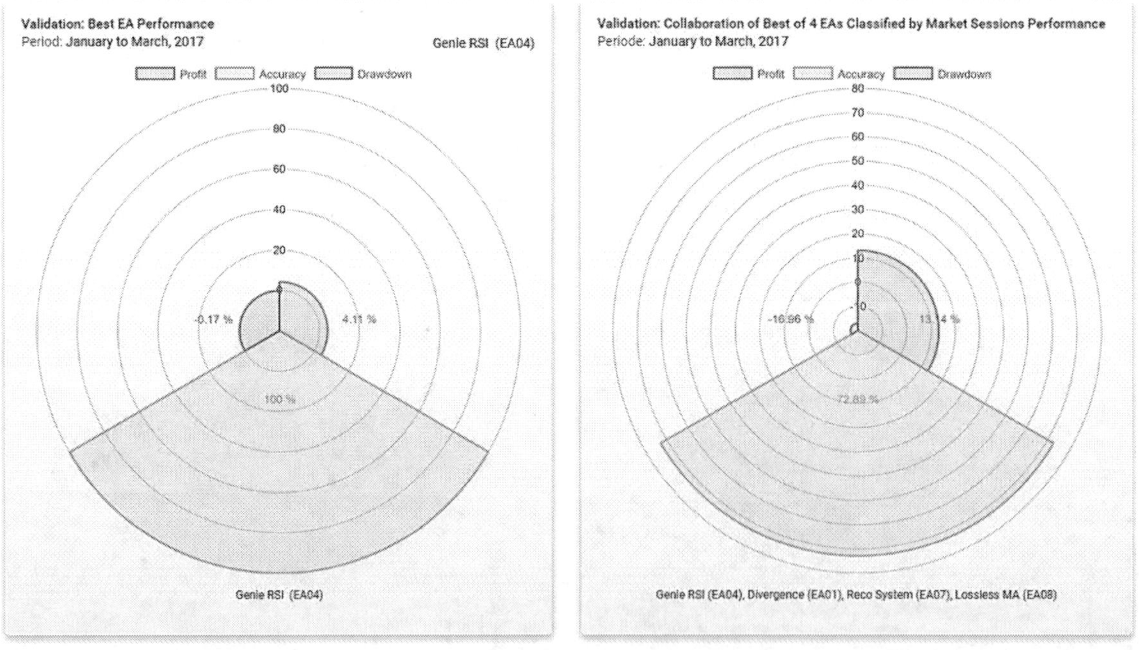

Fig. 5. EAs performance validation

4 Conclusion

The decision-making system with web-based ELECTRE method to determine the best collaboration EAs to be used based on its best market session performance, EAs performance are calculated by its profit, accuracy and drawdown criteria and the use of combination four EAs were best on its market session can increase profit consistently while maintaining accuracy and drawdown levels, this can be proven in the testing and validation processes with the profit increase level of 0.63 (9.20%) in the testing period and 9.03 (219.71%) in the validation period, and keep securing accuracy and drawdown levels when compared to other EAs.

References

1. T. Balabanov, I. Zankinski, N. Dobrinkova, Large-Scale Scientific Computing (LSSC) 2011, Vol 7**116**, 198-205 (2012).
2. R.M.C. Pinto, J.C.M. Silva, 2012 12th International Conference on Intelligent Systems Design and Applications (ISDA), 34-39 (2012).
3. P. Czekalski, M. Niezabitowski, R. Styblinski, 2015 20th International Conference on Control Systems and Science, 322-328 (2015).
4. J. Korczak, M. Hernes, M. Bac, 2015 Federated Conference on Computer Science and Information Systems (FedCSIS), 1367-1376 (2015).
5. K.K. Lai, L. Yu , S. Wang, Data Mining and Knowledge Management, Vol **3327**, 243-253 (2004).
6. D. Simić, J. Gajić, V. Ilin, V. Svirčević, S. Simić, *A* Hybrid Artificial Intelligent Systems, Vol **9648**, 283-295 (2016).
7. R.P. Barbosa, O. Belo, Advances in Data Mining, Medical Applications, E-Commerce, Marketing, and Theoretical Aspects, Vol **5077**, 389-403 (2008).
8. R.F.B. de Brito, A.L.I. Oliveira, 2014 International Joint Conference on Neural Networks (IJCNN), 4251-4258 (2014).
9. M.H. Eng, Y. Li, Q.G. Wang, T.H. Lee, 2008 International Conference on Information Management, Innovation Management and Industrial Engineering, Vol **1**, 279-282 (2008).
10. Y.L. Yong, Y. Lee, D. Nga, 2015 IEEE International Symposium on Robotics and Intelligent Sensors (IRIS), 313-317 (2015).
11. V.H.T. Kiet, R. 2005 IEEE International Conference on e-Technology, e-Commerce and e-Service, 518-521 (2005).
12. Z. Liu, D. Xiao, 2009 Sixth International Conference on Fuzzy Systems and Knowledge Discovery, Vol **3**, 239-243 (2009).

Simple Additive Weighting to Diagnose Rabbit Disease

Ramadiani[1*], *Dyna* Marissa[1], Muhammad *Labib* Jundillah[1], *Azainil*[2] and *Heliza* Rahmania Hatta[1]

[1]Faculty of Computer Science and Information Technology Mulawarman University, Samarinda - Indonesia
[2]Faculty of Teacher Training and Education Mulawarman University, Samarinda, Indonesia

Abstract. Rabbit is one of the many pets maintained by the general public in Indonesia. Like other pet, rabbits are also susceptible to various diseases. Society in general does not understand correctly the type of rabbit disease and the way of treatment. To help care for sick rabbits it is necessary a decision support system recommendation diagnosis of rabbit disease. The purpose of this research is to make the application of rabbit disease diagnosis system so that can help user in taking care of rabbit. This application diagnoses the disease by tracing the symptoms and calculating the recommendation of the disease using Simple Additive Weighting method. This research produces a web-based decision support system that is used to help rabbit breeders and the general public.

1 Introduction

Rabbits are funny and cute animals that can be nurtured and breed by everyone. Many people maintain rabbits because they have clean, smooth, soft feathers like carpets, in addition to low cost of feed and not a lot of noise. Other causes that make people interested in maintaining the rabbit because it is one source of low-cholesterol meat providers, which can be an alternative treatment for some people who have asthma and reproductive problems [1-2].

The type of rabbits varies depending on the origin of the country where the rabbit is alive. There are rabbits that have thick, long and smooth fur, but there are also thin, fluffy, soft rabbits such as carpets. There are rabbits that have long, wide and standing ears, but many rabbits are short, small and face down. There are rabbits whose body size reaches a weight of more than 5 kg, but there is also a type of rabbit whose body size is small and weighs not more than 3 kg. Each rabbit has its own uniqueness, such as feathers, eye color, ears, and body size. Some types of rabbits need special attention in their daily care [1].

Like other pets, rabbits are also susceptible to disease attacks. Unfortunately, not everyone understands correctly the various diseases of rabbits and his treatment [2]. If the rabbit is already affected by the disease, it is necessary to get special treatment. If we are slow in overcoming it, it will cause other rabbits infected by sick rabbits. In addition, the highest cause of rabbit death is disease. Therefore, prevention needs to be done early on by knowing the symptoms of rabbits affected by the disease. Good prevention is expected to reduce rabbit death rates.

The study aims to assist the determination of Rabbit disease type, by applying Simple Additive Weighting method. Based on the background that has been described then needed a system that can provide disease diagnosis in rabbits as a consideration for the care of the disease appropriately. This study uses SAW method because this method is one of multi-criteria settlement, where in diagnose rabbit disease there are many symptoms or criteria to be considered.

2 Literature Review

2.1 Decision Support System

Decision Support System (DSS) is a computer-based information system that approaches to produce various decision alternatives to assist certain parties in handling problems using data and various decision-making models. Decision-making model is adjusted to the number of variables, weighting, matrix multiplication and decision making flow. A decision making system only provides an alternative decision and then submitted to the user as the final decision maker [3].

The model that describes the decision-making process consists of four phases, namely [3]:

- Search (Intelligence)

This stage is the stage of defining the problem and the identification of the required information relating to the problems faced and the decisions to be taken.

- Design (Design)

This stage is a process to represent the model of the system to be built based on the assumptions that have been set. In this stage, a model of the problem is created, tested and validated.

- Choice (Choice)

* Corresponding author: ilkom.ramadiani@gmail.com

This stage is a process of testing and selecting the best decision based on certain criteria that have been determined and lead to the goal to be achieved.
- Implementation (Implementation)
This stage is the stage of implementation of the decisions that have been taken. At this stage it is necessary to prepare a series of planned actions so that the results of the decisions can be monitored and adjusted as necessary.

2.2 Simple Additive Weighting

It is a weighted sum method. The basic concept of the SAW method is to find the weighted sum of performance ratings on each alternative on all criteria [4]. The SAW method requires the process of normalizing the decision matrix (X) to a scale comparable to all existing alternative ratings. The completion step in using it is [4] [5] [6]:

1. Determine the alternative, Ai.
2. Determining the criteria that will be used as a reference in decision making, namely Cj.
3. Provide an alternative match rating on each criterion.
4. Determine the weight of preference or importance level (W) of each criterion.
   $$W = [\, W_1 \;\; W_2 \;\; W_3 \;\; \ldots \;\; W_j \,] \quad (1)$$
5. Create a match rating table of each alternative on each criteria.
6. Create a decision matrix formed from the match rating table of each alternative on each criterion. The value of x each alternative (Ai) on each criterion (Cj) has been determined.

$$X = \begin{bmatrix} x_{11} & x_{12} & \cdots & x_{1j} \\ \vdots & & & \vdots \\ x_{i1} & x_{i2} & \cdots & x_{ij} \end{bmatrix} \quad (2)$$

where i = 1,2, ... m and j = 1,2, ... n.

7. Perform normalize the decision X matrix by calculating the normalized performance x rating value (r_{ij}) of the alternative A_i on the criterion C_j.

$$r_{ij} = \begin{cases} \dfrac{x_{ij}}{Max_i\,(x_{ij})} & \text{If j is a benefit criterion} \\ \dfrac{Min_i\,(x_{ij})}{x_{ij}} & \text{If j is a cost criterion} \end{cases} \quad (3)$$

Information :

a. It says the profit criteria if the x_{ij} value provides benefits to the decision maker, otherwise the cost criterion if x_{ij} raises costs for the decision maker.
b. If it is a profit criterion, the value x_{ij} is divided by the value of $Max_i\,(x_{ij})$ of each column, while for the cost criterion, the value $Min_i\,(x_{ij})$ of each column is divided by the value x_{ij}.

8. The result of the normalized performance rating value (r_{ij}) forms a normalized matrix (R).

$$R = \begin{bmatrix} r_{11} & r_{12} & \cdots & r_{1j} \\ \vdots & & & \vdots \\ r_{i1} & r_{i2} & \cdots & r_{ij} \end{bmatrix} \quad (4)$$

9. The end result of the preference value (V_i) is derived from the sum of the matrix elements of the normalized matrix row (R) with the corresponding weight of preference (W) of the matrix column element (W).

$$V_i = \sum_{i=1}^{n} W_j \, r_{ij} \quad (5)$$

Greater Vi calculation results indicate that the alternative Ai is the best alternative.

2.3 Rabbit Disease

At first the rabbits are wild animals that live in Africa to mainland Europe. After humans immigrated to various parts of America, Australia, and Asia. In Indonesia, especially in Java, rabbits were brought by the Dutch as ornamental livestock in 1835 [1] [2]. Like other pets the rabbit is susceptible to various diseases, such as:

2.3.1. Diarrhea

This disease attacks the digestive tract, and is the most common cause of death in rabbits on farms. The victims are rabbit children who are still nursing. Children of rabbits who recover from this disease subsequent growth is less good, uneconomical if maintained.

2.3.2 Bloat (Bloated)

The disease is known by the name of bloating, diarrhea, and bloat. The cause of bloating can be due to humid air, wet, or exposed to the night wind directly, and bad weather. Bloating can also be caused by food, because the ratio of crude fiber, protein, and fat is not appropriate.

2.3.3 Young Doe Syndrome (Swollen Milk Glands)

This disease occurs in female rabbits in the first and second birth cohorts. The cause is septicemia due to mastitis resulting in swelling of the milk glands. *Staphyloccus aureus* germs enter the mammary glands through injuries to the nodes or nipples due to mastitis, heat body temperature, lack of appetite. The nipple is swollen and hard so the mother does not want to breastfeed her child.

2.3.4 Constipation

Constipation shows symptoms cannot defecate. Piss very little. Rabbit behavior is very restless. The cause of dry ration is less balanced with sufficient drinking water needs. Raw fiber intakes in dry rations with forage feeds

 https://doi.org/10.1051/e3sconf/20183110002

are less precise. Rabbits are less mobile because of narrow cages.

2.3.5 Cold

Colds cause people to sneeze. Symptoms nose issued lenders clear or cloudy color. The forelegs are always trying to scratch their noses. Feet and body hair come wet. The eyes are puffy, wet, watery. Causes of disease due to bacteria or viruses. The nature of the disease is highly contagious, invading the mucous membranes of the nasal passages. Infection can spread to the throat and lungs causing shortness of breath.

2.3.6 Scabies

Scabies cause itching. Parts of the body are attacked first of the head, then spread to the eyes, nose, legs, and then the whole body. The cause of mites Sarcoptes sacbiei so the disease is called scabesiosis scabies. Skin damage causes injury and itching. As a result of skin infections. Reddish skin, hair loss, accompanied by very tortious itching.

2.3.7 Favus (Infection of the skin)

Favus is an infection of the scalp caused by a fungus. Attacks on the scalp lead to flat round scales, red, and hard. The scalp looks chapped, and the fur fall off.

3. Result and Discussion

3.1 Data Analysis

After conducted a collection of research data on diseases that attack the rabbits and their symptoms. The diseases diagnosed in the system can be seen in Table 1.

Table 1. Various Kinds of Rabbit Disease.

Alternative	Type of Rabbit Disease
A1	Diarrhea
A2	Bloat
A3	Young Doe Syndrome (gland inflammation)
A4	Constipation
A5	Cold
A6	Scabies (Scabiosis)
A7	Favus (Skin Infections)

We can see the determine of disease symptoms in rabbits as in Table 2.

Table 2. Criteria for Rabbit Disease Symptoms

Criteria	Symptoms of Rabbit Disease
C1	Appetite
C2	Body temperature
C3	Sluggish symptoms
C4	Glazed eyes
C5	Diarrhea
C6	Skin scaly / ulcer
C7	Gas formation in the abdomen
C8	Nipple hard and hot
C8	Shortness of breath
C10	Aqueous nose
C11	Defects
C12	Feathers around the anus

Then determined the relation or relationship of each criterion on each alternative, according to Table 3. Where R = Low, N = Normal, and T = Height.

Table 3. Alternative Relations with Criteria

Symptoms Code	Disease Code						
	A1	A2	A3	A4	A5	A6	A7
C1	R	R	T	R	R	R	R
C2	T	T	T	N	N	N	N
C3	T	T	T	T	T	T	T
C4	T	T	T	T	T	T	T
C5	T	T	N	N	N	N	N
C6	N	N	N	N	N	T	T
C7	N	T	N	N	N	N	N
C8	N	N	T	N	N	N	N
C9	N	N	N	N	T	N	N
C10	N	N	N	N	T	N	N
C11	T	T	N	R	N	N	N
C12	T	R	N	N	N	N	N

3.2 System Design

First of all system design is a systematic plan in the process of system development after enough data to support the development activity of rabbit disease diagnosis system. In designing this system the author uses two diagrams Unified Modeling Language (UML) is a use case diagram and activity diagram.

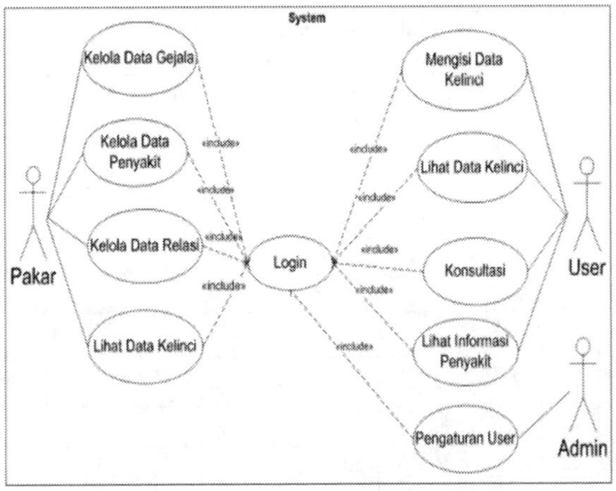

Fig. 1. Use case Diagram System

Figure 1 describes the activities performed by experts, users and admin on the system. Experts performed on this system started by logging into an expert account, after the expert successfully logged in, the expert can manage symptoms data, manage disease data, manage relation data between symptoms with illness, and can see rabbit data.

Activities performed by the user on this system start with login into user account, after user successfully login into account, user can fill rabbit data, see rabbit data, consultation, and see disease information. Activities performed by the admin also begins by logging into the admin account, after admin successfully logged in, the admin can manage the user according to its role.

3.3 System Implementation

This stage is important how the stages of the plan, stages of analysis and design stages can be applied to the program so that it can be used in accordance with the purpose of the researcher is making a decision support system to diagnose rabbit disease using Simple Additive Weighting (SAW) method.

3.3.1 Start Page

The start page is the initial view from when the user opens the app and the login page. The page for the user enter the system according to his / her interests

Fig. 2. Start Page

3.3.2 User Settings Menu

This menu is only open specifically for operator, where operator can specify user types based on user names that exist within the system.

3.3.3 Alternative Menu

This menu is specifically intended for experts in managing alternative data and their relation to criteria.

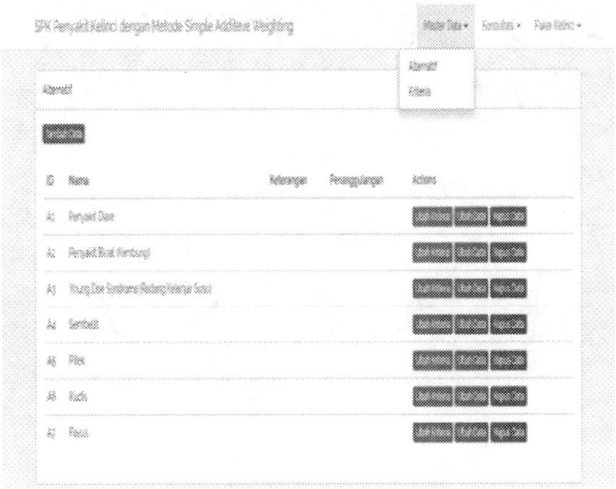

Fig. 3. Alternative Menu

3.3.4 Criteria Menu

This criteria menu is also specifically intended for experts to change the criteria data, add criteria data, delete criteria data and determine the weight value of each criterion.

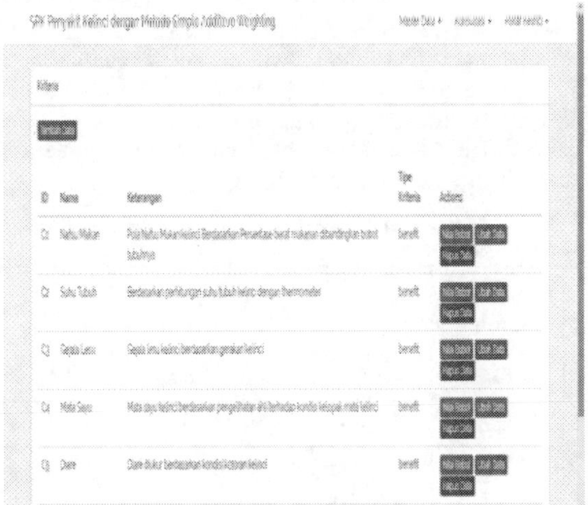

Fig. 4. Criteria Menu

3.3.5 Rabbit Data Menu

The rabbit data menu is the menu where the user enters the rabbit data to be diagnosed.

3.3.6 Rabbit Data Consultation Form

This form contains symptom data that must be filled by user based on rabbit circumstance, which will become the reference in generating the diagnosis. After all the symptom data have been filled to full, then pressed consultation button.

3.3.7 Decision Making Results

Results of decision-making show the results of the order of calculation using SAW method based on symptoms or criteria that have been entered by the user. Information on the cause and handling it by clicking for details.

Fig. 5. Results of Decision Making

3.4 System Testing

System testing is done to compare between the calculation of the system with manual calculation is appropriate. This manual calculation is performed by the user by inputting the values on the symptoms of the disease. The values are selected and entered by the user according to Table 4 and Table 5.

Table 4. Parameter Value Testing Criteria

Symptoms Code	Symptoms	Value Weight (*W*)
C1	<10% body weight (Low)	5
C2	38,5-40,1^0 C (Normal)	1
C3	Rabbits rarely move (Height)	5
C4	His eyes look glazed (High)	5
C5	Feces solid and fluid defecation (Normal)	1
C6	Skin looks healthy and clean (Normal)	1
C7	Abdomen looks healthy (Normal)	1
C8	Healthy nipple (Normal)	1
C9	Rabbit looks difficulty breathing (High)	5
C10	There is liquid around the nose (Height)	5
C11	Dispose of water smoothly (Normal)	1
C12	Net (Norm) 1	1

Table 5. Alternative Value

	Criteria											
	C1	C2	C3	C4	C5	C6	C7	C8	C9	C10	C11	C12
A1	5	5	5	5	5	1	1	1	1	1	5	5
A2	5	5	5	5	5	1	5	1	1	1	5	5
A3	1	5	5	5	1	1	1	5	1	1	1	1
A4	5	1	5	5	1	1	1	1	1	1	3	1
A5	5	1	5	5	1	1	1	1	5	5	1	1
A6	5	1	5	5	1	5	1	1	1	1	1	1
A7	5	1	5	5	1	5	1	1	1	1	1	1

Then form an X matrix corresponding to equation (2).

$$X=\begin{bmatrix} 5 & 5 & 5 & 5 & 5 & 1 & 1 & 1 & 1 & 1 & 5 & 5 \\ 5 & 5 & 5 & 5 & 5 & 1 & 5 & 1 & 1 & 1 & 5 & 5 \\ 1 & 5 & 5 & 5 & 1 & 1 & 1 & 5 & 1 & 1 & 1 & 1 \\ 5 & 1 & 5 & 5 & 1 & 1 & 1 & 1 & 1 & 1 & 3 & 1 \\ 5 & 1 & 5 & 5 & 1 & 1 & 1 & 1 & 5 & 5 & 1 & 1 \\ 5 & 1 & 5 & 5 & 1 & 5 & 1 & 1 & 1 & 1 & 1 & 1 \\ 5 & 1 & 5 & 5 & 1 & 5 & 1 & 1 & 1 & 1 & 1 & 1 \end{bmatrix}$$

Furthermore, normalize the decision matrix X with equation (3). So we get the value of matrix normalized R in Table 6.

Table 6. Normalized R.

$r_{11} = \frac{5}{Max\,(5;5;1;5;5;5;5)} = 1$	$r_{12} = \frac{5}{Max\,(5;5;5;1;1;1;1)} = 1$
$r_{13} = \frac{5}{Max\,(5;5;5;5;5;5;5)} = 1$	$r_{14} = \frac{5}{Max\,(5;5;5;5;5;5;5)} = 1$
$r_{15} = \frac{5}{Max\,(5;5;1;1;1;1;1)} = 1$	$r_{16} = \frac{1}{Max\,(1;1;1;1;1;5;5)} = 0.2$
$r_{17} = \frac{1}{Max\,(1;5;1;1;1;1;1)} = 0.2$	$r_{18} = \frac{1}{Max\,(1;1;5;1;1;1;1)} = 0.2$
$r_{19} = \frac{1}{Max\,(1;1;1;1;5;1;1)} = 0.2$	$r_{110} = \frac{1}{Max\,(1;1;1;1;5;1;1)} = 0.2$
$r_{111} = \frac{5}{Max\,(5;5;1;3;1;1;1)} = 1$	$r_{112} = \frac{5}{Max\,(5;3;1;1;1;1;1)} = 1$
$r_{21} = \frac{5}{Max\,(5;5;1;5;5;5;5)} = 1$	$r_{22} = \frac{5}{Max\,(5;5;5;1;1;1;1)} = 1$
$r_{23} = \frac{5}{Max\,(5;5;5;5;5;5;5)} = 1$	$r_{24} = \frac{5}{Max\,(5;5;5;5;5;5;5)} = 1$
$r_{25} = \frac{5}{Max\,(5;5;1;1;1;1;1)} = 1$	$r_{26} = \frac{1}{Max\,(1;1;1;1;1;5;5)} = 0.2$
$r_{27} = \frac{5}{Max\,(1;5;1;1;1;1;1)} = 1$	$r_{28} = \frac{1}{Max\,(1;1;5;1;1;1;1)} = 0.2$
$r_{29} = \frac{1}{Max\,(1;1;1;1;5;1;1)} = 0.2$	$r_{210} = \frac{1}{Max\,(1;1;1;1;5;1;1)} = 0.2$
$r_{211} = \frac{5}{Max\,(5;5;1;3;1;1;1)} = 1$	$r_{212} = \frac{3}{Max\,(5;3;1;1;1;1;1)} = 0.6$
$r_{31} = \frac{1}{Max\,(5;5;1;5;5;5;5)} = 0.2$	$r_{32} = \frac{5}{Max\,(5;5;5;1;1;1;1)} =$
$r_{33} = \frac{5}{Max\,(5;5;5;5;5;5;5)} = 1$	$r_{34} = \frac{5}{Max\,(5;5;5;5;5;5;5)} = 1$

$r_{35} = \frac{1}{Max(5;5;1;1;1;1;1)} = 0.2$	$r_{36} = \frac{1}{Max(1;1;1;1;1;5;5)} = 0.2$
$r_{37} = \frac{1}{Max(1;5;1;1;1;1;1)} = 0.2$	$r_{38} = \frac{5}{Max(1;1;5;1;1;1;1)} = 1$
$r_{39} = \frac{1}{Max(1;1;1;1;5;1;1)} = 0.2$	$r_{310} = \frac{1}{Max(1;1;1;1;5;1;1)} = 0.2$
$r_{311} = \frac{1}{Max(5;5;1;3;1;1;1)} = 0.2$	$r_{312} = \frac{1}{Max(5;3;1;1;1;1;1)} = 0.2$
$r_{41} = \frac{5}{Max(5;5;1;5;5;5;5)} = 1$	$r_{42} = \frac{1}{Max(5;5;5;1;1;1;1)} = 0.2$
$r_{43} = \frac{5}{Max(5;5;5;5;5;5;5)} = 1$	$r_{44} = \frac{5}{Max(5;5;5;5;5;5;5)} = 1$
$r_{45} = \frac{3}{Max(5;5;1;1;1;1;1)} = 0.6$	$r_{46} = \frac{1}{Max(1;1;1;1;1;5;5)} = 0.2$
$r_{47} = \frac{1}{Max(1;5;1;1;1;1;1)} = 0.2$	$r_{48} = \frac{1}{Max(1;1;5;1;1;1;1)} = 0.2$
$r_{49} = \frac{1}{Max(1;1;1;1;5;1;1)} = 0.2$	$r_{410} = \frac{1}{Max(1;1;1;1;5;1;1)} = 0.2$
$r_{411} = \frac{3}{Max(5;5;1;3;1;1;1)} = 0.6$	$r_{412} = \frac{1}{Max(5;3;1;5;1;1;1)} = 0.2$
$r_{51} = \frac{5}{Max(5;5;1;5;5;5;5)} = 1$	$r_{52} = \frac{1}{Max(5;5;5;1;1;1;1)} = 0.2$
$r_{53} = \frac{5}{Max(5;5;5;5;5;5;5)} = 1$	$r_{54} = \frac{5}{Max(5;5;5;5;5;5;5)} = 1$
$r_{55} = \frac{1}{Max(5;5;1;1;1;1;1)} = 0.2$	$r_{56} = \frac{1}{Max(1;1;1;1;1;5;5)} = 0.2$
$r_{57} = \frac{1}{Max(1;5;1;1;1;1;1)} = 0.2$	$r_{58} = \frac{1}{Max(1;1;5;1;1;1;1)} = 0.2$
$r_{59} = \frac{5}{Max(1;1;1;1;5;1;1)} = 1$	$r_{510} = \frac{5}{Max(1;1;1;1;5;1;1)} = 1$
$r_{511} = \frac{1}{Max(5;5;1;3;1;1;1)} = 0.2$	$r_{512} = \frac{1}{Max(5;3;1;1;1;1;1)} = 0.2$
$r_{61} = \frac{5}{Max(5;5;1;5;5;5;5)} = 1$	$r_{62} = \frac{1}{Max(5;5;5;1;1;1;1)} = 0.2$
$r_{63} = \frac{5}{Max(5;5;5;5;5;5;5)} = 1$	$r_{64}\ \frac{5}{Max(5;5;5;5;5;5;5)} = 1$
$r_{65} = \frac{1}{Max(5;5;1;1;1;1;1)} = 0.2$	$r_{66} = \frac{5}{Max(1;1;1;1;1;5;5)} = 1$
$r_{67} = \frac{1}{Max(1;5;1;1;1;1;1)} = 0.2$	$r_{68} = \frac{1}{Max(1;1;5;1;1;1;1)} = 0.2$
$r_{68} = \frac{1}{Max(1;1;5;1;1;1;1)} = 0.2$	$r_{610} = \frac{1}{Max(1;1;1;1;5;1;1)} = 0.2$
$r_{611} = \frac{1}{Max(5;5;1;3;1;1;1)} = 0.2$	$r_{612} = \frac{1}{Max(5;3;1;1;1;1;1)} = 0.2$
$r_{71} = \frac{5}{Max(5;5;1;5;5;5;5)} = 1$	$r_{72} = \frac{1}{ax(5;5;5;1;1;1;1)} = 0.2$
$r_{73} = \frac{5}{Max(5;5;5;5;5;5;5)} = 1$	$r_{74} = \frac{5}{Max(5;5;5;5;5;5;5)} = 1$
$r_{75} = \frac{1}{Max(5;5;1;1;1;1;1)} = 0.2$	$r_{76} = \frac{5}{Max(1;1;1;1;1;5;5)} = 1$
$r_{77} = \frac{1}{Max(1;5;1;1;1;1;1)} = 0.2$	$r_{78} = \frac{1}{Max(1;1;5;1;1;1;1)} = 0.2$
$r_{79} = \frac{1}{Max(1;1;1;1;5;1;1)} = 0.2$	$r_{710} = \frac{1}{Max(1;1;1;1;5;1;1)} = 0.2$
$r_{711} = \frac{1}{Max(5;5;1;3;1;1;1)} = 0.2$	$r_{712} = \frac{1}{Max(5;3;1;1;1;1;1)} = 0.2$

$$R = \begin{bmatrix} 1 & 1 & 1 & 1 & 1 & 0.2 & 0.2 & 0.2 & 0.2 & 0.2 & 1 & 1 \\ 1 & 1 & 1 & 1 & 1 & 0.2 & 1 & 0.2 & 0.2 & 0.2 & 1 & 0.6 \\ 0.2 & 1 & 1 & 1 & 0.2 & 0.2 & 0.2 & 1 & 0.2 & 0.2 & 0.2 & 0.2 \\ 1 & 0.2 & 1 & 1 & 0.2 & 0.2 & 0.2 & 0.2 & 0.2 & 0.2 & 0.6 & 0.2 \\ 1 & 0.2 & 1 & 1 & 0.2 & 0.2 & 0,2 & 0.2 & 1 & 1 & 0.2 & 0.2 \\ 1 & 0.2 & 1 & 1 & 0.2 & 1 & 0.2 & 0.2 & 0.2 & 0.2 & 0.2 & 0.2 \\ 1 & 0.2 & 1 & 1 & 0.2 & 1 & 0.2 & 0.2 & 0.2 & 0.2 & 0.2 & 0.2 \end{bmatrix}$$

Next is process of calculating preference (V) by using the following equation. The end result of the preference value (Vi) is derived from the sum of the matrix elements of the normalized matrix row (R) with the corresponding weight of preference (W) of the matrix column element (W). V larger value indicates that the alternative V5 is the best alternative, in other words, A5 were the best alternative decision using Simple Additive Weighting method (Table 7).

Table 7. The Preference Value Vi

No	(Vi)
1	[redacted] = (1x5)+(1x1)+(1x5)+(1x5)+(1x1)+(0.2x1)+ (0.2x1)+(0.2x1)+(0.2x5)+(0.2x5)+(1x1)+(1x1) = 5+1+5+5+1+0.2+0.2+0.2+1+1+1+1 = 21.6
2	[redacted] = (1x5)+(1x1)+(1x5)+(1x5)+(1x1)+(0.2x1)+ (1x1)+(0.2x1)+(0.2x5)+(0.2x5)+(1x1)+ (0.6x1) = 5+1+5+5+1+0.2+1+ 0.2+1+1+1+ 0.6 = 22
3	[redacted] = (0.2x5)+(1x1)+(1x5)+(1x5)+(0.2x1)+(0.2x1)+ (0.2x1)+(1x1)+(0.2x5)+(0.2x5)+(0.2x1)+(0.2x1) = 1+1+5+5+0.2+0.2+0.2+1+1+1+0.2+0.2 = 16
4	[redacted] = (1x5)+(0.2x1)+(1x5)+(1x5)+(0.2x1)+(0.2x1)+ (0.2x1)+(0.2x1)+(0.2x5)+(0.2x5)+(0.6x1)+ (0.2x1) = 5+0.2+5+5+0.2+0.2+0.2+0.2+1+1+0.6+0.2 = 18.8
5	V_5 = (1x5)+(0.2x1)+(1x5)+(1x5)+(0.2x1)+(0.2x1) + (0.2x1)+(0.2x1)+(1x5)+(1x5)+ (0.2x1)+(0.2x1) =5+0.2+5+5+0.2+0.2+0.2+0.2+5+5+0.2+0.2 = **26.4**
6	[redacted] = (1x5)+(0.2x1)+(1x5)+(1x5)+(0.2x1)+(1x1) + (0.2x1)+(0.2x1)+(0.2x5)+(0.2x5)+(0.2x1) + (0.2x1) = 5+0.2+5+5+0.2+1+0.2+0.2+1+1+0,2+0.2 = 19.2
7	[redacted] = (1x5)+(0.2x1)+(1x5)+(1x5)+(0.2x1)+(1x1) + (0.2x1)+(0.2x1)+(0.2x5)+(0.2x5)+(0.2x1)+(0.2x) = 5+0.2+5+5+0.2+1+0.2+0.2+1+1+0.2+0.2 = 19.2

Determine the final result of the preference value using equation (5). So we get the ordering of the calculation result is:

Cold Disease $V_5 = \mathbf{26.4}$
Bloating Disease $V_2 = 22$
Diarrhea Disease $V_1 = 21.6$
Mange Disease $V_6 = 19.2$
Favus disease $V_7 = 19.2$
Constipation Disease $V_4 = 18.4$

The result of the calculation is $V_5 = \mathbf{26.4}$ has the greatest value, so the result of diagnosis is Colds disease. System diagnostic results are: Rabbits have the possibility of having a cold with the largest preference value of 26.4. So the result of calculation system compared with manual calculation accordingly.

How to overcome disease in the system is step handling patients (rabbits) should be treated. The snot-nosed nose was sprayed with an antiseptic solution, a hardened crust cleansed with warm water. Treat the rabbit with a given antibiotic injection.

4. Conclusions

From the implementation that has been done, there are some conclusions that can be presented as a result of research. This research has produced a decision support system for the diagnosis of the rabbit disease on the web-based with Simple Additive weighting method (SAW). It can assist the public in providing a diagnosis on the sick rabbit.

The decision support system works based on symptoms that have been selected by the user and processed by the system so as to produce a diagnosis of rabbit disease. Based on the computation calculations performed manually and done with the system then it can be concluded from both the manual calculation and the system, yielding the same value with the same disease.

References

1. B. Sarwono, *Kelinci Potong dan Hias*, Jakarta : Agro Media Pustaka (2010).
2. R. Hustamin, *Panduan Memelihara Kelinci Hias,* Jakarta : Agro Media (2006).
3. E. Turban, *Decision Support System and Inteligent Systems.* Jilid1. Edisi 7. Yogyakarta: Penerbit Andi (2005).
4. Kusumadewi, Sri, dkk. *Fuzzy Multi Attribute Decision Making*. Yogyakarta (2006).
5. Nurmalini1, R. Robbi, IJSRST Vol. **3**: 541-544, March-April (2017).
6. Risawandi, R. Robbi, IJSRST Vol. **3**: 491-494, Nov-Dec (2016).

Humanoid Robot Control System Balance Dance Indonesia and Reader Filters Using Complementary Angle Values

Sholihin[1,*] , *Eka* Susanti[2]

[1]Department Polytechnic of Sriwijaya, Srijaya Negera Street, Bukit Besar, Palembang - Indonesia
[2]Department Polytechnic of Sriwijaya, Srijaya Negera Street, Bukit Besar, Palembang - Indonesia

Abstract. The development of increasingly advanced technology, make people want to be more developed and curiosity to know more to determine the development of advanced technology. Robot is a tool that can be used as a tool for people who have several advantages. Basically humanoid robot is a robot that resembles a human being with all the driving structure. In the application of this humanoid robot manufacture researchers use MPU6050 module which is an important component of the robot because it can provide a response to the angle reference axis X and Y reference axis, the reading corner still has noise if not filtered out beforehand. On the other hand the use of Complementary filters are the answer to reduce the noise. By arranging the filter coefficients and time sampling filter that affects the signal updates corner. The angle value will be the value of the sensor to the process to the PID system which generates output values that are integrated with the servo pulses. Researchers will test to get a reading of the most stable angle for this experiment is the "a" or the value of the filter coefficient = 0.96 and "dt" or the sampling time = 10 ms.

1 Introduction

The ability to be a technology developed in the modern era this point is crucial, in particular the application of robot technology. One application of the conducted research is the humanoid robot that its application to be able to dance to the theme of dance that has been determined, when the robot walks then it is also necessary to balance the robot so that the robot does not fall when the robot move or walk, and therefore need to be made of a system which can control the balance of the humanoid robot so that the robot has become more balanced and able to minimize interference from outside. The parameters that will be used to control systems that balance is the tilt position of the robot body to the reference axis X and the reference axis Y. The slope of the robot is obtained by what angle parameter robot will fall. To get the angles can be searched by the formula on MPU6050, but the angle with the formula it still has a readout noise so that the angle obtained is not so good. Complementary application of filter is one filter that can improve the reading corner robot movement.

2 Supporting theory

2.1 Robot

The robot is a mechanical device that can perform physical tasks, both in the monitoring and control of human, or use a program that has been defined in advance in artificial intelligence. Robots are usually used for heavy duty, dangerous and dirty jobs repeatedly. Usually most industrial robots used in manufacturing. The applicability of other robotic included for cleaning up toxic waste, the exploration of outer space and under water, mining, jobs "find and help" (search and rescue), and search for mines. Lately the robot began to enter the consumer market in the fields of entertainment, and a housekeeper tool, such as vacuum cleaners and lawn mowers [1,2].

2.2 Arduino Mega 2560

Arduino Mega 2560 is a microcontroller based development board arduino using ATmega chip 2560. This board has a pin I / O that is quite a lot, some 54 pieces of digital I / O pin (pin 15 of which is PWM), 16 analog input pin, 4 pin UART (serial port hardware). Arduino Mega 2560 is equipped with a 16 MHz oscillator, a USB port, DC power jack, ICSP header, and a reset button. This board is very complete, it has everything needed for a microcontroller [3].

* Corresponding author: hin00308@yahoo.com

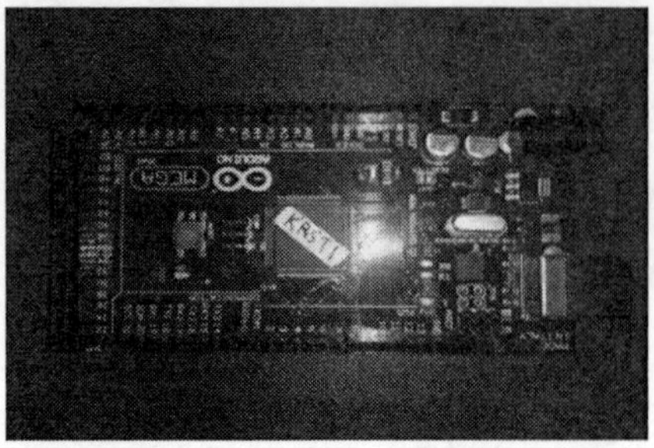

Fig. 1. Arduino Mega 2560

2.3 Sensor

The sensor is a component that is used to detect a physical quantity into electrical quantities that can be analyzed with a particular electrical circuit. Sensors very important role in the world of robotics which function as a robot navigational input, while the types of sensors that sound sensors, light sensors, pressure sensors fire, temperature sensors, humidity sensors, ultrasonic sensors, magnetic sensors [4,5].

2.4 MPU6050 module, gyrescope, and accelerometer

Module MPU6050 a balance sensor that can read the angular velocity and the angular velocity. MPU6050 an IC comprising digital accelerometer and gyroscope which each have a 3 axis, both accelerometer and gyroscope that exist in MPU6050 has a 16 bit output [9]. Decision accelerometer values is done by the following formula :
Pich = arc tan $(-x/z^2 + y^2)$ x (360/2pi)
Roll = arc tan $(y/z^2 + x^2)$ x (360/2pi)
MPU6050 module has a selection maximum scale gyroscope readings can be selected, which is 250 deg / sec, 500 deg / sec, 1000 deg / sec, 2000 deg / sec.

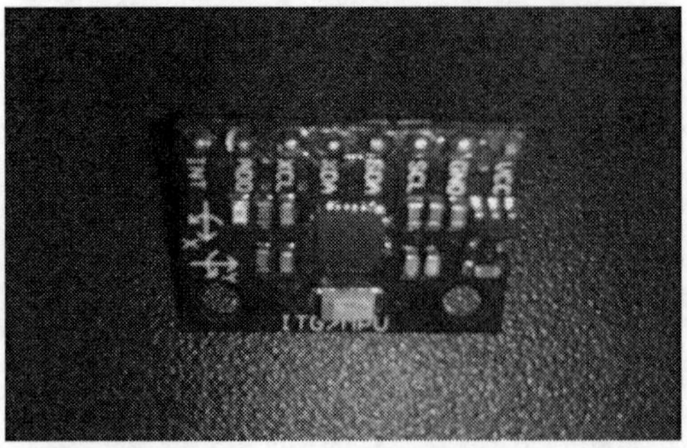

Fig. 2. MPU6050 Module

Sensor gyroscope module MPU6050 own digital outputs 16 bits, so as to perform the conversion to detection to the angular velocity, this output must be divided by the divisor factor that is different on each scale were selected. as shown in the table below ;

Table 1. Scale Factor Measurement and Dividers MPU6050

No.	Scale (deg/sec)	Dividing Factor (LSB/deg/sec)
1.	250	131
2.	500	65,5
3.	1000	32,8
4.	2000	6,4

That way, the gyroscope data conversion from 16 bit ADC output to units of degrees / second, so; ω = Output ADC divided by the divider factor [5,6].

2.5 complementary filter

Complementary Filter is a recursive digital filter to estimate the condition of any process very well. Complementary filter is used as an algorithm to provide a reliable estimate the condition of the state of the process. Complementary Filters are also used to control systems that are sensitive to noise from the environment because it minimizes the square error [4]. This filter can reduce noise from the measurement of the affected sensors before entering into the control system. In the algorithm is applied to the microcontroller with software, use discrete mathematical equations. The system to be measured must be modeled by a linear system [6].

3 Methods

3.1 Design mechanical robot

Robot skeleton is made using acrylic material and PLA filament, using a 3D printer to process PLA filament into line servo desired shape of the container.

Fig. 3. Physical form Mechanical Robot

3.2 Flowchart robot

Design flowchart robot is based on a balancing robot that uses an algorithm, complementary filters and PID, as follows:

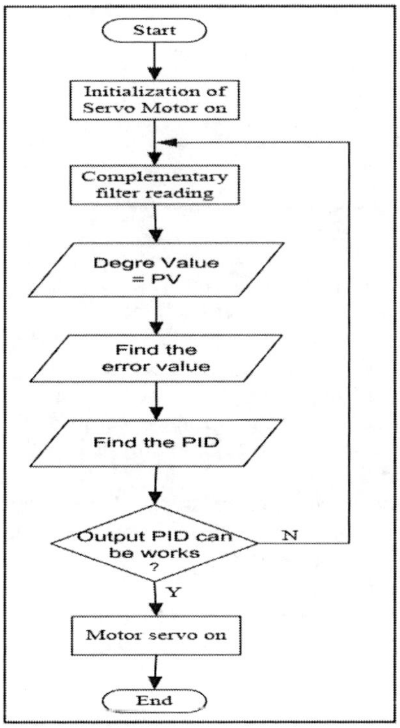

Fig. 4. Flowchart Robot

3.3 Design of complementary filter

On balance control processing using PID control, complementary filter function to filter the value point value or the value of the sensor modules that emit MPU6050 pitchdan roll angle. The following block diagram complementary filter design.

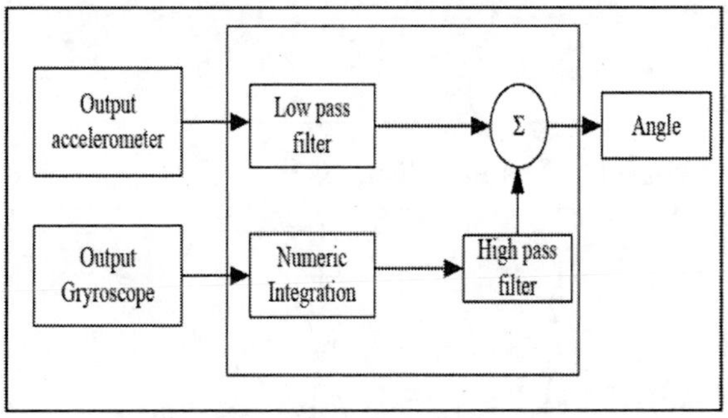

Fig. 5. Diagramm Block Complementary Filter

Using 2-axis accelerometer sensor to obtain the value of the tilt angle (θ), the axis X and Z. Then the angle of inclination of this accelerometer in to feed the low-pass filter to eliminate noise. While Gyroscope sensor output in the form of angular velocity (ω) must be integrated in advance to get the value slope. Then fed to a high-pass filter to eliminate the effect of bias on the gyroscope.

Complementary Filter is making the value of the angle in which the value of the output of the accelerometer is fed to the low pass filter is summed with the output value of the gyroscope is fed to a high pass filter which further if added together would produce a stable output value angle.

Meanwhile, the output value is stable gyroscope using the high pass filter to obtain the desired value of the speed through the corners. Here's the formula of the complementary filter.

Angle = (a)*(Angle + out_gyro*dt)+(1-a)*(Angle_acc) (1)

The output of the gyroscope that form the angular velocity is added to the output value of the angle complementary previous filter, the output angle before the angle now, separated by time sampling. Here the authors chose the sampling time of 0.01 seconds or 10 ms. This value is multiplied by a factor of the filter coefficients or symbolized by "a". here the author choose the value "a" of 0.96. That value is added to the value of output accelerometer that has not been filtered out earlier. Then tu value multiplied by the value of "1-a", so that the value of the multiplier for output accelerometer angle is 0.04. In the complementary filter there is time constant, namely the length of real time signal, with the following formula.

T = (a*dt)/(1-a) (2)

4 Current results

Complementary Filters used in digital filters or use the program. Complementary filter testing is done by doing the reading corner of the roll and pitch, as well as the results of calculations comparing data with the filter without filter. At the first measurement, the value of the filter coefficients (a) will be changed - a change of 0.6; 0.8; 0.96 as well as the sampling time (dt) fixed at 10 ms. Here is the response signal output from the complementary filter. Red line is the sensor output accelerometer, gyroscope sensor output is black and blue is the output signal of the complementary filter.

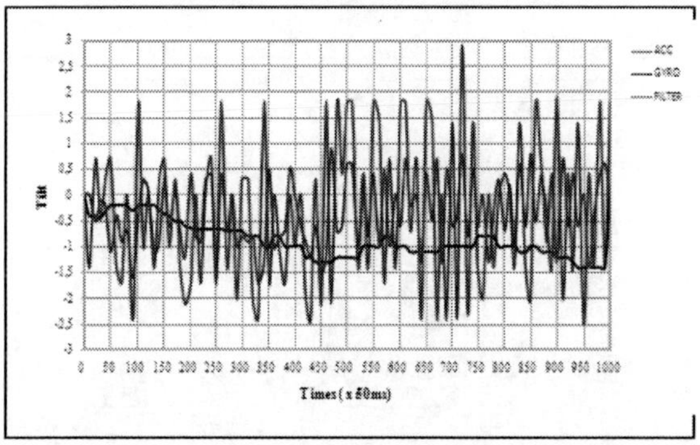

Fig. 6. Filter coefficient 0.6 was stationary non-rotating

https://doi.org/10.1051/e3sconf/20183110003

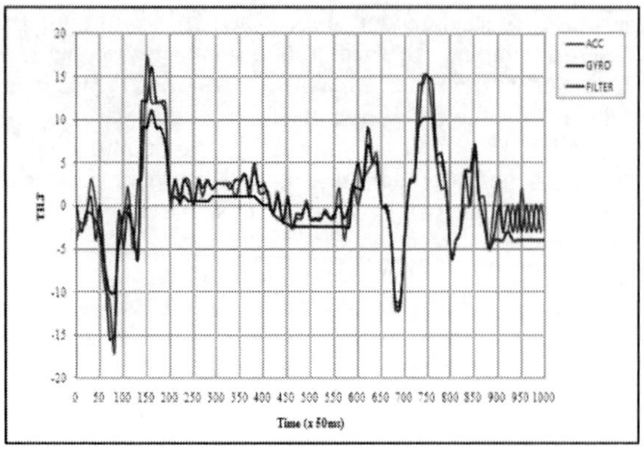

Fig. 7. Filter Cofficient 0,6 was Stationary Rotating

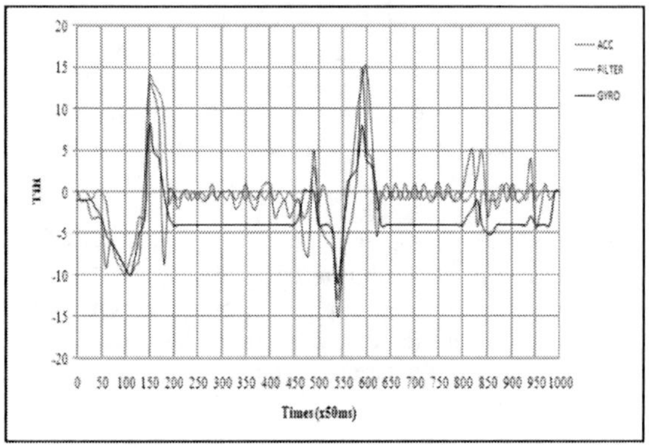

Fig. 8. Filter Coeffecint 0,8 being stationary non-Rotating

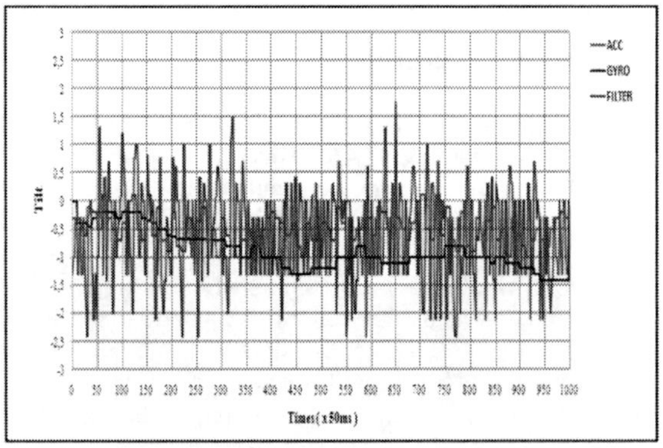

Fig. 9. Filter Coeffecient 0,8 being Stationary Rotating

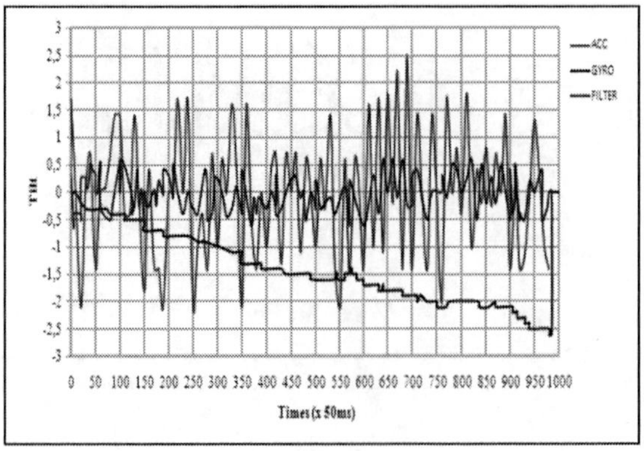

Fig. 10. Filter Coeffecient 0,96 no-Respon

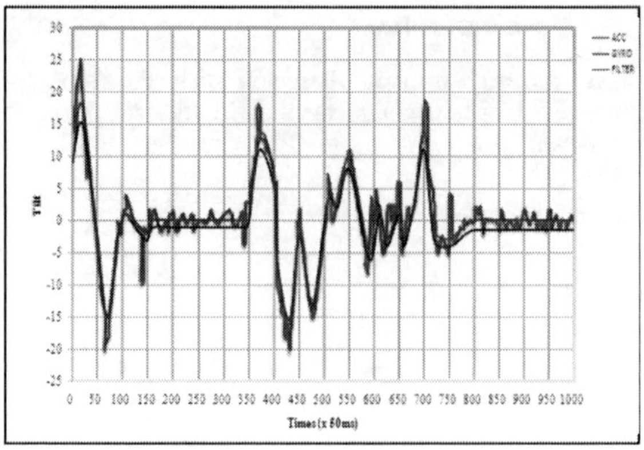

Fig. 11. Filter Coeffecinet 0,96 Respon

From some of the pictures above we can conclude the greater the value of the filter coefficients (a), the output signal will be more stable complementary filter less air noise) but will prolong the time constant resulting signal update time is getting longer. If we compare with the equation to find the value of time constant, then the above conclusions based testing will be in accordance with the equation.

Next was measured by changing change time sampling, the sampling time value will be changed to change the value of 0.1 seconds (10Hz), 0.05 sec (20Hz) and 0.01 second (100Hz).

Red line is the sensor output accelerometer, gyroscope sensor output is black and blue are complementary filter output signal.

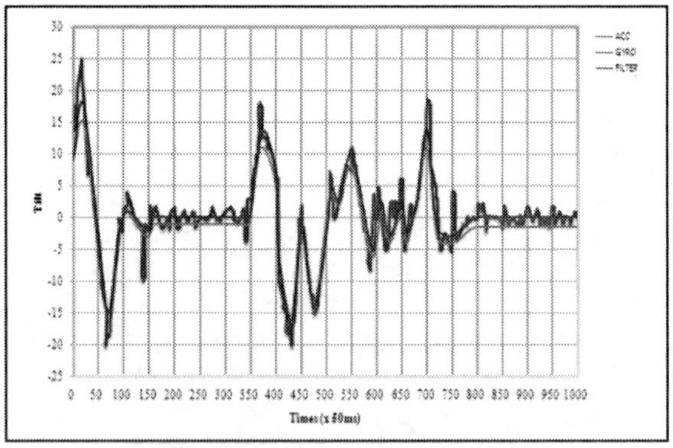

Fig. 12. Time Sampling with 0,01 second

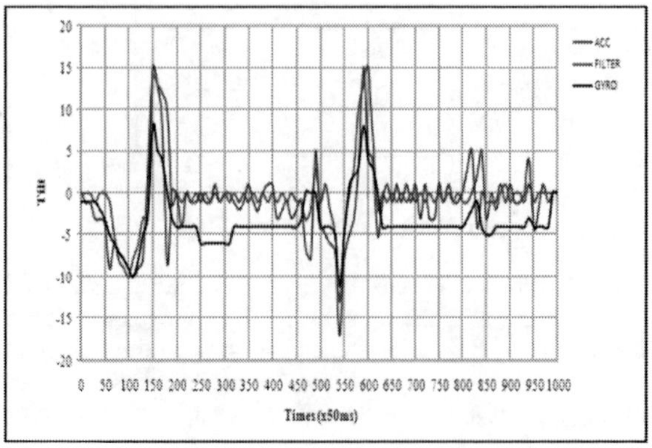

Fig. 13 Time Sampling with 0,05 second

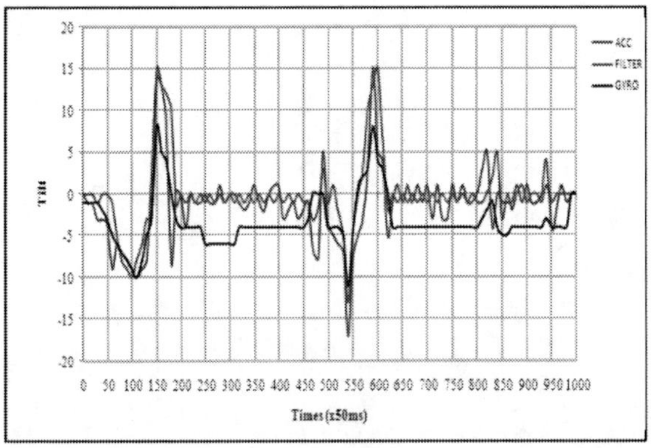

Fig. 14. Time Sampling with 0,1 second

It can be concluded that the faster sampling time used the faster time constant of the output signal complementary filter. Again if we look at the equation for calculating the time constant, the output signal complementary filter testing, will be in accordance with the theory of equations to calculate the value time constant.

Selection of the value of the filter coefficients (a) and the sampling time (dt) shall be adjusted to the response on the application you want to create. If the response is fast, it is advisable to accelerate the value of time constant, and if the response to the application system that we created is relatively slow, then the value of time Constans the complementary filter algorithm we can make relatively longer.

5 Conclusion

In making this humanoid robot dance filter coefficient values corresponding to the sensor MPU6050 is 0.96 because the robot should have a fast response on balance to keep from falling while dancing. If the sampling time accelerated the signal sampling time is faster and has a faster response. Selection of the value of the filter coefficients (a) and the sampling time (dt) must be adapted to respond to the application that you created. If the system response desired application quickly, it is advisable to accelerate the value of time constant, and if the system response to the application that you created relatively slowly, the time constant value at the algorithm complementary filter you can make relatively longer.

References

1. Asma Bait Fadhah, Athra' Al-Lawati, Sara Al-Maskari, Abderezak Touzene and Ahmed Al-Kindi, *Applications of Digital Information and Web Technologies*,p. 151-155, (2008)
2. W. Budiharto, and Nawalan, *Membuat Sendiri Robot Humanoid*. Jakarta: Elex Media Komputindo.
3. Eka, Yulius. 2011.*Buku Pintar Pemrograman C#.* Yogyakarta: Media Kom, (2009)
4. Gourab Sen Gupta, S.C. Mukhopadhyay, Matthew Finnie-" WiFi-based Control of a Robotic Arm with Remote Vision" I2MTC International Instrumentation and Measurement Technology Conference, Singapore, (2009)
5. Carlos Cardei and Jose Sa da Costa, in Industrial Electronics Societ, p. 1-6., (2005)
6. Johan, Pramudita, *Sistem Kontrol Keseimbangan Statis Robot Humanoid Joko Klana Berbasis Pengontrol PID.* Elektronika dan Instrumentasi. Universitas Gadja Mada. Yogyakarta, (2013)

I. Maulana, Susilo, A,*Robot Humanoid dan Kecerdasan Buatan*,, Jurusan Teknik Komputer, Fakultas Teknik, STIMIK-AKI, Semarang (2010)

7. Munir, Rinaldi, *Algoritma & Pemrograman Dalam Bahasa Pascal dan C*. Bandung: Elex Media Komputindo,(2009)
8. Nurfansyah, Rahadian, Estimasi Sudut Orientasi Menggunakan Sensor 6 DOF IMU dan Sensor Magnetometer 3 aksis, thesis of Dept Elctro, Universitas Diponegoro,Semarang. Volume **6**, No. 1. 3 Juli (2013).
9. Septian, Taufik dwi suyadh, *Buku Pintar Robotika*.Yogyakarta : Andi,(2010)
10. Setyono, Arif, Perancangan Perangkat Lunak Pendeteksi Posisi Benda 6 Derajat Kebebasan, Tesis of Dept Electro, Universitas Diponegoro, Semarang. Volume **3**, No. 2. 10 Juli (2016)

Production Planning and Planting Pattern Scheduling Information System for Horticulture

Tanhella Zein Vitadiar[1,*], *Farikhin* Farikhin[1,2], *Bayu* Surarso [1,2]

[1]Master Program of Information System, Diponegoro University, Semarang - Indonesia
[2]Faculty of Science and Mathematics, Diponegoro University, Semarang – Indonesia

Abstract. This paper present the production of planning and planting pattern scheduling faced by horticulture farmer using two methods. Fuzzy time series method use to predict demand on based on sales amount, while linear programming is used to assist horticulture farmers in making production planning decisions and determining the schedule of cropping patterns in accordance with demand predictions of the fuzzy time series method, variable use in this paper is size of areas, production advantage, amount of seeds and age of the plants. This research result production planning and planting patterns scheduling information system with the output is recommendations planting schedule, harvest schedule and the number of seeds will be plant.

1 Introduction

Production planning is one of a topic to which is of special attention for the researchers and practitioners academic industry, because it is a problem that was important in the production process, with planning production of good then can increase the company profitability [1]. Several factors that need to be considered in planning the production of is consumer demand, advantage to be achieved, a charge required during the production process, the number of resources that there was either natural resources or human resources and rescheduled, it is rescheduled planting patterns that is the determination of the schedule cropping and harvest the schedule so that it can boost the profitability of agricultural products [2].

In scheduling patterns of cropping farmers must respect all existing attributes, So as to maximize the profits of farmers, some attribute to note is, size of areas, production advantage, amount of seeds and age of the plant and production as well as the results of the life cycle of horticulture plant, in this paper study case using mushroom where mushrooms which have a life cycle as follows: mushroom life cycle averages 4 months, the process of harvesting the Mushrooms start 2 month until 4 month, harvest daily variety, mushroom harvest pattern starting from the initial harvest will always increase with each passing day until a phase peak at 2 month to 3 month and will decrease again after passing the peak phase.

For planning and scheduling patterns of cropping required a model that is able to perform calculations for planning the best among the possibilities of action that can be performed, a linear programming model is the right model because it can take decisions multi objective so different possibilities to obtain optimum resource utilization can be achieved [3].

Linear Programming is helpful in the decision for the farmers by way of the estimated number of farmer acceptance in various alternative in planning to achieve specific goals on a limited resource. These conditions encountered in agricultural commodities, such as the amount of land available, limited manpower, funding limitations and other supporting the factors such as agricultural tools are inadequate, so that Linear Programming can be used for two way i.e. minimizing costs in getting the total acceptance or total profit (minimization program) and maximize the total profit by the constraints of limited resources (maximization program) [4].

A linear Programming model is composed of the coefficients of the function of the purpose and function of the constraints which the objective function aims to describe a mathematical equation to goals are obtained while the function of mathematical equations to describe the constraints of any effect on production [5]. Then after describing the mathematical equations the next step is the calculation using the simplex method. In Linear programming, There is a sensitivity analysis. Because there is always a factor of uncertainty, sensitivity is enacting the resource parameter that is available on most small limits (Lower Limit) and limit most large (Upper Limit) it was allowed to change the parameter to a certain extents [6].

Development of cropping pattern scheduling starts with knowing the request data because the data request is the most urgent given the one key of a business is able to meet the demand of the consumer so that the consumer

* Corresponding author: tanhellavitadiar@gmail.com

satisfaction are meet, their relationship with consumers waking, addressing the existence of the advantages and disadvantages of stock is excessive, to know the consumer demand will then do the prediction using fuzzy time series forecasting method for the future which is done based on the value of the pattern or past data of a variable or an error in the past. The concept of fuzzy time series presented based on fuzzy set theory, fuzzy logic and approximate reasoning [7].

Some literature relating with the planning production, scheduling of planting patterns in plants horticultural: tomato : Rocco dan Morabito [1], citrus : Munhoz and Morabito [8], the difference with this paper is horticulture crops study case is mushroom and this paper use 2 method there is fuzzy time series method and linear programming. Fuzzy time series prediction is very useful, if there is no farm demand to data so that the prediction results in the form of demand data can be one of the constraints variable in linear programming.

Production planning is one of the activities conducted by the company, with production planning is good then it can control the production process so that it does not produce excessive production or do not comply with the request, due to the cropping pattern of scheduling information system then expected the company can meet the amount of production effectively and efficiently so that consumer can be meet, and corporate profits can be achieved.

2 Theoretical Framework

2.1 Production Planning

The production is a set of procedures or activities conducted to create a goods and services, in any production process required a planning process. Production planning is planning about what products and how that will be produced by the company in question in a period to come. Production planning is part of operational planning within the company. In the preparation of production planning things to consider is the optimization of production so as to be able to achieve the lowest cost level for the implementation of the production process [9]. The role of planning is merely intended to coordinate the activities of the sections that directly or indirectly in producing, planning, scheduling and controlling the production of the starting stages of raw materials, process, until the resulting output so that the company really can produce goods or services effectively and efficiently.

Some of the goals of production planning, namely as a preliminary step to determine the activity of production, as a reference to more detailed planning, input of resource plans resource planning so that it can be developed to support production planning, production and labor stability against fluctuations in demand.

2.2 Scheduling of Planting Pattern

The cropping pattern of scheduling is a plantings on a plot of land with a set order of the layout and sequence of the plant over a given period includes the period of tillage and the period didn't planted during a certain time period [10]. Plants in one area can be set according to its kind. Monoculture cropping pattern exists, namely the growing of plants similar to the one planting acreage. There is a mixed cropping pattern, i.e., plants are grown on a single area. There is also a growing pattern of passes, that is, planting crops in turn some kinds of plants at different time on the same acreage.

In planning the production agribusiness factor to note is the cropping pattern so that scheduling can increase the profits of agriculture [11]. In the preparation of the scheduling is expected to reduce the delays in the work, and the processing time, maximizing the available working tools and labor as well as reduce the time delay.

2.3 Information System

Information system is a tool to present information so beneficial to the recipient for decision making on planning, controlling, and organizational management of company, information systems are used to automate the performance of routine and management activities. The field of information technology is very special because always follow the dynamic changes that develop every year there is always a new version in terms of hardware and software [12]. the basic concepts of information.

With the information system is expected to simplify and automate all the performance in the field of any kind of health, education and agriculture.

2.4 Fuzzy Time Series

Fuzzy time series a concept based on fuzzy set theory and its linguistic connection and application variables, fuzzy time series are used to solve the problem of forecasting which historical data are linguistic values. For example, in the matter of forecasting, historical data is not in the form of real numbers, but in the form of linguistic data. In this case, there is no conventional time series model that can be applied, but the fuzzy time series models can be applied with more precise. The main difference between the conventional and the fuzzy time series time series in the value used in forecasting, which is a fuzzy set of the real numbers over the set of hosts is specified, fuzzy set is defined as a class of numbers with vague boundaries, in this paper, *fuzzy time series* applied to the sale of horticulture crops in this study is mushroom, *fuzzy time series* used to predict demand of next month, the parameter used are Januari 2012 until April 2017.

Steps by steps with the prediction method of Fuzzy Time Series [2].

1. Partition the universe of discourse U = [Dmin - D1, Dmax -[- D2] into even lengthy and equal length intervals ul, u2, ..., um, historical sales of mushroom data Januari 2012 - April 2017. (1)
2. Let A1,A2, ... ,A k be fuzzy sets which are linguistic values of the linguistic variable "en- rollments". Define fuzzy sets A1, A2 ..... Ak on the universe of discourse U. (2)

3. Divide the derived fuzzy logical relation- ships into groups based on the current states of the enrollments of fuzzy logical relationships, by using the formula :

$$r(MBF)^{w+1} = \frac{MBF_1 + MBF_2 + \dots + MBF_w}{W} \quad (3)$$

Dimana MBF_1, MBF_2, MBF_w is the membership value of the sales historical data.

4. Calculate the forecasted outputs use *Center of Grafity method* (COG). By using the formula :

$$y = \frac{(A1*c1+A2*c2+A3*c3 + \dots + A7*c7)}{A1 + A2 + A3 + \dots + A7} \quad (4)$$

$$v = r(NS) * A1 + r(NM) * A2 + \dots + r(PB) * A7$$

$$Fi = y + v$$

2.5 Linear Programming

Linear Programming is a general model that can be used to solving the problem of allocating limited resources optimally. The linear program includes planning activities to attain optimal results that is an outcome which reflects the achievement of a specific target is the best (according to a mathematical model) between the alternatives are possible by using a linear function, with the extant problems i.e. optimize a linear functions are limited by constraints in the form of the equation and a linear inequality [12].

In Linear Programming, there are two kinds of linear functions as follows :

a. The destination function (objective function) is a function that leads analysts to detect goal formulation problems.
b. The function of the constraints that is a direct function of the analyst to know the resources available and the demand for these resources.

Linear programming right used to potentially problematic agriculture horticulture because agricultural resources horticulture limited, a case study the research is plant of a mushroom where linear programming used to optimize advantage existing with problems such as demand was, the production not been able to meet the demand that is, the work force in need, their land less optimized, several variables need in optimize advantage is the data on demand, data production , the number of labor, data their land, data advantage production, the number of seeds and old data grow and inserted into the equation linear.

Common form of linear programming :

Maximize :

a) Objective Function: $Z = c_1x_1 + c_2x_2 + \dots + c_nx_n$ (6)

b) Constraint : (7)

Based on common form of linear programming , so can be concluded :

1. This is linear programming there must be function purpose (expressed with the equation a straight line function z or f (z)) namely benefit sales by must be agriculture.
2. Linear programming must constraints expressed with the equation a straight line , the input in function obstacles in the agricultural sector that affects sales have to less than equal to available resources on the farm.
3. Is that all value x is positively or equals 0. Or in other words , there should not be value of x negative .Thus hence the magnitude of the coefficients input output should not be negative.

Next phase is a method of settlement with simplex .Is a method of procedure simplex algebra that is spatially the iterative moving step by step toward the point of optimal , formulations of a method of simplex [5].

$x_1, x_2, \dots, x_n \geq 0$ is :

$$\begin{array}{lllll}
x_1 & & & + a_{1\cdot m+1} x_{m+1} + .. + a_{1j}x_j + .. + a_{1nxn} = & b_1 \\
& x_2 & & + a_{2\cdot m+1} x_{m+1} + .. + a_{2j}x_j + .. + a_{2nxn} = & b_2 \\
& . & & + \dots\dots + \dots + \dots + \dots + \dots = & \dots \\
& & x_m & + a_{m\cdot m+1} x_{m+1} + \dots + a_{mj}x_j + .. + a_{mnxn} = & b_m \\
& & z & + c_{m+1} x_{m+1} + \dots + c_jx_j + .. + c_{nxn} = & -z_0
\end{array}$$

In a method of simplex function problem with inequality ((≤) to be turned into equation (=), z is the variable the base while a1j, cj, b1 and z0 is the constant .To solution get a optimal then xn to be zero, in the form of canonical solution essentially as follows

$z = z_0$, $x_1 = b_1$, $x_2 = b_2$, ..., $x_m = b_m$, $x_{m+1} = x_{m+2}$, ... $= xn = 0$

In assigning value to optimal there should not be a negative value if there is to be multiplied by (-1) so that the $b_1, b_2, \dots, b_m \geq 0$.

3 Result and Discussion

The result of this research provides display system information on planning the production and scheduling planting patterns by applying a method of programming linear and fuzzy time series values is the selling data January 2012 - April 2017, so that obtained the results of optimize the production and scheduling of planting patterns. The results of data processing fuzzy time series of the results of a prediction demand for months later will be used as a function obstacles in the process of programming linear, by making formulation mathematics the determination of function the purpose and obstacles based on variables for which there are in order to obtain optimize the production and scheduling planting patterns to optimize advantage production with planted the seed mushroom in accordance with an estimated demand months later so mushroom produced according to demand that is.

Initial phase data processing started by noting the selling data, production, profit production , long planting , the amount of labor and there have in the agricultural sector mushroom the selling data January 2012 - April 2017 mixed with fuzzy time series so as to produce data demand for months later. Data demand, histori planting, the amount of labor and there have will be variable

$$\begin{array}{lllllll}
a_{11}x_1 & + & a_{21}x_2 & + \dots & + & a_{m1}x_m & \leq b_1 \\
a_{12}x_1 & + & a_{22}x_2 & + \dots & + & a_{m2}x_m & \leq b_2 \\
. & & . & . & & . & . \\
. & & . & . & & . & . \\
a_{1m}x_1 & + & a_2x_2 & + \dots & + & a_{nm}x_n & \leq b_m
\end{array}$$

$x_1, x_2, \dots, x_n \geq 0$

function obstacles data on advantage production will to

be functions of the purpose of which will formulated in the form of mathematics by using programming linear, of the results of data processing will get some of the scenarios with the lapse of time planting mushroom different, the lapse of time planting start of an interval planting one month to hose planting 4 months to produce sum advantage different also from here function of incorporation fuzzy time series and linear programming namely optimize advantage farmers a mushroom with planting seeds mushroom in accordance with an estimated demand months later so can produce mushroom in accordance with consumer demand.

The planning process the producing and scheduling of planting patterns commenced by determining prediction demand months later using data historical sales a few years back, the data used was the selling data from year January 2012 - April 2017, the process prediction demand by using the method fuzzy time series.

Steps in fuzzy time series is

1. Partition the universe of discourse U = [Dmin - D1, Dmax -[- D2] into even lengthy and equal length intervals ul, u2, ..., um (1)

Based on data of year January 2012 - April 2017, obtained universe of discourse as follows:

D_{min} = 133 and D_{max} = 598, Divided long intervals as many as 7 so obtained u1 – u7 such as table 1.

Table 1. Number of interval.

u1	133	199.4286	1
u2	199.4285714	265.8571	3
u3	265.8571429	332.2857	9
u4	332.2857143	398.7143	9
u5	398.7142857	465.1429	20
u6	465.1428571	531.5714	8
u7	531.5714286	598	12

2. Defines the set of fuzzy in of U with determine the value of nb (negative big), nm (negative medium), ns (negative small), z (zero), ps (positive small), pm (positive medium) and pb (positive big) a curve to describe variation from the sale, as in a picture 1.

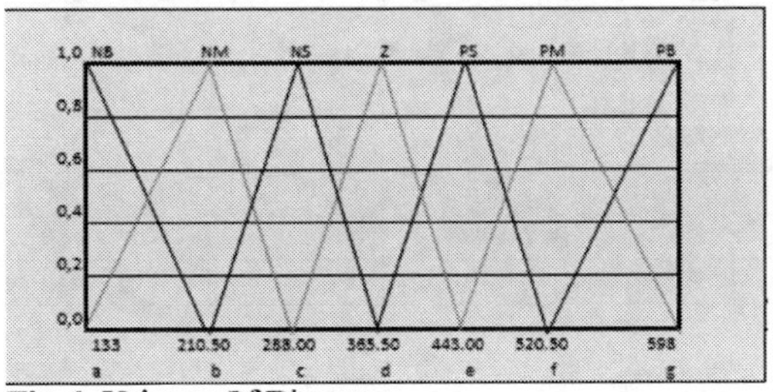

Fig. 1. Universe Of Discourse

Table 2. Fuzzification process

Universe of Discourse U	Fuzzy Set		Fuzzification			Calculation Fuzzy w=1		Defuzzificatio n	
Month	**Year**	**Actual Value**	**NB**	**NM**	**NS**	**Z**	**PS**	**PM**	**PB**
JAN	2017	485	0.00	0.00	0.00	0.00	0.458	0.542	0.00
FEB	2017	585	1.007	0.00	0.00	0.00	0.00	0.168	0.832
MARCH	2017	417	0.614	0.00	0.00	0.335	0.665	0.00	0.00
APRIL	2017	312	1.009	0.00	0.69	0.31	0.00	0.00	0.00

Table 3. Table the results of a prediction fuzzy

Universe of Discourse U		Fuzzy Set		Fuzzification		Calculation Fuzzy w=1		Defuzzification	
Month	**Year**	**Actual Value**	**Y**	**v**	**Prediction value / Fi**	**Error**	**MAD**	**MSE**	**MAPE %**
JAN	2017	485	1.01	386.038	387.048	97.952	97.952	9594.594	4.951
FEB	2017	585	1.007	485.005	486.012	98.988	9798.624	9798.624	5.91
MARCH	2017	417	0.614	584.98	585.594	-168.594	28423.937	28423.937	2.437
APRIL	2017	312	1.009	417.038	418.047	-106.047	11245.966	11245.966	2.942

3. Fuzzification historical data with the w = 1, in table 2 explained the process fuzzification the final period. Namely january 2012 - april 2017. (3)
4. Defuzzification

In table 3 explained the result of a prediction in the final period to be used as variable demand at calculation linear programming and value mape.(4)

After obtained the results of predictive value demand may as many as 418, so predictive value is used as variable consumer demand in function obstacles in calculation linear programming. Besides variable consumer demand that to be functions of obstacles is variable amount of sales, there have, working hours and the amount of labor (the coefficients) and variable advantage production to be functions of the purpose.

Besides the determination of function the purpose and obstacles is scenario optimize consisting of scenario one, two and three where scenario one distance planting mushrooms in one batch have hose planting one month, scenario two have hose planting two months and scenario of 3 have hose planting 3 months and scenario 4 have hose planting four months.

Calculation resolved with the methods simplex who has formulated in the form of mathematical linear model programming, the data is required for the planning the producting and scheduling of planting patterns is presented in Table 4.

Table 4. Data for a period

	Amount of Sales Mushrooms (kg)	**Profit (Rp)**	**Seed (Batch)**	**Coefficient**	**Demand (kg)**
X1	485	2 907 000	970	0.192	387
X2	585	3 527 000	1170	0.16	486
X3	417	2 485 400	834	0.222	586
X4	312	1 834 400	624	0.294	418

The next step is to formulating into mathematical form linear programming.

1. Formulation programming of linear programming scenario 1

Objective Fuction

Z = 2907000X1+3527000X2+2485000X3 + 1834000X4 (profit) (6)

Constraints (7)

0.19X1 + 0.16X2 + 0.22X3 + 0.29X4 <= 288 (koefisien)
970X1 + 1170X2 + 834X3 + 624X4 <= 5000 (bibit)
485X1 <= 387
1070X2 <= 486
1487X3 <= 586
1314X4 <= 418

2. Formulation programming of linear programming scenario 2

Objective Fuction

Z = 2907000X1+3527000X2+2485000X3 + 1834000X4 (profit) (6)

Constraints (7)

0.19X1 + 0.16X2 + 0.22X3 + 0.29X4 <= 288 (coefficient)
970X1 + 1170X2 + 834X3 + 624X4 <= 5000 (bibit)
485X1 <= 387
1070X2 <= 486
1002X3 <= 586
729X4 <= 418

3. Formulation programming of linear programming scenario 3

Objective Fuction

Z = 2907000X1+3527000X2+2485000X3 + 1834000X4 (profit) (6)

Constraints (7)

0.19X1 + 0.16X2 + 0.22X3 + 0.29X4 <= 288 (coefficient)
970X1 + 1170X2 + 834X3 + 624X4 <= 5000 (bibit)
485X1 <= 387
585X2 <= 486
417X3 <= 586
312X4 <= 418

4. Formulation programming of linear programming scenario 4

Objective Fuction

Z = 2907000X1+3527000X2+2485000X3 + 1834000X4 (profit) (6)

Constraints (7)

0.19X1 + 0.16X2 + 0.22X3 <= 288 (coefficient)
970X1 + 1170X2 + 834X3 <= 5000 (bibit)
485X1 <= 387
585X2 <= 486
417X3 <= 586

From scratch linear programming completed by method simplex in the system information planning production and scheduling planting patterns obtained a result of each scenario shown in table 5.

Table 5. Comparison every scenario (SC)

SC	Interval Planting (mount)	Number of seeds grown (mount)				Profit (Rp)
		1st	2nd	3rd	4th	
1	1	774	531	164	115	5.327.200
2	2	774	531	487	357	6.426.900
3	3	774	972	1172	836	11.200.000
4	4	774	972	1172	-	8.742.400

Calculation scenario 1, 2, 3, and 4 showed optimal results found in scenario 3 with a hose planting three months. In an analysis of solution optimal the number of planting in scenario 3 in proportion to the number of requests that is while in scenario 1 and 2 scenario the number of planting excess and also on scenario 4 the number of planting in 1 year only 3 cropping.

Results obtained in scenario 3 as much Rp 11.200.000,00 with the number of mushroom that must be produced in a period as many as x1: 775, x2: 972, x3: 1172 and x4: 836, recommendations the number of mushroom should be planted based on formulation programming linear and calculation with the methods simplex namely the results of multiplied by the number of seeds grown, planting patterns to be demonstrated in table 6.

Table 6. Planting patterns distance planting three months

	Month											
X1	**J**	**F**	**M**	**A**	**M**	**J**	**J**	**A**	**S**	**O**	**N**	**D**
X2	T:775	P	P	P								
X3					T:1173	P	P	P				
X4									T:837	P	P	P

https://doi.org/10.1051/e3sconf/20183110004

4 Conclusions

Based on the research done and discussion information system planning the production and scheduling planting patterns this can provide recommendations planting patterns and the number of mushroom should be planted in a period by farmers mushroom, the life cycle mushrooms in a period is four months, the number of planting adapted to the number of requests is there who had previously been predicted by using fuzzy time series, the results showed of four scenario hose planting obtained a value on first scenario distance planting a month Rp 5.327.266,00, on second scenario, The distance planting two months Rp 6.426.950,00 , value third scenario , the distance planting three months worth Rp 11.200.000,00 , and the distance planting four months worth Rp 8.742.400,00 based on the results of scenario one, two, three and four obtained value optimal in scenario to three Rp 11.200.000,00 with seeds planting mushroom do not all planted in early , but broken down with seeds planting next given a distance of three months as many as planting the first month 775, second its 972, third 1172, and fourth 836.

References

1. S. A. Raza, F. C. Abdullakutty, and S. Rathinam, Applied Mathematical Modelling **40**, 8446 –8463 (2016).
2. N. M. C. Garcia, A. G. B. Lozano, and Y. A. R. Solis, Computers and Electronics in Agriculture **107**, 20 – 28 (2015).
3. B. B. Pal, M. Kumar, and S. Sen, *Second International conference on Computing, Communication and Networking Technologies* (2010).
4. G. Dantzig, *Linier Programming and extensions*. The Rand Corporation (1963).
5. A. G. Martín, C. Pozo, and G. G. Gosalbez, Land Use Policy **48**, 515 – 524 (2015).
6. C. D. Rocco, and R. Morabito, Computers and Electronics in Agriculture **127**, 763 – 774 (2016).
7. Q. Song, and B., S. Chissom, In Fuzzy Sets and System **54**, 1 - 9 (1993).
8. J.R. Munhoz, R. Morabito, Comput. Electron. Agric **107**, 45-57. (2014).
9. J. Heizer, B. Render, *Operation Management Tenth edition*. US : Pearson (2005).
10. N. A. Campbell, *Biology Tenth edition*. US : Pearson (2002).
11. R. A. R. Junqueira, and R. Morabito, Computers and Electronics in Agriculture **84**, 100 – 110 (2012).
12. V. Democ, Z. Vyhnalikova, P. Alac, Procedia Economics and Finance **34**, 477 – 484 (2015).

Hybrid Method for Mobile learning Cooperative: Study of Timor Leste

Ofelia Cizela da Costa Tavares[1, 2,*], *Suyoto*[2, 1], and *Pranowo* [1]

[1]Magister Informatics Engineering, Universitas Atma Jaya Yogyakarta, Yogyakarta, Indonesia 55281
[2]Department of Computer Science, Faculty of Engineering and Science, Dili Institute of Technology, Dili – Timor Leste

Abstract. In the modern world today the decision support system is very useful to help in solving a problem, so this study discusses the learning process of savings and loan cooperatives in Timor Leste. The purpose of the observation is that the people of Timor Leste are still in the process of learning the use DSS for good saving and loan cooperative process. Based on existing research on the Timor Leste community on credit cooperatives, a mobile application will be built that will help the cooperative learning process in East Timorese society. The methods used for decision making are AHP (Analytical Hierarchy Process) and SAW (simple additive Weighting) method to see the result of each criterion and the weight of the value. The result of this research is mobile leaning cooperative in decision support system by using SAW and AHP method. Originality Value: Changed the two methods of mobile application development using AHP and SAW methods to help the decision support system process of a savings and credit cooperative in Timor Leste.

1 Introduction

In today's modern world, cooperatives are well-known in various countries, with the aim of helping to process fast and good savings and loans. As we have seen, cooperatives are an organization that has a business owned and operated by a person for the sake of the common good. Seeing that the people of Timor Leste have not understood well the process of a cooperative system with good savings and loans, the community still lends money and interest with manual processes without good administrative processes [1].

Community activities in Timor Leste's country in credit unions were not well understood by the newly independent Timor-Leste state in 2002, and the lending process that existed to the community was still through a manual process, and without going through better processes, resulting in a lack of community understanding Timor Leste in the use of cooperatives. So in this study, the author uses the merger of two methods such as SAW and AHP in the calculation process to get a decision on who will get the loan in the cooperative and cooperative learning contained in the mobile application [2]. Already we know that credit cooperatives are non-bank groups in various developed countries and together with the fact that credit cooperatives are an important part of the financial system in Timor Leste [3].

Based on an analysis of the simplified savings and loan cooperative business activities in Timor Leste, this research is how to change the understanding of the people of Timor Leste in the learning process of financial cooperatives with a credit decision support system with the help of combining methods such as SAW and AHP to decide a person Can receive loans in the cooperative [4][5][6]. Movement of cooperative learning changes from applications that will be run from the merger methods SAW and AHP is one of the means for the community can be well understood process of learning good savings and loan cooperatives. In this study the author also uses decision support system to decide who will receive the loan.

The results of the decision of this WEB application will help the process of inputting data of new cooperative members, and the results of its decisions through mobile applications in which in the form of information transactions and cooperative learning. Learning is an aid provided by educators to occur the process of acquiring knowledge and knowledge and mastery of skills and the formation of attitudes and beliefs in learners. In other words learning aims to help learners to learn well. So cooperative learning with mobile is assisted by decision support system using SAW and AHP method [7]. To make the learning process of credit cooperatives in Timor Leste, the results of the application using decision support system with SAW and AHP method, then the results of the calculation of both methods are included in web and mobile applications for cooperative learning process.

2 Literature Review

Previously there have been several studies conducted and cannot be separated from the results of earlier research from the topic of research on credit cooperatives:

* Corresponding author: ofeliacizela0305@gmail.com

There are several studies that have been conducted to discuss the impact of computerization on the efficiency of cooperatives [8], Using learning analysis to make changes to credit cooperative knowledge and this study also shows that the application of technology such as computers that encourage the efficiency of credit cooperatives and see the results of the implementation of the important community and efficiency.

Research conducted by David P. Ely and Kenneth J. Robinson is the banking industry is the chance that credit union is engaged in the loan business related to the activity and demand for small business loans in the local market [9]. In developing countries credit cooperatives and credit cooperatives are thriving with laws and branched cooperatives, so in the same way as the country's advanced experience, research conducted by Ryland A. Taylor on the role to be played by Community responsibility to the rapid economic development of credit cooperatives [10].

In this paper, the credit learning system in the proposed multi-stage stage for credit risk evaluation issues according to Yu and others discusses that the results obtained show that the proposed learning model can give promising solutions to the problem of credit risk evaluation that multi-type learning techniques Which are proposed to be of great potential use in the application of problem classification and so forth [11]. Research conducted by Begoña Gutiérrez-Nieto, Carlos Serrano-Cinca which says that In a bank system, a microfinance institution or a particular credit cooperative provides socially responsible loans, it can present a credit score system and model check the social and financial aspects of the borrower as well The loan proceeds will be managed through a system that is very advanced and accessible anywhere through the system that has been made [12].

In learning to consider person and group credit, research conducted by Erin Carrol, Robert L. Williams's in 2010 discusses cooperative learning in a State to increasing academic engagement and achievement, any cooperative learning procedure may be a special incollegecourses Very importantly related in the same way [13].

The SAW (Simple additive weighting) and AHP (analytical hierarchy Process) methods in decision-making are very helpful for the process of calculating the weighting of the value of the cucumber in the results of the credit decision, then the research conducted by Rizka ella Setyani in 2016 and Seyit Ali Erdogan in 2017 Discusses the aid of the SAW method that can see the summation of the criteria and the greatest value for the crediting decision [14] [15].

In a study [16] conducted by Idris and Bunyamin in 2013 said that the need to learn mobile learning is increasingly important. Its use is very easy and access from mobile makes them more important than ever, this study has benefited greatly in expanding the mobile learning needs and meeting new learning conditions.

3 Theoretical Basis

3.1 Mobile app

Mobile app is a system of application development process for handheld devices like PDA, which is also a company's digital system or mobile phone. In this application system is located on the fellow phone manufacturing and then downloaded by customers from the application and distribution tools of a mobile software [17].

In a faster mobile app development system there is the idea that the speed of mobile devices is said to be fast or slow depending on the comparison of development efforts from other applications. By using mobile applications, people can do various activities that can help their work, because viewing from the results of 70% presentation by phone users, with the features of games, music player to video player. From the application system contained in the phone proves that the use of mobile applications easy to use anytime and anywhere.

3.2 Learning Credit Cooperative

Learning credit itself is a process that must be followed by the community with the aim for the good of society. From this the author will do research to see that the people of Timor Leste today still need cooperative learning to make a good savings and loan process. This Web and Mobile application was created to help the East Timorese community about the credit cooperative [18].

3.3 Decision Support System

Decision support systems are part of computer-based information systems used to support decision-making within an organization or company. In this study we discussed the calculations shown to represent the thinking processes of decision-making often occur with multi objective decision support tools, and these calculations most often use weighting factors for each want or goal to combine utility values into a scale to allow alternative rankings. And illustrate how the choice of weighting factors is meant to be of equal value or weight [19].

4 Methodology and Implementation

4.1 Method

This research will have a concept to understand the learning process of credit cooperative through credit decision support system by using SAW (Simple Additive Weighting) and AHP (Analytical Hierarchy Process) method with the result of mobile application process on lending information process and learning process that exist in the system The. At this stage it is necessary to build a system with collaborative learning concepts that can help the community with the help of SAW and AHP methods. Introduction to credit cooperative learning

https://doi.org/10.1051/e3sconf/20183110005

system will be built in several stages in the form of Web system for cooperative and mobile cooperative part. In a study [20] [21] [22], which discusses that now this smartphone a good friend is preferred for users of the desktop or notebook because it sees that by using smartphones more easily understood and run by mobile users very well by the community especially in the learning process to be more effective? In a cooperative system only use the database on the web to input data cooperative members who make loans in the cooperative. The role of this system is how to introduce a mobile system that will be diverted to aid members of the cooperative process, which can be used by consumers to retrieve information about the cooperative's lending and learning. The system used with the aid of both methods in the information process of the mobile aims to find information from lending and cooperative learning.

Collaborative learning app process

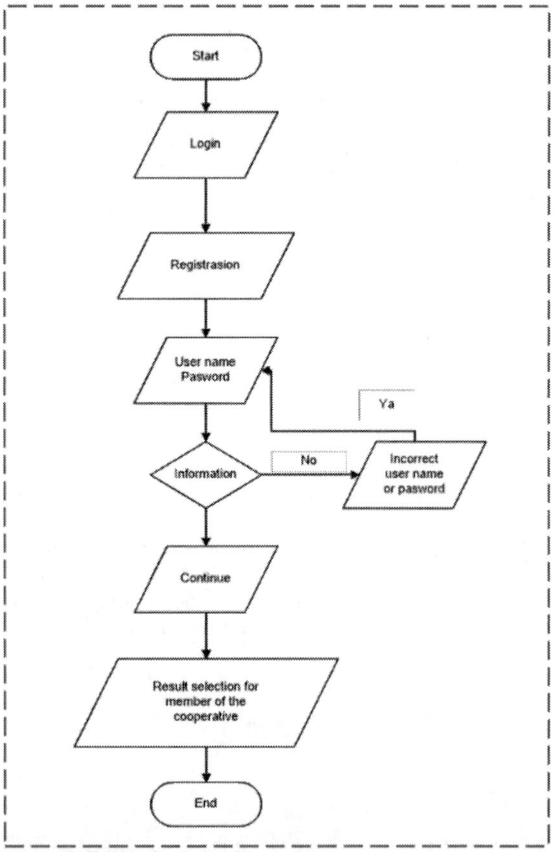

Fig. 1. Collaborative Learning app Process

The system used with the aid of both methods in the information process of the mobile aims to find information from lending and cooperative learning, so that consumers do lending (credit) but there is also cooperative credit learning as the best means of lending money and each consumer can use the calculation of existing methods on the system for decision-making of credit, and information that can be accessed through the phone of each user To be able to see the results of transactional information and cooperative learning, and thus the system to be used for decision-making is proposed for a good cooperative learning process for East Timor people.

Originality Value: Changed the two methods of mobile application development using AHP and SAW methods to help the decision support system process of a savings and credit cooperative in Timor Leste.

4.1 Decision Support System

In the decision support system that use to run on the learning process is a member of the cooperative that can make money or not, then the credit decision support system using AHP (Analytical Hierarchy Process) and SAW (Simple Additive Weighting) which is also a method that can solve complex situation that unstructured, determines which variable has the highest priority to influence the outcome of the situation [23], And look for the sum of weighted ratings on each alternative across all attributes and must the process of normalizing the decision matrix to a Scale that is comparable to all existing alternative ratings [24] [25] [26]. To process the calculation can be done through the existing criteria.

$$\mathrm{CI} = \frac{\lambda maks - n}{\mathrm{n} - 1}$$

Information:
n = the number of criteria or sub criteria
CI = consistent Index
RI = Random Index

$$CR = \frac{\mathrm{CI}}{\mathrm{RI}}$$

(AHP method)

$$Vi = \sum_{j+1}^{n} (Wj_x r_{ij})$$

(SAW methods)

5 Results and Implementation

5.1 The process of calculating SAW and AHP methods

In this process will show how one can receive credit results from cooperatives in Timor Leste. In this case some people will apply for credit in a cooperative, and the results of the calculation will decide someone will first receive a credit loan.

Table 1. Warranty Criteria

Warranty Ci	Value
$ 100.00	0.25
$ 1000.00	0.5
$ 2000.00	0.75
$ 3000.00	1

Table 2. Loans Criteria

Loans C2	Value
$ 10.00	0.25
$ 500.00	0.5
$ 2000.00	0.75
$ 5000.00	1

Table 3. Instalment Criteria

Instalments C3	Value
$ 10.00	0.25
$ 50.00	0.5
$ 150.00	0.75
$ 300.00	1

Table 4. Timeframe criterion

Time period C4	Value
3 Months	0.25
12 Months	0.5
24 Months	0.75
30 Months	1

Table 5. Age Criteria

Age C4	Value
> 15 years	0.25
> 20 years	0.5
> 30 years	0.75
> 40 years	1

Example Calculation
There are data of cooperative members who will take credit.

Francisco:
Warranty = $ 1000.00
Loans = $ 400.00
Instalment = $ 40.00
Time Period = 10 Months
Age = 25 Years

Efigenia:
Warranty = $ 1500.00
Loans = $ 600.00
Instalment = $ 60.00
Time Period = 12 Months
Age = 30 Years

Maria:
Warranty = $ 2000.00
Loans = $ 500.00
Instalment = $ 400.00
Time Period = 15 Months
Age = 20 Years

In this process will (table 1 – table 5) show how one can receive credit results from cooperative in Timor Leste.

Table 6. Table of credit applicants

alternative	Criteria				
	C1	C2	C3	C4	C5
Francisco	0.25	0.25	0.25	0.25	0.5
Efigenia	0.25	0.5	0.5	0.25	0.5
Maria	0.25	0.25	0.25	0.5	0.25

Calculation
V11 = 0.25/Max{0.25;0.25;0.25}= 0.25/0.25 =1
V12 = 0.25/Max {0.25;0.5;0.25} = 0.25/0.5=0.5
V13=0,25/Max{0,25;0,5;0,25} = 0,25/0,5 =0,5
V14=0,25/Max{0,25;0,25;0, = 0,25/0,5 =0,25
V15=PP0,5/Max{0,5:0,5;0,25} = 0,5/0,5 =1
V21=0,25/Max{0,25:0,25;0,25} = 0,25/0,25 =1
V22=0,5/Max{0,5:0,25;0,25} = 0,5/0,25 =2
V23=0,5/Max{0,25;0,5;0,25} = 0,5/0,5 =1
V24=0,25/Max{0,25;0,25;0,5} = 0,25/0,5 =0,5
V25=0,5/Max{0,5:0,5;0,25} = 0,5/0,5 =1
V31=0,25/Max{0,25:0,25;0,25} = 0,25/0,25 =1
V32=0,25/Max{0,25:0,5;0,25 = 0,25/0,5 =0,5
V33=0,25/Max{0,25;0,5;0,25} = 0,25/0,5 =0,5
V34=0,5/Max{0,25;0,25;0,5} = 0,5/0,5 =1
V35=0,25/Max{0,5:0,5;0,25} = 0,25/0,5 =0,5

From the above calculation can be formed matrix R
R=((1&0.5&0,5&0,25&1@1&2&1
&0,5&1@1&0,5&0,5&1&0,5))
W=((0,4&0,25&0,2&0,1&0,05))

Perform Process Ranking

V1 =(0,4)(1) + (0,25)(0,5) + (0,2)(0,5) + (0,1)(0,25) + (0,05)(1) = 0,4 + 0,125 + 0,1 + 0,025 + 0,05 = 0,7
V2 = (0,4)(1) + (0,25)(2) + (0,2)(1) + (0,1)(0,5) + (0,05)(1) = 0,4 + 0,5+ 0,2 + 0,05 + 0,05 = 1,2
V3 = (0,4)(1) + (0,25)(0,5) + (0,2)(0,5) + (0,1)(1) + (0,05)(0,5) = 0,4 + 0,125 + 0,1 + 0,1 + 0,025 = 0,75
The Greatest Value falls on V2 (Efigenia) is the alternative chosen as the best alternative.

From the alternative values that have been determined by the method then the credit granting process will be given to the results that have been determined.

Value of Originality: Change the two methods of mobile application development using AHP and SAW methods to aid the decision-making process of saving and loan cooperative system in Timor Leste.

5.2 The result of application implementation

As discussed earlier, at this stage is the system process used to register cooperative members through cooperative management for new cooperative member data. From members of the cooperative management in determining the calculation of SW and AHP methods so that it can be implemented.

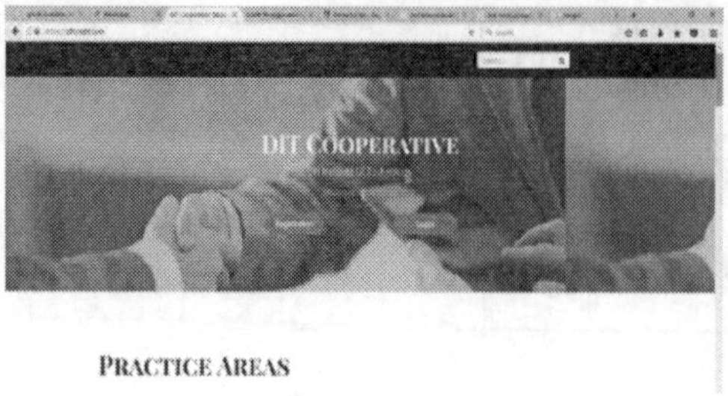

Fig. 2. Main Menu.

On the main page (Figure 2), this web application will discuss the form of filling to become a member of the cooperative and is entitled to get a cooperative loan. Prospective members of the cooperative will click the register button and will perform the registration process

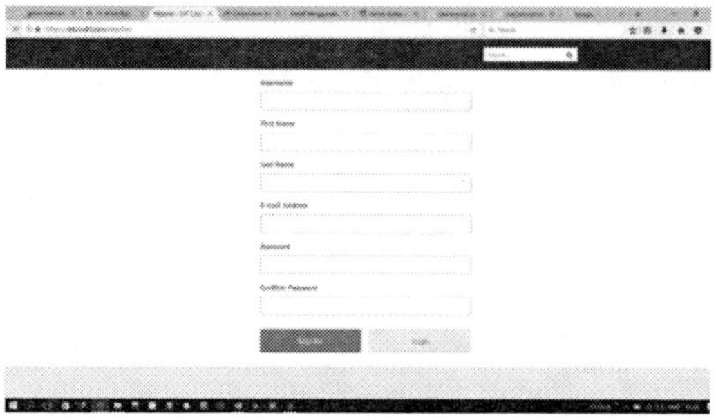

Fig.3. New cooperative member registration.

Ina Figure 3 shows new cooperative member registration. It discusses how a person wants to be a new member in a cooperative, at a stage someone will click on the register button and fill out the form provided by the cooperative management and create their own user name and password, for the form there are two languages, English and Tetun from Timor Leste. Fill out the form to see if the new member can apply for the cooperative ask.

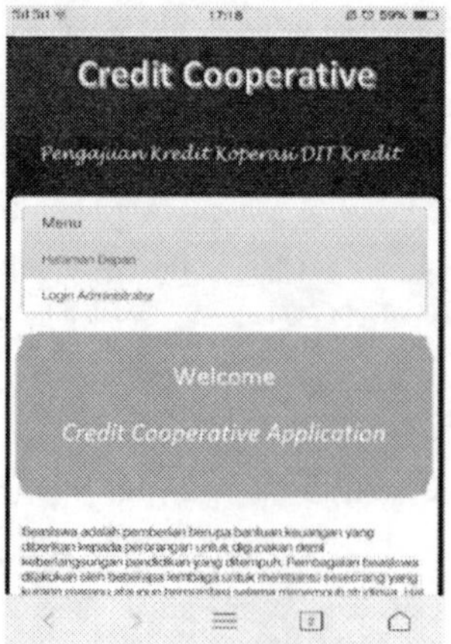

Fig. 4. Login in Mobile for Result of Cooperative Notification.

On this page (Figure 4), hold this new members will be informed about the credit application results from each criterion that has been applied by the cooperative. In this stage will decide the results of each criteria that has been automatically on the system executed to decide who is entitled to get the credit application first. So the results can be seen on the mobile application.

Normalization

NO	NAMA	C1 C2 C3 C4 C5
1	Efigenia da Costa	1 1 1 0.667
2	Francisco Vitor Tavares	0.660.330.250.333
3	Maria Celia	0.660.660.5 1 0.5
4	Ofelia Cizela da Costa Tavares	0.660.660.5 0.660.5

The Final Result

NO	NAMA	NILAI	RANK
1	Efigenia da Costa	12.001	1
2	Maria Celia	8.668	2
3	Ofelia Cizela da Costa Tavares	7.669	3

Fig. 5. Result Selection for member of the Cooperative.

In Figure 5 shows the final result of for member of the cooperative. This will discuss about the selection of criteria that have been determined by the cooperative with the calculation using two methods namely SAW method and AHP method to decide who is entitled to receive credit applications in a savings and loan cooperatives. The members of the cooperative will see the results are mobile applications with each user name and password.

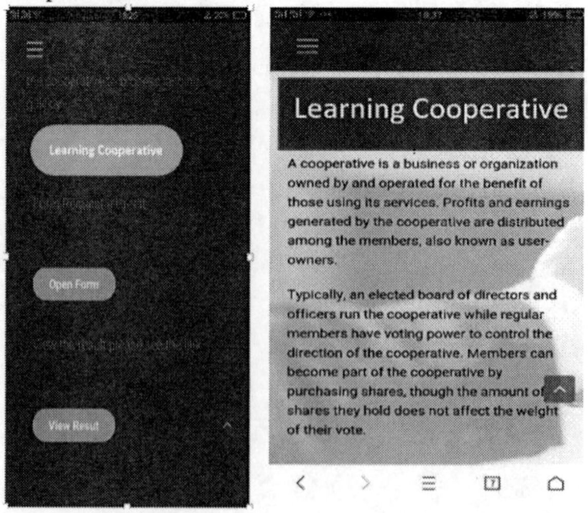

Fig. 6. Cooperative Learning.

On the third page (Figure 6) of the mobile app, This will showcase the cooperative learning process in mobile applications for all members of the cooperative, with the aim that each member is not only to lend to a cooperative but the people of Timor Leste can also understand well how a cooperative is in good savings and loan process Takes a long time, so that the learning process will change the changes in East Timorese society not to lend without proper administrative process.

6 Conclusions And Recommendation

In this paper, a lesson for the people of Timor-Leste to get a good lending process, such as credit cooperatives that use mobile apps for the transformation of East Timorese society, the transformation of East Timorese society to a safe and comfortable savings and loan process. There are learning methods that can help the decision support system process, and the method is a hybrid method consisting of SAW and AHP methods. This paper is assisted by applications that will be used by cooperative boards and members of the cooperative, so that the data included an analysis that can produce a review of the problems being faced by the people of Timor Leste on cooperative learning, the authors concluded by using information technology using decision support system Which is very supportive Helping the use of a good cooperative. With the expansion of cooperatives that have expanded with 60% presentation, learning cooperative savings and loan by using mobile applications. In cooperative learning with the help of SAW and AHP methods will make a change of understanding of cooperatives for the people of Timor Leste. In today's modern, with the view that smart phone is very used by the society around the world, the writer took the initiative to help the learning process of the people of East Timor by using mobile application well and virtual.

7 Contribution

In improving the quality of society in addition to improving the welfare of the quality of life of the people of Timor Leste. So the main contribution is to change the first mindset that comes to mind or the model of society about good saving and lending process. With these problems, the authors compose information communication technology is a cooperative learning process assisted by decision support system with SAW and AHP method for the process of calculating the assessment of the provision of cooperatives. Decision support systems also consist of computer systems designed to aid or affect the needs of nominal workers in the savings and loan sections of a cooperative, so that the people of East Timor will also experience major changes in cooperative learning as well.

Reference

1. A. Saretto and H. E. Tookes, Rev. Financ. Stud., vol. **26**, no. 5, pp. 1190–1247, (2013).
2. C. Serrano-cinca, B. Gutiérrez-nieto, and N. M. Reyes, J. Clean. Prod., (2015).
3. T. Harris, Expert Syst. Appl., vol. **42**, no. 2, pp. 741–750, (2015).
4. B. A. Alyoubi, Procedia Comput. Sci., vol. **65**, no. Iccmit, pp. 278–284, (2015).
5. S. S. Hosseinian, H. Navidi, and A. Hajfathaliha, Gr. Decis. Negot., vol. **21**, no. 3, pp. 233–254, (2012).
6. A. Afshari, M. Mojahed, and R. Yusuff, Int. J. Innov. Manag. Technol., vol. **1**, no. 5, pp. 511–515, (2010).
7. W. H. Wu, Y. C. Jim Wu, C. Y. Chen, H. Y. Kao, C. H. Lin, and S. H. Huang, Comput. Educ., vol. **59**, no. 2, pp. 817–827, (2012).
8. C. Nelson, Academe, vol. **96**, no. 1, pp. 10–14, (2010).
9. D. P. Ely and K. J. Robinson, J. Financ. Serv. Res., vol. **35**, no. 1, pp. 53–80, (2009).
10. R. A. Taylor, Ann. Public Coop. Econ., vol. **45**, no. 2, pp. 105–118, (2010).
11. L. Yu, W. Yue, S. Wang, and K. K. Lai, Expert Syst. Appl., vol. **37**, no. 2, pp. 1351–1360, (2010).
12. B. Gutiérrez-Nieto, C. Serrano-Cinca, and J. Camón-Cala, J. Bus. Ethics, vol. **133**, no. 4, pp. 691–701, (2016).
13. E. Carroll, R. L. Williams, and B. Hautau, J. Behav. Educ., vol. **15**, no. 4, pp. 191–202, (2010).
14. R. E. Setyani and R. Saputra, Procedia - Soc. Behav. Sci., vol. **227**, no. November 2015, pp. 378–386, (2016).
15. S. A. Erdogan, J. ?aparauskas, and Z. Turskis, Procedia Eng., vol. **172**, pp. 270–276, (2017).
16. I. Educational and T. Conference, vol. **103**, pp. 685–694, (2013).
17. Y. M. Huang, Y. L. Jeng, and T. C. Huang, Educ. Technol. Soc., vol. **12**, no. 2, pp. 163–175, (2009).
18. M. Bal, M. F. Amasyali, H. Sever, G. Kose, and A. Demirhan, Sci. World J., vol. **2014**, pp. 1–15, (2014).
19. R. L. Smith and G. J. Ruiz-Mercado, Clean Technol. Environ. Policy, vol. **16**, no. 4, pp. 749–755, (2014).
20. Z. W. Abas, T. Lim, H. Kaur, D. Singh, W. W. Shyang, and K. Lumpur, vol. **20**, no. 1, pp. 13–15, (2009).
21. Suyoto, Prasetyaningrum T., Gregorius R.M., In: Kim T. et al. (eds) *Multimedia, Computer Graphics and Broadcasting. Communications in Computer and Information Science,* vol **262**. Springer, Berlin, Heidelberg. 10.1007/978-3-642-27204-2_27. pp 217-226, (2011).
22. Suyoto, Suselo T., Dwiandiyanta Y., Prasetyaningrum T., In: Kim T. et al. (eds)

Multimedia, Computer Graphics and Broadcasting. Communications in Computer and Information Science, vol **262**. Springer, Berlin, Heidelberg. 10.1007/978-3-642-27204-2_28. pp 227-236, (2011).

23. A. Rezaei, M. Shayestehfar, H. Hassani, and M. R. T. Mohammadi, Environ. Earth Sci., vol. **74**, no. 4, pp. 3191–3205, (2015).
24. M. Danner et al., Int. J. Technol. Assess. Health Care, vol. **27**, no. 4, pp. 369–375, (2011).
25. Sharifah Nadiyah Razali,Faaizah Shahbodin,Mohd Hafiez Ahmad and Helmi Adli Mohd Nor, *International Journal on Advanced Science, Engineering and Information Technology*, vol. 7, no. 3, pp. 799-807, (2017). [Online]. Available: http://dx.doi.org/10.18517/ijaseit.7.3.1310.
26. Ahmad Nazari Mohd Rose,Mohd Isa Awang,Fadhilah Ahmad, Nurnadiah Zamri, Mohamad Afendee Mohamed and Mustafa Mat Deris, *International Journal on Advanced Science, Engineering and Information Technology*, vol. 7, no. 3, pp. 1032-1037, (2017). [Online].Available:http://dx.doi.org/10.18517/ijaseit.7.3.1610.

The Decision Support System (DSS) Application to Determination of Diabetes Mellitus Patient Menu Using a Genetic Algorithm Method

Nia Zuliyana[1,*], *Jatmiko* Endro Suseno[2], and *Kusworo* Adi[2]

[1]Master Program of Information System, School of Postgraduate Studies, Diponegoro University, Semarang - Indonesia
[2]Department of Physics, Faculty of Science and Mathematics, Diponegoro University, Semarang – Indonesia

Abstract. Composition of foods containing sugar in people with Diabetes Mellitus should be balanced, so an app is required for facilitate the public and nutritionists in determining the appropriate food menu with calorie requirement of diabetes patient. This research will be recommended to determination of food variation for using Genetic Algorithm. The data used is nutrient content of food obtained from *Tabel Komposisi Pangan Indonesia* (TKPI). The requirement of caloric value the patient can be used the PERKENI 2015 method. Then the data is processed to determine the best food menu consisting of energy (E), carbohydrate (K), fat (L) and protein (P) requirements. The system is comparised with variation of Genetic Algorithm parameters is the total of chromosomes, Probability of Crossover (Pc) and Probability of Mutation (Pm). Maximum value of the probability generation of crossover and probability of mutation will be the more variations of food that will come out. For example, patient with gender is women aged 61 years old, height 160 cm, weight 55 kg, will be resulted number of calories: (E=1621.4, K=243.21, P=60.80, L=45.04), with the gene=4, chromosomes=3, generation=3, Pc=0.2, and Pm=0.2. The result obtained is the three varians: E=1607.25, K=198.877, P=95.385, L=47.508), (E=1633.25, K=196.677, P=85.885, L=55.758), (E=1630.90, K=177.455, P=85.245, L=64.335).

1. Introduction

In 2005 there were 200 million people with diabetes and is expected to increase to 333 million people in 2025. Countries like India, China, the United States, Japan, Indonesia, Pakistan, Bangladesh, Italy, Russia and Brazil is the top 10 countries with the largest population of diabetes [1]. The development of Diabetes Mellitus disease causes concern throughout the world, especially people who have nutritional transition marked by reduced traditional food intake and increased intake of instant food or fast food.

In the American Diabetes Association (ADA) research concluded that there are no more effective nutrients than carbohydrate, protein and fat. The proportion of macro nutrition should be tailored to each individual need to meet metabolic goals in a person with Diabetes Mellitus [2]. Diabetes Mellitus occurs sustainably, nutritional therapy in Diabetes Mellitus is effective for improving blood sugar control and other metabolic functions. In general, the energy needs of individuals with Diabetes Mellitus is not different from normal people.

Paradigm that happened in society is patient with Diabetes Mellitus should not consume sugar, but sugar is very needed for body metabolism process. Metabolism that occurs in the body will produce energy that will be needed for the activity. Lack of intake of sugar in a person will result in the occurrence of hypoglycemia, a condition in which blood sugar levels dropped dramatically. It would be dangerous and even fatal. Diabetes Mellitus sufferers should continue to consume sugar in sufficient quantities, no more and no less.

Therefore, the determination of energy needs in people with Diabetes Mellitus is considered very important. How to determine it is by using the formula of the PERKENI 2015. The problem that exists today is that people still rarely use the formula and more often use the formula approximately. So the provision of diet to patients with Diabetes Mellitus is not accurate. Hospitals have set the PERKENI 2015 formula but when determining the portion of the menu still uses the approximate formula because it is too complex, and it takes time for the community or nutritionist to calculate the energy needs of each patient each day.

The Genetic Algorithm has been widely used in various optimizations, including numerical optimization and combinatorial optimization problems. The advantages of Algorithm methods from other methods are capable of handling various types of objective functions with constraints on discrete, continuous, or mixed search spaces. The advantages possessed by

* Corresponding author: zuliyanania@student.undip.ac.id

Genetic Algorithm are selection evolution operations that work probabilistically against several possible solutions to make genetic algorithm very effective in global optimization. The genetic algorithm has a fairly good flexibility and is sufficiently possible to combine it with other methods and make a more efficient implementation of a particular problem [3].

From the problems and conditions that have been discussed earlier, it is necessary an information system to make the determination of food menu of people with diabetes mellitus using Genetic Algorithm. Genetic Algorithm is used to randomize food based on the calories needed.

2. Theoretical Framework

2.1. Diabetes Mellitus

Diabetes Mellitus (DM) is a collection of symptoms that arise in someone who has increased blood glucose levels due to deficiency of insulin hormone in absolute or relative. The implementation of the diet should be accompanied by physical exercise and changes in food behavior. According to the consensus of Management of Diabetes Mellitus in Indonesia by the association of Endocrinology of Indonesia, Diabetes Mellitus disease is divided into 4 groups, namely: Type I and Diabetes Mellitus, Gestational Diabetes Mellitus, and other types [4].

Determination of food for people with Diabetes disease aims to maintain blood glucose levels to near normal, maintain normal serum lipid levels, and provide enough energy to maintain or achieve normal weight. The dietary requirements of inner Diabetes disease [4] are as follows:

a. Foods with sufficient energy correspond to the total energy required by the patient to achieve and maintain a normal weight.
b. Food with enough protein, ie 10-15% of total energy needs in a day. Fat is 20-25% of total energy needs in a day. And carbohydrates are the rest of the total energy requirement, which is 60-70%.
c. The use of pure sugar in beverages and foods is not allowed unless the amount is a bit as a spice. When the glucose level is controlled, it is allowed to consume pure sugar to 5% of total energy needs.
d. Vitamins and minerals.
e. Distribution of food in a day:
   1. Breakfast: 20% x Total Energy
   2. Food interlude morning: 10-15% x Total Energy
   3. Lunch: 30% x Total Energy
   4. Interlude afternoon snack: 10-15% x Total Energy
   5. Dinner: 25% x Total Energy

2.1.1 Food List and Nutrient Composition

Data of nutritional composition of food used in this research taken from the Indonesian Food Composition Table (TKPI) and report on the results of research on the composition of food nutrients carried out in the Center for Research and Development of Nutrition and Food Ministry of Health.

2.1.2 Consensus Formulas PERKENI 2015

Consensus formula 2015 is used to calculate the nutritional needs of patients with Diabetes Mellitus. The formula is as follows [5]:

1. Energy

The main components that determine energy demand are Basal Metabolic Rate (AMB) or Basal Metabolic Rate (BMR) influenced by age, gender, weight, and height. To determine BMR as follows:

$$\begin{aligned} &BMR\ for\ men = 30\ x\ Ideal\ weight \\ &BMR\ for\ women = 25\ x\ Ideal\ weight \end{aligned} \tag{1}$$

With the calculation of Ideal Weight using Broca's modified formula:

- For men again with Height ≥ 160 cm and women with Height ≥ 150 cm. The formula used is:

$$Ideal\ weight = 90\%\ x\ (Height - 100) \tag{2}$$

- For men again with Height <160 cm and women with Height <150 cm. The formula used is:

$$Ideal\ weight = Height - 100$$

$$\begin{aligned} Energy = (BMR + Activity\ Factor) - Age\ Factor + \\ Stress\ Factor +/- Body\ Correction\ Factor \end{aligned} \tag{3}$$

Table 1. Activity Factor

Activity Level	Formula
Bedrest	10% from BMR
Mild	20% from BMR
Medium	30% from BMR
Heavy	40% from BMR

Table 2. Stress Factor

Stress Factor	Value
No stress, normal nutritional status	1.2-1.3
Mild Stress: inflammation of the gastrointestinal tract, cancer, effective surgery, trauma, post minor surgery	1.3-1.4
Medium stress: sepsis, bone surgery, burns, liver disease, major surgery post	1.4-1.5
High Stress: HIV AIDS + complications, multisystem surgery, Pulmonary TB + complications	1.5-1.6
Very Heavy Stress: Severe head trauma	1.7

Table 3. Age Factor

Age	Formula
0-40 years old	0% from BMR
40-59 years old	5% from BMR
60-69 years old	10% from BMR
≥ 70 years old	15% from BMR

Table 4. Body Correction

Nutrition Status	Formula
Normal	0% from BMR
Malnutrition	+ 20% from BMR
Obesity	- 20% from BMR

With the formula as:

$$Nutritional\ Status = \frac{Actual\ Weight}{Ideal\ Weight} x100\% \quad (4)$$

Information:
Status result > 110 = Obesity
Status result 90-100 = Normal
Status result < 90 = Malnutrition

1. Protein

Protein requirement normally is 15% of total energy requirement. The protein equation is as follows:

$$Protein = \frac{15\%\ x\ Total\ Energy\ (TE)}{4} \quad (5)$$

2. Fat

While for normal fat requirement is 25% of total energy requirement. The fat equation on Harris Benedict method is as follows:

$$Fat = \frac{25\%\ x\ Total\ Energy\ (TE)}{9} \quad (6)$$

3. Carbohydrate

While for normal carbohydrate needs is 60% of total energy needs, or residual energy after being reduced by energy derived from protein and fat. The equation determines carbohydrates are as follows:

$$Carbohydrate = \frac{60\%\ x\ Total\ Energy\ (TE)}{4} \quad (7)$$

Diet used as part of Diabetes Mellitus management is controlled based on energy, protein, fat and carbohydrate content. As a guideline used 8 types of Diabetes Mellitus Diet as can be seen in Table 5 [4]:

Table 5. Type Diet Diabetes Melitus

Type Diet	Energy (Kcal)
I	1100
II	1300
III	1500
IV	1700
V	1900
VI	2100
VII	2300
VIII	2500

2.2. Genetic Algorithm

The genetic algorithm is a heuristic search algorithm based on the mechanism of biological evolution. The diversity in biological evolution is a variation of the chromosomes between individual organisms. The Genetic algorithm was developed by John Holland from Michigan University. John Holland says if any problem (natural or artificial) associate with adaptation can be formulated in genetic terms. The genetic algorithm is a simulation of the evolutionary process of Darwin and the genetic operation of the chromosomes.

In this algorithm, the search technique is done simultaneously on a number of solutions known as the population. Individual in one population are called chromosomes. This chromosomes is a solution that is still a symbol. The initial populations randomly constructed, while the next population is the result of chromosomes evolution through an iteration called by the term generation. In each generation, the chromosomes will go through the evaluating process using a measuring instrument called a fitness function.

The fitness value of a chromosomes will show the quality of the chromosomes in that population. The next generation known as term child (off-spring) formed from the combined 2 generation chromosomes now acting as the parent using a cross carrier (crossover), in addition to the cross carrier, a chromosomes may also be modified by a mutation operator. The population of new generation is formed by selecting the fitness value of the parent chromosomes and the fitness value of a child's chromosomes (offspring), as well as rejecting the other chromosomes so the population size (the number of chromosomes in a population) still constant. After going through several generations, this algorithm will converge to the best chromosomes. In general, a simple genetic algorithm flow diagram as shown in the figure 1:

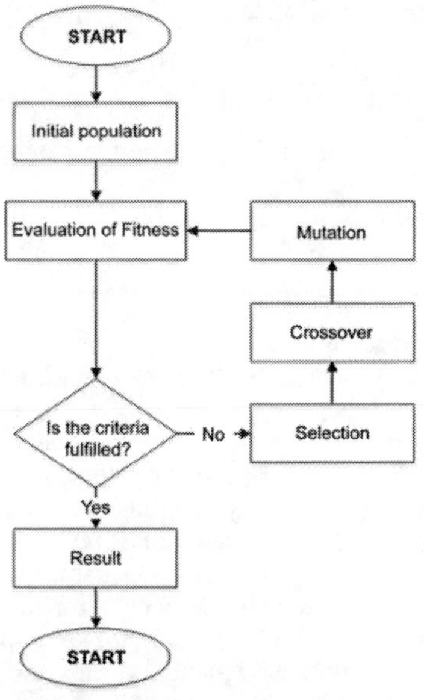

Fig 1. Genetic Algorithm Cycle

To implement genetic algorithm, the main components to do are:

1. Generate the initial population

Before generating the initial population, it must first determine the number of individuals in the population. For example the number of individuals is N. After that, it generates only initial populations that have individual beginnings at random.

2. Evaluation function

There are 2 things to do in evaluating chromosomes, that is: evaluate objective function (objective function) and conversion of objective function to fitness function. In general, the fitness function is derived from an objective function with a non-negative value.

3. Selection

Selection will determine which individuals will be selected for recombination and how offspring is formed from the selected individual. The first step in this selection is the fitness value search. Each individual in a selection vessel will accept the probability of reproduction that depends on its own objective value on the objective value on an individuals in the container. This fitness value will be used in the next selection stage.

A commonly used selection method is roulette-wheel. This selection aims to provide greater reproductive opportunities for members of high-fitness populations to reproduce. The workings of the roulette-wheel method are as follows:

a. Calculate the total fitness (F):

$$TotFitness = \sum F_k \,; k = 1,2,\ldots,popsize \quad (8)$$

b. Calculate the relative fitness of each individual:

$$P_k = F_k/TotFitness \quad (9)$$

c. Calculate cumulative fitness

$$Q_1 = P_1$$
$$Q_k = Q_{k-1} + P_k \,; k = 2,3,\ldots,popsize \quad (10)$$

d. Select the parent that will be the candidate for the crossover:

   1. Generate random r
   2. If Q_k £ r and $Q_{k+1} > r$, then select the chromosome into (k+1) as the parent candidate.

4. Crossover

Crossover (crossing) is done on two chromosomes to produce a child's chromosome (offspring). The child's chromosomes that are formed will inherit some of the parental chromosomal properties. In crossover there is one parameter that is very important that is probability of crossover (Pc). Probability of crossover show the ratio of the children generated in each generation to the size of the population. Example the population size (popsize = 100), while the probability of crossover (Pc = 0.25), means there is expected 25 chromosomes of 100 chromosomes in the population will be crossover.

5. Mutation

Mutation is the process of modifying a child's chromosomes randomly. Mutations will create new individuals by altering one or more genes present in a chromosome. Mutations play a role in replacing the missing genes of the population due to the selection process and allow for the emergence of genes not present in the initial population. In the mutation there is one parameter that is very important is the probability of mutation (Pm). Probability of mutation show the percentage of total genes in the population that will undergo a mutation. To make a mutation, we must first calculate the total number of genes in the population. Then generate a random number that will determine which position will be mutated. Example the population size (popsize = 100), each chromosome has a length of 20 genes, then the total gene is 100 x 20 = 2000 genes. If the mutation probability (Pm = 0.01), means that there is expected (1/100) x 2000 = 20 genes will mutate.

3. Methodology

This research uses Genetic Algorithm method to optimize calorie and calorie needs of food so get the appropriate menu variation. Food data used in this study is Indonesian Food Composition Table (TKPI) and report on the results of research on the composition of food nutrients carried out at the Center for Research and Development of Nutrition and Food Ministry of Health.

In an effort to obtain optimal results, this research is done through the stages to be done, namely:

1. Literature review.
   At this stage the process of collecting literature such as international and national journals in accordance with the topic of research, as well as articles relevant to the topic of research.
2. Interview.
   Observation is done by conducting interview process to nutritionist of General Hospital of Kudus.
3. Analysis and Design of Information System
   The design of this system is the conceptualization stage, which is a stage that requires analysis in the design of the system (software) so that the system is made according to the needs of the user. The diagram used in the design of this application is the Unified Language.
4. System implementation
   At this stage will implement the system into the application to be built using PHP and MySql database based web.

Flow Chart on the Implementation of Genetic Algorithm for Decision Support System Food Menu Determination of Diabetes Mellitus Patients is shown in figure 2.

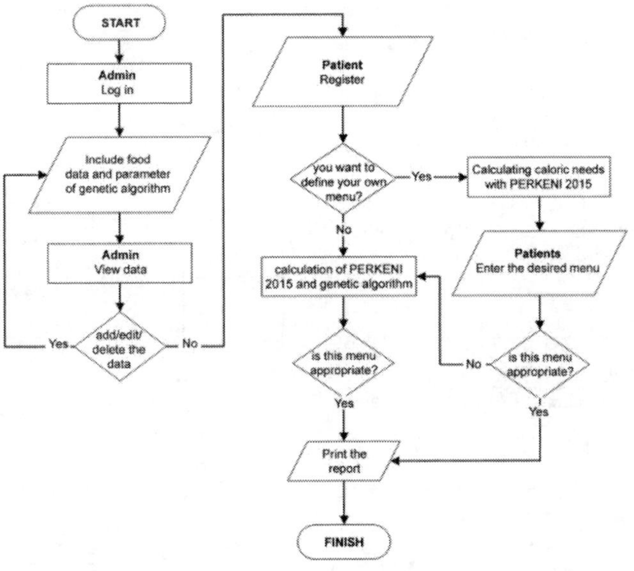

Fig 2. Flow Chart

4. Result and Discussion

4.1. Research result

The results of the research were conducted by comparing the parameters used in the Genetic Algorithm, so it can form a conclusion of the use of suitable parameters for a relatively large amount of data. Food menu determination system can provide information on which food menu is suitable to serve as a menu for people with diabetes mellitus with the calories needed.

4.2. Discussion

Before calculating the food menu using Genetic Algorithm, first calculate caloric requirement using PERKENI 2015.

Examples of patient test data used:

Name	: Siti Fatimah
Age	: 61 years old
Sex	: Women
Height	: 160 cm
Weight	: 55 kg
Activity	: Medium
Stress factor	: Medium

a. Calculate the Nutritional Needs

1. Energy

BMR = 25 x Ideal weight

Ideal weight = 90% x (TB-100)

= 90% x (160-100)

= 54

BMR = 25 x 54

= 1350

$$\text{Nutritional Status} = \frac{\textbf{Actual Weight}}{\textbf{Ideal Weight}} \; x\; 100\%$$

$$= \frac{55}{54} \; x\; 100\%$$

= 101 % = Normal

Energy = (BMR + Activity Factor) – Age Factor + Stress Factor +/- Body Correction Factor

= (1350 + (30% x 1350)) – (10% x 1350) + 1.4 – 0

= (1350 + 405) – 135 + 1.4 – 0

= 1621.4

2. Carbohydrate $= \frac{60\% \; x \; Total\ Energy}{4} = 243.21$

3. Protein $= \frac{15\% \; x \; Total\ Energy}{4} = 60.80$

4. Fat $= \frac{25\% \; x \; Total\ Energy}{9} = 45.04$

b. Process of Initialization Phase and Evaluation Function

The size of the population depends on the problem to be solved. After the population size is determined, then initialization of the chromosomes in the population is performed. Chromosome initiation is done randomly. Initial population by taking samples of 4 kinds of foods as genes and 3 menus as chromosomes to be done manual calculations.

There are two things to do in evaluating chromosomes, namely evaluating the value of objective function and conversion of objective function into fitness function. The number of calories here as an objective function.

The fitness function is lowered by a non-negative value. At the fitness value here, the minimum caloric difference is the optimization to be achieved in the determination of the food menu. So the smallest calorie difference will get the value of fitness with the greatest value. The initial population show in table 6, objective function show in table 7 and fitness value in table 8.

Table 6. Initial Population

Chromosome	Random Menu
K[1]	Rice, Sea Cucumber Dendeng, Fried Tempeh, Spinach Steamed
K[2]	Brown Rice, Rendang, Fried Tofu, Papaya Salad
K[3]	Savory Rice, Peda Banjar, Fried Tofu, Boiled Bean

Table 7. Objective Function

Chromosome	Total Calories of Each Food				Total of Calories
FO K[1]	135	231	175	30	571
FO K[2]	111.75	96,5	57.5	49	314.75
FO K[3]	142.5	156	57.5	30	386

Table 8. Fitness Value

Chromosome	Total of Calories	Deviation of Calories	Fitness Value (F)
Fitness K[1]	571	$\lvert 324,28 - 571 \rvert = 246,72$	$\frac{1}{246,72} = 0,0041$
Fitness K[2]	314.75	$\lvert 324,28 - 314,75 \rvert = 9.53$	$\frac{1}{9,53} = 0,1038$
Fitness K[3]	386	$\lvert 324,28 - 386 \rvert = 61.72$	$\frac{1}{61,72} = 0,0162$
Total Fitness			0.1241

c. Stage of Selection

After obtaining the value of the cultivated fitness (Q) and the random value (R), an approximation of R values to the Q value is altered to obtain a new chromosome. The following stage of selection in a table 9 and result of selection in table 10.

d. Stage of Crossover

Crossovers are performed on two chromosomes to produce a child's chromosome (offspring). The child's chromosomes that are formed will inherit some of the parental chromosomal properties. Probability of crossover (Pc) is 0.5 then the value below 0.5 will be selected to be the parent chromosome. The following stage of crossover in a table 11 and result of crossover in table 12.

Table 9. Stage of Selection

Chromosome	Fitness Relative (P)	Fitness Cumulative (Q)	Random Value (R)
K[1]	$\frac{0,0041}{0,1241} = 0,0330$	0.0330	0.3553
K[2]	$\frac{0,1038}{0,1241} = 0,8364$	$0,0330 + 0,8364 = 0,8694$	0.9198
K[3]	$\frac{0,0162}{0,1241} = 0,1305$	$0,8694 + 0,1305 = 0,9999$	0.5786

Table 10. Result of Selection

Random Value (R)	Fitness Cumulative (Q)	Comparative Result	New Chromosome
0.3553	0.0330	K[2] from the previous population	Red Rice, Rendang, Tahu goreng, Papaya Salad
0.9198	0.8694	K[3] from the previous population	Savory Rice, Peda banjar, Fried Tofu, Boiled Bean
0.5786	0.9999	K[2] from the previous population	Red Rice, Rendang, Fried Tofu, Papaya Salad

Random Value (R)
0.1197
0.4297
0.6950

Table 11. Stage of Crossover

Selected Chromosome	Population
K[1]	Brown Rice, Rendang, Fried Tofu, Papaya Salad
K[2]	Savory Rice, Peda banjar, Fried Tofu, Boiled Bean

Table 12. Result of Crossover

Chromosome	New Population from Crossover Outcome	Total Calories
K[1]	Brown Rice, Rendang, Fried Tofu, Boiled Bean	295.75
K[2]	Savory Rice, Peda Banjar, Fried Tofu, Papaya Salad	405
K[3]	Brown Rice, Rendang, Fried Tofu, Papaya Salad	314.75

e. Stage of Mutation

The number of chromosomes that have mutations in one population is determined by the probability of mutation (Pm). In this mutation it is possible to emerge a new chromosome that had not originally appeared in the initial population. Each chromosome has a length of 4 genes, then the total gene is 3x4 = 12 genes. If the probability of mutation (Pm = 0.5), the expected means is 0,5x12 = 6. It is expected that there are 6 genes that will have mutations. The following calculation of the mutation stage is shown in table 13 and result of mutation in table 14

Table 13. Stage of Mutation

Random from 1 to 20	Random Mutation
Random number = 2 1st Chromosome, 2nd index	Rendang → Fried Chicken
Random number = 4 1st Chromosome, 4th index	Boiled Bean → Cap cai
Random number = 5 Kromosom ke-2 index ke-1	Savory Rice → Rice
Random number = 6 2nd chromosome, 2nd	Peda Banjar → Fried Fish

Random from 1 to 20	Random Mutation
index	
Random number = 7 2nd chromosome, 3rd index	Fried Tofu → Fried Tempeh
Random number = 8 2nd chromosome, 4th index	Papaya salad → Boiled Cassava Leaves

Table 14. Result of Mutation

Chromosome	Result of Mutation	Calories	Fitness
K[1]	Brown Rice, Fried Chicken, Fried Tofu, Cap cai	365.25	0.02441
K[2]	Rice, Fried Fish, Fried Tempeh, Boiled Cassava Leaves	551.5	0.00440
K[3]	Brown Rice, Rendang, Fried Tofu, Papaya Salad	314.75	0.10493

So, one of selected food menu is K[3]. Calculated to the desired generation.

4.3. Testing System

Below is a picture of the system implementation:

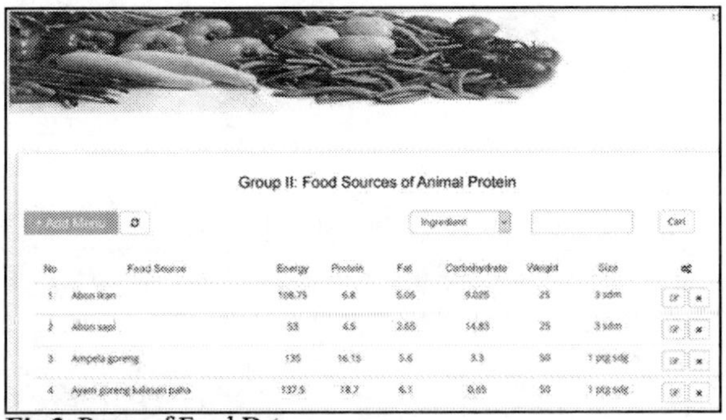

Fig 3. Pages of Food Data

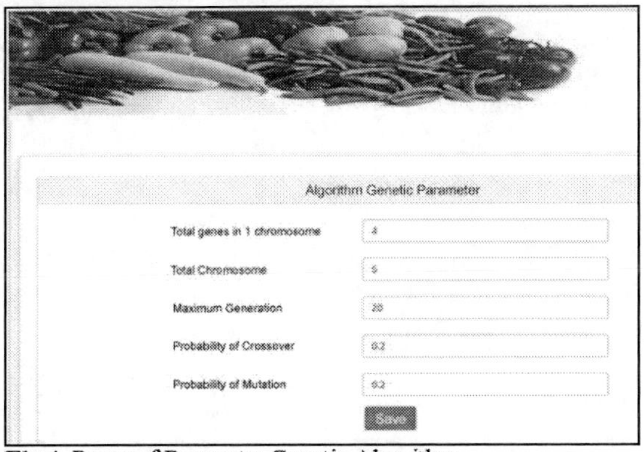

Fig 4. Pages of Parameter Genetic Algorithm

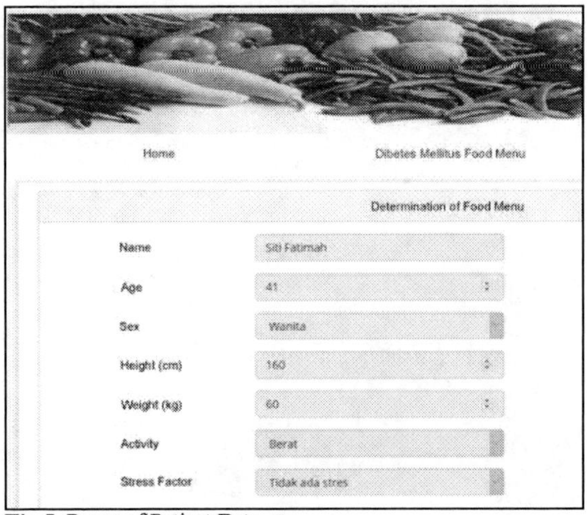

Fig 5. Pages of Patient Data

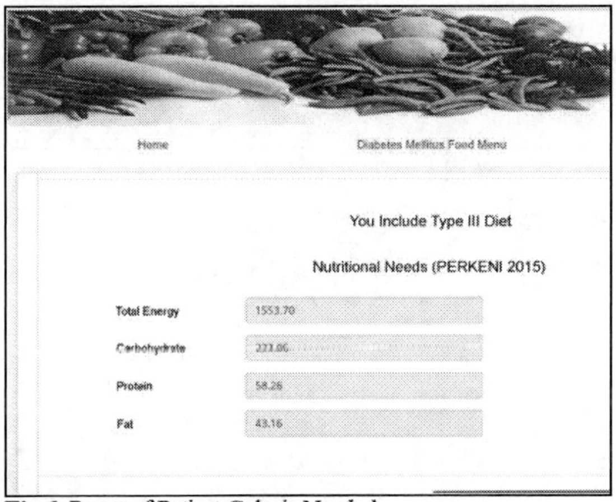

Fig 6. Pages of Patient Calorie Needed

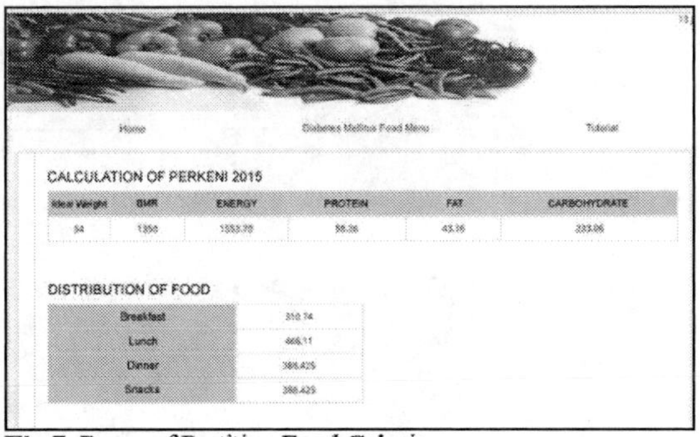

Fig 7. Pages of Partition Food Calorie

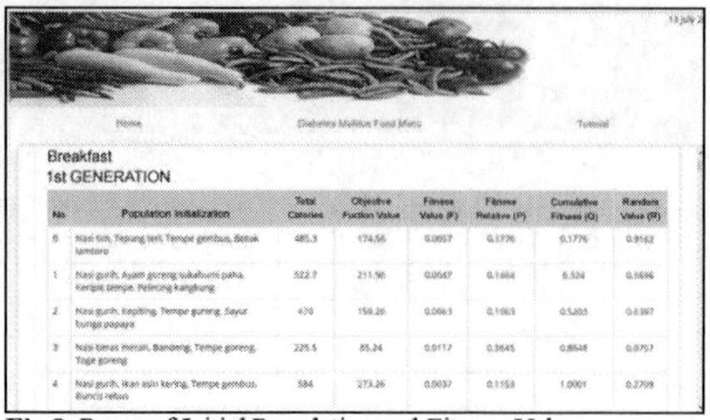

Fig 8. Pages of Initial Population and Fitness Value

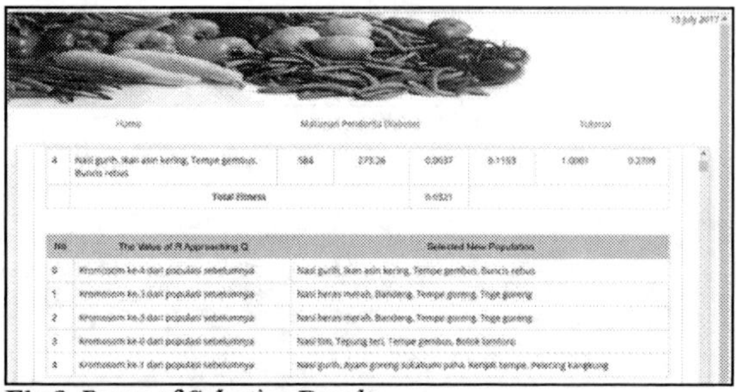

Fig 9. Pages of Selection Result

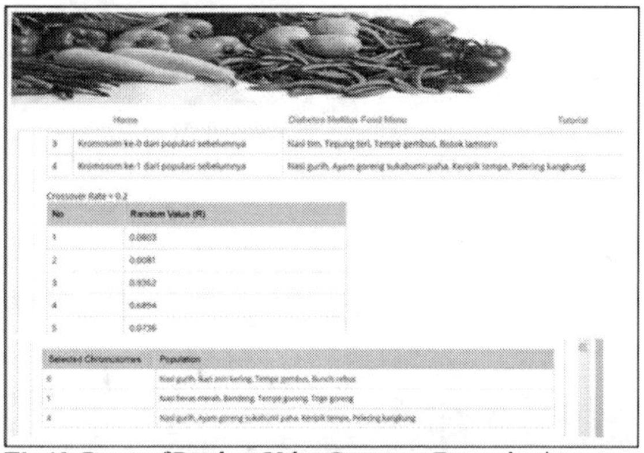

Fig 10. Pages of Random Value Crossover Determination

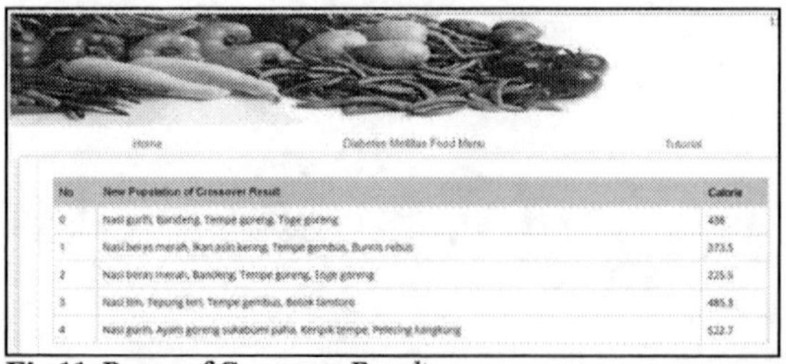

Fig 11. Pages of Crossover Result

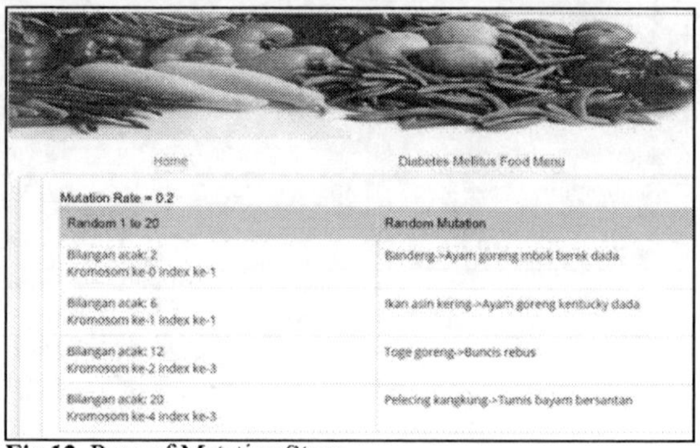

Fig 12. Page of Mutation Stage

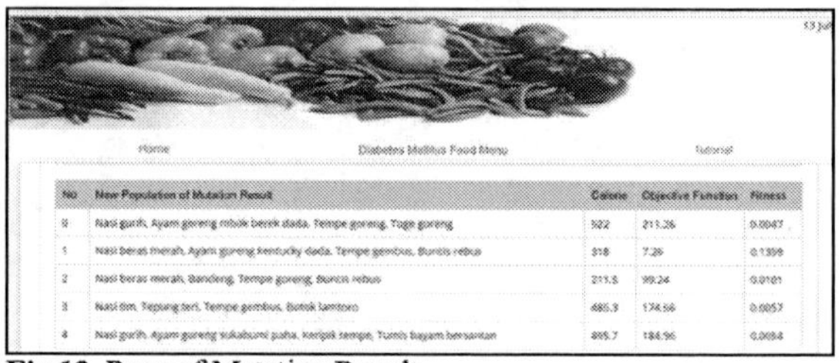

Fig 13. Page of Mutation Result

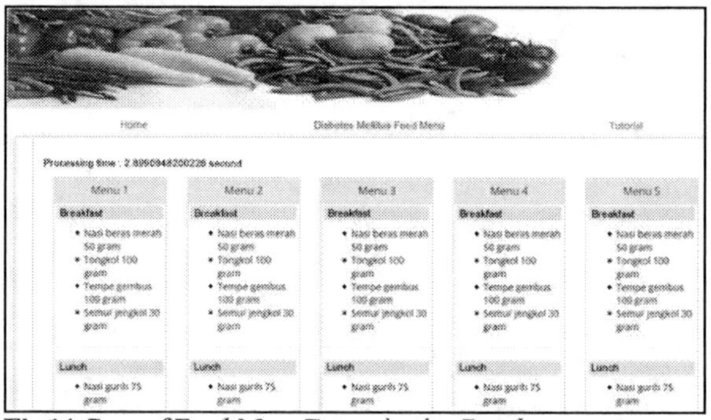

Fig 14. Page of Food Menu Determination Result

In testing this system using different genetic algorithm parameters. The following system test is shown in table 15.

Table 15. Testing System

Number	Total of Genes in 1 Chromosomes	Total of Chromosomes	Maximum of Generation	Probability of Crossover (Pc)	Probability of Mutation (Pm)	Attemps to -										Time
						1	2	3	4	5	6	7	8	9	10	
1	4	5	5	0.1	0.1	3	0	1	0	5	1	2	4	0	5	1 second
2	4	15	5	0.3	0.1	1	3	3	5	2	5	3	2	0	3	2 second
3	4	25	5	0.5	0.1	2	5	1	5	1	5	4	4	4	5	3 second

Number	Total of Genes in 1 Chromosomes	Total of Chromosomes	Maximum of Generation	Probability of Crossover (Pc)	Probability of Mutation (Pm)	Attemps to -										Time
						1	2	3	4	5	6	7	8	9	10	
4	4	5	10	0.1	0.3	10	10	7	0	7	6	7	9	5	5	2 second
5	4	15	10	0.3	0.3	9	6	7	8	9	7	7	8	7	7	4 second
6	4	25	10	0.5	0.3	6	9	7	6	9	7	5	6	5	4	7 second
7	4	5	15	0.1	0.5	9	7	14	2	10	11	11	5	9	3	3 second
8	4	15	15	0.3	0.5	13	11	10	12	13	15	14	12	10	12	7 second
9	4	25	15	0.5	0.5	13	12	14	15	14	10	14	13	13	11	11 second
10	4	5	20	0.1	0.1	9	9	8	11	7	8	2	10	7	11	3 second
11	4	15	20	0.3	0.1	8	10	9	13	8	8	10	2	4	14	8 second
12	4	25	20	0.5	0.1	13	16	5	9	9	6	14	7	7	6	12 second
13	4	5	25	0.1	0.3	21	18	16	15	18	15	17	9	11	1	4 second
14	4	15	25	0.3	0.3	10	17	16	16	9	14	10	14	10	8	10 second
15	4	25	25	0.5	0.3	11	6	8	9	13	18	5	11	19	11	15 second
16	4	5	30	0.1	0.5	23	18	19	20	22	15	16	23	20	22	5 second
17	4	15	30	0.3	0.5	25	28	29	27	29	25	29	28	23	24	12 second
18	4	25	30	0.5	0.5	26	28	29	30	28	26	25	27	27	26	21 second

4.4. Analysis result

The results of the analysis that can be taken from the results of system testing performed:

1. In the determination of diabetes mellitus food made comparison using genetic algorithm parameters namely Chromosome Number, Maximum Generation, Crossover Probability, and Probability of Mutation.
2. By entering the number of chromosomes a little for example entering the number of chromosomes 5 will use a fast time during the process. Whereas if entering chromosome more than 15 will use a lot of time.
3. Output results will be more optimal if the number of chromosomes are used slightly and the greater the value of Maximum Generation, Probability Crossover and Probability Mutations will be more varied food that will come out.
4. Testing the system performed to produce benefits in terms of users is that if the user is a nutritionist should be use total chromosome = 5, Maxium of Generation = 30, Probability of Crossover = 0,1 and Probability of Mutation = 0,5. Because, with the total of parameter will produce a stable variation and use shorted time, that is 4 second. If the user is a layman and want to choose a lot of dietary variations it is better to use the total chromosomes = 15, Maximum of Generation = 30, Probability of Crossover = 0,3, and Probability of Mutation = 0,5. Because, with the number of parameter will produce a lot of dietary variations and use shorted time to compared with other, that is 12 second.

5. Conclusion

Genetic Algorithm Method can be applied to decision support system of food menu selection for people with diabetes mellitus. In the results of research conducted by comparing the parameters of genetic algorithm. By entering a small number of chromosomes will use a fast time during the process. And the greater the value of Maximum Generation, Crossover Probability and Probability Mutation will be more variation of food that will come out. Using Number of Chromosomes = 5, Maximum Generation = 30, Crossover Probability = 0.1 and Probability Mutation = 0.5 using a not too long time of 5 seconds with little food variation. While using the Number of Chromosomes = 15, Maximum of 30, Probability Crossover = 0.3 and Probability Mutation = 0.5 using a slightly longer time of 12 seconds with many variations of food and by using Number of Chromosomes = 25, Maximum Generation = 30, Probability Crossover = 0.5 and Probability of Mutation = 0.5 using a very long time of 21 seconds with many food variations and relatively stable.

References

1. I. D. Federation, Brussel: Europan Diabetes (2005)
2. M. J. Franz, *Diabetes Education for Primary Health care Professionals*, 374–379, (2016)
3. S. Kusumadewi, S. Hartati, A. Harjoko, and R. Wardoyo, *Fuzzy Multi Attribute Decision Making (Fuzzy MADM),* Yogyakarta: Graha Ilmu, (2006)
4. S. Almatsier, *Penuntun Diet Edisi Baru Instalasi Gizi Perjan RS Dr. Cipto Mangunkusumo dan Asosiasi Dietisien Indonesia,* Jakarta: PT. Gramedia Pustaka Utama Jakarta, (2010)
5. S. A. Fajar, *Buku Saku Gizi, (*2017*)*

6. A. D. Association, *Nutrition Principle and Recommendations in Diabetes,* Brussels: Diabetes Care, (2004)
7. M. Fogue, J. A. Sanguesa, F. Naranjo, J. Gallardo, P. Garrido, and F. J. Martinez, *Expert Systems With Applications*, 262-271, (2016)
8. C. Groba, A. Sartal, and X. H. Varquez, *Computer & Operations Research*, 22-32, (2015)
9. M. Kusrini, *Konsep dan Aplikasi Sistem Pendukung Keputusan,* Yogyakarta: ANDI Yogyakarta, (2007)
10. S. Kusumadewi, and H. Purnomo, *Penyelesaian Masalah Optimasi dengan Teknik-teknik Heuristik,* Yogyakarta: Graha Ilmu, (2005)
11. L. Liu, H. Mu, X. Yang, R. He, and Y. Li, *Applied Soft Computing*, 506-515, (2012)
12. M. K. Mahmud, Hermana, N. A. Zulfianto, R. R. Apriyantono, I. Ngadiarti, B. Hartati, Tinexcelly, *Tabel Komposisi Pangan Indonesia,* Jakarta: PT Elex Media Komputindo, (2008)
13. T. Sutojo, E. Mulyanto, and V. Suhartono, *Kecerdasan Buatan.* Yogyakarta: ANDI, ((2011))
14. E. Turban, E. Aronson, J., and T.P. Liang, *Decision Support Systems and Intelligent Systems -7th Ed. Jilid 1 (Sistem Pendukung Keputusan dan Sistem Cerdas),* Yogyakarta: ANDI, (2005)
15. D. Whitham, *Canadian Journal of Diabetes*, 90-93, (2014)
16. L. Yang, X. Zhao, S. Peng, and X. Lic, Water quality assessment analysis by using combination of Bayesian and genetic algorithm approach in an, (2016)

Implementation of Rivest Shamir Adleman Algorithm (RSA) and Vigenere Cipher In Web Based Information System

Aryanti Aryanti [1,*], *Ikhthison* Mekongga[2]

[1] Telecommunication engineering State Polytechnic of Sriwijaya, Palembang - Indonesia
[2] Computer engineering State Polytechnic of Sriwijaya, Palembang - Indonesia

Abstract. Data security and confidentiality is one of the most important aspects of information systems at the moment. One attempt to secure data such as by using cryptography. In this study developed a data security system by implementing the cryptography algorithm Rivest, Shamir Adleman (RSA) and Vigenere Cipher. The research was done by combining Rivest, Shamir Adleman (RSA) and Vigenere Cipher cryptographic algorithms to document file either word, excel, and pdf. This application includes the process of encryption and decryption of data, which is created by using PHP software and my SQL. Data encryption is done on the transmit side through RSA cryptographic calculations using the public key, then proceed with Vigenere Cipher algorithm which also uses public key. As for the stage of the decryption side received by using the Vigenere Cipher algorithm still use public key and then the RSA cryptographic algorithm using a private key. Test results show that the system can encrypt files, decrypt files and transmit files. Tests performed on the process of encryption and decryption of files with different file sizes, file size affects the process of encryption and decryption. The larger the file size the longer the process of encryption and decryption.

1 Introduction

Data security and confidentiality are one of the most important aspects of information systems today. One attempt to secure data such as by using cryptography. Cryptography is the art or science used to maintain the security of information or messages by turning it into something that has no meaning. In cryptography, encryption and decryption, encryption is to convert the message into data that cannot be read or understood while the decryption is to restore the data as before so that data can be read properly [1].

Various kinds of cryptographic algorithms can be implemented to realize the data security system. Among them is Rivest Shamir Adleman (RSA) cryptography algorithm. RSA that use asymmetric algorithms have two different keys, called the key pair for the encryption and decryption process. The security level of the RSA encryption algorithm depends heavily on the key size of the password, because the smaller the key size, the greater the possible combination of locks can be broken by the method of examining the combination of one by one key or better known as Brute Force Attack. The prime numbers generated in the RSA algorithm affect the key size of the password [2].

Cryptography using the Vigenere Cipher Algorithm by adopting the operation mode of Cipher Block Chaining (CBC) operation is one of the methods of many data security methods. This application includes encryption and data decryption, created using Borland Delphi 6.0. The encrypted data will have an extension. Merging the Vigenere Chiper algorithm and the CBC mode of operation will produce a new method that researchers call Vigenere Chiper +, in this method the weaknesses of the Vigenere Chiper algorithm will be improved. As it extends the reach of 26 letters of the alphabet into 256 ASCII characters [3].

In this study developed a data security system by implementing the cryptography algorithm Rivest Shamir Adleman (RSA) and Vigenere Chiper. The research was done by combining Rivest Shamir Adleman (RSA) and Vigenere Cipher cryptographic algorithms to document file either word, excel, ppt and pdf. This application includes encryption and data decryption, created using PHP and my SQL. Data encryption is done on the send side through RSA cryptographic calculations first using public key, then proceed with Vigenere Cipher algorithm which also use a public key. As for the stage of the description side received by using the Vigenere Cipher algorithm first still use public key and then use the RSA cryptographic algorithm using a private key.

2 Research methods

2.1 The proposed method

Fig. 1. Design diagram of the algorithm

The method used in this research is by combining Rivest Shamir Adleman (RSA) cryptographic algorithm and Vigenere Cipher. Plain text or first original file will be encrypted using the RSA cryptographic algorithm by using public key. And it will generate RSA ciphertext. Then in encryption again using public key cryptography algorithm vigenere cipher. After going through a double locking process. Furthermore, the decryption process uses cryptography algorithm vigenere cipher by using public key. Because vigenere cipher is a symmetric algorithm which in the process of encryption and decryption using the same key. And finally the decryption process uses the private key RSA cryptographic algorithm. Then back to the plaintext.

2.2 Stages of key establishment

2.2.1 RSA lock generation stage

For the RSA key generation process in this test the following steps are performed [4]:

1 Determine the two prime numbers, with the names p and q. Suppose the value of p = 51 and q = 5.

2 Calculate the modulus value (n):

$$n = p * q \quad (1)$$

$n = 51 \text{ x } 5$

$n = 255$

3 Calculate the totient value n:

$$\phi(n) = (p\text{-}1) * (q\text{-}1) \quad (2)$$

$\phi(n) = (51\text{-}1) * (5\text{-}1)$

$\phi(n) = (50 * 4)$

$\phi(n) = 200$

4 Determine the value of e with the terms gcd (e, ϕ (n)) = 1
Where e = prime number, and 1 <e <ϕ (n).
Select public key e is 7 (relatively primed tor 200)

5 Looking for exponent deciphering value (d), then:

$$d = (1 + (k \text{ x } \phi(n)) / e) \quad (3)$$

$$d = (1 + (k \text{ x } 200)) / 7$$

The value of k is any number of searches until an integer or integer value is generated. By trying the value of k = 1,2, ..., to obtain the value of d that is round, that is d = 343.

6 From the previously described steps, the values n, e, and d have been obtained so that the key pair has been formed.
Public key pair (n, e) = (255, 7)
The secret key pair (n, d) = (255, 343)

2.2.2 RSA encryption process

For the encryption process uses the public key of the RSA that has been established previously e = 7 by using the formula $a^e \ mod \ n$ [5].
For example, ICENIS is taken as Plain text.
I have a value of 73 in the ASCII table
C has a value of 67 in the ASCII table
E has a value of 69 in the ASCII table
N has a value of 78 in the ASCII table
I have a value of 73 in the ASCII table
S has a value of 83 in the ASCII table
For the encryption process uses the public key from RSA, which has been formed before that is 7 pieces, using the formula $a^e mod \ n$
73^7 mod 255 = 112 on the ASCII table p
67^7 mod 255 = 118 on the ASCII table v
69^7 mod 255 = 69 on the ASCII table E
78^7 mod 255 = 192 on ASCII table └
73^7 mod 255 = 112 on the ASCII table p
83^7 mod 255 = 212 on ASCII table ╘

After the encryption process using RSA public key has obtained ciphertext which will be used as the plaintext for encryption process using a vigenere cipher.

2.2.3. The formation of vigenere cipher keys

In the first process is done by determining the length of key variables that will be used. In this study, the authors limit the length of values ranging from 0-96. In this study using ASCII code them for mod used mod 256. As for the encryption formula of vigenere cipher is Ci = (pi + ki) mod 256. Plain text used is the result Of the ciphertext contained in the RSA. IE pvE └p ╘. In this study, the authors use the POLSRI key as its public and private key

2.2.4. An Encryption process vigenere cipher

P has a value of 112 in the ASCII table
Shift: (P) -96
80-96 = -16
Ci: 112 + (- 16) mod 256
= 96 in the ASCII table = `
V has a value of 118 in the ASCII table
Shift: (O) -96

79-96 = -17
Ci: 118 + (- 17) mod 256
= 101 in ASCII table = e
E has a value of 69 in the ASCII table
Shift: (L) -96
76-96 = -20
Ci: 69 + (- 20) mod 256
= 49 in the ASCII table = 1
└ has a value of 192 in the ASCII table
Shift: (S) -96
83-96 = -13
Ci: 192 + (- 13) mod 256
= 179 in ASCII table = |
P has a value of 112 in the ASCII table
Shift: (R) -96
82-96 = -14
Ci: 112 + (- 14) mod 256
= 98 in ASCII table = ~
╘ has a value of 212 in the ASCII table
Shift: (I) -96
73-96 = -23
Ci: 212 + (- 23) mod 256
= 189 in ASCII table = ╜

2.2.5. The Decryption process vigenere cipher

For the decryption stage of Vigenere Cipher, the process is almost identical to the encryption process. It's just the mathematical process that distinguishes Ci = (pi-ki) mod 256. With the same key with encryption, as also at the time of decryption is POLSRI Has a value of 96 in the ASCII table

Shift: (P) -96
80-96 = -16

Ci: 112+(-16) mod 256
=96 in the ASCII table = `
v has a value of 118 in the ASCII table
Shift: (O) -96
79-96=-17
Ci : 118+(-17) mod 256
=101 in the ASCII table = e
E has a value of 69 in the ASCII table
Shift : (L)-96
76-96=-20
Ci : 69+(-20) mod 256
=49 in the ASCII table = └
└ has a value of 192 in the ASCII table
Shift : (S)-96
83-96=-13
Ci : 192+(-13) mod 256
=179 in the ASCII table = |
~ has a value of 98 in the ASCII table
Shift : (R)-96
82-96=-14
Ci : 98-(-14) mod 256
= 112 in the ASCII table =
Has a value of 189 in the ASCII table
Shift : (I)-96
73-96=-23
Ci : 189-(-23) mod 256
=212 in the ASCII table ╘

2.2.6. RSA decryption process

In the RSA decryption process, use the specified private key that is d = 343
112^{343} mod 255 = 73 on the ASCII table I
118^{343}mod 255 = 67 on the ASCII table C
69^{343}mod 255 = 69 on the ASCII table E
192^{343}mod 255 =78oin the ASCII table N
112^{343} mod 255 = 112 on the ASCII table I
212^{343} mod 255 =83 on the ASCII table S

3 Results and discussion

3.1 Program Implementation

3.1.1 Testing of encryption and decryption process

For testing, performed the process of encryption and decryption as shown below. In this menu will be the file encryption process. Before the process of uploading password input for the file can be maintained its security. Then click save, then the encryption process will run and will generate a random file

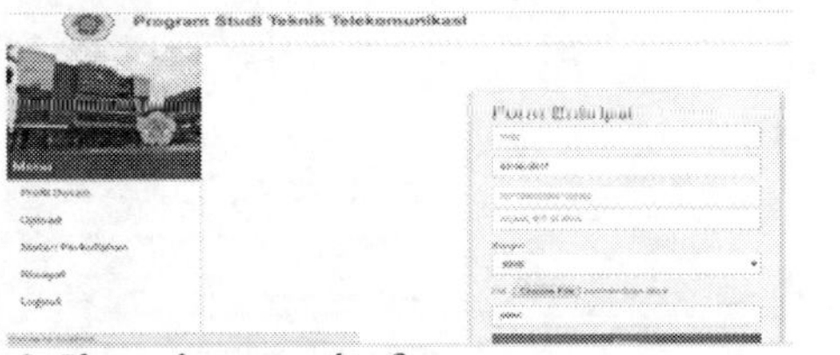

Fig. 2. Shows the encryption form

Files that have been successfully encrypted will appear on the encryption list.

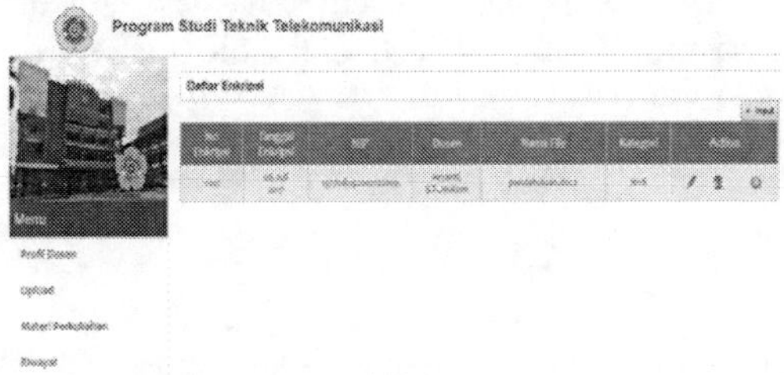

Fig. 3. Shows the encryption list

In the decryption process can be done in the file list. Encrypted files can be downloaded and use the same password during the encryption process. If the password differs between encryption and decryption, then the file will not be a decrypted file or an actual file.

Fig. 4. Display input decryption password

Figure 4 is a password input view. When the student encryption process first input password. If the password is correct then the downloaded file will be decrypted. But if the password is wrong the downloaded file is an encrypted file or a random file that cannot be read.

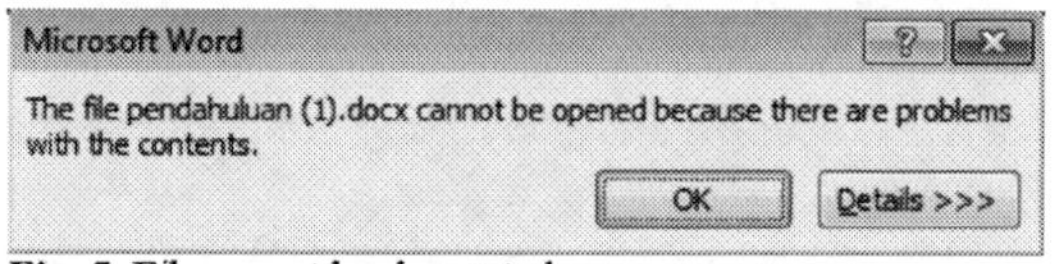

Fig. 5. File cannot be decrypted

Figure 5 incorrect password of the downloaded file is an encrypted file or a random file that cannot be opened.

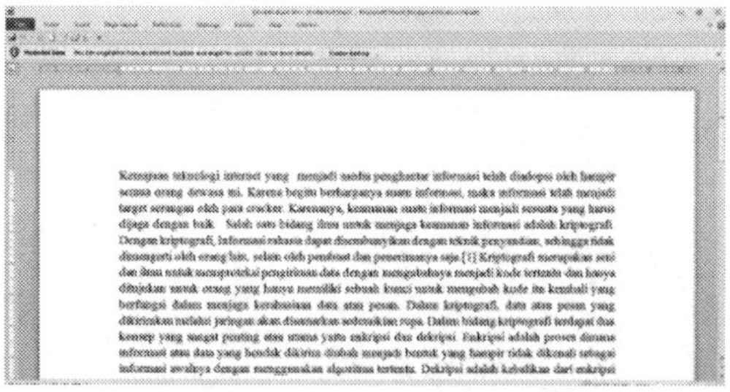

Fig 6. File Decryption

Figure 6 is the result of the file successfully decrypted. When downloading the file, first enter the password. The file will be decrypted if the password is correct.

3.2 Discussion

To analyze the security of files, it is necessary to check the file. Is the file perfectly encrypted so that information on the file cannot be accessed by unauthorized people.

In the tests that have been performed on the file "data2.docx", the original file structure or files that have not been decrypted can be viewed as follows. The encrypted file is named data2 (1). Here is an encrypted file structure in hexadecimal form.

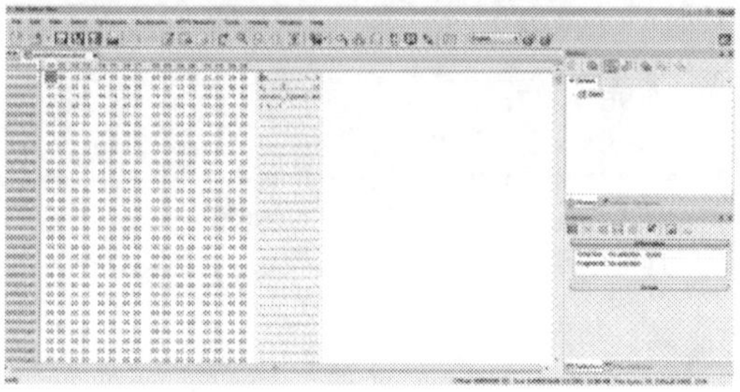

Fig 7. View Files"data2.docx in hexadecimal"

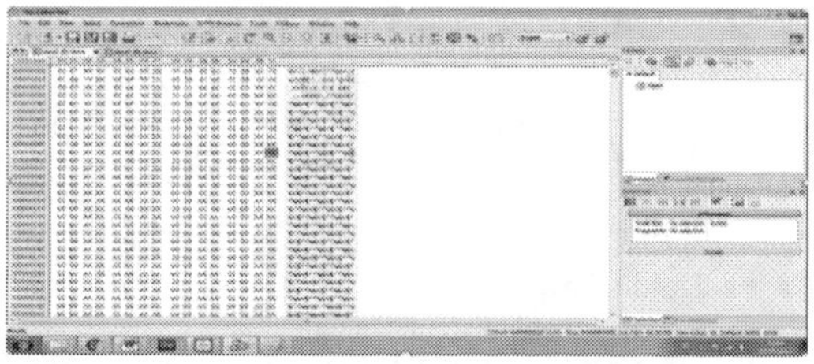

Fig 8. View Files "Data2 (1) .docx in hexadecimal"

In Figure 8 is the structure of the file "Data2 (1) .Docs" in the left side of the display shows the file structure in the form of hexadecimal, while on the right is the character form (ASCII) of hexadecimal. Can be seen in the picture above. Both of them have different results. This can prove that the "Design Diagram.docx" file is successfully encrypted. To prove that the application is running well, we can do the decryption process to file "Data2 (1) .Docs"

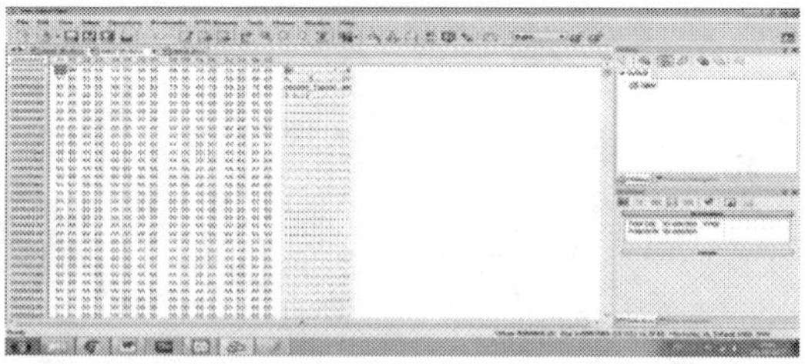

Fig 9. View Files "Data2 (2) .docs" in the decimal, hex

Figure 9 shows that the file decryption process is successful. Can be seen from its file structure.

4 Conclusions

From the research conducted can be seen that the implementation of RSA and Vigenere Chiper algorithm on the WEB-based information system is successfully encrypted and decrypt the file well. Test results show that the system can encrypt files, decrypt files and transmit files. Tests performed on the process of encryption and decryption of files with

 https://doi.org/10.1051/e3sconf/20183110007

different file sizes, file size affects the process of encryption and decryption. The larger the files size the longer the process of encryption and decryption.

References

1. JB, R. Kristoforus and BP. Stefanus Aditya. *Impl of Rijndael Algthm for Encrypt and Decrypt In Dgtl Img*. Yogyakarta (2012)
2. N. Aini, DA. Ningrum, Pradhipta M RSA *Algthm Impl for Encryp and Decrypt Using Java Progrm Language on Netbeans* (2012).
3. EK. Nurnawati, *National Appl Science and Tech. IST AKPRIND* Yogyakarta (2008)
4. *RSA*Algthm:www.dimgt.com.au/rsa_alg.html
5. Evgeny Milanov, *"The RSA Algthm"*, June. pp. 1-11(2009)

The Design of The Monitoring Tools Of Clean Air Condition And Dangerous Gas CO, CO_2 CH_4 In Chemical Laboratory By Using Fuzzy Logic Based On Microcontroller

Slamet Widodo [1,*], *M.Miftakul* Amin[2], *Adi* Sutrisman[3]

[1,2,3] Department of Computer Engineering, Politeknik Negeri Sriwijaya, Palembang, Indonesia

Abstract: There are many phenomena that human are exposed to toxins from certain types such as of CO2, CO2 and CH4 gases. The device used to detect large amounts of CO, CO2, and CH4 gas in air in enclosed spaces using MQ 135 gas sensors of different types based on the three sensitivity of the Gas. The results of testing the use of sensors MQ 135 on the gas content of CO, CO2 and CH4 received by the sensor is still in the form of ppm based on the maximum ppm detection range of each sensor. Active sensor detects CO 120 ppm gas, CO2 1600 ppm and CH4 1ppm "standby 1" air condition with intermediate rotary fan. Active sensor detects CO 30 ppm gas, CO2 490 ppm and CH4 7 ppm "Standby 2" with low rotating fan output. Fuzzy rulebase logic for motor speed when gas detection sensor CO, CO2, and CH4 output controls the motion speed of the fan blower. Active sensors detect CO 15 ppm, CO2 320 ppm and CH4 45 ppm "Danger" air condition with high fan spin fan. At the gas level of CO 15 ppm, CO2 390 ppm and CH4 3 ppm detect "normal" AC sensor with fan output stop spinning.

1 Introduction

Currently, a major concern of scientists is global warming, largely due to huge emissions of carbon dioxide (CO2). Considered as one of the main greenhouse gases inducing a warming climate, CO2 concentration in the atmosphere is under special scrutiny of many weather services in the world. The program involves capturing samples of atmospheric air at the Assekrem station twice a week, using 2 special bottles of 1.5 litres. These bottles are then sent to the NOAA laboratory (Boulder-USA) to determine concentrations of major greenhouse gases: CO_2 in ppm, CH_4 and CO in ppm.[1]

In human life can not be separated from the Gas carbon monoxide Gas CO, Gas Carbon dioxide CO_2, and Gas Methane is CH_4 hydrocarbons. The odorless CO gas is very dangerous, Methane (Methane) is the simplest hydrocarbon in the form of gas. Pure methane is odorless, colorless, extremely flammable, asphyxian, non toxic and non corrosive. Burning one molecule of methane with oxygen will release one molecule of CO_2 (carbon dioxide). [2]

At certain depths below the earth's surface there are dangerous gases such as Carbon Dioxide (CO_2) and Methane gas (CH_4). While carbon monoxide gas (CO) can be produced from leakage or emissions of fuel used as a source of propulsion power from the generator set and room temperature. Third gas is very dangerous if accumulate in the room without air circulation is not good. Hazardous gases that accumulate will often be inhaled by the workers who are in the room. [4]

Health problems will arise if inhaled methane gas in high concentrations. The symptoms are oxygen deprivation, rapid breathing, increased pulse rate, decreased muscle coordination, increased emotion, nausea, vomiting, loss of consciousness, respiratory failure, and death. [4]

To solve the above problem, it is necessary to design the toxic detector of toxic monoxide (CO) gas, Carbon Monoxide (CO_2) and Gas methane (CH_4) gas as human safety from poisoning. This research was built for gas leak detection tool CO, CO_2, and CH_4 which work using MQ-7 Sensor. The MQ-7 sensor is a gas sensor that can detect vehicle exhaust gases in air carbon monoxide (CO) gas. If the CO gas is detected, the MQ-7 sensor will provide input (insert) to the Arduino output of the microcontroller which has three types of outputs, Namely LCD display (Liquid Crystal Display), Motor Fan, and buzzer. LCD function information Gas CO, CO_2 and CH_4 in ppm quantities. The LCD is used to provide information on the occurrence of gas leak indoors when the sensor reads CO_2, and CH_4 through microcontroller. Exhaust Fan is used to neutralize air conditions in the room according to sensor readings against harmful gas gases in ppm. Fan motor rotation works based on fuzzy logic during low, middle, and high conditions. The larger the hazardous gas gases detected by the sensor in ppm size the microcontroller will send the information to the fan fan circuit with high conditions, as well as the low conditions the microcontroller will send the information to the fan at low speed. The fan driver is used to neutralize the condition of the room by removing air inside the room when it detects CO, CO_2 and CH_4 gas in the laboratory and workshop room in safe, standby, alert and danger conditions.

* Corresponding author: slametwidodo160573@gmail.com

2. Experimental details

Stages done in this research is by the method of planning and design. Here is the design of CO gas detector using ATMEGA 8535. [4]
This CO gas leak detector will work with the MQ-7 sensor and MQ 135 is a gas sensor that detects exhaust gases in the air of Carbon Monoxide (CO), Carbon Dioxide (CO_2) and Methane (CH_4) gas. If the CO, CO_2 and CH_4 gases have been detected then the MQ-7 sensor will give input (insert) to the microcontroller, then from the microcontroller has four outputs, LCD (Liquid Crystal Display), Driver Motor fan, and buzzer. LCD is used to display information readings of sensor detectors CO, CO_2, and CH_4. Driver The fan motor is used to drive the fan blades when it is detected CO, CO_2 and CH_4 gases inside the room during Safe, Standby1, Standby and danger so that space will be neutralized toxic gas content in ppm. LCD gives message information to the user that there is a toxic Gas leak.

The driver acts as an additional link to connect between the microcontroller with the buzzer and the Exhaust Fan. Explanation of the flow diagram of the CO gas detector in the laboratory room Figure 1. the following:

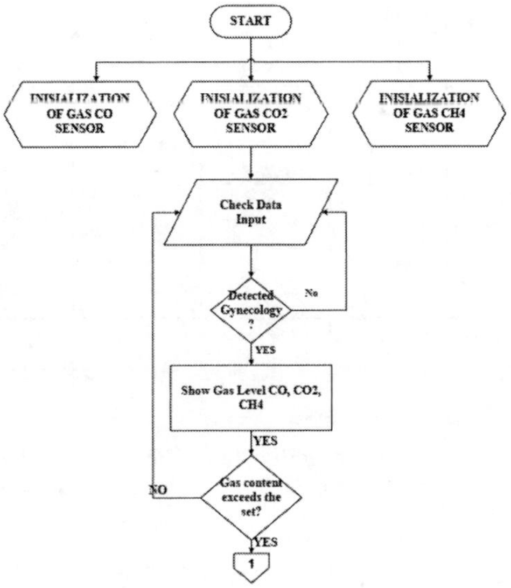

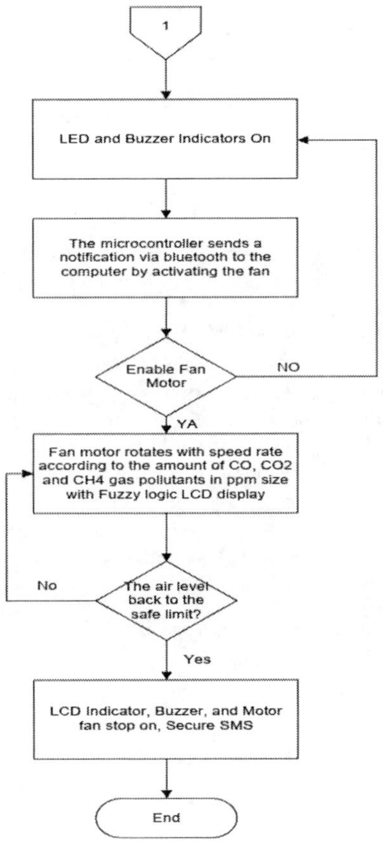

Fig. 1. Flowchart of How the Tool Works

The flow and work on the flow diagram is first to initialize the serial and then seen the input of the three gas sensors CO, CO_2, and CH_4 which will appear in gas levels. Then the sensor will work based on the smoke fumes detected by the MQ-7 and MQ-135 sensors. When the CO, CO_2 and CH_4 gases Active sensors detect CO 120 ppm, CO_2 1600 ppm and CH_4 1ppm air condition "standby 1" with medium spinning exhaust fan output. In active sensor detect CO 30 ppm gas, CO_2 490 ppm and CH_4 7 ppm "Standby 2" air condition with low rotating exhaust fan output. While the active sensor detects CO gas 15 ppm, CO_2 320 ppm and CH_4 45 ppm "Danger" air condition with High Spin Exhaust fan output. At the gas level of CO 15 ppm, CO_2 390 ppm and CH_4 3 ppm detected "normal" air conditioner sensors with the exhaust fan output stop spinning. To detect CO gas of 5 volt DC voltage sensor with ADC 0 - 1023 range for ADC value. For the detection of CO gas it is determined that the maximum value of the sensor to detect CO gas is connected to the microcontroller ADC at port D by 30 ppm [16]. From these provisions, the CO membership function for membership function is 0 - 30 ppm. So in order to scale the membership function accordingly ie 30 ppm then the ADC value must be divided by 34 which can be searched with the following equation: [5]

ADC divider value = $\frac{\textbf{ADC Value}}{\textbf{CO (ppm)}}$......[5]

Information :
CO (ppm) = Levels maximum desired CO gas
ADC value = 1023
So, can we enter the value
ADC divider value = 1023 / (30 ppm)
ADC divider value = 34

The divisor value of ADC = 343.2 Metode Fuzzy

Rule evaluation process In this process the specified rules will be applied. Or it could be said function is to find the value of the fuzzy output of fuzzy input. Rule made to control the work of the sensor in detecting sensor in order to generate value in accordance with the expected output. The process is a fuzzy input of the process fuzzyfication included in a rule that has been created to serve fuzzy output. There are several methods of decision making in fuzzy logic Mamdani among which methods Figure 2.[2]. The following picture:

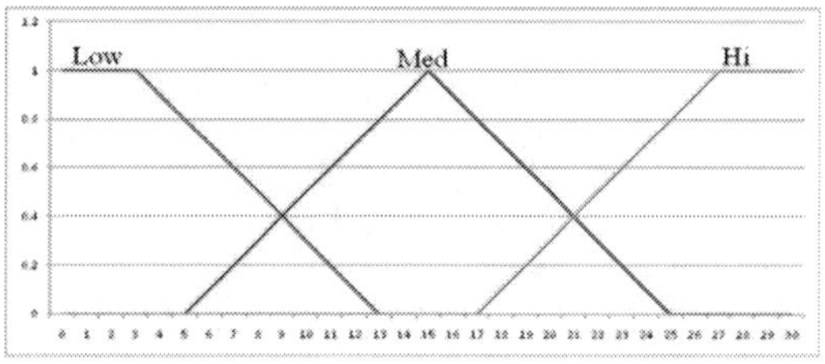

Fig. 2. Decision Methods Mamdani

Implications functions used in decision-making with Mamdani method by using MIN and in doing composition using MAX. The composition method is often called MAX-MIN. decision with Mamdani method. Here is a ten-rule evaluation that can be determined:
Table 1. Rule Evaluation
The following is stage of detection

1. If CO is low and CO_2 is low and CH_4 is low then condition is normal, blower OFF.
2. If CO is low and CO_2 is medium and CH_4 is low then blower is SPD_1.
3. If CO is medium and CO_2 is low and CH_4/CH_4 is low then blower is SPD_1.
4. If CO low and CO2 is low and CH_4/CH_4 medium then blower is SPD_1.
5. If CO is medium and CO_2 medium and CH_4/CH_4 is high then blower is SPD_2.
6. If CO is medium and CO_2 high and CH_4/CH_4 is medium then blower is SPD_2.
7. If CO high and CO_2 is medium and CH_4/CH_4 is medium then blower is SPD_2.
8. If CO is not low and CO_2 high and CH_4/CH_4 is high then blower is SPD_3.
9. If CO is high and CO_2 is not low and CH_4/CH_4 is high then blower is SPD_3.
10. If CO is high and CO_2 is high and CH_4/CH_4 is not low then blower is SPD_3.

Is not low , mean high or medium
SPD = Exhaust fan rte

3. Results and discussion

Based on tests conducted with test samples consisting of sources of pollutants ie gas originating from clean air, vehicle fumes, livestock manure and cigarette smoke for CO and CO_2 sensor testing, and gases derived from animal waste to compute the CH_4 sensor response, the following

Table 1. test for each input response.

No	Type of Gas	Clean air (ppm)	Vehicle emission (ppm)	Animal waste (ppm)	Cigarette smoke(ppm)
1	CO	2	1150	159	9206
		2	1132	156	9340
		3	1211	153	9604
		5	1256	154	9113
		7	1252	148	9480
Average value :		**3.8**	**1200.2**	**154**	**9348.6**
2	CO_2	320	1750	390	5200
		300	1671	387	5440
		295	1455	391	5105
		310	1723	387	5201
		312	1559	391	5130
Average value :		**370.4**	**1631.6**	**389.2**	**5215.2**
3	CH_4	1	90	70	25
		2	87	73	23
		1	89	71	24
		2	88	73	22
		1	98	72	21
Average value :		**1.4**	**90.4**	**71.8**	**23**

The result of the room condition test on the CO gas, CO_2 and CH_4 gases in the Graph using Fuzzy Logic Rulebase as described below:

1. High CO_2, CH_4 Low and CO High Standby conditions

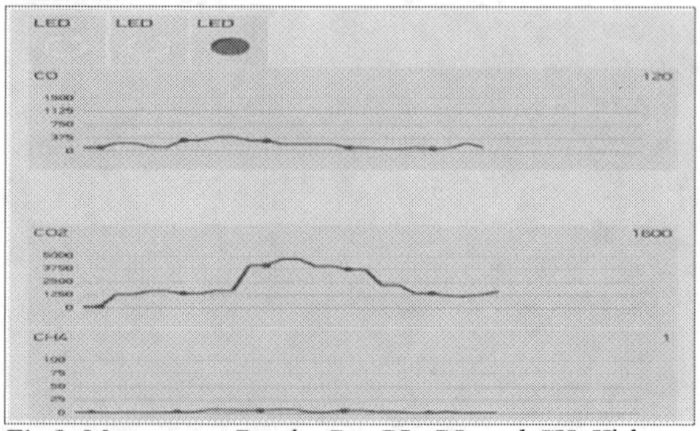

Fig 3. Measurement Results Gas CO, CO_2 and CH_4 High Standby conditions

2. Condition of CO_2 med, CH_4 Low and CO med standby output 2

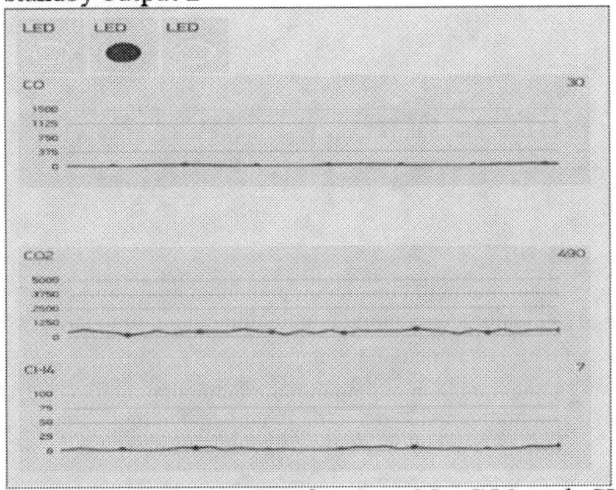

Fig 4. Measurement Results Gas CO, CO2 and CH4 Standby conditions

3. Condition of CO_2 med, CH_4 high and low CO output hazard

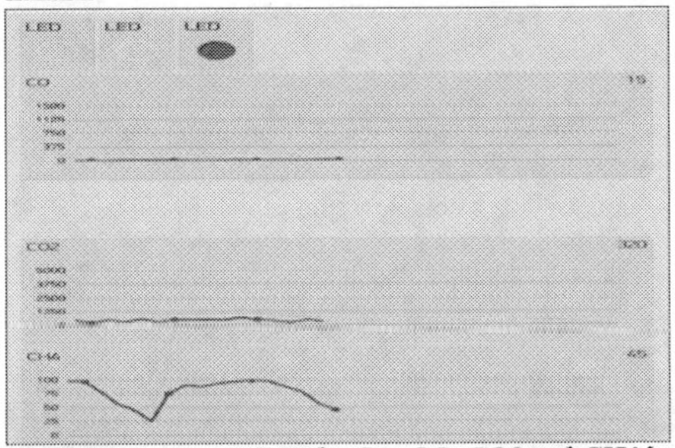

Fig 5. Measurement Results Gas CO, CO2 and CH4 hazard conditions

4. Low CO_2 conditions, Low CH_4 Low and low CO output are safe

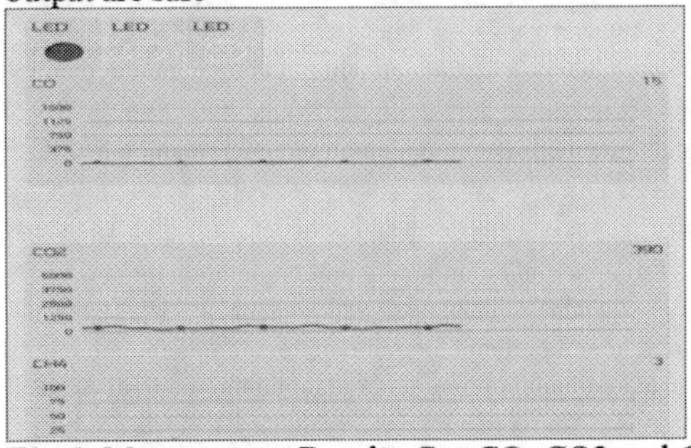

Fig 6. Measurement Results Gas CO, CO2 and CH4 Safe conditions

4 Conclusion

In conclusion, this study shows that the test results of the system when the active sensor detects CO 120 ppm gas, CO_2 1600 ppm and CH_4 1ppm air condition "standby 1" with a medium spinning exhaust fan output. In active sensor detect CO 30 ppm gas, CO_2 490 ppm and CH_4 7 ppm "Standby 2" air condition with low rotating exhaust fan output. While the active sensor detects CO gas 15 ppm, CO_2 320 ppm and CH_4 45 ppm "Danger" air condition with High Spin Exhaust fan output. At the gas level of CO 15 ppm, CO2 390 ppm and CH_4 3 ppm detected "normal" air conditioner sensors with the exhaust fan output stop spinning.

References

1. H. Abderrahim, M. Berrebia, A. Hamou, H. Kherief, Y. Zanoun, K. Zenata. J. Mater. Environ. Sci. **2** (2) 94-103 (2011)
2. Andrian Dendy, *Computer Engineering Department of the Polytechnic of Sriwijaya. Palembang,* 34-36. (2013)
3. Hapsari, Chrismalia; Wilujeng, Susi Agustina. *Study of Carbon Dioxide Emissions (CO_2) and Methane (CH_4) from Waste Reduction Activity in Southern Part of Surabaya.* Environmental Engineering, ITS. Surabaya. 2010
4. Aldo, Aziz Putra,. *Work Build Clean Air Monitoring Tool and Harmful Gas CO, CO_2 and CH4 Using Microcontroller.* (2016)
5. Sari Rahma, *Eka.Toxic Gas Leak Detection Equipment CO By Car Using Sensor Array with Fuzzy Controller.* Computer Engineering Department of the Polytechnic of Sriwijaya.Palembang, 36-37. (2012)
6. Indahwati, Elly; Nurhayati. *Design of Carbon Monoxide (CO) Gas Concentration Measurements Using Micro-Controller Based MQ-135 Gas Sensors With USART Serial Communications.* UNESA. Surabaya. (2012)
7. Nurrahmah; Prasetiawan; Suparno. *Gas Leak Detector System on LPG Tube.* UMS Surakarta. Surakarta. (2006)
8. Rofiq, Muhammad, Yusron, *M. Design of Lights Control and Monitoring Systems By Utilizing Bluetooth Technology on Android Smartphone.* STMIK ASIA. Malang. (2014)
9. Saputra, Andri, Febriansyah, Dwi, Kuswara, Haris. Volume **4**. Number 1. *Home Light Control Device Using Bluetooth Based Android.* TECHNOMETICS. (2014)
10. Siker Nas. *Carbon Dioxide Carbon Dioxide.* Badan POM RI. (2010)
11. Safrizal, Rino. *Global Warming.* Http://www.jejaringkimia.web.id/2010/12/mengenal-gas-metana.html (accessed on March 28, 2016) . (2010)
12. *Introducing Earth, Earth Atmosphere* http://www.trunitydemo3.net/boxford_classroom/view/article/171011/ (accessed on May 5, 2016).

Expert System Application of Forward Chaining and Certainty Factors Method for The Decision of Contraception Tools

Dwi Arief *Prambudi*[1,*], *Catur* Edi Widodo[2], and *Aris* Puji Widodo[3]

[1] Magister Program of Information System, School of Postgraduate Studies, Diponegoro University, Semarang – Indonesia 50242
[2] Department of Physics, Faculty of Science and Mathematics, Diponegoro University, Semarang – Indonesia 50275
[3] Department of Computer Science, Faculty of Science and Mathematics, Diponegoro University, Semarang – Indonesia 50275

Abstract. The choice of contraceptive tools is not an easy thing because the risks or effects will give impact on the body that never using it previously. in the other side , there is no contraception always suit for everybody because the circumstances of each body is different, so the extensive knowledge must be needed to know the advantages and disadvantages of each contraceptive tools then adjusted to the user's body.The expert system for contraceptive tools uses Forward Chaining search method combined with Certainty Factors Method. These method value the patient's indication. The Expert system gives the output data which define the kind of tool uses of the patient. the results obtained will be able to help people to find indications that lead to appropriate contraceptive tools and advice or suggestions about these tools. The success rate of the contraceptive tools decision experienced by experienced by the user by using forward chaining combined with the CF computation method is also influenced by the number of indication criteria selected by the user. Based on testing that has been done, expert system contraception tools has accuracy level equal to 75%.

1 Introduction

To suppress the rapid population growth, can be done with a variety of actions, such as promoting family planning programs to limit the number of children in a family to be built later, including in planning time and the number of children born to each child in the future [1]. The only way to set the number and time of pregnancy is through the use of contraceptives. Steps for setting up the Nativity or postpone pregnancy needed contraceptives in order toward the prosperous and happy families. Many of the undesirable pregnancy, unplanned pregnancy happen because these couples do not use preventive of pregnancy, and partly to use contraceptives but not correct usage.

Choosing birth control is not an easy thing because the risk or effects that affect to the body will not be known for have not used it. In addition there is no contraceptive are always suitable for everyone because of the situation and condition of the body is always different, so the extensive knowledge must be needed. Risk factors in the choice of contraceptives is very necessary to be considered. Factors that affect the frequency of sexual intercourse, such as ease of returns pregnant again, side effects to lactation, and the effects of the contraceptive in the future. In addition to the foregoing consideration of contraception also is based on the age, the use of contraceptives before, the period of use of contraception, effectiveness, security, frequency of use, menstrual periods, health disorders, breast feeding childbirth, types of contraceptives, as well as the willingness and ability to do a regular contraceptive. As an example of the use of the contraceptive pill, Mini-pills are highly effective (98.5%) for use in nursing mothers when used correctly and consistently. But the effectiveness of the contraceptive pill will be reduced when breast feeding, and the effectiveness of the mini pill would be lower if the use of the mini pill is consumed along with tuberculosis or epilepsy. The side effects caused from the use of the mini pill menstrual disorders such as increased weight gain, pain in the breast, nausea, dizziness, bloating and the incidence of dermatitis or Acne [2].

One of the researches associated with risk factors such as the analysis of the negative effects of the use of contraceptive implants on the lives of individuals [3]. There are many methods that can be used in the expert system, including method of forward chaining. Related research forward chaining has been widely performed by some researchers previously, one of which implementation method of forward chaining rule-based expert systems on classification of nutritional status toddler [4].

Other studies relating to contraception is using expert system to help diagnose ectopic pregnancy with the use of a factor of certainty to get quality answers from each patient [5]. The selection of efficient contraception can be judged from the cost of contraception in protecting pregnancy per year of its use from a pair. Then the election effectively contraceptives should be based on

* Corresponding author: dwiariefprambudi@gmail.com or dwiariefprambudi@student.undip.ac.id

consideration of the effectiveness and risks of each type of contraception is based on the level of its success.

2 Literature review

Development of expert system applied in many different areas of life. Several studies from different fields has ever done about expert system, introduce a visual basic expert system to help graduate students in the diagnosis and treatment. The system used to help graduate students in the diagnosis and treatment of oral ulcers are known to most common to the rarest [6] . In the health field has been widely developed an expert system to diagnose the disease diabetes complications mikrovaskular using cereal knowledge [7]. In the field of industry, expert system can be used to identify the location or any damage to the electronic systems, industrial systems, or on other systems. Expert system can also be used for optimization of the grinding force adjustments to get maximum results [8].

Related research Forward chaining rule-based pun has been much done by some earlier researchers, such as the implementation of Forward chaining on a complex interaction of social response on autistic children. In this study, a few examples of autistic children are taught how to receive the response in the process of sharing inducing [9].

There are also many research about birth control, such as a decision support system in the selection of an alternative birth control using fuzzy logic [10]. Other research trying to integrate expert system with analytical hierarchy process to determine the right birth control for each individual. The results in the form of information about contraception are recommended [11]. Research related to risk factors such as the analysis of the negative effects of the use of contraceptive implants on the life of the user [12]. research related to risk factors against the teenager on the use of contraception i.e. the existence of qualitative research that influences the selection of contraception in teen age [13]. The presence of risk factors posed on the use of oral contraception tool against cervical cancer diagnosis on the user [14]. The risk of the use of hormonal contraceptives against the incidence of cervical cancer suffered by users who performed research for a year from 2006 to 2007 [15].

2.1. Expert system

Expert system is a computer software that has a knowledge base for the specific domain and use reasoning inference resembles that of an expert in solving problems and finding solutions [16]. In General, the expert system is a system that attempted to adopt human knowledge to a computer that is designed to model the ability to resolve problems such as befits an expert. With this expert system, any layman can solve the problem or just simply looking for actual quality information can only be obtained with the help of experts in their field. Expert system will also be able to assist the activities of the experts as an experienced Assistant and has the required knowledge.

An expert with expert system has a lot of difference. Comparison of ability between an expert with an expert system as shown in table 1 [17].

Table 1. Comparison of experts and expert system (Durkin, 1994)

No	Factor	Human expert	Expert sistem
1	Time availability	workday	Anytime
2	Geographic condition	Local/ certain	Anywhere
3	perishable	Yes	No
4	Performance	Variable	Consistent
5	speed	Variable	Consistent
6	cost	High	Affordable

2.2 Forward chaining method

An inference that connects a multiplication problem with solution called chain. A chain is sought or is bypassed or crossed from a problem to obtain the solution referred to as Forward Chaining. Another way of describing the forward chaining this is by reasoning from facts leading to the conclusion that there is from the facts.

In the advanced rules of reasoning are tested one by one in a specific order. The sequence may be either a sequence of rules of incorporation into the base rules or also other sequence determined by the user. Each time a rule is tested, the expert system will evaluate whether conditions are true or false. If the condition is true, then the rule is kept then the next rule is tested. Conversely, if conditions are wrong, it is not stored and the next rule is tested. This process would be repeated until the entire base of the rules tested with a variety of conditions. Advanced reasoning work with problems that started with the recording of the initial information and the final settlement to be achieved, then the whole process will be done sequentially. Figure 1 shows the process forward chaining

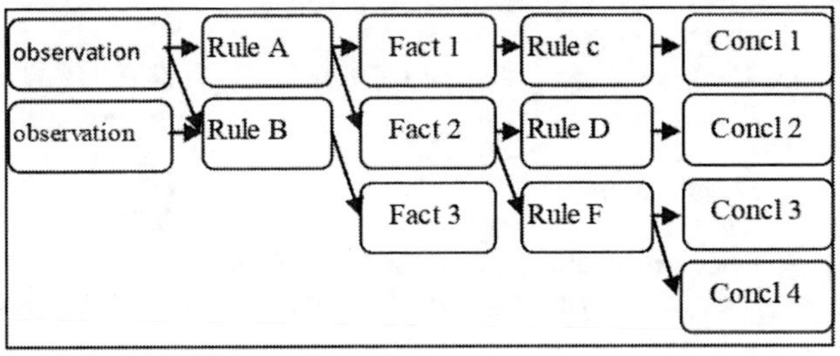

Fig. 1. Forward chaining method

2.3 Certainty factors

Expert system must be able to work in uncertainty [18] . Certainty factor is the way of the merger of trust and distrust in the single numbers. In certainty theory, qualitative data is represented as the degree of conviction. There are two steps in the qualitative data representations. The first step is the ability to express the degree of confidence in accordance with the methods.

The second step is the ability to place and the degree of confidence in combines expert system. In expressing a degree of certainty, belief theory using a value called the CF to assumes degrees of belief an expert against a data. The CF introduced the concept of belief and disbelief. This concept was later formulated in the following basic formula:

$$CF(H,E) = MB(H,E) - MD(H,E) \qquad (1)$$

Description:

CF (H, E): certainty factor of hypothesis H that is affected by symptoms
(evidence) E. CF Magnitude ranging between – 1 to 1. A value of – 1 indicates an absolute disbelief while a value of 1 indicates an absolute trust.

MB (H, E): the size of the increase in confidence (a Measure of Increased Belief) hypothesis H that is affected by symptoms of e.

MD (H, E): the size of the increase of unbelief (a Measure of Disbelief Increased) hypothesis H that is affected by symptoms of e.

CF network shows confidence in a hypothesis based on some fact or symptoms in medicine. CF-positive meaningful facts supporting the hypothesis because MB > MD.

CF = 1 means that the fact by definition proves a hypothesis. CF = 0 means one of two possibilities, namely the first CF = MB-MD = 0 both MB and MD is zero which means no facts. The second possibility is that the MD = MB and both are not the same as zero which means that trust is deleted or removed by distrust. CF negative means that the facts indicate negation of the hypothesis because MB < MD. In other words, it is more reasonable to express distrust toward the hypothesis than believe it.

Specify parallel CF obtained from some premise on a rule. The magnitude of the CF sequential influenced by the CF for each premise and the operators of the premises. The formula of each operator can be seen as follows.

CF (x and y) = Min (CF(x) , CF(y)) (2)
CF (x or y) = Max (CF(x) , CF(y)) (3)
CF (Not x) = - CF(x) (4)

The basic formula for the CF from the rule IF E THEN H is given by formula:

CF (H, E) = CF (E, e) * CF (H, e) (5)

Description:

CF (H, E): certainty factor hypotheses assuming the evidence uncertain, i.e. when the CF (E, e) = 1

CF (E, e): certainty factor evidence E influenced by evidence e

CF (H, e): certainty factor hypothesis was influenced by evidence e

If there are two rules that have the same hypothetical conclusions, then factor in the conviction of two rules combined with calculated confidence, namely:

CF CF (x) (y) (CF CF * (x) (y)), if (x) 0 CF and CF (y) 0 (6)

CF (x, y) = (CF CF (x) (y))/(((Min (CF (x) | ⊢| CF (y))))), one (CF, CF (x) (y)) 0 (7)

CF (x) (CF (y) * (1 CF (x))) if (x) 0 CF and CF (y) 0 (8)

CF is CF joint end of a prospective conclusions drawn. CF joint is required if a conclusion obtained from several rules at once. CF end of one rule with the other rules are combined to get the final value of the CF for the prospective conclusion.

3 Method

In this study were divided into several stages that are used to build an expert system application, starting from the stage of problem identification or analysis, knowledge acquisition and knowledge representation. In this study also used the method certainty factor to deal with the uncertainty of the results in the application of expert systems.

3.1 Identify the problem and knowledge

Making this expert system begins with determining the problem, in this case is to decide the appropriate contraception tools. It is very important to determine which will further the knowledge required in the system. The identification process begins from the knowledge acquisition of knowledge and continued with knowledge representation.

3.2 Knowledge acquisition

The acquisition of knowledge is a process to collect data knowledge will be a problem. In this study used two types of data obtained through different ways. The data used is primary data taken directly through the speakers in this case the experts that people be made the object of research to get information and data. Search primary data in this study will be conducted by interviewing an expert who was one of the doctors who are currently working in one hospital in Samarinda east Kalimantan. In this study also obtained information from the health books in the library, research journals and other literature that supports this research.

3.2 Knowledge representation

After the data collection process is completed, then the representation of data into the knowledge base and the rule base is then encoded, organized and described in draft form another so that a systematic form. The knowledge representation made into an expert system decides the appropriate contraception tools is to use the decision table and then formed a production rule.

4 Results

On the final assignment is examined about the use of contraceptives appropriate to each of its users. Basically any free user in selecting contraceptives. Users can use contraceptives is desirable. But any contraceptives have terms and conditions so that the tool is safe to use. The terms that will be used as factors in choosing contraception that is appropriate for each user

https://doi.org/10.1051/e3sconf/20183110009

Table 2 is a knowledge representation table of contracepstion tools.

Table 2. Table of knowledge representation.

no	Description Of Indication	Tools							
		P01	P02	P03	P04	P05	P06	P07	P08
1	In the reproductive age (20-35) = 0.85	X							
2	Blood pressure less than 180/110 mmHg = 0.80	X							
3	Smokers of all ages = 0.80	X							
4	want long-term contraception = 1:00	X	X	X	X	X	X	X	X
5	want a short-term contraception = 1:00		X	X	X	X		X	X
6	Not taking estrogen = 0.80	X							
7	breast cancer = 0.85	X							
8	impaired lung = 0.75	X							
9	Do not have a history of stroke = 0.80	X							
10	Do not take your medicine and epilepsy tbc = 0.85	X							
11	There is no inflammation of the genital = 0.95		X						
12	Requires protection against sexually transmitted diseases = 0.90		X						
13	No smoker = 0.90		X						
14	Do not have a history of allergy to rubber = 0.50		X						
15	Do not have a history of poisoning shock syndrome = 0.50		X						
16	No experience infection in the urethra (the tube pee) = 0.65		X						
17	Gender male = 0.85		X	X					
18	Age over 30 years = 0.85			X					
19	Children over 2 = 0.90			X					
20	Understand and agree to vasectomy procedure and the consequences = 0.65			X					
21	good psychological condition = 0.85			X					
22	Age over 26 years = 0.90				X				
23	Post-childbirth = 0.90				X				
24	Post-miscarriage = 0.90				X				
25	Understand and agree on the procedures and consequences tubektomi = 0.85				X				
26	obese more than 140/90 = 0.90				X				
27	hemoglobin less than 8 = 0.60				X				
28	There is no ovulation (fertilization in the uterus) = 0.85				X				
29	Do not want to be pregnant within a period of 1-5 years = 0.80					X			
30	The blood pressure <180/110 mmHg = 0.90					X			
31	suffered severe varicose = 0.65					X			
32	Do not have a history of jaundice = 0.60					X			
33	regular Haid = 0.60					X			
34	There are nursing = 0.90						X		
35	suffering from hiv / ims = 0.80						X		
36	suffer from pelvic inflammatory = 0.75						X		
37	infection in the genital = 0.85						X		
38	No bleeding in the vagina = 0.60						X		
39	anemic = 0.95							X	
40	had a history of heart disease = 0.80							X	
41	Not having migraine = 0.95							X	
42	Approaching the age of menopause = 0.85							X	
43	Do not want more children but do not want to use safe contraception = 0.95								X
44	Mother postpartum over 6 weeks = 0.85								X
45	Mothers who have a history of hypertension = 0.55								X

Using Forward Chaining and Certainty Factors then those factors will be made of the facts that are then processed by the existing rules so that it will yield a conclusion. Where is the conclusion in the form of contraceptive tools that can be used by any user. The results obtained by each user can be more than one type of contraception. So that the user can select the safe contraceptives to suit each user. Because according to the medical rules, users are welcome to choose which tools will be used. While only an expert help in choosing birth control.

The following list of contraceptives that are used:

1. P01The contraceptive pill
2. P02 Condoms
3. P03 Vasectomy
4. P04 Tubektomi
5 P05 norplant
6 P06 Iud.
7. P07 Contraception hypodermic/ injection
8. P07 Implants

4.1 Handling uncertainty

In this study used a method certainty factor (CF) for the treatment of the uncertainty of the expert system. In the processing of data after their parameters, the next step is to create a rule that will be used in the expert system, after making the rules, then making inference engine with forward chaining methods as search stages and methods of the CF as the stage of the weighting. Examples of the application of the method of certainty factor in a case where an existing assumptions

In this expert system, the value obtained from the certainty factor CF and CF expert users. CF value obtained when users perform diagnostics on a CF specialist symptoms while the value given to a disease in a rule (rule) by using the AND operator in any rule made. Note the State is represented by a variable number 4, 1, 2, 3, 6, 7, 8 (relation 1) assuming that all circumstances are met (experienced by the patient).

If known cf 4 = 1.00, 1 = 0.85 , 2 = 0.80 , 3 = 0.80 , 6 = 0.80, 7 = 0.85 , 8 = 0.75 cf expert relationship 1 = 0.80 then calculated the value of cf conclusion With the following steps:

1. Calculate the cf parallel relations 1

   CF parallel relation 1: max (cf 4, cf 1, cf 2, cf 3, cf 6, cf 7, cf 8)

   : max (1.00, 0.85, 0.80, 0.80, 0.80, 0.85, 0.75)

   : 1.00

2. Calculate the cf sequensial relation 1

   CF sequensial cf 1: parallel relations relations 1 * cf expert relationship 1

   : 1.00 * 0.80

   : 0.80

From the above calculation means of 0.80 percentage magnitude,, that means producing 80%.

4.2 Examination systems

System testing is performed to find out the extent success of expert system that has been built. The process of testing is done by evaluating the input and output that is generated by the application. This phase is carried out by field testing involving related experts (in this case the doctor or tocologist). Testing with experts performed a number of times with the intention of enabling an application able to be corrected as early as possible. Experts try to direct the system created by selecting symptoms from the system then check if diagnoze are produced accordingly and not distorted. After testing then proceed with validation as the process of testing the performance or accuracy against the example given during the process of testing is in progress. Stage of the validation process is carried out after the test was given, next the application is tested with another data. The purpose of validation is done in order to find out the extent of which the application is able to predict the values of the output and input values are given. The following equation validation.

$$\text{Validation } (\%) = \left(\frac{A}{B}\right) \text{x } 100\% \qquad (9)$$

From the equation 2, A is the amount of data the same estimation results with the target, and B is the number of target data.
The equation will be applied at the time of calculating the level of accuracy in deciding application of contraception tools.
The application level test results fault detection of the tools , there are 15 of the 20 testing appropriate test. So that the application has a success rate of the contraception tools decision by 75 %.

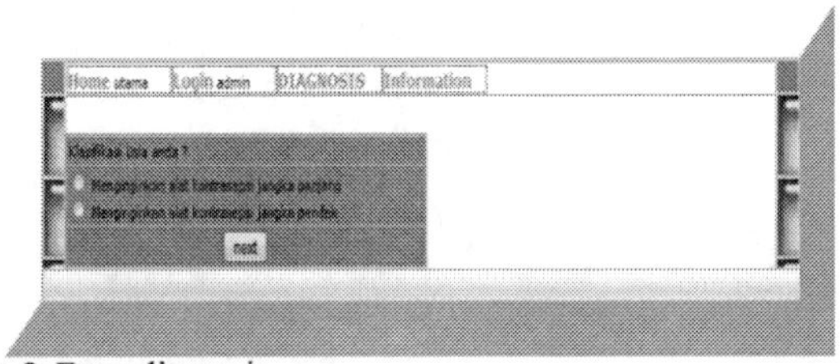

Fig. 2. Form diagnosis

Fig. 3. Form home

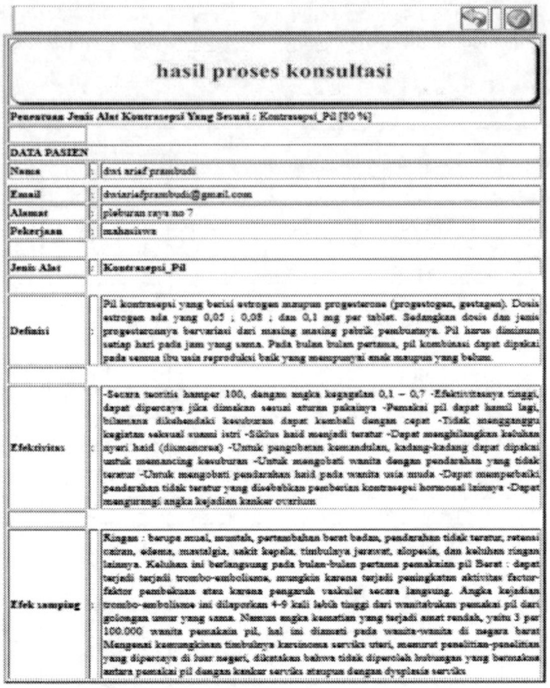

Fig. 4. Final calculation result

5 Conclusion

From a series of research methodology, analysis and discussion design of a system that has been done then it can be drawn the conclusion, has produced an application expert system for deciding the appropriate contraception tools is built using web-based method of forward chaining and certainty factors, using the programming language PHP and the MySQL database, which is a work based on user selected symptoms are then processed by the system so that it generates output in the form of the name tools, definition of tools, risk factors ,examples of tools pictures, the percentage of confidence handling solutions diagnosis and tools.

Based on the testing that has been done the application expert system of the decision of contraception tools has an accuracy rate of 75%.

References

1. Y. E. Chan, H. K. Ma, E. T. Chan, H. Y. Chen, and T. Y. Chen, (1995). International Journal of Computer Science Vol. **24**, 293-298, (1995)
2. S.N. Mulyani, *Keluarga Berencana dan Alat Kontrasepsi,* Nuha Medika ; Yogyakarta, (2013)
3. E . S. Watkins, Journal J Womes Hist, **88**-111, (2013).
4. N. Suhati, *Implementasi Metode Forward chaining Berbasis Aturan Pada Sistem Pakar Klasifikasi Status Gizi Balita,* thesis, Universitas Diponegoro, Semarang, (2015)

5. G. Astuti, *Sistem Pakar Pendiagnosis Kehamilan Ektopik,* thesis, Universitas Gajah Mada, Yogyakarta, (2004)
6. S. A. Ali, and H. I. Saudi, (2014), Tanta Dental Journal Vol. **11,** 42-46, (2014)
7. T.L.T. Nguyen, and N.V. Do, 2013, IJCSI Vol. **10,** 308-316, (2013)
8. L. Rubio, M. D. Sen, A. P. Longstaff, and S. Fletcher, (2013). Journal Expert Systems with Applications Vol. **40**, 2312-2322, (2013)
9. J. A. DeQuinzio, D. B. Townsend, and C. L. Poulson, (2008), Research in Autism Spectrum Disorders Vol. **2** 264-275, (2008)
10. D.R. Wardhani, *Sistem Pendukung Keputusan Dalam Penentuan Alat Kontrasepsi Untuk Keluarga Berencana Dengan Pemodelan Logika Fuzzy, seminar nasional,* Stimik Amikom, Yogyakarta, (2014)
11. W. Priharyanti, *Sistem Penunjang Keputusan Pemilihan Alat Kontrasepsi,* thesis, Universitas Indonesia, Depok, (2013)
12. E . S. Watkins, Journal J Womes Hist, 88-111, (2013).
13. A.J. Hoopes, K. Gilmore, and K.R. Ahrens,. Journal of Pediatric and Adolescent Gynecology, Vol. **10**, 164-259, (2015)
14. P . Lesmana, *Faktor Risiko Pemakaian Alat Kontrasepsi Oral, Merokok,Dan Karakteristik Penderita Terhadap Kejadian Kanker Leher Rahim Di Rsd Dr. Soebandi Jember,* Skripsi , Universitas Negeri Jember, Jember, (2011)
15. D . S. Tira, *Risiko Jumlah Perkawinan, Riwayat Abortus, Dan Pemakaian Alat Kontrasepsi Hormonal Terhadap Kejadian Kanker Serviks Di Rumah Sakit Pelamonia Makassar Tahun 2006 – 2007,* thesis , Universitas Nusa Cendana, Kupang, (2008)
16. F. Aziz, *Belajar Sendiri Pemrograman Sistem Pakar,* Elex Media Komputindo , Jakarta, (1994)
17. J. Durkin, *Expert Systems: Design and Development,* Macmillan Coll Div, , New York, (1994)
18. Kusrini, *Sistem Pakar Teori & Aplikasi.* Andi ,Yogyakarta, (2006)

https://doi.org/10.1051/e3sconf/20183110010

Paskibraka Member Selection Using A Combination Of AHP and TOPSIS Methods On The Office Of Youth And Sports Of Kutai Kartanegara Regency

Septya Maharani [1,*] , *Heliza* Rahmania Hatta [1] , *Afif* Nur Anzhari [1] , *Dyna* Marisa Khairina [1]

[1]Departemen of Computer Science. Faculty of Computer Science and Information Technology, Mulawarman Unversity, Samarinda-Indonesia

Abstract. Paskibraka as troops whose job is to flap the heritage duplicates flag. To become a Paskibraka a selection that participants are high school students are made. Because the number of participants of the selection of many support systems to facilitate the assessment process is made. This system uses Analytical Hierarchy Process (AHP) to determine the weight value criteria that comprise the value of the interview, health, physical, .height and value rules for marching as well as using Technique For Others Preference by Similarity to Ideal Solution (TOPSIS) methods to seek best alternative participants. The calculation results of 21 alternative names best male and female of the participants and their school origin. The system has also been tested by performing the calculations manually using Microsoft Excel (Ms.Excel) to calculate the calculation of the system using AHP and TOPSIS.

1 Introduction

In Paskibraka elections a several main criteria on which to base an assessment is conducted. The assessment is then processed manually in order to get some of the names of the members elected Paskibraka. With so many alternative candidate members, it will take a long time in the selection process. To find the optimal alternative of a number of alternatives and certain criteria can use a decision support system with several methods namely Analytical Hierarchy Process (AHP) and Technique For Order Preference By Similarity To Ideal Solution (TOPSIS).

AHP is a method that uses human perception as the primary input for use as a base to look for eigenvalues, then determine Consistency Ratio. TOPSIS method to determine the best alternative solution [1].

Based on the description that has been presented, then built a system to help determine the election decision Paskibraka in Kutai Kartanegara regency using a combination of AHP and TOPSIS to find the best alternative solution.

2 Literature review

2.1 Decision Support System

Decision support system is a computer-assisted interactive system that supports the user in the ease of access to data and decision models in an effort to help the decision-making process effective in solving problems that are semi-structured and unstructured [2].

Decision support system (DSS) is an interactive information system that provides information, modeling, and data manipulation. This system is used to assist decision making in semi-structured and unstructured situations [3].

From the above understanding, it can be concluded that Decision Support System (DSS) is a computer-based information system that approaches to generate various alternative decisions to assist certain parties in dealing with problems using data and models. A DSS only provides an alternative decision and then submitted to the user to make a decision.

Decision-making is the result of a selection process from a variety of alternative actions that may be chosen by a particular mechanism, with the aim of producing the best decision. Where the decision process is gradual, systematic, consistent and in every step from the beginning has included all parties, will give good results.

2.2 *Paskibraka*

Pasukan Pengibar Bendera Pusaka, or more commonly known as PASKIBRAKA, is an army in charge of raising the duplicate of the heirloom flag in the commemoration ceremony of the proclamation of Indonesian independence on 17 August held in three places, namely the Regency / Municipal level (Office of the Regent / Mayor) Provincial (Governor's Office), and National (State Palace) [4].

* Corresponding author: septyamaharani@yahoo.com

https://doi.org/10.1051/e3sconf/20183110010

2.3 Analytical Hierarchy Process (AHP)

Analytical Hierarchy Process (AHP) is one form of decision-making methods that use a functional hierarchy with the primary input is human perception. With hierarchy, something that is complex and unstructured broken into groups and then the group is set to be a form of hierarchy [5].

2.4 Technique for order preference by similarity to ideal solution (TOPSIS)

TOPSIS based on the concept that the best alternative was selected not only has the value of the shortest distance from the positive ideal solution, but it also has the longest distance value from the negative ideal solution. This concept is widely used in several models of Multi-Attribute Decision Making (MADM) to solve problems in a practical decision. This is due to the concept is simple and easy to understand, efficient computation, and has the ability to measure the relative performance of the alternatives in the form of a simple mathematical decision [6].

3 Research methodologies

3.1 Method of collecting data

1) *Literature:* The authors conducted a study of literature to find references related to the Decision Support System (DSS), Personal Home Page (PHP), My Structured Query Language (MySQL), books, text, journals and scientific works as a review of the literature to support research that does not deviate from the context.
2) Field Observation: Writers make a field trip to collect the necessary data is based on information FH Dian Suryadi, SE Paskibraka 2016 Selection Coordinator Kutai regency for participants judging criteria.
3) *Interview:* The authors conducted interviews to Mr. FH Dian Suryadi, SE to get the value criteria and the father Akhmad Dwi Ari Saputra, Amd.Kep to get the value of health.

3.2 Stage analysis

In the next stage of data analysis required data in the form of health values, interviews, physical ability, high body, and the value of the United Nations. The value is obtained from the recapitulation during selection done. The values will then be calculated using the AHP and TOPSIS methods.

3.3 Modeling of AHP-TOPSIS method

The initial phase of this system is a pad by clicking the input value comparison of each criterion by using the standard value comparison of AHP, then input the results are processed using AHP to obtain eigenvalues. Eigen value is then used to find the value of concistency ratio, if in accordance with the standard value of concistency ratio continued to the calculation process by using TOPSIS method, otherwise it will repeat again to input the value of each criteria comparison. At the stage of using TOPSIS method will find the best solution of each alternative. Alternate data of selection participants entered and then stored. These alternatives have data in the form of health values, interviews, physical ability, high body and value rules for marching which is then used to find the value of each alternative weight. The best preference value would be PASKIBRAKA member recommendation of 21 sons and 21 daughters.

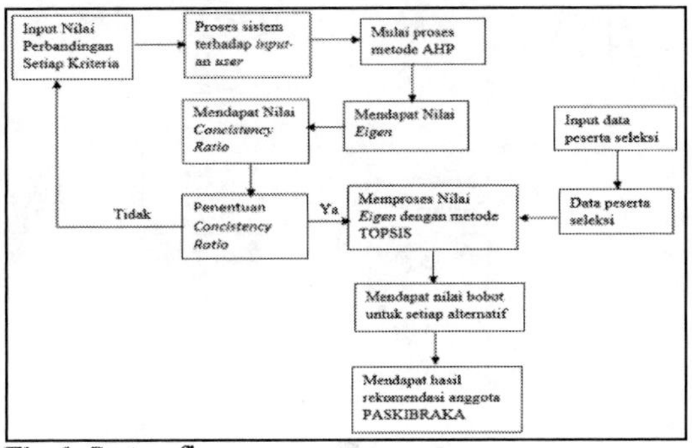

Fig. 1. System flow

4 Results and discussion

4.1 System analysis

AHP method is used as the first process base whose input value comes from admin and get the result of weight value criteria will be processed in the next calculation using TOPSIS method. The first stage in the system process is to include the comparative value of each criterion based on the standard AHP comparison value.

The second stage, the system will arrange the value of the comparison that has been inputted into a matrix of equation comparison. The third stage, after being a comparison matrix of equations, then the matrix is normalized. The fourth stage, the matrix normalization in total per line criteria provided then divided by 5 according to the number of criteria available produce eigen value / weight value.

The fifth stage, eigen values / weight value is then used to find the value consistency index, if the value CR of the calculation process AHP qualify CR (Consistency Ratio) received $CR \leq 0.1$ then the calculation process can be continued to methods TOPSIS, if No admin is required to enter again the criterion comparison value. The next stage is using TOPSIS method to search for the best alternative. The seventh stage, the system will store the critical weight value of the AHP method and then the admin will input the weight value of each alternative. The value of alternative weight is then made into matrix and normalized.

The eighth stage, after normalization of the alternative assessment continued with weighted normalization using the weighted value criteria that have been obtained in the AHP calculation process. The ninth stage computes the ideal positive and ideal solution that produces the positive and negative ideal spacing of each alternative at the tenth stage. The final stage calculates the preference value for each alternative followed by sorting the preference value.

The recommendations are the best preferences of 21 alternative male and 21 female alternatives.

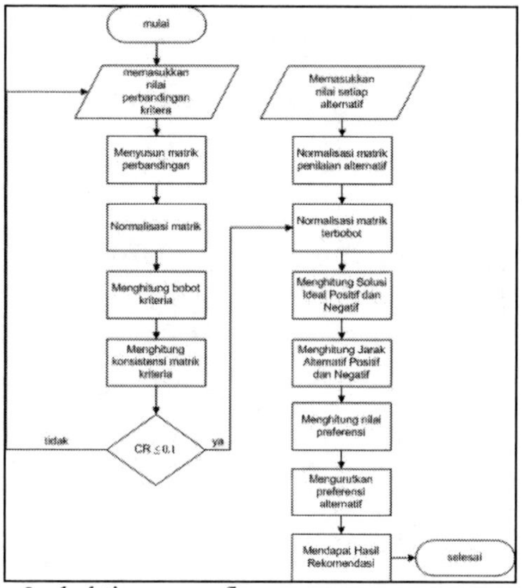

Fig. 2 calculation process flow system

4.2 System Implementation

Implementation system is a stage that is done after the system has been built based on the designs that have been made.

4.3 Home Page

Home Page contains an explanation of the method of AHP and TOPSIS on Selection Paskibraka and how to use this system. Homepage menu display can be seen in Figure 4.

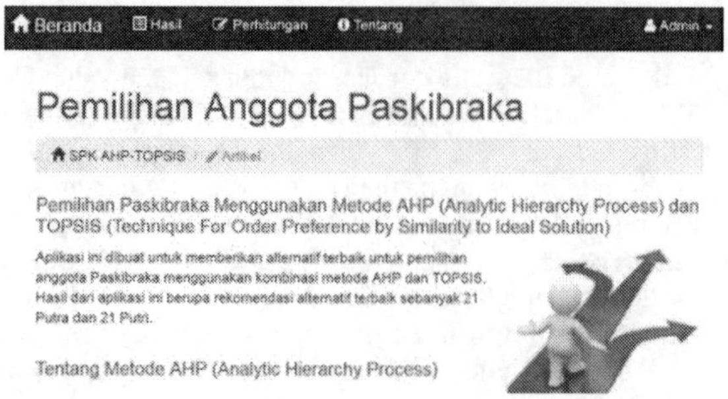

Fig. 3. Home Page

4.4 calculations page

On the Calculation page displays detailed calculations of AHP and TOPSIS methods starting from calculating the criteria weighted value to an alternative preference value. The Calculation page shows the entire system calculation from matrix comparison to alternative sorting based on value of preference / value of weighted criteria obtained. Calculation menu view can be seen in figure 4.

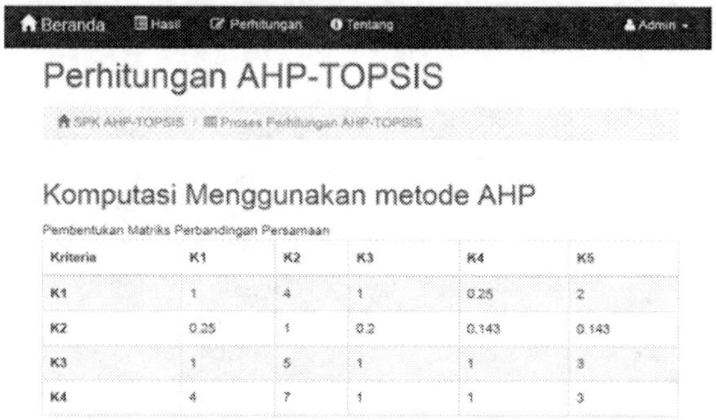

Kriteria	K1	K2	K3	K4	K5
K1	1	4	1	0.25	2
K2	0.25	1	0.2	0.143	0.143
K3	1	5	1	1	3
K4	4	7	1	1	3

Fig. 4. Calculation Page

4.5 Result Page

Display Results menu displays the results of system recommendations based on the value of weight that has been entered by the admin. The recommendations in the form of 21 male and female names along with the school of origin and final score calculations. result page display can be seen in Figure 5.

Fig. 5. Page Results

4.6 System testing

System testing is done to determine the extent to which the success of the calculation of decision support systems that have been built. At this stage, the system will be tested by comparing the calculation of AHP and TOPSIS system with AHP and TOPSIS calculations performed manually.

Table 2. *Sample* Selection Participants Paskibraka

NO	NAME	PARTICIPANTS'S NUMBER	WHICH SCHOOL ARE YOU FROM	SELECTION RESULT			Height	UN
				Health	Interview	Physical test		
1	MUHAMMAD ROJALI	002	SMAN 1 Ma.MUNTAI	S	A	76	171	70
2	YUDHA WAHYU ADRI SETIAWAN	278	SMAN 1 LOA KULU	S	B	71	173	70
3	THE RAHMAN HADI	251	SMK YPK TGR	S	A	70	172	90
4	ABDOEL MANAB	254	SMAN 1 Ma.WOOD	S	B	6 8	170	89
5	MUHAMMAD LUFIANTO	052	SMKN 1 TGR	S	B	70	168	80

Table 3. Selection Criteria Paskibraka

Criteria	Criteria code
Health	K1
Interview	K2
Physical test	K3
Height	K4
marching	K5

1. Calculation of AHP method

Table 4. Table *Pair Wise Comparisons*

Criteria	k1	k2	k3	k4	k5
k1	1	4	1	0.25	2
k2	0.25	1	0.2	0.143	0.143
k3	1	5	1	1	3
k4	4	7	1	1	3
k5	0.5	7	0.333	0.333	1
amount	6.75	24	3.533	2.726	9.143

Table 5 Table Normalized Matrix *Pair Wise Comparisons*

criteria	k1	k2	k3	k4	k5	Jumlah	*Average* / priority weight
k1	0.148	0.166	0.283	0.091	0.218	0.908	0.182
k2	0.037	0.041	0.056	0.052	0.015	0.203	0.041
k3	0.148	0.208	0.283	0.367	0.328	1.335	0.267
k4	0.592	0.291	0.283	0.367	0.328	1.862	0.372
k5	0.074	0.291	0.093	0.121	0.109	0.689	0.138

Table 6. Vector Weight Value

Vector of weight
0.981
0.212
1.439
2.066
0.727

Calculating CR value

$$\lambda_{max} = \frac{((0.981/0.182) + (0.212/0.041) + (1.439/0.267) + (2.066/0.372) + (0.727/0.138))}{5}$$

$$= 5.361$$

$$CI = \frac{5.361 - 5}{5-1} = 0.090$$

$$CR\ (Consistency\ Ratio) = \frac{0.090}{1.12} = 0.080$$

By *CR (Consistency Ratio)* generated is less than 0.1 or $CR \leq 0.1$, the final result can be justified and can proceed to the next step and conclude that the comparison between the criteria consistently.

2. Calculation of TOPSIS method

Table. 7 Alternative Assessments

Alternative/ No participants	Alternative Rating Matrix				
	K1	K2	K3	K4	K5
002	5	5	76	171	70
052	5	3	70	168	80
251	5	5	70	172	90
278	5	3	71	173	70
254	5	3	68	170	89

Table. 8 Normalized Matrices Weighted

Alternative/ No participants	Normalized Matrix Weighted				
	K1	K2	K3	K4	K5
002	0.081	0.023	0.127	0.166	0.053
052	0.081	0.013	0.117	0.163	0.061
251	0.081	0.023	0.117	0.167	0.069
278	0.081	0.013	0.119	0.168	0.053
254	0.081	0.013	0.114	0.165	0.068

Table 9 Positive Ideal Matrix Solutions & Ideal Negative Matrix Solutions

	K1	K2	K3	K4	K5
K^+	0.081	0.023	0.127	0.168	0.069
K^-	0.081	0.013	0.114	0.163	0.053

Table 10 Positive Distance Matter and Alternative Negative Distance

Alternative	Distance +	Distance -
2	0.0155	0.0166
52	0.016	0.008
251	0.010	0.0187
278	0.0198	0.007
254	0.0166	0.0147

$$V_2 = \frac{0.01663}{0.01663 + 0.0155} = 0.517$$

$$V_{52} = \frac{0.008}{0.008+0.016}$$

$= 0.337$

$$V_{251} = \frac{0.0187}{0.0187+0.010}$$

$= 0.648$

$$V_{278} = \frac{0.007}{0.007+0.0198}$$

$= 0.260$

$$V_{254} = \frac{0.0147}{0.0147+0.0166}$$

$= 0.469$

Table 11 Preference Value

Alternative	Preference	Alternate Sequence	Preference Sequence
2	0.517	251	0.648
52	0.337	2	0.517
251	0.648	254	0.469
278	0.2608	52	0.337
254	0.469	278	0.260

The order of selection of members Paskibraka recommendations were counted manually in the order recommendation Paskibraka members is calculated using the system seen in Figure 10 with the difference of each alternative 0001.

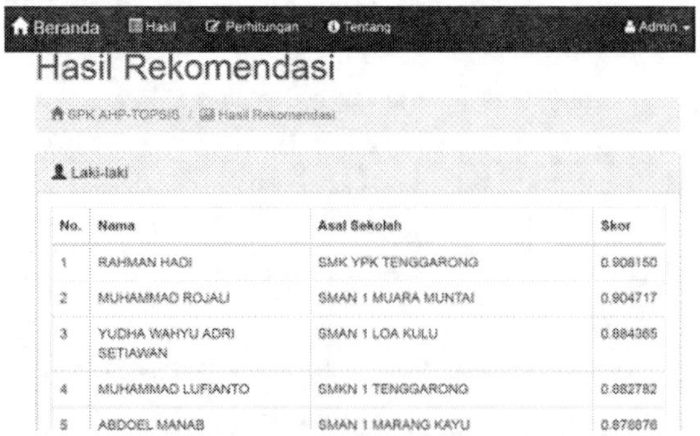

Fig. 6. System Calculation Results

4.7 Validation of results

After testing the calculation of the system run in accordance with the manual calculation then input all participants as much as 73 men and 69 women. All participants of the election Paskibraka alternative stored next step is to do a comparison of the value of the 5 existing criteria namely health, interview, physical test, height, and the value of the UN. Then the results of recommendations based on the system validated against the results of selection conducted directly by the Dispora Kutai Kartanegara.

The validation result between the recommendation of the system and the direct selection by the committee is 40.47% value less than 50% so it is assumed the result of less precise but precise calculation method. The result of validation is very less because there are several factors that can not be calculated by the system such as preferably the members paskibraka school and the appearance of the selection participants.

5 Conclusion

Analytical Hierarchy Process (AHP) and Technique For Order Preference By Similarity To Ideal Solution (TOPSIS) successfully applied in decision support system of Paskibraka selection in Youth and Sports Department of Kutai Kartanegara Regency. The criteria chosen in this research are health, interview, physical test, high body, and PBB can be processed and then produce recommendation through stages that have been specified in method AHP and TOPSIS

References

1. S. Maharani, HR. Hatta, G. Merdiko, " *Articles of Bali International Seminar On Science and Technology.* (2014)
2. D. U. Daihani, *Computerized Decision Making*. Bogor: Ghalia Indonesia. (2001)
3. Kusrini. *Concept and Application of Decision Support System.* Yogyakarta: Andi Offset. 2007.
4. *Regulation of the Minister of Youth and Sports of the Republic of Indonesia* Number 0065 of 2015.
5. Padmowati, RdLE, *Measurement Index Consistency In Decision Making Process Using AHP Method.* Semnas IF. 2009.
6. Kusumadewi, Sri, et al. Fuzzy *Multi Attribute Decision Making*. Yogyakarta: Graha Science. 2006.
7. A. Rasyid, S. Maharani, Journal of Informatics Mulawarman Vol. **11**, No. 2, September 2016. "
8. W Idodo, PP, and Herlawati. Using UML *(Unified Modeling Language.* Bandung: Informatics, 2011.
9. JL. Whiten, and LD. Bentley, *System Analysis and Design Methods* Seventh Edition. New York: The McGraw-Hills Company. 2007.

https://doi.org/10.1051/e3sconf/20183110011

Total Quality Management of Information System for Quality Assessment of Pesantren Using Fuzzy-SERVQUAL

Arbiati Faizah[1,*], *Wahyul* Amien Syafei[2], and *R. Rizal* Isnanto[3]

[1]Master of Information System, School of Postgraduate Studies, Diponegoro University, Semarang – Indonesia 50242
[2]Department of Electronic Engineering, Faculty of Engineering, Diponegoro University, Semarang – Indonesia 50275
[3]Department of Computer System, Faculty of Engineering, Diponegoro University, Semarang – Indonesia 50275

Abstract. This research proposed a model combining an approach of Total Quality Management (TQM) and Fuzzy method of Service Quality (SERVQUAL) to asses service quality. TQM implementation was as quality management orienting on customer's satisfaction by involving all stakeholders. SERVQUAL model was used to measure quality service based on five dimensions such as tangible, reliability, responsiveness, assurance, and empathy. Fuzzy set theory was to accommodate subjectivity and ambiguity of quality assessment. Input data consisted of indicator data and quality assessment aspect. Input data was, then, processed to be service quality assessment questionnaires of Pesantren by using Fuzzy method to get service quality score. This process consisted of some steps as follows : inputting dimension and questionnaire data to data base system, filling questionnaire through system, then, system calculated fuzzification, defuzzification, gap of quality expected and received by service receivers, and calculating each dimension rating showing quality refinement priority. Rating of each quality dimension was, then, displayed at dashboard system to enable users to see information. From system having been built, it could be known that tangible dimension had the highest gap, -0.399, thus it needs to be prioritized and gets evaluation and refinement action soon.

1 Introduction

Pesantren is the oldest education institution in Indonesia, in which this institution takes a part in giving education especially religion education to society [1]. Pesantren institution is demanded to implement quality control system to ensure performance achievement in product or merit service made.

Information system of Total Quality Management (TQM) is used as a tool to support management process so that the decision makers can arrange planning and make decision more efficiently [2]. Information system of TQM also can be used as benchmarks based on customers' satisfaction on merit service given by education institution. Customers meant in this case are those having relation with education institution such as students, teachers, parents, and society. Customers' satisfication on merit service quality can be measured by using SERVQUAL (Service Quality) model. This model measures service quality using five different dimensions which can be admitted as indicators making quality service received by customers [3].

In service satisfaction analysis, customers' judgment subjectivity has uncertain score of information characteristic given. Fuzzy set is used as representation and information processing influenced by some uncertainties because of natural language use. Fuzzy set can be used as a tool to calculate linguistic data and change it to be a numeric format so it is easily manipulated by computer [4].

Based on how important TQM implementation is in Pesantren, this research was hoped to be a solution in building information system of total quality management to measure achievement of pesantren quality by using fuzzy SERVQUAL. This system was applied online, so it enabled the decision makers to evaluate and make decision about quality refinement efficiently.

2 Total Quality Management (TQM) of Pesantren

In Pesantren education context, the starting point of quality is the strengthening of vision and mission directed to customers' satisfaction; it is satisfaction of education merit users (students' parents, working world) [5]. Criteria of qualified Pesantren are as follows:

1. Safe and orderly education environment.
2. Clear vision, mission and quality target forming.
3. Reliable education leadership.
4. High achievement expectation and output.
5. Qualification development and improvement.
6. Effective and efficient learning evaluation.
7. Interaction among institution, parents and society [6].

TQM is as an integrated organization strategy to increase qualified product and service as a helper or a solution for various problems faced by an organization,

* Corresponding author: arbiati_f@student.undip.ac.id

including organization performance improvement. Four main principles in TQM are as follows:

1. Customers' satisfaction.
2. Respect to everyone.
3. Fact based management.
4. Continuous refinement.

Quality management system always has something to do with cycle concept of PDCA (Plant-Do-Check_Act). Fig. 1 shows PDCA cycle in continuous refinement.

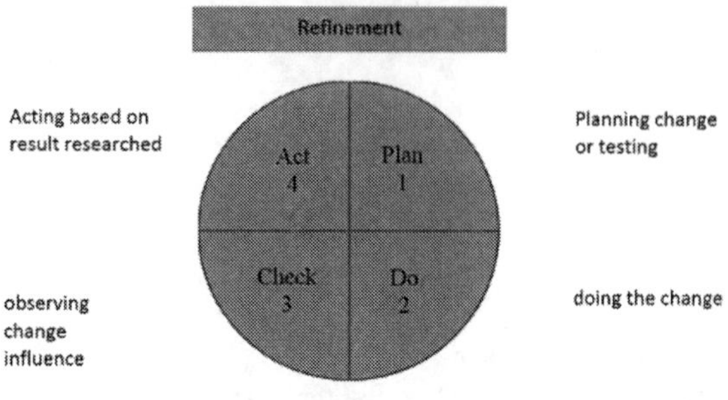

Fig. 1. PDCA Cycle [7]

The success benchmarks of quality management in Pesantren are measured from internal and external customers' satisfaction levels. Pesantren is said to be successful if they can give service as customers' expectation. Customers' satisfaction indicators on service given by Pesantren are as follows:

1. Students are satisfied with service of Pesantren.
2. Parents feel satisfied with service given to their children.
3. Users or receivers of alumni are satisfied because of receiving alumni with high quality and suitable with their expectation
4. Teachers and staffs are satisfied with service given by Pesantren [8].

3. Service Quality (SERVQUAL)

Quality in merit organization is a measurement of achievement range of service given to fulfill customers' expectation. Model of SERVQUAL is used to measure service quality received by customers. Merit quality dimensions are used in measuring service merit quality. The service quality dimensions consist of tangible, reliability, responsiveness, assurance, and empathy. The measuring of merit quality using model of SERVQUAL consists of score gap given by customers for statements relating to expectation and perception about merit received by customers. The calculation of score gap between expectation and perception using formula 1.

$$SQ = \sum_{i=1}^{k}(P_i - E_i) \qquad (1)$$

The assessment of service quality is calculated based on the number of attribute k. Then, it is continued by calculating score differences from perception scores of each attribute (P_i) and expectation scores of each attribute (E_i).

4 Fuzzy Set

Fuzzy compilation theory gives tool to present uncertainty and also becomes a good tool for uncertainty modeling relating to vagueness, incorrectness and lack of information about certain elements of problem faced. The power becoming base of Fuzzy compilation theory is that the use of linguistic variable is better than quantitative variable to present concept which is not precisive. Fuzzy compilation is a set containing elements having different member degrees and are very different from classic set (crisp).

Membership function can be expressed as triangular function, gaussian function, or trapeziodal function. Membership function on triangular curve can be expressed by using Triangular Fuzzy Number (TFN). TFN is fuzzy compilation expressed in the interval form used to represent subjective judgment of customers. Triangular curve is basically combination of lines shown in Fig. 2.

Membership function :

$$\mu[z] = \begin{cases} 0; & x \le a \text{ or } x \ge c \\ (x-a)/(b-a); & a \le x \le b \\ (b-x)/(c-b); & b \le x \le c \end{cases} \qquad (2)$$

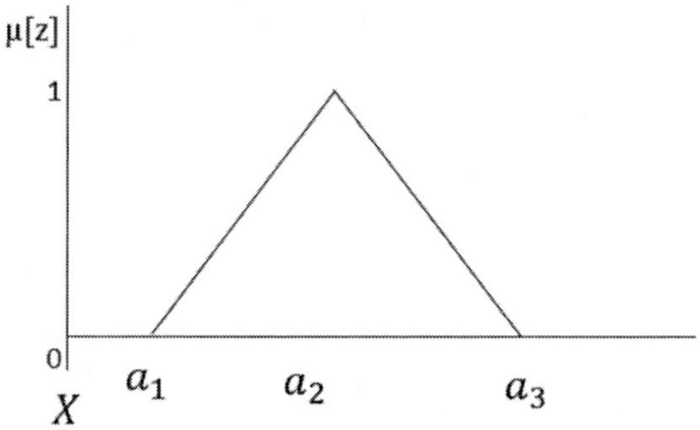

Fig. 2. Triangular curve [9]

Calculation of fuzzification on service quality assessment can be done by looking for score of a_1, a_2, a_3 for each aspect based on formula 3, 4, 5 [10].

Lower limit (a_1) =

$$\frac{b_{i1}*n_1 + b_{i1}*n_2 + b_{i2}*n_3 + \dots + b_{i(k-1)}*n_k}{n_1 + n_2 + n_3 + \dots + n_k} \qquad (3)$$

Middle Score (a_2) =

$$\frac{b_{i1}*n_1 + b_{i2}*n_2 + b_{i3}*n_3 + \dots + b_{ik}*n_k}{n_1 + n_2 + n_3 + \dots + n_k} \qquad (4)$$

Upper limit (a_3) =

$$\frac{b_{i2}*n_2 + b_{i3}*n_3 + \dots + b_i*n_{i(k-1)} + b_{ik}*n_k}{n_1 + n_2 + n_3 + \dots + n_{(k-1)} + n_{jk}} \qquad (5)$$

The next step is defuzzification. Defuzzification process is done to get a single score by using formula of Mean of the maximum as follows:

Deffuzification = ((a_3-a_1) + (a_2-a_1))/3 + a_1 (6)

5 Methodology

This research had some steps which had to be done in information system built. Those steps consisted of determining input, storing, processing, and output. Design form of system modelled is appropriate with this research purpose; which was building quality assessment system using method of fuzzy-SERVQUAL. In Pesantren, variables involved in quality assessment are : quality assurance unit, tutors, teachers/ustadz, staffs and *pengasuh*. Assessment indicators come from work planning and program of quality assurance. Attribute of questions are arranged from those indicators and classified into dimensions of SERVQUAL shown in picture 3.

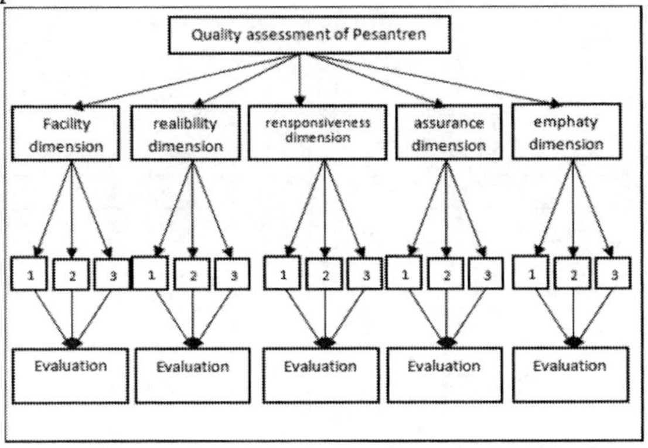

Picture 2. SERVQUAL Hierarchy Model of Pesantren quality assessment

Attribute of questions are used as criteria of quality assessment between expectation and perception. Perception meant in this discussion is customers' satisfaction on merit or service given by Pesantren. Input system consists of quality assessment by merit users. Attributes of questions of each dimension are shown in table 1. Result of this quality assessment is then stored at database. Assessment data got is then processed by using fuzzy method.

Calculation processes consist of fuzzification, and defuzzification of triangular fuzzy number. Fuzzy compilation data taken for fuzzification calculation were statement data on questionnaire about fact and expectation as shown in table 2.

Fuzzification calculation on service quality assessment can be done by looking for scores of a,b,c of each attribute by using formula 3,4,5. The next step is defuzzification. Defuzzification process is done to get single score by using formula 6. After single score is got, system, then, calculates gaps of each criteria based on indicators and quality of scores given by respondents.
Output is resulted from this process in the form of result data of quality achievement assessment of Pesantren. Those data can be used as materials of dimension evaluation and to know which aspect must be improved.

Table 1. Attributes and variables in SERVQUAL dimension

Dimensions	Attributes	Variables
Tangible	Completeness and readiness in teaching and learning process	X_{11}
	Availability of clean eating place	X_{12}
	Every day cleanliness and comfort of Pesantren	X_{13}
	Spatial arrangement in library and other supporting facilities	X_{14}
	Availability and capacity of parking area	X_{15}
Responsiveness	Teachers' capability which is professional and right	X_{21}
	Clear and correct explanation when there is students' or their parents' question	X_{22}
	Teachers' good capabilities in delivering material	X_{23}
	Security always works well	X_{24}
	Builders always do their tasks as their capacities	X_{25}
Assurance	Teachers' knowledge and capabilities in teaching	X_{31}
	Other skills of teachers and staffs in working	X_{32}
	Polite and friendly service	X_{33}
	Safety and comfort in teaching and learning process	X_{34}
	Students get assurance of caring facilities in Pesantren clinic	X_{35}
Emphaty	The same service is given without considering social status and others	X_{41}
	Staffs always give friendly and polite service to students	X_{42}
	There is communication among staffs, builders, and boards	X_{43}
	Boards always watch on the running of Pesantren program	X_{44}
	Boards pay attention to program service of Pesantren	X_{45}
Reliability	Discipline of time delivering material as what is determined on the schedule	X_{51}
	Easy administration service	X_{52}
	The improving of students' abilities in using foreign languages (Arabic and English)	X_{53}
	Technology based administration service	X_{54}
	Guidelines of ethic codes in Pesantren environment (students, teachers, staffs, builders)	X_{55}

Table 2. Fuzzy set Data set on quality assessment variables

Variable	Statement	Membership Function
Expectation	Very unimportant	(0,0,2)
	Unimportant	(0,2,4)
	Important enough	(2,4,6)
	Important	(4,6,8)
	Very Important	(6,8,8)
Perception/ Satisfaction	Very Bad	(0,0,2)
	Bad	(0,2,4)
	Good Enough	(2,4,6)
	Good	(4,6,8)
	Very Good	(6,8,8)

6 Result and Discussion

Implementation of total quality management of information system consisted of three user levels. In this research, system users cover : administrator, quality control, caregiver, teachers, staffs, and students.

1. Level one : administrator and quality control
   In this level, users have access rights to manage all contents existing in system and see report of quality assessment result from users.
2. Level two : *pengasuh*
   Pengasuh is as top management and decision maker. In this level, user has access right to see planning and quality aspects as well as report of quality assessment result.
3. Level three : teachers/ustadz, staffs, and students
   As the receivers of merit service given by Pesantren, this level has access to fill questionnaire list on the system and see a whole assessment result.

In this research, quality assessment was done by 100 respondents consisting of teachers/ustadz, staffs and students. Assessment was done by filling questionnair forms about users' perception and expectation on merit service given by Pesantren. Result of this assessment was then processed by system by using fuzzy method. Calculation result of quality assessment by users is shown in table 3 and 4.

Table 3. Result of perception fuzzification and defuzzification

Variable	Fuzzification			Defuzzifica-tion
	a	b	c	
X_{11}	4.392	6.392	7.629	6.137
X_{12}	4.722	6.722	7.711	6.385
X_{13}	4.887	6.887	7.835	6.536
X_{14}	4.598	6.598	7.711	6.302
X_{15}	4.309	6.309	7.670	6.096
X_{21}	5.010	7.010	7.835	6.619
X_{22}	4.598	6.598	7.670	6.289
X_{23}	4.845	6.845	7.711	6.467
X_{24}	4.309	6.309	7.546	6.055
X_{25}	4.722	6.722	7.670	6.371
X_{31}	4.887	6.887	7.794	6.522
X_{32}	4.928	6.928	7.876	6.577
X_{33}	5.052	7.052	7.753	6.619
X_{34}	4.722	6.722	7.794	6.412
X_{35}	5.175	7.175	7.918	6.756
X_{41}	4.433	6.392	7.629	6.151
X_{42}	4.515	6.515	7.629	6.220
X_{43}	4.474	6.474	7.629	6.192
X_{44}	4.763	6.763	7.835	6.454
X_{45}	4.928	6.928	7.835	6.564
X_{51}	4.598	6.598	7.670	6.289
X_{52}	4.722	6.722	7.794	6.412
X_{53}	3.814	5.773	7.175	5.588
X_{54}	4.515	6.515	7.711	6.247
X_{55}	4.804	6.804	7.876	6.495

Table 5. Result of expectation fuzzification and deffuzification

Variable	Fuzzification			Deffuzification
	a	b	c	
X_{11}	5.155	7.155	7.959	6.756
X_{12}	5.361	7.361	7.959	6.893
X_{13}	5.443	7.443	7.959	6.948
X_{14}	4.680	6.680	7.773	6.378
X_{15}	4.784	6.784	7.856	6.474
X_{21}	5.320	7.320	7.959	6.866
X_{22}	5.216	7.216	7.814	6.749
X_{23}	5.175	7.175	7.897	6.749
X_{24}	4.804	6.804	7.753	6.454
X_{25}	5.072	7.072	7.897	6.680
X_{31}	5.320	7.320	7.959	6.866
X_{32}	5.155	7.155	7.959	6.756
X_{33}	5.278	7.278	7.876	6.811
X_{34}	5.237	7.237	7.959	6.811
X_{35}	5.485	7.485	7.959	6.976
X_{41}	4.948	6.887	7.773	6.536
X_{42}	4.742	6.742	7.711	6.399
X_{43}	4.680	6.680	7.814	6.392
X_{44}	4.969	6.969	7.897	6.612
X_{45}	4.948	6.948	7.856	6.584
X_{51}	5.113	7.113	7.876	6.701
X_{52}	5.052	7.052	8.000	6.701
X_{53}	4.557	6.557	7.546	6.220
X_{54}	4.784	6.784	7.918	6.495
X_{55}	5.052	7.052	7.876	6.660

Deffuzification result shown in table 4 is higher than deffuzification result shown in table 3. It shows that expectation score is higher than perception score. This result shows that receivers of merit service want something better than this condition. Expectation score which is higher than perception score causes the existence of negative gap. It shows that the real condition has not fulfilled expectation wanted by merit service receivers yet in Pesantren. So there must be evaluation and follow up from decision makers.

Table 5. Gap Score of SERVQUAL

Variable	Perception Score	Expecta-tion Score	Gap
X_{11}	6.137	6.756	-0.619
X_{12}	6.385	6.893	-0.509
X_{13}	6.536	6.948	-0.412
X_{14}	6.302	6.378	-0.076

Variable	Perception Score	Expecta-tion Score	Gap
X_{15}	6.096	6.474	-0.378
X_{21}	6.619	6.866	-0.247
X_{22}	6.289	6.749	-0.460
X_{23}	6.467	6.749	-0.282
X_{24}	6.055	6.454	-0.399
X_{25}	6.371	6.680	-0.309
X_{31}	6.522	6.866	-0.344
X_{32}	6.577	6.756	-0.179
X_{33}	6.619	6.811	-0.192
X_{34}	6.412	6.811	-0.399
X_{35}	6.756	6.976	-0.220
X_{41}	6.151	6.536	-0.385
X_{42}	6.220	6.399	-0.179
X_{43}	6.192	6.392	-0.199
X_{44}	6.454	6.612	-0.158
X_{45}	6.564	6.584	-0.021
X_{51}	6.289	6.701	-0.412
X_{52}	6.412	6.701	-0.289
X_{53}	5.588	6.220	-0.632
X_{54}	6.247	6.495	-0.247
X_{55}	6.495	6.660	-0.165

Table 5 shows gap score of each attribute of perception and expectation. The biggest gap is on variable X_{53} which is the improving of students' abilities in using foreign languages, with gap score of -0.632. From this result, it can be known that students' foreign language abilities need to be improved and developed. Meanwhile, the smallest gap score is on variable X_{45} ; which is the attentions of the boards to program service of Pesantren. It means that service given by Pesantren has already been good even though it is very small and there is still gap.

Score of each attribute is then classified into each dimension to get SERVQUAL score for each dimension. This SERVQUAL score is then ranked to know which dimension have the highest gap level so that it can become material for quality refinement evaluation in the next. SERVQUAL score of each dimension is shown in table 6.

Table 6. SERVQUAL score of each quality assessment dimension

Dimensi on	Percepti on Score	Expecta-tion Score	GAP	Rank
Tangible	6.690	6.291	-0.399	1
Responsi veness	6.700	6.360	-0.340	3
Assuran ce	6.844	6.577	-0.267	4
Emphaty	6.504	6.316	-0.188	5
Reliabi lity	6.555	6.206	-0.349	2

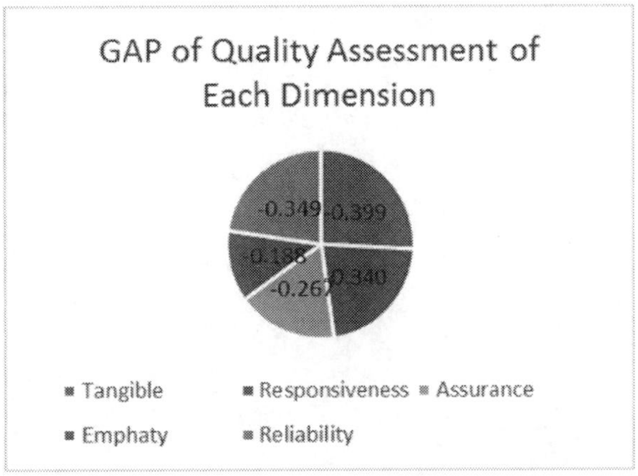

Fig. 3. GAP of each dimension

Fig. 3 tells that gap score of each dimension is less than 0, it means that expectation of service given is still higher than perception received by users. From those five dimensions, tangible dimension has the highest gap ; which is -0.399, and empathy dimension has the smallest gap, which is -0.188. It shows that there is a need to do refinement of service related to attributes existing on tangible dimension to fulfill standard of quality expected. The gap score of each dimension is shown in picture 3.

7 Conclusion

Total quality management of information system can be implemented to assess Pesantren quality based on merit service given. Assessment of quality given is based on dimensions existing on SERVQUAL, such as tangible, reliability, responsiveness, assurance, and empathy dimensions, then their attributes are arranged for each dimension. Assesment is processed by using fuzzy method to get the certainty of score. From research result, it can be got that the highest gap score was on attribute of the improving of students' foreign language abilities, with the gap score of -0.632, meanwhile for assessment of each dimension, tangible dimension has the highest gap, which is r -0.399. The result got is hoped to be recommendation for decision makers to evaluate Pesantren quality refinement.

References

1. Akhiruddin. Lembaga Pendidikan Islam Nusantara. Jurnal Tarbiya, 195-219.(2015)
2. S. Srima, P. Wannapiroon, & P. Nilsook. *Procedia Social and Behavioral Sciences* Vol. **174**. 2160 – 2165, (2014)
3. E. Akhlaghi, S. Amini, & H. Akhlaghi. *Procedia Social and Behavioral Sciences,* Vol. **46**. 5285 – 5289, (2012)
4. T. Lupo, *Journal of Expert Systems with Applications* Vol. **40**. 7096–7110, (2013)
5. Z. Dali, *Jurnal At-Ta'lim, Vol.12* No.1, (2013)

6. M.K. Fadillah, *Jurnal At-Ta'lim* Vol.**10** No.1,(2015)
7. M.N. Nasution, *Manajemen Mutu Terpadu (Total Quality Management)*, Ghalia Indonesia, Bogor. (2015
8. F. Mubarak, *Jurnal Management of Education* Vol. **1**, 10-18.(2015)
9. C.J. Chien, H.H. Tsai, *Fuzzy and Set Systems Journal. Vol.* ***116***, 289-300, (2000)
10. D.P. Maruvada, R.S. Bellamkonda, *International Journal of Innovation, Management and Technology*, Vol. **1**, 478-482, (2010).

 https://doi.org/10.1051/e3sconf/20183110012

Translation Analysis on Civil Engineering Text Produced by Machine Translator

Anam Sutopo,
Department of English Education, Faculty of Teacher Training and Education, Universitas Muhammadiyah Surakarta – Indonesia

Abstract. Translation is extremely needed in communication since people have serious problem in the language used. Translation activity is done by the person in charge for translating the material. Translation activity is also able to be done by machine. It is called machine translation, reflected in the programs developed by programmer. One of them is Transtool. Many people used Transtool for helping them in solving the problem related with translation activities. This paper wants to deliver how important is the Transtool program, how effective is Transtool program and how is the function of Transtool for human business. This study applies qualitative research. The sources of data were document and informant. This study used documentation and in dept-interviewing as the techniques for collecting data. The collected data were analyzed by using interactive analysis. The results of the study show that, first; Transtool program is helpful for people in translating the civil engineering text and it functions as the aid or helper, second; the working of Transtool software program is effective enough and third; the result of translation produced by Transtool is good for short and simple sentences and not readable, not understandable and not accurate for long sentences (compound, complex and compound complex) thought the result is informative. The translated material must be edited by the professional translator.

1. Introduction

Discussing translation means talking about communication. It is due to the fact that between translation and communication has the same focus, namely: message. The message or meaning is extremely important for communication and translation. Therefore, without considering meaning translation is nothing. A good translation result may be stated as a nice communication.

Translation activities have a great role in developing science and technology. The development of science and technology is only depending on the natural resource that from time to time has been reduced, but also influenced by the quality of human resources. There are many ways to improve the quality of human resources. One of them is by increasing their ability in understanding of the recent knowledge and technology. Meanwhile, it cannot be denied that most of the available books, relevant with knowledge and technology are written in English. There are 75 % of the available books, which are collected by Indonesian library, are written in English and there is only 5 % of the readers who understand them [1]. It means that the coming of translation becomes an alternative solution. In other words, we may say that translation can be a bridge in transferring the message from English to Indonesian. By translating books, readers who do not understand English can read and get the knowledge that they want.

Translation is not an easy task. It needs a serious attention and concentration. It is due to the fact that the core of translation is transferring 'meaning' (message). The translator must be careful in replacing meaning from source to target language [2]. It is also hard forbidden for translator to gain or loss the messages. The translator should be able to look for the closest meaning from source to target language. The translator must keep the excess and mission in the source language then transfer them to target one. If the translator is able to translate well semantically but he cannot move the closest mission, the result of translation is useless. The translator spends his time and energy carelessly. Therefore, to avoid this useless activity and to have a good work, the translator should master not only linguistics aspect and but also the materials that will be translated as well as the theory of translation [3].

There is nice phenomenon in translation result done by machine translator [4]. One of them is translation which is done using Transtool program.

* Corresponding author: anam.sutopo@ums.ac.id

This program has been used by many people. The following is the example of translation done by this program:

SL : *Over the past two decades concrete has enjoyed a renewed level of research and testing, resulting in the development of many new types of concrete.*

TL : *(Di) atas masa lalu dua dekade beton telah menikmati suatu [yang] diperbaharui tingkat riset dan pengujian, menghasilkan pengembangan dari banyak jenis baru beton.*

From the TL, it can be seen that the result translation done by Transtool above is not readable. It means that the language used is not easy to read and the content is difficult to understand. There are too many choices in the bracket and difficult to understand.

There are many definitions of translation. In analyzing translation, there are two languages involved in translation, they are the source language and the target language. Larson argues that translation is basically a change of form. When we speak of a language, we are referring to the actual words, phrases, clauses, sentences, paragraphs, et., which are spoken or written. These forms are referred to as the surface structure of a language[5]. She also says that the form which the translation is made will be called the source language and the form into which it is to be changed will be called the receptor language. Ttranslation is an operation performed on language: a process of substituting a text in one language for a text in other [6]. From those definition it can be drawn conclusion that translation as a process of replacing language and culture content from a Source Language into a Target Language.

There are two kind of translation, they are; form-based translation and meaning-based translation [7]. Literal translation is called as form-based translation. The literal translation is the form of the source language that transferred into the form of target language. Although the literal translation is useful for the purposes of the related study of the source language, it can help the speakers of target language who are interested for the meaning of the source language. The literal translation gives priority of form whether in words, clause or sentences make the result of the translation sound unnatural and has a little value of communication. The result of translation text becomes bad translation because the translator makes the use the words with the contextual meaning.

According to Jumpelt, there are three major steps when the process of translation happens[8]. First, the translator needs to discover the meaning of the message in the source language to be translated. Second is the process of determining of the meaning to gets related meaning for communicated the message in the target language. The last, the translators need to write or re-express the meaning or the message with the context in the target language using the translation equivalent.

The equivalence of the translation is very important things in transferring one language into other languages, getting relation between the source language and the target language. Nababan states equivalence of translation have some conditions translation equivalence occurs when a SL and a TL text are relatable to (at least some of) the same features of substance [9]. Based on the statement, it has meaning that when source language has some language relations for the relations for the features of substance with the target language, is possible for the translation equivalence to occurs. The translation cannot be based only for some meanings but it must be related to the some situations and contexts where the equivalence should be established. There are three aspects of the quality of the translation; they are readability, accuracy and acceptability of the translation [10]. Readability in translation is about how easily written materials can be read and understood. Readability is what makes some texts easier to read by readers. To maximize readability is a goal that every writer, technical or otherwise, should strive to achieve. Readability has two common meanings, one applying to document design, the other to language. Readability as it is applied to document design is concerned with such matters as line length, leading, white space, font type and the like. Readability as it is applied to language is concerned with the comprehensibility or understandability of a piece of written text. Readability level is about some standards criteria to get the readability score of a text. Readability level must be divided into several levels with different score of each.

The accuracy of translation is an important part of translation [11]. Accuracy is one of the factors which are used to determine the quality of translation. Accuracy also means that the meaning of the source text is transferred in the target text correctly. Basically, accuracy is related to choose the correct words in order to reveal the same idea in target language. Choosing the right word is very important because if the translators incorrectly choose the word, the text will not be accurate. Choosing the right word should consider some factors, such as the target readers, context of the text and the socio – cultural condition. Acceptability refers to how the translation product can be accepted by the readers [12]. The target text must be intended to be a text and accepted as such in order to be utilized in communicative interaction. Acceptability is a term firstly used the translator referring to a target–text-oriented approach to translation [13]. He concludes that a translation is assumed to be acceptable if it fulfills the requirements of 'reading as an original' written in the target language and sound natural for the target reader rather than that of 'reading as the original' in source language. The acceptability of translation is determined by the faithfulness to the linguistic and literary norms of target system [14].

Meanwhile, linguistics Unit is the one of the natural units into which linguistic messages can be analyzed. In English grammar we have five part of linguistics unit from the larger units being made up of smaller units. They form a scale of units at different ranks, also called as rank scale. We have five units; morpheme, word, phrase, clause, and sentence. The largest or highest on the rank scale is the sentence and the smallest on the rank scale is the morpheme.

Besides explanation on English grammatical/ linguistics unit, Indonesia grammatical needed to be explained as well. Actually there is bo difference between English and Indonesia linguistics unit. Indonesia also have five units of linguistic, morfem (morpheme), *kata* (word), *frasa* (phrase), *klausa* (clause) and *kalimat* (sentence) and this is the linguistics unit from English and Indonesia.

Sentence is a grammatically complete construction, it does not need the help of other construction to make its grammatical meaning clearer. Pym states that a sentence is an independent linguistic form, not included by virtue of any grammatical construction in any longer linguistic form[15]. According to its structure, sentence is divided into three classes, they are simple sentence, compound and complex sentences.

Simple sentence is the sentence that contains only one subject and one predicate, in other word the simple sentence contains no clauses. Compound sentence is a sentence containing more than one predication; the over-all construction is a coordination of two or more simple sentences. It means that two or more simple sentences (independent clauses) can be combined to form a compound sentence, frequently using coordinating conjunction[16]. Complex sentence is a sentence containing more than one predication, one of the predications is some kind of subordinate element – a modifier, a subject, or complement.

2. Research Method

The type of this study belongs to descriptive qualitative study. The aims of this research to know the function of Transtool program, to show how this program works and to describe the result of translation civil engineering text done by this program. The writer uses document (civil Engineering texts) and informant as the source of data while the techniques for collecting data used by the writer are documentation and interview. The researcher has following the procedures for this research, as follows: comparing the data between the source language and the target language, classifying them and then justifying the quality of the translation. To check the validity of data, the writer uses data and method triangulations. Then, he collected data are analyzed by using Interactive Analysis as seen in the following diagram:

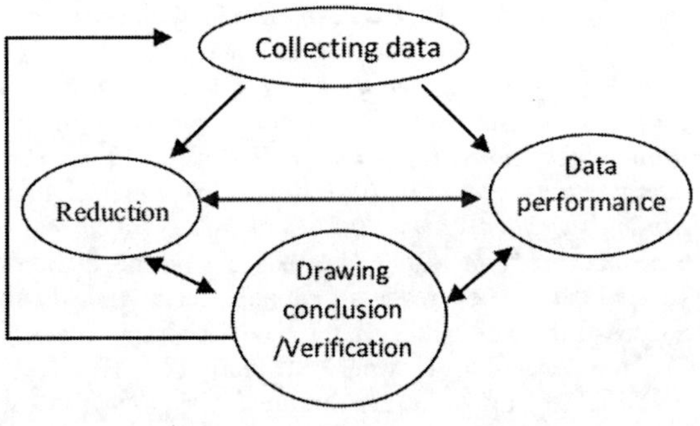

3. Result and Discussion

3.1 Helpful

Based on the collected information given by the informants, Transtool program is very helpful. This software can help many them in translating the civil Engineering text from English to Indonesian and from Indonesian to English. Most of them felt that Transtool program is always ready helping them in transferring the meaning from Indonesian to English.

At least, it was expressed by informant W. He said that when he wrote an article as the assignment from his lecturer in post graduate program, he got problem with reference. The solution is using Transtool program to translate the collected material by using Transtool software. He could find the main information from the translated material done by the Transtool program.

The same expression said by informants. She used Transtool program when she wrote not only article but also theses. As we know that writing theses is not easy. It needs some references and most of articles and references were written in English she said. She had no choice. If she wanted to hire translator, she would spend much money and time. Not all the result of translation was useful to completing writing her theses. The only Transtool program that helped her in finding some references. The similar argument also given by informant N and B.

After crosschecking the implementation of Transtool program as software, the writer may conclude that the Software of Transtool is helpful. It can support the everyone who has problem dealing with translation working. It may help them to change the language used from Indonesia to English and from English to Indonesian. It means, the writer concludes that the Transtool program is helpful for translating the Civil Engineering texts.

3.2. Effectiveness

In line with the effectiveness of this software, informant S said that as the Transtool program is effective enough seen from the time needed. The fact shows that Transtool program can help him in translating many pages of the text in short time. He just needed the time for copying and pasting. Copying the source language then transferring by Transtool program and finally pasting the result of translation into the new file.

This step is also done by other informants. They have similar argument that as the aid Transtool program just needs a few minutes for doing this translation work. From the consolidated information or statement above it can be concluded that the effectiveness of Transtool program is effective enough as the aid. It is as the helper to translate the provided articles.

3.3. Accuracy and Readability.

Although the Transtool program is useful and effective enough as the aid for translating the civil Engineering text both from English to Indonesian or the opposite one, the result of translation produced by this software is questionable. Even, the translation result produced by this program is broken. At least, it can be seen from the readability aspect. Yes, this machine translator can transferred the meaning from Bsu to Bsa quickly. It just needs a short time to do translation or to change the message from English to Indonesian.

Anyhow, when it is seen from the quality of result (readability), the Transtool program cannot do it well. The result of this software is broken, not good. Let's see the following example:

SL : Civil engineering is the term for the work of designing and building infrastructure.
TL : *Rancang-Bangun sipil adalah istilah untuk pekerjaan merancang dan membangun infrastruktur.*

This datum is belonging to simple or short sentence. From this datum it can be seen that the Trandtool program can change or transfer the meaning from SL to TL successfully. The phrase '*civil engineering* ' is transferred into *'rancang-bangun cipil'*. The phrase of *'in term of'* is translated into *'istilah'* and the phrase of *'building infrastructure'* is translated into *'membangun infrastruktur'*.

So, this translated sentence is easily to read the meaning is easy to understand, therefore this datum includes into the high readability level. In this case the writer concludes that the source language translated into the target language well and can be understood by the reader. This translation can be stated accurate because the machine translator has translate into the target language correctly.

SL : Civil engineering has many different areas or disciplines.
TL : *Rancang-Bangun sipil mempunyai banyak area berbeda atau disiplin.*

This datum is belonging to simple or short sentence. From this datum it can be seen that the Trandtool program can change or transfer the meaning from SL to TL successfully. The phrase '*civil engineering* ' is transferred into *'rancang-bangun cipil'*. The phrase of 'different areas' is translated into *'area berbeda'* and the phrase of 'disciplines' is translated into *'disiplin*.

So, this translated sentence is easily to read the meaning is easy to understand, therefore this datum includes into the high readability level. In this case the writer concludes that the source language translated into the target language well and can be understood by the reader. This translation can be stated accurate because the machine translator has translate into the target language correctly.

SL : Chartership shows employers and clients that they are both qualified and experienced in their jobs.
TL : *Pencarteran menunjukkan pemberi kerja dan klien yang mereka adalah kedua-duanya [yang] berpengalaman dan berkwalitas di (dalam) pekerjaan mereka.*

In this case the writer classifies that this datum belonging to medium readability level because the translator no needed to rewriting some words from source language to translates into the target language. Medium readability occurs if the translation makes the readers need to read more than one to catch the meaning. For example, it can be happened when the source language translated into the target language well and can be understood by the reader. It indicates that the phrase '*both qualified and experienced*' which is translated into '*kedua-duanya [yang] berpengalaman dan berkwalitas'* and the phrase 'in their jobs' which is translated into '*di (dalam) pekerjaan mereka'* .

This data also belongs to less accurate because the meaning or message from the source language and target language less inaccurate. Translation can be less accurate because the Transtool machine translator translates the word '*chartership'* into *'pencarteran'*. So, this translation categories into less accurately.

SL : Structural engineering is a subset of civil engineering dealing with the design and analysis of buildings and large non-building structures to withstand both the gravity and wind loads as well as natural disasters. Besides, it may also cover design of machinery, medical equipment, vehicles or any other objects where structural functionality or safety are involved. Structural engineers must ensure their designs satisfy building codes.
TL : *Rancang-Bangun struktural adalah suatu subset tehnik sipil berhadapan dengan disain dan analisa bangunan dan struktur tidak membangun besar ke withstand kedua-duanya gaya berat dan memutar beban seperti halnya bencana alami. Di samping, mungkin juga [meliput/tutup] perancangan permesinan, peralatan kedokteran, sarana (angkut) atau object lain [di mana/jika] keselamatan atau kemampuan struktural dilibatkan. Insinyur struktural harus memastikan disain mereka mencukupi membangun kode.*

This datum belonging to not readability and not accurate. It is due to the fact that the target language '*Rancang-Bangun struktural adalah suatu subset tehnik sipil berhadapan dengan disain dan analisa bangunan dan struktur tidak membangun besar ke withstand kedua-duanya gaya berat dan memutar beban seperti halnya bencana alami. Di samping, mungkin juga [meliput/tutup] perancangan permesinan, peralatan kedokteran, sarana (angkut) atau object lain [di mana/jika] keselamatan atau kemampuan struktural dilibatkan. Insinyur struktural harus memastikan disain mereka mencukupi membangun kode'* cannot be read easily and difficult understand. The readers cannot read this BSa successfully. The content is also not accurate. There is different meaning between SL and TL. This data belongs to not accurate because the meaning or message from the source language and target language is different.

SL : Entry-level structural engineers may design individual structural elements of a structure, for example, beams, columns, and floors of a building. More experienced engineers would be responsible for the structural design and integrity of an entire system, such as a building.

TL : *Entry-Level insinyur struktural boleh mendisain unsur-unsur [yang] struktural individu suatu struktur, sebagai contoh, [balok/berkas cahaya], kolom, dan lantai suatu bangunan. Insinyur [yang] lebih berpengalaman akan bersifat bertanggung jawab untuk integritas dan disain yang struktural dari suatu keseluruhan sistem, seperti suatu membangun.*

This datum belongs to not readability and not accurate. It is due to the fact that the target language '*Entry-Level insinyur struktural boleh mendisain unsur-unsur [yang] struktural individu suatu struktur, sebagai contoh, [balok/berkas cahaya], kolom, dan lantai suatu bangunan. Insinyur [yang] lebih berpengalaman akan bersifat bertanggung jawab untuk integritas dan disain yang struktural dari suatu keseluruhan sistem, seperti suatu membangun*' cannot be read easily and difficult understand. The readers cannot read this BSa successfully. The content is also not accurate. There is different meaning between SL and TL. This data belongs to not accurate because the meaning or message from the source language and target language is different.

4. Conclusion

Translation is not an easy task. It needs a serious attention and concentration. Translation is the process of transferring meaning or message from source into target language. There are three steps in translation process, namely (1) analyzing, (2) transferring and (3) restructuring. Translation can be done by the machine translator. One of them is Transtool software program. This study recommends that Transtool is a helpful software. It can support the everyone who has problem dealing with translation working. It may help them to change the language used from Indonesia to English and from English to Indonesian. It means, the writer concludes that the Transtool program is helpful for translating the Civil Engineering texts. The Transtool program is effective enough as the aid. The Transtool program helps people to translate the text in short time. It is for copying they material to be translated to the machine and pasting the result of translation to another program. It means that copying the source language then transferring by Transtool program and finally pasting the result of translation into the new file. Dealing with Quality of translation seen from readability and accuracy, for the simple sentences text the translated material is easy to read and understandable. For the long sentences (complex and compound complex) result of translation is not readable (the readability is low), not understandable and not accurate although the result above is informative.

References

1. Sadtono, F. *Pedoman Penerjemahan*. 141 (1985).
2. Miyanda, Fewdays. *Total Meaning and Equivalence in Translation*. *NAWA* Journal of Language and Communication. 56 (2007)
3. Sang, Jian dan Zhang, Grace. *Communication across languages and cultures: A perspective of brand name translation from English to Chinese.* Journal of Asian Pacific Communication ISSN 0957–6851/E-ISSN 1569–9838. 246. (2008)
4. Hui-juan, Ma. *Exploring the differences between Jin Di's translation theory and Eugene A. Nida's translation theory.* The Journal of *Babel.* ISSN 0521–9744/E-ISSN 1569–9668. 99. (2007)
5. Larson, M. *Meaning Based Translation, A Guide to Cross-Language Equivalent*. 15. (1991)
6. Catford, J.C. *A Linguistic Theory of Translation***.** 156. (1974.)
7. Hui-juan, Ma. (2007). *Exploring the differences between Jin Di's translation theory and Eugene A. Nida's translation theory.* The Journal of *Babel*. issn 0521–9744/e-issn 1569–9668. 101 (2007)
8. Jumpelt, B. *The Science of Translation*. 67. (1981)
9. Nababan M R. *Aspek Teori Penerjemahan dan Pengalihbahasaan*. Surakarta : PPS-UNS (1997)
10. Sutopo, Anam. *Terjemahan dan Penerjemahan: Kajian Lintas Pemahaman*. 17. (2014)
11. Sutopo, Anam. (2015). *Penerjemahan Naskah Resmi: Telaah Holistik Pidato Kenegaraan*. Solo: MUP Press UMS.
12. Newmark, P. *A Textbook of Translation*. 42 (1987)
13. Nida, E A. *Commponental Analysis of Meaning. An Introduction to Semantic Structures*.78 (2008)
14. Ordudary, M. *Translation Procedures, Strategy and Methods*. Translation Journal. Volume **11**, No 3, July 2007. 265 (2007).
15. Pym, Anthony. *Natural and directional equivalence in theories of Translation .Target* Volume 19 Issue 2. ISSN 0924–1884 / E-ISSN 1569–9986. 284 (2007).
16. Miyanda, Fewdays. *Total Meaning and Equivalence in Translation*. *NAWA* Journal of Language and Communication, . Volume 11, No 3, June 2007. (2007)

Effect of Value Congruence, Brand Distinctiveness, Brand Social, Brand Warmth, and Memorable Brand Experience on Customer-Brand Identification and Brand Loyalty (Case Study: Brand of ACER Laptop)

Aries Susanty[1,*], *Aprilia* Tresnaningrum[2]

[1,2]Program study of Industrial Engineering, Faculty of Engineering, Diponegoro University, Semarang - Indonesia

Abstract. This study has several purposes. First, this study aims to investigate the effect of consumer–brand value congruence, brand distinctiveness, brand social benefit, brand warmth, and memorable brand experience on customer-brand identification (CBI). We call all of those factors as the antecedent factor of CBI. Second, this study aims to investigate the effect of CBI on customer loyalty. Third, investigate the role of product involvement as a moderating variable of the relationship between brand distinctiveness, brand social benefit, brand warmth, memorable brand experience and CBI. This research used primary data collected through closed questionnaires using a Likert scale of 1 - 5. The total sample size was 273 respondents located in Semarang City who has or has been using Acer Laptop for minimal one year. This research was conducted using Partial Least Square (PLS) method through SmartPLS 3.0 software. The result of data processing indicated that all of the antecedent factors of CBI have the positive and significant effect on CBI of the user of Acer Laptop. In this case, among the five antecedent factors of CBI, value congruence has the greatest effect on CBI of the user of Acer Laptop. The result of data processing also indicated that CBI has the positive and significant effect on brand loyalty of user of Acer Laptop. This study fails to prove the role of product involvement as a moderating variable of the relationship between brand distinctiveness, brand social benefit, brand warmth, memorable brand experience and CBI of the user of Acer Laptop. Moreover, based on the result of hypothesis testing, this study gives some recommendation to Acer Laptop to develop or create some features which are match with the value of user of Laptop Acer in Semarang City.

1 Introduction

According to The American Marketing Association, brand related to the name, term, sign, symbol or design that identifies the products and services resulted by one seller or group of sellers and brand can be used to differentiate the products and services from the competitors. The brand concept is much more comprehensive than its graphical representation [1-7]. Brands have role to ensure the company's identity. Moreover, several researchers have concluded that brand which can create the real value in customer's mind is more important than the company's products and services [8]. Then, one of the important brand concept is customer-brand identification (CBI). The concept of CBI is important to understand about how, when, and why the consumer articulate their identities through the brand. This concept is important in the recent condition when the time of great customer skepticism toward brands, coupled with the fall in the value of traditional media in promoting brands [9]. This concept even more important today when there are too many choices available for customers, so the company should make the effort to work out what makes a particular brand specific [10] Moreover, the importance of the concept of CBI in building the brand of the products or services cannot be separated from the several impacts of this concept on individual consumer behavior including: consumer buying-related decisions [11], consumer satisfaction and a higher possibility of repurchase [12], positive word of mouth [12]. [13], [14], psychological sense of brand community and brand commitment [15], brand loyalty [14], [16], [17[, and brand preference [18]. Among the several impacts of CBI on individual customer behavior, this study will focus on the impact of CBI on customer or brand loyalty.

There were several definitions of CBI. According to, Bergami and Bagozzi [19] and Elbedweihy et al [20], CBI can be defined as the primary psychological substrate for deep, committed and meaningful relationships that marketers are increasingly seeking to build with their customers. CBI can be seen as a kind of cognitive state of self-categorization [21]. According to Wolter et al [22], CBI is a decision of consumer to define his or her "self" by including a specific brand into their self-concept. In the marketing literature, the process

* Corresponding author: ariessusanty@gmail.com

of CBI was influencing of several antecedent factors. Referring to the perspective of social identify, CBI is resulted when a brand satisfies self-consistency, self-enhancement, or self-differentiation motives [21], [23]. Theory suggests the symbolically-based antecedents of identification are prestige for the self-brand similarity for self-consistency motives, satisfaction of self-enhancement motives, and distinctiveness for self-differentiation motives [21]. On the simple way, Bachman and Wilkins [24] only use self-congruity as an antecedent factor of CBI. Bachman and Wilkins [24] proposed that the higher the perceived congruity between the self and a brand, the higher the positive evaluation of the brand. In this case, the process of associating the self -image owned by one person with the personality of the brand contributes to the state of consumer-brand identification. Different with Bachman and Wilkins [24], more comprehensive framework in describing the antecedent factor of CBI was proposed by Stokburger-Sauer et al [17]. According to Stokburger-Sauer et al [17], antecedent factors of the CBI can be differentiated into two categories, namely cognitive and affective factors. The cognitive factor includes the extent to which customers (i) have a personality that is similar to his or her own (i.e. brand–self similarity), (ii) to be unique or distinctive (brand distinctiveness), and (iii) and to be prestigious (brand prestige). Then, the affective factor includes the extent to which customers (i) feel that their interactions with a brand help them connect with important social others (brand social benefit), (ii) perceive a brand in warm, emotional terms rather than cold, rational ones (brand warmth), and (iii) have fond memories of brand consumption experience (memorable brand experience). Moreover, Stokburger-Sauer et al [17] proposed that consumer's involvement in the product category in which a brand belongs (product involvement) act as a moderating variable; it made the relationship between the antecedent factor and CBI stronger. Rather than six factors, this study only uses the five factors as an antecedent factor of CBI or exclude the brand prestige from the antecedent factor of CBI because this study will test the construct of antecedent factors of CBI and the effect of CBI on customer loyalty on one of the laptop brands (Acer Laptop) and this brand do not belong to luxury items. Referring to the result of hypothesis testing from Stokburger-Sauer et al [17], CBI is less sensitive to brand prestige in product categories such as soft drinks, supermarkets, and even athletic shoes than in the more conventionally status or luxury product categories in building the CBI. Besides that, rather to use brand–self similarity in describing the extent of customers have a personality that is similar to the brand, this study prefer to use consumer–brand value congruence (or value congruence for short) in building the CBI as Elbedweihy et al [20].

This study will use the brand of Acer Laptop as the object for testing the relationship between consumer–brand value congruence, brand distinctiveness, brand social benefit, brand warmth, memorable brand experience, product involvement, and the CBI and customer loyalty for the following reason. In Indonesia, Acer was awarded the Top Brand Award 2016 in the laptop / notebook category. This success reflects that Acer laptop/notebook products become the main choice of Indonesian society. Top Brand survey results obtained through data collection by interviewing directly to 14,000 respondents in 15 major cities of Indonesia. The results of this interview show that Acer has the highest index value, which is 34.7%. This result exceeds Asus and Lenovo which only have index values of 16.5% and 11.1%. Moreover, Acer Indonesia received an Indonesian Customer Service Award (ICSA) 2015 from the SWA magazine for our outstanding customer service quality. ICSA 2015 marked our achievements for 8 consecutive years in this annual awarding. It seems that Acer Laptop have good brand in the market of laptop and also achieve a loyalty from their customer. In the worldwide, Acer group was one of the top five vendors which had significant contribution towards worldwide personal computer shipments during 2014 and 2015. Acer group capture around 7 % market share towards worldwide personal computer shipments [25].

2 Literature Review

As stated in the introduction, basically, this study refers to the framework from Stokburger-Sauer et al [17] in building the CBI. Specifically, after removing the brand prestige as an antecedent factor of CBI and changing brand–self similarity with customer-brand value congruence, this study use consumer–brand value congruence, brand distinctiveness, brand social benefit, brand warmth, and memorable brand experience as an antecedent factor of CBI and product involvement as a moderating factor for the relationship between antecedent factor and CBI. Then, the CBI will impact the individual customer behavior through brand loyalty [14], [16], [17].

- Customer-brand value congruence or value congruence in short. Reference [20] among the first to introduce value congruence as an antecedent of CBI. According to self-congruity theory from Johar and Sirgy [26], value congruity can be defined as a mental evaluation that consumers make the comparison of the similarity or dissimilarity between their own set of values and the entity's values. Moreover, self-congruity theory describes customer behavior as partly determined by the congruence resulting from a psychological comparison between the consumer self-concept and the product user or brand image [27].
- Brand distinctiveness refers to "need for uniqueness". Brand distinctiveness can be defined as pursuit of individual's of differentness relative to others that is achieved through the acquisition, utilization, and disposition of consumer goods for the purpose of developing and enhancing one's personal and social identity [17]. Brand distinctiveness is significant for several reasons. Brand distinctiveness can provide an effective communication with consumers, help consumers to identify brands easily, and reduce consumer's cognitive effort. Moreover, brand distinctiveness is

a significant concept to help consumers identify a certain brand among other brands easily. The brand name, colors, logos, taglines, symbols/characters, celebrities and advertising styles are some important elements to achieve the brand distinctiveness [28].

- Brand social benefit refers to the condition that consumers are more likely to identify with brands that help them to connect with important others, groups, communities, or subcultures. In this case, the brand used by their reference groups become an important thing to gain or strengthen the membership of certain group [29]. Brand is the important tool for linking people to one another [30]. [31], [32]. Sometimes, consumers join into distinct subgroups of society on the basis of a shared commitment to a brand [33], [34].
- Brand warmth perceptions is a judgment of the brand's pro-sociality and trustworthiness, and its ability to achieve corporate and market goals. In this case, the brand's production and distribution practices, the way it treats its customers, suppliers, and employees, or the brand's financial performance can all be informed about the brand's warmth [35]. Brand warmth perception makes the consumer have more intense feelings about warmer brands and the role of those brands in their lives [36].
- A memorable brand experience can be defined as the extent to which consumers have positive, effectively charged memories of a certain brand. In this case, memorable brand experience is related to the ability of a brand to deliver the remarkable and vivid experiences creating from extraordinary consumption activities, regardless the frequency of use [37], [38]. The brands that offer memorable experiences are more likely to lead to individuals' intertwining of brand-related and self-related thoughts, thus contributing to CBI [17].
- Product involvement can be defined to a common level of concern about a product class [39]. Product involvement should be distinguished from product evaluation. Product involvement can be classified into two categories, high involvement and low involvement. The highly involved consumer is defined as one who is very interested in differences between particular brands and is willing to invest considerable energy in decision making; whereas, whereas, the low involvement is defined as one who is not very interested in differences between particular brands because it is related with something that are not very important to consumer and the customer only hold little perceived risk on the product [40]. According to Stokburger et al [17], the higher product involvement can moderate the relationship between the various antecedents and CBI because the knowledge structure of consumers regarding high involvement product categories are more likely to contain deeply processed and highly elaborated beliefs regarding brands' abilities to meet self-definitional needs.
- Brand loyalty

Reference [41] state that there is no simple one or two-dimensional concept to illustrate the contract of brand loyalty. The concept of brand loyalty is a complex construction which involves multiple dimension. Reference [41] also state that brand loyalty related with the action of the consumer regarding with their thoughts and feelings of the brand. In this way, reference [41] divide attitudinal loyalty into a simple two component in order to understand the concept of brand loyalty as a whole. The first component is cognitive loyalty and the second component is emotional loyalty. Cognitive loyalty can be described to the decision of the consumers to repurchase a certain brand after they analyse some factors such as attribute of the brand. Emotional loyalty can be described as an affective commitment to a brand involving of the positive feelings about and attached to purchasing a brand on the next purchase occasion. Related with CBI, Stokburger et al [17] argue that CBI is an significant predictor of brand loyalty.

So, based on the relationship between value congruence, brand distinctiveness, brand social benefit, brand warmth, memorable brand experience, product involvement, CBI, and brand loyalty which was expressed by the previous researcher, the conceptual model and the hypothesis of this study can be described as follow.

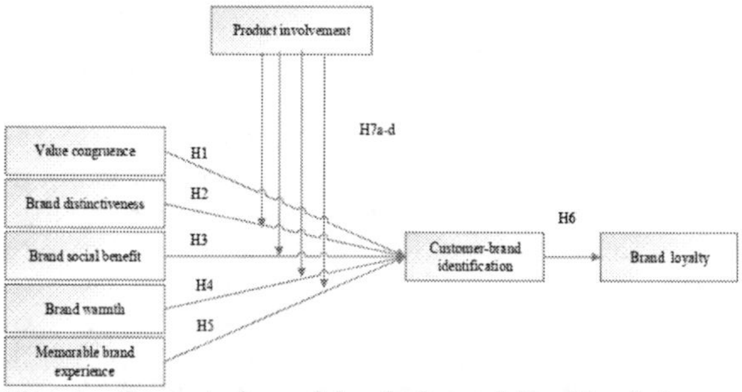

Fig. 1. Conceptual model of the relationship between antecedent factor of CBI, CBI, and brand loyalty

H1	Value congruence has a positive significant effect on customer-brand identification
H2	Brand distinctiveness has a positive significant effect on customer-brand identification
H3	Brand social benefit has a positive significant effect on customer-brand identification
H4	Brand warmth has a positive significant effect on customer-brand identification
H5	A memorable brand experience has a positive significant effect on customer-brand identification
H6	Customer-brand identification has a positive significant effect on customer loyalty
H7a-e	Product involvement has a positive significant effect in strengthening the relationship between (a) brand distinctiveness and CBI, (b) brand social value and CBI, (c) brand warmth and CBI, and (d) memorable brand experiences and CBI

3 Method of Research

3.1. Sample of Research

The research was done in Semarang, the capital of Central Java Province, Indonesia. There are 273 citizens of Semarang City chosen as the respondent of this study. In this study, the researcher cannot define the population frame during the period of data collection because the actual number of a user of Acer Laptop cannot determine exactly. So, the researcher had to give the questionnaire to the user of Laptop Acer who was available at the time questionnaire of data were distributed or the user of Laptop Acer who was available to fill the online questionnaire. The sampling technique used in this study was non-probability purposive sampling. In this technique, the selection of respondent in the sample was based upon certain appropriate characteristics, i.e. people of ages 18 to 50 who have used Laptop Acer at least one year. Characteristics of a respondent in a study can be seen in the following section.

3.2 Instrument and Measurement

Forty-eight items were used to test the relationship between value congruence, brand distinctiveness, brand social benefit, brand warmth, memorable brand experience, product involvement, CBI, and customer loyalty. Out of these forty-eight items, ten items were used for measuring value congruence, three items were used for measuring brand distinctiveness, four items were used for measuring brand social benefit, three items were used for measuring brand warmth, three items were used for measuring memorable brand experience, four items were used for measuring product involvement, seven items were used for measuring CBI, and four items were used for measuring customer loyalty. Except for value congruence, most of those items were adopted from Stokburger-Sauer et al. [17]. For measuring the CBI and customer loyalty, this study had adopted the items from Elbedweihy et al [20]. Items for measuring value congruence were adopted from the short version of Schwartz's value survey [42]. In detail, all the items used in this study can be seen in the next section (Table 1). Then, the condition of all of the items used in this study was measuring in Likert Scale (1= strongly disagree, 2=disagree, 3=neither agree nor disagree, 4=agree, and 5= strongly agree).

3.3 Data Analysis Tool

The study used the Partial Least Square (PLS) method through SmartPLS 3.0 software for processing the data obtained from the questionnaire.

4 Result and Discussion

4.1 Characteristic of Respondent

The characteristics of the respondent of this study can be explained as follow. The respondents consisted of 158 males (57.86%) and 115 females (42.12%) which indicated that the sample of research consisting more male than females. The majority of the respondent of this study have 23-30 years old (60.07%), followed by 15-22 years old (24.71%), 31-38 years old (8.43%), 39-46 years old (4.03%), and 47-54 years old (1.83%). The level of the majority of respondents has the highest level of education on bachelor degree (70.69%), followed by senior high school (16.12%), Magister degree (9.89%), and junior high school (3.29%). Then, the majority of respondent of this study have monthly as much as IDR 2,000,000-4,000,000 (42.12%), followed by IDR 4,000,000-6,000,000 (21.98%), less than IDR 2,000.000 (27.47%), and more than IDR 6,000,000 (8.43%). The majority of respondents have been using the Acer Laptop for 2-4 years (54.58%), followed by less than 2 years (29.67%), 4-6 years (10.99%), and more than 6 years (4.76%).

4.2 Evaluation of Outer Model Test

The result of evaluation of outer model test can be seen in Table 1.

Table 1. The result of outer model test.

Construct	Items	*Cross Loading*	*AVE/ Composite Reliability*
Value Congruence (VCO)	Brand X is ....; I am ...		0.592/ 0.935
	Power (VCO1)	0.753	
	Achievement (VCO2)	0.763	
	Hedonism (VCO3)	0.777	
	Stimulation (VCO4)	0.776	
	Self-Direction (VCO5)	0.736	
	Universalism (VCO6)	0.775	
	Benevolence (VCO7)	0.818	
	Tradition (VCO8)	0.761	
	Conformity (VCO9)	0.763	
	Security (VCO10)	0.767	
Brand Distinctiveness (BDT)	Acer brand has a distinctive identity (BDT1)	0.832	0.707/ 0.879
	Acer brand is unique (BDT2)	0.846	
	Acer brand stands out from its competitors (BDT3)	0.845	
Brand Social Benefit (BSB)	Acer brand offers me the opportunity to socialize. (BSB1)	0.790	0.657/ 0.885
	I feel a sense of kinship with other people who use Acer brand (BSB2)	0.841	
	I gain a lot from interactions with other customers/users of Acer brand (BSB3)	0.797	
	Being a customer of Acer brand makes me feel like I belong to a special (BSB4)	0.813	
Brand Warmth (BWM)	Acer brand creates warm feelings among its users (BWM1).	0.834	0.685/ 0.867
	Acer brand is very loveable (BWM2)	0.838	
	Acer brand is emotional rather than rational (BWM3)	0.811	
Memorable Brand Experience (MBE)	I have had a lot of memorable experiences with Acer brand (MBE1)	0.849	0.707/ 0.879
	Thinking of Acer brand brings back good memories (MBE2)	0.817	
	I have fond memories of Acer brand (MBE3)	0.856	
Product Involvement (PIN)	I am very interested in anything related to laptop (PIN1)	0.749	0.608/ 0.861
	Which brand of laptop I buy matters a lot (PIN2)	0.788	
	I value laptop as an important	0.772	

Construct	Items	*Cross Loading*	*AVE/ Composite Reliability*
	part of my life (PIN3)		
	Laptop mean a lot to me (PIN4)	0.809	
Consumer-Brand Identification (CBI)	I feel a strong sense of belonging to Acer brand (CBI1)	0.793	0.615/ 0.918
	I identify strongly with Acer brand (CBI2)	0.745	
	Acer brand embodies what I believe in (CBI3)	0.773	
	Acer brand is like a part of me (CBI4)	0.842	
	Acer brand has a great deal of personal meaning for me (CBI5)	0.755	
	The degree my self-image overlaps with Acer brand image (CBI6)	0.783	
	The degree my self-identity overlaps with Acer brand identity (CBI7)	0.797	
Brand Loyalty (BLO)	I will buy Acer brand the next time I buy (BLO1)	0.868	0.696/ 0.902
	I would be willing to pay a higher price for Acer brand over other brands of laptop (BLO2)	0.824	
	I stick with Acer laptop because I know it is the best for me (BLO3)	0.819	
	I intend to keep purchasing Acer laptop (BLO4)	0.825	

The outer model was evaluated by convergent validity and composite reliability. For convergent validity, the outer loading value of the indicator is considered sufficient on early stage if its value between 0.5 until 0.6 [43] or ≥ 0.5 [44] and each construct has an average variance extracted (AVE) value higher than 0.5 [45]. In the study, the result of convergent validity test indicates that all items have outer loading and AVE value higher than 0.5. Therefore, no indicator needs to be excluded from the construct. Then, the composite reliability is used to assess whether the sample is truly free from bias or if the responses –on the whole– are reliable [46]. Composite reliability coefficients between 0.60 and 0.70 are considered appropriate in exploratory studies, while coefficients of 0.70 and 0.90 are considered satisfactory for the other types of research [47]. Table 1 shows that CR value of each construct had the value higher than 0.7. Based on this condition, we can conclude that each construct exhibited satisfactory reliability.

4.2 Evaluation of Inner Model

The evaluation of inner model could be seen from the value of determinant coefficient (R^2), and the value of Goodness of Fit (GoF). The value of R^2 indicates the amount of variance in the dependent variables that are determined by the independent variables. According to reference [43], the R^2 values can be 0.19, 0.33, and 0.67. The value of 0.19 indicates weak levels of determination, the value of 0.33 indicates moderate levels of determination, and the value of 0.33 indicates substantial/strong levels of determination.. The value of R^2 for CBI was about 0.884 which indicated that CBI variables could be determined through value congruence, brand distinctiveness, brand social benefit, brand warmth, and memorable brand experience with its percentage of 88.4%; while the value of R^2 for the brand loyalty variables was 0.696 which indicated that the variables could be determined by CBI with percentage around 69.6%.

GoF suggested by reference [48] was used for assessing the global validity of PLS-based complex models. According to reference [48], the GoF values can be 0.1, 0.25, and 0.36. The value of GoF O.1 indicate the small global validity, the value of GoF 0.25 indicate the moderate global validity, and the value of GoF 0.36 indicate the large global validity of a PLS based complex model. The result of calculation indicated that GoF value from the model was 0.721, which exceeds the cut-off value of 0.36 for the large global validity of a PLS based complex model. It indicates that the model has a better prediction power in comparison with the baseline values.

4.3 Result of Hypothesis Testing

The hypothesis is accepted if the value of t statistic (t stat) larger than 1.96 (t critical) and value significance (p-value) less than 5% (0.05). The result of hypothesis testing can be seen in Table 2.

Table 2. The result of hypothesis testing

Hypothesis	*Path Coefficient*	*Standard Error*	*T-value (p-value)*	Conclusion
H1: VCO→ CBI	0.438	0.053	8.311 (p-value<0.05)	Accepted
H2: BDT→ CBI	0.112	0.033	3.401 (p-value<0.05)	Accepted
H3: BSB→ CBI	0.086	0.035	2.417 (p-value<0.05)	Accepted
H4: BMW→ CBI	0.078	0.036	2.134 (p-value<0.05)	Accepted
H5: MBE→ CBI	0.105	0.039	2.709 (p-value<0.05)	Accepted
H6: CBI→ BLO	0.834	0.021	39.095 (p-value<0.05)	Accepted
H7a: BDT * PIN → CBI	0,374	0,198	1,890 (p-value>0.05)	Rejected
H7b: BSB * PIN → CBI	-0,414	0,253	1,639 (p-value>0.05)	Rejected
H7c: BMW * PIN → CBI	0,084	0,228	0,371 (p-value>0.05)	Rejected
H7d: MBE * PIN → CBI	-0,135	0,292	0,461 (p-value>0.05)	Rejected

As seen in Table 2, the relationship between value congruence, brand distinctiveness, brand social benefit, brand warmth, memorable brand experience, and CBI had a path coefficient of 0.438, 0.112, 0.086, 0.078, 0.105 and the t-stat value of 8.311, 3.401, 2.417, 2.134, 2.709 (more than 1.96). Based on this result, value congruence, brand distinctiveness, brand social benefit, brand warmth, memorable brand experience have a significant positive relationship with CBU. Therefore, hypothesis 1 (H1) until hypothesis 5 (H5) were accepted. We can also see that the relationship between CBI and brand loyalty had a path coefficient of 0.834 and the t-stat value of 39.095 (more than 1.96). Based on this result, CBI has a significant positive relationship with brand loyalty. Therefore, hypothesis 6 (H6) was accepted. Moreover, this study failed to prove the effect of brand involvement in strengthening the relationship between brand distinctiveness and CBI, brand social value and CBI, brand warmth and CBI, and between memorable brand experiences and CBI. It could happen because a majority of respondents have an age between 23-30 years with the level of education bachelor degree which tends to use a laptop just as a complementary tool to help their work and as well as lifestyle. Evidence of involvement as a moderator does not exist and the situational brand choice of factors was a solely factors influence the decision of consumer in using a brand.

5 Conclusions

According to its purpose, our findings showed that value congruence, brand distinctiveness, brand social benefit, brand warmth, and memorable brand experience of the consumer have the positive significant effect on CBI. In this case, based on the value of path coefficient, the value congruence has the higher effect on CBI compared to the other factors. Brand distinctiveness is the second factor that has positive significant on CBI, followed by memorable brand experience, brand social benefit, and brand warmth. Then, our finding also showed that CBI has the positive significant effect on CBI. However, this study fails to prove the role of product involvement in strengthening the relationship between brand distinctiveness, brand social benefit, brand warmth, and memorable brand experience on CBI.

There are several limitations in this study which are wished to motivate forthcoming researcher in this field. The first limitation of this study is associated with the location of the respondent who became the sample of this study. The sample of the study was restricted to the customer of Acer laptop in Semarang city who have the willingness to participate in filling the questionnaire. So, the sample may do not represent all the users of Acer laptop. Regarding this limitation, the next study may increase the range of the data collection to another area in Central of Java and also to another area in Indonesia. Second, the impact of brand loyalty isn't observed in the context of the process of interaction between customer and brand, so this study cannot capture the indication about the dynamism of process of development of brand loyalty over the time. To capture the impact of brand loyalty as an essential of an interactive activity of marketing of the company, it would need to adopt the longitudinal study through more than one case study. In this case, it is better to use a longitudinal data and not a cross-sectional data to track the experience of the development of CBI and its impact on brand loyalty. Third, this study is using the Likert scale to measure a brand loyalty. It can be the source of biased in articulating the level of loyalty of customer to the Acer brand. Regarding this limitation, a future study can improve the scale of measurement of brand loyalty of the customer of Acer Brand by using a direct measure of how long the customer being the user of Acer laptop, how often the customer gives the other third party to buy Acer laptop, etc.

Acknowledgment

This study has been supported by the Faculty of Engineering, Diponegoro University through grand for "Penelitian Dasar". The authors wish to acknowledge to Dean of Faculty Engineering for giving the authors the opportunity to conduct this study. The authors also wish to acknowledge our appreciation and gratitude all the partners in grand of "Penelitian Dasar" for their support to the development of numerous ideas and concepts presented in this paper.

References

1. M. McDonald, E. Wallace, L. de Chernatony, *Creating powerful brands*. (Butterworth-Heinemann, 2013).
2. D.A.Aaker, *Building strong brands* (Simon and Schuster, 2012)
3. D. A. Aaker, E. Joachimsthaler, *Brand leadership* (Simon and Schuster, 2012)
4. N. A. Morgan, L. L. Rego, Journal of Marketing, **73**, 1 (2009)
5. K. L. Keller, D. R. Lehmann, Marketing science, **25**, (2006).
6. J. N. Kapferer, **2**. *Auflage, London* (1997)
7. K. L. Keller,. the Journal of Marketing, **57** (1993).
8. Shirazi, H. Z. Lorestani, A. K. Mazidi, Iranian Journal of Management Studies, **6**, 2 (2013).
9. U. Tuškej, U. Golob, K. Podnar, Journal of business research, **66**, 1 (2013).
10. J-N. Kapferer. *The new strategic brand management, creating and sustaining brand equity long term* (Kogan Page, London, 2008).
11. M. Ahearne, C.B. Bhattacharya, T.Gruen, 2005. J.Appl. Phys, **90** (3), 574 (2005)
12. S. Kuenzel, S. Vaux Halliday, Journal of Product & Brand Management, **17**, 5 (2008)
13. Belén del Río, R. Vazquez, V. Iglesias, V.. Journal of consumer marketing, **18**, 5 (2001)
14. C. K. Kim, D. Han, S. B. Park,. Japanese psychological research, **43**, 4, (2001).
15. L. V. Casaló, C. Flavián, M. Guinalíu, International Journal of Bank Marketing, **26**, 6 (2008).
16. S. C. Fang, Y. K. Liao, Management Review, **34**, (2015).
17. N. Stokburger-Sauer, S. Ratneshwar, S. Sen, International journal of research in marketing, **29**, 4, (2012)
18. F. Tildesley, L. V. Coote. Advances in Consumer Research, **36**, (2009)
19. M. Bergami, R.P. Bagozzi, British Journal of Social Psychology, **39**, 4 (2000).
20. M. Elbedweihy, C. Jayawardhena, M.H. Elsharnouby, T.H. Elsharnouby, Journal of Business Research, **69**, 8, (2016)
21. C.B. Bhattacharya, S. Sen, Journal of marketing, **67**, 2 (2003).
22. J. S. Wolter, S. Brach, J. J. Cronin, M. Bonn, Journal of Business Research, **69**, 2 (2016).
23. B.E. Ashforth, F. Mael, Acad Manage Rev, **14**, 1, (1989)

24. K. Bachman, S. Wilkins. *Research with Plymouth University* (2014)
25. Data Corporation (IDC), (January, 2016).
26. J. S. Johar, M. J. Sirgy, Journal of advertising, **20**, 3 (1991).
27. M. Hamilton, X. Sun, Actual self and ideal brand image: an application of self-congruity to brand image positioning. In *International Communication Association Annual Meeting, New York, NY* (2005).
28. J. Romaniuk, B. Sharp, A. Ehrenberg, Australasian Marketing Journal (AMJ), **15**, 2 (2007)
29. J. E. Escalas, J. R. Bettman, Handbook of Brand Relationships, ME Sharpe, **6581**, (2009)
30. M. Muniz, T. C. O'guinn, Journal of consumer research, **27**, 4 (2001).
31. T. C. O'Guinn, A. M. Jr. Muniz, Collective brand relationships. In D. J. MacInnis, C. W. Park, & J. R. Priester (Eds.), *Handbook of brand relationships* (pp. 173–194) (M.E. Sharpe, Armonk, New York, 2009).
32. N. Stokburger-Sauer, Psychology & Marketing, **27**, 4, (2010).
33. J. W. Schouten, J. H. McAlexander, Journal of consumer research, **22**, 1 (1995).
34. C. J. Thompson, A. Rindfleisch, Z. Arsel, Journal of marketing, **70**, 1, (2006)
35. B. Bratanova, N. Kervyn, O. Klein, Psychologica belgica, **55**, 2 (2015).
36. C. Whan Park, D. J. MacInnis, J. Priester, A. B. Eisingerich, D. Iacobucci, Journal of marketing, **74**, 6 (2010)
37. F. Davis,. *Yearning for yesterday: A sociology of nostalgia*. (Free Press, 2008).
38. E. S. Moore, W. L. Wilkie, We are who we were: Intergenerational influences in consumer behavior. In S. Ratneshwar, & D. G. Mick (Eds.), *Inside consumption* (pp. 208–232). (Routledge, London, New York, 2005).
39. N. Hupfer, D. Gardner, Differential involvement with products and issues: An exploratory study. In D. M. Gardner (Ed.), *Proc., association for consumer research College Park, MD: Association for Consumer Research* (1971).
40. L. G. Schiffman, L. L. Kanuk, *Consumer behavior (2nd ed.)* (New Jersey Prentice Hall, 1991).
41. S. Worthington, R. Russell-Bennett, C. Härtel, Journal of Brand Management, **17,** 4 (2010).
42. M. Lindeman, M. Verkasalo, Journal of personality assessment, **85**, 2 (2005)
43. W. W. Chin, Modern methods for business research, **295**, 2, (1998)
44. M. Igbaria, T. Guimaraes, G. B. Davis, Journal of management information systems, **11**, 4 (1995).
45. C. Fornell, D. F. Larcker, Journal of marketing research, **18,** (1981)
46. J. C. H. B. do Nascimento, M. A. da Silva Macedo, Revista de Educação e Pesquisa em Contabilidade, **10**, 3 (2016).
47. F. Hair Jr, J., Sarstedt, M.. L. Hopkins, G. V. Kuppelwieser, European Business Review, **26,** 2 (2014).
48. .M. Tenenhaus, V. E. Vinzi, Y. M. Chatelin, C. Lauro, Computational statistics & data analysis, **48**, 1, (2005)

Enterprise Architecture Planning in developing A planning Information System: a Case Study of Semarang State University

Kholiq Budiman[1,*], *Toni* Prahasto[2], and *Amie* Kusumawardhani[3]

[1]Magister Program of Information System, School of Postgraduate Studies, Diponegoro University, Semarang - Indonesia
[2]Department of Mechanical Engineering, Faculty of Engineering, Diponegoro University, Semarang – Indonesia
[3]Department of Accounting, Faculty of Economic and Business, Diponegoro University, Semarang - Indonesia

Abstract. This research has applied an integrated design and development of planning information system, which is been designed using Enterprise Architecture Planning. Frequent discrepancy between planning and realization of the budget that has been made, resulted in ineffective planning, is one of the reason for doing this research. The design using EAP aims to keep development aligned and in line with the strategic direction of the organization. In the practice, EAP is carried out in several stages of the planning initiation, identification and definition of business functions, proceeded with architectural design and EA implementation plan that has been built. In addition to the design of the Enterprise Architecture, this research carried out the implementation, and was tested by several methods of black box and white box. Black box testing method is used to test the fundamental aspects of the software, tested by two kinds of testing, first is using User Acceptance Testing and the second is using software functionality testing. White box testing method is used to test the effectiveness of the code in the software, tested using unit testing. Tests conducted using white box and black box on the integrated planning information system, is declared successful. Success in the software testing can not be ascertained if the software built has not shown any distinction from prior circumstance to the development of this integrated planning information system. For ensuring the success of this system implementation, the authors test consistency between the planning of data and the realization of prior-use of the information system, until after-use information system. This consistency test is done by reducing the time data of the planning and realization time. From the tabulated data, the planning information system that has been built reduces the difference between the planning time and the realization time, in which indicates that the planning information system can motivate the planner unit in realizing the budget that has been designed. It also proves that the value chain of the information planning system has brought implications for budget realization.

1 Introduction

Weak and not mature planning in a government budget process resulting in low absorption of the budget. The minimum absorption can be prevented by revising the budget during the unaccomplished budget period. However, because the delivery of a limited budget revisions make the realization of the budget in the unit not the same as planning revision resulted in the reporting process will be more difficult to do. If the revised budget reporting process is not easy, then it will take a long time to report budget changes, causing the budget too long to be arranged, it shows the weakness of budgeting [1]. With the help of information systems, budget report will be easier, because administrative data that is processed will be more accurate and human errors will be more easily prevented too.

The construction and development of information systems should be in harmony and in accordance with the organization's (Enterprise) strategic direction, many cases of failure of information systems management in achieving the objectives organization, because the way using and building the information system not in accordance with the direction and goals and needs [2]. Enterprise architecture is very important for the conversion of information systems and develop a new system that will optimize the company's mission. This is achieved in terms of logical or business (eg, objectives and business functions, information flow, and the environment outside the system) and technical (e.g., software, hardware, communications), and includes plans harmonization for the transition from basic environment to targeted e-government environment [3]. To the real needs that exist in the budget absorption effective enterprise architecture also did not excluded rules that are used as budgetary constraints.

2 Theory

2.1 Enterprise Architecture

Architecture is an area of science in information systems that basically meant to build the foundation of the

* Corresponding author: kholiq.budiman@student.undip.ac.id

structure, characteristics behaviors and principles as well as a guide, alteration and operation in the making and developing of the system in the long term [4].

Enterprise Architecture(EA) is a representation of the structure and behavior of a company's business processes. This is describe a system that currently exists and the system in the future. EA include:

- An insight into the utilization current information technology in business operations
- A vision for the future utilization of information technology in business operations, and
- A roadmap for the evolution of information technology landscape from the current state to the future state, along with the transient states in between [4].

2.2 Enterprise Architecture Planning

Enterprise Architecture Planning (Enterprise Architecture Planning, EAP) is a collection of architectural and strategic fields which include information, business systems, and architectural engineering [5]. EAP is a modern approach to the planning of the quality of data in order to achieve the mission of Information Systems. EAP is also a process of defining a number of architectures, that is: data architecture, application architecture, and architecture in using information technology to support the business. EAP has associated with how to align business strategy with IT strategy in which the organization's business strategy development will be the starting point for determining the IT strategy further. EAP will provide a map of the enterprise and enterprise plan to change the path of business and technology. The linkage between the existing architecture is essential for EAP [5]. It is therefore not developed in isolation enterprise, EAP should be looked at in the perspective of enterprise-wide. The steps in using Enterprise Architecture planning is shown in Fig. 1.

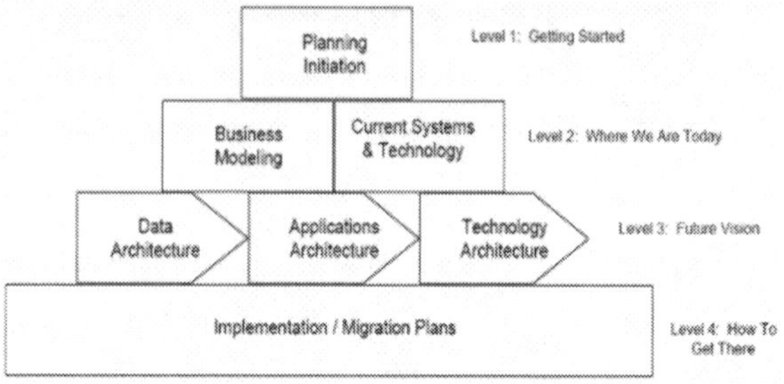

Fig. 1. Layers in the EAP [5]

2.3 Zachman Framework

Enterprise Architecture Framework is a set of assumptions, concepts, values and practices that constitute a realistic perspective of a company through the perspective of architectural models. So EA offers fixed structures and assist the designer in developing, maintaining and using EA [4]. Zachman Framework was originally an enterprise modeling tool. Basically Zachman is a 6x6 matrix that defines 6 levels that are relevant for any company. The arrangement provided by the Zachman framework placed on all relevant scales, as well as on all relevant aspects and also take into account all the conditions [6].

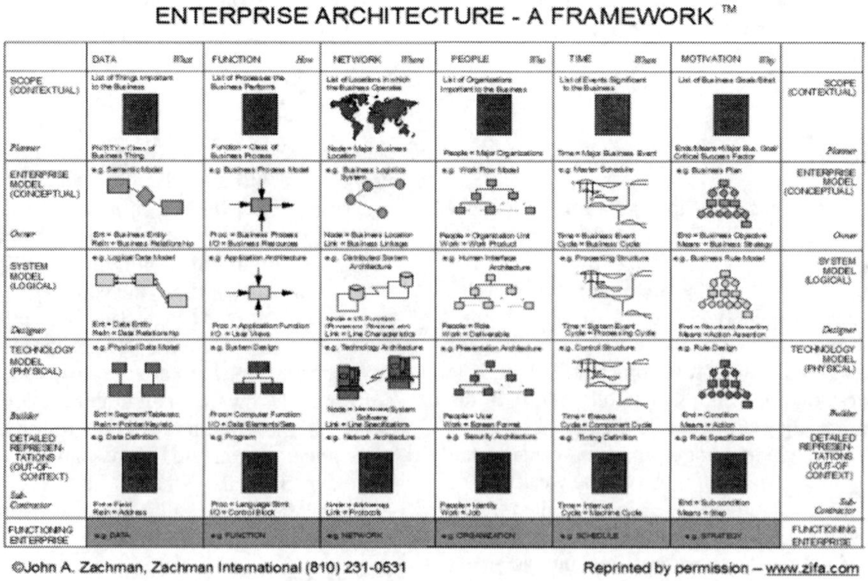

Fig. 2. Zachman Framework [7]

2.4 EAP & Zachman Framework

EAP is part of the Zachman Framework. in the Zachman Framework, EAP include first and second row of the first three columns as shown in Fig. 3.

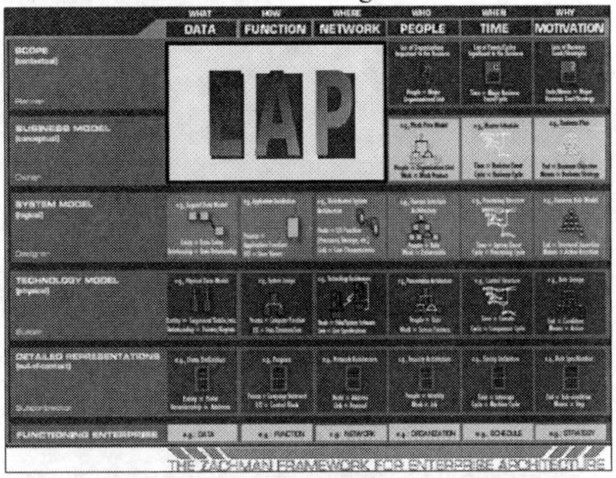

Fig. 3. Relations EAP & Zachman Framework

2.5 Value Chain Model (Value Chain)

Model of value chain consists of a series of activities that create and build a margin value that can generate added value for the organization [8]. The value chain provides a framework for identifying and inventorying areas of business functions is by grouping functional areas into core activities and supporting activities.

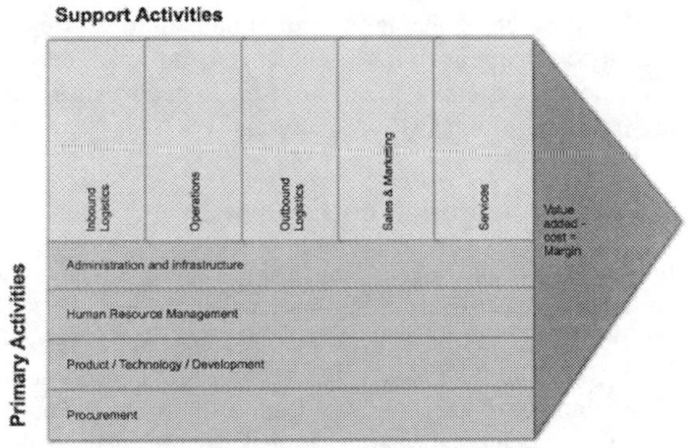

Fig. 4. The Value Chain Model [8]

2.6 Software Testing

Software testing used to identify errors of its information system, deficiencies or errors in the application code is must be corrected, this test has the main purpose is to fix errors to guarantee the quality, reliability estimation, validation and verification of software that is built [9]. There are three techniques to look for errors in the software it is White-box test, Black-box test and the Gray-box test [9]. Structural test (or so-called white-box tests and glass-box tests) found a bug in structural elements such low levels structure that occur at the level of code, database schemas, chips, subassemblies and interfaces [10]. Structural testing is based on how a system operates. Black box testing is test intended to test the functionality of the software. This black-box testing ignoring the effectiveness of the internal structure and software. Thus it is programmer or testers get a series of input conditions that fully utilizes all of the functional requirements for a program. While Gray box testing allows testers to focus on all layers of the software by combining Black Box and White Box [9]. By combining the two tests is expected to make the software more effective in its use because the code is structured and has been functioning effectively.

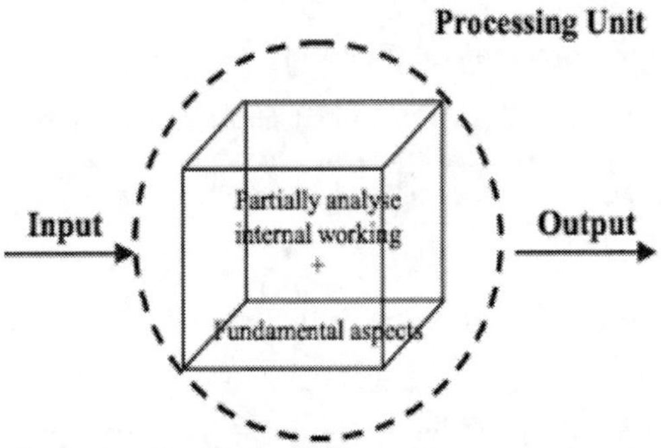

Fig. 5. Gray Box Testing [9]

2.7 Planning and Budget

Planning means to predefine possible activities and how to do it. Planning is an effort to anticipate before doing something so that what is done can be managed properly aim is to provide the process feed forward in order to give direction to every manager in decision making daily operations [11], the linkage between planning and budgeting described in the Fig. 6.

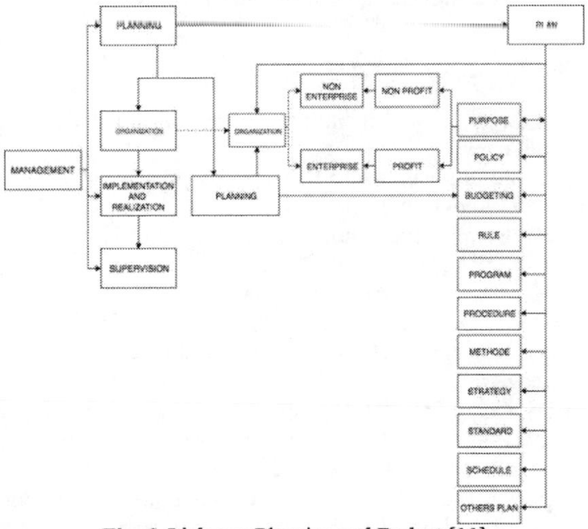

Fig. 6. Linkages Planning and Budget [11]

Budget is a plan of systematically arranged, which covers all activities of the company, which is expressed in units (entity) monetary and valid for a period of time (period) certain to come [12].

In the execution of the budget can be considered effective if it meets several criteria, including:

- Prediction activities
- Clear rules
- Accounting calculations are accurate, reliable and precise
- Appropriate information and can be easily understood
- Support at all levels of the organization both the most top level to the lowest level in the organization [1].

3 Research Methods

3.1 Initialize method Planning

The initiation phase is the initial phase of the study in which the author identifies the basic rules that exist within an existing budget budgeting especially those in the case study that is at the University.

3.2 Methods Identify and define business functions

At this stage, identification of key functions in budgeting. Based on observations made there are six major functions of budgeting in the University:

- Determining the purpose to determine the plans and strategies of each unit in the university.
- Analyzing the available resources, those in terms of revenue coming from routine income that is from students payment and non-routine income.
- Negotiating in the forecast budget components
- Coordinate and review the budgeting component
- Final approval from policy makers
- Distributing the budget approved

3.3 Architectural Design Method

3.3.1 Data Architecture

In designing data architecture steps that must be done include :

- Identifying candidates that are possible to support the business entity;
- Identifying modeling data such as hierarchical, network and relational model;
- Make a diagram of relationships between entities. Modeling that describes the relationship between entities can use the E-RD (Entity-Relationalship Diagram)
- Data entities relate to the identification of business functions that have been performed to determine create, read / reference and update the data so that it can be a reference to the development of applications, where entities that support the functions done first because it create a source of data / master data.

3.3.2 Application Architecture

In designing application architecture steps that must be done include :

- Identify candidate applications to be created and developed to match the needs of business functions are identified;
- Identify actors and designing use case useful for related actors to business functions that support applications that are built;
- Designing Data Flow Diagram to relate the data entities and business functions that will be accommodated applications.
- Designing Class diagram that contains objects that support business functions identified

3.3.3 Architecture Technology

In designing architecture technology steps that must be done include :

- Identify technologies that can support data architecture and application architecture which has been designed so that the use of appropriate technology will be possible;
- Describing the concept of technology is built into a conceptual-frame.

3.4 Methods of Implementation Plan Planning and Budget

Implementation plan is a plan for the realization of the system to be built consists of the selection of programming languages, system test planning, preparation of human resources and other needs that will be required in the field. This implementation plan is useful in the implementation guide data architecture and architectural applications made for formulated into an information system that can be used in accordance with the initial requirements

3.5 Implementation of the System

In implementing the system, the authors made the development of information systems in accordance with the priorities that have been planned before the priority is based on the system which is the holding data from the planning system itself and proceed with system priorities that are identified using matrix data flow entity relationship data against business functions.

3.6 Results Testing Method of Enterprise Architecture

Testing of Enterprise Architecture is intended to test the draft that has been created is already in line with user needs, the needs of companies and the rules that apply in the planning process. In this test done with black-box and white box testing to users. Black-box testing intended to test whether the system has been designed according to the EA can be accepted by the user by using UAT (User Acceptance Testing), in addition to testing with UAT, authors also take functional tests performed by the user, in this case is part of planning and information system at Semarang State University using functional testing, the writer can see if the system is running properly according to its function.

3.7 Analysis System Implemented

Analysis of integrated planning information system which has been applied to an object of research conducted to obtain valid data on the implementation of the information system itself or not. Valid data that you want to search from the analysis of implementing information systems integrated planning is in the form of data changes between planning data and data realization.

4 Results

4.1 Initialization and Identification Business Functions

Initiation by the author in the planning system takes some of the information system that supports the objectives of the budget or the planning itself, the purpose of planning and budget is categorized into a number of things which the authors make it as an Key performance Indicator. The following indicators of success outlined by the initiation of the identification of the rules and the important things that become the primary function of the planning itself, it is:

- Integrated planning information systems can be used in the planning process and strategic plan of a unit / sub-structured organization (Budget support planning function).
- Integrated Planning Information System can be used for coordination (Budget support coordination function).
- Terintegras Planning Information System can be used in communicating in the budgeting process (Budget support communication functions).
- Integrated Planning Information System to motivate the executive in carrying out the activities that have been budgeted and can achieve the goals (Budget As a function of motivation).
- Integrated Planning Information System can be used as a means of controlling the activities (Budget as a Control and Evaluation).
- Integrated Planning Information System can be used as a means of optimizing activities to achieve the objectives of the strategic plan of the unit.
- Planning Information System can be used as a tool for monitoring the activities, so that the activities carried out does not deviate from the plan and strategy and organizational units (Budget as Education).

Information system who identified by the author and required in enterprise architecture planning information system including:

- Budgeting Information System

Budgeting information system is the primary system, which includes the main function of planning is to determination of possible activities and how to carry out these activities.

- Systems Implementation Plan

This information system is used to clarify the budget that were previously defined in the budgeting information system, and the benefits of this system is to explain as clearly as possible the steps of each activity to defining the detailed step-activity (time, person, place, spending details) by defining the full implementation of the realization is also easier because there is a guide implementation of the plan.

- Operational-Reporting System

This reporting information system is a system used to make a proposal before the project is implemented and used too for reporting activities that have been realized with details of activities undertaken.

- LAKIP Information Systems

This system is used as a support in the planning system because the system is required to support the reporting function in the budget.

Of the few systems that are described by the authors above have their respective different destinations that exist in these systems support the value chain so as to form a complete integrated system planning. The value chain that will result from the integration of the system described in Fig. 7.

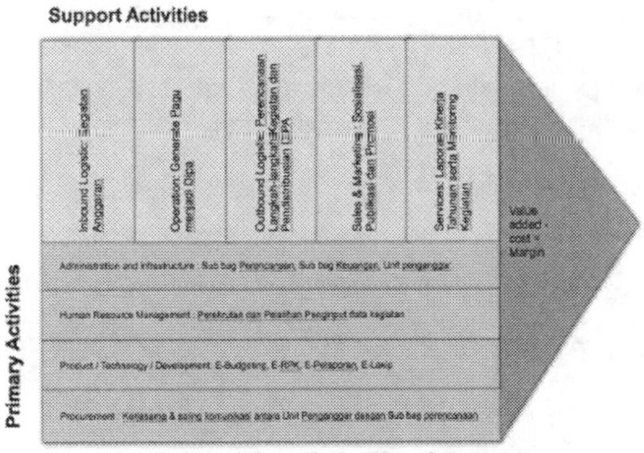

Fig. 7. Value Chain Planning System

4.2 Data Architecture

Based on the data entity identification by author on integrated planning information system generates the data entity relationships as shown in Fig. 8.

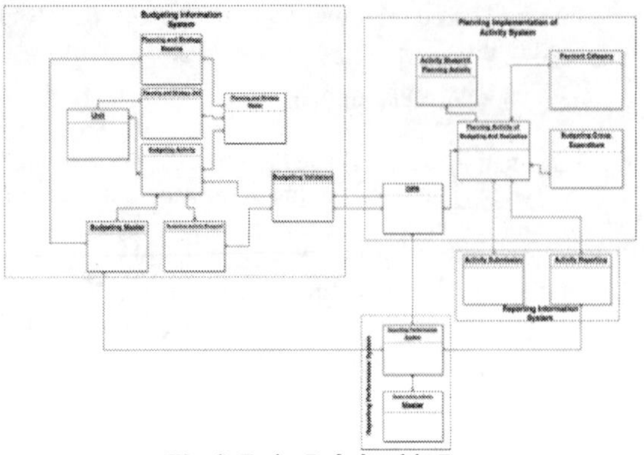

Fig. 8. Entity Relationship Data

4.3 Application Architecture

Products of this application architecture is use case, DFD, and also class diagram

4.3.1 Use case Diagram

In the integrated system described use case each actor to function. The planning information system is divided into 4 use case diagrams based on the function they have, the first diagram is used for the budgeting system in which the system is used to plan all budgeting within the university, this budgeting use case is shown on Fig. 9. Second is the use case used as the implementation plan of the activity, this use case explains and describes how the implementation of activities in the budgeting system will be implemented, this use case is shown on Fig. 10. Third is the use case used for reporting, this use case has a function to view the report between the budgeting system and its realization, the use case is shown on Fig. 11. The last use case used to see the performance of a unit seen from the achievement of the realized realization, the use case is shown on Fig. 12.

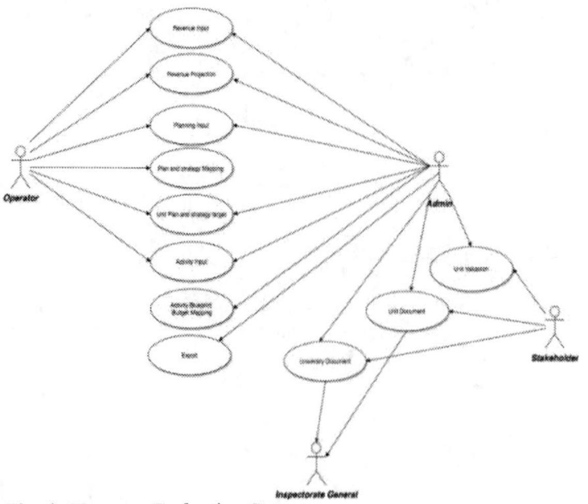

Fig. 9. Use case Budgeting System

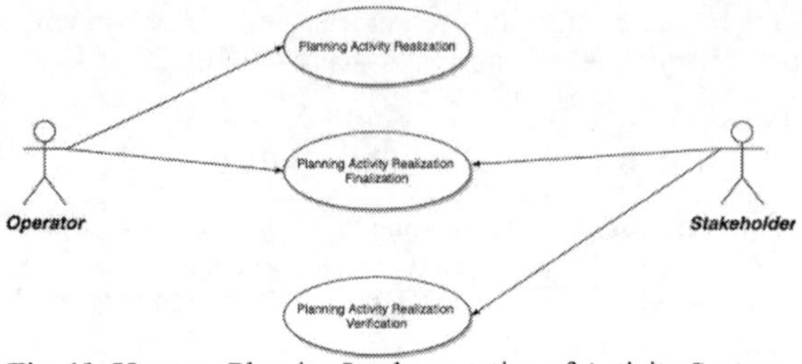

Fig. 10. Use case Planning Implementation of Activity System

Fig. 11. Use case Activity Reporting System

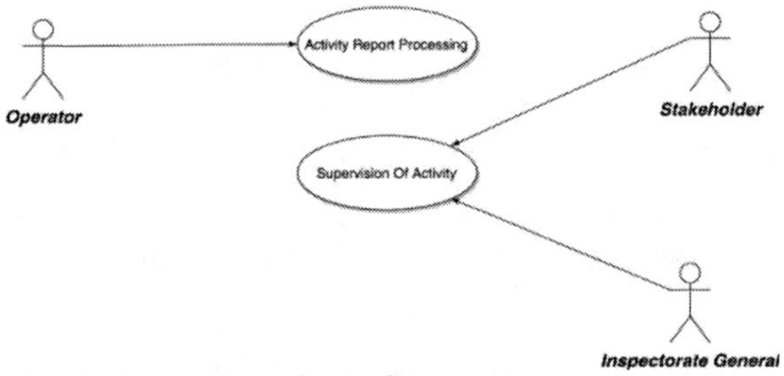

Fig. 12. Use case Reporting Performance System

4.3.2 Class Diagram Planning system

In facilitating the implementation of the system required class diagram that shows the relationship, dependency and the relationship between objects and classes that are contained in information systems planning. Class diagrams contained in the information system planning showed in Fig. 13.

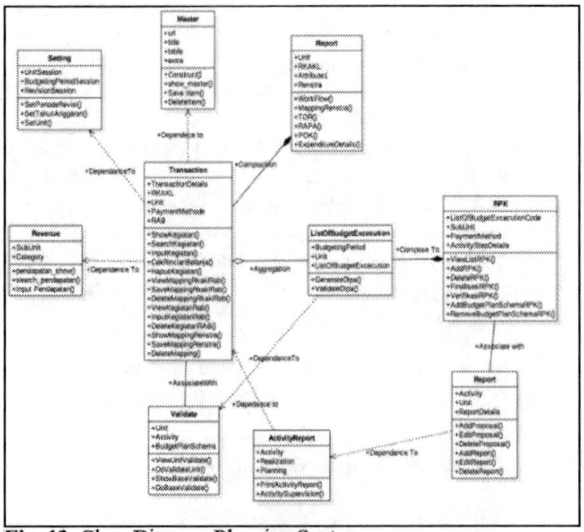

Fig. 13. Class Diagram Planning System

 https://doi.org/10.1051/e3sconf/20183111002

4.3.3 Architecture Technology

In designing this architecture technology the author describes into the conceptual network architecture shown in Fig. 14.

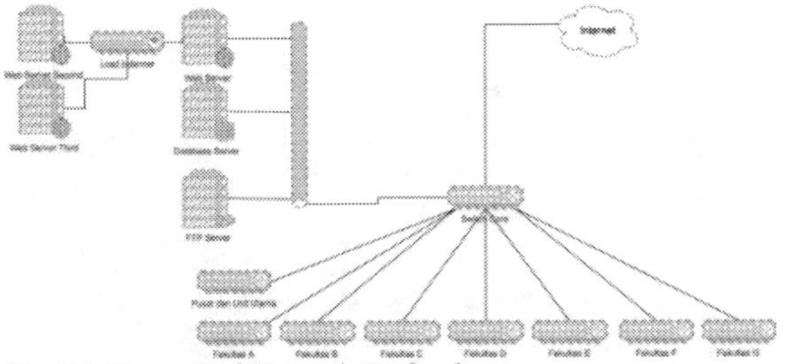

Fig. 14. Conceptual Network Technology

4.4 Implementation of Application

In the implementation, authors write the program code and create a database of architectural structure that has been designed. Then in the execution priorities author adjust to the importance of using entity relationship matrix data from flow data against business functions illustrated in Fig. 15, Due to the EAP approach has the principle application is made or form (create) done first than to other applications, so that the budgeting system has a priority to work earlier than other applications.

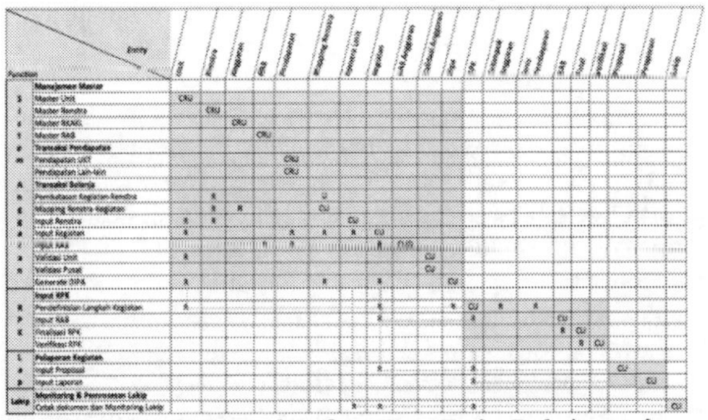

Fig. 15. Matrix Data Flow from Data Entity Relation to the Business function

4.5 Testing Applications

In testing this application the author use gray-box testing techniques, it is the merger between black box and white box testing techniques.

4.5.1 Black box Testing

Black box testing conducted by the author using two methods it is by User Acceptance Testing (UAT) and also functional test where the two tests are without looking at the structure of the programming code. UAT testing is done by using the instrument of acceptance taken from the success indicator that has been set and adjusted with the planning function. Here are the results of the UAT:

a. From the calculation results of questionnaires obtained from 45 respondents, the percentage of the value obtained from the ease and effectiveness of integrated planning information system that has been built has a percentage value of 168/45/5 x 100 = 75%.
b. system planning information can be used in the process of planning plans and strategies of a unit / sub-organizations are structured so that all the planned activities of a unit within the budget in accordance with the plans and strategies that are owned by a unit so this application supports the planning function has a percentage value of 186 / 45/5 x 100 = 83%.
c. Information Systems integrated planning can be used to coordinate the budget so that every unit that exist within an organization can coordinate well and does not cause a miss conception between unit and providers of funds so that the planning system supports support functions of coordination has the percentage of the 166/45/5 x 100 = 75 %.
d. Integrated Planning Information System can be used to communicate in the budgeting process so that everyone in the organization is responsible for the budget has been drawn up so that the system supports the planning information communication function has the percentage of 175/45/5 x 100 = 78%.
e. Analysis of integrated information systems planning can motivate implementers in activities already budgeted and can achieve the goals that support the functions of information systems planning motivation has a percentage of 183/45/5 x 100 = 81%.
f. Analysis of information systems integrated planning can be used as a means of controlling the activities because the approved budget is the commitment of the implementers who participate in the preparation of the budget so that the system supports the function of controlling and evaluation has the percentage of the 182/45/5 x 100 = 81% ,
g. Analysis of integrated planning information system can be used as a means of optimizing activities to achieve the objectives of the strategic plan unit has a percentage value of 185/45/5 x 100 = 82%.
h. Analysis of planning information systems can be used as a tool for monitoring the activities, so that the activities carried out does not deviate from the plan and strategy and organizational units so that the information system supporting the planning of educational functions has the percentage of the 161/45/5 x 100 = 72%.

As for the functional testing conducted by researchers from each of the test items in the form of business functions identified indicates success.

4.5.2 Whit box Testing

In the white-box testing these researchers used unit testing. Unit testing is code-based testing performed by the developer, this test is performed on each test individual units separately. White box testing results can be seen in Fig. 16.

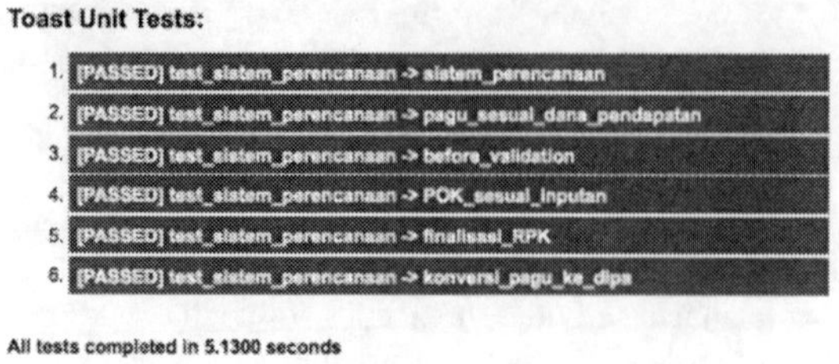

Fig. 16. White box Testing Results

4.6 Information Systems Analysis

The analysis was performed to determine that the system can control the activities of planning information to remain focused on the process of planning and implementation (realization), it is this which makes one of the added value (added value) of the integrated planning information systems that are built using EAP.

Analysis was performed using the planning and realization of the data prior to use integrated planning and information systems after usage of integrated information systems. Testing is done by finding the difference between the plan and the date of realization. If the difference is very high, it is indicates that the implementation or realization is not consistent and can be said to be planning to do less than the maximum. In the analysis of this system the authors saw that the time data from the planning and the time realized tends to retreat in its implementation. Writer screening data and finding the difference between the time data planning and implementation period. Data were analyzed by researchers tabulated on Table 1.The table shows that the average difference between before and after the use of information systems planning tends to fall.

Table 1. The average data tabulation Difference Planning and Realization

Difference	Month	Year	Information
15:48	January	2016	Before Use Information Systems
15:34	February	2016	
15:11	March	2016	
14.72	April	2016	
15:40	May	2016	
15:33	June	2016	
15:56	July	2016	
15:26	August	2016	
15:14	September	2016	
15.71	October	2016	
15:58	November	2016	
0:00	December	2016	Transition Year Budget
1.80	January	2017	After Usage Information System
1.80	February	2017	
1.78	March	2017	
1.75	April	2017	

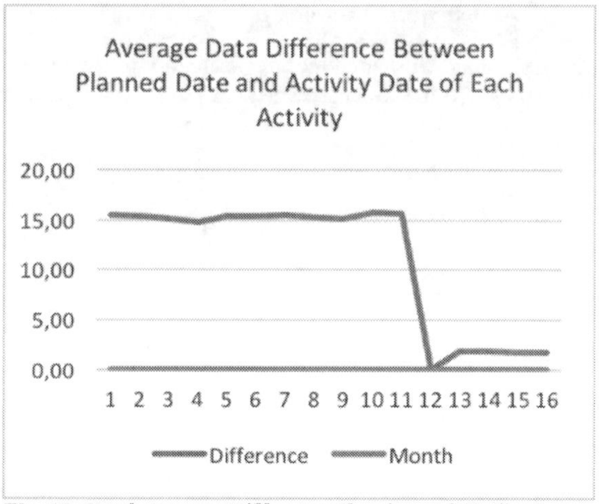

Fig. 17. Graph Average Difference Planning and Realization

5 Conclusions

Judging from the analysis of the User acceptance Test each question in User Acceptance Test has a percentage score above 70%. This shows that the integrated information system that is designed to use EAP is effective and easy to use and most importantly information system integrated planning is in accordance with the functions of planning / budgeting is the planning, the coordination function, the communication function, the function of motivation, controlling and evaluation function and educational function.

Integrated planning information system has been able to motivate the executor in carrying out the activities that already budgeted, can be seen in the graph the average difference between execution of the activities carried out by the time budgeted in the plan, after implementation information system planning applied the difference between planning and realization is becoming less. This shows absorption of the budgeting information system makes budget more realized timely..

From the analysis of users acceptance Testing, planning system integrated is already qualified as a system of communication and coordination system between planner unit and provider of funds, as well as the result of the difference between the schedule of implementation and budgeting more timely because of the planning information system is definitely a miss conception between planner unit and provider of funds.

References

1. 1 J.K. Shim, J.G. Siegel, *Budgeting Basics and Beyon*. (2005)
2. 2 B. Kurniawan, *Enterprise Architecture Planning Sistem Informasi Pada Perguruan Tinggi Swasta Dengan Zachman Framework*, **1**, 21 (2011)
3. *3 B. Bellman, R. Felix, Enterprise Architecture for e-Government **1**, 48 (2004)*
4. 4 S. Bente, B. Uwe, L. Shailendra, *Collaborative Enterprise Architecture* (2012)

https://doi.org/10.1051/e3sconf/20183111002

5. 5 S. Spewak, M. *Tiemann, Updating The Enterprise architecture planning model*,11 (2006)
6. 6 B. Dantu, S. Eric, *Budgeting Basics and Beyond*, **6**, 76 (2005)
7. 7 S. Spewak, *Enterprise Architecture Planning (Developing a Blueprint for Data, Application and Technology)*. (1992)
8. 8 M.E. Porter, *Competitive Advantage : Creating and Sustaining Superior Performance : with a new introduction*. (1985)
9. 9 M.E. Khan, F. Khan. International Journal of Advanced Computer Science and Applications, **6**, 12 (2012)
10. 10 R. Black, *Managing the Testing Process : Practical Tools and Techniques for Managing Hardware and Software Testing* **3**,(2009)
11. 11 M. Nafarin, *Penganggaran Perusahaan,* (2007)
12. 12 M. Munandar. *Budgeting: Perencanaan Kerja, Pengkoordinasian Kerja, Pengawasan Kerja. Cetakan Ketigabelas* (2000)

Consumer's Buying Decision-Making Process in E-Commerce

Nia Budi Puspitasari[1,*], *Susatyo* Nugroho W P[2], *Deya* Nilan Amyhorsea, and *Aries* Susanty[4]

[1,2,3,4] Department of Industrial Engineering, Faculty of Engineering, Diponegoro University, Semarang - Indonesia

Abstract. The e-commerce growth and development in Indonesia is very rapid as well as the internet grows, but it is not well-balanced with the number of online buying transaction which is still relatively low. Even the today's biggest B2C e-commerce people in Indonesia, Lazada, has continually decreased online purchasing. This research is aimed to describe factors affecting online buying decision- making in the e-commerce Lazada. The type of this research is confirmatory research. The variable used is following conceptual model i.e. Electronic Word of Mouth (EWOM), social identity, risk perception, trust, and purchase intention. The data were obtained through the questionnaire with Likert scale 1-5. There are 104 people researching sample who meets the criteria as Lazada consumer that, at least do a transaction in recent six months. Data analyzing were done using Structural Equation Modelling (SEM) method by Analysis of Moment Structures (AMOS) software. The results showed that the purchase intention has positively related to the purchase decision. Variable EWOM toward trust has positive relation, variable social identity and risk perception have no any significant relation to trust. Variable risk perception toward purchase intention has no significant relation, while the variable trust has significant positive relation to purchase intention.

1 Introduction

Internet as a cutting-edge information technology developed that now its use has penetrated into all sectors of life, could not be separated from people nowadays lifestyle. One of those opportunities that take advantage of internet technology is in the business sector, by using electronic commerce (e-commerce)as a trading medium which everyone can access anywhere anytime.

In Indonesia, the growth and development of e-commerce are the second largest in the world in the last four years based on SparksLabs, one of the global research company. The result of the research showed the growth in 2012 by 85%, 2013 by 71%, 2014 by 45%, and 2015 by 37%, while the each forecast for 2016 and 2017 is by 26% and 22% [1, 2]. Can be said that the growth of e-commerce in Indonesia is experiencing rapid growth. According to the Communication and Information Ministry of Republic of Indonesia (KEMKOMINFO), the growth in 2015 is driven by Internet users continues to grow as well. At 2015, Indonesian internet users reached 93,4 million that is increased compared with 2014 which is 88,1 million users. Around 77% of internet users searched product information and online shopped with the most popular products are clothes, shoes, bags, and phones [3].

Despite the growth and development of e-commerce always increased each year, but the conducted survey by Spire Research & Consulting in 2007-2012 stated that in Indonesia the online transaction value of e-commerce is still low if it is compared with the total value of the online and offline transaction. The percentage of online transaction in consecutive years 2007-2012 was recorded at 1,19%; 1,27%, 1,42%; 1,51%; 1,79% and 2,25% [4]. It can be said of that discovery that Indonesian online shopping transaction is still relatively low.

It is also supported by the comparative value of online transaction value data of Indonesia with other south east Asian country in 2014 which is taken from Euromonitor in [5]. Based on those data, the carrying value of transaction in Indonesia was recorded at USD 1.1 billion, which only 0.7% of the total value of retail trade in Indonesia. In South East Asian, Indonesian e-commerce transaction is still inferior to the other three countries namely Malaysia, Thailand, and Singapore with each percentage at 0.9%, 1.2%, and 3.4%.

E-commerce is an online business activity both in product offerings and services, and all business transactions that involve the interaction of various parties occurs electronically and not in direct contact [6]. According Waghmare in [7] the types of e-commerce are b2b (business to business), b2c (business to consumer), c2c (consumer to consumer) and others such as the g2g (government to government).The most developed type in Indonesia is type b2c [8].

Based on the five major b2c e-commerce traffic data in Indonesia, namely Lazada, Zalora, Bhinneka, Elevania, and Blibli in the period September 2015-September 2016; this period Lazada always ranked first compared with other e-commerce competitors. At the end of 2015, Lazada had a significant increase, although it was decreasing in early 2016 until mid-2016. Lazada which always occupied the first rank proves that Lazada is the biggest b2c e-commerce people today. Besides, in Alexa.com, a website visitor counter site, stated that in September 2016 Lazada was ranked out of the whole

* Corresponding author: niabudipuspitasari@gmail.com

website in Indonesia and placed itself on the first rank in category e-commerce website.

Based on the official website, Lazada Indonesia is part of Lazada Group operates in a number of countries in Southeast Asia, namely Indonesia, Malaysia, Philippines, Singapore, Thailand, and Vietnam. While the center of Lazada Group is in the Middle East, the Dubai, United Arab Emirates. The products were offered in the hundreds of thousands that are available for various categories, health and beauty, household appliances, mobile phones and tablets, electronics, and household appliances. The range of products are available from international brands to local brands. Completeness of the products offered have become one of the advantages of Lazada. The company gives consumers a chance to buy all kinds of products simply by accessing to the website that has been provided, then the product orders will be delivered to the homes of consumers with the courier services provided. Payment of the purchase transactions can be made either by cash or credit card.

Based on the theory of human behavior in the use of technology that is Theory Acceptance Model (TAM), which is the development of the Theory of Planned Behavior (TPB) has the assumption that a behavior of humans is preceded by intention or interest that is needed to be done or not to be done the behavior [9]. Thus, purchase intention and purchase actual decision have a proportional relationship.

The high interest of online shopping on e-commerce site can be seen on traffic data or traffic data web [10]. Based on traffic data, Lazada's decreasing traffic occurred in the early 2016 until mid-2016. This decrease is proportional with the online purchasing intention and purchase; thus it can be concluded that online purchasing intention and purchase also decreased. Online buying decision is affected by EWOM factors, social identity, risk perception, and trust [11].

2 Literary Reviews

2.1 Technology Acceptance Model (TAM)

Technology Acceptance Model (TAM) is one of the theories on the use of information technology systems that can be used to describe the individual acceptance of the use of information technology system. This influential theory was first introduced by Davis (1989), it is the development of Theory of Reasoned Action(TRA) by Ajzen and Fishbein (1975) in[9].

2.2 Purchase Decision

Consumer behavior will determine the decision-making process in their purchases. The decision-making process is a problem-solving approach that consists of five stages: problem recognition, information search, evaluation of alternatives, purchase decision and post-purchase behavior [12].

2.3 Purchase Intention

Purchase intention is a mental activity that precedes the act of a person before taking action using the product or service [13].

2.4 Electronic Word of Mouth (EWOM)

According to [14] Word of Mouth (WOM) can be described as an consumers' interpersonal communication about products or services, and in general now WOM plays a key role in influencing consumer attitudes, consumer trends and consumer behavior. WOM in this online system was called EWOM.

2.5 Social Identity

Identification with a social group is a psychological state that is very different from a specified social category and has important self-evaluative result. Overall, social identity can support when in a depressed state [15]. The reason why consumers actively participate in the online community is found by Dholakia et al (2004). She found that social identity motivates participation in online interaction through increased "we-intention", for example, is the commitment of the community participating in the joint action, and includes an agreement between the participants to engage in collective action. They added membership, frequency and level of participation that are driven by the will of choice [16].

2.6 Perceived Risk

According to Murphy and Enis (1986), quoted by [17] perception of risk is an unpredictable phenomenon encountered by consumers during the purchase process error caused by the consumer or the result of decisions that do not fit on the subjective assessment in the decision-making process. On consumer behavior and marketing literature, the perception of risk is an important concept and the various risks that have been identified [14].

2.7 Trust

[18] defined trust as the tendency of one party to a willingness to accept the attitude of the other party even though the first party is not protected by the second and failed to control the attitude of second parties.

3 Method

3.1 Research Model

Conceptual Research Model can be seen in Fig 1.

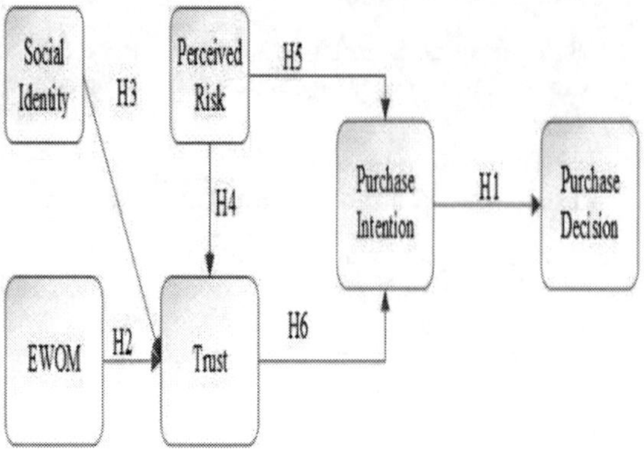

Fig. 1. Conceptual Research Model

Based on the conceptual model proposed by [11], the hypotheses will be arranged as follows:

- Hypothesis 1: Online consumer purchase intention has a positive relationship toward an actual purchase decision.
- Hypothesis 2: Electronic Word of Mouth(EWOM) has a positive relationship toward trust on e-commerce.
- Hypothesis 3: Social identity has a positive relationship toward trust on e-commerce.
- Hypothesis 4: Risk perception has a negative relationship toward trust on e-commerce.
- Hypothesis 5: Risk perception has a negative relationship toward online purchase intention on e-commerce.
- Hypothesis 6: Consumer trust has a positive relationship toward online purchase intention.

3.2 Respondent

This research was conducted in September 2016 using a questionnaire with Likert scale 1-5 (strongly disagree to strongly agree). The population in this research is Semarang people who within the last six months have been purchasing transaction on e-commerce Lazada. The determination of the number of samples for SEM method is based on [19], the sample size, which is recommended and generally accepted to obtain appropriate results is by using Maximum Likelihood numbered 100-200 people. Therefore, the number of samples taken in this research is 100 people and the respondents in this research are 104 people. Then, sampling technique that had been done in this research is non-probability sampling or in another word a non-random sampling. Nonrandom sampling or non-probability sampling give no any similar chance for each individual in the population to be chosen as samples [20].

3.3 Study Design

Research indicator for each six variables are can be seen in Table 1.

Table 1. Indicator for Each Variables

Construct	Variable	Indicator	Reference
Purchase	X1	Amount of Information Search	[21]
	X2	Consideration Set	
	X3	Purchase Decision	
Purchase Intention	X4	Transactional Intention	[22]
	X5	Preferential Intention	
	X6	Explorative Intention	
Trust	X7	Trustee's Ability	[14]
	X8	Trustee's Benevolence	
	X9	Trustee's Integrity	
Perceived Risk	X10	Financial Risk	[14]
	X11	Privacy Risk	
	X12	Product Risk	
Social Identity	X13	Cognitive (have a community or group of social network sites/social media)	[14]
	X14	Affective	
	X15	Evaluative	
Electronic Word of Mouth (EWOM)	X16	EWOM Intensity	[23]
	X17	Positive Valence EWOM	
	X18	Negative Valence EWOM	
	X19	EWOM Content	

4 Results

Data analyzing was performed by using software SPSS 22.00 to test the reliability and validity of the questionnaire and AMOS 22.00 to calculate using SEM.

4.1 Tests of Validity and Reliability of Questionnaire

A result of SPSS output validity test states that all indicators are valid, because it has met the requirements that r count is bigger than r tabulated. The number of r table 0,1466 is obtained from table of critical value (r) product moment with degrees of freedom (N-2), i.e. 125 and *p*-value 0.05. Meanwhile, obtained reliability test states that all variables can be said reliably because had met the requirements of cronbach's alphabigger than 0,6.

4.2 Structural Equation Modelling(SEM)

There are 6 variables and 19 indicators which are used in this research. However, in the SEM analyzing the 18th indicators must be dropped because it does not meet criteria on Confirmatory Factor Analysis (CFA) step. Obtained result of data analyzing could be seen in figure 2.

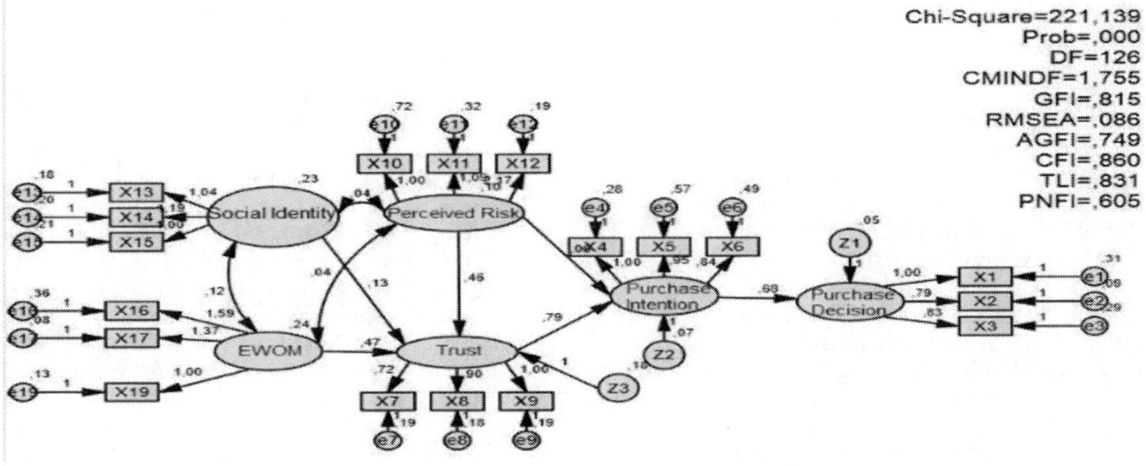

Fig. 2.AMOS Output

Based on AMOS output results, goodness of fit test can be said a good model. According to Ghozali, a model can be said good if two or more goodness of fit shows a good model fit [19] *Goodness of fit* test that are included in good fit are CMINDF and PNFI, it is by fullfilling required CMINDF value less than 2 and PNFI value more than 0.60.Meanwhile, another goodness of fit test shows fit index value with marginal and poor fit criteria still acceptable because the value is between 0 until 1, which is still acceptable.

Meanwhile, SEM assumption that must be fulfilled are normality, reliability test, and validity test. In normality, 23 data must be removed and only 104 data that met this assumption. Reliability and validity testing, all variables are valid (>0,700) and reliable (>0,500). Result of reliability test can be seen in table 2, while the result of validity test can be seen in table 3.

Table 2. Assumption Test Reliability

Variable	Sum of Const Reliability	Result
Purchase	0,660	Good Enough Reliability
Purchase Intention	0,617	Good Enough Reliability
Trust	0,781	Good Reliability
Perceived Risk	0,603	Good Enough Reliability
Social Identity	0,803	Good Reliability
EWOM	0,877	Good Reliability

Table 3. AssumptionAVE Test

Variable	AVE Score	Result
Purchase	0,398	Moderate
Purchase Intention	0,353	Moderate
Trust	0,544	Valid
Perceived Risk	0,365	Moderate
Social Identity	0,576	Valid
EWOM	0,704	Valid

Then, the result of hypothesis testing using SEM can be seen in table 4.

Table 4. Result of Hypothesis Testing

Hypothesis	C.R. (>1,96)	P (<0,05)	Result
H1	4,464	0,000	Accepted
H2	3,189	0,001	Accepted
H3	0,849	0,396	Rejected
H4	1,842	0,065	Rejected
H5	0,271	0,786	Rejected
H6	4,901	0,000	Accepted

5 Discussions

The comparison between this research result and previous journal are similar results for hypothesis 1, 2, 3, and 6. However, hypothesis 1, 2, and 6 stated result significantly affects whereas hypothesis 3 has no significant influence. There is a difference on hypothesis 4 and hypothesis 5 where the journal stated that the conclusion is accepted which means it has significant influence while the research result stated it has no significant influence.

Based on obtaining results, first hypothesis regarding online consumer purchase intention have a positive relationship toward a purchase decision, it has significant influence. Therefore, research result is appropriate to previous research conducted by [11] in Iran. This research result is also supported by [14] result that stated that consumer intention in performing a behavior is an important determinant of the next actual behavior. The result of second hypothesis is also supported by research that is conducted by [24] which stated that the more distributed information quantity and EWOM actuality, the bigger consumer trust on e-commerce. The third one is third hypothesis, social identity has a positive relationship with trust on e-commerce. As a result, it has no significant influence, similar to previous research by [11]. This hypothesis is unacceptable because the questionnaire might not be directed at appropriate consumer. It is caused by almost 50% of respondents has once purchase transaction in the last six months, which can identify the increased level of distrust of obtaining answers. Besides, can be caused by members of a group or a social network community have no close relationship.

Fourth hypothesis, risk perception has negative toward trust on e-commerce. Previous research concludes it is significant, which means higher risk perception, the lower trust will be. Meanwhile, this research shows another result, non-significant influence which is contrary to a study conducted by [14]. This phenomenon can be caused by respondents who have experience in performing online purchasing transaction. In the fifth hypothesis, although theory says high risk perception will decrease purchase intention, but this research says the contrary. The reason is because of their attractive promotions such as discount price given by Lazada even the risk perception is high, their free shipping for certain products also affects the purchase interest is still high. The sixth hypothesis is supported by [17] who stated that there is a positive relationship between trust and online purchase intention. The higher the trust, the higher purchase intention will be.

Improvements recommendation performed on the most influencing variable. Based on SEM analyzing on standardized total effects, trust variable has the biggest value. Therefore, this variable is the most influencing toward purchase intention. The recommendations are it should renew website design regularly, shorten link about product guarantee and its term and condition, use logo to protect personal data and the room, regularly report Lazada'strack record, fill Lazada blog content with profile video and press release, and maximize the use of social media official Lazada.

6 Conclusions

Based on the analysis that has been done, it can be concluded that:

1. Online purchase intention has a significant positive influence toward an actual decision on online purchase of e-commerce Lazada.
2. Electronic Word of Mouth(EWOM) has a positive effect toward trust in e-commerce Lazada purchase transactions while social identity and risk perception do not have a significant influence toward trust on e-commerce Lazada purchase transactions.
3. Risk perception has no significant relationship toward the online purchase intention in purchasing transactions of e-commerce Lazada but the trust has a significant positive influence on online purchase intention in e-commerce Lazada purchasing transactions.
4. The factors that most significantly affect online purchase intention in e-commerce Lazada is trust, because it has the greatest results in the processing of SEM. If Lazada can give trust to consumers and they feel it is high enough in Lazada, it can push them to perform online purchasing transactions at Lazada.
5. Recommendations or suggestion given to e-commerce Lazada is by increasing the trust which are composed of three indicators, namely the ability, the track record t, and the integrity of party that give trust.

References

1. M. P. AlHafiz, http://marketeers.com (2016)
2. A. M. A. Maulani, http://e27.co (2016)
3. Kominfo,*https://kominfo.go.id/index.php/content/detail/6385/Potensi+Besar+untuk+Pemasukan+Negara/0/infografis/*(2015)
4. Spireresearch,https://www.spireresearch.com/newsroom/media/ (2012)
5. Y Syahrul, A Daud, http://katadata.co.id/telaah/2015/12/18/indonesia-pasar-e-commerce-terbesar-di-asean-dengan-banyak-kendala (2015)
6. Z.R. Andam, *E-Commerce and E-Business*. (UNDP-APDIP, Kuala Lumpur, 2003)
7. D.K. Gangeshwer, I.J. of u- and e-Service, Science and Tech, **6(6)**, 187-194 (2013)
8. Clare, *http://www.prnewswire.com/news-releases.* (2016)
9. Fred D.Davis, R. P. Bagozzi, P. R. Warshaw. Mgt. Sc.,**35**, 982-1003(1989)
10. C. Andini, Claudia, M. Ariyanti, Sumrahadi. *e-Proceeding of Management*, **3** (2016)
11. M. Takhire, M. R Taghizadezh Joorshari. Int Journal of Managing Public Sector Informtand Comm Techs, **6**, 3 (2015)
12. Kotler and Armstrong. *Prinsip-Prinsip Pemasaran Jilid 1 Edisi Kedelapan* (2001)
13. N. D. Paramita, and N. N. Kerti Yasa, JMK, **17, 2,** 177-185, (2015)
14. D.J. Kim, D. L. Ferrin, dan H. R. Rao. DSS,**44,** 544-564 (2008)
15. D. Frings, Daniel, I. P. Albery, Addict Behav., **44**, 35-42 (2015)
16. U. M. Dholakia, R. P. Bagozzi, dan L. K. Pearo, IJRM, **21**, 241-263 (2004)

17. K. C. Ling, D. bin Daud, T. H. Piew, K. H. Keoy, P. Hassan, IJBM, **6**,167-182 (2011)
18. R. C. Mayer, J. H. Davis, dan F. D.Schoorman, Acad. Of Manag. Rev., **10**,709-734. (1995)
19. I. Ghozali, *Structural Equation Modelling Teori, Konsep dan Aplikasi* (Badan Penerbit Universitas Diponegoro, 2008)
20. Sugiyono. *Metode Penelitian Pendidikan* (Alfabeta, 2011)
21. G. Haubl, V. Trifts. MKSC, **19**, 4-21 (2000)
22. R. D. Agustin, S. Kumadji, E. Yulianto, J Adm Bis UB,**22**, 1-10 (2015)
23. I. Goyette, L. Richard, J. Bergeron, F. Marticotte, Can J Adm Sci, **27**, 5-23 (2010)
24. F. Xiaorong, Z. Bin, X. Qinghong, X. Liuli, C Yu, AJBR, **1**, 37-52 (2011)

The Impact of Work-Family Supportive Supervisors on Job Involvement and Job Satisfaction with Career Competencies as an Intervening Variable

Suharnomo[1,*], *Paguh* Raja Johnpray[2]

[1] Department of Management, Faculty of Economics and Business, Diponegoro University, Semarang - Indonesia
[2] Department of Management, Faculty of Economics and Business, Diponegoro University, Semarang - Indonesia

Abstract. This study was conducted to explore the relationship between work-family supportive supervisor, career competencies, job involvement, and job satisfaction. Data were collected from a sample of 162 respondents who worked as a nurse at hospital in Jakarta, Indonesia. This study used Structural Equation Modeling (SEM) to examine and estimate the relationship between the constructs. The study results indicate that work-family supportive supervisor has a positive and significant impact on career competencies, career competencies have a positive and significant impact on job involvement, and on job satisfaction. Work-family supportive supervisor has a positive and significant impact on job involvement and on job satisfaction. A total of 162 respondents with a percentage of 96.30% women showed that women need good support and policies in achieving a balance between work and family. Finally, this study highlights the importance of work-family supportive supervisor and career competencies for improving job involvement and job satisfaction. This study provides insights on how the human resources department can retain qualified staff and develop employee satisfaction.

1 Introduction

Unbalanced problem between life and work receives considerable attention from the human resources department, because the current workforce is owned by Gen-Y generation. Gen-Y's generation is the generation that considers quality of life as their top priority. The generation was born in the 1970s until the early 1990s. Research in 2013 in China showed employees above 30 years suffered from extreme stress, where the survey shows more than 64% of employees below 35 years get three main pressures which are the conflict with the balance of life and work, study pressure, and the pressure of competition [1]. In the western countries, reaching equality between life and work is used as a management tool of strategic human resources to maintain quality staff[2]. In addition, most of employees below 30 years and become the only child in their family, have an obligation to take care of their parents, especially when their parents get disease. As a result of marriage, work (job), and family, becomes a source of considerable pressure for employees below 30 years. These people need work-family-friendly policy and endowment.

Work-family supportive supervisor is a supervisor who supports subordinate to participate in formal role that existed in family, and gives information and suggestion about how a certain job could decrease a problem that related with family responsibilities. *Work-family supportive supervisor* is a signal to employee that proved they got an appreciation, an important coping source psychologist and also a buffer of tension, or a human resources protection. Similar with general supervisor's support, but supervisor whose support family more having positive relation with *in-role job performance* because of enhancement quality of human resource also contextual performance and obligation which perceived by an employee as an mutual relation. In this case, work-family supportive supervisor considered as a part of internal relation between employee, where supervisor should maintain good relationship with their subordinate in order to create favorable atmosphere in the working's life. Work-family supportive supervisor has a strong relationship with other variable such as career competencies [2].

Career competencies is a characteristic which can be measured, usually relative stable of individual whose can predict superior performance such as strategic orientation, business knowledge, technical ability, and teamwork. *Career competencies* is a figuration professional competency which owned by employee because of supervisor support in case of motivation and giving positive direction to the employee in case of working. *Career competencies* is becoming an important thing in working atmosphere, which is also getting more

* Corresponding author: suharnomo.undip@gmail.com; suharnomo@live.undip.ac.id

dynamic now on. *Career competencies* could be classified into 3 types, "knowing-why, knowing-whom, and knowing-how". Career competencies have an impact to job involvement and job satisfaction of an employee in company. The number of employee ,who less support from supervisor is getting increase more pressure in working life, then human resource party should retain qualified staff by knowing level of employee's work satisfaction to the job and employee's courage level to be involved in a certain job with their owned career competencies.

Job involvement as an internalization values about good jobs or importance job for existence of person. *Job involvement* is a personal level of job performance affects their dignity and level of physiology in case of identification their selves to the job or the importance of job in personal reflection. Personal who has high involvement is more identified their selves to the job and assume a job as the most important thing in their life. In the end, every effort which is done by human resources department in retaining qualified staff, especially employee' competency development, and applied a policy which is supporting to work-family conflict with supervisor who also supports that, will cause higher level of employee job satisfaction

Job satisfaction is an affective or emotional response reaction from a job, a condition which someone could feel a satisfaction in one and another aspects. It stated that *job satisfaction* is a positive feeling about a job which can be a result of evaluation from several job characteristic. One of many indicator of job satisfaction is supervisor's satisfaction. Research proved that level of satification can be diverse, depend on aspect that job satisfaction viewed.

Competition which made the employees to increase their capabilities and competencies, confiscate their time and the need of balance between work and family. Hospital as a place where a person needs careness and protection is now beginning to compete closely in the selection of employees, services provided, and the development of its employees. Nurse is an employee who must give a full attention to each condition and time, also require attention for its work. As the nurse provides care to patients, they also requires attention such as policy and regulatory support for any problems that occur both in work and family. Therefore, this study aimed to explore the effect of the work-family supportive supervisor at the career competencies and its influence on job involvement and job satisfaction.

2 Previous Researches

Work-family supportive supervisor is someone who sympathized with employee's desire to find equilibrium between work and family responsibilities[4]. Work-family supportive supervisor is an informal adjustment demands and work scheduling along with suggestions about coping strategies provided by supervisors who support families who allow employee to meet their family responsibilities without unduly limiting their opportunity to fulfill their work's need. According to Greenhouse et al in Kong [1] there are seven measurements that need to be considered in work-family supportive supervision, which are :

Work-family supportive supervisor are:

1. Family supportive supervision
2. Family-supportive organizational perceptions
3. Spousal support
4. Work interference with family/WIF
5. Family interference with work/FIW
6. Work-family balance
7. Control and presence of children

Based on considered measurements above, could identified that several factors related to supervisor supports will be able to affect the way emloyee fulfill their work responsibility and family, so will get better work performance. Work-family supportive supervisor has a strong relationship with the career competencies and job involvement. Therefore, it could be seen that work-family supportive supervision could affect career competencies[5].

Research conducted by Akkermans [3] concluded that the role of career competency role as a mediator. Results showed that career competencies positive associated with job resources and job involvement, but not (negatively) to the demands of work and emotional exhaustion. In addition, career competencies have mediated effect on the relationship between job demands and job involvement, and job resources have the effect of partially mediators in relations between career competencies and job involvement.

According to Akkermans et al. [3], career competencies is a knowledge, skills, and ability to develop career that could be affected by individual. It means that *career competencies* is getting more important and became requirement to navigate career. Career competencies defined as a valuable competencies for all employees, which that competencies become career development for themselves, regardless of certain jobs. *Career competencies* also refer to characteristic that describe behavior patterns, personal characteristics, self-concept, values, knowledge or expertise that brought by someone who performs superior (*superior performer*) in the workplace. This is about personality characteristic regarding their ability and the way they do it. Competencies is found in people whose classified as superior performers or effective. There are 6 important measurement which affect to career competencies, such as ;

1. *Reflection on motivation*
2. *Reflection on qualities*
3. *Networking*
4. *Self-Profiling*
5. *Work-exploration*
6. *Career control*

Research about work-family supportive supervisor also found that work-family supportive supervisor has a positive contribution to the career competencies, job involvement and job satisfaction. Moreover, this discovery indicates that work-family supportive supervisor not only helps to retain qualified staff but also increases employee's job involvement and job satisfaction. In the end, the manager is obliged to listen to employees and appreciate the differences and needs of employees. "Such an organizational climate interference would help employees balance between work and family domains, have high job satisfaction and well-being, and ultimately serve customers happy." The outline research above explained how a good climate organization can help frontline employees to resolve problems that occur in the balance of family and work, with its impact on job satisfaction as well as satisfaction of service perceived by the customer to the frontline employees [1].

Job satisfaction is a matter that can be said as the level of happiness or the extent to which employees feel satisfied with what he/she has resolved in his/her work. It depends on key factors in job satisfaction itself. One of the factors most often discussed is the supervisor. When employee work, employee work not only for himself, but for the company, and his supervisor. Employees are required to take full responsibility to supervisor. In this relation, the relationship between the employee and the supervisor will be established stronger. It means, the relationship shows a positive/strong inclination on the side or even to strain that occurs due to the lack of effective communication. Employees who are married, and have children of course also became a problem itself, well it needs the support of supervisors as motivation in the work.

Job involvement can be explained as involvement about internalization values regarding job righteousness or the importance of individual's work value. Most of researchers agreed, that the fact of job involvement is building a different construction related, encompass of: organization commitment, job satisfaction, and intrinsic motivation. Good relation from supervisor to employee will increase employee's willingness to be involved in a job, especially to employee that already had family where their concentration to work will be split to their family affairs. If there are good relation between supervisor and employee, so family affairs wouldn't be a significant constraint

Beside of research job involvement and satisfaction, the important of work family supportive supervisor also reflected in research about work family conflict. Work-family conflict is a form of inter-role conflict in which pressures from the work role are incompatible with pressures from the family role, and this conflict can have an important effect on the quality of both work and family domains [2] This happen because of role pressure incompatibility exists when participation in one role is made more difficult by virtue of participation in another role. As such, the requirements of one role make performance of the other role more difficult, so that commitment to succeed in one role may heighten the concern for succeeding in another role. This implies that the amount of time a workers devotes to progress and succeeding in a chosen career interferes with handling family-work-related responsibilities. Specifically, the additional family-role activities required for nurses who struggle to balance the demands of work and home life as they attempt to manage simultaneous work and family domain pressures. Good level of job satisfaction and job involvement will be able to reduce possibility of work family conflicts, which could degrade workers's performance

The importance of work-family supportive supervisor also consider in the policies making when the manager can identify the policy that support employee and family, it will increase and satisfaction and work-family enrichment, which can lead to performance.

3 The relationship between Work Family Supportive Supervisor and Career Competencies

Supervisor can maximize the success of career and career networking. Individuals who are devoted to the career management, finally reach the highest career competencies, because they have a high-quality family, and supportive management. Work-family supportive supervisor has a positive influence on the career competencies [1]

In addition, research found that work-family supportive supervisor has an influence on the career competencies. The presence of supervisors who support a family, and work (work-family supportive supervisor) is the organization's efforts to maintain their job responsibilities and families. Activities that support organizations reduce employee's stress experience of work-family conflicts, and contribute to the highest productivity of the employees [4].

Research found that work-family supportive supervisor has a positive influence on the career self-efficacy. Career self-efficacy is a person's belief in his ability in the process of managing, developing and completing a career, to achieve the goal of his career. There are several factors that influence the career self-efficacy, which one is the condition of the family, and personality. If an employee who has a dual role as workers and housewives, and she is not able to meet the demands of both roles, there will be conflict-called work-family conflict [5].

Research [6] also revealed that work-family supportive supervisor had a shape that is supportive policies (policy support). This policy needs to be applied with the basis of any form of appreciation and concerns the company/organization to its employees. The support of work-life policies were a form of policy and the attention given by the company to its employees to provide convenience and comfort within the scope of private life that have an impact on working life. So that, it reduces the various clashes and supports of all aspects on the work, and as a result it will increase the career competencies and work commitments [7].

Work-family supportive supervisor has a positive and significant impact on the career competencies. Supervisory support has a positive relation to job satisfaction of employees and families, with the help from a supervisor who supports the employees. It can balance the life and work, and concentrate on the competence and career management [2].

H1: There is a positive relationship between the work-family supportive supervisors with career competencies.

4 The Relationship between Career Competencies and Job Involvement.

Individuals with high competencies' career tend to achieve a high level of job involvement, because "knowing why" career competency can inspire the internal motivation of employees. Employees with a high enough internal motivation tend to be involved in their work, because they have the inner desire to devote more effort to their work. Results from this study showed career competencies have a positive influence on job involvement [1].

. In addition, individuals with high career competencies can rise to higher levels of performance work by proposing creative suggestions, develop additional skills, and introducing new practices. Employees with high scores in performance usually obtain a relatively higher level of assessment performance compared with colleagues from supervisors/managers of them. In turn, high appraisal can encourage employees to involve more in their job.

Career competencies have a positive influence on job involvement. Where an employee who knows what their value and strength, and better in making action plans and goals settings, can develop themselves and consequently, suffered in job involvement. They actively communicate with others and create opportunities associated with career where it creates a positive and energetic working environment that make employees engaged in their work.

When employees feel that their jobs can meet the need to acquire competence, it can encourage them a sense of satisfaction to exert more effort, and thereby it increases the involvement of their work (job involvement) [8].

H2: There is a positive relationship between career competencies and job involvement

5 The relationship between Career Competencies and Job Satisfaction

Research found that career competencies have influence on job satisfaction. Employees with high scores on the career competencies reached a high level of career satisfaction and job satisfaction. Employees who are satisfied with their careers can psychologically identify their career with their current job [1].

Career competencies have a positive influence on job satisfaction. People with high career competencies "know why, "know who", and" "know how" can set their career goals, and develop skills associated with the job. For example, people who get a high score on the " know why " competence, usually have realistic career expectations, knowledge of their strengths and weaknesses, and specific career goals. People who know themselves well showing a very strong career identity.

People who have a strong career identity and career competencies tend to seek regular training, acquiring portable skills, and building a network. Besides knowing how performance improves confidence and identity work. It can allow an individual to remodel his career. Given the unstable employment situation and the need to be always aware, is a way to develop new skills to set up. It is estimated that the "knowing how" competence is an important predictor of job satisfaction and career satisfaction [9].

H3: There is a positive relationship between career competencies and job satisfaction

6 The relationship between Work-Family Supportive Supervisor and Job Involvement and Job Satisfaction

Job satisfaction can be defined as a state affective resulting from the evaluation of a person from his job. When the hotel's employee found that their supervisor gives attention to their problems and provides sufficient support for the resolution to these problems, they become more satisfied in their work. These results found that there is a positive influence of the work-family supportive supervisor on job satisfaction [10].

Moreover, this research is also supported by research [11] where supportive supervisors has a positive and significant impact on job satisfaction of the server/ food waiters. Sergeant and Frenkel [12] found evidence in his research on call centers that work-family supportive supervisor has a positive and significant effect to job satisfaction frontline officer. Support of supervision has a significant and positive impact on job satisfaction frontline officer of the bank [13].

Work-family supportive supervisor has influence on the job involvement. When employees feel their supervisor supports their family, they tend to feel more comfortable psychologically, and devote themselves to their work. Psychological climate, such as management support became a positive contribution to job involvement and job satisfaction bank [1].

Based on the results of data analysis in research Siahaan (2011) obtained the value of r xy = 0395, R-square = 0156 ($p < 0.05$), which means that the quality of work life is proven to have a significant positive relationship to job involvement and accounted for 15, 6% of the increase in job involvement on the subject of research. The results also showed that the majority of research subjects have high levels of quality of work life and job involvement is classified in the category of medium. This proves that the strong support from the supervisor, and the policies of the pro-work-family issues is needed.

H4: There is a positive relationship between the work-family supportive supervisor with job involvement.

H5: There is a positive relationship between the work-family supportive supervisor with job satisfaction.

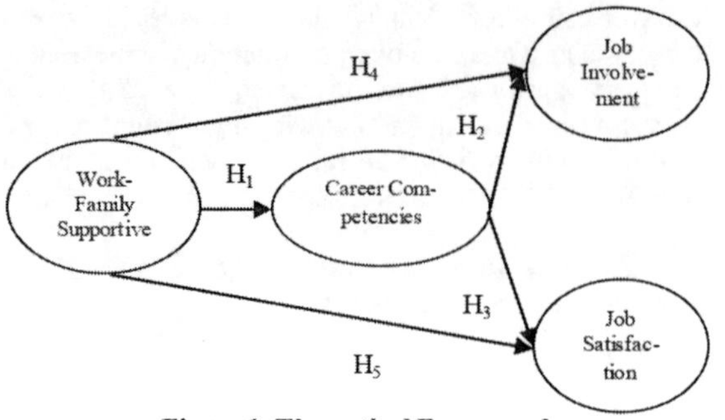

Figure 1. Theoretical Framework

7 Reserach Methods

7.1 Sample

Samples are a part of the number and characteristics owned by population [14]. There are some guidelines for determining the number of samples according to Ferdinand. One of them in the analysis of SEM (Structural Equation Modeling) requires a sample of at least 5 times the number of indicators used. This study uses 17 indicators. The sample selection was done because researchers have understood that the required information can be obtained from a specific target group that is able to provide the desired information because they have the information like that and they meet the criteria set by the researchers [15]. Type of purposive sampling is judgment sampling. These samples were selected by using certain considerations that are tailored to the purpose of research or research issues developed. Nurses are being sampled in this study were nurses PGI Cikini Hospital in Jakarta who still have family, this means that nurses still have relatives/parents /husband/wife and children.

Nurses who still have family are believed to still have a problem between work and family balance. In this study, nurses who still have family are believed to be able to support the research into work-family supportive supervisor and career competencies as variables that affect job involvement and job satisfaction.

A researcher must examine in detail of all members of the population. The next stage is to determine the number of samples of the population. The study population consisted of all nurses PGI Cikini Hospital in Jakarta. After known the population, the next is statistics determining (traditional statistical model), based on a statistical formula approach formula Yamane [15]. This study uses a slovin formula:

$$n = \frac{N}{1+N.d^2}$$

Specification:
N = Number of samples
N = Population size
D = Precision expected or percentage of leeway inaccuracy due to sampling error can be tolerated or desirable.
Amount of population is 272, and the desired error rate is 5%, then the number of samples used are:

$$N = 272/272\ (0.05^2)\ + 1 = 161 \text{ employees/nurses}$$

8 Methods

Methods of collecting primary data in this study was conducted using questionnaires. Questionnaires are a number of written questions to be answered by the survey respondents, so that researchers obtain empirical data to solve the problem of research and testing hypotheses that have been established. There are two types of questionnaires, which assists people and self-administered. Data analysis is the process of simplification of data into a form that is easier to read and implement. Data analysis was performed with the aim to present empirical findings in the form of descriptive statistics that describe the characteristics of respondents and inferential statistical analysis that used to test the hypothesis proposed research [15].

Confirmatory factor analysis was conducted to test the validity of a theoretical construct, or in other words, whether the indicators used are uni-dimension size of a latent constructs. In accordance with the model developed in this study, the data analysis technique used is Structural Equation Modeling which is operated by AMOS program 20. Researchers used SEM analysis tool because it can directly analyze the relationship between the dependent variables with several independent variables, so that it help to take a decision which will be applied in the future.

9 Results and Decision

Analysis of Structural Equation Modeling (SEM) with a full model had been done to test the model and hypothesis. SEM analysis results are as follows:

https://doi.org/10.1051/e3sconf/20183111004

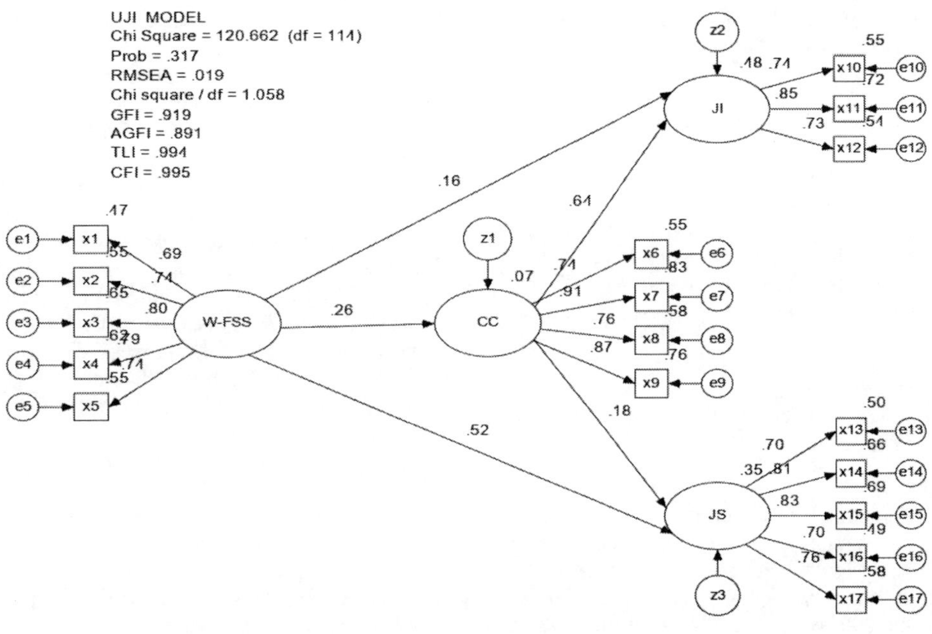

Figure.2 . Results of Testing Structural Equation Modelling (SEM) Full Model
Source: Processed Primary Data, 2016.

Test for the feasibility of SEM full model is tested using Chi-square, CFI, TLI, CMIN/DF and RMSEA which are within the expected range of values, although GFI and AGFI marginally acceptable, as described in following table:

Table 1. Test Results of Goodness-of-Fit

Goodness of Fit Index	Cut-off Value	Result	Evaluation Model
Chi – Square	Little	120 662	Good
Probability	≥ 0.05	0317	Good
RMSEA	≤ 0.08	0019	Good
Chi-square / df	≤ 2.00	1.0558	Good
GFI	≥ 0.90	0919	Good
AGFI	≥ 0.90	0891	Marginal
TLI	≥ 0.95	0994	Good
CFI	≥ 0.95	0995	Good

9.1 Effect of Work-Family Supportive Supervisor on Career Competencies

Parameter estimation of influence Work-Family Supportive Supervisor for Career Competencies showed coefficient values with the positive direction. Tests of significance influence values obtained CR = 2,913 with probability = 0.004 <0.05. Significance value less than

0.05 indicates, that the Work-Family Supportive Supervisor has a significant influence on career competencies. Thus the results of this statistical test successfully supported hypothesis 1.

9.2 Effect of Career Competencies on Job Involvement

Parameter estimation of the influence of career competencies on job involvement showed coefficient values with the positive direction. Tests of significance influence values obtained CR = 6,404 with probability = 0.000 <0.05. Significance value less than 0.05 indicated that the career competencies have a significant effect on job involvement. Thus the results of this statistical test successfully supported hypothesis 2.

9.3 Effect of Career Competencies on Job Satisfaction

Parameter estimation of the influence of career competencies on job satisfaction showed coefficient values with the positive direction. Tests of significance influence values obtained CR = 2,247 with probability = 0.025 <0.05. Significance value less than 0.05 indicated that the career competencies have a significant effect on job satisfaction. Thus the statistical test is successfully supported hypothesis 3.

9.4 Effect of Work Family Supportive Supervisor on Job Involvement

Parameter estimation of the influence of the work family supportive supervisor on job involvement showed coefficient values with the positive direction. Tests of significance influence values obtained CR = 1,983 with probability = 0.047 <0.05. Significance value less than 0.05 indicates that the work family supportive supervisor has a significant impact on job involvement. Thus the results of this statistical test successfully supported hypothesis 4.

9.5 Effect of Work Family Supportive Supervisor on Job Satisfaction

Parameter estimation of the influence of work-family supportive supervisor on job satisfaction showed coefficient values with the positive direction. Tests of significance influence values obtained CR = 5,302 with probability = 0.000 <0.05. Significance value less than 0.05 indicates that the work-family supportive supervisor has a significant impact on job satisfaction. Thus the results of this statistical test successfully supported hypothesis 5.

10 Conclusion

The result of this study showed that work-family supportive supervisor has a positive and significant impact on the nurse career competencies. Better work-family supportive supervisor will increase the career competencies of nurse at PGI Cikini Hospital Jakarta. Career competencies have a positive and significant impact on nurse's job involvement and job satisfaction. The better career competencies that employees have, will lead to greater job involvement that conducted by nurses. This also applied to job satisfaction of nurse obtained.

Work-Family Supportive Supervisor has a positive and significant impact on nurse job involvement and job satisfaction at PGI Cikini Hospital in Jakarta. The better work-family supportive supervisory gave to nurse, the greater job involvement will be conducted by nurses. Same with job involvement, the better Work-Family Supportive Supervisor gave to the employee, the greater job satisfaction obtained by nurses.

Based on result, 96.30% percentage of the sample who are women showed that they need a good support and policies to achieve a balance between work and family. When the support and policies are compliance, they will feel safe and satisfied to do their job, which lead to increase their performance.

If the employee get more support from supervisor, they will get other power resources and reduce the level of job responsibilities that involve and impact the family, thereby increasing job satisfaction. This results supported research done in tourist industry [16]. At the same time, organizations and supervisors will restore the psychological identity of their employees, which increase work engagement and organizational commitment [17].

In conclusion, the results of this study support the importance of family supportive supervisory behavior (FSSB) as a resource to assist employees in managing job demands and family. These findings suggest that not only do employees benefit from FSSB as a resource, but the organization also has a possibility to get benefit through positive outcomes such as affective commitment and job satisfaction.

Based on the theories and previous research, current research also convince the future researchers to understand helpfully the process where the FSSB can affect these work and health outcomes. This research will help others researchers and future managers understand the importance of FSSB. This research illustrate the model-based relationship system between FSSB, work-family conflicts and enrichment, and work-family results. The current study covers the aspect of the model and adds it by including additional variables and explanation processes. Significant direct and indirect effects are found, some supporting the proposed model and another research done in the same setting[18, 19]

11 Limitations and Suggestions

This study has limitations, where there are 38 questionnaires from a total of 200 questionnaires handicapped. It happened because due to lack of response from the respondents to fulfill the questionnaire research. So that proper questionnaire used in a further analysis in order to produce good results is only 162 questionnaires.

This quantitative research will be more fruitful, if it is accompanied by a qualitative approach. Which is, it still needed to dig how much enthusiastic respondents about every aspect in the variable, so that it will show more accurate results. The uses of open-ended questions are better need to be used to explore the indicators of each variable in depth needs to be done.

In some studies found an answer questionnaire/scale that is almost equal to the value of one indicator on all questionnaires. This can actually be said to be legitimate only. However, when it was discovered that the practice on the field do not match, then there is a possibility that respondents did not give an honest answer. In the future, studies that discuss work-family supportive supervisors should conduct in-depth study with the results from the matching the questionnaire with supervisors, members (employees), and their family members.

References

1. H. Kong, International Journal of Hospitality Management ,**33**, 304-309, (2013) doi:10.1016/j.ijhm.2012.09.006
2. L. T. Thomas, and D. C. Ganster, Journal of Applied Psychology, **80(1)**, 6-15, (1995) doi:10.1037//0021-9010.80.1.6

3. J. Akkermans, W. Schaufeli, V. Brenninkmeijer, and R. Blonk, Journal of Vocational Behavior,**83(3)**, 356-366, (2013) doi:10.1016/j.jvb.2013.06.011
4. T. J. Clifton, and E. Shepard, International Journal of Manpower,**25(8)**, 714-728, (2004) doi:10.1108/01437720410570036
5. Wulandari, *Hubungan Konflik Peran Ganda dengan Stres Kerja Karyawan Wanita di Pusat Administrasi Universitas Indonesia,* (Fakultas Ilmu Sosial Ilmu Politik Universitas Indonesia, Depok, 2012)
6. S. Wicaksono, H. A. Bieantri, and A. Nugroho, Jurnal Hospitality dan Manajemen Jasa, **1**, 1-20. (2014)
7. R. L. Mathis, and J. H. Jackson, *Human resource management.* (Thomson/South-Western, Australia, 2003)
8. S. P. Brown, Psychological Bulletin,**120(2)**, 235-255, (1996) doi:10.1037//0033-2909.120.2.235
9. Cheung C, TTRA Annual Conference Proceedings, (2009)
10. O. M. Karatepe, and H. Kilic, Tourism Management, **28(1)**, 238-252. (2007) doi:10.1016/j.tourman.2005.12.019
11. L. E. Ross, and J. S. Boles, International Journal of Hospitality Management,**13**(2), 155-171, (1994) doi:10.1016/0278-4319(94)90036-1
12. A. Sergeant, and S. Frenkel, Journal of Service Research, **3(1)**, 18-34, (2000) doi:10.1177/109467050031002
13. M. H. Yoon, J. H. Seo, and T. S. Yoon, Journal of Services Marketing, **18(5)**, 395-412, (2004) doi:10.1108/08876040410548311
14. Sugiyono, *Statistika untuk Penelitian.* (Alfabeta, Bandung, 2001)
15. T. Ferdinand Augusty *Metode Penelitian Manajemen: Pedoman Penelitian untuk Skripsi, Tesis dan Disertasi Ilmu Manajemen.* (Badan Penerbit Universitas Diponegoro, Semarang, 2006)
16. H. Kong, and Q. Yan, International Journal of Hospitality Management, **41**, 133-139, (2014) doi:10.1016/j.ijhm.2014.05.013
17. S. P. Robbins, and M. Coulter, *Management: fifth edition.* (Qing hua da xue chu ban she, Beijing, 1997).
18. I. B. Siahaan, Hubungan antara Quality of Work Life dengan Keterlibatan Kerja pada Karyawan (2011).
19. A. Zopiatis, P. Constanti, and A. L. Theocharous, Tourism Management, **41**, 129-140, (2014) doi:10.1016/j.tourman.2013.09.013

Prediction of Safety Stock Using Fuzzy Time Series (FTS) and Technology of Radio Frequency Identification (RFID) for Stock Control at Vendor Managed Inventory (VMI)

Chamdan Mashuri[1,3,*], *Suryono* Suryono[1,2], and *Jatmiko* Endro Suseno[2]

[1]Master of Information System, Graduate School, Diponegoro University Semarang Indonesia
[2]Department of Physics, Faculty of Science and Mathematics, Diponegoro University Semarang Indonesia.
[3]Department of Information System, Faculty of Information Technology, Hasyim Asy'ari University, Jombang - Indonesia

Abstract. This research was conducted by prediction of safety stock using Fuzzy Time Series (FTS) and technology of Radio Frequency Identification (RFID) for stock control at Vendor Managed Inventory (VMI). Well-controlled stock influenced company revenue and minimized cost. It discussed about information system of safety stock prediction developed through programming language of PHP. Input data consisted of demand got from automatic, online and real time acquisition using technology of RFID, then, sent to server and stored at online database. Furthermore, data of acquisition result was predicted by using algorithm of FTS applying universe of discourse defining and fuzzy sets determination. Fuzzy set result was continued to division process of universe of discourse in order to be to final step. Prediction result was displayed at information system dashboard developed. By using 60 data from demand data, prediction score was 450.331 and safety stock was 135.535. Prediction result was done by error deviation validation using Mean Square Percent Error of 15%. It proved that FTS was good enough in predicting demand and safety stock for stock control. For deeper analysis, researchers used data of demand and universe of discourse U varying at FTS to get various result based on test data used.

1 Introduction

Stock management is one of production process planning and controls whose purpose is to decrease total cost of stock material and stock level during lead time and acquisition cost. Management developing stock policy which can minimize operational total cost is the main purpose of planning and control. Stock management is an important factor in production process, one factor influencing stock management is demand prediction; demand fluctuation influences product stock and production activity greatly [1].

An important component of chained supply management is stock management. Stock management can spotlight prediction mistake and decision policy depending on demand having potential to prediction mistake. Prediction having the greatest influence to final user's decision can be used to develop demand prediction concept which can give significant influence to the improvement of a company profit [2].

Demand prediction in control management and production supply becomes interesting challenge to be researched because most of them work on data of time series as having been done to overcome problem prediction, like prediction in information system management, health care, economy prediction, selling prediction, budgeting analysis, stock exchange fluctuation, and business analysis, etc [3].

Fuzzy time series (FTS) can design problem of prediction having linguistic value with information having been long time. FTS also can use more observation in prediction having been applied to overcome non-linear. Based on theory of fuzzy compilation, FTS model came from Song and Chissom in 1993, FTS was used to predict the registration of Alabama University. Chen presents new model by using simple fuzzy relation and simple arithmetic calculation [4-5].

Fuzzy time series can predict product need for the next period and this prediction can be arranged based on time period needed. By integrating fuzzy time series to an information system to calculate ROP score of each product, the error average of ROP score got after being examined by using method of Average Forecasting Error Rate (AFER) was 7,13%. Fuzzy times series can predict the number of stock needed in stock room, report stock availability, and give goods stock information so high economy efficiency is got [6].

Time series is an ordered time series arranged from quantitative individual characteristics or collective phenomenon taken from time period successively. To understand time series characteristic, many researchers

* Corresponding author: chamdan.mashuri@gmail.com

have adopted, analyzed, and developed time series method whose final purpose is to find pattern or formula that can be used to predict the future [7].

Radio Frequency Identification (RFID) technology is one technology used in supply chain management using modern. By using wireless technology, a company can track RFID tags easily without physical contact. RFID technology has been proven to be very useful in planning of production, transportation, and warehousing [8]

RFID can integrate into company business process so that it is possible for every entity marked can communicate with all organization information infrastructure, so it can enhance information of supply chain. In business technology process, RFID shows that it can operate in small and middle retail industry and can describe effect of RFID in business operation [9]

Vendor Managed Inventory (VMI) has very significant benefit for supply chain and each company. VMI gives competitive profit to retailer related to higher product availability provided by suppliers with the chance to increase production and marketing efficiency. VMI can increase fulfillment frequency with a small number and decrease stock level for all involved in distribution and supply chain. VMI can optimize supply chain performance in which the producer is in charge to keep distributor's stock level. Producer has access to distributor's stock data and is in charge to order [10-11].

2 Method

2.1 Data Acquisition

This acquisition data process on application of safety stock prediction using Fuzzy Time Series (FTS) and Radio Frequency Identification (RFID) technology for stock control at Vendor Managed Inventory (VMI) applies RFID censor technology which is censor detecting id tags put on the goods using radio waves and analyzed to be data time series stored at local database on microprocessor by using internet network, data of goods demand history is sent to web server and stored at online database then predicted by using fuzzy time series and used to determine safety stock. Data acquisition route is shown in Fig 1.

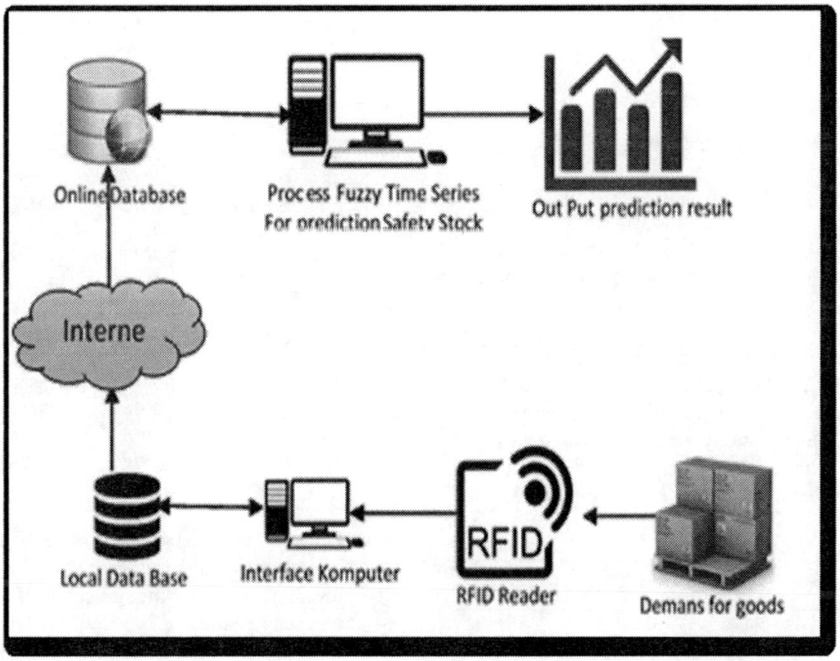

Fig. 1. Acquisition process route of goods demand data

2.2 Modelling by using fuzzy time series

Predicting by using fuzzy time series model is method of data prediction using principles of fuzzy whose base is catching formula of long time data then used to project the future data. Modelling of prediction by using fuzzy time series has some steps as follows :

1. Defining universe of discourse U until *fuzzy* set can be determined as U = [x, y].
2. Determining minimal and maximal score of actual history data (Xmin = x, Xmax = y).

3. Dividing universe of discourse U with some series of data u1, u2, ... un and determining linguistic score.
4. Doing fuzzification and fuzzy set from data of actual histories.
5. Calculating score of fuzzy data of actual history by using the following pattern :

$$\mu NB[x] = \frac{b-x}{b-a} 203659 \leq x \leq 471585 \quad (1)$$

6. Choosing basis of model W which is very appropriate and calculating fuzzy using the following pattern:

$$r(MBF)^{w+1} = \frac{MBF1 + MBF2 + \cdots + MBFw}{w} \quad (2)$$

7. Doing defuzzyfication of calculation result from the above step then, calculating prediction result by using the following pattern :

$$y = \frac{(A1 * c1 + A2 * c2)}{A1 + A2} \quad (3)$$

$$v = r(NB) * A1 + r(NM) * A2 + \cdots + r(PB) * A7 \quad (4)$$

$$Fi = y + v \quad (5)$$

3 Results and Discussion

Data acquisition process on application of safety stock prediction using Fuzzy Time Series (FTS) and Radio Frequency Identification (RFID) technology for stock control at Vendor Managed Inventory (VMI) has some steps as follows:

a. Acquisition process of goods demand data
History data used to predict by using fuzzy time series is data of every month-actual demand data from PT. Quindo food, period of 2012 – 2016, with 60 data as sample taken from scan tag id on goods using censor technology of RFID as in table 1.

Table 1. Data of actual demand

No	Month	Year	Demand Number	Fuzzification
1	January	2012	382635	A5
2	February	2012	355766	A4
3	March	2012	325994	A4
4	April	2012	344349	A4
5	May	2012	362127	A5
6	June	2012	332272	A4
7	July	2012	344733	A4
8	August	2012	262136	A2
9	September	2012	371755	A5
10	October	2012	344931	A4
11	November	2012	360428	A5
12	December	2012	305567	A3
13	January	2013	398608	A6
14	February	2013	401103	A6
15	March	2013	410591	A6
16	April	2013	391991	A5
17	May	2013	373435	A5
18	June	2013	390023	A5
19	July	2013	415428	A6
20	August	2013	294396	A3
21	September	2013	417544	A6
22	October	2013	439641	A7
23	November	2013	422857	A6
24	December	2013	342836	A4
25	January	2014	366797	A5
26	February	2014	423950	A6
27	March	2014	463070	A7
28	April	2014	445420	A7
29	May	2014	452353	A7
30	June	2014	471585	A7
31	July	2014	327364	A4
32	August	2014	388073	A5
33	September	2014	459309	A7
34	October	2014	452508	A7
35	November	2014	425409	A6
36	December	2014	375814	A5
37	January	2015	339850	A4
38	February	2015	376973	A5
39	March	2015	376571	A5
40	April	2015	371001	A5
41	May	2015	304900	A3
42	June	2015	361767	A5
43	July	2015	278754	A2
44	August	2015	430953	A6
45	September	2015	425458	A6
46	October	2015	453944	A7
47	November	2015	394726	A5
48	December	2015	338991	A4
49	January	2016	287776	A3
50	February	2016	362668	A5
51	March	2016	440171	A7
52	April	2016	348626	A4
53	May	2016	339128	A4
54	June	2016	380019	A5
55	July	2016	203659	A1
56	August	2016	388847	A5
57	September	2016	423256	A6
58	October	2016	446611	A7
59	November	2016	450331	A7
60	December	2016	309796	A3

b. Defining universe of discourse U until fuzzy set can be determined. After actual data was calculated, so minimal and maximal score of sample data was obtained (Xmin = 203659, Xmax = 471585). Based on that score difference, universe of discourse U can be defined as U = [203659,471585].

c. Dividing universe of discourse U with some data series u1, u2, …, un, and calculate linguistic score. Firstly, universe of discourse U was divided into ke 7 intervals which have the same size, using the following way: Xmin + interval length. Interval length= (Xmax – Xmin) / 7, for example, 153 + ((471585 – 203659) / 7) = 38275. u1 = [203659, 241934], u2 = [241934, 280209], u3 = [280209, 318484], u4 = [318484, 356760], u5 = [356760,

395035], u6 = [395035, 433310], u7= [433310, 471585]. Then, we admit them as 7 linguistic scores , such as (Negative Big), (Negative Medium), (Negative Small), (Zero), (Positive Small), (Positive Small), (Positive Big) to describe variance of selling number. Based on that definition, 7 fuzzy sets A1, A2, A3, A4, A5, A6, A7, in which A1= (Negative Big), A2= (Negative Medium), A3= (Negative Small), A4= (Zero), A5= (Positive Small), A6= (Positive Small), A7= (Positive Big). In universe of discourse U with the following scores:
A1=203659, A2=248313, A3=292968, A4=337622, A5=382276, A6=426931, A7=471585.

d. Fuzzification of score from history data. In the condition of membership functions (MBF) and fuzzy sets as illustrated in step 3, actual score of selling number can be fuzzified with the norm : "if actual score of selling number is p and score of p on the interval Uj, so p can be translated as Aj". Fuzzifying final score of selling number is based on the norms summarized in Table 1.
e. Calculating fuzzy score of selling number history data of product A, in each fuzzy set.
f. Choosing base of model w which is very appropriate and calculating fuzzy operation.
g. Defuzzification of calculation in step 5, and then, calculate prediction result. After the calculation of fuzzy was done, we need to translate fuzzy output, next, final prediction result was got. By using Center of Gravity (COG) method and equality of (3) (4) (5) so it got final calculation result as shown in table 2, and prediction result and graphic on application can be shown in Figure 2.

Table 2. Data of prediction result

Month	Year	Actual Score	Fi (Prediction Score)
January	2012	382635	0
February	2012	355766	382636
March	2012	325994	355767
April	2012	344349	325995
May	2012	362127	344350
June	2012	332272	362128
July	2012	344733	332273
August	2012	262136	344734
September	2012	371755	262137
October	2012	344931	371756
November	2012	360428	344932
December	2012	305567	360429
January	2013	398608	305568
February	2013	401103	398609
March	2013	410591	401104
April	2013	391991	410592
May	2013	373435	391992
June	2013	390023	373436
July	2013	415428	390024
August	2013	294396	415429
September	2013	417544	294397
October	2013	439641	417545
November	2013	422857	439642
December	2013	342836	422858
January	2014	366797	342837
February	2014	423950	366798
March	2014	463070	423951
April	2014	445420	463071
May	2014	452353	445421
June	2014	471585	452354
July	2014	327364	471586
August	2014	388073	327365
September	2014	459309	388074
October	2014	452508	459310
November	2014	425409	452509
December	2014	375814	425410
January	2015	339850	375815
February	2015	376973	339851
March	2015	376571	376974
April	2015	371001	376572
May	2015	304900	371002
June	2015	361767	304901
July	2015	278754	361768
August	2015	430953	278755
September	2015	425458	430954
October	2015	453944	425459
November	2015	394726	453945
December	2015	338991	394727
January	2016	287776	338992
February	2016	362668	287777
March	2016	440171	362669
April	2016	348626	440172
May	2016	339128	348627
June	2016	380019	339129
July	2016	203659	380020
August	2016	388847	203659
September	2016	423256	388848
October	2016	446611	423257
November	2016	450331	446612
December	2016	......	450332

3.1 Evaluation and validation of calculation result

From evaluation and validation of error deviation toward fuzzy time series above, error deviation also has been tested by using variance of number of universe of discourse starting from 3, 4, 5, 6 and 7 as well as data number starting from 12, 24, 36, 48 and 60 data so that the result got is shown in Figure 3.

4 Conclusion

Prediction by applying algorithm of Fuzzy time series done with variance of interval score toward universe of discourse and variance of data number can be applied to predict safety stock. It can be proven by testing result using data number of 60 and the average error score got was 15% measured by using method of Mean Absolute Percentage Error (MAPE). Prediction result accuracy is influenced by data number fluctuation, the size (small and big) of interval score of universe of discourse and minimal and maximal score of universe of discourse.

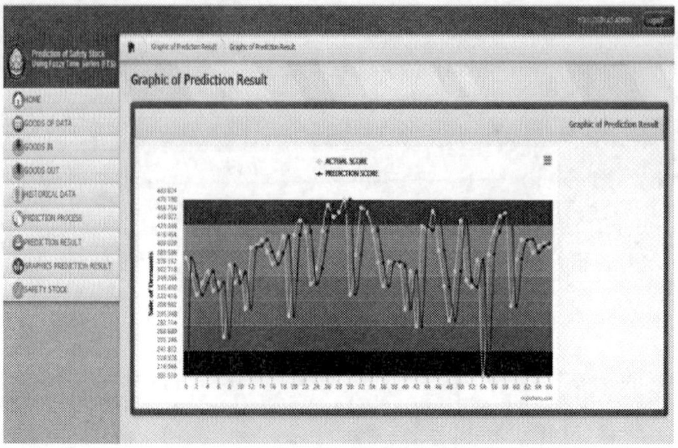

Fig. 2. Graphic of prediction result

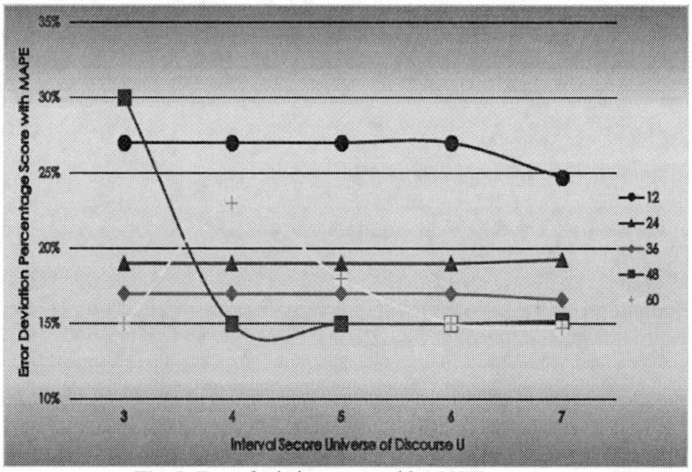

Fig. 3. Error deviation score with MAPE

References

1. T. Suesut, B. Mongkhoin, *Demand Forecasting Approach Inventory Control For CIMS,* 1869-1873 (2004)
2. J.D. Schwartz, M.R. Arahal, D.E. Rivera, *Control-relevant demand forecasting for management of a production-inventory system,* 4053-4058 (2008)
3. B. Garg, M.M.S. Beg, A.Q. Ansari, *Fuzzy Time Series Model to Forecast Rice Production,* 1-8 (2013)
4. N.S. Bajestani, A. Zare, *Application of Optimized Type 2 Fuzzy time series,* 1-6 (2009)
5. Y. Lin, Y. Yang, *Stock markets forecasting based on fuzzy time series model,* 782 – 786 (2009)
6. S. Suryono, J.E. Suseso, C. Mashuri, A.D. Sabila, J.A.M. Nugraha, M.H. Primasiwi, RFID Sensor for Automated Prediction of Reorder Point (ROP) Values in a Vendor Management Inventory (VMI) System Using Fuzzy Time Series **23,** 2398-2400 (2017)
7. S. Hansun, *Jakarta Stock Exchange (JKSE) forecasting using fuzzy time series,* 130 – 134 (2013)
8. T.M. Choi, Coordination and Risk Analysis of VMI Supply Chains With RFID Technology **3,** 497 – 504 (2011)
9. H. Dane, K. Michael, S.F. Wamba, *RFID enabled Inventory Control Optimization A Proof of Concept,* 1-10 (2010)
10. R.H. Rad, J. Razmi, M.S. Sangari, Z.F. Ebrahimi, Optimizing an integrated vendor-managed inventory system for a single-vendor two-buyer supply chain with determining weighting factor for vendor's ordering cost **153**, 295–308 (2014)
11. M.Xie, D. L. Olson, *Modeling and Values of Vendor managed inventory in the Retail Supply Chain,* . 465 – 470 (2006)

Automation Diagnosis of Skin Disease in Humans using Dempster-Shafer Method

Dyna Marisa Khairina[1,*], *Heliza Rahmania* Hatta[1], *Rustam* Rustam[1], and *Septya* Maharani[1]

[1]Department of Computer Science, Faculty of Computer Science and Information Technology, Mulawarman University, Samarinda - Indonesia

Abstract. Skin disease is an infectious disease that is common in people of all ages. Disorders of the skin often occur because there are factors, among others, are climate, environment, shelter, unhealthy living habits, allergies and others. Skin diseases in Indonesia are mostly caused by bacterial, fungal, parasitic, and allergies. The objective of the research is to diagnose skin diseases in humans by using the method of making decision tree then performing the search by forward chaining and calculating the probability value of Dempster-Shafer method. The results of research in the form of an automated system that can resemble an expert in diagnosing skin disease accurately and can help in overcoming the problem of skin diseases.

1 Introduction

The skin is the organ that lies outside and limits it from the human environment. The skin is an essential and vital organ and is a mirror of health and life. The skin is also very complex, elastic and sensitive, varying in climatic conditions, age, gender, race and also highly dependent on the location of the body [1].

Skin diseases in Indonesia are mostly caused by bacterial infections, fungi, parasites, and allergic basic diseases. This is different from Western countries that are more influenced by degenerative factors. Besides the difference in causes, other factors such as climate, habits and the environment also contributed to differences in clinical images of skin diseases [2].

Epidemiology data show that superficial dermatomycosis is a skin disease common in all societies, both rural and urban, not only in developing countries but also in developed countries. Although the disease is not fatal, but because it is often chronic, and not a few are resistant to anti-fungal drugs, the disease can cause disruption of comfort and reduce quality of life for the sufferer [3].

The Dempster-Shafer method is the representation, combination and propagation of uncertainty, in which this theory has some characteristics that are institutive in accordance with the way an expert thinks, but is based on a strong mathematical [4].

2 Uncertainties in Expert System

The construction of expert and other intelligent computer systems requires to be sophisticated mechanism for representing and reasoning with uncertain information. At least three forms of uncertainty can be identified as playing a significant role in these types of systems. The first of these possibility and uncertainty appears in situations where the value of a variable can only be narrowed down to a set of values one of which is the actual value of the variable. The second kind of uncertainty is related to situations in which there exists uncertainty as to satisfaction of a predicate by an element. This is manifested by concepts which have imprecise or gray boundaries. A very powerful tool for handling this type of uncertainty which also handles the first type of uncertainty is the fuzzy set. The third type of uncertainty is related to situations in which the value of a variable assumes can be modeled by the performance of a random experiment. [5]

3 Dempster-Shafer Method

The Dempster-Shafer method was first introduced by Dempster [6], which attempts to model the uncertainty with range probabilities rather than as a single probability. Then in 1976 Shafer published the Dempster theory in a book called Mathematical Theory Of Evident [7]. Dempster-Sahfer Theory of Evidence, suggests a way to give the weight of confidence according to the facts collected. In theory it can distinguish uncertainty and ignorance. The Dempster-Shafer Theory is a representation, combination and propagation of uncertainty, in which this theory has some characteristics that are institutive in accordance with the way an expert thinks, but a strong mathematical basis [4].

In general Dempster-Shafer theory is written in an interval: [*Belief, Plausibility*]. *Belief* (Bel) is a measure of the strength of evidence in favor of a set of propositions. If the value of 0 then indicates that there is no evidence, and if the value of 1 indicates a certainty. *Plausibility* (Pls) will reduce the level of certainty of the evidence. Plausibility is 0 to 1. If it is sure of X', then it

* Corresponding author: dyna.ilkom@gmail.com

can be said that Bell (X') = 1, so the above formula values from Pls (X) = 0.

According to Giarratano and Riley the Belief function can be formulated and shown in the equation:

$$Bel\ (X) = \sum_{Y \subseteq X} m(Y) \quad (1)$$

And Plausibility is denoted in the equation:

$$Pls\ (X) = 1 - Bel\ (X) = 1 - \sum_{Y \subseteq X} m(X) \quad (2)$$

Where:
Bel (X) : Belief (X)
Pls (X) : Plausibility (X)
m(X) : Mass function from (X)
m(Y) : Mass function from (Y)

The Dempster-Shafer theory states the existence of a frame of discrement denoted by a symbol (θ). Frame of discrement is the universe of speech of a set of hypotheses so often called the environment shown simultaneously: θ = {01, 02, ... 0N}
Where:
θ : Frame of discernment
01, 02, ... 0N : Elements in the environment

The environment contains elements that describe the possibility of an answer, and there is only one that will correspond to the required answer. This possibility in Dempster-Shafer theory is called the power set and denoted by P (θ), each element in this power set has an interval value of 0 to 1.

$$m : P(\theta) \rightarrow [0,1]$$

So it can be formulated in equations with:

$$\sum_{X \in P(\theta)} m(X) = 1 \quad (3)$$

Where:
P(θ) : Power set
m (X) : Mass function (X)

The mass function (m) in Dempster-shafer theory is the degree of confidence of an evidence, often called an evidence measure so that it is denoted by (m). The goal is to relate the trust size of the elements θ. Not all evidences directly support each element. For that we need probability density function (m). The m value not only defines the elements θ only, but also all the subset. So if θ contains n elements, then the subset θ is 2n. The number of sems in a subset θ equals 1. If there is no information whatsoever to select the hypothesis, then the value: m{θ} = 1,0

If X is a subset of θ, with m1 as its density function, and Y is also a subset of θ with m2 as its density function, a function of combinations m1 and m2 as m3 can be formed as shown in equation:

$$m_3(Z) = \frac{\sum_{X \cap Y = Z} m_1(X).m_2(Y)}{1 - \sum_{X \cap Y = \emptyset} m_1(X).m_2(Y)} \quad (4)$$

Where:
$m_3(Z)$: mass function from evidence (Z)
$m_1(X)$: mass function from evidence (X), which is obtained from the confidence value of an evidence multiplied by the disbelief value of the evidence.
$m_2(Y)$: mass function from evidence (Y), which is obtained from the confidence value of an evidence multiplied by the disbelief value of the evidence.
$\sum_{X \cap Y = Z} m_1(X).m_2(Y)$: is a strength value from the evidence Z obtained from a combination of confidence values set of evidence.

4. Research Method

4.1 Knowledge Acquisition

Initial identification process to obtain initial data and understanding for specific determination of the problem then conducted an interview with a skin specialist. Preliminary data and understanding obtained data of symptoms and illness then illustrated in a decision tree diagram. In the decision tree diagram of the skin disease diagnosis expert system, the symptoms are grouped and denoted by a combination of numeric and letter codes, then disease (diagnosis) and handling (solution) are denoted by numeric and letter codes. The decision tree is illustrated in Figure 1.

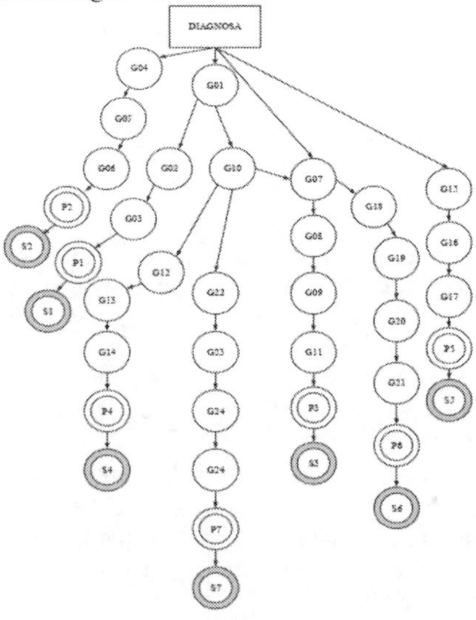

Fig.1. *Decision Tree* for Diagnosis of Skin Disease in Humans

Table 1 and Table 2 show the symptoms and illnesses of the decision tree in Figure 1.

Table 1. Description of the Symptom Code

Code	Symptoms
G01	Reddish patches on the skin
G02	The presence of plaques (skin lesions whose surface is elevated and flat top)
G03	Symptoms occur koebner phenomenon (skin disorder, where if healthy skin exposed to trauma / scratched, healthy skin will also be a disorder)
G04	The presence of a papula (solid protrusion bordered on a skin surface of <1 cm in diameter) of a small needle-sized head
G05	Papula grows prominent
G06	The surface of the skin becomes darker and hyperkeratosis (the skin becomes thick rough)
G07	Fever
G08	Rashes appear that become small bumps then turn into nodules with fluid
G09	Limp
G10	Itchy
G11	Appetite is gone
G12	The skin feels dry
G13	Skin thickened
G14	Skin scab
G15	The presence of macula hipopigmentasi (skin disorder where the white color is not rising flat compared to surrounding healthy skin) on the asymptomatic skin
G16	The appearance of white fine spots on the skin is not scaly
G17	Skin looks white and slippery
G18	Shivering
G19	Out of breath
G20	Pain or joint joints in one part of the body
G21	The appearance of reddish spots on the skin that eventually form a bubble of liquid
G22	Skin blisters
G23	Appears a red bump on the itchy part
G24	Itching is more intense especially at night
G25	Thickening and wrinkles on the skin are covered by yellowish-gray crust

Table 2. Decription of the Disease Code

Code	Disease
P1	Psoriasis
P2	Verruca
P3	Varicella
P4	Eczema
P5	Vitiligo
P6	Herpes
P7	Scabies

4.2 Knowledge Base

After the acquisition of knowledge then afterwards done representation of knowledge by making model of rule.

a. For example, for the first rule (R01) is:

IF G1
AND G2
AND G3
THEN Psoriasis

with the weighting value of each of the following symptoms:

G01 = 0.5 = P1 and P4
G02 = 0.8 = P1
G03 = 0.8 = P1

from the value of the symptom weights above, then determined initial density (m) value consisting of belief and plausibility.

Symptom 1: Reddish spots on the skin, then obtained :
$m_1\{P1,P4\} = 0.5$
The plausibility value of the formula is then determined
Pl=1 – Bel
$m_1\{P1,P4\} = 1 - 0.5 = 0.5$

Symptom 2: The presence of a plaque (skin lesions that surface is raised and flat top), then obtained: $m_1\{P1\} = 0.8$
The plausibility value of the formula is then determined
Pl=1 – Bel
$m_1\{P1\} = 1 - 0.8 = 0.2$

Based on the initial density determination on symptoms 1 and 2, the initial density of the next symptoms can be seen in Table 3.

Table 3. Determination of Initial Density (m)

No	Symptoms	Disease	Density (m) Belief	Density (m) Plausibility
1	Reddish patches on the skin	P1, P4	0.5	0.5
2	The presence of plaques (skin lesions whose surface is elevated and flat top)	P1	0.8	0.2
3	Symptoms occur koebner phenomenon (skin disorder, where if healthy skin exposed to trauma / scratched, healthy skin will also be a disorder)	P1	0.8	0.2

From Table 3 we can calculate the new density (m) value by creating a combination rule table first. Then the resulting combination will be used when indicating a new symptom.

Table 4. Combination Rule for m_3

Density 1 \ Density 2	m_2 {P1} 0,8	m_2 {θ} 0.2
m_1 {P1,P4} 0.5	{P1} 0.4	{P1,P4} 0.1
m_1 {θ} 0.5	{P1} 0.4	{θ} 0.1

Then calculated :

a) $m_3\ \{P1\} = \frac{0.4 + 0.4}{1 - 0} = 0.8$

b) $m_3\ \{P1,P4\} = \frac{0.1}{1 - 0} = 0.1$

c) $m_3\ \{\theta\} = \frac{0.1}{1 - 0} = 0.1$

Symptom 3: Symptoms of koebner phenomenon (skin disorder, where if healthy skin is affected by trauma / scratches, healthy skin will also be a disorder).

Based on the decision tree and the density value of the symptoms of the disease it is obtained: $m_4\ \{P1\} = 0,8$
Then plausibility value: $m_4\ \{\theta\} = 0,2$

Table 5. Combination Rule for m_5

Density 3 \ Density 4	m_4 {P1} 0,8	m_4 {θ} 0,2
m_3 {P1} 0.8	{P1} 0.64	{P1 } 0.16
m_3 {P1,P4} 0.1	{P1} 0.08	{P1,P4} 0.02
m_3 {θ}0.1	{P1} 0.08	{θ} 0.02

Then calculated :

a) $m_5\{P1\} = \frac{0.64+0.16+0.08+0.08}{1-0} = 0.96$

b) $m_5\{P1,P4\} = \frac{0.02}{1-0} = 0.02$

c) $m_5\{\theta\} = \frac{0.02}{1-0} = 0.02$

Then the magnitude of the possibility of users affected by Psoriasis disease is 96%.

b. For example, the rule for the second rule (R02) is:
IF G1
AND G10
AND G12
AND G13
AND G14
THEN Eczema

with the weighting value of each of the following symptoms:

G01 = 0,5 = P1 and P4
G10 = 0,5 = P3, P4 and P7
G12 = 0,7 = P4
G13 = 0,6 = P4
G14 = 0,6 = P4

from the value of the symptom weights above, then determined initial density (m) value consisting of belief and plausibility.

Symptom 1: Reddish patches on the skin, then obtained: $m_1\{P1,P4\} = 0.5$
The plausibility value of the formula is then determined
Pl=1 – Bel
$m_1\{P1,P4\} = 1 - 0.5 = 0.5$

Symptom 10: Itchy, then obtained: $m_1\{P3,P4,P7\} = 0.5$
The plausibility value of the formula is then determined
Pl=1 – Bel
$m_1\{P1\} = 1 - 0.5 = 0.5$

Table 6. Determination of Initial Density (m)

No	Symptoms	Disease	Density (m) Belief	Plausibility
1	Reddish patches on the skin	P1, P4	0.5	0.5
2	Itchy	P3, P4, P7	0.5	0.5
3	The skin feels dry	P4	0.7	0.3
4	Skin thickened	P4	0.6	0.4
5	Skin scab	P4	0.6	0.4

Based on the initial density determination on symptoms 1 and 10, the initial density of the next symptoms can be seen in Table 6.

From Table 6 we can calculate the new density (m) value by creating a combination rule table first. Then the resulting combination will be used when indicating a new symptom.

Table 7. Combination Rule for m_3

Density 1 \ Density 2	m_2 {P3,P4,P7} 0.5	m_2 {θ} 0.5
m_1 {P1,P4} 0.5	{P4} 0.25	{P1,P4} 0.25
m_1 {θ} 0.5	{ P3,P4,P7} 0.25	{θ} 0.25

Then calculated :

a) $m_3\{P4\} = \frac{0.25}{1-0} = 0.25$

b) $m_3\{P1,P4\} = \frac{0.25}{1-0} = 0.25$

c) $m_3\{P3,P4,P7\} = \frac{0.25}{1-0} = 0.25$

d) $m_3\{\theta\} = \frac{0.25}{1-0} = 0.25$

Symptom 12: The skin feels dry
Based on the decision tree and the density value of the symptoms of the disease it is obtained: $m_4\{P4\} = 0,7$
Then plausibility value: $m_4\{\theta\} = 0,3$

Table 8. Combination Rule for m_5

Density 3 \ Density 4	m_4 {P4} 0,7	m_4 {θ} 0,3
m_3 {P4} 0,25	{P4} 0,175	{P4} 0,075
m_3 {P1,P4} 0,25	{P4} 0,175	{P1,P4} 0,075
m_3 {P3,P4,P7} 0,25	{P4} 0,175	m_3 {P3,P4,P7} 0,075
m_3 {θ} 0,25	{P4} 0,175	{θ} 0,075

Then calculated:

a. $m_5\{P4\} = \frac{0.175+0.175+0.175+0.175+0.175}{1-0} = 0.775$

b. $m_5\{P1,P4\} = \frac{0.075}{1-0} = 0.075$

c. $m_5\{P3,P4,P7\} = \frac{0.075}{1-0} = 0.075$

d. $m_5\{\theta\} = \frac{0.075}{1-0} = 0.075$

Symptom 13: Skin thickened
Based on the decision tree and the density value of the symptoms of the disease it is obtained: $m_6\{P4\} = 0,6$
Then plausibility value: $m_6\{\theta\} = 0,4$

Table 9. Combination Rule for m_5

Density 5 \ Density 6	m_6 {P4} 0,6	m_6 {θ} 0,4
m_5 {P4} 0,775	{P4} 0,465	{P1} 0,31
m_5 {P1,P4} 0,075	{P4} 0,045	{P1,P4} 0, 03
m_5 {P3,P4,P7} 0,075	{P4} 0,045	{P3,P4,P7} 0,03
m_5 {θ} 0,075	{P4} 0,045	{θ} 0,03

Then calculated:

a. $m_7\{P4\} = \frac{0.465+0.045+0.045+0.045+0.031}{1-0} = 0.91$

b. $m_7\{P1,P4\} = \frac{0.03}{1-0} = 0.03$

c. m_7 {P3,P4,P7} $= \frac{0.03}{1 - 0} = 0.03$

d. m_7 {θ} $= \frac{0.03}{1 - 0} = 0.03$

Symptom 14: Skin scab

Based on the decision tree and the density value of the symptoms of the disease it is obtained: m_8 {P4} = 0,6 Then plausibility value: m_8 {θ} = 0,4

Table 10. Combination Rule for m_5

Density 7 \ Density 8	M_8 {P4 } 0,6	m_8 {θ} 0,4
m_7 {P4} 0,91	{P4} 0,546	{P4} 0,364
m_7 {P1,P4} 0,03	{P4} 0,018	{P1,P4} 0,012
m_7 {P3,P4,P7} 0,03	{P4} 0,018	{P3,P4,P7} 0,012
m_7 {θ}0,03	{P4} 0,018	{θ} 0,012

Then calculated:

a. m_9 {P4} $= \frac{0.546+0.0364+0.018+0.018+0.18}{1 - 0} = 0.964$

b. m_9 {P1,P4} $= \frac{0.012}{1 - 0} = 0.012$

c. m_9 {P3,P4,P7} $= \frac{0.012}{1 - 0} = 0.012$

d. m_9 {θ} $= \frac{0.012}{1 - 0} = 0.012$

Then the magnitude of the possibility of users affected by Eczema disease is 96%.

5. Results and Discussion

This system aims to create a model system that can effectively diagnose skin diseases automatically and work like an expert. Measurement of effectiveness is done by testing the accuracy of diagnostic results by using 20 data samples. Table 11 shows detailed diagnostic accuracy test data.

Table 11. Testing Accuracy of Diagnostic Results

No	Symptoms Suffered		System Diagnostic Results	Expert Diagnosis Results	Accuracy of Comparison Results
1	G08 G12 G17 G20 G21	Rashes appear that become small bumps then turn into nodules with fluid The skin feels dry Skin looks white and slippery Pain or joint joints in one part of the body The appearance of reddish spots on the skin that eventually form a bubble of liquid	Herpes	Herpes	Suitable
2	G01 G04 G07 G10 G13	Reddish patches on the skin The presence of a papula (solid protrusion bordered on a skin surface of <1 cm in diameter) of a small needle-sized head Fever Itchy Skin thickened	Eczema	Eczema	Suitable
3	G02 G03 G04 G05	The presence of plaques (skin lesions whose surface is elevated and flat top) Symptoms occur koebner phenomenon (skin disorder, where if healthy skin exposed to trauma / scratched, healthy skin will also be a disorder) The presence of a papula (solid protrusion bordered on a skin surface of <1 cm in diameter) of a small needle-sized head Papula grows prominent	Psoriasis	Veruca	Not Suitable
4	G06 G08 G09 G15	The surface of the skin becomes darker and hyperkeratosis (the skin becomes thick rough) Rashes appear that become small bumps then turn into nodules with fluid Limp The presence of macula hipopigmentasi (skin disorder where the white color is not rising flat compared to surrounding healthy skin) on the asymptomatic skin	Varicella	Varicella	Suitable
5	G09 G10 G11 G19 G20	Limp Itchy Appetite is gone Out of breath Pain or joint joints in one part of the body	Varicella	Varicella	Suitable
6	G04 G05 G06 G10 G13	The presence of a papula (solid protrusion bordered on a skin surface of <1 cm in diameter) of a small needle-sized head Papula grows prominent The surface of the skin becomes darker and hyperkeratosis (the skin becomes thick rough) Itchy Skin thickened	Veruca	Veruca	Suitable
7	G10 G12 G23 G24 G25	Itchy The skin feels dry Appears a red bump on the itchy part Itching is more intense especially at night Thickening and wrinkles on the skin are covered by yellowish-gray crust	Scabies	Scabies	Suitable

No		Symptoms Suffered	System Diagnostic Results	Expert Diagnosis Results	Accuracy of Comparis on Results
8	G12 G13 G15 G16 G17	The skin feels dry Skin thickened The presence of macula hipopigmentasi (skin disorder where the white color is not rising flat compared to surrounding healthy skin) on the asymptomatic skin The appearance of white fine spots on the skin is not scaly Skin looks white and slippery	Vitiligo	Vitiligo	Suitable
9	G07 G08 G10 G23 G25	Fever Rashes appear that become small bumps then turn into nodules with fluid Itchy Appears a red bump on the itchy part Thickening and wrinkles on the skin are covered by yellowish-gray crust	Varicella	Varicella	Suitable
10	G06 G09 G10 G16 G17 G23	The surface of the skin becomes darker and hyperkeratosis (the skin becomes thick rough) Limp Itchy The appearance of white fine spots on the skin is not scaly Skin looks white and slippery Appears a red bump on the itchy part	Vitiligo	Scabies	Not Suitable
11	G01 G02 G03	Reddish patches on the skin The presence of plaques (skin lesions whose surface is elevated and flat top) Symptoms occur koebner phenomenon (skin disorder, where if healthy skin exposed to trauma / scratched, healthy skin will also be a disorder)	Psoriasis	Psoriasis	Suitable
12	G03 G04 G05	Symptoms occur koebner phenomenon (skin disorder, where if healthy skin exposed to trauma / scratched, healthy skin will also be a disorder) The presence of a papula (solid protrusion bordered on a skin surface of <1 cm in diameter) of a small needle-sized head Papula grows prominent	Veruca	Veruca	Suitable
13	G04 G07 G08	The presence of a papula (solid protrusion bordered on a skin surface of <1 cm in diameter) of a small needle-sized head Fever Rashes appear that become small bumps then turn into nodules with fluid	Varicella	Varicella	Suitable
14	G08 G09 G10	Rashes appear that become small bumps then turn into nodules with fluid Limp Itchy	Varicella	Varicella	Suitable
15	G03 G04 G06 G07	Symptoms occur koebner phenomenon (skin disorder, where if healthy skin exposed to trauma / scratched, healthy skin will also be a disorder) The presence of a papula (solid protrusion bordered on a skin surface of <1 cm in diameter) of a small needle-sized head The surface of the skin becomes darker and hyperkeratosis (the skin becomes thick rough) Fever	Veruca	Veruca	Suitable
16	G10 G11 G12 G15	Itchy Appetite is gone The skin feels dry The presence of macula hipopigmentasi (skin disorder where the white color is not rising flat compared to surrounding healthy skin) on the asymptomatic skin	Eczema	Eczema	Suitable
17	G01 G08 G09 G10 G21	Reddish patches on the skin Rashes appear that become small bumps then turn into nodules with fluid Limp Itchy The appearance of reddish spots on the skin that eventually form a bubble of liquid	Varicella	Varicella	Suitable
18	G01 G10 G21	Reddish patches on the skin Itchy The appearance of reddish spots on the skin that eventually form a bubble of liquid	Varicella	Varicella	Suitable
19	G10 G11 G18 G20 G22	Itchy Appetite is gone Shivering Pain or joint joints in one part of the body Skin blisters	Herpes	Herpes	Suitable
20	G22 G23 G24 G25	Skin blisters Appears a red bump on the itchy part Itching is more intense especially at night Thickening and wrinkles on the skin are covered by yellowish-gray crust	Scabies	Scabies	Suitable

https://doi.org/10.1051/e3sconf/20183111006

From Table 11 we can calculate the accuracy value based on 20 tested data so that the percentage value obtained by 90% indicates that the accuracy of the diagnosis can work well according to the rules and the specialist skin diagnosis result so that the expert knowledge used has 90% value truth. The value of inaccuracy of 10% is likely due to errors in the assignment of the confidence of symptoms in each disease.

6. Conclusion

Based on the results of research and discussion it can be concluded that the automatic diagnosis of skin disease in humans requires accurate data and model rules are packed in the representation of knowledge as a basis for determining the final diagnosis. The sample simulation trials show that there is a 90% truth score that compares the diagnostic results based on expert knowledge. Value of accuracy results obtained by testing as many as 20 data samples so it can be said the accuracy obtained has a very good accuracy.

References

1. A. Djuanda, *Ilmu Penyakit Kulit dan Kelamin*, Ed.4. Jakarta: FKUI, (2005)
2. Siregar, *Atlas Berwarna Saripati Penyakit Kulit*, Jakarta: EGC, (2005)
3. Soebono, *Dermatomikosis Superfisialis,* Jakarta; Balai Penerbit FKUI, (2001)
4. E. G. Wahyuni, W. Prijodiprojo, IJCCS, **Vol. 7, No. 2**, (2013), pp. 133-144
5. R. Yager, International Joint Conference on Artificial Intelligence, (1985)
6. A. P. Dempster, Journal of the Royal Statistical Society, pp. 205-247, (1968)
7. G. Shafer, *A Mathematical Theory of Evidence*, Princeton University Press, New Jersey, (1976)
8. A. Maseleno, Md. M. Hasan, International Conference on Informatics for Development, pp. 93-98, (2011)
9. A. Maseleno, Md. M. Hasan, International Journal of Advanced Science and Technology, **Vol. 50**, pp. 111-120, (2013)
10. H. R. Hatta, F. Ulfah, D. M. Khairina, Hamdani, and S. Maharani, Journal of Theoretical and Applied Information Technology, **Vol. 95, No. 11**, (2017), pp. 2589-2599
11. S. Maharani, N. Dengen, G. Y. Saputra, D. M. Khairina, and H. R. Hatta, 2015 2nd International Conference on Information Technology, Computer, and Electrical Engineering ICITACEE, IEEE, pp. 87-91, (2015)

https://doi.org/10.1051/e3sconf/20183111007

Application Mail Tracking Using RSA Algorithm As Security Data and HOT-Fit a Model for Evaluation System

Ginanjar Setyo Permadi[1,*], *Kusworo* Adi[1,2], *Rahmad* Gernowo[1,2]

[1]Master Program of Information System, School of Postgraduate School, Universitas Diponegoro
[2]Physic department, Faculty of Science and Mathematics, Universitas Diponegoro

Abstract. RSA algorithm give security in the process of the sending of messages or data by using 2 key , namely private key and public key .In this research to ensure and assess directly systems are made have meet goals or desire using a comprehensive evaluation methods HOT-Fit system .The purpose of this research is to build a information system sending mail by applying methods of security RSA algorithm and to evaluate in uses the method HOT-Fit to produce a system corresponding in the faculty physics. Security RSA algorithm located at the difficulty of factoring number of large coiled factors prima, the results of the prime factors has to be done to obtain private key. HOT-Fit has three aspects assessment, in the aspect of technology judging from the system status, the quality of system and quality of service. In the aspect of human judging from the use of systems and satisfaction users while in the aspect of organization judging from the structure and environment. The results of give a tracking system sending message based on the evaluation acquired.

1 Introduction

Letter is communication in writing containing a information would be delivered or sent by one party to another individually or organization, company, and the office. The utilization of technology in assist with the send a message/ data is part of information system. System information can be defined as a means to present information in such manner that beneficial for recipients for the purpose present information to decision-making on planning, organizing, control operations subsystem an enterprise, and presenting synergy organization on a process [1].

Mail delivery system over the internet from various hardware and software components that include the sender's client computer, server, client and server computer with the required software and services are installed on each client [2]. The use of the mail delivery system is not only used to communicate between users but also between users and other organizations online services [3]. Organizations such as academic institutions or business companies typically manage their own mailing system, providing everyone with an address post mail in mail server by using the mailing list [4].

Mail Tracking is a way to monitor the delivery of a mail to the recipient or destination of the shipment. Most of the tracking technology using some form of digital records to reveal the exact time and date of the mail has been opened or received, as well as the IP address of the recipient [5]. In a mail tracking information system development needs must be viewed from a supporting factor for the creation by choosing the method of development of the system.

In this study add Algorithm Rivest Shamir Adleman (RSA) to provide security of a message or data in the process of delivery. RSA has two keys, a public key and a private key. RSA security located in the difficulty of a large number of factors being the prime factors [6]. With the addition of the RSA can give a sense of security in the process of sending data over the internet then it needs the addition of an algorithm or method in the construction of a data security system.

As for previous research about the implementation of digital signatures for security systems for authentication data mail tracking sender or signer of the letter, and based on one letter if signed by n sender verification should be done as much then n times. If all n Verify valid value means have penetrated the security tier n in terms of verification. Conversely, if one or more of the results invalid and verification or fail then the recipient can find out if the letter is received is not authentic and or one or more of the signatories of the letter was not the one who actually signed the letter. In the previous research for the correct letter, it sends the same letter maker directly without any changes from others.

* Corresponding author: ginanjar.s.permadi@gmail.com

 https://doi.org/10.1051/e3sconf/20183111007

To ascertain and assess directly whether the system has been made already achieve the goal or desire, in this research adds a method evaluation system. There are many methods to do the evaluation system, in this study using the method of human, organization and technology fit (HOT-fit). HOT-fit approach discusses the important components of the information system (SI), i.e., human, organizational and technological, and conformity between them using a comprehensive approach, sustainability and systematically to analyze the key components of the system, development, use, and net profit [7].

Based on government regulation republic of Indonesia No. 82 in 2012 about organizing systems and electronic transactions mentioned in article 1 paragraph 6 and 7 containing about understanding electronic information which is the set of electronic data (sounds, images, maps, photos, e-mail, electronic data Interchange, telegram) and understanding electronic documents that have a sense of any electronic information are made, forwarded, transmitted, received, or stored in the form of analog, digital, optical, electromagnetic, or similar ones, which can be seen, displayed, and/or heard through a computer or electronic systems. The use of e-mail has also stated in the Legislation of the Republic of Indonesia number 11 Year 2008 Electronic information and Transactions About article 5 and 6, which contains Information about electronic and/or electronic documents and/or the results of the print is legitimate and legal evidence of electronic information and/or electronic documents are considered valid throughout the information contained in it can be accessed, displayed, secured unity , and socially so that explains a situation.

With information from the Government of the Republic of Indonesia and the Republic of Indonesia that has been described above, making the system mail tracking in the systematic study of can help and contribute directly to provide information with a message or mail in send. This will give advantage in the proposed system can be run in an organization or company. So making the mail tracking system using RSA and HOT-Fit is expected to optimize the performance in terms of data transfer or messages, so that the information will be shared or posted can be easily on the monitor.

2 Theoretical Framework

2.1. Mail Tracking

Mail tracking is the way or method to monitor sending message to recipients sent in a system .The grooves of delivery systems message that is , messaging transmits a message be kept in a database mail server messaging and system continue the sending of messages to mail server the purpose of delivery , then the system give notice incoming messages to recipients of a message [2]. As for explanation structure / a groove system of process of shipping mail on Figure 1.

Fig 1. Flowchart of mail

Based on a groove system the e-mails and then in system that built this users can do tracking message sent have been opened, passed on, status of the message has been in verification, and status message on process. Messaging to know a response from recipients message to a message that which had been shipped. Figure 2. Drawing a groove delivery and response system.

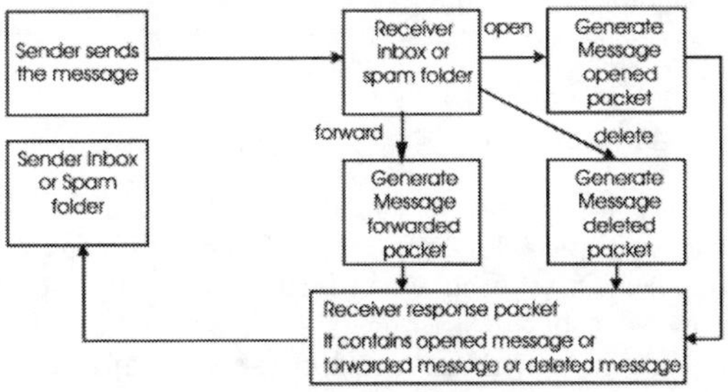

Fig. 2. Mail Message and Response Flowchart Diagram

2.2. Algorithm RSA

Including algorithm RSA algorithm asymmetry, namely the algorithms having 2 key, public key and private key. Security RSA algorithm located at the difficulty of factorization number of large coiled factors prima. Factorization is conducted to gather private key. During factorization number of large coiled factors prima has not been found the algorithms that good, so long as it security rsa fixed guaranteed [6]. Key to do in such a way that a decryption key may not be easy deduced of the encryption key public, as a in Figure 3.

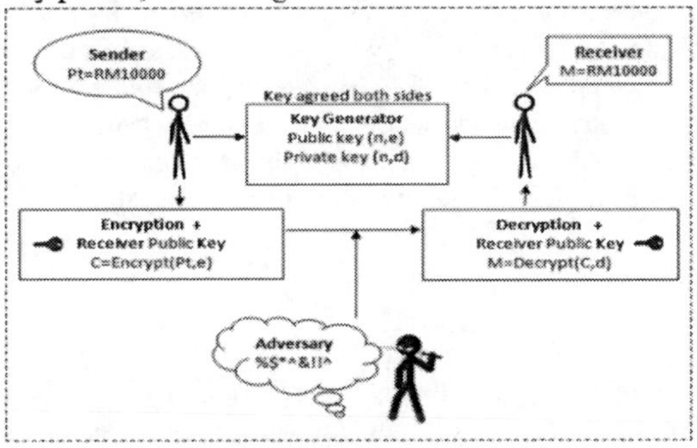

Fig. 3. RSA encryption algorithm

2.2.1 Key Generation Process

a. Select two parameters, namely p and q, these two parameters must be random prime number
b. Compute $n = p \times q$.
c. Compute $\Phi(n) = (p-1)(q-1)$.
d. Choose the exponent k if : GCD (h(id), phi (n)) = 1
Else, looping the process 1 to 6 and until k=1

e. Compute $d = e$ -1 mod $\Phi(n)$
f. Public key : (e, n)
g. Private key : (d, n)

2.2.2 Encryption Process

a. use public key (e, n).
b. Choose plainteks M, $0 \leq M \leq n - 1$.
c. Compute $C = M\ e$

2.2.3 Decription Process

a. Use private key (d, n)
b. Compute $M = C\ d \bmod n$.

2.2.4 Hash Function

Hash function is one of a mathematical function, who takes a long variable string input, called pre-image and convert them to a string output by long fixed and usually smaller consisting of letters and digits look random (binary data written in hexadecimal notation), called message digest [8].

Function hash one direction (one-way hash function) is hash function who works one direction, which is a hash function who can calculate message of pre-image digest, but is very difficult to count pre-image of message digest.

2.5 Evaluation System

Evaluation research is to collect , analyze , and presenting information which is useful about the object evaluation , rating them by comparing it with an indicator evaluation and the results should be used to evaluate and making decisions on the object evaluation.

Function evaluation system in this case to assess the implementation of delivery systems message may be accepted or need of improvements to systems. Methodology data collection using evaluation system with the spread of questioner directly based on an indicator variable model evaluation selected system.

2.5.1 HOT-fit Model

The framework of human, organization, technology-fit (HOT-fit) model was a development of success of the information system delone and mclean. Model evaluation explain all of the components that contained in information system, "human" assessment or in view of the use of (use of system) relating to the use, training, experience, knowledge, hope, the reception and rejects the. "organization" who feel system of organization the structure and organization environment associated with management planning, control system , support management, and financing. While in terms of "technology" can be seen from system quality, quality of information, and services [9].

HOT-fit has 3 aspects and different perspectives in every aspect. In aspect of technology, there are three dimensions: (1) system mention status; (2) the quality of information; (3) quality of service. In aspect of people, there are two dimensions: (1) use; system and (2) satisfaction users. In the organization, there are two dimensions: (1) structure; and (2) environment.

As for explanation three aspects that side in figure 5 the following [10].

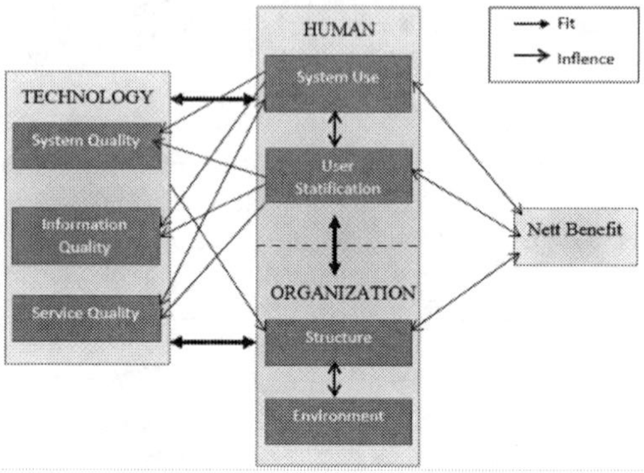

Fig. 5. HOT-Fit Evaluation Framework

3 Result and Discussion

3.1 Information System Sending Mail

The sending of messages information system is a system that provide the services the sending message a message the notification letter, the meeting, and the permission. In building information systems sending mail need data of mail template used at an object research. This system can do tracking message based on the number of letters and display information about a letter of the message timestamp opened, continued, verification status, and the status of being processed.

3.1.1 Algorithm

In this section, the method proposed is RSA algorithm. Figure 6 shows diagram a groove algorithm key power station. The first step in key generation is choose two random prime number for p and q, after that calculate n = p*q. Next step is calculate value of modulo phi(n)=(p-l)*(q-l). User input their email id as a parameter e, and get decimal value for email id using CRC hash function h(id)= hash (e). After that test either GCD (h(id),phi(n)) = \ or not. If value of GCD=\, so that email id can be a public key, and if GCD is not equal to one, the process will be looping from the first step to get new value of modulo until GCD equal to one.

https://doi.org/10.1051/e3sconf/20183111007

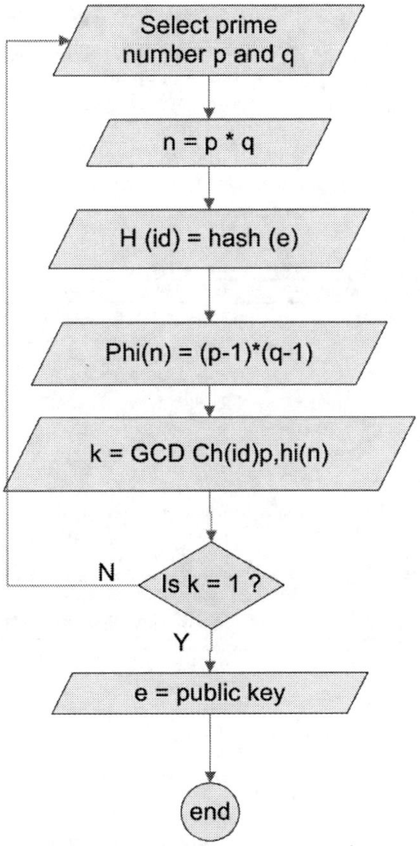

Fig. 6. Flow chart of i-RSA algorithm key generation

Table 1. Showing samples of data which implements RSA algorithm stored in a database.

Table 1. Applied algorithm RSA

Mail Number	Body	Publik Key	Mod Key
SJ/1234/1234	4521031425665177139 4748813461942348316 9508409774...	5	56147513 70746471
SJ/1234/1235	1816691989958088317 0032316292452398393 799954485 ...	5	54330613 89057521
SJ/1234/1236	751333.1506281076143 469.2696393490444714 .5754029...	5	86700716 34754457
SJ/1234/1237	4460713701637443034 1360827509712800700 3498064 11...	5	49342639 9904741
SJ/1234/1238	1741431677165025210 6022005741841916661 6584 14364...	5	25296861 766199

3.1.2 Mail Tracking

Figure 7. Showing menu tracking a message from information systems that built. The first step in do tracking message by inserting number a message to his search string, after that select the search button

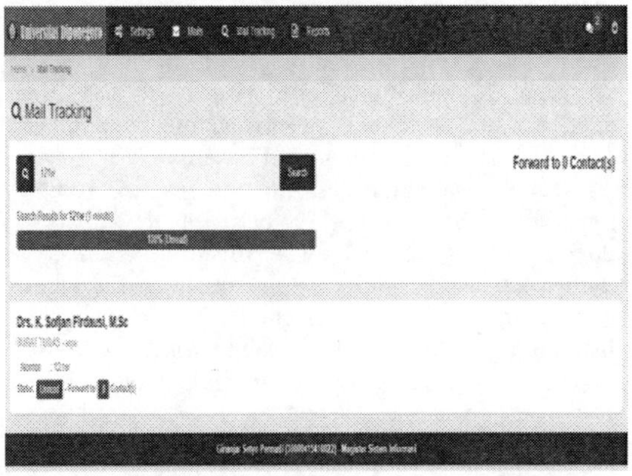

Fig. 7. Mail Tracking

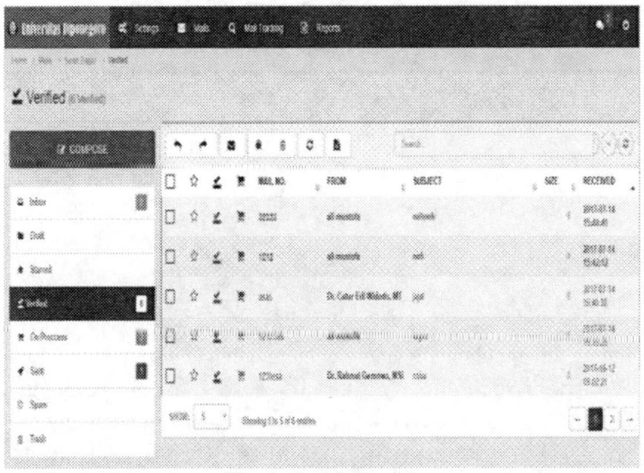

Fig. 8. Status mail

Figure 8. Showing the parts of a feature system can give status message sent. Status the message of status verification and status being processed. If the message has been in verification so menu the

3.2 Evaluation of Applied information system

The final stage is do the analysis afterwards evaluation of the implementation of a system of information the sending of messages which have been made .This research use the model hot fit developed by Yusof et al [7] to assess the success of the implementation of information system the sending of messages. Hot fit model is a model that complete and most according to the condition of the existing problems compared with a model that other.

Then the system will do the search based on the number the letter. The result of the process search provides information that of the message has been is read with details date and time opened, and information message has forward.

Verification will display messages that was in verification. Where as if message being processed so menu being processed will display messages that has been in the process.

Variable in this research consists of eight variable i.e. quality of a system, quality of information, quality of services, organizational of structure, and use of endogenous variable, satisfaction complainants, organizational environment and benefits.

Model evaluation information system portray a hypothesis the relationship between the factors that directly have an impact on information systems, on the model the evaluation was, the quality of system, the quality of information and quality of service used as a hypothesis to influence the assumption users against the system beneficial (net benefit and easy to use [6]. Hypothesized model developed for the evaluation of information system the sending of messages as follows :

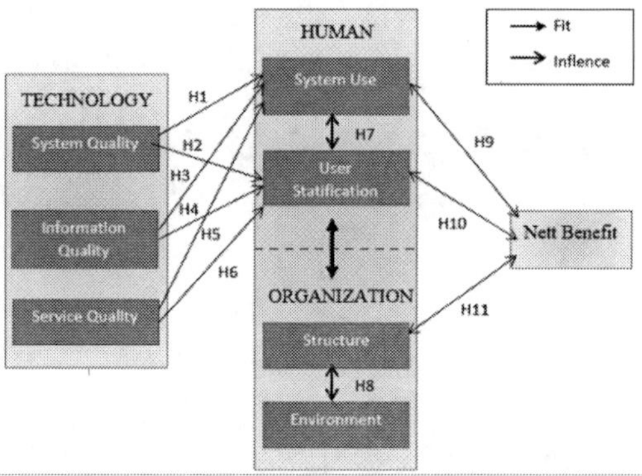

Fig. 7. Evaluation Hypothesis Model

The conceptual framework research can be seen on Figure 7. There are seventeen hypotheses in this research:

1. H1 : system quality has a significant positive effect towards system use.
2. H2 : system quality has a significant positive effect towards user satisfaction.
3. H3 : information quality has a significant positive effect towards system use.
4. H4 : information quality has a significant positive effect towards user satisfaction.
5. H5 : service quality has a significant positive effect towards system use.
6. H6 : service quality has a significant positive effect towards user satisfaction.
7. H7 : system use has a significant positive effect towards user satisfaction.
8. H8 : structure has a significant positive effect towards environment.
9. H9 : system use has a significant positive effect towards net benefits.
10. H10 : user satisfaction has a significant positive effect towards net benefits.
11. H11: environment has a significant positive effect towards net benefits

In evaluation system uses the data collected from questioner granted to information system sending of mail person as respondents. Users is lecturer and employees in the physics faculty Diponegoro University with the total respondents 35 people. Table 2. Show framework evaluation begins with the interpretation of the every aspect on the model hot-fit into one statement measurable, consisting of variable and indicators.

Table 2. Variables and indicator

No	Variables	Indicator
1	System Quality	System easy to use and user friendly
		System view is not confusing
		Security system
		System easy of learning
		The system rarely errors
2	Information Quality	Accuracy
		Availability
		Timeliness
		Completeness
		Information easy to read
3	Service Quality	Fast quality and responsive
		System easy to access
4	System Use	System use to easy for searching information
		System Use helping work every day
		System use can help in decision making
		Level of use
5	Use Stratification	Facilities and features as need of user
		Need system development
		Facilities and features according to needs
		Accurate information
		User satisfied with system view
		Over all system liked expectation
		Easy to use
6	Structure Organization	System is a strategy for performance improvement
		Always updating hardware or software
		System has been well planned
		Institution support implemented system
7	Environment	Assist in submitting letters
		Improve work efficiency
		Help make decision
		Help reach the goal effectively
		Improve communication between all parts of organization

Table 3. Test results statistic-t

	Original Sample (O)	Sample Mean (M)	Standard Deviation (STDEV)	Standard Error (STERR)	T Statistics (|O/STERR|)
KI -> KP	0,190418	0,183789	0,073820	0,073820	2,579470
KI -> PS	0,800221	0,811994	0,073040	0,073040	10,955888
KL -> KP	0,632795	0,645636	0,084925	0,084925	7,451253
KL -> PS	0,700640	0,696069	0,115518	0,115518	6,065214
KP -> NB	-0,084379	-0,043700	0,044658	0,044658	1,889464
KP -> PS	-0,526889	-0,534306	0,182610	0,182610	2,885326
KS -> KP	0,221444	0,213358	0,067641	0,067641	3,273805
KS -> PS	0,089517	0,088287	0,026991	0,026991	3,316587
LO -> NB	-0,690410	-0,660785	0,057967	0,057967	11,910355
PS -> NB	-1,114080	-1,001627	0,194129	0,194129	5,738872
SO -> LO	0,882306	0,889061	0,015202	0,015202	58,037257
SO -> NB	2,751230	2,580272	0,234069	0,234069	11,753918

In Table 3. It is a reject hypothesis or accepted based on value from t statistic compared with t table, in this

research have 95% (α =0,05) value t table with significant of degree 95% is 1,77.
Based on result of research t statistic, then can be determined hypothesis test in this research :

Table 4. Result of research hypothesis test

H1	system quality has a significant positive effect towards system use	Accepted
H2	system quality has a significant positive effect towards user satisfaction	Accepted
H3	information quality has a significant positive effect towards system use	Accepted
H4	information quality has a significant positive effect towards user satisfaction	Accepted
H5	service quality has a significant positive effect towards system use	Accepted
H6	service quality has a significant positive effect towards user satisfaction	Accepted
H7	system use has a significant positive effect towards user satisfaction	Accepted
H8	structure has a significant positive effect towards environment	Accepted
H9	system use has a significant positive effect towards net benefits	Accepted
H10	user satisfaction has a significant positive effect towards net benefits	Accepted
H11	environment has a significant positive effect towards net benefits	Accepted

4 Conclusion

In this research has been built information system sending mail by add RSA algorithm in the security of the mail sending process. The output of system is mail status report has been opened and forwarded. Result of the system is evaluated using HOTfit method based on the questionnaire of user system. Evaluation result show that the system has been good implemented in physics faculty Diponegoro University.

References

1. 1 J. J. Y. Chen, M. Z. Wu, *Integrating Extreme Programming with Software Engineering Education*. MIPRO (2015).
2. 2 J. Divya, D. Jagadeesan, G. Asha, *i-manager's Journal on Communication Engineering and Systems* (2016).
3. 3 I. Kounelis, SeadMuftic, J. Löschner, *Secure and Privacy-enhanced E-Mail System based on the Concept of Proxies.* (2014)
4. 4 H. Kim, S-Y. Shin, T. Motomichi, *Proceedings of the Seventh International Conference on Web-Age Information Management Workshops* (WAIMW'06) IEEE. (2006)
5. 5 G. Singh, M. Kaur, International Journal of Advanced Research in Computer Science and Software Engineering, Volume **5**, Issue 1. (2015)
6. 6 H. Delfs, and H. Knebl, *Introduction to Cryptography, Principles and Applications* Second Edition. Springer (2007).
7. 7 M. M. Yusof, HOT-fit Evaluation Framework: Validation Using Case Studies and Qualitative Systematic Review in Health Information Systems Evaluation Adoption. (2008)
8. 8 R. Munir, *Kryptografi,* Bandung: Informatika. (2004)
9. 9 I. P. Ramayasa, Journal of Theoretical and Applied Information Technology. (2015)
10. 10 L.M. Erlirianto, A. H. N. Ali, A. Herdiyanti, The Third Information Systems International Conference. Procedia Computer Science **72** 580 – 587 (2015).
11. H. A. *Fatta, Analisis dan Perancangan Sistem Informasi*. Yogyakarta : Andi (2007)
12. N. Muhammadi, J.M. Zaini, M. Y. M. Saman, Interational Conference on Control, Automation and Systems (ICCAS 2013) in Kimdaejung Convention Center, Gwangu, Korea. (2013)

Differentiation and Exploration of Model MACP for HE VER 1.0 on Prototype Performance Measurement Application for Higher Education

R. Reza El Akbar[1], *Muhammad Adi Khairul* Anshary[1] and *Dennis* Hariadi[2]

[1] Lecture in Department of Informatics Engineering, Siliwangi University, Tasikmalaya 46115 - Indonesia
[2] Student in Department of Informatics Engineering, Siliwangi University, Tasikmalaya 46115 - Indonesia

Abstract. Model MACP for HE ver.1. Is a model that describes how to perform measurement and monitoring performance for Higher Education. Based on a review of the research related to the model, there are several parts of the model component to develop in further research, so this research has four main objectives. The first objective is to differentiate the CSF (critical success factor) components in the previous model, the two key KPI (key performance indicators) exploration in the previous model, the three based on the previous objective, the new and more detailed model design. The final goal is the fourth designed prototype application for performance measurement in higher education, based on a new model created. The method used is explorative research method and application design using prototype method. The results of this study are first, forming a more detailed new model for measurement and monitoring of performance in higher education, differentiation and exploration of the Model MACP for HE Ver.1. The second result compiles a dictionary of college performance measurement by re-evaluating the existing indicators. The third result is the design of prototype application of performance measurement in higher education.

1 Introduction

Measurement of performance in every organization or institution is required, both governmental and private institutions. This is because the performance measurement related to the quality of the institution [1]. The quality of an institution that is measured both quantitatively and qualitatively based on supporting information, easy to describe and explain its achievement, will be beneficial to the institution in establishing, managing and monitoring its performance targets and achievement strategies. No exception for higher education institutions should always be consistent in maintaining and improving the quality of graduates. Various ways taken, one of which measures each performance on the parts of the process that exist in the institution, to then carried out monitoring, evaluation and improvement.

Based on the preliminary literature review, a performance measurement model found that simultaneously monitored performance. The model is Model Measurement and Controlling Performance for Higher Education Ver. 1.0 (MACP for HE Ver. 1.0) [2]. The model shows the existence of input transformation process and output in activities of higher education.

In the model has not shown in detail each of the translation from CSF to KPI. Further exploration for the model is required. In addition, the translation of each KPI need to be revised and adjusted as required. The description of the background is one reason why the study was conduct.

There are four main objectives of this study. First, perform Differentiation and Exploration of CSF components on references research models. The two conduct KPI Exploration on each CSF. Third design a new model that more detail and the last make prototype application of college performance measurement.

2 Experimental Details

The sequence of this research step follows four stages of research, which described as follows:

- Differentiation of CSF components by reviewing the related literature through literature study.
- The exploration of KPI components by literature search. Adjustment process is the next stage to understand the condition of the environment, especially the higher education environment in Indonesia.
- The design of the new model after the Differentiation and Exploration stage is completed. This stage adds and completes the model that becomes the reference model MACP for HE Ver. 1.0, so that model of Version 2 formed. At this stage the first step is to evaluate

* Corresponding author: reza@unsil.ac.id

the reference model, then modify the old model to produce a new model.

- The prototype design is the final step in this research to get an overview of the application. Prototype method consists of five stages, at this stage presented in this journal only the results of the design or modelling stage in the form of display interface application.

3 Results and Discussions

3.1 Differentiation of CSF components

At this stage, a review of each component of the CSF (Critical Success Factor) model is modelled base. In this section, CSF produced a new product. CSF defined as all products of research and other useful works and bias directly used in the community. This section includes research products, books, intellectual property rights and more [3]. Some CSF components incorporated in this section, such as CSF information in the input and output areas combined.

3.2 Exploration of KPI components

In this section the translation of each KPI (Key Performance Indicator) becomes more detailed, each KPI is grouped based on three areas namely input area, process area and output area.

All the elaborations in this section plotted on the KPI tables presented after the depiction of the new model presented at the design stage and the creation of a new model. In general, the KPI in this section divided into five parts of KPI namely:

1. KPI derived from CSF Area Input and given the symbol with I-KPI (KPI for Input Area).
2. KPI derived from CSF Process Area. In this section divided into three parts:
   a. KPI derived from CSF Information Flow and symbolized with P-KPI Information (KPI for Process Area Information);
   b. KPI derived from CSF Transformation Process and symbolized with P-KPI Transformation (KPI for Process Area Transformation);
   c. KPI derived from CSF Flow of Funds and given a symbol with P-KPI Flow of Funds (KPI for Process Flow Area Fund)
3. KPI derived from CSF Area Output and symbolized with O-KPI (KPI for Output Area).

3.3 New Model Creation

The old model that became the reference in this study, as shown in Figure 1 MACP Model for HE Ver. 1.0 describes the relationship between input, process and output.

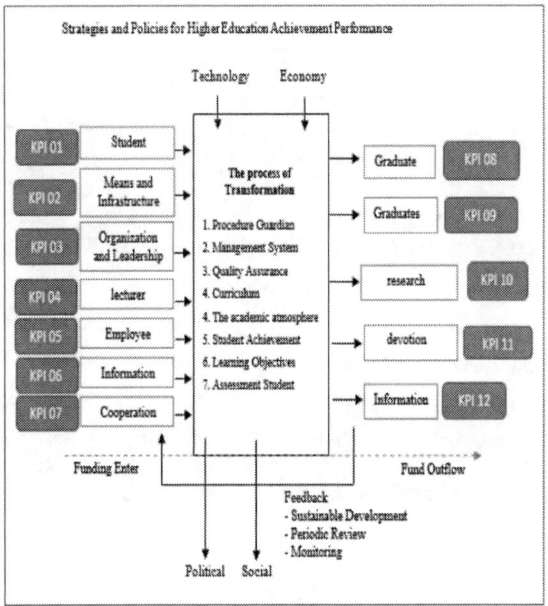

Fig. 1. Model MACP for HE ver 1.0 [2]

Based on the review and discussion by reviewing some parts of the model then there are some models revision or improvement. The following descriptions and improvements made, as a basis for the creation of new models.

1. The input, process and output reinforced as an effort to separate the areas.
2. Incoming and outgoing information on the model merged into a part of the process area
3. Based on the previous study that is differentiation and exploration, both CSF and KPI there is a new

KPI that is CSF product is new part of Area Output.
Based on the previous descriptions, a new model with the name of MACP for HE Ver. 0.2 as shown in Figure 2 below. The model shows that the area for Performance Measurement Model of Higher Education divided into three main areas namely Input Area, Process Area and Output Area.

Information and Financial or fund flows are in the process area and each has a KPI. There are five CSF groups: CSF in Input group / area, CSF group / process area of transformations, CSF group / Information process area, CSF group / financial process area, last CSF Group / output area.

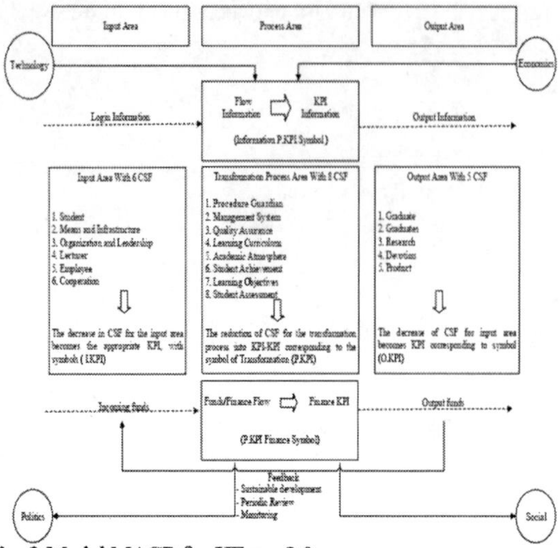

Fig. 2 Model MACP for HE ver 2.0

Based on the model then the following described some of the KPI to be calculated. The following presented in the form of a table, to facilitate which parts will be measured.

Table 1. KPI for CSF in Input Area

No.	CSF	KPI
1	College student	Number of scholarships available [4], Ratio of prospective students participating in selection and capacity [5]
2	Facilities and infrastructure	Number of Rooms available [4], Condition of learning support facilities [4], Number of library materials and other equipment [6]
3	Organizations and Leadership	Periodic evaluation of roles and functions of organization and leadership [7]
4	Lecturer	Level of education, academic position, certification, achievement level of learning evaluation result by lecturer [4]
5	Employee	Employee competency score [8],
6	Cooperation	Number of cooperation grouped by level [9]

Table 2. KPIs for CSFs in the Information Process Area

No.	CSF	KPI
1	Information	Total availability of information system all activities of each unit [10]

Table 3. KPIs for CSF in the Transformation Process Area

No.	CSF	KPI
1	Procedure Guardian	The level of satisfaction of the leadership of each unit [11]
2	Management System	Number of systems available for the management of each unit [11]
3	Quality assurance	Quality standard report [11]
4	Learning Curriculum	Percentage of curriculum completeness [11]
5	Academics Atmosphere	Presentation of comfort academic atmosphere [11]
6	Students Achievement	Total student achievement [11]
7	Learning objectives	Percentage of learning achievement [11]
8	Students Assessment	Average student score [11]

Table 4. KPIs for CSFs in the Financial Process Area

No.	CSF	KPI
1	Financial	Financial statements for cash inflows and outflows [11]

Table 5. KPI for CSF in Output Area

No.	CSF	KPI
1	Graduate	Length of waiting period, average IPK [10], [11]
2	Graduates	Work suitability, [12]
3	Research	Number of lecturers' research, student participation [13]
4	Devotion	Total devotion of lecturers, student participation [14]
5	Product	Number of commercial products [13]

3.4 Prototype Design Results

Based on the results of the application design obtained interface as shown below:

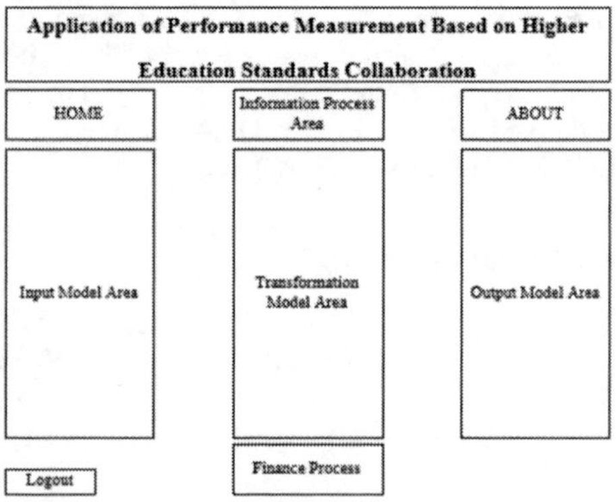

Fig. 3 User Interface Applications

4 Conclusion

After the research steps consisting of four research steps, the Measurement and Controlling Performance for Higher Educations (MACP for HE Ver. 2.0) model has successfully constructed from the previous version of the model. The details of each KPI can be trace further through the performance measurement table presented in this journal.

The prototype application design has been successfully created and presented, but has not been tested or tested thoroughly on prototype results of performance measurement applications in this college.

Further research needs conducting laboratory-scale trials. Research related to this topic that needs to be developed and research is the application of monitoring performance appraisal results and applications to evaluate the performance of universities.

Acknowledgments

The authors would like to thank to the Ministry of Technology, Research, Higher Educations of the Republic Indonesia for funding this project. Especially to LPPM *(Lembaga Penelitian dan Pengabdian Masyarakat)* Siliwangi University, to give authors funding of research innovation for the second year.

References

1. S. Luis, and P.A. Biromo, *Step by Step in Cascading Balanced Scorecard*, Jakarta: PT Gramedia Pustaka Utama (2008)
2. Akbar, R. Reza El; Anshary, Muhammad Adi Khairul., Volume **23**, Number 3, pp. 2354-2357(4), (2017)
3. *Panduan penelitian* SIMLITABMAS (2017)
4. SPM-ITB, *Standar Mutu Akademik dan Instrumen Pengukuran*, 4 (2015)
5. F.C. Saputro, W. Anggraeni, A. Mukhlason, *Pembuatan Dashboard Berbasis Web Sebagai Sarana Evaluasi Diri Berkala untuk Persiapan Penilaian Akreditasi Berdasarkan Standar Badan Akreditasi Nasional Perguruan Tinggi*, 1 (2012)
6. SPMI-UNDIP, *Standar Sarana Dan Prasarana Sistem Penjaminan Mutu Internal* Universitas Diponegoro, (2011)
7. A. Junaidi, *Standar Nasional Pendidikan Tinggi Permenristekdikti No. 44 Tahun* 2015, (2016)
8. Yodhia Antariksa. *KPI Bidang SDM,Marketing, IT Finance dan Produksi* Universitas Airlangga, (2010)
9. H. Nur Fatima, Alimuddin, D. Said. *Hubungan Antar key performance indicators* (2013)
10. F.Cahyo Saputro, W. Anggraeni dan A. Mukhlason. *Pembuatan Dashboard Berbasis Web* Institut Teknologi Sepuluh Nopember (2012)
11. Badan Akreditasi Nasional-Perguruan Tinggi;*Buku Ii Standar Dan Prosedur Akreditasi Program Studi Sarjana*;2008;Jakarta, Indonesia.
12. Syahru Rahmayuda. *Sistem Informasi Evaluasi Tracerstudy Untuk Pengukuran Key Performance Indicators Alumni Menggunakan Metode Fuzzy.* Program Pascasarjana Universitas Diponogoro Semarang,(2013)
13. I Made Suartika, *Perancangan Dan Implementasi Sistem Pengukuran Kinerja Dengan Metode Integrated Performance Measurement Systems* (Studi Kasus: Jurusan Teknik Mesin Universitas Mataram)
14. Jurnal Teknik ITS Vol. **1**, ISSN: 2301-9271 Jurusan Sistem Informasi, Fakultas Teknologi Informasi, Institut Teknologi Sepuluh Nopember (ITS), (2012)

Implementation of Online Promethee Method for Poor Family Change Rate Calculation

Dhady Lukito Aji[1,*], *Suryono* Suryono[1,2], *Catur* Edi Widodo[1,2]

[1]Master of Information System, School of Post-Graduate Studies, Diponegoro University, Semarang, Indonesia
[2]Departement of Physics, Faculty of Science and Mathematic, Diponegoro University, Semarang, Indonesia

Abstract. This research has been done online calculation of the rate of poor family change rate by using Preference Ranking Method of Organization Of Enrichment Evaluation (PROMETHEE) .This system is very useful to monitor poverty in a region as well as for administrative services related to poverty rate. The system consists of computer clients and servers connected via the internet network. Poor family residence data obtained from the government. In addition, survey data are inputted through the client computer in each administrative village and also 23 criteria of input in accordance with the established government. The PROMETHEE method is used to evaluate the value of poverty and its weight is used to determine poverty status. PROMETHEE output can also be used to rank the poverty of the registered population of the server based on the netflow value. The poverty rate is calculated based on the current poverty rate compared to the previous poverty rate. The rate results can be viewed online and real time on the server through numbers and graphs. From the test results can be seen that the system can classify poverty status, calculate the poverty rate change rate and can determine the value and poverty ranking of each population.

1 Preface

Poverty is a term used to denote absolute poverty wherein a person experiences a very hard life due to difficulties to meet their basic needs such as food, clothes, shelter and education [1]. The rate numbers of poverty according to the Regional Development Planning Agency (BAPPEDA) is the data rate of poor people who are in the process of monitoring the number of poverty to know the impact of Poverty alleviation and social assistance programs to reduce the number of poor families. Planning in poverty alleviation strategies and social assistance budget allocations requires a rate of change in a city to see the level of success and achievement of program objectives with *valid* data.

Method of Preference Ranking Organization Method for Enrichment Evaluations (abbreviated as PROMETHEE) online based is used to validate the proposed data of insufficient information application used by government in the city poverty data collection before doing the calculation of poverty rate change. Based on the test results, it can be seen that the system can classify poverty status, calculate the poverty rate change rate and can determine the value and poverty ranking of each population. The rate results can be viewed online and real time in the form of numbers and graphics on http://lajukemiskinankotasemarang.info page.

The *PROMETHEE* method is a well-known method and has many separate criteria [2]. The *PROMETHEE* method was first developed by Brans and published in 1982 at a conference hosted by Nadeau and Landry at the University of Laval, Quebec, Canada [3]. The PROMETHEE method is not only applicable for ranking but can also be used to evaluate the feasibility of program execution by determining its net flow number which can not be found in other methods. If a ranking is using PROMETHEE net flow results positive, then a new object deserves to be called good data. Conversely, if the net flow is negative, it is called unfavorable data. The example of using PROMETHEE method is that the method can be used in employee selection process at a company. Experts from the company are appointed to determine the company's required quality of criteria, the value obtained, then normalized and determining attributes. For negative net flow, the value is ignored and will not be processed at the next calculation. The results of the PROMETHEE process on the required quality of the existing criteria are then set as the best criterion alternative [4].

The use of the PROMETHEE method can be used in the selection of industrial equipment. The labor, materials, and equipment that will be used in the production process is a very important component in the production system. The used equipment is an important subject that greatly affects efficiency system, work effectiveness, and quality of product. Some mistakes in selecting equipment may have a negative effect on the production system [5]. The PROMETHEE method can be used to predict the bankruptcy of an institution based on harmony and discrepancy between the members of the organization. Relationships pattern between members

* Corresponding author: dhadylukitoaji@gmail.com

https://doi.org/10.1051/e3sconf/20183111009

can be used as a reference in making measurements and calculations [6]. The PROMETHEE method can be used to measure the performance quality of a school. By looking at achievements, student absenteeism, social activities and project criteria, the method can show that Junior High School performs better than senior high schools and schools in the cities are better performing than the village schools. There was also a strong relationship between attainment and absenteeism [7]. Comparison between PROMETHEE method and AHP method in selecting building structure system, and the result PROMETHEE were chosen as the right method to evaluate alternative solution in the process of building structural system selection. The PROMETHEE method is more consistent, easy to understand, and requires less interaction with decision makers. Linearity and other assumptions for preference functions can be accepted by the decision maker. By utilizing the PROMETHEE Method in analyzing a thing, the most effective criteria in the decision-making process can be determined. This is a unique feature of PROMETHEE that is not available in AHP, or other MCDM techniques such as the Elimination Et Choix Tradiniant la Realité (ELECTRE) and TOPSIS methods [7].

The rate of changed numbers calculation is an analysis technique that calculates the percentage change of the amount of data of an object from period to period obtained by way of survey so that the data can be observed trends or patterns of change for use in evaluation and input in making decision for a problem. The calculation of the rate change rate is the main input in the final conclusion to fill in the monitoring instrument and evaluation of the benefits of a program. The example of using the rate change calculation that the calculation can be used in the Evaluation of Program Benefits in Tourism Affairs in a city. The rate of change in the number of tourist arrivals in 2005 - 2008 was calculated at the rate of change of numbers and used as the input to fill monitoring instruments by the government [8].

2 Method

2.1 Materials and Research Tools

This study requires data based on the criteria. The survey data were obtained from questionnaires filled out by residents who proposed the Certificate of Poor (SKTM). The criteria used are relevant criteria based on Mayor Regulation on indicators, criteria and classification of urban poor. While the sub criteria, the value of the scale of each criterion is obtained from the interview to the Regional Development Planning Agency. The used tools are a set of client computers and servers and PHP software.

2.2 Architecture Network

This software will be used by two categories of users, namely Regional Development Planning Agency data center as server and urban village administrator as client. The proposed architecture network is a network connected to the internet and both computer servers and clients must have internet access in order to be connected to each other. The client-server system is the best system to use, the system is capable of generating robust database applications in terms of securities, as well as reducing network traffic density [9]. The architecture network is shown in Figure 1.

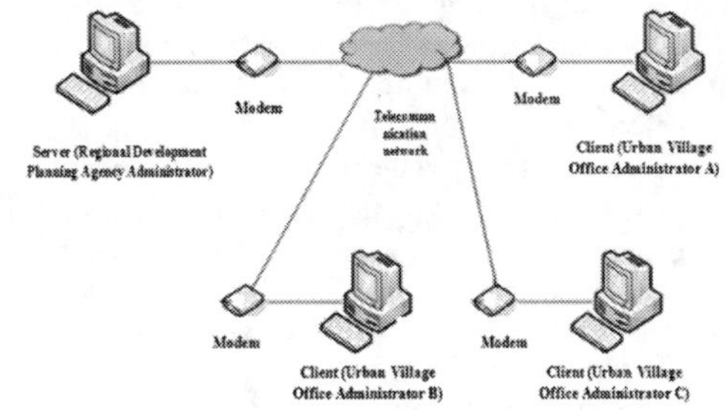

Fig. 1. Architecture network

2.3 Access Rights

In this system, there are two different access rights namely Regional Development Planning Agency data center as server and urban village administrator as client. The server has facilities to access the application form, user form, criteria form, family card form, application form of poor certificate, PROMETHEE forms, resident status report forms, poor certificate application forms, and poverty rate reporting forms. Meanwhile urban village administrator as client only has facility to access letter application form of poor certificate, PROMETHEE forms, resident status report forms, poor certificate entry application forms, and poverty rate reporting forms. The access rights above are shown in the use case shown in Figure 2.

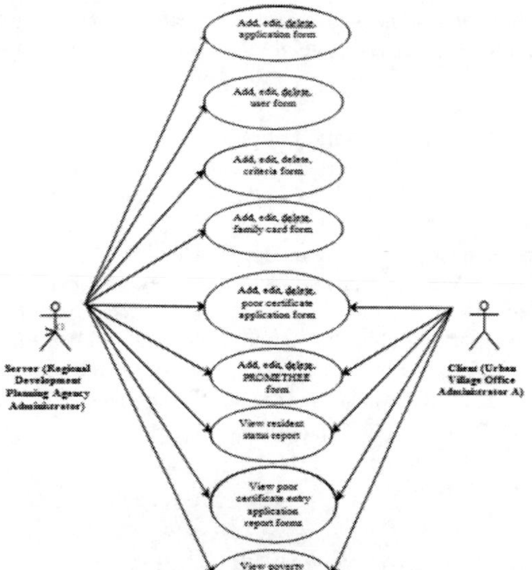

Fig. 2. Use case access rights

2.4. PROMETHEE Modeling

The use of the PROMETHEE model is accomplished by the following steps:

Step 1

Determine criteria and subcriteria.

The criteria used are 6 criteria, namely food, house, clothing, education, health, income and asset ownership. While the sub criteria is taken from the questionnaire questionnaires filled out by the poor applicants as many as 23 sub criteria such as how many times consuming the staple food in a day, the frequency of eating meat and or eggs eaten by the family in a week, the status of the house inhabited today, The floor of the house inhabited today, the type of floor of the house, the type of wall of the house, the availability of electricity, the average number of new clothes that can be purchased by the family in the latest year, whether each family member has different clothes for at home / work / school / travel, able to finance the family's education to what level, is there any family member who dropped out of school, if there is a sick family brought to primary health care, if there is family member Sick and require advanced health care / hospital, who finances him/her, whether family or family members are active smokers, if taken to health care facilities, to whom you do the treatment, whether there are family members who have disability (physical / mental), Availability of clean water, availability of toilet room, household income source, employment status of head of household, number of family member working, occupation of head of household, number of family member, amount of family income per month, and whether owning productive asset.

Step 2

Determine the value of the alternative.

The determination of the value of the alternative is obtained from questionnaires that are based on the criteria and scale that have been predetermined and filled by the applicant of Poor Certificate (SKTM) and submitted to the kelurahan. This calculation using the minimization and the weight ratio are equally important.

Step 3

Selection and calculation of preference values.

There are six forms of preference function in the Promethee, namely Type I or Usual Criterion, Type II or QuasiCriterion, Type III or Criteria with Linear Preferences, Type IV or Level Criterion, Type V or Criteria With Linear and Unbound Area and Type VI or Gaussian Criterion. The Usage of Usual Criterion is particularly suitable for equal importance of qualitative criteria. Examples of cases of usual criteria usage are yes / no answers or answers that have range of 0-5 [4]. The counter is using the equation as follow:

Usual Criterion

$$H(d)=\begin{cases}0 \text{ jika } & d=0\\ 1 \text{ jika } & d\neq 0\end{cases} \tag{1}$$

Notes :

1. H (d) : Function of cross-alternative differences
2. d : Difference in criterion value {d = f (a) – f (b)}

Function H (d) for this preference function is shown in Figure 3

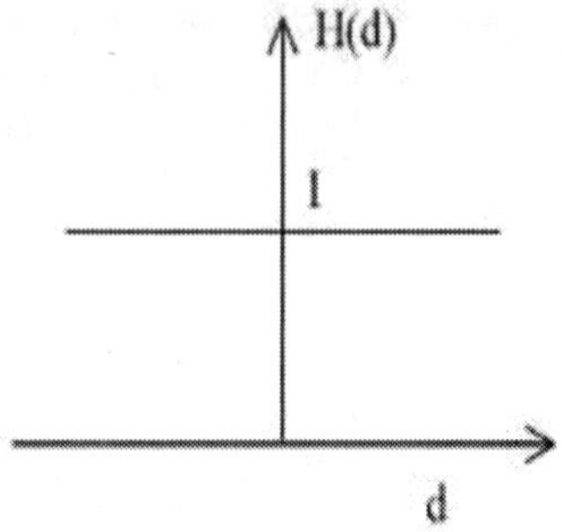

Fig.3. Usual Criterion [4]

Step 4

The calculation of preference index value.

The Calculation of Preference Index Value is a preference measure from one alternative into another, the closer to 1 (one), the greater the preference be. Therefore, the preference index is the intensity of the decision maker's preference which states that the other alternatives with simultaneous consideration of all criteria. The counter is using the equation as follow:

Preference Index Value

$$\wp(a,b)=\frac{1}{n}\sum_{i=1}^{n}P_i(a,b) \tag{2}$$

with

$\wp(a,b)$ Shows an a alternative preference index is better than b alternative

$P_i(a,b)$ Is an a alternative preference to b alternative

n Is the number of criteria

Step 5 : Calculating the value of leaving flow and entering flow.

The calculation of the value of *leaving flow* (Φ +), *entering flow* (Φ-) is the process of determining the priority relationship between one alternatives to others'. the counter is using the equation as follow:

Leaving Flow

$$\phi^{+}(a)=\frac{1}{n-1}\sum_{x\in A}^{\square}p(a,x) \tag{3}$$

Entering Flow

$$\phi^{-}(a)=\frac{1}{n-1}\sum_{x\in A}^{\square}p(x,a) \tag{4}$$

with

$\Phi^{+}(a)$ is *Leaving Flow* a alternative

ϕ (a) is *Entering Flow* a alternative
p (a,x) is the number of criteriais showing alternative preferences a more than x.
n is the number of criteria

Step 6

Net flow calculation.

Net flow calculation (Φ) is done by determining the differences of *Leaving Flow* and *Entering Flow* value and the result is dominance value of each alternative. If the netflow result is positive, then a new object can to be called as good data. For negative netflow, the value is ignored and will not be processed at the next calculation. The higher the netflow value, the higher the rank be.
The determination is using the equation as follow:

Net flow

$$\Phi(a) = \Phi^{+}(a) - \Phi^{-}(a) \quad (5)$$

With
Φ (a) is *Net Flow* a alternative
Φ^{+} (a) is *Leaving Flow* a alternative
Φ^{-} (a) is *Entering Flow* a alternative

2.5. The Rate Of Changed Numbers Calculation

The rate of changed numbers calculation is an analysis technique that calculates the percentage change of the number of an object which is observed by using *time series* data obtained by *survey*. The determination is using the equation as follow:

The rate of changed numbers calculation

$$P = \frac{A_{tn} - A_{to}}{A_{to}} \times 100\% \quad (6)$$

with
P is the percentage change in the number of alternative
A is the Number of alternative
tn is the current month
to was the last month

3 Results and Discussion

Stages of web based calculation of the poverty rate are done by doing with the following stages:

a. Data Retrieval
The data used are data from Regional Development Planning Agency in January - June 2016 a number of 150 Heads of Families include 350 residents. Out of 150 Heads of Families, 60 heads of families signed up for SKTM/Poor Certificate. While the manual calculation took 5 data of the heads of the families that is Family 1, Family 2, Family 3, Family 4, Family 5. The 5th family signed himself up for Poor Certificate.

b. Verification of poverty status data will be done by following steps :

The first step is to determine the value of alternatives. The determination of the value of the alternative is obtained from the questionnaire based on the criteria and scale that has been predetermined and filled by Poor Certificate and submitted to the urban village office. This calculation is using the minimization and weight ratio which are equally important.

The second step is the determination of criteria preference. In this study, the use of the usual criterion preference is very suitable to use. Because it is having the same important weight and the answer is yes / no and the answer is using a 0-5 choice scale.

The Third Step is the Calculation of Preference Index Value is a preference measure of one alternative to another, the closer to 1 (one), the greater the preference be. Thus, the preference index is the intensity of the decision maker's preference which states that the other alternatives are simultaneously consideration of all rituals. This calculation uses the usual criteria where the preference value is 0 if the difference between alternatives results 0 or less than 0 and the preference value is 1 if the difference between the alternatives is more than 1. The counter on the dashboard system showed in figure 4.

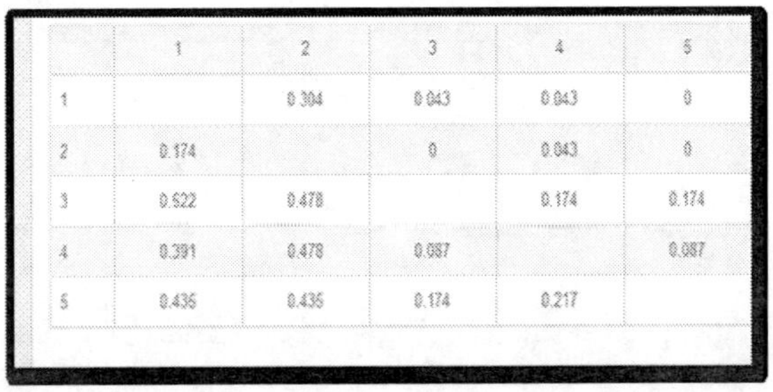

	1	2	3	4	5
1		0.304	0.043	0.043	0
2	0.174		0	0.043	0
3	0.522	0.478		0.174	0.174
4	0.391	0.478	0.087		0.087
5	0.436	0.435	0.174	0.217	

Fig. 4. Dashboard view of the preference index value

The view on the dashboard of preference index values is then verified by counting the manual preference index. Manual counter results are compared with the dashboard preference index table as shown in table 1.

Table 1. Manual Preference Index Verification

	Family 1	Family 2	Family 3	Family 4	Family 5	amount	verification result
Family 1		0.304	0.043	0.043	0.000	0.391	consistent
Family 2	0.174		0.000	0.043	0.000	0.217	consistent
Family 3	0.522	0.478		0.174	0.174	1.348	consistent
Family 4	0.391	0.478	0.087		0.087	1.043	consistent
Family 5	0.435	0.435	0.174	0.217		1.261	consistent
amount	1.522	1.391	0.261	0.435	0.261		consistent
verification result	consistent	consistent	consistent	consistent	consistent	consistent	

The fourth step is the calculation of the value of leaving flow (Φ +) and entering flow (Φ-). Arrangement of this domination relationship is the process of determining the priority relationship between one alternative to another. At this stage, the calculation steps consist of Leaving flow (Φ +) calculation which means the greater the value, the more dominance an alternative to other alternatives will be and the calculation of Entering flow (Φ-) where the smaller the value is, the more dominance an alternative to other alternatives will be. The counter on the dashboard is shown in figure 5.

NAMA	LEAVING FLOW	RANK	ENTERING FLOW	RANK
NAME	0.098	4	0.381	2
2	0.054	5	0.424	1
3	0.337	1	0.076	4
4	0.261	3	0.119	3
5	0.315	2	0.065	5

Fig. 5. The dashboard calculation view leaving flow and entering flow

Views on the dashboard leaving flow and entering flow value then the results are verified by performing manual calculations. The manual counter results are compared to the table leaving flow and entering flow dashboard as shown in table 2 and table 3.

Table 2. Manual Leaving Flow Verification

Name	Counter of Leaving Flow			Rank	Verivication
Family 1	1/(5-1)* ((Family 1,Family 2)+(Family 1,Family 3)+ (Family 1,Family 4)+(Family 1,Family 5))	1/4 * 0.391=	0.098	2	consistent
Family 2	1/(5-1)* ((Family 2,Family 1)+(Family 2,Family 5) +(Family 2,Family 4)+(Family 2,Family 3))	1/4 * 0.217=	0.054	1	consistent
Family 3	1/(5-1)* ((Family 3,Family 1)+(Family 3,Family 5) +(Family 3,Family 4)+(Family 3,Family 2))	1/4 * 1.348=	0.337	4	consistent
Family 4	1/(5-1)* ((Family 4,Family 1)+(Family 4,Family 5) +(Family 4,Family 3)+(Family 4,Family 2))	1/4 * 1.043=	0.261	3	consistent
Family 5	1/(5-1)* ((Family 5,Family 1)+(Family 5,Family 4) +(Family 5,Family 3)+(Family 5,Family 2))	1/4 * 1.261=	0.315	5	consistent

Table 3. Manual Entering Flow Verivication

Name	Counter of Entering Flow			Rank	Verivication
Family 1	1/(5-1)* ((Family 1,Family 2)+(Family 1,Family 3) +(Family 1,Family 4)+(Family 1,Family 5))	1/4 * 1.522=	0.380	2	consistent
Family 2	1/(5-1)* ((Family 2,Family 1)+(Family 2,Family 5) +(Family 2,Family 4)+(Family 2,Family 3))	1/4 * 1.391=	0.424	1	consistent
Family 3	1/(5-1)* ((Family 3,Family 1)+(Family 3,Family 5) +(Family 3,Family 4)+(Family 3,Family 2))	1/4 * 0.261=	0.076	4	consistent
Family 4	1/(5-1)* ((Family 4,Family 1)+(Family 4,Family 5) +(Family 4,Family 3)+(Family 4,Family 2))	1/4 * 0.435=	0.120	3	consistent
Family 5	1/(5-1)* ((Family 5,Family 1)+(Family 5,Family 4) +(Family 5,Family 3)+(Family 5,Family 2))	1/4 * 0.261=	0.065	5	consistent

The fifth step is to calculate the Net flow (Φ). Net flow calculation (Φ) Is the difference between Leaving Flow and Entering Flow value and the result is the dominance value for each alternative. If the net flow result is positive, then a new object deserves to be called good data. Evaluation and ranking are done by selecting the net flow result data. If the net flow result is positive, then a new object deserves to be called good data. In this case a family with a positive net flow can be called a poor family which is eligible to be given poor certificate. Conversely, if the family has a negative net flow then it is called a non-poor family that is not worthy to be given poor certificate. In the case of poor families, only families with positive net flow can be included in the rank, while families with negative net flow are not included in the rank. The higher the value of net flow, the poor of a family be. Which means the higher rank, they will get as shown in figure 6.

NO	NF	RANK	STATUS
1	0.261	1	Poor
2	0.25	2	Poor
3	0.142	3	Poor
4	-0.283	4	Not poor
5	-0.37	5	Not poor

Fig. 6. Netflow calculation dashboard view

After the PROMETHEE method is verified by manual counting, the numbers, rank and status appearing on the system and the manual numbers will be the same. Then the system can be said that it runs well and can be used to process data with a larger amount and then calculated its poverty rate. At this stage, it is using data from Regional Development Planning Agency in January - June 2016 a number of 150 heads of families include 350 residents. Out of 150 Heads of Families, 60 heads of household signed up for Poor Certificate. The applicants were evaluated using the PROMETHEE method and resulted in only 44 families who could be declared poor. 16 The rest are not eligible to be called poor families. the number of 44 heads of households, are further explained that the total population in which the male sex are poor as many as 71 people and the number of female population including the poor as many as 60 people.

MONTH	POOR FAMILY	RATE
Januari	7	0%
Februari	10	42.857142857143%
Maret	8	-20%
April	5	-37.5%
Mei	5	0%
Juni	9	80%
Juli	0	-100%

Fig.7. Dash board view of poverty rate calculation

The sixth step is the calculation of the poverty rate change. This calculation is done by using data that have been evaluated and declared as poor family for 6 month, then counted the percentage change of poverty number from one month to other months. The poverty table count results on the system dashboard as shown in Figure 7.

The display on the dashboard calculates the rate of poverty then the results derived. The results of manual count are compared to the dashboard view of poverty rate calculation as shown in Table 4.

Table 4. Calculation of Percentage of poverty rate

Month	Poor Family	Counter		RATE	Verivication
Januari	7	=poor family amount Januari-poor family amount before/poor family amount before	=(7-0)/0*100%	0%	Consistent
Februari	10	=poor family amount Februari-poor family amount Januari/Julah KK Januari	=(10-7)/7*100%	42.86%	Sesuai
Maret	8	=poor family amount Maret-poor family amount Februari/poor family amount Februari	=(8-10)/10*100%	-20%	Sesuai
April	5	=poor family amount April-poor family amount Maret/poor family amount Maret	=(5-8)/8*100%	-37.50%	Sesuai
Mei	5	=poor family amount Mei-poor family amount April/poor family amount April	=(5-5)/5*100%	0%	Sesuai
Juni	9	=poor family amount Juni-poor family amount Mei/Julah KK Mei	=(9-5)/5*100%	80%	Sesuai

From the poverty rate calculation data above, the figures are then shown in graphical form as shown in Figure 8.

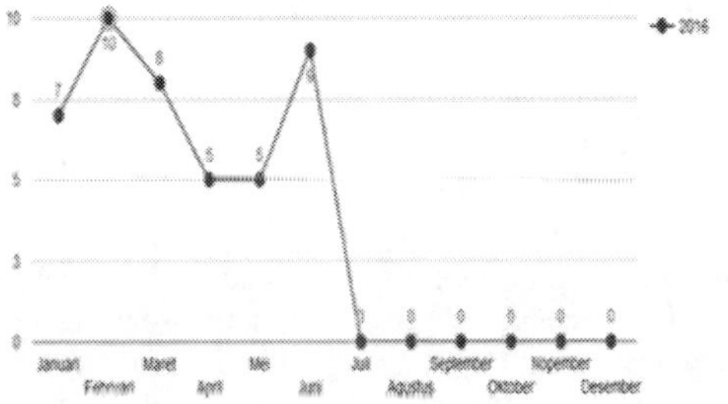

Fig. 8. Poverty rate

4 Conclusions

Based on the results of research and discussion on the system of calculating the poverty rate change, it results some conclusions that the system of calculating the poverty rate change rate, can evaluate the feasibility of determining the status of poverty. Out of the 60 residents who signed up for the granting of Poor Certificate, 16 were declared to be poor. The poverty rate change calculation system can also provide a percentage of poverty rate change data per month. The given percentage of rate poverty change data can be used for monitoring poverty rates. The development of a poverty rate, web-based poverty rate, the evaluation system, alternative rankings, and poverty percentages can be viewed online and in real time so that decisions can be made quickly and accurately.

References

1. B. Ikejiaku, *Journal of Sustainable Development* (2009)
2. S. Vinodh, and R.J. Girubha., *Applied Mathematical Modelling 36,* 5301-5308 (2012)
3. M. Behzadian, R.B. Kazemzadeb, A. Albadvi, M. Aghdasi, *European Journal of Operational Research 200,* 198-215 (2010)
4. T. Dereli, A. Durmusoglu, S.U. Seckiner, and N. Avlanmaz, *An OfficialTurkis Fuzzy System Association, Vol.1,No 2.,pp.* 126-140 (2010)
5. B. Yilmaz, and M. Dagdeviren, *Expert Systems with Applicatications 38,* 11641-11650. (2011)
6. 6. Y.C. Hu, and C.J. Chen, *Information Sciences* ***181***,4959-4968. (2011)
7. S. Murat, H. Kazan, and S. Coskun, *Procedia - Social and Behavioral Sciences* ***195*** 729 – 738 (2015)
8. V. Balali, B. Zahraie, A. Roozbahani, *American Journal of Civil Engineering and Architecture,* ***Vol. 2, No. 5,*** 149-159 (2014)
9. M. Muktiali, Riptek, **Vol.3, No.2**, Hal.: 11 – 20 (2009)
10. H.S. Oluwatosin, *IOSR Journal of Computer Engineering (IOSR-JCE, -ISSN: 2278-0661, p-ISSN: 2278-8727,* ***Volume 16, Issue 1, Ver. IX***), *PP 67-71)* (2014)
11. S.C. Deshmukh, *International Journal of Engineering Science Invention,* www.ijesi.org **Volume 2 Issue 11** PP.28-34 (2013)

Power Distribution Analysis For Electrical Usage In Province Area Using Olap (Online Analytical Processing)

Riza Samsinar[1,*], *Jatmiko* Endro Suseno [2], and *Catur* Edi Widodo [3]

[1] Master Program of Information Systems, School of Postgraduate Studies, Diponegoro University, Semarang - Indonesia
[2] Department of Physics, Faculty of Science & Mathematics, Diponegoro University, Semarang – Indonesia
[3] Department of Physics, Faculty of Science & Mathematics, Diponegoro University, Semarang - Indonesia

Abstract. The distribution network is the closest power grid to the customer Electric service providers such as PT. PLN. The dispatching center of power grid companies is also the data center of the power grid where gathers great amount of operating information. The valuable information contained in these data means a lot for power grid operating management. The technique of data warehousing online analytical processing has been used to manage and analysis the great capacity of data. Specific methods for online analytics information systems resulting from data warehouse processing with OLAP are chart and query reporting. The information in the form of chart reporting consists of the load distribution chart based on the repetition of time, distribution chart on the area, the substation region chart and the electric load usage chart. The results of the OLAP process show the development of electric load distribution, as well as the analysis of information on the load of electric power consumption and become an alternative in presenting information related to peak load.

1 Introduction

Network control center is the center of the grid operating data, a large amount of data that produced from SCADA, EMS, CPS and other systems contains a grid in the operation and safety of all information. However, the current dispatch center information system is mainly for data collection, storage and simple summary of daily operations query, and not specifically for a lot of historical information for effective management and correlation analysis, not effectively use the data from the valuable resource the data extract useful knowledge paired scheduling management [1]. Procedure to calculate total loss in the distribution transformer under non-linear distortion environment is proposed. The power factor capacitor performance under non-linear load conditions is also analyzed. The relation of total current harmonic distortion in the distribution system with load power factor, transformer losses, efficiency and maximum current delivered is also analyzed. The mitigation methods are proposed to minimize the non-linear load impact on the distribution transformer performance. Instead of K-factor transformer approach, a passive harmonic filter method is developed based on higher savings in energy losses [2].

Operational analysis and large data systems for data transmission in electrical power. Data processing techniques use online analytical system to manage and analyze large data capacity. The current center information delivery system is primarily for data collection, data overwriting and simple request summaries of day-to-day operations, and does not specifically have historical information used for category management and effective correlation analysis. Data collection techniques are large databases that extract previously unknown data information about the operation of the knowledge discovery process, including data warehouses, this processing is done with OLAP analysis, and can be visualized data from the latest information science technology [3]. The data cube generated by the data warehouse is processed using statistical analysis in the process of data mining. One of the methods used to perform the analysis is OLAP (OnLine Analytical Processing). The results of this simplification and summary are presented to users who are the basis of business decision making. Thus management can make decisions based on actual facts and not just rely on intuition and quantitative experience alone [4]. A data warehouse is a "subject-oriented, integrated, time varying, non-volatile collection of data that is used primarily in organizational decision making. Data warehouses provide on-line analytical processing (OLAP) tools for the interactive analysis of multidimensional data of varied granularities, which facilitates effective data mining [5]. The use of systems other than Data warehouses and OLAP as a tool to analyze and collect existing data into information that is very important. This OLAP implementation demonstrates excellence at the level of validation of information, so that the resulting information can be trusted. In addition, with OLAP systems can also analyze data from several experts [6].

* Corresponding author: riza.upi@student.undip.ac.id

2 Literature review

2.1 Electrical energy

Energy is the ability to perform work, stored energy work. Not much different from physics is as the ability to produce energy. The law of energy preservation states that energy can not be created and can not be destroyed. Energy can only be changed from one form to another. Similarly, electrical energy is the result of changes in mechanical energy (motion) into electrical energy. The existence of electrical energy can be utilized as much as possible. The use of electrical energy in everyday life is lighting, heating, electric motors and others. The energy used by the appliance is the rate of energy use (power) multiplied by the time during which the device is used. When power is measured in clock watts, then

$$w = p * t \text{ .... } (1)$$

With:
P = power in watts (W)
T = time in hours (h)
W = energy in watts clock (Wh)
Watt clock (wathour = Wh) is energy released if 1 watt is used for 1 hour.

Provision of electric power to customers, requires a variety of electrical equipment. These various electrical devices connected to each other are interrelated and overall form the power system. As is the power system here is a set of power centers and substations (load center) that are connected by the transmission network so that the unity of interconnection. The need for electricity from customers always increases over time. To keep serving the electricity needs of customers, the power system must be developed in line with the increasing demand of electricity from consumers. To do this as well as possible, the results of operations need to be analyzed and evaluated, among others, to determine.

Distribution system is one of the power system that has an important role because it deals directly with users of electrical energy, especially users of medium and low voltage electrical energy. Usually there is often an unbalanced load on the phase (distribution system is a phase 3 system) or overloaded due to the use of electrical equipment from consumers of electrical energy. A balance between phase loads is required for load equalization to minimize changes caused by full load. This is also important because it is useful in optimization techniques to produce a reliable and efficient system. The 1 phase configuration with 3 wires can be said to be unbalanced if neutral current is not zero. This happens because the connected load, between phases and neutrals are not the same

1.1 OLAP System

OLAP is an integrated decision-making approach for the multidimensional analysis of business data by operations such as Rollup, Drill-down and Pivot. The applications mostly involve business decision support, such as business intelligence, business performance management, financial reporting, knowledge discovery, data mining and other related areas. The advantage of the OLAP system is that it allows end-users to employ a multi-dimensional method for the analysis of ad-hoc data and efficiently provides them with reliable information. Based on these features, OLAP might be useful in other scientific and engineering fields that require multi-dimensional data analyses. It is said that OLAP as a supplement to create a more effective data mining tool for a gene expression database. It integrat the geographic information system and OLAP to facilitate geographic knowledge discovery and to analyze environmental health data. OLAP is employed to analyze the incidence of mealybugs on cotton crops. OLAP is also applicable to environmental problems. The application of OLAP in ecological areas is strongly needed because the data are usually analyzed along multiple dimensions. We therefore consider adopting OLAP for water quality analyses.

A data warehouse (or smaller-scale data mart) is a specially prepared repository of data designed to support decision making. The data comes from operational systems and external sources. To create the data warehouse, data are extracted from source systems, cleaned (e.g., to detect and correct errors), transformed (e.g., put into subject groups or summarized), and loaded into a data store (i.e., placed into a data warehouse). The data in a data warehouse have the following characteristics:

- Subject oriented — The data are logically organized around major subjects of the organization, e.g., around customers, sales, or items produced.
- Integrated — All of the data about the subject are combined and can be analyzed together.
- Time variant — Historical data are maintained in detail form.
- Nonvolatile — The data are read only, not updated or changed by users.

A data warehouse draws data from operational systems, but is physically separate and serves a different purpose. Operational systems have their own databases and are used for transaction processing; a data warehouse has its own database and is used to support decision making. Once the warehouse is created, users (e.g., analysts, managers) access the data in the warehouse using tools that generate SQL (i.e., structured query language) queries or through applications such as a decision support system or an executive information system. "Data warehousing" is a broader term than "data warehouse" and is used to describe the creation, maintenance, use, and continuous refreshing of the data in the warehouse.

 https://doi.org/10.1051/e3sconf/20183111010

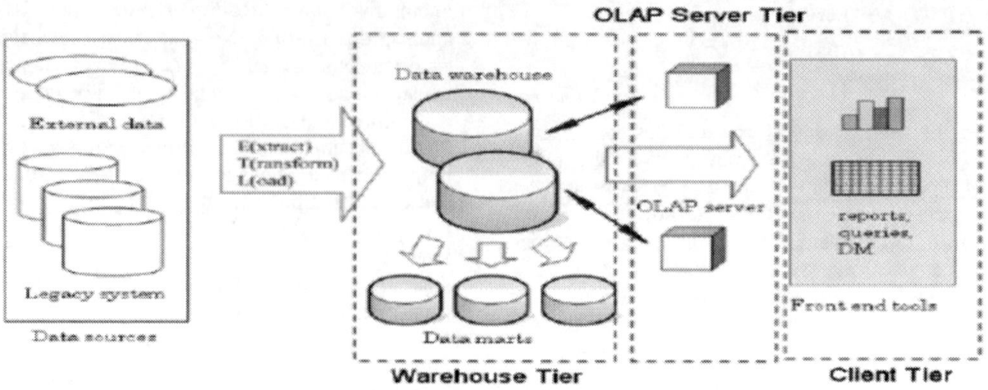

Fig. 1: Data Warehousing Architecture [7]

It includes tools for extracting data from multiple operational databases and external sources; for cleaning, transforming and integrating this data; for loading data into the data warehouse; and for periodically refreshing the warehouse to reflect updates at the sources and to purge data from the warehouse, perhaps onto slower archival storage. In addition to the main warehouse, there may be several departmental data marts. Data in the warehouse and data marts is stored and managed by one or more warehouse servers, which present multidimensional views of data to a variety of front end tools: query tools, report writers, analysis tools, and data mining tools. Finally, there is a repository for storing and managing metadata, and tools for monitoring and administering the warehousing system.

Reporting is public news that contains a collection of information. Business Reporting is public news that contains a collection of business process information and company information, or periodically providing information to decision makers within an organization to support them in their work. Reporting is the result of a process that has an effect on improving companies such as Business Intelligence and knowledge management. Often the implementation involves extract, transform, and load (ETL) in coordination with the data warehouse and then using reporting tools. While reporting can be distributed in print or over the web, they can also be accessed via corporate intranets.

3 Method

At this stage is to collect data related to the total load power distribution of the substation and the transformer special area of DKI Jakarta at PT. PLN (Persero) Distribution Jakarta Raya and Tangerang. The data will be analyzed with OLAP information system with excel data method that has been extracted from internal portal web PT. PLN APD (Persero) Distribution APD Jakarta. PT. PLN (Persero) Distribution Jakarta Raya and Tangerang, especially Jakarta Raya has 16 areas of distribution area, 51 substations and 1906 units of distribution. Incoming data on web portals is not real time, each substation has a shift operator in charge of loading a large repeater load on each substation with an hourly interval input. Then each distribution area has a load graph of the repeater load and monitor graph of the transformer load on web port APD PT. PLN (Persero) Distribution APD Jakarta

System analysis can be explained in the framework of data warehouse storage. Starting with the processed data phase, ETL, cube on OLAP to create an analytics dimension.

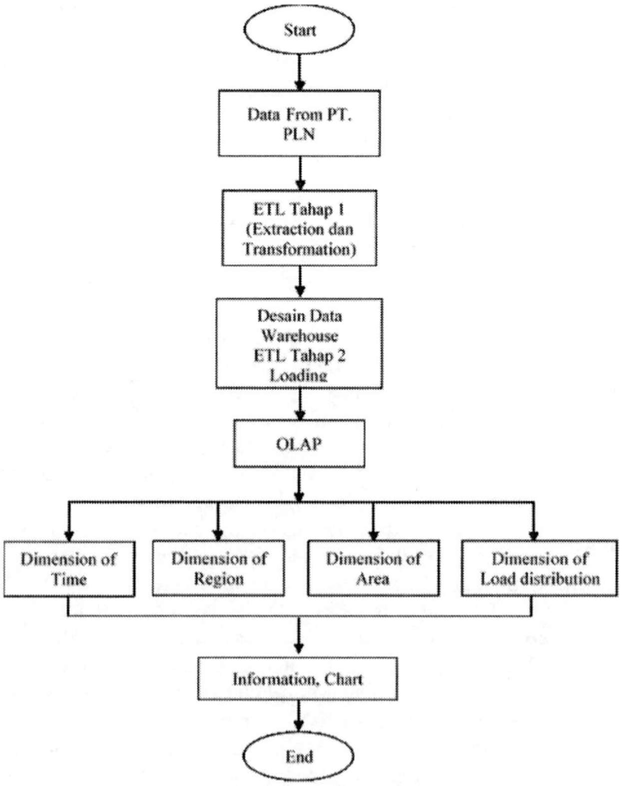

Fig. 2 Research flow

Figure 2 is a framework. that will be used to build a data warehouse for System Analysis Distribution of load power distribution information of PT. PLN (Persero).. ETL on loading step is to design a data model that will be used as a data storage that has been transformed, a model designed with Star Schema Model. Based on data obtained at PT. PLN (Persero), then the Star Schema data model to be built can be seen in Figure 3. In the process of loading data or data into the data warehouse can use the user interface built with programming language.

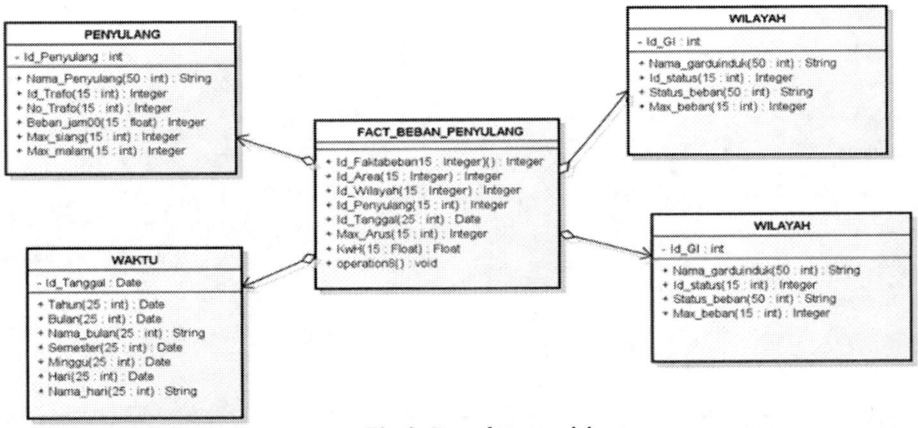

Fig. 3. Star schema model

4 Implementing OLAP system

This section discusses the system design and implementation phases. We begin with the system infrastructure and the database design. Then, we present a system output sample. Olap system with pentaho tools, reporting results with the dimensions of the area. Data that is displayed in accordance with the data warehouse displayed according to the date

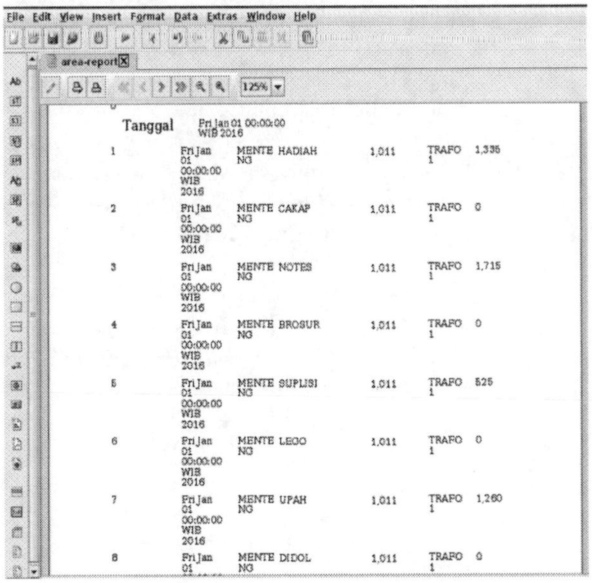

Fig.4. area dimensions

The output results that display the analysis of the use of power load based on no transformer on the substation.

View reporting with dimension of area in sorting based on date, area and main transformer as power distribution from substation.

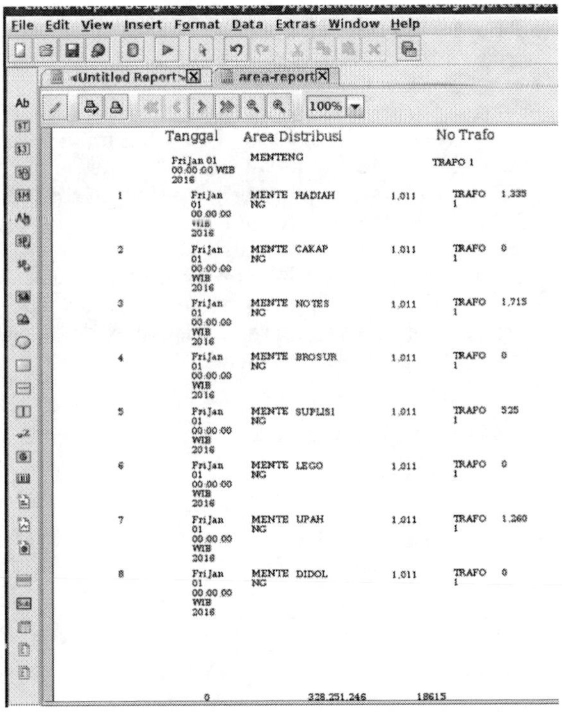

Fig. 5. Area dimensions by area and transformer

Reporting chart area as an example of one of the results of the chart display in other dimensions.

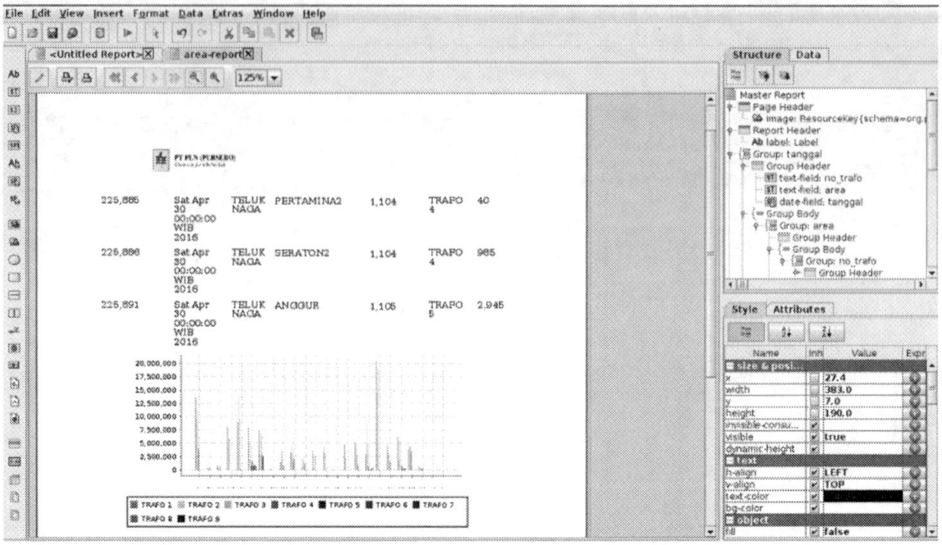

Fig. 6. Data and chart results

5 Conclusion

Information analysis systems can provide information relating to electrical power loads in a single province area. A system that can track information based on multidimensional with area dimension, area dimension, power load dimension and time dimension. The information system built with the olap system with pentaho tools provides load analysis results in the form of charts and queries. The results of its analysis can provide the minimum load, peak load and the amount of load in use on each substation distribution. Lack of information systems that are designed with olap still a little on the dimension side, while olap can be developed more multidimensional dimensions

References

1. H. Zhou, D. Liu, D. Li, G. Shao, Q. Li, Journal Energy and Power Engineering, Vol. **5**, 616-620. (2013)
2. A. D. Sanjay, and M. W. Laxman, International Journal of Emerging Tecnology and Advancedd Engineering, Vol. **1**. (2011)
3. A. Singh, D. C. Upadhyay, Yadav., Hemant, International Journal of Engineeringg Science and Technology (IJEST), Vol. **3**, No. 7. (2011)
4. F. Dehnea, Q. Kongb, C. Rau, H. Zaboli, R. Zhoua, Journal Parallel Distrkib. Computing. Vol. **79**, No. 3. (2014)
5. G. S. Reddy, R. Srinivasu, M. P. Chander Rao, and S. R. Rikkula, International Journal on Computer Science and Engineering. 2865-2873. (2010).
6. H. C. Wang, and J. L. Guo, *Journa*l of Cleaner Production. 40-45. (2013)
7. P. Ponniah, 2001. *Data warehouse Fundamental : A Comprehensive Guide for IT Profesionals*. John Wiley & Sons, Inc : NewYork.

Digital Divide Measurement in Lembata Regency Using SIBIS

Cecilia Dai Payon Binti Gabriel[1,*], *Djoko* Budiyanto Setyohadi[2], *Suyoto*[2]

[1] Informatics Engineering, STIMIKOM Stella Maris Sumba,Sumba Barat Daya NTT Indonesia, 85274
[2] Master Program of Informatics Engineering, Universitas Atma Jaya Yogyakarta, Yogyakarta Indonesia, 55281

Abstract. Along with technological development in Indonesia, digital divide occurs in various regions, which were behind in terms of information on how to use, access and utilize ICT in collecting information from internet. One of the regions is Lembata Regency in East Nusa Tenggara, where digital divide among the people should be measured. The purpose of this study was to determine the level of digital divide among the people of Lembata Regency. To determine the level of digital divide, we used SIBIS GPS (General Population Survey) method, which consisted of several indicators or aspect, i.e. internet usage behavior, internet utilization, and e-government. We also performed two tests, i.e. validity test and reliability test to obtain value of index of digital divide measurement among the people of Lembata Regency. The results of validity test which is processed using SPSS program are categorized valid for each variable indicator and the reliability test results show reliable status. According to the test results on digital discrepancy in Lembata people, the internet usage attitude indicator is categorized low which is 63.1%, the internet usage function indicator is categorized low which is 64%, and the digital discrepancy of e-government indicator is categorized medium which is 40.4%. Therefore, the result of this study because consideration for the government of Lembata Regency in improving ICT services in e-government and in distributing ICT access and ability equally to the people.

1 Introduction

Technological development greatly influences people's lives. This is due to a divide in obtaining information and accessing information via internet in the society. Digital divide is an individual or group's ability in using ICT by accessing and utilizing it [1]. Digital divide is defined as those who access information technology and those who can't access it in collecting information via internet and unable to use it [2]. Digital divide is a gap of internet access and usage, which is separated by social economic status, gender, lifestyle, ethnicity, and geographic location, which illustrates a divide between communities and individuals who have resources to participate in the information era and those who don't. This digital divide really happens across the globe [3]. Digital divide is a divide between individuals, households, businesses and geographic areas at different social economic levels, opportunity to access ICT and internet usage for various activities [4].

Lembata is one region of East Nusa Tenggara Province, which consists of nine sub-districts. The vision of Lembata region is that "the realization of Lembata Region which is independent and productive based on its potency in layout" [5]. In order to fulfill that vision, knowledge in science and technology is needed. However, not all of its areas have the proper access to and skill in information and communication technology and thus there is digital discrepancy between them.

The purpose of this paper was to discuss digital divide in the government of Lembata Regency. ICT utilization in governance is called e-government. E-government aims to help efficiently run a government system and use ICT for the people in Lembata Regency. However, not everyone in Lembata Regency has access and ability in using and utilizing ICT. It's due to difference in access and ability in utilizing ICT, creating digital divide among the people of Lembata Regency.

Digital divide among the people in Lembata Regency should be measured to contribute to understanding on digital divide along with various literatures and studies by the government of Lembata Regency. The contribution was used as consideration in creating policy strategy in public service and to be used by the government to spread ICT access and ability equally to the people in Lembata Regency. Digital divide is measured by SIBIS (Statistical Indicators Benchmarking the Information Society) method. SIBIS consists of several indicators, i.e. internet usage behavior, internet utilization and e-government.

The result of this study became a reference for the government of Lembata Regency in providing ICT services for the people by providing integrated services, training for internet access and information collection.

* Corresponding author: cecylgabriel266@gmail.com

Therefore, digital divide should be measured by adjusting with local condition in Lembata Regency.

2 Material and Method

2.1 Digital Divide

The definition of digital divide is a divide between those who own computers and can access internet and those who don't own computer and can't access information via internet [2]. Moreover, there is digital divide between the rich who can easily get information from internet and the poor who can't get information or access information via internet. [5]. Digital divide is also defined as a divide between individuals, households, businesses, geographies, economic levels in opportunity to get and access information by using internet. [6]. Digital divide is also called a divide in accessing computer and internet between men and women, among people with different social economic statuses, ages and demographics. [7]. Furthermore, digital divide isn't only about a divide of access to ICT but also a divide in ability in using ICT. Whether accessing internet or using internet, it can't be removed from individual's ability and skill [8].

Digital divide according to Presidential Instruction number 3 of 2003 on National Policy and Strategy for E-Government Development is defined as isolation from global development due to inability to utilize information. It's also mentioned that inability to adjust with global trend will lead Indonesia into a great digital divide, which is isolation from global development due to inability to utilize information [9]. E-Government itself is utilization of communication and information technology in governance process (e-government) to improve efficiency, effectiveness, transparence and accountability of governance. Article 5 of Presidential Instruction No.3 of 2003 on National Policy and Strategy for E-Government Development states that "Therefore, the government must immediately perform transformation process toward e-government. Through the transformation process, the government can optimize utilization of information technology development to eliminate dividers between bureaucratic organizations, and form management system network and work process which enable government agencies to work in integration to simplify access to all information and public services which must be provided by the government.

Therefore, all national institutions, the society, business world, and other stakeholders can use government information and services optimally anytime. It requires strong leadership in each institution or government unit so that the transformation process toward e-government can be performed optimally."

Studies on e-government to measure digital divide have been performed by other researchers. According to Chalita Srinuan, a measurement was performed among community groups and countries in Thailand using econometric method. The study showed that demand factors commonly found in United States and European Union are also found in Thailand. Digital divide in Thailand was created by interactions between factors of demand and offer. They are both required to promote benefits in adopting internet and to bridge digital divide [10]. Beside using econometric method, digital divide can be measured by a method used in European society, i.e. SIBIS (Statistic Indicators Benchmarking The Information Society). One of the studies using SIBIS model is the study by Alivia Yulfitri. The object of the study is the world of education by a case study in Senior High School Kotamadya in Bandung.

This study was performed on teachers in the school. The result of the research shows relation between availability of ICT access facility and mastery of ICT, availability of ICT access facility and level of ICT mastery, and availability of ICT access facility and utilization of ICT [11]. SIBIS method was also used in the study by Syarif Hidayatullah in South Tapanuli Regency in Department of Plantation and Animal Husbandry. The study was performed using SIBIS GPS (General Population Survey) method. The research result shows that the level of ICT access divide I high and the ICT skill divide is medium, while digital divide by age, income, education has quite great influence on digital divide among human resources and gender doesn't significantly influence the level of digital divide [12]. A study on digital divide in local government was performed by Ike Pertiwi Windasari and Kridanto Surendro. The study was performed in the local government of Semarang for equalizing ICT skill of human resources and to close the gap in competence in developing e-government system by training employees and recruiting new employees who can use SIBIS GPS and DIDIX instruments. The result of this study shows that the government of Semarang has obstacles in ICT adoption due to lack of training for the employees, low awareness of the employees to join training, and a lack of training by organization [13]. A study on digital divide in Pekalongan was performed by Dyah Listianing Tyas. The research result is used by the government of Pekalongan in spreading ICT access and skill by providing infrastructures and training programs to develop human resources [14].

Digital divide is caused by factors of internet access and usage, which are separated by social economic status, gender, ethnicity, lifestyle and geographic location [15]. Digital divide was focused on the following:

a. Infrastructure
   Defining difference between individual who owns ICT infrastructure and internet connection and individual who doesn't own ICT infrastructure and internet connection.
b. ICT skill achievement
   Defining individual who tries to have required ICT skill and individual who doesn't put effort in having the required ICT skill.
c. Resources utilization
   Defining individual's limitation in using resources available in internet media. Moreover, it defines individual who is unable to access information, knowledge and the latest information.

2.2 SIBIS

To measure the level of digital divide, SIBIS (Statistical Indicators Benchmarking the Information Society) method was used. It's a European commissioned project to analyze and compare various indicators of divide. The indicators used in measuring digital divide among the people of Lembata Regency are:

a. Internet usage behavior
b. Internet utilization
c. Demography
d. E-government.

The SIBIS method used in measuring digital divide in the people of Lembata Regency has the following strengths and weaknesses:

a. Strength
   i. Readiness of internet
   ii. Digital divide
   iii. Security of information
   iv. Immediate response to access
   v. Digital literacy, leaning and training
   vi. E-commerce, e-work, e-science, e-government, e-health
b. Weakness
   Indicator of digital divide doesn't properly emphasize social and economic gaps (Source: [16])

In SIBIS, digital divide of internet usage behavior includes computer usage, internet usage, internet access, digital divide index, internet usage utilization divide, intensity of internet usage, termination of internet usage, email usage and e-government divide, which includes avalaibility of e-government, e-government usage, and e-government assessment [17].

2.3 Research Method

The research is conducted in several steps simultaneously which starts from collecting until measuring data to process and to investigate the data. The data is divided into two which are primary data and secondary data. The primary data are gained through spreading queries while the secondary data are obtained through desk research method, any support is taken from books, journals and papers. The author uses Business Information System Method to investigate digital discrepancy in governmental level of Lembata Region when they plan the community service policy strategies regarding information and communication technology. Figure 1 is the flow diagram to measure digital discrepancy of Lembata people.

The research method applied in this study is started with literature study and problem formulation. Literature study and problem formulation are meant to gain knowledge in terms of data collecting process of this research. The next step is determining which method is appropriate to carry this research. If the method is decided, preparing the queries can be conducted for the queries is the base of this study. After that, the author uses the queries to collect data and the required information regarding digital discrepancy. Then, the collected data will be investigated using SPSS version 20 software. These steps end with query data processing and making conclusions from the results.

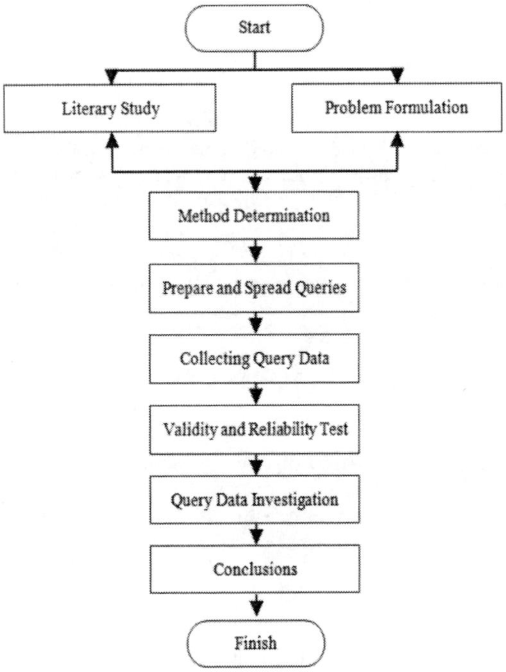

Fig. 1. Flow Diagram

In the present study, the researchers collected data through questionnaires distributed to the people of Lembata Regency. The population of Lembata Regency was 132,171. The data was collected in 2016 [18]. The detail of the population of each sub-district is below:

Table 1. The detail of the population of each sub-district

No	Sub-district	Population (people)
1	Nagawutung Sub-district	9368
2	Wulandoni Sub-district	8503
3	Atadei Sub-district	7568
4	Ile Ape Sub-district	12158
5	East Ile Ape Sub-district	5119
6	Lebatukan Sub-district	45585
7	Omesuri Sub-district	15548
8	Buyasuri Sub-district	19523

From the population of Lembata Regency, samples were collected using Slovin's formula to determine the number of sample [19]. The total sample was 133 respondents. The Slovin's formula is:

$$\mathbf{n} = \frac{\mathbf{N}}{\mathbf{1 + Ne^2}} \quad (1)$$

Note:

n : total sample
N : population size
e : percent of allowance of inaccuracy due to error

Furthermore, the sampling technique in determining population was proportionate stratified random sampling method [20]. The sample result was processed and tested by validity and reliability tests.

2.3.1 Validity Test

Validity test is used to the validity of questionnaire. If a questionnaire is valid, it can reveal what is measured by the questionnaire. Validity test is a measurement which showed how variable to be measured matches variable to be studied by researcher [21]. In the present study, validity test was validity of each question and validity index was found using product moment correlation formula :

$$r = \frac{N\sum XY - (\sum X \sum Y)}{\sqrt{[N\sum X^2 - (\sum X)^2][N\sum Y^2 - (\sum Y)^2]}} \qquad (2)$$

Note :

r = Correlation rate
N = Number of respondent
X = Score or question or statement
Y = Total Score of sub variable

2.3.2 Reliability Test

Reliability test is an index showing how far a measurement instrument can be trusted or relied on. In other words, the instrument has consistent result despite being used repeatedly at different times. Reliability test is performed by alpha cronbach technique. An instrument is reliable if it has reliability of alpha amounting to 0,6 or more. The technique uses the following formula:

$$\alpha = \left[\frac{K}{K-1}\right]\left[1\frac{\sum \sigma_i^2}{\sum \sigma_t^2}\right] \qquad (3)$$

Note :

α : Reliability value
k : Total question or statement item
$\sum \sigma_i^2$: Variance value of each item
$\sum \sigma_t^2$: Total value

This test processes and analyzes in descriptive statistic to determine the category of the level of digital divide among the people of Lembata Regency. The categories in the assessment of the digital divide were:

a. Index <20.00% : very high
b. 20.00% ≤ index < 40.00% : high
c. 40.00% ≤ index < 60.00% : medium
d. 60.00% ≤ index 80.00% : low
e. Index ≥80.00% : very low

3 Results and Discussion

In the discussion, validity test and reliability tests were performed using an application or program called SPSS. Validity test was performed using correlation, where correlation value (pearson correlation) must be positive to mean that the instrument is valid and if it's negative then the instrument is invalid. Meanwhile, reliability test was performed by alpha-cronbach test, in which if coefficient alpha value is bigger than 0,6, then it's reliable.

The research method applied in this study is started with literature study and problem formulation. Literature study and problem formulation are meant to gain knowledge in terms of data collecting process of this research. The next step is determining which method is appropriate to carry this research. If the method is decided, preparing the queries can be conducted for the queries is the base of this study. After that, the author uses the queries to collect data and the required information regarding digital discrepancy. Then, the collected data will be investigated using SPSS version 20 software. These steps end with query data processing and making conclusions from the results.

The following are the instruments which their validity and reliability are tested to determine whether or not they meet the analysis requirements. Validity test is used to examine whether the instruments are valid and can be used to measure. Reliable research instruments are the instruments used to examine the same research objects several times and produce same results.

3.1 Validity Test

The validity test is carried using correlation in SPSS version 20 software. Table 2 presents the validity test result on variable X_1 for internet usage attitude:

Table 2. Validity test result on variable X_1 for internet usage

Variable	r Count	r Table	Status
X1,1	0,36	0,1703	Valid
X1,2	0,226	0,1703	Valid
X1,3	0,282	0,1703	Valid
X1,4	0,394	0,1703	Valid
X1,5	0,308	0,1703	Valid
X1,6	0,241	0,1703	Valid
X1,7	0,344	0,1703	Valid
X1,8	0,422	0,1703	Valid
X1,9	0,341	0,1703	Valid
X1,10	0,313	0,1703	Valid
X1,11	0,305	0,1703	Valid
X1,12	0.405	0,1703	Valid
X1,13	0,371	0,1703	Valid
X1,14	0,368	0,1703	Valid
X1,15	0,232	0,1703	Valid
X1,16	0,175	0,1703	Valid
X1,17	0,222	0,1703	Valid
X1,18	0,281	0,1703	Valid
X1,19	0,245	0,1703	Valid

If r-count > r-table, thus it is categorized as valid and if r-count < r-table, thus it is categorized as invalid. r-Count value is gained from SPSS Reliability analysis in corrected item total correlation column. With 200 correspondents and 2 independents in which significance rate is at 0.05, thus r-table is acquired at 0.1703. The following is X_2 or internet usage function validity test (Table 3) result:

Table 3. Internet usage function validity test result

Variable	r Count	r Table	Status
X2,1	0,426	0,1703	Valid
X2,2	0,345	0,1703	Valid
X2,3	0,338	0,1703	Valid
X2,4	0,284	0,1703	Valid
X2,5	0,495	0,1703	Valid
X2,6	0,487	0,1703	Valid
X2,7	0,267	0,1703	Valid
X2,8	0,392	0,1703	Valid
X2,9	0,345	0,1703	Valid
X2,10	0,255	0,1703	Valid
X2,11	0,313	0,1703	Valid
X2,12	0,416	0,1703	Valid
X2,13	0,288	0,1703	Valid
X2,14	0,298	0,1703	Valid
X2,15	0,439	0,1703	Valid
X2,16	0,449	0,1703	Valid
X2,17	0,345	0,1703	Valid
X2,18	0,377	0,1703	Valid
X2,19	0,272	0,1703	Valid
X2,20	0,323	0,1703	Valid
X2,21	0,343	0,1703	Valid
X2,22	0,358	0,1703	Valid
X2,23	0,315	0,1703	Valid
X2,24	0,285	0,1703	Valid

The result of Y or e-government variable validity test is presented in table 4:

Table 4. The result of Y or e-government variable validity test

Variable	r Count	r Table	Status
Y,1	0,233	0,1703	Valid
Y,2	0,416	0,1703	Valid
Y,3	0,372	0,1703	Valid
Y,4	0,391	0,1703	Valid
Y,5	0,178	0,1703	Valid
Y,6	0,342	0,1703	Valid
Y,7	0,228	0,1703	Valid
Y,8	0,431	0,1703	Valid
Y,9	0,402	0,1703	Valid
Y,10	0,339	0,1703	Valid
Y,11	0,319	0,1703	Valid
Y,12	0,373	0,1703	Valid
Y,13	0,191	0,1703	Valid
Y,14	0,219	0,1703	Valid
Y,15	0,474	0,1703	Valid
Y,16	0,523	0,1703	Valid
Y,17	0,502	0,1703	Valid
Y,18	0,452	0,1703	Valid
Y,19	0,446	0,1703	Valid
Y,20	0,434	0,1703	Valid
Y,21	0,399	0,1703	Valid
Y,22	0,553	0,1703	Valid
Y,23	0,509	0,1703	Valid
Y,24	0,403	0,1703	Valid

According to the table, it is found that all tested indicators show r-count > table value which is 0.1703. Therefore, it is concluded that all research variable indicators are valid.

3.2 Reliability Test

To examine the reliability of each research variable, it uses Cronbach-Alpha test. The queries are declared reliable if they have alpha coefficient value bigger than 0.6. Here are the results:

Table 5. To examine the reliability

Variable	Alpha Value	Status
Internet usage behavior (X1)	0,735	Reliable
Internet usage utilization (X2)	0,814	Reliable
E-government (Y)	0,833	Reliable

Based on the table above, it is concluded that internet usage attitude variable (X1), internet usage function (X2) and e-government have reliable statuses because their alpha coefficient values are > 0.06.

Using SIBIS GPS method which consists of several indicators discussed above, a measurement should be performed on the people of Lembata Regency. The table of digital divide by indicator is shown below:

a. The level of digital divide by internet usage behavior

Table 6. The level of digital divide by internet usage behavior

Indicator	Sub-indicator	Percentage (%)
Internet usage behavior	In work	65,2
	Personal activity	63,7
	Ease of access	60,4
Total average		63,1

The test result on the digital divide among the people of Lembata Regency by internet usage behavior in SIBIS method in terms of sub-indicator internet usage in work of 133 respondents was 65.2%. It showed that the level of internet usage in work is low. Sub-indicator internet usage in personal activity among the people of Lembata Regency was 63.7%. It showed a divide in internet usage in personal activities, such as finishing tasks from teacher or lecturer or work. Sub-indicator ease in accessing internet among the people of Lembata Regency was 60.4%. It showed that that the divide among the people in being facilitated to access information was low. Therefore, it is concluded that the digital divide among the people of Lembata Regency in terms of internet usage behavior was 63.1%, showing low level of digital divide among the people of Lembata Regency.

b. The level of digital divide by internet usage utilization

The percentage of internet usage utilization in SIBIS method in terms of sub-indicator internet usage utilization to find information on internet of 133 respondents was 65.7%. It showed that the people of Lembata Regency lacked knowledge on information through internet usage. Meanwhile, sub-indicator internet usage to collect information online was 64.2%. It showed that the people of Lembata Regency didn't understand how to collect information online. Sub-indicator sending work data by email was 62,1%. It showed that e-mail usage was low and not understood by the general public of Lembata Regency. It's concluded that the level of digital divide of indicator internet usage utilization was 64%, and the three sub-indicators showed that the level of the divide among the people of Lembata Regency was low.

Table 7. The level of digital divide by internet usage utilization

Indicator	Sub-indicator	Percentage (%)
Internet usage utilization	Finding information via internet on daily basis	65,7
	Using internet to get information online	64,2
	Sending work data via e-mail	62,1
	Total average	64

c. The level of digital divide by e-government

Table 8. The level of digital divide by e-government

Indicator	Sub-indicator	Percentage (%)
E-government	Finding information via e-government service	40,3
	Ease of e-government service	37,4
	Ease of access	42,5
	Confidence in correctness of e-government information	41,4
	Total Average	40,4

The percentage of e-government behavior in SIBIS method in terms of sub-indicator search for information via e-government service of 133 respondents was 40.3%. It showed that people were able to use the application prepared by the government of Lembata Regency with e-government service well. Sub-indicator ease of e-government service was 37.4%. It showed that e-government service was easier to understand and easy to access by the people in Lembata Regency. Sub-indicator ease of internet access was 42.5%. Sub-indicator confidence in correctness of e-government information was 41.4%. It showed that the people in Lembata Regency were able to look for the correctness of information via the service provided by the local government. So, it's concluded that the level of digital divide of indicator e-government from four sub-indicators was 40.4%. It showed that the level of the divide for indicator e-government was medium.

The result above showed internet usage behavior, internet usage utilization and e-government according to SIBIS GPS [17]. It's concluded the government of Lembata Regency must consider a lot of things in providing ICT in public services.

Moreover, the result above showed that the people in Lembata Regency lacked experience and understanding on internet usage and how to access internet properly. So, the government of Lembata Regency must provide socialization to people to improve knowledge on internet usage in getting information.

It's concluded that the level of digital divide among the people of Lembata Regency was at medium level. It's evident in low internet usage behavior, low internet usage utilization and medium e-government. So, the government of Lembata Regency should consider reducing digital divide in the society.

4 Conclusion

The analysis result showed that SIBIS method described digital divide values among the people of Lembata Regency. The digital divide in terms of internet usage behavior was low (63.1%). Internet usage utilization low (64%), and e-government medium (40.4%)

The analysis result can be used as consideration for the government of Lembata Regency to improve public ICT service related with digital divide and reduce digital divide in Lembata Regency.

References

1. S. Baase, *Issues for Computing Technology 4th New Jersey* : Prentice Hall PTR, (2012).
2. E. Hargittai, Sociology Departemen Princeton University, pp. 1-22, (2003).
3. W. C. a. B. Wellman, *Charting and Bridging Digital Divides: Comparing Socio-economic. Gender, Life Stage, and Rural-Urban Internet Access and use in Eight Countries*, (2003).
4. OECD, *Understanding the Digital Divide*, Paris Cedex 16, France: OECD, (2001).
5. Administrator, *"Situs Resmi BPS Kabupaten Lembata,"* 2016. [Online]. Available: https://lembatakab.bps.go.id/. [Accessed 06 May 2016].
6. I. C. a. T. T. Kidd, Digital Divide Implications and Trends, Texas: IGI Global, pp. 130-132, (2008)
7. J. a. D. Preeti Dubey, BIJIT-BVICAM's Internasional Journal of Information Techonology, pp. **341-346**, July-December (2011).
8. S. J. Min, Journal of Information Techonology and Politics, pp. 1-15, (2010).
9. E. A.-S. a. R. Khasawneh, International Conference on Information Technology, vol. **6** th , pp. 1-8, 8 May (2013).
10. Inpres, *Kebijakan dan Strategi Nasional*, Jakarta, (2013).

11. P. S. E. B. Chalita Srinuan, Telematics and Informatics, pp. 254-262, (2012).
12. A. Yulfitri, *Konferensi dan Temu Nasional Teknologi Informasi dan Komunikasi untuk Indonesia*, pp. 1-5, 21-23 Mei (2008).
13. S. Hidayatullah, Universitas Negeri Padang, Padang, (2013).
14. K. S. Ike Pertiwi Windasari, Jurnal Sistem Komputer, Vol.**1**, No.2, pp. 71-75, 2011.
15. A. B. A. J. S. Dyah Listianing Tyas, SENTIKA 2016, pp. 590-598, 18-19 Maret (2016).
16. C. a. B. W. Wenhong, AMD Global Consumer Advisory Board, (2003).
17. K. Barzilai-Nahon, *Washington : The Information Society*: An International Journal, (2006).
18. SIBIS, Information Society Technologies and European Commission, Germany, (2003).
19. Martono, *Metode penelitian kualitatif*, Bandung: Alfabeta, (2012).
20. Sugiyono, *Metode Penelitian Bisnis,* Bandung: Alfabeta., (2010).
21. D. R. C. a. P. S. Schindler, McGraw-Hill Internation Edition, (2006).
22. 2. Yulfitri.
23. SIBIS, Information Society Technologies, Germany, (2003).
24. B. N. K, The Information Society, (2006).
25. E. Hargittai, The Digital Divide and What To Do About It, San Diego, CA: Sociology Department, Princeton University and New Economy Handbook edited by Derek C. Jones, (2003).

The Development of Mobile Application to Introduce Historical Monuments in Manado

Moshe Markhasi Rupilu[1,*], *Suyoto*[1], and *Albertus* Joko Santoso[1]

[1]Magister Informatics Engineering, Universitas Atma Jaya Yogyakarta, Yogyakarta, Indonesia 55281

Abstract. Learning the historical value of a monument is important because it preserves cultural and historical values, as well as expanding our personal insight. In Indonesia, particularly in Manado, North Sulawesi, there are many monuments. The monuments are erected for history, religion, culture and past war, however these aren't written in detail in the monuments. To get information on specific monument, manual search was required, i.e. asking related people or sources. Based on the problem, the development of an application which can utilize LBS (Location Based Service) method and some algorithmic methods specifically designed for mobile devices such as Smartphone, was required so that information on every monument in Manado can be displayed in detail using GPS coordinate. The application was developed by KNN method with K-means algorithm and collaborative filtering to recommend monument information to tourist. Tourists will get recommended options filtered by distance. Then, this method was also used to look for the closest monument from user. KNN algorithm determines the closest location by making comparisons according to calculation of longitude and latitude of several monuments tourist wants to visit. With this application, tourists who want to know and find information on monuments in Manado can do them easily and quickly because monument information is recommended directly to user without having to make selection. Moreover, tourist can see recommended monument information and search several monuments in Manado in real time.

1 Introduction

Manado is one of the most famous tourism destination in Indonesia and is located in Asia-Pacific, which is bordering on the Philippines, Republic of Palau and the Pacific Ocean [1]. It is nicknamed Maldives Van Celeb because this city has the beautiful National Park of Bunaken which is similar with Maldives (Maldives) in the north-western tip of Sulawesi [2]. Beside Bunaken, Manado also has many interesting destinations which are historical monuments that have religious history, cultural history, and even the history of Indonesian Independence struggle. Cultural diversity and monuments make Manado as a city that has so many tourism destinations. In fact, in 2017, there is government program which is MWTC (Manado as World Tourism Centre 2017) so that Manado becomes the destination of the International and Local tourists. Travelling is a necessity that cannot be separated from human life. The monuments are needed by tourists in order to experience adventure tour in Manado.

Access to information services are still inadequate and not proportional with many tourism destinations that are available, especially the monuments in Manado. Tourists get difficult to find the locations. They are not getting detailed information's about the destinations they want to visit, for example, they cannot get information about route which is short, safe, and free from congestion. So, the time wasted on the way to the location, even the tourists can have over budget to hire a tour guide who is not organized.

The development of information technology, especially mobile phones, becomes a facility that improves services in accessing the data needed by tourist to know the locations, distance, and description of the monuments in Manado. One result of the mobile technology development is the emergence of a mobile phone that has Android as its operation system. The term "Android" comes from the Greek word with the prefix – and that means "male" and the suffix -elides means "the same or similar", the full meaning is together "to be human". Android is a set of software for mobile device that has a relationship between one to another and becomes a set of system programs or a set of application programs that forms a complete system [3].

Android is meant to revolutionize the mobile phone market by bringing internet to mobile phone and enabling its operations just like operating a personal computer (PC), so with Android, consumers can easily get information's via smart applications on their phone [3]. An application that consists of information's about the historical monuments such as location, distance, and their description is needed for accuracy in determining which monuments are priority and can be visited by tourists.

Based on the problems described above, there is an application that can help visitors and tourists in finding information and location of the monument in Manado. The application is location-based service (LBS). LBS means mobile service providers facilitate their mobile phone with information about tourism destination that based on their users location.

* Corresponding author: mosherupilu@gmail.com

2 Literature Review

The development of mobile campus applications to guide students, parents, and/or visitors in finding places on campus. By using LBS (Location Based Service) and NFC (Near Field Communication) feature, phones will be connected automatically in close range with other NFC phones [4]. The increase of popular Smartphone also opens opportunities for mobile services to develop mobile tourism application to suggest tourism destinations based on context factors such as location, weather, and time available. This mobile tourism application is seen as the most efficient way to help travelers on their trip.

Location Based Service (LBS) is a service which is location-based, and will be accessed via mobile device (Smartphone, etc.) So that, it can display a map along with the location where the mobile device is located. Location-based service (LBS) provide customized information according to user characteristics, so it can be easily expanded for many other usages, such as guiding routes for transportation systems or for tourism destination [5].

Location-Based Services application provides a set of services for users who are from geographic location of mobile devices. Use of this service makes it possible for users to search and find other people, vehicles, resources and provide location-sensitive services, besides tracking their own location. Requests for location may come from mobile devices or other entities such as application or networks providers. It is possible to automatically trigger Location Based Services when the mobile device is in a particular location. In this study, it is discussed how to implement location-based services on Android [6].

It is possible to automatically trigger Location Based Service when the mobile device is in a particular Location. This service can also be derived from the user's mobile device to meet location-based requests such as finding areas of interest, checking traffic conditions, finding friends, vehicles, resources, machinery and emergency requests. This study will discuss how to implement location-based services on Android [7].

Developing Location Based Services applications uses data mining for mobile users. Data Mining aims to find interesting and useful knowledge from the database. Conventionally, the data were analyzed manually. Many useful relationships are not revealed and may not be user identity. Through data mining, it can extract interesting knowledge and regularity. Mobile application which is built using Data Mining approach can find the nearest and famous location around a certain area. By using the data mining approach, it extracts the database with the help of different data mining algorithms. The nearest location is determined by the wireless network. It provides data using open source Android. It provides a world-class platform for creating applications and games for android users everywhere, and open markets for immediate distribution to users [8].

Mobile computing has grown in such a way that users can access all the information on a single device where people are always moving with mobile devices like laptops, cell phones, tablets, etc. Using a user geographic location, many of the information's related to mobile device users can be collected. The location information of mobile users can improve the class of services and applications that can be provided to mobile device users. These application and service classes are called as location-based service. Location Based Services (LBS) is a type of service that helps obtaining user's geographic location and more useful information near the user's location. This location-based information can be obtained by various terms such as position, vicinity, proximity, context, maps, routes, places, and more [9].

Design of tour guide system is based on three layers of architecture. The architecture includes the browser layer, top layer and bottom layer. It uses KNN algorithms and collaborative filtering to calculate and recommend tourism information to users. Limitations found in this study are the application is not very efficient in providing information and predicting the right places with an affordable price [10].

Tour guide system uses web technology services and three layers architecture, by using web technology and three layers architecture that are browser layer, business logic layer and server layer. Lucene is used to create an index for data which is used to execute efficient requests. Limitations found in this study are the differences in the number of Geographic description details of the code at different locations, for example, one place has information and detail description in the form of building address, but the other place only has city name and zip code [11].

3 Research Methodology

In conducting this research there is a research methodology used to obtain information that is completely understood and the results in accordance with the expected results, as well as get the results of quality scientific work [12]. Therefore, the researcher uses methods in the form of: Methods of data collection uses interviews and field studies. Interviews were conducted with local tourism authorities to obtain information in monuments in Manado. In addition, researcher also uses sources from other existing research such as journals, papers and reference books. Research methods can be described as follows [13].

3.1. Observation Method

At this stage, researcher collects data by doing observation to monuments that are interesting.

3.1. Library Method

This method is data collection by studying books (literature study), journals and other references that are related to this research.

For developing monument recommendation system, it uses initial determination of KNN algorithm, grouping items with K-means and the process of filtering or evaluating items with collaborative filtering. K-Nearest Neighbor (KNN) algorithm is a method to classify objects based on learning that is very close to the object. The working principle of K-Nearest Neighbor (KNN) is to find the closest distance between the data that will be evaluated with the nearest K-Neighbor in the training data.

$$d_i = \sqrt{\sum_{i=1}^{p}(X_{2i} - X_{1i})^2} \quad (1)$$

One of partition algorithm is the K-means algorithm, it is algorithm based on the definition of initial centroid after determining the number of first group. The process is periodically used in the K-means algorithm to obtain a database cluster. The data items in this algorithm are grouped into a data set according to the closest distance to a cluster [14].

Filtering process or evaluating items which are based on the perceptions of other users becomes an important determinant of collaborative filtering algorithm. A profile is used to filter a number of items. Collection and profile development become a collaborative filtering technique, then determine the relationship between the suitability of model equations and data. The purpose of collaborative filtering is to filter number of items which users select, and provide recommendations to users with the remaining options [15].

4 Analysis

In developing this system, it is required a phase which is called analysis phase, and there will be need analysis in this phase. This phase is done because the system that is made sometimes is not working as what system maker expects. this problem happens because need analysis is not done correctly. The purpose of the phase is to examine and identify problems that happen to system users, so that systems which will be built and developed in accordance with what is expected by the user. Besides, disadvantages and errors in building the system can be minimized by making or adding minor changes to the built system.

Information about the monuments is the important thing that should be given to tourists, and one of the information that must be covered is the monuments location. There are still many tourists who do not know about the existence of monuments in Manado, they do not know about the location of each monument, address, short description, route location and related information needed. This obstacle happens, because not all tourists know the location and information related to the existing monument, especially for tourists who visit one of the monuments in Manado for the first time.

Tourists still search information manually by asking other people who they meet, besides, some of them find the information from brochures in the form of hardcopy, or find monument information via internet. However, the search for monument information by reading brochure is rarely encountered because of the globalization. Similarly, asking others is also rarely done now, because the information that they get is limited to the experience of person they meet. In addition, searching the information on the internet is not helpful enough because there is no special website that provides sufficient information's about the monuments. Therefore, it is necessary to develop a system that can help and facilitate the tourists to obtain the monument information's that are recommended in a real time.

As the result of this study, the proposed system is developing a system that can recommend the right monument to the users regarding the existence of monuments in Manado. Tourists can be facilitated by the recommendation about the monument they want to visit. The information displayed is recommended monument, brief description, map and route location to the monument using Smartphone device through mobile monument application.

5 Design

Design phase is the stage that describes the application design and develops the system structure. This stage includes modelling data, architecture, interface design and the implementation.

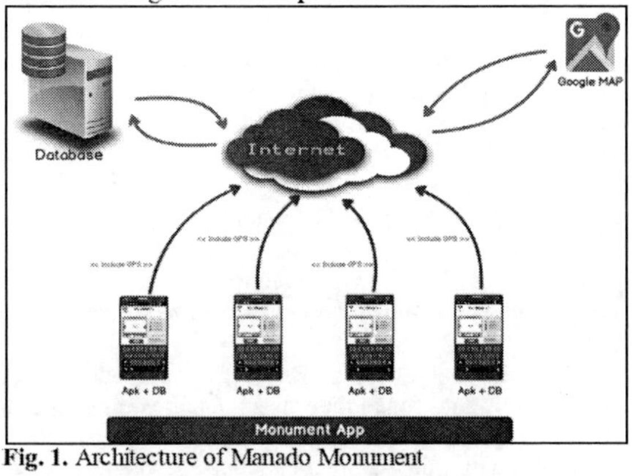

Fig. 1. Architecture of Manado Monument

This phase describes about design process that will be transformed into software design, which can be predicted before it is implemented into the program. The process will focus on determining the overall system architecture, the design phase will also be used to perform system-building activities. Then, in developing this system, use-case is used to explain what users can do to monument recommendation system using mobile device. Figure 2 shows the use-case diagram of the monument recommendation system.

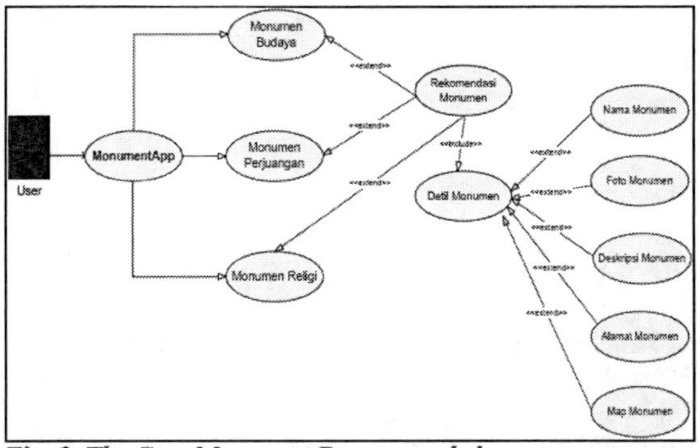

Fig. 2. The Case Monument Recommended

The following is the design of monument recommendation system architecture that will be shown in Figure 3 below.

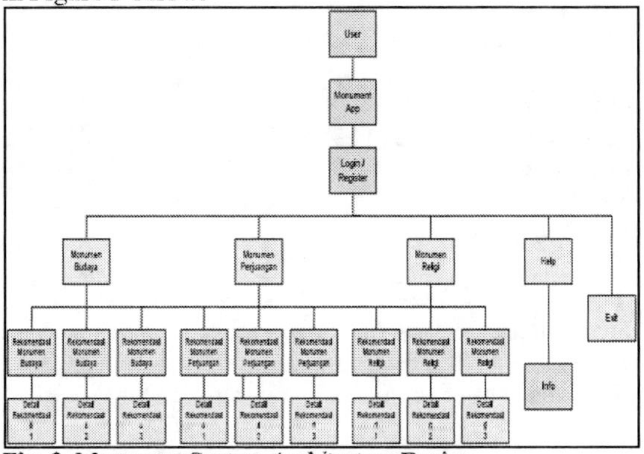

Fig. 3. Monument System Architecture Design

4 Results and Discussion

The monument recommendation system which is built uses four modules in its application. The first module is used for system authentication purposes, the second module is used to display the initial system interface, the third module is used to display monument recommendations to users and the fourth module is used to display details of the monuments.

In module 1, there is checking system about user registration that is done by monument recommendation system. If users have not registered, they can register in the system directly. After done registration, user will get authentication from the system, then they can access the system.

In module 2, system will display the monument categories which are historical, cultural, and religious monuments. The classification of monuments will use different icons in representing them, in order to ease users when they select category they want. Each recommendation of the monument given will be shown with the list of monuments based on the classification. Besides, there are also features to know information about the use of system and how to use it.

Module 3 is used to provide information about classification of the monuments. The information displayed is a list of monuments recommended. Furthermore, in Module 4, each monument classification will display detailed information, maps and location routes that are needed by the users. The maps and routes are displayed on Google Maps that is tracking users' location by utilizing GPS. GPS is a navigation system that gives information about the coordinate of users location. Then, the system will display some interesting locations or other places such as public facilities, cafes, airports and so forth.

Users can interact into the system using Smartphone devices to access GPS using monument applications, but GPS must be active to do it. Then, the system will respond command given by performing related processes. The process will provide the information needed according to the longitude and latitude to display the monument in interface system. In addition, the system also automatically reads the database related to some monuments through the information that has been built, and the monument will be displayed based on each choice according to wishes of the user. Monuments displayed include historical monuments, cultural monuments and religious monuments.

Information's displayed by Manado monument application are:

4.1 Cultural Monument

- *Kesatuan Bangsa* Park
- City Park Monument
- *Adipura* Monument
- Boboca Monument
- Toar Lumimuut Monument

4.2 Monuments of Struggle

- The monument of Worang Battalion Landing
- Monument of World War 2 (Sentrum Church)
- Sam Ratulangi Monument
- Zero Point Monument
- Moraya Castle

4.3 Religious Monument

- Statue of Blessing Jesus
- The Tree of Love
- Cross Warembungan Monument
- The Hill of Love
- Coral Triangle Monument

KNN algorithm is used to find and get the distance between items and users needs in cluster. Then, the classification can be grouped in neighboring categories. K-mean algorithm is used to observe and group items into the allocation of several sections to each cluster observation, according to the nearest mean, as well as the prototype cluster. Collaborative filtering is used to filter the users choice which then recommend some of remaining options to the users.

To calculate the suitability between tourist object criteria and tourists needs, similarity formula is used.

$$Similarity(P, C) = (s1 * w1 + s2 * w2 + ... + sn * wn) / (w1 + w2 + ... + wn) \quad (2)$$

Information:
P: Problem
C: case
S: similarity which is 1 (similar) 0 (different)
W = weight (weight given).

Case Study
Giving weight to tourist priority.

Main Priority = 5 (w)
Criteria of tourism destination that become main priority are:
1. Cultural Monument
2. Monument of Struggle
3. Religious monuments
4. Toilet
5. Cheap Price

Medium Priority = 3 (w)
The criteria of tourism places which become medium priority are:
1. Restaurant
2. Lodge
3. Medium price

Low priority = 1 (w)
Criteria of tourism place that become low priority are:
1. Souvenir
2. The expensive price

In this case study, there is a new user who will search for tourist attractions with the following criteria:
1. Museum of struggle
2. Medium Price
3. Lodge
4. Restaurant

The system already has the data of previous cases such as in table 1 below.

Table 1. The data of selecting tourist attractions

Case 1 **Criteria:** 1. Cultural Museum 2. Medium Price 3. Toilet 4. Souvenir 5. Restaurant	**Solution:** Toar Lumimuut Monument
Case 2 **Criteria:** 1. Museum of Struggle 2. Expensive Price 3. Toilet 4. Souvenir 5. Lodges 6. Restaurants	Solutions: Sam Ratulagi monument
Case 3 **Criteria:** 1. Religious Museum 2. Medium Price 3. Toilet 4. Restaurant	**Solution:** The Hill of Love

In order to make the system recommend users about which place to visit, so calculation and selecting one of the cases that is very close to user criteria need to be done.

Calculation 1 (case 1)
Cultural Museum = 5
Medium Price = 3
Toilet = 5
Souvenirs = 1
Restaurant = 3
Similarity
(X, 1) = (0*5+1*3+0*5+0*1+1*3)/(5+3+5+1+3)= 0.35

Calculation 2 (case 2)
Museum of Struggle = 5
Expensive Price = 1
Toilet = 5
Souvenirs = 1
Lodge = 3
Restaurant = 3
Similarity
(X, 1) =
(1*5+0*1+0*5+0*1+1*3+1*3)/(5+1+5+1+3+3)=0.64

Calculation 3 (case 3)
Religious Museum = 5
Medium Price = 3
Toilet = 5
Restaurant = 3
Similarity
(X, 1) = (0*5+1*3+0*5+1*3)/(5+3+5+3) = 0.37

From three cases that have occurred. The case that has a value close to 1 is case number 2. The solution in case 2 can be used as a recommendation to the new user.

The initial interface implementation is displayed when the user selects the monument icon. It can be shown in the following picture.

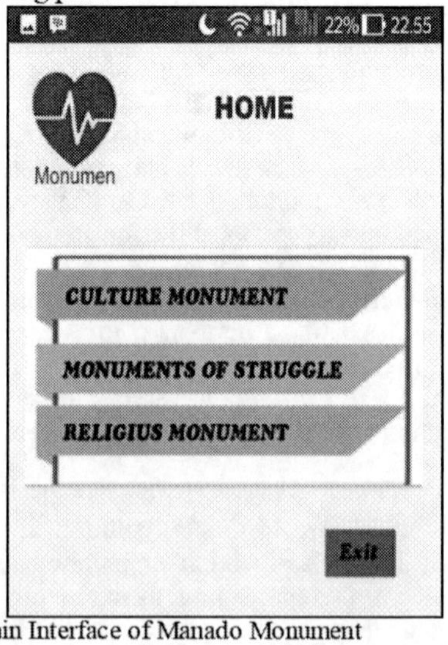

Fig. 4. Main Interface of Manado Monument

In the interface picture above, it shows the initial interface of monument recommendation system that consists of historical monuments, cultural monuments, and religious monuments. Each monument has been distinguished according to its category and lists. The lists are list of historical monuments, list of cultural monuments, and list of religious monuments. Through

https://doi.org/10.1051/e3sconf/20183111012

the monument lists, users can find detailed monument information or can find map of monument locations and monument location routes that can direct users to the monument they want to go.

Then, maps and routes to the monuments which are recommended by the system, will be shown in the following picture.

Fig. 5. Map and Monument Route Interface

The picture above shows map of the location and route of one monument that has been recommended by the system when the user selects the monument as he wishes. The system will respond and give route of the location to the monument based on the user starting position by using GPS support on the user's mobile device.

7 Conclusions

In this research, researcher presents a design for the monument recommendation system using mobile devices' location-based services (LBS), KNN, K-Nearst, and Collaborating Filtering. The monument recommendation system makes tourists get information about the monument in Manado more easily and efficiently. Monument location can be obtained using the directions menu on each monument that aims to facilitate the users in order to reach the location, because it displays the route that must be passed from users starting position. The combination of mobile applications using Location Based Services (LBS) and KNN is chosen because it is interactive. It means that there is a two-way interaction between users and the system, which the system will easily recommend interesting monuments that can be selected by tourists. So, tourists realize that monuments' attractiveness in Manado increases.

References

1. A. E. Tungka, A. A. Omran, A. O. Gebril, W. S. Wah, And A. B. Suprapti, "Manado Waterfront Development Concept As Sustainable City Of Tourism," *ACTA Tech. CORVINIENSIS – Bull. Eng.*, pp. 31–36, 2012.
2. Kholil and D. Tangian, Int. J. Dev. Sustain., vol. **1**, no. 2, pp. 391–401, 2012.
3. O. Uke and A. Shaikh, Int. J. Comput. Appl., vol. **2**, no. 2, pp. 543–549, 2016.
4. *S. Bhattacharya and M. B. Panbu, Int. J. Innov. Technol. Explor. Eng., vol. **2**, no. 3, pp. 25–29, 2013.*
5. J. M. Noguera, M. J. Barranco, R. J. Segura, and L. Martínez, *"A Mobile 3D-GIS Hybrid Recommender System For Tourism,"* pp. 37–52, 2012.
6. J. M. Noguera, M. J. Barranco, R. J. Segura, and L. Martínez, Inf. Sci. (Ny)., vol. **215**, pp. 37–52, 2012.
7. C. R. Rani, A. P. Kumar, D. Adarsh, K. K. Mohan, and K.V.Kiran, Int. J. Adv. Eng. Technol., vol. **3**, no. 1, pp. 209–220, 2012.
8. A. Malik and M. A. Kajala, Int. J. Recent Innov. Trends Comput. Commun., vol. **2**, no. 4, pp. 920–923, 2014.
9. P. Doshi, P. Jain, and A. Shakwala, Int. J. Eng. Comput. Sci., vol. **3**, no. 3, pp. 5072–5077, 2014.
10. Mrs.S.S.Pawar, M. A. S.Kadam, M. P. R. Chavhan, M. P. R. Ranjane, and M. A. S. Lohar, *Int. J. Eng. Technol. Manag. Appl. Sci.*, vol. **4**, no. 2, pp. 42–46, 2016.
11. B. S. N. Reddy and D. R. P. Sam, Int. J. Comput. Trends Technol., vol. **4**, no. 5, pp. 1287–1290, 2013.
12. V. S. Molchanova, Eur. J. Soc. Hum. Sci., vol. **2**, no. 2, pp. 75–80, 2014.
13. P. P. A. Manjare, M. P. V. Ninawe, M. M. L. Dabhire, M. R. S. Bonde, M. D. S. Charhate, and M. M. S. Gawande, Int. Res. J. Eng. Technol., vol. **3**, no. 4, pp. 877–881, 2016.
14. *U. R. Raval and C. Jani, Int. J. Comput. Sci. Mob. Comput., vol. **4**, no. 11, pp. 72–76, 2015.*
15. M. A. Hameed, O. Al Jadaan, and S. Ramachandram, Int. J. Comput. Sci. Eng., vol. **4**, no. 5, pp. 859–876, 2012.

Assessment of Information Security Management System based on ISO/IEC 27001:2013 On Subdirectorate of Data Center and Data Recovery Center in Ministry of Internal Affairs

Ari Kurnianto[1,*], *Rizal* Isnanto[2], and *Aris* Puji Widodo[3]

[1]Magister of Information System, Diponegoro, University, Semarang - Indonesia
[2]Department of Computer Systems Engineering, Diponegoro University, Semarang - Indonesia
[3]Physics Department of Science and Mathematics Faculty, Diponegoro University, Semarang – Indonesia

Abstract. Information security is a problem effected business process of an organization, so it needs special concern. Information security assessment which is good and has international standard is done using Information Security Management System (ISMS) ISO/IEC 27001:2013. In this research, the high level assessment has been done using ISO/IEC 27001:2013 to observe the strength of information secuity in Ministry of Internal Affairs. The research explains about the assessment of information security management which is built using PHP. The input data use primary and secondary data which passed observation. The process gets maturity using the assessment of ISO/IEC 27001:2013. GAP Analysis observes the condition now a days and then to get recommendation and road map. The result of this research gets all of the information security process which has not been already good enough in Ministry of Internal Affairs, gives recommendation and road map to improve part of all information system being running. It indicates that ISO/IEC 27001:2013 is good used to rate maturity of information security management. As the next analyzation, this research use Clause and Annex in ISO/IEC 27001:2013 which is suitable with condition of Data Center and Data Recovery Center, so it gets optimum result and solving problem of the weakness information security.

1 Preliminary

Management is the structure between connectivity and process which is directing and controlling the organization to achieve the goal by giving added value from utilization of information technology to balance and to compare between the risk and the result which is given by information technology and its process [1].

Application of IT management must suitable with business needs and the result which is usually happened in the risk management improvement, communication, and connectivity between business and IT related. Application of IT management is important thing of organization or company which gives direction and control to ensure the significant infestation bringing added value. IT resources become reference of responsibility and mitigation process risk [2].

ISO 27001:2013 is part of management system in an organization based on business risk approach that purposed to build, to implement, to operate, to observe, to maintain and to improve information security. Application of ISO/IEC 27001 allows the organization or company comparing the competition and giving relevant information about IT security [3].

ISO/IEC 27001 known as ISMS (Information Security Management System) is the standard which identifies technical security, but as reality it is a management system, it is not technical standard of information security. There are many organizations or companies assume ISMS implementation is needed by who run in data security only, so it cause lack of awareness for data security in all sectors. To protect information is not only adding device or application, but also all of the process in using application and device must be check, that is why ISMS implementation must be applied in organizations or companies (Broderick, 2006).

ISO 27001:2013 is standard which provides information security management system, the standard is used in the world by organization (commercials and government) as policy management and implementation of information security of organization. ISO/IEC 27001:2013 is designed flexible for stakeholders, internally and externally as the first standing, so the framework can be more focus [4].

* Corresponding author: ari.kurnianto07@gmail.com

The structure of ISO 27001:2013 use management principle which is called as "Deming Cycle" or more known as PDCA (Plan – Do – Check – Action) which is shown at the picture 2.7 [5,6].

2 Methods

2.1 Assessment

In data processing, the result of IT management questionnaire has connectivity to main clause process and annex. The respondents of analysis process are the stakeholders of DC and DRC in population administration information management directorate. The questionnaire calculation procedure is shown below:

1. Every assessment level has attribute processes. Attribute process consists of some criteria which must be fulfilled based on attribute process requirements.
2. Every criteria has assessment score from 1 to 5. The score presented achievement level which is achieved each criteria
3. The next step is accumulation of all the scores have been achieved in each level.
4. The average value must be found by Average the accumulation score.
5. The last step is the average value is divided by the biggest value, and then it is multiplied with 100. The result is categorized suitable with the rules: ISO 270001:2013 requirement haven't been established (not achieved, 0%), ISO 270001:2013 requirement have been established practically (practically achieved, 25%), ISO 270001:2013 requirement have been established overall (largely achieved 50%), ISO 270001:2013 requirement have been established implemented (implemented achieved, 75%) and ISO 270001:2013 requirement have been evaluated effectively (effectively achieved, 100%) [7].

To support IT management audit, the data from questionnaire will be processed and must be done:

1. Calculation of average to each process attribute of all respondent's answer.
2. The assessment is gotten by doing the calculation of all attribute average or processes
3. Existing condition of representation information technology.

The size in the model covers ordinal size and nominal size. The ordinal size is the number which means as level. Nominal size is the number which is used to sort the object from the lowest level to the highest level. The size is not giving absolute value to the object, but it is only giving sort of the level from the lowest to the highest. The next process is to implement between value and absolute value which is done by calculating to be index form using the formula which is shown below :

$$Indeks = \frac{\sum Answer\ the\ questionnaire}{\sum Questionnaire\ question} x\ 100\% \quad (1)$$

2.2 Analysis

Information security management system (ISMS) has plan model to assess, the steps to asses is shown at the Figure 1.

Fig. 1. Assessment process ISMS

Determination of assessment activities determine coverage of the thing will be assessed which covers:

1. Obedience to the vision and mission of the organization

2. All of the asset form are the technology form, process, system, service, application, network, and software.
3. The risk assessment covers asset identification, asset definition, asset classification, asset priority, asset determination, control determination, and monitoring.

Gap analysis compares existing condition, whether ISO/IEC 27001:2013 controls have been done well, including policy, procedures, work instruction, till to the documentation. Before doing analysis, first of all the assessment must be done with interview and observation using obedience list assessment direction which is suitable with ISO/IEC 27001:2013.

Table 1. Sensitivity analysis

	Sensitivity analysis	**Assessment point**		
		high (3)	**Medium (2)**	**Low (1)**
confidentiality	How big disadvantage will rise if the disturbance is happened?	if disadvantage is very big and disturb IT services which cause very confidential information leak.	If disadvantage is not significant	If disadvantage is small
Accuracy	How big disadvantage will rise when the disturbance is happened to IT service which is caused by miss using of the data?	if disadvantage is very big and disturb IT services which cause very confidential	If disadvantage is not significant	If disadvantage is small
Availability	How big disadvantage will rise when the disturbance is happened to IT service which is caused by unavailability?	if disadvantage is very big and disturb IT services which cause very confidential	If disadvantage is not significant	If disadvantage is small

The next step by doing risk assessment plan, this step consists of some risk assessment steps, such as:

1. Doing asset identification
   Asset in sub directorate of DC and DRC management in the ministry of internal affairs.
   a. Asset : OceanStor 18000 Series-Storage Rack Exp(4U)
      Classification : hardware
      Sub classification : Storage
      Serial Number : 210211431510FB000002
      Description : -
      Sub location : B1

   b. Asset : SAS Disk Unit(3.5")
      Classification : Hardware
      Sub classification : Hardisk
      Serial Number : 210235875110FB000230
      Description : 600GB 15K RPM
      Sub location : B1

   c. Asset : Disk Enclosure
      Classification : Hardware
      Sub classification : Hardisk area
      Serial Number : 210235877910FB000009
      Description : (4U,3.5",AC,DAE12435U4)
      Sub location : B1

   d. Asset : Disk Enclosure
      Classification : Hardware
      Sub classification : Hardisk area
      Serial Number : 210235877910F9000066
      Description : (4U,3.5",AC,DAE12435U4)
      Sub location : B1

2. Risk determination

   Risk determination of asset refers to confidentiality level, accuracy, and availability. Criteria of critical asset assessment is shown in table 2.

3. Determination of risk priority
   Risk assessment is done using possibility assessment criteria and, effect, and risk value calculation as shown in the table 2.

Table 2. Likelihood scale

Threats	(T1) threat of natutal disaster	(T2) threat of operational disturbance in weakness information security	(T3) threat of hardware	(T4) threat of software
Very seldom	>3 years probability <20%	Very minimum, more than 1 years	Very minimum, more than 1 years	Very minimum, more than 1 years
seldom	between 1- 3 years or probability 21% - 40%	Average is once a year	Average is once a year	Average is once a year
Can be	between 1- 3 years or probability 41% - 60%	Average is once per 6 month	Average is once per 6 month	Average is once per 6 month
often	In 1 years perhaps with probability 61% - 80 %	Average is once per 3 month	Average is once per 3 month	Average is once per 3 month
Very often	Very possible happened in 1 or probability 81%- 100%	Average is once a month	Average is once a month	Average is once a month

Table 3. Impact scale

	Effects	
	(E1) disadvantage of yearly income effect	(E2) reputation effect and law
Not significant	Don't give effect	Don't give effect
low	< (less than) 0.1 % or profit before bank interest, tax,depreciation and amortitsation <0.1%	Complain via communication media (telepon/email/ fax)
medium	between 0,1 % - 1% or disadvantage before bank interest, tax, depreciation, and amortization 0.1 - 1%	Bad reputation PIAK in user perspective, specially electronic identity.
high	between 1 % - 10% or disadvantage before bank interest, tax, depreciation, and amortization 1 - 10%	Bad publication in local media (news,news papper)
critical	More than 10% or disadvantage before bank interest, tax,depreciation, and amortisation >10% or income lost >2%	Bad publication in national media (news,news papper) or law inforcement

The formula of risk calculation criteria shown below:

$$Risk\ value = K \times D \qquad (2)$$

Risk value = the result of risk value
K = tendency or possibility
D = effect

Risk value criteria
1. Value 0-5 = low (L)
2. Value 6-15 = medium (M)
3. Value 16-25 = high (H)

The matrix which is used to see the risk value is divided into 3 level; high, medium, and low. The matrix is shown in the Table 4.

Table 4. Risk matrix

		impact				
		1	2	3	4	5
likelihood	1	L	L	L	L	L
	2	L	L	M	M	M
	3	L	M	M	M	M
	4	L	M	M	H	H
	5	L	M	H	H	H

4. Control development
   In the control development step, this step identify existing controls in information security assets, and it is developed to get control which will decrease the incident will be appear.

5. Making road map
   After getting risks which should be fixed, the next step is choosing control that will be used to decrease the risk. When determination of control, it is gotten vulnerability, threat, and effect.

Table 5. Determine of controls

Vulnerability	Control of ISO 27001:2013	Work planning	Code
Disobedience to all of information security process	Management Direction for Information Security	Making rule of implementation and period implementation review to information security policy.	T2
Vulnerability in new employee recruitment	Prior to Employment	Making procedure for new employee recruitment	T2
Minimum awareness of stakeholders and other related people	During Employment	Determine and define responsibility of awareness to information security	T2
Vulnerability of loosing asset	Responsibility for Asset Management	Making procedure for asset borrowing and pointed person in charge (PIC)	T2, T3
Vulnerability of malware attack to DC	Protection from malware	Making procedures to handle, and prevent from virus or malware.	T1, T2, T4, E1, E2
Vulnerability of disaster to DC	Backup	Making back up mechanism to related document and back up testing.	T1, T4, E1, E2

3 Result and Discussion

Data processing in application of information security management system based on ISO/IEC 27001:2013, its assessment process depend on the result of observation and evidence which is saved by the organization. Its process is shown at the Figure 2.

After assessment process has been finished, the result of main clause is gotten and the control has been determined shown at the Figure 3.

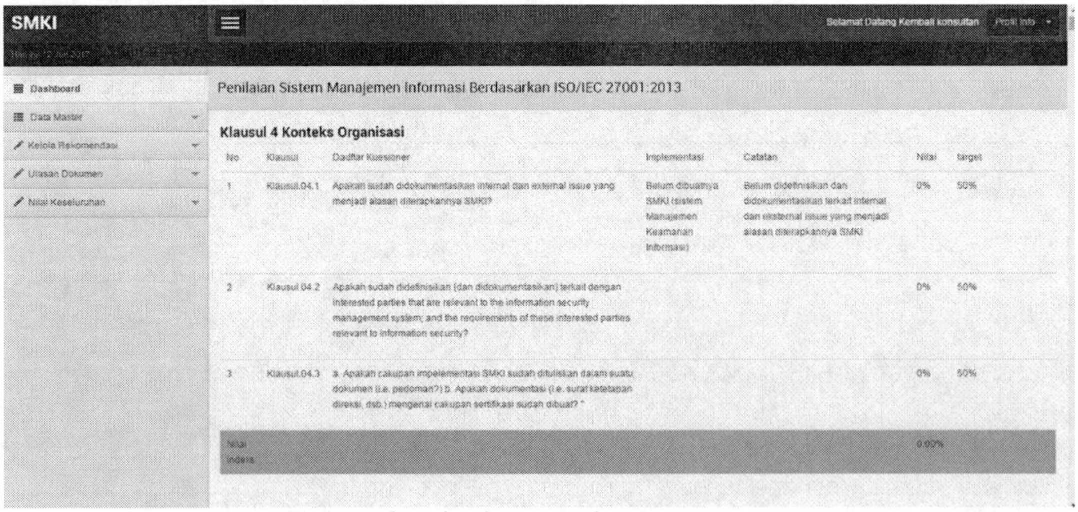

Fig. 2. Results review document and assessment

https://doi.org/10.1051/e3sconf/20183111013

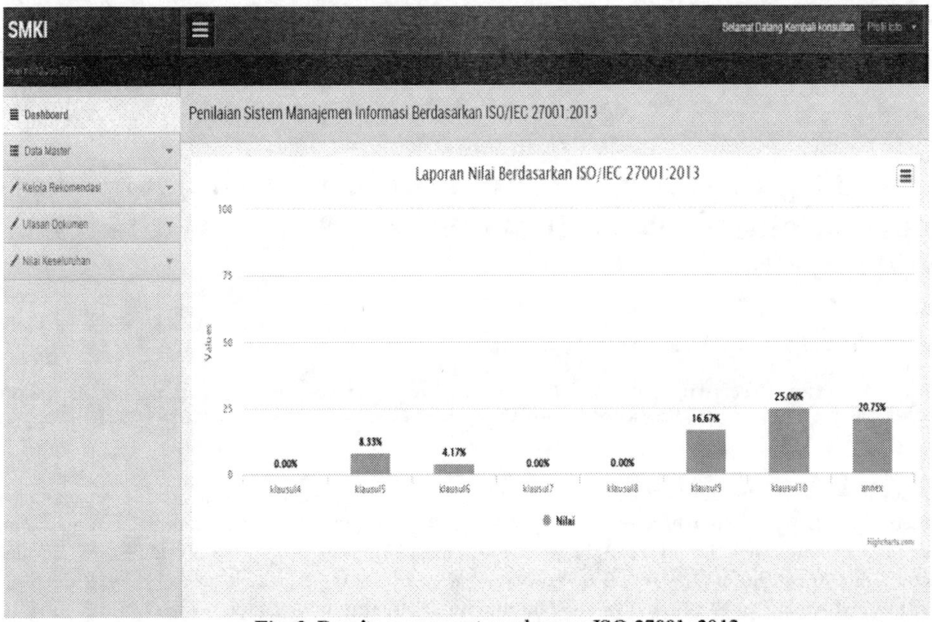

Fig. 3. Results assessment graph score ISO 27001: 2013

4 Conclusions

Assessment of information security management system using ISO 27001:2013 in data center and data recovery center become optimal by choosing control. It is proved with some main clause process and control which is located in sub directorate of management data center and data recovery center still has lack of information security and it must be improved periodically.

References

1. Andriaole, S.. IGTT Governance in the Trenhes. **57-62** (2016)
2. E. Bilbao, A. Bilbao , K. Pecina, and R. Estremera, Physical and Logical Security Management Organization Model Based On ISO 31000 and ISO 27001. IEEE, **1-5**, (2011).
3. E. Bilbao, A. Bilbao , K. Pecina, Physical Logical Security Risk Analysis Model. IEEE, **1-7**, (2011).
4. GEIT, Global Status Report on the Governance of Enterprise It. USA: ISACA, (2011).
5. M. Gerber, and Solms, R. v. Information Security Requirements - Interprenting at Legal Aspects. Computers & Security, **124-135**, (2008).
6. E. Humphreys, Information security management standards: Compliance,. Information Security technical Report , **247-255**, (2008).
7. C. Pelnekar, Planning for and Implementing ISO 27001. Journal ISACA, **1-8**, (2011).

Understanding customers' intention to use social network sites as complaint channel: an analysis of young customers' perspectives

Retno Agus Setiawan[1,*], *Djoko* Budiyanto Setyohadi[2], *Pranowo*[2]

[1]Master Program of Information Technology, School of Postgraduate Studies, Universitas Atma Jaya Yogyakarta, Yogyakarta - Indonesia
[2] Department of Information Technology, Faculty of Industrial Engineering, Universitas Atma Jaya Yogyakarta, Yogyakarta - Indonesia

Abstract. Social network sites (SNSs) have grown rapidly in recent years. More and more companies have used SNSs as part of their business strategy. SNSs offer numerous advantages, especially in enhancing communication. SNSs have a potential as a new complaint channel for young customers to file their complaints to companies. The objective of this study is to investigate the acceptance of SNSs as complaint channel based on TAM. A structured questionnaire was distributed to young participants, which collected 222 valid questionnaires. Furthermore, structural equation modeling was utilized to investigate the structural model. The results revealed that perceived ease of use and perceived usefulness have a positive correlation on the attitude towards SNSs. While the attitude plays an important role in understanding customers' intention to use SNSs to voice complaints. However perceived usefulness has no significant impact on intention to use. Limitations and further research were also discussed.

1 Introduction

Technology acceptance model (TAM) suggested by Davis [1], is an effort to explain the adoption process of any technology within organization. TAM has been extensively utilized to explain the determinants influencing users' acceptance of the technology [2]. TAM suggests that users' perceptions about usefulness and ease of use are two key elements that determine the adoption of various technologies. Conforming to Davis et al [3], the attitudes toward using technology can be described by perceived ease of use and usefulness, while attitudes will influence behavioral intention and usage behavior.

TAM is an essential model for understanding an individuals' attitude and trustworthy of use technology through the individuals' beliefs about a technology is useful and effortless [4]. When users are introduced with a new technology, perceived ease of use and usefulness are two significant determinants that affect their decision to use it. TAM has been examined with kind of technologies and populations and has been confirmed to be an important and effective model in predicting the acceptance of technologies [5].

In recent years, technology has grown rapidly, particularly social network sites (SNSs). The use of SNSs has increased and SNSs have grown very popular among internet users. As of January 2016, more than 79 million people in Indonesia use SNSs, accounting for 89% of its internet population [6]. In addition, Facebook has almost 12 million active users, Google+ has 9 million active users, and more than 170 minutes are spent on SNSs per day. There is no doubt that SNSs offer many advantages for its users. SNSs provide a huge amount of data [7]. SNSs may allow quick access to a huge audience and almost no cost [8]. Furthermore, SNSs also provide a broad way to communicate with anyone anywhere with no geographical limitations.

More and more companies have used SNSs as part of business strategy. In order to obtain the advantages of SNSs, companies need to understand the nature of SNSs and how it is used [5]. Generally, TAM is frequently utilized to understand the use of new technology. Few studies have successfully used TAM to explain the factors that affect the acceptance of SNSs [5][9][10]. Furthermore, in light of business tools, SNSs can be used as tools for customer engagement. According to Garding & Bruns [11], SNSs have the potential as a new complaint channel to accommodate any complaints voiced by customers. Customers may use SNSs to file their complaints to companies.

Mattila & Wirtz [12] defined complaint channel as the tool that allows customers to address their complaints about the products or services related to the company. Complaints are derived from customers provide an overview to analyze about issues and make the right decisions to make sure that issues do not occur again [13]. The adequacy of complaint channels is essential towards realizing a successful complaint management [14]. However, it should be noted that these channels are not all suitable from a customer's point of view. Thus, companies should think deeply in determining communication channels.

* Corresponding author: retnoagussetiawan@gmail.com

Mattila & Wirtz [12] examined the possibility of communication channels for customer to file complaints. They examined the traditional communication channels telephone, face to face, mail, and email. Generally, these communication channels are suitable for making a complaint [15]. Mattila & Wirtz [12] revealed that the choice of communication channels depends on customers' motivation when voicing a complaint. However, not all customer groups are willing to voice their complaints to companies. When customer groups "young non-complainers" decide to complain, they would choose SNSs to voice their complaints to companies [16]. SNSs can be implemented as complaint channel to motivate younger customer to voice their complaints.

Based on the above brief background of the problem, this study examined the acceptance of social network sites as complaint channel. This study aims to examine whether TAM can predict the use of SNSs in the context of complaint channel, especially from a younger customers' perspectives. Several studies have found that TAM is useful to explain SNSs adoption [5][9]. Therefore, this study used the basic framework of TAM to explain the acceptance of SNSs as complaint channel.

2 Literature review

2.1 Complaint management

In business activities, not all business deals lead to customer satisfaction [14]. Dissatisfaction may come with some reasons. For example, there is a difference between customers' expectations towards a product or service. When this happens, customers may voice their complaints to companies. Customer complaints provide an occasion to analyze about issues and take possible reaction to resolve service failures [13]. Companies can collect information from all complainants into account as sources of meaningful knowledge that can be used to improve their performance. A good quality of complaint handling can generate satisfied and loyal customers. Furthermore, a successful complaint management can provide myriad of benefits for companies.

The accessibility of complaint channels is a key determinant for realizing a successful complaint management. Mattila & Wirtz [12] defined complaint channel as the medium that customers use to file a complaint to companies. Companies also use these channel to respond to customer complaints. There are several popular communication channels like telephone, face to face, mail, and email, which commonly used for making a complaint. In addition, SNSs emerged as a modern communication channel which increasingly important [17]. SNSs may have capability as new complaint channel based on their unique characteristics.

Finally, complaint management is necessary for company's strategies. A successful complaint management may help companies to keep their market by maintain customer satisfaction. Furthermore, the success of complaint handling provides customer satisfaction and prevents negative behavior from customers [12]. A good complaint management can help to enhance the relation between companies and customers.

2.2 Technology acceptance model and social network sites

Introduced by Davis, technology acceptance model (TAM) was adapted from theory of planned behavior (TPB) by Ajzen [18] and theory of reasoned (TRA) by Fishbein & Ajzen [19]. TAM is a model that developed to predict and analyze the adoption of technology. Within TAM there are two main factors, perceived ease of use and perceived usefulness, which influence the adoption of technology. Perceived ease of use is defined as "the extent to which individuals believe that using a specific system will be effortless". Meanwhile, perceived usefulness is defined as "the extent to which individuals believe that using a specific system will enhance their work performance" [1]. Furthermore, Davis et al [3] argued that both variables may influence user attitude towards using a technology and then affect users' intention to use. TAM has grown into a major model in explaining the predictions of individuals' behavior towards the adoption or rejection of technology. The effectiveness of this model is verified by various researches that apply it extensively to numerous technologies [2].

In general, TAM has been commonly utilized and consistently explains the adoption of various technologies. Several researches have effectively utilized TAM to evaluate the adoption of numerous technologies, including Tablet PCs [20], Cloud Computing [21], Instant messaging [22], and Social network sites [5][10] [23][24][25]. Willis [5] investigated whether TAM is feasible to be utilized to explain the acceptance of SNSs technology which has a relationship-oriented. The results indicated that TAM deserves to be utilized to explain the acceptance of SNS technology. Furthermore, this results have been confirmed by Constantinides et al [9], which revealed that TAM can explain and predict the usage of SNSs.

Willis [5] applied TAM to explain the determinants influencing the acceptance of SNSs by presenting a new model that contains the influences of past experience. The results indicated that prior experience significantly correlated to intention to use, perceived ease of use, and perceived usefulness. However, the results revealed the negative effects of prior experience on subjective norm.

Pinho & Soares [23] identified the determinants that affecting the adoption of SNSs among students. The results demonstrated that perceived ease of use and usefulness positively influence individuals' intention to use SNSs. They believe that SNSs is easy to use, they quickly become expert at using SNSs. In addition, the attitude toward using SNSs significantly correlated to individuals' intention to use, as respondents who believe that SNSs are quite fun to use are willing to use these technologies. However, the results do not support the influences of perceived usefulness on intention to use.

Due to the fact that businesses are continuously interested in SNSs as a source of valuable data, Constantinides et al [9] identified the determinants that affecting the acceptance and use of SNSs as a business tool in marketing strategy. They revealed that ease of use significantly affects perceived usefulness of SNSs, both variables have an indirect influence on individuals' intention through the attitude and a direct and significant influence on intention to use. In addition, intention to use has a positive correlation on the level of use SNSs.

Chiu and Huang [24] attempted to identify the determinants that may influence individuals participation in SNSs. The results indicated that the attitudes towards SNSs have a positive influence on individual intention. Perceived ease of use and usefulness significantly affect behavior and attitude. In addition, it was shown that there is a significant correlation among use behavior and behavioral intention. However, the results noted that perceived usefulness has no significant influence on individual intention.

Yeh et al [10] applied TAM to investigate the adoption of SNSs in the context of human resource, with a particular focus on staffing activities within the organization. Subjective norm has been added to the research model with a belief that SNSs have a social aspect that may affect the adoption of these technologies. The results demonstrated that subjective norm, perceived ease of use and usefulness have a positive correlation on human resource practitioners' intentions to use SNSs in organizational staffing activities.

Furthermore, Yoon et al [22] identified online communication technology that have similar characteristics as SNSs. They identified the influence of social aspects on the adoption of new technologies. In addition to social aspects, the results indicated that perceived usefulness positively correlated to intention to use. However, the results additionally indicated that perceived ease of use has no significant correlation on perceived usefulness.

3 Research model and hypotheses

In this study, technology acceptance model was utilized as the major model to explain the adoption of SNSs in the context of complaint channel. The research model contains perceived ease of use, perceived usefulness, attitude, and intention to use.

3.1 Perceived usefulness, perceived ease of use, and attitude

Davis et al. [3] argue that individuals' motivation can be interpreted by three factors; attitude, perceived usefulness, and perceived ease of use. Perceived usefulness and perceived ease of use are key determinants affecting attitude, where perceived ease of use has a direct influence on perceived usefulness. In the context of SNSs, prior study has confirmed these relationships [9]. They revealed that ease of use and usefulness significantly correlated with user perceptions and attitude toward SNSs. Thus, this study suggests the following hypotheses:

H1: Perceived usefulness of SNSs as complaint channel has a positive and significant influence on the attitude toward using them.

H2: Perceived ease of use of SNSs as complaint channel has a positive and significant influence on the attitude toward using them

H3: Perceived ease of use of SNSs as complaint channel has a positive and significant influence on perceived usefulness of using them

3.2 Attitude and intention to use

The correlation among attitude and intention to use is essential and popular in behavioral models. Individuals' attitudes towards technology have a significant correlation on intention to use [9]. This correlation has been confirmed by several studies. They found a strong influence of attitude on individuals' intention to use SNSs [25]. Moreover, the following hypothesis is suggested:

H4: Attitude toward using SNSs as complaint channel has a positive and significant influence on intention to use them.

3.3 Perceived usefulness and intention to use

The findings have consistently shown that perceived usefulness is the key determinant that affect users' intention to use technology [2]. Previous researches have revealed that there is a significant correlation among perceived usefulness and intention to use [5] [9]. Furthermore, Constantinides et al [9] revealed that perceived usefulness positively affected intention to use. Then, the following hypothesis is proposed in this study:

H5: Perceived usefulness of SNSs as complaint channel has a positive and significant influence on intention to use them

4 Research methodology

This study used SPSS to analyze the data and applied structural equation modeling to investigate the hypothesized correlations in the research model using AMOS.

4.1 Sample and data collection

To achieve the goal of this study, a survey method was applied to gather data from participants. This study selected participants who have an active SNSs account. The target participants for this study were young consumers aged 18 and 30 years old. At first, this study organized a pre-test survey on 30 SNSs users to test the consistency of measurement items. In addition, this pre-test was conducted to ensure unambiguousness of each question and to calculate the time participants required to complete the questionnaire. Furthermore the questionnaire was distributed in two ways, paper and electronic questionnaire. 79 paper questionnaires and 196 electronic questionnaires were distributed. Overall, 275 questionnaires were collected. However, only 222 were included in the sample for analysis, 53 responses were deleted because of incompleteness and contained the same values for all questions. The profiles of participants are listed in Table 1.

Table 1. Demographic profile of participants (N=222)

	Frequency	Percentage
Gender		
Male	121	54.5
Female	101	45.5
Age		
18 – 24	181	81.5
25 – 30	41	18.5
Educational level		
Senior high school	92	41.4
3-year diploma	14	6.3
Bachelor	113	50.9
Postgraduate	3	1.4
Occupation		
Students	129	58.1
Private employees	50	22.5
Public servants	4	1.8
Others	39	17.6
Subscription period of the SNS		
< 6 months	1	0.5
6 months – 1 year	4	1.8
1 – 3 years	20	9.0
More than 3 years	197	88.7
Frequency of using SNS		
Few/barely	22	9.9
Frequently	64	28.8
Every day	136	61.3

4.2 Measurement items

This study adapted questionnaire items from prior researches and were all checked by 5-points Likert scale from strongly disagree (1) to strongly agree (5). Items of perceived usefulness were adapted from Davis [1], Constantinides et al [9], Kwon et al [25], Hung et al [26], Seol et al [27], Ernst [28], and Mouakket [29]. Items of perceived ease of use were adapted from Davis [1], Willis [5], Constantinides et al [9], Hung et al [26], Ernst [28], and Hsu et al [30]. Items of attitude were adapted from Constantinides et al [9], Kwon et al [25], Hung et al [26], and Hsu et al [30]. While items of intention to use were adapted from studies organized by Constantinides et al [9], Kwon et al [25], Hung et al [26], and Balakrishnan [31].

5 Data analysis and results

5.1 Measurement model

This study applied a confirmatory factor analysis (CFA) to examine the measurement model. AMOS software was utilized to perform the analysis. The results indicated that several model fit indices did not meet the standard basically. Thus, the model was optimized and several responses were deleted according to modification indices. As a result, all fit indices have better values. The *p-value* was 0.088 ($\chi^2 = 58.325$, df = 45), which greater than 0.05, indicating a good fit. The normed chi-square (χ^2/df) was 1.296, while the normed chi-square value should be less than 5 [32]. The GFI and AGFI were 0.941 and 0.898, respectively, showing a good fit [33]. Then, the RMSEA was 0.044, which is below the maximum value of 0.08 [34]. The other fit indices NFI, TLI, and CFI were 0.960, 0.990, and 0.986, respectively, all exceed the suggested value of 0.9 [35][36]. Therefore, the overall fit indices indicate an acceptable model fit.

Furthermore, this study investigated the reliability and validity of the measurement model. All the average variance extracted (AVE) and factor loadings must be exceed 0.5 [37]. As shown in Table 2, most loadings loaded highly above 0.70. The value of AVE for various construct is larger than 0.50. Then, the composite reliability (CR) must be larger than the AVE. The CR values for each construct varied between 0.802 and 0.917, which above the recommended value of 0.70 [37]. These results indicate good validity of all measurement items. Furthermore, the cronbach's alpha for various construct exceeded Nunnaly's [38] recommendation of 0.70, indicating internal consistency and good reliability.

Table 2. Factor loadings, AVE, CR, and Alpha values

Items	Factor loadings	AVE	CR	Alpha
Perceived usefulness				
PU1	0.81	0.79	0.92	0.91
PU2	0.96			
PU3	0.89			
Perceived ease of use				
PEOU2	0.75	0.67	0.86	0.88
PEOU3	0.80			
PEOU4	0.89			
Attitude				
AT1	0.67	0.68	0.86	0.84
AT2	0.86			
AT3	0.93			
Intention to use				
IU1	0.88	0.58	0.80	0.85
IU2	0.70			
IU3	0.69			

Source: Author calculations

Table 3. Structural model results

Fit indices	Measurement model	Structural model	Recommended values
Chi square/degrees freedom (χ^2/df)	1.296	1.270	<5.00 Hair et al [32]
Goodness-of-fit index (GFI)	0.941	0.941	>0.90 Baumgartner & Homburg [33]
Adjusted goodness-of-fit (AGFI)	0.898	0.900	>0.80 Baumgartner & Homburg [33]
Root mean square error of approximation (RMSEA)	0.044	0.042	<0.08 Hair et al [34]
Normalized fit index (NFI)	0.960	0.960	>0.90 Hu & Bentler [36]
Comparative fit index (CFI)	0.990	0.991	>0.90 Hu & Bentler [36]
Tucker Lewis index (TLI)	0.986	0.987	>0.90 Bagozzi & Yi [35]

Source: Author calculations

5.2 Structural model

This study also conducted a structural equation modeling to explain the hypothesized correlations among various constructs. As Table 3 shown, the overall fit indices for the research model suggest a very good model fit (*p-value*=0.104, χ^2/df=1.270, GFI=0.941, AGFI=0.900, RMSEA=0.042, NFI=0.960, CFI=0.991, and TLI=0.987).

Since the research model demonstrated a good fitness, this study examined the causal relationships between constructs. As Fig. 1 shown, the results indicate that perceived usefulness (β=0.260, *t*-value=1.972, p<0.05) and perceived ease of use (β=0.515, *t*-value=3.685, p<0.001) have a positive related to the attitude, confirming hypothesis H1 and H2. Then, perceived ease of use has a significant influence on perceived usefulness (β=0.810, *t*-value=10.148, p<0.001), supporting hypothesis H3. The results also found support for hypothesis H4, the attitude toward using SNSs has a strong correlation with intention to use (β=0.960, *t*-value=8.460, p<0.001). However, the perceived usefulness of SNSs is found to have no significant impact on intention to use (β=0.34, *t*-value=0.473, p<0.05). Thus, hypothesis H5 is not supported. In addition, based on squared multiple correlations (R^2), perceived ease of use explains about 65.6% of variance in perceived usefulness. Perceived usefulness and ease of use together explain 54.9% of variance in the attitude toward SNSs. Meanwhile, perceived usefulness and attitude jointly explain 96.7% variance in intention to use.

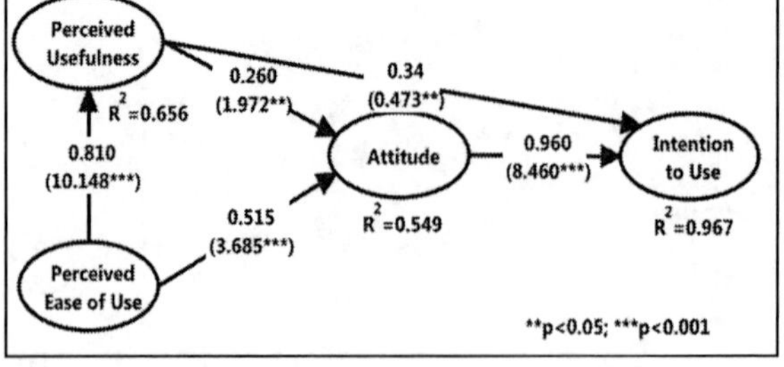

Fig. 1. Structural model results.

6 Discussions and conclusions

The use of SNSs has grown rapidly and very popular among young users. In addition, SNSs have been widely used by companies because of their clear benefits from a business point of view. SNSs offer two-way communication among the companies and customers. Customers may file their complaints to the company through SNSs quickly. Moreover, companies can easily gain feedback from customers. These possibilities have led to interest in evaluating the adoption of SNSs as complaint channel, particularly for young customers. Furthermore, this study confirms the explanatory power of technology acceptance model in the adoption of SNSs.

As listed in Table 4, most hypotheses were supported except the relationship among perceived usefulness and intention to use. The results indicate that ease of use positively influences perceived usefulness and both variables have an indirect influence on intention to use through the attitude toward SNSs. However, perceived usefulness of SNSs as complaint channel has no significant influence on intention to use them.

Table 4. Hypothesis testing results

Hypothesis	Std. Coefficient	t-value	Supported
H1: PU → AT	0.260	1.972**	Yes
H2: PEOU → AT	0.515	3.685***	Yes
H3: PEOU → PU	0.810	10.148***	Yes
H4: AT → IU	0.960	8.460***	Yes
H5: PU → IU	0.340	0.473**	No

Source: Author calculations
Notes: **p<0.05; ***p<0.001
Perceived usefulness (PU); Perceived ease of use (PEOU); Attitude (AT); Intention to use (IU).

The results indicate that perceived ease of use and perceived usefulness significantly and positively correlated with attitude, which is consistent with prior studies [9][10][23]. They revealed that perceived usefulness and ease of use positively correlated with attitude. Furthermore, perceived ease of use (β=0.515) has a greater effect on attitude than perceived usefulness (β=0.260). Similar to a prior study by Constantinides et al [9], the results show that perceived ease of use of SNSs has a direct impact on perceived usefulness of using them.

Perceived ease of use reflects the consumers believe that contacting the company will be effortless. They have no difficulty completing a complaint procedure. Furthermore, some participants believe that using SNSs will enhance their effectiveness in voicing their complaints to companies. SNSs allow the consumers to file their complaints quickly. The results suggest that maintaining easy and useful policies and procedures are essential prerequisites for delivering positive attitudes.

In light of attitudes, the attitude toward SNSs as complaint channel is a major determinant in understanding customers' intention to use them, as shown by the huge value of path coefficient for itself. This result is consistent with prior studies, which verified the role of attitude in user adoption [9][23][24]. Furthermore, Balakrishnan [31] stated that user who feels positive in using the tools for working may literally use them for working. Similarly, consumers who feel positively about using SNSs as complaint channel may actually take the initiative to use them for delivering complaints to companies.

However, it should be noted that individual attitudes may change immediately. Therefore, continuous supports should be granted to preserve the attitude. User-friendly, low-time consumption *(the time consumers needed to complete a complaint procedure*), and fast response should be given to maintain the positive attitudes. Furthermore, in the initial stage, complaints that are delivered through SNSs must be managed properly in order to eliminate prejudgments. In addition, companies must have the ability to adapt the procedures to fit the consumer needs. In the end, the results reject the hypothesis perceived usefulness significantly correlated to intention to use. The rejection of this correlation is consistent with prior study by Pinho & Soares [23], perceived usefulness does not have positive correlation with intention to use.

This study, however, has some limitations. First, this study was established in single country, Indonesia. The majority of the participants were from special region of Yogyakarta, where it is known as a centre of education and culture. Accordingly, the results of this study should be generalized carefully. Second, the present study focuses only on young adults who are member of social network sites as a target population. Participants in this age segment may be more active in using social network sites. Third, the influence of demographic factors such as gender, and occupation were not considered in this study. Thus, these factors should be examined in further research. Lastly, this study not focused on particular social network sites. Future studies could examine the particular SNSs such as Twitter and Facebook as complaint channel.

References

1. F.D. Davis, MIS Q., **13**, 319–340 (1989).
2. N. Marangunic and A. Granic, Univers. Access Inf. Soc., **14**, 81–95 (2015).
3. F.D. Davis, R.P. Bagozzi, and P.R. Warshaw, Manage. Sci., **35**, 982–1003 (1989).
4. Q. Zhao, C.-D. Chen, and J.-L. Wang, Telemat. Informatics, **33**, 959–972 (2016).
5. T.J. Willis, *An Evaluation of the Technology Acceptance Model as a Means of Understanding Online Social Networking Behavior*, in: Dr. Diss., University of South Florida, (2008).
6. S. Kemp, Digital in (2016), http://wearesocial.com/sg/special-reports/digital-2016.
7. A.J. Wicaksono, Suyoto, and Pranowo, *A proposed method for predicting US presidential election by analyzing sentiment in social media,* in: 2016 2nd Int. Conf. Sci. Inf. Technol., pp. 276–280, (2016),
8. P. Karampelas, *Techniques and Tools for Designing an Online Social Network Platform,* in: Springer Vienna, Vienna, pp. 3–13, (2013)
9. E. Constantinides, C. Lorenzo-Romero, and M.-C. Alarcón-del-Amo, *Social Networking Sites as Business Tool: A Study of User Behavior, in: Ed. by M. Glykas, Bus. Process Manag.* Theory Appl., Springer Berlin Heidelberg, Berlin, Heidelberg, pp. 221–240, (2013)
10. C.-C.R. Yeh, K.C. Gossmann, and Y.-H. Tao, *Acceptance of Online Social Networks as an HR Staffing Tool: Result from a Multi-country Sample, in: Ed. by L. Wang, S. Uesugi, I.-H. Ting, K. Okuhara, and K. Wang, Multidiscip. Soc. Networks* Res. Second Int. Conf. MISNC 2015, Matsuyama, Japan, Sept. 1-3, 2015. Proc., Springer Berlin Heidelberg, Berlin, Heidelberg, pp. 70–79, (2015)
11. S. Garding and A. Bruns, *Moving Towards Successful Complaint Management, in:* Complain. Manag. Channel Choice An Anal. Cust. Perceptions, Springer International Publishing, Cham, pp. 13–26, (2015)
12. A.S. Mattila and J. Wirtz, J. Serv. Mark., **18**, 147–155 (2004).
13. Y. Namkung, S. (Shawn) Jang, and S.K. Choi, Int. J. Hosp. Manag., **30**, 495–502 (2011).
14. S. Garding and A. Bruns, *Organisational Complaint Management, in: Complain. Manag. Channel Choice An Anal. Cust.* Perceptions, Springer International Publishing, Cham, pp. 1–12, (2015)
15. S. Garding and A. Bruns, *Analysis of Customers' Complaint Channel Choice and Complaint Behaviour, in: Complain. Manag. Channel Choice An Anal. Cust.* Perceptions, Springer International Publishing, Cham, pp. 35–74, (2015)
16. S. Garding and A. Bruns, *Conclusions for Organisational Complaint Management and Future Research, in: Complain. Manag. Channel Choice An Anal. Cust.* Perceptions, Springer International Publishing, Cham, pp. 75–82, (2015)
17. A.M. Kaplan and M. Haenlein, Bus. Horiz., **53**, 59–68 (2010).
18. I. Ajzen, Organ. Behav. Hum. Decis. Process., **50**, 179–211 (1991).

19. M. Fishbein and I. Ajzen, *An Introd. to Theory Res. Massachussets Addison-Wesley*, (1975).
20. E. Park and A.P. Del Pobil, Wirel. Pers. Commun., **73**, 1561–1572 (2013).
21. M. Eltayeb and M. Dawson, *Understanding User's Acceptance of Personal Cloud Computing: Using the Technology Acceptance Model*, in: Ed. by S. Latifi, Inf. Technol. New Gener. 13th Int. Conf. Inf. Technol., Springer International Publishing, Cham, pp. 3–12, (2016)
22. C. Yoon, C. Jeong, and E. Rolland, Inf. Technol. Manag., **16**, 139–151 (2015).
23. J.C.M.R. Pinho and A.M. Soares, J. Res. Interact. Mark., **5**, 116–129 (2011).
24. Y.-K. Chiu and C.-W. Huang, U*sing Behavior of Social Network Sites Based on Acceptance Model, in: Ed. by Y. Yang, M. Ma, and B. Liu, Inf. Comput.* Appl. 4th Int. Conf. ICICA 2013, Singapore, August 16-18, 2013 Revis. Sel. Pap. Part II, Springer Berlin Heidelberg, Berlin, Heidelberg, pp. 57–66, (2013)
25. S.J. Kwon, E. Park, and K.J. Kim, Soc. Sci. J., **51**, 534–544 (2014).
26. S.-Y. Hung, J.C.-A. Tsai, and S.-T. Chou, Inf. Manag., **53**, 698–716 (2016).
27. S. Seol, H. Lee, J. Yu, and H. Zo, *Inf. Manag.*, (2016).
28. H.C.-P. Ernst, *The Influence of Perceived Belonging on Social Network Site Usage BT - Factors Driving Social Network Site Usage,* in: Ed. by H.C.-P. Ernst, Springer Fachmedien Wiesbaden, Wiesbaden, pp. 29–44, (2015)\
29. S. Mouakket, Comput. Human Behav., **53**, 102–110 (2015).
30. C.-L. Hsu, C.-C. Yu, and C.-C. Wu, Inf. Syst. E-Bus. Manag., **12**, 139–163 (2014).
31. V. Balakrishnan, *Univers. Access Inf. Soc.*, 1–13 (2016).
32. J.F. Hair, R.E. Anderson, R.L. Tatham, and W.C. Black, *"Multivariate Data Analysis with Readings"*, 4th ed., Prentice-Hall, Englewood Cliffs (1995).
33. H. Baumgartner and C. Homburg, Int. J. Res. Mark., **13**, 139–161 (1996).
34. J.F. Hair, R.E. Anderson, R.L. Tatham, and W.C. Black, *"Multivariate Data Analysis"*, 5th ed., Prentice-Hall, Upper Saddle (1998).
35. R.P. Bagozzi and Y. Yi, J. Acad. Mark. Sci., **16**, 74–94 (1988).
36. L. Hu and P.M. Bentler, Struct. Equ. Model. A Multidiscip. J., **6**, 1–55 (1999).
37. J.F. Hair, W.C. Black, B.J. Babin, and R.E. Anderson, *"Multivariate Data Analysis"*, 7th ed., Prentice-Hall, Upper Saddle (2009).
38. J. Nunnally, *"Psychometric theory"*, 2nd ed., McGraw-Hill, New York, NY (1978).

https://doi.org/10.1051/e3sconf/20183111015

Inventory Control System by Using Vendor Managed Inventory (VMI)

Alzena Dona Sabila[1,*], *Mustafid* Mustafid[2], and *Suryono* Suryono[3]

[1]Master of Information System, School of Postgraduate Studies, Diponegoro University, Semarang – Indonesia 50242
[2]Department of Statistic, Faculty of Sains and Mathematic, Diponegoro University, Semarang – Indonesia 50275
[3]Department of Physic, Faculty of Sains and Mathematic, Diponegoro University, Semarang – Indonesia 50275

Abstract. The inventory control system has a strategic role for the business in managing inventory operations. Management of conventional inventory creates problems in the stock of goods that often runs into vacancies and excess goods at the retail level. This study aims to build inventory control system that can maintain the stability of goods availability at the retail level. The implementation of Vendor Managed Inventory (VMI) method on inventory control system provides transparency of sales data and inventory of goods at retailer level to supplier. Inventory control is performed by calculating safety stock and reorder point of goods based on sales data received by the system. Rule-based reasoning is provided on the system to facilitate the monitoring of inventory status information, thereby helping the process of inventory updates appropriately. Utilization of SMS technology is also considered as a medium of collecting sales data in real-time due to the ease of use. The results of this study indicate that inventory control using VMI ensures the availability of goods ± 70% and can reduce the accumulation of goods ± 30% at the retail level.

1 Introduction

Market changes demand complex business processes. Business complexity issued to answer questions about how businesses deal with daily changes, what business actions can predict and prepare businesses for change [1]. In line with technological developments with faster and cheaper information exchange, business processes especially for controlling inventory can grow rapidly.

The right inventory system can be used for tracking, recording, and providing information related to the product. The inventory system has an important role in monitoring the condition of the product [2]. One of the methods that governs inventory is the Vendor Managed Inventory (VMI). VMI grants full right to the suppliers to control inventory and manage amounts of products inventory for stores. VMI performs transparency of sales data and stock levels to the suppliers [3-4]. VMI offers benefits to the stores due to the high availability of the product and provides an opportunity for distributors to improve the efficiency of their product marketing, and is able to maintain the availability of fluctuating demand and demand uncertainty [5]. VMI responds quickly and efficiently in meeting customer needs as well as being a decisive factor for business survival and development of business prediction [6].

Scheduled inventory management is not based on the high sales of goods but is carried out with a fixed delivery period. This results in a shortage of inventory under high sales conditions and excess inventory under declining sales conditions at the retailer level. This researches one solution to solve business complexity to control inventory for distributors in the delivery of the appropriate goods. The inventory control system built can maintain the stability of goods availability that refers to the level of sales at the retailer level.

2 Vendor Managed Inventory (VMI)

Vendor Managed Inventory is an inventory and procurement approach, in which the supplier is responsible for managing and updating stock. This seems to contradict the principle of pull scheduling, since the previous process (production) deciding how much and when it will be sent to the store/retailer. In practice, the information reference lies with the store / retailer through sales information. Within the VMI, suppliers have to monitor sales and inventory, then information will be generated when the procurement is done [7].

The implementation of VMI in the supply chain simplifies the steps that occur when the purchase information made by the last consumer can be accessed directly by the distributor / supplier. Thus the feedback to determine the speed of the flow of goods can be known. The supply chain in the VMI is shown in Fig. 1, the firm line represents the material flow from the factory to the last consumer opposite the dashed line indicating the information flow. By applying VMI, the sales information at stores/retailers to suppliers, can be simplified, so that information can be immediately known.

* Corresponding author: alzena.dona@gmail.com

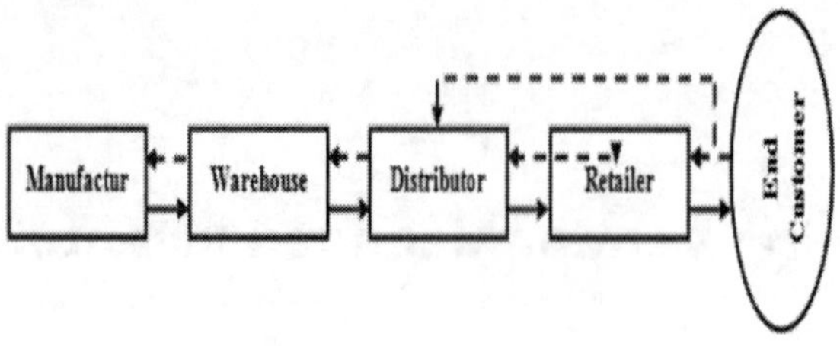

Fig. 1. Supply chain of VMI [8]

The material flow and information flow on the VMI supply chain is shown in Figure 2, the dotted line indicates how the VMI system controls the transactions in the store / retailer, procurement of goods is processed based on sales data input and stock levels. After that, the delivery of goods is set to ensure the availability of goods.

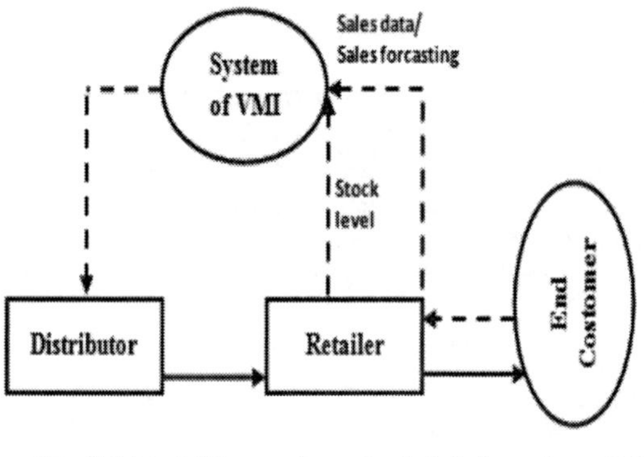

Fig. 2. Material flow and supply chain information of VMI [3]

VMI performs data transparency in the form of sales data and stock levels to suppliers [3]. Implementation of decision support systems on VMI, generating profit in the form of dramatic developments in customer service levels as well as a significant increase in inventory turnover [9].

3 Inventory Control

Inventory control is performed to maintain the stability of the availability of goods in accordance with the high low sales of goods. Inventory control is performed by passing the calculation of *safety stock* associated with time and amount of sales that are not constant. The estimation of sales variation in the N period is shown in Figure 3, so the modeling estimation of safety stock based on the standard deviation of sales as given in Figure 3 produces the calculation of sales standard deviation Equation 1. And *safety stock* in Equation 2.

$\sigma_t(i,j)$ is standard deviation of the sale of goods j in store j by walk t calculated in the period N, N is fixed in accordance with the period that has been running. In one distribution period it is estimated to be sold out within 14 days, thus using the period N = 14. Meanwhile, ST as *safety time* is the estimated time of safe distribution of the product up to the store in days and SS as the *safety stock* which is the estimated safe stock in the unit Pcs.

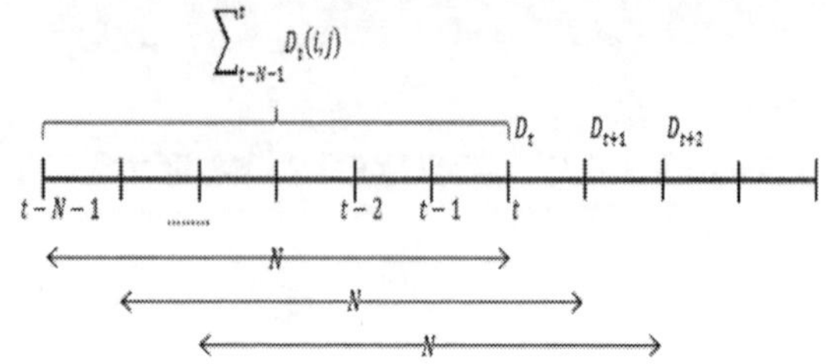

Figure. 3. Estimation of sales based on N period

$$\sigma_t(i,j) = \sqrt{\frac{\sum_{t-N-1}^{t} D_t{}^2(i,j) - \frac{\left(\sum_{t-N-1}^{t} D_t(i,j)\right)^2}{N}}{N-1}} \quad (1)$$

$$SS_t(i,j) = \sigma_t(i,j).ST \quad (2)$$

4 Estimation of Re-Stock Updates

The estimation of re-stock updates is performed to meet the need of demand for using *reorder point* (ROP). Stock changes will be made if the current stock position is smaller than the ROP, and then, the products will be sent on the difference between the current stock position and the required stock. The calculation of *reorder point* based on the average sales in a certain period of time plus *safety stock*, then stock renewal will be done based on the number of *level reorder* that must be achieved using Equation 3 and Equation 4.

$$r_t(i,j) = LT.\ \frac{\sum_{t-s-1}^{t} D_t(i,j)}{s} + + SS(i,j) \quad (3)$$

$$R_t(i,j) = \frac{\sum_{t-s-1}^{t} D_t(i,j)}{s} + + r_t(i,j)\ . \quad (4)$$

$$Q_t(i,j) = R_t(i,j) - IP_t(i,j). \quad (5)$$

The calculation of *reorder point* is performed monitoring periodically with period S, then R as the number of product that becomes the target number of product inventory j in store i, IP is the current stock position, and Q is the number of product j that will be sent back to shop j so the position Product inventory reaches level R [9].

5 Methodology

Research that has been conducted has stages to be done in the information system built. These stages start from the determination of input, storage, process, and output.

https://doi.org/10.1051/e3sconf/20183111015

The form of system design is modeled in accordance with the purpose of this research: to build inventory control system using VMI method shown in Figure 4. The system explains the existence of two current flows that is flow of material and information. Material flow in the form of a firm line from suppliers to stores through the process of delivery of goods, while the flow of information from stores to suppliers starts from the amount of goods sales, then control is performed on the amount of inventory that is important for the supplier in the delivery of goods.

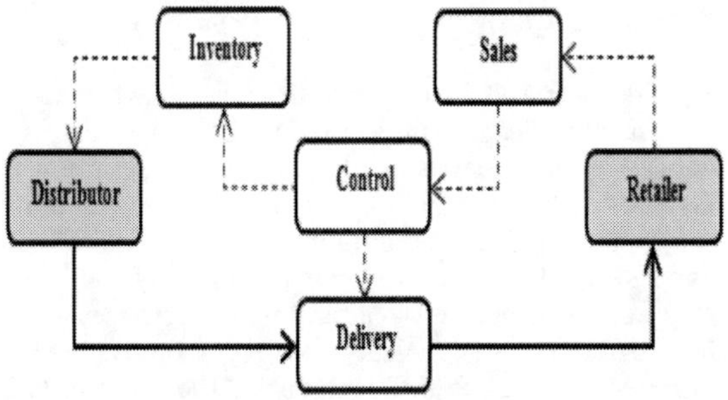

Fig. 4. Model of inventory control system

Changes in inventory due to a final buyer transaction on the store are monitored through a control system. Everyday, sales data in the form of goods code and the amount sold is sent by the shop owner via SMS facility on mobile devices. Data of the inventory changes and sales become system input that will be processed by the system.

Sales data stored on the system will be processed through the calculation of safety stock (SS) as in Equations (1) and (2) and ROP as in equation (3), and (4). The result of the calculation will be compared with the inventory quantity in the store using rule-based reasoning to determine the inventory status as shown in Table 1. Next, at the status of warning inventory, the amount of goods to be distributed is calculated as in (5). The output obtained in the form of product inventory status information in the store whether to re-distribute or not, so that it can be known immediately, when and the right amount for the goods distribution of each store.

Table 1. Rules for determining inventory status

No	Ruler	Status	Definition
1	IP < ROP	Secure	No delivery
2	SS<IP<ROP	Backup	Prepare of delivery
3	IP<SS	Alert	Immediate delivery of goods according R level

6 Results and Discussion

The monitoring of the store's stock of products is generated from the safe stock and ROP calculation through the implementation of Equation (2) resulting in the safe stock of the product, Equation (3) resulting in the number of product ROP and Equation (4) generating the number of goods to distribute. Based on product sales data, inventory control in the form of safe stock amount is obtained, ROP and level R as shown in Table 2.

Table 2. Total safe stock, ROP, and Level R

Date	IP	D	SS	ROP	R level
1-Jun	53	33	31	48	53
2-Jun	58	21	33	52	58
3-Jun	37	25	36	57	64
4-Jun	64	20	36	57	64
5-Jun	44	0	36	57	64
6-Jun	44	0	36	57	64
7-Jun	44	0	36	57	64
8-Jun	44	28	39	66	75
9-Jun	86	30	41	75	86
10-Jun	56	19	41	78	91
11-Jun	96	20	40	82	96
12-Jun	76	0	40	82	96
13-Jun	76	0	40	82	96
14-Jun	76	0	40	82	96
15-Jun	76	21	37	76	89
16-Jun	55	30	39	80	94
17-Jun	90	17	37	77	90
18-Jun	73	12	37	75	88
19-Jun	61	11	35	76	89
20-Jun	50	10	34	76	90

Inventory control is performed daily on the amount of inventory available in the store through the safe stock calculation and ROP in equation (2), (3), and (4) based on the input of goods sales data. Table 2 shows that on June 20, 2016 the number of safe stock was 34 pcs, ROP was 76 pcs and level R was 90 pcs which would be compared to the last inventory at the store.

The rules of inventory status determination in Table 1 provides convenience in monitoring the inventory status of the goods on the store as given in Figure 5. This provides information for suppliers to determine whether or not to re-ship the goods in controlling the store's inventory. The alert status is interpreted as a warning to immediately deliver the goods in accordance with the needs.

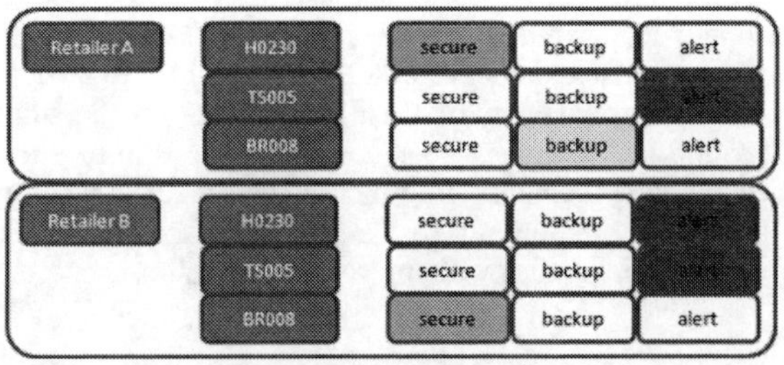

Fig. 5. Monitoring of goods inventory at stores

Based on the monitoring of goods performed by suppliers, the goods can be delivered in accordance with the store needs. Inventory control using VMI is shown in Figures 6 and 7 in the period from 1st to 20th of June 2016, the amount of inventory follows the high sales.

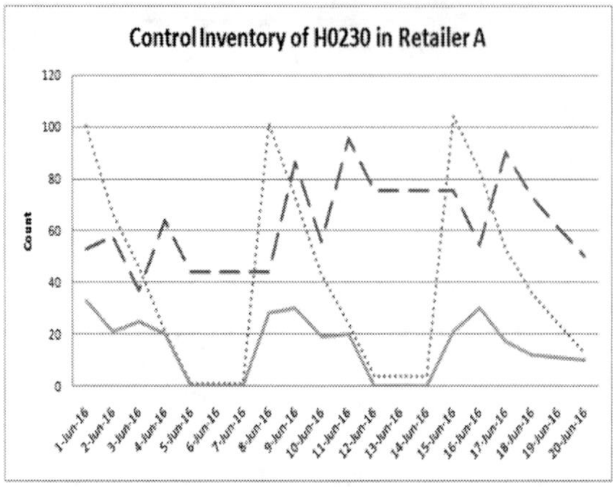

Fig. 6 Control Inventory of H0230 inventory in retailer A

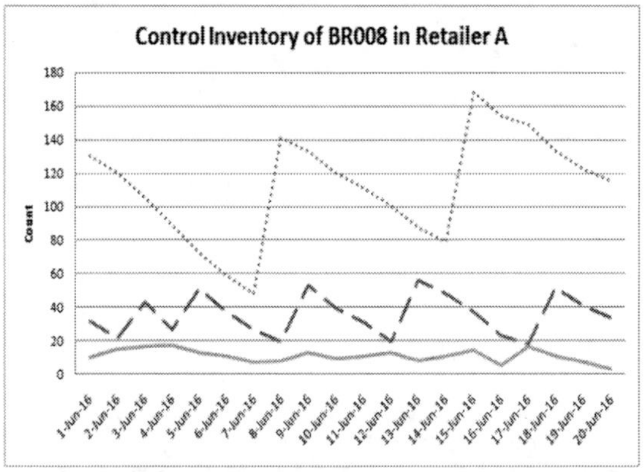

Fig. 7 Control Inventory of BR008 in retailer A

In Figure 6, a firm line shows the number of sales, the dotted line indicates the number of scheduled inventories and the dashed line represents the amount of inventory using the VMI. It is said that the level of availability of H0230 products on Store A through VMI's inventory control is more secure because of safe stock calculation and stock renewal so that stock can be stable. The stability of the goods inventory in the period 1st to 20th of June 2016 was fulfilled with the lower limit of 37 pcs and the upper limit of 96 pcs. This was not found in scheduled stocks because of the void of goods supply for several times. Safe stock estimation that is always updated daily allows sales transactions to be done more with inventory of existing products on the store. Monitoring the inventory status of the goods in the store can provid inventory information before the inventory amount is running out, so the delivery of goods can run again in accordance with the needs of the store.

Figure 7 shows that BR008 on Store A is experiencing low sales, but has a high inventory in scheduled stock. Inventory control using VMI follows the sale transaction of the goods. The sales transaction of the goods BR008 that is relatively low has resulted in low inventory of BR008 as well. The inventory status monitoring also provides information that the inventory is still sufficient so that no deliveries are returned to the store. This can reduce the amount of the product BR008 stock accumulation in Store A so that it can anticipate the return of large quantities of products to the suppliers.

7 Conclusion

The inventory control system that uses VMI equipped with sms technology can be used to monitor inventory of goods in accordance with the needs of retailers. Inventory control can be done through the calculation of safety stock and reorder point based on the data of daily good sales. The monitoring of inventory status is also given in detail because a similar item in different stores has different sales levels as well, moreover if the different items are at different stores. The result of using the system gives benefit for the store in the availability of goods as shown in the higher product availability level on H0230 product at shop A by ± 30%. The distribution is done based on the needs of the store by looking at the condition of the product inventory as in BR008 product in store A so as to reduce the risk of return of the product in the future by ± 70% due to the accumulation of unsold products in the store. Inventory management by using VMI maintains inventory stability by balancing the amount of supply with the sales.

Reference

1. B.S. Sahay, and J. Ranjan, *Information Management & Computer Security16*, 1 (2008)
2. A.A. Alamri, I. Harris, A.A. Syntetos, European Journal of Operational Research **254**, 1 (2016)
3. M. Gronalt, and P. Rauch, Silva Fennica **42**, 1, (2008)
4. R.H. Rad, J. Razmi, M.S. Sangari, and Z.F. Ebrahimi, International Journal of Production Economics **153** (2014)
5. K. Ryu, I. Moon, S. Oh, and M. Jung, International Journal Production Economics **143** (2013)
6. S. Mateen, and A.K. Chatterjee, International Journal of Decision Support Systems **70** (2015)
7. Horrison, Alan and van R. Hoek, 3rd, Essex: Pearson Education Limited (2008)
8. S.M. Disney, A.T. Potter, and B.M. Gardner, Transportation Research Part **E 39** (2003)
9. D. Achabal, S. McIntyre, S. Smith and K. Kalyanam, Journal of Retailing **76**, 4 (2000)
10. M. Rawat, and T. Altiok, International Journal of Production Research, (2008)

Rule Based System for Medicine Inventory Control Using *Radio Frequency Identification* (RFID)

Joanna Ardhyanti Mita Nugraha[1,*], *Suryono* Suryono[1, 2], dan *Jatmiko* Endro Suseno[2]

[1]Magister of Information System, Diponegoro University, Semarang - Indonesia
[2]Physics Department of Science and Mathematics Faculty, Diponegoro University, Semarang – Indonesia

Abstract. *Rule based system* is very efficient to ensure stock of drug to remain available by utilizing Radio Frequency Identification (RFID) as input means automatically. This method can ensure the stock of drugs to remain available by analyzing the needs of drug users. The research data was the amount of drug usage in hospital for 1 year. The data was processed by using ABC classification to determine the drug with fast, medium and slow movement. In each classification result, rule based algorithm was given for determination of safety stock and Reorder Point (ROP). This research yielded safety stock and ROP values that vary depending on the class of each drug. Validation is done by comparing the calculation of safety stock and reorder point both manually and by system, then, it was found that the mean deviation value at safety stock was 0,03 and and ROP was 0,08.

1 Introduction

Rule based system uses explicit knowledge and provide certain flexibility in terms of altering knowledge taken [1]. It was once used in ship design research and was able to create principles for machine layout architecture by concerning the relationship among equipment and ship-owner needs, insight, design, etc. In addition, rule based system that has been developed become more common and can be applied to a wider product [1], [2].

Rule based system is highly efficient and achieves an accuracy of up to 94.29% in research aiming at estimating tomato ripeness based on the color. The set of rules is generated automatically from a set of derived features using a decision tree. Expert knowledge is quickly converted into computer programs in the form of if-then-rules. Moreover, expert knowledge also reduces the search space while optimizing the system [3].

The use of instructions is needed in the pharmaceutical world, for example in the field of drug supply, which is the most important factor in the process of health care. Poor management of drugs will result in stagnant drug stocks (excess supply of medicines) and stockout (deficiency or void of drug supply). Stagnant drug supplies have risks in expiration and damage if they are not stored properly. The stagnant and stockout drugs will have an impact on health cares [4].

In inventory systems, the use of RFID technology is believed to have a major impact on revenue increase [5], such as in clothing retail industries, health industries, education, etc. The application of RFID technology used in healthcare apparel management research can also save up to 30% of total cloth spending. In addition, RFID has also been used in research to ensure that patients' beds and mattresses can be cleaned properly with efficient time and cost, and can shorten work time [5], [6].

Hospital is a health service institution that provides inpatient, outpatient and emergency care services. Hospitals together with their sub-organizations must be managed as well as possible in order to provide health services as much as possible. Pharmacy service is a supporting service and also a main revenue center. This is because more than 90% of health services in hospitals use pharmaceutical supplies, such as medicines, chemicals, medical devices, medical devices and medical gases [7].

The management of drugs in hospitals is routine, urgent and periodic activities which means they must always be available and should not be void. In other words, stockout can disrupt the operational activities of the hospital. In addition, the drug stockout may result in a loss of having to order and immediately deliver medication with higher price than buying from a supplier [7].

Some previously unknown information or knowledge can be found from the data. The data has the potential as supporting evidence to reflect some important patterns by using modeling techniques. Moreover, the model can be revised automatically because the database can be updated in real time when modeling techniques are used. Data can be represented as a rule. Basic rules can be used to manage regulations effectively and efficiently by noticing storage and retrieval. The most commonly used method for representing rules is the *rule based system*

* Corresponding author: joanna.nugraha@gmail.com

[8]. Always Better Control (ABC) analysis is one way of controlling inventory by sorting and classifying the type of goods. Research in India used ABC analysis and Vital Esensial Non-esensial (VEN) analysis for efficient drug management. The results obtained from ABC-VEN matrix analysis showed that only 21% of 1536 stock items of drugs belonging to class A require maximum attention [9].

From the above explanation, the *rule-based system* is very efficient and can help in the process of controlling the inventory system by combining ABC analysis in it. In addition, RFID is capable of performing the process of reading data automatically that can be used for the identification process of drug data. Therefore this research will apply *rule-based system* method to know safety stock of drug supplies using RFID input system.

2 Methods

Research has several stages to obtain maximum results, including use-value analysis using ABC classification, Safety Stock (SS) and Reorder Point (ROP) calculations, drug supply system analysis and drug supply system implementation.

2.1 Analysis of use-value by ABC classification

The ABC classification is performed by the analysis of drug use-value. The drug use-value is obtained from the amount of usage in one period, then sorted from the highest usage amount to the lowest usage amount. After the item data is sorted, the percentage of usage, then, is calculated with the following calculation [7].

$$\text{Percentage of usage}: \left(\frac{x}{\Sigma x}\right) \times 100\% \quad (1)$$

From the above formula, it is known that x is the amount of usage per item of drug and Σx is the total amount of drug usage. After the percentage of usage is calculated, the categories of preparations are made in accordance with the cumulative percentage of 70%, 20%, and 10% [20]. The steps used in the ABC classification are:

a. Counting the number of monthly usage in 2016 for each item.
b. Sorting usage values, ranging from the largest to the smallest, then a percentage of the value of usage is made.
c. Finding the cumulative value of usage by summing the value of the percentage of usage that has been ranked. Group A with a percentage of 70%, group B with a percentage of 20% and group C with a percentage of 10%.

2.2 Calculation of Safety Stock (SS) and Reorder Point (ROP)

The importance of calculating safety stocks due to frequent new orders after the lead time is too late (e.g. late travel due to traffic jams or hit by a disaster) and the frequent increase in production demand. This situation will result in the stock out which will further decrease the level of service at hospitals. Due to the amount of investment for safety stock inventory especially for the most frequently used drugs (class A) then safety stock is prioritized to vital medicines. The way to calculate the amount of safety stock by using VEN system guidance is shown in Table 2.1. [7]

Tabel 2.1 Determination of safety stock using VEN guideline

Groups of VEN	Safety Stock
Vital	20% of last usage
Essential	10% of last usage
Non essential	5% of last usage

Factors affecting ROP are lead time, average usage and safety stock, can be calculated by the formula [10]:

$$ROP = (LT \times d) + SS \quad (2)$$

With ROP as the reorder point, LT is the lead time which is the time required when the drug is ordered until the drug arrives at hospital. Then d is the average usage in a given period of time. And SS is safety stock which is an additional supply to protect and preserve the possibility of material deficiency [10].

2.3 Analysis of drug supply system

A shortage and excess supply of drugs commonly happen in hospitals. This is caused by supply planning that does not suit the patients' needs. By applying *rule-based system* method, drug supply planning is made based on the needs of patients.

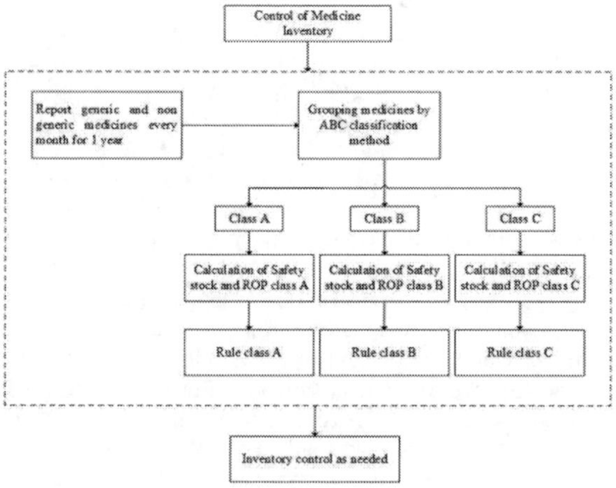

Fig. 1. Theoretical framework

The data used are reports of drug use which will be processed using ABC classification and *rule based system* method. RFID technology serves to accelerate the process of *input* and *output* data and avoid errors. RFID is used to *scan* every incoming and used drug, so the officers do not need to enter the drug data to accelerate the process of drug entry. There are two users involved in the system, in the warehouse and hospital pharmacy. The description of the research flow lies in the theoretical framework as shown in Figure 1.

2.4 Implementation of drug supply system

The result of data analysis using ABC classification is fast, moderate and slow moving drugs. After classification is done, next is to apply *rule based system* method to know safety stock and ROP from each type of drug based on fast moving, moderate and slow moving drug classification. The procedures for performing decision tree analysis are:

1. Identifying all decision points and other possibilities that will occur
2. Identifying decision alternatives for each decision point
3. Identifying what might happen from every decision
4. Creating a tree diagram showing the order of decisions and events that may occur

And the rule based will be shown with the decision tree in Figure 2.

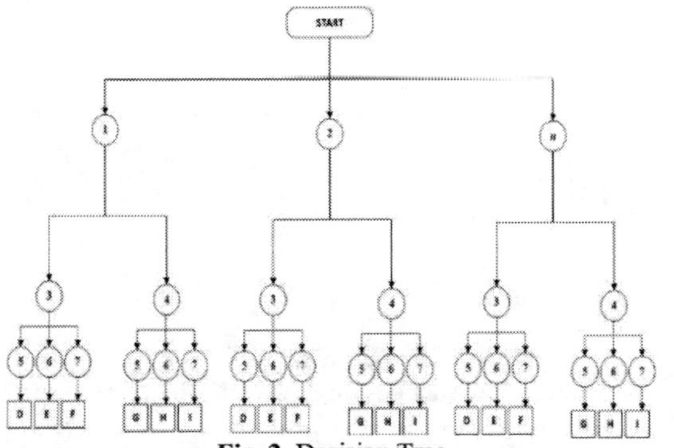

Fig. 2. Decision Tree

In the decision tree, it shows number 1, number 2 to n, where n = 12 which is the number of months of the year. Then number 3 is a generic drug and number 4 is a non generic drug. Then number 5 is the classification of fast moving drugs, number 6 is the classification of moderate drugs and number 7 is the classification of slow moving drugs. And D is the rule for fast moving drug classification, E is the rule for moderate drug classification and F is the rule for the classification of slow moving drugs. The rule description of the decision tree will be shown in Table 1.

Table 1. Explanation of rule of decision tree

Alphabet	Rule
D	**IF** Medicine out of January **AND** generic medicine **AND** fast moving **THEN** safety stock 20% of last usage
E	**IF** Medicine out of January **AND** generic medicine **AND** moderate **THEN** safety stock 10% of last usage
F	**IF** Medicine out of January **AND** generic medicine **AND** slow moving **THEN** safety stock 5% of last usage
G	**IF** Medicine out of January **AND** non-generic medicine **AND** fast moving **THEN** safety stock 20% of last usage
H	**IF** Medicine out of January **AND** non-generic medicine **AND** fast moving **THEN** safety stock 10% of last usage
I	**IF** Medicine out of January **AND** non-generic medicine **AND** fast moving **THEN** safety stock 5% of last usage

3 Results and Discussion

3.1 Calculation process of ABC classification

To calculate ABC classification based on use-value, the steps are as follows:

a. Calculating the monthly usage amount in 2016 for each drug item by using formula (1) as shown in Table 2.

Table 2. The calculation of the amount of usage

NO	MEDICINE NAME	VALUE	AMOUNT USAGE	AMOUNT USAGE (%)
1	ASAM MEFENAMAT TAB 500 MG	TABLET	2370	6,34%
2	METFORMIN TAB 500 MG	TABLET	2283	6,11%
3	PARACETAMOL TAB 500 MG	TABLET	2071	5,54%
4	RINGER LAKTAT LAR INF 500 ML	PLABOT	1913	5,12%
5	VITAMIN B COMPLEX	TABLET	1299	3,48%
6	OMEPRAZOL KAP 20 MG	CAPSULE	1102	2,95%
7	RANITIDIN TAB 150 MG	TABLET	1031	2,76%
8	CEFADROXIL KAP 500 MG	CAPSULE	943	2,52%
9	VITAMIN C 50 MG TAB	TABLET	834	2,23%
10	CIPROFLOXACIN TAB 500 MG	TABLET	795	2,13%
			...	
196	SALEP KULIT BASITRASIN 500 IU/G	TUBE	1	0,00%
	TOTAL		37353	

b. Sorting usage values, ranging from the largest to the smallest, then a percentage of usage values is made as shown in Table 3.

c. Finding the cumulative value of usage by summing the value of the percentage of usage that has been ranked. Class A with a percentage of 70%, class B with a percentage of 20% and class C with a percentage of 10%.

So from the classification of ABC, the drugs in class A include the vital group with the calculation of safety stock of 20% of the past use, class B includes the essential group with the calculation of safety stock of 10% of past use and class C includes non-essential group with the calculation Safety stock of 5% of the usage then shown in Table 3. For ROP calculation has been determined by the formula according to the average monthly usage and safety stock in each class, as shown in Table 4.

Table 3. Result of safety stock calculation in each class

NO	MEDICINE NAME	VALUE	AMOUNT USAGE	AMOUNT USAGE (%)	CLASS	SS
1	ASAM MEFENAMAT TAB 500 MG	TABLET	2370	6,34%	A	474
2	METFORMIN TAB 500 MG	TABLET	2283	6,11%	A	457
3	PARACETAMOL TAB 500 MG	TABLET	2071	5,54%	A	414
4	RINGER LAKTAT LAR INF 500 ML	PLABOT	1913	5,12%	A	383
5	DIAZEPAM 2 MG TABLET	TABLET	310	0,83%	B	31
6	CEFTRIAXON 1 GR INJEKSI	VIAL	308	0,82%	B	31
7	ONDANSETRON INJ 4 MG/2 ML	AMPUL	297	0,80%	B	30
8	LOPERAMID TAB 2 MG	TABLET	94	0,25%	C	5
9	LISINOPRIL TAB 10 MG	TABLET	90	0,24%	C	5
10	PAPAVERIN TAB	TABLET	90	0,24%	C	5

Table 4. Result of ROP calculation in each class

NO	MEDICINE NAME	VALUE	AMOUNT USAGE	AMOUNT USAGE (%)	CLASS	SS	d	ROP
1	ASAM MEFENAMAT TAB 500 MG	TABLET	2370	6,34%	A	474	79	632
2	METFORMIN TAB 500 MG	TABLET	2283	6,11%	A	457	76	609
3	PARACETAMOL TAB 500 MG	TABLET	2071	5,54%	A	414	69	552
4	RINGER LAKTAT LAR INF 500 ML	PLABOT	1913	5,12%	A	383	64	510
5	DIAZEPAM 2 MG TABLET	TABLET	310	0,83%	B	31	10	52
6	CEFTRIAXON 1 GR INJEKSI	VIAL	308	0,82%	B	31	10	51
7	ONDANSETRON INJ 4 MG/2 ML	AMPUL	297	0,80%	B	30	10	50
8	LOPERAMID TAB 2 MG	TABLET	94	0,25%	C	5	3	11
9	LISINOPRIL TAB 10 MG	TABLET	90	0,24%	C	5	3	11
10	PAPAVERIN TAB	TABLET	90	0,24%	C	5	3	11

3.2 Evaluation and validation of calculation result

Verification is done by comparing manual calculations to perform ABC classification with calculation result using system. The data used for verification are 20 random data of generic drug types taken from each class, as shown in Table 5.

Table 5. Verification result of ABC classification

No	MEDICINE NAME	Calculation Result of ABC Manual Classification			Calculation Result of ABC Classification Using System			Conformity
		Fast Moving	Moderate	Slow Moving	Fast Moving	Moderate	Slow Moving	
1	2	3	4	5	6	7	8	9
1	AMOXICILLIN TAB 500 MG	√			√			CORRESPONDING
2	RINGER LAKTAT LAR INF 500 ML	√			√			CORRESPONDING
3	RANITIDIN TAB 150 MG	√			√			CORRESPONDING
4	FUROSEMID TAB 40 MG		√			√		CORRESPONDING
5	RAMIPRIL TAB 10 MG			√			√	CORRESPONDING
6	AMLODIPIN TAB 5 MG	√			√			CORRESPONDING
7	IBUPROFEN TAB 400 MG	√			√			CORRESPONDING
8	ASAM TRANEKSAMAT TAB 500 mg		√			√		CORRESPONDING
9	COTRIMOXAZOLE TAB 480 MG		√			√		CORRESPONDING
10	ASAM FOLAT 1 MG TAB		√			√		CORRESPONDING
11	ANTASIDA DOEN TAB		√			√		CORRESPONDING
12	CIPROFLOXACIN INF 2 MG/ML			√			√	CORRESPONDING
13	PROPANOLOL TAB 40 MG			√			√	CORRESPONDING
14	CO-AMOXICLAV 625 MG			√			√	CORRESPONDING
15	BETAHISTIN TAB	√			√			CORRESPONDING
16	CEFIXIME KAP 100 MG	√			√			CORRESPONDING
17	AMITRIPTILIN TAB 25 MG			√			√	CORRESPONDING
18	OMEPRAZOL INJ 40 MG / VIAL			√			√	CORRESPONDING
19	FLUNARIZIN TAB 5 MG			√			√	CORRESPONDING
20	CIPROFLOXACIN TAB 500 MG	√			√			CORRESPONDING

Next, verification for safety stock and ROP calculation is performed. The verification is done by using the mean deviation formula by calculating the difference between system calculation and manual calculation, then searching for the average of data difference and dividing the amount of data used to perform the verification. The verification results show that the average deviation generated for the safety stock is 0.03 and for the ROP is 0.08 as shown in Table 6.

Table 6. Verification of safety stock and ROP calculation

No	MEDICINE NAME	Manual ABC Classification Calculation Result		Calculation Result of ABC Classification Using System		Difference	
		Safety Stock	ROP	Safety Stock	ROP	Safety Stok	ROP
1	2	3	4	5	6	7	8
1	AMOXICILLIN TAB 500 MG	106	142	107	143	1	1
2	RINGER LAKTAT LAR INF 500 ML	383	510	383	511	0	1
3	RANITIDIN TAB 150 MG	206	275	207	277	1	2
4	FUROSEMID TAB 40 MG	15	26	16	28	1	2
5	RAMIPRIL TAB 10 MG	2	4	2	4	1	1
6	AMLODIPIN TAB 5 MG	101	134	101	135	0	1
7	IBUPROFEN TAB 400 MG	92	123	93	125	1	2
8	ASAM TRANEKSAMAT TAB 500 m	23	39	24	40	1	1
9	COTRIMOXAZOLE TAB 480 MG	16	26	16	28	0	2
10	ASAM FOLAT 1 MG TAB	15	26	16	28	1	3
11	ANTASIDA DOEN TAB	13	22	13	23	0	1
12	CIPROFLOXACIN INF 2 MG/ML	5	11	4	10	1	1
13	PROPANOLOL TAB 40 MG	5	11	4	10	1	1
14	CO-AMOXICLAV 625 MG	4	10	4	10	0	0
15	BETAHISTIN TAB	71	95	72	96	1	1
16	CEFIXIME KAP 100 MG	85	113	85	115	0	2
17	AMITRIPTILIN TAB 25 MG	4	9	4	10	0	1
18	OMEPRAZOL INJ 40 MG / VIAL	4	8	4	10	0	2
19	FLUNARIZIN TAB 5 MG	3	7	4	10	1	3
20	CIPROFLOXACIN TAB 500 MG	159	212	159	213	0	1
					AVERAGE DEVIATION	0,03	0,08

4 Conclusions

The supply control system can identify drugs with fast, moderate and slow movements by applying the ABC classification method. From the results of experiments in this study, ABC classification generated from the system has been in accordance with the results of manual ABC classification calculation. The rule-based system method can be applied to know safety stock and ROP on each drug based on its class. This is evident from the test results indicating that the average deviation produced is relatively small, i.e. for safety stock with an average deviation of 0.03 and for ROP with an average deviation of 0.08.

References

1. C. A. Costa, M. A. Luciano, C. P. Lima, and R. I. Young, Advanced Engineering Informatics, 292–305 (2012)
2. K.-S. Kim, M.-l. Roh, and S. Ha, Expert Systems With Applications, 1-14 (2015)
3. N. Goel, and P. Sehgal, Applied Soft Computing, 45-56 (2015)
4. F. A. Rosmania, and S. Supriyanto, *Jurnal Administrasi Kesehatan Indonesia*, (2015)
5. S. Suryono, J.E.Suseno, C.Mashuri, A.D.Sabila, J.A.M.Nugraha, M.H.Primasiwi, RFID Sensor for Automated Prediction of Reorder Point (ROP) values in a Vendor Management Inventory (VMI) System Using Fuzzy Times Series **23,** 2398-2400 (2017)
6. H.-L. Chan, T.-M. Choi, and C.-L. Hui, Decision Support Systems, 803–811 (2012)
7. A. Ramadhan, *Jurlan Ilmiah Komputer dan Informatika (KOMPUTA)*, 1-12 (2013)

8. H. Liu, A. Gegov, and M. Cocea, *Rule Based Systems for Big Data*. London: Springer International Publishing Switzerland (2016)
9. M. S. Kumar, and B. Chakravarty, Medical Journal Armed Forces India, 1-4 (2014)
10. F. Rangkuti, *Manajement Persediaan*. Jakarta: PT Raja Grafindo Persada (1996)
11. R. Pressman, *Software Engineering: A Practitioner is Approach*. New York: McGraw- Hill (2001)

Comparison of Genetic Algorithm and Hill Climbing for Shortest Path Optimization Mapping

Mona Fronita [1,*], *Rahmat* Gernowo [2], and *Vincencius* Gunawan [2]

[1]Master of Information System, School of Postgraduate Studies, Diponegoro University, Semarang – Indonesia 50242
[2] Department of Physics, Faculty of Science and Mathematics, Diponegoro University, Semarang – Indonesia 50275

Abstract. Traveling Salesman Problem (TSP) is an optimization to find the shortest path to reach several destinations in one trip without passing through the same city and back again to the early departure city, the process is applied to the delivery systems. This comparison is done using two methods, namely optimization genetic algorithm and hill climbing. Hill Climbing works by directly selecting a new path that is exchanged with the neighbour's to get the track distance smaller than the previous track, without testing. Genetic algorithms depend on the input parameters, they are the number of population, the probability of crossover, mutation probability and the number of generations. To simplify the process of determining the shortest path supported by the development of software that uses the google map API. Tests carried out as much as 20 times with the number of city 8, 16, 24 and 32 to see which method is optimal in terms of distance and time computation. Based on experiments conducted with a number of cities 3, 4, 5 and 6 producing the same value and optimal distance for the genetic algorithm and hill climbing, the value of this distance begins to differ with the number of city 7. The overall results shows that these tests, hill climbing are more optimal to number of small cities and the number of cities over 30 optimized using genetic algorithms.

1 Introduction

Traveling Salesman Problem (TSP) is a complex combinatorial optimization problem, originally presented by Dantzig. TSP concept can be illustrated as follows: A salesman depart from a city, visiting all the cities in one trip without visiting the same city more than once and finally returned to the city he left early. The goal is to minimize the total distance and the cost of salesmen who have to travel to visit all the cities [1].

Some methods can be applied to solve the TSP is *Ant Colony Optimization* (ACO) [2], *Annealing Algorithm* [3], *Hill Climbing* [4] *Genetic algorithm* [5], *Greedy algorithm* [6].

From several methods of TSP completion, hill climbing algorithm has good performance in local searching. Starting from defining the initial group, deciding the better search area up to iterating from level to level, the results of each level is taken from best and then compared, so they can get more optimal [7]. This TSP method is used to determine the nodes that has been given the distance among other nodes by comparing the existing node based on selection of the shortest distance from the initial position [8].

Beside hill climbing to solve TSP problems, there is also genetic algorithm. Genetic algorithm is a powerful and flexible metaheuristic as well as the relatively new type of algorithm by adopting the idea of natural selection and genetic changes naturally. This algorithm is known as a tool that can solve combinatorial in optimization problems such as TSP [9]. GA worked by solving the problem of appropriate population selection (known as a chromosome) that will later perform iterations / generations by applying the three basic genetic operators, i.e. selection, crossover and mutation. By combining simple operator as a single operator, to create complex operator, which can accelerate the evolutionary process, so that it can improve the efficiency of GA [10].

In the problems of *Traveling Salesman Problem* is finding a solution to compute the shortest path or minimum travel route of all paths. Genetic algorithm and hill climbing has advantages and disadvantages. An algorithm has a different result, because an algorithm that has a high optimization for a case uncertainly having a high optimization also for other cases. Both of these algorithms depend on the number of cities to determine which is more efficient in calculating the optimum distance. If the number of cities visited by sales fewer so then it is obtained a shorter distance. However, it is still not assured that the total shortest travel distance is obtained when the number of cities which must be visited by sales is the fewest.

Thus, the research it will be carried out a comparison between genetic algorithm and hill climbing based on the calculation of the shortest path and computing time to see which one is more optimal to finish salesman traveling problems.

* Corresponding author: mona_fron@yahoo.co.id

2 Theoretical Framework

2.1. Traveling Salesman Problem (TSP)

The problem of Traveling Salesman Problem (TSP) is one of the most studied examples in combinatorial optimization. This problem is easy to declare but very difficult to solve. When completed in exact time the required computation will increase exponentially as the problem increases. TSP can be expressed as a problem in finding the minimum distance of a closed journey to a number of n cities where the cities are only visited once. TSP is represented by using a complete and weighted graph G = (V, E) with V set of vertices representing a set of points, and E is the set of edges. Each edge (r, s) ∈ E is the value (distance) rs which is the distance from city r to city s, with (r, s) ∈ V. In a symmetric TSP (distance from city r to point s is equal to distance from point S to point r), drs = drs for all edges (r, s) ∈ E . In a graph, the TSP is drawn as shown in Figure 1 below:

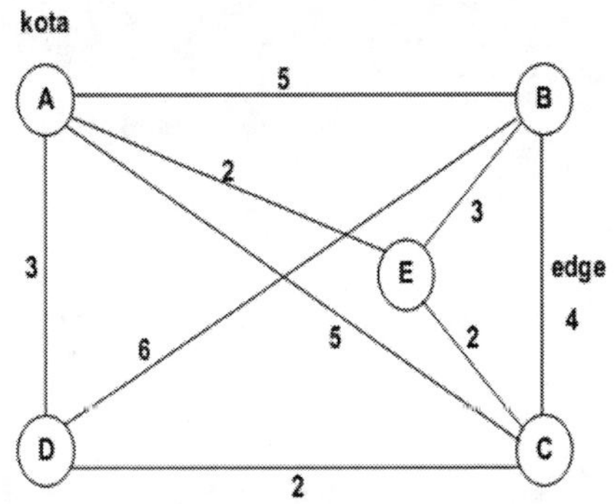

Fig 1 The illustration of TSP

2.2 Hill Climbing Method

The Simple Hill Climbing method is an offshoot of Hill Climbing method. The Simple Hill Climbing search movement starts from the leftmost position after the initial and pointed points are determined by comparing the current point state with a single point regardless of the next point at the same level, and a better first point being selected to the next. The move is done continuously until the final goal is obtained. Hill Climbing algorithm is as follows:

1. Determine the initial random trajectory and calculate the distance of the initial path, then tested by swapping each city. After testing if the initial path is the destination city, stop, and if the initial path is not a destination city continue with the current state as the initial path.
2. Do the following steps until the solution is found, or until no new point will be applied to the current state.
   A) Look for a point that has not been used in the current state, and use an exchange point to get a new state.
   B) Evaluation of the new situation;

If the new state is a destination city, then the search process stops,

If it is not a destination city, but its value is better or smaller than it is now, then make that new state into the present state.

If the new situation is no better than the current situation then continue the search, as in the previous step to get the destination city with the smallest weight. Hill climbing process flow can be seen on fig. 3 Hill Climbing flowchart.

There are several problems that may occur in Hill Climbing method:

A. The algorithm will stop if it has reached the local maximum value i.e. a current path that is considered to have the shortest distance from all previous trajectories, but not necessarily the current path is considered to be the shortest distance is the destination because there is still a track that has not been visited in the predefined iteration.
B. The sequence of points used to swap positions will greatly affect the discovery of solutions.
C. Hill Climbing passes a point of exchange that has been used in the previous step [11].

2.3 Genetic algorithms

Main components of Genetic Algorithm are Encoding, Function Evaluation, Selection, Crossover and Mutation. Encoding is a difficult process in the genetic algorithm. It because the encoding process for each problem is different because not all coding techniques suitable for each problem. The encoding process results a row is called chromosomes. Chromosomes consist of a set of bits known as genes. There are several kinds of coding techniques can be done in a genetic algorithm [12], including binary encoding, permutation encoding, value encoding, and tree encoding.

In TSP, used coding techniques are permutation encoding, where a chromosome represents a permutation of the city by serial numbering 1, 2, 3, ..., n. The evaluation process is a process of calculating fitness values expresses chromosomes quality level as a representation of problem solving. In this case, fitness function *eval (v)* cannot be defined directly from the objective function *f (v)* because TSP is the problem of finding the minimum value. The fitness function must be mapped from the objective function. The objective function is a calculation of all sides formed from existing genes on a chromosome. For example there are as many as *n* genes on a chromosome then *f (v)* of chromosomes is:

$$f(v) = D_{1\,2} + D_{2\,3} + \ldots + D_{n-1\,n} \qquad (1)$$

With $D_{1\,2}$ is the distance between city 1 to city 2, to city *n-1* to town *n*. The simplest way to determine the fitness function is mapping every addition value of objective function as a declining of the fitness function value with the following equation:

$$eval\,(v) = \frac{1}{\sum_{i=1}^{n} f(v)} \qquad (2)$$

Selection will determine which individuals are chosen to be recombined and how the *offspring* is made up from those elected individuals. Each individual in a selection forum will receive the reproduction probability depends on the objective value itself against the objective value of all individuals in those selection forum. A selection methods commonly used is the *roulette-wheel*. As the name implies, this method imitates the roulette-wheel game where each chromosome occupies a piece of the circle on the roulette wheel proportionally based on its fitness value. The chromosome which has greater fitness value occupies a piece of the larger circle than the low fitness valuable chromosome, How the roulette-wheel method works [15]:

1. Calculating the probability value of each chromosome

$$P[i] = \frac{eval\ (v)}{\sum_{i=1}^{n} eval\ (v)} \tag{3}$$

2. Generating random number with interval (0-1) for the value of R
3. Calculate the cumulative probability of each chromosome, defined

$$K[0]=0 \quad K[i]=K[i-1]+P[i] \tag{4}$$

Crossover (crossing) is carried on 2 chromosomes to produce children chromosome (*offspring*). Children chromosomes formed will inherit some of their parents chromosomes nature. In the crossover there is a very important parameter, namely the crossover opportunities (P_c). Crossover opportunities shows the ratio of children produced from each generation by the population size. For example, population size (pop size = 100), while the crossover opportunities (P_c = 0.25), it means that there are 25 chromosomes of 100 chromosomes which are hoped to be exist in the population and undergoing *crossover*. Crossing over of TSP can be implemented by *order crossover* scheme.

Mutation is the process of modifying the children chromosomes at random. Mutation will create new individual by altering one or more genes in one chromosome. Mutation role is to replace the missing genes from the population as a result of selection process and allow the emergence of genes that are not present in the initial population [13]. In TSP, the mutation operator is usually implemented by exchanging mutated genes with other genes selected randomly. For example, chromosome {2, 3, 4, 1, 5} can be mutated into a chromosome {4, 3, 2, 1, 5}. In this case, gene 1 and gene 3 are exchanged. This mutation scheme known as *swapping mutation*. Genetic algorithm process flow can be seen on fig. 2 Genetic algorithm flowchart.

3 Methodology

3.1 Analysis and Design of Information Systems

The shape of system design can be modeled as a chromosome made up from a number of genes in which each of these genes represent each city that will be passed by a salesman. The number of genes on one chromosome is the number of cities plus one city that will be the similar initial city with the final city which will be passed by the salesman so that the chromosome is a trip route that may be passed by the salesman with the arrangement of input and output process. In fig 2 and 3 there is a flow of performance from the algorithm and hill climbing. In this arrangement, it consists of administrator and user (salesman).

1. *Input* is the process of data input in the form of a series of activities performed by the administrator and salesman.
   Web Administrator, facilitating the entry of data master or data admin and performing management of items data input, city data in the form of latitude and longitude coordinate points, initialization parameters used in the calculation of genetic algorithm and hill climbing
   User, enter data cities to be visited to make delivery for determining the shortest track.

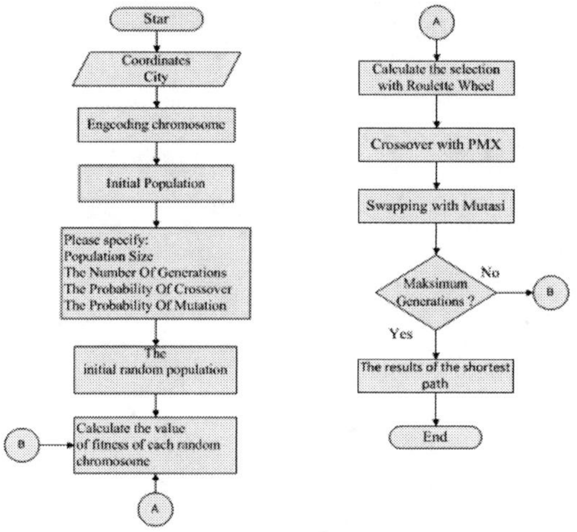

Fig.2 Genetic algorithm flowchart

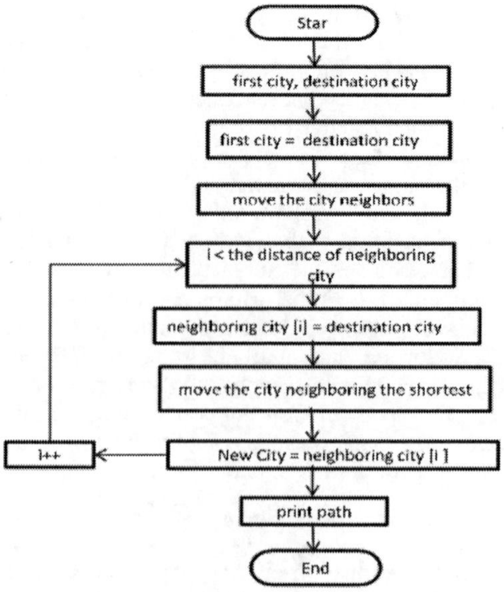

Fig. 3 Hill Climbing flowchart

2. *Process* is by creating coordinate location point by performing coordinate input in the form of latitude and longitude of JNE locations throughout the territory of Central Java which later became the master data that is stored in the database. Creating a track using Google Map API that connects coordinate points using the marker of inter-city destinations by salesmen that would be seeing a path to be passed by graph using a polyline, and then it will be sorted to obtain the optimal shortest track.The search process using a genetic algorithm for TSP and Hill Climbin.

4 Finding and Discussion

In the TSP process genetic algorithm and hill climbing is compared with some aspects such as the ratio of distance, time of execution, the generation / level and the passed track. Comparison of generation at the level of genetic algorithm and hill climbing, the genetic algorithm parameters to generation, the greater the number of generations, then the more optimal result will be got. While on the *hill climbing*, the repetition process will continue at the next level as far as finding the most optimum distance then the repetition stops. But, in both of these methods the optimal results are not necessarily in the last generation or level. Based on the calculation of the research finding, it is obtained the same and optimal distance by the testing 4, 5 and 6 cities either using genetic algorithm or *hill climbing*. If the number of cities inserted more than 7 cities producing a different city distance but optimal for distance and computing time such as shown in Table 1 and table 2

Table 1

Comparison of distance of the trajectory of the GA and HC

No	Number of cities	GA			HC		
		Average	The shortest time	The longest time	Average	The shortest time	The longest time
1	8	11112.1	553.5	559.9	11450	572.5	572.5
2	16	24217	1038.6	1319.9	26718	1335.9	1335.9
3	24	38650.6	1736.6	2087.2	43356	2167.8	2167.8
4	32	53539.04	115.136	2993.8	2889.1	2889.1	2889.1

Distance on the genetic algorithm varies while on his hill climbing, the distance is more consistent. Further different distance can be seen from the image of a distance testing with the number of different cities.

Distance from *hill climbing* is always similar proven from 20 times testing, while the genetic algorithm to get various distance. The test based on this distance can be seen from the optimal value of the shortest track on genetic algorithm and *hill climbing*. The results of distance comparison among the eight cities shows that the use of *hill climbing* is more optimal and always similar in fig 4.

Table 2

Comparison of the time path of GA dan HC

No	Number of cities	GA			HC		
		Average	The shortest time	The longest time	Average	The shortest time	The longest time
1	8	850.575	41.791	44.766	3.518	0.185	0.251
2	16	3101.754	131.99	236.338	47.833	2.013	3.292
3	24	3570.397	149.786	282.84	73.326	3.715	4.298
4	32	9845.27	108.23	2720.4	139.559	7.165	7.502

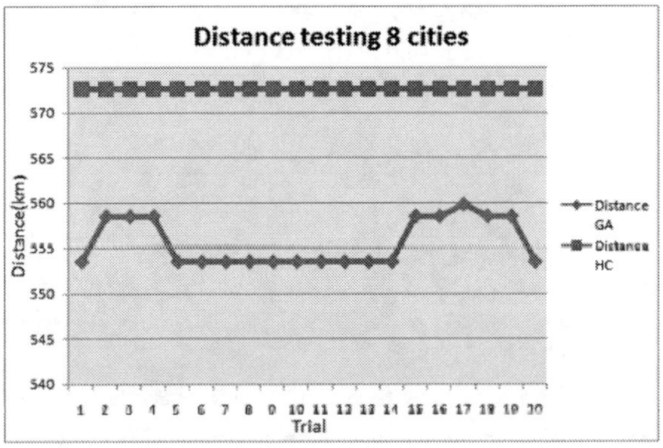

Fig. 4 Distance testing 8 cities

In fig 5, The distance using *hill climbing* with 16 testing of the city is similar, while the genetic algorithm produces various distances. The distance resulting from *hill climbing* is more optimal than the genetic algorithm. However, the genetic algorithm has a shorter distance.

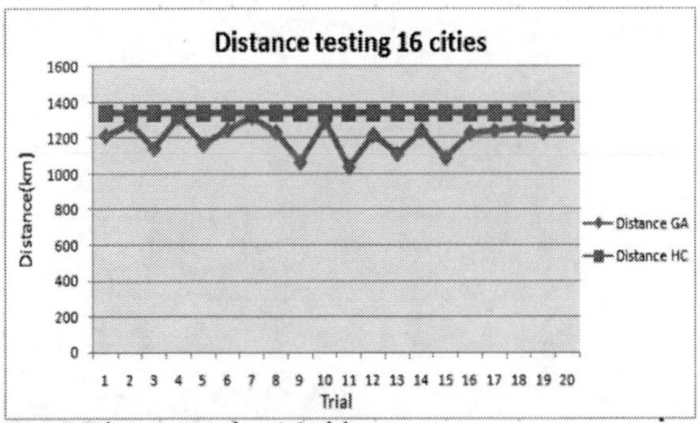

Fig. 5 Distance testing 16 cities

By testing within the 24 cities, in fig 6 hill climbing results are still more optimal than the genetic algorithm, but the results of the genetic algorithm distance does not differ much from *hill climbing*. On testing to 4 and 9 to produce the same distance.

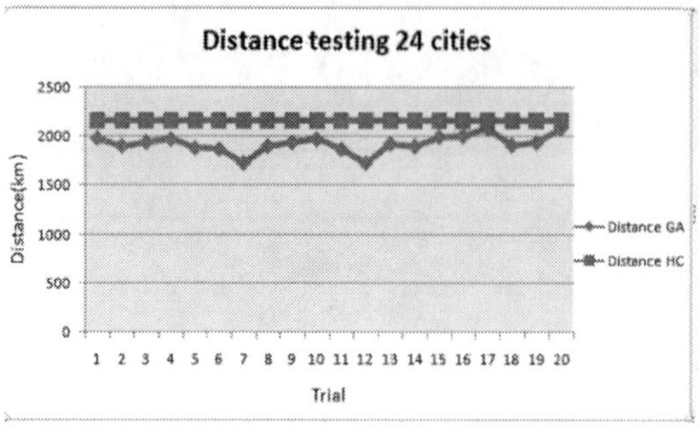

Fig. 6 Distance testing 24 cities

By testing of 32 cities, it is produced an optimal distance using genetic algorithm, even it is lower than the *hill climbing*, although the most optimal average distance is *hill climbing*. The optimal distance obtained from testing 1, 6 and 9.

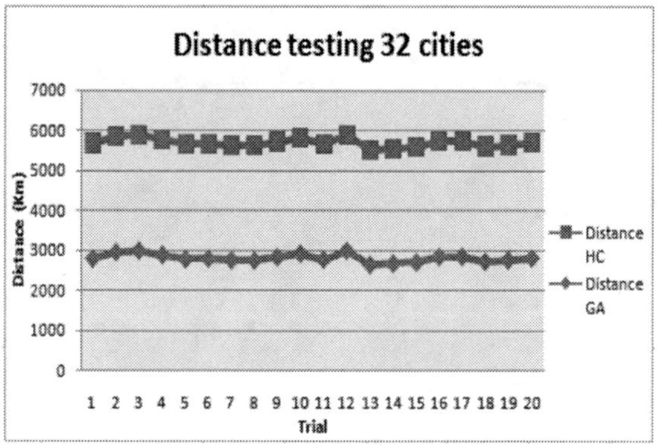

Fig. 7 Distance testing 32 cities

After testing the entire distance with the number 8, 16, 24 and 32 cities, it can be seen that the genetic algorithm has the long enough average distance in testing the amount of 8, 16 and 24 cities, in fig 7 with 32 cities which has relatively shorter distance. While *hill climbing* on testing the amount of 8, 16 and 24, the city which has shorter track distance, with the number of 32 cities which has longer track

5 Conclusion

Web-based system built by mapping optimization the shortest track with genetic comparisons and *hill climbing*. In a system of genetic algorithm, parameters to a total population of 10, 60% of the crossover probability value, mutation probability value of 50% and the number of generations are 200. Meanwhile, on *hill climbing* there is no input of certain parameters. The result of the system that will be compared based on distance, computation time, generation / level and track sequence. The results can be seen after 20 times of testing with the amount of city 8, 16, 24 and 32. Based on the overall computation of the fastest time using *hill climbing* is the shortest time, 57.753 seconds. Based on the relatively optimal distance of *hill climbing* with the number of cities which are relatively fewer, for cities over 30, it will be more optimal to use Genetic Algorithms.

References

1. Gutin, abd, G., and A. P. Punnen, *The traveling salesman problem and its varia-tions*. NewYork: Springer, (2007)
2. M. Mavrovouniotis, and Yang., Applied Soft Computin **13** 4023–4037, (2013)
3. X. Zheng, Z. Li, and M. Li, *Applying Simulated Annealing Algorithm*, (2010)
4. H. S. Jacobson, A. L. McLay, N. S. Hall, D. Henderson, E. D. Vaughan, International Journal of Mathematical And Computer Modelling. Vol **43** page 1061-1073, (2006)
5. Lim, European of operational reaserch **5**, 1459-1478, (2005)
6. S. Maity, A. Roy, and M. Maity, Journal Computer and Industrial Engineering Vol. **83**, *273-296*, (2015)
7. J. Brecklinghaus, and S. Hougardy, Operations Research Letters, **43** 259–261, (2015)
8. H. Junliang, M. Weijian, Y. Wei, *Journal of Shanghai Maritime University*, **28** (4),11, (2007)
9. J. Majumdar, and A,K. Bhunia, Journal of Computational and Applied Mathematics **235** page 3063–3078,(2011)
10. S. Kusumadewi, and H. Purnomo, *Penyelesaian Masalah Optimasi Dengan Teknik-Teknik Heuristik*. GRAHA ILMU : Yogyakarta, (2005)
11. Z. Zukhri, *Algoritma Genetika : Metode komputasi Evolusioner untuk Menyelesaikan Masalah Optimasi,* Penerbit Andi Yosgyakarta, (2014)
12. Suyanto, *Artificial Intelligence*. Informatika: Bandung, (2014)

Rule Based Expert System for Monitoring Real Time Drug Supply in Hospital Using Radio Frequency Identification Technology

Galih Driandanu[1*], *Bayu* Surarso [1], and *Suryono*[1]

[1]Master in Information System, School of Postgraduate Studies, Diponegoro University, Semarang - Indonesia

Abstract. A radio frequency identification (RFID) has obtained increasing attention with the emergence of various applications. This study aims to examine the implementation of rule based expert system supported by RFID technology into a monitoring information system of drug supply in a hospital. This research facilitates in monitoring the real time drug supply by using data sample from the hospital pharmacy. This system able to identify and count the number of drug and provide warning and report in real time. the conclusion is the rule based expert system and RFID technology can facilitate the performance in monitoring the drug supply quickly and precisely.

1 Introduction

Rule Based Expert System is a branch of artificial intelligence that was developed in 1960 by Newl and Simon. The name was derived from the *knowledge-based expert system*, in which the expert knowledge entered into a computer to resolve the problems. The expert system is a knowledge – based programme that provides expert-level solutions in solving specific problems [1]. The use of *Rule Based Expert System* becomes widespread increasingly, such as in mineral water identification which supported by *visual basic* for the implementation and *Microsoft access* for creating the *database*. This is driven by the insufficient mineral water experts who give the knowledge in the water resource area. Mineral water knowledge expert system can help cater the knowledge in the water resource areas [2]. The purpose of *Rule Based Expert System* application is providing fast response with the given rules based on the experts in their fields [3]. India has used the technology in handling crimes against women for better judiciary. The technology which combined with the *Rule Based Expert System* method helps in giving decisions based on the rules given to the legal experts system which matched by entering the data [4]. Judicial problem solution in India which makes use of *Rule Based Expert System* encourages the writer to develop in the pharmacy field by utilizing this same method for monitoring drug supplies supported by *Radio Frequency Identification* technology or known as RFID.

The RFID technology is used for system optimization and shrinkage control. Unlike barcode technology, RFID in pharmacy enables giving accurate and *real-time* information. Besides, RFID also has a bigger storage capacity compared to the barcodes. Counterfeiting is easier to do in this technology because the *barcode* is easily duplicated. While, information stored in RFID has been encrypted, therefore it makes more difficult to counterfeit the information. RFID also has better resistance than *barcode* (5).

The use of *Rule Based Expert System* and RFID technology will help the process of monitoring the *real-time* drug supply. The system which supported by RFID technology can ease the drug supply calculation manually. Besides, this technology also has the expert knowledge that was put into the system. Hence, it facilitates in monitoring the drug supply in a hospital obviously.

Rule Based Expert System compiled with the formula 1.1 as follows:

R1 : If (*stock free* > *Safety stock*) Then (1.1)
: safe stock order

R2 : If (*stock free* <= *Safety Stok*) Then
: unsafe stock order

The current stock is obtained from the existing ones in the pharmaceutical warehouse in a hospital. While, for the *safety stock* gained from the calculation containing some parameters, in which the first are *lead time*, drug usage in a hospital and also *service level*. The *free stock* is generated from the current inventory variable in the warehouse minus the ordering value. The secure and insecure messages status will be formed by comparing between the *free stock* and *safety stock*.

The status determination of the *service level* in a company is expressed as the percentage in each company itself, while the *service factor* used as a multiplier with the standard deviation to calculate a specific quantity to meet the specified service level. The Excel NORMSINV function can be used for converting the *service level* percentage to the *service factor*. Its percentage is shown in Table 1 [6].

* Corresponding author: Galih.driandanu@gmail.com

Tabel 1. *Service factor* based on *service level*

No.	Service Level	Service Factor	No.	Service Level	Service Factor
1	50.00%	0	17	90.00%	1.28
2	55.00%	0.13	18	91.00%	1.34
3	60.00%	0.25	19	92.00%	1.41
4	65.00%	0.39	20	93.00%	1.48
5	70.00%	0.52	21	94.00%	1.55
6	75.00%	0.67	22	95.00%	1.64
7	80.00%	0.84	23	96.00%	1.75
8	81.00%	0.88	24	97.00%	1.88
9	82.00%	0.92	25	98.00%	2.05
10	83.00%	0.95	26	99.00%	2.33
11	84.00%	0.99	27	99.50%	2.58
12	85.00%	1.04	28	99.60%	2.65
13	86.00%	1.08	29	99.70%	2.75
14	87.00%	1.13	30	99.80%	2.88
15	88.00%	1.17	31	99.90%	3.90
16	89.00%	1.23	32	99.99%	3.72

A company that has implemented *safety stock* is Danfoss company by entering 4.247 products as the research material to prove the *safety stock* method can decrease the cost up to 12%. The *safety stock* calculation conducted by Danfoss company as follows [7] :

a. Computing the forecast deviation for each month
b. Squaring each deviation
c. Computing the standard deviation by the formula shown in equation (1) which N is the sum of data and σ is standard deviation, d1 is the usage of –i and d is the usage average.

$$\sigma = \sqrt{\frac{\sum_{i=1}^{N} (d_1 - d)^2}{N-1}} \quad (1)$$

d. Computing the *safety stock* shown in formula (2) where SS is the *safety stock*, z is the value based on the *service level* and LT means the lead time

$$SS = z\sqrt{LT.\sigma^2} \quad (2)$$

The *safety stock* model can be applied in the supply chain to minimize the production cost, transportation and stock supply. It is done because consumer demand which based on the date is changing frequently. This model implements the objectivity function with the procedures applied to the intel dual core 3.0 Ghz and 8Gb memory PCs supported by VB.Net software whose it results can be exported into the spreadsheet [8].

2 The Problem

The increasing need of drugs nowadays makes the drug supply in a hospital being concerned. It is also triggered by the increasing number of people which contrasts with the good quality of life. Besides, drugs is the biggest income in the hospital which is very simple to calculate the profits gained each month. The drug turnover amount in the hospital reaches 50%-60% of the entire hospital budget. Besides being the biggest income in the hospital, drugs are also something that is needed by the patients. Therefore, the supply must be maintained so that there is no delay in giving the drugs to the patients whom needed [9]. Regular monitoring can be done in order to avoid the delay in the required drug supplies. During this time, the drug supply relies on the pharmacists who check regularly. On the manual checks, miscalculations are often happened due to the officers′ limitation such as inaccurate or miscalculate count.

3 The Idea for Solution

The main proposed idea is illustrated in Figure 1 which consists of three main parts: input, process, and output.

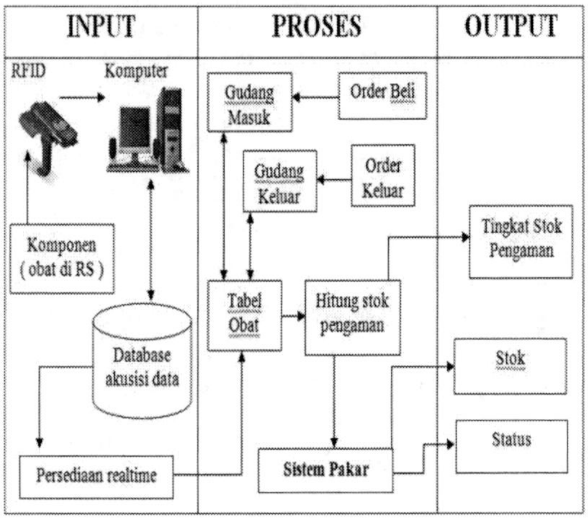

Figure 1 Drug Monitoring System Design

RFID technology is required in the INPUT to obtain, collect and prepare the data, until process to produce the data needed that is identifying the amount of the drug supplies. For example, RFID technology reading the RFID tag that has already installed in the drug wrapping so that it will get the drug data which includes the name, the number and also the supplier of the drug.

The next is the PROCESS section. This section processes the drug data which already received from the input section. The process section will manage the drug by calculating the supply or the minimum limit. By identifying the supply processed for entering into the warehouse that will increase the amount or processed for leaving the warehouse which will reduce the amount of the drug supply. The expert system in this system is used to avoid the void of the drug supply due to being late in ordering the drug to suppliers. The system will facilitate the process of monitoring the drug supply. The decision of making the minimum limit of the drug supply can be done only by the pharmacists who have the specific skill. The rules which are from the pharmaceutical experts are embedded into the system so that the drug monitoring process becomes real time. Therefore, there is no need to wait for the pharmacists to come for calculating the drug supply that are often happened within a drug period.

The next part is the OUTPUT section. This section provides the result consists of the amount, the status and the minimum quantity of the drug supply. The drug supply status obtained from the rules from the experts as in the formula (1.1), while for the minimum quantity taken from the formula (2) which is the calculation between the service level, lead time and the standard deviation. Based on the calculation and the rules from the experts, this system provides the minimum, the status, and the amount of the drug supply.

3 System Implementation

The implementation of the system is carried out in accordance with the design on the *rule based expert system* for monitoring the drug supply as seen in the figure 1.1. The system has five inputs and four processes to create the transaction report and the *safety stock* level and the safe or insecure status of the drug supply. The five inputs include the supplier, the drug master, the poly master, the *rule based* master and the *service level* input. Four processes that include drug ordering process to the supplier, drug entering process to the warehouse from the supplier, drug ordering process from the poly to the warehouse to deliver the drug needed and the last one is drug recording process from the warehouse to the poly which order some drugs needed.

The generated output in this drug monitoring system are the drug master report, the drug ordering report to the supplier, the drug report which going to the warehouse, the drug ordering report ordered by the poly to the warehouse, the drug report from the warehouse to the ordered poly and the final output is the *safety stock* report.

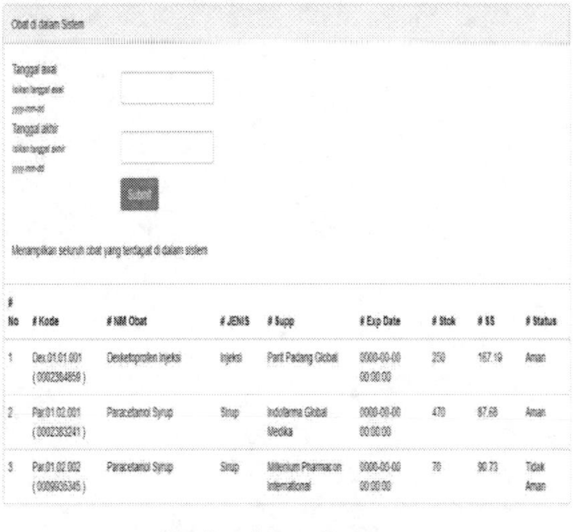

Fig. 2. *safety stock* report

The *safety stock* report is the main result from this system that provides the drug supply which can be monitored in *real time* by providing the data results in the form of the quantity, the minimum quantity and also the safe or unsafe status of the drug supply at that time as in Figure 2.

4 Discussion

The implementation of *Rule Based Expert System* for monitoring the *real time* drug supply by using rfid technology is a designed system to assist the process of the drug hospital warehouse. Hence, it is expected that the drug needed is already available and there is no vacuum of drug stocks due to manually calculation which causes miscalculation or misinformation of the available stock. The drug monitoring system can be accessed by online. It becomes an added value since for monitoring and deciding new policy of the drug supply can be done everywhere. This system will facilitate the process of monitoring the drug supply from the pharmaceutical staffs until its director who has the right access in the monitoring system. Besides, this monitoring system is a *real-time* system. The process of the drug supply will be seen directly and at that moment without waiting for the records which done in every single period to find out the remaining stock from the warehouse. This system has the ability of pharmaceutical expert to determine the status of the drug supply. The expert knowledge is applied in this system by providing the experts' decision on the drug supply status, either the stock is safe or unsafe. Therefore the drug should be ordered to the supplier so that the drug void wont be happened since it will be fatal for the patients who need the drug.

The study as in figure 4.1 uses sample data that are three kinds of drugs from different suppliers; dexketoprofen (injection) from the Parit Padang Global, paracetamol syrup from Indofarma Global Medika and paracetamol syrup from Millenium Pharmacon International. Besides, the *safety stock* calculation done from April 2016 until July 2016 with the *service level* 75% which had *service factor* value 0.67 from the hospital policy.

Information obtained from monitoring of the drug supply at the end July 2016 with 14days leadtime on each drug such as dexketoprofen injection from Parit Padang Global with the code dex.01.01.001 has the *safety stock* level of 167 tablets with the remaining stock 250 tablets. The rule based expert system of this drug is if it is less from 30% of the *safety stock*, the status of this drug will be unsafe. While, the recent stock is still more that 30% of the drug *safety stock* so the status of dexketoprofen injection from the Parit Padang Global supplier is safe.

Paracetamol syrup from the Indofarma Global Medika supplier with the code par.01.02.001 has the *safety stock* level from the calculation in formula 2.2 with the sevice level 75% which has *service factor* value 0.67 and 14days leadtime so that the *safety stock* produced is 87tablets with the available stock 470 tablets. *The rule based expert system* of this drug is if the drug stock less than 50% will be given unsafe status. In fact, the available drug stock is still more than 50% of the *safety stock*, so the paracetamol syrup from the Indofarma Global Medika supplier is safe.

For Paracetamol syrup from the Millenium Pharmacon International supplier with the code par.01.02.002 001 has the *safety stock* level from the calculation in formula 2.2 with the *sevice level* 75% which has service factor value 0.67 and 14days leadtime. Therefore the *safety stock* produced is 90tablets with the available stock 70 tablets. The *rule based expert system* of this drug is if the drug stock less than 30% of the safety stock will be given unsafe status. In fact, the available drug stock is less than 30% of the *safety stock*, so the paracetamol syrup from the Millenium Pharmacon International supplier is unsafe.

5 Conclusion and further research

After discussing *Rule Based Expert System* in monitoring the *real time* drug supply in the hospital using *radio frequency identification technology*, can be inferred that this system :

1. Facilitate the hospital in providing the drug supply information
2. Facilitate the pharmacy in monitoring the drug supply both in the warehouse and in the poly of the hospital itself.
3. Facilitate the drug data collection
4. Facilitate in deciding the order time to the supplier when the drug is in short supply.

For the further development of *Rule Based Expert System* in monitoring the *real time* drug supply in the hospital using *radio frequency identification technology* research, the writer gives some suggestions to help the hospitals and also the researchers, namely:

1. The need of additional computer equipment to support this system so that it can works as expected
2. The need of assistance and training to optimize the computers uses so that it can run properly and quickly.

6 Acknowledgements

Galih Driandanu thanked to Mr. Bayu Suroso and Mr Suryono who has guided in this research. As well as the RSUD K.R.M.T Wongsonegoro team who willing to help and provide the drug sample data needed in this study.

References

1. Luger, George F. and Stubblefield, William A, *AI : Structure And Strategic For Complex Problem Solving*, 2nd edition, (1993)
2. I.O, Folorunso., O.C. Abikoye, R.G. Jimoh, and K.S. Raji, Journal of Emerging Trends in Computing and Information Sciences, **vol. 3 no. 2**, 205 – 210, (2012)
3. R. Keller, Prentice Hall, Englewood Clifs, N.J, (1988)
4. J.S. Jadhav, K.M. Nalawade, and M.M. Bapat, In Indian Judicial System, IOSR Journal Of Humanities And Social Science, **vol.16 no. 6,** 19-22, (2013).
5. Cakici, O. Engin, H. Groenevelt, and Decision Support Systems, ***51***, 842–852, (2011)
6. O. Celik, school of industrial engineering and telecommunication university of Cantabria, (2013)
7. L.F. Tratar, 2009, economic and bussiness review, **vol. 11 no.2**, 109-117, (2009)
8. K. Funaki, 2010. int. J. Production Economics .**135**, 4 – 13, (2010)
9. L. Trisnantoro. *Aspek Strategi Manajemen Rumah Sakit*. Yogyakarta: Penerbit Andi, (2005)

Real-time Geographic Information System (GIS) for Monitoring the Area of Potential Water Level Using Rule Based System

Wirdah Anugrah[1,*], *Suryono* Suryono[1, 2], and *Jatmiko* Endro Suseno[2]

[1]Magister of Information System, Diponegoro University, Semarang - Indonesia
[2]Physics Department of Science and Mathematics Faculty, Diponegoro University, Semarang – Indonesia

Abstract. Management of water resources based on Geographic Information System can provide substantial benefits to water availability settings. Monitoring the potential water level is needed in the development sector, agriculture, energy and others. In this research is developed water resource information system using real-time Geographic Information System concept for monitoring the potential water level of web based area by applying rule based system method. GIS consists of hardware, software, and database. Based on the web-based GIS architecture, this study uses a set of computer that are connected to the network, run on the Apache web server and PHP programming language using MySQL database. The Ultrasound Wireless Sensor System is used as a water level data input. It also includes time and geographic location information. This GIS maps the five sensor locations. GIS is processed through a rule based system to determine the level of potential water level of the area. Water level monitoring information result can be displayed on thematic maps by overlaying more than one layer, and also generating information in the form of tables from the database, as well as graphs are based on the timing of events and the water level values.

1 Introduction

Management of water recourses which is carried out in real time is urgently needed due to its benefit in irrigation, food supply, transportation, recreation, climate, fishery, farms, energy resources, natural resources and many others. An error in managing the water resources can trigger disasters such as floods, tsunami, and landslides, so that it needs a system to monitor the level of water in a certain area to avoid those disasters. The information of Water Resources Management as a real-time monitoring will be useful to any departments that need water resource management and the society. The main concept of real-time water resource information system is to give any accurate and informative [1].

The monitoring process using rule-based system is very easy to be applied so that the system can work in a smart way [2]. Rule-based system is used in the region with small scale problems, and it can suit the data available. The available data will be processed in the monitoring system by using Rule-Based System. Rule-based system is applied in storing and manipulating the information to interpret the condition as effective and easy to understand so that ru-based system is often used in implementation handmade smart reviews and researches. The rule-based system is mostly applied in expert system using rules to have deductive conclusion or alternative choices, written in if-then style.

Based on the fact on the field, the management of water resources faces problems on the expensiveness of the instruments, real-time, data continuum, memory capacity, limitation of the human resource, the wide cope of the area, area disturbance, maintenance, calibration and spare part for any malfunctions. Monitoring water real-timely is based on computerized communication network and database with sophisticating monitoring instruments of a sensor to bring up online real-time monitoring [3].

From the explanation above, if monitoring the level of water is not carried out real-timely, it will give negative impacts, such as flood happening mostly in lowland. The information from real-time monitoring system can give predictive description about what is going to happen and what action taken will be more precise. The use of rule-based system method in the monitoring model of potential water level in a certain spotted area which is done real-timely, will be displayed in a form of GIS map. Using data of water level, time, geographical location and coordinate spot, potential water level monitoring can be done easily. The output of the system of using the result can be in form of web, application for smartphone or even as the computer desktop. It discusses any location in 5 spot, so it can represents relevant monitoring result with the location

* Corresponding author: wirdahanugrah1989@gmail.com

based system. GIS will give easy assess to its users or decision makers to make any policies. GIS has been widely used to assess any information about a certain location.

2 Geographic Information System (GIS)

GIS is an information system that shows information in a form of geographical map as interface data. GIS is set of concepts of layers and its relations [4]. The analyse used in GIS is overlay analyse as data integrated process among the different layers. The analyse needs more than one layer which is physically staged and will be analyzed visually. GIS also needs spatial analyse as a group of techniques which can be used as the data processing of GIS [5]. GIS is developed using internet technology called **web-based GIS** as shown in Figure 1.

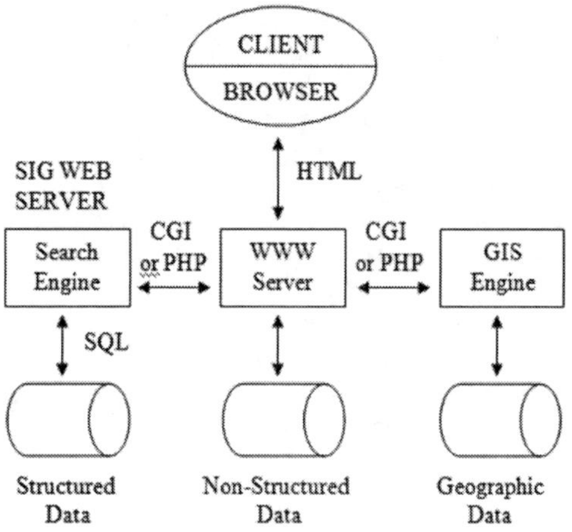

Fig. 1. Architecture of Web-based GIS [6]

Architecture of web-based GIS is a client-server structure. The use of this structure is in order to make easy access to the data and efficient in displaying information as it reduces the use of paper based data processing. This GIS structure consist of some main components, such as :

1. *Browser* as the media for the clients. They can use Mozilla Firefox.
2. *Web server* as the data provider that can be operated by any operator to share data in web, the web server use dis Apache.
3. *Programme language* used to process any request from server is PHP or CGI
4. The internet is used as the communication media between server and clients.
5. Engine used to process the structured and non structure data or geographycal data is map server. Map server is to read mapfile, map data and other dataas file.shp which produces map in .png.
6. Structured data is in form of database as the data storage in form of MySQL.

3 *Rule Based System Method*

Rule based system consists of *if-then* rules, a group of data and an interpreter as the controller of the rules applied in providing facts. Statements used of if-then rules to formulate the conditional reports consist of complete knowledge. Single *If-then* rules assume '*if* x is A so y is B' and *if* is a part of 'x rules is A it is called premise, and part of command 'y is B' is called conclusion [7]. *Rule based system* is a relative simple model that can be suited to many problems. *Rule based system* is a part of *Artificial Intelligent* (AI). *Rule based system* can only be used to problems with science and rules and in the rather small area [8]. *Rule* has distinctive form:

***If* <conditions>*then*<conclusion>;**

Rule based method fits to application domain which is easy, understandable, complete, small cope area and limited [9].

4 Research Methodology

The research carried out has stages that are done with the aim to get the maximum result done by using System Development Life Cycle (SDLC) development method with waterfall method, including system planning, system requirement analysis, system design, system implementation, system test, and System maintenance [9].

4.1 Planning of System

The first stage of planning is the activities undertaken before the research took place is the stages in conducting the collection of literature such as international and national journals that fit and support the research topic, as well as articles relevant to the research topic. Observations made include the contemporary problems, the current method will be applied as inputs in the design of the information system to be built.

4.2 Analyzing of System Requirement

Stages of system requirement analysis are the steps to determine the required variables such as water level data, coordinate point, time data, location data, and existing layers such as, river boundary map, Watershed (DAS) map, and regional map Flood with the system to be built. The system requirements analysis also addresses some of the needs and or requirements related to input, process, and output, as follows:

1. *Input*
   Is a data input process in the form of a series of activities performed by administrators and users (local government and the general public). The administrator (manager) manages the web of geographic information systems and facilitates added data, modify and delete. The data input is in the form of water level data acquisition, time, and location

coordinate point. Users (local governments and the general public) can only view information such as: maps, water level potential status, altitude values, sensor locations, and graphs based on time and water levels.

2. *Process*
   Represents processes that run on geographic information systems for monitoring the potential of the water level of the region using a rule based system with the step of making maps and overlays taken from the google map API. Creation of coordinate points of location by performing coordinate input in the form of latitude and longitude of locations with 5 sensors will be sent to the server to be stored in the database.
3. *Output*
   The output obtained is information in the form of maps, water level potential status, height values, sensor location, and graphs based on the time and value of water level that have been processed by using rule based system.

The analysis used in the Geographic Information System in monitoring the potential of the regional water level is spatial analysis. Analyzes require more than one layer to be physically compiled in order to be visually analyzed. Layers used in this GIS, namely: Google map as the base map, the sensor is a type of point layer, watershed, river border, and flooded areas in the form of polygon-type layer, and non-spatial data in the form of water level data in meters. Spatial analysis is done by calculating area based on spatial data (coordinate point). Spatial analysis in this study is the calculation of the area that appears for each polygon that is clicked. The method of rule base system is used in monitoring the potential of the water level of the area by applying the existing rules based on the regulation of the Head of National Disaster Management Agency Number 02 of 2012 on General Guidelines for Disaster Risk Assessment.
In the broad calculations that appear for each polygon that is clicked, as in the following calculation:

```
var polygon = features[i].getGeometry();

var sourceProj = map.getView().getProjection();

var geom=/**@type{ol.geom.Polygon}*/ (polygon.clone(). Transform
(sourceProj, 'EPSG:4326'));

var coordinates = geom.getLinearRing(0).getCoordinates();

var area_m = Math.abs(wgs84Sphere.geodesicArea(coordinates));

var area_km = area_m / 1000 / 1000;
```

4.3 Designing of System

Development of information systems begins with the determination of input (input) to be used, then proceed with the process as the implementation of the applied method and ends with the form of output (output) to be displayed. The design of information systems as shown Figure 2.

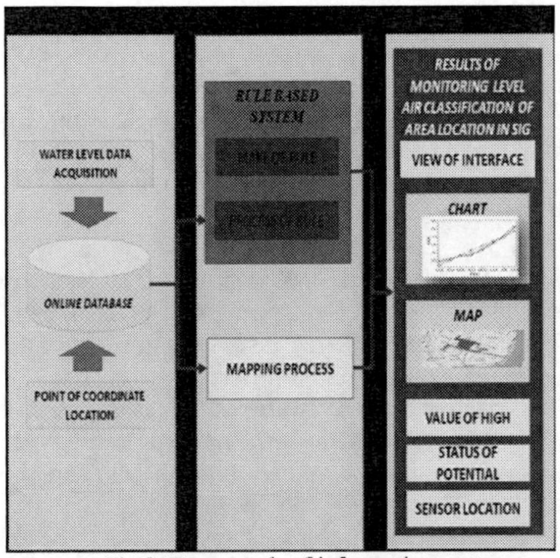

Fig.2. Framework of information system

As for the description of the flow of water level data acquisition starts from the data acquisition process using the sensor, then storage of historical data in the local database (local database) in the microcontroller, then by utilizing the request http (http request) and modem connection of historical data is sent to the web Server. The internet-connected web server accepts and stores all historical data of water level levels sent periodically from those sensors in the database (online database). Data stored on the database in the network (online database) is used as an analysis material as shown in Figure 3.

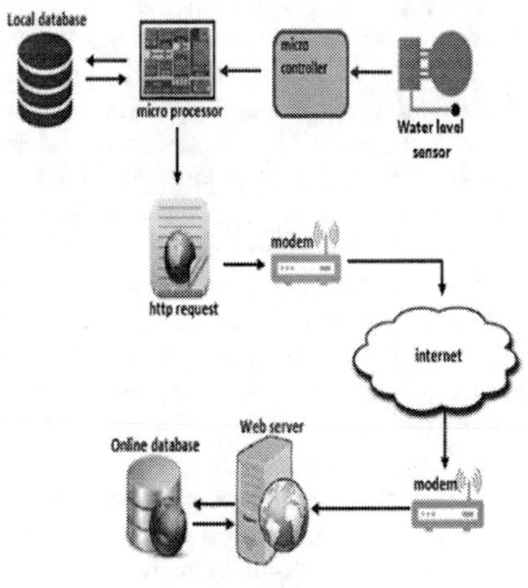

Figure 3. The process of acquisition and transmission of water level sensor data

4.4 Implementing of System

Implementation stage is the stage of programming. At this stage will implement the information system by applying the rule-based system method used in the process of monitoring the potential water level of the area into the application to be built using PHP and MySQL, then displayed in the form of Geographic Information System.

4.5 Coding of System

In the real-time application of Geographic Information System application for monitoring the potential of the water level of this area, to prove that the system can run well the testing method used is black box testing which means the test process is displayed in the form of a table which describes the description of test, initial precondition, Test procedures, input data used, expected outputs, evaluation criteria, results obtained from the test and the final conclusion of the test.

The result of GIS software testing that has been done shows that the system runs in accordance with the functions required by the user (user) and admin (manager). Admin can enter water level data that has been processed with rule based system also to be able to add, change, and delete data, then system can monitor potential of level air in accordance with water level data, so that obtained information accurate according to user requirement presented in the form map. And the information obtained also in the form of related layers, water level potential status, height values, sensor location, and graphs based on the time of occurrence and value of water level. The above is the advantage of this GIS application program, the presentation of information more clear and easy to understand and always updated and real-time.

5 Results and Discussion

The results of the implementation are generally described with screenshot images from the GIS in real time for monitoring the potential of the water level of the region by using the rule based system method. Under the following explanation there are some images of the results of the system screenshot. The results of system implementation on user pages (Local Government and Public) can only see information in the form of: maps, related layers, water level potential status, height values, sensor location, polygon area, and graphs based on time and water level . If one of the sensors is clicked it will display a description of the object identification indicating that the sensor is located on one of the existing layers, as shown Figure 4.

Users (Local Government and Public) can also see graphs from the monitoring results of potential water levels in the area. Monitoring results are shown in graphical form based on time and water level, as shown in Figure 5. In addition to the implementation results of the system there are also screenshot implementation of the GIS database. Implementation of the program by using PHP which is a programming language used for the creation of web applications with MySQL database, as shown Figure 6.

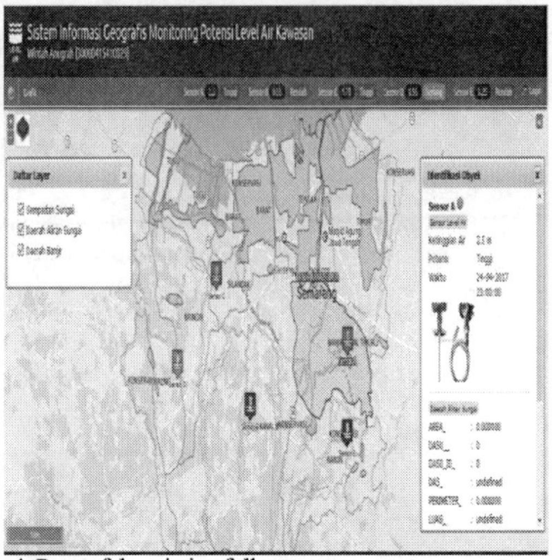

Fig. 4. Page of description full user

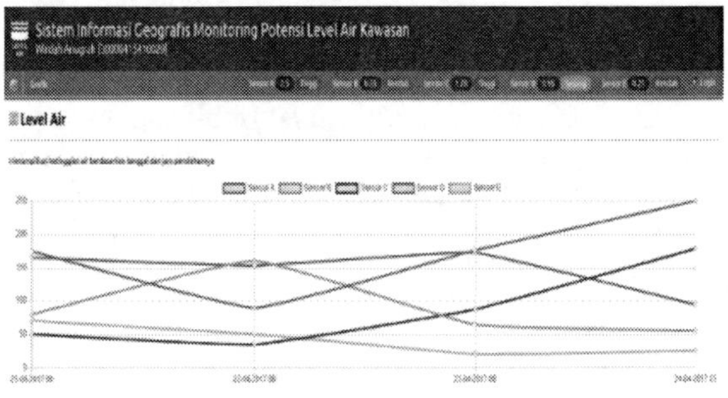

Fig. 5. Page of user graph

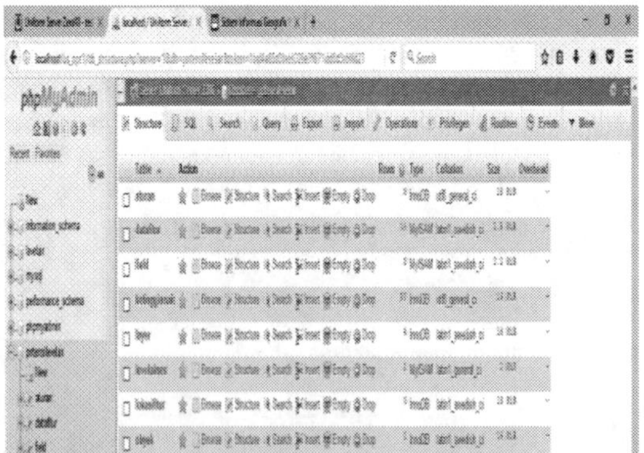

Fig. 6. View of database GIS

Water level data obtained from sensor acquisition results are stored into the database and then processed by using a rule-based system based on the Regulation of the Head of National Agency for Disaster Management Number 02 Year 2012 on General Guidelines for

Disaster Risk Assessment. In accordance with these rules can be determined the rules, as follows:

```
$ Level Air
  if Level Air < 0.76 m
        then potential of area = Low
  if Level Air > 0.76 m and Level Air <1.5 m
        then potential of area = Medium
  if Level Air > 1.5 m
        then potential of area = High
```

In the GIS process of data collection coordinates by mapping or retrieving from Google Map as much as 5 (five) locations in Semarang City, the data as input data for data processing which will be used for monitoring the potential of water level in the area that has been determined. The coordinate data can be seen in Table 1.

Table 1. Data of Coordinates

No.	Coordinates	Location	Sensor
1.	-7.030024, 110.464439	Banjir Kanal Timur	A
2.	-7.055238, 110.380669	Banjir Kanal Barat	B
3.	-7.004808, 110.352516	Silandak	C
4.	-7.038883, 110.319557	Plumbon	D
5.	-7.065460, 110.463753	Babon	E

6 Conclusions

GIS for monitoring the water level potential of the built area is able to behave intelligently by acquiring water level data, time data, and sensor location data which is then processed with rule based system whose source of knowledge is the regulation of Head of National Disaster Management Agency Number 02 of 2012 About General Guidelines Disaster Risk Assessment. In accordance with the regulation, the established rule if the water level is <0.76 m then the area is potentially low, if the Water level> 0.76 m to <1.5 m then the area is potentially moderate, and if the water level> 1.5 m then the area is potentially high.

Coordinate data on GIS was taken with mapping from google map, ie coordinate point -7.030024, 110.464439 at east flood canal location, -7.055238, 110.380669 at western canal flood location, -7.004808, 110.352516 at silandak location, -7.038883, 110.319557 at plumbon location , And -7.065460, 110.463753 at the baboon location. GIS is also processed through mapping which results in maps, layers, charts, altitude values, potential status, and location of sensors that are real-time.

References

1. M. Xiaocong, Q. X. Jiao, and S. Shaohong, *International Conference on Intelligent Human-Machine Systems and Cybernetics*, *Vol.2* , 365 – 368, *IEEE*, (2015)
2. K. Lukasiewicz, K. Teymourian, and A. Paschke, *The Semantic Web: ESWC 2014 Satellite Events - Lecture Notes in Computer Science 8798*, 401-406, (2014)
3. H. Yingzhan, H. Xuemei, and Y. Xingzhi, *International Workshop on Intelligent Systems and Applications (ISA)*, 1- 4, *IEEE*, (2011)
4. E. Prahasta, *Build a Web-based GIS Application with MapServer,* Publisher Informatics, Bandung, (2006)
5. E. Prahasta, *Geographic Information System: Basic Concepts (Geodesy & Geomatical Perspectives)*, Publisher Informatics, Bandung, (2005)
6. Riyanto, P. E. Putra, H. Indelarko, *Application Development of Geographic Information System,* Publisher Media Style, Yogyakarta, (2009)
7. A. Abraham, Rule-based Expert Systems, *Handbook of Measuring System Design,* Oklahoma State University, USA, (2005)
8. A. Sulistiyo, and S. Suryono, Youngster Physics Journal, Vol.**5**, No.2, 43 – 50, (2016)
9. R. Pressman, *Software Engineering: A Practitioner is Approach*, New York: McGraw-Hill, (2001)

Information Management of Web Application Based Environmental Performance Management in Concentrating Division of PTFI

Arif Susanto[1,2,*] and *Nur Budi* Mulyono[2]

[1]Doctoral Program in Environmental Science, School of Postgraduate Studies, Diponegoro University, Semarang - Indonesia
[2]School of Business and Management, Institut Teknologi Bandung, Bandung - Indonesia

Abstract. The changes of environmental management system standards into the latest version, i.e. ISO 14001:2015, may cause a change on a data and information need in decision making and achieving the objectives in the organization coverage. Information management is the organization's responsibility to ensure that effectiveness and efficiency start from its creating, storing, processing and distribution processes to support operations and effective decision making activity in environmental performance management. The objective of this research was to set up an information management program and to adopt the technology as the supporting component of the program which was done by PTFI Concentrating Division so that it could be in line with the desirable organization objective in environmental management based on ISO 14001:2015 environmental management system standards. Materials and methods used covered technical aspects in information management, i.e. with web-based application development by using usage centered design. The result of this research showed that the use of Single Sign On gave ease to its user to interact further on the use of the environmental management system. Developing a web-based through creating entity relationship diagram (ERD) and information extraction by conducting information extraction which focuses on attributes, keys, determination of constraints. While creating ERD is obtained from relational database scheme from a number of database from environmental performances in Concentrating Division.

1 Introduction

Change of environmental management system (EMS) standard into the latest version, i.e. ISO 14001:2015 where an organization is given a three-year transition period [1], can also cause a change on the need of data and information in decision making and achieving the objective of that organization scope [2-3]. There is enough evidence that explains that the success level of an organization is determined by how well information is managed, because information is one of the organization resources [4].

Information is some data that has been processed into a meaningful and useful form for receivers and possesses some real values, or is felt and can cause a change in decision making [5-6] and is communicated well and can be in a form of records, documents, data and files that are created and managed by the organization [7].

Information management is gathering and managing information from one or more resources and their distribution to those who deserve them [8]. Information management is a systematic information management that is imaginative and responsible in creating and using it, where this information will give strategic contribution for organization objective achievement and also ensure that groups and individuals have efficient access to and make effective use of required information to perform the task and to develop themselves [10]. Information management is a continuous activity cycle and is strongly related to the identification of information need, acquisition and information creation, information analysis and interpretation, information organization and storage, information access and dissemination, and information use [8].

Information management includes developing and applying information and strategy policy; data creation and management; information processing; storage and transfer; and information use [10]. Technically, information management covers all system and processes such as web content management; document management; record management; digital asset management; learning system management; technical infrastructure used to support information management programs [4]. In management point of view, information management is an organizational, social, cultural and strategic factor that needs to be paid attention to in improving information in organization [8].

Information management is the organization responsibility that needs to be handled and followed from top management for lower level employees to ensure the effectiveness and efficiency of creation, storage, processing and distribution of information to

* Corresponding author: arifssnt@yahoo.com

 https://doi.org/10.1051/e3sconf/20183112001

support operations and decision making activity in an organization [4]. There are five bases of competitive strategy given by information management, including cost leadership, product differentiation, innovation, growth and alliances [5].

Concentrating Division of PT Freeport Indonesia (PTFI) is a process business unit in mine mineral separation [11-12], and in its environmental management, this division has conducted a transition based on EMS ISO 14001:2015 standard. Therefore a change related to information management is needed, especially for environment performance implementation and achieving the objectives stated in the organization. This information management is to give accurate, real time and comprehensive information for effective decision making in environmental management in PTFI Concentrating Division; to give information and records with low cost and to give maximum service to customers; disposing records which are no longer needed in organization.

Furthermore, information management in Concentrating Division has objective for processing, recording information and distributing it to all stakeholders as effective and efficient as possible to ensure that the information value is identified, and used in order that the environmental management objective achievement occur maximally. In managerial function, the definition of information objective is for leadership, information policy development, allocating proper resources, trainings, and evaluating information and providing feedback [8]. Therefore, Concentrating Division has to be responsible for catching, processing, storing, sharing, preserving and conveying information properly and responsibly.

The strategy to support necessary information management program through digital interface will provide high quality services to the users. There is an important issue arises from this research, i.e. the need to arrange information management program and technology adoption as a support for the program components done by PTFI Concentrating Division in order to be in line with organization objective to be achieved in environmental management based on EMS ISO 14001:2015 standard.

2 Methods

In this paper, the method used covers technical aspects in the information management of environmental management performance in the framework of implementing environmental management system based on ISO 14001:2015, i.e. by conducting web-based application planning, designing and development with usage centered design.

This web-based application development is assessed and reviewed base on the process business unit in PTFI Concentrating Division that can be used as reference. Meanwhile, limitation and recommendation in implementing this framework is discussed from practical point of view.

3 Results and discussion

3.1 Policy

Information management evolution category consists of traditional era and technical era. Traditional era covers a period where information is managed manually by using human brain and hands, papers and pens or pencils. Technology era is a period where information technology is introduced into information program management to handle several problems attached in the manual system [13]. Effective information management involves three activities, i.e. planning method, procedure control and organization management. This emphasizes that the function of management, i.e. Planning-Organizing-Actuating-Controlling (POAC), is very important in every information management program.

There were two important elements in this research; the need for policy and strategy to manage information management program and the adoption of technology as a supporting component of information management program upon the implementation of EMS ISO 14001:2015 in PTFI Concentrating Division.

Similarly with all ISO management system standards, ISO 14001 also covers the need for organization system improvement as well as continuous approach for environmental problems. In EMS ISO 14001:2015, there are key elements related to information management stated in : clause 7.4–Communication (covering 7.4.1-7.4.3); clause 7.5–Documented Control (7.5.1-7.5.3); clause 8–Operation (covering sub clauses 8.1-8.2); clause 9–Performance Evaluation (covering sub clauses 9.1-9.3); and clause 10–Improvement (clause 10.1-10.2).

The need for policy and strategy to manage information management program upon the implementation of EMS 14001:2015 in PTFI Concentrating Division is the first step stated in environmental policy change. This environmental policy states information management created by the top management to be internalized into the organization where this policy ensures that environment is an integrated part that starts with planning, engineering and operations. This is in line with the function of management, i.e. POAC. The implementation of environmental management system was done sustainably and continuous improvement is done to achieve the goals based on valid and scientific data. Therefore information management is a very necessary element.

The need for technology adoption as a supporting component of information management program upon the implementation of EMS 14001:2015 in PTFI Concentrating Division is achieved by developing information management infrastructure in the form of web based application where environmental data can be processed into accessible information for all organization members internally and externally in PTFI Concentrating Division.

3.2 Environmental performance management

The stage of environmental performance data collection is done with direct observation on the field, interview with stakeholders, and searching for organization data sources to identify environmental aspects which will be *Key to Environment Performance Indicator* (KEPI), and questionnaire distribution to the people who are competent in production process.

Identifying KEPI by determining performance indicators starts with defining context or organization, listing all interested parties and scope of Environmental Management System. Afterwards, corporate *objectives* (goals) and environmental targets to achieve, where leadership, which acts as part of management system, shall confirm them through plan-do-check-action cycle, and through review on environmental policy. Then, the environmental aspects from corporate activities are identified to be given weighing based on the criteria of risk and opportunity assessment results.

Evaluation based on these criteria aims to know the environmental aspects of the production process that shall create quite big impacts so they need to be considered as environmental performance indicators. The environmental aspects considered significant as an impact are monitored and controlled so they keep fulfilling the compliances, thus operational control acts as a control in environmental performance achievement.

Environmental performance has a two-side meaning. On one side, it means all the negative effects on the environmental elements, i.e water, ground, air, etc., caused by activities of organization. And the other side, it can be defined as acts and their effects in connection with prevention and decreasing the charging of these elements [14]. Environmental performance evaluation is a management possibility for:

1. decreasing cost and expenditures,
2. increasing the market position and the value of the organization,
3. helping the observation of direction,
4. getting a higher motivation of leaders, managers and staff as well.

Designing Environmental Performance Measurement System uses integration from a model of environmental management system which follows *Plan-Do-Check-Act* (PDCA) with a concept of IEMPS (*Integrated Environment Performance Measurement System*). The objective of using IEMPS model is to obtain a structured process to achieve continuous improvement in environmental management, as illustrated in figure 1 below. Environmental performance assessment using IEPMS method shall consider two measures; quantitative and qualitative. Therefore the result obtained shall be more integrated, because it considers relevance, analysis accuracy, measurability and comparability [15]. To conduct KEPI designing, a comprehensive study on identifying organization activity based on life-cycle perspectives is conducted.

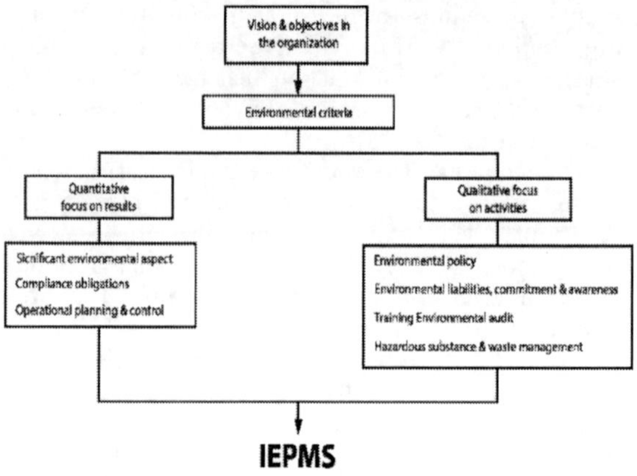

Fig. 1. Integrated Environment Performance Measurement System (IEPMS)

Furthermore, a model of environmental performance measurement [16] is illustrated in figure 2 as follows.

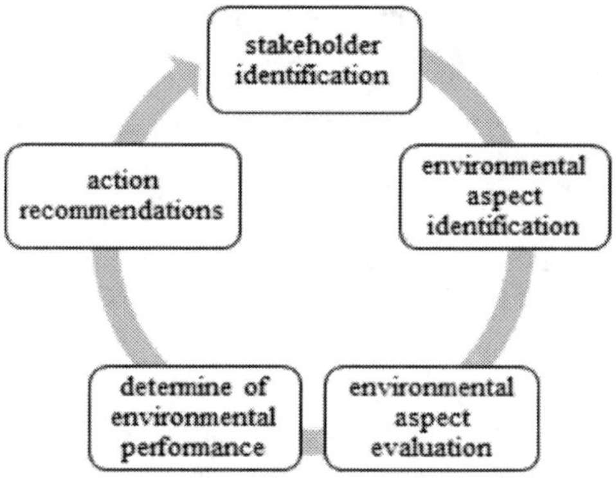

Fig. 2. Model of Environmental Performance Measurement

3.3 Control of Documented Information and Web Application

The major change to ISO 14001:2015 means that new standard is opportunity for Concentrating Division PTFI to align a strategic direction with PTFI wide environmental management system, increased focus on improving environmental performance. Mandatory documents required by ISO 14001:2015 depicted in table 1. A mandatory records depicted in table 2 as below.

In implementing environmental performance information management, web applications were used. This application planning aimed for the users to benefit all information on the implementation of environmental management system in Concentrating Division easily, started with access to enter the intranet network system by using usage centered design.

Usage centered design is a systematic process using abstract models to design the smallest, simplest system that fully and directly supports all the tasks users need to accomplish [17]. Its focus on usage (improved tools supporting ask accomplishment), driving by models and modelling, selective user involvement, design by modelling, systematic and fully specified process, and design by engineering.

Table 1. Mandatory documents required by ISO 14001:2015 in PTFI Concentrating Division

Documents	
Clause 4.3	Scope of the EMS
Clause 5.2	Environmental Policy
Clause 6.1.1	Risk and Opportunities
Clause 6.1.2	Criteria for evaluation of significant environmental aspects
Clause 6.1.2	Significant environmental aspects
Clause 6.2	Environmental objectives and plans for achieving them
Clause 8.1	Operational control
Clause 8.2	Emergency preparedness and response

Table 2. Mandatory records required by ISO 14001:2015 in PTFI Concentrating Division

Records	
Clause 6.1.3	Compliance obligations records
Clause 7.2	Records of training, skills, experience and qualifications
Clause 7.4	Evidence of communication
Clause 9.1.1	Monitoring and measurement results
Clause 9.2	Internal audit programs
Clause 9.2	Results of internal audits
Clause 9.3	Results of the management review
Clause 10.1	Results of corrective actions

Web single sign on (SSO) was used as a technique to authenticate the users, which later on, automatically enter the Concentrating Division EMS as shown in figure 3. SSO could increase the usability of a network as a whole and at the same time to manage the other environmental performance indicators. SSO as operating system login was made by inserting username in the form of 8-digit Universal Identity number of PTFI employees, and with a password of the last 4-digit of the UID number, so the users can enter the environmental web based (figure 4) and manage their environmental performance.

Single Sign-On (SSO) illustrated in figure 3 was used in the adoption of web-based application in ISO 14001:2015 implementation in PTFI Concentrating Division, where network users must do credential authentication with username and password for every services for registered users. Users authenticate once only for login to the system services to be used without further interaction manuals.

The web-based application developed in ISO 14001:2015 implementation uses usage-centered design, where this design focuses on usage (improved tools supporting task accomplishement); driven by model and modeling; selective user involvement; systematic and fully specified process and design by engineering.

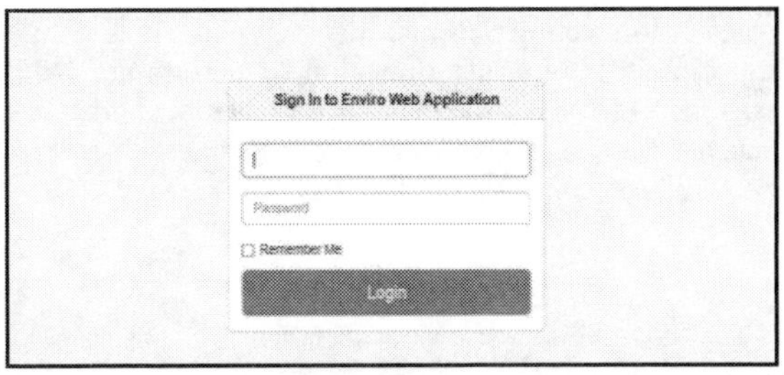

Fig. 3. Web Single Sign On to Environmental Web Application in PTFI Concentrating Division

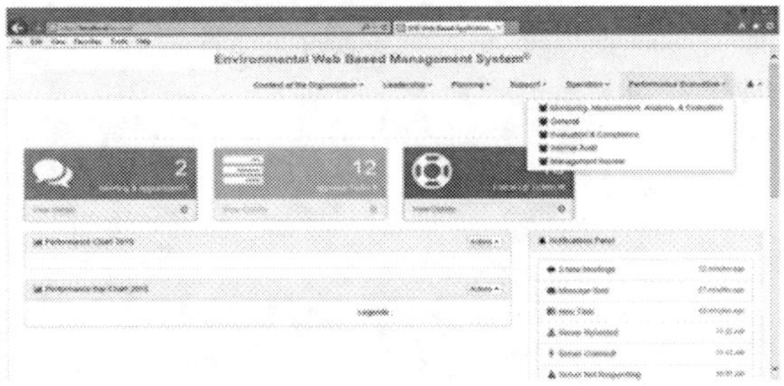

Fig. 4. View of Web Application in the Implementation of EMS in PTFI Concentrating Division

Usage centered web design projects a lightweight form of the process has evolved, takes certain modeling and design shortcuts, when practical, the time to gather, model and validate user requirements is further reduced by developing models collaboratively with end users. Agile usage centered design proceeds [18]:

1. Preliminary steps:
   a. Essential purpose and preconception with clarify business and user purposes then fantasize and set aside preconceptions of features, facilities, content and capabilities;
   b. Exploratory modelling: identify questions, ambiguities, and areas of risk and uncertainty.
2. First iteration:
   a. Role modelling with inventory and prioritize all user roles and select intially targeted subset;
   b. Task modelling with inventory and prioritize all tasks and select initially targeted subset;
   c. Task clustering with group all tasks by affinity and draft overall navigation architectures;
   d. Design draft visual and interaction scheme, review and revise aesthetic design;
   e. Abstract prototyping with develop content model for interaction contexts supporting selected subset of tasks;
   f. Design and develop detailed user interface design for selected interaction contexts;
   g. Construction program designed portions of user interface.

Methods consist of two stages, i.e. developing a web-based through creating entity relationship diagram (ERD) and information extraction (figure 5). The first stage is conducting information extraction which focuses on attributes, keys, determination of constraints. While creating ERD is obtained from relational database scheme from a number of database from environmental performances in Concentrating Division.

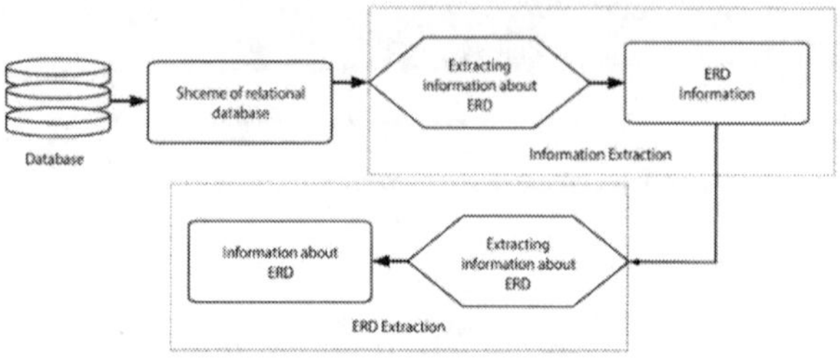

Fig. 5. ERD Diagram Database Scheme

Entity Relationship Diagram (ERD) is a common diagram, which aims to present the database structure in conceptualized form. To build good ERD and excellent schema depends on type of word (verb, noun, adjective, and gerund) [19]. It can be generated from database that make it understandable and simple diagram as depicted for stack emission test (figure 6).

Extracted ERD schema set from an operational relational database [20]. Designers must use tools to reach best implementation of these methodologies by ER-Draw to represent ERD and translated to schema of relational database [21].

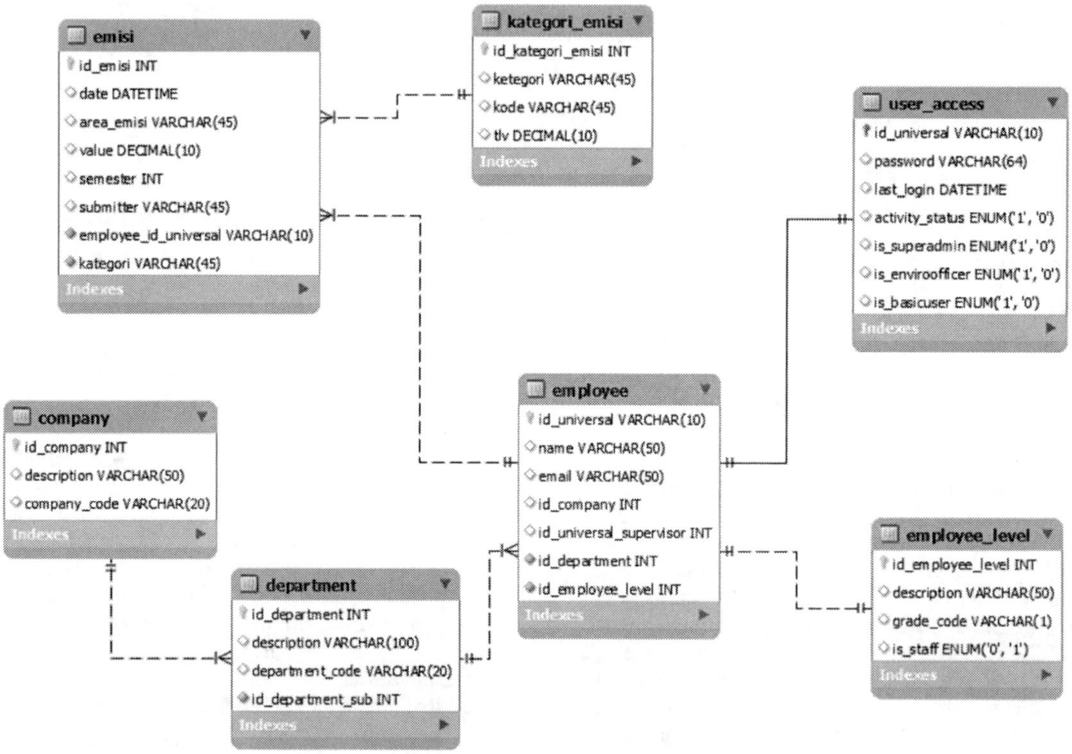

Fig. 6. ERD for Stack Emission Test

The use of web-based application in EMS ISO 14001:2015 starts with information extraction process by using relational database scheme, (sample for stack emission test and its depicted in figure 6) as main input, and depends on the available contextual data, or depends on database script code. In the first sight of the relational database scheme, several information can become clear and on a closer look, some important information can be defined. This information contains database about attributes, entities and relationship between these entities. Figure 6 presents ERD from relational database and validation of the extraction results used by more advanced rules and based on keywords, and depends on the table that contains data or script code of the database. The web application contains the management of compliance obligations records which are obtained from the implementation of the regulations and laws.

References

1. ISO. *Introduction ISO 14001:2015*. Ed. **1** (2015).
2. R. Saloojee, D. Groenwald, A.S. du Toit, S Afr J Inf Manag, **6**, 62-70 (2007).
3. A. Weintraub, L. Owens, E. Jedinak. Cambridge M.A. *USA: Forrester Research, Inc.* (2013).

4. G.S. Reddy, R. Srinivasu, S.R. Rikkula, V.S. Rao. International Journal of Reviews in Computing, **5**, 2076-3336 (2009).
5. J.A. O'Brien, and G.M. Marakas. E **14**. New York: McGraw-Hill (2008).
6. K.C. Laudon, and Z. Irani. Elsevier: Information & Management, 227-242, (2004).
7. M.F. Robek, G. F. Brown, and D.O. Stephens. E **4**. New York: McGraw-Hill, Columbus (1995).
8. J. Robertson. *Step Two Design* Pty. Ltd., (2005).
9. S. Henczel. *Worldwide Conference on Special Librarianship.* Brighton: Special Libraries Association (SLA), 210-226, (2000).
10. E. Maceviciute, and T.D. Wilson. Information Research, **7**, (2002).
11. A. Susanto, E.K. Putro, R. N. Wulan, and W.E. Yochu. Globally Harmonized System (GHS) Implementation and Management of Hazardous Substances. *1st International Conference on Sustainable Agriculture and Environmental (ICSAE),* Solo: Sebelas Maret University, 302-308 (2012).
12. A. Susanto. Thesis. School of Business and Management, Institut Teknologi Bandung, (2017).
13. M.J. Earl. Mayland Avenue: Prentice Hall International (UK) Ltd., (1989).
14. W.L. Kuhre. Prentice Hall, Inc., (1996).
15. M. Stutz, P. Singhal, M. Terho, S. Ahonen, G. Rice, H. van der Wel. LCA Info, (2004).
16. E. Gunther, and A. Sturm. Technische Universitat Dresden, (2000).
17. L.L. Constantine and L.A.D. Lockwood. MA: Addison-Wesley, (1999).
18. L.L. Constantine and L.A.D. Lockwood. MA: Addison-Wesley, (2001).
19. P.P.S. Chen. Information Sciences, **29**, 127-149 (1983).
20. S. Xu, L. Yu and S. Lu. Computers and Their Applications: 143-146, (2003).
21. N. Mfourga, N. Reverse Engineering, Proceeding of the Fourth Working Conference on IEEE, (1997).

Assessment of groundwater recharge potential zone using GIS approach in Purworejo regency, Central Java province, Indonesia

Daniel Eko Aryanto[1,*], *Gagoek* Hardiman[2]

[1]Master Program of Environmental Science, School of Postgraduate Studies, Diponegoro University, Semarang-Indonesia
[2]Faculty of Engineering, Diponegoro University, Semarang-Indonesia

Abstract. Floods and droughts in Purworejo regency are an indication of problems in groundwater management. The current development progress has led to land conversion which has an impact on the problem of water infiltration in Purworejo regency. This study aims to determine the distribution of groundwater recharge potential zones by using geographic information system as the basis for ground water management. The groundwater recharge potential zone is obtained by overlaying all the thematic maps that affect the groundwater infiltration. Each thematic map is weighted according to its effect on groundwater infiltration such as land-use - 25%, rainfall - 20%, litology - 20%, soil - 15%, slope - 10%, lineament - 5%, and river density - 5% to find groundwater recharge potential zones. The groundwater recharge potential zones thus obtained were divided into five categories, viz., very high, high, medium, low and very low zones. The results of this study may be useful for better groundwater planning and management.

1 Introduction

Ground water is the amount of water beneath the earth's surface that can be collected with wells, tunnels or drainage systems or by pumping. It can also be called a stream that naturally flows into the soil surface through jets or seepage [1]. The availability of ground water in a region is very important for the survival of the region.

Purworejo as one of the regencies in Central Java Province, based on the results of the research have alert status related to water balance and critical water resources per Regency in Central Java [1]. While another study concluded that by 2025 the number of districts with water deficits increased by 78.4 percent with deficits ranging from one to twelve months, or deficits throughout the year. This means that water availability in Purworejo regency needs to serious attention to its conservation.

The rapid development in Purworejo regency has caused land conversion which resulted in the decreasing of area for groundwater infiltration. The reduced water catchment area causes groundwater filling to be disrupted and result in a decrease in groundwater availability. In general, water infiltration process occurs through the filtration process (movement of water from the top into the soil surface) and percolation (ground water movement down from the unsaturated zone into the saturation zone) that occur sequentially. Infiltration rate will be equal to rain intensity if infiltration rate is still less than infiltration power. Percolation will not occur if porosity in the zone is not saturated yet optimally contain water [3].

Disaster occurring in Purworejo regency, the floods that occur almost every year that cause fatalities and property [4], as well as droughts that cause the residents difficulty to get clean water [5] is an indication of problems related to groundwater infiltration. Rainwater that is not maximally infiltrated into the ground causes such a disaster.

The absence of a comprehensive groundwater recharge potential zone map in Purworejo Regency has resulted in the development program not considering the catchment area as an area that should be protected and conserved as part of groundwater conservation. Groundwater recharge area is the area where the rain water infiltrated into the soil which then becomes ground water.

To determine the potential groundwater recharge area is to use mapping and processing through Geographic Information System. Geographic Information System (GIS) is a computer-based system or technology constructed for the purpose of collecting, storing, processing and analyzing, and presenting data and information of an object or phenomenon related to its location or existence Earth surface. Basically GIS can be broken down into several interrelated sub-systems that include data input, data management, data processing or analysis, reporting (output) and data results [6].

Mapping of potential groundwater recharge potential zones using Geographic Information System

* Corresponding author: dea.nuha@gmail.com

https://doi.org/10.1051/e3sconf/20183112002

has been done by many researchers, such as Manap et al. [7], Madani and Niyazi [8], Elbeih [9], Nagarajan [10], Oikonomidis, et al. [11], Senanayake, et al. [12]. Some steps taken to obtain the groundwater recharge potential map are: (1) identification of thematic map that influences groundwater infiltration. (2) weighting and scoring on each thematic map. (3) a thematic map application with a GIS program for spatial mapping.

2 Study Area

Purworejo is one of regency in Central Java province, located between 109° 47'28" to 110° 8'20" East Longitude and between 7° 32 to 7° 54" South Latitude. North of Purworejo borders Magelang and Wonosobo, South by the Indonesian Ocean. The west of Purworejo borders Kebumen Regency and the east borders of Yogyakarta Province precisely Kulonprogo Regency [13].

Topographical of Purworejo is wet tropical regions with temperatures between 19° C - 28° C while humidity between 70 percent until 90 percent.

Purworejo regency has an area of Groundwater Basin and Non Basin of Ground Water. The Groundwater Basin for Purworejo Regency is in accordance with Presidential Decree Number 26 of 2011 on the Establishment of Groundwater Basin [14], including in Kebumen-Purworejo Groundwater Basin with Longitude Coordinate: 109° 27' 18.50"- 110° 03' 36.50" and latitude coordinates : -07° 34' 14.61" - -07° 53' 59.96".

3 Methodology

Groundwater recharge refers to the entry of water from the saturated zone into unsaturated zone. Below the water table surface, together with the associated flow away from the water table within the saturated zone.

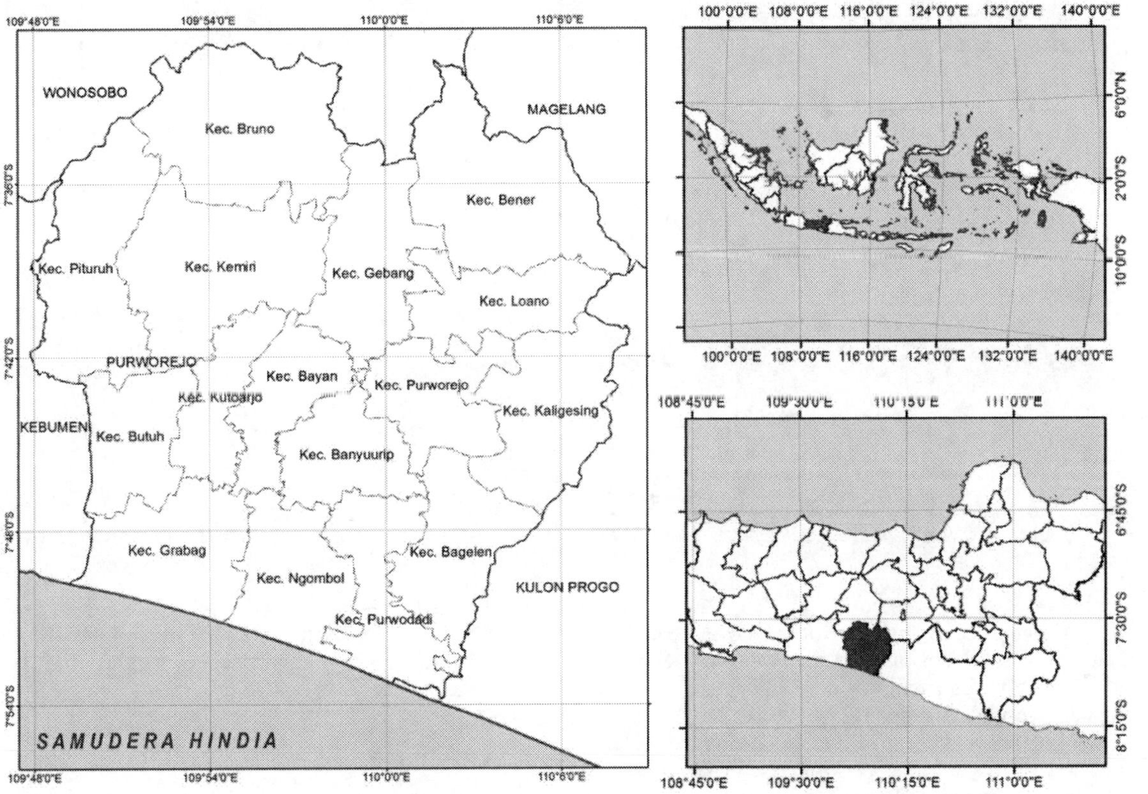

Fig. 1. Location map of the study area, Purworejo regency, Central Java province, Indonesia

Purworejo Regency is divided into 16 sub districts and 494 villages. Region of Purworejo in 2015 has vast 103,481 ha or about 3.18 percent of Jawa Tengah. Land area of 103,481 ha in Purworejo consists of 87,105 ha (84.18 percent) of agricultural land and 16,375 ha (15.82 percent) is non agricultural land. Agricultural land that is used as a wetland 30,225 ha (34,70 percent) and non wetland 56,880 ha (65.30 percent).

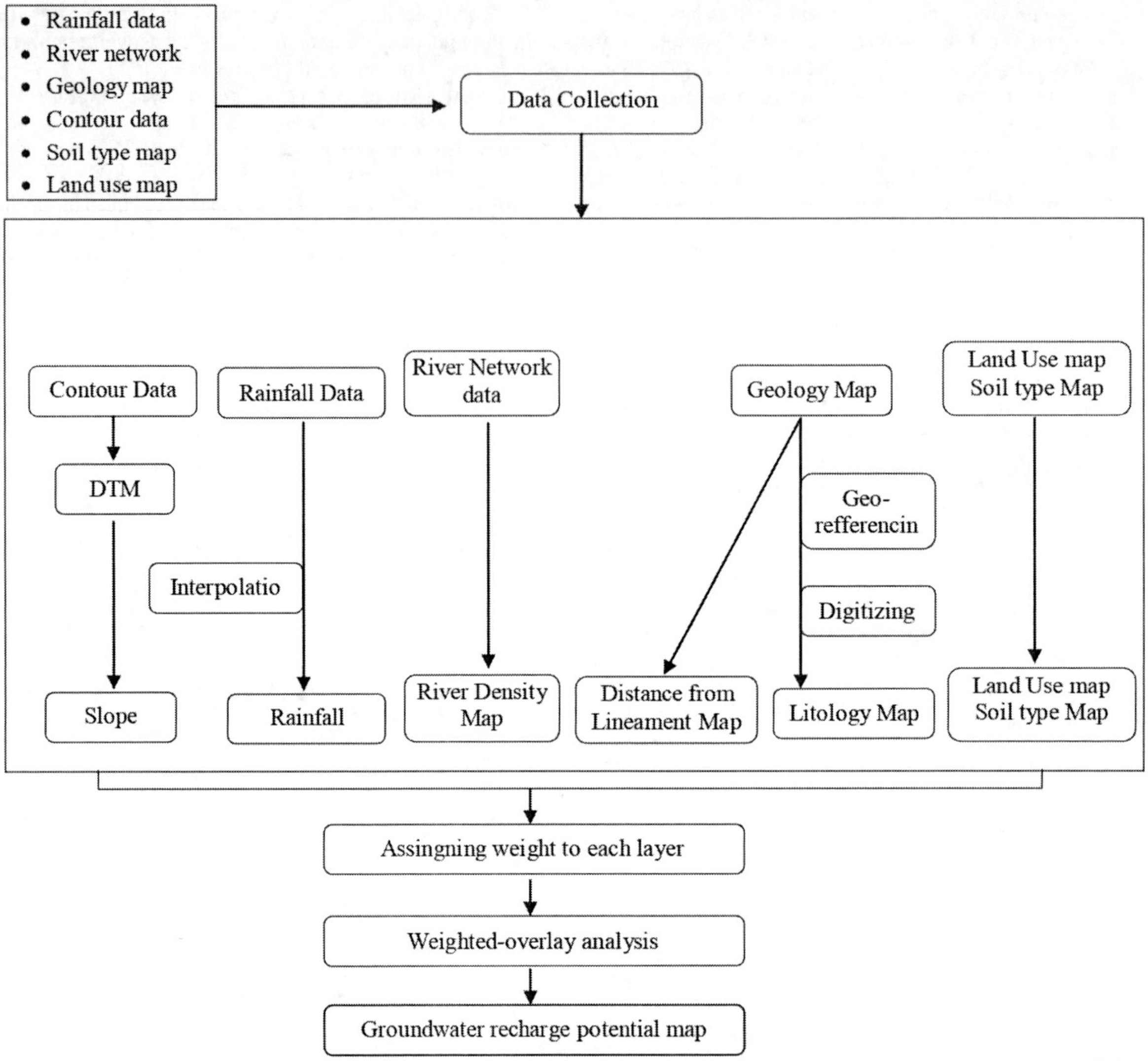

Fig. 2. Flow chart of the metodology employed in the study to delineate groundwater recharge potential of Purworejo regency.

Recharge occurs when water flows past the groundwater level and infiltrates into saturated zone. It is an extremely important water component of the circulation cycle in nature [15].

With the development of remote-sensing technology and its spatially accurate performance in georeference processes, the factors influential to groundwater prospect zonation can be more easily identified on a broad scale [16]. GIS approach were employed in this study to delineate the groundwater recharge potential of Purworejo regency based on time and cost effectiveness [10]. Identification of suitable zones for groundwater recharge was conducted through a knowledge-based factor analysis, using rainfall, land, use, lithology, soil type, slope, river density, and distance from lineament layers.

The methodology followed in this study to delineate suitable sites for groundwater recharging has been illustrated in Fig. 2.

3.1 Weightage

Weighting for identification of groundwater recharge [potential zones has been done by many researchers using geographic information systems. This research uses a subjective approach, where the final result depends on the signification level of the thematic map and the weighting factor used in this study. The integration of the layer factor that influences the groundwater recharge using overlay method in a GIS platform in this study is presented in Table 1.

Table 1. Rank and weight for factors used for groundwater potentiality mapping (modified after ; Wibowo, [3]; Madani & Niyazi, [8]; Elbeih, [9]; Nagarajan & Singh, [10]; Oikonomidis, Dimogianni, Kazakis, & Voudouris, [11]; Senanayake, Dissanayake, Mayadunna, & Weerasekera, [12];; Yeh, Cheng, Lin, & Lee, [15]; Ahmed, [17]; Magesh, Chandrasekar, & Soundranayagam, [18]; Sigit, [19]; Adibah, [20]).

Parameters	Class		Rank	Weight (%)
Land use	Very high	Forest	5	25
	High	Garden	4	
	Moderate	Shrubs	3	
	Low	Empty land. Vacant land,	2	
	Very Low	Settlement, Swamp, Rice field,	1	
Litology	High	Alluvium Deposits	5	20
	Moderate	Sandstone and Limestone	3	
	Low	Volcanic Breccia	2	
	Very Low	Andesit	1	
Rainfall	Very high	> 3000 mm/thn	5	20
	High	2000 – 3000	4	
Soil type	Very high	Regosol	5	15
	Moderate	Alluvial	3	
	Very Low	Litosol	1	
Slope	Very high	< 8 %	5	10
	High	8 – 15	4	
	Moderate	15 – 25	3	
	Low	25 – 40	2	
	Very Low	> 40	1	
River density	Very high	> 2.7 km/km^2	5	5
	High	2.12 – 2.7	4	
	Moderate	1.56 – 2.12	3	
	Low	0.95 – 1.56	2	
	Very Low	< 0.95	1	
Distance from Lineament	Very high	< 1.500 meter	5	5
	Very Low	> 1.500	1	

3.2 Land Use Layer

Land use maps in the study area were obtained from the Public Works and Spatial Planning Office of Purworejo regency. The use of space in Purworejo regency with total area of 107,980.67 ha consists of 8 types of land use covering Rice field (34.669,05 ha), Swamp (354,6 ha), Shrubs (5,350,31 ha), Empty land (554.44 ha), Vacant land (5,621.38 ha), Gardens (39,826.62 ha), Forest (1,339.33 ha) and Settlements (19,062.06 ha). Rice fields, gardens and settlements dominate land use types in the study area. Types of land use with large amounts of vegetation indicate a high groundwater infiltration potential compared to the type of land use that lacks vegetation. Land use maps are presented in Figure 3.

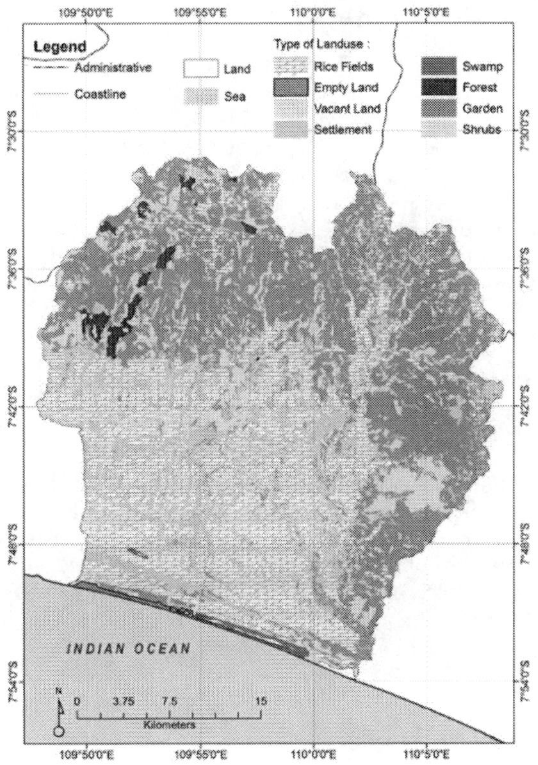

Fig. 3. Land use map of the study area.

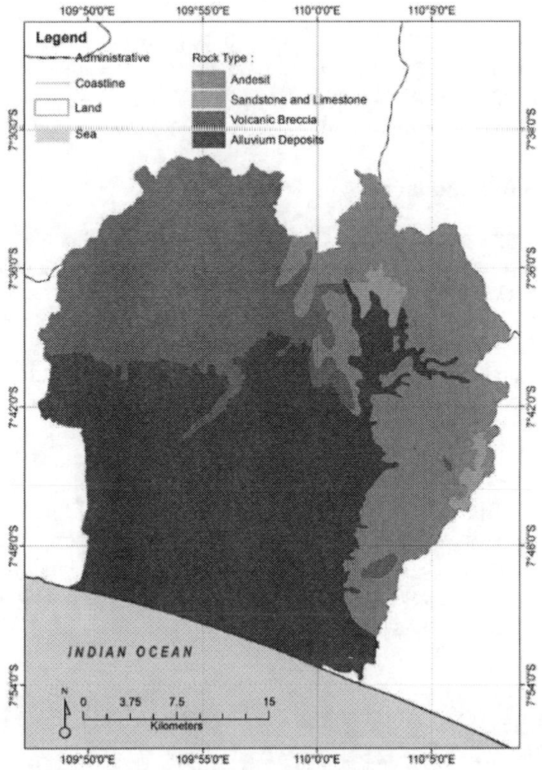

Fig. 4. Litology map of the study area.

3.3 Lithology layer

Lithology map of Purworejo regency was prepared by scanning, geo-referencing and digitizing the 1: 25.000 geology map of central java. The study area consists of Andesite rock types (26,580.56 ha), Volcanic Breccia (27,765.03 ha), Sandstone & limestone (1,812.39 ha), and Alluvium deposits (51,822.69 ha). The most potent Alluvium deposits types are potential for groundwater infiltration compared to other rock types. The study area lithology map is presented in Figure 4.

3.4 Rainfall layer

Rainfall in the study area is almost high seen from annual rainfall data. The rainfall map is obtained from average annual rainfall data from 2011 - 2016, then interpolated to get map of rainfall distribution. The level of rainfall in the study area is categorized Very high on the area of 60,531.13 ha and High area of 47,449.54 ha. Higher levels of rainfall are potentially infiltrates into the soil and stored as groundwater. The rainfall maps in the study area are presented in Figure 5.

3.5 Soil type layer

The soil type map in the study area is obtained from the spatial map of Purworejo regency, which consists of 3 types of land, Latosol (68,580,56 ha), Alluvial (34,439 ha), and Regosol (5,432 ha). Type of soil Regosol potentially infiltates groundwater better than alluvial and latosol soil types. The soil type charts in the study area are presented in Figure 6.

3.6 Slope layer

Slope is the rate of change of elevation, and is also a significant factor in identifying groundwater potential zones. Increased slope results in high runoff and erosion of surface soil [17]. The study area outline is lowland with a fairly sloping slope. Areas with a fairly steep slope exist only in the hills in the north and east. Study area with slope level of <8% covering 55,305,69 ha, 8 - 15% (19,579,02 ha), 15 - 25% (17,415,96 ha), 25-40% (11,252,22 ha), and> 40% (4,427,78 ha). The higher the slope of the land the run-off rate is greater, but the water absorption rate of the soil is smaller. In the area is actually suitable to be used as a conservation area, because the water level is increasingly large and the erosion rate is also getting bigger. The land slope map of the study area is presented in Figure 7.

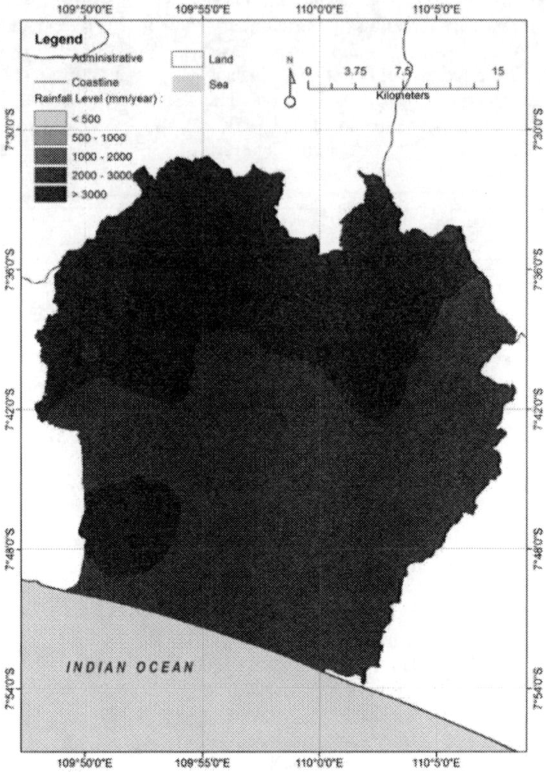

Fig. 5. Rainfall map of the study area.

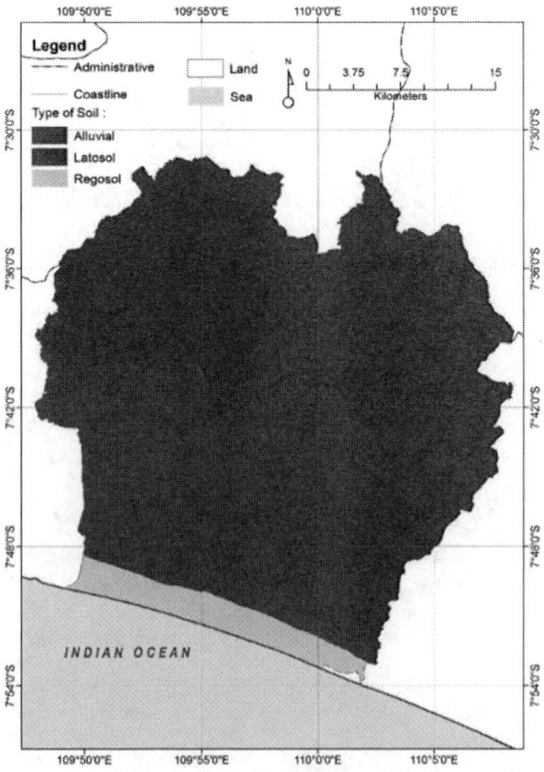

Fig. 6. Soil type map of the study area

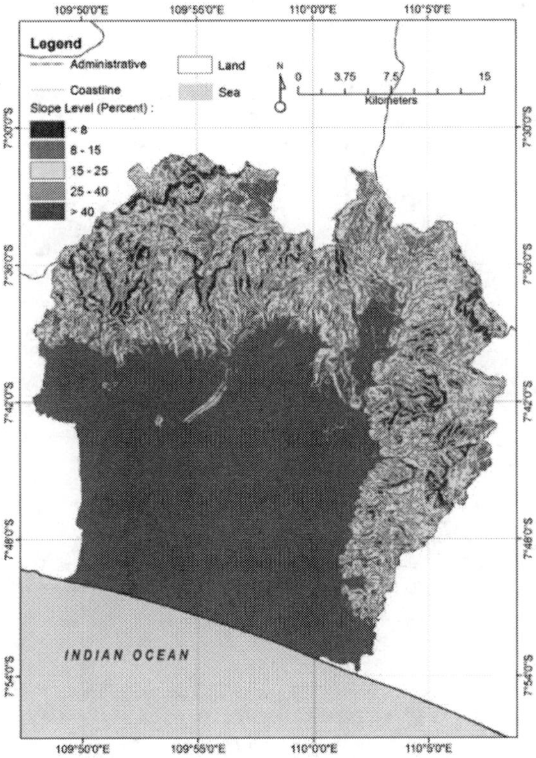

Fig. 7. Slope map of the study area

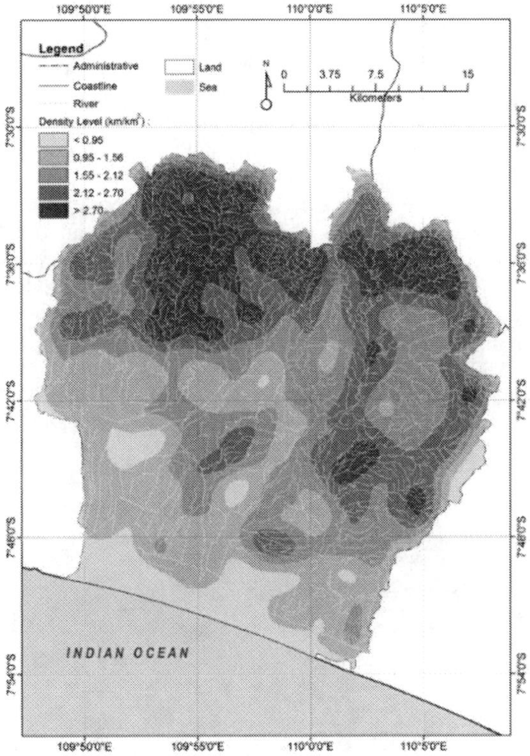

Fig.8. River density map of the study area

3.7 River density layer

River density levels indicate the potential for greater groundwater infiltration. In the study area, areas with large river densities are located in hilly areas. River density levels are divided into 5 categories, ie very high, high, medium, low, and very low. Areas with very high river density of 17,628 ha, High (24,160, 65 ha), Medium (28,671, 27 ha), Low (27,968.28 ha), and Very Low (9.662,29 ha). A map of river density in the study area is presented in Figure 8.

3.8 Distance from Lineament layer

Lineament is structurally controlled linear or curvilinear features, which are identified from the satellite imagery by their relatively linear alignment. Lineaments represent the zones of faulting and fracturing resulting in secondary porosity and permeability. These are are hydro-geologically very important as they provide the pathways for groundwater movement [18]. In this study used lineament in the fault form as a potential zones for the infiltrates of groundwater which we known as non basin groundwater area. Map of this fault is obtained from the regional geological map of Central Java. The proximity level of a zone on the lineament indicates more potential for flowing into the lineament and then become groundwater. The distance from lineament map is shown in Figure 9.

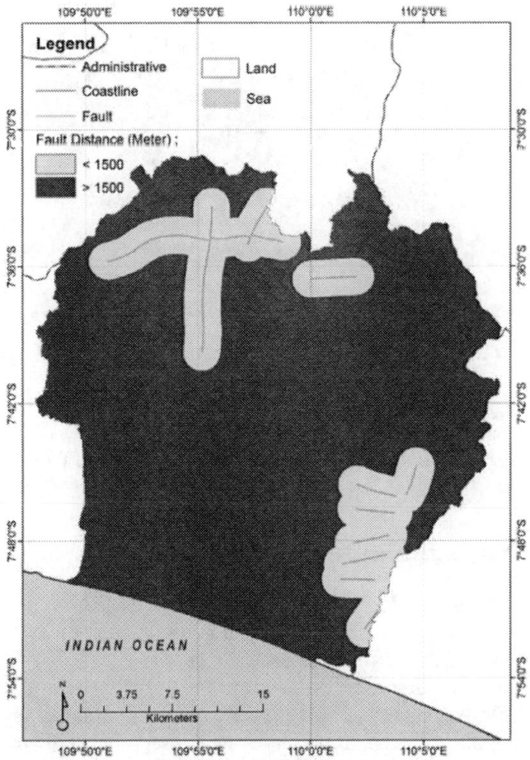

Fig. 9. Lineament map of the study area

4 Result

The groundwater potential zones for the study area were generated through the integration of various thematic maps viz., land use, lithology, rainfall, soil type, slope, river density, and distance from lineament using GIS techniques. The demarcation of groundwater potential zones for the study area was made by grouping of the interpreted layers through weighted factor and finally assigned different potential zones. The groundwater recharge potential zone divided into five classes from very low into very high. According to the final map, the areas of very high occurs an area of 4.597.56 ha, high (24.832,08 ha), moderate (38.675,79 ha), low (30.398,13 ha), and very low occurs in area of 9.477, 09 ha. Groundwater recharge potential zones map is shown in figure 10.

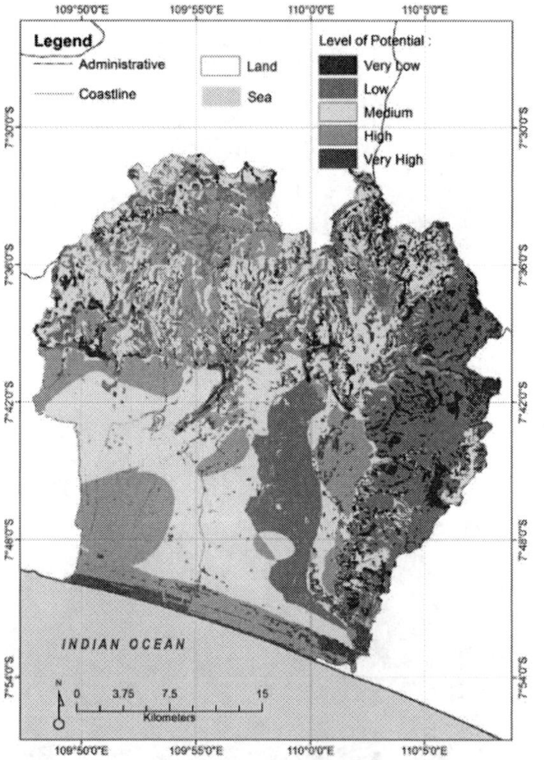

Fig. 10. Groundwater recharge potential zone map of the study area

5 Conclusions

Mapping groundwater recharge potential zones using Geographic Information System is one technique that is very effective and efficient. From the results of this study to describe the region with the potential to infiltrates rainwater into groundwater as consideration for policy makers to preserve and conserve groundwater. The results of this study can also be used as a reference in the revised spatial planning Purworejo.

Reference

1. Kodoatie JR. *Tata Ruang Air Tanah.*; (2012).
2. Nugroho SP. Evaluasi Keseimbangan Air di Provinsi Jawa Tengah, 3(2):175-182, (2007)
3. Wibowo M. Model Penentuan Kawasan Resapan Air untuk Perencanaan Tata Ruang Berwawasan Lingkungan, **1**(1):1-7, (2006)
4. Yuliana. (2016). Kesiapsiagaan Masyarakat dalam Menghadapi Bencana Banjir di Desa Bendungan, Kecamatan Grabag, Kabupaten Purworejo 4. Adi, H. P. (2011). Kondisi dan Konsep Penanggulangan Bencana Kekeringan Di Jawa Tengah. Seminar Nasional Mitigasi Dan Ketahanan Bencana, 1–10. http://doi.org/978-602-8420-85-3.
6. A. Ekadinata, S. Dewi, D. Hadi, D. Nugroho, and F. Johana, World Agroforestry Centre, Bogor, Indonesia, (2008)
7. M. A. Manap, H. Nampak, B. Pradhan, S. Lee, W. N. A. Sulaiman, and M. F. Ramli. Arabian Journal of Geosciences, 7(2), 711–724, (2014) https://doi.org/10.1007/s12517-012-0795-z
8. A. Madani, B. Niyazi, *Environ Earth Sci*.74(6):5129-5142, (2015) doi:10.1007/s12665-015-4524-2.
9. SF. Elbeih, *Ain Shams Eng J*, 6(1):1-15, (2015) doi:10.1016/j.asej.2014.08.008.
10. M. Nagarajan, S. Singh, J Indian Soc Remote Sens, 37(1):69-77, (2009) doi:10.1007/s12524-009-0012-z.
11. D. Oikonomidis, S. Dimogianni, N. Kazakis, K. Voudouris, *J Hydrol*. 525:197-208, (2015) doi:10.1016/j.jhydrol.2015.03.056.
12. IP. Senanayake, Dissanayake DMDOK, BB. Mayadunna, WL. Weerasekera. *Geosci Front*.7(1):115-124, (2016).
13. Badan Pusat Statistik. Kabupaten Purworejo dalam angka. [internet].(2016). Available from: https://www.purworejokab.bps.go.id/index.php/publikasi/index?Publikasi%5BtahunJudul%5D=&Publikasi%5BkataKunci%5D=purworejo+dalam+angka&yt0
14. Presiden P. KEPUTUSAN PRESIDEN REPUBLIK INDONESIA NOMOR 26 TAHUN 2011 TENTANG PENETAPAN CEKUNGAN AIR TANAH. 1999:1-16. http://www.komnasham.go.id/sites/default/files/dokumen/UU NO 39 TAHUN 1999 HAM_0.pdf.
15. H. Yeh, Y. Cheng, H. Lin, C. Lee. *Sustain Environ Res*, 26(1):33-43, (2016).
16. T. N. Adji, and S. P. Sejati, *Identification of groundwater potential zones within an area with various geomorphological units by using several field parameters and a GIS approach in Kulon Progo Regency , Java , Indonesia,* 161–172. https://doi.org/10.1007/s12517-012-0779-z16..
17. SA. Ahmed, *Egypt J Remote Sens Sp Sci*, 19(2):223-234, (2016) doi:10.1016/j.ejrs.2016.06.002.
18. NS. Magesh, N. Chandrasekar, JP. Soundranayagam, *Geosci Front*, 3(2):189-196, (2012) doi:10.1016/j.gsf.2011.10.007.
19. A. A. Sigit, *Pemanfaatan Teknologi Penginderaan Jauh dan Sistem Informasi Geografis untuk Pendugaan Potensi Peresapan Air DAS Wedi Kabupaten Klaten*-Boyolali, 25(1), (2011).
20. N. Adibah, S. Kahar, B. Sasmito, *Aplikasi Penginderaan Jauh dan Sistem Informasi Geografis untuk Analisis Daerah Resapan Air (Studi Kasus :*

Kota Pekalongan).2(April):141-153, (2013)
21. Bappenas, *Studi Prakarsa Strategis SDA untuk Mengatasi Banjir dan Kekeringan di Pulau Jawa.* Tidak Diterbitkan. Jakarta, (2005)

Assessing the Priority Area of Mountainous Tourism Using Geospatial Approach in Kendal Regency, Central Java

Riwayatiningsih[2,*] and *Hartuti* Purnaweni[1,2]

[1]Doctoral Program of Environmental Science, School of Postgraduate Studies, Diponegoro University, Semarang - Indonesia
[2]Master Program of Environmental Science, School of Postgraduate Studies, Diponegoro University, Semarang - Indonesia

Abstract. Kendal is one of 35 regencies in Central Java which has diverse topographies, from low land, hilly, to mountainous areas. Mountainous area of Kendal with numerous unique and distinct natural environments, supported by various unique and distinct culture of its community can be used for tourism activities. Kendal has natural and sociocultural resources for developing tourism that must be considered by the local government. Therefore, nature based tourism resources assessment is important in order to determine the appropriate area in the planning of sustainable tourism destination. The objectives of this study are to assess and prioritize the potential area of mountainous tourism object in Kendal using geospatial approach based on criteria attractiveness, accessibility and amenity of the tourism object. Those criteria are modification of ADO-ODTWA guidelines and condition of the study location. There are 16 locations of tourism object that will be assessed. The result will be processed using *ArcMap 10.3*. The result will show the most potential tourism object that could become priority for mountainous tourism development in Kendal.

1 Introduction

Kendal is one of 35 regencies in Central Java which has diverse topographies, from low land, hilly to mountainous areas. Low land area is located in the northern area with altitude between 0 up to 10 meters above sea level and temperature ranges around 27^{0}C. The hills are in the middle area, and the mountainous area are located in the most southern area with elevations up to 2.579 meters above sea level and temperature ranges around 25^{0}C. [1]

The area of Kendal regency presents numerous unique and distinct natural environment, supported by various unique and distinct culture of its community which could become tourism potency to attract visitors. However, many of these areas have not been well managed, despite the potencies found such as in Gonoharjo and Curugsewu rainfall, both are located in mountainous area [2, 3]

As a key factor in revenue, job creation, business and infrastructure development, tourism sector needs good planning to develop it. This tourism planning should not affect in environmental damage but instead must be planned and implemented to support sustainable development. [4-5]

Today, there is a growing demand for nature based tourism which becomes a new trend in the tourism industry. It was seen from a lot of documentation uploaded in social media used as self-actualization. This condition requires tourism actors, including in mountainous tourism area of Kendal Regency.

In order to develop the tourism potencies, a proper strategy is needed to manage them. By using ADO-ODTWA (Operational Area Analysis-Nature based Tourism Object and Attraction) Guidelines issued by the Directorate General of Forest Protection and Nature Conservation [6] Oktadiyani, et.al [7] analyzed the Kamojang Crater Nature Park of Bandung Regency, West Java Province using several criterias including attractiveness, relationship level, lodging condition, supporting facilities and infrastructure, clean water availability, and connection with other tourism objects. While another study held in Tanjung Belandang Beaches of Ketapang Regency used criteria of attractiveness, accessibility, socio-economic conditions, accommodation, supporting facilities and infrastructure, as well as clean water availability to assess the attractiveness and development of Tanjung Belandang beaches [8].

There was also another research conducted by Bunruamkaew and Murayama [9] which was aimed to identify and prioritize the potencies of ecotourism in Surat Thani Province, Thailand using GIS and AHP. Meanwhile Rahayuningsih, et.al [4] aimed to develop spatial model of natural tourism planning based on attraction criteria and accessibility of tourism object in Bogor.

Hai-ling [10] concluded that there was a power in integrating GIS for ecotourism applications. GIS is a useful tool to solve some semi-structured spatial decision problems that is encountered in the real world. To assess the effectiveness of this application, survey can be conducted in the future to evaluate whether the system improves the experience of ecotourism tourist. This study implies that the key process can be a resource and reference for other ecosystem services analysis.

Utilizing Geographic Information System (GIS) technology as geospatial approach, it is expected that

* Corresponding author: tinink@yahoo.com

 https://doi.org/10.1051/e3sconf/20183112003

analysis on the spatial aspects of the mountainous area of Kendal Regency can be done, therefore the existing tourism potencies can be optimally developed into object and tourism attraction.

This article contains result of a preliminary study using 2016 data from the Youth, Sports and Tourism Office of Kendal Regency, and analysis of the 2016 data using geospatial approach

2. Literature review

2.1 Tourism

Tourism is a prime sector that is one of the most important factors in the regional development and the community welfare improvement in a country, which has continuous expansion and diversification and become one of the largest growth sectors in the world.

Tourism is a variety of travel activities supported by various facilities and services provided by the community, industrialists, and government [11]. Marpaung [12] defined tourism as human temporary transfers in order to getting out from their home and routine jobs. While tourist is everyone who travels and settles for a while in a place that is not his/ her residence for any particular reason except looking for a job.

2.2 Tourism potency

Tourism potency is everything that is owned by a tourist attraction and useful for developing the tourism industry in that area [13]. There are three tourism potencies in this research, i.e. nature potency, culture potency and man-made potency. Based on these three kinds of tourism potencies, tourism object also divided into three, i.e. nature based tourism, culture based tourism and man-made based tourism.

In the Operational Area Analysis Nature Based Tourism Object and Attraction (ADO-ODTWA) guidelines[6], nature based tourism is defined as a travel activity or part of voluntary and temporary activity to enjoy the symptoms uniqueness and natural beauty of natural attraction, people's forest park and nature parks. Mountainous tourism is one of nature based tourism since the tourist object is nature, beauty and phenomenal.

2.3 Geospatial approach

Geospatial approach is another term of Geographic Information System (GIS) that is an integration between hardware, software, human data (brain ware) which cooperate in processing data starting from data management, manipulation and data analysis so as to produce output or final result that can be used as a reference in decision making to a problem [14,15,16].

GIS technology integrates common database operations, such as query and statistical analysis with maps. GIS manages location-based information and provides tools for the display and analysis of various statistics, including population characteristics, economic development opportunities, and vegetation types.

GIS allows users to connect between databases and maps to create a dynamic view. In addition, GIS also provides tools for displaying, querying, and uploading such databases in ways that traditional spreadsheets may not. This capability differentiates GIS from other information systems, and makes it valuable for government, public and private companies to explain events, predict outcomes, and planning strategies implementation.

Purnaweni [17] concluded that GIS is believed to be an important step towards more efficient and effective public policy making by an innovative government eventhough not all necessary data for formulating, planning and evaluating public policies can be provided by GIS.

2.4 ADO-ODTWA

Natural resources such as volcanoes, waterfalls, hot springs, craters, rivers, caves, lakes, coral reefs, and mangrove forests, are the potential objects and natural tourist attraction (ODTWA) that need serious handling in order to maintain its sustainability and existence [6].

Furthermore, the development of ODTWA requires criteria and indicators through determination of criteria elements, weighting, each sub elements calculation and the sum of all criteria element values. The objective is to determine the priority scale of ODTWA development and to intensify the utilization and development of an ODTWA.

Criteria of ODTWA assessment was issued by the Directorate General of Forest Protection and Nature Conservation in 2003, consists of attractiveness, market potency, accessibility, condition of area, visitor services, climate, accommodation, supporting facilities and infrastructure, clean water availability, connection with other tourism objects, safeness, etc. [6]

The criterias of ADO-ODTWA assessment used in this study were modified from ADO-ODTWA guidelines and Strategic Plan of The Indonesian Tourism Ministry 2015-2019. The assessment criterias are as follow

Table 1. Assessment criteria of nature based tourism in Kendal Regency, a modification of Rahayuningsih, Tri et.al [4], ADO-ODTWA Guidelines [6] and Strategic Plan of Indonesian Tourism Ministry [18].

No	Criteria	Elements	Sub elements
1	Attraction (weight : 6)	Landscape variation	Mountainous landscape, Forest landscape, Plantation landscape, Rice field landscape, Water body (lake, river, waterfall, etc)
		Tourism object variation	Natural beauty, Natural phenomenon, Water body (lake, river, waterfall, etc), Cultural attraction, History
		Tourism activity type	Tracking/ hiking, Religius tourism, Tubing, Camping, Education/ research, Enjoying scenery, Photography, Viewing cultural/ historical attraction, Fishing
		Location cleanliness	Natural factor, Industrial factor, Transportation, Settlements, Garbage, Graffiti
		Disaster safeness area	Landslide, Flash floods, Earthquakes, Mountain eruption, Fires
		Slope variation	45 % (very steep), 25-45 % (steep), 15-25 % (wavy), 8-15 % (rather flat), 0-8 % (flat)
		Contour variation	> 2.000 m asl, 1.000 – 2.000 m asl, 500 – 1.000 m asl, 100 -500 m asl, 15 – 100 m asl
2	Accessibility (weight : 5)	Estimated distance of tourism potency to the district center	0-30 km (Ring 1), 30-70 km (Ring 2), 70-100 km (Ring 3), >100 (Ring 4)
		Estimated distance of tourism potency to the provincial capital	0-30 km (Ring 1), 30-70 km (Ring 2), 70-100 km (Ring 3), >100 (Ring 4)
		Estimated distance of tourism potency to the nearest nearby tourist attraction	0-30 km (Ring 1), 30-70 km (Ring 2), 70-100 km (Ring 3), >100 (Ring 4)
3	Amenity (weight : 3)	Supporting facilities	Accommodation, Restaurant, Transportation, Souvenir shop, Worship facilities, Banking facilities, Healthcare facilities
		Infrastructures	Street, Bridge, Parking area, Electric network, Telecommunication network, Water supply network

3 Research Method

The location of this study is in Kendal Regency especially in the mountainous areas of Plantungan, Pageruyung, Sukorejo, Patean, Singorojo, Boja, and Limbangan. Below is the location of the study:

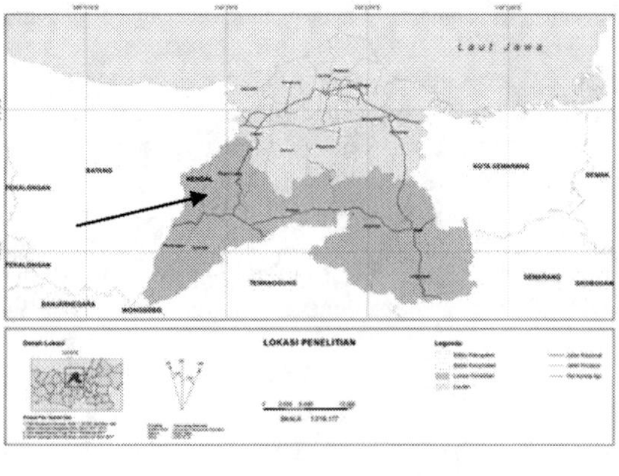

Fig. 1. Study location.

A modified method of Rahayuningsih, et.al. [4] using ADO-ODTWA (Operational Area Analysis - Nature based Tourism Object and Attraction) Guidelines [6] will be used in this study. The result of ADO-ODTWA assessment then would be integrated with Geographic Information System (GIS) by tabulating it into the database of *Arc Map 10.3*.

The phase of this study includes preliminary study in order to find criteria of resources assessment, data collection, and data analysis, which is presented in Fig.2.

3.1 Preliminary study

Literature study is the first step to obtain general illustration of the study location as administrative map with the geography and demography conditions. It is carried out by collecting and reviewing data of the study locations. It is also used to determine the assessment criteria of the study [4].

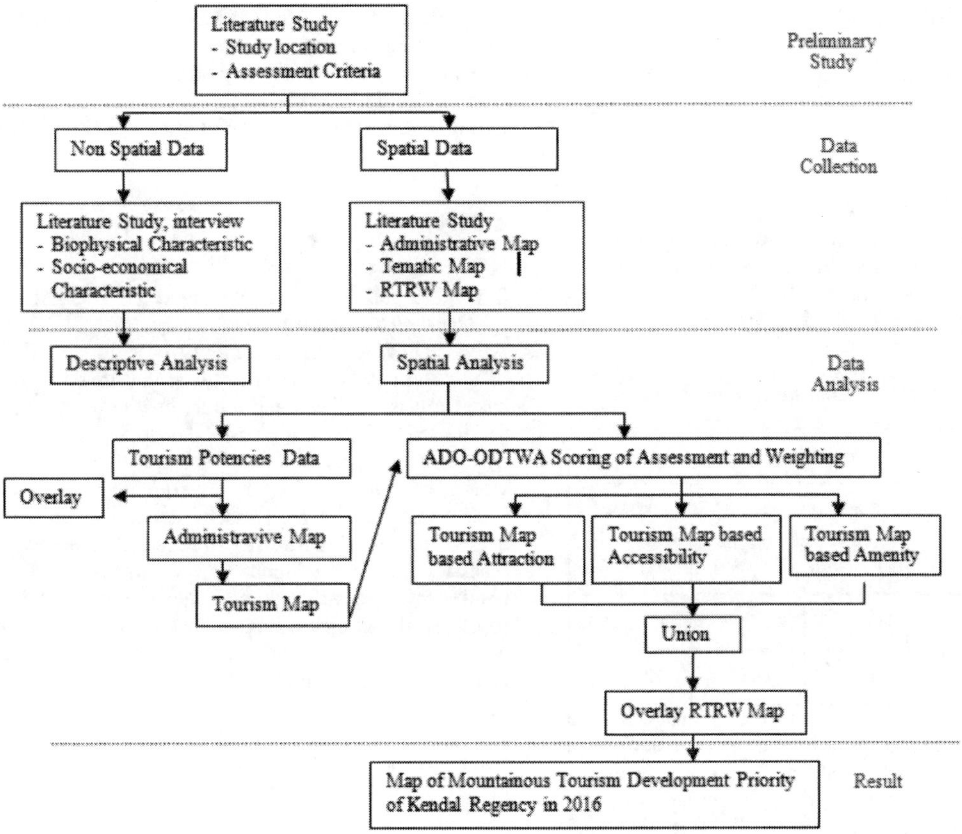

Fig. 2. Phase of Preliminary Study.

3.2 Data collection

Data collection phase needs a technique related with the process of data extracting. Data must be relevant with study needs and data resources [4] includes biophysical and socio-economic characteristics of the community. Stages of data collection conducted in this study include literature study and interview.

3.2.1 Literature study

Data from literature studies are often referred to as secondary data. Data for this article was obtained from the library of Postgraduate Program of Diponegoro University of Semarang, Kendal District Library; Youth, Sports and Tourism Department of Kendal Regency, and other institutions related to this research.

3.2.2 Interview

In this study, interviews were conducted with 2 (two) officers from the tourism division of the Youth, Sports and Tourism Office of Kendal Regency.

3.3 Data analysis

Data analysis technique is continuously done from the beginning when the data are started to be intensively collected until the research is finished.

Spatial data analysis was the main analysis in this research to produce map of the priority area of mountainous tourism in Kendal Regency in 2016, for the development of priority areas in 2017. The first step was to assess the tourism potency using ADO-ODTWA (Operational Area Analysis - Nature based Tourism Object and Attraction) Guidelines used by Rahayuningsih, et. al [4] modified with the Strategic Plan of Tourism Ministry of Republic Indonesia 2015-2019 on tourism product development strategy embodied in "3A", namely Attraction, Accessibility and Amenity. The criteria is shown in the Table 1.

The attraction criteria consisted of seven elements, i. e. landscape variation based on land cover, tourism object variation, tourism activities type, location cleanliness, disaster safeness area, slope and contour variation. The weight of attraction criteria is 6, the highest number because the attraction is the main capital that allows the visitors arrival.

Accessibility criteria consisted of three elements, i.e. the estimated distance of tourism potency to the district center, the provincial capital and the nearest nearby tourist attraction, with the weight of the assessment was 5. While the criteria of amenity consisted of 2 elements, i.e. supporting facilities and infrastructure of tourism activities by rating of weight was 3.

Scores of assessment of each criterion would be calculated using the following formula :

$$S = \Sigma \text{ (criteria value x weight)} \quad (1)$$

While classification of each tourism objects and attractions also would be calculated using the formula :

$$K = (Nt - Nr) : 3) \quad (2)$$

Nt = highest value and Nr = lowest value. The final assessment classification would be categorized into low, medium and high, on certain intervals (Table 2). The purpose of this assessment was to obtain description of tourism potency that should be put in high priority for the development [4].

Table 2. Classification of assessment.

No	Criteria	Assessment Classification		
		Low	**Med**	**High**
1	Attraction	42-448	449-854	855-1260
2	Accessibility	300-600	601-900	901-1200
3	Amenity	6-64	65-122	123-180

The next analysis used GIS as a tool of analysis. First step was to compose basic map of distribution of mountainous tourism obtained from overlay between the administrative map with the data of tourism potency. After that, integrated the result of ADO-ODTWA assessment into Arc Map 10.3 by using tabular data. The final spatial analysis was to overlay the result of integrated tabulation and map with tourism development area of Kendal Regency so it will produce the priority of mountainous tourism development map of Kendal Regency.

4. Results

4.1 ADO-ODTWA assessment of mountainous tourism potency

ADO-ODTWA is an instrument to obtain the suitability of a certain tourism potency to be developed as nature-based tourism objects and attractions. Criteria functions as the basis in planning the development of tourism potencies, by establishing the elements of criteria, weight and calculation of each sub elements [6, 7].

By using the 2016 secondary data from Youth, Sports and Tourism Department of Kendal Regency, the assessment of mountainous tourism was conducted. The result of the assessment can be seen in the Table 3 as the result of attraction assessment, Table 4 as the result of accessibility assessment and Table 5 as the result of amenity assessment in 2016..

Based on the assessment classification (Table 2), there were 2 (two) classes of mountainous tourism potency in Kendal Regency in the criteria of attraction, accessibility and amenity, i.e. medium and high classes.

Table 3. Score of attraction assessment in 2016

No	Tourism Potency	Landscape Variation	Tourism Object Variation	Tourism Activity Type	Location Cleanliness	Disaster Safeness Area	Slope Variation	Countour Variation	Scores Total
1	Agrowisata Plantera	20	15	20	20	25	25	10	810
2	Dewi Selo Arjuno	30	15	15	15	25	25	15	840
3	Dewi Blumah	30	20	20	15	25	20	15	870
4	Dewi Bendosari	30	20	20	15	25	20	10	840
5	Curug Terong	30	15	15	15	25	15	15	780
6	Curugsewu	25	30	20	15	25	20	10	870
7	Dewi Cacaban	30	25	20	15	25	25	10	900
8	Dewi Gondang	30	30	25	15	25	20	15	960
9	Dewi Ngesrepbalong	30	20	20	15	25	15	15	840
10	Dewi Cening	25	20	15	15	25	15	20	810
11	Dewi Sembrani/ Santosa Sable	30	25	25	20	25	20	15	960
12	Dewi Peron	30	30	20	15	25	20	20	960
13	Kebun Cengkeh Banaran	30	15	15	25	25	15	15	840
14	Kebun Teh Medini	25	10	20	20	25	20	15	810
15	Pemandian Air Panas Gonoharjo	25	20	20	20	20	20	20	870
16	River Tubing Genting	25	20	20	15	25	20	15	840

Table 4. Score **of accessibility assessment in 2016**

No	Tourism Potency	Estimated distance of tourism potency to the district center	Estimated distance of tourism potency to the provincial capital	Estimated distance of tourism potency to the nearest nearby tourist attraction	Scores Total
1	Agrowisata Plantera	60	60	60	900
2	Dewi Selo Arjuno	60	60	60	900
3	Dewi Blumah	60	40	60	800
4	Dewi Bendosari	60	40	60	800
5	Curug Terong	60	40	60	800
6	Curugsewu	60	60	60	900
7	Dewi Cacaban	60	60	60	900
8	Dewi Gondang	60	60	60	900
9	Dewi Ngesrepbalong	60	60	60	900
10	Dewi Cening	60	60	60	900
11	Dewi Sembrani/ Santosa Sable	60	80	80	1100
12	Dewi Peron	60	60	60	900
13	Kebun Cengkeh Banaran	60	40	60	800
14	Kebun Teh Medini	60	60	60	900
15	Pemandian Air Panas Gonoharjo	60	60	60	900
16	River Tubing Genting	60	60	60	900

Table 5. Score of amenity assessment in 2016

No	Tourism Potency	Supporting facilities	Infrastructure s	Scores Total
1	Agrowisata Plantera	20	30	150
2	Dewi Selo Arjuno	15	20	105
3	Dewi Blumah	15	25	120
4	Dewi Bendosari	15	25	120
5	Curug Terong	15	15	90
6	Curugsewu	25	30	165
7	Dewi Cacaban	20	25	135
8	Dewi Gondang	20	25	135
9	Dewi Ngesrepbalong	15	25	120
10	Dewi Cening	15	20	105
11	Dewi Sembrani/ Santosa Sable	20	30	150
12	Dewi Peron	15	25	120
13	Kebun Cengkeh Banaran	15	25	120
14	Kebun Teh Medini	15	25	120
15	Pemandian Air Panas Gonoharjo	15	30	135
16	River Tubing Genting	15	20	105

4.2 GIS analysis result

The first step done in GIS analysis was to compose basic map of distribution of mountainous tourism by overlaying between the administrative map with the data of tourism potency. Score of each elements of ADO-ODTWA assessment was then integrated into *Arc Map 10.3* by tabulating them. It would resulted in 3 (three) thematic maps, i.e. Tourism Map based on Attraction (Fig. 3), Tourism Map based on Accessibility (Fig. 4) and Tourism Map based on Amenity (Fig. 5).

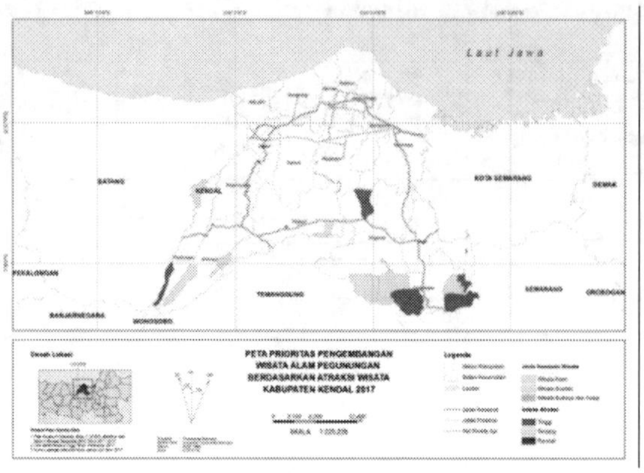

Fig. 3. Tourism map based on attraction in 2016

From Fig.3, it can be seen that there were 8 (eight) tourism potencies in 2016 which were then classified into 'high priority' based on attraction criteria, such *Dewi Blumah, Curugsewu, Dewi Cacaban, Dewi Peron, Dewi Gondang, Kebun Teh Medini, Pemandian Air Panas Gonoharjo* and *Dewi Sembrani/ Santosa Stable*. The other eight are belonged to 'middle priority' including *Dewi Bendosari, Curug Terong, Kebun Cengkeh Banaran, Agrowisata Plantera, River Tubing Genting, Dewi Selo Arjuno, Dewi Cening,* and *Dewi Ngesrepbalong*.

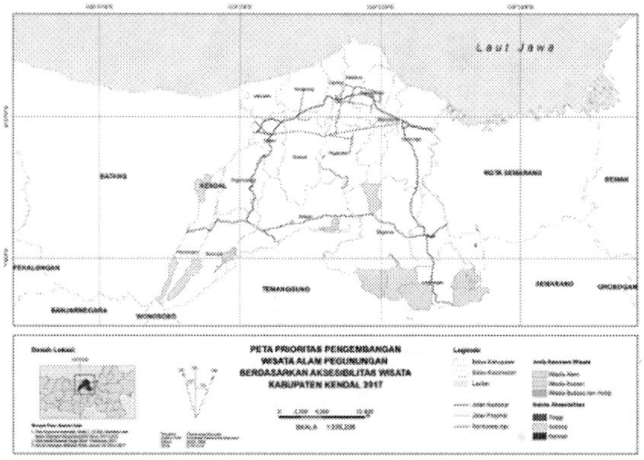

Fig. 4. Tourism map based on accessibility in 2016

While in Fig 4, almost all of the tourism potencies are included into 'middle priority' based on accessibility criteria, except *Dewi Sembrani/ Santosa Stable*. It is because the location of *Dewi Sembrani/ Santosa Stable* is near to the provincial capital of Central Java and also near to the nearest nearby tourist attraction.

Same as Fig 4, Fig 5 also shows that mostly of tourism potencies are included into 'middle priority' based on accessibility criteria, except *Dewi Sembrani/ Santosa Stable* that has more complete supporting facilities and infrastructures than others.

The final step did in spatial analysis was to overlay both of those thematic maps (tourism map based on attraction, accessibility, and amenity) with the map of tourism area development on spatial planning of Kendal Regency. So it would result the map of mountainous tourism development priority of Kendal Regency as shown on Fig. 6 below.

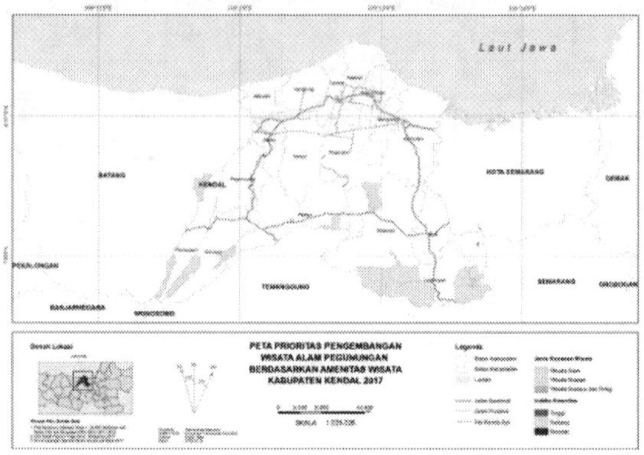

Fig. 5. Tourism map based on amenity in 2016

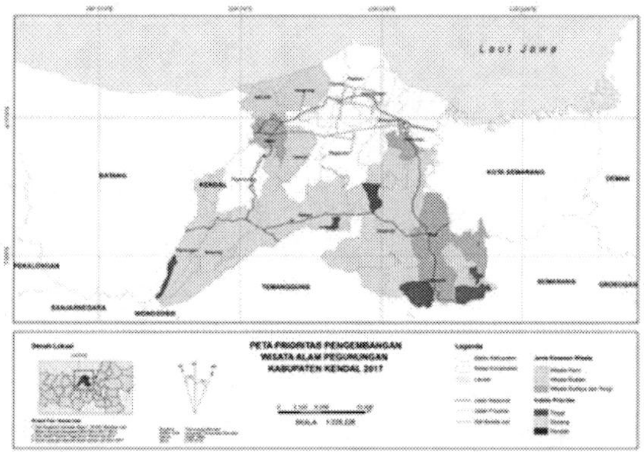

Fig. 6. Map of mountainous tourism development priority of Kendal Regency in 2016

5 Conclusion

The final result showed that there were 7 (seven) tourism potencies than could be prioritized as a mountainous tourism development in 2016, i.e. (1) *Dewi Blumah* in the Plantungan District, (2) *Curugsewu* and (3) *Agrowisata Plantera* in Patean District, (4) *Dewi Cacaban* in Singoojo District, (5) *Dewi Gondang,* (6) *Pemandian Air Panas Gonoharjo* and (7) *Kebun Teh Medini* in Limbangan District.

Eventhough *Dewi Sembrani/ Santosa Stable* was categorized into high priority in all criteria, it could not become a priority in the mountainous tourism development area because it was not included in the area of nature tourism development but rather in the cultural tourism development in the spatial planning of Kendal Regency.

Therefore, this result can be used as a guidance to determine the further priority area of mountainous tourism development in Kendal Regency, such as in 2017.

 https://doi.org/10.1051/e3sconf/20183112003

References

1. Kabupaten Kendal, B.P.S.K, *Kabupaten Kendal Dalam Angka 2015,* Kendal: Badan Pusat Statistik Kabupaten Kendal, (2015).
2. Sunardiono, *Upaya Pengembangan Obyek Wisata Gonoharjo Kecamatan Limbangan Kabupaten Kendal,* Skripsi, Universitas Negeri Semarang (2009)
3. S. Yuliadi, *Analisis Potensi Pengembangan Obyek Wisata Alam di Kabupaten Kendal,* Skripsi, Universitas Muhammadiyah Surakarta, (2011
4. T.Rahayuningsih, E.K.S.H. Muntasib ,& L. Budi, J.Proenv, **33**, 365-375, (2016)
5. JM. Tambunan, S. Anggoro & H. Purnaweni, Proceeding of National Conference on Natural Resources and Environment Management. *Kajian Kualitas Lingkungan dan Kesesuaian Wisata Pantai Tanjung Pesona Kabupaten Bangka*, 356-362. (2013)
6. Perlindungan Hutan dan Konservasi Alam (PHKA), *Pedoman Analisis Daerah Operasi Obyek dan Daya Tarik Wisata Alam (ADO-ODTWA),* Bogor : Direktorat Jenderal Perlindungan Hutan dan Konservasi Alam, (2003)
7. P. Oktadiyani, *Alternatif Strategi Pengelolaan Taman Wisata Alam Kawah Kamojang Kabupaten Bandung Provinsi Jawa Barat,* Bandung : Institut Pertanian Bogor, (n.d),
8. J.S. Muin and R.S. Wulandari, *Penilaian Daya Tarik dan Pengembangan Obyek Wisata Pantai Tanjung Belandang di Kabupaten Ketapang,* Pontianak : Universitas Tanjungpura, (n.d).
9. K. Bunruamkaew & Y. Murayama, J.Sbspro, **21**, 269-278, (2011)
10. G. Hai-Ling, et.al. J.Proeng, **15**,1988-1992, (2011)
11. Departemen Kebudayaan dan Pariwisata, *Undang-Undang Republik Indonesia No. 10 Tentang Kepariwisataan.*
12. H. Marpaung, *Pengetahuan Kepariwisataan.* Bandung : Alfabeta, (2000)
13. R.G. Soekadijo, *Anatomi Pariwisata.*Jakarta : PT. Gramedia Pustaka Utama, (2000)
14. P.A. Burrough, *Principles of GIS for Land Resources Survey,* New York : Oxford Science Publications, (1986)
15. S. Aronoff, *A Review of Geographic Information System : A Management Perspective,* Ottawa : WDL Publication, Pp. 294, (1989)
16. NR. Chrisman, *ExploringGeographic Information System,* John wilwy, 298 pp (first Ed), (1997)
17. H. Purnaweni, Proceeding of International Annual Conference of Public Policy, *Intensive Use of Geographic Information System for Public Policy,* 134-141. Malang : University of Brawijaya, (2012)
18. Kementrian Pariwisata, *Rencana Strategis : Pengembangan Destinasi dan Industri Pariwisata Kementrian Pariwisata Tahun 2015-2019,* Jakarta : Deputi Bidang Pengembangan Destinasi dan Industri Pariwisata Kementrian Pariwisata, (2015).

The Role of Spatial Analysis in Detecting the Consequence of the Factory Sites : Case Study of Assalaya Factory- Sudan

Amar Sharaf Eldin Khair[1,*], *Purwanto*[2], *Henna* RyaSunoko[3], *Omer* Adam Abdullah[4]

[1]Lecturer at Omdurman Islamic University, Geography Department, Omdurman city – Sudan
[2]School of Postgraduate Studies, Diponegoro University, Semarang-Indonesia
[3]Diponegoro University, Pharmacy Faculty, Semarang – Indonesia
[4]Dean of art faculty, Omdurman Islamic University, Omdurman city – Sudan

Abstract. Spatial analysis is considered as one of the most important science for identifying the most appropriate site for industrialization and also to alleviate the environmental ramifications caused by factories. This study aims at analyzing the Assalaya sugarcane factory site by the use of spatial analysis to determine whether it has ramification on the White Nile River. The methodology employed for this study is Global Position System (GPS) to identify the coordinate system of the study phenomena and other relative factors. The study will also make use Geographical Information System (GIS) to implement the spatial analysis. Satellite data (LandsatDem- Digital Elevation Model) will be considered for the study area and factory in identifying the consequences by analyzing the location of the factory through several features such as hydrological, contour line and geological analysis. Data analysis reveals that the factory site is inappropriate and according to observation on the ground it has consequences on the White Nile River. Based on the finding, the study recommended some suggestions to avoid the aftermath of any factory in general. We have to take advantage of this new technological method to aid in selecting most apt locations for industries that will create an ambient environment.

1 Introduction

The spatial analysis in this paper is emphasizes on the spatial elaboration for the factory site by several element using Global Position System (GPS) , Geographical Information System (GIS) and satellite data, making way for spatial interpretation of the factory potential zones. It has the ability to decide whether the factory location is suitable for the industrial process through special techniques. It will be elaborated in data analysis. It also has potential to find the affinity between Assalaya factory location and water resources–white Nile River and human settlement. The study will further illustrate the potential aftermath affected by the factory through analyzing the data by GIS technique, [1].
The GIS technique is considered as one of the important scientific technology that is recently in use as decision-maker for selecting compatible location for industries and has ability to predict the future ramification as well as the influence by the factories through special techniques because it can be one of the scientific technological
Innovation which has ability to put scientific research findings into practice,[2].

The use modern technology with different techniques like Spatial Analysis and Digital Elevation Model (DEM) is the best way of selecting the right position of factories to avoid consequences caused by the factory production which has massive impact on water resources. So if factories are well-sited will bring forth both economic and environmental benefit especially in recent case of rapid population growth,[3].

Digital Elevation Model (DEM) is suitable to exhibit the continuous change of the earth topography. It is the basic data source for terrain analysis and spatial applications. It can be used for studies that are related to science and engineering. The function of the DEM is supported by the widespread availability of digital topographic data,[4].

There are significant reasons for the selection of this topic and choosing Assalaya factory. In relation to the selection of this topic, it has been observed that factory wastewater is discharged into White Nile River. The White Nile River is considered the main branch of River Nile and the fundamental source of drinking water in Sudan in general. This being the specific area under study, there arise the need to emphasize on the role and ability of Spatial Analysis to select compatible location of factories to secure human life from the consequences from the factories and also to conserve the environment and the realization of the economic efficiency as well.

Assalaya Factory is being chosen for this study because the factory has been allocated in an environment that is inappropriate and complained by citizens that stay around this area, where some people have been suffering as a result of sugar cultivation production output. It has

* Corresponding author: amar77600@gmail.com

brutal consequences which affects the ecology, drinking water, air and the arable land. Furthermore, social impacts such as stinky smell, respiratory diseases, unsafe drinking water and the likes has caused the relocation of neighborhoods and villages.

The main purpose of the study is to eliminate and reduce the industrial wastewater which has significant role in polluting drinking water and endangering living organisms inside it. Through persuasion of the stakeholder and owners of industries and community participation, this cause can be properly addressed. This study also aims to emphasize the ability of Spatial Analysis in selecting the right position for any kind of industry through GIS software and arc toolbox window.

2 Material and Method

2.1 Study sites

Assalaya is located in Assalaya locality- White Nile State- Sudan between Longitude 33:33:10 West 32:58:52 East and, latitude 13:12:45 South 23:58:40 North. It is bordered by Aljazeera State and El jabalian in the north direction. To the south are Rabak and Knana cities. Sennar State, White Nile River and kosti city are to its west as it is seen in figure 1. While the Assalaya factory is sited between Longitude 32:43:15 West and 32:45 East either the latitude between 13:14:40 south and 13:15:50 north. Its sugarcane fields spread over from east to west in 25 km and 12 km from north to south. The distance between the factory and the White Nile River is 8,5 km and the distance to the residential community is 2km as seen in figure 1.

2.2 Study Type

The study topic analyzes the sugar factory location and identifies whether its location is appropriate in order to prevent its ramification on the nearby society. Geospatial Information System will be used through spatial analysis function for the study. This study uses Global Positioning System (Ground survey), interviews and observation to get primary data. Also data will be collected from the field and secondary data as Satellite data DEM of the study area, has also been included in this study.

2.3 Data Collection and Acquisition

The study requires two kinds of data to realize its target and it includes secondary data which are; satellite data image represented in land sate Dem and GIS software to analyze the factory location as hydrological, contour line, slopping and geological analysis. Primary data include the data collected from the following methods; Observation, Global Positioning System GPS and the use of interviews.

2.4 Data Analysis

This study relies on landsat DEM that has been processed by GIS software program to analyze the data and to achieve the results as obvious in the study title. The study is confined on specific tools in the GIS program found in arc tools box. Overall, the study will analyze the slope map to find the slope direction of the factory position and hydrological analysis to determine whether the factory location is in runoff of the water and also geological map to investigate the geological location of the factory area. Lastly, contour line map for the study area with two meter interval will be used to identify the factory land level.

3 Results:

Through analysis, the contour line which had been figured out by GIS software which is processed via Arc tools box window. The purpose of that is to find out how far the factory location is suitable for the industrial production from the river. In evaluating the consequences on the water resource (White Nile River), the study found that the Contour lines have retreated towards the Nile reaching to 909 in the study phenomena and it has further declined to779. This finding is based on analysis of the contour map of study area with interval 2 meters to determine the decline of contour line as seen in Figure 2.

Depending on the hydrological analysis for the factory location that has been analyzed based on land sate DEM and special techniques as identifying the study area, through the contour line and the hydrology, the study found that the factory location is located in water catchment area - valleys, lagoon- those are runoff toward White Nile River. Indicated in Figure 3.

Through analysis with the geological map of the study area, the study found that the industrial production site is within Ancient River Sedimentary adjacent to Modern River Sedimentary and Sediments of the Umm Ruwaba Formation. Meanwhile, the agriculture production the (sugarcane) is located in Ancient River Sedimentary, Umm Ruwaba Formation and the Nubian Sandstone as seen in Figure 4.

4 Discussion

The study through checking of the contour line indicates that the location of the factory is inappropriate because of the proximity in contour lines, where it declines subsequently in measures of909, 868,861, 807 till779. The contour indicates that, the smaller the vertical separator, the more the contour lines converge,[5]. Because contour successfully separate and identify the different objects and have been applied to many different applications, it is have many different approaches for detecting the location which have been developed and could be broadly used,[6].

 https://doi.org/10.1051/e3sconf/20183112004

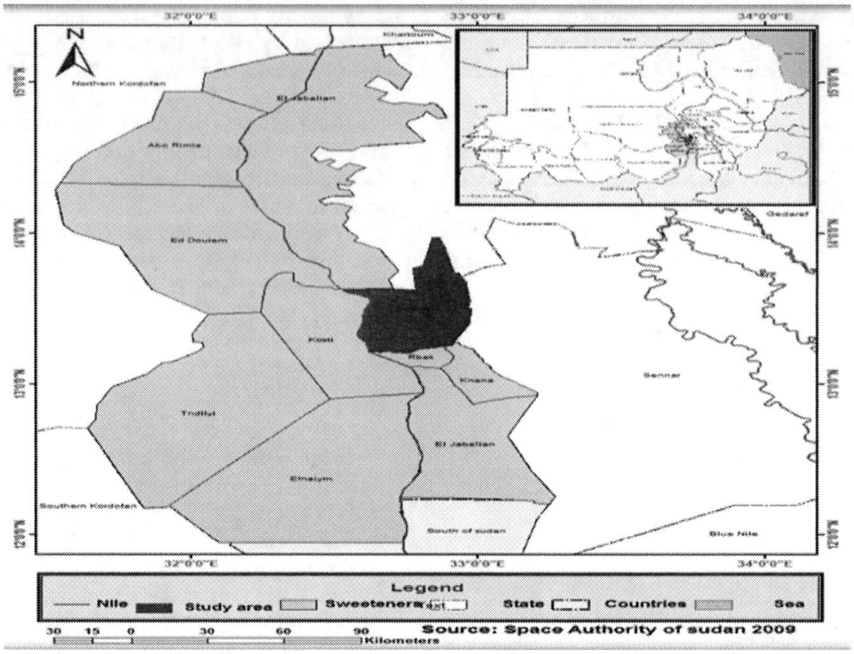

Fig. 1. Study sites

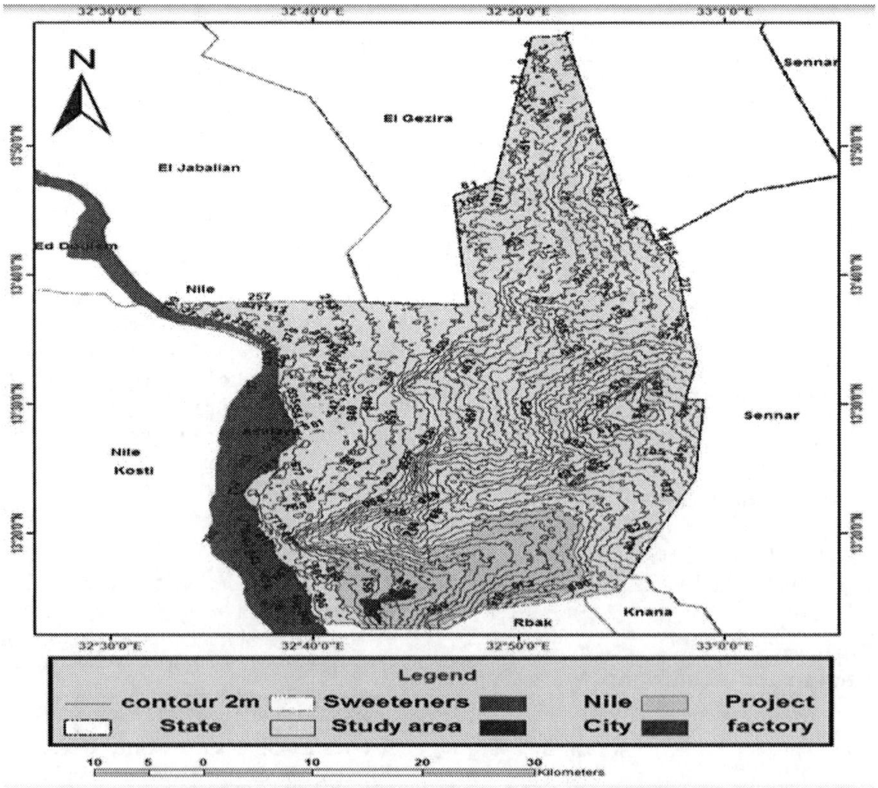

Fig. 2. Location Contour map

The direction of the valleys and water flow have been shown in fig. 3 . To emphasize and prove that the study has selected two-meter contour to investigate the magnitude of the convergence in recognizing the factual and authentic way of flow. So the retreat of the contour line indicates that the direction of flow of the river. Based on the Geometry studies. It is wrong to set up factories in inappropriate position, because it may aggravate the condition as it creates phenomenon such as water pollution particularly in the white Nile river for instance as we become conscious of the study area and Sudan in general will rely on it for daily drinking. Not only is the White Nile River considered as the main source of water supply to several countries after joining with Blue Nile at Khartoum city. At this point, the emphases on how important the water resource that has been disturbed by the sugar factory is really realised. To avoid the ramification of it, it is better we keep proper planning in implementing projects, especially projects that have aftermath on ambient environmental resources.It is better if we built our ideas on geometry and other modern science because geometry constitutes the shape of the object and Physics imposes the constraints in how the shape of the object can contrast over time,[7].

The Hydrological analysis is done via satellite data DEM by GIS processing using several step which are; first, fill the Watercourses till the study area is flat and facilitate the water flow direction calculations, all of these process are done by GIS Arc Map → Arc tools box → Spatial analysis tools → Hydrology → Fill. Secondly; cutting the basin area done by Arc tools box →Spatial analysis tools → extract → clip hydrological models has become one of the fundamental tools that is widely used to identify such matters [8], through both processes appears that the study area had been situated in the catchment area which descends from the north-east direction and crosses the study area westward way downstream into the White Nile River at South-west in tree-delta form which indicate the location is not fit for establishing any kind of factory due to the valleys and rivers that runoff on it , it is likely to pollute the Nile water by the agricultural and industrial waste (pollutant).

The geological result found confirms that part of the factory area is situated in Ancient River Sedimentary and other parts in Sediments of The Umm Ruwaba Formation which indicates that the factory area in Agricultural production belt as well as it's sediment area which has created several Lagoons and valleys as seen in hydrological Fig 3. Establishing factories in these sites may infiltrate and percolate the industrial wastewater into groundwater and thus definitely connected with the Nile River (approximately 2km between the river and the factory within Sedimentary area) therefore the factory needs industrial re-planning, it is better if it is set up in basement complex.

Overall, through analysis the collected data emphasizes that, it is inappropriate to establish any kind of industry in the research area. Besides, confirmation also comes from the conducted interviews at the study area, that the factory inflicts so much harm on the water resource which has changed the water quality by way of colour, bad smell, mutilation and malformation of some kind of fishes, [9].These effects have to a large extent had adverse bearing on many who reside near to the factory resettle to other places,[10].Through the observation, it turns out that the wastewater that comes from sugarcane irrigation and the operation have consequences on citizens'arable lands due to the slope, contour, hydrological and geological factors.

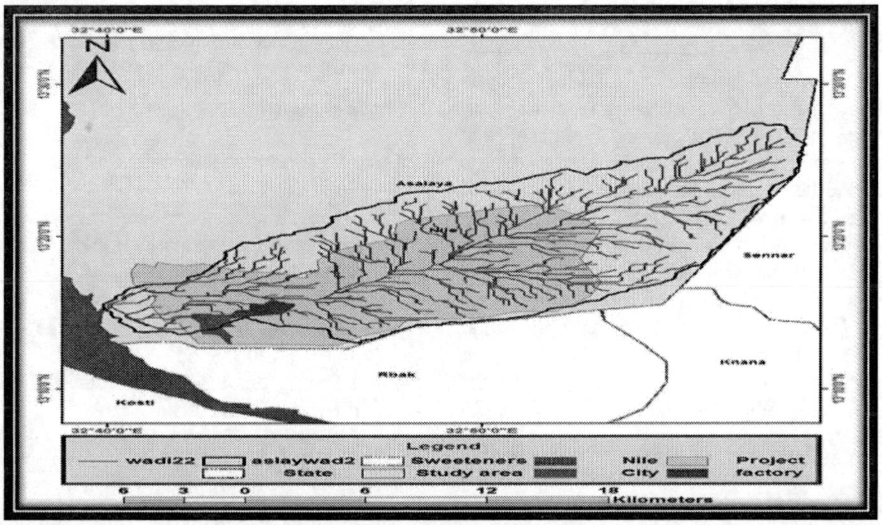

Fig.3. Hydrological map

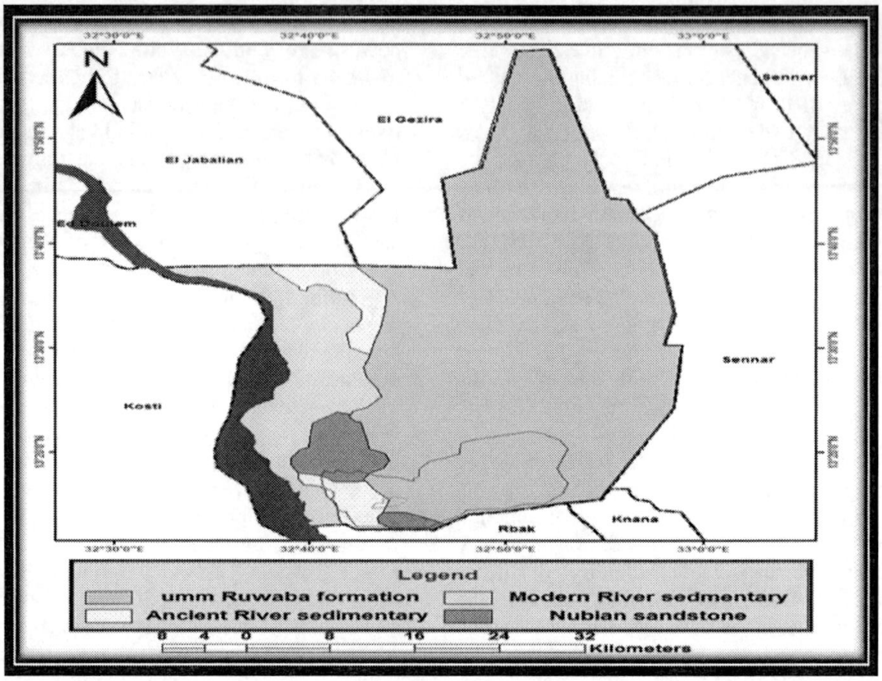

Fig.4. Geological map

5 Conclusion and Suggestion

The study concludes that all the study parameters that have been checked reveal that the factory location is inappropriate. The contour map data with Vertical separator2 Meters prove that via the retreat in contour lines subsequently 909, 868,861 , 807 and till779. The hydrological map confirms that the area is located in water catchment area. The geological map has the same logic because the area is still within the Ancient River Sedimentary, and other part in Sediments of The Umm Ruwaba Formation. Based on the conclusion, the study suggests some recommendations for establishing any kind of developmental projects. Siting of factories such as that of Assalaya should be properly planned and thought of before putting up its structures. The Sudanese government and other responsible stakeholders should make and implement entrenched laws that will ensure that an industry like the aforementioned is well planned and situated. It is better if we use and follow modern techniques and technology such as environmental impact assessment (EIA), strategic environmental assessment(SEA) and Remote sensing and geographic information systems (RS&GIS) in order to prevent the adverse effects but rather strengthen the scientific aspects.

6 Acknowledgments

First of all, the author would like to express heartfelt gratitude to his family for bringing him up from cradle to this stage. Secondly, he accords much appreciation to his supervisors namely Prof. Purwanto and Dr. Henna Rya Sunoko for guiding him through this dissertation. Also, a big thank you to the Sudanese Higher Education and Omdurman Islamic University for supporting this project. Sincere gratitude also goes to prof Saleh hasab Al Rasul, Dr. Omar Adam, Dr.Elyas Mansour, Mr. Abdullahabkar and Samuel Anderson for the incredible academic support. Thank you all of you for the input that contributed in making sure we are where we are today. Furthermore to the Lecturers at Doctoral Program of Environmental Science- School of Postgraduate Studies, Diponegoro University for endured the sacrifices in making everything is possible.

References

1. A.Wadood, A. Dewan, AOSI, RP. J. E **48**, 104–111, P3,(2016).
2. Taylor. Francis, *GIS for sustainable development*, International Standard Book Number-10: 0-8493-3051-3 (Hardcover), P 6-7, (2006).
3. S. Sánchez, D. Athanassiadis, C. Martínez, E. Tolosana, J. Majada, E. Canga, GISMFOL,JCLP 9412, J. E 0959-6526, P3, (2017).
4. J.Thompson, J.Bell, C.Butler, DEM, JGEM100 , J. E 67–89, P1, (2001).
5. J ISPRS **62** 93–103, P 94, (2007).
6. X. Wanga, C.Wei, K.Xianga, W. Chenb A Zhejiang, , J.VCIR, J. E 2022, P2, (2017).
7. R.Verdu´-Monedero, J. Morales-Sanchez, L. Weruaga, CAOAC,J.VCIR, J. E **26** 1118–1128, P 1119,(2008).
8. R. Turcotte, J. Fortin, A.Rousseau, S. Massicotte, J.Villeneuve Institut, DOFTDS, J.H, J. E 240 225–

https://doi.org/10.1051/e3sconf/20183112004

242,P3, (2001).
9. A. Ahmed, Interview with a fisherman, Al JazeeraAba city, (2017).
10. O. Mohammed , Interview with seiner citizen, Assalaya city, (2017).

Designing Web-based GIS Application by CSF Method: A Case Study in Boven Digoel Papua

Hendrykus Saritangdan Letsoin[1], *Albertus* Joko Santoso[1], and *Suyoto*[1]

[1]Magister Informatics Engineering. Universitas Atma Jaya Yogyakarta, Yogyakarta, Indonesia 55281

Abstract. Since Boven Digoel Regency was established in 2003, to this day its infrastructures are very limited, the region is still not developed, Human Resources (SDM) and Information and Communication Technology are still manual. The development of Geographic Information System was expected to contribute to ICT development in Papua, particularly Boven Digoel Regency, in displaying data of existing local potentials, e.g. tourism, plantation and forestry. Critical Success Factor Method was the right method in considering factors determining the success of the implementation of a strategy, in good information management and in providing data of local potentials of geographic information system for people in Boven Digoel Regency and the people of Indonesia in general to introduce Boven Digoel Regency, Papua. Framework Laravel with PHP programming language was expected to support Geographic Information System well to determine the distribution of local potentials in Boven Digoel Regency..

1 Introduction

As a new government center since 2003, Boven Digoel face several complex issues which are also experienced by newly formed regencies and. The main issue which becomes principle of policy implementation and government management is adjustment of local income (Locally-generated Revenue) and needs which emerge and must be fulfilled especially for the society as a form of public accountability. With the background of being a former prison camp, Boven Digoel is the only region in Papua or even Indonesia which has rapid development, growth and expansion in various sectors. With extensive local potentials, GIS application was made to help getting information to be processed, stored and displayed by location on the application in writing, color, picture and direction. Therefore, GIS application users can easily get information they need.

According to T. A. N. Xiaoshan and S. Yizhong (2010), geographic Information System is very important to get better display and also present map between user and computer. The Geographic Information System (GIS) in this study in synchronized in (GIS) map [1] in assessing a region for different energy potential affected by various technical, topographical and environmental variables. The study by A. Oudin (2017) discussed modeling of different obstacles for the latest energy potentials and result visualization on an interactive map. Geographic Information System (GIS) is used in various regions to assess a location and its situation [2]. According to L. Wang, et.al. (2010), a new geographic information technology application can be demonstrated in government using "Geo-Windows" as the Geographic Information System (GIS) software for E-Government to be applied in the network [3]. The study by A. V. Vitianingsih and D. Cahyono (2016) state that without adequate road construction, traffic will exceed road volume capacity. It will make roads difficult to pass, causing risk of congestion and traffic accident. This paper will discuss the usage of Geographic Information System (GIS) in analyzing spatial data and attribute date (geoprocessing layers) in mapping accident-prone roads and mapping heavy traffic roads as alternatives for constructing new roads [4]. Geographic Information System (GIS) substantially defied existing method by relational data since the navigational operation of GIS map (e.g. panning, zooming in / out) has their own unique access patterns vastly different from relational equivalents. This study presents an efficient spatial caching framework in handling spatial object adjusted to navigational operational map so that density is visible in the system or GIS application so that application user can travel to their destination [5].

The current development of internet technology, according to, N. Hakim, Maulana and R. Cahyana (2015), is very rapid so everyone should use it to access information, whether via internet media or other social media, e.g. Facebook, Instagram, Youtube, etc. Due to the importance of the information service, Geographic Information System should be developed to help users search social facilities and their specific locations in Garut. Web Modeling Language method is expected to help the development of web engineering which can run in web platform to display existing facilities as required, especially in map [6]. Each region in Indonesia has unique potential that should be analyzed to be considered in national development. Analysis was performed by categorizing data of potentials in Indonesia via PPODES

* Corresponding author: hendrik.letsoin@gmail.com

https://doi.org/10.1051/e3sconf/20183112005

census, thus consisting of two attributes, i.e. numeric and categorical [7]. In the study by B. T. Nguyen et al (2017) on prediction user movement from historical location data, Movement Predictor application was made to help collect location data from users registered by Global Positioning System (GPS). To analyze various types of feature in comparing three learning models [8]. E-Governance (E-Gov) is generally recognized as an innovation which encourages socio-economic development across the world. Therefore, developed and developing countries continue to try to perform E-Gov projects because they're proven to improve good governance and government services for citizens or society. Developed countries such as Republic of Korea, United States, Singapore, and Denmark have successfully transformed their economies by E-Gov. It's different from countries which have many failed E-Gov projects which didn't develop well [9].

Susanty, et. al. examine the influence of a company on the extent of the assimilation of internet technology into the company's daily operation. The study examined critical success factors adopted by internet technology and the effect of internet technology adoption on company performance [10]. Internet development and current technology offer new opportunities for the government to serve and interact with the public. Previously, E-Government was used to provide better relationship whether from the government or the society. Most governments have realized the importance of ICT and therefore may inhibit the role of E-Government by presenting some challenges by comparing with others, especially developing countries, thus affecting the success rate in E-Government in public service. A study in Kenya to develop success factor and adoption of E-Government from the society's perspective aims to identify factors and challenges to the success of the implementation of E-Government [11]. It's very important to get better understanding on what's effective in government project management and how to manage project effectively, thus clarifying the definition of government project management, effective government project management standard and the effect of success by critical factor in governance model. A questionnaire in China produces a list of success factors affecting the success in governance quantitatively [12]. According to Alfat, Triwiyatno and Isnanto reliable web-based development of monitoring system of online application media to simplify recapitulation and documentation of temperature and humidity, to accelerate web development, there should be web technology application of the framework. In developing the application, LARAVEL MVC framework is used as scripting method [13].

A study by Kainde, Santoso and Budiyanto shows that Bali is famous as a tourism city in some countries across the globe which makes Bali unique as an Indonesian icon. There are many tourism spots famous for diving which are favored by local and foreign tourists. Geographic Information System was developed to map diving locations in Bali [14][15]. Yogyakarta is also special in Java. Every year, tourists visit Yogyakarta during holiday to visit tourism spots, e.g. Borobudur Temple, so there should be alternative information in searching tourism location, i.e. mobile web-based Case Based Reasoning application, to determine tourist destination [15][14]. A study by Nugraha, et.al shows that internet and smartphone technology development has changed people's behavior in their daily interactions. People more easily ask for and obtain information via mobile device. This study is used to help tourists get location information in Surakarta. The research result shows that tourists easily and quickly get location information in tourism city Surakarta [16].

2 Material and method

Application is run by Framework and method as benchmarks in the current case study.

2.1 Geographic information system of Boven Digoel

Geographic Information System in Boven Digoel Regency was developed to realize the vision and missions of the Regent of Boven Digoel in ICT, particularly by application development by the Department of Communication and Informatics of Boven Digoel Regency. In general, GIS was created by many stages and data, and the reviewed location. The Department of Communication and Informatics of Boven Digoel Regency has worked with an educational institution in Yogyakarta to develop human resources and information technology. As a partner in human resources and information technology development, YOGYA EXECUTIVE SCHOOL "YES" has held various national information technology trainings and human resources development joined by representatives of local government agencies, universities, hospitals, schools, SMEs, NGOs and private companies since 2004.

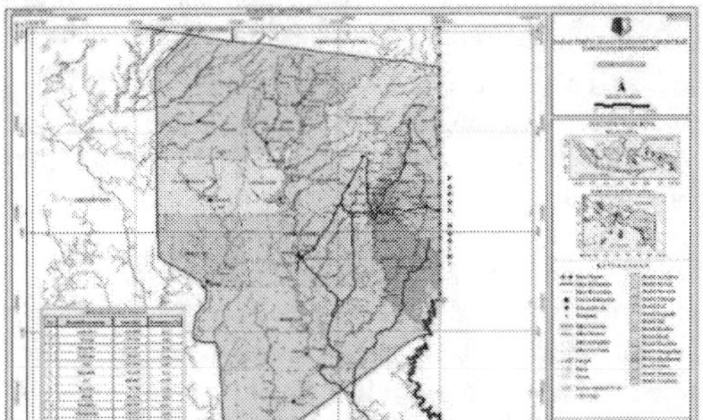

Fig. 1. Geographic Information System of Boven Digoel Regency
(Source : http://www.bovendigoelkab.go.id/halaman/peta)

2.2 Critical success factor

The method was Critical Success Factor which is run for Geographic Information System application of descriptive Data of Local Potentials of Boven Digoel Regency, whether tourism, plantation, and forestry potentials in information and location data of the latitude and longitude in the GIS application. This study was performed systematically, factually and accurately based on the case of the development of Geographic Information in Boven Digoel Regency. Critical Success Factor is an analysis method which considers several critical things in the environment to define factors affecting the success.

In the present study, the authors used Critical Success Factors methods to successfully develop geographic information system application of data of local potential of Boven Digoel Regency, especially in information and location, so that the result of the geographic referenced information can display the correct location to fulfill the needs for local information of Boven Digoel Regency.

2.3 Laravel

According to Komara, et. Al. developing generic software application built on web dynamic is complex and expensive because programmer must focus on certain platform, technology and architecture. The programmer should understand the behavior and environment of each platform, technology and architecture. This study was aimed to produce generic pattern generator for web dynamic application for three web dynamic platforms: Java EE, .NET, PHP.

We used MVC (Model-View-Controller) architecture to develop generic pattern. Every layer of MVC was analyzed by their special technology. The explored technologies were: layer view, Java, PHP [17]. Traditional framework design method to design web has many limitations, take a long time, etc. Therefore, this paper presents design and implementation of a web method based on LARAVEL framework. LARAVEL makes standard development process, processes several automatic non-business logical relations. This paper designed and implemented simple LARAVEL model which achieved automatic processing for part of the design, experimental and proven simulation, web design based on LARAVEL framework [18]. Parkar et al (2016) states that in today's era, Internet in widely known. Today, most business processes are online. Companies prefer to send job vacancies in their websites and receive CVs from qualified candidates. Millions of people post their CVs on the websites. The task of selecting and verifying the CVs are difficult, time consuming, and inaccurate. Therefore, computerization should be considered. In this system, a model or tool extracts information required from qualified candidates. The proposed model uses searching and sorting algorithms [19], evaluating procedure and PHP performance by CodeIgniter and LARAVEL framework. The study was performed because there is a lack of comparison between popular PHP frameworks such as LARAVEL, codeigniter by php. The results are analyzed and interpreted to expose the performance of the target frameworks [20]. According to Verma View controller (MVC) is architecture in standard design pattern used in website design or web application development. It classifies software applications into three interconnected parts so internal information is separated by information presented or received from users. The main component of the model consists of application data, business rule, logics and functions. The view would be information representation output, such as graph or diagram. Some views of the same information is possible, such as bar chart for management and table interface. The third part, controller, receives input and converts into command for MVC model of LARAVEL Framework which uses PHP programming language [21].

Developing a good application required some tools for support in developing Geographic Information System of data of local potentials of Boven Digoel Regency, whether tourism, plantation, or forestry potential. The framework was Laravel. The programming language was PHP hypertext preprocessor. The model, view and controller of each function are:

- Model
  The model serves to store data based on a list of name from database as required. The names which are used will be stored and processed by the system to call between view and controller consistent with application user's input.
- View
  The model serves to display HTML of web. As a service for user, view should be designed well to present information for users in terms of color, font, video or picture stored in database.
- Controller
  So that application can run in accordance with the procedure, controller function is used. If user wants to display the geography of a tourism potential, the system will display it by calling model and view from database and web interface in HTML.

2.4 Flowchart system

As contribution for the authors, data of local potentials are displayed in the developed geographic information system application consistent with the data from the Department of Communication and Informatics of Boven Digoel Regency. This application is based on web in general, providing information to users on data of tourism, plantation and forestry potentials, and the location of the potential. The method was expected to be suitable with the target of the success of the application in promoting its data content and application in the appropriate interface and location so that user can understand the location as an illustration. Figure 2 shows three existing potentials and each of them has different data content consistent with existing potentials in Boven Digoel Regency.

 https://doi.org/10.1051/e3sconf/20183112005

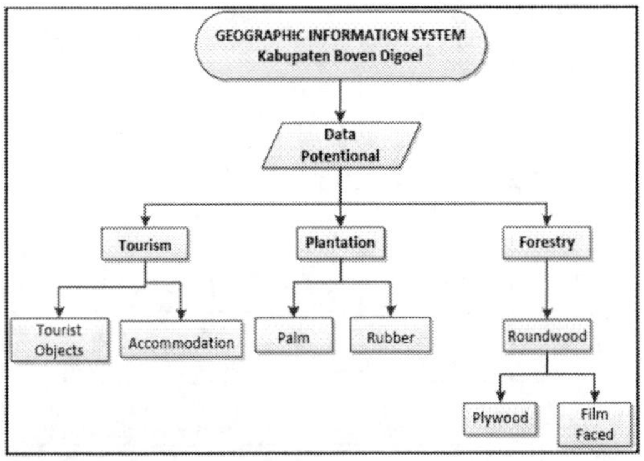

Fig. 2. Flowchart System

2.5 Use case Diagram

In this study, the authors' contribution in developing web-based application of data of local potentials is expected to run and be used by users. The system homepage of the application shows web home, tourism, plantation and tourism which contain information on existing potentials and their locations. For management, the application provides a database to store input data in the application so the data managed by the administrator is up to date.

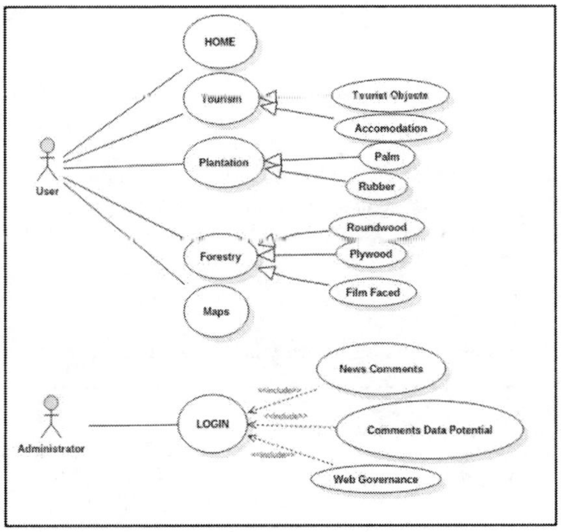

Fig. 3. Use case Diagram

2.5 Application architecture

In the architecture of the geographic information system, user can access the website by internet to get information and location. The displayed information is consistent with the data in the database managed by the administrator, so the data is appropriate for user and is published with corresponding and interesting domain. Application can be used using cellphone and computer as required.

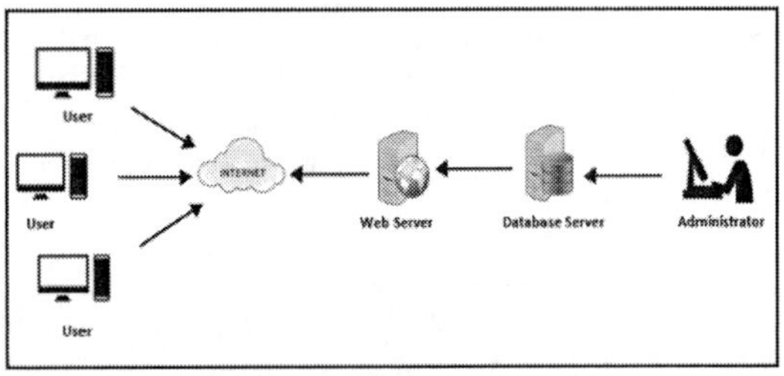

Fig. 4. Application Architecture

3 Result and discussion

The web-based geographic information system of data of local potentials of Boven Digoel Regency displays the homepage as the master, tourism objects as references and accommodation as details of tourism potential. The list in the menu bar contains 6 functions of each interface which shows information on the target vision and missions, gallery, map, archive, and contact. The geographic information system application uses CMS Content Management System function so that the display in the website contents of the geographic information system shows header, content and footer.

The header shows the logo and text of the Department of Communication and Informatics along with the moot of Boven Digoel Regency. The menu bar in the application has profile for information on the Department of Communication and Informatics, the government to show this vision and missions of Boven Digoel Regency related to the Department of Communication and Informatics of Boven Digoel, gallery to store pictures related with local potentials, whether tourism, plantation, or forestry. The gallery contains pictures and videos. Map displays the locations of existing potentials by location, archive displays news on local development, especially local potentials. The news are based on data stored in the database which is managed by administrator.

Fig. 5. Tourist Objects Interface

https://doi.org/10.1051/e3sconf/20183112005

This page displays tourism object. Every region has its own tourism objects. Boven Digoel Regency has unique tourism objects to visit. The regency is known to be a location for exile during Dutch colonization. The Proclamator was exiled in Tanah Merah city which is now known as Boven Digoel prison. The webpage in the application provides.

Fig. 5. Accommodation Interface

The accommodation interface displays hospitality services for tourists visiting Boven Digoel Regency as one of Indonesia's historical cities. This interface shows some accommodations services, e.g. hotel, motel, guesthouse etc. and displays gallery of the accommodations, as well as their locations. The information is shown as the information function of accommodation service.

Table 1. Accomodation in Boven Digoel Regency.

Business	Classification	Location	
		Lat.	Longi.
Idaman	Hotel	-6.1091481	140.324523
Mandiri	Hotel	-6.097116	140.304465
Boven Jaya	Lodging	-6,100487	140.309521
Pelangi	Lodging	-6.100871	140.303006
Ramayana	Losmen	-6.102172	140.306209

To support tourism and other activities in Boven Digoel Regency, there are some businesses which provide accommodation services. Some accommodation business services in Boven Digoel Regency are shown in the following table 1.

It is can be seen on tabel 2, that beside historical sites, Boven Digoel also has other tourist destinations, i.e. cultural, natural and religious tourist destination

Table 2. Objec Tourism in Boven Digoel Regency.

Tourist Objects	Kind of Tourism Object	Area	Location	
			Lat	Longi
Korowai Tribe's High House	Culture	Distrik Yaniruma	-5.328406	139.823793
Boven Digoel's Jail	Historical sites	Distrik Mandobo	-6.101513	140.299922
Pesta Ulat Sagu	Culture	Mindiptana	-6.013390	140.639266
Gua Maria	Religy tourism	Distrik Mandobo	-6.103926	140.312294

Table 3. Plantation crop potentials in Boven Digoel Regency

Plants	Area Of Land	Rubber Product	Year	Location	
				Lat	Longi
Rubber	1.813 ha	190 ton	2009	-6.421219	140.432 360
	1.913 ha	220.07 ton	2010		
	2.127 ha	273.69 ton	2011		
	2.616 ha	430.92 ton	2012		
	2.811 ha	383.34 ton	2013		

The plantation are dominated by rubber and palm oil. Rubber is people's plantation crops as most people in Boven Digoel Regency cultivate this plant, while palm oil is a large plantation crop managed by companies.

Table 4. The area for palm oil

Plants	Area Of Land	Palm Product	Year	Location	
				Lat	Longi
Palm	17.268 ha	22.842 ton	2009	-6.602175	140.5015 38
	17.268 ha	21.729 ton	2010		
	17.596 ha	42.307 ton	2011		
	17.648 ha	32.536 ton	2012		
	17.715 ha	40.197 ton	2013		

It can be seen in tabel 4 that the area for pal oil managed by major companies keeps increasing every year. In 2004, the area was 9.608 hectares, then in 2005 and 2006 it became 10.408 hectares.

Table 5. Product Timber Yearly in Boyen Digoel Regenct

Company	Round Timber Product	Year	Location	
			Lat	Longi
Round Timber IUPHHK & IPK	264,322.71	2009	-6.646954	140.7669763
	295,484.61	2010		
	275,617.19	2011		
	221,638.99	2012		
	121,310,01	2013		

And the top table is Boven Digoel Regency has large forest with various forest products which can be processed, increasing local income. One of the leading forest commodities is round wood. Below is a figure of the growth of round wood products by IUPHHK and IPK companies. The production of year to year increasing, caused by the businesses producing from the company except in 2013 experience reduction.

Table 6. Forest product yearly

Forest Product	Company	Sales Volume	Year
Plywood & Film Faced	IUPHHK & IPK	45600	2009
		89200	
		74513.19	2010
		51039.86	
		40701.8	2011
		120881.13	
		38107.56	2012
		97760.79	
		47174.85	2013
		87555.11	

Some forest products are processed and sold abroad, i.e. Plywood and Film Faced. Both forest products started to be sold abroad in 2007 and the sales volumes are shown in the table. In 2007, the sales volume of Plywood abroad was 72.607,01 m^3, then declined in 2008 to 45.588,18 m3.

The present study only displays data of tourism potentials as an example of application interface developed to contribute to the Department of Communication and Informatics of Boven Digoel Regency in developing local development applications for the general public and the people of Boven Digoel. The geographic information system is expected to run and be used.

4 Conclusion

The developed application design used Critical Success Factors method to clearly direct the success of geographic information system in displaying the data of potentials as local information, location and promotion for application users to attract people by information on the local potentials of Boven Digoel which is the main objective of the website application. The web-based geographic information system displays information data based on original data from the department of Communication and Informatics of Boven Digoel Regency. Boven Digoel Regency has many local potentials, i.e. agriculture, fishery, mining, and farming. In the present study, the authors only used the data of tourism, plantation and forestry potentials as contribution. It's expected that in the future all existing potentials can be inserted in the web-based geographic information system to be better and usable.

References

1. T. A. N. Xiaoshan and S. Yizhong, no. 40771164, pp. 183–186, (2010).
2. A. Oudin, D. Sprake, Y. Vagapov, and O. Simonova, pp. 513–517, (2017).
3. L. Wang, Y. Wang, Y. Li, A. Qiu, and K. Tao, 2010 Int. Conf. Web Inf. Syst. Min., pp. 138–142, (2010).
4. A. V. Vitianingsih and D. Cahyono, pp. 0–4,(2016).
5. M. R. Vieira, P. Bakalov, E. Hoel, and V. J. Tsotras, pp. 89–98, (2012).
6. N. Hakim, Maulana and R. Cahyana, J. Algoritm., pp. 1–7, (2015).
7. H. Prasetyo and A. Purwarianti, 2014 *Int. Conf. Inf. Technol. Syst. Innov.* ICITSI 2014 - Proc., no. November, pp. 13–18, (2014).
8. B. T. Nguyen, N. V Nguyen, N. T. Nguyen, and M. H. T. *Tran, Seventh Int. Conf. Inf. Sci. Technol.,* pp. 1–6, (2017).
9. S. Hatsu and E. K. Ngassam, pp. 1–8, (2015).
10. A. Susanty, D. P. Sari, and D. Anastasia, *2016 2nd Int. Conf. Sci. Inf. Technol.*, pp. 35–40, (2016).
11. I. Otieno and E. Omwenga, pp. 1–9, (2015).
12. W. Xiang, Y. Li, and Y. Shou, no. 71072119, pp. 405–409, (2013).
13. L. Alfat, A. Triwiyatno, and R. R. Isnanto, ICITACEE 2015 - *2nd Int. Conf. Inf. Technol. Comput. Electr. Eng. Green Technol. Strength.* Inf. Technol. Electr. Comput. Eng. Implementation, Proc., pp. 46–51, (2016).
14. H. V. F. Kainde, A. J. Santoso, and D. Budiyanto, J. Sist. dan Inform., vol. **11**, no. 1, pp. 173–180, (2017).
15. E. K. Dewi, Suyoto, and K. Anindito, *Journal in Indonesian*, **vol. 20**, no. semnasIF, pp. 33–40, (2012).
16. B. Nugraha, Nur, Suyoto, and Pranowo, "*Mobile Application Development For Smart Tourist Guide,*" pp. 2475–2481, (2017).
17. H. Komara, B. Hendradja, and P. Saptawati, pp. 1–5, (2016).
18. H. E. R. E. N. Yu, *Int. Conf. Comput. Sci. Electron. Technol. (ICCSET 2014)*, no. Iccset 2014, pp. 301–304, (2015).
19. V. V Parkar, P. P. Shinde, S. C. Gadade, and P. M. Shinde, *IOSR J. Comput. Eng.*, pp. 36–41, (2016).
20. R. Das and L. P. Saikia, Int. J. Curr. Trends Eng. Res., **vol. 2**, no. 6, pp. 42–48, (2016).

21 A. Verma, Indian J. Comput. Sci. Eng., **vol. 5**, no. 5, pp. 196–198, 2014.

E3S Web of Conferences **31**, 12006 (2018) https://doi.org/10.1051/e3sconf/20183112006
ICENIS 2017

The *Decision Making Trial and Evaluation Laboratory* (Dematel) and *Analytic Network Process* (ANP) for *Safety Management System* Evaluation Performance

Lisa Rolita[1,*], *Bayu* Surarso[2], *Rahmat* Gernowo[3]

[1]Master Program of Information System, School of Postgraduate Studies, Diponegoro University, Semarang - Indonesia
[2]Department of Mathematics, Faculty of Science and Mathematics, Diponegoro University, Semarang – Indonesia
[3]Department of Physics, Faculty of Science and Mathematics, Diponegoro University, Semarang - Indonesia

Abstract. In order to improve airport safety management system (SMS) performance, an evaluation system is required to improve on current shortcomings and maximize safety. This study suggests the integration of the DEMATEL and ANP methods in decision making processes by analyzing causal relations between the relevant criteria and taking effective analysis-based decision. The DEMATEL method builds on the ANP method in identifying the interdependencies between criteria. The input data consists of questionnaire data obtained online and then stored in an online database. Furthermore, the questionnaire data is processed using DEMATEL and ANP methods to obtain the results of determining the relationship between criteria and criteria that need to be evaluated. The study cases on this evaluation system were Adi Sutjipto International Airport, Yogyakarta (JOG); Ahmad Yani International Airport, Semarang (SRG); and Adi Sumarmo International Airport, Surakarta (SOC). The integration grades SMS performance criterion weights in a descending order as follow: safety and destination policy, safety risk management, healthcare, and safety awareness. Sturges' formula classified the results into nine grades. JOG and SMG airports were in grade 8, while SOG airport was in grade 7.

1 Introduction

Evaluation systems have a vital role in assessing and improving quality service by certain criteria. This allows an organization to take decision or to formulate a policy that control, monitor, and improve the quality of service in the future. The Decision Making Trial and Evaluation Laboratory (DEMATEL) and Analytic Network Process (ANP) methods allow the modeling of the problems in a evaluation system. They also excel at measuring the interdependencies among criteria in a hierarchy or network. The results can be more objective than those of the other conventional Multiple Criteria Decision Making (MCDM) methods which assume independence among criteria. The DEMATEL and ANP methods in an evaluation system enable various trouble-shooting, such as supplier performance, learning materials, information security risks, company performance, etc [1]. The results of the criteria analyses and grading using these methods show that the graded qualities of each criterion affect the result of an evaluation.

ANP grading presents the interdependencies among criteria and sub-criteria, giving a more general framework for decision making without assuming independency among elements [2]. The ANP method, on the other hand, is unable to identify interdependencies among individual criteria; to do so requires in-depth discussion and expert opinions. The ANP method needs an additional instrument to present such interdependencies among criteria, such as the DEMATEL method. Additionally, the DEMATEL method makes apparent the interactions in a decision model, grades the weights of criteria, supports ANP decision making [1], and finds and analyses the dominant criterion in a system [3].

The implementation of DEMATEL and ANP aims at improving decision making processes by taking into account interdependencies among criteria [4]. The DEMATEL method delineates the relations between various perspectives and helps the understanding of complex performance-related problems. It is these relations among the criteria acquired by the DEMATEL method that will be used to grade each criterion. The grading itself employs the ANP method that more effective results can be expected. The ANP method utilizes pair-wise comparison and offers a more accurate quantitative model of decision [5].

A performance evaluation system integrating the DEMATEL and ANP methods can be applied in aviation security management to minimize the loss of life in a situation, aviation property damage, and license revocation; to improve efficiency in capital outflow; and to reduce human and environmental externalities. The Safety Management System (SMS) is a systematic framework for safety management: it integrates performance-based approach to the regulation-based

* Corresponding author: lisarolita@student.undip.ac.id

https://doi.org/10.1051/e3sconf/20183112006

approach. Airport safety analysis is crucial in complying with the certification of aviation standards [6].

The airport SMS can help improve on existing shortcomings of an airport or provide quality and safe aviation training [7]. The aviation evaluation system has been used in airport safety evaluation by analyzing relevant factors using the Fuzzy method [8], as well as in a study on airport SMS evaluation using the ANP and Fuzzy TOPSIS methods [9]. The study highlights the need for an evaluation system in enhancing safety and in maintaining quality service to the highest aviation standards.

The current study will integrate the DEMATEL and ANP methods in evaluating airport SMS in its efforts to improve airport safety quality based on relevant factors. The result will grade the SMS-es in three Indonesian airports, namely Adi Sutjipto International Airport, Yogyakarta (JOG); Ahmad Yani International Airport, Semarang (SMG); and Adi Sumarmo International Airport, Surakarta (SOC). The grades rank the airport SMS-es that will motivate further improvements.

2 Theoretical framework

2.1 Safety management system

According to the International Civil Aviation Organization, SMS is an organized approach for safety management, which covers structured organization, accountability, policy, and procedures. SMS as a management instrument makes use of proactive components on safety performance, with a focus on the process and the identification of the risks of reducing the components [10]. The Federal Aviation Administration, on the other hand, regards SMS as a formal, top-down business-like approach to risk management [11].

In Indonesia, safety management system is regulated in Act 1, 2009 on aviation and in Transportation Ministerial Regulation KM.20, 2009 on the creation of safety management system for airport operation: a safety management system has to cater to the size, nature, and complexity of the airport as well as the safety risks. Civil aviation safety is regulated in the Transportation Ministerial Regulation PM.55, 2015, chapter 139 on aeorodrome. As for safety management system operation, it is regulated in Air Transportation Directorate Regulation KP.622, 2015 on the directives on technical safety regulations of civil aviation, chapter 139-08. Four components from these regulations will be treated as the criteria for SMS airport evaluation, as shown in Figure 1.

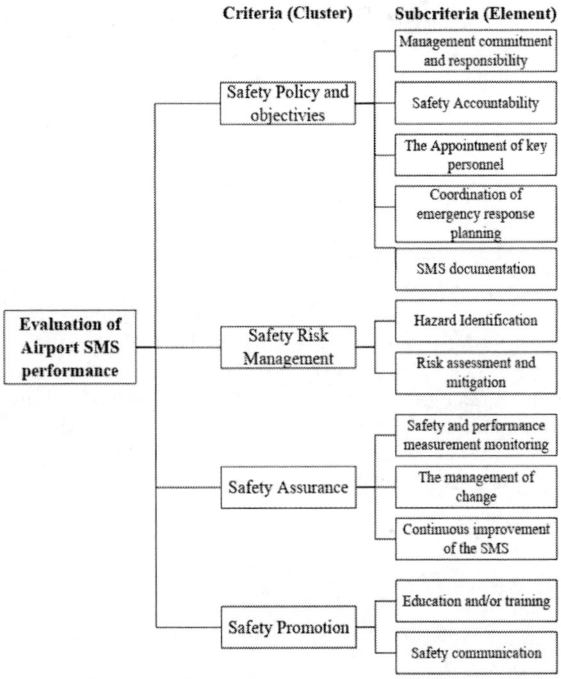

Fig. 1. Criteria and subcriteria on airport SMS evaluation system.

2.2 DEMATEL

The DEMATEL method was developed by the Science and Human Affairs Program of the Battele Memorial Institute of Geneva between 1972 to 1976. It was intended to trouble-shoot and solve complex and interconnected groups systems [12]. The DEMATEL method is a comprehensive method for building and analyzing structural models involving relations among factors [13].

The method aims at discovering immediate or immediate relations (dependency) in a system of variables. Steps in the DEMATEL method [1]:

1. Identifying variable and determining measurement scale Measurement scale presents levels of significance from 0-4 [14, 15].
2. Building immediate relation matrix $A = [a_{ij}]_{nxn}$.

The a_{ij} value represents the influence of the ai criteria or cluster against the aj criteria or cluster and n represents the number of criteria.

$$A = \begin{bmatrix} 0 & a_{12} & \cdots & a_{1n} \\ a_{21} & 0 & \cdots & a_{2n} \\ \vdots & \vdots & \ddots & \vdots \\ a_{n1} & a_{n2} & \cdots & 0 \end{bmatrix} \quad (1)$$

3. The calculation of normalization of the immediate relation matrix uses the normalization factor S in equation (2).

$$S = Min\left[\frac{1}{Max_{1\le i\le n}\left(\sum_{j=1}^{n} aij\right)}, \frac{1}{Max_{1\le j\le n}\left(\sum_{i=1}^{n} aij\right)}\right] \quad (2)$$

Immediate relation matrix normalization is signified with X and is taken from the multiplication immediate relation matrices A and S as in equation (3).

$$X = S\,x\,A \tag{3}$$

4. Calculation total relation matrix (T) uses equation (4).

$$T = \lim_{k\to\infty}(X + X^2 + \cdots + X^k) = X(I - X)^{-1} \tag{4}$$

I is the identity matrix. An essential feature of the DEMATEL method is the visualization of the most influenced and influential factor. After calculating the total relation matrix T, the number of rows and columns are available. The sum of the values in a row is signified with D, as in equation (5). The sum of the values in a column is signified with R, as in equation (6).

$$D = [\textstyle\sum_{j=1}^{n} t_{ij}]_{nx1}, (i = 1,2,\cdots,n) \tag{5}$$

$$R = [\textstyle\sum_{i=1}^{n} t_{ij}]_{nx1}, (j = 1,2,\cdots,n) \tag{6}$$

This figure shows the overall influence of other factors on the particular factor. Additionally, a visual causal diagram can be represented by the calculation of (D+R) and (D-R). The (D+R) values are on axis x, showing how important a criterion is, whereas the (D-R) values are placed on axis y and are termed relations. The relation values activate factors to be classified into cause and effect groups. If a relation value is positive, then it is classified into the cause group, and into the effect group otherwise. To integrate the DEMATEL method to the ANP method, several additional steps are taken, as follow:

5. Creating a total relation matrix using equation (7).

$$T_D = \begin{bmatrix} t_{11}^D & \cdots & t_{1j}^D & \cdots & t_{1n}^D \\ \vdots & & \vdots & & \vdots \\ t_{i1}^D & \cdots & t_{ij}^D & \cdots & t_{in}^D \\ \vdots & & \vdots & & \vdots \\ t_{n1}^D & \cdots & t_{nj}^D & \cdots & t_{nn}^D \end{bmatrix} \tag{7}$$

T_D is the total relation matrix for the criteria and t_{ij}^D is the level of influence criterion i to criterion j.

6. Summing up all row values on the total relation matrix using equation (8).

$$T_D = \begin{bmatrix} t_{11}^D & \cdots & t_{1j}^D & \cdots & t_{1n}^D \\ \vdots & & \vdots & & \vdots \\ t_{i1}^D & \cdots & t_{ij}^D & \cdots & t_{in}^D \\ \vdots & & \vdots & & \vdots \\ t_{n1}^D & \cdots & t_{nj}^D & \cdots & t_{nn}^D \end{bmatrix} \rightarrow d_i = \textstyle\sum_{j=1}^{n} t_{ij}^D \tag{8}$$

7. Normalizing the total relation matrix using equation (9).

8. Transposing the matrix T_D^α. The transposed matrix will be used as an input to the supermatrix in the ANP method.

$$T_D^\alpha = \begin{bmatrix} t_{11}^D/d_1 & \cdots & t_{1j}^D/d_1 & \cdots & t_{1n}^D/d_1 \\ \vdots & & \vdots & & \vdots \\ t_{i1}^D/d_i & \cdots & t_{ij}^D/d_i & \cdots & t_{in}^D/d_i \\ \vdots & & \vdots & & \vdots \\ t_{n1}^D/d_n & \cdots & t_{nj}^D/d_n & \cdots & t_{nn}^D/d_n \end{bmatrix} \tag{9}$$

2.3 ANP

The Analytic Network Process (ANP) was introduced by Saaty as an improvement of the Analytical Hierarchy Process (AHP) [2]. ANP builds on a shortcoming of AHP and introduces the ability to accommodate interdependencies among criteria or alternatives. There are two kinds of interdependencies in the ANP method, namely innerdependence (the interdependency elements of the same set) and outerdependence (the interdependency between different elements). These interdependencies make ANP more complex than AHP.

ANP is a general theory of relative measurement that can reduce the priority ratio from individual ratio scales. It reflects relative measurements of the influence of interacting elements against the control criteria. ANP provides a general framework for decision making without the assumption of independence of higher order elements [2]. Steps in the ANP method are as follow:

1. Creating decision network hierarchy of decision factors.
2. Creating pair-wise comparison matrix.

The comparison of significance among all criteria is transformed into the form of matrix A. The value a_{ij} represents the relative level of significance of the element on row *i* on the element of column *j*. For example, $a_{ij} = \frac{wi}{wj}$. If there are an *n* number of element to compare, then the comparison matrix A is defined in equation (10). The relative significance values are represented in a ration scale from 1 to 9, as proposed by Saaty.

$$A = \begin{bmatrix} \frac{w_1}{w_1} & \frac{w_1}{w_2} & \cdots & \frac{w_1}{w_n} \\ \frac{w_2}{w_1} & \frac{w_2}{w_2} & \cdots & \frac{w_2}{w_n} \\ \vdots & \cdots & \cdots & \vdots \\ \frac{w_n}{w_1} & \frac{w_n}{w_2} & \cdots & \frac{w_n}{w_2} \end{bmatrix} = \begin{bmatrix} 1 & a_{12} & \cdots & a_{1n} \\ a_{21} & 1 & \cdots & a_{2n} \\ \vdots & \cdots & \cdots & \vdots \\ a_{n1} & a_{n2} & \cdots & 1 \end{bmatrix} \tag{10}$$

3. Measuring element weights. Calculation of the priority vector value *w* is the eigen vector, uses equation (11).

$$A \cdot w = \lambda_{maks} \cdot W \tag{11}$$

A is a pair-wise comparison matrix, whereas w is the eigen vector and the maximum eigen value is represented with λ_{maks}.

4. Calculating consistency index using equation (12).

$$\mathrm{CI} = \frac{\lambda maks - n}{n-1} \tag{12}$$

CI is the consistency index and n represents the number of elements. By comparing the consistency index and the random index (RI), a threshold value becomes available to determine the consistency level of a matrix. This is called consistency ration (CR). CR is calculated using equation (13).

$$CR = \frac{CI}{IR} \quad (13)$$

Random index (RI) is the value of a matrix that measures the consistency of the results acquired up to this step. If the pair-wise comparison matrix has a value of less than 0.10, then the inconsistency of the opinions of decision makers is to be expected; if not, data assessment in the decision making process needs to be improved. If CR equals zero, the result is consistent, and if it is less than or equal to 0.1, the result is consistent enough.

5. Creating the supermatrix.

Supermatrix is a matrix of the results of the priority vector from pair-wise comparison between clusters, elements, and alternatives. There are three components of a supermatrix, namely:

a. Unweighted supermatrix, acquired from the comparisons among clusters, criteria, and alternatives with the eigen vector column in the matrix.
b. Weighted supermatrix, acquired by multiplying all elements in the unweighted supermatrix by the value of the transposed matrix T_D^{α} in the corresponding DEMATEL operation.
c. Limiting supermatrix, acquired by multiplying the supermatrix values with themselves until each row has the same value.

3 Methodologies

Data collection was through surveys with questioners on SMS performance. The study cases on this evaluation system were Adi Sutjipto International Airport, Yogyakarta (JOG); Ahmad Yani International Airport, Semarang (SRG); and Adi Sumarmo International Airport, Surakarta (SOG). The questioners were responded to by the SMS head officers and SMS officers. The questioner had three components, namely the levels of significance of criteria, levels of significance of sub-criteria, and airport SMS evaluation. The criteria for the indicators in the evaluation process were based on the Air Transportation Directorate General Regulation KP.622, 2015 on the operations of airport safety management system. The following indicaters were considered: safety and destination policy, safety risk management, healthcare, and safety awareness. The steps in the evaluation process of airport SMS are shown in Figure 1.

4 Results and discussion

The DEMATEL method managed to improve the ANP method in identifying the interdependency relations of each criterion. In view of this, the DEMATEL and ANP methods can be integrated in modeling the problems in an evaluation system, with especial benefits in measuring the relations among criteria in a hierarchy or network. This integration can be implemented effectively on airport SMS evaluation system as it can determine the value and direction of the interdependencies among the criteria. This study results in an evaluation system that can be used to assess airport SMS performance and to improve it.

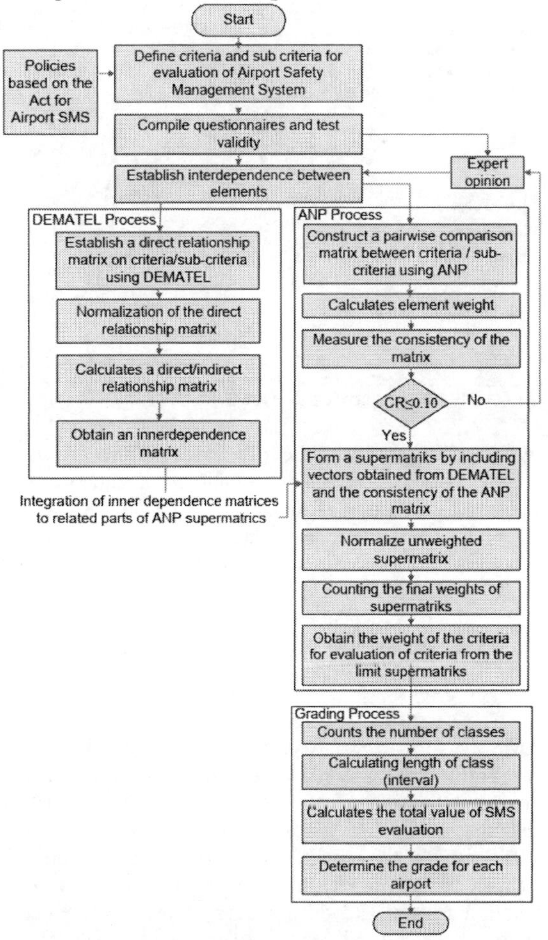

Fig. 2. The steps in the evaluation process of SMS.

This evaluation system presents relevant information on the grades of the criteria to be used in evaluating the SMS. The grades can in turn be evaluated and be used by stakeholders or the management in implementing airport SMS performance. This system also grades the results of the evaluation of SMS performance, to which the current airport SMS performance can be compared.

Assessments of the levels of influence and levels of significance by the SMS head officers were conducted based on the proposed criteria and subcriteria. The results from the assessments were further analyzed using the DEMATEL and ANP methods. The DEMATEL method sees relations among criteria as shown by the Network Relation Map (NRM). The NRM reflecting the assessments is shown in Figure 2.

Fig. 3. NRM of Airport SMS using DEMATEL

The NRM indicates the following relations:.

1. C1 has innerdependent relations: it interacts and gives and receives feedback within a cluster.
2. C1 has outerdependent relations: it interacts with and gives and receives feedback to and from clusters C2, C3, and C4.
3. C2 has outerdependent relations to C1, C3, and C4.
4. C3 has outerdependent relations to C1, C2, and C4.
5. C4 has outerdependent relations to C1, C2 dan C3.

The suggested model indicates that the criteria for SMS performance measurement were interdependent. The influencing and influenced criteria were taken from the total relation matrix from the DEMATEL method calculation. To find those criteria, the dispatcher vector (D) and the receiver vector (R) were calculated. Dispatcher vector is the number of rows in a total relation matrix, and the receiver vector the number of columns in a total relation matrix. Next, the (D+R) values are placed on the x axis of a impact-digraph map, while the (D-R) values on the y axis. Positive (D-R) values also put criteria as criteria dispatchers and negative (D-R) values put them as criteria receivers. D and R vectors are shown in Table 1.

Table 1. Vector *dispatcher* and vector *receiver*

Criteria	D	R	D+R	D-R
C1	8.567	8.008	16.575	0.559
C2	7.444	7.624	15.068	-0.180
C3	7.641	8.016	15.657	-0.375
C4	7.620	7.624	15.244	-0.004

After determining the (D+R) and (D-R) values, the impact-digraph map can be created. The direction of the arrow (following the row-to-column direction) is represented on the matrix with values exceeding the threshold value. The threshold value qualifies a parameter forming a network. This value is the mean of the total matrix minus 0.1; in this case, the threshold value (α) was 1.8545. The resulting impact-digraph map is in Figure 3.

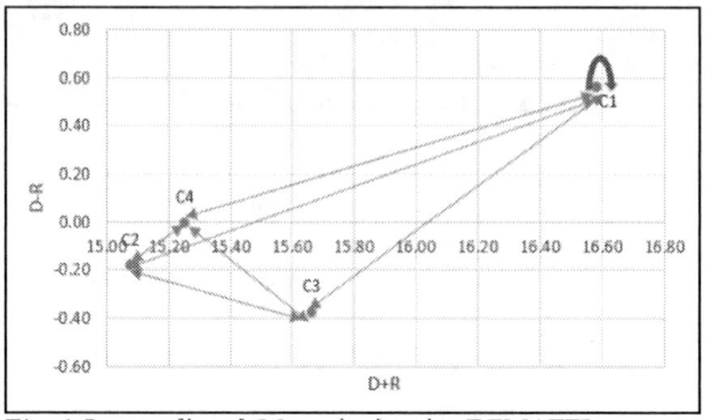

Fig. 4. Impact-digraph Map criteria using DEMATEL

To integrate the DEMATEL method to ANP, the values of the criteria matrix, or the Tαd matrix, were calculated The transposed results for Tαd matrix were used in calculating the supermatrix values in ANP, as shown in Table 2.

Table 2. Transpose Matriks Tαd

Criteria	C1	C2	C3	C4
C1	0.23	0.34	0.35	0.34
C2	0.26	0	0.32	0.32
C3	0.26	0.34	0	0.34
C4	0.25	0.32	0.33	0

After the levels of influence of the four criteria were calculated, the levels of significance among the sub-criteria were calculated. The ANP method determined the levels of significance of the sub-criteria. The initial stage determined the pair-wise comparison matrix, giving the weight of each sub-criteria. Weight measurement was is intended to determine the dominant sub-criteria to be improved on in the SMS performance evaluation process.

The CR value, 0.0661, was acquired from the pair-wise grade matrix. As the value is less than 0.10, the matrix is regarded as consistent. The implication is that the opinions gathered during the ANP-method assessment should be regarded. The final stage of ANP is the creation of the supermatrix. This creation is comprised of three components, namely the unweighted supermatrix, the weighted supermatrix, and the limiting supermatrix. The creation of the supermatrix is integrated into the transposed matrix Tαd (total relations) from the DEMATEL method. The unweighted supermatrix is shown in Table 3, the weighted supermatrix on Table 4, and the limiting supermatrix on Table 5. In calculating the supermatrices, an alternative ranking of weights is acquired from the sum of each row in the limiting supermatrix. This alternative weight is shown on Table 6.

Table 3. Unweighted supermatrix

		C1					C2		C3			C4	
		e11	**e12**	**e13**	**e14**	**e15**	**e21**	**e22**	**e31**	**e32**	**e33**	**e41**	**e42**
C1	e11	0.029	0.009	0.017	0.008	0.019	0.020	0.016	0.013	0.009	0.012	0.008	0.009
	e12	0.009	0.003	0.005	0.003	0.006	0.007	0.005	0.004	0.003	0.004	0.003	0.003
	e13	0.017	0.005	0.010	0.005	0.011	0.012	0.009	0.008	0.005	0.007	0.005	0.005
	e14	0.008	0.003	0.005	0.002	0.005	0.006	0.005	0.004	0.003	0.003	0.002	0.003
	e15	0.019	0.006	0.011	0.005	0.012	0.013	0.011	0.009	0.006	0.008	0.005	0.006
C2	e21	0.020	0.007	0.012	0.006	0.013	0.000	0.000	0.009	0.006	0.008	0.006	0.006
	e22	0.016	0.005	0.009	0.005	0.011	0.000	0.000	0.008	0.005	0.007	0.005	0.005
C3	e31	0.013	0.004	0.008	0.004	0.009	0.009	0.008	0.000	0.000	0.000	0.004	0.004
	e32	0.009	0.003	0.005	0.003	0.006	0.006	0.005	0.000	0.000	0.000	0.003	0.003
	e33	0.012	0.004	0.007	0.003	0.008	0.008	0.007	0.000	0.000	0.000	0.003	0.004
C4	e41	0.008	0.003	0.005	0.002	0.005	0.006	0.005	0.004	0.003	0.003	0.000	0.000
	e42	0.009	0.003	0.005	0.003	0.006	0.006	0.005	0.004	0.003	0.004	0.000	0.000

Table 4. Weighted supermatrix

		C1					C2		C3			C4	
		e11	**e12**	**e13**	**e14**	**e15**	**e21**	**e22**	**e31**	**e32**	**e33**	**e41**	**e42**
C1	e11	0.007	0.002	0.004	0.002	0.004	0.007	0.006	0.005	0.003	0.004	0.003	0.003
	e12	0.002	0.001	0.001	0.001	0.001	0.002	0.002	0.002	0.001	0.001	0.001	0.001
	e13	0.004	0.001	0.002	0.001	0.003	0.004	0.003	0.003	0.002	0.002	0.002	0.002
	e14	0.002	0.001	0.001	0.001	0.001	0.002	0.002	0.001	0.001	0.001	0.001	0.001
	e15	0.004	0.001	0.003	0.001	0.003	0.005	0.004	0.003	0.002	0.003	0.002	0.002
C2	e21	0.005	0.002	0.003	0.001	0.003	0.000	0.000	0.003	0.002	0.003	0.002	0.002
	e22	0.004	0.001	0.002	0.001	0.003	0.000	0.000	0.002	0.002	0.002	0.001	0.002
C3	e31	0.003	0.001	0.002	0.001	0.002	0.003	0.003	0.000	0.000	0.000	0.001	0.001
	e32	0.002	0.001	0.001	0.001	0.002	0.002	0.002	0.000	0.000	0.000	0.001	0.001
	e33	0.003	0.001	0.002	0.001	0.002	0.003	0.002	0.000	0.000	0.000	0.001	0.001
C4	e41	0.002	0.001	0.001	0.001	0.001	0.002	0.001	0.001	0.001	0.001	0.000	0.000
	e42	0.002	0.001	0.001	0.001	0.001	0.002	0.002	0.001	0.001	0.001	0.000	0.000

Table 5. Limiting supermatrix

		C1					C2		C3			C4	
		e11	**e12**	**e13**	**e14**	**e15**	**e21**	**e22**	**e31**	**e32**	**e33**	**e41**	**e42**
C1	e11	0.185	0.185	0.185	0.185	0.185	0.185	0.185	0.185	0.185	0.185	0.185	0.185
	e12	0.061	0.061	0.061	0.061	0.061	0.061	0.061	0.061	0.061	0.061	0.061	0.061
	e13	0.108	0.108	0.108	0.108	0.108	0.108	0.108	0.108	0.108	0.108	0.108	0.108
	e14	0.052	0.052	0.052	0.052	0.052	0.052	0.052	0.052	0.052	0.052	0.052	0.052
	e15	0.121	0.121	0.121	0.121	0.121	0.121	0.121	0.121	0.121	0.121	0.121	0.121
C2	e21	0.104	0.104	0.104	0.104	0.104	0.104	0.104	0.104	0.104	0.104	0.104	0.104
	e22	0.084	0.084	0.084	0.084	0.084	0.084	0.084	0.084	0.084	0.084	0.084	0.084
C3	e31	0.072	0.072	0.072	0.072	0.072	0.072	0.072	0.072	0.072	0.072	0.072	0.072
	e32	0.049	0.049	0.049	0.049	0.049	0.049	0.049	0.049	0.049	0.049	0.049	0.049
	e33	0.065	0.065	0.065	0.065	0.065	0.065	0.065	0.065	0.065	0.065	0.065	0.065
C4	e41	0.047	0.047	0.047	0.047	0.047	0.047	0.047	0.047	0.047	0.047	0.047	0.047
	e42	0.052	0.052	0.052	0.052	0.052	0.052	0.052	0.052	0.052	0.052	0.052	0.052

The results of the ANP calculation will also determine the weight of each sub-criterion to be used in the SMS performance evaluation. These results will determine the dominant criteria and sub-criteria or the priority to be improved in airport SMS performance, as represented with the lowest rank in the alternative weight ranking is shown in Table 6. It can be seen that the safety and destination policy has been graded with the maximum average value of 0.1053, while management commitment and responsibility 0.1852.

The average assessment figures from the evaluations by the SMS head officers and SMS officers in the three airports studied were multiplied with the weight of each sub-criterion. The results are as follow: the SMG airport was given a score of 4.1487, the JOG airport 4.3996, and the SOG airport 3.9851.

The results of the SMS performance evaluation were then classified using Sturges formula into nine grades. This classification would hopefully motivate each airport to improve the safety system performance. The Sturges's formula determined the number of classes in the classification.

Table 6. Alternative ranking weights

		Sub-criteria Weight	Percen-tage of Sub-criteria weight	Rate	Average Criteria Weight
C1	e11	0.1852	18.52%	1	0.1053
	e12	0.0607	6.07%	8	
	e13	0.1079	10.79%	3	
	e14	0.0522	5.22%	9	
	e15	0.1206	12.06%	2	
C2	e21	0.1041	10.41%	4	0.0939
	e22	0.0838	8.38%	5	
C3	e31	0.0718	7.18%	6	0.0620
	e32	0.0493	4.93%	11	
	e33	0.0650	6.50%	7	
C4	e41	0.0473	4.73%	12	0.0496
	e42	0.0520	5.20%	10	
Total		1	100%		

The formula yielded nine classes from observation data of all 298 airports in Indonesia. The between-class interval values of the nine classes (grades) are shown in Table 7. The classification of the three airports in the current study is shown in Table 8.

Table 7. Interval value between class of evaluation result

Grade	Interval		
1	1.0000	-	1.4444
2	1.4445	-	1.8888
3	1.8889	-	2.3332
4	2.3333	-	2.7776
5	2.7777	-	3.2220
6	3.2221	-	3.6664
7	3.6665	-	4.1108
8	4.1109	-	4.5552
9	4.5553	-	5.0000

Table 8. Classification of the results class of Airport SMS evaluation

Airport Name	Total evaluation value	Grade
Ahmad Yani International Airport	4.1487	8
Adi Stjucipto International Airport	4.3996	8
Adi Sumarmo International Airport	3.9851	7

5 Conclusions

The DEMATEL and ANP methods can be integrated in modeling the problems of an evaluation system with the advantage on measuring the interdependencies among criteria in a hierarchy or network. The current study investigates the airport SMS performance evaluation system using those methods. It employs the DEMATEL and ANP methods in determining the priority of airport SMS performance improvement areas. This study yields several conclusions. First, the criteria used in the evaluation system were interdependent in terms of the levels of influence among criteria with the safety and destination criterion being innerdependent. Second, the airport SMS performance system can rank criteria and sub-criteria to be multiplied. The safety and destination policy has been graded with the maximum average value of 0.1053, while management commitment and responsibility 0.1852. Third, with the weight measurement by the ANP method, the results of SMS evaluation of the three airports are as follow: Adi Sutjipto International Airport, Yogyakarta, was graded 4.3995, Ahmad Yani International Airport, Semarang, 4.1487, and Adi Sumarmo International Airport, Surakarta, 3.9851. Regarding the nine classes acquired from the Sturges formula, these results put the Yogyakarta and Semarang airports in grade 8, while the Surakarta airport in grade 7.

References

1. I. Golcuk, A. Baykasoglu, Expert Systems With Applications **46**, 346-366 (2016)
2. T.L. Saaty, Pittsburgh: RWS Publications (2005)
3. Y.C. Chou, C.C. Sun, H.Y. Yen, Applied Soft Computing **12**, 64-71 (2012)
4. J.J.H. Liou, Y. T. Chuang, Expert System with Applications **37**, 3755-3761 (2010)
5. C.C. Hsu, J.J.H. Liou, Journal of Air Transport Management **28**, 40-46 (2013)
6. C. Kang, L. Zhong, Z. Haijun, Procedia Engineering **99**, 180-184 (2015)
7. Y. Yan, *Safe, Smart, and Sustainable Multimodal Transportation Systems* ASCE, 62-67 (2014)
8. X.F. Lu, S.G. Huang, *International Conference on Fuzzy Theory and Its Applications*, 144-149 (2012)
9. Y.H. Chang, P.C. Shao, H.J. Chen, Safety Science **75**, 71-86 (2015)
10. E. Gerede, Safety Science **73**, 106-116 (2015)
11. J.C. Price, J.S. Forrest, *Practical Airport Operation, Safety, and Emergency Management*. Butterworth Heineman, (2016)
12. G.H. Tzeng, C.H. Chiang, C.W. Li, Expert Systems with Applications **32**, 1028-1044 (2007)
13. G. Buyukozkan, G. Cifci, Expert System with Application **39**, 3000-3011 (2012)
14. I.S. Chen, Journal of the Operational Research Society **63**, 478-488 (2012)
15. J.L. Yang, G.H. Tzeng, Expert Systems with Applications **38**, 1417–1424 (2011)

Development of Airport Noise Mapping using Matlab Software (Case Study: Adi Soemarmo Airport – Boyolali, Indonesia)

Pertiwi Andarani[1,*], *Haryono* Setiyo Huboyo[2], *Diny* Setyanti[1], and *Wiwik* Budiawan[2]

[1]Department of Environmental Engineering, Faculty of Engineering, Diponegoro University, Semarang – Indonesia
[2]Department of Industrial Engineering, Faculty of Engineering, Diponegoro University, Semarang – Indonesia

Abstract. Noise is considered as one of the main environmental impact of Adi Soemarmo International Airport (ASIA), the second largest airport in Central Java Province, Indonesia. In order to manage the noise of airport, airport noise mapping is necessary. However, a model that requires simple input but still reliable was not available in ASIA. Therefore, the objective of this study are to develop model using Matlab software, to verify its reliability by measuring actual noise exposure, and to analyze the area of noise levels.. The model was developed based on interpolation or extrapolation of identified Noise-Power-Distance (NPD) data. In accordance with Indonesian Government Ordinance No.40/2012, the noise metric used is WECPNL (Weighted Equivalent Continuous Perceived Noise Level). Based on this model simulation, there are residence area in the region of noise level II (1.912 km^2) and III (1.16 km^2) and 18 school buildings in the area of noise levels I, II, and III. These land-uses are actually prohibited unless noise insulation is equipped. The model using Matlab in the case of Adi Soemarmo International Airport is valid based on comparison of the field measurement (6 sampling points). However, it is important to validate the model again once the case study (the airport) is changed.

1 Introduction

Adi Sumarmo International Airport (ASIA) is the second largest airport in Central Java Province, Indonesia. It is located in Boyolali, around 14 km away from Surakarta City. It served 1,412,638 passengers in 2015 [1]. Some extra flights is always added during Hajj season and Ramadhan season. Different with Achmad Yani International Airport (AYIA) - Semarang, ASIA is located on land far from the sea. It is built near Waduk Cengklik (Cengklik Reservoir). Due to the airport operation, noise cannot be avoided, particularly from the aircrafts departures and landing (approaches).

Many researches have proved that noise from the airport could bring negative impact, such as annoyance [2], sleep disturbance [3], insomnia and daytime hypersomnia [4], myocardial infarctions [5], children's development [6], [7], psychosomatic effects [8], chronic noise stress and hypertension [9], and immune system [10]. In the vicinity of ASIA, it is reported that 65% of the inhabitants in the region suffered from sleep disturbance, hearing impairment, and annoyance based on Hartono's research in 2006 [10]. The aircraft noise impact on immune system has been also investigated by Hartono [10] that there was a significant effect of the aircraft noise level to the number of Natural Killer Cell (CD56+CD16+CD3-) among people in the area of Adi Sumarmo Airport Boyolali.

The scientific evidence of aforementioned negative impacts due to aircraft noise exposure must be avoided by properly managing the airport operation. In order to identify whether there is higher noise exposure than the standard limit around the airport, airport noise mapping is necessary. The airport noise mapping can be estimated by a model. The FAA (Federal Aviation Administration) has developed INM (Integrated Noise Model) and AEDT (Aviation Environmental Design Tool). However, these tools require advanced understanding on aircraft specification and operation. Therefore, another model using simplified input data was needed to be developed, particularly adjusted to the condition of airports in Indonesia. The objective of this study are to develop model using Matlab and to verify its reliability by measuring actual noise exposure.

2 Methods

2.1. Airport specification

Adi Sumarmo International Airport (ASIA) is one of the airport operated by PT Angkasa Pura I (Persero) with service area of Surakarta City, Boyolali, Sukoharjo, Karanganyar, Sragen, Klaten, Wonogiri, a part of districts in East Java Province (Ngawi, Magetan, Pacitan, and Ponorogo District). ASIA is located at coordinate 07 31' 04"S 110 45'18"E with relatively flat topography (1-2%) and elevation of 128 meters above sea level. The temperature at ASIA is relatively constant, the minimum average temperature annually is from 20,6°C to 22,8°C,

* Corresponding author: andarani@ft.undip.ac.id

while maximum is from 31,7°C to 32,8°C. The maximum wind velocity monthly is 5 knot (8 km/hr) while minimum 2 knot (3.2 km/hr) with prevailing wind N240E.

At ASIA, there are two runways, i.e. Runway-08 (7°31'06.4"S 110°44'47.4"E) and Runway-26 (7°30'48.6"S 110°46'07.1"E), with dimension of 2600 m x 45 m. This airport also serves extra flights to Mecca and Jeddah during Hajj season. In this study, the busiest schedule during regular time was selected as case study, i.e. September 15th, 2016 (22 approaches and 22 departures). The operational hours during the day was from 06.00 – 20:40 Western Indonesia Time (GMT+07h).

2.2 Model Development

The first step in this modeling was to determine the layout area and grids in the layout so that the coordinate of each grid can be determined. Coordinates denoted by x, y, and z. Coordinate-x is the coordinate in the direction of the runway, coordinate-y represents the coordinate of perpendicular to the runway, and coordinate-z shows the coordinates of the height of the surface (ground level). Furthermore, flight tracks and flight profile were defined. Flight track data obtained from ASIA were used to determine the position of the aircraft (the sound source). Meanwhile, the flight profiles were obtained from the flight profile, i.e. Aircraft Noise Performance (ANP) database. Flight profile data shows the relationship between altitude of aircraft with aircraft horizontal distance from the runway during landing or take-off.

The calculation step was started from the altitude of aircraft during landing (approach) or take-off (departure). To calculate the altitude at each grid from flight profile data, interpolation or extrapolation is necessary [11].

$$z(x) = z(x_i) + \frac{z(x_{i+1}) - z(x_i)}{x_{i+1} - x_i} . (x - x_i) \qquad (1)$$

where, $z(x)$: aircraft altitude at the observed grid point; $z(x_i)$: aircraft altitude at lower grid point; $z(x_{i+1})$: aircraft altitude at higher grid point; x: aircraft distance from the runway at the observed grid point; x_i: aircraft distance from the runway at closer grid point; x_{i+1}: aircraft distance from the runway at further grid point.

The next step was to calculate the distance between aircraft and each grid. The distance is the closest distance between line on flight profile with grid.

$$d = \sqrt{(x_a - x_o)^2 + (y_a - y_o)^2 + (z_a - z_o)^2} \qquad (2)$$

where, d: distance between aircraft and observed grid; x_a: coordinate-x of the aircraft; x_o: coordinate-x of the observed grid; y_a: coordinate-y of the aircraft; y_o: coordinate-y of the observed grid; z_a: coordinate-z of the aircraft; z_o: coordinate-z of the observed grid.

After the calculation of the distance in each grid has been done, the noise level of each grid was calculated using Noise-Power-Distance (NPD) data. Interpolation or extrapolation at logarithmic scale was undertaken to calculate the noise level of the grid that previously unknown. According to ICAO [11], to determine the noise level at grid between two grids that have previously known, the interpolation could be calculated as equation (3). Meanwhile, extrapolation should be done when the observed grid was beyond the other two grid that have been known its noise levels. If the distance (d) is shorter than the shortest distance in NPD data (d_i), then the equation (4) should be done, while if (d) is larger than the largest distance in NPD data (d_i), then the calculation was done by equation (5).

$$L(d) = L(d_i) + \frac{L(d_{i+1}) - L(d_i)}{\log d_{i+1} - \log d_i} \cdot (\log d - \log d_i) \qquad (3)$$

$$L(d) = L(d_2) + \frac{L(d_1) - L(d_2)}{\log d_2 - \log d_1} \cdot (\log d_2 - \log d) \qquad (4)$$

$$L(d) = L(d_{i-1}) + \frac{L(d_{i-1}) - L(d_i)}{\log d_i - \log d_{i-1}} \cdot (\log d - \log d_{i-1}) \qquad (5)$$

where, $L(d)$: noise level at the observed grid point; $L(d_i)$: noise level at lower grid point than the observed grid point; $L(d_{i+1})$: noise level at higher grid point than the observed grid point; $L(d_1)$: noise level at the first point in NPD data; $L(d_2)$: noise level at the second point in NPD data; $L(d_i)$: noise level at the last point in NPD data; $L(d_{i-1})$: noise level at the second latest in NPD data; d: distance from grid point to aircraft in the observed grid point; d_i: distance from grid point to aircraft that shorter than the observed grid point; d_{i+1}: distance from grid point to aircraft that further than the observed grid point; d_1: distance from the aircraft at the first point in NPD data; d_2: distance from the aircraft to the second point n NPD data; d_i: distance from the aircraft to the last point in NPD data; d_{i-1}: distance from the grid point to the second latest in NPD data.

After all of noise levels of grids has been calculated, WECPNL (Weighted Equivalent Continuous Perceived Noise Level) as the noise metric should be used in accordance with Indonesian Government Ordinance No.40/2012. In this study, a day simulation (September 15th, 2016) was undertaken. WECPNL can be calculated as the following equation based on Indonesian National Standard, SNI 8150:2015:

$$\mathrm{WECPNL} = \overline{\mathrm{dB(A)}} + 10 \log \mathrm{N} - 27 \qquad (6)$$

$$\overline{\mathrm{dB(A)}} = 10 \log [(\tfrac{1}{n}) \textstyle\sum_{i=1}^{n} 10^{L_i/10}] \qquad (7)$$

$$\mathrm{N} = \mathrm{N2} + 3\,\mathrm{N3} + 10\,(\mathrm{N1} + \mathrm{N4}) \qquad (8)$$

where, $\overline{\mathrm{dB(A)}}$ means the energy averaged maximum A-weighted sound levels (L_i) of noise events N. The noise events include the number of approaches and departures of aircrafts. N_1, N_2, N_3, and N_4 mean the numbers of noise events observed during the night-early morning

(00:00-07:00), the day-time (07:00-19:00), evening (19:00-22:00), night-time (22:00-00:00), respectively. Calculating dB(A) requires n and Li observed during 24 hours. The total number of departures and approaches during 24 hours is denoted by n while Li means the maximum measured noise level (A-weighted) during one noise event.

2.3 Programming steps

The program was developed in Bahasa Indonesia (Indonesian Language) instead of English so that Indonesian could easily understand how to input the data needed. The initial step in programming using Matlab is by designing function m-file to read or process data used in the simulation, then the process of calculating and creating the contours and the output process both the noise contour map and the noise levels at the determined points.

The input data includes length of runway, number of flights, operation type (approach or depart), flight direction (right or left), and flight time schedule. The calculation and creating the contour process were done in m-file. The contour map can be saved as *.jpg format. All of these processes are integrated in GUI (Graphical User Interface) program.

2.4 Validation of the model

This study used ASIA as a case study, hence, validation of the model was undertaken using the data from the airport. In this case, the WECPNLs generated from the model were compared to the WECPNL measured in the field. WECPNL needs Li or the maximum A-weighted sound levels (L_i) of each noise event such as take-off (departure) or landing (approach) of an aircraft. There were 6 locations of sampling point in the vicinity of ASIA, e.g. 4 points crossing the runway (residence area) and 2 points in school area (see Fig. 1). The model is considered to be accurate if >75% of sampling points have ΔWECPNL<5.

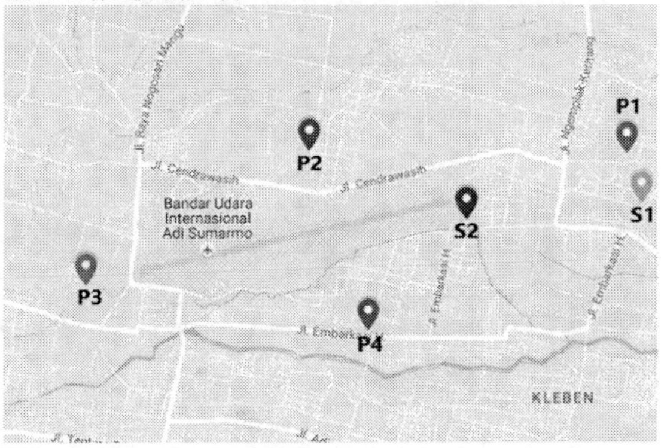

Fig. 1. Location of noise sampling points

In this study, the background noise levels (A-weighted) during noise sampling were also measured then Leq was used as the noise metric. The method was based on the Indonesian Ministry of Environment Decree No. 48/Menlh/11/1996 regarding Noise Level Standard. The measurement was done using Sound Level Meter Extech RS232 (at S2), Lutron-4010 (at P1 and P4), Benetech GM1351 (at P2, P3, and S1). All of the equipment have been checked its accuracy. Leq calculation is as follows:

$$L_{eq} = 10\log\left(\sum_{i=1}^{n}\left(fi.10^{Li/10}\right)\right) \quad (9)$$

where fi = time fraction during measurement, Li = dBA value, n = total measurement during one period.

3 Results and discussion

3.1 Simulations of the model

The simulation was based on the data during busiest schedule in September 2016. There were 22 departures and 22 approaches, starting from 06.00 to 20.40 GMT+7h. GUI for inputting the data can be seen in Fig. 2.

According to the Indonesian Government Ordinance No.40/2012, the evaluation of noise levels in the vicinity of airport is based on the area in each category level noise region, namely region level I, II, and III. The total area of this noise region is 12.8 km^2. The area of noise region level I, II, and III are 6.4, 2.27, 4.13 km2, respectively. Directions distribution of the noise region extends toward Runway-26 as far as 10.34 km and 8.71 km from Runway-08 and 0.92 km from each runway side. For details, the noise contour map can be seen in Fig. 3.

Noise region level I can be used for various types of activities and / or building unless the type of building hospitals and schools. Noise region level II can be used for various types of activities and / or building unless the type of building hospitals, schools, and residence area. Meanwhile, the noise region level III can only be used to build airport facilities which are equipped with sound insulation and can be used also as a green belt or a means of controlling the environment and agriculture which are not inviting birds.

However, based on this model simulation, there are residence area in the region of noise level II (1.912 km^2) and III (1.16 km^2) as well as the school buildings in the area of noise levels I, II, and III. There are 18 schools in total, namely 8 schools in level I, 5 schools in level II, and 5 schools in level III. Unfortunately, schools and residence area are still not insulated from aircraft noise exposure properly.

3.2 Validation of the model

It is important to validate the model, how far the model is accurate. In this case, 6 sampling points were selected to identify whether the WECPNL in the model is accurate compared to the actual noise measurement. Table I represents the results. It is clear that only P3 has ΔWECPNL>5. Therefore, the model is considered valid

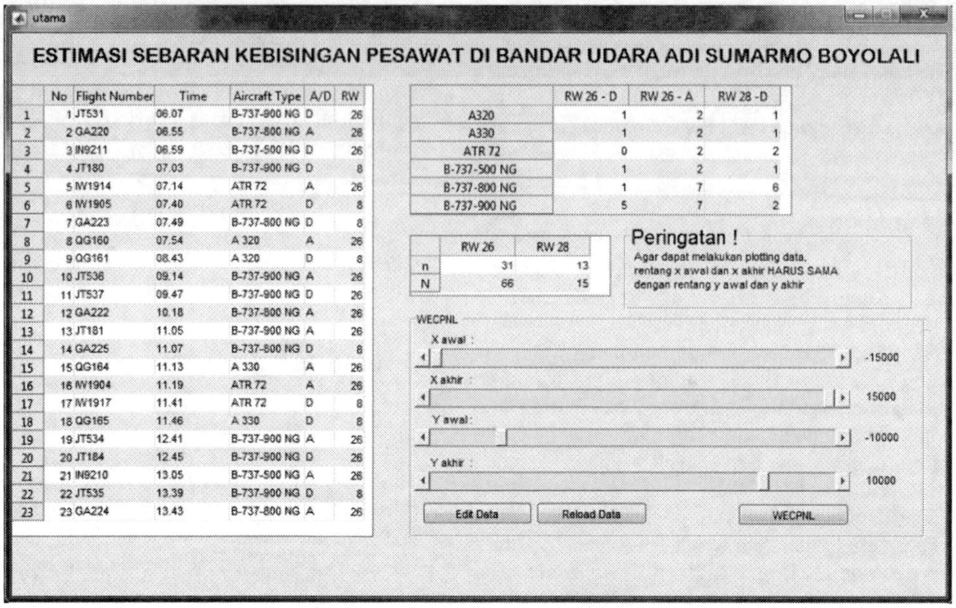

Fig. 2. Graphical User Interface (GUI) developed in Indonesian Language for data input

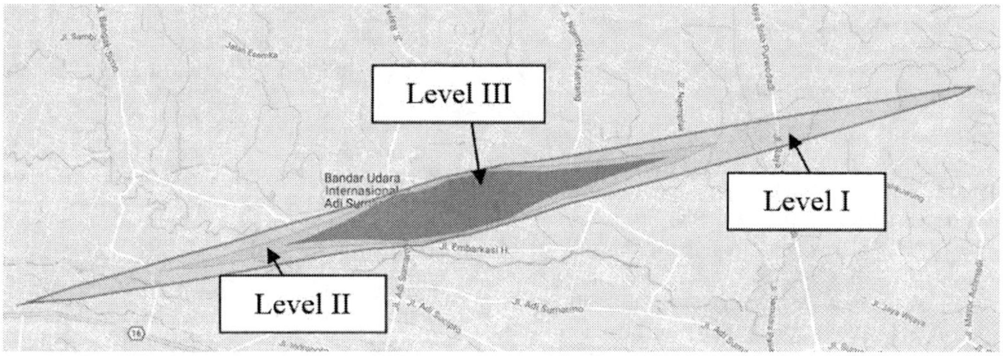

Fig. 3. Noise contour map of Adi Soemarmo International Airport

in this case. However, if the model is to be implemented at other airports, the model should be re-checked to make sure that the error is within the acceptable limit. Because, in other case, such as Husein Sastranegara International Airport – Bandung, the model using Matlab was not considered accurate because several aspects such as bank angel, atmospheric, absorption adjustment, acoustic impedance adjustment, noise fraction adjustment, duration adjustment, lateral attenuation adjustment, and line-of sight blockage adjustment, were not considered properly [12].

Table 1. Comparison between model and field measurement

No	Point	Background Noise (Leq) during 15 hours of sampling dB(A)	WECPNL		ΔWECPNL
			Model	Field measurement	
1.	P1	59,70	83,25	80,93	2,31
2.	P2	59,29	69,25	64,28	4,96
3.	P3	67,46	87,31	79,0359	8,28
4.	P4	54,31	65,71	65,27	0,44
5.	S1	58,93	74,69	74,54	0,14
6.	S2	48,49	82,54	83,49	-0,95

4. Conclusions

The total area of this noise region is 12.8 km^2. The area of noise region level I, II, and III are 6.4, 2.27, and 4.13

km^2, respectively. Directions distribution of the noise region extends toward Runway-26 as far as 10.34 km and 8.71 km from Runway-08, and 0.92 km from each runway side. The model using Matlab in the case of Adi Soemarmo International Airport is valid based on comparison of the field measurement (6 sampling points). However, it is important to validate the model once the case study (the airport) is changed. For example, the model developed in this study was specified for Adi Soemarmo International Airport, then it will not appropriate to model the noise in other airports. Therefore, in the next study, the model could be developed more detail to consider the neglected factors and user input for adjustment of the airport specification.

References

1. BPS (Statistics Indonesia in Surakarta City), "*Statistics of Air Transportation at Adi Sumarmo Airport - Surakarta in 2015,*" Surakarta, 2016, <https://surakartakota.bps.go.id> accessed on July 14, 2016.
2. W. Babisch, H. Danny, P. Göran, C. Ennio, K. Klea, V.Manolis, Environ. Int., **35** (8), 1169–1176 (2009)
3. S.J. Kim, S.K. Chai, K.W. Lee, J.B. Park, K.B. Min, H.G. Kil, C. Lee, K.J. Lee, Osong Public Heal. Res. Perspect., **5** (2), 110–114 (2014)
4. K.M. Kwak, Y.S. Ju, Y.J. Kwon, Y.K. Chung, B.K. Kim, H.J. Kim, K.W. Youn, Ann. Occup. Environ. Med., **28** (1), 38, (2016).
5. P. J. Wolfe, J. L. Kramer, and S. R. H. Barrett, J. Air Transp. Manag., **58**, 91–99 (2017).
6. L. M. Schell and Y. Ando, J. Sound Vib., **151** (3), 371–382 (1991)
7. M. M. Haines, S. A. Stansfeld, S. Job, and J. Head, Int J Epidemiol., **30** (4), 839–845 (2001).
8. K. Hiramatsu, T. Yamamoto, and K. Taira, J. Sound Vib., **205** (4), 451–460 (1997).
9. D. A. Black, J. A. Black, T. Issarayangyun, and S. E. Samuels, J. Air Transp. Manag., **13** (5), 264–276, (2007)
10. Hartono, J. Ekosains, **II** (3), 43–50 (2010).
11. ICAO, *Recommended Method for Computing Noise Contours Around Airports,* Volume 9911 of Doc (International Civil Aviation Organization, 2008).
12. Ahyani, *"Assessment of Noise Control at Husein Sastranegara Airport - Bandung,"* Master Thesis, Faculty of Civil and Environmental Engineering, Institut Teknologi Bandung, 2013.

Integration of Remote Sensing Technology Using Sentinel-2A Satellite images For Fertilization and Water Pollution Analysis in Estuaries Inlet of Semarang Eastern Flood Canal

Sawitri Subiyanto[1]*, *Zainab* Ramadhanis[1], *Aditya* Hafidh Baktiar[1]
[1]Department of Geodesy Engineering, Faculty of Engineering, Diponegoro University, Semarang

Abstract. One of the waters that has been contaminated by industrial waste and domestic waste is the waters in estuaries inlet of Semarang Eastern Flood Canal which is the estuary of the river system, which passes through the eastern city of Semarang which is dense with residential and industrial. So it is necessary to have information about the assessment of water quality in Estuaries Inlet of Semarang Eastern Flood Canal. Remote sensing technology can analyze the results of recording the spectral characteristics of water with water quality parameters. One of the parameters for assessing water quality is Chlorophyll-a and Total Suspended Solid, can be estimated through remote sensing technology using multispectral Sentinel-2A Satellite images. In this research there are 3 algorithms that will be used in determining the content of chlorophyll a, and for determining TSS. Image accuracy test is done to find out how far the image can give information about Chlorophyll-a and TSS in the waters. The results of the image accuracy test will be compared with the value of chlorophyll-a and TSS that have been tested through laboratory analysis. The result of this research is the distribution map of chlorophyll-a and TSS content in the waters.

1 Introduction

One of the waters that has been contaminated by industrial waste and domestic waste is the waters of the Semarang Eastern Flood Canal which is the estuary of the river system of Eastern Flood Canal, Tambak Lorok (Kali Banger) and Kali Tenggang. The East Flood Canal River crossing the eastern city of Semarang is dense residential and industrial. Many industrial activities around this watershed (DAS). Among other things are the textile industry, foodstuff, plastic, karoseri, printing, pharmaceutical and herbal medicine, paint, furniture, lubricating oil, workshop, even there is a fish auction. These waters become domestic and urban waste disposal sites and industrial wastes generated by activities around the river basin [1].

With the amount of waste discharged along the river of East Canal Flood, Tambak Lorok and Kali Tenggang is so that there is need for information on the assessment of water quality in the estuary of the Eastern Flood Canal. One of the parameters for assessing water quality is Chlorophyll-a and Total Suspended Solid. Chlorophyll-a is a parameter used to determine the level of water fertility while Total Suspended Solid (TSS) is one of the parameters used to determine the level of water pollution. Because the Eastern Flood Canal is one of the estuary from the rivers system as dumping industrial waste. By knowing the content of Chlorophyll-a and Total Suspended Solid (TSS) in the waters of this Semarang Eastern Flood Canal will be obtained information about water quality.

Distribution of Chlorophyll-a and Total Suspended Solid (TSS) can be estimated through remote sensing technology, such as from Sentinel-2A satellite imagery from remote sensing technology. Remote sensing technology can analyze the results of recording the spectral characteristics of water with water quality parameters. Remote sensing technology using satellite imagery offers a variety of facilities for obtaining data on fertility and water pollution compared to using more complex and costly laboratory tests. One of the benefits of using remote sensing technology is its wide coverage area, having an uncomplicated processing and low cost.

In Sentinel-2A satellite image processing to obtain information on fertility distribution data and water pollution, an algorithm is required to obtain suitable distribution value between image data and laboratory test data so that the selection of this algorithm is expected to get the value of distribution of fertility and water contamination matching In the study area. In this research there is one algorithms that will be used in determining the content of chlorophyll. The algorithm for chlorophyll use [2] algorithm and algorithms for determining TSS concentration is Sturn algorithm.

After obtaining the proper algorithm in determining the content of Chlorophyll-a and TSS concentration, then satellite image analysis was done to obtain real-time Chlorophyll-a and TSS distribution profile. Image accuracy test is done to find out how far the image can give information about Chlorophyll-a and TSS in the

* Corresponding author: sawitrisubiyanto66@gmail.com

waters. The results of the image accuracy test will be compared with the value of Chlorophyll-a and TSS that have been tested through laboratory analysis.

2 Literature Review

2.1 Water Pollution

Definition of water pollution is the entry or inclusion of living creatures, substances, energy, and / or other components into the water by human activities, so that water quality decreases to some extent that causes water to not function according to its designation [3].

Water pollution caused by humans can arise from a variety of activities, whether intentionally or not, and in general greatly affect the environment resulting from pollution by living things. Pollution if not prevented or reduced is essentially harmful and detrimental to human beings in terms of health as well as in terms of social life or the survival of living things [4].

Water pollution affects organisms and plants in water bodies. In many cases this effect destroys not only individual populations and species but also natural biological communities. Water pollution is a global problem that requires immediate evaluation and water source policy at all levels. It is considered also the leading cause of illness and death [4].

2.2 Phytoplankton

Phytoplankton are tiny creatures that float on the surface of the water [5]. According to [6], phytoplankton is a plant that is often found throughout the mass of water in the euphotic zone, microscopic and has chlorophyll to form organic substances from inorganic substances through photosynthesis. Phytoplankton as an autotrophic organism produces oxygen to be utilized by other organisms, so that phytoplankton have an important role in supporting aquatic productivity. The existence of phytoplankton can be seen based on its abundance in waters, which is influenced by environmental parameters [7]. Apart from being a primary producer, phytoplankton is also a producer of dissolved oxygen in the waters for other organisms [8].

Phytoplankton is a biological parameter that can be used as an indicator to evaluate the quality and level of fertility of a water. Phytoplankton is also the largest contributor of oxygen in a waters. The importance of the role of phytoplankton as the initial binding of solar energy makes phytoplankton play an important role for aquatic life.

As a water organism, phytoplankton has many advantages as a biological benchmark that is able to show the level of ecological instability and evaluate various forms of pollution. Each type of phytoplankton differs physiologically and behaviorally to changes in environmental quality [10].

2.3 Chlorophyll-a

Chlorophyll comes from the Greek, which consists of two syllables, chloros which means green and phylum which means leaf. Chlorophyll captures life force or solar energy and is used to split H_2O molecules into elements of H and O_2, then combine them between elements H with CO_2 gas and produce sugar or carbohydrates. From the process of photosynthesis is also produced byproducts of gas O_2 [11].

Chlorophyll is better known as green leaf substance which is a pigment found in a producer organism that functions as a converter of carbon dioxide into carbohydrates, through photosynthesis. Chlorophyll has a chemical formula C55H72O5N4Mg with Mg atoms as its center. Chlorophyll-a is one of the most decisive parameters of primary productivity in the ocean. The high distribution of low chlorophyll-a concentration is closely related to oceanographic condition of a waters. Some physical-chemical parameters that control and influence the distribution of chlorophyll-a are light intensity, nutrients (especially nitrates, phosphates and silicates). These physical-chemical parameter differences are directly the cause of the variation in primary productivity in some places at sea. In addition, "grazing" also has a major role in controlling the concentration of chlorophyll-a at sea [12]. Chlorophyll-a is an important component that is supported by phytoplankton and water plants which are both a source of natural food for fish. Chlorophyll-a is an active pigment in plant cells that play an important role in the ongoing process of photosynthesis [13].

2.4 Trophic Water Status

Eutrophication is a water enrichment with nutrients in the form of inorganic materials needed by plants and leads to an increase in primary productivity of waters. Nutrients in question are nitrogen and phosphorus. Eutrophication is classified into two, namely artificial or cultural eutrophication and natural eutrophication. Eutrophication is classified as artificial (cultural) eutrophication if nutrient uptake in waters is caused by human activity; And are classified as natural eutrophication if nutrient uptake in waters is not caused by human activity but natural activity [14].

Eutrophication is classified into four categories of trophic status [15], namely:

a. Oligotrophf; Is trophic status of lake water and / or reservoir containing low nutrient content, this status indicates that water quality is still unspoiled nature from nutrient sources N and P.
b. Mesotroph; Is the trophic status of lake water and / or reservoir containing nutrients with moderate levels, this status indicates an increase in levels of N and P, but still within the tolerance limit because it has not shown any indication of water pollution.
c. Eutrophic; Is the trophic status of lake water and / or reservoir containing high levels of nutrients, this status indicates that water has been contaminated by increased levels of N and P.

d. Hypereutrophic; Is the trophic status of lake water and / or reservoir containing very high levels of nutrients, this status indicates that water has been heavily polluted by increasing levels of N and P.

Next on the criteria of trophic water status according to [16] are listed in the following Table 1.

Table 1. Water Status Trophic Criteria

Trophic Status	Average Chlorophyll a (μg/l)
Oligotroph	<2,6
Mesotroph	2,6-7,3
Eutroph	7,3-56
Hypereutroph	>56

2.5 Total Suspended Solid

TSS are suspended materials (diameter> 1 μm) retained on a sieve of miliopore with a pore diameter of 0.45 μm. TSS consists of mud and fine sand and microorganisms. The cause of TSS in the main waters is the soil erosion or soil erosion carried to the water body. TSS concentrations if too high will inhibit the penetration of light into the water and result in disruption of the process of photosynthesis. The spread of TSS in coastal waters and estuaries is influenced by several physical factors such as wind, rainfall, waves, currents, and tides [17].

The TSS concentrations in waters generally consist of phytoplankton, zooplankton, human waste, animal waste, sludge, crop residues and animals, and industrial waste [18]. Suspended materials in natural waters are not toxic, but if excessive amounts can increase the turbidity value which further inhibits the penetration of sunlight into the water column [17].

TSS is closely related to soil erosion and erosion of river channels. TSS is highly variable, ranging from less than 5 $mg.L^{-1}$ is the most extreme 30.000 $mg.L^{-1}$ in some rivers. TSS is not only an important measure of erosion in the river channel, it is also closely related to transport through the river system of nutrients (especially phosphorus), metals, and various industrial and agricultural chemicals [19].

2.6 Pollution Index Determination

According with the [20], University of Texas, A.S., proposed an index relating to a significant pollutant compound for a designation. This index is expressed as the Pollution Index used to determine the level of pollution relative to permitted water quality parameters [20]. This index has a different concept with the Water Quality Index.

The pollution index (PI) is determined for a designation, then it can be developed for some designation for all parts of the water body or part of a river. Water quality management on the basis of pollution index (PI) can provide input to decision makers in order to assess the quality of water bodies for a designation and take action to improve quality in case of quality deterioration due to the presence of pollutant compounds. PI encompasses a variety of independent and meaningful quality parameter groups [21].

Results of image data processing on Total Suspended Solid (TSS) concentration and subsequent laboratory test results are calculated by pollution index in accordance with Decree of State Minister of Environment No.115 Year 2003 about Water Quality Status as comparison with quality standard to know pollution level at point location The sampling. The formula for calculating the pollution index is as follows (1):

$$PI = Ci/Lij \quad (1)$$

Where :

Ci: concentration of water quality parameters obtained from laboratory test results (TSS).

Lij: concentration of water quality parameters specified in the Water Quality Standard.

Evaluation of PI values are:

$0 \leq PI \leq 1.0$: meet the quality standard

$1.0 < PI \leq 5.0$: lightly contaminated

$5.0 < PI \leq 10$: moderately polluted

PI> 10: heavily polluted

2.7 Cholorophyll-a Calculation Algorithm

According to equation (1), algorithms that will be used in this research with the aim to get the best algorithm that can be used as formula of chlorophyll-a content processing in research area and can be used by multitemporal data of multilayered area study.

Nuriya Algorithm et a

The algorithm used for the determination of chlorophyll-a content using the Nuriya algorithm. The band ratio used to match the Sentinel 2A wavelength is Band Band, Band 5, and Band 4 band ratios. So in the processing of Sentinel 2A image data the algorithm is used as follows (2):

$$C = 0,2818 \times \left(\frac{B5+B6}{B4}\right)^{3,497} \quad (2)$$

Information:

C: Chlorophyll-a concentration (mg / l)

B4: The reflectance value of channel 4

B5: Reflectance value of channel 8

B6: The channel reflectance value of 6

Because the algorithm uses the reflectance value, then the image data must be extracted first from the value of DN (digital number) to the value radiance. Then from the value of radiance to the value of Reflectance.

2.8 TSS Concentration Determination Algorithm

There are one algorithms that will be used in this research with the aim to get the best algorithm that can be used as a formula for processing Total Suspended Solid (TSS) concentration in research area and can be used by multitemporal data of Sentinel-2A study area.

Sturn (1988) Algorithms

In this algorithm Sturn (1988) mapping suspended material. The Algorithm of TSS decrease as follows (3):

$$TSS\ (mg/l) = 0{,}4*(Rad_{b1} - Rad_{b2}) - 0{,}88 \qquad (3)$$

Where :

TSS = Total Suspended Solid
Refb2 = Band Reflectant Value 2

2.9 Data Collection

Data to be used in this research shown in Table 2.

Table 2. Data and Data Source

No	Data	Source	Acquisition Time
1.	Sentinel-2A	https://scihub.copernicus.eu/dhus/#/home	April 2017
2.	Map of Semarang City RBI Scale 1: 25,000	BAPPEDA Kota Semarang	2001
3.	Chlorophyll-a and TSS Concentration	Field Survey Results	April 13, 2017

In this study satellite image used is Sentinel-2A must be free from cloud. Map of Semarang City RBI Scale 1: 25,000 are used to extract basic elements of roads, rivers and administrative boundaries. In-situ measurements of Chlorophyll-a and TSS in the waters of the East Canal Flooding use randomly sampled.

The Chlorophyll-a and TSS concentration data from the field survey as shown in the Table 3 and Table 4 were obtained by sampling in the waters of the East Canal Flooding of Semarang City. Samples taken are at a depth of 25 cm from sea level. At the same time the sampling coordinates were also taken using GPS Handheld with accuracy of ± 2 m. Table 3 is shown the result of Chlorophyll-a laboratory test result in sample locations.

Table 3. Content of Chlorophyll-a Laboratory Test Result

Date	Name Point	Coordinates (m) E	Coordinates (m) N	Chlorophyll-a (mg/l)
April 13, 2016	P1	437862	9232628	1,817540
	P2	437811	9232623	1,740550
	P3	438927	9238418	0,086445
	P4	439123	9233220	0,463155
	P5	438675	9233356	0,155235
	P6	440472	9234578	0,121995
	P7	440182	9235726	0,035430

Description: UTM Projection, Datum WGS 84, Zone 49 S

To select the appropriate algorithm used determination method using regression analysis between TSS result of algorithm processing with TSS value taken from field sample result. Table 4 is shown the result of TSS Concentration of Laboratory Test Results in sample locations.

Table 4. TSS Concentration of Laboratory Test Results

Date	Name Point	Coordinates (m) E	Coordinates (m) N	Total Suspended Solid Concentration (mg/l)
April 13, 2017	P1	437862	9232628	460
	P2	436624	9233616	500
	P3	437733	9232880	580
	P4	437941	9234370	520
	P5	439123	9233220	420
	P6	438675	9233356	760
	P7	440472	9234578	460

Description: UTM Projection, Datum WGS 84, Zone 49 S

4 Results and Discussion

4.1 Results Chlorophyll-a Algorithm and Total Suspended Solid Algorithm

Based on the processing of each algorithm obtained regression and RMSe results as Table 5.

Table 5. Regression Result and RMSe on each Chlorophyll-a Algorithm and Total Suspended Solid (TSS)

Results	chlorophyll-a (Nuriya et al Algorithm)	Total Suspended Solid (TSS) (Sturn Algorithm)
Regression	93%	85%
RMSe	0,1992 mg/l	41.03 mg/l

From result of regression and RMSe obtained that best Suspended Solid algorithm is Sturn with 85% regression and RMSe 41.03 mg/l. For clorofil-a algorithm with regression value of 93% and RMSe of 0,1992 mg/l. From RMSE these two algorithms meet the standards for the use of water pollution index determination.

4.2. Analysis of Water Fertility Distribution

Results of satellite image processing to determine the content of chlorophyll a using the algorithm Nuriya et al and classify using the trophic status of waters performed on sentinel years 2017 get the following results as Figure 1.

Table 6. Water Fertility Classification April 2017

Class	Value	Area (ha)	Percentage
Oligotroph	<2,6	5274,04	88,83%
Mesotroph	2,6-7,3	265,33	4,47%
Eutroph	7,3-56	248,56	4,19%
Hypeutroph	>56	149,40	2,52%
Total (ha)		5937,33	100,00%

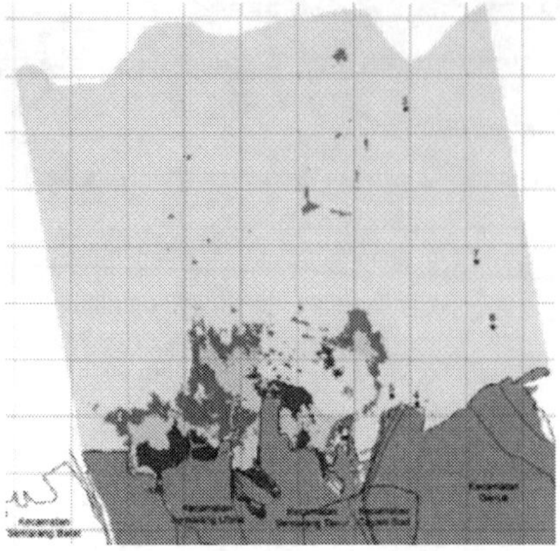

Figure 1. Distribution of Water Fertility in April 2017

From result of processing of chlorophyll concentration a using algorithm Nuriya et al in Table 6 got result of area of waters of flood waters of east canal of city of Semarang with classification of trophic status of waters, while result from classification is obtained oligotroph class of 5274,04 hectares occupy 88,83% From the research area, this oligotrophic class is spread evenly in all research areas, including river mouth, beach and loose seas. Mesotroph class of 265.33 hectares occupying 4.47% of the research area, this mesotrophic class is widely dispersed in the western seas of the harbor and is scattered over the loose seas. The eutrophic class of 248.56 hectares occupying 4.19% of the research area, this eutrophic class is spread over harbor, beach and ocean areas. While the hypereutrophic class of 149.40 hectares occupies 2.52% of the research area, this hypereutrophic class is spread over the estuary of the river, coastal and harbor areas.

4.3 Analysis of Multipurpose Water Pollution

After the processing is done on the reflectance value to the total suspended solid (TSS) concentration value using, the process of classifying the Total Suspended Solid (TSS) value to the pollution index value is calculated, The extent of the pollution area in accordance with the pollution index to obtain the following results is shown in Figure 2.

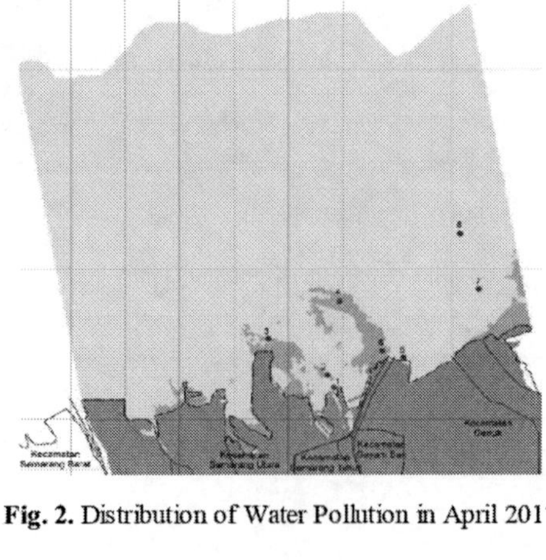

Fig. 2. Distribution of Water Pollution in April 2017

From the result of the processing of total suspended solid concentration in Table 7 is obtained the result of the extent of the flood waters of east canal flooding of Semarang city with the classification of pollution index, while the result of the classification is obtained lightly contaminated class of 5721,29 hectares occupying 96.18% From all the waters of the east canal flood that became the research site. While the polluted class was 227.52 hectares occupying 3.82% of the research area. The mild contaminated dispersion dominates the research area and it can be seen that the distribution of mild contamination lies almost in all research areas such as coastal areas and loose seas, whereas the polluted distribution is at the estuary of the rivers, beaches and harbor areas.

Table 7. Classification of Water Pollution April 2017

Class	Value	Area (ha)	Percentage
Mild Polluted	1<PI≤5	5721,29	96,18%
Medium Polluted	5<PI≤10	227,52	3,82%
Total (ha)		5948,81	100,00%

TSS is one element to determine the level of water pollution in a waters because it affects the entry of sunlight into the sea water. One of the factors used to determine water pollution is TSS which has a quality standard that has been established in [22] regarding the quality standards of seawater for harbor waters and sea water quality standards for marine biota, especially mangrove, because the area of this research area Harbor and planting mangrove plants so that the quality standard taken is the standard of seawater quality for port waters and marine biota. This further becomes the basis for classification based on the pollution index.

5 Conclusions and Recommendations

5.1 Conclusion

From a series of processes and analyzes in the previous chapter we can draw some conclusions as follows:
The content of chlorophyll a in the waters of the East Canal Flood viewed from the trophic status of the waters more dominated in the range 0-2.6 mg / l, spread almost throughout the study area ranging from the coast, river mouth and open seas. For the smallest trophic status respectively were eutrophs (7.3-56 mg / l), mesotrophs (2.6-7.3 mg / l), hypereutroph (> 56 mg / l), Spread over the estuary of the river and beach. As for the concentration of Total Suspended Solid (TSS) has a concentration of TSS dominated by the range of TSS concentration of 84 - 504 mg / l, spread almost in all research areas. For the smallest concentration of TSS is in the range 504 -5048 mg / l located at the estuary of the river and the port area. The suitable algorithm for the distribution of chlorophyll content in Eastern Channel Flood waters is Nuriya et al (2010) algorithm with determination coefficient of 93% And RMSe of 0.1992 mg/l. While the best suitable algorithm for the distribution of Total Suspended Solid (TSS)

concentration in the waters of East Canal Flood Sturn (1988) algorithms with determination coefficient Of 85% and RMSe of 41.03 mg / l. Based on the results of the analysis it is known that the distribution of water fertility in the waters of the East Canal Flood in 2017 the trophic oligotroph status dominated the study area with a percentage of 88.83% and the lowest trophic status was in trophic hypereutroph status with a percentage of 2.52%. While the distribution of water pollution with the category of contaminated light has 5721,29 hectares.

Bibliography

1. S.Y. Wulandari. Buletin Oseanografi Marina. **1(1-7)** (2012)
2. H. Nuriya, Z. Hidayah, W.A. Nugraha. Jurnal Kelautan. **3**(1):60-66 (2010)
3. Permenneg LH. Nomor 01 Tahun 2010. *Tentang Laksana Pengendalian Pencemaran Air.* (2010)
4. Dix, H. M. *Environmental Pollution*. John Willey and Sons Inc. Chischester. New York. (1981)
5. Yatim, W. 2003. *Kamus Biologi*. Yayasan Obor Indonesia. Jakarta.
6. Nontji, *A. Plankton*. Lembaga Ilmu Pengetahuan Indonesia-Pusat Penelitian Oseanografi. Jakarta. (2006)
7. Lukman, Sulastri, D.S. Said, T. Tarigan, and T. Widiyanto. *Prosiding Seminar Nasional Limnologi 2006.* Pusat Penelitian Limnologi-LIPI. Bogor.(2006)
8. Kamali. *Komunitas Fitoplankton. Institut Pertanian Bogor.* Bogor. (2004)
9. A. Astirin, A.D. Seryawan, and M. Harini, Jurnal. Biodiversitas, **3(2),** 236-241 (2002)
10. A. Astirin, A.D. Seryawan, and M. Harini, Jurnal. Biodiversitas, **3(2),** 236-241 (2002)
11. Sinurat, *Studi Tentang Nilai Produktivitas Primer di Pangururan Perairan Danau Toba*. Skripsi. Universitas Sumatera Utara. Medan. (2009)
12. Hatta, *Hubungan Antara Klorofil-a dan Ikan Pelagis.* Institut Pertanian Bogor. Bogor. (2002)
13. LIMNOTEK (2010) **17 (2)** : 171-180. (2010)
14. Effendi, Hefni. *Telaah Kualitas Air Bagi Pengelolaan Sumber Daya dan Lingkungan*. (2003)
15. Peraturan Menteri Negara Lingkungan Hidup. Nomor 28 Tahun 2009. *Tentang Daya Tampung Beban Pencemaran Air Danau Dan/Atau Waduk.* (2009)
16. R.E Carlson, and J. Simpson. *North American Lake Management Society.* **96** pp. (1996)
17. Effendi, *Telaahan Kualitas Air Bagi Pengelolaan Sumberdaya dan Lingkungan Perairan.* Jurusan Manajemen Sumberdaya Perairan. Fakultas Perikanan dan Ilmu Kelautan.Institut Pertanian Bogor.hal 12-18. (2000)
18. US. Wiradisastra, *delineasi Agro-ecological Zone. Bahan Kuliah Pelatihan Apresiasi Metodologi Delineasi Agroekologi. Bogor, 8-17 Januari 1996.* Kerjasama Proyek Pembinaan Kelembagaan Penelitian dan Pengembangan Pertanian/AMRP dengan Fakultas Pertanian-IPB. Bogor. (1996)
19. Bird, Tony. *Kimia Fisik untuk Universitas*. Jakarta: Gramedia. (1987)
20. Nemerow, N.L.Scientific *Stream Pollution Analysis.* McGraw-Hill Book Co. (1974)
21. Keputusan Menteri Negara Lingkungan Hidup Nomor : 115 Tahun 2003 *Tentang Pedoman Penentuan Status Mutu Air,* Menteri Negara Lingkungan Hidup, Jakarta, 2003
22. MENLH. Keputusan Menteri Lingkungan Hidup Nomor: 51/MENLH/2004 Tahun 2004. *Tentang Penetapan Baku Mutu Air Laut Dalam Himpunan Peraturan di Bidang Lingkungan Hidup*. Jakarta. (2004)

Risk Analysis of Coastal Disaster of Semarang City, Indonesia

Sunaryo Sunaryo[1]*, *Ambariyanto* Ambariyanto[2,3], *Denny* Nugroho Sugianto[2,3], *Muhammad* Helmi[2,3], *Awaluddin* Halirin Kaimuddin[3], *Agus* Indarjo[2]

[1]Marine and Fisheries Affairs Office of Semarang City, Jl.Pemuda 175 Semarang Indonesia 50139
[2]Faculty of Fisheries and Marine Science, Diponegoro University, Jl. Prof.Soedarto SH, TembalangCampus,Semarang, Indonesia 50275
[3]Coastal Disaster Rehabilitation and Mitigation Center, Integrated Laboratory, Diponegoro University, Jl. Prof. H. Soedarto, SH, Tembalang, Semarang. 50275

Abstract. Coastal areas are highly vulnerable to disasters, as they are affected by events occurring both on land and at sea. In the development of cities in these areas, information on vulnerability levels is needed as a consideration in determining policy. This study aims to identify potential vulnerability of Semarang city, and to investigate the potential of hazard and disaster risk levels of the city. The study was conducted in 17 villages in the northern coastal area of Semarang. The assessment approach used was score analysis to some variables, i.e geomorphology, erosion, coastal slope, waves and socio-economic aspects. The research showed that the highest level of coastal vulnerability in the Semarang city were at three villages i.e.Terboyo Kulon, Terboyo Wetan and Trimulyo with value of 4.5; while the lowest level were at the Maron beach, Marina beach and Tanah Mas with value of 2.8. The highest potential coastal hazard in the Semarang city were at two villages i.e. Tanjung Mas and Tambak Lorok with value of 4.5; and the lowest value were in the village of Randugarut, Karanganyar, Tugurejo, Marina beach and Tanah Mas with value of 1.25.The highest level of coastal risk in the Semarang city were in the Trimulyo village with value of 3.13; and the lowest level were in the Marina beach and Tanah mas with value of 1.32.

1 Introduction

Climate change has slowly happening within long time period which hard to be avoided. It closely related to temperature rise and season sifts which have impact on various human livelihoods [1, 2] as well as biodiversity [3]. Based on the report of intergovernmental panel on climate change [4], it gives significant effect, since 1850 had been recorded 11 of the hottest temperature happened in the past 12 years. The increasing temperatures in the period of 1850-1899 to 2001-2005 was 0,76°C.This incident has caused the sea water mass swell and sea level rise. The global mean sea level have increased about 1,8 mm/yearin between 1961 to 2003.The recorded sea level rise in the 20 century is about 0,17 m [4].

Semarang city is one of coastal city at the northern coast of Java island, which is directly influenced by ocean. This condition causes various components including ecosystem, community and socioeconomic to get the risk pressure from marine dynamics affected by climate change and classified as vulnerable. Vulnerability is a function magnitude of change and impact as well as climate change variation [5].

Along with the cause of sea level rise as one indicator of climate change in coastal areas of Semarang city that achieve 0.8 mm/year and the mean land subsidence ranging from 6-19 cm/year[5], causing some coastal communities as well as agricultural areas, in which their social economy depend on the resources, are becoming increasingly unreliable. Land subsidence has been reported to influence vulnerability [6,7]. The study on the effects of sea level rise becomes very important to understand how these impacts influencing the coastal cities. The impact of climate change through sea level rise has a tendency for coastal areas will be flooded, consequently, the coastal areas balance will also disturbed.

In this regard, this research attempts to identify potential risk and vulnerability of coastal area due to climate change impact; to assess potential hazard level and to assess disaster level on the coastal areas based on its' potential risk level.

* Corresponding author: sunaryosuling@gmail.com

2 Research method

The quantitative research method was used by undertaking scoring analysis on the specific research objects to get the real image of mapping pattern with different risk level on the coastal city of Semarang. Observation strategy (grand tour) and grouping strategy were applied to achieve the research objectives. Grand tour strategy was conducted to recognize the research areas, observation process was developed as an effort to search risk phenomena of coastal disaster. In addition, observation was also used to support grouping strategy in accordance to the characteristic of coastal areas.

Primary data were collected by interviewing respondent of fishermen and farmers, field observation and documentation; whereas secondary data collected from related institutions, such asthe Fisheries and Marine Board of Semarang city, the Board of Meteorology, Climatology dan Geophysics, Statistic Centerboards of Central Java Province, Development Planning Agency – Bappeda, and Office of Public Work - DPU. Data from 17 stations at Semarang coastal areas were collected both primary and secondary data. Disaster potential, coastal vulnerability and risk analysis were done based on collected data.

2.1 Analysis of disaster potential

Coastal disaster potential was determined using scoring model (see **Table 1**)

2.2 Analysis of coastal vulnerabilitylevel

Coastal vulnerability level was determined using scoring model based on the guide of disaster potential area analysis and climate change from the Ministry for Environment of New Zealand [8] (see **Table 2**).
The formula of coastal vulnerability as follows:

$$\overline{K} = \frac{(S1 + S2 + S3 + S4 + S5 + S6 + I + E)}{8}$$

$\overline{K}$ = coastal vulnerability value

2.3 Analysis of risk potential

Risk potential was calculated by using guide of coastal vulnerability level toward sea level rise [1] using formula:

$$Risk = \frac{\sqrt{\text{disaster potential x Coastal vulnerability}}}{2}$$

The risk potential value of the research areas is classified as follows:

a. Low level : 0,1 – 0,7
b. Medium level : 0,8 – 1,4
c. High level : 1,5 – 2,1

Table 1. Scoring of Coastal Disaster Potential Level

Variable	Very low	Low	Medium	High	Very high
	1	2	3	4	5
Geomorphology	Rocky shore	Stepping coast	Low lying coast, alluvial land	Estuary, laguna, beach	Sandy beach, bay muddy beach, swamp, delta, mangrove, coral reefs
Coast line erosion/accretion (m/year)	> 2.0	1.0 – 2.0	-1.0 – 1.0	-2.0 – 1.0	< -2.0
Coastal slope (%)	> 1.9	1.9 – 1.3	1.3 – 0.9	0.9 – 0.6	< 0.6
Sea level rise (mm/year)	< 1.8	1.8 – 2.5	2.5 – 3.0	3.0 – 3.4	> 3.4
Mean wave high (m)	< 1.1	1.1 – 2.0	2.0 – 2.25	2.25 – 2.6	> 2.6
Mean tidal high (m)	> 6.0	4.0 – 6.0	2.0 – 4.0	1.0 – 2.0	< 1.0

Table 2. Coastal Vulnerability Level

Acceptance Compliant	Impacts				
	Insignificant	Low	Medium	Significant	High
	1	2	3	4	5
Total household migration – S1	< 10 (short term)	10 – 50 (couple days)	50 – 100 (weeks /year)	100 – 200 (destruct housing permanently)	> 200 (massive destruction housing permanently)
Resident (total victim – S2	< 5	1 – 10	10 – 25	25 – 50	> 50
Resident (total fatalistic) – S3	0	0	1	< 5	> 5
Economy Impact – S4	Very low impact	Medium impact to several	Medium impact to many entrepreneur	Big impact to many individu and several	Very big impact to many individu and entrepreneur and/or local government

Acceptance Compliant	Impacts				
	Insignificant	**Low**	**Medium**	**Significant**	**High**
	1	**2**	**3**	**4**	**5**
		entrepreneur		entrepreneur	
Important services – S5	short term	In a day	For several day – a week	Long term	Long term impact to broader services
Commercial services – S6	short term	In a day	For several day – a week	Long term	Long term impact to broader services
Infrastructure - I	short term	In a day	For several day – a week	Need development /restoration	Important infrastructure need development /restoration/relocation
Ecosystem – E	short term	to some natural resources	to several natural resources	to several natural resources	to all natural resources

3 Result

Semarang as one of Indonesia big cities located at coastal area. Geographically, it lays at 06^{0}50’ – 7^{0}10” LSand 109^{0}35’ - 110^{0}50’ BT with area of 37,360.947 ha[9] with coastline of 36.63 km[10].

The coastal area of Semarang has elevation of 0 – 0.75 m SWL with dominant slope of 0 – 2%is about 92% within 43% of coastal areas. The area nearby the sea has same elevation to the sea level and therefore often inundated due to tide, land subsidence and sea level rise. Most of these areas are used as brackish ponds operating by the tidal phenomena.

Coastal abrasion and accretion occur on the north coast of Central Java. These two natural phenomena have change coast line in which coastal abration affects accretion in other area because of long shore current activities. The worst abration was found at Genuk sub-district while worst accretion was at West Semarang [11]. See **Table 3.**

Analysis of coastal vulnerability at 17 stations show that station 8, 9, 10, 11, 12 and 13 have medium level with value of 2.8. While high level occurred at station 4, 5, and 6 with value of 3.6. Very high level were found at station of1, 2, 3, 14, 15, 16 and 17 with value of4.1 – 4.5.

Disaster potential level in coastal area of Semarang within 17 locations shown that the station of 7, 8, 9, 10, 11 and 12 has medium level with value of 2.8, high level occurred in the station of 4, 5, 6 and 7 with value of 3.8; while the very high level occurred the station of1, 2, 3, 13, 14, 15, 16 and 17 with value of 4.2 – 4.6.

Table 3. Abrasion and accretion inSemarang

No	Sub-district	Abration (ha)	Accretion (ha)	Alteration (ha)
1	Genuk	153.47	5.12	148.35
2	North Semarang	15.32	46.43	- 31.11
3	West Semarang	4.59	48.89	- 44.30

Tabel 4. Alignment between vulnerability, hazard and risks Semarang coastal

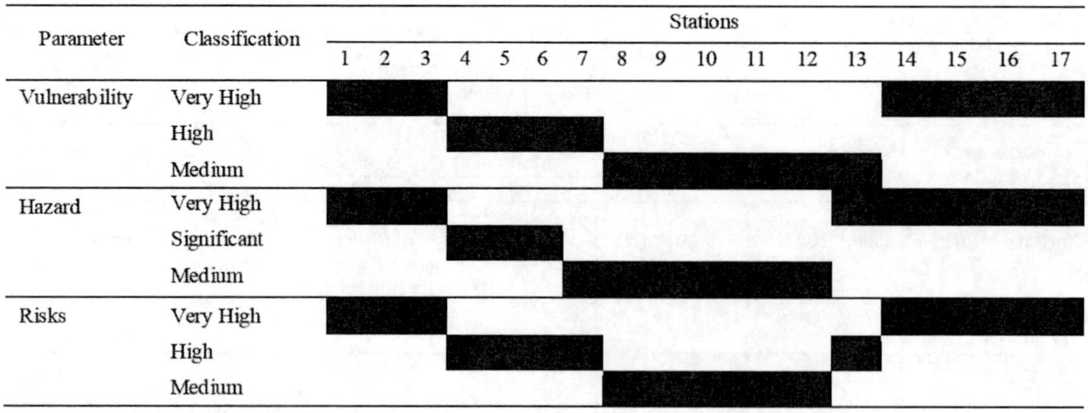

The results of the alignment analysis between vulnerability, hazards and risk indicate suitability in the study area and are correlative. **Table 4.** The higher the vulnerability demonstrated high potential of hazard, which caused increase the level of risks. Furthermore, if assessed from the aspect or determinant components of this alignment then there is a variable that significantly influence it i.e. the wave.

4 Discussion

Coastal environment characterized by two different sites, the terrestrial which is static and the sea which is dynamic. The nature of cohesion between the two has eliciting a very specific condition [12]. Land is a contributor of material and energy to the marine ecosystems, subsequently the sea will assemble into a specific ecosystem. In addition, commonly there are three typical coastal ecosystems i.e. mangroves, sea grass beds and coral reefs.The next relations could be interpreted as reverse interaction, a place of processing land material to be a source of energy to the waters' environment, while those three ecosystems can be useful as a beach protection.

However cohesive relation between land and sea more often was suffering interruption from various factors. Antrophogenic factors regarded as a dominant factor that can cause disruption in between the two, which finally have impact against the static system of terrestrial [13]. The distractions are, for example, abrasion and accretion along the coast. There are some variables that related closely and by function become important factors, i.e. physical variables of tidal, elevation and slope, andsediment structure; social variables of population; and economic variables of land use [14].

When assessed of the environmental conditions of the study areas, the eastern and western regions of Semarang receive the load of energy from incoming wave. The amount of energy was different from one time to time in the annual period. Large load in eastern and western areas probably due in two ways, the first formed by natural and the next formed by artificially. By nature area of Mangunharjo region, Mangkang Wetan and Randugarut as the representation of aside point 1, 2 and 3 (S1, S2 and S3) in the west, while Tambak Lorok, Terboyo Kulon and Terboyo Wetan and Trimulyo as the representation of sampling points 14 (S- 14), S-15, S-16 and S-17 in the eastern region more developed towards the north compared with the S-4 to S-13 which are the representation of middle area. Furthermore, the middle region (ranging from Karanganyar, tapak, TugurejoMaron Beach, Coastal Marina to Tambakrejo) has changed considerably by land conversion and has relatively lower from the pressure wave.

Another impact is tide which cause increased of *rob* (water inundation) at Semarang coastal areas. This event is worsenby land subsidence. The rate of land subsidence in Semarang in general higher towards to the northern area, since the conditions in Semarang southern region is highland. It is also influenced by the type of soil in the coastal area which is young soft soil layers so they are still continue to compression/ consolidation. Furthermore, Semarang coastal area is very dense with industrial activities, building construction, reclamation, hoarding land and groundwater extraction which accelerates the rate of soil degradation. Land subsidence caused by groundwater for residential, and industrial, building increasing load, groundwater is increasing from year to year. Indications even this have of synergism with the pattern of use of space along the coastal Semarang.

Accumulative influence of wave, tidal and land subsidence is a determining factor to the increasing flood of in Semarang. Observations estimated that the increase in sea level a result of increasing temperature was around 1 mm / year in the last decade. Deep sea waves are generally no direct effect on coastal area, while near the coast waves, especially in the area of the surf (surf zone) has a great energy which influence the formation of coastal morphology like a drag sediment (sediment-sized sand and gravel) are located on the seabed transported and dumped in the form sandbars (sand bar) [15]. Some reports show the influence of sediment and waves on the coastal line [16, 17].

Based on this fact and when linked with land use along the coast Semarang indicate that sedimentation potentially occur in areas in the western region (representation S-1 to S-4) and the eastern region (representation S-13 and S-17).The effects associated with it was the high abrasion. Factors that increase the influence of waves on the shore abrasion can occur with increasing accumulation with the high tides [12]. This indication appears in the areas of development, especially in the eastern region of Semarang, where its influence is very strong in the region of Sayung. This condition has been aggravated by the development of inland ponds and thinness a mangrove forest, which cause flood in the eastern region.

Therefore, the land management along the coastal area of Semarang indicates potential risk from the west to the east. The impact of this regard is, for example, high abrasion in Genuk region approximately153.47 ha. Waves could increase abrasion on the beach and occur synergistically in accumulation of the tide [12]. This indication happened on the development areas, particularly in Sayung. This condition is worsen by brackish water ponds as well as decreasing mangrove areas that enlarge inundation areas in the eastern part of Semarang city.

5 Conclusion

The highest vulnerability level at Semarang city are in the villages of Terboyo Kulon, Terboyo Wetan and Trimulyo with value of 4.5; while the lowest level are in the Maron beach, Marina beach and Tanah Mas with value of 2.8. The highest potential coastal danger in the Semarang city were at the Tanjung Mas and TambakLorok with value of 4.5; and the lowest value were in the village of Randugarut, Karanganyar,

Tugurejo, Marina beach and Tanah Mas with value of 1.25.The highest level of coastal risk in the Semarang city were in the Trimulyo village with value of 3.13; and the lowest level were in the Marina beach and Tanah mas with value of 1.32.

References

1. S. Diposaptono, Budiman, Agung, F. Penerbit Buku Ilmiah Populer, Bogor,359 pp. (2009)
2. O. Hoegh-Guldberg, H. Hoegh-Guldberg, Veron, J.E.N., A. Green, E. D. Gomez, J. Lough, M. King, A. Ambariyanto, L. Hansen, J. Cinner, G. Dews, G. Russ, H. Z. Schuttenberg, E.L. Peñaflor, C. M. Eakin, T. R. L. Christensen, M. Abbey, F. Areki, R. A. Kosaka, A. Tewfik, J. Oliver, WWF Australia. (2009)

a. Ambariyanto, In IOP Conference Series: Earth and Environmental Science, **55**, 1: 012002. IOP Publishing. (2017)

3. IPCC.In:Solomon, S., D. Qin, M. Manning, Z. Chen, M. Marquis, K.B. Averyt, M.Tignor and H.L. Miller (eds.). Cambridge University Press, Cambridge, United Kingdom and New York, USA.(2007)
4. H.M. Füssel, and R.J. Klein, Climatic change, **75**,3:301-329 (2006)
5. P.L.A. Wibowo, A. Hartoko, and A. Ambariyanto Ilmu Kelautan, **20**,3:127-134 (2015)
6. H.Z. Abidin, H. Andreas, I. Gumilar, T.P. Sidiq, and Y. Fukuda, Geomatics, Natural Hazards and Risk, **4**,3: 226-240(2013)
7. Ministry for Environment – New Zealand Government. (2008)
8. BPS Kota Semarang. Balai Pusat Statistik Kota Semarang. (2015)
9. DKP. Laporan Dinas Kelautan dan Perikanan, Kota Semarang. (2010)
10. DKP. Laporan Dinas Kelautan dan Perikanan, KotaSemarang. (2011)
11. A.S. Trenhaile, Clarendon Press, Oxford, United Kingdom,366 pp (1997)
12. R.J. Nicholls, and J. Branson.Geographical Journal, **164**,3:255–258 (1998)
13. G. Ozyurt, and A. Ergin. Journal of Coastal Research. Lisbon Portugal. (2009)
14. R. Dahuri, J.Rais, S.P.Ginting and M.J. Sitepu. PT. Pradnya Paramita,Jakarta.(1996)
15. Rifardi, R. Ilmu Kelautan. **13**, 3: 147-152 (2010)
16. Rositasari, R.Ilmu Kelautan. **12**, 3:146-156(2007)
17. A.D. Siswanto, W.A. Pratikto, and S., Suntoyo, Ilmu Kelautan **15**, 4: 221-230(2012)

Landslides susceptibility mapping at Gunung Ciremai National Park

Faizin[1], *Bambang* Azis Nur[2]

[1]Master Program in Environmental Science, Diponegoro University, Semarang - Indonesia
[2]Faculty of Marine and Fisheries, Diponegoro University, Semarang - Indonesia

Abstract. In addition to agriculture, tourism became one of primary economic income for communities around Mount Ciremai, West, Java. Unfortunately, the landscape of West Java has many potential causes to disasters, mainly landslides. Mapping of disaster susceptibility area is needed as a consideration of tourism planning. The study was conducted in Gunung Ciremai National Park, West Java. This paper propose a methodology to map landslides susceptibilities based on spatial data. Using Geographic Information System tools, several environmental parameters such as slope, land use, elevation, and lithology are scored to build a landslide susceptibility map. Then, susceptibility map is overlaid with Utilization Zone.

1 Background

As declared by Minister of Tourism, Dr. Ir. Arief Yahya, M.Sc, tourisms a main contributor to Indonesia GDP, devisa resource and provide easiest and cheapest occupations [1]. Tourism proved can support the economic growth and activities of the community [2].

The change of tourism trends from conventional (mass tourism) to alternative tourism encourages the creation of a tourism product based on natural resources and managed by conservation principles as known as ecotourism [3, 4].

Indonesia has a lot of natural beauty, so that the beauty of nature becomes one of the main stay of Indonesian tourism. Unfortunately, Indonesia's geographical conditions provide great opportunities for natural disasters, such as volcanic eruptions, landslides and floods. So the tourism area is also not separated from the disaster aspect.

Not all disasters will bad affect tourism activities. Even an active volcano can be developed for tourism [5]. Yogyakarta has developed a lava tour that utilizes this disaster phenomenon.

Landslides *(landslides)* associated with the soil mass motion caused by gravity [6]. Landslide is one of the most devastating natural disasters [7]. In Indonesia, West Java is the province with the most landslide events [6]. Many factor triggering landslides. Usually landslides occur in areas with steep slopes. Areas that fall into the landslide zone are generally located in areas of corrugated hills and steep hills [8]. In various models of erosion and landslide determination, slope factors are included in the main elements [9].

Rain fall is one of the major climatic elements that role against landslide and erosion events [10]. The rain has the potential to cause an avalanche depending on the intensity and duration. Raindrops destroy the soil aggregates and cause landslides. The rain water infiltrated into the soil pores that escaped the water softens the soil so that the soil loses its carrying capacity of the slope.

Geological factors affect the landslide potential based on the arrangement and lithology of rocks. Rock properties are determined by the origin of rocks and mineral composition that affect the sensitivity to erosion and landslides. In mountainous areas, soil materials dominated by solid rocks from volcanic, sedimentary, and metamorphic rocks. Soils formed from sedimentary rocks, especially clay, calcareous limestone or marl and limestone, are relatively sensitive to erosion and landslides. Volcanic rocks are generally resistant to erosion and landslides. The scoring of geological factors is distinguished by a geological arrangement or consisting of one or more textural properties.

The land cover contributed to the landslide. Low land cover is suspected as one of the factors that cause landslide prone [11]. Buchori asserted that land cover is influenced by human intervention factor in fulfilling the necessities of life [10].

Landslide vulnerability can be interpreted as landslides possibilities to occur in an area based on local terrain conditions [12], vulnerability maps are usually obtained by cartography [13]. This thermal so describes the speed and intensity of landslides that have occurred or potentially occur. But the map of vulnerability does not consider the aspect of time [12].

Researchers try to understand the phenomenon of landslide disaster as a stub for mitigation, minimize the casualties and losses from the impact [14]. The assessment of landslide vulnerability is also required in regional development considerations. The assessment of landslide vulnerability can be considered as a first step for hazard management and landslide risk [15].

* Corresponding author: ojinskw3@gmail.com

From several methods of landslide vulnerability assessment, broadly classified into two namely quantitative and qualitative. Qualitative methods depend on expert opinion [13, 16, 17]. Sabokbar declare there are three method, quantitative, qualitative and hybrid [18]. Medium quantitative methods have less dependence on human opinion (experts).

2 Methodology

The research was held in July 2017 laboratory of computation, Diponegoro University. Mapping of landslide susceptibility in this study use quantitative method which the proposed by Wati with a slight modification [19]. The data were collected by literature study, mainly from official maps, field observation, documentation and study of literature related to the research topic.

The creation of landslides susceptibility map was built by simple overlay with weighted scores. The determination of score and weighting is based on literature study. The score range is 1-4, which the higher score mean higher effect on the landslide event. In operation, the making susceptibility class is technically done with the ArcGis 10.3 tools.

Four of the five parameters are used in susceptibility analysis, i.e. slopes, lithology, soil type, and land use. Rainfall parameters are not included in the calculations because the rainfalls in TNGC area are relatively same. Determination of weights and scores of each parameter as in table 2 using the following assumptions:

1. slope; slope to an related with Voltage sliding more high. Meaning increasingly big slope, the possibility damage more Great.
2. Lithology; Volcanic generally more solid and persistent more persistent with erosion.
3. Soil types; soil have each texture and permeability. Soil with soft texture keep water and release it more slowly release so that more susceptible on landslides. Low permeability increasing high possibility landslides occurrences.
4. Land use. Land use not directly indicator of vulnerability, but densely of vegetation can increasing the mass of soil. Paddy field with it terrace plantation reducing possibility of landslides.

Table 1 Weights and scores of each parameter

Parameters and weighted factors	Score
Slope (weight: 40.8)	
Flat and undulating (0- 8%)	1
Moderately sloping (8-15%)	2
Hilly and moderately steep (15-45%)	3
Steep (> 45%)	4
Lithology (weight: 24.2)	
Andesite	1
Lava rock	2
Breccia	3
volcanic rock	4
The soil type (weight: 19.2)	
Alluvial,	1
Andosol, latosol	2
Regosol, Litosol	3
Brown forestsoil	4
Land use (weight: 15.8)	
Area limestone, wetland, paddy field, pine plantation	1
Shrubs and bush, mixed garden, forests	2
Mixed paddy fields and vegetable garden	3
Settlement, vegetable garden, sparse vegetation in forest region	4

Source: Wati [19], with modification

Classes of vulnerability are classified into five levels i.e. very low, low, medium, high and very high based on the accumulation of weighted scores. The class of was done by natural break classification.

3. Result ad Discussion

Landslides Susceptibility Map (LSM) built based on four parameters, i.e. slope, lithology, soil type, and land use. Each parameter is represent by map. Slope (*peta kelerengan*), lithology (litologi), Soil type (*jenis tanah)* land use (*penutupan lahan*) are obtained from RBI countur map, geological map, soil map and RBI map, respectively. Each map is clipped by border map of TNGC. After each map is weighted on each field attribute, then it is over laid by union fiture on ArcGIS. Then, susceptibility map is over laid with zonation map on TNGC.

Land slide susceptibility in this research is categorized in to five classes utilizing natural break classification based on accumulative weight and score.

3.1. Slope

Slope map obtained from Digital Elevation Model (DEM) is processed from SRTM imagery at Kuningan Regency, which downloaded from earth explorer.usgs.glovis. The resulted slope map is raster data. Through field calculator in ArcGIS, the score of 1, 2, 3 and 4 area are 1872, 2739, 4109, 86 Ha, respectively

3.2.Lithology

The lithology here refers to the formation of the soil-forming rock obtained from the Geological map of the system, Djawasheet Ardjawinangun sheet 1309-01 from the Directorate of Geology. The resulting geologic maPpis raster. Through field calculator obtained area with a score of 1 is 1880 Ha, score 2 is 3200 Ha, score 3 is 9 Ha and score 4 is 3689 Ha.

https://doi.org/10.1051/e3sconf/20183112010

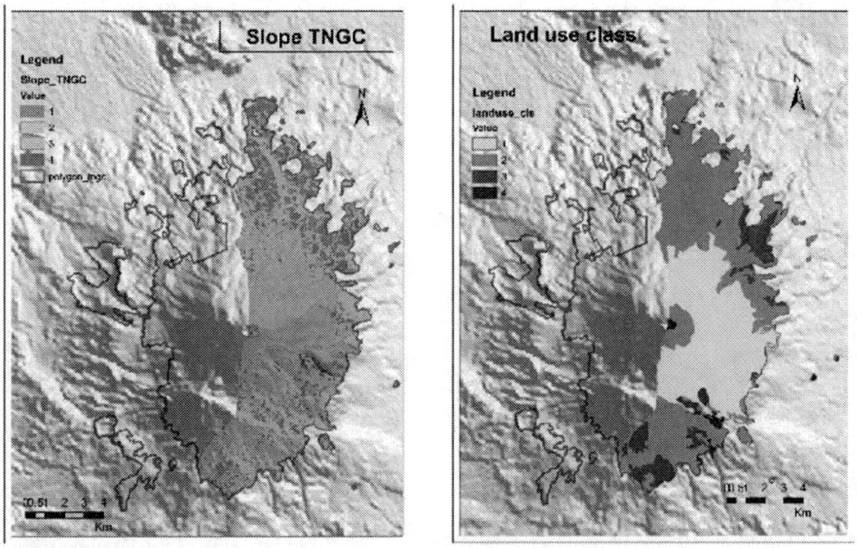

(a) (b)

Fig. 1. Map of (a) Slope TNGC, (b) Land use class

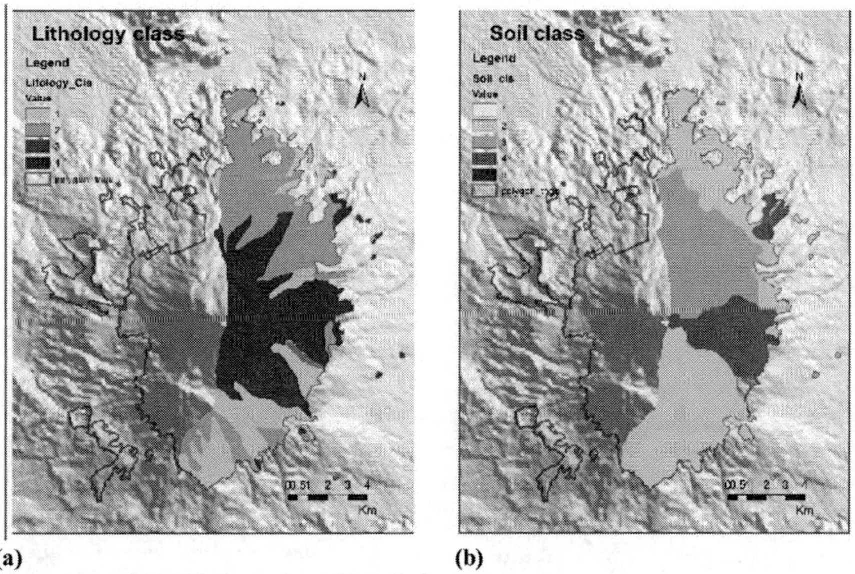

(a) (b)

Fig. 2. Map of (a) Lithology class, (b) Soil class

3.3.Soil Type

The soil type is obtained from the map of the *Puslittanak* land system. Land class of vulnerability 1 is 3 hectare, class of vulnerability 2 is 4656 Ha, class of vulnerability of land 3 is 2528 Ha, and class of vulnerability 4 is 1517 Ha.

3.4. Land Use

The use is the result of processed land cover map from *Balai* TNGC. The result of land use land use map shows that land cover is 3305 Ha land-use prone class, 4479 Ha of land use class 2, land use class 3 area of 925 Ha and 4 ha of land use class of 86 Ha.

3.5. Map of Landslide Hazard TNGC

In making the landslide susceptibility map, the four variables were processed with Raster Calculator (Spatial Analyze) tools with the formula (40.8 * slope) + (24.2 * lytology) + (19,2 * ground) + (15,8 * land use) With the result shown in Figure 5 below. Very low vulnerability is 1882 Ha, low vulnerability 1980 Ha, vulnerability is 1356 Ha high vulnerability 1308 Ha and vulnerability very high 2001 Ha.

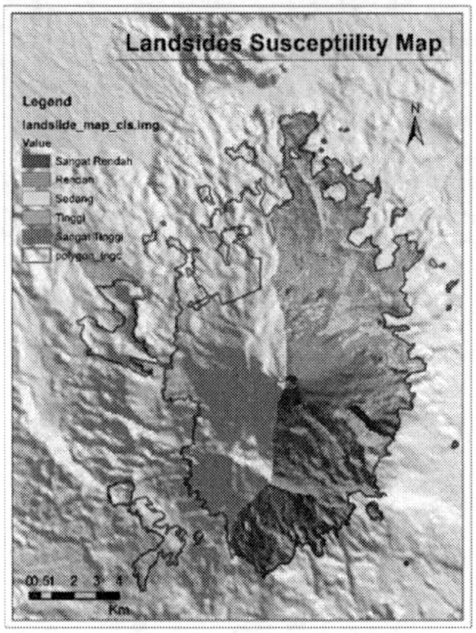

Fig. 2. Map of Landslides Susceptility

4 Conclusion

1. It can be said that susceptibility area spread on middle area of TNGC.
2. Elevation factor not so influential, but slope factor maybe more influents.
3. Area with highly susceptibility covering rehabilitation zone and utilization area.

References

1. Alamsjah, M. I. (2016) *'The Ministry of Tourism Exposure on KIDi 2016'*. Yogyakarta.
2. Ioan, S. (2013) 'Ecological Tourism and Public Administration in Romania', *Procedia - Social and Behavioral Sciences*. **81**, pp. 235–240.
3. Damanik, J. and Weber, H. F. (2006) *Ecotourism Planning: from Theory to Practice.* Yogyakarta: CV. Andi Offset.
4. Latupapua, Y. T. (2011) 'Public Perception of Potential Coastal Tourism *Objects* in Kei Kecil Subdistrict, Southeast Maluku District, *Jurnal Agroforestri*, 7(2), pp. 92–102.
5. Erfurt-cooper, P. (2015) *'Volcanoes and Tourism', in The Encyclopedia of Volcanoes.* Second Edi. Elsevier Inc., pp. 1295–1311.
6. PVMBG (2006) *Introduction of Land Movements. Center for Volcanology and Geological Hazard Mitigation.*
7. Lin, G., Chang, M., Huang, Y. and Ho, J. (2017) Engineering Geology. **224**(September 2016), pp. 62–74.
8. Lukmantara, L. I. (2007). Scientific Contribution, **5**(2), pp. 105–110.
9. Guillard, C. and Zezere, J. Environmental Management, **50**(4), pp. 721–735.
10. Buchori, I. and Susilo, J. (2012). TATALOKA, **14**(4), pp. 282–294.
11. Atmojo, S. W. (2008) 'The role of Agroforestry in Tackling Flood and Landslide', in *Seminar Nasional Prndidikan* Agroforestry *Sebagai Strategi Menghadapi Pemanasan Global*. Solo: Universitas Negeri Surakarta, pp. 1–15.
12. Fell, R., Corominas, J., Bonnard, C., Cascini, L., Leroi, E. and Savage, W. Z. (2008). Engineering Geology. **102**(3–4), pp. 85–98.
13. Yalcin, A. (2008). *CATENA*, **72**, pp. 1–12.
14. Feizizadeh, B. and Blaschke, T. (2011) Journal of Earth Science and Engeneering, **1**(June 2017), pp. 66–71.
15. Leonardi, G., Palamara, R. and Cirianni, F. (2016) 'Landslide Susceptibility Mapping Using a Fuzzy Approach', *Procedia Engineering*. 161, pp. 380–387.
16. Ercanoglu, M. (2008) *'An Overview on the Landslide Susceptibility Assessment Techniques'*, in The 1st WSEAS International Conference *on Environmental and Geological and Engineering* pp. 131–134.
17. Ayalew, L., Yamagishi, H., Marui, H. and Kanno, T. (2005) *'Landslides in Sado Island of Japan : Part II* . GIS-based susceptibility mapping with comparisons of results from two methods and verifications', 81, pp. 432–445.
18. Sabokbar, H. F., Roodposhti, M. S. and Tazik, E. (2014) Geomorphology. **226**, pp. 15–24.
19. Wati, S. E., Hastuti, T., Widjojo, S., Pinem, F. and Information, W. (2010) *'Landslides Susceptibility Mapping with Heuristic Approach in Mountainous Area a Case Study in Tawang Mangu Subsdistrict, Central Java, Indonesia'*, in International Archives of the Photogrammetry, Remote Sensing and Spatial Information Science. Kyoto, pp. 248–253.

AUTHOR INDEX

AUTHOR INDEX

AUTHOR INDEX

AUTHOR INDEX

AUTHOR INDEX